D0138435

MATHEMATICAL PRACTICES

Mathematics for Teachers

Activities, Models, and Real-Life Examples

Ron Larson
The Pennsylvania State University
The Behrend College

Robyn Silbey
Montgomery County Public Schools

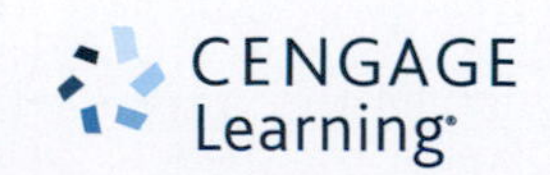

Australia • Brazil • Mexico • Singapore • United Kingdom • United States

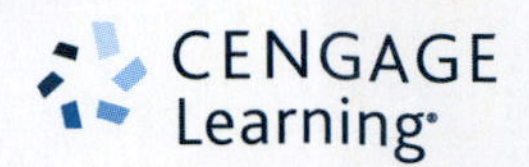

Mathematical Practices, Mathematics for Teachers: Activities, Models, and Real-Life Examples

Ron Larson
Robyn Silbey

Product Director: Liz Covello
Product Manager: Richard Stratton
Senior Content Developer: Laura Wheel
Product Assistant: Danielle Hallock
Media Developer: Andrew Coppola
Content Project Manager: Jill Quinn
Manufacturing Planner: Doug Bertke
Rights Acquisition Specialist: Shalice Shah-Caldwell
Production Service: Larson Texts, Inc.
Text and Cover Designer: Larson Texts, Inc.
Cover Image: ©iStockphoto.com/porcorex
Compositor: Larson Texts, Inc.

Library of Congress Control Number: 2013953540

Student Edition:
ISBN-13: 978-1-285-44710-0
ISBN-10: 1-285-44710-7

Cengage Learning
200 First Stamford Place, 4th Floor
Stamford, CT 06902
USA

Printed in the United States of America
1 2 3 4 5 6 7 17 16 15 14 13

Contents

Preface

Welcome to *Mathematical Practices, Mathematics for Teachers: Activities, Models, and Real-Life Examples*. Through reliable, research-based pedagogical techniques, this book will help you master the concepts needed to excel at teaching mathematics. In presenting the content, special care has been taken to help you become familiar and comfortable with the Common Core State Standards for Mathematics. Our goal is to help you become a great math teacher. We are happy to present this textbook as a key part of your preparation for teaching.

Features of Mathematical Practices

Activities

Each section begins with an activity providing you an opportunity to discover the concepts of the section. Many of these activities are adapted from the Big Ideas Math *Middle School* series. While completing each activity as part of your coursework, keep in mind how you will use it in your future classroom! You can watch Robyn Silbey review and discuss these activities at MathematicalPractices.com.

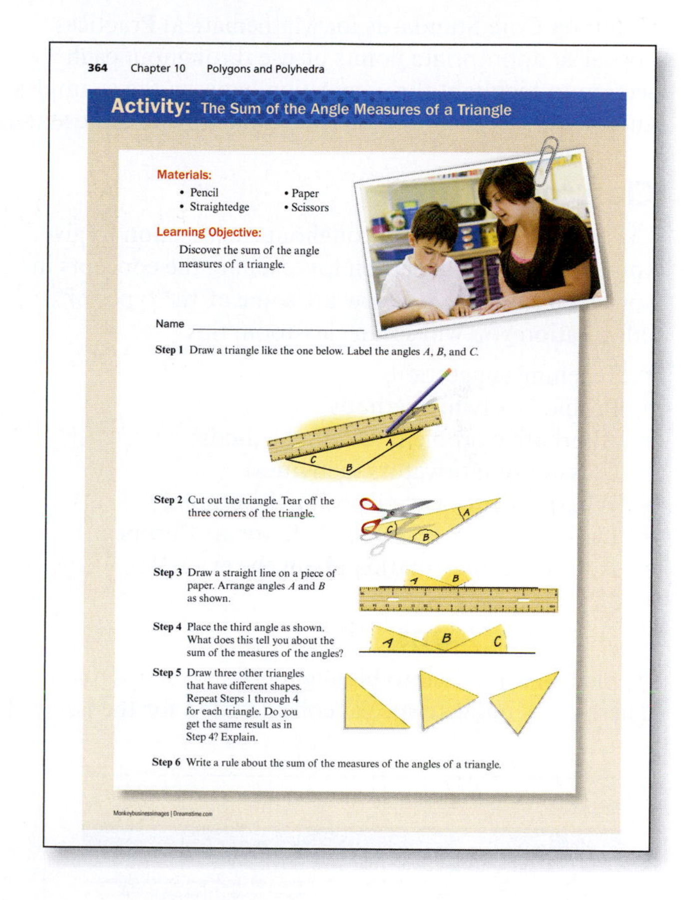

364 Chapter 10 Polygons and Polyhedra

Activity: The Sum of the Angle Measures of a Triangle

Materials:
- Pencil
- Straightedge
- Paper
- Scissors

Learning Objective:
Discover the sum of the angle measures of a triangle.

Name ____________

Step 1 Draw a triangle like the one below. Label the angles A, B, and C.

Step 2 Cut out the triangle. Tear off the three corners of the triangle.

Step 3 Draw a straight line on a piece of paper. Arrange angles A and B as shown.

Step 4 Place the third angle as shown. What does this tell you about the sum of the measures of the angles?

Step 5 Draw three other triangles that have different shapes. Repeat Steps 1 through 4 for each triangle. Do you get the same result as in Step 4? Explain.

Step 6 Write a rule about the sum of the measures of the angles of a triangle.

Monkeybusinessimages | Dreamstime.com

Standards

The standards listed at the beginning of each section are the primary Common Core State Standards related to the section. These are included to show you the relevance of the section topics to the pedagogical responsibilities you will have as a teacher. Standards are grouped according to the grade levels K–2, 3–5, and 6–8.

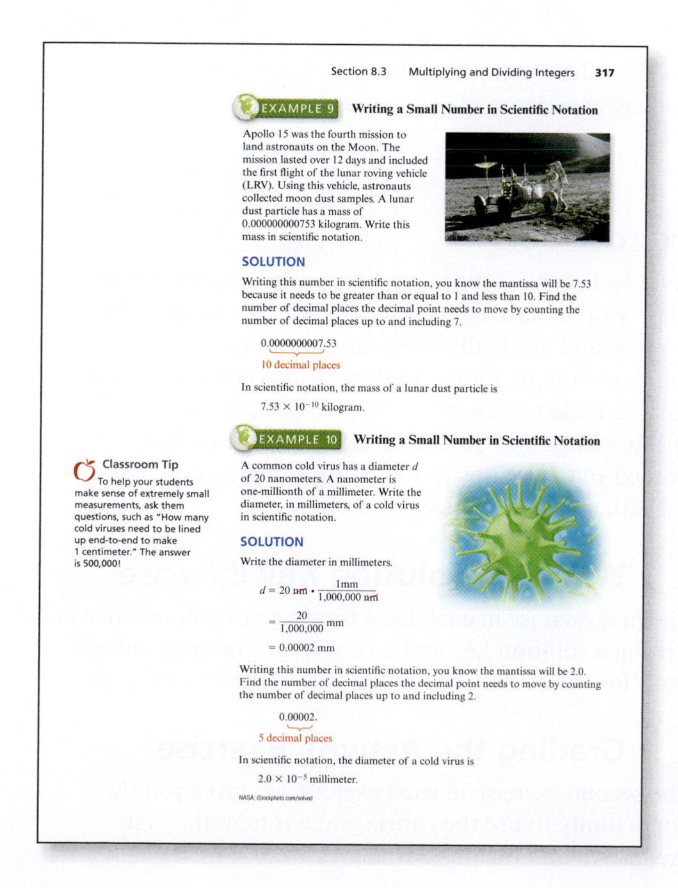

Section 8.3 Multiplying and Dividing Integers 317

EXAMPLE 9 Writing a Small Number in Scientific Notation

Apollo 15 was the fourth mission to land astronauts on the Moon. The mission lasted over 12 days and included the first flight of the lunar roving vehicle (LRV). Using this vehicle, astronauts collected moon dust samples. A lunar dust particle has a mass of 0.000000000753 kilogram. Write this mass in scientific notation.

SOLUTION

Writing this number in scientific notation, you know the mantissa will be 7.53 because it needs to be greater than or equal to 1 and less than 10. Find the number of decimal places the decimal point needs to move by counting the number of decimal places up to and including 7.

$\underbrace{0.0000000007}_{\text{10 decimal places}}.53$

In scientific notation, the mass of a lunar dust particle is

7.53×10^{-10} kilogram.

EXAMPLE 10 Writing a Small Number in Scientific Notation

A common cold virus has a diameter d of 20 nanometers. A nanometer is one-millionth of a millimeter. Write the diameter, in millimeters, of a cold virus in scientific notation.

SOLUTION

Write the diameter in millimeters.

$$d = 20 \text{ nm} \cdot \frac{1 \text{mm}}{1{,}000{,}000 \text{ nm}}$$

$$= \frac{20}{1{,}000{,}000} \text{ mm}$$

$$= 0.00002 \text{ mm}$$

Writing this number in scientific notation, you know the mantissa will be 2.0. Find the number of decimal places the decimal point needs to move by counting the number of decimal places up to and including 2.

$\underbrace{0.00002}_{\text{5 decimal places}}.$

In scientific notation, the diameter of a cold virus is

2.0×10^{-5} millimeter.

Classroom Tip: To help your students make sense of extremely small measurements, ask them questions, such as "How many cold viruses need to be lined up end-to-end to make 1 centimeter." The answer is 500,000!

NASA; iStockphoto.com/solvod

Section Objectives

A bulleted list of learning objectives provides you with the opportunity to preview what will be presented in the section.

Examples

The examples in each section teach you the concepts of .the section and show you how to solve key types of problems, step-by-step. A mix of skill examples, real-life examples, and modeling examples will give you a stimulating and pedagogically-sound learning experience.

Real-Life Examples

The wide variety of real-life examples present you opportunities to see how the math of the section is used in engaging and relevant ways. These examples are indicated by a green globe and example box.

Modeling Examples

Modeling examples show you how to learn, model, and teach mathematical concepts in the classroom by using manipulatives, fraction strips, number line models, area models, set models, and many other types of models. These examples are indicated by orange example boxes.

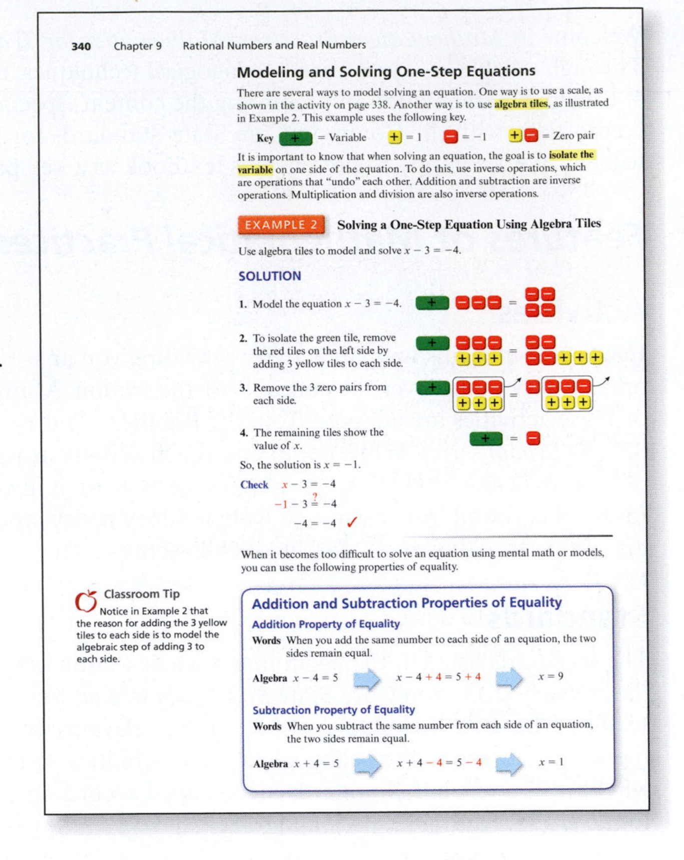

340 Chapter 9 Rational Numbers and Real Numbers

Modeling and Solving One-Step Equations

There are several ways to model solving an equation. One way is to use a scale, as shown in the activity on page 338. Another way is to use **algebra tiles**, as illustrated in Example 2. This example uses the following key.

It is important to know that when solving an equation, the goal is to **isolate the variable** on one side of the equation. To do this, use inverse operations, which are operations that "undo" each other. Addition and subtraction are inverse operations. Multiplication and division are also inverse operations.

EXAMPLE 2 **Solving a One-Step Equation Using Algebra Tiles**

Use algebra tiles to model and solve $x - 3 = -4$.

SOLUTION

1. Model the equation $x - 3 = -4$.
2. To isolate the green tile, remove the red tiles on the left side by adding 3 yellow tiles to each side.
3. Remove the 3 zero pairs from each side.
4. The remaining tiles show the value of x.

So, the solution is $x = -1$.

Check $x - 3 = -4$

$-1 - 3 \stackrel{?}{=} -4$

$-4 = -4$ ✓

When it becomes too difficult to solve an equation using mental math or models, you can use the following properties of equality.

Classroom Tip Notice in Example 2 that the reason for adding the 3 yellow tiles to each side is to model the algebraic step of adding 3 to each side.

Addition and Subtraction Properties of Equality

Addition Property of Equality

Words When you add the same number to each side of an equation, the two sides remain equal.

Algebra $x - 4 = 5$ → $x - 4 + 4 = 5 + 4$ → $x = 9$

Subtraction Property of Equality

Words When you subtract the same number from each side of an equation, the two sides remain equal.

Algebra $x + 4 = 5$ → $x + 4 - 4 = 5 - 4$ → $x = 1$

Mathematical Practices

Common Core Standards for Mathematical Practices appear at appropriate points of use throughout each section to highlight the connection between the examples and the processes and proficiencies identified by the standards.

Classroom Tips

Classroom tips appear throughout each section to give you additional information for teaching the concepts in your own classroom. Below are some of the types of information you will see in classroom tips.

- Teaching suggestions
- Problem-solving methods
- Alternative problem-solving methods
- Classroom motivators (activities)
- What students learn in doing the example
- Differing definitions (i.e., U.S. versus European)
- Additional information about the example

Apple Icons

Apple icons are used to highlight classroom tips and exercises that involve applications for the classroom. These icons indicate special consideration for the issues that you may experience in your own classroom.

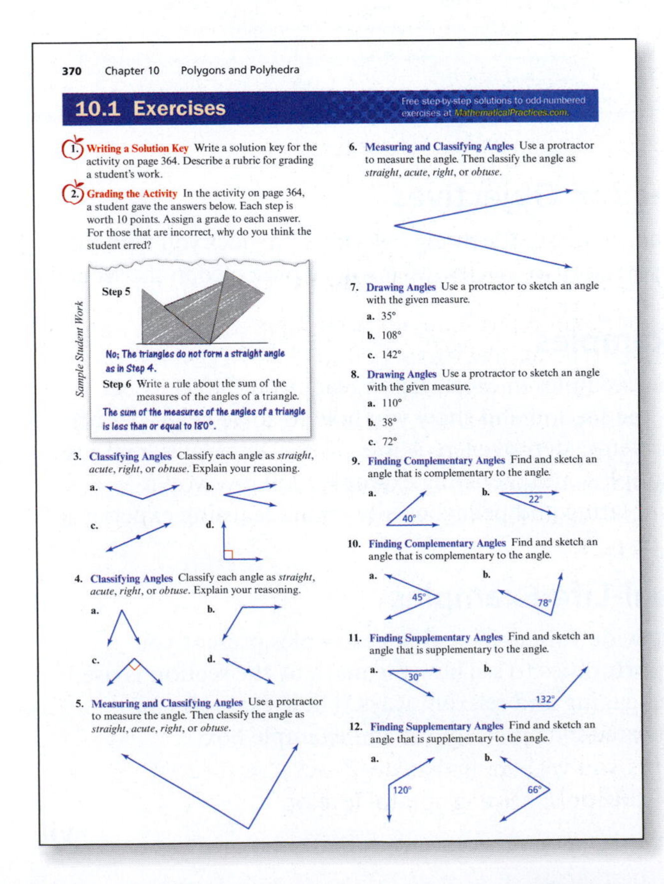

370 Chapter 10 Polygons and Polyhedra

10.1 Exercises

Free step-by-step solutions to odd-numbered exercises at MathematicalPractices.com.

1. **Writing a Solution Key** Write a solution key for the activity on page 364. Describe a rubric for grading a student's work.
2. **Grading the Activity** In the activity on page 364, a student gave the answers below. Each step is worth 10 points. Assign a grade to each answer. For those that are incorrect, why do you think the student erred?

3. **Classifying Angles** Classify each angle as *straight*, *acute*, *right*, or *obtuse*. Explain your reasoning.
   a. b. c. d.
4. **Classifying Angles** Classify each angle as *straight*, *acute*, *right*, or *obtuse*. Explain your reasoning.
   a. b. c. d.
5. **Measuring and Classifying Angles** Use a protractor to measure the angle. Then classify the angle as *straight*, *acute*, *right*, or *obtuse*.
6. **Measuring and Classifying Angles** Use a protractor to measure the angle. Then classify the angle as *straight*, *acute*, *right*, or *obtuse*.
7. **Drawing Angles** Use a protractor to sketch an angle with the given measure.
   a. 35°
   b. 108°
   c. 142°
8. **Drawing Angles** Use a protractor to sketch an angle with the given measure.
   a. 110°
   b. 38°
   c. 72°
9. **Finding Complementary Angles** Find and sketch an angle that is complementary to the angle.
   a. 40° b. 22°
10. **Finding Complementary Angles** Find and sketch an angle that is complementary to the angle.
    a. 45° b. 78°
11. **Finding Supplementary Angles** Find and sketch an angle that is supplementary to the angle.
    a. 30° b. 132°
12. **Finding Supplementary Angles** Find and sketch an angle that is supplementary to the angle.
    a. 120° b. 66°

Exercise Sets

Each section has four pages of exercises organized to guide you to mastery. The set begins with basic skills practice and gradually progresses to exercises that help you explain, apply, analyze, and connect concepts. Solving these exercises will help you attain a depth of knowledge and build confidence. You can view worked-out solutions to the odd-numbered exercises at MathematicalPractices.com.

Writing a Solution Key Exercise

The first exercise in each exercise set gives you practice in writing a solution key and a rubric for grading students' work for the activity presented before the lesson.

Grading the Activity Exercise

The second exercise in each exercise set gives you the opportunity to use the rubric you wrote in the first exercise to grade sample student work.

 In Your Classroom Exercises

These exercises involve the types of real-life applications and concept exercises found in classroom textbooks. They include a connection to teaching the concept in the classroom.

 Grading Student Work Exercises

Each section includes grading student work exercises to give you practice in reviewing student solutions to identify errors, and suggesting topics for review based on the errors.

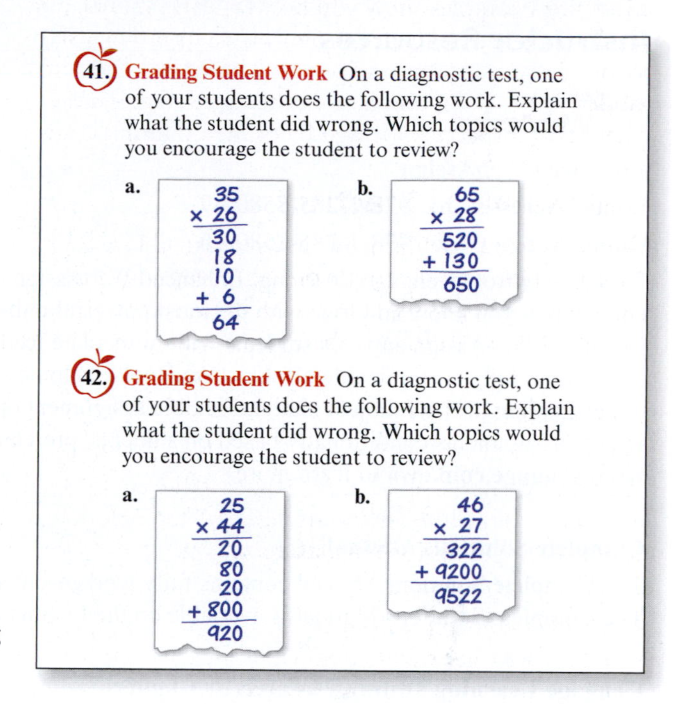

41. Grading Student Work On a diagnostic test, one of your students does the following work. Explain what the student did wrong. Which topics would you encourage the student to review?

a.
35
× 26
30
18
10
+ 6
64

b.
65
× 28
520
+ 130
650

42. Grading Student Work On a diagnostic test, one of your students does the following work. Explain what the student did wrong. Which topics would you encourage the student to review?

a.
25
× 44
20
80
20
+ 800
920

b.
46
× 27
322
+ 9200
9522

Writing Exercises

There are writing exercises in every section to give you practice in describing, explaining, and analyzing mathematical concepts in your own words.

 My Dear Aunt Sally Exercises

My Dear Aunt Sally exercises appear in sections that include topics relating to basic operations or order of operations. These exercises encourage the development of problem-solving skills and number sense. You can find more problems of this type at **AuntSally.com.**

Chapter Summary

Each chapter summary includes key vocabulary, section learning objectives, and key learning concepts and mathematical formulas.

Review Exercises

Three pages of review exercises give you extra practice and a review of the important exercise types from the chapter.

Chapter Test

Chapter test exercises are written in the style and format of state certification exams. These tests will aid in your preparation for state certification exams by familiarizing you with their form.

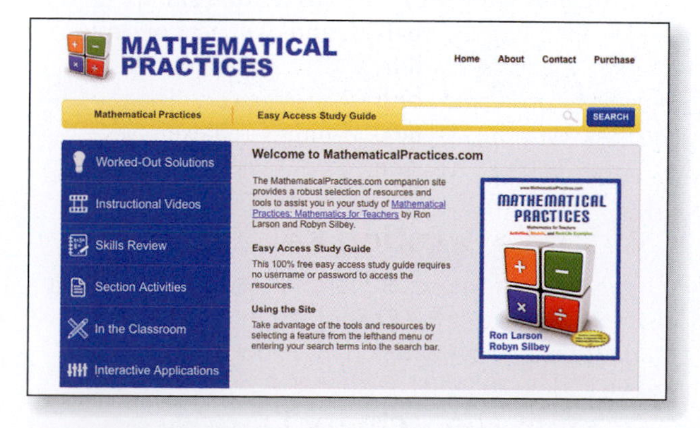

MathematicalPractices.com

The companion website offers multiple tools and resources to supplement your learning. Access to these features is free. Watch Robyn Silbey discussing the section activities, use tools and resources developed by Big Ideas Math for a middle school program, view worked-out solutions to odd-numbered exercises, watch students learn math concepts in a classroom setting, and much more.

Enhanced WebAssign combines exceptional Math Practices content with WebAssign, the most powerful online homework solution. Enhanced WebAssign engages you with immediate feedback, rich tutorial content and interactive, fully customizable eBooks (YouBook) helping you to develop a deeper conceptual understanding of the subject matter.

Supplements

Instructor Resources

ENHANCED WebAssign

Enhanced WebAssign

Printed Access Card: 978-1-285-85802-9

Online Access Code: 978-1-285-85803-6

Exclusively from Cengage Learning, Enhanced WebAssign combines the exceptional mathematics content that you know and love with the most powerful online homework solution, WebAssign. Enhanced WebAssign engages students with immediate feedback, rich tutorial content, and interactive, fully customizable eBooks (YouBook), helping students to develop a deeper conceptual understanding of their subject matter. Flexible assignment options give instructors the ability to release assignments conditionally based on students' prerequisite assignment scores. Visit us at **www.cengage.com/ewa** to learn more.

Complete Solutions Manual

The Complete Solutions Manual contains fully worked-out solutions for all exercises in the text. The Complete Solutions Manual is available on the Instructor Companion Site.

Cengage Learning Testing Powered by Cognero

Cengage Learning Testing Powered by Cognero is a flexible, online system that allows you to author, edit, and manage test bank content from multiple Cengage Learning solutions; create multiple test versions in an instant; and deliver tests from your LMS, your classroom or wherever you want. Access to Cognero is available on the Instructor Companion Site.

Instructor Companion Site

Everything you need for your course in one place! This collection of book-specific lecture and class tools is available online via **www.cengage.com/login.** Access and download the Complete Solutions Manual, Cognero, videos and more.

Student Resources

Student Solutions Manual

ISBN-13: 978-1-285-44714-8

Giving you more in-depth explanations, this insightful resource includes fully worked-out solutions for selected exercises in the textbook.

Enhanced WebAssign

Printed Access Card: 978-1-285-85802-9

Online Access Code: 978-1-285-85803-6

Enhanced WebAssign (assigned by the instructor) provides you with instant feedback on homework assignments. This online homework system is easy to use and includes helpful links to textbook sections, video examples, and problem-specific tutorials.

CengageBrain.com

To access additional course materials and companion resources, please visit www.cengagebrain.com. At the CengageBrain.com home page, search for the ISBN of your title (from the back cover of your book) using the search box at the top of the page. This will take you to the product page where free companion resources can be found.

Acknowledgements

To those who asked us to write this book, thank you for the inspiration and encouragement to do so. Our goal is to make mathematics understandable to elementary school teachers, and in turn, make it understandable to their students.

Reviewers

Jim Brandt, *Southern Utah State University*
Jill Drake, *University of West Georgia*
Thomas Fisher, *Frontrange Community College*
Linda Fitzpatrick, *Western Kentucky University*
Marie Franzosa, *Oregon State University*
Melinda Gann, *Mississippi College*
Sue Glascoe, *Mesa Community College*
Jerrold Grossman, *Oakland University*
Cheryl Herrmann, *California State Univerity-San Marcos*
Peter Johnson, *Eastern Connecticut State University*
Blair Madore, *The State University of New York-Potsdam*
Maria Mitchell, *Central Connecticut State University*
Perla Myers, *University of San Diego*
Connie Schrock, *Emporia State University*

We would also like to thank the staff of Larson Texts, Inc., who assisted in preparing the manuscript, rendering the art package, and typesetting and proofreading the pages and the supplements. On a personal level, we are grateful to our spouses, Deanna Gilbert Larson and Sam Silbey, for their love, patience, and support. Also, a special thanks goes to R. Scott O'Neil.

If you have suggestions for improving this text, please feel free to write to us. We value your comments very much.

Ron Larson
Professor of Mathematics
Penn State University
www.RonLarson.com

Robyn Silbey
Montgomery County Public Schools
www.RobynSilbey.com

1 Problem Solving

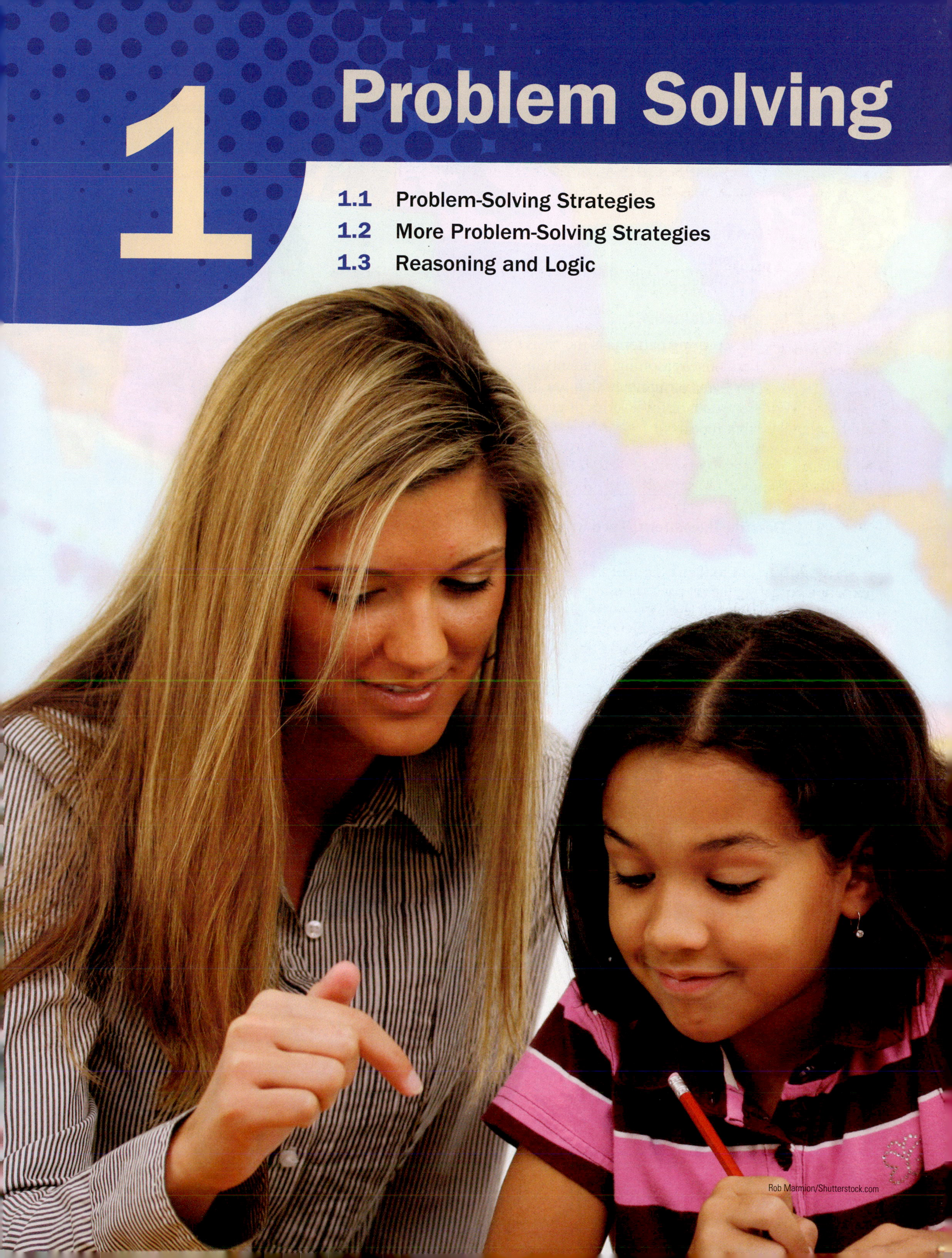

Rob Marmion/Shutterstock.com

Activity: Describing Patterns

Materials:

- Pencil
- Colored pencils

Learning Objective:

Discovering patterns is the basis for inductive reasoning. Students look for items or numbers that are repeated, or for a series of events that repeats.

Name ______________________________

Describe the pattern. Then write or draw the next three numbers, letters, or figures.

Pattern Description

1. 1, 3, 5, 7, ___, ___, ___ ______________________________

2. 2, 3, 5, 8, 12, ___, ___, ___ ______________________________

3. 1, 4, 9, 16, ___, ___, ___ ______________________________

4. 20, 17, 14, ___, ___, ___ ______________________________

5. A, a, C, c, ___, ___, ___ ______________________________

6. Z, X, V, T, ___, ___, ___ ______________________________

7. ■, ●, ■, ●, ___, ___, ___ ______________________________

8. ■, ■, ■, ■, ___, ___, ___ ______________________________

9. ●, •, •, ●, ___, ___, ___ ______________________________

10. 1, 3, 9, 27, ___, ___, ___ ______________________________

iofoto/Shutterstock.com

1.1 Problem-Solving Strategies

Standards

Grades K–2
Measurement and Data
Students should represent and interpret data.

Grades 3–5
Operations and Algebraic Thinking
Students should identify, generate, and analyze patterns.

- Solve a problem by looking for a pattern.
- Solve a problem by reading or making a table or list.
- Solve a problem by reading or making a graph.

Looking for a Pattern

The book *How to Solve It*, written by mathematician George Pólya, describes the following four-step approach to problem solving.

Four-Step Approach to Problem Solving

1. **Understand the problem.** Carefully read and reread the problem. Try to state the problem in your own words. What information is given? What information is missing? What is being asked?
2. **Make a plan** for solving the problem. This plan may involve one or more of the problem-solving strategies discussed in this section and in the following section.
3. Carry out the plan and **solve the problem.** Implement the strategy in Step 2. Check each step of the work carefully for accuracy and correctness.
4. **Look back** and examine the results. Check that the results make sense in the statement of the original problem. In some cases, you may want to consider other possible approaches.

Learning how to use this approach is an important first step in becoming an efficient problem solver.

EXAMPLE 1 **Strategy: Looking for a Pattern**

Describe the pattern. Then write the next three numbers.

3, 12, 48, 192, , ,

SOLUTION

1. **Understand the Problem** You need to find and describe the pattern. Then use the pattern to find the next three numbers.
2. **Make a Plan** Look for a pattern by comparing each number to the previous number. Then use the pattern to find the next three numbers.
3. **Solve the Problem** Each given number is 4 times the previous number.

   3 12 48 192

   ×4 ×4 ×4

   Continue the pattern by multiplying by 4 to find the next number.

   3 12 48 192 768 3072 12,288

4. **Look Back** Check your result by dividing each of the three new numbers by the previous number. The result should always be 4.

   $\frac{768}{192} = 4$ ✓ $\frac{3072}{768} = 4$ ✓ $\frac{12,288}{3072} = 4$ ✓

Classroom Tip

Here is a fun pattern that you can try with your students.

O, T, T, F, F, S, S, E, ?, ?

It is tempting to say that the next letter is "E." However, what is the pattern? What would the next letter be? The pattern is the first letters of the digits 1, 2, 3, 4, 5, 6, 7, and 8 written in words. So, the next two letters are "N" for 9 and "T" for 10.

There are many problem-solving strategies. You will learn about the most common strategies in this section and in the following section. The problem-solving strategy in Examples 1 and 2 is called *Look for a Pattern.*

Classroom Tip

Our ability to recognize patterns is one of the primary characteristics that make us human. For example, early humans recognized patterns for the length of a day and the length of a year. Ask your students to think about other patterns that early humans may have recognized.

Strategy: Look for a Pattern

Looking for a Pattern is a problem-solving strategy in which students look for patterns in the given information. Students look for items or numbers that are repeated, or for a series of events that repeats. Other problem-solving strategies can be combined with this strategy. For example, recognizing a pattern can be easier when you make a table, draw a graph, or draw a diagram.

EXAMPLE 2 Strategy: Looking for a Pattern

The triangle below is called *Pascal's Triangle*, named after the French mathematician Blaise Pascal (1623–1662). Find the pattern in the sums of the numbers in the rows of Pascal's Triangle.

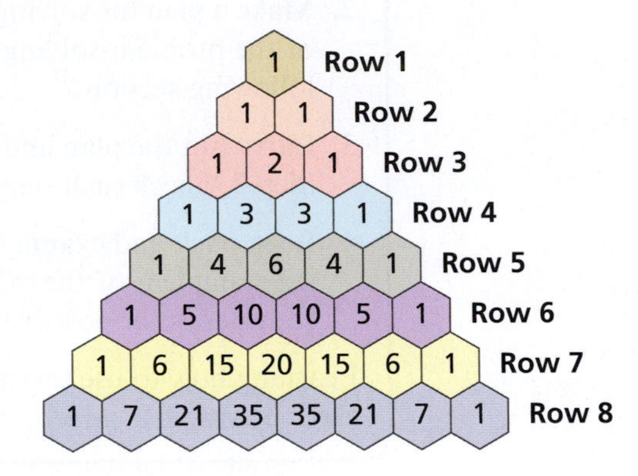

SOLUTION

1. **Understand the Problem** You need to describe the pattern in the sums of the numbers in the rows of Pascal's Triangle.
2. **Make a Plan** Add the numbers in each row and look for a pattern. Organize the sums in a table.
3. **Solve the Problem** The sums are shown in the table.

Row	1	2	3	4	5	6	7	8
Sum	1	2	4	8	16	32	64	128

×2 ×2 ×2 ×2 ×2 ×2 ×2

Each sum is twice the previous sum.

4. **Look Back** Check your answer by dividing each sum in the table by the previous sum.

$\frac{2}{1} = 2$ ✓ $\frac{4}{2} = 2$ ✓ $\frac{8}{4} = 2$ ✓ $\frac{16}{8} = 2$ ✓

$\frac{32}{16} = 2$ ✓ $\frac{64}{32} = 2$ ✓ $\frac{128}{64} = 2$ ✓

Notice that making a table in Step 3 helps you see the pattern in the data.

Reading or Making a Table or List

Strategy: Read or Make a Table or List

Reading or Making a Table or List is a problem-solving strategy that students use to solve mathematical word problems by writing the given information in an organized format. This strategy allows students to discover relationships and patterns. It encourages students to organize information in a logical way and to look critically at the information to find patterns and develop a solution.

EXAMPLE 3 Strategy: Making a Table

You make \$16 per hour, plus time and a half for each hour over 40 hours that you work in a week. How many hours do you need to work in a week to earn \$808?

SOLUTION

1. **Understand the Problem** You need to find the number of hours you need to work in a week to earn \$808.
2. **Make a Plan** Make a table showing your earnings as the number of work hours increases.
3. **Solve the Problem** To find your earnings, multiply the amount you earn per hour by the number of hours worked. Begin with 40 hours. Then increase the number of hours worked until you reach \$808. Note that "time and a half" for \$16 per hour is $1.5 \cdot 16 = \$24$ per hour.

Hours	40	41	42	43	44	45	46	47
Earnings	\$640	\$664	\$688	\$712	\$736	\$760	\$784	\$808

You need to work 47 hours.

4. **Look Back** Check your answer by multiplying \$16 per hour by 40 hours and adding 7 hours of overtime pay at \$24 per hour.

$$\frac{\$16}{\cancel{h}} \cdot 40\,\cancel{h} + \frac{\$24}{\cancel{h}} \cdot 7\,\cancel{h} = \$640 + \$168$$
$$= \$808 \checkmark$$

Mathematical Practices

Reason Abstractly and Quantitatively

MP2 Mathematically proficient students create a coherent representation of the problem at hand, considering the units involved, attending to the meaning of quantities, and knowing and flexibly using different properties of operations and objects.

Classroom Tip

Always encourage your students to write the units when solving word problems. Then show them how to use unit analysis. For instance, in Step 4 of Example 3, you can use unit analysis to conclude that "dollars per hour" times "hours" is equal to "dollars."

An alternative way to solve the problem in Example 3 is to use algebra.

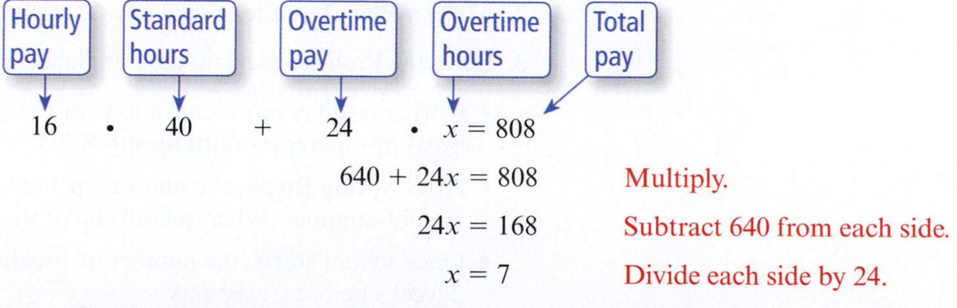

$$16 \cdot 40 + 24 \cdot x = 808$$
$$640 + 24x = 808 \quad \text{Multiply.}$$
$$24x = 168 \quad \text{Subtract 640 from each side.}$$
$$x = 7 \quad \text{Divide each side by 24.}$$

Note that this value does not represent the solution to the problem. It represents the number of additional hours you must work over 40 hours. So, you need to work $40 + 7 = 47$ hours to earn \$808.

Classroom Tip

Here is a historical note that your students may find interesting. The Russian novelist Ivan Turgenev wrote (in *Fathers and Sons*, 1862), "A picture shows me at a glance what it takes dozens of pages of a book to expound." This statement is especially true for data sets consisting of hundreds of numbers, as in Example 4.

Reading or Making a Graph

Strategy: Read or Make a Graph

Reading or Making a Graph is a problem-solving strategy in which students use a graph to represent information visually. Representing information using pictures makes it easier to visualize results, identify relationships, describe trends, make comparisons, and draw conclusions. Different types of graphs include line graphs, bar graphs, picture graphs, and circle graphs.

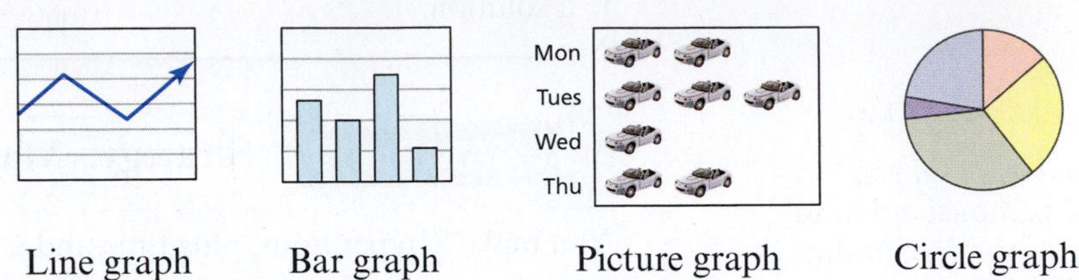

Line graph | Bar graph | Picture graph | Circle graph

Strategy: Reading a Graph

The graph below is taken from *The Visual Miscellaneum* by David McCandless. Describe the patterns in the graph. (The data were collected from Facebook status updates.)

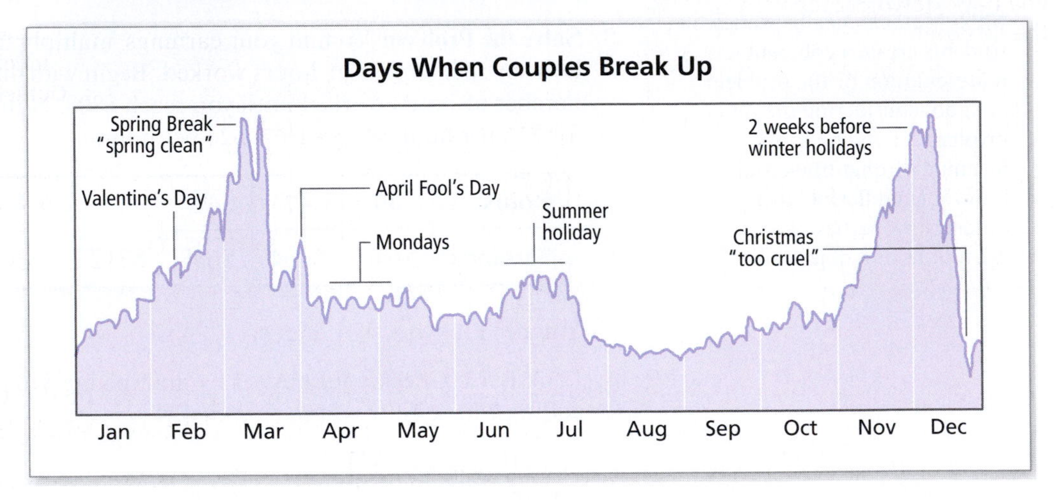

SOLUTION

1. **Understand the Problem** You need to describe the patterns in the information represented by the line graph of when people break up.
2. **Make a Plan** Use the special days that seem to stand out in the graph.
3. **Solve the Problem** Several days are labeled in the graph.
   - Christmas Day represents a low for the year. After that, the number of breakups increases until Spring Break.
   - After Spring Break, the number of breakups is relatively steady until the end of summer. When school starts, the number of breakups decreases.
   - Once school starts, the number of breakups increases to a yearly high 2 weeks before Christmas.
4. **Look Back** Do you see other patterns in the graph? For example, if the graph were larger, you would notice that more breakups are posted on Mondays than any other day of the week.

EXAMPLE 5 Comparing Two Graphs

Use the graphs to compare the weights of the players on the two teams.

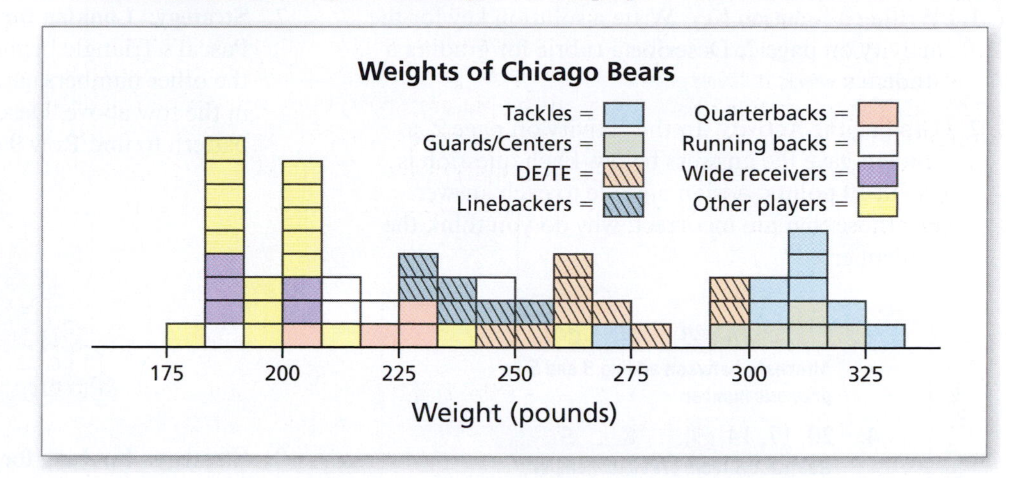

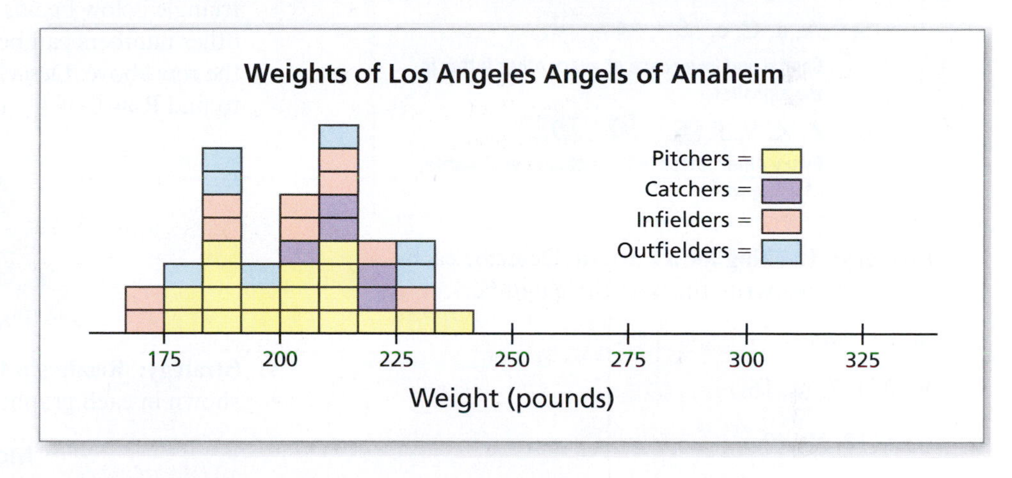

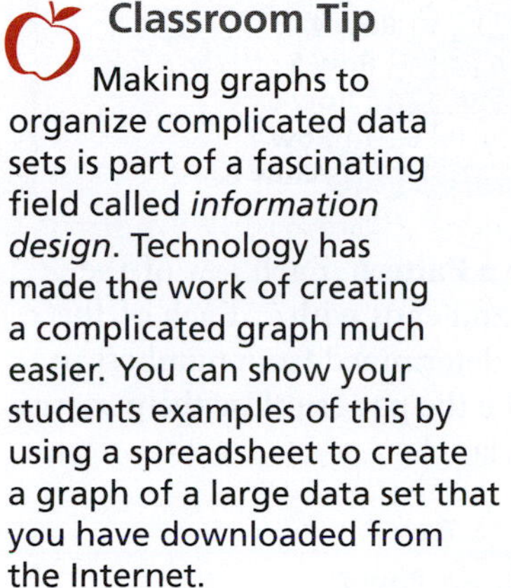

Classroom Tip

Making graphs to organize complicated data sets is part of a fascinating field called *information design*. Technology has made the work of creating a complicated graph much easier. You can show your students examples of this by using a spreadsheet to create a graph of a large data set that you have downloaded from the Internet.

SOLUTION

1. **Understand the Problem** You need to compare the weights of the players on the Chicago Bears and on the Los Angeles Angels of Anaheim.
2. **Make a Plan** Use the color codes to draw conclusions about the weights of the players on each of the two teams.
3. **Solve the Problem** The distributions of the weights are quite different for the two teams.
   - The weights of the Chicago Bears are more spread out than the weights of the Los Angeles Angels of Anaheim.
   - The Chicago Bears is a football team. From the graph, it appears that there is a relationship between a player's weight and the player's position. For example, defensive ends and tight ends (DE/TE), linebackers, quarterbacks, running backs, and receivers have lesser weights. Tackles, guards, and centers have greater weights.
   - The Los Angeles Angels of Anaheim is a baseball team. From the graph, there appears to be no relationship between the weight of a baseball player and the player's position.
4. **Look Back** Check that your answer is reasonable. If you have seen a football team and a baseball team, does your answer agree with what you saw?

1.1 Exercises

Free step-by-step solutions to odd-numbered exercises at *MathematicalPractices.com*.

1. Writing a Solution Key Write a solution key for the activity on page 2. Describe a rubric for grading a student's work.

2. Grading the Activity In the activity on page 2, a student gave the answers below. Each question is worth 10 points. Assign a grade to each answer. For those that are incorrect, why do you think the student erred?

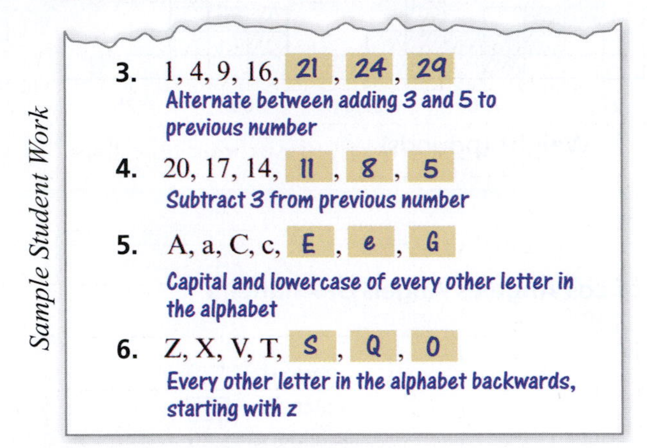

3. Strategy: Looking for a Pattern Describe each pattern. Then write the next three numbers.

a. 98, 90, 82, 74, ▭, ▭, ▭

b. 2, 4, 7, 11, 16, ▭, ▭, ▭

c. 5, 10, 20, 40, ▭, ▭, ▭

d. 100, 81, 64, 49, ▭, ▭, ▭

4. Strategy: Looking for a Pattern Describe each pattern. Then write the next three numbers.

a. 37, 43, 49, 55, ▭, ▭, ▭

b. 30, 29, 27, 24, 20, ▭, ▭, ▭

c. 2, 6, 18, 54, ▭, ▭, ▭

d. 1, 8, 27, 64, ▭, ▭, ▭

5. Strategy: Looking for a Pattern Describe each pattern. Then draw the next figure.

a. 

b.

6. Strategy: Looking for a Pattern Describe each pattern. Then draw the next figure.

a.

b. 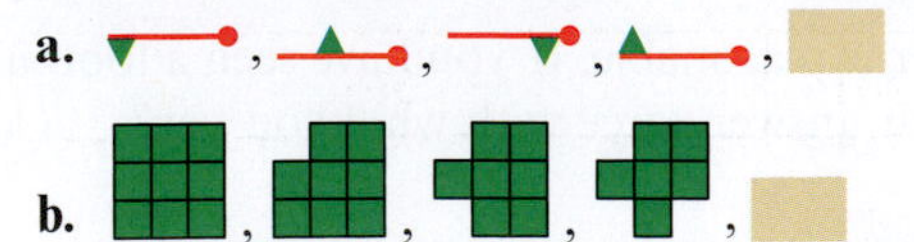

7. Strategy: Looking for a Pattern Each row of Pascal's Triangle begins and ends with 1. Each of the other numbers can be determined from numbers in the row above. Describe the pattern. Use the pattern to find Row 9 of Pascal's Triangle.

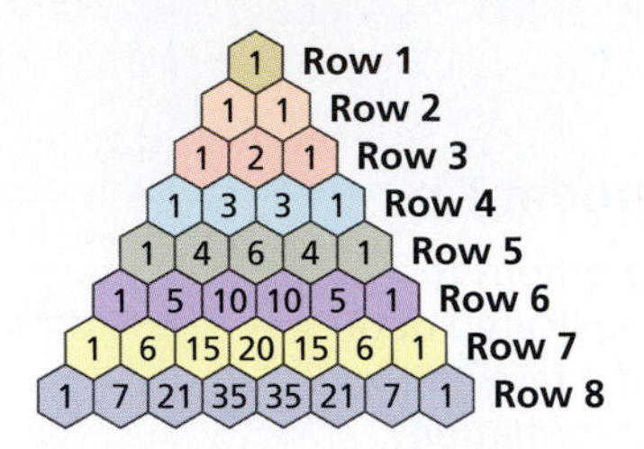

8. Strategy: Looking for a Pattern Each row of the triangle below begins and ends with 2. Each of the other numbers can be determined from numbers in the row above. Describe the pattern. Use the pattern to find Row 6 of the triangle.

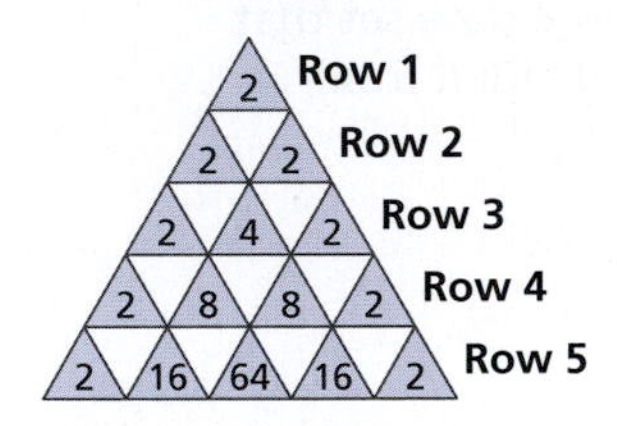

9. Strategy: Reading a Graph Describe the pattern(s) shown in each graph.

a.

b.

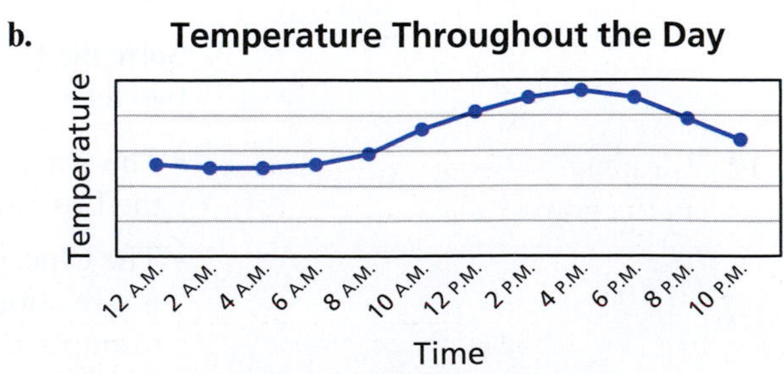

c. **Age of Graduate Students at a College**

50–59, 60 or older, 40–49, 30–39, 20–29

10. **Strategy: Reading a Graph** Describe the pattern(s) shown in each graph.

a.

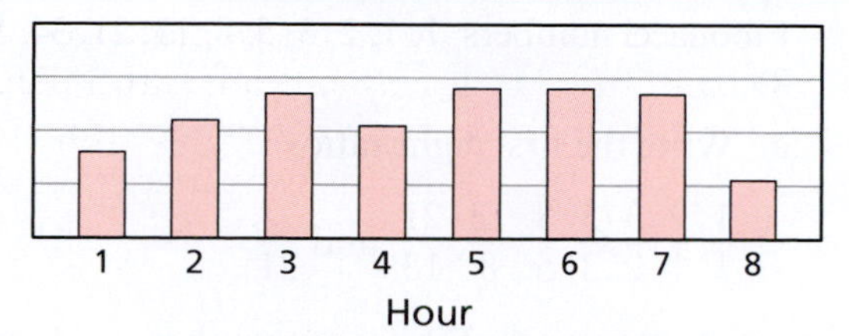

b.

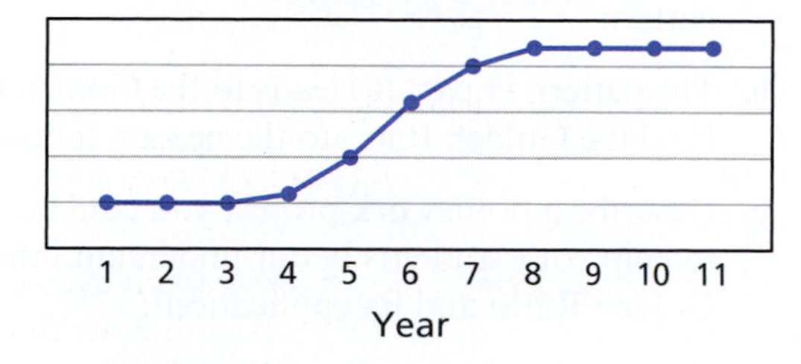

c.

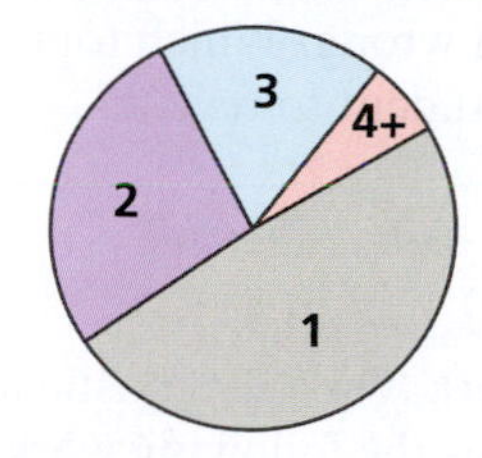

11. **Strategy: Reading a Graph** Describe the pattern(s) in the graph.

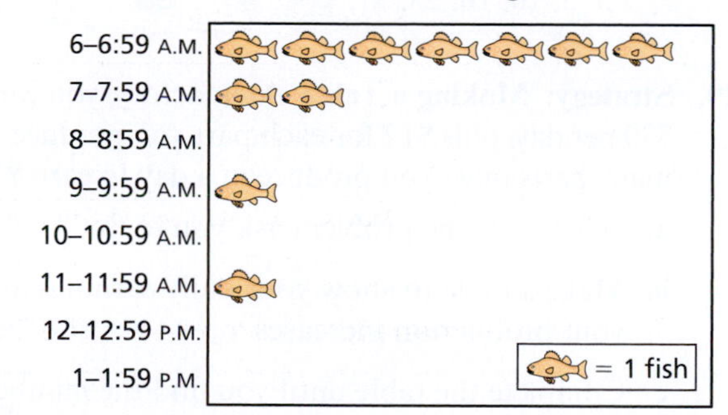

12. **Strategy: Reading a Graph** Describe the pattern(s) in the graph.

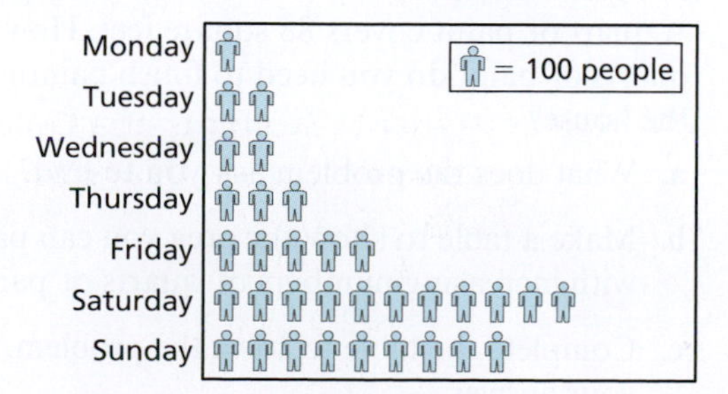

13. **Strategy: Making a Graph** Make a bar graph to display the data in the table. Describe the pattern(s).

Day	Text messages
Monday	62
Tuesday	29
Wednesday	24
Thursday	33
Friday	54

14. **Strategy: Making a Graph** Make a bar graph to display the data in the table. Describe the pattern(s).

Week	Profit (thousands of dollars)
1	22
2	16
3	27
4	11
5	18

15. **Strategy: Making a Graph** Make a line graph to display the data in the table. Describe the pattern(s).

Month	Savings account balance
July	$60
August	$80
September	$100
October	$160
November	$240
December	$180

16. **Strategy: Making a Graph** Make a line graph to display the data in the table. Describe the pattern(s).

Year	Salary (thousands of dollars)
2009	34
2010	38
2011	44
2012	49
2013	52
2014	56

17. **Strategy: Making a Graph** Real estate agency A sold 7 homes, agency B sold 4 homes, agency C sold 5 homes, and agency D sold 2 homes. Make a picture graph to display the data. Describe the pattern(s).

18. **Strategy: Making a Graph** The class grades on a math test are 6 "A"s, 5 "B"s, 7 "C"s, 6 "D"s, and 3 "F"s. Make a picture graph to display the data. Describe the pattern(s).

19. **Strategy: Looking for a Pattern** You are watching a pitcher who throws two types of pitches, a fastball (F) and a curveball (C). The order of the pitches thrown so far is F, C, F, F, C, F, F, F, C. Assuming that the pattern continues, predict the next five pitches. Explain.

20. **Strategy: Looking for a Pattern** Stopping a car in an emergency depends both on the driver's reaction time and the braking distance of the car. The table shows typical stopping distances at speeds of 20, 30, 40, and 50 miles per hour. Assuming that the pattern continues, find typical stopping distances at speeds of 60 and 70 miles per hour. Explain.

Speed (miles per hour)	20	30	40	50
Stopping distance (feet)	40	75	120	175

21. **Writing** Write a real-life application that involves a pattern of numbers in which each number is 8 more than the previous number.

22. **Writing** Write a real-life application that involves a pattern of numbers in which each number is 5 less than the previous number.

23. **Writing** Write a real-life application that involves a pattern of numbers in which each number is 3 times the previous number.

24. **Writing** Write a real-life application that involves a pattern of numbers in which each number is one-half the previous number.

25. **Fibonacci Numbers: In Your Classroom** The Fibonacci numbers 1, 1, 2, 3, 5, 8, 13, 21, 34, 55, 89,. . . appear in many patterns in nature, such as the growth of a nautilus shell.

a. Describe the pattern of the Fibonacci numbers.

b. Write the next three Fibonacci numbers.

c. Describe a homework project you could assign that applies patterns of Fibonacci numbers.

jordache/Shutterstock.com

26. **The Golden Ratio: In Your Classroom** The *Golden Ratio* is a special relationship found in nature, art, and architecture. The Golden Ratio can be approximated as a ratio of two consecutive Fibonacci numbers 1, 1, 2, 3, 5, 8, 13, 21, 34, 55, 89, ...

a. Write the first eight ratios

$$\frac{1}{1}, \frac{2}{1}, \frac{3}{2}, \frac{5}{3}, \frac{8}{5}, \frac{13}{8}, \frac{21}{13}, \text{ and } \frac{34}{21}$$

of consecutive Fibonacci numbers as decimals rounded to five decimal places. Describe the pattern.

b. The pattern in part (a) leads to the Golden Ratio. Find the Golden Ratio to the nearest thousandth.

c. Describe a homework project you could assign to help your students better understand the Golden Ratio and its applications.

27. **Grading Student Work** On a diagnostic test, one of your students does the following work. Explain what the student did wrong. Which topics would you encourage the student to review?

5, 10, 20, 40, 60, 80, 100, . . .

28. **Grading Student Work** On a diagnostic test, one of your students does the following work. Explain what the student did wrong. Which topics would you encourage the student to review?

5, 10, 16, 23, 31, 39, 47, 55, . . .

29. **Strategy: Making a Table** At a factory, you earn $80 per day, plus $12 for each part you produce. How many parts must you produce in a day to earn $140?

a. What does the problem ask you to find?

b. Make a table to show your daily earnings as your production increases.

c. Complete the table until you find the number of parts you must produce. Check your answer.

30. **Strategy: Making a Table** You are painting a house that has a surface area of 2200 square feet. You have five gallons of paint that will cover 1760 square feet. A quart of paint covers 88 square feet. How many quarts of paint do you need to finish painting the house?

a. What does the problem ask you to find?

b. Make a table to show the area you can paint with increasing numbers of quarts of paint.

c. Complete the table to solve the problem. Check your answer.

31. **Strategy: Reading a Graph** The graph shows the net profits for a company each year from 2002 through 2011. Describe the patterns in the graph.

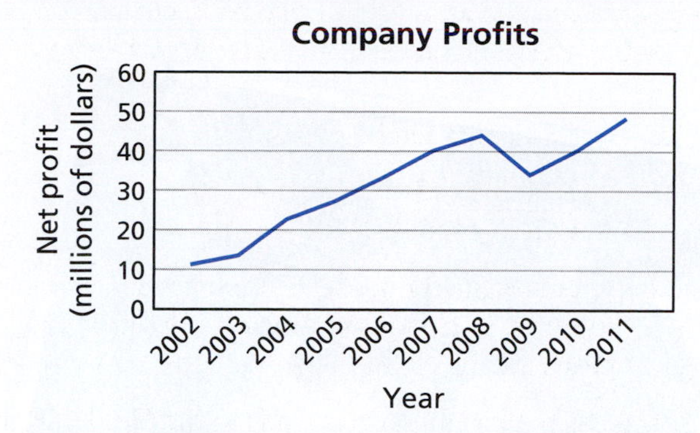

32. **Strategy: Reading a Graph** The graph shows the numbers of votes cast in U.S. presidential elections from 1960 through 2012. Describe the patterns in the graph.

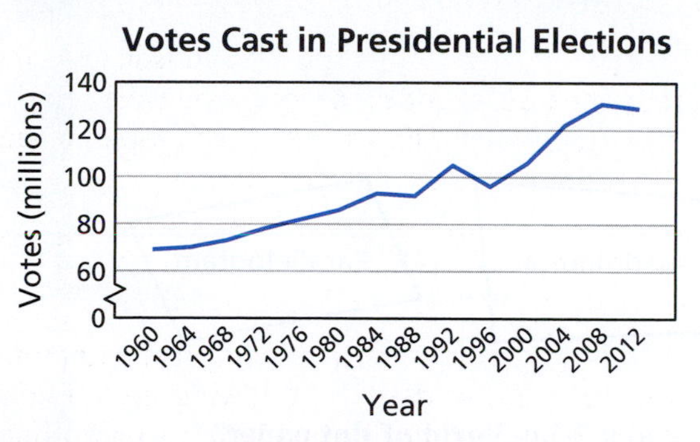

33. **Reading a Graph: In Your Classroom** Which bar graph shows steadily increasing yearly sales? How would you explain this concept to your students?

a.

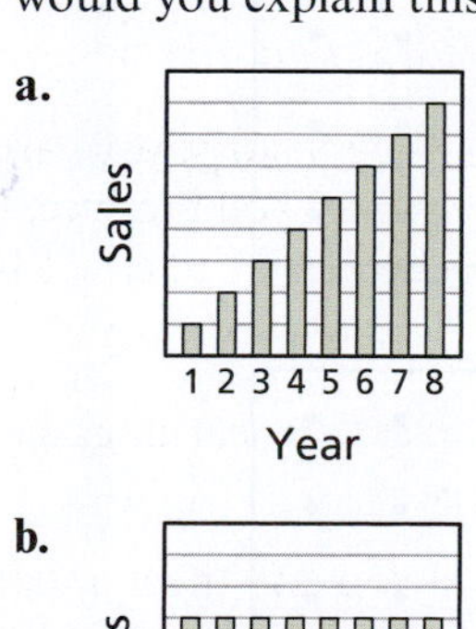

b.

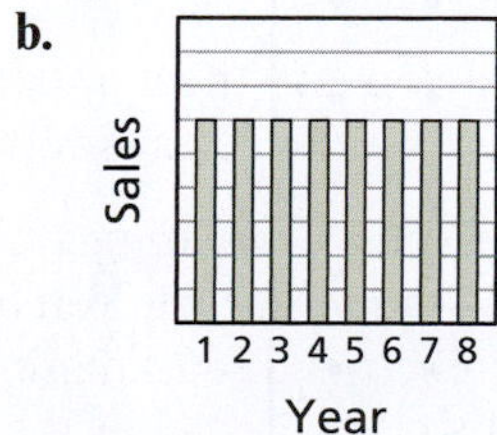

c.

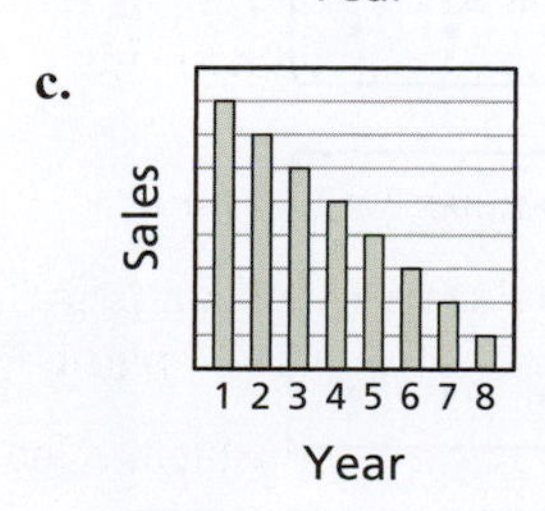

34. **Reading a Graph: In Your Classroom** Which line graph shows steadily decreasing yearly profits? How would you explain this concept to your students?

a.

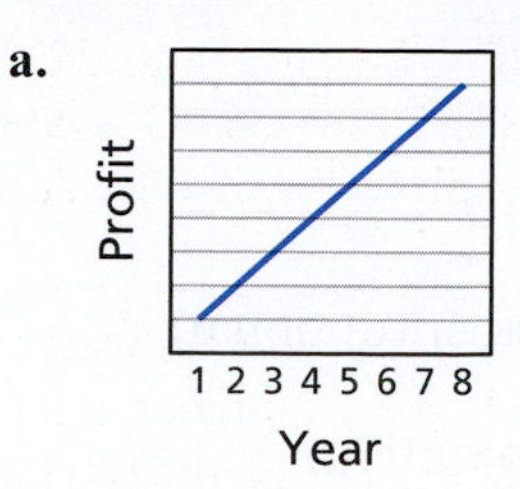

b.

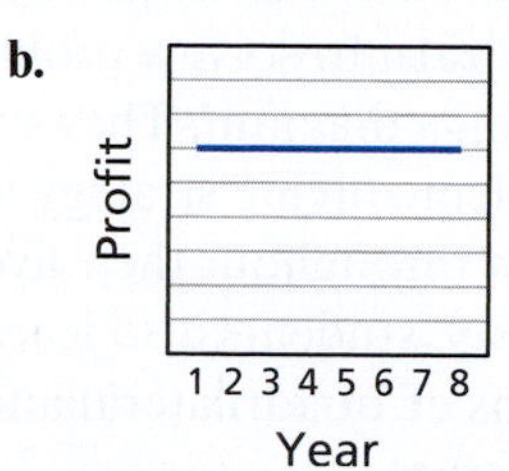

c.

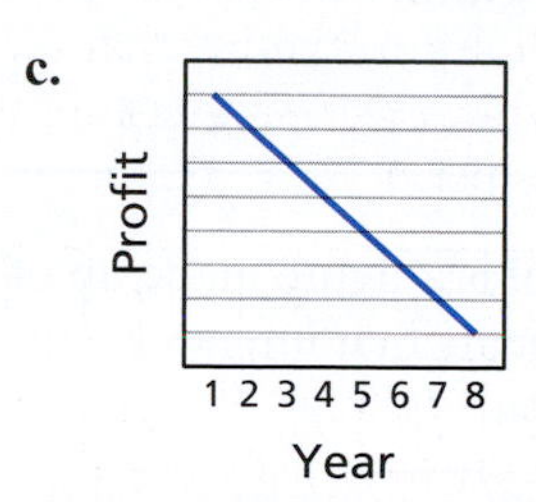

35. **Using Two Graphs** Use the graphs below to compare the heights of the players on each team by position.

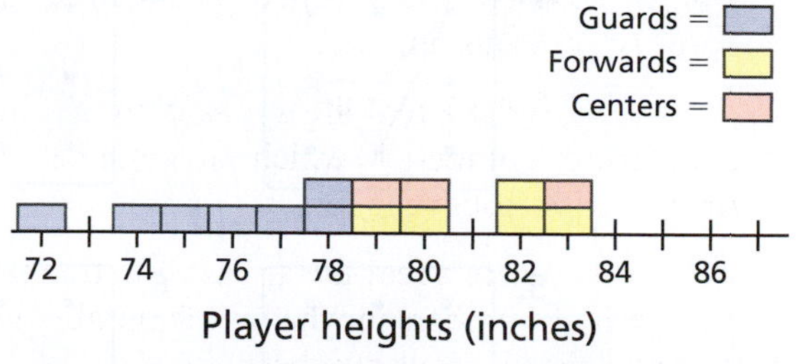

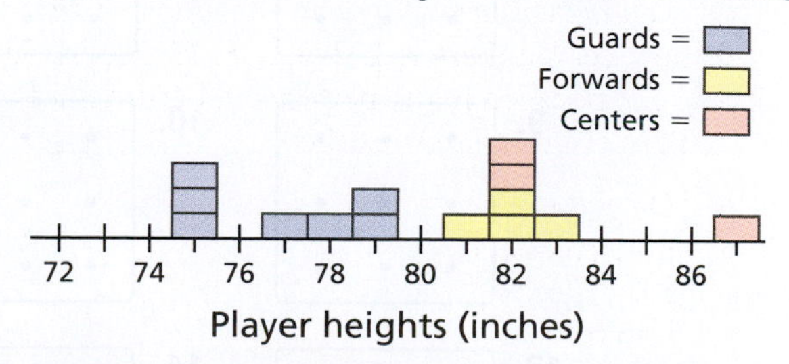

36. **Comparing Two Graphs** Use the graphs above to compare the heights of the San Antonio Spurs players to the heights of the Oklahoma City Thunder players.

Activity: Drawing Diagrams

Materials:

- Pencil
- Dot paper (optional)

Learning Objective:

Students learn to solve a problem by drawing a diagram. They will use this problem-solving strategy again and again throughout their lives. In this activity, students also learn the definitions of quadrilateral and parallelogram.

Name ______________________________

A **quadrilateral** is a figure made up of four line segments that intersect only at their endpoints.

A **parallelogram** is a quadrilateral that has two pairs of parallel sides.

Quadrilateral

Parallelogram

How many different quadrilaterals can you draw in a 3-by-3 grid of dot paper? Of these, how many are parallelograms?

1. 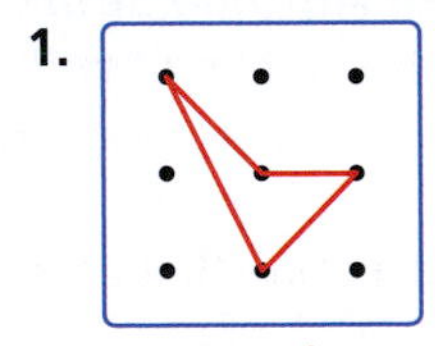
Sample

2. 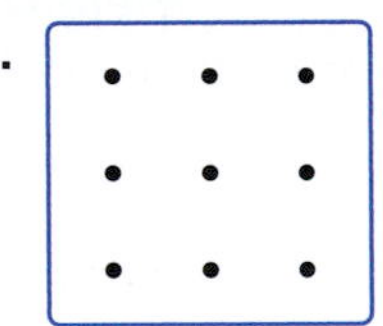

3. 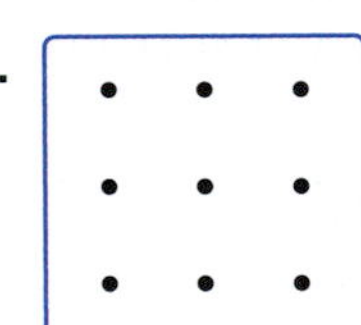

4. 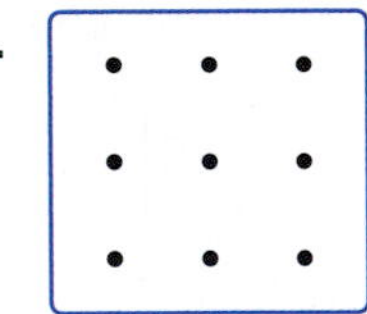

5. 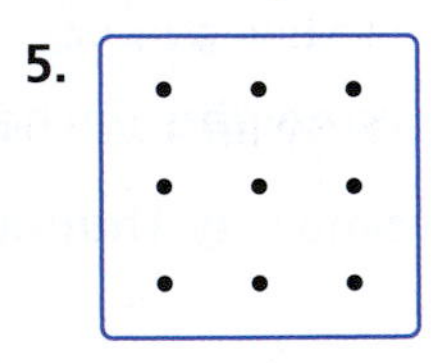

6. 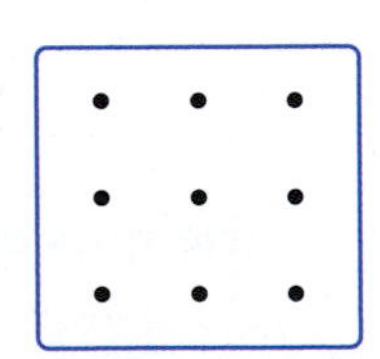

7. 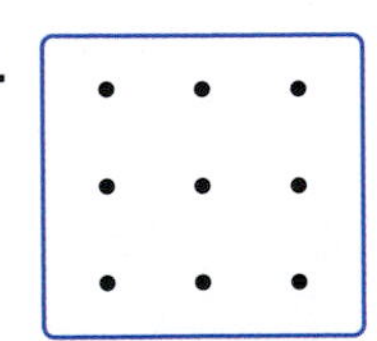

8. 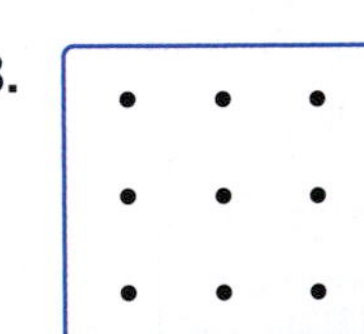

9. 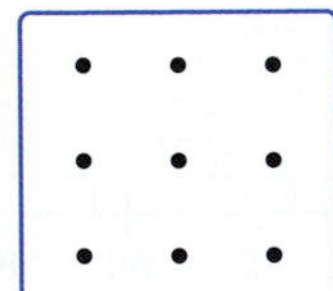

10. 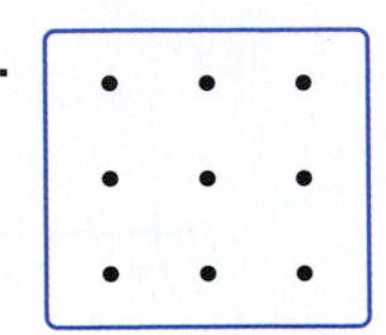

11. 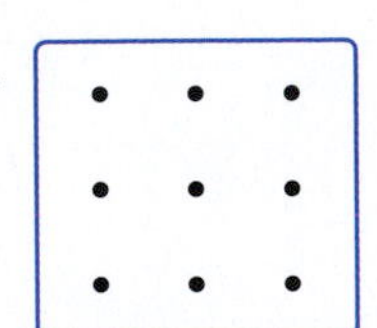

12. 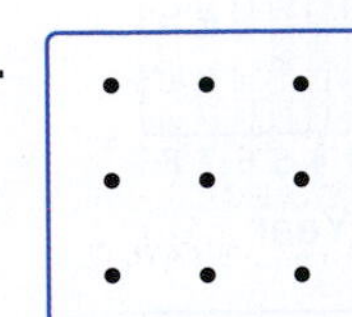

13. 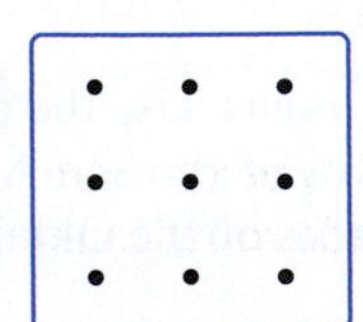

14. 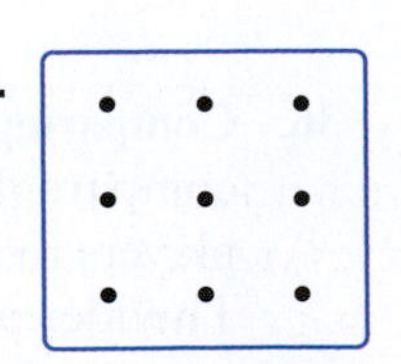

15. 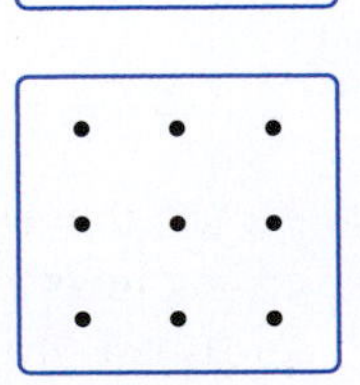

16.

Lisa F. Young/Shutterstock.com

1.2 More Problem-Solving Strategies

Standards

Grades K–2
Numbers and Operations in Base Ten
Students should use place value understanding and properties of operations to add and subtract.

Grades 3–5
Numbers and Operations in Base Ten
Students should use place value understanding and properties of operations to perform multi-digit arithmetic.

Grades 6–8
Expressions and Equations
Students should solve real-life mathematical problems using numerical and algebraic expressions and equations.

- Solve a problem by drawing a diagram.
- Solve a problem by working backwards.
- Solve a problem by solving a simpler problem.
- Solve a problem by using *Guess, Check, and Revise.*

Drawing a Diagram

Strategy: Draw a Diagram

Drawing a Diagram is a problem-solving strategy in which students make a visual representation of a problem. Drawing a diagram is a good starting point for solving many types of word problems. By representing units of measurement and other objects visually, students can begin to think about the problem mathematically.

EXAMPLE 1 **Strategy: Drawing a Diagram**

Can a parade cross all 7 bridges without crossing a bridge twice?

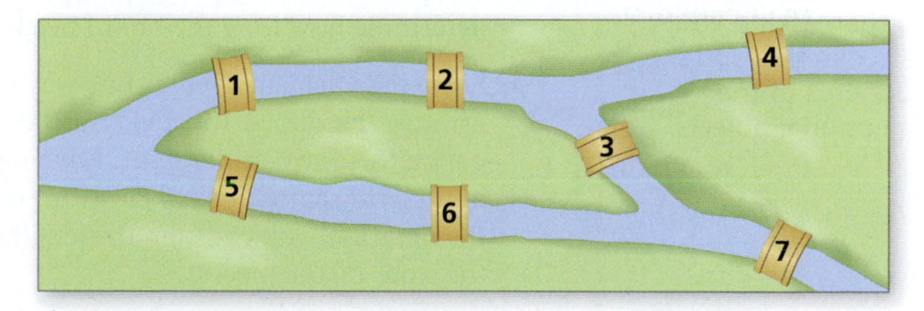

SOLUTION

1. **Understand the Problem** You must try to find a route that crosses each bridge only once. No doubling back is allowed.
2. **Make a Plan** Make copies of the map and try to find a route.
3. **Solve the Problem** Try drawing parade routes. Here are three trial routes.

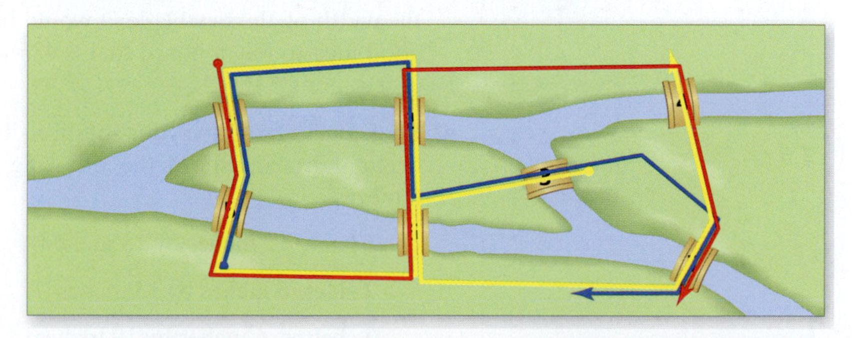

Notice that each parade route will either omit one or more of the bridges or cross one of the bridges twice. In fact, a parade route cannot cross all 7 bridges without crossing a bridge twice.

4. **Look Back** Try several different trial routes on your own.

Classroom Tip
The problem in Example 1 is called *The Seven Bridges of Königsberg*. It was solved by the well-known mathematician Leonhard Euler in 1736. He proved that the answer is "no." This famous problem shows your students that sometimes problems in mathematics have *no solution.*

EXAMPLE 2 Strategy: Drawing a Diagram

You are building a storage box for your video game collection and download the diagram shown at the right. You plan to cut the pieces from a 1-by-12 board, which is a board that is 1 inch thick and 12 inches wide. How long does the board need to be?

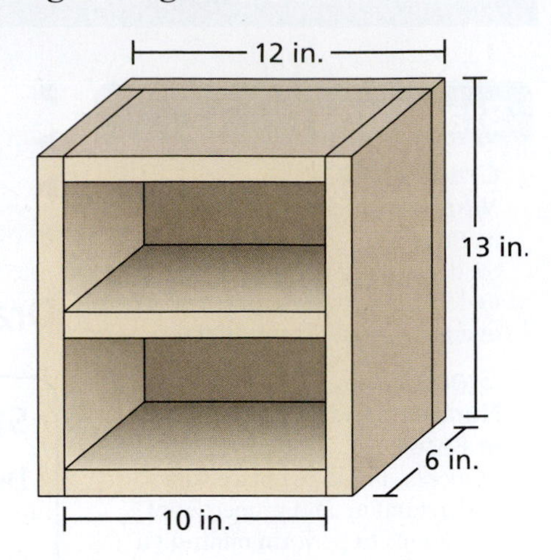

Classroom Tip

Some students have a difficult time visualizing three-dimensional drawings and translating them into two-dimensional diagrams. You can help your students by bringing blueprint drawings of three-dimensional objects to class, and then asking your students to describe and sketch the components of the three-dimensional objects.

SOLUTION

1. **Understand the Problem** You need to determine how long the 1-by-12 board needs to be to build the storage box.

2. **Make a Plan** Draw a diagram of a board that is 12 inches wide. On the diagram, draw and label each piece that is shown in the completed drawing of the storage box. Then determine how long the board needs to be.

3. **Solve the Problem** Here is one way to arrange the pieces. It helps to use grid paper, as shown at the right. Each unit in the grid represents 1 inch. After drawing the diagram, you can sum the lengths of the sides of the pieces.

$$\begin{array}{r} 13 \text{ inches} \\ 13 \text{ inches} \\ 10 \text{ inches} \\ +\ 6 \text{ inches} \\ \hline 42 \text{ inches} \end{array}$$

So, the board needs to be at least 42 inches long. Remember that in order to allow for the width of the saw blade, it would be a good idea to buy a board that is 48 inches (4 feet) long.

4. **Look Back** Examine your solution. The original drawing shows the following 6 pieces.

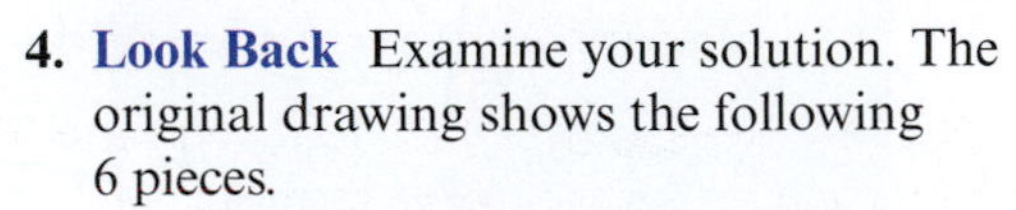

1 back 12 inches by 13 inches

2 sides 6 inches by 13 inches

3 shelves 6 inches by 10 inches

The pieces are shown in the diagram at the right.

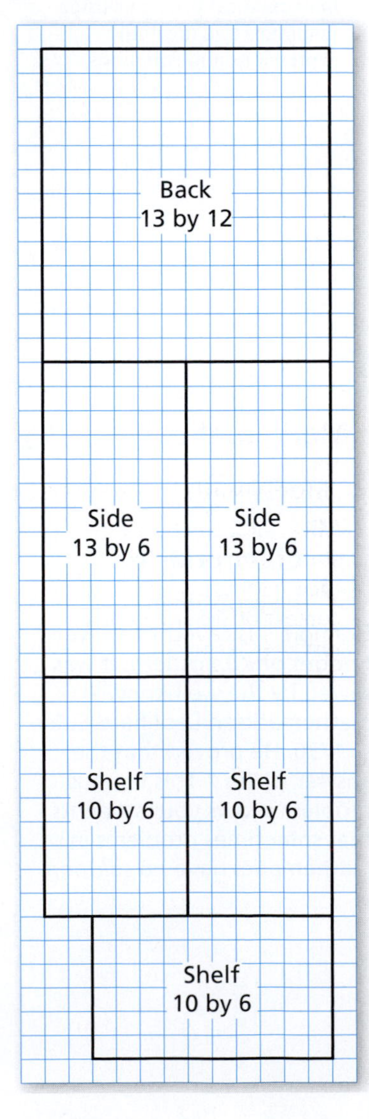

Working Backwards

Strategy: Work Backwards

Working Backwards is a problem-solving strategy involving a sequence of events in which students are given a final event and are asked to find an original event. Students begin at the end, with the final event, and work through a process in reverse order to establish what happened in the original situation. Other strategies can be combined with this strategy.

EXAMPLE 3 Strategy: Working Backwards

A video game console costs $325. Each individual game costs about $30. You have $500 to buy the console and some games. How many games can you buy?

SOLUTION

1. **Understand the Problem** You are given the prices of the video game console and the individual games. You need to find the numbers of possible games you can buy with the amount of money you have.
2. **Make a Plan** Decide to work backwards. Write the given information.
   - The video game console costs $325.
   - Each individual game costs $30.
   - You have $500 to spend on the console and games.
3. **Solve the Problem** Begin with the amount you have, $500. Then subtract the cost of the video game console to get $500 − $325 = $175. Make a table that shows the amount of money you have left after buying various numbers of video games.

Number of video games	Amount of money left
0	$175
1	$175 − $30 = $145
2	$145 − $30 = $115
3	$115 − $30 = $85
4	$85 − $30 = $55
5	$55 − $30 = $25 ← $25 left

Because you have only $25 left, you cannot buy another $30 video game. So, you can buy up to 5 video games.

4. **Look Back** Check your answer by making a table. Begin with the cost of the video game console. Then increase the number of video games until you reach or exceed $500.

Video games	0	1	2	3	4	5	6
Total cost	$325	$355	$385	$415	$445	$475	$505

Mathematical Practices

Model with Mathematics

MP4 Mathematically proficient students routinely interpret their mathematical results in the context of the situation and the results make sense.

Solving a Simpler Problem

Some problems may seem difficult because they contain large numbers or they appear to require many steps.

Strategy: Solve a Simpler Problem

Instead of solving a given problem, students can **Solve a Similar, but Simpler, Problem.** Look for easier numbers, patterns, and relationships. Then use previously learned concepts to solve the original problem.

EXAMPLE 4 Strategy: Solving a Simpler Problem

You are teaching a class of 20 students. Each day, you want a different pair of students to present the solution of a homework problem. There are 150 days in which students will present problems. Are there enough days for you to use all possible pairs of students?

SOLUTION

1. **Understand the Problem** You need to find how many different pairs of students you can choose from 20 students.

2. **Make a Plan** Listing all possible pairs seems cumbersome. So, you decide to solve a similar, but simpler, problem and look for a pattern.

3. **Solve the Problem** You cannot form any pairs with only 1 student. So, begin with 2 students. Then increase to 3 students. Then 4 students, and so on.

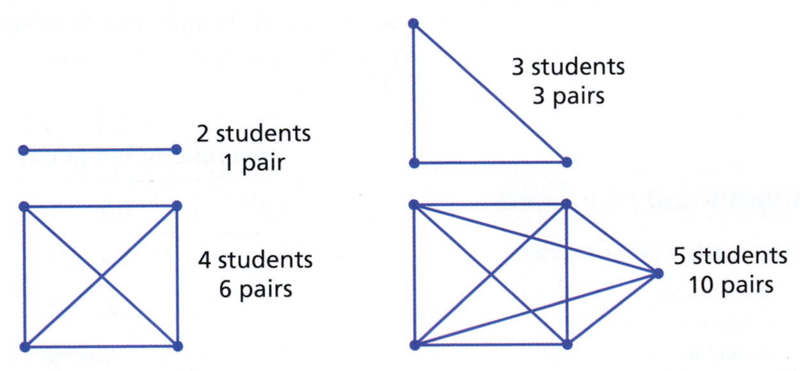

The following pattern appears.

1 student	0 pairs	
2 students	1 pair	$+1$
3 students	$1 + 2 = 3$ pairs	$+2$
4 students	$1 + 2 + 3 = 6$ pairs	$+3$
5 students	$1 + 2 + 3 + 4 = 10$ pairs	$+4$

By continuing this pattern, you can conclude that with 20 students you have

20 students $1 + 2 + 3 + 4 + \cdots + 18 + 19 = 190$ pairs.

You only have 150 days in which students will present homework problems. Because you plan to use a different pair of students each day, there will not be enough days to use all possible pairs of students.

4. **Look Back** You can use a calculator to verify that your answer is correct.

Classroom Tip

There is a wonderful story about the famous mathematician Carl Friedrich Gauss (1777–1855). In school, at the age of 10, Carl's teacher asked each student to add the numbers from 1 through 100. In a flash, Carl came up with 5050. Young Carl had noticed a pattern.

$$\left.\begin{array}{r} 1 + 100 = 101 \\ 2 + 99 = 101 \\ 3 + 98 = 101 \\ \vdots \\ 50 + 51 = 101 \end{array}\right\} \frac{100}{2} \cdot 101 = 5050$$

This is the same pattern that occurs in Example 4.

$$1 + 2 + 3 + 4 + \cdots + 19 = \frac{19}{2} \cdot 20 = 190$$

Using Guess, Check, and Revise

Strategy: Guess, Check, and Revise

Guess, Check, and Revise is a problem-solving strategy in which students begin by making a reasonable guess about the solution. After checking the guess in the statement of the problem, students adjust the guess by making a new, refined guess. This process continues until the correct answer is found.

EXAMPLE 5 Strategy: Guess, Check, and Revise

The floor of a living room is square and has an area of 450 square feet. What are the dimensions of the floor to the nearest tenth of a foot?

s ft Area = 450 ft^2
s ft

SOLUTION

1. **Understand the Problem** You are given that the floor is square and its area is 450 square feet. Because the floor is square, you know that each side has the same length. You need to find the length of each side s.

2. **Make a Plan** You can use algebra to solve the equation $s^2 = 450$. However, if you are teaching at a grade level in which square roots and square root equations have not been introduced, use *Guess, Check, and Revise* to find a number whose square is 450.

3. **Solve the Problem** A reasonable first guess is $s = 20$.

Guess	Check	Revise
$s = 20$	$s^2 = (20)^2 = 400$	Because $400 < 450$, make guess greater.
$s = 21$	$s^2 = (21)^2 = 441$	Because $441 < 450$, make guess greater.
$s = 22$	$s^2 = (22)^2 = 484$	Because $484 > 450$, make guess lesser.
$s = 21.5$	$s^2 = (21.5)^2 = 462.25$	Because $462.25 > 450$, make guess lesser.
$s = 21.1$	$s^2 = (21.1)^2 = 445.21$	Because $445.21 < 450$, make guess greater.
$s = 21.2$	$s^2 = (21.2)^2 = 449.44$	Because $449.44 <450$, make guess greater.
$s = 21.3$	$s^2 = (21.3)^2 = 453.69$	Stop, because 449.44 is closer to 450 than 453.69.

So, the length of each side of the floor is about 21.2 feet.

4. **Look Back** Look at the differences between the values of the checks and the area of the living room. Determine which value of s gives an area that is the closest to 450 square feet.

$450 - 400 = 50$ $\qquad 450 - 441 = 9$

$484 - 450 = 34$ $\qquad 462.25 - 450 = 12.25$

$450 - 445.21 = 4.29$ $\qquad 450 - 449.44 = 0.56$

$453.69 - 450 = 3.69$

The least difference is 0.56 between 450 and 449.44. So, you know that your answer of 21.2 feet is reasonable.

Classroom Tip
In Example 5, notice that the final answer to the question is written as a complete sentence, including the units of measure: "The length of each side of the room is about 21.2 feet." Encourage your students to develop the habit of writing out solutions as complete sentences. Writing the answer as simply 21.2 does not sufficiently communicate the solution to the problem.

1.2 Exercises

Free step-by-step solutions to odd-numbered exercises at *MathematicalPractices.com.*

1. **Writing a Solution Key** Write a solution key for the activity on page 12. Describe a rubric for grading a student's work.

2. **Grading the Activity** In the activity on page 12, a student gave the answers below. Each question is worth 10 points. Assign a grade to each answer. For those that are incorrect, why do you think the student erred?

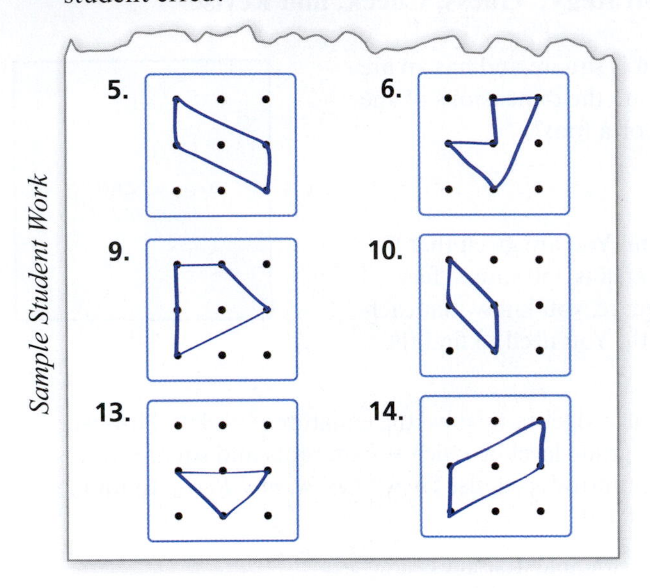

3. **Strategy: Guess, Check, and Revise** Use the symbols +, −, ×, and ÷ to make each statement true. (A symbol may be used more than once.)

   a. $4 \square 3 \square 2 = 5$

   b. $5 \square 4 \square 3 = 12$

   c. $2 \square 5 \square 4 = 6$

4. **Strategy: Guess, Check, and Revise** Use the symbols +, −, ×, and ÷ to make each statement true. (A symbol may be used more than once.)

   a. $2 \square 3 \square 4 = 2$

   b. $3 \square 4 \square 5 = 17$

   c. $4 \square 5 \square 6 = 3$

5. **Strategy: Drawing a Diagram** Using four straight cuts, can the dessert be cut into 8 equal pieces?

   a.

   b.

bonchan/Shutterstock.com; Treb999 | Dreamstime.com

6. **Strategy: Drawing a Diagram** Using five straight cuts, can the pizza be cut into 10 equal pieces?

   a.

   b.

7. **Strategy: Drawing a Diagram** Passengers enter an elevator of a building. The elevator moves up 6 floors, stops, moves down 5 floors, stops, and then moves up 7 floors and stops. The elevator is now 2 floors from the top floor. Find the number of floors in the building when the passengers enter the elevator on the given floor.

   a. Bottom floor **b.** Middle floor

8. **Strategy: Drawing a Diagram** Passengers enter an elevator of a building. The elevator moves down 8 floors, stops, moves up 3 floors, stops, and then moves down 12 floors and stops. The elevator is now 2 floors from the bottom floor. Find the number of floors in the building when the passengers enter the elevator on the given floor.

   a. Top floor **b.** Middle floor

images.etc/Shutterstock.com; iStockphoto.com/LauriPatterson
Natykach Nataliia/Shutterstock.com

9. **Solving a Puzzle** Place the numbers in the circles so that no two numbers joined by a line differ by 1.

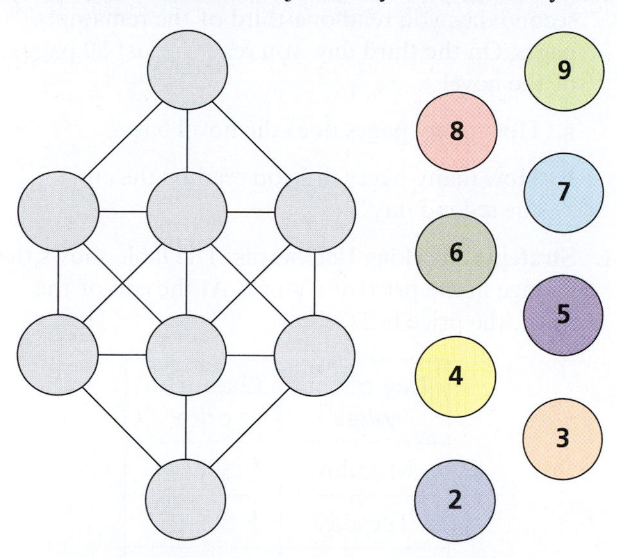

10. **Solving a Puzzle** Place the 4s and 5s in the empty circles so that no three numbers adjacent to each other have a sum that is divisible by three.

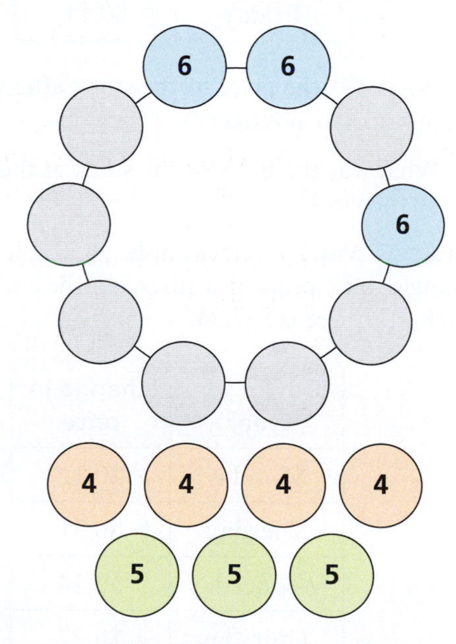

11. **Strategy: Drawing a Diagram** The dimensions of wrapping paper you need to wrap one gift box is given. Find the smallest dimensions of a single sheet of wrapping paper you need to wrap 10 gift boxes.

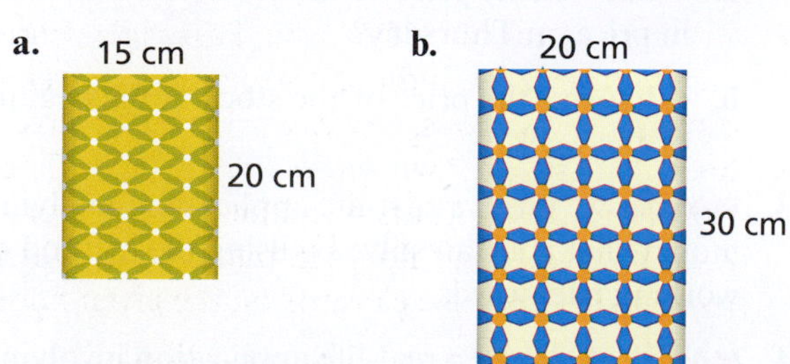

12. **Strategy: Drawing a Diagram** You are building a storage box for your movie collection. You plan to cut the pieces from a 1-by-10 board, which is a board that is 1 inch thick and 10 inches wide. How long does the board need to be?

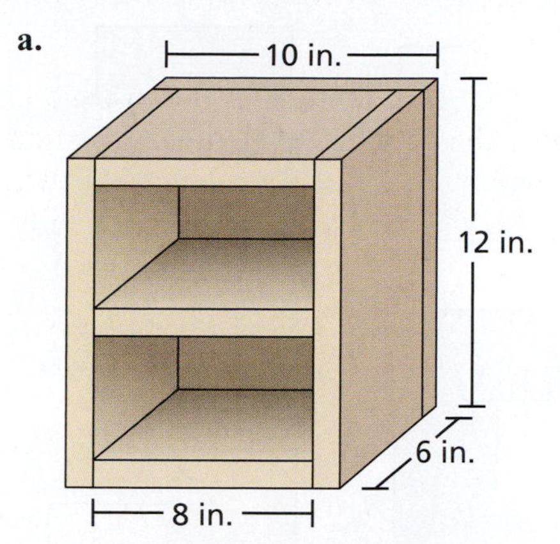

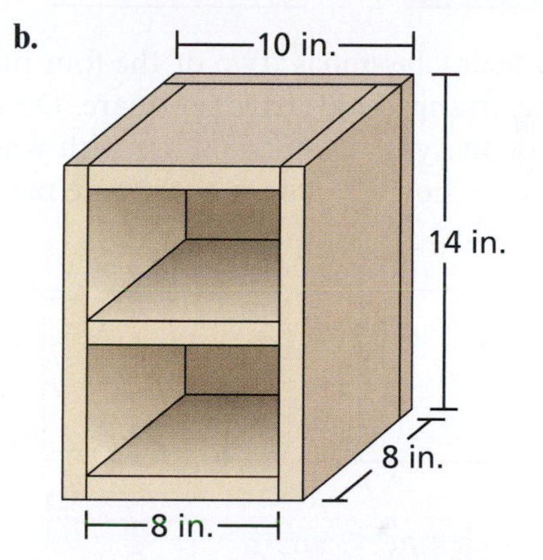

13. **Using a Geometric Model** Use the diagram of the cube.

a. How many squares are on the green side of the cube?

b. How many squares are on all sides of the cube?

14. **Using a Geometric Model** A large triangle is made up of smaller triangles. How many triangles are in the figure? Describe the method you used to find the answer.

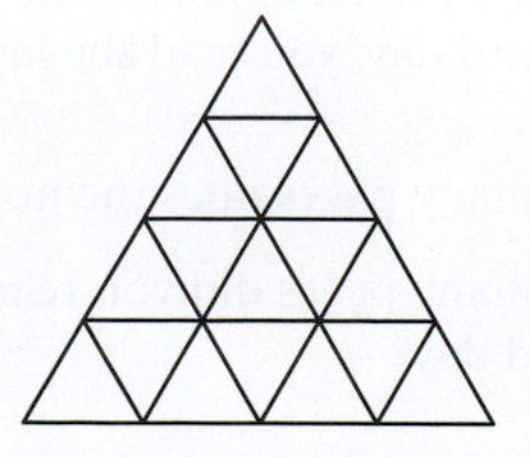

15. **Activity: In Your Classroom** Two of the four puzzle pieces can be arranged to form one square. Design a classroom activity that you could use with your students to show how to arrange the puzzle pieces to form the square.

16. **Activity: In Your Classroom** Two of the four puzzle pieces can be arranged to form one square. Design a classroom activity that you could use with your students to show how to arrange the puzzle pieces to form the square.

17. **Strategy: Working Backwards** At the end of the day, you have \$18. You remember taking \$15 out of the bank, spending \$5 on lunch, and spending \$21 on a new phone case. How much money did you have at the beginning of the day?

18. **Strategy: Working Backwards** You need to be at dance class at 3:00 P.M. It takes 25 minutes to drive to the studio. You want to watch a 1.5 hour movie before leaving for dance class. When should you start watching the movie so you can get to class on time?

19. **Strategy: Working Backwards** On the first day of vacation, you read one-quarter of a novel. On the second day, you read half of the remaining pages. On the third day, you read the last 114 pages of the novel.

a. How many pages does the novel have?

b. How many pages did you read by the end of the second day?

20. **Strategy: Working Backwards** On the first day of vacation, you read one-third of a novel. On the second day, you read one-third of the remaining pages. On the third day, you read the last 80 pages of the novel.

a. How many pages does the novel have?

b. How many pages did you read by the end of the second day?

21. **Strategy: Working Backwards** The table shows the change in the price of a stock. At the end of the week, the price is \$28.17.

Day of the week	Change in price
Monday	↑ \$0.21
Tuesday	↑ \$0.09
Wednesday	↓ \$0.33
Thursday	↓ \$0.02
Friday	↑ \$0.11

a. What was the price of the stock after the change in price on Thursday?

b. What was the price of the stock at the beginning of the week?

22. **Strategy: Working Backwards** The table shows the change in the price of a stock. At the end of the week, the price is \$32.24.

Day of the week	Change in price
Monday	↓ \$0.12
Tuesday	↑ \$0.31
Wednesday	↑ \$0.14
Thursday	↓ \$0.27
Friday	↓ \$0.06

a. What was the price of the stock after the change in price on Thursday?

b. What was the price of the stock at the beginning of the week?

23. **Writing** Write a real-life application involving money that you can solve by using the method of working backwards.

24. **Writing** Write a real-life application involving distance that you can solve by using the method of working backwards.

25. **Strategy: Solving a Simpler Problem** A diameter of a circle is a segment that passes through the center of the circle and has endpoints on the circle. How many sections are made when you draw 20 diameters of a circle? Describe the simpler problem you used to solve this problem.

26. **Strategy: Solving a Simpler Problem** A line is a geometrical object that extends to infinity in both directions. How many sections are made when you draw 30 lines through the center of a square? Describe the simpler problem you used to solve this problem.

27. **Strategy: Solving a Simpler Problem** A restaurant has 10 square tables that can seat one person on each side. The tables are joined together in a row to form one long table. How many people can be seated at the long table? Describe the simpler problem you used to solve this problem.

28. **Strategy: Solving a Simpler Problem** A restaurant has 10 rectangular tables that can seat one person on each short side and two people on each long side. The tables are joined on the shorter sides to form one long table. How many people can be seated at the long table? Describe the simpler problem you used to solve this problem.

29. **Grading Student Work** On a diagnostic test, one of your students does the following work. Explain what the student did wrong. Which topics would you encourage the student to review?

> What is the seventh entry in the list below?
> 2, 4, 6, 8, . . . 16

30. **Grading Student Work** On a diagnostic test, one of your students does the following work. Explain what the student did wrong. Which topics would you encourage the student to review?

> What is the eighth entry in the list below?
> 3, 6, 9, 12, . . . 21

31. **Strategy: Guess, Check, and Revise** The shape of a vegetable garden is square. What is the length of a side of the vegetable garden to the nearest tenth of a foot?

Area = 677 ft^2

32. **Strategy: Guess, Check, and Revise** The floor of a classroom is square. What is the length of a side of the floor to the nearest tenth of a foot?

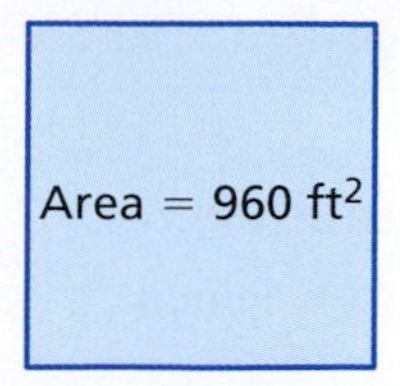

33. **Strategy: Guess, Check, and Revise** You are saving money for the smartphone shown below You already have \$70. You can save \$20 a week. How long will it take you to save enough money for the smartphone?

34. **Strategy: Guess, Check, and Revise** You are saving money for the smartphone shown below. You already have \$55. You can save \$25 a week. How long will it take you to save enough money for the smartphone?

35. **Problem Solving: In Your Classroom** Use the numbers 4, 5, 6, 7, 8, and 9 to find the largest possible sum. (Each number can be used only once.)

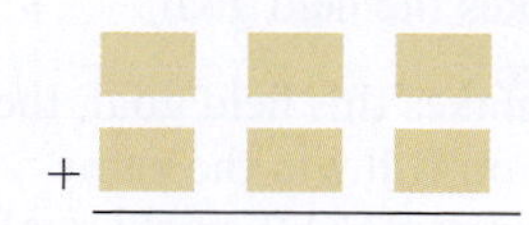

36. **Problem Solving: In Your Classroom** Use the numbers 4, 5, 6, 7, 8, and 9 to find the smallest possible positive difference. (Each number can be used only once.)

Janfilip/Shutterstock.com

Activity: Logical Reasoning

Materials:

- Pencil

Learning Objective:

Students look at a logical argument to decide whether the reasoning is valid or flawed. Students should only use the *given information*. They should not try to reason whether the conclusion is true or false in real life. The goal is to decide whether the conclusion follows logically from the given information.

Name ________________________________

Use the given information to decide whether the conclusion is *valid* or *flawed*. Circle the correct answer. Explain your reasoning.

	Given information	Conclusion		
1.	Everyone in Tyrone's class got an A on the math test. Diane is in Tyrone's class.	Therefore, Diane got an A on the math test.	Valid	Flawed
2.	Everyone in Tyrone's class got an A on the math test. James is not in Tyrone's class.	Therefore, James did not get an A on the math test.	Valid	Flawed
3.	All rectangles are parallelograms. All parallelograms are quadrilaterals.	Therefore, all rectangles are quadrilaterals.	Valid	Flawed
4.	All rectangles are parallelograms. All squares are quadrilaterals.	Therefore, all rectangles are quadrilaterals.	Valid	Flawed
5.	If Robert makes this field goal, then Robert's team will win the game. Robert makes the field goal.	Therefore, Robert's team wins the game.	Valid	Flawed
6.	If Robert makes this field goal, then Robert's team will win the game. Robert did not make the field goal.	Therefore, Robert's team did not win the game.	Valid	Flawed
7.	Some horses are paints. Toby is a paint.	Therefore, Toby is a horse.	Valid	Flawed
8.	All paints are horses. Toby is a paint.	Therefore, Toby is a horse.	Valid	Flawed

1.3 Reasoning and Logic

Standards

Grades K–2
Measurement and Data
Students should represent and interpret data.

Grades 3–5
Operations and Algebraic Thinking
Students should analyze patterns and relationships.

- Use statements, negations, and truth tables.
- Use compound statements and De Morgan's Laws.
- Use direct and indirect reasoning.

Statements, Negations, and Truth Tables

Logic is a tool used in mathematical thinking that allows a student to analyze an argument and determine whether or not the conclusion is valid. There are two basic types of logical reasoning: *deductive reasoning* and *inductive reasoning*. This section contains a brief introduction to deductive reasoning.

Statements and Negations

The building blocks of a logical argument are statements. A **statement** is a sentence that is either true (T) or false (F), but not both. The **negation** of a statement is a sentence that is false when the statement is true and is true when the statement is false.

Classroom Tip

The phrasing of the negation of a statement can be done in many ways. For instance, each of the following sentences is a correct negation of the statement "I am happy."

"I am not happy."

"It is not true that I am happy."

"I am unhappy."

EXAMPLE 1 **Statements and Negations**

	Statement	Negation
a.	Fluffy is a cat.	Fluffy is not a cat.
b.	$2 + 3 + 4 = 9$	$2 + 3 + 4 \neq 9$
c.	I am happy.	I am not happy.

The following are not statements.

	Non-statement	
d.	Are you going to town?	This is a question, not a statement.
e.	The UFO incident near Roswell, New Mexico	This is a phrase, not a statement.
f.	$x + 7 = 10$	There is no value assigned to x. So, you do not know whether this is true (T) or false (F).

EXAMPLE 2 **Writing the Negation of a Statement**

Write the negation of each statement.

a. I received an A on the mid-term exam.

b. Randy was elected class president.

c. A vanda orchid is a type of flower.

SOLUTION

a. I did not receive an A on the mid-term exam.

b. Randy was not elected class president.

c. It is not true that a vanda orchid is a type of flower.

Some statements contain **quantifiers** that give additional information about the existence of something. Typical quantifiers are *all*, *some*, *every*, *no*, and *there exists*. Euler diagrams are often used to illustrate statements containing quantifiers. An **Euler diagram** (usually circles or ovals) shows the relationship between two or more collections of objects.

Classroom Tip

When presenting Euler diagrams to students, remember that the goal is to help students investigate the truth of a statement. Adding illustrations to a diagram can be helpful, as shown below.

All canaries are birds.

EXAMPLE 3 Quantifiers and Euler Diagrams

Use a quantifier to write a statement that represents the information given by each Euler diagram.

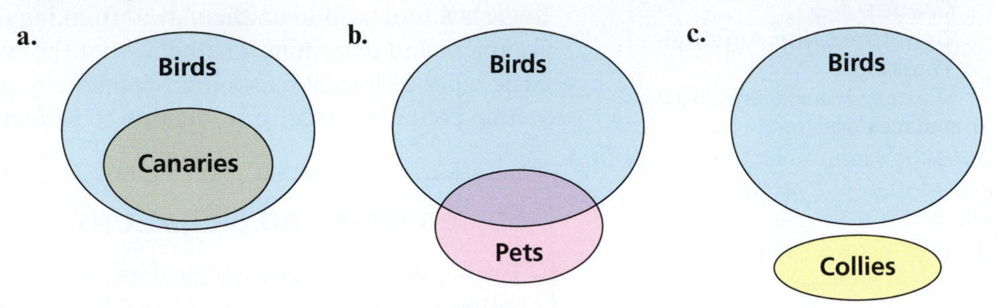

SOLUTION

a. Because the region representing canaries lies entirely within the region representing birds, you can reason that "*All* canaries are birds."

b. The region representing pets intersects the region representing birds. So, one possible statement for this situation is "*Some* pets are birds."

c. The region representing collies has no points in common with the region representing birds. So, one possible statement for this situation is "*No* collies are birds."

Other statements are also possible. For example, the Euler diagram in part (b) can also be interpreted as "*Some* birds are pets."

Statements can be represented symbolically by lowercase letters, such as p, q, and r. The negation of a statement p can be written as Not(p). A **truth table** shows all possible true-false patterns for statements and their components. Truth tables are commonly used in the study of logic.

EXAMPLE 4 Making a Simple Truth Table

Let p represent the statement "It is raining." Construct a truth table that shows the relationship between p and its negation, Not(p).

SOLUTION

One possible negation of p is "It is not raining." So, let Not(p) represent the statement "It is not raining." Construct a table with two columns and three rows. Label one column "p" and label the other column "Not(p)," as shown.

p	Not(p)
T	F
F	T

When the statement p is true, Not(p) is false.

When the statement p is false, Not(p) is true.

The truth table above shows that the statement p and its negation have opposite truth values. This is true for all possible cases.

Compound Statements and De Morgan's Laws

A **compound statement** involves the connection of two or more simpler statements using the connectors "*and*" or "*or*."

Conjunction: p **and** q — True if and only if both p and q are true.

Disjunction: p **or** q — False if and only if both p and q are false.

> **Classroom Tip**
>
> There are two common uses of "or" in everyday language: the exclusive "or" and the inclusive "or."
>
> **Exclusive** "I will have iced tea *or* I will have milk."
>
> Both parts of the statement cannot be true at the same time.
>
> **Inclusive** "I will put sugar in my iced tea or I will put lemon in my iced tea."
>
> This statement means I will either put sugar, lemon, or both sugar *and* lemon in my iced tea. It allows for both parts of the statement to be true. In logic, the inclusive "or" is used.

EXAMPLE 5 Making a Compound Truth Table

Let p represent the statement "I will do my chores" and let q represent the statement "I will go shopping." Construct a truth table for the compound statement (p or q).

SOLUTION

Construct a table with three columns labeled "p," "q," and "p or q." Notice that there are four combinations of true and false. So, there are four distinct possibilities for the truth value of (p or q), "I will do my chores or I will go shopping."

p	q	p or q	
T	T	T	I do my chores and I go shopping.
F	T	T	I do not do my chores but I go shopping.
T	F	T	I do my chores but I do not go shopping.
F	F	F	I do not do my chores and I do not go shopping.

Two statements are **logically equivalent** if and only if they have the same truth values in every possible situation. This is denoted by the symbol ≡.

> **Classroom Tip**
>
> The laws shown at the right are named after British logician Augustus De Morgan (1806–1871). When teaching De Morgan's Laws, try to stay away from just the symbolic representations. Instead, emphasize specific examples. For instance, the *Declaration of Independence* states that all people have certain rights, and "that among these are Life, Liberty, and the pursuit of Happiness." This statement is true only if people are given the right to all three things: life and liberty and the pursuit of happiness.

De Morgan's Laws

1. The negation of a conjunction of two statements is logically equivalent to the disjunction of the negations of the statements.

   Not(p and q) ≡ Not(p) or Not(q)

2. The negation of a disjunction of two statements is logically equivalent to the conjunction of the negations of the statements.

   Not(p or q) ≡ Not(p) and Not(q)

EXAMPLE 6 Verifying One of De Morgan's Laws

The truth table verifies the second of De Morgan's Laws.

p	q	p or q	Not(p or q)	Not(p)	Not(q)	Not(p) and Not(q)
T	T	T	F	F	F	F
F	T	T	F	T	F	F
T	F	T	F	F	T	F
F	F	F	T	T	T	T

The fourth and seventh columns have the same truth values. So, it follows that Not(p or q) ≡ Not(p) and Not(q).

Using Direct and Indirect Reasoning

Conditional Statements

A **conditional statement** (or implication) has the following form.

Algebra If p, then q. **Example** If it rains, then the grass will get wet.

(Algebra: p = Hypothesis, q = Conclusion. Example: "it rains" = Hypothesis, "then the grass will get wet" = Conclusion.)

A conditional statement has three related statements as shown below.

Statement If p, then q. **Converse** If q, then p.

Contrapositive If Not(q), then Not(p). **Inverse** If Not(p), then Not(q).

Every statement is logically equivalent to its contrapositive. Similarly, the converse and inverse of any statement are logically equivalent. The conjunction of a statement and its converse gives a **biconditional statement**. A biconditional statement has the following form.

Algebra p if and only if q.

Example You will win the lottery if and only if your number is drawn.

Classroom Tip

Symbols may be used to represent the negation of a statement, connectors, a conditional statement, and a biconditional statement.

Negation of p: $\sim p$

p and q: $p \wedge q$

p or q: $p \vee q$

If p, then q: $p \rightarrow q$

p if and only if q: $p \leftrightarrow q$

EXAMPLE 7 Making a Truth Table for a Conditional Statement

Your friend says "If you do this magic trick, then I will give you $1." Construct the truth table for this conditional statement.

SOLUTION

Let p represent the statement "you do this magic trick" and let q represent the statement "I will give you a dollar."

p	*q*	If *p*, then *q*.	
T	T	T	You do the magic trick; your friend gives you $1.
F	T	T	You do not do the magic trick; your friend gives you $1.
T	F	F	You do the magic trick; your friend does not give you $1.
F	F	T	You do not do the magic trick; your friend does not give you $1.

EXAMPLE 8 Verifying a Contrapositive Statement

The truth table shows that the statement in Example 7 is logically equivalent to its contrapositive "If I do not give you $1, then you do not do this magic trick."

p	*q*	If *p*, then *q*.	Not(*q*)	Not(*p*)	If Not(*q*), then Not(*p*).
T	T	T	F	F	T
F	T	T	F	T	T
T	F	F	T	F	F
F	F	T	T	T	T

The third and sixth columns have the same truth values. So, it follows that the statement and its contrapositive are logically equivalent.

Direct and Indirect Reasoning

1. **Direct reasoning** has the following form.
   - If p, then q.
   - p is true.
   - Therefore, q.
2. **Indirect reasoning** has the following form.
   - If p, then q.
   - Not(q) is true.
   - Therefore, Not(p).

You can use the principles of direct and indirect reasoning to determine whether the conclusions in the activity on page 22 are valid or flawed. Example 9 shows you how this can be done.

EXAMPLE 9 Using Direct and Indirect Reasoning

Decide whether each conclusion is *valid* or *flawed.* If it is valid, state whether it shows direct reasoning or indirect reasoning.

	Given information	Conclusion
a.	If Robert makes this field goal, then Robert's team will win the game. Robert makes the field goal.	Therefore, Robert's team wins the game.
b.	If Robert makes this field goal, then Robert's team will win the game. Robert did not make the field goal.	Therefore, Robert's team did not win the game.
c.	If Robert makes this field goal, then Robert's team will win the game. Robert's team did not win.	Therefore, Robert did not make the field goal.

Mathematical Practices

Construct Viable Arguments and Critique the Reasoning of Others

MP3 Mathematically proficient students compare the effectiveness of two plausible arguments, distinguish correct logic or reasoning from that which is flawed and explain the flaw if one exists.

SOLUTION

a. The conclusion is valid. This is an example of direct reasoning.

- If Robert makes this field goal, Robert's team will win the game. — If p, then q.
- Robert makes the field goal. — p is true.
- Therefore, Robert's team wins the game. — Therefore, q.

b. This reasoning is faulty, so the conclusion is flawed.

- If Robert makes this field goal, then Robert's team will win the game. — If p, then q.
- Robert did not make the field goal. — p is not true.
- Therefore, Robert's team did not win the game. — Therefore, Not(q).

c. The conclusion is valid. This is an example of indirect reasoning.

- If Robert makes this field goal, then Robert's team will win the game. — If p, then q.
- Robert's team did not win. — q is not true.
- Therefore, Robert did not make the field goal. — Therefore, Not(p).

Classroom Tip

Thinking that the reasoning in Example 9(b) is valid is one of the most common errors in logic. To convince your students that this is faulty reasoning, point out that the field goal attempt might have occurred in the last second of the game. Suppose that Robert's team is already winning. Robert's team would win regardless of whether Robert makes the field goal or not.

1.3 Exercises

Free step-by-step solutions to odd-numbered exercises at *MathematicalPractices.com.*

1. **Writing a Solution Key** Write a solution key for the activity on page 22. Describe a rubric for grading a student's work.

2. **Grading the Activity** In the activity on page 22, a student gave the answers below. Each question is worth 10 points. Assign a grade to each answer. For those that are incorrect, why do you think the student erred?

Sample Student Work

Given information	Conclusion
1. Everyone in Tyrone's class got an A on the math test. Diane is in Tyrone's class.	Therefore, Diane got an A on the math test.

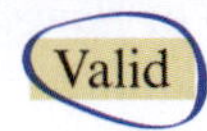

Flawed

Everyone in Tyrone's class got an A.

Given information	Conclusion
2. Everyone in Tyrone's class got an A on the math test. James is not in Tyrone's class.	Therefore, James did not get an A on the math test.

Flawed

Because James in not in Tyrone's class, you can assume that James did not get an A.

Given information	Conclusion
7. Some horses are paints. Toby is a paint.	Therefore, Toby is a horse.

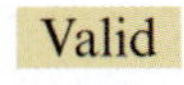

It is not stated that all paints are horses.

3. **Identifying Statements** Identify each part as a statement or non-statement. If it is a statement, write the negation. If it is a non-statement, explain why.
   a. Traveling at a high rate of speed
   b. George Washington was a United States president.
   c. $3 + 5 = 8$
   d. $5x = 10$

4. **Identifying Statements** Identify each part as a statement or non-statement. If it is a statement, write the negation. If it is a non-statement, explain why.
   a. A marine biologist studies ocean life.
   b. Choose running shoes with good support.
   c. $x > 3$
   d. $6 - 2 = 4$

5. **Validity of Statements** Tell whether each statement is *true* or *false*.
   a. A mouse is an animal.
   b. A canary is a reptile.
   c. $3 + 2 = 6$
   d. $5 - 1 < 10$

6. **Validity of Statements** Tell whether each statement is *true* or *false*.
   a. Five billion people live in Canada.
   b. People sometimes buy used cars.
   c. $12 - 3 = 9$
   d. 5 is an even number.

7. **Writing the Negations of Statements** Write the negation of each statement.
   a. Patience is important in teaching.
   b. The gold medalist had her fastest time ever.
   c. The shoes are uncomfortable.
   d. $4 + 1 = 5$

8. **Writing the Negations of Statements** Write the negation of each statement.
   a. The instructions are helpful.
   b. Beans are nutritious.
   c. The results were inconceivable.
   d. $3.5 \div 3 > 1$

9. **Quantifiers and Euler Diagrams** Use a quantifier to write a statement that represents the information given by the Euler diagram for each specified relationship.

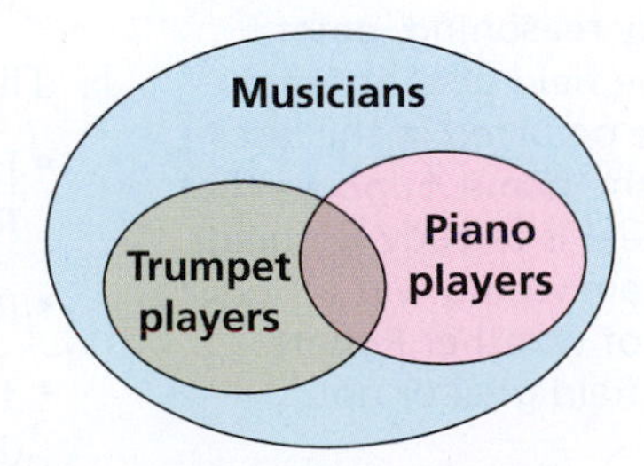

   a. Piano players that are musicians
   b. Trumpet players that are piano players
   c. Musicians that are trumpet players
   d. Trumpet players that are not musicians

10. Quantifiers and Euler Diagrams Use a quantifier to write a statement that represents the information given by the Euler diagram for each specified relationship.

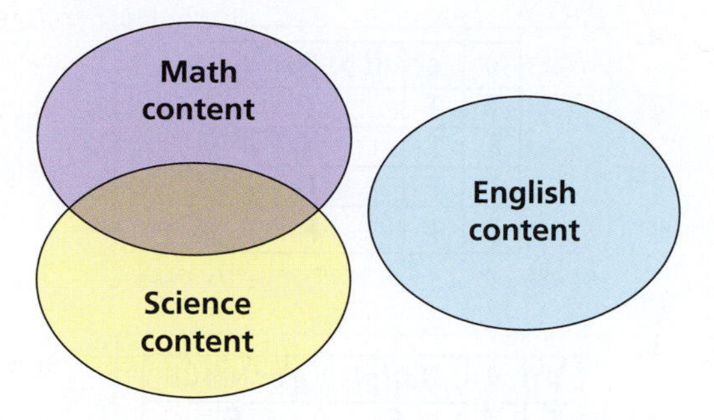

a. Math content that is also science content

b. Science content that is not math content

c. English content that is math content

d. English content that is not science content

11. Making a Truth Table Let p represent the statement "The Reds win." Complete the truth table to show the truth values for Not(p). Write the statement for Not(p).

p	Not(*p*)
T	
F	

12. Making a Truth Table Let p represent the statement "The light is on." Complete the truth table to show the truth values for Not(p) and Not[Not(p)]. Write the statements for Not(p) and Not[Not(p)]. Are p and Not[Not(p)] logically equivalent? Explain.

p	Not(*p*)	Not[Not(*p*)]
T		
F		

13. Making a Compound Truth Table Complete the truth table to show the truth values for the conjunction (p and q).

p	*q*	*p* and *q*
T	T	
F	T	
T	F	
F	F	

14. Making a Compound Truth Table Complete the truth table to show the truth values for the disjunction [p or Not(q)].

p	*q*	Not(*q*)	*p* or Not(*q*)
T	T		
F	T		
T	F		
F	F		

15. Verifying De Morgan's Laws Construct a truth table to verify the first of De Morgan's Laws.

16. Verifying Equivalence Construct a truth table to verify the equivalence statement below.

p and $q \equiv$ Not[Not(p) or Not(q)]

17. Verifying Equivalence Construct a truth table to verify that the inverse and the converse of a conditional statement are logically equivalent.

18. Verifying Equivalence Construct a truth table to verify that the biconditional statement

p if and only if q

is logically equivalent to the statement

(p and q) or [Not(p) and Not(q)].

19. Using Direct and Indirect Reasoning Decide whether the conclusion is *valid* or *flawed*. If it is valid, state whether it shows direct or indirect reasoning.

	Given information	Conclusion
a.	If Cassie goes to the party, then she will meet Chris. Cassie does not go to the party.	Therefore, Cassie will not meet Chris.
b.	If Cassie goes to the party, then she will meet Chris. Cassie goes to the party.	Therefore, Cassie will meet Chris.

20. Using Direct and Indirect Reasoning Decide whether the conclusion is *valid* or *flawed*. If it is valid, state whether it shows direct or indirect reasoning.

	Given information	Conclusion
a.	If the puppies are well-fed, then they will grow fast. The puppies grow fast.	Therefore, the puppies are well-fed.
b.	If the puppies are well-fed, then they will grow fast. The puppies do not grow fast.	Therefore, the puppies are not well-fed.

21. Quantifiers and Euler Diagrams The Euler diagram represents information about candidates for the office of President of the United States in past elections.

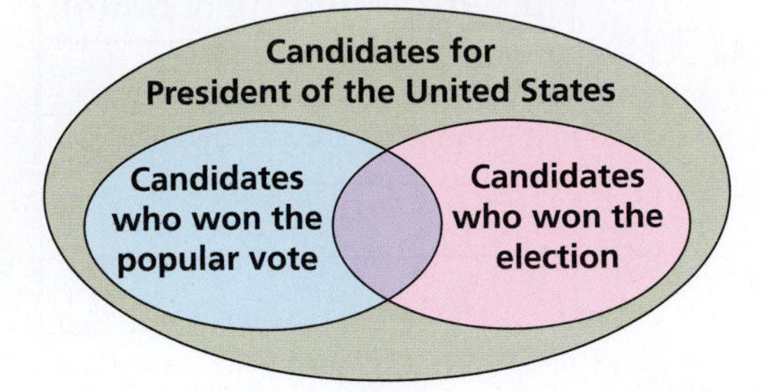

a. Use a quantifier to complete the statement.

▭ candidates won both the popular vote and the election.

b. Use a quantifier to complete the statement.

▭ people who won the election were candidates.

c. Use quantifiers to write two statements about candidates who did not win the popular vote.

22. Quantifiers and Euler Diagrams The Euler diagram represents information about United States presidents, vice presidents, and first ladies.

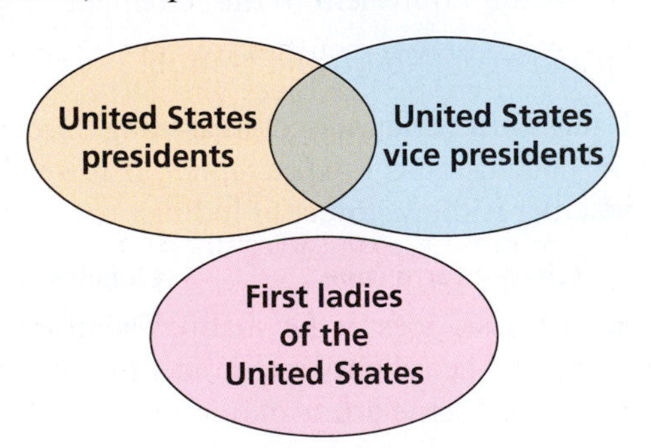

a. Use a quantifier to complete the statement.

▭ presidents were vice presidents.

b. Use a quantifier to complete the statement.

▭ first ladies were presidents.

c. Use quantifiers to write two statements about vice presidents.

23. Writing Write a real-life application that involves a conjunction of two simple statements p and q.

24. Writing Write a real-life application that involves a disjunction of two simple statements p and q.

25. Writing Write a real-life application that involves a conditional statement.

26. Writing Write a real-life application that involves a biconditional statement.

27. Grading Student Work On a diagnostic test, one of your students does the following work. Explain what the student did wrong. Which topics would you encourage the student to review?

a.

p	q	If p, then q.
T	T	T
F	T	F
T	F	F
F	F	F

b.

p	q	Not(p)	q or Not(p)
T	T	F	F
F	T	T	T
T	F	F	F
F	F	T	F

28. Grading Student Work On a diagnostic test, one of your students does the following work. Explain what the student did wrong. Which topics would you encourage the student to review?

a.

p	q	If q, then p.
T	T	T
F	T	T
T	F	F
F	F	T

b.

p	q	Not(p)	q and Not(p)
T	T	F	T
F	T	T	T
T	F	F	F
F	F	T	T

29. Writing Statements Write two statements using quantifiers that represent the information given by each Euler diagram.

a.

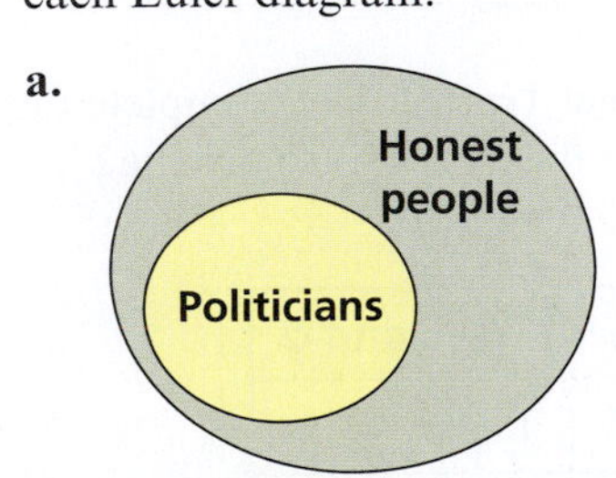

b.

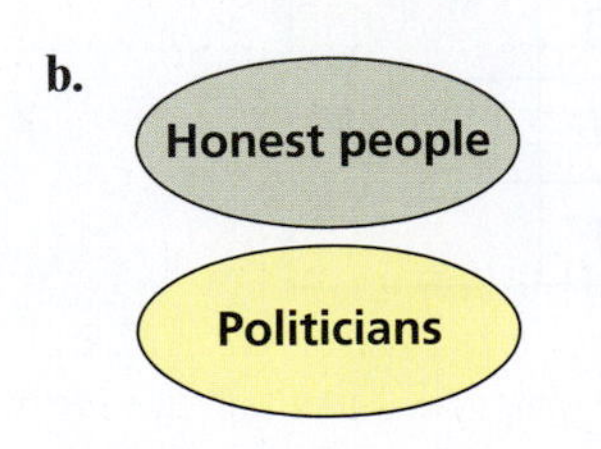

30. Writing Statements Write two statements using quantifiers that represent the information given by each Euler diagram.

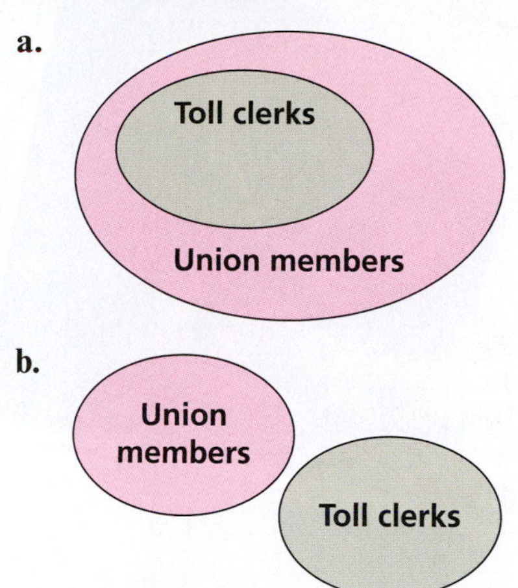

31. Using De Morgan's Laws The Euler diagram represents information about amphibians.

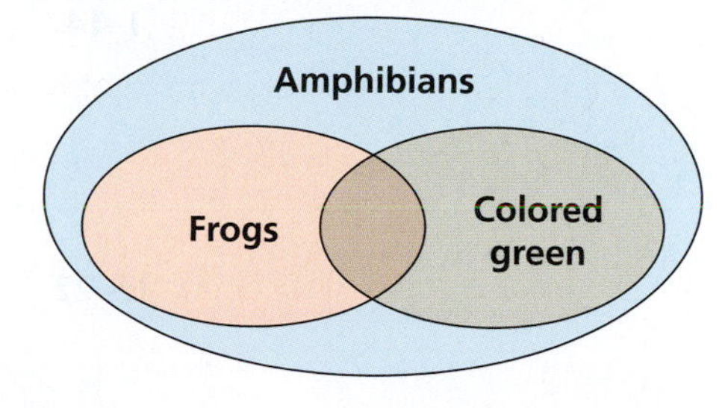

a. Use the first of De Morgan's Laws to write the negation of the compound statement "The amphibian is a frog and the amphibian is green" in the form of a disjunction.

b. Use the second of De Morgan's Laws to write the negation of the compound statement "The amphibian is a frog or the amphibian is green" in the form of a conjunction.

32. Using De Morgan's Laws The Euler diagram represents information about pizza toppings.

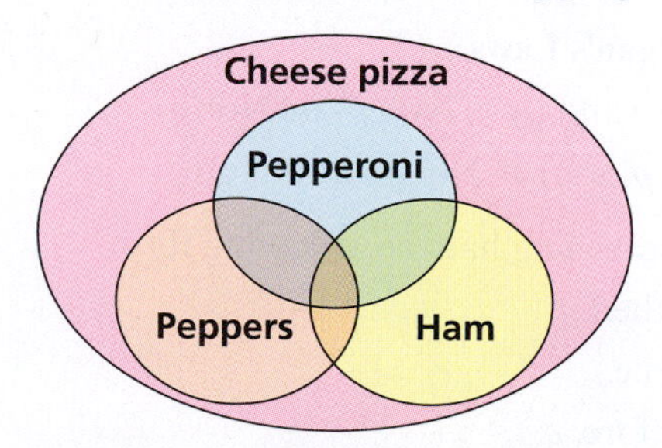

a. Use the second of De Morgan's Laws to write the negation of the compound statement "The pizza has pepperoni or ham" in the form of a conjunction.

b. Use the first of De Morgan's Laws to write the negation of the compound statement "The pizza has peppers, pepperoni, and ham" in the form of a disjunction.

33. Direct or Indirect Reasoning: In Your Classroom If a public policy treats people unequally, then the policy is considered to be unconstitutional. In the court case *United States v. Virginia*, the Supreme Court decided that Virginia Military Institute's male-only admission policy treated women unfairly.

a. What conclusion can you make about the constitutionality of the policy?

b. Did you use direct or indirect reasoning to make the conclusion in part (a)?

c. Describe a project that you could use in your classroom that involves direct or indirect reasoning and the constitution.

34. Direct or Indirect Reasoning: In Your Classroom If sufficient consent is given by residents of a house, then it is constitutionally legal for police to search the house without a warrant. In the court case *Georgia v. Randolph*, the Supreme Court ruled it unconstitutional for police to search a house without a warrant if one resident consents but another resident objects.

a. What can you conclude about the standards of sufficient consent for a search without a warrant?

b. Did you use direct or indirect reasoning to make the conclusion in part (a)?

c. Describe a project that you could use in your classroom that involves direct or indirect reasoning and the constitution.

35. Problem Solving If a person who is not a U.S. citizen has been a permanent resident (green card holder) for at least 3 years, has been the spouse of the same U.S. citizen for that period of time, and meets other eligibility requirements, then the person may qualify for naturalization under Section 319(a) of the Immigration and Nationality Act.

a. Can a person who just arrived in the United States qualify for naturalization under these conditions? Explain.

b. An unmarried immigrant is granted naturalization. Were the conditions above used? Explain.

36. Problem Solving There are two levels of teaching certification in Pennsylvania. A Level I certificate is valid for 6 years of service. After 6 years of service, you can remain teaching if and only if you upgrade to a Level II certificate. If you teach in Pennsylvania on a Level I certificate for 3 to 6 years and earn 24 post-baccalaureate credits, then you may be eligible to upgrade to a Level II certificate.

a. Can a teacher possess a valid Level I certificate after 8 years of teaching in Pennsylvania? Explain.

b. How many years of in-state teaching are required to obtain a Level II certificate? Explain.

Chapter Summary

Chapter Vocabulary

Looking for a Pattern *(p. 4)*
Reading or Making a Table or List *(p. 5)*
Reading or Making a Graph *(p. 6)*
Drawing a Diagram *(p. 13)*
Working Backwards *(p. 15)*
Solve a Similar, but Simpler, Problem *(p. 16)*
Guess, Check, and Revise *(p. 17)*
logic *(p. 23)*
statement *(p. 23)*
negation *(p. 23)*
quantifiers *(p. 24)*
Euler diagram *(p. 24)*
truth table *(p. 24)*
compound statement *(p. 25)*
logically equivalent *(p. 25)*
conditional statement *(p. 26)*
converse *(p. 26)*
contrapositive *(p. 26)*
inverse *(p. 26)*
biconditional statement *(p. 26)*
direct reasoning *(p. 27)*
indirect reasoning *(p. 27)*

Chapter Learning Objectives — Review Exercises

1.1 Problem-Solving Strategies *(page 3)* — 1–14

- Solve a problem by looking for a pattern.
- Solve a problem by reading or making a table and list.
- Solve a problem by reading or making a graph.

1.2 More Problem-Solving Strategies *(page 13)* — 15–22

- Solve a problem by drawing a diagram.
- Solve a problem by working backwards.
- Solve a problem by solving a simpler problem.
- Solve a problem by using *Guess, Check, and Revise.*

1.3 Reasoning and Logic *(page 23)* — 23–40

- Use statements, negations, and truth tables.
- Use compound statements and De Morgan's Laws.
- Use direct and indirect reasoning.

Important Concepts and Formulas

Problem-Solving Strategies

1. Looking for a Pattern
2. Reading or Making a Table or List
3. Reading or Making a Graph
4. Drawing a Diagram
5. Working Backwards
6. Solve a Similar, but Simpler, Problem
7. Guess, Check, and Revise

A **statement** is a sentence that is either true or false, but not both. The **negation** of a statement is a sentence that is false when the statement is true and is true when the statement is false.

De Morgan's Laws

1. Not(p and q) $\equiv$ Not(p) or Not(q)
2. Not(p or q) $\equiv$ Not(p) and Not(q)

Direct reasoning has the following form:

- If p, then q.
- p is true.
- Therefore, q.

Indirect reasoning has the following form:

- If p, then q.
- Not(q) is true.
- Therefore, Not(p).

dotshock/Shutterstock.com

Review Exercises

Free step-by-step solutions to odd-numbered exercises at MathematicalPractices.com.

1.1 Problem-Solving Strategies

Strategy: Looking for a Pattern Describe the pattern. Then write the next three numbers.

1. 12, 22, 32, 42, ___, ___, ___
2. 2, 4, 8, 16, ___, ___, ___
3. 76, 69, 62, 55, ___, ___, ___
4. 64, 32, 16, 8, ___, ___, ___
5. 3, 9, 27, 81, ___, ___, ___
6. 99, 90, 81, 72, ___, ___, ___

Strategy: Looking for a Pattern Each row of the triangle begins and ends with the same number. Every other number can be determined from the numbers in the row above. Describe the pattern. Use the pattern to find Row 6 of the triangle.

7.

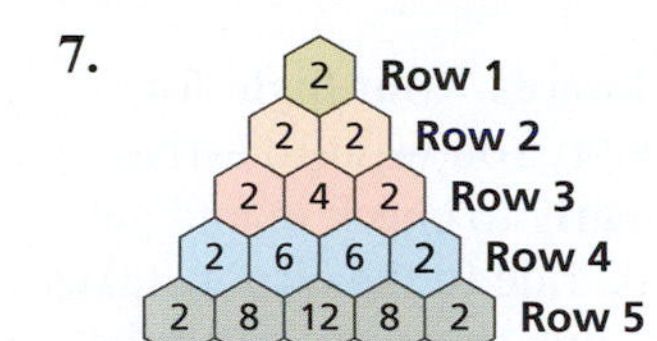

8.

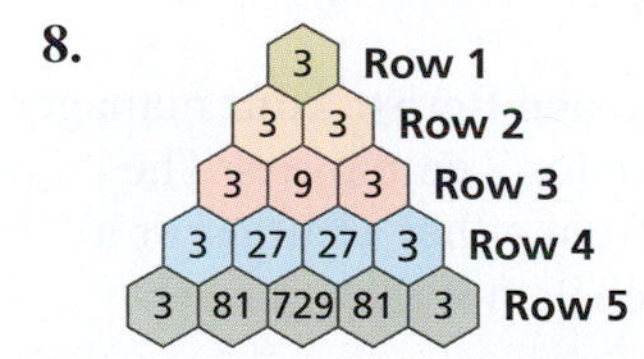

Strategy: Reading a Graph Describe the pattern(s) shown in the graph.

9.

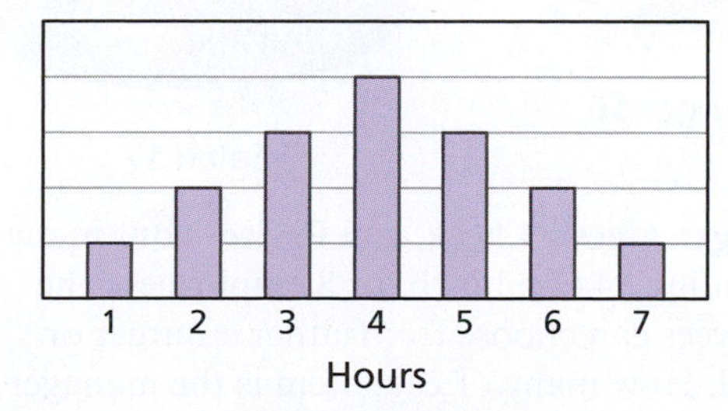

10.

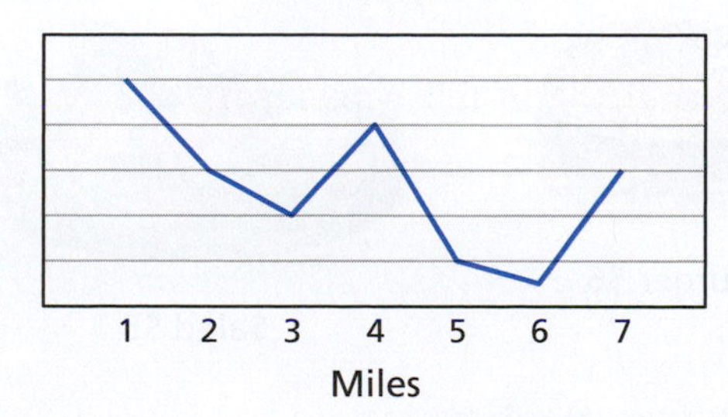

11. **Strategy: Making a Table** You make $12 per hour, plus time and a half for each hour over 40 hours that you work in a week. How many hours do you need to work in a week to earn $588?

 a. Make a table showing your weekly earnings as the number of work hours increases.

 b. Complete the table until you find the number of hours you need to work. Check your answer.

12. **Strategy: Making a Table** You are planning on filling a 5-gallon cooler with fruit punch. You have 2 packets of drink mix that will make 1 gallon of fruit punch. How many more packets do you need to fill the cooler?

 a. Make a table showing the numbers of gallons as the numbers of packets increase.

 b. Complete the table until you find the number of packets you need to fill the cooler. Check your answer.

Comparing Two Graphs The graphs show the distributions of AP exam scores by grade.

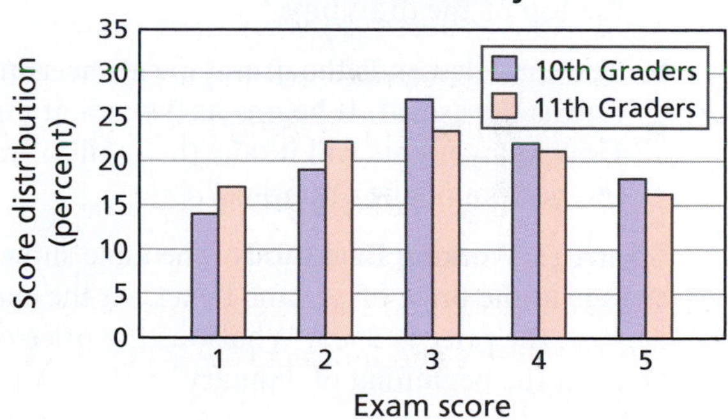

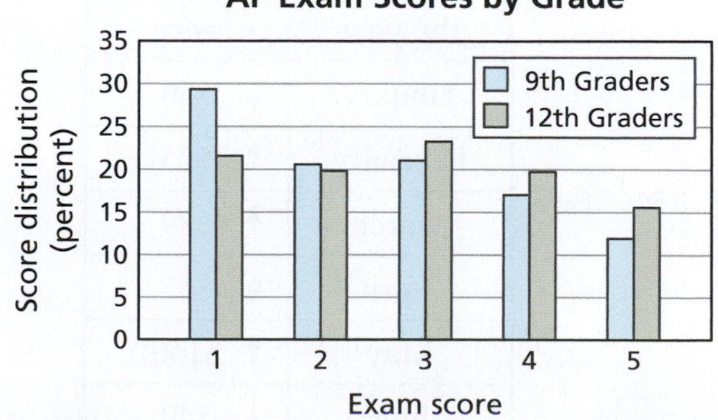

13. Compare the exam scores of 9th graders to those of 10th graders.

14. Compare the exam scores of 10th and 11th graders to those of 9th and 12th graders.

1.2 More Problem-Solving Strategies

15. **Strategy: Drawing a Diagram** A snail is climbing up a 40-meter high drainpipe. Each night the snail climbs 7.5 meters. Every day when it rains the snail slips down the drainpipe 1.5 meters.

a. How many nights will it take the snail to reach the top of the drainpipe?

b. The snail descends the drainpipe at the same rate as the ascent. It begins its descent at night. How many nights will it take the snail to reach the bottom of the drainpipe?

16. **Strategy: Drawing a Diagram** A snail is climbing up a 36-meter high drainpipe. Each night the snail climbs 8 meters. Every day when it rains the snail slips down the drainpipe 1 meter.

a. How many nights will it take the snail to reach the top of the drainpipe?

b. The snail descends the drainpipe at the same rate as the ascent. It begins its descent at night. How many nights will it take the snail to reach the bottom of the drainpipe?

17. **Strategy: Working Backwards** The table shows the change in the price of a plane ticket. At the end of August, the price is $505. What was the price of the ticket at the beginning of January?

Month of the year	Change in price
January	↓ $50
February	↑ $25
March	↑ $89
April	↓ $57
May	↑ $150
June	↓ $40
July	↓ $20
August	↑ $110

18. **Strategy: Working Backwards** The table shows the change in the price of a plane ticket. At the end of May, the price is $599. What was the price of the ticket at the beginning of January?

Month of the year	Change in price
January	↑ $75
February	↓ $33
March	↑ $27
April	↓ $20
May	↑ $110

19. **Strategy: Working Backwards** Your flight for Miami leaves at 5:15 P.M. You want to arrive at the airport 2 hours early to check in and get through security. The taxi ride to the airport takes about 30 minutes. What time should you ask the taxi driver to pick you up?

20. **Strategy: Working Backwards** Your flight for Houston leaves at 6:45 P.M. You want to arrive at the airport 1.5 hours early to check in and get through security. The taxi ride to the airport takes about 37 minutes. What time should you ask the taxi driver to pick you up?

21. **Strategy: Guess, Check, and Revise** Your manager is spending $45 on lunch for 7 employees. The employees can choose from either a burger or a salad. How many of each item is the manager buying?

Burger $6

Salad $7

22. **Strategy: Guess, Check, and Revise** Your manager is spending $44 on lunch for 8 employees. The employees can choose from either a burger or a salad. How many of each item is the manager buying?

Burger $6

Salad $5

1.3 Reasoning and Logic

Identifying a Statement Identify the words or symbols as a statement or non-statement. If it is a statement, write the negation. If it is a non-statement, explain why.

23. Do you want to go see a movie?

24. Mathematicians study math.

25. $x + 5 = 7$

26. $9 - 1 = 8$

Writing the Negation of a Statement Write the negation of the statement.

27. I am home.

28. $3 = 9 + 2 - 8$

29. I enjoy science class.

30. This book is heavy.

Quantifiers and Euler Diagrams Use a quantifier to write a statement that represents the information given by the Euler diagram for the specified relationship.

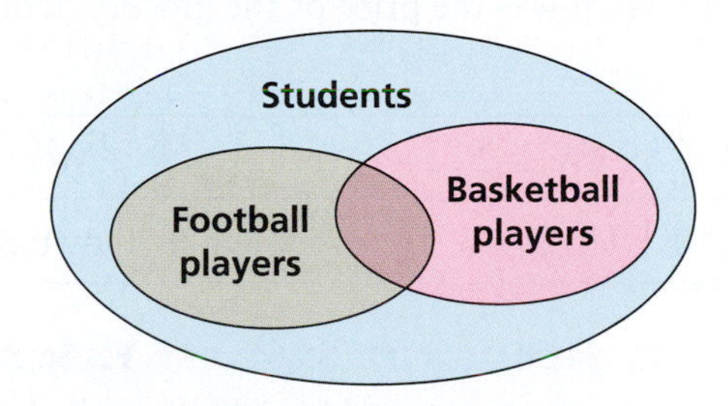

31. Basketball players that are students

32. Football players that are basketball players

Quantifiers and Euler Diagrams Use a quantifier to write a statement that represents the information given by the Euler diagram for the specified relationship.

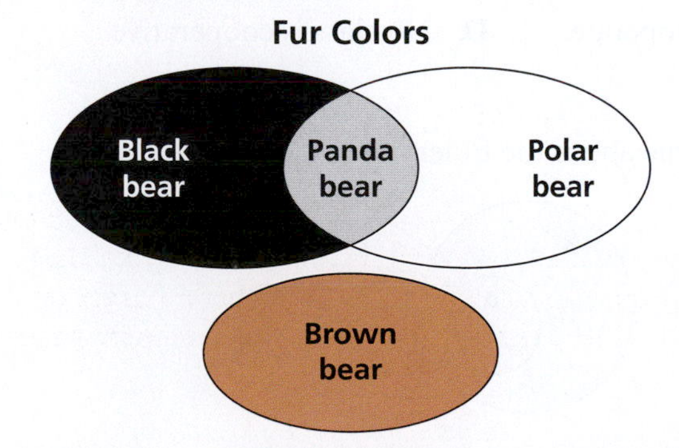

33. Panda bears that share fur colors with polar and black bears

34. Brown bears that share fur color with polar bears

35. Polar bears that share fur color with brown bears

36. Black bears that share fur color with panda bears

37. **Making a Simple Truth Table** Let p represent the statement "Dubstep is a genre of music." Complete the truth table to show the truth values for Not(p). Write the statement for Not(p).

p	Not(p)
T	
F	

38. **Making a Compound Truth Table** Complete the truth table to show the truth values for the conjunction [p and Not(q)].

p	q	Not(q)	p and Not(q)
T	T		
F	T		
T	F		
F	F		

39. **Using Direct and Indirect Reasoning** Decide whether the conclusion is *valid* or *flawed*. If it is valid, state whether it shows direct or indirect reasoning.

	Given information	Conclusion
a.	If Ben gets an A, then he can go to the party. Ben gets an A.	Therefore, Ben can go to the party.
b.	If Ben gets an A, then he can go to the party. Ben does not get an A.	Therefore, Ben can not go to the party.
c.	If Ben gets an A, then he can go to the party. Ben cannot go to the party.	Therefore, Ben does not get an A.

40. **Using Direct and Indirect Reasoning** Decide whether the conclusion is *valid* or *flawed*. If it is valid, state whether it shows direct or indirect reasoning.

	Given information	Conclusion
a.	If Sarah goes skiing, then she will see her cousin. Sarah does not go skiing.	Therefore, Sarah will not see her cousin.
b.	If Sarah goes skiing, then she will see her cousin. Sarah will not see her cousin.	Therefore, Sarah does not go skiing.
c.	If Sarah goes skiing, then she will see her cousin. Sarah goes skiing.	Therefore, Sarah will see her cousin.

Chapter Test

The questions on this practice test are modeled after questions on state certification exams for K–8 teaching.

Test Directions: Each of the questions is followed by five choices. Choose the *best* response to each question.

1. Find the next three terms in this sequence: 1, 4, 10, 22, …

A. 22, 24, 30 **B.** 24, 30, 38 **C.** 38, 46, 94
D. 46, 94, 190 **E.** 48, 96, 192

2. You make $20 per hour, plus time and a half for each hour over 40 hours that you work in a week. How many hours do you need to work in a week to earn $980?

A. 44 hours **B.** 45 hours **C.** 46 hours
D. 48 hours **E.** 49 hours

3. John and Kevin have a total of $87. John has $11 more than Kevin. How much money does Kevin have?

A. $30 **B.** $34 **C.** $38 **D.** $40 **E.** $42

4. The table show the change in the price of a grocery item. At the end of the fourth week, the price is $4.99. What was the price of the grocery item at the beginning of the first week?

Day of the week	Week 1	Week 2	Week 3	Week 4
Change in price	Down $0.50	Up $0.35	Down $0.40	Down $0.10

A. $3.94 **B.** $4.34 **C.** $4.54 **D.** $5.64 **E.** $6.34

5. The sum of 8 and twice a number is 32. Find the number.

A. 12 **B.** 14 **C.** 16 **D.** 18 **E.** 20

6. Which statement is not a negation of the statement "The class is cooperative"?

A. The class is not cooperative. **B.** The class is uncooperative.
C. The class does not cooperate. **D.** No class is cooperative.
E. None of the above.

7. Which statement is not true about the Euler diagram below?

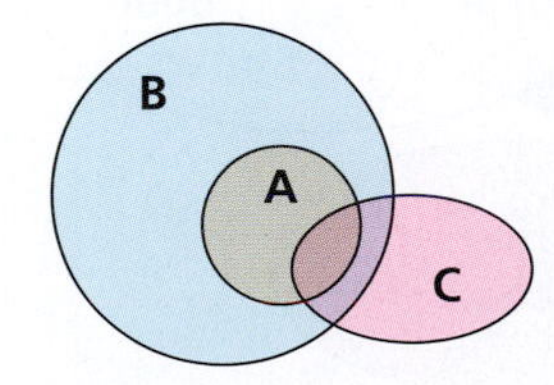

A. No B are C. **B.** All A are B. **C.** Some C are B.
D. Some B are A. **E.** Some C are A.

8. Assume that one pig eats 5 pounds of food each week. How much food do 10 pigs eat in a week?

A. 20 lb **B.** 40 lb **C.** 50 lb **D.** 500 lb **E.** 2,080 lb

2 Sets and Numeral Systems

iStockphoto.com/Mark Bowden

Activity: Matching Sets and Whole Numbers

Materials:

- Pencil

Learning Objective:

Learn to count the number of elements in a set and match the total with a whole number from 0 through 20. Use deductive reasoning to determine which whole numbers are not used.

Name ______________________________

Count the number of elements in the set. Then write the total.

1. { } ▭
2. { } ▭
3. { } ▭
4. { } ▭
5. { } ▭
6. { } ▭
7. { } ▭
8. { } ▭
9. { } ▭
10. { } ▭
11. { } ▭
12. { } ▭
13. { } ▭
14. { } ▭
15. { } ▭
16. {8 8 8 8 8 8 8 8
8 8 8 8 8 8 8 8} ▭

17. Which whole numbers from 0 through 20 are not used as totals?

2.1 Sets

Standards

Grades K–2

Counting and Cardinality

Students should be able to count to tell the number of objects and compare numbers.

- Describe a set using the listing method and set-builder notation, and identify when sets are equal.
- Develop the concept of cardinality of a set and use it to describe sets of numbers.
- Compare the equality and equivalence of different sets.
- Use the concept of subset to develop the concept of inequality.

Describing a Set

Set theory was developed by the German mathematician Georg Cantor (1845–1918). The basic principles of set theory can be taught to very young children, as shown in the activity on page 38.

Definition of a Set

A **set** is a collection of objects. The **elements** (or members) of a set can be numbers, people, letters of the alphabet, other sets, and so on. Sets are usually denoted by capital letters.

There are two common ways to describe a set.

$C = \{1, 2, 3, 4, 5\}$ Listing method

$C = \{x \mid x \text{ is a whole number from 1 through 5.}\}$ Set-builder notation

The set of all x such that x is a whole number from 1 through 5.

The **null set**, denoted by $\{\ \}$ or $\varnothing$, is empty and has no elements. Sets A and B are **equal** if and only if they have the same elements.

Example $A = \{a, b, c, d, e, f\}$

$B = \{d, f, e, a, c, b\}$

$A = B$. The order of the elements does not matter.

EXAMPLE 1 **Equality of Sets**

Write each set using the listing method. Then decide which of the sets are equal.

$A = \{x \mid x \text{ is an odd number between 2 and 10.}\}$

$B = \{x \mid x \text{ is a prime number between 2 and 10.}\}$

$C = \{x \mid x \text{ is a whole number between 2 and 10 and is not divisible by 2.}\}$

SOLUTION

Use the listing method to describe the elements of each set.

$A = \{3, 5, 7, 9\}$ A consists of the odd numbers between 2 and 10.

$B = \{3, 5, 7\}$ Note that 9 is not prime because $9 = 3 \cdot 3$.

$C = \{3, 5, 7, 9\}$ Odd numbers are not divisible by 2.

Sets A and C have exactly the same elements. So, it follows that

$A = C$. Two sets are equal.

Classroom Tip

Emphasize to your students that describing the elements of a set is an exercise in precise communication. If your younger students struggle using set-builder notation, then encourage them to list the elements of a set.

The Cardinality of a Set and Whole Numbers

The words *number* and *numeral* tend to be used interchangeably. Technically, however, there is a difference. A **number** is an abstract concept, while a **numeral** is a symbol used to represent a number. For example, "3," "three," and "III" are all numerals that represent the same number (or the concept of "threeness").

Cardinality of a Set

The **cardinality of a set** is the number of unique elements of the set. A **finite set** is a set that has a limited number of elements. An **infinite set** is a set that has an unlimited number of elements. The null set is considered to be finite with a cardinality of 0.

Classroom Tip

The natural numbers represent the oldest number concepts of humans. Primitive humans developed the concept of 1, 2, 3, and so on. However, it took thousands of years after humans developed the concept of numbers before they learned to write symbols (numerals) to represent these numbers. The concept of counting may have started when one primitive man wanted to trade an animal he had slain for another man's two spears. Ask your students to think about other situations in which primitive humans might have needed to use numbers.

Early man knew "one" and "two." Some primitive humans used "many" to represent "three."

iStockphoto.com/megasquib

EXAMPLE 2 Finding the Cardinality of a Set

Tell whether the set is *finite* or *infinite*. If the set is finite, then find its cardinality.

a. $A = \{a, b, c, d\}$

b. $B = \{2, 7, 12, 14, 16, 20\}$

c. $C = \varnothing$

d. $D = \{x \mid x \text{ is an odd number.}\}$

SOLUTION

a. A has 4 elements so it is a finite set. Its cardinality is 4.

b. B has 6 elements so it is a finite set. Its cardinality is 6.

c. C is the null set. It is finite and its cardinality is 0.

d. D is an infinite set because there are infinitely many odd numbers.

Natural Numbers and Whole Numbers

The set of **natural numbers** is the set of all cardinalities of *nonempty* finite sets. The set of natural numbers is also called the set of **counting numbers**.

Natural numbers $\{1, 2, 3, 4, 5, 6, \ldots\}$

The set of **whole numbers** is the set of all cardinalities of finite sets.

Whole numbers $\{0, 1, 2, 3, 4, 5, 6, \ldots\}$

EXAMPLE 3 Comparing Natural Numbers and Whole Numbers

What is the difference between the set of natural numbers and the set of whole numbers?

SOLUTION

The only difference is that the set of whole numbers contains the number zero, while the set of natural numbers does not.

Comparing Equality and Equivalence of Sets

Equivalent Sets

Two sets A and B are **equivalent**, written as $A \sim B$, when they have the same cardinality. When two sets are equivalent, their elements can be put into a **one-to-one correspondence** such that each element in set A corresponds to exactly one element in set B, and vice versa.

Example $A = \{a, b, c, d, e\}$

$B = \{1, 2, 3, 4, 5\}$

(with arrows $a \leftrightarrow 1$, $b \leftrightarrow 2$, $c \leftrightarrow 3$, $d \leftrightarrow 4$, $e \leftrightarrow 5$)

Sets A and B are equivalent because they have the same number of elements. The sets are *not* equal because they do not contain exactly the same elements.

EXAMPLE 4 Comparing Equality and Equivalence

Tell which sets are *equal* and which sets are *equivalent*. Explain your reasoning.

$A = \{2, 3, 5, 7\}$ $\quad B = \{\text{dog, cat, rat, owl}\}$ $\quad C = \{a, e, i, o, u\}$

$D = \{\text{April, May, June, July, August}\}$ $\quad E = \{x \mid x \text{ is a vowel.}\}$

SOLUTION

Two of the sets, A and B, have a cardinality of 4. So, these two sets are equivalent.

$A = \{2, 3, 5, 7\}$

$B = \{\text{dog, cat, rat, owl}\}$ $\qquad A \sim B$

The sets C and E are equal because each consists of the five vowels in the alphabet. Because sets C and E are equal, it follows that they are also equivalent.

$C = \{a, e, i, o, u\} = E$ $\qquad C = E$ and $C \sim E$.

Three of the sets, C, D, and E, have a cardinality of 5. So, these three sets are equivalent.

$C = \{a, e, i, o, u\}$

$\qquad C \sim D$

$D = \{\text{April, May, June, July, August}\}$

$\qquad D \sim E$

$E = \{a, e, i, o, u\}$

So, $C \sim D \sim E$.

Classroom Tip

As a classroom activity, ask your students to think of a situation in which they know that two sets have the same cardinality, without actually knowing how many elements are in each set. Here is an example. You could take a stack of cookies and a bunch of paper bags. If there are exactly enough bags to put one cookie in each bag, then you know the two sets (cookies and bags) must have the same cardinality.

Two sets can have several possible one-to-one correspondences as shown.

$A = \{2, 3, 5, 7\}$ $\qquad A = \{2, 3, 5, 7\}$

$B = \{\text{dog, cat, rat, owl}\}$ $\qquad B = \{\text{dog, cat, rat, owl}\}$

Classroom Tip

Here is a thought-provoking activity for your class. Ask students to find a pattern for the number of subsets of a set with n elements.

1 Element $A = \{a\}$

$\{\ \}, \{a\}$ 2 subsets

2 Elements $B = \{a, b\}$

$\{\ \}, \{a\},$

$\{b\}, \{a, b\}$ 4 subsets

3 Elements $C = \{a, b, c\}$

$\{\ \}, \{a\}, \{b\}, \{c\}, \{a, b\},$

$\{a, c\}, \{b, c\}, \{a, b, c\}$

8 subsets

4 Elements $D = \{a, b, c, d\}$

$\{\ \}, \{a\}, \{b\}, \{c\}, \{d\}, \{a, b\},$

$\{a, c\}, \{a, d\}, \{b, c\}, \{b, d\},$

$\{c, d\}, \{a, b, c\}, \{a, b, d\},$

$\{a, c, d\}, \{b, c, d\},$

$\{a, b, c, d\}$ 16 subsets

Do you see the pattern? In general, a set with n elements has 2^n subsets.

Subsets and Inequalities

Definition of a Subset

Set A is a **subset** of set B, written as $A \subseteq B$, if and only if every element in A is also an element in B.

Example $A = \{a, b, c\}$

$B = \{a, b, c, d, e\}$

$A \subseteq B$

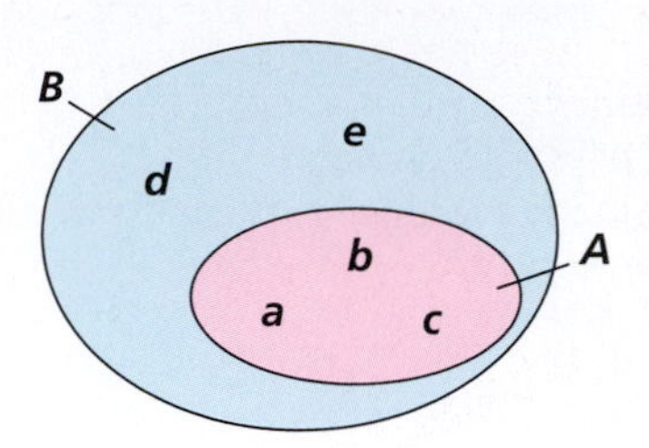

The definition of a subset also allows A to be equal to B. If $A \subseteq B$, and B contains at least one element that is not in A, then A is called a **proper subset** of B, written as $A \subset B$. The null set $\{\ \}$ is a subset of every set.

EXAMPLE 5 Determining Subsets and Proper Subsets

Tell whether the statement is *true* or *false*. Explain your reasoning.

a. The set of natural numbers is a subset of the set of whole numbers.

b. The set of natural numbers is a proper subset of the set of whole numbers.

c. The set of odd numbers is a subset of the set of natural numbers.

d. The set of whole numbers is a subset of the set of natural numbers.

e. The set of the squares of the whole numbers is a subset of the set of even numbers.

SOLUTION

a. This statement is true.

Natural numbers **Whole numbers**

$\{1, 2, 3, 4, 5, 6, \ldots\} \subseteq \{0, 1, 2, 3, 4, 5, 6, \ldots\}$

b. This statement is true. Notice that 0 is a whole number, but it is not a natural number.

Natural numbers **Whole numbers**

$\{1, 2, 3, 4, 5, 6, \ldots\} \subset \{0, 1, 2, 3, 4, 5, 6, \ldots\}$

c. This statement is true.

Odd numbers **Natural numbers**

$\{1, 3, 5, 7, 9, \ldots\} \subseteq \{1, 2, 3, 4, 5, 6, 7, 8, 9, \ldots\}$

d. This statement is false.

Whole numbers **Natural numbers**

$\{0, 1, 2, 3, 4, 5, 6, \ldots\} \not\subseteq \{1, 2, 3, 4, 5, 6, \ldots\}$

e. This statement is false.

Squares of the whole numbers **Even numbers**

$\{0, 1, 4, 9, 16, 25, \ldots\} \not\subseteq \{0, 2, 4, 6, 8, 10, 12, \ldots\}$

You can use the concept of subset to help define the concept of "less than" among whole numbers.

Definition of Inequality for Whole Numbers

Let A and B be two finite sets. Let m be the cardinality of A and let n be the cardinality of B. If $A \subseteq B$, then m is **less than or equal to** n, which is written as $m \leq n$. If $A \subset B$, then m is **less than** n, which is written as $m < n$.

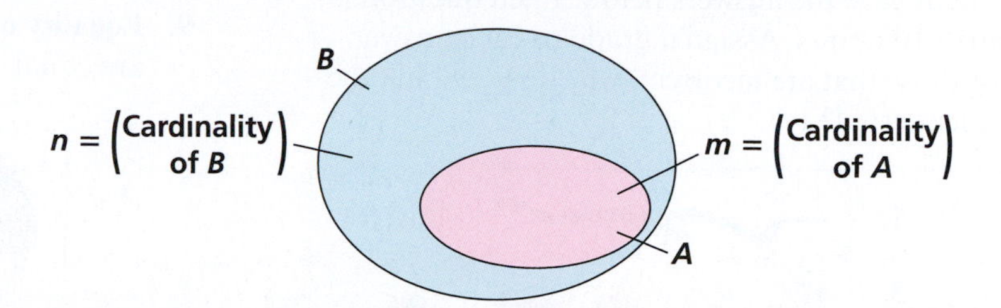

The concepts **greater than or equal to** and **greater than** can be defined in a similar manner.

Classroom Tip

You can help students remember which way an inequality sign points by remembering that the small end always points to the lesser number.

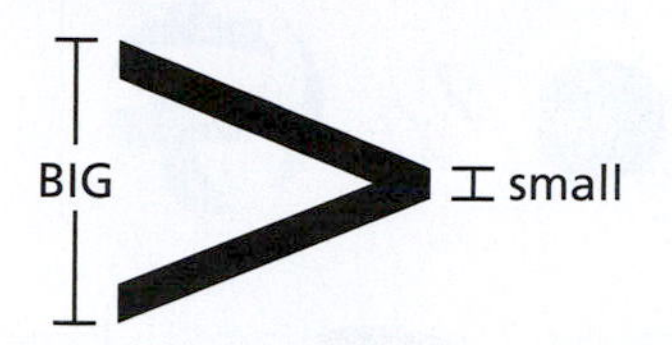

EXAMPLE 6 Comparing Two Numbers

Use sets to determine which number is greater, 3 or 7.

SOLUTION

Let $A = \{a, b, c\}$ and $B = \{a, b, c, d, e, f, g\}$. You can match the three elements in set A with a proper subset of set B. Because there are unmatched elements in set B, $3 < 7$ and $7 > 3$.

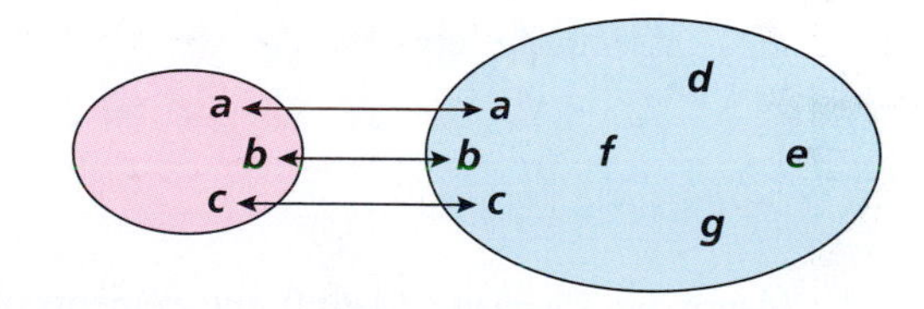

EXAMPLE 7 Using Inequalities to Define Sets

Each set is defined using set-builder notation or the listing method. Write each set using the other form.

a. $A = \{0, 1, 2, 3, 4, 5, 6, 7\}$ **b.** $B = \{1, 2, 3, 4, 5, 6, 7\}$

c. $C = \{x \mid x \text{ is a whole number, and } x \leq 8.\}$

d. $D = \{x \mid x \text{ is a whole number, and } 7 < x \text{ and } x < 14.\}$

e. $E = \{x \mid x \text{ is a whole number, and } 7 \leq x \text{ and } x \leq 14.\}$

SOLUTION

a. $A = \{x \mid x \text{ is a whole number, and } x < 8.\}$ — This set consists of whole numbers that are less than 8.

b. $B = \{x \mid x \text{ is a natural number, and } x < 8.\}$ — This set consists of natural numbers that are less than 8.

c. $C = \{0, 1, 2, 3, 4, 5, 6, 7, 8\}$ — 0 is a whole number, and $8 \leq 8$.

d. $D = \{8, 9, 10, 11, 12, 13\}$ — Does not include 7 or 14.

e. $E = \{7, 8, 9, 10, 11, 12, 13, 14\}$ — Does include 7 and 14.

Mathematical Practices

Attend to Precision

MP6 Mathematically proficient students state the meaning of the symbols they choose, including the equal sign, consistently and appropriately.

2.1 Exercises

Free step-by-step solutions to odd-numbered exercises at *MathematicalPractices.com.*

1. **Writing a Solution Key** Write a solution key for the activity on page 38. Describe a rubric for grading a student's work.

2. **Grading the Activity** In the activity on page 38, a student gave the answers below. Each question is worth 10 points. Assign a grade to each answer. For those that are incorrect, why do you think the student erred?

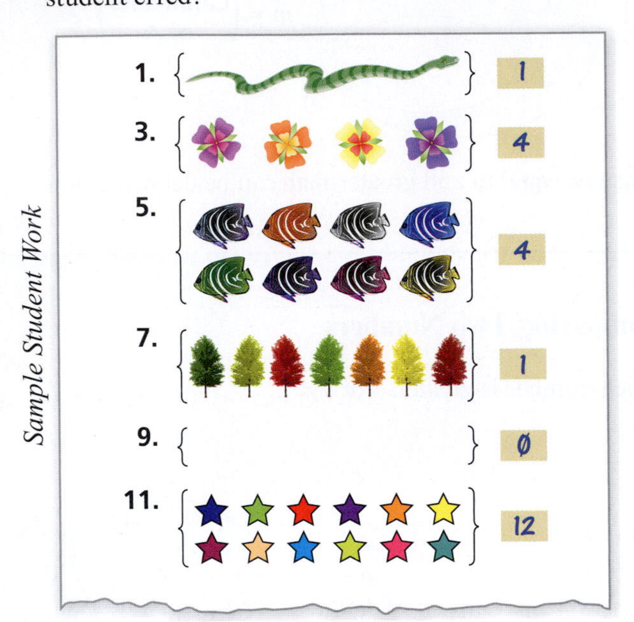

3. **Identifying Elements** Identify the elements of each set.
   - **a.** $\{1, 2, 3, 4, 5\}$
   - **b.** $\{0\}$

4. **Identifying Elements** Identify the elements of each set.
   - **a.** $\{a, b, c, d\}$
   - **b.** $\{\ \}$

5. **Using the Listing Method** Write each set using the listing method.
   - **a.** The set of integers from 1 through 9
   - **b.** The set of letters in the word *set*

6. **Using the Listing Method** Write each set using the listing method.
   - **a.** The set of even integers from 10 through 16
   - **b.** The set of letters in the words *math practices*

7. **Using Set-Builder Notation** Write each set using set-builder notation.
   - **a.** The set of odd integers from 1 through 7
   - **b.** The set of letters in the word *math*

8. **Using Set-Builder Notation** Write each set using set-builder notation.
   - **a.** The set of prime numbers from 3 through 11
   - **b.** The set of letters in the words *common core*

9. **Equality of Sets** Decide which of the following sets are equal. Explain your reasoning.

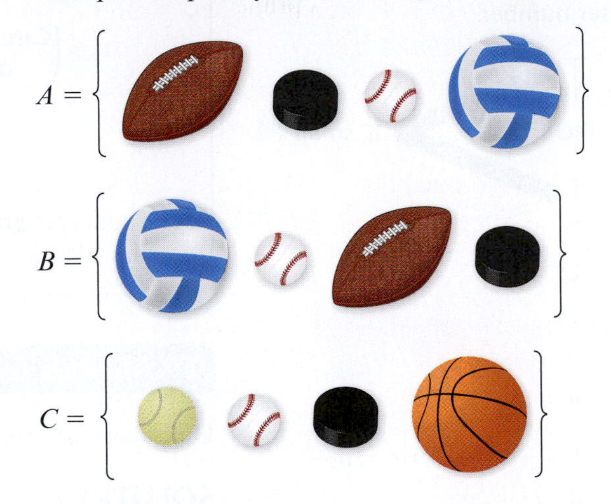

10. **Equality of Sets** Decide which of the following sets are equal. Explain your reasoning.

$A =$

$B =$

$C =$

11. **Equality of Sets** Write each set using the listing method, if necessary. Then decide which of the sets are equal. Explain your reasoning.
   - **a.** $A = \{2, 4, 6, 10\}$

     $B = \{x \mid x$ is an even number from 2 through 10.$\}$

     $C = \{x \mid x$ is a number from 2 through 10 and is divisible by 2.$\}$
   - **b.** $A = \{x \mid x$ is a composite number from 10 through 20.$\}$

     $B = \{x \mid x$ is a number from 10 through 20 and is divisible by 2 or 3.$\}$

     $C = \{x \mid x$ is an even number from 10 through 20.$\}$

12. Equality of Sets Write each set using the listing method, if necessary. Then decide which of the sets are equal. Explain your reasoning.

a. $A = \{a, b, c, d, e, f\}$

$B = \{a, e, c, g, b, d\}$

$C = \{b, e, f, d, a, c\}$

b. $A = \{\ \}$

$B = \{x \mid x$ is an even number from 6 through 16.$\}$

$C = \{x \mid x$ is an even prime number from 5 through 16.$\}$

13. Finding Cardinality Find the cardinality of each set.

a. $\{n, u, m, b, e, r\}$ **b.** $\{g, r, a, d, e\}$

14. Finding Cardinality Find the cardinality of each set.

a. $\{\ \}$

b. $\{x \mid x$ is a negative integer from -15 through -1.$\}$

15. Finding Cardinalities of Sets Tell whether each set is *finite* or *infinite*. If the set is finite, find its cardinality.

a. $A = \{r, s, t\}$ **b.** $B = \{3, 6, 9, 12, 15\}$

c. $C = \{x \mid x$ is an even number.$\}$

d. $D = \{1\}$

16. Finding Cardinalities of Sets Tell whether each set is *finite* or *infinite*. If the set is finite, find its cardinality.

a. $A = \{x \mid x$ is an odd number from 0 through 100.$\}$

b. $B = \{x \mid x$ is a prime number from 1 through positive infinity.$\}$

c. $C = \{x \mid x$ is a consonant from the alphabet.$\}$

d. $D = \{b, a, s, i, c\}$

17. One-to-One Correspondence Determine whether the elements in the sets can be put in a one-to-one correspondence. Explain your reasoning.

a. $A = \{9, 8, 7\}$

$B = \{6, 5, 4\}$

b. $A = \{$balloon, kite, tent, dollar, dime$\}$

$B = \{2, 4, 6, 8\}$

18. One-to-One Correspondence Determine whether the elements in the sets can be put in a one-to-one correspondence. Explain your reasoning.

a. $A = \{1, 2, 3, 4, 5, 6, 7, 8, 9\}$

$B = \{x \mid x$ is a letter from the word *practices*.$\}$

b. $A = \{\$, ♫, ♣, ®\}$

$B = \{0, 1, 2, 3\}$

19. Comparing Equality and Equivalence Tell which sets are *equal* and which sets are *equivalent*. Explain your reasoning.

a. $A = \{v, w, x, y, z\}$

$B = \{2, 3, 5, 7\}$

$C = \{x \mid x$ is a prime number from 1 through 10.$\}$

$D = \{+, -, \times, \div, =\}$

b. $A = \{x \mid x$ is a prime number from 24 through 26.$\}$

$B = \{$red, white, blue$\}$

$C = \{0\}$

$D = \{\ \}$

20. Comparing Equality and Equivalence Tell which sets are *equal* and which sets are *equivalent*. Explain your reasoning.

a. $A = \{$cow, horse, chicken, pig$\}$

$B = \{0, 1, 2, 3\}$

$C = \{$black, pink, white, blue, purple, orange, green$\}$

$D = \{<, >, \approx, \neq\}$

$E = \{$purple, green, black, orange, white, pink, blue,$\}$

b. $A = \{r, b, e, g, l, a\}$

$B = \{0, 1, 2, 3, 4, 5, 6, 7\}$

$C = \{$Monday, Tuesday, Wednesday, Thursday, Friday, Saturday, Sunday$\}$

$D = \{l, g, e, b, r, a\}$

$E = \{0, 1, 2, 3, 4, 5, 6\}$

21. One-to-One Correspondence How many one-to-one correspondences are there between two sets with two elements each?

22. One-to-One Correspondence How many one-to-one correspondences are there between two sets with three elements each?

23. Finding Subsets Find the number of subsets of each set. List the subsets and identify any proper subsets.

a. $A = \{\ \}$

b. $B = \{\ \ \}$

c. $C = \{$north, south, east, west$\}$

Africa Studio/Shutterstock.com

24. Finding Subsets Find the number of subsets of each set. List the subsets and identify any proper subsets.

a. $E = \{10, 100, 1000\}$ **b.** $F = \{1\}$

c. $G = \{$ [two coins] $\}$

25. Subsets and Proper Subsets Tell whether each statement is *true* or *false*. Explain your reasoning.

a. The set of integers is a subset of the set of whole numbers.

b. The set of positive integers is a subset of the set of whole numbers.

c. The set of positive integers is a proper subset of the set of natural numbers.

d. The set of natural numbers is a proper subset of the set of integers.

26. Subsets and Proper Subsets Tell whether each statement is *true* or *false*. Explain your reasoning.

a. The set $\{0\}$ is a proper subset of the set of rational numbers.

b. The set of irrational numbers is a subset of the set of rational numbers.

c. The set of integers is a proper subset of the set of rational numbers.

d. The set $\{0\}$ is not a subset of the set of real numbers.

27. Comparing Two Sets Write an inequality to show the relationship between the cardinality of the sets A and B.

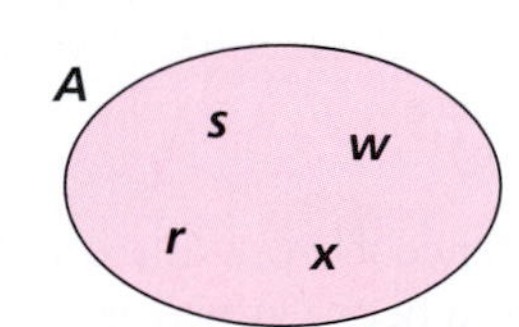

28. Comparing Two Sets Write an inequality to show the relationship between the cardinality of the sets A and B.

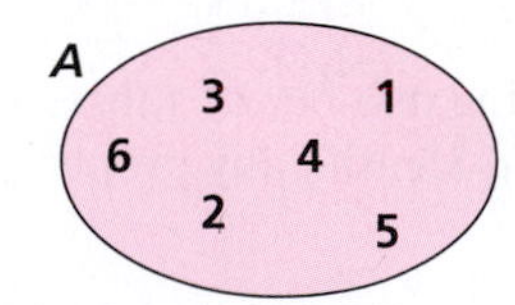

29. Comparing Two Numbers Use a diagram to determine which number is greater. Write your answer as an inequality.

a. 4, 8 **b.** 9, 5

30. Comparing Two Numbers Use a diagram to determine which number is greater. Write your answer as an inequality.

a. 1, 7 **b.** 5, 3

31. Using Inequalities to Define Sets Write each set using the listing method.

a. $A = \{x \mid x \text{ is a positive integer, and } x < 12.\}$

b. $B = \{y \mid y \text{ is a natural number, and } y \leq 9.\}$

c. $C = \{s \mid s \text{ is a negative integer, and } s \geq -4.\}$

d. $D = \{t \mid t \text{ is a positive integer, and } t > 7.\}$

32. Using Inequalities to Define Sets Write each set using the listing method.

a. $Q = \{x \mid x \text{ is a positive integer, and } 2 < x \text{ and } x < 12.\}$

b. $R = \{y \mid y \text{ is a negative integer, and } -15 < y \text{ and } y < -6.\}$

c. $S = \{z \mid z \text{ is a whole number, and } 0 \leq z \text{ and } z \leq 4.\}$

d. $T = \{w \mid w \text{ is a natural number, and } 1 \leq w.\}$

33. Writing Sets Using Set-Builder Notation Write each set using set-builder notation and inequalities.

a. $A = \{1, 2, 3, 4, 5, 6, \ldots\}$

b. $B = \{2, 4, 6, 8, 10, 12, 14, 16, 18, 20\}$

c. $C = \{-1, -2, -3, -4, -5\}$

34. Writing Sets Using Set-Builder Notation Write each set using set-builder notation and inequalities.

a. $A = \{7, 8, 9\}$

b. $B = \{2, 3, 5, 7, 11\}$

c. $C = \{-10, -11, -12, -13, \ldots\}$

35. Grading Student Work On a diagnostic test, one of your students made the following statement about the diagram of sets A and B. Explain what the student did wrong. Which topics would you encourage the student to review?

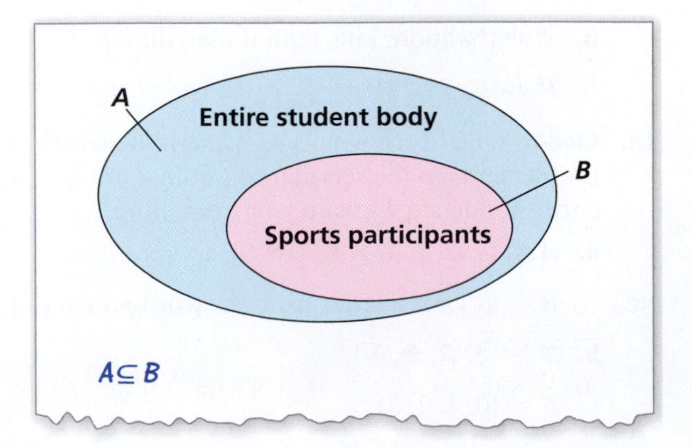

cl2004lhy/Shutterstock.com

36. **Grading Student Work** On a diagnostic test, one of your students made the following statement about the diagram of sets A and B. Explain what the student did wrong. Which topics would you encourage the student to review?

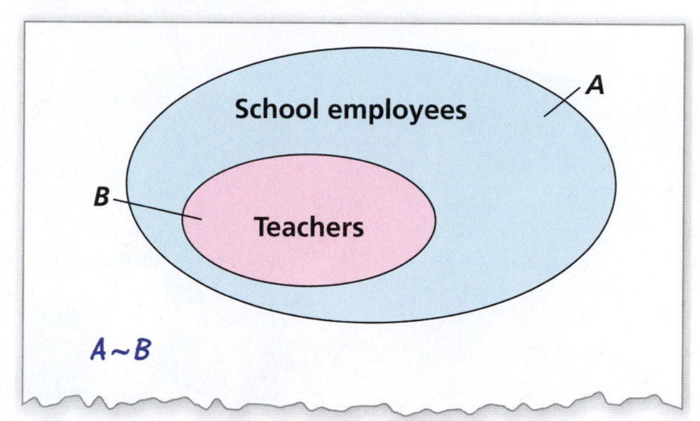

37. **Finding Subsets: In Your Classroom** Consider set X shown below.

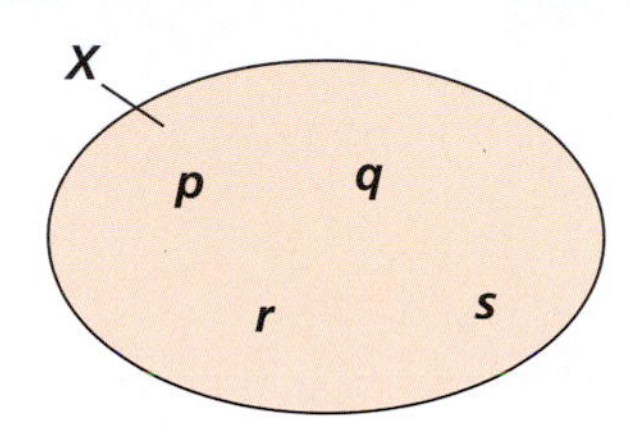

a. How many subsets does X have?

b. How many proper subsets does X have?

c. How many subsets does X have that include the elements p and q?

d. Describe a demonstration you could use to build students' knowledge about finding subsets.

38. **Finding Subsets: In Your Classroom** Consider set X shown below.

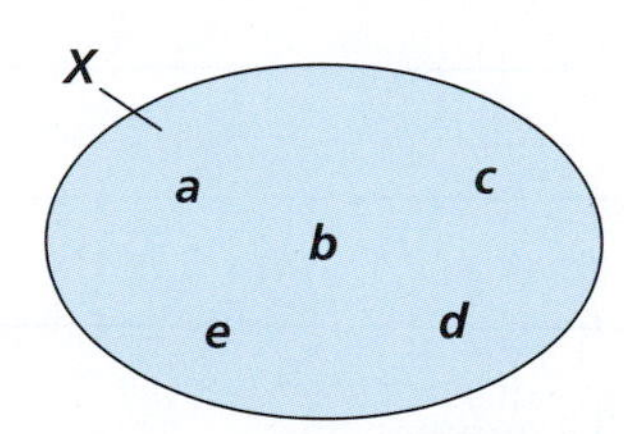

a. How many subsets does X have?

b. How many proper subsets does X have?

c. How many subsets does X have that include the elements a and b?

d. Describe a demonstration you could use to build students' knowledge about finding subsets.

39. **Analyzing Sets** Tell whether each statement is *true* or *false*. Explain your reasoning.

a. When $G \subseteq H$, it is always true that $G = H$.

b. When $G = H$, it is always true that $G \subseteq H$.

40. **Analyzing Sets** Tell whether each statement is *true* or *false*. Explain your reasoning.

a. When $G \subseteq H$, it is always true that $G \subset H$.

b. When $G \subset H$, it is always true that $G \subseteq H$.

41. **Discussion: In Your Classroom** Let $E \subseteq F$ and $F \subseteq G$. Discuss whether $E \subseteq G$.

42. **Discussion: In Your Classroom** Let $X \subseteq Y$.

a. What is the minimum number of elements in set X?

b. Can set Y be the null set? If so, give an example of sets X and Y.

43. **Problem Solving** There are four fruits available at a salad bar. How many ways can you sample two different fruits?

44. **Problem Solving** There are five vegetables available at a salad bar. How many ways can you sample two different vegetables?

45. **Writing** Can the null set be a proper subset of every nonempty set? Explain your reasoning.

46. **Writing** Describe two sets from real-life situations where the sets are equivalent. Show this using one-to-one correspondence.

47. **Problem Solving** There are six members on a student council. Three of these members will serve on a spring formal committee. How many possible spring formal committees are there?

48. **Problem Solving** There are seven members on a student council. Three of these members will serve on a prom committee. How many possible prom committees are there?

Activity: Whole Numbers and Place Value

Materials:

- Pencil

Learning Objective:

Understand the base-ten place system for whole numbers. Write the word name and base-ten numeral for whole numbers.

Name ______________________

Write the word name for the whole number.

1. 23,500 ______________________
2. 650,000 ______________________
3. 540 ______________________
4. 381 ______________________
5. 10,947 ______________________

Write the whole number for the word name.

6. Nine hundred twenty-two thousand, four ______________________
7. Five hundred thirty-seven ______________________
8. Ten thousand, four hundred ______________________
9. Three hundred six thousand, seventeen ______________________
10. Four hundred nineteen thousand, fifty-nine ______________________

11. A mystery number has:
    1 in the ten thousands place
    6 in the hundred thousands place
    4 in the tens place
    2 in the ones place
    0 in the hundreds place
    3 in the thousands place
    What is the mystery number?
    ____

12. A mystery number has:
    0 in the ones place
    5 in the hundred thousands place
    7 in the tens place
    8 in the ten thousands place
    4 in the thousands place
    2 in the hundreds place
    What is the mystery number?
    ____

2.2 Whole Numbers

Standards

Grades K–2
Number and Operations in Base Ten
Students should understand place value.

Grades 3–5
Number and Operations in Base Ten
Students should understand the place value system and use place-value understanding to round numbers.

- Understand and use the properties of equality.
- Understand the base-ten system for whole numbers.
- Round whole numbers to given place values.

Properties of Equality

As you progress through this course, you will study various sets of numbers such as whole numbers, integers, fractions, decimals, and real numbers. These sets of numbers share the properties that are presented here for the set of whole numbers.

Properties of Equality for Whole Numbers

Reflexive Property

Words A whole number equals itself.

Algebra $a = a$ **Example** $3 = 3$

Symmetric Property

Words The order of the equality of whole numbers does not matter.

Algebra If $a = b$, then $b = a$. **Example** If $4 = x$, then $x = 4$.

Transitive Property

Words Two whole numbers that are equal to the same whole number are equal to each other.

Algebra If $a = b$ and $b = c$, then $a = c$.

Example If $x = 5$ and $5 = y$, then $x = y$.

Classroom Tip

To help your students remember the Transitive Property of Equality, tell them that this property allows them to eliminate the "middle man."

If $a = b$ and $b = c$, then $a = c$.

"middle man"

EXAMPLE 1 **Applying a Property of Equality**

One of your students solves an equation, as follows.

$7 + x = 3x - 5$ Original equation

$+5 \qquad +5$ Add 5 to each side.

$12 + x = 3x$ Simplify.

$-x \quad -x$ Subtract x from each side.

$12 = 2x$ Simplify.

$\frac{12}{2} = \frac{2x}{2}$ Divide each side by 2.

$6 = x$ Simplify.

Your student is unsatisfied with the solution because the variable is on the right side. What would you say to your student?

SOLUTION

Tell your student that the variable can also be on the left side. Include a last step in the solution and justify the step using the Symmetric Property of Equality.

$x = 6$ Symmetric Property of Equality

The Base-Ten System for Whole Numbers

Classroom Tip

In the base-ten system, it is important that your students can identify each position (or place). Here is an example.

The number 721,040 has 7 in the hundred thousands place, 2 in the ten thousands place, 1 in the thousands place, 4 in the tens place, and 0 in both the hundreds place and the ones place.

At one time in history, there were dozens of different systems for writing numerals. In the next section, you will get a brief introduction to a few of these systems. Today, the *base-ten system*, or *decimal system,* has become almost universal.

The Base-Ten System for Whole Numbers

A **positional numeral system** is a system in which the value of a digit depends on the position or place it occupies in a numeral. The **base-ten system** (or **decimal system**) is a positional numeral system that is based on powers of ten. It uses **place value** to assign a value to a digit in a numeral.

Position	Place Value
1	Ones (units)
10	Tens
100	Hundreds
1000	Thousands
10,000	Ten thousands
100,000	Hundred thousands
1,000,000	Millions
1,000,000,000	Billions
1,000,000,000,000	Trillions

The **expanded form** of a number is the sum of the product of each digit times its place value position.

Example $2083 = 2 \times 1000 + 8 \times 10 + 3 \times 1 = 2000 + 80 + 3$

Mathematical Practices

Look for and Make Use of Structure

MP7 Mathematically proficient students look closely to discern a pattern or structure.

EXAMPLE 2 Finding the Pattern Behind Base Ten

Describe the pattern in the position values of the base-ten system.

SOLUTION

1. **Understand the Problem** You need to describe the pattern in the position values of the base-ten system.

2. **Make a Plan** Make a table that shows the place values, the position values, and the position values' relationship to the number 10.

3. **Solve the Problem** The place values and the position values are shown below.

Place Value	Position Value	Pattern
Ones	1	
Tens	10	1×10
Hundreds	100	10×10
Thousands	1000	100×10
Ten thousands	10,000	1000×10

Each position value is 10 times the previous position value.

4. **Look Back** Check your answer by dividing each position value by the previous position value.

$$\frac{10}{1} = 10 \ ✓ \quad \frac{100}{10} = 10 \ ✓ \quad \frac{1000}{10} = 10 \ ✓ \quad \frac{10{,}000}{10} = 10 \ ✓$$

Physically manipulating objects is an important technique when learning basic mathematical principles. Throughout this text, you will be introduced to several of the more commonly used mathematical manipulatives.

Model for the Base-Ten Numeral System

Base-ten blocks can help visualize the base-ten numeral system.

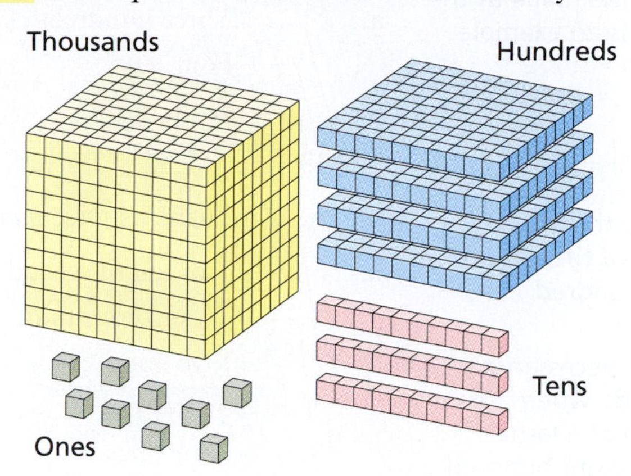

There are four base-ten block sizes.

Ones 1 *small cube*; also called a *unit cube*

Tens 10 small cubes stacked in a column; also called a *long* or *rod*

Hundreds 100 small cubes forming a 10-by-10 square; also called a *flat*

Thousands 1000 small cubes forming a 10-by-10-by-10 *big cube*

Classroom Tip

Base-ten blocks are available from manufacturers of mathematics manipulatives. They are made in various colors of plastic.

EXAMPLE 3 Using Base-Ten Blocks and Expanded Form

Write the whole number represented by the base-ten blocks.

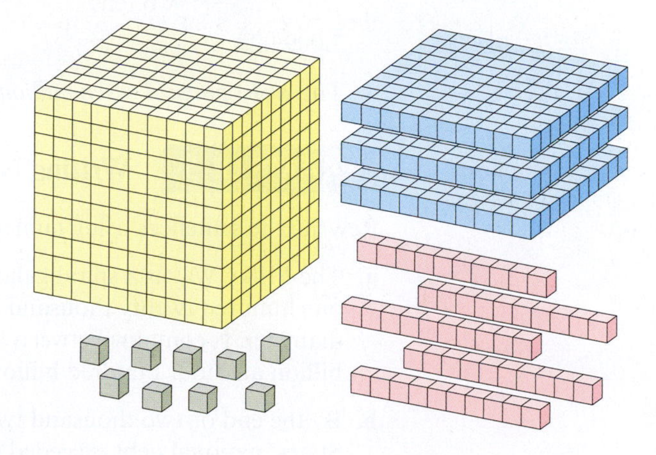

Classroom Tip

As a classroom discussion, ask your students how the base-ten blocks are related to each other.

1. How many small cubes are needed to make a long?
2. How many longs are needed to make a flat?
3. How many flats are needed to make a big cube?
4. If there was a next size block, what would it be? How could you make it?

SOLUTION

In the figure, there is 1 big cube, 3 flats, 5 longs, and 9 small cubes.

In expanded form, the whole number represented by these base-ten blocks is

$$\begin{aligned} &1(\text{thousand}) + 3(\text{hundreds}) + 5(\text{tens}) + 9(\text{ones}) \\ &= 1 \times 1000 + 3 \times 100 + 5 \times 10 + 9 \times 1 \\ &= 1000 + 300 + 50 + 9 \\ &= 1359. \end{aligned}$$

Classroom Tip

When you are teaching students to read the word name for a large number, instruct them to look at the commas, say the name of each three-digit number between the commas, and then include the name of the place value. Here is an example.

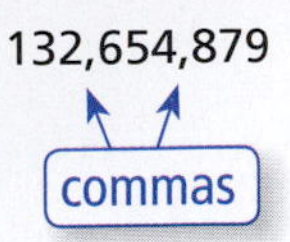

Say "one hundred thirty-two million, six hundred fifty-four thousand, eight hundred seventy-nine."

Note that it is not necessary to use the word "and" when reading the name of a large number, such as "eight hundred *and* seventy-nine."

EXAMPLE 4 Writing Word Names for Numerals

Write the word name for the numeral.

a. 348 **b.** 23,569 **c.** 145,000 **d.** 7,009,020

SOLUTION

a.

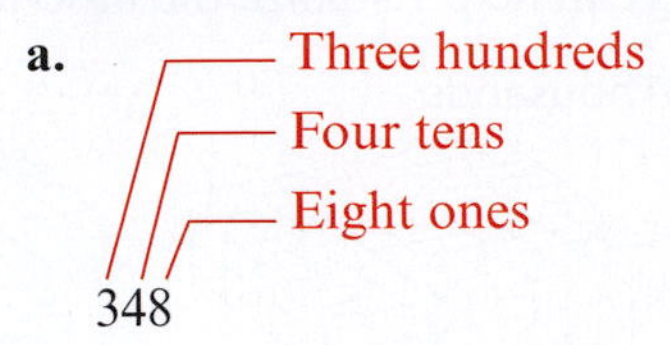

The word name is *three hundred forty-eight*.

b.

Two ten thousands
Three thousands
Five hundreds
Six tens
Nine ones
23,569

The word name is *twenty-three thousand, five hundred sixty-nine*.

c.

One hundred thousand
Four ten thousands
Five thousands
145,000

The word name is *one hundred forty-five thousand*.

d.

Seven millions
Nine thousands
Two tens
7,009,020

The word name is *seven million, nine thousand, twenty*.

EXAMPLE 5 Writing Numerals for Word Names

Rewrite the sentences using numerals instead of word names.

a. The Milky Way is a spiral galaxy that is about one hundred twenty thousand light-years in diameter. It contains between two hundred billion and four hundred billion stars.

b. By the end of two thousand twelve, the United States' national debt exceeded sixteen trillion dollars. This implied that each U.S. citizen's share of the national debt was about fifty-one thousand dollars.

SOLUTION

a. The Milky Way is a spiral galaxy that is about 120,000 light-years in diameter. It contains between 200,000,000,000 and 400,000,000,000 stars.

b. By the end of 2012, the United States' national debt exceeded \$16,000,000,000,000. This implied that each U.S. citizen's share of the national debt was about \$51,000.

NASA

Classroom Tip

To help your students remember the rules for rounding, consider the following. To round 391 to the nearest ten, draw a hill with 390 at the bottom left, 395 at the top, and 400 at the bottom right.

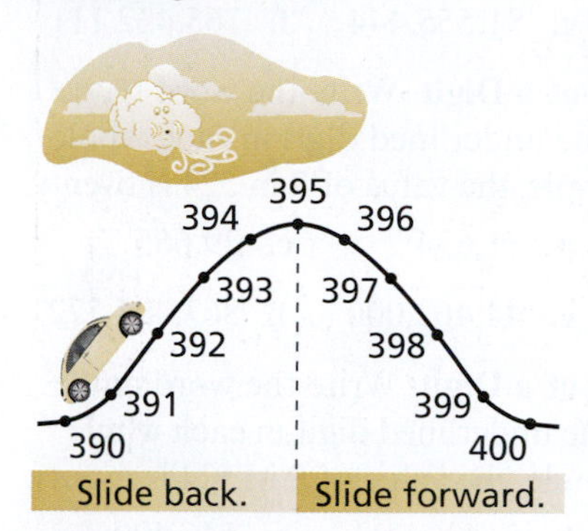

A wind is blowing at the top of the hill. So, at 395, the wind blows the car forward.

Rounding Whole Numbers

Rounding Whole Numbers

A rounded number has *about the same value* as the original number, but is less exact. To round a number, first circle the place value to be rounded. The decision digit is the digit to the right of this place value. Underline this digit.

Round up When the decision digit is 5, 6, 7, 8, or 9, round up by adding 1 to the circled place value. All digits to the right of the circled place value become 0.

Example Rounded to the nearest hundred, 3 9 1 is 400.

Round down When the decision digit is 0, 1, 2, 3, or 4, round down by leaving the circled place value as is. All digits to the right of the circled place value become 0.

Example Rounded to the nearest ten, 3 9 1 is 390.

EXAMPLE 6 **Rounding Whole Numbers**

Round the whole number to the nearest ten, hundred, and thousand.

a. 4865 **b.** 6239 **c.** 18,007

SOLUTION

a. Rounded to the nearest ten, 4 8 6 5 is 4870. Round up.

Decision digit is 5.

Rounded to the nearest hundred, 4 8 6 5 is 4900. Round up.

Decision digit is 6.

Rounded to the nearest thousand, 4 8 6 5 is 5000. Round up.

Decision digit is 8.

b. Rounded to the nearest ten, 6 2 3 9 is 6240. Round up.

Decision digit is 9.

Rounded to the nearest hundred, 6 2 3 9 is 6200. Round down.

Decision digit is 3.

Rounded to the nearest thousand, 6 2 3 9 is 6000. Round down.

Decision digit is 2.

c. Rounded to the nearest ten, 1 8, 0 0 7 is 18,010. Round up.

Decision digit is 7.

Rounded to the nearest hundred, 1 8, 0 0 7 is 18,000. Round down.

Decision digit is 0.

Rounded to the nearest thousand, 1 8, 0 0 7 is 18,000. Round down.

Decision digit is 0.

2.2 Exercises

Free step-by-step solutions to odd-numbered exercises at MathematicalPractices.com.

1. **Writing a Solution Key** Write a solution key for the activity on page 48. Describe a rubric for grading a student's work.

2. **Grading the Activity** In the activity on page 48, a student gave the answers below. Each question is worth 10 points. Assign a grade to each answer. For those that are incorrect, why do you think the student erred?

Sample Student Work

> **4.** 381   *Three hundred eighty-one*
>
> **5.** 10,947   *Ten and nine hundred forty-seven*
>
> **6.** Nine hundred twenty-two thousand, four
> *900,022,004*
>
> **7.** Five hundred thirty-seven   *537*
>
> **11.** A mystery number has:
> 1 in the ten thousands place
> 6 in the hundred thousands place
> 4 in the tens place
> 2 in the ones place
> 0 in the hundreds place
> 3 in the thousands place
> What is the mystery number?
> *164,203*

3. **Applying Properties of Equality** Complete each statement to illustrate the indicated property.
   a. Reflexive Property: $16 =$ ____.
   b. Symmetric Property: If $23 = x$, then ____.
   c. Transitive Property: If $a = b$, and $b = 7$, then ____.

4. **Applying Properties of Equality** Complete each statement to illustrate the indicated property.
   a. Reflexive Property: $x =$ ____.
   b. Symmetric Property: If $y = 47$, then ____.
   c. Transitive Property: If $a = 6$, and $b = 6$, then ____.

5. **Finding Place Values** State the place value of the underlined digit in each whole number.
   a. $24\underline{1}$ b. $5\underline{1}6$ c. $\underline{1}549$
   d. $\underline{1}65{,}342$ e. $\underline{1}2{,}000{,}000$ f. $322{,}122{,}\underline{1}44$

6. **Finding Place Values** State the place value of the underlined digit in each whole number.
   a. $42\underline{1}6$ b. $7\underline{1}55$ c. $\underline{1}4{,}557$
   d. $3\underline{1}{,}549$ e. $8\underline{1}{,}555{,}444$ f. $765{,}432{,}11\underline{1}$

7. **Finding the Value of a Digit** Write the word name for the value of the underlined digit in each whole number. For example, the value of $\underline{2}$ in $5\underline{2}4$ is twenty.
   a. $9\underline{4}51$ b. $71{,}6\underline{3}4$ c. $\underline{2}9{,}665$
   d. $4{,}\underline{7}54{,}112$ e. $44{,}4\underline{6}2{,}000$ f. $\underline{8}67{,}494{,}772$

8. **Finding the Value of a Digit** Write the word name for the value of the underlined digit in each whole number. For example, the value of $\underline{2}$ in $5\underline{2}4$ is twenty.
   a. $524\underline{6}$ b. $31\underline{4}3$ c. $5\underline{4}6{,}400$
   d. $66\underline{2}{,}479$ e. $\underline{9}9{,}777{,}202$ f. $115{,}\underline{4}16{,}882$

9. **Using Base-Ten Blocks and Expanded Form** Write the whole number represented by the base-ten blocks.

a.

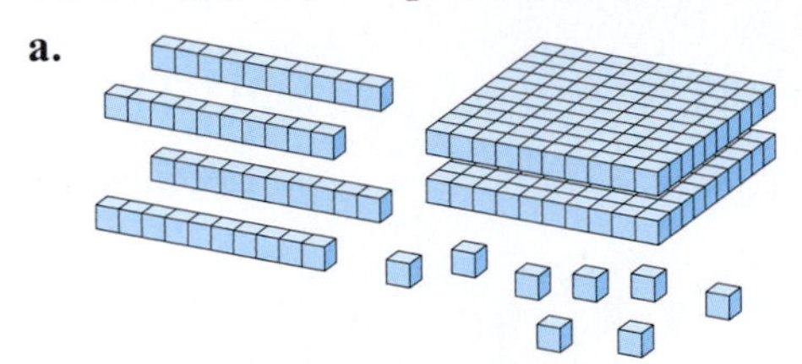

b.

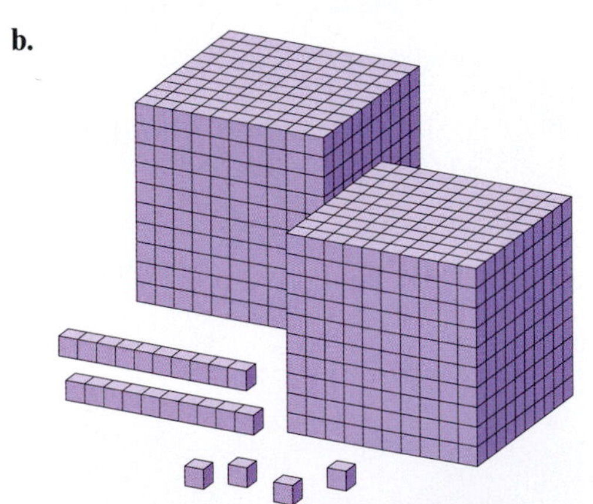

c.

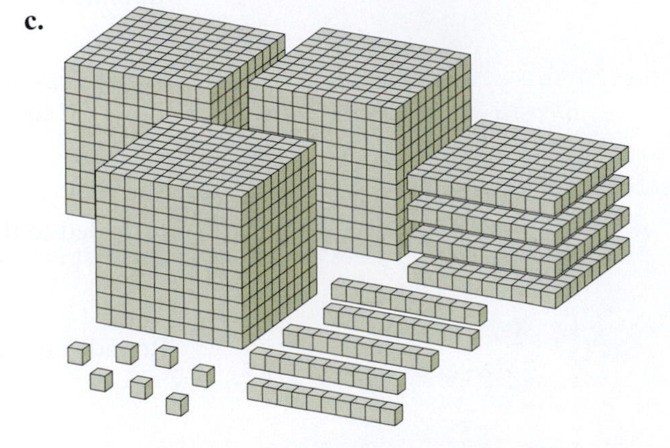

10. Using Base-Ten Blocks and Expanded Form Write the whole number represented by the base-ten blocks.

a.

b.

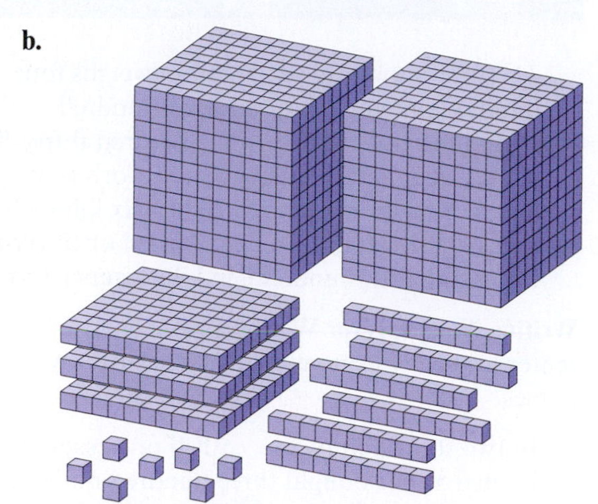

c.

11. Writing Word Names for Numerals Write the word name for each numeral.

a. 691 **b.** 98,435

c. 606,001 **d.** 2,900,450

12. Writing Word Names for Numerals Write the word name for each numeral.

a. 786 **b.** 30,501

c. 565,030 **d.** 23,465,000

13. Writing Numerals for Word Names Write the numeral for each word name.

a. Seven hundred twenty-three

b. Eight thousand, six hundred four

c. Two hundred thousand, seventy-six

d. Sixteen million, eight hundred twenty-four thousand, seven hundred eighty-six

14. Writing Numerals for Word Names Write the numeral for each word name.

a. Three hundred ninety-seven

b. One thousand, two hundred seventy-two

c. Thirteen thousand, five hundred sixty-seven

d. Four million, three hundred thousand, nine

15. Rounding Whole Numbers Round each whole number to the nearest ten.

a. 75 **b.** 413

c. 1058 **d.** 12,672

16. Rounding Whole Numbers Round each whole number to the nearest ten.

a. 41 **b.** 625

c. 4118 **d.** 41,009

17. Rounding Whole Numbers Round each whole number to the nearest hundred.

a. 723 **b.** 2067

c. 4350 **d.** 22,499

18. Rounding Whole Numbers Round each whole number to the nearest hundred.

a. 571

b. 5475

c. 9451

d. 45,902

19. Rounding Whole Numbers Round each whole number to the nearest thousand.

a. 9235

b. 78,500

c. 112,901

d. 2,000,843

20. Rounding Whole Numbers Round each whole number to the nearest thousand.

a. 7599

b. 32,855

c. 300,500

d. 42,616,099

21. **Properties of Equality: In Your Classroom** Let n represent the amount of money Nick has, and let r represent the amount of money Rochelle has.

a. Nick has \$100. Rochelle has \$100. By what property can you say that $100 = 100$?

b. So, $n = 100$ and $r = 100$. What property can you use to rewrite $r = 100$ as $100 = r$?

c. Explain how you would use the information in part (b) to show your students how to apply the Transitive Property of Equality.

22. **Properties of Equality: In Your Classroom** Let m represent the number of electoral votes for Minnesota, and let w represent the number of electoral votes for Wisconsin.

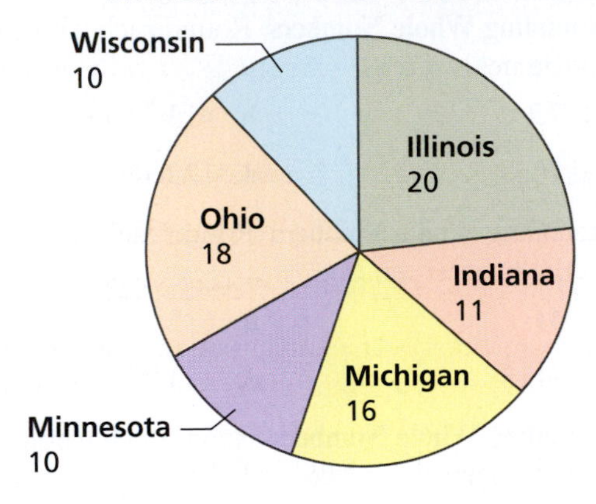

a. Minnesota has 10 electoral votes. Wisconsin has 10 electoral votes. By what property can you say that $10 = 10$?

b. So, $m = 10$ and $w = 10$. What property can you use to rewrite $w = 10$ as $10 = w$?

c. Explain how you would use the information in part (b) to show your students how to apply the Transitive Property of Equality.

23. **Writing Word Names for Numerals** Rewrite the sentences below using word names instead of numerals.

The total national health expenditures in 2011 were \$2,700,700,000,000. Based on an estimated population of 311,141,000, the per capita national health expenditures were about \$8680.

24. **Writing Word Names for Numerals** Rewrite the sentences below using word names instead of numerals.

The total amount spent in the United States on health insurance in 2011 was \$1,960,100,000,000. This includes \$896,300,000,000 for private health insurance, \$554,300,000,000 for Medicare, and \$407,700,000,000 for Medicaid.

25. **Writing Numerals for Word Names** Rewrite the sentences below using numerals instead of word names.

The International Space Station weighs four hundred nineteen thousand, six hundred kilograms and contains nine hundred thirty-five cubic meters of habitable space. It orbits at a distance of three hundred eighty-six kilometers above the Earth, moving at a speed of thirty-two thousand, four hundred ten kilometers per hour.

26. **Writing Numerals for Word Names** Rewrite the sentences below using numerals instead of word names.

In two thousand twelve, catfish processors in the United States bought three hundred million, one hundred fifty-one thousand pounds of catfish at an average price of ninety-eight cents per pound. They sold one hundred sixty-one million, four hundred thirty-four thousand pounds of processed catfish at an average price of three hundred eight cents per pound.

27. **Grading Student Work** On a diagnostic test, one of your students does the following work. Explain what the student did wrong. Which topics would you encourage the student to review?

> 1,014,000
>
> The word name is one thousand fourteen thousand.

28. **Grading Student Work** On a diagnostic test, one of your students does the following work. Explain what the student did wrong. Which topics would you encourage the student to review?

> Four hundred thirty-four thousand, two
>
> The numeral is 434,200.

NASA

29. *Writing* Write a real-life application that involves rounding whole numbers to the nearest thousand.

30. *Writing* Write a real-life application that involves rounding whole numbers to the nearest million.

31. **Rounding Whole Numbers: In Your Classroom** You are teaching your students to round whole numbers to the nearest ten using base-ten blocks.

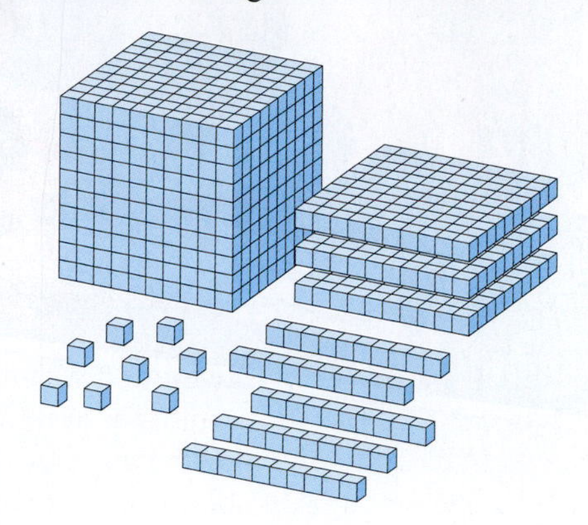

a. To round the whole number represented by the base-ten blocks to the nearest ten, what block(s) need to be removed?

b. To round the whole number represented by the base-ten blocks to the nearest ten, what block(s) need to be added?

c. Write the steps that you would teach to your students for rounding a whole number to the nearest ten using base-ten blocks.

32. **Rounding Whole Numbers: In Your Classroom** You are teaching your students to round whole numbers to the nearest hundred using base-ten blocks.

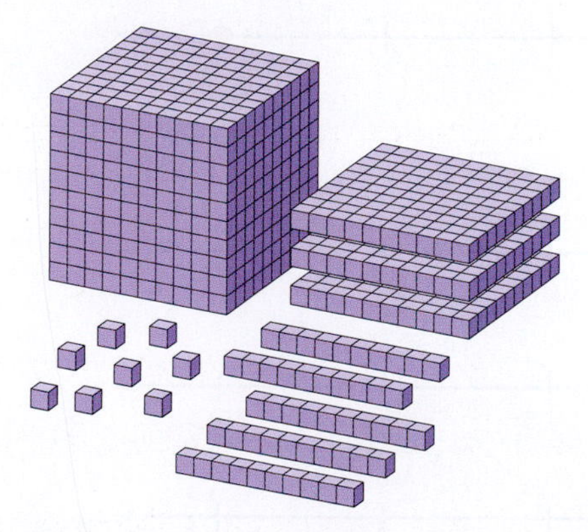

a. To round the whole number represented by the base-ten blocks to the nearest hundred, what block(s) need to be removed?

b. To round the whole number represented by the base-ten blocks to the nearest hundred, what block(s) need to be added?

c. Write the steps that you would teach to your students for rounding a whole number to the nearest hundred using base-ten blocks.

33. **Interpreting Rounded Numbers** The estimated populations of several U.S. cities are shown. Each population is rounded to the nearest thousand. Write the range of whole numbers that each rounded population could represent.

	City	2010 Population
a.	Baton Rouge, LA	802,000
b.	Columbus, OH	1,837,000
c.	Laredo, TX	250,000
d.	Oklahoma City, OK	1,253,000

34. **Interpreting Rounded Numbers** The estimated populations of several countries are shown. Each population is rounded to the nearest hundred thousand. Write the range of whole numbers that each rounded population could represent.

	Country	2010 Population
a.	The Bahamas	300,000
b.	Canada	33,800,000
c.	Mexico	112,500,000
d.	United States	310,200,000

35. **Problem Solving** Female antelopes form different sized herds at different times of the year. A park ranger is patrolling a nature reserve to count the numbers of females in the herds.

Herd counts
63
77
62
96
43
48

a. Estimate the number of female antelopes accounted for on the card by rounding each number to the nearest ten and adding.

b. At last estimate, the park had about 450 female antelopes. Based on your estimate in part (a), should the ranger keep looking for herds? Explain.

36. **Problem Solving** You are shopping for furniture for a home remodeling project. You have chosen pieces that cost \$119, \$485, \$324, \$198, \$339, \$665, \$138, and \$324.

a. Estimate the total cost of the furniture you have chosen so far by rounding each price to the nearest \$100 and adding.

b. Your spending limit is \$2500. Should you trust your estimate and buy the furniture that you have chosen? Explain.

Activity: Placing Whole Numbers on a Number Line

Materials:

- Pencil

Learning Objective:

Determine the scale of a number line and understand that the numbers increase from left to right. Students should be able to graph whole numbers on a number line.

Name ____________________________

Write the number on the number line that corresponds to ●.

1.
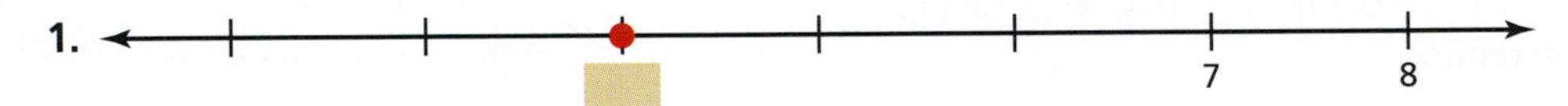

2.
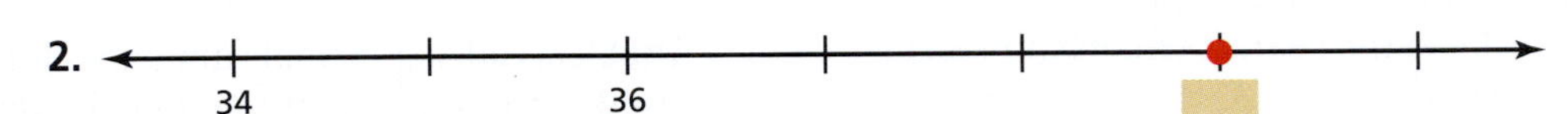

3.
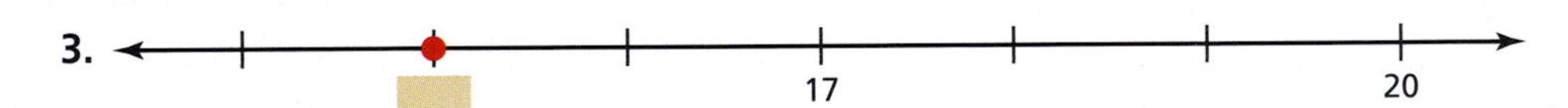

4.
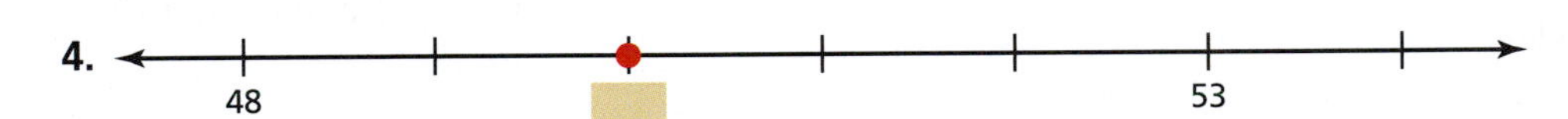

5.
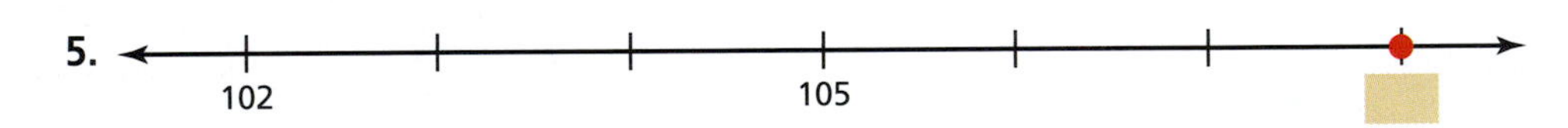

Fill in the missing labels of the number line.

6.
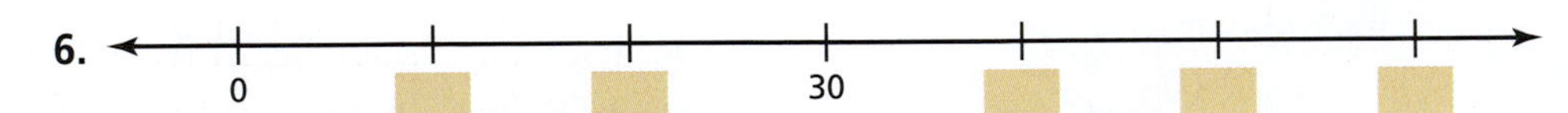

7.
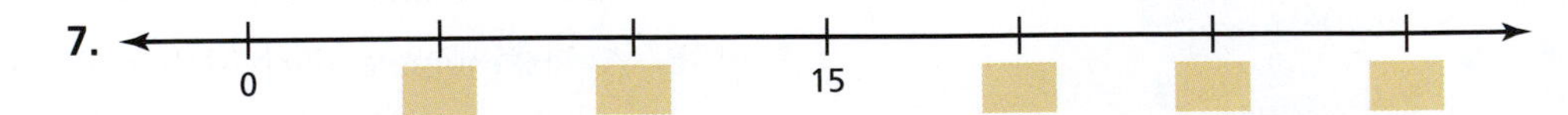

8.
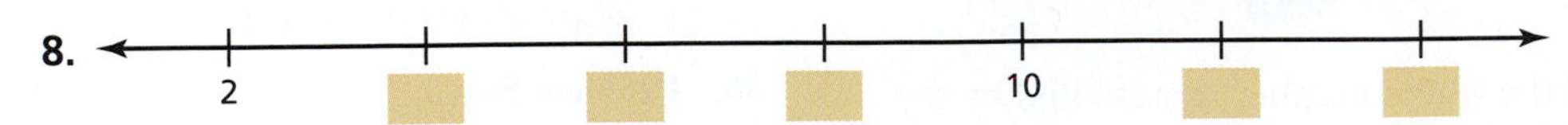

9.
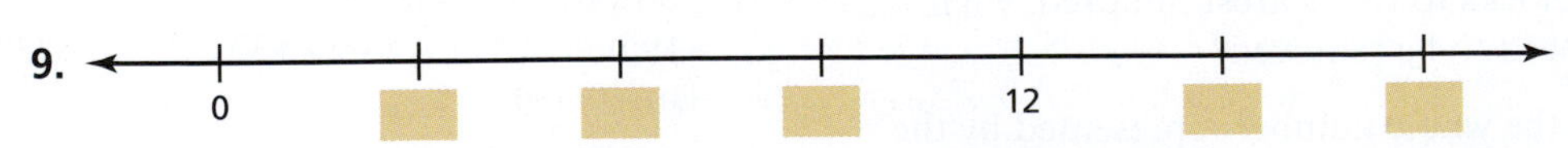

10.
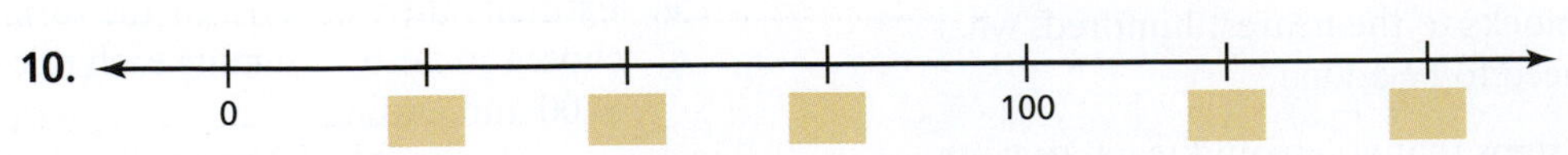

Kzenon/Shutterstock.com

2.3 Number Lines and Numeral Systems

Standards

Grades K–2
Number and Operations in Base Ten
Students should understand place value.

Grades 3–5
Number and Operations in Base Ten
Students should generalize place-value understanding for multi-digit whole numbers.

- Understand and use the properties of the number line for whole numbers.
- Understand the concept of base for a positional numeral system, and represent numbers using alternative numeral systems.

The Number Line for Whole Numbers

It is uncertain exactly when humans first learned the concept of numbers. Whenever it was, it was some time later when humans assigned names to different numbers. And it was even later when humans assigned symbols or *numerals* to those numbers.

By the time the early civilizations developed, humans had established numeral systems. Within these systems, they could write, add, subtract, multiply, and divide numbers. They also knew that whole numbers could be ordered. In ancient Egypt, construction guidelines were labeled at regular intervals using whole numbers.

The Number Line for Whole Numbers

A **number line** is a line on which numbers are marked at intervals.

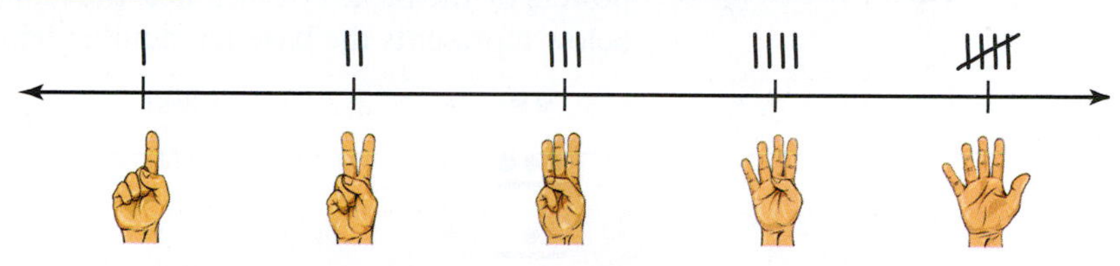

The concept that numbers are ordered is independent of the symbols or numerals used to represent the numbers. Below are some properties of the number line that are important for elementary school students to know. Students can use a number line to compare and order whole numbers.

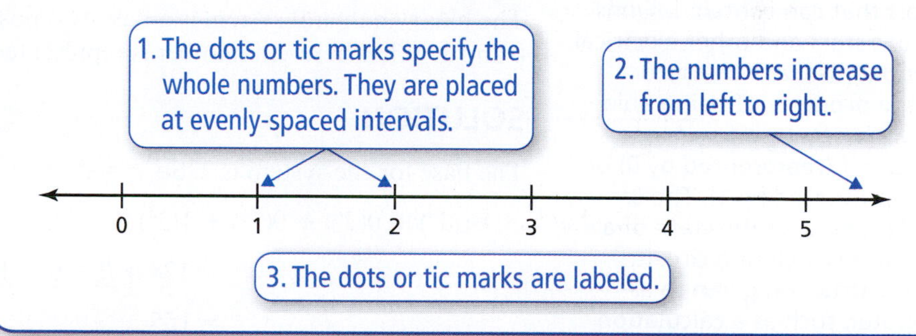

EXAMPLE 1 **Comparing Whole Numbers Using a Number Line**

Compare 7 and 3.

SOLUTION

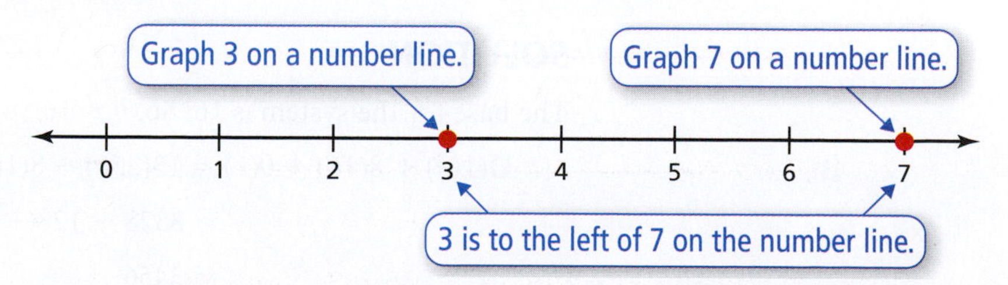

So, 3 is less than 7, or $3 < 7$.

Bases for Positional and Nonpositional Systems

Classroom Tip

As a classroom activity for your students, ask them to invent their own numeral systems. Require that it not be base ten and that it be positional.

For example, suppose that a cartoon character, who only has 8 fingers and thumbs instead of 10, invents a numeral system.

Bases for Numeral Systems

Let b be a natural number that is greater than 1. A **base-b numeral system** is based on powers of b. Typically, the numeral system has a different symbol for each power of b.

$$\ldots,\quad b^5,\quad b^4,\quad b^3,\quad b^2,\quad b,\quad 1$$

Recall from page 50 that a *positional numeral system* is a system in which the value of a digit depends on the position or place it occupies in a numeral.

EXAMPLE 2 **The Mayan Numeral System**

The Mayan numeral system is a base-twenty positional system. The numbers are written vertically with the lowest digit representing ones, and the higher digits representing more powers of the base. For instance, the numeral below represents the base-ten number 966.

$$\rightarrow \quad 2(20^2) \quad \rightarrow 800$$
$$\rightarrow \quad 8(20) \quad \rightarrow 160$$
$$\rightarrow \quad 6(1) \quad \rightarrow +\ 6$$
$$966$$

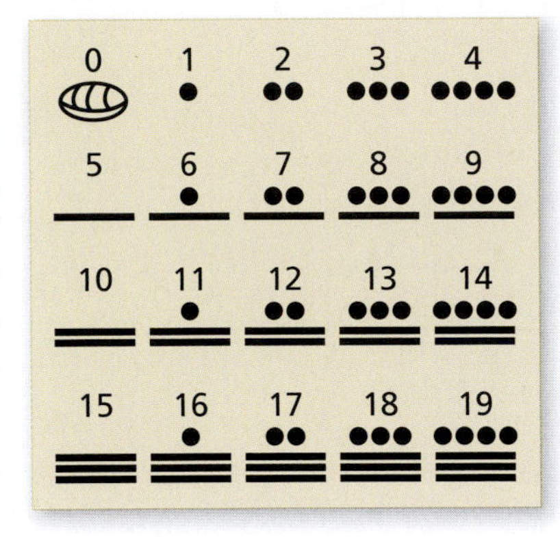

Classroom Tip

Students might be interested to know that computers store data using integrated circuits (chips) that can contain billions of transistors and other electrical components. Each transistor has the property of being able to switch between two binary states: off (represented by 0) or on (represented by 1). The 0s and 1s represent the state of a transistor switch or a circuit. Every instruction given to a computer, such as a calculation, is represented using the binary numeral system.

EXAMPLE 3 **The Binary Numeral System**

The binary numeral system is a base-two positional system that uses the symbols 0 and 1. What base-ten number is represented by the binary number 10011011?

SOLUTION

The base for the system is 2. So, $b = 2$.

$$\begin{aligned} 1(2^7) + 0(2^6) + 0(2^5) + 1(2^4) + 1(2^3) + 0(2^2) + 1(2) + 1(1) &= 128 + 0 + 0 + 16 + 8 + 0 + 2 + 1 \\ &= 155 \end{aligned}$$

EXAMPLE 4 **The Hexadecimal Numeral System**

The hexadecimal numeral system is a base-sixteen positional system that uses the symbols 0, 1, 2, 3, 4, 5, 6, 7, 8, 9, A, B, C, D, E, F for the numbers 0 through 15. What base-ten number is represented by the hexadecimal number D80?

SOLUTION

The base for the system is 16. So, $b = 16$.

$$\begin{aligned} D(16^2) + 8(16) + 0(1) &= 13(256) + 8(16) + 0 \\ &= 3328 + 128 + 0 \\ &= 3456 \end{aligned}$$

EXAMPLE 5 The Egyptian Numeral System

The ancient Egyptian numeral system was base ten. This numeral system was not positional. The hieroglyphs (artistic pictures or symbols) representing the powers of 10 could be placed in any order, horizontally or vertically.

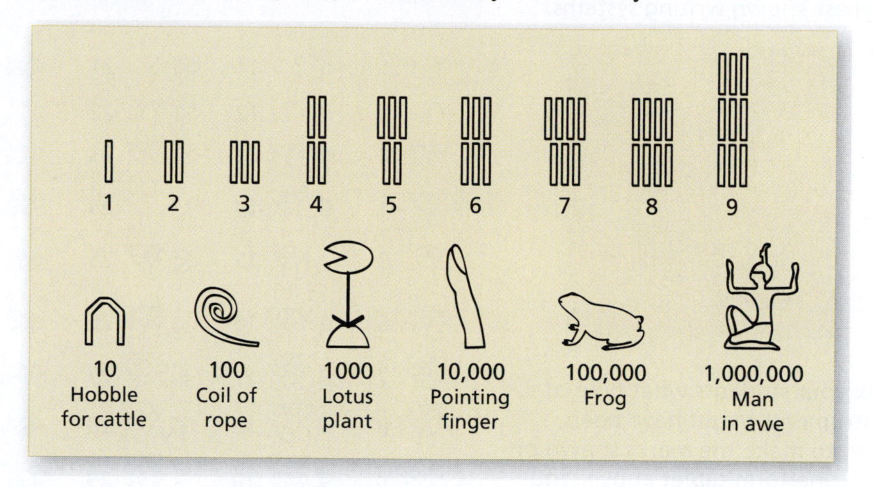

What number is represented below?

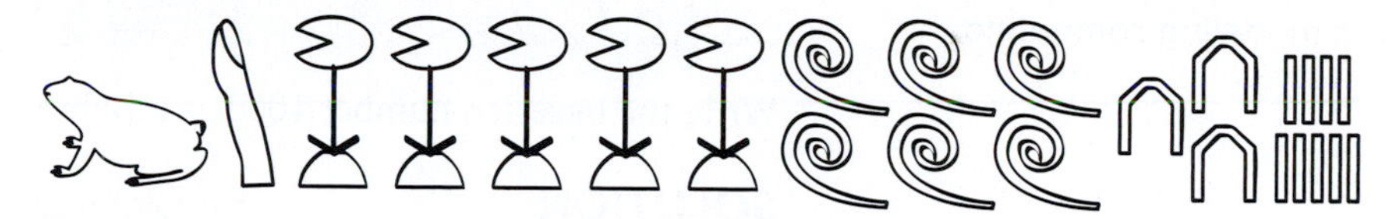

SOLUTION

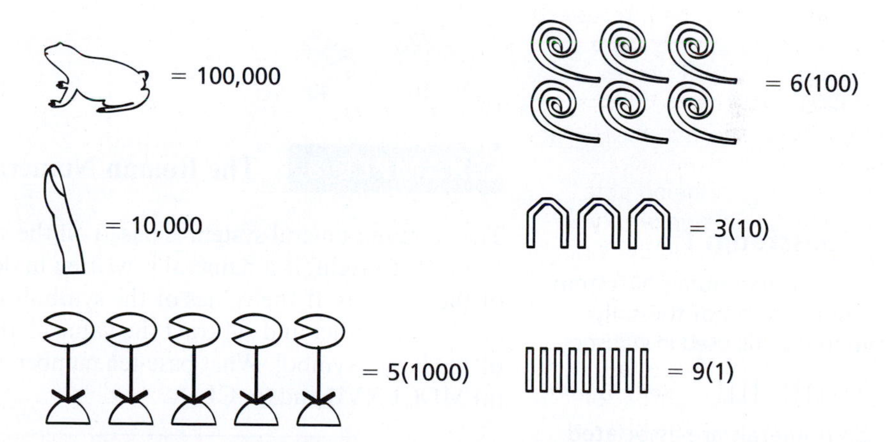

The number is 100,000 + 10,000 + 5(1000) + 6(100) + 3(10) + 9(1) = 115,639.

EXAMPLE 6 The Chinese Numeral System

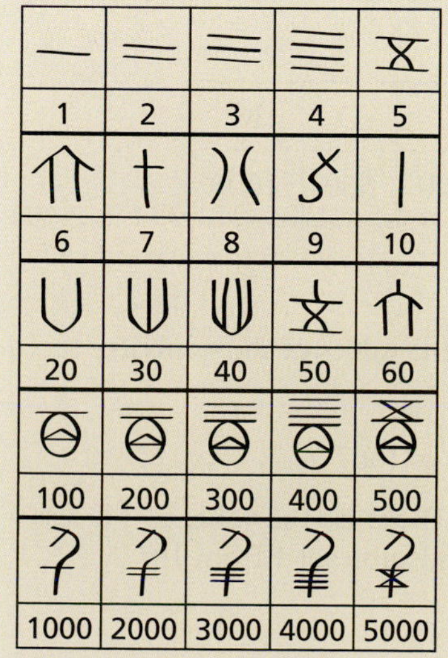

In 1899, archaeologists made a discovery in China that dated back to 1400 B.C. Items found contained inscriptions that had numerical information detailing various events. The table at the left shows some of the numerals found. What is the base for this numeral system? Is this numeral system positional?

SOLUTION

The system is base ten. Notice that it has no single symbols for the numbers 11 through 19. The system is nonpositional. If it were, then it would have no need for special symbols for 20, 30, 40, 50, 60, 200, 300, 400, 500, 2000, 3000, 4000, and 5000.

EXAMPLE 7 The Babylonian Numeral System

The ancient Babylonian numeral system was a base-sixty positional system. In this system, there was no symbol for zero. Zero was indicated by a blank space in the number being written. As a result, cuneiform writing was very difficult to learn.

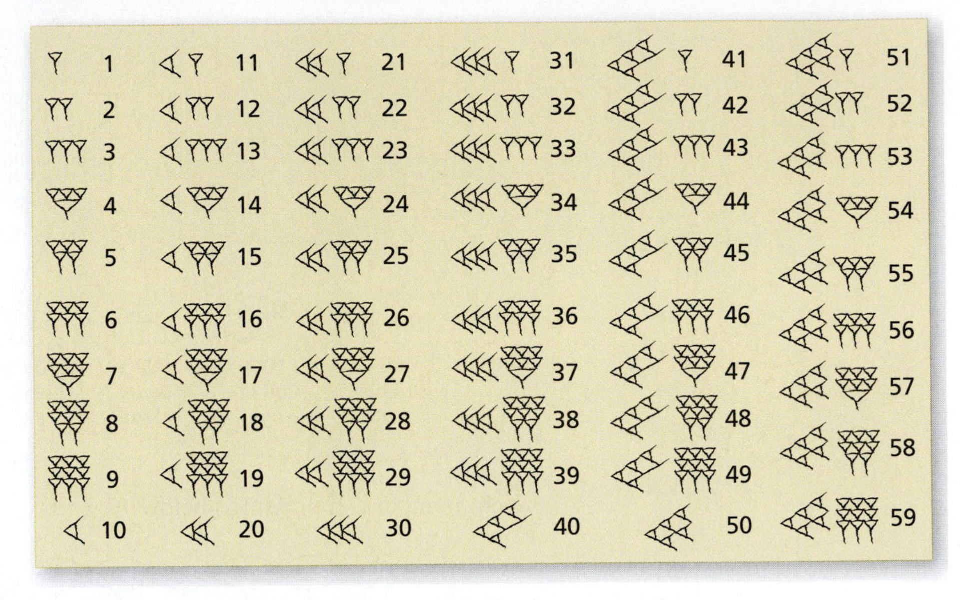

Write the base-ten number 1000 as a Babylonian numeral.

SOLUTION

Because $1000 = 16(60) + 40$, the Babylonian numeral can be written as shown.

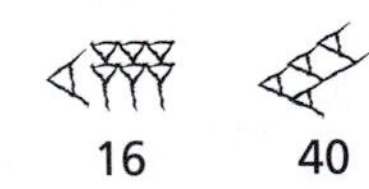

EXAMPLE 8 The Roman Numeral System

The Roman numeral system is based on the symbols in the table below. Reading from left to right, if a numeral is written in decreasing order, then add the values of the symbols. If the values of the symbols increase, then group the pair of symbols together and subtract the value of the smaller symbol from the value of the larger symbol. What base-ten number is represented by the Roman numeral (a) MDCLXVII and (b) CLIX?

Symbol	I	V	X	L	C	D	M
Value	1	5	10	50	100	500	1000

SOLUTION

a. The numeral is written in decreasing order. Add the values of the symbols.

M D C L X V I I

$$1000 + 500 + 100 + 50 + 10 + 5 + 1 + 1 = 1667$$

b. The numeral is not written in decreasing order. Because the value of I is less than the value of X, group I and X together and subtract 1 from 10.

C L IX

$$100 + 50 + (10 - 1) = 159$$

Classroom Tip

The ancient Babylonian system of writing is called *cuneiform*. It was written on wet clay tablets and is one of the earliest known writing systems.

Ask your students what type of instruments might have been used to make the marks shown on the cuneiform tablet above. You can even have your students try to duplicate some of the symbols using modeling compound.

Classroom Tip

The Roman numeral system is a sophistication of the tally system that is still used in statistics.

9 = IIII IIII Tally system

Roman numerals are associated with hand signals. For instance, the V for 5 comes from the V-shape between a person's thumb and index finger.

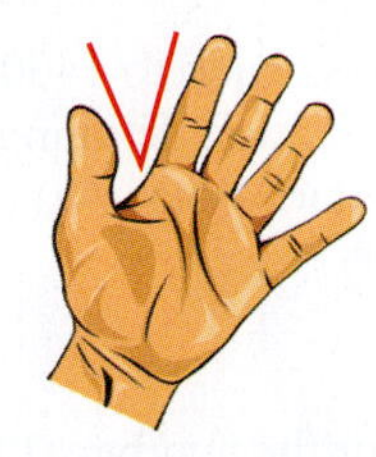

Ask your students where Roman numerals are still used today.

Wikimedia Commons; iStockphoto.com/susaro

EXAMPLE 9 An Octal Numeral System

Some Native American tribes used an octal numeral system because they counted on the spaces between their fingers rather than on their fingers. What is the base for this system?

SOLUTION

The numeral system is base eight. Notice that there are 8 spaces between the digits of a person's two hands.

EXAMPLE 10 The Arabic-Hindu Numeral System

The Arabic-Hindu numeral system is a base-ten positional system that was developed over many centuries. Its symbols have been traced to the Hindus of India as far back as 200 B.C. The system was adopted by Arab mathematicians around A.D. 800. Eventually, the symbols were transported to Spain, where a late 10th-century version looked like the numerals shown at the left. As books and printing become more standardized, the system spread throughout Europe and today it is the dominant system used throughout the world. Even so, the symbols used for the 10 digits in the system still vary.

European	0	1	2	3	4	5	6	7	8	9
Arabic-Indic	٠	١	٢	٣	٤	٥	٦	٧	٨	٩
Eastern Arabic-Indic (Persian and Urdu)	۰	۱	۲	۳	۴	۵	۶	۷	۸	۹
Devanagari (Hindi)	०	१	२	३	४	५	६	७	८	९

Early versions of the Hindu numeral system did not have a symbol for zero. Explain why it is necessary to have a symbol for zero in a positional numeral system.

SOLUTION

A positional numeral system must have a symbol for zero to distinguish between base-ten numbers such as $15 = 1(10) + 5(1)$ and $105 = 1(100) + 0(10) + 5(1)$.

Mathematical Practices

Look for and Express Regularity in Repeated Reasoning

MP8 Mathematically proficient students continually evaluate the reasonableness of intermediate results.

The table shows the different representations among the numeral systems.

Mayan	𝋠	𝋡	𝋢	𝋣	𝋤	𝋥	𝋦	𝋧	𝋨	𝋩	𝋪
Hexadecimal	0	1	2	3	4	5	6	7	8	9	A
Egyptian		𓏺	𓏻	𓏼	𓏽	𓏾	𓏿	𓐀	𓐁	𓐂	𓎆
Chinese		一	二	三	亖	〤	〥	〦	〧	〨	〡
Babylonian		𒁹	𒈫	𒐈	𒐉	𒐊	𒐋	𒑂	𒑄	𒑆	𒌋
Roman		I	II	III	IV	V	VI	VII	VIII	IX	X
European	0	1	2	3	4	5	6	7	8	9	10

iStockphoto.com/susaro

2.3 Exercises

Free step-by-step solutions to odd-numbered exercises at *MathematicalPractices.com.*

1. Writing a Solution Key Write a solution key for the activity on page 58. Describe a rubric for grading a student's work.

2. Grading the Activity In the activity on page 58, a student gave the answers below. Each question is worth 10 points. Assign a grade to each answer. For those that are incorrect, why do you think the student erred?

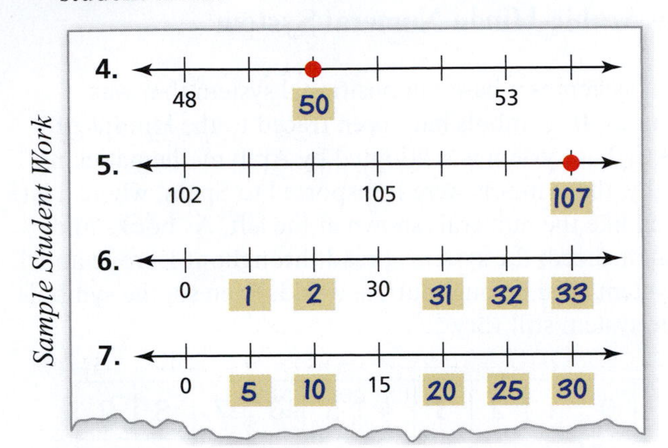

3. Comparing Whole Numbers Using Number Lines Use $<$ or $>$ to compare the numbers that correspond to ● on each number line.

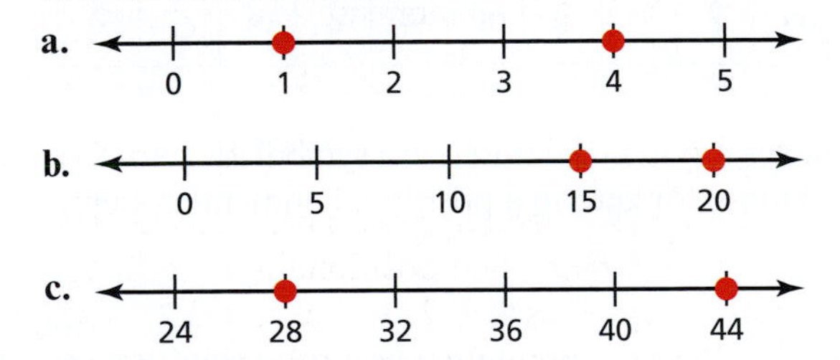

4. Comparing Whole Numbers Using Number Lines Use $<$ or $>$ to compare the numbers that correspond to ● on each number line.

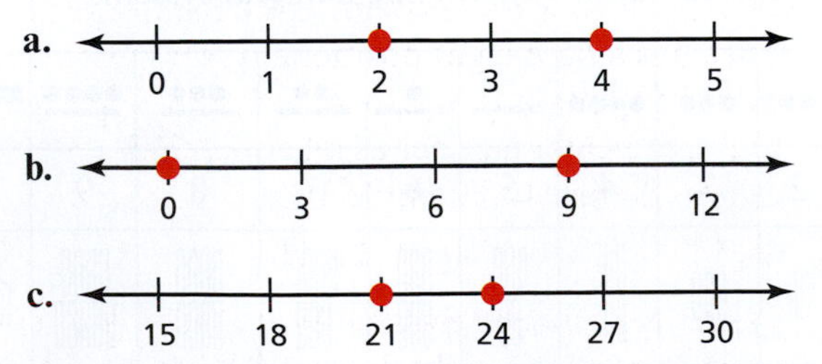

5. Comparing Whole Numbers Using Number Lines Graph each pair of numbers on a number line. Then compare the numbers.

a. 4 and 2 **b.** 3 and 8

c. 20 and 10 **d.** 22 and 26

6. Comparing Whole Numbers Using Number Lines Graph each pair of numbers on a number line. Then compare the numbers.

a. 5 and 1 **b.** 6 and 9

c. 15 and 10 **d.** 39 and 43

7. The Mayan Numeral System Write each Mayan number as a base-ten number.

a. **b.**

c. **d.**

8. The Mayan Numeral System Write each Mayan number as a base-ten number.

a. **b.**

c. **d.**

9. The Binary Numeral System Write each binary number as a base-ten number.

a. 101 **b.** 1101

c. 10001 **d.** 111111

10. The Binary Numeral System Write each binary number as a base-ten number.

a. 110 **b.** 1011

c. 11011 **d.** 101101

11. The Hexadecimal Numeral System Write each hexadecimal number as a base-ten number.

a. B **b.** 1C

c. 1036 **d.** 21AF

12. The Hexadecimal Numeral System Write each hexadecimal number as a base-ten number.

a. D **b.** AB

c. 104E **d.** 31B8

13. The Egyptian Numeral System Write each Egyptian number as a base-ten number.

a. 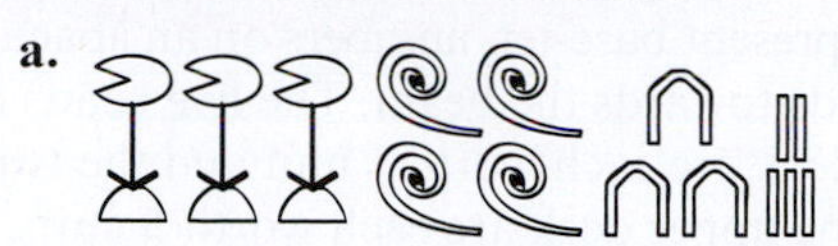

b. 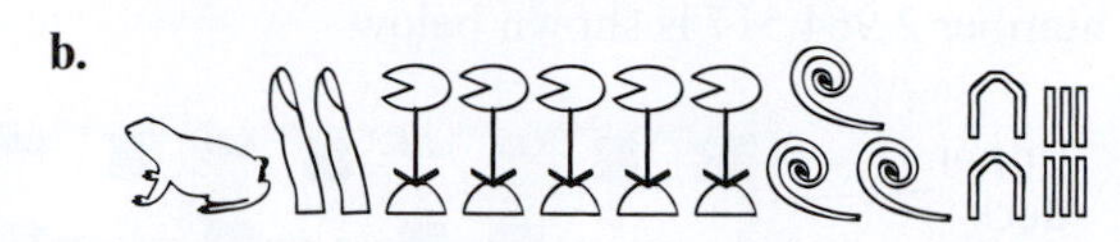

c. 

d. 

14. The Egyptian Numeral System Write each Egyptian number as a base-ten number.

a.

b.

c.

d.

15. The Chinese Numeral System The Chinese numeral system on page 61 has both additive and multiplicative properties. Explain how one or both of these properties are used to form each number using two or more Chinese numerals.

a. 80:

b. 33:

c. 221:

16. The Chinese Numeral System The Chinese numeral system on page 61 has both additive and multiplicative properties. Explain how one or both of these properties are used to form each number using two or more Chinese numerals.

a. 800:

b. 52:

c. 6060:

17. The Babylonian Numeral System Write each base-ten number as a Babylonian number.

a. 61 b. 70

c. 130 d. 602

18. The Babylonian Numeral System Write each base-ten number as a Babylonian number.

a. 63 b. 80

c. 150 d. 1220

19. The Roman Numeral System Write each Roman numeral as a base-ten number.

a. XVII b. DLXXVI

c. XCIX d. MCLXXIV

20. The Roman Numeral System Write each Roman numeral as a base-ten number.

a. XXVI b. MMDLXI

c. XLIX

d. DCCXXXIX

21. The Roman Numeral System Write each base-ten number as a Roman numeral.

a. 19 b. 43

c. 334 d. 1567

22. The Roman Numeral System Write each base-ten number as a Roman numeral.

a. 38 b. 144

c. 799 d. 2008

23. The Octal Numeral System The octal numeral system is a base-eight positional system that uses the symbols 0, 1, 2, 3, 4, 5, 6, and 7. Write each base-eight number as a base-ten number. (*Hint:* Similar to Example 3)

a. 11 b. 115

24. The Octal Numeral System The octal numeral system is a base-eight positional system that uses the symbols 0, 1, 2, 3, 4, 5, 6, and 7. Write each base-eight number as a base-ten number. (*Hint:* Similar to Example 3)

a. 12 b. 117

25. The Octal Numeral System Write each base-ten number as a base-eight number. (*Hint:* Similar to Example 7)

a. 19 b. 70

26. The Octal Numeral System Write each base-ten number as a base-eight number. (*Hint:* Similar to Example 7)

a. 26 b. 68

27. Comparing Whole Numbers: In Your Classroom There is a Transitive Property of Inequalities for Whole Numbers, and it has two parts. One part is stated below for three whole numbers, a, b, and c.

"If $a < b$ and $b < c$, then $a < c$."

a. Explain how to demonstrate this property to your students by graphing the whole numbers a, b, and c on a number line.

b. Do you think that there is a reflexive property for inequalities? Explain your reasoning.

28. Comparing Whole Numbers: In Your Classroom There is a Transitive Property of Inequalities for Whole Numbers, and it has two parts. One part is stated below for three whole numbers, a, b, and c.

"If $a > b$ and $b > c$, then $a > c$."

a. Explain how to demonstrate this property to your students by graphing the whole numbers a, b, and c on a number line.

b. Do you think that there is a symmetric property for inequalities? Explain how you can use a number line to demonstrate your reasoning to your students.

29. A Chinese Numeral System A Chinese numeral system from the 4th century B.C. used the two sets of symbols shown below for the numbers 1 through 9.

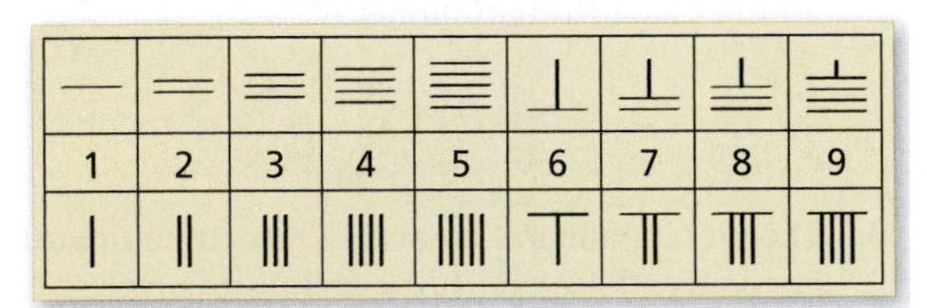

1	2	3	4	5	6	7	8	9

To form a number, the two types of symbols were alternately placed in squares along a row of a counting board. The base-ten number 2,058,731 in this numeral system is shown below.

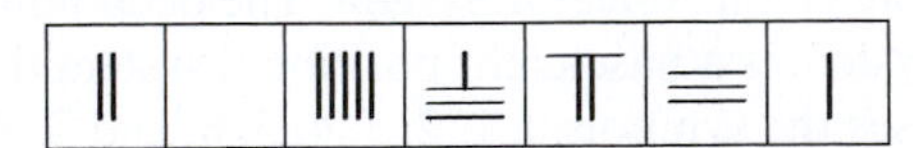

a. What is the base for this numeral system?

b. Is this numeral system positional? Explain.

c. How is zero represented in this system?

d. What base-ten number is represented by the Chinese number below?

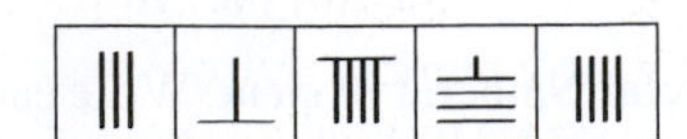

30. The Chinese Abacus The Chinese have used the abacus for counting and arithmetic for centuries. You can represent base-ten numbers on an abacus by sliding beads towards the beam. The five beads on the lower deck are each worth 1 unit and the two beads on the upper deck are each worth 5 units. The number 2,964,517 is shown below.

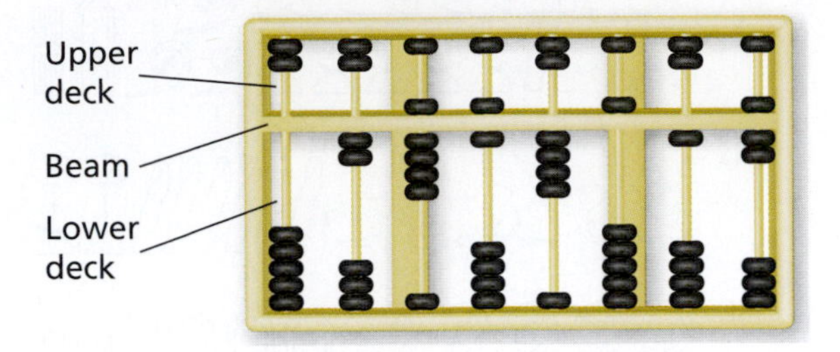

a. Is this numeral system positional? Explain.

b. How is a zero represented in this system?

c. What base-ten number is represented below?

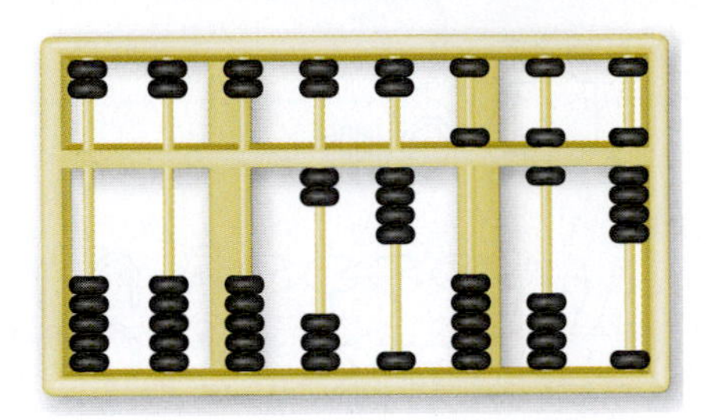

31. Modeling Binary Numbers Using Base-Two Blocks There are four base-two block sizes.

Position	Place Value	Block Model
1	Ones (units)	
10	Twos	
100	Fours	
1000	Eights	

a. Write the binary number represented by the base-two blocks.

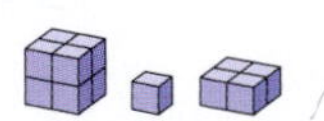

b. What base-ten number is represented by the binary number in part (a)?

c. What is the largest binary number that can be modeled using the four sizes of base-two blocks? What base-ten number is represented by this binary number?

d. Describe the set of base-ten numbers whose binary representations can be modeled with the four sizes of base-two blocks.

32. Modeling Octal Numbers Using Base-Eight Blocks There are four base-eight block sizes.

Position	Place Value	Block Model
1	Ones (units)	
10	Eights	
100	Sixty-fours	
1000	Five hundred twelves	

a. Write the octal number represented by the base-eight blocks.

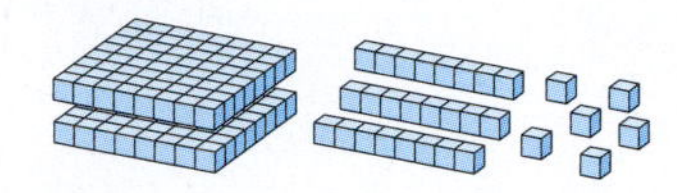

b. What base-ten number is represented by the octal number in part (a)?

c. What is the largest octal number that can be modeled using the four sizes of base-eight blocks? What base-ten number is represented by this octal number?

d. Describe the set of base-ten numbers whose octal representations can be modeled with the four sizes of base-eight blocks.

33. Writing Is the Roman numeral system a positional numeral system? Explain.

34. Writing Is the Roman numeral system a base-ten system? Explain.

35. Grading Student Work On a diagnostic test, one of your students does the following work. Explain what the student did wrong. Which topics would you encourage the student to review?

a.

> Write the hexadecimal number E31 as a base-ten number.
>
> $16(14^2) + 16(3) + 1 = 3136 + 48 + 1$
> $= 3185$

b.

> Write the Roman numeral DCXL as a base-ten number.
>
> $500 + 100 + 10 + 50 = 660$

36. Grading Student Work On a diagnostic test, one of your students does the following work. Explain what the student did wrong. Which topics would you encourage the student to review?

a.

> Write the binary number 101 as a base-ten number.
>
> $1(2^3) + 0(2^2) + 1(2) = 8 + 0 + 2$
> $= 10$

b.

> Write the Roman number CMX as a base-ten number.
>
> $100 + 1000 + 10 = 1110$

37. Problem Solving You can also use an abacus to represent hexadecimal numbers. The five beads on the lower deck are each worth 1 unit and the two beads on the upper deck are each worth 5 units. To represent the hexadecimal digit F (decimal 15), slide all the beads in one column toward the beam.

a. What hexadecimal number is represented below?

b. What base-ten number is represented by the hexadecimal number in part (a)?

38. Problem Solving Use the capacity conversions shown below.

1 cup = 16 tablespoons

1 pint = 2 cups

1 quart = 2 pints

1 gallon = 4 quarts

a. Write the binary number that represents the number of cups in 1 gallon.

b. Write the base-four number that represents the number of tablespoons in 1 gallon.

c. There are 2 tablespoons in 1 fluid ounce. Write the binary number that represents the number of fluid ounces in 1 gallon.

Chapter Summary

Chapter Vocabulary

set *(p. 39)*
elements *(p. 39)*
null set *(p. 39)*
equal *(p. 39)*
number *(p. 40)*
numeral *(p. 40)*
cardinality of a set *(p. 40)*
finite set *(p. 40)*
infinite set *(p. 40)*
natural numbers *(p. 40)*
counting numbers *(p. 40)*
whole numbers *(p. 40)*
equivalent *(p. 41)*
one-to-one correspondence *(p. 41)*
subset *(p. 42)*
proper subset *(p. 42)*
less than or equal to *(p. 43)*
less than *(p. 43)*
greater than or equal to *(p. 43)*
greater than *(p. 43)*
Reflexive Property *(p. 49)*
Symmetric Property *(p. 49)*
Transitive Property *(p. 49)*
positional numeral system *(p. 50)*
base-ten system *(p. 50)*
decimal system *(p. 50)*
place value *(p. 50)*
expanded form *(p. 50)*
base-ten blocks *(p. 51)*
number line *(p. 59)*
base-*b* numeral system *(p. 60)*

Chapter Learning Objectives — Review Exercises

2.1 Sets *(page 39)* — 1–26

- Describe a set using the listing method and set-builder notation, and identify when sets are equal.
- Develop the concept of cardinality of a set and use it to describe sets of numbers.
- Compare the equality and equivalence of different sets.
- Use the concept of subset to develop the concept of inequality.

2.2 Whole Numbers *(page 49)* — 27–68

- Understand and use the properties of equality.
- Understand the base-ten system for whole numbers.
- Round whole numbers to given place values.

2.3 Number Lines and Numeral Systems *(page 59)* — 69–112

- Understand and use the properties of the number line for whole numbers.
- Understand the concept of base for a positional numeral system, and represent numbers using alternate numeral systems.

Important Concepts and Formulas

Definition of a Set

A **set** is a collection of objects. The **elements** (or members) of a set can be numbers, people, letters of the alphabet, others sets, and so on. The **null set**, denoted by $\{\ \}$ or $\varnothing$, is empty and has no elements.

Definition of a Subset

Set A is a **subset** of set B, written as $A \subseteq B$, if and only if every element in A is also an element in B.

Reflexive Property $a = a$

Symmetric Property If $a = b$, then $b = a$.

Transitive Property If $a = b$ and $b = c$, then $a = c$.

The Number Line for Whole Numbers

A **number line** is a line on which numbers are marked at intervals.

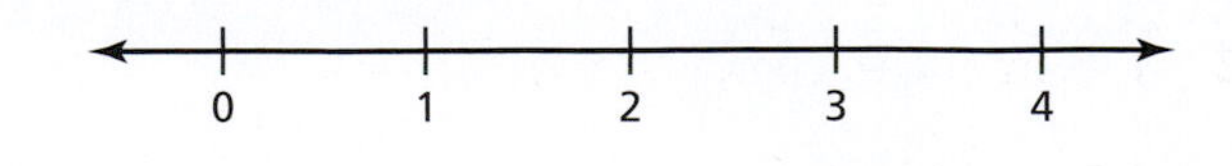

Review Exercises

Free step-by-step solutions to odd-numbered exercises at MathematicalPractices.com.

2.1 Sets

Using the Listing Method Write the set using the listing method.

1. The set of odd integers from 21 through 29
2. The set of letters in the word *science*

Using Set-Builder Notation Write the set using set-builder notation.

3. The set of odd integers from 17 through 21
4. The set of letters in the word *geology*

Equality of Sets Write the set using the listing method, if necessary. Then decide which of the sets are equal. Explain your reasoning.

5. 

6. $A = \{x \mid x$ is a number from 1 through 15 and is divisible by 3.$\}$

   $B = \{1, 3, 5, 7, 9\}$

   $C = \{x \mid x$ is an odd number from 1 through 10.$\}$

Finding the Cardinality of a Set Tell whether the set is *finite* or *infinite*. If the set is finite, find its cardinality.

7. $A = \{a, b, c, d\}$
8. $B = \{x \mid x$ is an odd number from 0 through positive infinity.$\}$
9. $C = \{6, 7, 8\}$
10. $D = \{x \mid x$ is a real number from 0 through 1.$\}$

One-to-One Correspondence Determine whether the elements in the sets can be put in a one-to-one correspondence. Explain your reasoning.

11. $A = \{21, 31, 41\}$

    $B = \{x, y, z\}$

12. $A = \{$math, science, history, English$\}$

    $B = \{1, 2, 3\}$

Comparing Equality and Equivalence Tell which sets are *equal* and which sets are *equivalent*. Explain your reasoning.

13. $A = \{9999, 9, 999\}$

    $B = \{$basketball, softball, football, volleyball$\}$

    $C = \{9, 99, 999, 9999\}$

    $D = \{$pen, pencil, marker$\}$

    $E = \{999, 99, 9, 9999\}$

14. $A = \{\ \}$

    $B = \{8, 9, 10, 11, 12, 13, 14\}$

    $C = \{x \mid x$ is a letter from the alphabet.$\}$

    $D = \{s, o, l, v, i, n, g\}$

    $E = \{x \mid x$ is an integer from 1 through 26.$\}$

Subsets and Proper Subsets Tell whether the statement is *true* or *false*. Explain your reasoning.

15. The set of integers is a proper subset of the set of real numbers.
16. The set of negative integers is a subset of the set of integers.
17. The set $\{0\}$ is a subset of the set of positive integers.
18. The set of positive integers is a proper subset of the set of integers.

Comparing Two Numbers Use a diagram to determine which number is greater. Write your answer as an inequality.

19. 10, 7
20. 3, 11

Using Inequalities to Define a Set Write the set using the listing method.

21. $A = \{x \mid x$ is a whole number, and $4 < x$ and $x < 11.\}$
22. $B = \{x \mid x$ is an integer, and $-6 < x$ and $x < 3.\}$
23. $C = \{x \mid x$ is a natural number, and $1 \le x$ and $x \le 3.\}$
24. $D = \{x \mid x$ is a negative integer, and $x \le -10.\}$
25. **Problem Solving** There are six condiments available for a sandwich. How many ways can you make a sandwich using two different condiments?
26. **Problem Solving** There are five toppings available for a pizza. How many ways can you make a pizza using three different toppings?

2.2 Whole Numbers

Applying a Property of Equality Complete the statement to illustrate the indicated property.

27. Reflexive Property: $21 =$ ______.
28. Symmetric Property: If $17 = x$, then ______.
29. Transitive Property: If $a = b$ and $b = 5$, then ______.
30. Reflexive Property: $33 =$ ______.
31. Symmetric Property: If $9 = x$, then ______.
32. Transitive Property: If $a = b$ and $b = 4$, then ______.

Finding a Place Value State the place value of the underlined digit in the whole number.

33. $30\underline{2}$
34. $20\underline{5},023,154$
35. $75,\underline{3}34$
36. $10\underline{4},760$
37. $28\underline{5}9$
38. $\underline{5}36,412$

Finding the Value of a Digit Write the word name for the value of the underlined digit in the whole number. For example, the value of $\underline{3}$ in $\underline{3}65$ is three hundred.

39. $9\underline{4},875$
40. $4\underline{5}0,621,950$
41. $65\underline{1}9$
42. $5\underline{2}3,045$

Using Base-Ten Blocks and Expanded Form Write the whole number represented by the base-ten blocks.

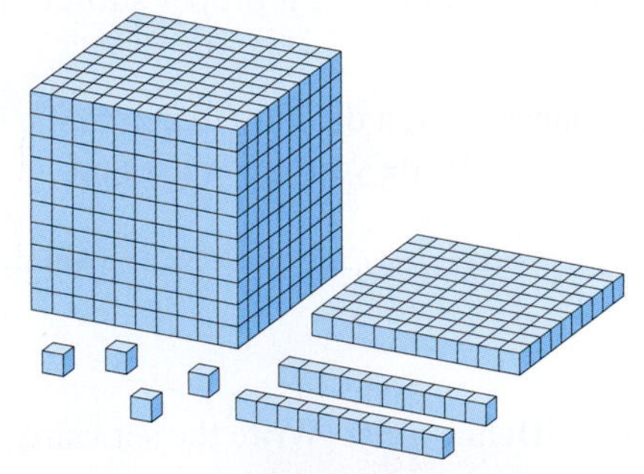

44.

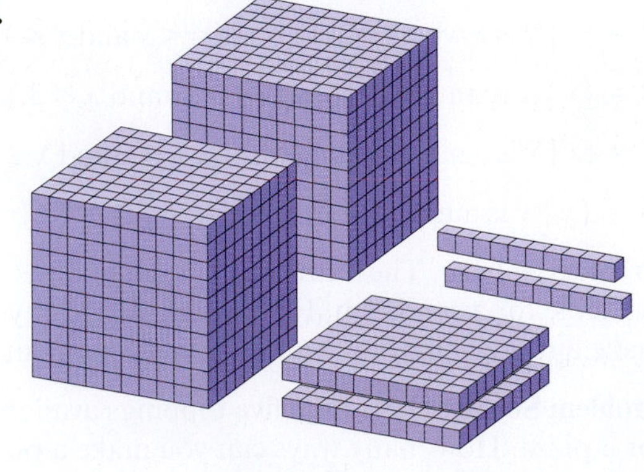

Writing the Word Name for a Numeral Write the word name for the numeral.

45. 431
46. 60,250
47. 123,462
48. 207,345,000

Writing the Numeral for a Word Name Write the numeral for the word name.

49. Three hundred twenty-seven
50. One thousand, fifty-four
51. Eighteen thousand, two hundred fourteen
52. Six hundred three thousand, ninety-one
53. Two million, five hundred fifty-one thousand, one hundred sixty-five
54. Nine hundred million, eight hundred thirty-one thousand

Rounding a Whole Number Round the whole number to the nearest ten.

55. 86
56. 231
57. 2507
58. 10,212

Rounding a Whole Number Round the whole number to the nearest hundred.

59. 456
60. 9023
61. 78,270
62. 102,881

Rounding a Whole Number Round the whole number to the nearest thousand.

63. 999
64. 2001
65. 55,555
66. 651,900

67. **Writing Word Names for Numerals** Rewrite the sentences below using word names instead of numerals.

    A school's French club would like to help raise \$1565 to help pay for a class trip overseas. They plan to sell magazine subscriptions for \$22, coupon booklets for \$15, and raise at least \$350 at a car wash.

68. **Writing Numerals for Word Names** Rewrite the sentences below using numerals instead of word names.

    In two thousand thirteen, a company's profit exceeded four million five hundred fifty thousand dollars. The company plans to distribute two hundred twenty-seven thousand five hundred dollars in bonuses to its employees.

2.3 Number Lines and Numeral Systems

Comparing Whole Numbers Using a Number Line Use < or > to compare the numbers that correspond to • on the number line.

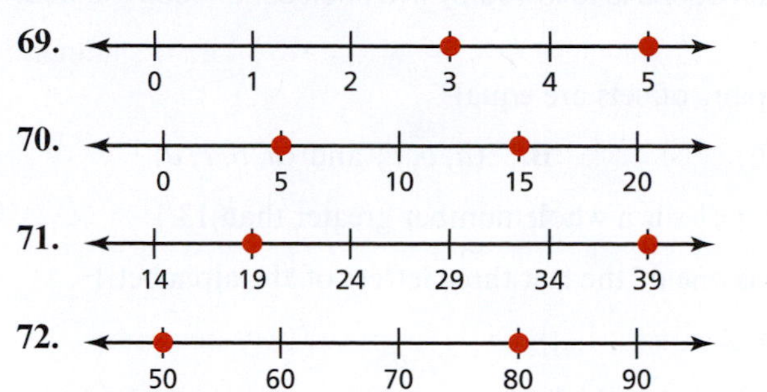

The Mayan Numeral System Write the Mayan number as a base-ten number.

73.

74.

75.

76.

The Binary Numeral System Write the binary number as a base-ten number.

77. 110

78. 1011

79. 10101

80. 111000

The Hexadecimal Numeral System Write the hexadecimal number as a base-ten number.

81. F **82.** 1D

83. 145 **84.** AC

The Egyptian Numeral System Write the Egyptian number as a base-ten number.

85.

86.

87.

88.

The Chinese Numeral System The Chinese numeral system on page 61 has both additive and multiplicative properties. Explain how one or both of these properties are used to form the number below using two or more Chinese numerals.

89. 60:

90. 207:

91. 19:

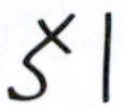

92. 50:

The Babylonian Numeral System Write the base-ten number as a Babylonian number.

93. 63 **94.** 65

95. 72 **96.** 140

The Roman Numeral System Write the Roman numeral as a base-ten number.

97. XVI

98. LII

99. MDLX

100. DCCXV

The Roman Numeral System Write the base-ten number as a Roman numeral.

101. 11 **102.** 40

103. 950 **104.** 2013

The Octal Numeral System The octal numeral system is a base-eight positional system that uses the symbols 0, 1, 2, 3, 4, 5, 6, and 7. Write the base-eight number as a base-ten number.

105. 12 **106.** 105

107. 36 **108.** 412

The Octal Numeral System Write the base-ten number as a base-eight number.

109. 20 **110.** 42

111. 80 **112.** 279

Chapter Test

The questions on this practice test are modeled after questions on state certification exams for K–8 teaching.

Test Directions: Each of the questions is followed by five choices. Choose the *best* response to each question.

1. Which of the following pairs of sets are equal?

A. $\{1, 3, 5\}$ and $\{2, 4, 6\}$ **B.** $\{a, b, c\}$ and $\{a, b, c, d\}$

C. $\{14, 16, 18, 20\}$ and $\{x \mid x$ is a whole number greater than 13.$\}$

D. $\{x, y, z\}$ and $\{w \mid w$ is one of the last three letters of the alphabet.$\}$

E. {$, ♫, ♠} and {♫, ♣, $}

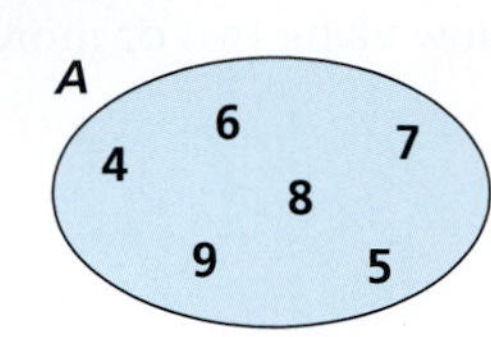

2. What is the cardinality of set A at the left?

A. 2 **B.** 4 **C.** 6 **D.** 7 **E.** 8

3. Which contains all the subsets of $\{w, y, z\}$?

A. $\{w\}, \{y\}, \{z\}, \{w, y\}, \{w, z\}, \{y, z\}, \{\ \}$

B. $\{w\}, \{y\}, \{z\}, \{w, y\}, \{w, z\}, \{y, z\}, \{w, y, z\}$

C. $\{w, y\}, \{w, z\}, \{y, z\}, \{w, y, z\}$

D. $\{w, y, z\}, \{y\}, \{w, z\}, \{z\}, \{w, y\}, \{\ \}, \{y, z\}, \{w\}$

E. $\{\ \}, \{w\}, \{y\}, \{z\}, \{w, y\}, \{y, z\}, \{w, y, z\}$

4. In the number 82,304.61, what is the place value of the 2?

A. Thousands **B.** Millions **C.** Ten thousands

D. Tenths **E.** Thousandths

5. If 406,832 is rounded to the nearest ten thousands place, the number is

A. 406,800 **B.** 406,830 **C.** 407,000

D. 410,000 **E.** 500,000

6. What is the word name for 45,720?

A. Forty-five hundreds, seventy-two

B. Forty-five, seven hundred twenty

C. Forty-five thousand, seven hundred

D. Forty-five million, seven hundred twenty

E. Forty-five thousand, seven hundred twenty

7. Which of the following is not a proper subset of $\{2, 4, 6, 8\}$?

A. $\{2, 4\}$ **B.** $\{2, 4, 6\}$ **C.** $\{2, 8\}$

D. $\{4, 6, 8\}$ **E.** $\{2, 9\}$

8. Which of the following statements includes a cardinal number?

A. There are 20 volumes in the set of periodicals.

B. I received my 9th volume recently.

C. The students meet in Room 304.

D. My phone number is 412-555-3425.

E. Art lives on 1892 Sheffield Avenue.

3 Operations with Whole Numbers

iStockphoto.com/monkeybusinessimages

Activity: Using Base-Ten Blocks to Add Whole Numbers

Materials:

- Pencil
- Base-ten blocks

Learning Objective:

Learn to use base-ten blocks to add multi-digit whole numbers, including the concepts of regrouping 10 ones for 1 ten and 10 tens for 1 hundred.

Name ______________________________

Write the addition problem, and find the sum.

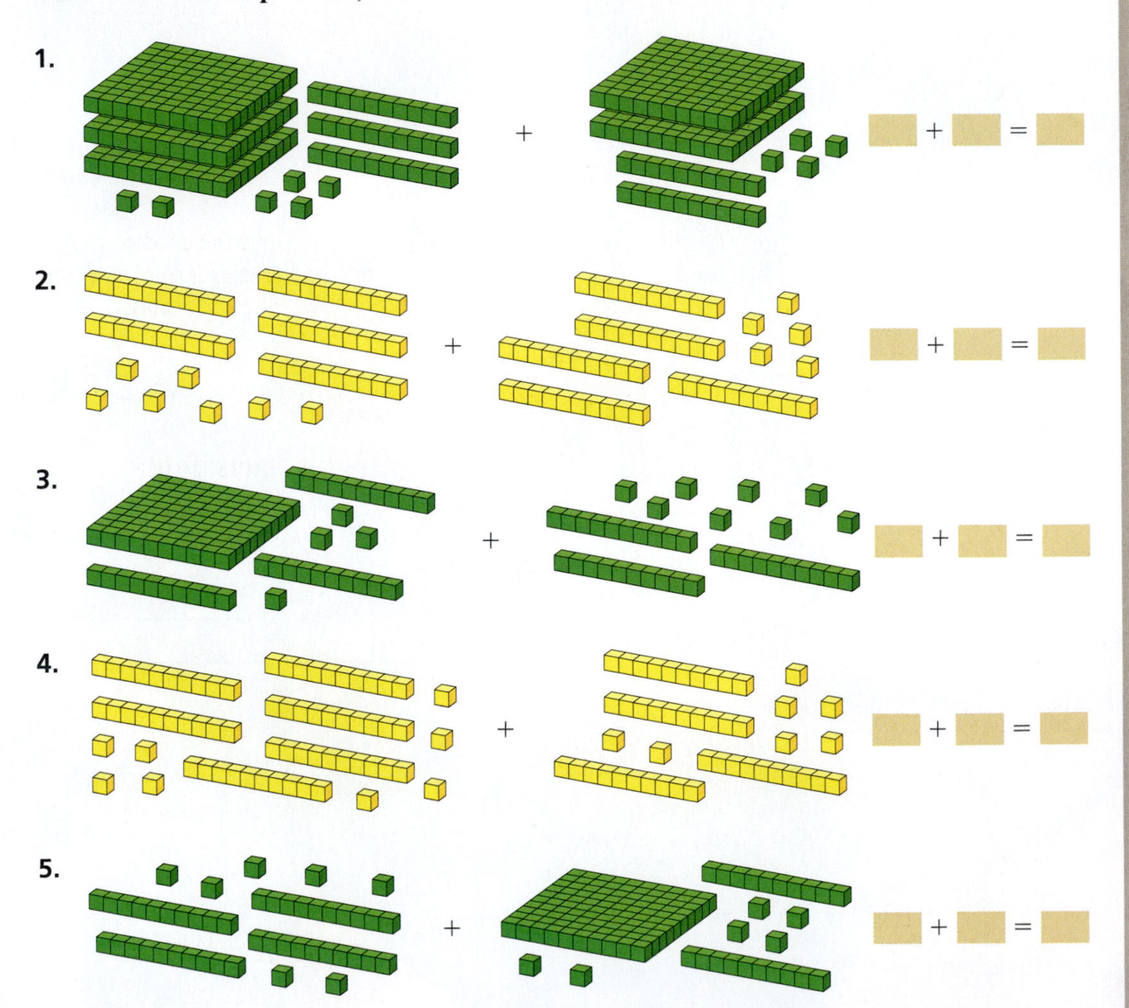

3.1 Adding Whole Numbers

Standards

Grades K–2

Numbers and Operations in Base Ten

Students should use place value understanding and properties of operations to add.

Grades 3–5

Numbers and Operations in Base Ten

Students should use place value understanding and properties of operations to perform multi-digit arithmetic.

- Understand and use the definition of the sum of two whole numbers.
- Understand and use the properties of addition for whole numbers.
- Use models to add whole numbers.
- Use algorithms to add whole numbers.

Defining the Sum of Two Whole Numbers

The mastery of the material in this chapter is critical for students in elementary grades. It forms the foundation for most of the mathematics they will learn in subsequent grades.

Sum of Two Whole Numbers

Let A and B be two sets that have no elements in common. The **union** of A and B, written as $A \cup B$, is the set whose elements consist of precisely the elements of A and B.

A **set model** is one way to represent whole number addition.

Example

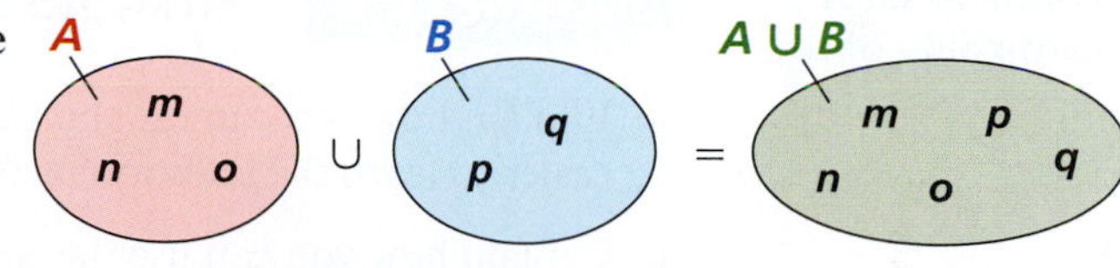

If the cardinality of A is the whole number a and the cardinality of B is the whole number b, then the **sum** of a and b, written as $a + b$, is the cardinality of $A \cup B$. The numbers a and b are the **addends** of the sum.

Numbers $3 + 2 = 5$ "The sum of 3 and 2 is 5" or "3 plus 2 equals 5."

Classroom Tip

As indicated by the cover of this book, there are four basic number operations in mathematics: addition, subtraction, multiplication, and division. Students spend much of the first six years of their education (K–5) learning how to perform these operations with various types of numbers.

iStockphoto.com/porcorex

EXAMPLE 1 Creating an Addition Facts Table

Create an addition facts table that shows the sum of any two whole numbers from 0 through 10.

SOLUTION

There are several ways to do this. One way is shown at the right. For young children who are just learning addition facts, there are dozens of learning aids they can use, including flash cards and games.

+	0	1	2	3	4	5	6	7	8	9	10
0	0	1	2	3	4	5	6	7	8	9	10
1	1	2	3	4	5	6	7	8	9	10	11
2	2	3	4	5	6	7	8	9	10	11	12
3	3	4	5	6	7	8	9	10	11	12	13
4	4	5	6	7	8	9	10	11	12	13	14
5	5	6	7	8	9	10	11	12	13	14	15
6	6	7	8	9	10	11	12	13	14	15	16
7	7	8	9	10	11	12	13	14	15	16	17
8	8	9	10	11	12	13	14	15	16	17	18
9	9	10	11	12	13	14	15	16	17	18	19
10	10	11	12	13	14	15	16	17	18	19	20

Properties of Addition for Whole Numbers

Properties of Addition for Whole Numbers

Commutative Property of Addition

Words The sum of two whole numbers does not depend on how the addends are ordered.

Numbers $1 + 4 = 4 + 1$ **Algebra** $a + b = b + a$

Associative Property of Addition

Words The sum of whole numbers does not depend on how the addends are grouped.

Numbers $(3 + 2) + 8 = 3 + (2 + 8)$ **Algebra** $(a + b) + c = a + (b + c)$

Additive Identity Property

Words The sum of any whole number and the **additive identity**, 0, is the whole number.

Numbers $7 + 0 = 7$ **Algebra** $a + 0 = a, 0 + a = a$

Classroom Tip

You can justify the *Additive Identity Property* by using the definition of the sum of two whole numbers. Notice that when you find the union of a set *A* (with cardinality *a*) and the null set, you get the original set *A*. Recall that the cardinality of the null set is 0.

$$A \cup \{\ \} = A$$
$$\downarrow \quad \downarrow \quad \downarrow \quad \downarrow \quad \downarrow$$
$$a + 0 = a$$

EXAMPLE 2 Strategies for Learning Addition Facts

a. Explain how you can use the *Commutative Property of Addition* to make it easier to learn the addition facts shown in Example 1.

b. Explain how you can use the *Additive Identity Property* to make it easier to learn the addition facts shown in Example 1.

SOLUTION

a. As shown in Example 1, the addition facts table contains 121 sums. By teaching students the *Commutative Property of Addition*, they can reduce the number of addition facts to 66.

b. The *Additive Identity Property* states that the sum of any whole number and 0 is the whole number. Due to simplicity, it is not necessary to learn this property for each whole number and 0. This reduces the number of addition facts to 55.

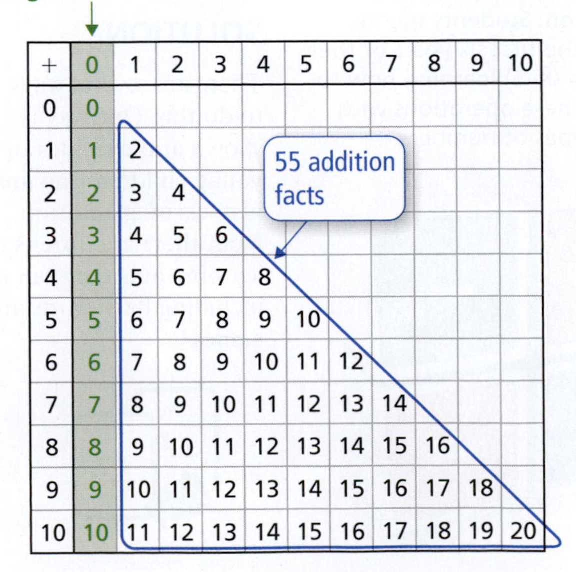

+	0	1	2	3	4	5	6	7	8	9	10
0	0										
1	1	2									
2	2	3	4								
3	3	4	5	6							
4	4	5	6	7	8						
5	5	6	7	8	9	10					
6	6	7	8	9	10	11	12				
7	7	8	9	10	11	12	13	14			
8	8	9	10	11	12	13	14	15	16		
9	9	10	11	12	13	14	15	16	17	18	
10	10	11	12	13	14	15	16	17	18	19	20

Mathematical Practices

Look For and Make Use of Structure

Mathematically proficient students look closely to discern a pattern or structure.

EXAMPLE 3 More Strategies for Learning Addition Facts

a. Explain how you can use the sum of 1 and any whole number to make it easier to learn the addition facts shown in Example 1.

b. Explain how you can use the sum of 10 and any whole number from 0 through 9 to make it easier to learn the addition facts shown in Example 1.

c. Explain how you can use *doubles* (when two addends are the same) to make it easier to learn the addition facts shown in Example 1.

d. Explain how you can use the *Associative Property of Addition* to add 9 and a whole number.

SOLUTION

a. The sum of 1 and any whole number can be found by simply counting on from the whole number. For instance, the sum $7 + 1$ can be found by starting at 7 and counting on to 8. This reduces the number of addition facts to 45.

b. The sum of 10 and any whole number from 0 through 9 can be found by simply putting a 1 in front of the whole number. For instance, $10 + 8 = 18$. This reduces the number of addition facts to 37.

c. Doubles such as $1 + 1 = 2$, $2 + 2 = 4$, $3 + 3 = 6$, and so on, are easily learned as a result of counting by twos: 2, 4, 6, 8, 10, ... This reduces the number of addition facts to 28.

d. To add 9 and a whole number, use the *Associative Property of Addition* as shown.

$9 + 8 = 9 + (1 + 7)$ Think of 8 as $1 + 7$.

$= (9 + 1) + 7$ Regroup using the *Associative Property*.

$= 10 + 7$ Add 9 and 1.

$= 17$ Use the strategy for adding 10 and a whole number.

This reduces the number of addition facts to 21.

Adding 1 and a number

+	0	1	2	3	4	5	6	7	8	9	10
0											
1		2									
2		3	4								
3		4	5	6							
4		5	6	7	8						
5		6	7	8	9	10					
6		7	8	9	10	11	12				
7		8	9	10	11	12	13	14			
8		9	10	11	12	13	14	15	16		
9		10	11	12	13	14	15	16	17	18	
10		11	12	13	14	15	16	17	18	19	20

21 addition facts

Adding 9 and a number

Adding 10 and a number

Doubles

Classroom Tip

Do you see the advantages of teaching children strategies when learning facts in mathematics? By learning the strategies described in Examples 2 and 3, students can reduce the number of addition facts that have to be "memorized" from 121 to only 21.

Using Models to Add Whole Numbers

Adding Whole Numbers Using a Number Line

Use a **number line model** to add *small* whole numbers. For instance, to model the sum of 4 and 5, use two arrows: one that is 4 units long and one that is 5 units long. Start at 0. Place the arrows end to end with no overlaps as shown.

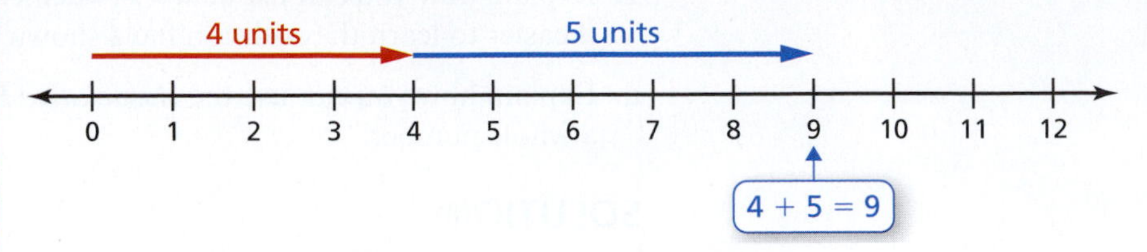

The location on the number line where the second arrow ends represents the sum. The second arrow ends at 9, so the sum is 9. A number line model is sometimes called a **measurement model**.

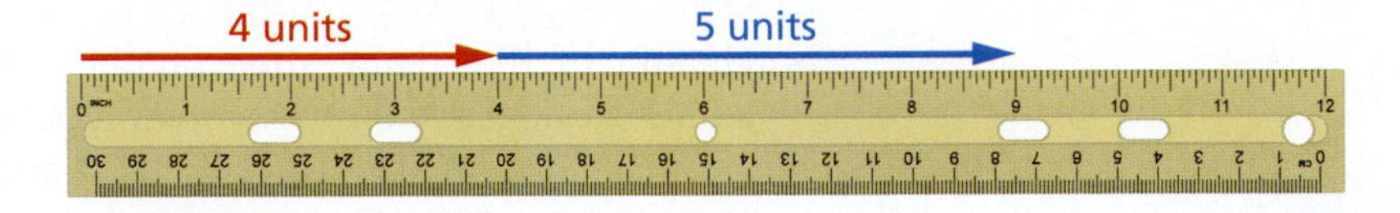

EXAMPLE 4 **Adding Whole Numbers Using a Number Line**

Use a number line to find (a) $3 + 5$, (b) $2 + 9$, and (c) $5 + 6$.

SOLUTION

a. Start at 0. Place two arrows of lengths 3 units and 5 units end to end.

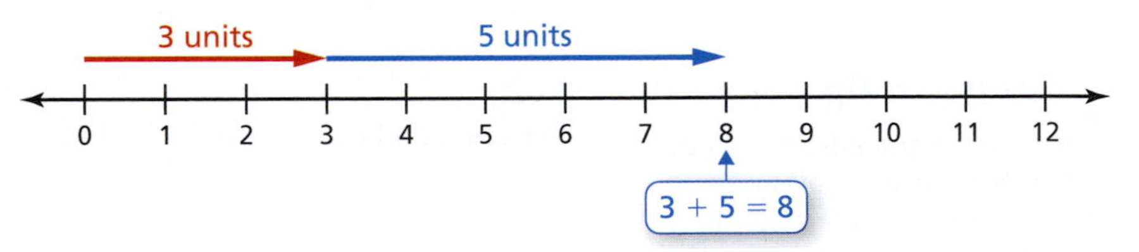

You can see that the *sum of 3 and 5* is 8.

b. Start at 0. Place two arrows of lengths 2 units and 9 units end to end.

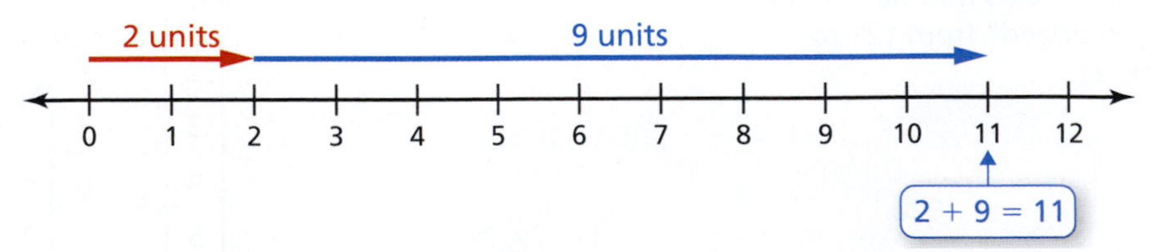

You can see that the *sum of 2 and 9* is 11.

c. Start at 0. Place two arrows of lengths 5 units and 6 units end to end.

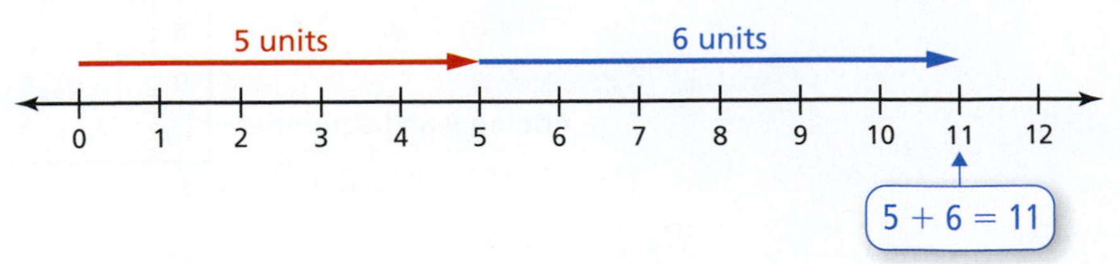

You can see that the *sum of 5 and 6* is 11.

Classroom Tip

When you are teaching young children addition facts, it is often helpful to use real-life situations. For instance, in Example 4(a), ask your students to think of a real-life situation that may involve the given sum. Here is an example.

"I slept for 3 hours. Then a noise woke me up. After I went back to bed, I slept for 5 more hours. How many total hours did I sleep?"

Adding Whole Numbers Using Base-Ten Blocks

Use **base-ten blocks** to add *large* whole numbers. For instance, model the sum of 45 and 67 as shown.

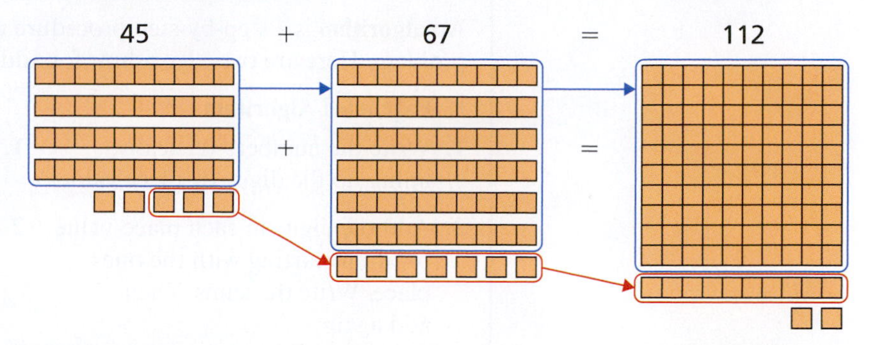

Notice how 10 ones are **regrouped** as 1 ten, and 10 tens are **regrouped** as 1 hundred.

Solving a Real-Life Problem Using Base-Ten Blocks

You have a part-time job waiting tables at a restaurant. On three nights, you earn \$37, \$52, and \$46 in tips. How much do you earn in tips for the three nights?

SOLUTION

1. **Understand the Problem** You need to find the total amount you earn in tips for the three nights.
2. **Make a Plan** Use base-ten blocks to find the total.
3. **Solve the Problem** Use base-ten blocks to model each number.

   37 + 52 + 46

   Join all of the blocks into a single group. For each group of 10 blocks of the same size, *regroup* using the next larger size block.

   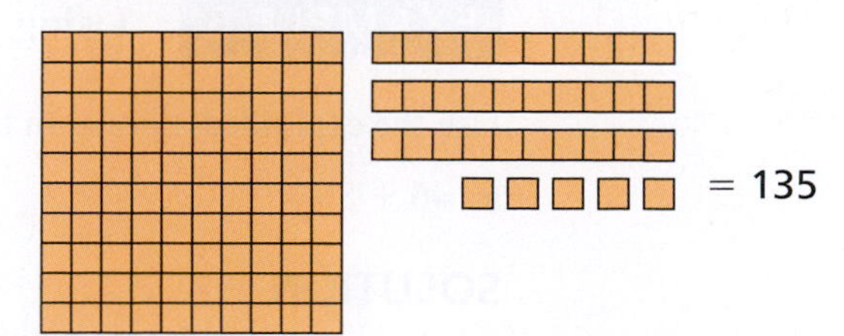

   You earn \$135 in tips.

4. **Look Back** You can check the reasonableness of your answer by rounding each addend to the nearest ten, and then finding the sum.

$$\begin{array}{ccccccc} \$37 & + & \$52 & + & \$46 & = & \$135 \\ \downarrow & & \downarrow & & \downarrow & & \downarrow \\ \$40 & + & \$50 & + & \$50 & = & \$140 \end{array} \quad ✓$$

Using Algorithms to Add Whole Numbers

Algorithms for Adding Whole Numbers

An **algorithm** is a step-by-step procedure used to find an answer to a problem. Here are two algorithms for adding whole numbers.

Expanded Algorithm

1. Write the numbers vertically, lining up the digits by place value.
2. Add the digits in each place-value position starting with the ones place. Write the sums. Then add again.

Standard Algorithm

1. Write the numbers vertically, lining up the digits by place value.
2. Add the digits in each place-value position starting with the ones place. Regroup mentally as you add.

Example $34 + 27 = 61$

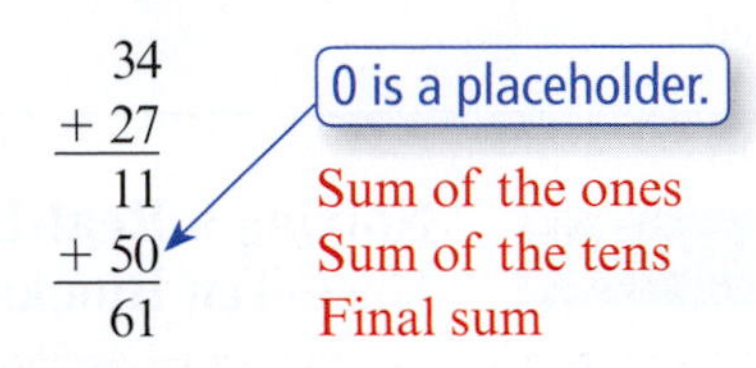

1
34
+ 27
61

Mentally regroup the sum of the ones as $11 = 1(10) + 1$.

Notice how the sums 11 and 50 in the expanded algorithm are shown in the base-ten block model below.

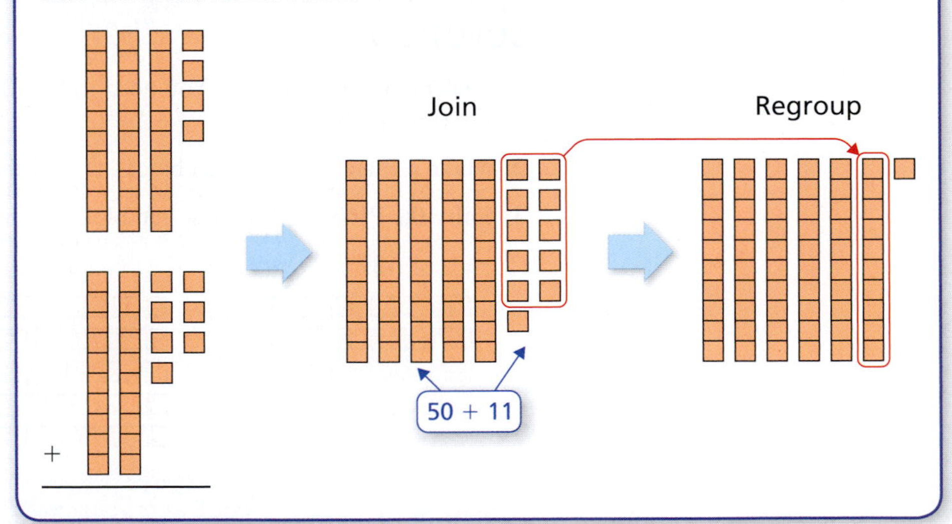

Classroom Tip

The expanded algorithm and the standard algorithm are essentially the same. The difference is that the standard (or traditional) algorithm uses mental math and regrouping.

EXAMPLE 6 Using the Expanded Algorithm

Use the expanded algorithm to find each sum.

a. $46 + 85$

b. $237 + 669$

SOLUTION

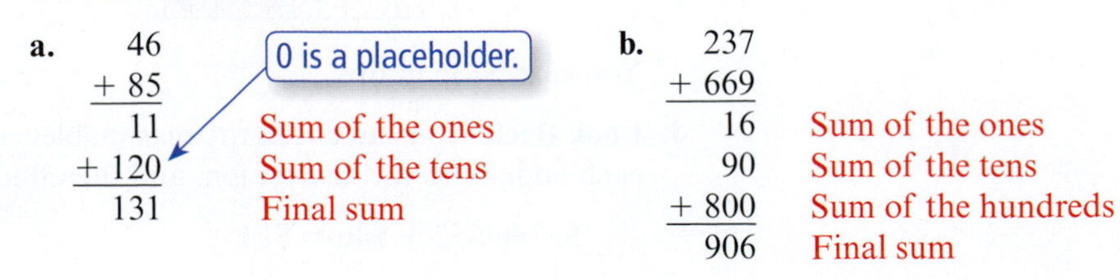

EXAMPLE 7 Using the Standard Algorithm

Use the standard algorithm to find each sum.

a. $46 + 85$ **b.** $237 + 669$

SOLUTION

a.
$$\begin{array}{r} \scriptstyle 1 \\ 46 \\ +\ 85 \\ \hline 131 \end{array}$$

Mentally regroup the sum of the ones as $11 = 1(10) + 1$.

b.
$$\begin{array}{r} \scriptstyle 11 \\ 237 \\ +\ 669 \\ \hline 906 \end{array}$$

Mentally regroup the sum of the ones as $16 = 1(10) + 6$.

Mentally regroup the sum of the tens as $100 = 1(100) + 0(10)$.

EXAMPLE 8 Solving a Real-Life Addition Problem

You work as a park ranger and are counting the numbers of sea turtle nests along five different stretches of a beach. Your counts are 147, 204, 158, 324, and 92. What is the total number of nests along the five stretches of beach?

SOLUTION

1. **Understand the Problem** You need to find the total number of nests along the five stretches of beach.
2. **Make a Plan** Use the standard algorithm to add the five numbers.
3. **Solve the Problem** Write the five numbers vertically, lining up the digits by place value.

$$\begin{array}{r} \scriptstyle 22 \\ 147 \\ 204 \\ 158 \\ 324 \\ +\ 92 \\ \hline 925 \end{array}$$

Mentally regroup the sum of the ones as $25 = 2(10) + 5$.

Mentally regroup the sum of the tens as $220 = 2(100) + 2(10)$.

There are 925 nests along the five stretches of beach.

4. **Look Back** You can use technology to check your result.

Use a calculator.

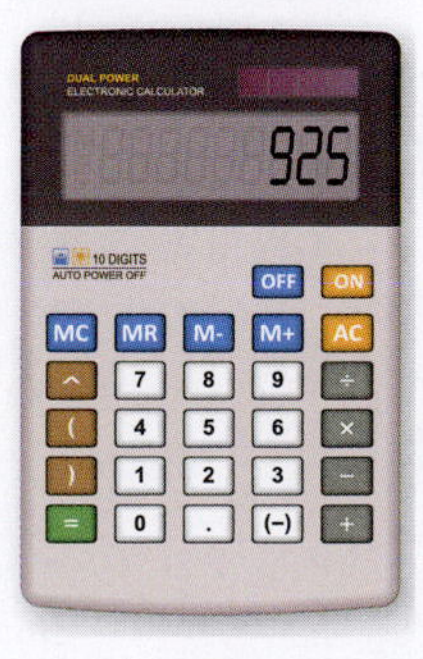

Use a spreadsheet.

	A	
1	147	
2	204	
3	158	
4	324	
5	92	
6	925	

3.1 Exercises

Free step-by-step solutions to odd-numbered exercises at MathematicalPractices.com.

1. **Writing a Solution Key** Write a solution key for the activity on page 74. Describe a rubric for grading a student's work.

2. **Grading the Activity** In the activity on page 74, a student gave the answers below. Each question is worth 10 points. Assign a grade to each answer. For those that are incorrect, why do you think the student erred?

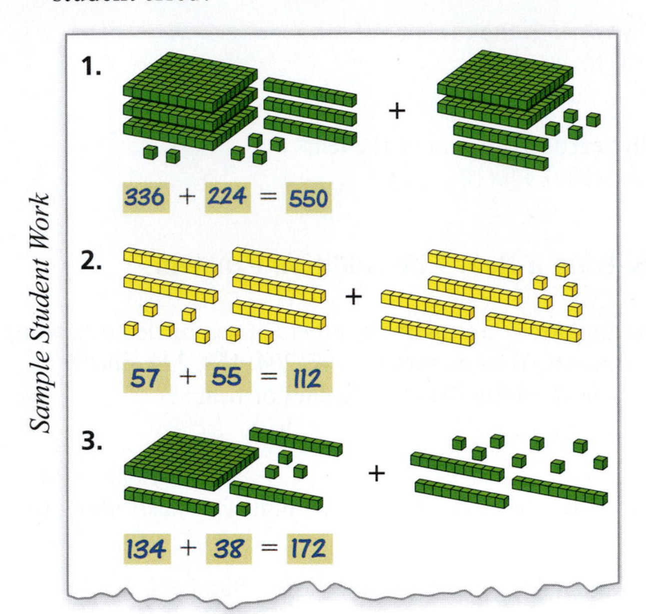

3. **Using Vocabulary** Identify the addends of each addition problem. Then find the sum.
   a. $6 + 5$
   b. $7 + 9$

4. **Using Vocabulary** Identify the addends of each addition problem. Then find the sum.
   a. $8 + 3$
   b. $2 + 4$

5. **Using the Definition of Sum** Write the addition problem represented by the set model. Explain how the model represents the addends and the sum.

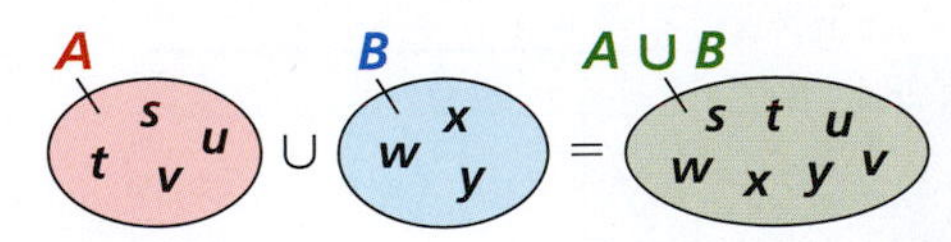

6. **Using the Definition of Sum** Write the addition problem represented by the set model. Explain how the model represents the addends and the sum.

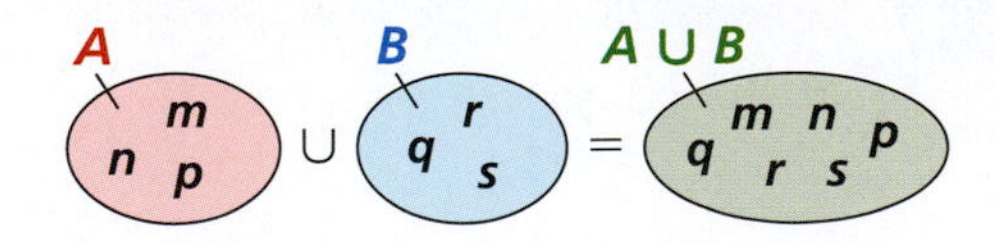

7. **Identifying Properties of Addition** Identify the property of addition that makes each equation true.
   a. $2 + x = x + 2$
   b. $5 + 0 = 5$
   c. $(2 + 9) + 1 = 2 + (9 + 1)$

8. **Identifying Properties of Addition** Identify the property of addition that makes each equation true.
   a. $x = 0 + x$
   b. $(8 + 6) + 4 = 8 + (6 + 4)$
   c. $17 + 86 = 86 + 17$

9. **Using Properties of Addition** Use the given property of addition to complete each equation.
   a. Commutative Property: $7 + 8 =$ ▢
   b. Associative Property: $4 + (6 + 7) =$ ▢
   c. Additive Identity Property: $4 + 0 =$ ▢

10. **Using Properties of Addition** Use the given property of addition to complete each equation.
   a. Commutative Property: $2 + a =$ ▢
   b. Associative Property: $(4 + 9) + 1 =$ ▢
   c. Additive Identity Property: $x + 0 =$ ▢

11. **Using Addition Strategies** Let a be any whole number from 0 through 9. Describe a strategy for adding each pair of numbers.
   a. $0 + a$ **b.** $1 + a$ **c.** $10 + a$

12. **Using Addition Strategies** Let b be any whole number from 1 through 9. Describe a strategy for adding each pair of numbers.
   a. $9 + b$ **b.** $b + 10$ **c.** $b + b$

13. **Adding Whole Numbers Using Number Lines** Describe how each number line model is used to add two numbers. Then write and solve the addition problem represented by the model.

a.

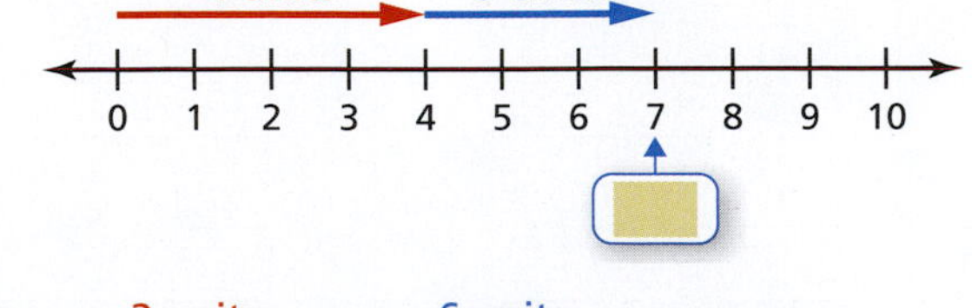

b.

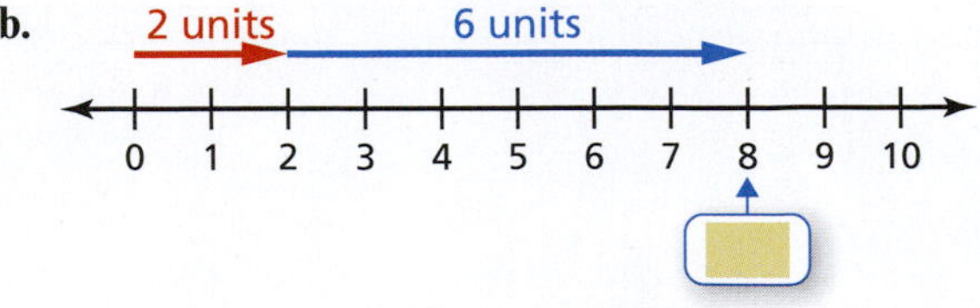

14. Adding Whole Numbers Using Number Lines
Describe how each number line model is used to add two numbers. Then write and solve the addition problem represented by the model.

a.

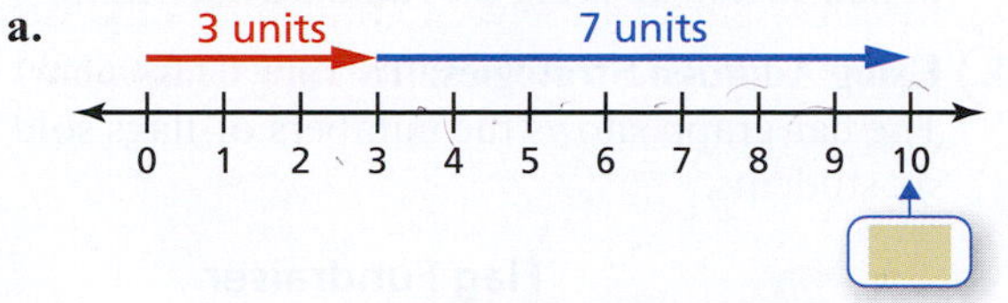

b.

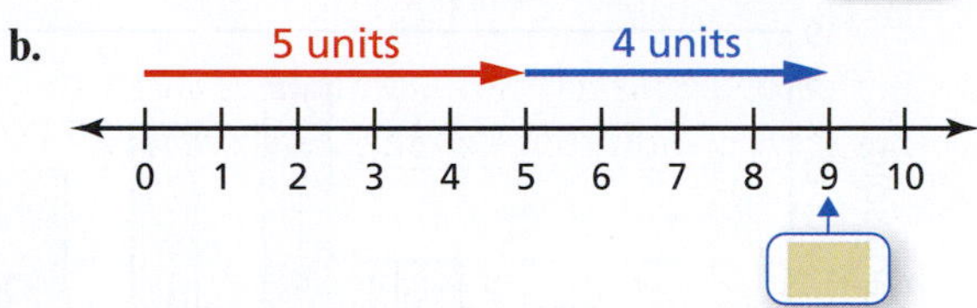

15. Adding Whole Numbers Using Number Lines
Draw and use a number line to find each sum.

a. 6 + 3 **b.** 4 + 7

c. 2 + 5 + 3 **d.** 4 + 2 + 7

16. Adding Whole Numbers Using Number Lines
Draw and use a number line to find each sum.

a. 8 + 2 **b.** 3 + 9

c. 3 + 3 + 6 **d.** 8 + 7 + 3

17. Adding Whole Numbers Using Base-Ten Blocks
Describe how to use the base-ten blocks to add two numbers. Then model the sum. Finally, write and solve the addition problem represented by the model.

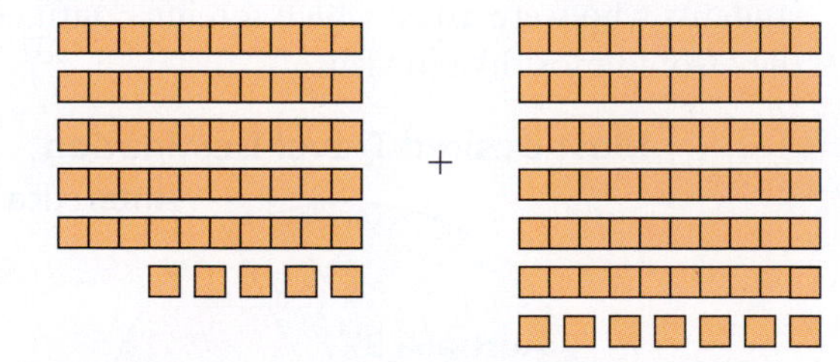

18. Adding Whole Numbers Using Base-Ten Blocks
Describe how to use the base-ten blocks to add two numbers. Then model the sum. Finally, write and solve the addition problem represented by the model.

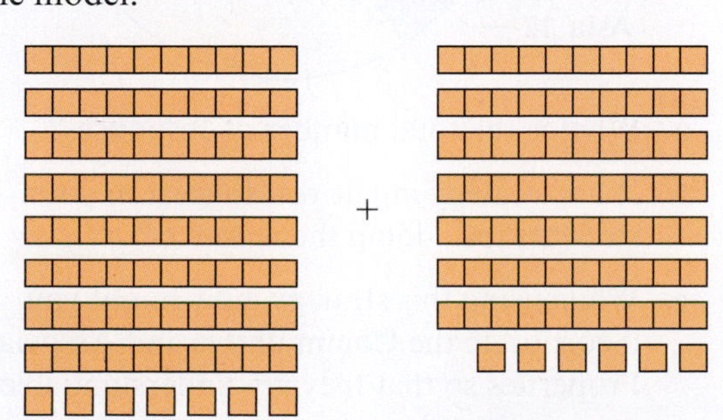

19. Adding Whole Numbers Using Base-Ten Blocks
Use base-ten blocks to find each sum. Justify your answer with a sketch of the model.

a. 28 + 36 **b.** 33 + 48

c. 59 + 54 **d.** 162 + 79

20. Adding Whole Numbers Using Base-Ten Blocks
Use base-ten blocks to find each sum. Justify your answer with a sketch of the model.

a. 25 + 35 **b.** 42 + 29

c. 67 + 46 **d.** 277 + 68

21. Using the Expanded Algorithm Use the expanded algorithm to find each sum.

a. $\begin{array}{r} 66 \\ +\ 38 \\ \hline \end{array}$ **b.** $\begin{array}{r} 638 \\ +\ 493 \\ \hline \end{array}$

c. $\begin{array}{r} 497 \\ +\ 73 \\ \hline \end{array}$ **d.** $\begin{array}{r} 1057 \\ +\ 296 \\ \hline \end{array}$

e. 59 + 73 **f.** 598 + 664

g. 17 + 23 + 65 **h.** 114 + 247 + 56

22. Using the Expanded Algorithm Use the expanded algorithm to find each sum.

a. $\begin{array}{r} 74 \\ +\ 97 \\ \hline \end{array}$ **b.** $\begin{array}{r} 579 \\ +\ 661 \\ \hline \end{array}$

c. $\begin{array}{r} 562 \\ +\ 59 \\ \hline \end{array}$ **d.** $\begin{array}{r} 2558 \\ +\ 644 \\ \hline \end{array}$

e. 94 + 78 **f.** 333 + 778

g. 43 + 36 + 49 **h.** 637 + 284 + 29

23. Using the Standard Algorithm Use the standard algorithm to find each sum.

a. $\begin{array}{r} 29 \\ +\ 77 \\ \hline \end{array}$ **b.** $\begin{array}{r} 489 \\ +\ 433 \\ \hline \end{array}$

c. $\begin{array}{r} 2648 \\ +\ 2777 \\ \hline \end{array}$ **d.** $\begin{array}{r} 896 \\ +\ 4598 \\ \hline \end{array}$

e. 6825 + 329 **f.** 53 + 7179

g. 87 + 44 + 37

h. 226 + 467 + 51

24. Using the Standard Algorithm Use the standard algorithm to find each sum.

a. $\begin{array}{r} 86 \\ +\ 65 \\ \hline \end{array}$ **b.** $\begin{array}{r} 276 \\ +\ 775 \\ \hline \end{array}$

c. $\begin{array}{r} 6677 \\ +\ 2356 \\ \hline \end{array}$ **d.** $\begin{array}{r} 567 \\ +\ 5675 \\ \hline \end{array}$

e. 1896 + 654 **f.** 7268 + 34

g. 49 + 96 + 28

h. 541 + 629 + 73

25. **Adding Whole Numbers** Find the total cost of the handbag and the shoes using the expanded algorithm and the standard algorithm.

26. **Adding Whole Numbers** Find the total cost of both wedding bands using the expanded algorithm and the standard algorithm.

27. **Adding Whole Numbers** Is the total weight of the motorcycle and the rider greater than 900 pounds? Explain your reasoning.

28. **Adding Whole Numbers** Is the total weight of the truck, boat, and trailer greater than 8900 pounds? Explain your reasoning.

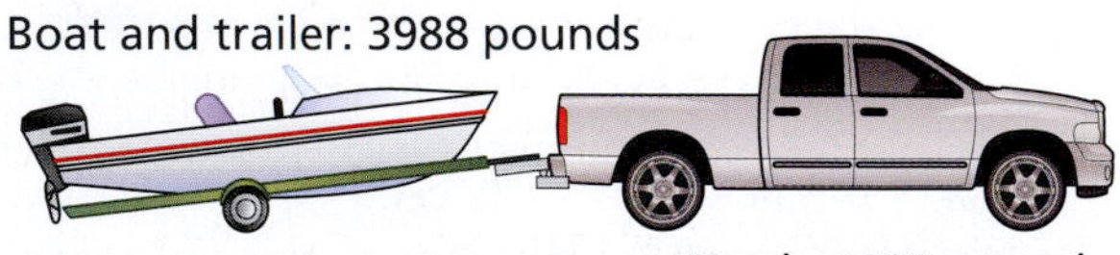

29. **My Dear Aunt Sally** Arrange the numbers 1, 3, 4, and 6 so that the equation is true.

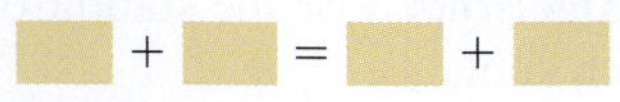

For more practice with this type of reasoning, go to *AuntSally.com.*

30. **My Dear Aunt Sally** Arrange the numbers 3, 4, 8, and 9 so that the equation is true.

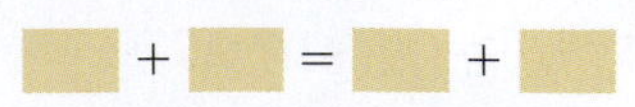

For more practice with this type of reasoning, go to *AuntSally.com.*

yourmedia/Shutterstock.com
Yaviki/Shutterstock.com; Aleksei Makarov/Shutterstock.com

31. **Writing** Write a real-life application that involves adding four whole numbers to find a total cost.

32. **Writing** Explain why you need to regroup mentally to add 48 and 29 using the standard algorithm.

33. **Using Addition Strategies: In Your Classroom** The bar graph shows the numbers of flags sold by six students.

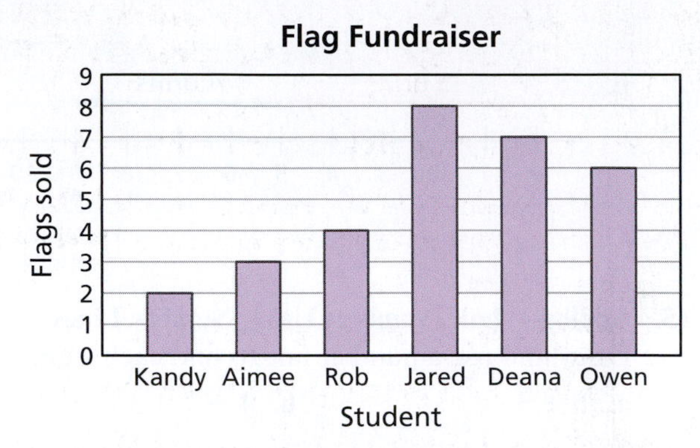

a. What is the total number of flags sold by the students?

b. What strategy might you suggest to your students for adding the numbers?

c. While using this strategy, how might you incorporate the Commutative and Associative Properties so that they are understandable to your students?

34. **Using Addition Strategies: In Your Classroom** The circle graph shows the responses of a group of students who were asked which foreign continent they would most like to visit.

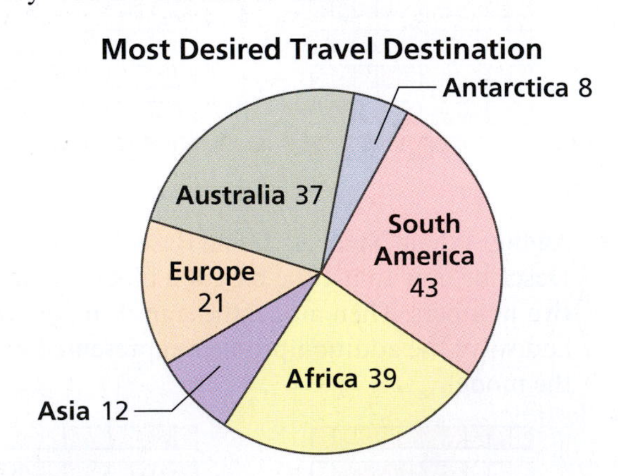

a. What is the total number of responses?

b. What strategy might you suggest to your students for adding the numbers?

c. While using this strategy, how might you incorporate the Commutative and Associative Properties so that they are understandable to your students?

35. Strategies for Adding Whole Numbers Fill in the steps of the addition problem and justify each step. Describe the strategy in words.

$$\begin{aligned} 49 + 45 &= 49 + (1 + \square) && \square \\ &= (49 + 1) + \square && \square \\ &= 50 + \square && \square \\ &= \square && \square \end{aligned}$$

36. Strategies for Adding Whole Numbers Fill in the steps of the addition problem and justify each step. Describe the strategy in words.

$$\begin{aligned} 390 + 440 &= 390 + (10 + \square) && \square \\ &= (390 + 10) + \square && \square \\ &= 400 + \square && \square \\ &= \square && \square \end{aligned}$$

37. Grading Student Work On a diagnostic test, one of your students does the following work. Explain what the student did wrong. Which topics would you encourage the student to review?

a. $9 + 24 = 10 + 25 = 35$

b.

$$\begin{array}{r} 76 \\ +\ 48 \\ \hline 114 \end{array}$$

38. Grading Student Work On a diagnostic test, one of your students does the following work. Explain what the student did wrong. Which topics would you encourage the student to review?

a. $79 + 33 = 80 + 30 = 110$

b.

$$\begin{array}{r} 1 \\ 37 \\ +\ 94 \\ \hline 11 \\ +\ 130 \\ \hline 141 \end{array}$$

39. Magic Square Starting with the numbers shown, fill in the rest of the numbers from 1 through 9 to form a *magic square* in which the sum of each row, column, and diagonal is 15. Describe your strategy.

		6
3		
	9	

40. Magic Square Starting with the numbers shown, fill in the rest of the numbers from 1 through 16 to form a *magic square* in which the sum of each row, column, and diagonal is 34. Describe your strategy.

16			13
		11	
	6		
	15		

41. Problem Solving The table shows the number of years of experience of each teacher at an elementary school.

Teacher	Experience (years)
Ms. Bledsoe	11
Mr. Chang	8
Mrs. Duncan	43
Mrs. Fetzner	34
Mr. Harris	16
Mr. Johnson	12
Miss Jones	2
Miss Lopez	5
Ms. Nguyen	27
Mr. Sanders	3

a. How many total years of experience do the teachers have?

b. Mrs. Duncan retires and is replaced by a new teacher with no experience. Describe two ways to find the new total number of years of experience.

42. Problem Solving How many laps do you take around the trail to run approximately 4 miles? Explain.

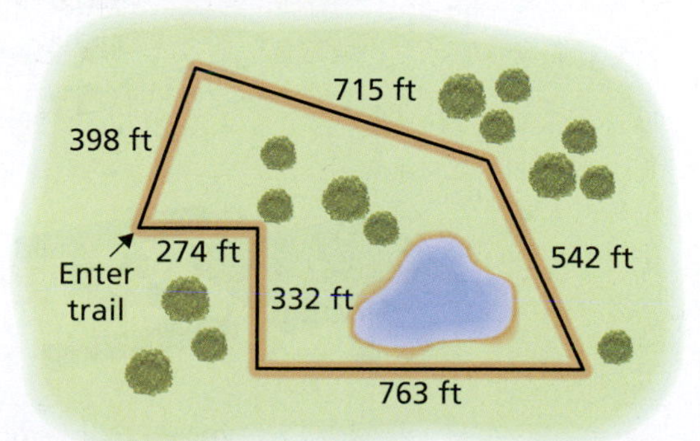

Activity: Using Base-Ten Blocks to Subtract Whole Numbers

Materials:

- Pencil
- Base-ten blocks

Learning Objective:

Learn to use base-ten blocks to subtract multi-digit whole numbers, including the concepts of regrouping 1 ten for 10 ones and 1 hundred for 10 tens.

Name ______________________________

Write the subtraction problem, and find the difference.

1. 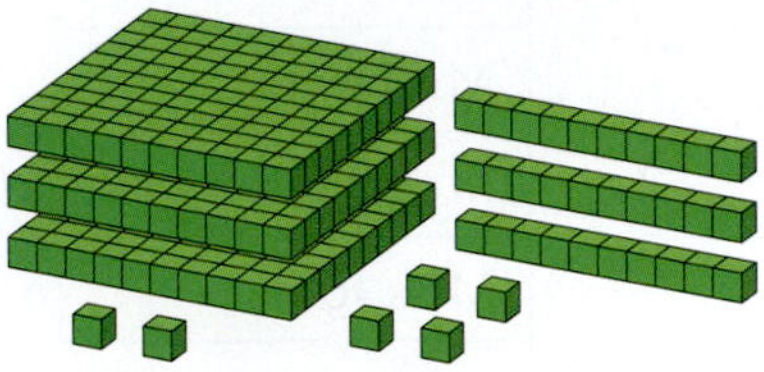 − ▭ − ▭ = ▭

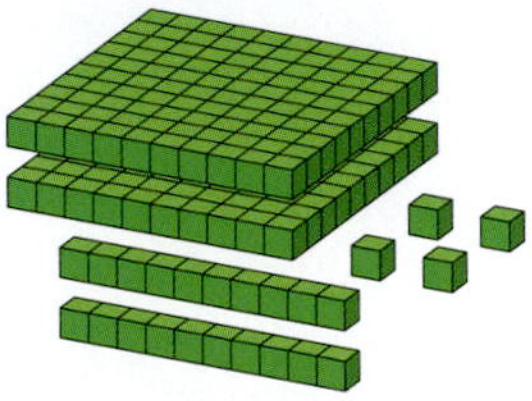

2. 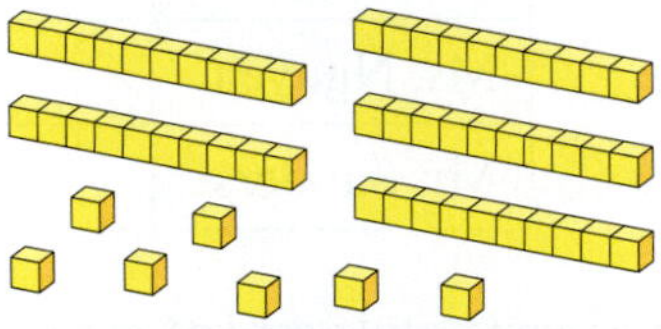 − ▭ − ▭ = ▭

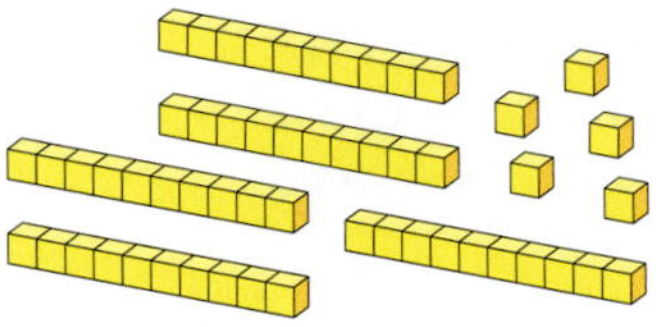

3.  − ▭ − ▭ = ▭

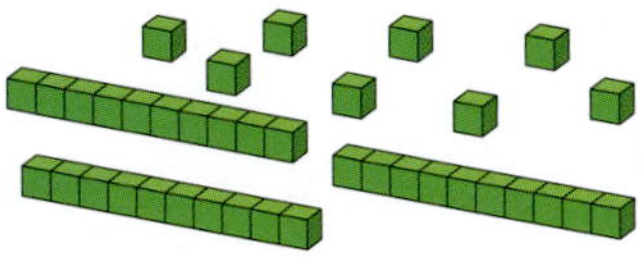

4. 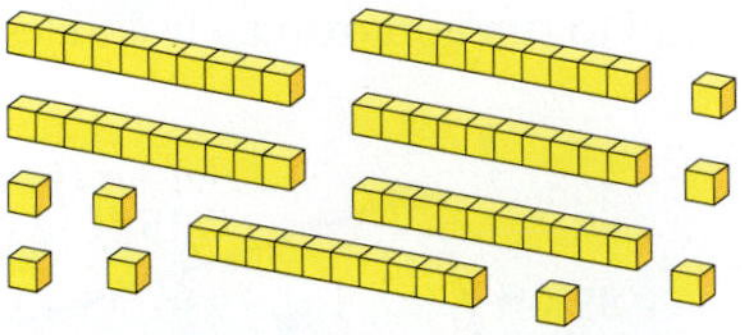 − ▭ − ▭ = ▭

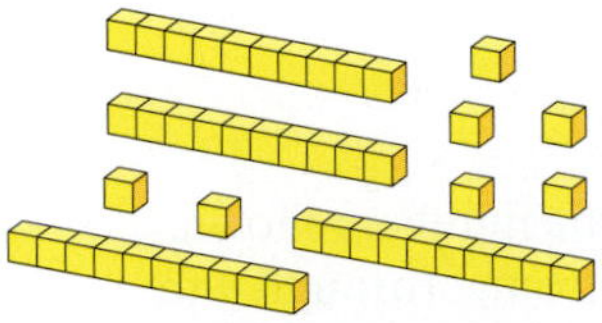

5.  − ▭ − ▭ = ▭

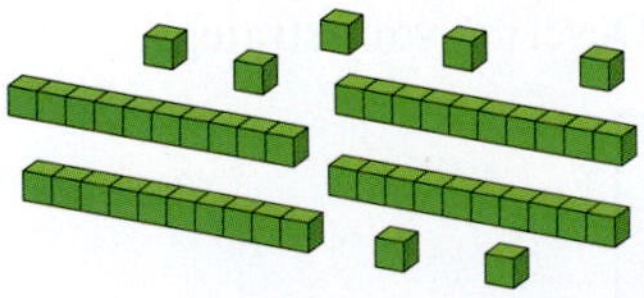

Goodluz/Shutterstock.com

3.2 Subtracting Whole Numbers

Standards

Grades K–2
Numbers and Operations in Base Ten
Students should use place value understanding and properties of operations to subtract.

Grades 3–5
Numbers and Operations in Base Ten
Students should use place value understanding and properties of operations to perform multi-digit arithmetic.

- Understand and use the definition of the difference of two whole numbers.
- Use models to subtract whole numbers.
- Use algorithms to subtract whole numbers.

Defining the Difference of Two Whole Numbers

In Section 3.1, you learned that addition is commutative, associative, and has 0 as the additive identity. Subtraction is quite different because it is neither commutative nor associative. Addition and subtraction are **inverse operations**, which means that the two operations "undo" each other.

Difference of Two Whole Numbers

Missing Addend Method

Let a and b be two whole numbers such that $a \geq b$. The **difference** of a and b, written as $a - b$, is the whole number such that

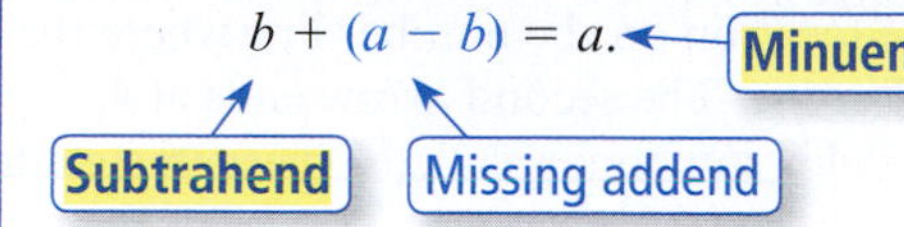

What can you add to b to get a?

Take-Away Method

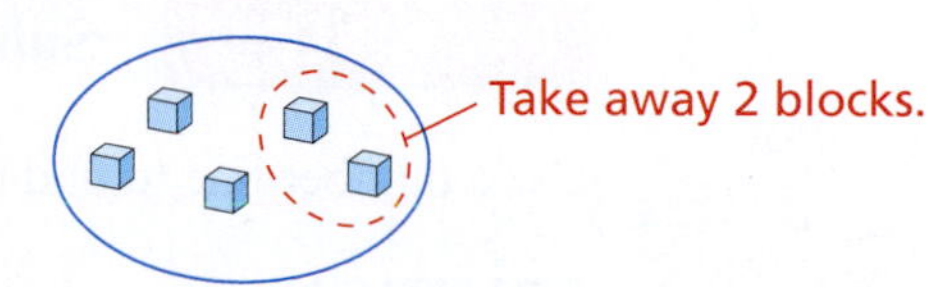

(5 elements) − (2 elements) = (3 elements)

"The difference of 5 and 2 is 3" or "5 minus 2 equals 3."

Classroom Tip

Teach young children to think of subtraction as a question. For instance, to subtract 4 from 7, use the missing addend method to write the following.

$4 + \boxed{?} = 7$

"What number can you add to 4 to get 7?"

This approach also prepares students for algebra. The addition facts table below may help students better understand the relationship between addition and subtraction facts.

+	0	1	2	3	4	5
0						
1						
2						
3						
4				7		
5						

$4 + 3 = 7$
$3 + 4 = 7$
$7 - 4 = 3$
$7 - 3 = 4$

EXAMPLE 1 **Using the Definition of Difference**

Use the missing addend method and the take-away method to find each difference.

a. $9 - 5$ **b.** $10 - 3$

SOLUTION

Missing addend method

a. $5 + \square = 9$ What can you add to 5 to get 9?

$5 + 4 = 9$ The answer is 4.

So, $9 - 5 = 4$.

b. $3 + \square = 10$ What can you add to 3 to get 10?

$3 + 7 = 10$ The answer is 7.

So, $10 - 3 = 7$.

Take-away method

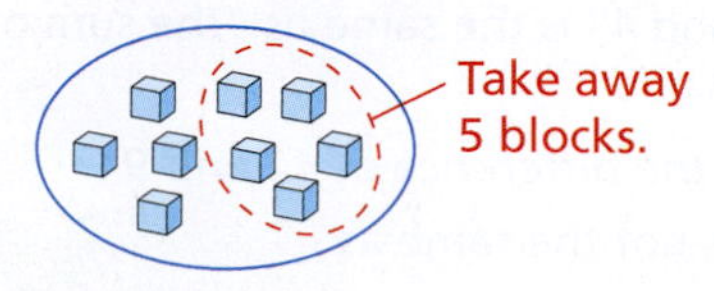

So, $9 - 5 = 4$.

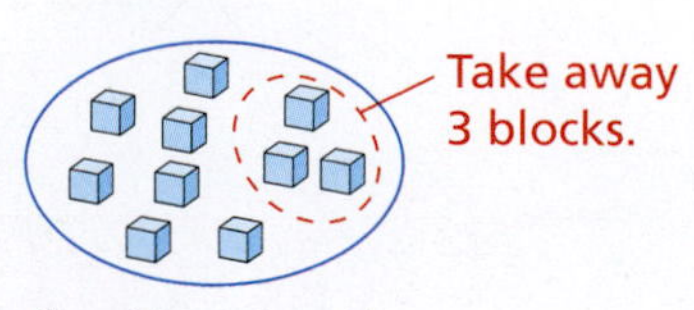

So, $10 - 3 = 7$.

Using Models to Subtract Whole Numbers

Subtracting Whole Numbers Using a Number Line

Use a **number line model** to subtract *small* whole numbers. For instance, to model the difference of 9 and 5, use two directed arrows: one that is 9 units long and one that is 5 units long. Start at 0. Place the arrow that represents the minuend. Place the arrow that represents the subtrahend below where the first arrow ends and count back to subtract as shown.

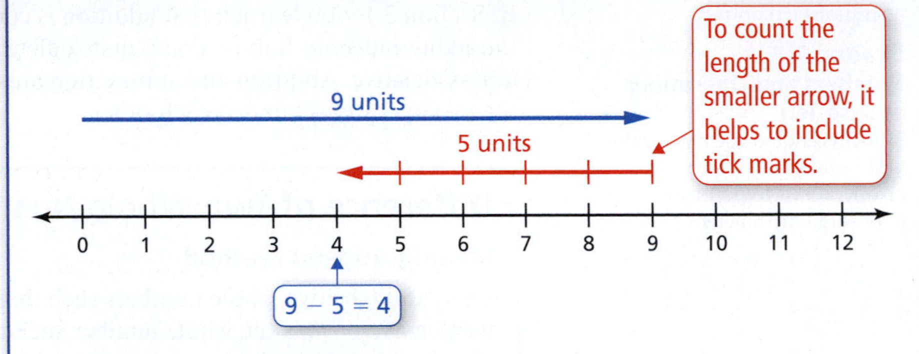

The location on the number line where the second arrow ends represents the difference. The second arrow ends at 4, so the difference is 4. A number line model is sometimes called a **measurement model**.

EXAMPLE 2 **Subtracting Whole Numbers Using a Number Line**

Use a number line to find (a) 12 − 9 and (b) 10 − 3.

SOLUTION

a. Start at 0. Place an arrow of length 12 units. Place another arrow of length 9 units below the first arrow and count back.

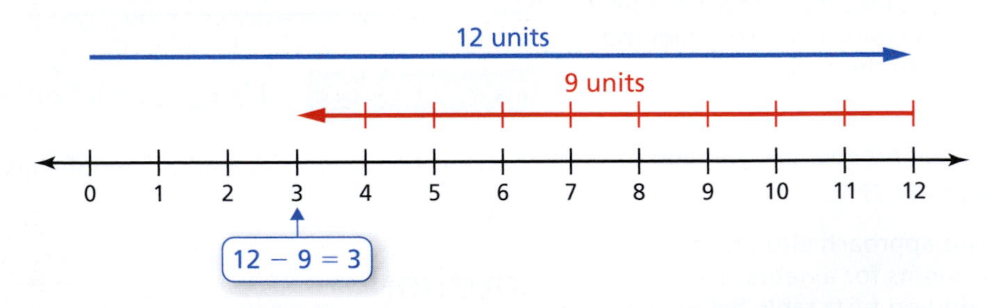

You can see the *difference of 12 and 9* is 3.

b. Start at 0. Place an arrow of length 10 units. Place another arrow of length 3 units below the first arrow and count back.

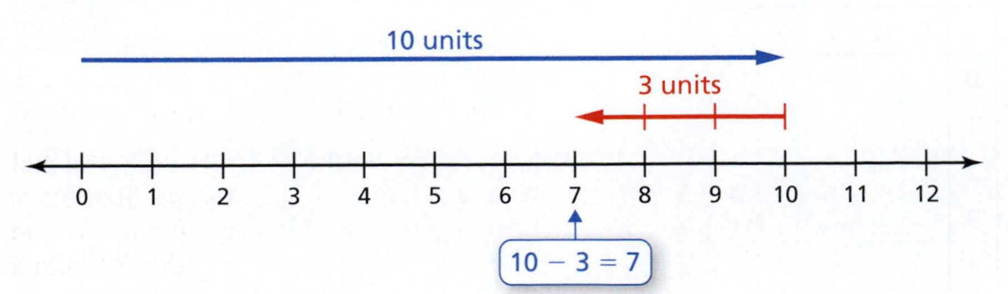

You can see that the *difference of 10 and 3* is 7.

Classroom Tip

When you are teaching addition and subtraction of whole numbers, remind students to be careful how they phrase subtraction problems. Because subtraction is not commutative, the order of the numbers is important. While "the sum of 3 and 4" is the same as "the sum of 4 and 3,"

"the difference of 12 and 9"

is not the same as

"the difference of 9 and 12."

Subtracting Whole Numbers Using Base-Ten Blocks

Use **base-ten blocks** to subtract *large* whole numbers. For instance, you can model the difference of 57 and 23 as shown.

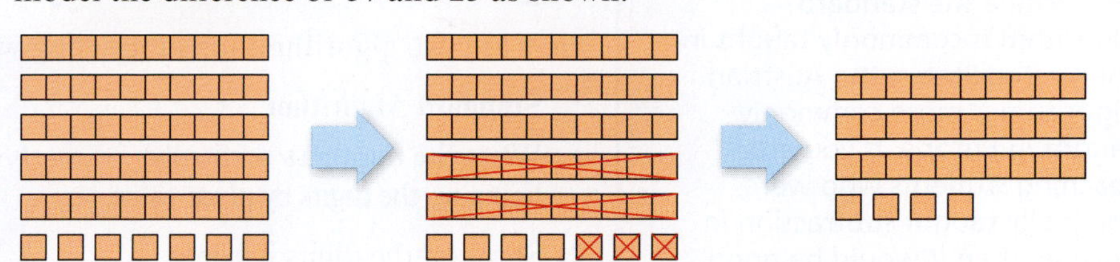

First, represent 57 using 5 tens and 7 ones.

Take away 23, or 2 tens and 3 ones.

That leaves 3 tens and 4 ones, or 34.

EXAMPLE 3 Subtracting Whole Numbers Using Base-Ten Blocks

Use base-ten blocks to find each difference.

a. 345 − 242 **b.** 113 − 96

SOLUTION

a. First, represent 345. Because you can directly take away 2 ones from 5 ones, 4 tens from 4 tens, and 2 hundreds from 3 hundreds, the use of base-ten blocks is straightforward.

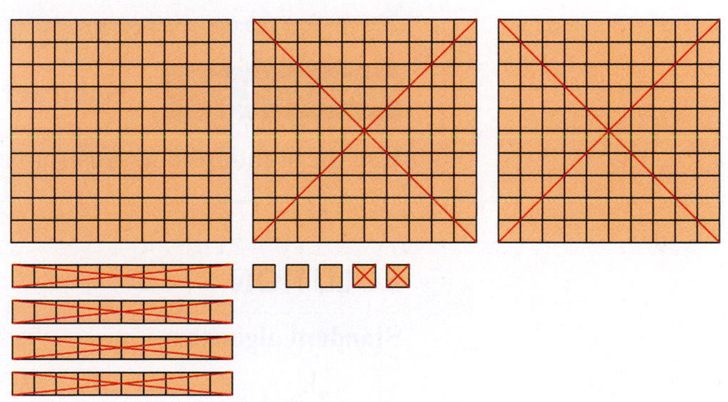

That leaves 1 hundred and 3 ones, or 103. So, 345 − 242 = 103.

b.

First, represent 113 using 1 hundred, 1 ten, and 3 ones.

You cannot directly take away 6 ones. So, *regroup* 1 ten as 10 ones. You must also regroup 1 hundred as 10 tens.

Take away 96, or 9 tens and 6 ones.

That leaves 1 ten and 7 ones, or 17. So, 113 − 96 = 17.

Using Algorithms to Subtract Whole Numbers

Classroom Tip

While the standard algorithm is commonly taught in the United States, the Austrian algorithm is more commonly taught in Europe. If you are teaching students who were originally taught subtraction in Europe, then it would be good for you to be familiar with this algorithm.

Consider teaching both algorithms to your students and ask them which algorithm they prefer. People who prefer the Austrian algorithm often cite the fact that it does not have so many messy cross-outs.

Algorithms for Subtracting Whole Numbers

Here are two algorithms for subtracting whole numbers.

Standard Algorithm

1. Write the numbers vertically, lining up the digits by place value.
2. Subtract the digits in each place-value position starting with the ones place. When you do not have enough ones, tens, or hundreds to subtract the bottom number from the top number, regroup and then subtract.

Austrian Algorithm

1. Write the numbers vertically, lining up the digits by place value.
2. Subtract the digits in each place-value position starting with the ones place. When you need to regroup, write a 1 between the columns. The 1 has two meanings, as shown below.

Example $61 - 34 = 27$

$$\begin{array}{r} \overset{5}{\cancel{6}}\,{}^{1}1 \\ -\ 3\,4 \\ \hline 2\,7 \end{array}$$

Regroup 61 as 5(10) + 11(1).

11 − 4 = 7

5 − 3 = 2

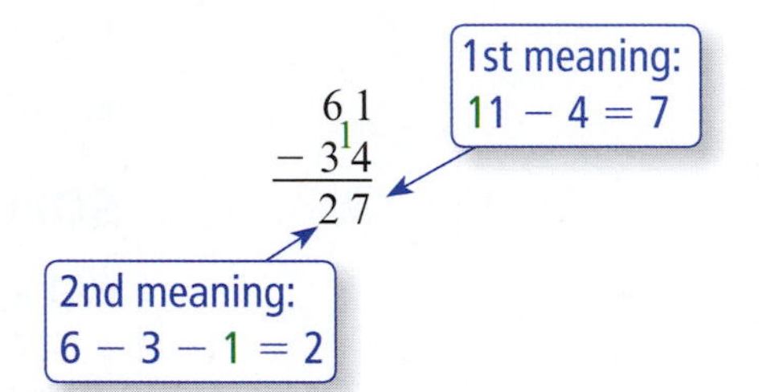

EXAMPLE 4 **Subtracting Using Algorithms**

Use the standard and Austrian algorithms to find the difference

331 − 172.

SOLUTION

Standard algorithm

$$\begin{array}{r} \overset{\overset{1}{2}}{\cancel{3}}\,\overset{2}{\cancel{3}}\,{}^{1}1 \\ -\ 1\,7\,2 \\ \hline 1\,5\,9 \end{array}$$

11 − 2 = 9

12 − 7 = 5

2 − 1 = 1

Austrian algorithm

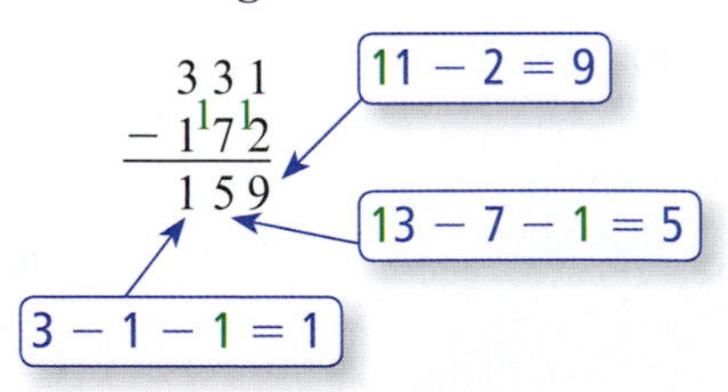

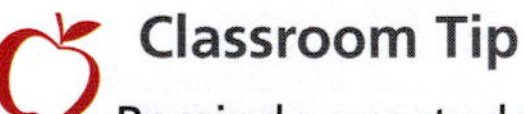

Classroom Tip

Remind your students that subtraction is the inverse operation of addition. Therefore, your students can check each subtraction problem using addition. You can check Example 4, as shown.

$$\begin{array}{r} {}^{1\,1} \\ 1\,5\,9 \\ +\ 1\,7\,2 \\ \hline 3\,3\,1 \end{array}\ \checkmark$$

Check your result using base-ten blocks. Regroup as needed.

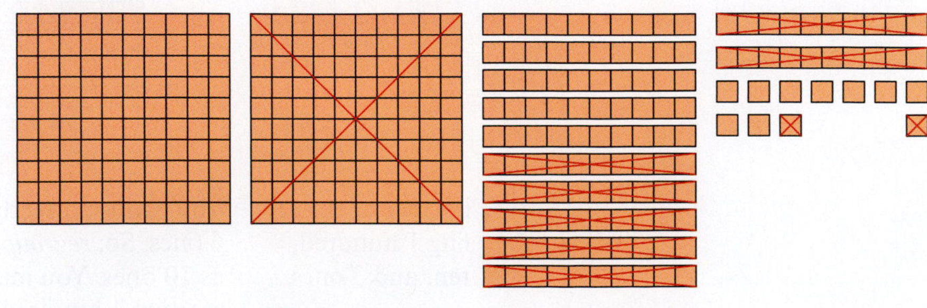

You are left with 1 hundred, 5 tens, and 9 ones, or 159. ✓

Classroom Tip

As well as checking an answer using the inverse operation, encourage your students to check the reasonableness of the answer using estimation.

```
  7000
- 4500
  2500 ✓
```

EXAMPLE 5 Subtracting Using Algorithms

Use the standard and Austrian algorithms to find the difference

7132 − 4568.

SOLUTION

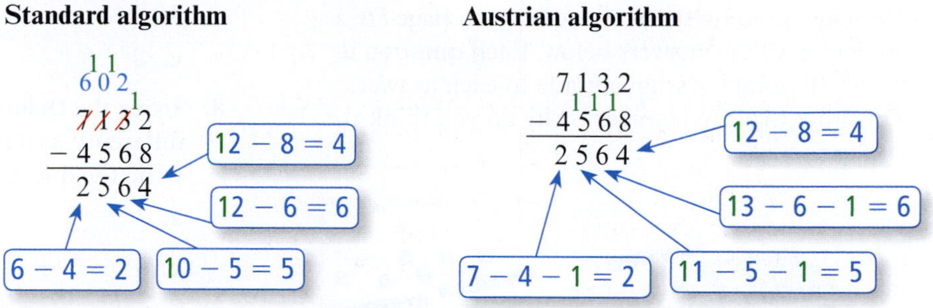

Standard algorithm

```
   1 1
   6 0 2
       1
   7 1 3 2
 − 4 5 6 8
   2 5 6 4
```

12 − 8 = 4

12 − 6 = 6

10 − 5 = 5

6 − 4 = 2

Austrian algorithm

```
   7 1 3 2
     1 1 1
 − 4 5 6 8
   2 5 6 4
```

12 − 8 = 4

13 − 6 − 1 = 6

11 − 5 − 1 = 5

7 − 4 − 1 = 2

Check your result using addition.

2564 + 4568 = 7132 ✓

EXAMPLE 6 Solving a Real-Life Subtraction Problem

You write a check to buy some clothes. After subtracting the amount from your account, your balance does not seem correct. What is the correct balance?

CHECK NUMBER	DATE	DESCRIPTION OF TRANSACTION	PAYMENT, FEE OR WITHDRAWAL (-)	T	DEPOSIT OR INTEREST (+)	$ 300
1291		Clothing Store	67			243

Mathematical Practices

Model with Mathematics

Mathematically proficient students can apply the mathematics they know to solve problems arising in everyday life.

SOLUTION

1. **Understand the Problem** You need to find the correct balance in your checking account.
2. **Make a Plan** Subtract the amount of the check from your current balance using the standard algorithm.
3. **Solve the Problem** Subtract 67 from 300.

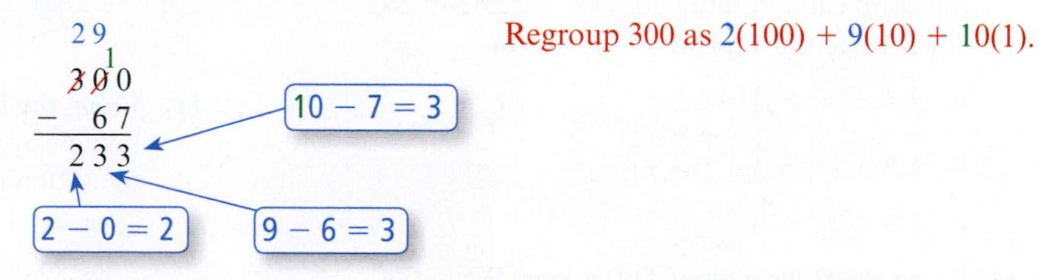

The correct balance is $233.

4. **Look Back** You can check your answer using addition.

```
   1 1
  $2 3 3      Write the new balance.
 + $6 7       Add amount of check.
  $3 0 0 ✓    Obtain original balance.
```

3.2 Exercises

Free step-by-step solutions to odd-numbered exercises at *MathematicalPractices.com.*

1. Writing a Solution Key Write a solution key for the activity on page 86. Describe a rubric for grading a student's work.

2. Grading the Activity In the activity on page 86, a student gave the answers below. Each question is worth 10 points. Assign a grade to each answer. For those that are incorrect, why do you think the student erred?

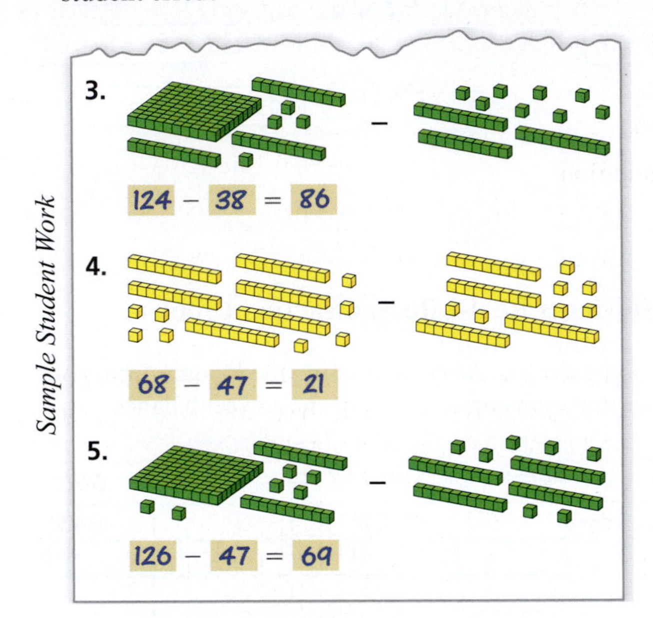

3. Using Vocabulary Identify the minuend, subtrahend, and difference in each problem.

a. $10 - 7 = 3$ **b.** $2 + (9 - 2) = 9$

4. Using Vocabulary Identify the minuend, subtrahend, and difference in each problem.

a. $8 - 3 = 5$ **b.** $4 + (12 - 4) = 12$

5. Using the Definition of Difference Write the question each missing addend statement asks. Then write the subtraction problem.

a. $9 + \square = 11$

b. $4 + \square = 7$

c. $2 + \square = 8$

6. Using the Definition of Difference Write the question each missing addend statement asks. Then write the subtraction problem.

a. $3 + \square = 9$

b. $6 + \square = 10$

c. $5 + \square = 12$

7. Using the Definition of Difference Write each difference as a missing addend statement. Then write the question the statement asks.

a. $9 - 7$ **b.** $11 - 1$

c. $8 - 6$ **d.** $6 - 1$

8. Using the Definition of Difference Write each difference as a missing addend statement. Then write the question the statement asks.

a. $12 - 7$ **b.** $15 - 10$

c. $13 - 7$ **d.** $11 - 10$

9. Using the Definition of Difference Use each take-away diagram to write the subtraction problem.

a.

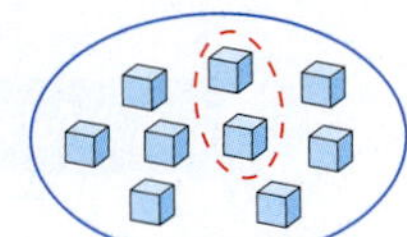

b.

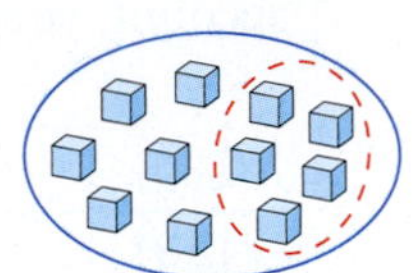

10. Using the Definition of Difference Use each take-away diagram to write the subtraction problem.

a.

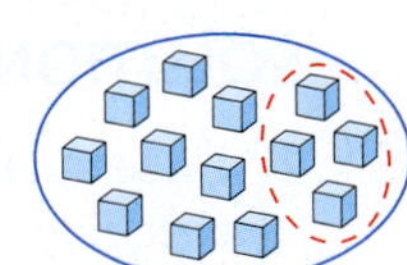

b.

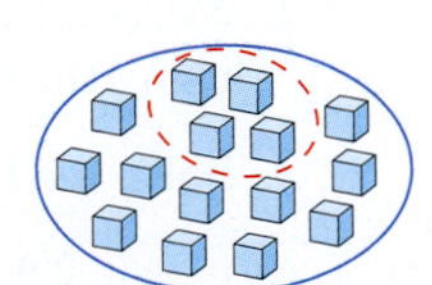

11. Using the Definition of Difference Draw a take-away diagram for each difference. Then find the difference.

a. $4 - 1$ **b.** $9 - 2$

c. $3 - 2$ **d.** $10 - 5$

12. Using the Definition of Difference Draw a take-away diagram for each difference. Then find the difference.

a. $11 - 6$ **b.** $8 - 5$

c. $9 - 3$ **d.** $15 - 10$

13. Subtracting Whole Numbers Using Number Lines Describe how each number line model is used to subtract two numbers. Then write and solve the subtraction problem represented by the model.

a.

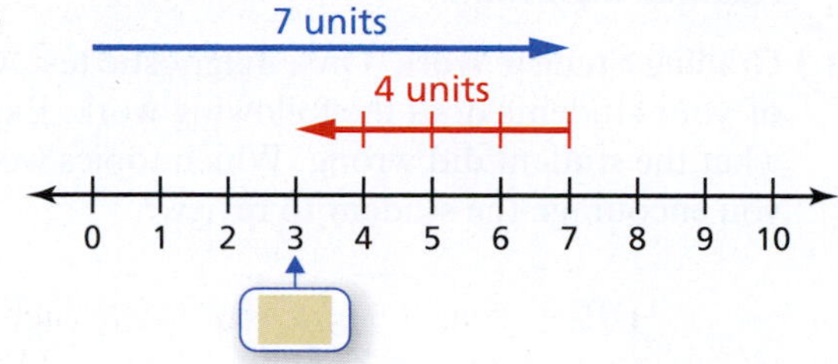

b.

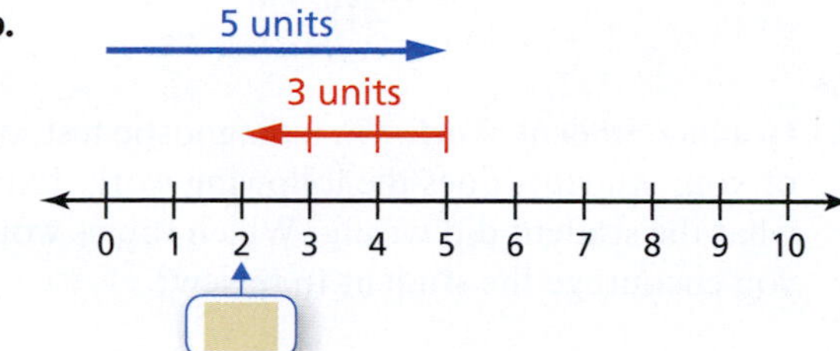

c.

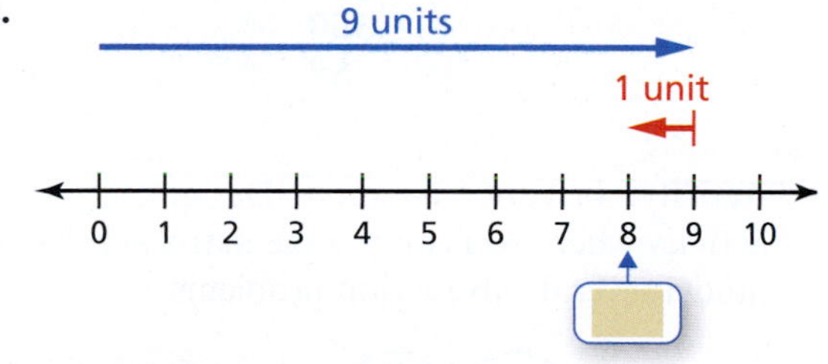

14. Subtracting Whole Numbers Using Number Lines Describe how each number line model is used to subtract two numbers. Then write and solve the subtraction problem represented by the model.

a.

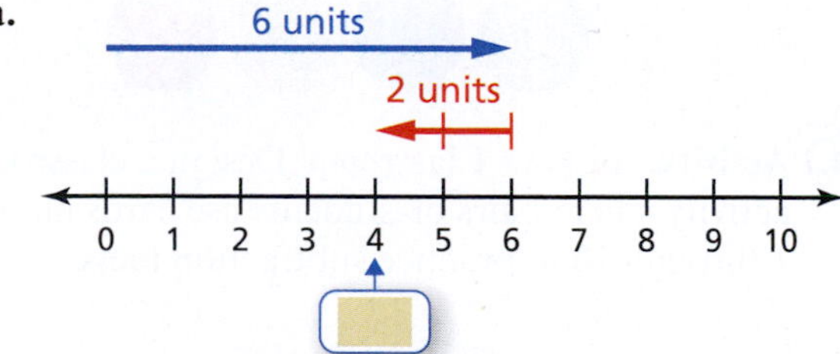

b.

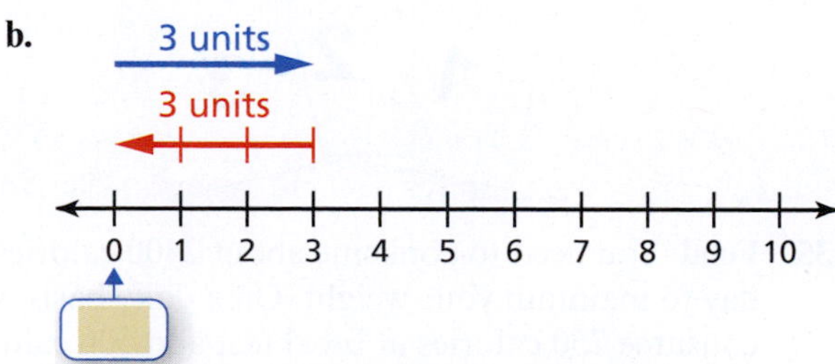

c.

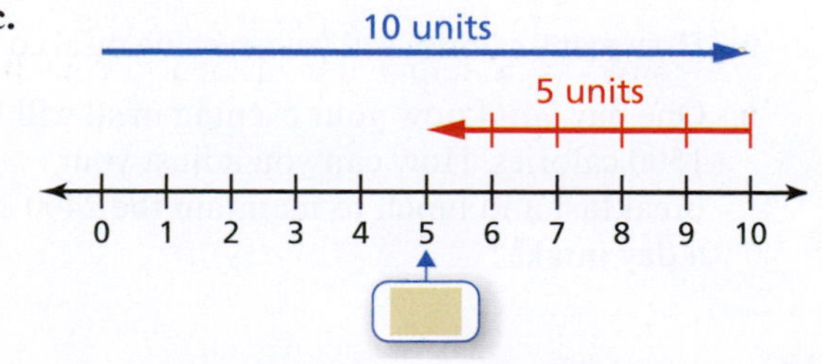

15. Subtracting Whole Numbers Using Number Lines Draw and use a number line to find each difference.

a. $12 - 2$ **b.** $7 - 1$

c. $2 - 2$ **d.** $9 - 4$

16. Subtracting Whole Numbers Using Number Lines Draw and use a number line to find each difference.

a. $6 - 4$ **b.** $11 - 4$

c. $15 - 6$ **d.** $20 - 10$

17. Subtracting Whole Numbers Using Base-Ten Blocks Describe how to use the base-ten blocks to subtract two numbers. Then write and solve the subtraction problem represented by the model.

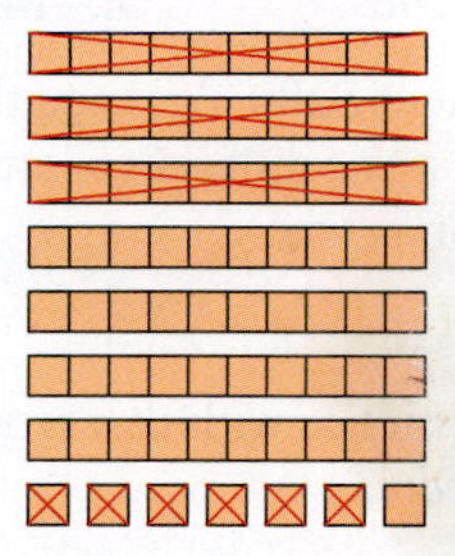

18. Subtracting Whole Numbers Using Base-Ten Blocks Describe how to use the base-ten blocks to subtract two numbers. Then write and solve the subtraction problem represented by the model.

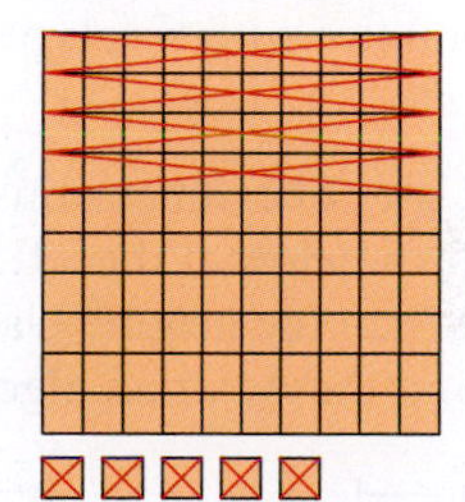

19. Subtracting Whole Numbers Using Base-Ten Blocks Use base-ten blocks to find each difference. Justify your answer with a sketch of the model.

a. $100 - 60$ **b.** $250 - 120$

c. $90 - 33$ **d.** $105 - 55$

20. Subtracting Whole Numbers Using Base-Ten Blocks Use base-ten blocks to find each difference. Justify your answer with a sketch of the model.

a. $117 - 45$ **b.** $88 - 17$

c. $330 - 275$ **d.** $289 - 61$

21. Using the Standard Algorithm Use the standard algorithm to find each difference.

a. $54 - 23$

b. $88 - 44$

c. $5321 - 412$

d. $7264 - 358$

22. Using the Standard Algorithm Use the standard algorithm to find each difference.

a. 995 − 794 **b.** 367 − 255

c. 3059 − 97 **d.** 9616 − 784

23. Using the Austrian Algorithm Use the Austrian algorithm to find each difference.

a. 2340 − 120 **b.** 1274 − 141

c. 1325 − 470 **d.** 2565 − 879

24. Using the Austrian Algorithm Use the Austrian algorithm to find each difference.

a. 7729 − 508 **b.** 4818 − 315

c. 1010 − 190 **d.** 1433 − 652

25. My Dear Aunt Sally Arrange the numbers 2, 5, 6, and 9 so that the differences are whole numbers and the equation is true.

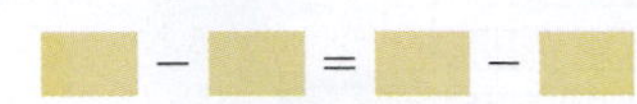

For more practice with this type of reasoning, go to *AuntSally.com*.

26. My Dear Aunt Sally Arrange the numbers 4, 5, 7, and 8 so that the differences are whole numbers and the equation is true.

For more practice with this type of reasoning, go to *AuntSally.com*.

27. Protective Case The dimensions of a smartphone case are shown below at the left. How much shorter is the width than the height? Use the standard algorithm and the Austrian algorithm.

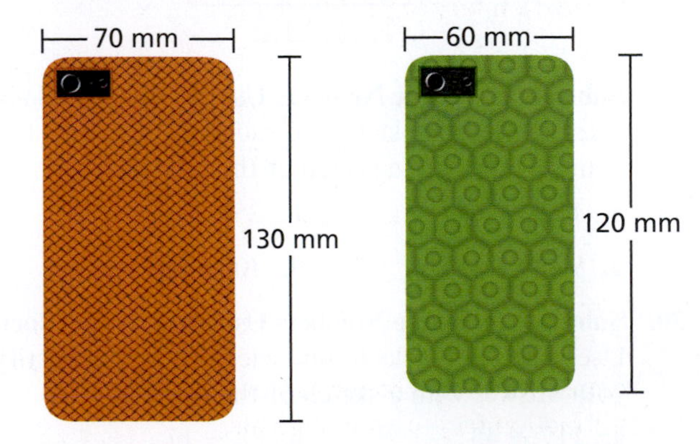

28. Protective Case The dimensions of a smartphone case are shown above at the right. How much shorter is the width than the height? Use the standard algorithm and the Austrian algorithm.

29. Aquatic Park On Friday, there were 674 people at an aquatic park. On Saturday, there were 806 people. How many more people were at the aquatic park on Saturday? Use the standard algorithm and the Austrian algorithm.

Egorich/Shutterstock.com

30. Comparing Weights A male walrus weighs 1410 kilograms. A female walrus weighs 1130 kilograms. How much more does the male walrus weigh? Use the standard algorithm and the Austrian algorithm.

31. Grading Student Work On a diagnostic test, one of your students does the following work. Explain what the student did wrong. Which topics would you encourage the student to review?

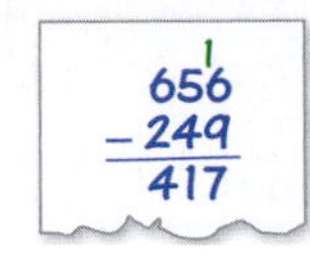

32. Grading Student Work On a diagnostic test, one of your students does the following work. Explain what the student did wrong. Which topics would you encourage the student to review?

33. Activity: In Your Classroom Design a classroom activity where you can use the marbles below to model several subtraction problems.

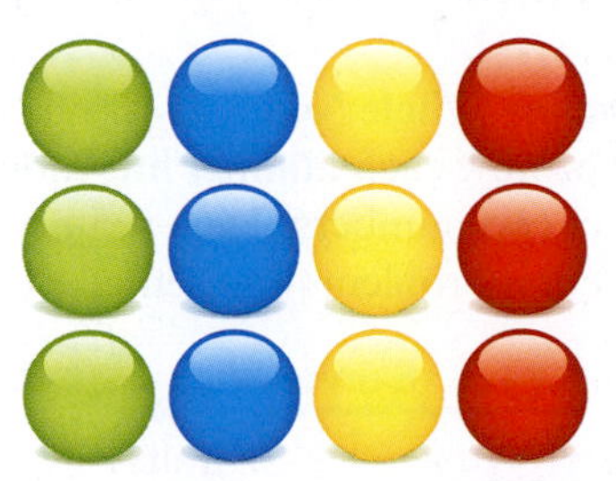

34. Activity: In Your Classroom Design a classroom activity where pairs of students use cards numbered 1 through 10 to practice subtraction facts.

35. Food You need to consume about 2400 calories a day to maintain your weight. On a daily basis, you consume 750 calories at breakfast and 500 calories at lunch.

a. How many calories can your evening meal contain?

b. One day you know your evening meal will be 1500 calories. How can you adjust your breakfast and lunch to maintain the 2400 calorie a day intake?

36. Budget You budget $1700 for bills each month. You budget $575 for rent and utilities and $725 for groceries.

a. How much money do you have left to spend on entertainment expenses?

b. One month you have an additional car expense of $200. How can you adjust your monthly expenses to maintain the $1700 budget?

37. Writing Write a real-life application that involves subtracting two whole numbers.

38. Writing How are addition and subtraction related? Explain.

39. Mental Math Find the missing digit in each subtraction problem. Explain which operation, addition or subtraction, you used to find the answer.

a. $\begin{array}{r} 7? \\ -\ 61 \\ \hline 18 \end{array}$ **b.** $\begin{array}{r} ?7 \\ -\ 43 \\ \hline 54 \end{array}$

c. $\begin{array}{r} 64 \\ -\ 5? \\ \hline 6 \end{array}$ **d.** $\begin{array}{r} 82 \\ -\ ?7 \\ \hline 35 \end{array}$

40. Mental Math Find the missing digit in each subtraction problem. Explain which operation, addition or subtraction, you used to find the answer.

a. $\begin{array}{r} 65? \\ -\ 247 \\ \hline 410 \end{array}$ **b.** $\begin{array}{r} 4?6 \\ -\ 245 \\ \hline 191 \end{array}$

c. $\begin{array}{r} 206 \\ -\ 1?9 \\ \hline 47 \end{array}$ **d.** $\begin{array}{r} 951 \\ -\ ?62 \\ \hline 89 \end{array}$

41. Using a Table The table shows the prices of several electronic devices.

Electronic device	Cost
Plasma TV	$1499
Printer	$129
Cellular phone	$289
Game console	$179
DVD player	$45

a. You purchase a cellular phone and a DVD player. How much more is your purchase than the cost of the game console?

b. You purchase a plasma TV. Your friend purchases a game console and a printer. Who pays more? How much more?

42. Making a Score Sheet The Bobcats and the Beavers are playing a basketball game. Use the information below to complete the score sheet showing the number of points scored each quarter and the final score of the game.

	1	2	3	4	Final
Bobcats					
Beavers					

- The Bobcats scored 12 points in the first quarter.
- The Bobcats were 2 points ahead at the end of the first quarter.
- The Bobcats scored 7 more points in the second quarter than they did in the first quarter.
- The Beavers scored 11 points less than the Bobcats in the second quarter.
- The Beavers scored as many points in the third quarter as the Bobcats did in the first quarter.
- The Bobcats scored as many points in the third quarter as the Beavers did in the first and third quarters combined.
- The Beavers scored a total of 52 points in the game.
- The Bobcats outscored the Beavers by 4 points in the fourth quarter.

43. Fact Families: In Your Classroom Explain how you can use the model below to show each addition and subtraction fact to your students.

a. $2 = 7 - 5$ **b.** $5 = 7 - 2$

c. $5 + 2 = 7$ **d.** $2 + 5 = 7$

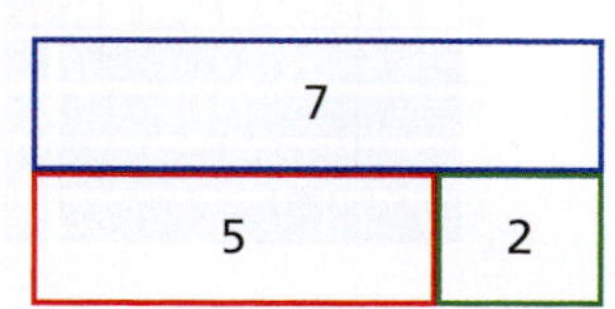

44. Fact Families: In Your Classroom Explain how you can use the model below to show each addition and subtraction fact to your students.

a. $3 = 12 - 9$ **b.** $9 = 12 - 3$

c. $9 + 3 = 12$ **d.** $3 + 9 = 12$

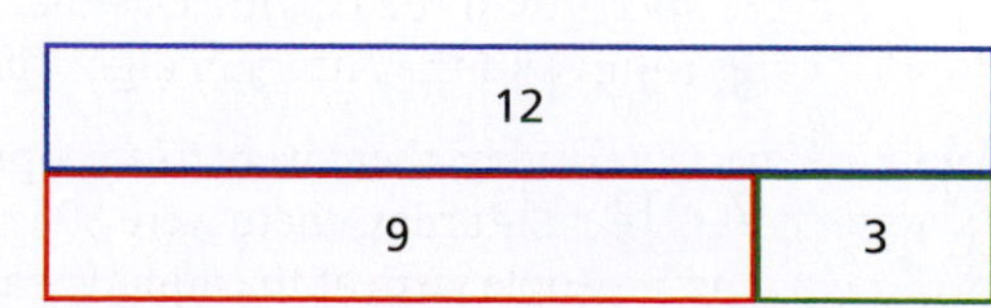

Activity: Using Models to Multiply Whole Numbers

Materials:

- Pencil
- Base-ten blocks

Learning Objective:

Learn to use base-ten blocks to multiply multi-digit whole numbers. Understand the concept of an area model as it is used to represent multiplication.

Name ______________________________

Write the multiplication problem, and find the product.

1.

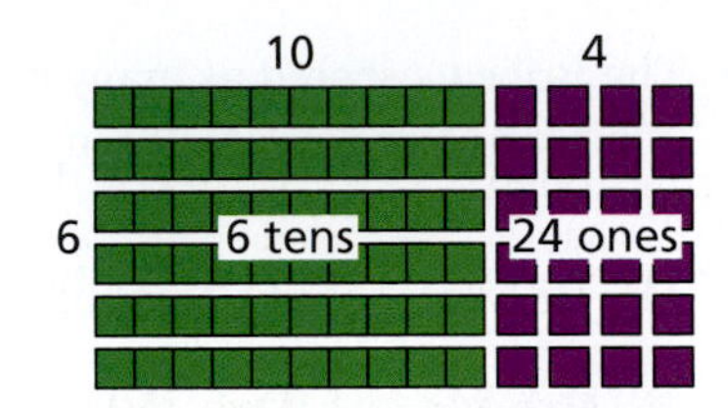

___ × ___ = ___

2.

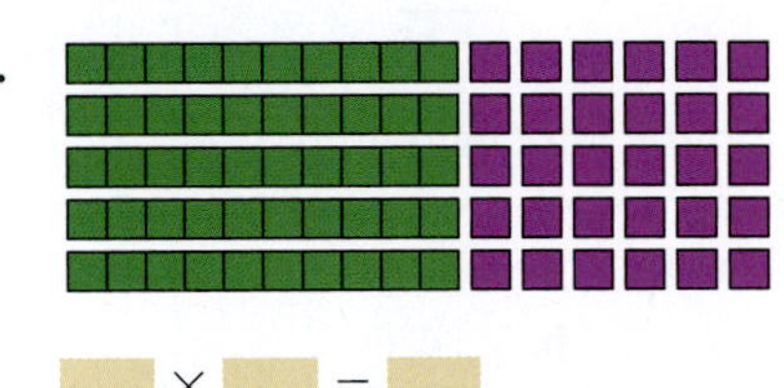

___ × ___ = ___

3.

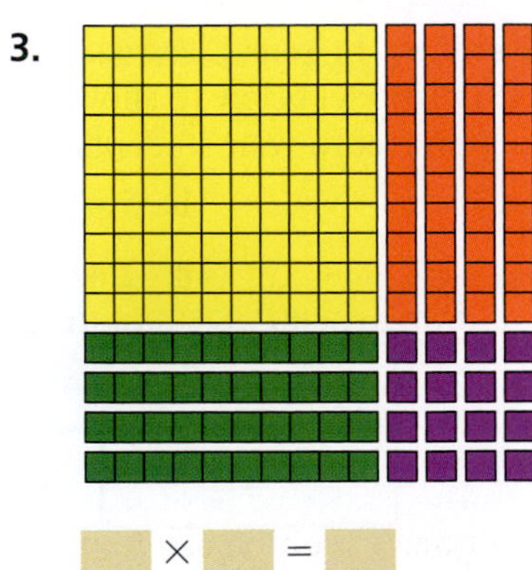

___ × ___ = ___

4.

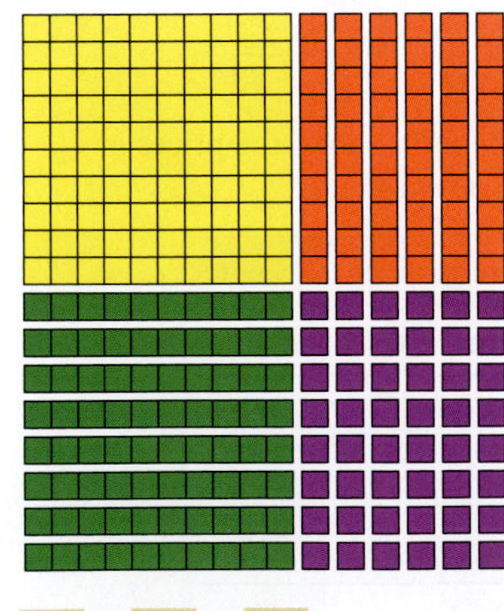

___ × ___ = ___

Draw a model using base-ten blocks to find the product. Explain your reasoning.

5. 7×13

6. 17×9

7. 12×15

8. 18×13

iStockphoto.com/kali9

3.3 Multiplying Whole Numbers

Standards

Grades K–2
Operations and Algebraic Thinking
Students should work with equal groups of objects to gain foundations for multiplication.

Grades 3–5
Numbers and Operations in Base Ten
Students should use place value understanding and properties of operations to perform multi-digit multiplication.

- Understand and use the definition of the product of two whole numbers.
- Understand and use the properties of multiplication for whole numbers.
- Use models to multiply whole numbers.
- Use algorithms to multiply whole numbers.
- Understand and use the Distributive Properties.

Defining the Product of Two Whole Numbers

There is more than one way to define the product of two whole numbers. The most common definition that is given to younger children is to define multiplication as *repeated addition*.

Product of Two Whole Numbers

Let n and a be two whole numbers such that $n \neq 0$. The **product** of n and a, written as $n \cdot a$, is the whole number such that

$$n \cdot a = \underbrace{a + a + a + \cdots + a}_{n \text{ addends}}. \qquad \text{Repeated addition}$$

The numbers, n and a, being multiplied are the **factors** of the product.

EXAMPLE 1 **Finding the Product of Two Whole Numbers**

Use the definition of the product of two whole numbers to find each product.

a. $8 \cdot 11$

b. The product of 9 and 5

c. Seven times four

d. Ten multiplied by nine

Classroom Tip
Example 1 illustrates an important concept in the teaching and learning of mathematics. Mathematical expressions, such as the product of two whole numbers, can be written using symbols. But, expressions can also be written using different word phrases. When you are introducing young students to mathematical concepts, you should make an effort to not always give expressions or instructions in the same manner. Your students will benefit from reading and hearing expressions and instructions given in a variety of ways.

SOLUTION

a. Using the repeated addition definition of multiplication, the product is

$$8 \cdot 11 = \underbrace{11 + 11 + 11 + 11 + 11 + 11 + 11 + 11}_{8 \text{ addends}} = 88.$$

b. Using the repeated addition definition of multiplication, the product of 9 and 5 is

$$9 \cdot 5 = \underbrace{5 + 5 + 5 + 5 + 5 + 5 + 5 + 5 + 5}_{9 \text{ addends}} = 45.$$

c. "Seven times four" can mean either $7 \cdot 4$ or $4 \cdot 7$. On the next page, you will learn that multiplication is *commutative*. So, you obtain the same result with either meaning. You usually interpret "7 times 4" to mean 7 groups of 4.

$$7 \cdot 4 = 4 + 4 + 4 + 4 + 4 + 4 + 4 = 28$$

d. "Ten multiplied by nine" is generally interpreted as $10 \cdot 9$.

$$10 \cdot 9 = 9 + 9 + 9 + 9 + 9 + 9 + 9 + 9 + 9 + 9 = 90$$

Properties of Multiplication for Whole Numbers

Classroom Tip

There are several ways to indicate multiplication. Three of the most common ways are shown below.

7 • 3	Dot
7(3)	Parentheses
7 × 3	Times symbol

Properties of Multiplication for Whole Numbers

Commutative Property of Multiplication

Words The product of two whole numbers does not depend on their order.

Numbers $3 \cdot 4 = 4 \cdot 3$ **Algebra** $a \cdot b = b \cdot a$

Associative Property of Multiplication

Words The product of whole numbers does not depend on how the factors are grouped.

Numbers $(2 \cdot 3) \cdot 6 = 2 \cdot (3 \cdot 6)$ **Algebra** $(a \cdot b) \cdot c = a \cdot (b \cdot c)$

Multiplicative Identity Property

Words The product of any whole number and the **multiplicative identity**, 1, is the whole number.

Numbers $7 \cdot 1 = 7$ **Algebra** $a \cdot 1 = a, 1 \cdot a = a$

Zero Multiplication Property

Words The product of 0 and any whole number is 0.

Numbers $7 \cdot 0 = 0$ **Algebra** $a \cdot 0 = 0, 0 \cdot a = 0$

Classroom Tip

There are many strategies and "tricks" for remembering multiplication facts. Here is one for multiplying a whole number by 9.

Example 9 × 7

1. Hold both palms facing you.

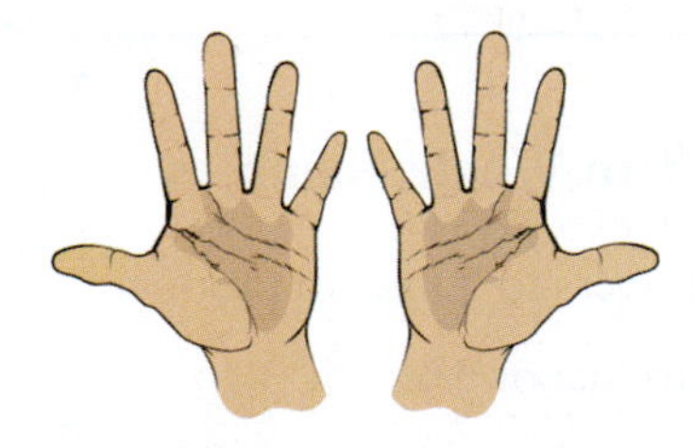

2. From left to right, hold down the finger that represents 7.
3. The number of fingers to the left of the down finger is the tens digit. The number of fingers to the right is the ones digit.

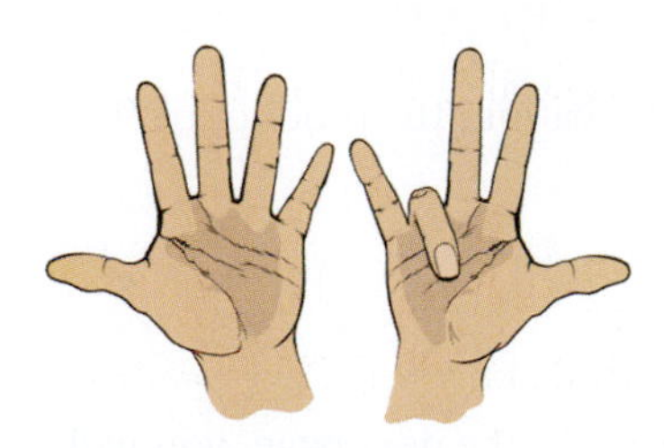

So, 9 × 7 = 63.

Whenever you teach "tricks" like this one, be sure to ask your students why they think the method works.

EXAMPLE 2 Creating a Multiplication Facts Table

Create a multiplication facts table that shows the product of any two whole numbers from 0 through 10.

SOLUTION

There are several ways to do this. One way is shown below.

•	0	1	2	3	4	5	6	7	8	9	10
0	0	0	0	0	0	0	0	0	0	0	0
1	0	1	2	3	4	5	6	7	8	9	10
2	0	2	4	6	8	10	12	14	16	18	20
3	0	3	6	9	12	15	18	21	24	27	30
4	0	4	8	12	16	20	24	28	32	36	40
5	0	5	10	15	20	25	30	35	40	45	50
6	0	6	12	18	24	30	36	42	48	54	60
7	0	7	14	21	28	35	42	49	56	63	70
8	0	8	16	24	32	40	48	56	64	72	80
9	0	9	18	27	36	45	54	63	72	81	90
10	0	10	20	30	40	50	60	70	80	90	100

Even numbers → (row 2)

There are many memorization aids for multiplication tables. For instance, notice that the row showing the product of 2 and a whole number is a subset of the *even numbers*.

EXAMPLE 3 Strategies for Learning Multiplication Facts

Explain how you can use (a) the *Zero Multiplication Property*, (b) the *Multiplicative Identity Property*, (c) the *Commutative Property of Multiplication*, and (d) multiplication by 10 to make it easier to learn the multiplication facts shown in Example 2.

SOLUTION

a. By applying the *Zero Multiplication Property*, students do not have to memorize the first row or the first column.

b. By applying the *Multiplicative Identity Property*, students do not have to memorize the second row or the second column.

c. By applying the *Commutative Property of Multiplication*, students do not have to memorize the product $b \cdot a$ once they know the product $a \cdot b$.

d. Multiplication by 10 is straightforward because students can simply put a zero digit to the right of the whole number. For instance, $7 \cdot 10 = 70$.

Classroom Tip In Example 3(d), when you are teaching students how to multiply by 10, you can justify "adjoining a zero" by using base-ten blocks or by resorting to the definition of the base-ten positional numeral system.

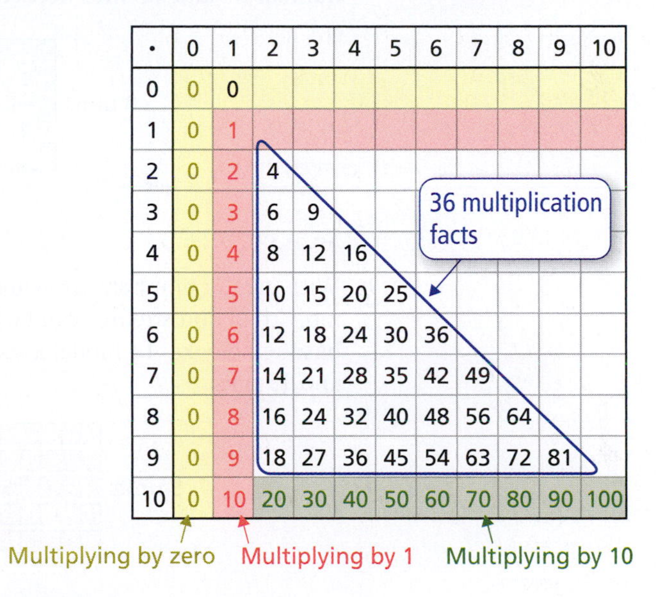

These four strategies reduce the number of multiplication facts from 121 to 36.

Classroom Tip In Examples 3 and 4, you can see that many of the 121 multiplication facts shown in the table in Example 2 can be reasoned out using properties of multiplication or addition facts.

EXAMPLE 4 More Strategies for Learning Multiplication Facts

a. Multiplying a whole number by 2 is the same as "doubling the whole number." In other words,

$2 \cdot a = a + a.$ Multiplying by 2

Because students have already learned addition facts, they can use this prior knowledge to determine the product of 2 and a whole number. This reduces the number of multiplication facts from 36 to 28.

b. By the time students learn multiplication facts, they generally have learned to *skip count* (counting by a number other than 1) by fives:

5, 10, 15, 20, 25,. . .

The products $5 \cdot 1$ and $5 \cdot 2$ have already been eliminated. So, this reduces the number of multiplication facts from 28 to 21.

Using Models to Multiply Whole Numbers

Multiplying Whole Numbers Using an Area Model

An **area model** is a pictorial way to multiply two whole numbers. In this model, the length and width of a rectangle represent the factors, and the area represents the product of the length and width.

EXAMPLE 5 Multiplying Whole Numbers Using an Area Model

Use an area model to find each product.

a. $4 \cdot 6$ **b.** $5 \cdot 13$ **c.** $12 \cdot 11$

SOLUTION

a. To model the product $4 \cdot 6$, draw a rectangle that is 4 units by 6 units. The number of unit squares needed to fill the rectangle represents the product.

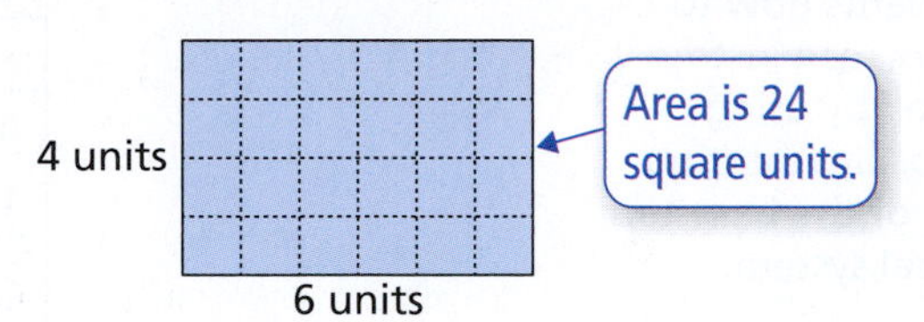

So, $4 \cdot 6 = 24$.

b. With greater numbers, drawing a rectangle and finding the number of individual unit squares can be tedious. To model the product $5 \cdot 13$, use base-ten blocks. To model a rectangle that is 5 units by 13 units, use 5 tens and 15 ones.

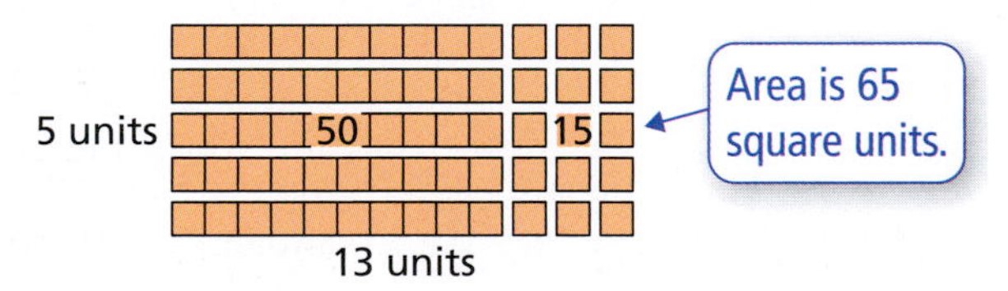

The two portions of the rectangle are 50 and 15 square units. So, the area of the rectangle is $50 + 15 = 65$ square units. So,

$$5 \cdot 13 = 65.$$

c. To model a rectangle that is 12 units by 11 units, use 1 hundred, 3 tens, and 2 ones.

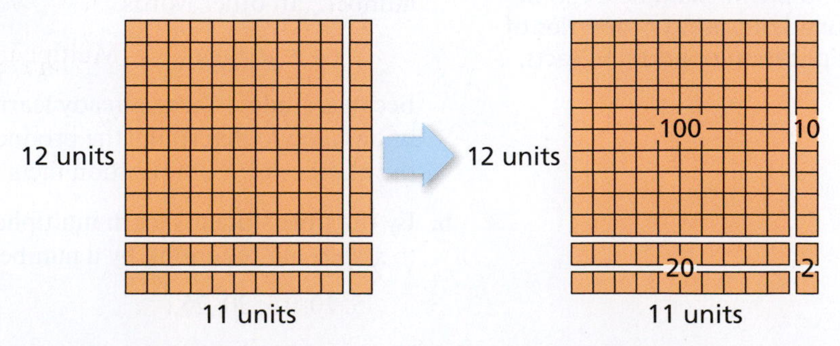

The four portions of the rectangle are 100, 20, 10, and 2 square units. So, the area of the rectangle is $100 + 20 + 10 + 2 = 132$ square units. So,

$$12 \cdot 11 = 132.$$

Using Algorithms to Multiply Whole Numbers

Algorithms for Multiplying Whole Numbers

Here are two algorithms for multiplying whole numbers.

Expanded Algorithm

1. Write the numbers vertically, lining up the digits by place value.
2. Multiply by place value. Then add the products.

Standard Algorithm

1. Write the numbers vertically, lining up the digits by place value.
2. Multiply the first number by each digit of the second number. Regroup if necessary. Then add the products.

Example $19 \cdot 16 = 304$

Expanded:

$$\begin{array}{rl} 19 & \\ \times\ 16 & \\ \hline 54 & 6 \times 9 \\ 60 & 6 \times 10 \\ 90 & 10 \times 9 \\ +\ 100 & 10 \times 10 \\ \hline 304 & \end{array}$$

Standard:

$$\begin{array}{r} \scriptstyle 5 \\ 19 \\ \times\ 16 \\ \hline 114 \\ +\ 190 \\ \hline 304 \end{array}$$

Using 0 as a placeholder helps keep the digits properly aligned.

Notice how the products being added in the expanded algorithm are shown in the base-ten block model below.

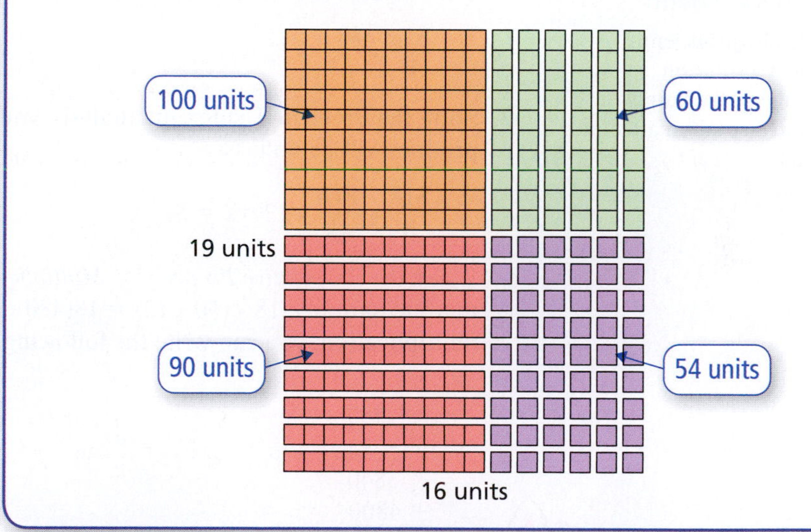

Classroom Tip In the standard algorithm shown at the far right, the 190 is often written as 19 tens, as follows.

$$\begin{array}{r} \scriptstyle 5 \\ 19 \\ \times\ 16 \\ \hline 114 \\ +\ 19 \\ \hline 304 \end{array}$$

EXAMPLE 6 **Using the Expanded Algorithm**

Use the expanded algorithm to find each product.

a. $85 \cdot 46$ **b.** $237 \cdot 60$

SOLUTION

a.

$$\begin{array}{rl} 85 & \\ \times\ 46 & \\ \hline 30 & 6 \times 5 \\ 480 & 6 \times 80 \\ 200 & 40 \times 5 \\ +\ 3200 & 40 \times 80 \\ \hline 3910 & \end{array}$$

b.

$$\begin{array}{rl} 237 & \\ \times\ 60 & \\ \hline 0 & 0 \times 237 \\ 420 & 60 \times 7 \\ 1{,}800 & 60 \times 30 \\ +\ 12{,}000 & 60 \times 200 \\ \hline 14{,}220 & \end{array}$$

EXAMPLE 7 Using the Standard Algorithm

Use the standard algorithm to find each product.

a. 85 • 46 **b.** 237 • 60

SOLUTION

a.

$$\begin{array}{r} {}^{2\,3}\\ 85 \\ \times\ 46 \\ \hline 510 \\ +\ 3400 \\ \hline 3910 \end{array}$$

0 is a placeholder.

Multiply 85 by 6.
Multiply 85 by 4.

b.

$$\begin{array}{r} {}^{2\,4}\\ 237 \\ \times 60 \\ \hline 0 \\ +\ 14{,}220 \\ \hline 14{,}220 \end{array}$$

0 is a placeholder.

Multiply 237 by 0.
Multiply 237 by 6.

EXAMPLE 8 Solving a Real-Life Multiplication Problem

You earn $18 per hour at a summer job. How much will you earn working 40 hours per week for 12 weeks?

SOLUTION

Mathematical Practices

Make Sense of Problems and Persevere in Solving Them

Mathematically proficient students plan a solution pathway rather than simply jumping into a solution attempt.

1. **Understand the Problem** You need to find the total amount you will earn working for 12 weeks.

2. **Make a Plan** Write an expression, including units, for your earnings. Your answer should be in *dollars*.

$$\frac{\$18}{\text{h}} \cdot \frac{40\text{ h}}{\text{wk}} \cdot 12\text{ wk}$$

3. **Solve the Problem** Using unit analysis, you can see that the units of the product will be dollars.

$$\frac{\$18}{\cancel{\text{h}}} \cdot \frac{40\ \cancel{\text{h}}}{\cancel{\text{wk}}} \cdot 12\ \cancel{\text{wk}} = \$\square$$

Using the *Associative Property of Multiplication* to group two factors and mental math to simplify, 18 • (40 • 12) = 18(480). Using the standard algorithm for multiplication, you can write the following.

$$\begin{array}{r} {}^{6}\\ 480 \\ \times\ 18 \\ \hline 3840 \\ +\ 4800 \\ \hline 8640 \end{array}$$

You will earn $8640 for working 12 weeks.

4. **Look Back** You can use a calculator to verify that your answer is correct.

You can also check the reasonableness of your answer by rounding each factor to the nearest ten and then finding the product.

18 • 40 • 12 = 8640

↓ ↓ ↓

20 • 40 • 10 = 8000 ✓

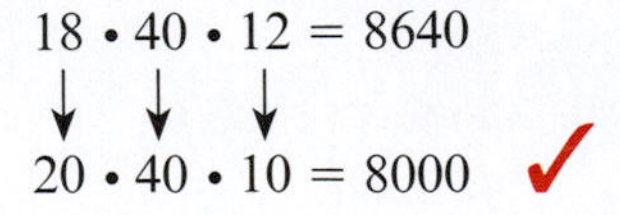

The Distributive Properties

A property that connects addition and multiplication is called the *Distributive Property of Multiplication Over Addition*. This property can be illustrated by the following area model, which shows that $4 \cdot (5 + 3) = 4 \cdot 5 + 4 \cdot 3$. Note that there is also a subtraction version of the Distributive Property.

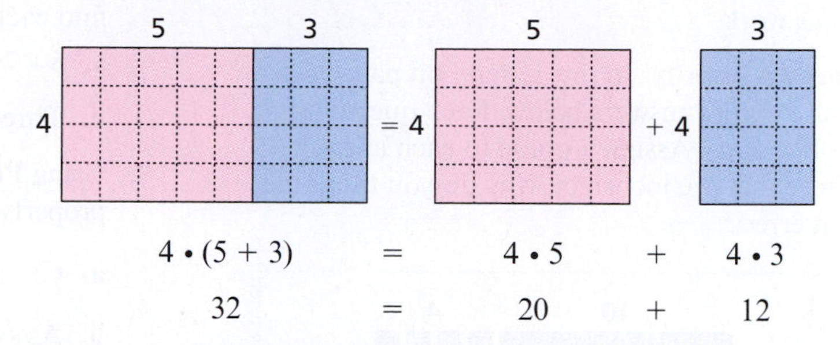

$4 \cdot (5 + 3) = 4 \cdot 5 + 4 \cdot 3$

$32 = 20 + 12$

Distributive Properties

Let a, b, and c be whole numbers.

Distributive Property of Multiplication Over Addition

Numbers $5(2 + 4) = 5(2) + 5(4)$ **Algebra** $a \cdot (b + c) = a \cdot b + a \cdot c$

Distributive Property of Multiplication Over Subtraction

Numbers $2(7 - 3) = 2(7) - 2(3)$ **Algebra** $a \cdot (b - c) = a \cdot b - a \cdot c$

EXAMPLE 9 Using the Distributive Properties

Use one of the Distributive Properties to rewrite each expression. Then evaluate.

a. $5 \cdot (12 + 4)$ **b.** $9 \cdot 11 + 9 \cdot 3$

c. $8 \cdot (21 - 4)$ **d.** $6(7 + 3 + 2)$

SOLUTION

a. $5 \cdot (12 + 4) = 5(12) + 5(4)$
$= 60 + 20$
$= 80$

b. $9 \cdot 11 + 9 \cdot 3 = 9(11 + 3)$
$= 9(14)$
$= 126$

c. $8 \cdot (21 - 4) = 8(21) - 8(4)$
$= 168 - 32$
$= 136$

d. $6(7 + 3 + 2) = 6(7) + 6(3) + 6(2)$
$= 42 + 18 + 12$
$= 72$

EXAMPLE 10 Using a Distributive Property

Use one of the Distributive Properties to evaluate $6 \cdot 317$.

SOLUTION

The Distributive Property presents an efficient method for evaluating this product.

$6 \cdot 317 = 6 \cdot (300 + 10 + 7)$ Write 317 in expanded form.

$= 1800 + 60 + 42$ Multiply using the *Distributive Property.*

$= 1902$ Add.

3.3 Exercises

Free step-by-step solutions to odd-numbered exercises at *MathematicalPractices.com.*

1. **Writing a Solution Key** Write a solution key for the activity on page 96. Describe a rubric for grading a student's work.

2. **Grading the Activity** In the activity on page 96, a student gave the answers below. Each question is worth 10 points. Assign a grade to each answer. For those that are incorrect, why do you think the student erred?

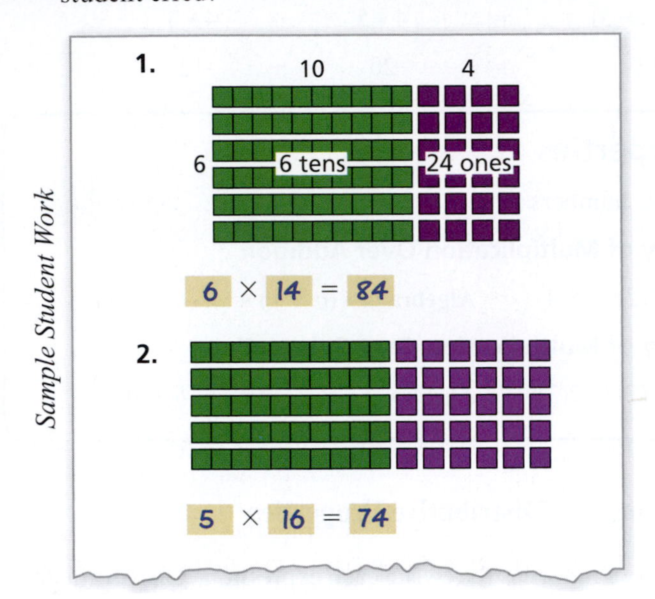

3. **Using the Definition of Product** Rewrite each expression as the product of two whole numbers.
   a. $4 + 4 + 4$ b. $7 + 7 + 7 + 7 + 7$

4. **Using the Definition of Product** Rewrite each expression as the product of two whole numbers.
   a. $3 + 3 + 3 + 3$
   b. $6 + 6 + 6 + 6 + 6 + 6 + 6$

5. **Using Vocabulary** Identify the factors of each multiplication problem. Then find the product.
   a. $4 \cdot 5$ b. $7 \cdot 2$

6. **Using Vocabulary** Identify the factors of each multiplication problem. Then find the product.
   a. $3 \cdot 8$ b. $6 \cdot 4$

7. **Finding Products of Two Whole Numbers** Use the definition of the product of two whole numbers to find each product.
   a. $6 \cdot 5$ b. The product of 4 and 8
   c. Five times seven
   d. Nine multiplied by three

8. **Finding Products of Two Whole Numbers** Use the definition of the product of two whole numbers to find each product.
   a. $8 \cdot 7$ b. The product of 6 and 8
   c. Three times eight d. Nine multiplied by four

9. **Using Properties of Multiplication** Use the given property of multiplication to complete each equation.
   a. Commutative Property: $7 \cdot 9 =$
   b. Associative Property: $4 \cdot (5 \cdot 7) =$
   c. Multiplicative Identity Property: $9 \cdot 1 =$
   d. Zero Multiplication Property: $7 \cdot 0 =$

10. **Using Properties of Multiplication** Use the given property of multiplication to complete each equation.
   a. Commutative Property: $x \cdot 8 =$
   b. Associative Property: $6 \cdot (2 \cdot x) =$
   c. Multiplicative Identity Property: $1 \cdot x =$
   d. Zero Multiplication Property: $0 \cdot x =$

11. **Using Multiplication Strategies** Let a be a one-digit whole number. Describe a strategy for multiplying each pair of numbers.
   a. $1 \cdot a$ b. $2 \cdot a$
   c. $10 \cdot a$

12. **Using Multiplication Strategies** Let b be a one-digit whole number. Describe a strategy for multiplying each pair of numbers.
   a. $0 \cdot b$
   b. $5 \cdot b$
   c. $b \cdot 3$ (given that $3 \cdot b = 27$)

13. **Multiplying Whole Numbers Using Models** Describe how each model is used to multiply two numbers. Then write and solve the multiplication problem represented by the model.
   a.
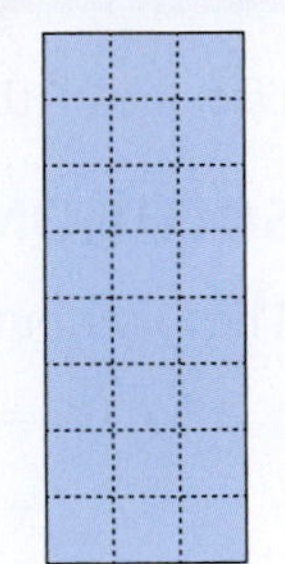
   b.

14. Multiplying Whole Numbers Using Models Describe how each model is used to multiply two numbers. Then write and solve the multiplication problem represented by the model.

a.

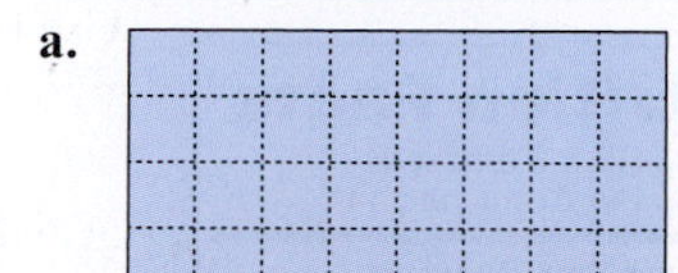

b.

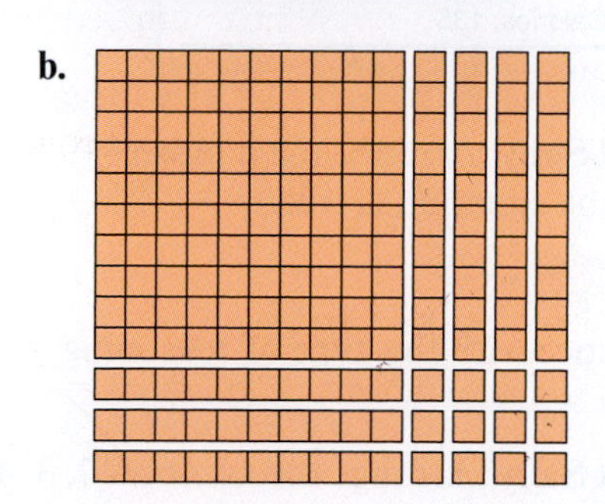

15. Identifying the Products in the Expanded Algorithm Identify the numbers that are multiplied at each step in the expanded algorithm solution of each problem.

a. $\begin{array}{rl} 47 & \\ \times 16 & \\ \hline 42 & 6 \times 7 \\ 240 & \square \\ 70 & \square \\ +\ 400 & \square \\ \hline 752 & \end{array}$

b. $\begin{array}{rl} 326 & \\ \times 42 & \\ \hline 12 & \square \\ 40 & \square \\ 600 & \square \\ 240 & \square \\ 800 & \square \\ +\ 12{,}000 & \square \\ \hline 13{,}692 & \end{array}$

16. Identifying the Products in the Expanded Algorithm Identify the numbers that are multiplied at each step in the expanded algorithm solution of each problem.

a. $\begin{array}{rl} 36 & \\ \times 24 & \\ \hline 24 & 4 \times 6 \\ 120 & \square \\ 120 & \square \\ +\ 600 & \square \\ \hline 864 & \end{array}$

b. $\begin{array}{rl} 531 & \\ \times 27 & \\ \hline 7 & \square \\ 210 & \square \\ 3{,}500 & \square \\ 20 & \square \\ 600 & \square \\ +\ 10{,}000 & \square \\ \hline 14{,}337 & \end{array}$

17. Using the Expanded Algorithm Use the expanded algorithm to find each product.

a. $\begin{array}{r} 78 \\ \times 40 \\ \hline \end{array}$

b. $\begin{array}{r} 39 \\ \times 75 \\ \hline \end{array}$

c. 32×63

d. 70×418

e. 24×363

f. 300×227

18. Using the Expanded Algorithm Use the expanded algorithm to find each product.

a. $\begin{array}{r} 68 \\ \times 90 \\ \hline \end{array}$

b. $\begin{array}{r} 57 \\ \times 76 \\ \hline \end{array}$

c. $29 \cdot 89$

d. $50 \cdot 782$

e. $77 \cdot 385$

f. $220 \cdot 330$

19. Using the Standard Algorithm Use the standard algorithm to find each product.

a. $\begin{array}{r} 62 \\ \times 21 \\ \hline \end{array}$

b. $\begin{array}{r} 77 \\ \times 72 \\ \hline \end{array}$

c. $\begin{array}{r} 149 \\ \times 40 \\ \hline \end{array}$

d. $\begin{array}{r} 376 \\ \times 37 \\ \hline \end{array}$

e. $200 \cdot 476$

f. $220 \cdot 193$

g. $303 \cdot 246$

h. $341 \cdot 761$

20. Using the Standard Algorithm Use the standard algorithm to find each product.

a. $\begin{array}{r} 48 \\ \times 32 \\ \hline \end{array}$

b. $\begin{array}{r} 87 \\ \times 59 \\ \hline \end{array}$

c. $\begin{array}{r} 362 \\ \times 60 \\ \hline \end{array}$

d. $\begin{array}{r} 679 \\ \times 43 \\ \hline \end{array}$

e. 600×986

f. 630×724

g. 407×239

h. 651×677

21. Using the Distributive Properties Use one of the Distributive Properties to rewrite each expression. Then evaluate.

a. $6 \cdot (9 + 5)$

b. $7 \cdot 13 + 7 \cdot 7$

c. $9 \cdot (20 - 3)$

d. $13(27) - 13(17)$

e. $4(9 + 7 + 6)$

f. $8(6) + 8(3) + 8(9)$

22. Using the Distributive Properties Use one of the Distributive Properties to rewrite each expression. Then evaluate.

a. $8 \cdot (10 + 7)$

b. $9 \cdot 7 + 9 \cdot 13$

c. $7 \cdot (30 - 4)$

d. $14(34) - 14(24)$

e. $9(8 + 8 + 7)$

f. $7(12) + 7(14) + 7(24)$

23. Using the Distributive Properties Use one of the Distributive Properties to evaluate each expression.

a. $8 \cdot 94$

b. $4 \cdot 246$

24. Using the Distributive Properties Use one of the Distributive Properties to evaluate each expression.

a. $9 \cdot 87$ **b.** $6 \cdot 345$

25. Using the Distributive Properties Evaluate $7 \cdot 73$ in two ways using the Distributive Properties.

26. Using the Distributive Properties Evaluate $8 \cdot 65$ in two ways using the Distributive Properties.

27. Modeling Properties: In Your Classroom You can use base-ten blocks to model a product of three numbers as shown below.

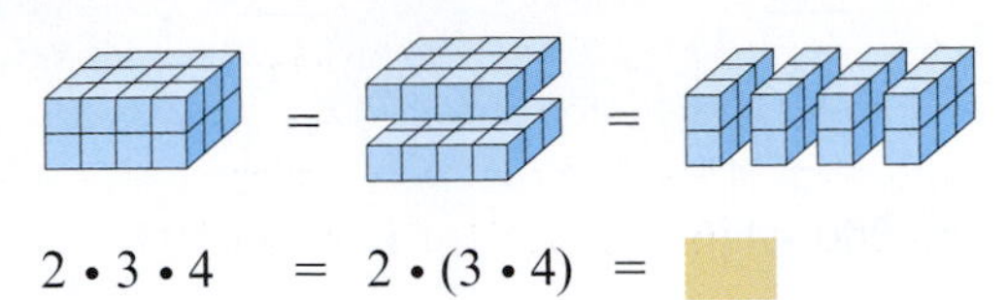

$2 \cdot 3 \cdot 4 \quad = \quad 2 \cdot (3 \cdot 4) \quad = \quad$ ____

a. What property is shown?

b. Write the missing expression.

c. Explain to your students how to use base-ten blocks to model other properties of multiplication of whole numbers.

28. Modeling Properties: In Your Classroom You can use an area model to represent one of the Distributive Properties.

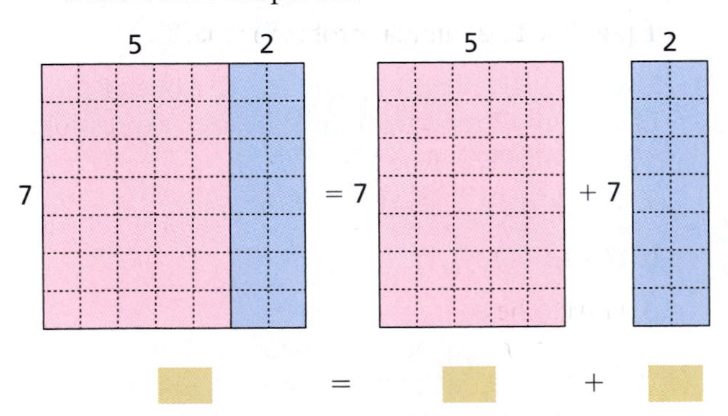

____ = ____ + ____

a. Which Distributive Property is represented by the area model?

b. Complete the equation represented by the area model.

c. Explain to your students how to use an area model to represent the Distributive Property of Multiplication Over Subtraction.

29. Multiplying Whole Numbers The nutritional information below is from a box of cereal. How many calories are there in the full box of cereal?

Nutrition Facts
Serving Size: 1 cup
Serving Per Container: 12
Amount Per Serving
Calories: 190

30. Multiplying Whole Numbers The nutritional information below is from a box of crackers. How many calories are there in the full box of crackers?

Nutrition Facts
Serving Size: 6 crackers
Serving Per Container: 14
Amount Per Serving
Calories: 135

31. My Dear Aunt Sally Arrange the numbers 1, 2, 4, and 8 so that the equation is true.

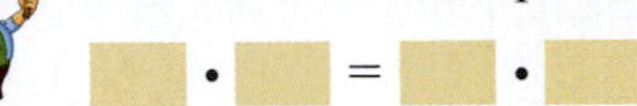

For more practice with this type of reasoning, go to *AuntSally.com.*

32. My Dear Aunt Sally Arrange the numbers 2, 3, 4, and 6 so that the equation is true.

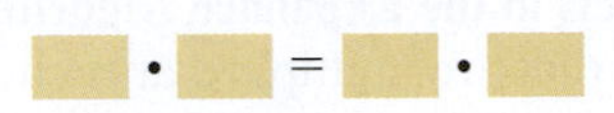

For more practice with this type of reasoning, go to *AuntSally.com.*

33. My Dear Aunt Sally Arrange the numbers 1, 2, 3, 3, 4, and 8 so that the equation is true.

For more practice with this type of reasoning, go to *AuntSally.com.*

34. My Dear Aunt Sally Arrange the numbers 3, 4, 5, 5, 6 and 8 so that the equation is true.

For more practice with this type of reasoning, go to *AuntSally.com.*

35. Finding Cost An ad in a local newspaper costs \$32 per column inch. A column inch is a unit of space that is 1 column wide and 1 inch high as shown below.

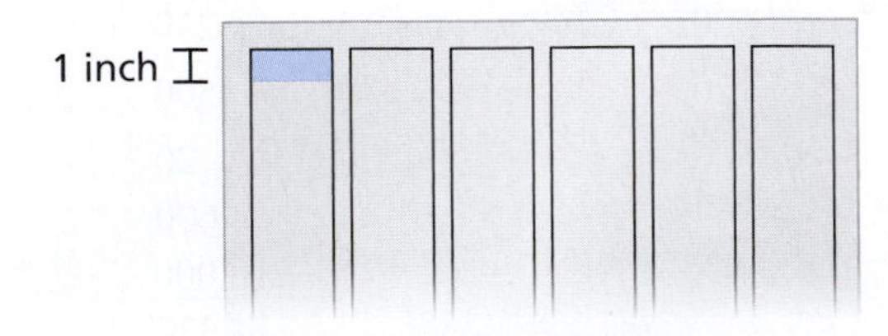

a. Find the cost of an ad that is 1 column wide and 4 inches high.

b. Write an expression for the cost of an ad that is 4 columns wide and 5 inches high. Then find the cost of the ad.

36. Finding Cost The monthly cost of a digital billboard ad is $3878 per 4-second spot in a 64-second loop.

a. Find the monthly cost of a 12-second spot.

b. Find the total monthly cost of an 8-second spot and a 20-second spot.

37. Writing Explain how the Zero Multiplication Property helps students to learn multiplication facts.

38. Writing Write a real-life application that uses the Distributive Property to estimate annual profit based on revenues and expenses for one month.

39. Finding Area You cut 2 yards of fabric to reduce the length of the fabric by 18 inches.

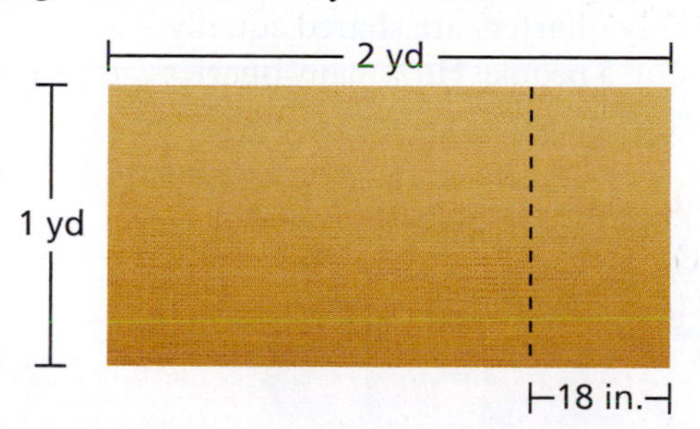

a. What is the area, in square inches, of the 2 yards of fabric?

b. Write an expression that can be evaluated using the Distributive Property to find the area of the shortened piece of fabric. Then find the area.

40. Finding Area The football field is 60 feet shorter without its end zones.

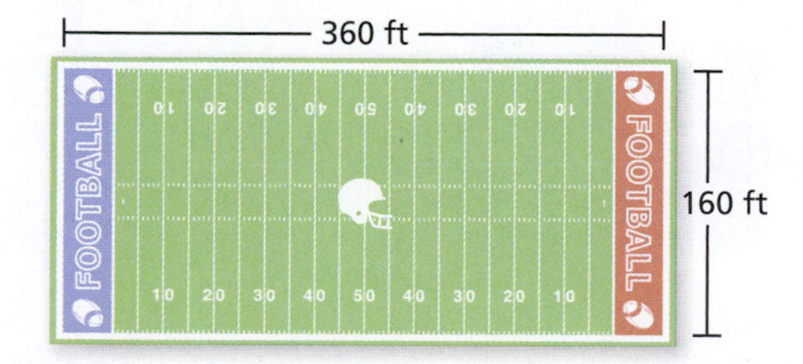

a. What is the total area of the football field including the end zones?

b. Write an expression that can be evaluated using the Distributive Property to find the area of the football field without its end zones. Then find the area.

iStockphoto.com/Jason Ganser

41. Grading Student Work On a diagnostic test, one of your students does the following work. Explain what the student did wrong. Which topics would you encourage the student to review?

a.

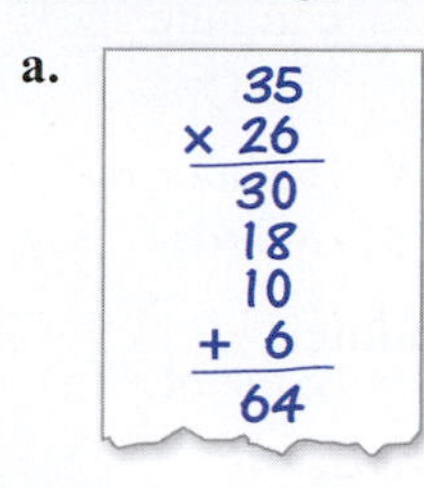

b.

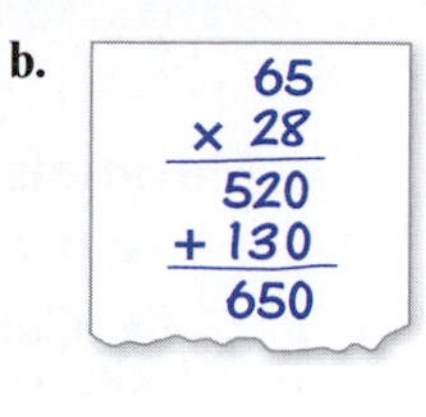

42. Grading Student Work On a diagnostic test, one of your students does the following work. Explain what the student did wrong. Which topics would you encourage the student to review?

a.

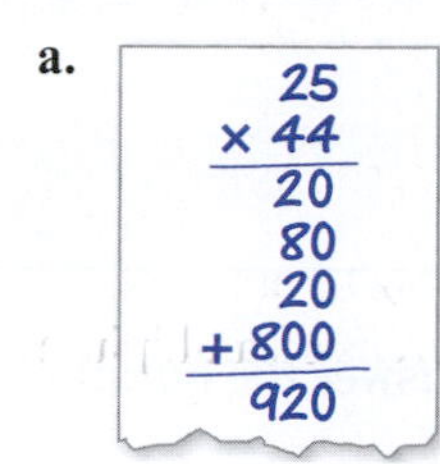

b.

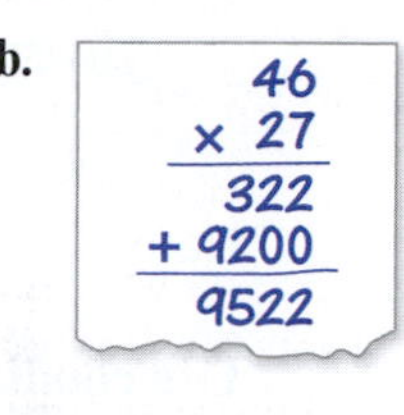

43. Problem Solving The range of possible monthly revenues and the monthly expenses for an employment agency are shown. Find the range of possible total annual profits for the agency.

Monthly Finances	
Revenues	$45,000–$65,000
Expenses	$42,000

44. Problem Solving A family needs nursing care for a relative. The charges and financial assistance available to the patient at two nursing homes are shown. The patient's family must pay the difference between the charges and the financial assistance.

Elder Care Lodge	
Daily charges	$244
Daily assistance	$187

Golden Ages Village	
Monthly charges	$6824
Monthly assistance	$5278

a. Which nursing home would cost the patient's family more money for a 21-day stay, given that Golden Ages Village will charge for a full month? How much more?

b. Which nursing home would cost the patient's family more money annually? How much more?

Activity: Modeling Division of Whole Numbers

Materials:

- Pencil
- Colored counters

Learning Objective:

Learn to connect division with making equal groups of items. Understand the concept of a set model as it is used to represent division.

Name ______________________________

Use counters to answer the question. Write the division problem, and find the quotient.

1. Twenty quarters are shared equally among 4 people. How many quarters does each person get?

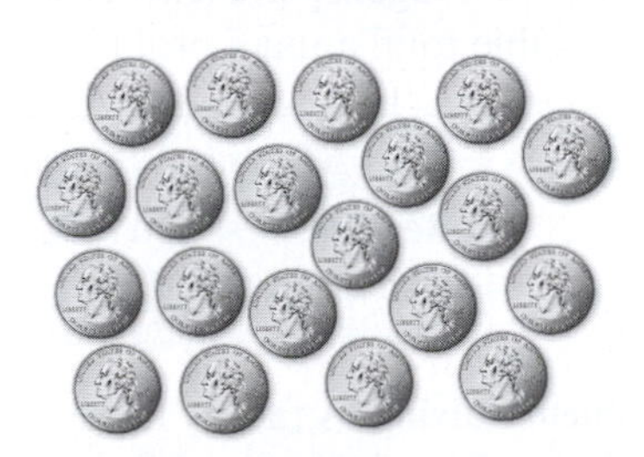

___ ÷ ___ = ___

2. Twenty quarters are shared equally among 5 people. How many quarters does each person get?

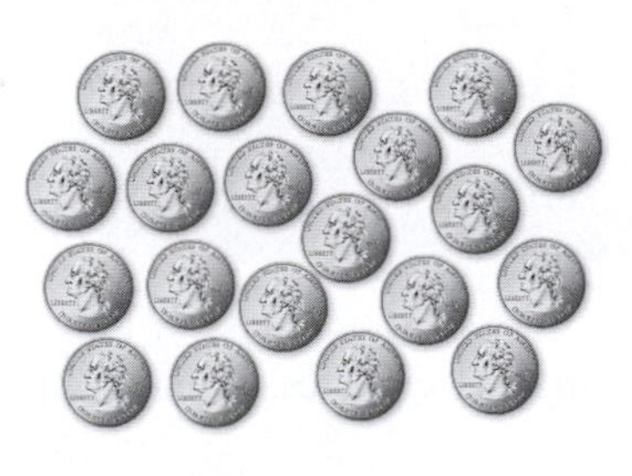

___ ÷ ___ = ___

3. Eighteen dimes are distributed among 6 people. Each person receives the same number of dimes. How many dimes does each person get?

___ ÷ ___ = ___

4. Eighteen dimes are distributed among 3 people. Each person receives the same number of dimes. How many dimes does each person get?

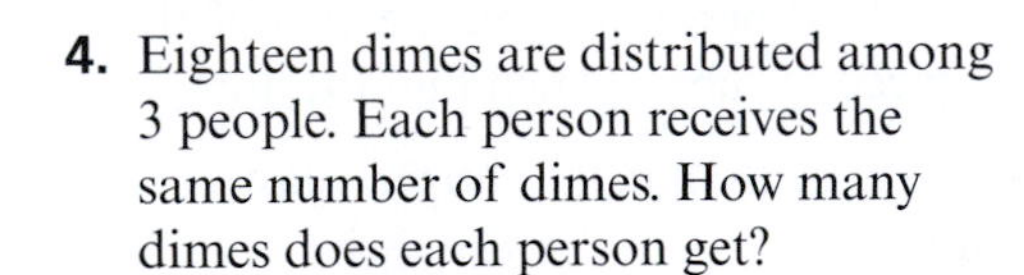

___ ÷ ___ = ___

5. Three jars contain a total of 24 nickels. Each jar has the same number of nickels. How many nickels are in each jar?

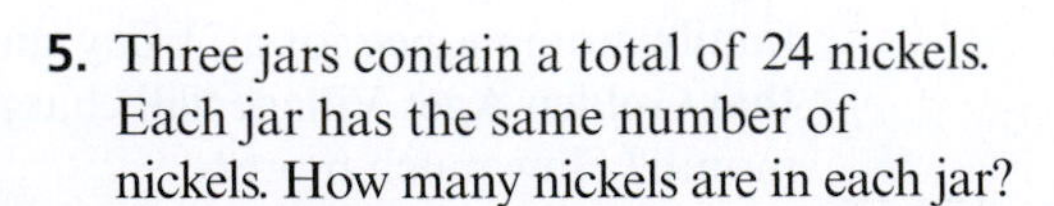

6. Eight jars contain a total of 24 nickels. Each jar has the same number of nickels. How many nickels are in each jar?

iStockphoto.com/CEFutcher

3.4 Dividing Whole Numbers

Standards

Grades 3–5

Operations and Algebraic Thinking

Students should represent and solve problems involving division.

Grades 6–8

The Number System

Students should compute fluently with multi-digit numbers.

- Understand and use the definition of the quotient of two whole numbers.
- Use models to divide whole numbers.
- Use algorithms to divide whole numbers.

Defining the Quotient of Two Whole Numbers

Just as subtraction is defined as the inverse operation of addition, **division** can be defined as the inverse operation of multiplication. The *missing factor method* helps define division of whole numbers in terms of multiplication.

Quotient of Two Whole Numbers

Missing Factor Method

Let a and b be two whole numbers, where $b \neq 0$. The **quotient** of a and b, written as $a \div b$, is the whole number such that

$$b \cdot (a \div b) = a.$$

What can you multiply b by to get a?

Divisor Quotient Dividend

The quotient $a \div b$ is also written as $\frac{a}{b}$ and $b\overline{)a}$.

Examples $10 \div 5 = 2$, $\frac{10}{5} = 2$, $5\overline{)10} = 2$, or $5\overline{)10}$ with quotient 2

Set (Partition) Method

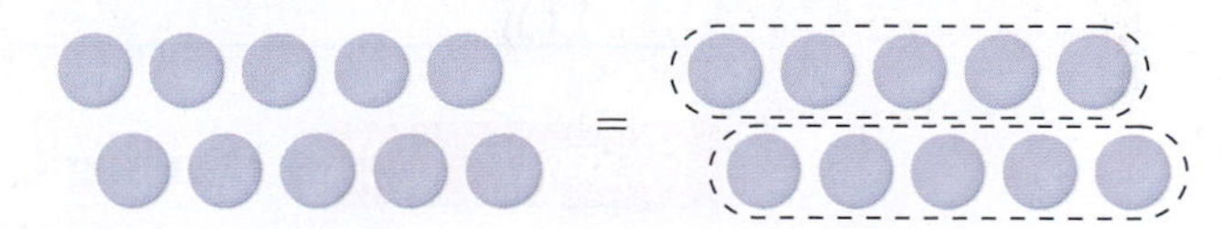

$10 \div 5 = 2$ "The quotient of 10 and 5 is 2" or "10 divided by 5 equals 2."

Classroom Tip To help students better understand the relationship between multiplication and division, use the missing factor method to write

$10 \div 2 \cdot 2 = 10.$

Dividing by 2, then multiplying by 2 gives back the original number. The multiplication facts table below also shows this relationship.

+	0	1	2	3	4	!
0						
1						
2						
3						
4						
5			10			

$5 \cdot 2 = 10$
$2 \cdot 5 = 10$
$10 \div 5 = 2$
$10 \div 2 = 5$

EXAMPLE 1 **Using the Definition of Quotient**

Answer each question. Then write the division problem.

a. What can you multiply 3 by to get 12?

b. Thirty-six people are grouped into 4 committees. Each committee has the same number of people. How many people are on each committee?

SOLUTION

a. You can multiply 3 by 4 to get 12.

$12 \div 3 = 4$ The quotient of 12 and 3 is 4.

b. Each committee has 9 people.

$36 \div 4 = 9$ The quotient of 36 and 4 is 9.

Using Models to Divide Whole Numbers

Classroom Tip

You have now seen the four basic operations with whole numbers. When teaching these operations, it is important to display and use all of the vocabulary words.

Addition: add, sum, addend, plus, total

Subtraction: subtract, difference, minuend, subtrahend, take away, minus

Multiplication: multiply, product, factor, times, multiplied by

Division: divide, quotient, divisor, dividend, divided by, divided into, remainder

Dividing Whole Numbers Using an Array

An **array** is a display of objects in a rectangular arrangement. The array at the right is formed with cereal. It has 3 rows, with each row having 4 pieces of cereal. It can be used to model the division problem

$$12 \div 4 = 3.$$

When modeling division of whole numbers, it is possible that the quotient of two whole numbers is not a whole number. In such cases, there is a **remainder**. For instance, to model $29 \div 6$, use 29 counters. Arrange them in rows of 6 counters each until you use all of the counters. Then count the number of full rows. When there is a partial row, the number of counters in the partial row is the remainder.

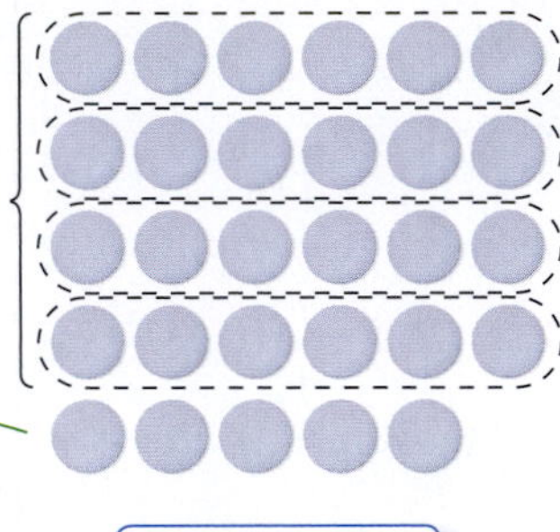

$29 \div 6 = 4R5$

EXAMPLE 2 Dividing Using an Array

Use an array to find (a) $30 \div 5$ and (b) $\frac{44}{8}$.

SOLUTION

a. Use 30 counters. Arrange them in rows of 5 until you use all of the counters. There are 6 full rows with no counters left over.

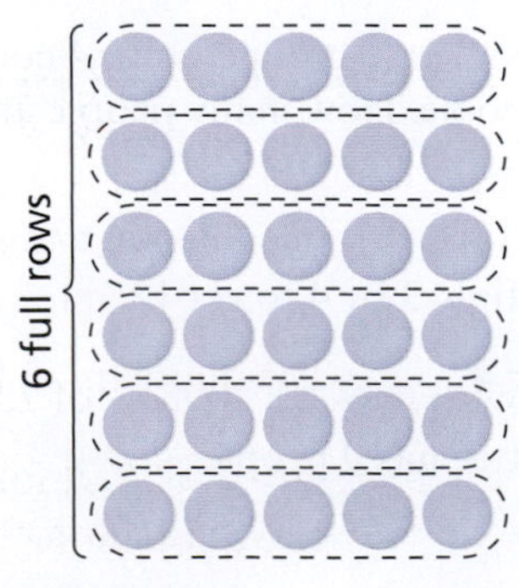

So, $30 \div 5 = 6$.

b. Use 44 counters. Arrange them in rows of 8 until you use all of the counters. There are 5 full rows. The sixth row is a partial row with 4 counters.

So, $\frac{44}{8} = 5R4$.

Dividing Whole Numbers Using a Number Line

Use a **number line model** to divide small whole numbers. For instance, to model 12 divided by 3, place an arrow that is 12 units long on a number line. Then count the number of arrows of length 3 units that are in the arrow representing 12 units.

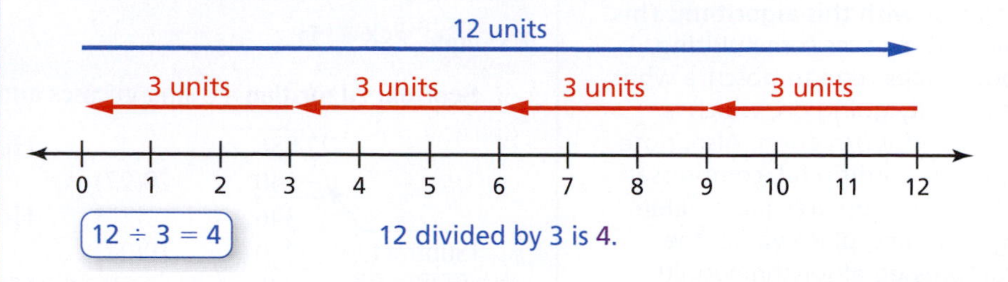

Here is an example of using a number line model to divide whole numbers and getting a remainder.

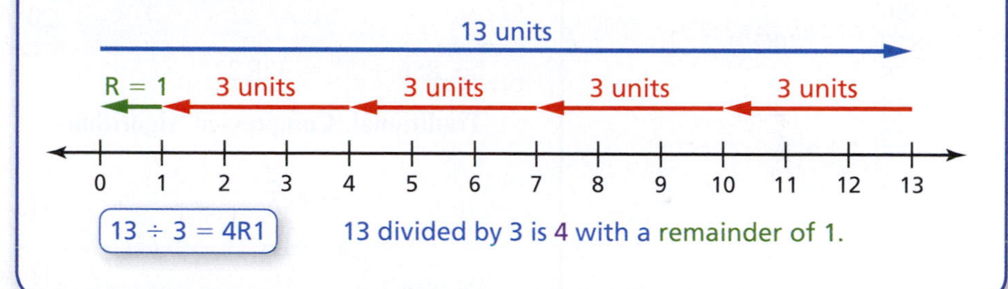

Classroom Tip

The number line model helps show how division can be viewed as *repeated subtraction*. For instance, you can divide 12 by 3 by repeatedly subtracting 3.

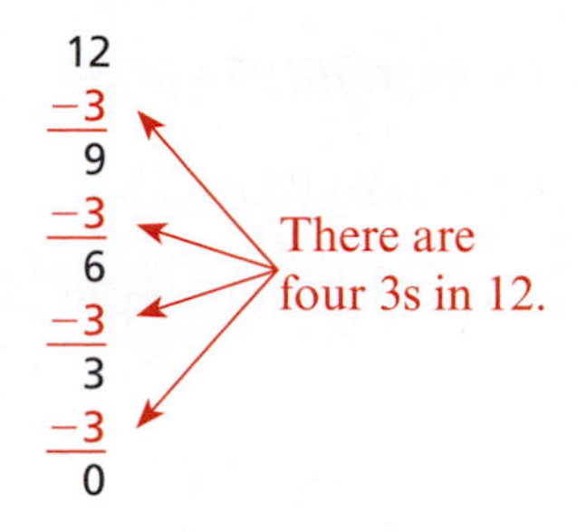

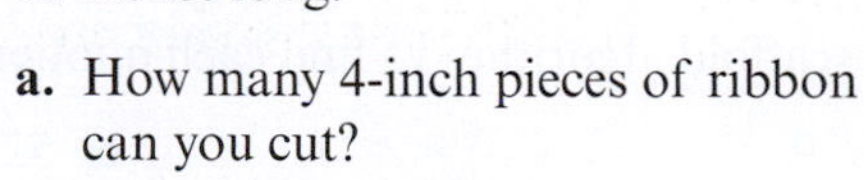

Dividing Using a Number Line Model

You are using pieces of ribbon in an art project. You cut 4-inch pieces of ribbon from a ribbon that is 38 inches long.

a. How many 4-inch pieces of ribbon can you cut?

b. Interpret the remainder.

SOLUTION

You can use a number line model.

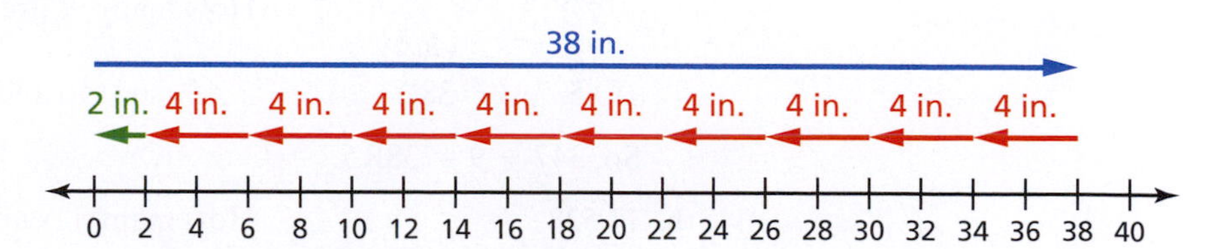

a. From the number line model, you can see that

$38 \div 4 = 9R2.$ 38 divided by 4 is 9, with a remainder of 2.

So, you can cut nine 4-inch pieces of ribbon.

b. Because there is a remainder of 2, it follows that you will have a 2-inch piece of ribbon left over. The leftover piece is not long enough to make a tenth 4-inch piece of ribbon.

Classroom Tip

When you are teaching mathematics, remember that "hands-on" activities have been proven to help students learn abstract concepts. For instance, in Example 3, students will benefit from actually measuring and cutting a piece of ribbon (or string or paper) and counting the number of pieces they obtain.

Using Algorithms to Divide Whole Numbers

Classroom Tip

The traditional, compressed algorithm has been taught for many years. Most parents are familiar with this algorithm. This algorithm saves some writing, but it does tend to obscure what is actually going on, which is *repeated subtraction.* Also, note that the traditional, compressed algorithm obscures place value. By including place value, the compressed algorithm would appear as follows.

```
      32
  27)886
    -810        30(27)
      76
     -54         2(27)
      22
```

Algorithms for Dividing Whole Numbers

There are many algorithms for dividing whole numbers. Here are two commonly used algorithms.

Example $886 \div 27$

Scaffold Algorithm (Using guesses and repeated subtraction)

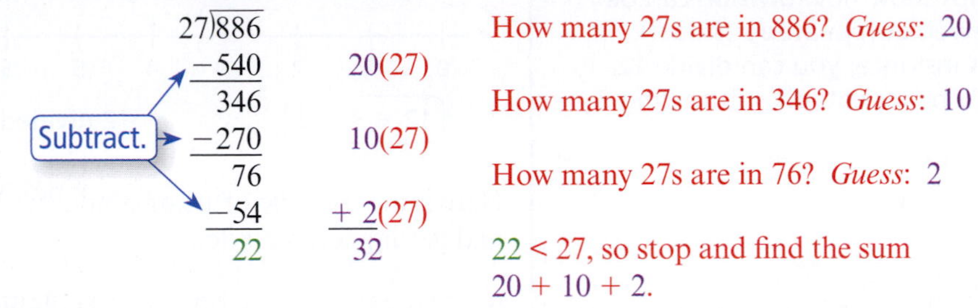

So, $886 \div 27 = 32R22$.

Traditional, Compressed Algorithm

```
      32
  27)886        How many 27s are in 88? There are 3.
     -81        3(27)
      76        How many 27s are in 76? There are 2.
     -54        2(27)
      22
```

(Subtract.)

So, $886 \div 27 = 32R22$.

EXAMPLE 4 **Using the Scaffold Algorithm**

Use the scaffold algorithm to find each quotient.

a. $347 \div 9$ **b.** $528 \div 11$

SOLUTION

```
a.  9)347                  How many 9s are in 347? Guess: 30
     -270        30(9)
       77                  How many 9s are in 77? Guess: 8
      -72       +8(9)
        5        38        5 < 9, so stop and find the sum 30 + 8.
```

So, $347 \div 9 = 38R5$.

```
b.  11)528                 How many 11s are in 528? Guess: 40
      -440       40(11)
        88                 How many 11s are in 88? Guess: 8
       -88      +8(11)
         0       48        0 < 11, so stop and find the sum 40 + 8.
```

So, $528 \div 11 = 48$.

In either of the algorithms, the key to success is to be familiar with multiplication facts. For instance, when deciding how many 9s are in 347, you can mentally reason that $40 \cdot 9 = 360$, which is greater than 347. So, the guess can be revised to $30 \cdot 9 = 270$.

EXAMPLE 5 Using the Traditional, Compressed Algorithm

Use the traditional, compressed algorithm to divide 2245 by 38.

SOLUTION

```
      59
  38)2245
    −190      5(38)     How many 38s are in 224? There are 5.
      345
     −342     9(38)     How many 38s are in 345? There are 9.
        3
```

So, $2245 \div 38 = 59\text{R}3$.

EXAMPLE 6 Solving a Real-Life Division Problem

On a car trip across the United States, you drive 2927 miles and use 125 gallons of gasoline. How many miles per gallon did your car average?

SOLUTION

1. **Understand the Problem** You need to find the miles per gallon that your car averaged for the trip.
2. **Make a Plan** To find the miles per gallon, divide the total miles driven by the total gallons used.

   2927 mi ÷ 125 gal = ☐ mi per gal

3. **Solve the Problem** Divide 2927 by 125 using the traditional, compressed algorithm.

```
        23
  125)2927
     −250      2(125)     How many 125s are in 292? There are 2.
       427
      −375     3(125)     How many 125s are in 427? There are 3.
        52
```

   Your car averaged about 23 miles per gallon.

4. **Look Back** You can check the reasonableness of your answer by rounding 2927 to the nearest thousand and then finding the quotient.

   $2927 \div 125 = 23\text{R}52 \rightarrow 3000 \div 125 = 24$ ✓

> **Mathematical Practices**
>
> **Reason Abstractly and Quantitatively**
>
> **MP2** Mathematically proficient students bring two complementary abilities to bear on problems involving quantitative relationships: The ability to *decontextualize* and the ability to *contextualize*.

3.4 Exercises

Free step-by-step solutions to odd-numbered exercises at MathematicalPractices.com.

1. **Writing a Solution Key** Write a solution key for the activity on page 108. Describe a rubric for grading a student's work.

2. **Grading the Activity** In the activity on page 108, a student gave the answers below. Each question is worth 10 points. Assign a grade to each answer. For those that are incorrect, why do you think the student erred?

Sample Student Work

> **1.** Twenty quarters are shared equally among 4 people. How many quarters does each person get? *4 quarters*
>
> 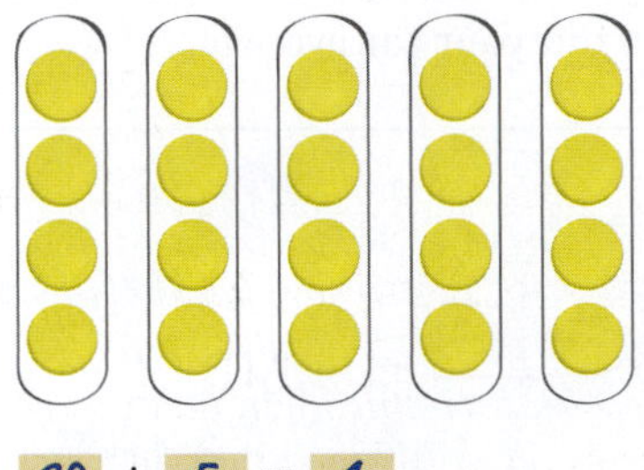
>
> 20 ÷ 5 = 4
>
> **3.** Eighteen dimes are distributed among 6 people. Each person receives the same number of dimes. How many dimes does each person get? *3 dimes*
>
> 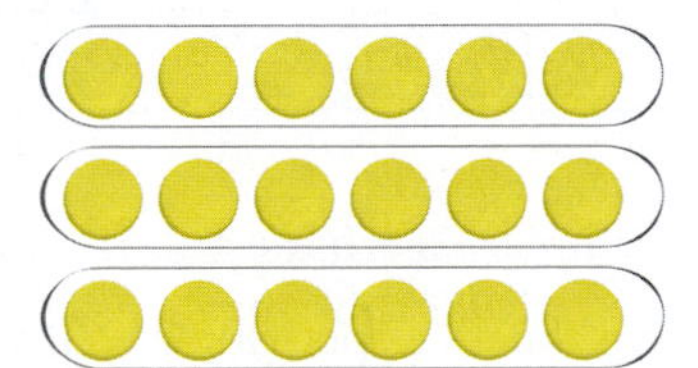
>
> 18 ÷ 6 = 3
>
> **5.** Three jars contain a total of 24 nickels. Each jar has the same number of nickels. How many nickels are in each jar? *8 nickels*

3. **Using Vocabulary** Identify the divisor, dividend, and quotient in $8 \div 2 = 4$.

4. **Using Vocabulary** Identify the divisor, dividend, and quotient in $6 \div 3 = 2$.

5. **Using the Definition of Quotient** Answer each question. Then write the division problem.
   a. What can you multiply by 5 to get 30?
   b. Twenty-eight students are divided into four groups. How many students are in each group?

6. **Using the Definition of Quotient** Answer each question. Then write the division problem.
   a. What can you multiply by 7 to get 42?
   b. Four cups of a breakfast cereal contain a total of 72 grams of sugar. How much sugar is in a one-cup serving?

7. **Dividing Whole Numbers** Write a division problem that can be answered using each set (partition) model.

a.

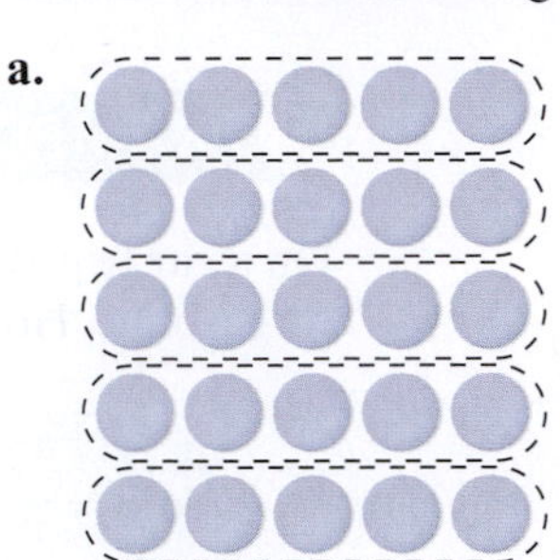

☐ ÷ ☐ = ☐

b.

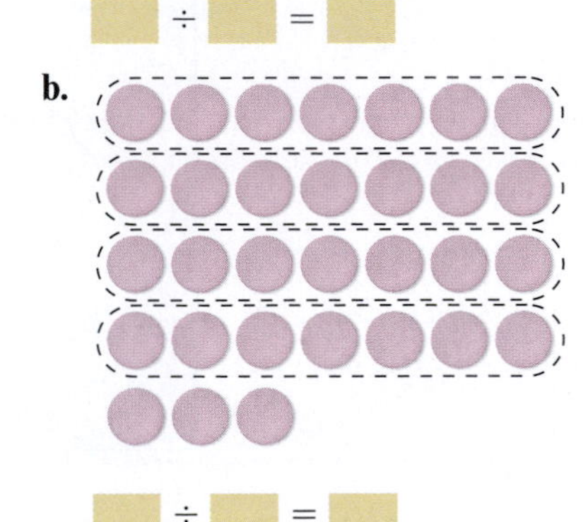

☐ ÷ ☐ = ☐

8. **Dividing Whole Numbers** Write a division problem that can be answered using each set (partition) model.

a.

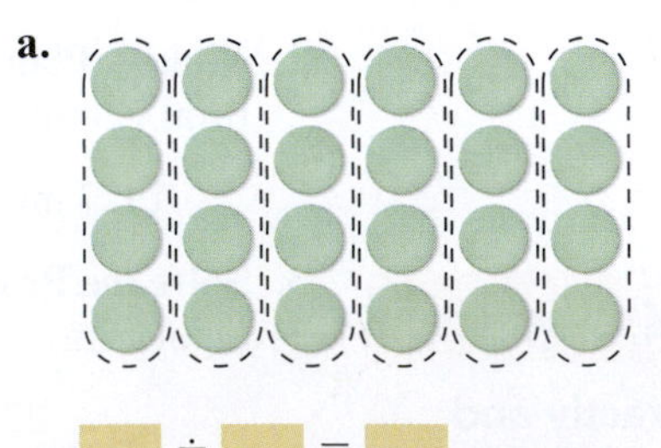

☐ ÷ ☐ = ☐

b.

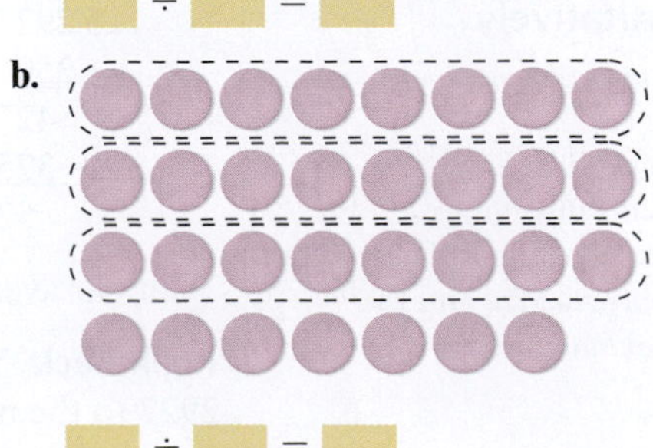

☐ ÷ ☐ = ☐

9. Dividing Whole Numbers Using Models Draw a set (partition) model that can be used to find each quotient. Then find the quotient.

a. $18 \div 3$ **b.** $\frac{20}{4}$ **c.** $35 \div 8$ **d.** $\frac{32}{6}$

10. Dividing Whole Numbers Using Models Draw a set (partition) model that can be used to find each quotient. Then find the quotient.

a. $42 \div 6$ **b.** $\frac{48}{4}$ **c.** $54 \div 7$ **d.** $\frac{47}{8}$

11. Dividing Whole Numbers Write the division problem shown by each number line model.

a.

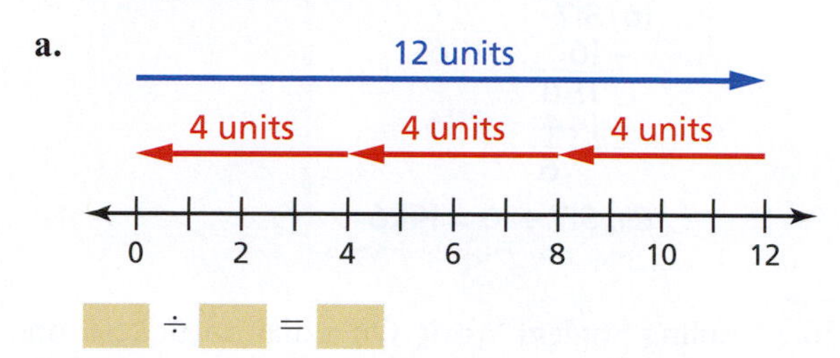

b.

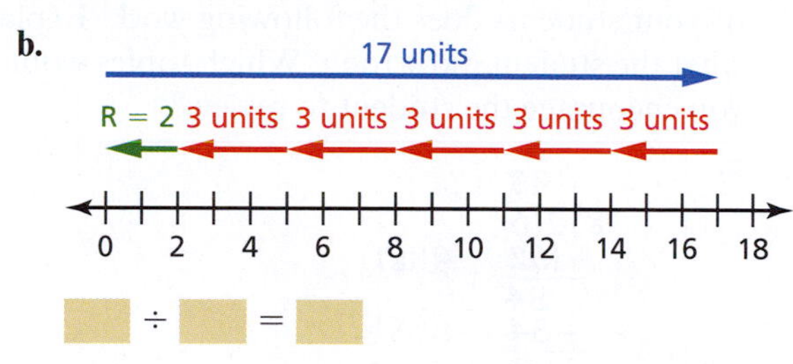

12. Dividing Whole Numbers Write the division problem shown by each number line model.

a.

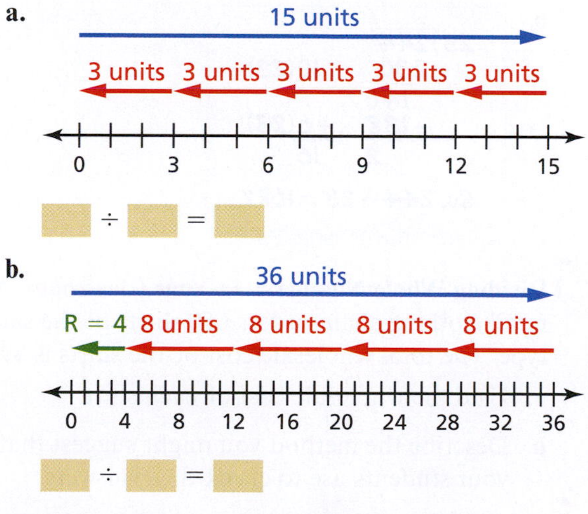

13. Dividing Whole Numbers Using Number Lines Draw a number line that can be used to find each quotient. Then find the quotient.

a. $18 \div 2$ **b.** $\frac{16}{4}$

c. $14 \div 4$ **d.** $\frac{27}{5}$

14. Dividing Whole Numbers Using Number Lines Draw a number line that can be used to find each quotient. Then find the quotient.

a. $36 \div 6$ **b.** $\frac{27}{9}$

c. $43 \div 8$ **d.** $\frac{39}{12}$

15. Using Division Algorithms State whether each solution uses the *scaffold algorithm* or the *traditional, compressed algorithm*. Then fill in the missing numbers in the solution, and find the quotient.

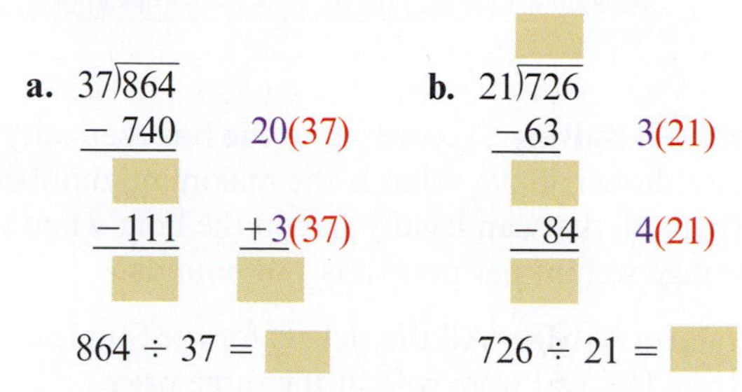

16. Using Division Algorithms State whether each solution uses the *scaffold algorithm* or the *traditional, compressed algorithm*. Then fill in the missing numbers in the solution, and find the quotient.

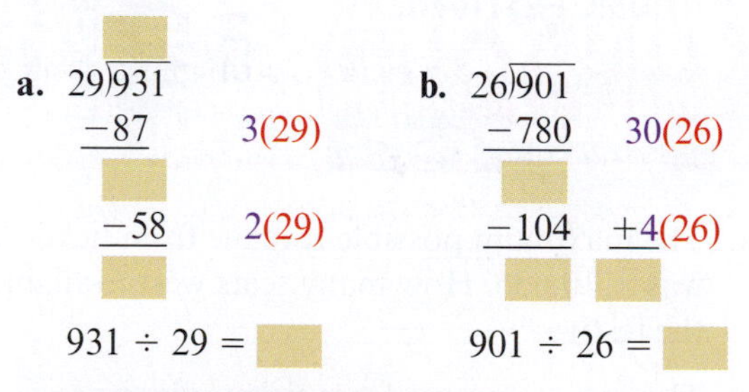

17. Using the Scaffold Algorithm Use the scaffold algorithm to find each quotient.

a. $427 \div 9$ **b.** $393 \div 12$ **c.** $741 \div 13$

d. $828 \div 36$ **e.** $2640 \div 47$ **f.** $5005 \div 56$

18. Using the Scaffold Algorithm Use the scaffold algorithm to find each quotient.

a. $363 \div 8$ **b.** $415 \div 17$ **c.** $912 \div 16$

d. $957 \div 33$ **e.** $5390 \div 82$ **f.** $5995 \div 71$

19. Using the Traditional, Compressed Algorithm Use the traditional, compressed algorithm to find each quotient.

a. $893 \div 19$ **b.** $782 \div 23$ **c.** $800 \div 27$

d. $3596 \div 58$ **e.** $2414 \div 46$ **f.** $2200 \div 115$

20. Using the Traditional, Compressed Algorithm Use the traditional, compressed algorithm to find each quotient.

a. $946 \div 22$ **b.** $929 \div 27$ **c.** $990 \div 61$

d. $2438 \div 53$ **e.** $3499 \div 75$ **f.** $16{,}850 \div 351$

21. Problem Solving The boat capacity plate shown below gives both the maximum number of people and the maximum total weight of people that can legally ride in the boat. Find the average weight per person at the maximum capacity.

22. Problem Solving According to the boat capacity plate shown above, what is the maximum number of people that can legally ride in the boat when the average weight per person is 180 pounds?

23. Problem Solving All the tickets for the Sundown Music Festival were sold at the same price.

a. The maximum possible revenue from ticket sales was \$110,636. How many seats were available for the festival?

b. The actual revenue from ticket sales was \$101,762. How many tickets were sold?

24. Problem Solving The poster shows the cost of general admission to an auto show.

a. The general admission revenue from adults was \$32,396. How many adults attended?

b. The general admission revenue from children was \$11,022. How many children attended?

25. Grading Student Work On a diagnostic test, one of your students does the following work. Explain what the student did wrong. Which topics would you encourage the student to review?

a.

```
32)697
  −640    20(32)
    57
   −32   +1(32)
    25     21
So, 697 ÷ 32 = 25R21
```

b.

```
     19
16)317
  −16      1(16)
   150
  −144     9(16)
     6
So, 317 ÷ 16 = 19R6
```

26. Grading Student Work On a diagnostic test, one of your students does the following work. Explain what the student did wrong. Which topics would you encourage the student to review?

a.

```
     93
18)216
  −162     9(18)
    54
   −54     3(18)
     0
So, 216 ÷ 18 = 93
```

b.

```
23)244
  −230    10(23)
   140
  −138    +6(23)
     2     16
So, 244 ÷ 23 = 16R2
```

27. Dividing Whole Numbers: In Your Classroom A retail clothing chain orders 650 shirts of the same type. The total wholesale cost of the shirts is \$9100.

a. Find the wholesale cost of each shirt.

b. Describe the method you might suggest that your students use to check their answers.

28. Dividing Whole Numbers: In Your Classroom A shoe store orders 124 pairs of shoes at the same cost per pair. The total wholesale cost of the shoes is \$3844.

a. Find the wholesale cost of each pair of shoes.

b. Describe the method you might suggest that your students use to check their answers.

29. **My Dear Aunt Sally** Arrange the numbers 1, 2, 3, and 6 so that the quotients are whole numbers and the equation is true.

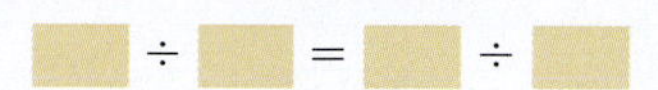

For more practice with this type of reasoning, go to *AuntSally.com.*

30. **My Dear Aunt Sally** Arrange the numbers 2, 4, 4, and 8 so that the quotients are whole numbers and the equation is true.

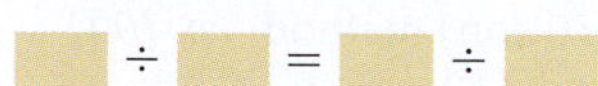

For more practice with this type of reasoning, go to *AuntSally.com.*

31. **Writing** Explain and give an example of how division and repeated subtraction are related.

32. **Writing** Explain and give an example of how an array of objects can be used to model division.

33. **Dividing Whole Numbers** Cans of iced tea in the 12-fluid-ounce size are commonly sold by the case.

Case of Iced Tea
24 cans = 288 fluid ounces

a. How many cases can be produced from 12,960 fluid ounces of iced tea?

b. How many cases can be produced from 1000 cans of iced tea?

c. Interpret the remainder in part (b).

34. **Dividing Whole Numbers** Bushels are commonly used to measure a quantity of apples.

1 Bushel of Medium Apples
Weight: 48 pounds
Number: about 126 apples

a. How many bushels do you get from 3120 pounds of medium apples?

b. How many bushels do you get from 5000 medium apples?

c. Interpret the remainder in part (b).

35. **Dividing Whole Numbers** The water in a hot tub weighs 4100 pounds. One cubic foot of water weighs about 62 pounds. Approximately how many cubic feet of water does the hot tub contain?

36. **Dividing Whole Numbers** Sixteen thousand pounds of concrete are poured to form a concrete patio. The concrete weighs 144 pounds per cubic foot. Approximately how many cubic feet of concrete are poured to form the patio?

37. **Problem Solving: In Your Classroom** The table below shows the numbers of calories a person who weighs 100 pounds will burn while doing various activities for one hour.

Activity (1 hour)	Calories burned
Bicycling, 6 mi/h	160
Jogging, 7 mi/h	610
Jumping rope	500
Swimming, 25 yd/min	185
Tennis, singles	265

a. How many calories does the person burn by jumping rope for 3 minutes?

b. Explain how to show your students a way to find the answer to part (a) by dividing whole numbers twice.

c. Describe other questions you might ask your students using the table to give them practice with dividing whole numbers.

38. **Problem Solving: In Your Classroom** The table below shows typical maximum speeds of various creatures.

Creature	Speed (miles per hour)
Pronghorn antelope	61
Frigatebird	95
Kangaroo	44
Lion	50
Sailfish	68

a. How far can a sailfish travel in 15 minutes at its maximum speed?

b. Explain how to show your students a way to find the answer to part (a) by dividing whole numbers twice.

c. Describe other questions you might ask your students using the table to give them practice with dividing whole numbers.

Chapter Summary

Chapter Vocabulary

union *(p. 75)*
set model *(p. 75)*
sum *(p. 75)*
addends *(p. 75)*
Commutative Property of Addition *(p. 76)*
Associative Property of Addition *(p. 76)*
Additive Identity Property *(p. 76)*
additive identity *(p. 76)*
number line model *(pp. 78, 88, 111)*
measurement model *(pp. 78, 88)*
base-ten blocks *(pp. 79, 89)*
regrouped *(p. 79)*
algorithm *(p. 80)*
expanded algorithm *(pp. 80, 101)*
standard algorithm *(pp. 80, 90, 101)*
inverse operations *(p. 87)*
missing addend method *(p. 87)*
difference *(p. 87)*
subtrahend *(p. 87)*
minuend *(p. 87)*
take-away method *(p. 87)*
Austrian algorithm *(p. 90)*
product *(p. 97)*
factors *(p. 97)*
Commutative Property of Multiplication *(p. 98)*
Associative Property of Multiplication *(p. 98)*
Multiplicative Identity Property *(p. 98)*
multiplicative identity *(p. 98)*
Zero Multiplication Property *(p. 98)*
area model *(p. 100)*
Distributive Property of Multiplication Over Addition *(p. 103)*
Distributive Property of Multiplication Over Subtraction *(p. 103)*
division *(p. 109)*
missing factor method *(p. 109)*
quotient *(p. 109)*
divisor *(p. 109)*
dividend *(p. 109)*
set (partition) method *(p. 109)*
array *(p. 110)*
remainder *(p. 110)*
scaffold algorithm *(p. 112)*
traditional, compressed algorithm *(p. 112)*

Chapter Learning Objectives / Review Exercises

3.1 Adding Whole Numbers *(page 75)* — Review Exercises 1–32

- Understand and use the definition of the sum of two whole numbers.
- Understand and use the properties of addition for whole numbers.
- Use models to add whole numbers.
- Use algorithms to add whole numbers.

3.2 Subtracting Whole Numbers *(page 87)* — Review Exercises 33–64

- Understand and use the definition of the difference of two whole numbers.
- Use models to subtract whole numbers.
- Use algorithms to subtract whole numbers.

3.3 Multiplying Whole Numbers *(page 97)* — Review Exercises 65–98

- Understand and use the definition of the product of two whole numbers.
- Understand and use the properties of multiplication for whole numbers.
- Use models to multiply whole numbers.
- Use algorithms to multiply whole numbers.
- Understand and use the Distributive Properties.

3.4 Dividing Whole Numbers *(page 109)* — Review Exercises 99–126

- Understand and use the definition of the quotient of two whole numbers.
- Use models to divide whole numbers.
- Use algorithms to divide whole numbers.

Review Exercises

Free step-by-step solutions to odd-numbered exercises at *MathematicalPractices.com*.

3.1 Adding Whole Numbers

Using Properties of Addition Use the given property of addition to complete the equation.

1. Commutative Property: $6 + 9 =$ ____
2. Commutative Property: $14 + d =$ ____
3. Associative Property: $(67 + 9) + 1 =$ ____
4. Associative Property: $3 + (3 + x) =$ ____
5. Additive Identity Property: $12 + 0 =$ ____
6. Additive Identity Property: $0 + a + 1 =$ ____

Adding Whole Numbers Using a Number Line Describe how the number line model is used to add two numbers. Then write and solve the addition problem represented by the model.

7.

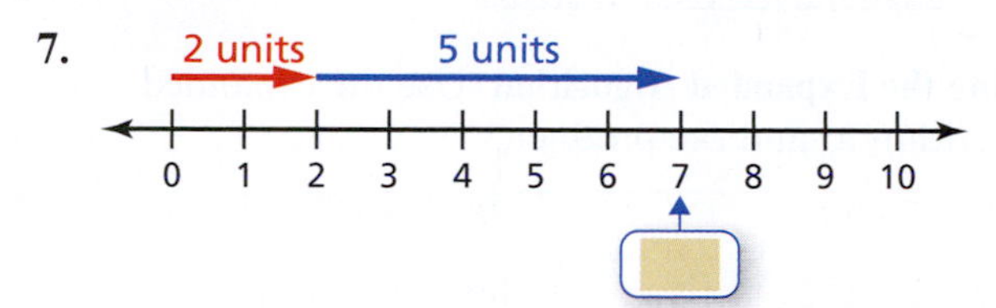

8.

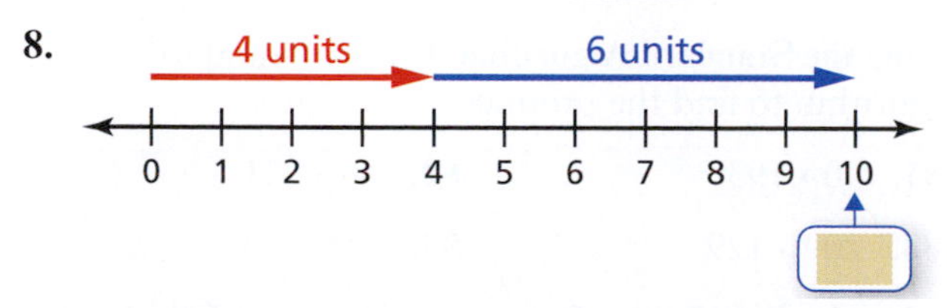

Adding Whole Numbers Using a Number Line Draw and use a number line to find the sum.

9. $7 + 6$
10. $8 + 7$
11. $2 + 4 + 5$
12. $3 + 2 + 9$

Adding Whole Numbers Using Base-Ten Blocks Describe how to use the base-ten blocks to add two numbers. Then model the sum. Finally, write and solve the addition problem represented by the model.

13.

14.

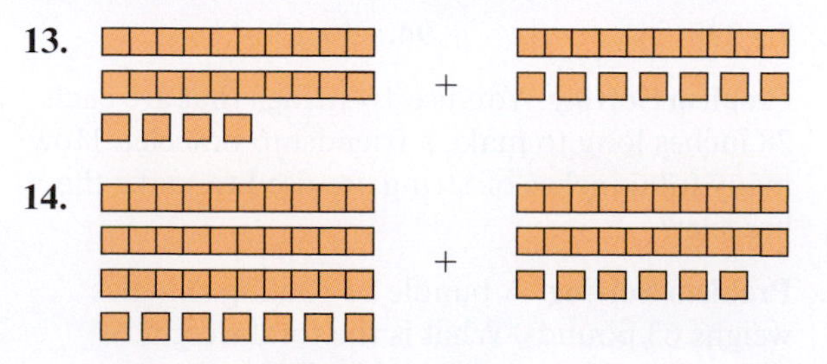

Adding Whole Numbers Using Base-Ten Blocks Use base-ten blocks to find the sum. Justify your answer with a sketch of the model.

15. $29 + 44$
16. $37 + 56$
17. $48 + 59$
18. $64 + 89$

Using the Expanded Algorithm Use the expanded algorithm to find the sum.

19. $77 + 84$
20. $87 + 698$
21. $379 + 569$
22. $7454 + 697$
23. $15 + 25 + 56$
24. $233 + 187 + 65$

Using the Standard Algorithm Use the standard algorithm to find the sum.

25. $6365 + 239$
26. $2053 + 79$
27. $1896 + 654$
28. $7268 + 9534$
29. $86 + 44 + 537$
30. $229 + 458 + 4951$

31. **Problem Solving** In one year, an insurance agent earns a base pay of \$33,544, commissions of \$22,690, and a sales bonus of \$6600. What are the total earnings of the insurance agent? What algorithm did you use and why?

32. **Problem Solving** A blue whale calf has a mass of 19,958 kilograms. The calf gains 2722 kilograms one month and 3016 kilograms the next month. What is the weight of the calf after two months?

3.2 Subtracting Whole Numbers

Using the Definition of Difference Write the question the missing addend statement asks. Then write the subtraction problem.

33. $8 +$ ____ $= 19$
34. $5 +$ ____ $= 13$
35. $6 +$ ____ $= 23$
36. $9 +$ ____ $= 25$

Using the Definition of Difference Write the difference as a missing addend statement. Then write the question the statement asks.

37. $24 - 22$
38. $18 - 15$
39. $18 - 6$
40. $29 - 7$

Using the Definition of Difference Use the take-away diagram to write the subtraction problem.

41.

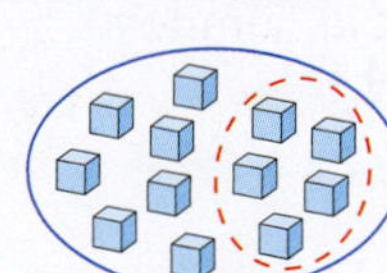

42.

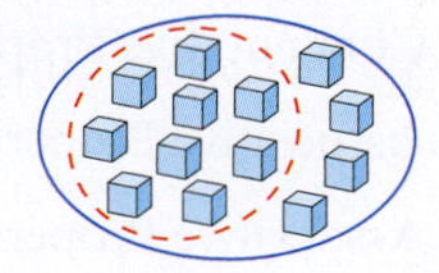

Using the Definition of Difference Draw a take-away diagram for the difference. Then find the difference.

43. $10 - 5$ **44.** $11 - 4$

45. $12 - 2$ **46.** $14 - 6$

Subtracting Whole Numbers Using a Number Line Draw and use a number line to find the difference.

47. $5 - 3$ **48.** $8 - 4$

49. $11 - 9$ **50.** $13 - 6$

Subtracting Whole Numbers Using Base-Ten Blocks Use base-ten blocks to find the difference. Justify your answer with a sketch of the model.

51. $87 - 33$ **52.** $72 - 28$

53. $224 - 146$ **54.** $343 - 179$

Using the Standard Algorithm Use the standard algorithm to find the difference.

55. $532 - 325$ **56.** $742 - 68$

57. $2315 - 1557$ **58.** $4911 - 945$

Using the Austrian Algorithm Use the Austrian algorithm to find the difference.

59. $244 - 163$ **60.** $412 - 58$

61. $2426 - 1445$ **62.** $7232 - 836$

63. **Satellite Television** A satellite TV technician installed 68 new systems two months ago and 87 new systems last month. How many more systems did the technician install last month than two months ago? Use the standard algorithm and the Austrian algorithm.

64. **Shopping** You have $76 to spend at the mall. After buying a shirt for $27 and sandals for $29, how much money do you have left to spend? Describe your strategy for solving this problem.

3.3 Multiplying Whole Numbers

Finding the Product of Two Whole Numbers Use the definition of the product of two whole numbers to find the product.

65. $7 \cdot 9$ **66.** The product of 9 and 8

67. Four times thirteen **68.** Six multiplied by sixteen

Using Properties of Multiplication Use the given property of multiplication to complete the equation.

69. Commutative Property: $6 \cdot 9 = \square$

70. Commutative Property: $d \cdot 34 = \square$

71. Associative Property: $(7 \cdot 16) \cdot 4 = \square$

72. Associative Property: $3 \cdot (7 \cdot x) = \square$

73. Multiplicative Identity Property: $22 \cdot 1 = \square$

74. Zero Multiplication Property: $0 \cdot 341 = \square$

Multiplying Whole Numbers Using Models Describe how the model is used to multiply two numbers. Then write and solve the multiplication problem represented by the model.

75.

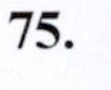

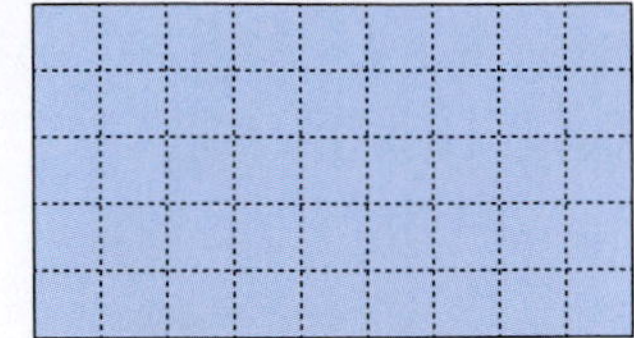

76.

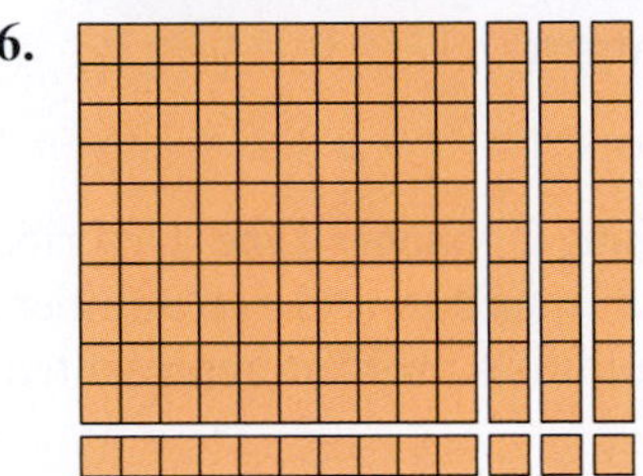

Using the Expanded Algorithm Use the expanded algorithm to find the product.

77. $45 \cdot 59$ **78.** $63 \cdot 78$

79. $23 \cdot 152$ **80.** $402 \cdot 617$

Using the Standard Algorithm Use the standard algorithm to find the product.

81. $40 \cdot 193$ **82.** $36 \cdot 321$

83. $209 \cdot 129$ **84.** $551 \cdot 146$

Using the Distributive Properties Use one of the Distributive Properties to rewrite the expression. Then evaluate.

85. $6 \cdot (10 + 8)$ **86.** $6 \cdot 7 + 6 \cdot 23$

87. $9 \cdot (40 - 7)$ **88.** $22(26) - 22(16)$

89. $4(7 + 8 + 9)$

90. $7(22) + 7(13) + 7(15)$

Using the Distributive Properties Use one of the Distributive Properties to evaluate the expression.

91. $8 \cdot 87$ **92.** $6 \cdot 325$

93. $7 \cdot 297$ **94.** $9 \cdot 1206$

95. **Problem Solving** You use 16 strings that are each 28 inches long to make a friendship bracelet. How many total inches of string are used to make the bracelet?

96. **Problem Solving** A bundle of roofing shingles weighs 63 pounds. What is the total weight of 42 bundles of the roofing shingles?

97. Finding Cost An ad in a local newspaper costs \$28 per column inch. A column inch is a unit of space that is 1 column wide and 1 inch high as shown below.

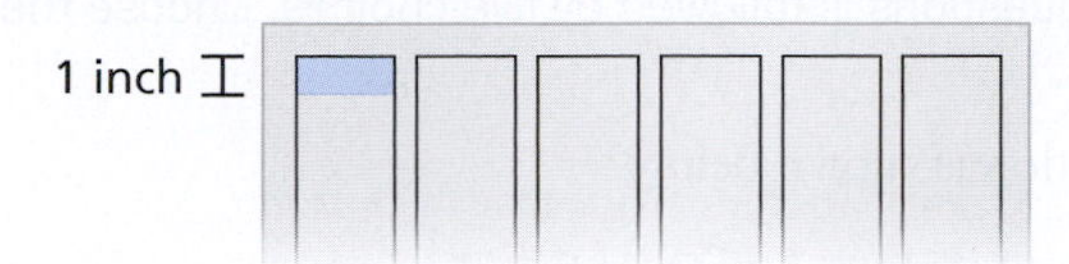

a. Find the cost of an ad that is 6 inches high and 1 column wide.

b. Write an expression for the cost of an ad that is 6 columns wide and 4 inches high. Then find the cost of the ad.

98. Finding Area You cut 3 yards of fabric to reduce the length of the fabric by 18 inches.

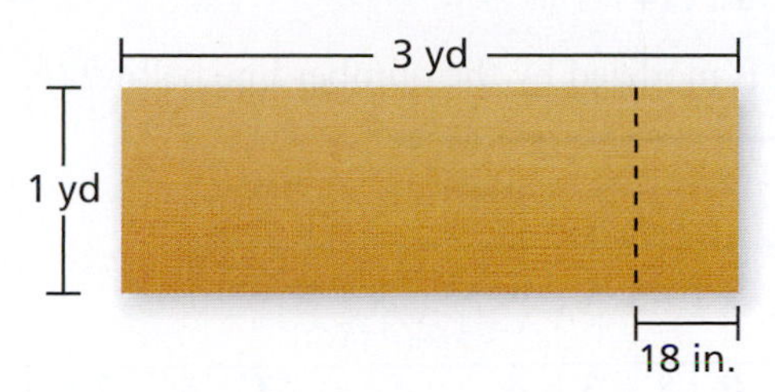

a. What is the area, in square inches, of the 3 yards of fabric?

b. Write an expression that can be evaluated using the Distributive Property to find the area of the shortened piece of fabric. Then find the area.

3.4 Dividing Whole Numbers

Using the Definition of Quotient Answer the question. Then write the division problem.

99. What can you multiply by 6 to get 42?

100. Twenty-four students are divided into six groups. How many students are in each group?

101. A total of 32 tennis balls come in 8 containers. How many tennis balls come in each container?

102. You swim 4 laps for a total distance of 80 meters. How far is each lap?

Dividing Whole Numbers Write a division problem that can be answered using the set (partition) model.

103.

104.

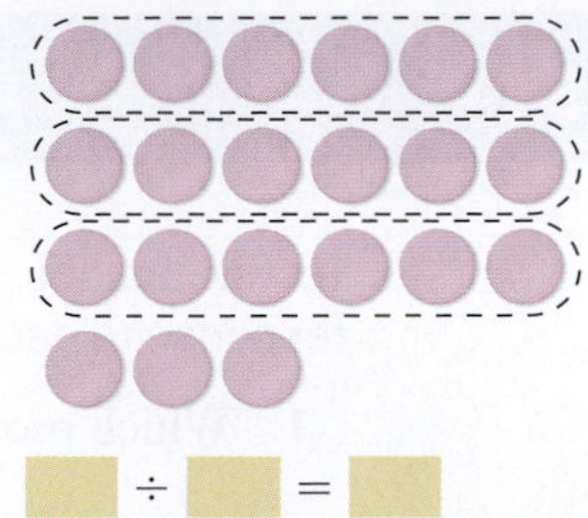

Dividing Whole Numbers Using a Model Draw a set (partition) model that can be used to find the quotient. Then find the quotient.

105. $36 \div 9$

106. $35 \div 7$

107. $39 \div 13$

108. $48 \div 12$

Dividing Whole Numbers Using a Number Line Draw a number line that can be used to find the quotient. Then find the quotient.

109. $10 \div 2$

110. $24 \div 4$

111. $21 \div 7$

112. $44 \div 11$

Using the Scaffold Algorithm Use the scaffold algorithm to find the quotient.

113. $336 \div 8$

114. $319 \div 11$

115. $521 \div 14$

116. $798 \div 33$

117. $3381 \div 51$

118. $5374 \div 63$

Using the Traditional, Compressed Algorithm Use the traditional, compressed algorithm to find the quotient.

119. $748 \div 17$

120. $962 \div 26$

121. $1307 \div 22$

122. $1760 \div 38$

123. $2800 \div 43$

124. $14{,}365 \div 221$

125. Problem Solving An oil storage tank holds 1980 pounds of home heating oil. A cubic foot of the oil weighs 54 pounds. How many cubic feet of heating oil does the tank hold?

126. Problem Solving How far can the helicopter fly in 4 minutes?

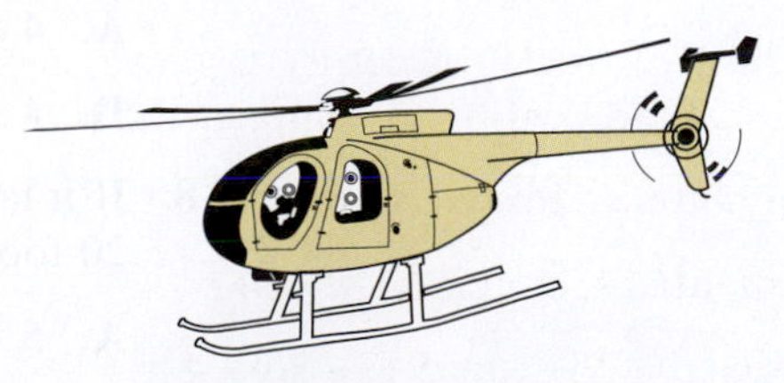

Maximum speed: 285 km/h

Chapter Test

The questions on this practice test are modeled after questions on state certification exams for K–8 teaching.

Test Directions: Each of the questions is followed by five choices. Choose the *best* response to each question.

1. Which property of addition is shown below?

$$(x + y) + z = x + (y + z)$$

A. Commutative **B.** Associative **C.** Additive Identity

D. Multiplicative **E.** None of the above

2. Find the sum.

$$5 + 2 + 7$$

A. 7 **B.** 9 **C.** 11

D. 12 **E.** 14

3. Which sum or difference is modeled by the number line below?

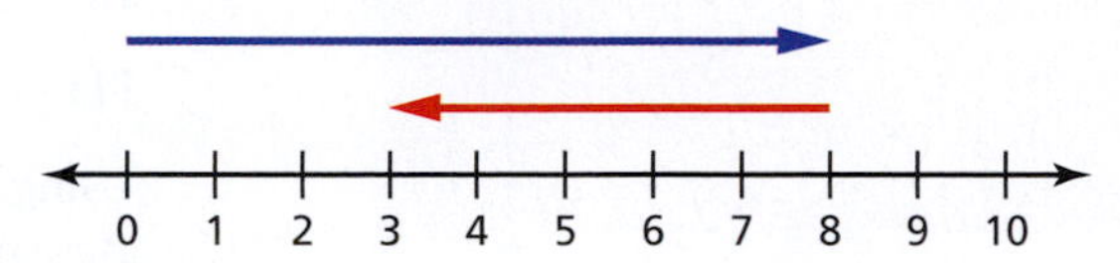

A. $8 - 3 = 5$ **B.** $5 + 3 = 8$ **C.** $8 - 5 = 3$

D. $3 + 5 = 8$ **E.** None of the above

4. Evaluate the expression.

$$6 \cdot (5 + 2)$$

A. 12 **B.** 30 **C.** 32

D. 42 **E.** 49

5. You have $175 to spend at a shopping center. You purchase the headphones at the left. How much money do you have left?

A. $141 **B.** $144 **C.** $145

D. $146 **E.** $149

6. There were 21 hot dogs for ten people on a picnic. How many whole hot dogs were there for each person if they were divided equally?

A. 1 **B.** 2 **C.** 3 **D.** 4 **E.** 5

7. Mrs. Wall has $200,000. She wishes to give each of her 4 children an equal amount of her money. Which of the following methods will result in the amount that each child is to receive?

A. $4 \times 200,000$ **B.** $4 \div 200,000$ **C.** $200,000 \div 4$

D. $4 - 200,000$ **E.** $200,000 - 4$

8. If it takes one minute per cut, how long will it take, in minutes, to cut a 20-foot-long timber into 20 equal pieces?

A. 5 **B.** 10 **C.** 14

D. 19 **E.** 20

4 Exponents and Estimation

iStockphoto.com/kali9

Activity: Performing Whole Number Operations Using Mental Math

Materials:

- Pencil

Learning Objective:

Learn to add, subtract, multiply, and divide whole numbers using mental math. Be able to explain why various mental math strategies are valid.

Name ______________________________

Addition Find the sum using mental math. Explain your reasoning.

1. $37 + 43$

2. $26 + 26$

3. $77 + 25$

4. $29 + 32$

Subtraction Find the difference using mental math. Explain your reasoning.

8. $200 - 55$

7. $100 - 36$

6. $79 - 29$

5. $87 - 35$

Multiplication Find the product using mental math. Explain your reasoning.

9. 32×5

10. 26×10

11. 19×8

12. 46×2

Division Find the quotient using mental math. Explain your reasoning.

14. $105 \div 3 =$ ____

13. $210 \div 10 =$ ____

16. $438 \div 2 =$ ____

15. $85 \div 5 =$ ____

4.1 Mental Math and Estimation

Standards

Grades K–2
Numbers and Operations in Base Ten
Students should use place value understanding and properties of operations to add and subtract.

Grades 3–5
Numbers and Operations in Base Ten
Students should use place value understanding and properties of operations to perform multi-digit arithmetic.

- Use mental math strategies to add and subtract whole numbers.
- Use mental math strategies to multiply and divide whole numbers.
- Use rounding to estimate sums, differences, products, and quotients.

Mental Math Strategies for Adding and Subtracting

When you use **mental math** to do a calculation, you find an answer without using any computational aids. There are many strategies for doing this type of math "in your head." Some of the more common strategies include the use of **compatible numbers**, which are numbers whose sums, differences, products, or quotients are easy to compute mentally. Here are some examples.

$$\begin{array}{r} 26 \\ +\ 24 \\ \hline 50 \end{array} \qquad \begin{array}{r} 33 \\ +\ 33 \\ \hline 66 \end{array} \qquad \begin{array}{r} 100 \\ -\ 25 \\ \hline 75 \end{array} \qquad \begin{array}{r} 500 \\ -\ 230 \\ \hline 270 \end{array}$$

Mental Math Strategy for Adding and Subtracting

Compensation is a mental math strategy in which you create compatible numbers by adjusting the actual numbers in a sum or difference.

To find a sum, add a number to one addend to make the addition easier. Then compensate by subtracting that same number from the other addend.

Example

$$\begin{array}{r} 28 \\ +\ 25 \\ \hline 53 \end{array} \rightarrow \begin{array}{r} 30 \\ +\ 23 \\ \hline 53 \end{array}$$

Think: $28 + 2 = 30$. So, $25 - 2 = 23$.

To find a difference, add or subtract the same number from *both* numbers.

In the examples shown, you are making one of the numbers a multiple of 10.

Example

$$\begin{array}{r} 81 \\ -\ 29 \\ \hline 52 \end{array} \rightarrow \begin{array}{r} 82 \\ -\ 30 \\ \hline 52 \end{array}$$

Think: $81 + 1 = 82$. So, $29 + 1 = 30$.

EXAMPLE 1 Using Mental Math in Real Life

You purchase 3 different-sized lag bolts. The price of each bolt is shown below. You give the cashier \$2. How much change do you receive?

49¢ 69¢ 79¢

SOLUTION

One way to find the total cost mentally is to use compensation. Add 49, 69, and 79 by increasing each amount by 1 cent. Then subtract 3 to maintain the same sum.

Add 1 to each addend.

$$\begin{aligned} 49 + 69 + 79 &= 50 + 70 + 80 - 3 && \text{Subtract 3 to } \textit{compensate}. \\ &= 200 - 3 && \text{Add.} \\ &= 197 && \text{Subtract.} \end{aligned}$$

So, the total amount of your purchase is 197 cents. Because you give the cashier \$2, or 200 cents, your change is $200 - 197 = 3$ cents.

Classroom Tip
As your students develop cognitively using language and symbols to manipulate concepts, their mathematical abilities also develop. Some of the most critical cognitive abilities for learning math are memory, language skills, and the ability to do mental math. When you are teaching mental math, be sure to ask your students to explain why the various mental math strategies work. This "thinking about thinking" is called *metacognition.*

Mental Math Strategies for Multiplying and Dividing

Mental Math Strategies for Multiplying

Rearranging the factors is a mental math strategy in which you create compatible numbers.

Example

$2 \cdot 16 \cdot 8 \cdot 5 = (2 \cdot 5) \cdot (16 \cdot 8)$ Commutative and Associative Properties

$= 10 \cdot 128$ Multiply.

$= 1280$ Adjoin a zero digit to the right of 128.

Front-end multiplying is a mental math strategy in which you multiply each place digit separately and then add the results.

Example

$30 \cdot 5 = 150$ $7 \cdot 5 = 35$

$37 \cdot 5 = 150 + 35$

$= 185$

Use the Distributive Property mentally: $37 \cdot 5 = (30 + 7)5$.

Thinking money is a mental math strategy in which you think that one of the factors of the product is money.

Examples

$56 \times 5 = 280$ Think of the product as 56 nickels, which is 28 dimes, or 280 cents.

$56 \times 50 = 2800$ Think of the product as 56 half dollars, which is 28 dollars, or 2800 cents.

Classroom Tip

When multiplying by 8, students can imagine that they are doubling 3 times.

$16 \cdot 8 = 16 \cdot 2 \cdot 2 \cdot 2$

EXAMPLE 2 Using Mental Math in Real Life

How many square feet are in the house represented by the blueprint at the right?

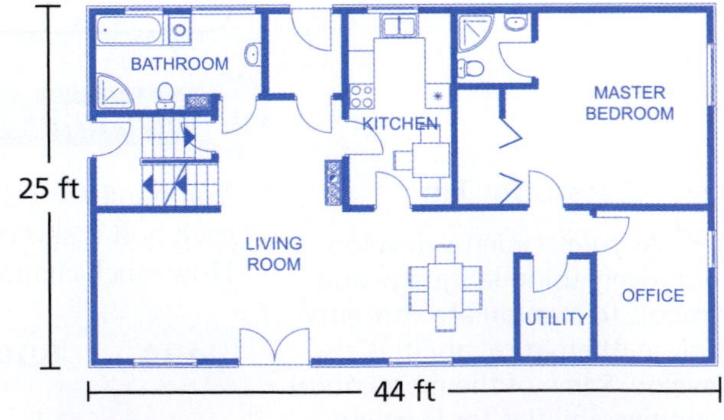

SOLUTION

To find the number of square feet, use the formula for the area of a rectangle.

$\text{Area} = (\text{length})(\text{width})$

$= (44 \text{ ft})(25 \text{ ft})$

One way to find this product using mental math is to use the thinking money strategy.

$44 \times 25 = 1100$ Think of the product as 44 quarters, which is 22 half dollars or 11 dollars. So, you have 1100 cents.

The blueprint represents 1100 square feet. You can check your answer using the Distributive Property.

$44 \cdot 25 = (40 + 4)(25)$ Think of 44 as 40 + 4.

$= 1000 + 100$ Multiply.

$= 1100$ ✓ Add.

Mathematical Practices

Model with Mathematics

MP4 Mathematically proficient students routinely interpret their mathematical results in the context of the situation and the results make sense.

Mental Math Strategy for Dividing

Rewrite the dividend to create compatible numbers.

Example Divide 312 by 8 using compatible numbers.

$320 \div 8 = 40$ $\quad 8 \div 8 = 1$

$$8\overline{)312} = 8\overline{)320 - 8} \quad \text{(quotient } 40 - 1\text{)}$$

Think of 312 as 320 − 8.

$$= 40 - 1$$

$$= 39$$

Example Divide 4254 by 6 using compatible numbers.

$4200 \div 6 = 700$ $\quad 54 \div 6 = 9$

$$6\overline{)4254} = 6\overline{)4200 + 54} \quad \text{(quotient } 700 + 9\text{)}$$

Think of 4254 as 4200 + 54.

$$= 700 + 9$$

$$= 709$$

EXAMPLE 3 Using Mental Math in Real Life

Lions belong to social groups called *prides*. A typical pride consists of about 15 lions. A nature preserve has a total population of 330 lions. About how many prides are in the nature preserve?

SOLUTION

1. **Understand the Problem** You need to approximate the number of prides.
2. **Make a Plan** Divide the total lion population by the typical pride size.
3. **Solve the Problem** You can use mental math by rewriting the dividend.

$$15\overline{)330} = 15\overline{)300 + 30} \quad \text{(quotient } 20 + 2\text{)}$$

Think of 330 as 300 + 30.

$$= 20 + 2$$

$$= 22$$

There are about 22 prides in the nature preserve.

4. **Look Back** Another problem-solving strategy students can use to find the number of prides in the nature preserve is *Guess, Check, and Revise*. A reasonable first guess is 20.

Guess	Check	Revise
20	$20 \cdot 15 = 300$	Because $300 < 330$, make a greater guess.
22	$22 \cdot 15 = 330$	This equals the total lion population.

So, there are about 22 prides in the nature preserve.

Estimating Sums, Differences, Products, and Quotients

In Section 2.2, you used estimation by rounding whole numbers. Now, you will round the numbers in a calculation.

EXAMPLE 4 Estimating by Rounding

Use rounding to estimate each answer.

a. Sum $408 + 873$

b. Difference $946 - 253$

c. Product 31×391

d. Quotient $12{,}082 \div 585$

SOLUTION

a. Round each number in the sum to the nearest hundred.

$400 + 900 = 1300$

Round 408 to 400. Round 873 to 900.

So, a reasonable answer for this sum is 1300.

b. Round each number in the difference to the nearest ten.

$950 - 250 = 700$

Round 946 to 950. Round 253 to 250.

So, a reasonable answer for this difference is 700.

c. Round 31 to the nearest ten and round 391 to the nearest hundred.

$30 \times 400 = 12{,}000$

Round 31 to 30. Round 391 to 400.

So, a reasonable answer for this product is 12,000.

d. Round 12,082 to the nearest thousand and round 585 to the nearest hundred.

$12{,}000 \div 600 = 20$

Round 12,082 to 12,000. Round 585 to 600.

So, a reasonable answer for this quotient is 20.

EXAMPLE 5 Using Estimation in Real Life

You have been given a bid for a website design. Estimate the total cost.

$1525 frames
$3875 website design
$4435 development
$4125 code integration

SOLUTION

You can round each amount to the nearest $500 to estimate the total.

$1525 → $1500
$3875 → $4000
$4435 → $4500
+ $4125 → + $4000

$1500 + $4500 = $6000
$4000 + $4000 = $8000
$6000 + $8000 = $14,000

The bid is about $14,000.

Classroom Tip

Be sure your students understand that rounding is dependent upon the context of a real-life problem. In Example 5, it is reasonable to round to the nearest $500. However, in estimating a weekly paycheck, for example, it may be unreasonable to round to the nearest $500.

Classroom Tip

When you are teaching front-end estimation, you may want to include a refined estimation technique called *front-end estimation with adjustment*. This technique uses the next group of digits to "adjust" the estimate as shown.

$$\begin{array}{r} 351 \\ 429 \\ 852 \\ +\ 354 \\ \hline 1800 \end{array} \qquad \begin{array}{l} 351, 429 \to 50 + 30 = 80 \\ 852, 354 \to 50 + 50 = 100 \end{array}$$

Front-end estimation with adjustment gives 1800 + 80 + 100 = 1980, which is even closer to the actual sum of 1986.

Other Estimation Techniques

Front-end estimation is an estimation technique that considers only the *leading digits*, which are the digits in the left-most place value.

Example	Actual		Estimate
	351		300
	429	→	400
	852		800
	+ 354		+ 300
			1800

Front-end estimation always produces a *low estimate* for sums and products. With **range estimation**, increase each leading digit by 1. This gives a *high estimate* for sums and products.

Example	Actual		Estimate
	351		400
	429	→	500
	852		900
	+ 354		+ 400
			2200

The actual sum, 1986, is between 1800 and 2200.

EXAMPLE 6 Using Estimation in Real Life

You are considering two job offers. One job offer pays $21 per hour. The other job offer pays $38,000 per year. Which job offer will pay you more money?

SOLUTION

1. **Understand the Problem** You need to determine which of the two job offers pays more money.
2. **Make a Plan** Use front-end estimation to approximate the annual salary of the job offer that pays $21 per hour.
3. **Solve the Problem** There are 52 weeks in one year. Assume that the number of work hours in a typical week is 40.

$$\textbf{Actual} \quad 52\ \cancel{wk} \cdot \frac{40\ \cancel{h}}{\cancel{wk}} \cdot \frac{\$21}{\cancel{h}}$$

$$\textbf{Estimate} \quad 50\ \cancel{wk} \cdot \frac{40\ \cancel{h}}{\cancel{wk}} \cdot \frac{\$20}{\cancel{h}} = 50 \cdot 40 \cdot \$20$$

The product of the leading digits is $5 \cdot 4 \cdot 2 = 40$. So, earning $21 per hour is about the same as earning $40,000 per year. This is a *low* estimate. So, you can conclude that the job offer that pays $21 per hour will pay you more money.

4. **Look Back** To get a more accurate comparison, use a calculator to find the actual product.

Earning $21 per hour is the same as earning $43,680 per year.

Classroom Tip

Because of the nature of multiplication, using range estimation to estimate a product may give a very wide range. For instance, in Example 6, increasing the leading digits by one for 52 and 21 gives a high estimate of

$$60 \cdot 40 \cdot 30 = 72{,}000.$$

So, the actual product is between 40,000 and 72,000.

4.1 Exercises

Free step-by-step solutions to odd-numbered exercises at MathematicalPractices.com.

1. **Writing a Solution Key** Write a solution key for the activity on page 124. Design a rubric for grading a student's work.

2. **Grading the Activity** In the activity on page 124, a student gave the answers below. Each question is worth 10 points. Assign a grade to each answer. For those that are incorrect, why do you think the student erred?

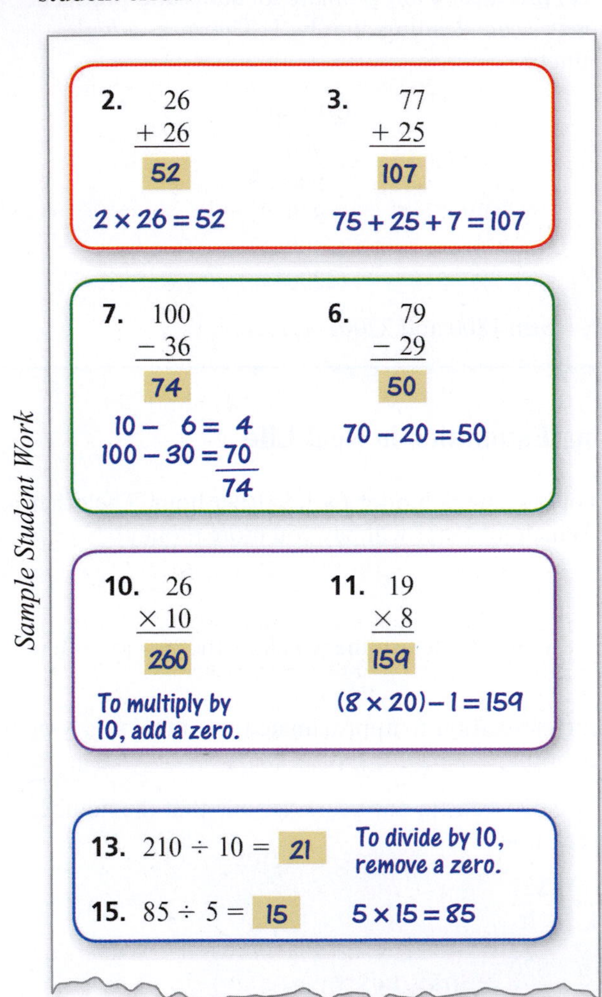

3. **Identifying Compatible Numbers** Decide whether each sum or difference involves compatible numbers.
   a. 77 + 26 b. 80 + 22
   c. 125 + 75 d. 68 − 30
   e. 100 − 75 f. 74 − 26

4. **Identifying Compatible Numbers** Decide whether each sum or difference involves compatible numbers.
   a. 50 + 23 b. 46 + 27
   c. 75 + 325 d. 125 − 75
   e. 50 − 35 f. 53 − 38

5. **Using Mental Math** Use mental math to find each sum or difference. Describe your reasoning.
   a. 48 + 33 b. 67 + 23
   c. 220 + 140 d. 72 − 29
   e. 102 − 77 f. 400 − 260

6. **Using Mental Math** Use mental math to find each sum or difference. Describe your reasoning.
   a. 59 + 44 b. 76 + 47
   c. 230 + 320 d. 66 − 22
   e. 128 − 78 f. 540 − 200

7. **Using Mental Math** Use mental math to find the total cost of the items represented by each set of prices. Explain your reasoning.
   a. 28¢ 59¢
   b. \$61 \$72
   c. 19¢ 28¢ 47¢
   d. \$52 \$41 \$32

8. **Using Mental Math** Use mental math to find the total cost of the items represented by each set of prices. Explain your reasoning.
   a. 48¢ 28¢
   b. \$81 \$93
   c. 68¢ 29¢ 39¢
   d. \$61 \$42 \$33

9. **Rearranging Factors** Rewrite each product by rearranging the factors to create compatible numbers. Then find the product.
   a. $2 \cdot 6 \cdot 5 \cdot 3$ b. $4 \cdot 7 \cdot 25 \cdot 5$
   c. $2 \cdot 13 \cdot 50 \cdot 3$ d. $250 \cdot 7 \cdot 4$
   e. $5 \cdot 17 \cdot 20$ f. $200 \cdot 23 \cdot 5$

10. Rearranging Factors Rewrite each product by rearranging the factors to create compatible numbers. Then find the product.

a. $7 \cdot 5 \cdot 9 \cdot 2$ b. $6 \cdot 25 \cdot 9 \cdot 4$

c. $5 \cdot 7 \cdot 200 \cdot 3$ d. $4 \cdot 11 \cdot 250$

e. $7 \cdot 20 \cdot 3 \cdot 5$ f. $500 \cdot 13 \cdot 2$

11. Front-End Multiplying Use front-end multiplying to find each product.

a. $53 \cdot 5$ b. $46 \cdot 6$ c. $624 \cdot 7$

12. Front-End Multiplying Use front-end multiplying to find each product.

a. $67 \cdot 4$ b. $94 \cdot 6$ c. $732 \cdot 7$

13. Thinking Money Think of one of the factors as money to find each product. Explain your reasoning.

a. $34 \cdot 5$ b. $36 \cdot 25$ c. $18 \cdot 50$

14. Thinking Money Think of one of the factors as money to find each product. Explain your reasoning.

a. $68 \cdot 5$ b. $52 \cdot 25$ c. $82 \cdot 50$

15. Using Mental Math Use mental math to find each product. Explain your strategy.

a. $36 \cdot 4$ b. $50 \cdot 28$

c. $4 \cdot 43 \cdot 25$ d. $78 \cdot 6$

16. Using Mental Math Use mental math to find each product. Explain your strategy.

a. $2 \cdot 37 \cdot 50$ b. $25 \cdot 84$

c. $56 \cdot 7$ d. $68 \cdot 5$

17. Identifying Compatible Numbers Decide whether the rewritten dividend uses compatible numbers.

a. $9\overline{)351} = 9\overline{)360 - 9}$

b. $9\overline{)351} = 9\overline{)342 + 9}$

18. Identifying Compatible Numbers Decide whether the rewritten dividend uses compatible numbers.

a. $14\overline{)308} = 14\overline{)350 - 42}$

b. $14\overline{)308} = 14\overline{)280 + 28}$

19. Rewriting Dividends Rewrite each dividend using compatible numbers. Then find the quotient.

a. $9\overline{)261}$ b. $8\overline{)2464}$ c. $45\overline{)495}$ d. $12\overline{)372}$

20. Rewriting Dividends Rewrite each dividend using compatible numbers. Then find the quotient.

a. $6\overline{)474}$ b. $7\overline{)4865}$ c. $25\overline{)575}$ d. $16\overline{)864}$

21. Estimating Sums Use rounding to estimate each sum.

a. $\begin{array}{r} 214 \\ 331 \\ +\ 490 \\ \hline \end{array}$ b. $\begin{array}{r} 468 \\ 728 \\ 211 \\ +\ 196 \\ \hline \end{array}$

22. Estimating Sums Use rounding to estimate each sum.

a. $\begin{array}{r} 646 \\ 321 \\ +\ 459 \\ \hline \end{array}$ b. $\begin{array}{r} 661 \\ 243 \\ 367 \\ +\ 492 \\ \hline \end{array}$

23. Estimating Differences Use rounding to estimate each difference.

a. $980 - 712$ b. $487 - 362$ c. $588 - 214$

24. Estimating Differences Use rounding to estimate each difference.

a. $676 - 411$ b. $208 - 147$ c. $846 - 221$

25. Estimating Products Use rounding to estimate each product.

a. $16 \cdot 435$ b. $38 \cdot 216$ c. $62 \cdot 553$

26. Estimating Products Use rounding to estimate each product.

a. $24 \cdot 372$ b. $28 \cdot 421$ c. $56 \cdot 724$

27. Estimating Quotients Use rounding to estimate each quotient.

a. $3054 \div 482$ b. $24{,}362 \div 8021$

c. $15{,}671 \div 516$ d. $452{,}667 \div 2233$

28. Estimating Quotients Use rounding to estimate each quotient.

a. $6124 \div 527$ b. $62{,}868 \div 7114$

c. $56{,}222 \div 785$ d. $623{,}118 \div 1617$

29. Estimating Sums Estimate the total cost of the items represented by the set of prices using (a) rounding, (b) front-end estimation, (c) range estimation, and (d) front-end estimation with adjustment.

\$668 \$374 \$529 \$212

30. Estimating Sums Estimate the total cost of the items represented by the set of prices using (a) rounding, (b) front-end estimation, (c) range estimation, and (d) front-end estimation with adjustment.

\$4399 \$8289 \$3778 \$2134

31. Estimating Differences Use front-end estimation to estimate each difference.

a. $84 - 29$ b. $123 - 72$ c. $583 - 364$

32. Estimating Differences Use front-end estimation to estimate each difference.

a. $67 - 19$ b. $226 - 92$ c. $577 - 314$

33. **Grading Student Work** On a diagnostic test, one of your students does the following work. Explain what the student did wrong. Which topics would you encourage the student to review?

a. $68 + 34 = 70 + 36$
$= 106$

b. $63 \cdot 6 = 36 + 18$
$= 54$

34. **Grading Student Work** On a diagnostic test, one of your students does the following work. Explain what the student did wrong. Which topics would you encourage the student to review?

a. $72 - 47 = 70 - 49$
$= 21$

b. $24 \times 5 = 480$ Think of the product as 24 nickels, which is 48 dimes, or 480 cents.

35. **Using Mental Math in Real Life** Use mental math to find the total cost of each set of items. Explain your strategy.

a. 99¢ 49¢ 79¢ 88¢

b. 9 items at 89¢ each

c. 5 cases of 24 items that cost \$2 per item

36. **Using Mental Math in Real Life** Use mental math to find the total cost of each set of items. Explain your strategy.

a.

b. 11 items at 48¢ each

c. 5 cases of 18 items that cost \$4 per item

37. **Writing** Explain how using compensation to add two numbers is different from using compensation to subtract two numbers.

38. **Writing** Explain the difference between *front-end multiplying* and *front-end estimating* for the product 31×9.

39. **Using Mental Math** A school with 352 students has an average class size of 16 students. How many classes does the school have? Rewrite the dividend using compatible numbers to find your answer.

40. **Using Mental Math** A soccer field has a width of 75 yards and an area of 7875 square yards. Find the length of the soccer field using compatible numbers.

41. **Using Mental Math or Estimation** The table shows the seating capacities of several basketball arenas. Use mental math or estimation to decide whether each statement below is *true* or *false*. State your reasoning.

Arena	Seating capacity
Rupp Arena	23,500
Thompson-Boling Arena	24,535
Viejas Arena	12,414
Comcast Center (MD)	17,950
Quicken Loans Arena	20,562

a. Thompson-Boling Arena has more than twice the capacity of Viejas Arena.

b. Thompson-Boling Arena holds over 1000 more people than Rupp Arena.

c. Rupp Arena holds over 3000 more people than Quicken Loans Arena.

d. The Maryland Comcast Center holds over 5500 more people than Viejas Arena.

42. **Using Mental Math or Estimation** The table shows the odometer readings of five vehicles on a used car lot. Use mental math or estimation to decide whether each statement below is *true* or *false*. State your reasoning.

Vehicle	Odometer (miles)
Compact	36,458
Sport utility	24,783
Convertible	9326
Minivan	16,772
Luxury sedan	12,981

a. The compact vehicle has more than four times as many miles as the convertible.

b. The sport utility vehicle has about twice the number of miles as the luxury sedan.

c. The compact vehicle has over 20,000 more miles than the minivan.

d. The compact vehicle has more miles than the sport utility vehicle and the luxury sedan combined.

43. **Activity: In Your Classroom** Without calculating, determine whether each pair of products is equal. Explain your reasoning. Then design a classroom activity for your students to help them to discover the strategy of halving and doubling different factors of a product to make calculations easier.

a. 12 • 16 and 6 • 32

b. 24 • 18 and 12 • 9

c. 18 • 23 and 9 • 46

44. **Activity: In Your Classroom** Without calculating, determine whether each pair of products is equal. Explain your reasoning. Then design a classroom activity for your students to help them to discover the strategy of multiplying and dividing different factors of a product by the same number to make calculations easier.

a. 16 • 11 and 4 • 44

b. 12 • 22 and 4 • 66

c. 25 • 12 and 5 • 48

45. **Using Cluster Estimation** When addends of a sum are clustered around a number, use **cluster estimation** to estimate the sum. For example, an estimate of the sum 288 + 309 + 304 is 3 • 300 = 900. Use cluster estimation to complete each sum.

a. 92 + 111 + 108 + 93

b. 244 + 239 + 260 + 255

c. 581 + 611 + 604 + 592 + 608

46. **Using Cluster Estimation** When addends of a sum are clustered around a number, use **cluster estimation** to estimate the sum. For example, an estimate of the sum 288 + 309 + 304 is 3 • 300 = 900. Use cluster estimation to complete each sum.

a. 189 + 209 + 206 + 193

b. 434 + 409 + 395 + 372 + 392

c. 354 + 346 + 363 + 367 + 341

47. **Analyzing a Method** Do you think cluster estimation is a good method to estimate each sum? If not, what method would you use? Explain your reasoning.

a. 345 + 1434 + 2197 + 867

b. 2472 + 2634 + 2388 + 2576 + 2481

c. 314 + 281 + 302 + 886 + 909 + 921

48. **Analyzing a Method** Do you think cluster estimation is a good method to estimate each sum? If not, what method would you use? Explain your reasoning.

a. 9301 + 8712 + 8852 + 9014 + 9109

b. 3766 + 2216 + 8454 + 975

c. 696 + 711 + 688 + 418 + 388 + 492

49. **Analyzing Estimates** Without calculating the answer, state whether each estimate is *less than* or *greater than* the actual answer. Explain your reasoning.

a. 4 • 399, estimate: 1600

b. 1000 ÷ 26, estimate: 40

c. 899 ÷ 29, estimate: 30

50. **Analyzing Estimates** Without calculating the answer, state whether each estimate is *less than* or *greater than* the actual answer. Explain your reasoning.

a. 12 • 202, estimate: 2400

b. 2500 ÷ 24, estimate: 100

c. 3201 ÷ 81, estimate: 40

51. **Reasoning** Under what circumstances is an estimate of the sum of several numbers using rounding the same as an estimate using front-end estimation? Give an example.

52. **Reasoning** Under what circumstances is an estimate of the sum of several numbers using rounding the same as an estimate obtained by considering only the leading digits increased by 1? Give an example.

53. **Problem Solving** A factory worker can earn an annual bonus by producing 15,000 parts in one year. Use front-end estimation to determine whether the worker will earn the bonus by producing 11 parts per hour, 40 hours per week, for 48 weeks. Explain your reasoning.

54. **Problem Solving** You are considering two jobs. One job pays $32 per hour, 40 hours per week. The other job pays $60,000 per year. Use front-end estimation to approximate the annual pay for the job that pays $32 per hour. Which job pays more money? Explain your reasoning.

55. **Problem Solving** Your checking account balance is $1476. You deposit checks for $407 and $1239. Use estimation or mental math to decide whether there is enough money to cover a check for $3000. Explain your reasoning.

56. **Problem Solving** Use estimation or mental math to decide which parking lot has a greater area. Explain your reasoning.

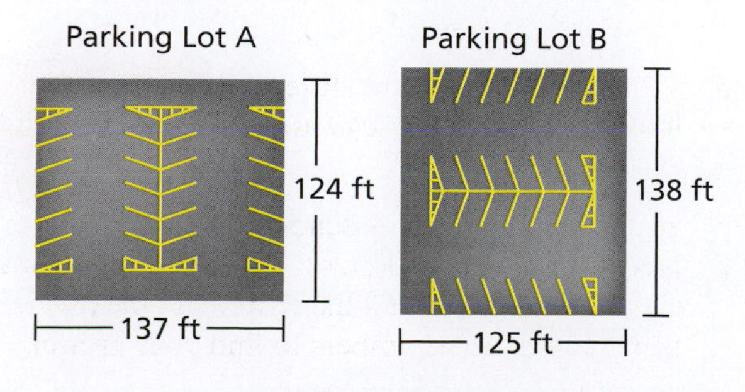

Activity: Writing Areas of Squares and Volumes of Cubes

Materials:

- Pencil
- 100 small base-ten cubes

Learning Objective:

Learn to write the area of a square using exponents. Learn to write the volume of a cube using exponents.

Name ______________________

1. Using up to 100 small base-ten cubes, how many different-sized squares can you make? ▭

2. Sketch each square in Exercise 1. Then write the area of each square's *upper surface* using exponential form. The upper surface of each square is the portion you see looking directly down on the square.

Sample 3-by-3 square

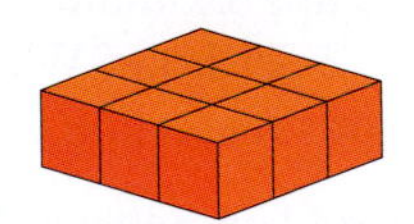

Area of upper surface

$A = 3 \times 3 = 3^2$ square units

3. Using up to 100 small base-ten cubes, how many different-sized cubes can you make? ▭

4. Sketch each cube in Exercise 3. Then write the volume of each cube using exponential form.

Sample 3-by-3-by-3 cube

Volume of cube

$V = 3 \times 3 \times 3 = 3^3$ cubic units

4.2 Exponents

Standards

Grades 6–8

Expressions and Equations

Students should apply and extend previous understandings of arithmetic to algebraic expressions and work with integer exponents.

- Use the definition of exponential form to write and evaluate powers.
- Use the properties of exponents involving the same base.
- Discover and use other properties of exponents.

Exponents

Exponents are a shorthand method of writing repeated multiplication.

Definition of Exponents

Let a and m be natural numbers. The expression that has m factors of a is written in **exponential form** as follows.

$$a^m = \underbrace{a \cdot a \cdot a \cdot \cdots \cdot a}_{m \text{ factors}}$$

In the expression a^m, a is called the **base**, and m is called the **exponent**. The expression a^m is called a **power**, which is read as "a to the mth."

Example $4 \times 4 \times 4 \times 4 \times 4 = 4^5$ "Four to the fifth"

$a^1 = a$ is read as "a to the first."

$a^2 = a \cdot a$ is read as "a to the second" or "a **squared**."

$a^3 = a \cdot a \cdot a$ is read as "a to the third" or "a **cubed**."

EXAMPLE 1 Finding the Volume of Cubes

Write the volume of each cube in exponential form. Then find the volume.

a.

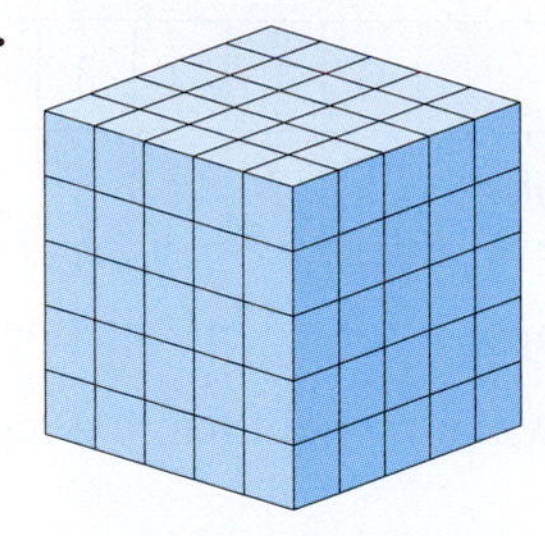

b.

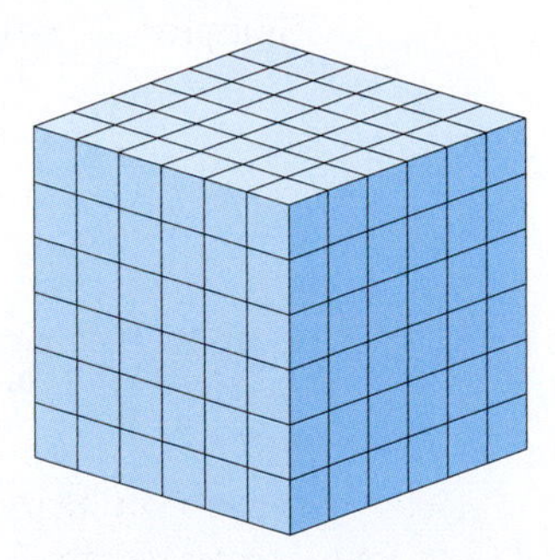

Classroom Tip

When you are teaching the vocabulary of exponents, be sure to point out to your students that the words *squared* and *cubed* have connections to geometry.

Area of a square with side length s: s^2

Volume of a cube with edge length s: s^3

SOLUTION

a. The dimensions of the cube are 5 units by 5 units by 5 units. Its volume is

$$\begin{aligned} V &= \text{(length)(width)(height)} \\ &= 5 \cdot 5 \cdot 5 \\ &= 5^3 \\ &= 125 \text{ cubic units.} \end{aligned}$$

b. The dimensions of the cube are 6 units by 6 units by 6 units. Its volume is

$$\begin{aligned} V &= \text{(length)(width)(height)} \\ &= 6 \cdot 6 \cdot 6 \\ &= 6^3 \\ &= 216 \text{ cubic units.} \end{aligned}$$

Classroom Tip
The expression 0^0 is undefined. Restricting the base a to be a natural number avoids this special case, allowing students to focus on the properties and how to use them.

Definition of Zero Exponent

For all natural numbers a,

$a^0 = 1.$

EXAMPLE 2 Using Exponents to Write a Pattern

Start with a single piece of paper of any size. Fold it in half several times.

a. Describe the pattern shown below.

Original: 1 paper thick — 1 fold: 2 papers thick — 2 folds: 4 papers thick

b. Use exponents to write the thickness of the folded paper in terms of the number of folds.

c. Use the expression you wrote in part (b) to find the thickness of a piece of paper that has been folded in half 10 times.

SOLUTION

a. The table shows the pattern for the thickness of the folded paper after each successive folding.

Folds	0	1	2	3	4	5	6	7	8
Thickness	1	2	4	8	16	32	64	128	256

Each time the paper is folded, its thickness doubles.

b. The pattern shown in the table can be represented using exponential form, as follows.

Folds	Thickness
0	$2^0 = 1$ paper thick
1	$2^1 = 2$ papers thick
2	$2^2 = 4$ papers thick
3	$2^3 = 8$ papers thick
4	$2^4 = 16$ papers thick

In general, a paper that has been folded in half n times is 2^n papers thick.

c. After folding a piece of paper in half 10 times, it is

$$2^{10} = 2 \cdot 2 \cdot 2 \cdot 2 \cdot 2 \cdot 2 \cdot 2 \cdot 2 \cdot 2 \cdot 2$$
$$= 1024 \text{ papers thick.}$$

Classroom Tip
Your students might like to try the paper-folding experiment in Example 2. The world record for folding a single piece of paper in half is 12 folds.

Properties of Exponents Involving the Same Base

There are several properties of exponents. Some of these properties can be easily explained, even to young children. For instance, suppose you multiply the two powers a^m and a^n.

$$a^m \cdot a^n = \underbrace{a \cdot a \cdot a \cdot \cdots \cdot a}_{m \text{ factors}} \cdot \underbrace{a \cdot a \cdot a \cdot \cdots \cdot a}_{n \text{ factors}}$$

$$= \underbrace{a \cdot a \cdot a \cdot \cdots \cdot a}_{m + n \text{ factors}}$$

You can see that the product has $m + n$ factors.

Properties of Exponents Involving the Same Base

Product of Powers Property

Let a be a natural number and m and n be whole numbers.

Words To multiply two powers having the same base, keep the base and *add* the exponents.

Numbers $2^2 \cdot 2^3 = 2^{2+3} = 2^5$ **Algebra** $a^m \cdot a^n = a^{m+n}$

Quotient of Powers Property

Let a be a natural number and m and n be whole numbers with $m \geq n$.

Words To divide two powers having the same base, keep the base and *subtract* the exponents.

Numbers $\frac{3^7}{3^2} = 3^{7-2} = 3^5$ **Algebra** $\frac{a^m}{a^n} = a^{m-n}$

Classroom Tip

Notice the restriction $m \geq n$ for the *Quotient of Powers Property*. When $m = n$, you have the following.

$$\frac{a^m}{a^m} = a^{m-m} = a^0 = 1$$

EXAMPLE 3 **Product and Quotient of Powers**

Simplify each expression *without* using the properties shown above. Then use the above properties to justify your answers.

a. $10^3 \cdot 10^4$ **b.** $\frac{8^6}{8^2}$

SOLUTION

a. $10^3 \cdot 10^4 = (10 \cdot 10 \cdot 10) \cdot (10 \cdot 10 \cdot 10 \cdot 10)$ — 7 factors of 10

$= 10^7$ — 10 to the seventh

Using the Product of Powers Property,

$10^3 \cdot 10^4 = 10^{3+4}$ — Product of Powers Property

$= 10^7.$ — Simplify.

b. $\frac{8^6}{8^2} = \frac{\cancel{8} \cdot \cancel{8} \cdot 8 \cdot 8 \cdot 8 \cdot 8}{\cancel{8} \cdot \cancel{8}}$ — Divide out common factors.

$= 8^4$ — 8 to the fourth

Using the Quotient of Powers Property,

$\frac{8^6}{8^2} = 8^{6-2}$ — Quotient of Powers Property

$= 8^4.$ — Simplify.

Classroom Tip

Notice the teaching strategy in Example 3. Even though there are "rules" for multiplying and dividing powers having the same base, it is still important for your students to perform the calculations using reasoning. Once they obtain the answers, they can check to see that the "new rules" work.

More Properties of Exponents

Property of Exponents Involving Different Bases

Power of a Product Property

Let a and b be natural numbers and m be a whole number.

Words To find a power of a product, find the power of each factor and multiply.

Numbers $(2 \cdot 5)^3 = 2^3 \cdot 5^3$

Algebra $(a \cdot b)^m = a^m \cdot b^m$

EXAMPLE 4 Using the Power of a Product Property

You show your students a drawing of a cube that has an edge length of 2 feet. You ask them to find the volume of the cube in cubic inches. Two of your students turn in the following solutions. Which, if either, is correct?

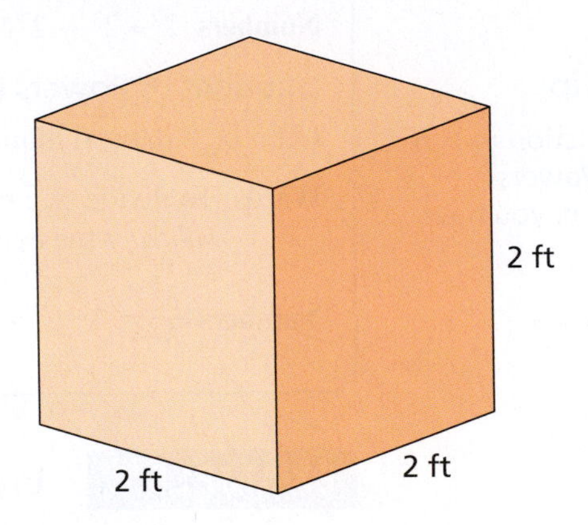

Solution 1 One student reasons that there are $2 \times 2 \times 2$ cubic feet in the cube. Each cubic foot has $12 \times 12 \times 12$ cubic inches. So, the volume of the cube is

$$V = 2^3 \cdot 12^3 \text{ cubic inches.}$$

Number of cubic feet (2^3)

Number of cubic inches in a cubic foot (12^3)

Solution 2 Another student reasons that the cube has an edge length of 24 inches. So, the volume of the cube is

$$V = 24^3 \text{ cubic inches.}$$

SOLUTION

Both of your students are correct. The equivalence of their solutions is justified by the *Power of a Product Property*.

$2^3 \cdot 12^3 = (2 \cdot 12)^3$ Power of a Product Property

$= 24^3$ Multiply.

$= 13{,}824$ Use a calculator.

So, the volume of the cube is 13,824 cubic inches.

Classroom Tip

As shown in Example 4, it is possible to obtain the correct answer using two different methods *that are both correct*! When your students solve problems in different ways, applaud them. Ask them to share their thinking with the class. Realizing that there are multiple ways to arrive at a correct answer is essential for a student's mastery of mathematics. It is possible for your students to obtain the correct answer using incorrect reasoning. When they do this, point out that their reasoning is incorrect.

Discovering another property of exponents on their own is within the grasp of most of your students. The next example shows how you can guide them.

EXAMPLE 5 **Discovering a Property**

Ask your students to discover a rule for finding a power of a power.

$(a^m)^n = \square$

SOLUTION

> **Mathematical Practices**
>
> **Look For and Make Use of Structure**
>
> Mathematically proficient students are able to see complicated things as single objects or as being composed of several objects.

Depending on the ages and thought processes of your students, some might use inductive reasoning, while others might use deductive reasoning.

Inductive reasoning To find the *Power of a Power Property* using inductive reasoning, a student can consider several examples, recognize the pattern, and then formulate a rule that represents the pattern.

$(3^2)^2 = (3 \cdot 3) \cdot (3 \cdot 3) = 3^4$ Note that $2 \cdot 2 = 4$.

$(3^2)^3 = (3 \cdot 3) \cdot (3 \cdot 3) \cdot (3 \cdot 3) = 3^6$ Note that $2 \cdot 3 = 6$.

$(3^3)^2 = (3 \cdot 3 \cdot 3) \cdot (3 \cdot 3 \cdot 3) = 3^6$ Note that $3 \cdot 2 = 6$.

$(3^4)^2 = (3 \cdot 3 \cdot 3 \cdot 3) \cdot (3 \cdot 3 \cdot 3 \cdot 3) = 3^8$ Note that $4 \cdot 2 = 8$.

$(3^2)^4 = (3 \cdot 3) \cdot (3 \cdot 3) \cdot (3 \cdot 3) \cdot (3 \cdot 3) = 3^8$ Note that $2 \cdot 4 = 8$.

From these examples, it appears that to find a power of a power, you multiply the exponents. So, $(a^m)^n = a^{m \cdot n}$.

Deductive reasoning Deductive reasoning is the use of logic and previously proven rules to deduce new rules.

$$(a^m)^n = \overbrace{\underbrace{(a \cdot a \cdot \cdots \cdot a)}_{m \text{ factors}}\underbrace{(a \cdot a \cdot \cdots \cdot a)}_{m \text{ factors}} \cdot \cdots \cdot \underbrace{(a \cdot a \cdot \cdots \cdot a)}_{m \text{ factors}}}^{n \text{ times}}$$

$$= \underbrace{a \cdot a \cdot a \cdot \cdots \cdot a}_{mn \text{ factors}}$$

$$= a^{m \cdot n}$$

Power of a Power Property

Let a be a natural number and m and n be whole numbers.

Words To find a power of a power, multiply the exponents.

Numbers $(4^3)^2 = 4^{3 \cdot 2} = 4^6$ **Algebra** $(a^m)^n = a^{m \cdot n}$

EXAMPLE 6 **Using the Power of a Power Property**

Write each expression as a single power. Then evaluate the power.

a. $(2^4)^2$ **b.** $(8^3)^0$

SOLUTION

a. $(2^4)^2 = 2^8$ Power of a Power Property

$= 256$ Evaluate the power.

b. $(8^3)^0 = 8^0$ Power of a Power Property

$= 1$ Evaluate the power.

4.2 Exercises

Free step-by-step solutions to odd-numbered exercises at *MathematicalPractices.com.*

1. **Writing a Solution Key** Write a solution key for the activity on page 134. Describe a rubric for grading a student's work.

2. **Grading the Activity** In the activity on page 134, a student gave the answers below. Each question is worth 10 points. Assign a grade to each answer. For those that are incorrect, why do you think the student erred?

Sample Student Work

> **3.** Using up to 100 small base-ten cubes, how many different-sized cubes can you make? 4
>
> **4.** Sketch each cube in Exercise 3. Then write the volume of each cube using exponential form.

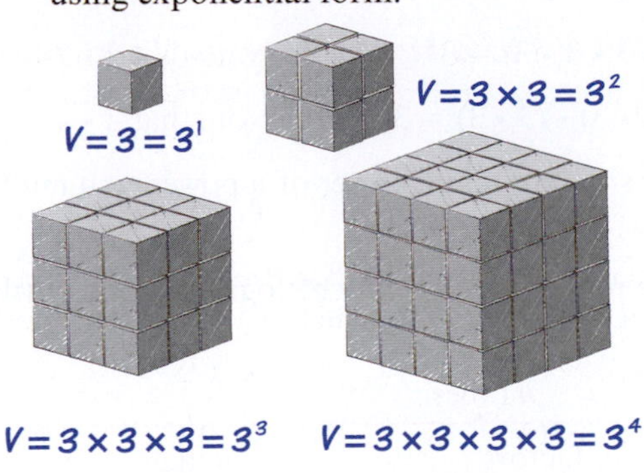

3. **Writing in Exponential Form** Write each expression in exponential form.
   - **a.** $17 \cdot 17$
   - **b.** $31 \cdot 31 \cdot 31$
   - **c.** $4 \cdot 4 \cdot 4 \cdot 4 \cdot 4$
   - **d.** $x \cdot x \cdot x \cdot x \cdot x \cdot x$

4. **Writing in Exponential Form** Write each expression in exponential form.
   - **a.** $12 \cdot 12 \cdot 12 \cdot 12$
   - **b.** $51 \cdot 51 \cdot 51$
   - **c.** $7 \cdot 7 \cdot 7 \cdot 7 \cdot 7$
   - **d.** $a \cdot a \cdot a \cdot a \cdot a \cdot a$

5. **Writing in Exponential Form** Use exponents to write each expression in simpler form.
   - **a.** $6 \cdot 6 \cdot 6 \cdot 6 \cdot 6 \cdot x \cdot x$
   - **b.** $15 \cdot 15 \cdot y \cdot 15 \cdot y \cdot y \cdot 7 \cdot 7$

6. **Writing in Exponential Form** Use exponents to write each expression in simpler form.
   - **a.** $5 \cdot 5 \cdot 5 \cdot 5 \cdot x \cdot x \cdot x \cdot x \cdot x$
   - **b.** $11 \cdot 11 \cdot a \cdot a \cdot 11 \cdot a \cdot 3 \cdot 3 \cdot a$

7. **Writing Powers as Products** Write each expression as a product.
   - **a.** 13^4
   - **b.** 9^6
   - **c.** y^7
   - **d.** b^8

8. **Writing Powers as Products** Write each expression as a product.
   - **a.** 19^5
   - **b.** 8^5
   - **c.** x^3
   - **d.** c^{10}

9. **Finding the Volume of a Cube** Write the volume of the cube in exponential form. Then find the volume.

10. **Finding the Volume of a Cube** Write the volume of the cube in exponential form. Then find the volume.

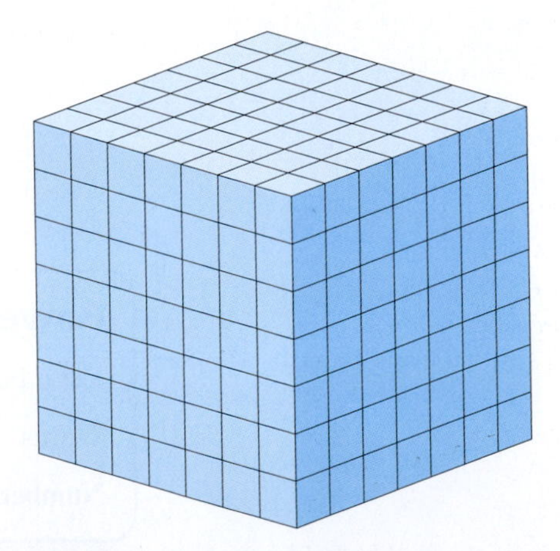

11. **Finding the Area of a Square** Write the area of the square in exponential form. Then find the area.

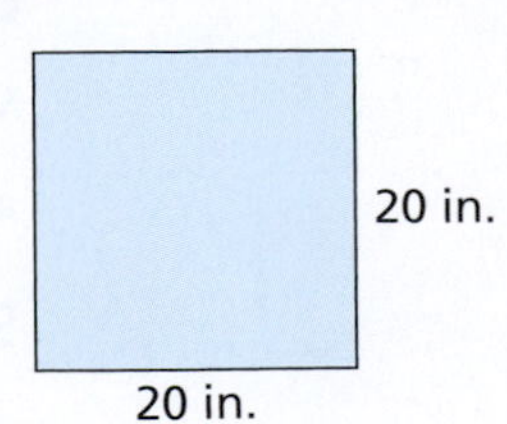

12. **Finding the Area of a Square** Write the area of the square in exponential form. Then find the area.

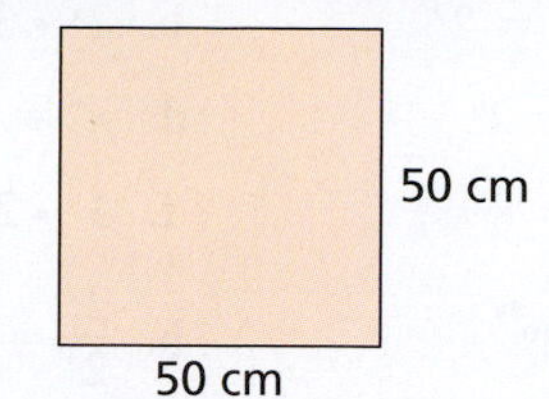

13. **Evaluating Powers** Evaluate each power.

a. 2^5 **b.** 3^4

c. 16^0 **d.** x^0

14. **Evaluating Powers** Evaluate each power.

a. 2^6 **b.** 5^4

c. 29^0 **d.** a^0

15. **Using Exponents to Write Patterns** Write each sequence of numbers using powers.

a. 1, 3, 9, 27, 81, 243, . . .

b. 1, 10, 100, 1000, 10,000, 100,000, . . .

16. **Using Exponents to Write Patterns** Write each sequence of numbers using powers.

a. 1, 5, 25, 125, 625, 3125, . . .

b. 1, 20, 400, 8000, 160,000, 3,200,000, . . .

17. **Using the Definition of Exponents** Write each expression as a single power *without* using the Product of Powers Property or the Quotient of Powers Property. Then use one of the properties to justify your answer.

a. $2^3 \cdot 2^4$ **b.** $\frac{6^7}{6^2}$

18. **Using the Definition of Exponents** Write each expression as a single power *without* using the Product of Powers Property or the Quotient of Powers Property. Then use one of the properties to justify your answer.

a. $7^5 \cdot 7^6$ **b.** $\frac{13^8}{13^3}$

19. **Properties of Exponents Involving the Same Base** Write each expression as a single power. Identify the property that you used.

a. $11^4 \cdot 11^5$

b. $\frac{23^7}{23^4}$

c. $\frac{19^6}{19^2}$

d. $x^6 \cdot x^7 \cdot x^2$

20. **Properties of Exponents Involving the Same Base** Write each expression as a single power. Identify the property that you used.

a. $\frac{8^5}{8}$ **b.** $\frac{y^{19}}{y^7}$

c. $12^4 \cdot 12^3$ **d.** $g^5 \cdot g^4 \cdot g^3$

21. **Using Properties of Exponents** Write each expression as a single power.

a. $3^5 \cdot 4^5$ **b.** $6^3 \cdot 5^3$

c. $5^4 \cdot 7^4$ **d.** $6^9 \cdot 7^9$

e. $8^7 \cdot 3^7 \cdot 5^7$ **f.** $9^3 \cdot 4^3 \cdot 3^3$

22. **Using a Property of Exponents** Write each expression as a single power.

a. $3^6 \cdot 5^6$ **b.** $2^2 \cdot 19^2$

c. $6^3 \cdot 8^3$ **d.** $9^8 \cdot 5^8$

e. $2^4 \cdot 6^4 \cdot 5^4$ **f.** $7^{11} \cdot 6^{11} \cdot 3^{11}$

23. **Using a Property of Exponents** Write each expression as a product of two or more powers.

a. $(3x)^6$ **b.** $(5y)^8$

c. $(4xy)^3$ **d.** $(6xy)^2$

24. **Using a Property of Exponents** Write each expression as a product of two or more powers.

a. $(5a)^3$ **b.** $(7c)^6$

c. $(6ab)^5$ **d.** $(9cd)^7$

25. **Using a Property of Exponents** Write each expression as a single power. Then evaluate the power.

a. $(2^3)^2$ **b.** $(5^2)^3$

c. $(11^4)^0$ **d.** $(10^3)^3$

26. **Using a Property of Exponents** Write each expression as a single power. Then evaluate the power.

a. $(3^2)^3$ **b.** $(16^0)^5$

c. $(2^4)^2$ **d.** $(10^4)^2$

27. **Using a Property of Exponents** Write each expression as a product of two powers.

a. $(2^3a)^2$

b. $(5c^3)^4$

28. **Using a Property of Exponents** Write each expression as a product of two powers.

a. $(3^4a)^3$

b. $(7c^2)^5$

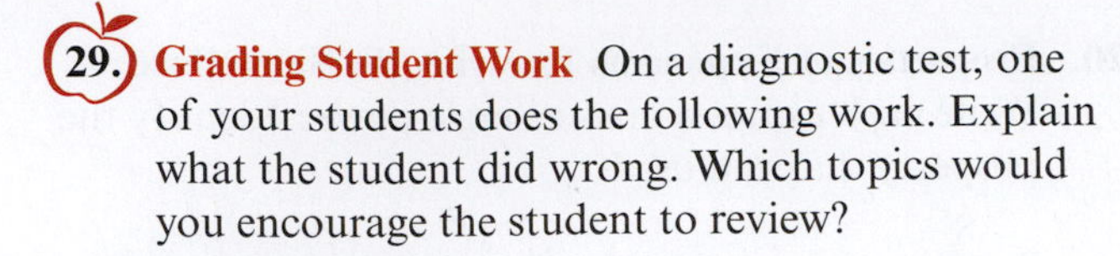

29. Grading Student Work On a diagnostic test, one of your students does the following work. Explain what the student did wrong. Which topics would you encourage the student to review?

a. $8^2 \cdot 8^3 = 8^{2 \cdot 3} = 8^6$

b. $3^3 \cdot 4^3 = 12^{3+3} = 12^6$

30. Grading Student Work On a diagnostic test, one of your students does the following work. Explain what the student did wrong. Which topics would you encourage the student to review?

a. $\frac{7^9}{7^3} = 7^{9 \div 3} = 7^3$

b. $(6^3)^2 = 6^{(3^2)} = 6^9$

31. Explaining a Property: In Your Classroom The description below is from the top of page 137.

$$a^m \cdot a^n = \underbrace{a \cdot a \cdot a \cdot \cdots \cdot a}_{m \text{ factors}} \cdot \underbrace{a \cdot a \cdot a \cdot \cdots \cdot a}_{n \text{ factors}}$$

$$= \underbrace{a \cdot a \cdot a \cdot \cdots \cdot a}_{m + n \text{ factors}}$$

a. Which property of exponents is described?

b. Explain how you might help your students relate the description above with the property in words.

c. Does the description above use *inductive reasoning* or *deductive reasoning*?

d. Explain how you could use the *other* type of reasoning mentioned in part (c) to give your students another description of the property.

32. Explaining a Property: In Your Classroom

a. Explain how you could use inductive reasoning to explain the Quotient of Powers Property to your students.

b. Explain how you could use deductive reasoning to explain the Quotient of Powers Property to your students.

33. Inductive Reasoning Does the inductive reasoning in Example 5 show that the Power of a Power Property is always true? Explain.

34. Deductive Reasoning Does the deductive reasoning in Example 5 show that the Power of a Power Property is always true? Explain.

Maksim Chuvashov/Shutterstock.com

35. Using Properties of Exponents Find the value(s) of x that make the equation true.

a. $2^3 \cdot 2^x = 2^9$ **b.** $2^x \cdot 3^x = 6^6$

c. $(2^x)^3 = 2^9$ **d.** $x^5 = 2^{10}$

e. $(2^x)^3 = 8^x$ **f.** $3^5 \cdot 2^x = 6^x$

g. $\frac{2^x}{2^4} = 2^6$ **h.** $\frac{2^x}{2^3} = 8$

36. Using Properties of Exponents Find the value(s) of x that make the equation true.

a. $3^{15} = 3^x \cdot 3^5$ **b.** $5^x \cdot 4^7 = 20^x$

c. $(5^3)^x = 125^x$ **d.** $x^4 = 2^{12}$

e. $(4^5)^x = 4^{10}$ **f.** $12^x = 4^x \cdot 3^5$

g. $\frac{6^{24}}{6^x} = 6^4$ **h.** $\frac{3^8}{3^x} = 9$

37. Using Properties of Exponents Tell whether each statement is *true* or *false*. Explain your reasoning.

a. $\frac{x^{k+1}}{x} = x^k$

b. $(a^n)^k = (a^k)^n$

38. Using Properties of Exponents Tell whether each statement is *true* or *false*. Explain your reasoning.

a. $y^{3k} \cdot y^k = y^{4k}$

b. $(a^n b^n)^k = (ab)^{2nk}$

39. Area of a Painting The square painting shown below has a side length of 4 feet.

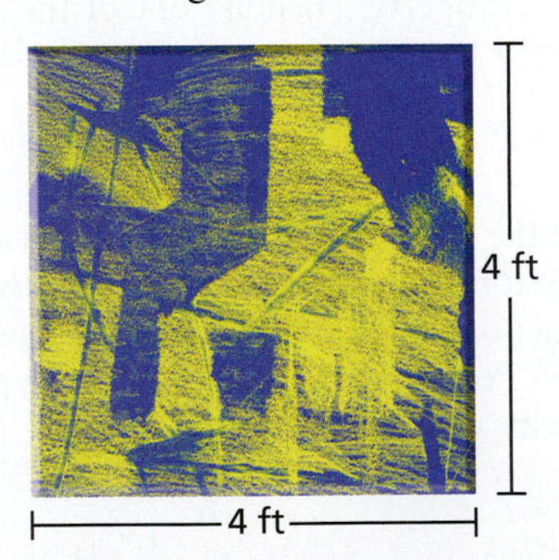

a. Use exponents to write an expression for the area of the painting (in square inches) as the product of the number of square feet of the painting and the number of square inches in a square foot.

b. Use exponents to write an expression for the area of the painting (in square inches) based on its side length in inches.

c. Use the Power of a Product Property to show that the expressions in parts (a) and (b) give the same answer.

40. Volume of a Cube A cube has an edge length of 3 meters.

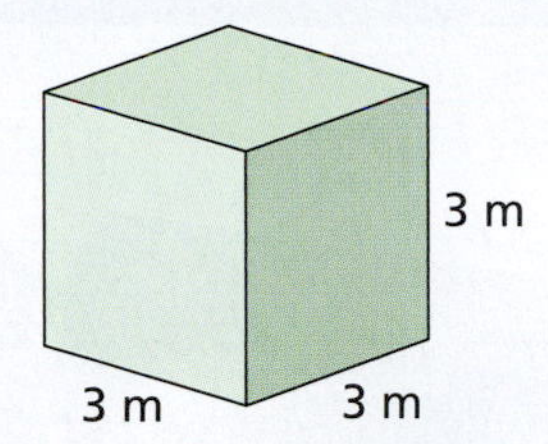

a. Use exponents to write an expression for the volume of the cube (in cubic centimeters) as the product of the number of cubic meters in the cube and the number of cubic centimeters in a cubic meter.

b. Use exponents to write an expression for the volume of the cube (in cubic centimeters) based on its edge length in centimeters.

c. Use the Power of a Product Property to show that the expressions in parts (a) and (b) give the same answer.

41. Writing Is *inductive* or *deductive* reasoning used to reach the conclusion in part (b) of Example 2? Explain.

42. Writing Explain how the Product of Powers Property is different from the Power of a Product Property.

43. Comparing Areas of Squares Three squares with side lengths of x, $2x$, and $3x$ are shown.

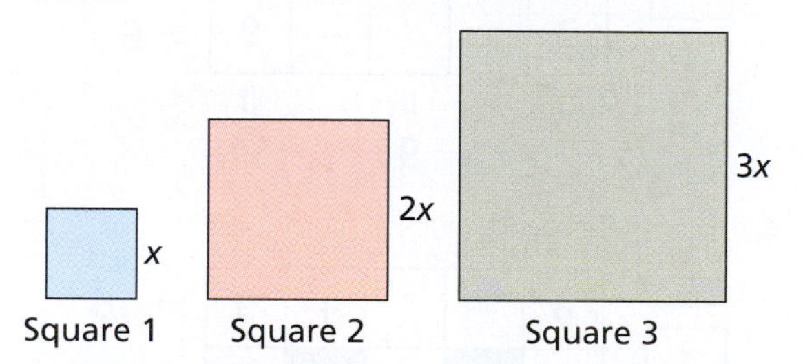

a. Write an exponential expression for the area of each square.

b. Rewrite the area of Square 2 as the product of two powers. Identify the property that you used.

c. How many times greater is the area of Square 2 than the area of Square 1? Explain.

d. Rewrite the area of Square 3 as a product of two powers. Identify the property that you used.

e. How many times greater is the area of Square 3 than the area of Square 1?

f. Explain how multiplying the side length of a square by a number n affects the area of the square.

44. Comparing Volumes of Cubes Three cubes with edge lengths of x, $2x$, and $3x$ are shown.

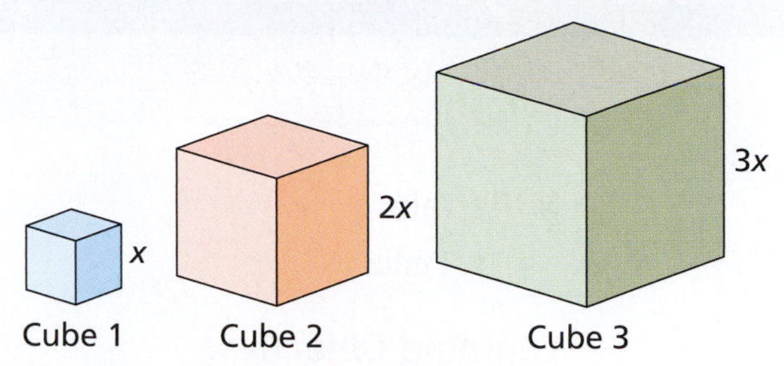

a. Write an exponential expression for the volume of each cube.

b. Rewrite the volume of Cube 2 as the product of two powers. Identify the property that you used.

c. How many times greater is the volume of Cube 2 than the volume of Cube 1? Explain.

d. Rewrite the volume of Cube 3 as a product of two powers. Identify the property that you used.

e. How many times greater is the volume of Cube 3 than the volume of Cube 1?

f. Explain how multiplying the edge lengths of a cube by a number n affects the volume of the cube.

45. Problem Solving Use part (f) of Exercise 43 to determine the side length of a square that has an area 64 times greater than the area of a square with a side length of 3 feet.

46. Problem Solving Use part (f) of Exercise 44 to determine the edge length of a cube that has a volume 64 times greater than the volume of a cube with an edge length of 3 feet.

47. Problem Solving You cut a long piece of string in half. You place the two resulting pieces of string together and then cut them in half. You repeat the process until you have made eight cuts. Write an exponential expression for the number of pieces of string you end up with. Then determine the number.

48. Problem Solving You place 4 bacteria in a petri dish. The number of bacteria quadruples every 20 minutes.

a. Write exponential expressions for the numbers of bacteria after 1, 2, 3, and 4 hours.

b. Assume that the reproduction rate continues. Write an exponential expression for the number of bacteria after n hours. What type of reasoning did you use to make your conclusion? Explain.

Activity: Please Excuse My Dear Aunt Sally

Materials:

- Pencil

Learning Objective:

Explore the order in which you should perform operations by *evaluating*, or finding the value of, expressions in different ways.

Name ______________________________

Magic Operation Square **Use the numbers in the boxes to fill in the magic operation square.**

1.

Numbers:  4  2 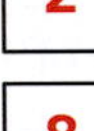 8  3  5

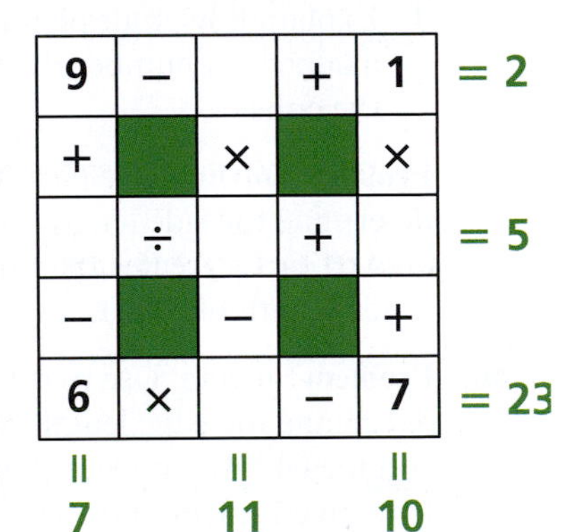

9	−		+	1	= 2
+		×		×	
	÷		+		= 5
−		−		+	
6	×		−	7	= 23
= 7		= 11		= 10	

2.

Numbers: 6, 4, 1, 5, 8

2	−		+	7	= 8
+		×		×	
	÷		+		= 8
−		+		+	
3	×		−	9	= 6
= 7		= 9		= 51	

3.

Numbers:  3, 4, 6,  9,  1

2	×		+	5	= 13
×		×		×	
	−		+		= 4
−		÷		+	
8	+		−	7	= 4
= 10		= 8		= 12	

4.

Numbers:  3,  1, 4,  6, 2

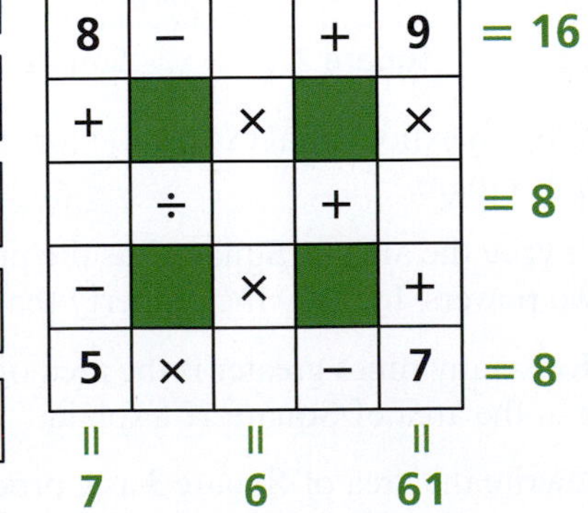

8	−		+	9	= 16
+		×		×	
	÷		+		= 8
−		×		+	
5	×		−	7	= 8
= 7		= 6		= 61	

Evaluate each expression by performing the circled operation first. Which result do you think is correct? Explain your reasoning. Use a calculator to check your results.

5. a. $20 \ominus 5 \times 4$ **b.** $20 - 5 \otimes 4$

6. a. $18 \oplus 9 \div 3$ **b.** $18 + 9 \odot 3$ (circled ÷)

7. a. $9 \times 4 \oplus 3$ **b.** $9 \otimes 4 + 3$

8. a. $24 \div 6 \ominus 2$ **b.** $24 \odot 6 - 2$ (circled ÷)

iStockphoto.com/Alina555

4.3 Order of Operations

Standards

Grades 3–5
Operations and Algebraic Thinking
Students should write and interpret numerical expressions.

Grades 6–8
Expressions and Equations
Students should apply and extend previous understandings of arithmetic to algebraic expressions, and solve real-life problems using numerical and algebraic expressions.

- Use the order of operations to evaluate numerical expressions.
- Evaluate algebraic expressions.
- Use the order of operations to evaluate algebraic expressions.

Order of Operations

Mathematics is a language. In this language, the numbers can be thought of as the nouns. The four basic math operations ($+$, $-$, $\times$, $\div$) and the exponential operations can be thought of as the verbs. The *order of operations* prioritizes these operations.

Order of Operations

A **numerical expression** contains only numbers and operations. When *evaluating* a numerical expression that has two or more operations, perform the operations using the **order of operations**. The order of operations is the set of rules that tells you which operations should be performed first.

1. Perform operations in Parentheses.
2. Evaluate numbers with Exponents.
3. Multiply or Divide from left to right.
4. Add or Subtract from left to right.

A common mnemonic for remembering this order is "Please Excuse My Dear Aunt Sally."

EXAMPLE 1 **Using Order of Operations**

Use the order of operations to evaluate each expression.

a. $17 - 4 \times 2$ **b.** $12 + 12 \div 3$ **c.** $13 - (4 - 3)$

d. $5 \times 11 - 2$ **e.** $9^2 - (1 + 4)$ **f.** $28 \div (7 - 5)$

SOLUTION

a. $17 - 4 \times 2 = 17 - 8$ Multiply.
$= 9$ Then subtract.

b. $12 + 12 \div 3 = 12 + 4$ Divide.
$= 16$ Then add.

c. $13 - (4 - 3) = 13 - 1$ Subtract inside the parentheses.
$= 12$ Then subtract.

d. $5 \times 11 - 2 = 55 - 2$ Multiply.
$= 53$ Then subtract.

e. $9^2 - (1 + 4) = 9^2 - 5$ Add inside the parentheses.
$= 81 - 5$ Evaluate the power.
$= 76$ Then subtract.

f. $28 \div (7 - 5) = 28 \div 2$ Subtract inside the parentheses.
$= 14$ Then divide.

Classroom Tip
Ask your students to evaluate $(12 + 12) \div 3$ in part (b) to see that the order in which you perform the operations is important.

Evaluating Algebraic Expressions

A collection of letters, or **variables**, and numbers combined by operations is an **algebraic expression**. To *evaluate an algebraic expression*, substitute a number for each variable. This changes the algebraic expression into a numerical expression.

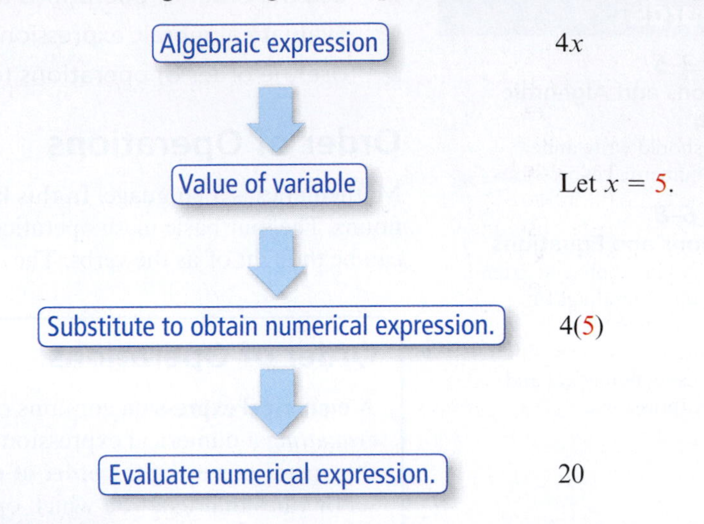

EXAMPLE 2 Evaluating Algebraic Expressions

a. Evaluate $3x$ when $x = 6$.

b. Evaluate p^3 when $p = 5$.

c. Evaluate $q + 12$ when $q = 13$.

d. Evaluate $m - 8$ when $m = 20$.

e. Evaluate $y \div 4$ when $y = 24$.

f. Evaluate 7^n when $n = 3$.

SOLUTION

a. $3x = 3(6)$ Substitute 6 for x.

$= 18$ Multiply.

The expression $3x$ is equal to 18 when $x = 6$.

b. $p^3 = 5^3$ Substitute 5 for p.

$= 125$ Evaluate the power.

The expression p^3 is equal to 125 when $p = 5$.

c. $q + 12 = 13 + 12$ Substitute 13 for q.

$= 25$ Add.

The expression $q + 12$ is equal to 25 when $q = 13$.

d. $m - 8 = 20 - 8$ Substitute 20 for m.

$= 12$ Subtract.

The expression $m - 8$ is equal to 12 when $m = 20$.

e. $y \div 4 = 24 \div 4$ Substitute 24 for y.

$= 6$ Divide.

The expression $y \div 4$ is equal to 6 when $y = 24$.

f. $7^n = 7^3$ Substitute 3 for n.

$= 343$ Evaluate the power.

The expression 7^n is equal to 343 when $n = 3$.

EXAMPLE 3 Comparing Problem-Solving Approaches

You give the following problem to your class.

You have a part-time job that pays $16 per hour. You receive a paycheck with a gross amount of $512. How many hours did you work?

Describe some problem-solving approaches.

SOLUTION

You can represent the gross amount of your paycheck by using the algebraic expression $16n$.

$$\text{Paycheck amount} = \frac{\$16}{\text{hour}} \cdot n \text{ hours}$$

- **Use a table** One approach is to use a table.

Hours, n	0	4	8	12	16	20	24	28	32
Paycheck amount	0	$64	$128	$192	$256	$320	$384	$448	$512

From the table, you see that you worked 32 hours.

- **Use Guess, Check, and Revise** A reasonable first guess is $n = 20$.

Guess	Check	Revise
$n = 20$	$16 \cdot 20 = \$320$	Because $320 < $512, make guess greater.
$n = 30$	$16 \cdot 30 = \$480$	Because $480 < $512, make guess greater.
$n = 34$	$16 \cdot 34 = \$544$	Because $544 > $512, make guess lesser.
$n = 32$	$16 \cdot 32 = \$512$	Correct ✓

So, you see that you worked 32 hours.

- **Use algebra** You can write an equation and solve it.

$16n = 512$ Set the expression $16n$ equal to the total amount earned.

$\frac{16n}{16} = \frac{512}{16}$ Divide each side by 16.

$n = 32$ Simplify.

So, you see that you worked 32 hours.

- **Use a spreadsheet** This is similar to using a table, but involves technology. From the spreadsheet, you see that you worked 32 hours.

	A	B	
1	Hours	Paycheck amount	
2	28	$448	
3	29	$464	
4	30	$480	
5	31	$496	
6	32	$512	
7	33	$528	
8	34	$544	
9	35	$560	
10			

Classroom Tip

When you are teaching real-life problem solving, use the guide "Quality is better than quantity." For instance, in Example 3, taking the time to go over several problem-solving strategies is a far better use of classroom time than quickly going through half a dozen similar real-life problems.

Mathematical Practices

Construct Viable Arguments and Critique the Reasoning of Others

Mathematically proficient students compare the effectiveness of two plausible arguments, distinguish correct logic or reasoning from that which is flawed and explain the flaw if one exists.

Order of Operations and Algebraic Expressions

Each of the algebraic expressions on pages 146 and 147 involves only one operation. When an algebraic expression involves two or more operations, you need to use the order of operations.

EXAMPLE 4 Evaluating Algebraic Expressions

a. Evaluate $3x + 4$ when $x = 6$.

b. Evaluate $3(x + 4)$ when $x = 6$.

c. Evaluate $12 + n^2$ when $n = 3$.

d. Evaluate $(12 + n)^2$ when $n = 3$.

e. Evaluate $3y \div 4 - 9$ when $y = 12$.

f. Evaluate $3(y \div 4) - 9$ when $y = 12$.

SOLUTION

a. $3x + 4 = 3(6) + 4$ Substitute 6 for x.

$= 18 + 4$ Multiply.

$= 22$ Add.

The expression $3x + 4$ is equal to 22 when $x = 6$.

b. $3(x + 4) = 3(6 + 4)$ Substitute 6 for x.

$= 3(10)$ Add inside parentheses.

$= 30$ Multiply.

The expression $3(x + 4)$ is equal to 30 when $x = 6$.

c. $12 + n^2 = 12 + 3^2$ Substitute 3 for n.

$= 12 + 9$ Evaluate the power.

$= 21$ Add.

The expression $12 + n^2$ is equal to 21 when $n = 3$.

d. $(12 + n)^2 = (12 + 3)^2$ Substitute 3 for n.

$= 15^2$ Add inside parentheses.

$= 225$ Evaluate the power.

The expression $(12 + n)^2$ is equal to 225 when $n = 3$.

e. $3y \div 4 - 9 = 3(12) \div 4 - 9$ Substitute 12 for y.

$= 36 \div 4 - 9$ Multiply.

$= 9 - 9$ Divide.

$= 0$ Subtract.

The expression $3y \div 4 - 9$ is equal to 0 when $y = 12$.

f. $3(y \div 4) + 9 = 3(12 \div 4) - 9$ Substitute 12 for y.

$= 3(3) - 9$ Divide inside parentheses.

$= 9 - 9$ Multiply.

$= 0$ Subtract.

The expression $3(y \div 4) - 9$ is equal to 0 when $y = 12$.

EXAMPLE 5 Solving a Real-Life Problem

Make a table that compares temperatures in degrees Celsius C and degrees Fahrenheit F using the formula below. Then describe the pattern.

$$F = \frac{9C}{5} + 32$$

SOLUTION

1. **Understand the Problem** You need to make a table that compares Celsius and Fahrenheit temperatures. Then find and describe any patterns.

2. **Make a Plan** The range of numbers you should include in the table is not given. Should you use common air temperatures? Should you use temperatures ranging from freezing to boiling? You decide to use common air temperatures ranging between 0°C and 30°C, in increments of 5°.

3. **Solve the Problem** Use the formula to make a table. Substitute the values for C and evaluate. Notice when the Celsius temperature increases by 5 degrees, the Fahrenheit temperature increases by 9 degrees.

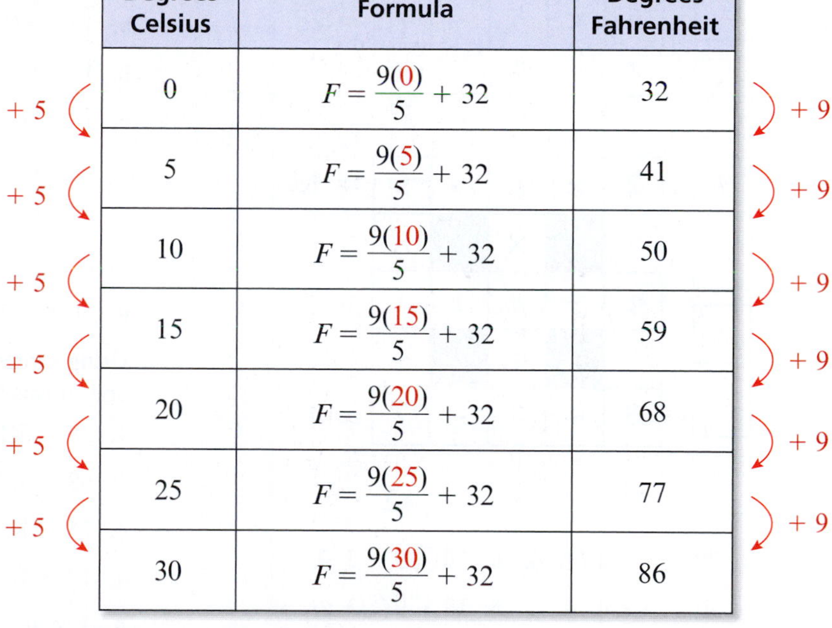

Degrees Celsius	Formula	Degrees Fahrenheit
0	$F = \frac{9(0)}{5} + 32$	32
5	$F = \frac{9(5)}{5} + 32$	41
10	$F = \frac{9(10)}{5} + 32$	50
15	$F = \frac{9(15)}{5} + 32$	59
20	$F = \frac{9(20)}{5} + 32$	68
25	$F = \frac{9(25)}{5} + 32$	77
30	$F = \frac{9(30)}{5} + 32$	86

4. **Look Back** You can use a thermometer that displays both Celsius and Fahrenheit temperature readings to compare the temperatures shown in the table. If you ever travel to a country that uses the Celsius scale, you may notice the following.

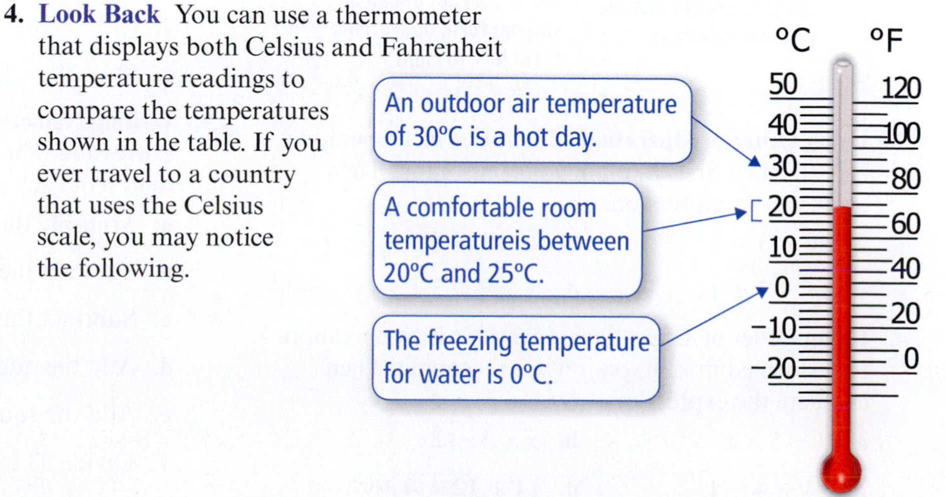

Classroom Tip

In Example 5, ask your students to describe the advantage of using Celsius temperature increments of 5°. The answer is that 5 will divide evenly into the numerator resulting in a whole number for each Fahrenheit temperature.

Classroom Tip

An easy estimate for changing temperatures from degrees Celsius to degrees Fahrenheit is to "double C and then add 30." It is not exact, but it gives a good estimate.

4.3 Exercises

Free step-by-step solutions to odd-numbered exercises at *MathematicalPractices.com.*

1. Writing a Solution Key Write a solution key for the activity on page 144. Describe a rubric for grading a student's work.

2. Grading the Activity In the activity on page 144, a student gave the answers below. Each question is worth 10 points. Assign a grade to each answer. For those that are incorrect, why do you think the student erred?

Sample Student Work

1.

Number tiles: 4, 2, 8, 3, 5

9	−	8	+	1	= 2
+		×		×	
4	÷	2	+	3	= 5
−		−		+	
6	×	5	−	7	= 23
= 7		= 11		= 10	

3.

Number tiles: 3, 4, 6, 9, 1

2	×	4	+	5	= 13
×		×		×	
9	−	4	+	1	= 4
−		÷		+	
8	+	3	−	7	= 4
= 10		= 8		= 12	

5. **a.** $20 \ominus 5 \times 4$ 60 **6.** **a.** $18 \oplus 9 \div 3$ 9

b. $20 - 5 \otimes 4$ 0 **b.** $18 + 9 \,(\div)\, 3$ 21

(b) is correct because you multiply first.

(a) is correct because you perform operations from left to right.

3. Using Order of Operations Identify which operation is performed first. Explain your reasoning. Then evaluate the expression.

a. $26 - 3 \times 8$ **b.** $4 \times 3 + 7$

c. $14 \div 7 + 11$ **d.** $5 - 9 \div 3$

4. Using Order of Operations Identify which operation is performed first. Explain your reasoning. Then evaluate the expression.

a. $9 + 5 \times 6$ **b.** $5 \times 3 - 2$

c. $24 \div 4 - 1$ **d.** $14 + 12 \div 4$

5. Using Order of Operations List the operations in the order they are performed to evaluate each expression. Explain your reasoning. Then evaluate the expression.

a. $3^2 + 4 - 2$ **b.** $(6 - 3) \times 3 + 1$

c. $3 \times 5^2 - 10$ **d.** $7 + 3 \times 4 - 5$

6. Using Order of Operations List the operations in the order they are performed to evaluate each expression. Explain your reasoning. Then evaluate the expression.

a. $5 - 3 + 4^2$ **b.** $4 + 3(7 - 5)$

c. $18 \div 3^2 + 5$

d. $12 - 2 \times 5 + 4$

7. Using Order of Operations Use the order of operations to evaluate each expression.

a. $18 - 3(5 - 3)$

b. $13 + 6(10 \div 2)$

c. $16 - 8 - 2 \times 3$

d. $6 \times 9 + 5^2$

e. $4 \times 3^2 + 4$

f. $(2^2)^3 \div 8 - 5 + 7 \times 3$

g. $3(18 - 16)^2 \div 6(27 \div 3^3)$

8. Using Order of Operations Use the order of operations to evaluate each expression.

a. $28 - 5(1 + 3)$

b. $68 - 4(24 \div 6)$

c. $25 - 12 + 9 \times 5$

d. $12 \times 10 - 3^4$

e. $3 \times 4^2 - 18$

f. $(2^3)^2 \div 16 - 3 + 12 \times 4$

g. $(8 \div 2)^4 - 10^2 - (160 \div 2^3)$

9. Writing Numerical Expressions Write a numerical expression that uses the order of operations to represent each problem.

a. Multiply the sum of 3 and 4 by 5.

b. Subtract the product of 3 and 6 from 23.

c. Subtract the difference of 6 and 4 from 12.

d. Add the quotient of 14 and 7 to 10.

e. Add the square of 7 to the sum of 4 and 6.

f. Divide 33 by the sum of 8 and 3.

10. Writing Numerical Expressions Write a numerical expression that uses the order of operations to represent each problem.

a. Multiply the difference of 8 and 5 by 7.

b. Add the product of 5 and 7 to 9.

c. Add the difference of 13 and 4 to 20.

d. Subtract the quotient of 50 and 10 from 26.

e. Add the square of 6 to the difference of 9 and 2.

f. Divide 63 by the difference of 32 and 11.

11. Substituting for a Variable Substitute 9 for x to rewrite the algebraic expression $18 - x$ as a numerical expression.

12. Substituting for a Variable Substitute 39 for n to rewrite the algebraic expression $n \div 13$ as a numerical expression.

13. Evaluating Algebraic Expressions Evaluate each expression for the given value of the variable.

a. $4x, x = 9$ **b.** $s^5, s = 2$

c. $n + 27, n = 12$ **d.** $m - 26, m = 30$

e. $z \div 7, z = 56$ **f.** $4^a, a = 3$

14. Evaluating Algebraic Expressions Evaluate each expression for the given value of the variable.

a. $11t, t = 6$ **b.** $u^4, u = 3$

c. $16 + b, b = 21$ **d.** $67 - d, d = 25$

e. $48 \div x, x = 12$ **f.** $7^y, y = 2$

15. Writing Algebraic Expressions Write an algebraic expression that represents each quantity.

a. The amount earned for working h hours at \$22 per hour

b. Your checking account balance when w dollars is withdrawn from the previous balance of \$160

c. The number of students in each study group when n students are divided into 8 groups of equal size

d. The area of a square in terms of its side length s

16. Writing Algebraic Expressions Write an algebraic expression that represents each quantity.

a. The total cost of x shirts at \$12 each

b. Your new body weight after losing 12 pounds of your previous w-pound body weight

c. The number of plastic beads each of n children receives when they divide 120 plastic beads evenly

d. The volume of a cube in terms of its edge length s

17. Evaluating Algebraic Expressions Evaluate each expression for the given value of the variable.

a. $6x + 3, x = 4$ **b.** $6(x + 3), x = 4$

c. $16 \div 2 + c, c = 6$ **d.** $16 \div (2 + c), c = 6$

e. $8 + a^2, a = 4$ **f.** $(8 + a)^2, a = 4$

g. $4b \div 3 - 7, b = 15$

h. $4(b \div 3) - 7, b = 15$

18. Evaluating Algebraic Expressions Evaluate each expression for the given value of the variable.

a. $7y - 2, y = 6$ **b.** $7(y - 2), y = 6$

c. $24 \div p - 4, p = 6$ **d.** $24 \div (p - 4), p = 6$

e. $14 - n^2, n = 3$ **f.** $(14 - n)^2, n = 3$

g. $5m \div 6 + 5, m = 12$

h. $5(m \div 6) + 5, m = 12$

19. Using a Table Use the table to solve $13n = 208$.

n	0	4	8	12	16	20
$13n$						

a. Complete the table by evaluating the expression $13n$ for the given values of n.

b. Use the table to determine the solution of $13n = 208$.

c. Check your solution by using algebra to solve the equation.

20. Using a Table Use the table to solve $7(n + 11) = 161$.

n	0	3	6	9	12	15	18
$7(n + 11)$							

a. Complete the table by evaluating the expression $7(n + 11)$ for the given values of n.

b. Use the table to determine the solution of $7(n + 11) = 161$.

c. Check your solution by using algebra to solve the equation.

21. Using Guess, Check, and Revise Use *Guess, Check, and Revise* to find the value of n for which $8n = 136$. Then use a spreadsheet to check your result.

22. Using Guess, Check, and Revise Use *Guess, Check, and Revise* to find the value of n for which

$$\frac{621}{n} = 23.$$

Then use a spreadsheet to check your result.

23. **Using Order of Operations** Match each expression with its value.

Expression	Value
a. $(8 + 3) \times (6 + 4)$	**i.** 70
b. $(8 + 3) \times 6 + 4$	**ii.** 30
c. $8 + 3 \times 6 + 4$	**iii.** 38
d. $8 + 3 \times (6 + 4)$	**iv.** 110

24. **Using Order of Operations** Match each expression with its value.

Expression	Value
a. $2 + 3^2 \times 4 - 1$	**i.** 29
b. $(2 + 3)^2 \times 4 - 1$	**ii.** 37
c. $2 + 3^2 \times (4 - 1)$	**iii.** 99
d. $(2 + 3)^2 \times (4 - 1)$	**iv.** 75

25. **Grading Student Work** On a diagnostic test, one of your students does the following work. Explain what the student did wrong. Which topics would you encourage the student to review?

$12(8 - 6) = 12 \times 8 - 6$
$= 96 - 6$
$= 90$

26. **Grading Student Work** On a diagnostic test, one of your students does the following work. Explain what the student did wrong. Which topics would you encourage the student to review?

$8^2 - (2 + 5) = 64 - 2 + 5$
$= 62 + 5$
$= 67$

27. **Using Parentheses** Evaluate each expression. Then insert parentheses into the expression so that it has a value of 60.

a. $10 + 5 \times 4$ **b.** $83 - 29 - 6$

c. $46 + 24 \div 2 + 25$ **d.** $11 + 4 \times 5 - 1$

e. $14 + 7 \times 3 - 5 - 2$

f. $2 + 4^2 + 6 + 2 \times 3$

28. **Using Parentheses** Evaluate each expression. Then insert parentheses into the expression so that it has a value of 84.

a. $2 + 4 \times 14$ **b.** $97 - 18 - 5$

c. $98 - 14 \div 2 + 42$ **d.** $5 + 3 \times 8 + 3 - 4$

e. $65 + 4 + 76 \div 4 - 1$

f. $5 + 3^2 + 3 + 2 \times 4$

29. **Writing** Describe the difference in evaluating the expression $(3x)^2$ and the expression $3x^2$ when $x = 4$.

30. **Writing** Explain how the mnemonic phrase "Please Excuse My Dear Aunt Sally" can help you to remember the order of operations.

31. **My Dear Aunt Sally** Arrange the numbers 1, 3, and 5 so that the equation is true.

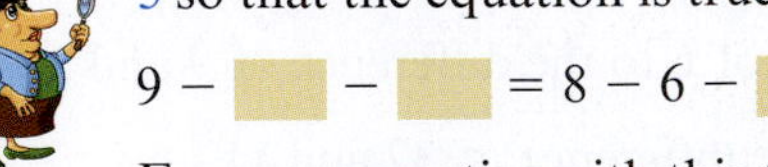

$9 - \square - \square = 8 - 6 - \square$

For more practice with this type of reasoning, go to *AuntSally.com.*

32. **My Dear Aunt Sally** Arrange the numbers 3, 4, and 7 so that the equation is true.

$8 + \square - \square = 9 - \square + 5$

For more practice with this type of reasoning, go to *AuntSally.com.*

33. **My Dear Aunt Sally** Arrange the numbers 1, 2, 4, and 6 so that the equation is true.

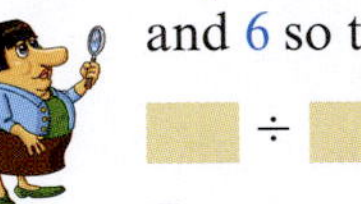

$\square \div \square + \square = 5 \times \square + 2$

For more practice with this type of reasoning, go to *AuntSally.com.*

34. **My Dear Aunt Sally** Arrange the numbers 0, 0, 9, and 9 so that the equation is true. (*Note*: The symbol ^ means to raise to an exponent. So, a ^ $b = a^b$.)

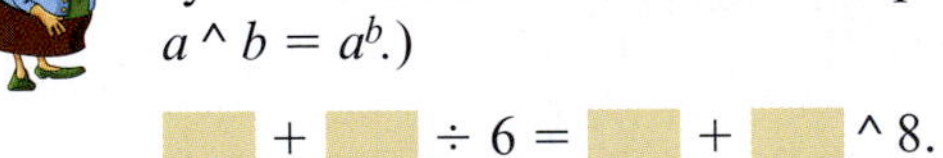

$\square + \square \div 6 = \square + \square$ ^ 8.

For more practice with this type of reasoning, go to *AuntSally.com.*

35. **Comparing Problem-Solving Approaches** A class sells candles as a fundraiser. The class earns a profit of \$9 for each candle sold. Use each method to determine the number of candles the class must sell to earn a profit of \$720.

a. Use a table.

b. Use *Guess, Check, and Revise.*

c. Use algebra.

d. Use a spreadsheet.

36. **Comparing Problem-Solving Approaches** An estimate for painting a house shows a total paint cost of \$336 at \$28 per gallon. Use each method to determine the estimated number of gallons of paint.

a. Use a table.

b. Use *Guess, Check, and Revise.*

c. Use algebra.

d. Use a spreadsheet.

37. Writing an Expression Think of a number. Multiply it by 6. Add 6. Divide the total by 6. Subtract 1. How is the result related to the original number? Write a numerical expression that uses the order of operations to represent the operations you performed. Evaluate the expression to show that it gives the same answer.

38. Writing an Expression Think of a number. Multiply it by 7. Add 21. Divide the total by 7. Subtract 3. How is the result related to the original number? Write a numerical expression that uses the order of operations to represent the operations you performed. Evaluate the expression to show that it gives the same answer.

39. Order of Operations: In Your Classroom Describe the process used in each solution of $4(8 + 2)$. Determine whether the answer is correct. How might you use the results to discuss with your students the relationship between using the Distributive Property and using the order of operations?

a. $4(8 + 2) = 32 + 2 = 34$

b. $4(8 + 2) = 4 \times 10 = 40$

c. $4(8 + 2) = 4 \times 8 + 4 \times 2 = 32 + 8 = 40$

40. Order of Operations: In Your Classroom One of your students rewrites $(x + y)^n$ as $x^n + y^n$. Design an activity to help your students see that the two expressions are not generally equal.

41. Finding Area Write an expression for the area of the shaded rectangle. Evaluate the expression when $x = 4$, first by using the order of operations, and then by using the Distributive Property. Compare the results.

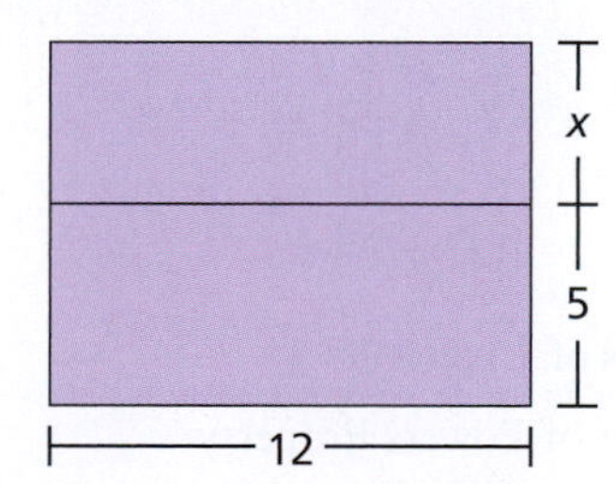

42. Finding Area Write an expression for the area of the shaded rectangle. Evaluate the expression when $a = 6$, first by using the order of operations, and then by using the Distributive Property. Compare the results.

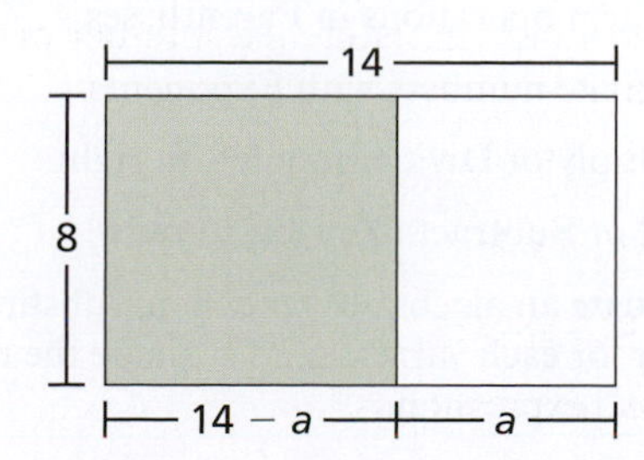

43. Finding Area The shaded region is composed of four smaller regions.

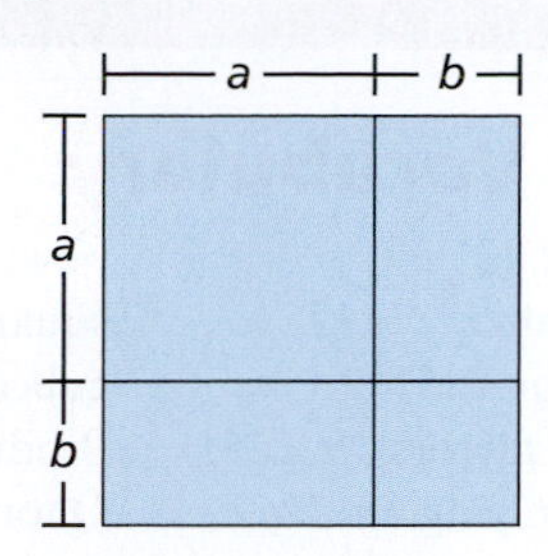

a. Write an expression for the area of the shaded region as the area of a square of side length $a + b$.

b. Write an expression for the area of the shaded region as the sum of the areas of the four smaller regions.

c. Evaluate each expression for the area of the shaded region when $a = 6$ and $b = 3$. Compare the results.

44. Finding Area The shaded region is composed of four rectangular regions.

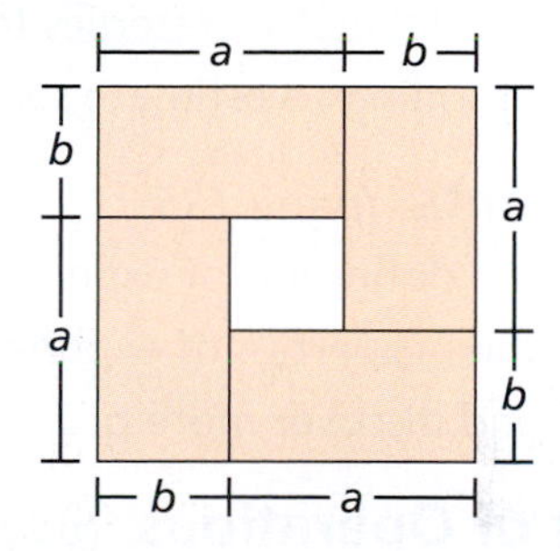

a. Write an expression for the area of the shaded region as the difference of the areas of two squares with the side lengths $a + b$ and $a - b$.

b. Write an expression for the area of the shaded region as the sum of the areas of the four rectangular regions.

c. Evaluate each expression for the area of the shaded region when $a = 8$ and $b = 4$. Compare the results.

45. Problem Solving Use the formula below to make a table that shows the annual operating costs C (in dollars) of a commercial van driven m miles annually, for various values of m. Then describe the pattern.

$$C = 113m \div 200 + 2500$$

46. Problem Solving You travel along a zip line. Use the formula below to make a table that compares your height h (in feet) above the ground with the distance d (in feet) you travel. Then describe the pattern.

$$h = 30 - 2d \div 25$$

Chapter Summary

Chapter Vocabulary

mental math *(p. 125)*
compatible numbers *(p. 125)*
compensation *(p. 125)*
rearranging the factors *(p. 126)*
front-end multiplying *(p. 126)*
thinking money *(p. 126)*
rewrite the dividend *(p. 127)*
front-end estimation *(p. 129)*
range estimation *(p. 129)*
exponential form *(p. 135)*
base *(p. 135)*
exponent *(p. 135)*
power *(p. 135)*
squared *(p. 135)*
cubed *(p. 135)*
Product of Powers Property *(p. 137)*
Quotient of Powers Property *(p. 137)*
Power of a Product Property *(p. 138)*
Power of a Power Property *(p. 139)*
numerical expression *(p. 145)*
order of operations *(p. 145)*
variables *(p. 146)*
algebraic expression *(p. 146)*

iStockphoto.com/monkeybusinessimages

Chapter Learning Objectives — Review Exercises

4.1 Mental Math and Estimation *(page 125)* — 1–56

- Use mental math strategies to add and subtract whole numbers.
- Use mental math strategies to multiply and divide whole numbers.
- Use rounding to estimate sums, differences, products, and quotients.

4.2 Exponents *(page 135)* — 57–100

- Use the definition of exponential form to write and evaluate powers.
- Use the properties of exponents involving the same base.
- Use and discover more properties of exponents.

4.3 Order of Operations *(page 145)* — 101–144

- Use the order of operations to evaluate numerical expressions.
- Evaluate algebraic expressions.
- Use the order of operations to evaluate algebraic expressions.

Important Concepts and Formulas

Mental Math Strategies

- Compensation (adding or subtracting)
- Rearranging the factors (multiplying)
- Front-end multiplying (multiplying)
- Thinking money (multiplying)
- Rewrite the dividend (dividing)

Estimation Techniques

- Estimation by rounding
- Front-end estimation
- Front-end estimation with adjustment
- Range estimation

Properties of Exponents

- Product of Powers Property
- Quotient of Powers Property
- Power of a Product Property
- Power of a Power Property

Order of Operations

1. Perform operations in Parentheses.
2. Evaluate numbers with Exponents.
3. Multiply or Divide from left to right.
4. Add or Subtract from left to right.

To **evaluate** an algebraic expression, substitute a number for each variable and evaluate the resulting numerical expression.

Review Exercises

Free step-by-step solutions to odd-numbered exercises at *MathematicalPractices.com.*

4.1 Mental Math and Estimation

Identifying Compatible Numbers Decide whether the sum or difference involves compatible numbers.

1. 37 + 59 **2.** 58 + 22

3. 42 − 18 **4.** 83 − 50

Using Mental Math Use mental math to find the sum or difference. Describe your reasoning.

5. 11 + 31 + 71 **6.** 19 + 49 + 89

7. 66 + 14 **8.** 91 − 50

9. 600 − 360 **10.** 410 − 270

Rearranging Factors Rewrite the product by rearranging the factors to create compatible numbers. Then find the product.

11. 2 • 3 • 7 • 5 **12.** 4 • 6 • 25 • 7

13. 4 • 13 • 250 **14.** 50 • 9 • 4 • 2

15. 7 • 20 • 8 • 5 **16.** 5 • 16 • 200 • 3

Front-End Multiplying Use front-end multiplying to find the product.

17. 47 • 6 **18.** 72 • 5

19. 430 • 6 **20.** 312 • 7

Thinking Money Think of one of the factors as money to find the product. Explain your reasoning.

21. 5 • 38 **22.** 44 • 25

23. 76 • 50 **24.** 43 • 5

Using Mental Math Use mental math to find the product. Explain your strategy.

25. 47 • 7 **26.** 4 • 2 • 8 • 5

27. 54 • 50 **28.** 59 • 9

Rewriting a Dividend Rewrite the dividend using compatible numbers. Then find the quotient.

29. $9\overline{)369}$ **30.** $7\overline{)2849}$

31. $35\overline{)3570}$ **32.** $14\overline{)2842}$

Estimating a Sum Use rounding to estimate the sum.

33.
$$\begin{array}{r} 518 \\ +\ 394 \\ \hline \end{array}$$

34.
$$\begin{array}{r} 2033 \\ +\ 2975 \\ \hline \end{array}$$

35.
$$\begin{array}{r} 506 \\ 621 \\ 213 \\ +\ 192 \\ \hline \end{array}$$

36.
$$\begin{array}{r} 272 \\ 334 \\ 109 \\ 631 \\ +\ 268 \\ \hline \end{array}$$

Estimating a Difference Use rounding to estimate the difference.

37. 698 − 403 **38.** 978 − 372

39. 267 − 189 **40.** 558 − 216

Estimating a Product Use rounding to estimate the product.

41. 19 • 207 **42.** 51 • 191

43. 82 • 301 **44.** 74 • 668

Estimating a Quotient Use rounding to estimate the quotient.

45. 2452 ÷ 623 **46.** 27,486 ÷ 3208

47. 49,221 ÷ 720 **48.** 242,517 ÷ 608

Estimating a Sum Estimate the total cost of the items with the given prices using (a) rounding, (b) front-end estimation, (c) range estimation, and (d) front-end estimation with adjustment.

49. $132 $388 $472 $641

50. $418 $343 $229 $382

51. $749 $238 $519 $488

52. $2240 $9413 $2679 $5898

53. **Problem Solving** You can type at a rate of 43 words per minute. Use front-end estimation to approximate how many words you can type in 106 minutes.

54. **Problem Solving** A factory has a contract to produce 150,000 parts in one year. Use front-end estimation to determine whether the factory can produce the parts at a rate of 81 parts per hour, 40 hours per week, for 52 weeks. Explain your reasoning.

Problem Solving Use estimation or mental math to answer the question. Explain your reasoning.

55. Your checking account balance is $331. You deposit checks of $1211 and $443. Is there enough money to cover a check for $1900?

56. You tie ropes of lengths 29 feet, 68 feet, 47 feet and 49 feet together. Will you have at least 200 feet of rope?

4.2 Exponents

Writing in Exponential Form Write the expression in exponential form.

57. $12 \cdot 12 \cdot 12$

58. $113 \cdot 113$

59. $6 \cdot 6 \cdot 6 \cdot 6 \cdot 6$

60. $b \cdot b \cdot b \cdot b \cdot b \cdot b$

Writing in Exponential Form Use exponents to write the expression in simpler form.

61. $2 \cdot 2 \cdot 2 \cdot x \cdot x$

62. $9 \cdot 9 \cdot y \cdot y \cdot 5 \cdot y \cdot 5 \cdot y \cdot 5$

Writing a Power as a Product Write the expression as a product.

63. 14^3 **64.** 5^7

65. x^5 **66.** a^8

67. Finding the Area of a Square Write the area of the square in exponential form. Then find the area.

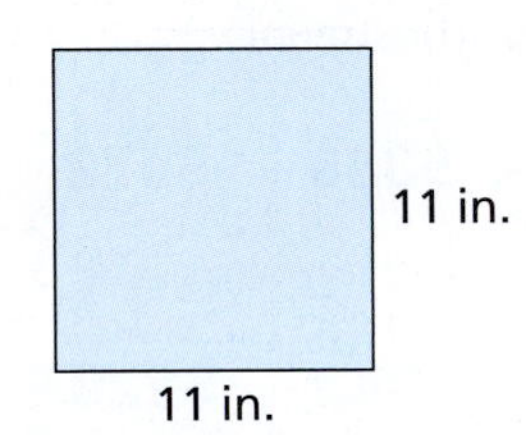

68. Finding the Volume of a Cube Write the volume of the cube in exponential form. Then find the volume.

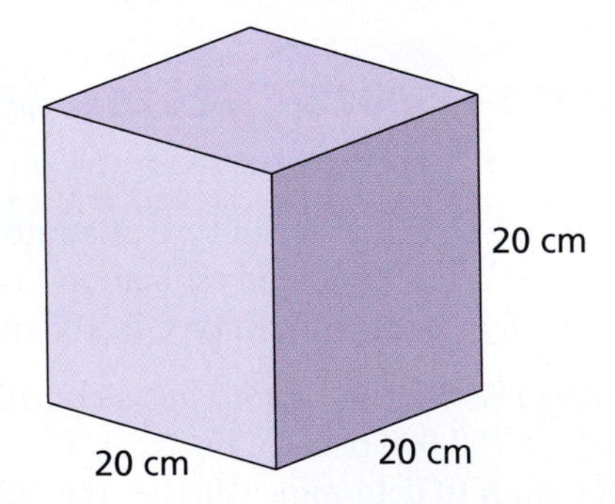

Evaluating a Power Evaluate the power.

69. 3^5

70. 6^4

71. 8^0

72. a^0

Using Exponents to Write a Pattern Write the sequence of numbers using powers.

73. 1, 7, 49, 343, 2401, . . .

74. 1, 6, 36, 216, 1296, . . .

Using Properties of Exponents Write the expression as a single power. Identify the property that you used.

75. $11^5 \cdot 11^4$ **76.** $7^5 \cdot 4^5$

77. $5^8 \cdot 5^4$ **78.** $3 \cdot 3^3$

79. $3^3 \cdot 4^3 \cdot 5^3$ **80.** $a^2 \cdot a^3 \cdot a^6$

81. $\frac{5^4}{5}$ **82.** $\frac{6^6}{6^2}$

83. $\frac{3^{12}}{3^3}$ **84.** $\frac{z^7}{z^6}$

Using a Property of Exponents Write the expression as a single power. Then evaluate the power.

85. $(10^3)^2$ **86.** $(10^5)^2$

87. $(2^2)^4$ **88.** $(5^2)^0$

Using a Property of Exponents Write the expression as a product of two or more powers.

89. $(3a)^2$ **90.** $(5ab)^3$

91. $(7a^2b)^3$ **92.** $(11a^3b^5)^2$

Using Properties of Exponents Find the value of x that makes the equation true.

93. $3^4 \cdot 3^x = 3^8$ **94.** $3^x \cdot 7^x = 21^5$

95. $(7^x)^4 = 7^8$

96. $x^4 = 3^8$

97. $\frac{5^x}{5^3} = 5^4$

98. $\frac{2^x}{2^2} = 8$

99. Comparing Areas of Squares Determine how many times greater the area of the large square is than the area of the small square. Explain your reasoning.

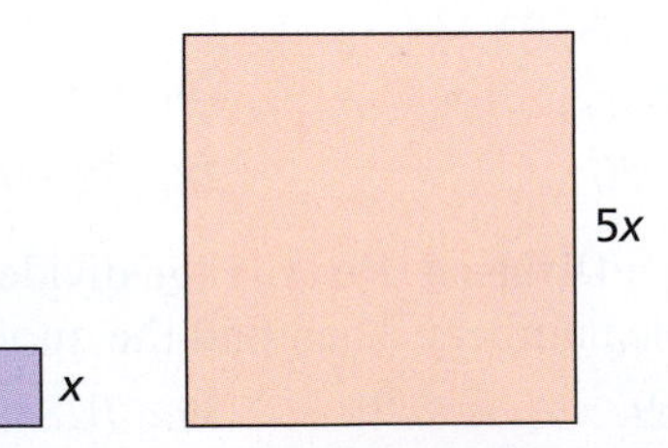

100. Comparing Volumes of Cubes Determine how many times greater the volume of the large cube is than the volume of the small cube. Explain your reasoning.

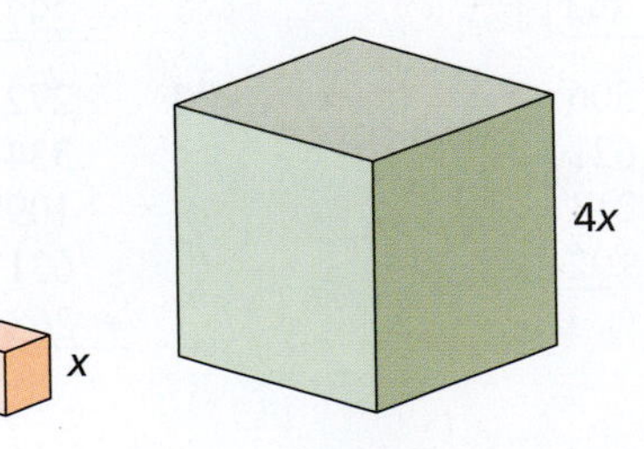

4.3 Order of Operations

Using Order of Operations List the operations in the order they are performed to evaluate each expression. Explain your reasoning. Then evaluate the expression.

101. $5^2 - 10 + 4$ **102.** $(7 - 2) \times 4 - 3$

103. $4 \times 3^2 + 4$ **104.** $16 - 3 \times 4 + 8$

Using Order of Operations Use the order of operations to evaluate the expression.

105. $21 - 4(6 - 4)$ **106.** $38 - 5(36 \div 12)$

107. $66 - 24 - 4 \times 3$ **108.** $5 \times 7 - 4^2$

109. $3 \times 4^2 - 20$

110. $2(19 - 15)^2 - 3^2 - 2(72 \div 6^2)$

Writing a Numerical Expression Write a numerical expression that uses the order of operations to represent the problem.

111. Multiply the difference of 6 and 2 by 3.

112. Subtract the product of 5 and 6 from 35.

113. Add the difference of 18 and 14 to 11.

114. Subtract the quotient of 24 and 6 from 29.

115. Add the square of 5 to the sum of 18 and 7.

116. Divide 80 by the difference of 34 and 24.

Evaluating an Algebraic Expression Evaluate the expression for the given value of the variable.

117. $7s$, $s = 6$ **118.** a^4, $a = 3$

119. $x + 13$, $x = 17$ **120.** $99 - n$, $n = 66$

121. $m \div 6$, $m = 48$ **122.** 9^y, $y = 2$

Writing an Algebraic Expression Write an algebraic expression that represents the quantity.

123. The amount a supermodel earns for working h hours at \$10,000 per hour

124. The amount of money you have left after spending \$28 of the x dollars you had originally

125. The number of baseballs each team receives when x baseballs are divided equally among 11 teams

126. The fourth power of a number n

Evaluating an Algebraic Expression Evaluate the expression for the given value of the variable.

127. $4y - 3$, $y = 10$ **128.** $4(y - 3)$, $y = 10$

129. $20 \div 4 + z$, $z = 6$ **130.** $20 \div (4 + z)$, $z = 6$

131. $(3 + s)^2$, $s = 7$

132. $3 + s^2$, $s = 7$

133. $4x \div 7 + 11$, $x = 14$

134. $4(x \div 7) + 11$, $x = 14$

Using Parentheses Evaluate the expression. Then insert parentheses into the expression so that it has a value of 48.

135. $4 \times 16 - 4$ **136.** $60 - 10 + 2$

137. $72 \div 8 + 28 + 46$ **138.** $2 + 6 \times 5 + 8$

139. $70 - 4 + 44 \div 2 + 2$ **140.** $8 - 2^2 + 2^3 + 2^2$

141. **Finding Area** Write an expression for the area of the shaded rectangle. Evaluate the expression when $x = 5$, first by using the order of operations, and then by using the Distributive Property. Compare the results.

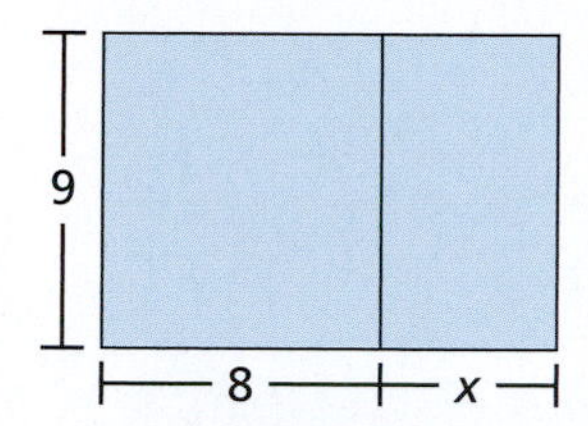

142. **Finding Area** The shaded region is composed of three smaller regions.

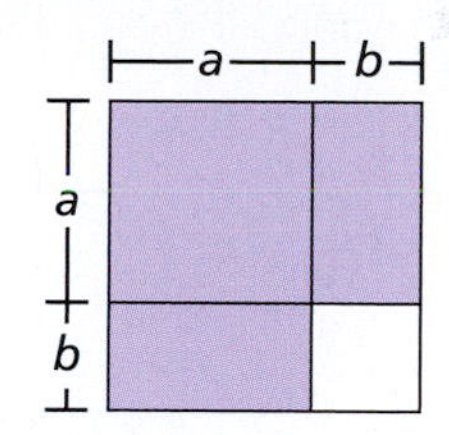

a. Write an expression for the area of the shaded region as the difference of the areas of two square regions.

b. Write an expression for the area of the shaded region as the sum of the areas of the three smaller regions.

c. Evaluate each expression for the area of the shaded region when $a = 5$ and $b = 2$. Compare the results.

143. **Comparing Problem-Solving Approaches** An estimate for painting a retail store shows a total paint cost of \$792 at \$24 per gallon. Use each method to determine the estimated number of gallons of paint.

a. Use a table.

b. Use *Guess, Check, and Revise.*

c. Use algebra.

d. Use a spreadsheet.

144. **Problem Solving** A hot air balloon rises from a height of 25 feet. Use the formula below to make a table that compares the balloon's height h (in feet) to the time s (in seconds) since the balloon started to rise. Then describe the pattern.

$h = 20s \div 3 + 25$

Chapter Test

The questions on this practice test are modeled after questions on state certification exams for K–8 teaching.

Test Directions: Each of the questions is followed by five choices. Choose the *best* response to each question.

1. Use front-end estimation to estimate the sum.

$$\begin{array}{r} 324 \\ 231 \\ 610 \\ +157 \\ \hline \end{array}$$

A. 900 **B.** 1,000 **C.** 1,100 **D.** 1,200 **E.** 1,500

2. Round the following number to the nearest tens place: 293,326.

A. 290,000 **B.** 293,000 **C.** 293,300 **D.** 293,400 **E.** 300,000

3. What would be a reasonable estimate for the answer to the following problem?

$$480 \cdot 12$$

A. 48 **B.** 480 **C.** 4,800 **D.** 48,000 **E.** 480,000

4. Rewrite the expression as a product.

$$2x^2y^3$$

A. $2 \cdot 2 \cdot x \cdot x \cdot y \cdot y \cdot y$

B. $2 \cdot x \cdot x \cdot y \cdot y \cdot y$

C. $2 + x + x + y + y + y$

D. $2 + x^2 + y^2$

E. None of the above

5. Simplify the following expression.

$$\frac{x^3x^6}{x}$$

A. x^6 **B.** x^7 **C.** x^8 **D.** x^{10} **E.** x^{13}

6. Use the order of operations to evaluate the expression.

$$3 + 7(4 \div 2)^2 - 4$$

A. 9 **B.** 10 **C.** 13 **D.** 27 **E.** 40

7. Evaluate the expression when $x = 1$.

$$(4 - x)^2 + 5$$

A. 8 **B.** 14 **C.** 21 **D.** 30 **E.** 35

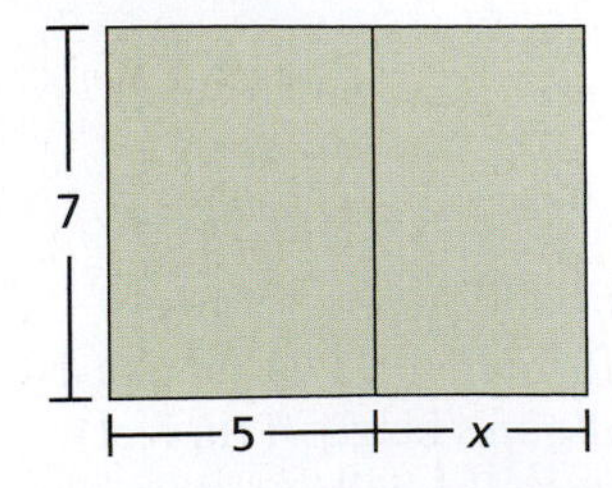

8. Evaluate the expression $7(x + 5)$ for the area of the shaded rectangle at the left when $x = 4$.

A. 7 **B.** 28 **C.** 33 **D.** 35 **E.** 63

5 Number Theory

iStockphoto.com/digitalskillet

Activity: Describing Divisibility Rules

Materials:

- Pencil

Learning Objective:

Decide whether given whole numbers are divisible by 2, 3, 4, 5, and 10. Then describe a divisibility test for each of the numbers 2, 3, 4, 5, and 10.

Name ______________________________

1. **Divisibility by 2** Circle each whole number that is evenly divisible by 2.

6 13 14 18 25 32 50 77

Write a rule that tells you when a whole number is evenly divisible by 2.

____

2. **Divisibility by 5** Circle each whole number that is evenly divisible by 5.

15 18 30 36 45 74 75 90

Write a rule that tells you when a whole number is evenly divisible by 5.

____

3. **Divisibility by 10** Circle each whole number that is evenly divisible by 10.

15 18 30 36 45 74 75 90

Write a rule that tells you when a whole number is evenly divisible by 10.

____

4. **Divisibility by 4** Circle each whole number that is evenly divisible by 4.

13 24 30 76 148 220 504 618

Write a rule that tells you when a whole number is evenly divisible by 4.

____

5. **Divisibility by 3** Circle each whole number that is evenly divisible by 3.

6 18 26 27 30 44 58 72

Write a rule that tells you when a whole number is evenly divisible by 3.

____

5.1 Divisibility Tests

Standards

Grades 3–5
Operations and Algebraic Thinking
Students should gain familiarity with factors and multiples, and generate and analyze patterns.

Grades 6–8
The Number System
Students should compute fluently with multi-digit numbers and find common factors and multiples.

- Use divisibility tests to decide whether a whole number is divisible by 2, 5, or 10.
- Use divisibility tests to decide whether a whole number is divisible by 4 or 8.
- Use divisibility tests to decide whether a whole number is divisible by 3, 6, or 9.

Divisibility Tests for 2, 5, and 10

Definition of "Divides"

Let a and b be whole numbers such that $a \neq 0$. a **divides** b if and only if there is a whole number n such that $b = an$. The notation $a \mid b$ is read as "a divides b." The notation $a \nmid b$ is read as "a does *not* divide b."

Example 3 divides 12, written as $3 \mid 12$, because $12 = 3 \cdot 4$.

The statement "a divides b" can also be written as a is a **factor** of b, b is a **multiple** of a, b is **divisible** by a, and b is **evenly divisible** by a.

The following divisibility tests are well known.

Classroom Tip
By the time students are in sixth grade, they usually have an understanding of the divisibility rules.

Divisibility Tests for 2, 5, and 10

Test for 2 A whole number is divisible by 2 if and only if its ones digit is an even number (0, 2, 4, 6, or 8).

Test for 5 A whole number is divisible by 5 if and only if its ones digit is 0 or 5.

Test for 10 A whole number is divisible by 10 if and only if its ones digit is 0.

EXAMPLE 1 **Comparing Two Definitions of Kilobyte**

The term *kilobyte* has two commonly used meanings.

a. A kilobyte is 1024 bytes. **b.** A kilobyte is 1000 bytes.

Tell whether each whole number is divisible by 2, 5, or 10.

SOLUTION

a. The number 1024 is divisible by 2 because its ones digit is 4.

$$1024 = 2 \cdot 512$$

1024 is not divisible by 5 or 10.

b. The number 1000 is divisible by 2, 5, and 10 because its ones digit is 0.

$$1000 = 2 \cdot 500 \qquad 1000 = 5 \cdot 200 \qquad 1000 = 10 \cdot 100$$

Although 1 kilobyte = 1024 bytes is the traditional definition, the definition 1 kilobyte = 1000 bytes is becoming more common because 1000 has more divisors and the prefix "kilo" actually means 1000.

Divisibility Tests for 4 and 8

Properties of Divisibility

Let a, b, m, and n be whole numbers such that $n \neq 0$.

1. If $n \mid a$ and $n \mid b$, then $n \mid (a + b)$. Sum Property
2. If $n \mid a$ and $n \mid b$, then $n \mid (a - b)$ for $a \geq b$. Difference Property
3. If $n \mid a$, then $n \mid am$. Product Property

The Properties of Divisibility can be used to develop divisibility tests for 4 and 8. For instance, consider the whole number 348. You can write this number as $348 = 3 \cdot 100 + 48$. Because 100 is divisible by 4, it follows that $3 \cdot 100$ is divisible by 4 by the Product Property of Divisibility. So, $3 \cdot 100 + 48$ is divisible by 4 if and only if 48 is divisible by 4. Because 48 is also divisible by 4, 348 is divisible by 4. A similar reasoning follows for divisibility by 8.

Divisibility Tests for 4 and 8

Test for 4 A whole number is divisible by 4 if and only if the number represented by its last two digits is divisible by 4.

Test for 8 A whole number is divisible by 8 if and only if the number represented by its last three digits is divisible by 8.

EXAMPLE 2 Using Divisibility Tests for 4 and 8

Tell whether each whole number is divisible by 4 or 8.

a. 1450 **b.** 2728

SOLUTION

a. 14(50) — $50 \div 4 = 12\text{R}2$

Because 50 is *not* divisible by 4, 1450 is *not* divisible by 4.

1(450) — $450 \div 8 = 56\text{R}2$

Because 450 is *not* divisible by 8, 1450 is *not* divisible by 8.

b. 27(28) — $28 \div 4 = 7$

Because 28 is divisible by 4, 2728 is divisible by 4.

2(728) — $728 \div 8 = 91$

Because 728 is divisible by 8, 2728 is divisible by 8.

Classroom Tip

Encourage students to use the Properties of Divisibility to help them divide. For instance, in Example 2(b), you can think of 728 as 720 + 8.

$$728 = 720 + 8$$
$$= 8(90) + 8$$

Because 8 divides both 720 and 8, it divides 728.

In Example 2(a), notice that when a number is not divisible by 4, it follows that the number is not divisible by 8. To see why this is true, consider the following.

$4n$ Divisible by 4 $4(2m)$ Divisible by 8

From these two forms, you can see that a number that is not divisible by 4 cannot have 4 as a factor and is therefore not divisible by 8.

EXAMPLE 3 Using a Divisibility Test in Real Life

A school has 376 students. Is it possible to divide the students into groups of 4 so that each student is in exactly one group?

SOLUTION

1. **Understand the Problem** You need to determine whether it is possible to divide the number of students in the school into groups of 4 so that each student is in exactly one group.
2. **Make a Plan** Essentially, you need to determine whether 376 is divisible by 4. Apply the divisibility test for 4.
3. **Solve the Problem**

   376 —— $76 \div 4 = 19$

   Because 76 is divisible by 4, 376 is divisible by 4. So, it is possible to divide the students into groups of 4 with each student in exactly one group.
4. **Look Back** Use a calculator to check that 4 divides evenly into 376.

   $376 \div 4 = 94$ ✓

EXAMPLE 4 Using a Divisibility Test in Real Life

New lights are to be installed 8 meters apart along the roadways of three bridges. The table shows the lengths of the bridges. Which of the bridges will have lights at the beginning and at the end of its roadway?

Bridge	Length (meters)
Bay Bridge	1800
East Bridge	2088
West Bridge	1012

SOLUTION

1. **Understand the Problem** You know the length of each bridge. You need to determine whether each bridge can have a light at the beginning and end of its roadway, and have 8 meters between lights.
2. **Make a Plan** Essentially, you need to determine whether the length of each bridge is divisible by 8. Apply the divisibility test for 8.
3. **Solve the Problem**

   Bay Bridge 1800 —— $800 \div 8 = 100$

   Because 800 is divisible by 8, 1800 is divisible by 8.

   East Bridge 2088 —— $88 \div 8 = 11$

   Because 88 is divisible by 8, 2088 is divisible by 8.

   West Bridge 1012 —— $12 \div 8 = 1R4$

   Because 12 is *not* divisible by 8, 1012 is *not* divisible by 8. So, the Bay Bridge and the East Bridge will have lights at the beginning and at the end of their roadways and the West Bridge will not.
4. **Look Back** Use a calculator to check whether 8 divides evenly into each length.

   $1800 \div 8 = 225$ ✓ 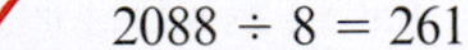 $2088 \div 8 = 261$ ✓ $1012 \div 8 = 126.5$ ✗

Mathematical Practices

Reason Abstractly and Quantitatively

MP2 Mathematically proficient students make sense of quantities and their relationships in problem situations, and bring the abilities to *decontextualize* and *contextualize* while solving a problem.

Divisibility Tests for 3, 6, and 9

The Properties of Divisibility can be used to develop divisibility tests for 3 and 9. For instance, consider the whole number 741. Expand this number as follows.

$$\begin{aligned}741 &= 7(100) + 4(10) + 1\\ &= 7(99 + 1) + 4(9 + 1) + 1\\ &= 7(99) + 7(1) + 4(9) + 4(1) + 1\\ &= \underbrace{7(99) + 4(9)}_{\text{This number is divisible by both 3 and 9.}} + (7 + 4 + 1)\end{aligned}$$

So, 741 is divisible by 3 and 9 if and only if the sum of the digits of 741, $7 + 4 + 1$, is divisible by 3 and 9.

Divisibility Tests for 3 and 9

Test for 3 A whole number is divisible by 3 if and only if the sum of its digits is divisible by 3.

Test for 9 A whole number is divisible by 9 if and only if the sum of its digits is divisible by 9.

EXAMPLE 5 Using Divisibility Tests for 3 and 9

Tell whether the whole number is divisible by 3 or 9.

a. 1375 **b.** 2094 **c.** 3744

SOLUTION

a. The sum of the digits of 1375 is

$1 + 3 + 7 + 5 = 16.$ $16 \div 3 = 5R1$ $16 \div 9 = 1R7$

Because 16 is *not* divisible by 3 or 9, 1375 is *not* divisible by 3 or 9.

b. The sum of the digits of 2094 is

$2 + 0 + 9 + 4 = 15.$ $15 \div 3 = 5$ ✓ $15 \div 9 = 1R6$

Because 15 is divisible by 3, 2094 is divisible by 3. Because 15 is *not* divisible by 9, 2094 is *not* divisible by 9.

c. The sum of the digits of 3744 is

$3 + 7 + 4 + 4 = 18.$ $18 \div 3 = 6$ ✓ $18 \div 9 = 2$ ✓

Because 18 is divisible by both 3 and 9, 3744 is divisible by both 3 and 9.

In Example 5(a), notice that when a number is not divisible by 3, it follows that the number is not divisible by 9. To see why this is true, consider the following.

$3n$ Divisible by 3 $3(3m)$ Divisible by 9

From these two forms, you can see that a number that is not divisible by 3 cannot have 3 as a factor and is therefore not divisible by 9.

Divisibility Test for 6

A whole number is divisible by 6 if and only if it is divisible by both 2 and 3.

EXAMPLE 6 Using the Divisibility Test for 6

Tell whether the whole number is divisible by 6.

a. 2645 **b.** 1750 **c.** 4728

SOLUTION

a. Because 2645 is not an even number, it is not divisible by 2. So, 2645 cannot be divisible by 6.

b. Because 1750 has a ones digit of 0, it is divisible by 2. The sum of the digits of 1750 is

$1 + 7 + 5 + 0 = 13.$ $13 \div 3 = 4R1$

Because 13 is *not* divisible by 3, 1750 is *not* divisible by 3. So, 1750 is *not* divisible by 6.

c. Because 4728 has a ones digit of 8, it is divisible by 2. The sum of the digits of 4728 is

$4 + 7 + 2 + 8 = 21.$ $21 \div 3 = 7$

Because 21 is divisible by 3, 4728 is divisible by 3. 4728 is divisible by both 2 and 3, so 4728 is divisible by 6.

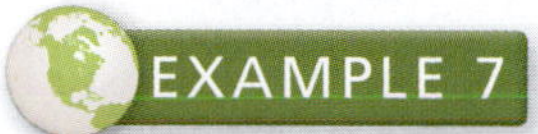

EXAMPLE 7 Using a Divisibility Test in Real Life

A biologist is studying the muscle structure of ant legs. A collection of ants has been scanned by a computer. The scan counts 946 ant legs. Each ant in the collection has 6 legs. Is the scan count correct?

SOLUTION

1. **Understand the Problem** The scan count is 946 ant legs. You need to determine whether this count is correct.
2. **Make a Plan** Essentially, you need to determine whether the scan count is divisible by the number of legs on each ant, 6. Apply the divisibility test for 6.
3. **Solve the Problem** 946 has a ones digit of 6, so, it is divisible by 2. The sum of the digits of 946 is

   $9 + 4 + 6 = 19.$ $19 \div 3 = 6R1$

   Because 19 is *not* divisible by 3, 946 is *not* divisible by 3. So, 946 is not divisible by 6. The total number of ant legs should have been a multiple of 6. So, it follows that the scan count is not correct.
4. **Look Back** Use a calculator to check that 6 does not divide evenly into 946.

   $946 \div 6 = 157R4$ 

Fotocrisis/Shutterstock.com

5.1 Exercises

Free step-by-step solutions to odd-numbered exercises at *MathematicalPractices.com.*

1. **Writing a Solution Key** Write a solution key for the activity on page 160. Describe a rubric for grading a student's work.

2. **Grading the Activity** In the activity on page 160, a student gave the answers below. Each question is worth 10 points. Assign a grade to each answer. For those that are incorrect, why do you think the student erred?

Sample Student Work

3. Divisibility by 10 Circle each whole number that is evenly divisible by 10.

15 18 (30) 6 45 74 75 (90)

Write a rule that tells you when a whole number is evenly divisible by 10.

A whole number is divisible by 10 when its ones digit is 0.

4. Divisibility by 4 Circle each whole number that is evenly divisible by 4.

13 (24) 30 76 148 220 (504) 618

Write a rule that tells you when a whole number is evenly divisible by 4.

A whole number is divisible by 4 when its ones digit is 4.

5. Divisibility by 3 Circle each whole number that is evenly divisible by 3.

(6) (18) 26 (27) (30) 44 58 (72)

Write a rule that tells you when a whole number is evenly divisible by 3.

A whole number is divisible by 3 when the sum of its digits is divisible by 3.

3. **Identifying Factors** Identify all factors of each whole number.
   a. 60
   b. 100

4. **Identifying Factors** Identify all factors of each whole number.
   a. 88 **b.** 95

5. **Identifying Multiples** Identify two multiples of each whole number.
   a. 3 **b.** 5
   c. 6 **d.** 10

6. **Identifying Multiples** Identify two multiples of each whole number.
   a. 2 **b.** 4
   c. 12 **d.** 16

7. **Analyzing Divisibility Statements** Tell whether each statement is *true* or *false*. Explain your reasoning.
   a. 4 is a factor of 20.
   b. $20 \mid 4$
   c. 20 is a multiple of 4.
   d. $4 \nmid 20$
   e. 4 is a divisor of 20.

8. **Analyzing Divisibility Statements** Tell whether each statement is *true* or *false*. Explain your reasoning.
   a. 30 is a factor of 5.
   b. $5 \mid 30$
   c. 5 is a multiple of 30.
   d. $30 \nmid 5$
   e. 5 is a divisor of 30.

9. **Using the Divisibility Test for 2** Tell whether each whole number is divisible by 2. Explain your reasoning.
   a. 6 **b.** 10
   c. 61 **d.** 97
   e. 254 **f.** 782

10. **Using the Divisibility Test for 2** Tell whether each whole number is divisible by 2. Explain your reasoning.
    a. 34 **b.** 78
    c. 19 **d.** 101
    e. 1056 **f.** 1780

11. **Using Divisibility Tests for 5 and 10** Tell whether each whole number is divisible by 5 or 10. Explain your reasoning.
    a. 20 **b.** 35
    c. 116 **d.** 402
    e. 1400 **f.** 1995

12. Using Divisibility Tests for 5 and 10 Tell whether each whole number is divisible by 5 or 10. Explain your reasoning.

a. 40 **b.** 95

c. 345 **d.** 888

e. 2000 **f.** 3005

13. Applying Properties of Divisibility Complete each statement to illustrate the indicated property of divisibility.

a. Sum Property:

2 | 6 and 2 | 12, so 2 | .

b. Difference Property:

5 | 105 and 5 | 65, so 5 | .

c. Product Property:

4 | 32, so 4 | 5 • ____.

14. Applying Properties of Divisibility Complete each statement to illustrate the indicated property of divisibility

a. Sum Property:

2 | 18 and 2 | 4, so 2 | ____.

b. Difference Property:

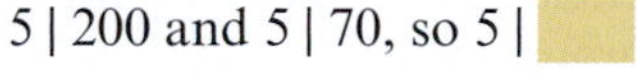
5 | 200 and 5 | 70, so 5 | ____.

c. Product Property:

4 | 40, so 4 | 5 • ____.

15. Using Divisibility Tests for 4 and 8 Tell whether each whole number is divisible by 4 or 8. Explain your reasoning.

a. 1056 **b.** 2460

c. 1519 **d.** 3536

e. 7530 **f.** 8676

16. Using Divisibility Tests for 4 and 8 Tell whether each whole number is divisible by 4 or 8. Explain your reasoning.

a. 1492 **b.** 1152

c. 2538 **d.** 3965

e. 4858 **f.** 9816

17. Using Divisibility Tests for 3 and 9 Tell whether each whole number is divisible by 3 or 9. Explain your reasoning.

a. 243 **b.** 783

c. 1857 **d.** 2058

e. 3253 **f.** 4259

18. Using Divisibility Tests for 3 and 9 Tell whether each whole number is divisible by 3 or 9. Explain your reasoning.

a. 234 **b.** 801

c. 1860 **d.** 2127

e. 3671 **f.** 4492

19. Using the Divisibility Test for 6 Tell whether each whole number is divisible by 6. Explain your reasoning.

a. 474 **b.** 547

c. 864 **d.** 913

e. 2460 **f.** 3289

20. Using the Divisibility Test for 6 Tell whether each whole number is divisible by 6. Explain your reasoning.

a. 468 **b.** 559

c. 858 **d.** 931

e. 2478 **f.** 3704

21. Using Divisibility Tests Complete each blank with the greatest digit that makes the statement true.

a. 2 | 10_

b. 3 | 4_9

c. 8 | 5_68

d. 6 | 71_

22. Using Divisibility Tests Complete each blank with the greatest digit that makes the statement true.

a. 5 | 55_

b. 9 | 610_

c. 4 | 21,5_6

d. 3 | _269

23. True or False? Tell whether each statement is *true* or *false*. Explain your reasoning.

a. When a natural number is divisible by 8, it is also divisible by 4.

b. When a natural number is divisible by 3, it is also divisible by 9.

24. True or False? Tell whether each statement is *true* or *false*. Explain your reasoning.

a. When a natural number is divisible by 6, it is also divisible by 2 and 3.

b. When a natural number is divisible by 2 and 3, it is also divisible by 6.

25. **Reasoning** The whole number 32 divides the whole number m. What other numbers must divide m?

26. **Reasoning** The whole number 24 divides the whole number m. What other numbers must divide m?

27. **Using a Divisibility Test for 7** Use the divisibility test for 7 to determine whether each whole number is divisible by 7.

Test Subtract 2 times the last digit from the remaining digits of the number. Continue until you can determine whether the difference is or is not divisible by 7. For example, 17,948 is divisible by 7.

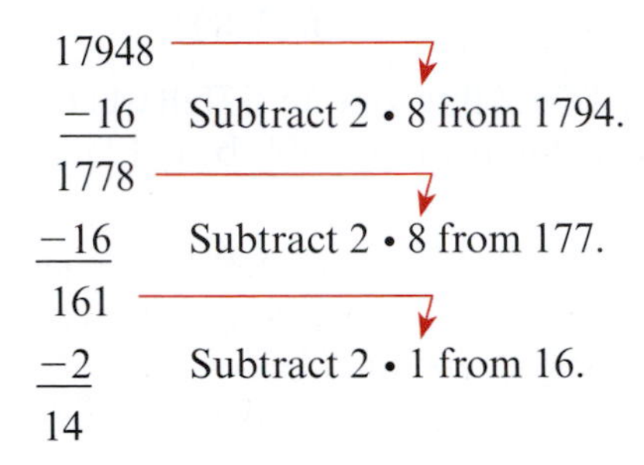

a. 10,801

b. 13,536

c. 15,071

d. 19,254

28. **Using a Divisibility Test for 11** Use the divisibility test for 11 to determine whether each whole number is divisible by 11.

Test A whole number is divisible by 11 when the sum of the digits in the places that are even powers of 10 minus the sum of the digits in the places that are odd powers of 10 is divisible by 11. For example, 190,938 is divisible by 11.

$$(8 + 9 + 9) - (3 + 0 + 1) = 26 - 4$$
$$= 22$$

a. 71,654

b. 12,784

c. 81,917

d. 70,454

29. **Reasoning** When using the divisibility test for 7, the final difference is 0. Is the original number divisible by 7?

30. **Reasoning** When using the divisibility test for 11, the final difference is 0. Is the original number divisible by 11?

istockphoto.com/Christopher Futcher

31. **Grading Student Work** On a diagnostic test, one of your students does the following work. Explain what the student did wrong. Which topics would you encourage the student to review?

> Tell whether 2576 is divisible by 4.
>
> 257(6) ⟶ 6 ÷ 4 = 1R2
>
> Because 6 is not divisible by 4, 2576 is not divisible by 4.

32. **Grading Student Work** On a diagnostic test, one of your students does the following work. Explain what the student did wrong. Which topics would you encourage the student to review?

> Tell whether 438 is divisible by 6.
>
> 4 + 3 + 8 = 15 15 ÷ 6 = 2R3
>
> Because 15 is not divisible by 6, 438 is not divisible by 6.

33. **Using a Divisibility Test in Real Life** A school has 490 students. Is it possible to evenly divide the students into groups of 4?

34. **Using a Divisibility Test in Real Life** A company has 366 employees. Is it possible to evenly divide the employees into groups of 6?

35. **Writing** When $n \mid (a + b)$, is it always true that $n \mid a$ and $n \mid b$? Explain your reasoning.

36. **Writing** When $n \mid am$, is it always true that $n \mid a$? Explain your reasoning.

37. **Using a Divisibility Test in Real Life** There are 261 children attending a summer soccer program. Each practice team needs 9 children. Is it necessary that a practice team has fewer than 9 children?

38. Using a Divisibility Test in Real Life There are 452 children attending a summer camp program. Each cottage can sleep 8 children. Is it necessary that a cottage sleeps fewer than 8 children?

39. Nursery A nursery employee plants seedlings 8 feet apart. The seedlings will be planted in a straight row that is 371 yards in length.

a. Can seedlings be planted at the beginning and end of the row and still be planted 8 feet apart? If so, how many seedlings are planted?

b. Can the employee plant the seedlings a different uniform whole number of feet apart and still plant a seedling at the beginning and end of the row? Explain your reasoning.

40. Wedding Planner A wedding planner wants twinkling light arrangements placed every 8 feet around the perimeter of a reception hall. The dimensions of the reception hall are 97 feet by 63 feet.

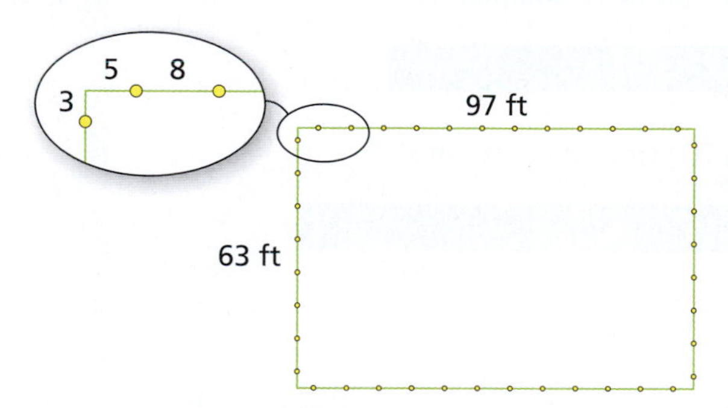

a. Can the arrangements be placed 8 feet apart without overlapping? If so, how many arrangements are needed?

b. Can the wedding planner place the arrangements a different uniform whole number of feet apart? Explain your reasoning.

41. Paper Towels A convenience store marks down the price of a pack of paper towels from \$3.00 but keeps the price above \$2.50. A customer buys several packs for a total cost of \$24.66. How many packs did the customer buy?

42. Online Order A teacher orders erasers. The total bill is \$3.08. The teacher notices that the actual price of each eraser is higher than the advertised price of \$0.25.

a. What are the possible prices of each eraser?

b. Which of the prices in part (a) is most reasonable? Explain.

43. Activity: In Your Classroom Design a classroom activity that you can use with your students to show that the area of the rectangle is divisible by 4.

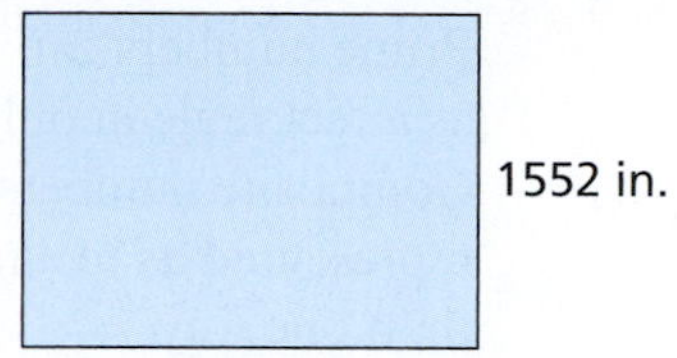

44. Activity: In Your Classroom Design a classroom activity that you can use with your students to show that the area of the rectangle is divisible by 4.

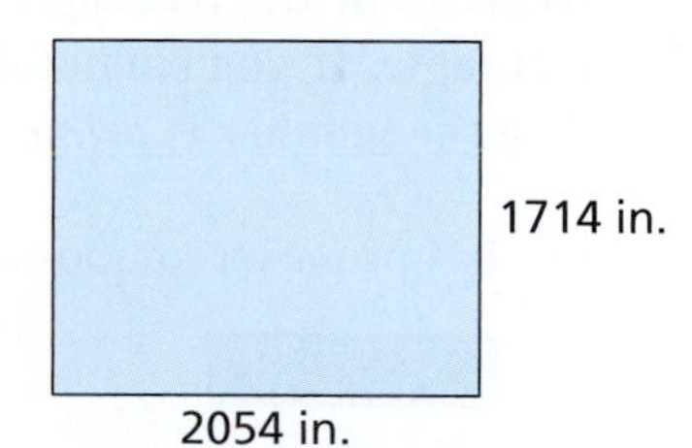

45. Grocery Shopping A customer orders the items on the list below. The total bill is \$4.69. The customer knows the bill is incorrect. Explain how the customer figures out the error.

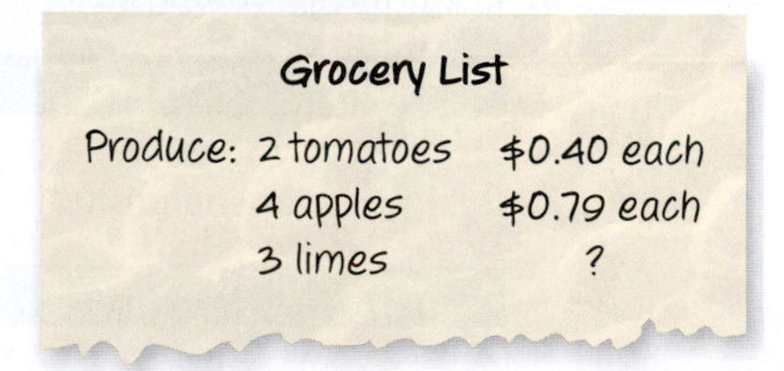

46. Grocery Shopping A customer orders the items on the list below. The total bill is \$6.26. The customer knows the bill is incorrect. Explain how the customer figures out the error.

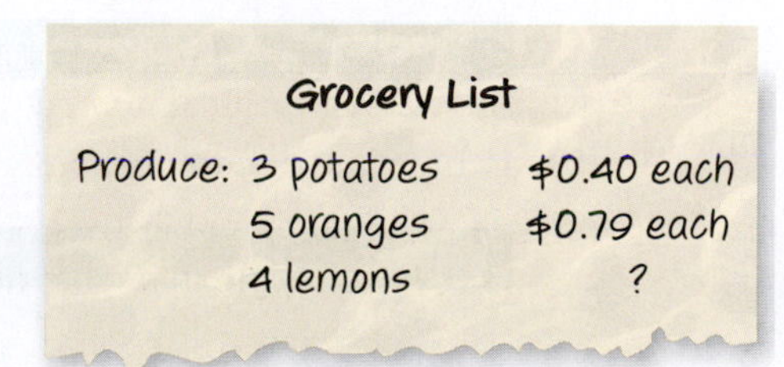

Activity: Discovering Primes and Composites

Materials:

- Pencil
- Square tiles

Learning Objective:

Prime numbers can be represented as a rectangle in only one way. Composite numbers can be represented as rectangles in more than one way.

Name ______________________________

Draw all of the rectangles that have exactly the same number of tiles as the given rectangle. If you cannot draw another rectangle that has the same number of tiles, then the number is *prime*. If you can, then the number is *composite*.

1. Is 4 prime or composite?

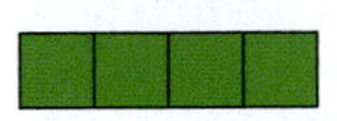

2. Is 6 prime or composite?

3. Is 7 prime or composite?

4. Is 8 prime or composite?

5. Is 9 prime or composite?

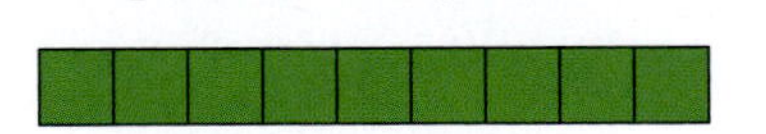

6. Is 10 prime or composite?

7. Is 11 prime or composite?

8. Is 12 prime or composite?

9. Is 13 prime or composite?

10. Is 14 prime or composite?

5.2 Primes and Composites

Standards

Grades 3–5
Operations and Algebraic Thinking
Students should gain familiarity with factors and multiples.

- Decide whether a natural number is prime or composite.
- Use a factor tree to write a composite number as the product of primes.
- Use prime factors to determine whether a large number is prime.

Prime and Composite Numbers

Definitions of Prime Number and Composite Number

A natural number is a **prime number** (or simply a **prime**) when it has exactly two factors: 1 and itself.

Example The only factors of 5 are 1 and 5. So, 5 is a prime number.

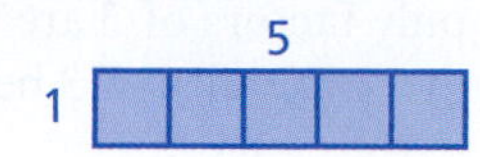

5 can only be factored as 1 • 5.

A natural number is a **composite number** (or simply a **composite**) when it has more than two factors.

Example The natural number 8 has 1, 2, 4, and 8 as factors. So, 8 is a composite number.

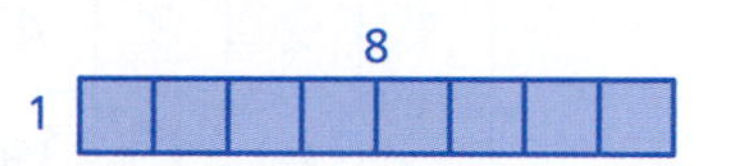

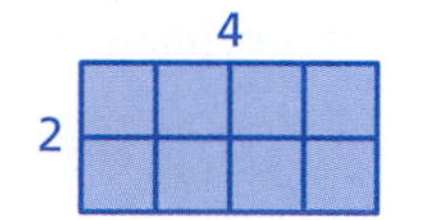

8 can be factored as 1 • 8 and as 2 • 4.

The natural number 1 has only one factor. So, it is neither prime nor composite.

Classroom Tip Be sure that your students understand that the natural number 1 is not a prime number because it cannot be written as the product of two unique factors. The least prime number is 2. It is the only even prime number. All other prime numbers are odd.

EXAMPLE 1 **Deciding Whether a Number Is Prime**

Which of the natural numbers from 20 to 29 are prime?

SOLUTION

Number	Factors	Conclusion
20	1, 2, 4, 5, 10, 20	Composite
21	1, 3, 7, 21	Composite
22	1, 2, 11, 22	Composite
23	1, 23	Prime ✓
24	1, 2, 3, 4, 6, 8, 12, 24	Composite
25	1, 5, 25	Composite
26	1, 2, 13, 26	Composite
27	1, 3, 9, 27	Composite
28	1, 2, 4, 7, 14, 28	Composite
29	1, 29	Prime ✓

There are only 2 primes from 20 to 29. They are 23 and 29.

Classroom Tip

The Sieve of Eratosthenes, named after the ancient Greek mathematician Eratosthenes (276 B.C.–195 B.C.), is one of the most efficient ways to find all of the small primes. Eratosthenes invented the modern discipline of geography, and he is thought to be the first person to have calculated the circumference of Earth.

The **Sieve of Eratosthenes** is an algorithm that identifies all of the prime numbers from 1 through a given natural number n. To make the Sieve of Eratosthenes, begin with a table listing the first n natural numbers.

EXAMPLE 2 Making the Sieve of Eratosthenes

Use the Sieve of Eratosthenes to identify the number of prime numbers from 1 through 100.

SOLUTION

Step 1 Begin by making a 10-by-10 table. Then label the cells in the table from 1 through 100.

Step 2 *Cross out 1*, because it is neither prime nor composite.

Step 3 The only factors of 2 are 1 and 2, so it is a prime. *Keep 2*. Then cross out all other multiples of 2 because they cannot be prime.

Step 4 The only factors of 3 are 1 and 3, so it is a prime. *Keep 3*. Then cross out all other multiples of 3 because they cannot be prime.

Step 5 The next prime is 5. So, *keep 5*. Then cross out all other multiples of 5 because they cannot be prime.

Step 6 The next prime is 7. So, *keep 7*. Then cross out all other multiples of 7 because they cannot be prime.

~~1~~	2	3	~~4~~	5	~~6~~	7	~~8~~	~~9~~	~~10~~
11	~~12~~	13	~~14~~	~~15~~	~~16~~	17	~~18~~	19	~~20~~
~~21~~	~~22~~	23	~~24~~	~~25~~	~~26~~	~~27~~	~~28~~	29	~~30~~
31	~~32~~	~~33~~	~~34~~	~~35~~	~~36~~	37	~~38~~	~~39~~	~~40~~
41	~~42~~	43	~~44~~	~~45~~	~~46~~	47	~~48~~	~~49~~	~~50~~
~~51~~	~~52~~	53	~~54~~	~~55~~	~~56~~	~~57~~	~~58~~	59	~~60~~
61	~~62~~	~~63~~	~~64~~	~~65~~	~~66~~	67	~~68~~	~~69~~	~~70~~
71	~~72~~	73	~~74~~	~~75~~	~~76~~	~~77~~	~~78~~	79	~~80~~
~~81~~	~~82~~	83	~~84~~	~~85~~	~~86~~	~~87~~	~~88~~	89	~~90~~
~~91~~	~~92~~	~~93~~	~~94~~	~~95~~	~~96~~	97	~~98~~	~~99~~	~~100~~

Look at the remaining numbers. The next prime is 11. Because $10 \cdot 10 = 100$, you do not have to check any of these remaining numbers. If any of the remaining numbers are divisible by a prime greater than 10, then they must also be divisible by a prime less than 10. All primes less than 10 have already been identified. The smallest composite number with prime factors greater than 10 is $11 \cdot 11 = 121$. Because $121 > 100$, it follows that the remaining numbers are primes. So, there are 25 prime numbers from 1 through 100:

2, 3, 5, 7, 11, 13, 17, 19, 23, 29, 31, 37, 41,
43, 47, 53, 59, 61, 67, 71, 73, 79, 83, 89, and 97.

Prime Factorization of a Number

One of the most important theorems in arithmetic is the *Fundamental Theorem of Arithmetic*. (Algebra also has a fundamental theorem, as does calculus.)

The Fundamental Theorem of Arithmetic

Every natural number greater than 1 is a prime number or can be written as the product of prime factors in exactly one way (disregarding order of factors).

So, every composite number can be written as a product of prime factors.

Prime Factorization

The product of prime factors for a given composite number is called a **prime factorization**.

Examples $15 = 3 \cdot 5$ $24 = 2 \cdot 2 \cdot 2 \cdot 3$

One way to determine the prime factors of a composite number is to use a **factor tree**.

Examples

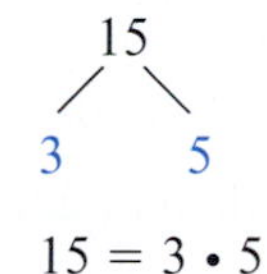

$15 = 3 \cdot 5$

24
3 8
2 4
2 2

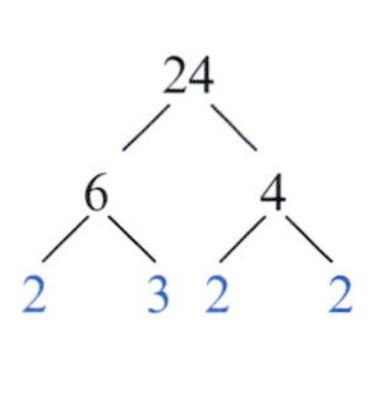

$24 = 2 \cdot 2 \cdot 2 \cdot 3$, or $2^3 \cdot 3$

> **Classroom Tip**
>
> When you are teaching students how to construct factor trees, it does not matter which number you factor out first. The point of the Fundamental Theorem of Arithmetic is that you will always end up with the same prime factors. For instance, suppose that in Example 3(a) you factor 3 out first and obtain the following factor tree.
>
> 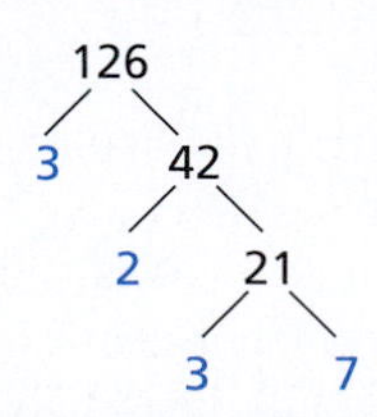
>
>
> Even though the factor tree is different, the prime factors are the same.

EXAMPLE 3 **Writing the Prime Factorization of a Number**

Write the prime factorization of each number.

a. 126 **b.** 225

SOLUTION

a.

126
2 63

126
2 63
7 9

126
2 63
7 9
3 3

The prime factorization of 126 is $2 \cdot 3 \cdot 3 \cdot 7$, or $2 \cdot 3^2 \cdot 7$.

b.

225
5 45

225
5 45
5 9

225
5 45
5 9
3 3

The prime factorization of 225 is $3 \cdot 3 \cdot 5 \cdot 5$, or $3^2 \cdot 5^2$.

Classroom Tip

In the Sieve of Eratosthenes on page 172, the table included the natural numbers from 1 through 100. Because the square root of 100 is 10, it was only necessary to cross out multiples of primes that were less than or equal to 10. If your table included more than 100 numbers, then you might have to use additional primes. For instance, if your table contained 225 natural numbers, then you would have to cross out multiples of primes that are less than or equal to 15.

Using Prime Factors of a Number

When you are looking for prime factors of a large number n, you can use a strategy that uses the *square root* of n. The **square root** of a number n is the number whose square is equal to n.

Examples The square root of 9 is 3, because $3^2 = 9$.

The square root of 36 is 6, because $6^2 = 36$.

Testing for Prime Factors

To test for prime factors of a number n, you need only check for prime factors that are less than or equal to the square root of n.

EXAMPLE 4 Determining Prime and Composite Numbers

Tell whether the number is *prime* or *composite*.

a. 901 **b.** 223

SOLUTION

a. The square root of 901 is about 30 because $30^2 = 900$. So, you need only test for prime factors that are less than or equal to 30.

Primes less than or equal to 30: 2, 3, 5, 7, 11, 13, 17, 19, 23, 29

$901 \div 2 = 450R1$ 2 is not a factor of 901.

$901 \div 3 = 300R1$ 3 is not a factor of 901.

$901 \div 5 = 180R1$ 5 is not a factor of 901.

$901 \div 7 = 128R5$ 7 is not a factor of 901.

$901 \div 11 = 81R10$ 11 is not a factor of 901.

$901 \div 13 = 69R4$ 13 is not a factor of 901.

$901 \div 17 = 53$ 17 is a factor of 901. ✓

So, $901 = 17 \cdot 53$, which means that 901 is composite. Notice that once you find a prime that is a factor of 901, you have shown that 901 is composite by the Fundamental Theorem of Arithmetic. So, it is unnecessary to continue checking primes that are greater than 17.

b. The square root of 223 is about 15 because $15^2 = 225$. So, you need only test for prime factors that are less than or equal to 15.

Primes less than or equal to 15: 2, 3, 5, 7, 11, 13

$223 \div 2 = 111R1$ 2 is not a factor of 223.

$223 \div 3 = 74R1$ 3 is not a factor of 223.

$223 \div 5 = 44R3$ 5 is not a factor of 223.

$223 \div 7 = 31R6$ 7 is not a factor of 223.

$223 \div 11 = 20R3$ 11 is not a factor of 223.

$223 \div 13 = 17R2$ 13 is not a factor of 223.

None of the primes less than or equal to 15 are factors of 223. So by the Fundamental Theorem of Arithmetic, 223 is prime.

It may be interesting for students to know that not all problems in mathematics have been solved. *Goldbach's Conjecture*, named after the German mathematician Christian Goldbach, is a problem that was stated in 1742 and has never been proven to be true, nor has it ever been proven to be false.

EXAMPLE 5 An Unsolved Problem in Number Theory

Goldbach's Conjecture states the following:

Every even integer greater than 2 can be written as the sum of two primes.

Check that the conjecture is true for all even numbers from 4 through 28.

Classroom Tip

After completing Example 5, ask your students whether they believe that *Goldbach's Conjecture* is true. Most people who have looked into the conjecture believe that it is true. And yet, no one has ever been able to prove it!

SOLUTION

One approach to this problem is to create a diagram that shows how the even numbers from 4 through 28 can be written as the sum of two primes.

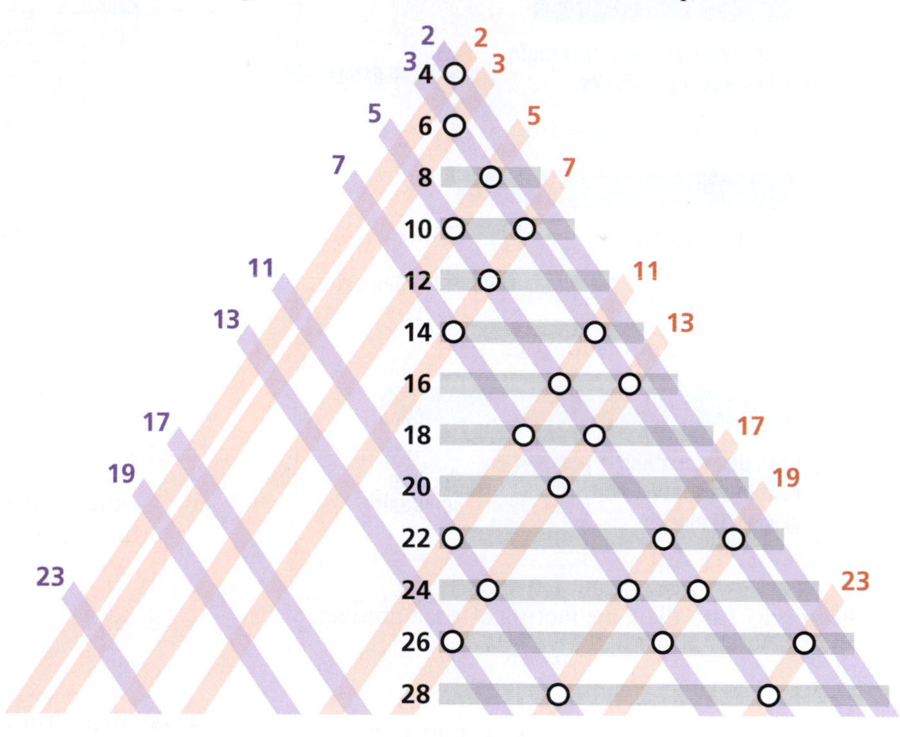

From the diagram, you can read the following sums.

Number	Sums
4	2 + 2
6	3 + 3
8	3 + 5
10	5 + 5, 3 + 7
12	5 + 7
14	7 + 7, 3 + 11
16	3 + 13, 5 + 11
18	5 + 13, 7 + 11
20	3 + 17, 7 + 13
22	3 + 19, 5 + 17, 11 + 11
24	5 + 19, 7 + 17, 11 + 13
26	3 + 23, 7 + 19, 13 + 13
28	5 + 23, 11 + 17

Mathematical Practices

Look For and Express Regularity in Repeated Reasoning

MP8 Mathematically proficient students notice when calculations are repeated, and look both for general methods and for shortcuts.

5.2 Exercises

Free step-by-step solutions to odd-numbered exercises at *MathematicalPractices.com.*

1. **Writing a Solution Key** Write a solution key for the activity on page 170. Describe a rubric for grading a student's work.

2. **Grading the Activity** In the activity on page 170, a student gave the answers below. Each question is worth 10 points. Assign a grade to each answer. For those that are incorrect, why do you think the student erred?

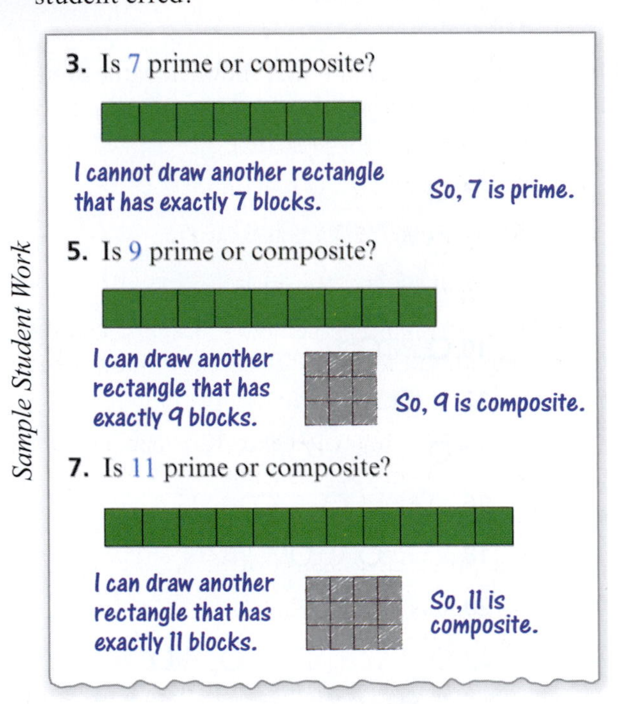

3. **Factoring** List all of the factors of each number.

a. 9 **b.** 16
c. 51 **d.** 63
e. 125 **f.** 144

4. **Factoring** List all of the factors of each number.

a. 72 **b.** 81
c. 62 **d.** 54
e. 326 **f.** 242

5. **Determining Prime and Composite Numbers** Tell whether each number is *prime* or *composite*.

a. 7 **b.** 3 **c.** 9
d. 12 **e.** 14 **f.** 11

6. **Determining Prime and Composite Numbers** Tell whether each number is *prime* or *composite*.

a. 2 **b.** 6 **c.** 15
d. 17 **e.** 19 **f.** 10

7. **Deciding Whether a Number Is Prime** List the natural numbers from 30 through 39 with their factors. Which of the numbers are prime?

8. **Deciding Whether a Number Is Prime** List the natural numbers from 40 through 49 with their factors. Which of the numbers are prime?

9. **Extending the Sieve of Eratosthenes** Continue the Sieve of Eratosthenes below through 120. Then list the prime numbers from 101 through 120.

71	~~72~~	73	~~74~~	~~75~~	~~76~~	~~77~~	~~78~~	79	~~80~~
~~81~~	~~82~~	83	~~84~~	~~85~~	~~86~~	~~87~~	~~88~~	89	~~90~~
~~91~~	~~92~~	~~93~~	~~94~~	~~95~~	~~96~~	97	~~98~~	~~99~~	~~100~~
101	102	103	104	105	106	107	108	109	110
111	112	113	114	115	116	117	118	119	120
121	122	123	124	125	126	127	128	129	130
131	132	133	134	135	136	137	138	139	140
141	142	143	144	145	146	147	148	149	150

10. **Extending the Sieve of Eratosthenes** Continue the Sieve of Eratosthenes above through 150. Then list the prime numbers from 121 through 150.

11. **Writing Prime Factorizations of Numbers** Write the prime factorization of each number.

a. 12 **b.** 25
c. 66 **d.** 18
e. 145 **f.** 100

12. **Writing Prime Factorizations of Numbers** Write the prime factorization of each number.

a. 16 **b.** 27
c. 52 **d.** 75
e. 112 **f.** 125

13. **Using Factor Trees** Fill in the missing numbers in the factor tree.

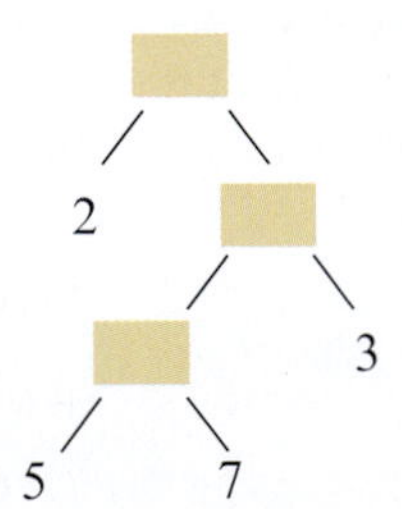

Is it possible to find the top number using only the given numbers? Explain.

14. Using Factor Trees Fill in the missing numbers in the factor tree.

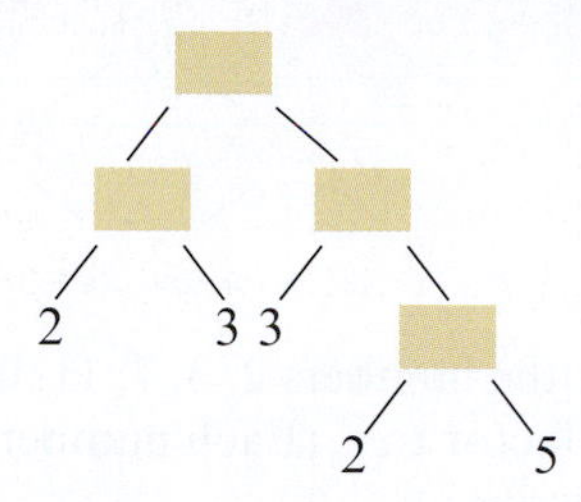

Is it possible to find the top number using only the given numbers? Explain.

15. Writing Square Roots of Numbers Write the square root of each number.

a. 4 **b.** 25

c. 121 **d.** 169

16. Writing Square Roots of Numbers Write the square root of each number.

a. 16 **b.** 81

c. 100 **d.** 144

17. Determining Prime and Composite Numbers Tell whether each number is *prime* or *composite*.

a. 319 **b.** 683

c. 1009 **d.** 1091

e. 19,017 **f.** 30,537

18. Determining Prime and Composite Numbers Tell whether each number is *prime* or *composite*.

a. 417 **b.** 729

c. 1571 **d.** 4587

e. 35,721 **f.** 87,451

19. Determining Prime Factorizations of Numbers Tell whether each number is *prime* or *composite*. If composite, write the prime factorization of the number.

a. 419 **b.** 363

c. 1369 **d.** 1319

e. 19,019 **f.** 15,827

20. Determining Prime Factorizations of Numbers Tell whether each number is *prime* or *composite*. If composite, write the prime factorization of the number.

a. 971 **b.** 847

c. 2197 **d.** 1323

e. 22,287 **f.** 66,339

21. Writing Prime Factorizations of Numbers Write the prime factorization of each number as a product of powers.

a. 352 **b.** 2160

c. 3773 **d.** 5929

e. 3500 **f.** 2025

22. Writing Prime Factorizations of Numbers Write the prime factorization of each number as a product of powers.

a. 432 **b.** 1568

c. 2079 **d.** 6318

e. 6048 **f.** 8281

23. True or False? Tell whether each statement is *true* or *false*. Explain your reasoning.

a. All prime numbers are odd.

b. All multiples of 5 are composite.

c. The prime factors of 30 are 2, 3, and 5.

d. To determine whether 127 is prime, you only need to test for prime factors less than or equal to 7.

24. True or False? Tell whether each statement is *true* or *false*. Explain your reasoning.

a. All even numbers are composite.

b. The prime factors of 60 are 2, 5, and 6.

c. The product of any two prime numbers is always odd.

d. To determine whether 1393 is prime, you only need to test for prime factors less than or equal to 37.

25. Problem Solving At soccer practice, the coach divides the team into groups of equal size to run drills. There are 28 players on the team. What are the possible group sizes?

26. Problem Solving You divide your students into groups of equal size for a project. There are 24 students in your class. What are the possible group sizes?

27. Problem Solving There are 133 members in a marching band. Is it possible for the band director to divide the band into equal-sized groups for a new formation? Explain.

28. Problem Solving A little league program has 216 players signed up to participate. Is it possible to divide the players into teams so that each team has the same number of players? Explain.

29. **Activity: In Your Classroom** Elementary school teachers often use Cuisenaire rods as a hands-on method for teaching math concepts. The figure shows the rods with their unit lengths and colors. A row of rods that are all the same color is called a *one-color train*.

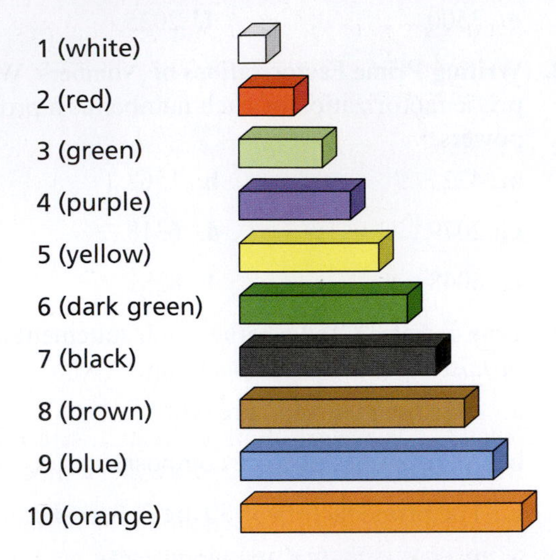

a. How many green rods are in a one-color train representing 15?

b. Which rods can be used to make a one-color train representing 30? Identify the rod length and the number of rods needed.

c. Explain how colored rods can be used to demonstrate factors.

30. **Activity: In Your Classroom** Refer to the Cuisenaire rods described in Exercise 29.

a. Which rods can be used to make a one-color train representing 7? Identify the rod length and the number of rods needed.

b. How can Cuisenaire rods be used to demonstrate prime numbers?

31. **Using Factor Trees: In Your Classroom** Refer to the factor tree below.

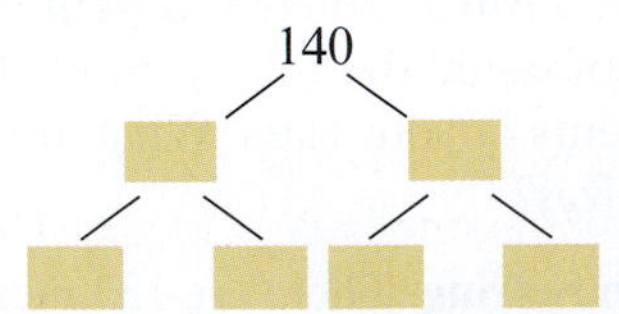

a. Use the numbers 2, 2, 5, 7, 10, and 14 to fill in the factor tree. (Each number can be used only once.)

b. Is this the only factor tree that can be used to find the prime factorization of 140? Explain.

c. Design a classroom activity that can be used to demonstrate several techniques for finding the prime factorization of a number.

32. **Using Factor Trees: In Your Classroom** Refer to the factor tree below.

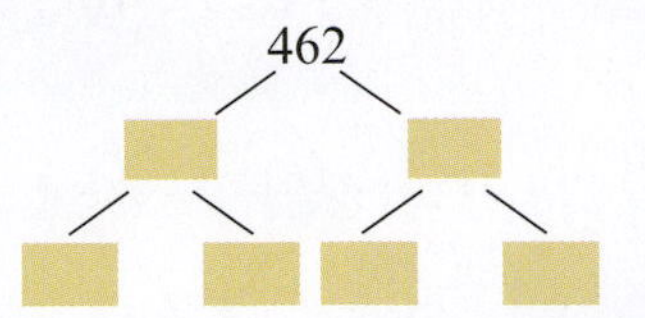

a. Use the numbers 2, 3, 7, 11, 21, and 22 to fill in the factor tree. (Each number can be used only once.)

b. Is this the only factor tree that can be used to find the prime factorization of 462? Explain.

c. Design a classroom activity that can be used to explain that the prime factorization of a number is unique.

33. **Deciding Whether a Number Is Prime** What is the greatest prime number you need to consider when determining whether 1367 is prime? Explain your reasoning.

34. **Deciding Whether a Number Is Prime** What is the greatest prime number you need to consider when determining whether 2213 is prime? Explain your reasoning.

35. **Deciding Whether a Number Is Prime** Determine whether 1749 is a prime number. What is the greatest prime number you test? Explain your reasoning.

36. **Deciding Whether a Number Is Prime** Determine whether 5083 is a prime number. What is the greatest prime number you test? Explain your reasoning.

37. **Determining the Prime Factorization of a Number** Explain why $2^4 \cdot 3 \cdot 5^{12} \cdot 8^{15}$ is not the prime factorization of a number. Write the prime factorization of the number.

38. **Determining the Prime Factorization of a Number** Explain why $5^{16} \cdot 9^{25} \cdot 19^{12} \cdot 23^3$ is not the prime factorization of a number. Write the prime factorization of the number

39. **Writing Prime Factorizations of Numbers** Write the prime factorization of each number as a product of powers.

a. $8^{10} \cdot 64^6 \cdot 121^4$

b. $9^{12} \cdot 49^8 \cdot 144^3$

40. **Writing Prime Factorizations of Numbers** Write the prime factorization of each number as a product of powers.

a. $12^9 \cdot 36^{15} \cdot 169^5$

b. $16^{13} \cdot 14^7 \cdot 81^{14}$

41. **Grading Student Work** On a diagnostic test, one of your students does the following work. Explain what the student did wrong. Which topics would you encourage the student to review?

> Use a factor tree to write the prime factorization of 28.
>
> 28
> 2 14
> 1 2 2 7
>
> $28 = 1 \cdot 2^2 \cdot 7$

42. **Grading Student Work** On a diagnostic test, one of your students does the following work. Explain what the student did wrong. Which topics would you encourage the student to review?

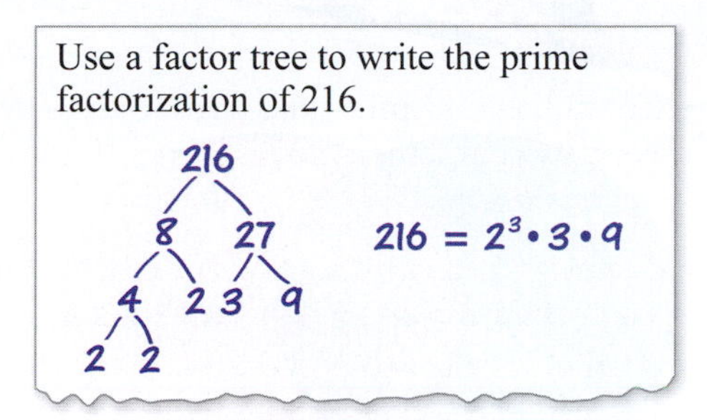

43. **Writing** Write the prime factorization of 48. Then list all of the composite factors of 48 and their prime factorizations. Describe the relationship that exists between the factorizations.

44. **Writing** Explain why the least whole number divisor, greater than 1, of a composite number must be prime.

45. **Classifying Numbers** A *perfect number* is a whole number n whose factors sum to $2n$. For example, consider the whole number 6. The factors of 6 are 1, 2, 3, and 6.

$1 + 2 + 3 + 6 = 12$

$2 \cdot 6 = 12$

So, 6 is a perfect number. Tell whether each number is perfect.

a. 28 **b.** 225 **c.** 496

46. **Classifying Numbers** A *deficient number* is a whole number n whose factors sum to less than $2n$. For example, consider the whole number 2. The factors of 2 are 1 and 2.

$1 + 2 = 3$

$2 \cdot 2 = 4$

So, 2 is a deficient number. Tell whether each number is deficient.

a. 36 **b.** 45 **c.** 50

47. **Finding Different Types of Prime Numbers** A *semiprime* is a number that is the product of two primes. For example, $2 \cdot 2 = 4$ is the product of two primes, both of which are 2. So, 4 is a semiprime. List all of the semiprimes between 5 and 30.

48. **Finding Different Types of Prime Numbers** *Twin primes* are pairs of prime numbers that differ by 2. For example, 3 and 5 are twin primes. List the first 10 pairs of twin primes.

49. **Patterns Involving Prime Numbers** The following are sums that yield prime numbers.

$19 + 4 = 23$

$23 + 6 = 29$

$29 + 8 = 37$

$31 + 10 = 41$

Identify the pattern. What is the first sum following this pattern that yields a composite number?

50. **Patterns Involving Prime Numbers** A *factorial* is defined for a natural number n as the product $n! = n \cdot (n - 1) \cdot \cdots \cdot 2 \cdot 1$. For example, 3! (read as "three factorial") is found by evaluating the product $3 \cdot 2 \cdot 1$. The following are expressions involving factorials that yield primes.

$3! - 2! + 1! = 5$

$4! - 3! + 2! - 1! = 19$

$5! - 4! + 3! - 2! + 1! = 101$

Identify the pattern. What is the first expression following this pattern that yields a composite number?

51. **An Unsolved Problem in Number Theory** *Waring's Prime Number Conjecture*, named after the English mathematician Edward Waring, states the following:

Every odd integer greater than 1 is a prime or can be written as a sum of three primes.

Check that the conjecture is true for all odd integers from 7 through 31.

52. **An Unsolved Problem in Number Theory** *Legendre's Conjecture*, named after Adrien-Marie Legendre, states the following:

For every positive integer, there exists a prime number between the square of the integer and the square of the sum of the integer and 1.

For example, consider the integer 1. The square of 1 is $1^2 = 1$ and the square of the sum of 1 and 1 is $(1 + 1)^2 = 4$. The prime numbers 2 and 3 are between 1 and 4. Check that the conjecture is true for all integers from 2 through 16.

Activity: Using Diagrams to Find GCF and LCM

Materials:

- Pencil

Learning Objective:

Learn how to use prime factorizations and diagrams to find the greatest common factor and the least common multiple of two whole numbers.

Name ________________________

The diagram at the right shows the prime factors of $a = 30$ (the blue region) and $b = 24$ (the red region). The product of the common factors is the *greatest common factor* (*GCF*). The product of all the factors is the *least common multiple* (*LCM*).

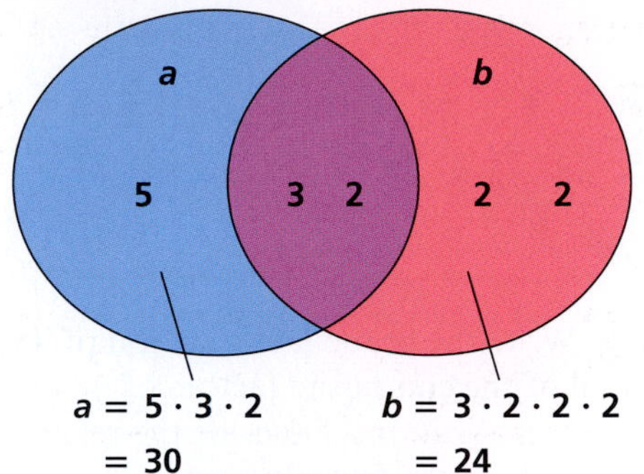

$\text{GCF} = 3 \cdot 2 = 6$, $\text{LCM} = 5 \cdot 3 \cdot 2 \cdot 2 \cdot 2 = 120$

Use the diagram to find the numbers.

1.

a
b
2
3 2
5

$a =$ ___, $b =$ ___,

GCF = ___, LCM = ___

2.

a
b
2 5
2 3
3

$a =$ ___, $b =$ ___,

GCF = ___, LCM = ___

3.

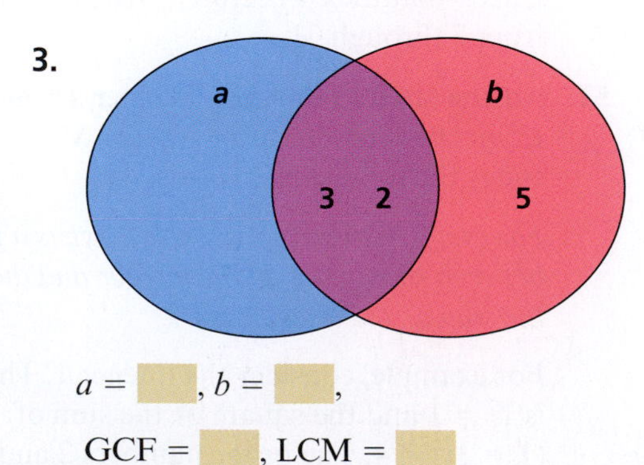

$a =$ ___, $b =$ ___,

GCF = ___, LCM = ___

4.

a
b
5
2 7
3 3 2

$a =$ ___, $b =$ ___,

GCF = ___, LCM = ___

ZouZou/Shutterstock.com

5.3 Greatest Common Factor and Least Common Multiple

Standards

Grades 3–5
Operations and Algebraic Thinking
Students should gain familiarity with factors and multiples.

Grades 6–8
The Number System
Students should find common factors and multiples.

- Find the greatest common factor of two or more whole numbers.
- Use algorithms to find the greatest common factor of two whole numbers.
- Find the least common multiple of two or more nonzero whole numbers.

The Greatest Common Factor

Definition of Greatest Common Factor (GCF)

The **greatest common factor** (GCF) of two or more whole numbers is the greatest whole number that is a factor of both or all of the whole numbers. The notation for the greatest common factor of a and b is written as **GCF(a, b)**. To find the GCF(a, b), write the prime factorizations of a and b. The greatest common factor is the product of the common factors.

Example $a = 20, b = 12$

$$a = 2 \cdot 2 \cdot 5$$

$$b = 2 \cdot 2 \cdot 3$$

$$\text{GCF}(20, 12) = 2 \cdot 2 = 4$$

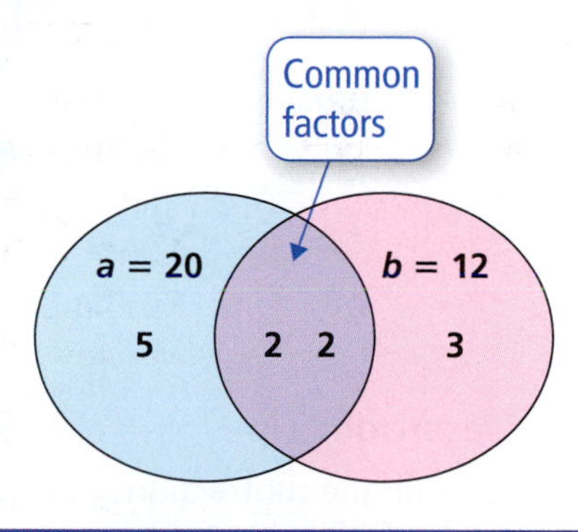

If a divides b, then GCF(a, b) = a. If b divides a, then GCF(a, b) = b.

Example GCF(3, 12) = 3

EXAMPLE 1 **Finding the Greatest Common Factor**

Find the greatest common factor of 16 and 24.

SOLUTION

Begin by writing the prime factorization of each number.

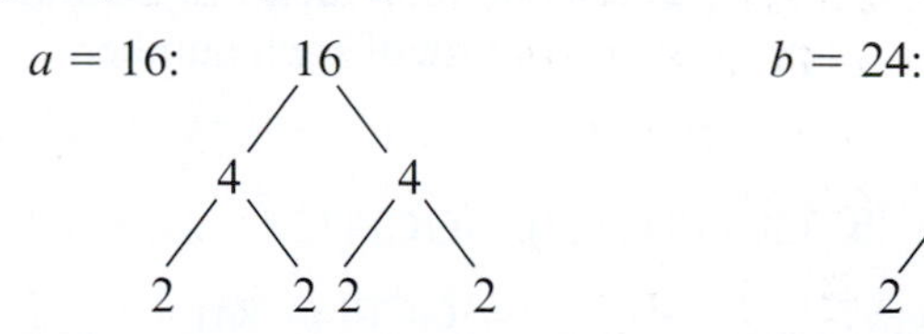

So, $16 = 2 \cdot 2 \cdot 2 \cdot 2$ and $24 = 2 \cdot 2 \cdot 2 \cdot 3$. Next, use a diagram to organize the prime factors of a and b.

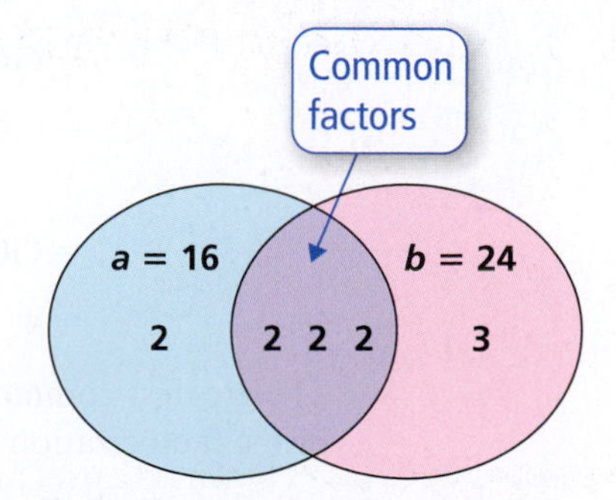

The greatest common factor of 16 and 24 is $2 \cdot 2 \cdot 2 = 8$.

Classroom Tip

When you are teaching strategies for finding the GCF of two whole numbers, remind your students to first think about the numbers.

1. If both are primes, then the GCF is 1.

   GCF(3, 5) = 1

2. If one number is a multiple of the other, then the GCF is the lesser of the two numbers.

   GCF(3, 15) = 3

3. If one of the numbers is 0, then the GCF is the greater of the two numbers because 0 is divisible by any nonzero number.

   GCF(3, 0) = 3

Algorithms for Finding the GCF

On page 162, one of the Properties of Divisibility states that when a number divides each of two numbers, it also divides their difference. So, if c is a common factor of a and b ($a \geq b$), then c is also a common factor of b and $a - b$. Because every common factor of a and b is also a common factor of b and $a - b$, the pairs (a, b) and $(a - b, b)$ have the same common factors. This leads to the following algorithm.

Subtraction Algorithm for Finding the GCF

If a and b are two whole numbers where $a \geq b$, then

$$\text{GCF}(a, b) = \text{GCF}(a - b, b).$$

To use the subtraction algorithm, continue subtracting the lesser number from the greater number until one number is a multiple of the other number.

Example $\text{GCF}(8, 6) = \text{GCF}(2, 6)$ $8 - 6 = 2$

$= 2$ 6 is a multiple of 2.

EXAMPLE 2 Using the Subtraction Algorithm to Find the GCF

Find the greatest common factor of each pair of whole numbers.

a. 36, 75 **b.** 84, 124

Classroom Tip

Although the subtraction algorithm for GCF(a, b) specifies that $a \geq b$, you can apply the algorithm for any two numbers. All you have to remember is to subtract the lesser number from the greater number, as shown in Example 2(a).

SOLUTION

a. $\text{GCF}(36, 75) = \text{GCF}(75 - 36, 36)$ — Apply subtraction algorithm.

$= \text{GCF}(39, 36)$ — Subtract: $75 - 36 = 39$.

$= \text{GCF}(39 - 36, 36)$ — Because 39 is not a multiple of 36, apply subtraction algorithm again.

$= \text{GCF}(3, 36)$ — Subtract: $39 - 36 = 3$.

$= 3$ — 36 is a multiple of 3.

The greatest common factor of 36 and 75 is 3. Check this result using the prime factorization of each number.

$36 = 2 \cdot 2 \cdot 3 \cdot 3$ $75 = 3 \cdot 5 \cdot 5$ ✓

b. $\text{GCF}(84, 124) = \text{GCF}(124 - 84, 84)$ — Apply subtraction algorithm.

$= \text{GCF}(40, 84)$ — Subtract: $124 - 84 = 40$.

$= \text{GCF}(84 - 40, 40)$ — Because 84 is not a multiple of 40, apply subtraction algorithm again.

$= \text{GCF}(44, 40)$ — Subtract: $84 - 40 = 44$.

$= \text{GCF}(44 - 40, 40)$ — Because 44 is not a multiple of 40, apply subtraction algorithm again.

$= \text{GCF}(4, 40)$ — Subtract: $44 - 40 = 4$.

$= 4$ — 40 is a multiple of 4.

The greatest common factor of 84 and 124 is 4. Check this result using the prime factorization of each number.

$84 = 2 \cdot 2 \cdot 3 \cdot 7$ $124 = 2 \cdot 2 \cdot 31$ $\longrightarrow$ $2 \cdot 2 = 4$ ✓

Because division can be represented as repeated subtraction, division can help shorten the process of finding the GCF of large numbers.

Classroom Tip

The Euclidean algorithm is named after the famous Greek mathematician Euclid, who lived around 300 B.C. Euclidean geometry is also named after Euclid.

Euclidean Algorithm for Finding the GCF

If a and b are two whole numbers where $a \geq b$ and the remainder obtained by dividing a by b is r, then

$$\text{GCF}(a, b) = \text{GCF}(b, r).$$

To use the Euclidean algorithm, continue dividing until the remainder is 0. The divisor that gives a zero remainder is the GCF.

Example GCF(8, 6): $6\overline{)8}$ = 1R2 → $2\overline{)6}$ = 3R0 (The GCF is 2.)

EXAMPLE 3 Using the Euclidean Algorithm to Find the GCF

Find the greatest common factor of 54 and 414.

SOLUTION

$54\overline{)414}$ = 7R36 → $36\overline{)54}$ = 1R18 → $18\overline{)36}$ = 2R0

The greatest common factor of 54 and 414 is 18. Check this result using the prime factorization of each number.

$54 = 2 \cdot 3 \cdot 3 \cdot 3$ $414 = 2 \cdot 3 \cdot 3 \cdot 23$ → $2 \cdot 3 \cdot 3 = 18$ ✓

Notice how using the Euclidean algorithm in Example 3 is more efficient than using the subtraction algorithm and repeatedly subtracting 54.

EXAMPLE 4 Using the GCF in Real Life

You are making identical balloon arrangements for a party. You have 32 red balloons, 24 white balloons, and 16 blue balloons. What is the greatest number of arrangements you can make if you use every balloon?

SOLUTION

One way to begin is to write the prime factorization of each number.

$32 = 2 \cdot 2 \cdot 2 \cdot 2 \cdot 2$ $24 = 2 \cdot 2 \cdot 2 \cdot 3$ $16 = 2 \cdot 2 \cdot 2 \cdot 2$

From these factorizations, you can see that the GCF of 32, 24, and 16 is $2 \cdot 2 \cdot 2 = 8$. So, you can make 8 identical arrangements. You can check this by drawing a diagram of the arrangements.

The Least Common Multiple

Definition of Least Common Multiple (LCM)

The **least common multiple** (LCM) of two or more natural numbers is the least natural number that is a multiple of both or all of the natural numbers. The notation for the least common multiple of a and b is written as **LCM(a, b)**. To find the LCM(a, b), write the prime factorizations of a and b using two intersecting sets. The least common multiple is the product of the factors in all three regions.

Example $a = 20, b = 12$

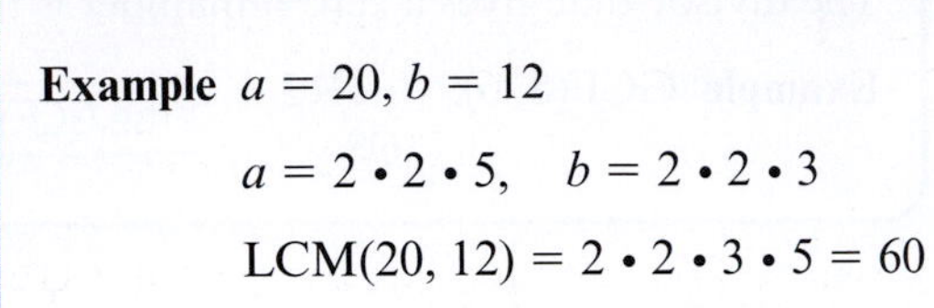

$a = 2 \cdot 2 \cdot 5, \quad b = 2 \cdot 2 \cdot 3$

$\text{LCM}(20, 12) = 2 \cdot 2 \cdot 3 \cdot 5 = 60$

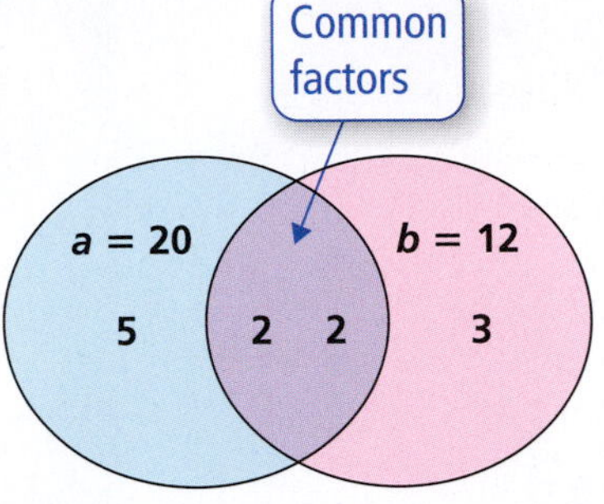

If the only common factor of a and b is 1, then the $\text{LCM}(a, b) = a \cdot b$.

Example $\text{LCM}(3, 5) = 3 \cdot 5 = 15$

Classroom Tip

The least common multiple of 1 and any nonzero whole number is simply the whole number.

EXAMPLE 5 Finding the Least Common Multiple

Find the least common multiple of each pair of whole numbers.

a. 6, 10 **b.** 18, 20

SOLUTION

a. Begin by writing the prime factorization of each number.

$a = 6$: 6 → 2, 3 $b = 10$: 10 → 2, 5

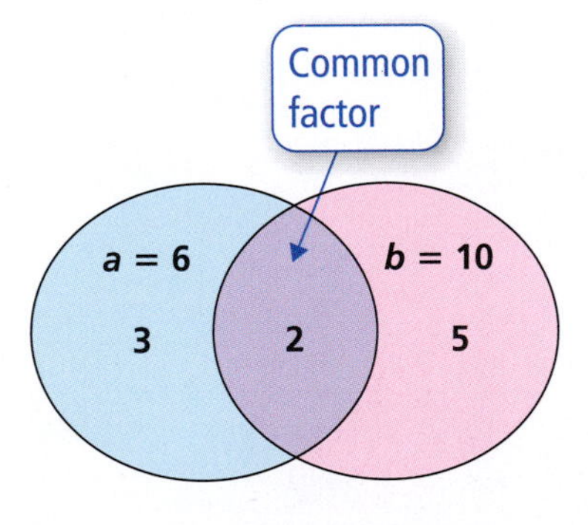

So, $6 = 2 \cdot 3$ and $10 = 2 \cdot 5$. Next, use a diagram to organize the prime factors of a and b. The least common multiple of 6 and 10 is

$2 \cdot 3 \cdot 5 = 30.$

b. Begin by writing the prime factorization of each number.

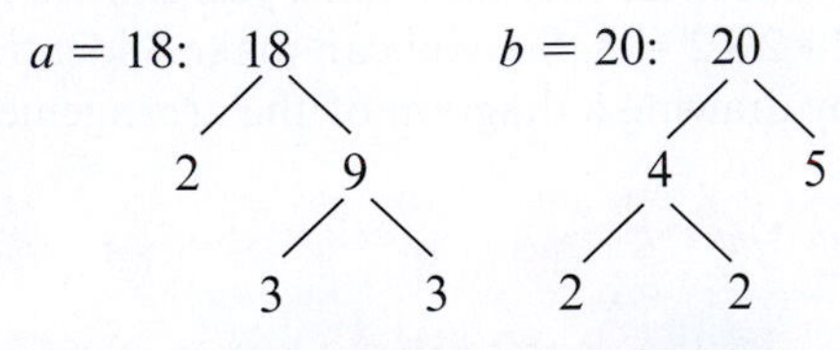

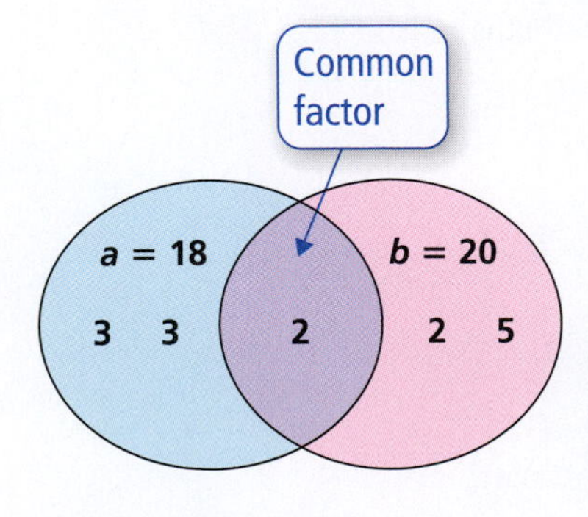

So, $18 = 2 \cdot 3 \cdot 3$ and $20 = 2 \cdot 2 \cdot 5$. Next, use a diagram to organize the prime factors of a and b. The least common multiple of 18 and 20 is

$2 \cdot 2 \cdot 3 \cdot 3 \cdot 5 = 180.$

Product Property of GCF and LCM

Let a and b be nonzero whole numbers.

Words The product of a and b is equal to the product of their GCF and LCM.

Algebra $a \cdot b = \text{GCF}(a, b) \cdot \text{LCM}(a, b)$

Numbers $a = 4$ $\quad$ $\text{GCF}(4, 6) = 2$

$b = 6$ $\quad$ $\text{LCM}(4, 6) = 12$

$a \cdot b = 4 \cdot 6 = 24$ $\quad$ $\text{GCF}(4, 6) \cdot \text{LCM}(4, 6) = 2 \cdot 12 = 24$

Classroom Tip

Another method for finding the LCM of two or more nonzero numbers is to make a list of the multiples of each number. For instance, the LCM of 6 and 8 is as follows.

Multiples of 6:
6, 12, 18, (24), 30, 36, 42, 48

Multiples of 8:
8, 16, (24), 32, 40, 48, 56, 64

You compare the two lists and circle the least common multiple. This method can be easier at times for students because they can clearly see all of the multiples and which is the least one.

EXAMPLE 6 Using the LCM in Real Life

You are buying hot dogs and hot dog buns for a picnic. The hot dogs come in packages of 6. The buns come in packages of 8. What is the least number of packages of hot dogs and buns you can buy to have exactly the same number of hot dogs as buns?

6 hot dogs per package

8 buns per package

SOLUTION

1. **Understand the Problem** You need to find the least number of packages of hot dogs and buns you can buy to have exactly the same number of hot dogs and buns.
2. **Make a Plan** Essentially, the question boils down to finding the least common multiple of 6 and 8, and then dividing to find the number of each type of package.
3. **Solve the Problem** You can use a diagram to find the LCM.

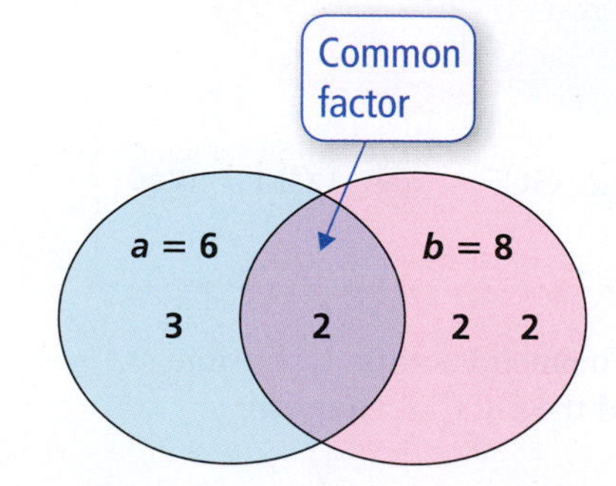

The least common multiple of 6 and 8 is $2 \cdot 2 \cdot 2 \cdot 3 = 24$. So, you need to buy $24 \div 6 = 4$ packages of hot dogs and $24 \div 8 = 3$ packages of buns.

4. **Look Back** Another way you can solve the problem is to use the Product Property of GCF and LCM. From the diagram, you know that the only common factor of 6 and 8 is 2. So, the GCF is 2. Because $6 \cdot 8 = 48$, you can use mental math to determine that $2 \cdot 24 = 48$. So, the LCM is 24.

Mathematical Practices

Reason Abstractly and Quantitatively

MP2 Mathematically proficient students make sense of quantities and their relationships in problem situations.

5.3 Exercises

Free step-by-step solutions to odd-numbered exercises at MathematicalPractices.com.

1. **Writing a Solution Key** Write a solution key for the activity on page 180. Describe a rubric for grading a student's work.

2. **Grading the Activity** In the activity on page 180, a student gave the answers below. Each question is worth 10 points. Assign a grade to each answer. For those that are incorrect, why do you think the student erred?

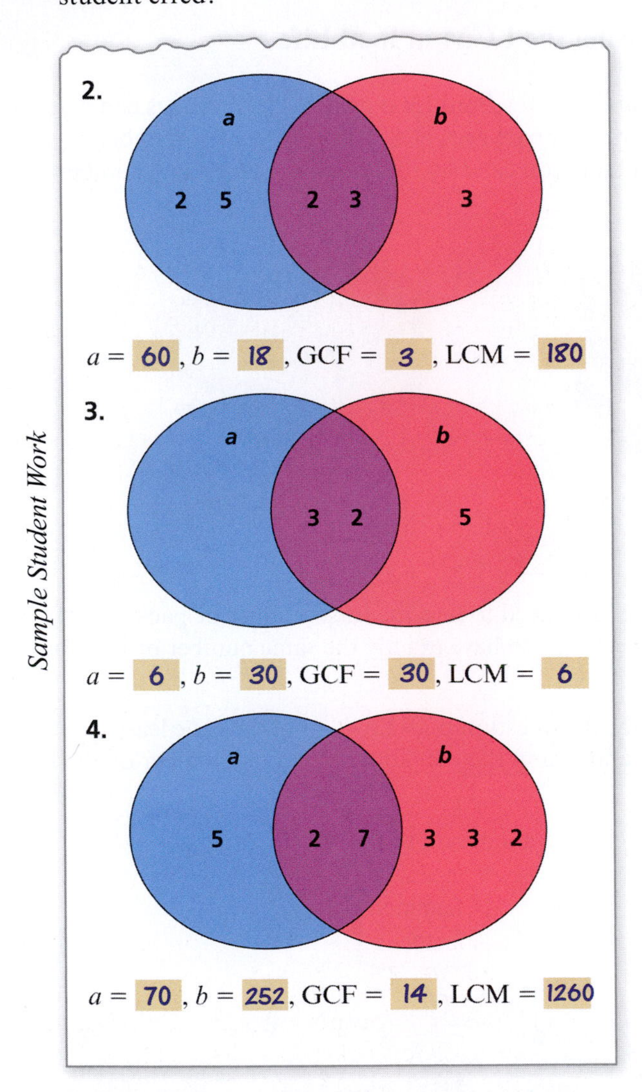

3. **Finding Greatest Common Factors** Use prime factorization to find the GCF of each pair of whole numbers.

   a. 4, 10 b. 3, 14
   c. 6, 12 d. 15, 33
   e. 30, 45 f. 150, 200

4. **Finding Greatest Common Factors** Use prime factorization to find the GCF of each pair of whole numbers.

   a. 9, 15 b. 13, 20
   c. 6, 8 d. 21, 63
   e. 36, 54 f. 100, 360

5. **Using the Subtraction Algorithm** Use the subtraction algorithm to find the GCF of each pair of whole numbers.

   a. 14, 42 b. 24, 36
   c. 60, 70 d. 48, 60
   e. 76, 86 f. 54, 89

6. **Using the Subtraction Algorithm** Use the subtraction algorithm to find the GCF of each pair of whole numbers.

   a. 6, 26 b. 8, 28
   c. 40, 56 d. 35, 42
   e. 34, 85 f. 32, 55

7. **Using the Euclidean Algorithm** Use the Euclidean algorithm to find the GCF of each pair of whole numbers.

   a. 90, 150 b. 126, 180
   c. 65, 182 d. 198, 216
   e. 32, 40 f. 120, 960
   g. 45, 825 h. 850, 918

8. **Using the Euclidean Algorithm** Use the Euclidean algorithm to find the GCF of each pair of whole numbers.

   a. 84, 216 b. 120, 192
   c. 63, 84 d. 98, 140
   e. 45, 90 f. 160, 800
   g. 56, 588 h. 544, 935

9. **Finding Least Common Multiples** Use prime factorization to find the LCM of each pair of whole numbers.

   a. 10, 12 b. 4, 6
   c. 7, 21 d. 7, 8
   e. 32, 80 f. 20, 28

10. Finding Least Common Multiples Use prime factorization to find the LCM of each pair of whole numbers.

a. 3, 8 **b.** 9, 12
c. 12, 36 **d.** 2, 9
e. 24, 45 **f.** 15, 25

11. Using Number Lines to Find Least Common Multiples Use a number line to find the LCM of each pair of whole numbers. The number line below shows the LCM of 2 and 3 is 6.

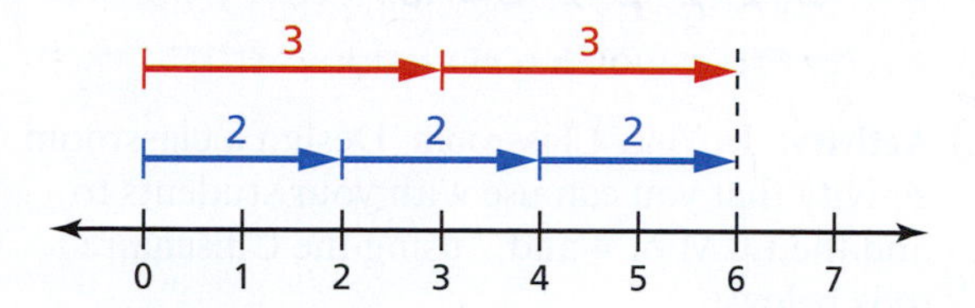

a. 2, 4
b. 5, 6
c. 3, 7

12. Using a Number Line to Find the Least Common Multiple Use a number line to find the LCM of each pair of whole numbers. The number line below shows the LCM of 3 and 4 is 12.

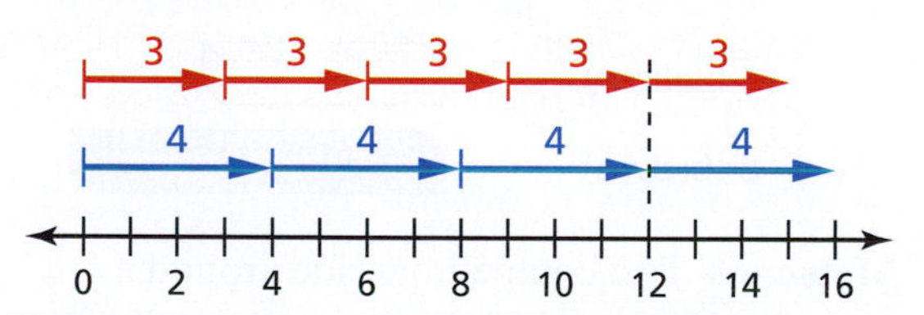

a. 3, 6
b. 4, 5
c. 2, 7

13. Finding the GCF and LCM For whole numbers a and b, use the diagram below to organize the factors of a and b. Then find the GCF and LCM of a and b.

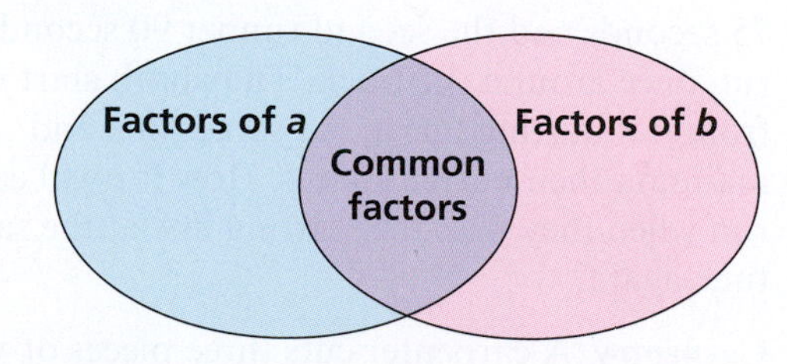

a. $a = 10, b = 45$
b. $a = 12, b = 96$
c. $a = 24, b = 36$

14. Finding the GCF and LCM For whole numbers a and b, use the diagram below to organize the factors of a and b. Then find the GCF and LCM of a and b.

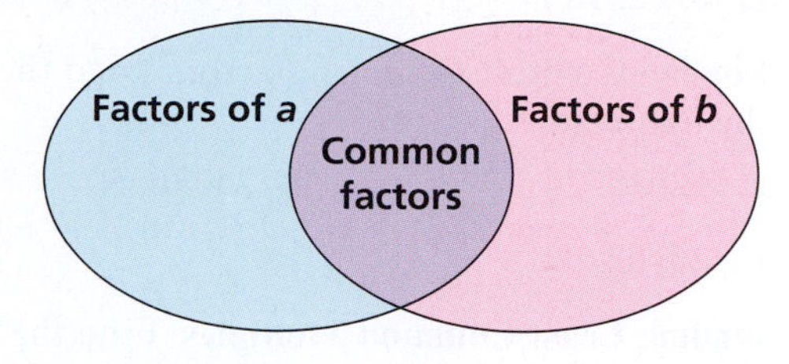

a. $a = 15, b = 20$
b. $a = 50, b = 100$
c. $a = 24, b = 30$

15. GCF vs. LCM Which is greater, the GCF of 48 and 72 or the LCM of 48 and 72? Explain your reasoning.

16. GCF vs. LCM Let a and b represent two different natural numbers. Which is greater, the LCM of a and b or the GCF of a and b? Explain your reasoning.

17. Product Property of GCF and LCM Show that the *Product Property of GCF and LCM* is true for each pair of whole numbers.

$a \cdot b = \text{GCF}(a, b) \cdot \text{LCM}(a, b)$

a. $a = 5, b = 15$
b. $a = 7, b = 21$
c. $a = 8, b = 20$

18. Product Property of GCF and LCM Show that the *Product Property of GCF and LCM* is true for each pair of whole numbers.

$a \cdot b = \text{GCF}(a, b) \cdot \text{LCM}(a, b)$

a. $a = 6, b = 8$
b. $a = 12, b = 36$
c. $a = 32, b = 80$

19. Party Balloons You are making identical balloon arrangements for a party. You have 12 pink balloons and 16 white balloons. What is the greatest number of arrangements you can make if you use every balloon?

20. Gift Bags You are making identical gift bags for a party. You have 18 candles and 24 gift cards. What is the greatest number of gift bags you can make if you use every candle and gift card?

21. Finding Greatest Common Factors Find the GCF of the numbers.

a. 6, 12, 20 b. 30, 45, 75

c. 6, 12, 18

22. Finding Greatest Common Factors Find the GCF of the numbers.

a. 12, 18, 24 b. 24, 36, 60

c. 26, 52, 78

23. Finding Least Common Multiples Find the LCM of the numbers.

a. 2, 3, 6 b. 3, 5, 10

c. 8, 12, 15

24. Finding Least Common Multiples Find the LCM of the numbers.

a. 4, 6, 12 b. 5, 16, 20

c. 24, 36, 48

25. Gas Mileage Your car gets 39 miles per gallon and your friend's car gets 30 miles per gallon. Both cars use a whole number of gallons of gasoline when traveling from Fresno, California, to Las Vegas, Nevada.

a. What is the minimum distance each car traveled?

b. What is the minimum number of gallons of gasoline each car used during the trip?

26. Business A crafter at a craft fair sells small change purses for \$16 each and large purses for \$45 each. At the end of the fair, the crafter earns the same amount of revenue from both types of purses.

a. What is the minimum number of purses of each type sold?

b. What is the minimum revenue earned at the craft fair?

27. Grading Student Work On a diagnostic test, one of your students does the following work. Explain what the student did wrong. Which topics would you encourage the student to review?

> Find the greatest common factor of 36 and 90.
>
> GCF (36, 90) = GCF(90 − 36, 36)
> = GCF(54, 36)
> = GCF(54 − 36, 36)
> = GCF(18, 36)
>
> Because 36 is a multiple of 18, the GCF of 36 and 90 is 36.

28. Grading Student Work On a diagnostic test, one of your students does the following work. Explain what the student did wrong. Which topics would you encourage the student to review?

> Find the least common multiple of 8 and 12.
>
> LCM (8, 12)   a = 8: 8 → 2, 4; 4 → 2, 2   b = 12: 12 → 2, 6; 6 → 2, 3
>
> The least common multiple of 8 and 12 is 2 • 2 • 2 • 2 • 2 • 3 = 96.

29. Activity: In Your Classroom Design a classroom activity that you can use with your students to find the LCM of 4 and 7 using the Cuisenaire rods below.

4 (purple)

7 (black)

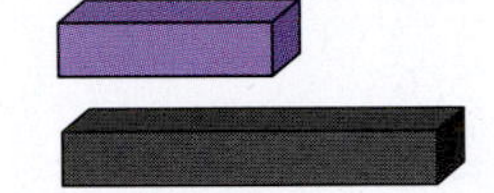

30. Activity: In Your Classroom Design a classroom activity that you can use with your students to find the LCM of 5 and 8 using the Cuisenaire rods below.

5 (yellow)

8 (brown)

31. Motocross Two competitors ride around a motocross track. The first competitor completes one round in 360 seconds and the second competitor completes one round in 420 seconds. They both start from the beginning of the track at the same time and maintain their current speeds. After how many minutes will they pass the beginning of the track together?

32. Track Team Two track team members are running laps on a quarter-mile track. It takes the first runner 75 seconds and the second runner 90 seconds to run once around the track. They both start running from the starting line at the same time and maintain their current paces. How far will each have run when they pass the starting line at the same time again?

33. Carpentry A carpenter cuts three pieces of wood that measure 36 inches, 45 inches, and 81 inches into smaller pieces of equal length. What is the longest possible length of the smaller pieces when all of the wood is used?

34. Florists A florist is making bouquets from the orchids, lilies, and tulips shown below. What is the greatest number of bouquets the florist can make when no flowers are left over?

180 orchids **240 lilies** **300 tulips**

35. Gears How many revolutions must each gear make before the arrows line up again?

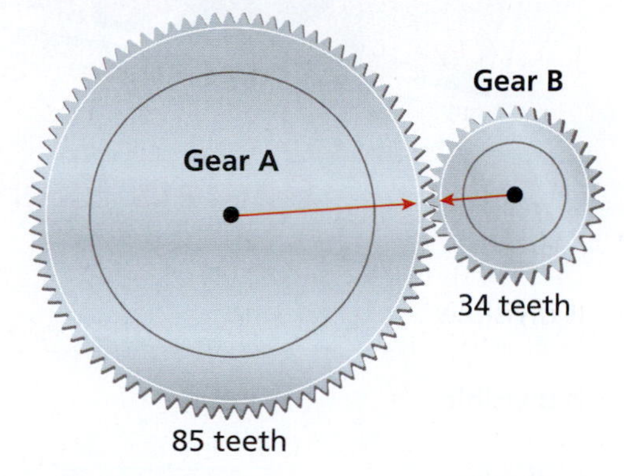

36. Gears How many revolutions must each gear make before the arrows line up again?

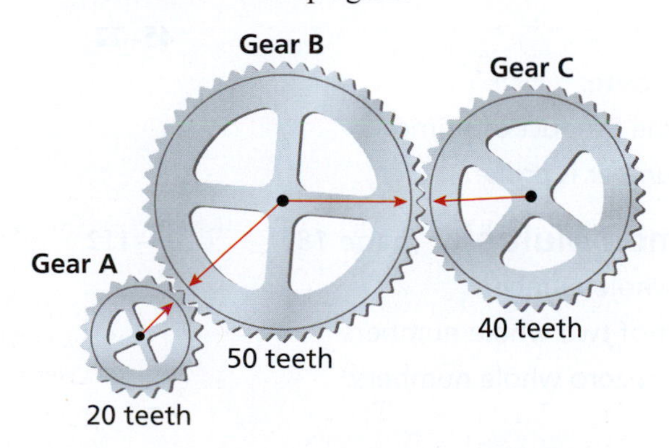

37. Math Class A student's class schedule changes on a three-day rotation. The student's math class is the last period of every third day. This week, the student's math class is on Friday. The student expects to have math class again on Friday in two weeks. Is the student correct or incorrect? Explain your reasoning.

38. Giveaways A movie and game rental store is giving away a free game rental to every ninth customer and a free movie rental to every twelfth customer. The store manager expects the forty-eighth customer to be the first person to win both rentals at the same time. Is the manager correct or incorrect? Explain your reasoning.

39. Picnic Supplies A party supply store sells packs of napkins, paper plates, and plastic cups in the quantities shown below. What is the least number of packs of each item you need to buy to have the same number of napkins, paper plates, and plastic cups?

Product	Quantity
Napkins	200
Paper plates	100
Plastic cups	24

40. Hardware Supplies A hardware store sells packs of nuts, bolts, and washers in the quantities shown below. What is the least number of packs of each item you need to buy to have the same number of nuts, bolts, and washers?

Product	Quantity
Nuts	100
Bolts	60
Washers	20

41. True or False? Tell whether the statement below is *true* or *false*. Let a and b be whole numbers. Explain your reasoning.

If $\text{GCF}(a, b) = 1$, then neither a nor b is even.

42. True or False? Tell whether the statement below is *true* or *false*. Let a and b be whole numbers. Explain your reasoning.

If $\text{GCF}(a, b) = 2$, then a and b are even.

43. Writing Explain whether the following statement is true.

$a \cdot b \cdot c = \text{GCF}(a, b, c) \cdot \text{LCM}(a, b, c)$

44. Writing When is the least common multiple of two numbers less than the product of the two numbers?

45. Problem Solving You have fewer than 100 trading cards. You try to sort these into groups of twos, threes, fours, and fives. Each time you have 1 card left over. How many trading cards do you have?

46. Problem Solving Two numbers less than 50 have a GCF of 5. When you increase each number by 5, the GCF doubles, but the LCM decreases by 15. Find the two numbers.

Chapter Summary

Chapter Vocabulary

divides *(p. 161)*
factor *(p. 161)*
multiple *(p. 161)*
divisible *(p. 161)*
evenly divisible *(p. 161)*
prime number *(p. 171)*
prime *(p. 171)*
composite number *(p. 171)*
composite *(p. 171)*
Sieve of Eratosthenes *(p. 172)*
Fundamental Theorem of Arithmetic *(p. 173)*
Prime Factorization *(p. 173)*
factor tree *(p. 173)*
square root *(p. 174)*
greatest common factor *(p. 181)*
GCF(a, b) *(p. 181)*
least common multiple *(p. 184)*
LCM(a, b) *(p. 184)*

Chapter Learning Objectives

Review Exercises

5.1 Divisibility Tests *(page 161)* — 1–44

- Use divisibility tests to decide whether a whole number is divisible by 2, 5, or 10.
- Use divisibility tests to decide whether a whole number is divisible by 4 or 8.
- Use divisibility tests to decide whether a whole number is divisible by 3, 6, or 9.

5.2 Primes and Composites *(page 171)* — 45–72

- Decide whether a natural number is prime or composite.
- Use a factor tree to write a composite number as the product of primes.
- Use prime factors to determine whether a large number is prime.

5.3 Greatest Common Factor and Least Common Multiple *(page 181)* — 73–112

- Find the greatest common factor of two or more whole numbers.
- Use algorithms to find the greatest common factor of two whole numbers.
- Find the least common multiples of two or more nonzero whole numbers.

Important Concepts and Formulas

Properties of Divisibility

1. If $n \mid a$ and $n \mid b$, then $n \mid (a + b)$.
2. If $n \mid a$ and $n \mid b$, then $n \mid (a - b)$ for $a \geq b$.
3. If $n \mid a$, then $n \mid am$.

Fundamental Theorem of Arithmetic

Every natural number greater than 1 is a prime number or can be written as the product of prime numbers in exactly one way (disregarding the order of the factors).

The product of prime factors for a given composite number is called a **prime factorization**.

To **test for prime factors** of a number n, you need only check for prime factors that are less than or equal to the square root of n.

To find the **GCF(a, b)**, write the prime factorizations of a and b. The greatest common factor is the product of the common factors.

Algorithms for finding the GCF(a, b)

1. Subtraction Algorithm
2. Euclidean Algorithm

To find the **LCM(a, b)**, write the prime factorizations of a and b using two intersecting sets. The least common multiple is the product of the factors in all three regions.

Monkey Business Images/Shutterstock.com

Review Exercises

Free step-by-step solutions to odd-numbered exercises at *MathematicalPractices.com.*

5.1 Divisibility Tests

Using the Divisibility Test for 2 Tell whether the whole number is divisible by 2. Explain your reasoning.

1. 328 **2.** 549

3. 5793 **4.** 7186

5. 9412 **6.** 8760

Using Divisibility Tests for 5 and 10 Tell whether the whole number is divisible by 5 or 10. Explain your reasoning.

7. 45 **8.** 950

9. 7485 **10.** 3029

11. 5900 **12.** 8000

Using Divisibility Tests for 4 and 8 Tell whether the whole number is divisible by 4 or 8. Explain your reasoning.

13. 5972 **14.** 1096

15. 1223 **16.** 1820

17. 9384 **18.** 2217

Using Divisibility Tests for 3 and 9 Tell whether the whole number is divisible by 3 or 9. Explain your reasoning.

19. 173 **20.** 201

21. 7254 **22.** 9362

23. 9027 **24.** 2178

Using the Divisibility Test for 6 Tell whether the whole number is divisible by 6. Explain your reasoning.

25. 24 **26.** 245

27. 1839 **28.** 8316

29. 3012 **30.** 7182

Applying Properties of Divisibility State the property illustrated by the statement. Explain your reasoning.

31. $6 \mid 96$ and $6 \mid 12$, so $6 \mid 84$.

32. $5 \mid 15$, so $5 \mid 105$.

33. $8 \mid 48$, so $8 \mid 336$.

34. $3 \mid 9$ and $3 \mid 24$, so $3 \mid 33$.

35. $4 \mid 572$ and $4 \mid 112$, so $4 \mid 460$.

36. $9 \mid 36$ and $9 \mid 540$, so $9 \mid 576$.

37. Reasoning The whole number 26 divides the whole number m. What other numbers must divide m?

38. Reasoning The whole number 42 divides the whole number m. What other numbers must divide m?

39. Using a Divisibility Test in Real Life A summer softball program has 104 players signed up to participate. Is it possible to divide the players into 8 teams with an equal number of players on each team?

40. Using a Divisibility Test in Real Life A summer soccer program has 126 players signed up to participate. Is it possible to divide the players into 9 teams with an equal number of players on each team?

41. Boardwalk New light posts are installed along a boardwalk that is 621 feet long. The distance between light posts is 9 feet.

a. Can the boardwalk have light posts at the beginning and end of its walkway and still have light posts 9 feet apart?

b. Can the light posts be installed along the boardwalk at a different uniform whole number of feet apart and have light posts at the beginning and end of its walkway? Explain your reasoning.

42. Landscaping Landscapers plant evergreen bushes along the edge of a parking lot which measure 192 yards in length. The evergreen bushes are planted 5 yards apart.

a. Can the landscapers plant bushes at each corner of the edge of the lot and still plant the bushes 5 yards apart?

b. Can the bushes be planted along the edge of the lot at a different uniform whole number of yards apart and still be planted at each end? Explain your reasoning.

43. Problem Solving The price of a package of construction paper is marked down from \$5.00 but is more than \$4.25. You buy several packages totalling \$18.72. How many packages of construction paper do you buy?

44. Problem Solving You order new posters to decorate your classroom. The total bill is \$17.85. You notice that each poster is more expensive than the original price of \$2.35.

a. What are the possible prices of each poster?

b. Which of the prices in part (a) is most reasonable? Explain.

5.2 Primes and Composites

Determining Prime and Composite Numbers Tell whether the number is *prime* or *composite*.

45. 42 **46.** 31

47. 53 **48.** 65

49. 39 **50.** 27

51. Using Factor Trees Fill in the missing numbers in the factor tree.

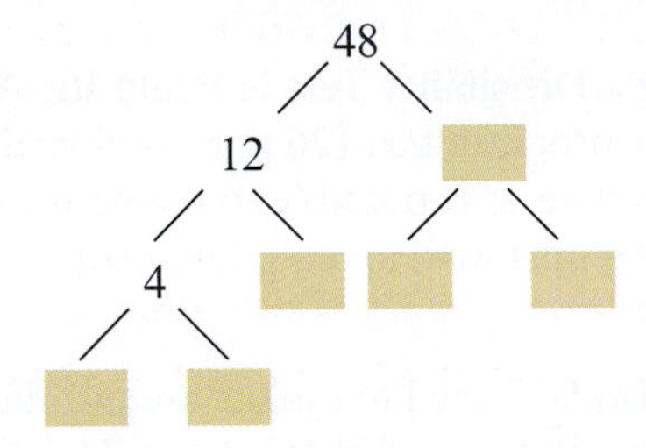

52. Using Factor Trees Fill in the missing numbers in the factor tree.

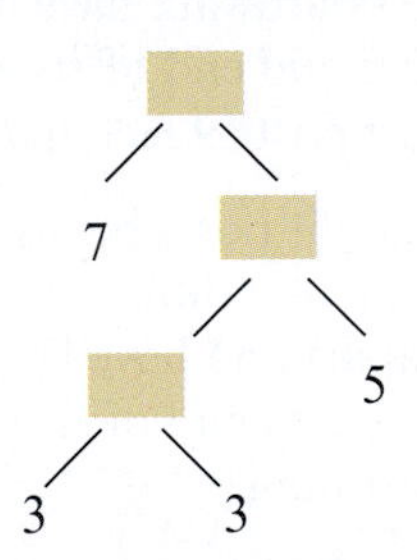

53. Deciding Whether a Number Is Prime What is the greatest prime number you need to test when determining whether 2371 is prime? Explain your reasoning.

54. Deciding Whether a Number Is Prime What is the greatest prime number you need to test when determining whether 3011 is prime? Explain your reasoning.

Determining the Prime Factorization of a Number Tell whether the number is *prime* or *composite*. If composite, write the prime factorization of the number.

55. 56 **56.** 73

57. 95 **58.** 65

59. 196 **60.** 257

61. 1257 **62.** 3791

63. 1691 **64.** 2153

Writing the Prime Factorization of a Number Write the prime factorization of the number as a product of powers.

65. $15^3 \cdot 33^2 \cdot 225^4$ **66.** $24^4 \cdot 49^{15} \cdot 155^3$

67. Problem Solving You want to rearrange 30 desks in your classroom into equal groups. One possibility is to make 6 groups of 5 desks as shown below. What other group sizes are possible?

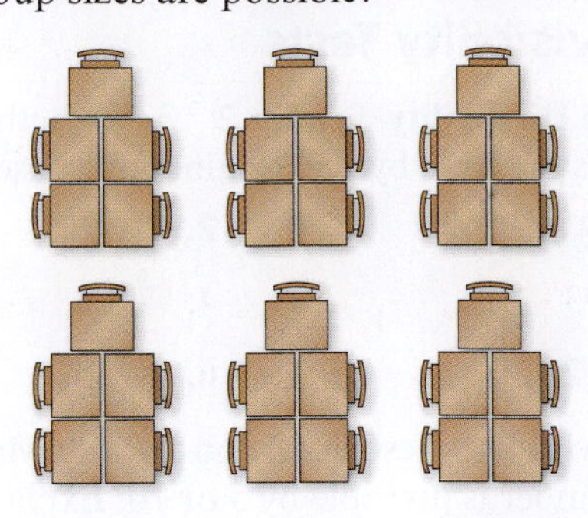

68. Problem Solving You take your class on a field trip to the zoo. You need to divide the students into equal groups to tour various exhibits. There are 20 students in your class. What are the possible group sizes?

69. Problem Solving There are 139 campers signed up for the summer art camp. Is it possible to divide the campers into groups so that each group has the same number of campers? Explain.

70. Problem Solving Volunteers to clean up local highways need to be separated into groups to cover a larger area. There are 2717 volunteers involved in the project. Is it possible to divide the volunteers into groups so that each group has the same number of volunteers? Explain.

71. Chen Primes A *Chen prime* is a prime number for which the sum of the prime number and 2 is either a prime or a semiprime. A *semiprime* is a number that is the product of two primes. For example, 2 is a prime number and $2 + 2 = 4$ is a semiprime. So, 2 is a Chen prime. List the rest of the Chen primes between 0 and 15.

72. An Unsolved Problem in Number Theory A *prime quadruplet* consists of four prime numbers beginning with the prime number p. The other three prime numbers in a prime quadruplet are then $p + 2$, $p + 6$, and $p + 8$. For example, consider the prime number 5.

$5 + 2 = 7$

$5 + 6 = 11$

$5 + 8 = 13$

So, the prime numbers 5, 7, 11, and 13 make up a prime quadruplet. Mathematicians have been unable to determine if there are infinitely many prime quadruplets. What are the next two prime quadruplets?

5.3 Greatest Common Factor and Least Common Multiple

Finding the Greatest Common Factor Use prime factorization to find the GCF of the whole numbers.

73. 18, 27
74. 8, 32
75. 12, 16
76. 24, 30

Using the Subtraction Algorithm Use the subtraction algorithm to find the GCF of the whole numbers.

77. 24, 120
78. 9, 84
79. 12, 54
80. 15, 70

Using the Euclidean Algorithm Use the Euclidean algorithm to find the GCF of the whole numbers.

81. 6, 482
82. 54, 369
83. 57, 152
84. 63, 294

Finding the Least Common Multiple Use prime factorization to find the LCM of the whole numbers.

85. 5, 45
86. 9, 16
87. 12, 16
88. 18, 27

Finding the Greatest Common Factor and Least Common Multiple For whole numbers a and b, use the diagram below to organize the factors of a and b. Then find the GCF and LCM of a and b.

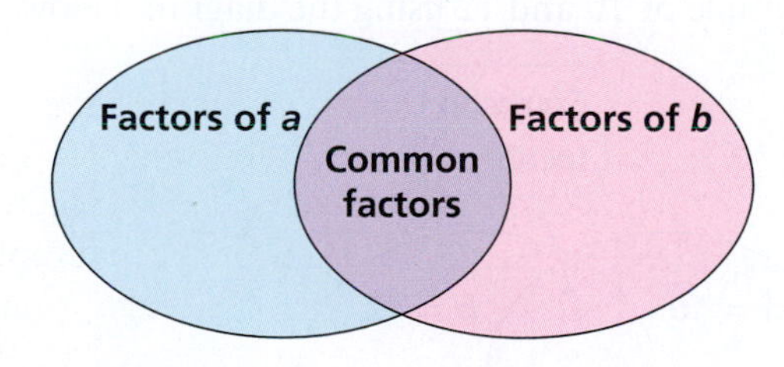

89. $a = 60, b = 84$

90. $a = 21, b = 90$

Finding the Greatest Common Factor Find the GCF of the whole numbers.

91. 4, 16, 68
92. 42, 63, 105
93. 54, 63, 72
94. 53, 79, 97
95. 51, 68, 85
96. 16, 48, 136

Finding the Least Common Multiple Find the LCM of the whole numbers.

97. 3, 13, 19
98. 15, 27, 45
99. 6, 54, 81
100. 9, 13, 39
101. 12, 40, 56
102. 10, 14, 45

Product Property of GCF and LCM Show that the *Product Property of GCF and LCM* is true for each pair of whole numbers.

$$a \cdot b = \text{GCF}(a, b) \cdot \text{LCM}(a, b)$$

103. $a = 9, b = 27$

104. $a = 15, b = 20$

105. $a = 8, b = 10$

106. $a = 14, b = 21$

107. **Marathon Training** Two runners are training for a marathon by running laps around a one-mile course. It takes the first runner 8 minutes and the second runner 9 minutes to complete one lap. Both runners begin at the same time and place and maintain their current paces. After how many minutes will the runners pass the beginning of the course together?

108. **Bus Routes** Two buses pass by your house along different routes. It takes the first bus 24 minutes and the second bus 40 minutes to complete their respective routes. Both buses pass your house at the same time. After how many minutes will both buses pass by your house at the same time again?

109. **Prize Bags** You are putting together prize bags to give your students. You have 78 stickers and 104 shaped erasers. What is the greatest number of identical prize bags you can make without having any items left over?

110. **Swimming** The swim team has 16 new swimmers and 24 returning swimmers. The coach wants to form practice teams with each team having the same number of new and returning swimmers. What is the greatest number of practice teams the coach can form using every swimmer?

111. **Giveaways** The local minor league baseball team is having a giveaway for children who attend the next game. They have 160 T-shirts, 240 souvenir baseballs, and 480 glow sticks. Children arriving at the game will receive identical prizes until the items are gone. What is the greatest number of children who receive a prize if there are no items left over?

112. **Craft Supplies** Your craft project requires one craft stick, one paper plate, and one sheet of construction paper. Craft sticks are sold in packs of 75 sticks. Paper plates are sold in packs of 100 plates. Construction paper is sold in packs of 50 sheets. How many packs of each item do you need to buy in order to have the same number of craft sticks, paper plates, and sheets of construction paper?

Chapter Test

The questions on this practice test are modeled after questions on state certification exams for K–8 teaching.

Test Directions: Each of the questions is followed by five choices. Choose the *best* response to each question.

1. Which whole number is divisible by 3 and 9?

A. 68 **B.** 289 **C.** 1,645
D. 2,439 **E.** 3,515

2. Which whole number is divisible by 4 and 8?

A. 54 **B.** 347 **C.** 1,544
D. 2,391 **E.** 3,154

3. Which whole number is prime?

A. 6 **B.** 18 **C.** 24
D. 35 **E.** 47

4. What is the prime factorization of 48?

A. $2 \cdot 2 \cdot 2 \cdot 2 \cdot 3$ **B.** $2 \cdot 2 \cdot 3$ **C.** $2^3 \cdot 3$
D. 2^4 **E.** $2 \cdot 3$

5. Find the greatest common factor of 36 and 48.

A. 2 **B.** 3 **C.** 6
D. 12 **E.** 48

6. Find the least common multiple of 10 and 12 using the diagram below.

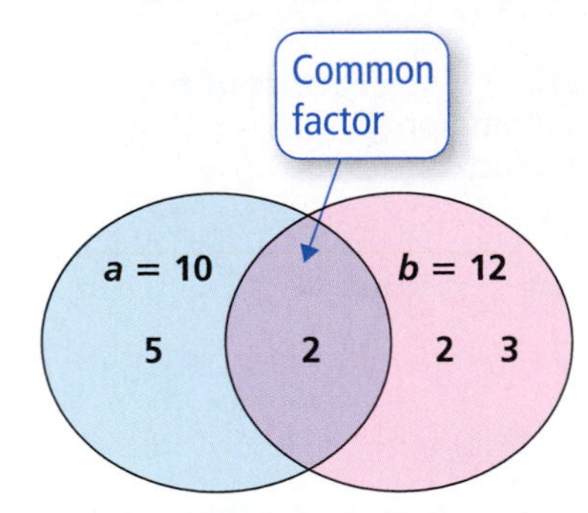

A. 2 **B.** $2^2 \cdot 3 \cdot 5$ **C.** $2 \cdot 3$
D. $2 \cdot 3 \cdot 5$ **E.** $3 \cdot 5$

7. You want to divide your soccer team into groups of equal size for a practice. You coach 36 players. Which is not one of the possible group sizes?

A. 3 **B.** 6 **C.** 8
D. 12 **E.** 18

8. You are making identical fruit baskets. You have 24 oranges, 36 apples, and 60 bananas. What is the greatest number of fruit baskets you can make if you use every piece of fruit?

A. 6 **B.** 12 **C.** 18
D. 24 **E.** 36

6 Fractions

iofoto/Shutterstock.com

Activity: Modeling Fractions with Pattern Blocks

Materials:

- Pattern blocks
- Pencil

Learning Objective:

Students recognize how many parts are in a whole. Then they use the result to write a fraction.

Name ______________________

1. How many are in ? If = 1, then = ——.

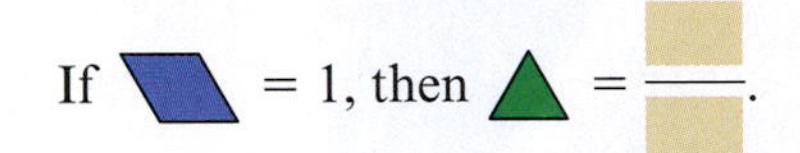

2. How many are in ? If = 1, then = ——.

3. How many are in ? If = 1, then = ——.

4. How many are in ? If = 1, then = ——.

5. How many are in ? If = 1, then = ——.

6. How many are in ? If = 1, then = ——.

7. How many are in ? If = 1, then = ——.

8. How many are in ? If = 1, then = ——.

9. How many are in ? If = 1, then = ——.

10. How many are in ? If = 1, then = ——.

Dmitriy Shironosov | Dreamstime.com

6.1 Fractions

Standards

Grades K–2
Geometry
Students should reason with shapes and their attributes.

Grades 3–5
Numbers and Operations–Fractions
Students should develop understanding of fractions as numbers, and extend understanding of fraction equivalence and ordering.

- Understand fractions and mixed numbers.
- Recognize equivalent fractions and find simplest form.
- Use a real number line to graph and order fractions.

Fractions and Mixed Numbers

Definition of Fraction

A **fraction** is a number of the form $\frac{a}{b}$ $(b \neq 0)$ where a is called the **numerator** and b is called the **denominator**.

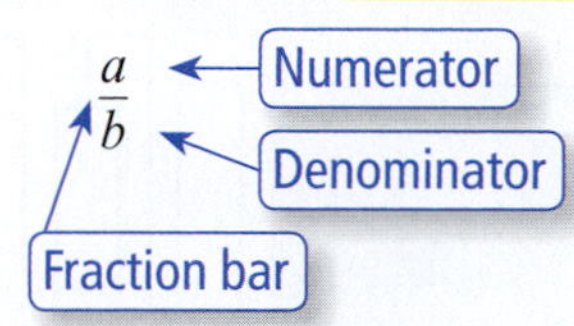

EXAMPLE 1 **Real-Life Examples of Fractions**

Classroom Tip Some students think of a fraction as two numbers separated by a line. Be sure that you help students understand that *a fraction is a single number*.

a. You ate $\frac{1}{6}$, or *one-sixth*, of a pizza.

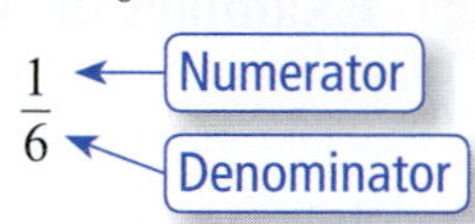

Five-sixths of the pizza is left.

b. *Five-sixths*, or $\frac{5}{6}$, of the pizza is left.

c. The recipe for frosting uses $\frac{2}{3}$ cup of butter.

d. The recipe for cupcakes uses $\frac{1}{2}$ cup of butter.

e. The caterpillar is $\frac{7}{8}$ inch long.

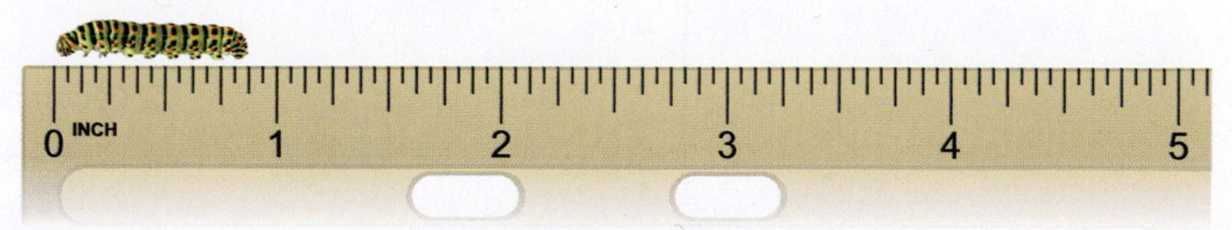

Africa Studio/Shutterstock.com; M. Unal Ozmen/Shutterstock.com; arka38/Shutterstock.com

Modeling fractions using concrete materials and drawings helps students develop an understanding of fractions. Three common fraction models are: *area models*, *set models*, and *linear models*.

Classroom Tip

Elementary school students may confuse the concepts of length and area. You can introduce area models with a simple demonstration, using two identical square slices of cheese. Cut one in half vertically (forming rectangles), and cut the other in half diagonally (forming triangles). Then ask the students, "Which piece is larger?" Although the shapes and dimensions are different, each piece represents the same area—one half of the whole slice of cheese.

Area Models for Fractions

In an **area model**, a geometric shape represents the whole unit where the whole unit is divided into fractional parts of the whole.

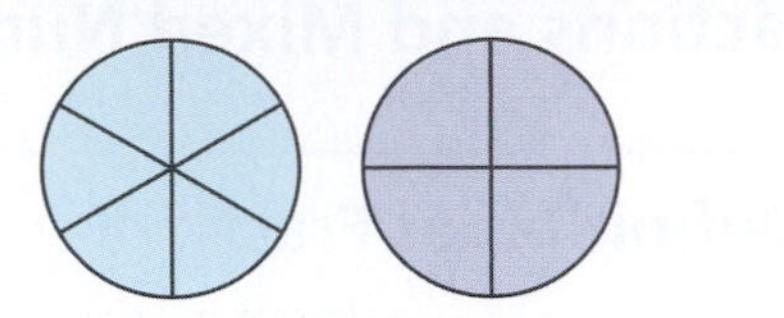
Fraction circles

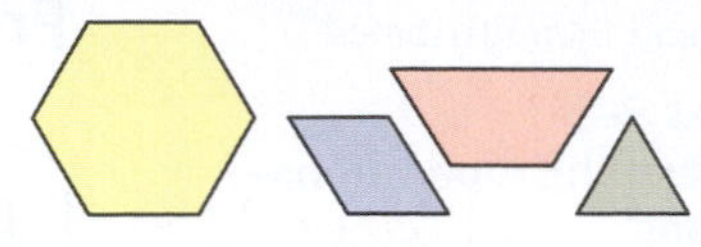
Pattern blocks

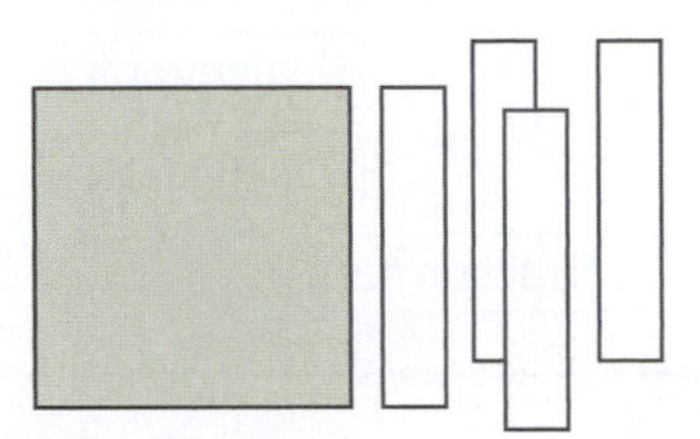
Fraction rectangles

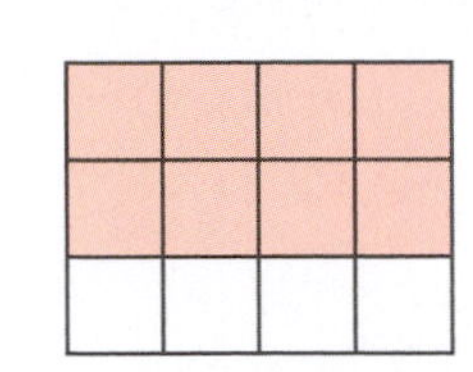
Square tiles

Classroom Tip

Students may need extensive practice reading fractions. For example, the fraction $\frac{2}{5}$ is read as "*two-fifths*" and not as "*two over five*."

EXAMPLE 2 Examples of Area Models for Fractions

a. $\frac{4}{6}$ $\frac{1}{4}$

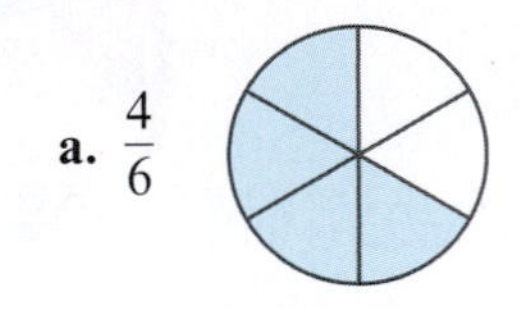

$\frac{2}{5}$

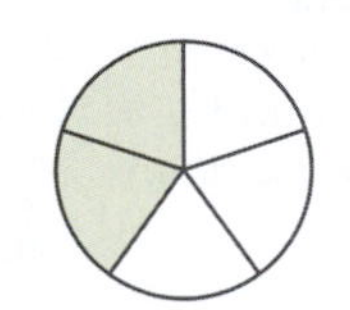

$\frac{2}{3}$

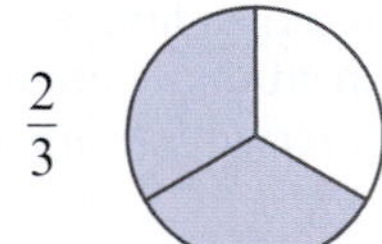

b. $\frac{4}{9}$ $\frac{7}{12}$ $\frac{8}{20}$

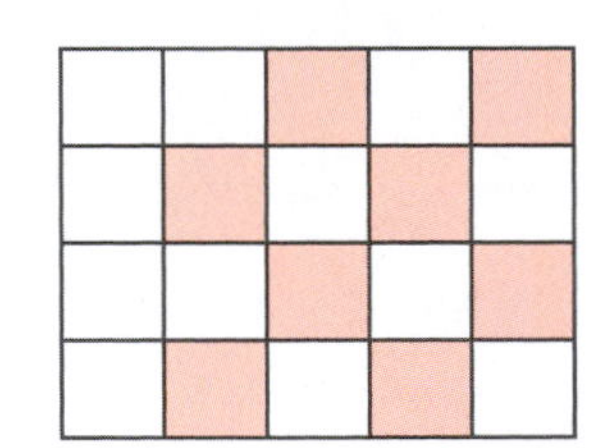

Set Models for Fractions

In a **set model**, a collection of objects represents the whole unit and a subset of the collection represents a fractional part of the whole.

Real objects

Counters

Square tiles

EXAMPLE 3 **Examples of Set Models for Fractions**

$\frac{2}{10}$ of the crayons are yellow.

$\frac{3}{10}$ of the crayons are green.

$\frac{5}{10}$ of the crayons are blue.

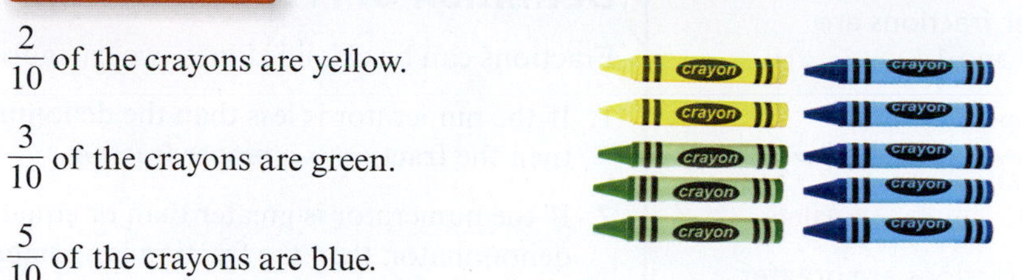

Classroom Tip

The terms *area model*, *set model*, and *linear model* are not terms intended for students in elementary school. Students should, however, be able to distinguish between area and length.

Linear Models for Fractions

In a **linear model**, a length represents the whole unit where the whole unit is divided into fractional parts of the whole.

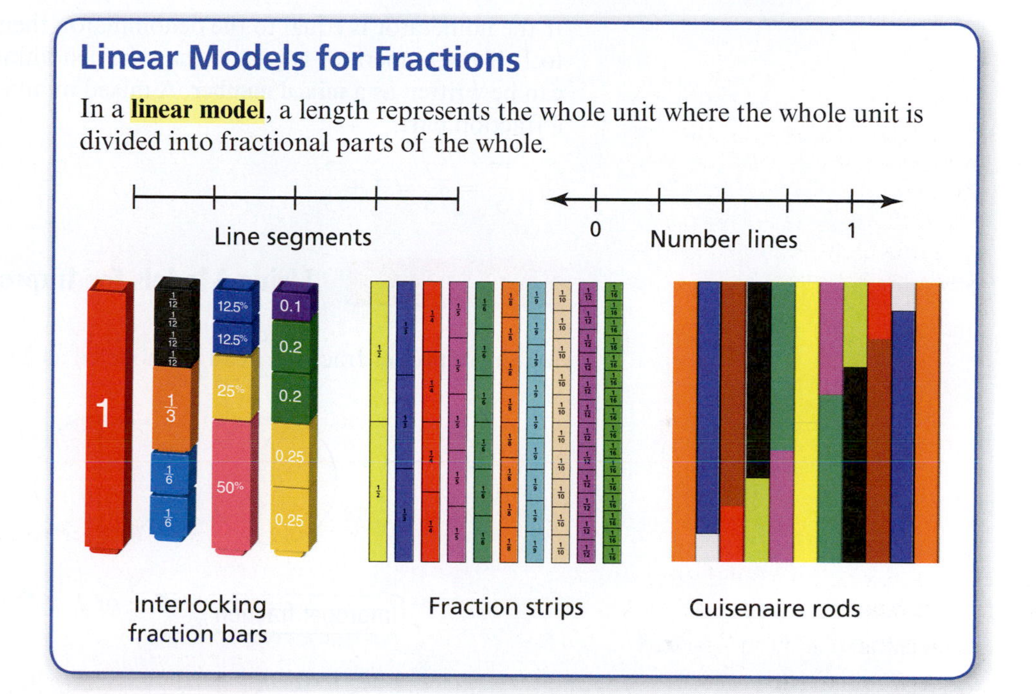

EXAMPLE 4 **Examples of Linear Models for Fractions**

Fraction strips can show students how different fractions represent the same length. Cut 3 paper strips that have the same length. This length represents the unit number 1. Divide one strip into two equal parts. Each part represents $\frac{1}{2}$.

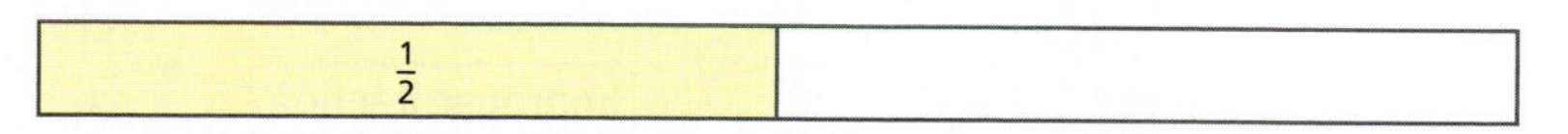

Divide the second strip into 4 equal parts. Each part represents $\frac{1}{4}$.

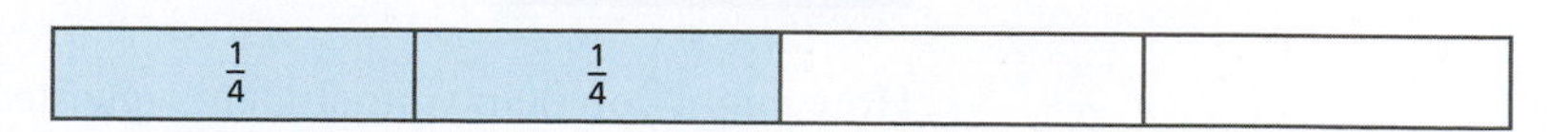

Divide the third strip into 8 equal parts. Each part represents $\frac{1}{8}$.

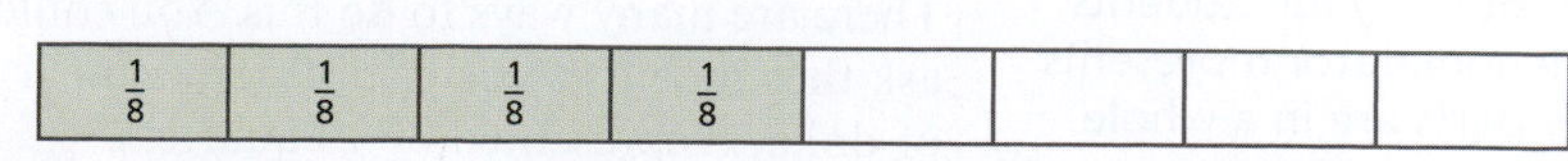

Each fraction strip has the same amount shaded. So, the fractions represent the same length.

Mathematical Practices

Use Appropriate Tools Strategically

MP5 Mathematically proficient students consider the tools available when solving a mathematical problem and are sufficiently familiar with tools appropriate for their grade to make sound decisions about when each of these tools might be helpful, recognizing both the insight to be gained and their limitations.

Classroom Tip
Proper fractions are between 0 and 1.

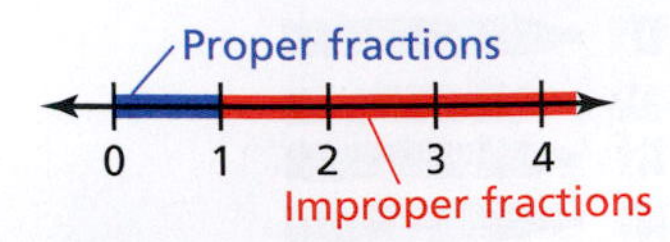

Improper fractions are greater than or equal to 1.

Definition of Proper and Improper Fractions

Fractions can be classified into two categories.

Examples

1. If the numerator is less than the denominator, then the fraction is a **proper fraction**. $\frac{2}{3}, \frac{1}{5}, \frac{2}{4}$
2. If the numerator is greater than or equal to the denominator, then the fraction is an **improper fraction**. $\frac{3}{2}, \frac{5}{3}, \frac{7}{2}, \frac{3}{3}$

If the numerator is equal to the denominator, then the improper fraction is equal to 1. If the numerator is greater than the denominator, then the improper fraction can be written as a **mixed number**. A mixed number has a whole number part and a fraction part.

$$\frac{17}{5} = 3\frac{2}{5} = 3 + \frac{2}{5}$$

EXAMPLE 5 Using Models for Improper Fractions

a. The improper fraction $\frac{9}{4}$ can be modeled as 2 wholes plus $\frac{1}{4}$.

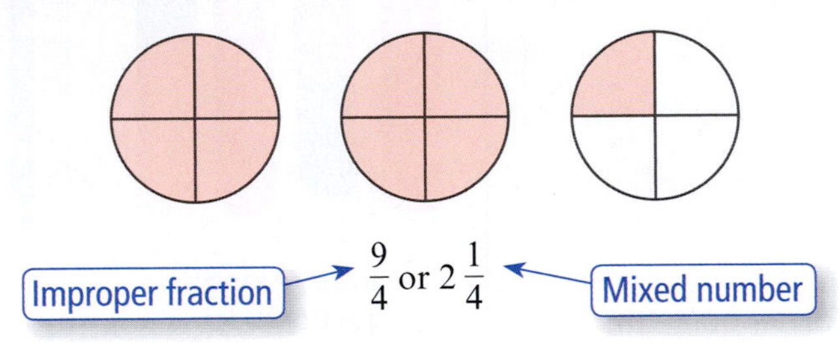

Improper fraction → $\frac{9}{4}$ or $2\frac{1}{4}$ ← Mixed number

b. The improper fraction $\frac{11}{3}$ can be modeled as 3 wholes plus $\frac{2}{3}$.

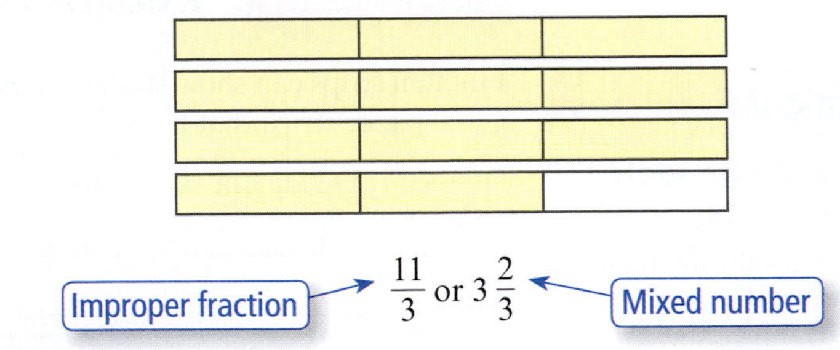

Improper fraction → $\frac{11}{3}$ or $3\frac{2}{3}$ ← Mixed number

EXAMPLE 6 Helping Students Write Mixed Numbers

How might you help your students rewrite $\frac{15}{4}$ as a mixed number?

SOLUTION

There are many ways to do this. You could ask the students to determine the number of dollars represented by 15 quarters.

$$15 \text{ quarters (fourths)} = 3\frac{3}{4} \text{ dollars}$$

Classroom Tip
Point out to your students that the denominator represents how many parts are in a whole.

Equivalent Fractions and Simplest Form

Two fractions are **equivalent** if they represent the same quantity.

EXAMPLE 7 **Recognizing Equivalent Fractions**

On each school day, you start with a whole round of cheddar cheese and cut it into pieces of equal size. On which days do you eat the same amount of cheese?

Whole

Classroom Tip
The concept of equivalent fractions should be developed by problem-solving rather than by procedure. Simply telling students to "multiply the numerator and denominator by the same number" does little to help their understanding of fractions or of equivalence.

Monday

Cut 2 pieces.
Eat 1 piece.

Tuesday

Cut 6 pieces.
Eat 3 pieces.

Wednesday

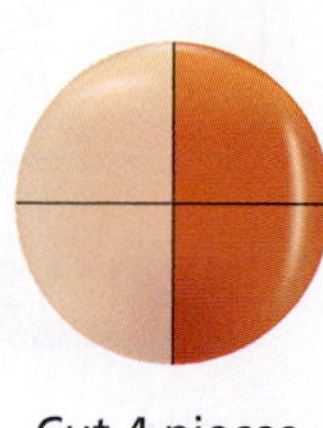

Cut 4 pieces.
Eat 2 pieces.

Thursday

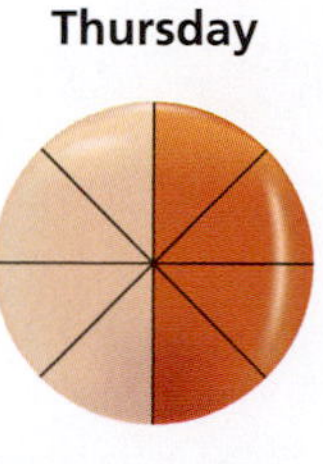

Cut 8 pieces.
Eat 4 pieces.

Friday

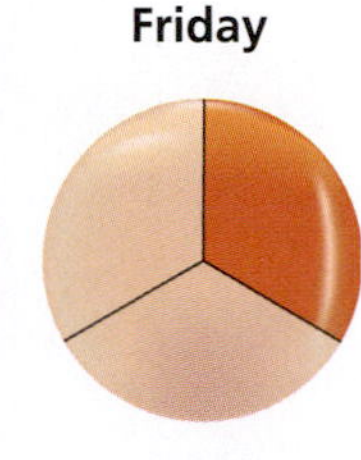

Cut 3 pieces.
Eat 2 pieces.

SOLUTION

On Monday, Tuesday, Wednesday, and Thursday, you eat the same amount of cheese. On each of these days, you eat half of the whole round.

$$\frac{1}{2} = \frac{3}{6} = \frac{2}{4} = \frac{4}{8} \qquad \text{Equivalent fractions}$$

On Friday, you eat more than half of the round.

> **Definition of Simplest Form**
>
> A fraction is in **simplest form** when its numerator and denominator have no common factors other than 1.

EXAMPLE 8 **Recognizing Fractions in Simplest Form**

a. When you factor the numerator and denominator of the fraction $\frac{9}{4}$, you can see that there are no common factors other than 1. So, the simplest form is $\frac{9}{4}$.

$$\frac{9}{4} = \frac{3 \cdot 3}{2 \cdot 2} \qquad \text{Simplest form is } \frac{9}{4}.$$

b. When you factor the numerator and denominator of the fraction $\frac{2}{6}$, you can see that 2 is a common factor. When there are no common factors other than 1 in the numerator and the denominator, the resulting fraction is in simplest form.

$$\frac{2}{6} = \frac{\overset{1}{\cancel{2}} \cdot 1}{\underset{1}{\cancel{2}} \cdot 3} = \frac{1}{3} \qquad \text{Simplest form is } \frac{1}{3}.$$

Classroom Tip
Simplified fractions are sometimes called *reduced fractions*. This term, however, can be misleading because students can interpret it to mean that the fraction is lesser.

Classroom Tip

Equivalence (or equality) is a core concept in mathematics. It is represented by an equal sign (=). For example, the fractions (or *numerals*) 1/2, 2/4, 3/6, 5/10, and 8/16 are all equivalent. Each represents the *number* "one-half."

Tests for Equivalent Fractions

Let $\frac{a}{b}$ and $\frac{c}{d}$ be any two fractions, where $b \neq 0$ and $d \neq 0$. Here are two tests for determining whether the fractions are equivalent (or equal).

1. **Compare simplified fractions** Write $\frac{a}{b}$ and $\frac{c}{d}$ in simplest form. If the two simplified fractions are the same, then the two original fractions are equivalent.
2. **Compare cross products** Find the two **cross products** by multiplying the denominator of each fraction by the numerator of the other fraction. This is called **cross-multiplication**. If the cross products are equal, then the two original fractions are equivalent.

$\frac{a}{b} \stackrel{?}{=} \frac{c}{d}$ — $b \cdot c$ Cross product; $a \cdot d$ Cross product

EXAMPLE 9 Helping Students Understand Equivalence

In how many ways can you help students show that $\frac{7}{15}$ and $\frac{2}{4}$ are not equivalent?

SOLUTION

Here are four ways to show that the fractions are not equivalent.

Compare linear models Be sure the models have the same total length.

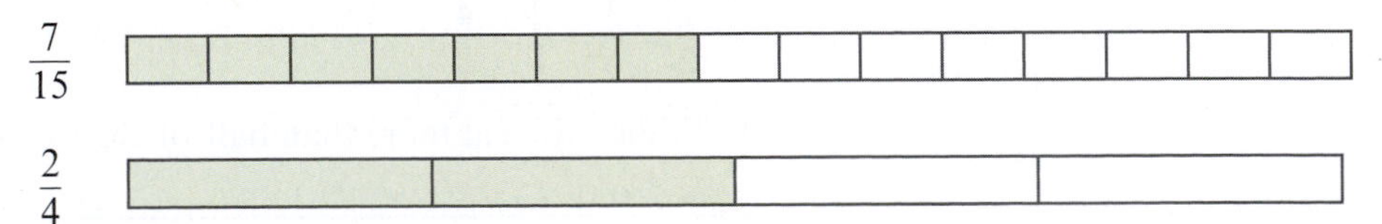

Because the shaded parts represent different lengths, the fractions are *not* equivalent.

Compare cross products Find the cross products.

$\frac{7}{15} \stackrel{?}{=} \frac{2}{4}$ — $15 \cdot 2 = 30$ Cross product; $7 \cdot 4 = 28$ Cross product

Because the cross products are not equal, the fractions are *not* equivalent.

Compare simplified fractions Write each fraction in simplest form.

$\frac{7}{15}$ Fraction is already in simplest form.

$\frac{2}{4} = \frac{1}{2}$ Fraction simplifies to $\frac{1}{2}$.

Because the simplified fractions are not the same, the fractions are *not* equivalent.

Compare area models Be sure the models have the same total area.

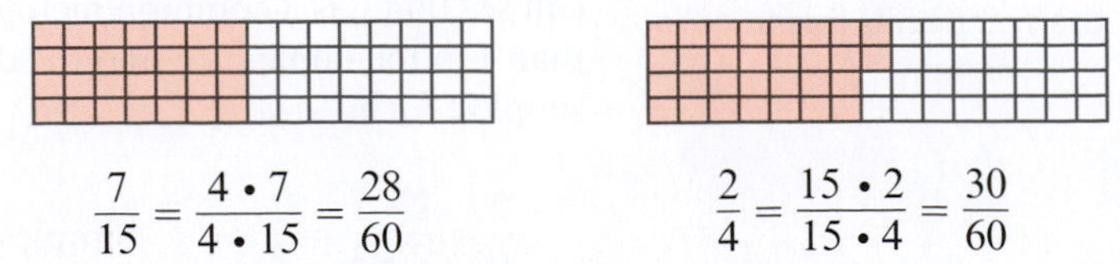

$$\frac{7}{15} = \frac{4 \cdot 7}{4 \cdot 15} = \frac{28}{60} \qquad \frac{2}{4} = \frac{15 \cdot 2}{15 \cdot 4} = \frac{30}{60}$$

Because the shaded parts represent different areas, the fractions are *not* equivalent.

Ordering Fractions

Ordering Fractions Using a Number Line

To order any two fractions, graph them on a number line. The fraction that is to the left **is less than** the fraction that is to the right.

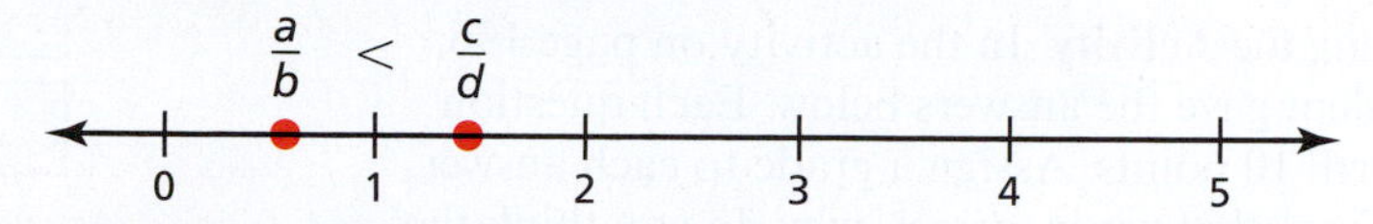

Classroom Tip

When two fractions have the same denominator, the lesser fraction is the one with the lesser numerator.

When two fractions have the same numerator, the lesser fraction is the one with the greater denominator.

EXAMPLE 10 **Ordering Fractions Using a Number Line**

a.

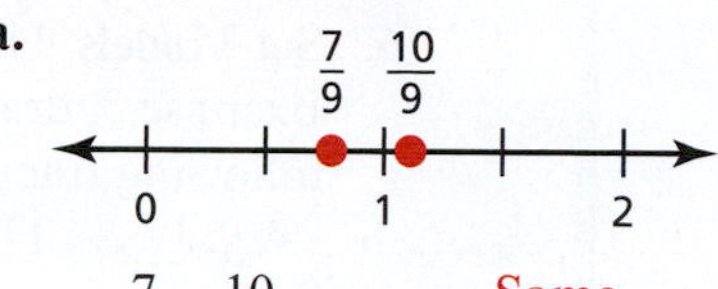

So, $\frac{7}{9} < \frac{10}{9}$. Same denominator

b.

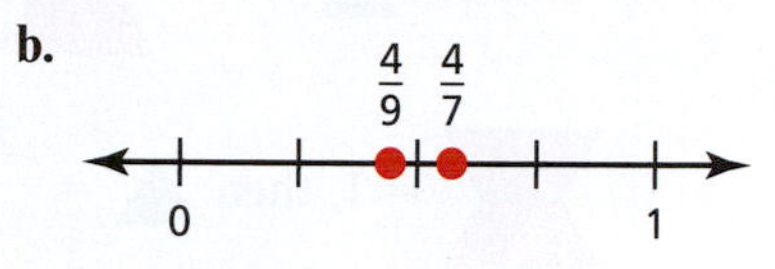

So, $\frac{4}{9} < \frac{4}{7}$. Same numerator

Sometimes it is possible to order proper fractions by rounding them to convenient **benchmarks**, such as 0, $\frac{1}{2}$, or 1.

- If the numerator is close to 0, then the fraction is close to the benchmark 0.
- If the numerator is close to half the denominator, then the fraction is close to the benchmark $\frac{1}{2}$.
- If the numerator is close to the denominator, then the fraction is close to the benchmark 1.

Classroom Tip

When students are first introduced to the concept of fractions and order, it is often difficult for them to understand that "larger denominators" tend to produce "smaller fractions." For example, it is common for students to think that $\frac{1}{9}$ is greater than $\frac{1}{6}$ because 9 is greater than 6.

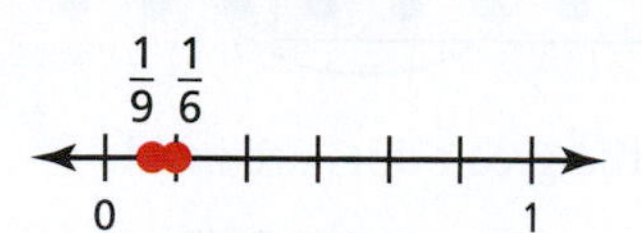

EXAMPLE 11 **Ordering Fractions Using Benchmarks**

Order the fractions $\frac{7}{8}$, $\frac{7}{12}$, and $\frac{9}{20}$ from least to greatest using benchmarks.

SOLUTION

- $\frac{7}{8}$: Because 7 is close to 8, $\frac{7}{8}$ is close to 1.
- $\frac{7}{12}$: Because 7 is close to half of 12, $\frac{7}{12}$ is close to $\frac{1}{2}$.
- $\frac{9}{20}$: Because 9 is close to half of 20, $\frac{9}{20}$ is close to $\frac{1}{2}$.

To compare fractions that are close to $\frac{1}{2}$, compare the numerator of each fraction to half of its denominator.

- $\frac{7}{12}$: Because half of 12 is 6, and $7 > 6$, $\frac{7}{12} > \frac{1}{2}$.
- $\frac{9}{20}$: Because half of 20 is 10, and $9 < 10$, $\frac{9}{20} < \frac{1}{2}$.

So, the fractions in order from least to greatest are $\frac{9}{20}$, $\frac{7}{12}$, and $\frac{7}{8}$.

6.1 Exercises

Free step-by-step solutions to odd-numbered exercises at *MathematicalPractices.com.*

1. **Writing a Solution Key** Write a solution key for the activity on page 196. Describe a rubric for grading a student's work.

2. **Grading the Activity** In the activity on page 196, a student gave the answers below. Each question is worth 10 points. Assign a grade to each answer. For those that are incorrect, why do you think the student erred?

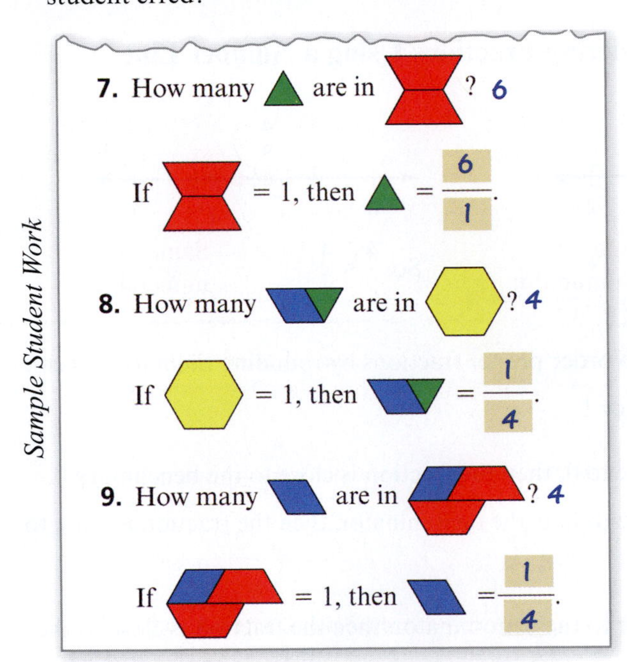

3. **Area Models** Write the fraction represented by the shaded portion of each figure. Identify the numerator and denominator.

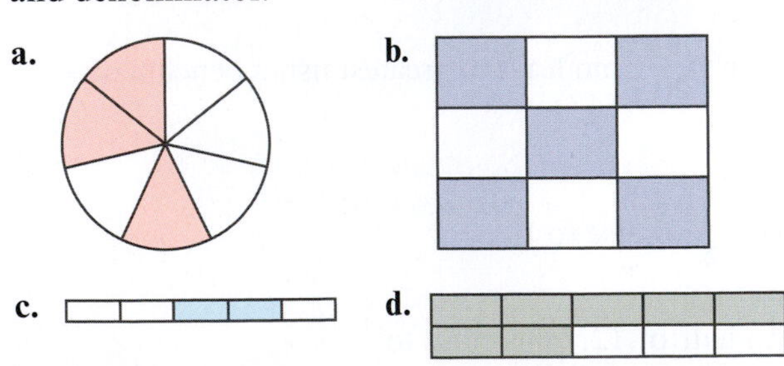

4. **Area Models** Write the fraction represented by the shaded portion of each figure. Identify the numerator and denominator.

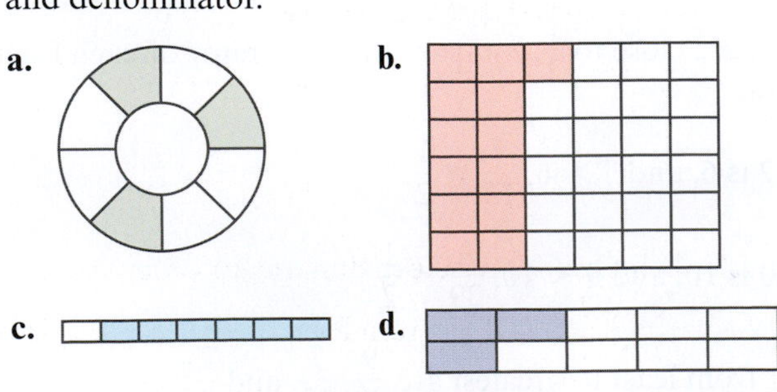

5. **Set Models** There are 12 polygon tiles. Answer the questions about the tiles.

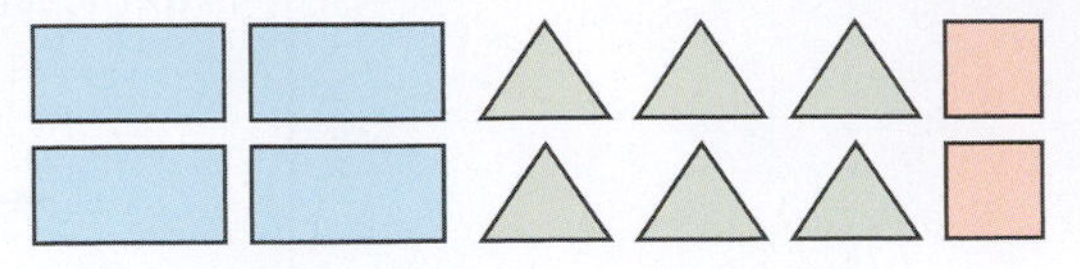

a. What fraction of the tiles are squares?

b. What fraction of the tiles are triangles?

c. What fraction of the tiles are quadrilaterals?

6. **Set Models** Using some or all of the tiles in Exercise 5, draw a set model for each of the following fractions.

a. $\frac{3}{5}$ b. $\frac{4}{7}$

c. $\frac{1}{3}$ d. $\frac{4}{6}$

7. **Set Models** Use the diagram.

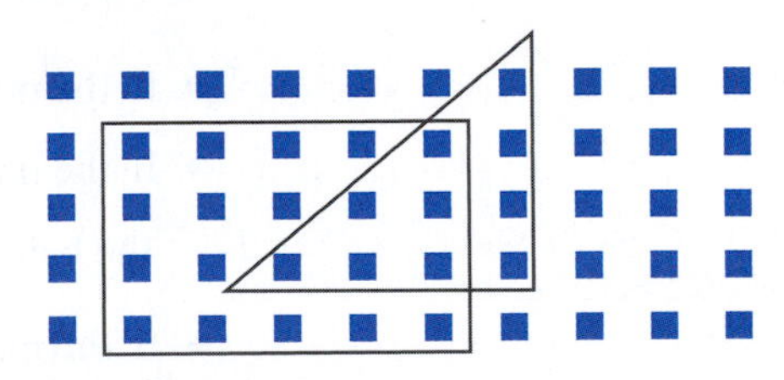

a. What fraction of the blue squares are inside the rectangle?

b. What fraction of the blue squares are inside the triangle?

c. What fraction of the blue squares are inside both the rectangle *and* the triangle?

8. **Set Models** Use the diagram.

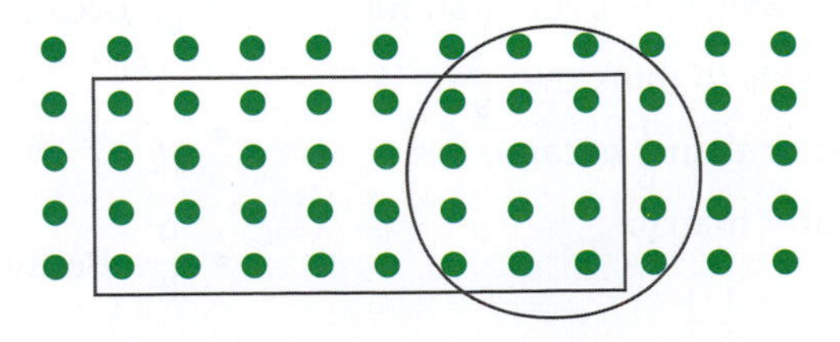

a. What fraction of the green dots are inside the circle?

b. What fraction of the green dots are inside the rectangle?

c. What fraction of the green dots are inside both the circle *and* the rectangle?

9. Linear Models Draw a number line like the one shown below. Use a dot ● on the number line to represent each fraction.

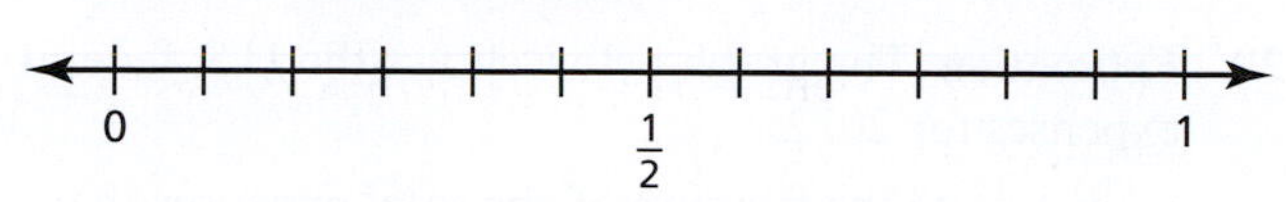

a. $\frac{1}{4}$ **b.** $\frac{2}{3}$ **c.** $\frac{3}{4}$ **d.** $\frac{11}{12}$

10. Linear Models Draw a number line like the one shown below. Use a dot ● on the number line to represent each fraction.

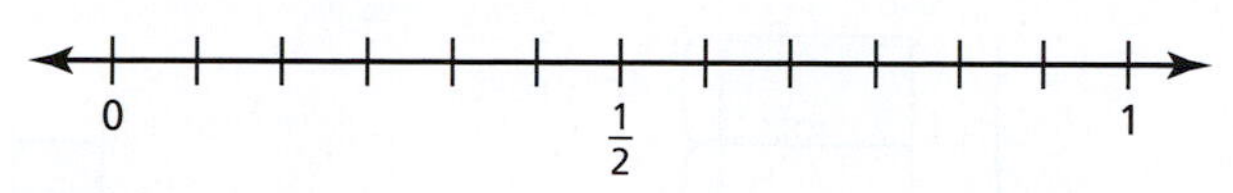

a. $\frac{3}{5}$ **b.** $\frac{7}{8}$ **c.** $\frac{1}{6}$ **d.** $\frac{7}{10}$

11. Improper Fractions and Mixed Numbers Draw a ruler like the one shown below. Represent each improper fraction or mixed number on your ruler.

a. $\frac{5}{4}$ **b.** $\frac{5}{2}$ **c.** $3\frac{1}{8}$ **d.** $2\frac{3}{4}$

12. Improper Fractions and Mixed Numbers Draw an area model to represent each improper fraction or mixed number.

a. $\frac{11}{6}$ **b.** $\frac{7}{3}$ **c.** $3\frac{5}{6}$ **d.** $2\frac{1}{3}$

13. Writing Simplified Fractions Estimate what fraction of the 60-mile trip the car has traveled. Write your answer in simplest form.

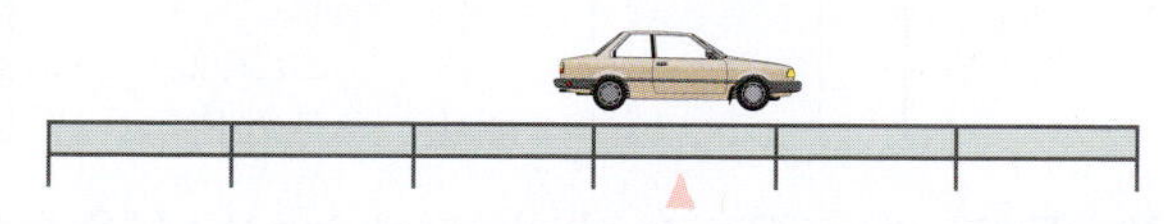

14. Writing Simplified Fractions Estimate what fraction of the 300-mile trip the car has traveled. Write your answer in simplest form.

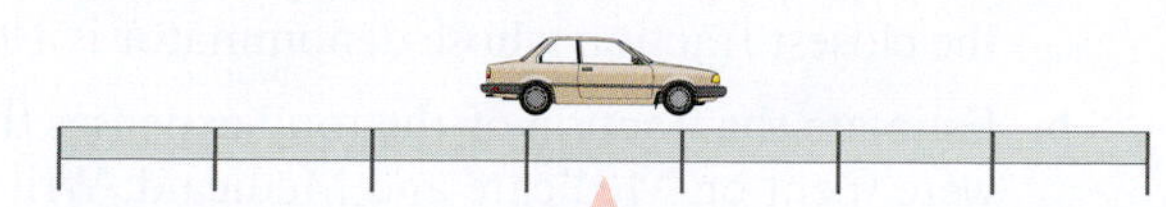

15. Real-Life Use of Fractions Use a fraction to fill in each blank. Write the fraction in simplest form.

a. 20 minutes is ________ of an hour.

b. 250 meters is ________ of a kilometer.

c. 8 ounces is ________ of a pound.

d. 8 inches is ________ of a foot.

16. Real-Life Use of Fractions Use a fraction to fill in each blank. Write the fraction in simplest form.

a. 1 foot is ________ of a yard.

b. 45 seconds is ________ of a minute.

c. 800 grams is ________ of a kilogram.

d. 6 hours is ________ of a day.

17. Modeling Equivalence Decide whether the two fractions are equivalent by using an area model.

a. $\frac{1}{3}$ and $\frac{2}{6}$ **b.** $\frac{2}{3}$ and $\frac{3}{4}$

c. $\frac{1}{2}$ and $\frac{6}{14}$ **d.** $\frac{3}{5}$ and $\frac{6}{10}$

18. Modeling Equivalence Decide whether the two fractions are equivalent by using an area model.

a. $\frac{1}{4}$ and $\frac{2}{7}$ **b.** $\frac{2}{8}$ and $\frac{1}{4}$

c. $\frac{1}{5}$ and $\frac{1}{4}$ **d.** $\frac{3}{8}$ and $\frac{6}{16}$

19. Determining Equivalence Decide whether the two fractions are equivalent by writing them in simplest form.

a. $\frac{9}{54}$ and $\frac{17}{102}$ **b.** $\frac{18}{54}$ and $\frac{19}{56}$

c. $\frac{15}{103}$ and $\frac{7}{49}$ **d.** $\frac{270}{1080}$ and $\frac{26}{104}$

20. Determining Equivalence Decide whether the two fractions are equivalent by writing them in simplest form.

a. $\frac{42}{70}$ and $\frac{51}{83}$ **b.** $\frac{38}{10}$ and $\frac{76}{20}$

c. $\frac{339}{501}$ and $\frac{61}{100}$ **d.** $\frac{81}{243}$ and $\frac{45}{135}$

21. Comparing Fractions Place the correct symbol (<, =, or >) between the two fractions.

a. $\frac{1}{3} \square \frac{1}{4}$ **b.** $\frac{1}{6} \square \frac{1}{5}$

c. $\frac{5}{6} \square \frac{10}{12}$ **d.** $\frac{3}{13} \square \frac{3}{14}$

22. Comparing Fractions Place the correct symbol (<, =, or >) between the two fractions.

a. $\frac{1}{5} \square \frac{1}{4}$ **b.** $\frac{5}{6} \square \frac{4}{5}$

c. $\frac{3}{7} \square \frac{9}{21}$ **d.** $\frac{4}{7} \square \frac{4}{9}$

23. Benchmarks Use benchmarks to order the fractions $\frac{8}{17}$, $\frac{5}{9}$, $\frac{11}{12}$, and $\frac{1}{16}$ from least to greatest. Explain.

24. Benchmarks Use benchmarks to order the fractions $\frac{11}{20}$, $\frac{1}{10}$, $\frac{5}{11}$, and $\frac{13}{14}$ from least to greatest. Explain.

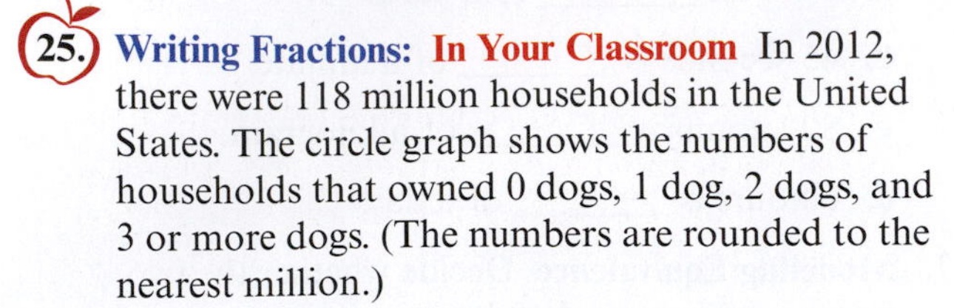

25. Writing Fractions: In Your Classroom In 2012, there were 118 million households in the United States. The circle graph shows the numbers of households that owned 0 dogs, 1 dog, 2 dogs, and 3 or more dogs. (The numbers are rounded to the nearest million.)

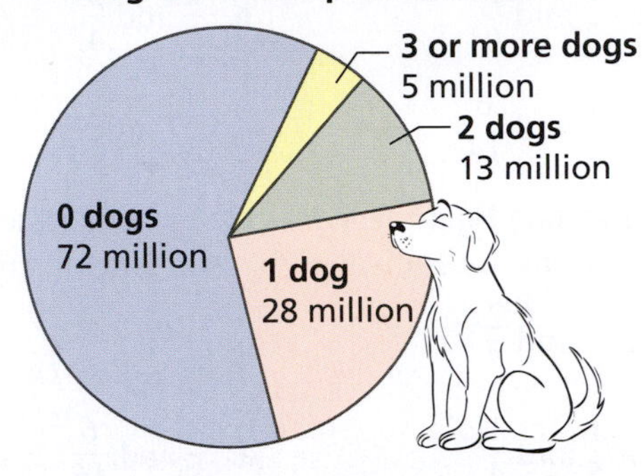

a. Write the fractions of U.S. households that owned 0 dogs, 1 dog, 2 dogs, and 3 or more dogs.

b. Describe a homework project that you could assign to your students that would involve data about pet ownership. Include a description of how you would ask your students to write and model fractions.

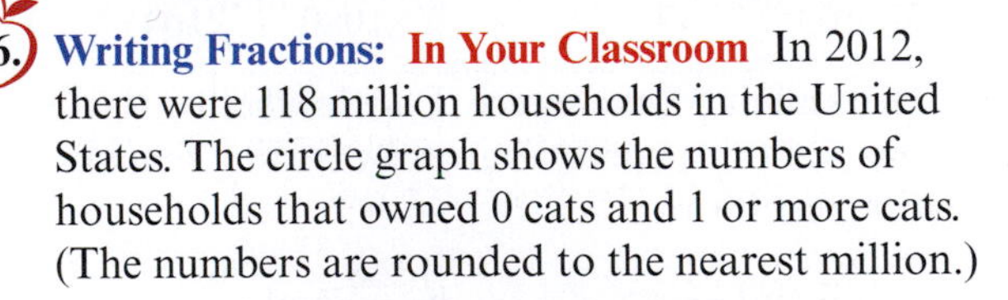

26. Writing Fractions: In Your Classroom In 2012, there were 118 million households in the United States. The circle graph shows the numbers of households that owned 0 cats and 1 or more cats. (The numbers are rounded to the nearest million.)

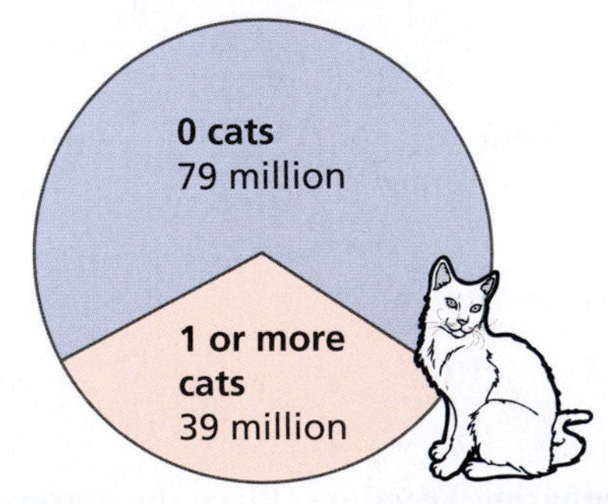

a. Write the fractions of U.S. households that owned 0 cats and 1 or more cats.

b. Describe a homework project that you could assign to your students that would involve data about pet ownership. Include a description of how you would ask your students to write and model fractions.

27. Writing Explain how to write a fraction in simplest form.

Andre Adams/Shutterstock.com; Petrovic Igor/Shutterstock.com

28. Writing Describe a way to use cross products to determine whether two mixed numbers are equivalent.

29. Estimation The graph below shows the U.S. federal expenses for 2012.

a. Estimate the fraction of the total expenses that were spent on defense. Write your answer as the closest fraction whose denominator is 100.

b. Estimate the fraction of the total expenses that were spent on Social Security. Write your answer as the closest fraction whose denominator is 100.

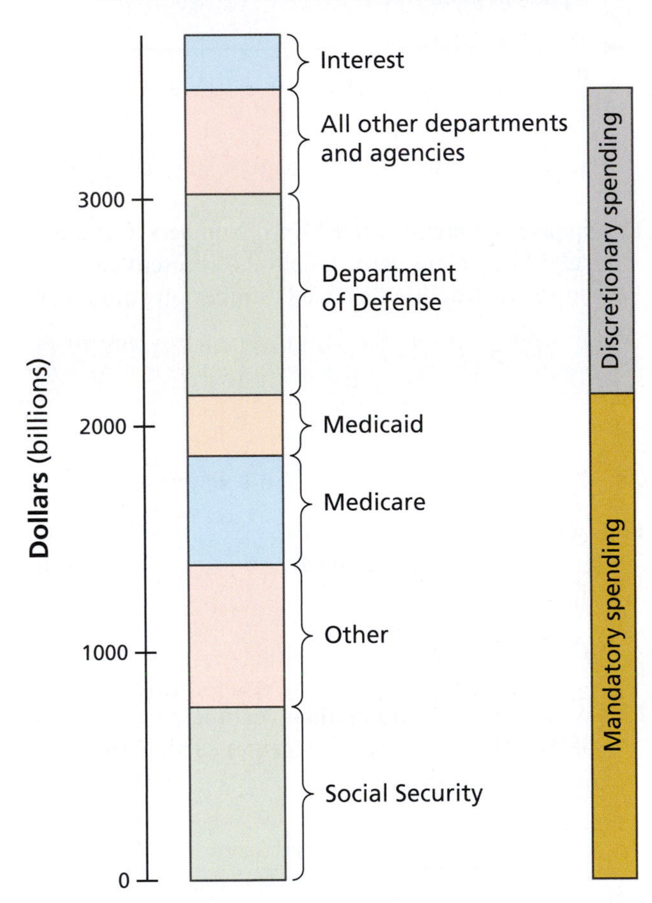

30. Estimation The graph above shows the U.S. federal expenses for 2012.

a. Estimate the fraction of the total expenses that were spent on Medicare. Write your answer as the closest fraction whose denominator is 100.

b. Estimate the fraction of the total expenses that were spent on Medicare and Medicaid. Write your answer as the closest fraction whose denominator is 100.

31. Grading Student Work On a diagnostic test, one of your students does the following work. Explain what the student did wrong. Which topics would you encourage the student to review?

$$\frac{42}{118} = \frac{\cancel{2}\cdot 3\cdot \cancel{7}}{\cancel{2}\cdot 7\cdot \cancel{7}} = \frac{3}{7}$$

32. **Grading Student Work** On a diagnostic test, one of your students does the following work. Explain what the student did wrong. Which topics would you encourage the student to review?

$$\frac{72}{104} = \frac{\cancel{2} \cdot \cancel{2} \cdot \cancel{2} \cdot 2 \cdot 3}{\cancel{2} \cdot \cancel{2} \cdot \cancel{2} \cdot 13} = \frac{6}{13}$$

33. **Number Sense** The circle graph shows the five things that are used to determine a person's credit score. Use your knowledge of circle graphs to find the fraction for payment history. Explain your reasoning.

Factors Used to Determine Credit Score

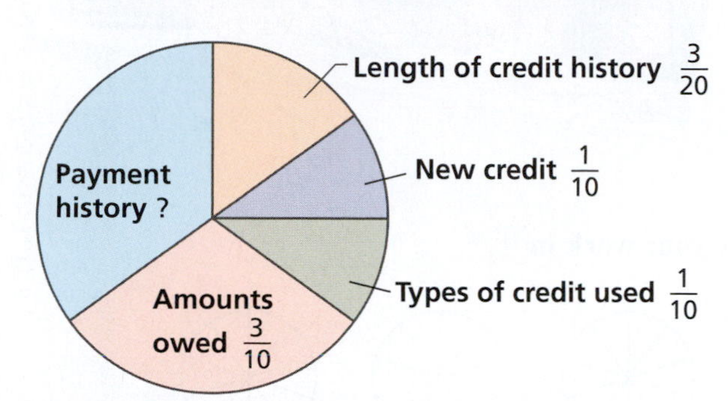

34. **Number Sense** The circle graph shows seven things that make up the total cost of fires in the United States each year. Estimate the fraction that represents the sum of fire insurance premiums and the two types of fire departments. Explain your reasoning.

Total Annual Cost of Fires in the United States

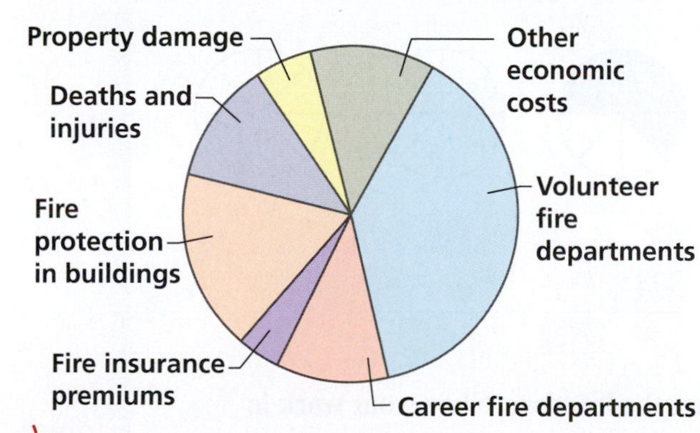

35. **Number Line Sense: In Your Classroom** Explain how you might teach your students to find a fraction that is between $\frac{1}{5}$ and $\frac{1}{4}$. Use a number line as part of your explanation.

36. **Number Line Sense: In Your Classroom** Explain how you might teach your students to find a fraction that is between $\frac{3}{5}$ and $\frac{3}{4}$. Use a number line as part of your explanation.

37. **Number Line Sense** Find a fraction that is less than one-hundredth.

38. **Number Line Sense** Find a fraction that is less than one-thousandth.

39. **Area Model Sense** The circle is divided into 12 equal parts. Determine whether you can represent each fraction by shading 1 or more of the 12 parts. Explain your reasoning.

a. $\frac{1}{1}$ **b.** $\frac{1}{2}$ **c.** $\frac{1}{3}$ **d.** $\frac{1}{4}$ **e.** $\frac{1}{5}$ **f.** $\frac{1}{6}$

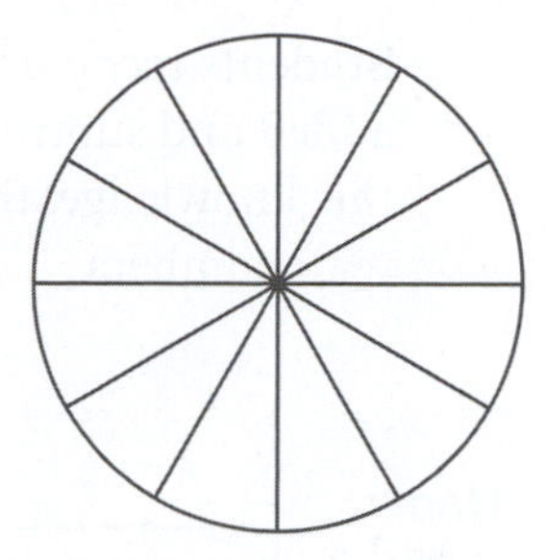

40. **Area Model Sense** The rectangle is divided into 10 equal parts. Determine whether you can represent each fraction by shading 1 or more of the 10 parts. Explain your reasoning.

a. $\frac{1}{1}$ **b.** $\frac{1}{2}$ **c.** $\frac{1}{3}$ **d.** $\frac{1}{4}$ **e.** $\frac{1}{5}$ **f.** $\frac{1}{6}$

41. **Using Fraction Notation** In printing, there are 3 ways that fractions are typeset.

Built up	Case	Shilling
$\frac{3}{4}$	$\tfrac{3}{4}$	3/4

In which way does the recipe for corn muffins write fractions? Why do you think the recipe was written with this style of fraction?

Corn Muffins

* 1 1/4 cup flour * 3/4 cup corn meal
* 2 tsp baking powder
* 1/4 tsp salt
* 1 cup milk
* 1/4 cup olive oil
* 1 egg

42. **Problem Solving** Use a number line to estimate the total number of cups in the recipe. If each muffin uses one-fourth cup of batter, about how many muffins does the recipe make? Explain your reasoning.

Activity: Using Area Models to Add and Subtract Fractions

Materials:

- Crayons or markers
- Pencil

Learning Objective:

Students recognize how area can be added and subtracted. Then, using this knowledge, they generalize with numbers.

Name ______________________________

Shade the sum. Then add the fractions. Show your work in ▭.

1.

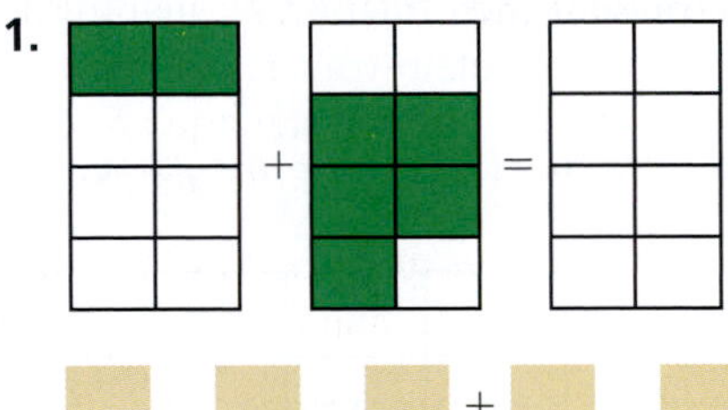

2.

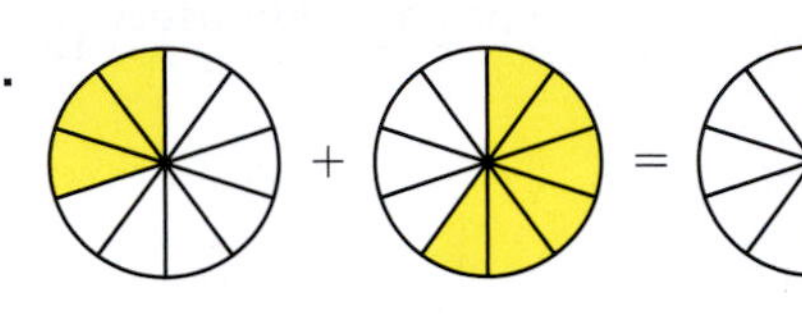

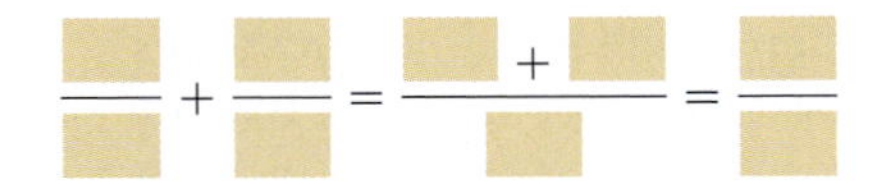

Shade the difference. Then subtract the fractions. Show your work in ▭.

3.

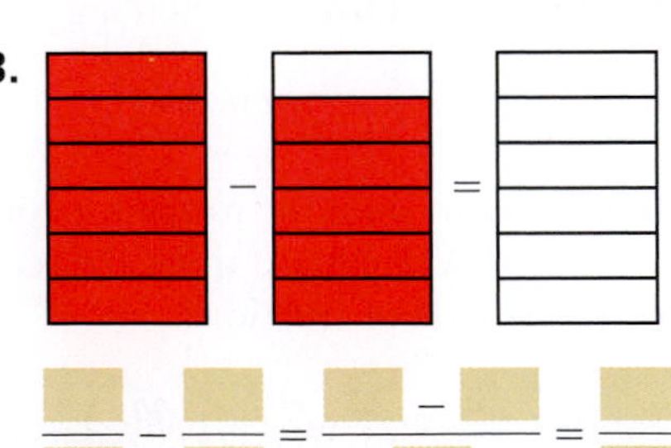

4.

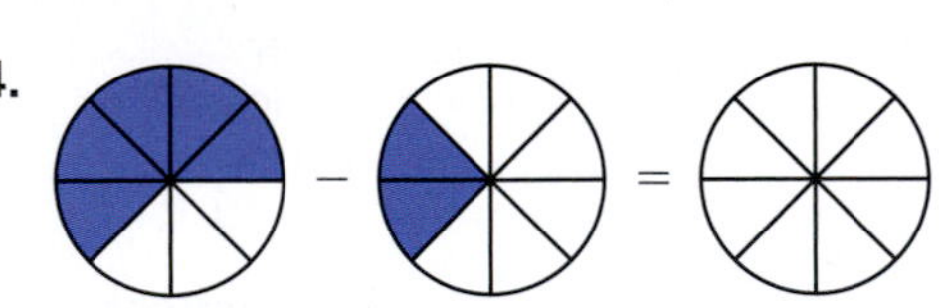

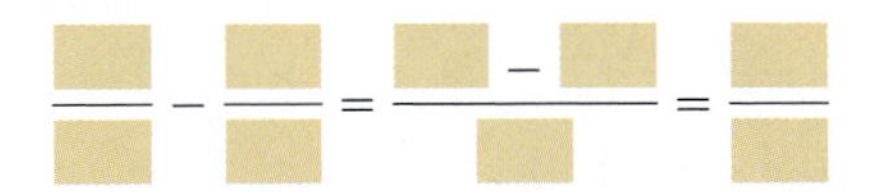

Shade the sum or difference. Then add or subtract the fractions. Show your work in ▭.

5.

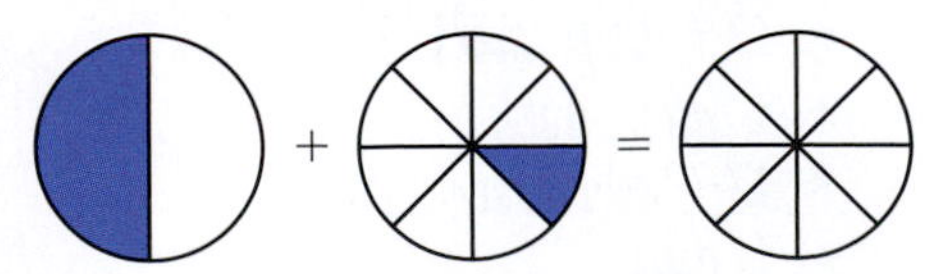

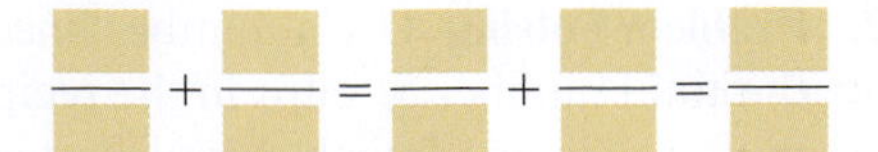

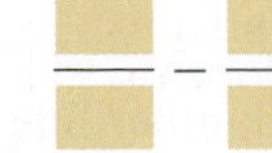

6.

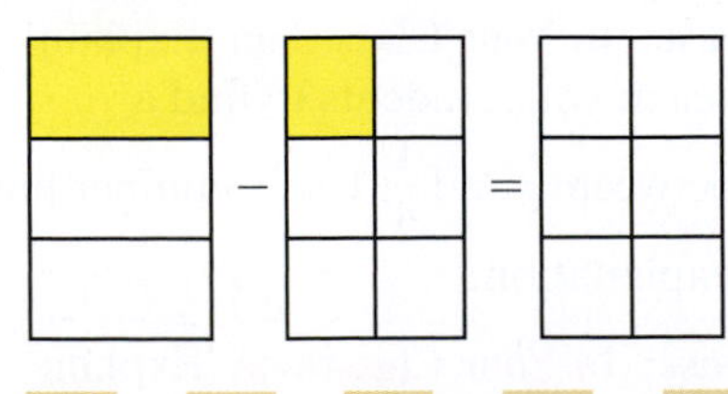

6.2 Adding and Subtracting Fractions

Standards

Grades 3–5
Numbers and Operations–Fractions
Students should build fractions from unit fractions by applying and extending previous understandings of operations on whole numbers, and use equivalent fractions as a strategy to add and subtract fractions.

- Add fractions with like denominators.
- Subtract fractions with like denominators.
- Add and subtract fractions with unlike denominators.

Adding Fractions with Like Denominators

Adding Fractions with Like Denominators

Words To add two fractions with like (or common) denominators, write the sum of the numerators over the like denominator.

Numbers $\frac{1}{5} + \frac{2}{5} = \frac{3}{5}$ (Like denominator)

Algebra $\frac{a}{c} + \frac{b}{c} = \frac{a+b}{c}$, $c \neq 0$. (Like denominator)

EXAMPLE 1 **Real-Life Examples of Fraction Addition**

a. A single serving oatmeal recipe calls for one-quarter cup of milk and three-quarters cup of water. How many cups of liquid are in the recipe?

b. You run two and one-half miles each weekday morning. You take the weekend off. How many miles do you run in one week?

c. Tile that is five-eighths inch thick is laid on top of plywood that is seven-eighths inch thick. How thick is the flooring?

Classroom Tip

Any of the fraction models you studied in Section 6.1 (area, set, and linear) can be used to visualize fraction addition and subtraction. Here is an example using a number line.

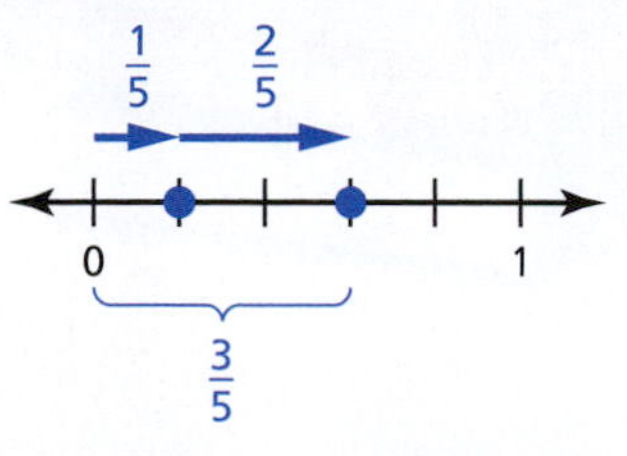

Use a variety of models with your students to build an intuitive understanding. Simply knowing the rules will not produce the confidence that is necessary for mastery.

SOLUTION

a. $\frac{1}{4} + \frac{3}{4} = \frac{1+3}{4} = \frac{4}{4} = 1$ Add numerators and simplify.

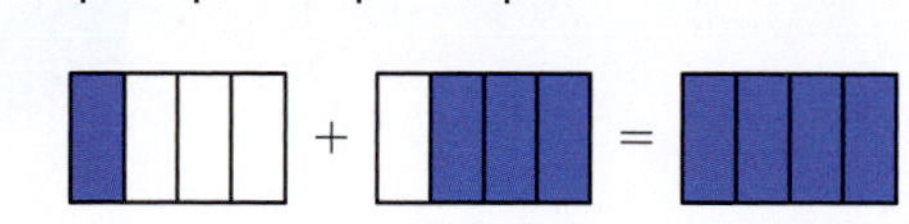

Use a visual model when possible.

There is 1 cup of liquid in the recipe.

b. Begin by writing the mixed number $2\frac{1}{2}$ as an improper fraction.

Mon	Tue	Wed	Thu	Fri

$$\frac{5}{2} + \frac{5}{2} + \frac{5}{2} + \frac{5}{2} + \frac{5}{2} = \frac{5+5+5+5+5}{2} = \frac{25}{2}$$ Add numerators.

You run $\frac{25}{2} = 12\frac{1}{2}$ miles in one week.

c. $\frac{5}{8} + \frac{7}{8} = \frac{5+7}{8} = \frac{12}{8} = \frac{3}{2}$ Add numerators and simplify.

The flooring is $\frac{3}{2} = 1\frac{1}{2}$ inches thick.

All of the properties of whole number addition studied in Section 3.1 are also true for fraction addition. For example, the Commutative and Associative Properties are shown below.

Classroom Tip

A common error when adding two fractions is adding the numerators *and* adding the denominators. When your students make this error, point out that the answer is not reasonable. For instance, $\frac{1}{2} + \frac{3}{2}$ is certainly not $\frac{4}{4}$ because $\frac{3}{2}$ is greater than $\frac{4}{4}$, or 1.

Properties of Fraction Addition

Commutative Property

Words The sum of two fractions does not depend on their order.

Numbers $\frac{1}{3} + \frac{4}{3} = \frac{4}{3} + \frac{1}{3}$

Algebra $\frac{a}{b} + \frac{c}{d} = \frac{c}{d} + \frac{a}{b}$, where $b, d \neq 0$

Associative Property

Words The sum of fractions does not depend on how the fractions are grouped.

Numbers $\left(\frac{1}{2} + \frac{3}{2}\right) + \frac{5}{2} = \frac{1}{2} + \left(\frac{3}{2} + \frac{5}{2}\right)$

Algebra $\left(\frac{a}{b} + \frac{c}{d}\right) + \frac{e}{f} = \frac{a}{b} + \left(\frac{c}{d} + \frac{e}{f}\right)$, where $b, d, f \neq 0$

EXAMPLE 2 **Addition Properties and Mental Math**

How tall is the stack of books?

SOLUTION

Adding the book thicknesses in the order that they are stacked produces a cumbersome addition problem. By reordering and grouping, you can show students that the total height can be found using mental math. Just look for pairs that add up to whole numbers. Begin by writing the mixed number $1\frac{3}{16}$ as $\frac{19}{16}$.

$$\text{Total height} = \left(\frac{8}{16} + \frac{8}{16}\right) + \left(\frac{19}{16} + \frac{13}{16}\right) + \left(\frac{10}{16} + \frac{6}{16}\right) + \left(\frac{9}{16} + \frac{7}{16}\right)$$

$$= \frac{16}{16} + \frac{32}{16} + \frac{16}{16} + \frac{16}{16}$$

$$= 1 + 2 + 1 + 1$$

$$= 5$$

The stack of books is 5 inches tall.

Mathematical Practices

Look for and Make Use of Structure

MP1 Mathematically proficient students try simpler forms of the original problem in order to gain insight into the solution.

Subtracting Fractions with Like Denominators

Classroom Tip

The word *common* has more than one meaning. Point out to students that in the context of fractions, *common* means *like*. An example is "Maria and James have common interests." For instance, they might both enjoy baseball and movies.

Subtracting Fractions with Like Denominators

Words To subtract two fractions with like (or common) denominators, write the difference of the numerators over the like denominator.

Numbers $\frac{3}{5} - \frac{2}{5} = \frac{1}{5}$ — Like denominator

Algebra $\frac{a}{c} - \frac{b}{c} = \frac{a-b}{c},\ c \neq 0$ — Like denominator

EXAMPLE 3 Real-Life Examples of Fraction Subtraction

a. A poll asks 700 people whether they approve of the president's job performance. The circle graph shows the results. What fraction of the respondents disapprove of the president's job performance?

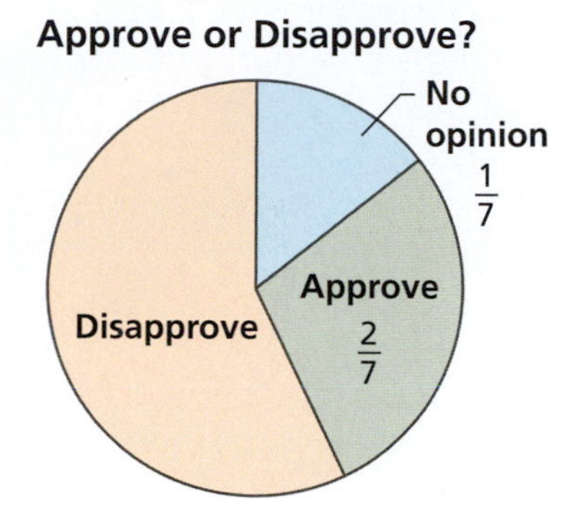

b. A survey asks 100 students which category from the table best describes the number of hours they spend texting each day. The table at the left shows the results. What fraction of the students spend more than 5 hours texting each day?

Hours per day texting	Fraction of students
0–1	5/10
2–3	2/10
4–5	1/10
More than 5	?

SOLUTION

a. The sum of the fractional parts of a circle graph is always 1.

$$\text{Disapprove} = 1 - \left(\frac{1}{7} + \frac{2}{7}\right)$$ Write equation.

$$= 1 - \frac{3}{7}$$ Add numerators.

$$= \frac{7}{7} - \frac{3}{7}$$ Rewrite 1 as a fraction.

$$= \frac{4}{7}$$ Subtract numerators.

So, four-sevenths of the respondents disapproved.

b. In a survey, the sum of the fractions of respondents is always 1.

$$\text{More than 5 hours} = 1 - \left(\frac{5}{10} + \frac{2}{10} + \frac{1}{10}\right)$$ Write equation.

$$= 1 - \frac{8}{10}$$ Add numerators.

$$= \frac{10}{10} - \frac{8}{10}$$ Rewrite 1 as a fraction.

$$= \frac{1}{5}$$ Subtract numerators and simplify.

So, one-fifth of the students spend more than 5 hours texting each day.

Adding and Subtracting Fractions with Unlike Denominators

Classroom Tip

Some texts emphasize using the *least common denominator (LCD)* when adding or subtracting fractions. This is a good method, but it is not necessary. Any common denominator, such as *bd*, will work, as shown in the algorithms here.

Algorithm for Adding and Subtracting Fractions with Unlike Denominators

Words To find the sum or difference of $\frac{a}{b}$ and $\frac{c}{d}$, where $b, d \neq 0$:

1. Rewrite the fractions using a *common denominator*.
2. Add or subtract the numerators.
3. Write the result over the common denominator.
4. Simplify if possible.

Numbers		Algebra	
$\frac{1}{2}+\frac{2}{3}=\frac{3+4}{6}=\frac{7}{6}$	Sum	$\frac{a}{b}+\frac{c}{d}=\frac{ad+bc}{bd}$	Sum
$\frac{3}{4}-\frac{2}{3}=\frac{9-8}{12}=\frac{1}{12}$	Difference	$\frac{a}{b}-\frac{c}{d}=\frac{ad-bc}{bd}$	Difference

EXAMPLE 4 Adding Fractions with Unlike Denominators

Find the sum $\frac{2}{5}+\frac{1}{3}$.

SOLUTION

$$\frac{2}{5}+\frac{1}{3}=\frac{6}{15}+\frac{5}{15}$$ Rewrite fractions using a common denominator of $5 \cdot 3 = 15$.

$$=\frac{6+5}{15}$$ Add numerators.

$$=\frac{11}{15}$$ Simplify.

Whenever your students apply an algorithm, encourage them to check the reasonableness of their answer using estimation or a visual such as a number line.

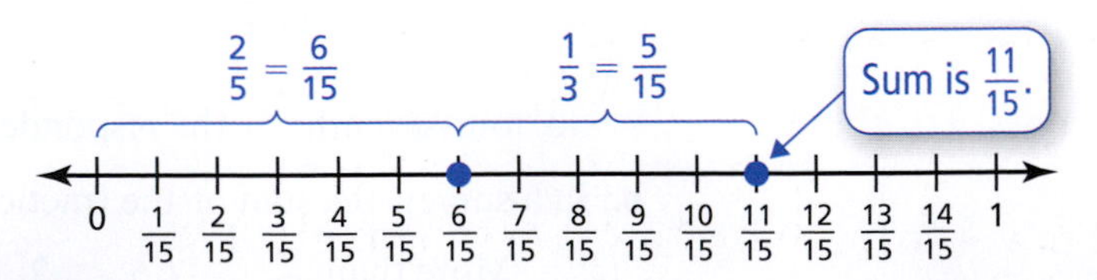

EXAMPLE 5 Subtracting Fractions with Unlike Denominators

Find the difference $\frac{1}{2}-\frac{1}{6}$.

SOLUTION

$$\frac{1}{2}-\frac{1}{6}=\frac{6}{12}-\frac{2}{12}=\frac{6-2}{12}=\frac{4}{12}=\frac{1}{3}$$ Apply algorithm.

EXAMPLE 6 Modeling Fraction Subtraction

Classroom Tip

There are many kinds of templates for fraction strips. Two are shown below.

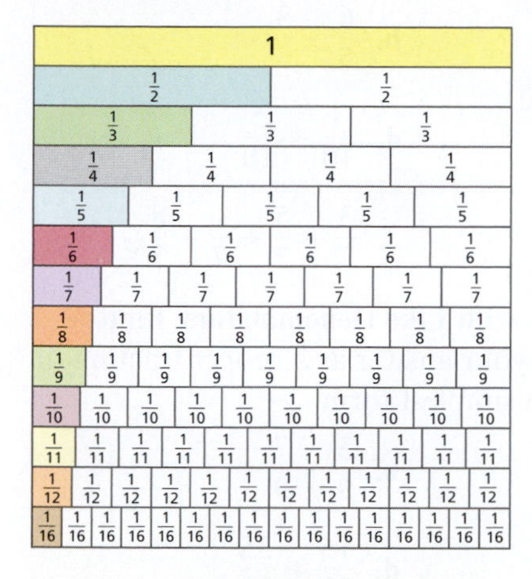

A few common templates are available online at *MathematicalPractices.com*.

Consider the difference $\frac{5}{6} - \frac{1}{3}$.

a. Use fraction strips to find the difference.

b. Use an algorithm to find the difference.

SOLUTION

a. Choose fraction strips for "sixths" and for "thirds."

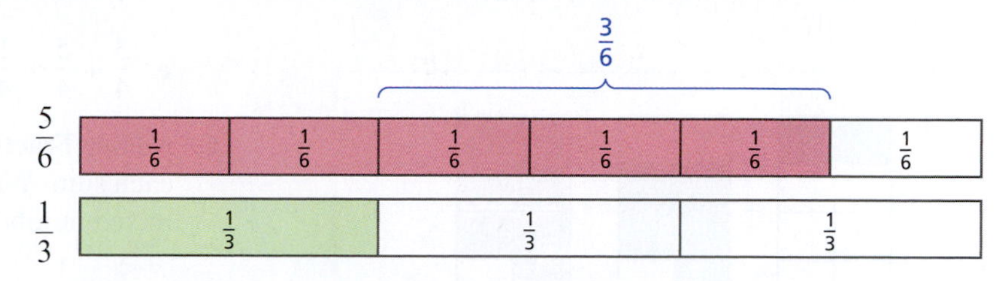

You can see that the difference is $\frac{3}{6}$, or $\frac{1}{2}$.

b. Using the subtraction algorithm on page 212, you get the same result.

$\frac{5}{6} - \frac{1}{3} = \frac{15}{18} - \frac{6}{18}$ Rewrite fractions using a common denominator of $6 \cdot 3 = 18$.

$= \frac{15 - 6}{18}$ Multiply.

$= \frac{9}{18}$ Subtract numerators.

$= \frac{1}{2}$ Simplify.

EXAMPLE 7 Using Models to Estimate a Difference

Consider the difference $\frac{11}{16} - \frac{2}{3}$.

a. Use fraction strips to *estimate* the difference.

b. Use an algorithm to find the difference.

SOLUTION

a. Using fraction strips, you can see that the difference is less than $\frac{1}{16}$.

b. Using the algorithm on page 212, you can find the *exact* difference.

$\frac{11}{16} - \frac{2}{3} = \frac{33}{48} - \frac{32}{48}$ Rewrite fractions using a common denominator of $16 \cdot 3 = 48$.

$= \frac{1}{48}$ Subtract numerators.

6.2 Exercises

Free step-by-step solutions to odd-numbered exercises at *MathematicalPractices.com.*

1. **Writing a Solution Key** Write a solution key for the activity on page 208. Describe a rubric for grading a student's work.

2. **Grading the Activity** In the activity on page 208, a student gave the answers below. Each question is worth 10 points. Assign a grade to each answer. For those that are incorrect, why do you think the student erred?

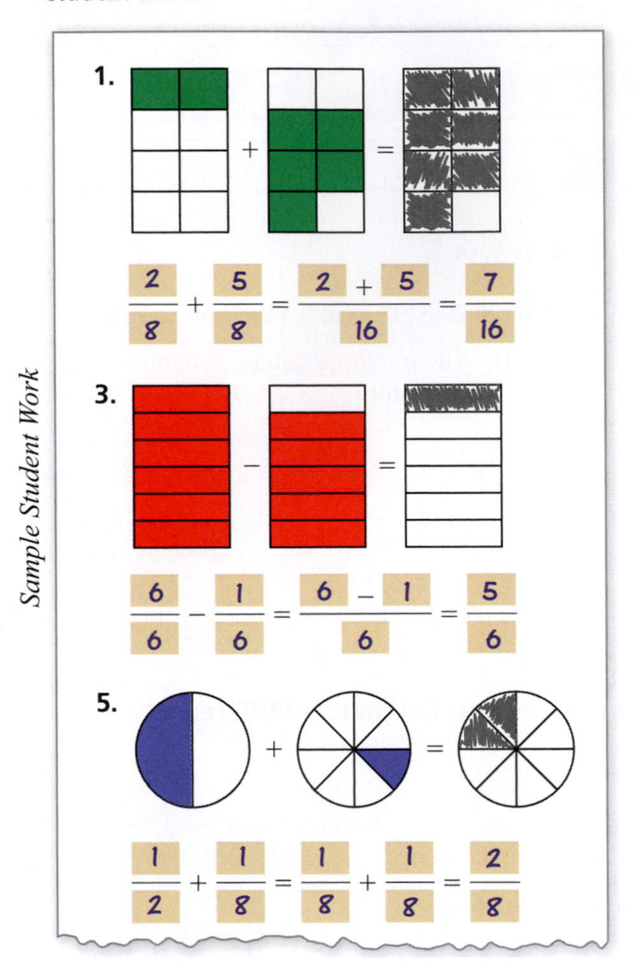

3. **Addition Models** Write the addition problem represented by each model. Solve the problem. Write your answer in simplest form.

a. b.

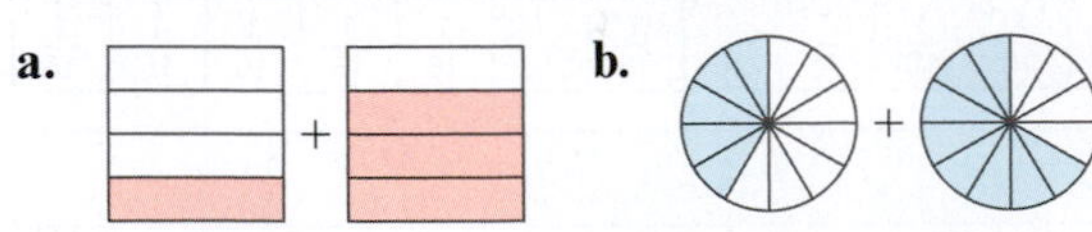

4. **Addition Models** Write the addition problem represented by each model. Solve the problem. Write your answer in simplest form.

a. b.

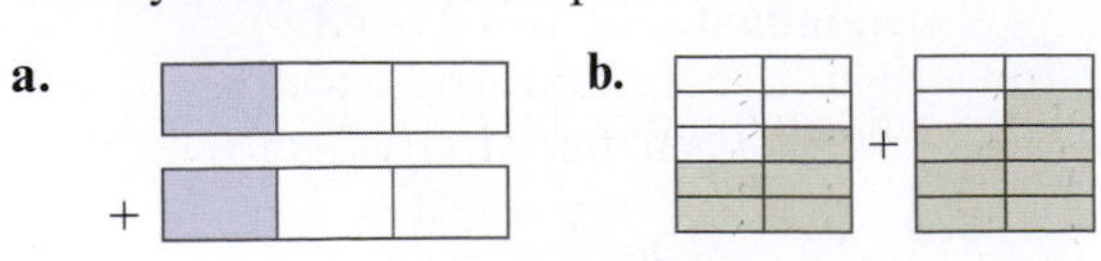

5. **Adding Fractions with Like Denominators** Find each sum. Write your answer as a mixed number in simplest form.

   a. $\frac{2}{3} + \frac{2}{3}$ b. $\frac{6}{5} + \frac{3}{5}$

   c. $\frac{5}{9} + \frac{8}{9}$ d. $\frac{7}{10} + \frac{5}{10}$

   e. $\frac{3}{4} + \frac{5}{4} + \frac{1}{4}$ f. $\frac{3}{7} + \frac{5}{7} + \frac{1}{7} + \frac{8}{7}$

6. **Adding Fractions with Like Denominators** Find each sum. Write your answer as a proper fraction or mixed number in simplest form.

   a. $\frac{5}{8} + \frac{1}{8}$ b. $\frac{5}{9} + \frac{7}{9}$

   c. $\frac{8}{11} + \frac{15}{11}$ d. $\frac{15}{20} + \frac{35}{20}$

   e. $\frac{5}{3} + \frac{5}{3} + \frac{5}{3}$ f. $\frac{3}{8} + \frac{11}{8} + \frac{9}{8} + \frac{7}{8}$

7. **Adding Mixed Numbers** Find each sum. Write your answer as a mixed number in simplest form.

   a. $2\frac{1}{2} + 3\frac{1}{2}$ b. $1\frac{3}{8} + \frac{7}{8}$ c. $3\frac{4}{5} + \frac{3}{5} + 2\frac{1}{5}$

8. **Adding Mixed Numbers** Find each sum. Write your answer as a mixed number in simplest form.

   a. $4\frac{2}{3} + 1\frac{2}{3}$ b. $1\frac{2}{9} + 2\frac{5}{9}$ c. $\frac{1}{4} + 5\frac{3}{4} + 3\frac{1}{4}$

9. **Number Line Models** Write the addition problem represented by each model. Solve the problem. Write your answer in simplest form.

a. b.

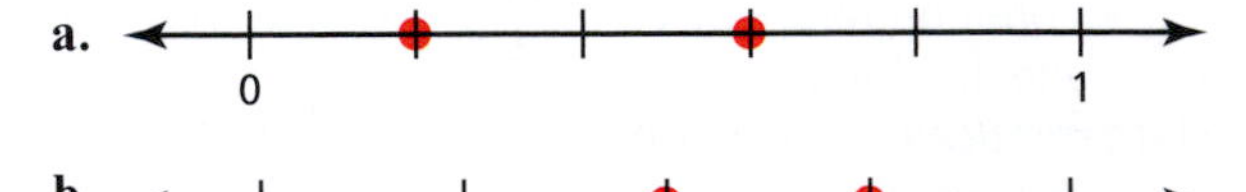

c.

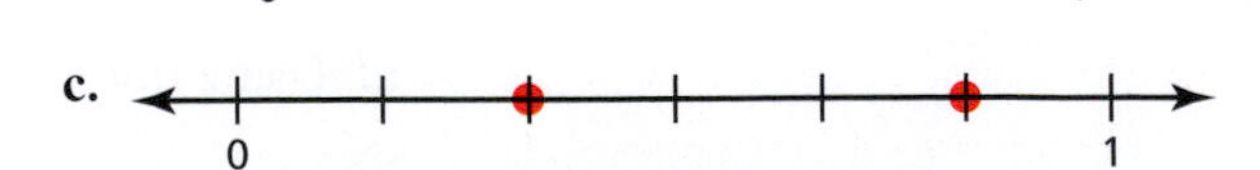

10. **Number Line Models** Write the addition problem represented by each model. Solve the problem. Write your answer in simplest form.

a.

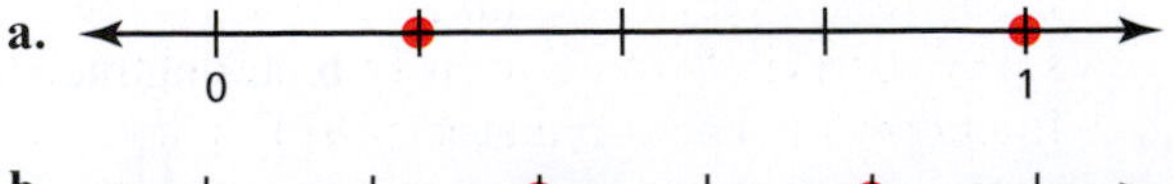

b. c.

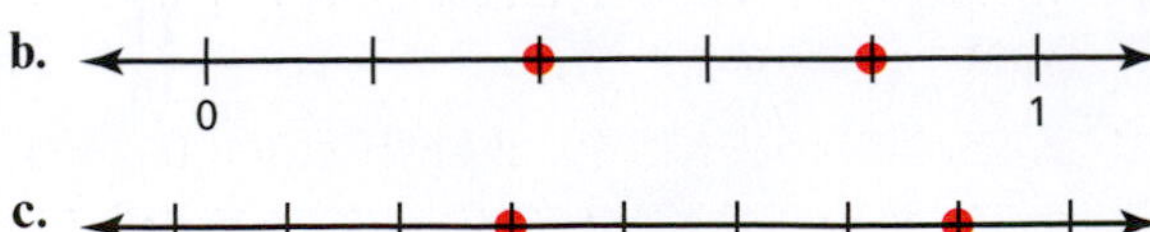

11. Subtraction Models Write the subtraction problem represented by each model. Solve the problem. Write your answer in simplest form.

a. 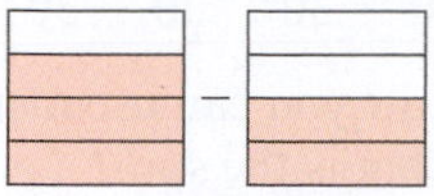 b.

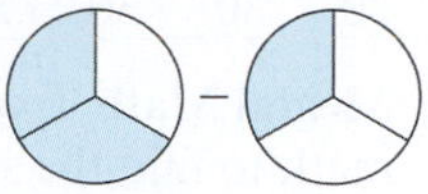

12. Subtraction Models Write the subtraction problem represented by each model. Solve the problem. Write your answer in simplest form.

a. 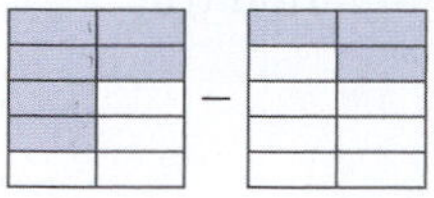 b.

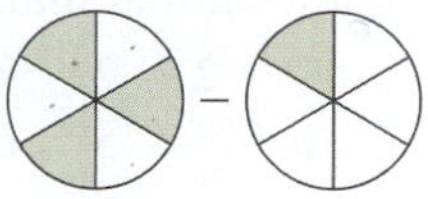

13. Subtracting Fractions with Like Denominators Find each difference. Write your answer in simplest form.

a. $\frac{5}{7} - \frac{4}{7}$ b. $\frac{8}{9} - \frac{5}{9}$ c. $\frac{13}{12} - \frac{7}{12}$ d. $\frac{9}{15} - \frac{9}{15}$

14. Subtracting Fractions with Like Denominators Find each difference. Write your answer in simplest form.

a. $\frac{9}{10} - \frac{4}{10}$ b. $\frac{6}{5} - \frac{3}{5}$ c. $\frac{11}{9} - \frac{8}{9}$ d. $\frac{5}{16} - \frac{1}{16}$

15. Subtracting Mixed Numbers Find each difference. Write your answer as a mixed number in simplest form.

a. $4\frac{1}{2} - 3$ b. $2\frac{3}{5} - 1\frac{1}{5}$ c. $3\frac{1}{5} - 1\frac{3}{5}$

16. Subtracting Mixed Numbers Find each difference. Write your answer as a mixed number in simplest form.

a. $2\frac{2}{3} - 1\frac{1}{3}$ b. $9\frac{7}{10} - 5\frac{3}{10}$ c. $4\frac{4}{5} - 2\frac{3}{5}$

17. Using Area Models Represent each addition problem with an area model. Then find the sum and write it in simplest form.

a. $\frac{1}{4} + \frac{2}{4}$ b. $\frac{1}{6} + \frac{3}{6}$ c. $\frac{3}{8} + \frac{3}{8}$ d. $\frac{3}{7} + \frac{2}{7}$

18. Using Area Models Represent each subtraction problem with an area model. Then find the difference and write it in simplest form.

a. $\frac{4}{12} - \frac{2}{12}$ b. $\frac{5}{6} - \frac{1}{6}$ c. $\frac{5}{9} - \frac{5}{9}$ d. $\frac{9}{10} - \frac{5}{10}$

19. Using Area Models Would it be easier to use a square or a circle to draw an area model for a fraction with a denominator of 9? Explain your reasoning.

20. Using Area Models Would it be easier to use a circle or a rectangle to draw an area model for a fraction with a denominator of 15? Explain your reasoning.

21. Addition and Subtraction Models Write the addition or subtraction problem represented by each model. Solve the problem. Write your answer as a proper fraction or mixed number in simplest form.

a. 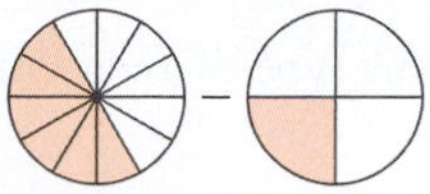 b.

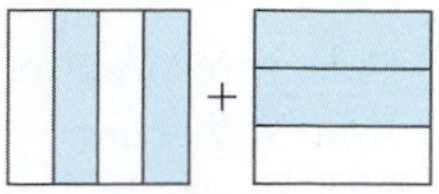

22. Addition and Subtraction Models Write the addition or subtraction problem represented by each model. Solve the problem. Write your answer as a proper fraction or mixed number in simplest form.

a. 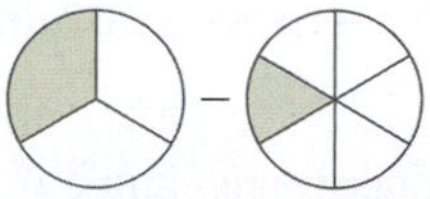 b.

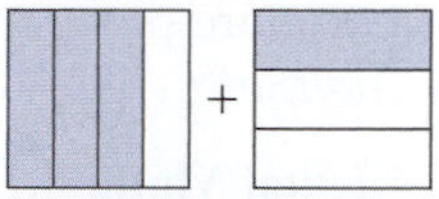

23. Adding and Subtracting with Unlike Denominators Find each sum or difference. Write your answer as a proper fraction or mixed number in simplest form.

a. $\frac{3}{4} - \frac{5}{12}$ b. $\frac{5}{6} + \frac{1}{2}$ c. $\frac{6}{7} + \frac{3}{14}$ d. $\frac{1}{2} - \frac{3}{10}$

24. Adding and Subtracting with Unlike Denominators Find each sum or difference. Write your answer as a proper fraction or mixed number in simplest form.

a. $\frac{4}{5} + \frac{1}{10}$ b. $\frac{7}{6} - \frac{1}{3}$ c. $\frac{5}{4} + \frac{1}{6}$ d. $\frac{3}{8} - \frac{1}{4}$

25. Adding or Subtracting Mixed Numbers Find each sum or difference. Write your answer in simplest form.

a. $1\frac{1}{2} - 1\frac{1}{3}$ b. $2\frac{3}{4} - 1\frac{1}{8}$ c. $\frac{1}{5} + 1\frac{1}{10}$

26. Adding or Subtracting Mixed Numbers Find each sum or difference. Write your answer in simplest form.

a. $4\frac{1}{2} + 2\frac{1}{5}$ b. $3\frac{1}{2} - 1\frac{1}{6}$ c. $4\frac{3}{10} - 3\frac{3}{5}$

27. Using Fraction Strips Use fraction strips to find each difference. Write your answer in simplest form.

a. $\frac{15}{16} - \frac{7}{8}$ b. $\frac{2}{5} - \frac{2}{15}$ c. $\frac{7}{9} - \frac{2}{3}$ d. $\frac{4}{5} - \frac{3}{10}$

28. Using Fraction Strips Use fraction strips to find each difference. Write your answer in simplest form.

a. $\frac{5}{12} - \frac{1}{3}$ b. $\frac{3}{4} - \frac{3}{8}$ c. $\frac{7}{12} - \frac{1}{6}$ d. $\frac{2}{3} - \frac{4}{9}$

29. Estimation Use fraction strips to estimate each sum or difference. Then, find the exact answer.

a. $\frac{11}{12} - \frac{3}{4}$ b. $\frac{3}{5} + \frac{11}{20}$ c. $\frac{1}{2} - \frac{5}{13}$

30. Estimation Use fraction strips to estimate each sum or difference. Then, find the exact answer.

a. $\frac{15}{16} - \frac{3}{4}$ b. $\frac{3}{7} + \frac{3}{10}$ c. $\frac{3}{5} - \frac{51}{100}$

31. **My Dear Aunt Sally** Arrange the numbers $\frac{1}{4}$, $\frac{3}{8}$, $\frac{3}{8}$, and $\frac{1}{2}$ so that the equation is true.

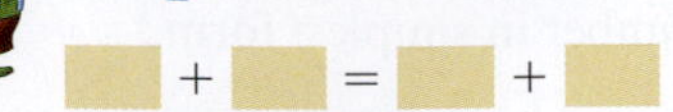

For more practice with this type of reasoning, go to *AuntSally.com.*

32. **My Dear Aunt Sally** Arrange the numbers $\frac{0}{1}$, $\frac{1}{4}$, $\frac{1}{2}$, and $\frac{3}{4}$ so that the equation is true.

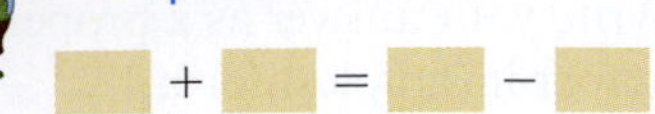

For more practice with this type of reasoning, go to *AuntSally.com.*

33. **Mental Math: In Your Classroom** The bar graph shows the distances (in kilometers) a person swam each day for one week.

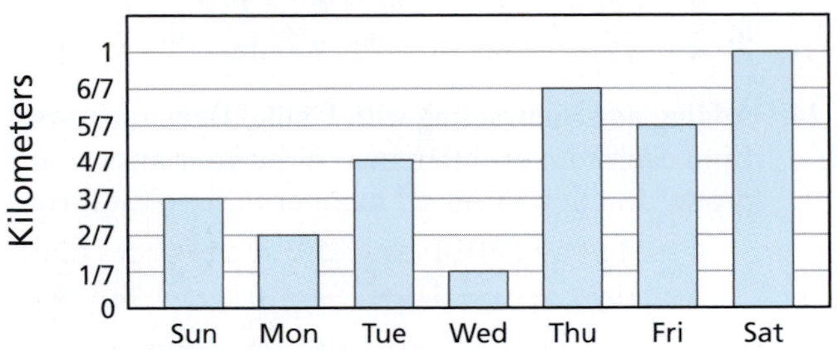

a. What is the total swimming distance in the week?

b. What mental math strategy might you identify for your students to solve part (a)?

c. In part (b), how will you incorporate the Commutative and Associative Properties so that they are understandable to your students?

34. **Mental Math: In Your Classroom** The circle graph shows the fractions of a monthly budget. Use mental math to verify that the fractions add up to 1. What are some reasons for students to use the Commutative and Associative Properties in their computations?

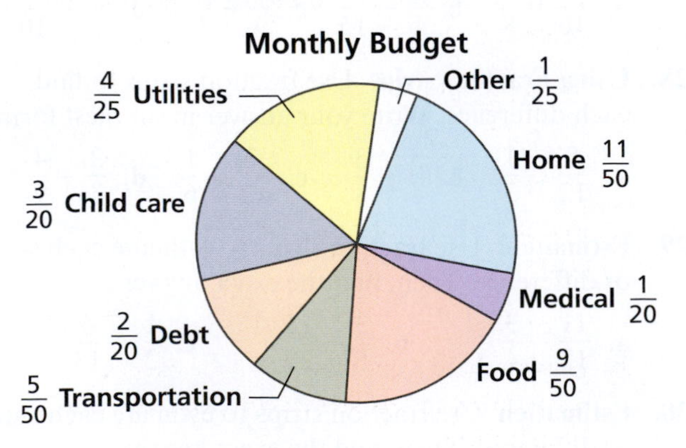

35. **Mental Math** Explain how you can use mental math to find the sum. What is the sum?

$$\frac{17}{30} + \frac{40}{30} + \frac{9}{30} + \frac{13}{30} + \frac{37}{30} + \frac{20}{30} + \frac{23}{30} + \frac{21}{30}$$

36. **Mental Math** Explain how you can use mental math to find the sum. What is the sum?

$$\frac{23}{50} + \frac{31}{50} + \frac{64}{50} + \frac{19}{50} + \frac{27}{50} + \frac{36}{50}$$

37. **Properties of Addition** Identify the property of addition that makes the equation true.

a. $\frac{7}{6} + \frac{2}{5} = \frac{2}{5} + \frac{7}{6}$

b. $\left(\frac{1}{4} + \frac{2}{7}\right) + \frac{3}{8} = \frac{1}{4} + \left(\frac{2}{7} + \frac{3}{8}\right)$

c. $\frac{0}{2} + \frac{3}{2} = \frac{3}{2}$

38. **Properties of Addition** Identify the property of addition that makes the equation true.

a. $\frac{7}{5} + \frac{0}{5} = \frac{7}{5}$

b. $\frac{9}{10} + \frac{7}{10} = \frac{7}{10} + \frac{9}{10}$

c. $\frac{3}{7} + \left(\frac{2}{9} + \frac{1}{5}\right) = \left(\frac{3}{7} + \frac{2}{9}\right) + \frac{1}{5}$

39. **Rewriting Expressions**

a. Use the Associative and Commutative Properties to rewrite the expression so the computation is simpler.

$$\left(\frac{3}{8} + \frac{4}{3}\right) + \left(\frac{2}{3} + \frac{4}{7}\right) + \left(\frac{5}{8} + \frac{10}{7}\right)$$

b. Use the result to find the sum.

40. **Rewriting Expressions**

a. Use the Associative and Commutative Properties to rewrite the expression so the computation is simpler.

$$\left(\frac{5}{6} + \frac{2}{5}\right) + \left(\frac{3}{10} + \frac{5}{9}\right) + \left(\frac{3}{5} + \frac{13}{9}\right) + \left(\frac{7}{10} + \frac{4}{6}\right)$$

b. Use the result to find the sum.

41. **Grading Student Work** On a diagnostic test, one of your students does the following work. Explain what the student did wrong. Which topics would you encourage the student to review?

$$\frac{2}{3} + \frac{1}{4} = \frac{2+1}{3+4} = \frac{3}{7}$$

42. Grading Student Work On a diagnostic test, one of your students does the following work. Explain what the student did wrong. Which topics would you encourage the student to review?

$$\frac{7}{10} - \frac{1}{5} = \frac{7-1}{10} = \frac{6}{10}$$

43. Writing Why do you sometimes rewrite two fractions to add them?

44. Writing Explain how to subtract a fraction from a whole number.

45. Activity: In Your Classroom The puzzle pieces can be arranged to form a square. Design a classroom activity that you could use with your students to show that the sum of the fractions representing the puzzle pieces is equal to 1.

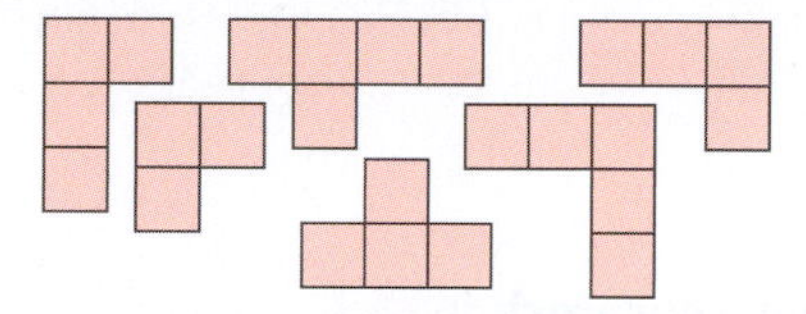

46. Activity: In Your Classroom The puzzle pieces can be arranged to form a square. Design a classroom activity that you could use with your students to show that the sum of the fractions representing the puzzle pieces is equal to 1.

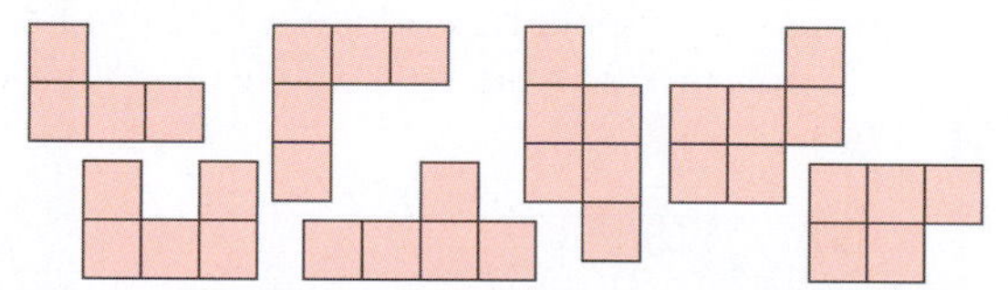

47. Poll: In Your Classroom A poll asked people to rate a brand of cereal. The circle graph shows the results.

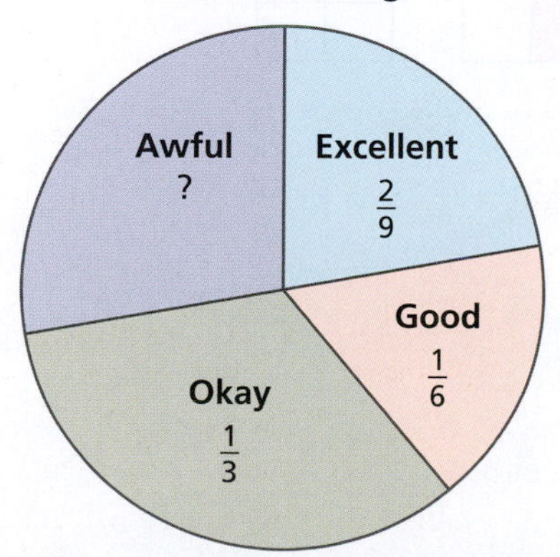

a. What fraction of respondents said awful?

b. Describe a similar poll you could use as a classroom activity with your students.

48. Survey: In Your Classroom A survey asked students which category best describes the number of hours they study each day. The table shows the results.

Hours spent studying per day	Fraction of students
0–1	1/5
2–3	7/20
4–5	3/10
More than 5	?

a. What fraction of the students spend more than 5 hours studying each day?

b. Describe a similar survey you could use as a classroom activity with your students.

49. Problem Solving The bar graph shows the scores of 6 students who took two math tests, one with preparation and one without preparation.

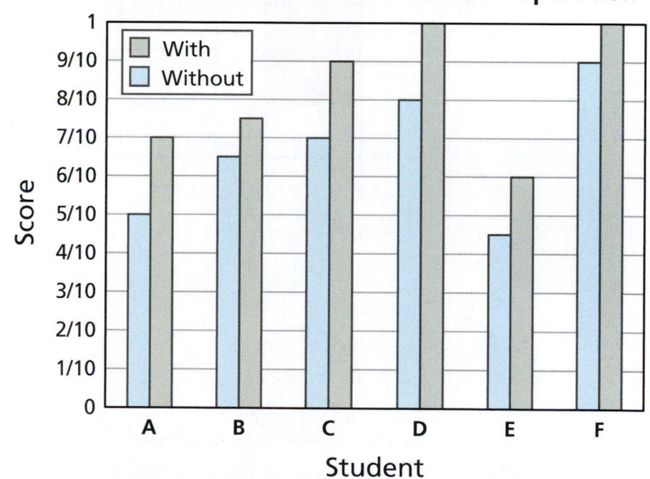

a. Find the sum of the test scores with and without preparation.

With preparation:

$$\frac{7}{10} + \frac{15}{20} + \frac{9}{10} + 1 + \frac{6}{10} + 1$$

Without preparation:

$$\frac{5}{10} + \frac{13}{20} + \frac{7}{10} + \frac{8}{10} + \frac{9}{20} + \frac{9}{10}$$

b. Explain how to use the result of part (a) to find the mean test scores with and without preparation.

Activity: Using Area Models to Multiply Fractions

Materials:

- Crayons or markers
- Pencil

Learning Objective:

Students use the meaning of fractions and multiplication to explain why the procedure for multiplying fractions makes sense. They understand that the word "of" in real-life problems is often interpreted as multiplication.

Name ______________________________

Shade the product. Then multiply the fractions. Show your work in ▭.

1. $\frac{1}{2}$ of ▭ = ▭

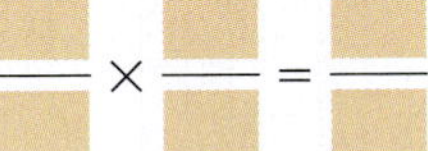

2. $\frac{2}{3}$ of

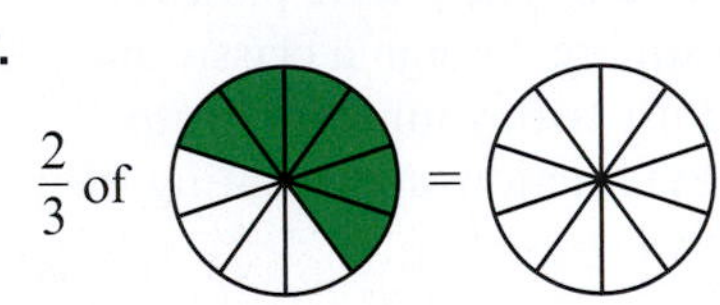

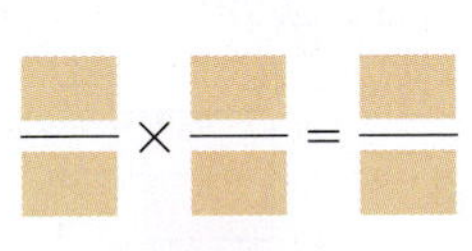

3. $\frac{2}{5}$ of ◯ = ◯

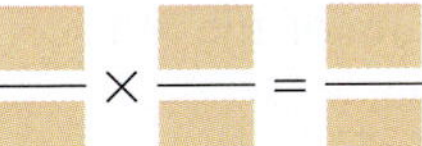

4. $\frac{1}{4}$ of

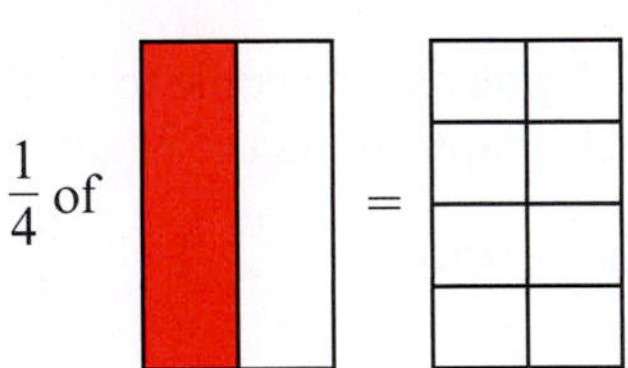

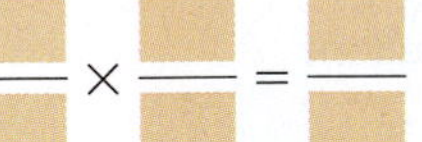

5. **Recipe** A scrambled eggs recipe calls for three-fourths cup of milk. You are making one-third of the recipe. How much milk should you use?

$\frac{\square}{\square}$ cup

6.3 Multiplying and Dividing Fractions

Standards

Grades 3–5

Numbers and Operations–Fractions

Students should build fractions from unit fractions by applying and extending previous understandings of operations on whole numbers, and apply and extend previous understandings of multiplication and division to multiply and divide fractions.

Grades 6–8

The Number System

Students should apply and extend previous understandings of multiplication and division to divide fractions by fractions.

- Multiply fractions, whole numbers, and mixed numbers.
- Divide fractions, whole numbers, and mixed numbers.
- Estimate products and quotients of fractions.

Multiplying Fractions, Whole Numbers, and Mixed Numbers

Multiplying Fractions

Words To multiply two fractions, multiply the numerators and multiply the denominators.

Numbers $\frac{2}{3} \cdot \frac{4}{5} = \frac{2 \cdot 4}{3 \cdot 5} = \frac{8}{15}$ **Algebra** $\frac{a}{b} \cdot \frac{c}{d} = \frac{a \cdot c}{b \cdot d}$, where $b, d \neq 0$.

Product of numerators

Product of denominators

EXAMPLE 1 Real-Life Examples of Fraction Multiplication

Classroom Tip

Encourage your students to read a fraction multiplication problem like $\frac{1}{2} \cdot \frac{2}{3}$ as "What is one-half of two-thirds?" This prepares them for many applications of fraction multiplication in real life.

a. Two-thirds of a litter of six kittens are females. How many of the kittens are female?

b. A soap bubble recipe calls for 1 cup warm water, 1/4 cup dishwashing liquid, and 1/2 teaspoon salt. How much of each ingredient is needed for one-half of the recipe?

c. Healthy fingernails normally grow about three-halves inch per year. How much do healthy fingernails grow in a month?

SOLUTION

a. $\frac{2}{3} \cdot 6 = \frac{2}{3} \cdot \frac{6}{1} = \frac{2 \cdot 6}{3 \cdot 1} = \frac{12}{3} = 4$

Four of the kittens are female. Notice that to multiply a whole number and a fraction, you can multiply the numerator of the fraction by the whole number.

b. **Water** $\frac{1}{2} \cdot 1 = \frac{1}{2}$ cup **Dishwashing liquid** $\frac{1}{2} \cdot \frac{1}{4} = \frac{1}{8}$ cup **Salt** $\frac{1}{2} \cdot \frac{1}{2} = \frac{1}{4}$ tsp

One-half of one-quarter is one-eighth.

c. One month is one-twelfth of a year.

$\frac{1}{12} \cdot \frac{3}{2} = \frac{1 \cdot 3}{12 \cdot 2} = \frac{3}{24} = \frac{1}{8}$ Multiply numerators and denominators.

Healthy fingernails grow about one-eighth inch in a month.

EXAMPLE 2 Modeling Fraction Multiplication

a. Repeated Addition You can model the product of a whole number and a fraction as repeated addition.

$$4 \cdot \frac{2}{5} = \frac{2}{5} + \frac{2}{5} + \frac{2}{5} + \frac{2}{5} = \frac{8}{5} \quad \text{Repeated addition of } \frac{2}{5}$$

Note that using the rule for multiplying fractions yields the same result.

$$4 \cdot \frac{2}{5} = \frac{4}{1} \cdot \frac{2}{5} = \frac{4 \cdot 2}{1 \cdot 5} = \frac{8}{5} \quad \text{Multiply numerators and denominators.}$$

b. Number Line Model You can use a number line to model fraction multiplication. For instance, you can model the product of $\frac{2}{3}$ and $4\frac{1}{2}$ as follows.

- Divide the length for $4\frac{1}{2}$ into thirds so that each third is $1\frac{1}{2}$.

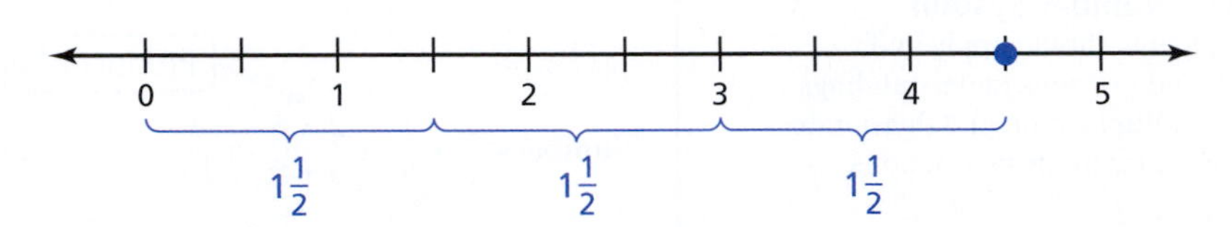

- Add two of the thirds to obtain

$$1\frac{1}{2} + 1\frac{1}{2} = 3. \quad \text{Number line model}$$

Note that using the rule for multiplying fractions yields the same result.

$$\frac{2}{3} \cdot 4\frac{1}{2} = \frac{2}{3} \cdot \frac{9}{2} = \frac{2 \cdot 9}{3 \cdot 2} = \frac{18}{6} = 3 \quad \text{Multiply numerators and denominators.}$$

Mathematical Practices

Reason Abstractly and Quantitatively

MP2 Mathematically proficient students create a coherent representation of the problem at hand, considering the units involved, and knowing and flexibly using different properties of operations.

c. Area Model You can use area models to represent fraction multiplication. For instance, you can model the product of $\frac{3}{4}$ and $\frac{2}{5}$ as follows.

- Begin by drawing an area model for $\frac{2}{5}$ using five vertical parts. Shade 2 of them.

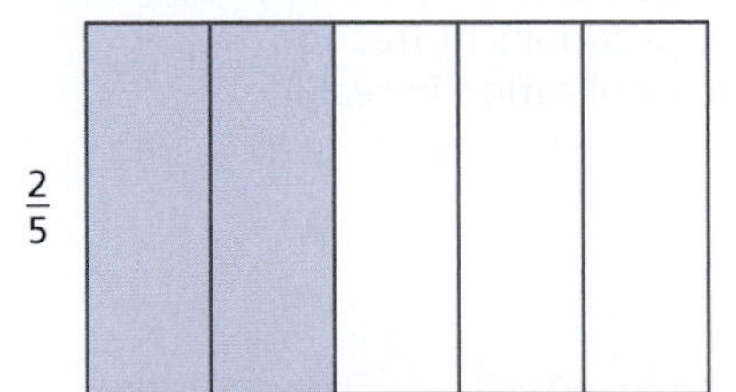

- Divide the rectangle into 4 equal horizontal parts and shade 3 of them.

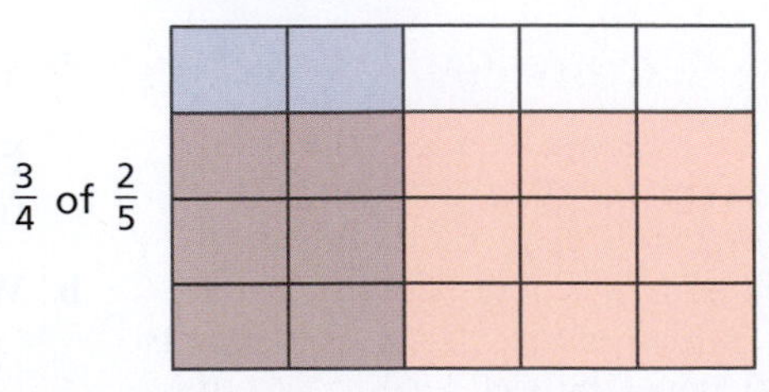

- The rectangle now has 20 equal parts. Six of these parts are shaded with both colors. So, the product of $\frac{3}{4}$ and $\frac{2}{5}$ is $\frac{6}{20}$, or $\frac{3}{10}$.

Note that using the rule for multiplying fractions yields the same result.

$$\frac{3}{4} \cdot \frac{2}{5} = \frac{3 \cdot 2}{4 \cdot 5} = \frac{6}{20} = \frac{3}{10} \quad \text{Multiply numerators and denominators.}$$

All of the properties of whole number multiplication studied in Section 3.3 are also true for fraction multiplication. The Commutative Property, Associative Property, and Multiplicative Identity Property of multiplication of fractions are shown below. There is also an additional property called the Multiplicative Inverse Property.

Properties of Fraction Multiplication

Let a, b, c, d, e, and f be any numbers such that $a, b, d, f \neq 0$.

Commutative Property

Words The product of two fractions does not depend on their order.

Numbers $\frac{1}{7} \cdot \frac{5}{2} = \frac{5}{2} \cdot \frac{1}{7}$ **Algebra** $\frac{a}{b} \cdot \frac{c}{d} = \frac{c}{d} \cdot \frac{a}{b}$

Associative Property

Words The product of fractions does not depend on how the factors are grouped.

Numbers

$$\left(\frac{3}{4} \cdot \frac{5}{8}\right) \cdot \frac{6}{7} = \frac{3}{4} \cdot \left(\frac{5}{8} \cdot \frac{6}{7}\right)$$

Algebra

$$\left(\frac{a}{b} \cdot \frac{c}{d}\right) \cdot \frac{e}{f} = \frac{a}{b} \cdot \left(\frac{c}{d} \cdot \frac{e}{f}\right)$$

Multiplicative Identity Property

Words The product of any fraction and 1 is equal to the fraction.

Numbers $\frac{3}{8} \cdot 1 = \frac{3}{8}$ **Algebra** $\frac{a}{b} \cdot 1 = \frac{a}{b}$

$1 \cdot \frac{5}{7} = \frac{5}{7}$ $1 \cdot \frac{a}{b} = \frac{a}{b}$

Multiplicative Inverse Property

Words Every nonzero fraction has a **multiplicative inverse**, or **reciprocal**, such that the product of the fraction and its multiplicative inverse is 1.

Numbers $\frac{3}{5} \cdot \frac{5}{3} = 1$ (Reciprocal of $\frac{3}{5}$) **Algebra** $\frac{a}{b} \cdot \frac{b}{a} = 1$ (Reciprocal of $\frac{a}{b}$)

EXAMPLE 3 **Using Multiplication Properties in Algebra**

$\frac{3}{4}x = \frac{3}{10}$ Original equation

$\frac{4}{3}\left(\frac{3}{4}x\right) = \frac{4}{3} \cdot \frac{3}{10}$ Multiply each side by the reciprocal of $\frac{3}{4}$.

$\left(\frac{4}{3} \cdot \frac{3}{4}\right)x = \frac{4}{3} \cdot \frac{3}{10}$ Associative Property of Multiplication

$(1)x = \frac{12}{30}$ Use the Multiplicative Inverse Property. Multiply $\frac{4}{3} \cdot \frac{3}{10}$.

$x = \frac{2}{5}$ Use the Multiplicative Identity Property. Simplify $\frac{12}{30}$.

Notice that this equation represents the same multiplication problem illustrated in part (c) of Example 2.

Classroom Tip

Using a multiplicative inverse, or reciprocal, is an important skill for algebra work in middle school. The foundation is laid in earlier grades as students develop their understanding of whole numbers, fractions, and the properties of multiplication and division.

The Distributive Properties studied in Section 3.3 are also valid for fractions.

Distributive Properties

Let a, b, c, d, e, and f be any numbers such that $b, d, f \neq 0$.

Distributive Property of Fraction Multiplication over Addition

Numbers

$$\frac{2}{3}\left(\frac{2}{5}+\frac{1}{5}\right)=\frac{2}{3}\cdot\frac{2}{5}+\frac{2}{3}\cdot\frac{1}{5}$$

Algebra

$$\frac{a}{b}\left(\frac{c}{d}+\frac{e}{f}\right)=\frac{a}{b}\cdot\frac{c}{d}+\frac{a}{b}\cdot\frac{e}{f}$$

Distributive Property of Fraction Multiplication over Subtraction

Numbers

$$\frac{2}{3}\left(\frac{2}{5}-\frac{1}{5}\right)=\frac{2}{3}\cdot\frac{2}{5}-\frac{2}{3}\cdot\frac{1}{5}$$

Algebra

$$\frac{a}{b}\left(\frac{c}{d}-\frac{e}{f}\right)=\frac{a}{b}\cdot\frac{c}{d}-\frac{a}{b}\cdot\frac{e}{f}$$

EXAMPLE 4 Real-Life Example of the Distributive Property

You and your spouse are filing your federal income tax return. You have $54,000 in taxable income and your spouse has $46,000. Each of you has a tax liability of two-fifths of your taxable income. Does it matter whether you file separately or jointly? Explain your reasoning.

SOLUTION

Method 1 First combine your two taxable incomes and then multiply by two-fifths.

$$\frac{2}{5}(54{,}000+46{,}000)=\frac{2}{5}(100{,}000) \quad \text{Add the two incomes.}$$

$$=\frac{2(100{,}000)}{5} \quad \text{Rewrite.}$$

$$=\frac{200{,}000}{5} \quad \text{Multiply.}$$

$$=40{,}000 \quad \text{Simplify.}$$

Method 2 First multiply each taxable income by two-fifths and then add the results.

$$\frac{2}{5}(54{,}000+46{,}000)=\frac{2}{5}\cdot 54{,}000+\frac{2}{5}\cdot 46{,}000 \quad \text{Distributive Property}$$

$$=\frac{2(54{,}000)}{5}+\frac{2(46{,}000)}{5} \quad \text{Rewrite.}$$

$$=\frac{108{,}000}{5}+\frac{92{,}000}{5} \quad \text{Multiply.}$$

$$=21{,}600+18{,}400 \quad \text{Divide.}$$

$$=40{,}000 \quad \text{Add.}$$

It does not matter how you file. Whether filing jointly or separately, the result is the same. You will owe $40,000 in federal tax.

In Example 4, notice that real-life applications of the Distributive Property relate to the order in which actions are performed. Do you combine first and then compute the tax? Or do you compute the tax first and then combine? With the Distributive Property, the result is the same with either order.

Dividing Fractions, Whole Numbers, and Mixed Numbers

Dividing Fractions

Words To divide by a nonzero fraction, multiply by its reciprocal.

Multiply by reciprocal.

Numbers $\frac{4}{3} \div \frac{2}{3} = \frac{4}{3} \cdot \frac{3}{2} = 2$ **Algebra** $\frac{a}{b} \div \frac{c}{d} = \frac{a}{b} \cdot \frac{d}{c}$, where $b, c, d \neq 0$

EXAMPLE 5 Real-Life Examples of Fraction Division

Classroom Tip
Continue to reinforce with your students that real-life applications of division often answer the question, "How many are in?" In Example 5(a), you can build your students' intuition by bringing a 2-cup measure and a quarter-cup measure and letting your students pour water to verify that there are six quarter cups in a cup and a half.

a. How many quarter cups are in a cup and a half?

b. You get paid twice per month. How many times do you get paid per year?

c. You earn \$8 for two-thirds of an hour of work. What is your hourly wage?

SOLUTION

a. $1\frac{1}{2} \div \frac{1}{4} = \frac{3}{2} \div \frac{1}{4}$

$= \frac{3}{2} \cdot \frac{4}{1}$

$= \frac{12}{2}$

$= 6$

There are 6 quarter cups in a cup and a half.

b. **Use Multiplication** You get paid twice per month. So, one way to answer the question is to multiply the number of months in one year by 2.

Number of paychecks $= 2 \cdot 12$

$= 24$

Use Division Another way to answer the question is to think, "How many half months are in twelve months?"

Number of paychecks $= 12 \div \frac{1}{2}$

$= 12 \cdot 2$

$= 24$

Classroom Tip
The application in Example 5(c) is often difficult for students. This common application of division is called a *rate problem*. Understanding the fact that "per" means "divide" can help students solve rate problems. You will study rates in more detail in Section 6.4.

c. Hourly wage $= \$8 \div \frac{2}{3}$ h Rate is "dollars per hour."

$= 8 \cdot \frac{3}{2}$

$= 12$

Your hourly wage is \$12 per hour.

Design56/Dreamstime.com

Definition of Complex Fraction

A **complex fraction** is a fraction that contains at least one fraction or mixed number in its numerator or denominator.

Numbers

$$\frac{\frac{2}{3}}{\frac{4}{5}}, \quad \frac{\frac{3}{4}}{1\frac{1}{8}}, \quad \frac{\frac{9}{11}}{2}, \quad \frac{4}{\frac{4}{7}+\frac{5}{7}}$$

Algebra

$$\frac{\frac{a}{b}}{\frac{c}{d}}, \text{ where } b, c, d \neq 0$$

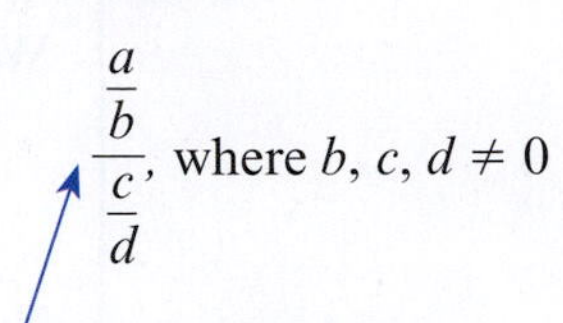

EXAMPLE 6 Evaluating Complex Fractions

Classroom Tip

Encourage students to write the major fraction bar of a complex fraction so that it is clearly longer than any fraction bars in the numerator or denominator.

Good $\dfrac{\frac{2}{3}}{\frac{4}{5}}$

Poor $\begin{matrix} 2 \\ \hline 3 \\ \hline 4 \\ \hline 5 \end{matrix}$

a. $\dfrac{\frac{2}{3}}{\frac{4}{5}} = \dfrac{2}{3} \div \dfrac{4}{5}$ Rewrite complex fraction using division sign.

$= \dfrac{2}{3} \cdot \dfrac{5}{4}$ Multiply by the reciprocal of $\dfrac{4}{5}$.

$= \dfrac{5}{6}$ Simplify.

b. $\dfrac{\frac{3}{4}}{1\frac{1}{8}} = \dfrac{\frac{3}{4}}{\frac{9}{8}}$ Rewrite denominator as improper fraction.

$= \dfrac{3}{4} \div \dfrac{9}{8}$ Rewrite complex fraction using division sign.

$= \dfrac{3}{4} \cdot \dfrac{8}{9}$ Multiply by the reciprocal of $\dfrac{9}{8}$.

$= \dfrac{2}{3}$ Simplify.

c. $\dfrac{\frac{9}{11}}{2} = \dfrac{9}{11} \div 2$ Rewrite complex fraction using division sign.

$= \dfrac{9}{11} \cdot \dfrac{1}{2}$ Multiply by reciprocal of 2.

$= \dfrac{9}{22}$ Multiply fractions.

d. $\dfrac{4}{\frac{6}{7}+\frac{5}{7}} = \dfrac{4}{\frac{11}{7}}$ Add fractions in denominator.

$= 4 \div \dfrac{11}{7}$ Rewrite complex fraction using division sign.

$= 4 \cdot \dfrac{7}{11}$ Multiply by the reciprocal of $\dfrac{11}{7}$.

$= \dfrac{28}{11}$ Multiply fractions.

Estimating Products and Quotients of Fractions

Some real-life applications may not necessitate exact computations. In such cases, you can use whole numbers or simple fractions to estimate the answer.

EXAMPLE 7 Estimating Products and Quotients in Real Life

a. You buy $9\frac{9}{10}$ gallons of gasoline for $\$4\frac{449}{1000}$ per gallon. Estimate the total cost.

b. You drive $243\frac{7}{10}$ miles on $10\frac{2}{10}$ gallons of gasoline. Estimate your gas mileage.

c. Use the figure to estimate the area of the living room floor.

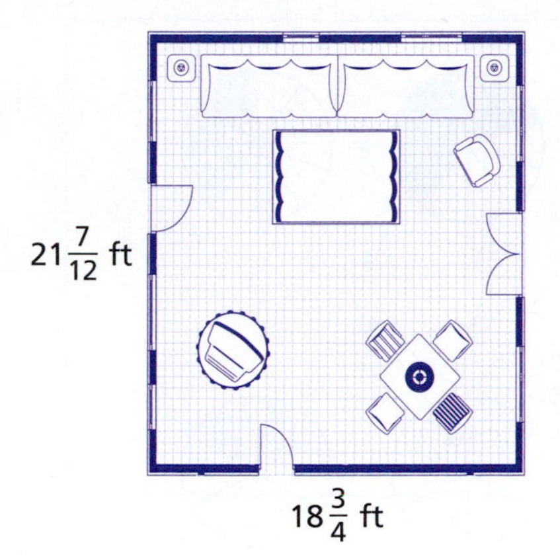

SOLUTION

a. The total cost is about

$$10 \text{ gallons} \cdot 4\frac{1}{2}\,\frac{\text{dollars}}{\text{gallon}} = \$45.$$

Estimate for $9\frac{9}{10}$ — Estimate for $4\frac{449}{1000}$

So, you pay about \$45 for gasoline.

b. Your gas mileage is about

$$240 \text{ miles} \div 10 \text{ gallons} = 24 \text{ miles per gallon}.$$

Estimate for $243\frac{7}{10}$ — Estimate for $10\frac{2}{10}$

So, your gas mileage is about 24 miles per gallon.

c. Roughly, the area of the living room floor is about

$$20 \text{ feet} \cdot 20 \text{ feet} = 400 \text{ square feet}.$$

Estimate for $21\frac{7}{12}$ — Estimate for $18\frac{3}{4}$

So, the living room floor has an area of about 400 square feet.

In the solution to part (c), notice that 20 feet is a low estimate for the length but a high estimate for the width. If 20 feet had been low (or high) for *both* the length and the width, then the resulting estimate for the area may not have been as good.

Classroom Tip

Make sure your students realize the importance of considering the real-life context when solving problems. In part (c) of Example 7, when you multiply the exact length and width, you obtain an area of $404\frac{11}{16}$ square feet. However, most situations involving the square footage of a room do not require this level of accuracy.

6.3 Exercises

Free step-by-step solutions to odd-numbered exercises at *MathematicalPractices.com*.

1. **Writing a Solution Key** Write a solution key for the activity on page 218. Describe a rubric for grading a student's work.

2. **Grading the Activity** In the activity on page 218, a student gave the answers below. Each question is worth 10 points. Assign a grade to each answer. For those that are incorrect, why do you think the student erred?

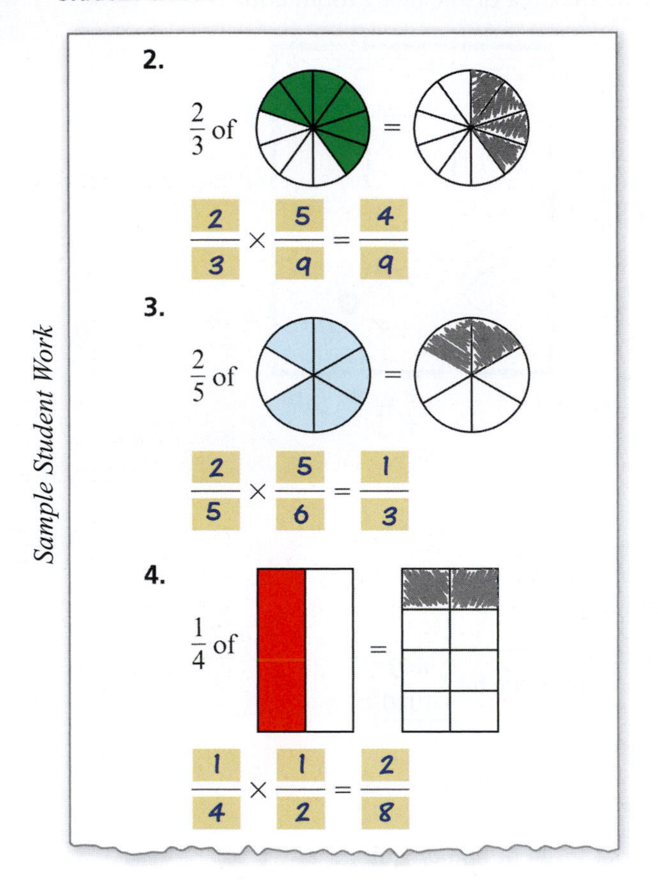

3. **Repeated Addition** Rewrite each addition problem as a multiplication problem. Then find the product. Write your answer in simplest form.

   a. $\frac{2}{5}+\frac{2}{5}+\frac{2}{5}+\frac{2}{5}+\frac{2}{5}+\frac{2}{5}$ **b.** $\frac{1}{8}+\frac{1}{8}+\frac{1}{8}$

   c. $\frac{3}{4}+\frac{3}{4}+\frac{3}{4}+\frac{3}{4}+\frac{3}{4}+\frac{3}{4}+\frac{3}{4}$

4. **Repeated Addition** Rewrite each addition problem as a multiplication problem. Then find the product. Write your answer in simplest form.

   a. $\frac{1}{6}+\frac{1}{6}+\frac{1}{6}+\frac{1}{6}+\frac{1}{6}$

   b. $\frac{3}{4}+\frac{3}{4}+\frac{3}{4}+\frac{3}{4}+\frac{3}{4}+\frac{3}{4}+\frac{3}{4}+\frac{3}{4}$

   c. $\frac{5}{8}+\frac{5}{8}+\frac{5}{8}+\frac{5}{8}+\frac{5}{8}+\frac{5}{8}+\frac{5}{8}+\frac{5}{8}$

5. **Number Line Models** Use a number line to model each problem.

   a. $\frac{1}{5}\cdot 2\frac{1}{2}$ **b.** $\frac{2}{3}\cdot 1\frac{1}{2}$ **c.** $\frac{3}{4}\cdot 1\frac{3}{5}$

6. **Number Line Models** Use a number line to model each problem.

   a. $\frac{1}{3}\cdot 4\frac{1}{2}$ **b.** $\frac{3}{7}\cdot 1\frac{2}{5}$ **c.** $\frac{6}{5}\cdot 1\frac{2}{3}$

7. **Area Models** Write the fraction multiplication problem shown by each area model.

   a.  **b.**  **c.** 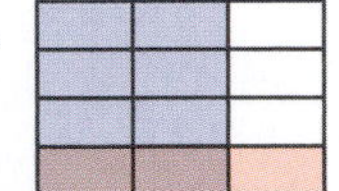

8. **Area Models** Write the fraction multiplication problem shown by each area model.

   a. 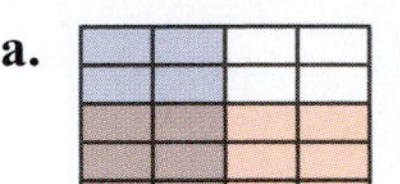 **b.** 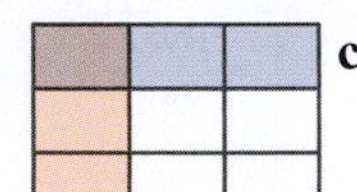 **c.** 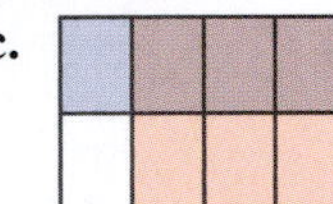

9. **Multiplying Fractions** Find each product. Write your answer in simplest form.

   a. $\frac{1}{5}\cdot 4$ **b.** $3\cdot\frac{5}{6}$ **c.** $\frac{4}{7}\cdot\frac{1}{2}$ **d.** $\frac{5}{8}\cdot\frac{4}{15}$

   e. $\frac{3}{5}\cdot 2\frac{1}{2}$ **f.** $3\frac{1}{3}\cdot\frac{1}{2}$ **g.** $1\frac{1}{2}\cdot 1\frac{1}{2}$ **h.** $0\cdot\frac{3}{11}$

10. **Multiplying Fractions** Find each product. Write your answer in simplest form.

    a. $\frac{4}{9}\cdot 3$ **b.** $4\cdot\frac{3}{2}$ **c.** $\frac{2}{5}\cdot\frac{5}{2}$ **d.** $\frac{2}{7}\cdot\frac{14}{3}$

    e. $\frac{2}{3}\cdot 7\frac{1}{2}$ **f.** $1\frac{4}{5}\cdot\frac{1}{3}$ **g.** $2\frac{2}{3}\cdot 0$ **h.** $0\cdot\frac{3}{11}$

11. **Reciprocals** Fill in the blanks to demonstrate the Multiplicative Inverse Property.

    a. $\frac{2}{3}\cdot$ ___ = ___ **b.** $\frac{5}{4}\cdot$ ___ = ___

    c. $3\frac{1}{2}\cdot$ ___ = ___

12. **Reciprocals** Fill in the blanks to demonstrate the Multiplicative Inverse Property.

    a. $\frac{4}{7}\cdot$ ___ = ___ **b.** $\frac{8}{3}\cdot$ ___ = ___

    c. $4\frac{1}{3}\cdot$ ___ = ___

13. **Order of Operations** Evaluate each expression.

    a. $2\cdot\frac{5}{2}+\frac{3}{2}$ **b.** $\frac{3}{2}\cdot\frac{1}{3}+\frac{1}{2}$ **c.** $5\cdot\frac{1}{2}-\frac{1}{4}$

14. Order of Operations Evaluate each expression.

a. $\frac{1}{4} \cdot \frac{1}{2} + \frac{3}{8}$ **b.** $3 \cdot \frac{1}{3} + \frac{4}{3}$ **c.** $\frac{3}{2} \cdot \frac{2}{3} - 1$

15. Multiplication Properties in Algebra Complete the solution of the equation. Then justify each step in the solution.

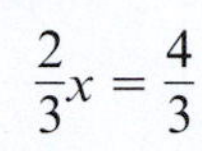

$\frac{2}{3}x = \frac{4}{3}$ Write original equation.

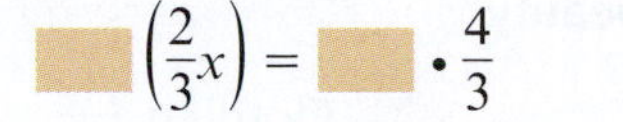

$\square\left(\frac{2}{3}x\right) = \square \cdot \frac{4}{3}$ $\square$

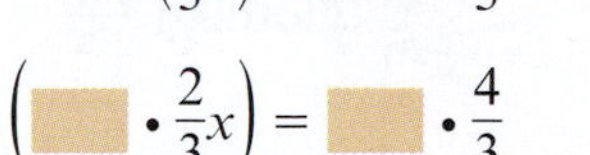

$\left(\square \cdot \frac{2}{3}x\right) = \square \cdot \frac{4}{3}$ $\square$

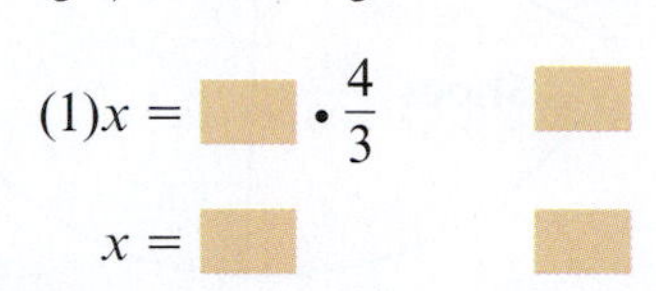

$(1)x = \square \cdot \frac{4}{3}$ $\square$

$x = \square$ $\square$

16. Multiplication Properties in Algebra Complete the solution of the equation. Then justify each step in the solution.

$\frac{4}{3}x = \frac{1}{6}$ Write original equation.

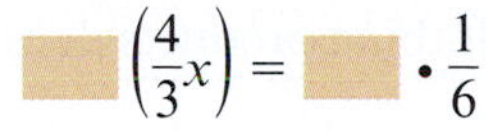

$\square\left(\frac{4}{3}x\right) = \square \cdot \frac{1}{6}$ $\square$

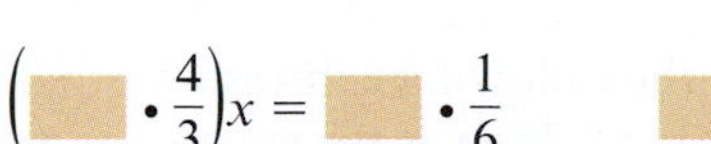

$\left(\square \cdot \frac{4}{3}\right)x = \square \cdot \frac{1}{6}$ $\square$

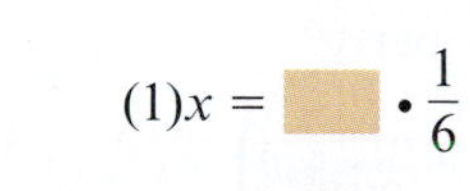

$(1)x = \square \cdot \frac{1}{6}$ $\square$

$x = \square$ $\square$

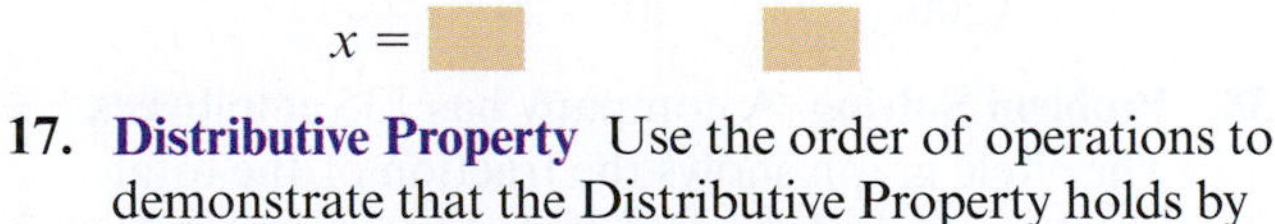

17. Distributive Property Use the order of operations to demonstrate that the Distributive Property holds by showing that both sides of each equation are equal.

a. $\frac{4}{3}\left(\frac{2}{5} + \frac{1}{5}\right) = \frac{4}{3} \cdot \frac{2}{5} + \frac{4}{3} \cdot \frac{1}{5}$

b. $\frac{4}{3}\left(\frac{2}{5} - \frac{1}{5}\right) = \frac{4}{3} \cdot \frac{2}{5} - \frac{4}{3} \cdot \frac{1}{5}$

18. Distributive Property Use the order of operations to demonstrate that the Distributive Property holds by showing that both sides of each equation are equal.

a. $\frac{2}{5}\left(\frac{4}{3} + \frac{1}{3}\right) = \frac{2}{5} \cdot \frac{4}{3} + \frac{2}{5} \cdot \frac{1}{3}$

b. $\frac{2}{5}\left(\frac{4}{3} - \frac{1}{3}\right) = \frac{2}{5} \cdot \frac{4}{3} - \frac{2}{5} \cdot \frac{1}{3}$

19. Dividing Fractions Find each quotient. Write your answer in simplest form.

a. $3 \div \frac{2}{3}$ **b.** $\frac{2}{3} \div 3$ **c.** $1 \div \frac{2}{3}$

20. Dividing Fractions Find each quotient. Write your answer in simplest form.

a. $4 \div \frac{1}{2}$ **b.** $\frac{1}{2} \div 4$ **c.** $\frac{4}{5} \div 1$

21. Dividing Fractions Write the question as a division problem. Then solve the division problem.

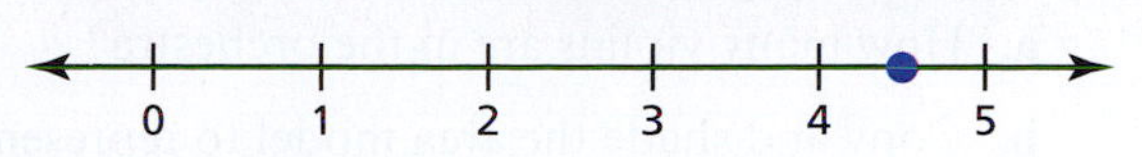

How many halves are in four and one-half?

22. Dividing Fractions Write the question as a division problem. Then solve the division problem.

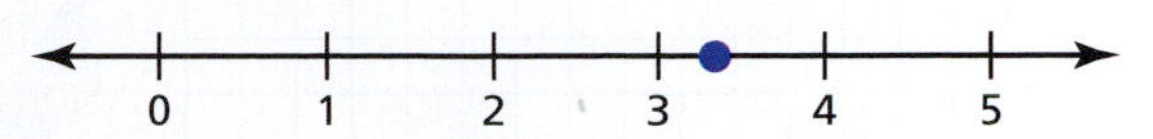

How many thirds are in three and one-third?

23. Dividing Mixed Numbers Find each quotient. Write your answer in simplest form.

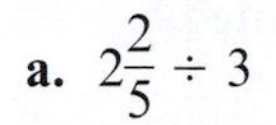

a. $2\frac{2}{5} \div 3$ **b.** $1\frac{2}{3} \div \frac{5}{6}$ **c.** $2\frac{4}{5} \div 1\frac{3}{5}$

24. Dividing Mixed Numbers Find each quotient. Write your answer in simplest form.

a. $5 \div 1\frac{7}{8}$ **b.** $\frac{7}{9} \div 4\frac{2}{3}$ **c.** $5\frac{1}{3} \div 1\frac{1}{9}$

25. Area Models Write the fraction multiplication problem shown by the area model. Then write two division problems that can be represented by the model.

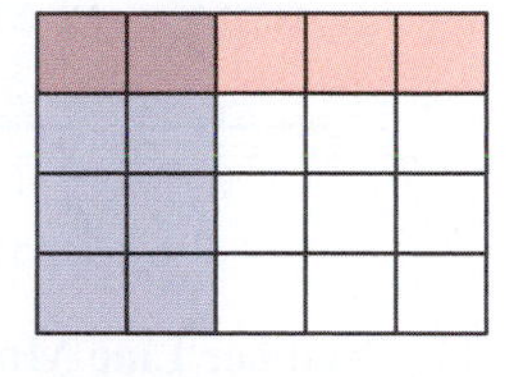

26. Area Models Write the fraction multiplication problem shown by the area model. Then write two division problems that can be represented by the model.

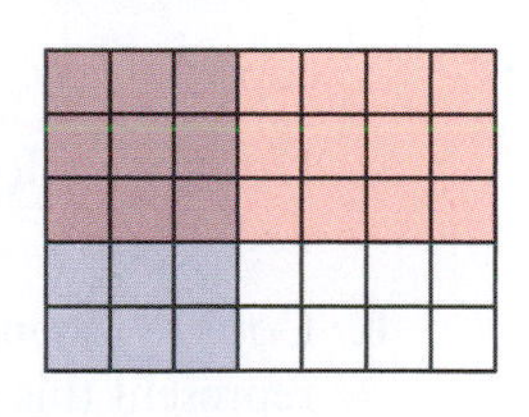

27. Evaluating Complex Fractions Evaluate each complex fraction. Write your answer in simplest form.

a. $\dfrac{\frac{3}{5}}{\frac{1}{3}}$ **b.** $\dfrac{\frac{1}{2}}{1\frac{1}{2}}$ **c.** $\dfrac{\frac{3}{10}}{4}$ **d.** $\dfrac{5}{\frac{1}{2} + \frac{1}{2}}$

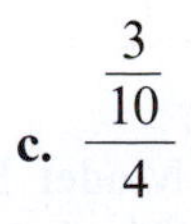

28. Evaluating Complex Fractions Evaluate each complex fraction. Write your answer in simplest form.

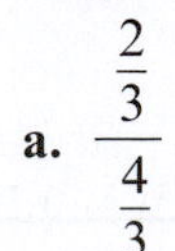

a. $\dfrac{\frac{2}{3}}{\frac{4}{3}}$ **b.** $\dfrac{\frac{3}{2}}{1\frac{1}{2}}$ **c.** $\dfrac{\frac{7}{15}}{7}$ **d.** $\dfrac{3}{\frac{2}{3} + \frac{4}{3}}$

29. Estimation Estimate each product or quotient.

a. $4\frac{8}{9} \cdot 11\frac{7}{8}$ **b.** $10\frac{1}{8} \div 4\frac{11}{12}$ **c.** $47\frac{4}{5} \div 11\frac{11}{16}$

30. Estimation Estimate each product or quotient.

a. $6\frac{1}{20} \cdot 14\frac{8}{9}$ **b.** $24\frac{1}{7} \div 5\frac{14}{15}$ **c.** $99\frac{3}{4} \div 25\frac{1}{3}$

31. **Area Model** An orchestra has 98 instruments. Of these, one-seventh are violins.

a. How many violins are in the orchestra?

b. Copy and shade the area model to represent the portion of the instruments that are violins.

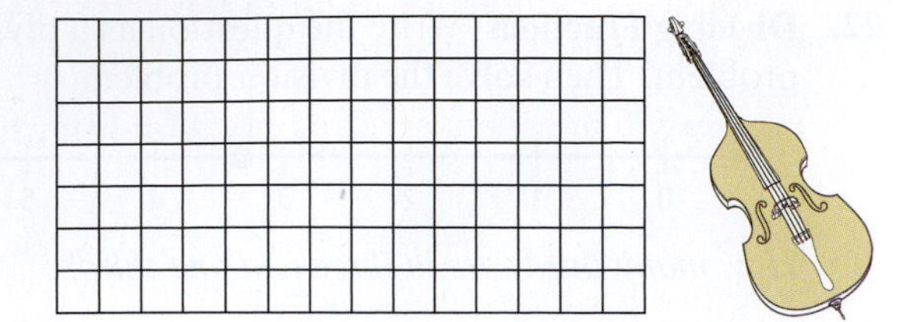

32. **Number Line Model** A clothing store advertises a sale in which all clothing is discounted by one-fourth up to one-third.

a. What is the most that you will pay for a jacket that originally had a price of $150?

b. What is the least that you will pay?

c. Copy and shade the number line model to represent the range of prices for the jacket.

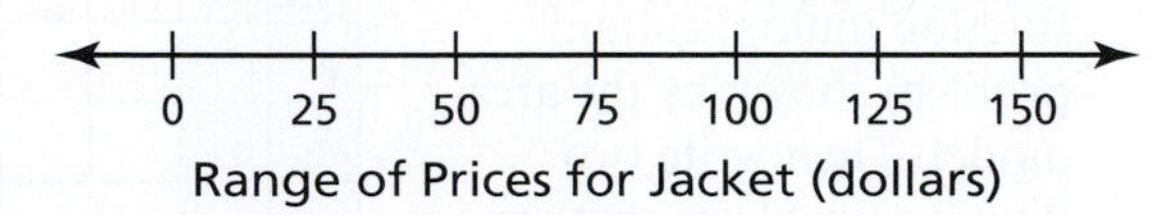

33. **Number Line Model** A recipe for cookies calls for $2\frac{1}{4}$ cups of flour. The recipe makes 3 dozen cookies.

a. How much flour do you need to make 8 dozen cookies?

b. Copy and complete the number line model to represent this problem.

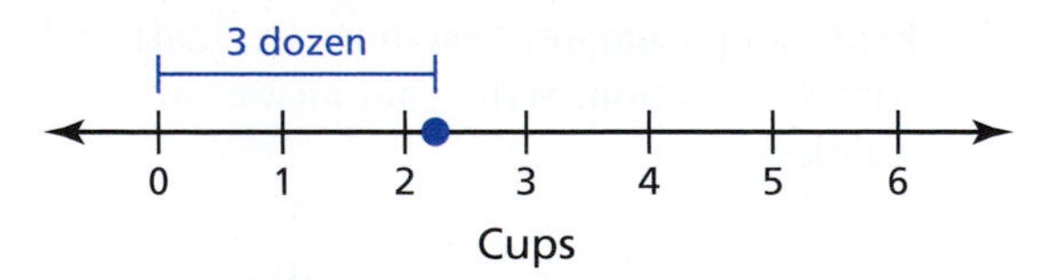

34. **Number Line Model** Human hair grows at a rate of about one-half inch per month.

a. How much does human hair grow in $3\frac{1}{2}$ months?

b. Copy and complete the number line model to represent this problem.

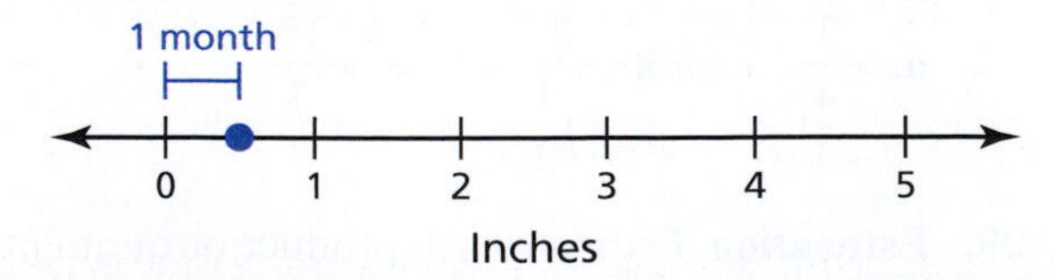

35. *Writing* Explain when the product of a fraction and a whole number will be greater than the whole number.

36. *Writing* Explain how to find the multiplicative inverse of a fraction.

37. **Problem Solving** A store sells 400 different products. The circle graph shows the fraction of the total number of product types in each department.

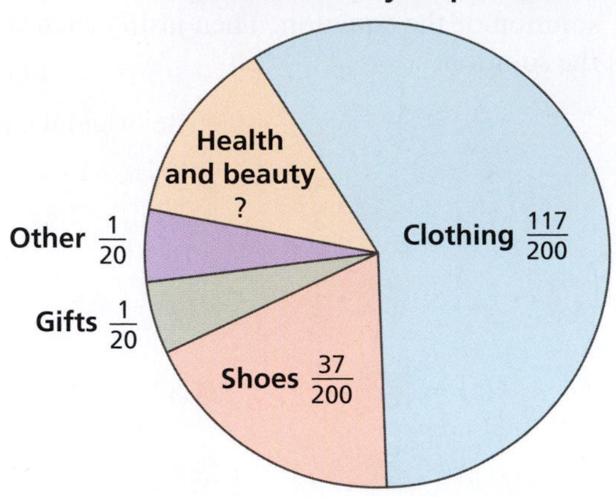

a. What fraction of the product types are health and beauty products?

b. How many types of health and beauty products does the store sell?

c. How many types of clothing products does the store sell?

d. Explain how the following expression is related to the circle graph. How is the expression related to the Distributive Property?

$$400\left(\frac{117}{200} + \frac{37}{200} + \frac{1}{20} + \frac{1}{20} + ?\right)$$

38. **Problem Solving** A company has 135 employees. The circle graph shows the fraction of the total number of employees who are in each age group.

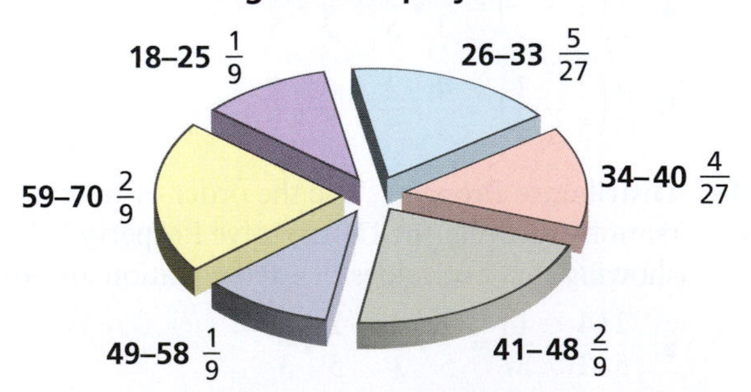

a. Make a table that shows the number of employees in each age group.

b. Add the numbers of employees in your table. Does your answer help to verify your results in part (a)? Explain.

c. Explain how the following expression is related to the circle graph. How is the expression related to the Distributive Property?

$$135\left(\frac{1}{9} + \frac{5}{27} + \frac{4}{27} + \frac{2}{9} + \frac{1}{9} + \frac{2}{9}\right)$$

39. My Dear Aunt Sally Arrange the fractions $\frac{2}{9}$, $\frac{1}{3}$, $\frac{2}{7}$, and $\frac{3}{7}$ so that the equation is true.

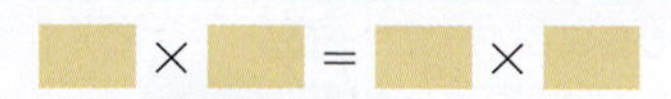

For more practice with this type of reasoning, go to *AuntSally.com.*

40. My Dear Aunt Sally Arrange the fractions $\frac{2}{9}$, $\frac{3}{8}$, $\frac{4}{9}$, and $\frac{3}{4}$ so that the equation is true.

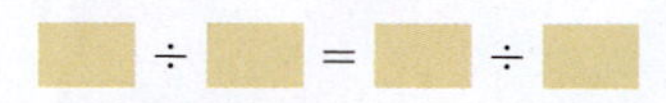

For more practice with this type of reasoning, go to *AuntSally.com.*

41. Grading Student Work On a diagnostic test, one of your students does the following work. Explain what the student did wrong. Which topics would you encourage the student to review?

a.

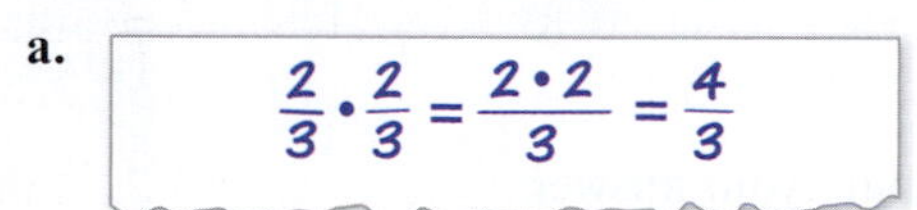

b. $\frac{3}{4} \div \frac{1}{4} = \frac{3}{4} \cdot \frac{1}{4} = \frac{3}{16}$

42. Grading Student Work On a diagnostic test, one of your students does the following work. Explain what the student did wrong. Which topics would you encourage the student to review?

a. $\frac{1}{3} \cdot \frac{3}{5} = \frac{3 \cdot 3}{1 \cdot 5} = \frac{9}{5}$

b. $\frac{3}{2} \div \frac{1}{2} = \frac{2}{3} \cdot \frac{1}{2} = \frac{1}{3}$

43. Measurement: In Your Classroom A piece of ribbon is three and a half inches long.

a. How many quarter-inch pieces can you cut from the ribbon?

b. Use the drawing below to explain how you can show your students that this question can be answered using division.

c. Explain how you can use mental math to answer the question.

44. Measurement: In Your Classroom

a. A one-cup jelly jar is one-sixteenth of a gallon. How many one-cup jelly jars can be filled with two and a half gallons of jelly?

b. Describe a demonstration you can use for your students to build their intuition for this type of measurement question.

c. Explain how to answer the question using mental math.

d. Suppose each jelly jar is filled only seven-eighths full. How many jars are needed?

45. *Writing* Explain how you can divide a number by a fraction using the multiplicative inverse of the fraction.

46. *Writing* Describe one way to evaluate a complex fraction.

47. Rate It takes two and a half minutes for an assembly line worker to sew the hem on a skirt. At this rate, how many hems can the worker sew in one hour?

48. Rate You earn twenty and a half dollars for two and a quarter hours of work. What is your hourly wage?

49. Estimating a Product

a. You buy $11\frac{9}{10}$ gallons of diesel fuel at $\$4\frac{489}{1000}$ per gallon. Estimate the total cost of the fuel.

b. Explain how you can check that your answer is reasonable.

50. Estimating a Product

a. Estimate the area of the bathroom.

b. Use fraction multiplication to find the exact area. How accurate is your estimate?

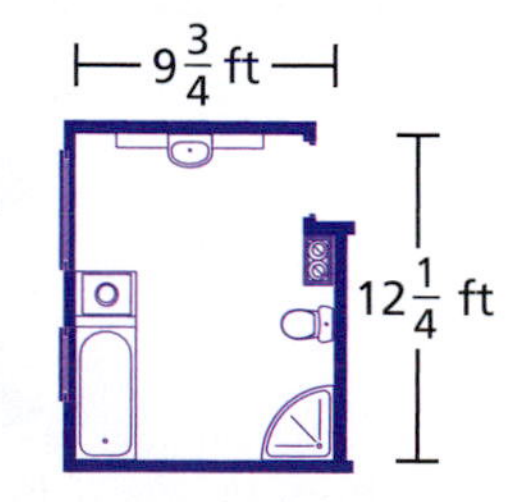

51. Estimating a Quotient

a. You drive $213\frac{3}{10}$ miles and use $11\frac{1}{4}$ gallons of gasoline. Estimate your gas mileage.

b. Use fraction division to find the exact mileage. How accurate is your estimation?

52. Estimating a Quotient

a. Your dog eats $1\frac{1}{4}$ pounds of dog food each day. Estimate how many days a 40-pound bag of dog food will last.

b. Use fraction division to find the exact answer. How accurate is your estimation?

Activity: Understanding Ratio and Proportion

Materials:

- Pencil

Learning Objectives:

Students understand that a ratio is a comparison of *part-to-part*, *part-to-whole*, or *whole-to-part*. Learn to write a ratio. Learn that some ratios compare quantities that have the same units, while others (rates) compare quantities that have different units.

Name ______________________

1. Write the ratio of football helmets to baseball caps. Simplify your answer.

Ratio = $\frac{\square}{\square}$

Simplified ratio = $\frac{\square}{\square}$

2. Write the ratio of army helmets to *all* hats. Simplify your answer.

Ratio = $\frac{\square}{\square}$

Simplified ratio = $\frac{\square}{\square}$

Complete the statement with = or ≠. Circle the correct answer.

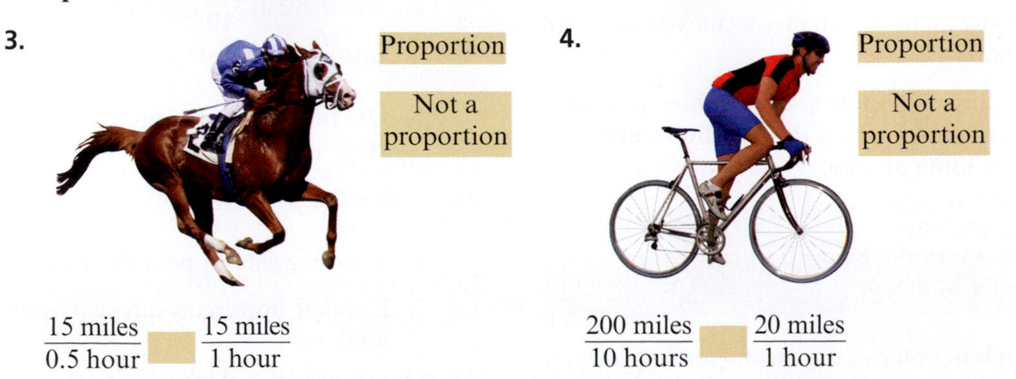

3. Proportion / Not a proportion

$\frac{15 \text{ miles}}{0.5 \text{ hour}} \ \square \ \frac{15 \text{ miles}}{1 \text{ hour}}$

4. Proportion / Not a proportion

$\frac{200 \text{ miles}}{10 \text{ hours}} \ \square \ \frac{20 \text{ miles}}{1 \text{ hour}}$

6.4 Ratios and Proportions

Standards

Grades 6–8
Ratios and Proportional Relationships
Students should understand ratio concepts and use ratio reasoning to solve problems, and analyze proportional relationships and use them to solve real-world and mathematical problems.

- ▶ Write a ratio as a part-to-part, a part-to-whole, or a whole-to-part comparison.
- ▶ Write and simplify rates.
- ▶ Decide whether two ratios or two rates are proportional.

Ratios

Ratios

Words The **ratio** of a number a to a nonzero number b is the quotient when a is divided by b. A ratio can be a **part-to-part**, a **part-to-whole**, or a **whole-to-part** comparison.

Ratio of blue to red marbles

Numbers $\frac{2}{4} = \frac{\cancel{2}^{1}}{2 \cdot \cancel{2}_{1}} = \frac{1}{2}$, 1 : 2, or 1 to 2

Ratio of a to b

Algebra $\frac{a}{b}$, $a : b$, or a to b, where $b \neq 0$

EXAMPLE 1 **Real-Life Examples of Ratios**

The string section of an orchestra consists of the instruments below.

- 18 violins
- 8 violas
- 6 cellos
- 4 double basses

a. Part-to-part What is the ratio of violas to violins?

b. Part-to-whole What is the ratio of violins to all stringed instruments?

Classroom Tip
Although the term *ratio* is often not introduced until Grade 6, students learn how to use fractions to describe "part-to-whole" relationships at a much earlier age.

Classroom Tip
In Example 1, note that the units "instruments" is the same in the numerator and denominator. In such cases, it is normal to not list the units in the final answer. However, encourage your students to always include the units during computation, as shown here.

SOLUTION

a. $$\frac{\text{Number of violas}}{\text{Number of violins}} = \frac{8 \text{ instruments}}{18 \text{ instruments}} = \frac{\cancel{2}^{1} \cdot 4 \cancel{\text{ instruments}}}{\cancel{2}_{1} \cdot 9 \cancel{\text{ instruments}}} = \frac{4}{9}$$

The ratio of violas to violins is 4 to 9.

b. $$\frac{\text{Number of violins}}{\text{All instruments}} = \frac{18 \text{ instruments}}{36 \text{ instruments}} = \frac{\cancel{18 \text{ instruments}}^{1}}{2 \cdot \cancel{18 \text{ instruments}}_{1}} = \frac{1}{2}$$

The ratio of violins to all stringed instruments is 1 to 2.

EXAMPLE 2 Classroom Activity for Teaching Ratios

Ask your students to sit in groups of 2 or 3. For each group, lay a collection of counting chips on the table. Here is an example.

Ask your students to describe and calculate all the possible ratios they see in the collection of counting chips.

SOLUTION

Part-to-whole ratios There are several possible ratios. Here are some "part-to-whole" ratios.

> **Classroom Tip**
> Notice how a simple concept, such as the ratio of one color of counters to another, can produce a rich problem.

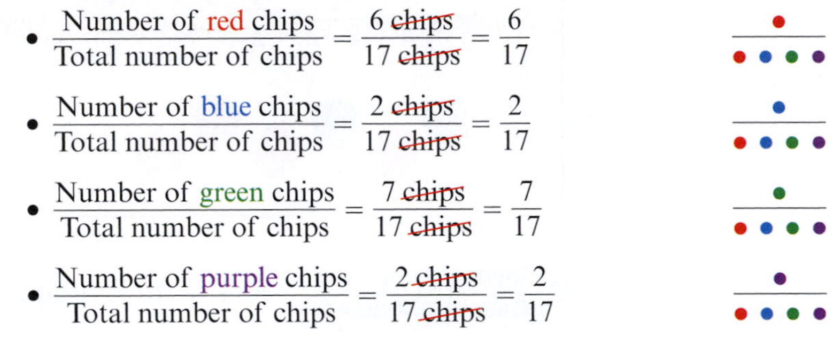

- $\dfrac{\text{Number of red chips}}{\text{Total number of chips}} = \dfrac{6\ \cancel{\text{chips}}}{17\ \cancel{\text{chips}}} = \dfrac{6}{17}$
- $\dfrac{\text{Number of blue chips}}{\text{Total number of chips}} = \dfrac{2\ \cancel{\text{chips}}}{17\ \cancel{\text{chips}}} = \dfrac{2}{17}$
- $\dfrac{\text{Number of green chips}}{\text{Total number of chips}} = \dfrac{7\ \cancel{\text{chips}}}{17\ \cancel{\text{chips}}} = \dfrac{7}{17}$
- $\dfrac{\text{Number of purple chips}}{\text{Total number of chips}} = \dfrac{2\ \cancel{\text{chips}}}{17\ \cancel{\text{chips}}} = \dfrac{2}{17}$

Part-to-part ratios There are many possible ratios. Here are some "part-to-part" ratios.

- $\dfrac{\text{Number of red chips}}{\text{Number of blue chips}} = \dfrac{6\ \cancel{\text{chips}}}{2\ \cancel{\text{chips}}} = \dfrac{6}{2} = \dfrac{3}{1}$
- $\dfrac{\text{Number of red chips}}{\text{Number of green chips}} = \dfrac{6\ \cancel{\text{chips}}}{7\ \cancel{\text{chips}}} = \dfrac{6}{7}$
- $\dfrac{\text{Number of red chips}}{\text{Number of purple chips}} = \dfrac{6\ \cancel{\text{chips}}}{2\ \cancel{\text{chips}}} = \dfrac{6}{2} = \dfrac{3}{1}$
- $\dfrac{\text{Number of blue chips}}{\text{Number of red chips}} = \dfrac{2\ \cancel{\text{chips}}}{6\ \cancel{\text{chips}}} = \dfrac{2}{6} = \dfrac{1}{3}$
- $\dfrac{\text{Number of blue chips}}{\text{Number of green chips}} = \dfrac{2\ \cancel{\text{chips}}}{7\ \cancel{\text{chips}}} = \dfrac{2}{7}$
- $\dfrac{\text{Number of blue chips}}{\text{Number of purple chips}} = \dfrac{2\ \cancel{\text{chips}}}{2\ \cancel{\text{chips}}} = \dfrac{1}{1}$

The list is not yet complete. You still have to use the green and yellow chips in the numerator. Moreover, your students might come up with more combinations, such as the following.

- $\dfrac{\text{Number of red and blue chips}}{\text{Number of blue chips}} = \dfrac{8\ \cancel{\text{chips}}}{2\ \cancel{\text{chips}}} = \dfrac{8}{2} = \dfrac{4}{1}$

With combinations such as this, there are literally dozens of additional ratios that your students can create.

Rates

Rates

Words A **rate** is a ratio of two quantities that have *different* units.

Numbers $\dfrac{130 \text{ mi}}{2 \text{ h}} = \dfrac{\cancel{2}^{1} \cdot 65}{\cancel{2}_{1}} = \dfrac{65 \text{ mi}}{1 \text{ h}}$ ← Unit rate

Algebra $\dfrac{a \text{ units of one type}}{b \text{ units of another type}}$, where $b \neq 0$

A rate that has a denominator of 1 is called a **unit rate**. The fraction bar in a rate is usually read as "per." For instance, the above unit rate is read as 65 miles *per* hour.

Classroom Tip

In real-life examples, teach your students to always list the units of the problem. This emphasis on units and unit analysis is a valuable gift you can give your students and will prepare them for success in algebra, precalculus, and calculus.

EXAMPLE 3 **Real-Life Examples of Rates**

a. Estimate the conversion between centimeters and inches. Write your result as a unit rate.

b. You earn \$41,600 per year. Write this as an hourly rate. (Assume a 40-hour work week.)

SOLUTION

a. Place a centimeter ruler and an inch ruler side by side. You can see that there are about $10\frac{2}{10}$ centimeters in 4 inches.

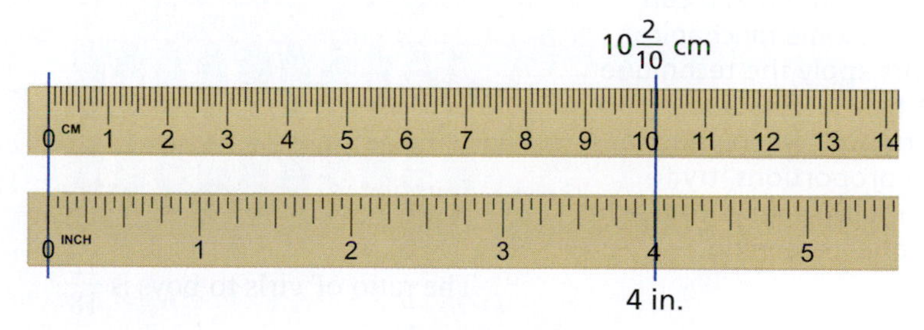

$$\text{Rate} = \frac{10\frac{2}{10} \text{ cm}}{4 \text{ in.}} = \frac{\frac{102}{10} \text{ cm}}{4 \text{ in.}} = \frac{\frac{51}{5} \text{ cm}}{4 \text{ in.}} = \frac{\frac{51}{20} \text{ cm}}{1 \text{ in.}} \qquad \text{Unit rate}$$

So, there are about $\frac{51}{20}$ centimeters per inch.

Classroom Tip

In Example 3(a), the estimate of $\frac{51}{20}$ centimeters per inch is close to the exact rate of 2.54 centimeters per inch. In the next chapter on decimals, you will learn how to write fractions such as $\frac{51}{20}$ as decimals.

In Example 3(b), notice that the key to changing from one unit of measure to another is multiplying by a convenient form of the multiplicative identity "1."

b. You need to change the denominator from years to hours. You can do this by using **unit analysis**, the process of changing from one unit of measure to another.

$$\text{Rate} = \frac{\$41{,}600}{1 \text{ yr}} \qquad \text{Rate is "per year."}$$

$$= \frac{\$41{,}600}{1 \cancel{\text{yr}}} \cdot \frac{1 \cancel{\text{yr}}}{52 \cancel{\text{wk}}} \cdot \frac{1 \cancel{\text{wk}}}{40 \text{ h}}$$

Multiplicative identities

$$= \frac{\$41{,}600}{2080 \text{ h}}$$

$$= \frac{\$20}{1 \text{ h}} \qquad \text{Rate is "per hour."}$$

You earn \$20 per hour.

Proportions

Proportions

Words A **proportion** is an equation stating that two ratios or rates are equivalent.

Numbers $\frac{5 \text{ mi}}{3 \text{ h}} = \frac{20 \text{ mi}}{12 \text{ h}}$

Algebra $\frac{a}{b} = \frac{c}{d}$, where $b, d \neq 0$

When two ratios form a proportion, they are said to be **proportional**.

Cross Products Property

Words The cross products of a proportion are equal.

Numbers

$\frac{5 \text{ mi}}{3 \text{ h}} = \frac{20 \text{ mi}}{12 \text{ h}}$ $3 \cdot 20 = 60$, $5 \cdot 12 = 60$

Algebra

If $\frac{a}{b} = \frac{c}{d}$, then $ad = bc$, where $b, d \neq 0$

EXAMPLE 4 **Testing for Proportions in Real Life**

Classroom Tip

Using cross products to check for proportionality is efficient, and middle school students tend to be able to master the technique. It can, however, become mechanical if students apply the technique without true understanding. For this reason, when you are teaching proportions, try to think of multiple ways to teach proportional reasoning.

a. Decide whether the two ratios are proportional.

Ratio 1 Of the 30 students in the classroom, 12 are girls.

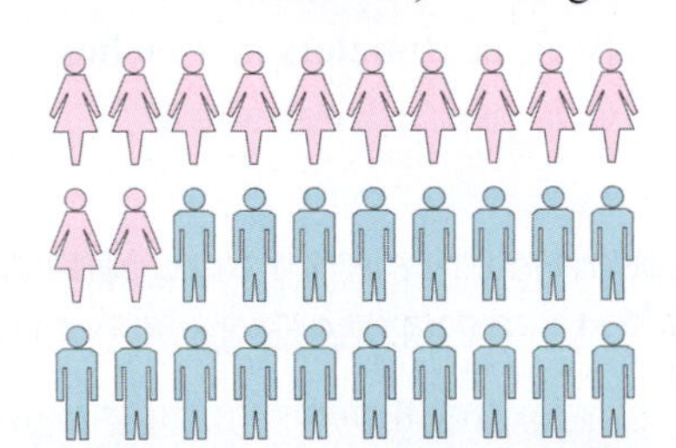

The ratio of girls to boys is $\frac{12}{18}$.

Ratio 2 Of the 25 students in the classroom, 10 are girls.

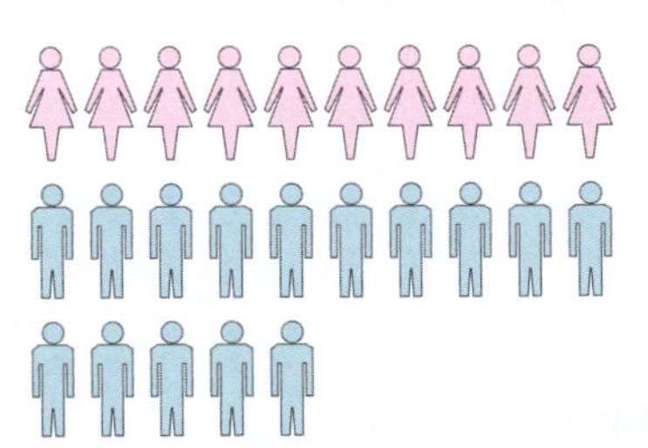

The ratio of girls to boys is $\frac{10}{15}$.

b. Decide whether the two rates are proportional.

Rate 1 A canoe travels 5 miles in 3 hours. The rate is 5 mi/3 h.

Rate 2 A canoe travels 16 miles in 10 hours. The rate is 16 mi/10 h.

SOLUTION

a. Here are two ways to conclude that the two ratios are proportional.

- You can check the cross products to see that they are equal.

$\frac{12 \text{ girls}}{18 \text{ boys}} = \frac{10 \text{ girls}}{15 \text{ boys}}$ $18 \cdot 10 = 180$, $12 \cdot 15 = 180$ Cross products are equal.

- A second way to conclude that the two ratios are proportional is to notice that each ratio simplifies to $\frac{2}{3}$.

b. These two rates are *not* proportional because the cross products are not equal.

$\frac{5 \text{ mi}}{3 \text{ h}} \stackrel{?}{=} \frac{16 \text{ mi}}{10 \text{ h}}$ $3 \cdot 16 = 48$, $5 \cdot 10 = 50$

EXAMPLE 5 Real-Life Example of Proportional Reasoning

Mathematical Practices

Model with Mathematics

MP4 Mathematically proficient students routinely interpret their mathematical results in the context of the situation, and the results make sense.

In *Gulliver's Travels* by Jonathan Swift, Gulliver's height and the height of a Lilliputian are "in the proportion of twelve to one." A Lilliputian is 6 inches tall. How tall is Gulliver?

SOLUTION

This problem can be solved using algebra by letting Gulliver's height be represented by x and by solving the following proportion for x.

$$\frac{\text{Gulliver's height}}{\text{Lilliputian's height}} = \frac{12}{1} = \frac{x \text{ in.}}{6 \text{ in.}}$$

For middle school students, this method can be too mechanical. A better method is to write the proportion without the variable x, as shown.

$$\frac{\text{Gulliver's height}}{\text{Lilliputian's height}} = \frac{12}{1} = \frac{? \text{ in.}}{6 \text{ in.}}$$

Then ask students to simply reason that Gulliver must be 12 times taller. So, Gulliver's height must be 6 • 12 or 72 inches.

EXAMPLE 6 Real-Life Example of Proportional Reasoning

You are reducing the size of a photograph. The photograph is 4800 pixels by 3000 pixels. You are proportionally reducing the photograph to 1600 pixels wide. What will the height be?

SOLUTION

To begin, write the following proportion.

$$\frac{3000 \text{ pixels}}{4800 \text{ pixels}} = \frac{? \text{ pixels}}{1600 \text{ pixels}}$$

Ask students to use proportional reasoning to visualize that when the width is reduced by one third, the height is also reduced by one third. So, the height of the reduced photograph will be $\frac{1}{3}$ • 3000 or 1000 pixels.

6.4 Exercises

Free step-by-step solutions to odd-numbered exercises at MathematicalPractices.com.

1. **Writing a Solution Key** Write a solution key for the activity on page 230. Describe a rubric for grading a student's work.

2. **Grading the Activity** In the activity on page 230, a student gave the answers below. Each question is worth 10 points. Assign a grade to each answer. For those that are incorrect, why do you think the student erred?

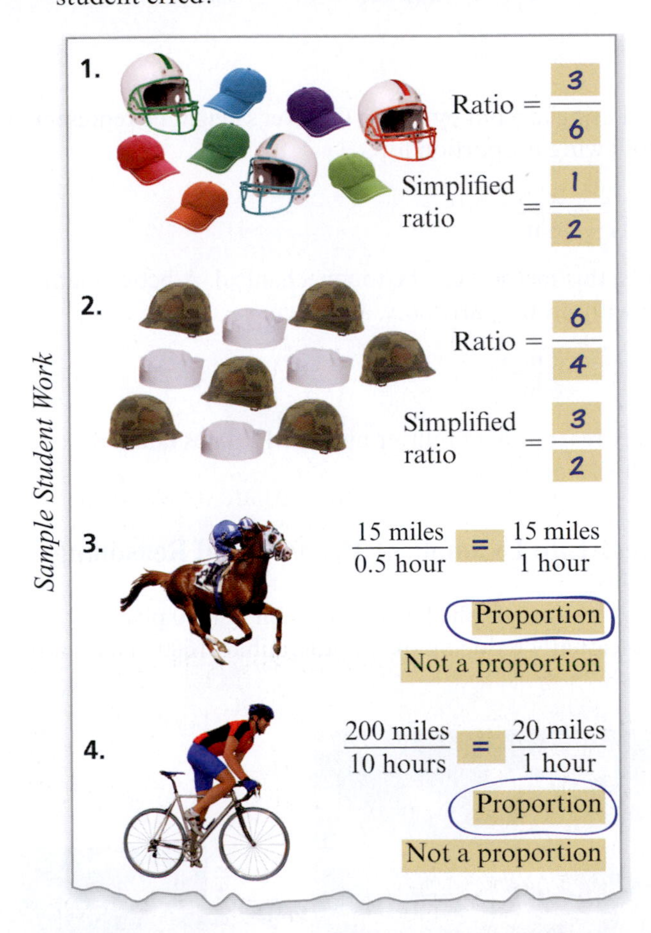

3. **Finding Ratios**

a. Write the ratio of red marbles to blue marbles. Simplify your answer.

b. Write the ratio of blue marbles to red marbles. Simplify your answer.

c. Write the ratio of red marbles to *all* marbles. Simplify your answer.

d. Write the ratio of blue marbles to *all* marbles. Simplify your answer.

Nicholas Piccillo/Shutterstock.com; jamalludin/Shutterstock.com; EchoArt/Shutterstock.com; Dani Simmonds/Shutterstock.com; Benjamin F. Haith/Shutterstock.com; Vaclav Volrab/Shutterstock.com

4. **Finding Ratios**

a. Write the ratio of red marbles to blue marbles. Simplify your answer.

b. Write the ratio of blue marbles to green marbles. Simplify your answer.

c. Write the ratio of green marbles to *all* marbles. Simplify your answer.

d. Write the ratio of red marbles to *all* marbles. Simplify your answer.

5. **Finding Ratios** Consider the set of whole numbers from 1 through 18.

{1, 2, 3, 4, 5, 6, 7, 8, 9, 10, 11, 12, 13, 14, 15, 16, 17, 18}

a. Write the ratio of even numbers to odd numbers. Simplify your answer.

b. Write the ratio of prime numbers to composite numbers. Simplify your answer. (Remember that 1 is neither prime nor composite.)

6. **Finding Ratios** Consider the set of whole numbers from 1 through 16.

{1, 2, 3, 4, 5, 6, 7, 8, 9, 10, 11, 12, 13, 14, 15, 16}

a. Write the ratio of even numbers to odd numbers. Simplify your answer.

b. Write the ratio of prime numbers to composite numbers. Simplify your answer. (Remember that 1 is neither prime nor composite.)

7. **Ratio in Geometry** The approximate circumferences and diameters of two circles are shown. For each circle, write the ratio of the circumference to the diameter. What can you conclude?

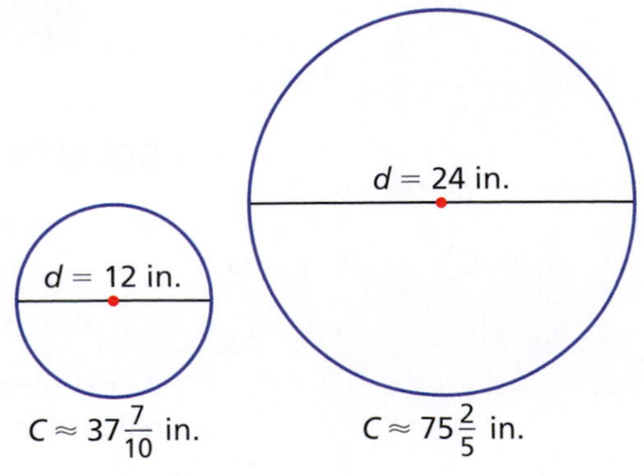

iStockphoto.com/sweetym

8. **Famous Ratio** The ratio in Exercise 7 is famous. What is it called? Research and write a brief history of this ratio.

9. **Unit Rates** Write the unit rate for each speed.

a. An automobile travels 165 miles in 3 hours.

b. A train travels 135 miles in $1\frac{1}{2}$ hours.

c. A jet travels 1250 miles in $2\frac{1}{2}$ hours.

10. **Unit Rates** Write the unit rate for each speed.

a. A wildebeest travels 4 miles in 5 minutes.

b. An elk travels 2 miles in 3 minutes.

c. A hyena travels 2 miles in 4 minutes.

11. *Writing* Simplify the expression. Then write a real-life problem that involves the expression.

$$\frac{15\text{ mi}}{1\text{ h}} \cdot \frac{1\text{ h}}{60\text{ min}} \cdot \frac{5280\text{ ft}}{1\text{ mi}}$$

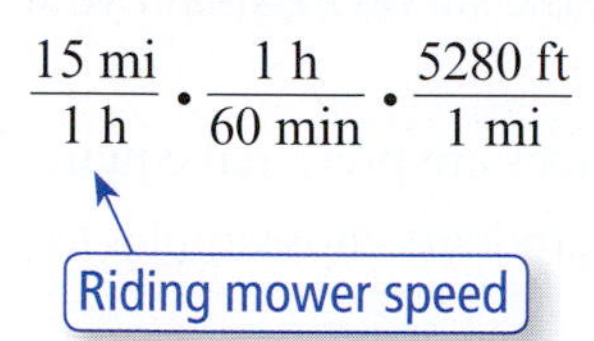

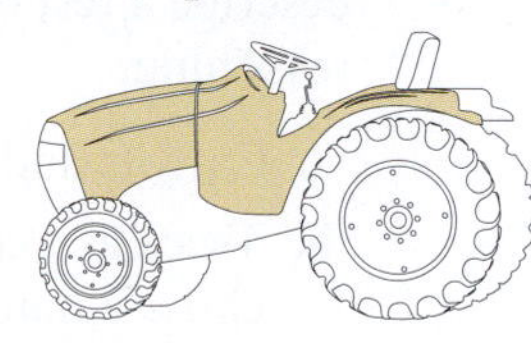

12. *Writing* Simplify the expression. Then write a real-life problem that involves the expression.

$$\frac{1100\text{ ft}}{1\text{ sec}} \cdot \frac{60\text{ sec}}{1\text{ min}} \cdot \frac{60\text{ min}}{1\text{ h}} \cdot \frac{1\text{ mi}}{5280\text{ ft}}$$

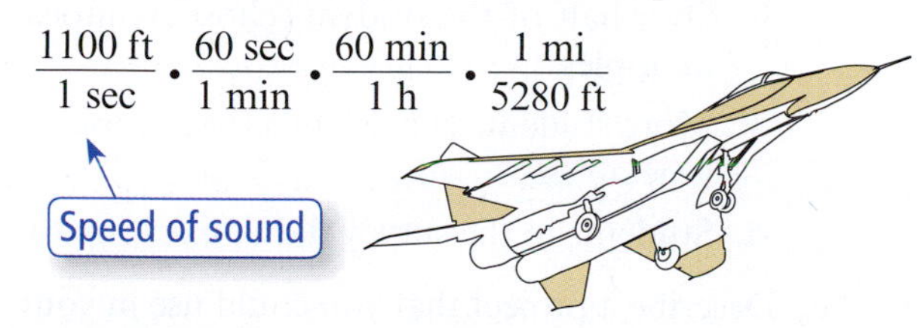

13. **Rate and Unit Analysis** A fast runner can run 1 mile in 4 minutes. Complete the work to find this rate in miles per hour.

$$\frac{1\text{ mi}}{4\text{ min}} \cdot \frac{\square\text{ min}}{\square\text{ h}} = \frac{\square\text{ mi}}{1\text{ h}}$$

14. **Rate and Unit Analysis** A parachutist glides to the ground at a rate of 18 feet per second. Complete the work to find this rate in miles per hour.

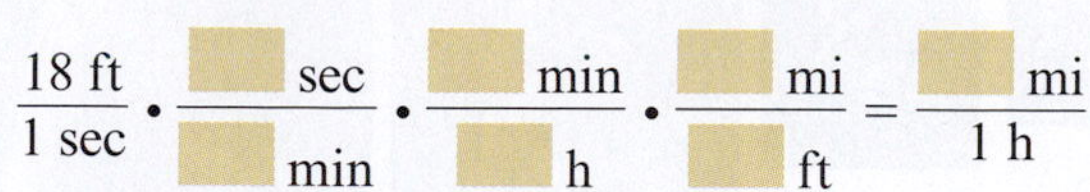

$$\frac{18\text{ ft}}{1\text{ sec}} \cdot \frac{\square\text{ sec}}{\square\text{ min}} \cdot \frac{\square\text{ min}}{\square\text{ h}} \cdot \frac{\square\text{ mi}}{\square\text{ ft}} = \frac{\square\text{ mi}}{1\text{ h}}$$

15. **Testing for Proportions** Decide whether the two ratios are proportional.

a. $\frac{3}{17}$ and $\frac{15}{85}$ **b.** $\frac{13}{14}$ and $\frac{41}{42}$ **c.** $\frac{2}{5}$ and $\frac{26}{65}$

16. **Testing for Proportions** Decide whether the two ratios are proportional.

a. $\frac{3}{10}$ and $\frac{15}{30}$ **b.** $\frac{4}{21}$ and $\frac{16}{84}$ **c.** $\frac{7}{13}$ and $\frac{28}{52}$

17. **Ratios and Proportion** Find the two ratios that are equal, and use them to write a proportion.

$\frac{3}{4}, \frac{12}{18}, \frac{18}{24}, \frac{20}{25}$

18. **Ratios and Proportion** Find the two ratios that are equal, and use them to write a proportion.

$\frac{2}{3}, \frac{12}{18}, \frac{18}{24}, \frac{18}{36}$

19. **Proportion** Two classrooms are shown. Is the ratio of boys to girls the same for each classroom? Explain your reasoning.

Classroom A

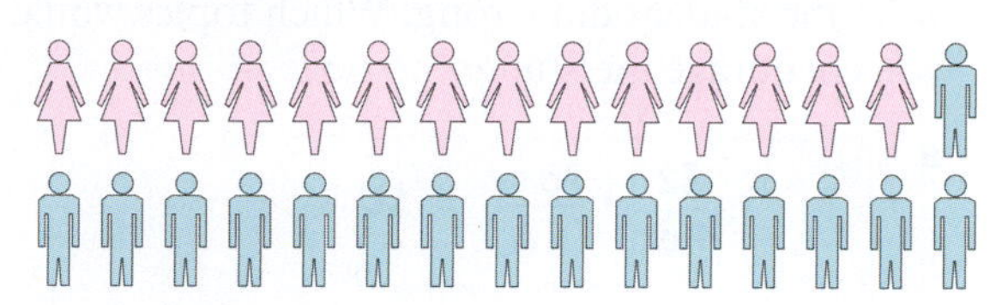

Classroom B

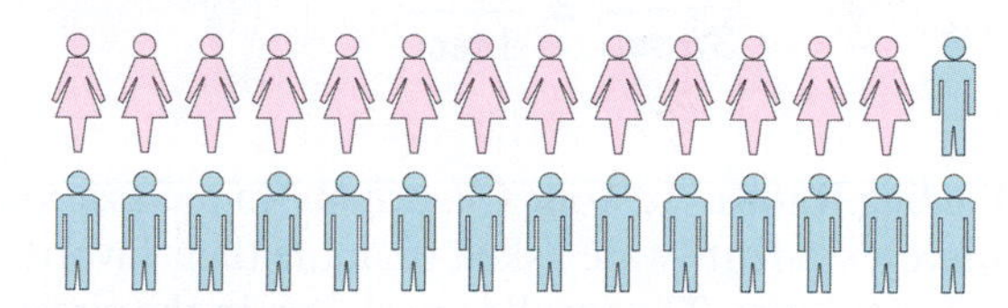

20. **Proportion** Two parking lots are shown. Is the ratio of cars to trucks the same for each parking lot? Explain your reasoning.

Parking lot A

Parking lot B

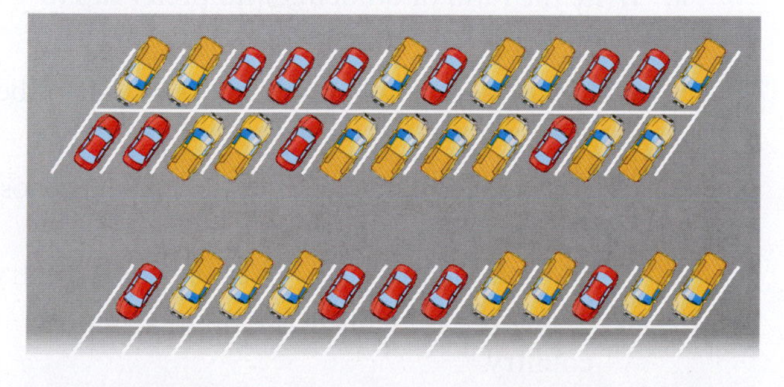

21. **Rate and Unit Analysis** An attorney charges \$350 per hour. What is this rate in dollars per year? (Assume a 40-hour work week.)

22. **Rate and Unit Analysis** A hotel housekeeper earns an hourly wage of \$10 per hour. What is this wage in dollars per year? (Assume a 40-hour work week.)

23. **Grading Student Work** On a diagnostic test, one of your students does the following work. Explain what the student did wrong. Which topics would you encourage the student to review?

a. $\frac{\$32}{4\text{ h}} \cdot \frac{40\text{ h}}{1\text{ wk}} = \frac{\$300}{1\text{ wk}}$

b. $\frac{120\text{ mi}}{2\frac{1}{2}\text{ h}} = \frac{60\text{ mi}}{1\text{ h}}$

24. **Grading Student Work** On a diagnostic test, one of your students does the following work. Explain what the student did wrong. Which topics would you encourage the student to review?

a. $\frac{\$2}{4\text{ oz}} \cdot \frac{16\text{ oz}}{1\text{ lb}} = \$8$

b. $\frac{42\text{ ft}}{3\frac{1}{2}\text{ sec}} = \frac{14\text{ ft}}{1\text{ sec}}$

25. **Finding Ratios: In Your Classroom** In a classroom survey, students were asked to name their favorite type of music. The results are shown in the circle graph.

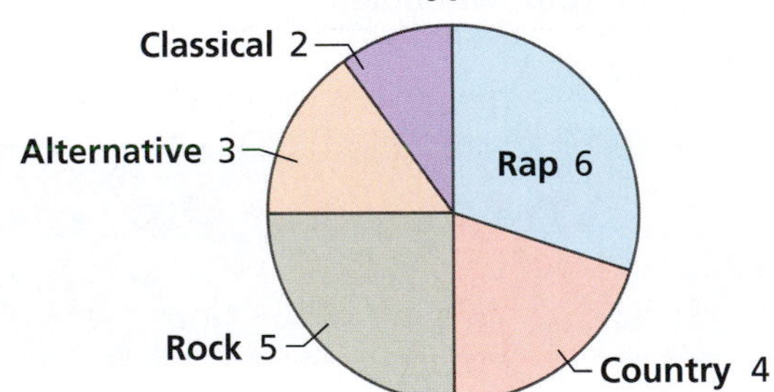

a. Write the ratio of students who prefer rap to students who prefer country.

b. Which of the following statements best describes the ratio in part (a)? Explain your reasoning.

- **i.** One and a half times as many students chose rap as chose country.
- **ii.** More students chose rap than chose country.
- **iii.** Two more students chose rap than chose country.
- **iv.** Half of the students chose rap or country.

c. Describe a project that you could use in your classroom that deals with ratios and favorite types of music.

26. **Finding Ratios: In Your Classroom** In a classroom survey, students were asked to name their favorite fruit or vegetable. The results are shown in the circle graph.

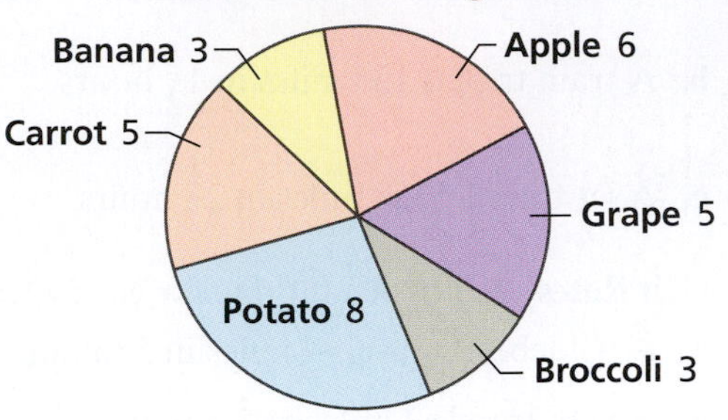

a. Write the ratio of students who prefer a fruit to students who prefer a vegetable.

b. Which of the following statements correctly describe a result of the survey? Explain your reasoning.

- **i.** Carrots and grapes are preferred equally.
- **ii.** Twice as many students chose apples as chose bananas.
- **iii.** The ratio of carrots to broccoli is the same as the ratio of grapes to bananas.
- **iv.** Over half of the students chose potatoes or apples.
- **v.** More students chose fruits than chose vegetables.
- **vi.** Students in the survey didn't like tomatoes.

c. Describe a project that you could use in your classroom that deals with ratios and favorite types of fruits or vegetables.

27. **Proportional Reasoning** The four most common sizes of photographic prints are shown below. Are any of these sizes proportional to each other? Explain your reasoning.

4″ by 6″ 5″ by 7″ 8″ by 10″ 11″ by 17″

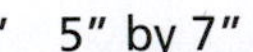

Makarova Viktoria/Shutterstock.com

28. **Proportional Reasoning** The four most common sizes of photographic prints are shown in Exercise 27. Less common sizes are shown below. Are any of the sizes below proportional to each other? Explain your reasoning.

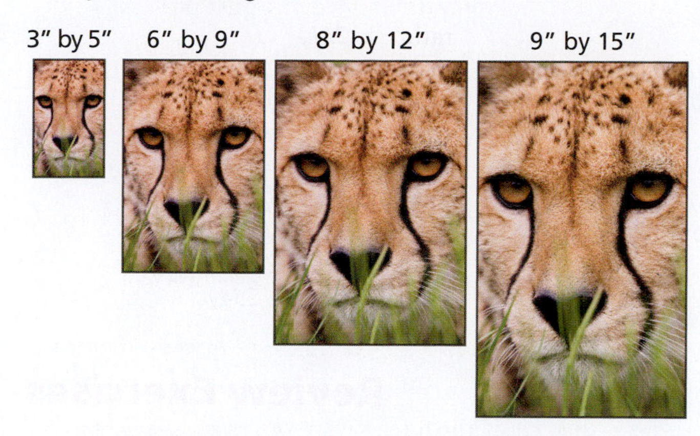

29. **Writing** Write a real-life application that involves a part-to-part ratio.

30. **Writing** Explain why a rate is a special type of ratio. Describe the parts of a rate.

31. **Scale Drawing (Length)** In the drawing below, each square is one-quarter inch long and represents one foot in real life (the actual measures in feet are shown). What is the ratio of lengths in the drawing to lengths in the actual room?

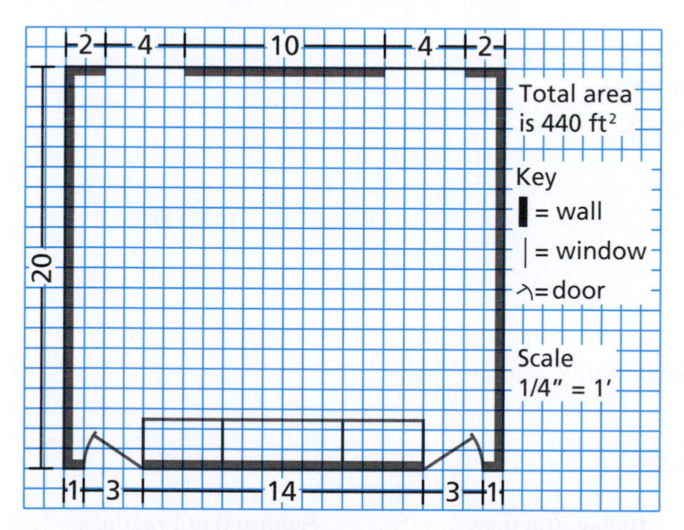

32. **Scale Drawing (Area)** What is the ratio of areas in the drawing above to areas in the actual room?

33. **Proportional Reasoning** Use proportional reasoning to solve each proportion.

a. $\dfrac{20 \text{ ft}}{3 \text{ ft}} = \dfrac{\square \text{ ft}}{9 \text{ ft}}$

b. $\dfrac{\$13}{1 \text{ h}} = \dfrac{\$104}{\square \text{ h}}$

Chaser/Shutterstock.com

34. **Proportional Reasoning** Use proportional reasoning to solve each proportion.

a. $\dfrac{8 \text{ gal}}{1 \text{ min}} = \dfrac{\square \text{ gal}}{15 \text{ min}}$ b. $\dfrac{\$36}{2 \text{ h}} = \dfrac{\$54}{\square \text{ h}}$

35. **Proportional Reasoning: In Your Classroom** In a children's story, a man grows to giant height and shrinks back to normal height, repeatedly. The ratio of his giant height to his normal height is 20 : 9.

a. Make a reasonable estimate of the man's normal height.

b. Use your estimate in part (a) to write and solve a proportion to find the giant's height.

c. Describe a project involving children's literature in which your students can use proportional reasoning.

36. **Proportional Reasoning: In Your Classroom** HO gauge model trains are built on a ratio of 1 : 87 with respect to actual trains.

a. Estimate the length of the HO gauge model train engine.

b. Use your estimate in part (a) to write and solve a proportion to find the length of an actual train engine.

c. Describe a project involving a type of model (other than model trains) in which your students can use proportional reasoning.

Ron and Joe/Shutterstock.com; David W Hughes/Shutterstock.com

Chapter Summary

Chapter Vocabulary

fraction *(p. 197)*
numerator *(p. 197)*
denominator *(p. 197)*
area model *(p. 198)*
set model *(p. 198)*
linear model *(p. 199)*
proper fraction *(p. 200)*
improper fraction *(p. 200)*
mixed number *(p. 200)*
equivalent *(p. 201)*
simplest form *(p. 201)*
cross products *(p. 202)*
cross-multiplication *(p. 202)*
is less than *(p. 203)*
benchmarks *(p. 203)*
multiplicative inverse *(p. 221)*
reciprocal *(p. 221)*
complex fraction *(p. 224)*
ratio *(p. 231)*
part-to-part *(p. 231)*
part-to-whole *(p. 231)*
whole-to-part *(p. 231)*
rate *(p. 233)*
unit rate *(p. 233)*
unit analysis *(p. 233)*
proportion *(p. 234)*
proportional *(p. 234)*

Chapter Learning Objectives — Review Exercises

6.1 Fractions *(page 197)* — 1–38

- Understand fractions and mixed numbers.
- Recognize equivalent fractions and find simplest form.
- Use a real number line to graph and order fractions.

6.2 Adding and Subtracting Fractions *(page 209)* — 39–84

- Add fractions with like denominators.
- Subtract fractions with like nominators.
- Add and subtract fractions with unlike denominators.

6.3 Multiplying and Dividing Fractions *(page 219)* — 85–132

- Multiply fractions, whole numbers, and mixed numbers.
- Divide fractions, whole numbers, and mixed numbers.
- Estimate products and quotients of fractions.

6.4 Ratios and Proportion *(page 231)* — 133–146

- Write a ratio as a part-to-part, a part-to-whole, or a whole-to-part comparison.
- Write and simplify rates.
- Decide whether two ratios or two rates are proportional.

Important Concepts and Formulas

Simplified Fraction

A fraction is in simplest form when its numerator and denominator have no common factors other than 1.

Adding Fractions

$$\frac{a}{c}+\frac{b}{c}=\frac{a+b}{c}$$

Common denominator

Adding Fractions

$$\frac{a}{b}+\frac{c}{d}=\frac{ad+bc}{bd}$$

Unlike denominators

Subtracting Fractions

$$\frac{a}{c}-\frac{b}{c}=\frac{a-b}{c}$$

Common denominator

Subtracting Fractions

$$\frac{a}{b}-\frac{c}{d}=\frac{ad-bc}{bd}$$

Unlike denominators

Multiplying Fractions

$$\frac{a}{b}\cdot\frac{c}{d}=\frac{a\cdot c}{b\cdot d}$$

Dividing Fractions

$$\frac{a}{b}\div\frac{c}{d}=\frac{a}{b}\cdot\frac{d}{c}$$

Multiply by reciprocal.

Proportion

If $\frac{a}{b}=\frac{c}{d}$, then $ad=bc$.

Review Exercises

Free step-by-step solutions to odd-numbered exercises at MathematicalPractices.com.

6.1 Fractions

Area Models Write the fraction represented by the shaded portion of the figure. Identify the numerator and denominator.

1. 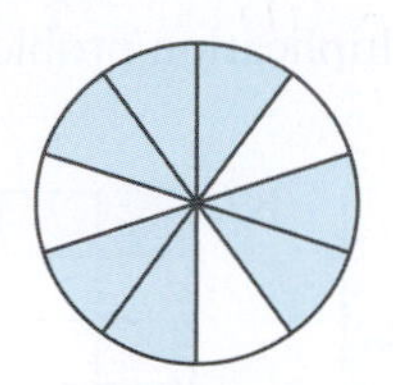

2. 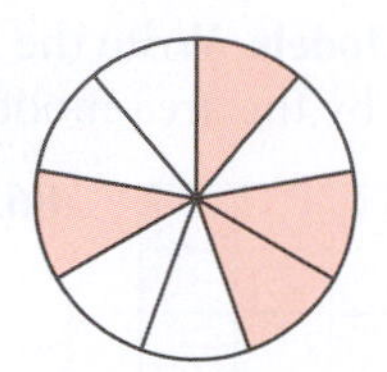

3. 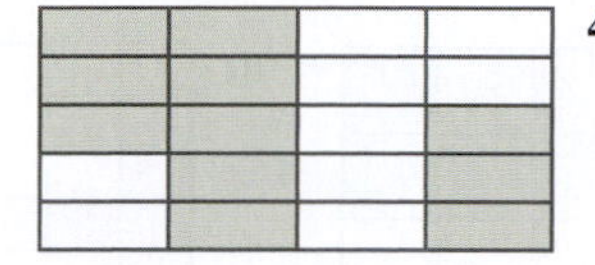

4. 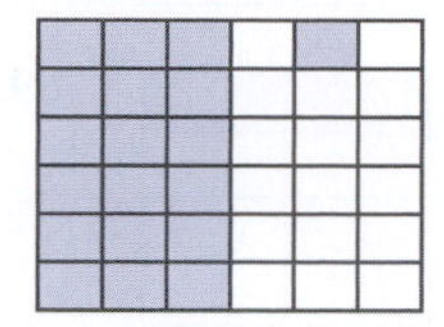

5. 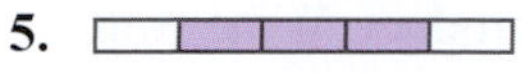

6. 

Linear Models Draw a number line like the one shown below. Use a dot • on the number line to represent the fraction.

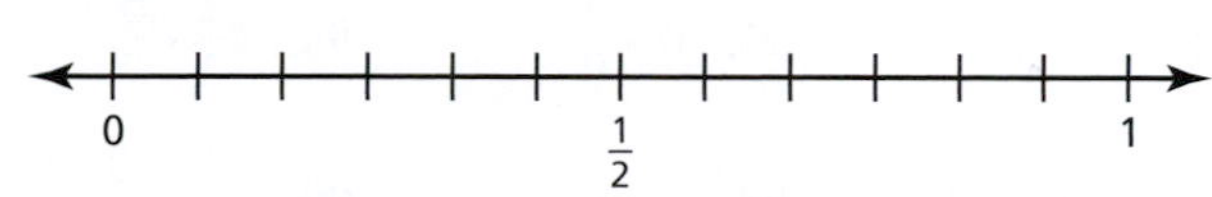

7. $\frac{1}{6}$ **8.** $\frac{1}{3}$ **9.** $\frac{5}{6}$ **10.** $\frac{7}{12}$

Improper Fractions and Mixed Numbers Draw a ruler like the one shown below. Represent the improper fraction or mixed number on your ruler.

11. $\frac{7}{2}$ **12.** $\frac{7}{4}$ **13.** $2\frac{3}{8}$ **14.** $3\frac{1}{2}$

Modeling Equivalence Decide whether the two fractions are equivalent. If they are, use an area model to demonstrate their equivalence.

15. $\frac{2}{8}$ and $\frac{3}{12}$ **16.** $\frac{5}{6}$ and $\frac{6}{5}$ **17.** $\frac{3}{7}$ and $\frac{9}{21}$

18. $\frac{9}{8}$ and $\frac{18}{16}$ **19.** $\frac{3}{5}$ and $\frac{30}{50}$ **20.** $\frac{6}{11}$ and $\frac{16}{21}$

Determining Equivalence Decide whether the two fractions are equivalent by writing them in simplest form.

21. $\frac{8}{48}$ and $\frac{2}{12}$ **22.** $\frac{13}{26}$ and $\frac{14}{28}$ **23.** $\frac{18}{99}$ and $\frac{2}{11}$

24. $\frac{12}{60}$ and $\frac{3}{15}$ **25.** $\frac{12}{48}$ and $\frac{25}{100}$ **26.** $\frac{16}{48}$ and $\frac{15}{60}$

Comparing Fractions Place the correct symbol (<, =, or >) between the two fractions.

27. $\frac{1}{4} \quad \frac{1}{5}$ **28.** $\frac{1}{11} \quad \frac{1}{9}$ **29.** $\frac{7}{8} \quad \frac{3}{4}$

30. $\frac{4}{9} \quad \frac{1}{3}$ **31.** $\frac{14}{28} \quad \frac{21}{42}$ **32.** $\frac{9}{36} \quad \frac{7}{28}$

Benchmarks Use benchmarks to order the fractions from least to greatest. Explain your reasoning.

33. $\frac{9}{10}, \frac{7}{15}, \frac{5}{8}, \frac{1}{17}$ **34.** $\frac{19}{36}, \frac{1}{19}, \frac{6}{14}, \frac{8}{9}$

Problem Solving In 2011, about \$51 billion was spent on pets in the United States. The circle graph shows how the money was spent.

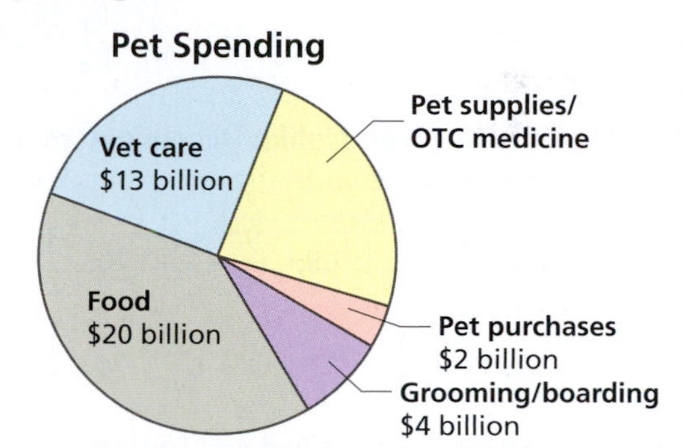

35. What fraction of the money was spent on vet care? On pet supplies and OTC medicine?

36. Find the total amount of money spent on vet care, and pet supplies and OTC medicine.

37. Write a fraction that represents the portion spent on vet care, and pet supplies and OTC medicine.

38. Is the fraction in Exercise 37 less than or greater than one-half? Explain your reasoning using benchmarks.

6.2 Adding and Subtracting Fractions

Adding Fractions with Like Denominators Find the sum. Write your answer as a proper fraction or mixed number in simplest form.

39. $\frac{2}{5} + \frac{1}{5}$ **40.** $\frac{3}{4} + \frac{3}{4}$

41. $\frac{3}{7} + \frac{4}{7}$ **42.** $\frac{6}{5} + \frac{4}{5}$

43. $\frac{7}{4} + \frac{7}{4}$ **44.** $\frac{2}{3} + \frac{5}{3}$

45. $\frac{3}{10} + \frac{9}{10}$ **46.** $\frac{5}{12} + \frac{1}{12}$

Adding Mixed Numbers Find the sum. Write your answer as a mixed number in simplest form.

47. $2\frac{1}{3} + 3\frac{2}{3}$ **48.** $1\frac{5}{8} + \frac{7}{8}$ **49.** $2\frac{1}{10} + \frac{3}{10} + 3\frac{7}{10}$

50. $3\frac{1}{2} + 3\frac{1}{2}$ **51.** $4\frac{3}{8} + \frac{9}{8}$ **52.** $4\frac{4}{5} + \frac{2}{5} + 7\frac{3}{5}$

Subtracting Fractions with Like Denominators Find the difference. Write your answer in simplest form.

53. $\frac{5}{8} - \frac{3}{8}$ **54.** $\frac{7}{9} - \frac{4}{9}$ **55.** $\frac{17}{10} - \frac{17}{10}$ **56.** $\frac{11}{4} - \frac{11}{4}$

57. $\frac{12}{5} - \frac{2}{5}$ **58.** $\frac{31}{7} - \frac{3}{7}$ **59.** $\frac{18}{7} - \frac{4}{7}$ **60.** $\frac{21}{12} - \frac{15}{12}$

Subtracting Mixed Numbers Find the difference. Write your answer as a mixed number in simplest form.

61. $5\frac{3}{4} - 3$ **62.** $8\frac{6}{7} - 1\frac{2}{7}$ **63.** $3\frac{3}{5} - 1\frac{3}{5}$

64. $3\frac{2}{3} - 2\frac{1}{3}$ **65.** $8\frac{3}{7} - 5\frac{6}{7}$ **66.** $6\frac{1}{5} - 2\frac{3}{5}$

Adding and Subtracting with Unlike Denominators Find the sum or difference. Write your answer in simplest form.

67. $\frac{5}{6} - \frac{5}{12}$ **68.** $\frac{4}{9} + \frac{2}{3}$ **69.** $\frac{9}{10} + \frac{4}{5}$ **70.** $\frac{3}{2} - \frac{3}{4}$

71. $\frac{3}{4} - \frac{7}{12}$ **72.** $\frac{7}{6} + \frac{1}{2}$ **73.** $\frac{6}{7} + \frac{5}{14}$ **74.** $\frac{1}{2} - \frac{1}{10}$

Adding or Subtracting Mixed Numbers Find the sum or difference. Write your answer as a whole number or mixed number in simplest form.

75. $3\frac{2}{3} - 2\frac{2}{3}$ **76.** $5\frac{7}{8} - 3\frac{1}{4}$ **77.** $\frac{7}{8} + 2\frac{5}{6}$

78. $4\frac{1}{2} - 1\frac{1}{3}$ **79.** $3\frac{3}{4} - 1\frac{1}{8}$ **80.** $\frac{1}{5} + 5\frac{1}{10}$

Problem Solving The circle graph shows the results of a poll in which people were asked to rate a new product.

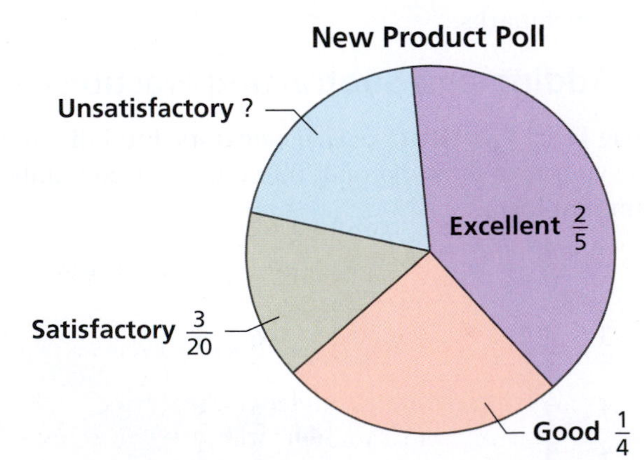

81. What fraction said excellent or good?

82. What fraction said unsatisfactory?

Mental Math Explain how you can use mental math to find the sum. What is the sum?

83. $\frac{6}{40} + \frac{50}{40} + \frac{34}{40} + \frac{28}{40} + \frac{9}{40} + \frac{12}{40} + \frac{31}{40} + \frac{30}{40}$

84. $\frac{8}{18} + \frac{21}{18} + \frac{30}{18} + \frac{15}{18} + \frac{10}{18} + \frac{6}{18}$

6.3 Multiplying and Dividing Fractions

Area Models Write the fraction multiplication problem shown by the area model.

85. 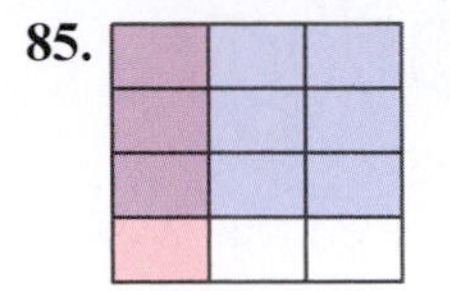

86. 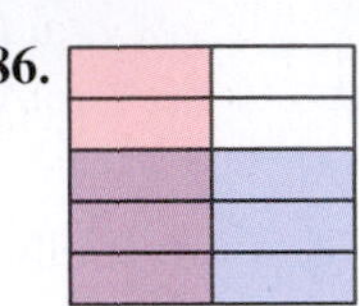

87. 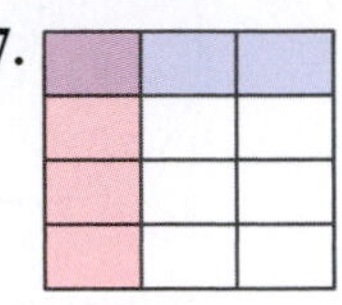

88. 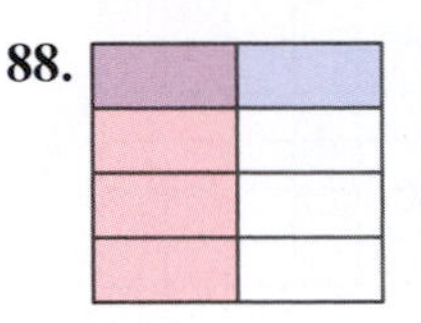

89. 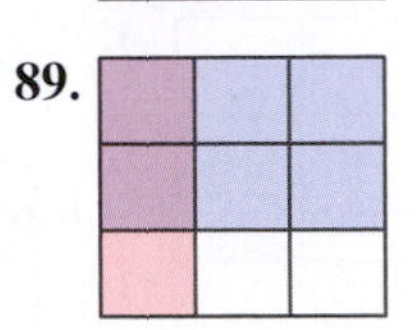

90. 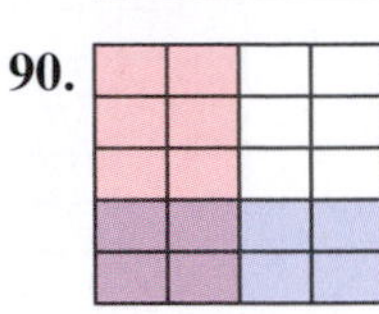

Multiplying Fractions Find the product. Write your answer in simplest form.

91. $\frac{2}{3} \cdot 5$ **92.** $8 \cdot \frac{3}{16}$ **93.** $\frac{5}{9} \cdot \frac{3}{2}$

94. $\frac{1}{4} \cdot 3\frac{1}{5}$ **95.** $0 \cdot \frac{4}{9}$ **96.** $4\frac{2}{3} \cdot \frac{5}{7}$

97. $\frac{3}{8} \cdot 2$ **98.** $\frac{3}{10} \cdot \frac{10}{3}$ **99.** $2\frac{1}{6} \cdot \frac{3}{13}$

100. $6\frac{1}{7} \cdot 0$ **101.** $\frac{2}{15} \cdot \frac{5}{8}$ **102.** $\frac{2}{3} \cdot 10\frac{1}{2}$

Dividing Fractions Find the quotient. Write your answer in simplest form.

103. $\frac{5}{6} \div 6$ **104.** $6 \div \frac{5}{6}$ **105.** $\frac{3}{16} \div \frac{7}{8}$

106. $4 \div \frac{3}{4}$ **107.** $\frac{3}{4} \div 4$ **108.** $\frac{2}{3} \div \frac{4}{21}$

Dividing Mixed Numbers Find the quotient. Write your answer in simplest form.

109. $4\frac{1}{2} \div \frac{1}{4}$ **110.** $3\frac{2}{3} \div \frac{1}{3}$ **111.** $\frac{11}{12} \div 2\frac{1}{5}$

112. $2\frac{3}{5} \div 13$ **113.** $5\frac{1}{4} \div 3\frac{1}{2}$ **114.** $2\frac{4}{7} \div 3$

Evaluating Complex Fractions Evaluate the complex fraction. Write your answer in simplest form.

115. $\dfrac{\frac{1}{4}}{\frac{5}{6}}$ **116.** $\dfrac{\frac{3}{8}}{1\frac{1}{5}}$ **117.** $\dfrac{\frac{4}{7}}{2\frac{2}{3}}$ **118.** $\dfrac{\frac{4}{5}}{\frac{7}{5}}$

119. $\dfrac{\frac{3}{5}}{12}$ **120.** $\dfrac{\frac{7}{8}}{2}$ **121.** $\dfrac{6}{\frac{1}{8}+\frac{5}{8}}$ **122.** $\dfrac{3}{\frac{9}{10}+\frac{3}{10}}$

Estimation Estimate the product or quotient.

123. $5\frac{7}{8} \cdot 2\frac{1}{32}$ **124.** $20\frac{1}{16} \div 1\frac{15}{16}$ **125.** $9\frac{9}{10} \cdot 4\frac{1}{25}$

126. $13\frac{8}{9} \div 2\frac{1}{20}$ **127.** $2\frac{1}{50} \cdot 3\frac{24}{50}$ **128.** $33\frac{7}{8} \div 17\frac{1}{10}$

129. **Measurement** How many quarter-inch pieces can you cut from a ribbon that is four and a half inches long?

130. **Measurement** A one-cup jelly jar is one-sixteenth of a gallon. How many one-cup jelly jars can be filled with one and a half gallons of jelly?

131. **Estimating a Product** A bathroom is $13\frac{1}{4}$ feet long and $9\frac{7}{8}$ feet wide.

a. Estimate the area of the bathroom.

b. Use fraction multiplication to find the exact area. How accurate is your estimate?

132. **Estimating a Quotient** A bathroom is $11\frac{3}{4}$ feet long and has an area of about 120 square feet. Estimate the width of the bathroom. Explain how you made your estimate.

6.4 Ratios and Proportions

Finding Ratios

133. Write the ratio of red marbles to blue marbles. Simplify your answer.

134. Write the ratio of blue marbles to red marbles. Simplify your answer.

135. Write the ratio of red marbles to *all* marbles. Simplify your answer.

136. Write the ratio of green marbles to *all* marbles. Simplify your answer.

Unit Rates Write the unit rate for the speed.

137. An automobile travels 135 miles in 3 hours.

138. A train travels 210 miles in $1\frac{1}{2}$ hours.

139. A jet travels 1750 miles in $2\frac{1}{2}$ hours.

140. A python travels 1 mile in 12 hours.

141. **Rate and Unit Analysis** A surgeon charges $450 per hour. What is this rate in dollars per year? (Assume a 40-hour work week.)

142. **Rate and Unit Analysis** A food server earns $15 per hour on the average. What is this wage in dollars per year? (Assume a 40-hour work week.)

Ratios and Proportions Find the two ratios that are equal, and use them to write a proportion.

143. $\frac{3}{4}, \frac{12}{18}, \frac{18}{24}, \frac{20}{25}$ **144.** $\frac{2}{3}, \frac{12}{18}, \frac{18}{24}, \frac{18}{36}$

145. **Proportional Reasoning** In a survey, students were asked to name their favorite class. The results are shown in the circle graph.

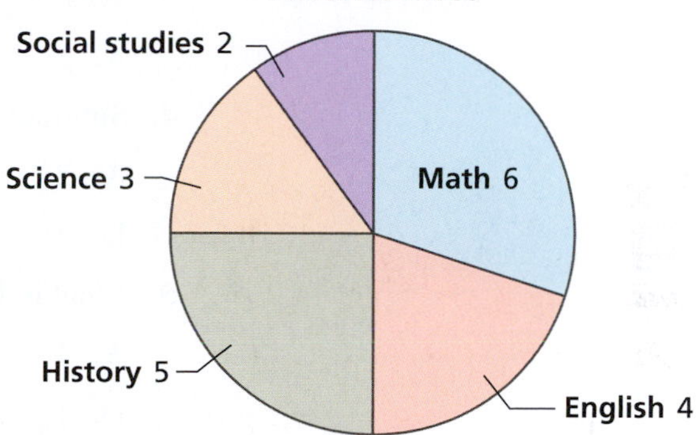

a. Write the ratio of students who prefer math to students who prefer science.

b. How many students were surveyed?

c. Suppose the same survey is given to a group of 200 students. How many students would you expect to prefer history? Explain your reasoning.

146. **Proportional Reasoning** In a survey, students were asked to name their favorite U.S. president. The results are shown in the circle graph.

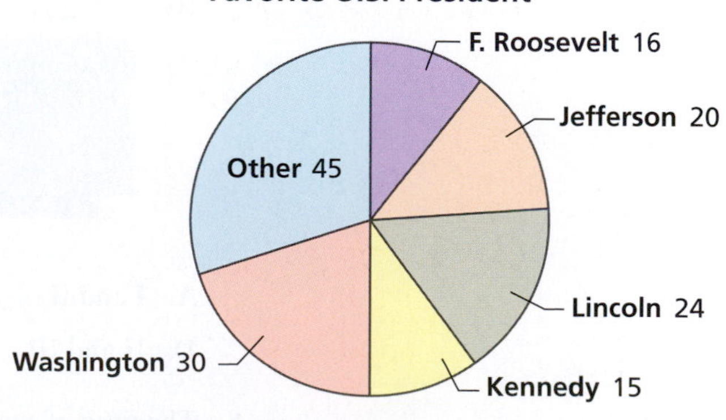

a. Write the ratio of students who prefer Washington to students who prefer Kennedy.

b. Suppose the same survey is given to a group of 300 students. How many students would you expect to prefer Lincoln? Explain your reasoning.

iStockphoto.com/sweetym

Chapter Test

The questions on this practice test are modeled after questions on state certification exams for K–8 teaching.

Test Directions: Each of the questions is followed by five choices. Choose the *best* response to each question.

1. List the fractions shown below from least to greatest.

$$\frac{1}{9}, \frac{2}{15}, \frac{3}{21}$$

A. $\frac{1}{9}, \frac{2}{15}, \frac{3}{21}$ **B.** $\frac{2}{15}, \frac{3}{21}, \frac{1}{9}$ **C.** $\frac{3}{21}, \frac{1}{9}, \frac{2}{15}$
D. $\frac{1}{9}, \frac{3}{21}, \frac{2}{15}$ **E.** $\frac{2}{15}, \frac{1}{9}, \frac{3}{21}$

2. What is the least common denominator of $\frac{2}{15}, \frac{1}{21}$, and $\frac{4}{35}$?

A. 35 **B.** 105 **C.** 175 **D.** 415 **E.** 735

3. What is $\frac{1}{2} + \frac{1}{3}$?

A. $\frac{1}{6}$ **B.** $\frac{1}{5}$ **C.** $\frac{2}{6}$ **D.** $\frac{2}{5}$ **E.** $\frac{5}{6}$

4. Subtract $4\frac{1}{3} - 1\frac{5}{6}$.

A. $2\frac{1}{6}$ **B.** $2\frac{1}{2}$ **C.** $3\frac{1}{2}$
D. $3\frac{2}{3}$ **E.** None of these

5. What is $4 \times \frac{2}{4} + \frac{5}{6}$?

A. 1 **B.** $\frac{3}{2}$ **C.** $\frac{5}{3}$
D. $1\frac{5}{6}$ **E.** $\frac{17}{6}$

6. Divide $3\frac{1}{5}$ by $1\frac{1}{3}$.

A. $2\frac{2}{5}$ **B.** $3\frac{1}{15}$ **C.** $3\frac{3}{5}$ **D.** $4\frac{4}{15}$ **E.** 8

7. Four LCD television screen sizes are shown. Which two screen sizes are proportional?

I. 28 cm, 22 cm

II. 30 cm, 24 cm

III.
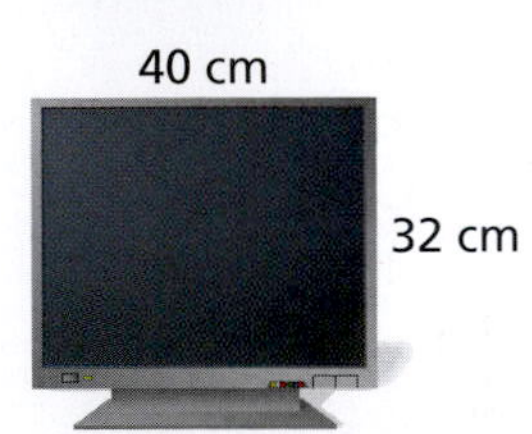

IV.
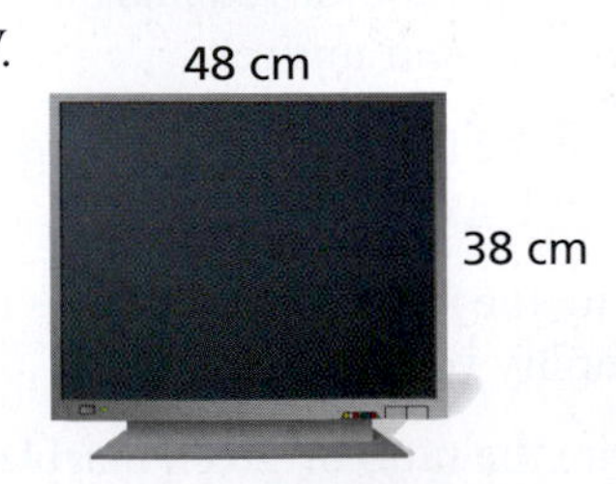

A. I and II **B.** I and III **C.** I and IV
D. II and III **E.** III and IV

8. The ratio of men to women at a college is 3 : 7. If there are 6,153 women at this college, how many men are at this college?

A. 879 **B.** 1,895 **C.** 2,051
D. 2,637 **E.** 14,357

7 Decimals and Percents

iStockphoto.com/JoseGirarte

Activity: Use Money to Review Decimals and Fractions

Materials:

- Pencil

Learning Objective:

Use money to review the relationship between decimal form and fraction form.

Name ______________________________

Write each amount of money in decimal form. Then write the amount of money in fraction form.

Sample

$2.86

Decimal form

$\$2\,\frac{86}{100}$

Fraction form

1.

Decimal form

Fraction form

2.

Decimal form

Fraction form

3.

Decimal form

Fraction form

4.

Decimal form

Fraction form

7.1 Decimals

Standards

Grades 3–5
Numbers and Operations–Fractions
Students should understand decimal notation for fractions, and compare decimal fractions.

Grades 6–8
The Number System
Students should apply and extend previous understandings of operations with fractions.

- Read and write decimals, and convert between decimal form and fraction form.
- Recognize which fractions can be written as terminating decimals.
- Order terminating decimals.

Converting Between Decimal and Fraction Form

Decimal Notation

A **decimal** (or *decimal fraction*) represents a fraction whose denominator is a power of 10. A decimal is represented without a denominator. A **decimal point** between the ones place and the tenths place separates the whole-number part of the decimal (to the left) from the fractional part of the decimal (to the right).

Example $3\frac{16}{100} = 3.16$

Fractional part
Whole-number part
Decimal point

Decimals are part of the base-ten positional system for writing numbers (as was introduced on page 50).

Example 1565.125 One thousand, five hundred sixty-five and one hundred twenty-five thousandths

1	5	6	5.	1	2	5
Thousands	Hundreds	Tens	Ones	Tenths	Hundredths	Thousandths

Classroom Tip

When teaching your students to read decimals: (1) Say the name of the whole number to the left of the decimal point. (2) Read the decimal point as "and." (3) Then say the name of the number to the right of the decimal point, followed by the name of the least place value. For instance, 4.753 is read as

"four *and* seven hundred fifty-three thousandths."

EXAMPLE 1 **Writing Numbers in Decimal Form**

a. $\frac{9}{10} = 0.9$ Nine tenths

b. $4\frac{7}{10} = 4.7$ Four and seven tenths

c. $12\frac{37}{100} = 12.37$ Twelve and thirty-seven hundredths

EXAMPLE 2 **Writing Numbers in Fraction Form**

a. $52.3 = 52\frac{3}{10}$ Fifty-two and three tenths

b. $800.095 = 800\frac{95}{1000}$ Eight hundred and ninety-five thousandths

In Section 2.2 (on page 51), you learned how base-ten blocks can model whole numbers. Base-ten blocks can also model decimals.

Classroom Tip

There is more than one way to use base-ten blocks to represent decimals. The value of each base-ten block depends on which block you use to represent 1. For instance, to represent a decimal that includes thousandths, you can let a big cube represent 1, a flat represent 0.1, a long represent 0.01, and a small cube represent 0.001.

Base-Ten Blocks as Models for Decimals

Base-ten blocks can model decimals, as shown at the right. Each small cube can represent one hundredth.

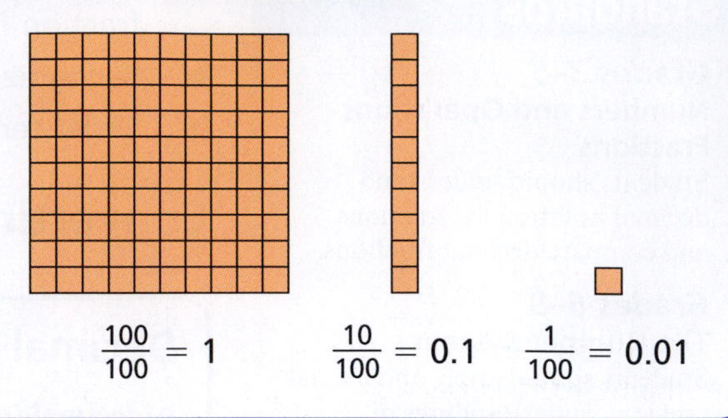

EXAMPLE 3 Modeling Decimals

Use base-ten blocks to model 2.34.

SOLUTION

Writing the decimal in expanded form helps to identify the model.

$$2.34 = 2 + \frac{3}{10} + \frac{4}{100}$$

From the expanded form, you can see that the model is as follows.

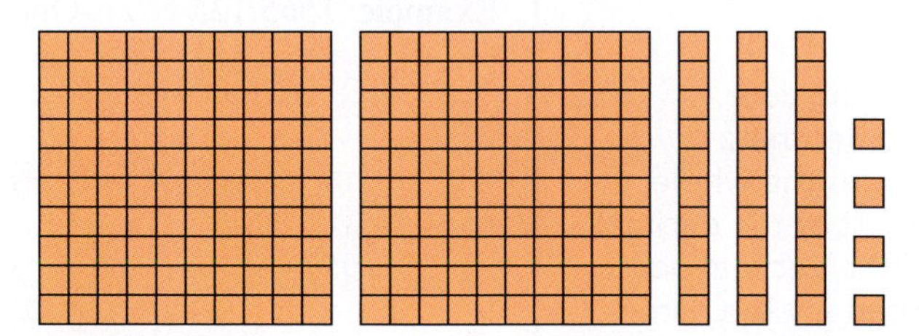

EXAMPLE 4 A Real-Life Example of Decimals

Write the gasoline prices in words.

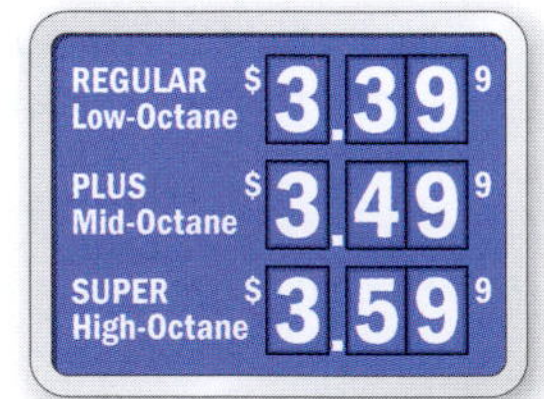

SOLUTION

As written, the price \$3.399 can be read as three and three hundred ninety-nine thousandths dollars. This, however, is not the common way to say a gasoline price. The common way is as follows.

Classroom Tip

Although it is not strictly necessary, teach your students to include a leading zero when writing a decimal whose whole-number part is 0. For instance, writing 0.25 is better than writing .25. Including a leading zero makes it clear that the whole-number part of the decimal is 0.

Regular $\$3.399 = \$3.39\frac{9}{10}$

This is read as "three dollars and thirty-nine and nine-tenths cents."

Plus $\$3.499 = \$3.49\frac{9}{10}$

This is read as "three dollars and forty-nine and nine-tenths cents."

Super $\$3.599 = \$3.59\frac{9}{10}$

This is read as "three dollars and fifty-nine and nine-tenths cents."

Terminating Decimals

Some fractions can be written as decimals that end, or terminate. The decimal forms of other fractions do not terminate and have infinitely many decimal places.

Terminating decimals	Nonterminating decimals
$\frac{1}{10} = 0.1$	$\frac{1}{3} = 0.333333\ldots$
$\frac{1}{8} = 0.125$	$\frac{1}{12} = 0.083333\ldots$

Classroom Tip

Point out to your students that the method at the right only works for fractions written in simplest form. For a fraction such as $\frac{3}{6}$, it might appear that the fraction cannot be written as a terminating decimal because the denominator contains the prime factor 3. However, after simplifying, the denominator contains only the prime number 2 and the fraction clearly can be written as a terminating decimal.

$$\frac{3}{6} = \frac{1}{2} = 0.5$$

Terminating Decimals

A fraction, written in simplest form, can be written as a **terminating decimal** if and only if the prime factorization of the denominator contains no primes other than 2 or 5. There are two common ways to write a fraction as a decimal.

1. Rewrite the fraction so that its denominator is a power of 10.

Example $\frac{1}{4} = \frac{1 \cdot 25}{4 \cdot 25} = \frac{25}{100} = 0.25$

2. Divide the numerator by the denominator.

Example $4\overline{)1.00}$ with quotient 0.25. So, $\frac{1}{4} = 0.25$. See Section 7.3 on dividing decimals.

EXAMPLE 5 Writing Terminating Decimals

Determine whether each fraction can be written as a terminating decimal. If so, write the terminating decimal.

a. $\frac{3}{4}$ **b.** $\frac{1}{6}$

c. $\frac{1}{50}$ **d.** $\frac{3}{8}$

SOLUTION

a. The denominator factors as $2 \cdot 2$. So, this fraction can be written as a terminating decimal.

$$\frac{3}{4} = \frac{3 \cdot 25}{4 \cdot 25} = \frac{75}{100} = 0.75$$

Three-fourths is equal to seventy-five hundredths.

b. The denominator factors as $2 \cdot 3$. Because 3 is one of its prime factors, this fraction *cannot* be written as a terminating decimal.

c. The denominator factors as $2 \cdot 5 \cdot 5$. So, this fraction can be written as a terminating decimal.

$$\frac{1}{50} = \frac{1 \cdot 2}{50 \cdot 2} = \frac{2}{100} = 0.02$$

One-fiftieth is equal to two hundredths.

d. The denominator factors as $2 \cdot 2 \cdot 2$. So, this fraction can be written as a terminating decimal.

$$\frac{3}{8} = \frac{3 \cdot 125}{8 \cdot 125} = \frac{375}{1000} = 0.375$$

Three-eighths is equal to three hundred seventy-five thousandths.

Ordering Terminating Decimals

Classroom Tip

If you have students who studied outside of the United States, then you might find that they use a comma as a "decimal point." For instance, 4.37 is usually written as 4,37 in Europe.

Ordering Decimals on a Number Line

Use a number line model to order decimals.

1. Draw a number line that includes all of the decimals to be ordered.
2. Graph each decimal on the number line.
3. Read the decimals from left to right to order them from least to greatest.

Example Order 0.34, 0.4, 0.21, and 0.58 from least to greatest.

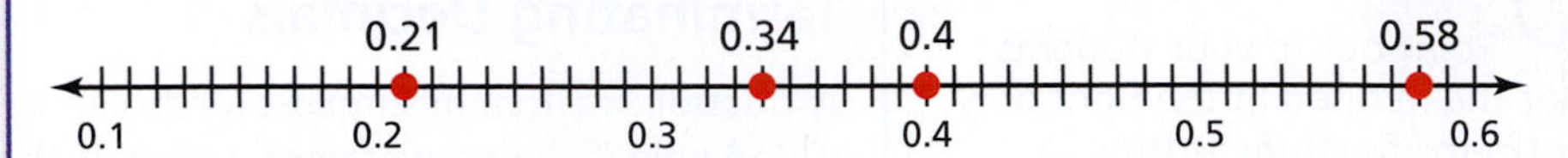

From the number line, you can see that the decimals in order from least to greatest are 0.21, 0.34, 0.47, and 0.58.

EXAMPLE 6 **A Real-Life Example of Ordering Decimals**

You are marking distances on a map from your home to several locations. Use a number line to decide which location is closest to your home.

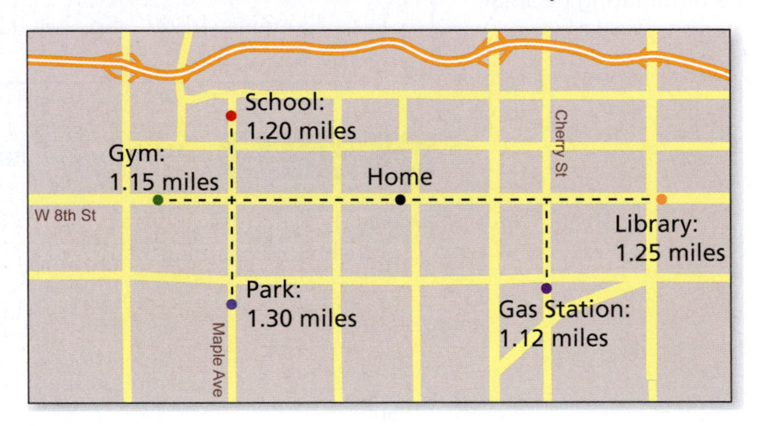

SOLUTION

1. **Understand the Problem** You are given the distances from your home to five locations. Basically, you need to order the five distances from least to greatest.
2. **Make a Plan** Use a number line model that includes all five distances. Then graph the decimal that represents each distance on the number line.
3. **Solve the Problem** The number line is shown below.

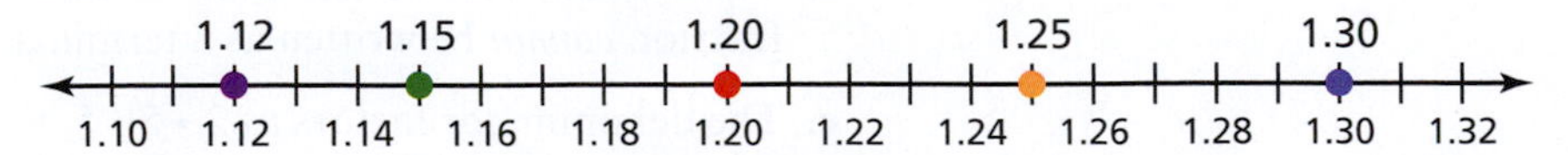

Reading from left to right on the number line, the distances from least to greatest are 1.12, 1.15, 1.20, 1.25, and 1.30. So, the location that is closest to your home is the gas station, which is 1.12 miles away.

4. **Look Back** Because the distances are between 1 and 2 miles away from your home, you can check your answer by looking at only the fractional parts of the distances as whole numbers. Of 15, 20, 30, 12, and 25, it is clear that 12 is the least.

EXAMPLE 7 Ordering Terminating Decimals

Order the decimals 0.3, 0.209, 0.25, and 0.2003 from least to greatest.

SOLUTION

Some students think that $0.3 < 0.25$ because $3 < 25$. To counteract this error, rewrite each decimal, annexing zeros, so that each has the same number of decimal places.

$0.3 = 0.3000$ 3000 ten thousandths

$0.209 = 0.2090$ 2090 ten thousandths

$0.25 = 0.2500$ 2500 ten thousandths

$0.2003 = 0.2003$ 2003 ten thousandths

In this form, it is easier to order the decimals from least to greatest because the reasoning is similar to ordering whole numbers. The order of the decimals from least to greatest is 0.2003, 0.209, 0.25, and 0.3.

Classroom Tip

When your students are first learning to order decimals, it helps to use money. The reason for this is that prices almost always have the same number of decimal places. Ordering decimals with different numbers of decimal places can be more difficult for students.

EXAMPLE 8 A Real-Life Example of Ordering Decimals

The results of the men's 100-meter dash final at the 2012 Summer Olympics are shown below. Who won the gold medal, silver medal, and bronze medal?

LANE	RUNNER	COUNTRY	TIME (seconds)
2	RICHARD THOMPSON	TTO	9.98
3	ASAFA POWELL	JAM	11.99
4	TYSON GAY	USA	9.80
5	YOHAN BLAKE	JAM	9.75
6	JUSTIN GATLIN	USA	9.79
7	USAIN BOLT	JAM	9.63
8	RYAN BAILEY	USA	9.88
9	CHURANDY MARTINA	NLD	9.94

Mathematical Practices

Attend to Precision

MP4 Mathematically proficient students analyze relationships mathematically to draw conclusions.

SOLUTION

The final times, from fastest to slowest, are shown below.

Lane	Runner	Country	Time (seconds)
7	Usain Bolt	Jamaica	9.63
5	Yohan Blake	Jamaica	9.75
6	Justin Gatlin	United States	9.79
4	Tyson Gay	United States	9.80
8	Ryan Bailey	United States	9.88
9	Churandy Martina	Netherlands	9.94
2	Richard Thompson	Trinidad and Tobago	9.98
3	Asafa Powell	Jamaica	11.99

Usain Bolt won the gold medal, Yohan Blake won the silver medal, and Justin Gatlin won the bronze medal.

7.1 Exercises

Free step-by-step solutions to odd-numbered exercises at MathematicalPractices.com.

1. **Writing a Solution Key** Write a solution key for the activity on page 246. Describe a rubric for grading a student's work.

2. **Grading the Activity** In the activity on page 246, a student gave the answers below. Each question is worth 10 points. Assign a grade to each answer. For those that are incorrect, why do you think the student erred?

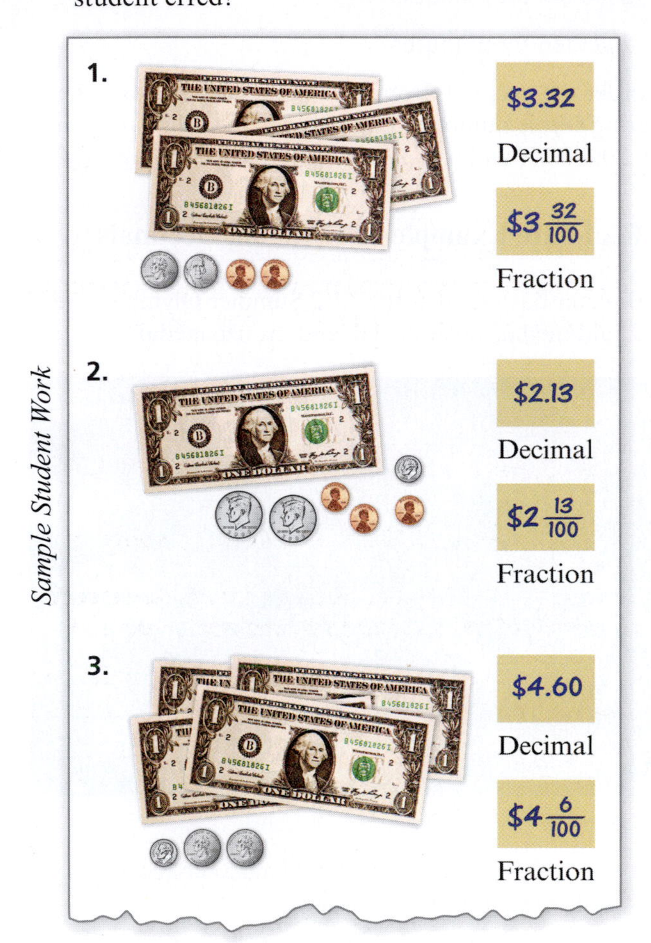

3. **Identifying Place Value** Identify the place value of each underlined digit.
   a. $1.\underline{4}3$ **b.** $\underline{6}08.29$ **c.** $21.7\underline{5}6$

4. **Identifying Place Value** Identify the place value of each underlined digit.
   a. $3\underline{2}.3$ **b.** $8.66\underline{1}$ **c.** $159.\underline{9}6$

5. **Writing Decimal Forms** Write each number in decimal form.
   a. $\frac{3}{10}$ **b.** $8\frac{29}{100}$
   c. $13\frac{9}{100}$ **d.** $\frac{741}{1000}$

6. **Writing Decimal Forms** Write each number in decimal form.
   a. $5\frac{1}{10}$ **b.** $\frac{81}{100}$
   c. $1\frac{177}{1000}$ **d.** $24\frac{15}{1000}$

7. **Writing Fraction Forms** Write each number in fraction form.
   a. 9.4 **b.** 3.612
   c. 2.01

8. **Writing Fraction Forms** Write each number in fraction form.
   a. 750.3 **b.** 4.820
   c. 6.038

9. **Identifying a Decimal** Identify the decimal represented by the base-ten blocks. A small cube represents one hundredth.

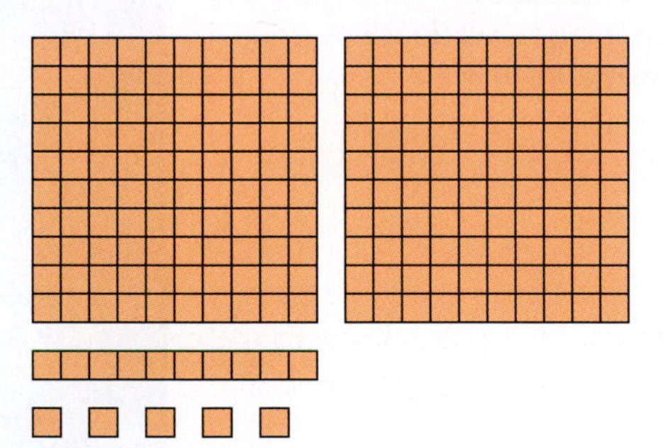

10. **Identifying a Decimal** Identify the decimal represented by the base-ten blocks. A small cube represents one hundredth.

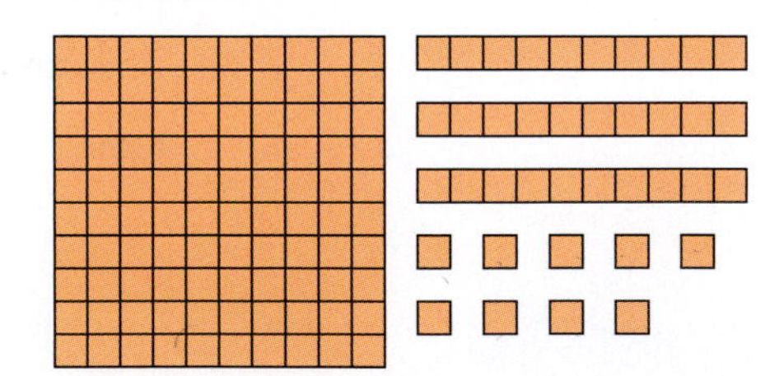

11. **Modeling Decimals** Write each decimal in expanded form. Model the decimal using base-ten blocks and draw a sketch of the model.
    a. 1.8 **b.** 3.19 **c.** 4.73

12. **Modeling Decimals** Write each decimal in expanded form. Model the decimal using base-ten blocks and draw a sketch of the model.
    a. 3.2 **b.** 2.88 **c.** 6.54

13. Writing Decimals Write the decimal for each word name.

a. Forty-five and eight-tenths

b. Nine and twenty-seven hundredths

c. Three and seven-hundredths

14. Writing Decimals Write the decimal for each word name.

a. Eight and nine-tenths

b. Sixty-three and fifty-five hundredths

c. Ten and twenty-four thousandths

15. Multiple Representations The career goals against average for an NHL goalie is shown below.

a. Write the decimal in expanded form.

b. Use base-ten blocks to model the decimal.

c. Write the decimal in words.

16. Multiple Representations The average number of rebounds per game for a basketball player in a recent season is shown below.

a. Write the decimal in expanded form.

b. Use base-ten blocks to model the decimal.

c. Write the decimal in words.

17. Writing Terminating Decimals Decide whether each fraction can be written as a terminating decimal. If so, write the terminating decimal. If not, explain why.

a. $\frac{1}{25}$ **b.** $\frac{5}{9}$ **c.** $\frac{5}{14}$ **d.** $\frac{21}{30}$

BrunoRosa/Shutterstock.com; Aspen Photo/Shutterstock.com

18. Writing Terminating Decimals Decide whether each fraction can be written as a terminating decimal. If so, write the terminating decimal. If not, explain why.

a. $\frac{5}{6}$ **b.** $\frac{7}{8}$ **c.** $\frac{4}{22}$ **d.** $\frac{11}{20}$

19. Ordering Terminating Decimals Graph the numbers on a number line. Then use the number line to decide which phone has the fastest upload speed (in millions of bits per second).

1.74 mbps 1.35 mbps 1.89 mbps 1.82 mbps

20. Ordering Terminating Decimals Graph the numbers on a number line. Then use the number line to decide which phone takes the shortest time (in seconds) to download an email.

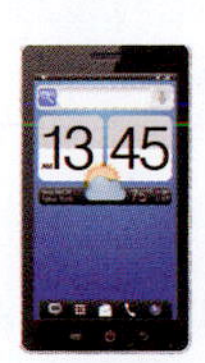 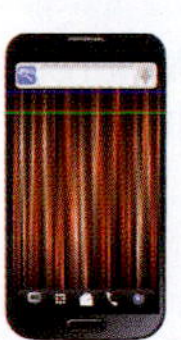  

0.28 sec 0.53 sec 0.48 sec 0.25 sec

21. Ordering Terminating Decimals Order the decimals 0.498, 0.55, 0.6, and 0.508 from least to greatest.

22. Ordering Terminating Decimals Order the decimals 0.73, 0.712, 0.706, and 0.7109 from least to greatest.

23. Ordering Numbers Order the numbers 0.835, $\frac{17}{20}$, $\frac{13}{16}$, and 0.9 from least to greatest.

24. Ordering Numbers Order the numbers 0.48, $\frac{7}{16}$, $\frac{19}{40}$, and $\frac{5}{10}$ from least to greatest.

25. Ordering Decimals The grade point averages (GPAs) of the top students in a senior class are shown below. Who are the valedictorian (ranked first) and salutatorian (ranked second) of the class?

Name	GPA	Name	GPA
Charlie	3.89	Cecilia	3.92
Benjamin	3.9	Jack	3.86
Angela	3.85	Alyssa	3.99
Oscar	3.97	Julia	3.8
Kelly	3.84	Gabrielle	3.94
Colin	3.95	Sean	3.88

cla78/Shutterstock.com; cobalt88/Shutterstock.com; losw/Shutterstock.com

26. Ordering Decimals The finishing times (in seconds) of drivers in a quarter-mile drag race are shown below. Which drivers won first place, second place, and third place?

Driver	Time	Driver	Time
1	11.74	5	13.6
2	13.1	6	11.48
3	12.45	7	12.9
4	12.7	8	12.63

27. Grading Student Work On a diagnostic test, one of your students does the following work. Explain what the student did wrong. Which topics would you encourage the student to review?

0.05 = *Five-tenths*

28. Grading Student Work On a diagnostic test, one of your students does the following work. Explain what the student did wrong. Which topics would you encourage the student to review?

0.702 = *Seven hundred and two thousandths*

29. Writing Decimals Rewrite each statement using a decimal.

a. A lap around a track is five-twentieths of a mile.

b. The atomic mass of Silicon (in atomic mass units) is approximately twenty-eight and forty-three five-hundredths.

c. A fossil shows that an ancient ant was six ninety-sixths of a yard long.

30. Writing Decimals Rewrite each statement using a decimal.

a. A piece of copy paper is about one two-hundred fiftieth of an inch thick.

b. The atomic mass of Magnesium (in atomic mass units) is approximately twenty-four and sixty-one two-hundredths.

c. It takes you twenty-two sixty-fourths of a second to blink your eye.

31. Writing Two fractions with like denominators are written as decimals. Is it possible that one decimal is nonterminating and the other is terminating? Explain.

32. Writing Write a real-life application for your students asking them to order six decimals composed of the digits 0, 2, 3, 6, and 7.

33. School Transportation: In Your Classroom The table below shows the fraction of students in each class who ride the bus to school.

Class	Fraction of students who ride the bus
1st grade class 1	$\frac{21}{24}$
1st grade class 2	$\frac{22}{25}$
2nd grade class 1	$\frac{21}{30}$
2nd grade class 2	$\frac{21}{28}$
3rd grade class 1	$\frac{17}{25}$
3rd grade class 2	$\frac{18}{24}$

a. Decide which 1st grade class has a greater fraction of students who ride the bus to school.

b. Describe a homework project that you could assign that would involve data about bus riding. Explain how the project would involve decimals.

34. School Transportation: In Your Classroom The table below shows the fraction of students in each class who walk to school.

Class	Fraction of students who walk
4th grade class 1	$\frac{6}{30}$
4th grade class 2	$\frac{7}{32}$
4th grade class 3	$\frac{6}{32}$
5th grade class 1	$\frac{7}{28}$
5th grade class 2	$\frac{5}{32}$
5th grade class 3	$\frac{6}{30}$

a. Decide which 5th grade class has the greatest fraction of students who walk to school.

b. Describe a homework project that you could assign that would involve data about walking to school. Explain how the project would involve decimals.

35. **Comparing Techniques** Write the fraction $\frac{4}{25}$ as a terminating decimal by (a) rewriting the fraction so that its denominator is a power of 10 and (b) dividing the numerator by the denominator. (c) Which technique do you prefer? Explain your reasoning.

36. **Comparing Techniques** Write the fraction $\frac{28}{70}$ as a terminating decimal by (a) rewriting the fraction so that its denominator is a power of 10 and (b) dividing the numerator by the denominator. (c) Which technique do you prefer? Explain your reasoning.

37. **Writing Terminating Decimals** Decide whether the heights (in feet) can be written as terminating decimals. If so, write the terminating decimal(s).

38. **Writing Terminating Decimals** Decide whether the prices can be written as terminating decimals. If so, write the terminating decimal(s).

39. **Number Line Sense** Use the two decimals graphed on the number line.

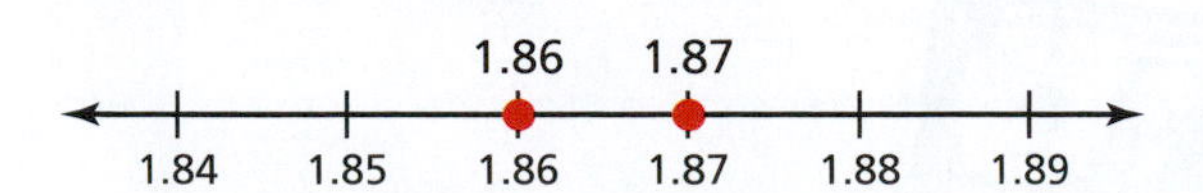

a. Find a decimal less than both decimals.

b. Find a decimal greater than both decimals.

c. Find a decimal between the two decimals.

Nevena Radonja/Shutterstock.com; jonson/Shutterstock.com; ecco/Shutterstock.com; nndrin/Shutterstock.com; spline_x/Shutterstock.com; Michael Dykstra | Dreamstime.com; prapass/Shutterstock.com; Africa Studio/Shutterstock.com

40. **Number Line Sense** Use the two decimals graphed on the number line.

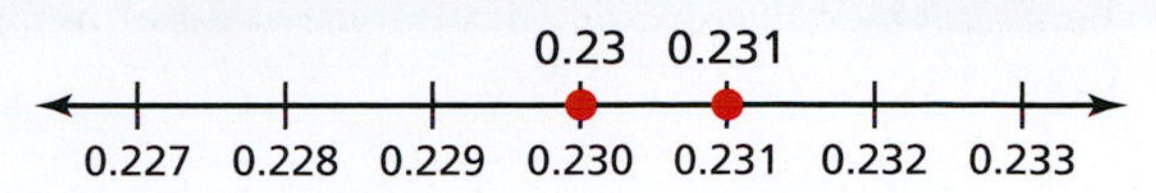

a. Find a decimal less than both decimals.

b. Find a decimal greater than both decimals.

c. Find a decimal between the two decimals.

41. **Number Sense: In Your Classroom** Explain how you would teach your students that, given any two terminating decimals, there is a terminating decimal between them.

42. **Number Sense: In Your Classroom** Explain how you would teach your students that, given any terminating decimal and any whole number, there is a terminating decimal between them.

43. **Reasoning** What is the greatest possible decimal you can write by filling in the spaces below with the digits 1, 2, 3, and 4? What is the least possible decimal?

0.__ __ __ __

44. **Reasoning** What is the greatest possible decimal you can write by filling in the spaces below with the digits 5, 6, 7, 8, and 9? What is the least possible decimal?

0.__ __ __ __ __

45. **Reasoning** Use the factored form of the fraction below.

$$\frac{7}{2 \cdot 2 \cdot \_\_ \cdot 5}$$

a. Fill in the missing digit so that the expression can be written as a terminating decimal.

b. How many different digits work for part (a)?

46. **Reasoning** Use the factored form of the fraction below.

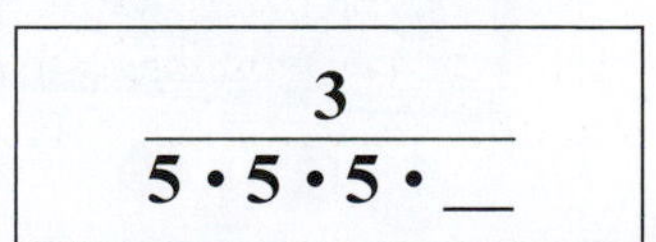

a. Fill in the missing digit so that the expression can be written as a terminating decimal.

b. How many different digits work for part (a)?

Activity: Using Money to Add and Subtract Decimals

Materials:

- Pencil

Learning Objective:

Use money to help understand how to add and subtract decimals.

Name ________________________________

Write each amount of money in decimal form. Then write the sum or difference. Explain how you found your answer.

1.

 + 

2.

 + 

3.

 − 

4.

 − 

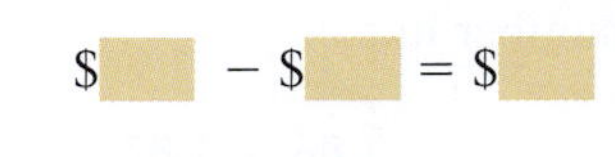

iofoto/Shutterstock.com; 2happy/Shutterstock.com; iStockphoto.com/Joe Potato Photo

7.2 Adding and Subtracting Decimals

Standards

Grades 3–5
Numbers and Operations in Base Ten
Students should perform operations with decimals to hundredths.

Grades 6–8
The Number System
Students should compute fluently with multi-digit numbers.

- Add and subtract decimals using models and algorithms.
- Add and subtract decimals using mental math.
- Round decimals.

Adding and Subtracting Decimals

Besides using algorithms, a common way to teach students to add and subtract decimals is to use base-ten blocks.

EXAMPLE 1 **Using Base-Ten Blocks to Add and Subtract Decimals**

Use base-ten blocks to find each sum or difference.

a. 1.3 + 1.57 **b.** 3.29 − 2.16

SOLUTION

a. Use base-ten blocks to model each number.

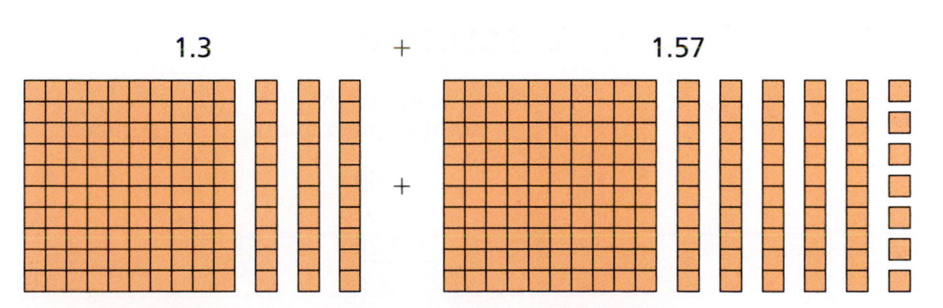

Put all of the blocks into a single group and regroup, if necessary.

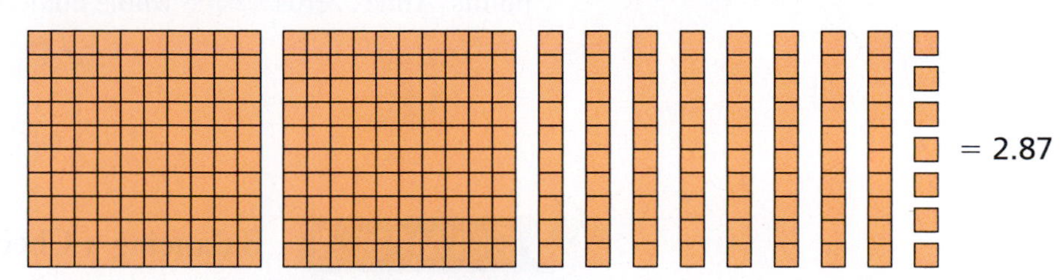

So, 1.3 + 1.57 = 2.87.

b. First, represent 3.29. Because you can directly take away 2 ones from 3 ones, 1 tenth from 2 tenths, and 6 hundredths from 9 hundredths, the use of base-ten blocks is straightforward.

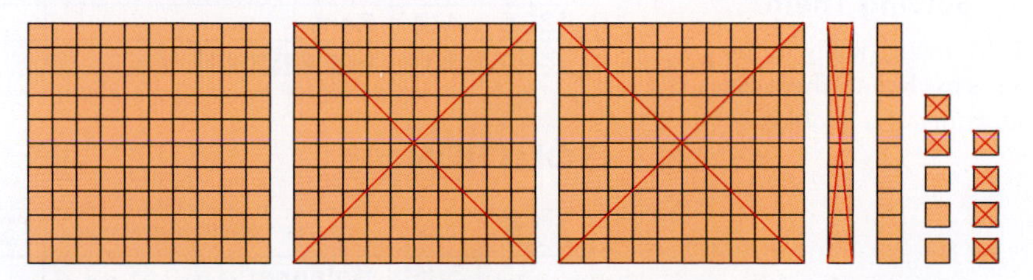

That leaves 1 one, 1 tenth, and 3 hundredths. So, 3.29 − 2.16 = 1.13.

Classroom Tip

Check your results by writing the decimals in fraction form. Then add or subtract.

a. $1.3 + 1.57 = 1\frac{30}{100} + 1\frac{57}{100}$

$= 2\frac{87}{100}$

$= 2.87$ ✓

b. $3.29 - 2.16 = 3\frac{29}{100} - 2\frac{16}{100}$

$= 1\frac{13}{100}$

$= 1.13$ ✓

Algorithm for Adding and Subtracting Decimals

To add or subtract decimals using a vertical algorithm, use the following steps.

1. Write the numbers vertically and line up the decimal points. Annex zeros so that the numbers have the same number of decimal places.
2. Add or subtract as with whole numbers.
3. Insert the decimal point in the sum or difference immediately below the decimal points in the numbers being added or subtracted.

Example 12.34 + 3.015

1. Line up the decimal points. Annex a zero.

$$\begin{array}{r} 12.340 \\ +\ 3.015 \\ \hline \end{array}$$

2. Add or subtract as with whole numbers.

$$\begin{array}{r} 12.340 \\ +\ 3.015 \\ \hline 15\ 355 \end{array}$$

3. Insert the decimal point in the sum.

$$\begin{array}{r} 12.340 \\ +\ 3.015 \\ \hline 15.355 \end{array}$$

Classroom Tip

Although not strictly necessary, it may help students to annex zeros as needed when using the vertical algorithm.

$$\begin{array}{r} 12.340 \\ +\ 3.015 \\ \hline 15.355 \end{array}$$

Annex a zero.

This is especially helpful when subtracting.

EXAMPLE 2 Adding and Subtracting Decimals

Use the vertical algorithm to find (a) 2.405 + 13.01 + 7.2 and (b) 5.3 − 1.255.

SOLUTION

a. Line up the decimal points. Annex zeros.

$$\begin{array}{r} 2.405 \\ 13.010 \\ +\ 7.200 \\ \hline \end{array}$$

Add as with whole numbers.

$$\begin{array}{r} 2.405 \\ {}^{1}13.010 \\ +\ 7.200 \\ \hline 22\ 615 \end{array}$$

Insert the decimal point in the sum.

$$\begin{array}{r} 2.405 \\ 13.010 \\ +\ 7.200 \\ \hline 22.615 \end{array}$$

b. Line up the decimal points. Annex zeros.

$$\begin{array}{r} 5.300 \\ -\ 1.255 \\ \hline \end{array}$$

Subtract as with whole numbers.

$$\begin{array}{r} {}^{2\,9\,1} \\ 5.\not{3}\not{0}0 \\ -\ 1.255 \\ \hline 4\ 045 \end{array}$$

Insert the decimal point in the difference.

$$\begin{array}{r} {}^{2\,9\,1} \\ 5.\not{3}\not{0}0 \\ -\ 1.255 \\ \hline 4.045 \end{array}$$

EXAMPLE 3 Balancing a Checkbook

Find the balance after each transaction.

CHECK NUMBER	DATE	DESCRIPTION OF TRANSACTION	PAYMENT, FEE OR WITHDRAWAL (-)	T	DEPOSIT OR INTEREST (+)	BALANCE
	3/31	Balance				356.78
	4/1	Deposit			802.56	
1295	4/2	Rent	875.00			
1296	4/4	Car payment	224.90			

Mathematical Practices

Make Sense of Problems and Persevere in Solving Them

MP1 Mathematically proficient students explain to themselves the meaning of a problem.

SOLUTION

CHECK NUMBER	DATE	DESCRIPTION OF TRANSACTION	PAYMENT, FEE OR WITHDRAWAL (-)	T	DEPOSIT OR INTEREST (+)	BALANCE
	3/31	Balance				356.78
	4/1	Deposit			802.56	1159.34
1295	4/2	Rent	875.00			284.34
1296	4/4	Car payment	224.90			59.44

Adding and Subtracting Decimals Using Mental Math

In Section 4.1, you learned strategies for using mental math to add and subtract whole numbers. These strategies can be used to add and subtract decimals. **Compatible numbers** are numbers whose sums, differences, products, or quotients are easy to compute mentally. Here are some examples involving decimals.

$$\begin{array}{r} 2.6 \\ +\ 2.4 \\ \hline 5.0 \end{array} \qquad \begin{array}{r} 0.75 \\ +\ 0.25 \\ \hline 1.00 \end{array} \qquad \begin{array}{r} 4.5 \\ -\ 0.5 \\ \hline 4.0 \end{array} \qquad \begin{array}{r} 3.90 \\ -\ 3.00 \\ \hline 0.90 \end{array}$$

Mental Math Strategy for Adding and Subtracting Decimals

Compensation is a mental math strategy in which you create compatible numbers by adjusting the actual numbers in a sum or difference.

To find a sum, add a number to one addend to make the addition easier. Then compensate by subtracting that same number from the other addend.

Example

$$\begin{array}{r} 2.8 \\ +\ 2.5 \\ \hline 5.3 \end{array} \begin{array}{c} \rightarrow \\ \rightarrow \\ \\ \end{array} \begin{array}{r} 3.0 \\ +\ 2.3 \\ \hline 5.3 \end{array}$$

Think $2.8 + 0.2 = 3.0$. So, $2.5 - 0.2 = 2.3$.

To find a difference, add or subtract the same number from *both* numbers.

Example

$$\begin{array}{r} 8.1 \\ -\ 2.9 \\ \hline 5.2 \end{array} \begin{array}{c} \rightarrow \\ \rightarrow \\ \\ \end{array} \begin{array}{r} 7.2 \\ -\ 2.0 \\ \hline 5.2 \end{array}$$

Think $2.9 - 0.9 = 2.0$. So, $8.1 - 0.9 = 7.2$.

Using Mental Math in Real Life

You purchase 5 packs of batteries online. The prices are \$9.97, \$11.97, \$5.97, \$6.97, and \$18.72. You pay using a \$75 gift card. How much credit is left on the gift card?

SOLUTION

First find the total cost of the batteries. Notice that four of the prices end in 0.97. Increase each of these prices by 0.03 and find the sum.

$$\begin{array}{r} 9.97 \\ 11.97 \\ 5.97 \\ +\ 6.97 \\ \\ \end{array} \begin{array}{c} \rightarrow \\ \rightarrow \\ \rightarrow \\ \rightarrow \\ \\ \end{array} \begin{array}{r} \overset{1}{1}0.00 \\ 12.00 \\ 6.00 \\ +\ 7.00 \\ \hline 35.00 \end{array}$$

Think $9.97 + 0.03 = 10.00$.
Think $11.97 + 0.03 = 12.00$.
Think $5.97 + 0.03 = 6.00$.
Think $6.97 + 0.03 = 7.00$.

To *compensate*, subtract the following sum from the remaining price of 18.72.

$$\begin{array}{r} 0.0\overset{1}{3} \\ 0.03 \\ 0.03 \\ +\ 0.03 \\ \hline 0.12 \end{array} \qquad \begin{array}{r} 18.72 \\ -\ 0.12 \\ \hline 18.60 \end{array} \qquad \begin{array}{r} \overset{1}{3}5.00 \\ +\ 18.60 \\ \hline 53.60 \end{array}$$

Subtract 0.12 to *compensate*.

Add 18.60 and the sum above.

So, you spend \$53.60 for batteries. To find the amount of credit left on the gift card, subtract the amount you spend from the amount of the gift card.

$$\begin{array}{r} 7\overset{4}{\not{5}}.\overset{1}{0}0 \\ -\ 53.60 \\ \hline 21.40 \end{array}$$

So, after paying for the batteries, \$21.40 is left on the gift card.

Classroom Tip The rules for rounding decimals are similar to the rules for rounding whole numbers shown in Section 2.2.

Rounding Decimals

Rounding Decimals

To round a decimal, circle the place value to be rounded. The decision digit is the digit to the right of this place value. Underline this digit.

Round up When the decision digit is 5, 6, 7, 8, or 9, round up by adding 1 to the circled place value. All digits to the right of the circled place value become 0.

Example Rounded to the nearest tenth, 2.(3)8 4 is 2.4.

Round down When the decision digit is 0, 1, 2, 3, or 4, round down by leaving the circled place value as is. All digits to the right of the circled place value become 0.

Example Rounded to the nearest hundredth, 1.0(5)4 is 1.05.

EXAMPLE 5 **Rounding Decimals**

Round each decimal to the nearest tenth, hundredth, and thousandth.

a. 4.8764 **b.** 23.0158 **c.** 0.1009

SOLUTION

a. Rounded to the nearest tenth, 4.(8)7 6 4 is 4.9. Round up.

Decision digit is 7.

Rounded to the nearest hundredth, 4. 8(7)6 4 is 4.88. Round up.

Decision digit is 6.

Rounded to the nearest thousandth, 4. 8 7(6)4 is 4.876. Round down.

Decision digit is 4.

b. Rounded to the nearest tenth, 23.(0)1 5 8 is 23.0. Round down.

Decision digit is 1.

Rounded to the nearest hundredth, 23. 0(1)5 8 is 23.02. Round up.

Decision digit is 5.

Rounded to the nearest thousandth, 23. 0 1(5)8 is 23.016. Round up.

Decision digit is 8.

c. Rounded to the nearest tenth, 0.(1)0 0 9 is 0.1. Round down.

Decision digit is 0.

Rounded to the nearest hundredth, 0. 1(0)0 9 is 0.10. Round down.

Decision digit is 0.

Rounded to the nearest thousandth, 0. 1 0(0)9 is 0.101. Round up.

Decision digit is 9.

Classroom Tip In Example 5(b), point out to your students that writing a rounded number as 23.0 is not technically the same as writing it as 23. Writing the answer as 23.0 means that, to the nearest tenth, you know the closest tenth is 0. Writing the answer as 23 means that you are not sure of the tenths digit.

EXAMPLE 6 Listing Reasonable Digits in Real Life

Rewrite each statement by rounding the decimal to a number that is reasonable in the real-life context. Justify your reasoning.

a. A typical weight of an adult Percheron horse is 1873.4 pounds.

b. Adult giraffes generally stand between 4.572 and 6.096 meters tall.

SOLUTION

a. It is unrealistic to measure the typical weight of *all* Percherons to the nearest pound or even to the nearest ten pounds. Estimate the typical weight as 1900 pounds.

A typical weight of an adult Percheron horse is 1900 pounds.

b. As in part (a), measuring the typical height of *all* giraffes to the nearest thousandth or to the nearest hundredth of a meter seems unreasonable. Estimate the typical height to the nearest tenth.

Adult giraffes generally stand between 4.6 and 6.1 meters tall.

> **Classroom Tip**
> Example 6 illustrates an important point for students to learn. Students need to understand that listing the reasonable digits in an answer is an important skill. Listing too many digits can be as misleading as not listing enough digits.

EXAMPLE 7 A Real-Life Example of Operations with Decimals

The first- and second-place winning times for the women's downhill Alpine skiing at a recent Winter Olympics were 1 minute, 44.19 seconds and 104.75 seconds, respectively. How much faster is the first-place skier?

SOLUTION

1. **Understand the Problem** You are given the first- and second-place finishing times. You need to find how much faster the first-place time is than the second-place time.
2. **Make a Plan** Find the difference of the two times. Change the first-place winning time to seconds and subtract.
3. **Solve the Problem** There are 60 seconds in one minute.

   1st place 1 minute and 44.19 seconds
   $$\begin{array}{r} 60.00 \\ +\ 44.19 \\ \hline 104.19 \end{array}$$

   Find the difference of 104.75 and 104.19.
   $$\begin{array}{r} \overset{6\,1}{104.\not{7}5} \\ -\ 104.19 \\ \hline 0.56 \end{array}$$

   The first-place skier is 0.56 second faster than the second-place skier.
4. **Look Back** You can check the reasonableness of your answer by rounding each addend to the nearest tenth, and then finding the difference.

   $$\begin{array}{r} 104.\textcircled{7}\underline{5} \\ -\ 104.\textcircled{1}\underline{9} \\ \hline 0.56 \end{array} \rightarrow \begin{array}{r} 104.8 \\ -104.2 \\ \hline 0.6 \end{array} \checkmark$$

   Decision digit is 5. Round up.
   Decision digit is 9. Round up.

> **Classroom Tip**
> Remind your students that subtraction is the inverse operation of addition. Your students can also check a subtraction problem using addition. You can check Example 7 as shown.
> $$\begin{array}{r} \overset{1}{0.56} \\ +\ 104.19 \\ \hline 104.75 \end{array} \checkmark$$

Eric Isselee/Shutterstock.com

7.2 Exercises

Free step-by-step solutions to odd-numbered exercises at *MathematicalPractices.com*.

1. **Writing a Solution Key** Write a solution key for the activity on page 256. Describe a rubric for grading a student's work.

2. **Grading the Activity** In the activity on page 256, a student gave the answers below. Each question is worth 10 points. Assign a grade to each answer. For those that are incorrect, why do you think the student erred?

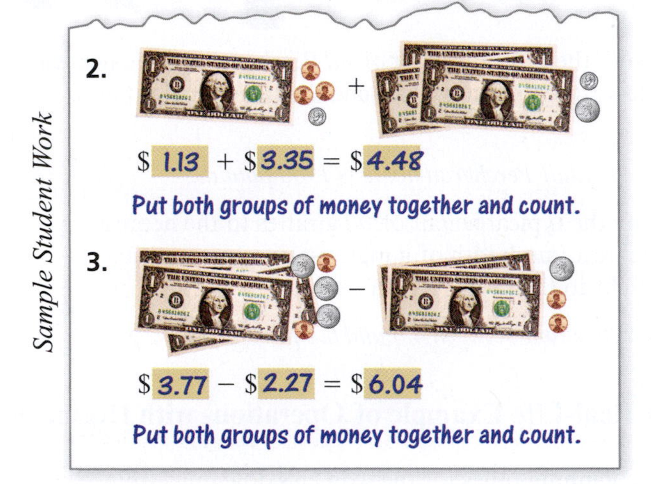

3. **Using Base-Ten Blocks to Add Decimals** Write and solve the addition problem represented by each model. A small cube represents one hundredth.

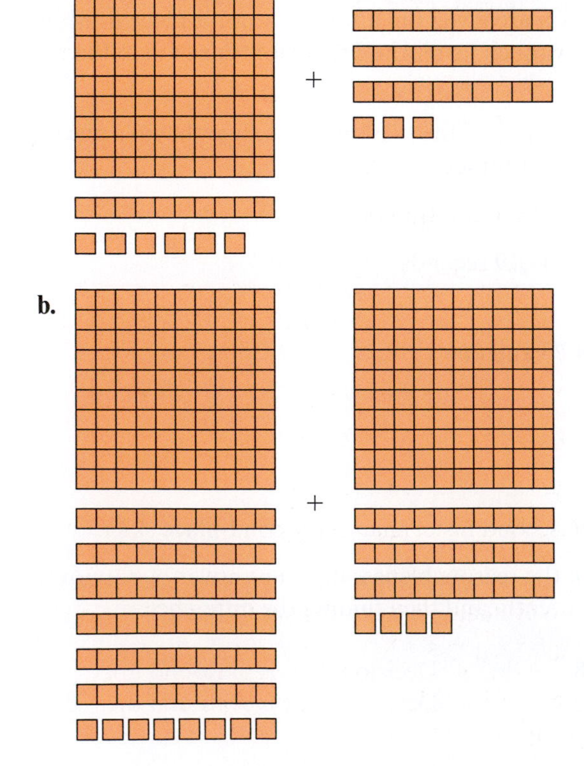

4. **Using Base-Ten Blocks to Add Decimals** Write and solve the addition problem represented by each model. A small cube represents one hundredth.

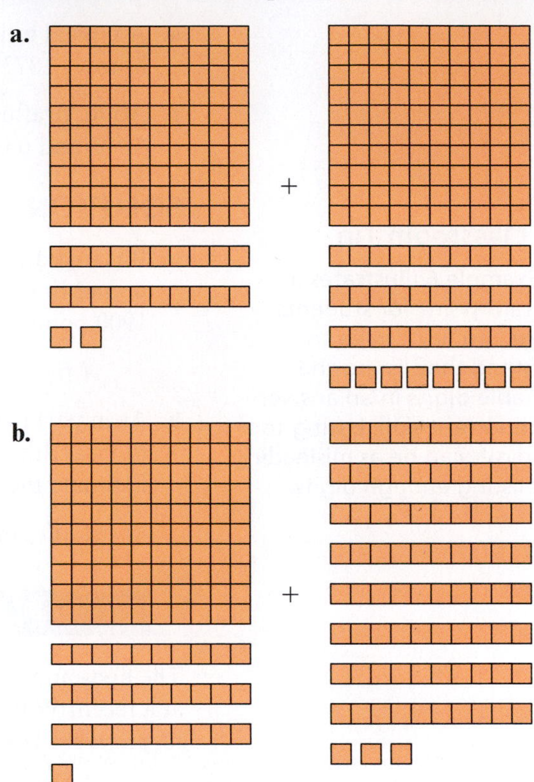

5. **Using Base-Ten Blocks to Subtract Decimals** Write and solve the subtraction problem represented by each model. A small cube represents one hundredth.

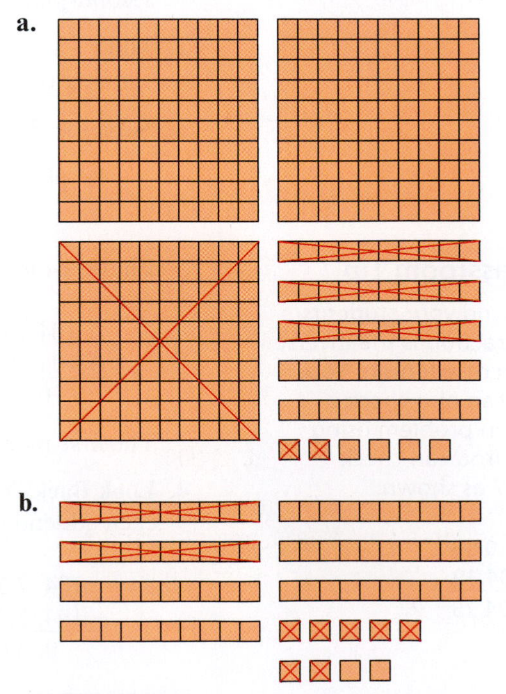

2happy/Shutterstock.com

6. **Using Base-Ten Blocks to Subtract Decimals** Write and solve the subtraction problem represented by each model. A small cube represents one hundredth.

a.

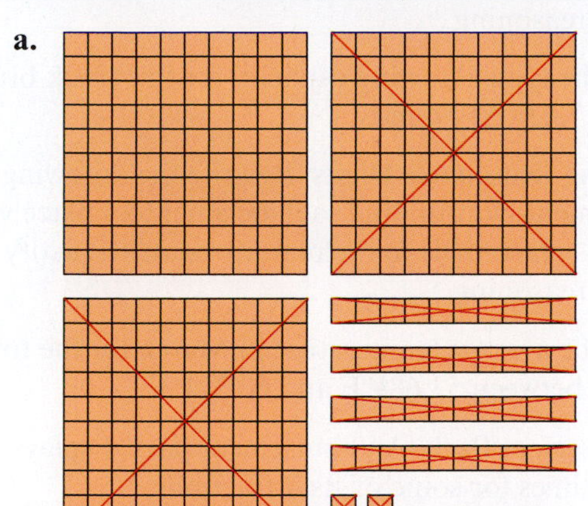

b.

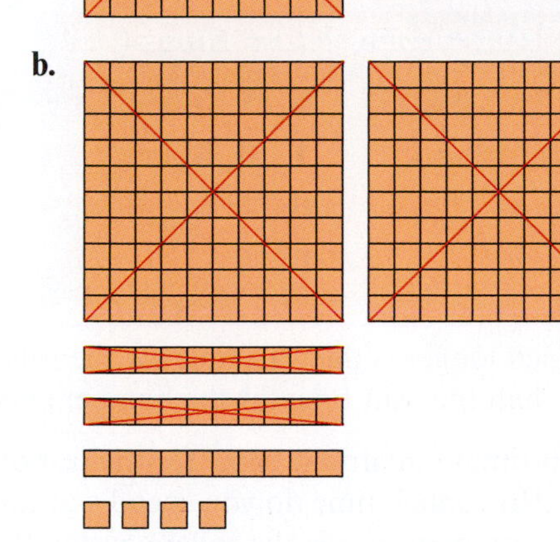

7. **Adding and Subtracting Decimals** Use base-ten blocks to find each sum or difference. Justify your answer with a sketch of the model.

a. 2 + 1.58 **b.** 1.82 + 1.4

c. 1.45 + 2.08 **d.** 3.28 − 1.06

e. 2.89 − 0.68 **f.** 1.94 − 1.03

8. **Adding and Subtracting Decimals** Use base-ten blocks to find each sum or difference. Justify your answer with a sketch of the model.

a. 1 + 2.42 **b.** 1.2 + 1.38

c. 2.17 + 1.43 **d.** 2.56 − 0.42

e. 3.71 − 1.61 **f.** 3 − 1.52

9. **Rounding Decimals** Round each decimal to the nearest tenth, hundredth, and thousandth.

a. 6.2536 **b.** 0.0862

c. 44.6981 **d.** 5.0008

10. **Rounding Decimals** Round each decimal to the nearest tenth, hundredth, and thousandth.

a. 3.5892 **b.** 18.2004

c. 9.0148 **d.** 7.5749

11. **Adding Decimals** Use the vertical algorithm to find each sum. Use estimation to see if your answer is reasonable.

a. 5.4 + 3.8 **b.** 2.5 + 4.03

c. 3.553 + 1.74 **d.** 14.927 + 1.064

e. 8.46 + 3 + 16.582

f. 1.85 + 4.221 + 0.826

12. **Adding Decimals** Use the vertical algorithm to find each sum. Use estimation to see if your answer is reasonable.

a. 4.23 + 6 **b.** 1.08 + 5.6

c. 0.858 + 12.21 **d.** 9.147 + 2.004

e. 17.201 + 1.15 + 2.4

f. 6.08 + 1.597 + 8.432

13. **Subtracting Decimals** Use the vertical algorithm to find each difference. Use estimation to see if your answer is reasonable.

a. 9 − 0.89 **b.** 4.86 − 2.1

c. 8.87 − 1.48 **d.** 15.094 − 5.18

e. 7.455 − 0.823 **f.** 42 − 6.08

14. **Subtracting Decimals** Use the vertical algorithm to find each difference. Use estimation to see if your answer is reasonable.

a. 5.17 − 2.8 **b.** 6 − 3.41

c. 31.545 − 4.06 **d.** 7.82 − 0.233

e. 18 − 4.588 **f.** 10.479 − 0.713

15. **Adding and Subtracting Decimals** Evaluate each expression.

a. 6.35 + 2.1 − 4.2

b. 5.631 − 3.12 + 0.844

c. 3.4 + 8.108 − 3.502

d. 19 − 5.42 + 1.532

16. **Adding and Subtracting Decimals** Evaluate each expression.

a. 5.3 + 1.87 − 2.66

b. 4.321 − 2.67 + 12.89

c. 7 − 6.33 + 0.315

d. 26.53 + 3.141 − 11.307

17. **Adding and Subtracting Decimals** Use mental math to evaluate each expression. Describe your reasoning.

a. 5.6 + 7.5 **b.** 6.2 − 3.4

c. 10.4 − 4.7 **d.** 1.98 + 3.96 + 2.93

18. **Adding and Subtracting Decimals** Use mental math to evaluate each expression. Describe your reasoning.

a. $13.7 + 4.2$ **b.** $8.4 - 1.3$

c. $9.6 - 5.9$ **d.** $2.03 + 1.01 + 8.05$

19. *Writing* Explain the importance of lining up the decimal points when adding decimals or subtracting decimals.

20. *Writing* Explain the steps in evaluating $8 - 6.321$ using the vertical algorithm.

21. **My Dear Aunt Sally** Arrange the numbers 0.2, 0.4, 0.6, and 0.8 so that the equation is true.

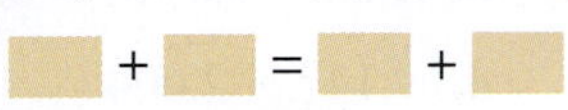

For more practice with this type of reasoning, go to *AuntSally.com*.

22. **My Dear Aunt Sally** Arrange the numbers 0.1, 0.3, 0.7, and 0.9 so that the equation is true.

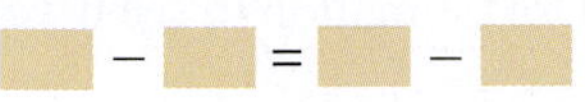

For more practice with this type of reasoning, go to *AuntSally.com*.

23. **Discus Throw** A discus thrower has a goal to throw a distance of 38 meters. Her farthest throw is 36.86 meters. How much farther must she throw to reach her goal?

24. **Balancing a Checkbook** The balance in your checking account is \$264.85. What is the balance after you deposit a check for \$54.50?

25. **Road Trip** You and your friend go on a road trip. The distances you and your friend drive are shown below.

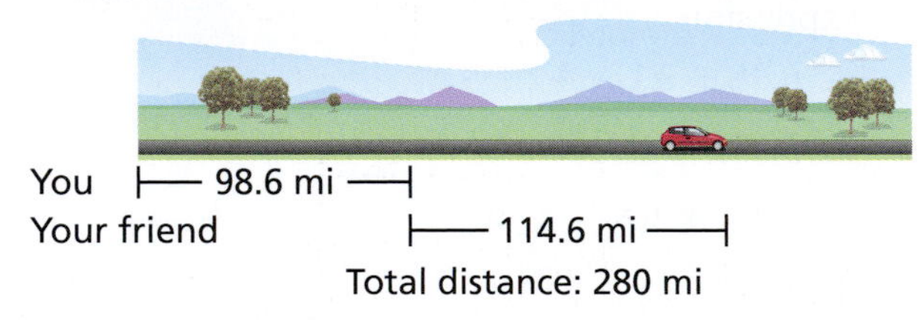

a. How many miles have you and your friend traveled so far?

b. How many miles are there left to travel?

26. **Balancing a Checkbook** You deposit a check for \$95.60 in your checking account and withdraw \$30.00. The balance in your checking account at the end of the day is \$180.70.

a. What was the beginning balance?

b. What is the change in the balance from the beginning to the end of the day?

27. **Listing Reasonable Digits** Rewrite the following statement by rounding the decimal to a place value that is reasonable in the real-life context. Justify your reasoning.

The typical towing capacity for this truck brand is 7,438.53 pounds.

28. **Listing Reasonable Digits** Rewrite the following statement by rounding each decimal to a place value that is reasonable in the real-life context. Justify your reasoning.

The average temperature in March for the town is between 53.689°F and 60.321°F.

29. **Amusement Park** An amusement park displays the wait times for some of its attractions.

a. How much longer is the wait time for the roller coaster than the wait time for the bumper cars?

b. The wait times remain the same for the rest of the day. How much time do you spend waiting in line if you plan to ride the roller coaster, the giant wheel, and the sky ride?

30. **Scholarship Money** You receive a scholarship for \$500 to purchase books for the semester. The prices of the books are \$68.90, \$78.85, \$98.80, \$124.95, and \$159.75. Is your scholarship enough to pay for all of the books? If not, how much more money do you need?

31. **My Dear Aunt Sally** Arrange the numbers 0.1, 0.2, 0.3, 0.5, 0.6, and 0.9 so that the equation is true.

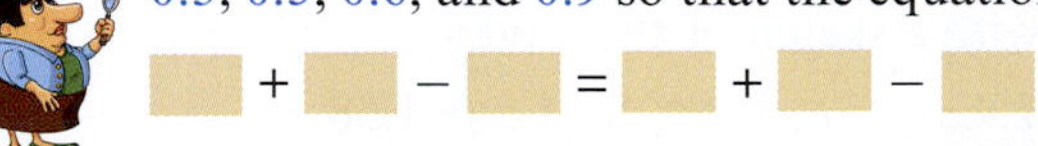

For more practice with this type of reasoning, go to *AuntSally.com*.

32. **My Dear Aunt Sally** Arrange the numbers 0.3, 0.4, 0.8, 0.9, 1.0, and 1.2 so that the equation is true.

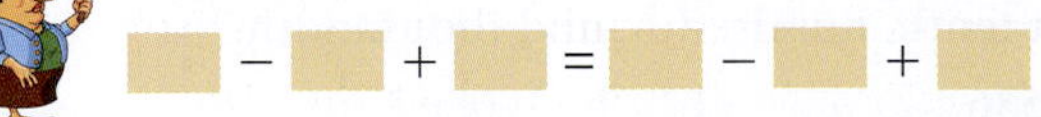

For more practice with this type of reasoning, go to *AuntSally.com*.

33. *Writing* Describe the similarities when evaluating $2.53 - 1.80$ and $253 - 180$. Describe the differences.

34. *Writing* Explain the steps in evaluating 1.85 + 3.42 using fractions.

35. *Writing* Write a real-life application involving distances represented by decimals in which you would solve using mental math.

36. *Writing* Write a real-life application involving temperatures represented by decimals in which you would solve using mental math.

37. **Grading Student Work** On a diagnostic test, one of your students does the following work. Explain what the student did wrong. Which topics would you encourage the student to review?

```
  7.520
  18.31
+  6.15
  99.66
```

38. **Grading Student Work** On a diagnostic test, one of your students does the following work. Explain what the student did wrong. Which topics would you encourage the student to review?

```
  4.5
-3.288
 1.388
```

39. **Comparing Perimeters: In Your Classroom** A school plans to build a playground. There are two options for locations: one triangular region and one rectangular region. Each location would require fencing around the entire playground.

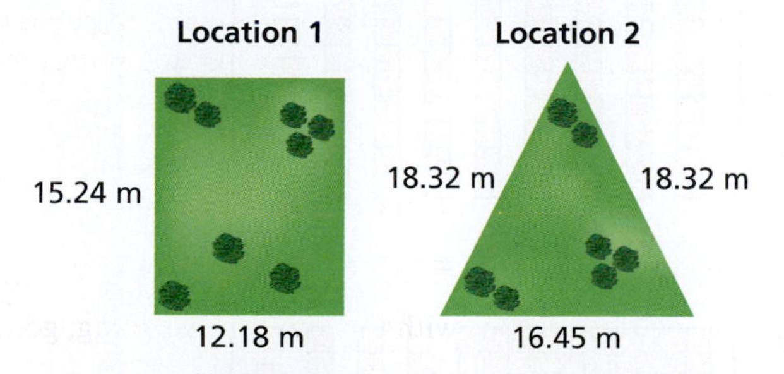

a. Find the perimeter of the rectangular region and the triangular region.

b. Which playground requires less fencing? How much more fencing is needed for the other playground?

c. Design a hands-on classroom activity for your students that would involve using decimals to calculate the perimeters of regions in your school.

40. **Comparing Perimeters: In Your Classroom** An artist must choose one of two paintings to frame for an art show. The artist will frame the one that requires less framing material.

Painting 1

121.92 cm

81.28 cm

Painting 2

96.52 cm

91.44 cm

a. Find the perimeter of each painting.

b. Which painting requires less framing material? How much more framing material is needed for the other painting?

c. Design a hands-on classroom activity for your students that would involve using decimals to calculate the perimeters of objects in your classroom.

41. **Magic Square** Fill in the magic square so that each row, column, and diagonal has the same sum.

1.8		3.6
	4.5	
5.4		

42. **Magic Square** Fill in the magic square so that each row, column, and diagonal has the same sum.

8.96		
	5.6	
	10.08	2.24

svetera/Shutterstock.com

Activity: Multiplying and Dividing Decimals by Whole Numbers

Materials:

- Pencil
- Base-ten blocks

Learning Objective:

Use base-ten blocks to multiply a decimal by a whole number. Use base-ten blocks to divide a decimal by a whole number.

Name ______________________

Use base-ten blocks to find the product or quotient. Explain your procedure. A small cube represents one hundredth.

1. $3 \cdot$ 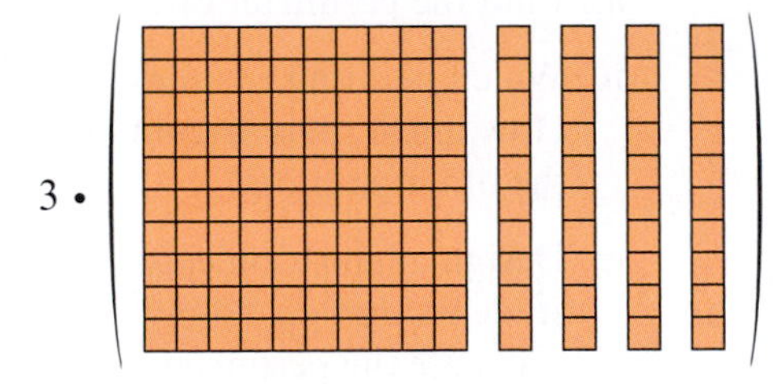

$3 \cdot$ ___ $=$ ___

2. $4 \cdot$ 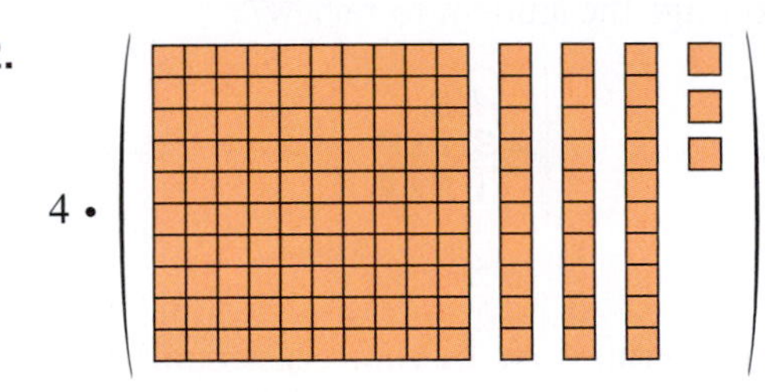

$4 \cdot$ ___ $=$ ___

3. $4 \cdot$ 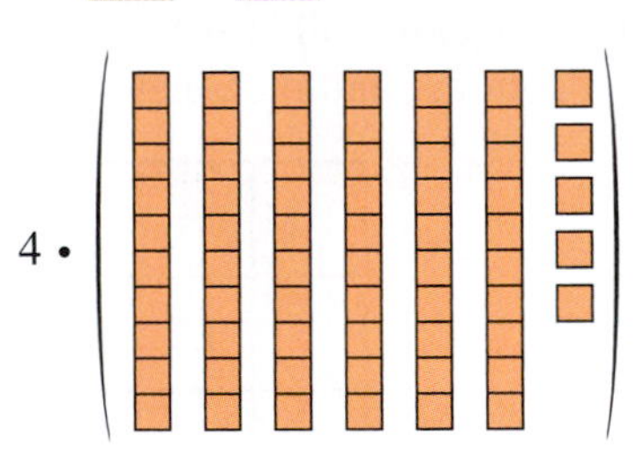

$4 \cdot$ ___ $=$ ___

4. 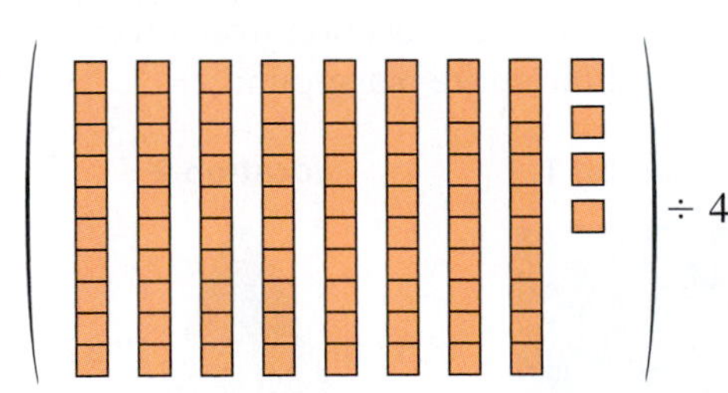 $\div 4$

___ $\div 4 =$ ___

5. 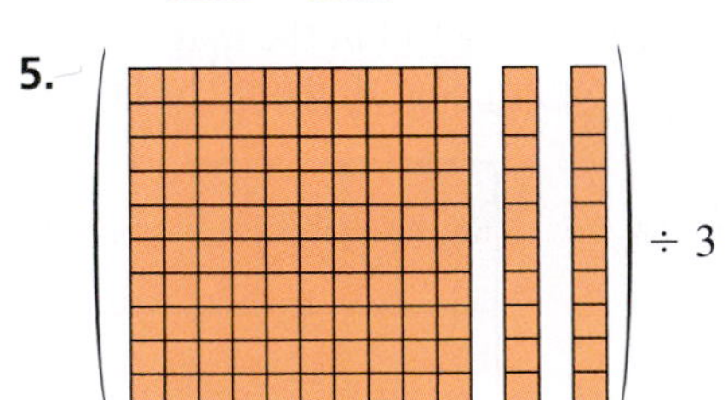 $\div 3$

___ $\div 3 =$ ___

6. 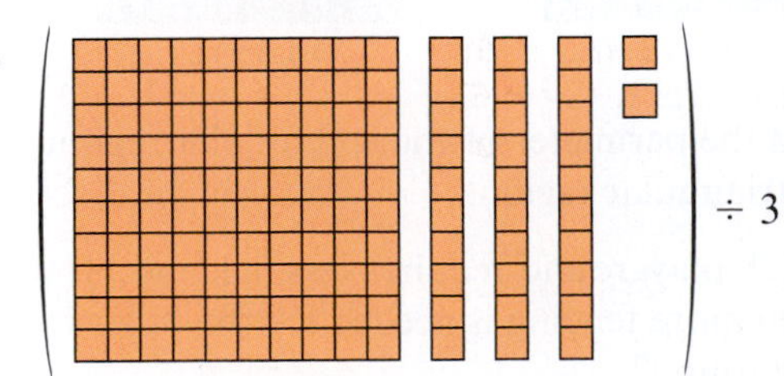 $\div 3$

___ $\div 3 =$ ___

7. $2 \cdot 1.26 =$ ___

8. $4 \cdot 2.31 =$ ___

9. $6.39 \div 3 =$ ___

10. $6.1 \div 5 =$ ___

Monkey Business Images/Shutterstock.com

7.3 Multiplying and Dividing Decimals

Standards

Grades 3–5
Numbers and Operations in Base Ten
Students should perform operations with decimals to hundredths.

Grades 6–8
The Number System
Students should compute fluently with multi-digit numbers.

- Multiply and divide decimals using fractions and algorithms.
- Write fractions as terminating or repeating decimals.
- Use scientific notation to represent large numbers.

Multiplying and Dividing Decimals

EXAMPLE 1 Multiplying Decimals Using Fractions

Find (a) $3 \cdot 2.1$ and (b) $2.5 \cdot 1.34$.

SOLUTION

a. $3 \cdot 2.1 = 3 \cdot 2\frac{1}{10} = 3 \cdot \frac{21}{10} = \frac{63}{10} = 6\frac{3}{10} = 6.3$

b. $2.5 \cdot 1.34 = 2\frac{5}{10} \cdot 1\frac{34}{100} = \frac{25}{10} \cdot \frac{134}{100} = \frac{3350}{1000} = 3\frac{350}{1000} = 3.35$

Algorithm for Multiplying Decimals

To multiply decimals using a vertical algorithm, use the following steps.

1. Multiply as with whole numbers.
2. Count the number of digits to the right of the decimal point in each of the two factors. Suppose this is m digits and n digits.
3. Insert the decimal point in the product so that there are $m + n$ digits to the right of the decimal point.

Example $0.26 \cdot 4.8$

1. Multiply as with whole numbers.

$$\begin{array}{r} 0.26 \\ \times\ 4.8 \\ \hline 208 \\ 1040 \\ \hline 1248 \end{array}$$

2. Insert the decimal point in the product.

$$\begin{array}{rl} 0.26 & \text{2 digits to the right of decimal point} \\ \times\ 4.8 & \text{1 digit to the right of decimal point} \\ \hline 208 & \\ 1040 & \\ \hline 1.248 & 2 + 1 = 3 \text{ digits to the right of decimal point} \end{array}$$

Classroom Tip

One way students can check their results in Example 2 is to use estimation. The product $3 \cdot 2.1$ is about $3 \cdot 2$, or 6. After multiplying as you would with whole numbers, the result is 63. Because the estimate is 6, you know that the decimal point must go after the 6.

This type of check can be used when dividing decimals as well.

EXAMPLE 2 Multiplying Decimals Using an Algorithm

Find each product.

a. $3 \cdot 2.1$

b. $2.5 \cdot 1.34$

SOLUTION

a.
$$\begin{array}{r} 2.1 \\ \times\ 3 \\ \hline 6.3 \end{array}$$

b.
$$\begin{array}{r} 1.34 \\ \times\ 2.5 \\ \hline 670 \\ 2680 \\ \hline 3.350 \end{array}$$

EXAMPLE 3 **Dividing Decimals Using Fractions**

Find (a) 2.4 ÷ 8 and (b) 7.5 ÷ 1.25.

SOLUTION

a. $2.4 \div 8 = 2\frac{4}{10} \div 8$

$= \frac{24}{10} \div 8$

$= \frac{\overset{3}{\cancel{24}}}{10} \cdot \frac{1}{\underset{1}{\cancel{8}}}$

$= \frac{3}{10}$

$= 0.3$

b. $7.5 \div 1.25 = 7\frac{5}{10} \div 1\frac{25}{100}$

$= 7\frac{1}{2} \div 1\frac{1}{4}$

$= \frac{15}{2} \div \frac{5}{4}$

$= \frac{\overset{3}{\cancel{15}}}{\underset{1}{\cancel{2}}} \cdot \frac{\overset{2}{\cancel{4}}}{\underset{1}{\cancel{5}}}$

$= 6$

Algorithm for Dividing Decimals

To divide decimals using long division, use the following steps.

1. Multiply the divisor and the dividend by the power of 10 necessary to make the divisor a whole number. Annex zeros to the dividend as needed.
2. Divide as with whole numbers.
3. Insert the decimal point in the quotient, above the decimal point in the dividend.

Example 9 ÷ 1.25

1. Multiply the divisor and the dividend by 100 annexing two zeros to the dividend.

   1.25.)9.00.

   This makes the divisor a whole number.

2. Divide as with whole numbers.

```
      7 2
125)900.0  ← Annex a zero.
    875
     25 0
     25 0
        0
```

3. Insert the decimal point in the quotient.

```
      7.2
125)900.0
    875
     25 0
     25 0
        0
```

Classroom Tip

In the long division algorithm at the right, multiplying the divisor and the dividend by the necessary power of 10 is equivalent to annexing zeros to the dividend, if needed, and "moving the decimal point" the necessary number of places (in both the divisor and the dividend) to make the divisor a whole number. Be sure your students understand that "moving the decimal point" to the right in the dividend often requires that they annex the appropriate number of zeros.

EXAMPLE 4 **Dividing Decimals Using an Algorithm**

Find (a) 2.4 ÷ 8, (b) 7.5 ÷ 1.25, and (c) 2.5875 ÷ 2.5.

SOLUTION

a.

```
   0.3
 8)2.4
 - 2.4
     0
```

b.

```
         6.
 1.25.)7.50.
     - 7 50
          0
```

c.

```
        1.035
 2.5.)2.5.875
    - 2.5
        0 8
      - 0
        87
      - 75
        125
      - 125
          0
```

Writing Fractions as Decimals

In Section 7.1 (on page 249), you learned that there are two common methods to write a fraction as a decimal.

1. Rewrite the fraction so that its denominator is a power of 10.
2. Divide the numerator by the denominator.

You have already studied the first method. The following example illustrates the second method.

EXAMPLE 5 Writing Fractions as Decimals

Write each fraction as a decimal. Tell whether the decimal terminates or repeats.

a. $\frac{3}{8}$ **b.** $\frac{2}{3}$

SOLUTION

a.
$$\begin{array}{r} 0.375 \\ 8\overline{)3.000} \\ -2\,4 \\ \hline 60 \\ -56 \\ \hline 40 \\ -40 \\ \hline 0 \end{array}$$

The decimal terminates.

So, $\frac{3}{8} = 0.375$.

b.
$$\begin{array}{r} 0.666\ldots \\ 3\overline{)2.000} \\ -1\,8 \\ \hline 20 \\ -18 \\ \hline 20 \\ -18 \\ \hline 2 \end{array}$$

The decimal repeats.

So, $\frac{2}{3} = 0.666\ldots$

Mathematical Practices

Look For and Express Regularity in Repeated Reasoning

MP8 Mathematically proficient students notice if calculations are repeated, and look for both general methods and for shortcuts.

In part (b) of Example 5, notice that no matter how long you continue the division process, the remainder will never be zero. Every remainder is 2. This type of quotient is called a *repeating decimal.*

Terminating Decimals and Repeating Decimals

Terminating Decimals A fraction, written in simplest form, can be written as a **terminating decimal** if and only if the prime factorization of the denominator contains no primes other than 2 or 5.

Repeating Decimals If the denominator of a fraction, written in simplest form, contains primes other than 2 or 5, then its decimal form is a *repeating decimal.* A **repeating decimal** is a type of nonterminating decimal where one or more digits repeat without end. A horizontal bar denotes a repeating digit or digits. This bar is placed above the first occurrence of repeating digits, as shown below.

Examples

$\frac{2}{3} = 0.666\ldots = 0.\overline{6}$

$\frac{1}{11} = 0.090909\ldots = 0.\overline{09}$

$\frac{2}{9} = 0.222\ldots = 0.\overline{2}$

$\frac{1}{7} = 0.142857142857\ldots = 0.\overline{142857}$

$\frac{1}{30} = 0.0333\ldots = 0.0\overline{3}$

$\frac{1}{99} = 0.010101\ldots = 0.\overline{01}$

Scientific Notation for Large Numbers

It is convenient to write very large numbers in *scientific notation*. This involves positive powers of 10. In Section 8.3, you will see that very small decimals can also be written in scientific notation using negative powers of 10.

Classroom Tip

If your students have access to calculators that display scientific notation, point out that most calculators do not display the power of 10. For instance, 1.35×10^7 may be displayed on a calculator as

1.35 E7.

This is also called *exponential notation*.

Scientific Notation for Large Numbers

A decimal is written in **scientific notation** when it is expressed in the form

$$a \times 10^n$$

where $1 \le a < 10$ and n is a whole number. The number a is called the **mantissa**. The number n is called the **characteristic**.

Example $13{,}500{,}000 = 1.35 \times 10{,}000{,}000$

$= 1.35 \times 10^7$ Scientific notation

Standard notation

EXAMPLE 6 Writing Numbers in Scientific Notation

Rewrite each statement using scientific notation.

a. The average distance from Earth to the Sun is about 150,000,000 kilometers. Astronomers call this distance 1 astronomical unit (or AU).

b. The distance from Earth to the nearest star, Proxima Centauri, is about 40,000,000,000,000 kilometers.

SOLUTION

a. $150{,}000{,}000 = 1.5 \times 10^8$ ← The exponent is 8.

Move decimal point 8 places to the left.

The average distance from Earth to the Sun is about 1.5×10^8 kilometers. Astronomers call this distance 1 astronomical unit (or AU).

b. $40{,}000{,}000{,}000{,}000 = 4.0 \times 10^{13}$ ← The exponent is 13.

Move decimal point 13 places to the left.

The distance from Earth to the nearest star, Proxima Centauri, is about 4.0×10^{13} kilometers.

EXAMPLE 7 Writing Numbers in Standard Form

Write each number in standard form.

a. 1.245×10^9 **b.** 3.06×10^5

SOLUTION

a. $1.245 \times 10^9 = 1{,}245{,}000{,}000$ Move decimal point 9 places to the right.

b. $3.06 \times 10^5 = 306{,}000$ Move decimal point 5 places to the right.

You can use the rules of exponents from Section 4.2 to help multiply and divide numbers written in scientific notation.

EXAMPLE 8 Dividing Numbers in Scientific Notation

Use the information from Example 6 to determine how many times greater the distance from Earth to Proxima Centauri is than the distance from Earth to the Sun.

SOLUTION

To find *how many times greater* the distance is from Earth to Proxima Centauri, divide the distances.

$$\frac{4.0 \times 10^{13}}{1.5 \times 10^{8}} = \frac{4.0}{1.5} \times \frac{10^{13}}{10^{8}}$$ Rewrite as a product of fractions.

$$\approx 2.67 \times \frac{10^{13}}{10^{8}}$$ Divide.

$$= 2.67 \times 10^{5}$$ Quotient of Powers Property

$$= 267{,}000$$ Write the number in standard form.

The distance from Earth to Proxima Centauri is about 267,000 times greater than the distance from Earth to the Sun.

EXAMPLE 9 Multiplying Numbers in Scientific Notation

There are about 1.2×10^{8} households in the United States. The mean income per household is about $\$7.0 \times 10^{4}$. What is the total household income in the United States? Write your answer in standard notation.

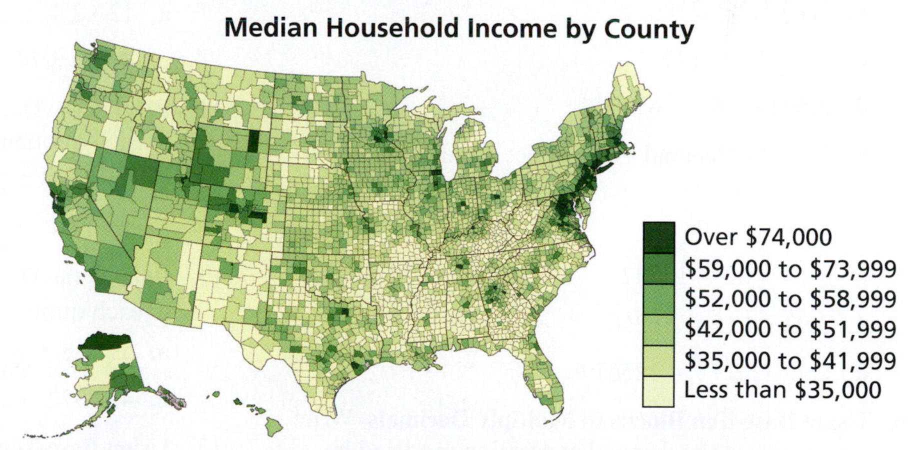

"Median Household Income by County" adapted from "US county household median income 2009" by Vikjam available at http://commons.wikimedia.org/wiki/File:US_county_household_median_income_2009.png#file. CC BY-SA 3.0.

SOLUTION

Multiply the number of households by the mean income per household.

$$(1.2 \times 10^{8})(7.0 \times 10^{4}) = (1.2)(7.0) \times 10^{12}$$

$$= 8.4 \times 10^{12}$$

$$= 8{,}400{,}000{,}000{,}000$$

The total household income in the United States is about $8.4 trillion.

Classroom Tip

When you teach your students to multiply and divide numbers written in scientific notation, remind them of the properties for multiplying and dividing powers with the same base. For instance, in Example 9, notice how the property for multiplying works.

$$10^{8} \times 10^{4} = 10^{8+4}$$

$$= 10^{12}$$

7.3 Exercises

Free step-by-step solutions to odd-numbered exercises at MathematicalPractices.com.

1. **Writing a Solution Key** Write a solution key for the activity on page 266. Describe a rubric for grading a student's work.

2. **Grading the Activity** In the activity on page 266, a student gave the answer below. The question is worth 10 points. Assign a grade to the answer. Why do you think the student erred?

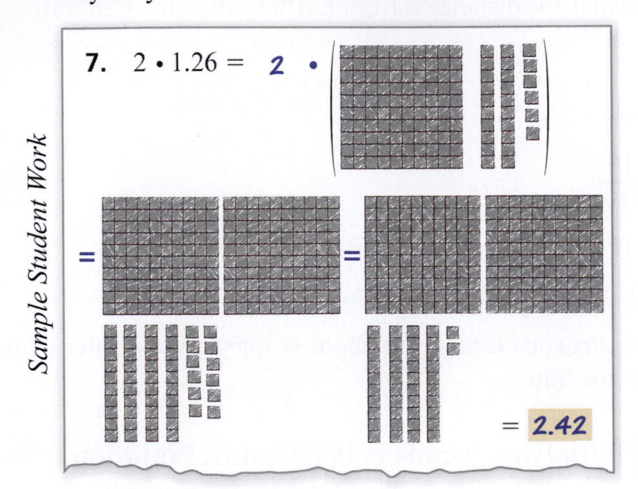

3. **Placing the Decimal Point** Place the decimal point in each product.
   - **a.** 6.9 • 3 = 207
   - **b.** 1.4 • 1.7 = 238
   - **c.** 7.5 • 2.6 = 195
   - **d.** 2.894 • 5.6 = 162064

4. **Placing the Decimal Point** Place the decimal point in each product.
   - **a.** 4 • 2.7 = 108
   - **b.** 3.1 • 4.62 = 14322
   - **c.** 3.92 • 4.15 = 16268
   - **d.** 2.43 • 5.172 = 1256796

5. **Using Base-Ten Blocks to Multiply Decimals** Write and evaluate the decimal expression modeled by the base-ten blocks. One small cube represents one hundredth.

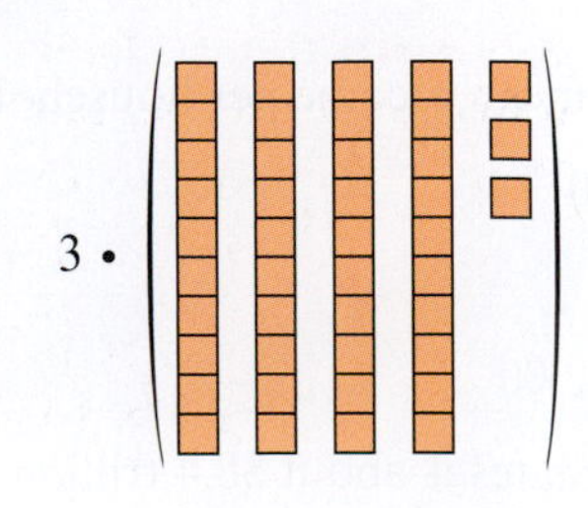

6. **Using Base-Ten Blocks to Multiply Decimals** Write and evaluate the decimal expression modeled by the base-ten blocks. One small cube represents one hundredth.

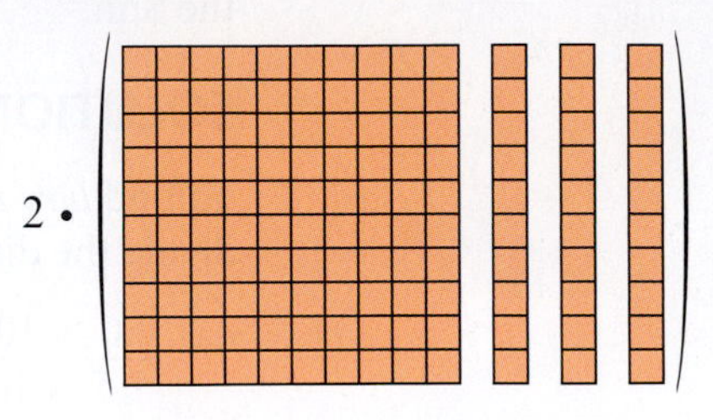

7. **Multiplying Decimals Using Base-Ten Blocks** Use base-ten blocks to find each product. Sketch the model.
   - **a.** 3 • 2.2
   - **b.** 5 • 0.97

8. **Multiplying Decimals Using Base-Ten Blocks** Use base-ten blocks to find each product. Sketch the model.
   - **a.** 2 • 1.46
   - **b.** 2.8 • 4

9. **Multiplying Decimals** Find each product.
   - **a.** 1.4 • 9
   - **b.** 4 • 6.25
   - **c.** 4.7 • 3.32
   - **d.** 5.38 • 1.51

10. **Multiplying Decimals** Find each product.
    - **a.** 12 • 2.4
    - **b.** 6.7 • 3.4
    - **c.** 4.15 • 8.78
    - **d.** 3.6 • 2.529

11. **Placing the Decimal Point** Place the decimal point in each quotient.
    - **a.** 27.5 ÷ 4.4 = 625
    - **b.** 2.96 ÷ 2 = 148
    - **c.** 16.6791 ÷ 5.3 = 3147

12. **Placing the Decimal Point** Place the decimal point in each quotient.
    - **a.** 59.888 ÷ 8 = 7486
    - **b.** 20.64 ÷ 9.6 = 215
    - **c.** 42.6972 ÷ 2.73 = 1564

13. **Using Base-Ten Blocks to Divide Decimals** Write and evaluate the decimal expression modeled by the base-ten blocks. One small cube represents one hundredth.

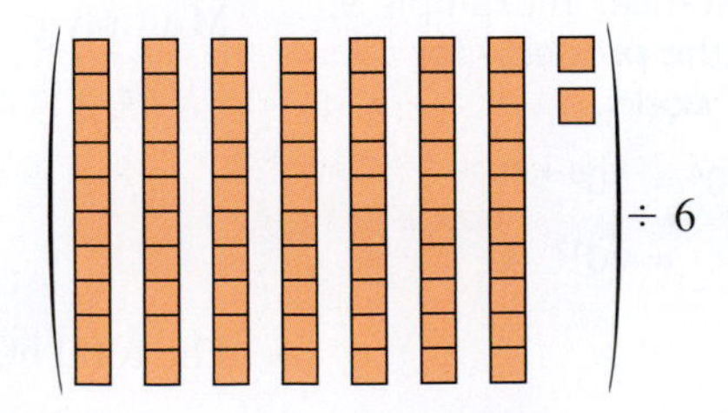

14. Using Base-Ten Blocks to Divide Decimals Write and evaluate the decimal expression modeled by the base-ten blocks. One small cube represents one hundredth.

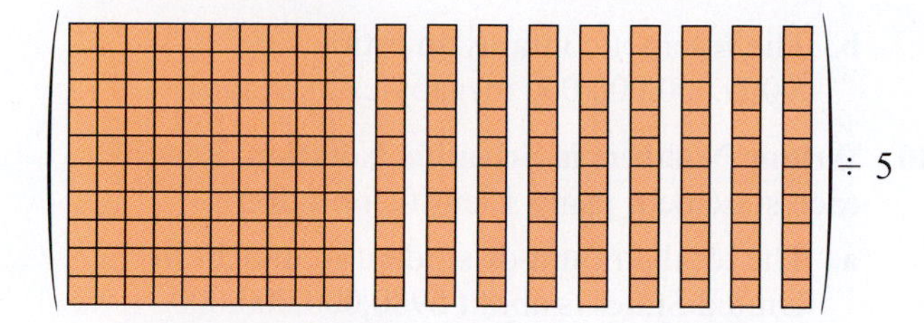

15. Dividing Decimals Using Base-Ten Blocks Use base ten blocks to find each quotient. Sketch the model.

a. $3.6 \div 3$ **b.** $1.76 \div 4$

16. Dividing Decimals Using Base-Ten Blocks Use base ten blocks to find each quotient. Sketch the model.

a. $2.7 \div 3$ **b.** $4.18 \div 2$

17. Dividing Decimals Find each quotient.

a. $3.6 \div 0.4$ **b.** $13.4 \div 2$

c. $10.8 \div 2.25$ **d.** $4.125 \div 1.5$

18. Dividing Decimals Find each quotient.

a. $17.4 \div 12$ **b.** $31.36 \div 7.84$

c. $7.9625 \div 4.9$ **d.** $21.2 \div 6.625$

19. Terminating and Repeating Decimals Decide whether each decimal is *terminating*, *repeating*, or *neither*.

a. 1.888. . . **b.** 3.75

c. 1.414213. . . **d.** $0.41\overline{6}$

20. Terminating and Repeating Decimals Decide whether each decimal is *terminating*, *repeating*, or *neither*.

a. 9.24 **b.** 3.14159. . .

c. $1.\overline{27}$ **d.** 0.1666

21. Writing Fractions as Decimals Write each fraction as a decimal. If the decimal repeats, then write the decimal using a horizontal bar over the first occurrence of the repeating digits.

a. $\frac{7}{8}$ **b.** $\frac{5}{9}$ **c.** $\frac{1}{25}$ **d.** $\frac{4}{15}$

22. Writing Fractions as Decimals Write each fraction as a decimal. If the decimal repeats, then write the decimal using a horizontal bar over the first occurrence of the repeating digits.

a. $\frac{1}{6}$ **b.** $\frac{3}{16}$ **c.** $\frac{7}{12}$ **d.** $\frac{47}{99}$

23. Identifying Scientific Notation Decide whether each number is written in scientific notation. If not, explain why and write it in scientific notation.

a. 3.9×10^5 **b.** 26.5×10^8

c. 1.98×20^2 **d.** 4.0876×10^2

24. Identifying Scientific Notation Decide whether each number is written in scientific notation. If not, explain why and write it in scientific notation.

a. 0.72×10^{14} **b.** 9.125×10^7

c. 8.1×100^3 **d.** 1.03×10^{15}

25. Scientific Notation Identify the mantissa and characteristic. Then write the number in standard form.

a. 1.5×10^7 **b.** 3.28×10^2

c. 9.96×10^6 **d.** 8.875×10^{10}

26. Scientific Notation Identify the mantissa and characteristic. Then write the number in standard form.

a. 5.6×10^3 **b.** 2.25×10^8

c. 4.73×10^1 **d.** 1.922×10^{12}

27. Writing Numbers in Scientific Notation Write each number in scientific notation.

a. 46 **b.** 9500

c. 204,250 **d.** 68,000,000,000

28. Writing Numbers in Scientific Notation Write each number in scientific notation.

a. 539 **b.** 60,130

c. 76,815,000 **d.** 120,040,000,000

29. Multiplying Numbers in Scientific Notation Evaluate each expression. Write your answer in standard form.

a. $(4.6 \times 10^3)(2.0 \times 10^4)$

b. $(7.5 \times 10^2)(5.4 \times 10^8)$

30. Multiplying Numbers in Scientific Notation Evaluate each expression. Write your answer in standard form.

a. $(8.25 \times 10^3)(3.8 \times 10^9)$

b. $(6.25 \times 10^5)^2$

31. Dividing Numbers in Scientific Notation Evaluate each expression. Write your answer in standard form.

a. $\dfrac{6.0 \times 10^{10}}{2.0 \times 10^8}$ **b.** $\dfrac{9.0 \times 10^9}{1.5 \times 10^3}$

c. $\dfrac{1.6 \times 10^5}{4.0 \times 10^2}$ **d.** $\dfrac{1.12 \times 10^6}{3.2 \times 10^5}$

32. Dividing Numbers in Scientific Notation Evaluate each expression. Write your answer in standard form.

a. $\dfrac{9.0 \times 10^{13}}{4.5 \times 10^{10}}$

b. $\dfrac{7.5 \times 10^{6}}{1.25 \times 10^{6}}$

c. $\dfrac{5.44 \times 10^{13}}{6.4 \times 10^{7}}$

d. $\dfrac{2.73 \times 10^{9}}{1.625 \times 10^{4}}$

33. My Dear Aunt Sally Arrange the numbers 0.6, 0.4, 0.9, and 0.6 so that the equation is true.

For more practice with this type of reasoning, go to *AuntSally.com*.

34. My Dear Aunt Sally Arrange the numbers 0.5, 2.8, 3.5, and 2.5 so that the equation is true.

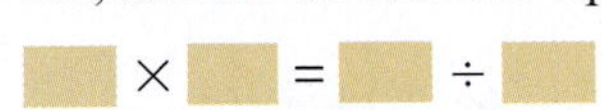

For more practice with this type of reasoning, go to *AuntSally.com*.

35. Average Cost An electronics store has a sale offering one free Blu-ray Disc™ when you buy two Blu-ray Discs for $14.97 each. What is the average cost per Blu-ray Disc?

36. Cost per Person You and four friends take a trip to an amusement park. Tickets for the park cost $64.35 each and one ticket is free. Hotel accommodations cost $96.90. You and your friends split the total cost of the amusement park passes and the hotel evenly. What is the cost per person?

37. Writing a Repeating Decimal You use a calculator to divide two integers and obtain the answer below. Write the quotient by placing a horizontal bar over the first occurrence of the repeating digits.

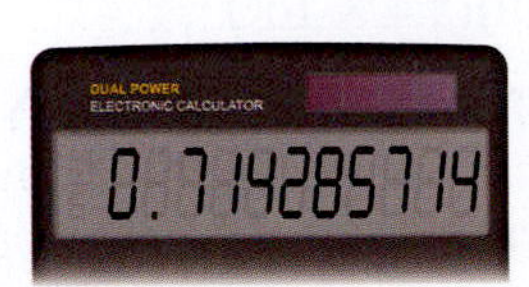

38. Writing a Repeating Decimal You use a calculator to divide two integers and obtain the answer below. Write the quotient by placing a horizontal bar over the first occurrence of the repeating digits.

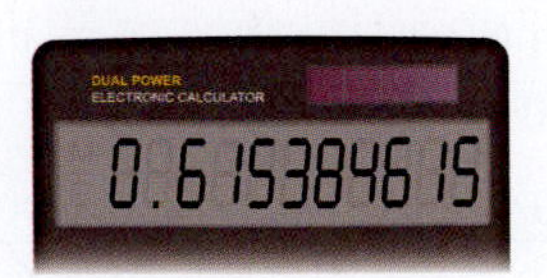

39. Writing Numbers in Scientific Notation Rewrite each statement using scientific notation.

a. During one month, Google had approximately 175,000,000 unique visitors.

b. The Internet contains an estimated 5,000,000,000,000 megabytes of data.

40. Writing Numbers in Scientific Notation Rewrite each statement using scientific notation.

a. The total amount of student loan debt in the United States is about $980,000,000,000.

b. The total amount of personal debt in the United States is about $15,999,000,000,000.

41. Writing A student claims that 3.24×50 has the same answer as 324×0.5. Is the student correct? Explain.

42. Writing A student claims that 80.4×25 has the same answer as 804×0.25. Is the student correct? Explain.

43. Social Networking In one month, the total amount of time spent on Facebook by 1.53×10^{8} unique visitors was 1.0251×10^{9} hours. What was the average amount of time spent on the website per person?

44. National Debt The population of the United States is about 3.16×10^{8}, and the amount of U.S. debt per person is about $\$1.88 \times 10^{5}$. What is the total U.S. national debt? Write your answer in standard form.

45. Comparing Techniques Find the product $1.8 \cdot 3.9$ by (a) rewriting the decimals as fractions and (b) using the algorithm for multiplying decimals. (c) Which technique do you prefer? Explain your reasoning.

46. Comparing Techniques Find the quotient $6.825 \div 5.25$ by (a) rewriting the decimals as fractions and (b) using the algorithm for dividing decimals. (c) Which technique do you prefer? Explain your reasoning.

47. Grading Student Work On a diagnostic test, one of your students does the following work. Explain what the student did wrong. Which topics would you encourage the student to review?

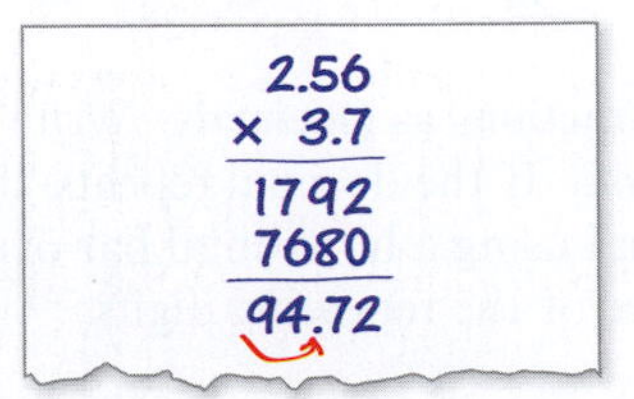

48. **Grading Student Work** On a diagnostic test, one of your students does the following work. Explain what the student did wrong. Which topics would you encourage the student to review?

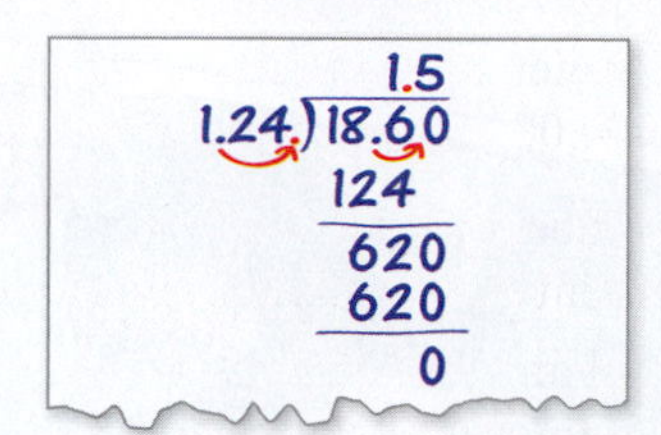

49. **Commutative Property** Use the Commutative Property and mental math to find the product $2.5 \times 7.9 \times 4$. Explain your reasoning.

50. **Associative Property** Use the Associative Property and mental math to find the product $40 \times (1.5 \times 0.75)$. Explain your reasoning.

51. **The Distributive Property: In Your Classroom** A gas station sells gasoline for $3.55 per gallon.

a. Use the algorithm for multiplying decimals to find the cost of 12 gallons of gasoline.

b. Write the cost of gasoline as $3 + 0.5 + 0.05$ dollars per gallon. Then use mental math to find the cost of 12 gallons.

c. Design an activity where students use mental math to find the cost of buying a large quantity of an item. Explain to your students how the Distributive Property can be used to make the computation easier.

52. **The Distributive Property: In Your Classroom** A meal at a fast food restaurant costs $9.38.

a. Use the algorithm for multiplying decimals to find the cost of buying four of the meals.

b. Write the cost of the meal as the sum of three terms. Then use mental math to find the cost of four meals.

c. Design an activity where students use mental math to find the cost of buying meals for a group. Explain to your students how the Distributive Property can be used to make the computation easier.

53. **Area** Find the area of the cooking grid.

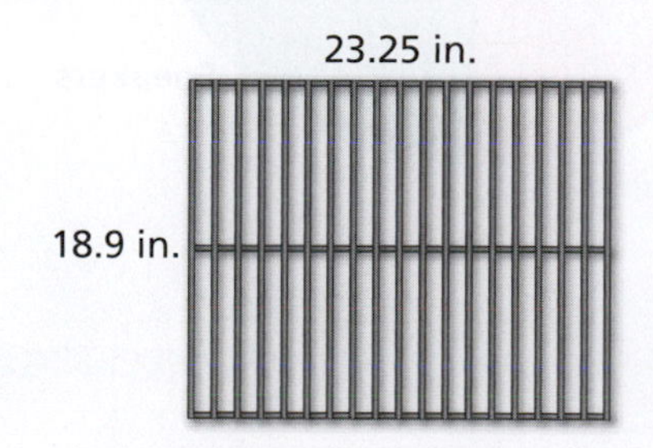

54. **Perimeter** Find the perimeter of the picture frame.

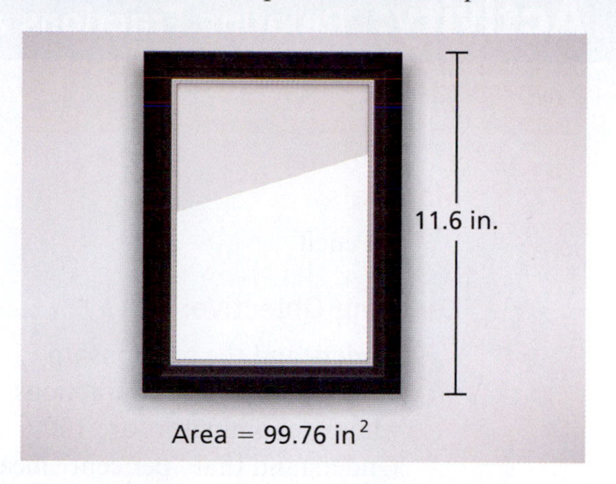

55. **My Dear Aunt Sally** Arrange the numbers 0.7, 2.5, 0.5, 1.5, 0.6, and 1.4 so that the equation is true.

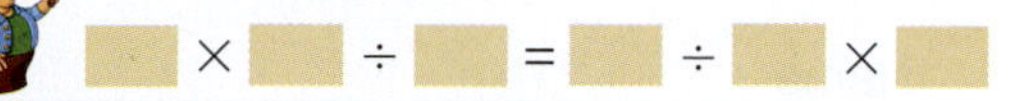

For more practice with this type of reasoning, go to *AuntSally.com*.

56. **My Dear Aunt Sally** Arrange the numbers 0.5, 1.8, 2.5, 0.6, 4.0, and 0.6 so that the equation is true.

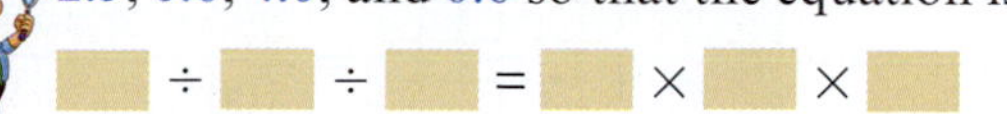

For more practice with this type of reasoning, go to *AuntSally.com*.

57. **Reasoning** What is the greatest possible product you can write by filling in the spaces below with the digits 1, 2, and 3? What is the least possible product?

0.__ __ × 0.__

58. **Reasoning** What is the greatest possible quotient you can write by filling in the spaces below with the digits 1, 2, and 3? What is the least possible quotient?

__ ÷ __.__

59. **Reasoning** Two numbers written in scientific notation are multiplied, as shown below. Describe the range of possible values of ab. Explain your reasoning.

$$(a \times 10^m)(b \times 10^n) = ab \times 10^{m+n}$$

60. **Reasoning** Two numbers written in scientific notation are divided, as shown below. Describe the range of possible values of $\frac{a}{b}$. Explain your reasoning.

$$\frac{(a \times 10^m)}{(b \times 10^n)} = \frac{a}{b} \times 10^{m-n}$$

Activity: Relating Fractions and Percents

Materials:
- Pencil

Learning Objective:

Understand the relationship between percents and fractions whose denominators are 100. Understand that "per cent" means "per 100."

Name ______________________________

Represent the shaded area as a fraction and as a percent.

1. 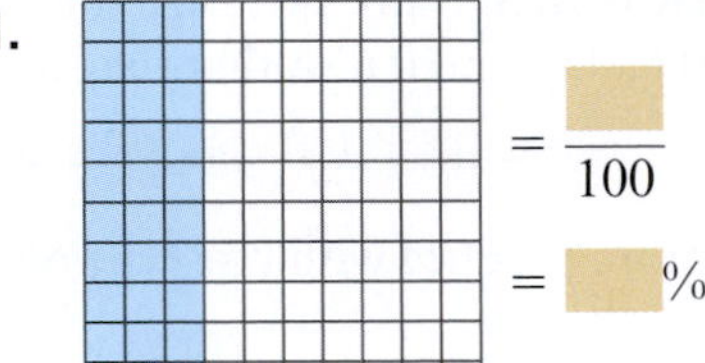

2. 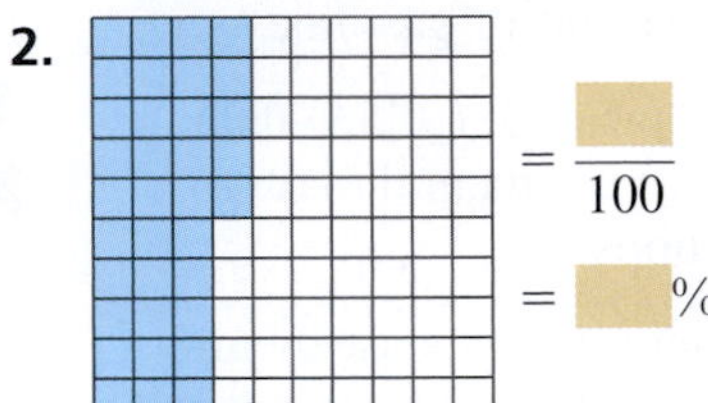

3. 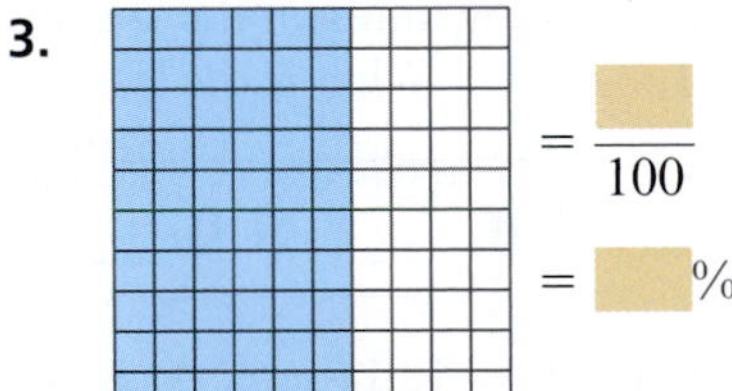

4. 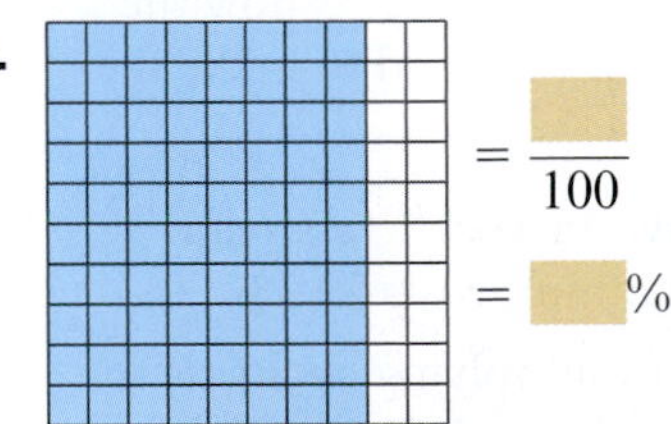

5. **Footwear** One hundred people were asked to name their favorite type of shoe. Estimate the percentage for each category.

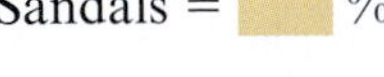

Sandals = ☐ %

Boots= ☐ %

Slippers = ☐ %

Sneakers= ☐ %

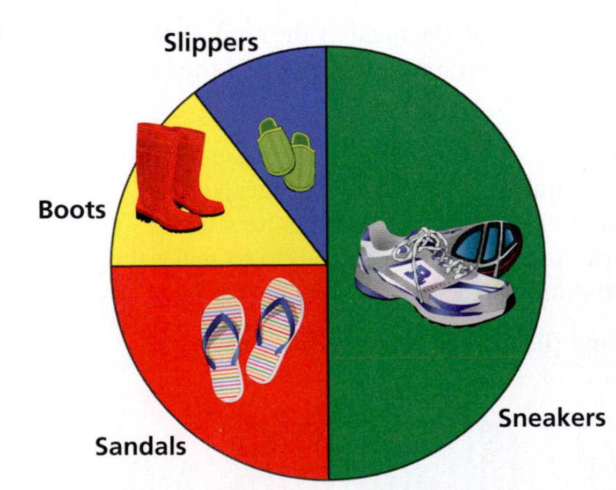

Lisa F. Young/Shutterstock.com

7.4 Percents

Standards

Grades 3–5
Numbers and Operations in Base Ten
Students should perform operations with multi-digit whole numbers and with decimals through hundredths.

Grades 6–8
Ratios and Proportional Relationships
Students should understand ratio concepts and use ratio reasoning to solve problems, and analyze proportional relationships and use them to solve real-world and mathematical problems.

- Convert between fractions, decimals, and percents.
- Solve percent problems using a model.
- Solve percent problems using the percent equation.

Percents

In Latin, "centum" means "one hundred." So, "per cent" means "per one hundred." You can see the use of the word cent in *centurion* (commander of 100 soldiers) and in *century* (100 years).

Definition of Percent

A **percent** is a part-to-whole ratio where the whole is 100. So, a percent can be written as a fraction with a denominator of 100. Percents are denoted by the **percent symbol %**.

Example $30\% = 30 \text{ out of } 100 = \frac{30}{100}$

30 — Per — Cent

There are two common ways to write fractions and decimals as percents.

1. For fractions, write the equivalent fraction with a denominator of 100. Then write the numerator with a percent symbol.
2. For decimals, multiply by 100. Then add a percent symbol.

Classroom Tip

The strategies shown in Examples 1 and 2 emphasize the understanding of the definition of percent as "per one hundred." Once your students understand this concept, you can also teach them the common shortcuts for converting between decimals and percents.

Decimal to percent

Move decimal point 2 places to the right.

0.35. = 35%

This is equivalent to multiplying by 100.

Percent to decimal

Move decimal point 2 places to the left.

45% = 0.45.

This is equivalent to dividing by 100.

EXAMPLE 1 Writing Fractions and Decimals as Percents

a. $\frac{1}{2} = \frac{50}{100} = 50\%$

b. $\frac{1}{4} = \frac{25}{100} = 25\%$

c. $\frac{5}{4} = \frac{125}{100} = 125\%$

d. $0.35 = 0.35 \cdot 100\% = 35\%$ Think $0.35 \cdot \frac{100}{100} = \frac{35}{100} = 35\%$.

e. $0.4 = 0.4 \cdot 100\% = 40\%$ Think $0.4 \cdot \frac{100}{100} = \frac{40}{100} = 40\%$.

f. $2.5 = 2.5 \cdot 100\% = 250\%$ Think $2.5 \cdot \frac{100}{100} = \frac{250}{100} = 250\%$.

EXAMPLE 2 Writing Percents as Fractions or Decimals

a. $28\% = \frac{28}{100} = \frac{7}{25}$

b. $150\% = \frac{150}{100} = \frac{3}{2}$

c. $45\% = 0.45$ Think $45\% = \frac{45}{100} = 0.45$.

d. $137\% = 1.37$ Think $137\% = \frac{137}{100} = 1.37$.

Finding the Percent of a Number

Finding the Percent of a Number Using a Model

Use a 10-by-10 grid to find the percent of a number.

1. Draw a 10-by-10 grid to represent the number. Divide the number by 100 to determine how much each square in the grid represents.
2. Shade the appropriate number of squares to represent the percent.
3. Multiply the number of shaded squares by how much each square represents.

Example Find 7% of \$2.

1. \$2 ÷ 100 = \$0.02
   So, □ = \$0.02.
2. Shade 7% of the grid.
3. 7 • \$0.02 = \$0.14

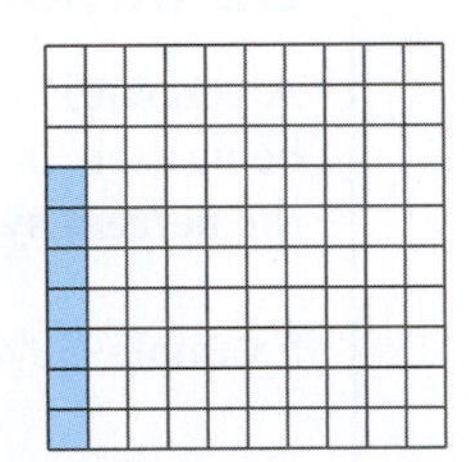

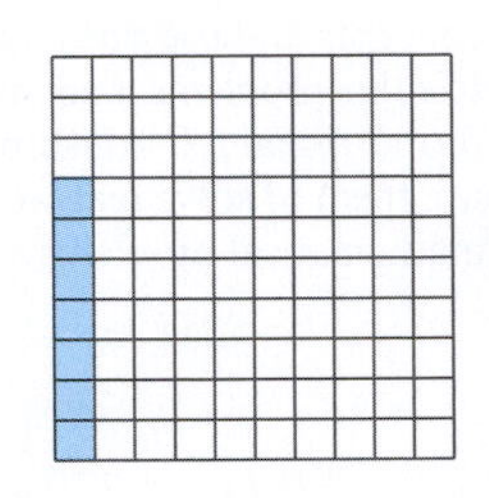

Classroom Tip

To help your students understand the 10-by-10 grid, point out that each square in the grid represents one hundredth of the number. For instance, each square in the grid at the right represents one hundredth of \$2.

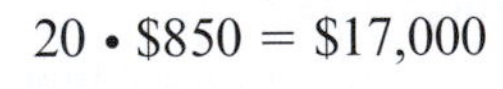

EXAMPLE 3 **Finding the Percent of a Number**

A couple is buying a house that costs \$85,000. They make a down payment that is 20% of the cost of the house. How much is the down payment?

SOLUTION

1. **Understand the Problem** You need to find the amount of the down payment, which is 20% of the cost of the house.
2. **Make a Plan** Use a 10-by-10 grid to help find 20% of \$85,000.
3. **Solve the Problem**

   \$85,000 ÷ 100 = \$850
   So, □ = \$850.

   Shade 20% of the grid.

   20 • \$850 = \$17,000

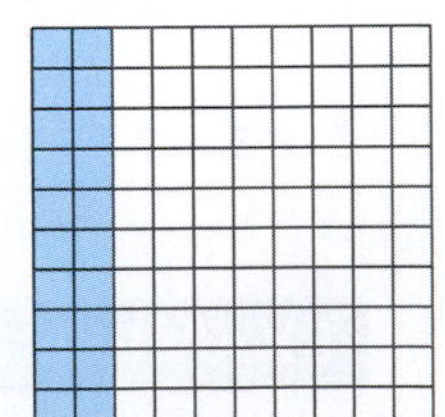

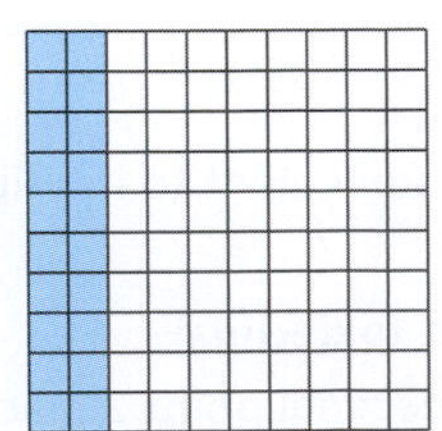

   The down payment is \$17,000.

4. **Look Back** A nice way to check a percent problem is to compare the answer to a common percent that is easy to work with, such as 10%. 10% of 85,000 is one tenth of \$85,000, which is

$$\frac{1}{10} \cdot \$85{,}000 = \$8500.$$

20% is twice this amount. So, the down payment is 2 • \$8500 = \$17,000. 

Classroom Tip

Finding 20% of a number is a common application because it is a typical amount of a tip. A shortcut for finding 20% of a number using mental math is shown below.

1. Move the decimal point one place to the left.
2. Multiply by 2.

For instance, 20% of 135 is

13.5 • 2 = 27.

Solving Percent Problems

Classroom Tip

The three types of percent problems at the right are listed in order of increasing difficulty for most students. The first type is the most straightforward. To find p percent of b, write p as a decimal and then multiply. The third type is the most difficult. Often, it is not studied until students are in an algebra class.

Using the Percent Equation

Let a and b be numbers and let p be a percent. The statement "a is p percent of b" is summarized by the following **percent equation**.

$$a = p \cdot b$$

(a: Part of the whole; p: Percent; b: Whole)

There are 3 common types of percent problems. Each involves the percent equation.

	Given	Find	Question	Strategy
1.	p and b	a	What is p percent of b?	Write p in decimal form. Then multiply b by p.
	Example		What is 12% of \$30?	$0.12 \cdot \$30 = \$3.60$
2.	a and b	p	a is what percent of b?	Divide a by b. Then rewrite the quotient as a percent.
	Example		6 is what percent of 8?	$\frac{6}{8} = 0.75 = 75\%$
3.	a and p	b	a is p percent of what?	Write p in decimal form. Then divide a by p.
	Example		5 is 25% of what?	$\frac{5}{0.25} = 20$

EXAMPLE 4 Finding the Percent of a Number

Classroom Tip

Notice that, although the method is different, the answer is the same as in Example 3.

Redo Example 3 using the percent equation.

SOLUTION

To find 20% of \$85,000, multiply \$85,000 by 0.2.

$$20\% \text{ of } \$85{,}000 = 0.2 \cdot \$85{,}000 \qquad a = p \cdot b$$
$$= \$17{,}000$$

EXAMPLE 5 Finding the Percent of a Number

According to the circle graph, how many hours does a typical teenager sleep each day?

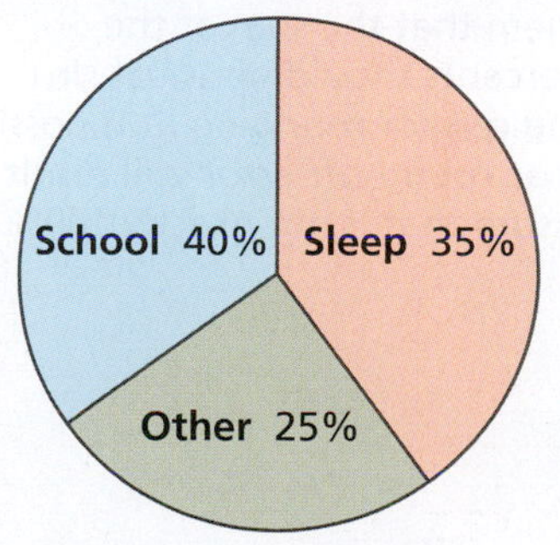

SOLUTION

A typical teenager sleeps 35% of a 24-hour day.

$$35\% \text{ of } 24 = 0.35 \cdot 24 \qquad a = p \cdot b$$
$$= 8.4$$

A typical teenager sleeps 8.4 hours each day. In other words, a typical teenager sleeps about eight and a half hours each day.

EXAMPLE 6 Finding a Percent p

You conduct a survey asking students in your class about their favorite type of pet. You use a spreadsheet program to create a circle graph showing the results. Write the percent for each of the six categories in the circle graph. Round each percent to the nearest whole percent.

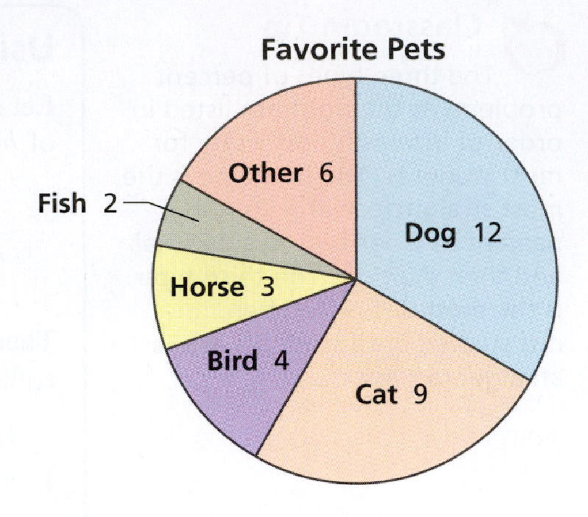

SOLUTION

To begin, add the numbers of responses in the circle graph to find the number of students surveyed.

$$12 + 9 + 4 + 3 + 2 + 6 = 36 \text{ students}$$

To find each percent, divide the number in each category by 36.

Dog $\frac{12}{36} = 0.333\ldots \approx 33\%$

Cat $\frac{9}{36} = 0.25 = 25\%$

Bird $\frac{4}{36} = 0.111\ldots \approx 11\%$

Horse $\frac{3}{36} = 0.08333\ldots \approx 8\%$

Fish $\frac{2}{36} = 0.0555\ldots \approx 6\%$

Other $\frac{6}{36} = 0.1666\ldots \approx 17\%$

You can write these percents for the six categories in the circle graph.

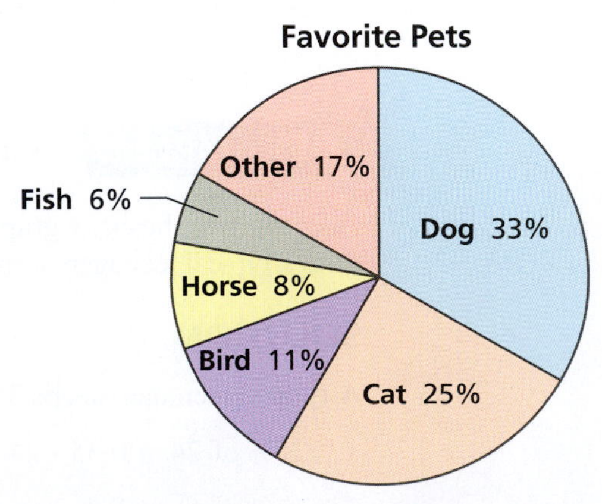

One way you can check your circle graph is to find the sum of the percents and see whether it equals 100%.

$$33\% + 25\% + 11\% + 8\% + 6\% + 17\% = 100\%$$ 

Mathematical Practices

Attend to Precision

MP6 Mathematically proficient students calculate accurately and efficiently and express numerical answers with a degree of precision appropriate for the problem context.

Classroom Tip

When your students are creating circle graphs, remind them that the sum of the percents should be 100%. In the case of rounding, it is possible that *round-off error* will result in a sum that is not exactly 100%.

EXAMPLE 7 Finding b in the Percent Equation

Classroom Tip
As shown in Examples 7 and 8, it is a good idea to teach students to use algebra to solve the third type of percent problem described on page 279. This promotes understanding and eliminates the need for memorizing a strategy. With this approach, they only need to remember the percent equation $a = p \cdot b$.

Your friend works in San Francisco. She pays a total of 45% in income taxes, which amounts to $54,000. What is her total taxable income?

SOLUTION

1. **Understand the Problem** You need to find your friend's total taxable income.
2. **Make a Plan** Using the percent equation $a = p \cdot b$, you are given a and p, and need to find b. This is the third type of percent problem. Rather than trying to remember the strategy described on page 279, you decide to use algebra.
3. **Solve the Problem**

$a = p \cdot b$	Write the percent equation.
$54{,}000 = 0.45b$	Substitute 54,000 for a and 0.45 for p.
$\frac{54{,}000}{0.45} = \frac{0.45b}{0.45}$	Divide each side by 0.45.
$\frac{54{,}000}{0.45} = b$	Simplify.
$120{,}000 = b$	Divide.

Your friend's total taxable income is $120,000.

4. **Look Back** You can use estimation to check the reasonableness of your answer. If your friend paid 50% of $120,000, then she would have paid

$$50\% \text{ of } \$120{,}000 = \frac{1}{2} \cdot \$120{,}000$$
$$= \$60{,}000.$$

Because $\$54{,}000 \approx \$60{,}000$, you know that your answer is reasonable. ✓

EXAMPLE 8 Finding b in the Percent Equation

The top 1% of income earners in the United States pay a total of about $355 billion in federal income taxes each year. This represents about 37% of all the federal income tax that is paid each year. Estimate the total federal income tax that is paid each year in the United States.

SOLUTION

The question you need to answer is "$355 billion is 37% of what?"

$a = p \cdot b$	Write the percent equation.
$355 = 0.37b$	Substitute 355 for a and 0.37 for p.
$\frac{355}{0.37} = \frac{0.37b}{0.37}$	Divide each side by 0.37.
$\frac{355}{0.37} = b$	Simplify.
$959 \approx b$	Divide.

The total federal income tax paid by income earners each year is about $959 billion.

7.4 Exercises

Free step-by-step solutions to odd-numbered exercises at MathematicalPractices.com.

1. **Writing a Solution Key** Write a solution key for the activity on page 276. Describe a rubric for grading a student's work.

2. **Grading the Activity** In the activity on page 276, a student gave the answers below. Each question is worth 10 points. Assign a grade to each answer. For those that are incorrect, why do you think the student erred?

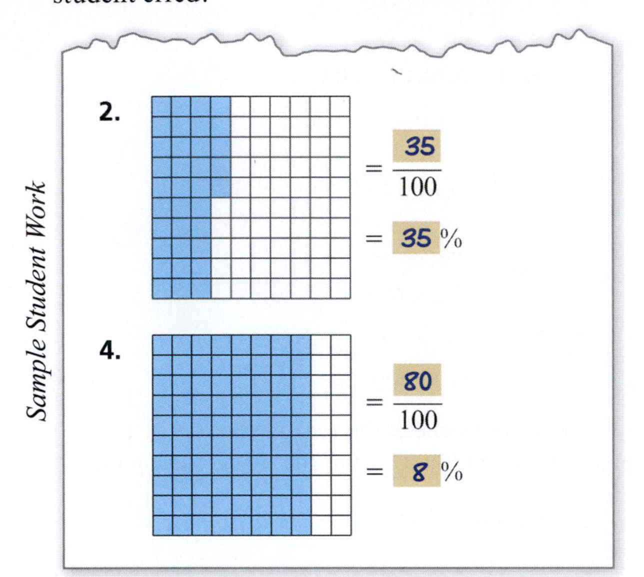

3. **Using Models to Represent Percents** Shade each grid to represent the given percent. Then write the percent as a decimal.

   a. 40% b. 73%

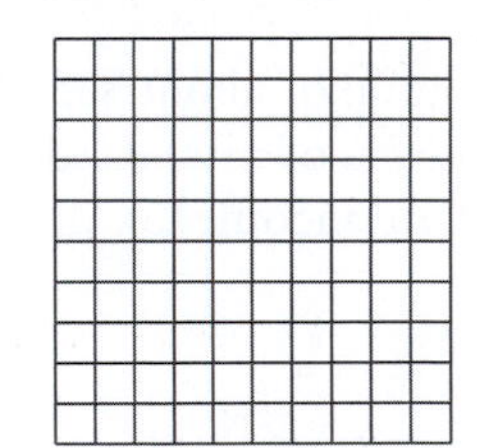
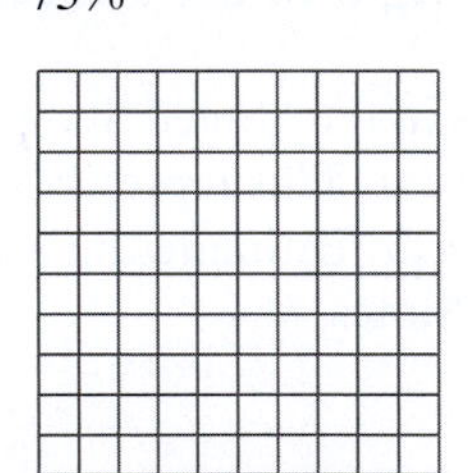

4. **Using Models to Represent Percents** Shade each grid to represent the given percent. Then write the percent as a decimal.

   a. 90% b. 26%

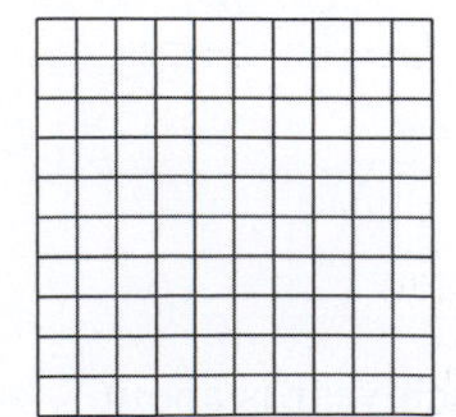
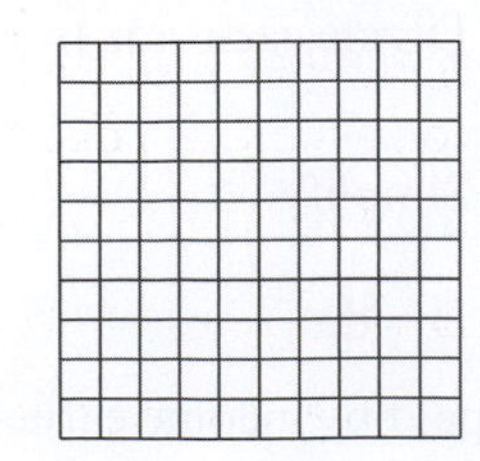

5. **Writing Fractions and Decimals as Percents** Write each fraction or decimal as a percent.

   a. $\frac{7}{50}$ b. 0.92

   c. $\frac{13}{10}$ d. 0.00004

6. **Writing Fractions and Decimals as Percents** Write each fraction or decimal as a percent.

   a. $\frac{13}{20}$ b. 0.27

   c. $\frac{37}{25}$ d. 0.0008

7. **Writing Percents as Fractions or Decimals** Write each percent as a decimal and as a fraction in simplest form.

   a. 30% b. 6% c. 224% d. 4.5%

8. **Writing Percents as Fractions or Decimals** Write each percent as a decimal and as a fraction in simplest form.

   a. 15% b. 7% c. 176% d. 18.4%

9. **Converting Fractions, Decimals, and Percents** Complete the table.

Fraction	Decimal	Percent
	0.8	
1/4		
		38%
	0.005	
		1%

10. **Converting Fractions, Decimals, and Percents** Complete the table.

Fraction	Decimal	Percent
	0.75	
2/5		
		84%
	0.003	
		22.5%

11. Comparing Fractions, Decimals, and Percents Order the numbers from least to greatest.

a. $\frac{7}{16}$, 0.45, 43.6%, $\frac{21}{50}$

b. 1.23, $\frac{31}{25}$, 125%, 1.225

12. Comparing Fractions, Decimals, and Percents Order the numbers from least to greatest.

a. 13.55%, 0.136, $\frac{33}{250}$, 12.9%

b. $\frac{77}{20}$, 382.5%, $\frac{191}{50}$, 3.86

13. Finding the Percent of a Number Write the percent represented by the shaded portion of each grid. Then find that percent of the given number.

a. 34,000 **b.** 2600

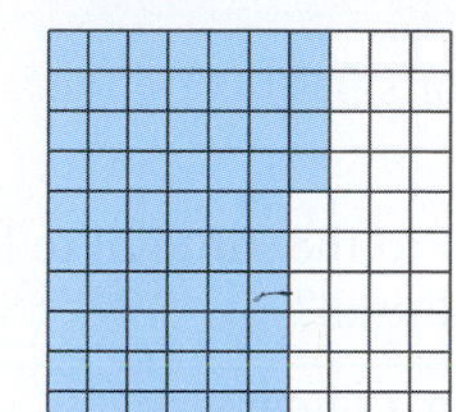

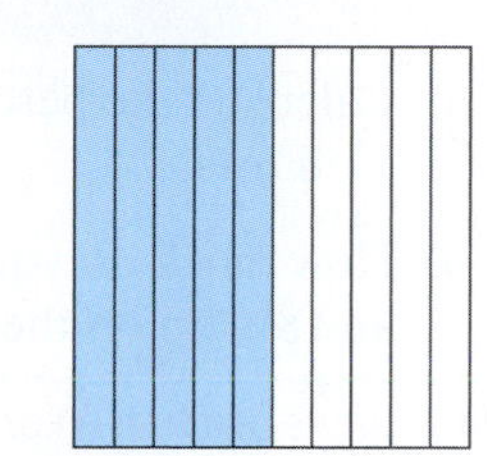

14. Finding the Percent of a Number Write the percent represented by the shaded portion of each grid. Then find that percent of the given number.

a. 62,800 **b.** 5420

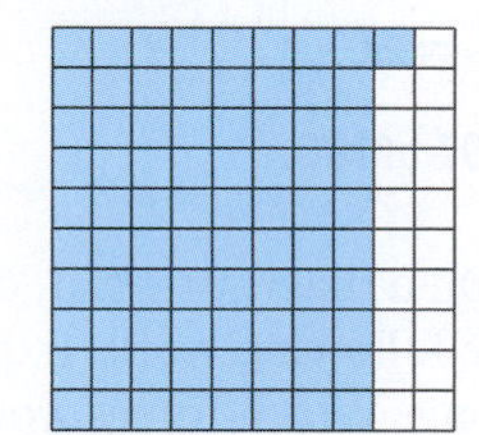

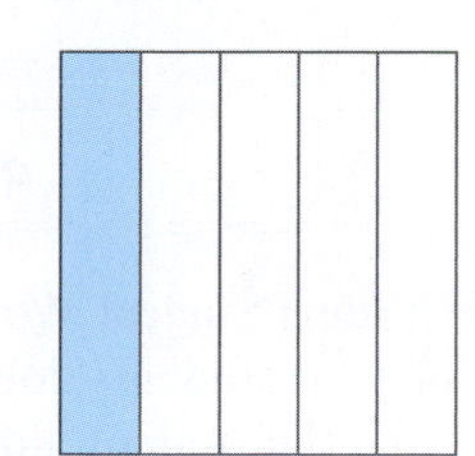

15. Finding the Percent of a Number Use a grid to find the percent of the number.

a. 14% of 50 **b.** 85% of 840

c. 45% of 91,000 **d.** 120% of 7160

16. Finding the Percent of a Number Use a grid to find the percent of the number.

a. 55% of 80 **b.** 6% of 6200

c. 32% of 104,500 **d.** 160% of 5060

17. Solving Percent Problems Match each question with the expression that would be used to answer the question.

a. What is 20% of 160? **i.** 20 ÷ 160

b. 20 is what percent of 160? **ii.** 160 ÷ 0.20

c. 160 is 20% of what? **iii.** 0.20 × 160

18. Solving Percent Problems Match each question with the expression that would be used to answer the question.

a. 280 is 140% of what? **i.** 1.40 × 280

b. 140 is what percent of 280? **ii.** 280 ÷ 1.40

c. What is 140% of 280? **iii.** 140 ÷ 280

19. Finding the Percent of a Number Use the percent equation to answer each question.

a. What is 18% of 600?

b. What is 62% of 90?

20. Finding the Percent of a Number Use the percent equation to answer each question.

a. What is 7% of 75?

b. What is 151% of 800?

21. Finding a Percent *p* Use the percent equation to answer each question.

a. 536 is what percent of 670?

b. 12 is what percent of 32?

22. Finding a Percent *p* Use the percent equation to answer each question.

a. 8 is what percent of 40?

b. 378 is what percent of 210?

23. Finding *b* in the Percent Equation Use the percent equation to answer each question.

a. 12 is 6% of what?

b. 168 is 48% of what?

24. Finding *b* in the Percent Equation Use the percent equation to answer each question.

a. 3680 is 92% of what?

b. 444 is 120% of what?

25. Solving Percent Problems Use the percent equation to answer each question.

a. 13 is what percent of 104?

b. 68 is 32% of what?

c. What is 175% of 820?

d. 58 is 40% of what?

26. Solving Percent Problems Use the percent equation to answer each question.

a. What is 15% of 7100?

b. 21 is what percent of 420?

c. 6300 is 210% of what?

d. 315 is what percent of 52,500?

27. **Solving Percent Problems** The circle graph shows the results of a survey asking recent college graduates how they spend their monthly income.

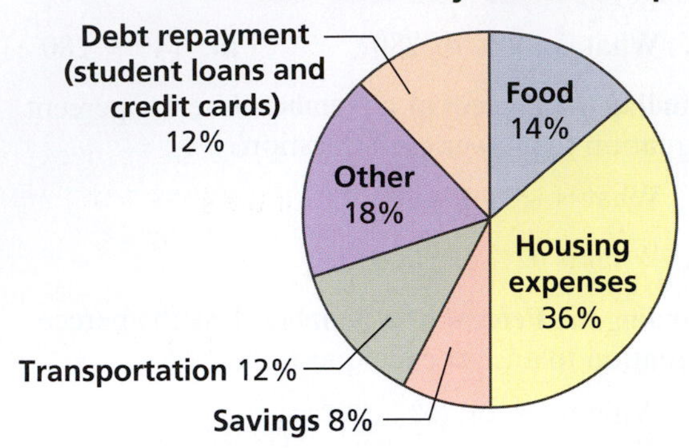

a. A recent college graduate earns a monthly income of \$4080. How much can you expect the graduate to spend on monthly housing expenses?

b. A recent college graduate earns an annual income of \$52,500. How much per month can you expect the graduate to save?

28. **Solving Percent Problems** The circle graph shows the distribution of enrolled students in different areas of study offered at a college.

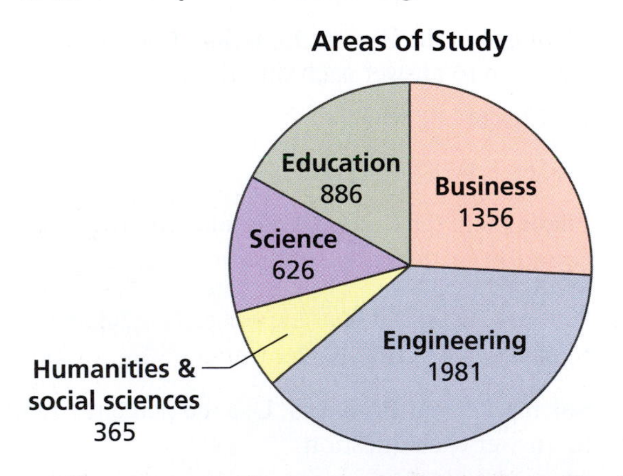

a. How many students are enrolled in the college?

b. What percent of students are in each area of study? Round your answers to the nearest whole percent.

29. **Writing** If 45% of a number is 920, is the number greater than or less than 920? Explain.

30. **Writing** If 135% of a number is 68, is the number greater than or less than 68? Explain.

31. **Writing** Can 25% of your paycheck be greater than 80% of your friend's paycheck? Explain.

32. **Writing** What does it mean to have reached 130% of your fundraising goal?

33. **Solving Percent Problems** An investor earns 11.5% on investments in one year, which amounts to \$6440. What is the amount invested?

34. **Solving Percent Problems** The receipt shows your total food cost of a meal and the sales tax.

40th Street Cafe

DATE: JUNE09'14 7:38PM
TABLE: 29
SERVER: APRIL

Food Total	**24.50**
Tax	**1.96**
Subtotal	**26.46**

TIP: ______________

TOTAL: ______________

Thank You

a. Calculate the percent of sales tax on the food total.

b. How much do you add to the subtotal to leave an 18% tip on the food total?

35. **Grading Student Work** On a diagnostic test, one of your students does the following work. Explain what the student did wrong. Which topics would you encourage the student to review?

> 9 is what percent of 180?
>
> $\frac{180}{9} = 20\%$
>
> 9 is 20% of 180.

36. **Grading Student Work** On a diagnostic test, one of your students does the following work. Explain what the student did wrong. Which topics would you encourage the student to review?

> What is 30% of 50?
>
> $30 \cdot 50 = 1500$
>
> 30% of 50 is 1500.

37. **Percent of Increase** A percent of change is the percent that a quantity changes from the original amount. When the original amount increases, the percent of change is called a *percent of increase* and can be calculated using the formula

$$\text{Percent of increase} = \frac{\text{new amount} - \text{original amount}}{\text{original amount}}.$$

a. What is the percent of increase from 60 to 84?

b. A class size increases from 26 students to 29 students. Calculate the percent of increase. Round your answer to the nearest tenth of a percent.

38. Percent of Decrease A percent of change is the percent that a quantity changes from the original amount. When the original amount decreases, the percent of change is called a *percent of decrease* and can be calculated using the formula

$$\text{Percent of decrease} = \frac{\text{original amount} - \text{new amount}}{\text{original amount}}.$$

a. What is the percent of decrease from 72 to 18?

b. What is the percent of decrease in the student's test scores? Round your answer to the nearest tenth of a percent.

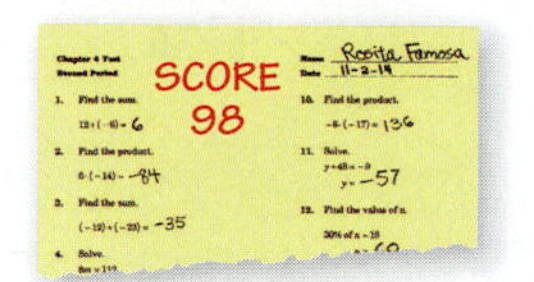

Pre-test

Post-test

39. Writing Refer to the definitions in Exercises 37 and 38. Can a percent of increase ever be greater than 100%? Can a percent of decrease ever be greater than 100%? Explain your reasoning.

40. Writing Refer to the definitions in Exercises 37 and 38. A number increases by 20% and then decreases by 20%. Will the resulting number be *less than*, *greater than*, or *equal to* the original number? Explain.

41. Simple Interest Simple interest is the money paid or earned only on the principal (the original amount of money that is borrowed or deposited). The formula used to calculate simple interest is

$$I = Prt$$

where I is the simple interest, P is the principal, r is the annual interest rate (in decimal form), and t is the time (in years).

a. Calculate the interest earned on a $500 deposit after 4 years at an annual simple interest rate of 4%.

b. You borrow $3000 at an 8% annual simple interest rate. What is the amount needed to pay off the entire loan after 2 years?

42. The Percent Proportion Another way to solve for unknowns in a percent problem is to use proportions. In the *percent proportion*

$$\frac{a}{w} = \frac{p}{100}$$

a represents the part, w represents the whole, and p represents the percent in percent form. Use the percent proportion to solve Exercise 34. What are the similarities between using the percent proportion and using the percent equation? What are the differences?

43. Markups: In Your Classroom To make a profit, a store needs to sell products for more than what it pays for the products. This increase from what the store pays to the selling price is called a *markup*.

a. A store pays $25 for a game. The percent of markup is 30%. Write the percent equation for this situation. What is the dollar amount of the markup?

b. To find the selling price of an item, the markup amount is added to the cost the store pays. What is the selling price of the game in part (a)?

c. Explain to your students how to find the selling price of a product with a markup using the percent equation.

44. Discounts: In Your Classroom A *discount* is a decrease in the original selling price of an item. A store advertises the following discounts.

a. Write the percent equation to find the sale price of the pair of sneakers. What is the sale price?

b. For the pair of sandals, what percent of the original price is the sale price? Use this percent to write the percent equation to find the original price of the pair of sandals. What is the original price?

c. Explain to your students how to find the sale price or original price of a discounted product using the percent equation.

d. Design a classroom project using local advertisements to help students understand percents and the percent equation.

45. Number Sense What whole number can you substitute for x in each list of numbers, so that the numbers are ordered from least to greatest?

a. $\frac{1}{x}, \frac{x}{5}, 68\%, \frac{x}{4}$

b. $\frac{x}{20}, \frac{2}{x}, 39\%, \frac{x}{8}$

Chapter Summary

Chapter Vocabulary

decimal *(p. 247)*
decimal point *(p. 247)*
terminating decimal *(p. 249, 269)*
compatible numbers *(p. 259)*
compensation *(p. 259)*
repeating decimal *(p. 269)*
scientific notation *(p. 270)*
mantissa *(p. 270)*
characteristic *(p. 270)*
percent *(p. 277)*
percent symbol % *(p. 277)*
percent equation *(p. 279)*

Chapter Learning Objectives — Review Exercises

7.1 Decimals *(page 247)* — 1–36

- Read and write decimals, and convert between decimal form and fraction form.
- Recognize which fractions can be written as terminating decimals.
- Order terminating decimals.

7.2 Adding and Subtracting Decimals *(page 257)* — 37–66

- Add and subtract decimals using models and algorithms.
- Add and subtract decimals using mental math.
- Round decimals.

7.3 Multiplying and Dividing Decimals *(page 267)* — 67–108

- Multiply and divide decimals using fractions and algorithms.
- Write fractions as terminating or repeating decimals.
- Use scientific notation to represent large numbers.

7.4 Percents *(page 277)* — 109–128

- Convert between fractions, decimals, and percents.
- Solve percent problems using a model.
- Solve percent problems using the percent equation.

Important Concepts and Formulas

Algorithm for Adding and Subtracting Decimals

1. Write the numbers vertically and line up the decimal points.
2. Add or subtract as with whole numbers.
3. Insert the decimal point in the sum or difference immediately below the decimal points in the numbers being added or subtracted.

The Percent Equation

The statement "a is p percent of b" is summarized by the percent equation.

$$a = p \cdot b$$

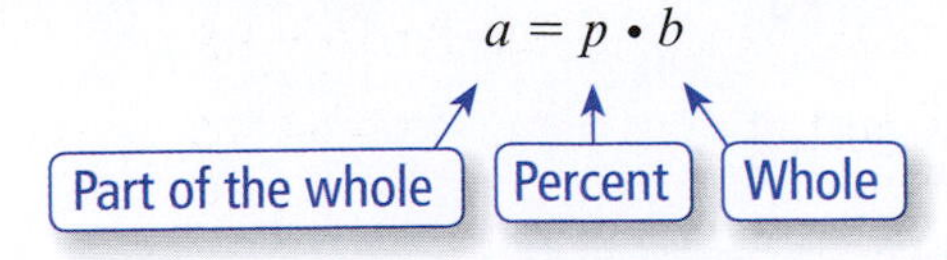

Algorithm for Multiplying Decimals

1. Multiply as with whole numbers.
2. Count the number of digits to the right of the decimal point in each of the two factors. Suppose this is m digits and n digits.
3. Insert the decimal point in the product so that there are $m + n$ digits to the right of the decimal point.

Algorithm for Dividing Decimals

1. Multiply the divisor and dividend by the power of 10 necessary to make the divisor a whole number.
2. Divide as with whole numbers.
3. Insert the decimal point in the quotient, above the decimal point in the dividend.

Review Exercises

Free step-by-step solutions to odd-numbered exercises at MathematicalPractices.com.

7.1 Decimals

Identifying Place Value Identify the place value of the underlined digit.

1. $7.5\underline{9}$ 2. $1\underline{6}8.95$ 3. $3.\underline{4}89$
4. $45.11\underline{5}$ 5. $\underline{2}32.641$ 6. $9\underline{6}.082$

Writing Decimal Form Write the number in decimal form.

7. $4\frac{3}{10}$ 8. $10\frac{63}{100}$
9. $82\frac{9}{100}$ 10. $\frac{724}{1000}$
11. $19\frac{7}{1000}$ 12. $308\frac{47}{1000}$

Writing Fraction Form Write the number in fraction form.

13. 15.8 14. 8.36 15. 2.73
16. 0.655 17. 7.032 18. 600.002

Multiple Representations Write the decimal in expanded form and in words.

19. 4.89 20. 0.54 21. 13.76
22. 26.21 23. 2.029 24. 80.955

Writing Terminating Decimals Decide whether the number can be written as a terminating decimal. If so, write the terminating decimal. If not, explain why.

25. $\frac{3}{10}$ 26. $\frac{12}{65}$ 27. $6\frac{31}{70}$
28. $18\frac{17}{26}$ 29. $42\frac{24}{30}$ 30. $\frac{63}{72}$

31. **Ordering Decimals** Order the decimals 0.14, 0.1392, 0.139, 0.1, and 0.145 from least to greatest.

32. **Ordering Numbers** Order the numbers 2.6, $2\frac{16}{25}$, 2.65, $2\frac{7}{10}$, and 2.642 from least to greatest.

Writing Decimals Rewrite the statement using a decimal.

33. A fluid ounce is equal to two-sixteenths of a cup.
34. The atomic mass of Neon (in atomic mass units) is approximately twenty and eighteen-hundredths.
35. Earth's distance from the Sun is about one hundred ninety-two thousandths of Jupiter's distance from the Sun.

36. **Ordering Numbers** The table below shows the fraction of students in each class who pack a lunch for school. Decide which classroom has the least fraction of students who pack a lunch.

Class	Fraction of students who pack lunch
3rd grade class 1	$\frac{15}{24}$
3rd grade class 2	$\frac{11}{25}$
4th grade class 1	$\frac{12}{30}$
4th grade class 2	$\frac{14}{28}$
5th grade class 1	$\frac{18}{30}$
5th grade class 2	$\frac{12}{32}$

7.2 Adding and Subtracting Decimals

Adding and Subtracting Decimals Use base-ten blocks to find the sum or difference. Justify your answer with a sketch of the model.

37. 1 + 1.36 38. 2.4 + 1.29
39. 1.07 + 1.86 40. 1.62 − 1.4
41. 3.1 − 2.04 42. 2.55 − 0.16

Rounding Decimals Round the decimal to the nearest tenth, hundredth, and thousandth.

43. 2.2481 44. 16.0287
45. 0.3968 46. 28.5483

Adding Decimals Use the vertical algorithm to find the sum. Use estimation to see if your answer is reasonable.

47. 5.74 + 1.41 48. 2.008 + 15.215
49. 10.43 + 8.2 + 0.868
50. 2.601 + 17.357 + 8.014

Subtracting Decimals Use the vertical algorithm to find the difference. Use estimation to see if your answer is reasonable.

51. 2.87 − 1.06 52. 8.562 − 4.5
53. 14.674 − 0.139 54. 52 − 24.367

Adding and Subtracting Decimals Evaluate the expression.

55. 3.55 + 1.48 − 2.6

56. 12 − 4.3 + 6.87

57. 27.845 + 13.22 − 5.606

58. 34 − 23.91 + 0.5714

Adding and Subtracting Decimals Use mental math to evaluate the expression. Describe your reasoning.

59. 8.4 + 4.8

60. 19.3 − 2.4

61. 5.87 − 3.12

62. 6.04 + 10.73

63. **Luggage** To check luggage for no additional fee, the weight of the luggage must not exceed 50 pounds. Your luggage has a weight of 28.62 pounds. How much more weight can you add to your luggage and still have no additional fee?

64. **Temperature** Yesterday's high temperature was 4.8°F warmer than the high temperature two days ago. Today's high temperature is 2.5°F cooler than the high temperature yesterday. Today's high temperature is 68.7°F. What was the high temperature two days ago?

65. **Balancing a Checkbook** The balance in your checking account is $245.86. Your paycheck is for $890.48. You write checks to pay $540.00 for rent, $79.90 for cable, and $130.68 for electricity. After you deposit your paycheck and pay your bills, what is the balance in your checking account?

66. **Walking** You walk the entire perimeter of the park. How many miles do you walk?

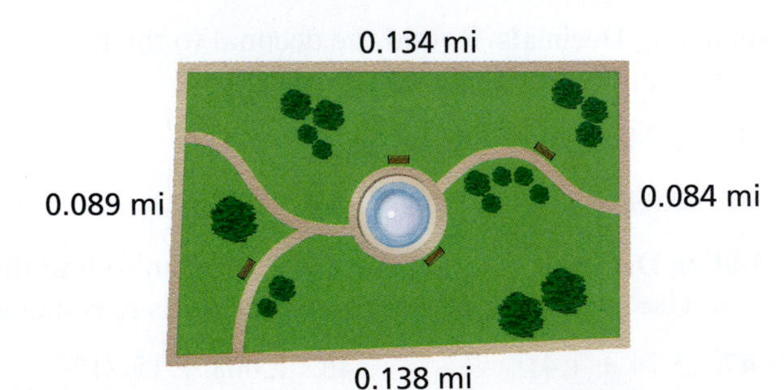

7.3 Multiplying and Dividing Decimals

Placing the Decimal Point Place the decimal point in the product or quotient.

67. 2.7 • 3.4 = 918

68. 6.922 ÷ 4 = 17305

69. 78.95 ÷ 12.5 = 6316

70. 5.2 • 4.81 = 25012

Multiplying Decimals Using Base-Ten Blocks Use base-ten blocks to find the product. Sketch the model.

71. 3 • 1.8 **72.** 0.35 • 5

73. 2.6 • 2 **74.** 4 • 1.67

Multiplying Decimals Find the product.

75. 14 • 8.5 **76.** 7.03 • 4.2

77. 3.9 • 5.244 **78.** 10.72 • 6.18

Dividing Decimals Using Base-Ten Blocks Use base-ten blocks to find the quotient. Sketch the model.

79. 2.4 ÷ 6 **80.** 4.5 ÷ 9

81. 5.46 ÷ 3 **82.** 3.84 ÷ 4

Dividing Decimals Find the quotient.

83. 28 ÷ 1.25 **84.** 12.48 ÷ 2.6

85. 8.48 ÷ 3.3125 **86.** 9.324 ÷ 1.11

Writing Fractions as Decimals Write the fraction as a decimal. If the decimal repeats, then write the decimal using a horizontal bar over the first occurrence of the repeating digits.

87. $\frac{8}{11}$ **88.** $\frac{6}{25}$

89. $\frac{1}{8}$ **90.** $\frac{5}{6}$

Identifying Scientific Notation Decide whether the number is written in scientific notation. If not, explain why and write it in scientific notation.

91. 0.45×10^{14} **92.** 9.2×10^{5}

93. 3.72×10^{8} **94.** 6.52×100^{11}

Scientific Notation Identify the mantissa and characteristic. Then write the number in standard form.

95. 2.9×10^{6} **96.** 7.46×10^{1}

97. 1.08×10^{10} **98.** 4.205×10^{4}

Writing Numbers in Scientific Notation Write the number in scientific notation.

99. 7820 **100.** 506,000

101. 3,421,000,000 **102.** 80,090,000

Multiplying Numbers in Scientific Notation Evaluate the expression. Write your answer in standard form.

103. $(7.8 \times 10^{2})(2.0 \times 10^{5})$

104. $(9.34 \times 10^{6})(1.6 \times 10^{3})$

Dividing Numbers in Scientific Notation Evaluate the expression. Write your answer in standard form.

105. $\frac{9.0 \times 10^{7}}{3.0 \times 10^{4}}$ **106.** $\frac{4.9 \times 10^{13}}{1.4 \times 10^{8}}$

107. Fuel Consumption The scale on a map is 1 inch to 53.9 miles. Your car gets 24.5 miles to the gallon and you travel a straight road measuring 6.75 inches on the map. How many gallons of gasoline do you need?

108. Volume of a Lake One cubic meter contains about 4.227×10^3 cups. The volume of water in a lake is about 1.3×10^{13} cubic meters. About how many cups of water are in the lake?

7.4 Percents

Using Models to Represent Percents Shade the grid to represent the given percent. Then write the percent as a decimal.

109. 60%

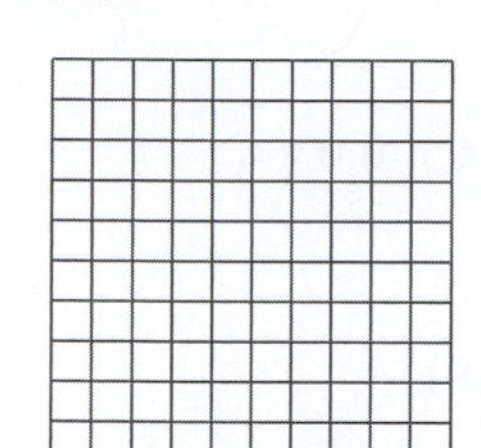

110. 34%

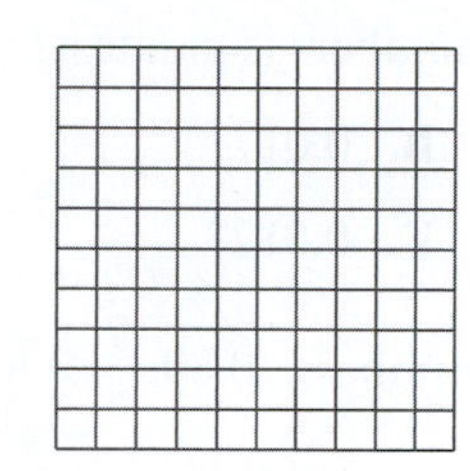

111. Converting Fractions, Decimals, and Percents Complete the table.

Fraction	Decimal	Percent
		42%
$\frac{5}{8}$		
	0.09	
		0.08%

112. Converting Fractions, Decimals, and Percents Complete the table.

Fraction	Decimal	Percent
$\frac{1}{5}$		
	0.16	
		79%
	0.0045	

Comparing Fractions, Decimals, and Percents Order the numbers from least to greatest.

113. 87.7%, $\frac{22}{25}$, 0.884, $\frac{17}{20}$

114. $\frac{61}{25}$, 240%, 2.368, $\frac{19}{8}$

Finding the Percent of a Number Write the percent represented by the shaded portion of the grid. Then find that percent of the given number.

115. 79,400

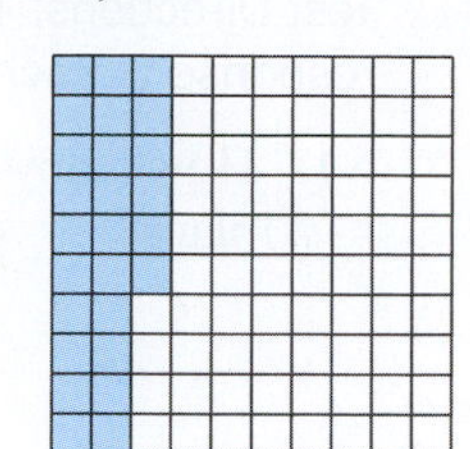

116. 4800

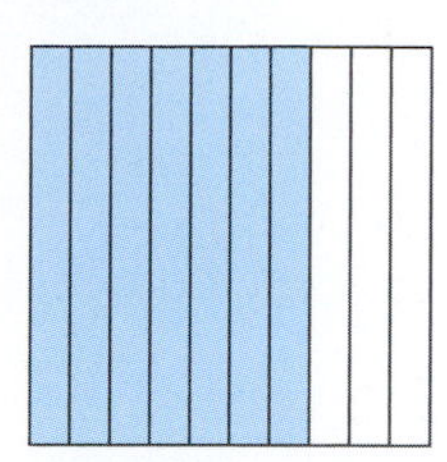

Solving Percent Problems Use the percent equation to answer the question.

117. 36 is what percent of 48?

118. What is 74% of 400?

119. 162 is what percent of 360?

120. 53 is 25% of what?

121. What is 230% of 70?

122. 144 is what percent of 80?

123. 81 is 135% of what?

124. What is 36% of 45?

125. 15 is 0.05% of what?

126. Field Trips In a survey, teachers were asked what type of field trip they prefer for their students. Of the 153 teachers surveyed, about how many prefer a museum?

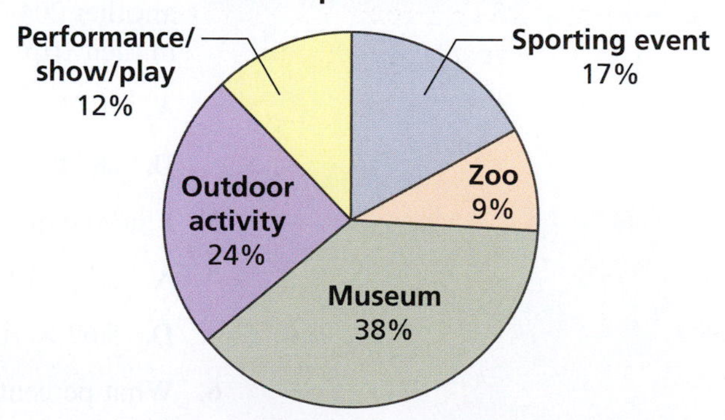

127. Winning Percentage A baseball team wins 38 out of 97 games. What is the team's winning percentage? Round your answer to the nearest tenth of a percent.

128. Student Survey The number of students who prefer having their teachers use an interactive white board instead of a chalkboard is 272. This represents 85% of the entire student body. What is the total number of students who attend the school?

Chapter Test

The questions on this practice test are modeled after questions on state certification exams for K–8 teaching.

Test Directions: Each of the questions is followed by five choices. Choose the *best* response to each question.

1. How many of the six numbers in the table have a 6 in the tens or the tenths place?

543.463	461.008	345.649
624.547	68.084	86.43

A. One **B.** Two **C.** Three
D. Four **E.** Five

2. Which of the following numbers is greatest?

A. 0.0036 **B.** 0.017 **C.** 0.0366
D. 0.0363 **E.** 0.0177

3. Consider the following numbers: $0.89, \frac{7}{8}, 0.\overline{8}, \frac{13}{16}, 0.81$

Which answer below shows the numbers arranged in ascending order?

A. $0.81, \frac{13}{16}, \frac{7}{8}, 0.\overline{8}, 0.89$ **B.** $0.89, \frac{7}{8}, 0.\overline{8}, \frac{13}{16}, 0.81$

C. $0.\overline{8}, \frac{13}{16}, 0.81, 0.89, \frac{7}{8}$ **D.** $0.89, 0.\overline{8}, \frac{7}{8}, \frac{13}{16}, 0.81$

E. $\frac{13}{16}, 0.81, 0.89, \frac{7}{8}, 0.\overline{8}$

4. A family needs to travel a total of 447.9 miles to arrive at their vacation destination. They drive 135.5 miles before making their first stop. They drive another 204.6 miles before stopping again. How many miles do they have left in their trip?

A. 107.8 **B.** 108.79 **C.** 243.3
D. 287.8 **E.** 378.8

5. Find the quotient $(8.37 \times 10^{12}) \div (2.7 \times 10^{4})$.

A. 3.1×10^3 **B.** 5.67×10^3 **C.** 3.1×10^8
D. 5.67×10^8 **E.** 31×10^8

6. What percent of 420 is 21?

A. 0.05% **B.** 0.5% **C.** 5% **D.** 20% **E.** 50%

7. A desktop computer sells for $1,800 to the general public. If you purchase one through the university, the price is reduced by 20 percent. What is the sale price of the computer?

A. $360 **B.** $1,000 **C.** $1,300
D. $1,440 **E.** $2,180

8. The enrollment at Eastern High School is 1,150. If the average daily attendance is 96 percent, how many students were absent daily?

A. 35 **B.** 46 **C.** 420 **D.** 987 **E.** 1,104

8 Integers

Activity: Comparing Integers on a Number Line

Materials:

- Pencil

Learning Objective:

Understand how to use a number line to compare integers. Recognize how integers can be used in real life to represent temperatures.

Name ______________________________

Write the integers shown on the number line in ▭ to make the inequality true.

1.

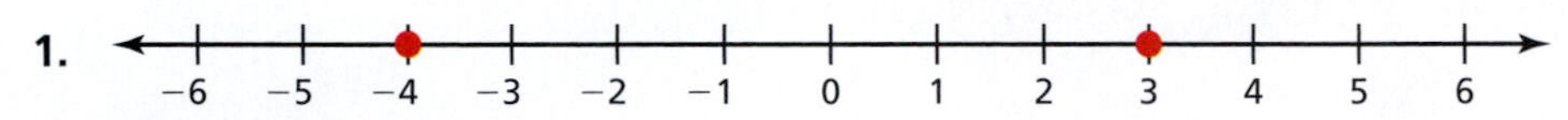

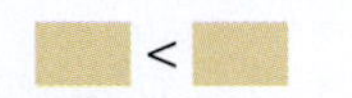

2.

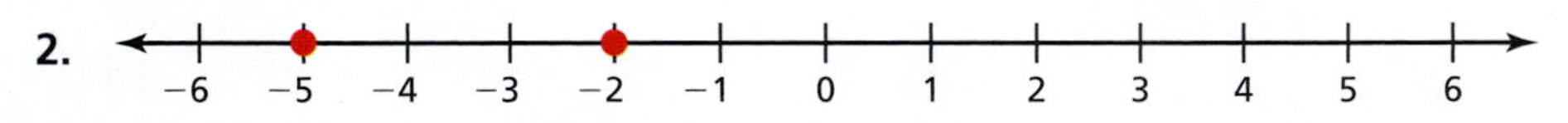

3.

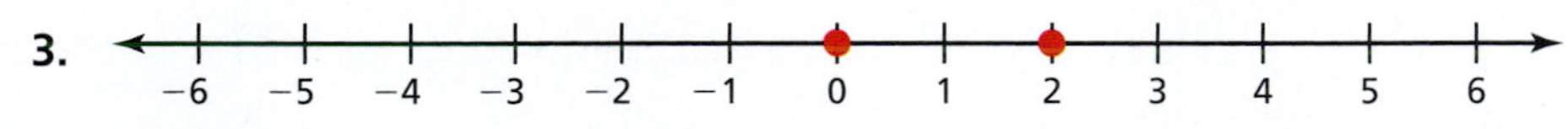

4.

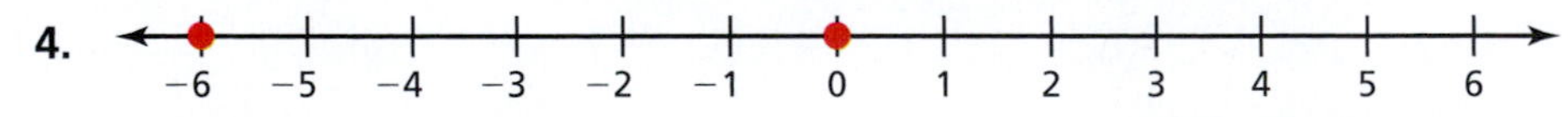

5. What is the temperature in Fairbanks?

▭ degrees Fahrenheit

6. What is the temperature in Anchorage?

▭ degrees Fahrenheit

7. Circle the city with the colder temperature. Explain your reasoning.

Fairbanks   Anchorage

Fairbanks Anchorage

8.1 Integers

Standards

Grades 6–8
The Number System
Students should apply and extend previous understandings of numbers to the system of rational numbers.

- Understand the definition of integers, and graph integers on an integer number line.
- Use an integer number line to order integers.
- Find the absolute value of an integer.
- Recognize real-life problems that can be represented by integers.

Integers

Definition of Integers

Integers are the set of whole numbers and their **opposites**. Two numbers are opposites when they are the same distance from 0 on a number line.

$\ldots -3, -2, -1, 0, 1, 2, 3, \ldots$ Integers

The *integer number line* is a measurement model that represents integers. Integers to the right of zero on the integer number line are called **positive integers**. Integers to the left of zero are called **negative integers**. The integer zero is neither positive nor negative.

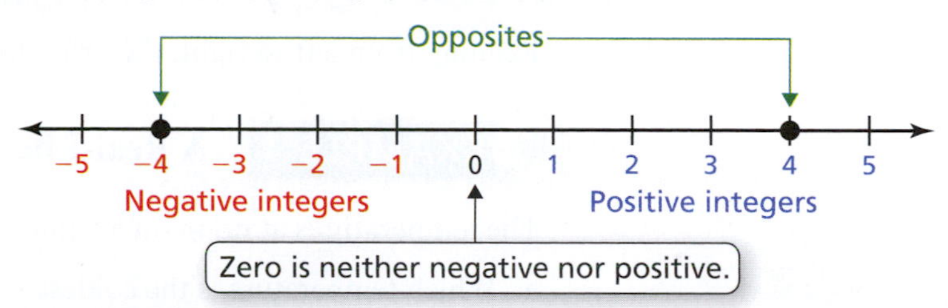

Locating the point that represents an integer on the integer number line is called **graphing an integer**.

Classroom Tip

You may hear students refer to the integer −5 as "minus five." In mathematics, *minus* represents subtraction, whereas *negative* describes a number that is to the left of zero on the integer number line. The integer −5 should be read as "negative five."

EXAMPLE 1 **Graphing Integers**

Graph the integers −5, −1, 0, and 4.

SOLUTION

EXAMPLE 2 **Graphing an Integer and Its Opposite**

Graph the integer 3 and its opposite.

SOLUTION

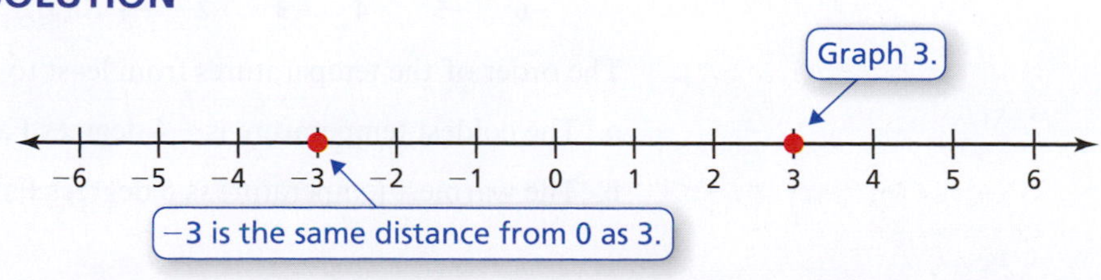

Comparing and Ordering Integers

Classroom Tip

Some students may think that $-1 < -4$ because $1 < 4$. Be sure to point out that it is the position on the integer number line that determines the order of two integers, not their magnitudes.

Comparing Integers

Let a and b be integers. ***a* is less than *b***, written as $a < b$, if and only if a is to the left of b on the integer number line.

Example $-4 < -1$, because -4 is to the left of -1 on the integer number line.

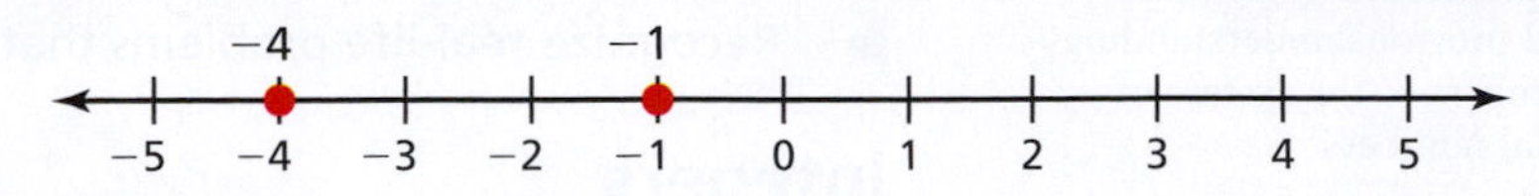

EXAMPLE 3 Ordering Integers

Order the integers 3, 0, -2, 4, and -5 from least to greatest.

SOLUTION

Graph the integers on the integer number line.

Reading from left to right, the order from least to greatest is -5, -2, 0, 3, and 4.

EXAMPLE 4 A Real-Life Example of Ordering Integers

The temperatures at noon on a winter day in four cities in Montana are shown.

a. Which temperature is the coldest? **b.** Which temperature is the warmest?

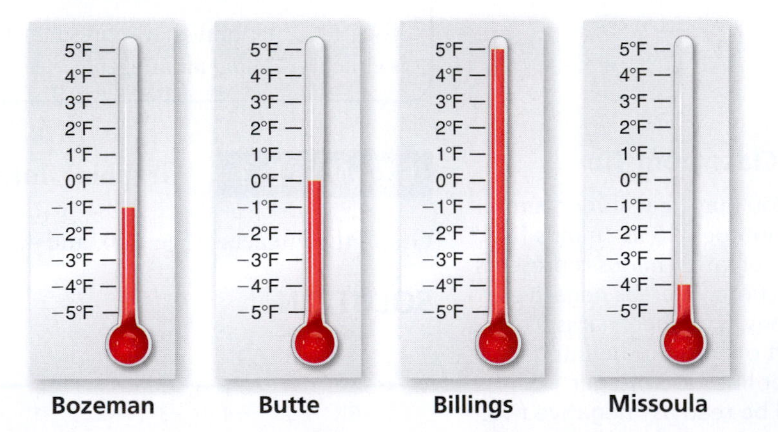

SOLUTION

Graph the integers on the integer number line.

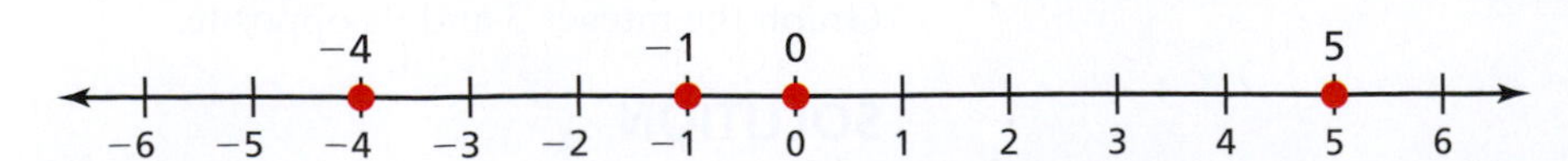

The order of the temperatures from least to greatest is -4, -1, 0, and 5.

a. The coldest temperature is -4 degrees Fahrenheit in Missoula.

b. The warmest temperature is 5 degrees Fahrenheit in Billings.

Absolute Value

Definition of the Absolute Value of an Integer

The **absolute value** of an integer is the distance between the integer and 0 on a number line. Let a be a whole number. The absolute values of a and $-a$ are denoted by the **absolute value bars** $|\quad|$ and are defined as follows.

$|a| = a$ The absolute value of the positive integer a is a.

$|0| = 0$ The absolute value of 0 is 0.

$|-a| = a$ The absolute value of the negative integer $-a$ is a.

Example $|4| = 4$

$|-4| = 4$

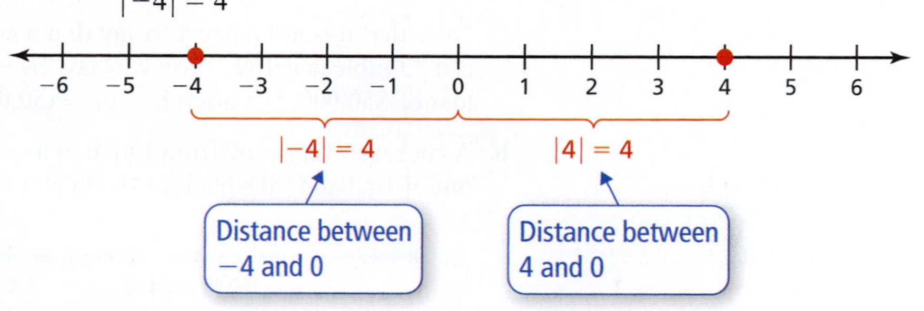

Classroom Tip

Note that the integers −4 and 4 are the same distance from 0. So, it makes sense that they have the same absolute value.

EXAMPLE 5 Comparing Integer Expressions

Complete the statement using the correct symbol: <, > , or =.

a. $|-9|$ ☐ $|9|$

b. 5 ☐ $|-3|$

c. 0 ☐ $|-7|$

d. -4 ☐ $-|-4|$

e. $|12|$ ☐ $|-15|$

SOLUTION

a. Because $|-9| = 9$ and $|9| = 9$,

$|-9| = |9|$. Both −9 and 9 are the same distance from 0.

b. Because $|-3| = 3$,

$5 > |-3|$. 5 is greater than 3.

c. Because $|-7| = 7$,

$0 < |-7|$. 0 is less than 7.

d. Because $-|-4| = -4$,

$-4 = -|-4|$. −4 is equal to −4.

e. Because $|12| = 12$ and $|-15| = 15$,

$|12| < |-15|$. 12 is less than 15.

Using Integers in Real Life

There are many examples of using negative integers in real life. Some are shown in Example 6.

Using Integers in Real Life

a. A company has a profit of \$60,000 in April and a profit of −\$50,000 (a loss) in May.

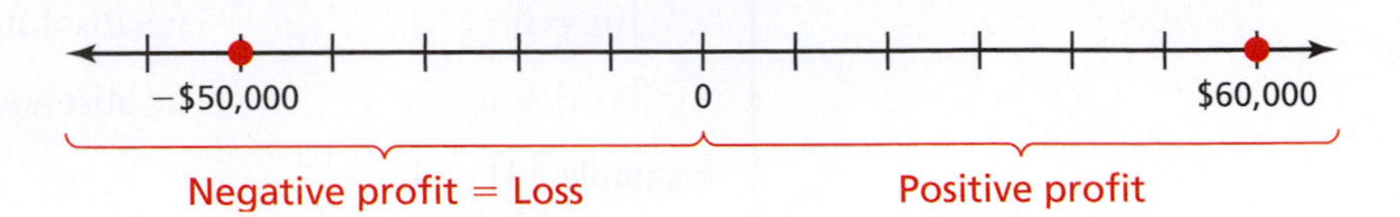

Note that it is not correct to say that a company has a loss of −\$50,000. This is a "double negative." You can say the company has a profit of −\$50,000 or a loss of \$50,000, but not a loss of −\$50,000.

b. A rocket travels away from Earth at a velocity of 500 feet per second. It runs out of fuel and falls back to Earth at a velocity of −300 feet per second.

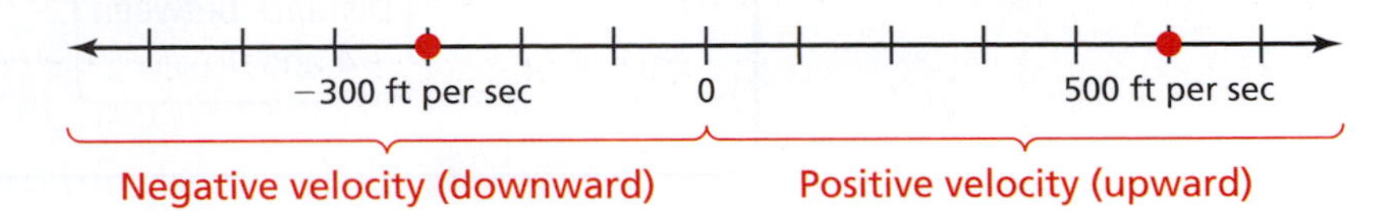

Note that in physics and engineering, velocity is not the same as speed. Velocity has both speed and direction (often upward or downward). Speed is the absolute value of velocity.

c. The balance in a checking account is \$350. In error, you write a check for \$400. The bank covers your check, but you now have a balance of −\$50. Moreover, the bank charged you an overdraft fee of \$35, which makes your balance −\$85.

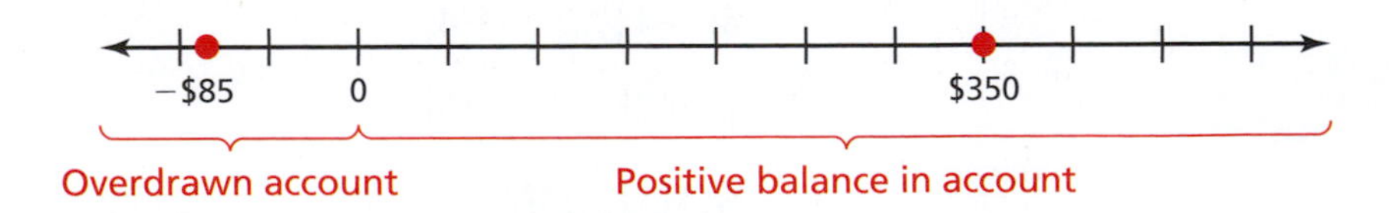

d. The greatest elevation in North America is the summit of Mount McKinley in Alaska. Its elevation is 20,320 feet. The least elevation in North America is Death Valley in California. Its elevation is −282 feet.

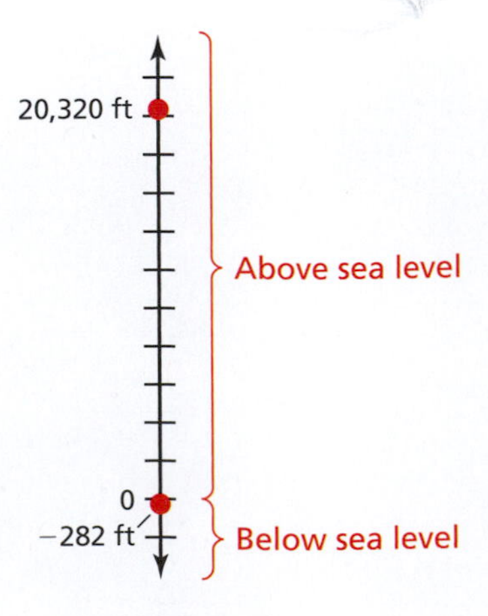

Budget Variance

Mathematical Practices

Model with Mathematics

MP4 Mathematically proficient students identify important quantities in practical situations and map their relationships using such tools as: diagrams, two-way tables, graphs, flowcharts, and formulas.

The spreadsheet shows some of the expenses for a small company. The variance for each expense is the difference of the actual expense and the budgeted expense. (You will study the subtraction of integers in Section 8.2.)

Variance = Actual − Budgeted

	A	B	C	D	E
1	Category	Actual expense	Budgeted expense	Variance	Absolute value of variance
2	Travel	$13,650	$13,500	$150	
3	Wages	$64,780	$63,000	$1780	
4	Utilities	$15,470	$16,000	-$530	
5					

The analysis program flags each variance whose absolute value is greater than $500.

a. Which variances should be flagged?

b. What does it mean when the variance is a positive number?

c. What does it mean when the variance is a negative number?

SOLUTION

a. Begin by completing the last column of the spreadsheet. Find the absolute value of each variance.

$|150| = 150 \qquad |1780| = 1780 \qquad |-530| = 530$

	A	B	C	D	E
1	Category	Actual expense	Budgeted expense	Variance	Absolute value of variance
2	Travel	$13,650	$13,500	$150	$150
3	Wages	$64,780	$63,000	$1780	$1780 ⬅
4	Utilities	$15,470	$16,000	-$530	$530 ⬅
5					

Because $1780 > $500, the absolute value of the variance for wages is greater than $500. So, it should be flagged. Because $530 > $500, the absolute value of the variance for utilities is greater than $500. So, it should be flagged.

b. When the variance is a positive number, the actual expense is greater than the budgeted expense. This occurs in two categories: travel and wages.

	Actual	**Budgeted**	
Travel	$13,650	$13,500	Actual expense is greater.

	Actual	**Budgeted**	
Wages	$64,780	$63,000	Actual expense is greater.

c. When the variance is a negative number, the actual expense is less than the budgeted expense. This occurs in only one category: utilities.

	Actual	**Budgeted**	
Utilities	$15,470	$16,000	Budgeted expense is greater.

8.1 Exercises

Free step-by-step solutions to odd-numbered exercises at MathematicalPractices.com.

1. **Writing a Solution Key** Write a solution key for the activity on page 292. Describe a rubric for grading a student's work.

2. **Grading the Activity** In the activity on page 292, a student gave the answers below. Each question is worth 10 points. Assign a grade to each answer. For those that are incorrect, why do you think the student erred?

Sample Student Work

3. 0 > 2

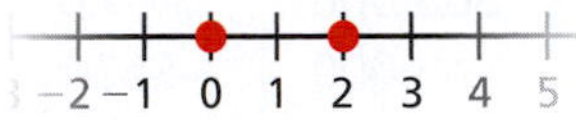

4. −6 < 0

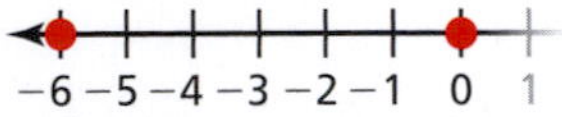

5. What is the temperature in Fairbanks?
 0 degrees Fahrenheit

6. What is the temperature in Anchorage?
 3 degrees Fahrenheit

7. Circle the day with the colder temperature. Explain your reasoning.
 (Fairbanks) Anchorage

 0 is less than 3. So, the temperature in Fairbanks is less than the temperature in Anchorage.

3. **Graphing Integers** Graph each integer.
 a. 8 b. −14
 c. −13 d. 22

4. **Graphing Integers** Graph each integer.
 a. −6 b. 11
 c. −19 d. 13

5. **Identifying Opposites** Write the opposite of each integer.
 a. −64 b. −18
 c. 25 d. −29

6. **Identifying Opposites** Write the opposite of each integer.
 a. 34 b. −96
 c. −41 d. 56

7. **Graphing an Integer and Its Opposite** Graph each integer and its opposite.
 a. 2 b. −12
 c. 42 d. 9
 e. −16 f. 20

8. **Graphing an Integer and Its Opposite** Graph each integer and its opposite.
 a. −7 b. 15
 c. −10 d. −21
 e. 26 f. 17

9. **Comparing Integers** Complete each statement using the correct symbol: < or >.
 a. 3 ___ 7 b. 0 ___ −2
 c. −5 ___ −1 d. 0 ___ 12
 e. 17 ___ −4 f. −27 ___ −33

10. **Comparing Integers** Complete each statement using the correct symbol: < or >.
 a. −10 ___ 1 b. 16 ___ 22
 c. −29 ___ −15 d. −2 ___ −3
 e. −37 ___ −33 f. 52 ___ −43

11. **Ordering Integers** Use a number line to order the integers from least to greatest.
 a. 22, −11, −7, 15, 4
 b. −3, 0, 5, 23, −15
 c. 2, 18, −10, −12, −9
 d. 16, 32, −30, −1, −21

12. **Ordering Integers** Use a number line to order the integers from least to greatest.
 a. 19, −13, −6, 7, 0
 b. −8, 14, −25, 32, −20
 c. −17, 24, −4, 9, −18
 d. 0, −7, 23, −9, −21

13. **Evaluating Absolute Values** Evaluate each expression.
 a. $|-5|$ b. $|0|$
 c. $-|9|$ d. $-|25|$
 e. $|17|$ f. $-|-21|$

14. **Evaluating Absolute Values** Evaluate each expression.
 a. $|-42|$ b. $-|-6|$
 c. $|11|$ d. $-|13|$
 e. $-|-23|$ f. $|-15|$

15. Comparing Integer Expressions Complete each statement using the correct symbol: <, >, or =.

a. $|-14| \square 15$

b. $-|3| \square |-3|$

c. $11 \square -|-12|$

d. $|-18| \square 18$

e. $-|22| \square -22$

f. $|-20| \square -19$

16. Comparing Integer Expressions Complete each statement using the correct symbol: <, >, or =.

a. $-13 \square |-13|$

b. $-|-27| \square -27$

c. $-|-10| \square -13$

d. $|-9| \square |-11|$

e. $|-26| \square |26|$

f. $|-2| \square 0$

17. Ordering Integer Expressions Use a number line to order the integer expressions from least to greatest.

a. $3, -11, |-17|, -9, -|-13|$

b. $0, -|5|, -7, 4, |-8|$

c. $-|-2|, 14, -22, |-10|, |1|$

d. $|-6|, -24, 0, 13, |29|$

e. $-|5|, 0, |-11|, 12, -16$

18. Ordering Integer Expressions Use a number line to order the integer expressions from least to greatest.

a. $|-1|, 11, -16, -|21|, -|-3|$

b. $-31, |-27|, 10, 0, |6|$

c. $28, -19, -|-14|, |-32|, 5$

d. $-|33|, 18, |-21|, 42, -21$

e. $-|-6|, |-9|, -11, 24, |13|$

19. True or False? Tell whether each statement is *true* or *false*. Explain your reasoning.

a. The opposite of a positive integer is always less than 0.

b. The absolute value of an integer is always the opposite of the integer.

c. The number 0 is a positive integer.

20. True or False? Tell whether each statement is *true* or *false*. Explain your reasoning.

a. The absolute value of a negative integer is greater than 0.

b. A negative integer is always greater than its opposite.

c. The absolute values of an integer and its opposite are equal.

21. Using Integers in Real Life Write the integer that represents each statement.

a. The temperature is 3°F below zero.

b. Last week, you weighed yourself and you lost 5 pounds. This week, you weighed yourself and you gained 3 pounds.

22. Using Integers in Real Life Write the integer that represents each statement.

a. The lowest elevation in the world is the Dead Sea at 1360 feet below sea level.

b. Yesterday, you deposited a check for $150 into your bank account. Today, you withdrew $50 from your bank account.

23. Using Integer Expressions in Real Life Write the speed at each velocity. (Remember, in physics and engineering, speed is the absolute value of velocity.)

a. 750 feet per second

b. -672 feet per second

c. -543 feet per second

d. 495 feet per second

24. Using Integer Expressions in Real Life Write the speed at each velocity. (Remember, in physics and engineering, speed is the absolute value of velocity.)

a. -35 miles per hour

b. 62 miles per hour

c. -46 miles per hour

d. -22 miles per hour

25. Graphing Integers Your checking account has a balance of $430. In error, you write a check for $450. This brings your balance to −$20. Your bank then charges you an overdraft fee of $30, bringing your balance to −$50. Graph the integers that represent your beginning and ending balances.

26. Graphing Integers A stunt man for a movie jumps off a cliff that is 18 feet above sea level and dives 11 feet into the water. Graph the integers that represent his elevations before the jump and at the deepest point of the dive.

27. Comparing Integers The average winter temperature in Juneau, Alaska, is 29°F above zero. The average winter temperature in Fairbanks, Alaska, is 7°F below zero. Write the integers that represent the temperatures. Which city has the warmer average winter temperature?

28. Comparing Integers The freezing point of Hexane is approximately 139°F below zero. The freezing point of Butane is approximately 217°F below zero. Write the integers that represent the freezing points. Which liquid has a lower freezing point?

29. Ordering Integers The table shows the counts for a few of the activities that occur during a rocket launch. "T−90 seconds" means 90 seconds before liftoff.

Count	Activity
T−90 seconds	Launch control system enabled
T+6 seconds	Rocket clears launch pad tower
T−16 seconds	Launch verification
T−3 seconds	Main engines start
T−0 seconds	Boosters ignite

a. Write the activities in the order in which they occur.

b. Which activity occurs last?

30. Ordering Integers Five friends play a 9-hole round of miniature golf. The table below shows each golfer's score at the end of the round. In golf, the lowest score wins.

Golfer	Score
Golfer A	−6
Golfer B	−2
Golfer C	3
Golfer D	−1
Golfer E	1
Golfer F	−4

a. Order the scores from best to worst.

b. Which golfer won the round?

31. Scuba Diving The depths of two scuba divers are shown below.

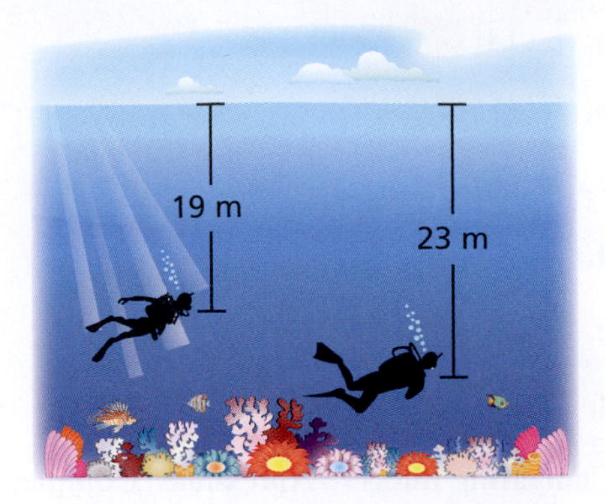

Write integers to represent the divers' positions relative to sea level. Which diver is farther from sea level?

32. Archaeology Archaeologists find two artifacts at a dig site.

Write integers to represent the artifacts' positions relative to ground level. Which artifact is closer to ground level?

33. Activity: In Your Classroom War is a card game in which the deck is divided evenly between two players. Each player places his or her stack face down. The players then place their top cards face up. The player with the higher value card keeps both cards, placing them at the bottom of his or her stack. If there is a tie, each player places three more cards face down and a new card face up. The player with the higher value new card keeps all of the cards in play and places them at the bottom of his or her stack. The game continues until one player has all of the cards. This player wins.

a. How could this game be altered so it can be used as an activity in the classroom to demonstrate comparing integers?

b. What is another game that could be used to demonstrate comparing integers? Describe the game.

34. **Activity: In Your Classroom** A common children's game involves tossing a bean bag at a target. The child whose bean bag lands closest to the bull's-eye wins.

a. How could this game be altered so it can be used as an activity in the classroom to demonstrate absolute values?

b. What is another game that could be used to demonstrate absolute values? Describe the game.

35. **Boiling Points** You are heating a solution that contains many liquids. The table below shows the liquids and their boiling points.

Liquid	Boiling point (°C)
Water	$-(-100)$
Ethanol	$\lvert -78 \rvert$
Propanol	97
Methanol	$\lvert -65 \rvert$

a. Order the liquids from lowest boiling point to highest boiling point.

b. While heating the solution, which liquid would you expect to evaporate last? Explain.

36. **Quarterly Earnings** The bar graph shows the quarterly earnings of a small company for one year.

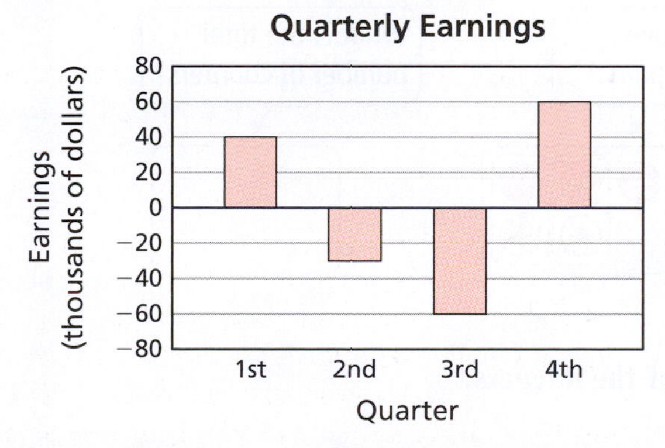

a. Which quarters show gains? Which quarters show losses?

b. Estimate the earnings for each quarter.

c. Did the company have a profit or a loss for the year? Explain.

37. **Writing** Write a list of six integers and order them from least to greatest. Then write the opposites of the integers and order them from least to greatest. Describe any patterns you observe.

38. **Writing** Let a and b be integers, and let $|a| < |b|$. Describe the possible relationships between a and b.

39. **Problem Solving** Use the following clues to order the variables a, b, c, d, and e from least to greatest.

- a is a positive integer.
- b is between a and 0.
- c and a are opposites.
- d lies to the left of c on the number line.
- e is the absolute value of d.

Which variables are positive? Which variables are negative?

40. **Problem Solving** Use the following clues to order the variables v, w, x, y, and z from least to greatest.

- v is a negative integer.
- w lies to the left of v on the number line.
- x is between v and w.
- y and x are opposites.
- z is the absolute value of v.

Which variables are positive? Which variables are negative?

41. **Grading Student Work** On a diagnostic test, one of your students does the following work. Explain what the student did wrong. Which topics would you encourage the student to review?

> Order the integers from least to greatest.
>
> -2, -5, -10, 0, 3

42. **Grading Student Work** On a diagnostic test, one of your students does the following work. Explain what the student did wrong. Which topics would you encourage the student to review?

> Order the integers from least to greatest.
>
> -8, |-4|, -1, 13, |16|

43. **Reasoning** An integer lies between -13 and -20 on the integer number line. What is the least possible value of this integer? What is the greatest possible value of this integer?

44. **Reasoning** Point A is halfway between -14 and 10 on the integer number line. Point B is halfway between point A and 0. What integer does point B represent?

Activity: Adding Integers Using Counters and Using a Number Line

Materials:

- Pencil
- Integer counters

Learning Objective:

Understand how to use positive (+) and negative (−) integer counters and how to use an integer number line to add integers.

Name ______________________________

Use integer counters to find the sum of the integers.

1.

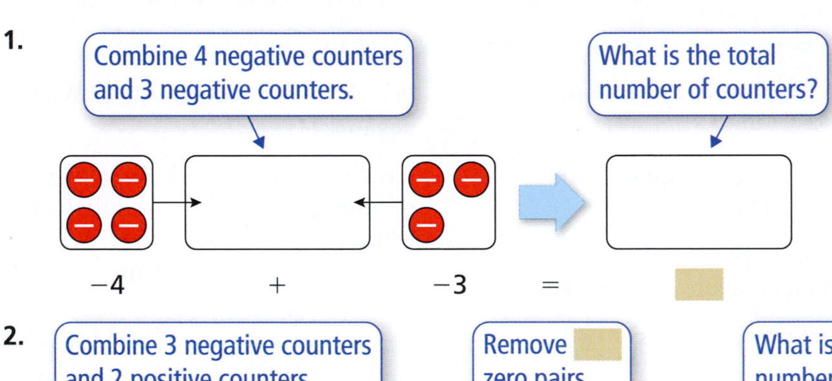

2.

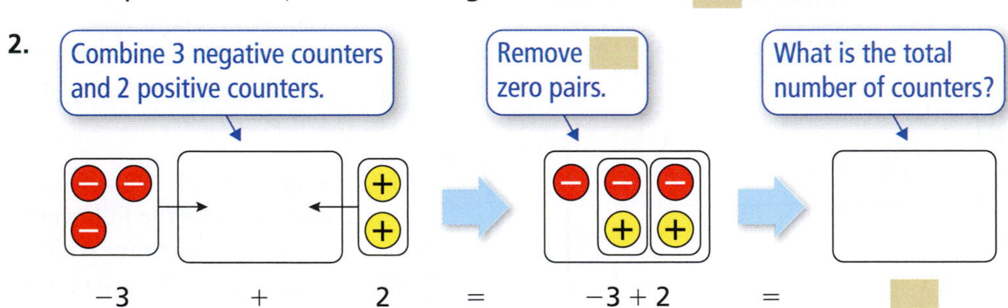

Use the integer number line to find the sum of the integers.

3. $5 + (-3) =$ ____

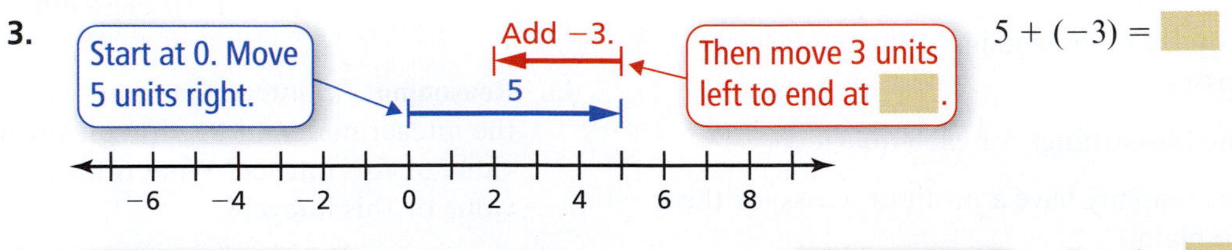

4. $-7 + 6 =$ ____

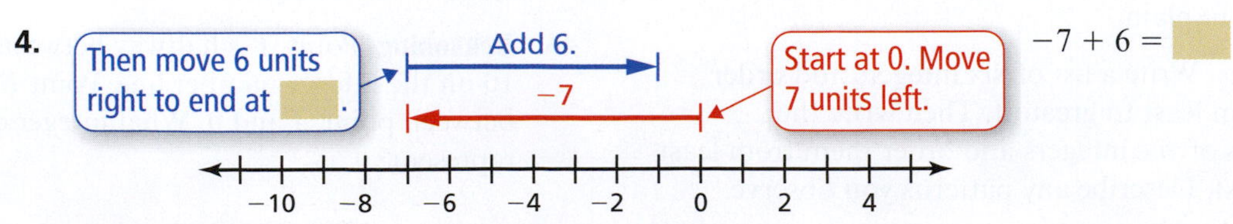

Monkey Business Images/Shutterstock.com

8.2 Adding and Subtracting Integers

Standards

Grades 6–8

The Number System

Students should apply and extend previous understandings of operations with fractions to add, subtract, multiply, and divide rational numbers.

- Understand how to add integers.
- Use models to add integers.
- Understand how to subtract integers.
- Use models to subtract integers.

Adding Integers

Adding Integers

Adding integers with the same sign To add integers with the *same* sign (both positive or both negative), add the absolute values of the integers. Then use the common sign.

Adding integers with different signs To add integers with *different* signs, subtract the lesser absolute value from the greater absolute value. Then use the sign of the integer with the greater absolute value.

Additive Inverse Property

The sum of an integer and its **additive inverse**, or opposite, is 0.

Numbers $6 + (-6) = 0,\ -3 + 3 = 0$ **Algebra** $a + (-a) = 0$

EXAMPLE 1 Adding Integers

Find each sum.

a. $5 + (-10)$ **b.** $-3 + 7$

c. $-12 + 12$ **d.** $-8 + (-14)$

SOLUTION

a. $5 + (-10) = -5$ The integers have different signs and $|-10| > |5|$. So, subtract $|5|$ from $|-10|$.

Use the sign of -10.

The sum is -5.

b. $-3 + 7 = 4$ The integers have different signs and $|7| > |-3|$. So, subtract $|-3|$ from $|7|$.

Use the sign of 7.

The sum is 4.

c. $-12 + 12 = 0$ The sum is 0 by the Additive Inverse Property.

-12 and 12 are opposites.

The sum is 0.

d. $-8 + (-14) = -22$ The integers have the same sign. So, add the absolute values $|-8|$ and $|-14|$.

Use the common sign.

The sum is -22.

Classroom Tip

The rules for adding integers apply to 3 or more integers. For instance,

$$-3 + (-2) + 4 = -1$$

and

$$-5 + (-1) + (-2) = -8.$$

Using Models to Add Integers

> **Classroom Tip**
> The following properties of whole number addition in Section 3.1 are also valid for the addition of integers.
>
> Commutative Property
> Associative Property
> Additive Identity Property

Using Models to Add Integers

Integer counters Model each integer using the appropriate number of positive or negative counters. Place them in a common area. Remove any *zero pairs* (⊖⊕). Then count the remaining counters to find the sum.

Integer number line Start at 0 and move the appropriate number of units for the first integer (move right for positive integers and move left for negative integers). From the ending position of the first integer, move the appropriate number of units for the second integer. The sum is the final location on the number line.

EXAMPLE 2 **Using Models to Find a Sum**

There are two ways to model the sum $4 + (-3)$.

Method 1

Method 2

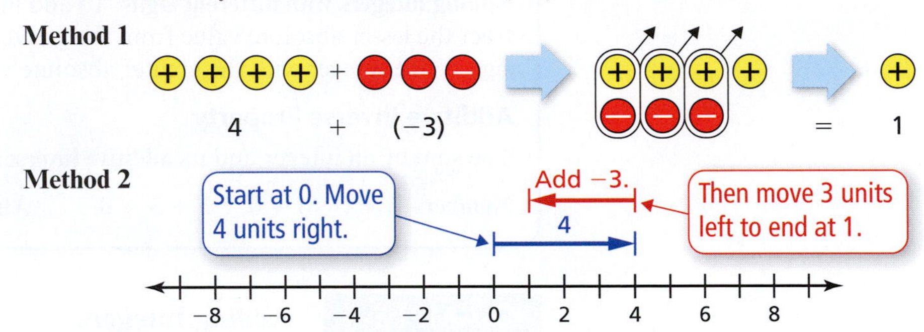

From the models, $4 + (-3) = 1$.

EXAMPLE 3 **Adding Integers in Real Life**

A football team has the following gains and losses for four downs. Does the team make a first down? (Note: A first down is a gain of 10 or more yards in four downs.)

1st down: Gains 3 yards 2nd down: Loses 4 yards
3rd down: Gains 7 yards 4th down: Gains 4 yards

SOLUTION

Use an integer number line to find the total yards that the team gains or loses. A positive integer represents a gain and a negative integer represents a loss.

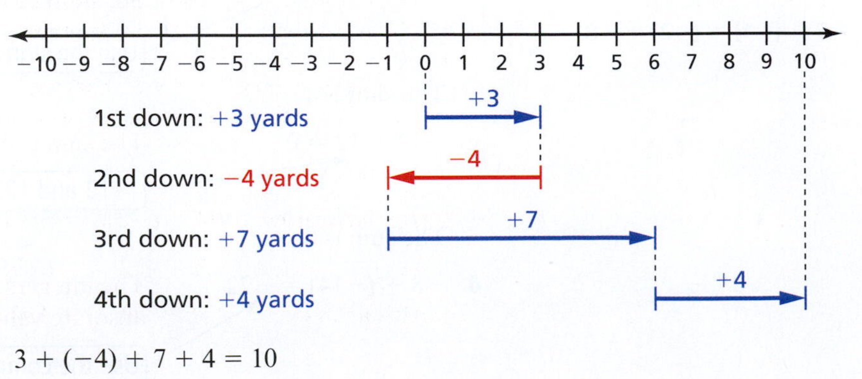

$3 + (-4) + 7 + 4 = 10$

The team gains 10 yards in four downs. So, the team makes a first down.

Subtracting Integers

Subtracting Integers

Words Let a and b be integers. To subtract b from a, add the opposite of b to a.

Numbers $4 - 5 = 4 + (-5) = -1$

Algebra $a - b = a + (-b)$

EXAMPLE 4 Subtracting Integers

Find each difference.

a. $3 - 12$

b. $-8 - (-13)$

c. $5 - (-4)$

d. $-20 - 7$

SOLUTION

a. $3 - 12 = 3 + (-12)$ Add the opposite of 12.

$= -9$ Add.

The difference is -9.

b. $-8 - (-13) = -8 + 13$ Add the opposite of -13.

$= 5$ Add.

The difference is 5.

c. $5 - (-4) = 5 + 4$ Add the opposite of -4.

$= 9$ Add.

The difference is 9.

d. $-20 - 7 = -20 + (-7)$ Add the opposite of 7.

$= -27$ Add.

The difference is -27.

EXAMPLE 5 Subtracting Integers in Real Life

Which continent has a greater range in elevations?

	North America	Africa
Highest elevation	6194 meters	5895 meters
Lowest elevation	−86 meters	−156 meters

SOLUTION

North America $6194 - (-86) = 6194 + 86$

$= 6280$ Range

Africa $5895 - (-156) = 5895 + 156$

$= 6051$ Range

Because $6280 > 6051$, North America has a greater range in elevations.

Using Models to Subtract Integers

> **Classroom Tip**
> When you are using integer counters to teach the subtraction of integers, be sure your students understand that "subtract" means to "take away" the appropriate number of counters. Often, this cannot be accomplished until the appropriate number of zero pairs have been added.

Using Models to Subtract Integers

Integer counters Model the first integer using the appropriate number of positive or negative counters. Remove the counters representing the second integer from the first integer. When the first integer does not have enough of the appropriate counters, add enough *zero pairs* (⊖⊕) so that you can remove the appropriate number of counters. Then count the remaining counters to find the difference.

Integer number line Start at 0 and move the appropriate number of units for the first integer (move right for positive integers and move left for negative integers). From the ending position of the first integer, move the appropriate number of units for the second integer (move left for positive integers and move right for negative integers). The difference is the final location on the number line.

EXAMPLE 6 Using Integer Counters to Find Differences

Find (a) $4 - 7$ and (b) $-2 - (-6)$.

SOLUTION

a. Model 4 using four positive counters. There are not enough positive counters to remove 7 of them. So, make a representation of 4 that uses 7 positive counters. To do this, add 3 zero pairs.

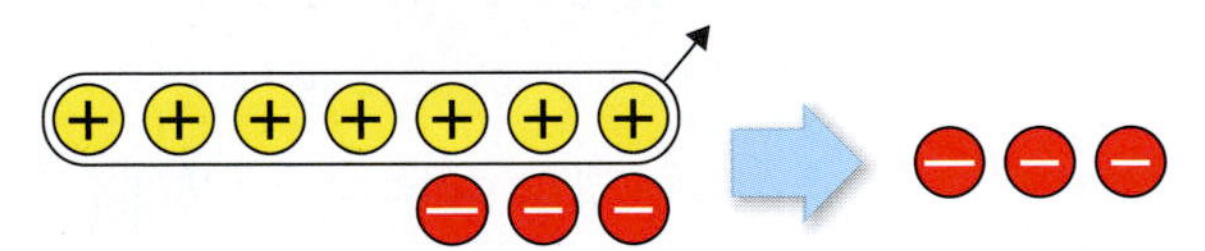

There are 3 negative counters left. So, $4 - 7 = -3$.

b. Model -2 using two negative counters. There are not enough negative counters to remove 6 of them. So, make a representation of -2 that uses 6 negative counters. To do this, add 4 zero pairs.

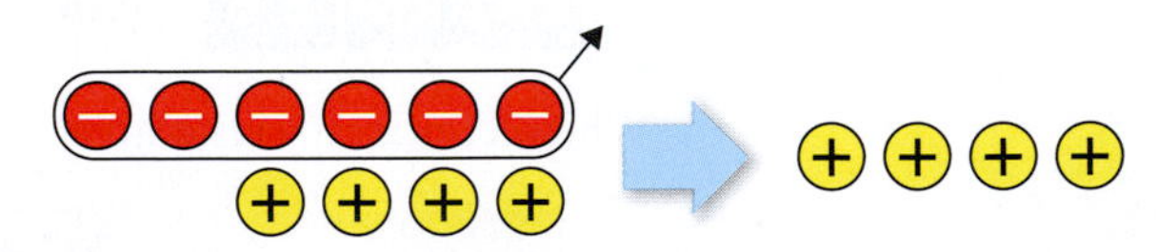

There are 4 positive counters left. So, $-2 - (-6) = 4$.

> **Classroom Tip**
> Notice that the movement on an integer number line for subtraction is the opposite of addition. This is because "when you subtract, you add the opposite."
>
> **For integers that are added**
> Move *right* for positive integers.
> Move *left* for negative integers.
>
> **For integers that are subtracted**
> Move *left* for positive integers.
> Move *right* for negative integers.

EXAMPLE 7 Using an Integer Number Line to Find a Difference

Find the difference $4 - 7$.

SOLUTION

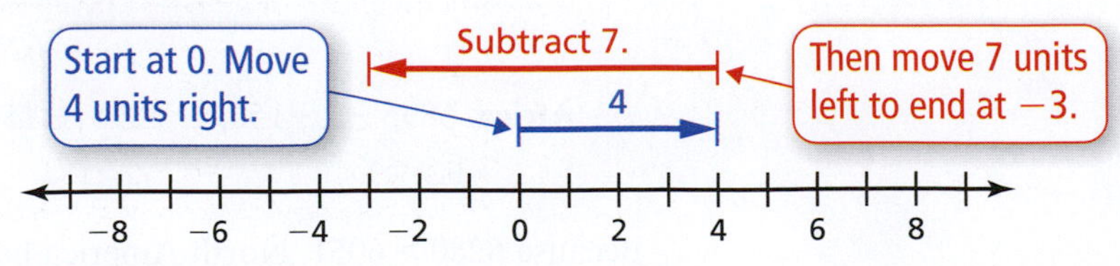

So, $4 - 7 = -3$.

EXAMPLE 8 Subtracting Integers in Real Life

The bottom of the shallow end of a swimming pool is represented by -3 feet. The bottom of the deep end is 9 feet deeper. Write an expression that represents the bottom of the pool at the deep end. What is the depth at the deep end?

SOLUTION

The bottom of the pool at the deep end can be represented by the expression

$-3 - 9 = -12.$ 9 feet deeper than -3 feet is represented by -12 feet.

The depth at the deep end of the pool is 12 feet.

EXAMPLE 9 Subtracting Integers in Real Life

A porpoise is 80 feet below sea level. It swims up and jumps out of the water to a height of 15 feet. Write a subtraction expression for the vertical distance the porpoise travels. What is the vertical distance?

SOLUTION

Note that the vertical distance should be positive. So, an expression is

$15 - (-80) = 15 + 80 = 95.$ Vertical distance

The porpoise travels a vertical distance of 95 feet.

Mathematical Practices

Reason Abstractly and Quantitatively

MP2 Mathematically proficient students make sense of quantities and their relationships in problem situations.

EXAMPLE 10 Subtracting Integers in Real Life

The table shows the record monthly high and low temperatures in Anchorage, Alaska. Which month has the greatest range?

	Jan	Feb	Mar	Apr	May	Jun
High (°F)	49	51	56	65	77	86
Low (°F)	−39	−28	−23	−3	17	34

	Jul	Aug	Sep	Oct	Nov	Dec
High (°F)	86	82	74	63	55	53
Low (°F)	40	32	−3	−11	−23	−34

SOLUTION

For each month, subtract the low temperature from the high temperature.

	Jan	Feb	Mar	Apr	May	Jun
Range (°F)	88	79	79	68	60	52

	Jul	Aug	Sep	Oct	Nov	Dec
Range (°F)	46	50	77	74	78	87

The greatest range is 88 degrees Fahrenheit in January.

8.2 Exercises

Free step-by-step solutions to odd-numbered exercises at *MathematicalPractices.com.*

1. **Writing a Solution Key** Write a solution key for the activity on page 302. Describe a rubric for grading a student's work.

2. **Grading the Activity** In the activity on page 302, a student gave the answers below. Each question is worth 10 points. Assign a grade to each answer. For those that are incorrect, why do you think the student erred?

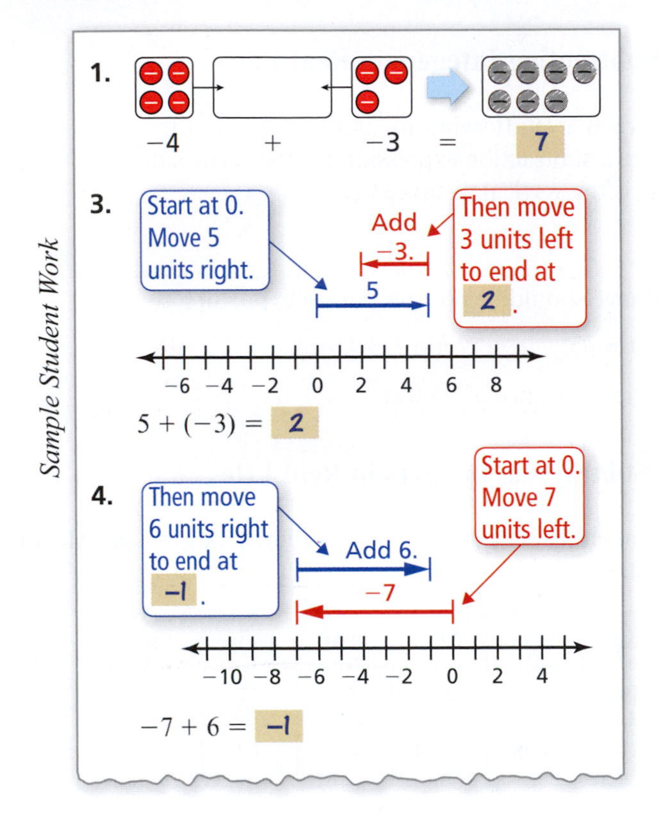

3. **Determining Sign** Tell whether each sum is negative or positive. Explain your reasoning.

   a. $3 + (-1)$ b. $-5 + (-2)$

   c. $-9 + 6$ d. $4 + (-7)$

4. **Determining Sign** Tell whether each sum is negative or positive. Explain your reasoning.

   a. $-8 + 10$ b. $2 + (-4)$

   c. $-6 + (-3)$ d. $7 + (-5)$

5. **Applying the Additive Inverse Property** Complete each statement to illustrate the Additive Inverse Property.

   a. $16 + \square = 0$

   b. $\square + (-8) = 0$

   c. $\square + 7 = 0$

   d. $-12 + \square = 0$

6. **Applying the Additive Inverse Property** Complete each statement to illustrate the Additive Inverse Property.

   a. $10 + \square = 0$ b. $\square + (-1) = 0$

   c. $\square + 5 = 0$ d. $-17 + \square = 0$

7. **Using Integer Counters** Describe how each model is used to add two integers. Then write and solve the addition problem represented by the model.

   a. ⊖ ⊖ ⊖ + ⊕ ⊕

   b. ⊕ ⊕ + ⊖

8. **Using Integer Counters** Describe how each model is used to add two integers. Then write and solve the addition problem represented by the model.

   a. ⊖ ⊖ ⊖ ⊖ ⊖ + ⊕ ⊕

   b. ⊕ ⊕ + ⊖ ⊖

9. **Using Integer Counters** Find each sum using integer counters.

   a. $-2 + 4$ b. $-3 + 9$

   c. $6 + (-11)$ d. $8 + (-1)$

   e. $-7 + (-5)$ f. $-9 + (-2)$

10. **Using Integer Counters** Find each sum using integer counters.

    a. $-7 + 10$ b. $-1 + 8$

    c. $3 + (-4)$ d. $12 + (-3)$

    e. $-4 + (-9)$ f. $-6 + (-13)$

11. **Using an Integer Number Line** Describe how the integer number line is used to add two integers. Then write and solve the addition problem represented by the model.

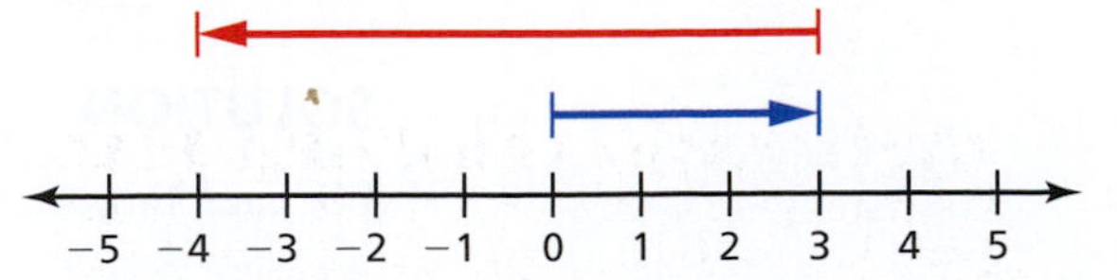

12. **Using an Integer Number Line** Describe how the integer number line is used to add two integers. Then write and solve the addition problem represented by the model.

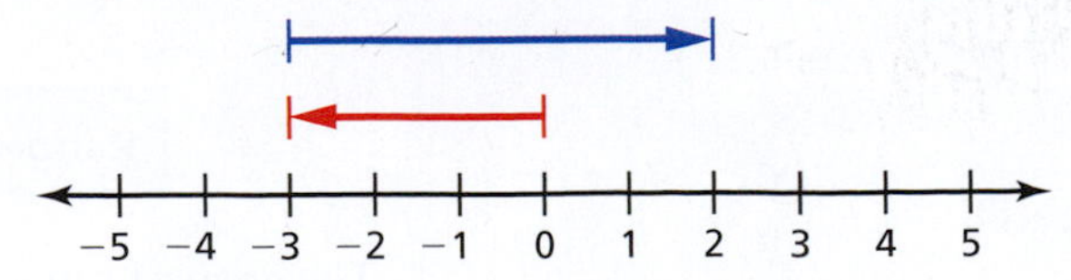

13. **Using an Integer Number Line** Find each sum using an integer number line.

a. $4 + (-1)$ b. $1 + (-4)$

c. $-8 + 2$ d. $9 + (-9)$

14. **Using an Integer Number Line** Find each sum using an integer number line.

a. $-1 + 2$ b. $-7 + 3$

c. $-5 + 5$ d. $8 + (-8)$

15. **Using Properties of Addition** Use the given property of addition to rewrite each expression.

a. Commutative Property: $3 + (-2) =$ ▭

b. Associative Property: $(-3 + 2) + 1 =$ ▭

c. Additive Identity Property: $-5 + 0 =$ ▭

16. **Using Properties of Addition** Use the given property of addition to rewrite each expression.

a. Commutative Property: $-7 + 2 =$ ▭

b. Associative Property: $9 + [(-4) + 6] =$ ▭

c. Additive Identity Property: $0 + (-10) =$ ▭

17. **Adding Integers** Find each sum.

a. $-3 + (-8)$ b. $-6 + (-9)$

c. $-3 + 15 + (-7)$ d. $-2 + 19 + (-13)$

e. $16 + (-17) + 2$ f. $19 + (-21) + (-4)$

18. **Adding Integers** Find each sum.

a. $-7 + (-1)$ b. $-3 + (-6)$

c. $-2 + 14 + (-5)$ d. $-10 + 12 + (-15)$

e. $11 + (-13) + 4$ f. $18 + (-22) + (-10)$

19. **Using Integer Counters** Which subtraction expression is represented by the integer counters model below?

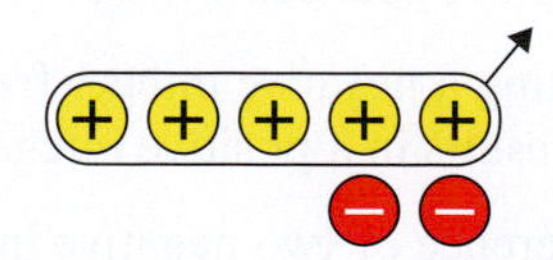

i. $5 - 2$ ii. $3 - 5$

iii. $2 - 5$ iv. $5 - 3$

20. **Using Integer Counters** Which subtraction expression is represented by the integer counters model below?

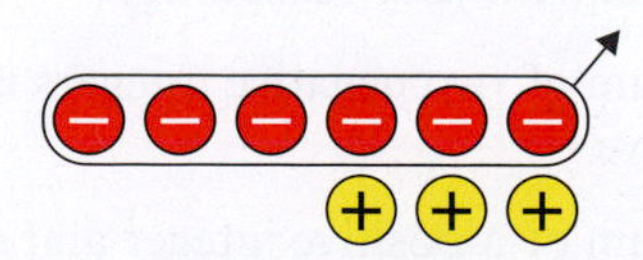

i. $-6 - 3$ ii. $3 - (-6)$

iii. $-3 - (-6)$ iv. $-6 - (-3)$

21. **Using Integer Counters** Find each difference using integer counters.

a. $7 - 14$ b. $5 - 10$

c. $-7 - 2$ d. $-1 - 3$

e. $3 - (-4)$ f. $-6 - (-3)$

22. **Using Integer Counters** Find each difference using integer counters.

a. $6 - 9$ b. $4 - 8$

c. $-9 - 4$ d. $-5 - 3$

e. $7 - (-2)$ f. $-4 - (-2)$

23. **Using an Integer Number Line** Describe how an integer number line is used to subtract two integers. Then write and solve the subtraction expression represented by the model.

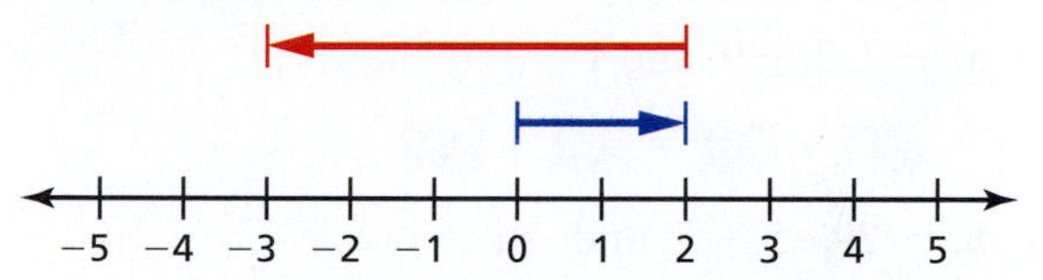

24. **Using an Integer Number Line** Describe how an integer number line is used to subtract two integers. Then write and solve the subtraction expression represented by the model.

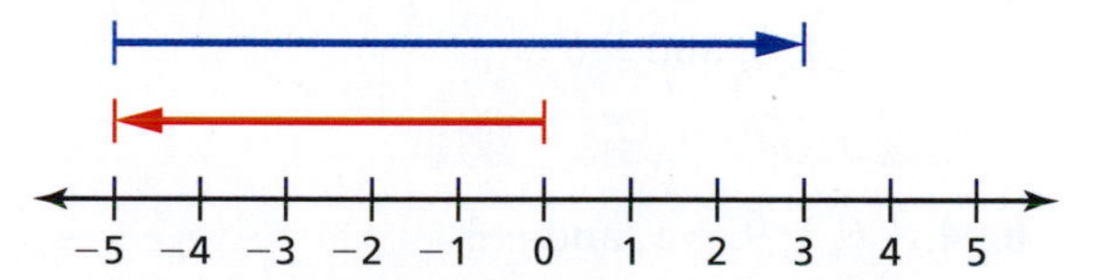

25. **Using an Integer Number Line** Find each difference using an integer number line.

a. $6 - 12$ b. $1 - (-2)$

c. $-12 - 16$ d. $-14 - (-3)$

26. **Using an Integer Number Line** Find each difference using an integer number line.

a. $3 - 9$ b. $-7 - 12$

c. $-5 - (-5)$ d. $-8 - (-1)$

27. **Subtracting Integers** Find each difference.

a. $0 - (-1)$ b. $0 - (-5)$

c. $-4 - 3 - 2$ d. $-7 - (-9) - 8$

e. $11 - 13 - (-6)$ f. $-15 - (-10) - 2$

28. **Subtracting Integers** Find each difference.

a. $0 - (-8)$ b. $0 - (-9)$

c. $4 - 5 - 12$

d. $-1 - (-7) - 14$

e. $-17 - (-2) - (-11)$

f. $6 - 13 - (-4)$

29. Adding and Subtracting Integers Evaluate each expression.

a. $12 + (-7) - 8$ b. $-19 - (-19) + 18$

c. $-17 - (-8) + (-9)$ d. $-13 + 10 - (-5)$

e. $|-4| + (-16) - 2$ f. $11 - |-9 + (-3)|$

30. Adding and Subtracting Integers Evaluate each expression.

a. $-12 + (-3) - (-15)$

b. $-8 - (-8) + 14$

c. $7 - 8 + (-1)$ d. $9 + (-6) - 11$

e. $-14 + |-9| - (-5)$ f. $|17 + (-19)| - 4$

31. My Dear Aunt Sally Arrange the numbers so that the equation is true. For more practice with this type of reasoning, go to *AuntSally.com*.

a. $-7, 3, -9$, and 1

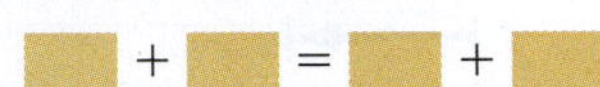

$\square + \square = \square + \square$

b. $-2, -5, -7$, and -4

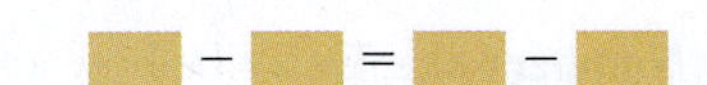

$\square - \square = \square - \square$

32. My Dear Aunt Sally Arrange the numbers so that the equation is true. For more practice with this type of reasoning, go to *AuntSally.com*.

a. $-7, -4, 5$, and -6

$\square + \square = \square - \square$

b. $4, -6, 1, 9, -1$, and -3

$\square + \square - \square = \square + \square - \square$

33. Problem Solving Write an expression that represents each real-life situation. Then evaluate the expression and answer the question.

a. The temperature is -10°C and is expected to rise 5°C by noon. What will the temperature be at noon?

b. In four downs, a football team lost 3 yards, gained 8 yards, gained 0 yards, and lost 7 yards. What is the total gain or loss?

34. Problem Solving Write an expression that represents each real-life situation. Then evaluate the expression and answer the question.

a. A plane is flying at 4000 feet and drops 600 feet. What is the new altitude of the plane?

b. In four bank transactions, you deposit \$130, withdraw \$70, withdraw \$35, and deposit \$90. What is the total debit or credit to your account?

35. Stocks A given stock drops 4 points and then gains 7 points. What is the net change in the stock's worth?

36. Elevators In an elevator, the ground floor is labeled Floor 0. The floors above and below ground level are labeled using positive integers and negative integers, respectively. You enter the elevator on the 5th floor, go down 7 floors and exit the elevator. Which floor are you on when you exit the elevator?

37. Scuba Diving A scuba diver dives to a depth of 50 feet, then rises 20 feet, dives 15 feet, and then rises 25 feet. What is the depth of the scuba diver relative to sea level?

38. Bird A bird dives 10 feet, then rises 12 feet, dives 8 feet, and then rises 3 feet. What is the change in altitude of the bird?

39. Grading Student Work On a diagnostic test, one of your students does the following work. Explain what the student did wrong. Which topics would you encourage the student to review?

a. $10 + (-15) = -25$

b. $3 - (-6) = -3$

40. Grading Student Work On a diagnostic test, one of your students does the following work. Explain what the student did wrong. Which topics would you encourage the student to review?

a. $-7 + |-13| = -6$

b. $|-11 - 2| = 9$

41. True or False? Tell whether each statement is *true* or *false*. Explain your reasoning.

a. Subtracting a negative integer from a positive integer results in a positive integer.

b. The difference of two negative integers is always negative.

c. To add two integers with different signs, you add their absolute values. Then use the sign of the integer with the greater absolute value.

42. True or False? Tell whether each statement is *true* or *false*. Explain your reasoning.

a. The sum of two negative integers is always negative.

b. The sum of a positive integer and a negative integer is never positive.

c. To subtract a from b, you add the absolute value of b to a.

43. Golf Scores Golf scores are measured by the number of strokes over or under par on each hole. For a nine-hole round of golf you score 1 birdie, 3 pars, 4 bogeys, and 1 double bogey. Use the table below to determine how your total score compares with par. Explain your reasoning.

Score	Compared with par
Birdie	1 under
Par	Even
Bogey	1 over
Double bogey	2 over

44. Missing Check You begin with \$150 in your checking account. You write 3 checks. One check is for \$45 and another is for \$61. You forget to record the amount of the third check. You receive a statement from the bank showing that your account is overdrawn by \$35. What was the amount of the third check?

45. Writing You choose an integer. You subtract 9 from the integer, find the opposite of the result, add -5, and then find the opposite of the new result. You end up with 1. What integer did you originally choose? Explain how you found the answer.

46. Writing You choose an integer. You add -6 to the integer, subtract 11, find the opposite of the result, and then find the opposite of the new result. You end up with -25. What integer did you originally choose? Explain how you found the answer.

47. Activity: In Your Classroom Pyramids like the one below can be used as a classroom activity for integer addition. Students fill in the pyramid so that the integer in each square is the sum of the pair of integers in the squares beneath it. Create a similar pyramid for your students to fill in, where the top square contains a negative integer.

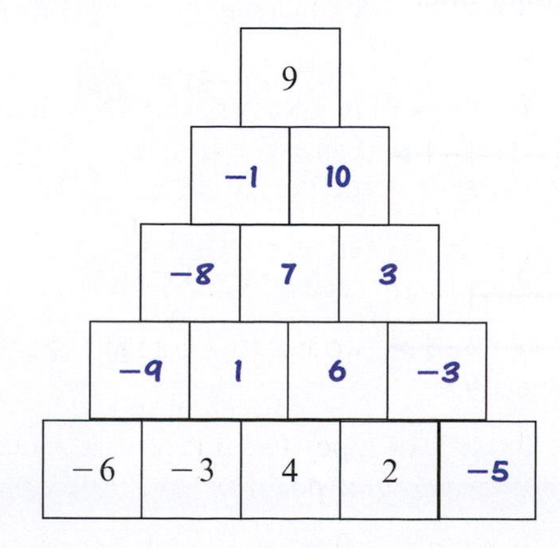

48. Activity: In Your Classroom Pyramids like the one below can be used as a classroom activity for integer subtraction. Students fill in the pyramid so that the integer in each square is the difference of the pair of integers in the squares beneath it. Create a similar pyramid for your students to fill in, where the integers in the bottom row are positive. Be sure that the completed pyramid includes some negative integers as well.

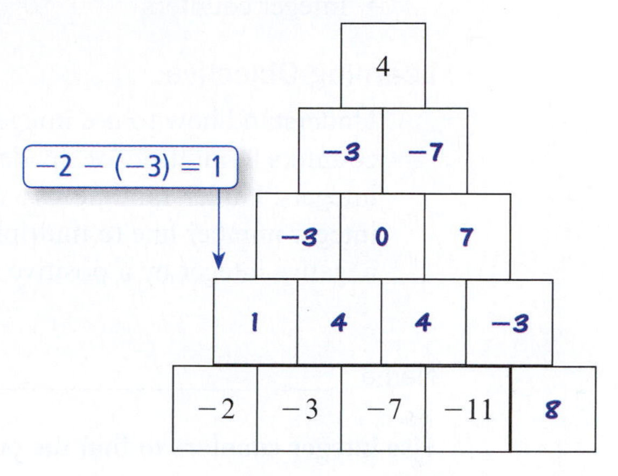

49. Equations Use mental math to find the value of x that makes each statement true.

a. $-x + 8 = -3$ **b.** $4 - x = -18$

50. Equations Use mental math to find the value of x that makes each statement true.

a. $-x - 7 = -11$ **b.** $-10 + x = -2$

51. Magic Square Use the integers -23, -11, 7, 13, and 25 to complete the *magic square* in which the sum of the entries in each row, column, and diagonal is the same.

−5		−17
	1	
19		

52. Magic Square Use the remaining integers from -8 through 6 to complete the *magic square* in which the sum of the entries in each row, column, and diagonal is the same.

6			
	0		
		−3	2
−6			−9

Activity: Multiplying and Dividing Integers Using Models

Materials:

- Pencil
- Integer counters

Learning Objective:

Understand how to use integer counters to multiply or divide integers. Understand how to use an integer number line to multiply a negative integer by a positive integer.

Name ______________________________

Use integer counters to find the product of the integers.

1. 3 · (⊖ ⊖ ⊖) = ▭

3 • ▭ = ▭

2. 4 · (⊕ ⊕ ⊕) = ▭

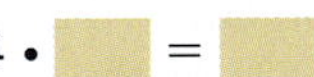

4 • ▭ = ▭

Use integer counters to find the quotient of the integers.

3.

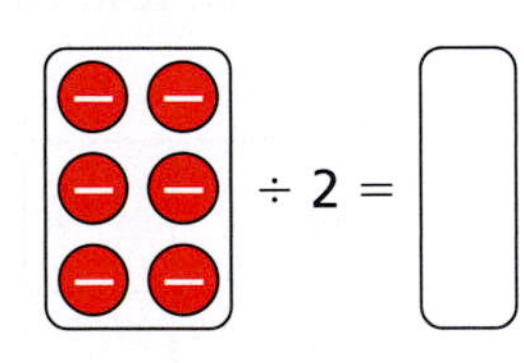

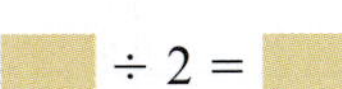

▭ ÷ 2 = ▭

4.

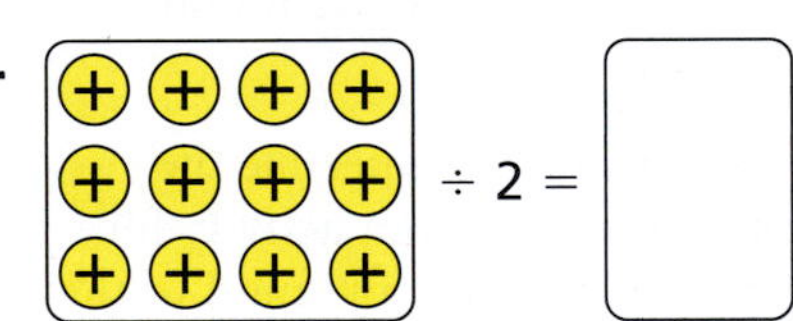

▭ ÷ 2 = ▭

Find the product that is modeled by the integer number line.

5.

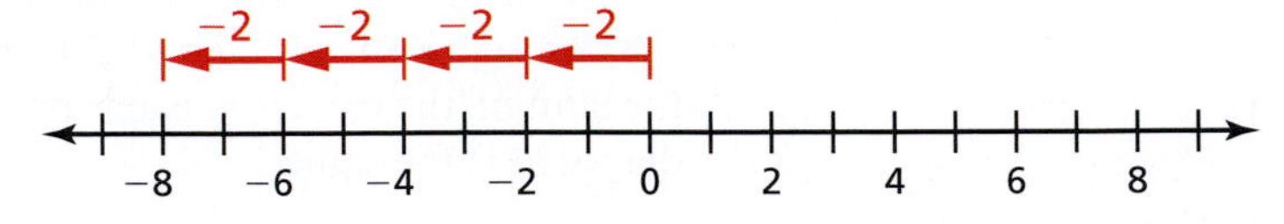

▭ • (−2) = ▭

6.

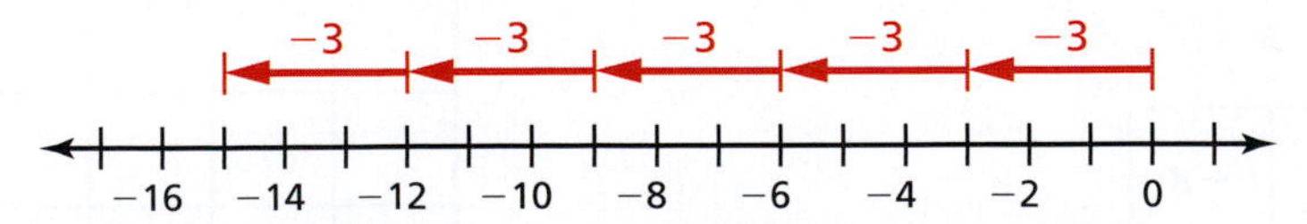

▭ • (−3) = ▭

8.3 Multiplying and Dividing Integers

Standards

Grades 6–8
The Number System
Students should apply and extend previous understandings of operations with fractions to add, subtract, multiply, and divide rational numbers.

- Understand how to multiply integers and to use properties of multiplication.
- Understand how to divide integers.
- Use negative integer exponents and scientific notation.

Multiplying Integers

In Section 3.3, you learned that the product of two whole numbers can be defined as repeated addition. Example 1 uses the same concept.

Classroom Tip
Multiplying and dividing integers is similar to multiplying and dividing whole numbers. When working with integers, pay close attention to the signs of the integers.

EXAMPLE 1 Multiplying Integers Using Repeated Addition

A company posts a loss of \$500 for each of its 4 quarters in a year. How much does the company lose for the entire year?

SOLUTION

Use repeated addition. The integer -500 represents a loss of \$500. Because the company *loses* \$500 each quarter for 4 quarters, add -500 four times.

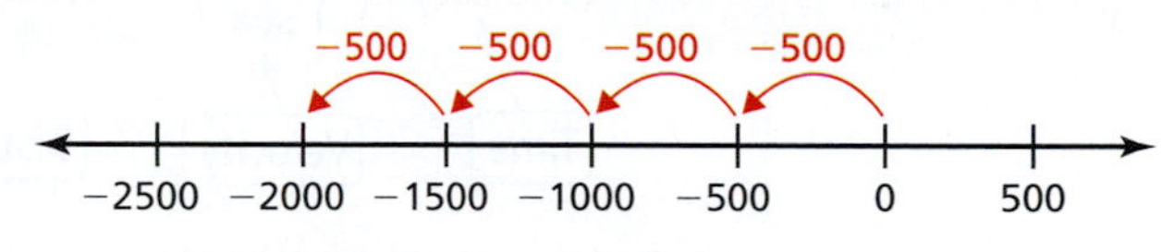

$$4 \cdot (-500) = -500 + (-500) + (-500) + (-500) = -2000$$

So, the company loses \$2000 for the year.

Multiplying Integers

Multiplying integers with the same sign The product of two integers with the *same* sign is positive.

Multiplying integers with different signs The product of two integers with *different* signs is negative.

Classroom Tip
The following properties of whole number multiplication in Section 3.3 are also valid for the multiplication of integers.
- Commutative Property
- Associative Property
- Multiplicative Identity Property
- Zero Multiplication Property
- Distributive Property of Multiplication Over Addition
- Distributive Property of Multiplication Over Subtraction

To better understand why the product of two *negative* integers is a positive integer, look at the product $(-1)(-1) = 1$. To show that this product is indeed equal to 1, assume that $(-1)(-1) = 1$ represents a true statement. In Section 8.2, you learned the Additive Inverse Property. So, you can write $1 + (-1) = 0$. Because you are assuming that $(-1)(-1) = 1$ is true, substitute $(-1)(-1)$ for 1 to get the equation

$$(-1)(-1) + (-1) = 0.$$

You can show this statement is true using the steps below.

$(-1)(-1) + (-1) = (-1)(-1) + 1(-1)$	Multiplicative Identity Property
$= (-1)[(-1) + 1]$	Distributive Property
$= (-1)0$	Additive Inverse Property
$= 0$	Zero Multiplication Property

Because $(-1)(-1) + (-1) = 0$, you know by the Additive Inverse Property that $(-1)(-1)$ is equal to the opposite of -1. So, $(-1)(-) = 1$.

Mathematical Practices

Look For and Make Use of Structure

MP7 Mathematically proficient students look closely to discern a pattern or structure.

EXAMPLE 2 Multiplying Integers

Find each product.

a. $-2 \cdot 8$

b. $-4 \cdot (-5)$

c. $-11 \cdot 0$

d. $9(-7)$

SOLUTION

a. $-2 \cdot 8 = -16$ — The integers have different signs. So, the product is negative.

b. $-4 \cdot (-5) = 20$ — The integers have the same sign. So, the product is positive.

c. $-11 \cdot 0 = 0$ — The product of any integer and 0 is 0.

d. $9(-7) = -63$ — The integers have different signs. So, the product is negative.

EXAMPLE 3 Multiplying Integers in Real Life

A parachutist falls 15 feet per second for 100 seconds. Write and evaluate an expression to find how far the parachutist falls in 100 seconds.

SOLUTION

$$(100 \text{ sec}) \cdot \left(\frac{-15 \text{ ft}}{1 \text{ sec}}\right) = -1500 \text{ ft}$$

Time — Velocity — Distance

The integers, 100 and -15, have different signs. So, the product is negative.

The parachutist falls 1500 feet in 100 seconds.

EXAMPLE 4 Using the Distributive Property

Use one of the Distributive Properties to evaluate

$-8 \cdot 113.$

SOLUTION

$-8 \cdot 113 = -8 \cdot (100 + 10 + 3)$ — Write 113 in expanded form.

$= -8(100) + (-8)(10) + (-8)(3)$ — Distributive Property

$= -800 + (-80) + (-24)$ — Multiply.

$= -904$ — Add.

Classroom Tip

When you are teaching the properties of multiplication and opposites, ask your students how these properties connect addition and multiplication. Here is one possible answer.

"Recall that another name for opposite is *additive inverse*. The first property at the right shows that you can obtain the additive inverse of *a* by multiplying *a* by −1."

The two properties below are important in algebra because they connect the two operations of addition and multiplication.

Properties of Multiplication and Opposites

Let a and b be any two integers.

1. **Words** The product of -1 and a is equal to the opposite of a.

   Numbers $-1 \cdot 3 = -3$ **Algebra** $(-1)a = -a$

2. **Words** The product of $-a$ and b is equal to the opposite of ab.

   Numbers $-2 \cdot 5 = -10$ **Algebra** $(-a)b = -(ab)$

Dividing Integers

Dividing Integers

Dividing integers with the same sign The quotient of two integers with the *same* sign is positive.

Dividing integers with different signs The quotient of two integers with *different* signs is negative.

EXAMPLE 5 Dividing Integers

a. $-12 \div 4 = -3$ The integers have different signs. So, the quotient is negative.

b. $-18 \div (-3) = 6$ The integers have the same sign. So, the quotient is positive.

c. $7 \div (-1) = -7$ The integers have different signs. So, the quotient is negative.

EXAMPLE 6 Dividing Integers in Real Life

The table shows the noon temperatures for a week in Duluth, Minnesota.

	Sun	Mon	Tues	Wed	Thu	Fri	Sat
Temperature (°F)	−10	−13	−8	−6	−2	0	4

What is the mean noon temperature for the week?

SOLUTION

To find the mean, find the sum of the temperatures. Then divide the sum by the number of days in a week, 7.

$$\text{Mean} = \frac{-10 + (-13) + (-8) + (-6) + (-2) + 0 + 4}{7}$$

$$= -\frac{35}{7}$$ Add the integers in the numerator.

$$= -5$$ The integers have different signs. So, the quotient is negative.

The mean noon temperature is -5°F.

Cancellation Property of Multiplication

Let a, b, and c be integers, where $c \neq 0$. If $ac = bc$, then $a = b$.

EXAMPLE 7 Using the Cancellation Property of Multiplication

Solve the equation $3x = -6$.

SOLUTION

$$3x = 3 \cdot (-2)$$ Rewrite -6.

$$x = -2$$ Cancellation Property of Multiplication

The solution is $x = -2$.

Negative Integer Exponents and Scientific Notation

In Section 4.2, you studied the Quotient of Powers Property. This property states that when dividing two powers that have the same base, keep the base and subtract the exponents.

$$\frac{a^m}{a^n} = a^{m-n}, \text{ where } a, m, \text{ and } n \text{ are nonzero and } m \geq n.$$

Now, after introducing negative integers, you can remove the requirement that $m \geq n$.

Negative Integer Exponents

Words For any integer n and any nonzero number a, a^{-n} is equal to 1 divided by a^n.

Numbers $2^{-3} = \frac{1}{2^3} = \frac{1}{8}$ **Algebra** $a^{-n} = \frac{1}{a^n}$, where $a \neq 0$

EXAMPLE 8 **Using Negative Exponents**

a. $4^{-2} = \frac{1}{4^2}$ Definition of negative integer exponent

$= \frac{1}{16}$ Evaluate 4^2.

b. $10^{-1} = \frac{1}{10^1}$ Definition of negative integer exponent

$= \frac{1}{10}$ Evaluate 10^1.

c. $(-2)^{-3} = \frac{1}{(-2)^3}$ Definition of negative integer exponent

$= \frac{1}{-8}$ Evaluate $(-2)^3$.

$= -\frac{1}{8}$ Rewrite.

In Section 7.3, you learned how to use scientific notation to represent large numbers. Recall that a decimal written in **scientific notation** is expressed in the form $a \times 10^n$, where $1 \leq a < 10$ and n is a whole number. Now, with the introduction of negative integer exponents, you can expand the definition so that n can be *any* integer. Below are some examples.

$$1.25 \times 10^{-2} = 1.25 \times \frac{1}{10^2} = \frac{1.25}{100} = 0.0125$$

Decimal point moves 2 decimal places to the left.

$$1.25 \times 10^{-5} = 1.25 \times \frac{1}{10^5} = \frac{1.25}{100{,}000} = 0.0000125$$

Decimal point moves 5 decimal places to the left.

$$1.25 \times 10^{-8} = 1.25 \times \frac{1}{10^8} = \frac{1.25}{100{,}000{,}000} = 0.0000000125$$

Decimal point moves 8 decimal places to the left.

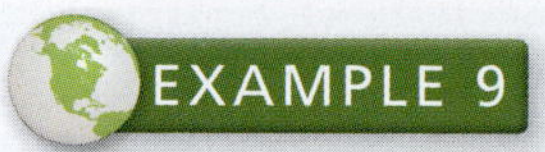

EXAMPLE 9 Writing a Small Number in Scientific Notation

Apollo 15 was the fourth mission to land astronauts on the Moon. The mission lasted over 12 days and included the first flight of the lunar roving vehicle (LRV). Using this vehicle, astronauts collected moon dust samples. A lunar dust particle has a mass of 0.000000000753 kilogram. Write this mass in scientific notation.

SOLUTION

Writing this number in scientific notation, you know the mantissa will be 7.53 because it needs to be greater than or equal to 1 and less than 10. Find the number of decimal places the decimal point needs to move by counting the number of decimal places up to and including 7.

$$0.\underbrace{0000000007}_{\text{10 decimal places}}.53$$

In scientific notation, the mass of a lunar dust particle is

7.53×10^{-10} kilogram.

EXAMPLE 10 Writing a Small Number in Scientific Notation

Classroom Tip

To help your students make sense of extremely small measurements, ask them questions, such as "How many cold viruses need to be lined up end-to-end to make 1 centimeter." The answer is 500,000!

A common cold virus has a diameter d of 20 nanometers. A nanometer is one-millionth of a millimeter. Write the diameter, in millimeters, of a cold virus in scientific notation.

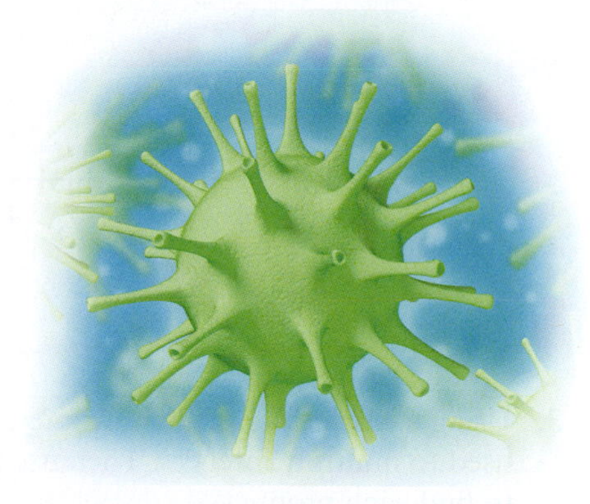

SOLUTION

Write the diameter in millimeters.

$$d = 20\ \cancel{\text{nm}} \cdot \frac{1\text{mm}}{1{,}000{,}000\ \cancel{\text{nm}}}$$

$$= \frac{20}{1{,}000{,}000}\ \text{mm}$$

$$= 0.00002\ \text{mm}$$

Writing this number in scientific notation, you know the mantissa will be 2.0. Find the number of decimal places the decimal point needs to move by counting the number of decimal places up to and including 2.

$$0.\underbrace{00002}_{\text{5 decimal places}}.$$

In scientific notation, the diameter of a cold virus is

2.0×10^{-5} millimeter.

NASA; iStockphoto.com/solvod

8.3 Exercises

Free step-by-step solutions to odd-numbered exercises at MathematicalPractices.com.

1. **Writing a Solution Key** Write a solution key for the activity on page 312. Describe a rubric for grading a student's work.

2. **Grading the Activity** In the activity on page 312, a student gave the answers below. Each question is worth 10 points. Assign a grade to each answer. For those that are incorrect, why do you think the student erred?

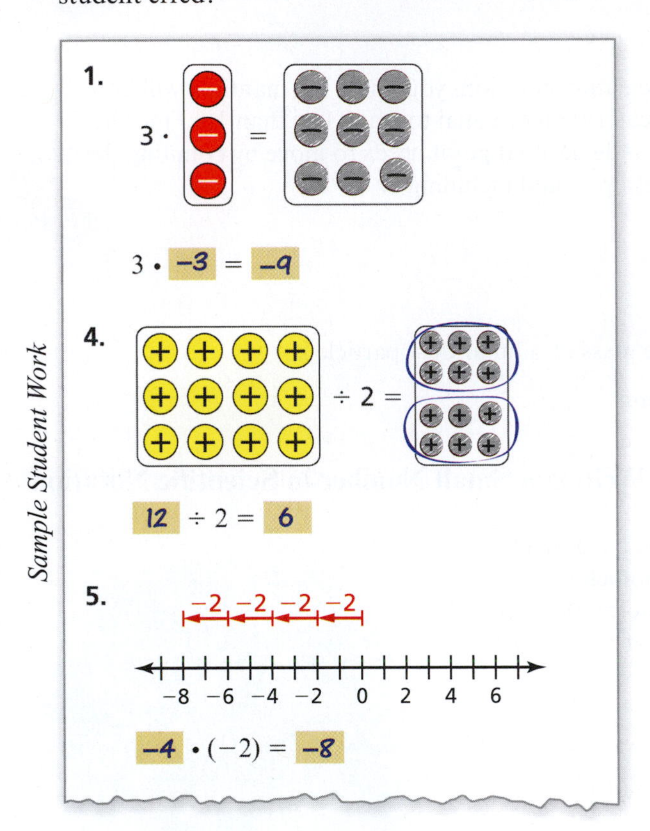

3. **Using a Model** Describe how integer counters are used to multiply integers. Then use integer counters to find each product.
   a. $-5 \cdot 2$ b. $-7 \cdot 6$
   c. $3 \cdot (-3)$ d. $1 \cdot (-12)$

4. **Using a Model** Describe how an integer number line is used to multiply integers. Then choose an integer number line to find each product.
   a. $-3 \cdot 4$ b. $-2 \cdot 8$
   c. $11 \cdot (-2)$ d. $7 \cdot (-1)$

5. **Multiplying Integers** Find each product.
   a. $-12 \cdot (-5)$ b. $-1 \cdot (-9)$
   c. $11 \cdot (-2)$ d. $8 \cdot (-20)$

6. **Multiplying Integers** Find each product.
   a. $-4 \cdot (-12)$ b. $-8 \cdot (-3)$
   c. $6 \cdot (-2)$ d. $3 \cdot (-7)$

7. **Using Properties of Multiplication** Use the given property of multiplication to rewrite each expression.
   a. Commutative Property: $-2 \cdot 8 = \square$
   b. Associative Property: $4 \cdot [(-5) \cdot 7] = \square$
   c. Multiplicative Identity Property: $-9 \cdot 1 = \square$
   d. Zero Multiplication Property: $-7 \cdot 0 = \square$

8. **Using Properties of Multiplication** Use the given property of multiplication to rewrite each expression.
   a. Commutative Property: $x \cdot (-7) = \square$
   b. Associative Property: $-6 \cdot (2 \cdot x) = \square$
   c. Multiplicative Identity Property: $-x \cdot 1 = \square$
   d. Zero Multiplication Property: $-x \cdot 0 = \square$

9. **Using the Distributive Properties** Use one of the Distributive Properties to find each product.
   a. $-6 \cdot 15$ b. $12 \cdot (-42)$

10. **Using the Distributive Properties** Use one of the Distributive Properties to find each product.
   a. $-9 \cdot 84$ b. $11 \cdot (-135)$

11. **Using Properties of Multiplication and Opposites** Use the properties of multiplication and opposites to find each product. Explain your process.
   a. $-1 \cdot 21$ b. $-3 \cdot 5$

12. **Using Properties of Multiplication and Opposites** Use the properties of multiplication and opposites to find each product. Explain your process.
   a. $x \cdot (-1)$ b. $x \cdot (-4)$

13. **Multiplying Integers** Find each product.
   a. $-3 \cdot (-8) \cdot 0$ b. $-2 \cdot (-5) \cdot (-9)$
   c. $7 \cdot (-6) \cdot (-1)$ d. $4 \cdot 1 \cdot (-5)$

14. **Multiplying Integers** Find each product.
   a. $-9 \cdot (-4) \cdot 0$ b. $-7 \cdot (-6) \cdot (-5)$
   c. $3 \cdot (-2) \cdot 1$ d. $8 \cdot (-1) \cdot (-7)$

15. Using Integer Counters Find each quotient using integer counters.

a. $14 \div (-7)$ b. $-16 \div 4$

c. $-8 \div 2$ d. $30 \div (-3)$

16. Using Integer Counters Find each quotient using integer counters.

a. $6 \div (-2)$ b. $-12 \div 3$

c. $-11 \div 1$ d. $10 \div (-10)$

17. Dividing Integers Find each quotient.

a. $-36 \div (-2)$ b. $-63 \div (-7)$

c. $-\frac{-72}{4}$ d. $36 \div (-12)$

e. $\frac{-44}{-4}$ f. $12 \div (-4)$

18. Dividing Integers Find each quotient.

a. $-24 \div (-12)$ b. $-56 \div (-7)$

c. $22 \div (-11)$ d. $98 \div (-14)$

e. $\frac{85}{-5}$ f. $-\frac{-81}{3}$

19. Cancellation Property of Multiplication Solve each equation using the Cancellation Property of Multiplication.

a. $-2x = 16$

b. $3x = -21$

c. $-6x = -36$

20. Cancellation Property of Multiplication Solve each equation using the Cancellation Property of Multiplication.

a. $-5x = 35$

b. $4x = -24$

c. $-10x = -150$

21. Using Negative Exponents Evaluate each expression.

a. 2^{-2} b. $(-3)^{-2}$

c. 6^{-3} d. $(-1)^{-1}$

22. Using Negative Exponents Evaluate each expression.

a. 5^{-1} b. $(-4)^{-2}$

c. 2^{-6} d. $(-6)^{-3}$

23. Using Properties of Exponents Evaluate each expression.

a. $3^{-4} \cdot 3^2$ b. $5^{-3} \cdot 5^3$

c. $(4^{-2})^3$ d. $(-2^2)^{-2}$

e. $\frac{6^4}{6^7}$ f. $\frac{4^{-4}}{4^{-5}}$

24. Using Properties of Exponents Evaluate each expression.

a. $6^1 \cdot 6^{-2}$ b. $2^2 \cdot 2^{-5}$

c. $(7^{-1})^2$ d. $(-1^8)^{-5}$

e. $\frac{3^2}{3^4}$ f. $\frac{5^{-2}}{5^{-3}}$

25. Writing Numbers in Standard Form Write each number in standard form.

a. 2.35×10^{-2} b. 1.11×10^{-4}

c. 7.51×10^{-7} d. 5.7×10^{-10}

26. Writing Numbers in Standard Form Write each number in standard form.

a. 9.45×10^{-3} b. 4.02×10^{-5}

c. 3.63×10^{-8} d. 8.2×10^{-9}

27. Writing Numbers in Scientific Notation Write each number in scientific notation.

a. 0.0264

b. 0.000587

c. 0.0000012

d. 0.0000000895

28. Writing Numbers in Scientific Notation Write each number in scientific notation.

a. 0.00671

b. 0.0000456

c. 0.00000065

d. 0.0000000052

29. Using Order of Operations Use the order of operations to evaluate each expression.

a. $(-2)^2 \cdot 3 \div 4$

b. $-24 - 18 \div (-3) + 2$

c. $-16 - (-3 - 1 \times 2)$

d. $-1 + 3[7(-6 \div 2)]$

e. $-6 + \frac{16 - 4}{2^2 + 2}$

30. Using Order of Operations Use the order of operations to evaluate each expression.

a. $-16 - 32 \div 2^2$

b. $-13 + 2 \times 3^2 \div (-3)$

c. $-(-5 - 10 \div 5) + (-8)$

d. $6 - 2[8 - (-3 \times 2)]$

e. $\frac{-19 + (-2)}{6^2 - 29}$

31. Seal A seal dives at a speed of 2 meters per second. Write and evaluate an expression to represent the seal's change in position after 30 seconds.

32. Airplane An airplane descends 6 feet per second prior to landing. Write and evaluate an expression to represent the change in the altitude of the airplane after 1 minute.

33. Temperature The low temperatures during a 5-day period were −20°F, −15°F, −8°F, 12°F, and 16°F. What is the mean low temperature for the 5 days?

34. Profit The table below gives the annual earnings of a company during a 6-year period. What is the average annual earnings of the company over these 6 years?

Year	Earnings (dollars)
2009	247,200
2010	185,300
2011	−81,500
2012	−227,600
2013	−109,700
2014	98,800

35. Oxygen The mass of an oxygen atom is approximately 0.0000000000000000000000266 gram. Write the mass in scientific notation.

36. Hydrogen The mass of a hydrogen atom is approximately 0.00000000000000000000000167 gram. Write the mass in scientific notation.

37. Hair The diameter of a strand of human hair is 99 micrometers. A micrometer is one millionth of a meter. Write the diameter, in meters, of a human hair in scientific notation.

38. Sand The diameter of a sand particle is approximately 0.25 millimeters. A millimeter is one thousandth of a meter. Write the diameter, in meters, of a grain of sand in scientific notation.

39. Grading Student Work On a diagnostic test, one of your students does the following work. Explain what the student did wrong. Which topics would you encourage the student to review?

a. $-10 \cdot (-5) = -50$

b. $3^{-3} = -27$

40. Grading Student Work On a diagnostic test, one of your students does the following work. Explain what the student did wrong. Which topics would you encourage the student to review?

a. $-1 \div (-1) = -1$

b. $-3^{-2} = \frac{1}{9}$

41. My Dear Aunt Sally Arrange the numbers −2, 2, and −3 so that the equation is true.

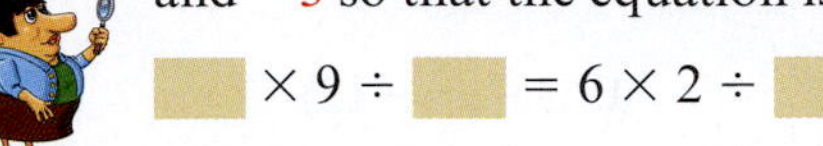

$\square \times 9 \div \square = 6 \times 2 \div \square$

For more practice with this type of reasoning, go to *AuntSally.com*.

42. My Dear Aunt Sally Arrange the numbers 4, −5, 2, and 6 so that the equation is true. (Note: The symbol ^ represents a power. So, $a \wedge b = a^b$.)

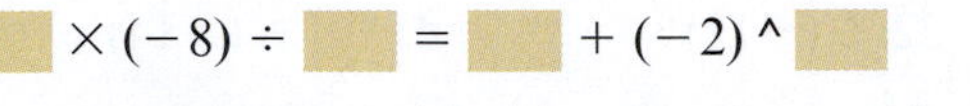

$\square \times (-8) \div \square = \square + (-2) \wedge \square$

For more practice with this type of reasoning, go to *AuntSally.com*.

43. Analysis Let x be an integer, where $x \neq 0$. Describe the set of integers that makes the expression positive. Then describe the set of integers that makes the expression negative. Explain your reasoning.

a. $-x$ **b.** x^2 **c.** x^3

44. Analysis Let x be an integer, where $x \neq 0$. Describe the set of integers that makes the expression positive. Then describe the set of integers that makes the expression negative. Explain your reasoning.

a. $(-x)^3$ **b.** $-x^2$ **c.** $(-x)^4$

45. Card Game You are playing a card game where you gain points by playing cards and lose points for cards you do not play. The face cards are each worth 10 points. The other cards are each worth 5 points. At the end of the game, you have earned 35 points but still have a king, 2 jacks, a 3, a 9, and an 8 in your hand.

a. Write an expression that represents your score for the game. Explain your reasoning.

b. What is your score for the game?

46. Smartphone You are playing an agility game on your smartphone. You begin the game with 400 points. Each time your character falls off a log, you lose 125 points.

a. Write an expression that represents your score for the game.

b. Your character falls off a log 4 times. What is your final score?

47. **Activity: In Your Classroom** Pyramids like the one below can be used as a classroom activity for integer multiplication. Students fill in the pyramid so that the integer in each square is the product of the pair of integers in the squares beneath it.

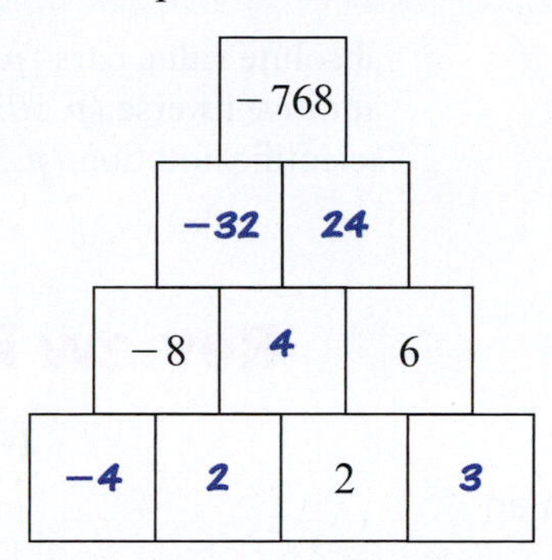

a. Is the solution given above unique? If not, give another solution.

b. Create a similar pyramid for your students to fill in, where the top square contains a positive integer. Be sure the completed pyramid includes some negative integers as well.

48. **Activity: In Your Classroom** Pyramids like the one below can be used as a classroom activity for integer division. Students fill in the pyramid so that the integer in each square is the quotient of the pair of integers in the squares beneath it.

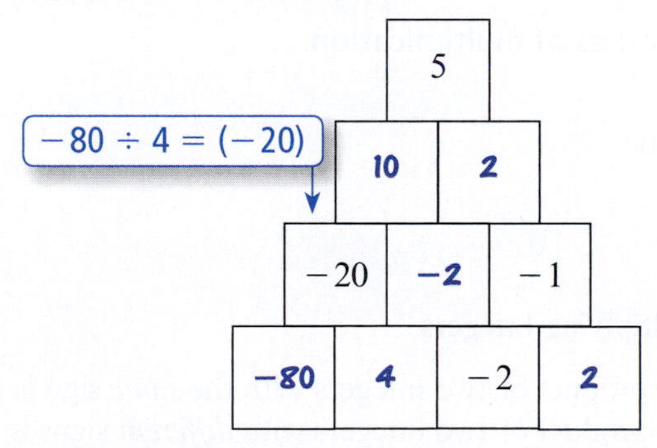

a. Is the solution given above unique? If not, give another solution.

b. Create a similar pyramid for your students to fill in, where the top square contains a negative integer.

49. **Problem Solving** The water level of a pond is measured once a week for 4 weeks. Each measurement is taken with respect to the normal water level. The mean of the 4 measurements is −2 inches. The first three measurements are shown below. What is the fourth measurement?

−2 inches, −5 inches, 1 inches

50. **Problem Solving** The mean earnings for a small business over a single year is $15,750. The earnings for the first three quarters are listed below. What are the company's earnings for the fourth quarter?

$43,000, −$17,000, $59,000

51. **Using Properties of Exponents** Evaluate each expression.

a. $\frac{3^3 \cdot 3^{-2}}{3^{-4}}$

b. $\frac{6^2}{(2^{-3})^2 \cdot 4^4}$

52. **Using Properties of Exponents** Evaluate each expression.

a. $\frac{(5^{-2})^{-3}}{5^7}$

b. $\frac{3^3 \cdot 2^{-2}}{-4^{-2}}$

53. **Chemistry** In a chemistry lab, one solution is heated to 37°C and then cooled at a rate of 7°C per minute. A second solution is cooled to −23°C and then heated at a rate of 5°C per minute.

a. Write an expression that represents the temperature of the first solution each minute.

b. Write an expression that represents the temperature of the second solution each minute.

c. Make a table that shows the temperature of each solution for the first 10 minutes.

d. After how many minutes will the temperature of both solutions be the same?

54. **Checking Account Balances** You have two checking accounts. In one account, you are overdrawn by $185. In the second account, you have a balance of $475. You deposit $35 daily into the first account but withdraw $25 daily from the second account.

a. Write an expression that represents the daily balance of the first account.

b. Write an expression that represents the daily balance of the second account.

c. Make a table that shows the daily balance of each account for the first two weeks.

d. After how many days will the balance in each account be the same?

55. **Writing** Explain how to use addition to find the product of 5 and −7.

56. **Writing** Evaluate $(-10)^1$, $(-10)^2$, $(-10)^3$, $(-10)^4$, and $(-10)^5$. How is the exponent related to the sign of the power?

57. **Reasoning** Let x and y be integers, and let $x < y$. Is it always true that $x^2 < y^2$? Explain your reasoning.

58. **Reasoning** Let x, y, and z be integers, and let $x < y$. Is it always true that $z - y < z - x$? Explain your reasoning.

Chapter Summary

Chapter Vocabulary

integers *(p. 293)*
opposites *(p. 293)*
positive integers *(p. 293)*
negative integers *(p. 293)*
graphing an integer *(p. 293)*
a is less than b *(p. 294)*
absolute value *(p. 295)*
absolute value bars *(p. 295)*
additive inverse *(p. 303)*
scientific notation *(p. 316)*

Chapter Learning Objectives | Review Exercises

8.1 Integers *(page 293)* — 1–36

- Understand the definition of integers, and graph integers on an integer number line.
- Use a number line to order integers.
- Find the absolute value of an integer.
- Recognize real-life problems that can be represented by integers.

8.2 Adding and Subtracting Integers *(page 303)* — 37–90

- Understand how to add integers.
- Use models to add integers.
- Understand how to subtract integers.
- Use models to subtract integers.

8.3 Multiplying and Dividing Fractions *(page 313)* — 91–146

- Understand how to multiply integers and to use properties of multiplication.
- Understand how to divide integers.
- Use negative integer exponents and scientific notation.

Important Concepts and Formulas

Absolute Value

Let a be a whole number.

$|a| = a$

$|0| = 0$

$|-a| = a$

Adding Integers

To add integers with the *same* sign, add the absolute values of the integers. Then use the common sign. To add integers with *different* signs, subtract the lesser absolute value from the greater absolute value. Then use the sign of the integer with the greater absolute value.

Additive Inverse Property

$a + (-a) = 0$

Subtracting Integers

Let a and b be integer. To subtract b from a, add the opposite of b to a.

Multiplying Integers

The product of two integers with the *same* sign is positive.
The product of two integers with *different* signs is negative.

Properties of Multiplication and Opposites

Let a and b be any two integers.

$(-1)a = -a$

$(-a)b = -(ab)$

Dividing Integers

The quotient of two integers with the *same* sign is positive.
The quotient of two integers with *different* signs is negative.

Cancellation Property of Multiplication

Let a, b, and c be integers, where $c \neq 0$.
If $ac = bc$, then $a = b$.

Negative Integer Exponents

$a^{-n} = \frac{1}{a^n}$, where $a \neq 0$

Review Exercises

Free step-by-step solutions to odd-numbered exercises at MathematicalPractices.com.

8.1 Integers

Graphing an Integer Graph the integer.

1. 5 **2.** 6

3. -8 **4.** -22

Identifying an Opposite Write the opposite of the integer.

5. -19 **6.** 54

7. 83 **8.** -35

Graphing an Integer and Its Opposite Graph the integer and its opposite.

9. -24 **10.** 32

11. 53 **12.** -40

Comparing Integers Complete the statement using the correct symbol: < or >.

13. -3 ▢ 0 **14.** -18 ▢ -21

15. 7 ▢ -4 **16.** -12 ▢ -23

Ordering Integers Use a number line to order the integers from least to greatest.

17. 0, -7, 5, 19, -13

18. -25, -17, 9, 1, -2

19. -12, 15, 0, -8, -30

20. 29, 14, -26, 11, -4

Evaluating an Absolute Value Evaluate the expression.

21. $|-51|$ **22.** $|27|$

23. $-|16|$ **24.** $-|-40|$

Comparing Integer Expressions Complete the statement using the correct symbol: <, >, or =.

25. $|-31|$ ▢ -31

26. $|-15|$ ▢ $|-16|$

27. $-|-9|$ ▢ -9

28. $|-35|$ ▢ 35

Ordering Integer Expressions Use a number line to order the integer expressions from least to greatest.

29. $|-28|$, 13, $-|-7|$, $-|9|$, 0

30. $|-13|$, -32, $-|-29|$, 33, -6

31. -22, $|-19|$, -8, $-|5|$, $|34|$

32. 37, $-|-39|$, -46, $-|2|$, $|-3|$

33. Comparing Integers Alaska's record low temperature during December is 72°F below zero. Alaska's record low temperature during January is 80°F below zero. Write the integers that represent the temperatures. Which month has the colder record low temperature?

34. Ordering Integers Five friends play a 9-hole round of golf. The table below shows each golfer's score at the end of the round. In golf, the lowest score wins.

Golfer	Score
Golfer A	2
Golfer B	-5
Golfer C	-1
Golfer D	0
Golfer E	-3

a. Order the scores from best to worst.

b. Which golfer won the round?

35. Guessing Game A game at the carnival involves guessing the number of marbles in a jar. You guess 125 marbles over the actual amount, and your friend guesses 76 marbles under the actual amount. Which guess is closer to the actual amount?

36. Quarterly Earnings The bar graph shows the quarterly earnings of a small company for one year.

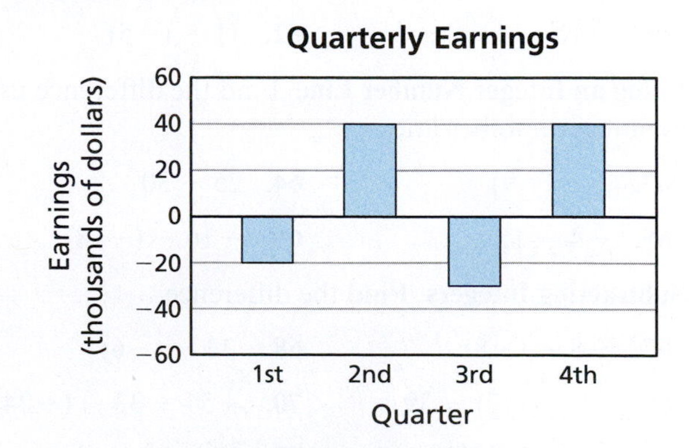

a. Which quarters show gains? Which quarters show losses?

b. Estimate the earnings for each quarter.

c. Did the company have a profit or a loss for the year? Explain.

8.2 Adding and Subtracting Integers

Applying the Additive Inverse Property Complete the statement to illustrate the Additive Inverse Property.

37. $15 + \square = 0$ **38.** $\square + (-13) = 0$

39. $-2 + \square = 0$ **40.** $\square + 21 = 0$

Using Integer Counters Find the sum using integer counters.

41. $-10 + 6$ **42.** $-5 + (-3)$

43. $4 + (-2)$ **44.** $-7 + 12$

Using an Integer Number Line Find the sum using an integer number line.

45. $-6 + 6$ **46.** $9 + (-11)$

47. $-18 + (-4)$ **48.** $17 + (-13)$

Identifying Properties of Addition State the property illustrated by the statement. Explain your reasoning.

49. $-12 + 22 = 22 + (-12)$

50. $-43 + 0 = -43$

51. $-17 + (9 + 11) = (-17 + 9) + 11$

52. $35 + (-35) = 0$

Adding Integers Find the sum.

53. $25 + (-16)$ **54.** $-22 + 9$

55. $-14 + (-5) + 23$ **56.** $13 + (-30) + 6$

57. $12 + (-23) + 9$ **58.** $-19 + 32 + (-4)$

Using Integer Counters Find the difference using integer counters.

59. $6 - 13$ **60.** $-10 - (-6)$

61. $-12 - 4$ **62.** $11 - (-5)$

Using an Integer Number Line Find the difference using an integer number line.

63. $2 - (-9)$ **64.** $25 - 30$

65. $-4 - 15$ **66.** $-16 - (-18)$

Subtracting Integers Find the difference.

67. $-8 - (-8)$ **68.** $32 - (-6)$

69. $47 - (-2) - 39$ **70.** $-54 - 33 - (-24)$

71. $-21 - (-30) - 5$ **72.** $22 - 28 - (-9)$

Adding and Subtracting Integers Evaluate the expression.

73. $15 + (-3) - 11$ **74.** $7 - 9 + 12$

75. $-11 + 3 - 4$ **76.** $5 - (-18) + (-24)$

77. $9 - (-3) + 6$ **78.** $-17 - (-10) + 2$

79. $-13 + (-8) - 1$ **80.** $-12 - 7 + (-5)$

81. **Problem Solving** You buy 5 trading cards. You trade 2 of the cards to a friend for 4 of his cards. You then trade 3 cards to another friend for 1 of her cards.

a. Write an expression that represents the real-life situation.

b. How many trading cards are you left with?

82. **Problem Solving** In four bank transactions, you withdraw \$75, deposit \$50, withdraw \$60, and withdraw \$35.

a. Write an expression that represents the real-life situation.

b. What is the total debit or credit to your account?

83. **Chemistry** In a lab, a chemical solution is heated to a temperature of 23°C. The chemical solution is checked 3 hours later and has cooled 30°C. What is the temperature of the solution when it is checked?

84. **Elevators** In an elevator, the ground floor is labeled Floor 0. The floors above and below ground level are labeled using positive integers and negative integers, respectively. You enter the elevator on the ground floor. The elevator goes up 5 floors, down 6 floors, up 2 floors, and down 3 floors, at which point you exit the elevator. What floor are you on when you exit the elevator?

85. **Classroom Supplies** A school agrees to reimburse a teacher up to \$85 for classroom supplies. The teacher wants to buy the supplies listed below. Subtract the total cost of the supplies from the amount of the reimbursement, and interpret the result.

Item	Cost
Chalk	\$4
Chalkboard erasers	\$7
Bulletin board	\$14
Craft supplies	\$72

86. **School Supplies** You receive a \$500 scholarship to purchase books for school. You purchase 5 books that cost \$125, \$82, \$154, \$53, and \$78. Subtract the total scholarship amount from the total cost of the books, and interpret the result.

Equations Use mental math to find the value of x that makes the statement true.

87. $x + 6 = -2$

88. $-5 - x = 4$

89. $12 + x = -1$

90. $-x - 9 = -13$

8.3 Multiplying and Dividing Integers

Multiplying Integers Find the product.

91. $-5 \cdot 42$ **92.** $13 \cdot (-6)$

93. $-12 \cdot (-7)$ **94.** $-62 \cdot 2$

95. $3 \cdot (-8) \cdot (-4)$ **96.** $-15 \cdot 5 \cdot (-1)$

97. $-9 \cdot (-2) \cdot (-3)$ **98.** $8 \cdot (-10) \cdot 0$

Identifying Properties of Multiplication State the property illustrated by the statement. Explain your reasoning.

99. $-5 \cdot 1 = -5$

100. $-3 \cdot 0 = 0$

101. $(-4 \cdot 6) \cdot 8 = -4 \cdot (6 \cdot 8)$

102. $7 \cdot (-2) = -2 \cdot 7$

Using the Distributive Properties Use one of the Distributive Properties to evaluate the expression.

103. $3 \cdot (200 - 90)$

104. $-7 \cdot (-50 + 6)$

105. $14 \cdot (-120)$

106. $-6 \cdot 93$

Dividing Integers Find the quotient.

107. $25 \div (-5)$ **108.** $-48 \div (-4)$

109. $-2 \div 2$ **110.** $-6 \div (-1)$

111. $-\dfrac{-27}{3}$ **112.** $\dfrac{-34}{-2}$

113. $\dfrac{72}{-9}$ **114.** $-\dfrac{56}{-7}$

Cancellation Property of Multiplication Solve the equation using the Cancellation Property of Multiplication.

115. $9x = -27$

116. $-12x = -48$

117. $-6x = 54$

118. $-3x = 45$

Using Negative Exponents Evaluate the expression.

119. 3^{-4} **120.** -9^{-2}

121. $(-5)^{-4}$ **122.** $(-2)^{-5}$

Using Properties of Exponents Evaluate the expression.

123. $9^3 \cdot 9^{-1}$ **124.** $(-2)^{-7} \cdot (-2)^4$

125. $(8^2)^{-1}$ **126.** $(-3^{-2})^2$

127. $\dfrac{(-5)^7}{(-5)^4}$ **128.** $\dfrac{12^{-4}}{12^{-6}}$

129. $\dfrac{4^2 \cdot 4^{-4}}{4^{-5}}$ **130.** $\dfrac{3^{-3}}{3^{-4} \cdot 3^{-2}}$

Writing Numbers in Standard Form Write the number in standard form.

131. 1.57×10^{-6} **132.** 3.24×10^{-9}

133. 6.4×10^{-2} **134.** 8.1×10^{-11}

Writing Numbers in Scientific Notation Write the number in scientific notation.

135. 0.00269

136. 0.000586

137. 0.000000073

138. 0.000000418

Using Order of Operations Use the order of operations to evaluate the expression.

139. $(-3)^{-2} \cdot 9 + (-17)$

140. $-8^2 - 4 \cdot 5 \div 2$

141. $6 - \dfrac{2(7 + 9 \div 3)}{-4}$

142. $12 + \dfrac{[-2 + (-1)]^3}{9}$

143. **Temperatures** At 8:00 A.M., the temperature is −3°F. The temperature is expected to rise 2°F each hour. Predict the temperature at noon.

144. **Scuba Diving** The scuba diver shown begins to ascend at a rate of 7 meters per minute. What is the diver's position relative to sea level after 3 minutes?

145. **Golf** A golfer has a mean score of −2 after four rounds. The golfer's scores in the first three rounds were −3, 1, and −2. What was the golfer's score in the fourth round?

146. **Atoms** The atomic radius of helium is about 31 picometers. A picometer is one billionth of a millimeter. Write the atomic radius, in millimeters, of helium in scientific notation.

Chapter Test

The questions on this practice test are modeled after questions on state certification exams for K–8 teaching.

Test Directions: Each of the questions is followed by five choices. Choose the *best* response to each question.

1. Consider the following integers $-16, 6, 7, -3, -7$. Which answer below shows the numbers arranged in descending order?

A. $-3, -7, -16, 6, 7$ **B.** $-16, -7, -3, 6, 7$ **C.** $7, 6, -3, -7, -16$

D. $-16, 7, -3, 6, -7$ **E.** $-7, -16, -3, 6, 7$

2. How many negative integers are between -10 and 4?

A. 6 **B.** 8 **C.** 9

D. 10 **E.** 13

3. Which expression represents the model below?

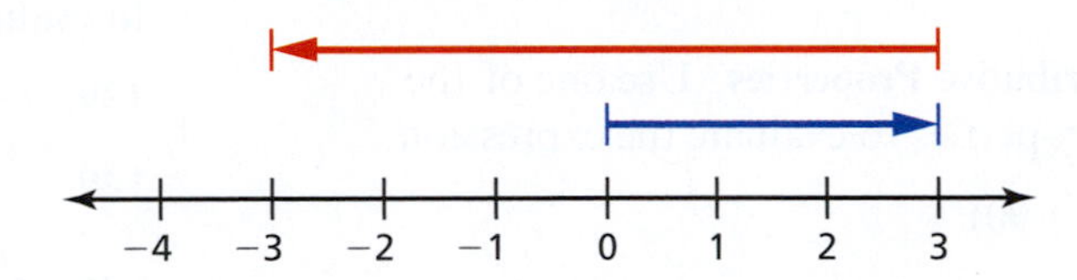

A. $3 + (-3)$ **B.** $-3 - 6$ **C.** $0 - 6$

D. 12 **E.** $3 - 6$

4. If $x = -2$, then find the value of $-x^2 + 3x$.

A. -10 **B.** -3 **C.** 3

D. 6 **E.** 15

5. On the number line below, point T (not shown) is located 3 units from point S, and point U (not shown) is located 2 units from point T. Which of the following could be the coordinate of point U?

A. -5 **B.** -3 **C.** 1

D. 4 **E.** 5

6. Simplify $-47 + (-13) - 99 - (-75)$.

A. -234 **B.** -208 **C.** -140

D. -84 **E.** -58

7. Calculate the expression shown below and write the answer in scientific notation.

$$0.002 \times 1.15$$

A. 2.3 **B.** 0.23×10^{-2} **C.** 0.0023×10^{2}

D. 2.3×10^{-3} **E.** 2.3×10^{3}

8. The width of a dust mite is 0.00032 meter. What is the width of the dust mite in scientific notation?

A. 0.32×10^{-3} m **B.** 3.2×10^{4} m **C.** 3.2×10^{3} m

D. 3.2×10^{-4} m **E.** 32.0×10^{4} m

9 Rational Numbers and Real Numbers

iStockphoto.com/monkeybusinessimages

Activity: Identifying and Comparing Rational Numbers

Materials:

- Pencil

Learning Objective:

Understand how to identify and compare rational numbers on a number line. Identify the opposite of a rational number.

Name ______________________________

Identify the rational number and write the number in ☐. Explain your reasoning.

1. $a =$ ☐

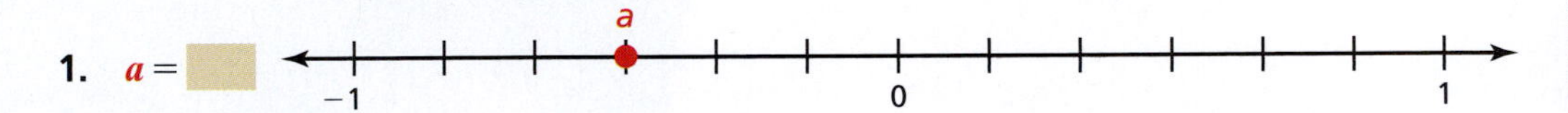

2. $b =$ ☐

b
−1 0 1

3. $c =$ ☐

c
−1 0 1

4. $d =$ ☐

d
−1 0 1

Write the rational numbers on the number line in ☐ to make the inequality true.

5. ☐ < ☐

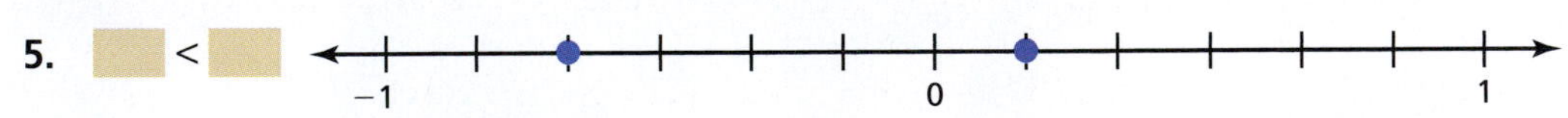

6. ☐ > ☐

−1 0 1

Write the opposite of the rational number. Then use the numbers to complete the addition problem.

7. $-a =$ ☐

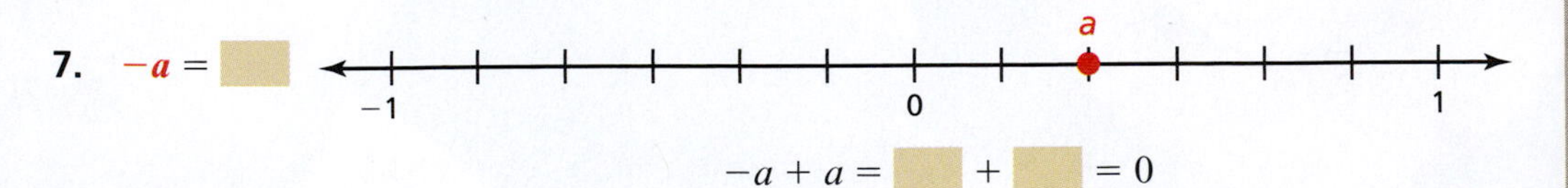

$-a + a =$ ☐ + ☐ $= 0$

8. $-b =$ ☐

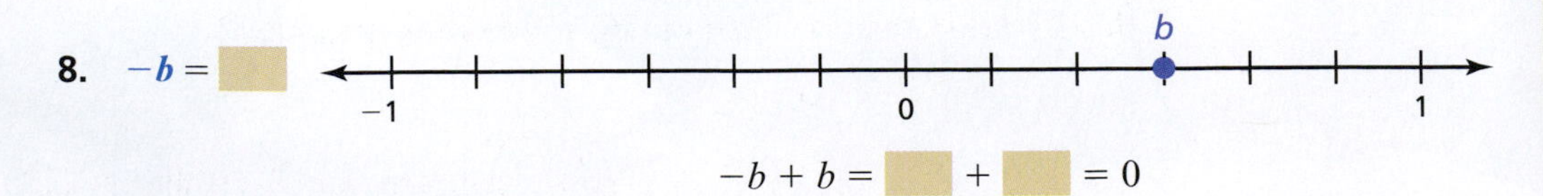

$-b + b =$ ☐ + ☐ $= 0$

9.1 Rational Numbers

Standards

Grades 6–8

The Number System

Students should apply and extend previous understandings of numbers to the system of rational numbers, and apply and extend previous understandings of operations with fractions to add, subtract, multiply, and divide rational numbers.

- Understand the definition of a rational number, and graph rational numbers on a number line.
- Use a number line to compare and order rational numbers.
- Understand properties of rational numbers and perform operations.

Rational Numbers

Definition of a Rational Number

A **rational number** is a number that can be written as

$$\frac{a}{b} \qquad \text{Ratio of } a \text{ to } b$$

where a and b are integers and $b \neq 0$.

Equality of Rational Numbers Two rational numbers, $\frac{a}{b}$ and $\frac{c}{d}$, are equal if and only if their cross products are equal.

$$\frac{a}{b} = \frac{c}{d} \qquad \text{if and only if} \qquad ad = bc.$$

Example $\frac{1}{-2} = \frac{-2}{4}$ because $(1)(4) = (-2)(-2)$.

The rational number $-\frac{1}{2}$ is read as "negative one-half." Locating the point that represents $-\frac{1}{2}$ on the number line is called **graphing a rational number**.

Negative one-half

−2 −1 0 1 2

Classroom Tip

Although the definition of a rational number allows for a negative integer in the numerator or denominator, it is more common to write the negative sign in front of the number. For instance,

$$\frac{1}{-2} \text{ and } \frac{-1}{2}$$

are usually written as $-\frac{1}{2}$.

EXAMPLE 1 Identifying Rational Numbers on a Number Line

Identify the rational numbers on the number line.

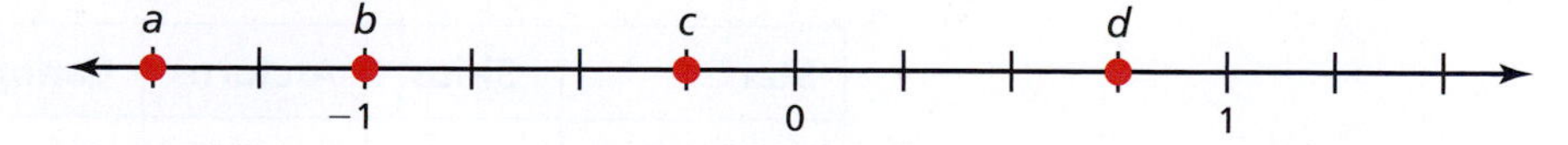

SOLUTION

$$a = -\frac{3}{2} \qquad b = -1 \qquad c = -\frac{1}{4} \qquad d = \frac{3}{4}$$

EXAMPLE 2 Justifying a Statement

Justify the statement "Every integer is a rational number."

SOLUTION

Let a be any integer. Because a can be written as $a = \frac{a}{1}$, it follows that a is a rational number.

Comparing and Ordering Rational Numbers

Comparing Rational Numbers

Let $\frac{a}{b}$ and $\frac{c}{d}$ be rational numbers. **$\frac{a}{b}$ is less than $\frac{c}{d}$**, written as $\frac{a}{b} < \frac{c}{d}$, if and only if $\frac{a}{b}$ is to the left of $\frac{c}{d}$ on a number line.

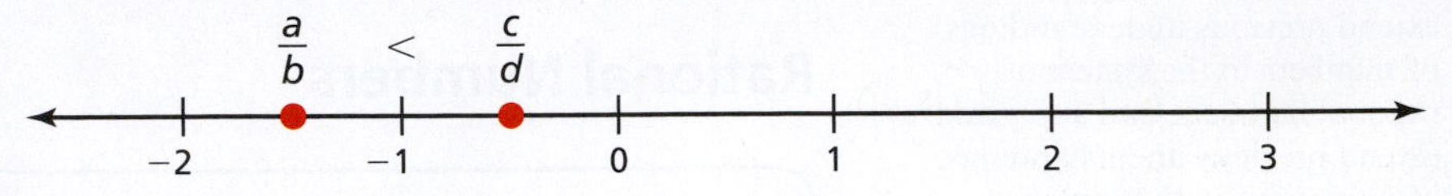

The inequality $\frac{a}{b} < \frac{c}{d}$ can also be written as $\frac{c}{d} > \frac{a}{b}$, read as **$\frac{c}{d}$ is greater than $\frac{a}{b}$**.

EXAMPLE 3 **Ordering Rational Numbers**

Order the rational numbers $\frac{3}{4}$, $-\frac{5}{4}$, $-\frac{1}{2}$, -1, and 0 from least to greatest.

SOLUTION

Graph the rational numbers on a number line.

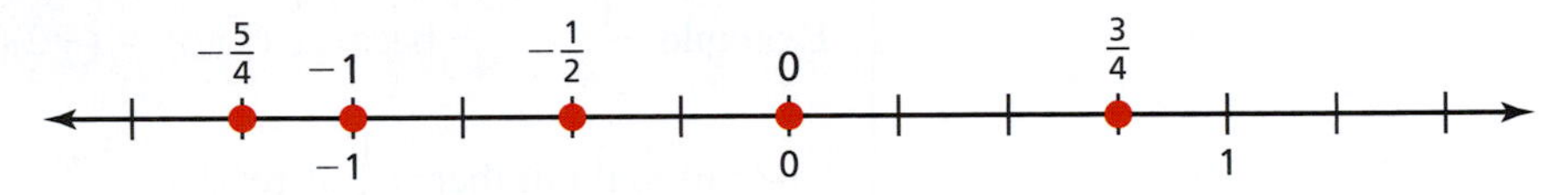

Reading from left to right, the order from least to greatest is $-\frac{5}{4}$, -1, $-\frac{1}{2}$, 0, and $\frac{3}{4}$.

EXAMPLE 4 **A Real-Life Example of Ordering Rational Numbers**

The apparent magnitude of a star is a number that indicates how bright the star appears as seen from Earth. The lesser the number, the brighter the star. The table shows the approximate apparent magnitudes of several stars. Which star is the brightest? Which star is the dimmest?

Star	Sirius	Arcturus	Canopus	Rigel	Rigil Kentaurus
Apparent magnitude	$-1\frac{1}{2}$	$-\frac{1}{25}$	$-\frac{7}{10}$	$\frac{1}{10}$	$-\frac{3}{10}$

SOLUTION

Graph the rational numbers on a number line.

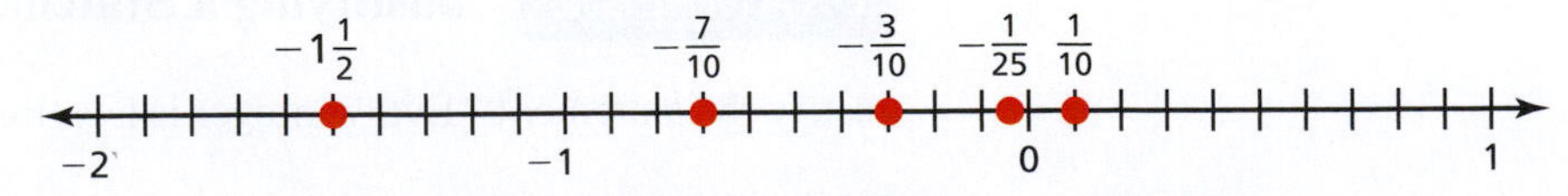

The order of the apparent magnitudes from least to greatest is

$$-1\frac{1}{2}, -\frac{7}{10}, -\frac{3}{10}, -\frac{1}{25}, \text{ and } \frac{1}{10}.$$

So, of these stars, the brightest is Sirius and the dimmest is Rigel.

Rational Numbers and Operations

Throughout this book, the number sets have been built in the same order as they are built in the K–8 curriculum. The natural, or counting, numbers are introduced first. The number 0 is included to form the set of whole numbers. To describe the parts of a whole, fractions are introduced. All fractions can be written as terminating or repeating decimals. To describe amounts less than zero, integers are introduced. Every fraction and every integer is a rational number.

Classroom Tip

Notice that the term *fraction* is generally used to imply only nonnegative numbers. In contrast, the term *rational number* implies that the number can be positive, zero, or negative.

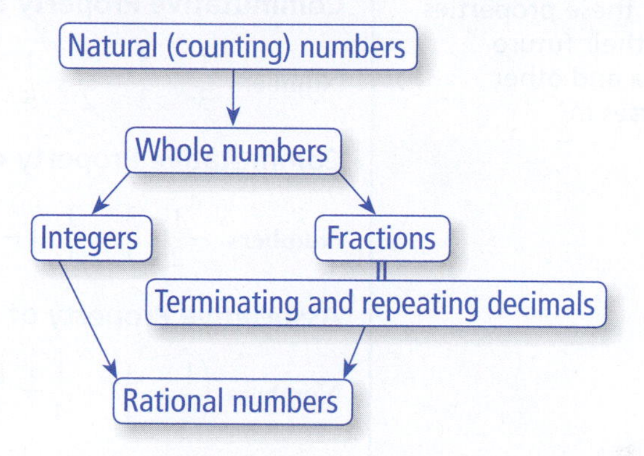

Definition of a Closed Set of Numbers

A set of numbers is **closed** under a mathematical operation if, when you apply the operation to two elements of the set, the result is also an element of the set.

Closure of the Set of Rational Numbers

The set of rational numbers is closed under all four operations: addition, subtraction, multiplication, and division.

Mathematical Practices

Attend to Precision

MP6 Mathematically proficient students use clear definitions in discussion with others and in their own reasoning.

EXAMPLE 5 Describing the Non-closure of Sets of Numbers

a. The set of natural numbers, or counting numbers, is closed under addition and multiplication, but *not* under subtraction and division.

Examples $2 - 2 = 0$ 0 is *not* a natural number.

$1 \div 5 = \frac{1}{5}$ $\frac{1}{5}$ is *not* a natural number.

b. The set of whole numbers is closed under addition and multiplication, but *not* under subtraction and division.

Examples $2 - 3 = -1$ -1 is *not* a whole number.

$3 \div 2 = \frac{3}{2}$ $\frac{3}{2}$ is *not* a whole number.

c. The set of integers is closed under addition, subtraction, and multiplication, but *not* under division.

Example $3 \div 2 = \frac{3}{2}$ $\frac{3}{2}$ is *not* an integer.

d. The set of (positive) fractions is closed under addition, multiplication, and division, but *not* under subtraction.

Example $\frac{1}{2} - 1 = -\frac{1}{2}$ $-\frac{1}{2}$ is *not* a (positive) fraction.

In addition to being a closed set, rational numbers also share several of the properties discussed in earlier chapters of this book.

Classroom Tip
It would be difficult to overstate the importance of the properties summarized at the right. Making sure your students are familiar with these properties will help ensure their future success in algebra and other higher-level courses in mathematics.

Properties of Rational Numbers

Let $\frac{a}{b}$, $\frac{c}{d}$, and $\frac{e}{f}$ be any rational numbers.

Commutative Property of Addition

Numbers $\frac{1}{3} + \frac{5}{2} = \frac{5}{2} + \frac{1}{3}$ **Algebra** $\frac{a}{b} + \frac{c}{d} = \frac{c}{d} + \frac{a}{b}$

Commutative Property of Multiplication

Numbers $-\frac{1}{2} \cdot \frac{3}{4} = \frac{3}{4} \cdot \left(-\frac{1}{2}\right)$ **Algebra** $\frac{a}{b} \cdot \frac{c}{d} = \frac{c}{d} \cdot \frac{a}{b}$

Associative Property of Addition

Numbers $\left(\frac{1}{2} + \frac{3}{4}\right) + \frac{1}{4} = \frac{1}{2} + \left(\frac{3}{4} + \frac{1}{4}\right)$ **Algebra** $\left(\frac{a}{b} + \frac{c}{d}\right) + \frac{e}{f} = \frac{a}{b} + \left(\frac{c}{d} + \frac{e}{f}\right)$

Associative Property of Multiplication

Numbers $\left(\frac{1}{2} \cdot \frac{1}{3}\right) \cdot \frac{3}{2} = \frac{1}{2} \cdot \left(\frac{1}{3} \cdot \frac{3}{2}\right)$ **Algebra** $\left(\frac{a}{b} \cdot \frac{c}{d}\right) \cdot \frac{e}{f} = \frac{a}{b} \cdot \left(\frac{c}{d} \cdot \frac{e}{f}\right)$

Distributive Property of Multiplication Over Addition and Subtraction

Numbers $\frac{2}{3}\left(\frac{2}{5} + \frac{1}{5}\right) = \frac{2}{3} \cdot \frac{2}{5} + \frac{2}{3} \cdot \frac{1}{5}$ **Algebra** $\frac{a}{b}\left(\frac{c}{d} + \frac{e}{f}\right) = \frac{a}{b} \cdot \frac{c}{d} + \frac{a}{b} \cdot \frac{e}{f}$

Numbers $\frac{2}{3}\left(\frac{2}{5} - \frac{1}{5}\right) = \frac{2}{3} \cdot \frac{2}{5} - \frac{2}{3} \cdot \frac{1}{5}$ **Algebra** $\frac{a}{b}\left(\frac{c}{d} - \frac{e}{f}\right) = \frac{a}{b} \cdot \frac{c}{d} - \frac{a}{b} \cdot \frac{e}{f}$

Additive Identity Property

Numbers $-\frac{1}{2} + 0 = -\frac{1}{2}$ **Algebra** $\frac{a}{b} + 0 = \frac{a}{b}, 0 + \frac{a}{b} = \frac{a}{b}$

Additive Inverse Property

Numbers $-\frac{1}{2} + \frac{1}{2} = 0$ **Algebra** $\frac{a}{b} + \left(-\frac{a}{b}\right) = 0$

Multiplicative Identity Property

Numbers $\frac{3}{8} \cdot 1 = \frac{3}{8}$ **Algebra** $\frac{a}{b} \cdot 1 = \frac{a}{b}, 1 \cdot \frac{a}{b} = \frac{a}{b}$

Multiplicative Inverse Property

Numbers $\frac{3}{5} \cdot \frac{5}{3} = 1$ **Algebra** $\frac{a}{b} \cdot \frac{b}{a} = 1, a \neq 0, b \neq 0$

Reciprocal of $\frac{3}{5}$ Reciprocal of $\frac{a}{b}$

Zero Multiplication Property

Numbers $\frac{3}{2} \cdot 0 = 0$ **Algebra** $\frac{a}{b} \cdot 0 = 0, 0 \cdot \frac{a}{b} = 0$

The rules for adding, subtracting, multiplying, and dividing rational numbers are the same as the rules for operations with fractions (Chapter 6), combined with the rules for operations with integers (Chapter 8).

Classroom Tip

When you are teaching operations with rational numbers, ask students to carefully write out all of the steps, as shown in Examples 6–9. Later, as they become more proficient, they can omit some of the steps by performing mental math. For instance, in Example 6(a), as students master operations with rational numbers, they should be able to use mental math to conclude that

$$-\frac{3}{2}+\frac{1}{2}=-1.$$

EXAMPLE 6 Adding Rational Numbers

a. $-\frac{3}{2}+\frac{1}{2}=\frac{-3}{2}+\frac{1}{2}$

$=\frac{-3+1}{2}$

$=\frac{-2}{2}$

$=-1$

b. $-\frac{1}{3}+\frac{4}{3}+\frac{1}{6}=\frac{-2}{6}+\frac{8}{6}+\frac{1}{6}$

$=\frac{-2+8+1}{6}$

$=\frac{7}{6}$

$=1\frac{1}{6}$

EXAMPLE 7 Subtracting Rational Numbers

a. $-\frac{3}{2}-\frac{1}{2}=\frac{-3}{2}-\frac{1}{2}$

$=\frac{-3-1}{2}$

$=\frac{-4}{2}$

$=-2$

b. $-\frac{1}{3}-\left(-\frac{1}{6}\right)=-\frac{1}{3}+\frac{1}{6}$

$=\frac{-2}{6}+\frac{1}{6}$

$=\frac{-2+1}{6}$

$=\frac{-1}{6}$

$=-\frac{1}{6}$

EXAMPLE 8 Multiplying Rational Numbers

a. $-\frac{3}{2}\cdot\frac{1}{2}=\frac{-3}{2}\cdot\frac{1}{2}$

$=\frac{-3\cdot 1}{2\cdot 2}$

$=\frac{-3}{4}$

$=-\frac{3}{4}$

b. $-\frac{1}{3}\cdot\left(-\frac{3}{4}\right)=\frac{-1}{3}\cdot\left(\frac{-3}{4}\right)$

$=\frac{(-1)(-3)}{3\cdot 4}$

$=\frac{(-1)(\overset{-1}{\cancel{-3}})}{\underset{1}{\cancel{3}}\cdot 4}$

$=\frac{1}{4}$

EXAMPLE 9 Dividing Rational Numbers

a. $-\frac{3}{2}\div\frac{1}{2}=\frac{-3}{2}\cdot\frac{2}{1}$

$=\frac{-3\cdot 2}{2\cdot 1}$

$=\frac{-3\cdot\overset{1}{\cancel{2}}}{\underset{1}{\cancel{2}}\cdot 1}$

$=-3$

b. $-\frac{1}{3}\div\frac{1}{6}=\frac{-1}{3}\cdot\frac{6}{1}$

$=\frac{-1\cdot 6}{3\cdot 1}$

$=\frac{-1\cdot\overset{2}{\cancel{6}}}{\underset{1}{\cancel{3}}\cdot 1}$

$=-2$

9.1 Exercises

Free step-by-step solutions to odd-numbered exercises at MathematicalPractices.com.

1. **Writing a Solution Key** Write a solution key for the activity on page 328. Describe a rubric for grading a student's work.

2. **Grading the Activity** In the activity on page 328, a student gave the answers below. Each question is worth 10 points. Assign a grade to each answer. For those that are incorrect, why do you think the student erred?

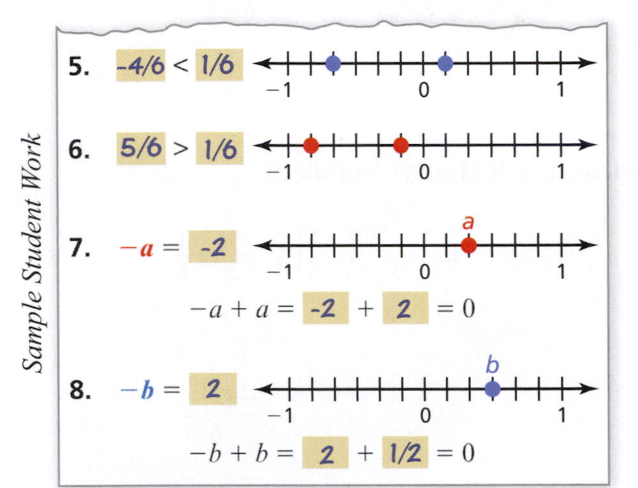

3. **Identifying Rational Numbers** Identify the rational numbers on the number line.

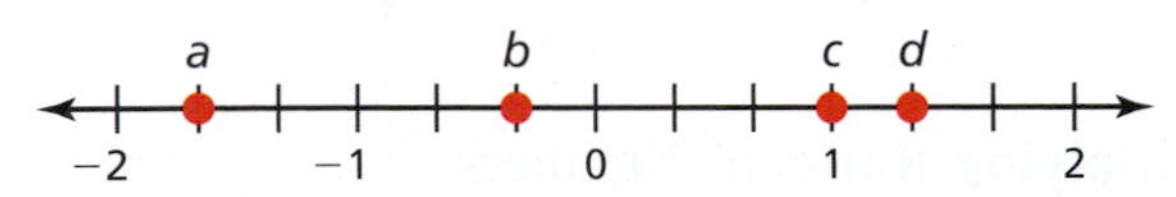

4. **Identifying Rational Numbers** Identify the rational numbers on the number line.

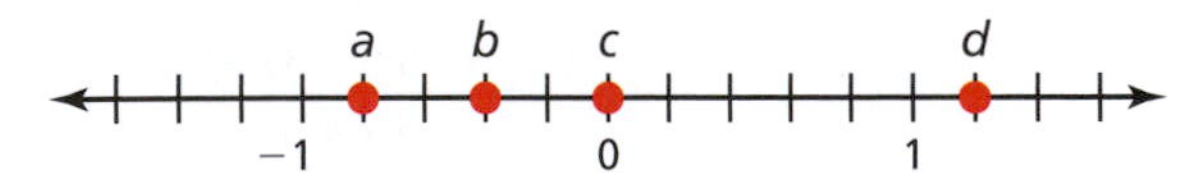

5. **Graphing Rational Numbers** Graph each rational number on a number line.

   a. $\frac{1}{4}$ **b.** $-\frac{3}{8}$ **c.** $-\frac{5}{4}$ **d.** $\frac{3}{2}$

6. **Graphing Rational Numbers** Graph each rational number on a number line.

   a. $\frac{7}{6}$ **b.** $-\frac{2}{3}$ **c.** 0 **d.** $-\frac{11}{6}$

7. **Definition of Rational Numbers** Show that each number is rational by writing it in the form $\frac{a}{b}$, where a and b are integers and $b \neq 0$.

   a. 1.5 **b.** -4 **c.** $1\frac{3}{4}$ **d.** $-0.\overline{3}$

8. **Definition of Rational Numbers** Show that each number is rational by writing it in the form $\frac{a}{b}$, where a and b are integers and $b \neq 0$.

   a. -1.9 **b.** 0 **c.** $2.\overline{6}$ **d.** $-3\frac{1}{5}$

9. **Writing Rational Numbers** Which of the following rational numbers are equal to -5?

   $\frac{1}{5}, \frac{-5}{1}, \frac{5}{-1}, \frac{-5}{-1}, -\frac{5}{1}, -\frac{-5}{1}, -\frac{5}{-1}, -\frac{-5}{-1}$

10. **Writing Rational Numbers** Which of the following rational numbers are equal to $\frac{1}{4}$?

    $\frac{-1}{4}, \frac{1}{-4}, \frac{-1}{-4}, -\frac{1}{4}, -\frac{-1}{4}, -\frac{1}{-4}, -\frac{-1}{-4}$

11. **Cross Products** Use cross products to determine whether the rational numbers are equal.

    a. $\frac{5}{8}$ and $\frac{7}{12}$ **b.** $\frac{-1}{6}$ and $\frac{-7}{42}$

    c. $\frac{20}{-36}$ and $\frac{-5}{9}$ **d.** $\frac{3}{-13}$ and $\frac{9}{-36}$

12. **Cross Products** Use cross products to determine whether the rational numbers are equal.

    a. $\frac{-7}{21}$ and $\frac{6}{-18}$ **b.** $\frac{3}{15}$ and $\frac{5}{20}$

    c. $\frac{-12}{26}$ and $\frac{-18}{39}$ **d.** $\frac{33}{-55}$ and $\frac{-6}{11}$

13. **Comparing Rational Numbers** Place the correct symbol (<, =, or >) between the two rational numbers.

    a. $-\frac{2}{3}$ ▢ $-\frac{3}{5}$ **b.** $-\frac{1}{4}$ ▢ $\frac{1}{6}$

    c. $\frac{-4}{6}$ ▢ $\frac{-8}{12}$ **d.** $\frac{8}{7}$ ▢ $-\frac{8}{7}$

14. **Comparing Rational Numbers** Place the correct symbol (<, =, or >) between the two rational numbers.

    a. $\frac{-5}{6}$ ▢ $\frac{1}{6}$ **b.** $\frac{-15}{40}$ ▢ $\frac{6}{-16}$

    c. $-\frac{5}{9}$ ▢ $-\frac{4}{7}$ **d.** $\frac{8}{-11}$ ▢ $\frac{11}{-14}$

15. **Ordering Rational Numbers** Use a number line to order the rational numbers $0, -\frac{5}{2}, -2, \frac{3}{2}$, and $\frac{5}{4}$ from least to greatest.

16. Ordering Rational Numbers Use a number line to order the rational numbers $\frac{2}{3}, -\frac{1}{3}, -\frac{5}{3}, -\frac{5}{6}$, and $\frac{1}{6}$ from least to greatest.

17. Finding Inverses Find the additive and multiplicative inverses of each rational number.

a. $\frac{1}{5}$ **b.** $-\frac{7}{3}$ **c.** 6 **d.** $-\frac{13}{16}$

18. Finding Inverses Find the additive and multiplicative inverses of each rational number.

a. -4 **b.** $\frac{9}{10}$ **c.** $\frac{15}{8}$ **d.** $-\frac{1}{12}$

19. Identifying Properties of Rational Numbers Identify the property of rational numbers that makes each equation true.

a. $\frac{3}{2} + 0 = \frac{3}{2}$ **b.** $-4 \cdot -\frac{1}{4} = 1$

c. $\left(-\frac{1}{4} \cdot \frac{3}{4}\right) \cdot \frac{2}{5} = -\frac{1}{4} \cdot \left(\frac{3}{4} \cdot \frac{2}{5}\right)$

d. $-\frac{1}{2}\left(\frac{2}{9} + \frac{4}{3}\right) = -\frac{1}{2} \cdot \frac{2}{9} + \left(-\frac{1}{2}\right) \cdot \frac{4}{3}$

20. Identifying Properties of Rational Numbers Identify the property of rational numbers that makes each equation true.

a. $-\frac{1}{7} \cdot 0 = 0$ **b.** $\frac{5}{8} + \left(-\frac{3}{8}\right) = -\frac{3}{8} + \frac{5}{8}$

c. $\frac{3}{11} + \left(-\frac{3}{11}\right) = 0$ **d.** $\frac{9}{5} \cdot 1 = \frac{9}{5}$

21. Using Properties of Rational Numbers Use the given property of rational numbers to complete each equation.

a. Associative Property of Addition

$\left(\frac{1}{3} + \frac{1}{5}\right) + \frac{3}{13} = $ ▭

b. Zero Multiplication Property

$\frac{9}{2} \cdot 0 = $ ▭

c. Commutative Property of Addition

$-\frac{4}{6} + \frac{2}{5} = $ ▭

d. Additive Identity Property

$-\frac{10}{11} + 0 = $ ▭

e. Commutative Property of Multiplication

$-\frac{6}{12} \cdot \frac{4}{3} = $ ▭

22. Using Properties of Rational Numbers Use the given property of rational numbers to complete each equation.

a. Additive Inverse Property

$-\frac{5}{12} + \frac{5}{12} = $ ▭

b. Distributive Property of Multiplication Over Subtraction

$\frac{3}{4}\left(\frac{1}{8} - \frac{1}{9}\right) = $ ▭

c. Multiplicative Identity Property

$-\frac{1}{14} \cdot 1 = $ ▭

d. Associative Property of Multiplication

$\frac{5}{7} \cdot \left[\left(-\frac{2}{8}\right) \cdot \frac{1}{2}\right] = $ ▭

e. Multiplicative Inverse Property

$\frac{11}{3} \cdot \frac{3}{11} = $ ▭

23. Operations with Rational Numbers Evaluate each expression. Write your answer in simplest form.

a. $\frac{1}{4} + \frac{7}{4}$ **b.** $-\frac{3}{5} - \frac{4}{5}$

c. $-\frac{2}{3} + \frac{1}{9}$ **d.** $\frac{7}{2} + \frac{3}{4} - \frac{5}{2}$

e. $\frac{5}{6} \cdot \frac{1}{2}$ **f.** $\frac{1}{4} \div \frac{3}{8}$

g. $-\frac{5}{3} \div \frac{7}{6}$ **h.** $\frac{1}{12} \cdot \frac{3}{2} \cdot \left(-\frac{4}{5}\right)$

24. Operations with Rational Numbers Evaluate each expression. Write your answer in simplest form.

a. $-\frac{2}{9} + \frac{4}{9}$ **b.** $\frac{8}{3} - \left(-\frac{5}{3}\right)$

c. $\frac{7}{10} - \frac{4}{15}$ **d.** $-\frac{3}{4} + \frac{1}{6} + \frac{5}{8}$

e. $\frac{3}{5} \div \left(-\frac{1}{10}\right)$ **f.** $-\frac{7}{3} \cdot \frac{2}{9}$

g. $\frac{5}{12} \div \frac{3}{4}$ **h.** $-\frac{10}{27} \cdot \frac{9}{5} \div \left(-\frac{4}{7}\right)$

25. True or False? Tell whether each statement is *true* or *false*. Justify your answer.

a. Every natural number is a rational number.

b. Every rational number can be written as a terminating decimal.

c. Every rational number can be written as the ratio of an integer to an integer.

26. True or False? Tell whether each statement is *true* or *false*. Justify your answer.

a. Every whole number is a rational number.

b. Every rational number can be written as a terminating or repeating decimal.

c. Every rational number can be written as the ratio of a whole number to a whole number.

27. Algebra Complete the solution of the equation. Then write the property that justifies each step in the solution.

Step	Reason
$\frac{6}{5}x = -3$	Write original equation.
$\square \cdot \left(\frac{6}{5}x\right) = \square \cdot (-3)$	Multiply each side by the multiplicative inverse of $\frac{6}{5}$.
$\left(\square \cdot \frac{6}{5}x\right) = \square \cdot (-3)$	$\square$
$(1)x = \square \cdot (-3)$	$\square$
$x = \square$	$\square$

28. Algebra Complete the solution of the equation. Then write the property that justifies each step in the solution.

Step	Reason
$\frac{3}{8} + x = -\frac{1}{8}$	Write original equation.
$\frac{3}{8} + x + \square = -\frac{1}{8} + \square$	Add the additive inverse of $\frac{3}{8}$ to each side.
$\frac{3}{8} + \square + x = -\frac{1}{8} + \square$	$\square$
$0 + x = -\frac{1}{8} + \square$	$\square$
$x = \square$	$\square$

29. Closure Under Subtraction Show that the set of natural numbers is not closed under subtraction by finding two natural numbers whose difference is not a natural number. Which number set(s) contain all possible differences of natural numbers?

30. Closure Under Subtraction Show that the set of positive fractions is not closed under subtraction by finding two positive fractions whose difference is not a positive fraction. Which number set(s) contain all possible differences of positive fractions?

31. Closure Under Division Show that the set of whole numbers is not closed under division by finding two whole numbers whose quotient is not a whole number. Which number set(s) contain all possible quotients of whole numbers?

32. Closure Under Division Show that the set of integers is not closed under division by finding two integers whose quotient is not an integer. Which number set(s) contain all possible quotients of integers?

33. River Depth: In Your Classroom The table shows the changes (in feet) of the depth of a river over five weeks.

Week	Change (feet)
1	$3\frac{5}{6}$
2	$-1\frac{1}{3}$
3	$-\frac{5}{6}$
4	$-\frac{7}{12}$
5	$2\frac{1}{6}$

a. During which week did the depth increase the most?

b. During which week did the depth decrease the most?

c. Describe a homework project that would involve data about river depth. Include a description of how you would ask your students to order and perform operations with rational numbers.

34. Snow Depth: In Your Classroom The table shows the changes (in inches) of the depth of snow in a town over five days.

Day	Change (inches)
1	$1\frac{3}{8}$
2	$6\frac{1}{4}$
3	$-\frac{3}{4}$
4	$1\frac{1}{2}$
5	$-1\frac{1}{4}$

a. On which day did the depth increase the most?

b. On which day did the depth decrease the most?

c. Describe a homework project that would involve data about snow depth. Include a description of how you would ask your students to order and perform operations with rational numbers.

35. Snorkeling The initial elevation of a snorkeler is $-4\frac{3}{5}$ meters. The snorkeler dives to three times the initial elevation and then swims $8\frac{1}{2}$ meters upward. What is the snorkeler's final elevation?

36. Giant Squid The initial elevation of a giant squid is $-\frac{7}{12}$ kilometer. The squid swims upward to $\frac{2}{3}$ of its initial elevation and then swims $\frac{1}{2}$ kilometer downward. What is the squid's final elevation?

37. Mental Math Use the properties of rational numbers and mental math to evaluate each expression. List the properties that you used to make calculations simpler.

a. $\frac{6}{7} + \left(-\frac{1}{4}\right) + \frac{1}{7}$ **b.** $\frac{2}{5} \cdot \left(-\frac{9}{8}\right) \cdot \frac{5}{2}$

c. $\frac{1}{6} \cdot \frac{5}{8} + \frac{1}{6} \cdot \frac{3}{8}$ **d.** $-\frac{1}{12} + \left(\frac{5}{12} + \frac{2}{3}\right)$

38. Mental Math Use the properties of rational numbers and mental math to evaluate each expression. List the properties that you used to make calculations simpler.

a. $\frac{4}{5} \cdot \left[\frac{5}{8} \cdot \left(-\frac{12}{13}\right)\right]$ **b.** $\left(\frac{7}{9} - 7\right)\frac{9}{7}$

c. $\frac{10}{3} + \frac{1}{3} + \left(-\frac{10}{3}\right)$ **d.** $\frac{4}{15} \cdot \frac{1}{6} + \left(-\frac{4}{15}\right) \cdot \frac{1}{6}$

39. *Writing* Find the sum of two arbitrary rational numbers, $\frac{a}{b}$ and $\frac{c}{d}$. How does this show that the set of rational numbers is closed under addition?

40. *Writing* Explain why it is easier to show that a set of numbers *is not* closed under an operation than it is to show that a set of numbers *is* closed under an operation. Give an example of each case.

41. Grading Student Work On a diagnostic test, one of your students does the following work. Explain what the student did wrong. Which topics would you encourage the student to review?

$\frac{4}{6}$ ▭ $\frac{3}{2}$

$4 \cdot 3 \stackrel{?}{=} 6 \cdot 2$

$12 = 12 \Rightarrow \frac{4}{6} = \frac{3}{2}$

42. Grading Student Work On a diagnostic test, one of your students does the following work. Explain what the student did wrong. Which topics would you encourage the student to review?

$\frac{10}{13}$ ▭ $\frac{7}{9}$

$10 > 7 \Rightarrow \frac{10}{13} > \frac{7}{9}$

43. Reasoning What rational number is its own additive inverse? Justify your answer.

44. Reasoning What two rational numbers are their own multiplicative inverses? Justify your answer.

45. Profit The bar graph shows the profit, in hundreds of thousands of dollars, of a new business for each of its first six years.

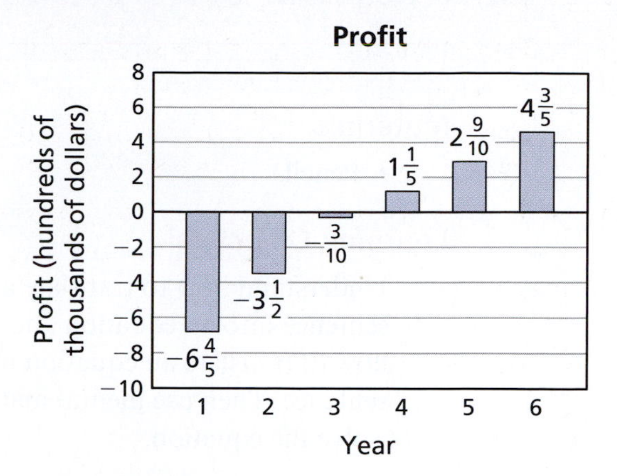

a. What was the difference in profit from Year 3 to Year 4?

b. How many times greater was the Year 1 profit than the Year 3 profit?

c. What is the total profit over the first six years?

46. Equity The bar graph shows the equity, in thousands of dollars, of a used car for each of the first five years after its purchase.

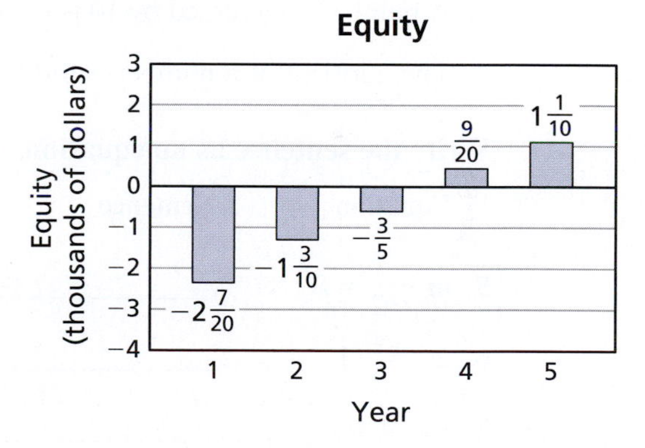

a. What was the difference in equity from Year 1 to Year 3?

b. How many times greater was the Year 5 equity than the Year 4 equity?

c. On average, how much did the equity increase each year?

47. Problem Solving A man started college $\frac{1}{4}$ of the way through his life, was unmarried for the next $\frac{1}{6}$ of his life, and then had a child six years after his marriage. The child's birth coincided with the half-way point of his life. How old did the man live to be?

48. Problem Solving A woman earned a master's degree $\frac{5}{16}$ of the way through her life, worked professionally for the next half of her life, and then spent the last 15 years of her life in retirement. How old did the woman live to be?

Activity: Writing and Solving Equations

Materials:

- Pencil

Learning Objective:

Understand how to translate a sentence into an equation and how to translate an equation into a sentence. Then use mental math to solve the equation.

Name ______________________________

Write the sentence as an equation. Then use mental math to solve the equation.

Sentence	Equation	Solution
1. 9 less than a number b equals 2.	____________	$b =$ ☐
2. The product of a number n and 5 is 30.	____________	$n =$ ☐
3. A number x increased by 10 is the same as 24.	____________	$x =$ ☐
4. The quotient of a number y and 4 is 12.	____________	$y =$ ☐

Write the sentence as an equation. Then use mental math to solve the equation.

Equation	Sentence	Solution
5. $m - \frac{1}{2} = 0$	________________________________	$m =$ ☐
6. $2x = -1$	________________________________	$x =$ ☐
7. $n \div 4 = \frac{3}{4}$	________________________________	$n =$ ☐
8. $d + \frac{1}{4} = \frac{5}{4}$	________________________________	$d =$ ☐

Write the equation that represents the scale. Then use mental math to solve the equation.

9.

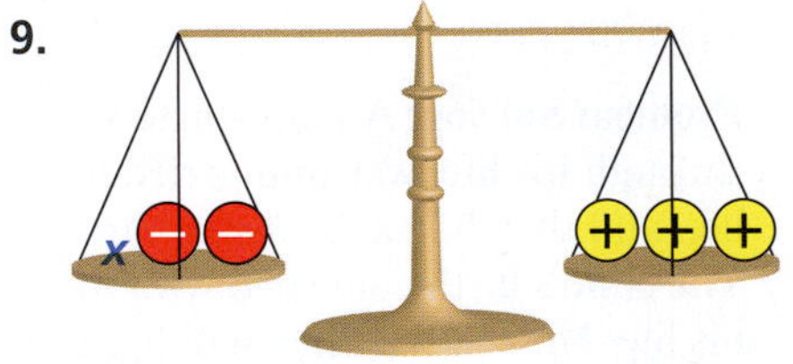

Equation ____________ Solution $x =$ ☐

10.

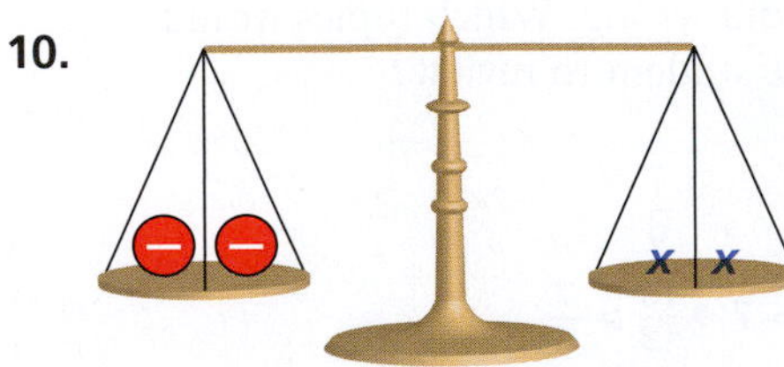

Equation ____________ Solution $x =$ ☐

Goodluz/Shutterstock.com

9.2 Solving Equations Using Rational Numbers

Standards

Grades 6–8
Expressions and Equations
Students should analyze and solve linear equations.

- Translate a sentence into an equation and vice versa.
- Solve one-step equations.
- Solve two-step equations.

Writing and Solving Equations

In Section 4.3, you studied algebraic expressions, such as

$2x + 3$ and $n - 4$. Algebraic expressions

When two expressions are set equal to each other, the resulting mathematical sentence is called an *equation*.

Definition of Equation

An **equation** is an algebraic statement that equates two expressions. When the equation has only one variable, it is called an **equation in one variable**.

Example $2x - 3 = 5$ Equation in one variable

Equations may be true for some values and false for others. A value of the variable that makes an equation true is called a **solution of the equation**.

Example $x = 4$ is a solution of $2x - 3 = 5$ because $2(4) - 3 = 5$.

Finding the solution (or solutions) of an equation is called **solving the equation**. For a one-step equation, it is possible to use mental math to solve the equation.

EXAMPLE 1 Using Mental Math to Solve One-Step Equations

Use mental math to solve each equation.

a. $x + 5 = 0$ **b.** $2n = 1$

c. $3 - m = -1$ **d.** $\frac{y}{8} = 4$

SOLUTION

	Equation	Think	Solution	Check
a.	$x + 5 = 0$	What number can be added to 5 to obtain 0?	$x = -5$	$-5 + 5 = 0$ ✓
b.	$2n = 1$	What number can be multiplied by 2 to obtain 1?	$n = \frac{1}{2}$	$2\left(\frac{1}{2}\right) = 1$ ✓
c.	$3 - m = -1$	What number can be subtracted from 3 to obtain −1?	$m = 4$	$3 - 4 = -1$ ✓
d.	$\frac{y}{8} = 4$	What number can be divided by 8 to obtain 4?	$y = 32$	$\frac{32}{8} = 4$ ✓

Notice that the final step in solving an equation is to **check the solution**. To check your solution, substitute it for the variable in the original equation.

Classroom Tip
Your students will benefit greatly if you make the final step of checking the solution a requirement. It is worth putting a poster in your classroom that states "Always Check Your Solutions."

Modeling and Solving One-Step Equations

There are several ways to model solving an equation. One way is to use a scale, as shown in the activity on page 338. Another way is to use **algebra tiles**, as illustrated in Example 2. This example uses the following key.

Key [+] = Variable [+] = 1 [−] = −1 [+][−] = Zero pair

It is important to know that when solving an equation, the goal is to **isolate the variable** on one side of the equation. To do this, use inverse operations, which are operations that "undo" each other. Addition and subtraction are inverse operations. Multiplication and division are also inverse operations.

EXAMPLE 2 Solving a One-Step Equation Using Algebra Tiles

Use algebra tiles to model and solve $x - 3 = -4$.

SOLUTION

1. Model the equation $x - 3 = -4$.
2. To isolate the green tile, remove the red tiles on the left side by adding 3 yellow tiles to each side.
3. Remove the 3 zero pairs from each side.
4. The remaining tiles show the value of x.

So, the solution is $x = -1$.

Check $x - 3 = -4$

$-1 - 3 \stackrel{?}{=} -4$

$-4 = -4$ ✓

When it becomes too difficult to solve an equation using mental math or models, you can use the following properties of equality.

Classroom Tip Notice in Example 2 that the reason for adding the 3 yellow tiles to each side is to model the algebraic step of adding 3 to each side.

Addition and Subtraction Properties of Equality

Addition Property of Equality

Words When you add the same number to each side of an equation, the two sides remain equal.

Algebra $x - 4 = 5$ → $x - 4 + 4 = 5 + 4$ → $x = 9$

Subtraction Property of Equality

Words When you subtract the same number from each side of an equation, the two sides remain equal.

Algebra $x + 4 = 5$ → $x + 4 - 4 = 5 - 4$ → $x = 1$

EXAMPLE 3 Solving One-Step Equations

Solve (a) $x - \frac{1}{2} = \frac{3}{2}$ and (b) $m + \frac{1}{6} = \frac{1}{3}$. Show your work and justify each step.

SOLUTION

a. $x - \frac{1}{2} = \frac{3}{2}$ Write the original equation.

$x - \frac{1}{2} + \frac{1}{2} = \frac{3}{2} + \frac{1}{2}$ Add $\frac{1}{2}$ to each side.

$x = 2$ Simplify.

Check

$2 - \frac{1}{2} \stackrel{?}{=} \frac{3}{2}$

$\frac{3}{2} = \frac{3}{2}$ ✓

b. $m + \frac{1}{6} = \frac{1}{3}$ Write original equation.

$m + \frac{1}{6} - \frac{1}{6} = \frac{1}{3} - \frac{1}{6}$ Subtract $\frac{1}{6}$ from each side.

$m = \frac{1}{6}$ Simplify.

Check

$\frac{1}{6} + \frac{1}{6} \stackrel{?}{=} \frac{1}{3}$

$\frac{1}{3} = \frac{1}{3}$ ✓

Classroom Tip

Some of your students may find a vertical format easier to use than the horizontal format shown in Example 3.

Vertical format

$$\begin{array}{r} x - \frac{1}{2} = \frac{3}{2} \\ + \frac{1}{2} = \frac{1}{2} \\ \hline x \quad = 2 \end{array}$$

Multiplication and Division Properties of Equality

Multiplication Property of Equality

Words When you multiply each side of an equation by the same nonzero number, the two sides remain equal.

Algebra $\frac{1}{2}x = 4$  $2 \cdot \frac{1}{2}x = 2 \cdot 4$  $x = 8$

Division Property of Equality

Words When you divide each side of an equation by the same nonzero number, the two sides remain equal.

Algebra $3x = -6$  $\frac{3x}{3} = \frac{-6}{3}$  $x = -2$

EXAMPLE 4 Solving One-Step Equations

Solve (a) $-3 = \frac{n}{2}$ and (b) $5b = -20$. Show your work and justify each step.

SOLUTION

a. $-3 = \frac{n}{2}$ Write the original equation.

$2(-3) = 2\left(\frac{n}{2}\right)$ Multiply each side by 2.

$-6 = n$ Simplify.

Check

$-3 \stackrel{?}{=} \frac{-6}{2}$

$-3 = -3$ ✓

b. $5b = -20$ Write original equation.

$\frac{5b}{5} = \frac{-20}{5}$ Divide each side by 5.

$b = -4$ Simplify.

Check

$5(-4) \stackrel{?}{=} -20$

$-20 = -20$ ✓

Solving Two-Step Equations

As the name "two-step" implies, a two-step equation involves two different operations. For instance, the equation $2x - 3 = -5$ involves multiplication and subtraction. To solve two-step equations, use inverse operations to isolate the variable.

EXAMPLE 5 **Solving a Two-Step Equation Using Algebra Tiles**

Use algebra tiles to model and solve

$$2x - 3 = -5.$$

SOLUTION

1. Model the equation $2x - 3 = -5$.
2. To isolate the green tiles, remove the red tiles on the left side by adding 3 yellow tiles to each side.
3. Remove the 3 zero pairs from each side.

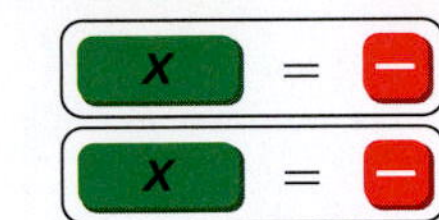

4. Because there are 2 green tiles, divide the red tiles into 2 equal groups.
5. Keep one of the groups. This shows the value of x.

So, the solution is $x = -1$.

Check $2(-1) - 3 \stackrel{?}{=} -5$

$-2 - 3 \stackrel{?}{=} -5$

$-5 = -5$ ✓

Classroom Tip

Notice in Examples 5 and 6 that "add 3 to each side" is the first step, before dividing each side by 2. If you divide each side by 2 first, the steps to your solution may look like this.

$$\frac{2x-3}{2} = \frac{-5}{2}$$

$$x - \frac{3}{2} = \frac{-5}{2}$$

$$x - \frac{3}{2} + \frac{3}{2} = \frac{-5}{2} + \frac{3}{2}$$

$$x = -1$$

EXAMPLE 6 **The Math Behind the Tiles**

Solve

$$2x - 3 = -5$$

without using algebra tiles. Describe each step.

SOLUTION

$2x - 3 = -5$ Write the original equation.

$2x - 3 + 3 = -5 + 3$ Add 3 to each side.

$2x = -2$ Simplify.

$\frac{2x}{2} = \frac{-2}{2}$ Divide each side by 2.

$x = -1$ Simplify.

The solution is $x = -1$.

EXAMPLE 7 Solving a Real-Life Problem

A taxi fare has an initial charge of $4.70 and $2.25 for each mile driven. You take a taxi to the airport and owe the driver $40.70. How many miles do you travel in the taxi?

SOLUTION

1. **Understand the Problem** You need to find the number of miles that you travel in the taxi.

2. **Make a Plan** Use the given information to write an equation. Then solve the equation.

3. **Solve the Problem** Begin by writing a verbal model. Then translate the verbal model into an algebraic equation. Let *m* represent the number of miles driven.

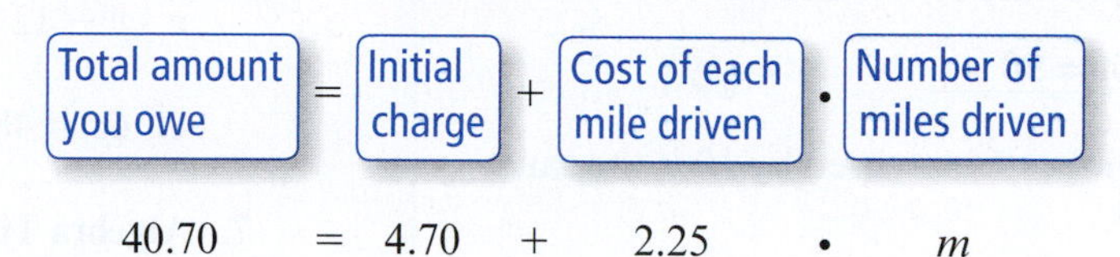

$$40.70 \quad = \quad 4.70 \quad + \quad 2.25 \quad \cdot \quad m$$

Finally, solve this two-step equation.

$40.70 = 4.70 + 2.25m$	Write the original equation.
$40.70 - 4.70 = 4.70 - 4.70 + 2.25m$	Subtract 4.70 from each side.
$36 = 2.25m$	Simplify.
$\frac{36}{2.25} = \frac{2.25m}{2.25}$	Divide each side by 2.25.
$16 = m$	Simplify.

So, you travel 16 miles in the taxi.

Check $40.70 \stackrel{?}{=} 4.70 + 2.25(16)$

$40.70 \stackrel{?}{=} 4.70 + 36$

$40.70 = 40.70$ ✓

4. **Look Back** Another way to check your answer is to create a spreadsheet showing the total amount you owe for different numbers of miles driven.

	A	B
1	**Number of miles driven**	**Total amount you owe**
2	0	$4.70
3	1	$6.95
4	2	$9.20
	⋮	⋮
16	14	$36.20
17	15	$38.45
18	16	$40.70 ✓

To check the reasonableness of your answer, use compatible numbers. The taxi fare has an initial charge of about $5 and charges about $2 for each mile driven. Because 5 + 2(16) = 37, and $37 is close to $40.70, you know that your answer is reasonable.

Mathematical Practices

Reason Abstractly and Quantitatively

MP2 Mathematically proficient students bring two complementary abilities to bear on problems involving quantitative relationships: the ability to *decontextualize* and the ability to *contextualize.*

Classroom Tip

When you are solving a real-life problem, be sure you see that checking the solution in the original equation is not the same as "Looking Back At Your Solution." You need two checks. First you need to check that 16 is the correct solution of the equation that *you wrote*. Then, to make sure that the equation you wrote is representative of the original problem, you need to check that 16 makes sense as the answer to the original problem.

9.2 Exercises

Free step-by-step solutions to odd-numbered exercises at *MathematicalPractices.com.*

1. **Writing a Solution Key** Write a solution key for the activity on page 338. Describe a rubric for grading a student's work.

2. **Grading the Activity** In the activity on page 338, a student gave the answers below. Each question is worth 10 points. Assign a grade to each answer. For those that are incorrect, why do you think the student erred?

Sample Student Work

2. The product of a number n and 5 is 30.

$5n = 30$ $n =$ 6

3. A number x increased by 10 is the same as 24.

$x = 10 + 24$ $x =$ 34

4. The quotient of a number y and 4 is 12.

$\frac{4}{y} = 12$ $y =$ 1/3

6. $2x = -1$

The product of 2 and a number x is -1.

$x =$ -1/2

7. $n \div 4 = \frac{3}{4}$

The number n divided by 4 is 3/4.

$n =$ 1/3

8. $d + \frac{1}{4} = \frac{5}{4}$

The number d plus 1/4 is 5/4.

$d =$ 3/2

3. **Expressions vs. Equations** Determine whether each of the following is an expression or an equation.
   a. $3x + 9$ b. $4x = 8$
   c. $-2 + y = 6$ d. $-7 - y$

4. **Expressions vs. Equations** Determine whether each of the following is an expression or an equation.
   a. $0 = 6x - 10$
   b. $5x$
   c. $12y + 8$
   d. $15y = 0$

5. **Using Mental Math** Use mental math to solve each equation.
   a. $x + 2 = 1$ b. $-4 + x = 9$
   c. $3 - y = 10$ d. $y - 8 = 13$
   e. $5z = 15$ f. $\frac{z}{6} = 3$

6. **Using Mental Math** Use mental math to solve each equation.
   a. $x + 8 = 19$ b. $3 + x = 13$
   c. $y - 12 = 5$ d. $y - 6 = -9$
   e. $6z = 48$ f. $\frac{z}{5} = -4$

7. **Algebra Tiles** Write the equation modeled by the algebra tiles. Then solve the equation.
   a.
   b. 

8. **Algebra Tiles** Write the equation modeled by the algebra tiles. Then solve the equation.
   a.
   b.

9. **Using Algebra Tiles** Use algebra tiles to model and solve each equation.
   a. $x + 7 = 14$ b. $9 + x = 12$
   c. $y - 5 = 10$ d. $-2 + y = 15$
   e. $4z = 16$ f. $3z = 21$

10. **Using Algebra Tiles** Use algebra tiles to model and solve each equation.
   a. $6 + x = 7$ b. $x + 9 = 18$
   c. $20 + y = 3$ d. $-4 + y = -10$
   e. $6z = 12$ f. $10z = 10$

11. **Properties of Equality** Decide which property of equality you will use to solve each equation.
   a. $x + 26 = 12$ b. $y - 5 = 14$
   c. $3m = 24$ d. $\frac{1}{4}n = 7$

12. Properties of Equality Decide which property of equality you will use to solve each equation.

a. $x + 10 = 8$ **b.** $y - 1 = -2$

c. $9m = 18$ **d.** $\frac{3}{2}n = 11$

13. Checking a Solution Check whether the given value is a solution of the equation.

a. $x + 9 = 17; x = 8$

b. $6x - 7 = 29; x = 5$

c. $18 - 4y = 6; y = 3$

d. $2.5y + 10 = 16; y = 4$

e. $\frac{1}{2}z - 1 = 5; z = 12$

f. $4z - 5 = 12; z = 1$

14. Checking a Solution Check whether the given value is a solution of the equation.

a. $4 - x = 8; x = 4$

b. $x + 1 = 7; x = -7$

c. $-2y + 9 = 7; y = 1$

d. $-5 + 2.2y = 18; y = 4$

e. $-1 + \frac{1}{4}z = -2; z = -4$

f. $-z + 4 = -3; z = 1$

15. Solving One-Step Equations Solve each equation. Show your work and justify each step.

a. $14 + x = 31$ **b.** $a - \frac{3}{4} = \frac{1}{8}$

c. $9 = \frac{y}{12}$ **d.** $91 = 13b$

e. $y - 0.5 = 2.80$ **f.** $\frac{u}{3} = 4.5$

16. Solving One-Step Equations Solve each equation. Show your work and justify each step.

a. $s - 13 = 19$ **b.** $\frac{2}{3} + x = \frac{5}{6}$

c. $-\frac{y}{7} = 12$ **d.** $-8z = 104$

e. $6.25 = -3.4 + t$ **f.** $12.46 = 9.7 + h$

17. Algebra Tiles Write the equation modeled by the algebra tiles. Then solve the equation.

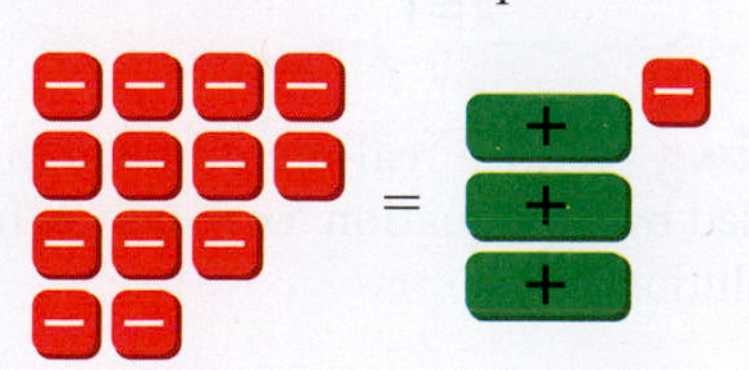

18. Algebra Tiles Write the equation modeled by the algebra tiles. Then solve the equation.

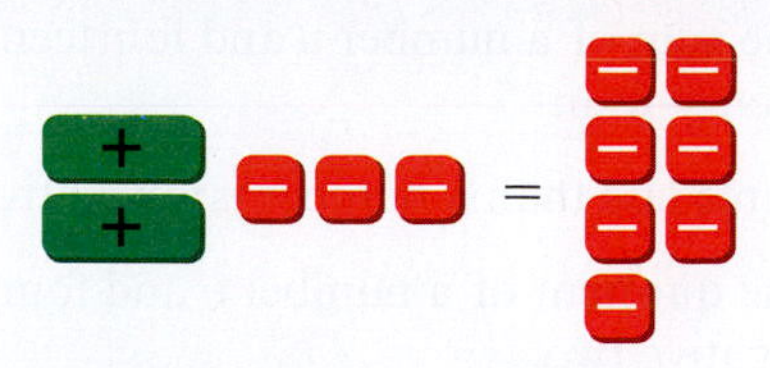

19. Using Algebra Tiles Use algebra tiles to model and solve each equation.

a. $6x + 1 = -11$

b. $8x - 16 = 0$

c. $3y - 15 = 15$

d. $-10 = 3 + 13y$

20. Using Algebra Tiles Use algebra tiles to model and solve each equation.

a. $2x + 5 = -15$

b. $3x + 21 = 0$

c. $4y - 10 = 6$

d. $1 = -1 + 2y$

21. Solving Two-Step Equations Solve each equation. Show your work and justify each step.

a. $4x + 3 = 11$ **b.** $7d - 1 = 13$

c. $10 = 7 - m$ **d.** $\frac{a}{3} + 4 = 6$

e. $16 - 2.4d = -8$

f. $\frac{c}{5.3} + 8.3 = 11.3$

g. $\frac{4}{5} - 2g = \frac{1}{10}$

h. $7 = \frac{5}{6}c - 8$

22. Solving Two-Step Equations Solve each equation. Show your work and justify each step.

a. $7y - 3 = 25$ **b.** $3x + 7 = 19$

c. $11 = 12 - q$ **d.** $17 = \frac{w}{5} + 13$

e. $14.4m - 5.1 = 2.1$

f. $3.2 + \frac{x}{2.5} = 4.6$

g. $-4j + \frac{1}{2} = \frac{7}{8}$

h. $10 = \frac{2}{7}n + 4$

23. Modeling Write and solve an equation to model each statement.

a. The sum of a number n and fourteen is twenty-four.

b. Three less than a number n is negative fifteen.

c. The quotient of a number n and four is negative two.

d. Twelve times a number n is thirty-six.

e. Nine less than the product of three and a number n is fifteen.

24. Modeling Write and solve an equation to model each statement.

a. Six more than a number n is two.

b. The difference between a number n and negative five is zero.

c. A number n divided by five is twenty.

d. The product of a number n and five-sixths is negative fifteen.

e. Fourteen more than the quotient of a number n and seven is twenty-five.

25. Cliff A cliff has a height of 1500 feet. You climb to a height of 675 feet. Write and solve an equation to find how much farther you have to climb to reach the top.

26. School Supplies You are buying school supplies that cost \$48.95. After sales tax is added, the total cost is \$51.89. Write and solve an equation to determine how much you pay in sales tax.

27. Bowling Your friend's bowling score is 105 pins. Your bowling score is 14 pins less than your friend's score. Write and solve an equation to find your score.

28. Tickets A discounted concert ticket is \$14.75 less than the original price. You pay \$54 for a discounted ticket. Write and solve an equation to find the original price of the ticket.

29. Park You clean a community park for 5.5 hours. You earn \$45.65. Write and solve an equation to find your hourly pay rate.

30. Theater A movie theater has 1960 seats. Each row has 40 seats. Write and solve an equation to find how many rows of seats are in the theater.

31. Dance A dance studio charges a one-time registration fee of \$15 and \$11 per class. A student has paid the studio \$92. How many classes has the student taken?

32. Party You need 124 plastic spoons for a party. A store has 1 box of 60 spoons and several boxes of 8 spoons available. How many boxes of 8 spoons should you buy along with the box of 60 spoons?

33. Activity: In Your Classroom Design a classroom activity in which students write and represent real-life situations involving one-step equations. Specify the steps that the students are to use to write an equation representing each real-life situation.

34. Activity: In Your Classroom Design a classroom activity in which students write and represent real-life situations involving two-step equations. Specify the steps that the students are to use to write an equation representing each real-life situation.

35. Grading Student Work On a diagnostic test, one of your students does the following work. Explain what the student did wrong. Which topics would you encourage the student to review?

a.
$-1.5 + k = 8.2$
$k = 8.2 + (-1.5)$
$k = 6.7$

b.
$7 - 3x = 12$
$4x = 12$
$\frac{4x}{4} = \frac{12}{4}$
$x = 3$

36. Grading Student Work On a diagnostic test, one of your students does the following work. Explain what the student did wrong. Which topics would you encourage the student to review?

a.
$s + \frac{3}{5} = -\frac{2}{5}$
$s + \frac{3}{5} + \frac{3}{5} = -\frac{2}{5} + \frac{3}{5}$
$s = \frac{1}{5}$

b.
$-6 + 2x = -10$
$-6 + \frac{2x}{2} = -\frac{10}{2}$
$-6 + x = -5$
$x = 1$

37. Writing Write a real-life problem that can be modeled by the equation $5x - 6 = 9$. Interpret the solution.

38. *Writing* Explain how you can use the Distributive Property to solve the equation below.

$3x + 4x = 21$

39. **Geometry** Write and solve equations to find the values of the variables in each rectangle.

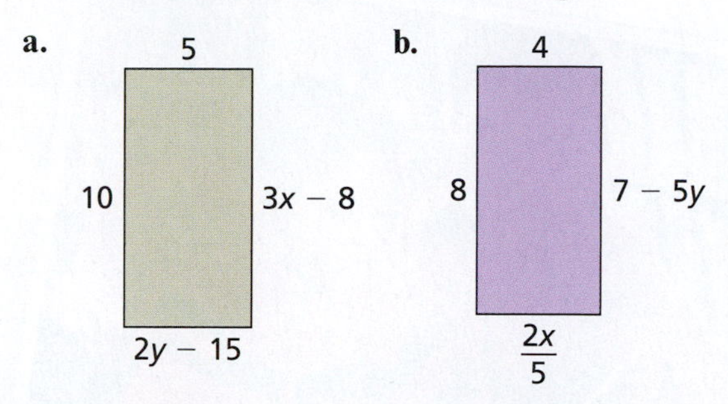

40. **Geometry** Write and solve equations to find the values of the variables in each rectangle.

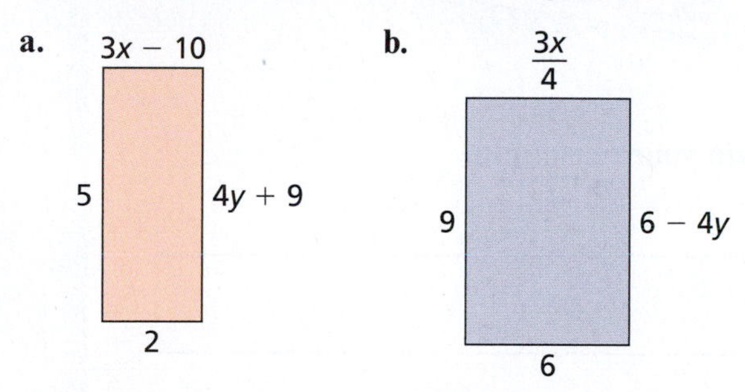

41. **Missing Information** The following real-life problem is missing information. Determine what information you need to solve the problem.

One day, you jog for 7 minutes and then run for 15 minutes. Each day after the first day, you jog for 7 minutes, but increase the time that you run. After how many days will you be exercising for a total of 90 minutes per day?

42. **Missing Information** The following real-life problem is missing information. Determine what information you need to solve the problem.

You have a job in which you earn \$7 per hour plus tips. You earned a total of \$34 yesterday. How much did you earn in tips?

43. **Rentals** The table shows costs to rent x videos and buy \$4 worth of snacks. Write a verbal model for the situation. Then translate the verbal model into an algebraic equation. Use the equation to find the cost to rent 9 videos and buy \$4 worth of snacks.

Videos, x	1	3	5
Cost	\$7	\$13	\$19

44. **Rentals** The table shows hourly costs to rent a boat including \$20 for gasoline. Write a verbal model for the situation. Then translate the verbal model into an algebraic equation. Use the equation to find the cost to rent the boat for 8 hours including \$20 for gasoline.

Hours, x	Cost
1	\$38.99
3	\$76.97
5	\$114.95
7	\$152.93

45. **School Pictures** One-fourth of the girls and one-eighth of the boys in a grade retake their school pictures. The photographer retakes pictures for 16 girls and 7 boys. How many students are in the grade?

46. **Sick Day** One-sixth of the fourth graders and one-third of the fifth graders miss school on Friday. The absentee sheet for Friday shows the names of 8 fourth graders and 18 fifth graders. How many fourth and fifth graders are enrolled in the school?

47. **Clothing Sale** At a sale, you buy 2 pairs of pants for \$24.95 each, 3 belts for \$12.99 each, and 3 shirts for the same price at a 20% discount. Your total cost (before sales tax) is \$132.07.

a. Write and solve an equation to find the discounted price of each shirt.

b. Write and solve an equation to find the original price of each shirt.

48. **Study Time** You have 28 hours available to study for 5 exams. You spend $6\frac{1}{2}$ hours, $5\frac{3}{4}$ hours, and $4\frac{3}{4}$ hours studying for 3 of the exams. Write and solve an equation to find equal amounts of time available to study for the 2 remaining exams.

49. **Reasoning** You have twice as many bus tokens as your colleague. You give your colleague several of your tokens. Each of you now has 120 bus tokens. How many bus tokens did you give your colleague? Explain your reasoning.

50. **Reasoning** You are 10 years older than your cousin. Two years ago, you were 3 times as old as your cousin is now. How old are you and your cousin now?

Activity: Calculating Square Roots

Materials:

- Pencil
- Calculator

Learning Objective:

Determine whether a decimal represents an approximation or the exact value of a square root. The *square root* of a number n is the number whose square is equal to n.

Name ______________________________

Complete the statement using = or ≈. Explain your reasoning.

1. $\sqrt{1}$ ▭ 1.00000 ______________________________

2. $\sqrt{2}$ ▭ 1.41421 ______________________________

3. $\sqrt{4}$ ▭ 2.00000 ______________________________

4. $\sqrt{2.25}$ ▭ 1.50000 ______________________________

5. $\sqrt{3}$ ▭ 1.73205 ______________________________

6. $\sqrt{100}$ ▭ 10.0000 ______________________________

7. $\sqrt{10}$ ▭ 3.16227 ______________________________

8. $\sqrt{0.36}$ ▭ 0.60000 ______________________________

9. The first 301 digits of $\sqrt{2}$ are shown. Do you see a pattern for the digits? If so, describe the pattern.

1.41421 35623 73095 04880 16887 24209 69807 85696 71875 37694 80731 76679
73799 07324 78462 10703 88503 87534 32764 15727 35013 84623 09122 97024
92483 60558 50737 21264 41214 97099 93583 14132 22665 92750 55927 55799
95050 11527 82060 57147 01095 59971 60597 02745 34596 86201 47285 17418
64088 91986 09552 32923 04843 08714 32145 08397 62603 62799 52514 07989

____

9.3 Real Numbers

Standards

Grades 6–8

The Number System

Students should apply and extend previous understandings of numbers to the system of rational numbers, and know that there are numbers that are not rational, and approximate them by rational numbers.

- Classify real numbers as rational or irrational.
- Understand the completeness of the real number line.
- Become familiar with some famous irrational numbers.

Classifying Real Numbers

In Section 7.3, you learned that any number written as $\frac{a}{b}$, where a and b are integers and $b \neq 0$, can be written as a terminating decimal or a repeating decimal.

$\frac{1}{4} = 0.25$ Terminating decimal $\qquad -\frac{1}{3} = -0.333\ldots$ Repeating decimal

There are other types of numbers whose decimal representations do not terminate or repeat.

$\sqrt{2} = 1.41421\ldots$ Nonrepeating decimal

Together, these three types of decimals help make up the set of *real numbers.*

Definition of the Set of Real Numbers

The set of **real numbers** is the set of all numbers that can be written in decimal form, including those numbers that require an infinite decimal expansion. If the decimal form terminates or repeats, then the real number is a *rational number*. If the decimal form neither terminates nor repeats, then the real number is an **irrational number**. Recall that a rational number is a number that can be written as the ratio of two integers. This implies that an irrational number cannot be written as the ratio of two integers.

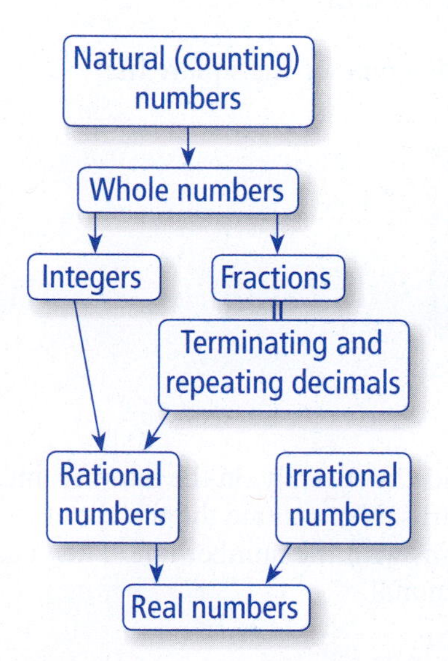

EXAMPLE 1 Showing that $\sqrt{2}$ Is Irrational

Use *proof by contradiction* to show that $\sqrt{2}$ is irrational.

SOLUTION

To begin, assume that $\sqrt{2}$ is rational and can be written as the ratio of two integers (in simplest form).

$\sqrt{2} = \frac{a}{b}$ Assumption: $\sqrt{2}$ can be written as a rational number $\frac{a}{b}$.

$2 = \frac{a^2}{b^2}$ Square each side.

$2b^2 = a^2$ Multiply both sides by b^2.

If $2b^2 = a^2$, then the prime factorizations of $2b^2$ and a^2 are the same by the Fundamental Theorem of Arithmetic. So the factor 2 must appear the same number of times in each prime factorization of $2b^2$ and a^2. Because squares have prime factors that occur in pairs, a^2 and b^2 have an even number of 2s. But for $2b^2$, there is another factor of 2, so there is an odd number of 2s in the prime factorization of $2b^2$. Because $2b^2 = a^2$, the factor 2 appears an odd number of times on the left side of the equal sign and an even number of times on the right side of the equal sign. This is a contradiction. Thus, the initial assumption that $\sqrt{2}$ is rational must be false. So, $\sqrt{2}$ is irrational.

Classroom Tip

The proof in Example 1 is usually considered beyond the scope of K–8 instruction. Even so, the proof is included in this book because of its historical importance in the study of mathematics.

Completeness of the Real Number Line

Each of the number lines you have studied so far has holes or gaps. For instance, the rational number line has a gap at the point corresponding to $\sqrt{2}$.

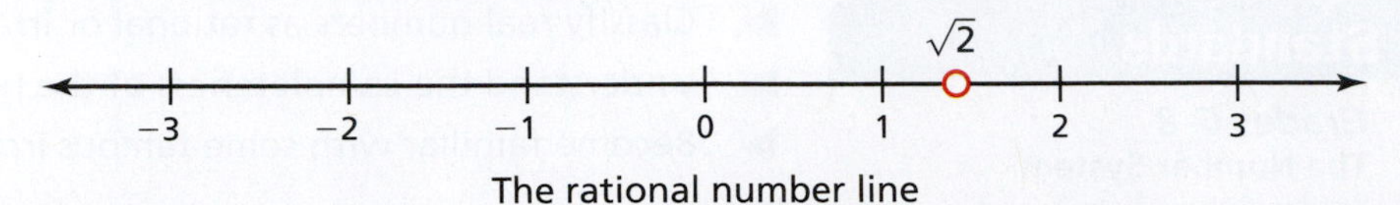

The rational number line

In fact, the rational number line has infinitely many gaps because there are infinitely many irrational numbers. The real number line, on the other hand, has no holes or gaps. In this sense, the set of real numbers is called "complete" because there are no other numbers that can be "squeezed" onto the real number line.

Completeness of the Real Number Line

The real number line is called **complete** because it has no holes or gaps. This implies the following two statements.

1. Every decimal representation (terminating, repeating, or nonrepeating) corresponds to exactly one point on the real number line.
2. Every point on the real number line corresponds to exactly one decimal representation.

EXAMPLE 2 Solving a Real-Life Problem

Use a number line to estimate the length of the diagonal of the square tile.

1 ft

Classroom Tip

A more straightforward approach is to use the Pythagorean Theorem that you studied in high school geometry. This states that $a^2 + b^2 = c^2$ for a right triangle shown below.

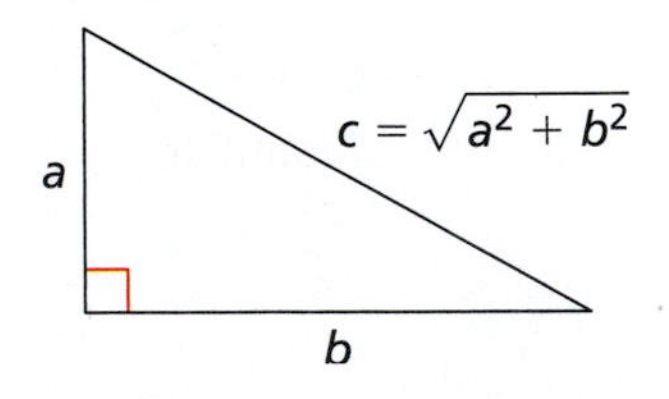

Using the dimensions of the tile in Example 2, you can find the exact length.

$$\begin{aligned}\text{Diagonal} &= \sqrt{1^2 + 1^2}\\ &= \sqrt{1 + 1}\\ &= \sqrt{2}\end{aligned}$$

So, the length of the diagonal of the tile is about 1.414 feet.

SOLUTION

Place the tile on a number line so that a 1-foot side lies exactly on the number line, as shown. Be sure to use the appropriate tick marks. Next, rotate the tile about zero on the number line so that the diagonal aligns with the number line. Then use the number line to estimate the length of the diagonal.

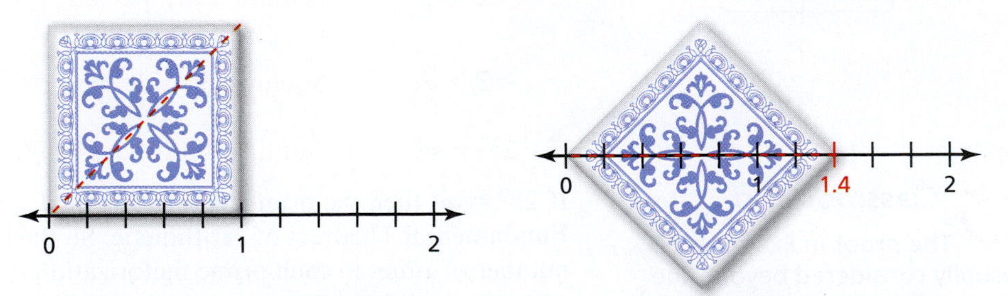

So, the length of the diagonal of the square tile is about 1.4 feet.

Note that the exact length of the diagonal is on the number line, but it is not practical to accurately determine the length past one or two decimal places.

Famous Irrational Numbers

One of the oldest known irrational numbers is the number π. The earliest written approximations of π were found in Egypt and Babylon, both within 1% of the exact value.

Classroom Tip

Sometimes students incorrectly think that $\frac{22}{7}$ is the exact value of π. When you are teaching this concept, be sure that your students understand that both $\frac{22}{7}$ and 3.14 are only approximations of π.

Definition of π (Pi)

The distance around a circle is called the **circumference**. The ratio of the circumference of a circle to its diameter, $\frac{\text{circumference}}{\text{diameter}}$, is the same for every circle. This ratio is represented by the Greek letter π, called pi.

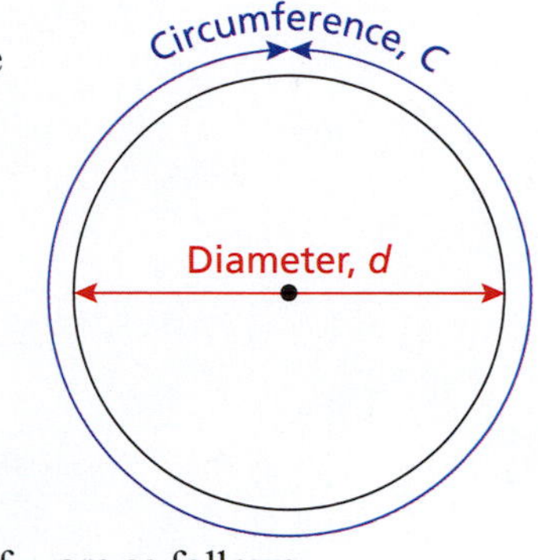

$$\pi = \frac{C}{d}$$

where C is the circumference and d is the diameter of the circle. π is an irrational number. Common rational approximations of π are as follows.

$$\pi \approx \frac{22}{7} \approx 3.1429 \qquad \pi \approx 3.14 \qquad \pi \approx \frac{355}{113} \approx 3.14159$$

The value of π to 8 decimal places is

$$\pi \approx 3.14159265.$$

Solving a Real-Life Problem

The diameter of Earth at the equator is about 7925 miles. Approximate the circumference of Earth at the equator. Use 3.14 for π.

SOLUTION

1. **Understand the Problem** You need to approximate the circumference of Earth at the equator.
2. **Make a Plan** Begin by writing the definition of π as the ratio of the circumference to the diameter. Substitute for π and d. Then solve for C.
3. **Solve the Problem**

$$\pi = \frac{C}{d} \qquad \text{Definition of } \pi$$

$$3.14 \approx \frac{C}{7925} \qquad \text{Substitute 3.14 for } \pi \text{ and 7925 for } d.$$

$$7925(3.14) \approx 7925\left(\frac{C}{7925}\right) \qquad \text{Multiply each side by 7925.}$$

$$24{,}884.5 \approx C \qquad \text{Simplify.}$$

At the equator, Earth's circumference is about 24,880 miles.

4. **Look Back** One way to check the reasonableness of your answer is to estimate. The circumference of a circle is roughly 3 times its diameter ($3.14 \approx 3$). Because the diameter of Earth is about 8000 miles, it is reasonable that the circumference is about $3 \times 8000 = 24{,}000$ miles.

Definition of Golden Ratio

If a line is divided into two parts so that the longer part divided by the shorter part is also equal to the sum of the parts divided by the longer part, then the resulting ratio is the **Golden Ratio**, which is denoted by the Greek letter phi, ϕ.

$$\phi = \frac{a}{b} = \frac{a + b}{a}$$

a b

$a + b$

The Golden Ratio is an irrational number.

$$\phi = \frac{1 + \sqrt{5}}{2} \approx 1.61803399$$

Rectangles whose sides are in the proportion of the Golden Ratio are considered in art and architecture to be pleasing to the human eye and are called **Golden Rectangles**.

Golden Rectangle

Real-Life Examples of the Golden Rectangle

Classroom Tip

As a classroom activity, you might consider asking your students to measure the dimensions of everyday rectangular objects such as index cards, photographs, picture frames, textbooks, door frames, computer screens, cell phones, and television screens. Ask them to determine which of the objects are Golden Rectangles.

Here are three examples of Golden Rectangles.

Front of the Parthenon in Athens, Greece

Mona Lisa's face

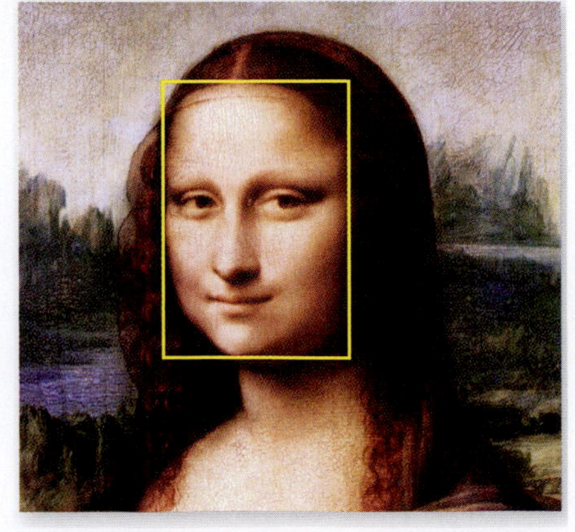

iStockphoto.com/MinistryOfJoy; J. Gatherum/Shutterstock.com; World History Archive/Image Asset Management Ltd. / Alamy

The final famous irrational number that we will study is called *Euler's number*, named after Swiss mathematician Leonard Euler (pronounced "oiler"). This number is critical in the study of calculus.

Definition of Euler's Number

Euler's number, denoted by the letter e, is given by the following infinite sum.

$$e = 1 + \frac{1}{1} + \frac{1}{1 \cdot 2} + \frac{1}{1 \cdot 2 \cdot 3} + \frac{1}{1 \cdot 2 \cdot 3 \cdot 4} + \frac{1}{1 \cdot 2 \cdot 3 \cdot 4 \cdot 5} + \cdots$$

e is an irrational number. The value of e to 8 decimal places is

$$e \approx 2.71828183.$$

EXAMPLE 5 Continuously-Compounded Interest

Classroom Tip

When an account earns simple interest, it earns money *only* on the principal (original amount of the deposit). When an account earns compound interest, it earns money on the principal as well as the interest that has already been earned.

The balance A in an account that earns interest that is compounded continuously is given by

$$A = Pe^{rt}$$

where P is the initial principal, r is the annual interest rate (in decimal form), and t is the time in years. You deposit \$10,000 into an account that earns 5% annual interest compounded continuously. What will the account balance be in 20 years? Use 2.72 for e.

SOLUTION

1. **Understand the Problem** You need to find the balance in the account at the end of 20 years.

2. **Make a Plan** Use the formula for continuously-compounded interest.

3. **Solve the Problem**

$$A = Pe^{rt} \quad \text{Write formula.}$$

$$\approx 10{,}000(2.72)^{0.05(20)} \quad \text{Substitute given values.}$$

$$= 10{,}000(2.72)^{1} \quad \text{Simplify exponent.}$$

$$= 27{,}200 \quad \text{Multiply.}$$

The account balance will be about \$27,200.

4. **Look Back** One way to check the reasonableness of your answer is to create a spreadsheet. You can approximate the final balance by compounding annually, instead of continuously.

Mathematical Practices

Make Sense of Problems and Persevere in Solving Them

MP1 Mathematically proficient students monitor and evaluate their progress and change course if necessary.

	A	B
1	Year	Balance
2	0	\$10,000.00
3	1	\$10,500.00
4	2	\$11,025.00
	⋮	⋮
21	19	\$25,269.60
22	20	\$26,532.98 ✓

The spreadsheet gives the balance for compounding annually (once a year). Continuous compounding is equivalent to compounding every moment which is why its balance is greater.

9.3 Exercises

Free step-by-step solutions to odd-numbered exercises at MathematicalPractices.com.

1. **Writing a Solution Key** Write a solution key for the activity on page 348. Describe a rubric for grading a student's work.

2. **Grading the Activity** In the activity on page 348, a student gave the answers below. Each question is worth 10 points. Assign a grade to each answer. For those that are incorrect, why do you think the student erred?

Sample Student Work

5. $\sqrt{3}$ ≈ 1.73205
   $(1.73205)^2$ is approximately 3.

6. $\sqrt{100}$ = 10.0000
   $(10)^2$ is exactly 100.

7. $\sqrt{10}$ = 3.16227
   $(3.16227)^2$ is exactly 10.

8. $\sqrt{0.36}$ ≈ 0.60000
   $(0.6)^2$ is approximately 0.36.

3. **Classifying Numbers** Place check marks to indicate the sets of numbers to which each number belongs.

		Natural	Integer	Rational	Irrational	Real
a.	5.4					
b.	-7					
c.	$\sqrt{11}$					
d.	$\frac{2}{5}$					
e.	71					

4. **Classifying Numbers** Place check marks to indicate the sets of numbers to which each number belongs.

		Natural	Integer	Rational	Irrational	Real
a.	$-4.\overline{5}$					
b.	$\sqrt{10}$					
c.	$\sqrt{16}$					
d.	$2\frac{8}{9}$					
e.	$\frac{6}{11}$					

5. **True or False?** Tell whether each statement is *true* or *false*. Explain your reasoning.
   a. The set of terminating decimals is a subset of the real numbers.
   b. The set of rational numbers is a subset of the irrational numbers.

6. **True or False?** Tell whether each statement is *true* or *false*. Explain your reasoning.
   a. The set of irrational numbers is a subset of the real numbers.
   b. The set of fractions is a subset of the irrational numbers.

7. **Rational and Irrational Numbers** Simplify each expression if possible. Then decide whether the number is rational or irrational.
   a. $\frac{1}{2}$ b. 0.4
   c. $\sqrt{5}$ d. $\sqrt{64}$
   e. $2 \cdot \sqrt{5}$ f. $5 + \sqrt{4}$

8. **Rational and Irrational Numbers** Simplify each expression if possible. Then decide whether the number is rational or irrational.
   a. $-\frac{2}{7}$ b. 0.9
   c. $\sqrt{3}$ d. $\sqrt{36}$
   e. $7 \div \sqrt{6}$ f. $11 - \sqrt{9}$

9. Ordering Real Numbers Order the real numbers from least to greatest.

a. $\frac{1}{2}$, $0.\overline{5}$, $0.\overline{54}$, $0.5\overline{54}$, $0.5\overline{4}$, $0.\overline{454}$, $\sqrt{0.54}$

b. $\frac{3}{4}$, $0.7\overline{5}$, $0.\overline{75}$, $0.\overline{755}$, 0.7, $\sqrt{0.75}$, $\sqrt{0.7}$

10. Ordering Real Numbers Order the real numbers from least to greatest.

a. $\frac{1}{4}$, $0.\overline{2}$, $0.\overline{25}$, $0.2\overline{25}$, $0.2\overline{5}$, $0.\overline{225}$, $\sqrt{0.25}$

b. $\frac{4}{7}$, 0.4, $0.\overline{47}$, $0.\overline{477}$, 0.47, $\sqrt{0.47}$, $\sqrt{0.4}$

11. Finding Irrational Numbers Find two irrational numbers between each pair of real numbers.

a. 2, 3

b. $\frac{3}{5}$, $\frac{9}{10}$

c. 0.1, $0.\overline{6}$

d. $\sqrt{2}$, $\sqrt{10}$

12. Finding Irrational Numbers Find two irrational numbers between each pair of real numbers.

a. 1, 6

b. $\frac{3}{10}$, $\frac{4}{5}$

c. $0.\overline{12}$, $0.\overline{8}$

d. $\sqrt{1}$, $\sqrt{8}$

13. Using the Pythagorean Theorem Use the Pythagorean Theorem to find or approximate the missing length of each triangle.

a.

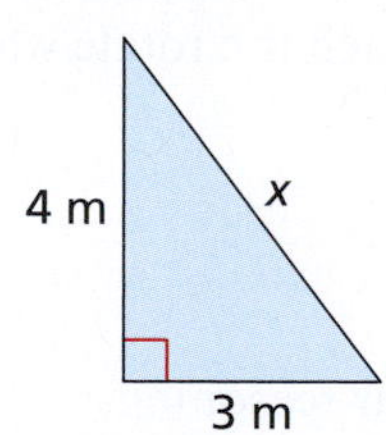

b.

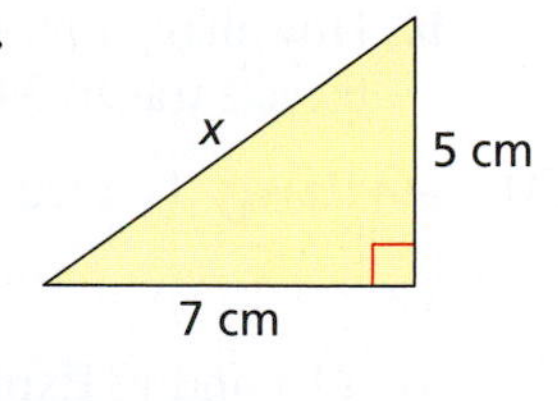

14. Using the Pythagorean Theorem Use the Pythagorean Theorem to approximate the missing length of each triangle.

a.

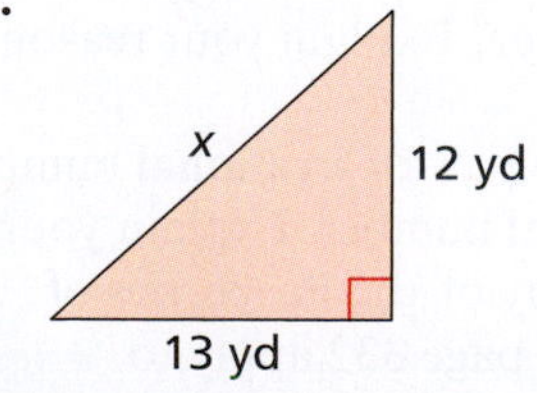

b.

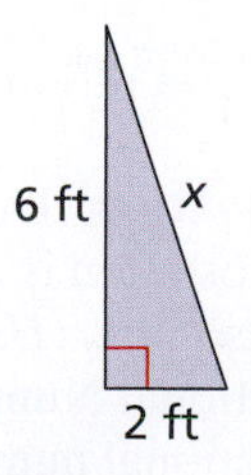

15. Finding the Circumference of a Circle Approximate the circumference of each circle. Use $\frac{22}{7}$ for π.

a.

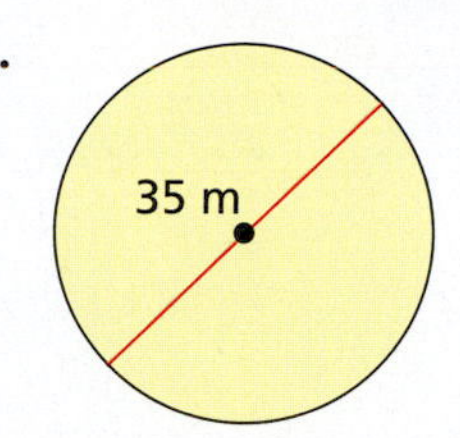

b.

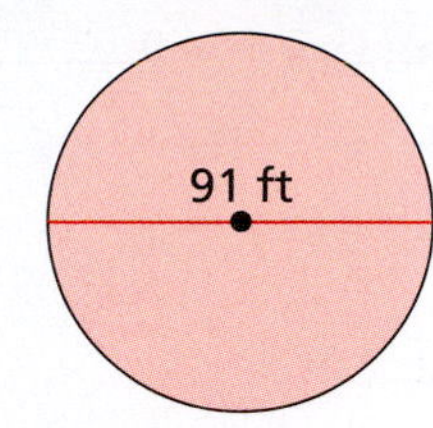

16. Finding the Circumference of a Circle Approximate the circumference of each circle. Use 3.14 for π.

a.

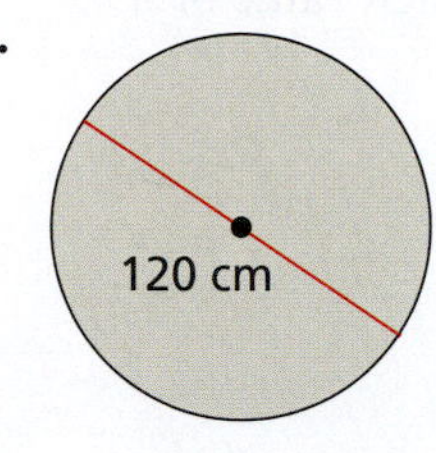

b.

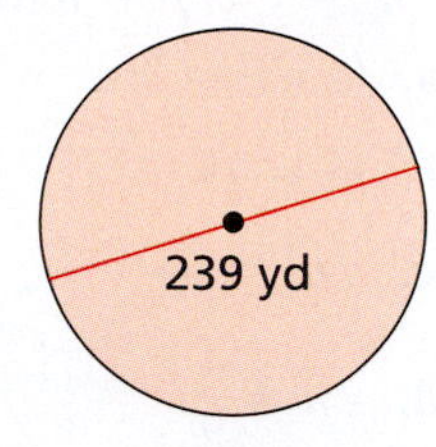

17. The Golden Ratio Determine whether the two parts of the line approximate the Golden Ratio.

a. 34 m 21 m

b.

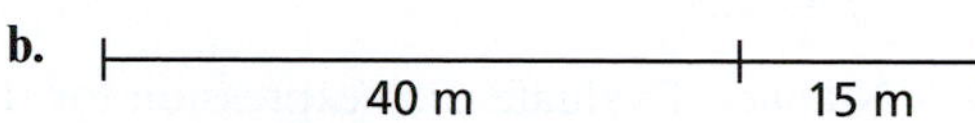

18. The Golden Ratio Determine whether the two parts of the line approximate the Golden Ratio.

a. 120 cm 113 cm

b. 144 cm 89 cm

19. The Golden Rectangle Determine whether each rectangle approximates a Golden Rectangle.

a.

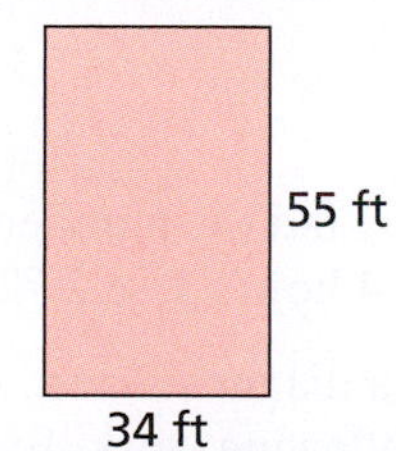

b.

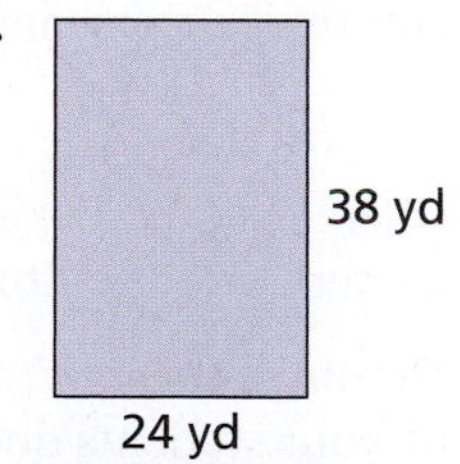

20. The Golden Rectangle Determine whether each rectangle approximates a Golden Rectangle.

a.

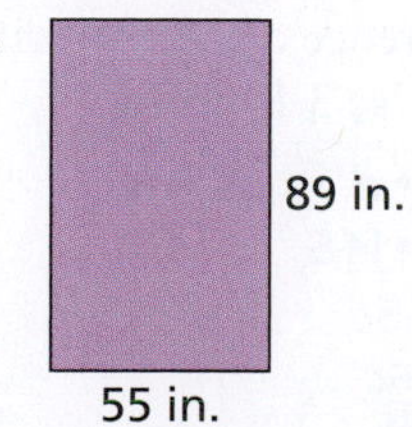

b.

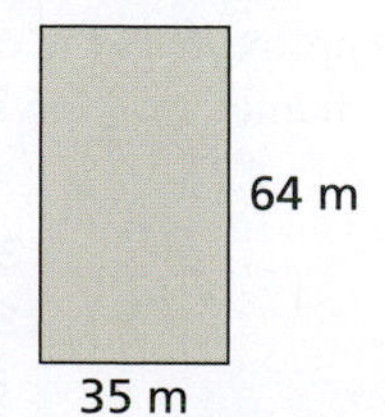

21. Using Properties of Squares and Square Roots Let n be a positive real number. Use the fact that $\sqrt{n^2} = (\sqrt{n})^2 = n$ to find each value of n.

a. $(\sqrt{n})^2 = 9$

b. $n^2 = 16$

c. $\sqrt{n^2} = 256$

d. $\sqrt{n} = \frac{1}{4}$

22. Using Properties of Squares and Square Roots Let n be a positive real number. Use the fact that $\sqrt{n^2} = (\sqrt{n})^2 = n$ to find each value of n.

a. $(\sqrt{n})^2 = 36$

b. $n^2 = 49$

c. $\sqrt{n^2} = 64$

d. $\sqrt{n} = \frac{4}{9}$

23. Euler's Number Evaluate each expression for the given value of x. Use 2.72 for e.

a. e^{2x}, $x = 4$

b. e^{-4x}, $x = 0.5$

24. Euler's Number Evaluate each expression for the given value of x. Use 2.72 for e.

a. $4e^{x}$, $x = -2$

b. e^{-5x}, $x = 0.6$

25. Bank Account You deposit \$8000 into an account that earns 5% annual interest compounded continuously. What will the account balance be in 40 years? Use 2.72 for e.

26. Bacteria Culture The number N of bacteria in a culture is represented by

$N = 100e^{0.25t}$

where t is the time in hours. Find the number of bacteria in the culture after 4 hours. Use 2.72 for e.

27. Grading Student Work On a diagnostic test, one of your students does the following work. Explain what the student did wrong. Which topics would you encourage the student to review?

> Approximate the circumference of a circle with a diameter of 145 inches. Use 3.14 for π.
>
> $\pi = C \cdot d$
>
> $3.14 \approx C \cdot 145$
>
> $3.14 \approx 145C$
>
> $\frac{3.14}{145} \approx \frac{145C}{145}$
>
> $0.02 = C$

28. Grading Student Work On a diagnostic test, one of your students does the following work. Explain what the student did wrong. Which topics would you encourage the student to review?

> You deposit \$5000 into an account that earns 4% annual interest compounded continuously. What will the account balance be in 5 years?
>
> $A = Pe^{rt}$
>
> $\approx 5000(2.72)^{0.4(5)}$
>
> $= 5000(2.72)^2$
>
> $= 36{,}992$
>
> The account balance will be about \$36,992.

29. Flying Disc A plastic flying disc has a circular hole in the middle as shown below. The diameter of the outer edge of the disc is 13 inches and the width of the disc is 1.5 inches. How much greater is the circumference of the outer edge than the circumference of the inner edge of the disc?

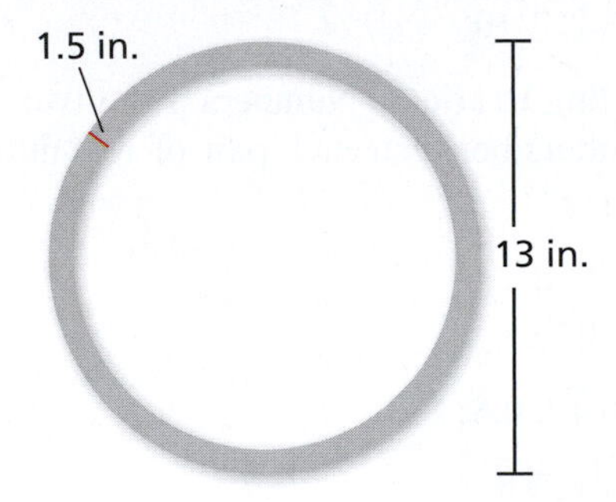

30. Bicycle Each tire on a bicycle has a diameter of 20 inches.

a. What is the circumference of each tire?

b. How many times does each tire rotate when the bicycle travels 340 inches?

31. Writing Is it true that

$\sqrt{x + y} = \sqrt{x} + \sqrt{y}$

for all x and y? Explain your reasoning.

32. Writing You can write $\sqrt{5}$ as

$\frac{\sqrt{5}}{1}$.

Is $\frac{\sqrt{5}}{1}$ a rational number? Explain your reasoning.

33. Writing Describe a pair of irrational numbers whose sum is a rational number. Explain your reasoning. (*Hint:* Many of the Properties of Rational Numbers on page 332 apply to irrational numbers.)

34. ***Writing*** Describe a pair of irrational numbers whose product is a rational number. Explain your reasoning. (*Hint:* Consider how the square root of a number is defined.)

35. **The Wheel of Theodorus** The *Wheel of Theodorus* is a figure formed by a chain of right triangles, with consecutive triangles sharing a common side. The hypotenuse of one triangle becomes a leg of the next, as shown below.

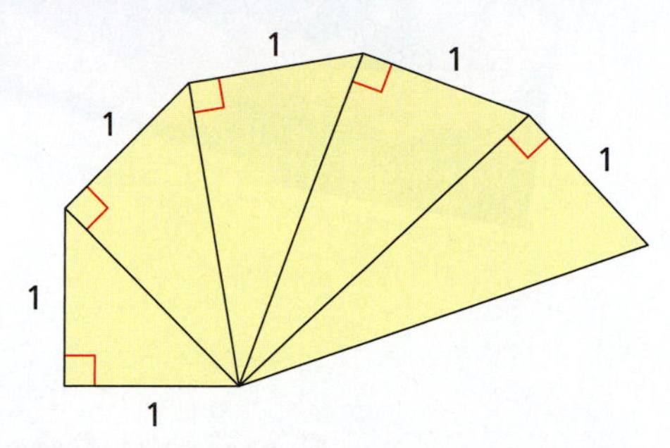

a. Find the length of the longest hypotenuse in the figure. Use only exact answers in your calculations.

b. Add two more triangles to the figure. Find the length of the longest hypotenuse. Use only exact answers in your calculations.

36. **Pythagorean Triple** A *Pythagorean triple* is a group of positive integers a, b, and c that represent the side lengths of a right triangle. For example, the integers 3, 4, and 5 form a Pythagorean triple because $3^2 + 4^2 = 5^2$.

a. Choose any two positive integers m and n such that $m < n$.

b. Find a, b, and c as follows: $a = n^2 - m^2$, $b = 2mn$, and $c = n^2 + m^2$.

c. Show that the numbers you generated in part (b) form a Pythagorean triple.

37. **Activity: In Your Classroom** Design a classroom activity that your students can use to discover the Pythagorean Theorem by working with several Pythagorean triples which you will give them. Describe the steps you will have your students perform. Some students may notice that some right triangles have side lengths that *do not* form Pythagorean triples. What will you tell your students about these triangles?

38. **Activity: In Your Classroom** Design a classroom activity that your students can use to discover the value of π. Describe the steps that students will perform to find several approximations of π. Include a description of any items or tools they will use, figures they will draw, or measurements they will make.

39. **Problem Solving** Fold an $8\frac{1}{2}$-inch by 11-inch sheet of paper along a diagonal as shown below. Use a ruler to measure the length of the diagonal in inches. Then use the Pythagorean Theorem to find the length of the diagonal. Compare your results.

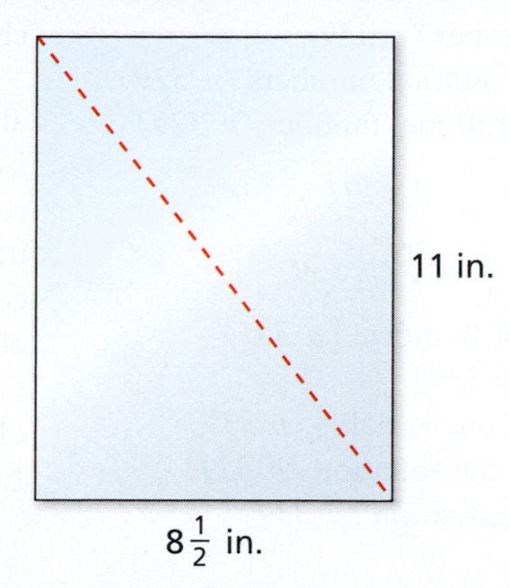

40. **Problem Solving** Cut an $8\frac{1}{2}$-inch by 11-inch sheet of paper in half, and then fold it along a diagonal as shown below. Use a ruler to measure the length of the diagonal in inches. Then use the Pythagorean Theorem to find the length of the diagonal. Compare your results.

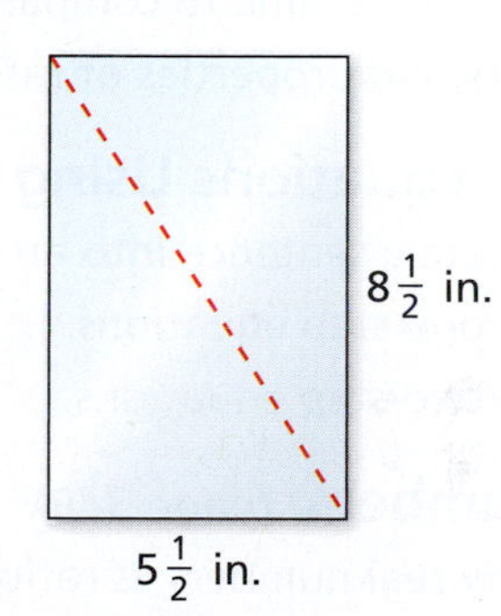

41. **Reasoning** The edge length of the cube shown below is 7 feet.

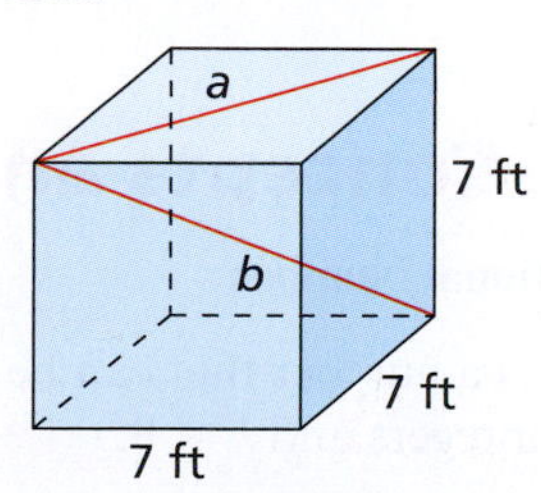

a. Find diagonal length a.

b. Find diagonal length b.

42. **Problem Solving** An events coordinator plans to use a square section of a park for a small outdoor concert. The section will have an area of 1500 square feet. There is 150 feet of rope available. Is there enough rope to surround the section? Explain your reasoning.

Chapter Summary

Chapter Vocabulary

rational number *(p. 329)*
equality of rational numbers *(p. 329)*
graphing a rational number *(p. 329)*
$\frac{a}{b}$ is less than $\frac{c}{d}$ *(p. 330)*
$\frac{c}{d}$ is greater than $\frac{a}{b}$ *(p. 330)*
closed set of numbers *(p. 331)*
equation *(p. 339)*
equation in one variable *(p. 339)*
solution of the equation *(p. 339)*
solving the equation *(p. 339)*
check the solution *(p. 339)*
algebra tiles *(p. 340)*
isolate the variable *(p. 340)*
real numbers *(p. 349)*
irrational number *(p. 349)*
complete *(p. 350)*
circumference *(p. 351)*
Golden Ratio *(p. 352)*
Golden Rectangles *(p. 352)*
Euler's number *(p. 353)*

Chapter Learning Objectives | Review Exercises

9.1 Rational Numbers *(page 329)* — 1–30

- Understand the definition of a rational number, and graph rational numbers on a number line.
- Use a number line to compare and order rational numbers.
- Understand properties of rational numbers and perform operations.

9.2 Solving Equations Using Rational Numbers *(page 339)* — 31–62

- Translate a sentence into an equation and vice versa.
- Solve one-step equations.
- Solve two-step equations.

9.3 Real Numbers *(page 349)* — 63–90

- Classify real numbers as rational or irrational.
- Understand the completeness of the real number line.
- Become familiar with some famous irrational numbers.

Important Concepts and Formulas

Definition of a Rational Number

A **rational number** is a number that can be written as $\frac{a}{b}$, where a and b are integers and $b \neq 0$.

Equations

Addition Property of Equality When you add the same number to each side of an equation, the two sides remain equal.

Subtraction Property of Equality When you subtract the same number from each side of an equation, the two sides remain equal.

Multiplication Property of Equality When you multiply each side of an equation by the same nonzero number, the two sides remain equal.

Division Property of Equality When you divide each side of an equation by the same nonzero number, the two sides remain equal.

Completeness of the Real Number Line

The real number line is called **complete** because it has no holes or gaps. This implies the following two statements.

1. Every decimal representation (terminating, repeating, or nonrepeating) corresponds to exactly one point on the real number line.
2. Every point on the real number line corresponds to exactly one decimal representation.

Review Exercises

Free step-by-step solutions to odd-numbered exercises at *MathematicalPractices.com.*

9.1 Rational Numbers

Identifying Rational Numbers Identify the rational numbers on the number line.

1.

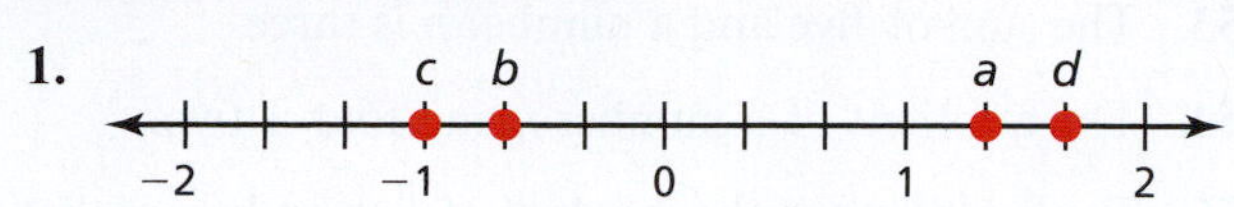

2.

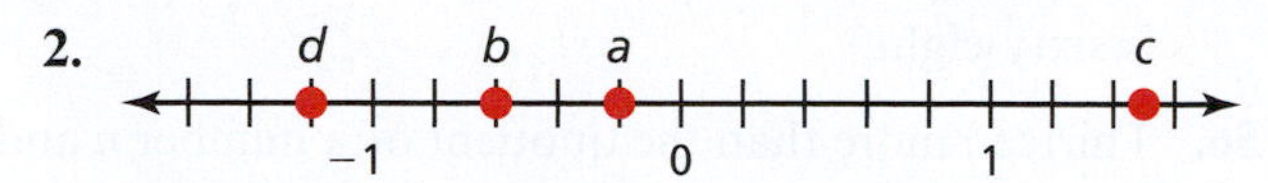

Graphing Rational Numbers Graph the rational number on a number line.

3. $\frac{1}{2}$ **4.** $-\frac{1}{4}$

5. $\frac{2}{7}$ **6.** $-\frac{3}{10}$

Definition of Rational Numbers Show that the number is rational by writing it in the form $\frac{a}{b}$, where a and b are integers and $b \neq 0$.

7. 0.5

8. -7

9. $2\frac{1}{6}$

10. $1\frac{2}{3}$

11. Ordering Rational Numbers Use a number line to order the rational numbers -1, $-\frac{1}{10}$, $-\frac{6}{10}$, $\frac{1}{3}$, and $\frac{3}{5}$ from least to greatest.

12. Ordering Rational Numbers Use a number line to order the rational numbers -2, 2, $\frac{8}{3}$, $-\frac{15}{6}$, and $-\frac{7}{3}$ from least to greatest.

Identifying Properties of Rational Numbers Identify the property of rational numbers that makes the equation true.

13. $-\frac{5}{6} + \frac{5}{6} = 0$

14. $\left(\frac{7}{9} \cdot \frac{1}{8}\right) \cdot \frac{6}{7} = \frac{7}{9} \cdot \left(\frac{1}{8} \cdot \frac{6}{7}\right)$

15. $\frac{2}{9} + \frac{1}{5} = \frac{1}{5} + \frac{2}{9}$

16. $-\frac{6}{5} \cdot 1 = -\frac{6}{5}$

Using Properties of Rational Numbers Use the given property of rational numbers to complete the equation.

17. Associative Property of Addition

$\left(\frac{2}{5} + \frac{4}{7}\right) + \frac{1}{15} = \square$

18. Zero Multiplication Property

$\frac{9}{2} \cdot 0 = \square$

19. Additive Identity Property

$\frac{4}{5} + 0 = \square$

20. Multiplicative Inverse Property

$\frac{10}{11} \cdot \frac{11}{10} = \square$

Operations with Rational Numbers Evaluate the expression. Write your answer in simplest form.

21. $\frac{5}{12} + \left(-\frac{3}{8}\right)$ **22.** $-\frac{5}{6} + \frac{5}{9}$

23. $\frac{1}{2} - \frac{5}{8}$ **24.** $-\frac{3}{4} - \left(-\frac{5}{6}\right)$

25. $\frac{3}{8} \cdot \frac{1}{2}$

26. $\frac{5}{12} \cdot \left(-\frac{3}{10}\right)$

27. $\frac{3}{8} \div \left(-\frac{9}{10}\right)$

28. $-\frac{2}{15} \div \frac{3}{5}$

29. Dog Food You feed your dog $\frac{3}{4}$ cup of dog food every morning and $\frac{1}{3}$ cup every evening. You buy a bag of dog food that contains 30 cups. How many days will the bag last?

30. Costumes You buy $37\frac{1}{2}$ feet of material to make costumes for a school play. Each costume needs $1\frac{1}{4}$ feet of material for a hat and $3\frac{1}{3}$ feet of material for a cape. How many costumes can you make?

9.2 Solving Equations Using Rational Numbers

Properties of Equality Determine which property of equality is used to solve the equation.

31. $z - 21 = 1$

32. $w + 2 = -13$

33. $\frac{1}{5}t = 9$

34. $2s = -10$

Algebra Tiles Write the equation modeled by the algebra tiles. Then solve the equation.

35.

36.

Solving a One-Step Equation Solve the equation. Show your work and justify each step.

37. $38 + x = -17$

38. $a - 53 = 28$

39. $119 = -7z$

40. $-114 = 6d$

41. $\frac{u}{3.1} = -7.2$

42. $\frac{1}{23}y = -8$

Algebra Tiles Write the equation modeled by the algebra tiles. Then solve the equation.

43.

44.

Solving a Two-Step Equation Solve the equation. Show your work and justify each step.

45. $-4y + 15 = 15$

46. $-5x - 30 = 0$

47. $\frac{1}{3}m - 1 = 5$

48. $6 = 4 + \frac{2}{5}t$

49. $-3.6 - 2j = -5.4$

50. $6.6 = -3r + 2.1$

Modeling Write and solve an equation to model the statement.

51. The difference between a number n and fifteen is seven.

52. The product of seven and a number n is negative twenty-one.

53. The sum of five and a number n is three.

54. The quotient of a number n and four is two.

55. Twelve less than the product of five and a number n is sixty-eight.

56. Thirteen more than the quotient of a number n and forty-eight is twenty-one.

Geometry Write and solve equations to find the values of the variables in the rectangle.

57.

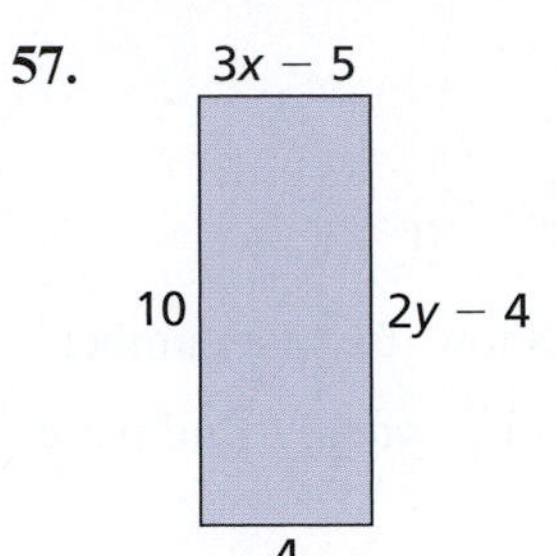

58.

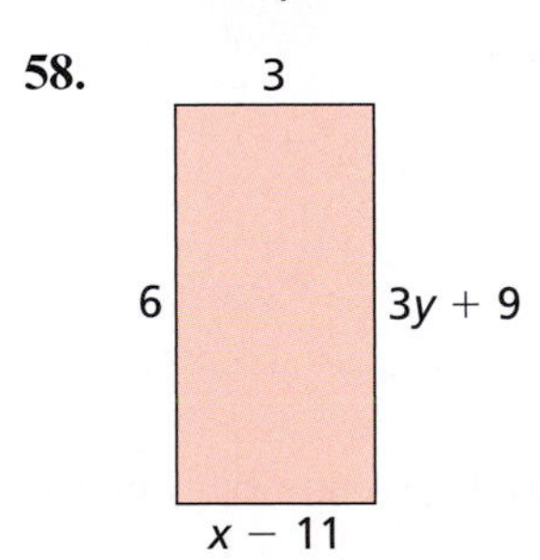

59. Flagpole A flagpole has a height of 40 feet. You raise a flag to a height of 27.5 feet. Write and solve an equation to find how much further you have to raise the flag to reach the top.

60. Sale A DVD player that has a regular price of \$74.58 is on sale for \$54.99. Write and solve an equation to determine how much the price of the DVD player is marked down.

61. Trampoline Dimensions A rectangular trampoline has an area of 196 square feet. The length of the trampoline is 17.5 feet. What is the width of the trampoline?

62. Carnival Package A carnival charges an entrance fee of \$18 and \$2.50 per ride. A student pays \$40.50. How many rides will the student be able to go on?

9.3 Real Numbers

Classifying a Number Place check marks to indicate the sets of numbers to which the number belongs.

		Natural	Integer	Rational	Irrational	Real
63.	$\sqrt{5}$					
64.	$\frac{3}{7}$					
65.	-2					
66.	3.9					

Rational and Irrational Numbers Simplify the expression if possible. Then decide whether the number is rational or irrational.

67. 0.3

68. -2.5

69. $\sqrt{\frac{1}{25}}$

70. $\sqrt{4} - \sqrt{9}$

71. $\sqrt{7}$

72. $\sqrt{11}$

Ordering Real Numbers Order the real numbers from least to greatest.

73. $\frac{3}{8}, 0.38, 0.3, 0.\overline{38}, 0.3\overline{8}, \sqrt{0.3}, \sqrt{0.38}$

74. $0.6\overline{64}, 0.\overline{6}, \sqrt{0.64}, 0.\overline{464}, 0.6, 0.6\overline{4}, 0.\overline{64}$

Using the Pythagorean Theorem Use the Pythagorean Theorem to approximate the missing length of the triangle.

75.

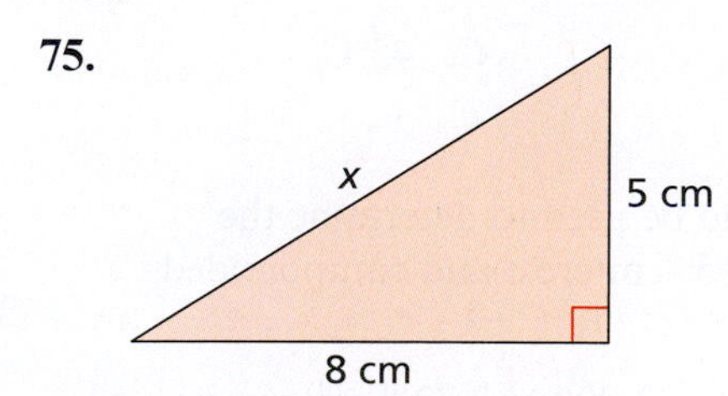

76.

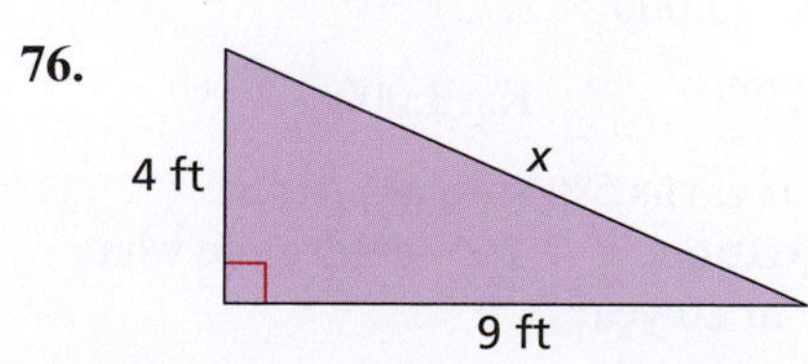

Finding the Circumference of a Circle Approximate the circumference of the circle. Use 3.14 or $\frac{22}{7}$ for π.

77. **78.**

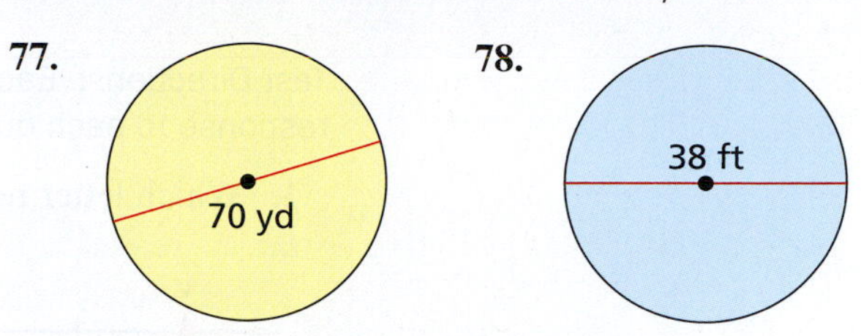

The Golden Ratio Determine whether the two parts of the line approximate the Golden Ratio.

79. **80.**

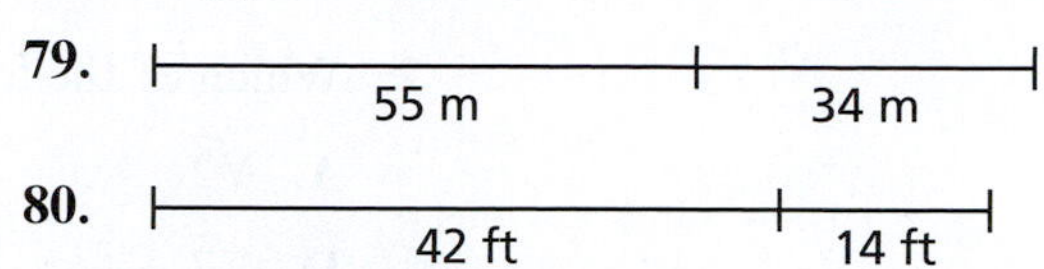

The Golden Rectangle Determine whether the rectangle approximates a Golden Rectangle.

81. **82.**

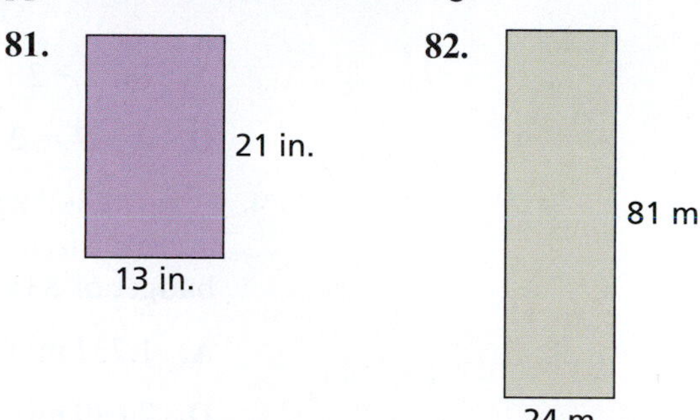

Using Properties of Squares and Square Roots Let n be a positive real number. Use the fact that $\sqrt{n^2} = (\sqrt{n})^2 = n$ to find the value of n.

83. $(\sqrt{n})^2 = 12$

84. $n^2 = 100$

85. $\sqrt{n^2} = 27$

86. $\sqrt{n} = \frac{1}{9}$

Euler's Number Evaluate the expression for the given value of x. Use 2.72 for e.

87. $2e^x, x = -3$

88. $e^{5x}, x = 0.4$

89. **Bank Account** You deposit $7500 into an account that earns 4% annual interest compounded continuously. What will the account balance be in 25 years? Use 2.72 for e.

90. **Inflatable Ball** An inflatable ball has a diameter of 20 centimeters. The ball rolls 300 centimeters down a hallway. How many times does the ball rotate?

Chapter Test

The questions on this practice test are modeled after questions on state certification exams for K–8 teaching.

Test Directions: Each of the questions is followed by five choices. Choose the *best* response to each question.

1. Which letter represents where $\frac{2}{7}$ is graphed on the number line?

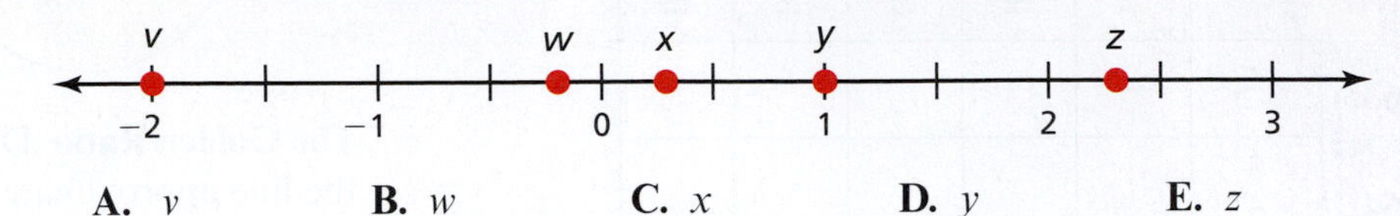

A. v **B.** w **C.** x **D.** y **E.** z

2. Which of the following is an example of a rational number?

A. $\sqrt{2}$ **B.** $6\sqrt{3}$ **C.** $4\frac{1}{2}$

D. $\sqrt{7} + \frac{3}{2}$ **E.** $2 - \frac{3}{19}$

3. Write an equation to model the following statement.

"The number n less than seven is negative two."

A. $7n = -2$ **B.** $7 - n = -2$ **C.** $7 - n = 2$

D. $n - 7 = 2$ **E.** $-2 = n - 7$

4. A car rental agency charges \$139 per week plus \$0.08 per mile for an average-sized car. How far can you travel to the nearest mile on a maximum budget of \$350?

A. 1,737 mi **B.** 2,110 mi **C.** 2,637 mi

D. 2,640 mi **E.** 4,375 mi

5. The area of a parking lot is 4,693 square feet. The length of the parking lot is 65 feet. What is the width of the lot?

A. 65 ft **B.** 72.2 ft **C.** 74.1 ft

D. 82 ft **E.** 87.5 ft

6. The formula relating the Celsius (C) and Fahrenheit (F) scales of temperature is shown below. Find the temperature in the Celsius scale when the temperature is 113°F.

$$F = \frac{9}{5}C + 32$$

A. 25°C **B.** 30°C **C.** 45°C

D. 203.4°C **E.** 235.4°C

7. Which of the following expressions could be used to determine the approximate value of \$3,000 invested at 5% interest and compounded continuously for 4 years?

A. $3,000 \times 2.72 \times 5 \times 4$ **B.** $(3,000 \times 2.72)^{0.05(4)}$

C. $3,000(2.72)^{5(4)}$ **D.** $3,000(2.72)^{0.05(4)}$ **E.** $3,000(3)^{0.05(4)}$

8. Luke deposits \$6,000 into an account that earns 5% annual interest compounded continuously. Using the equation, $A = Pe^{rt}$, determine what the approximate account balance will be in 20 years.

A. \$6,631.03 **B.** \$9,122.47 **C.** \$12,000.00

D. \$15,919.79 **E.** \$16,309.69

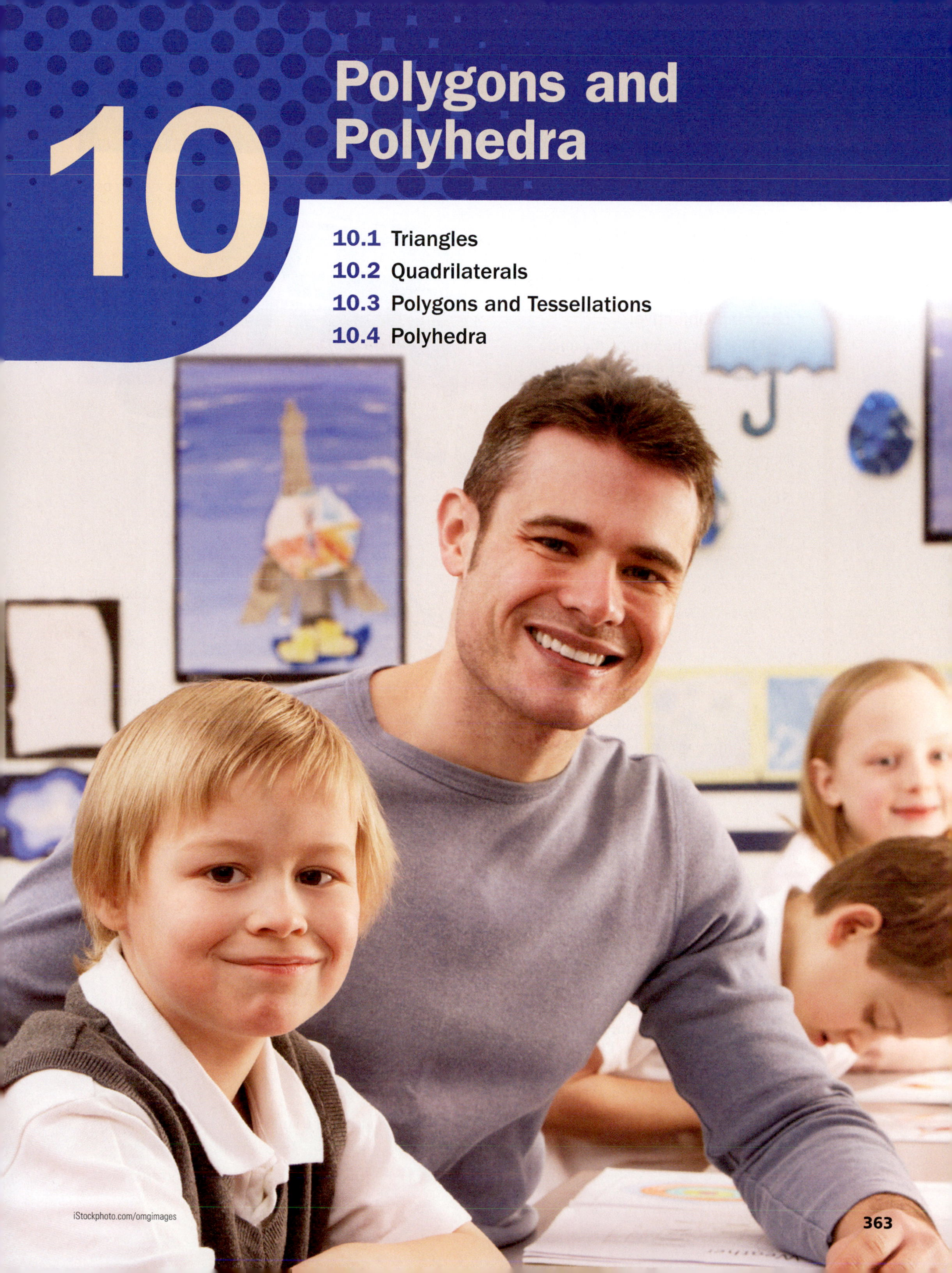

10 Polygons and Polyhedra

iStockphoto.com/omgimages

Activity: The Sum of the Angle Measures of a Triangle

Materials:

- Pencil
- Straightedge
- Paper
- Scissors

Learning Objective:

Discover the sum of the angle measures of a triangle.

Name ______________________________

Step 1 Draw a triangle like the one below. Label the angles A, B, and C.

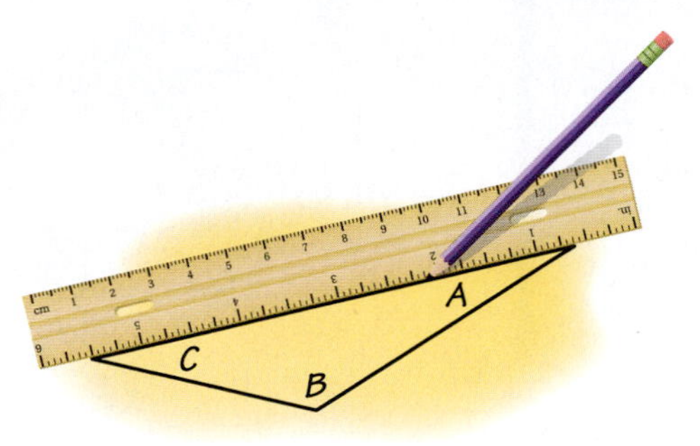

Step 2 Cut out the triangle. Tear off the three corners of the triangle.

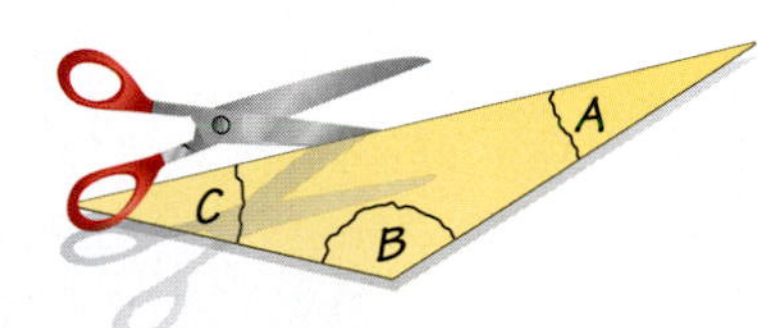

Step 3 Draw a straight line on a piece of paper. Arrange angles A and B as shown.

Step 4 Place the third angle as shown. What does this tell you about the sum of the measures of the angles?

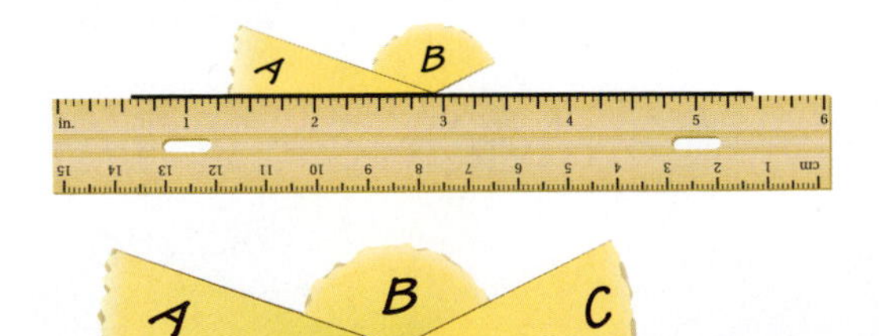

Step 5 Draw three other triangles that have different shapes. Repeat Steps 1 through 4 for each triangle. Do you get the same result as in Step 4? Explain.

Step 6 Write a rule about the sum of the measures of the angles of a triangle.

10.1 Triangles

Standards

Grades K–2
Geometry
Students should identify, describe, and reason with shapes and their attributes.

Grades 3–5
Measurement and Data
Students should understand concepts of angle and measure angles.

Grades 6–8
Geometry
Students should solve real-life and mathematical problems involving angle measure.

- Draw, measure, and classify angles.
- Classify triangles by their angles.
- Classify triangles by their sides.

Drawing, Measuring, and Classifying Angles

Drawing and Measuring Angles Using a Protractor

An **angle** consists of two rays that share the same endpoint. The rays are the **sides** of the angle. The endpoint is the **vertex** of the angle. A **protractor** is a tool you can use to draw and measure angles. Angles are measured in units called **degrees (°)**. The **degree measure** of a protractor ranges from 0° to 180°. The angle shown measures 70°.

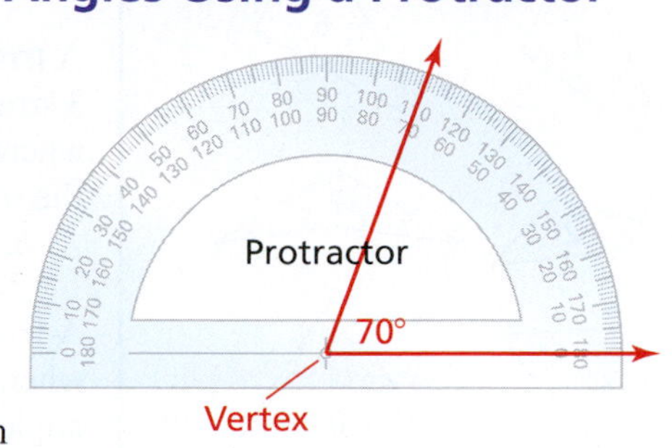

Classifying Angles

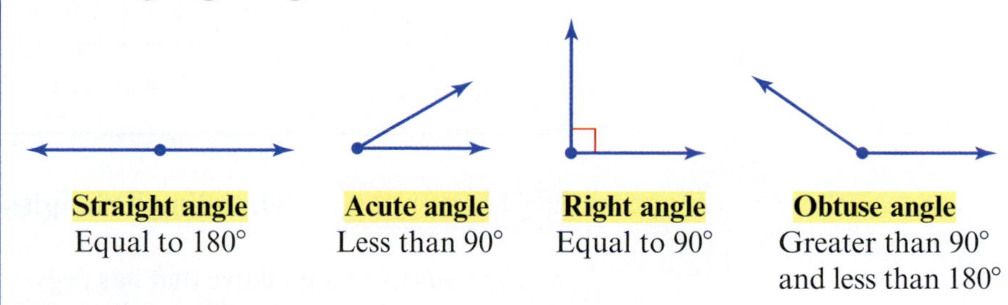

Two angles are **complementary** when the sum of their measures is 90°.
Two angles are **supplementary** when the sum of their measures is 180°.

Classroom Tip

When teaching students how to read an angle measure on a protractor, make sure they notice that the inner and outer measures are supplementary. Reading the angle measure depends on how the protractor is placed on the angle. For instance, on the protractor at the right, the angle measure is 70° and the angle measure of the supplement is 110°.

EXAMPLE 1 Finding Complementary and Supplementary Angles

a. Find and sketch an angle that is complementary to the angle shown.

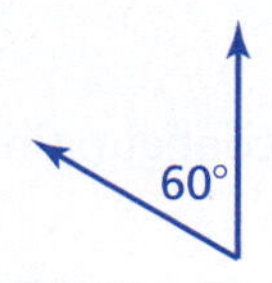

b. Find and sketch an angle that is supplementary to the angle shown.

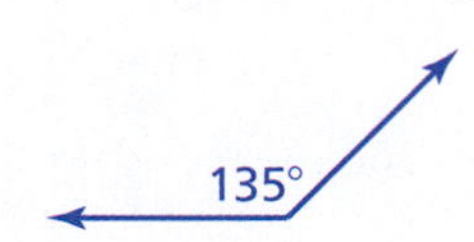

SOLUTION

a. An angle that is complementary to the angle has a measure of $90° - 60° = 30°$.

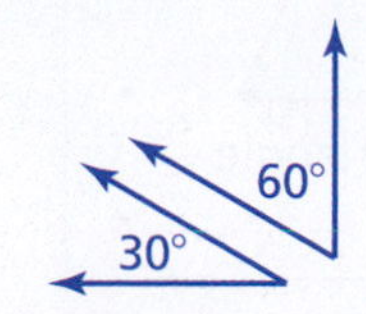

b. An angle that is supplementary to the angle has a measure of $180° - 135° = 45°$.

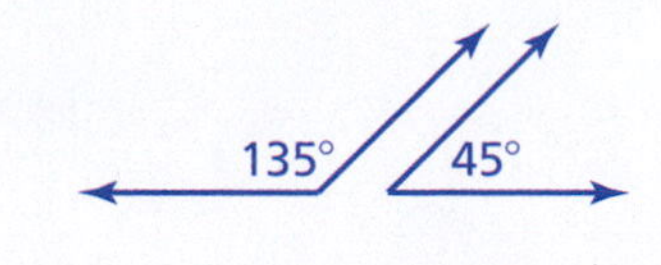

Classifying Triangles by Their Angles

As the name implies, a *tri-angle* has 3 angles. Because it is a **closed figure**, it also has 3 sides (another name could be *tri-lateral*).

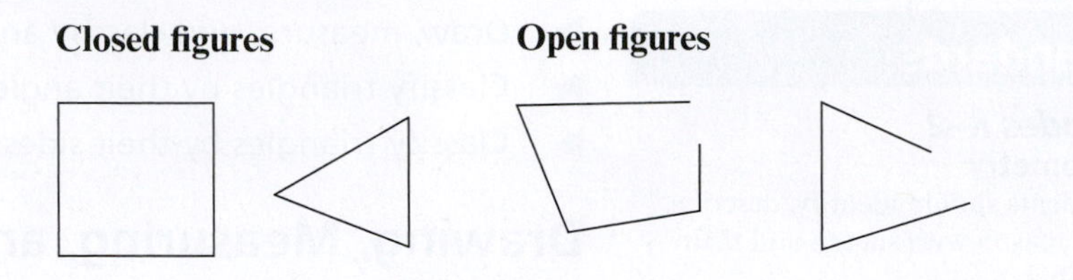

Classroom Tip

When you are teaching your students about triangles, you can assess their understanding by showing them the collection of figures shown below. You might be surprised at how your students classify the shapes (as triangles or not triangles).

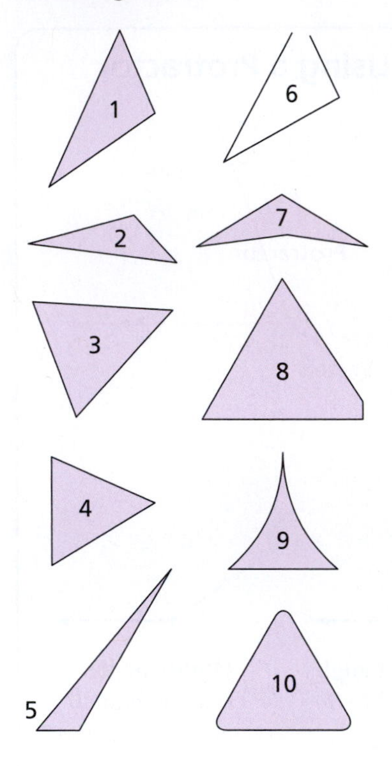

Classifying Triangles by Angles

A **triangle** is a closed figure that has 3 straight sides and 3 angles. The points where the sides meet are the **vertices** of the triangle. A triangle with vertices A, B, and C is written as $\triangle ABC$.

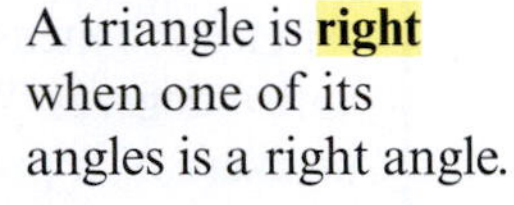
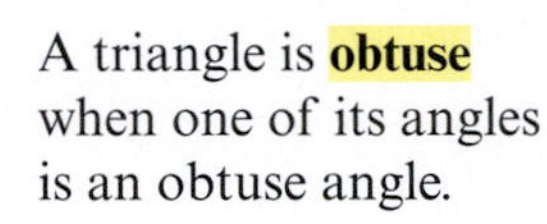

A triangle is **acute** when all three of its angles are acute angles.

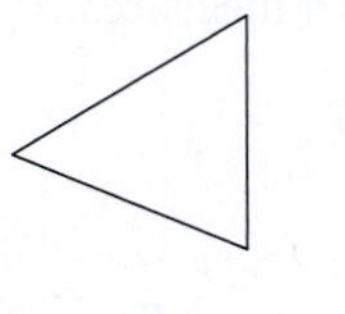

A triangle is **right** when one of its angles is a right angle.

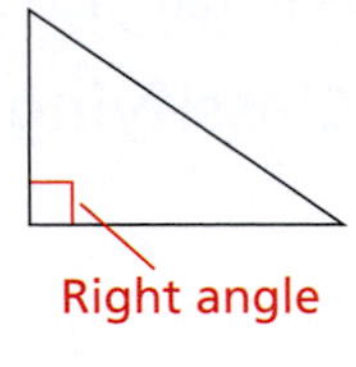

A triangle is **obtuse** when one of its angles is an obtuse angle.

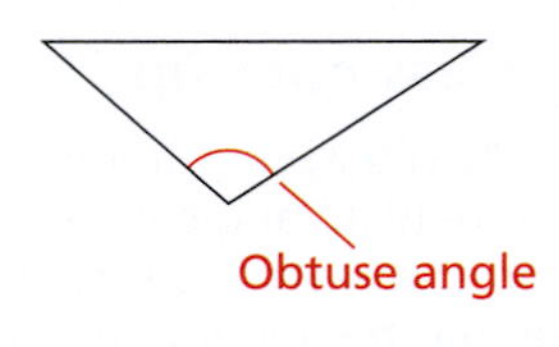

EXAMPLE 2 **Modeling Triangles with a Geoboard**

A *geoboard* is a manipulative that has pegs arranged in either rectangular or triangular arrays. Students use rubber bands to create various closed figures on geoboards. The geoboard at the right shows a right triangle.

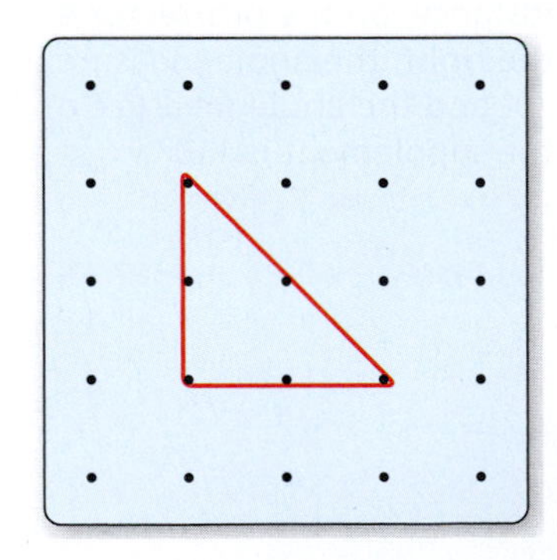

a. Move one of the vertices to create an acute triangle.

b. Move one of the vertices to create an obtuse triangle.

SOLUTION

Here is one possible configuration for each triangle.

a.

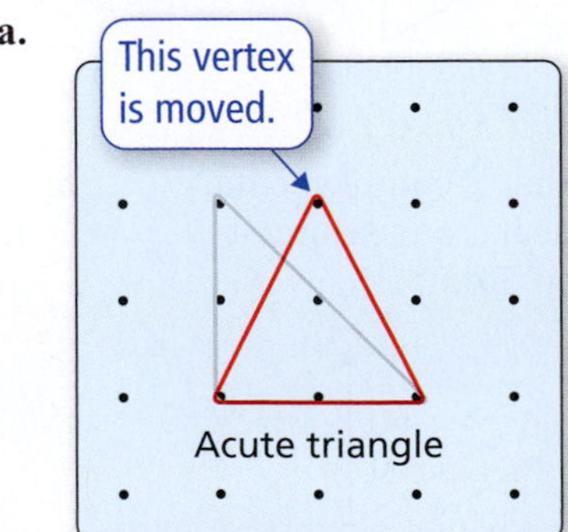

b.

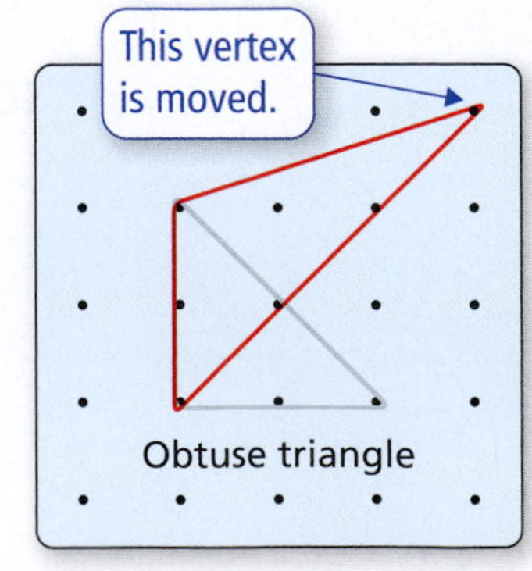

Classroom Tip

Notice that the activity on page 364 guides students to the rule at the right. The activity shows that the 3 angles of a triangle can be torn off and arranged to form a straight angle.

Sum of Angle Measures of a Triangle

Words The sum of the angle measures of a triangle is 180°.

Algebra $x + y + z = 180$

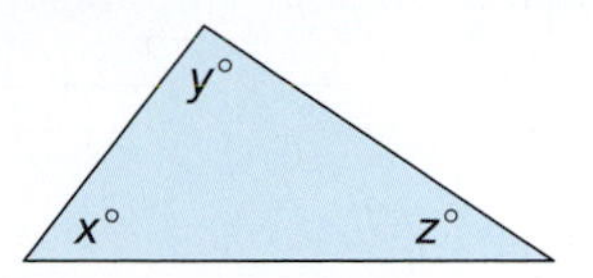

EXAMPLE 3 Finding Angle Measures

Find each value of x. Then classify each triangle by its angles.

a.

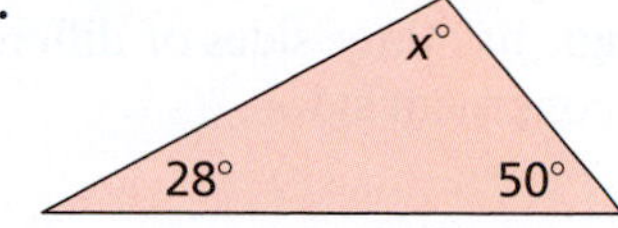

b.

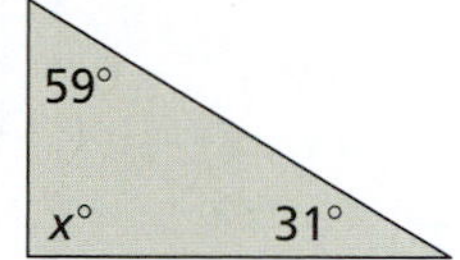

SOLUTION

Use the fact that the sum of the angle measures of a triangle is 180° to write an equation. Then solve the equation for x.

a. $x + 28 + 50 = 180$

$x + 78 = 180$

$x = 102$

The value of x is 102. The triangle has an obtuse angle. So, it is an obtuse triangle.

b. $x + 59 + 31 = 180$

$x + 90 = 180$

$x = 90$

The value of x is 90. The triangle has a right angle. So, it is a right triangle.

EXAMPLE 4 Real-Life Example of Angle Measures

An airplane leaves Miami and travels around the Bermuda Triangle.

a. Find the missing angle measure.

b. Classify the triangle.

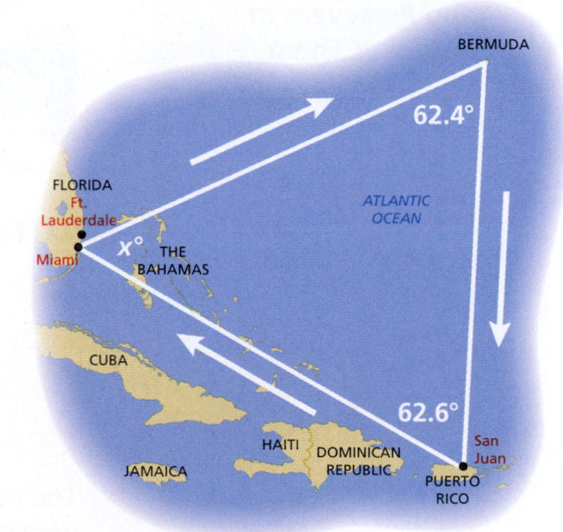

SOLUTION

a. Use the fact that the sum of the angle measures of a triangle is 180° to write an equation. Then solve the equation for x.

$x + 62.4 + 62.6 = 180$ Write equation.

$x + 125 = 180$ Add.

$x = 55$ Subtraction Property of Equality

The missing angle measure at Miami is 55°.

b. All three angles of the Bermuda Triangle are acute. So, the Bermuda Triangle is an acute triangle.

Classroom Tip

It used to be common to further divide degrees into minutes and seconds. Each degree was divided into 60 minutes and each minute was divided into 60 seconds. For instance, 2 degrees 5 minutes 30 seconds was written as 2° 5′ 30″. Today, it is more common to measure a portion of a degree using a decimal, as shown in Example 4.

Classifying Triangles by Their Sides

Two line segments are **congruent** when they have the same length. Tick marks on the segments indicate congruent line segments.

Congruent line segments **Noncongruent line segments**

Classifying Triangles by Sides

A **scalene triangle** has three sides of different lengths, or no congruent sides.

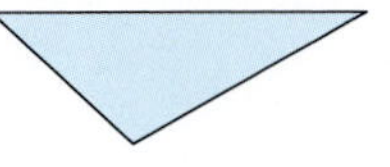

An **isosceles triangle** has at least two sides that are congruent.

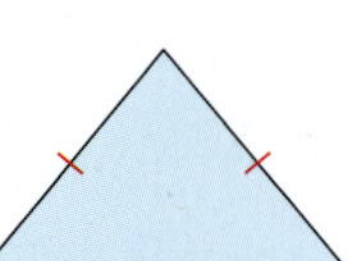

An **equilateral triangle** has three congruent sides. An equilateral triangle is also **equiangular** (three congruent angles).

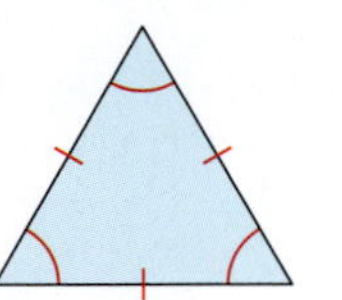

Classroom Tip

Just as tick marks indicate congruent sides, arcs indicate congruent angles. Congruent angles have the same measure.

All equilateral triangles are isosceles triangles. However, triangles with three congruent sides are usually only classified as equilateral when classifying by sides.

Mathematical Practices

Make Sense of Problems and Persevere in Solving Them

MP1 Mathematically proficient students look for entry points to a solution.

EXAMPLE 5 Classifying Triangles in Real Life

Find each value of x. Then classify each triangle by its sides.

a. Flag of Jamaica

b. Flag of Cuba

SOLUTION

a. $x + x + 128 = 180$ Write equation.

$2x + 128 = 180$ Add.

$2x = 52$ Subtraction Property of Equality

$x = 26$ Division Property of Equality

The value of x is 26. Two of the sides are congruent. So, it is an isosceles triangle.

b. $x + x + 60 = 180$ Write equation.

$2x + 60 = 180$ Add.

$2x = 120$ Subtraction Property of Equality

$x = 60$ Division Property of Equality

The value of x is 60. All three angles are congruent. So, the triangle is equiangular, and is therefore an equilateral triangle.

The **median** of a triangle is the line segment that connects a vertex to the *midpoint* of the opposite side, as shown in $\triangle ABC$. The **midpoint** of a line segment divides the segment into two congruent segments. A triangle has three medians.

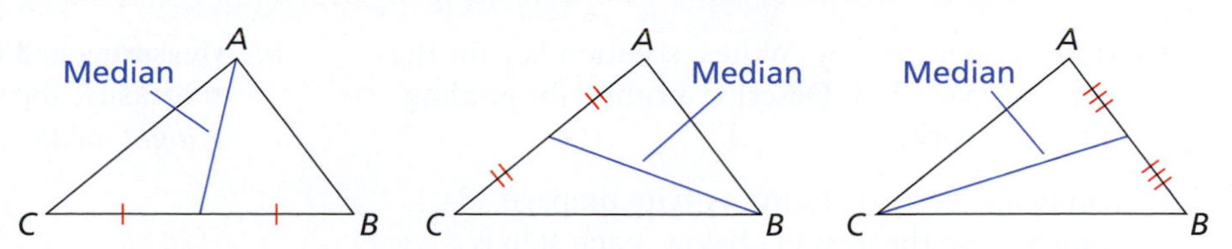

Paper folding is an excellent way to have students model and discover theorems in geometry. The example below shows how to use paper folding to discover a theorem about the medians of a triangle.

EXAMPLE 6 Exploring Medians of Triangles

As a classroom discovery project, perform the following steps.

- Draw and cut out a large scalene triangle. Label the vertices of the triangle A, B, and C.
- Fold the triangle so that vertices B and C coincide. Crease the edge of the paper to mark the midpoint of the side. Then fold the triangle along the median connecting the midpoint of the side and vertex A.
- Repeat the folding process to find the other two medians of the triangle.

What do you observe about the three medians?

Classroom Tip

The paper folding activity shown at the right can be duplicated using three other types of line segments.

a. The *altitudes* of a triangle intersect at a point called the *orthocenter* of the triangle.

b. The *perpendicular bisectors of the sides* of a triangle intersect at a point called the *circumcenter* of the triangle.

c. The *angle bisectors* of a triangle intersect at a point called the *incenter* of the triangle.

SOLUTION

After folding the triangle, you should discover that the three medians meet at a single point. This point is called a **point of concurrency**. The point of concurrency of the medians of a triangle is called the **centroid** of the triangle.

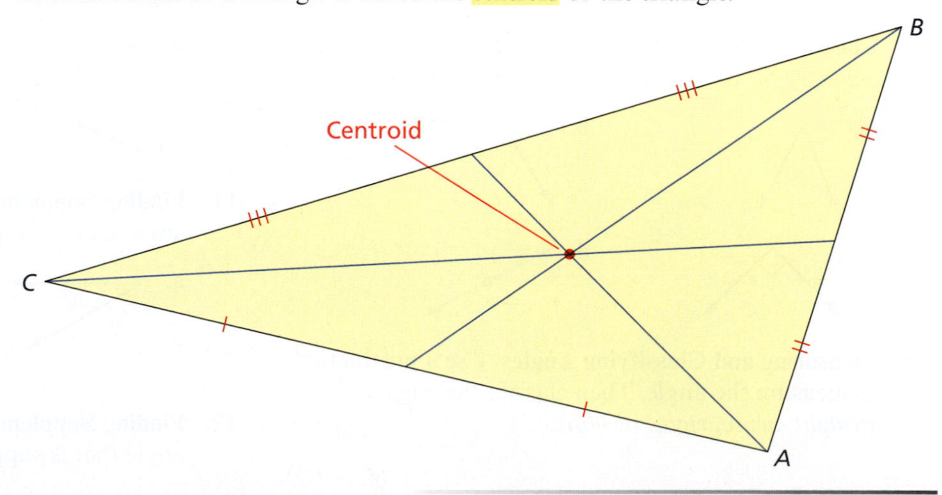

If you use Example 6 as an activity, it is a good idea to ask some students to draw scalene, isosceles, and equilateral triangles. When all of your students arrive at the same conclusion, you will have a good example of *inductive reasoning*. In other words, your students can inductively conclude that no matter which type of triangle is used, the medians of a triangle intersect at a single point.

10.1 Exercises

Free step-by-step solutions to odd-numbered exercises at *MathematicalPractices.com.*

1. Writing a Solution Key Write a solution key for the activity on page 364. Describe a rubric for grading a student's work.

2. Grading the Activity In the activity on page 364, a student gave the answers below. Each step is worth 10 points. Assign a grade to each answer. For those that are incorrect, why do you think the student erred?

Sample Student Work

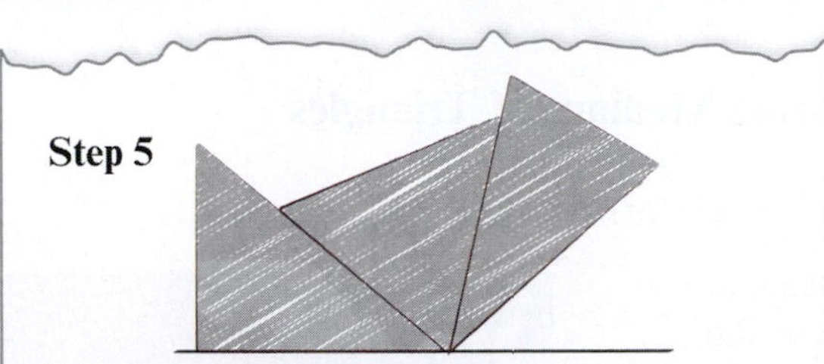

No; The triangles do not form a straight angle as in Step 4.

Step 6 Write a rule about the sum of the measures of the angles of a triangle.

The sum of the measures of the angles of a triangle is less than or equal to 180°.

3. Classifying Angles Classify each angle as *straight*, *acute*, *right*, or *obtuse*. Explain your reasoning.

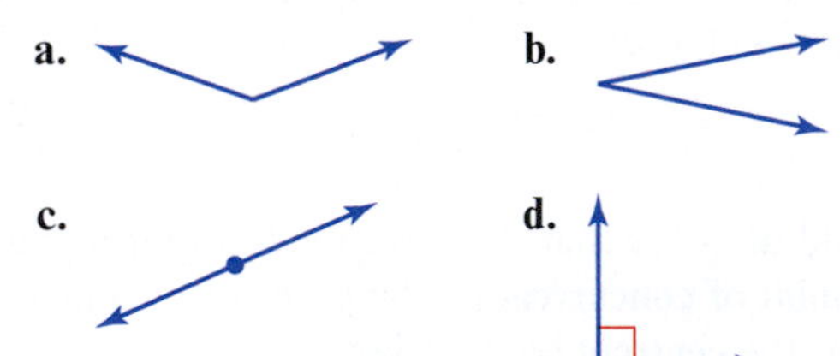

4. Classifying Angles Classify each angle as *straight*, *acute*, *right*, or *obtuse*. Explain your reasoning.

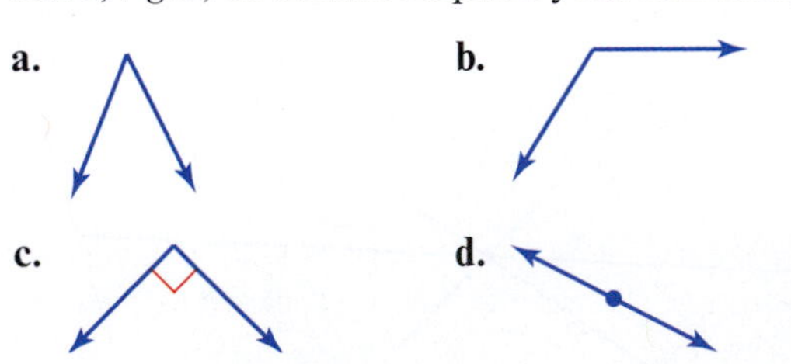

5. Measuring and Classifying Angles Use a protractor to measure the angle. Then classify the angle as *straight*, *acute*, *right*, or *obtuse*.

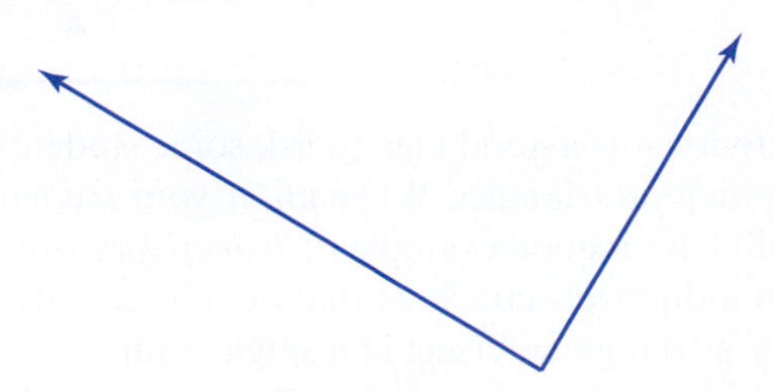

6. Measuring and Classifying Angles Use a protractor to measure the angle. Then classify the angle as *straight*, *acute*, *right*, or *obtuse*.

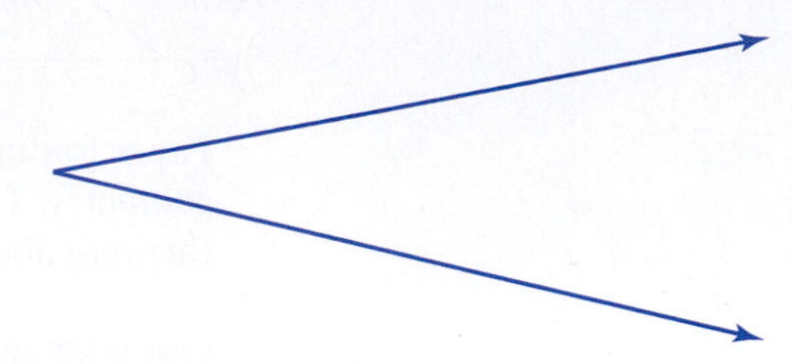

7. Drawing Angles Use a protractor to sketch an angle with the given measure.

a. 35°

b. 108°

c. 142°

8. Drawing Angles Use a protractor to sketch an angle with the given measure.

a. 110°

b. 38°

c. 72°

9. Finding Complementary Angles Find and sketch an angle that is complementary to the angle.

10. Finding Complementary Angles Find and sketch an angle that is complementary to the angle.

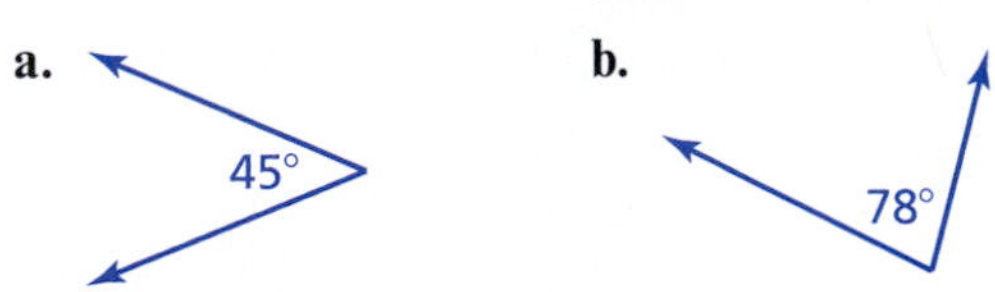

11. Finding Supplementary Angles Find and sketch an angle that is supplementary to the angle.

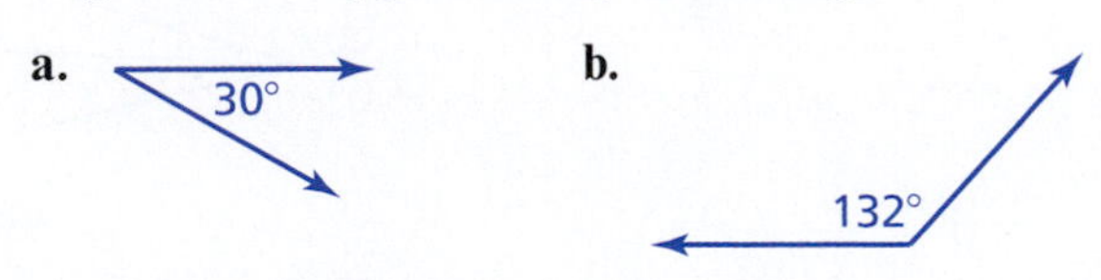

12. Finding Supplementary Angles Find and sketch an angle that is supplementary to the angle.

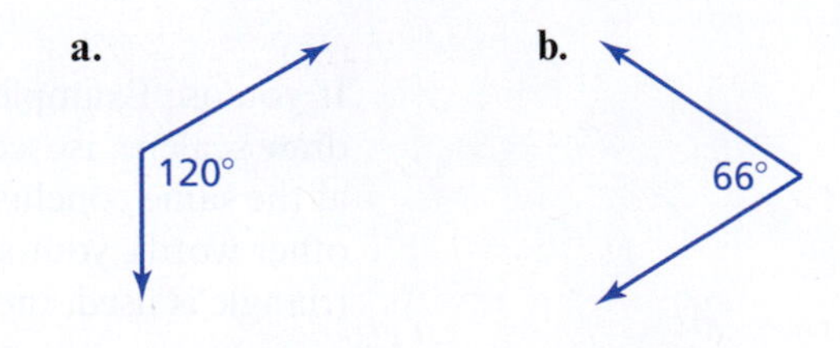

13. **Identifying Triangles** State whether the figure is a triangle. Explain your reasoning.

a.  **b.** 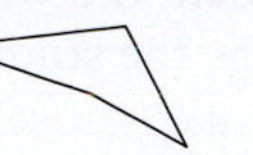 **c.** 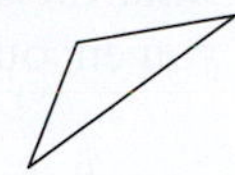

14. **Identifying Triangles** State whether the figure is a triangle. Explain your reasoning.

a. 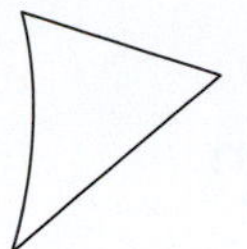 **b.** 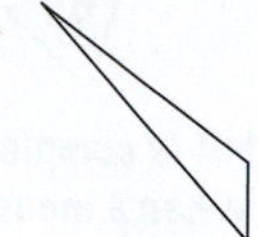 **c.** 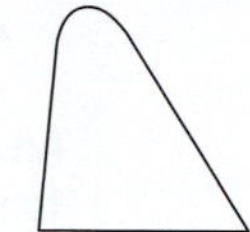

15. **Modeling Triangles with a Geoboard** Use a geoboard and vertices A and B, as shown below, to create $\triangle ABC$ for each type of triangle. Make a sketch of your answer with the vertices labeled.

a. Scalene **b.** Obtuse **c.** Right

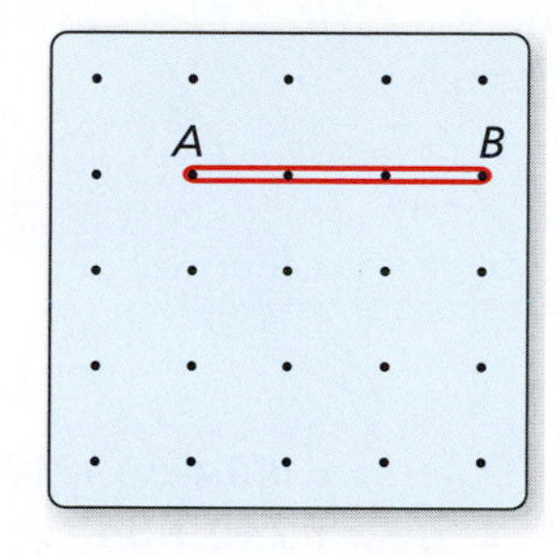

16. **Modeling Triangles with a Geoboard** Use a geoboard and vertices A and B, as shown below, to create $\triangle ABC$ for each type of triangle. Make a sketch of your answer with the vertices labeled.

a. Scalene **b.** Obtuse **c.** Right

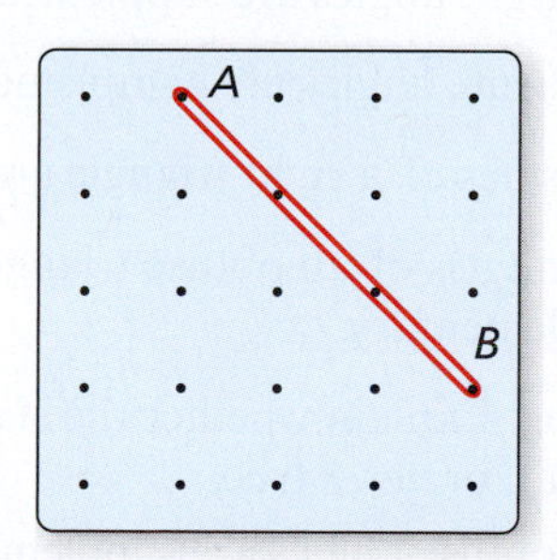

17. **Finding Angle Measures** Find the value of x for each triangle. Then classify the triangle by its angles.

a. 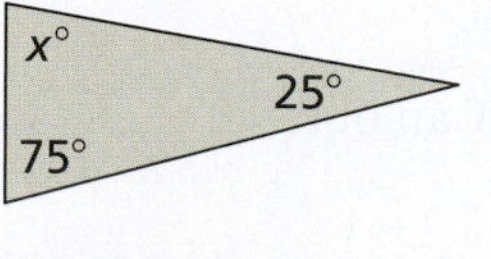

b. 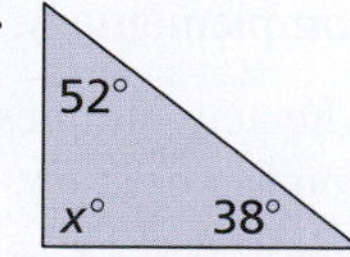

c. 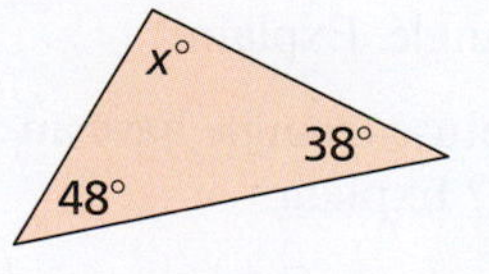

d. 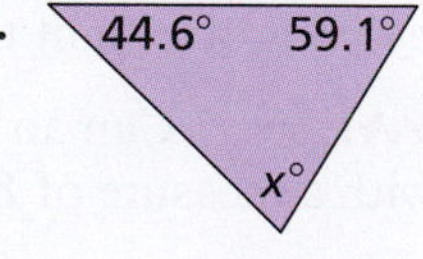

18. **Finding Angle Measures** Find the value of x for each triangle. Then classify the triangle by its angles.

a. 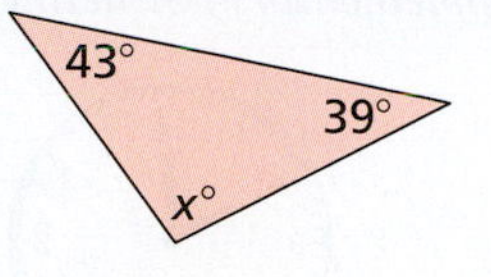

b. 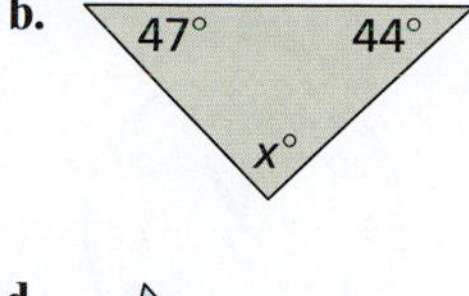

c. 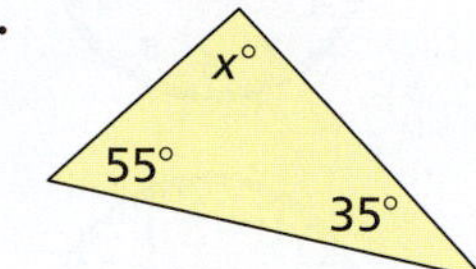

d. 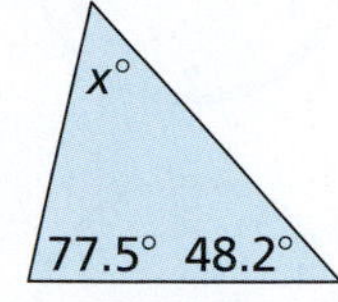

19. **Identifying Triangles** Can a triangle have the given angle measures? If not, change the measure of the first angle so that a triangle is possible.

a. 30°, 68°, 82° **b.** 54°, 58°, 59°

20. **Identifying Triangles** Can a triangle have the given angle measures? If not, change the measure of the first angle so that a triangle is possible.

a. 42°, 72°, 76° **b.** 24°, 63°, 93°

21. **Sketching a Triangle** Sketch an isosceles triangle. Use tick marks to indicate the congruent sides.

22. **Sketching a Triangle** Sketch an equilateral triangle. Use tick marks and arcs to indicate the congruent sides and angles.

23. **Classifying Triangles** Find the value of x in each triangle. Then classify the triangle in as many ways as possible.

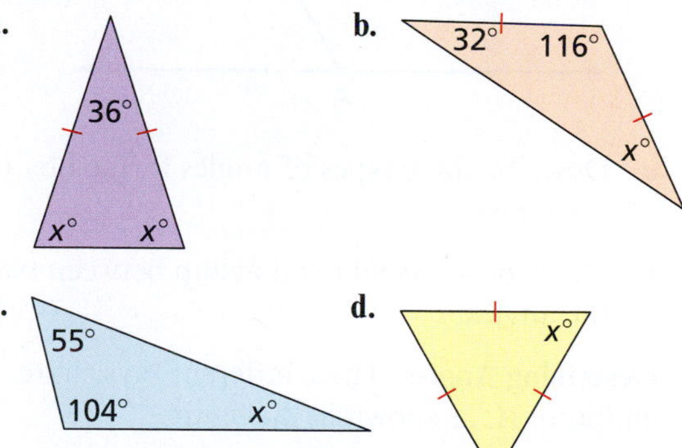

24. **Classifying Triangles** Find the value of x in each triangle. Then classify the triangle in as many ways as possible.

a. 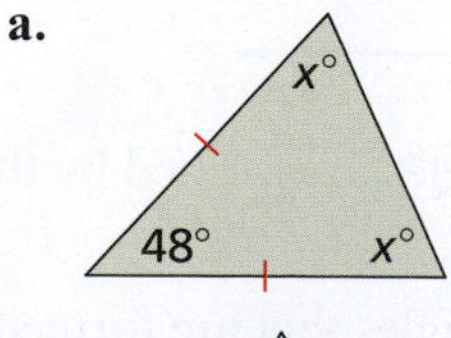

b. 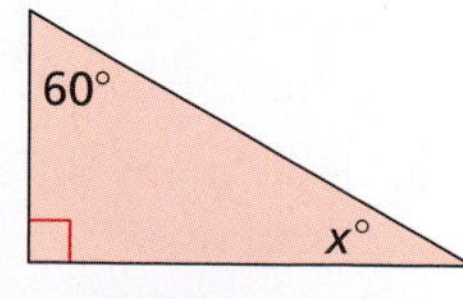

c. 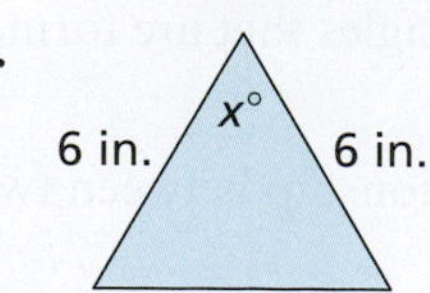

d. 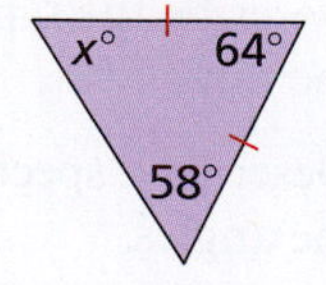

25. Angle Relationships Determine whether the two angles formed by the hands of each pair of clocks are *complementary*, *supplementary*, or *neither*.

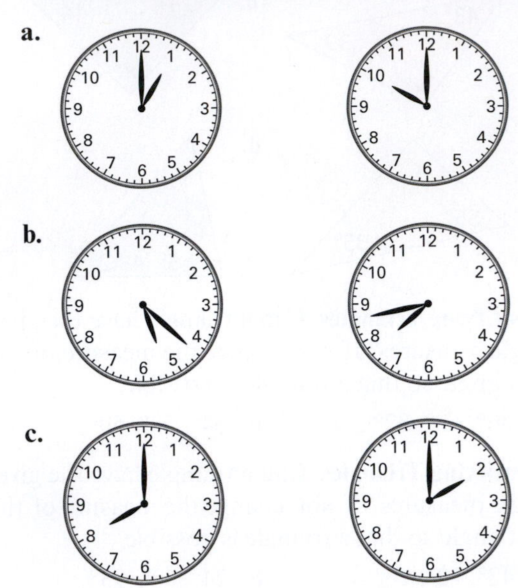

26. Angles in Triangles Is it possible for an obtuse angle to be complementary to another angle? Explain your reasoning.

27. Describing Angles Three different rays share endpoint A, as shown in the figure.

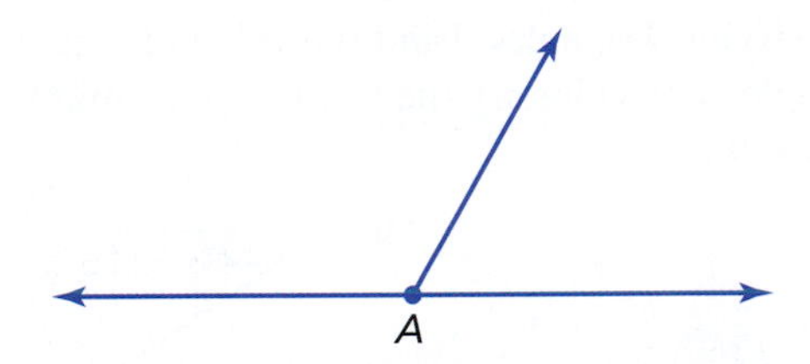

a. Describe three types of angles formed by the three rays.

b. Describe a special relationship between two of the angles.

28. Describing Angles Three different rays share endpoint A, as shown in the figure.

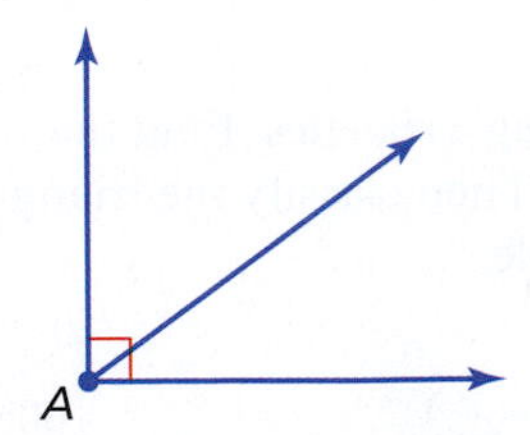

a. How many different angles are formed by the three rays?

b. Describe the types of angles that are formed by the three rays.

c. Describe a special relationship between two of the angles.

29. Grading Student Work On a diagnostic test, one of your students does the following work. Explain what the student did wrong. Which topics would you encourage the student to review?

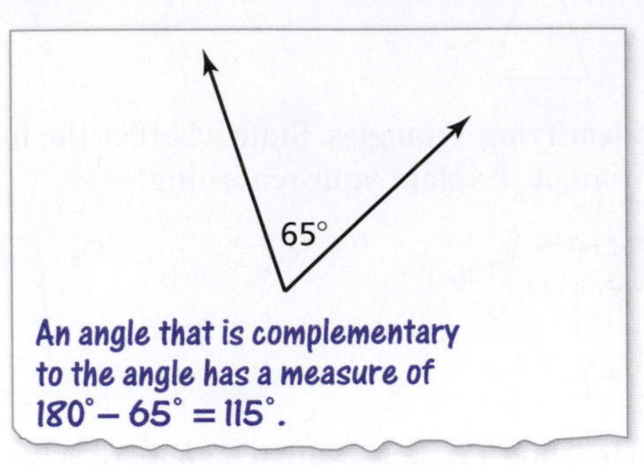

30. Grading Student Work On a diagnostic test, one of your students does the following work. Explain what the student did wrong. Which topics would you encourage the student to review?

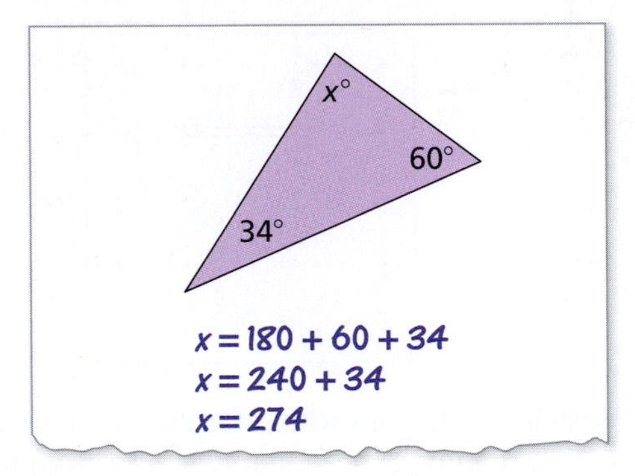

31. Reasoning Explain whether the statement is *always*, *sometimes*, or *never* true.

a. Two right angles are supplementary.

b. Two acute angles are complementary.

c. Two sides of a right triangle are congruent.

d. Two angles of an obtuse triangle are supplementary.

32. Reasoning Explain whether the statement is *always*, *sometimes*, or *never* true.

a. Two obtuse angles are supplementary.

b. One of two supplementary angles is obtuse.

c. The acute angles of a right triangle are complementary.

d. The acute angles of an obtuse triangle are complementary.

33. Writing Describe all the possible classifications by sides of a right triangle. Explain.

34. Writing Can an obtuse triangle have an angle with a measure of 89°? Explain.

35. Modeling Triangles with a Geoboard List each type of triangle classified by sides and by angles that can be modeled as $\triangle ABC$ on the geoboard using vertices A and B as shown. Sketch an example of each type.

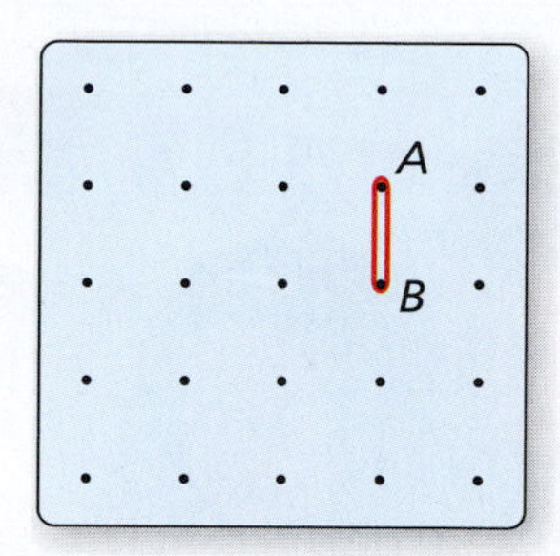

36. Modeling Triangles with a Geoboard List each type of triangle classified by sides and by angles that can be modeled as $\triangle ABC$ on the geoboard using vertices A and B as shown. Sketch an example of each type.

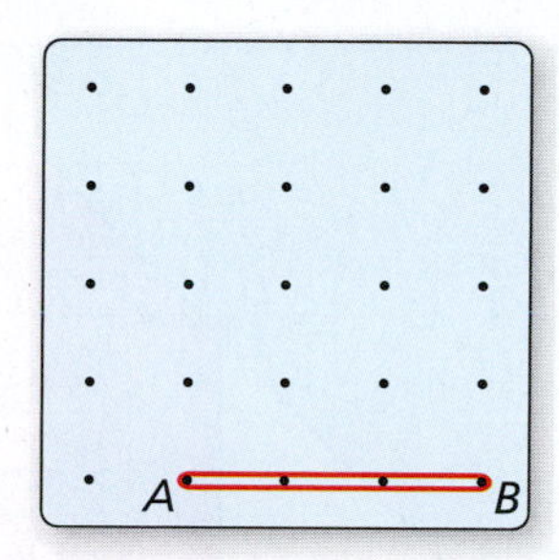

37. Making a Circle Graph The table shows the results of a public opinion poll about a luxury tax on hotels. Consider displaying the data in a circle graph.

Opinion	Results
For the tax	25%
Against the tax	50%
Don't care	10%
Undecided	15%

a. Describe how to find the angle that represents a given percent in a circle graph.

b. Which opinion can you represent in a circle graph using a right angle?

c. Which opinion can you represent in a circle graph using a straight angle?

d. Which opinions can you represent in a circle graph using acute angles?

e. Draw a circle graph for the data.

f. Draw another circle graph for the data, this time combining "Don't care" and "Undecided" into one category. What type of angle represents this category?

38. Finding Angle Measures in Real Life Use the angle relationships shown in the roof truss to find the values of x, y, and z. Classify the types of triangles formed.

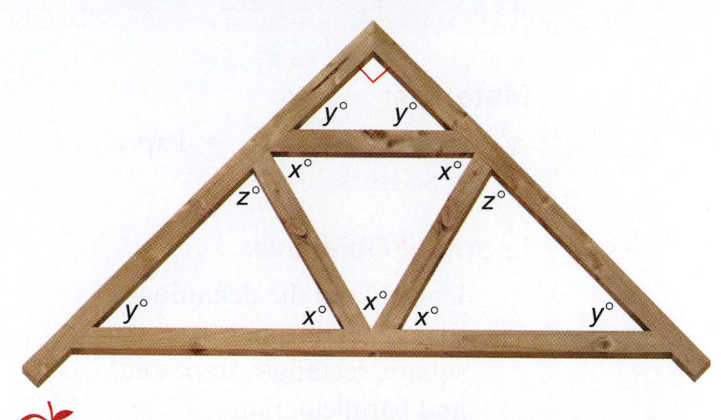

39. Activity: In Your Classroom Trace the figure onto your paper. Use the relationships shown.

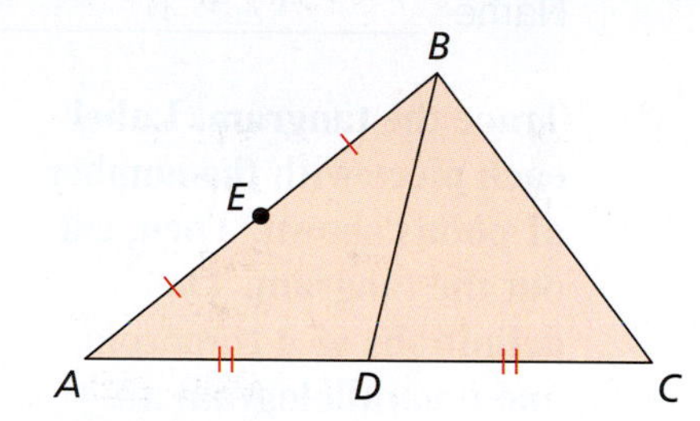

a. Identify the midpoint of the side connecting vertices A and B.

b. What type of line segment is drawn from vertex B to point D? Explain.

c. Draw the median from vertex C to the opposite side of the triangle.

d. Label the centroid of $\triangle ABC$. How do you know where the centroid is?

e. Describe a way to draw the median from vertex A without measuring the opposite side to find its midpoint. Then draw the median.

f. Describe the concepts that your students will practice and discover by doing this activity.

40. Activity: In Your Classroom Use the figure below to design an activity that you can use with your students. Have students discover how to use a straightedge to construct the midpoint of one side of a triangle given the midpoints of the other two sides. Describe the concepts that your students will practice by doing this activity.

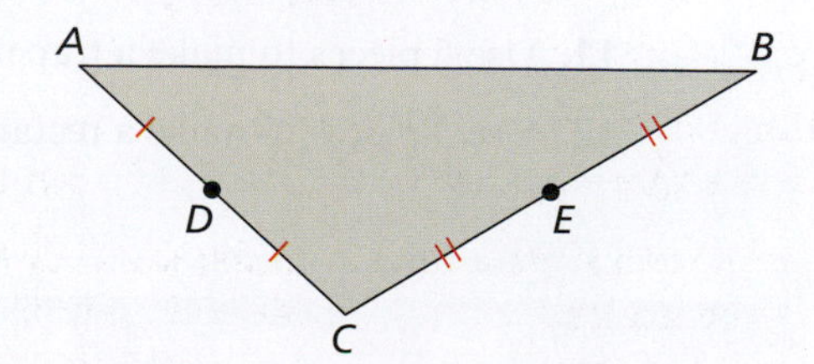

Activity: Reviewing Shapes of Quadrilaterals

Materials:

- Pencil
- Scissors
- Paper

Learning Objective:

Understand the definitions of four types of quadrilaterals: square, rectangle, trapezoid, and parallelogram.

Name ______________________________

Trace the tangram. Label each piece with the number of points shown. Then, cut out the tangram. The definitions of a trapezoid and a parallelogram are reviewed at the lower right side of this activity.

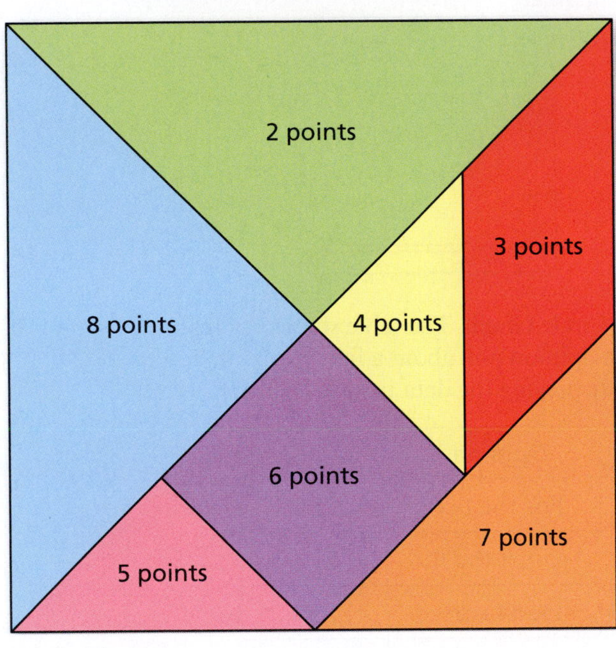

1. Use 2 pieces to make a triangle worth 10 points.
2. Use 2 pieces to make a square worth 9 points.
3. Use 2 pieces to make a trapezoid worth 10 points.
4. Use 3 pieces to make a rectangle worth 12 points.
5. Use 3 pieces to make a square worth 16 points.
6. Use 3 pieces to make a parallelogram worth 15 points.
7. Use 4 pieces to make a square worth 14 points.
8. Use 4 pieces to make a rectangle worth 18 points.
9. Use 4 pieces to make a square worth 20 points.
10. Use 5 pieces to make a parallelogram worth 25 points.
11. Use 5 pieces to make a trapezoid worth 25 points.
12. Use 5 pieces to make a rectangle worth 22 points.

A *trapezoid* is a quadrilateral that has exactly one pair of opposite sides that are parallel.

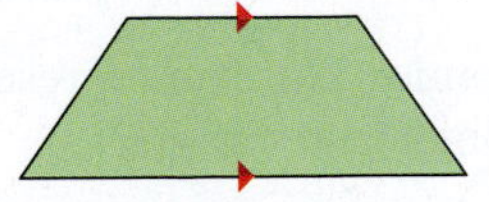

A *parallelogram* is a quadrilaterial that has two pairs of opposite sides that are parallel.

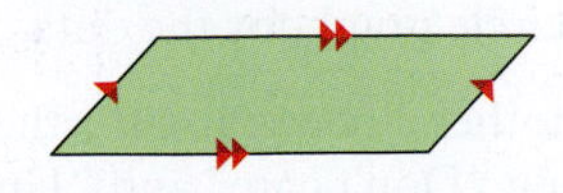

10.2 Quadrilaterals

Standards

Grades K–2
Geometry
Students should identify, describe, and reason with shapes and their attributes.

Grades 3–5
Geometry
Students should classify two-dimensional shapes into categories based on their properties.

- Understand and use the definitions of a rectangle and a square.
- Understand and use the definition of a parallelogram.
- Understand and use the definitions of a trapezoid and a rhombus.
- Understand and use the definitions of a convex figure and a kite.

Rectangles and Squares

A **quadrilateral** is a closed figure that has four straight sides. Two types of quadrilaterals are rectangles and squares.

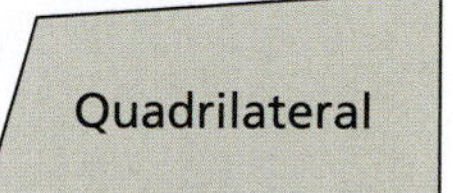

Definitions of a Rectangle and a Square

A **rectangle** is a quadrilateral that has four right angles.

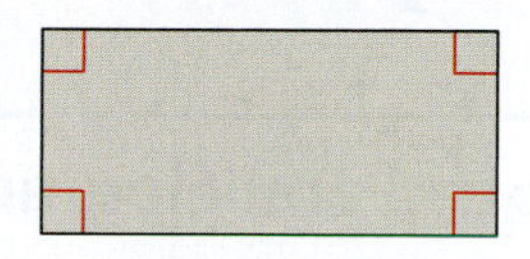

A **square** is a rectangle that has four congruent sides.

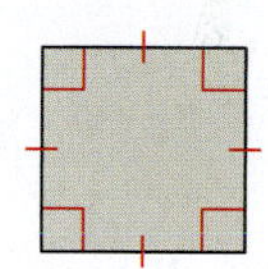

From the definitions above, notice that every rectangle is a quadrilateral and every square is a rectangle. These relationships can be illustrated by an Euler diagram, as shown at the right.

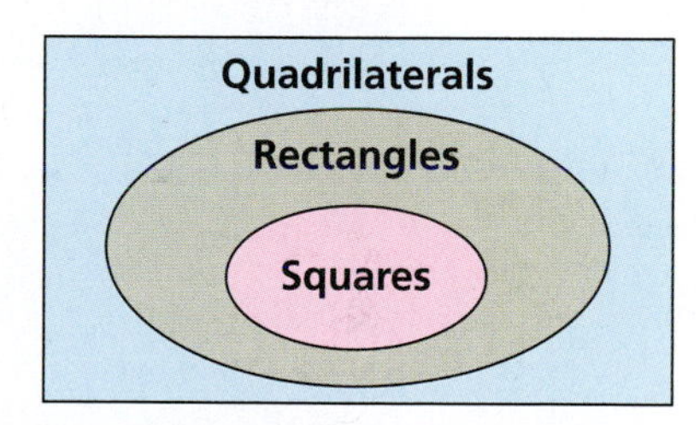

EXAMPLE 1 **Discovering a Property of Rectangles**

Use a geoboard to create several rectangles. Is it possible to create a rectangle in which the opposite sides are not congruent? What can you conclude?

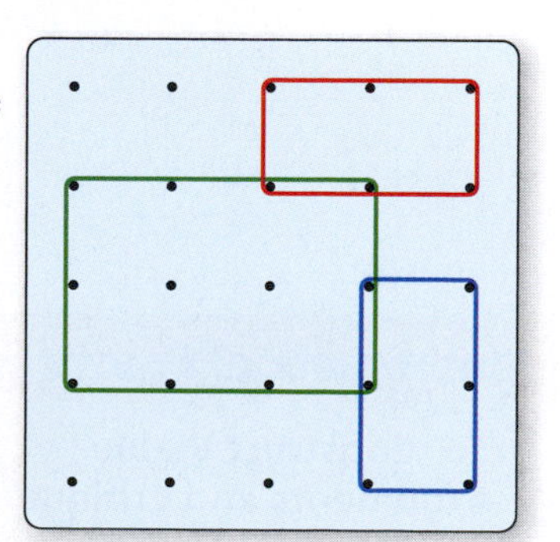

SOLUTION

In each case, when you make a quadrilateral that has four right angles, you end up with a figure whose opposite sides are congruent. From this, you can inductively conclude that the opposite sides of a rectangle are congruent. In a formal course in geometry, you can prove this property deductively.

Classroom Tip

The Euler diagram and the example at the right show the two types of reasoning: inductive and deductive. Inductive reasoning constructs general propositions from observations. In Example 1, you use an activity to inductively conclude a statement. Deductive reasoning constructs general propositions based on previously established definitions, axioms, or theorems. To make the Euler diagram, you use definitions to deduce that every rectangle is a quadrilateral, and every square is a rectangle.

Property of Rectangles

The opposite sides of a rectangle are congruent.

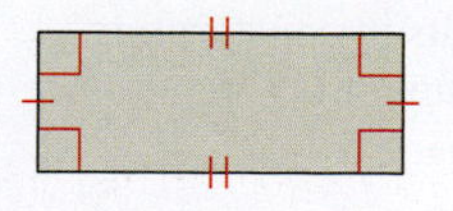

Parallelograms

Two lines in the same plane are **parallel** when they have no point of intersection. Two line segments are parallel when they are segments taken from parallel lines. Arrows on lines or segments indicate that they are parallel.

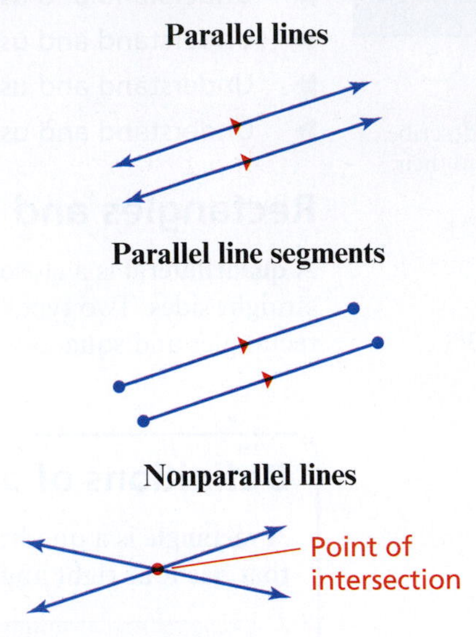

Definition of a Parallelogram

A **parallelogram** is a quadrilateral that has two pairs of opposite sides that are parallel.

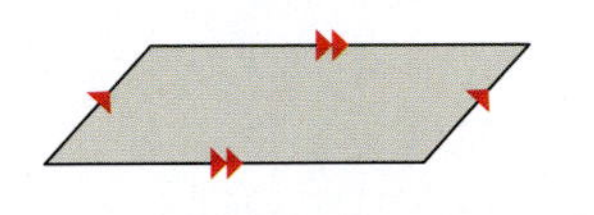

EXAMPLE 2 **Modeling Parallelograms with a Geoboard**

a. Use a geoboard to create several rectangles. Is it possible to create a rectangle that is not a parallelogram? What can you conclude?

b. Use a geoboard to create several parallelograms. Is it possible to create a parallelogram in which the opposite sides are not congruent? What can you conclude?

SOLUTION

a. From the geoboard below, it appears that it is not possible to create a rectangle that is not a parallelogram.

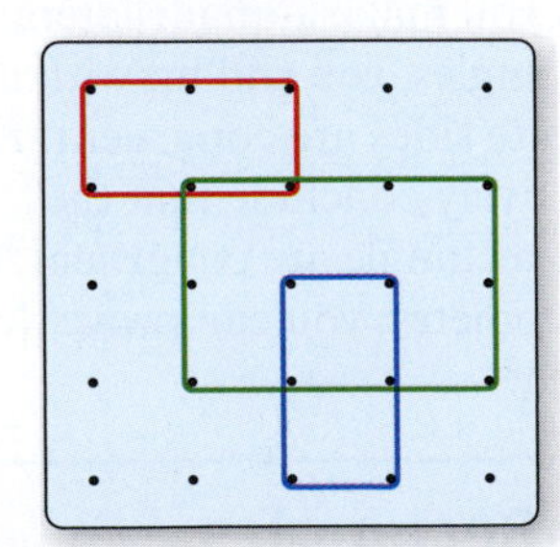

Conclusion Every rectangle is a parallelogram.

b. From the geoboard below, it appears that it is not possible to create a parallelogram in which the opposite sides are not congruent.

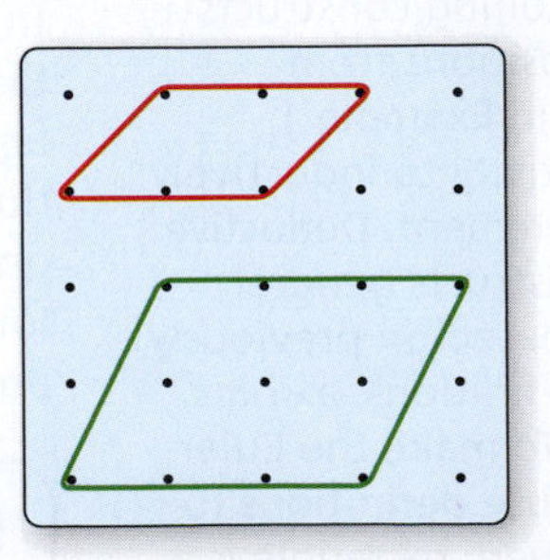

Conclusion Opposite sides of a parallelogram are congruent.

Mathematical Practices

Construct Viable Arguments and Critique the Reasoning of Others

MP3 Mathematically proficient students understand and use stated assumptions, definitions, and previously established results in constructing arguments.

EXAMPLE 3 Sum of the Angle Measures of a Quadrilateral

The activity on page 364 shows that the sum of the angle measures of a triangle is 180°. Create an activity to find the sum of the angle measures of a quadrilateral.

SOLUTION

Here is one possible activity.

- Draw a quadrilateral similar to the one below. Label the vertices of the quadrilateral A, B, C, and D.

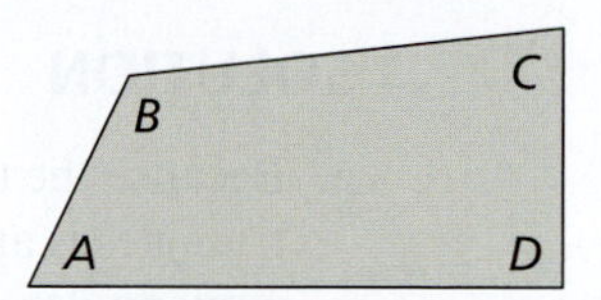

- Make 3 copies of the quadrilateral.

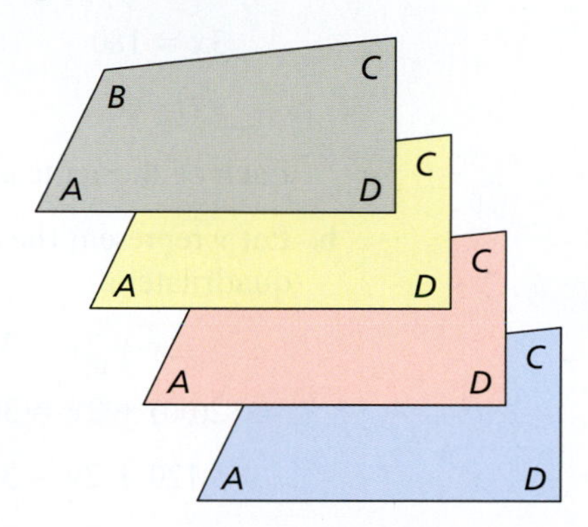

- Arrange four vertices, one from each quadrilateral so that they meet at a point without overlapping. Notice that the four angles form a circle. Because 360° represents a full circle, the sum of the angle measures of a quadrilateral is 360°.

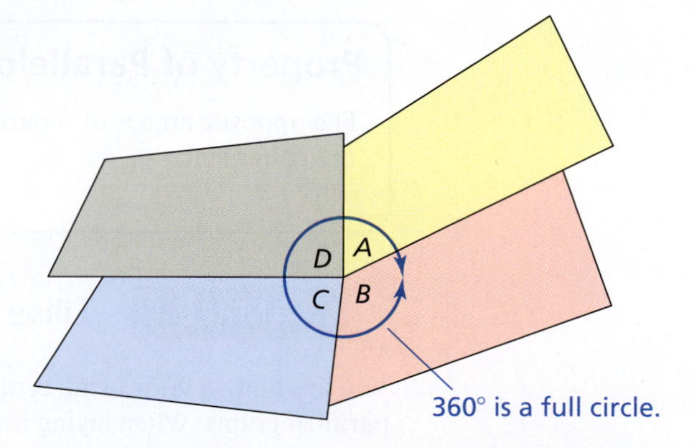

Sum of Angle Measures of a Quadrilateral

Words The sum of the angle measures of a quadrilateral is 360°.

Algebra $w + x + y + z = 360$

EXAMPLE 4 Finding the Angle Measures of a Parallelogram

Use 5 toothpicks to form a parallelogram, as shown at the right.

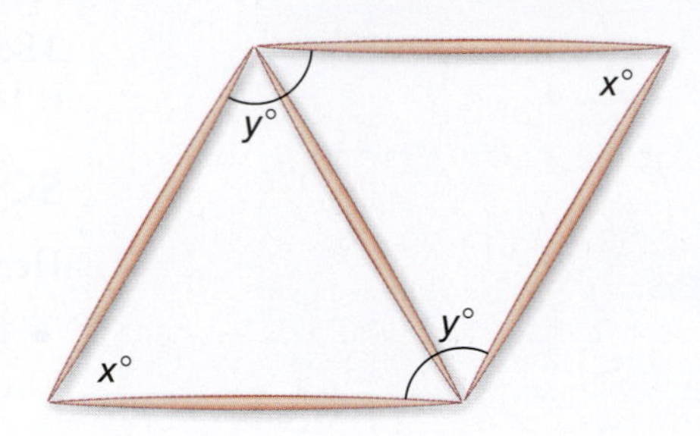

a. Find the measures of the two acute angles.

b. Find the measures of the two obtuse angles.

SOLUTION

a. Because the toothpicks are the same length, you can conclude that each of the triangles is an equilateral triangle. This implies that the two triangles are also equiangular. Let x represent the angle measure of each of the acute angles of the triangles.

$3x = 180$ Sum of the angle measures of a triangle

$x = 60$ Division Property of Equality

Each of the acute angles has a measure of 60°.

b. Let y represent the angle measure of each of the obtuse angles of the quadrilateral.

$2x + 2y = 360$ Sum of the angle measures of a quadrilateral

$2(60) + 2y = 360$ Substitute 60 for x.

$120 + 2y = 360$ Multiply.

$2y = 240$ Subtraction Property of Equality

$y = 120$ Division Property of Equality

Each of the obtuse angles has a measure of 120°.

Property of Parallelograms

The opposite angles of a parallelogram are congruent.

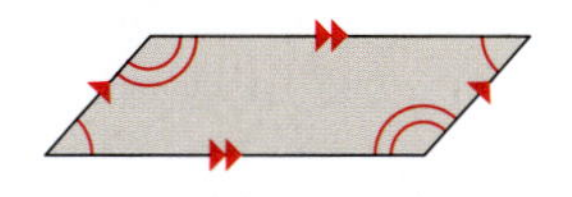

EXAMPLE 5 Tiling a Floor

You are tiling a floor using ceramic tiles that are congruent parallelograms. When laying a tile, does it matter if you turn the tile 180°? Explain your reasoning.

SOLUTION

It does not matter because the opposite angles of the tile are congruent. So, when you turn a tile 180°, it will still fit. The tiles have a design, however, so the design might not look right.

Trapezoids and Rhombuses

Classroom Tip

Definitions of mathematical concepts may differ from country to country. For instance, in European countries, it is common to refer to a trapezoid as a "trapezium" and define it as a quadrilateral that has *at least* one pair of opposite sides that are parallel. By this European definition, a parallelogram is considered a trapezoid. By the U.S. definition shown at the right, a parallelogram is not considered a trapezoid.

Definitions of a Trapezoid and a Rhombus

A **trapezoid** is a quadrilateral that has exactly one pair of opposite sides that are parallel.

A trapezoid that has two adjacent right angles is called a **right trapezoid**.

A trapezoid whose nonparallel sides are congruent is called an **isosceles trapezoid**.

A **rhombus** is a parallelogram that has four congruent sides. The plural of rhombus is either rhombuses or rhombi.

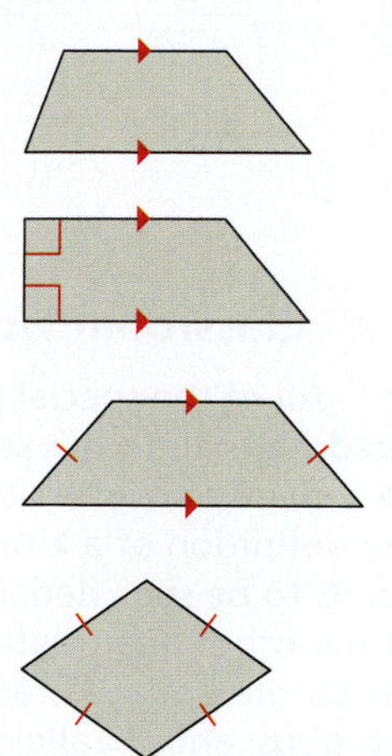

EXAMPLE 6 **Drawing an Euler Diagram**

Draw an Euler diagram that shows the relationships among quadrilaterals, rectangles, squares, parallelograms, trapezoids, and rhombuses.

SOLUTION

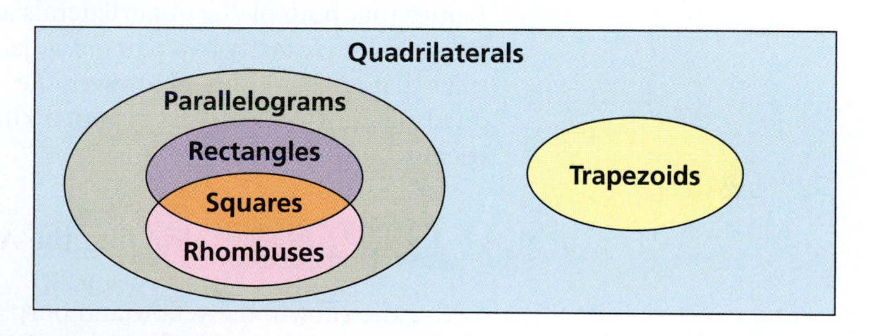

EXAMPLE 7 **Finding the Angle Measures of a Rhombus**

Find the angle measures of the rhombus portion of one of the earrings.

SOLUTION

A rhombus is a parallelogram, so the opposite angles are congruent. Use a protractor to determine that two of the angle measures are about 62°. Let y represent the angle measure of each of the other two angles.

$$2(62) + 2y = 360$$
$$124 + 2y = 360$$
$$2y = 236$$
$$y = 118$$

The angle measures are 62°, 62°, 118°, and 118°.

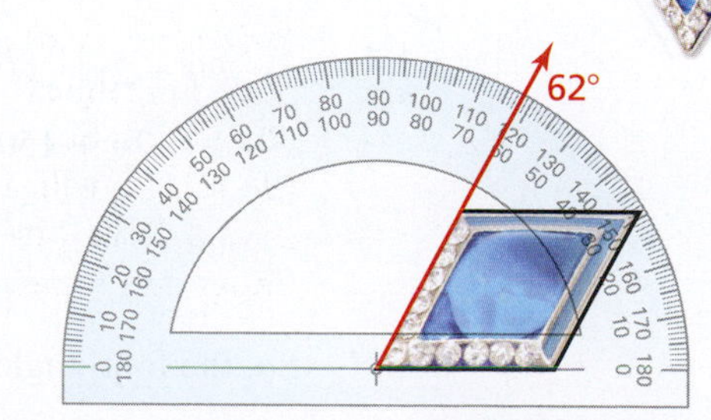

loraart8/Shutterstock.com

Convex Figures and Kites

A closed figure in a plane is **convex** when every line segment connecting any two points lies entirely inside the figure. A closed figure in a plane is **concave** when at least one line segment connecting any two points lies outside the figure.

Convex figures

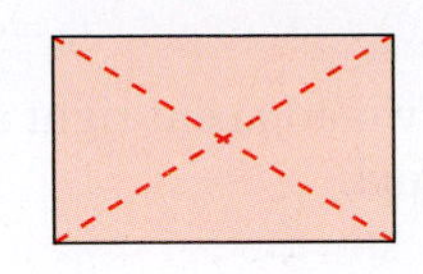
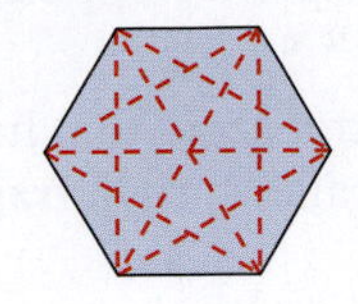

Concave figures

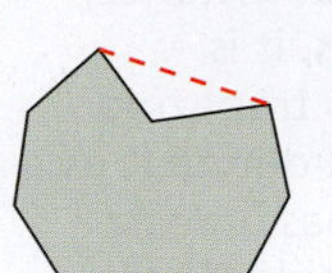
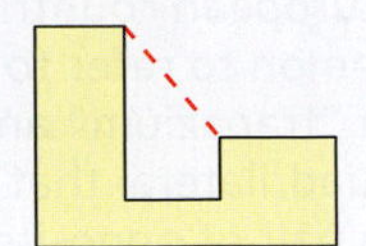

Classroom Tip

All of the special types of quadrilaterals in this section must be convex. However, it is only in the definition of a kite that this needs to be specified. For each of the other quadrilaterals (rectangle, square, trapezoid, rhombus, and parallelogram), the given definition can be used to prove that the quadrilateral must be convex.

Definition of a Kite

A **kite** is a *convex* quadrilateral that has exactly two pairs of adjacent sides that are congruent.

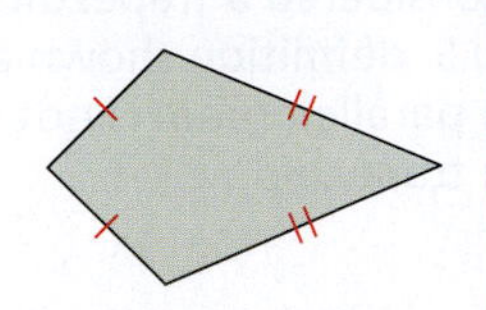

EXAMPLE 8 Drawing a Concave Quadrilateral

Draw a quadrilateral that is not a kite, but still has exactly two pairs of adjacent sides that are congruent.

SOLUTION

Notice that both of the quadrilaterals at the right have exactly two pairs of adjacent sides that are congruent. However, the quadrilateral at the far right is not a kite because it is concave.

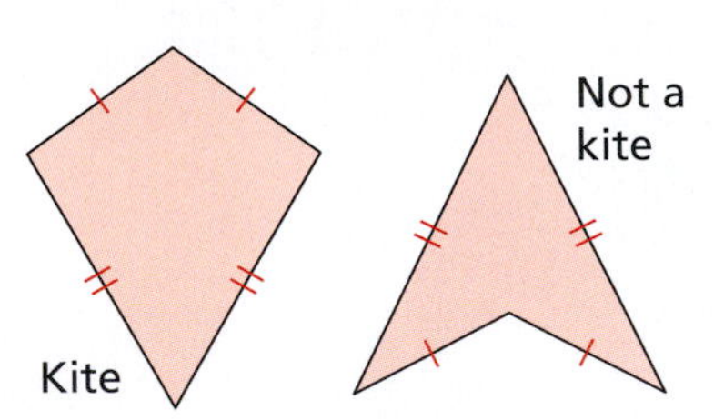

EXAMPLE 9 Finding the Angle Measures of a Kite

In the kite shown, the yellow and purple triangles are isosceles right triangles. The blue and green triangles are right triangles, in which the least angle measure of each is 30°. Find the four angle measures of the kite.

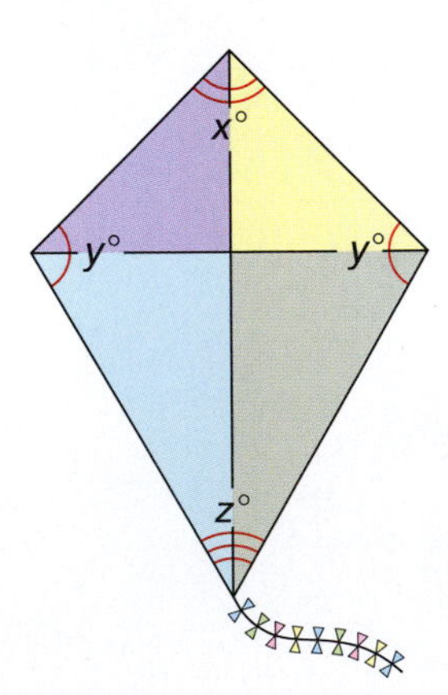

SOLUTION

An isosceles right triangle has angle measures of 45°, 45°, and 90°. This implies that $x = 45 + 45 = 90$. Because the least angle measure of each of the blue and green triangles is 30°, it follows that $z = 30 + 30 = 60$. To find y, use the fact that the sum of the angle measures of a quadrilateral is 360° to write and solve an equation.

$x + 2y + z = 360$ Sum of the angle measures of a quadrilateral

$90 + 2y + 60 = 360$ Substitute 90 for x and 60 for z.

$2y + 150 = 360$ Add.

$2y = 210$ Subtraction Property of Equality

$y = 105$ Division Property of Equality

So, the four angle measures of the kite are 90°, 105°, 105°, and 60°.

There are many properties of quadrilaterals. The example below uses paper folding to investigate properties about the diagonals of special types of quadrilaterals.

EXAMPLE 10 Exploring the Diagonals of Quadrilaterals

Use paper folding to investigate whether the diagonals of rectangles, squares, parallelograms, and rhombuses intersect at right angles.

SOLUTION

Begin by drawing a representative quadrilateral and its two diagonals. Then cut out the quadrilateral and fold it along one of the diagonals. Determine whether the diagonals intersect at right angles.

Classroom Tip There are other properties of the diagonals of quadrilaterals. For instance, the diagonals of each of the four quadrilaterals shown at the right *bisect* each other.

Quadrilateral | **Fold on Diagonal**

Rectangle

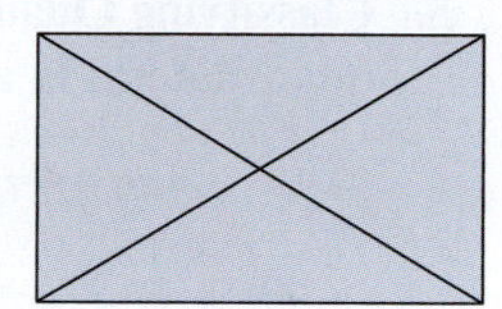

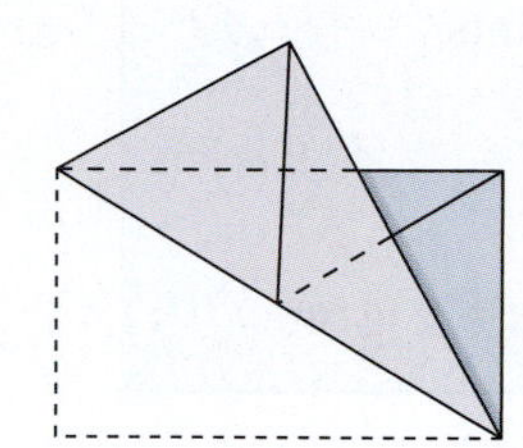

Conclusion The diagonals of a rectangle do not necessarily intersect at right angles.

Square

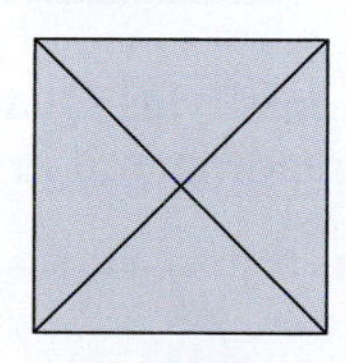

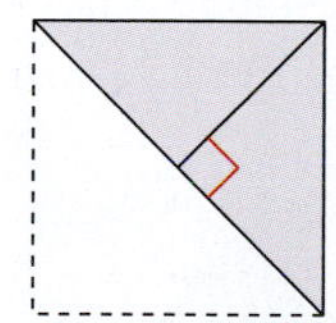

Conclusion The diagonals of a square intersect at right angles.

Parallelogram

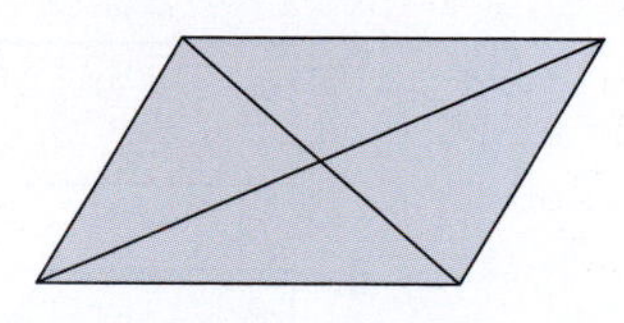

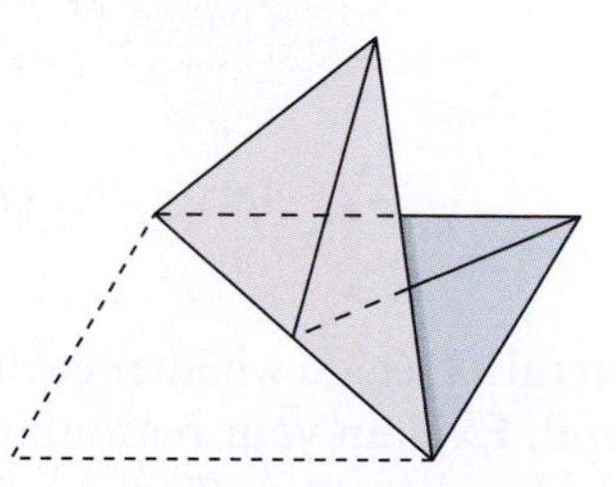

Conclusion The diagonals of a parallelogram do not necessarily intersect at right angles.

Rhombus

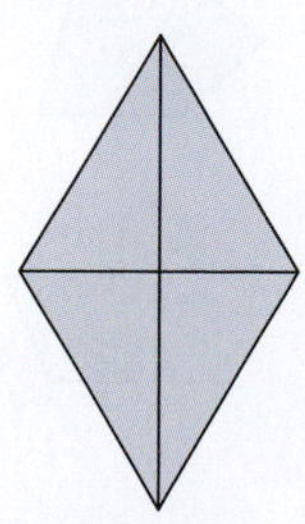

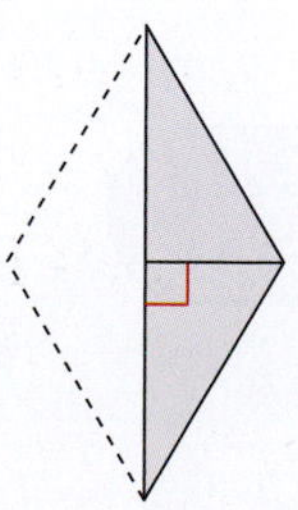

Conclusion The diagonals of a rhombus intersect at right angles.

10.2 Exercises

Free step-by-step solutions to odd-numbered exercises at *MathematicalPractices.com.*

1. **Writing a Solution Key** Write a solution key for the activity on page 374. Describe a rubric for grading a student's work.

2. **Grading the Activity** In the activity on page 374, a student gave the answers below. Each question is worth 10 points. Assign a grade to each answer. For those that are incorrect, why do you think the student erred?

Sample Student Work

1. Use 2 pieces to make a triangle worth 10 points.

$8 + 2 = 10$

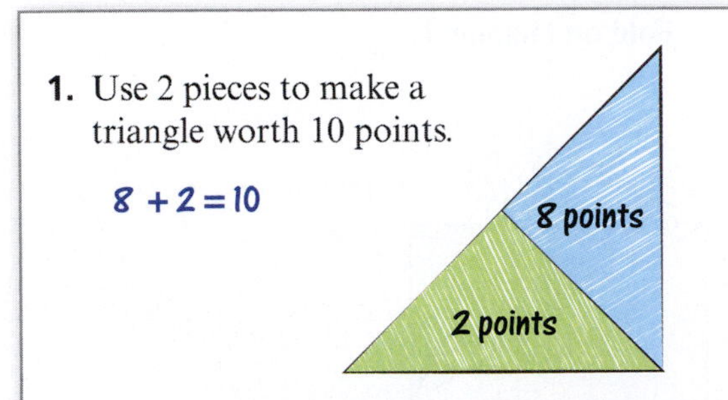

2. Use 2 pieces to make a square worth 9 points.

$5 + 4 = 9$

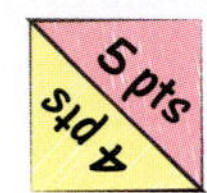

3. Use 2 pieces to make a trapezoid worth 10 points.

$8 + 2 = 10$

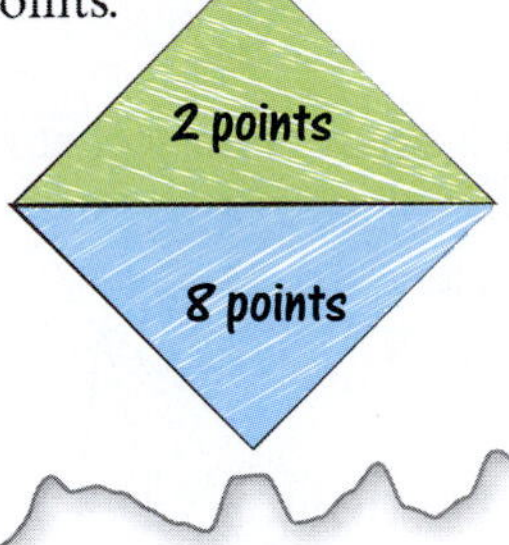

3. **Identifying Quadrilaterals** Decide whether each figure is a quadrilateral. Explain your reasoning.

a. 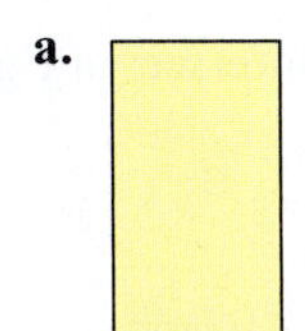

b.

c. 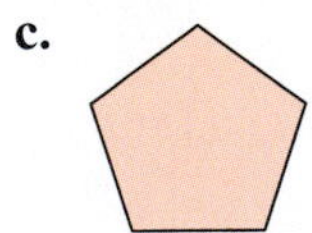

d. 

4. **Identifying Quadrilaterals** Decide whether each figure is a quadrilateral. Explain your reasoning.

a. 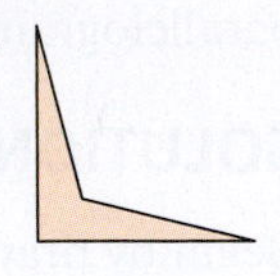

b. 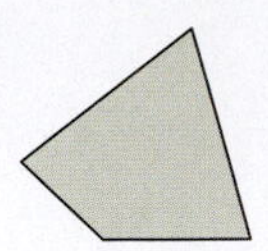

c.

d. 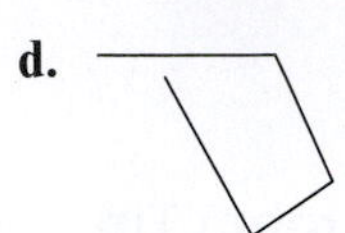

5. **Classifying Quadrilaterals** Classify each quadrilateral in as many ways as possible.

a. 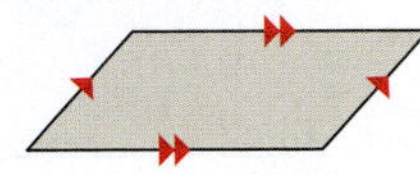

b. 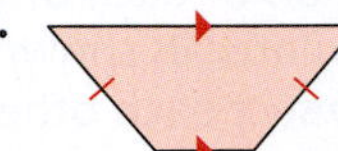

c. 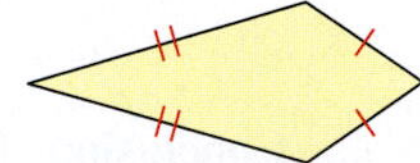

d. 

6. **Classifying Quadrilaterals** Classify each quadrilateral in as many ways as possible.

a. 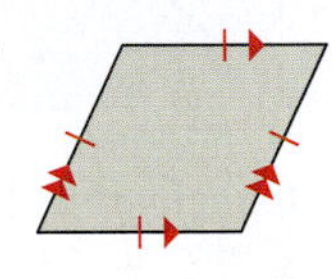

b. 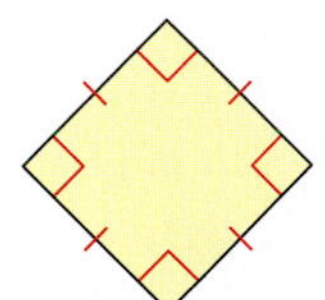

c. 

d.

7. **Convex and Concave Figures** Tell whether each figure is *convex* or *concave*.

a. 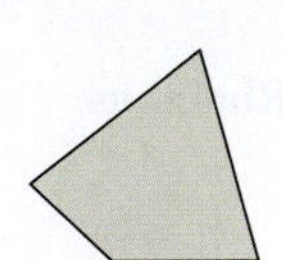

b. 

c. 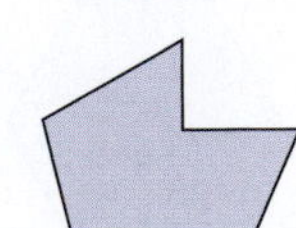

d. 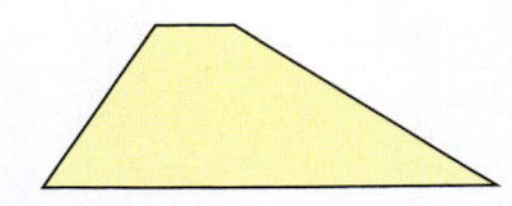

8. Convex and Concave Figures Tell whether each figure is *convex* or *concave*.

a.

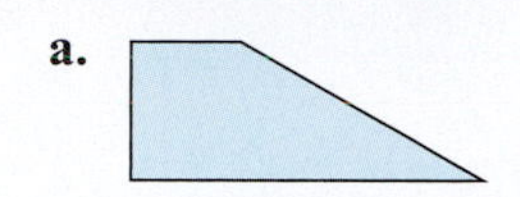

b.

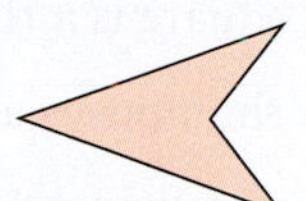

c.

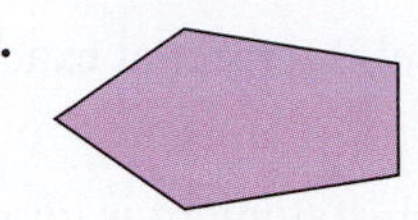

d.

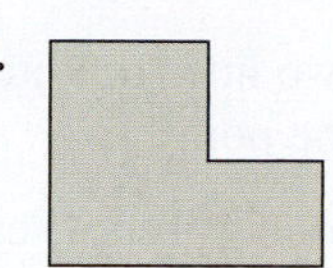

9. Rectangles Find the value of x for each rectangle.

a.

12

x

b.

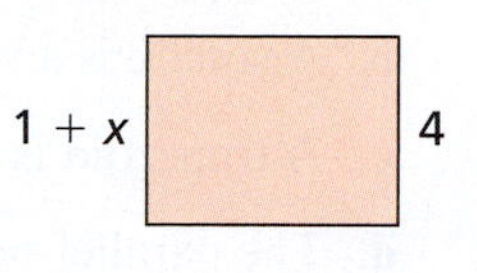

c.

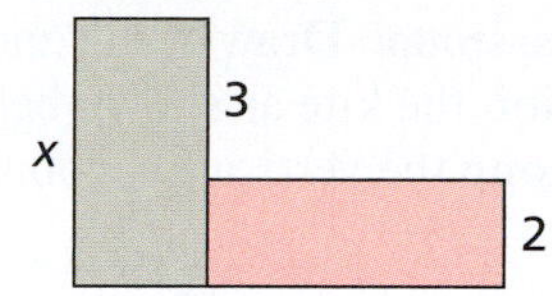

10. Rectangles Find the value of x for each rectangle.

a.

x 2

b.

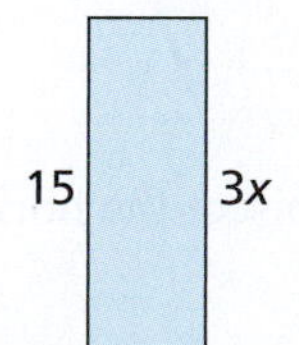

c.

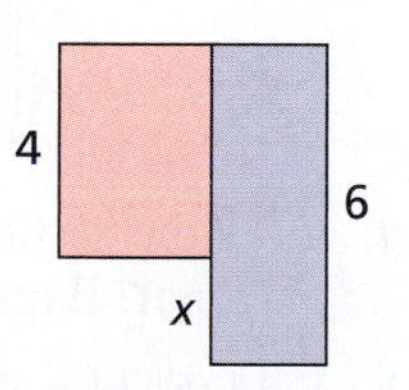

11. Angle Measures of Quadrilaterals The angle measures of three angles of a quadrilateral are given. Find the measure of the fourth angle of the quadrilateral.

a. 65°, 90°, 125°

b. 60°, 90°, 90°

c. 56°, 67°, 124°

d. 60°, 98°, 120°

12. Angle Measures of Quadrilaterals The angle measures of three angles of a quadrilateral are given. Find the measure of the fourth angle of the quadrilateral.

a. 62°, 95°, 135°

b. 54°, 68°, 100°

c. 65°, 90°, 146°

d. 50°, 50°, 130°

13. Angle Measures of Parallelograms Find the value of x for each parallelogram.

a.

47°

x°

b.

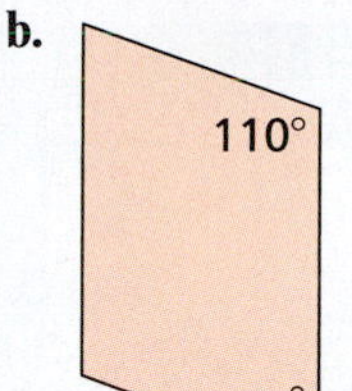

c.

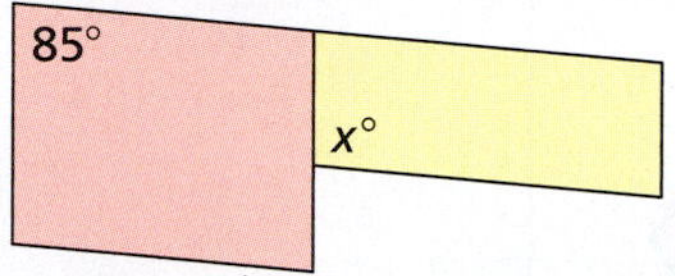

14. Angle Measures of Parallelograms Find the value of x for each parallelogram.

a.

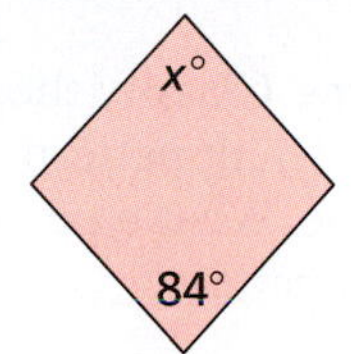

b.

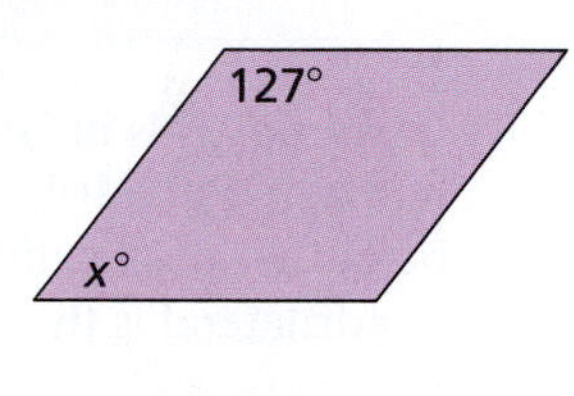

c.

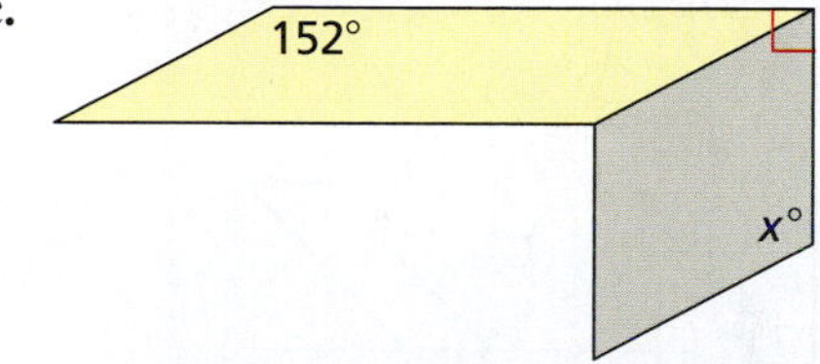

15. Identifying Quadrilaterals in Real Life Classify each quadrilateral as specifically as possible.

a.

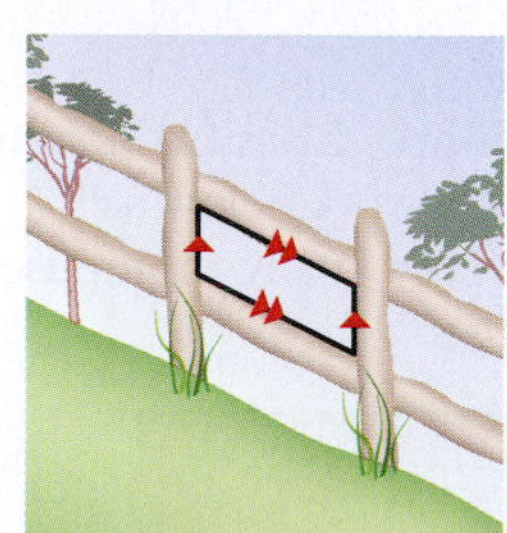

b.

16. Identifying Quadrilaterals in Real Life Classify each quadrilateral as specifically as possible.

a.

b.

17. Quadrilaterals in Quilt Patterns Classify the numbered quadrilaterals in the quilt pattern below as specifically as possible. What type of quadrilateral is the entire quilt pattern?

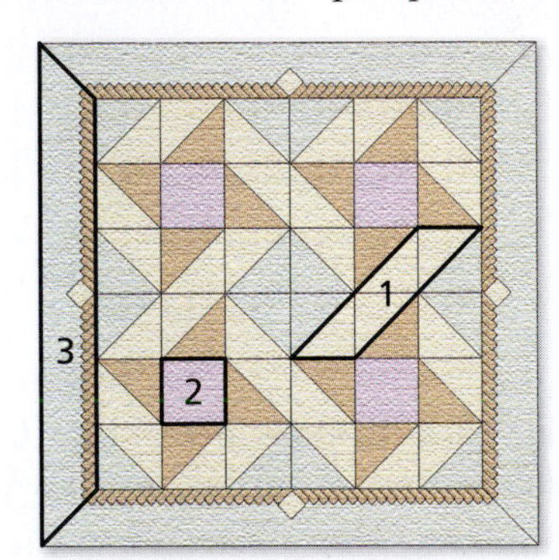

18. Quadrilaterals in Woodwork Classify the numbered quadrilaterals in the woodwork below as specifically as possible. How many rhombuses that are not squares do you see?

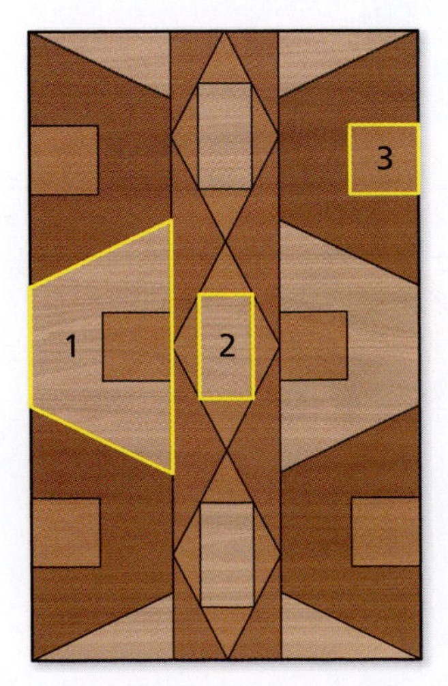

19. Reasoning Tell whether each statement is *always*, *sometimes*, or *never* true. Explain your reasoning.

a. A square is a rhombus.

b. A rhombus is a kite.

c. A parallelogram is a rectangle.

d. Two lines that are parallel intersect at exactly one point.

20. Reasoning Tell whether each statement is *always*, *sometimes*, or *never* true. Explain your reasoning.

a. A rectangle is a rhombus.

b. A square is a right trapezoid.

c. A trapezoid is convex.

d. The parallel sides of an isosceles trapezoid are congruent.

21. Activity: In Your Classroom Draw a kite and its two diagonals. Position the kite as shown below. Then fold the kite along the vertical diagonal.

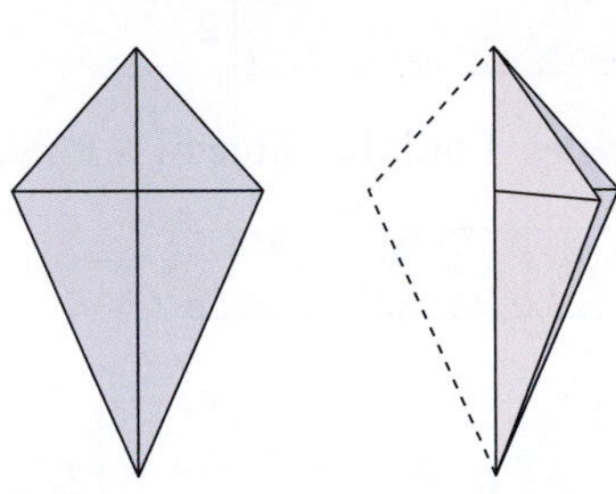

a. Does the vertical diagonal bisect the horizontal diagonal? Explain.

b. Fold the kite along the horizontal diagonal. Does the horizontal diagonal bisect the vertical diagonal?

c. Design an activity involving paper folding that can be used to demonstrate that the diagonals of rectangles, squares, parallelograms, and rhombuses bisect each other.

22. Activity: In Your Classroom Consider the parallelogram below.

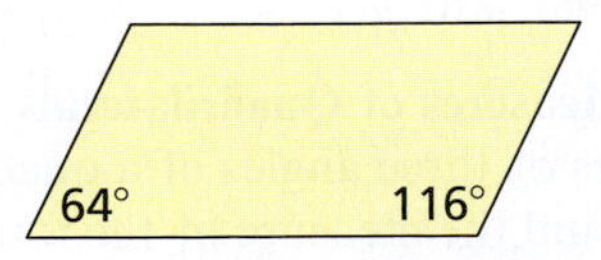

a. Tell whether the adjacent angles of the parallelogram are *complementary*, *supplementary*, or *neither*.

b. Does this property hold for all parallelograms? Explain.

c. Design an activity similar to the one in Example 3 to demonstrate this concept to students.

23. Parking Lot A parking lot is in the shape of a trapezoid. The two angles bordering the road each measure 60°. Another angle measures 120°.

a. What is the measure of the fourth angle?

b. Classify the trapezoid.

24. Fencing A farmer has 1100 feet of fencing to enclose a parallelogram-shaped pasture. The shorter sides of the pasture are each 180 feet long.

a. How much fencing will be used for each of the longer sides?

b. Draw and label a diagram of the pasture.

25. Problem Solving The figure below is a rhombus.

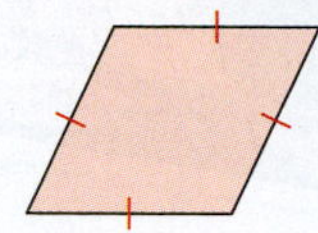

a. Draw one line through the rhombus to make two triangles.

b. Classify the triangles as *equilateral*, *isosceles*, or *scalene*.

26. Problem Solving The figure below is a right trapezoid.

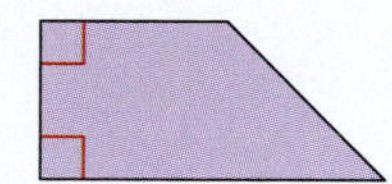

a. Draw one line perpendicular to the base through the right trapezoid to make a triangle and a quadrilateral.

b. Classify the triangle as *acute*, *right*, or *obtuse*.

c. Classify the quadrilateral as specifically as you can.

27. Writing Is it possible for a quadrilateral to have three 60° angles? Explain.

28. Writing What is the greatest number of obtuse angles a quadrilateral can have? Explain.

29. Writing Is it possible for a quadrilateral that is not a rectangle to have two right angles? three right angles? Explain your reasoning.

30. Writing Is it possible to find the remaining angle measures of a rhombus if you are given one angle measure? Explain your reasoning.

31. Reasoning Does the figure provide you with enough information to determine whether it is a parallelogram? Explain.

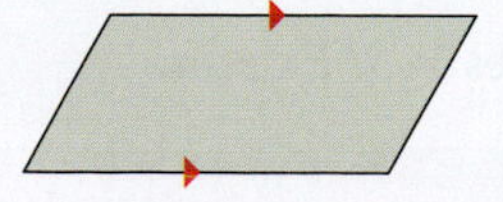

32. Reasoning What additional information do you need to determine whether the figure is a rhombus? Explain.

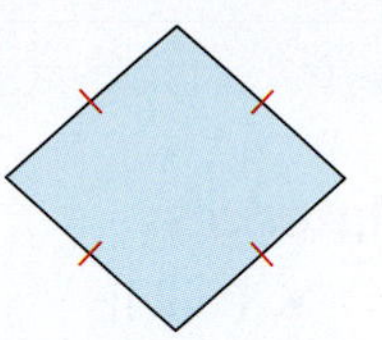

33. Problem Solving The ratios of the corresponding side lengths of the rectangles are equal. Find the value of x.

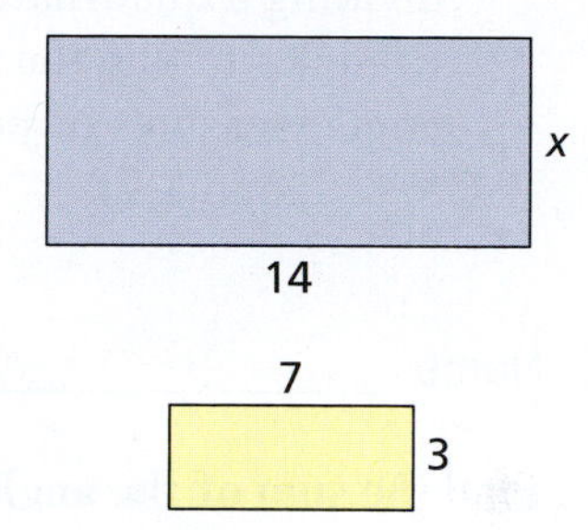

34. Problem Solving The ratios of the corresponding side lengths of the isosceles trapezoids are equal. Find the value of x.

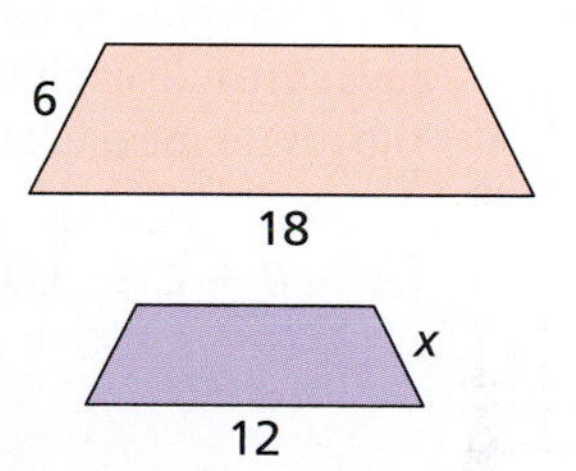

35. Grading Student Work On a diagnostics test, one of your students does the following work. Explain what the student did wrong. Which topics would you encourage the student to review?

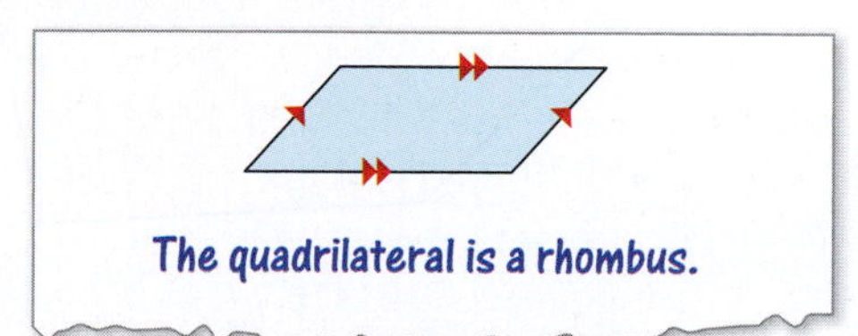

36. Grading Student Work On a diagnostics test, one of your students does the following work. Explain what the student did wrong. Which topics would you encourage the student to review?

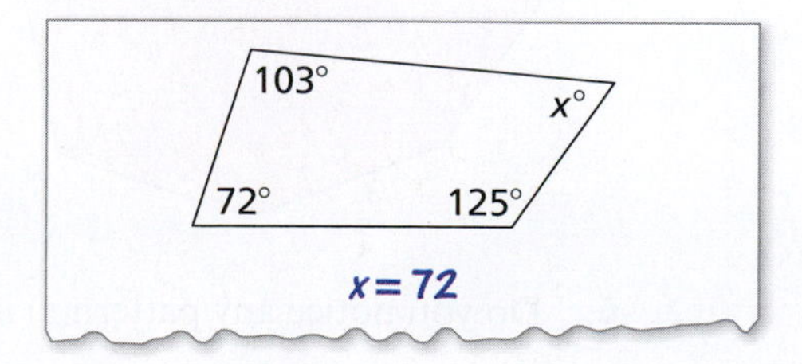

Activity: The Sums of the Angle Measures of Polygons

Materials:

- Pencil

Learning Objective:

Generalize the pattern given for dividing a quadrilateral into two triangles to find the sums of the angle measures of polygons with 5, 6, 7, and 8 sides.

Name ______________________________

Find the sum of the angle measures of the polygon.

1. Quadrilateral

Draw a line that divides the quadrilateral into two triangles. Because the sum of the angle measures of each triangle is 180°, the sum of the angle measures of a quadrilateral is

$(A + B + C) + (D + E + F) =$ ____ $+$ ____

$=$ ____.

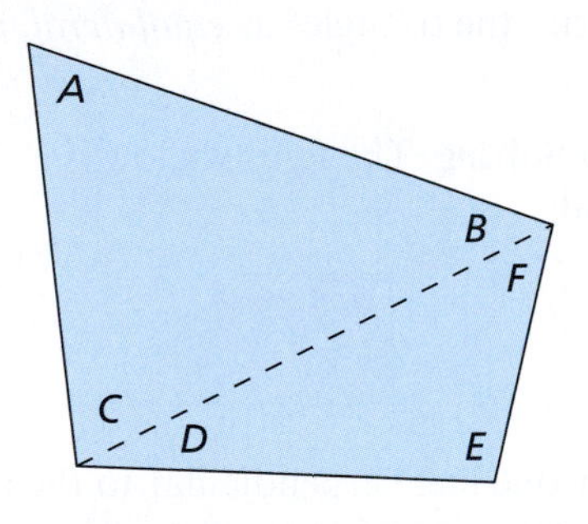

2. Pentagon

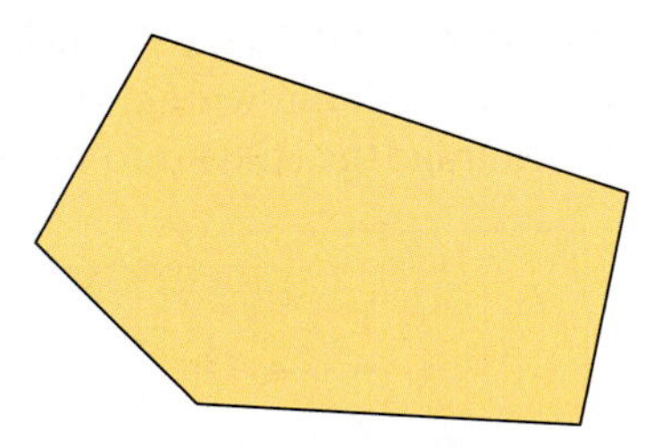

3. Hexagon

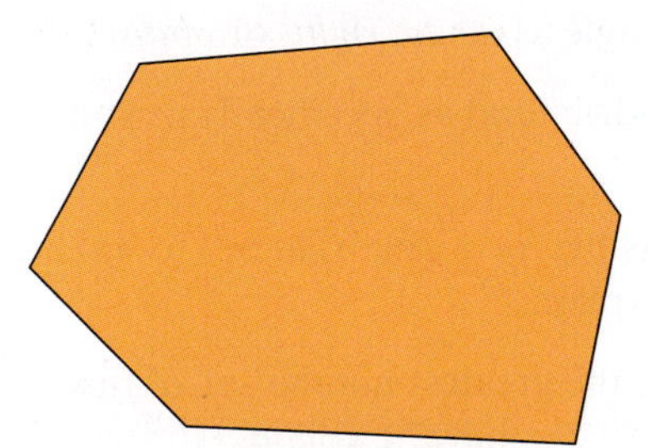

4. Heptagon (or Septagon)

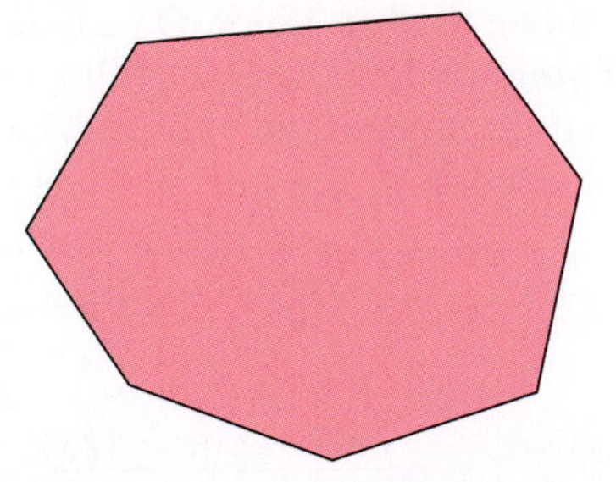

5. Octagon

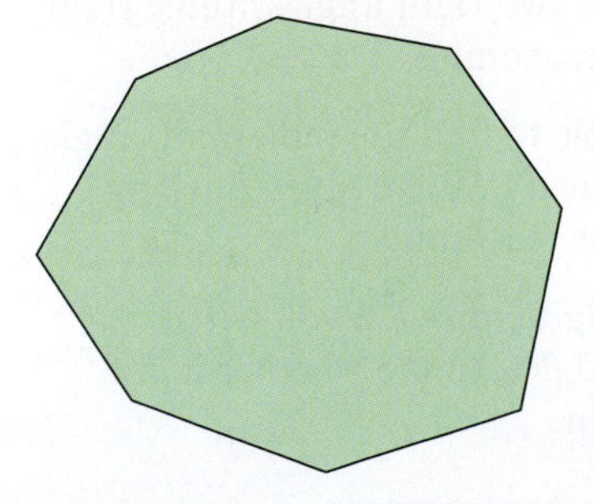

6. Do you notice any pattern(s) in your results from Exercises 1–5? Explain.

iStockphoto.com/Alina555

10.3 Polygons and Tessellations

Standards

Grades K–2
Geometry
Students should analyze, compare, create, and compose shapes.

Grades 3–5
Geometry
Students should draw and identify lines and angles, and classify shapes by properties of their lines and angles.

Grades 6–8
Geometry
Students should solve real-life and mathematical problems involving angle measure.

- Use the formula for the sum of the angle measures of a polygon.
- Understand the properties of regular polygons.
- Use regular polygons to create tessellations.

Sum of the Angle Measures of a Polygon

The word *polygon* comes from the Greek words poly meaning *many* and gonia meaning *angle*. So, a polygon is a figure that has many angles.

Definition of a Polygon

A **polygon** is a closed two-dimensional figure that has 3 or more straight sides. In general, a polygon with n sides is called an ***n*-gon**.

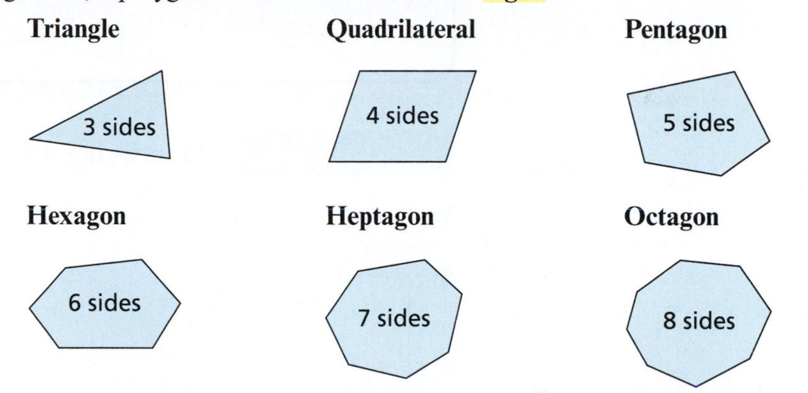

Sum of the Interior Angle Measures of a Polygon

An **interior angle**, or **vertex angle**, is formed by a vertex and the two sides that have the vertex as an endpoint. The sum of the interior angle measures of an n-gon is $180°(n - 2)$.

Classroom Tip

The formula for the sum of the interior angle measures of a polygon can be used for concave polygons. To do this, however, requires introducing angles whose degree measures are greater than 180°. For instance, the concave polygon below has an interior angle that is greater than 180°.

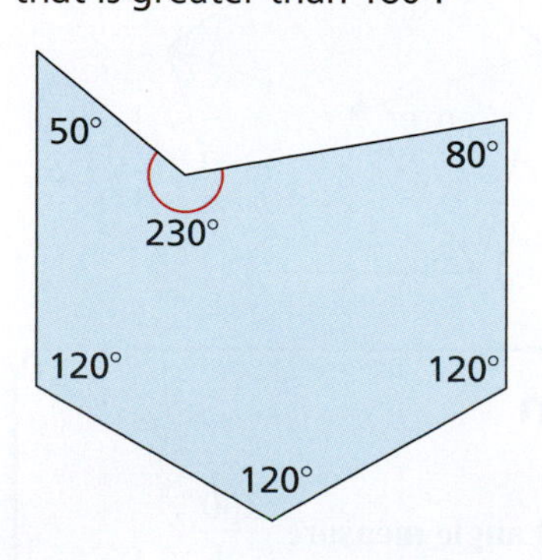

EXAMPLE 1 **The Sum of the Angle Measures of a Polygon**

The figure at the right is called a *buckyball*. It is not a sphere. Identify the polygons that make up the buckyball. Find the sum of the interior angle measures of each polygon.

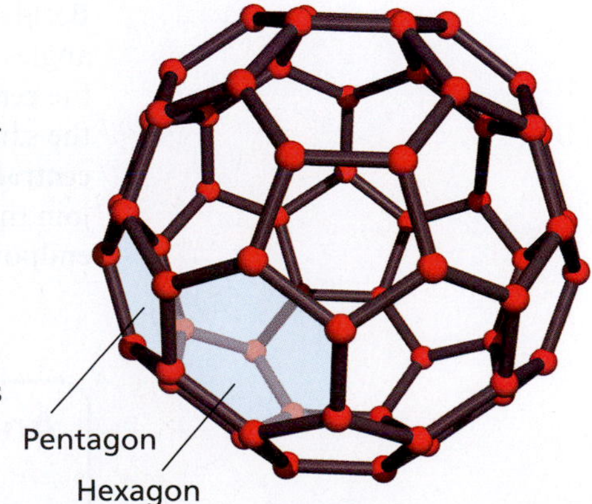

SOLUTION

The buckyball consists of two polygons: pentagons and hexagons. The sum of the interior angle measures for each pentagon is

$$180°(n - 2) = 180°(5 - 2)$$
$$= 540°.$$

The sum of the interior angle measures for each hexagon is

$$180°(n - 2) = 180°(6 - 2) = 720°.$$

Friedrich Saurer/ImageBroker/Glow Images

Regular Polygons

Classroom Tip

Notice that a regular triangle is also called an equilateral triangle and a regular quadrilateral is also called a square.

Definition of a Regular Polygon

A polygon is **regular** when all of its sides are congruent and all of its interior angles are congruent. A regular polygon is both equilateral and equiangular.

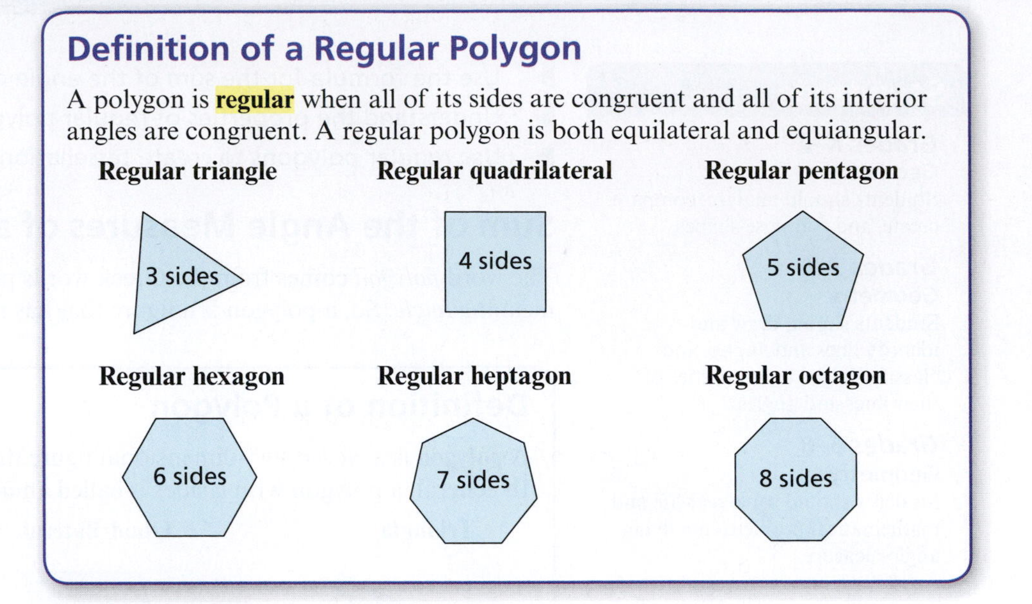

EXAMPLE 2 **Identifying a Regular Polygon**

Tell whether the polygon is regular. Explain your reasoning.

a.

b.

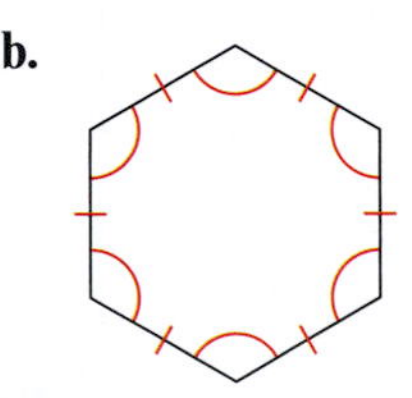

SOLUTION

a. All of the sides of the polygon are congruent. But, not all of the interior angles are congruent. So, the polygon is *not* regular.

b. All of the sides and all of the interior angles of the polygon are congruent. So, the polygon is regular.

Besides interior angles, there are other interesting angles in a regular n-gon. One such angle involves the **center of the polygon**, which is the point that is the same distance from all of the vertices. A **central angle** is formed by the line segments that join the center of the polygon with the two endpoints of one of the sides.

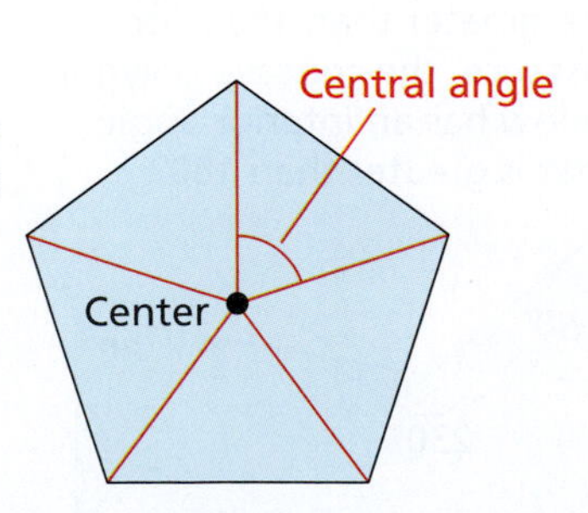

Angle Measures of a Regular n-gon

Interior angle measure $\dfrac{180°(n-2)}{n}$ **Central angle measure** $\dfrac{360°}{n}$

EXAMPLE 3 Drawing a Regular Polygon

Use a protractor and a ruler to draw a regular pentagon. Check that each of the interior angles has the correct measure.

SOLUTION

1. For a regular pentagon, the central angle has a measure of

$$\frac{360°}{n} = \frac{360°}{5} = 72°.$$

Use a protractor to measure 72°. Then draw the angle.

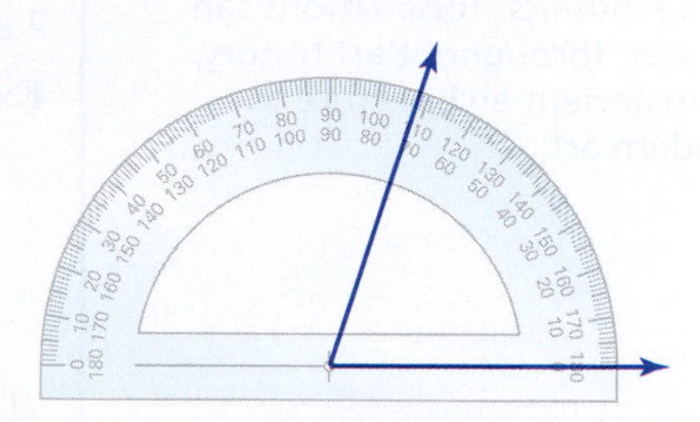

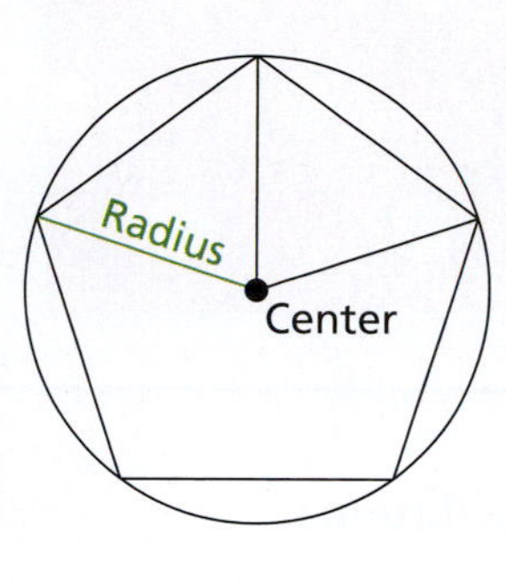

2. Choose a radius for the pentagon. Mark the radius on each ray of the angle.

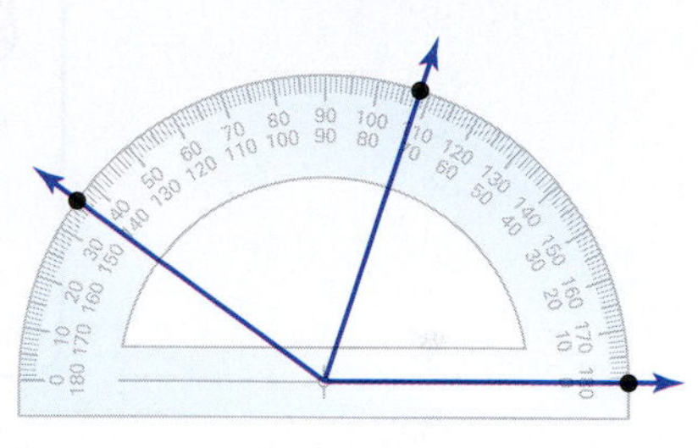

3. Continue this process until all five rays are drawn.

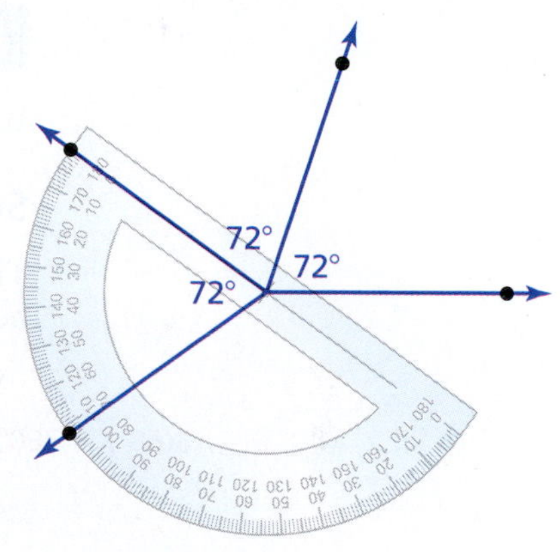

4. Connect the ends of the radii. The result is a regular pentagon.

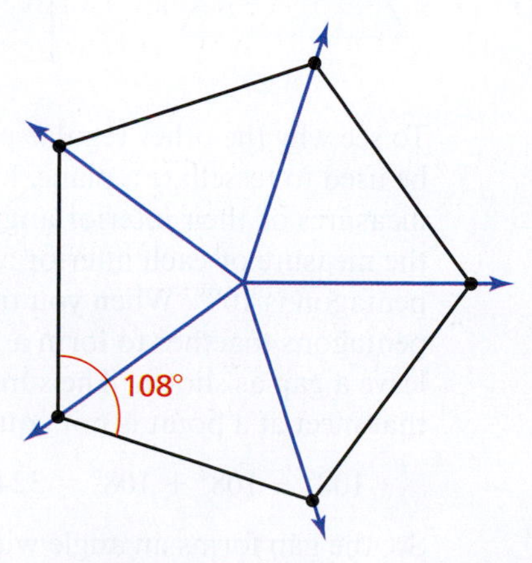

Each of the angles of the regular pentagon has a measure of

$$\frac{180°(n-2)}{n} = \frac{180°(5-2)}{5} = \frac{540°}{5} = 108°.$$

You can use a protractor to check each interior angle measure.

Tessellations

Classroom Tip

The Latin word *tessella* refers to a small block used to make mosaics. Tessellations can be seen throughout art history, from ancient architecture to modern art.

Definition of a Tessellation

A **tessellation** is a covering of a two-dimensional plane using the repetition of a geometric shape with no overlaps or gaps.

Example

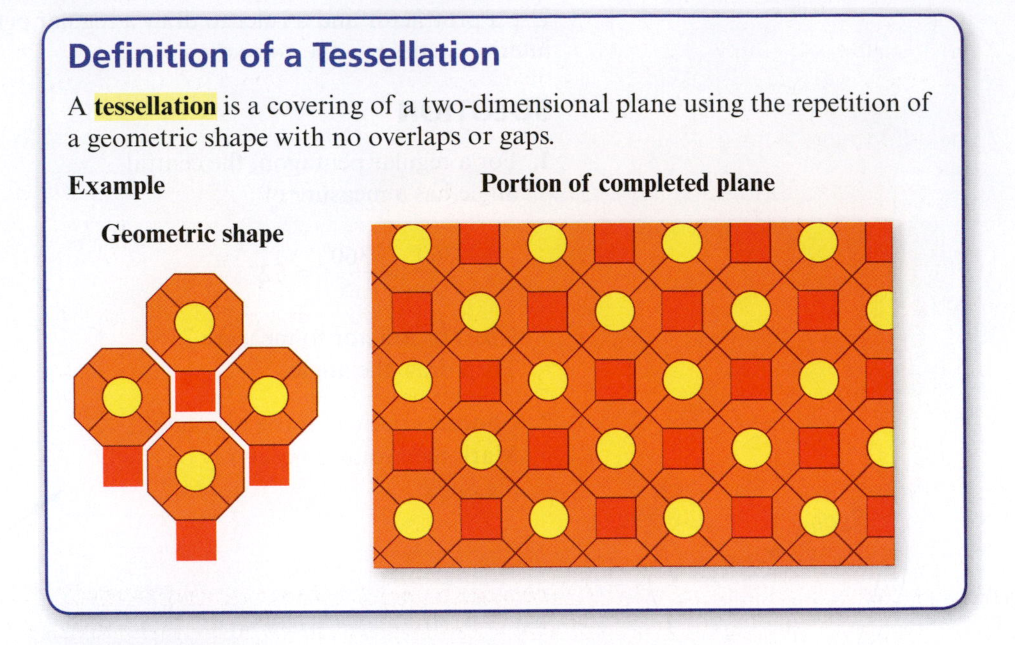

EXAMPLE 4 Regular Polygons and Tessellations

Which of the regular polygons can be used to tessellate a plane?

SOLUTION

There are only 3 regular polygons: equilateral triangles, squares, and regular hexagons.

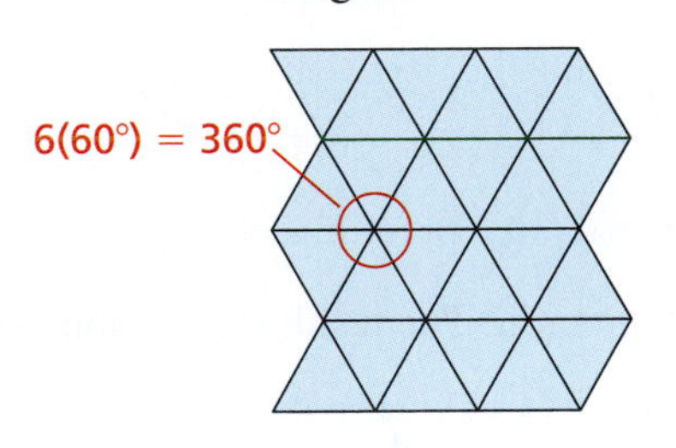

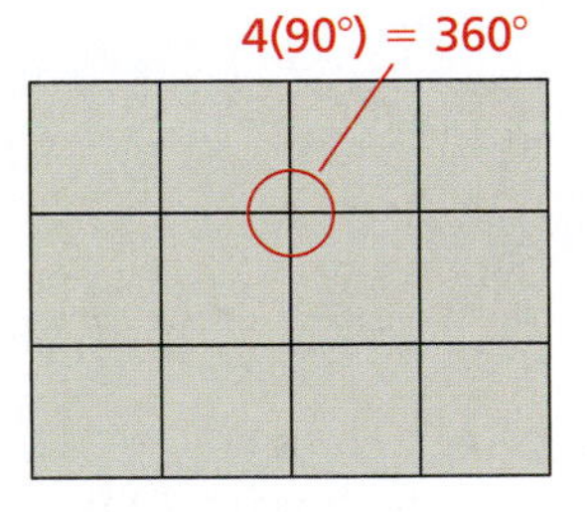

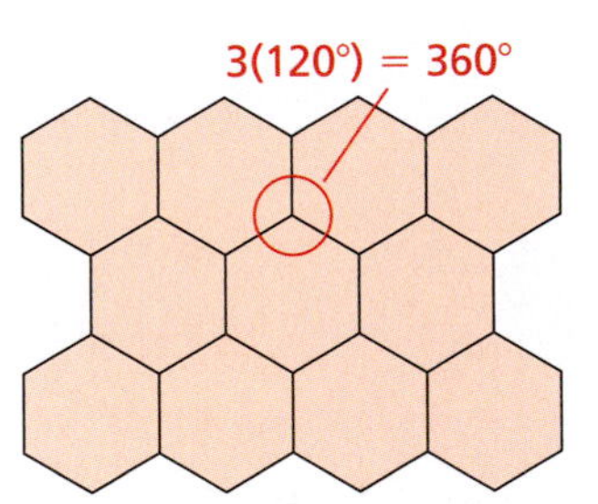

To see why the other regular polygons cannot be used to tessellate a plane, look at the measures of their interior angles. For instance, the measure of each interior angle of a regular pentagon is 108°. When you try to put regular pentagons together to form a tessellation, you leave a gap as shown. The sum of the angles that meet at a point is not equal to 360°.

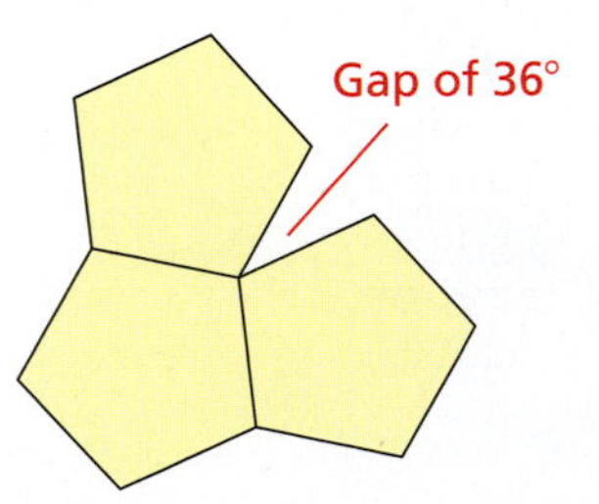

$$108° + 108° + 108° = 324°$$

So, the gap forms an angle with a measure of $360° - 324° = 36°$.

Notice that the regular polygons that tessellate have interior angle measures that divide 360°. That is why the other regular polygons shown at the top of page 388 do not tessellate; their angle measures are not divisors of 360°.

Classroom Tip

Although Example 5 only shows regular polygons as tiles for tessellations, nonregular polygons can also work. For instance, *any* parallelogram can be used to tessellate a plane.

EXAMPLE 5 Designing Floor Tiles

There are hundreds of different tessellations that can be made using combinations of regular polygons. Design several tessellations that can be used as floor tilings.

SOLUTION

a. Triangles and hexagons

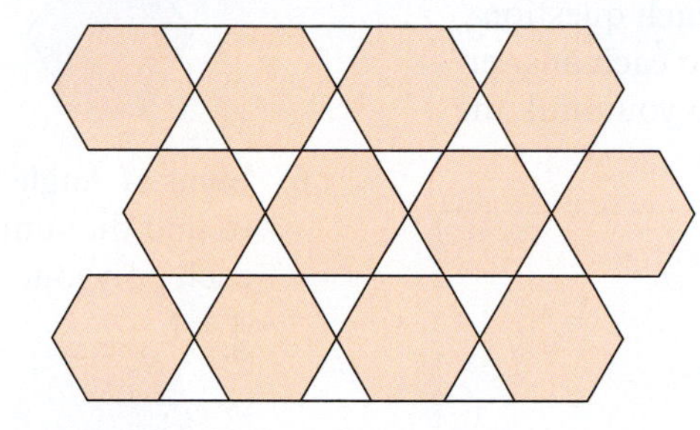

b. Squares and octagons

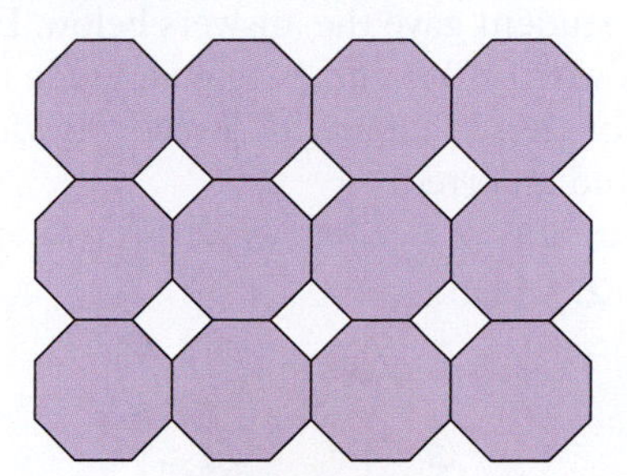

One of the more famous people to use tessellations in art is Dutch artist M.C. Escher (1898–1972). He is known for his mathematically inspired woodcuts, lithographs, and mezzotints which often incorporate tessellations like the one shown.

EXAMPLE 6 Tessellations in Art

Create an Escher-style tessellation art print.

SOLUTION

There are many ways to do this. One way is to start with a polygon that tessellates, such as a square. Cut two irregular shapes out of two of the sides, and tape each shape to its opposite side as shown at the right. Use the taped shape to make a tessellation. Use different colors to enhance your art piece.

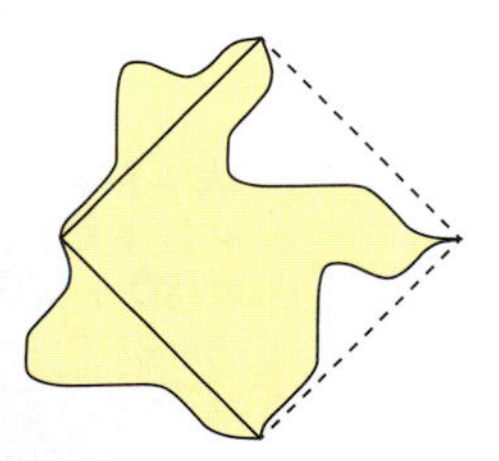

Mathematical Practices

Use Appropriate Tools Strategically

MP5 Mathematically proficient students consider available tools, such as dynamic geometry software, when solving a mathematical problem. They are able to use technological tools to explore and deepen their understanding of concepts.

10.3 Exercises

Free step-by-step solutions to odd-numbered exercises at *MathematicalPractices.com.*

1. **Writing a Solution Key** Write a solution key for the activity on page 386. Describe a rubric for grading a student's work.

2. **Grading the Activity** In the activity on page 386, a student gave the answers below. Each question is worth 10 points. Assign a grade to each answer. For those that are incorrect, why do you think the student erred?

Sample Student Work

2. Pentagon

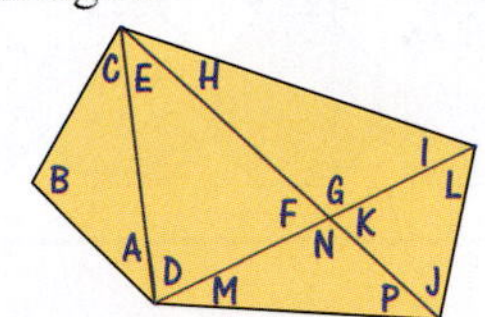

(A + B + C) + (D + E + F) + (G + H + I)
+ (J + K + L) + (M + N + P)
= 180 + 180 + 180 + 180 + 180 = 900°

3. Hexagon

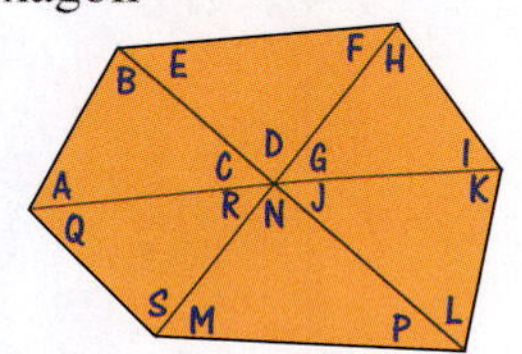

(A + B + C) + (D + E + F) + (G + H + I)
+ (J + K + L) + (M + N + P) + (Q + R + S)
= 180 + 180 + 180 + 180 + 180 + 180 = 1080°

4. Heptagon (or Septagon)

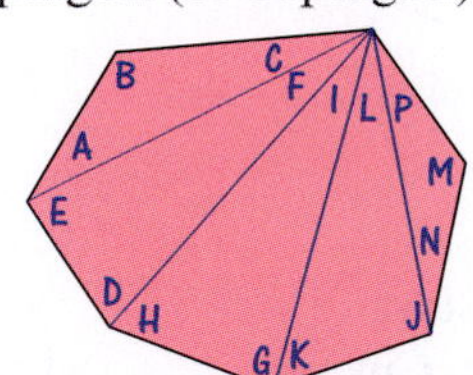

(A + B + C) + (D + E + F) + (G + H + I)
+ (J + K + L) + (M + N + P)
= 180 + 180 + 180 + 180 + 180 = 900°

3. **Identifying Polygons** Decide whether each figure is a polygon. Explain your reasoning.

a. 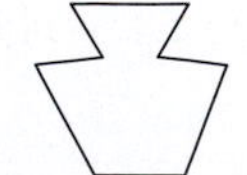 b. 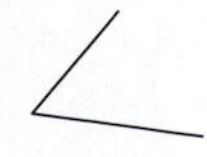 c. 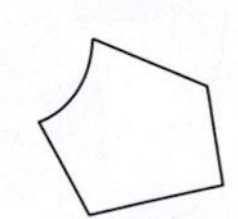

4. **Identifying Polygons** Decide whether each figure is a polygon. Explain your reasoning.

a. 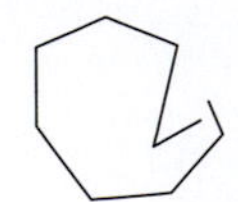 b. 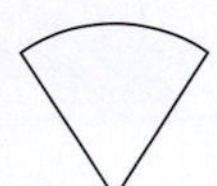 c. 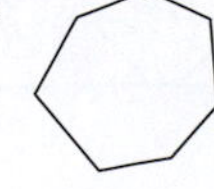

5. **Sums of Angle Measures of Polygons** Use triangles to find the sum of the interior angle measures of each polygon.

a. 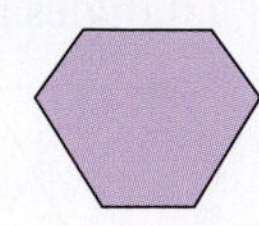 b. 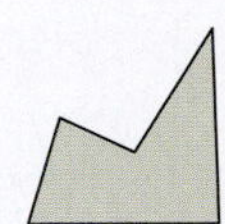

6. **Sums of Angle Measures of Polygons** Use triangles to find the sum of the interior angle measures of each polygon.

a. 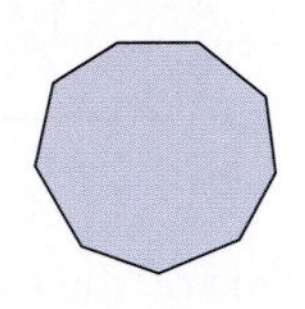 b. 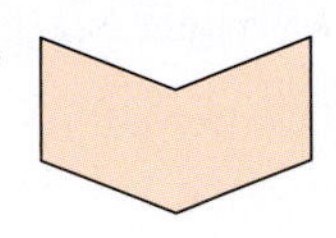

7. **Sums of Angle Measures of Polygons** Use a formula to find the sum of the interior angle measures of each polygon.

a. 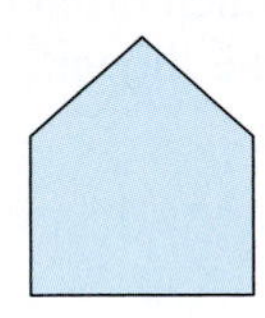 b. 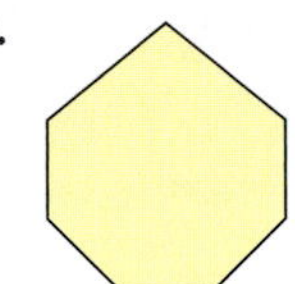

8. **Sums of Angle Measures of Polygons** Use a formula to find the sum of the interior angle measures of each polygon.

a. 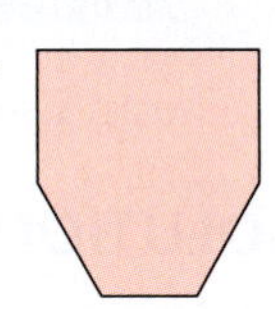 b. 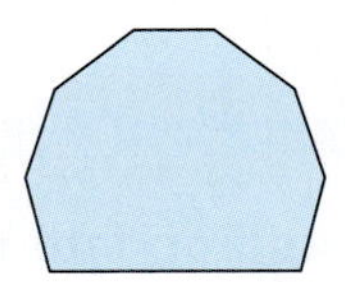

9. **Identifying Regular Polygons** Decide whether each polygon is regular. Explain your reasoning.

a. 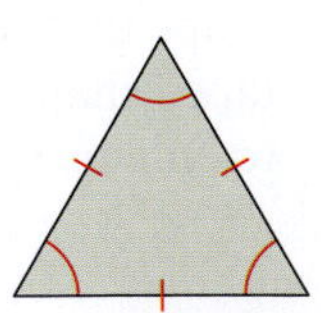 b. 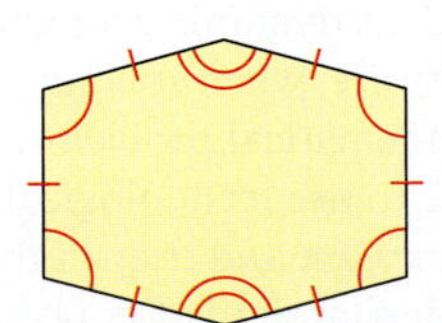

c.

7 m
6 m 6 m
6 m 6 m
7 m

d. 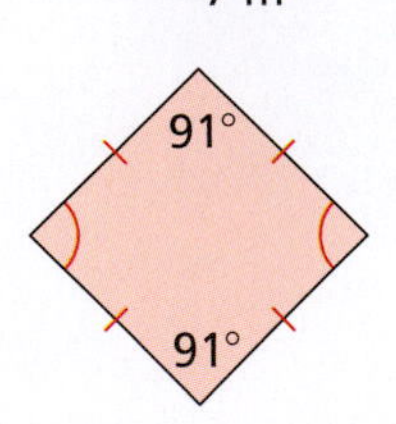

10. Identifying Regular Polygons Decide whether each polygon is regular. Explain your reasoning.

a.

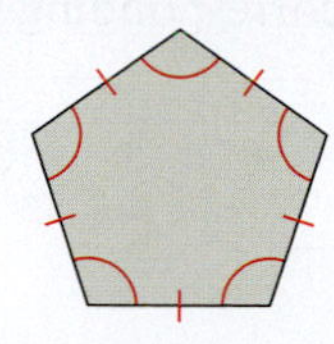

b.

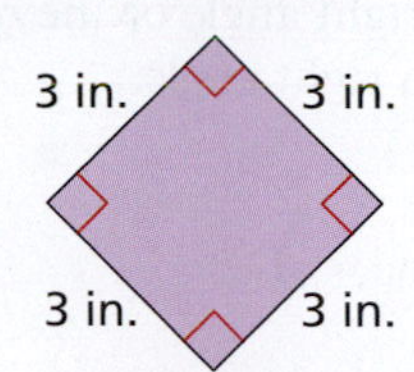

c.

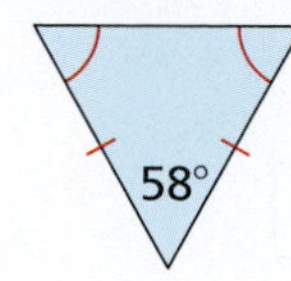

d.

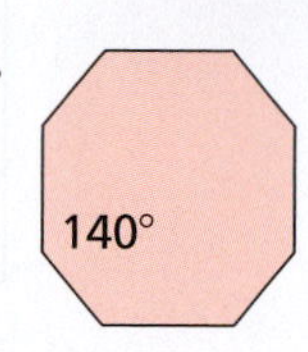

11. Finding Angle Measures Find the measures of an interior angle and a central angle of each regular polygon.

a. Equilateral triangle

b. Regular octagon

12. Finding Angle Measures Find the measures of an interior angle and a central angle of each regular polygon.

a. Regular hexagon **b.** Regular heptagon

13. Finding Angle Measures Find the value of x for each polygon.

a.

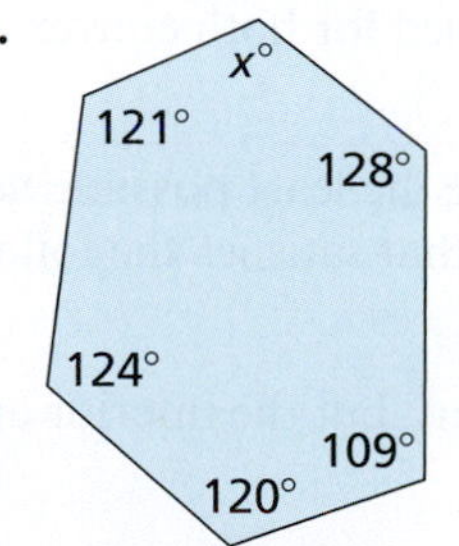

b.

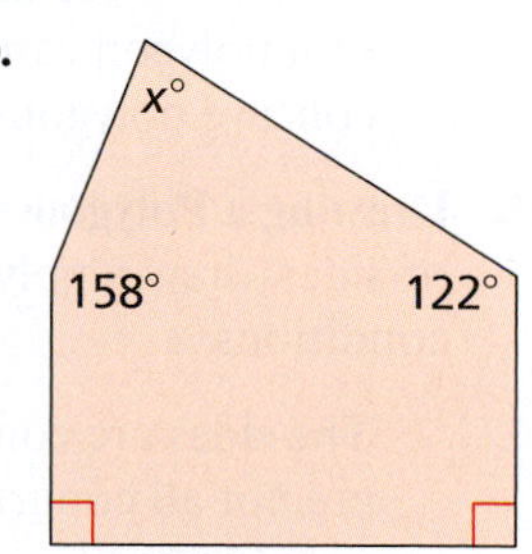

14. Finding Angle Measures Find the value of x for each polygon.

a.

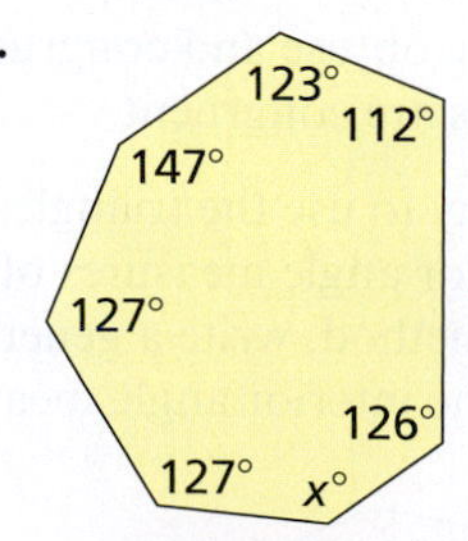

b.

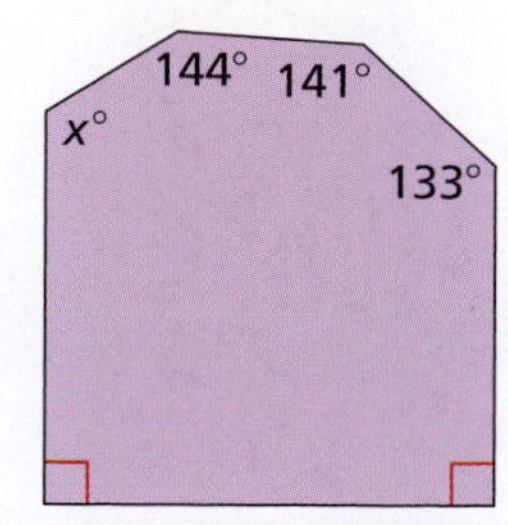

15. Drawing Regular Polygons Use a protractor and a ruler to draw the polygon. Check that each of the interior angles has the correct measure.

a. Equilateral triangle

b. Regular hexagon

c. Regular decagon (10-gon)

16. Drawing Regular Polygons Use a protractor and a ruler to draw the polygon. Check that each of the interior angles has the correct measure.

a. Square

b. Regular octagon

c. Regular nonagon (9-gon)

17. Tessellation Use dot paper to create a tessellation based on the figure shown.

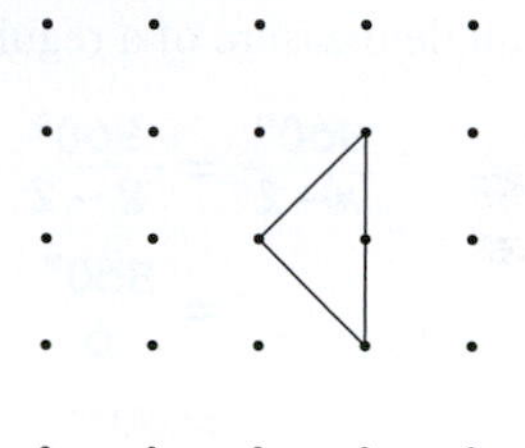

18. Tessellation Use dot paper to create a tessellation based on the figure shown.

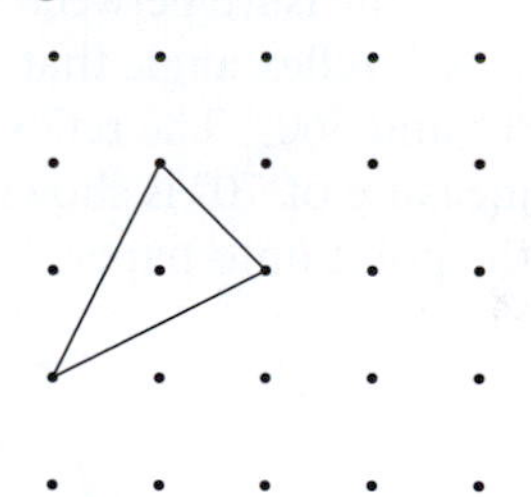

19. Tessellation Use dot paper to create a tessellation based on the figure shown.

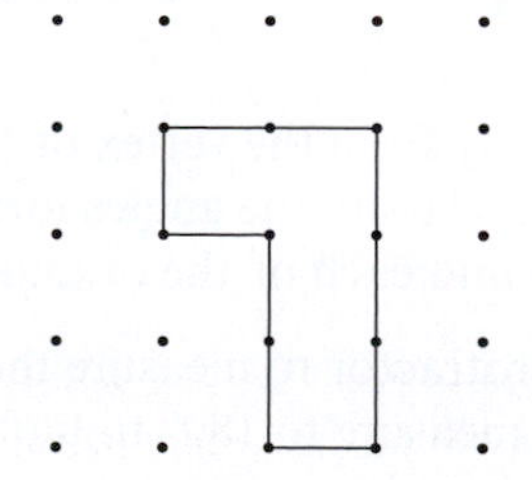

20. Tessellation Use dot paper to create a tessellation based on the figure shown.

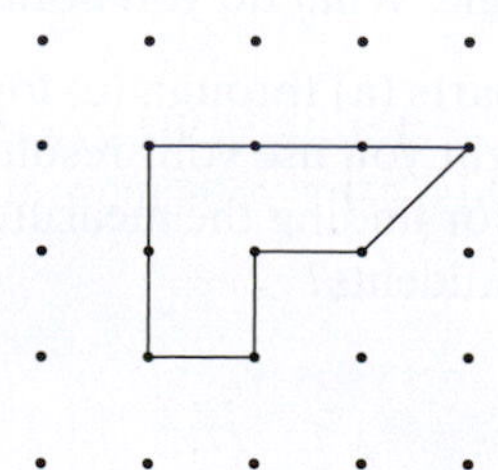

21. **Grading Student Work** On a diagnostic test, one of your students does the following work. Explain what the student did wrong. Which topics would you encourage the student to review?

> Interior angle measure of a regular heptagon:
>
> $$\frac{180° \cdot n}{n-2} = \frac{180° \cdot 7}{7-2}$$
> $$= \frac{1260°}{5}$$
> $$= 252°$$

22. **Grading Student Work** On a diagnostic test, one of your students does the following work. Explain what the student did wrong. Which topics would you encourage the student to review?

> Central angle measure of a regular octagon:
>
> $$\frac{360°}{n-2} = \frac{360°}{8-2}$$
> $$= \frac{360°}{6}$$
> $$= 60°$$

23. **Measuring Reflex Angles: In Your Classroom** An angle that has a measure between 0° and 180° corresponds to a **reflex angle** that has a measure between 180° and 360°. The reflex angle for an angle that has a measure of 70° is shown below. Trace the angle and the point onto paper.

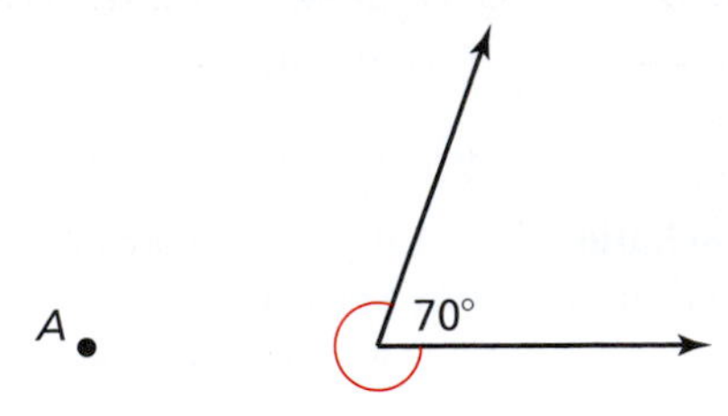

a. Draw a ray from the vertex of the angle through point A. Classify the angles formed between the new ray and each of the original rays.

b. Use a protractor to measure the obtuse angle. Add its measure to 180° to find the measure of the reflex angle.

c. Add the measures of the original angle and the reflex angle. What do you notice?

d. Repeat parts (a) through (c) for different angles. How might you use your results to describe a method for finding the measure of a reflex angle to your students?

24. **Activity: Concave Polygons: In Your Classroom** Each interior angle of the concave figure is either a right angle or the *reflex angle* corresponding to a right angle.

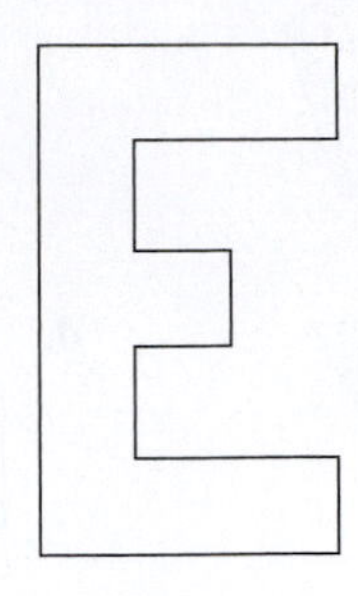

a. Determine the measure of the reflex angle corresponding to a right angle. (See Exercise 23.)

b. How many sides does the concave figure have? Make a sketch of the figure. Label the measure of each interior angle.

c. Find the sum of all the interior angle measures.

d. Use the formula on page 387 to find the sum of the angle measures of the polygon. Is your result the same as your sum in part (c)?

e. Design an activity that students can use to discover inductively that the formula for finding the sum of the measures of the interior angles of a polygon can be used for both convex and concave polygons.

25. **Drawing a Polygon** Using the least possible number of sides, draw a polygon that satisfies the following conditions.

The sides are congruent, but the interior angles are not all congruent.

26. **Drawing a Polygon** Using the least possible number of sides, draw a polygon that satisfies the following conditions.

The interior angles are obtuse and congruent, but not all of the sides are congruent.

27. **Writing** Describe a way to use the triangles to find the sum of the interior angle measures of the hexagon. Based on this method, write a general formula for the sum of the interior angle measures of a polygon.

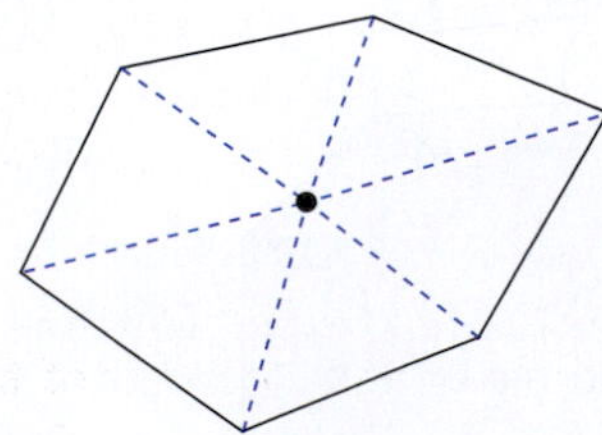

28. *Writing* Describe a way to use the triangles in the figure to find the sum of the interior angle measures of the hexagon. Explain how this method is related to the formula $180°(n - 2)$. Explain why this method gives the same result as the method in Exercise 27.

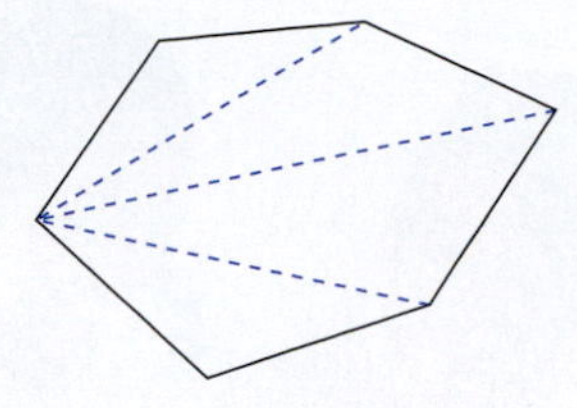

29. **Finding an Angle Measure** The interior reflex angles (see Exercise 23) of the polygon are congruent. The interior acute angles are also congruent. The measure of each interior acute angle is 30°. Find the measure of each interior reflex angle.

30. **Finding an Angle Measure** The interior reflex angles (see Exercise 23) of the polygon are congruent. The interior acute angles are also congruent. The measure of each interior reflex angle is 252°. Find the measure of each interior acute angle.

31. **Tessellations in Real Life** Describe the shape that tessellates in the brick wall. There are a few other ways the shape can be arranged to form tessellations. Why is the arrangement below generally used for a brick wall?

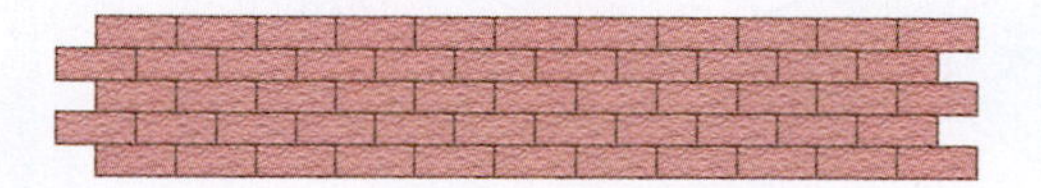

32. **Tessellations in Real Life** Make a sketch that shows a different way to tessellate the shape in the figure above. Describe a real-life application that might involve a tessellation like your sketch.

33. **Regular Polygons and Tessellations** Is it possible to arrange three regular heptagons so that all three share one vertex and each pair of heptagons shares exactly one side? Explain why or why not.

34. **Regular Polygons and Tessellations** Are there any regular polygons with more than six sides that can be used to tessellate a plane? Explain.

35. **Tessellations** Can the figure be used to tessellate a plane? If so, make a drawing to show how. If not, explain why not.

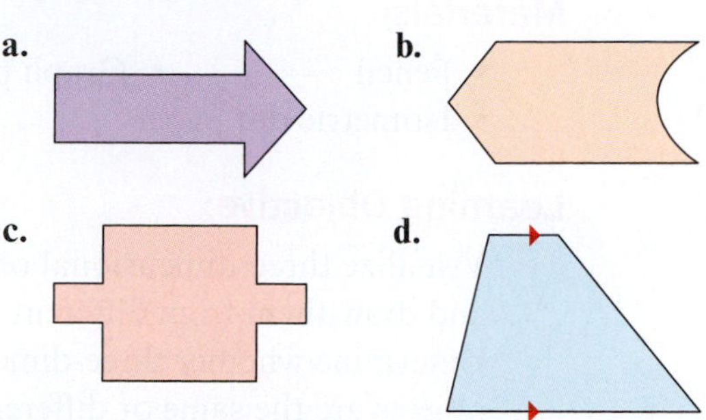

36. **Tessellations in Art** Trace the regular hexagon below onto paper and use the method of Example 6 to create an Escher-style tessellation art piece. Can you cut irregular shapes out of three sides and tape each shape to its opposite side and still have a shape that tessellates? Explain.

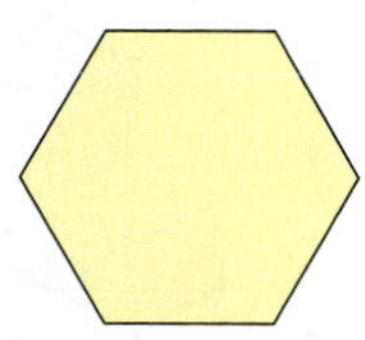

37. **Tessellations in Art** Trace the isosceles triangle below onto paper and use the method of Example 6 to create an Escher-style tessellation art piece. Explain how you can change the sides of the isosceles triangle to form a new shape that tessellates.

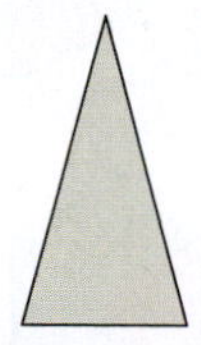

38. **Tessellations with Regular Pentagons** What is the measure of the marked obtuse angle in the figure? Is it possible to use a regular pentagon and one other regular polygon to tessellate a plane? Explain.

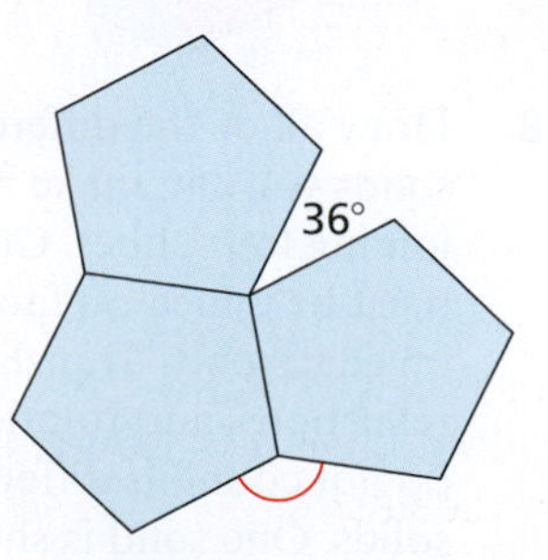

Activity: Visualizing Three-Dimensional Objects

Materials:

- Pencil
- Graph paper
- Isometric dot paper

Learning Objective:

Visualize three-dimensional objects and draw them from different views. Determine whether three-dimensional objects are the same or different.

Name ______________________________

Draw the top, front, and side views of the stack of cubes.

1.

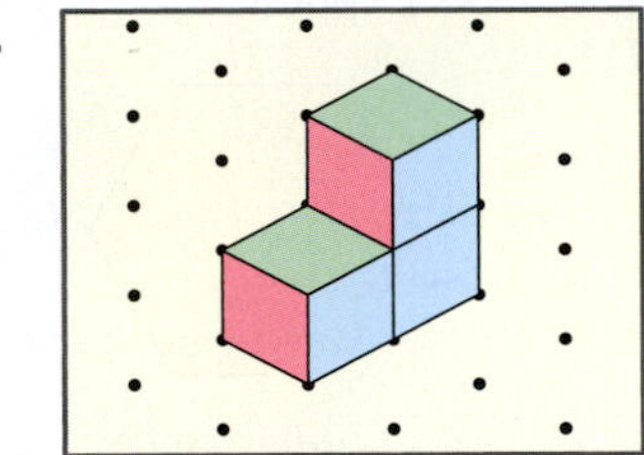

Top Front Side

2.

3.

4.

5.

6.

7.

8. Draw all of the different solids you can make by joining four cubes. Cubes must be joined on faces, not on edges only. Translations, reflections, and rotations do not count as different solids. One solid is shown at the right.

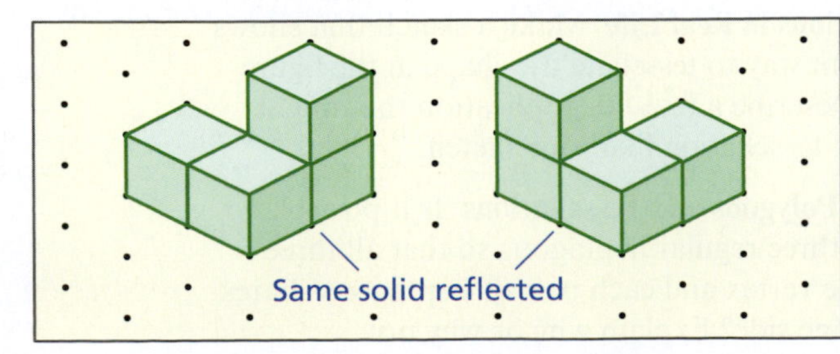

10.4 Polyhedra

Standards

Grades K–2
Geometry
Students should identify and describe shapes, and analyze, compare, create, and compose shapes.

- Classify prisms and pyramids.
- Use Euler's Formula to determine the number of vertices, edges, and faces for a polyhedron.
- Classify regular and semiregular polyhedra.

Prisms and Pyramids

Definitions of Prisms and Pyramids

A **three-dimensional figure**, or solid, has length, width, and depth. A **polyhedron** is a three-dimensional solid whose **faces** are all polygons. The plural of polyhedron is *polyhedra*.

A **prism** is a polyhedron that has two parallel, congruent **bases**. The **lateral faces** are parallelograms.

A **pyramid** is a polyhedron that has one **base**. The **lateral faces** are triangles.

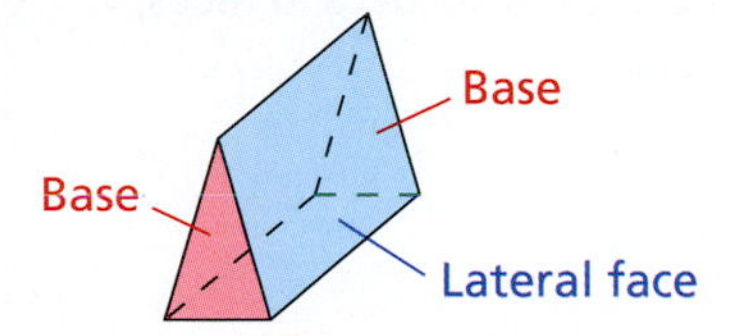

Triangular prism

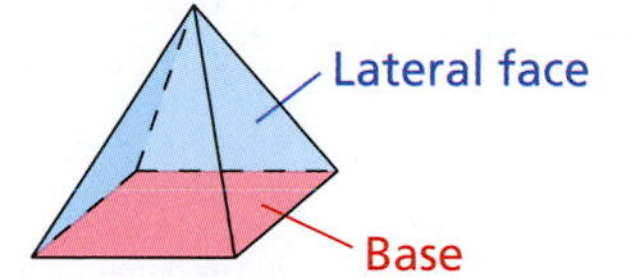

Rectangular pyramid

The shape of the base tells the name of the prism or the pyramid.

EXAMPLE 1 **Identifying Prisms and Pyramids**

Which of the wooden objects are prisms? Which are pyramids? Which are other types of solids?

SOLUTION

Three of the objects are prisms or pyramids. The other three objects are a cone, a cylinder, and a sphere.

Rectangular prism	**Triangular prism**	**Square pyramid**
2 rectangular bases	2 triangular bases	1 square base
4 rectangular lateral faces	3 rectangular lateral faces	4 triangular lateral faces

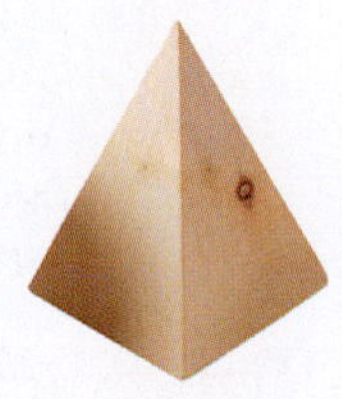

Classroom Tip

To make the drawings of polygons and polyhedra easier, you can use dot paper, either isometric or rectangular.

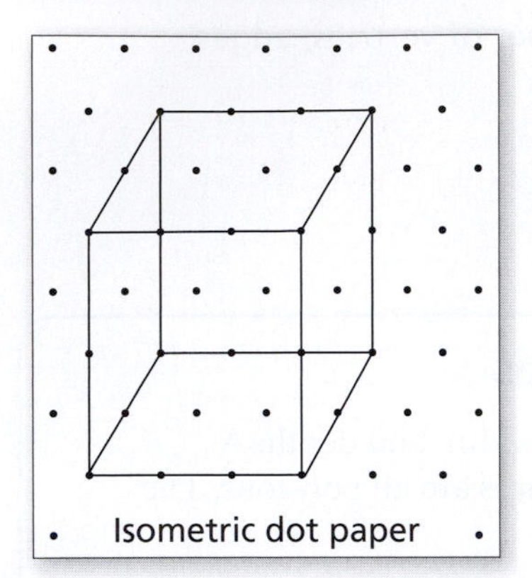
Isometric dot paper

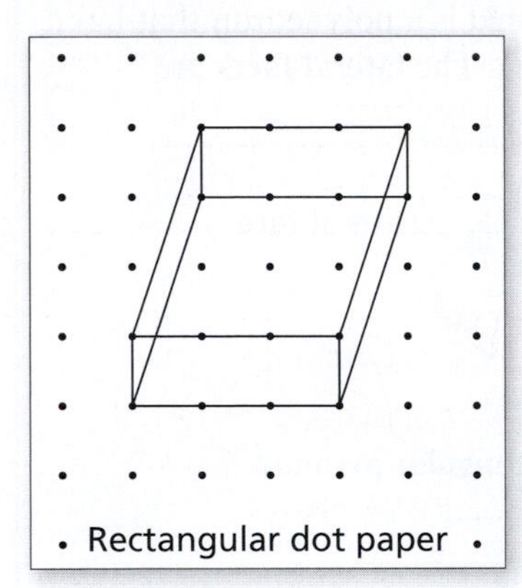
Rectangular dot paper

Using Euler's Formula

EXAMPLE 2 Drawing a Rectangular Prism

a. Draw a rectangular prism.

b. Count the numbers of vertices, faces, and edges of the prism.

SOLUTION

a. There are several ways to do this. One method is shown.

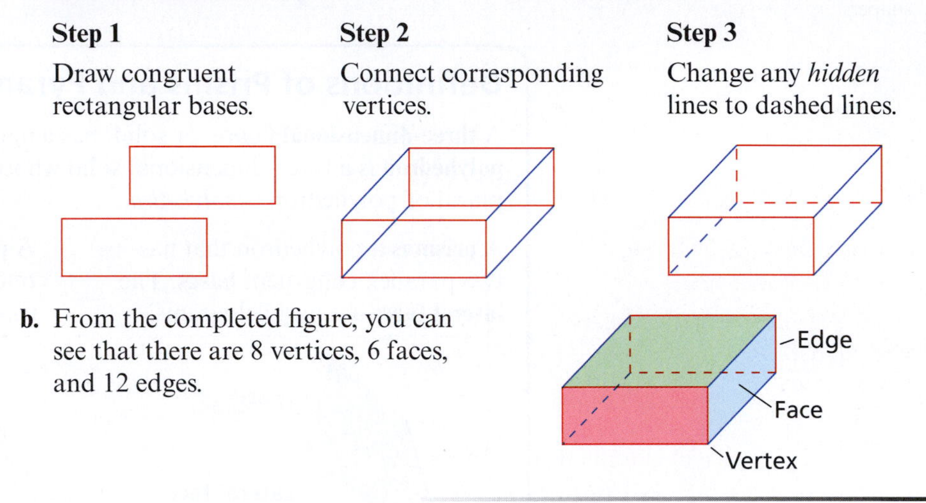

Step 1 Draw congruent rectangular bases.

Step 2 Connect corresponding vertices.

Step 3 Change any *hidden* lines to dashed lines.

b. From the completed figure, you can see that there are 8 vertices, 6 faces, and 12 edges.

The famous Swiss mathematician Leonhard Euler noticed that for any polyhedron, the sum of the numbers of vertices and faces was always two more than the number of edges. The formula for this relationship bears his name.

Euler's Formula

For any polyhedron, the sum of the numbers of vertices and faces is two more than the number of edges.

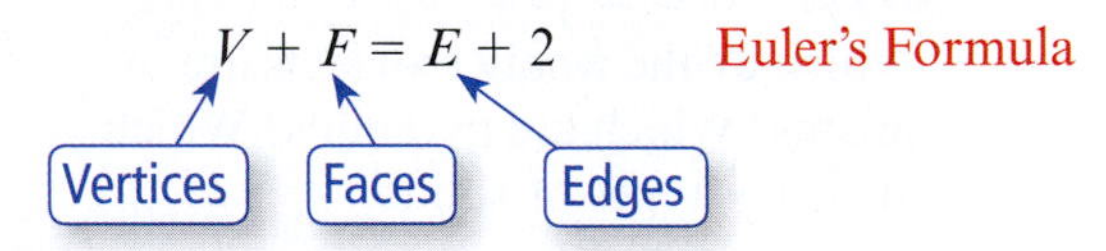

$V + F = E + 2$ Euler's Formula

EXAMPLE 3 Using Euler's Formula

The hexagonal prism at the right has 6 lateral faces and 12 vertices. How many edges does it have?

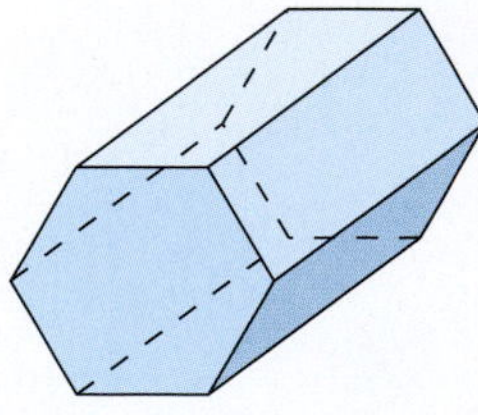

SOLUTION

You are given the number of vertices. The prism has 6 lateral faces and 2 bases for a total of $6 + 2 = 8$ faces.

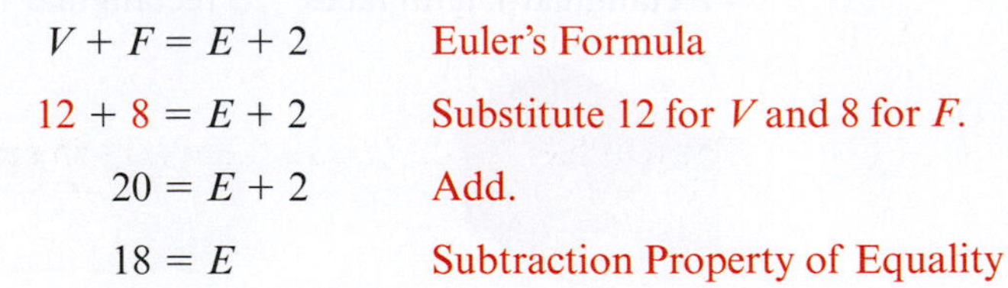

$V + F = E + 2$	Euler's Formula
$12 + 8 = E + 2$	Substitute 12 for V and 8 for F.
$20 = E + 2$	Add.
$18 = E$	Subtraction Property of Equality

The hexagonal prism has 18 edges.

EXAMPLE 4 Verifying Euler's Formula

Your class makes paper models of four different pyramids. Check that Euler's Formula applies for all four pyramids.

Triangular Square Pentagonal Hexagonal

Mathematical Practices

Look for and Express Regularity in Repeated Reasoning

MP8 Mathematically proficient students notice if calculations are repeated and look both for general methods and for shortcuts.

SOLUTION

Pyramid	Vertices, V	Faces, F	Edges, E	$V + F = E + 2$
Triangular	4	4	6	$4 + 4 = 6 + 2$ ✓
Square	5	5	8	$5 + 5 = 8 + 2$ ✓
Pentagonal	6	6	10	$6 + 6 = 10 + 2$ ✓
Hexagonal	7	7	12	$7 + 7 = 12 + 2$ ✓

A Real-Life Example of a Polyhedron

Count the numbers of vertices, faces, and edges in the polyhedral tent.

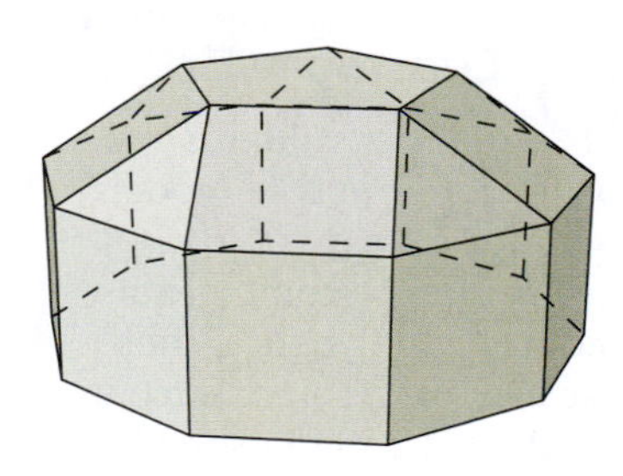

SOLUTION

- The base or floor of the tent is a decagon. It has 10 sides and 10 vertices.
- The sides of the tent are made up of 10 vertical rectangular faces, which includes 10 more vertices.
- The ceiling portion of the tent is made up of 5 rectangular faces, 5 triangular faces, and 1 pentagonal face. So, the ceiling has a total of 11 faces. It also has 5 more vertices that have not been counted.

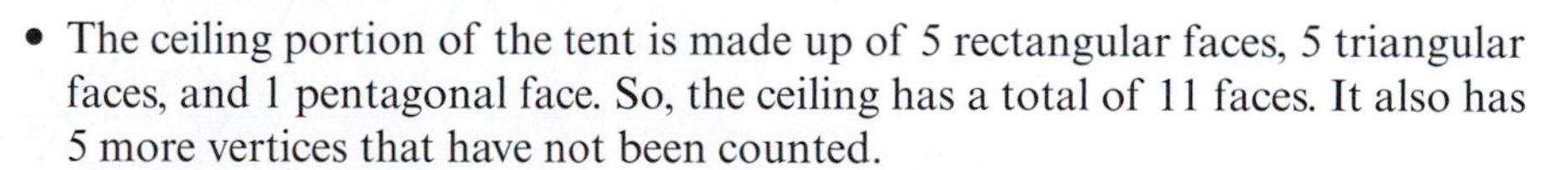

So, the tent has $10 + 10 + 5 = 25$ vertices and $1 + 10 + 5 + 5 + 1 = 22$ faces. Use Euler's formula to find the number of edges.

$V + F = E + 2$	Euler's Formula
$25 + 22 = E + 2$	Substitute 25 for V and 22 for F.
$47 = E + 2$	Add.
$45 = E$	Subtraction Property of Equality

The tent has 25 vertices, 22 faces, and 45 edges.

Regular and Semiregular Polyhedra

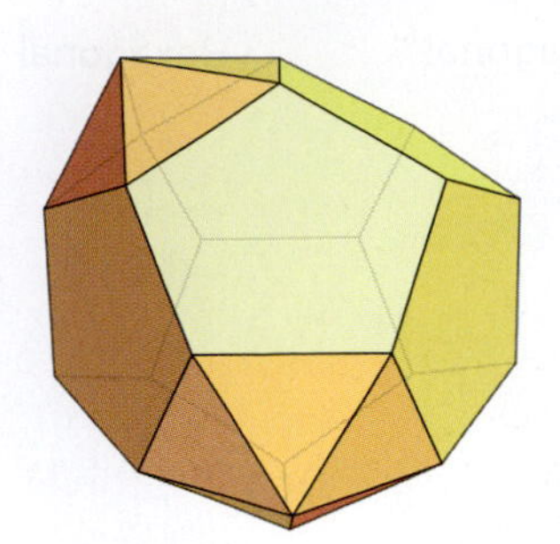

Polyhedron that is not regular

Definition of a Regular Polyhedron

A polyhedron is **regular** when all of its faces are congruent regular polygons and each of its vertices is formed by the same number of faces. A regular polyhedron is called a *Platonic solid*. There are five possible regular polyhedra.

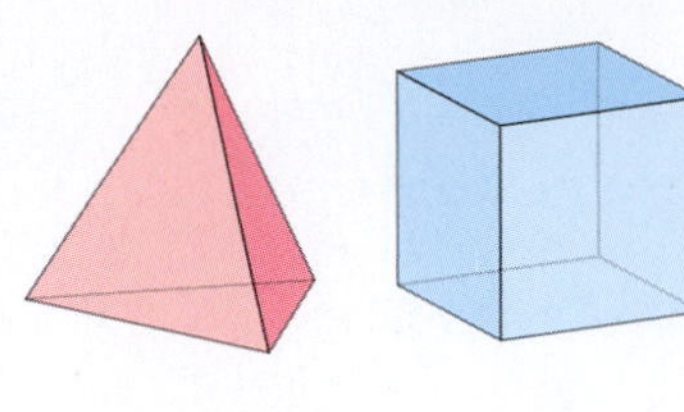

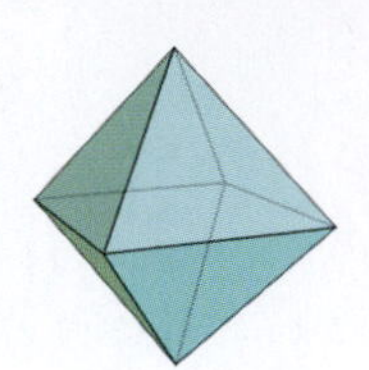

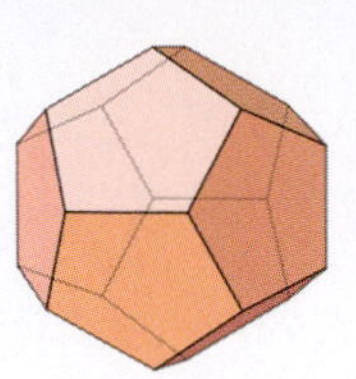

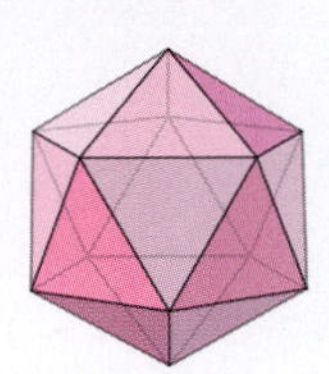

Tetrahedron (4 faces)	Cube (6 faces)	Octahedron (8 faces)	Dodecahedron (12 faces)	Icosahedron (20 faces)

Although the five Platonic solids are named after the Greek philosopher Plato, they were known long before his time. They are named after Plato because they were featured prominently in his philosophy regarding the five basic elements of the world: earth, air, water, fire, and universe.

EXAMPLE 6 **Folding a Net**

A net is a pattern that can be cut out and folded to make a polyhedron. What type of polyhedron does each net make?

SOLUTION

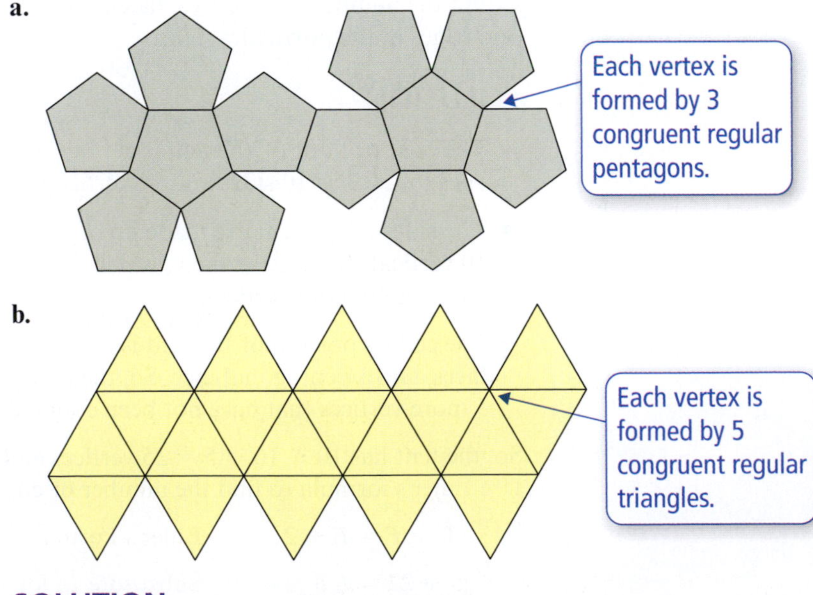

Classroom Tip

Younger students benefit from tactile models of polyhedra. If you do not have access to commercial models of polyhedra, then use everyday objects such as toothpicks and marshmallows to construct simple polyhedra.

SOLUTION

a. Dodecahedron

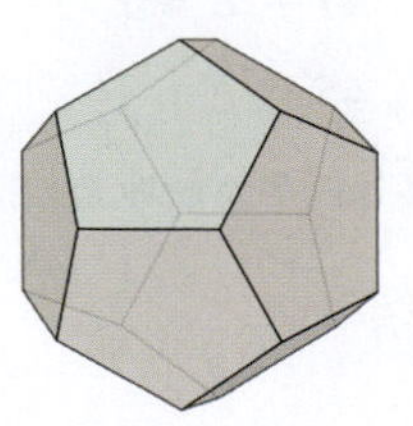

b. Icosahedron

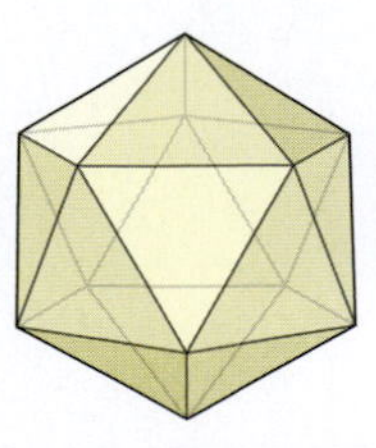

Definition of a Semiregular Polyhedron

A polyhedron is **semiregular** when all of its faces are regular polygons and each of its vertices is formed by the same types of faces. A semiregular polyhedron is called an *Archimedean solid*. There are 13 possible semiregular polyhedra.

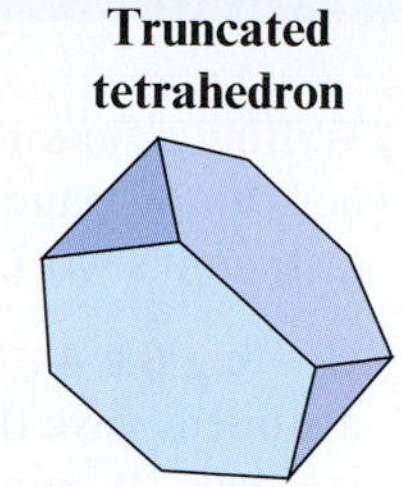

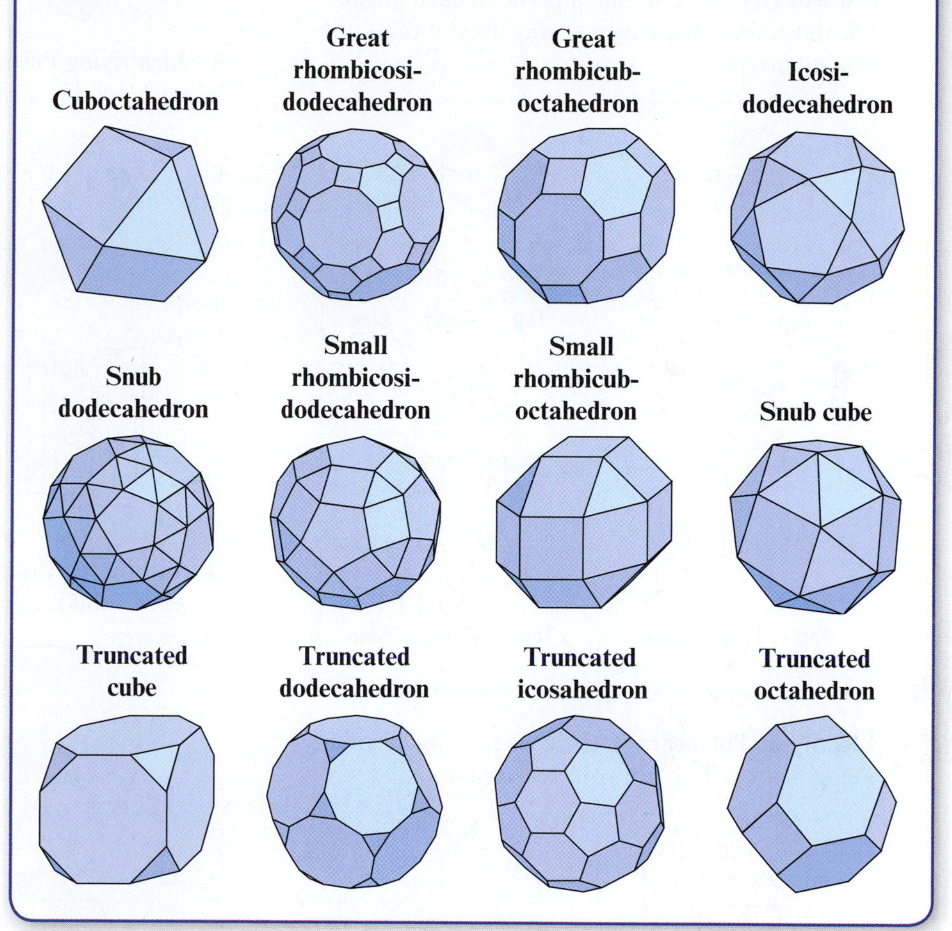

EXAMPLE 7 A Real-Life Semiregular Polyhedron

A soccer ball is spherical, so it is not a polyhedron. However, if it contained flat faces, which of the semiregular polyhedra would it be?

SOLUTION

The polyhedron would be a *truncated icosahedron*. There are 12 (black) regular pentagonal faces and 20 (white) regular hexagonal faces. Each of its vertices would be formed by one pentagon and two hexagons.

10.4 Exercises

Free step-by-step solutions to odd-numbered exercises at *MathematicalPractices.com.*

1. Writing a Solution Key Write a solution key for the activity on page 396. Describe a rubric for grading a student's work.

2. Grading the Activity In the activity on page 396, a student gave the answers below. Each question is worth 10 points. Assign a grade to each answer. For those that are incorrect, why do you think the student erred?

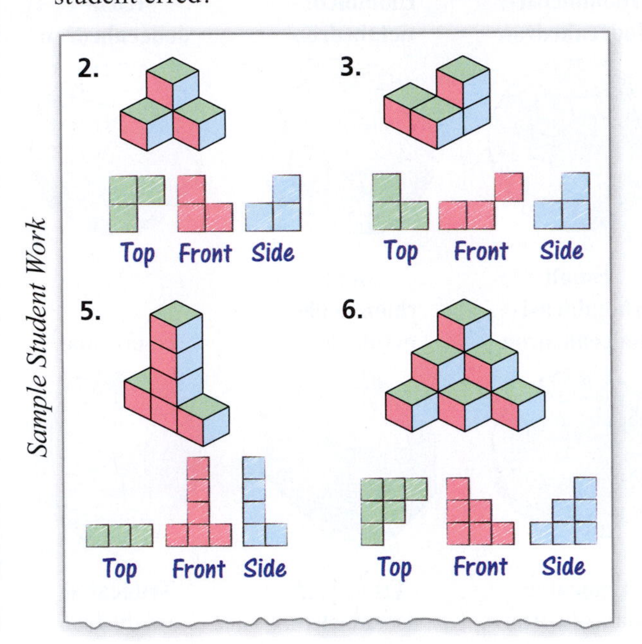

3. Identifying Polyhedra Decide whether each solid is a polyhedron. Explain your reasoning.

a.
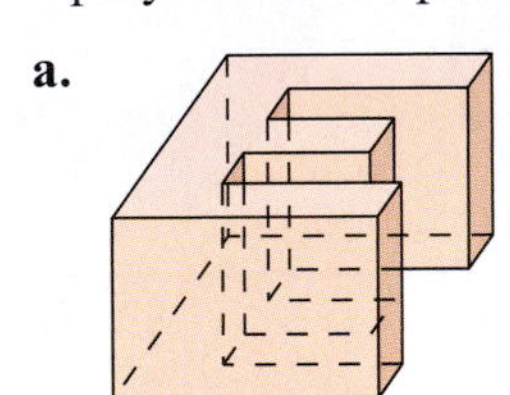

b.
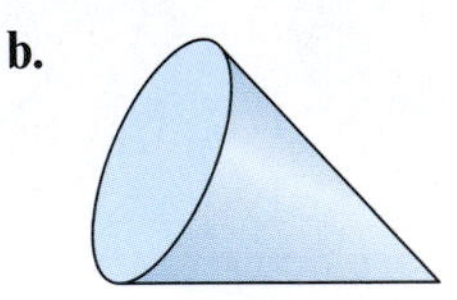

c.
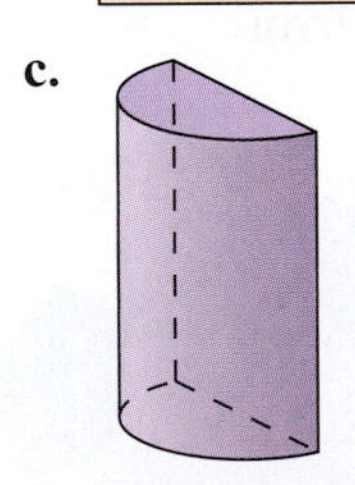

d.
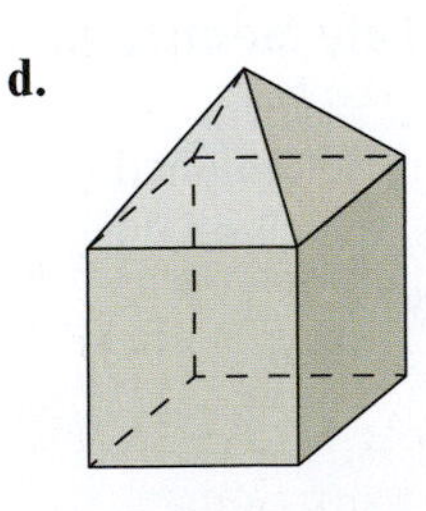

4. Identifying Polyhedra Decide whether each solid is a polyhedron. Explain your reasoning.

a.
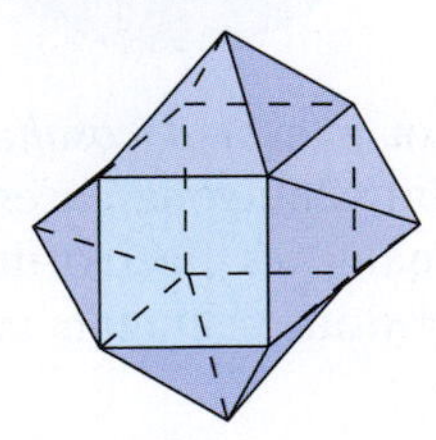

b.
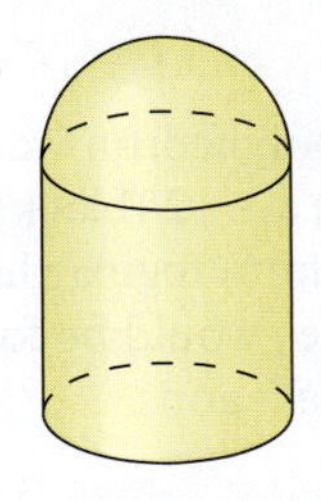

c.
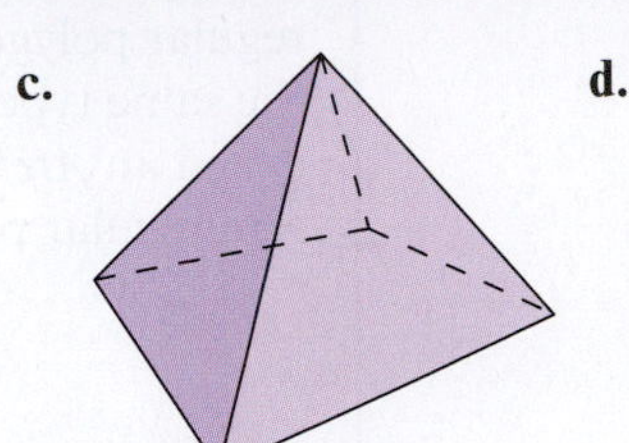

d.

5. Identifying Prisms and Pyramids Decide whether each solid is a *prism*, *pyramid*, or *neither*.

a.
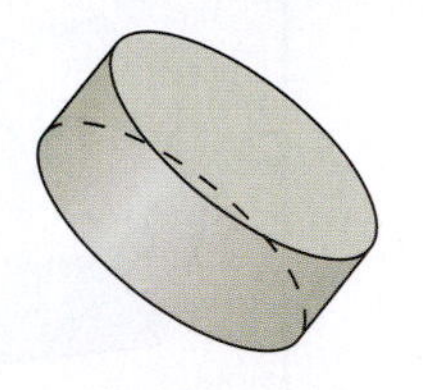

b.
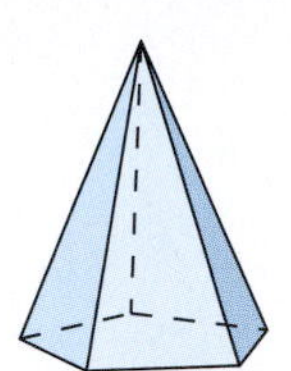

c.

d.
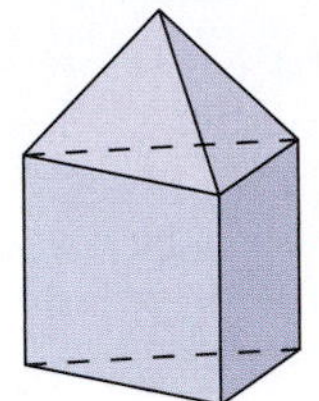

6. Identifying Prisms and Pyramids Decide whether each solid is a *prism*, *pyramid*, or *neither*.

a.
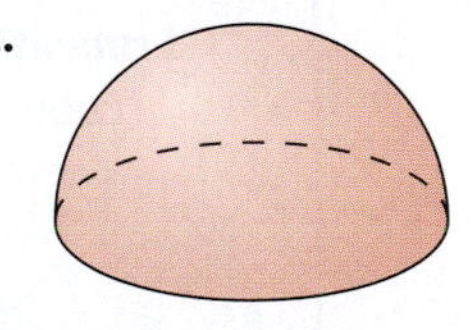

b.
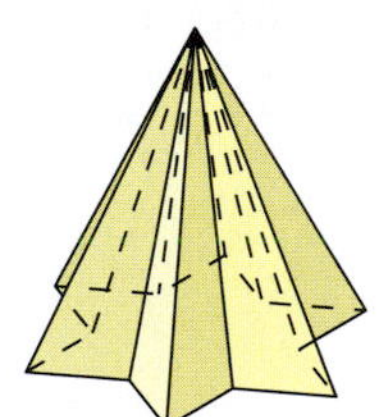

c.
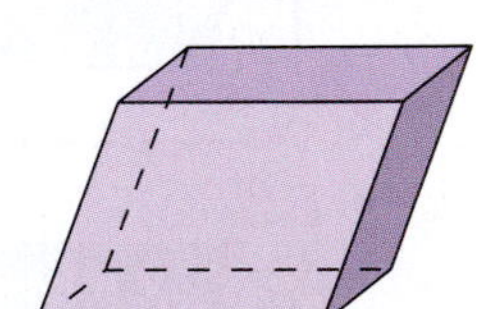

d.
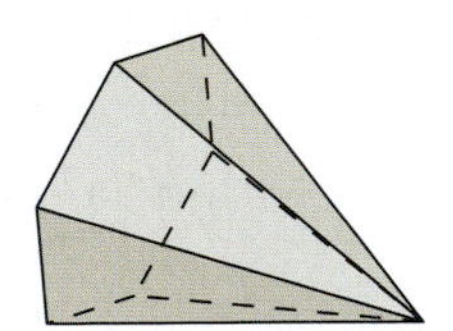

7. Naming Prisms and Pyramids Name each prism or pyramid.

a.
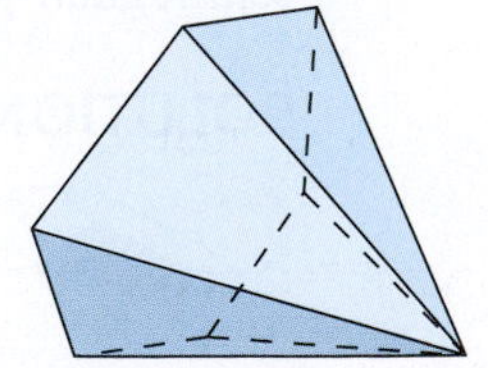

b.
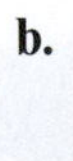
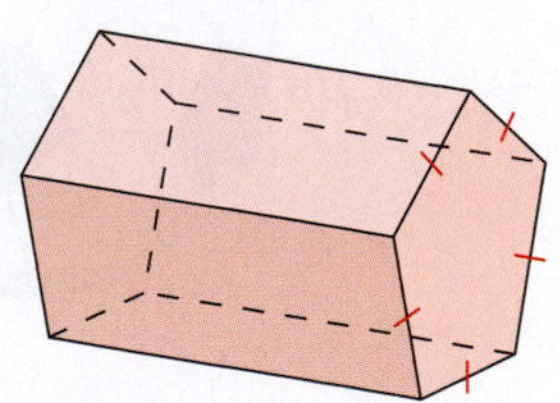

8. Naming Prisms and Pyramids Name each prism or pyramid.

a.

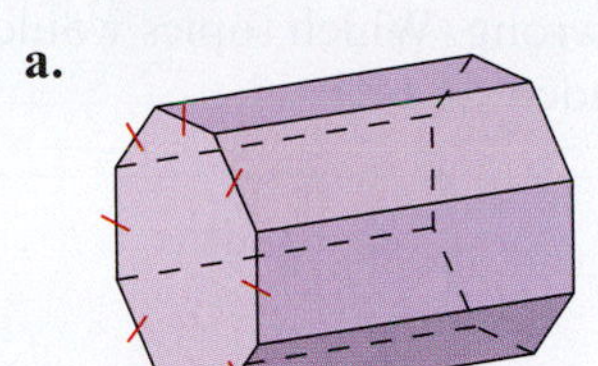

b.

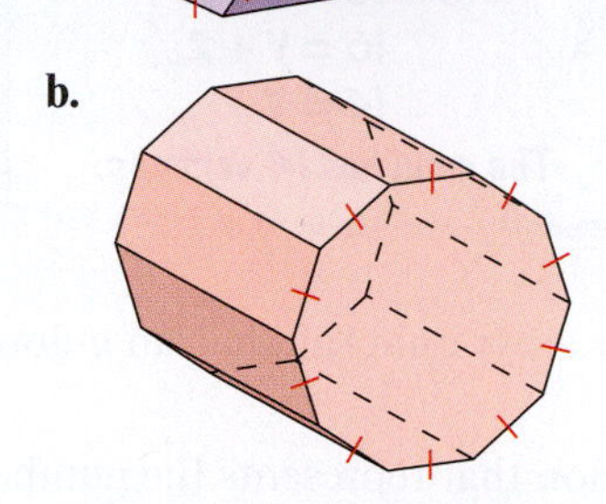

9. Drawing a Prism Draw a hexagonal prism with irregular bases. Then count the number of vertices, faces, and edges of the prism.

10. Drawing a Prism Draw a prism with bases that are trapezoids. Then count the number of vertices, faces, and edges of the prism.

11. Using Euler's Formula Find the missing number of vertices, faces, or edges for each polyhedron.

a. Pentagonal pyramid

5 lateral faces, 6 vertices, ____ edges

b. Octagonal prism

8 lateral faces, ____ vertices, 24 edges

c. Polyhedron

____ faces, 20 vertices, 30 edges

12. Using Euler's Formula Find the missing number of vertices, faces, or edges for each polyhedron.

a. Decagonal prism

10 lateral faces, 20 vertices, ____ edges

b. Pyramid with a base that is a kite

4 lateral faces, ____ vertices, 8 edges

c. Polyhedron

____ faces, 10 vertices, 18 edges

13. Archimedian Solids Refer to the Archimedian solids on page 401.

a. How many solids have faces that are triangles?

b. Name the solids from part (a).

14. Archimedian Solids Refer to the Archimedian solids on page 401.

a. How many solids have faces that are hexagons?

b. Name the solids from part (a).

15. Folding Nets Classify the polyhedron made by each net as *regular* or *semiregular*. Name the polyhedron.

a.

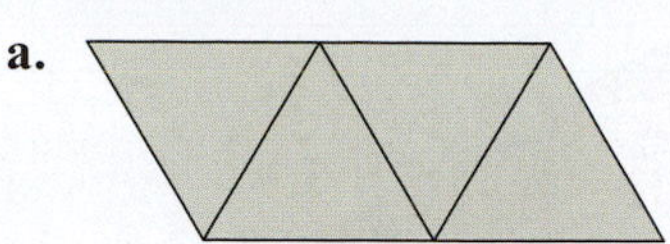

b.

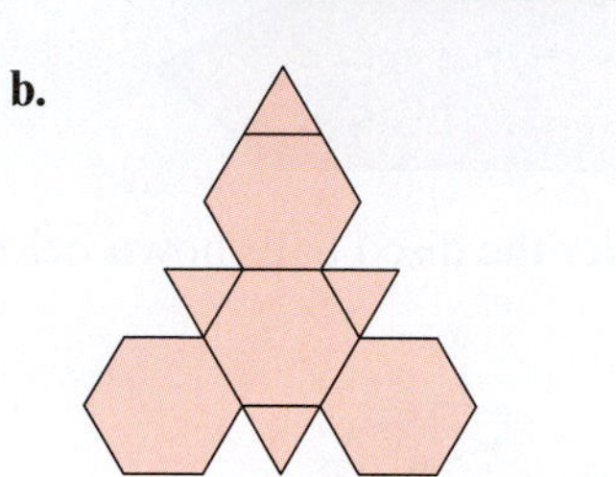

16. Folding Nets Classify the polyhedron made by each net as *regular* or *semiregular*. Name the polyhedron.

a.

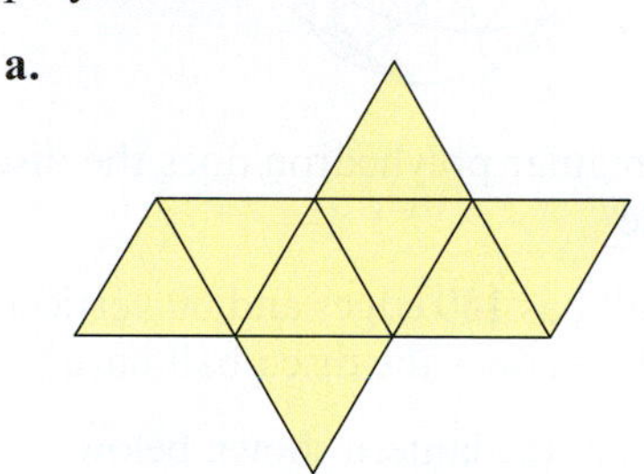

b.

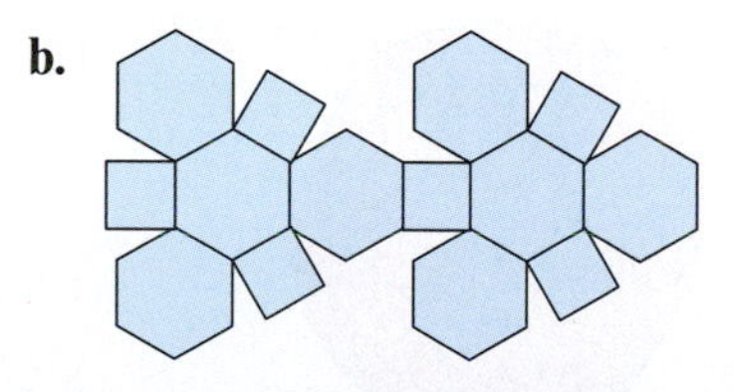

17. True or False? Decide whether each statement is *true* or *false*. Explain your reasoning.

a. A cube has 12 congruent faces.

b. A triangular prism has 9 edges.

c. The net of an icosahedron can be tessellated.

d. Any face of a prism has an opposite face that is parallel to it.

18. True or False? Decide whether each statement is *true* or *false*. Explain your reasoning.

a. A triangular prism has five triangular faces.

b. An octahedron is composed of two square pyramids.

c. The net of a dodecahedron can be tessellated.

d. Every tetrahedron has 6 congruent edges.

19. Ice Cube Trays An ice cube tray makes ice cubes whose lateral faces are slanted outward on four sides for easier removal from the tray. Are the ice cubes *prisms*, *pyramids*, or *neither*? Explain your reasoning.

20. Skateboarding What type of solid does the skateboarding ramp shown below resemble? How many faces, edges, and vertices does the ramp have?

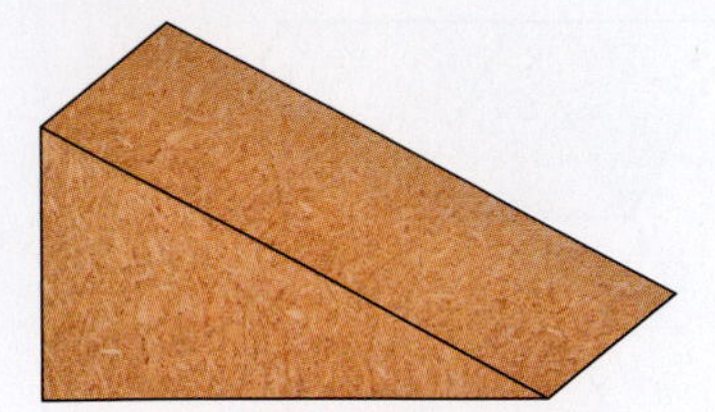

21. Disco Balls Consider the disco ball shown below.

a. Which semiregular polyhedron does the disco ball resemble?

b. The disco ball has 150 edges and 60 vertices. How many faces does the disco ball have?

22. Lanterns Consider the lantern shown below.

a. Which semiregular polyhedron does the lantern resemble?

b. The lantern has 90 edges and 60 vertices. How many faces does the lantern have?

23. Writing Explain how a prism and a pyramid with congruent bases are alike and how they are different.

24. Writing Explain how a rectangular pyramid and a triangular pyramid are alike and how they are different.

25. Grading Student Work On a diagnostics test, one of your students does the following work. Explain what the student did wrong. Which topics would you encourage the student to review?

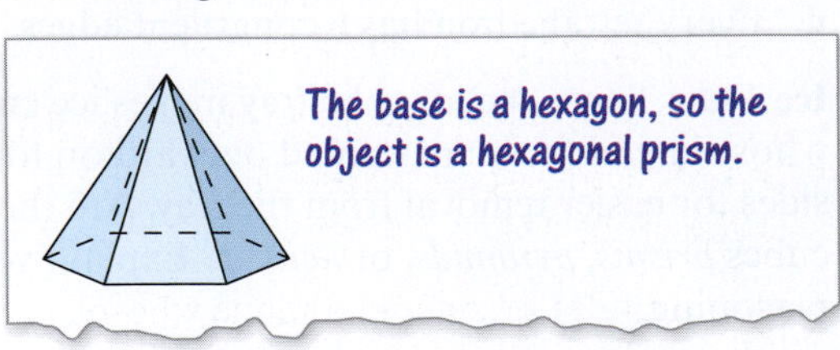

26. Grading Student Work On a diagnostics test, one of your students does the following work. Explain what the student did wrong. Which topics would you encourage the student to review?

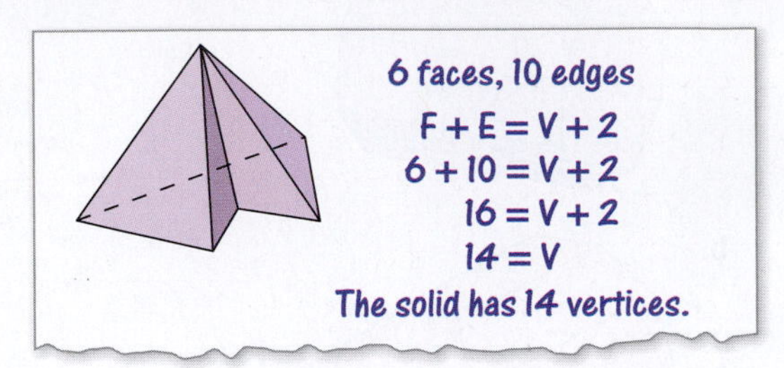

27. Reasoning Consider a pyramid that has an n-gon as a base.

a. Write an expression that represents the number of faces the pyramid has.

b. Write an expression that represents the number of vertices the pyramid has.

c. Write an expression that represents the number of edges the pyramid has.

d. Use your answers in parts (a), (b), and (c) to verify Euler's Formula for all pyramids.

28. Reasoning Consider a prism that has n-gons as bases.

a. Write an expression that represents the number of faces the prism has.

b. Write an expression that represents the number of vertices the prism has.

c. Write an expression that represents the number of edges the prism has.

d. Use your answers in parts (a), (b), and (c) to verify Euler's Formula for all prisms.

29. Activity: In Your Classroom Use the net below.

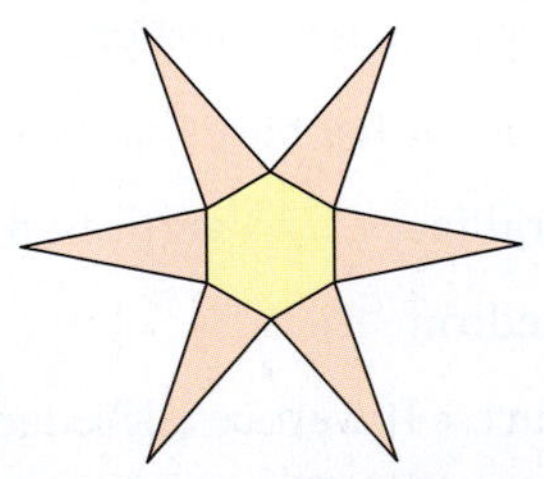

a. Trace the net onto a piece of plain paper. Then cut out the net. Count the number of faces.

b. Fold the paper to make a polyhedron. Which polyhedron is made?

c. Count the number of vertices and edges. Does Euler's Formula apply to this polyhedron?

d. Describe the concepts your students will practice and discover by doing this activity.

30. **Activity: In Your Classroom** Many common items can be used to construct polyhedra. For instance, in the image below, plastic drinking straws were used as the edges of a cuboctahedron.

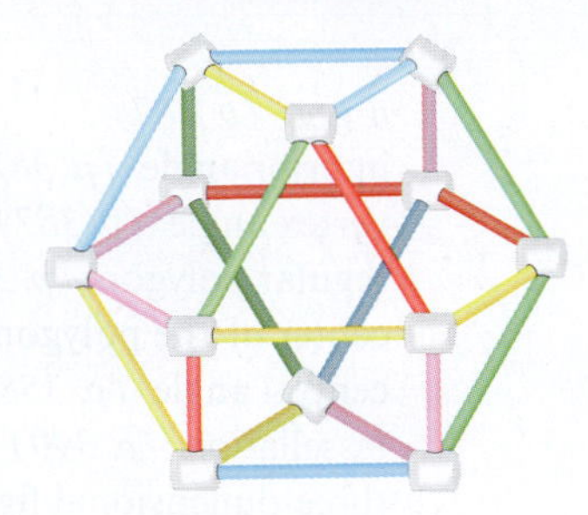

a. List two other common objects that could be used as edges in the construction of a polyhedron.

b. Describe the concepts your students will practice and discover by constructing a polyhedron in this way.

31. **Writing** Explain why there are only five possible regular polyhedra.

32. **Writing** Describe how you could divide a cube into six congruent square pyramids.

33. **Folding a Net** Which solid is made when the net is folded?

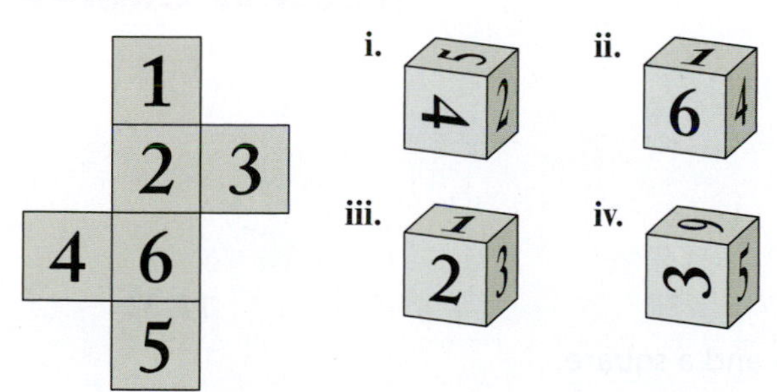

34. **Folding a Net** Which solid is made when the net is folded?

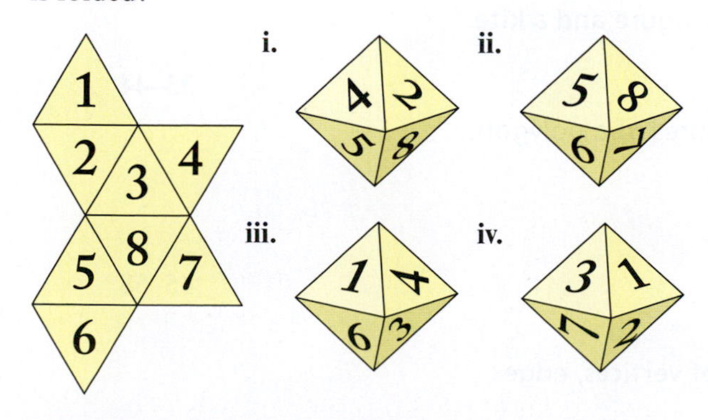

35. **Sketching a Solid** Sketch the solid that has the given views. Then name the solid.

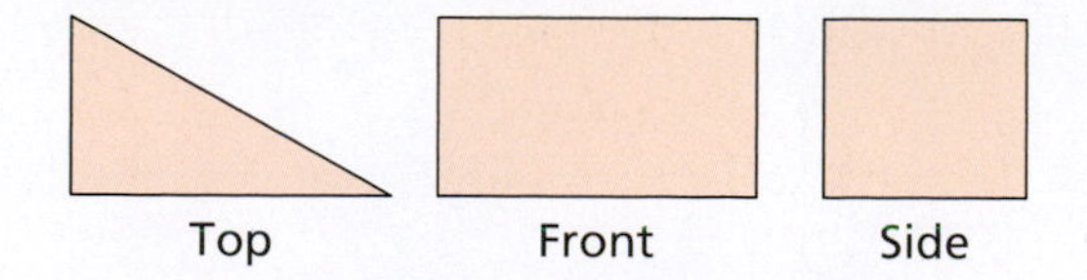

36. **Sketching a Solid** Sketch the solid that has the given views. Then name the solid.

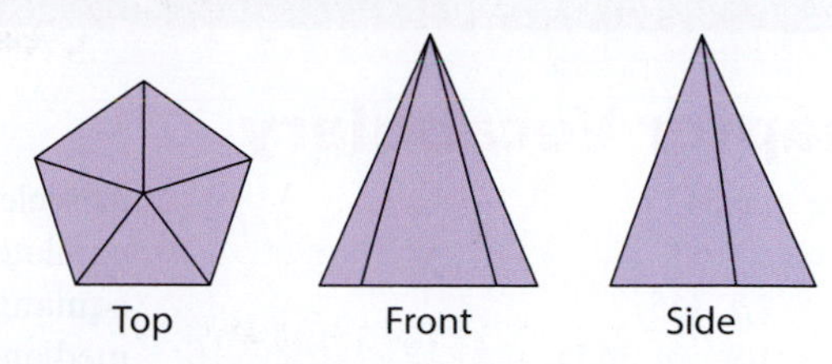

37. **Sketching Solids** Sketch the top, front, and side views of each solid.

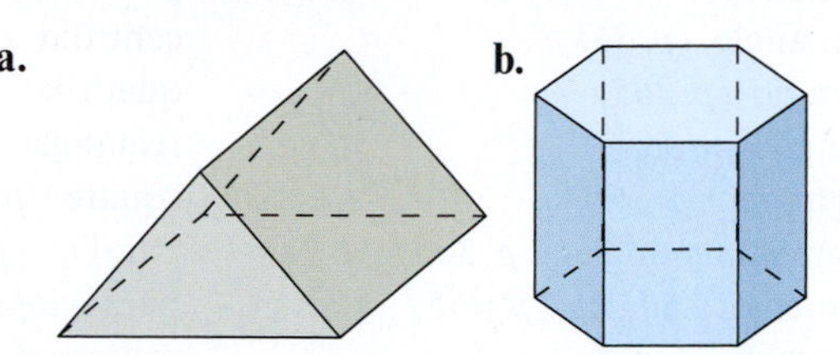

38. **Sketching Solids** Sketch the top, front, and side views of each solid.

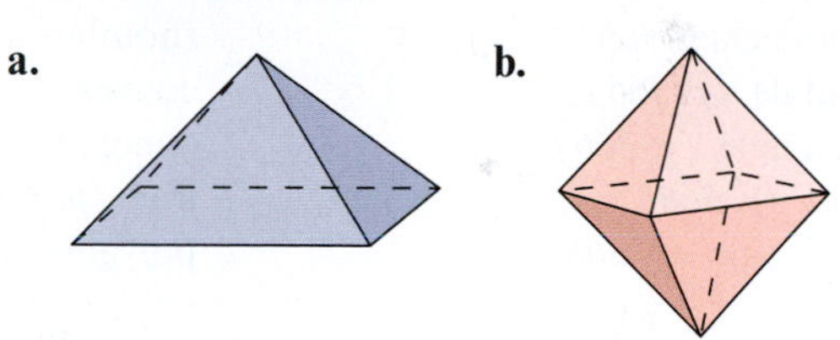

39. **Polyhedra and Planes** Classify each polygon formed by the intersection of a polyhedron and a plane.

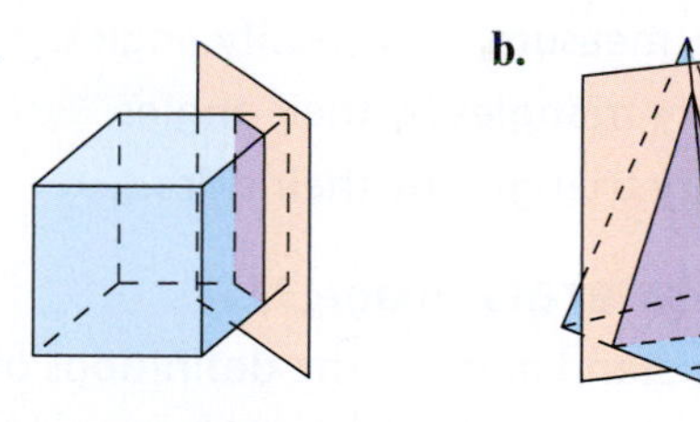

40. **Polyhedra and Planes** Classify each polygon formed by the intersection of a polyhedron and a plane.

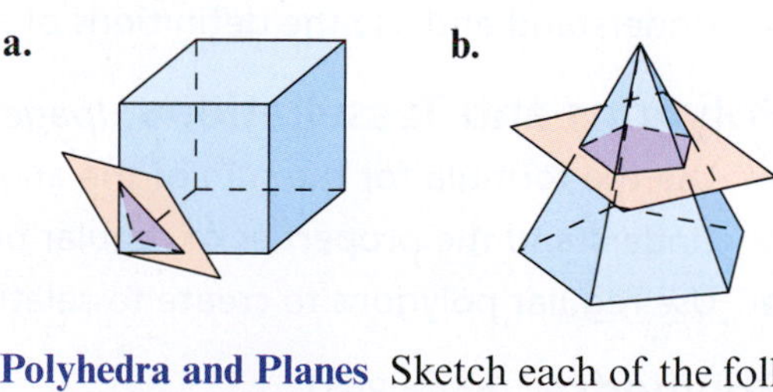

41. **Polyhedra and Planes** Sketch each of the following.

a. A square pyramid and a plane whose intersection is a trapezoid.

b. A triangular prism and a plane whose intersection is a pentagon.

42. **Polyhedra and Planes** Sketch each of the following.

a. A cube and a plane whose intersection is a hexagon.

b. A triangular prism and a plane whose intersection is a trapezoid.

Chapter Summary

Chapter Vocabulary

angle *(p. 365)*
sides *(p. 365)*
vertex *(p. 365)*
protractor *(p. 365)*
degrees (°) *(p. 365)*
degree measure *(p. 365)*
straight angle *(p. 365)*
acute angle *(p. 365)*
right angle *(p. 365)*
obtuse angle *(p. 365)*
complementary angles *(p. 365)*
supplementary angles *(p. 365)*
closed figure *(p. 366)*
triangle *(p. 366)*
vertices *(p. 366)*
acute triangle *(p. 366)*
right triangle *(p. 366)*
obtuse triangle *(p. 366)*
congruent *(p. 368)*
scalene triangle *(p. 368)*
isosceles triangle *(p. 368)*
equilateral triangle *(p. 368)*
equiangular *(p. 368)*
median *(p. 369)*
midpoint *(p. 369)*
point of concurrency *(p. 369)*
centroid *(p. 369)*
quadrilateral *(p. 375)*
rectangle *(p. 375)*
square *(p. 375)*
parallel *(p. 376)*
parallelogram *(p. 376)*
trapezoid *(p. 379)*
right trapezoid *(p. 379)*
isosceles trapezoid *(p. 379)*
rhombus *(p. 379)*
convex *(p. 380)*
concave *(p. 380)*
kite *(p. 380)*
polygon *(p. 387)*
n-gon *(p. 387)*
interior angle *(p. 387)*
vertex angle *(p. 387)*
regular polygon *(p. 388)*
center of the polygon *(p. 388)*
central angle *(p. 388)*
tessellation *(p. 390)*
three-dimensional figure *(p. 397)*
polyhedron *(p. 397)*
faces *(p. 397)*
prism *(p. 397)*
bases of a prism *(p. 397)*
lateral faces of a prism *(p. 397)*
pyramid *(p. 397)*
base of a pyramid *(p. 397)*
lateral faces of a pyramid *(p. 397)*
regular polyhedron *(p. 400)*
semiregular polyhedron *(p. 401)*

Chapter Learning Objectives — Review Exercises

10.1 Triangles *(page 365)* — 1–16

- Draw, measure, and classify angles.
- Classify triangles by their angles.
- Classify triangles by their sides.

10.2 Quadrilaterals *(page 375)* — 17–32

- Understand and use the definitions of a rectangle and a square.
- Understand and use the definition of a parallelogram.
- Understand and use the definitions of a trapezoid and a rhombus.
- Understand and use the definitions of a convex figure and a kite.

10.3 Polygons and Tessellations *(page 387)* — 33–44

- Use the formula for the sum of the angle measures of a polygon.
- Understand the properties of regular polygons.
- Use regular polygons to create tesselations.

10.4 Polyhedra *(page 397)* — 45–58

- Classify prisms and pyramids.
- Use Euler's formula to determine the number of vertices, edges, and faces for a polyhedron.
- Classify regular and semiregular polyhedra.

Review Exercises

Free step-by-step solutions to odd-numbered exercises at *MathematicalPractices.com.*

10.1 Triangles

Classifying an Angle Classify the angle as *straight*, *acute*, *right*, or *obtuse*. Explain your reasoning.

1.

2.

3. 

4. 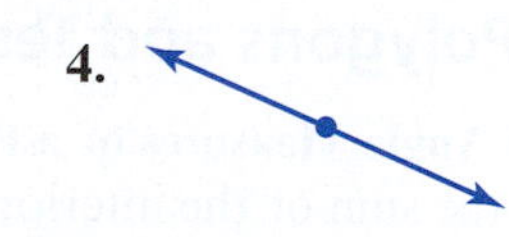

Drawing an Angle Use a protractor to sketch an angle with the given measure.

5. 15°

6. 122°

7. 173°

8. 81°

9. **Finding a Complementary Angle** Use a protractor to measure the angle. Then find the complement of the angle.

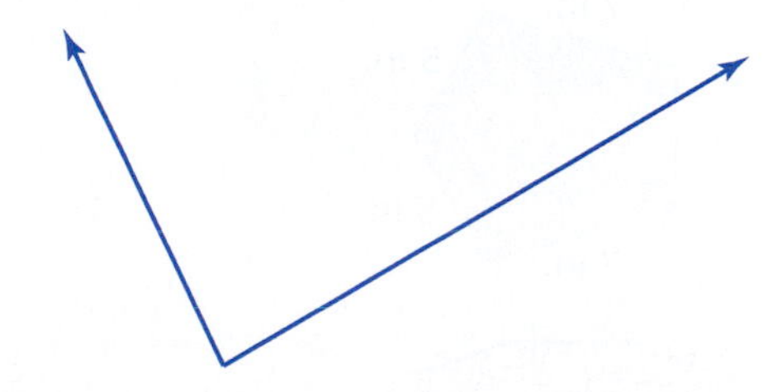

10. **Finding a Supplementary Angle** Use a protractor to measure the angle. Then find the supplement of the angle.

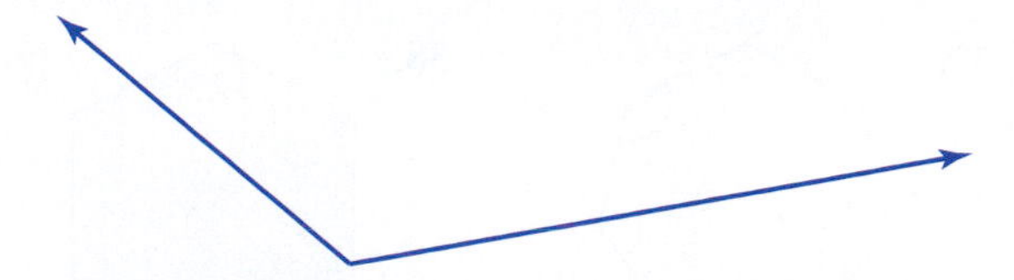

Classifying a Triangle Find the value of x for the triangle. Then classify the triangle in as many ways as possible.

11. 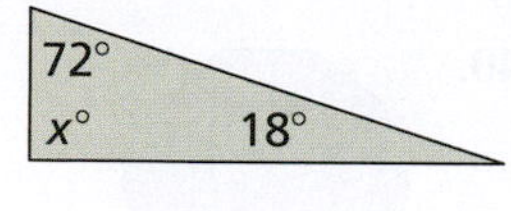

12. 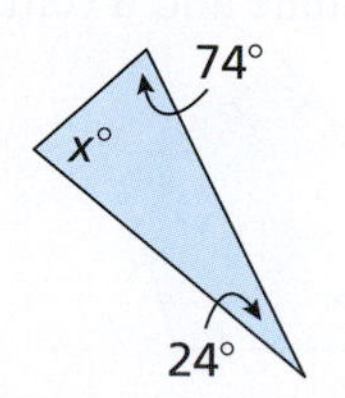

13. 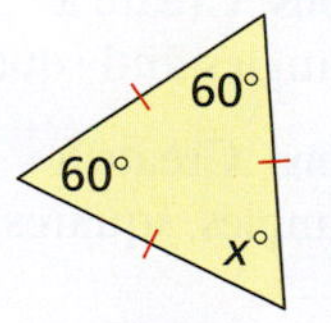

14. 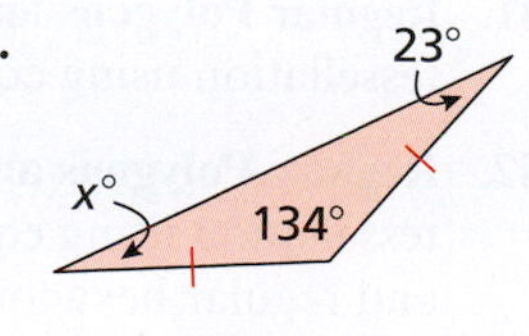

15. **Telling Time** The clock below reads exactly 20 seconds past 6 o'clock.

a. Classify the angle between the minute hand and the second hand.

b. Classify the angle between the hour hand and the second hand.

c. What special relationship exists between the two angles in parts (a) and (b)?

16. **Finding Angle Measures in Real Life** A cruise ship sails out of Fort Lauderdale, Florida, and docks at Freeport for two days. The ship then sails to Nassau, where it docks for two more days before returning to Fort Lauderdale.

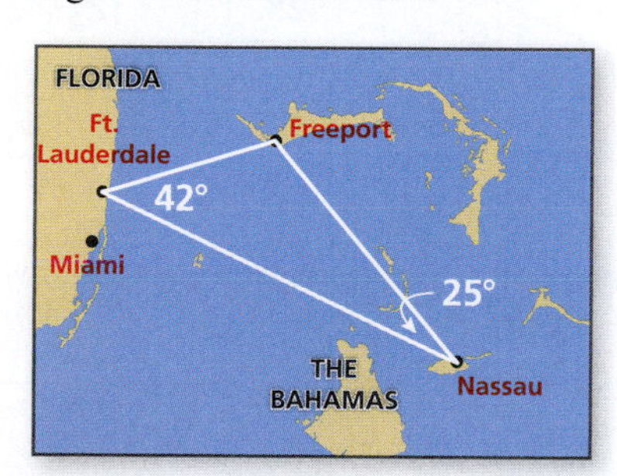

a. Find the missing angle measure.

b. Classify the triangle.

10.2 Quadrilaterals

Classifying a Quadrilateral Classify the quadrilateral in as many ways as possible.

17. 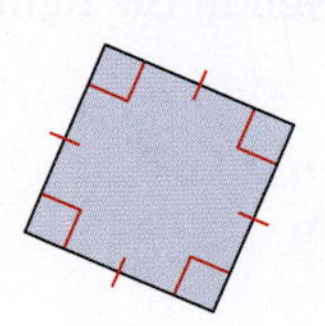

18. 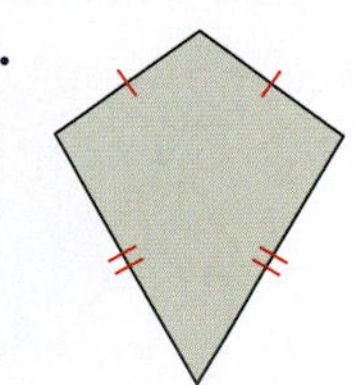

19. 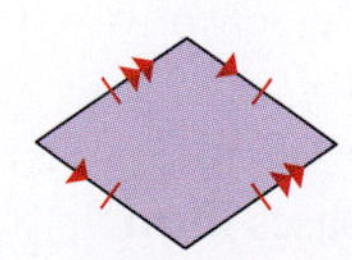

20. 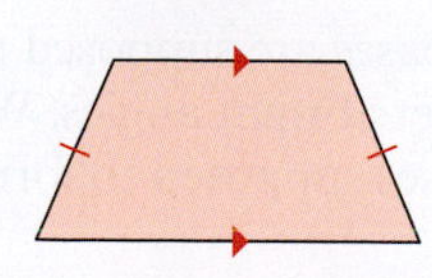

Convex or Concave? Tell whether the figure is *concave* or *convex*.

21.

22.

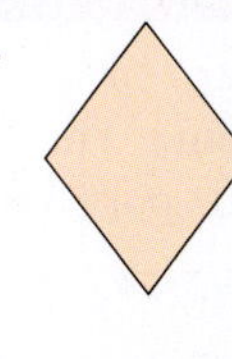

Rectangles Find the value of x for the rectangle.

23.

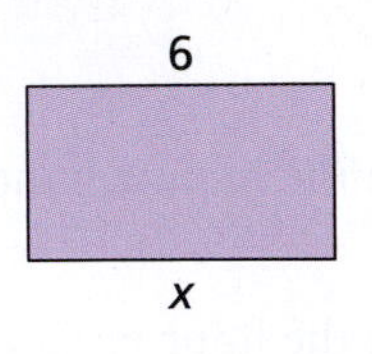

24.

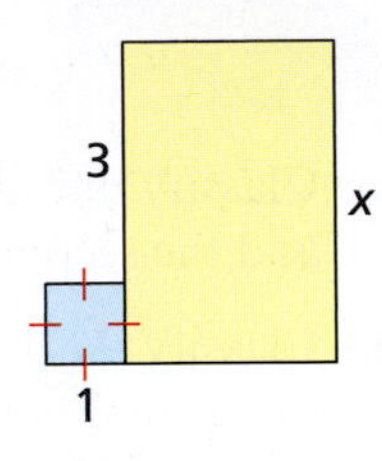

Angle Measures of a Quadrilateral Find the value of x for the quadrilateral.

25.

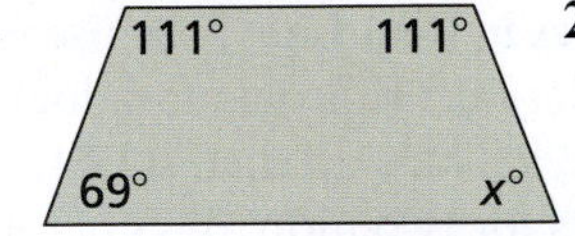

26.

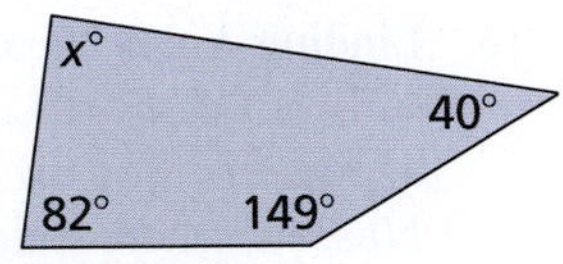

27.

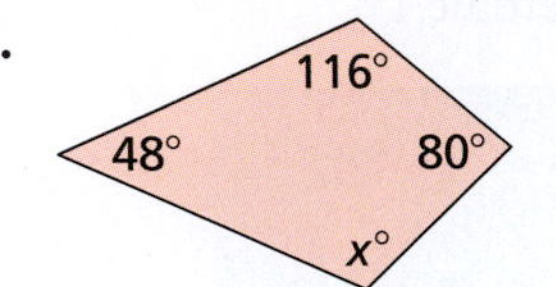

28.

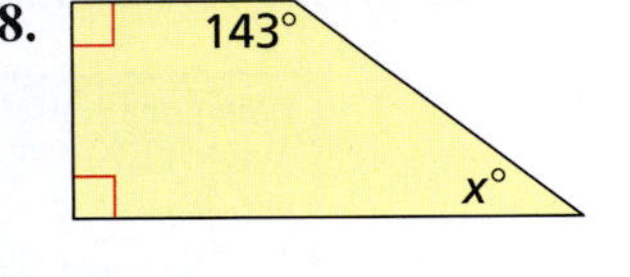

Angle Measures of a Parallelogram Find the value of x for the parallelogram.

29.

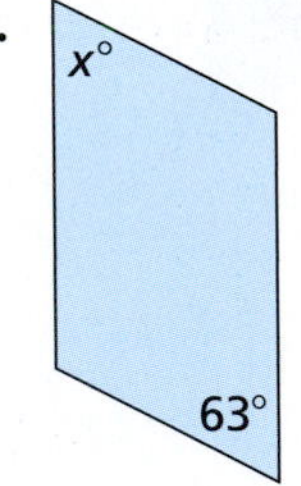

30.

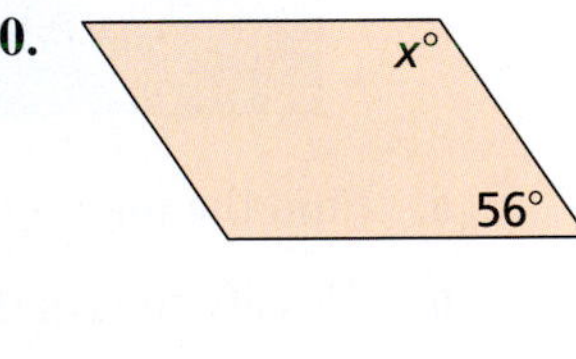

31. Sandlot Baseball The distances between bases on a sandlot baseball field are given in the figure.

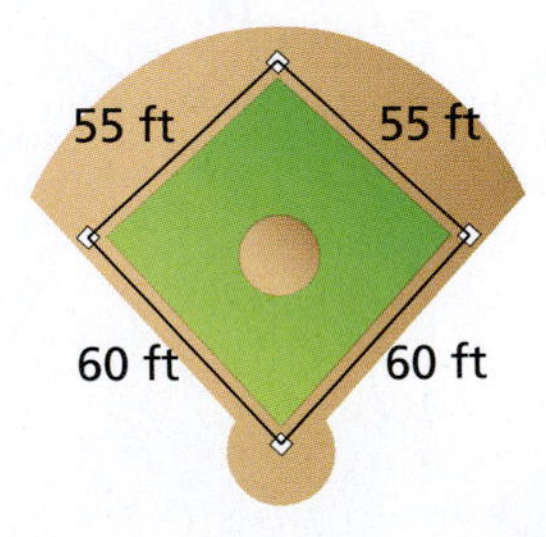

a. What quadrilateral do the bases form?

b. The distances between bases are supposed to be equal and should meet at right angles. What quadrilateral are the bases supposed to form?

32. Swimming Pool Design The bottom of an in ground swimming pool descends gradually as shown in the cross-section of the pool.

167°

a. Classify the quadrilateral formed by the cross section.

b. Find the missing angle measure.

10.3 Polygons and Tessellations

Sum of Angle Measures of a Polygon Use a formula to find the sum of the interior angle measures of the polygon.

33.

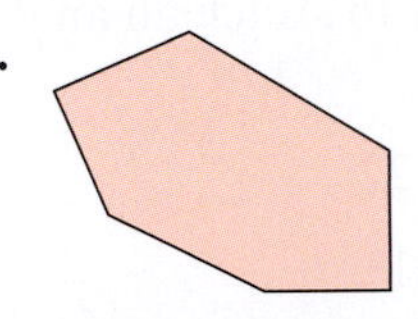

34.

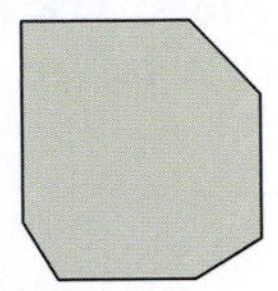

Identifying a Regular Polygon Decide whether the polygon is regular. Explain your reasoning.

35.

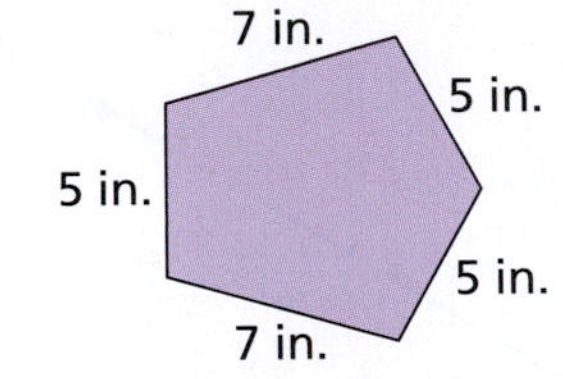

36.

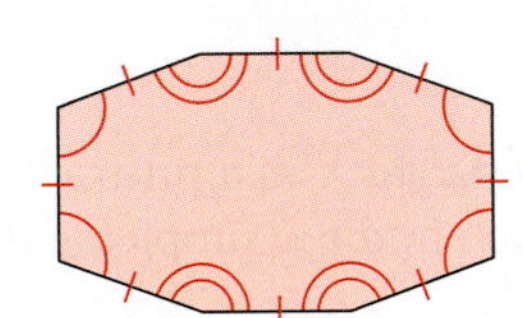

37.

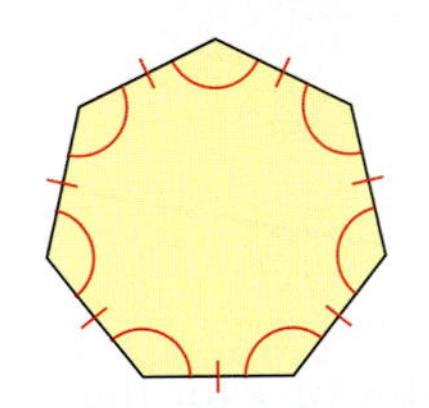

38.

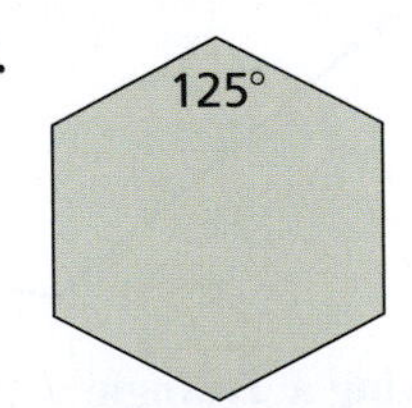

Finding an Angle Measure Find the measures of an interior angle and a central angle of the regular polygon.

39.

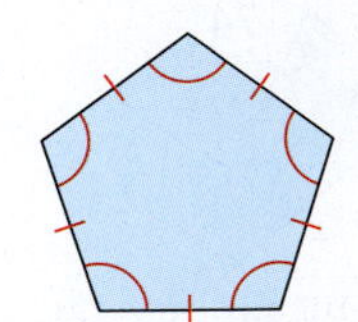

40.

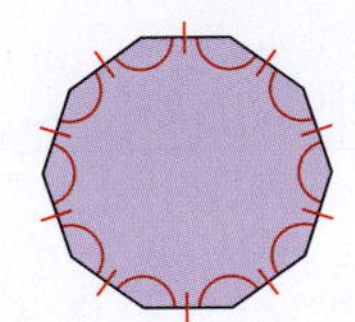

41. Regular Polygons and Tessellations Create a tessellation using equilateral triangles and squares.

42. Regular Polygons and Tessellations Create a tessellation using equilateral triangles, squares, and regular hexagons.

43. Drawing a Polygon Use a protractor and a ruler to draw a regular dodecagon (12-gon). Check that each of the interior angles has the correct measure.

44. Tessellations in Nature Honeycombs, like the one shown below, are examples of tessellations in nature.

a. Identify the polygon in the tesselation.

b. Give another example of a tessellation in nature.

10.4 Polyhedra

Identifying Prisms and Pyramids Use the solids below.

i.

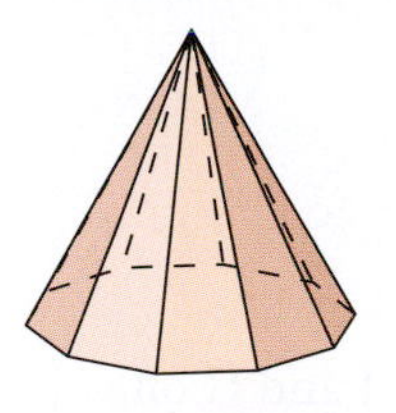

ii.

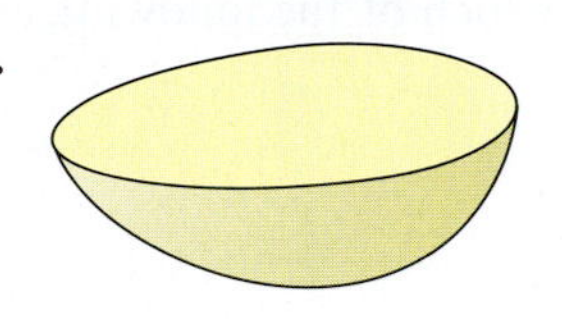

iii.

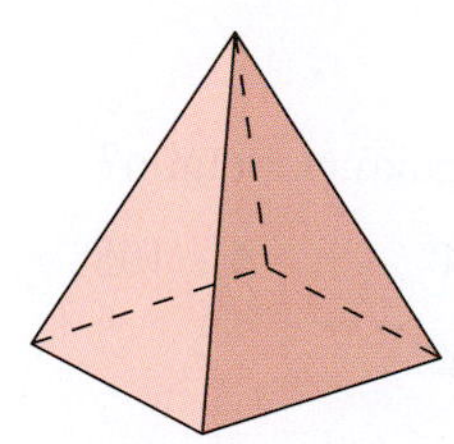

iv.

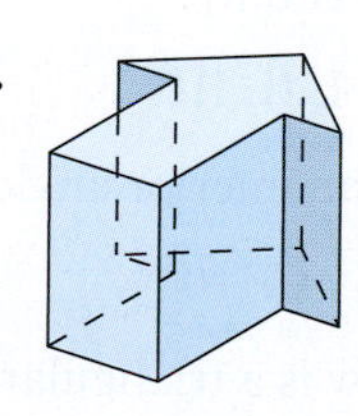

45. Which of the solids are prisms?

46. Which of the solids are pyramids?

Naming a Prism or a Pyramid Name the prism or pyramid.

47.

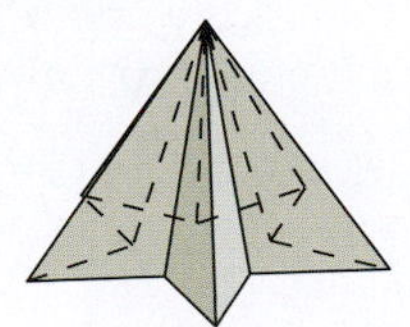

48.

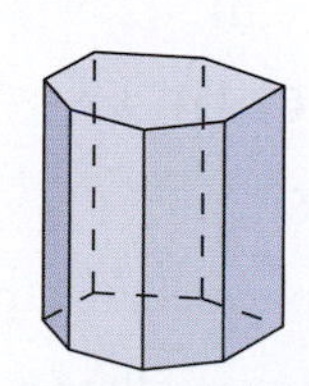

Using Euler's Formula Find the missing number of vertices, faces, or edges for the polyhedron.

49. Nonagonal pyramid

9 lateral faces, 10 vertices, ____ edges

50. Elongated square dipyramid

12 faces, 10 vertices, ____ edges

51. Heptagonal prism

7 lateral faces, ____ vertices, 21 edges

52. Elongated triangular dipyramid

9 faces, ____ vertices, 15 edges

53. 11-gonal pyramid

____ lateral faces, 12 vertices, 22 edges

54. Pentagonal dipyramid

____ faces, 7 vertices, 15 edges

Folding a Net Classify the polyhedron made by the net as *regular*, *semiregular*, or *neither*. Name the polyhedron.

55.

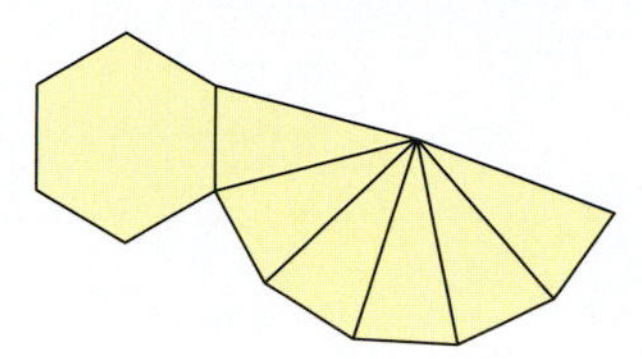

56.

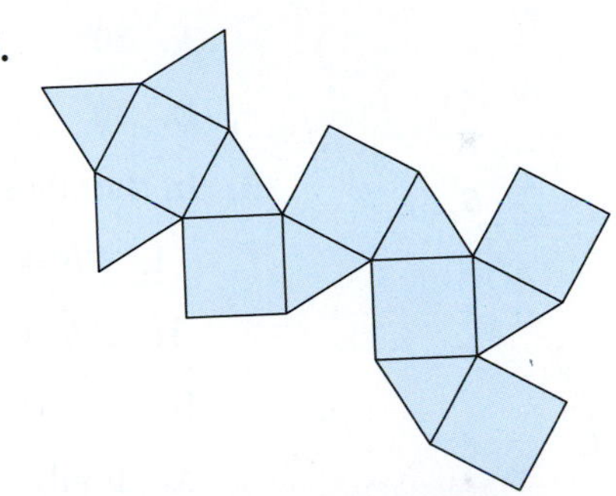

57. Board Games Board games use a variety of differently-shaped dice. For instance, consider the die shown below.

a. What regular polyhedron does the die resemble?

b. How many faces, edges, and vertices does the die have?

58. Ornaments Consider the ornament shown below.

a. What semiregular polyhedron does the ornament resemble?

b. The ornament has 60 edges and 24 vertices. How many faces does the ornament have?

StudioSmart/Shutterstock.com

Chapter Test

The questions on this practice test are modeled after questions on state certification exams for K–8 teaching.

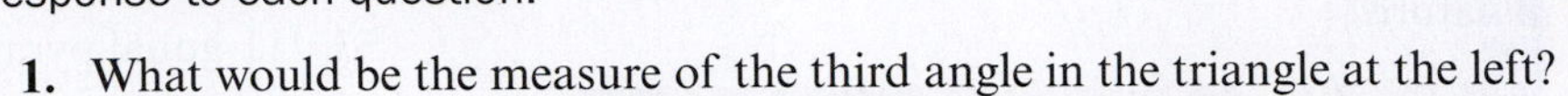

Test Directions: Each of the questions is followed by five choices. Choose the *best* response to each question.

1. What would be the measure of the third angle in the triangle at the left?

   A. 40° **B.** 45° **C.** 60°
   D. 70° **E.** 110°°

2. Referring to the following figure, if the measure of $\angle C$ is 50° and the measure of $\angle ABD$ is 130°, then what is the measure of $\angle A$?

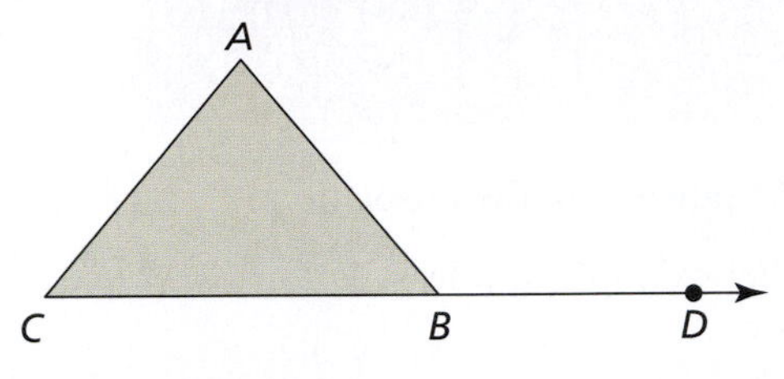

   A. 50° **B.** 60° **C.** 70°
   D. 80° **E.** 100°

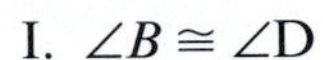

3. In rhombus $ABCD$, which of the following are true?

   I. $\angle B \cong \angle D$
   II. $\angle BAC \cong \angle BCA$
   III. $\angle DAC \cong \angle DCA$

   A. I only **B.** II only **C.** I and II only
   D. I and III only **E.** I, II, III

4. What is the measurement of an interior angle of a regular pentagon?

   A. 72° **B.** 84° **C.** 100° **D.** 104° **E.** 108°

5. Which of the polyhedra below is a triangular pyramid?

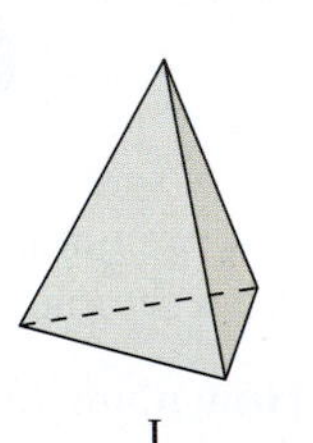
I

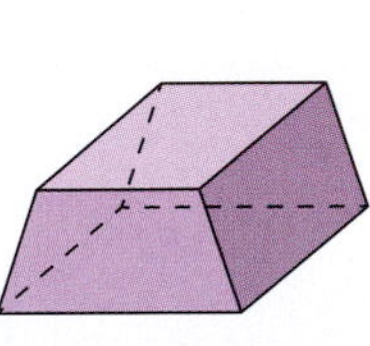
II

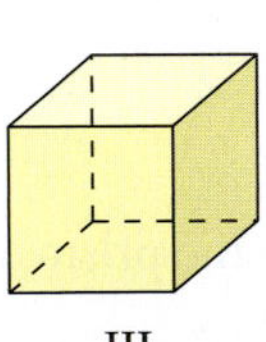
III

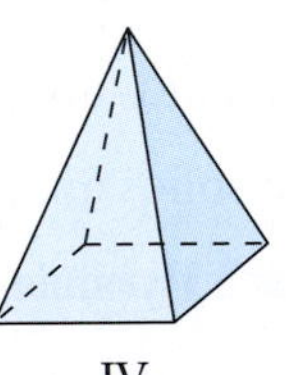
IV

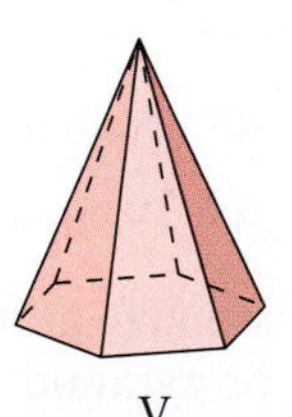
V

   A. I **B.** II **C.** III **D.** IV **E.** V

6. Which of the following are true?

   I. All triangles can be used to tessellate a plane.
   II. No hexagons can be used to tessellate a plane.
   III. All squares can be used to tessellate a plane.

   A. I only **B.** II only **C.** I and II only
   D. I and III only **E.** I, II, III

7. How many vertices does a triangular pyramid have?

   A. 3 **B.** 4 **C.** 5 **D.** 6 **E.** 7

11 Measurement

iStockphoto.com/kali9

Activity: Measuring Using Nonstandard Units

Materials:

- Pencil
- Jelly beans
- Paper clips (1.25 in. size)

Learning Objective:

Understand how any object, such as a paper clip or a jelly bean, can be used as a nonstandard unit of measure.

Name ______________________

Measure the object to the closest half paper clip. Then complete the sentence.

1. The pencil has a length of ▭ paper clips.

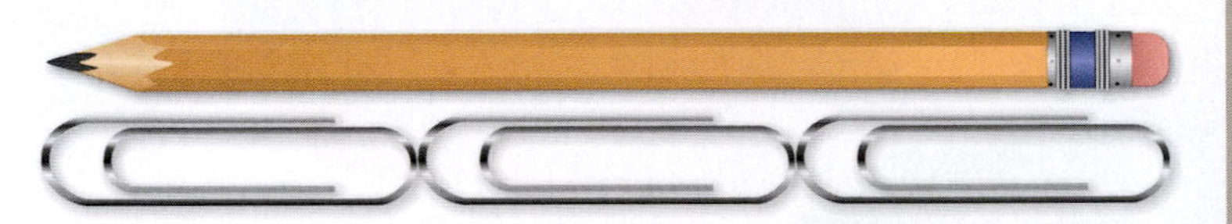

2. The highlighter has a length of ▭ paper clips.

3. The pair of scissors has a length of ▭ paper clips and a width of ▭ paper clips.

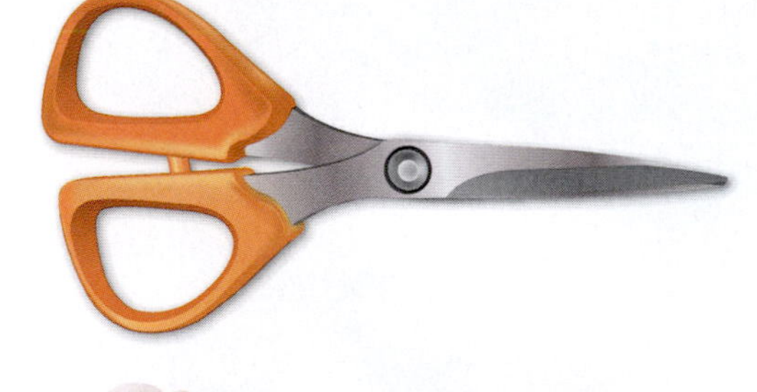

4. The athletic shoe has a length of ▭ paper clips and a height of ▭ paper clips.

Find an object in your classroom that has the given dimension.

5. Length of about 8 jelly beans

6. Width of about 3 paper clips

11.1 Standard and Nonstandard Units

Standards

Grades K–2
Measurement and Data
Students should measure and estimate lengths in standard units.

Grades 3–5
Measurement and Data
Students should convert like measurement units within a given measurement system.

Grades 6–8
Ratios and Proportional Relationships
Students should understand ratio concepts and use ratio reasoning to solve problems.

- Find lengths, weights, and volumes using the customary system.
- Find lengths, weights, and volumes using the metric system.
- Convert between the customary and metric systems.

The Customary System of Measure

The customary system of lengths used in the United States was derived over centuries and can be traced back to ancient Rome.

The U.S. Customary System of Lengths

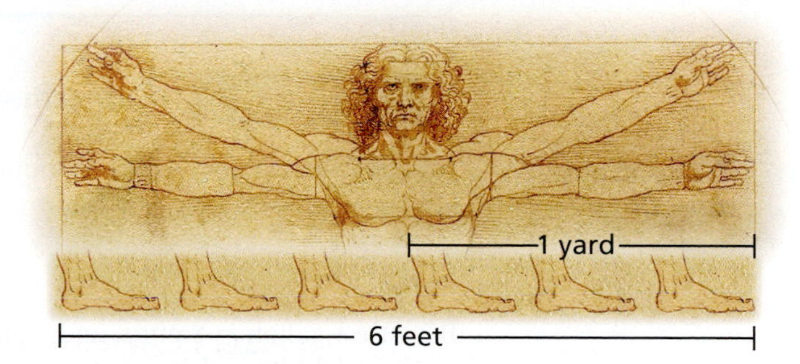

Lengths in the **U.S. customary system** of measurement are based on human lengths. Here are the historical beginnings of the units.

1 inch = width of an adult male's thumb

1 foot = length of an adult male's foot

1 yard = distance from an adult male's nose to his index finger

1 mile = distance a Roman soldier walked in 1000 paces (1 pace is about 2 steps)

Here are the standardized units of measure.

1 foot = 12 inches

1 yard = 3 feet = 36 inches

1 mile = 5280 feet = 1760 yards

Classroom Tip
Most people are familiar with "the rule of thumb." It means that you can estimate the length of an object by marking off the number of thumb widths of the object.

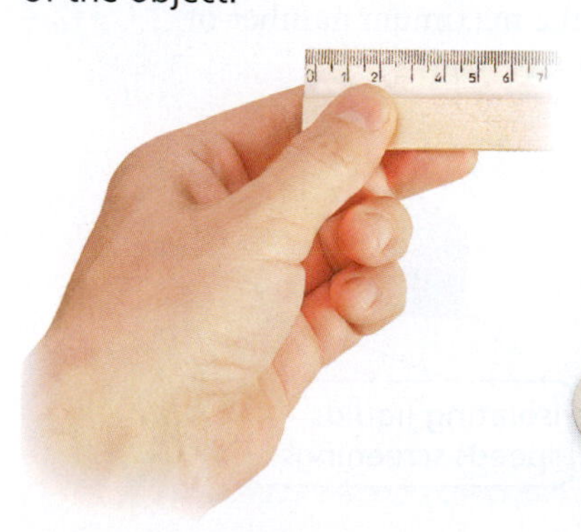

EXAMPLE 1 Converting Units in Real Life

a. The height of a basketball player is 6'11". Find the height in inches.

b. You are 1 mile from an exit on a highway. How many inches are you from the exit?

SOLUTION

a. $6'\,11'' = 6 \text{ ft} + 11 \text{ in.} = 6\cancel{\text{ft}}\left(\dfrac{12\text{in.}}{1\cancel{\text{ft}}}\right) + 11 \text{ in.} = 72 \text{ in.} + 11 \text{ in.} = 83 \text{ in.}$

b. $1 \text{ mi} = 1\cancel{\text{mi}}\left(\dfrac{5280\cancel{\text{ft}}}{1\cancel{\text{mi}}}\right)\left(\dfrac{12 \text{ in.}}{1\cancel{\text{ft}}}\right) = 63{,}360 \text{ in.}$

Furtseff | Dreamstime.com; Dennis Hallinan/Alamy

Classroom Tip

The abbreviation for pound, "lb," is an abbreviation of the Latin word *libra*. *Libra* is a shortened form of the Latin phrase *libra pondo*, which means "pound weight."

The abbreviation for ounce, "oz," is an abbreviation of the Italian word *onza*, which means "ounce."

The U.S. Customary System of Weights and Volumes

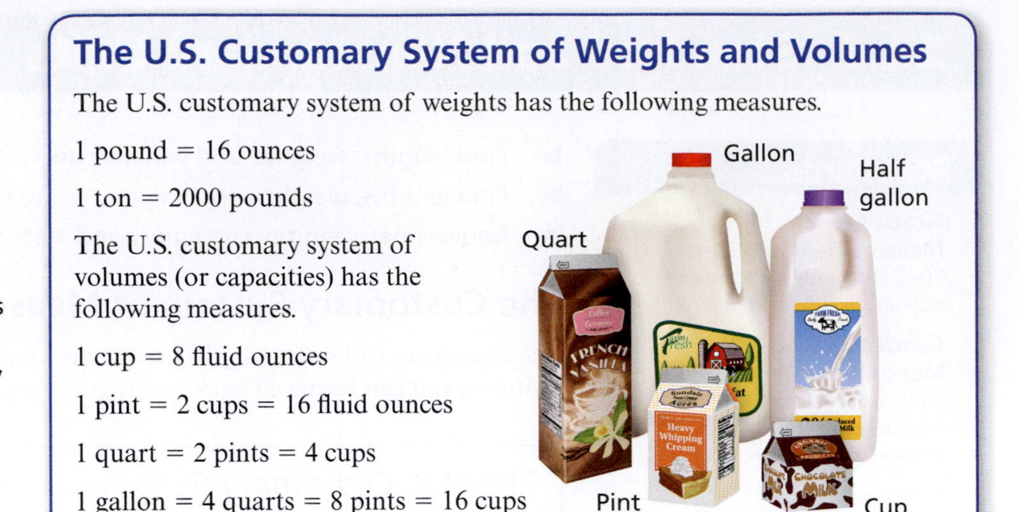

The U.S. customary system of weights has the following measures.

1 pound = 16 ounces

1 ton = 2000 pounds

The U.S. customary system of volumes (or capacities) has the following measures.

1 cup = 8 fluid ounces

1 pint = 2 cups = 16 fluid ounces

1 quart = 2 pints = 4 cups

1 gallon = 4 quarts = 8 pints = 16 cups

EXAMPLE 2 Converting from Volume to Weight

An old saying states "A pint's a pound the world around." Based on this saying, how much does a gallon of water weigh?

SOLUTION

Because 1 gallon = 8 pints, and 1 pint weighs 1 pound, it follows that 1 gallon of water weighs 8 pounds.

EXAMPLE 3 Estimating the Number of Containers

The Transportation Security Administration's (TSA) 3-1-1 rule regulates the amount of liquid passengers can have in their carry-on luggage when they board an aircraft. Each liquid container can be no greater than 3.4 fluid ounces. All liquid containers must be placed (and fit) into a 1-quart, clear plastic, zip-top bag. There is a limit of 1 bag per passenger. What is the maximum number of 3.4-fluid-ounce containers that can fit into a 1-quart bag?

Mathematical Practices

Model with Mathematics

MP4 Mathematically proficient students analyze relationships mathematically to draw conclusions.

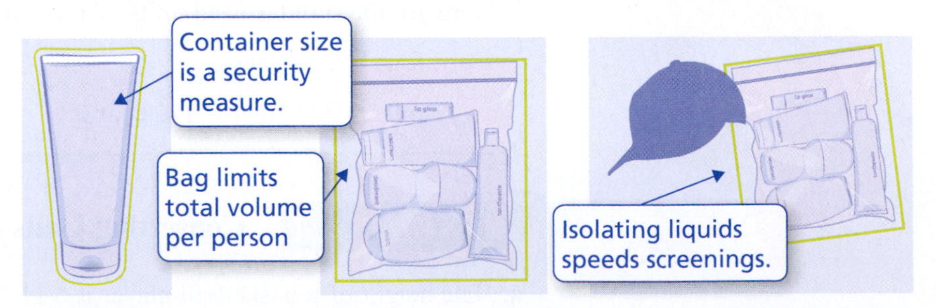

SOLUTION

A 1-quart bag contains

$$1 \text{ qt} = 1 \cancel{\text{qt}}\left(\frac{4 \cancel{c}}{1 \cancel{\text{qt}}}\right)\left(\frac{8 \text{ fl oz}}{1 \cancel{c}}\right) = 32 \text{ fl oz}.$$

To find the maximum number of 3.4-fluid-ounce containers that can fit into the 1-quart bag, divide 32 fluid ounces by 3.4 fluid ounces.

$$32 \text{ fl oz} \div 3.4 \text{ fl oz} \approx 9.4$$

So, the maximum number of 3.4-fluid-ounce containers that can fit into a 1-quart bag is 9.

The Metric System of Measure

In 1790, the National Assembly of France instructed the French Academy of Sciences to develop a new system of measurement. This system was to have a decimal base and be based on an unchanging standard found in nature. To determine a base unit of length for this new system, the academy chose the term *meter*, after the Greek word *metron*, meaning measure. The length of a meter would be precisely one ten-millionth of an imaginary arc that began at the North Pole, went through Paris, and ended at the equator.

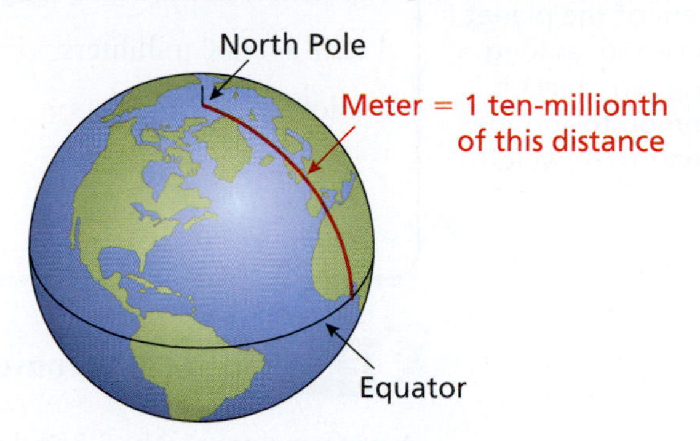

The Metric System of Lengths

Lengths in the **metric system** of measurement are based on the meter.

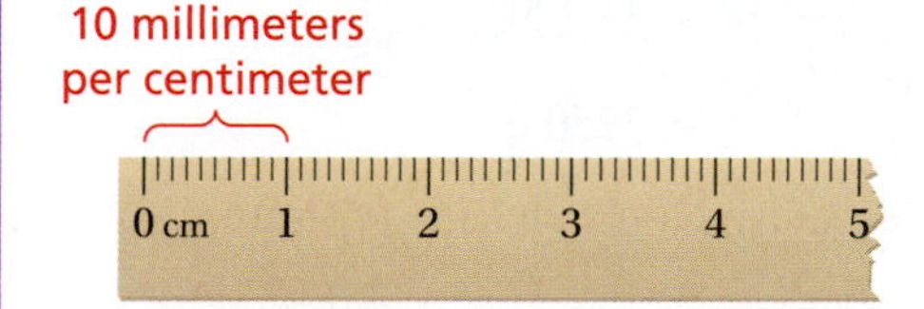

1 centimeter = 10 millimeters

1 meter = 100 centimeters

1 kilometer = 1000 meters

Classroom Tip

Most countries use the metric system. The United States is one of the few countries that continues to use the customary system of measure.

Converting Units in Real Life

The length of an alligator is 210 centimeters. Find the length in meters.

SOLUTION

$$210 \text{ cm} = 210 \cancel{\text{cm}} \left(\frac{1 \text{ m}}{100 \cancel{\text{cm}}}\right) = 2.1 \text{ m}$$

Converting Units in Real Life

Find the circumference of Earth in kilometers.

SOLUTION

A meter represents 1 ten-millionth of the arc length from the North Pole to the equator. So, the arc length measures 10,000,000 meters. The circumference of Earth is made up of 4 of these arc lengths. So, the circumference of Earth is

$$4 \cdot 10{,}000{,}000 \text{ m} = 40{,}000{,}000 \text{ m}.$$

Convert 40,000,000 meters to kilometers.

$$40{,}000{,}000 \cancel{\text{m}} \left(\frac{1 \text{ km}}{1000 \cancel{\text{m}}}\right) = 40{,}000 \text{ km}$$

The circumference of Earth is 40,000 kilometers.

Classroom Tip

Technically, kilograms are units of mass rather than units of weight. The difference between mass and weight is that the mass of an object is independent of the gravitational system of the planet the object is on. Even so, as long as it is assumed that an object is on Earth, it is common to consider kilograms as a measure of weight.

The Metric System of Weights and Volumes

The metric system of weights has the following measures.

1 kilogram = 1000 grams

1 metric ton = 1000 kilograms

The metric system of volumes (or capacities) has the following measures.

1 liter = 1000 milliliters

1 kiloliter = 1000 liters

EXAMPLE 6 Converting from Kilograms to Grams

A watermelon weighs 3.25 kilograms. How many grams does it weigh?

SOLUTION

$$3.25 \text{ kg} = 3.25 \cancel{\text{kg}}\left(\frac{1000 \text{ g}}{1 \cancel{\text{kg}}}\right)$$

$$= 3250 \text{ g}$$

The watermelon weighs 3250 grams.

EXAMPLE 7 Converting from Metric Tons to Kilograms

An automobile weighs 1.5 metric tons. How many kilograms does it weigh?

SOLUTION

$$1.5 \text{ t} = 1.5 \cancel{\text{t}}\left(\frac{1000 \text{ kg}}{1 \cancel{\text{t}}}\right)$$

$$= 1500 \text{ kg}$$

The automobile weighs 1500 kilograms.

EXAMPLE 8 Converting from Liters to Kiloliters

A swimming pool is 25 meters long and 9 meters wide. Its depth is 1.8 meters at the deep end and 1 meter at the shallow end. With these dimensions, the pool contains 315,000 liters of water. What is the volume of the swimming pool in kiloliters?

SOLUTION

$$315{,}000 \text{ L} = 315{,}000 \cancel{\text{L}}\left(\frac{1 \text{ kL}}{1000 \cancel{\text{L}}}\right)$$

$$= 315 \text{ kL}$$

The volume of the swimming pool is 315 kiloliters.

Martin Darley/Shutterstock.com; Maksim Toome/Shutterstock.com

Converting Between Measurement Systems

The Internet makes it easy to convert between U.S. customary and metric units. When you do not have the Internet or a convenient conversion chart, it may be helpful to know the common conversions below.

U.S. Customary and Metric Conversions

1 inch = 2.54 centimeters, 1 centimeter ≈ 0.39 inch

1 mile ≈ 1.61 kilometers, 1 kilometer ≈ 0.62 mile

1 foot ≈ 0.3 meter, 1 meter ≈ 3.28 feet

1 pound ≈ 0.45 kilogram, 1 kilogram ≈ 2.2 pounds

1 gallon ≈ 3.79 liters, 1 liter ≈ 0.26 gallon

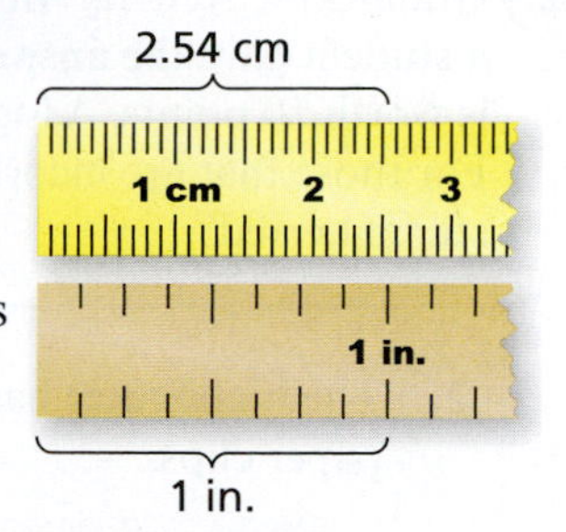

EXAMPLE 9 Converting from Kilograms to Pounds

Your friend weighs 90 kilograms. What is this weight in pounds?

SOLUTION

$$90 \text{ kg} \approx 90 \text{ kg}\left(\frac{2.2 \text{ lb}}{1 \text{ kg}}\right) = 198 \text{ lb}$$

Your friend weighs about 198 pounds.

EXAMPLE 10 Speed Limit in Canada

You are driving in Canada and see the speed limit sign shown.

a. What is the speed limit in miles per hour?

b. Explain how to estimate the speed limit in miles per hour using mental math.

SOLUTION

a. $\frac{100 \text{ km}}{1 \text{ h}} \approx \frac{100 \text{ km}}{1\text{h}}\left(\frac{0.62 \text{ mi}}{1 \text{ km}}\right) = \frac{62 \text{ mi}}{1\text{h}}$

The speed limit is about 62 miles per hour.

b. Because 1 kilometer is about 0.6 mile, 100 kilometers, or 1(100), is about 0.6(100) or 60 miles. So, the speed limit is about 60 miles per hour.

EXAMPLE 11 Price of Gasoline in Canada

You are traveling in Canada. Gasoline costs $1.32 per liter. What is the price in dollars per gallon? Assume that U.S. and Canadian dollars are of equal value.

SOLUTION

$$\frac{\$1.32}{1 \text{ L}} \approx \frac{\$1.32}{1 \text{ L}}\left(\frac{3.79 \text{ L}}{1 \text{ gal}}\right) \approx \frac{\$5.00}{1 \text{ gal}}$$

The gasoline costs about $5 per gallon.

11.1 Exercises

Free step-by-step solutions to odd-numbered exercises at *MathematicalPractices.com.*

1. **Writing a Solution Key** Write a solution key for the activity on page 412. Describe a rubric for grading a student's work.
2. **Grading the Activity** In the activity on page 412, a student gave the answers below. Each question is worth 10 points. Assign a grade to each answer. For those that are incorrect, why do you think the student erred?

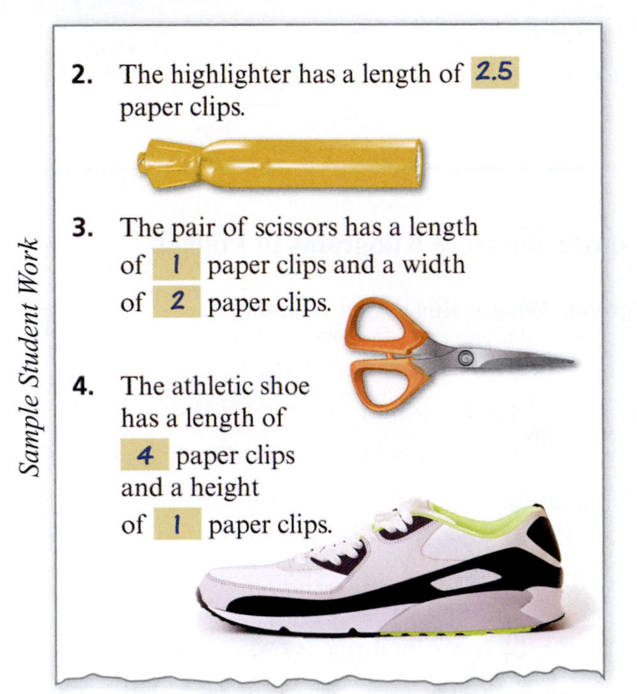

3. **Classifying Units** Tell whether each unit is a measure of *length*, *weight*, or *volume*.
   a. Yards **b.** Ounces **c.** Kilograms
4. **Classifying Units** Tell whether each unit is a measure of *length*, *weight*, or *volume*.
   a. Fluid ounces **b.** Centimeters **c.** Metric tons
5. **Measuring Length** How long is the pen in inches?

6. **Measuring Length** How long is the eraser in centimeters?

7. **Measuring Weight** What is the weight of the box in pounds?

8. **Measuring Weight** What is the weight of the block in grams?

9. **Measuring Volume** How many fluid ounces of water are in the measuring cup?

10. **Measuring Volume** How many milliliters of water are in the measuring cup?

11. **Converting Customary Units** Complete each statement.
    a. 3 ft = ___ in. **b.** 0.7 mi = ___ in.
    c. 22 oz = ___ lb **d.** 0.2 ton = ___ oz
    e. 19 fl oz = ___ c **f.** 1.2 gal = ___ pt

12. Converting Customary Units Complete each statement.

a. 7850 ft = ____ mi b. 9.2 yd = ____ in.

c. 5.6 lb = ____ oz d. 9000 lb = ____ ton

e. 3.4 pt = ____ c f. 76 c = ____ gal

13. Converting Metric Units Complete each statement.

a. 12 mm = ____ cm b. 9.25 km = ____ m

c. 0.003 t = ____ kg d. 7240 g = ____ kg

e. 7.2 L = ____ mL f. 573 L = ____ kL

14. Converting Metric Units Complete each statement.

a. 0.2 m = ____ cm

b. 0.001 km = ____ mm

c. 0.05 kg = ____ g

d. 0.6 t = ____ g

e. 0.0097 kL = ____ mL

f. 8000 mL = ____ L

15. Converting Between Measurement Systems Complete each statement.

a. 9 in. = ____ cm b. 12 mi ≈ ____ km

c. 9.2 m ≈ ____ ft d. 27 in. = ____ m

e. 0.33 oz ≈ ____ g f. 190 g ≈ ____ lb

g. 0.03 kL ≈ ____ gal

h. 320 mL ≈ ____ pt

16. Converting Between Measurement Systems Complete each statement.

a. 0.2 km ≈ ____ ft b. 15 in. = ____ mm

c. 3.7 m ≈ ____ yd d. 80 oz ≈ ____ kg

e. 490 kg ≈ ____ ton f. 52 c ≈ ____ L

g. 7 mL ≈ ____ fl oz h. 1.5 kL ≈ ____ qt

17. Comparing Measurements Complete each statement with the correct symbol: $<$ or $>$.

a. 2 in. ____ 54 mm b. 9 kg ____ 160 oz

c. 3 L ____ 2 qt d. 1 c ____ 500 mL

e. 25 mi ____ 0.02 m f. 3 g ____ 0.02 oz

18. Comparing Measurements Complete each statement with the correct symbol: $<$ or $>$.

a. 3 yd ____ 2 m b. 22 mL ____ 3 c

c. 2 t ____ 4500 lb d. 2 ft ____ 40 cm

e. 65 fl oz ____ 1 L f. 1 ton ____ 1 t

19. Choosing Units Choose an appropriate U.S. customary unit and metric unit to measure each item. Explain your reasoning.

a. Diameter of a quarter

b. Capacity of a contact lens case

c. Mass of a bicycle

20. Choosing Units Choose an appropriate U.S. customary unit and metric unit to measure each item. Explain your reasoning.

a. Distance of a marathon

b. Amount of water in a bird bath

c. Mass of a leaf

21. Converting Units Using Proportions Proportions can also be used to convert units. For example, to convert 250 centimeters to meters, you set up the proportion

$$\frac{1\text{ m}}{100\text{ cm}} = \frac{x\text{ m}}{250\text{ cm}}.$$

Solving this proportion, you obtain $x = 2.5$. So, 250 centimeters = 2.5 meters.

a. Use a proportion to convert 33 liters to milliliters.

b. Use a proportion to convert 33 liters to kiloliters.

22. Converting Units Using Proportions Use proportions as described in Exercise 21 to convert each of the following.

a. Convert 16 feet to inches.

b. Convert 16 feet to yards.

23. Reasoning Determine whether each measurement describes a model car or an actual car. Explain your reasoning.

a. The weight of the car is about 105 grams.

b. The height of the car is about 5 feet.

c. The length of the car is about 21 centimeters.

24. Reasoning Determine whether each measurement describes a bathtub or a swimming pool. Explain your reasoning.

a. The average depth is 4 feet.

b. The capacity is 5189.5 liters.

c. The length is 153.6 centimeters.

25. Converting Units in Real Life An aluminum can holds 12 fluid ounces of liquid. Find the capacity in cups.

26. Converting Units in Real Life A baseball weighs about 5 ounces. Find the weight in grams.

27. **Record Heights** In 1940, Robert Wadlow, who was approximately 8 feet and 11 inches tall, was declared the tallest man ever. In 2012, Chandra Bahadur Dangi, who was approximately 21.5 inches tall, was declared the shortest man ever. How much taller than Dangi was Wadlow? Give your answer in feet and inches.

28. **Groceries** Use the given prices to determine whether it is less expensive to buy a gallon of milk in a gallon container or in quart containers. How much money do you save?

29. **Animal Weights** The average adult blue whale can weigh up to 300,000 pounds. The average adult African elephant can weigh up to 7.5 tons. How many times greater is the weight of an adult blue whale than the weight of an adult African elephant?

30. **Baking** You want to bake muffins for a bake sale. Each batch of muffins calls for $\frac{1}{4}$ cup milk. How many batches of muffins can you make with $\frac{3}{4}$ gallon of milk?

31. **Comparing Speeds** A car is driving 70 miles per hour down the highway. A jet is flying above the car at 920 kilometers per hour. How much faster, in miles per hour, is the jet moving?

32. **Circulation** The figure below shows a human heart at rest. During intense exercise, the human heart may pump 27 liters of blood per minute. How much more blood, in quarts, does a human heart pump per minute during intense exercise than at rest?

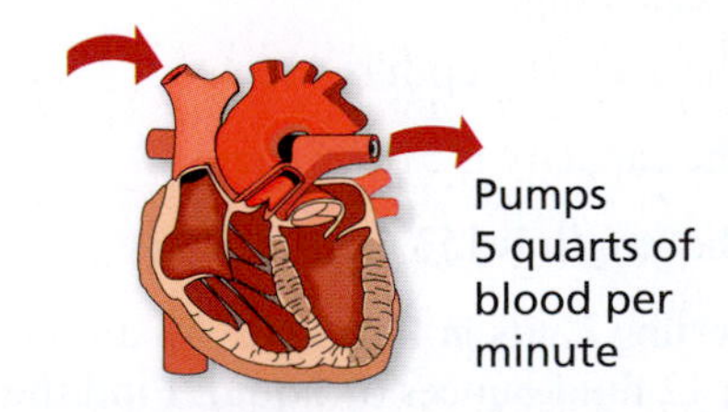

33. **Shampoo** Your shampoo bottle is about 60% full. The total capacity of the shampoo bottle is 24 fluid ounces. How many liters of shampoo are left in the bottle?

34. **Water** Your water bottle is about 30% empty. The total capacity of the bottle is 1.05 pints. How many milliliters of water are left in the bottle?

35. **Grading Student Work** On a diagnostic test, one of your students does the following work. Explain what the student did wrong. Which topics would you encourage the student to review?

> Choose an appropriate metric unit to measure the amount of liquid medicine to give a toddler.
>
> *Fluid ounces, because the toddler would need a smaller dose of the medicine.*

36. **Grading Student Work** On a diagnostic test, one of your students does the following work. Explain what the student did wrong. Which topics would you encourage the student to review?

$$6\text{ kg} \cdot \frac{0.45\text{ lb}}{1\text{ kg}} = 2.7\text{ lb}$$

37. **Prefixes** Knowing the definitions of certain prefixes can help you to determine the values of various measures.

Millimeter	
Centimeter	
Meter	1
Kilometer	

a. Fill in the table above so that all of the values in the right column represent the same length. What pattern do you notice?

b. What do you think the prefix "centi-" means?

c. The same prefixes that are used in the metric system can be applied to units of time. What fraction of a second is a millisecond?

38. **Prefixes** The prefixes "deci-," "deka-," and "hecto-" mean "one tenth part of," "ten," and "hundred," respectively.

a. Add decimeter, dekameter, and hectometer to the table in Exercise 37.

b. Use the definitions of "milli-" and "deci-" to determine how many millidecimeters are in one meter.

c. What is an equivalent name for a hectodekameter?

39. Mixing Paint You want to paint your room a specific shade of purple. To get this shade, you must mix the quantities of red, blue, and yellow paint shown below. How many total quarts of purple paint do you make?

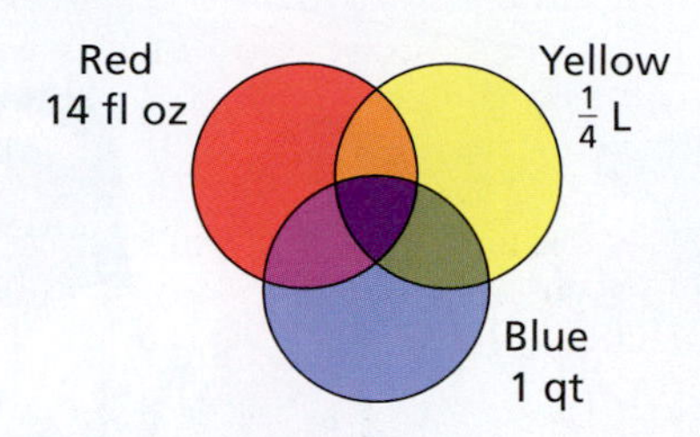

40. Making Punch You make Floating Island Punch for a party by following the recipe below.

Recipe for: Floating Island Punch
From: Mom

2 cups water	2 cups sugar
1 L ginger ale	1 L carbonated water
1 pt orange sherbet	4 cups ice
$1\frac{1}{2}$ cups frozen lemonade concentrate	$1\frac{1}{2}$ cups frozen orange juice concentrate

a. Do you need a 4-liter or a 6-liter punch bowl to hold the entire recipe? Explain.

b. You need to store the punch until the party. How many containers will you need to use if each container has a capacity of 3500 cubic centimeters? (*Hint*: 1 milliliter is equal to 1 cubic centimeter.)

41. Ancient Rome Ancient Romans used the *talent* and the *mina* as measures of weight. Use the conversions below to determine the conversion between minas and talents.

1 lb = 0.8 mina

1 talent = 75 lb

42. Using Currency Conversions The table shows the currencies of four countries and their exchange rates. How much of each currency would you receive in exchange for 25 U.S. dollars?

Country	Currency	Value in U.S. dollars
United States	Dollar	$1
Japan	Yen	$0.01
Great Britain	Pound	$1.52
Spain	Euro	$1.30

43. The Speed of Light The speed of light is about 300,000 kilometers per second.

a. Determine the length of one *light-year* by calculating the distance, in miles, light will travel in a year.

b. Proxima Centauri, the closest star to Earth, is approximately 4.22 light-years away. How far, in miles, will light from Proxima Centauri travel before it reaches Earth?

44. Thunderstorms You can estimate how far away a storm is by counting the seconds between the time you see a lightning strike and the time you hear thunder. The speed of sound is about 340 meters per second.

a. How many kilometers away is a storm when 24 seconds pass between a lightning strike and thunder?

b. How many seconds will pass between a lightning strike and thunder of a storm that is 1 mile away?

45. Activity: In Your Classroom Any object can be used as a nonstandard unit of measure. For instance, a horse's height is measured in hands by counting the number of hand widths from the ground to the horse's shoulder when standing.

a. Measure the height of a desk in hands.

b. How do you think your measurement will compare to those of your classmates? Explain your answer.

c. Explain how an activity such as this one helps students understand the importance of standard units of measure when accuracy is a factor.

46. Activity: In Your Classroom You fill an empty pint jug with water and pour it into a gallon jug, and repeat this process until the gallon jug is full.

a. How many times will you pour water into the gallon jug?

b. Why might an activity such as this be beneficial to students?

c. Create a similar activity to demonstrate conversions within metric units of weight.

47. Writing Use the units you chose in Exercise 19 to estimate the measure of each item. Explain your reasoning.

48. Writing A shelf in your entertainment system is 20 centimeters tall. To the nearest centimeter, your electronic gaming system is also 20 centimeters tall. If the shelf is wide enough and deep enough, can you be sure the gaming system will fit completely on the shelf? Explain.

Activity: Finding the Perimeter of a Polygon

Materials:

- Pencil
- Dot paper

Learning Objective:

Understand the concept of the perimeter of a polygon. Be able to find the perimeter of a polygon. Draw a polygon that has a given perimeter.

Name ______________________________

The perimeter of a polygon is the distance around the edges of the polygon.

The distance between 2 horizontal dots or 2 vertical dots is 1 unit. Find the perimeter of the polygon.

1. 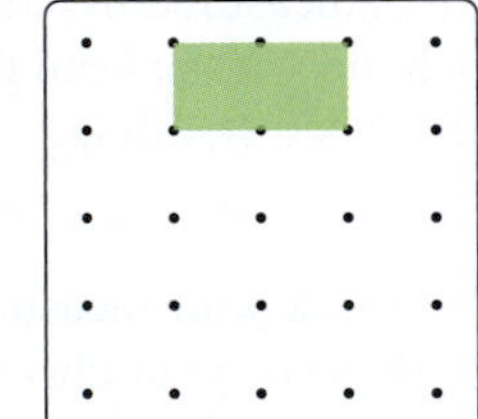 **2.**  **3.** 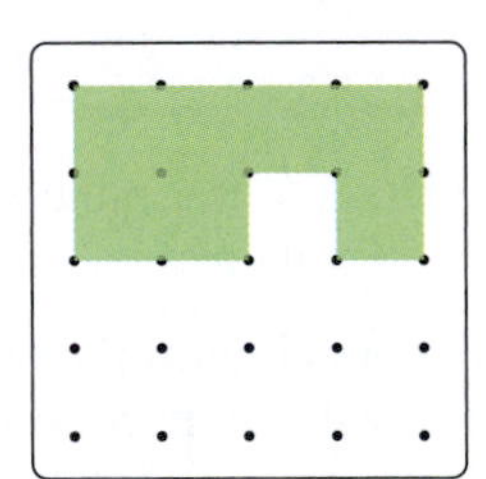

The distance between 2 horizontal dots or 2 vertical dots is 1 unit. Draw a polygon that has the given perimeter.

4. Perimeter = 12

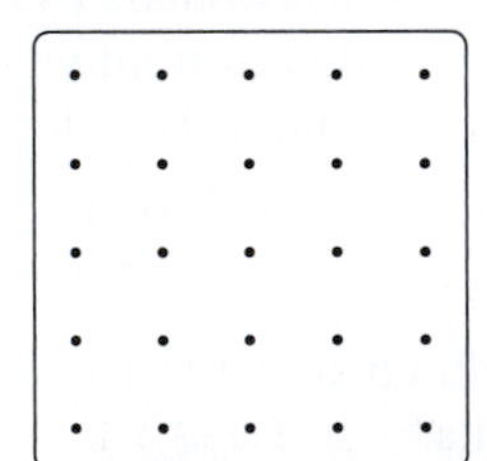

5. Perimeter = 16

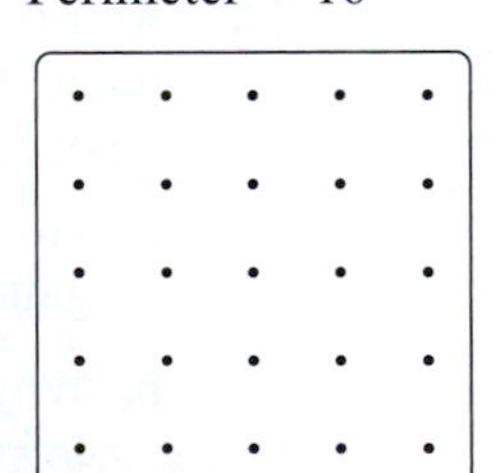

6. Perimeter = 18

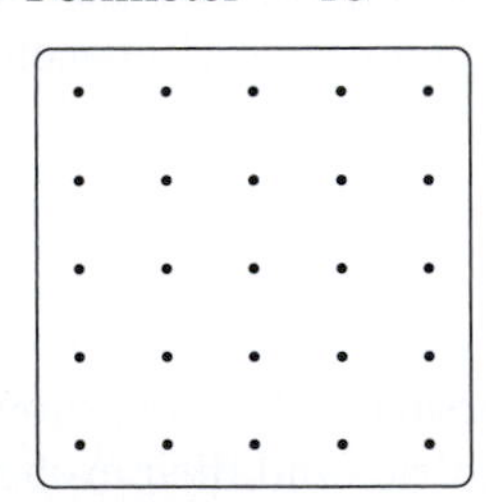

The red right triangle has sides of lengths 3 units, 4 units, and 5 units. Find the perimeter of the entire polygon.

7. 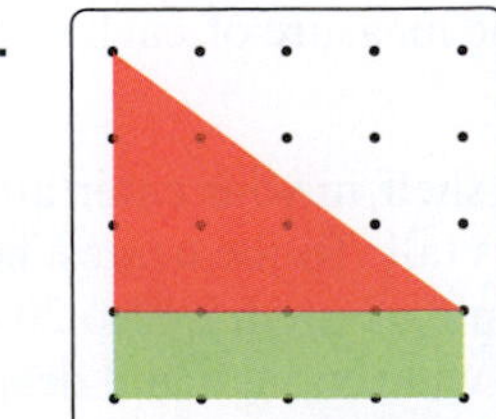 **8.** 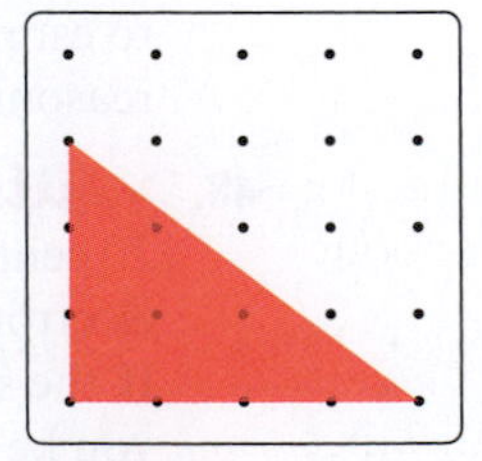 **9.** 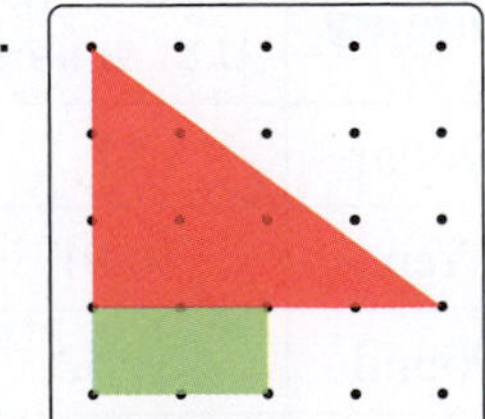

11.2 Perimeters and Areas of Polygons

Standards

Grades 3–5
Measurement and Data
Students should understand concepts of area, and recognize perimeter as an attribute of plane figures and distinguish between linear and area measures.

Grades 6–8
Geometry
Students should solve real-life and mathematical problems involving area.

- Find the perimeter of a polygon.
- Find the area of a quadrilateral.
- Find the area of a triangle.

Finding the Perimeter of a Polygon

Perimeter of a Polygon

The **perimeter** P of a polygon is the sum of the lengths of its sides. Perimeter is measured in **linear units** such as inches, feet, or centimeters.

Example $P = 2 + 4 + 4 + 5$

$= 15$ cm

The perimeter of a rectangle is given by

$P = 2(\text{width}) + 2(\text{length})$.

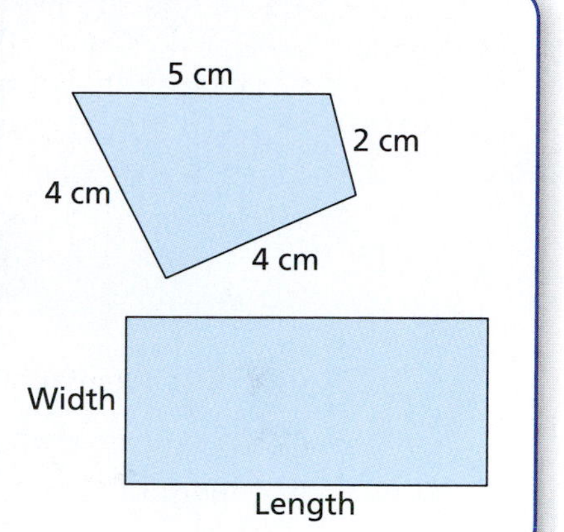

EXAMPLE 1 **Finding the Perimeters of Polygons**

Find the perimeter of each polygon.

a.

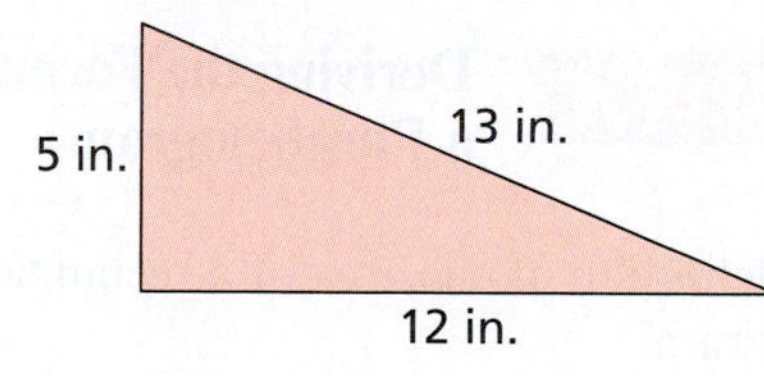

b.

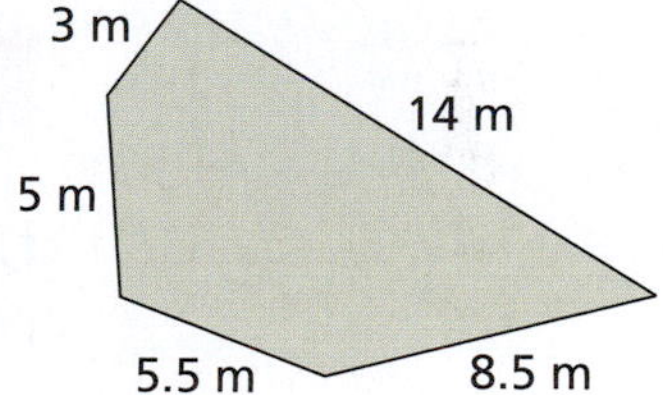

SOLUTION

a. $P = 5 + 12 + 13 = 30$

The perimeter is 30 inches.

b. $P = 3 + 5 + 5.5 + 8.5 + 14 = 36$

The perimeter is 36 meters.

EXAMPLE 2 **Finding a Perimeter in Real Life**

Central Park in New York City is rectangular with a width of about 800 meters and a length that is 5 times its width. You walk along the perimeter of the park. Estimate the distance you walk.

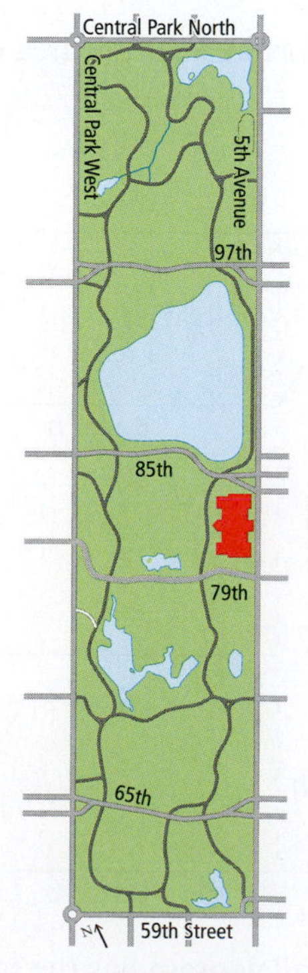

Central Park

SOLUTION

The length is 5 times the width. So, the length is about $5(800) = 4000$ meters. The perimeter is about

$$P = 2(\text{width}) + 2(\text{length}) = 2(800) + 2(4000) = 9600.$$

So, you walk a distance of about 9600 meters.

Rainer Lesniewski/Shutterstock.com

Finding the Area of a Quadrilateral

Classroom Tip

The formula for the area of a rectangle can be used to derive the other formulas for area, including triangles, circles, and even irregular areas in calculus.

Area of a Rectangle

The **area** A of a rectangle is the product of its length and width. Area is measured in **square units**, such as square inches, square feet, or square centimeters.

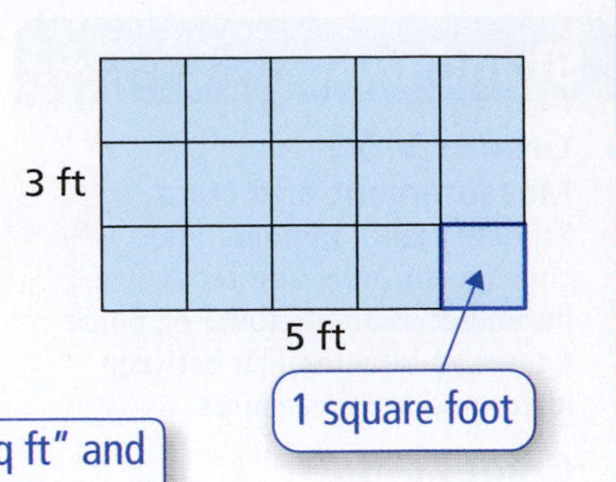

Example

$$A = (\text{length})(\text{width})$$
$$= (5 \text{ ft})(3 \text{ ft})$$
$$= 15 \text{ sq ft}$$

The abbreviations "sq ft" and "ft^2" represent square feet.

EXAMPLE 3 Finding an Area in Real Life

Find the area of the screen for the Global Positioning System (GPS) receiver.

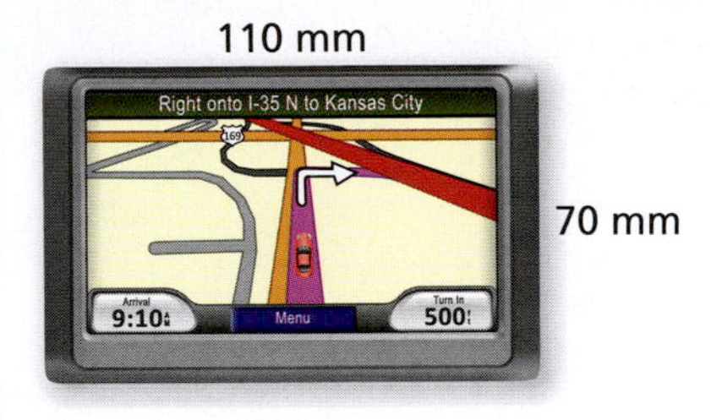

SOLUTION

$$A = (\text{length})(\text{width})$$
$$= (110)(70)$$
$$= 7700$$

The screen has an area of 7700 square millimeters.

EXAMPLE 4 Deriving the Formula for the Area of a Parallelogram

Use the definition of the area of a rectangle to derive a formula for the area of a parallelogram.

SOLUTION

Begin by drawing a representative parallelogram. Label the length of its **base** b. Label its **height** h.

Cut out the triangular portion of the parallelogram to the left of its height.

Rearrange the two pieces so that they form a rectangle whose length is b and height is h.

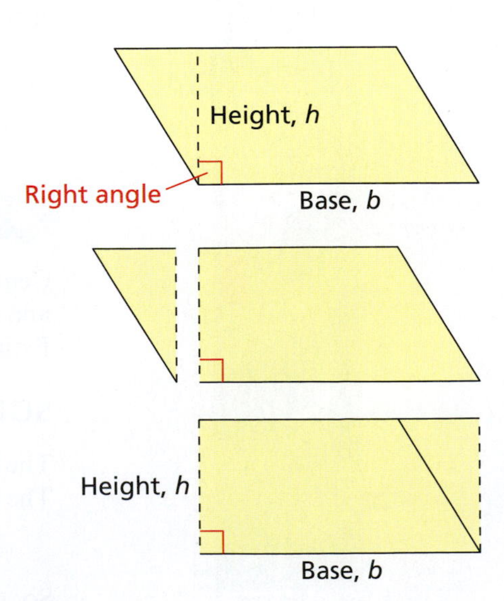

The area of the rectangle is $A = (\text{base})(\text{height})$. The parallelogram has the same area. So, its area is also $A = (\text{base})(\text{height})$.

Classroom Tip

The derivation in Example 4 is valid because of the following two properties of parallelograms.

1. Opposite sides of a parallelogram are congruent.
2. Opposite angles of a parallelogram are congruent.

You can extend the procedure shown in Example 4 to derive the formula for the area of a trapezoid. The formulas for the areas of four types of quadrilaterals are shown below.

Areas of Quadrilaterals

Rectangle $A = \ell w$

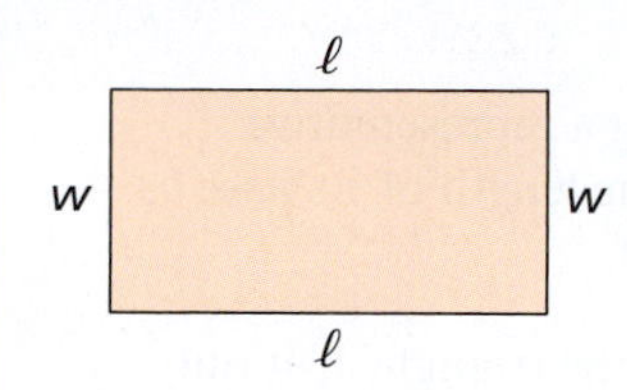

Square $A = s^2$

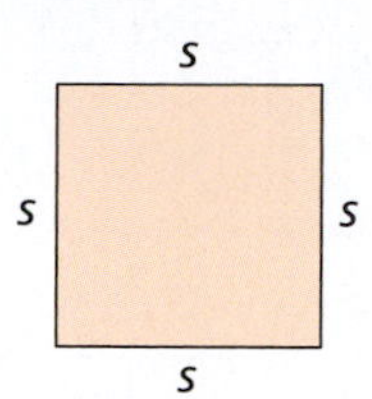

Parallelogram $A = bh$

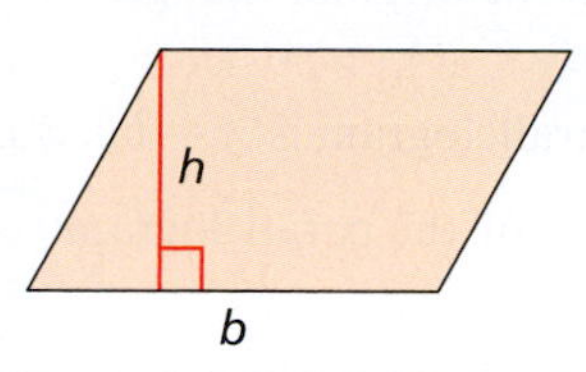

Trapezoid $A = \frac{1}{2}h(b + B)$

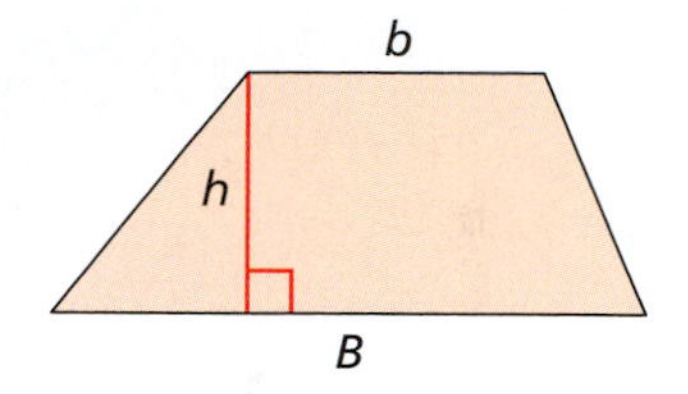

Classroom Tip

One way to understand the formula for the area of a trapezoid is to think of it as the height of the trapezoid multiplied by the mean of the lengths of the two bases.

$$A = h \cdot \frac{1}{2}(b + B)$$

Mathematical Practices

Reason Abstractly and Quantitatively

MP2 Mathematically proficient students create a coherent representation of the problem at hand, considering the units involved, attending to the meaning of quantities, not just how to compute them, and knowing and flexibly using different properties of operations.

EXAMPLE 5 Finding an Area in Real Life

A trapezoid can be used to approximate the shape of Scott County, Virginia. The population of Scott County is about 22,850. Estimate the population density of Scott County.

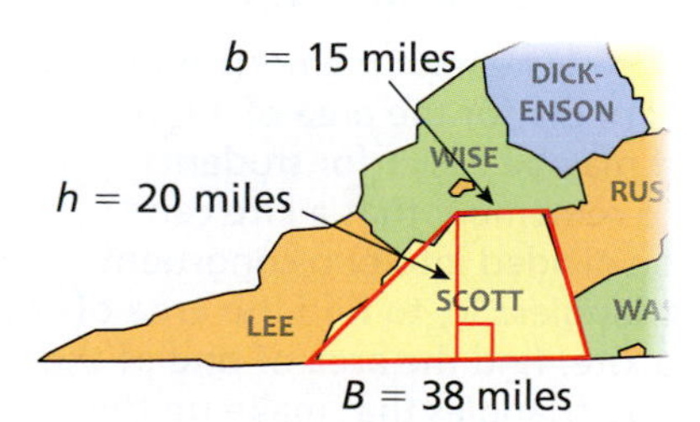

SOLUTION

Begin by estimating the area of Scott County.

$$A = \frac{1}{2}h(b + B) \quad \text{Write formula for area of a trapezoid.}$$

$$= \frac{1}{2}(20)(15 + 38) \quad \text{Substitute for } h, b, \text{ and } B.$$

$$= \frac{1}{2}(20)(53) \quad \text{Add.}$$

$$= 530 \quad \text{Multiply.}$$

The area of Scott County is about 530 square miles. To find the population density, divide the population by the area.

$$\text{Population density} = \frac{\text{Population}}{\text{Area}}$$

$$\approx \frac{22{,}850 \text{ people}}{530 \text{ mi}^2}$$

$$\approx 43 \text{ people/mi}^2$$

The population density is about 43 people per square mile.

Finding the Area of a Triangle

EXAMPLE 6 **Deriving the Formula for the Area of a Triangle**

Use the definition of the area of a parallelogram to derive a formula for the area of a triangle.

SOLUTION

Begin by drawing a representative triangle. Label the length of its **base** b. Label its **height** h.

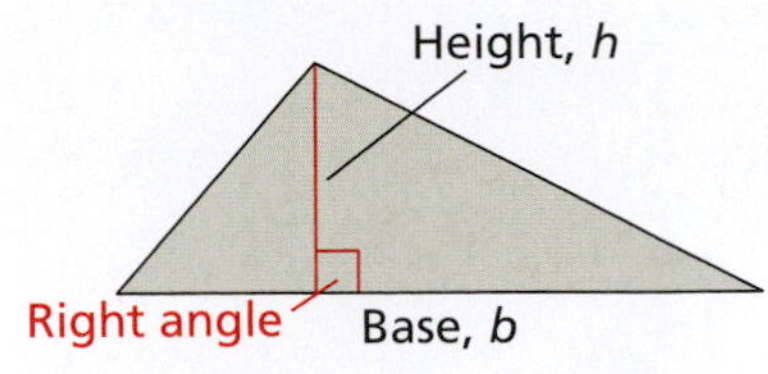

Make a copy of the triangle. Cut out the original triangle and the copy. Rotate the copy 180° so that it forms a parallelogram with the original triangle.

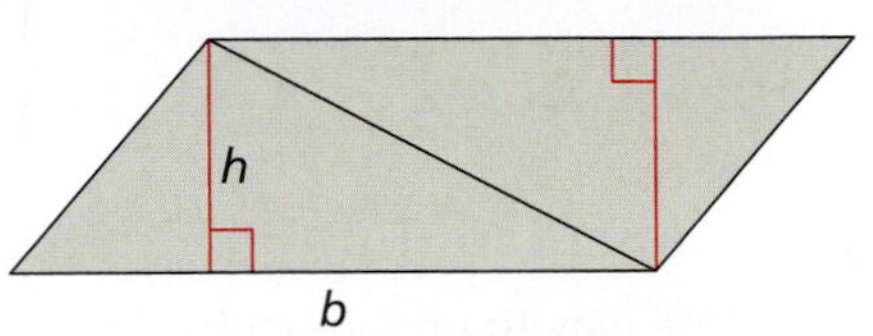

The area of the parallelogram is $A = bh$. So, the area of the original triangle is one-half of the area of the parallelogram: $A = \frac{1}{2}bh$.

Area of a Triangle and Area of a Kite

Triangle $A = \frac{1}{2}bh$

Words The area of a triangle is one-half the product of its base and its height.

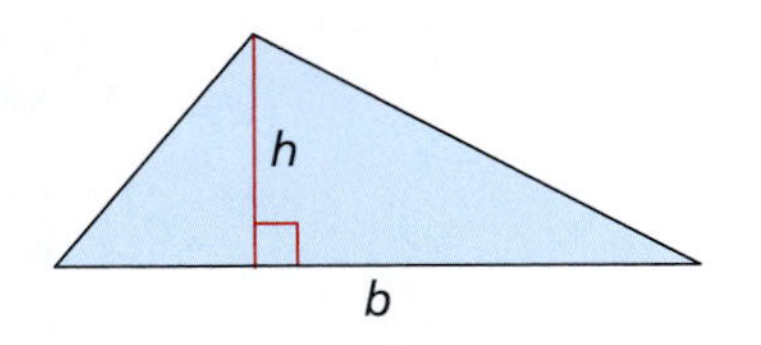

Kite $A = \frac{1}{2}ab$

Words The area of a kite is one-half the product of the lengths of its two diagonals.

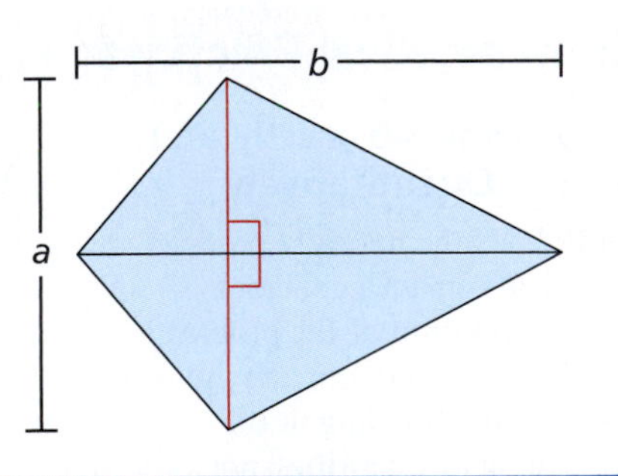

Classroom Tip

Instead of memorizing the formula for the area of a kite, it may be easier for students to remember that a kite can be divided into two congruent triangles. So, to find the area of a kite, find the area of one of the two triangles that make up the kite. Then multiply the area by 2.

EXAMPLE 7 **Finding the Area of a Hang Gliding Sail**

The wingspan of the triangular sail of a hang glider is 30 feet. How much fabric do you need to make the sail?

SOLUTION

The area of the sail is

$$A = \frac{1}{2}bh = \frac{1}{2}(30)(9) = 135.$$

You need 135 square feet of fabric to make the sail. Counting the hems, you would need even more fabric.

EXAMPLE 8 Finding the Area of a Kite

Find the area of the kite.

SOLUTION

The area of the kite is

$$A = \frac{1}{2}ab \qquad \text{Write formula for area of a kite.}$$

$$= \frac{1}{2}(10)(12) \qquad \text{Substitute for } a \text{ and } b.$$

$$= 60. \qquad \text{Multiply.}$$

The area of the kite is 60 square centimeters.

To find the area of a **composite figure**, which can be made up of triangles, squares, rectangles, and other two-dimensional figures, split the composite figure up into figures whose areas you know how to find.

EXAMPLE 9 Finding the Area of a Composite Figure

Find the area of the polygon. Each square on the grid is 1 inch by 1 inch.

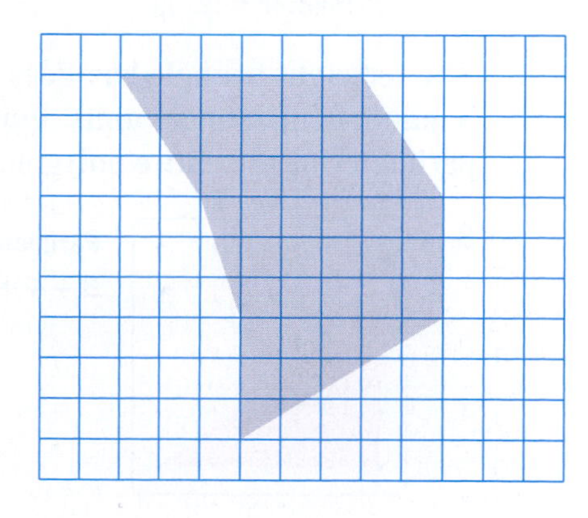

SOLUTION

Split up the polygon into three figures: a parallelogram, a trapezoid, and a triangle, as shown.

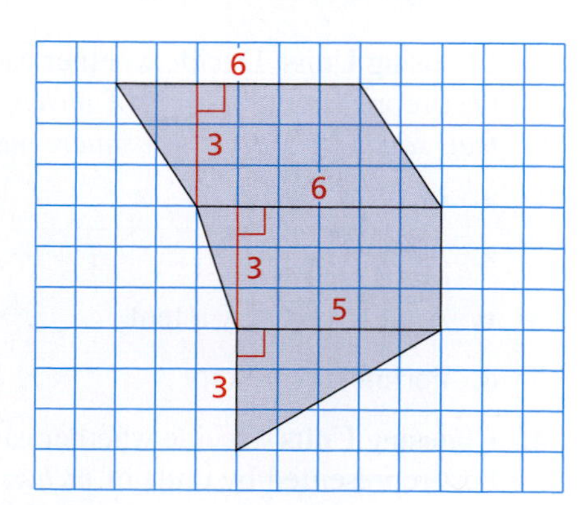

Area of parallelogram

$$A = bh$$

$$= 6(3)$$

$$= 18$$

Area of trapezoid

$$A = \frac{1}{2}h(b + B)$$

$$= \frac{1}{2}(3)(5 + 6)$$

$$= 16.5$$

Area of triangle

$$A = \frac{1}{2}bh$$

$$= \frac{1}{2}(5)(3)$$

$$= 7.5$$

The area of the polygon is the sum of the three areas.

Area of polygon $= 18 + 16.5 + 7.5 = 42$ square inches

Classroom Tip

When your students are finding the area of a polygon, like the polygon in Example 9, encourage them to check the reasonableness of the result. For instance, in Example 9, the polygon covers roughly the same area as a rectangle that is 5 inches by 9 inches. So, 42 square inches is a reasonable area for the polygon.

11.2 Exercises

Free step-by-step solutions to odd-numbered exercises at *MathematicalPractices.com.*

1. Writing a Solution Key Write a solution key for the activity on page 422. Describe a rubric for grading a student's work.

2. Grading the Activity In the activity on page 422, a student gave the answers below. Each question is worth 10 points. Assign a grade to each answer. For those that are incorrect, why do you think the student erred?

Sample Student Work

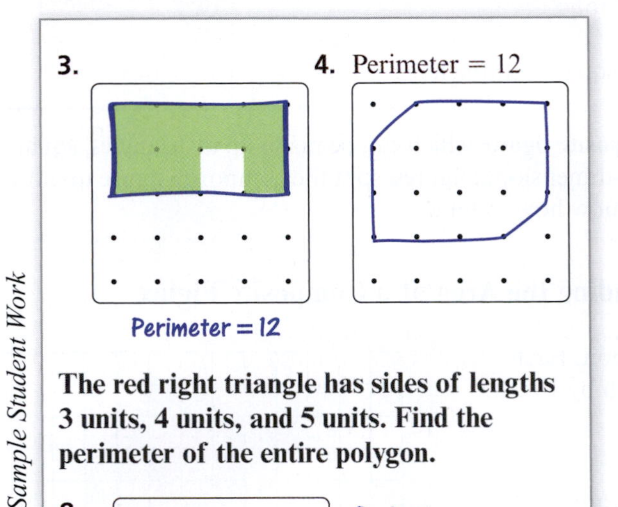

The red right triangle has sides of lengths 3 units, 4 units, and 5 units. Find the perimeter of the entire polygon.

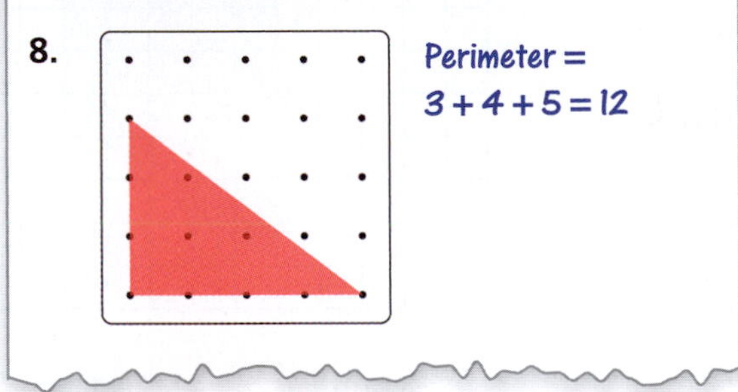

3. Choosing Units Decide whether each measure is best represented by units of *inches, square inches, feet, square feet, miles,* or *square miles.*

a. Length of a pencil

b. Area of a state

c. Area covered by a tent

d. Perimeter of a city

4. Choosing Units Decide whether each measure is best represented by units of *inches, square inches, feet, square feet, miles,* or *square miles.*

a. Length of a county

b. Area of a room

c. Area of the screen of a cell phone

d. Perimeter of a house

5. Perimeters of Polygons Find the perimeter of each polygon.

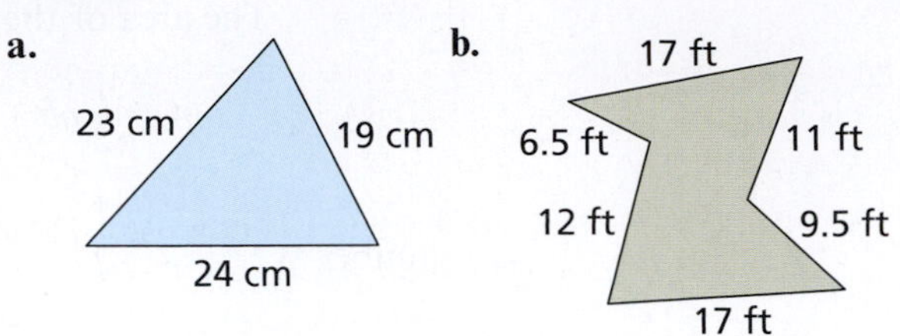

6. Perimeters of Polygons Find the perimeter of each polygon.

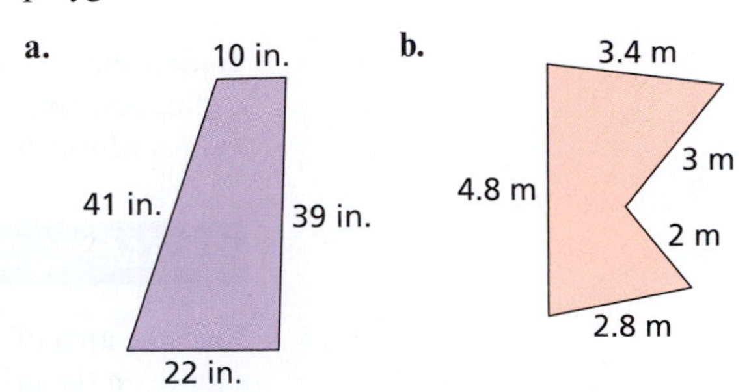

7. Perimeter and Area of a Rectangle The distance between 2 horizontal dots or 2 vertical dots is 1 unit. Find the perimeter and area of the rectangle by using formulas. Count the numbers of linear units and square units to check your answers.

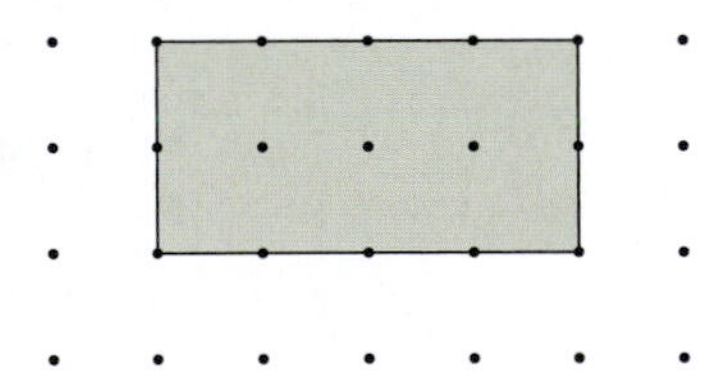

8. Perimeter and Area of a Rectangle The distance between 2 horizontal dots or 2 vertical dots is 1 unit. Find the perimeter and area of the rectangle by using formulas. Count the numbers of linear units and square units to check your answers.

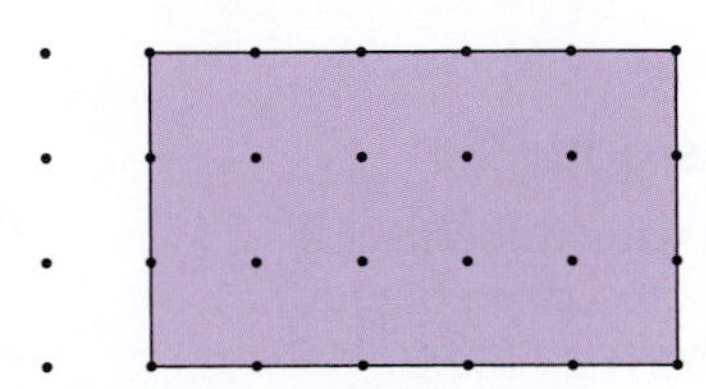

9. Perimeter and Area of a Rectangle Find the perimeter and area of the rectangle.

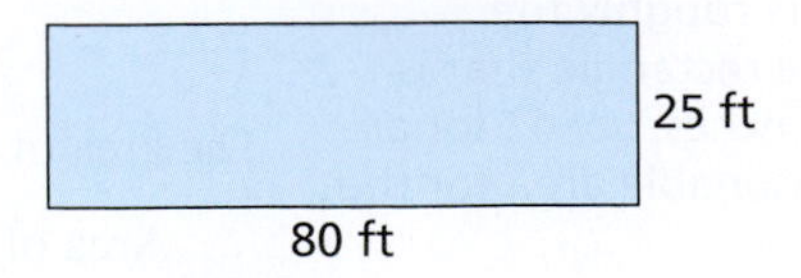

10. Perimeter and Area of a Rectangle Find the perimeter and area of the rectangle.

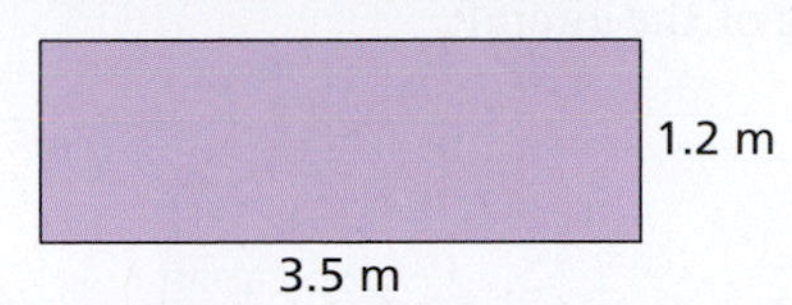

11. Area of a Parallelogram The distance between 2 horizontal dots or 2 vertical dots is 1 unit. Find the area of the parallelogram by using a formula, and then by counting the number of square units. Explain how you counted the number of square units.

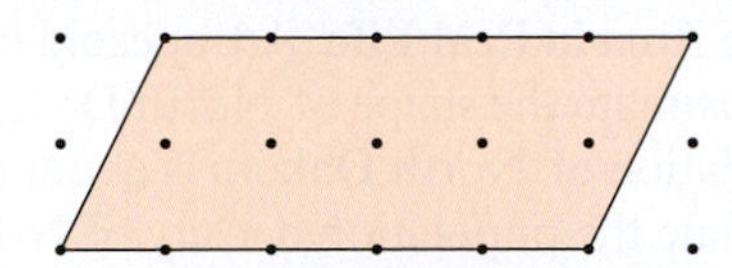

12. Area of a Trapezoid The distance between 2 horizontal dots or 2 vertical dots is 1 unit. Find the area of the trapezoid by using a formula, and then by counting the number of square units. Explain how you counted the number of square units.

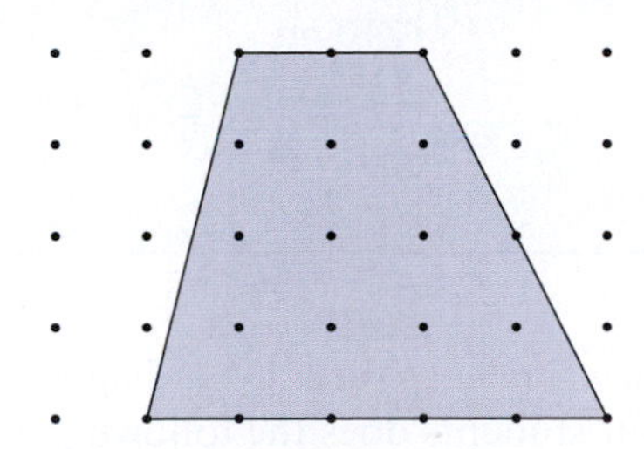

13. Area of a Parallelogram Find the area of the parallelogram.

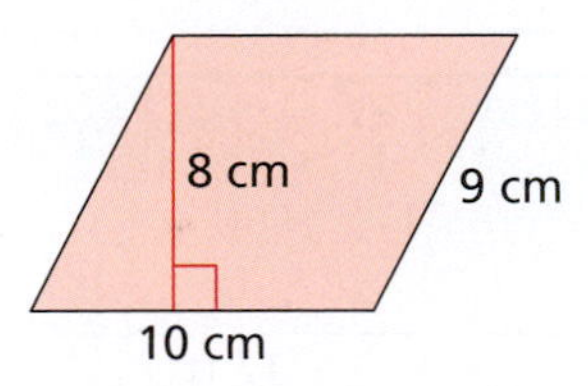

14. Area of a Parallelogram Find the area of the parallelogram.

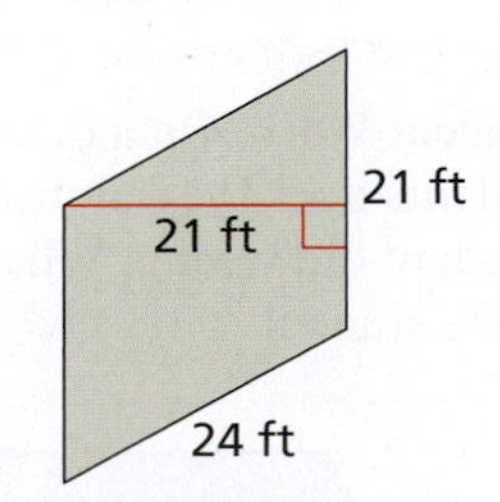

15. Area of a Trapezoid Find the area of the trapezoid.

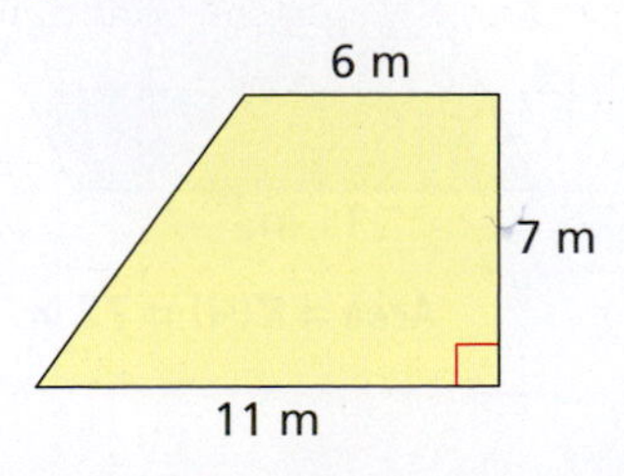

16. Area of a Trapezoid Find the area of the trapezoid.

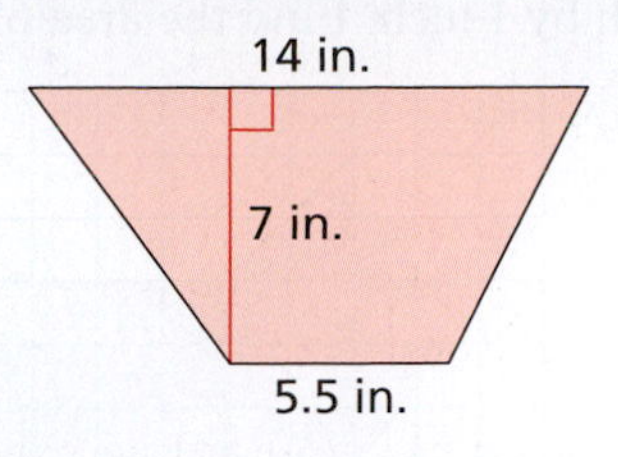

17. Area of a Triangle Find the area of the triangle.

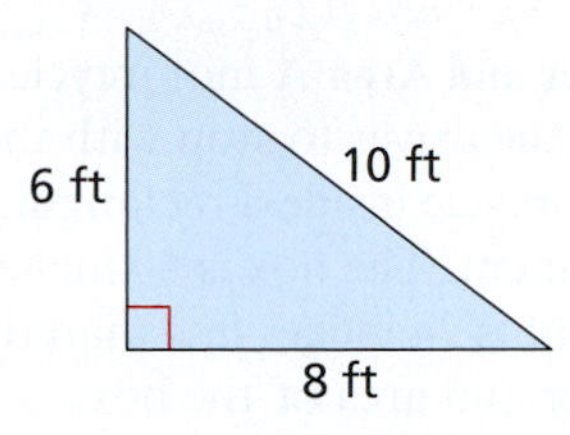

18. Area of a Triangle Find the area of the triangle.

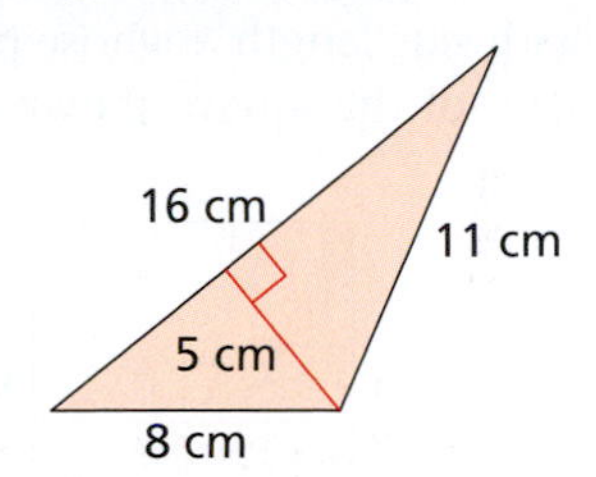

19. Area of a Kite Find the area of the kite.

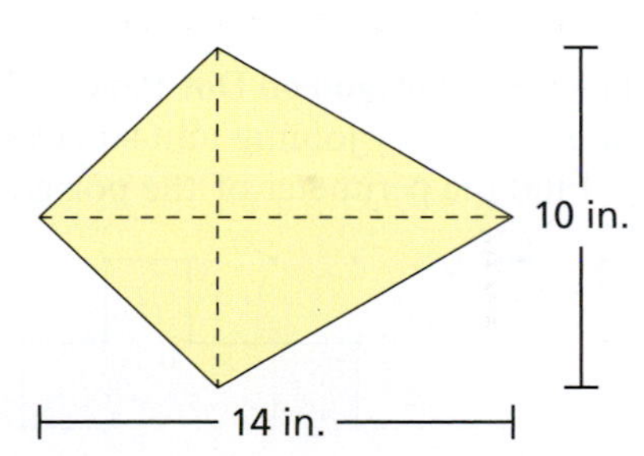

20. Area of a Kite Find the area of the kite.

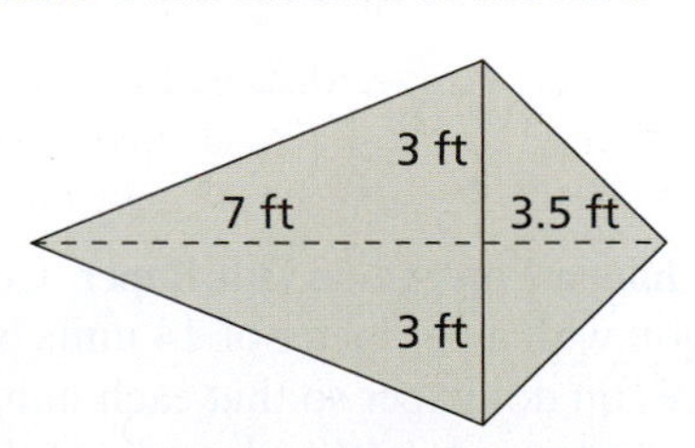

21. Area of a Composite Figure Each square on the grid is 1 inch by 1 inch. Find the area of the polygon.

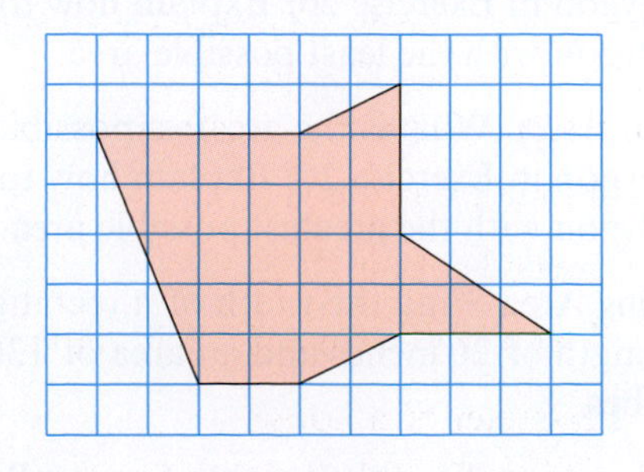

22. Area of a Composite Figure Each square on the grid is 1 inch by 1 inch. Find the area of the polygon.

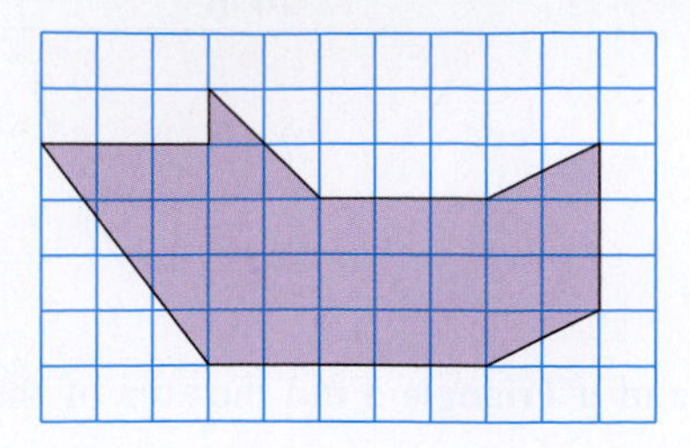

23. Perimeter and Area A motorcycle driving test requires the driver to stop with the front wheel of the motorcycle inside a rectangular box drawn on the pavement. The box is 60 inches long and has a width that is 24 inches less than the length. Find the perimeter and area of the box.

24. Perimeters of Regular Polygons Sketch a regular polygon with side length x whose perimeter is twice the perimeter of the square shown below.

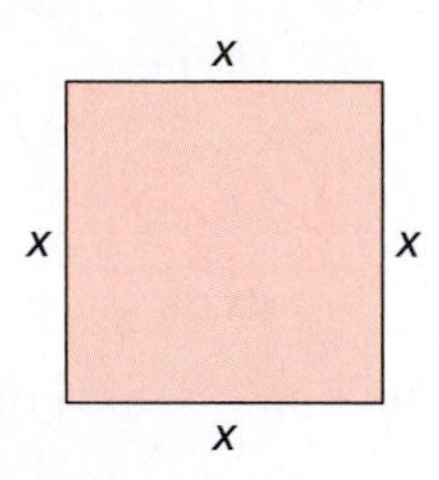

25. Perimeter of a Polygon on Dot Paper The polygon shown is formed by joining unit squares on dot paper. Find the perimeter of the polygon.

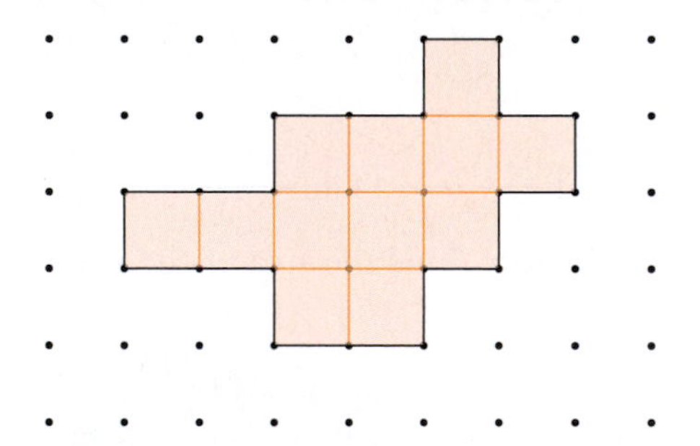

26. Sketching a Polygon on Dot Paper Construct a polygon with a perimeter of 14 units by joining unit squares on dot paper so that each unit square shares one complete side with at least one other unit square.

27. *Writing* What is the least possible area of the polygon in Exercise 26? Explain how to sketch a polygon with the least possible area.

28. *Writing* What is the greatest possible area of the polygon in Exercise 26? Explain how to sketch a polygon with the greatest possible area.

29. Using Area Find the width of a rectangle with a length of 26 inches and an area of 130 square inches.

30. Using Area Find the height of a parallelogram with a base of 8 feet and an area of 56 square feet.

31. Finding Area in Real Life Use the measurements shown to approximate the area of the top of one wing of the aircraft.

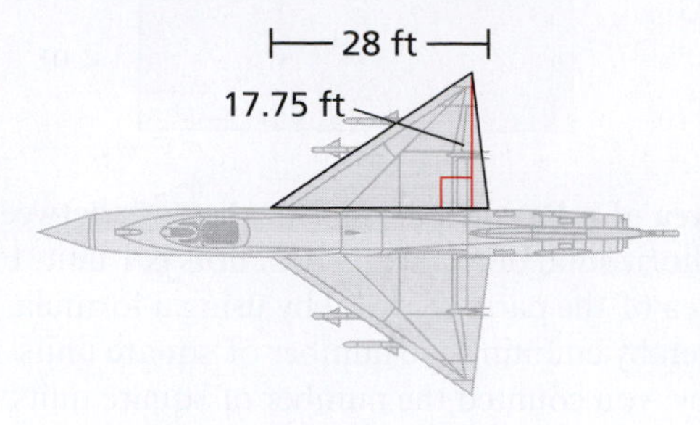

32. Using Area in Real Life A trapezoid can be used to approximate the shape of North Dakota. The population of North Dakota is about 672,000. Estimate the population density of North Dakota (to the nearest whole number of people per square mile).

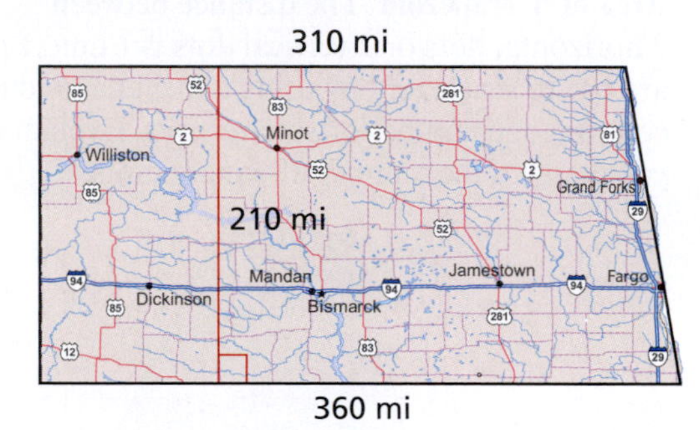

33. Grading Student Work On a diagnostic test, one of your students does the following work. Explain what the student did wrong. What topics would you encourage the student to review?

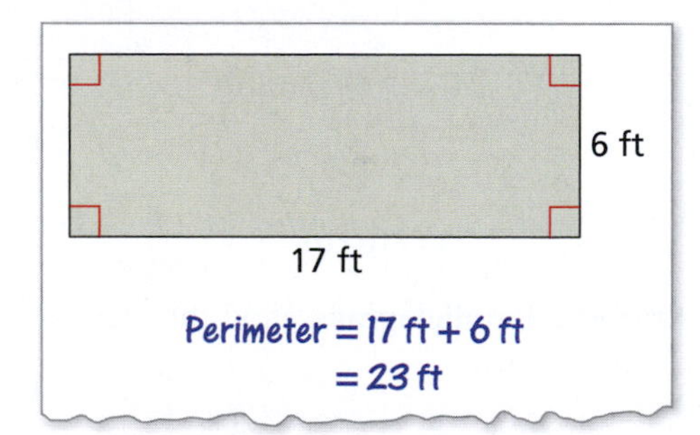

34. Grading Student Work On a diagnostic test, one of your students does the following work. Explain what the student did wrong. What topics would you encourage the student to review?

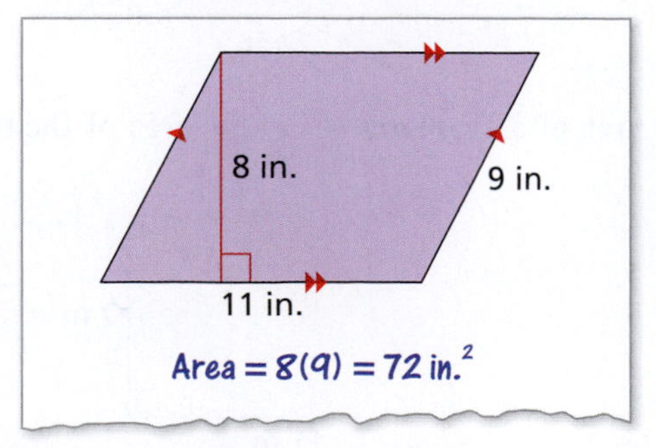

35. **Area of a Kite: In Your Classroom** Use the figure to explain how you can derive the formula for the area of a kite for your students by rearranging the parts of the kite to form a rectangle.

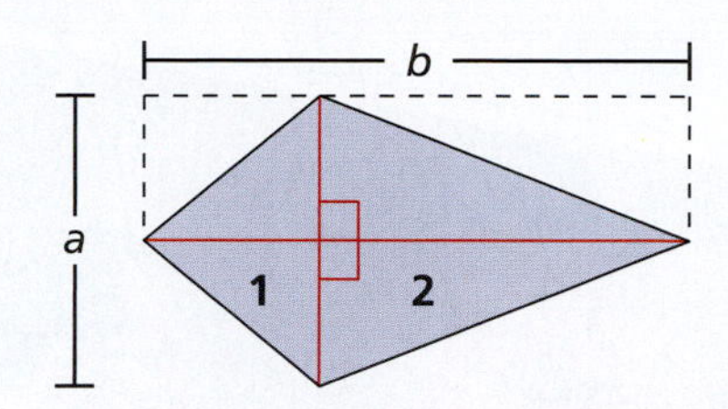

36. **Area of a Trapezoid: In Your Classroom** Use the figure to explain how you can derive the formula for the area of a trapezoid for your students based on the area of a parallelogram.

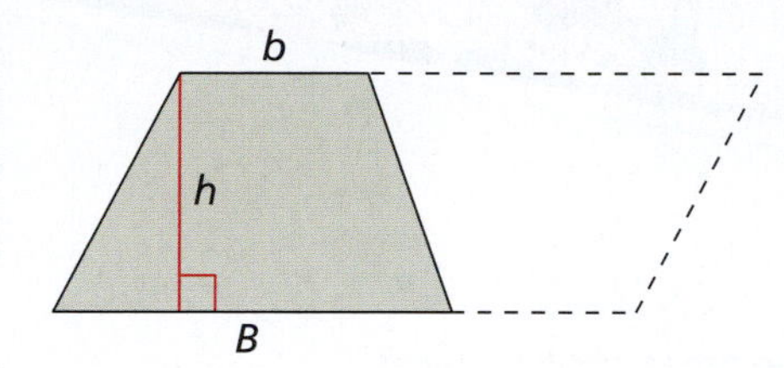

37. **Area of a Rectangle: In Your Classroom** Consider the areas of rectangles that have perimeters of 20 units.

a. Copy and complete the table to show the lengths, widths, and areas of rectangles with perimeters of 20 units.

Length, ℓ	9	8	7	6	5
Width, w					
Area, A					

b. What might you suggest to your students about finding the dimensions of a rectangle with the greatest area for a given perimeter?

c. Explain how your answer in part (b) is related to your answer in Exercise 28.

38. **Activity: In Your Classroom** Design an activity that you could use with your students to show that the greatest possible area for a parallelogram with the side lengths shown occurs when the height h is 6 units. What special type of parallelogram has the greatest area?

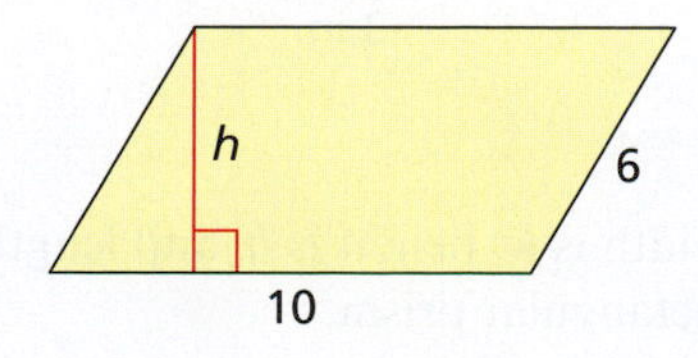

39. **Finding Area** Find the area of the shaded figure.

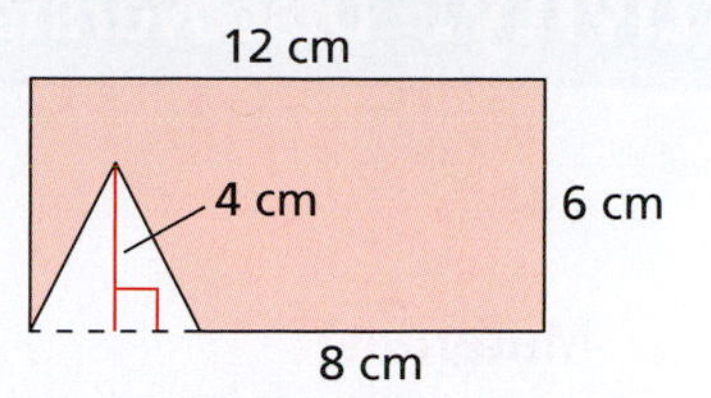

40. **Finding Area** Find the area of the shaded figure.

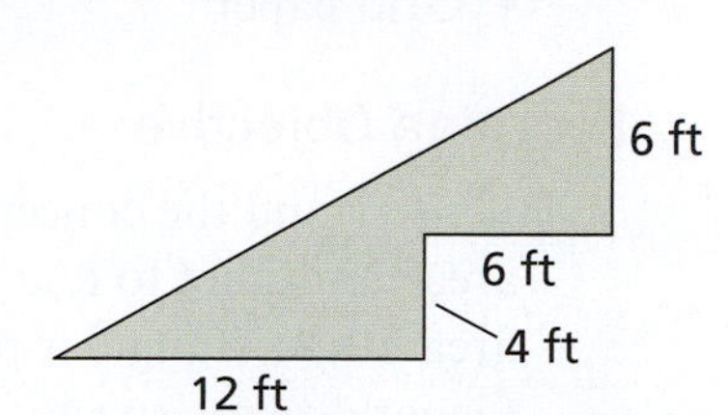

41. **Area in Envelope Design** Each figure shows a design for a 5-inch-by-7-inch envelope. Find the area of the paper used to make each envelope.

a.

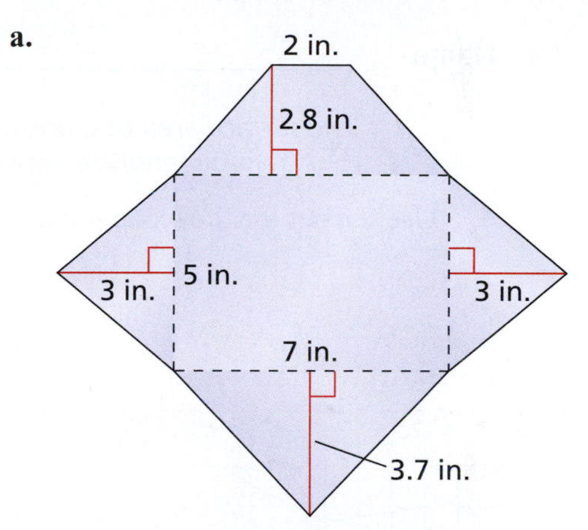

b.

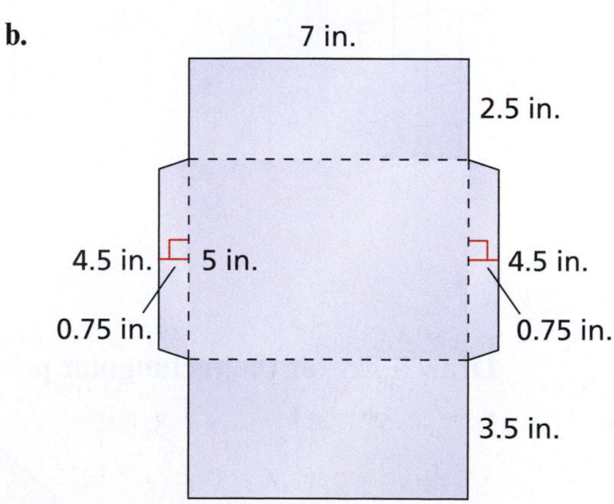

42. **Problem Solving** Answer the following questions about envelopes produced using the envelope designs in Exercise 41.

a. Which finished envelope contains less paper?

b. How much less paper, in square feet, is there in 5000 envelopes of the design mentioned in part (a) than in 5000 envelopes of the other design?

Activity: Discovering a Formula for the Surface Area of a Prism

Materials:

- Pencil
- Grid paper

Learning Objective:

Understand the concept of surface area. Use a net to find the surface area of a rectangular prism. Generalize the specific results to find a formula for the surface area of a rectangular prism.

Name ______________________________

The *surface area* of a prism is the sum of the areas of all its faces.
A two-dimensional representation of a solid is called a *net.*

1. Use the net for the rectangular prism to find its surface area.

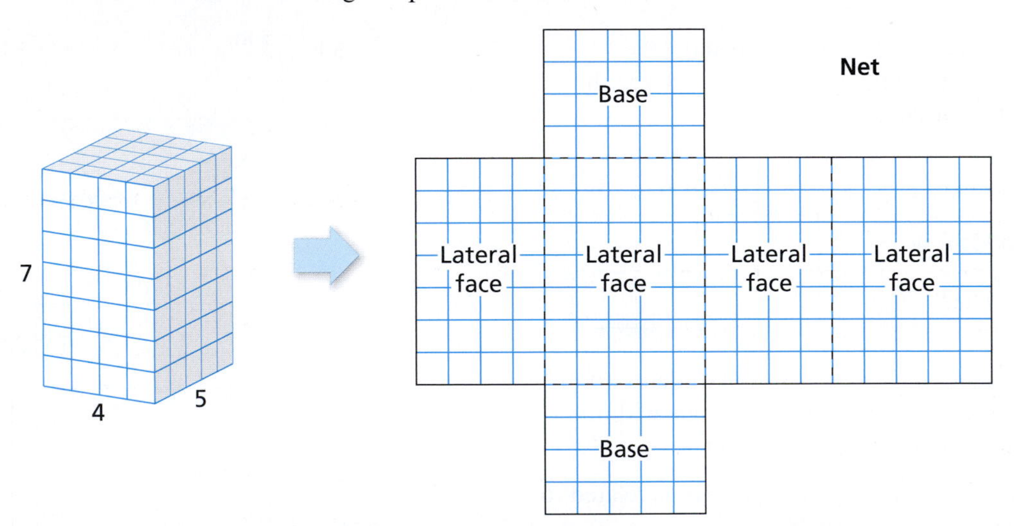

Draw a net for the rectangular prism. Then find the surface area of the prism.

2.

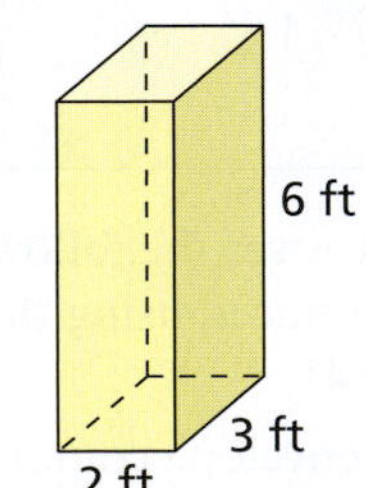

3.

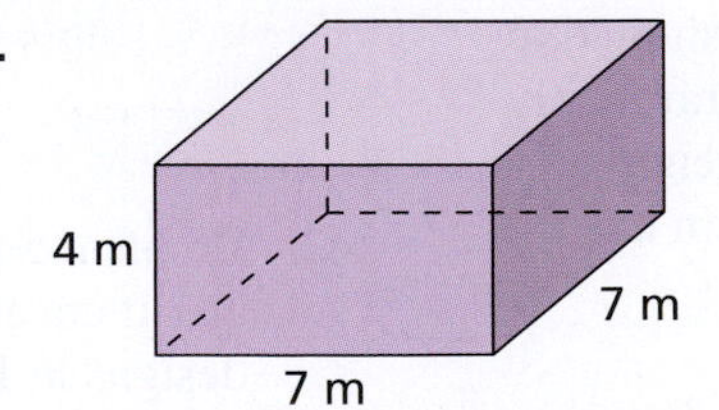

4.

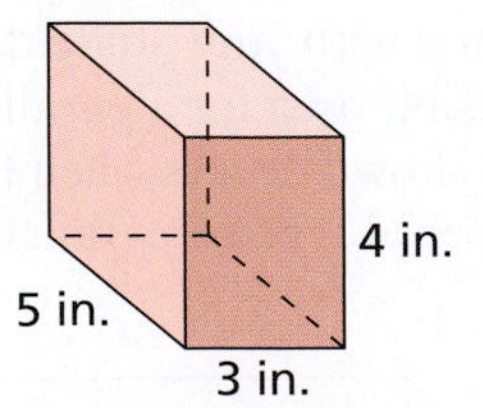

5. Draw a net for the general rectangular prism whose width is w, height is h, and length is ℓ. Then write a formula for the surface area of a rectangular prism.

11.3 Surface Areas of Polyhedra

Standards

Grades 6–8
Geometry
Students should solve real-world and mathematical problems involving surface area.

- Find the surface area of a rectangular prism.
- Find the surface area of a cube.
- Find the surface area of a pyramid.

Finding the Surface Area of a Rectangular Prism

Surface Area of a Rectangular Prism

Words The **surface area** S of a rectangular prism is the sum of the areas of the bases and the lateral faces.

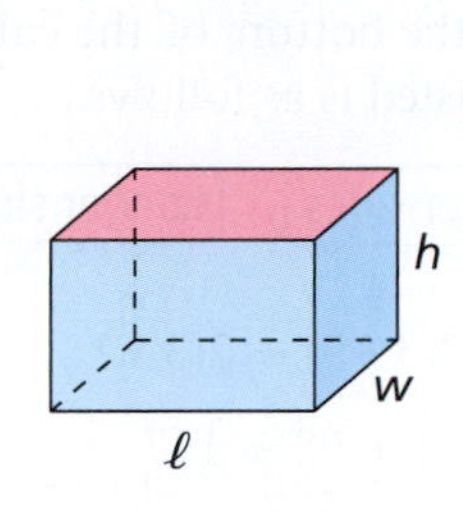

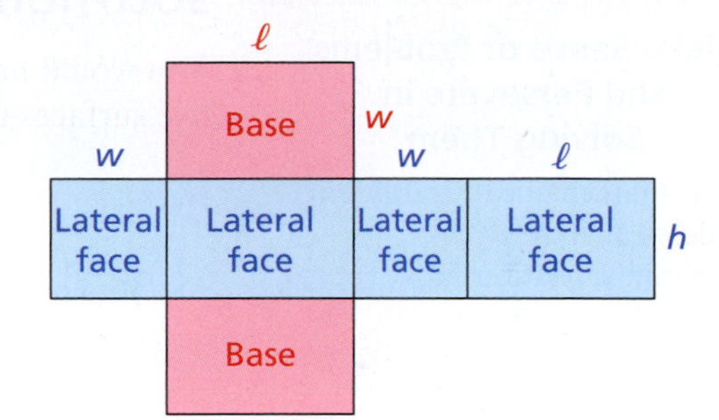

Algebra $S = 2\ell w + 2\ell h + 2wh$

Area of bases ($2\ell w$); Area of lateral faces ($2\ell h + 2wh$)

EXAMPLE 1 **Finding the Surface Area of a Rectangular Prism**

Find the surface area of the rectangular prism.

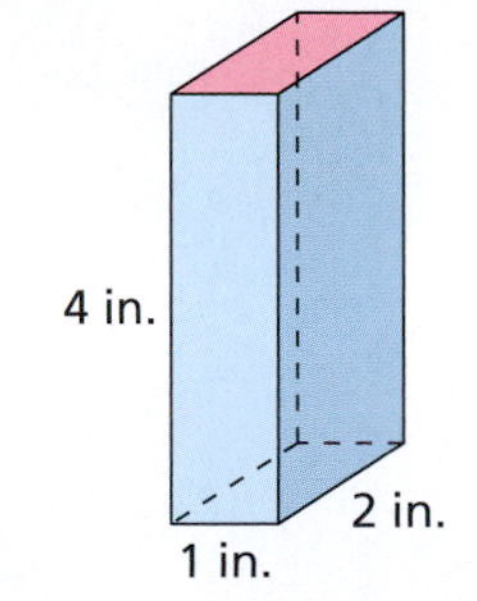

SOLUTION

$$S = 2\ell w + 2\ell h + 2wh$$
$$= 2(2)(1) + 2(2)(4) + 2(1)(4)$$
$$= 4 + 16 + 8$$
$$= 28$$

The surface area is 28 square inches. Use a net to check your result. Using grid paper, let each square represent 1 square inch.

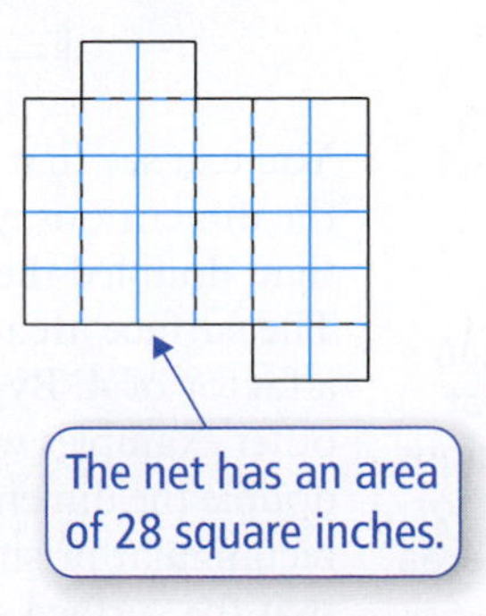

Classroom Tip

When introducing students to the concept of surface area, be sure to give them plenty of hands-on experience before providing them with a formula. In addition to using nets, you can also use building blocks to help build your students' intuitions.

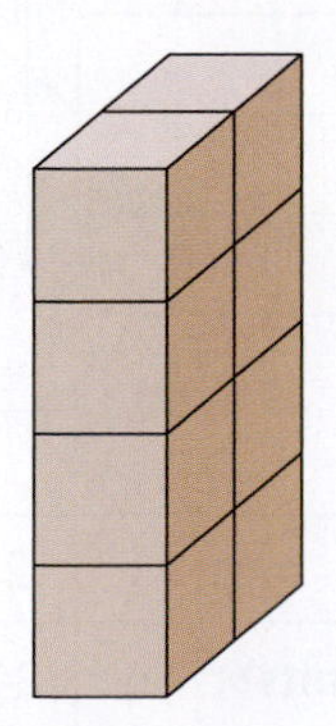

EXAMPLE 2 Finding a Surface Area in Real Life

One can of frosting covers about 280 square inches of cake. Is one can of frosting enough to cover the cake shown below? Explain your reasoning.

SOLUTION

You would not frost the bottom of the cake. So, the sum of the areas of the other five surfaces to be frosted is as follows.

Shorter sides | Longer sides | Top

$$\text{Frosted areas} = 2(3 \cdot 9) + 2(3 \cdot 13) + 9 \cdot 13$$
$$= 54 + 78 + 117$$
$$= 249$$

Because 249 in.^{2} < 280 in.^{2}, there is enough frosting to cover the cake.

Mathematical Practices

Make Sense of Problems and Persevere in Solving Them

MP1 Mathematically proficient students analyze givens, constraints, relationships, and goals.

EXAMPLE 3 Doubling the Dimensions of a Prism

When the dimensions of a rectangular prism are doubled, does the surface area of the prism double? Justify your answer.

SOLUTION

Use the problem-solving strategy *Solve a Similar, but Simpler, Problem*. Start with a cube.

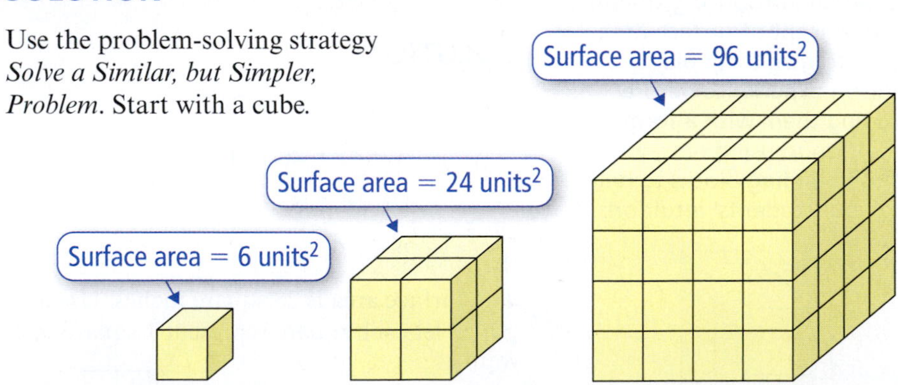

You can see that by doubling the dimensions, you have more than doubled the surface area. The surface area increases by a factor of 4. By trying several other examples where you double the dimensions of a rectangular prism, you will notice that the surface area *always* increases by a factor of 4.

Edge length of cube	1	2	4
Surface area (square units)	6	24	96

(Edge length: ×2, ×2; Surface area: ×4, ×4)

Classroom Tip

The result in Example 3 can be generalized to conclude that when you increase the dimensions of a rectangular prism by a factor of n, you increase the surface area by a factor of n^2.

Finding the Surface Area of a Cube

In Section 10.4, you learned that a cube is a regular polyhedron. Because a cube has 6 congruent square faces, the formula for its surface area is easier to understand.

Surface Area of a Cube

Words The **surface area** S of a cube is the product of 6 and the area of one of its square faces.

Algebra $S = 6s^2$

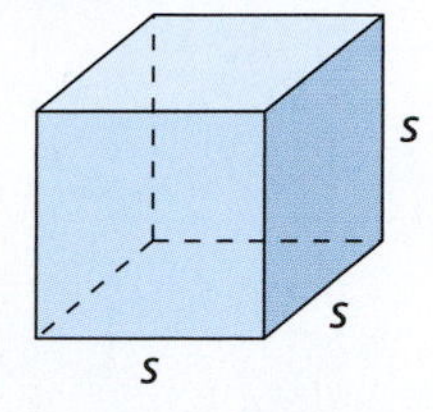

EXAMPLE 4 **Finding the Surface Area of a Cube**

Find the surface area of the cube.

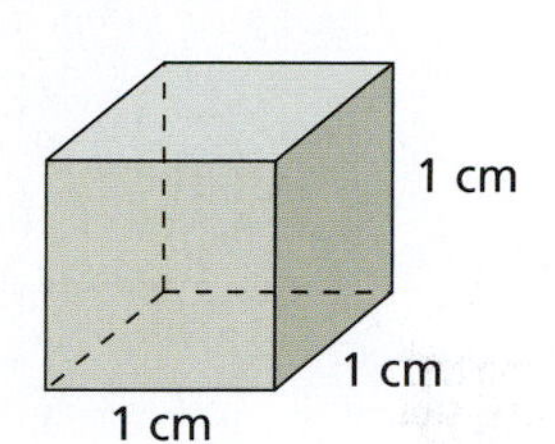

SOLUTION

$S = 6s^2$ Write the formula for surface area of a cube.

$= 6(1)^2$ Substitute 1 for s.

$= 6(1)$ Evaluate the power.

$= 6$ Multiply.

The surface area is 6 square centimeters.

> **Classroom Tip**
> In Section 11.4, you will learn that the cube in Example 4 has a volume of 1 cubic centimeter. One cubic centimeter is equivalent to one milliliter. This is an intentional connection between length and volume in the metric system.

EXAMPLE 5 **Finding a Surface Area in Real Life**

When comparing ice blocks with the same volume, the greater the surface area of the ice block, the faster it will melt. An ice sculptor has an ice block that measures 3 feet by 1 foot by 1 foot, as shown.

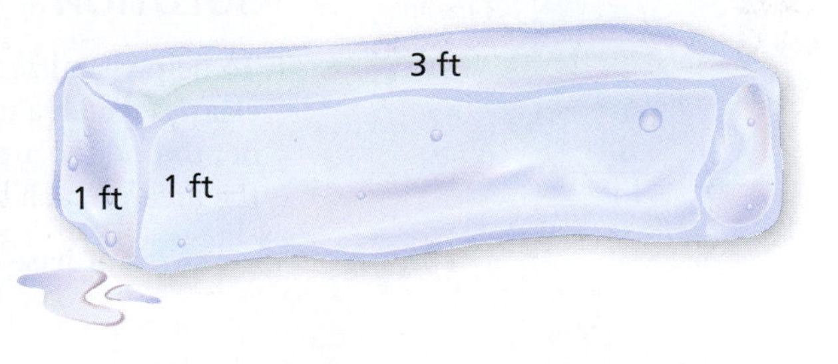

The sculptor cuts the block into three cubes, as shown. Will the three smaller blocks melt faster than the original ice block? Explain your reasoning.

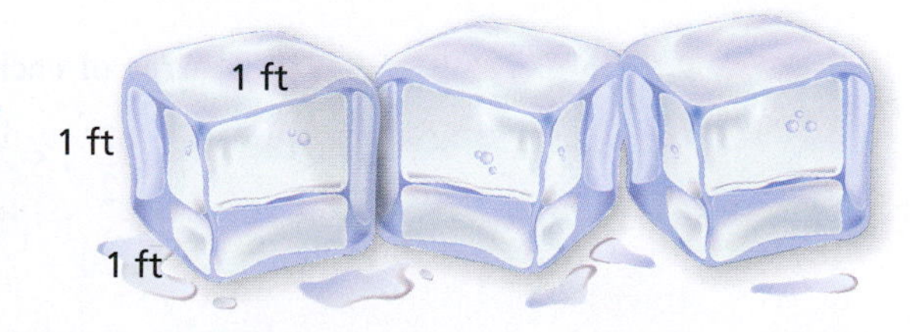

SOLUTION

Surface area of original ice block

$S = 2(3 \cdot 1) + 2(3 \cdot 1) + 2(1 \cdot 1)$

$= 6 + 6 + 2$

$= 14$

Surface area of three smaller ice blocks

$S = 3 \cdot 6(1)^2$

$= 3 \cdot 6$

$= 18$

Because 18 in.^{2} > 14 in.^{2}, the three smaller ice blocks have a greater surface area. So, they will melt faster than the original ice block.

Finding the Surface Area of a Pyramid

A **regular pyramid** is a pyramid whose base is a *regular polygon* and whose lateral faces are congruent triangles. The **slant height** of a regular pyramid is the height of the triangular lateral face.

Surface Area of a Pyramid

Words The **surface area** S of a pyramid is the sum of the area of the base and the areas of the lateral faces.

$S =$ area of base + areas of lateral faces

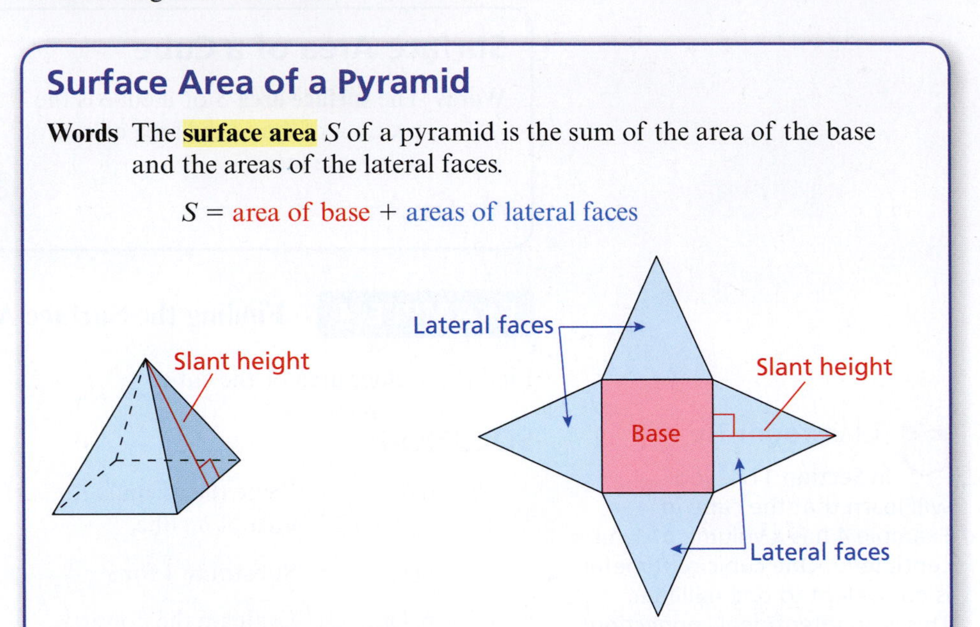

EXAMPLE 6 **Finding the Surface Area of a Pyramid**

Find the surface area of the regular pyramid shown at the left.

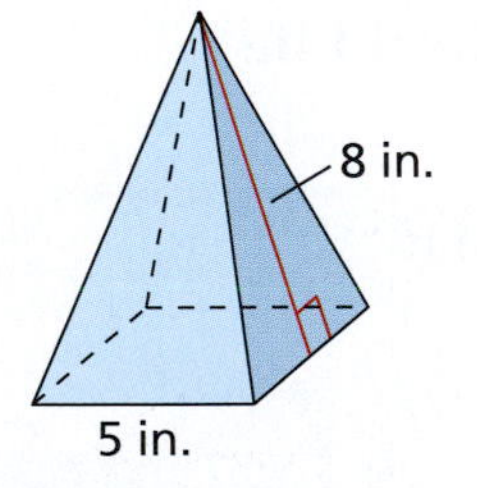

SOLUTION

To help visualize the surface area clearly, sketch a net. Then use the net to find the area of the base and the area of each lateral face.

Area of base

$A = 5 \cdot 5$

$= 25$

Area of each lateral face

$A = \frac{1}{2} \cdot 5 \cdot 8$

$= 20$

Finally, sum the areas to find the surface area.

$S =$ area of base + areas of lateral faces

$= 25 + 20 + 20 + 20 + 20$

$= 105$

The pyramid has four congruent lateral faces. Count the area 4 times.

The surface area is 105 square inches.

EXAMPLE 7 Finding the Surface Area of a Pyramid

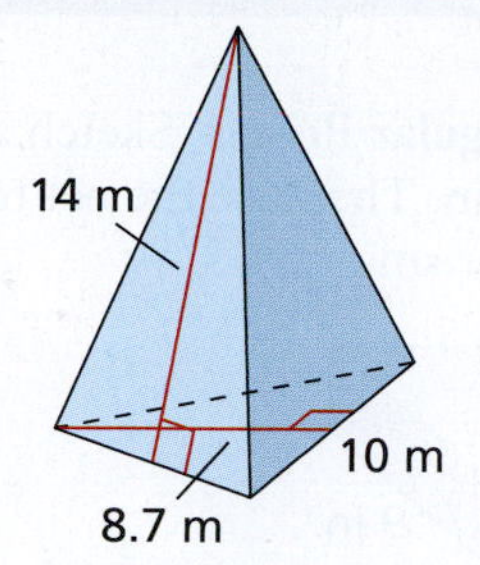

Find the surface area of the regular pyramid shown at the left.

SOLUTION

To help visualize the surface area clearly, sketch a net. Then use the net to find the area of the base and the area of each lateral face.

Area of base

$$A = \frac{1}{2} \cdot 10 \cdot 8.7$$

$$= 43.5$$

Area of each lateral face

$$A = \frac{1}{2} \cdot 10 \cdot 14$$

$$= 70$$

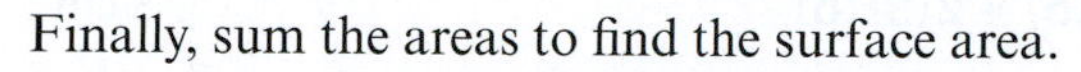

Finally, sum the areas to find the surface area.

$$S = \text{area of base} + \text{areas of lateral faces}$$

$$= 43.5 + 70 + 70 + 70$$

$$= 253.5$$

The pyramid has three congruent lateral faces. Count the area 3 times.

The surface area is 253.5 square meters.

EXAMPLE 8 Finding a Surface Area in Real Life

You are putting a new roof on the building shown. The roof is shaped like a square pyramid. One bundle of shingles covers 25 square feet. How many bundles should you buy to cover the roof?

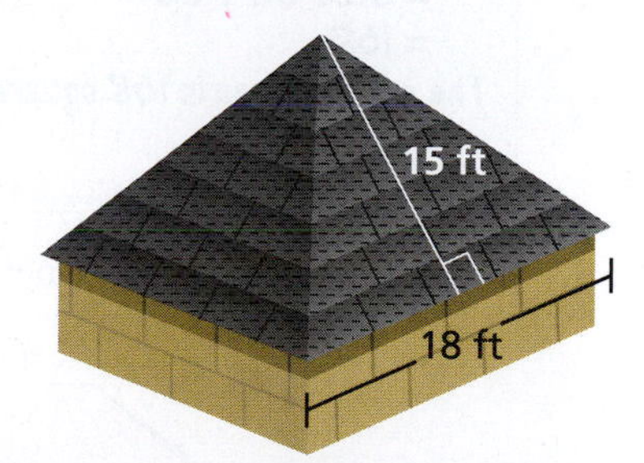

SOLUTION

The parts of the roof that require shingles are the lateral faces of the pyramid. So, find the sum of the areas of the lateral faces.

Area of each lateral face

$$A = \frac{1}{2} \cdot 18 \cdot 15$$

$$= 135$$

There are four congruent lateral faces. So, the sum of the areas of the lateral faces is

$$135 + 135 + 135 + 135 = 540.$$

Because one bundle of shingles covers 25 square feet, it will take $540 \div 25 = 21.6$ bundles to cover the roof. This means that you should buy 22 bundles. To account for waste and ensure you have enough shingles, you would probably want to buy 23 or 24 bundles.

11.3 Exercises

Free step-by-step solutions to odd-numbered exercises at MathematicalPractices.com.

1. **Writing a Solution Key** Write a solution key for the activity on page 432. Describe a rubric for grading a student's work.

2. **Grading the Activity** In the activity on page 432, a student gave the answers below. Each question is worth 10 points. Assign a grade to each answer. For those that are incorrect, why do you think the student erred?

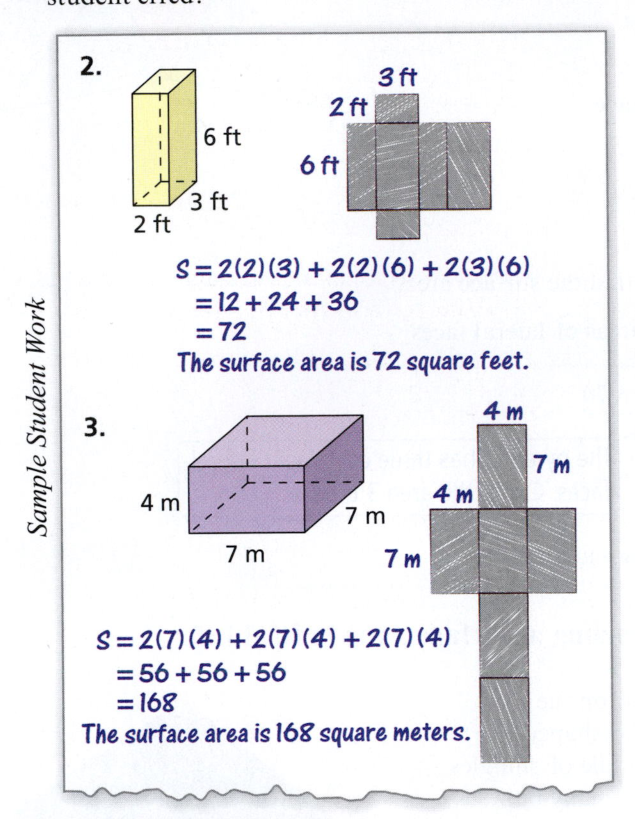

3. **Sketching a Net** Sketch a net of the polyhedron.

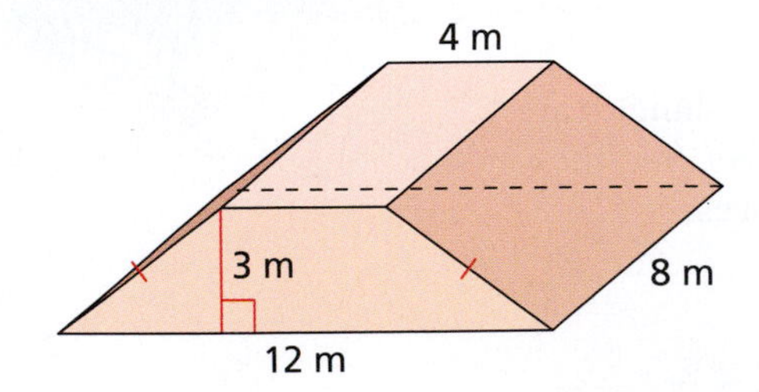

4. **Sketching a Net** Sketch a net of the regular pyramid.

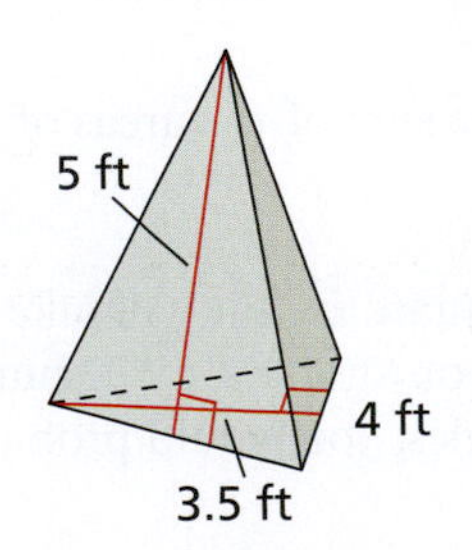

5. **Surface Areas of Rectangular Prisms** Sketch a net of each rectangular prism. Then use the net to find the surface area of the prism.

a.

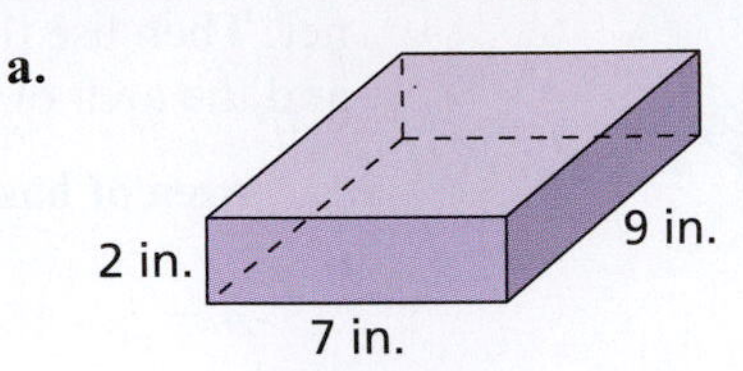

b.

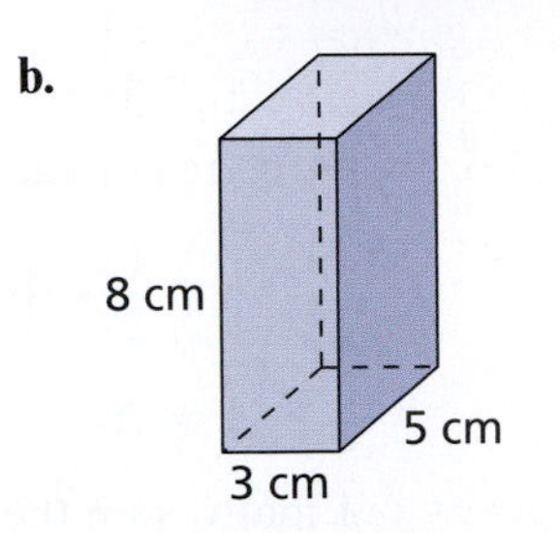

6. **Surface Areas of Rectangular Prisms** Sketch a net of each rectangular prism. Then use the net to find the surface area of the prism.

a.

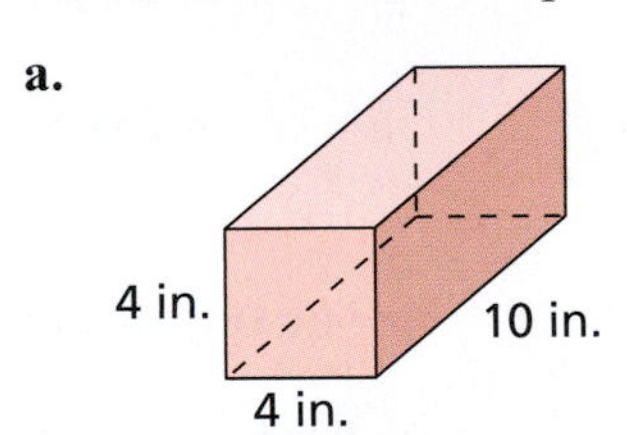

b.

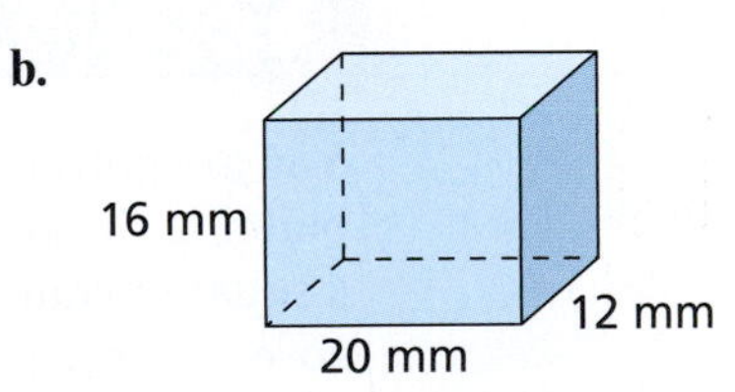

7. **Surface Area of a Cube** Sketch a net of the cube. Then use the net to find the surface area of the cube.

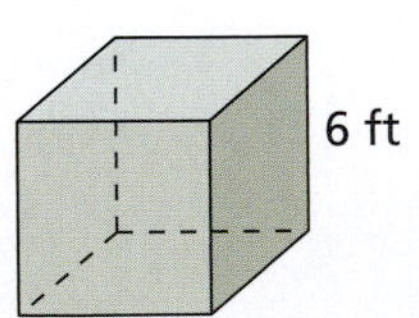

8. **Surface Area of a Cube** Sketch a net of the cube. Then use the net to find the surface area of the cube.

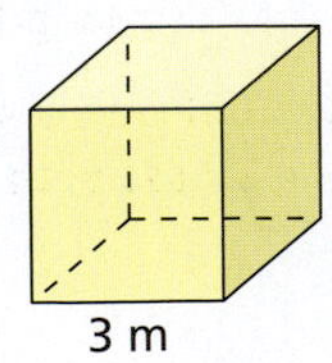

9. Surface Areas of Pyramids Sketch a net of each regular pyramid. Then use the net to find the surface area of the pyramid.

a.

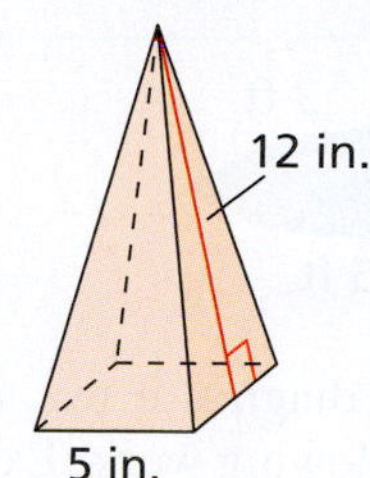

b.

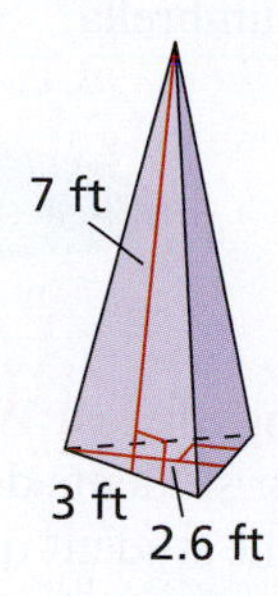

10. Surface Areas of Pyramids Sketch a net of each regular pyramid. Then use the net to find the surface area of the pyramid.

a.

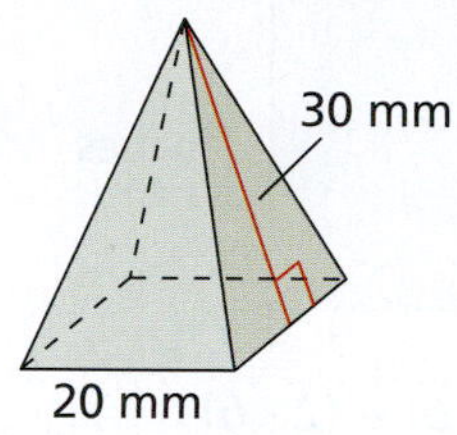

b.

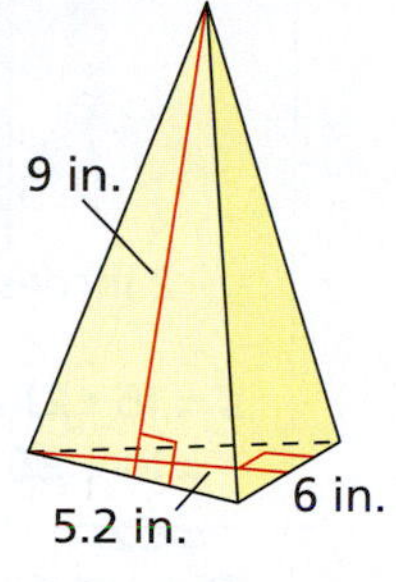

11. Surface Areas of a Prism Sketch a net of the prism. Then use the net to find the surface area of the prism by finding the sum of the areas of its faces.

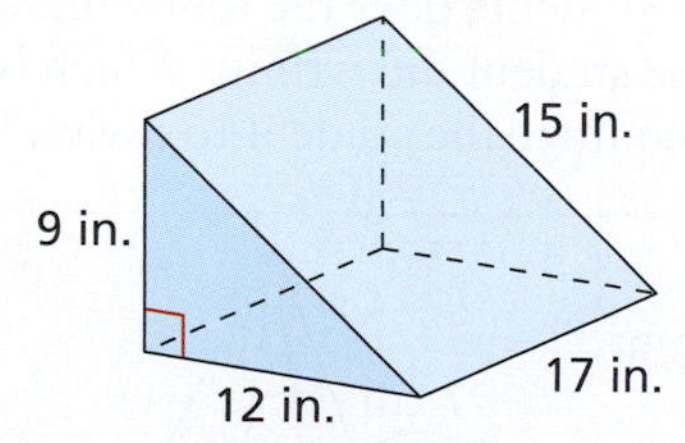

12. Surface Areas of a Prism Sketch a net of the prism. Then use the net to find the surface area of the prism by finding the sum of the areas of its faces.

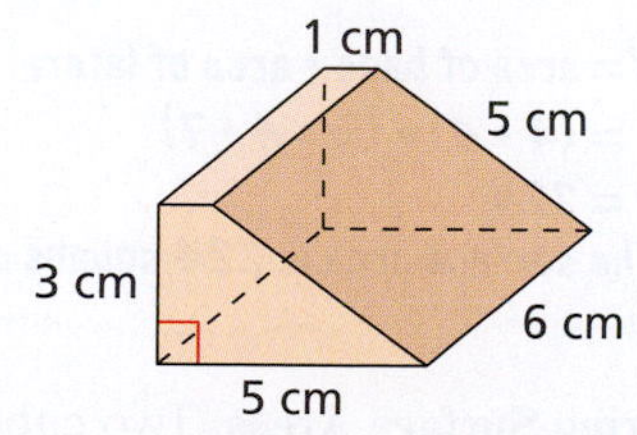

13. Surface Areas of Polyhedra Use a formula to find the surface area of each polyhedron described.

a. A cube with an edge length of 4 meters

b. A square pyramid with a slant height of 5.5 inches and a base with a side length of 2 inches

14. Surface Areas of Polyhedra Use a formula to find the surface area of each polyhedron described.

a. A rectangular prism that is 4 feet wide, as tall as it is wide, and twice as long as it is tall

b. A regular hexagonal pyramid with a base area of 93.5 square centimeters, a base side length of 6 centimeters, and a slant height of 6 centimeters

15. True or False? Tell whether each statement is *true* or *false*. Explain your reasoning.

a. The surface area of a polyhedron equals the area of its net.

b. When you triple the edge length of a cube, the surface area is also tripled.

16. True or False? Tell whether each statement is *true* or *false*. Explain your reasoning.

a. The surface area of a pyramid equals the sum of the areas of its lateral faces.

b. When the dimensions of a rectangular prism are halved, the surface area decreases by a factor of 4.

17. Finding Surface Areas Each solid is composed of rectangular prisms. Find the surface area of the solid.

a.

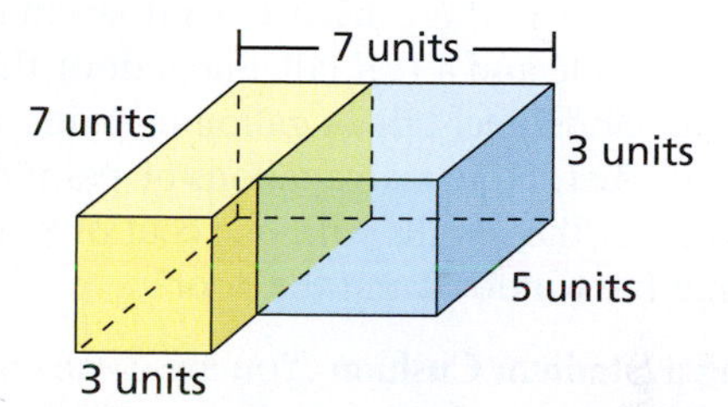

b.

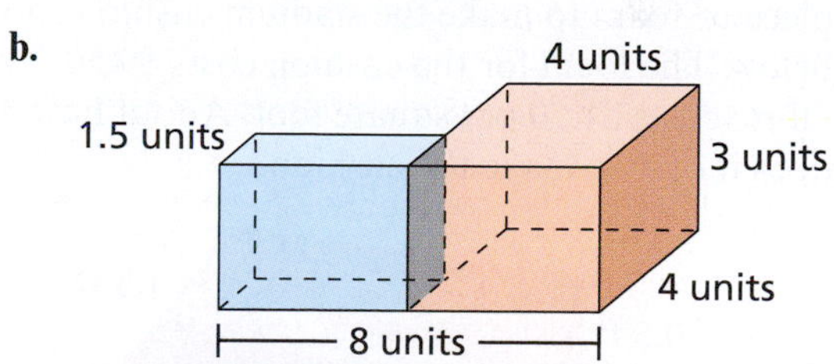

18. Finding Surface Areas Each solid is composed of a rectangular prism and a regular pyramid. Find the surface area of the solid.

a.

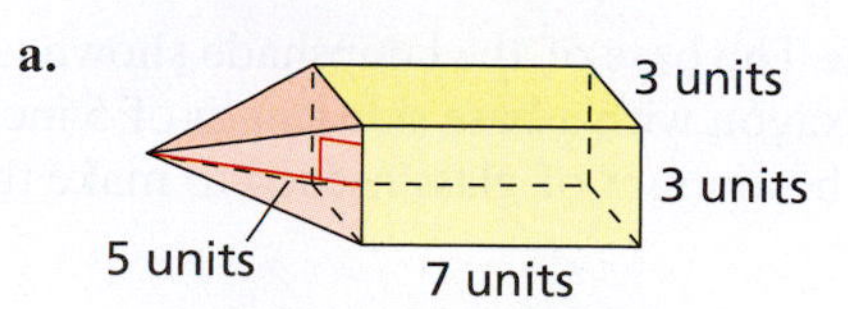

b.

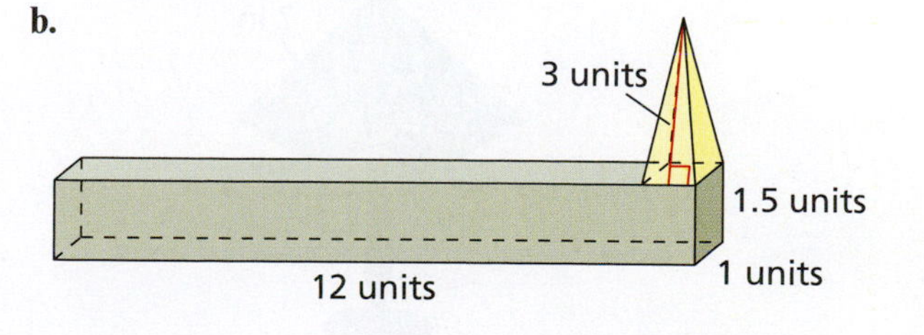

19. **Problem Solving** The surface area of a cube is 216 square centimeters. Find the length of each edge of the cube.

20. **Problem Solving** The surface area of a square pyramid is 176 square inches. The side length of the base is 8 inches. What is the slant height?

21. **Wrapping a Package** A cube-shaped package has an edge length of 1.5 feet. What is the least amount of wrapping paper needed to cover the package?

22. **Making a Tent** What is the least amount of fabric needed to make the tent shown below? (*Hint*: The floor of the tent should also be fabric.)

23. **Painting a Room** You are painting a room that is 14 feet long, 10 feet wide, and 8 feet high. On the walls of the room, there are two windows that are each 2 feet wide and 4 feet tall, and a door that is 4 feet wide and 7 feet tall. A gallon of paint covers 350 square feet. How many gallons of paint do you need to cover the 4 walls with one coat of paint, not including the windows and the door?

24. **Making a Stadium Cushion** You are covering a piece of foam to make the stadium cushion shown below. The foam for the cushion costs \$9.50. The fabric costs \$1.50 per square foot. About how much does it cost to make the cushion?

25. **Lampshade** The base of the lampshade shown is a regular hexagon with a base side length of 5 inches. Estimate the amount of glass needed to make the lampshade.

26. **Beach Umbrella** The beach umbrella shown is shaped like an octagonal pyramid. Estimate the amount of fabric that is needed to make the beach umbrella.

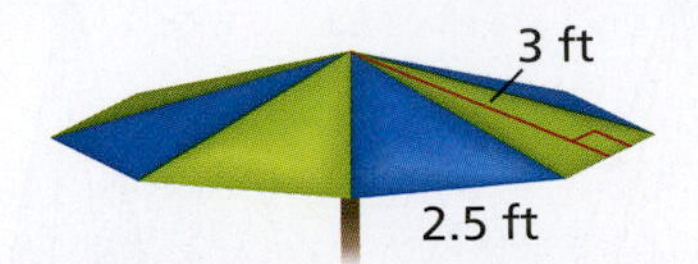

27. **Grading Student Work** On a diagnostic test, one of your students does the following work. Explain what the student did wrong. Which topics would you encourage the student to review?

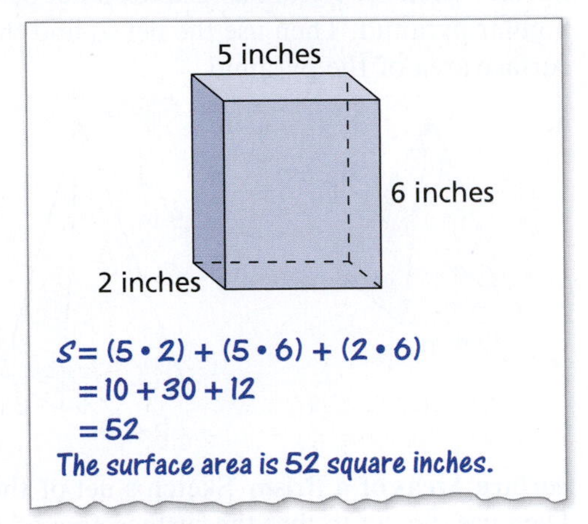

28. **Grading Student Work** On a diagnostic test, one of your students does the following work. Explain what the student did wrong. Which topics would you encourage the student to review?

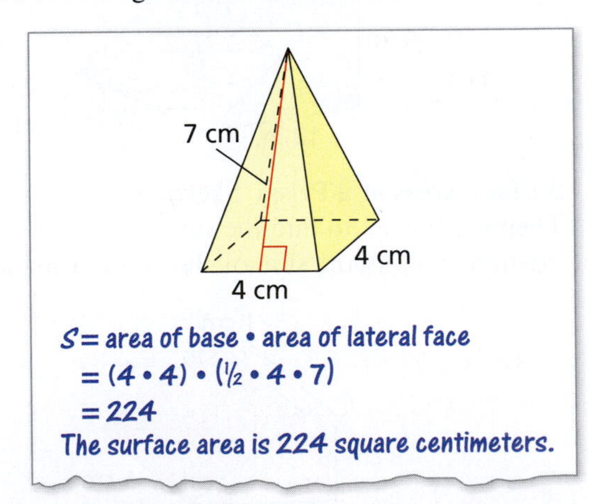

29. **Comparing Surface Areas** Two cubes have edge lengths of 6 inches and 8 inches. What is the ratio of the surface area of the smaller cube to the surface area of the larger cube?

30. **Comparing Surface Areas** Two cubes have surface areas of 96 square millimeters and 150 square millimeters. What is the ratio of the edge length of the smaller cube to the edge length of the larger cube?

31. **Pasta Boxes** A small box of pasta is made out of approximately 135 square inches of cardboard. The dimensions of a small box of pasta are doubled to make a large box of pasta. About how much cardboard do you need to make a large box of pasta?

32. **Cereal Boxes** The dimensions of a value-size box of cereal are 1.5 times the dimensions of a regular box of cereal. Approximately 3375 square centimeters of cardboard are used to make a value-size box of cereal. About how much cardboard is used to make a regular box of cereal?

33. **Problem Solving** A cube with a surface area of 6 square inches is removed from the corner of a cube with a surface area of 96 square inches, as shown.

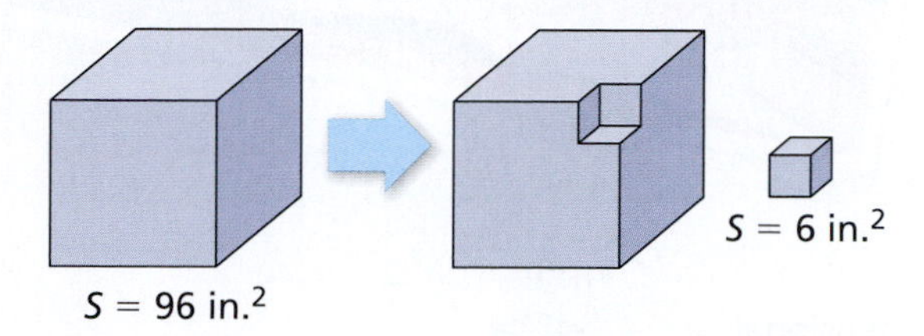

a. What effect do you think removing the smaller cube from the larger cube has on its surface area? Explain your reasoning.

b. Find the surface area of the new solid.

c. Was your answer in part (a) correct? Explain.

34. **Problem Solving** A solid is formed by removing a rectangular prism from the edge of a larger rectangular prism, as shown.

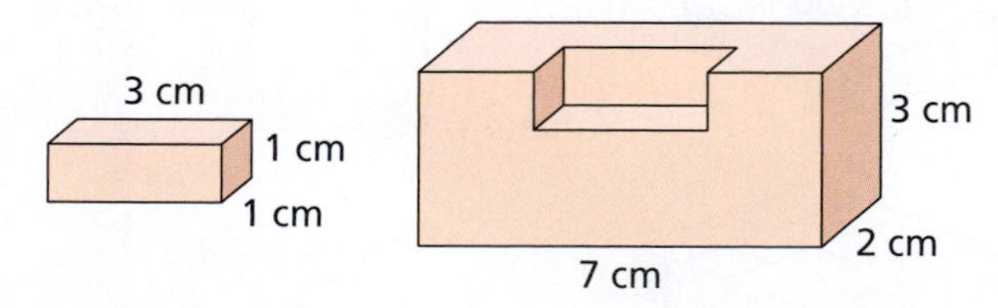

a. Find the original surface area of the larger prism.

b. Find the surface area of the solid formed by removing the smaller prism from the larger prism.

c. What effect does removing the smaller prism have on the surface area?

d. Why do you think the effect is different here than in Exercise 33?

35. **Problem Solving** A smaller rectangular prism is cut out of a larger prism as shown. What is the surface area of the new solid?

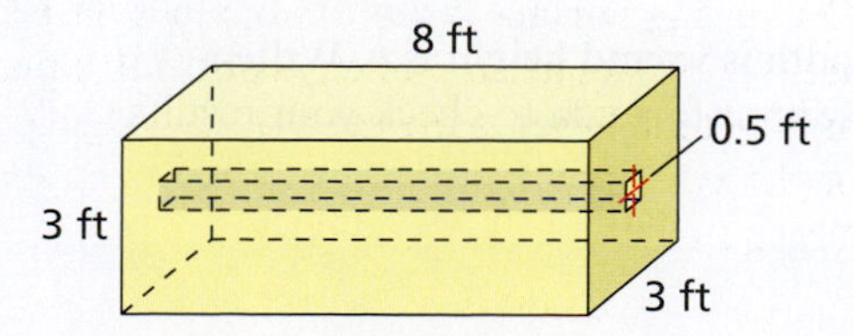

36. **Problem Solving** The surface area of a cube is 294 square inches. A hole is cut out of the cube as shown. What is the surface area of the new solid?

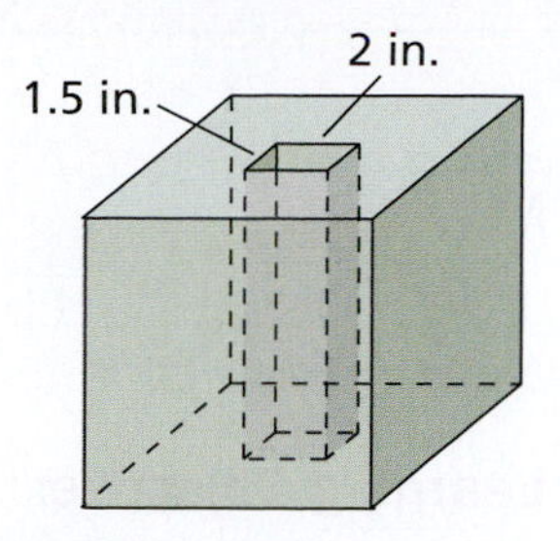

37. **Writing** The *lateral surface area* of a three-dimensional figure is the surface area of the figure excluding the area of its bases. Explain how you can find the lateral surface area of a prism using the height and the perimeter of the base.

38. **Writing** Is the total area of the lateral faces of a pyramid *greater than*, *less than*, or *equal to* the area of the base? Explain.

39. **Building Blocks: In Your Classroom** You have 24 cube-shaped building blocks like the one below.

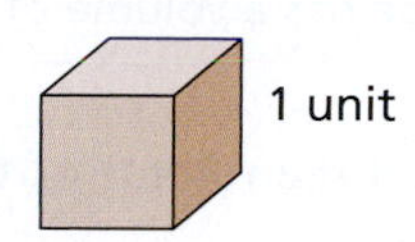

a. You create a rectangular prism by placing all 24 blocks in a single row. What is the surface area of the prism?

b. How could you arrange the blocks so they are in the shape of a rectangular prism with a surface area of 70 square units?

c. What arrangement of blocks gives you a rectangular prism with the least surface area? What is the surface area of this arrangement?

d. Why might this type of activity be beneficial to your students?

40. **Building Blocks: In Your Classroom** You have 24 rectangular-shaped building blocks like the one below.

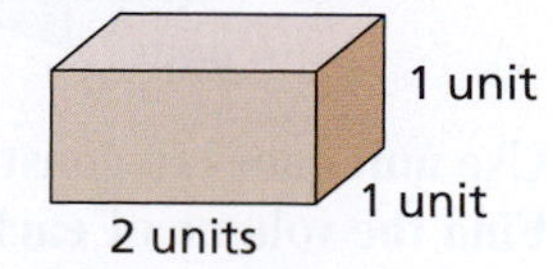

a. How many different rectangular prisms can you create by placing all 24 blocks in a single row? Explain.

b. Which arrangement has the greatest surface area?

c. How would you explain your conclusion in part (b) to your students?

Activity: Find the Volume of a Prism by Counting

Materials:

- Pencil
- Unit cubes

Learning Objective:

Understand the concept of volume. Use unit cubes to find the volume of a rectangular prism. Generalize the specific results to find a formula for the volume of a rectangular prism.

Name ______________________

The *volume* of a prism is the amount of three-dimensional space that the prism occupies. In the exercises below, each cube has a volume of 1 cubic unit.

1 cubic unit

Find the volume of the rectangular prism.

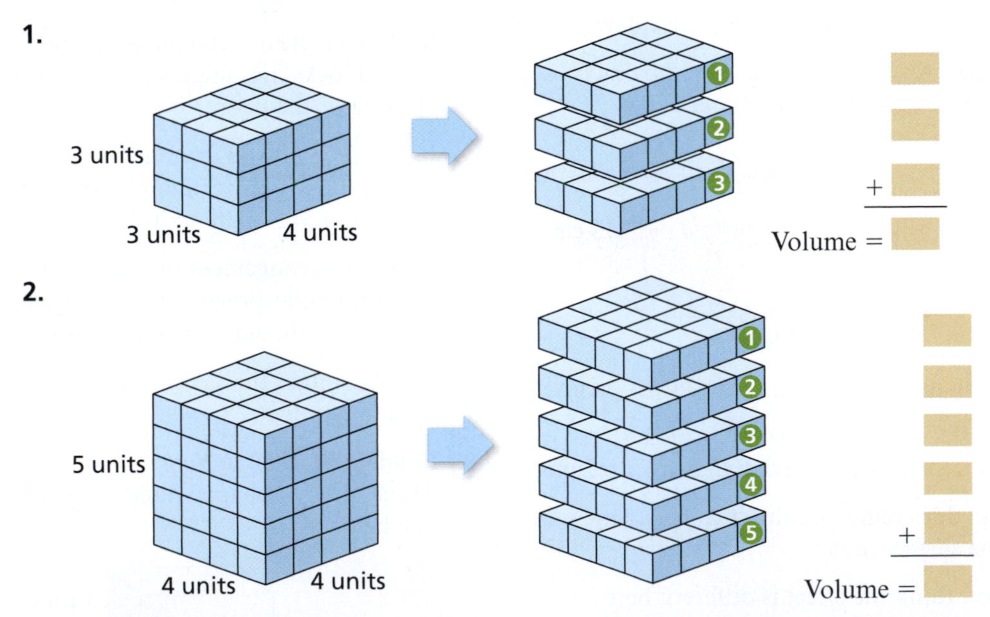

Use unit cubes to construct a rectangular prism that has the given dimensions. Find the volume of each prism by counting the number of unit cubes.

3. 2 by 2 by 2 **4.** 2 by 3 by 3 **5.** 2 by 3 by 4

6. Consider a rectangular prism whose length is ℓ, width is w, and height is h. Write a formula for the volume of a rectangular prism. Use your formula to check your results in Exercises 1–5.

Wavebreakmedia/Shutterstock.com

11.4 Volumes of Polyhedra

Standards

Grades 3–5
Measurement and Data
Students should understand concepts of volume and relate volume to multiplication and to addition.

Grades 6–8
Geometry
Students should solve real-life and mathematical problems involving volume.

- Find the volume of a prism.
- Find the volume of a pyramid.
- Find the volume of a composite solid.

Finding the Volume of a Prism

The **volume** of a three-dimensional solid is a measure of the amount of space that it occupies. Volume is measured in **cubic units**, such as cubic inches or cubic centimeters.

Volume of a Prism

Words The volume V of a prism is the product of the area of the base and the height of the prism.

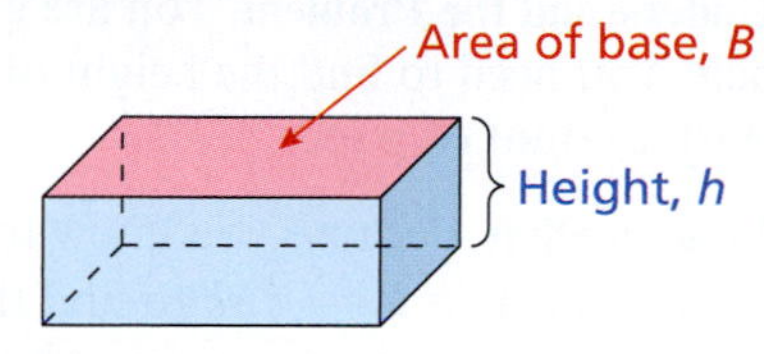

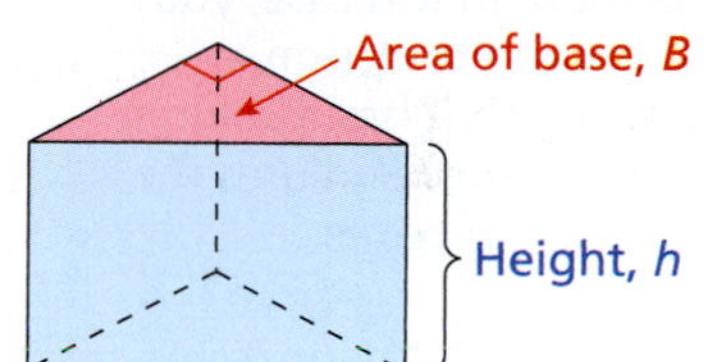

Algebra $V = Bh$

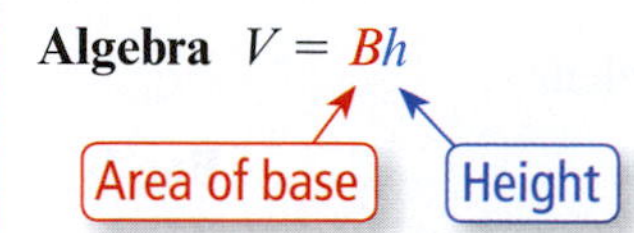

For a rectangular prism, the formula $V = Bh$ can be written as

$V = \text{(length)(width)(height)}$.

EXAMPLE 1 **Finding the Volumes of Prisms**

Find the volume of each prism.

a.

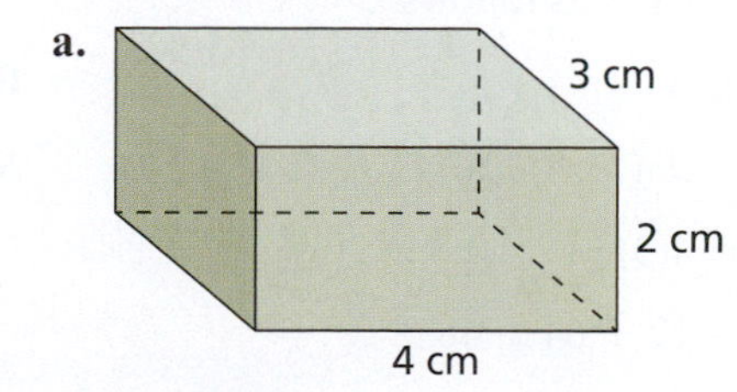

b.

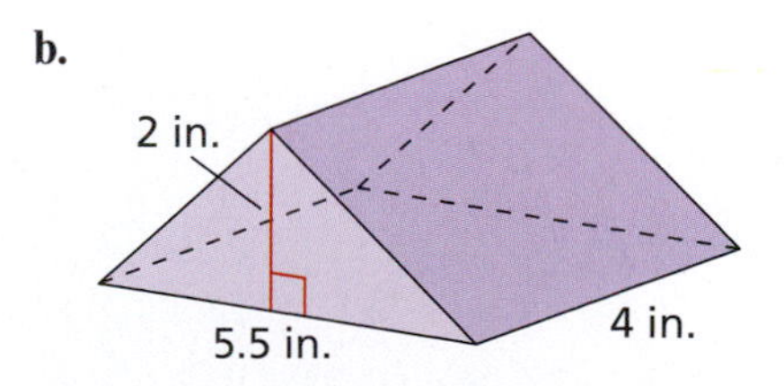

Classroom Tip
In Example 1(b), be sure you see that the base of a prism is not necessarily positioned so that it is on "the bottom of the prism." When you are teaching about prisms, be sure to include examples in which the base is not on the bottom.

SOLUTION

a. $V = Bh$

$= (4 \cdot 3) \cdot 2$

The base is a rectangle. So, its area is the product of the length and the width.

$= 24$

The volume is 24 cubic centimeters.

b. $V = Bh$

$= \left(\frac{1}{2} \cdot 5.5 \cdot 2\right) \cdot 4$

The base is a triangle. So, its area is one-half the product of its base and its height.

$= 22$

The volume is 22 cubic inches.

EXAMPLE 2 Using Volume in Real Life

A movie theater designs three bags of popcorn. Each bag holds 96 cubic inches of popcorn. Find the height of each bag. Which bag uses the least amount of paper?

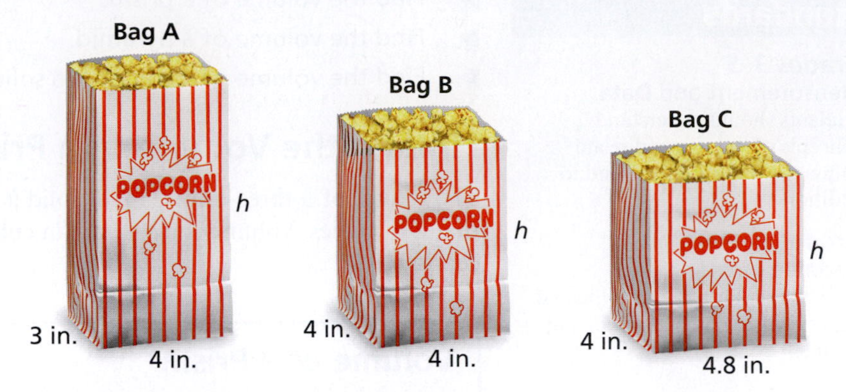

Classroom Tip

Example 2 is similar to a classic type of problem called *optimization*. In this case, you could be asked to design a popcorn bag that has a volume of 96 cubic inches and uses the least amount of paper.

SOLUTION

1. **Understand the Problem** You are given the dimensions of the base for each bag. You need to find the height of each bag and determine which bag uses the least amount of paper.
2. **Make a Plan** Because you know the length and width of the base for each bag, use the formula $V = \ell wh$ to find the height of each bag. Then find the surface area of each bag to determine which one uses the least amount of paper.
3. **Solve the Problem**

Bag A	Bag B	Bag C
$V = \ell wh$	$V = \ell wh$	$V = \ell wh$
$96 = 4(3)h$	$96 = 4(4)h$	$96 = 4.8(4)h$
$96 = 12h$	$96 = 16h$	$96 = 19.2h$
$8 = h$	$6 = h$	$5 = h$

So, Bag A has a height of 8 inches, Bag B has a height of 6 inches, and Bag C has a height of 5 inches. Because none of the bags has a top, the surface area of each bag is as follows.

Bag A

$$\begin{aligned} S &= \ell w + 2\ell h + 2wh \\ &= 4(3) + 2(4)(8) + 2(3)(8) \\ &= 12 + 64 + 48 \\ &= 124 \end{aligned}$$

Bag B

$$\begin{aligned} S &= \ell w + 2\ell h + 2wh \\ &= 4(4) + 2(4)(6) + 2(4)(6) \\ &= 16 + 48 + 48 \\ &= 112 \end{aligned}$$

Bag C

$$\begin{aligned} S &= \ell w + 2\ell h + 2wh \\ &= 4.8(4) + 2(4.8)(5) + 2(4)(5) \\ &= 19.2 + 48 + 40 \\ &= 107.2 \end{aligned}$$

Because $107.2 \text{ in.}^2 < 112 \text{ in.}^2 < 124 \text{ in.}^2$, Bag C uses the least amount of paper.

4. **Look Back** One way to check your answer is to draw a net that represents each bag. Then find the area of each net to determine which bag uses the least amount of paper.

Finding the Volume of a Pyramid

EXAMPLE 3 **Finding a Formula Experimentally**

As a classroom discovery project, perform the following steps to find the formula for the volume of a pyramid.

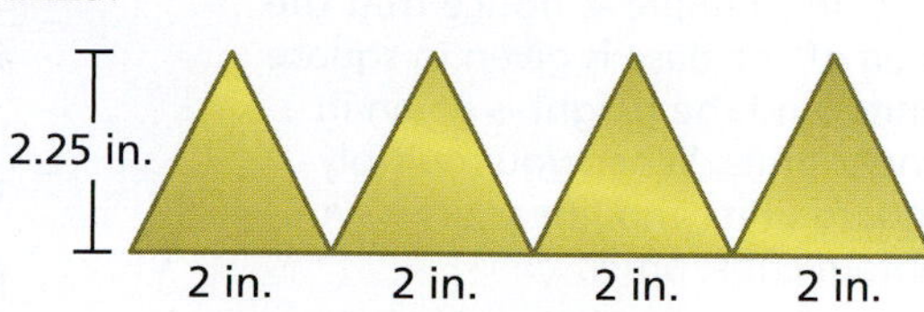

- Draw the two nets shown at the right on cardboard and cut them out.
- Fold and tape the nets to form an open square box and an open square pyramid.
- Fill the pyramid with rice. Then pour the rice into the box. Repeat this until the box is full. How many pyramids does it take to fill the box?

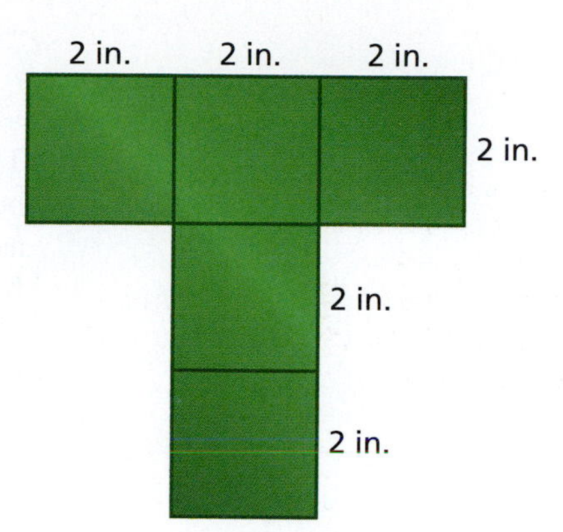

Use your result to find a formula for the volume of a pyramid.

SOLUTION

After making the open square pyramid and the open square box, you should discover that it takes 3 pyramids full of rice to fill the open box. The volume of the open box is

$$V = Bh.$$

Because it took 3 pyramids to fill the box, the volume of the open pyramid is one-third of the volume of the open box, or

$$V = \frac{1}{3}Bh. \qquad \text{Volume of a pyramid}$$

Mathematical Practices

Construct Viable Arguments and Critique the Reasoning of Others

MP3 Mathematically proficient students make conjectures and build a logical progression of statements to explore the truth of their conjectures.

Volume of a Pyramid

Words The volume V of a pyramid is one-third the product of the area of the base and the height of the pyramid. The **height of a pyramid** is the perpendicular distance from the base to the vertex that is shared by the lateral faces of the pyramid.

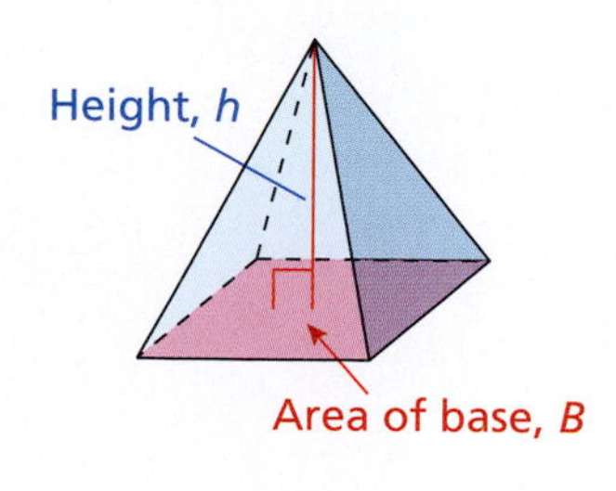

Algebra $V = \frac{1}{3}Bh$

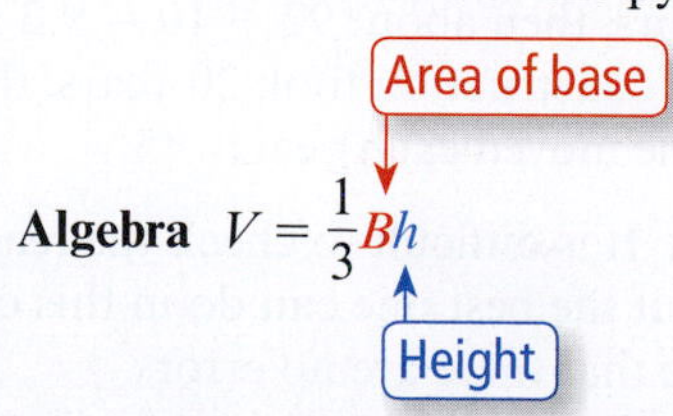

Classroom Tip

For a rectangular pyramid, the formula for the volume can be written as

$$V = \frac{1}{3}\ell wh$$

where ℓw is the area of the base.

EXAMPLE 4 Finding the Volume of a Pyramid

Find the volume of the pyramid.

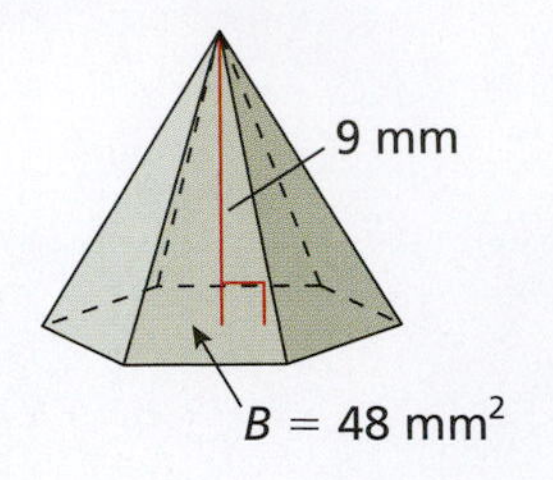

SOLUTION

$$V = \frac{1}{3}Bh$$

$$= \frac{1}{3}(48)(9)$$

$$= 144$$

The volume is 144 cubic millimeters.

> **Classroom Tip**
> In Example 4, notice that the area of the base is given in square units, and the height is given in linear units. When you multiply square units and linear units, you obtain cubic units.
>
> $\text{units}^2 \cdot \text{units} = \text{units}^3$

EXAMPLE 5 Finding a Volume in Real Life

The Great Pyramid of Giza was built over a period of 10 to 20 years as a tomb for Egyptian Pharaoh Khufu. The original height was 481 feet. Each side of its square base has a length of 756 feet. Estimate the volume of stone that was moved each year during the construction of the pyramid.

SOLUTION

1. **Understand the Problem** You are given the period of time it took to construct the pyramid, the side length of the square base, and the height of the pyramid. You need to estimate the volume of stone that was moved each year during construction.
2. **Make a Plan** Find the volume of the pyramid. Then divide the volume by the number of years it took to construct the pyramid.
3. **Solve the Problem**

$V = \frac{1}{3}Bh$ Write the formula for volume of a pyramid.

$= \frac{1}{3}(756^2)(481)$ Substitute 756^2 for B and 481 for h.

$= \frac{1}{3}(571{,}536)(481)$ Evaluate the power.

$= 91{,}636{,}272$ Multiply.

The volume of the pyramid is about 92 million cubic feet. If the construction took 10 years, then about $92 \div 10 = 9.2$ million cubic feet of stone moved each year. If the construction took 20 years, then about $92 \div 20 = 4.6$ million cubic feet of stone moved each year.

4. **Look Back** It is difficult to check the reasonableness of numbers that are so large. About the best one can do in this case is to go back and double check the steps to see that there are no errors.

The Great Pyramid consists of about 2.3 million blocks. If it took 20 years to construct, then this would have required moving about 13 blocks into place each hour—day and night.

Finding the Volume of a Composite Solid

A **composite solid** is a solid that is made up of two or more solids. To find the volume of a composite solid, decompose it into solids whose volumes you can find.

EXAMPLE 6 Finding the Volume of a Composite Solid

The composite solid is made up of a rectangular prism and a square pyramid. Find its volume.

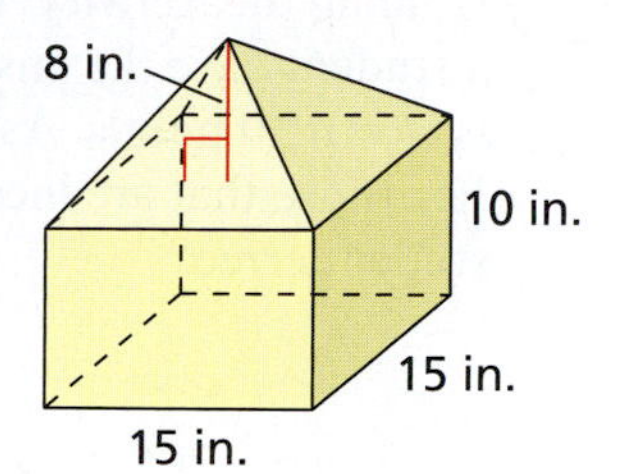

SOLUTION

Find the volumes of the prism and pyramid separately.

Prism

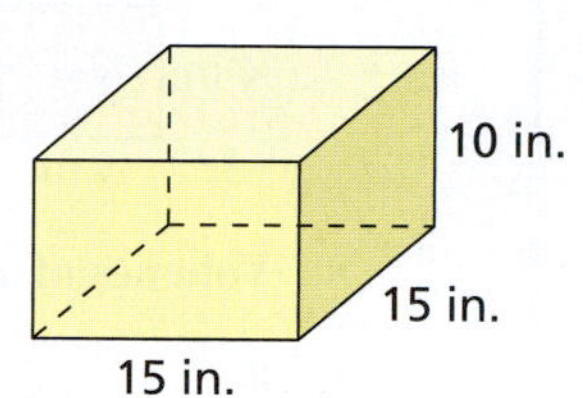

$V = Bh$

$= (15)(15) \cdot 10$

$= 2250$

Pyramid

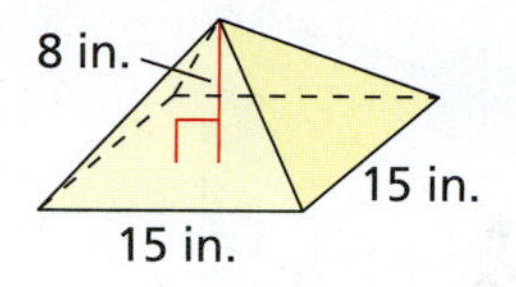

$V = \frac{1}{3}Bh$

$= \frac{1}{3}(15)(15) \cdot 8$

$= 600$

The volume of the composite solid is $2250 + 600 = 2850$ cubic inches.

EXAMPLE 7 Real-Life Example of a Composite Solid

The green hexagonal part of the bird feeder has a base area of 20 square inches. Estimate the amount of birdseed the feeder can hold. Assume that the feeder can hold birdseed in the pyramid portion as well.

SOLUTION

Decompose the bird feeder into a hexagonal prism and a hexagonal pyramid.

Volume of prism $= Bh$

$= 20(6)$

$= 120$

The volume of the prism is 120 cubic inches. For the pyramid portion, assume the base area to be 20 square inches as well.

$$\text{Volume of pyramid} = \frac{1}{3}Bh = \frac{1}{3}(20)(3) = 20$$

The volume of the pyramid is 20 cubic inches. The volume of the bird feeder is about $120 + 20 = 140$ cubic inches. So, it holds about 140 cubic inches of birdseed.

11.4 Exercises

Free step-by-step solutions to odd-numbered exercises at *MathematicalPractices.com.*

1. **Writing a Solution Key** Write a solution key for the activity on page 442. Describe a rubric for grading a student's work.

2. **Grading the Activity** In the activity on page 442, a student gave the answers below. Each question is worth 10 points. Assign a grade to each answer. For those that are incorrect, why do you think the student erred?

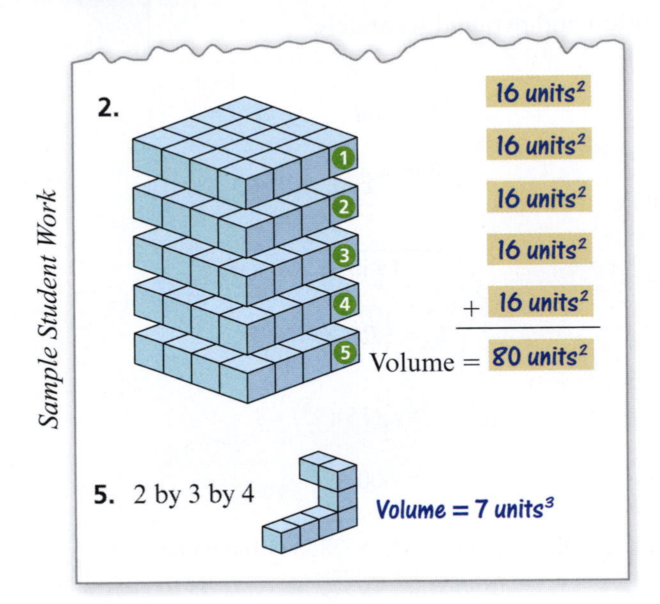

3. **Constructing Prisms** Give the dimensions of the different rectangular prisms that can be constructed using exactly 8 unit cubes. What is the volume of each prism?

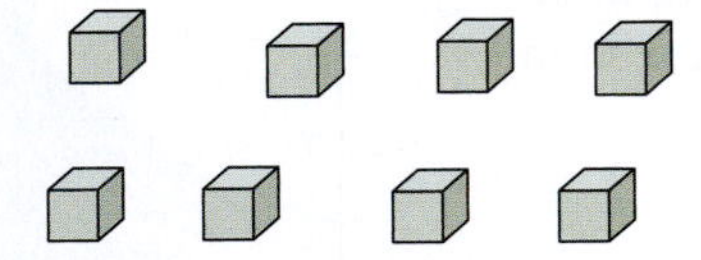

4. **Constructing Prisms** Give the dimensions of the different rectangular prisms that can be constructed using exactly 12 unit cubes. What is the volume of each prism?

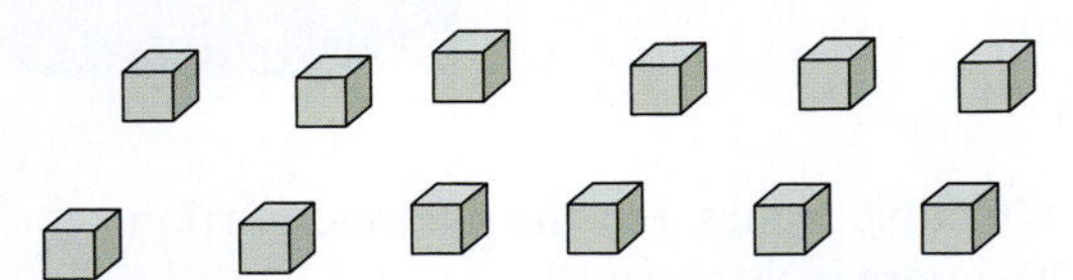

5. **Volume of a Prism** Find the volume of the prism by counting the number of unit cubes.

6. **Volume of a Prism** Find the volume of the prism by counting the number of unit cubes.

7. **Volumes of Prisms** Find the volume of each prism.

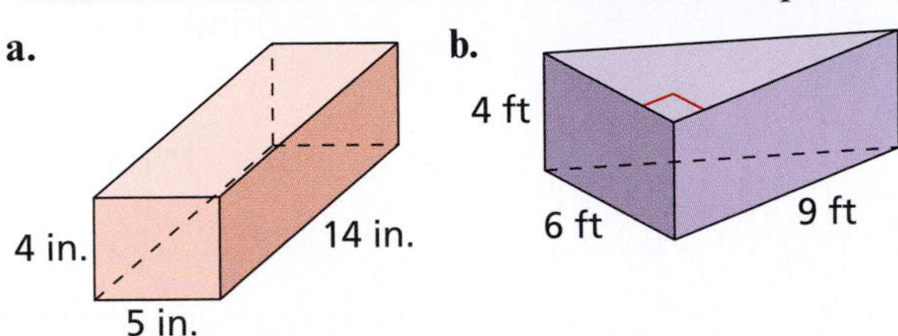

8. **Volumes of Prisms** Find the volume of each prism.

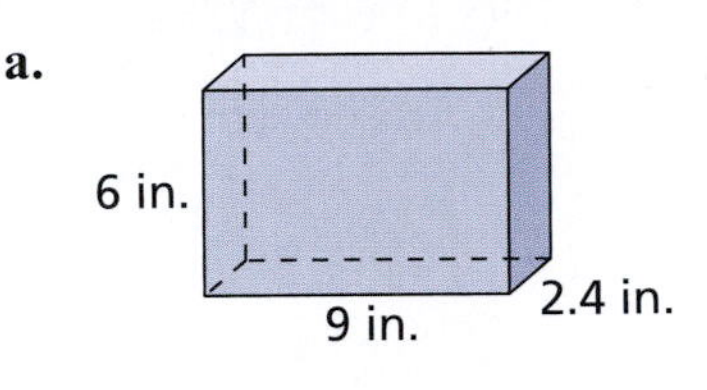

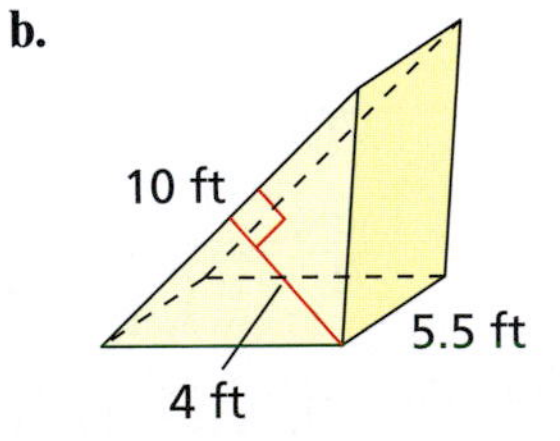

9. **Using the Volume** The prism has a volume of 160 cubic inches. Find the height h.

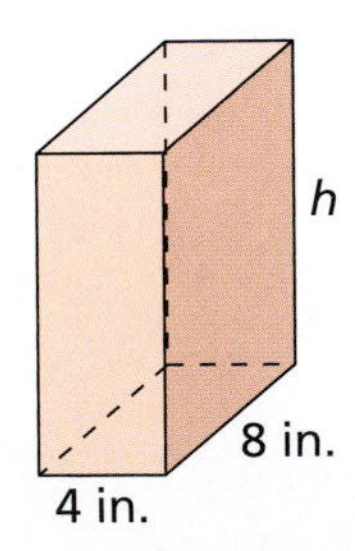

10. **Using the Volume** The prism has a volume of 1296 cubic centimeters. Find the height h.

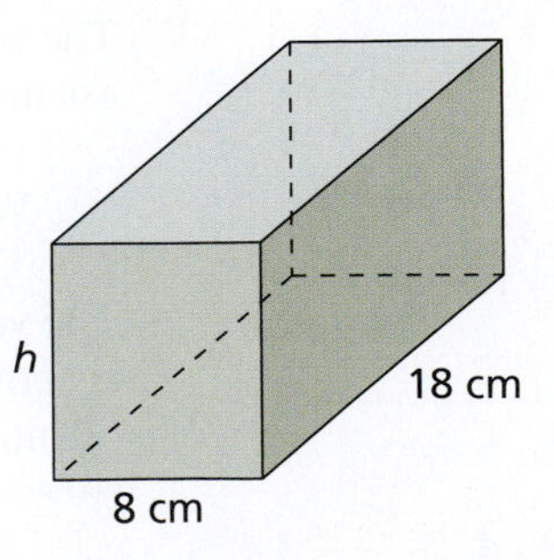

11. Using the Base Area to Find the Volume The base of a rectangular prism has an area of 102 square centimeters. The height of the prism is 9 centimeters. Find the volume of the prism.

12. Using the Base Area to Find the Volume The base of a triangular prism has an area of 38 square feet. The height of the prism is 3.5 feet. Find the volume of the prism.

13. Volumes of Pyramids Find the volume of each pyramid.

a.

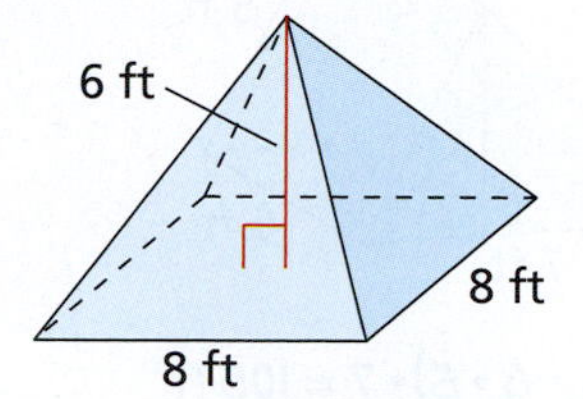

b.

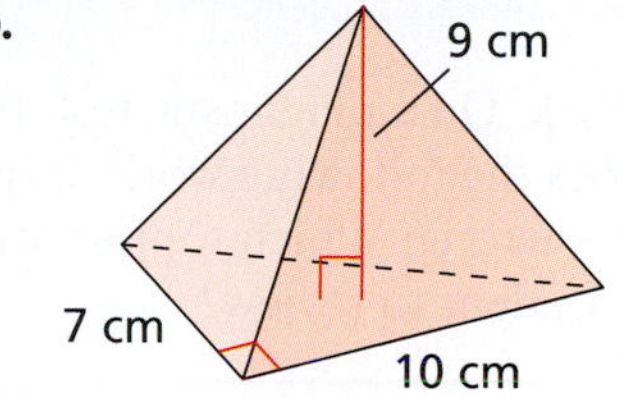

c.

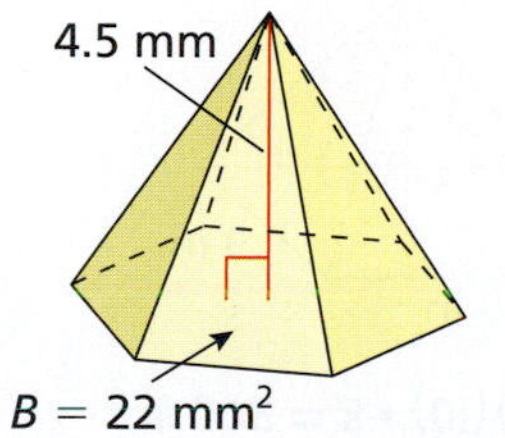

14. Volumes of Pyramids Find the volume of each pyramid.

a.

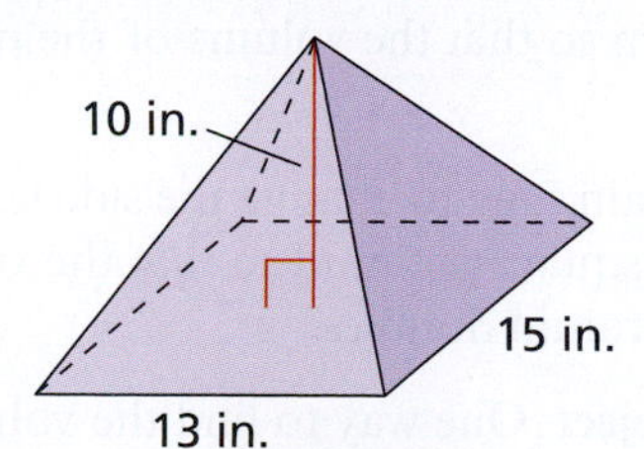

b.

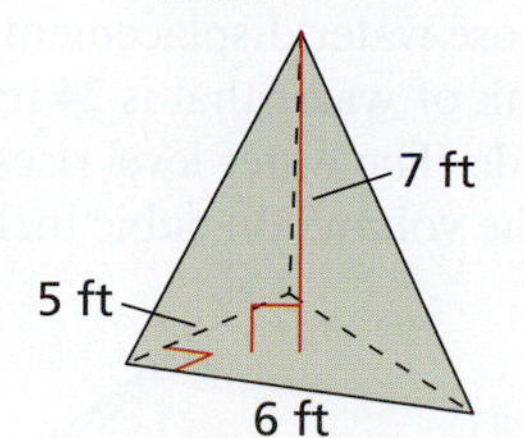

c.

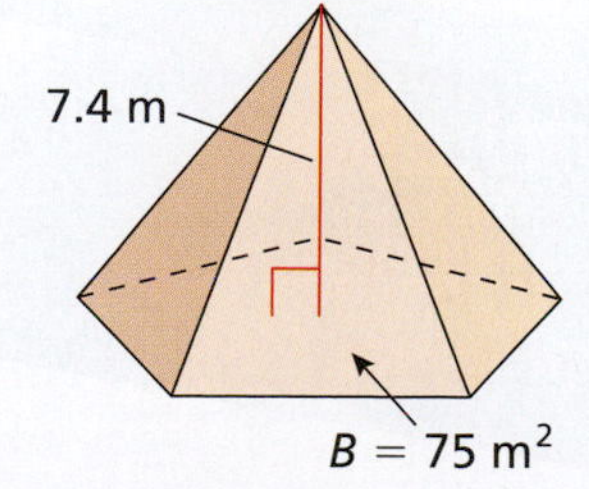

15. Finding Volumes Find the volume of each solid.

a.

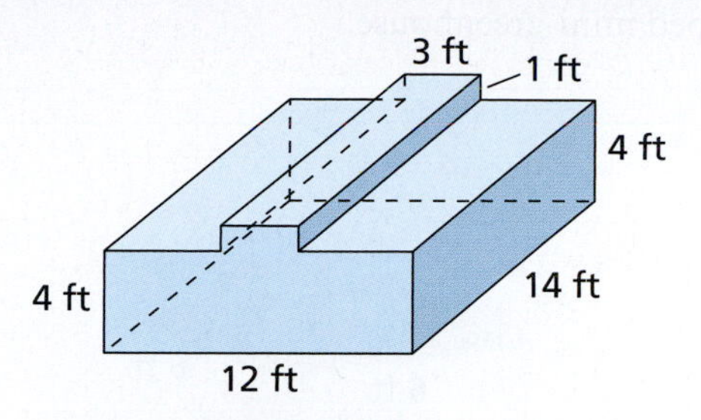

b.

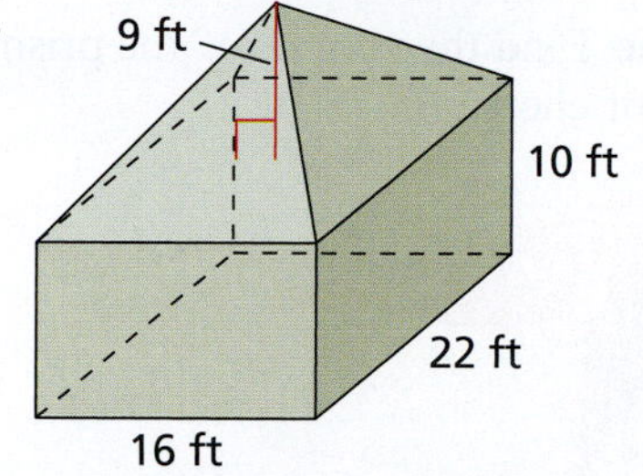

c.

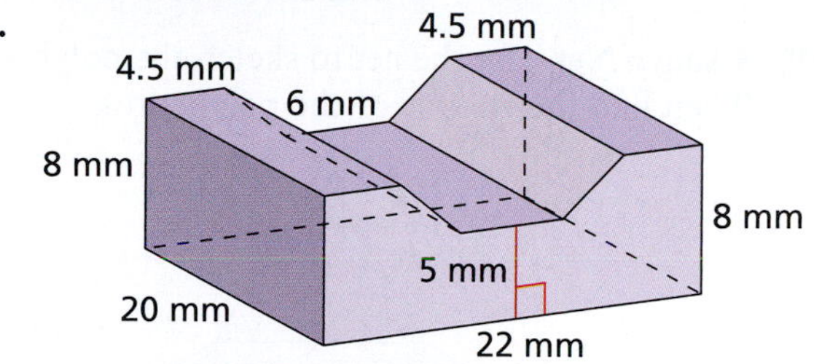

16. Finding Volumes Find the volume of each solid.

a.

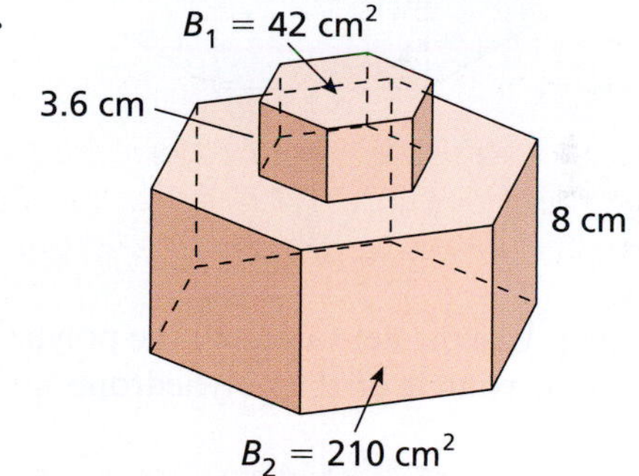

b.

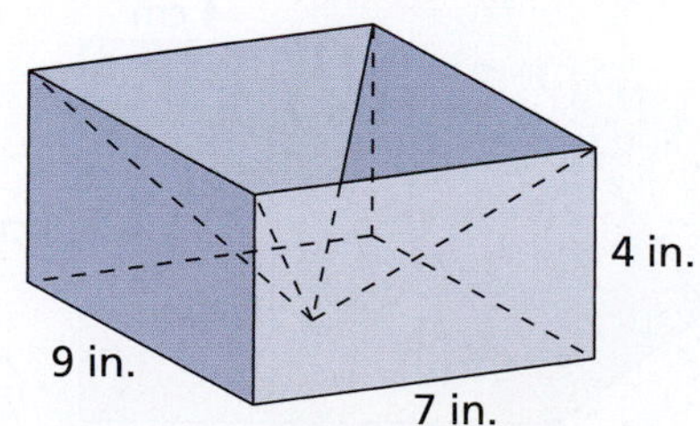

c.

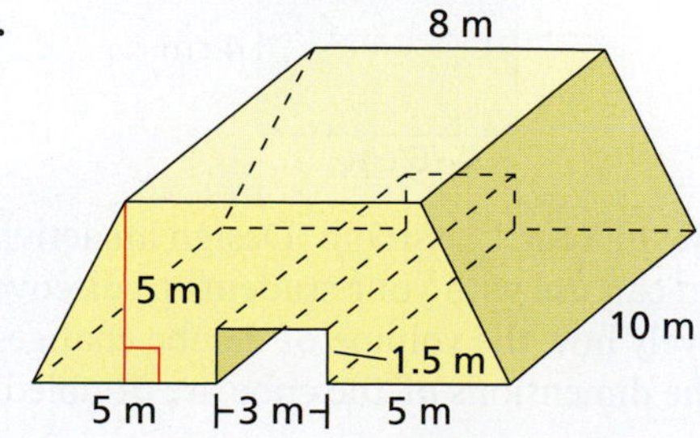

17. Finding Volume Find the volume of the pyramid-shaped mini-greenhouse.

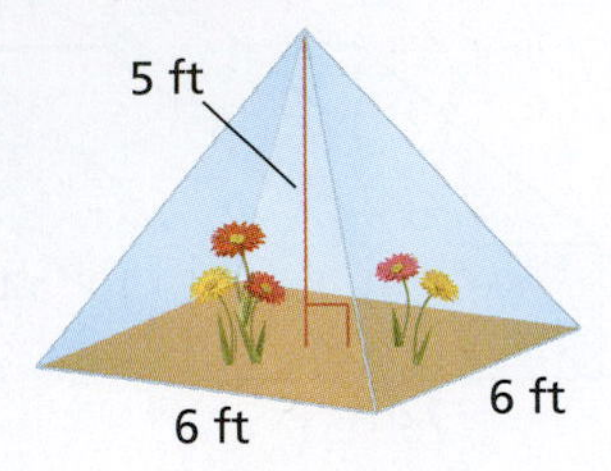

18. Finding Volume Find the volume of the prism-shaped block of cheese.

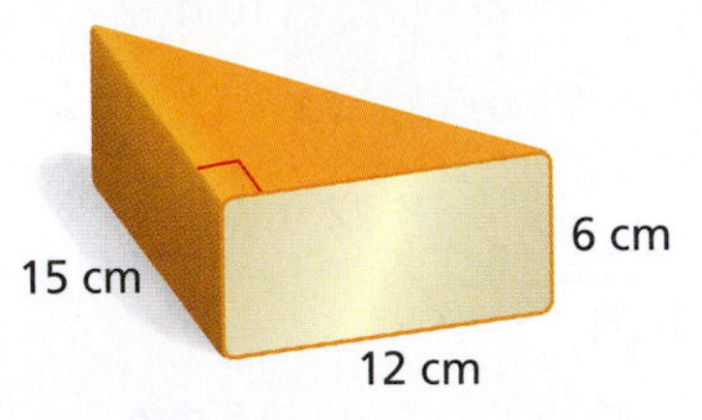

19. Using a Net Use the net to sketch the polyhedron. Then find the volume of the polyhedron.

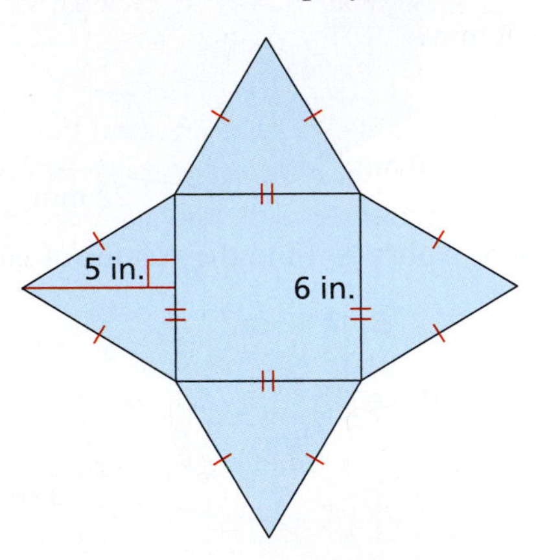

20. Using a Net Use the net to sketch the polyhedron. Then find the volume of the polyhedron.

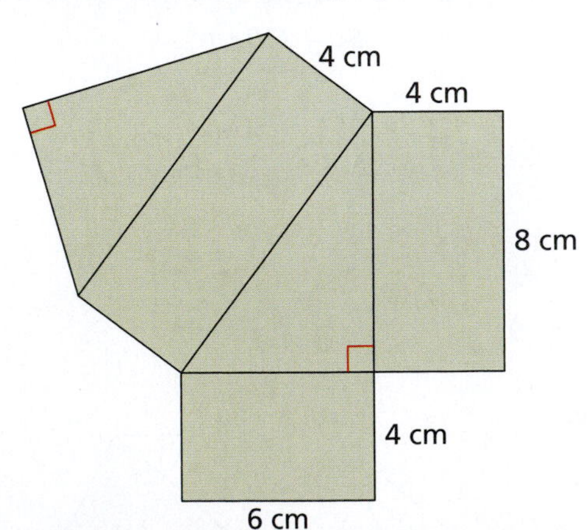

21. Activity: In Your Classroom Design an activity that you can use with your students to discover inductively how the volume of a cube changes when the dimensions of the cube are doubled.

22. Doubling Dimensions: In Your Classroom Describe a way that your students can show deductively how the volume of a cube changes when the dimensions of the cube are doubled.

23. Grading Student Work On a diagnostic test, one of your students does the following work. Explain what the student did wrong. Which topics would you encourage the student to review?

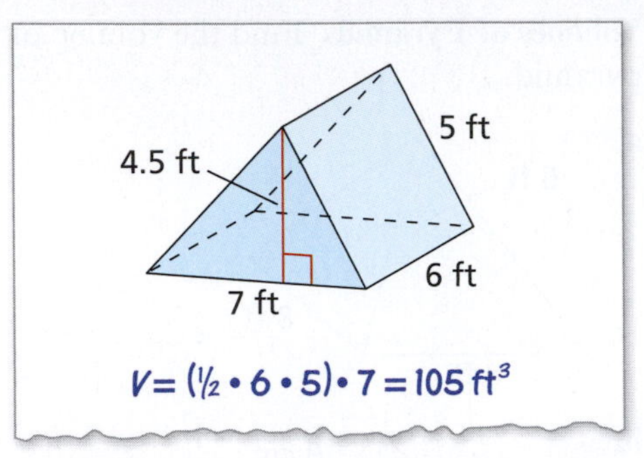

24. Grading Student Work On a diagnostic test, one of your students does the following work. Explain what the student did wrong. Which topics would you encourage the student to review?

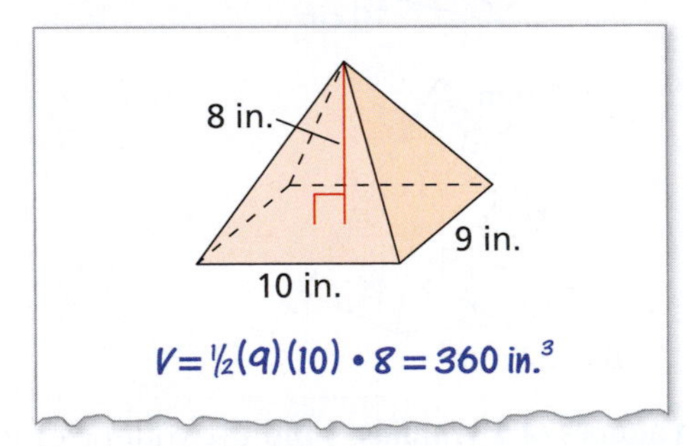

25. Writing Explain how to change the length of a rectangular prism so that the volume of the prism doubles.

26. Writing Explain how to change the side lengths of the base of a square pyramid so that the volume of the square pyramid doubles.

27. Volume of an Object One way to find the volume of a solid object is to use water displacement. You place a rock into a tank of water that is 24 inches long and 12 inches wide. The water level rises 1.25 inches. What is the volume (in cubic inches) of the rock?

28. Volume of a Gas Tank A gas tank for a boat has the shape of a rectangular prism. One gallon of gasoline has a volume of 231 cubic inches.

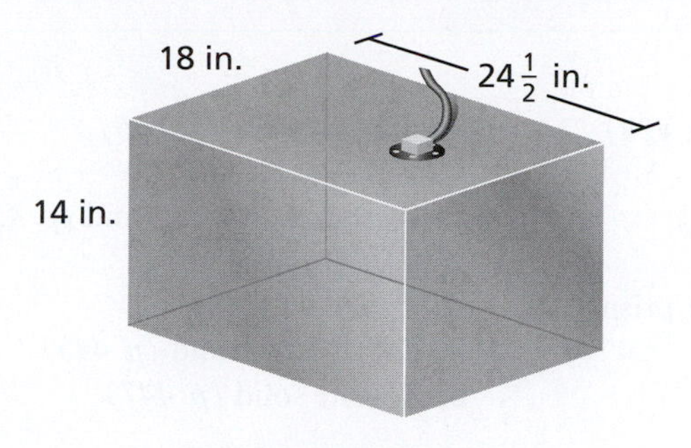

a. Use the dimensions shown to find the volume of the tank in cubic inches.

b. How many inches high is the fuel level when the tank contains 21 gallons of gasoline?

29. Volume and Surface Area The bottoms of three gift boxes are shown. Each has a volume of 285 cubic inches. Which size uses the least amount of cardboard? Which size uses the greatest amount of cardboard?

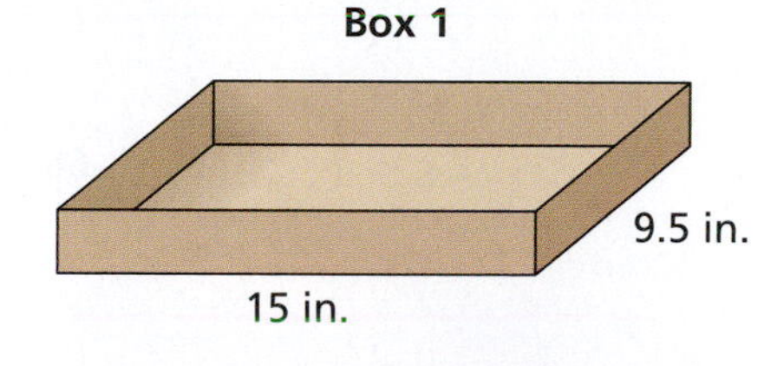

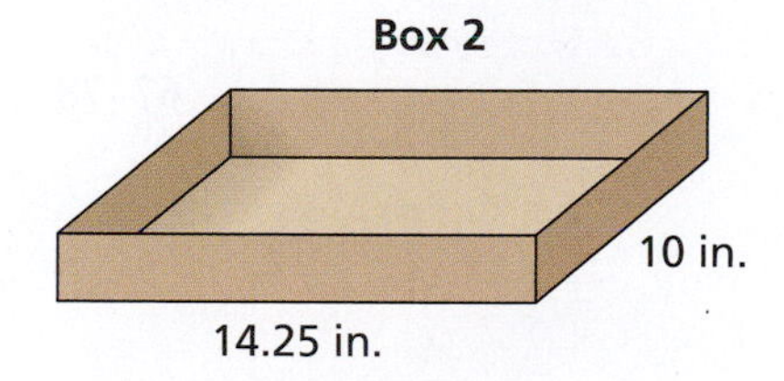

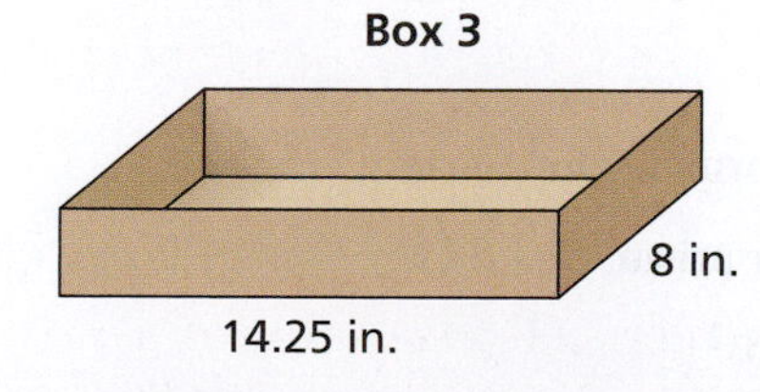

30. Volume and Surface Area One cup of frosting is equal to a volume of approximately 14 cubic inches. Are two cups of frosting enough to cover the top and sides of the square cake shown at a thickness of 1/8 inch? Explain.

31. Finding Volume Find the volume of the concrete used to make the concrete block.

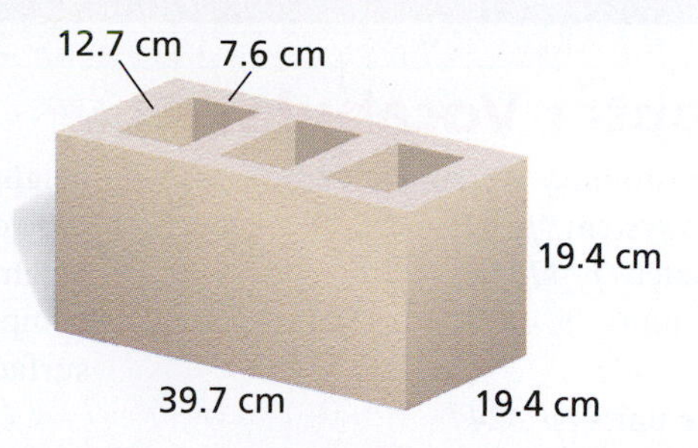

32. Finding Volume The dimensions of a paint tray are shown. A volume of 57.75 cubic inches is equal to one quart. What is the greatest whole number of quarts of paint that the paint tray can hold?

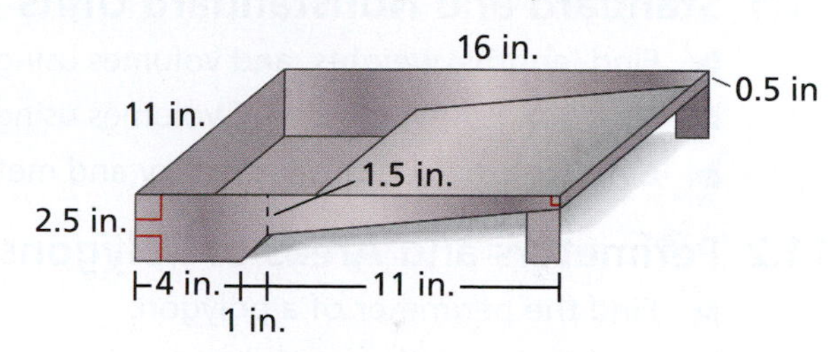

33. Problem Solving The volume of the pole barn is 10,440 cubic feet. Find the total area of the roof. (The front and back halves are the same size.)

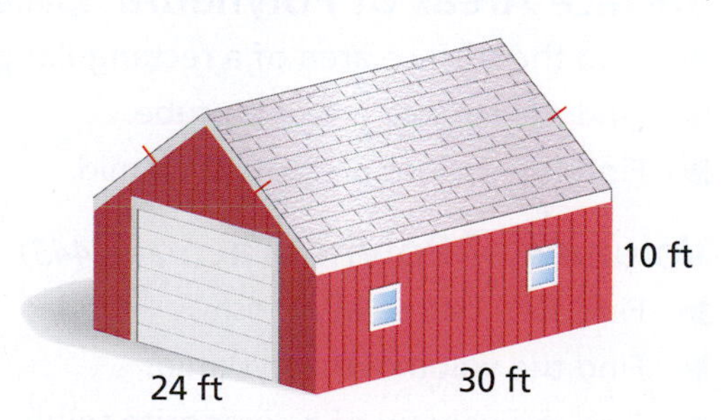

34. Problem Solving A cardboard pet carrier has the dimensions shown.

a. Make a sketch that divides the composite solid into a rectangular prism, a triangular prism, and two rectangular pyramids. (*Hint*: Each pyramid has a vertical lateral face.)

b. Find the volume of the composite solid.

Chapter Summary

Chapter Vocabulary

U.S. customary system *(p. 413)*
metric system *(p. 415)*
perimeter *(p. 423)*
linear units *(p. 423)*
area *(p. 424)*
square units *(p. 424)*
base of a parallelogram *(p. 424)*
height of a parallelogram *(p. 424)*
base of a triangle *(p. 426)*
height of a triangle *(p. 426)*
composite figure *(p. 427)*
surface area of a rectangular prism *(p. 433)*
surface area of a cube *(p. 435)*
regular pyramid *(p. 436)*
slant height *(p. 436)*
surface area of a pyramid *(p. 436)*
volume *(p. 443)*
cubic units *(p. 443)*
height of a pyramid *(p. 445)*
composite solid *(p. 447)*

Chapter Learning Objectives

Review Exercises

11.1 Standard and Nonstandard Units *(page 413)* — 1–36

- Find lengths, weights, and volumes using the customary system.
- Find lengths, weights, and volumes using the metric system.
- Convert between the customary and metric systems.

11.2 Perimeters and Areas of Polygons *(page 423)* — 37–50

- Find the perimeter of a polygon.
- Find the area of a quadrilateral.
- Find the area of a triangle.

11.3 Surface Areas of Polyhedra *(page 433)* — 51–66

- Find the surface area of a rectangular prism.
- Find the surface area of a cube.
- Find the surface area of a pyramid.

11.4 Volumes of Polyhedra *(page 443)* — 67–78

- Find the volume of a prism.
- Find the volume of a pyramid.
- Find the volume of a composite solid.

Important Concepts and Formulas

U.S. Customary and Metric Conversions

1 inch = 2.54 centimeters, 1 centimeter ≈ 0.39 inch
1 mile ≈ 1.61 kilometers, 1 kilometer ≈ 0.62 mile
1 foot ≈ 0.3 meter, 1 meter ≈ 3.28 feet
1 pound ≈ 0.45 kilograms, 1 kilogram ≈ 2.2 pounds
1 gallon ≈ 3.79 liters, 1 liter ≈ 0.26 gallon

Common Formulas for Area

Rectangle $A = \ell w$ **Square** $A = s^2$

Parallelogram $A = bh$ **Trapezoid** $A = \frac{1}{2}h(b + B)$

Triangle $A = \frac{1}{2}bh$ **Kite** $A = \frac{1}{2}ab$

Common Formulas for Surface Area

Rectangular prism $S = 2\ell w + 2\ell h + 2wh$

Cube $S = 6s^2$

Pyramid S = area of base + areas of lateral faces

Common Formulas for Volume

Prism $V = Bh$

Rectangular prism $V = \ell wh$

Pyramid $V = \frac{1}{3}Bh$

Review Exercises

Free step-by-step solutions to odd-numbered exercises at *MathematicalPractices.com*.

11.1 Standard and Nonstandard Units

Converting Customary Units Complete the statement.

1. 32 in. = ____ ft
2. 5500 yd = ____ mi
3. 2.4 ton = ____ lb
4. 60 oz = ____ lb
5. 7 qt = ____ pt
6. 80 c = ____ qt

Converting Metric Units Complete the statement.

7. 1.5 km = ____ cm
8. 9000 mm = ____ m
9. 1800 kg = ____ t
10. 0.08 t = ____ g
11. 7200 mL = ____ L
12. 0.3 kL = ____ L

Converting Between Measurement Systems Complete the statement.

13. 2 yd ≈ ____ m
14. 4500 m ≈ ____ mi
15. 26 oz ≈ ____ kg
16. 0.001 t ≈ ____ lb
17. 37 pt ≈ ____ L
18. 80 gal ≈ ____ kL

Comparing Measurements Complete the statement with the correct symbol: $<$ or $>$.

19. 40 mm ____ 0.5 yd
20. 1 t ____ 500 lb
21. 7 pt ____ 2 L
22. 0.02 km ____ 200 ft
23. 320 oz ____ 2 kg
24. 0.03 fl oz ____ 0.5 mL

Choosing Units Choose an appropriate U.S. customary unit and metric unit to measure the item. Explain your reasoning.

25. Mass of an eraser
26. Length of a clarinet
27. Capacity of a tanker truck
28. Capacity of a bottle of nail polish
29. Mass of a bowling ball
30. Length of a river
31. **Converting Units in Real Life** A soccer ball weighs about 15 ounces. Find its weight in grams.
32. **Converting Units in Real Life** The Mackinac Bridge in Michigan has a length of 26,372 feet. What is the length in kilometers?
33. **Comparing Units in Real Life** Can you pour 2 liters of water into a 2-quart container without overflowing the container? Explain.
34. **Converting Units in Real Life** You have a poster that has a width of 15 inches. You have a space on your wall to hang the poster that is 1.2 feet wide. Will the poster fit? Explain.
35. **Lemonade Mix** You are comparing lemonade mixes. Lemonade mix A makes four 2-liter pitchers of lemonade. Lemonade mix B makes twenty-five 500-milliliter glasses of lemonade. Which lemonade mix makes more lemonade?
36. **Reading a Map** On a map, 1 centimeter represents a distance of 300 meters. The distance between two cities on the map is 12.3 centimeters. How many miles apart are the cities?

11.2 Perimeters and Areas of Polygons

Perimeter of a Polygon Find the perimeter of the polygon.

37. 27 ft, 13 ft, 30 ft

38.

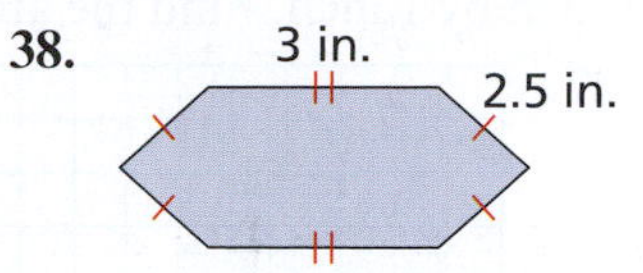

39. 0.5 cm, 7 cm, 3.5 cm, 3 cm, 1 cm, 6.5 cm

40.

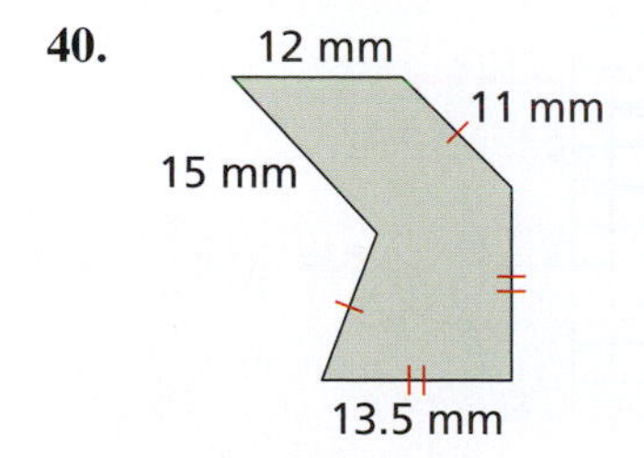

Perimeter and Area of a Rectangle Find the perimeter and area of the rectangle.

41. 42.

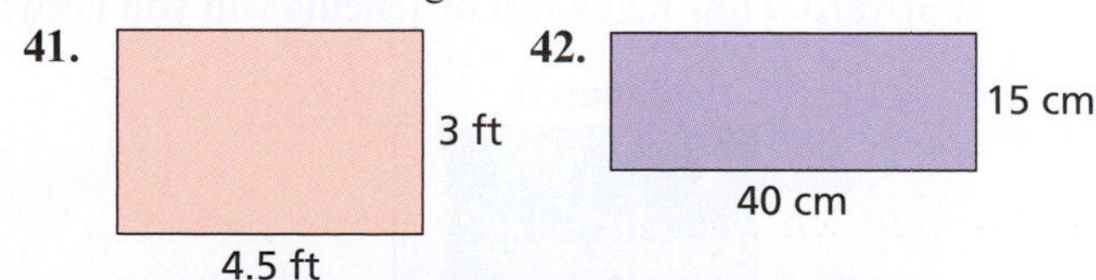

Area of a Quadrilateral Find the area of the quadrilateral.

43. Parallelogram

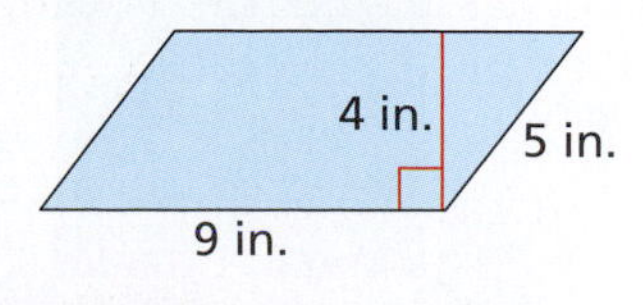

44. Trapezoid

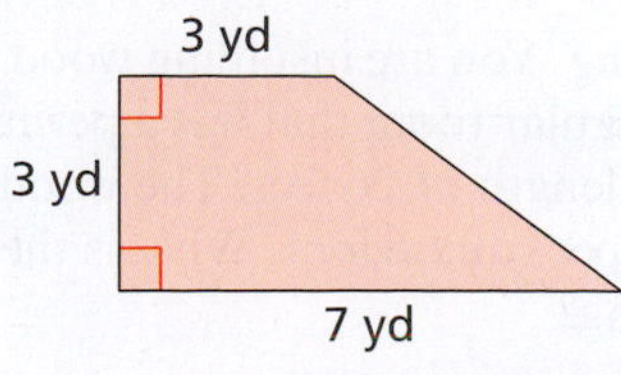

Area of a Kite and a Triangle Find the area of the kite or triangle.

45.

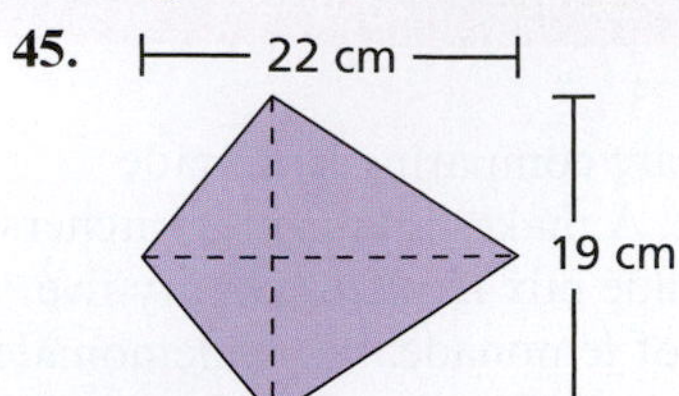

46.

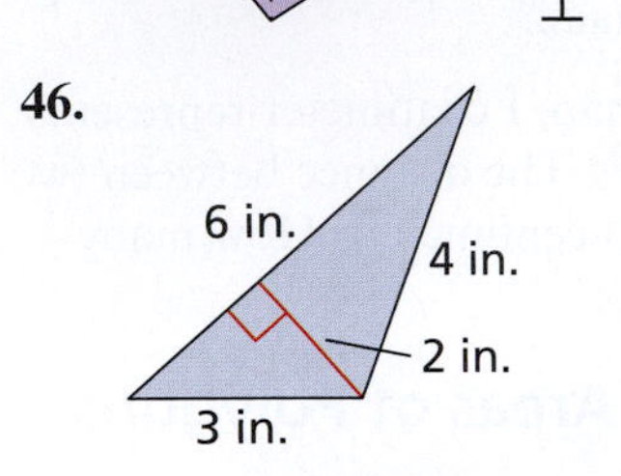

Area of a Composite Figure Each square on the grid is 1 inch by 1 inch. Find the area of the polygon.

47.

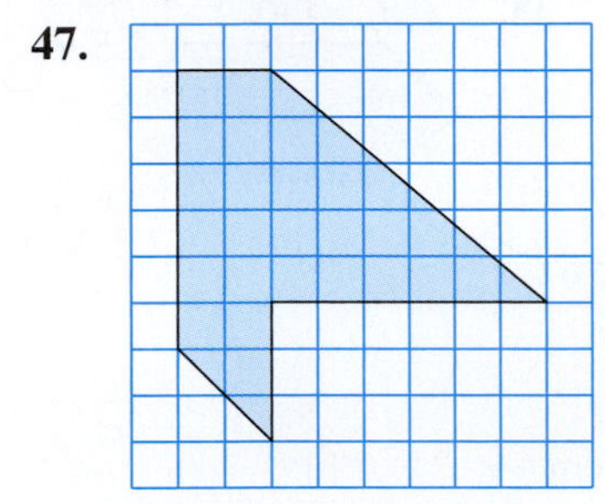

48.

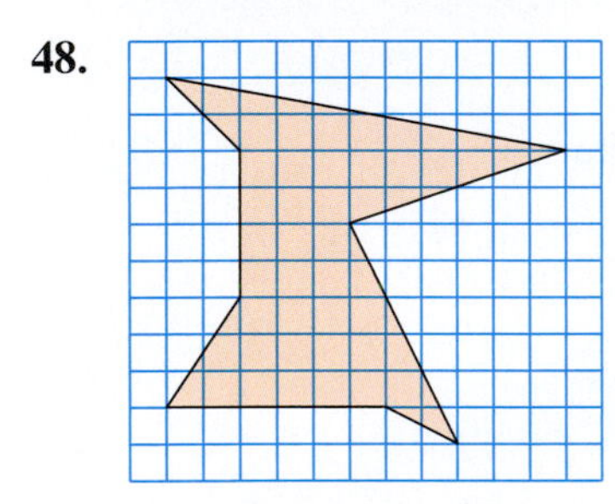

49. **Fencing** You want to install a privacy fence around your yard. How many feet of fencing will you need?

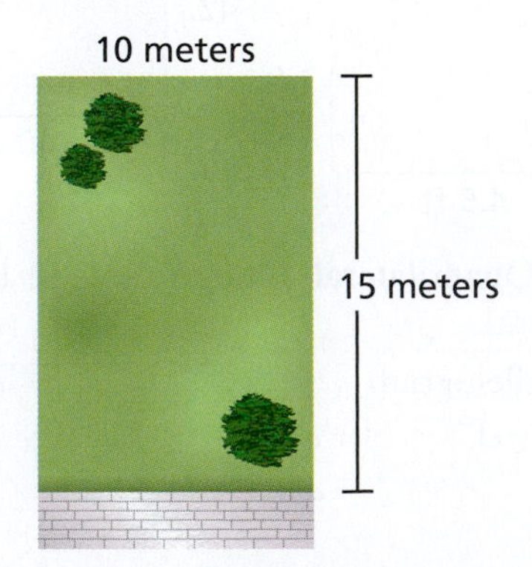

50. **Flooring** You are installing wood flooring in a rectangular room that has a perimeter of 116 feet and a length of 26 feet. The wood flooring costs $5.50 per square foot. What is the total cost of the flooring?

11.3 Surface Areas of Polyhedra

Surface Area of a Rectangular Prism Sketch a net of the rectangular prism. Then use the net to find the surface area of the prism.

51. 52.

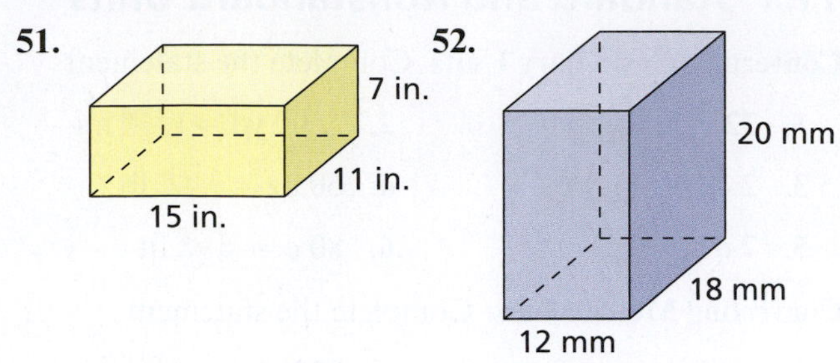

Surface Area of a Cube Sketch a net of the cube. Then use the net to find the surface area of the cube.

53. 54.

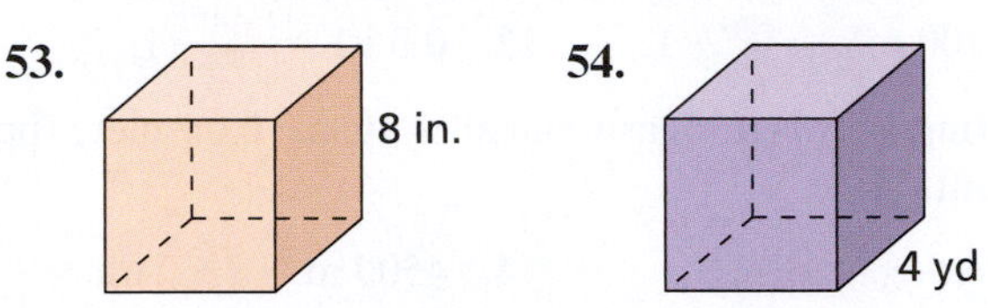

Surface Area of a Pyramid Sketch a net of the regular pyramid. Then use the net to find the surface area of the pyramid.

55. 56.

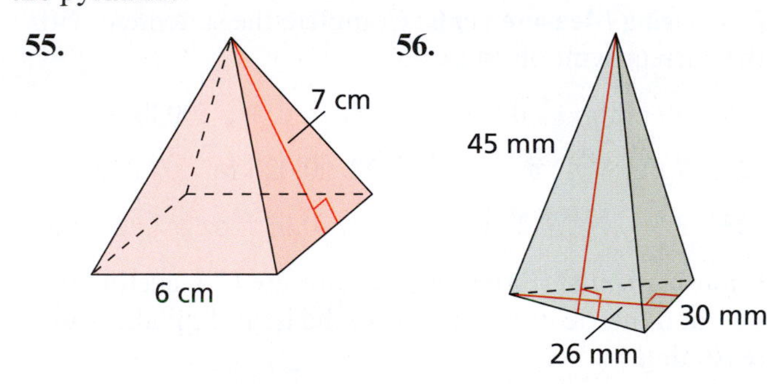

Surface Area of a Polyhedron Use a formula to find the surface area of the polyhedron described.

57. A cube with a height of 9 inches

58. A rectangular prism that is 2 centimeters tall, 10 centimeters wide, and twice as long as it is wide

59. A square pyramid with a slant height of 25 millimeters, and a base side length of 20 millimeters

60. A regular pentagonal pyramid with a slant height of 6 feet, a base area of about 27.5 square feet, and a base side length of 4 feet

61. **Problem Solving** The surface area of a cube is 150 square feet. Find the length of each edge of the cube.

62. **Problem Solving** The surface area of a square pyramid is 48,600 square meters. The side length of the base is 80 meters. What is the slant height?

63. Comparing Surface Areas One cube has an edge length of 4 inches and a second cube has an edge length of 5 inches. What is the ratio of the surface area of the first cube to the surface area of the second cube?

64. Comparing Surface Areas One cube has a surface area of 24 square inches and a second cube has a surface area of 216 square inches. What is the ratio of the edge length of the first cube to the edge length of the second cube?

65. Gift Wrapping You need to wrap the box shown below. What is the least amount of wrapping paper you need?

66. Prop Design You are building a cardboard square pyramid as a prop for a play that takes place in Egypt. The base of the pyramid must have side length of 3 feet, and the slant height must be 6 feet. How much cardboard do you need if you do not need to include the base?

11.4 Volumes of Polyhedra

Volume of a Polyhedron Find the volume of the polyhedron.

67.

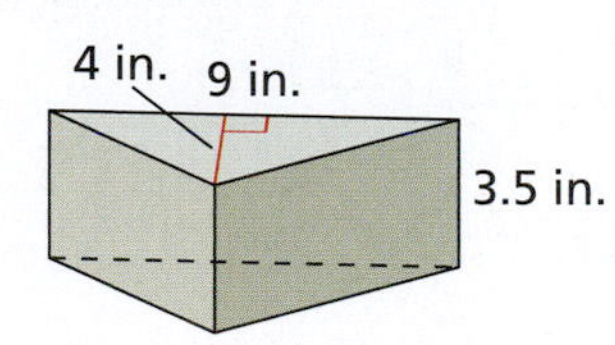

68.

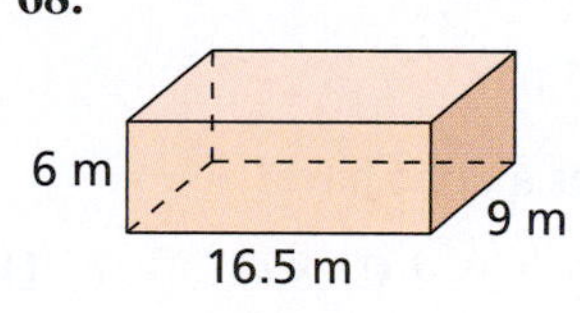

69.

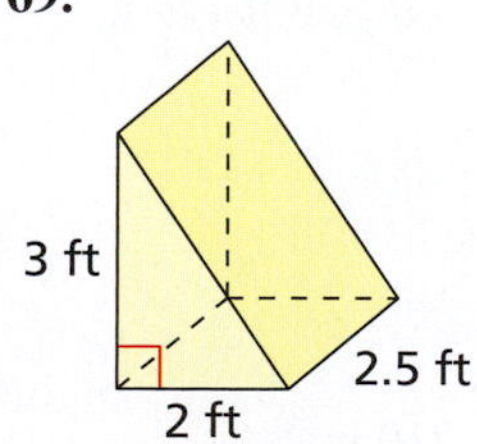

70.

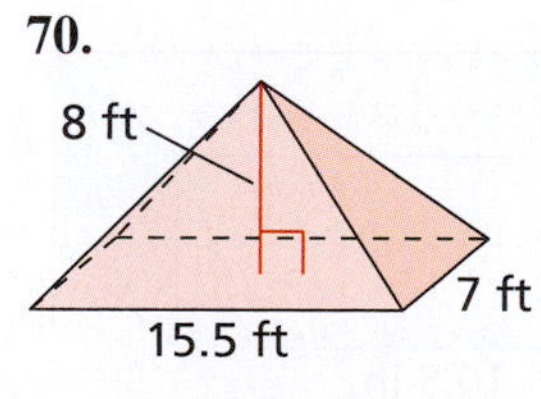

71.

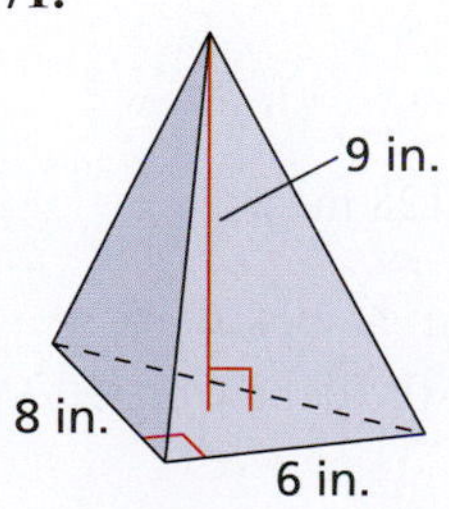

72.

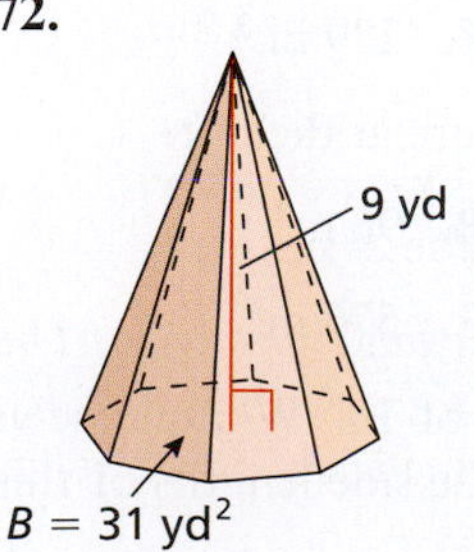

Finding Volume Find the volume of the solid.

73.

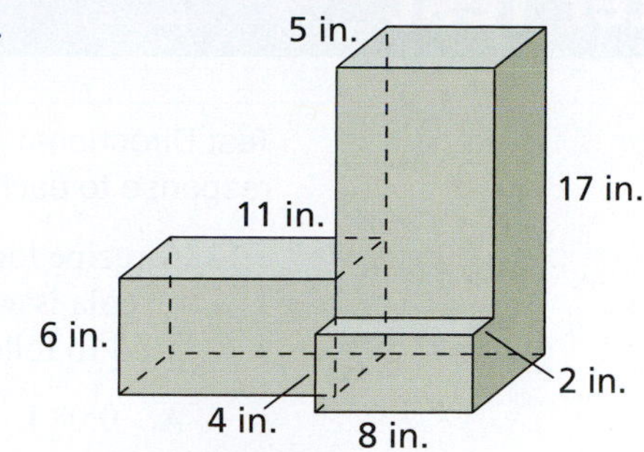

74.

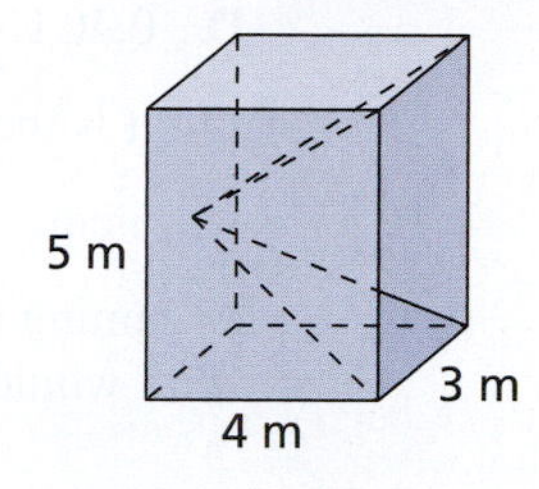

75.

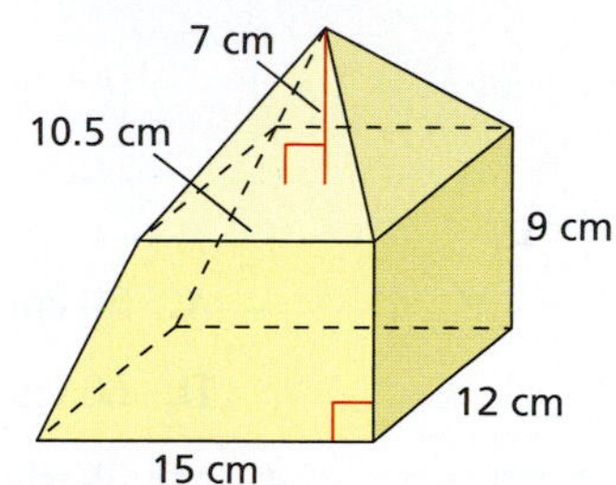

76.

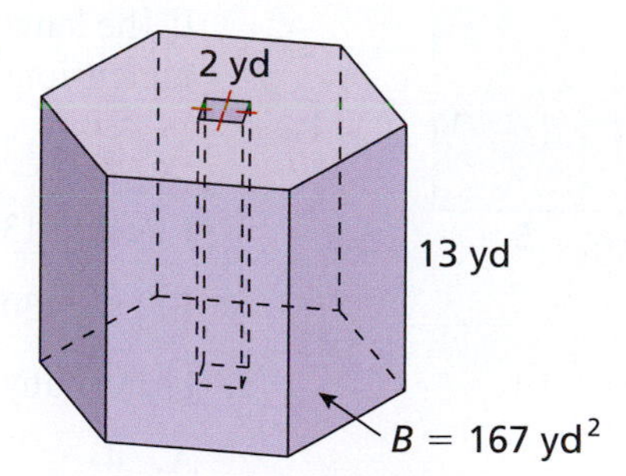

77. Sand Box How much sand will the sand box hold?

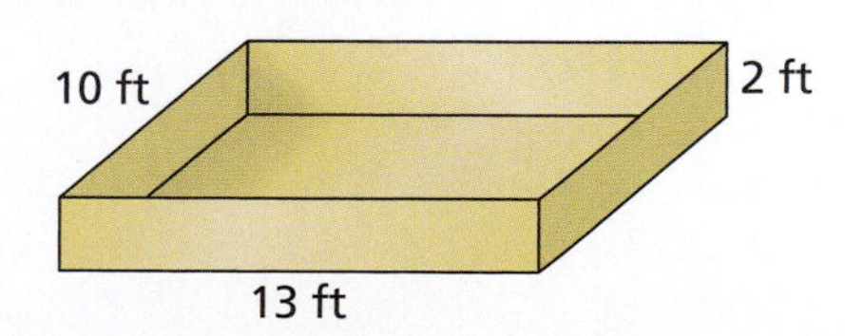

78. Swimming Pool Water flows from the hose at a rate of 10 gallons per minute. How long does it take to fill the swimming pool shown? (*Hint*: It takes about 7.5 gallons of water to fill 1 cubic foot.)

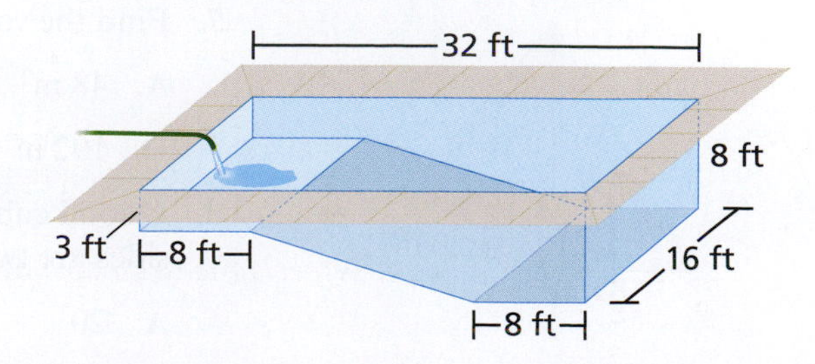

Chapter Test

The questions on this practice test are modeled after questions on state certification exams for K–8 teaching.

Test Directions: Each of the questions is followed by five choices. Choose the *best* response to each question.

1. A recipe for a beef roast uses a 12-fluid-ounce can of cola. A one-liter bottle of cola is available. Approximately how many liters of the bottle must be used to follow the recipe?

A. 0.08 L **B.** 0.20 L **C.** 0.36 L

D. 0.40 L **E.** 0.43 L

4 cm 3 cm

6 cm

2. What is the perimeter of the triangle at the left?

A. 6 cm **B.** 11 cm **C.** 12 cm **D.** 13 cm **E.** 15 cm

3. Assuming that the quadrilateral in the following figure is a parallelogram, what would be its area?

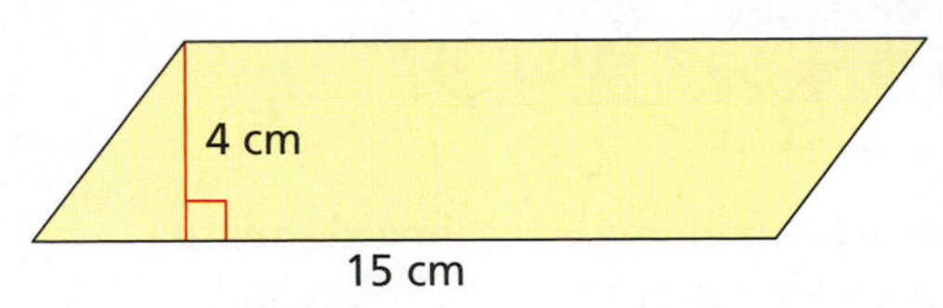

A. 30 cm **B.** 38 cm **C.** 38 cm^2

D. 60 cm **E.** 60 cm^2

w

$\ell = w + 5$

4. Mr. Wesley needs 65 meters of fencing to enclose his rectangular property. If the length of the property is 5 meters more than the width, what are the dimensions of his property? Note the figure at the left.

A. $w = 13.75$ m; $\ell = 18.75$ m **B.** $w = 12.25$ m; $\ell = 20.25$ m

C. $w = 13.0$ m; $\ell = 19.5$ m **D.** $w = 12.5$ m; $\ell = 20.0$ m

E. $w = 30.0$ m; $\ell = 35.0$ m

5. How many corners does a cube have?

A. 4 **B.** 6 **C.** 8 **D.** 12 **E.** 16

6. Find the surface area of the prism below.

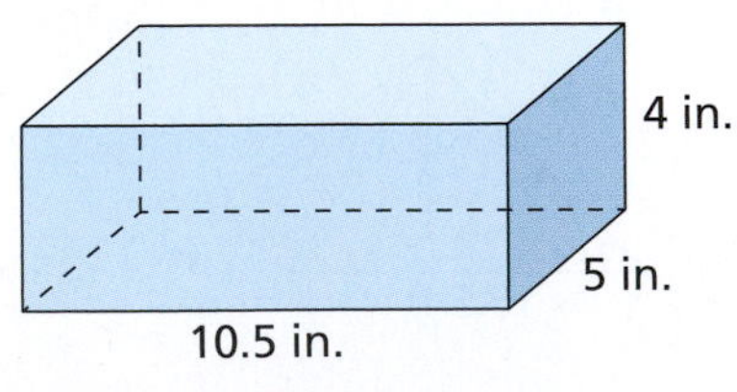

A. 114.5 $in.^2$ **B.** 210 $in.^2$ **C.** 210 $in.^3$

D. 229 $in.^2$ **E.** 229 $in.^3$

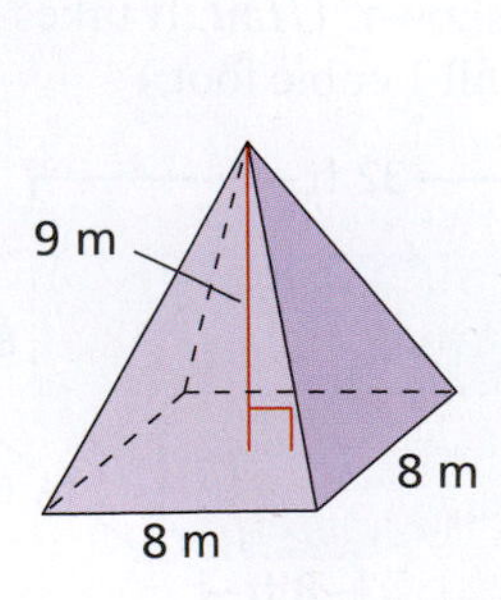

7. Find the volume of the figure at the left.

A. 48 m^2 **B.** 96 m^2 **C.** 128 m^3

D. 192 m^3 **E.** 576 m^3

8. A solid cube has a volume of 10. What is the volume of the cube whose sides are twice as long as the side lengths of this cube?

A. 20 **B.** 40 **C.** 60 **D.** 80 **E.** 100

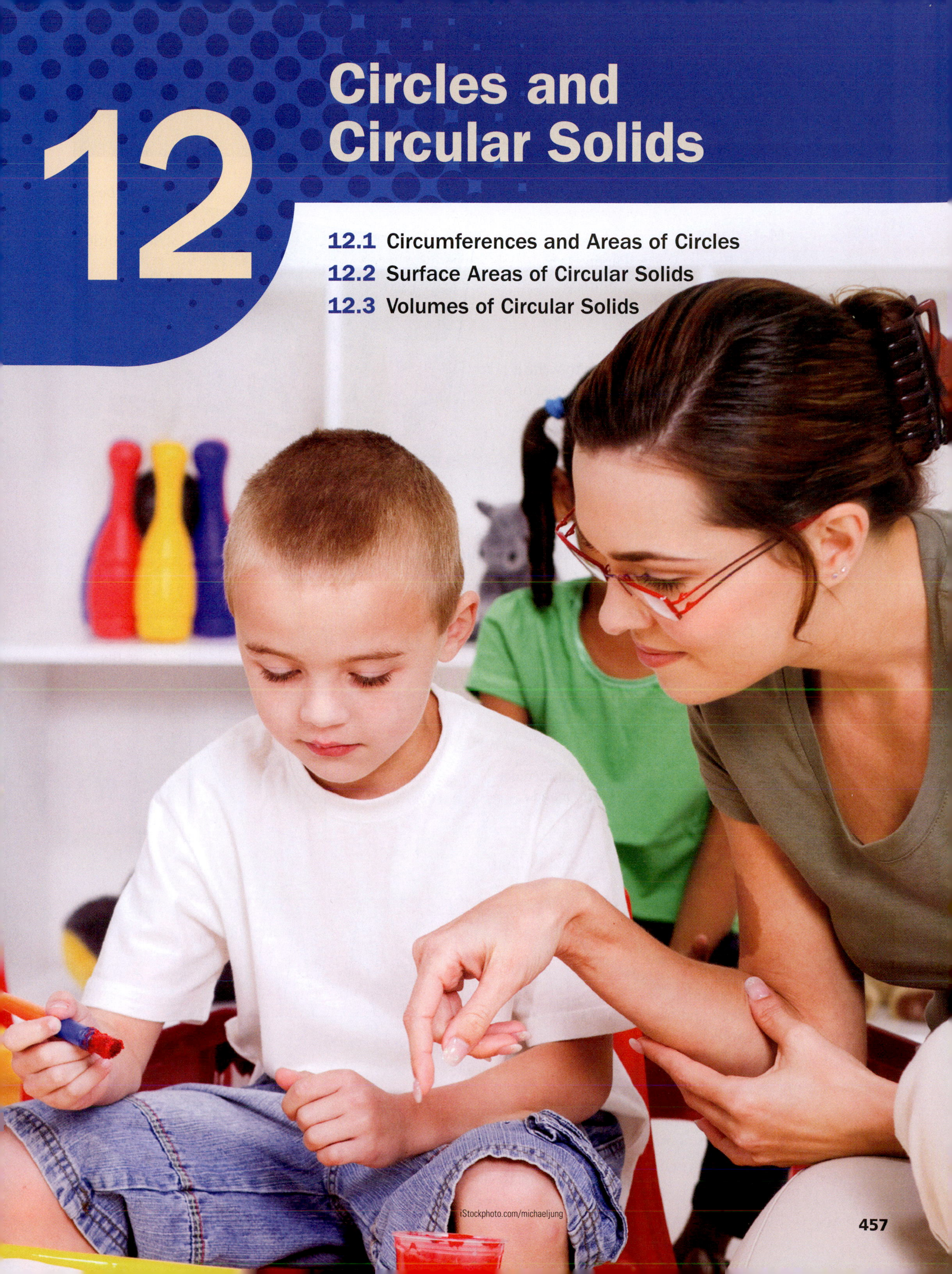

12 Circles and Circular Solids

Activity: Discovering a Formula for the Area of a Circle

Materials:
- Pencil
- Scissors
- Ruler
- Paper
- Compass

Learning Objective:

Use the definition of π to write a formula for the circumference of a circle. Write a formula for the area of a circle.

Name ______________________________

Recall that the real number π is defined as the ratio of the circumference of a circle C to its diameter d.

$$\pi = \frac{C}{d}$$

1. Draw a circle. Label its diameter d. Label its radius r.

2. Write the diameter d of the circle in terms of its radius r.

   $d =$ ☐

3. Write a formula for the circumference C of the circle in terms of π and r.

   $C =$ ☐

4. Divide the circle into 24 equal sections.

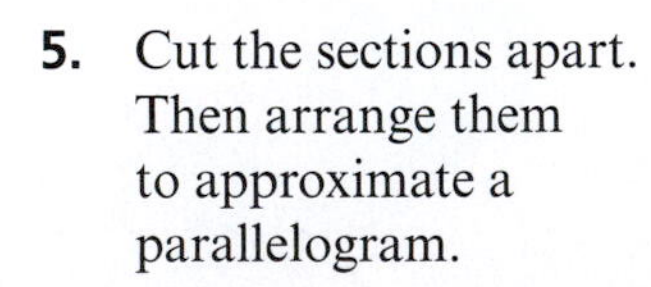

5. Cut the sections apart. Then arrange them to approximate a parallelogram.

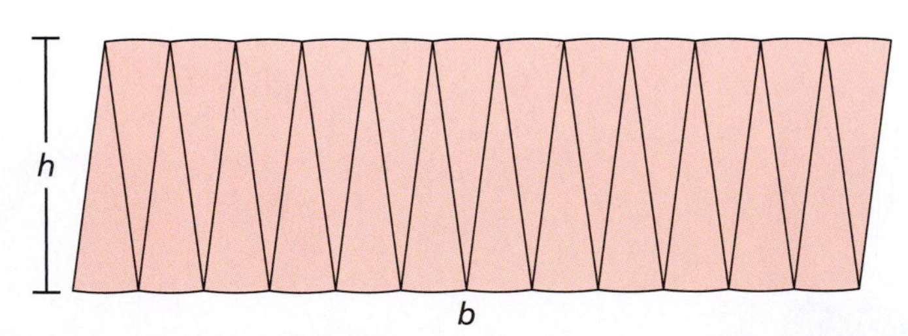

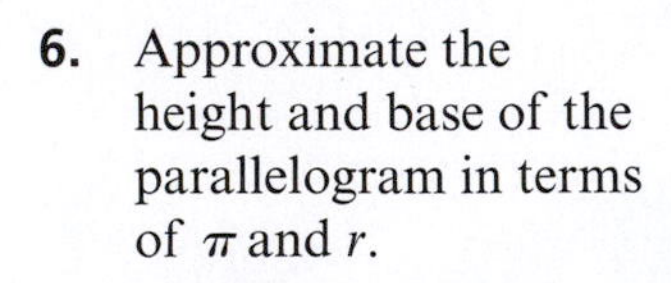

6. Approximate the height and base of the parallelogram in terms of π and r.

   $h \approx$ ☐ $b \approx$ ☐

7. Write a formula for the area of the parallelogram in terms of π and r. What can you conclude?

   $A \approx$ ☐

Dmitriy Shironosov | Dreamstime.com

12.1 Circumferences and Areas of Circles

Standards

Grades K–2
Geometry
Students should identify and describe shapes.

Grades 3–5
Measurement and Data
Students should understand concepts of angles and angle measurement.

Grades 6–8
Geometry
Students should solve real-life and mathematical problems involving area.

- Find the diameter and radius of a circle.
- Find the circumference of a circle.
- Find the area of a circle.

Parts of a Circle

Definition of a Circle

A **circle** is the set of all points in a plane that are the same distance from a point called the **center**.

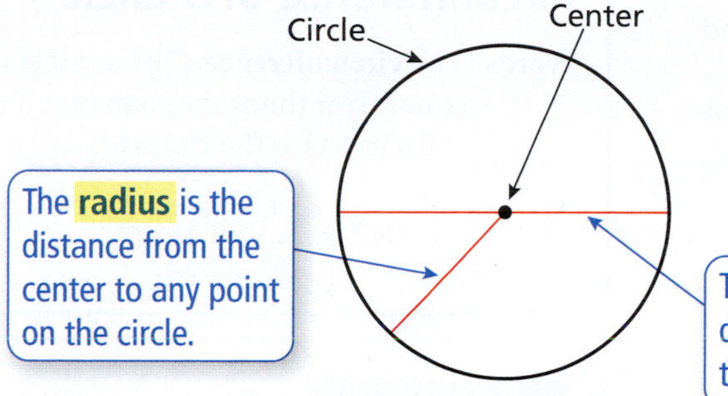

The **diameter** is the distance across the circle through the center.

Radius and Diameter of a Circle

Words The diameter d of a circle is twice the radius r. The radius r of a circle is one-half the diameter d.

Algebra Diameter: $d = 2r$ Radius: $r = \frac{d}{2}$

Classroom Tip

The tool used to draw a circle is called a *compass*. To use a compass, place the point at the center of a circle. Adjust the radius. Then rotate the compass so that the pencil draws the circle.

Lipskiy/Shutterstock.com

EXAMPLE 1 Finding a Diameter and a Radius

a. The diameter of a circle is 12 feet. Find the radius.

12 ft

b. The radius of a circle is 4.5 meters. Find the diameter.

4.5 m

SOLUTION

a. $r = \frac{d}{2}$ Radius of a circle

$= \frac{12}{2}$ Substitute 12 for d.

$= 6$ Divide.

The radius is 6 feet.

b. $d = 2r$ Diameter of a circle

$= 2(4.5)$ Substitute 4.5 for r.

$= 9$ Multiply.

The diameter is 9 meters.

Finding the Circumference of a Circle

You learned in Section 9.3 that the *circumference* of a circle is the distance around the circle. You also learned that the ratio of the circumference of a circle to its diameter can be represented by the Greek letter π, called **pi**.

$$\pi = \frac{C}{d}$$

where C is the circumference and d is the diameter of the circle.

Common approximations of π are 3.14 and $\frac{22}{7}$.

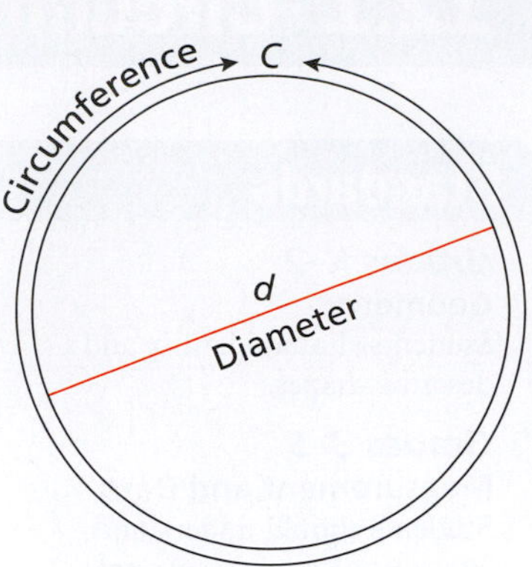

Classroom Tip

In most cases, 3.14 is used to approximate π. When the radius or diameter is a multiple of 7, it is easier to use $\frac{22}{7}$ as the approximation of π.

Circumference of a Circle

Words The **circumference** C of a circle is equal to π times the diameter d or π times twice the radius r.

Algebra $C = \pi d$ or $C = 2\pi r$

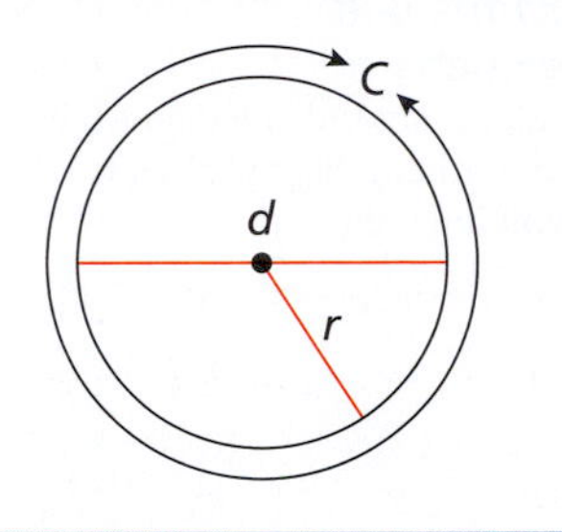

EXAMPLE 2 Finding a Circumference in Real Life

a. Find the circumference of the blade of the circular saw.

b. The blade has two teeth per inch. How many teeth are on the blade?

SOLUTION

a. The circumference of the blade is

$$C = \pi d \approx \frac{22}{7}(7) = 22 \text{ inches.}$$

b. The number of teeth on the blade is

$$(22 \text{ in.})\left(\frac{2 \text{ teeth}}{1 \text{ in.}}\right) = 44 \text{ teeth.}$$

Classroom Tip

To help give students a better understanding of the relationship between the circumference and the diameter of a circle, bring several sizes of cans to class. Ask students to measure the circumference and the diameter of each can and then divide to find the ratio of C to d.

EXAMPLE 3 Comparing Circumferences in Real Life

How many times does the smaller gear rotate each time the larger gear rotates once?

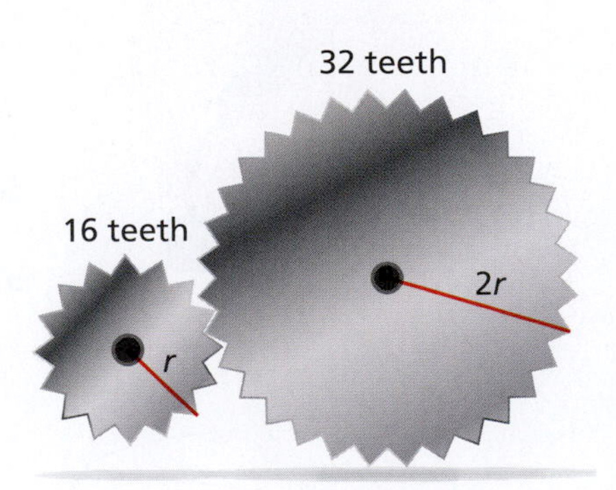

SOLUTION

The circumference of the smaller gear is $2\pi r$. The circumference of the larger gear is $2\pi(2r) = 4\pi r$. So, the smaller gear rotates twice each time the larger gear rotates once.

Finding a Perimeter in Real Life

A *semicircle* is one-half of a circle. Estimate the perimeter of the semicircular window.

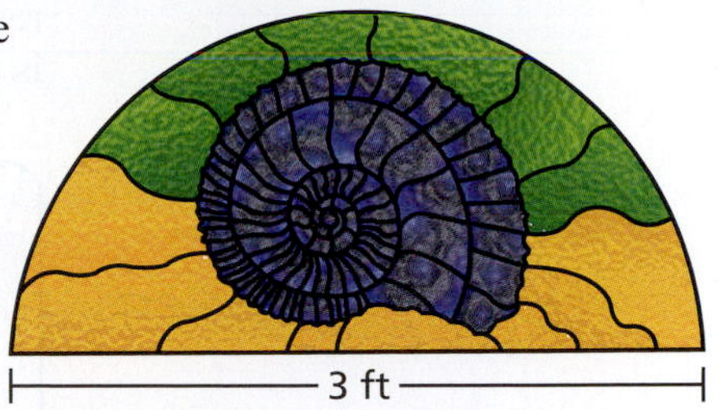

SOLUTION

The straight part of the window is 3 feet. The curved part is half the circumference C of a circle with a diameter of 3 feet.

$$\text{Perimeter} = \frac{\pi d}{2} + 3$$ The perimeter is the sum of the two parts.

$$\approx \frac{(3.14)(3)}{2} + 3$$ Substitute 3.14 for π and 3 for d.

$$= 7.71$$ Simplify.

The perimeter of the window is about 7.7 feet.

Classroom Tip

The formula at the right is derived using ratios and proportions that you studied in Section 6.4. Because the central angle of the arc is $a°$, the ratio of the length of the arc L to the entire circumference C is proportional to the ratio of $a°$ to 360°. So,

$$\frac{L}{C} = \frac{a°}{360°}.$$

Definition of a Circular Arc

A **circular arc** is a portion of the circumference of a circle C.

Length of a Circular Arc

When the central angle of the arc is $a°$, the length of the arc L is

$$L = \left(\frac{a}{360}\right)C \quad \text{or} \quad L = \frac{\pi a r}{180}.$$

A semicircle, determined by a diameter, is a special case of a circular arc, in which $a = 180$.

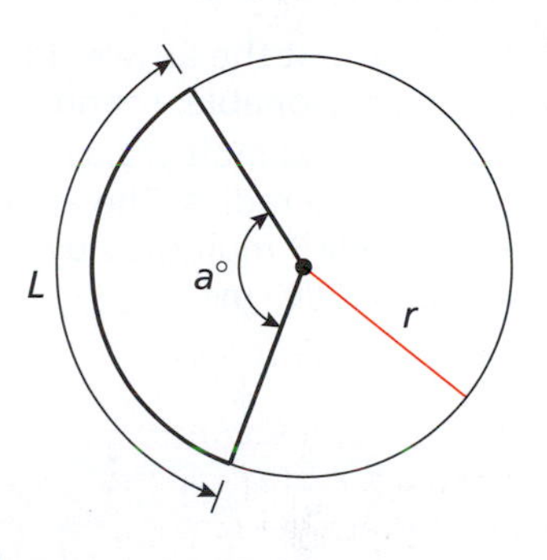

EXAMPLE 5

Finding the Length of a Circular Arc

The radius of the clock face is 10 inches. What is the length of the circular arc between the minute hand and the hour hand at 1:30?

SOLUTION

The angle between any two consecutive hours is $360° \div 12 = 30°$. So, the hour hand is at a 45° angle from noon, and the minute hand is at a 180° angle from noon. The central angle of the arc between the two hands is $180° - 45° = 135°$. The length of the arc is

$$L = \frac{\pi a r}{180}$$ Write formula for the length of a circular arc.

$$= \frac{\pi(135)(10)}{180}$$ Substitute 135 for a and 10 for r.

$$= 7.5\pi$$ Simplify.

$$\approx 23.55.$$ Multiply.

The length of the arc is about 23.6 inches.

Mathematical Practices

Construct Viable Arguments and Critique the Reasoning of Others

MP3 Mathematically proficient students construct arguments using concrete referents, such as objects, drawings, diagrams, and actions.

Finding the Area of a Circle

In the activity on page 458, a circle is divided into 24 equal sections and rearranged to approximate a parallelogram. The conclusion of the activity is that the area of a circle is $A = \pi r^2$.

Area of a Circle

Words The **area** A of a circle is equal to π times the square of the radius r.

Algebra $A = \pi r^2$

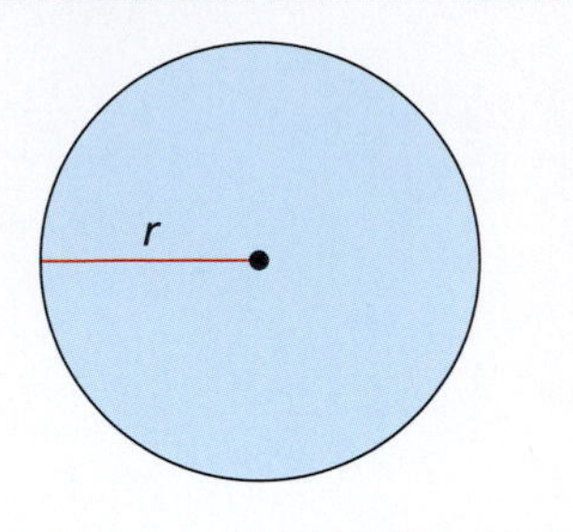

EXAMPLE 6 Finding the Area of a Circle

Find the area of the circle.

20 cm

SOLUTION

The radius is $20 \div 2 = 10$ centimeters.

$$A = \pi r^2 \quad \text{Write formula for area.}$$
$$\approx (3.14)(10)^2 \quad \text{Substitute 3.14 for } \pi \text{ and 10 for } r.$$
$$= (3.14)(100) \quad \text{Evaluate the power.}$$
$$= 314 \quad \text{Multiply.}$$

The area of the circle is about 314 square centimeters.

Classroom Tip

To check that the answer in Example 6 is reasonable, round the value of π and multiply by the square of the radius. The area is approximately 3 multipled by 10 squared, or 300 cm^2.

EXAMPLE 7 Finding an Area in Real Life

How much more area does the Hillsboro Lighthouse light up than the Jupiter Lighthouse?

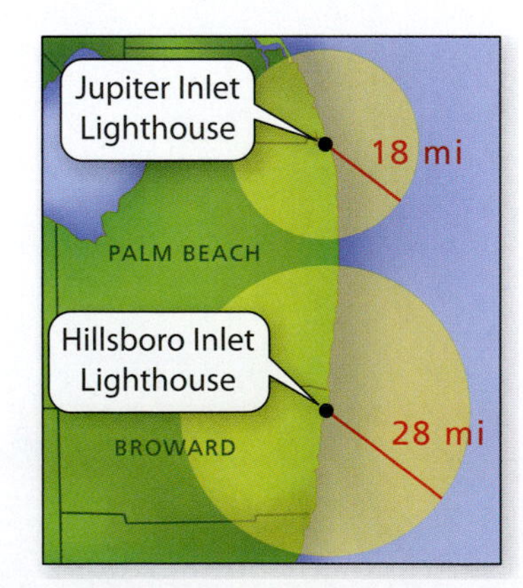

SOLUTION

The Jupiter Lighthouse lights up an area of

$$A = \pi r^2$$
$$= \pi(18)^2$$
$$= 324\pi \text{ square miles.}$$

The Hillsboro Lighthouse lights up an area of

$$A = \pi r^2$$
$$= \pi(28)^2$$
$$= 784\pi \text{ square miles.}$$

Subtract the area that the Jupiter Lighthouse lights up from the area that the Hillsboro Lighthouse lights up.

$$784\pi - 324\pi = 460\pi \approx 460(3.14) = 1444.4$$

So, the Hillsboro Lighthouse lights up about 1444.4 square miles more than the Jupiter Lighthouse.

Definition of a Sector

A **sector** of a circle is a pie-shaped region enclosed by two radii and their intercepted arc.

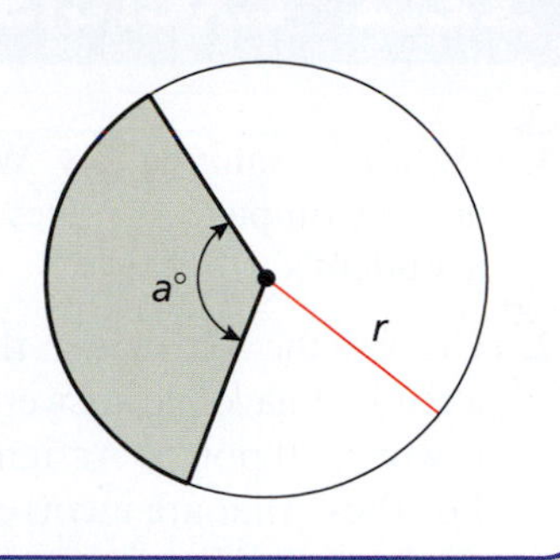

Area of a Sector

When the central angle of the sector is $a°$, the area A of the sector is

$$A = \left(\frac{a}{360}\right)\pi r^2.$$

EXAMPLE 8 Finding the Area of a Sector

a. Find the area of the sector.

b. What percent of the entire circle does the sector represent?

45°
12 in.

SOLUTION

a. $A = \left(\frac{a}{360}\right)\pi r^2.$ Write formula for the area of a sector.

$= \left(\frac{45}{360}\right)\pi(12)^2$ Substitute 45 for a and 12 for r.

$= 18\pi$ Simplify.

≈ 56.52 Multiply.

The area of the sector is about 56.52 square inches.

b. The sector makes up $\frac{45}{360} = \frac{1}{8}$ of the circle. So, the sector represents $\frac{1}{8} = 0.125 = 12.5\,\%$ of the entire circle.

EXAMPLE 9 Finding the Area of a Composite Figure

Find the area of the portion of the basketball court.

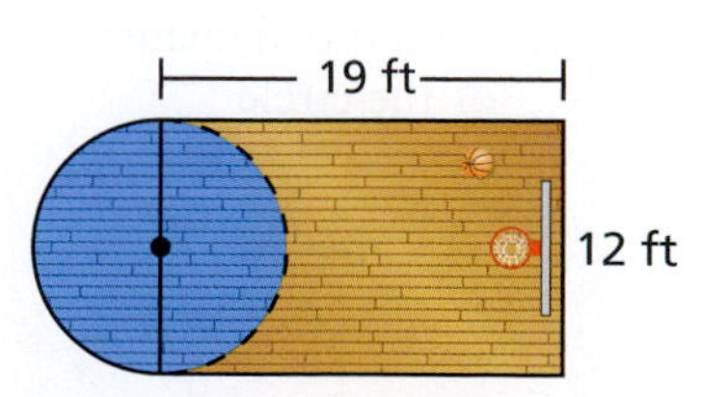

SOLUTION

The figure consists of a rectangle and a semicircle. You can find the area of the entire figure by finding the sum of the areas of these two figures.

Area of rectangle	**Area of semicircle**
$A = (\text{length})(\text{width})$	$A = \left(\frac{a}{360}\right)\pi r^2$
$= (19)(12)$	$= \left(\frac{180}{360}\right)\pi(6)^2$
$= 228$	$= 18\pi$
	≈ 56.52

The semicircle has a radius of $\frac{12}{2} = 6$ feet.

So, the area of the portion of the basketball court is about $228 + 56.52 = 284.52$ square feet.

12.1 Exercises

Free step-by-step solutions to odd-numbered exercises at *MathematicalPractices.com.*

1. Writing a Solution Key Write a solution key for the activity on page 458. Describe a rubric for grading a student's work.

2. Grading the Activity In the activity on page 458, a student gave the answers below. Each question is worth 10 points. Assign a grade to each answer. For those that are incorrect, why do you think the student erred?

Sample Student Work

6. Approximate the height and base of the parallelogram in terms of π and r.

$h \approx$ r $\qquad b \approx$ 2r

7. Write a formula for the area of the parallelogram in terms of π and r. What can you conclude?

$A \approx$ 2r²

The area of the circle is twice the square of the radius.

3. Finding the Diameter and the Radius of a Circle

a. Find the diameter of the circle.

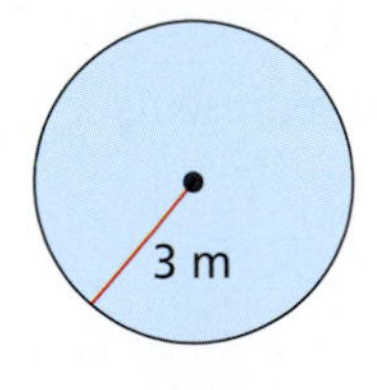

b. Find the radius of the circle.

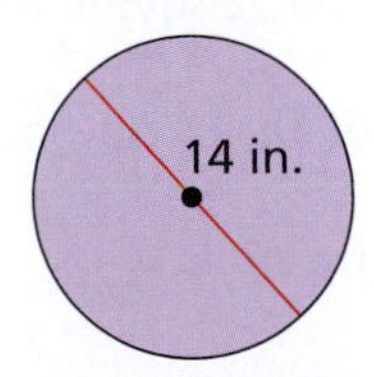

4. Finding the Diameter and the Radius of a Circle

a. Find the diameter of the circle.

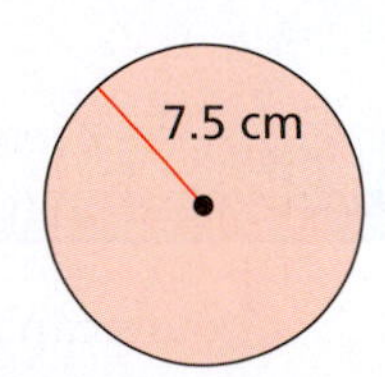

b. Find the radius of the circle.

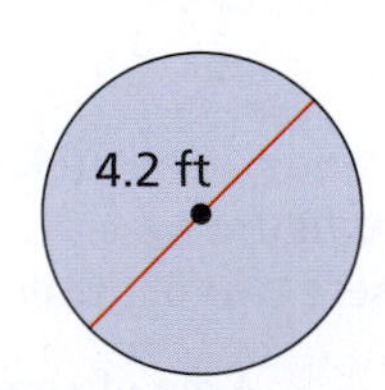

5. Finding Circumferences of Circles Find the circumference of each circle given its radius or diameter.

a. $d = 15$ mm **b.** $r = 5.5$ ft

c.

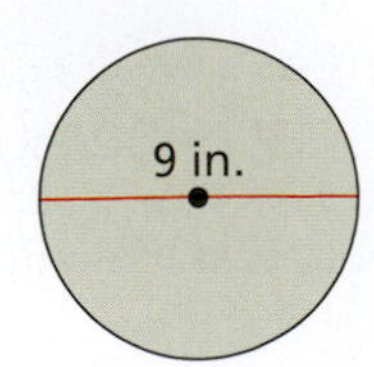

d.

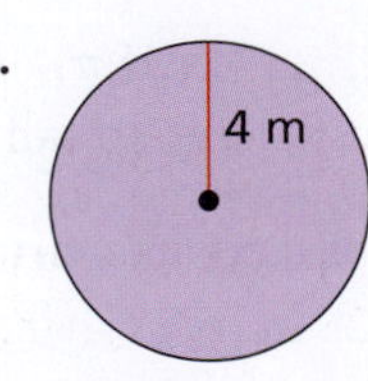

6. Finding Circumferences of Circles Find the circumference of each circle given its radius or diameter.

a. $d = 16$ in. **b.** $r = 9$ m

c.

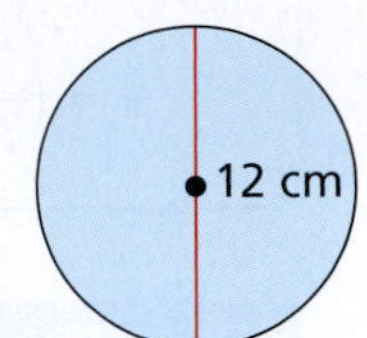

d.

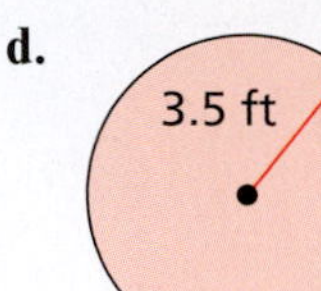

7. Finding Perimeters of Semicircles Find the perimeter of each semicircle given its radius or diameter.

a. $d = 5$ m **b.** $r = 16.5$ mm

c.

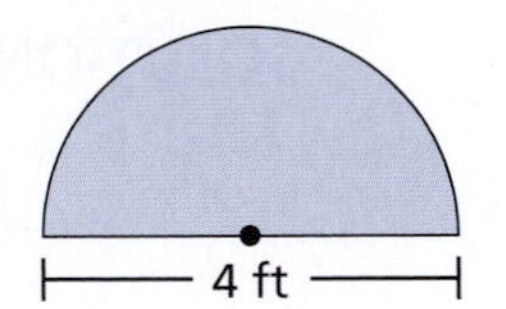

d.

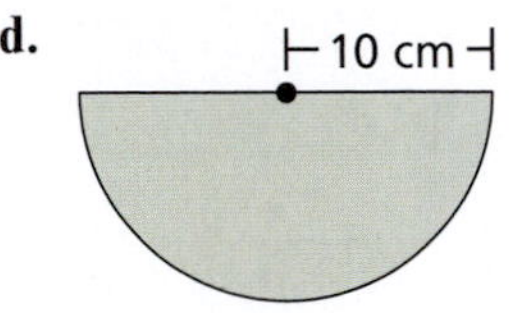

8. Finding Perimeters of Semicircles Find the perimeter of each semicircle given its radius or diameter.

a. $d = 13$ mm **b.** $r = 3$ ft

c. 7 m

d.

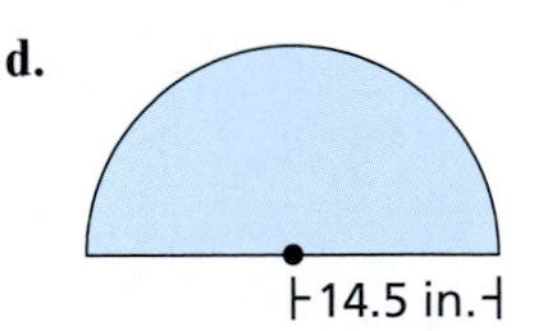

9. Finding Lengths of Circular Arcs Use the given dimensions to find the length of each circular arc.

a. $a = 60$, $C = 24$ in. **b.** $a = 40$, $r = 9$ m

c.

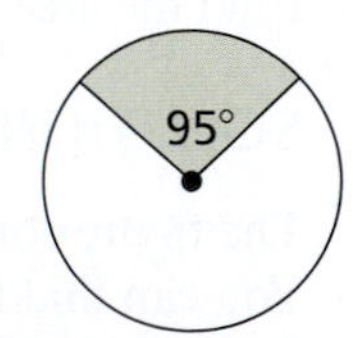

$C = 63$ mm

d.

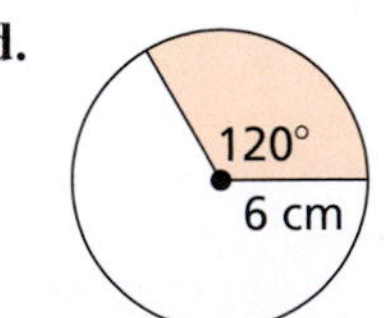

10. Finding Lengths of Circular Arcs Use the given dimensions to find the length of each circular arc.

a. $a = 30$, $C = 36$ cm **b.** $a = 72$, $r = 2$ ft

c.

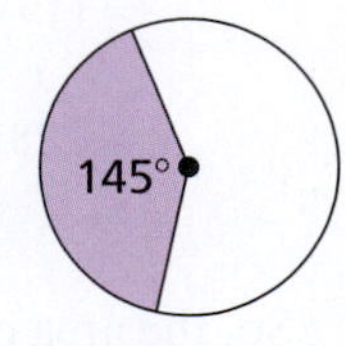

$C = 54$ in.

d.

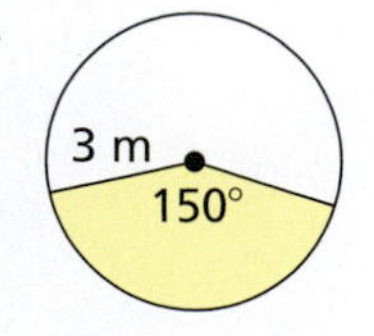

11. Finding Areas of Circles Find the area of each circle given its radius or diameter.

a. $r = 12$ ft **b.** $d = 26$ in.

c.

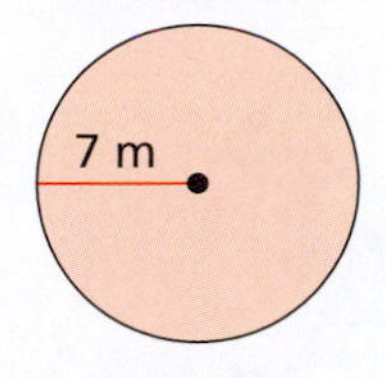

d.

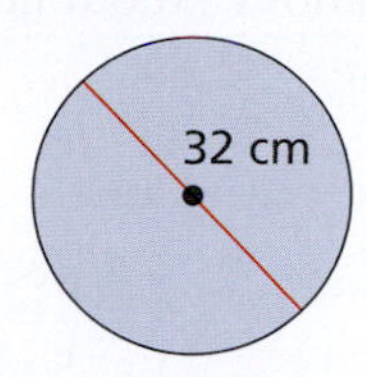

12. Finding Areas of Circles Find the area of each circle given its radius or diameter.

a. $r = 5$ m **b.** $d = 18$ ft

c.

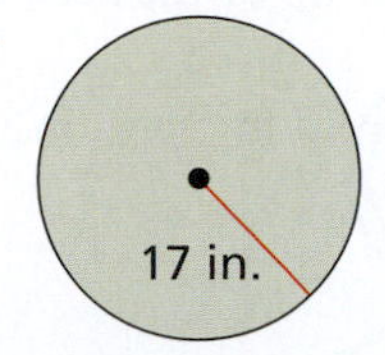

d.

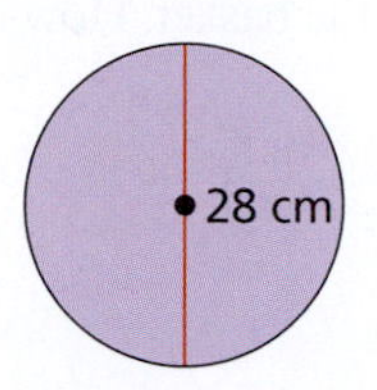

13. Finding Areas of Sectors Use the given dimensions to find the area of each sector.

a.

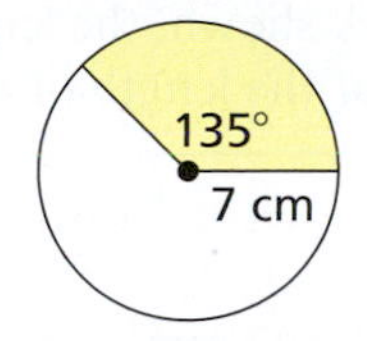

b.

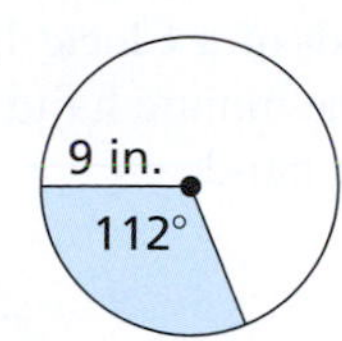

c. $a = 55$
$d = 12$ in.

d. $a = 63$
$d = 8$ m

14. Finding Areas of Sectors Use the given dimensions to find the area of each sector.

a.

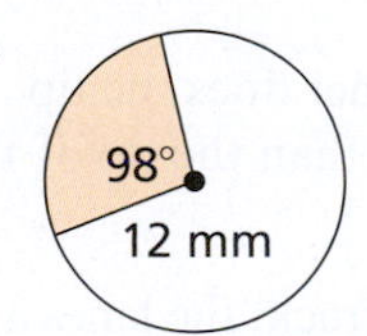

b.

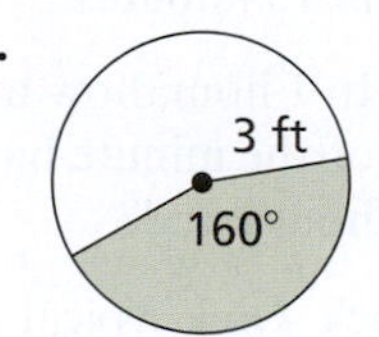

c. $a = 27$
$d = 10$ cm

d. $a = 80$
$d = 30$ in.

15. Finding Areas of Composite Figures Find the area of each composite figure. All curves are circular arcs with centers shown.

a.

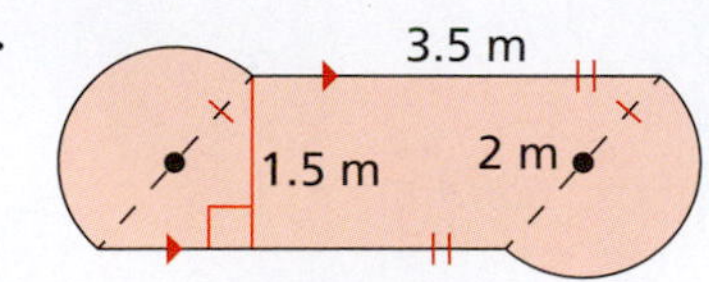

b.

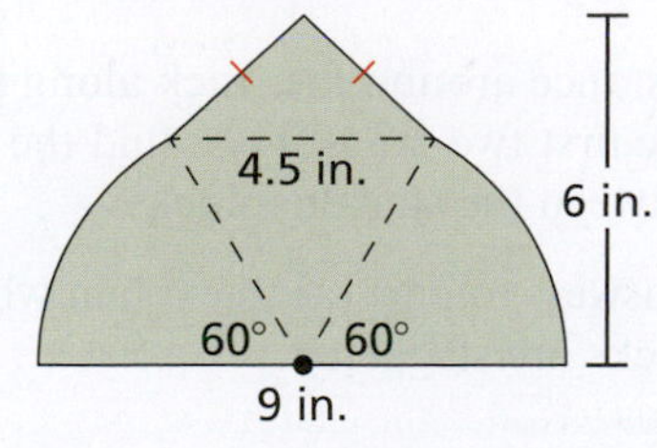

16. Finding Areas of Composite Figures Find the area of each composite figure. All curves are circular arcs with centers shown.

a.

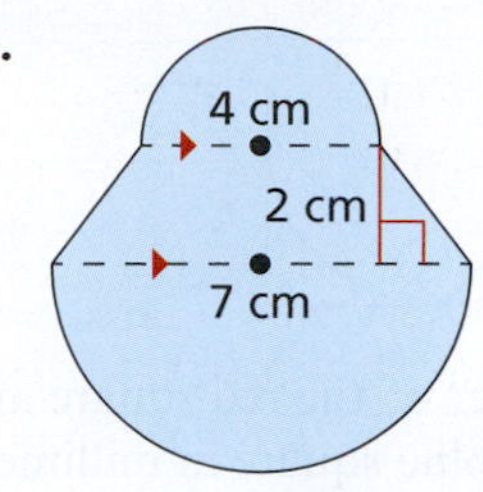

b.

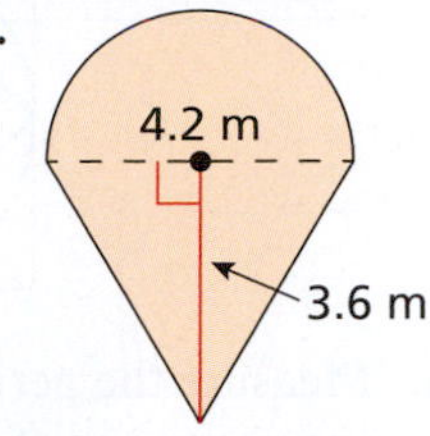

17. Finding Perimeters of Composite Figures Find the perimeter of each composite figure. All curves are circular arcs with centers shown.

a. **b.**

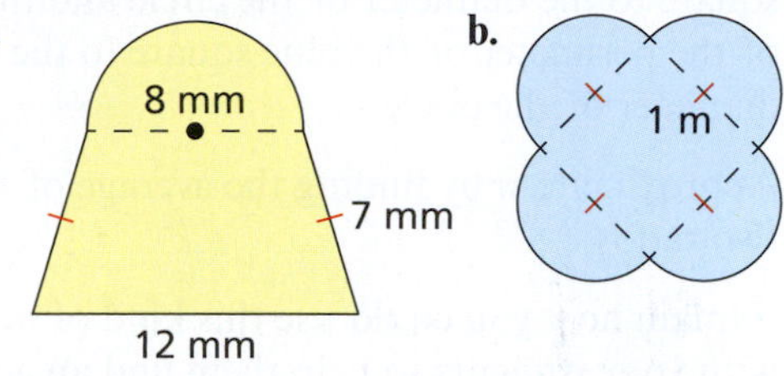

18. Finding Perimeters of Composite Figures Find the perimeter of each composite figure. All curves are circular arcs with centers shown.

a.

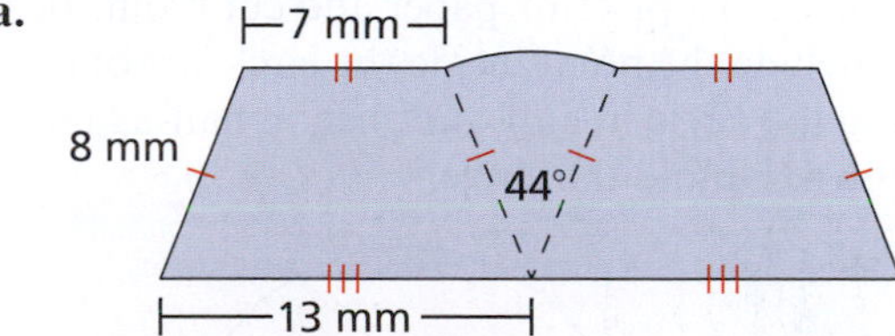

b.

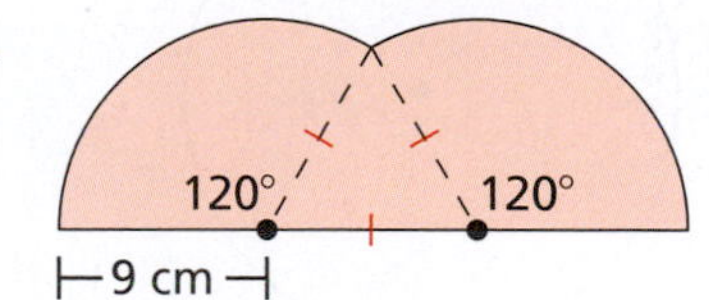

19. Finding a Circumference in Real Life A 12-inch round pizza has a diameter of 12 inches. What is the circumference of the pizza to the nearest inch?

20. Finding a Circumference in Real Life You want to make a skirt to go around the bottom of a round chair with a radius of 1 foot. What is the minimum length of fabric needed for the skirt to wrap the entire way around the chair?

21. Finding an Area in Real Life A circular logo centered at the 50-yard line of a football field has a radius of about 4 yards. What is the area of the logo to the nearest square yard?

22. Finding an Area in Real Life You have determined that you can install a round swimming pool that has an area that is less than 700 square feet. Are you able to install a swimming pool that has a diameter of 28 feet? Explain.

23. **Approximating Pi: In Your Classroom** Refer to the figure below.

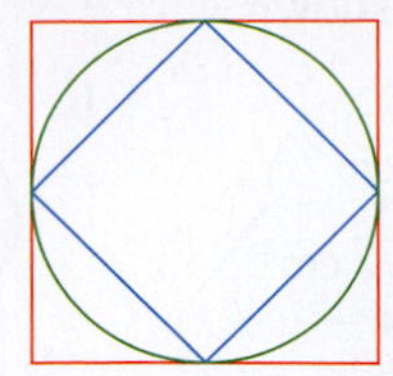

a. Measure the perimeter of the red square and the perimeter of the blue square in millimeters. Then measure the diameter of the circle in millimeters.

b. Calculate the ratio of the perimeter of the red square to the diameter of the circle and the ratio of the perimeter of the blue square to the diameter of the circle.

c. Approximate π by finding the average of these two ratios.

d. Explain how you could use this kind of activity with your students to help them find an accurate approximation of π.

24. **Determining the Center of a Circle: In Your Classroom** Refer to the figure below. Trace the circle onto a piece of paper and cut it out. Be sure to include the point inside the circle in your trace. Fold the circle in half and then in half again as shown. Unfold the circle.

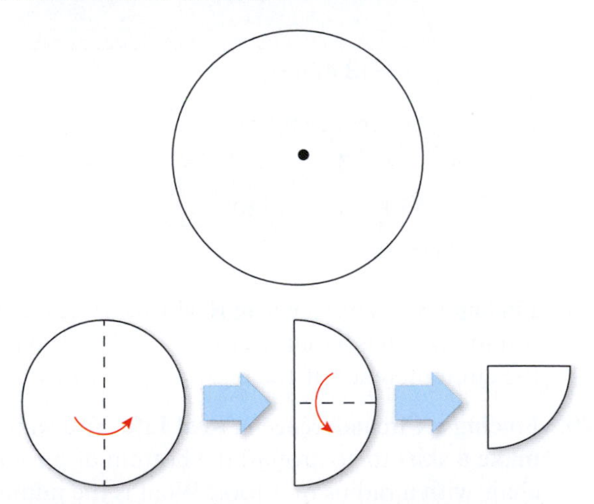

a. Is the point inside the circle at the center of the circle? Explain your reasoning.

b. Describe a similar activity you could use with your students to determine whether a line is a diameter of a circle.

25. **Tires of a Car** The tire of a car has a diameter of 28 inches. About how many rotations will the tire make when the car travels 7 miles?

26. **Ferris Wheel** The Ferris wheel shown travels at a rate of 157 meters per minute. Throughout the duration of one ride, the Ferris wheel makes 15 full rotations. About how long is each ride?

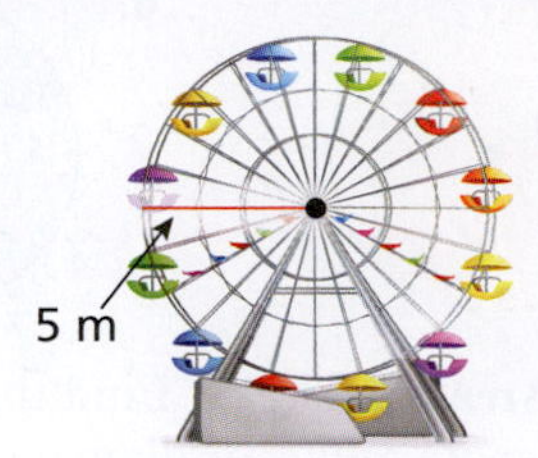

27. **Weighing Groceries** You place 2 pounds of bananas in the basket. How far does the tip of the needle move?

28. **Hands of a Clock** In the clock shown, the length of the minute hand is 150% of the length of the hour hand.

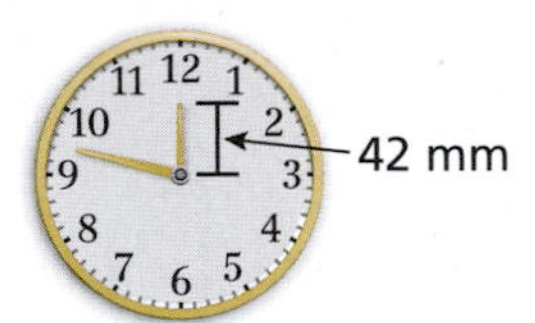

a. How far will the tip of the minute hand move in 15 minutes?

b. In 1 hour, how much farther does the tip of the minute hand move than the tip of the hour hand?

29. **Track** On a typical outdoor track, the lanes are about 1.2 meters wide and numbered beginning with the innermost lane. For a race that involves one complete lap around the track, the starting blocks are staggered as shown below.

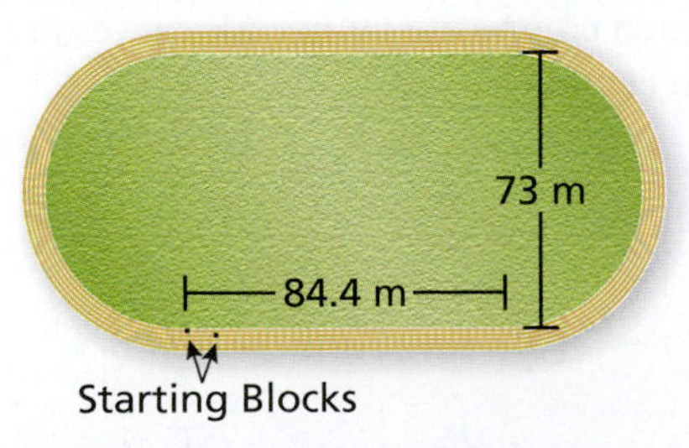

a. Find the distance around the track along the inside of the first two lanes. Then find the distance between the starting blocks.

b. Use your answers to part (a) to explain why the starting blocks are staggered.

30. Fountain The fountain shown below is made up of two semicircles and one-fourth of a circle. Find the perimeter and the area of the fountain.

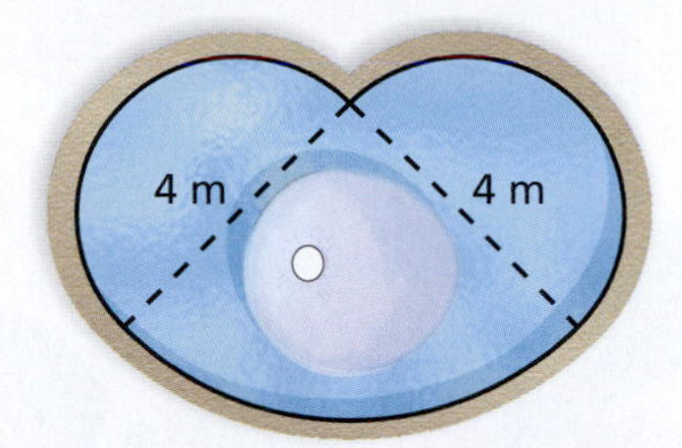

31. Problem Solving Find the area of each shaded region.

a.

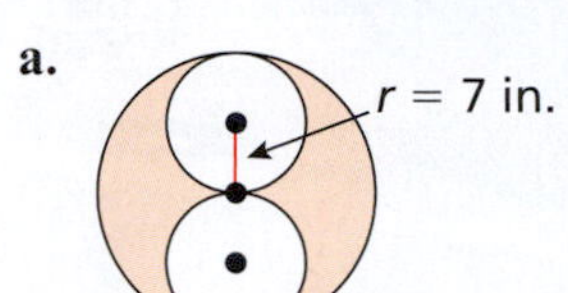

b.

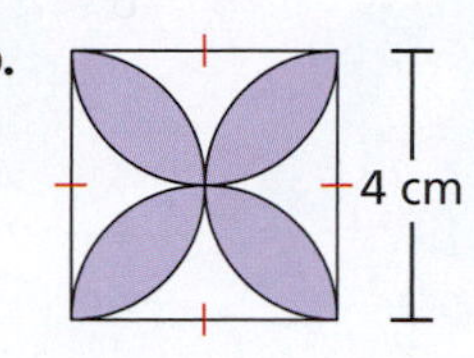

32. Problem Solving The diameter of the circle shown is 12 inches. What is the area of the yellow region?

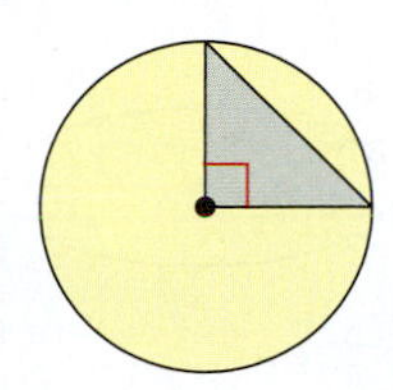

33. Problem Solving In the figure shown below at the left, two semicircles separate a circle into two regions. Two of the ends of the semicircles meet at the center of the circle, and the other ends lie on a diameter of the circle. Find the length of the curve that separates the regions. Then find the area of the shaded region.

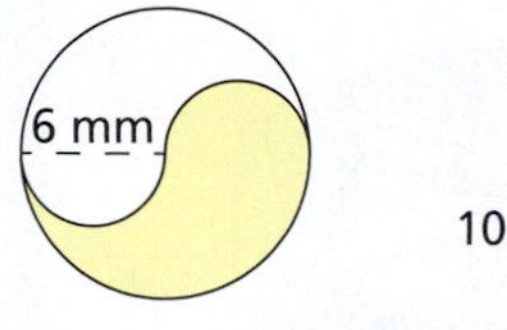

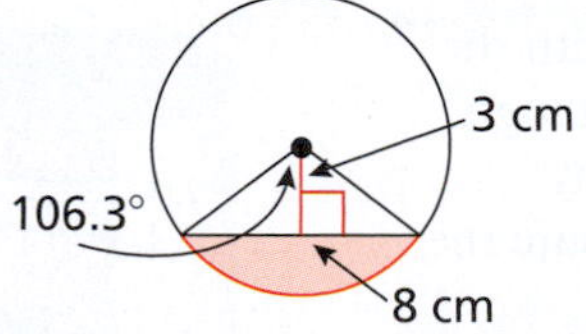

34. Problem Solving The radius of the circle, shown above at the right, is 5 centimeters. Find the length of the red circular arc. Then find the area of the shaded region.

35. Grading Student Work On a diagnostic test, one of your students does the following work. Explain what the student did wrong. Which topics would you encourage the student to review?

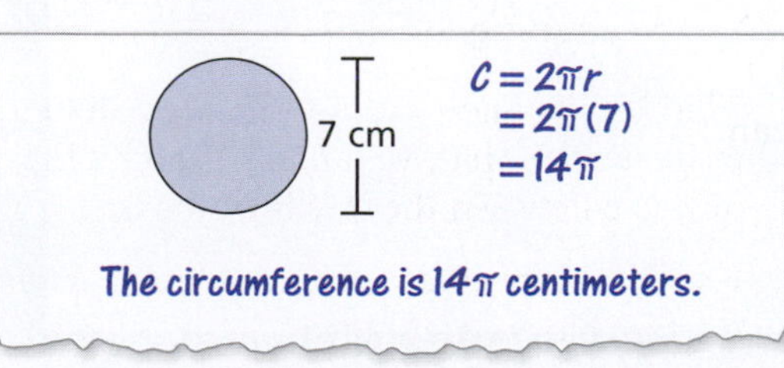

36. Grading Student Work On a diagnostic test, one of your students does the following work. Explain what the student did wrong. Which topics would you encourage the student to review?

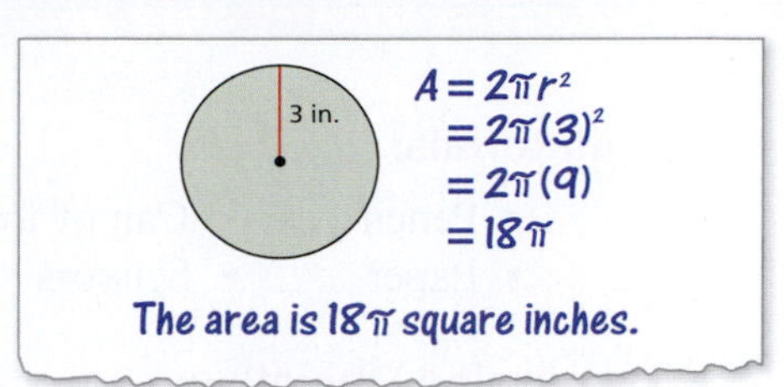

37. Irrigation A farmer has two options for irrigating his square field. Option A includes one large circular irrigation system, and Option B includes four smaller circular irrigation systems.

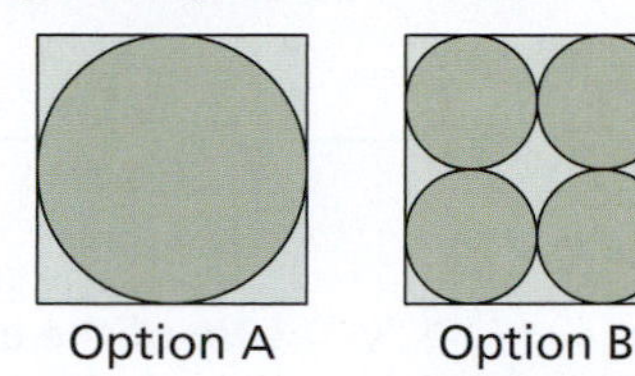

a. Without doing any calculations, which option do you think will irrigate more land? Explain your reasoning.

b. What percentage of the field will be irrigated in Option A?

c. What percentage of the field will be irrigated in Option B?

d. Do your answers to parts (b) and (c) support your answer to part (a)? Explain.

38. Darts In the dartboard below, the radius of the red circle is 1 inch. The width of each ring is also 1 inch.

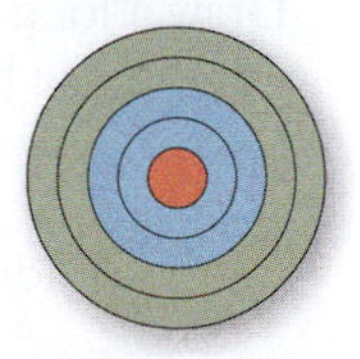

a. How does the area of the outermost ring compare to the sum of the areas of the blue rings and the red circle?

b. Use the areas of the red, blue, and green regions to describe a fair way to score the board when playing a game in which the first player to reach 100 points wins.

39. Writing Because the ratio of the circumference to the diameter is the same for every circle, is the ratio of the circumference to the radius the same for every circle? Explain your reasoning.

40. Writing Is the area of a semicircle with a radius of x *greater than*, *less than*, or *equal to* the area of a circle with a radius of $\frac{1}{2}x$? Explain your reasoning.

Activity: Discovering a Formula for the Surface Area of a Cylinder

Materials:

- Pencil
- Paper
- Can of food
- Scissors

Learning Objective:

Make a net for a circular cylinder. Use the net to write a formula for the surface of a circular cylinder.

Name ______________________________

The *surface area* of a circular cylinder is the sum of the areas of the two bases and the lateral surface.

1. Use a can of food. Trace the top and bottom of the can on paper. Cut out the two circles. Measure the diameter and find the radius r of each circle.

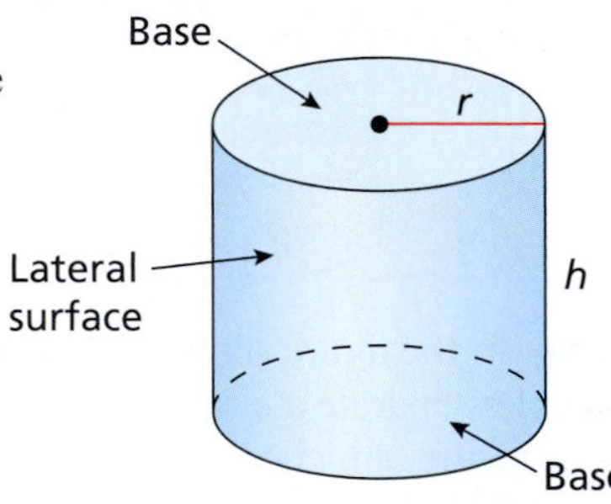

$r =$ ▭

Are the circles congruent?

2. Draw a long rectangle on paper. Make the width the same as the height of the can. Cut out the rectangle. Wrap the rectangle around the can. Cut off any excess paper so that the edges just meet. Measure the width w and length ℓ of the rectangle.

$w =$ ▭ $\quad \ell =$ ▭

What is the relationship between the length of the rectangle and the circumference of the base of the can?

3. Make a net for the can.

4. Use the net to write a formula for the surface area of a circular cylinder.

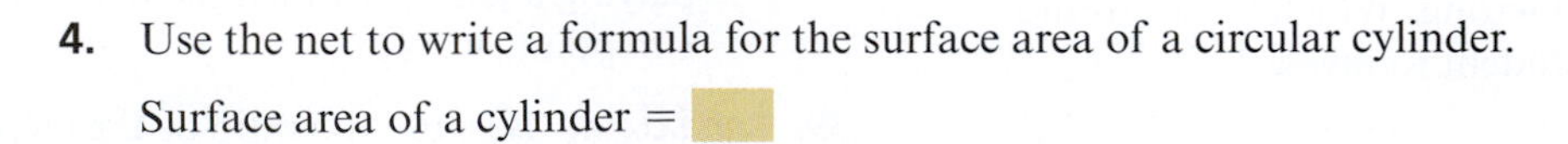

Surface area of a cylinder = ▭

5. Use the formula to find the surface area of the can.

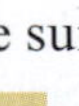

Surface area of a can = ▭

Monkey Business Images/Shutterstock.com

12.2 Surface Areas of Circular Solids

Standards

Grades K–2
Geometry
Students should identify and describe shapes.

Grades 6–8
Geometry
Students should solve real-life and mathematical problems involving area and surface area.

- Find the surface area of a right circular cylinder.
- Find the surface area of a right circular cone.
- Find the surface area of a sphere.

Finding the Surface Area of a Right Circular Cylinder

A **right circular cylinder** is a cylinder with circular bases that are perpendicular to the side of the cylinder. The **lateral surface** is the surface of the side of the cylinder.

Surface Area of a Right Circular Cylinder

Words The **surface area** S of a right circular cylinder is the sum of the areas of the two bases and the lateral surface.

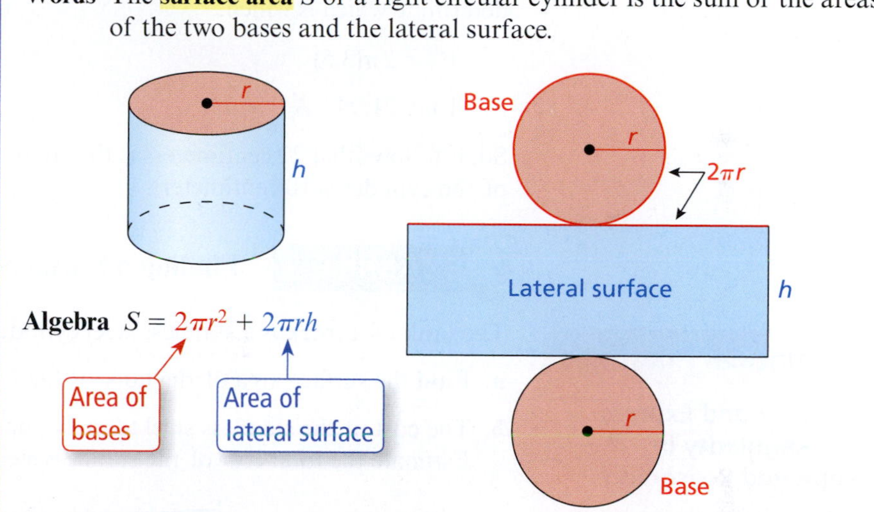

Algebra $S = 2\pi r^2 + 2\pi rh$

Classroom Tip

There are many types of cylinders. Their bases can be ellipses, and their lateral surfaces can meet the base at angles other than right angles. The formula for the surface area of a right circular cylinder does not apply to elliptical or oblique cylinders. In this book, only right cylinders that have circular bases are discussed.

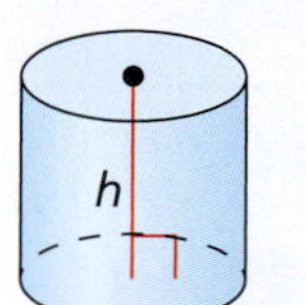

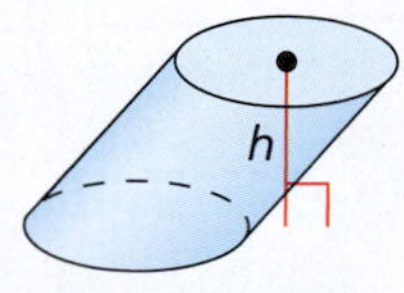

Right cylinder **Oblique cylinder**

EXAMPLE 1 **Finding the Surface Area of a Cylinder**

Find the surface area of the cylinder. Round your answer to the nearest tenth.

4 mm
3 mm

SOLUTION

Begin by drawing a net of the cylinder.

$$\begin{aligned} S &= 2\pi r^2 + 2\pi rh \\ &= 2\pi(4)^2 + 2\pi(4)(3) \\ &= 32\pi + 24\pi \\ &= 56\pi \\ &\approx 175.84 \end{aligned}$$

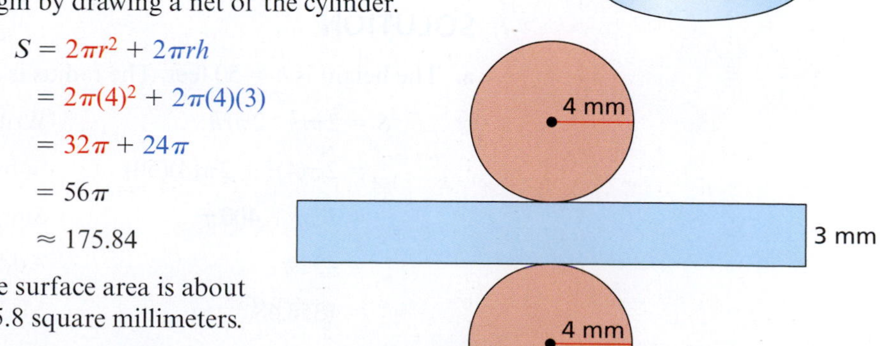

The surface area is about 175.8 square millimeters.

EXAMPLE 2 Identifying the Dimensions of a Cylinder

A student is constructing a cylinder, as shown. What is the height of the cylinder that the student is constructing? Explain your reasoning.

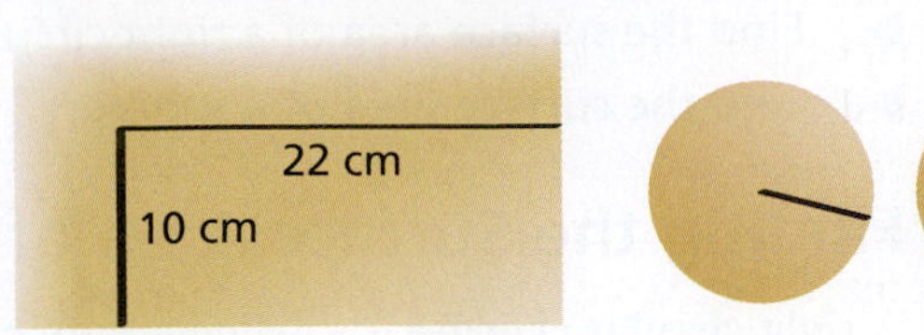

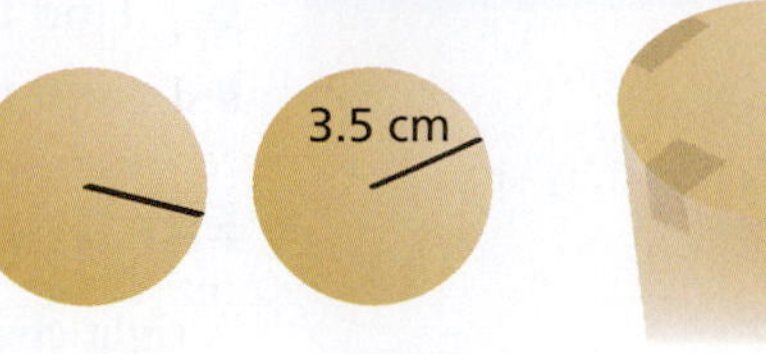

SOLUTION

The radius of each circle is 3.5 centimeters. Looking at the dimensions of the rectangle, one dimension is the height of the cylinder, and the other is the circumference of each base. Use the formula for circumference $C = 2\pi r$ to determine which is which.

$10 \stackrel{?}{=} 2\pi(3.5)$ $\qquad$ $22 \stackrel{?}{=} 2\pi(3.5)$

$10 \not\approx 21.98$ ✗ $\qquad$ $22 \approx 21.98$ ✓

So, it follows that 22 centimeters is the circumference. This implies that the height of the cylinder is 10 centimeters.

EXAMPLE 3 Finding a Surface Area in Real Life

Mathematical Practices

Look For and Express Regularity in Repeated Reasoning

MP8 Mathematically proficient students maintain oversight of the process when working on a problem.

The tank of a truck is a stainless steel cylinder.

a. Find the surface area of the tank. Round your answer to the nearest tenth.

b. The cost of the stainless steel to make the tank is \$60 per square foot. Estimate the total cost of the stainless steel.

SOLUTION

a. The height is $h = 50$ feet. The radius is $r = 4$ feet.

$S = 2\pi r^2 + 2\pi rh$	Write formula for surface area of a cylinder.
$= 2\pi(4)^2 + 2\pi(4)(50)$	Substitute 4 for r and 50 for h.
$= 32\pi + 400\pi$	Simplify.
$= 432\pi$	Add.
≈ 1356.48	Multiply.

The surface area is about 1356.5 square feet.

b. At \$60 per square foot, the total cost of the stainless steel is

$$\frac{\$60}{1\ \cancel{\text{sq ft}}} \cdot 1356.5\ \cancel{\text{sq ft}} = \$81{,}390.$$

iStockphoto.com/macroworld

Finding the Surface Area of a Right Circular Cone

To make a cone out of paper, begin with a circle. Cut out a sector of the circle. Then bend the paper to form the cone, as shown at the right. The greater the angle of the sector cut out, the more pointed the cone will be.

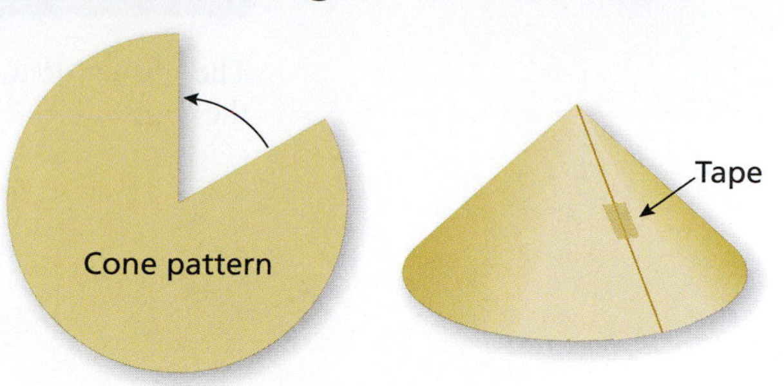

Surface Area of a Right Circular Cone

Words The **surface area** S of a right circular cone is the sum of the areas of the base and the lateral surface. The distance from the vertex of the cone to any point on the edge of its base is the **slant height** of the cone.

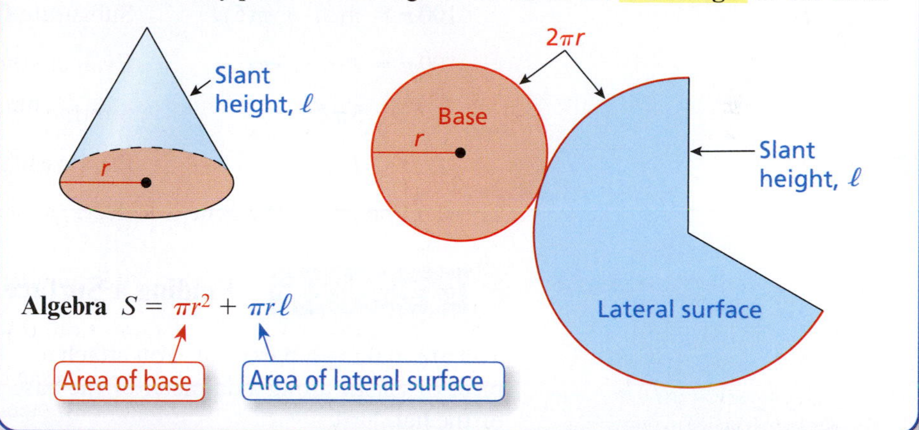

Algebra $S = \pi r^2 + \pi r \ell$

Area of base — Area of lateral surface

Classroom Tip

As is true of cylinders, not all cones are right circular cones. Some have bases that are ellipses, and some are oblique. The formula for the surface area of a right circular cone does not apply to elliptical or oblique cones.

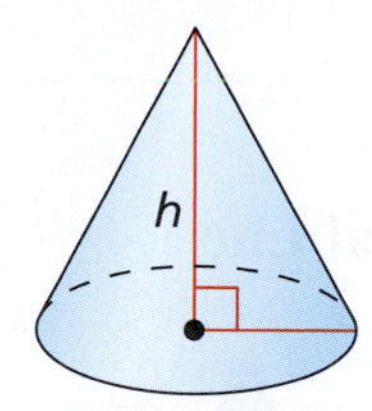

Right cone

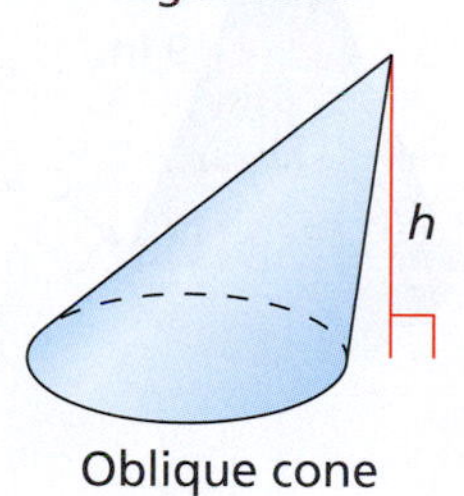

Oblique cone

EXAMPLE 4 Finding the Surface Area of a Cone

Find the surface area of the cone. Round your answer to the nearest tenth.

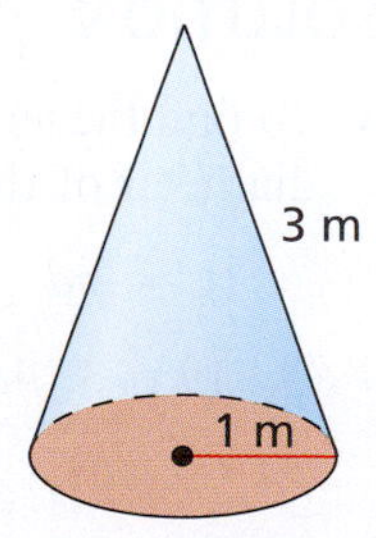

SOLUTION

Begin by drawing a net of the cone.

$$S = \pi r^2 + \pi r \ell$$
$$= \pi(1)^2 + \pi(1)(3)$$
$$= \pi + 3\pi$$
$$= 4\pi$$
$$\approx 12.56$$

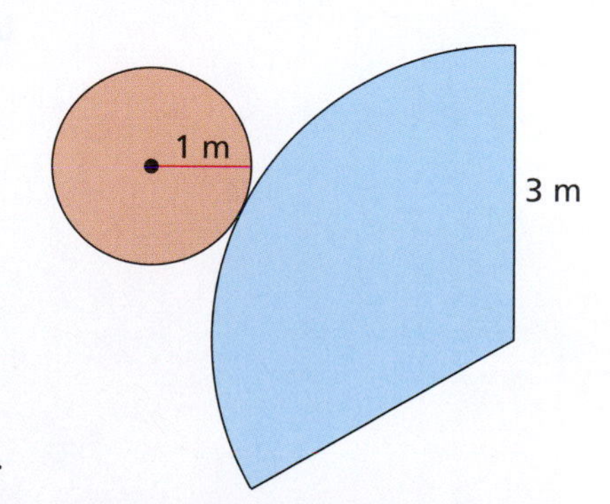

The surface area is about 12.6 square meters.

EXAMPLE 5 Finding the Slant Height of a Cone

The surface area of the cone is 100π square feet. What is the slant height of the cone?

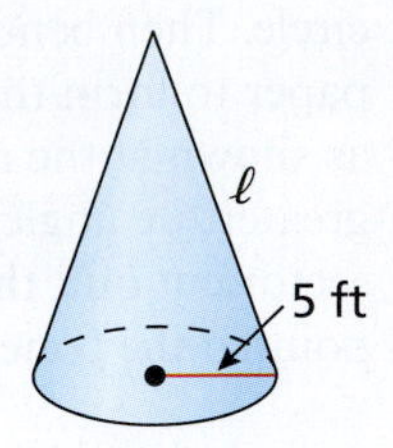

SOLUTION

To find the slant height, start with the formula for the surface area and substitute the known values. Then solve for the slant height.

$S = \pi r^2 + \pi r \ell$	Write formula for surface area of a cone.
$100\pi = \pi(5)^2 + \pi(5)\ell$	Substitute 100π for S and 5 for r.
$100\pi = 25\pi + 5\pi\ell$	Evaluate the power.
$75\pi = 5\pi\ell$	Subtract 25π from each side.
$15 = \ell$	Divide each side by 5π.

The slant height of the cone is 15 feet.

EXAMPLE 6 Finding a Surface Area in Real Life

You are making a party hat. You attach a piece of elastic along a diameter of the base of the hat.

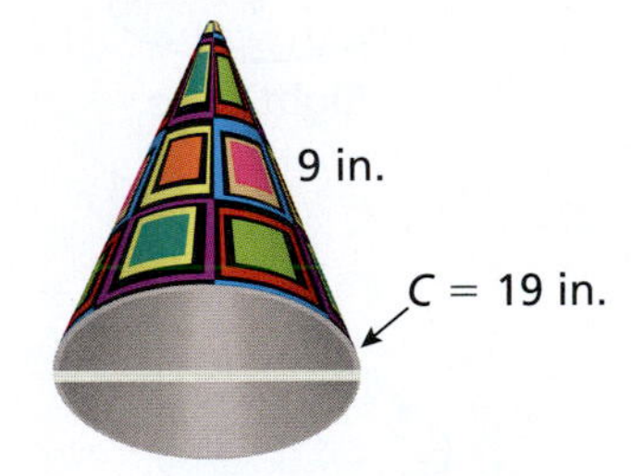

a. How long is the elastic?

b. How much paper do you need to make the hat?

SOLUTION

a. To find the length of the elastic, find the diameter of the base.

$C = \pi d$	Write formula for circumference.
$19 \approx 3.14d$	Substitute 19 for C and 3.14 for π.
$6 \approx d$	Divide each side by 3.14.

The elastic is about 6 inches long.

b. To find how much paper you need, find the lateral surface area. Because the diameter is 6 inches, the radius is 3 inches.

$S = \pi r \ell$	Write formula for the area of the lateral surface.
$= \pi(3)(9)$	Substitute 3 for r and 9 for ℓ.
$= 27\pi$	Multiply.
≈ 84.78	Multiply.

You need about 85 square inches of paper to make the hat.

Finding the Surface Area of a Sphere

A **sphere** is the set of all points in space that are the same distance from a point called the **center**.

Surface Area of a Sphere

Words The **surface area** S of a sphere is four times the area of a circular *cross section* that contains the center of the sphere. A **cross section** is the two-dimensional shape that is the result of the intersection of a plane with a three-dimensional solid.

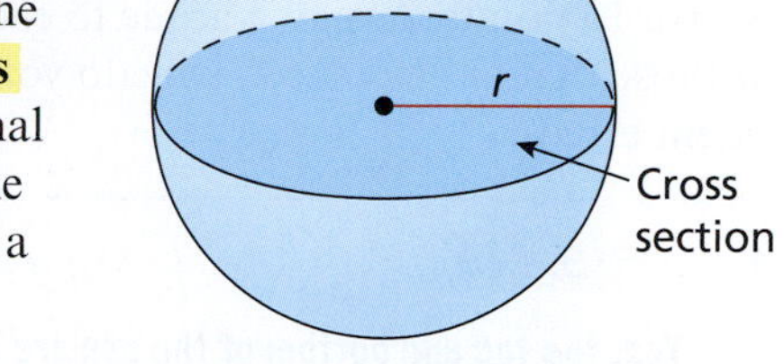

Algebra $S = 4\pi r^2$

Justifying the Surface Area of a Sphere

To show students that the formula for the surface area of a sphere is reasonable, bring two hand-sewn baseballs to class. Cut the stitches on one and unfold the two pieces of leather. You can see that the amount of leather necessary to cover the surface of the baseball is roughly equal to four cross sections of the baseball.

Finding a Surface Area in Real Life

The circumference of a basketball is 29.5 inches. What is its surface area?

SOLUTION

Begin by finding the radius of the basketball.

$C = 2\pi r$	Write formula for circumference.
$29.5 = 2\pi r$	Substitute 29.5 for C.
$4.7 \approx r$	Solve for r.

Find the surface area.

$S = 4\pi r^2$	Write formula for surface area of a sphere.
$= 4\pi(4.7)^2$	Substitute 4.7 for r.
$= 88.36\pi$	Simplify.
≈ 277.5	Multiply.

The surface area is about 277.5 square inches.

12.2 Exercises

Free step-by-step solutions to odd-numbered exercises at *MathematicalPractices.com.*

1. **Writing a Solution Key** Write a solution key for the activity on page 468. Describe a rubric for grading a student's work.

2. **Grading the Activity** In the activity on page 468, a student gave the answers below. Each question is worth 10 points. Assign a grade to each answer. For those that are incorrect, why do you think the student erred?

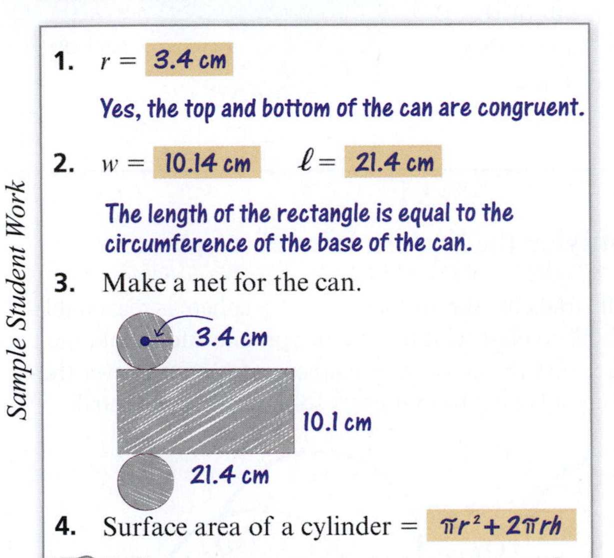

3. **Finding Surface Area Using a Net** The net for a cylinder is shown. Find the surface area of the cylinder.

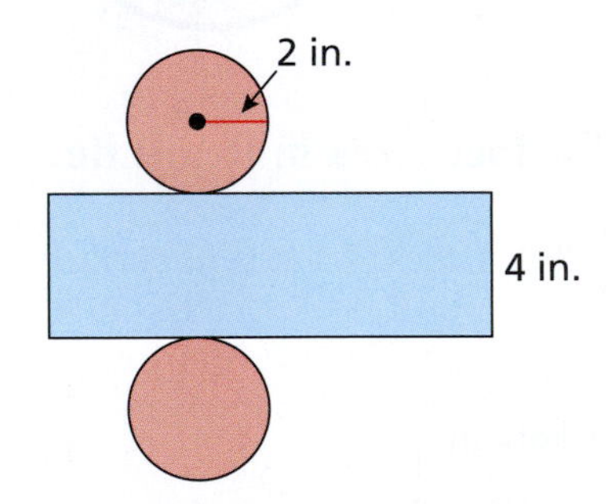

4. **Finding Surface Area Using a Net** The net for a cylinder is shown. Find the surface area of the cylinder.

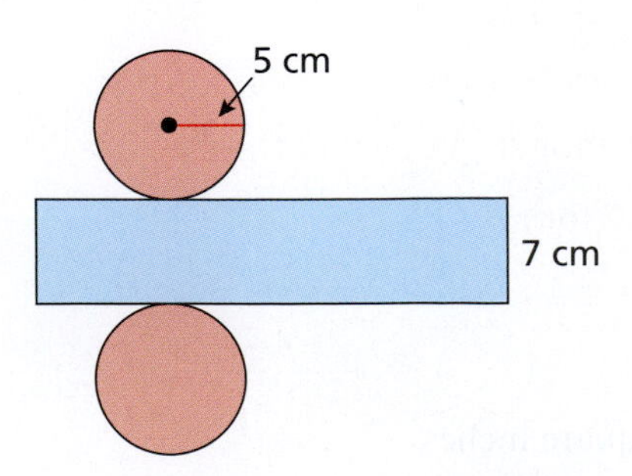

5. **Finding Surface Areas of Cylinders** Sketch a net of each cylinder. Then find its surface area.

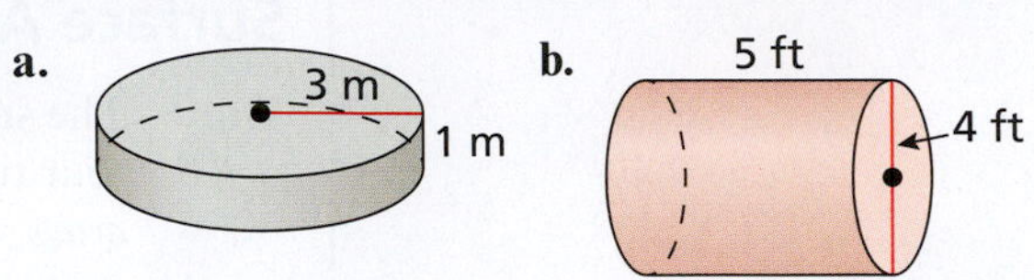

6. **Finding Surface Areas of Cylinders** Sketch a net of each cylinder. Then find its surface area.

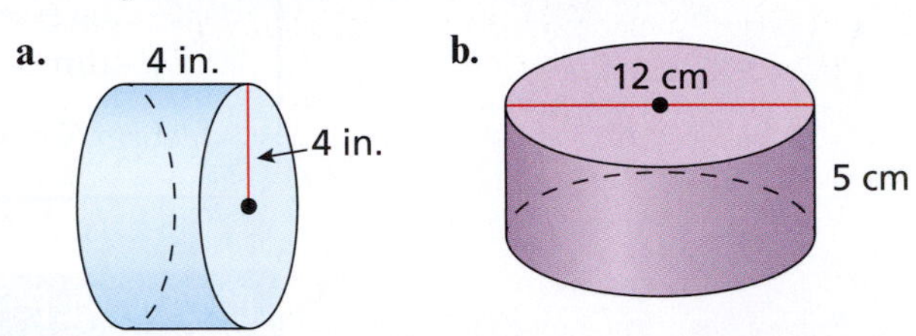

7. **Finding the Height of a Cylinder** The lateral surface and the bases of a cylinder are shown. Find the height of the cylinder.

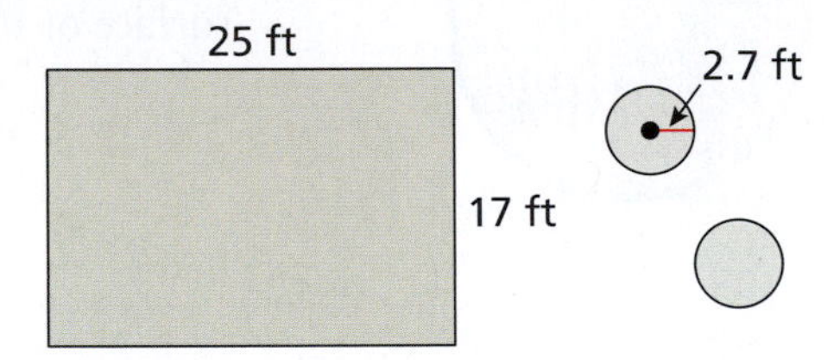

8. **Finding the Height of a Cylinder** The lateral surface and the bases of a cylinder are shown. Find the height of the cylinder.

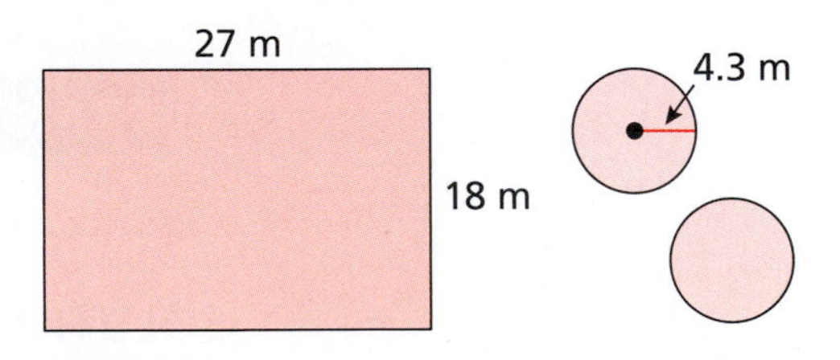

9. **Finding Surface Areas of Cones** Find the surface area of each cone with the given dimensions.

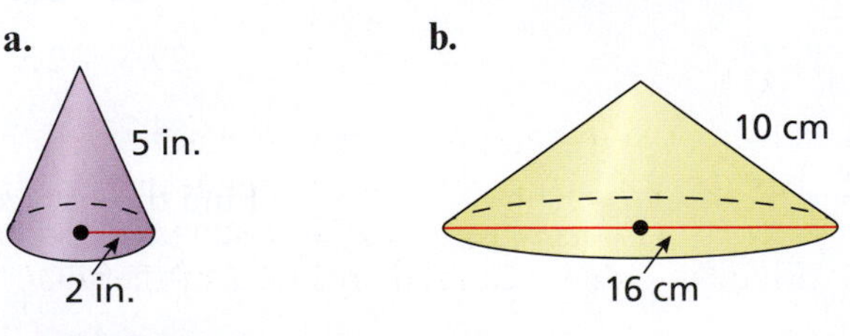

c. Radius: 7 ft
Slant height: 12 ft

d. Diameter: 22 m
Slant height: 15 m

10. Finding Surface Areas of Cones Find the surface area of each cone with the given dimensions.

a. **b.**

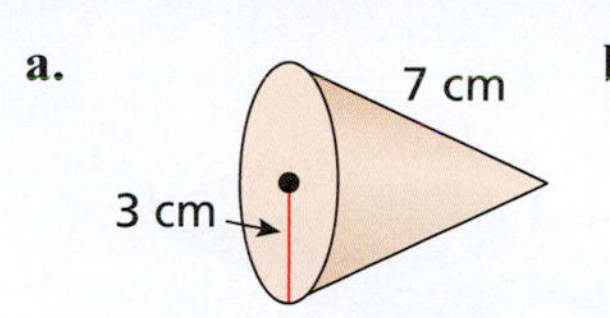

c. Radius: 4 yd
Slant height: 6.5 yd

d. Diameter: 14 ft
Slant height: 9.2 ft

11. Finding the Slant Height of a Cone The surface area of the cone is 300π square inches. Find the slant height of the cone.

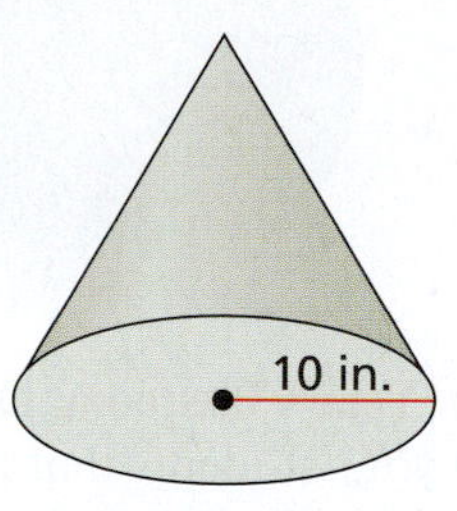

12. Finding the Slant Height of a Cone The surface area of the cone is 105π square feet. Find the slant height of the cone.

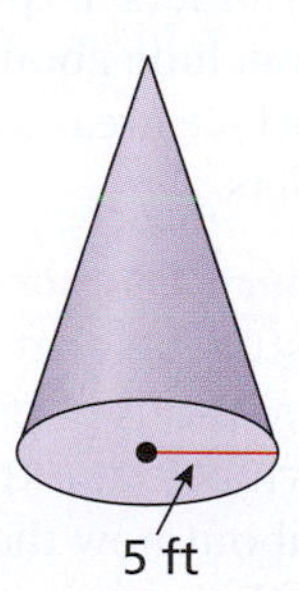

13. Finding the Dimensions of a Cone The area of the base of the cone is 16π square meters, and the area of the lateral surface is 24π square meters. Find the radius r and slant height ℓ of the cone.

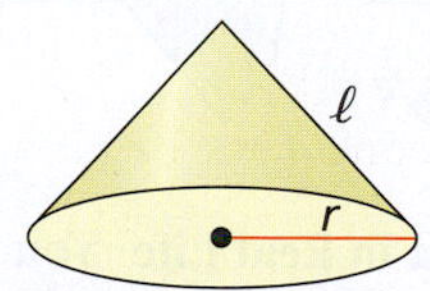

14. Finding the Dimensions of a Cone The area of the base of the cone is 81π square millimeters, and the area of the lateral surface is 135π square millimeters. Find the radius r and slant height ℓ of the cone.

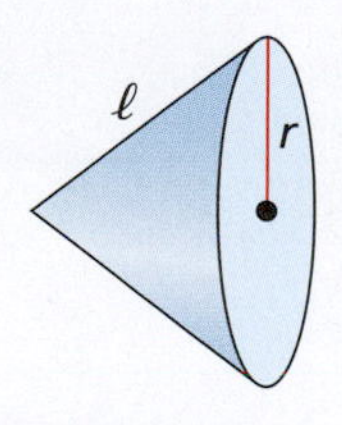

15. Finding the Surface Area of a Sphere The area A of a cross section through the center of a sphere is given. Find the surface area of the sphere.

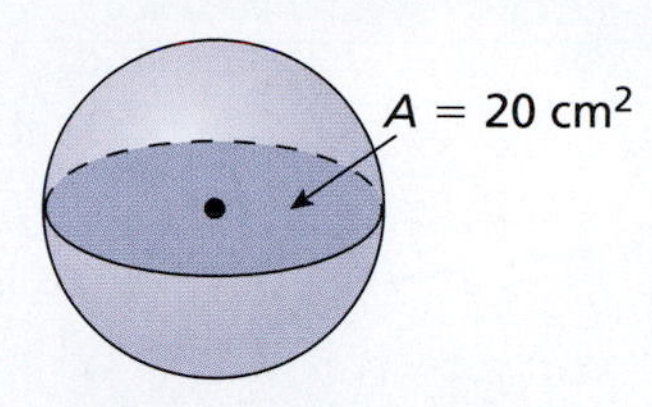

16. Finding the Surface Area of a Sphere The area A of a cross section through the center of a sphere is given. Find the surface area of the sphere.

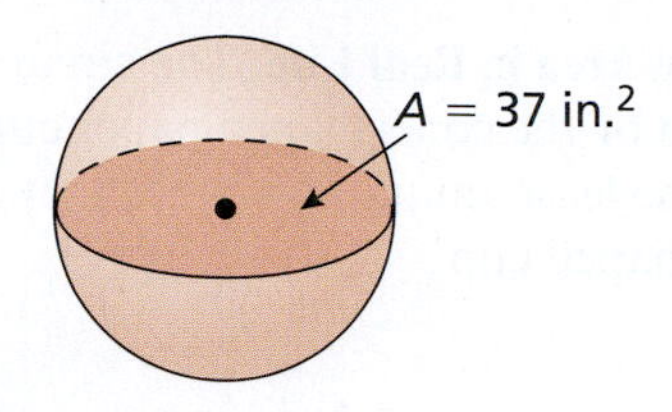

17. Finding Surface Areas of Spheres Find the surface area of each sphere.

a. **b.**

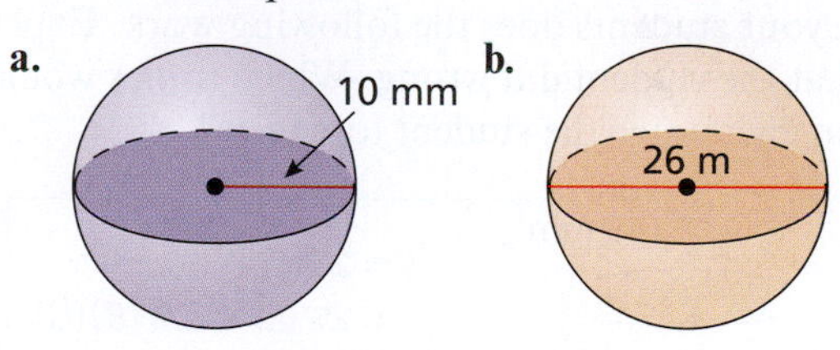

18. Finding Surface Areas of Spheres Find the surface area of each sphere.

a. **b.**

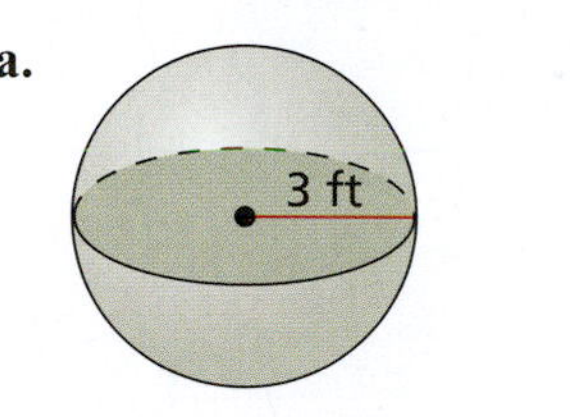

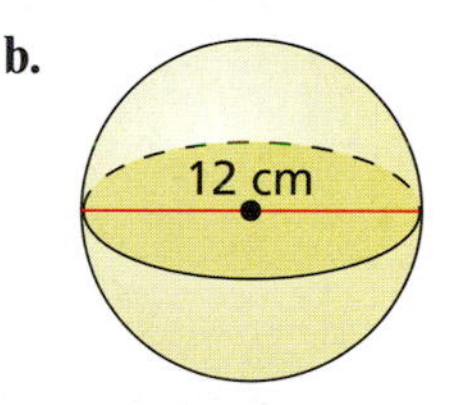

19. Finding the Surface Area of a Sphere The circumference C of a sphere is given. Find the surface area of the sphere.

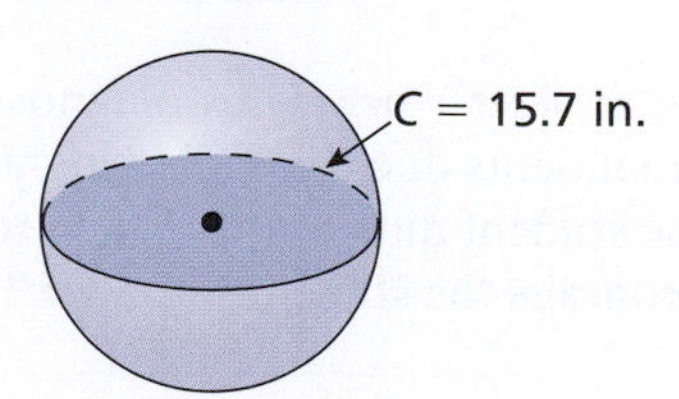

20. Finding the Surface Area of a Sphere The circumference C of a sphere is given. Find the surface area of the sphere.

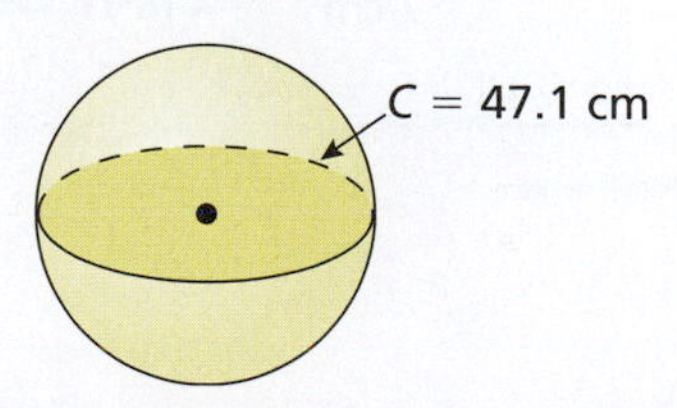

21. Surface Area in Real Life Find the surface area of the round bale of hay.

22. Surface Area in Real Life The circumference of the top rim of the cone-shaped paper cup is 7.85 inches. Find the least amount of paper that can form the cone-shaped cup.

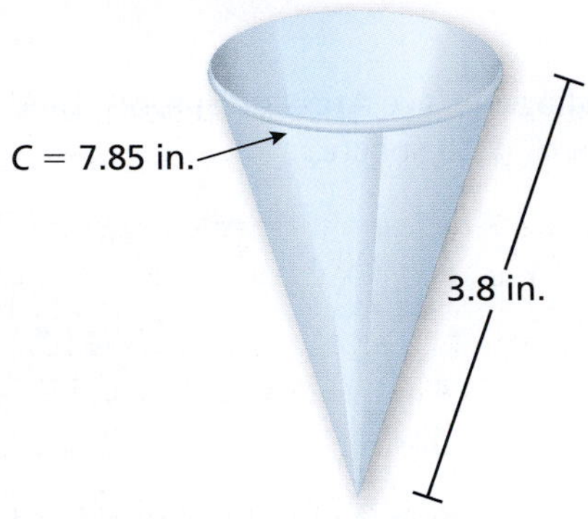

23. Grading Student Work On a diagnostic test, one of your students does the following work. Explain what the student did wrong. Which topics would you encourage the student to review?

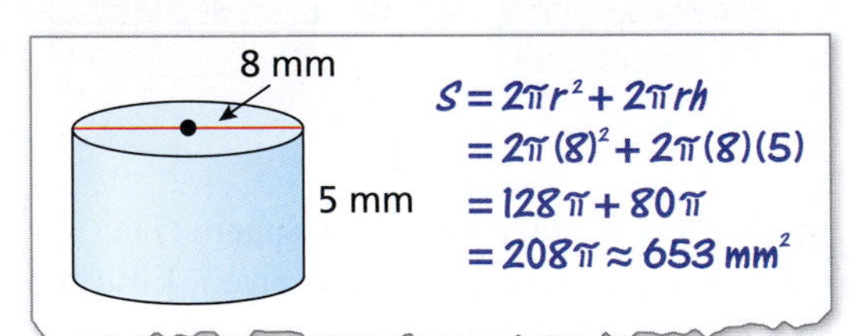

24. Grading Student Work On a diagnostic test, one of your students does the following work. Explain what the student did wrong. Which topics would you encourage the student to review?

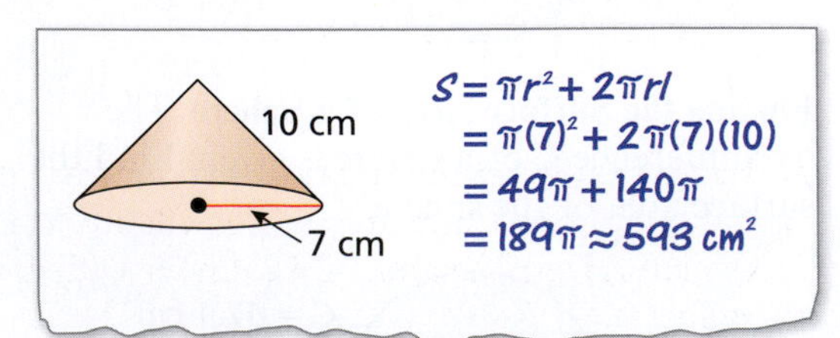

Boris15/Shutterstock.com

25. Surface Area in Real Life A cone-shaped grinding head is shown. Find the lateral surface area of the cone.

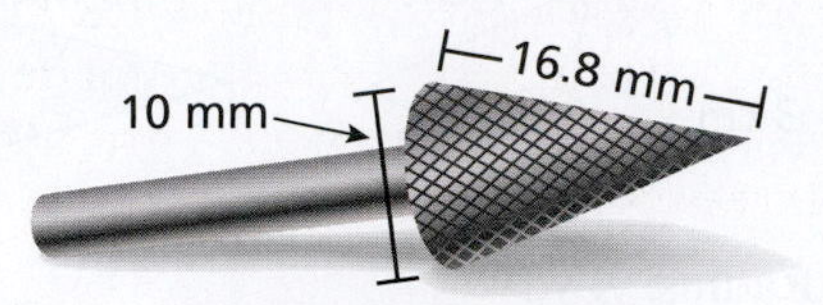

26. Surface Area in Real Life The circumference of the softball is 12 inches. Find the surface area of the softball.

27. Activity: In Your Classroom Design an activity to use with your students that involves using two 8.5-inch by 11-inch sheets of paper to form the lateral surfaces of two cylinders with different heights. Use additional sheets of paper to make bases for the cylinders. Explain what your students can conclude about the relationship between the surface areas and lateral surface areas of cylinders.

28. Activity: In Your Classroom Design an activity to use with your students that involves making cones from different sections of circles with the same radius, as shown below. Explain what your students can conclude about how the cones compare in terms of slant height, base circumference, and how tall they are.

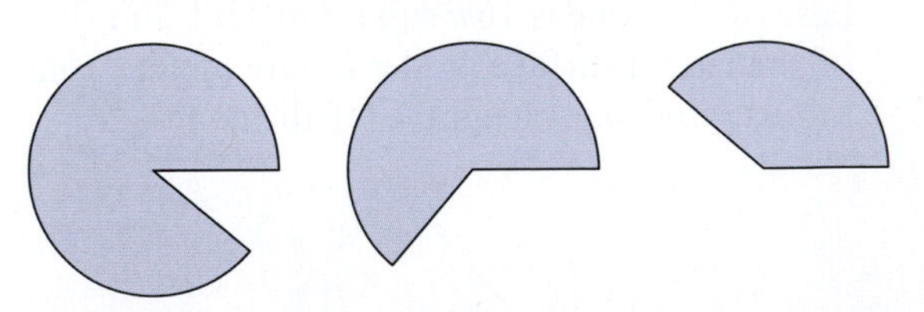

29. Surface Area in Real Life You cut a styrofoam ball in half for a project. Find the surface area of each half of the styrofoam ball.

30. **Surface Area in Real Life** A cylindrical piece of ice with a hole in the middle and an ice cube are shown. Both contain about the same volume of ice. Find the surface area of each piece of ice. Round each answer to the nearest tenth of a square inch. Which shape will chill a drink faster? Explain your reasoning.

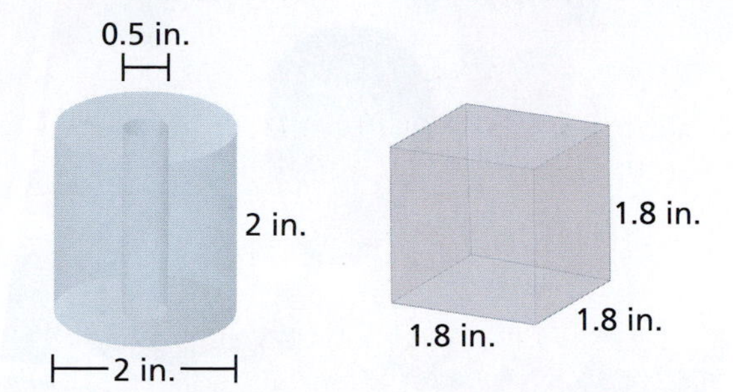

31. **Surface Area in Real Life** The exterior surface (except for the bottom) of the grain silo shown below needs to be painted. The silo has a circumference of 94.2 feet, and its top has the shape of a hemisphere. One gallon of paint covers 350 square feet of surface area. How many gallons of paint are needed to cover the silo?

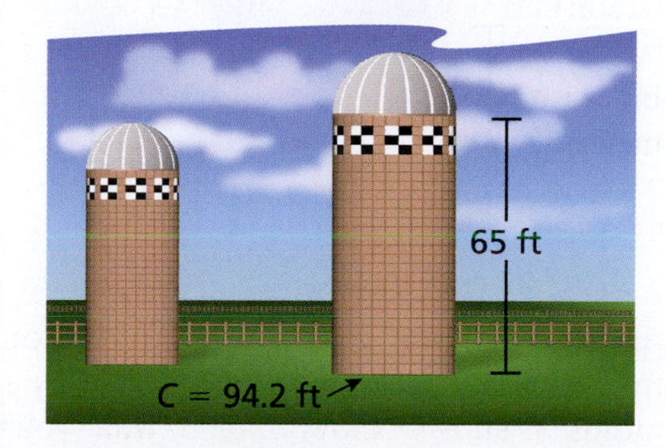

32. **Surface Area in Real Life** The bottom of the cone-shaped paper cup in Exercise 22 is cut off, and a flat circular base is added, as shown below. Find the least amount of paper that can form the cup.

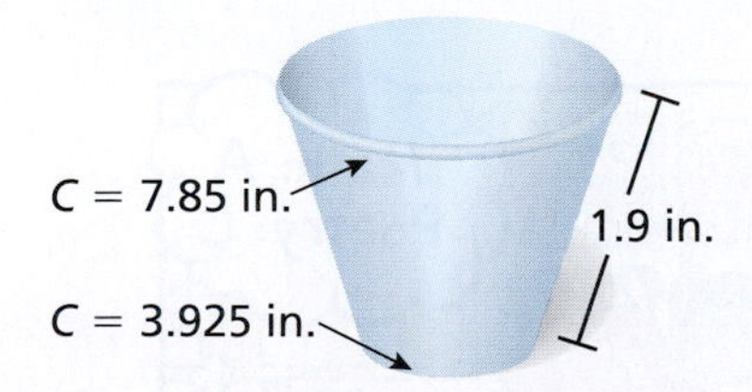

33. **Writing** You cut a cylindrical log in half, forming two congruent cylindrical logs. Does each of these two logs have exactly half as much surface area as the original log? Explain.

34. **Writing** A cone is contained in the smallest cylinder that can contain the cone. Which solid has a greater surface area? Explain.

35. **Writing** The slant height of one cone is twice the slant height of another cone. The bases of the two cones are congruent. Does the cone with twice the slant height have twice as much surface area as the other cone? Explain.

36. **Writing** The radius of one sphere is twice the radius of another sphere. How many times greater is the surface area of the larger sphere than the surface area of the smaller sphere? Explain your reasoning.

37. **Comparing Surface Areas** Consider the smallest possible cylinder that can contain a sphere of radius r.

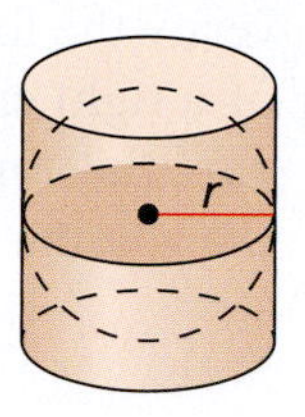

a. Give the height of the cylinder and the radius of the base of the cylinder in terms of r.

b. Write expressions in terms of r for the surface area of the sphere and the surface area of the cylinder.

c. Find the ratio of the surface area of the sphere to the surface area of the cylinder.

38. **Activity: In Your Classroom** A cylinder and a sphere have the same radius. Design an activity that you might use with your students to help them discover inductively how the height of the cylinder is related to the radius when the sphere and the cylinder have the same surface area. Explain what your students can conclude. Then show deductively that this conclusion is true.

39. **Problem Solving** The *Torrid Zone* is a region of Earth between the Tropic of Cancer and the Tropic of Capricorn that forms a 3250-mile-wide belt around the equator.

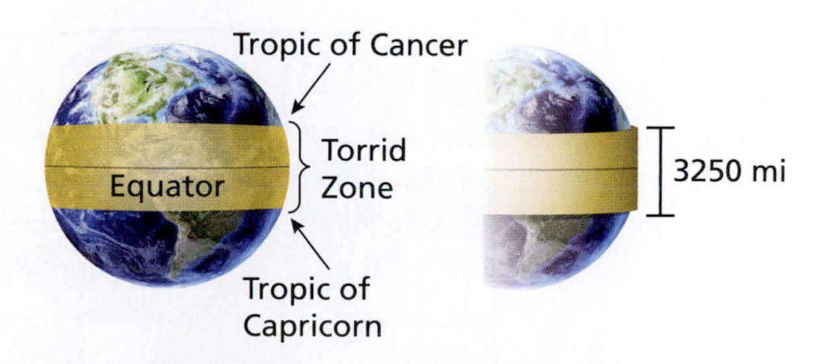

a. Earth's radius at the equator is about 3963 miles. Estimate the surface area of the Torrid Zone as the lateral surface area of a cylinder with this radius.

b. Approximately what percent of Earth's surface area is in the Torrid Zone?

Activity: Verifying a Formula for the Volume of a Cylinder

Materials:

- Pencil

Learning Objective:

Use the relationship between milliliters and cubic centimeters to verify the formula for the volume of a right circular cylinder. Compare this formula to the formula for the volume of a prism.

Name ______________________________

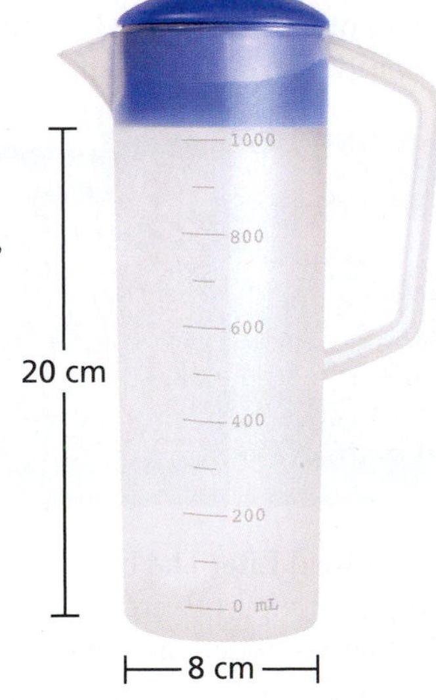

1. What is the volume V, in milliliters, of the metric measuring pitcher shown at the right?

$V =$ ▭ mL

2. In the metric system, 1 milliliter is equal to 1 cubic centimeter. What is the volume V, in cubic centimeters, of the measuring pitcher?

$V =$ ▭ cm^3

3. The diameter of the base is 8 centimeters. What is the area A of the base? Use 3.14 for π.

$A \approx$ ▭

4. For a prism, you can find the volume by multiplying the area of the base by the height. Is this true for the metric measuring pitcher? Explain your reasoning.

"Here's how I remember how to find the volume of any prism or cylinder."

"Base times tall, will fill 'em all."

12.3 Volumes of Circular Solids

Standards

Grades 6–8
Geometry
Students should solve real-life and mathematical problems involving volume of cylinders, cones, and spheres.

- Find the volume of a cylinder.
- Find the volume of a cone.
- Find the volume of a sphere.

Finding the Volume of a Cylinder

Volume of a Cylinder

Words The **volume** V of a cylinder is the product of the area of the base and the height of the cylinder.

Algebra $V = Bh$

Area of base (B), Height (h)

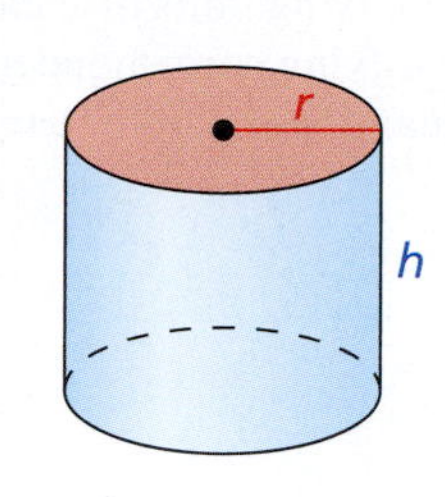

For a right circular cylinder, the volume is given by

$V = \pi r^2 h.$

Classroom Tip

In Section 12.2, it was mentioned that the formula for the surface area of a right circular cylinder did not apply to elliptical or oblique cylinders. For volume, however, the general formula $V = Bh$ applies to all cylinders. This is conveyed in the cartoon on page 478.

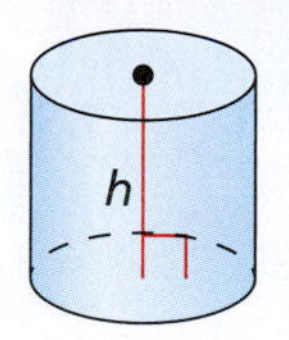

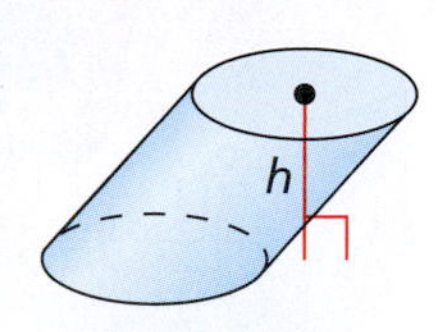

Right cylinder **Oblique cylinder**

EXAMPLE 1 **Finding the Volume of a Cylinder**

Find the volume of the cylinder. Round your answer to the nearest tenth.

4 mm
3 mm

SOLUTION

$V = Bh = \pi r^2 h$ — Write formula for volume of a cylinder.

$= \pi(4)^2(3)$ — Substitute 4 for r and 3 for h.

≈ 150.72 — Multiply.

The volume is about 150.7 cubic millimeters.

EXAMPLE 2 **Comparing Volumes in Real Life**

Which can holds more juice?

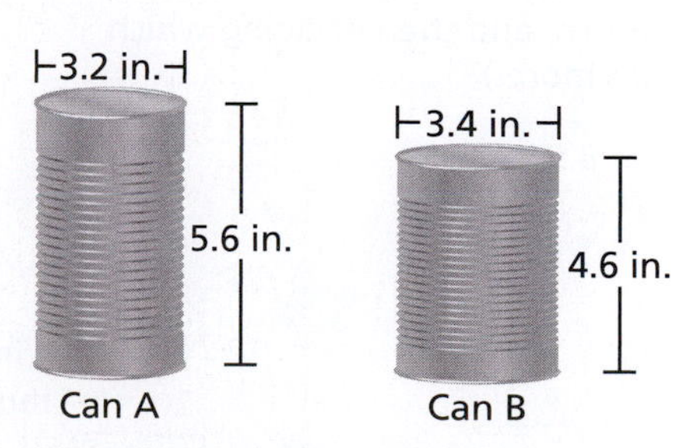

SOLUTION

The diameter of Can A is 3.2 inches, so the radius is 1.6 inches. The diameter of Can B is 3.4 inches, so the radius is 1.7 inches. Now find the volume of each can.

Can A $V = \pi r^2 h$
$= \pi(1.6)^2(5.6)$
$\approx 14.3\pi$

Can B $V = \pi r^2 h$
$= \pi(1.7)^2(4.6)$
$\approx 13.3\pi$

Because $14.3\pi > 13.3\pi$, Can A holds more juice.

EXAMPLE 3 Finding the Volume of a Cylinder

About how many gallons of water does the water cooler contain? One cubic foot is about 7.5 gallons.

1.7 ft

1 ft

Mathematical Practices

Attend to Precision

MP6 Mathematically proficient students are careful about specifying units of measure and clarifying correspondence with quantities in a problem.

SOLUTION

Begin by finding the volume of the cooler. The diameter is 1 foot. So, the radius is 0.5 foot.

$V = Bh = \pi r^2 h$ Write formula for volume of a cylinder.

$= \pi(0.5)^2(1.7)$ Substitute 0.5 for r and 1.7 for h.

$= 0.425\pi$ Simplify.

≈ 1.3345 Multiply.

The cooler contains about 1.3345 cubic feet of water. To find the number of gallons it contains, multiply by 7.5 gallons per 1 cubic foot.

$$1.3345 \text{ ft}^3 \times \frac{7.5 \text{ gal}}{1 \text{ ft}^3} \approx 10 \text{ gal}$$

The water cooler contains about 10 gallons of water.

EXAMPLE 4 Finding the Weight of a Cylinder

The tank on the road roller is a cylinder that has a height of 6 feet and a radius of 2 feet. The tank is completely filled with water to make it heavy. One cubic foot of water weighs 62.4 pounds. Find the weight of the water in the tank in tons.

Classroom Tip

To help your students build their intuitions for the volume of a right circular cylinder, ask them to try the following: Take an 8.5-inch by 11-inch sheet of paper. Roll it two different ways to produce right circular cylinders. Ask your students which way produces a greater volume. (They can verify which has a greater volume by filling each with rice or popcorn, and then deciding which holds more.)

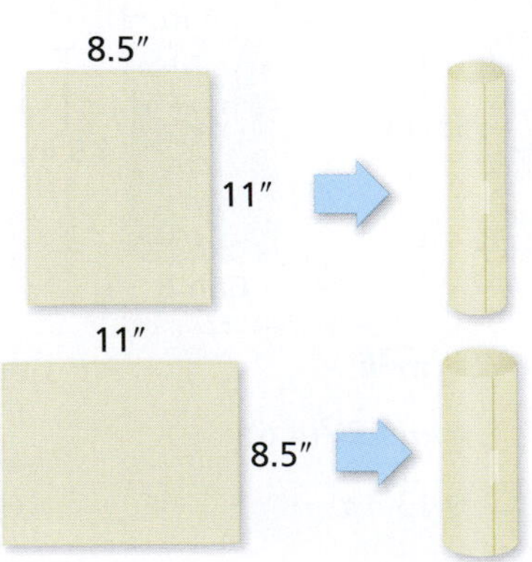

SOLUTION

Begin by finding the volume of the tank.

$V = Bh = \pi r^2 h$ Write formula for volume of a cylinder.

$= \pi(2)^2(6)$ Substitute 2 for r and 6 for h.

$= 24\pi$ Simplify.

≈ 75.36 Multiply.

The tank contains about 75.36 cubic feet of water. To find the weight, multiply by 62.4 pounds per 1 cubic foot.

$$75.36 \text{ ft}^3 \times \frac{62.4 \text{ lb}}{1 \text{ ft}^3} \approx 4702 \text{ lb}$$

The tank weighs about 4702 pounds.

$$4702 \text{ lb} \times \frac{1 \text{ ton}}{2000 \text{ lb}} \approx 2.4 \text{ tons}$$

So, the tank weighs about 2.4 tons.

Finding the Volume of a Cone

In Section 11.4, you verified that the volume of a pyramid is one-third the volume of a prism having the same base and height. You can use the same approach to show that the volume of a cone is one-third the volume of a cylinder having the same base and height.

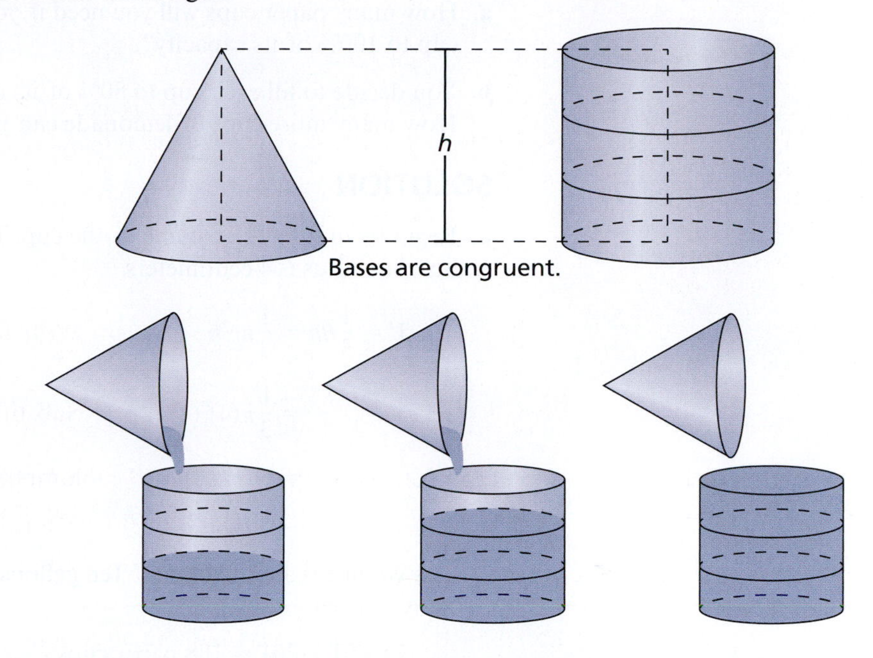

Classroom Tip

The formula for the volume of a cone applies to all cones, including oblique cones and cones with noncircular bases.

Volume of a Cone

Words The **volume** V of a cone is one-third the product of the area of the base and the height of the cone.

Algebra $V = \frac{1}{3}Bh$

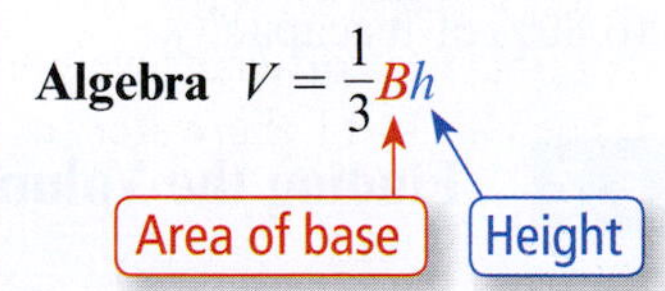

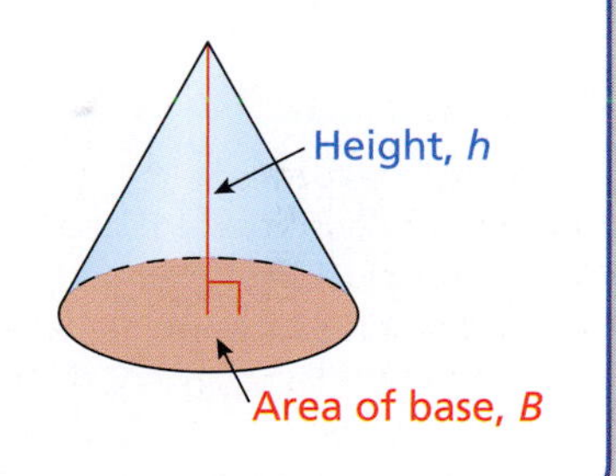

EXAMPLE 5 **Finding the Volume of a Cone**

Find the volume of the cone. Round your answer to the nearest tenth.

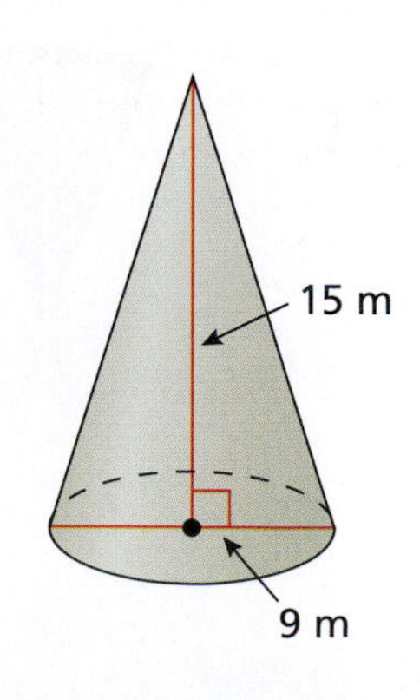

SOLUTION

The diameter is 9 meters. So, the radius is 4.5 meters.

$$V = \frac{1}{3}Bh = \frac{1}{3}\pi r^2 h \quad \text{Write formula for volume of a cone.}$$

$$= \frac{1}{3}\pi(4.5)^2(15) \quad \text{Substitute 4.5 for } r \text{ and 15 for } h.$$

$$= 101.25\pi \quad \text{Simplify.}$$

$$\approx 317.925 \quad \text{Multiply.}$$

The volume is about 317.9 cubic meters.

EXAMPLE 6 Finding the Volume of a Cone

You sell 10 gallons of lemonade in the paper cup shown at the right. One gallon is about 3785 cubic centimeters.

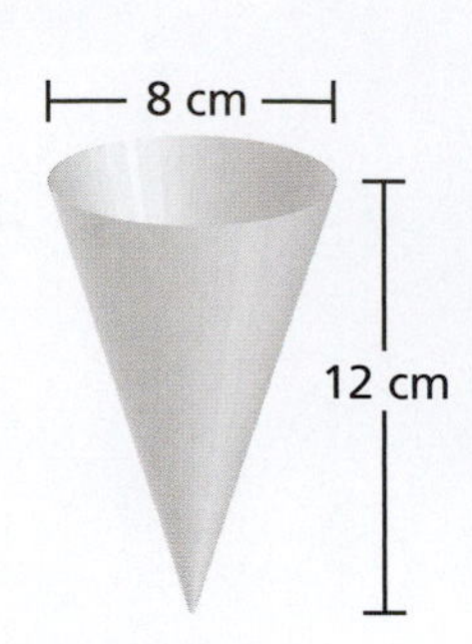

a. How many paper cups will you need if you fill each cup to 100% of its capacity?

b. You decide to fill each cup to 80% of its capacity. How many more cups of lemonade can you sell?

SOLUTION

a. Begin by finding the volume of the cup. The diameter is 8 centimeters. So, the radius is 4 centimeters.

$$V = \frac{1}{3}Bh = \frac{1}{3}\pi r^2 h \quad \text{Write formula for volume of a cone.}$$

$$= \frac{1}{3}\pi(4)^2(12) \quad \text{Substitute 4 for } r \text{ and 12 for } h.$$

$$= 64\pi \quad \text{Simplify.}$$

$$\approx 201 \quad \text{Multiply.}$$

The volume is about 201 cm^3. Ten gallons is about 10×3785 cm^3 = 37,850 cm^3. So, you need

$$37{,}850 \div 201 \approx 188 \text{ paper cups.}$$

b. When you fill each paper cup to 80% of its capacity, each cup will hold 80% of 201, or about 161 cm^3 of lemonade. In this case, you need

$$37{,}850 \div 161 \approx 235 \text{ paper cups.}$$

So, you can sell about $235 - 188 = 47$ more cups of lemonade when you fill each cup to 80% of its capacity.

EXAMPLE 7 Finding the Volume of a Cone

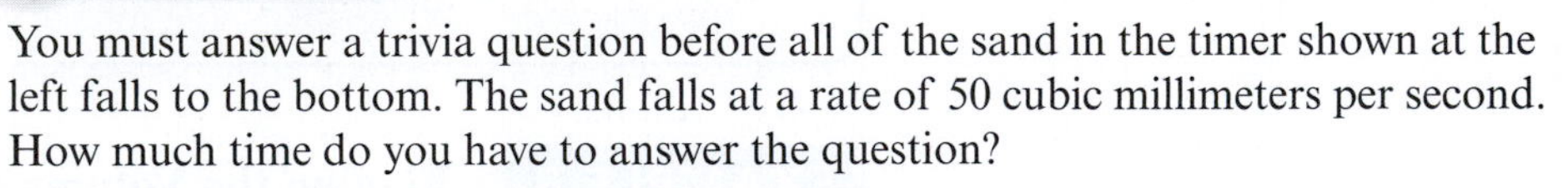

You must answer a trivia question before all of the sand in the timer shown at the left falls to the bottom. The sand falls at a rate of 50 cubic millimeters per second. How much time do you have to answer the question?

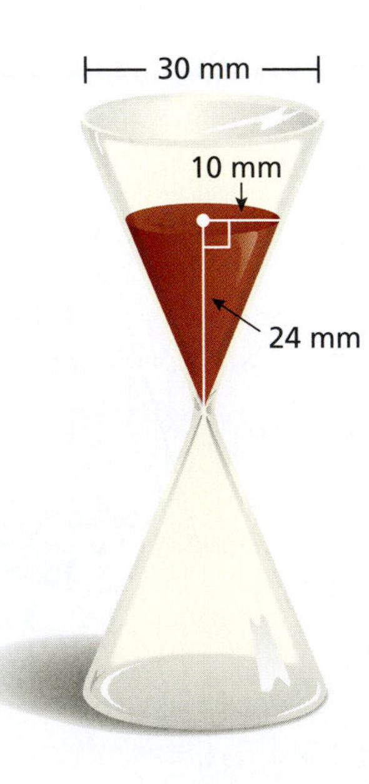

SOLUTION

Begin by finding the volume of the sand.

$$V = \frac{1}{3}Bh = \frac{1}{3}\pi r^2 h \quad \text{Write formula for volume of a cone.}$$

$$= \frac{1}{3}\pi(10)^2(24) \quad \text{Substitute 10 for } r \text{ and 24 for } h.$$

$$= 800\pi \quad \text{Simplify.}$$

$$\approx 2512 \quad \text{Multiply.}$$

The volume of the sand is about 2512 cubic millimeters. This means that the time you have is

$$2512 \cancel{\text{mm}^3} \times \frac{1 \text{ sec}}{50 \cancel{\text{mm}^3}} = 50.24 \text{ seconds.}$$

So, you have about 50 seconds to answer the question.

Finding the Volume of a Sphere

Volume of a Sphere

Words The **volume** V of a sphere is four-thirds the product of π and the cube of the radius of the sphere.

Algebra $V = \frac{4}{3}\pi r^3$ ← Cube of radius of sphere

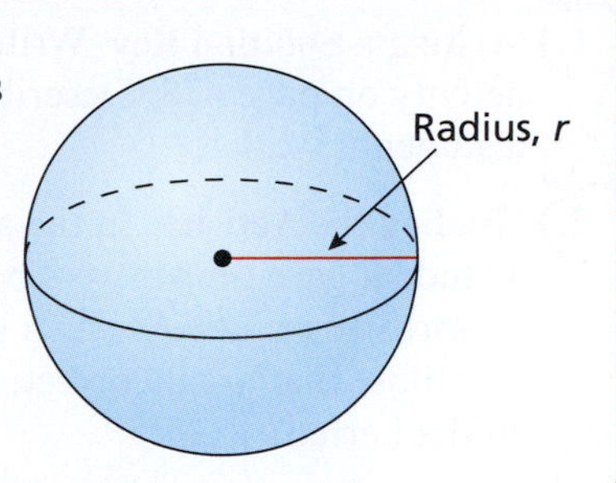

Classroom Tip

The point of Example 8 is that the volumes of the three circular solids having the same radius and height are related. The ratio of volumes (cylinder : cone : sphere) is

$1 : \frac{1}{3} : \frac{2}{3}$ or $3 : 1 : 2$.

EXAMPLE 8 Comparing Volumes of Solids

Compare the volumes of the three circular solids. Note that the "height" of the sphere is also $2r$.

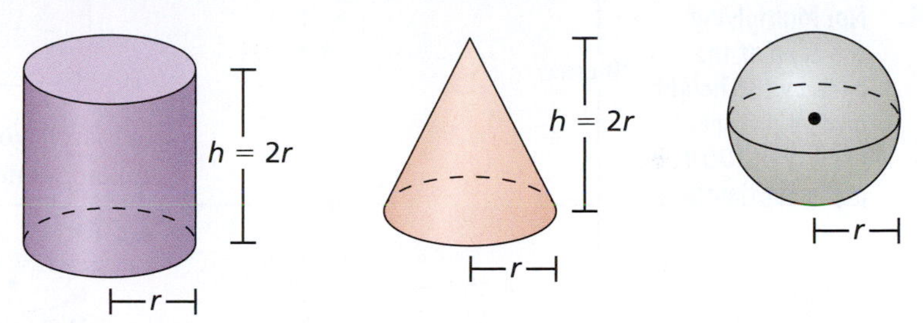

SOLUTION

Cylinder $V = Bh = \pi r^2 h = \pi r^2(2r) = 2\pi r^3$

Cone $V = \frac{1}{3}Bh = \frac{1}{3}\pi r^2 h = \frac{1}{3}\pi r^2(2r) = \frac{2}{3}\pi r^3$ $\quad \left(\frac{1}{3}\right)(2\pi r^3)$

Sphere $V = \frac{4}{3}\pi r^3$ $\quad \left(\frac{2}{3}\right)(2\pi r^3)$

The volume of the cone is $\frac{1}{3}$ the volume of the cylinder, and the volume of the sphere is $\frac{2}{3}$ the volume of the cylinder.

EXAMPLE 9 Finding the Volume of a Composite Solid

The ice cream cone is full of ice cream. Including the scoop on top, find the total volume of ice cream in the cone. Assume that the scoop on top of the cone forms a *hemisphere*, which is one-half of a sphere.

SOLUTION

The solid consists of a cone and a hemisphere. You can find the volume of the entire solid by finding the volume of each of these two solids.

Volume of cone $V = \frac{1}{3}\pi r^2 h = \frac{1}{3}\pi(1.5)^2(4) = 3\pi$

Volume of hemisphere $V = \frac{1}{2} \cdot \frac{4}{3}\pi r^3 = \frac{1}{2} \cdot \frac{4}{3}\pi(1.5)^3 = 2.25\pi$

So, the volume of ice cream in the cone, including the scoop on top, is $3\pi + 2.25\pi = 5.25\pi$, or about 16.5 cubic inches.

bonchan/Shutterstock.com

12.3 Exercises

Free step-by-step solutions to odd-numbered exercises at *MathematicalPractices.com.*

1. Writing a Solution Key Write a solution key for the activity on page 478. Describe a rubric for grading a student's work.

2. Grading the Activity In the activity on page 478, a student gave the answers below. Each question is worth 10 points. Assign a grade to each answer. For those that are incorrect, why do you think the student erred?

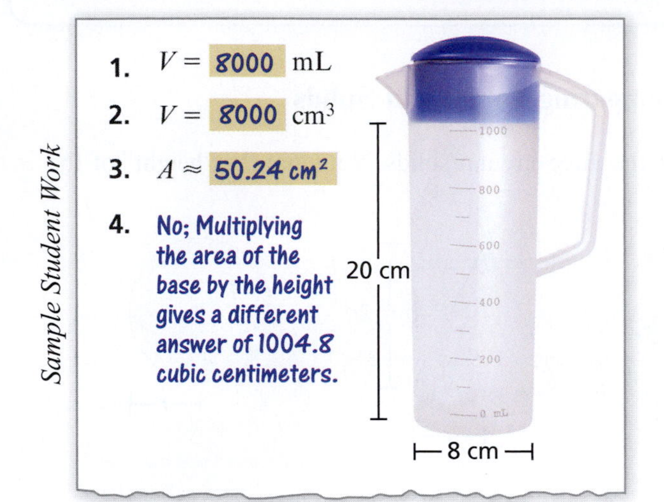

3. Finding Volume Using the Area of a Base

a. Find the area.

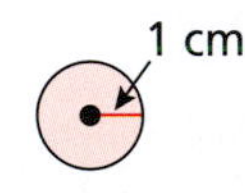

b. Find the volume.

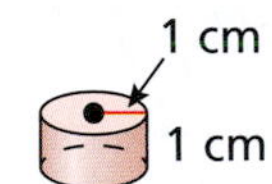

c. Find the volume.

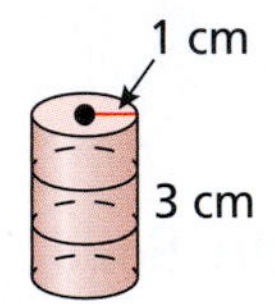

4. Finding Volume Using the Area of a Base

a. Find the area.

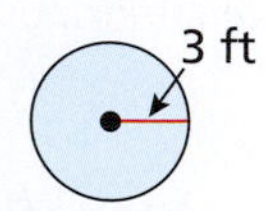

b. Find the volume.

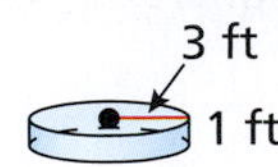

c. Find the volume.

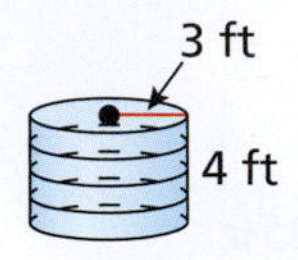

5. Finding Volumes of Cylinders Find the volume of each cylinder.

a. 4 ft, 6 ft

b.

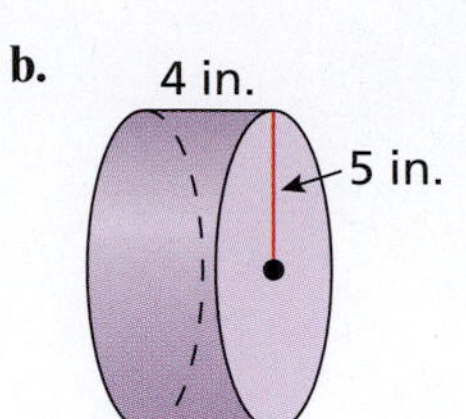

c. 12 mm, 3 mm

d.

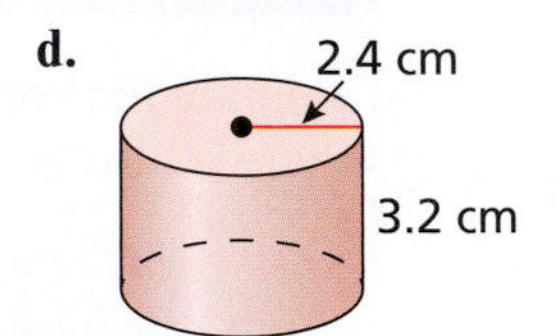

6. Finding Volumes of Cylinders Find the volume of each cylinder.

a. 5 cm, 7 cm

b.

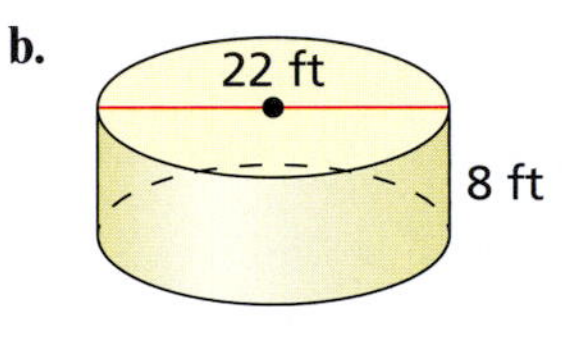

c.

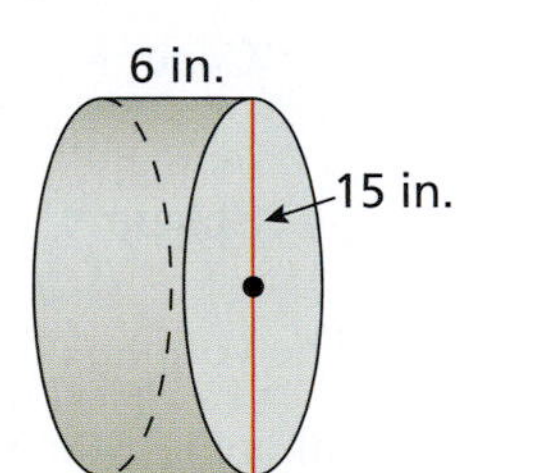

d.

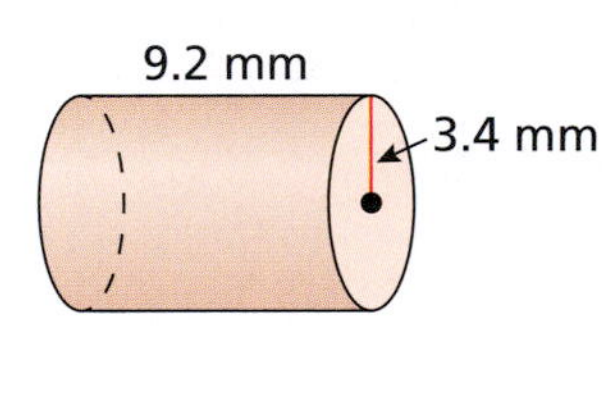

7. Finding Volumes of Cones Find the volume of each cone.

a. 7 in., 5 in.

b.

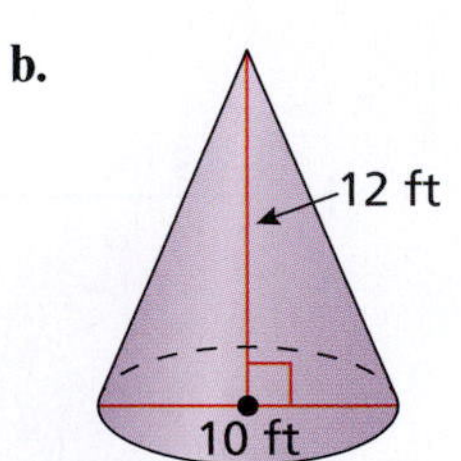

c.

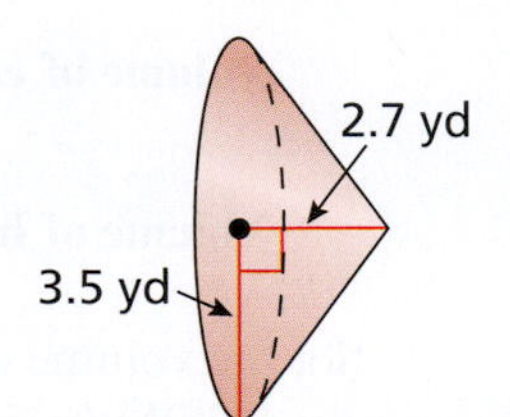

d.

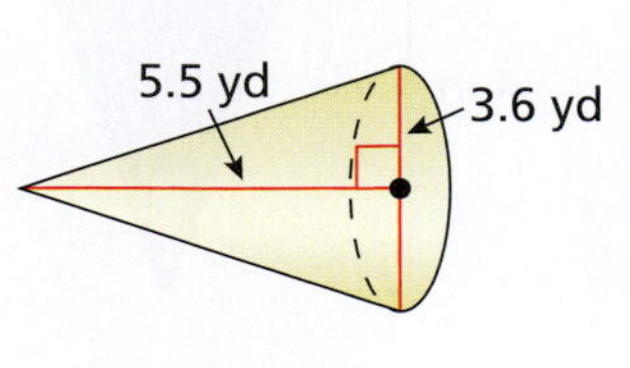

8. Finding Volumes of Cones Find the volume of each cone.

a.

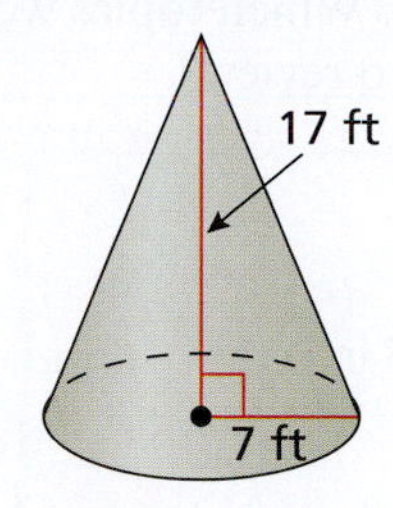

b.

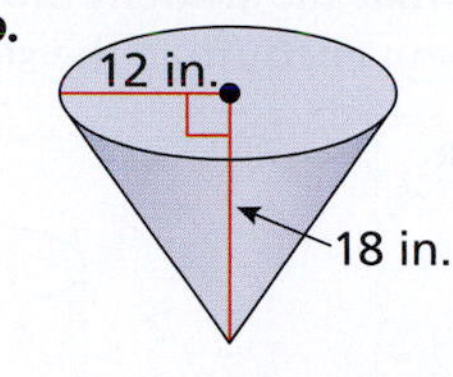

c.

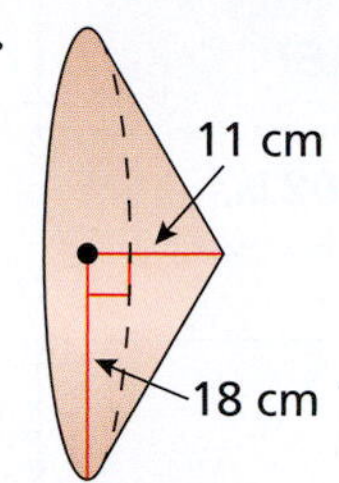

d.

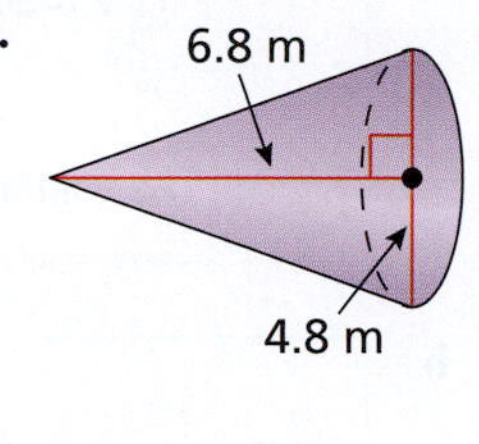

9. Finding Volumes of Spheres Find the volume of each sphere.

a.

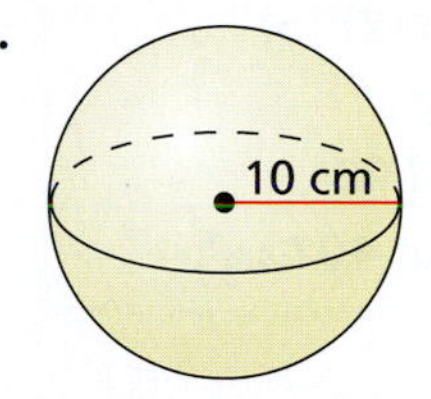

b.

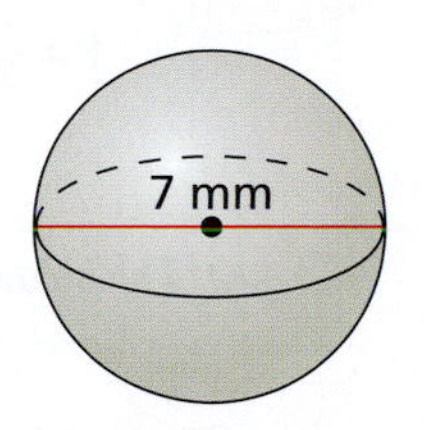

10. Finding Volumes of Spheres Find the volume of each sphere.

a.

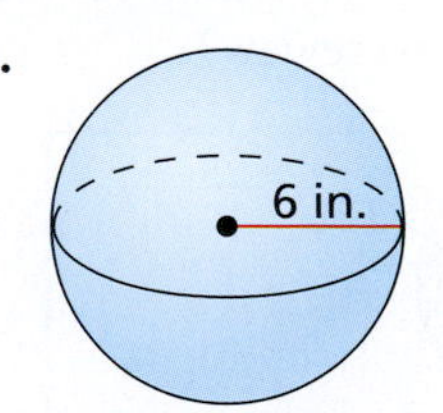

b.

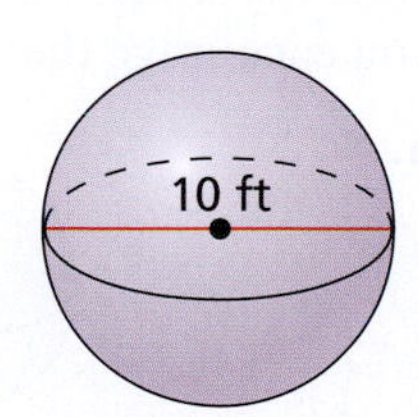

11. Finding the Height of a Cylinder or Cone Find the height h of each cylinder or cone.

a. Volume = 904 ft^3

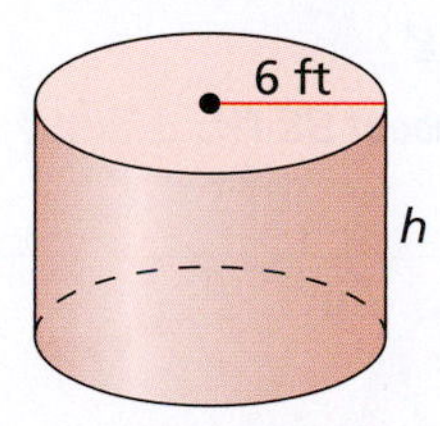

b. Volume = 737 cm^3

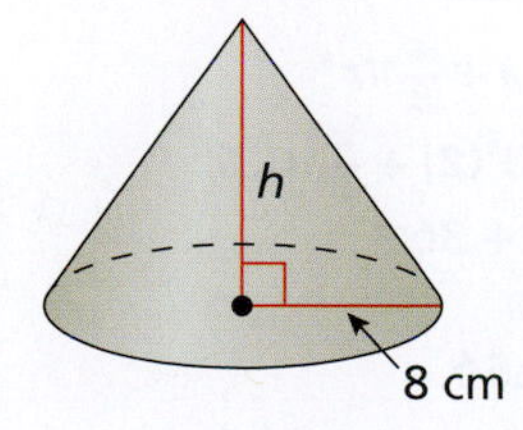

12. Finding the Height of a Cylinder or Cone Find the height h of each cylinder or cone.

a. Volume = 314 in.^3

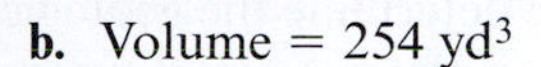
b. Volume = 254 yd^3

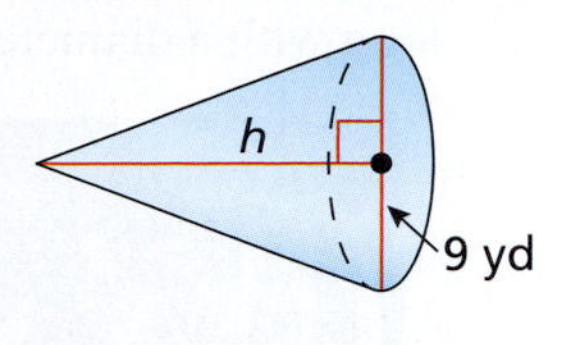

13. Finding the Volume of a Composite Solid The solid contains a cone and a hemisphere. Find the volume of the composite solid.

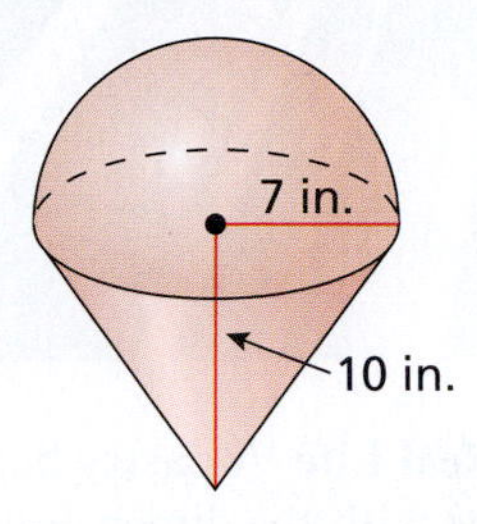

14. Finding the Volume of a Composite Solid The solid contains a cone and a cylinder. Find the volume of the composite solid.

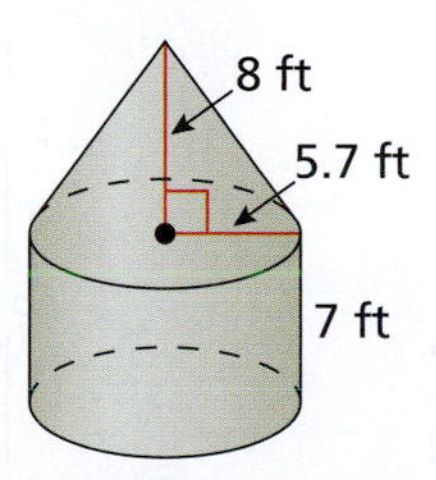

15. Finding the Volume of a Composite Solid The solid contains a cylinder and a hemisphere. Find the volume of the composite solid.

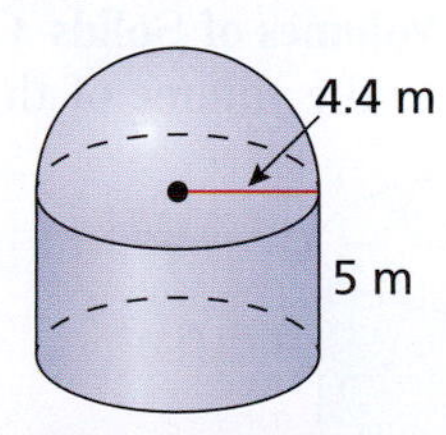

16. Finding the Volume of a Solid The solid is a cylinder with a hemisphere removed. Find the volume of the solid.

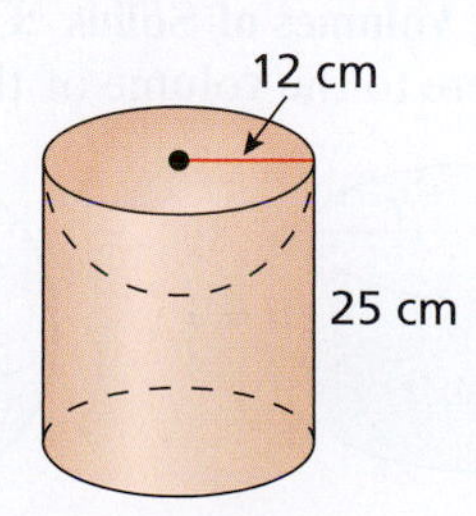

17. **Volume in Real Life** An oil exploration company takes cylindrical core samples from the ground to determine the geological features of a region. What is the volume of a core sample that is 100 meters long with a diameter of 6 centimeters?

18. **Volume in Real Life** A pastry bag full of frosting forms a cone with the dimensions shown. Find the volume of frosting in the pastry bag.

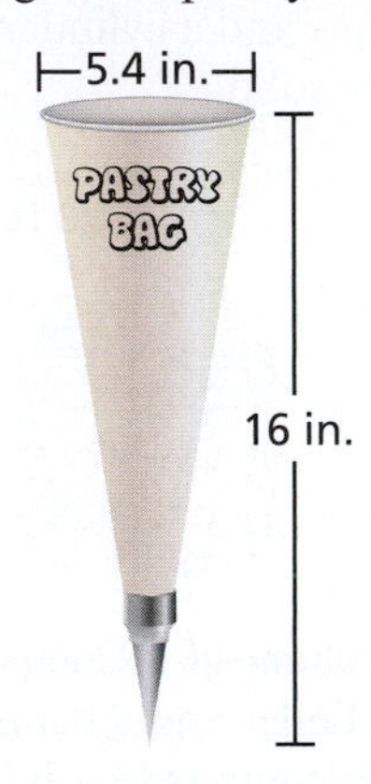

19. **Comparing Volumes of Solids** Compare the volume of the cone to the volume of the cylinder.

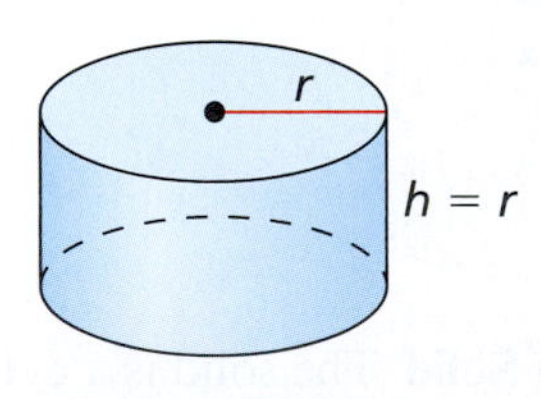

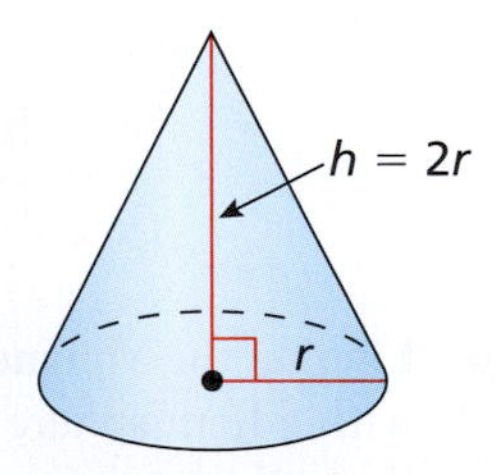

20. **Comparing Volumes of Solids** Compare the volume of the sphere to the volume of the cylinder.

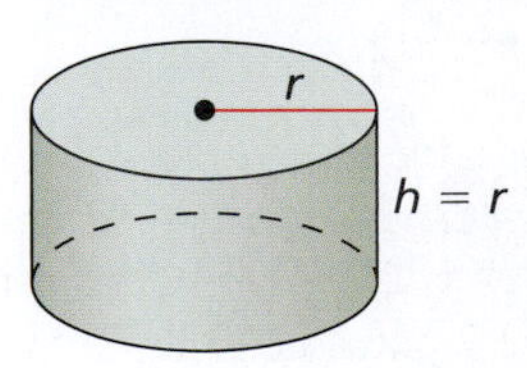

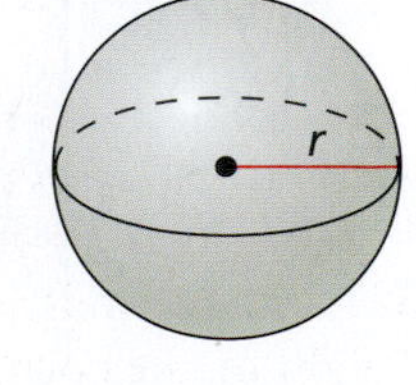

usbr.gov

21. **Grading Student Work** On a diagnostic test, one of your students does the following work. Explain what the student did wrong. Which topics would you encourage the student to review?

a.

4 in.

5 in.

$V = Bh = 2\pi r^2 h$
$= 2\pi(4)^2(5)$
≈ 502.4

The volume is about 502 in.^{3}

b.

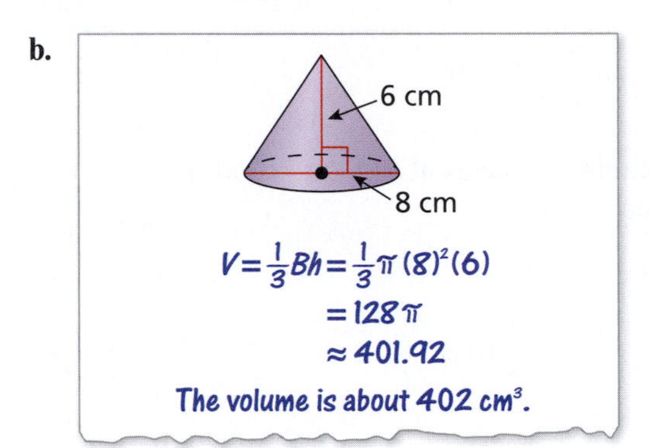

22. **Grading Student Work** On a diagnostic test, one of your students does the following work. Explain what the student did wrong. Which topics would you encourage the student to review?

a.

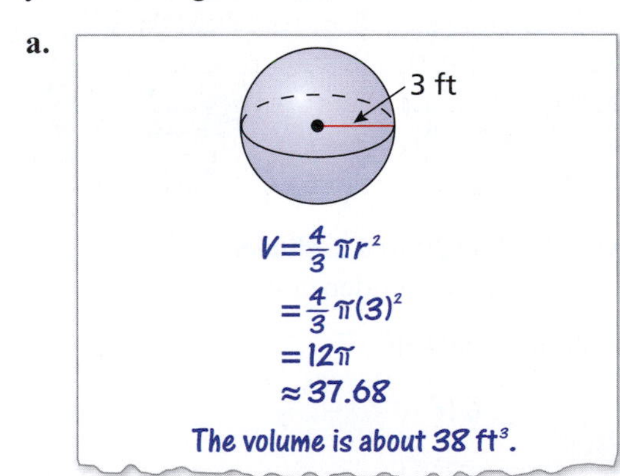

b.

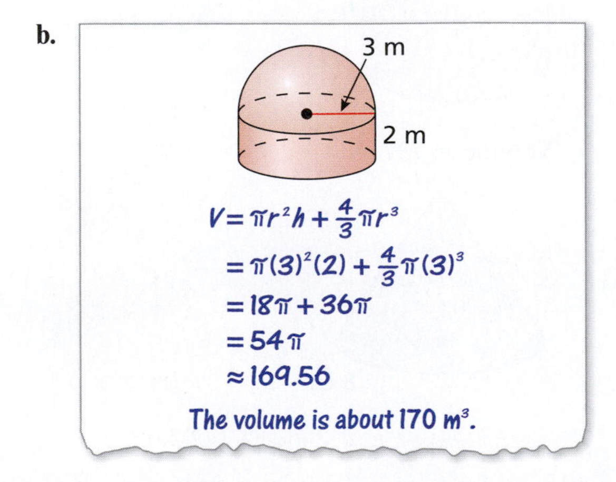

23. **Sphereing** A sphereing ball contains a sphere within a sphere. The diameter of the inner sphere is 1.9 meters. Find the volume of the inner sphere.

24. **Sphereing** The outer sphere of the sphereing ball in Exercise 23 has a diameter of 2.6 meters. Find the volume within the outer sphere but outside the inner sphere.

25. **Price Comparison** Which is the better buy? Explain.

26. **Volume in Real Life** Four tennis balls are packed in a cylindrical container that has a diameter of 2.6 inches and a height of 10.4 inches.

a. The diameter of each tennis ball is 2.6 inches. Find the volume of a tennis ball.

b. What percent of the space in the can is not occupied by the tennis balls?

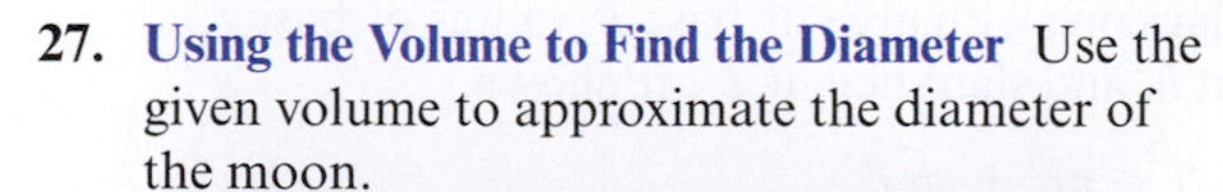

27. **Using the Volume to Find the Diameter** Use the given volume to approximate the diameter of the moon.

Volume $\approx$ 22,000,000,000 km^3

Chris Turner/Shutterstock.com; Klagyivik Viktor/Shutterstock.com

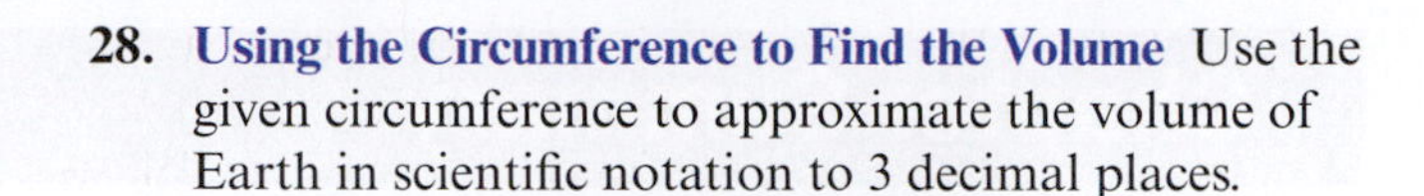

28. **Using the Circumference to Find the Volume** Use the given circumference to approximate the volume of Earth in scientific notation to 3 decimal places.

Equatorial Circumference $\approx$ 40,000 km

29. **Activity: In Your Classroom** Design an activity for your students that involves inflating balloons to discover how the volume of a sphere changes when its radius is doubled. Describe the conclusions you can expect your students to make.

30. **Activity: In Your Classroom** Design an activity for your students to discover how the volume of a cone changes when you double its height, its radius, or both. Describe the conclusions you can expect your students to make.

31. **Writing** Identify the types of solids to which the volume formula $V = Bh$ applies. Then describe the relationship between the bases and any cross section parallel to the bases in each of these solids.

32. **Writing** Explain how to double the volume of a cylinder by changing its height. Then explain how to double the volume of a cylinder by changing its radius.

33. **Problem Solving** You hit the target with a softball to dunk a man in the dunking booth. The man's floating body makes the water level in the tank rise 3.4 inches. One cubic foot of the water weighs 62.2 pounds. What is the man's weight?

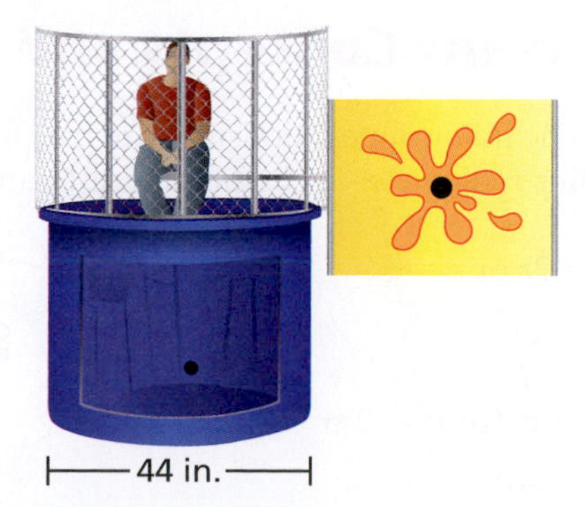

34. **Problem Solving** You place several ears of corn into a cylindrical pot of water. The water level in the 12-inch diameter pot rises 2.4 inches. One cubic foot of the water weighs 62.2 pounds. What is the total weight of the corn?

leonello calvetti/Shutterstock.com

Chapter Summary

Chapter Vocabulary

circle *(p. 459)*
center of a circle *(p. 459)*
radius *(p. 459)*
diameter *(p. 459)*
π (pi) *(p. 460)*
circumference *(p. 460)*
circular arc *(p. 461)*
area of a circle *(p. 462)*
sector of a circle *(p. 463)*
right circular cylinder *(p. 469)*
lateral surface of a right circular cylinder *(p. 469)*
surface area of a right circular cylinder *(p. 469)*
surface area of a right circular cone *(p. 471)*
slant height of a right circular cone *(p. 471)*
sphere *(p. 473)*
center of a sphere *(p. 473)*
surface area of a sphere *(p. 473)*
cross section *(p. 473)*
volume of a cylinder *(p. 479)*
volume of a cone *(p. 481)*
volume of a sphere *(p. 483)*

Chapter Learning Objectives — Review Exercises

12.1 Circumferences and Areas of Circles *(page 459)* — 1–26

- Find the diameter and radius of a circle.
- Find the circumference of a circle.
- Find the area of a circle.

12.2 Surface Areas of Circular Solids *(page 469)* — 27–44

- Find the surface area of a right circular cylinder.
- Find the surface area of a right circular cone.
- Find the surface area of a sphere.

12.3 Volumes of Circular Solids *(page 479)* — 45–58

- Find the volume of a cylinder.
- Find the volume of a cone.
- Find the volume of a sphere.

Important Concepts and Formulas

Circles The following relationships are true for a circle with radius r, diameter d, circumference C, and area A.

- $d = 2r$ or $r = \frac{d}{2}$
- $\pi = \frac{C}{d}$
- $C = \pi d$ or $C = 2\pi r$
- $A = \pi r^2$

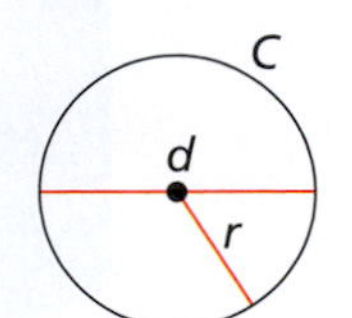

Cylinders The surface area S and volume V of a right circular cylinder with area of base B, radius of base r, and height h are shown.

- $S = 2\pi r^2 + 2\pi rh$
- $V = Bh$ or $V = \pi r^2 h$

Cones The surface area S and volume V of a right circular cone with area of base B, radius of base r, height h, and slant height ℓ are shown.

- $S = \pi r^2 + \pi r\ell$
- $V = \frac{1}{3}Bh$ or $V = \frac{1}{3}\pi r^2 h$

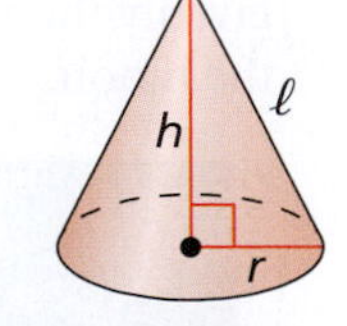

Spheres The surface area S and volume V of a sphere with radius r are shown.

- $S = 4\pi r^2$
- $V = \frac{4}{3}\pi r^3$

Review Exercises

Free step-by-step solutions to odd-numbered exercises at *MathematicalPractices.com*.

12.1 Circumferences and Areas of Circles

Finding the Radius of a Circle Find the radius of the circle.

1.

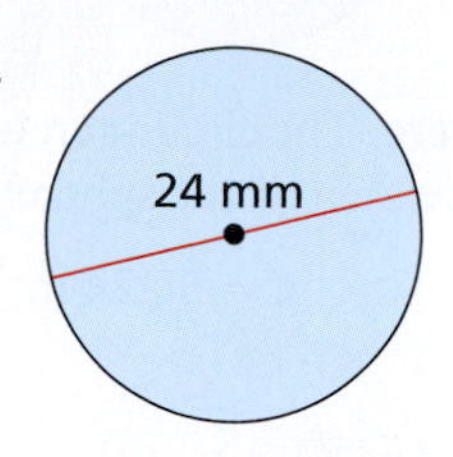

2.

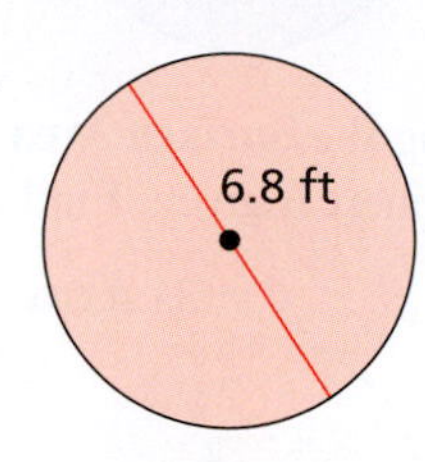

Finding the Diameter of a Circle Find the diameter of the circle.

3.

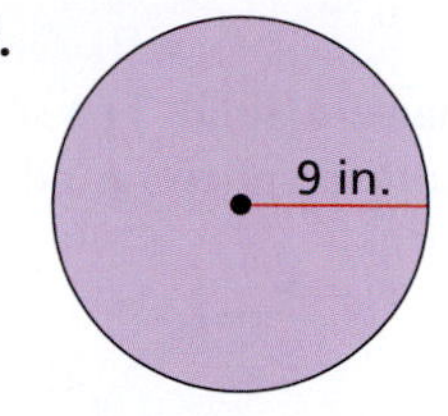

4.

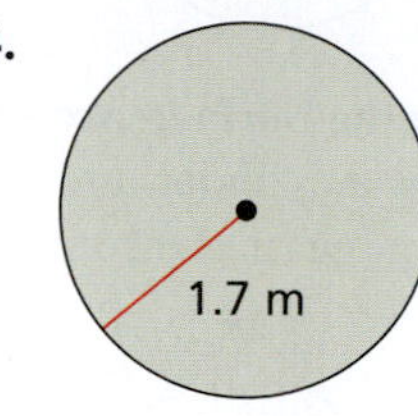

Finding the Circumference of a Circle Find the circumference of the circle given its radius or diameter.

5. $r = 8$ yd

6. $d = 20$ cm

7. $d = 13$ mm

8. $r = 21.5$ ft

Finding the Perimeter of a Semicircle Find the perimeter of the semicircle.

9.

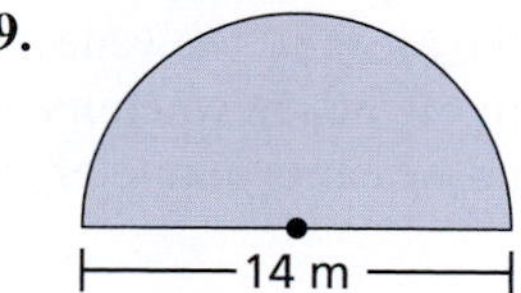

10.

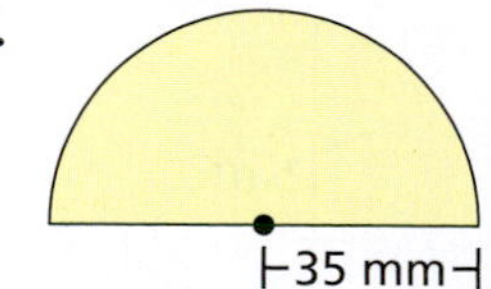

Finding the Length of a Circular Arc Use the given dimensions to find the length of the circular arc.

11.

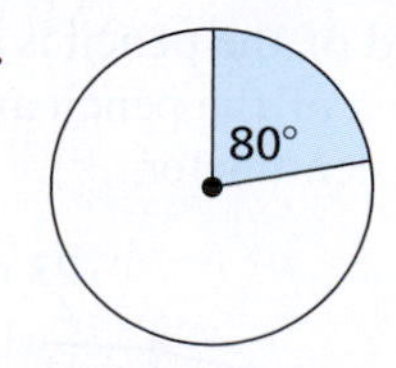

$C = 90$ mm

12.

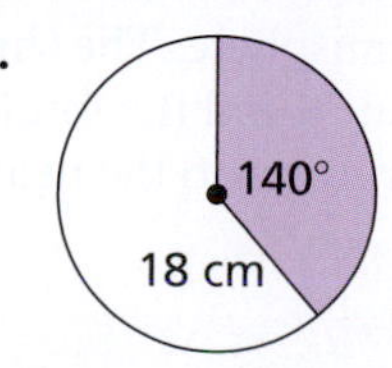

13. $a = 210$
$r = 9$ in.

14. $a = 10$
$r = 252$ m

Finding the Area of a Circle Find the area of the circle given its radius or diameter.

15.

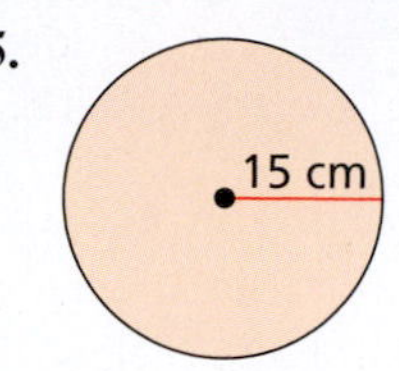

16.

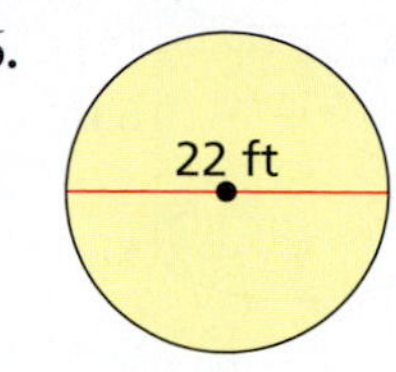

17. $r = 14$ mm

18. $d = 50$ in.

Finding the Area of a Sector Use the given dimensions to find the area of the sector.

19.

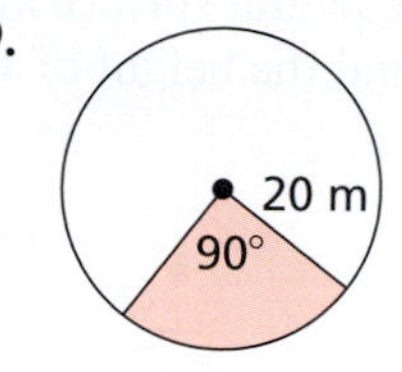

20.

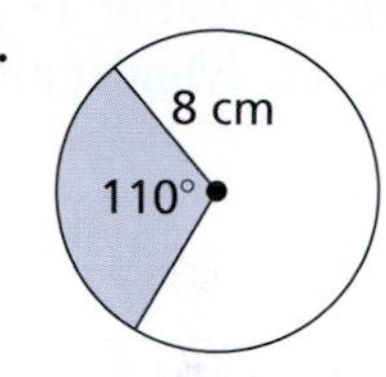

21. $a = 48$
$d = 28$ in.

22. $a = 16$
$d = 100$ km

Finding the Area of a Composite Figure Find the area of the composite figure. All curves are circular arcs with centers shown.

23.

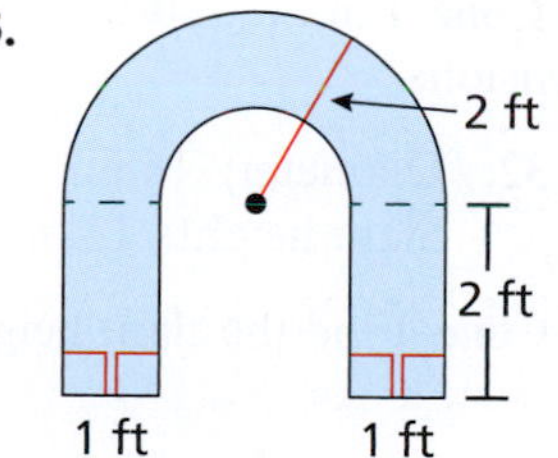

24.

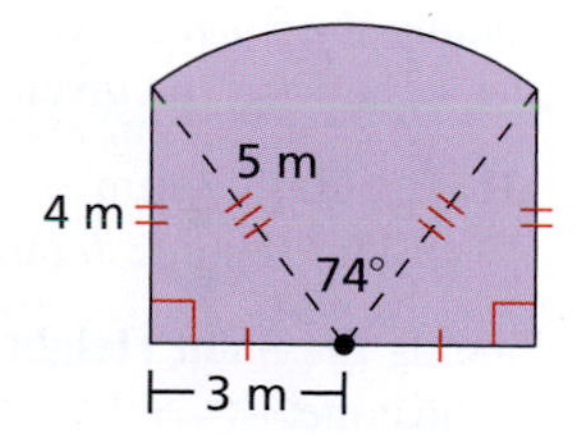

25. **Pizza Servings** The figure shows how a 10-inch pizza and a 12-inch pizza are sliced. Each pizza has the same thickness. Which has more pizza per slice? Explain.

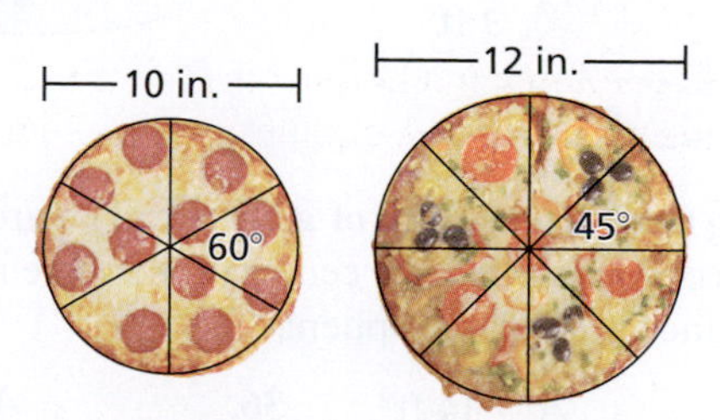

26. **Bicycling** The winner of a bicycle race finished 0.1 second before the runner-up. The winner's 26-inch diameter wheels were turning at a rate of 7.72 revolutions per second. How far in front of the runner-up did the winner finish?

12.2 Surface Areas of Circular Solids

Finding the Surface Area of a Cylinder Sketch a net of the cylinder. Then find its surface area.

27.

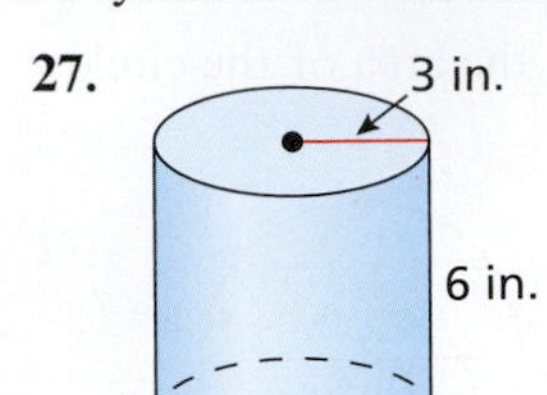

28.

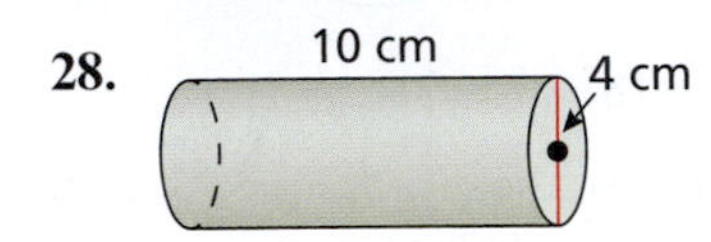

Finding the Height of a Cylinder The lateral surface and the bases of a cylinder are shown. Find the height of the cylinder.

29. **30.**

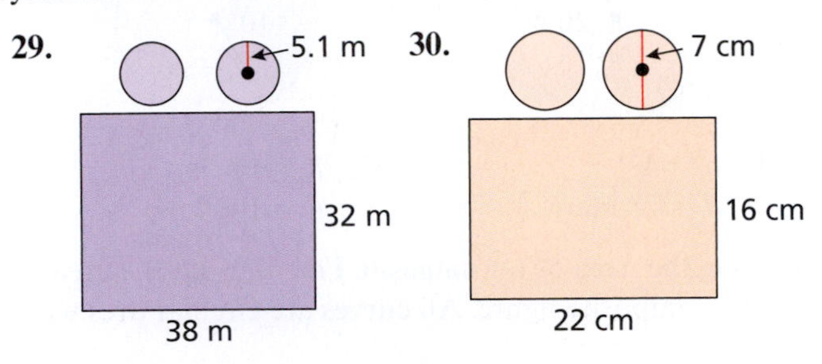

Finding the Surface Area of a Cone Find the surface area of a cone with the given dimensions.

31. Radius: 3 mm
Slant height: 6 mm

32. Diameter: 16 in.
Slant height: 11 in.

Finding the Slant Height of a Cone Find the slant height of the cone.

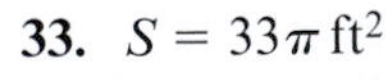

33. $S = 33\pi \text{ ft}^2$

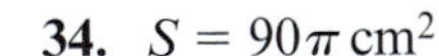

34. $S = 90\pi \text{ cm}^2$

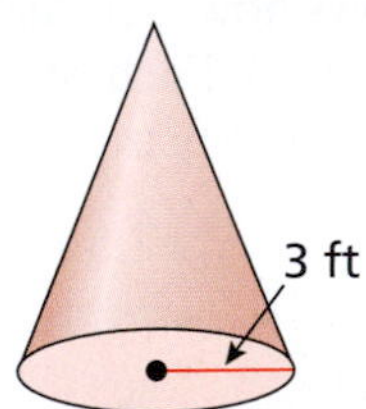

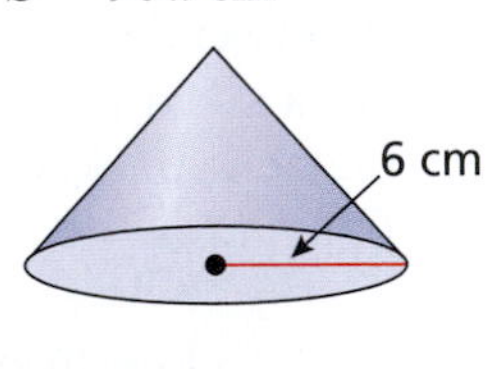

Finding the Surface Area of a Sphere The area A of a cross section through the center of a sphere is given. Find the surface area of the sphere.

35.

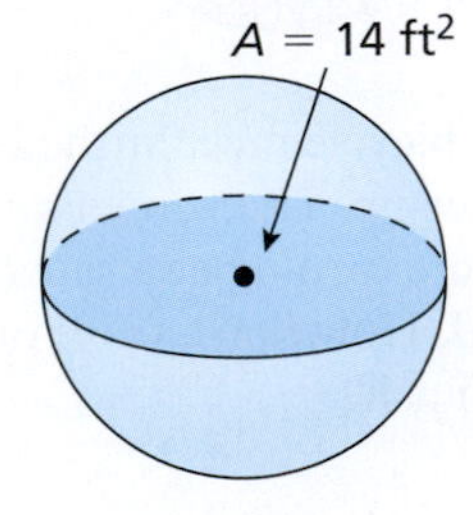

36.

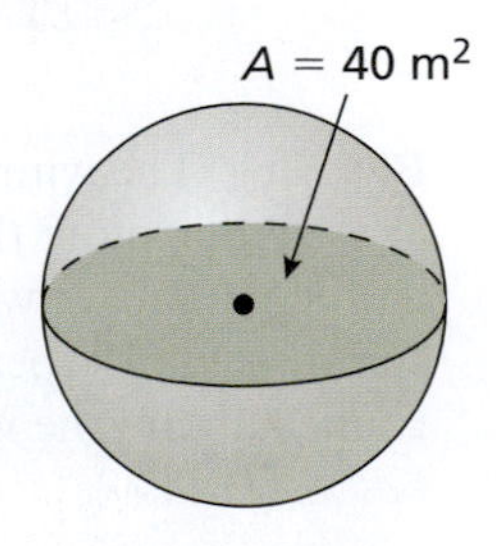

Finding the Surface Area of a Sphere Find the surface area of the sphere.

37.

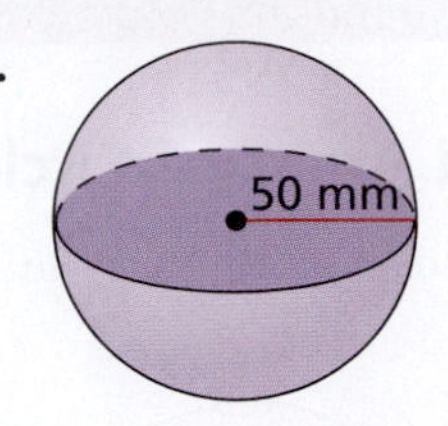

38.

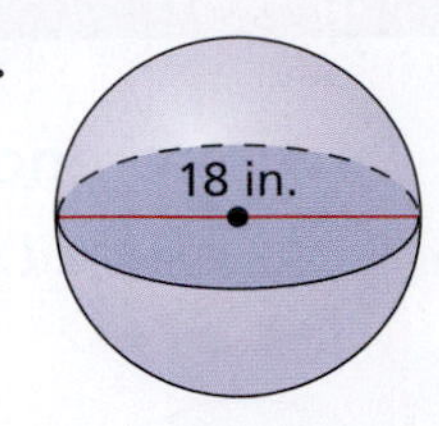

Finding the Surface Area of a Sphere The circumference C of a sphere is given. Find the surface area of the sphere.

39.

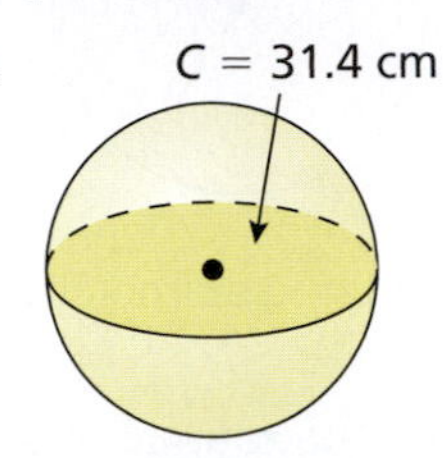

40.

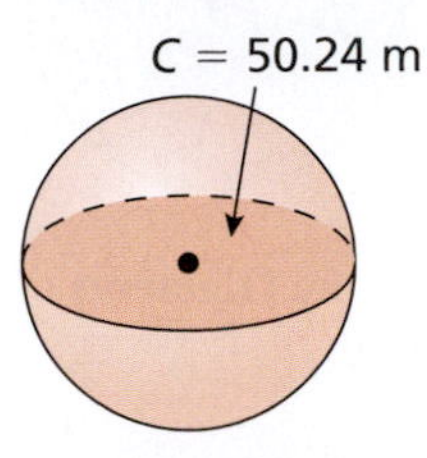

Finding the Surface Area of a Composite Solid The solid contains a cylinder and a cone or a hemisphere. Find the surface area of the composite solid.

41.

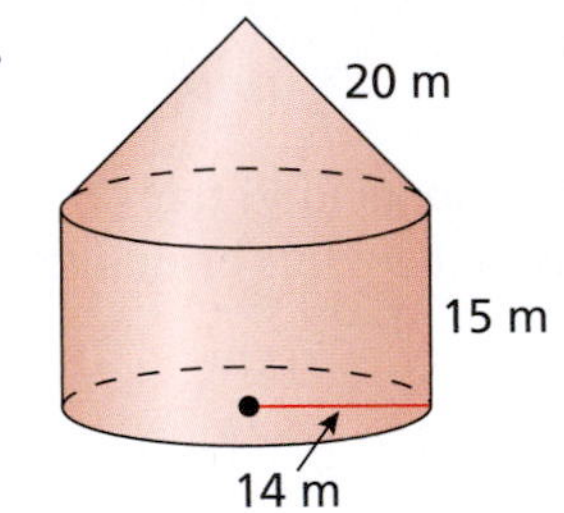

42.

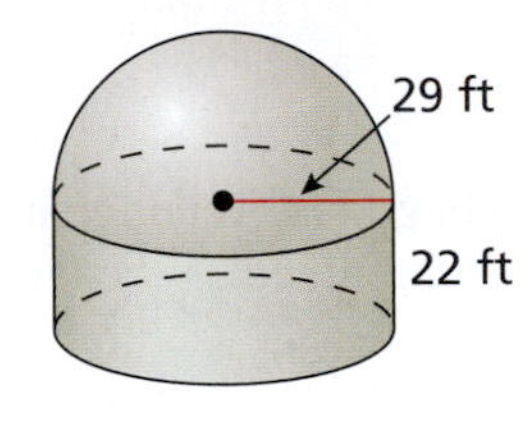

43. Wood Veneer A log that is 6 feet long is rotated on a lathe to slice off a thin, rectangular sheet of wood called a veneer. Compare the areas of the veneer produced from one rotation of the log when its diameter is $d = 8$ inches and when its diameter is $d = 4$ inches.

44. Pencil The eraser of the pencil shown is a hemisphere. The sharpened end of the pencil is a cone. Find the total surface area of the pencil and its eraser to the nearest square centimeter.

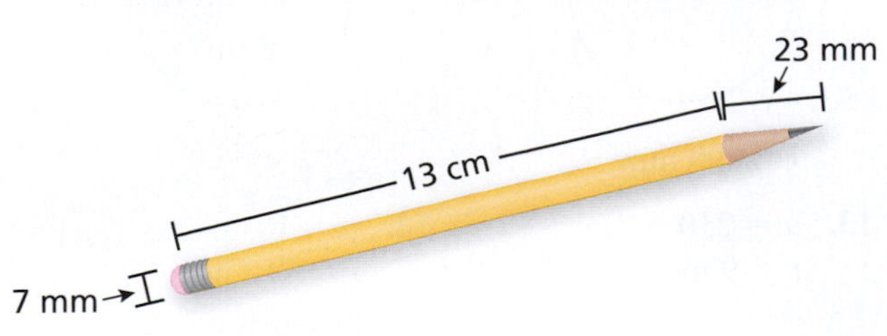

12.3 Volumes of Circular Solids

Finding the Volume of a Cylinder Find the volume of the cylinder.

45.

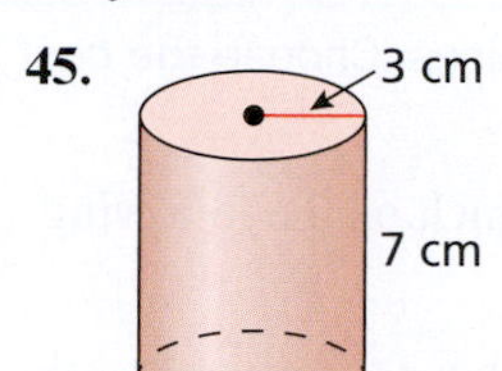

46.

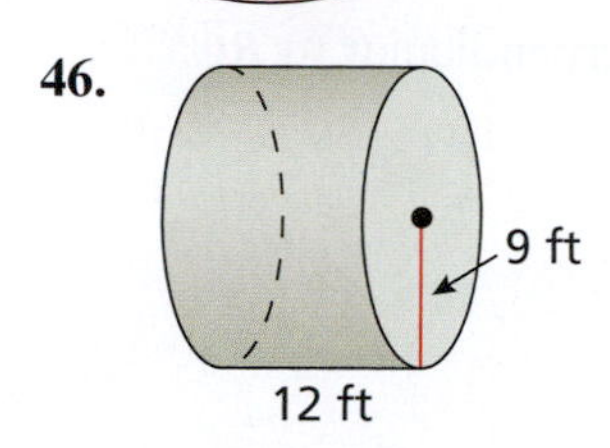

Finding the Volume of a Cone Find the volume of the cone.

47.

12 ft

6 ft

48.

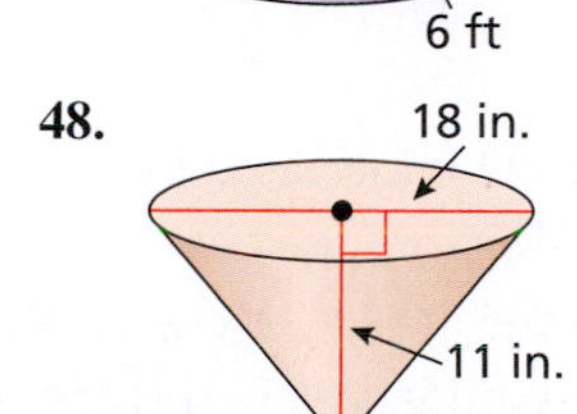

Finding the Volume of a Sphere Find the volume of the sphere.

49.

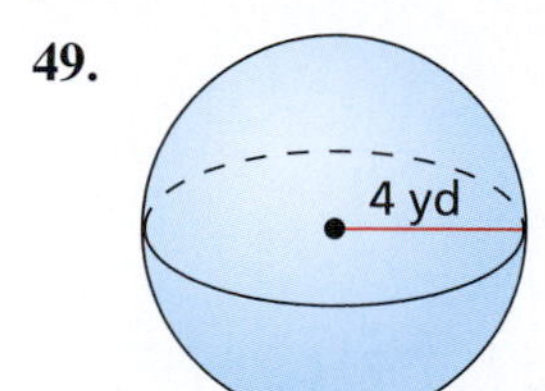

48.

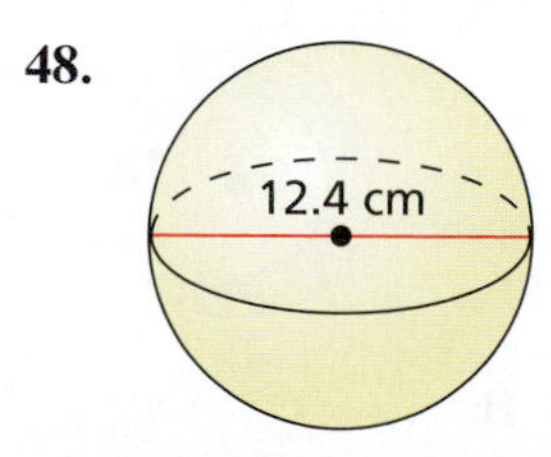

51. Finding the Volume of a Sphere Find the volume of a sphere with a diameter of 2 kilometers.

52. Finding the Volume of a Sphere Find the volume of a sphere with a radius of 21 centimeters.

Finding the Height of a Cylinder or Cone Find the height h of the cylinder or cone.

53. Volume $= 5087 \text{ in.}^3$

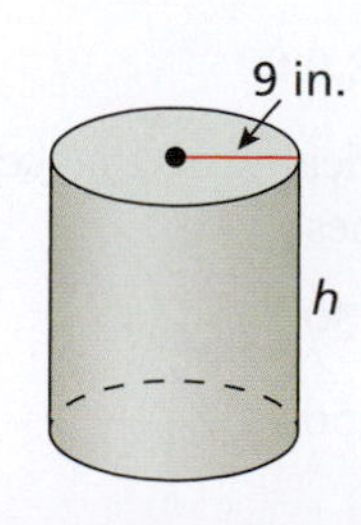

54. Volume $= 830 \text{ mm}^3$

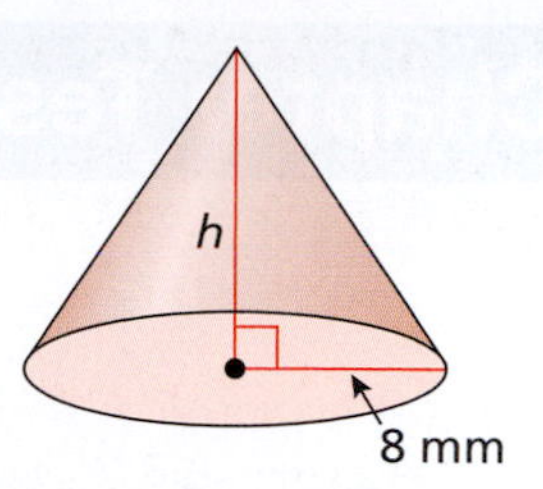

Finding the Volume of a Composite Solid The solid contains two cones, or a cylinder and a hemisphere. Find the volume of the composite solid.

55.

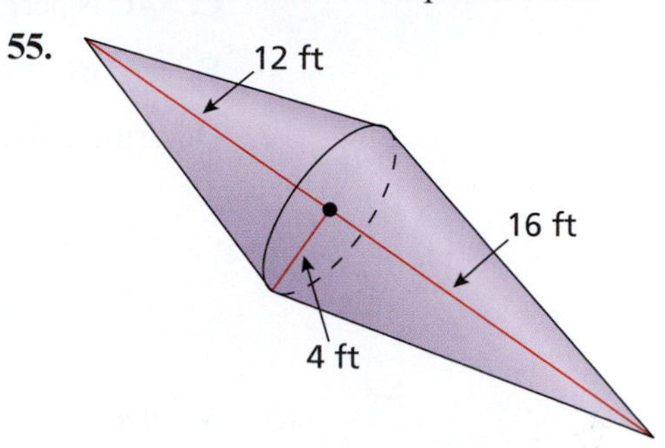

56.

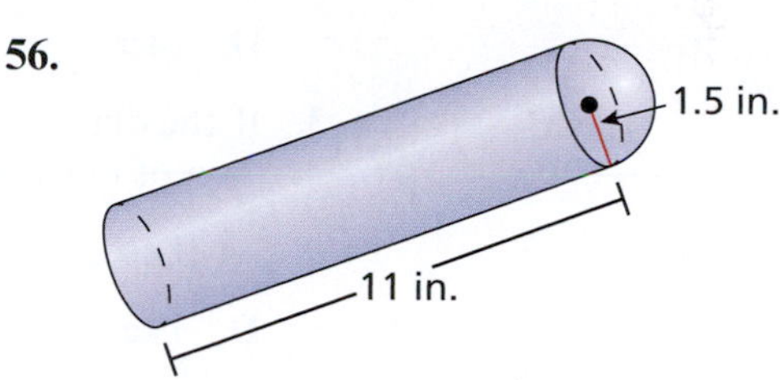

57. Finding the Volume of a Composite Solid The figure shows the dimensions of a round tent.

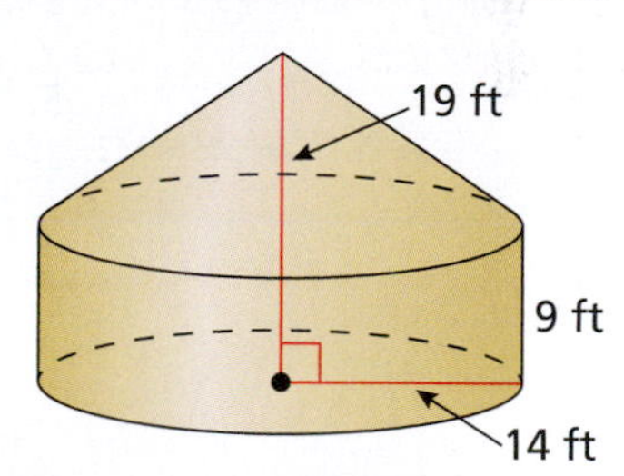

a. Find the volume of the tent.

b. What percent of the space in the tent is within the cone-shaped ceiling?

58. Problem Solving Twelve people enter the pool and start floating. The water level rises 0.28 feet. The weight of the water is 62.2 pounds per cubic foot. What is the average weight per person?

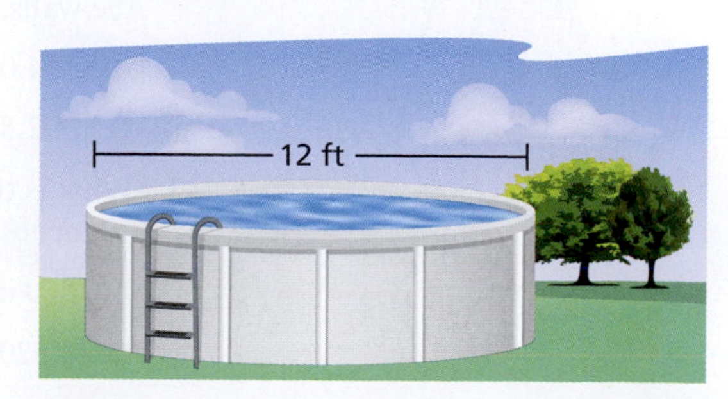

Chapter Test

The questions on this practice test are modeled after questions on state certification exams for K–8 teaching.

Test Directions: Each of the questions is followed by five choices. Choose the *best* response to each question.

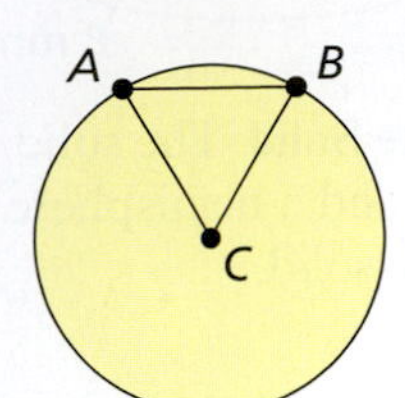

1. In the figure at the left, C is the center of the circle. Which of the following must be true?

   A. $\overline{AC}$ and $\overline{BC}$ have the same length. **B.** $\overline{AB}$ and $\overline{BC}$ have the same length.

   C. $\overline{AC}$ is perpendicular to $\overline{AB}$. **D.** $\overline{AC}$ is perpendicular to $\overline{BC}$.

   E. $\triangle ABC$ is equilateral.

2. Determine the area (in square inches) of the sector.

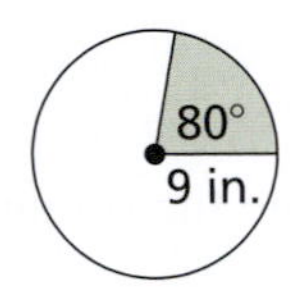

   A. 9π **B.** 18π **C.** 20π

   D. 24π **E.** 28π

3. If the diameter of circle A is half that of circle B, what is the ratio of the area of circle A to the area of circle B?

   A. 1 to 2 **B.** 2 to 1 **C.** 1 to 4

   D. 1 to π **E.** 1 to 8

4. The area of the circle at the left is 81π square feet. If $\overline{OA}$ is increased by 3 feet, what is the area of the new circle (in square feet)?

   O A

   A. 12π **B.** 100π **C.** 121π **D.** 144π **E.** 169π

5. Two concentric circles are shown in the figure below. The smaller circle has radius $OA = 2$ and the larger circle has radius $OB = 3$. Find the area of the shaded region.

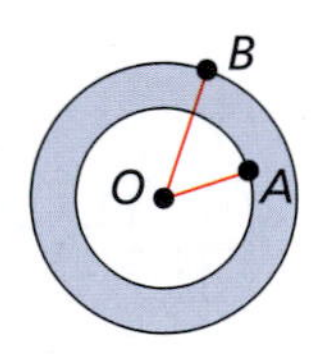

   A. π **B.** 4π **C.** 5π

   D. 8π **E.** 9π

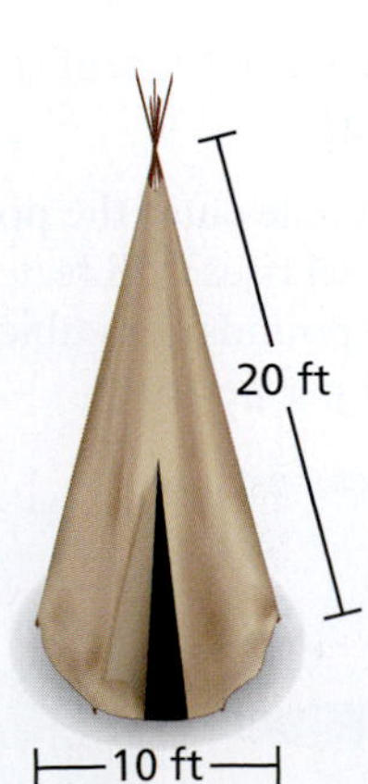

6. A scout troop is constructing a canvas recreation of a Native American teepee. The teepee will be approximately conical with a base diameter of 10 feet and a slant height of 20 feet. How much canvas is needed to form the walls of the teepee?

   A. 314.0 ft^2 **B.** 314.0 ft^3 **C.** 392.5 ft^2

   D. 392.5 ft^3 **E.** 628.0 ft^3

7. What is the capacity of a cylindrical waste basket that has a diameter of 12 inches and a height of 30 inches?

   A. 360 in.^3 **B.** 1,130 in.^3 **C.** 3,391 in.^3

   D. 4,320 in.^3 **E.** 13,565 in.^3

13 Congruence and Similarity

mangostock/Shutterstock.com

Activity: Discovering the SSS Congruence Postulate

Materials:

- Pencil
- Scissors
- Protractor
- Straws
- Ruler

Learning Objective:

Use straws to investigate and draw conclusions about triangles that have the same side lengths.

Name ______________________________

1. Start with three straws. Measure and cut one straw to a length of 11 centimeters. Measure and cut the second straw to a length of 8 centimeters. Measure and cut the third straw to a length of 5 centimeters.

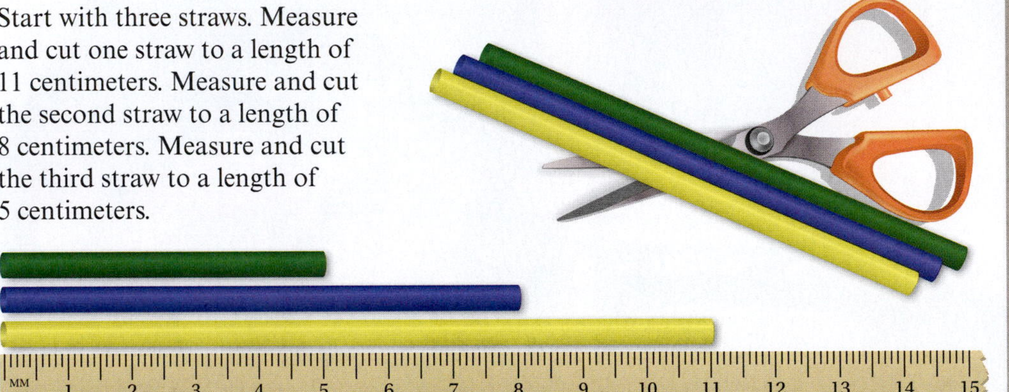

2. Arrange the straws to form a triangle. Use a protractor to approximate the angle measures of the triangle.

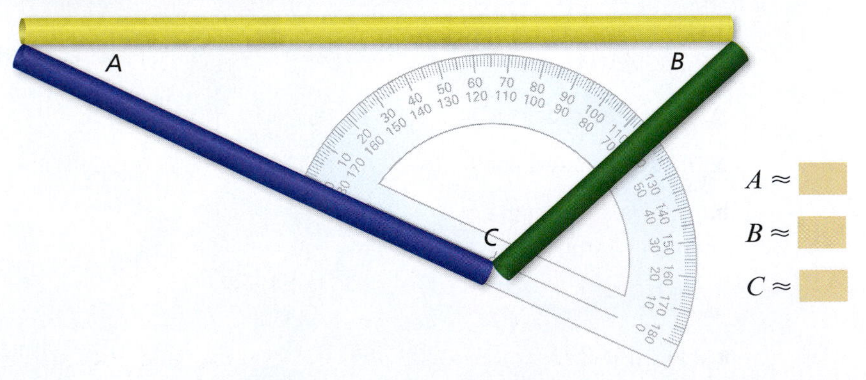

3. Try to rearrange the straws so that they form a triangle that has different angle measures. Is it possible? Explain your reasoning.

4. Two triangles are *congruent* when they have the same side lengths and the same angle measures. Is knowing that two triangles have the same side lengths enough to conclude that they are congruent? Explain your reasoning.

dotshock/Shutterstock.com

13.1 Congruence of Triangles

Standards

Grades 3–5
Geometry
Students should classify shapes by properties of their lines and angles.

Grades 6–8
Geometry
Students should describe the relationships between geometrical figures.

- Understand the definition of congruent triangles.
- Use the SSS, SAS, and ASA Congruence Postulates.
- Use congruent triangles to explore properties of parallelograms.

Congruent Triangles

Recall from Chapter 10 that line segments are congruent when they have the same length, and angles are congruent when they have the same measure. **Congruent figures** have the same size and shape.

Definition of Congruent Triangles

Two triangles are **congruent triangles** when their *corresponding sides* are congruent and their *corresponding angles* are congruent. In the figures below, triangle ABC is congruent to triangle DEF.

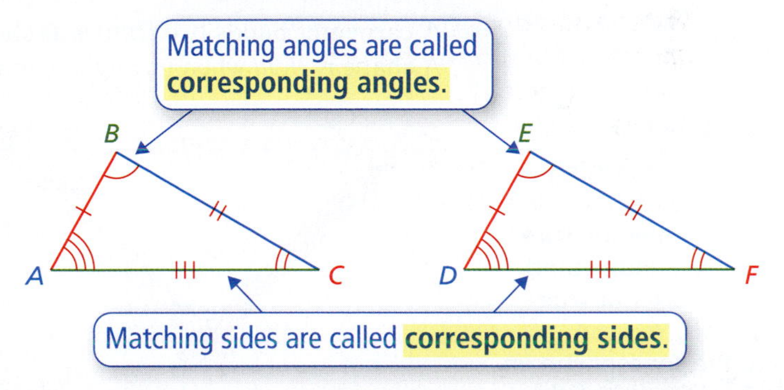

$\triangle ABC \cong \triangle DEF$ The symbol $\cong$ means "is congruent to."

When writing statements such as $\triangle ABC \cong \triangle DEF$, you can start at any vertex. Just be sure to list the vertices of the triangles in corresponding order.

EXAMPLE 1 Using the Definition of Congruent Triangles

$\triangle JKL$ is congruent to $\triangle XYZ$.

a. Find the length of $\overline{JK}$.

b. Find the measures of angles J and L.

L Z 65° 75° J K Y 10 cm X

SOLUTION

a. Because $\triangle JKL \cong \triangle XYZ$, the corresponding sides $\overline{JK}$ and $\overline{XY}$ are congruent. So, the length of $\overline{JK}$ is the same as the length of $\overline{XY}$.

$JK = XY = 10$ cm Corresponding sides

b. The corresponding angles J and X are congruent, and the corresponding angles L and Z are congruent. So, the measure of angle J, which can be written as $m\angle J$, is the same as the measure of angle X, and the measure of angle L is the same as the measure of angle Z.

$m\angle J = m\angle X = 75°$ Corresponding angles

$m\angle L = m\angle Z = 65°$ Corresponding angles

SSS, SAS, and ASA Congruence Postulates

In the activity on page 494, you learned that there is only one way to form a triangle with three given side lengths. This suggests the following *postulate*. A **postulate** is a rule that is accepted without justification.

Side-Side-Side (SSS) Congruence Postulate

If three sides of one triangle are congruent to three sides of another triangle, then the two triangles are congruent.

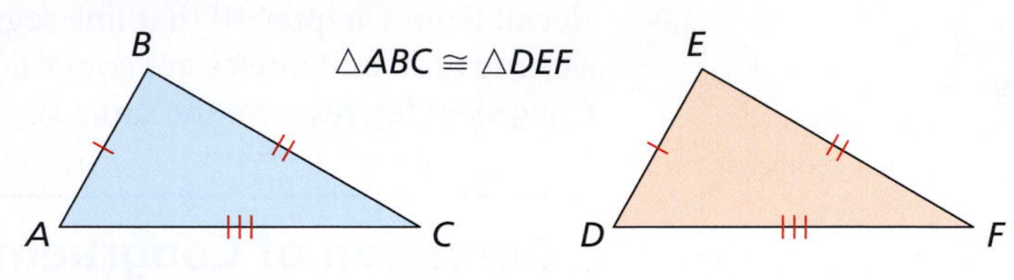

EXAMPLE 2 Side-Side-Side Congruence and Carpentry

A carpenter builds two frames as shown below. Tell whether each frame is rigid. A shape is *rigid* when its angles cannot be changed.

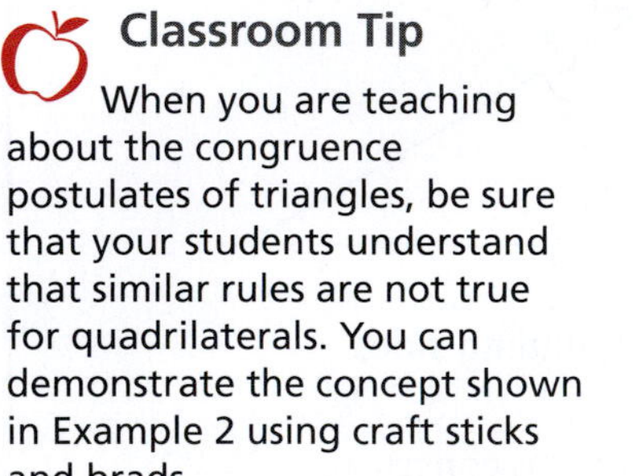

Classroom Tip

When you are teaching about the congruence postulates of triangles, be sure that your students understand that similar rules are not true for quadrilaterals. You can demonstrate the concept shown in Example 2 using craft sticks and brads.

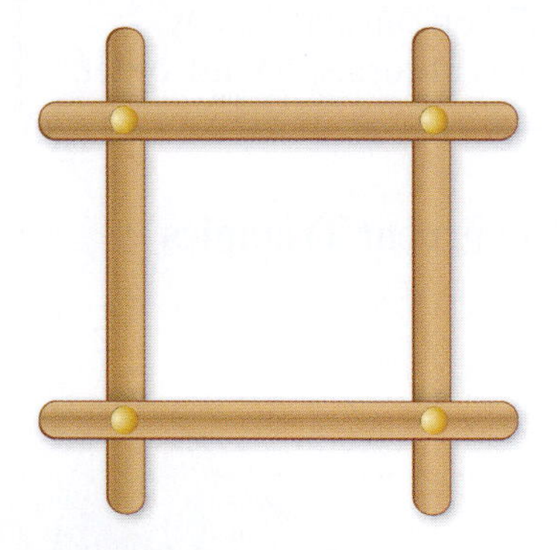

SOLUTION

As every carpenter knows, triangles are rigid. This is due to the SSS Congruence Postulate. Quadrilaterals do not share this attribute. You can take a rectangle, like the one shown above, and push and pull the vertices so that it is no longer a rectangle. This occurs even though the lengths of the sides are still the same.

A carpenter creates a triangular support for a gate because it makes the construction more rigid. Bridges also have triangular supports.

kezza/Shutterstock.com; Mazzzur/Shutterstock.com

Classroom Tip

In a triangle (or any polygon), the *included angle* is the angle formed by two adjacent sides. The vertex of the angle is the shared endpoint of the two sides.

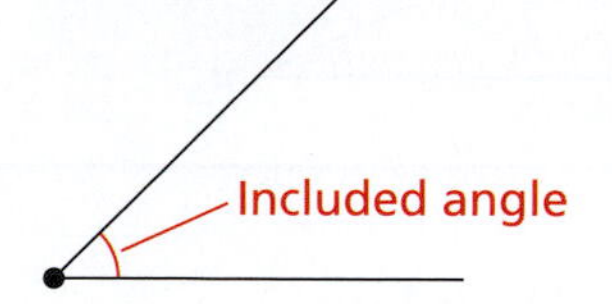

Side-Angle-Side (SAS) Congruence Postulate

If two sides and the included angle of one triangle are congruent to two sides and the included angle of another triangle, then the two triangles are congruent.

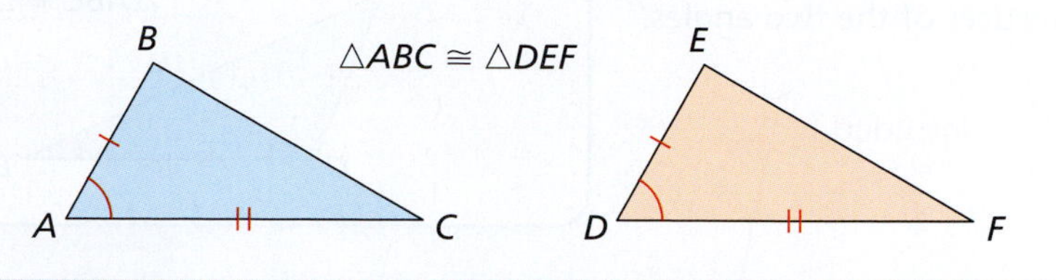

EXAMPLE 3 Verifying the SAS Congruence Postulate

Describe an activity that you can use with your students to show them that the SAS Congruence Postulate is valid.

SOLUTION

Here is one activity.

1. Use a straightedge to draw a triangle on a piece of paper.

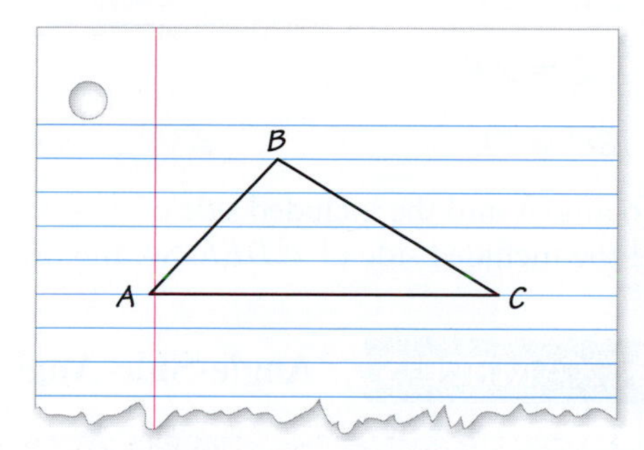

2. Use tracing paper to trace two sides and the included angle of the triangle.

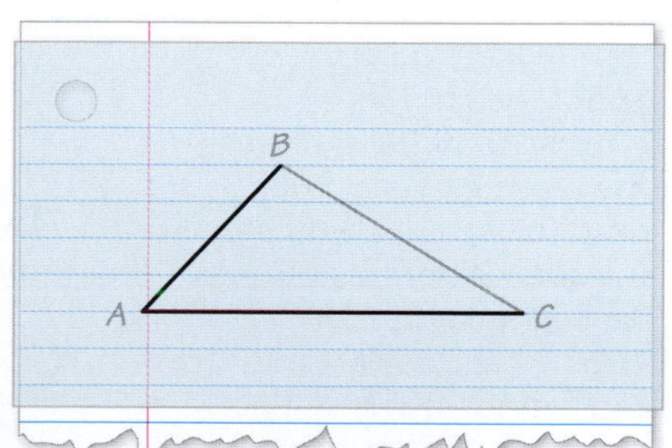

3. Ask your students to determine the number of ways to complete the triangle on the tracing paper. It should not take them long to determine that there is only one way.

4. Finally, take the triangle drawn on the tracing paper and place it on top of the original triangle. Your students can conclude that the two triangles are congruent.

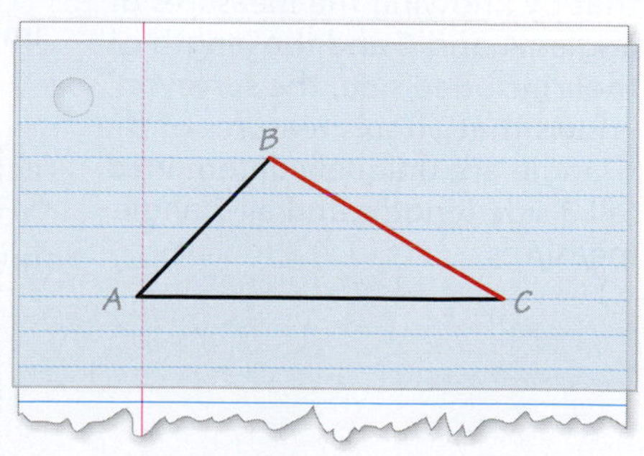

Angle-Side-Angle (ASA) Congruence Postulate

If two angles and the included side of one triangle are congruent to two angles and the included side of another triangle, then the two triangles are congruent.

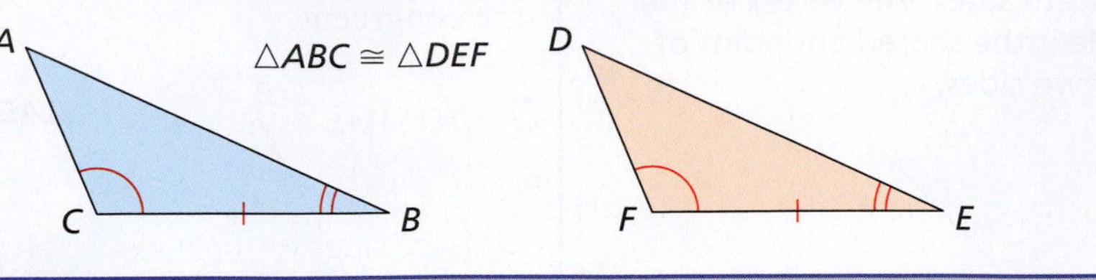

Classroom Tip

In a triangle, the *included side* is the side whose endpoints are the vertices of the two angles.

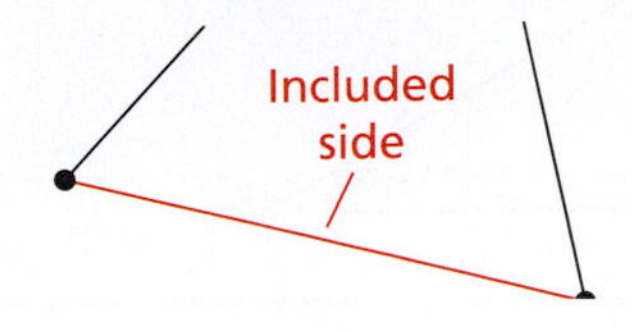

EXAMPLE 4 **Using the ASA Congruence Postulate**

Is there enough information given to conclude that the two triangles shown at the right are congruent? Explain your reasoning.

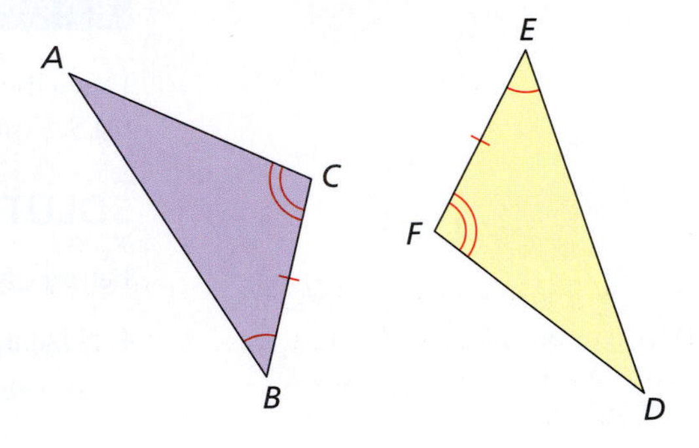

SOLUTION

From the triangles shown,

$\angle B \cong \angle E$,

$\angle C \cong \angle F$, and

$\overline{BC} \cong \overline{EF}$.

Two angles and the included side of $\triangle ABC$ are congruent to two angles and the included side of $\triangle DEF$. So, the triangles are congruent.

EXAMPLE 5 **Angle-Side-Angle Congruence and Surveying**

A surveyor is trying to determine the distance across the river. Explain why the surveyor can determine this distance by measuring the two indicated angles and the distance.

SOLUTION

The surveyor can determine the distance across the river because the Angle-Side-Angle Congruence Postulate implies that knowing two angles and the included side of a triangle uniquely determine the lengths of the other two sides.

To actually determine the length of $\overline{AB}$, which represents the distance across the river, the surveyor needs to use other geometric properties. You will be able to do this after the properties of similar triangles are introduced in Section 13.2.

Classroom Tip

The point of Example 5 is that by knowing the measures of angles A and C and the length of their included side, the surveyor knows that all six measures of the triangle are uniquely determined (all 3 side lengths and all 3 angle measures).

Using Congruent Triangles

In Section 10.2, you learned that opposite sides of a parallelogram are congruent and opposite angles of a parallelogram are congruent. Using the congruence postulates for congruent triangles, you can justify these statements.

Opposite Sides and Opposite Angles Property of a Parallelogram

Opposite sides of a parallelogram are congruent. Opposite angles of a parallelogram are also congruent.

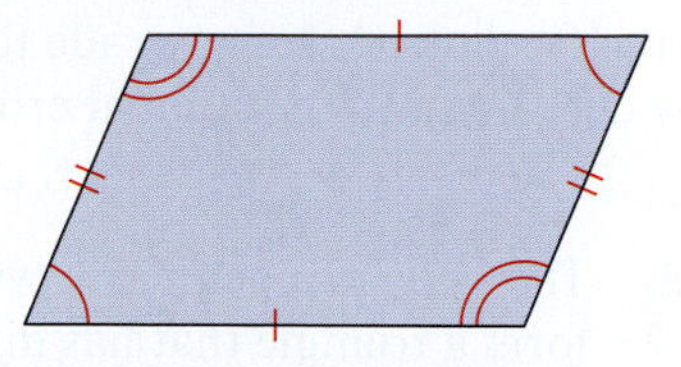

EXAMPLE 6 **Exploring Properties of Parallelograms**

Explain how to use the theorem given below and the ASA Congruence Postulate to show that opposite sides and opposite angles of a parallelogram are congruent.

When a transversal intersects two parallel lines, alternate interior angles are congruent.

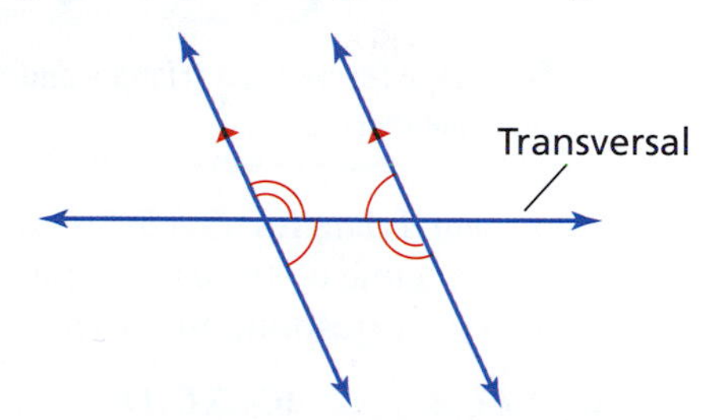

SOLUTION

To begin, draw the parallelogram given above. Recall from Section 10.2, that by definition, a parallelogram has two pairs of opposite sides that are parallel. Draw one of the diagonals of the parallelogram as shown.

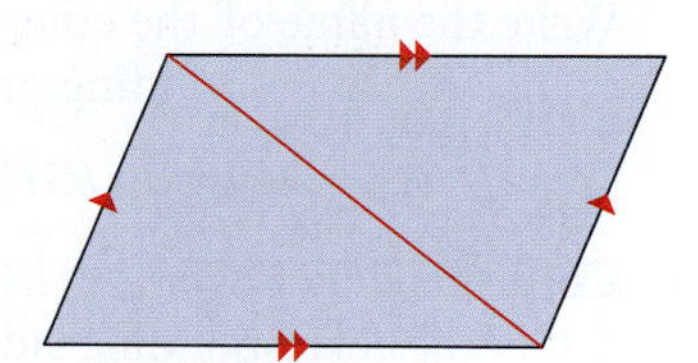

Using the theorem given above, mark the alternate interior angles that are congruent.

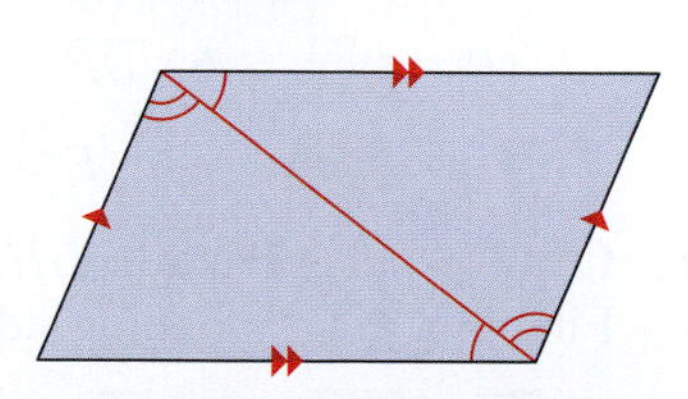

Cut the parallelogram along the diagonal. The result is two triangles that have two angles and the included side that are congruent.

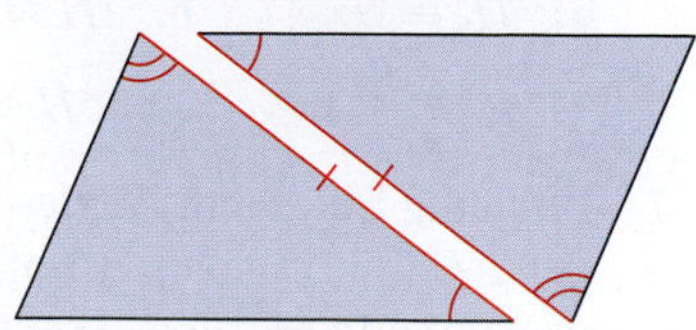

So, by the Angle-Side-Angle Congruence Postulate, you can conclude that the two triangles are congruent. From this, it follows that the opposite sides of the parallelogram are congruent and that the opposite angles of the parallelogram are congruent.

Classroom Tip

A *theorem* is a statement that can be proven, whereas a *postulate* is accepted without justification.

Mathematical Practices

Make Sense of Problems and Persevere in Solving Them

MP1 Mathematically proficient students make conjectures about the form and meaning of a situation.

13.1 Exercises

1. **Writing a Solution Key** Write a solution key for the activity on page 494. Describe a rubric for grading a student's work.

2. **Grading the Activity** In the activity on page 494, a student gave the answer below. The question is worth 10 points. Assign a grade to the answer. Why do you think the student erred?

Sample Student Work

> **3.** Try to rearrange the straws so that they form a triangle that has different angle measures. Is it possible? Explain your reasoning.
>
> 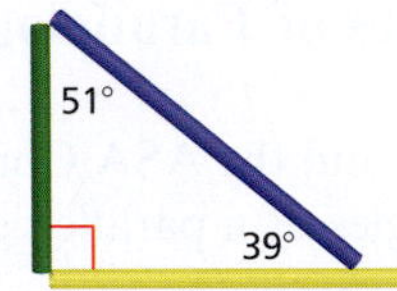
>
>
> The shape formed is a triangle and has different angle measures.

3. **Congruent Triangles** $\triangle ABC$ is congruent to $\triangle DEF$. Write the name of the congruent triangle listing the vertices in corresponding order.

a. $\triangle BCA$ **b.** $\triangle CAB$ **c.** $\triangle FED$

4. **Congruent Triangles** $\triangle GHI$ is congruent to $\triangle JKL$. Write the name of the congruent triangle listing the vertices in corresponding order.

a. $\triangle HIG$ **b.** $\triangle IGH$ **c.** $\triangle LKJ$

5. **Corresponding Parts** $\triangle ABC$ is congruent to $\triangle DEF$. Fill in the corresponding side or angle.

a. $\overline{AB} \cong$ ▭ **b.** $\overline{DF} \cong$ ▭ **c.** $\overline{CA} \cong$ ▭

d. $\angle D \cong$ ▭ **e.** $\angle E \cong$ ▭ **f.** $\angle C \cong$ ▭

6. **Corresponding Parts** $\triangle GHI$ is congruent to $\triangle JKL$. Fill in the corresponding side or angle.

a. $\overline{JK} \cong$ ▭ **b.** $\overline{HI} \cong$ ▭ **c.** $\overline{JL} \cong$ ▭

d. $\angle G \cong$ ▭ **e.** $\angle H \cong$ ▭ **f.** $\angle L \cong$ ▭

7. **Labeling Sides and Angles** $\triangle ABC$ is congruent to $\triangle DEF$. Find the length of each side or the measure of each angle.

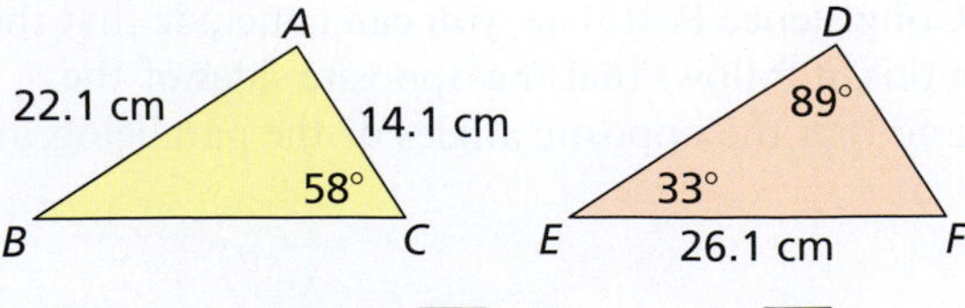

a. $\overline{DE}$ **b.** $\overline{DF}$ **c.** $\overline{BC}$

d. $\angle A$ **e.** $\angle B$ **f.** $\angle F$

8. **Labeling Sides and Angles** $\triangle GHI$ is congruent to $\triangle JKL$. Find the length of each side or the measure of each angle.

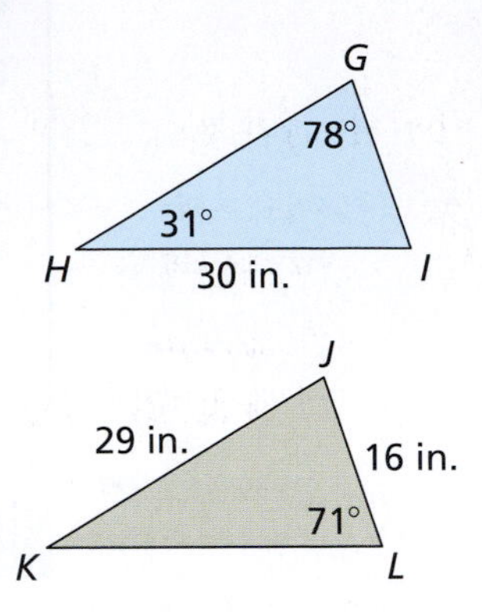

a. $\overline{GH}$ **b.** $\overline{GI}$ **c.** $\overline{KL}$

d. $\angle J$ **e.** $\angle K$ **f.** $\angle I$

9. **Naming Angles** Name the included angle between each pair of given sides.

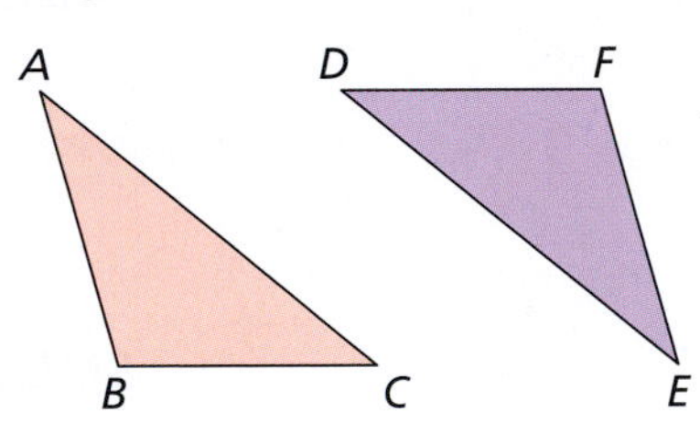

a. $\overline{AB}$ and $\overline{BC}$

b. $\overline{DF}$ and $\overline{FE}$

c. $\overline{BA}$ and $\overline{AC}$

10. **Naming Angles** Name the included angle between each pair of given sides.

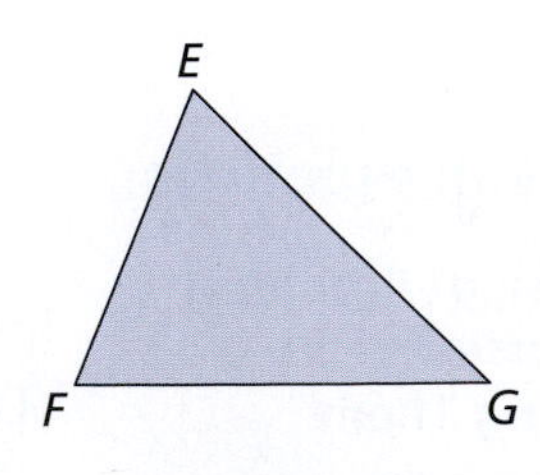

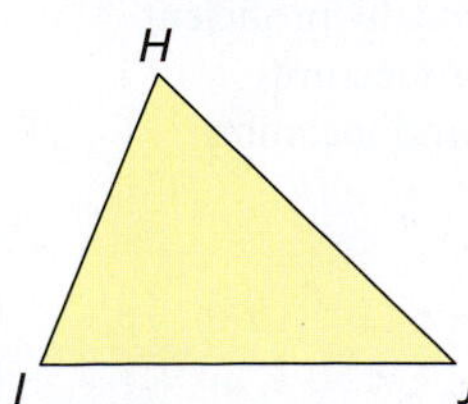

a. $\overline{HJ}$ and $\overline{JI}$

b. $\overline{FE}$ and $\overline{EG}$

c. $\overline{JI}$ and $\overline{IH}$

11. Congruent Triangles Decide whether each pair of triangles is congruent. Explain your reasoning.

a.

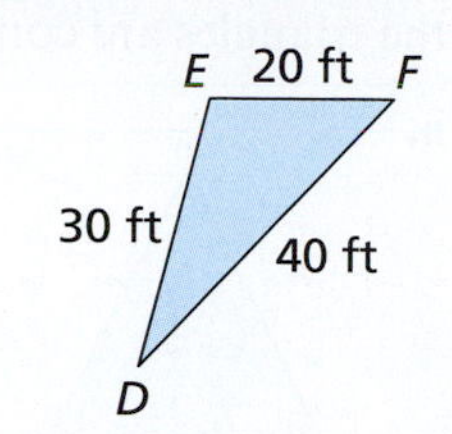

b.

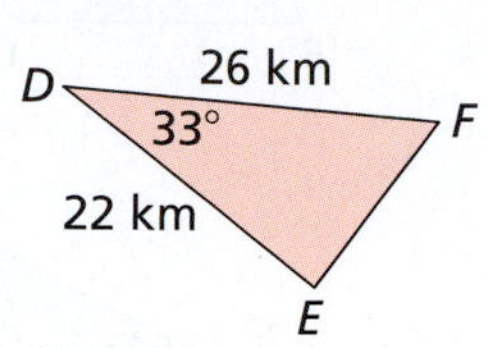

c.

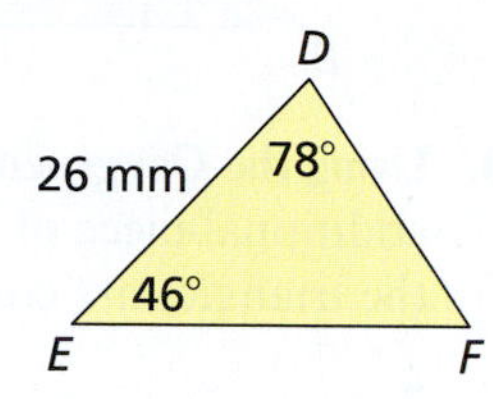

12. Congruent Triangles Decide whether each pair of triangles is congruent. Explain your reasoning.

a.

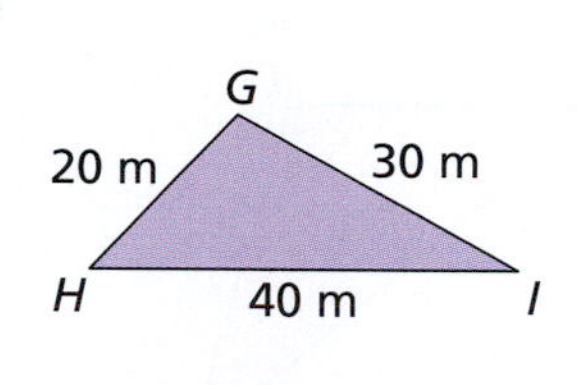

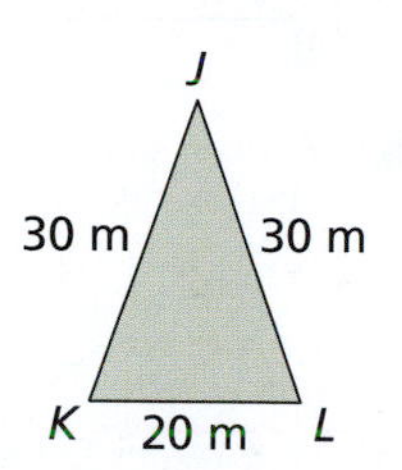

b.

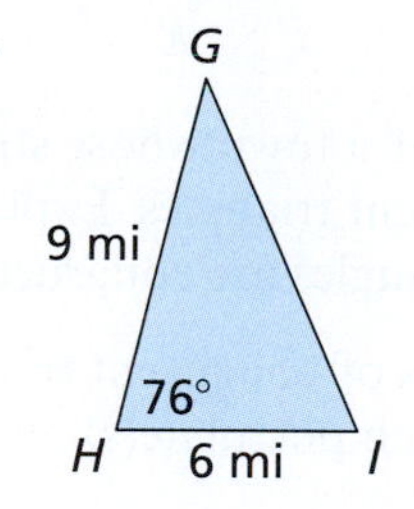

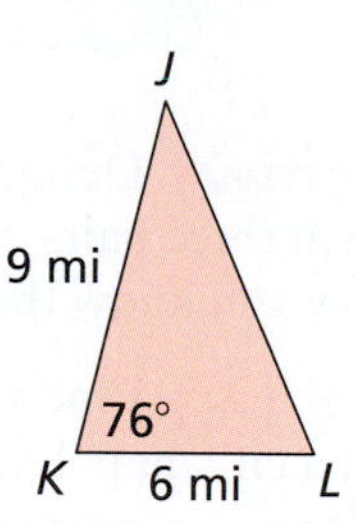

c.

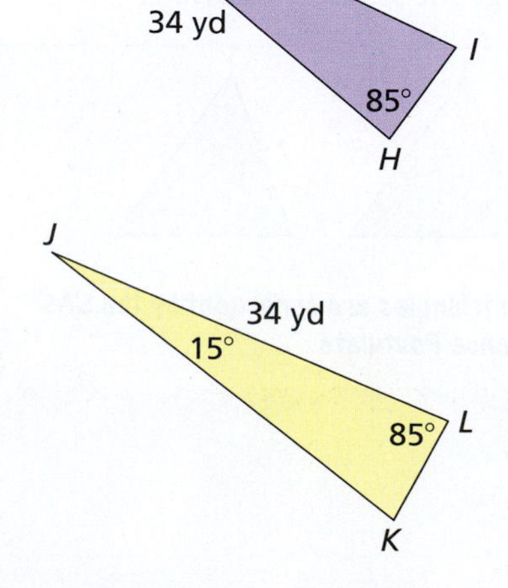

13. Corresponding Parts and Congruent Triangles Write the congruence statements needed to show $\triangle ABC \cong \triangle DEF$ using each postulate.

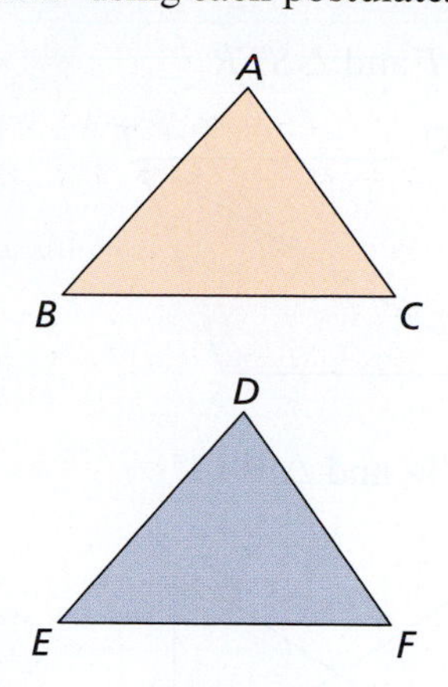

a. SAS Congruence Postulate

b. SSS Congruence Postulate

14. Corresponding Parts and Congruent Triangles Write the congruence statements needed to show $\triangle GHJ \cong \triangle KLM$ using each postulate.

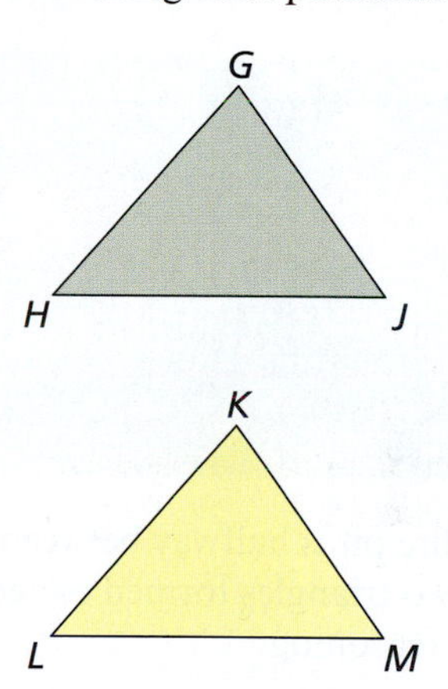

a. ASA Congruence Postulate

b. SAS Congruence Postulate

15. Using the Congruence Postulates Determine whether there is enough information to state that the triangles are congruent. Explain your reasoning.

a. $\triangle ABC$ and $\triangle ADC$

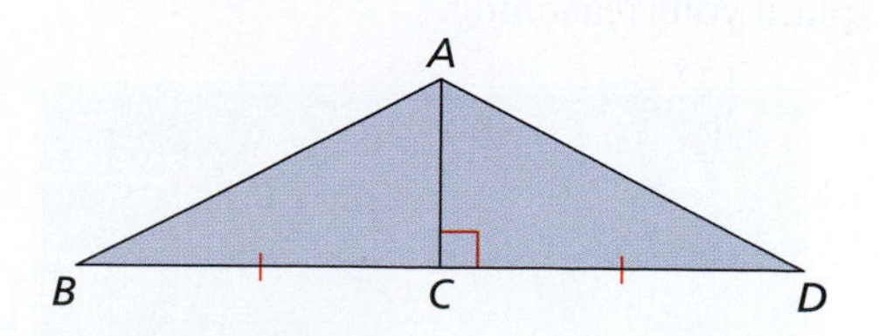

b. $\triangle EFG$ and $\triangle IHG$

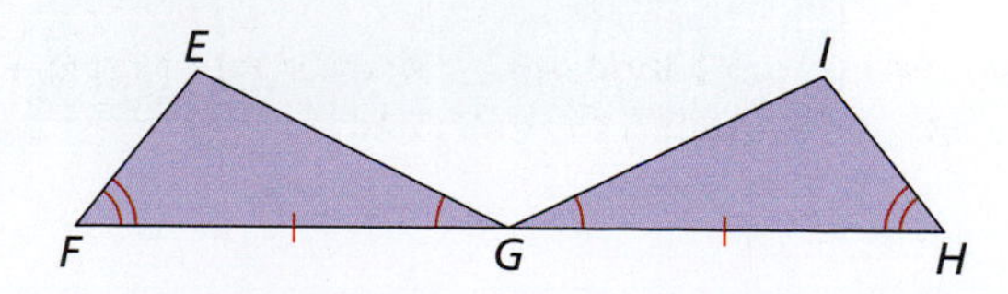

16. Using the Congruence Postulates Determine whether there is enough information to state that the triangles are congruent. Explain your reasoning.

a. $\triangle QRT$ and $\triangle STR$

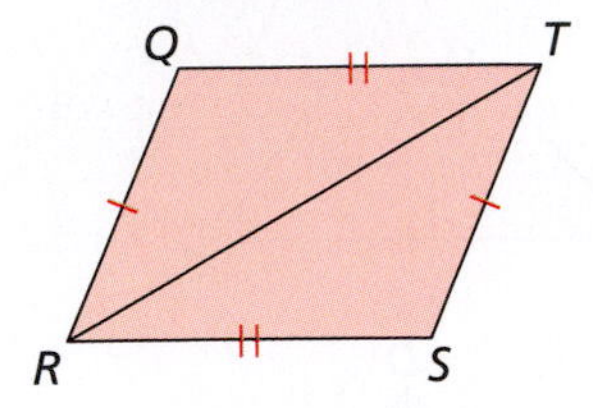

b. $\triangle UVW$ and $\triangle XWV$

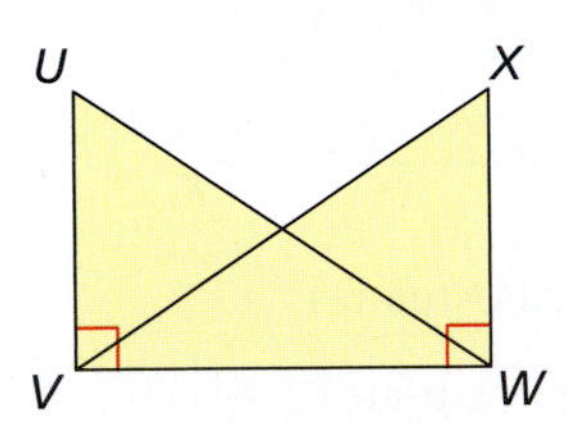

17. Using a Congruence Postulate to Measure Distance You and your friend set up a tent and a fire pit along the edge of a canyon as shown in the figure.

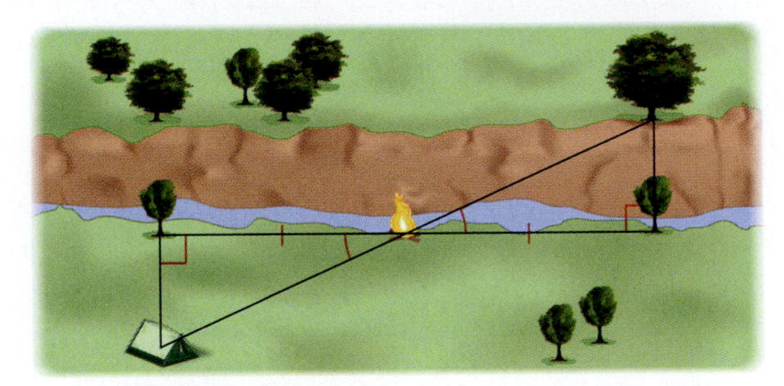

a. The fire pit is halfway between the trees. Are the two triangles formed congruent? Explain your reasoning.

b. Can you use the information you have to measure the distance across the canyon? Explain your reasoning.

18. Using a Congruence Postulate to Measure Distance The figure shows the positions of you, your teammate, the soccer ball, and two opponents on a soccer field. Are the triangles shown congruent? Explain your reasoning.

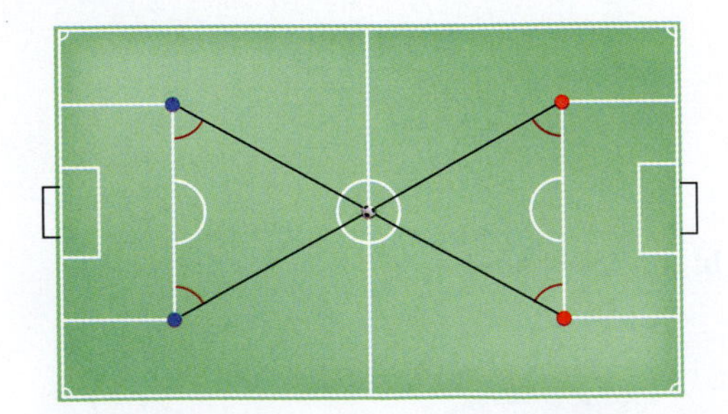

19. Using the Congruence Postulates Identify an additional piece of information needed to show that the triangles are congruent. Explain your reasoning.

a.

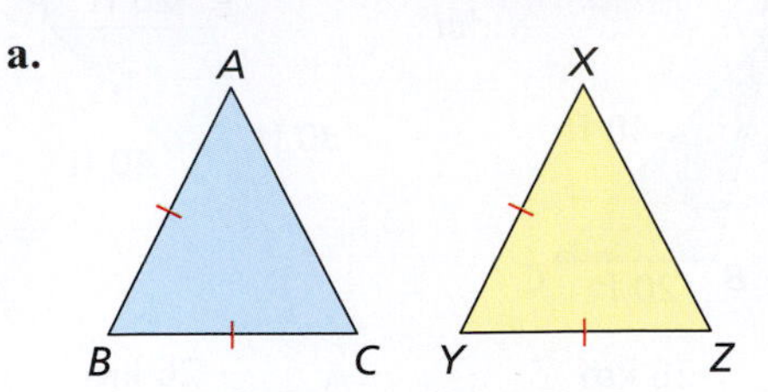

b.

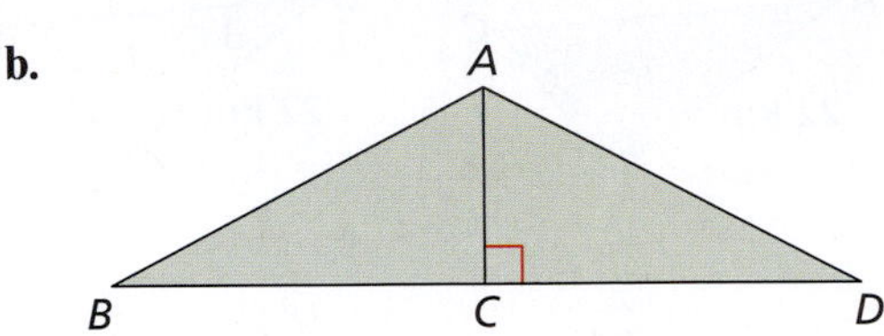

20. Using the Congruence Postulates Identify an additional piece of information needed to show that the triangles are congruent. Explain your reasoning.

a. **b.**

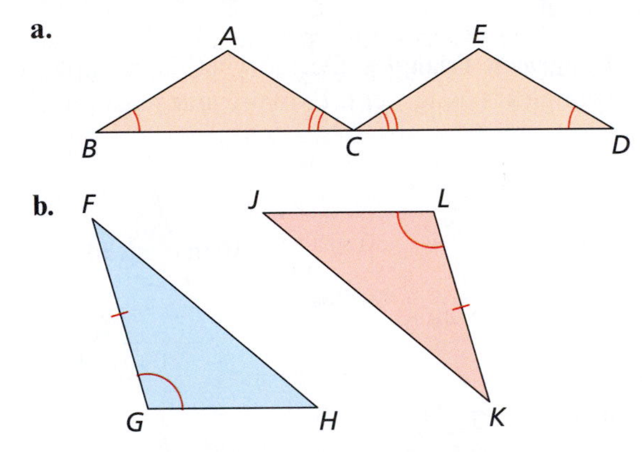

21. Writing Design a map of a town whose streets form three pairs of congruent triangles. Explain how you know that the triangles are congruent.

22. Writing Find three pairs of congruent triangles on a city map. Explain which postulate(s) make the triangles congruent.

23. Grading Student Work On a diagnostic test, one of your students does the following work. Explain what the student did wrong. Which topics would you encourage the student to review?

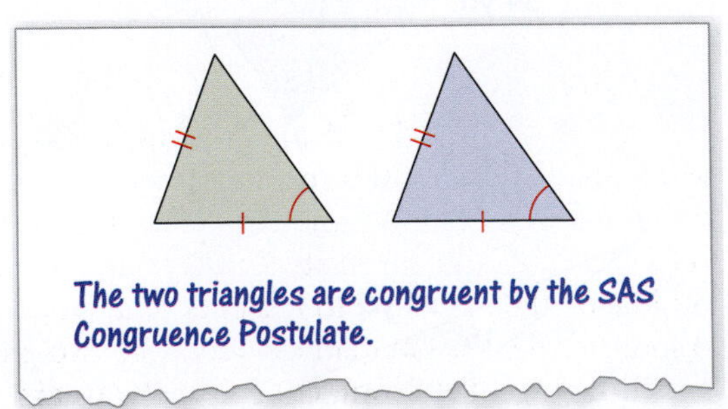

24. **Grading Student Work** On a diagnostic test, one of your students does the following work. Explain what the student did wrong. Which topics would you encourage the student to review?

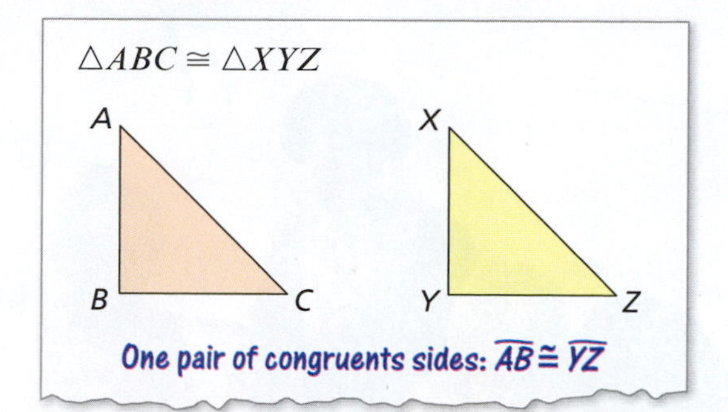

25. **Activity: In Your Classroom** Design an activity you could use to show your students that having two pairs of congruent sides is not enough to say that two triangles are congruent.

26. **Activity: In Your Classroom** Design an activity you could use to show your students that having two pairs of congruent angles is not enough to say that two triangles are congruent.

27. **Using a Congruence Postulate** When opposite angles of a quadrilateral are congruent, the quadrilateral is a parallelogram. In the figure, $\triangle EFH \cong \triangle GHF$. Determine whether quadrilateral $EFGH$ is a parallelogram.

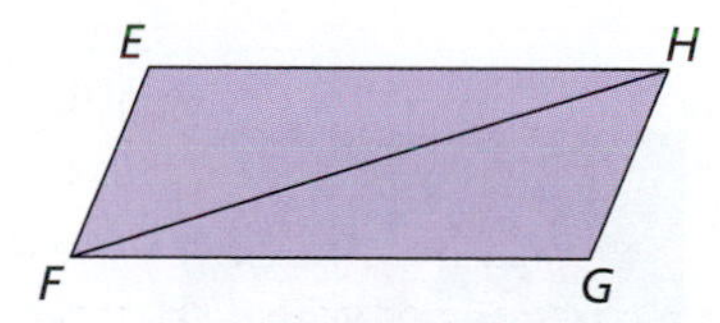

28. **Using a Congruence Postulate** Show that the opposite angles of the quadrilateral are congruent. (*Hint*: First, draw diagonal $\overline{BD}$, and then draw diagonal $\overline{AC}$).

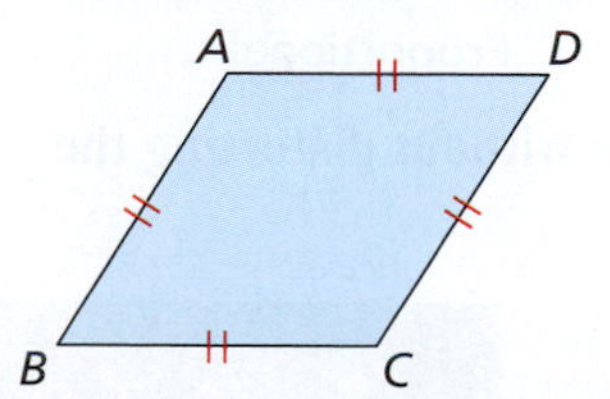

29. **Properties of Triangles** A student has a piece of paper that is in the shape of an isosceles triangle with angles of 75°, 75°, and 30°. To divide the paper into two equal parts, the student cuts down the middle of the 30° angle to the middle of the opposite side. The student says that the two pieces are congruent. Is the student correct? Explain.

30. **Finding Distance with Triangles** You stand on one side of a riverbank looking straight across. You tip the visor of your hat until the farthest thing you see is the opposite bank. You turn to the right without moving the position of your head or your visor and look down the riverbank. Your friend marks the place on the bank in line with your eyes and the visor of your hat. (See figure.) Is the distance on your side of the bank equal to the distance across the river?

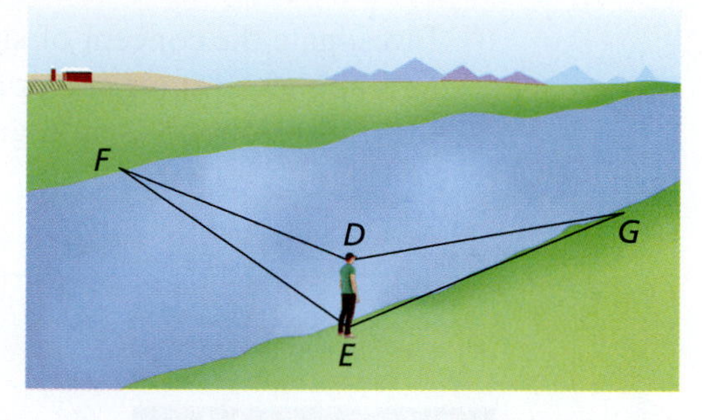

31. **Using a Congruence Postulate** A telephone pole on level ground has three guy wires anchored at the same height and the same distance from the pole, as shown in the figure. What do you know about the lengths of the three guy wires? Explain your reasoning.

32. **Using a Congruence Postulate** A telephone pole on the side of a hill has two guy wires anchored at the same height and the same distance from the pole, as shown in the figure. Are the two guy wires the same length? Explain your reasoning.

Activity: Investigating Similar Rectangles

Materials:

- Pencil

Learning Objective:

Investigate the concept of similarity as it applies to enlarging or reducing the dimensions of a photograph.

Name ______________________________

Original Photo

In a computer graphics program, when you push or pull on a side of a photograph, you distort it. When you push or pull on a corner of a photograph, it remains proportional to the original photograph.

Distorted

Distorted

Proportional

Can you resize the photograph to the indicated dimensions without distorting the photograph? Explain your reasoning.

1.

6 in. (height), 6 in. (width)

a. 5 in. by 5 in.
b. 5 in. by 6 in.
c. 8 in. by 8 in.

2.

6 in. (height), 4 in. (width)

a. 5 in. by 3 in.
b. 3 in. by 2 in.
c. 9 in. by 6 in.

3.

6 in. (height), 8 in. (width)

a. 3 in. by 4 in.
b. 4 in. by 6 in.
c. 8 in. by 10 in.

13.2 Similarity

Standards

Grades 3–5
Geometry
Students should classify shapes by properties of their lines and angles.

Grades 6–8
Geometry
Students should understand congruence and similarity using physical models, transparencies, or geometry software.

- Understand the definition of similar triangles.
- Use the AA Similarity Postulate.
- Use the SAS and SSS Similarity Theorems.
- Understand the definition of similarity as it applies to other polygons.

Similar Triangles

Similar figures have the same shape but not necessarily the same size.

Definition of Similar Triangles

Two triangles are **similar triangles** when their corresponding side lengths are proportional and their corresponding angles are congruent. In the figures below, triangle ABC is similar to triangle DEF.

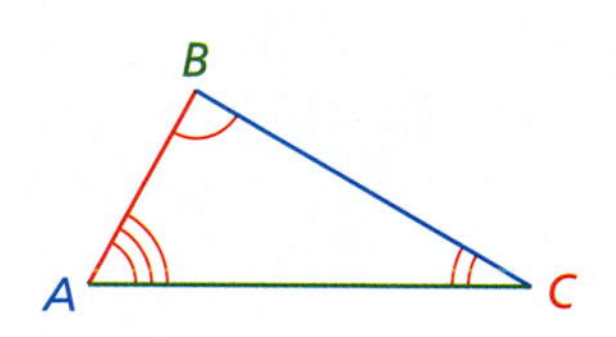

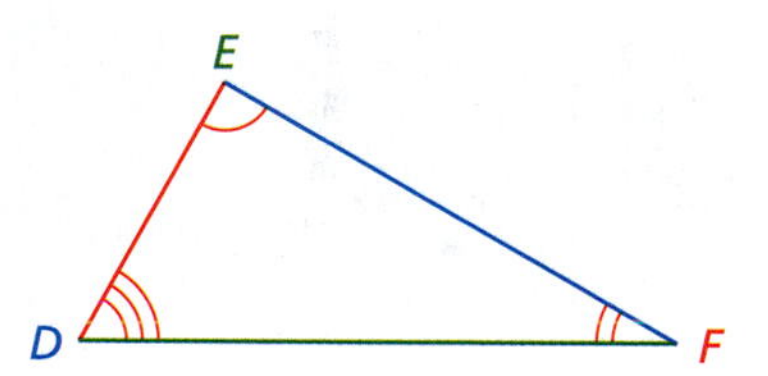

Triangles $\triangle ABC \sim \triangle DEF$

The symbol ~ means "is similar to."

Side Lengths $\dfrac{AB}{DE} = \dfrac{BC}{EF} = \dfrac{AC}{DF}$

Angles $A \cong D$, $B \cong E$, $C \cong F$

EXAMPLE 1 Using the Definition of Similar Triangles

$\triangle ABC$ is similar to $\triangle JKL$. Find the measure of (a) $\angle B$ and (b) $\angle L$.

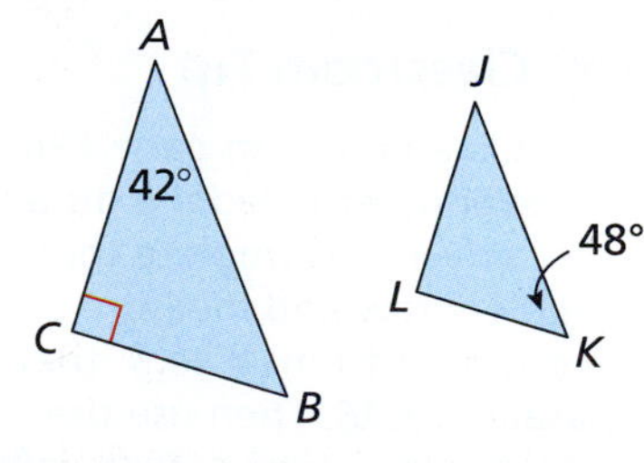

SOLUTION

a. Angle B and angle K are corresponding angles. So, the measure of angle B is 48°.

b. Angle L and angle C are corresponding angles. So, the measure of angle L is 90°.

EXAMPLE 2 Using the Definition of Similar Triangles

$\triangle FGH$ is similar to $\triangle PQR$. Find the value of x.

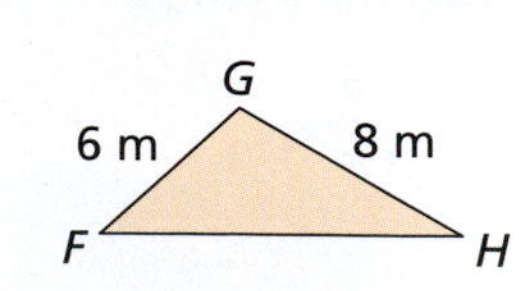

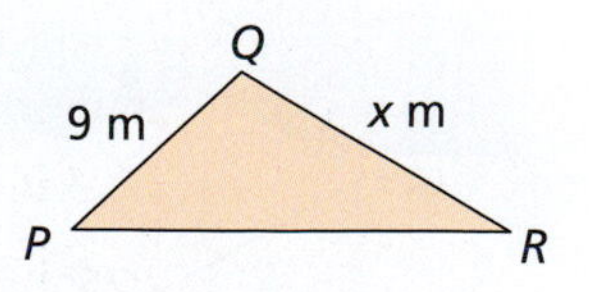

SOLUTION

Use a proportion to find x.

$\dfrac{6}{9} = \dfrac{8}{x}$ Write a proportion.

$6x = 72$ Use the Cross Products Property.

$x = 12$ Division Property of Equality

So, x is 12 and the length of the side is 12 meters.

AA Similarity Postulate

Angle-Angle (AA) Similarity Postulate

If two angles of one triangle are congruent to two angles of another triangle, then the two triangles are similar.

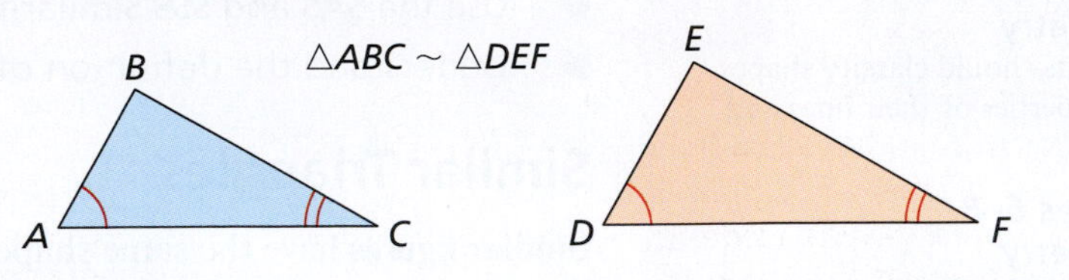

EXAMPLE 3 **Comparing Two Drafting Triangles**

Are the two drafting triangles similar? Explain your reasoning.

45°-45°-90° Drafting triangle

30°-60°-90° Drafting triangle

SOLUTION

Both triangles have a right angle. However, none of the other corresponding angles are congruent. So, the two triangles are not similar.

EXAMPLE 4 **Measuring Indirectly**

The sun casts shadows (with parallel sun rays), as shown at the right. The woman is 5 feet tall. What is the height of the tree?

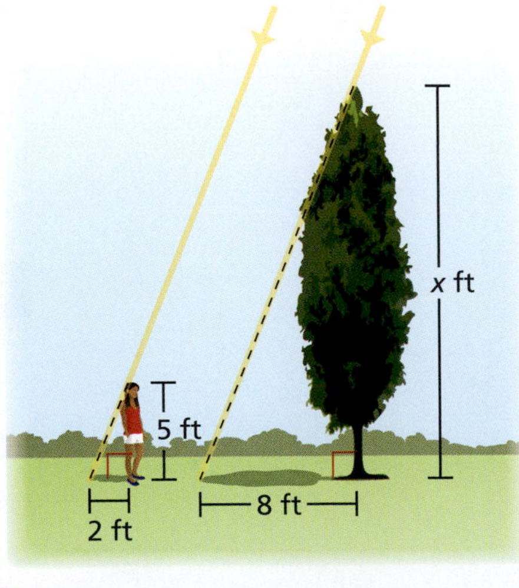

SOLUTION

Both triangles shown in the figure are right triangles. Besides the right angles, there is another pair of congruent angles. So, by the Angle-Angle Similarity Postulate, the two right triangles are similar. Write and solve a proportion.

$\frac{2}{8} = \frac{5}{x}$ Write a proportion.

$2x = 40$ Use the Cross Products Property.

$x = 20$ Division Property of Equality

The height of the tree is 20 feet.

Classroom Tip

In Example 5 on page 498, a surveyor determined the distance across a river. To complete that example, draw and measure a line segment from A' to B' that is parallel to AB. Then use the Angle-Angle Similarity Postulate to conclude that $\triangle ABC \sim \triangle A'B'C$. From this, you can write and solve a proportion to find the distance across the river.

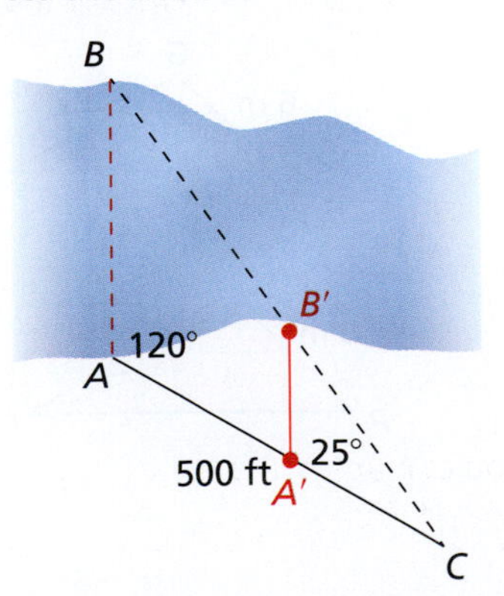

SAS and SSS Similarity Theorems

Side-Angle-Side (SAS) Similarity Theorem

If two side lengths of one triangle are proportional to two side lengths of another triangle, and the included angles are congruent, then the two triangles are similar.

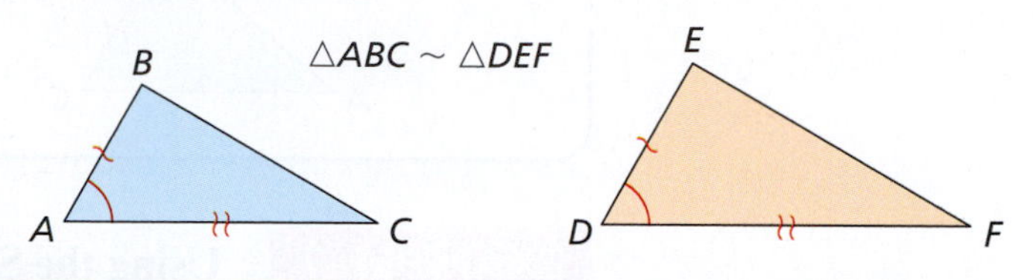

Classroom Tip

Just as tick marks indicate congruent sides, the marks shown at the right indicate when sides of triangles are proportional to each other.

EXAMPLE 5 Comparing Wings

How can you use the SAS Similarity Theorem to show that the triangles formed from the main wing and the tail wing of an F-16 jet are similar?

SOLUTION

There is more than one way to do this. One way would be to check that both triangles are right triangles. Next, measure the lengths of x, y, z, and w and check that

$$\frac{x}{z} = \frac{y}{w}.$$ Corresponding side lengths are proportional.

Then use the SAS Similarity Theorem to conclude that the two triangles are similar.

EXAMPLE 6 Using the SAS Similarity Theorem

Is $\triangle ABC$ similar to $\triangle A'BC'$? Explain your reasoning.

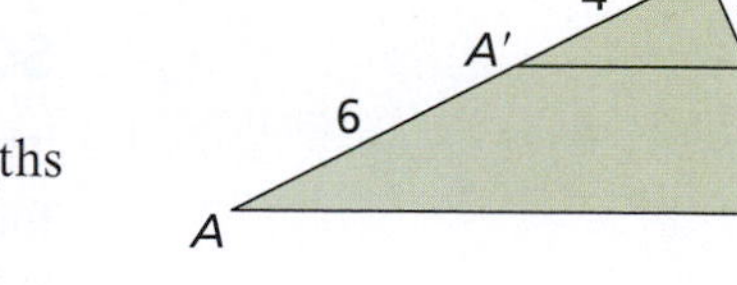

SOLUTION

Check whether the corresponding side lengths are proportional.

$$\frac{BA'}{BA} \stackrel{?}{=} \frac{BC'}{BC}$$ Write a proportion.

$$\frac{4}{4+6} \stackrel{?}{=} \frac{2}{2+3}$$ Substitute values.

$$\frac{4}{10} = \frac{2}{5}\ \checkmark$$ Corresponding side lengths are proportional.

Because the included angle of each triangle is B, you can apply the SAS Similarity Theorem to conclude that $\triangle ABC \sim \triangle A'BC'$.

mobil11/Shutterstock.com

Side-Side-Side (SSS) Similarity Theorem

If all three side lengths of one triangle are proportional to the corresponding side lengths of another triangle, then the two triangles are similar.

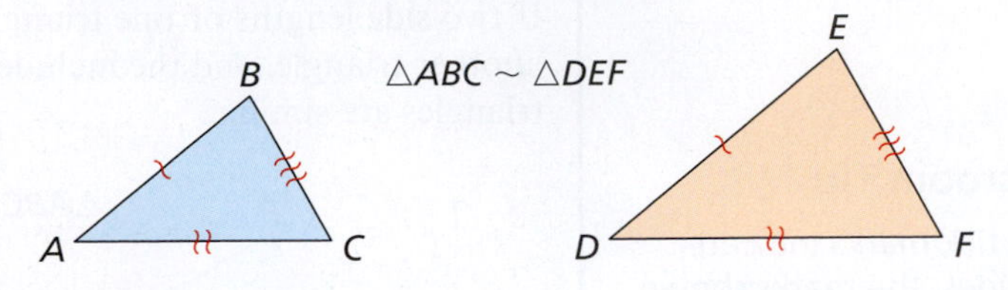

EXAMPLE 7 Using the SSS Similarity Theorem

You build a tetrahedron as shown at the right using a magnetic construction system. Is the larger triangle, $\triangle ABC$, similar to the smaller triangle, $\triangle DEF$? Explain your reasoning. Assume each rod has a length of 1 unit.

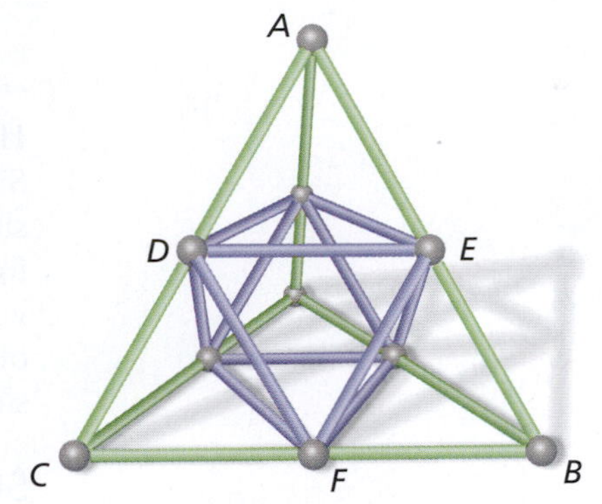

SOLUTION

The sides of $\triangle ABC$ are each 2 units long. The sides of $\triangle DEF$ are each 1 unit long. So, all three side lengths of $\triangle ABC$ are proportional to the corresponding side lengths of $\triangle DEF$. You can use the Side-Side-Side Similarity Theorem to conclude that the two triangles are similar.

EXAMPLE 8 Using the SSS Similarity Theorem

Is $\triangle ABC$ similar to $\triangle DEF$? Explain your reasoning.

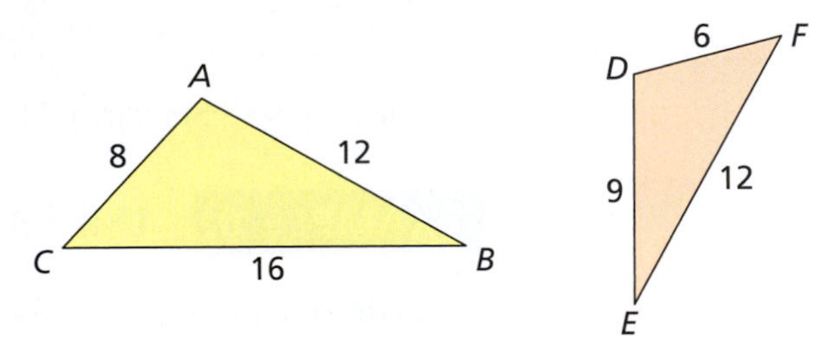

SOLUTION

The corresponding side lengths of two triangles must be proportional for the two triangles to be similar. Find the ratios of the corresponding side lengths: shortest to shortest, longest to longest, and so on.

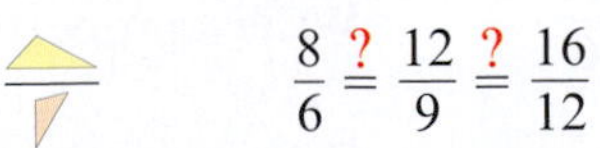

$$\frac{8}{6} \stackrel{?}{=} \frac{12}{9} \stackrel{?}{=} \frac{16}{12}$$

Each ratio simplifies to $\frac{4}{3}$. You can conclude that all three pairs of side lengths are proportional. So, it follows from the SSS Similarity Theorem that $\triangle ABC \sim \triangle DEF$.

Similar Polygons

Definition of Similar Polygons

Two polygons are **similar polygons** when their corresponding side lengths are proportional and their corresponding angles are congruent.

EXAMPLE 9 Determining Whether Two Polygons are Similar

Is $ABCD$ similar to $WXYZ$? Explain your reasoning.

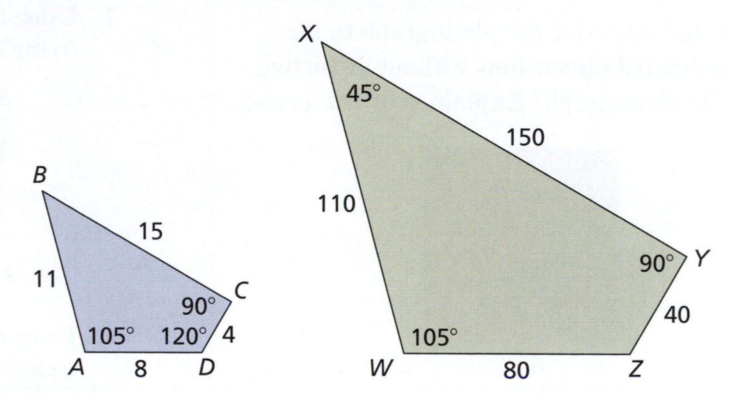

Mathematical Practices

Reason Abstractly and Quantitatively

MP2 Mathematically proficient students make sense of quantities and their relationships in problem situations.

SOLUTION

Recall from Section 10.2 that the sum of the angle measures of a quadrilateral is 360°. Using this fact, you can determine that the missing angle measures are 45° for angle B and 120° for angle Z. So, the corresponding angles of the quadrilaterals are congruent. Also, each side length of the larger quadrilateral is 10 times longer than the corresponding side length of the smaller quadrilateral. So, the side lengths are proportional.

Because the corresponding angles are congruent and the corresponding side lengths are proportional, $ABCD \sim WXYZ$.

EXAMPLE 10 Comparing Constellations

Are the quadrilateral portions of the Big Dipper and the Little Dipper constellations similar? Explain your reasoning.

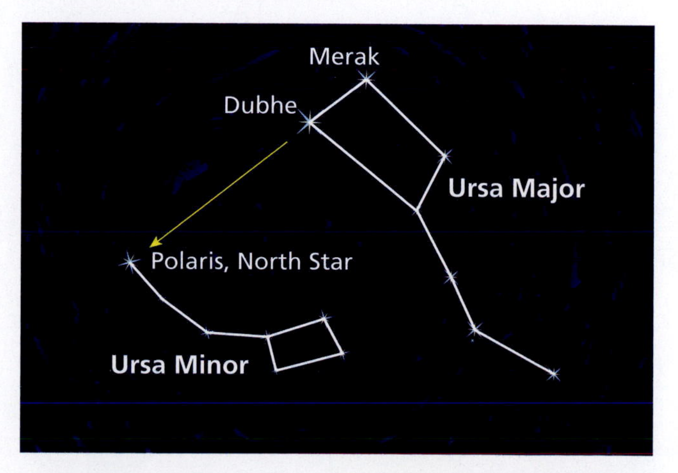

SOLUTION

Trace the quadrilateral portion of each constellation in the drawing and orient them so that corresponding sides align as shown. From this, you can see that the corresponding angles of the quadrilaterals are not congruent. So, the quadrilateral portions of the constellations are not similar.

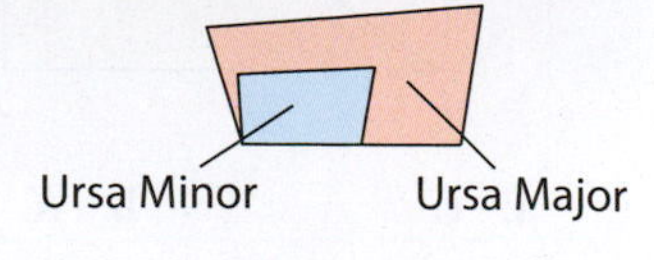

13.2 Exercises

Free step-by-step solutions to odd-numbered exercises at *MathematicalPractices.com*.

1. Writing a Solution Key Write a solution key for the activity on page 504. Describe a rubric for grading a student's work.

2. Grading the Activity In the activity on page 504, a student gave the answer below. The question is worth 10 points. Assign a grade to the answer. Why do you think the student erred?

Sample Student Work

Can you resize the photograph to the indicated dimensions without distorting the photograph? Explain your reasoning.

1.

6 in.

6 in.

a. 5 in. by 5 in.

b. 5 in. by 6 in.

c. 8 in. by 8 in.

(a) is not distorted because the sides are proportional. (b) and (c) are distorted because the sides are not proportional.

3. Using the Definition of Similar Triangles $\triangle ABC$ is similar to $\triangle DEF$. Find the measure of each angle.

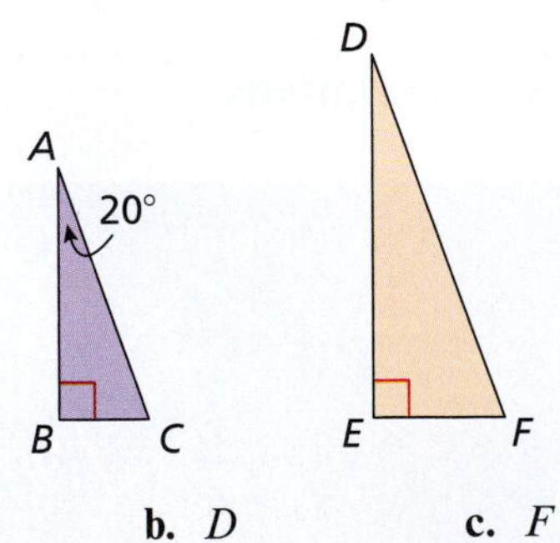

a. C **b.** D **c.** F

4. Using the Definition of Similar Triangles $\triangle GHJ$ is similar to $\triangle KLM$. Find the measure of each angle.

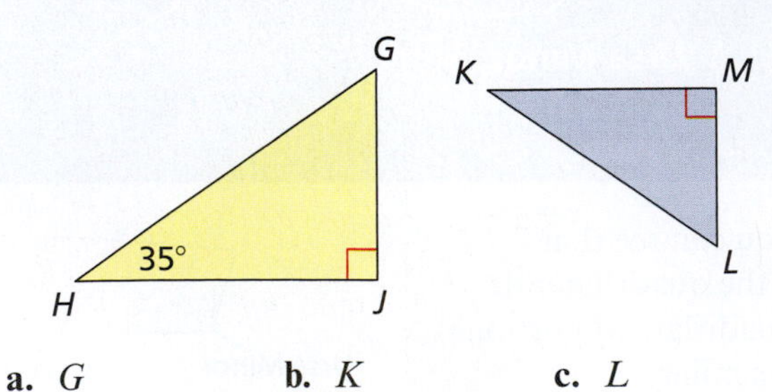

a. G **b.** K **c.** L

5. Solving Proportions Solve each proportion for x.

a. $\frac{3}{4} = \frac{x}{16}$ **b.** $\frac{x}{7} = \frac{4}{28}$

6. Solving Proportions Solve each proportion for x.

a. $\frac{3}{12} = \frac{5}{x}$ **b.** $\frac{5}{x} = \frac{40}{8}$

7. Using the Definition of Similar Triangles The triangles are similar. Find x and y.

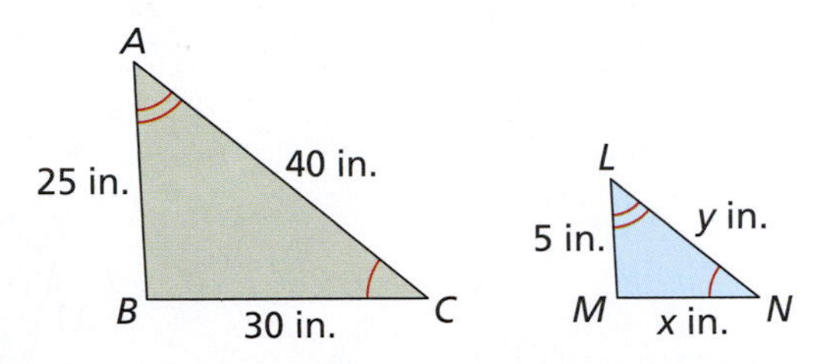

8. Using the Definition of Similar Triangles The triangles are similar. Find x and y.

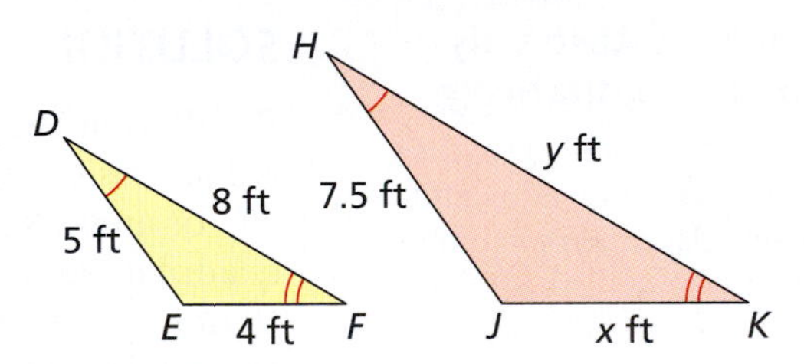

9. Similar Triangles Decide whether the triangles are similar. Explain your reasoning.

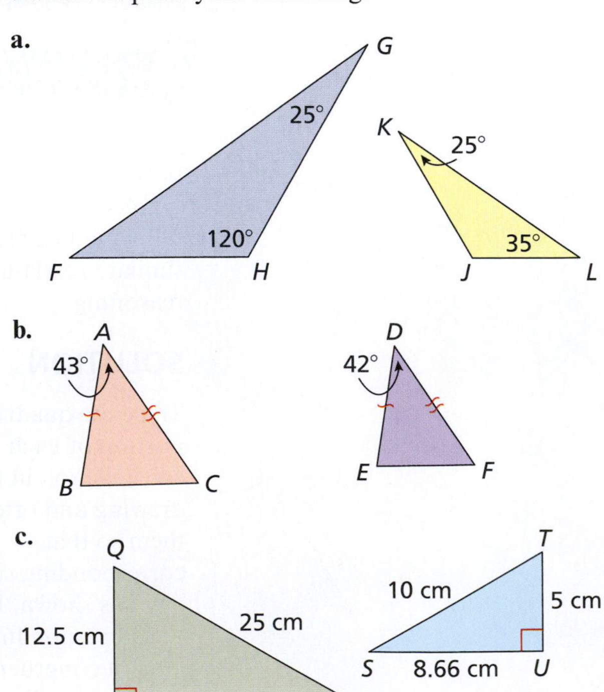

10. **Similar Triangles** Decide whether the triangles are similar. Explain your reasoning.

a.

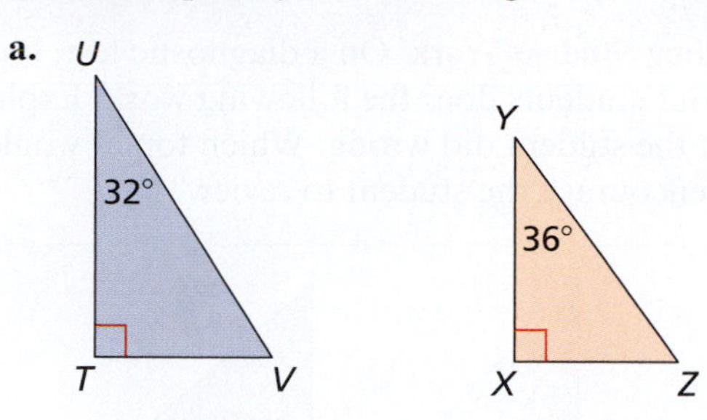

b.

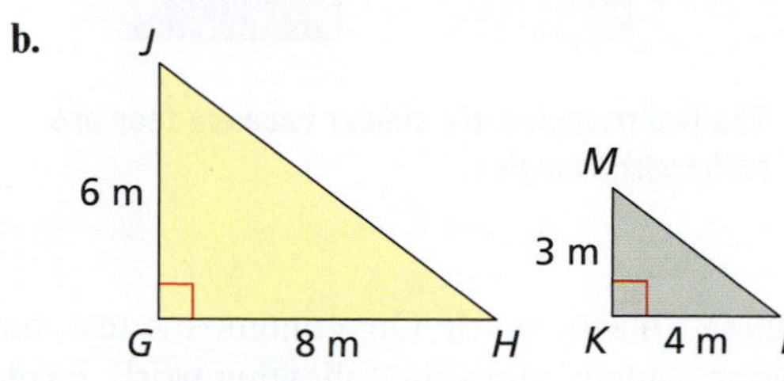

c.

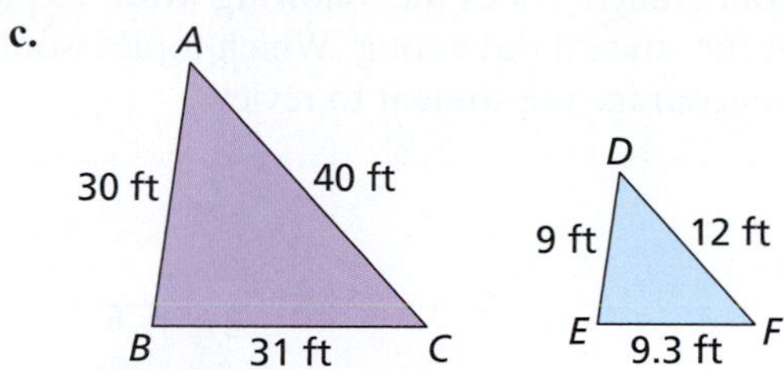

11. **Similar Polygons** Decide whether *ABCD* is similar to *EFGH*. Explain your reasoning.

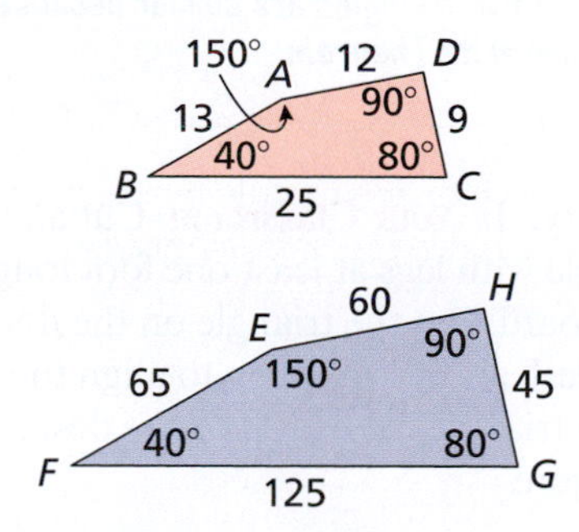

12. **Similar Polygons** Decide whether *JKLM* is similar to *NPRS*. Explain your reasoning.

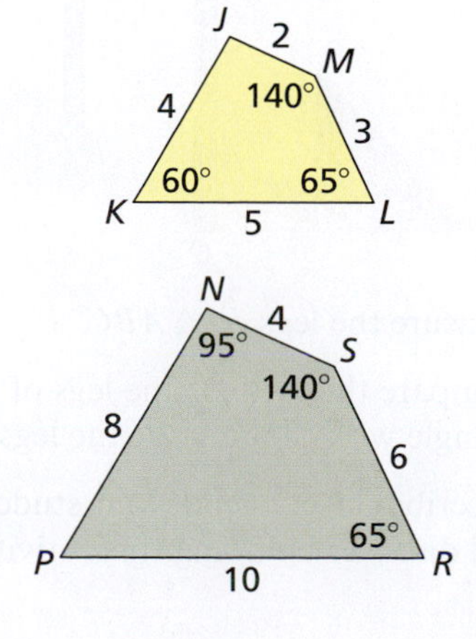

13. **Similar Polygons** *ABCD* is similar to *EFGH*. Find the length of each side.

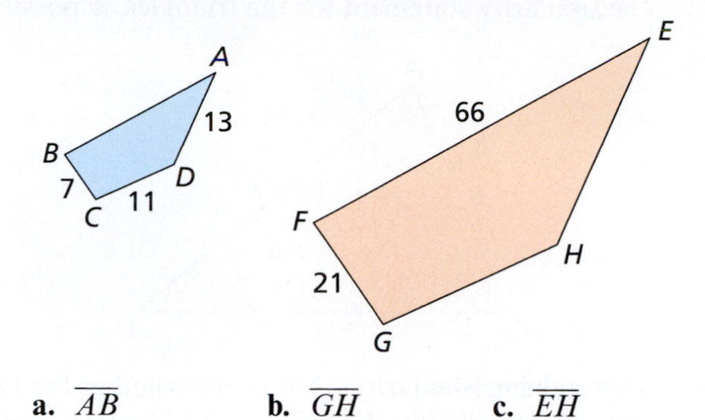

a. $\overline{AB}$ **b.** $\overline{GH}$ **c.** $\overline{EH}$

14. **Similar Polygons** *JKLM* is similar to *NPRS*. Find the length of each side.

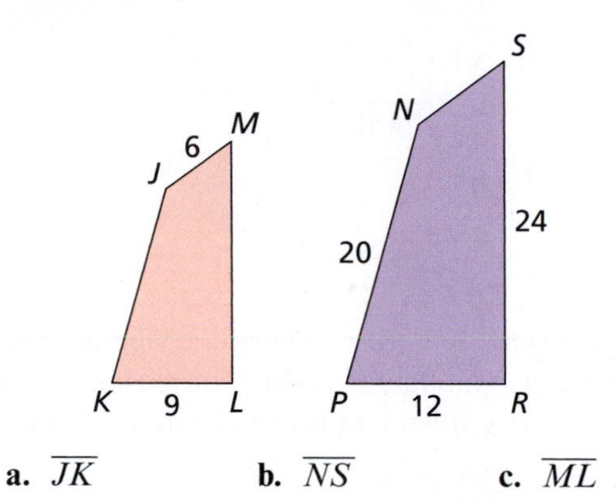

a. $\overline{JK}$ **b.** $\overline{NS}$ **c.** $\overline{ML}$

15. **Using a Property of Similar Triangles** $\triangle RST$ is similar to $\triangle NLT$. Solve for *x*.

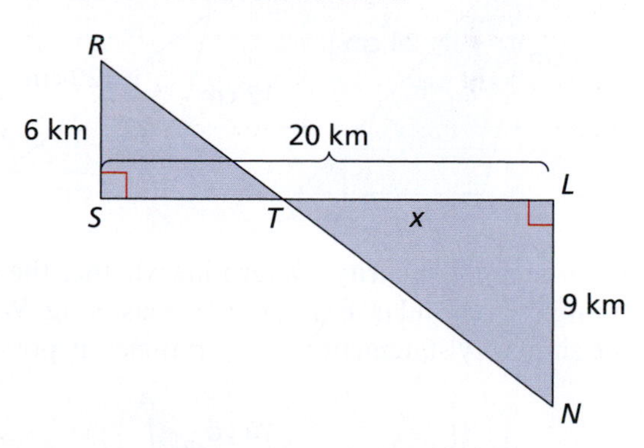

16. **Using a Property of Similar Triangles** $\triangle EDC$ is similar to $\triangle ABC$. Solve for *x*.

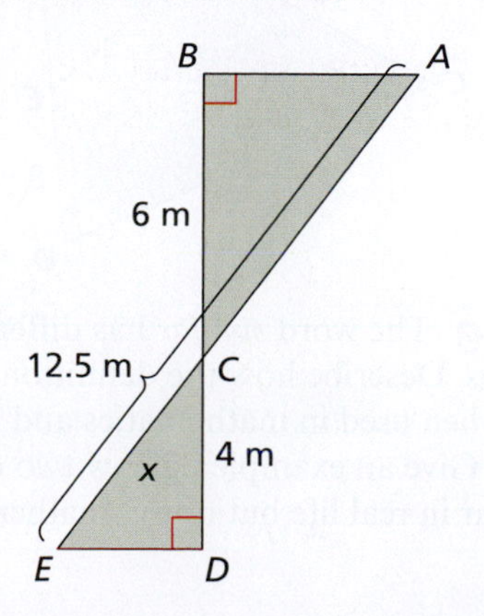

17. Determining Similarity Determine whether the two triangles are similar. Explain your reasoning. Write the similarity statement for the triangles, if possible.

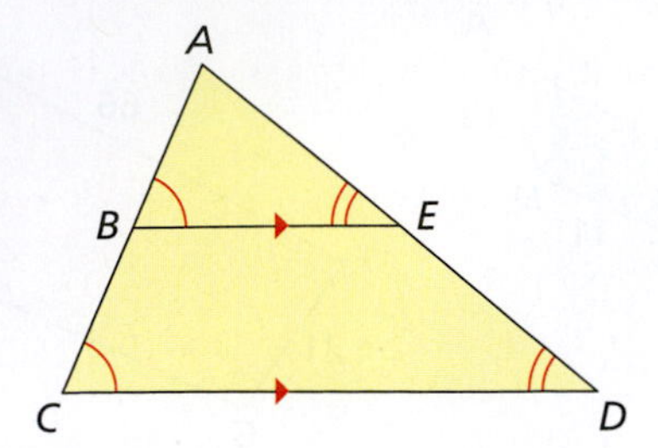

18. Determining Similarity Determine whether the two triangles are similar. Explain your reasoning. Write the similarity statement for the triangles, if possible.

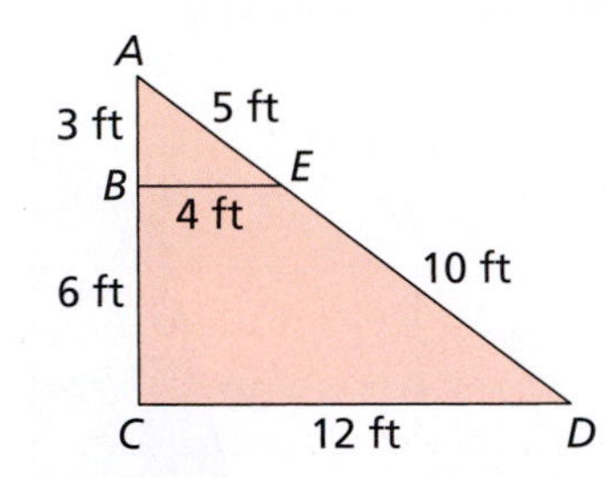

19. Determining Similarity Determine whether the two triangles are similar. Explain your reasoning. Write the similarity statement for the triangles, if possible.

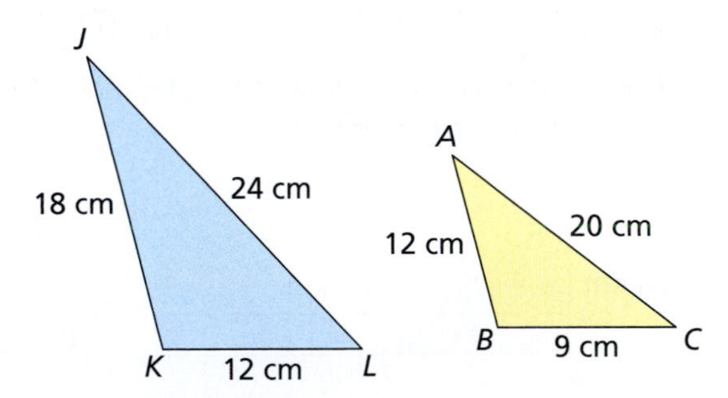

20. Determining Similarity Determine whether the two triangles are similar. Explain your reasoning. Write the similarity statement for the triangles, if possible.

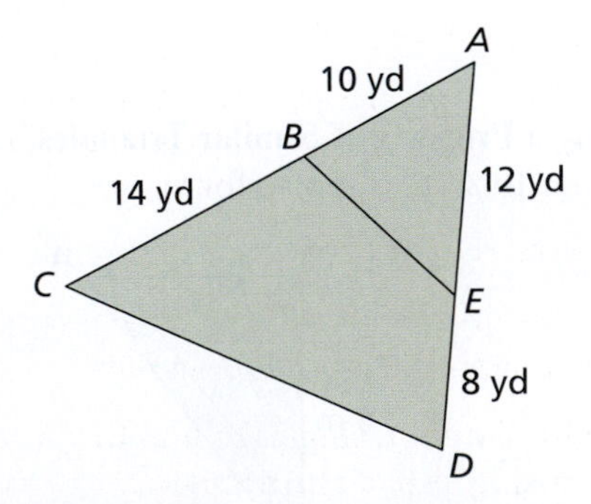

21. Writing The word *similar* has different meanings. Describe how the definition of similar differs when used in mathematics and when used in real life. Give an example of how two objects might be similar in real life but not in mathematics.

22. Writing Identify two careers that use similar figures. Explain how similar figures are used.

23. Grading Student Work On a diagnostic test, one of your students does the following work. Explain what the student did wrong. Which topics would you encourage the student to review?

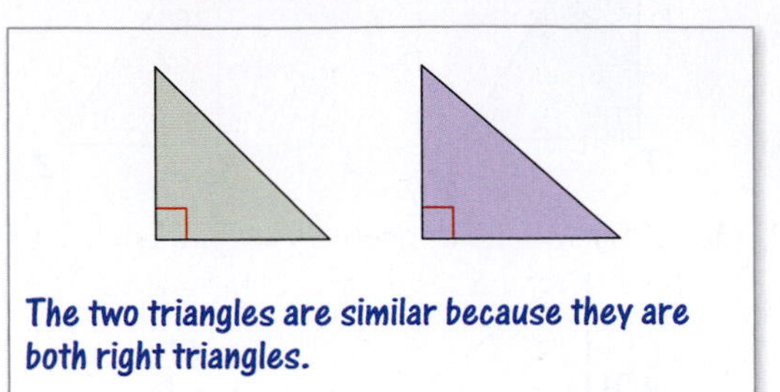

24. Grading Student Work On a diagnostic test, one of your students does the following work. Explain what the student did wrong. Which topics would you encourage the student to review?

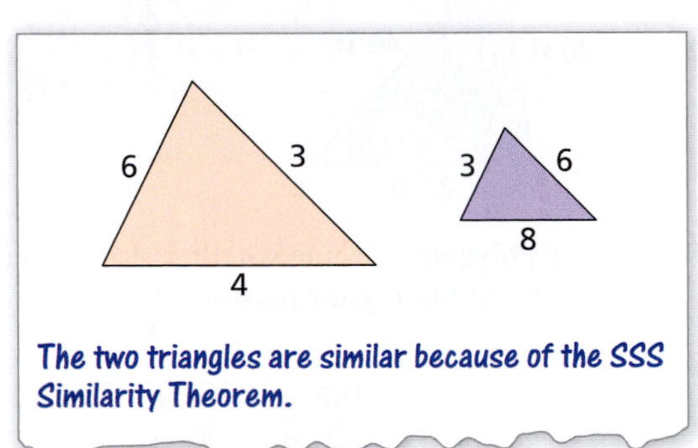

25. Activity: In Your Classroom Cut an isosceles right triangle with legs at least one foot long out of cardboard. Put the triangle on the floor in front of a desk and use a string to align the hypotenuse of the triangle to the top of the desk, as shown in the figure.

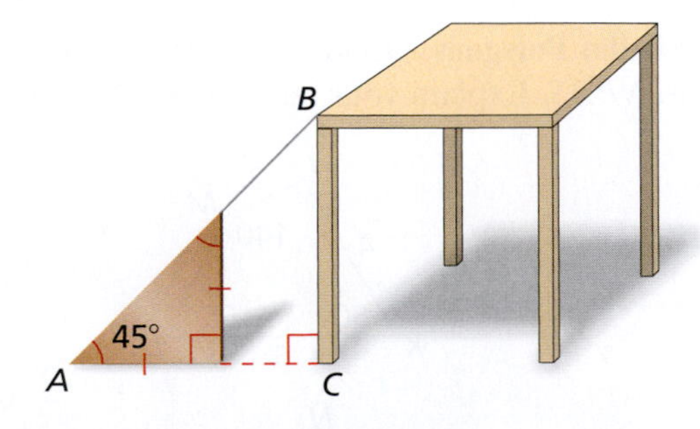

a. Measure the legs of $\triangle ABC$.

b. Compare the ratio of the legs of the cardboard triangle with the ratio of the legs of $\triangle ABC$.

c. Describe the concepts your students will practice and discover by doing this activity.

26. **Activity: In Your Classroom** To find the height of your classroom wall using the triangle and desk in Exercise 25, put the triangle on the desk. Use a line of sight to align the hypotenuse of the triangle with the top of the wall.

a. Measure the distance from the vertex of the triangle to the wall.

b. What other measurement is needed to find the height of the classroom wall? Explain your reasoning.

c. Describe the concepts your students will practice and discover by doing this activity.

27. **Measuring Indirectly** To find out how tall your house is, you stand next to your house so the tip of your shadow coincides with the tip of the shadow of your house.

a. Are the two triangles similar? Explain.

b. How tall is your house?

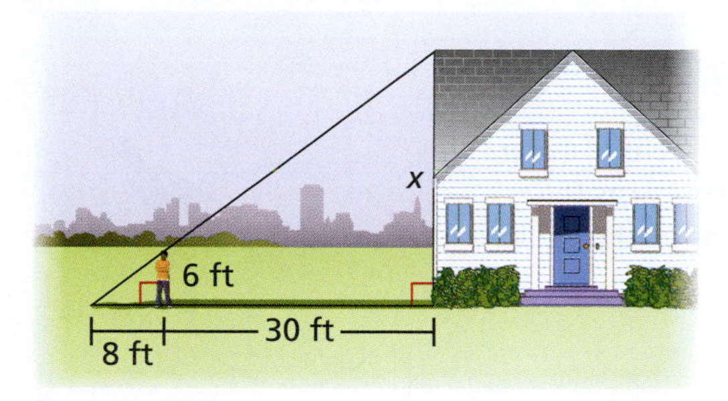

28. **Comparing Distance** When the sun is directly overhead, a leaning tree casts a shadow 10 feet long. The base of the tree is 21 feet away from the house. The homeowner takes a yardstick and touches one end to the tree and the other end to the ground and sees that the yardstick is 18 inches from the base of the tree.

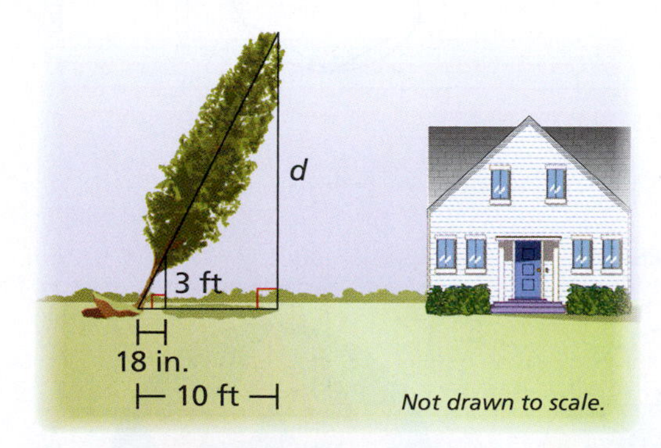

a. Find the distance d from the top of the tree to the ground.

b. Use the Pythagorean Theorem to find the length of the tree. If the tree falls, will it hit the house?

29. **Another Way to Measure Distance** An architect places a mirror 25 feet from the base of a building and stands back until the top of the building is visible in the mirror.

a. Are the two triangles formed similar? Explain your reasoning.

b. Find the height of the building.

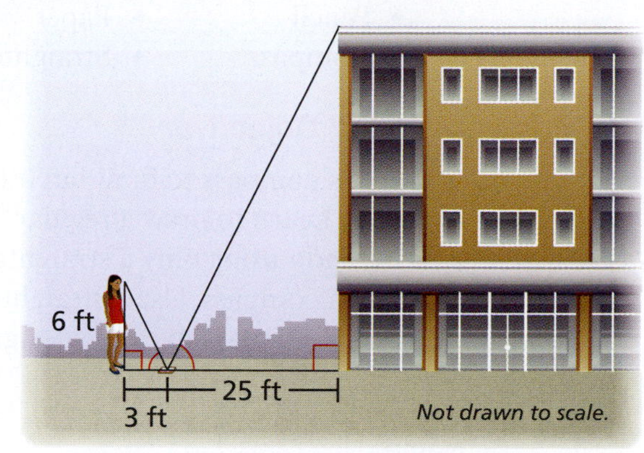

30. **Another Way to Measure Distance** A zip line runs from the top of a platform on the edge of a pond to the ground across the pond. A camp counselor places a mirror on the edge of the pond across from the platform and stands back until the top of the platform is visible in the mirror.

a. Are the two triangles formed similar? Explain your reasoning.

b. Find the length of the zip line.

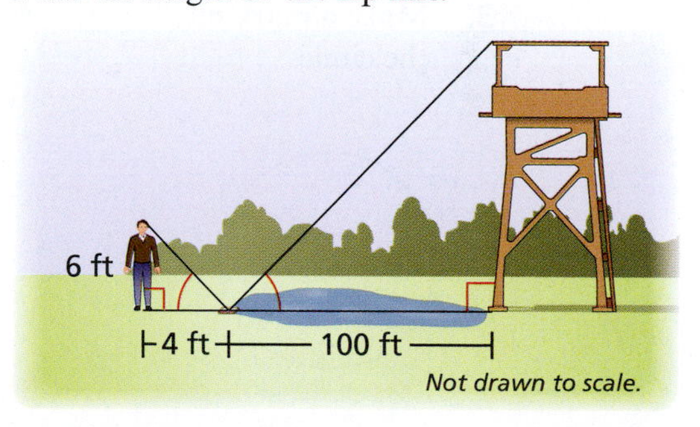

31. **Comparing Distance** You can ride your bike from your house to your friend's house on a straight path or on the road. To ride on the road, you ride 2 miles south, 3 miles east, then 4 miles south again.

a. Draw a map of the path and roads.

b. How do you know the triangles formed by the path and the roads are similar?

c. Use similar triangles to find the distances of the missing legs of the triangles.

d. How much shorter is it to ride on the path than to ride on the road?

Activity: Making Designs with a Straightedge and a Compass

Materials:

- Pencil
- Compass
- Paper
- Straightedge

Learning Objective:

Use a compass to draw circles and arcs. Learn to draw a regular hexagon using only a straightedge and a compass. Use a straightedge and a compass to create designs.

Name ______________________________

Use a straightedge and a compass to draw a regular hexagon.

1. Choose the length of each side of the hexagon. Set the compass to that length.

2. Draw a circle using the length as the radius of the circle.

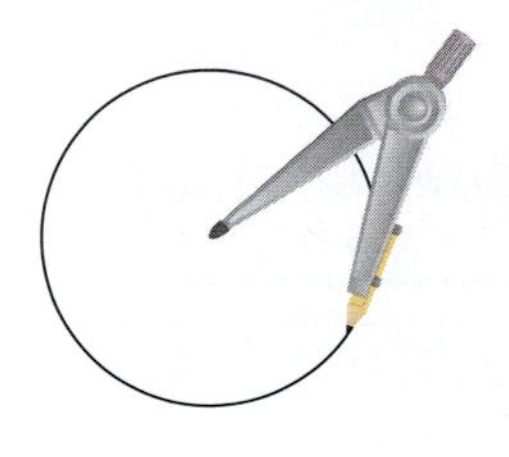

3. Make a mark on the circle.

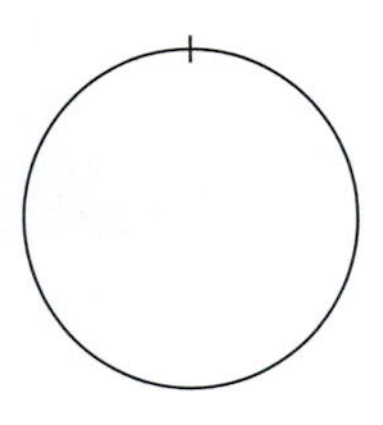

4. Start at the mark and make a series of arcs around the circle.

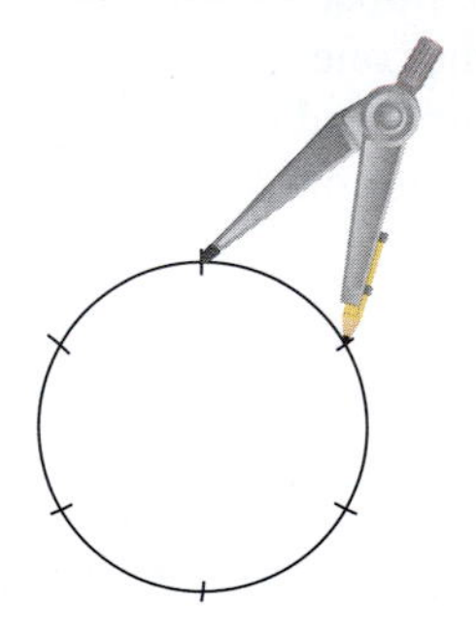

5. Connect the arcs using a straightedge.

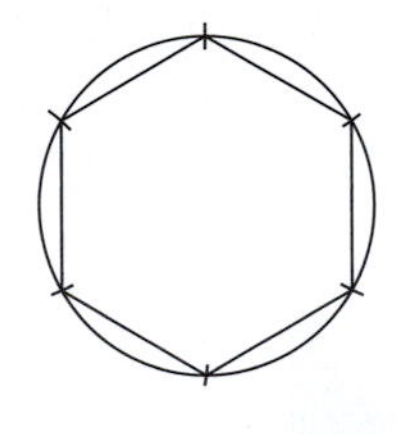

Use a straightedge and compass to create the design.

6.

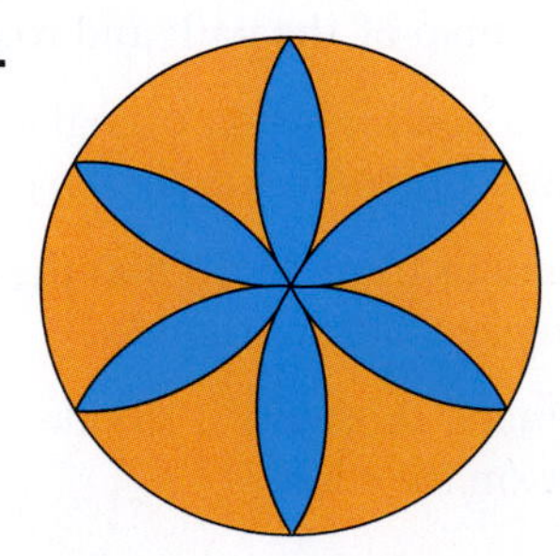

7.

8.

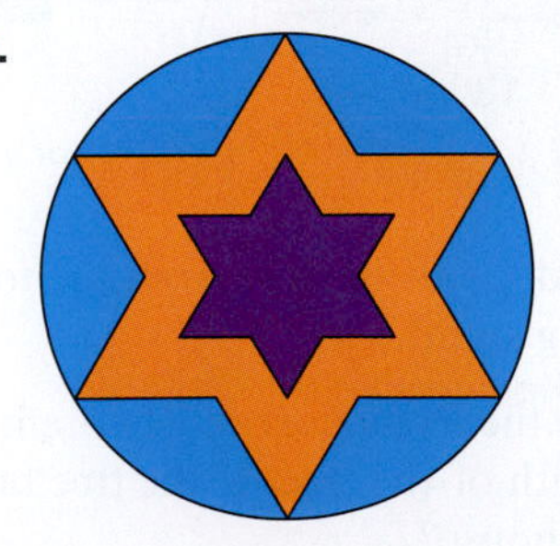

13.3 Construction Techniques

Standards

Grades 3–5

Geometry

Students should draw and identify lines and angles.

Grades 6–8

Geometry

Students should draw and construct geometrical figures.

- Use a straightedge and a compass to copy a line segment and an angle.
- Use a straightedge and a compass in parallel constructions.
- Use a straightedge and a compass in bisector constructions.

Copying Segments and Angles

A **construction** is a geometric drawing made with only a compass and a straightedge, no protractors or rulers.

EXAMPLE 1 Constructing a Copy of a Line Segment

1. To use a compass and a straightedge to construct a copy of $\overline{AB}$, draw a ray, beginning at C.

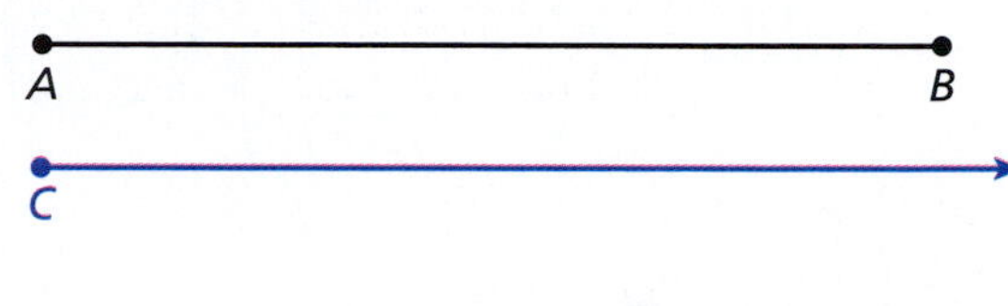

2. Use the compass to measure the length of $\overline{AB}$.
3. Transfer the length of $\overline{AB}$ to the ray by placing the needle on C and marking D.

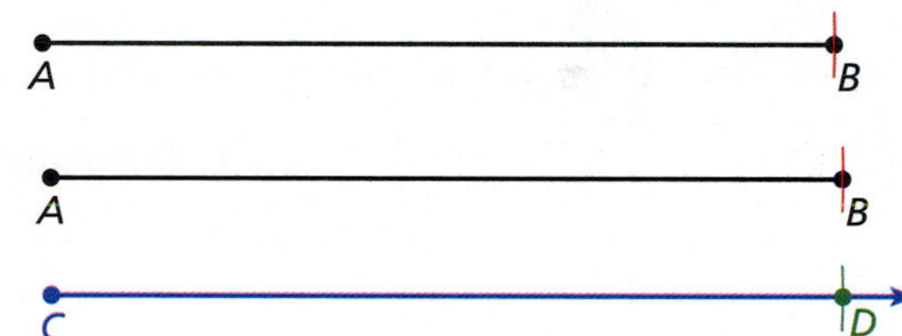

EXAMPLE 2 Constructing a Copy of an Angle

1. To use a compass and a straightedge to construct a copy of angle E, draw a ray, beginning at G.

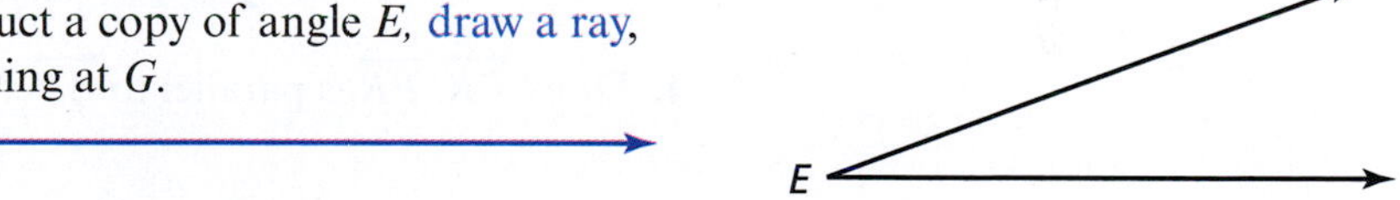

2. Use the compass to draw an arc with center at E. Copy the arc with center at G.

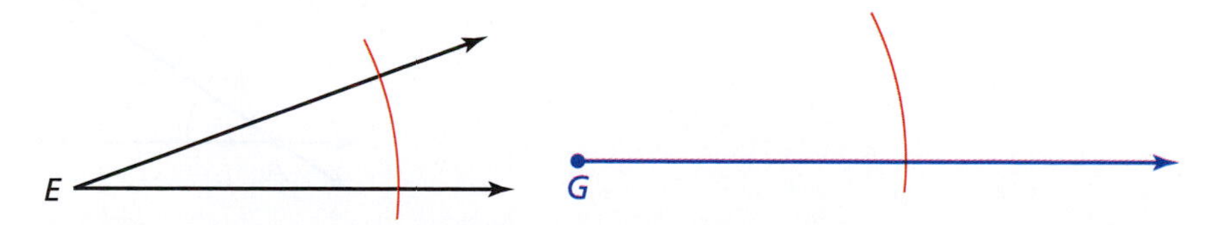

3. Measure the length between the sides of angle E where it intercepts the arc. Use this length to draw an arc with center at H.

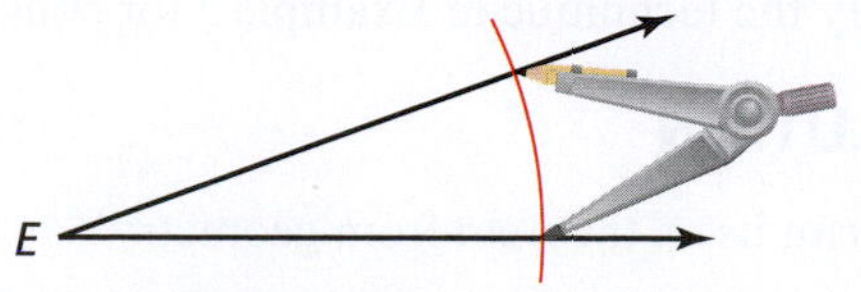

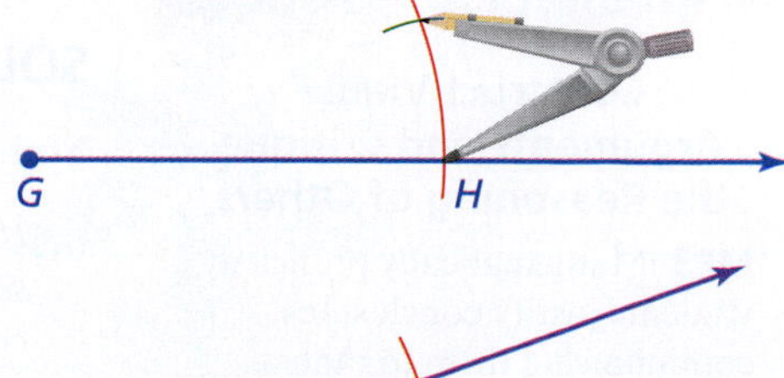

4. Use the straightedge to draw a ray from G through the intersection of the arcs.

G H

Classroom Tip

There are two types of compasses. A traditional compass has a needle (for the center of the arc or circle) and a pencil to draw the arc or circle.

Traditional compass

For teachers who are concerned about students having pointed objects, there are compasses that do not have needles.

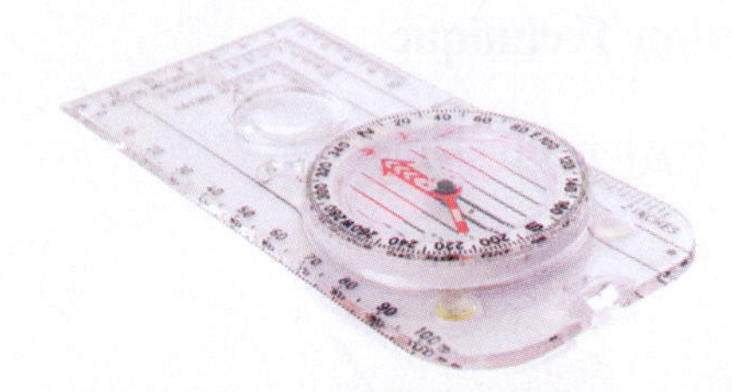

Needleless compass

Constructions Involving Parallel Lines

EXAMPLE 3 Constructing a Parallel Line

1. To use a compass and a straightedge to construct a line that is parallel to $\overleftrightarrow{AB}$ through point P, draw a line through P that intersects $\overleftrightarrow{AB}$ at A.

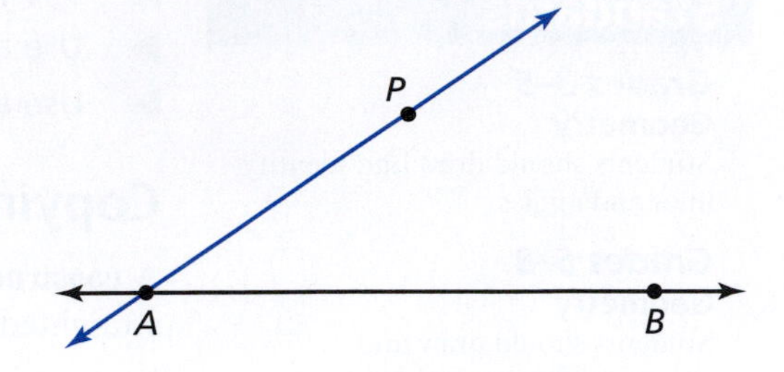

2. Use the compass to draw an arc with center at A so that it crosses both lines.

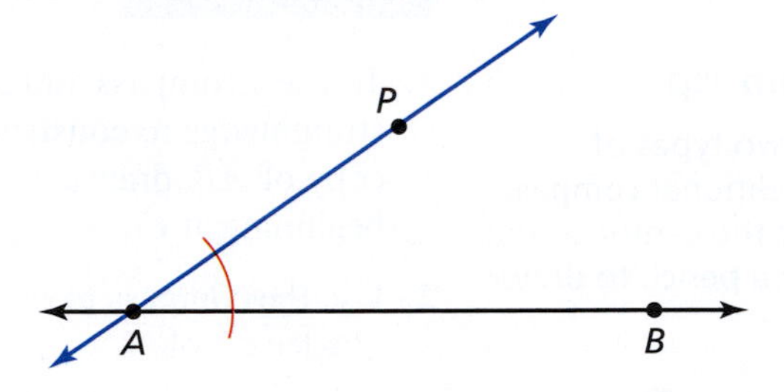

3. Copy angle PAB on $\overleftrightarrow{AP}$. Label the new angle.

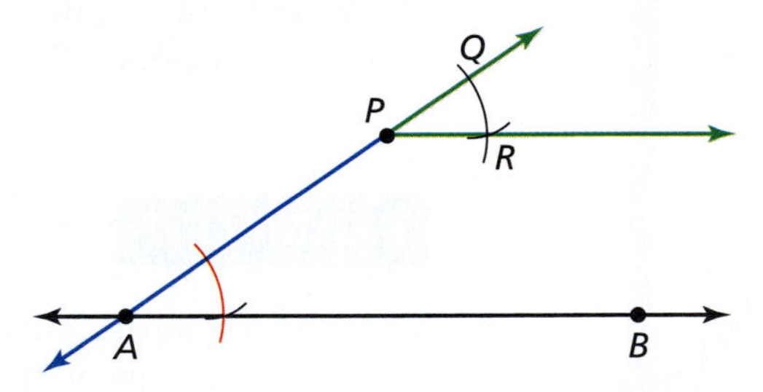

4. Draw $\overleftrightarrow{PR}$. $\overleftrightarrow{PR}$ is parallel to $\overleftrightarrow{AB}$.

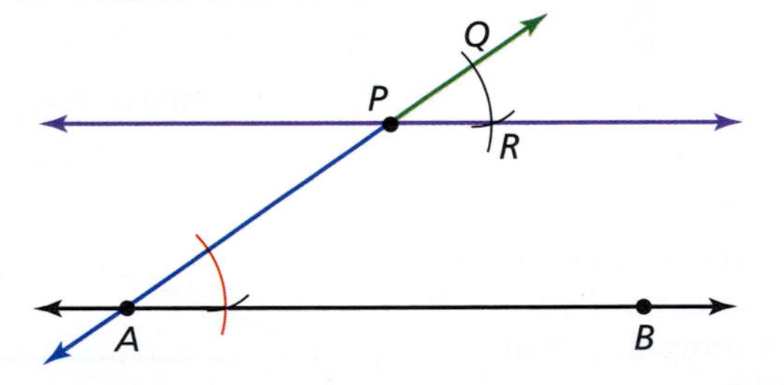

EXAMPLE 4 Justifying a Construction Technique

Justify the technique in Example 3 for constructing a parallel line.

SOLUTION

You can use a theorem from geometry.

If two lines are crossed by a transversal so that the corresponding angles are congruent, then the two lines are parallel.

From this theorem, you can see that the construction technique is valid because the original angle and the copied angle are congruent.

B
A
Corresponding angles

Mathematical Practices

Construct Viable Arguments and Critique the Reasoning of Others

MP3 Mathematically proficient students justify conclusions, communicate them to others, and respond to the arguments of others.

EXAMPLE 5 Constructing a Parallelogram

1. To use a compass and a straightedge to construct a parallelogram, draw an angle. Label it as shown below.

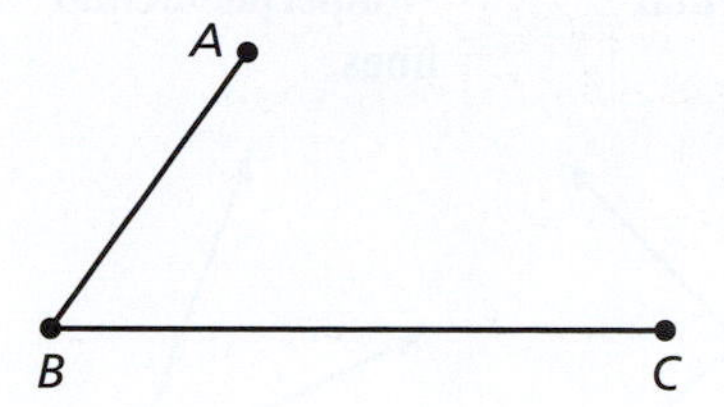

2. Use the compass to measure the length of $\overline{BC}$. Use this length to draw an arc with center at A.

3. Use the compass to measure the length of $\overline{AB}$. Use this length to draw an arc with center at C.

4. Label the intersection of the two arcs as D. Draw $\overline{AD}$ and $\overline{CD}$. $ABCD$ is a parallelogram.

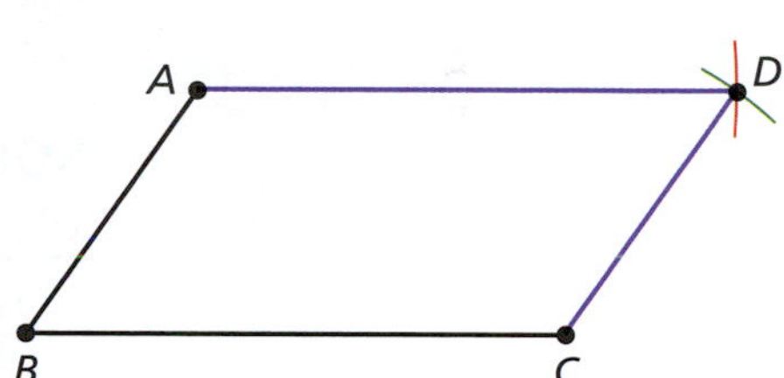

EXAMPLE 6 Justifying a Construction Technique

In Section 13.1, it was stated that opposite sides of a parallelogram are congruent. The converse of this statement is also true.

If the opposite sides of a quadrilateral are congruent, then the quadrilateral is a parallelogram.

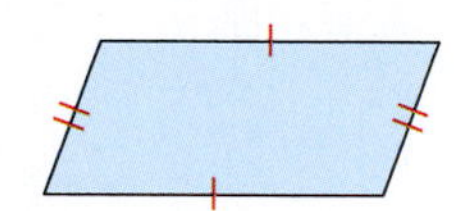

From this statement, the construction technique in Example 5 is valid because the construction produced a quadrilateral whose opposite sides are congruent.

EXAMPLE 7 Parallelogram in Real Life

A storm damages a rectangular door frame as shown. Is the resulting quadrilateral still a parallelogram? Explain your reasoning.

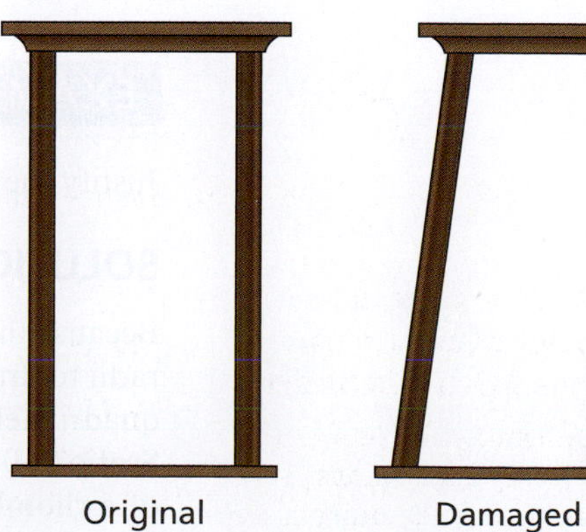

SOLUTION

The opposite sides of the original door frame are congruent. After the storm damage, the sides of the frame shift, but the lengths remain the same. So, the resulting quadrilateral is still a parallelogram.

Constructions: Line Segment and Angle Bisectors

Two lines are **perpendicular** when they intersect at right angles. A line is a **perpendicular bisector** of a line segment when it is perpendicular to the line segment and divides the line segment into two congruent parts.

Perpendicular lines

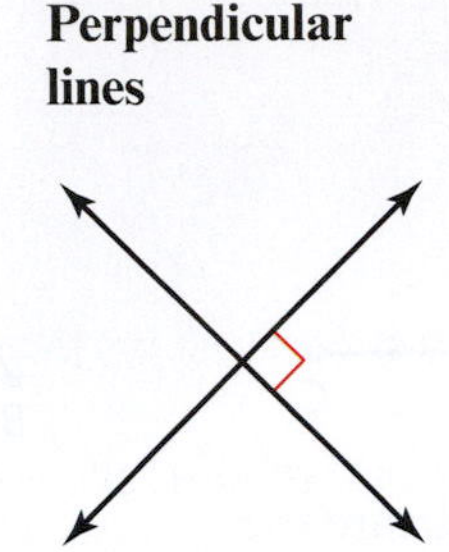

Nonperpendicular lines

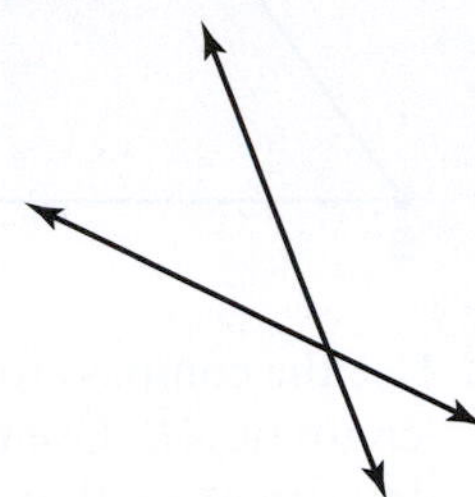

Perpendicular bisector

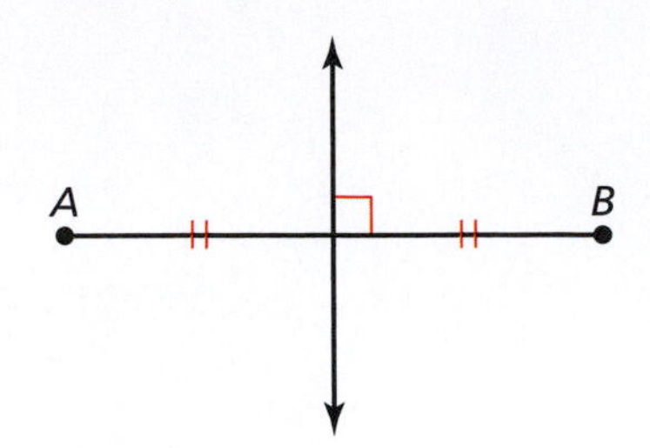

EXAMPLE 8 Constructing a Perpendicular Bisector

Use a compass and a straightedge to construct the perpendicular bisector of $\overline{AB}$.

SOLUTION

1. Set the radius of the compass so that it is greater than one-half of the length of $\overline{AB}$. Draw arcs with centers at A above and below $\overline{AB}$.

2. Using the same radius as in Step 1, draw arcs with centers at B above and below $\overline{AB}$.

3. Draw the line that passes through the intersections of the arcs. This line is the perpendicular bisector of $\overline{AB}$.

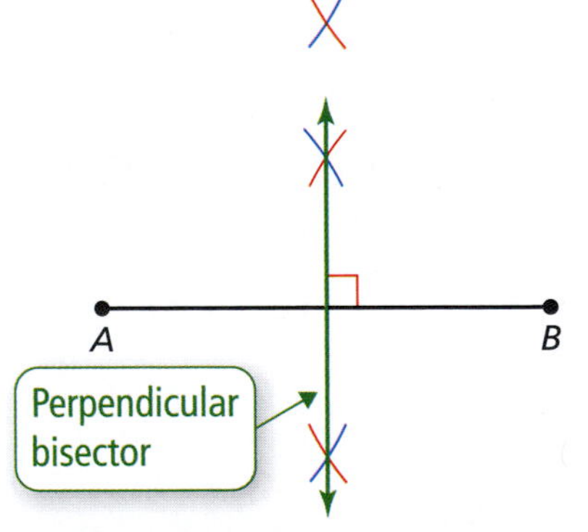

EXAMPLE 9 Justifying a Construction Technique

Justify the technique in Example 8 for constructing a perpendicular bisector.

SOLUTION

Because the construction technique uses equal radii to draw the arcs, you can conclude that quadrilateral $ACBD$ is a rhombus. In Section 10.2, you learned that the diagonals of a rhombus intersect at right angles and they bisect each other. So, it follows that $\overleftrightarrow{CD}$ is the perpendicular bisector of $\overline{AB}$.

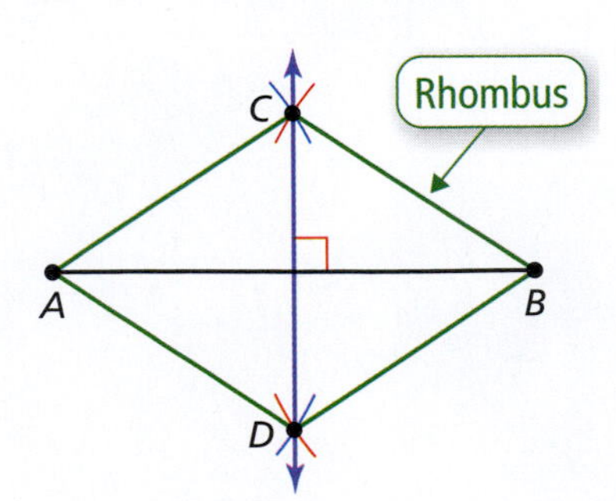

The **angle bisector** of an angle is the ray that divides the angle into two congruent angles.

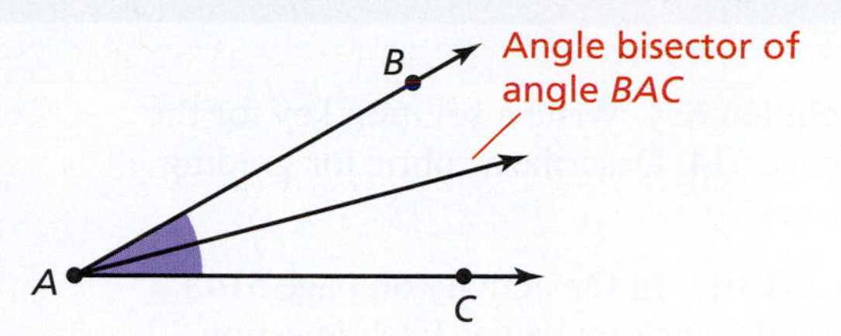

EXAMPLE 10 Constructing an Angle Bisector

Use a compass and a straightedge to construct the angle bisector of angle J.

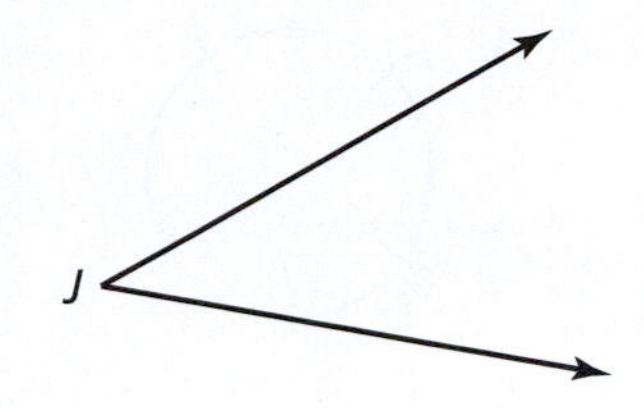

SOLUTION

1. Draw an arc with center at J. Label the intersections of the arc and the rays of the angle as K and L.
2. Draw an arc with center at K. Then use the same radius to draw an arc with center at L.

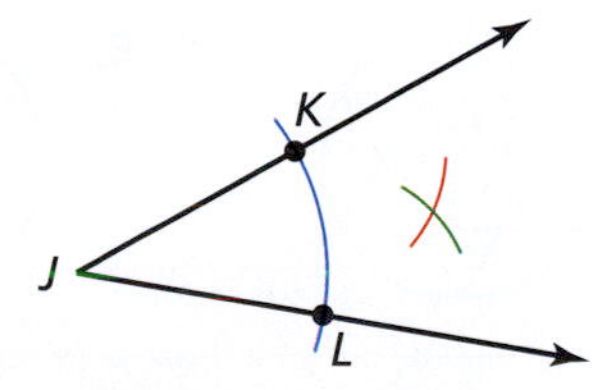

3. Label the point of intersection of the two arcs from Step 2 as M. Draw the ray from J through M. This ray is the angle bisector of angle KJL.

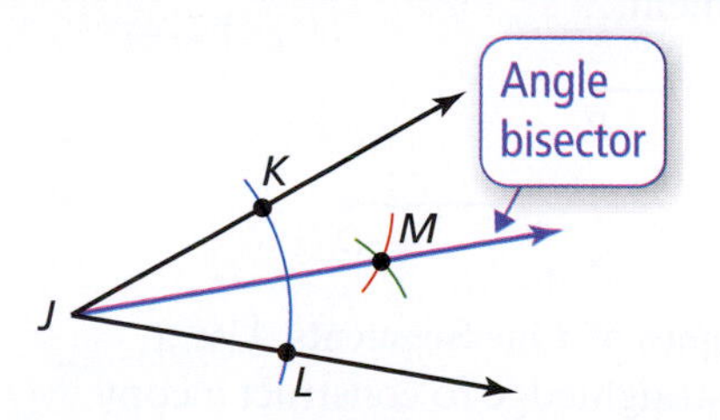

EXAMPLE 11 Justifying a Construction Technique

Justify the technique in Example 10 for constructing an angle bisector.

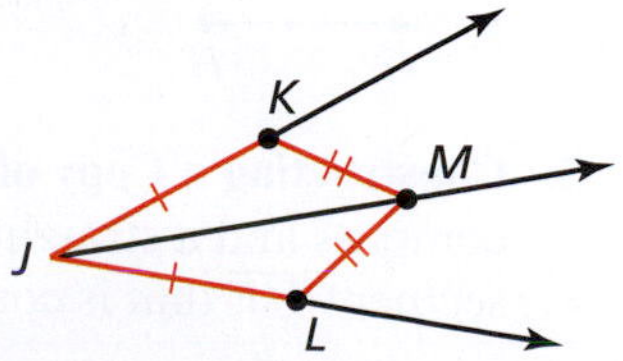

SOLUTION

Because the construction technique uses equal radii to draw the second and third arcs, you can conclude that quadrilateral $JKML$ is a kite. Using the definition of a kite and the SSS Congruence Postulate, you can conclude that $\triangle JKM$ and $\triangle JLM$ are congruent. Because the triangles are congruent, the corresponding angles KJM andLJM are congruent. So, $\overrightarrow{JM}$ bisects angle KJL.

13.3 Exercises

Free step-by-step solutions to odd-numbered exercises at MathematicalPractices.com.

1. **Writing a Solution Key** Write a solution key for the activity on page 514. Describe a rubric for grading a student's work.

2. **Grading the Activity** In the activity on page 514, a student gave the answers below. Each question is worth 10 points. Assign a grade to each answer. For those that are incorrect, why do you think the student erred?

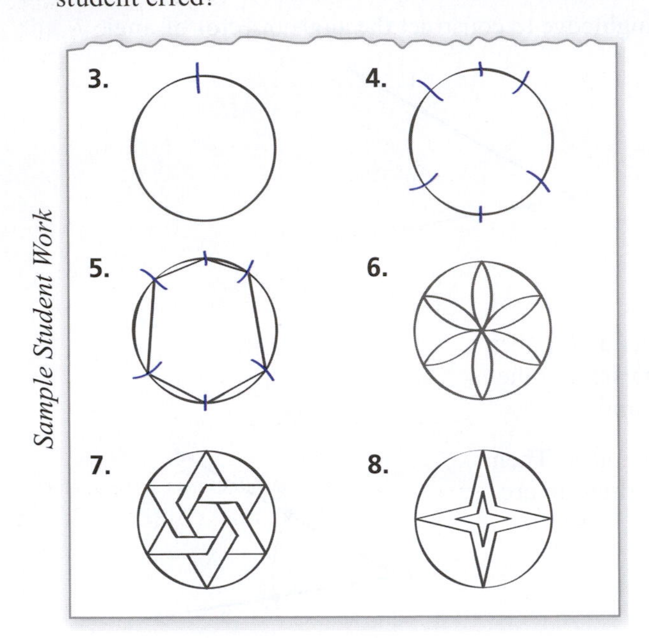

3. **Constructing Copies of Line Segments** Use a compass and a straightedge to construct a copy of each line segment.

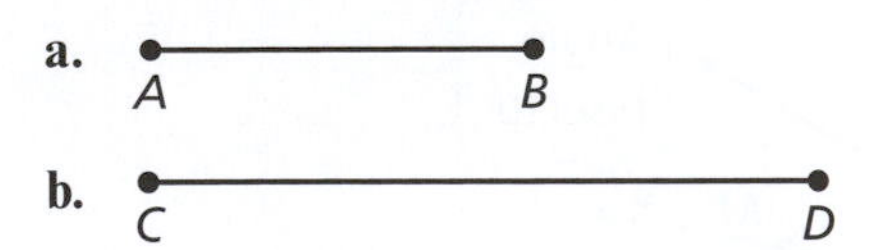

4. **Constructing Copies of Line Segments** Use a compass and a straightedge to construct a copy of each line segment.

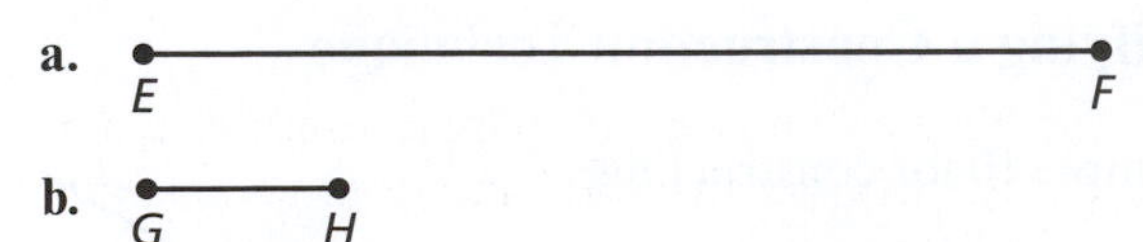

5. **Constructing a Copy of a Line Segment** Use a compass and a straightedge to construct a line segment $\overline{UV}$ that is congruent to $\overline{YZ}$.

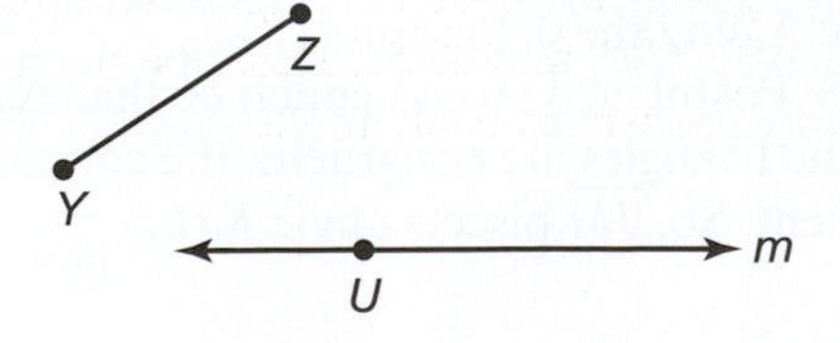

6. **Constructing a Copy of a Line Segment** Use a compass and a straightedge to construct a line segment $\overline{WX}$ that is congruent to $\overline{JK}$.

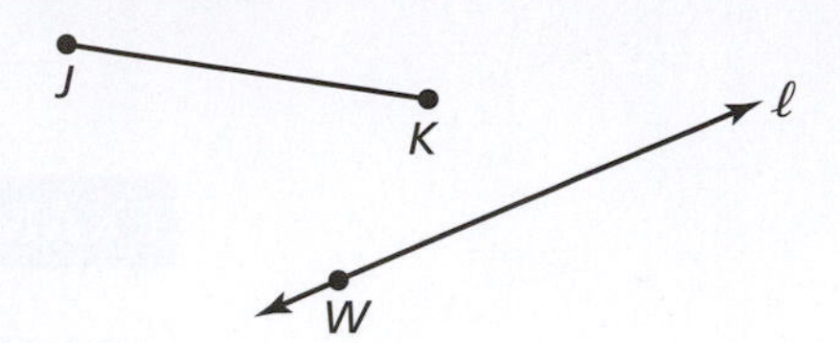

7. **Constructing Copies of Angles** Use a compass and a straightedge to construct a copy of each angle.

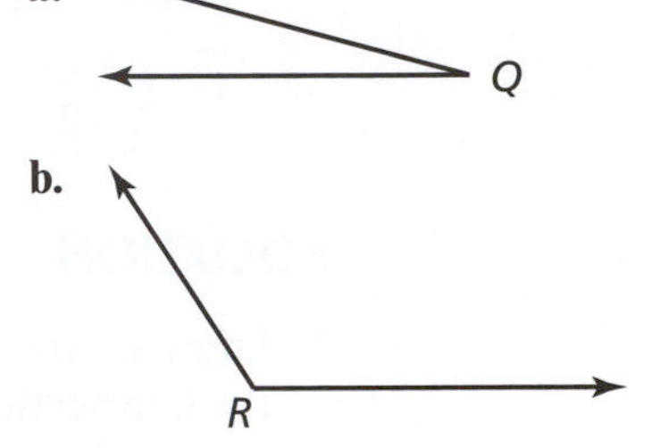

8. **Constructing Copies of Angles** Use a compass and a straightedge to construct a copy of each angle.

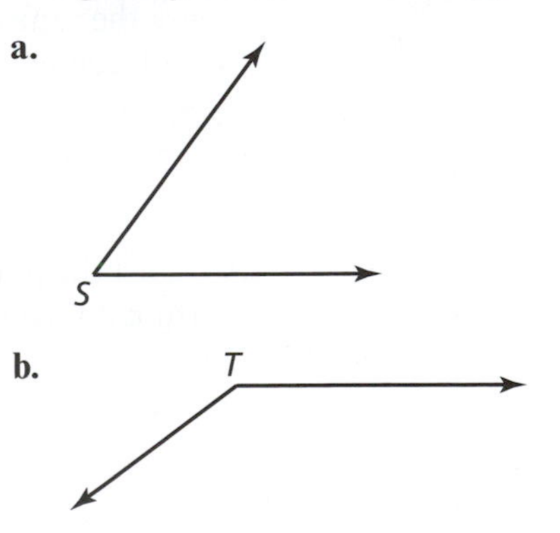

9. **Constructing a Copy of an Angle** Use a compass and a straightedge to construct an angle F that is congruent to angle D.

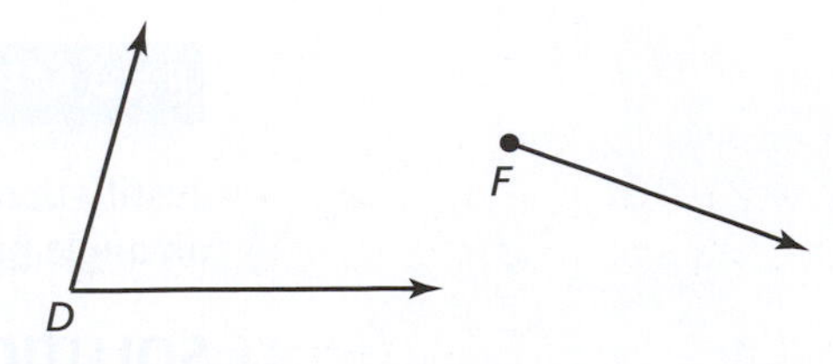

10. **Constructing a Copy of an Angle** Use a compass and a straightedge to construct an angle M that is congruent to angle N.

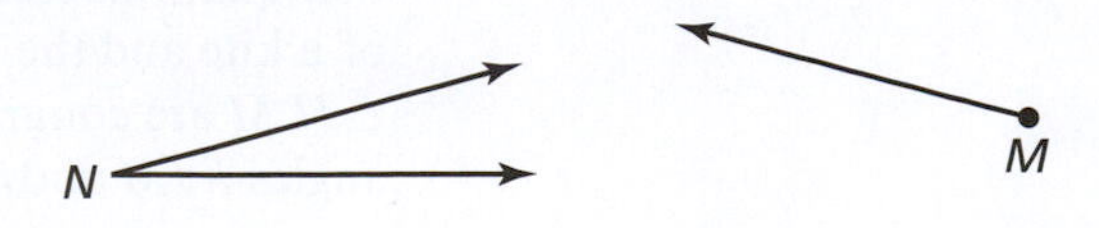

11. Constructing Parallel Lines Use a compass and a straightedge to construct a line that is parallel to the given line through point P.

a. P J K

b.

12. Constructing Parallel Lines Use a compass and a straightedge to construct a line that is parallel to the given line through point P.

a. Z Y P

b.

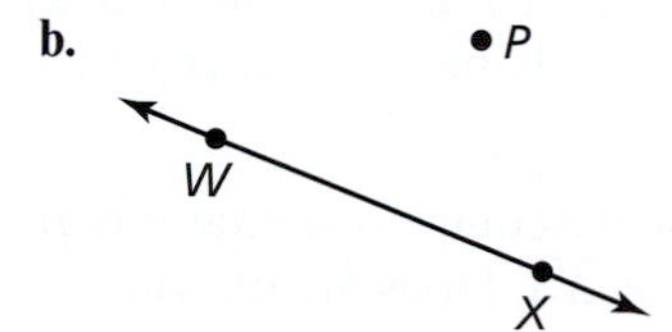

13. Constructing a Parallelogram Use a compass and a straightedge to construct a parallelogram using $\angle NML$.

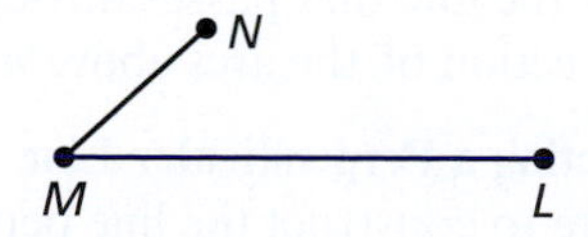

14. Constructing a Parallelogram Use a compass and a straightedge to construct a parallelogram using $\angle UVW$.

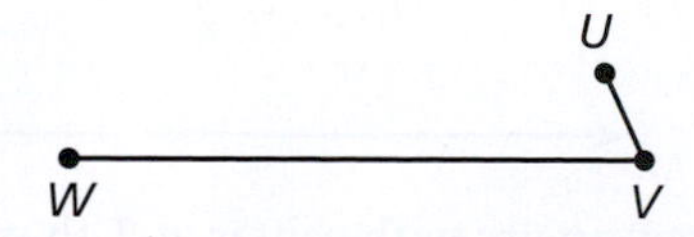

15. Constructing Perpendicular Bisectors Use a compass and a straightedge to construct the perpendicular bisector of each line segment.

a. S R

b.

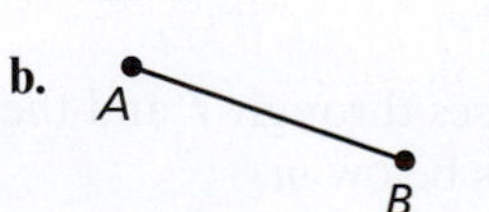

16. Constructing Perpendicular Bisectors Use a compass and a straightedge to construct the perpendicular bisector of each line segment.

a. F G

b. L K

17. Constructing Angle Bisectors Use a compass and a straightedge to construct the angle bisector of each angle.

a.

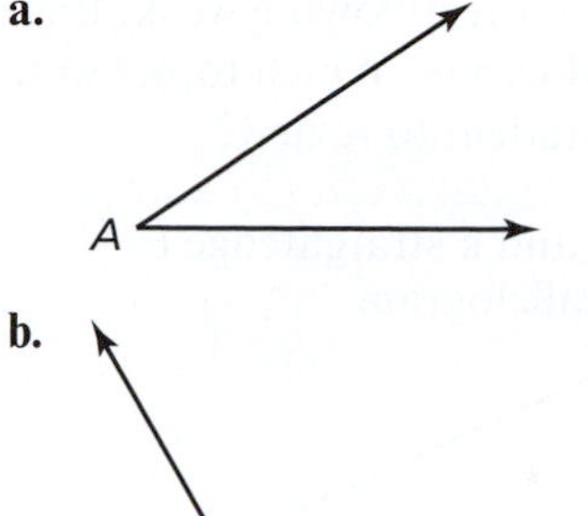

b. B

18. Constructing Angle Bisectors Use a compass and a straightedge to construct the angle bisector of each angle.

a. C

b.

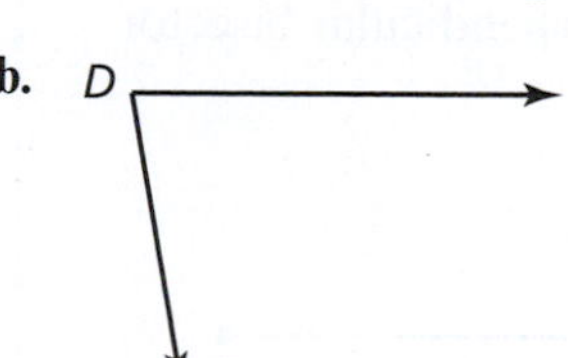

19. Finding a Midpoint Use a compass and a straightedge to find the midpoint M of $\overline{LN}$.

L N

20. Dividing a Line Segment Use a compass and a straightedge to divide $\overline{PQ}$ into four congruent line segments.

P Q

21. Constructing Angles Use a compass and a straightedge to construct angles with measures of 90°, 45°, 135°, and 67.5°. Describe the process you used to construct each angle.

22. **Constructing Angles** Use the angles below to construct an angle with each measure. Describe the process you used to construct each angle.

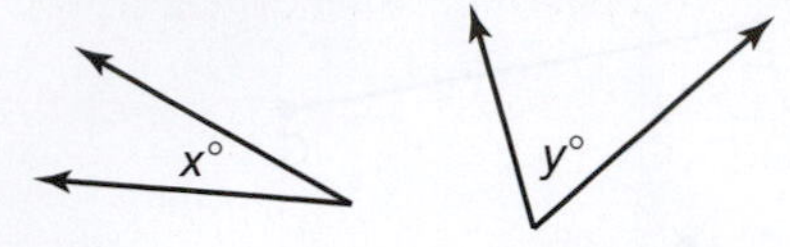

a. $(y - x)^\circ$

b. $(3x + y)^\circ$

c. $\left(\dfrac{x + y}{2}\right)^\circ$

23. **Grading Student Work** On a diagnostic test, one of your students does the following work. Explain what the student did wrong. Which topics would you encourage the student to review?

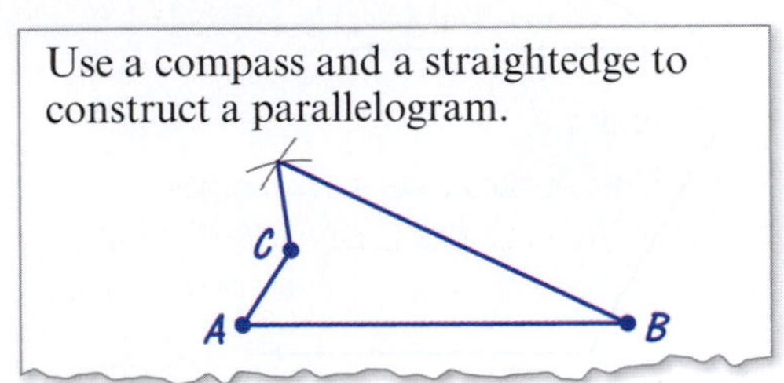
Use a compass and a straightedge to construct a parallelogram.

24. **Grading Student Work** On a diagnostic test, one of your students does the following work. Explain what the student did wrong. Which topics would you encourage the student to review?

Use a compass and a straightedge to construct the perpendicular bisector of $\overline{AB}$.

A B

25. **Creating Designs** Use a compass and a straightedge to create the design on each stained glass window.

a.

b.

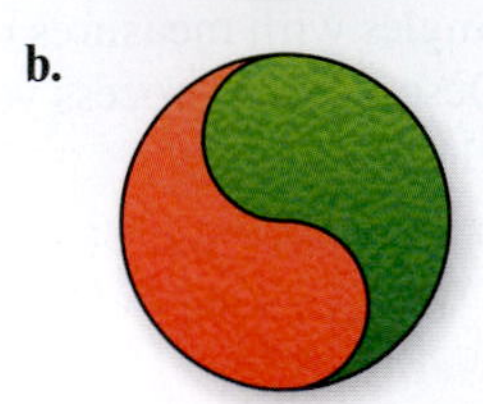

26. **Creating Designs** Use a compass and a straightedge to create the design on each stained glass window.

a.

b.

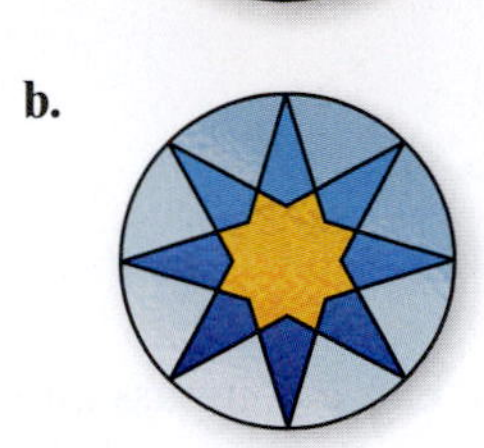

27. **Constructing a Perpendicular Line** Follow the procedure to construct the line perpendicular to line n through point S using only a compass and a straightedge.

i. Draw an arc with center at S that intersects n at two different points. Label these intersections as X and Y.

ii. Set the radius of the compass so that it is greater than the length of $\overline{XS}$. Draw an arc with center at X above n.

iii. Using the same radius as in Step (ii), draw an arc with center at Y above n.

iv. Draw the line that passes through S and the intersection of the arcs above n.

28. **Constructing a Perpendicular Line** Follow the procedure to construct the line perpendicular to line m through point T using only a compass and a straightedge.

i. Draw an arc with center at T that intersects m at two different points. Label these intersections as U and V.

ii. Using the same radius as in Step (i), draw an arc with center at U below m.

iii. Using the same radius as in Step (i), draw an arc with center at V below m.

iv. Draw the line that passes through T and the intersection of the arcs below m.

29. **Writing** Justify the technique in Exercise 27 for constructing a perpendicular line.

30. *Writing* Justify the technique in Exercise 28 for constructing a perpendicular line.

31. **Constructing Perpendicular and Parallel Lines** Refer to the figure below.

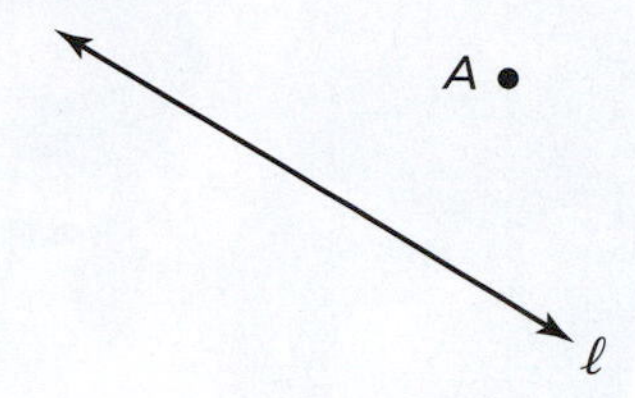

a. Use the procedure described in Exercise 28 to construct the line perpendicular to line ℓ through point A. Label this line j.

b. Construct a line parallel to ℓ through A. Label this line k.

c. Describe the relationship between lines j and k. Explain your reasoning.

32. **Constructing Perpendicular and Parallel Lines** Refer to the figure below.

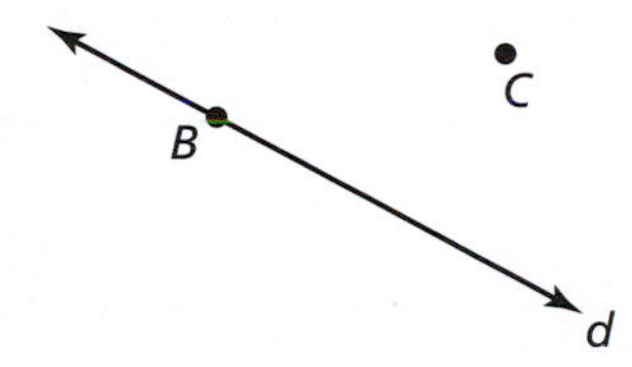

a. Use the procedure described in Exercise 27 to construct the line perpendicular to line d through point B. Label this line e.

b. Use the procedure described in Exercise 28 to construct the line perpendicular to e through point C. Label this line f.

c. Describe the relationship between lines d and f. Explain your reasoning.

33. **Constructing a Square** Use a compass and a straightedge to construct a square with diagonal length equal to the length of $\overline{EF}$. Explain your construction technique.

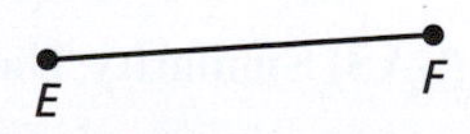

34. **Constructing a Triangle** Use a compass and a straightedge to construct an equilateral triangle with side lengths equal to the length of $\overline{JK}$. Explain your construction technique.

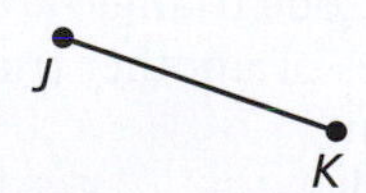

35. *Writing* Describe how to use a compass and a straightedge to construct a 60° angle. Then justify your technique.

36. *Writing* Describe how you can use a compass and a straightedge to draw a single line that divides an equilateral triangle into two congruent right triangles. Then justify your technique.

37. **Copying a Triangle: In Your Classroom** Refer to the figure below.

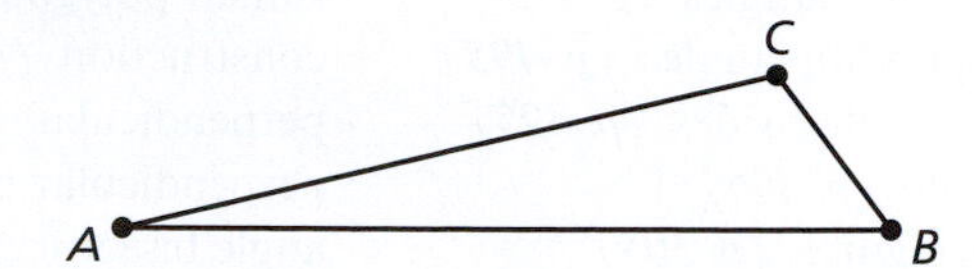

a. Follow the procedure to construct a copy of $\triangle ABC$.

i. Draw a ray, beginning at X.

ii. Construct a line segment $\overline{XY}$ on the ray such that $\overline{XY}$ is congruent to $\overline{AB}$.

iii. Set the radius of the compass to the length of $\overline{AC}$ and draw an arc with center at X above $\overline{XY}$.

iv. Set the radius of the compass to the length of $\overline{BC}$ and draw an arc with center at Y above $\overline{XY}$. Label the intersection of the arcs as Z.

v. Draw $\overline{XZ}$ and $\overline{YZ}$.

b. Which congruence postulate justifies this technique for constructing a copy of a triangle?

c. Write procedures for constructing a copy of $\triangle ABC$ that you could use with your students to reinforce the other congruence postulates.

38. **Constructing an Octagon: In Your Classroom** A polygon is *inscribed* in a circle when every vertex of the polygon is on the circle.

a. Explain why the endpoints of diameters $\overline{AB}$ and $\overline{DE}$ in the figure are vertices of a square. Then construct the inscribed square.

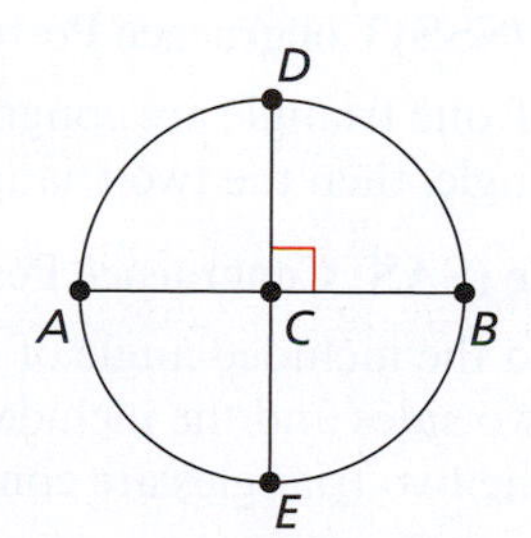

b. Construct the perpendicular bisector of $\overline{DB}$. Label the intersection of the perpendicular bisector and the circle as F.

c. Draw $\overline{BF}$. Explain why $\overline{BF}$ is the side of a regular octagon inscribed in the circle.

d. Describe the instructions you would give your students to construct a regular octagon using a compass and a straightedge.

Chapter Summary

Chapter Vocabulary

congruent figures *(p. 495)*
congruent triangles *(p. 495)*
corresponding angles *(p. 495)*
corresponding sides *(p. 495)*
postulate *(p. 496)*
similar figures *(p. 505)*
similar triangles *(p. 505)*
similar polygons *(p. 509)*
construction *(p. 515)*
perpendicular *(p. 518)*
perpendicular bisector *(p. 518)*
angle bisector *(p. 519)*

Chapter Learning Objectives

Review Exercises

13.1 Congruence of Triangles *(page 495)* — 1–24

- Understand the definition of congruent triangles.
- Use the SSS, SAS, and ASA Congruence Postulates.
- Use congruent triangles to explore properties of parallelograms.

13.2 Similarity *(page 505)* — 25–42

- Understand the definition of similar triangles.
- Use the AA Similarity Postulate
- Use the SAS and SSS Similarity Theorems.
- Understand the definition of similarity as it applies to other polygons.

13.3 Construction Techniques *(page 515)* — 43–60

- Use a straightedge and a compass to copy a line segment and an angle.
- Use a straightedge and a compass in parallel constructions.
- Use a straightedge and a compass in bisector constructions.

Important Concepts and Formulas

Side-Side-Side (SSS) Congruence Postulate

If three sides of one triangle are congruent to three sides of another triangle, then the two triangles are congruent.

Side-Angle-Side (SAS) Congruence Postulate

If two sides and the included angle of one triangle are congruent to two sides and the included angle of another triangle, then the two triangles are congruent.

Angle-Side-Angle (ASA) Congruence Postulate

If two angles and the included side of one triangle are congruent to two angles and the included side of another triangle, then the two triangles are congruent.

Opposite Sides and Opposite Angles Property of a Parallelogram

Opposites sides of a parallelogram are congruent. Opposite angles of a parallelogram are congruent.

Angle-Angle (AA) Similarity Postulate

If two angles of one triangle are congruent to two angles of another triangle, then the two triangles are similar.

Side-Angle-Side (SAS) Similarity Theorem

If two sides of one triangle are proportional to two sides of another triangle, and the included angles are congruent, then the two triangles are similar.

Side-Side-Side (SSS) Similarity Theorem

If all three sides of one triangle are proportional to the corresponding sides of another triangle, then the two triangles are similar.

Basic Constructions Using a Straightedge and a Compass

- Copy of a line segment
- Copy of an angle
- Parallel line
- Parallelogram
- Perpendicular bisector
- Angle bisector

Review Exercises

Free step-by-step solutions to odd-numbered exercises at *MathematicalPractices.com*.

13.1 Congruence of Triangles

Corresponding Parts $\triangle ABC$ is congruent to $\triangle JKL$. Fill in the corresponding side or angle.

1. $\overline{BC} \cong$ ▭
2. $\overline{JK} \cong$ ▭
3. $\overline{AC} \cong$ ▭
4. $\angle L \cong$ ▭
5. $\angle A \cong$ ▭
6. $\angle K \cong$ ▭

Labeling Sides and Angles $\triangle DEF$ is congruent to $\triangle GHI$. Find the length of the side or the measure of the angle.

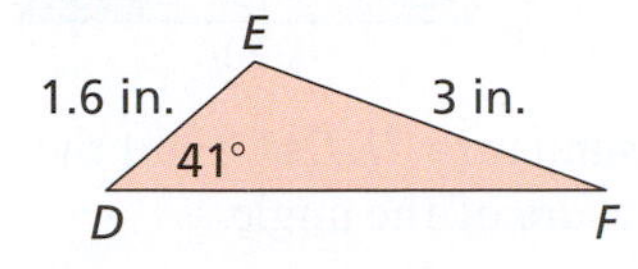

7. $\overline{GH}$
8. $\overline{DF}$
9. $\overline{HI}$
10. $\angle G$
11. $\angle E$
12. $\angle I$

Congruent Triangles Decide whether the triangles are congruent. Explain your reasoning.

13. 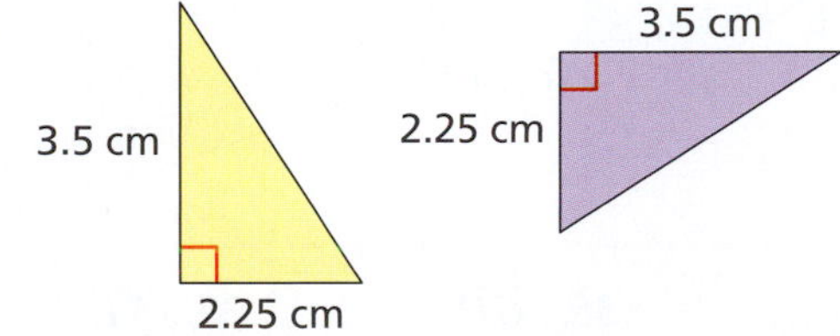

14. 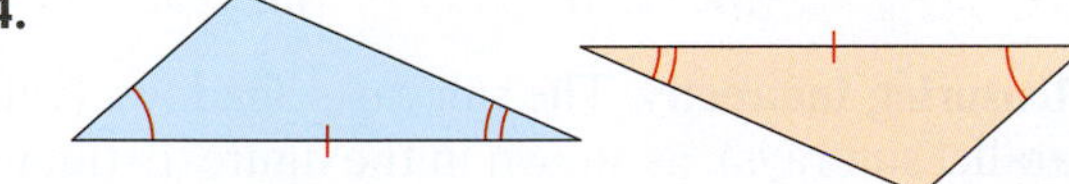

15. 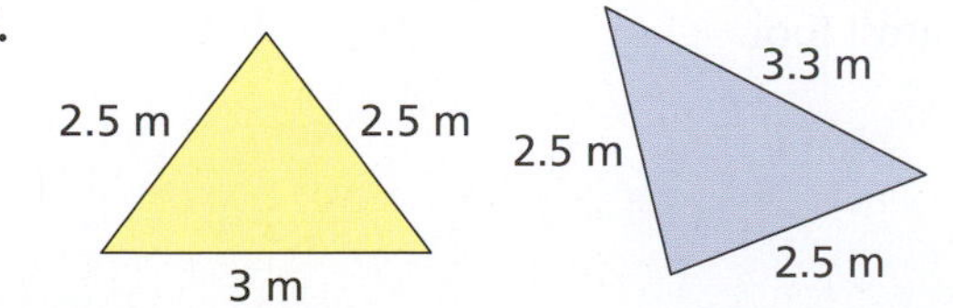

16. 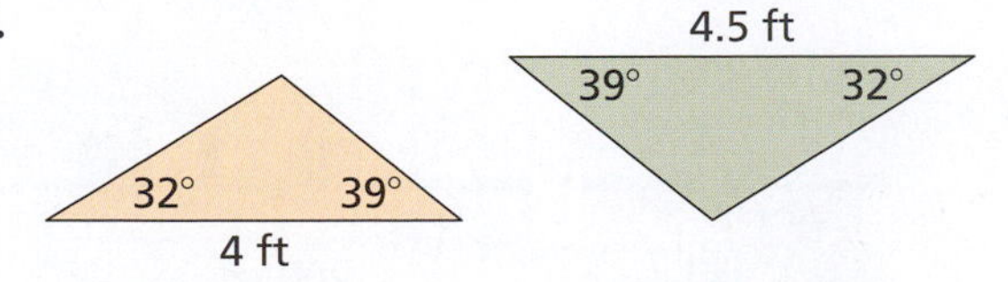

17. 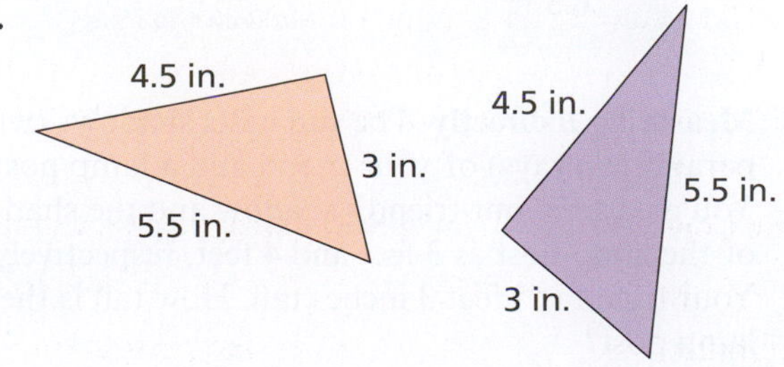

18. 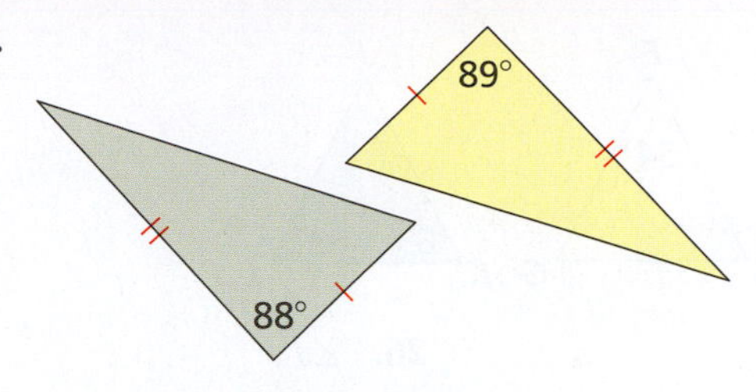

Using the Congruence Postulates Determine whether there is enough information to state that the triangles are congruent. Explain your reasoning.

19. $\triangle QRT$ and $\triangle SRT$

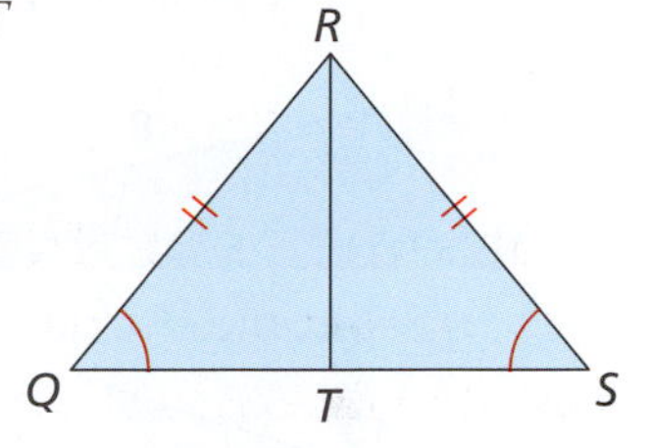

20. $\triangle HIJ$ and $\triangle HKJ$

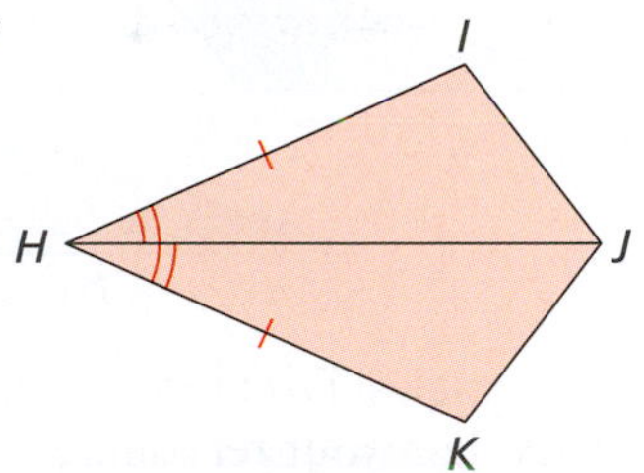

21. $\triangle LMP$ and $\triangle NMP$

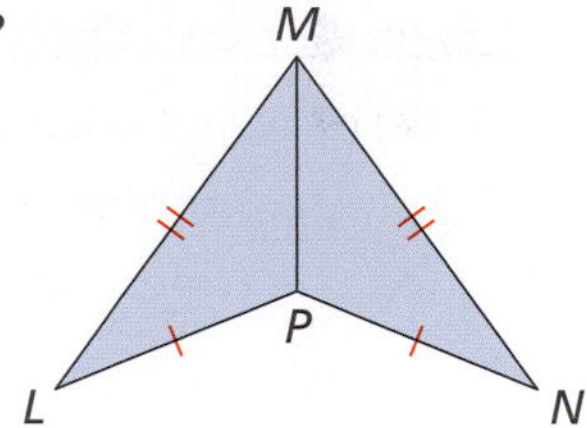

22. $\triangle ABC$ and $\triangle EDC$

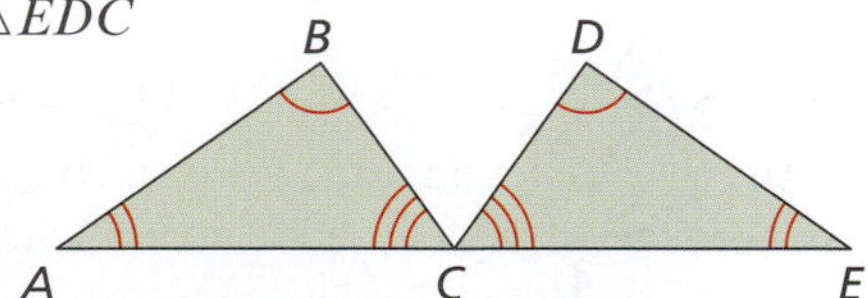

23. **Parallelograms** In the figure, $\triangle WVX$ is congruent to $\triangle YZX$. Explain how you can use this congruency to show that diagonals of a parallelogram bisect each other.

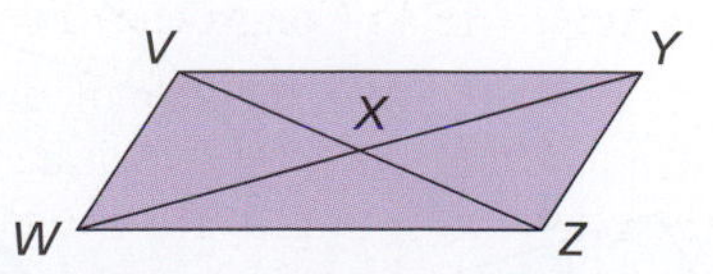

24. **Architecture** The roof of a tower is a rectangular pyramid. The faces of the roof are congruent triangles. Explain how you know that the base of the roof is a square.

13.2 Similarity

Using the Definition of Similar Triangles $\triangle EFG$ is similar to $\triangle JKL$. Find the measure of the angle.

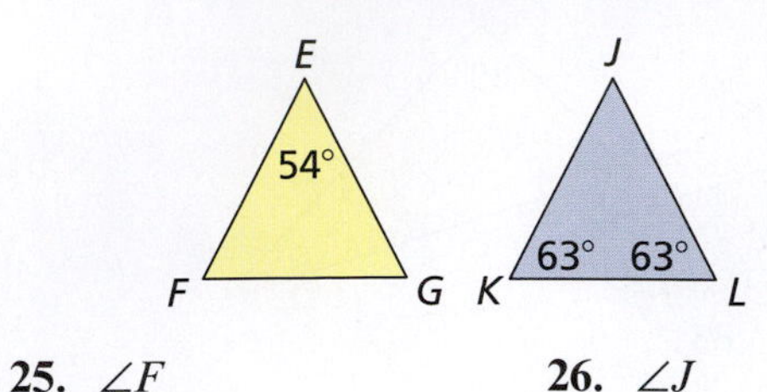

25. $\angle F$ **26.** $\angle J$

Using the Definition of Similar Triangles The triangles are similar. Find a and b.

27. **28.**

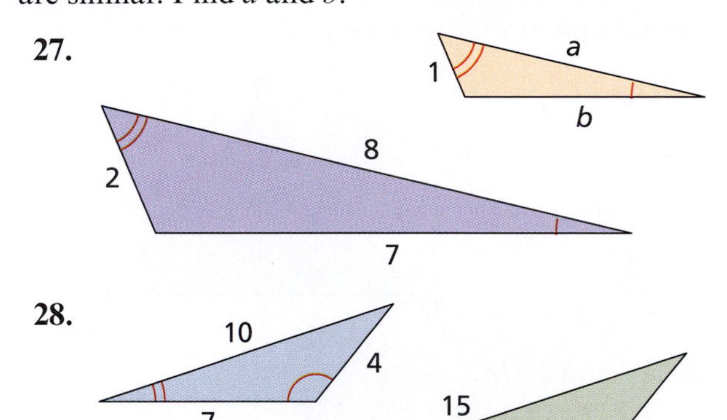

Similar Triangles Decide whether the triangles are similar. Explain your reasoning.

29. **30.** **31.** **32.**

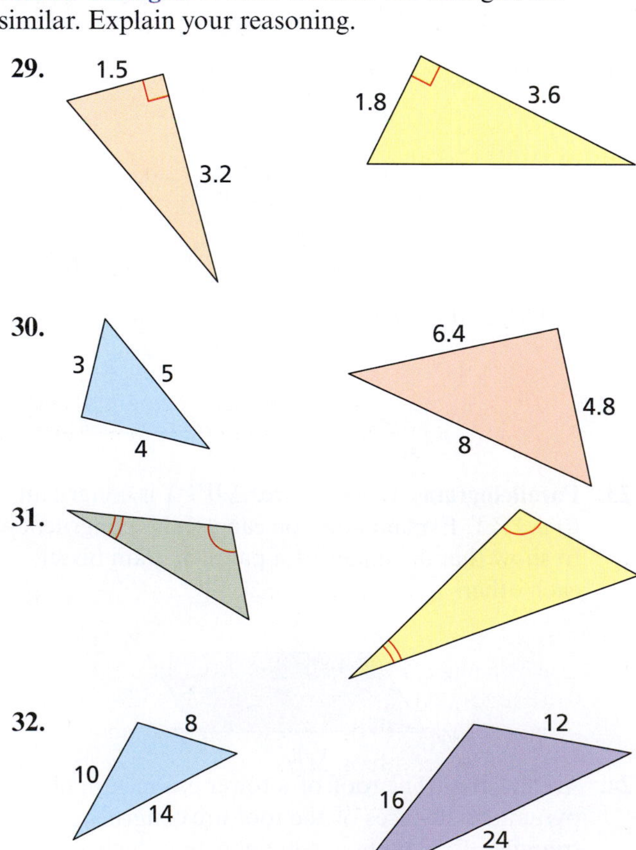

Similar Polygons Decide whether the polygons are similar. Explain your reasoning.

33. **34.**

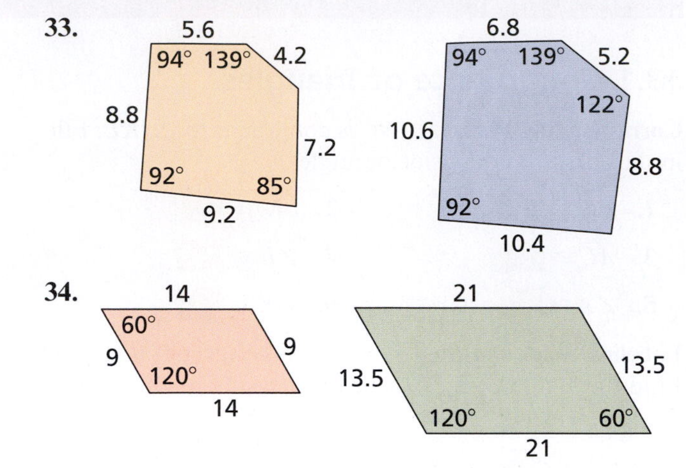

Similar Polygons $EFGH$ is similar to $JKLM$. Find the length of the side or the measure of the angle.

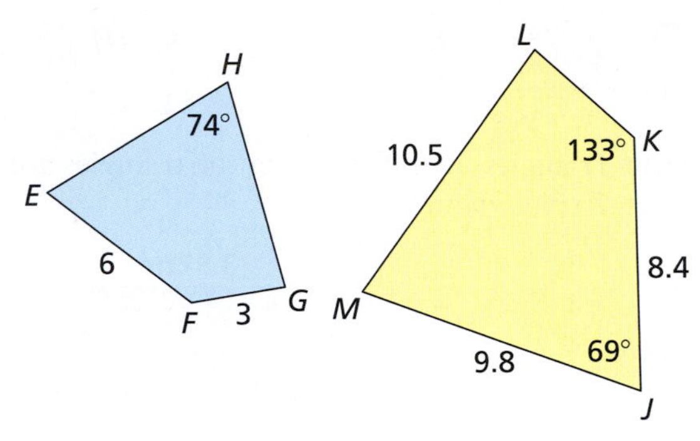

35. $\overline{EH}$ **36.** $\overline{GH}$ **37.** $\overline{KL}$

38. $\angle F$ **39.** $\angle M$ **40.** $\angle L$

41. Measuring Indirectly The sun casts shadows (with parallel sun rays), as shown in the figure. Estimate the height of the Washington Monument to the nearest foot.

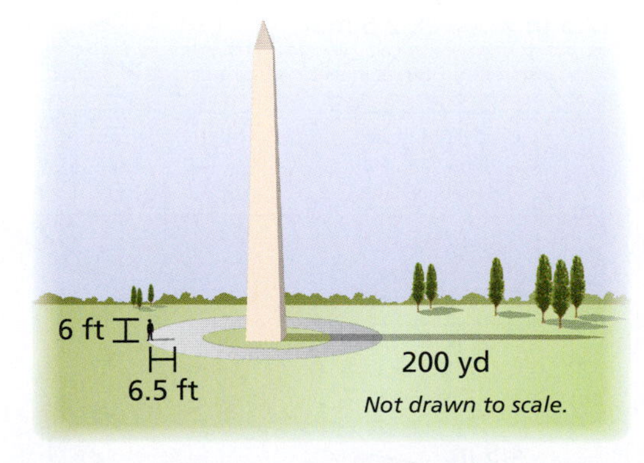

42. Measuring Indirectly The sun casts shadows (with parallel sun rays) of your friend and a lamp post. You measure your friend's shadow and the shadow of the lamp post as 3 feet and 4 feet, respectively. Your friend is 5 feet 3 inches tall. How tall is the lamp post?

13.3 Construction Techniques

Constructing a Copy of a Line Segment Use a compass and a straightedge to construct a copy of the line segment.

43.

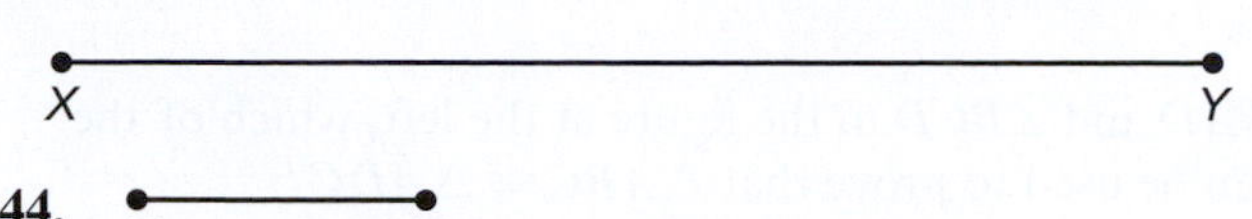

44. U V

Constructing a Copy of an Angle Use a compass and a straightedge to construct a copy of the angle.

45.

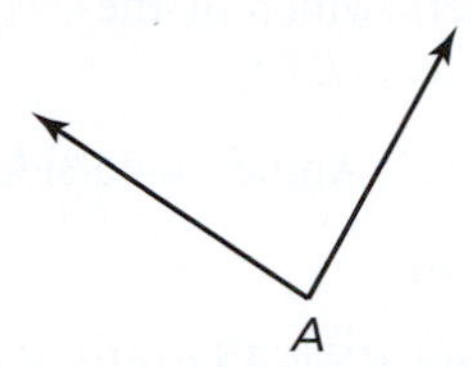

46.

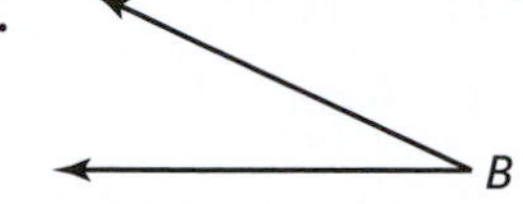

47. Constructing a Copy of a Line Segment Use a compass and a straightedge to construct a line segment $\overline{NP}$ that is congruent to $\overline{LM}$.

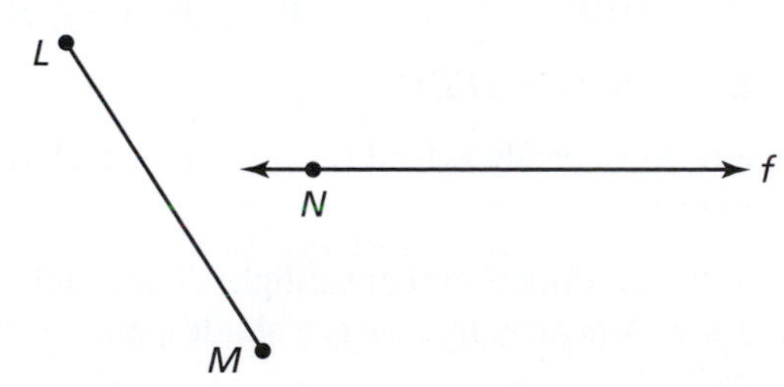

48. Constructing a Copy of an Angle Use a compass and a straightedge to construct an angle R that is congruent to angle Q.

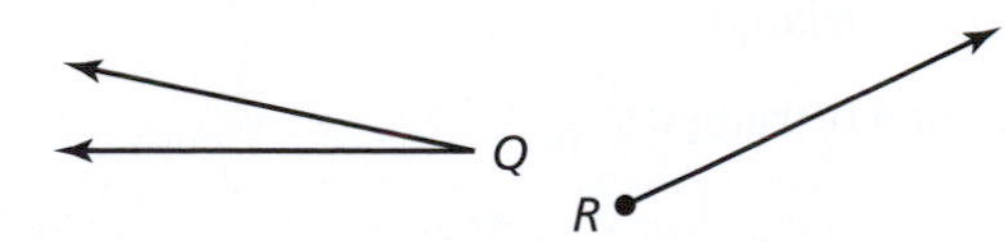

Constructing a Parallel Line Use a compass and a straightedge to construct a line that is parallel to the given line through point P.

49.

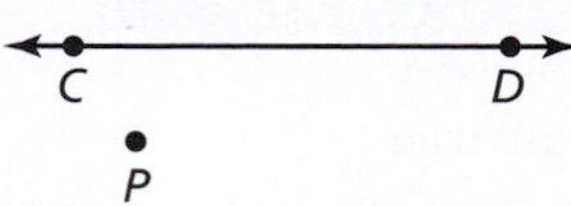

50.

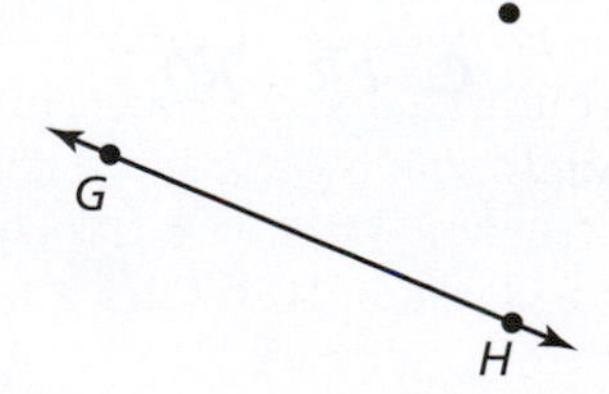

Constructing a Perpendicular Bisector Use a compass and a straightedge to construct the perpendicular bisector of the line segment.

51.

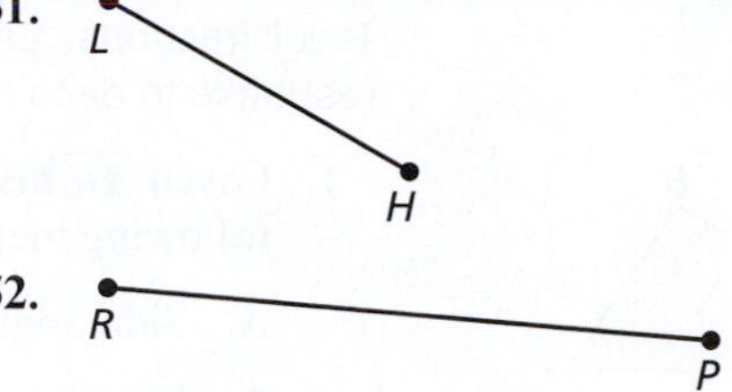

52. R P

Constructing an Angle Bisector Use a compass and a straightedge to construct the angle bisector of the angle.

53.

54.

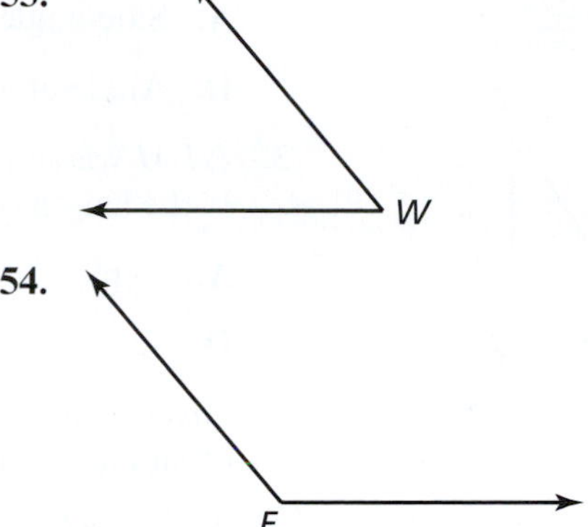

Constructing an Angle Use the angles to construct an angle with the given measure. Describe the process you used to construct the angle.

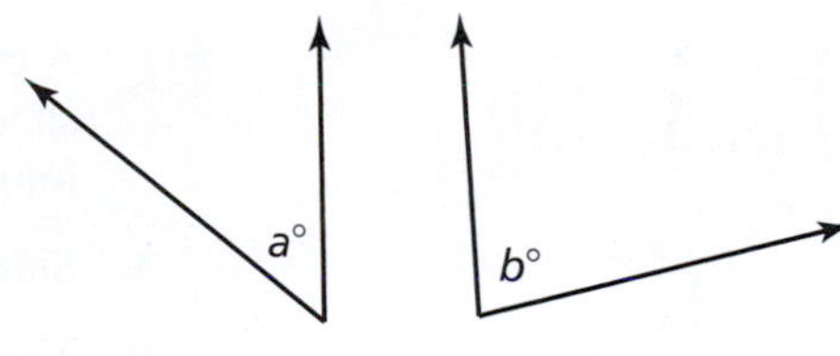

55. $(3a)°$

56. $(2a - b)°$

57. $(2a + b)°$

58. $\left(\frac{1}{4}b\right)^{\circ}$

59. Constructing a Parallelogram Use a compass and a straightedge to construct a parallelogram using $\angle LKJ$.

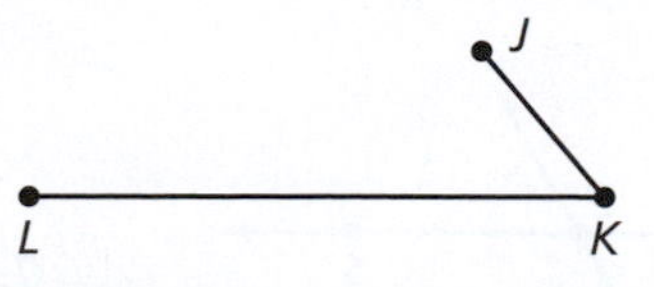

60. Constructing a Rectangle Use a compass and a straightedge to construct a rectangle with two sides equal to the length of $\overline{ST}$ and diagonal length equal to the length of $\overline{MN}$.

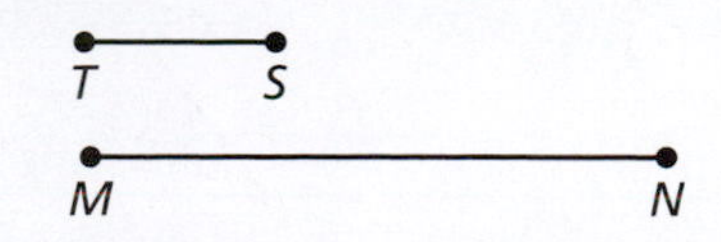

Chapter Test

The questions on this practice test are modeled after questions on state certification exams for K–8 teaching.

Test Directions: Each of the questions is followed by five choices. Choose the *best* response to each question.

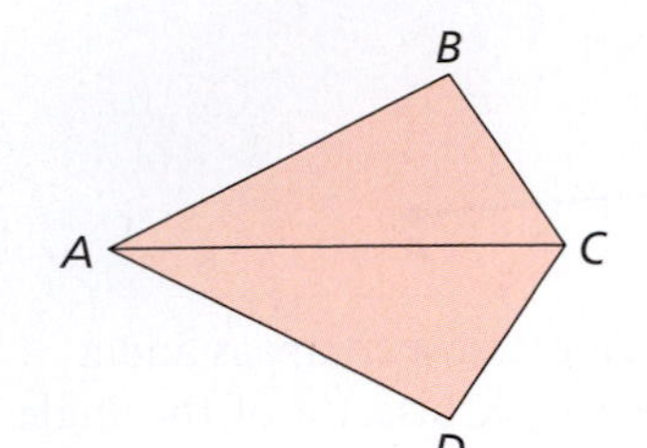

1. Given $\overline{AC}$ bisects $\angle BAD$ and $\angle BCD$ in the figure at the left, which of the following methods can be used to prove that $\triangle ABC \cong \triangle ADC$?

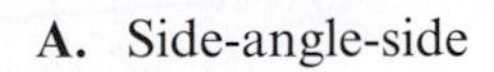

A. Side-angle-side **B.** Angle-side-angle **C.** Angle-angle-side

D. Angle-angle **E.** Not enough information

2. Given $\overline{AB} \cong \overline{DE}$ and $\angle A \cong \angle D$ in the figure at the left, which of the following methods can be used to prove that $\triangle ABC \cong \triangle DEC$?

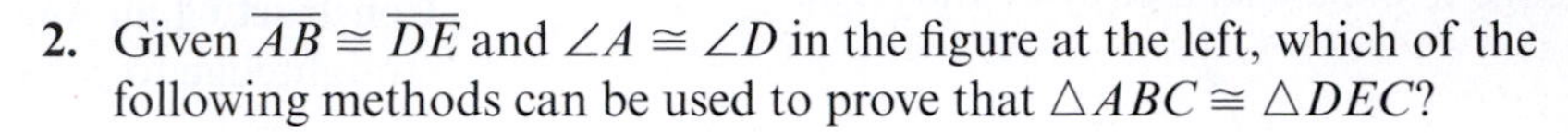

A. Side-angle-side **B.** Angle-side-angle **C.** Angle-angle-side

D. Angle-angle **E.** Not enough information

3. $\triangle LMN$ is similar to $\triangle RST$. LM is 10 centimeters and RS is 15 centimeters. If the length of $\overline{MN}$ is 6 cm, what is the length of $\overline{ST}$?

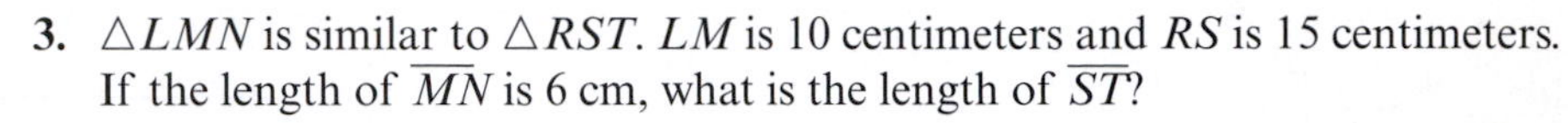

A. 4 cm **B.** 6 cm **C.** 9 cm

D. 12 cm **E.** 18 cm

4. Given that $\overline{EF}$ is parallel to $\overline{GH}$ in the figure at the left, write down the pair of similar (~) triangles.

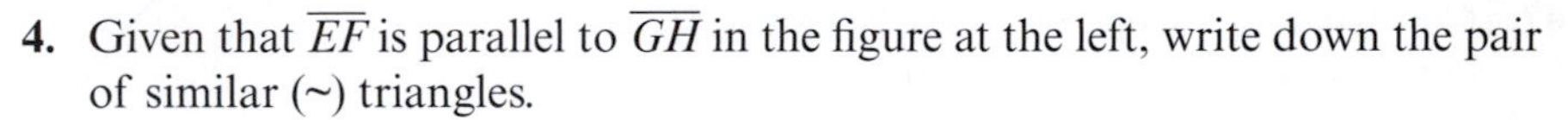

A. $\triangle DEF \sim \triangle DGH$ **B.** $\triangle DEF \sim \triangle DEH$ **C.** $\triangle DEF \sim \triangle DHG$

D. $\triangle DEF \sim \triangle EDF$ **E.** $\triangle DGH \sim \triangle FEH$

5. The following problem can be most easily solved by applying which of the following properties of triangles?

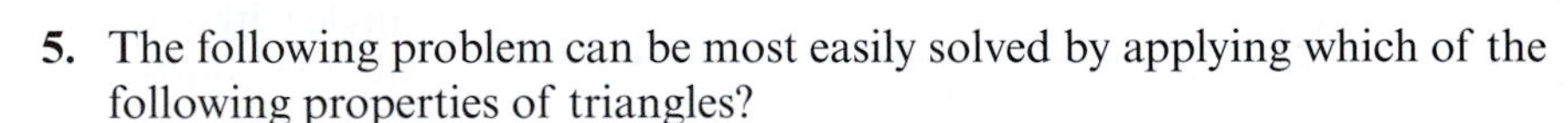

A child who is 4 feet tall is standing in direct sunlight. The child casts a shadow that is 10 feet long. A nearby tree casts a shadow that is 25 feet long. How tall is the tree?

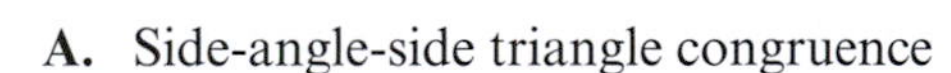

A. Side-angle-side triangle congruence

B. The square of the hypotenuse of a right triangle

C. The sum of the angles in a triangle

D. Proportional sides of similar triangles

E. None of the above

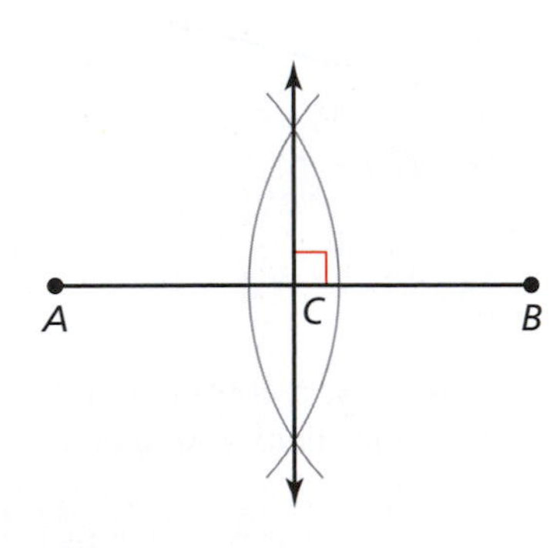

6. The diagram at the left illustrates the construction of the perpendicular bisector of $\overline{AB}$. Which statement is true?

A. $AC = CB$ **B.** $CB = \frac{1}{2}AB$ **C.** $2AC = AB$

D. $AC + CB = AB$ **E.** All of the above are true

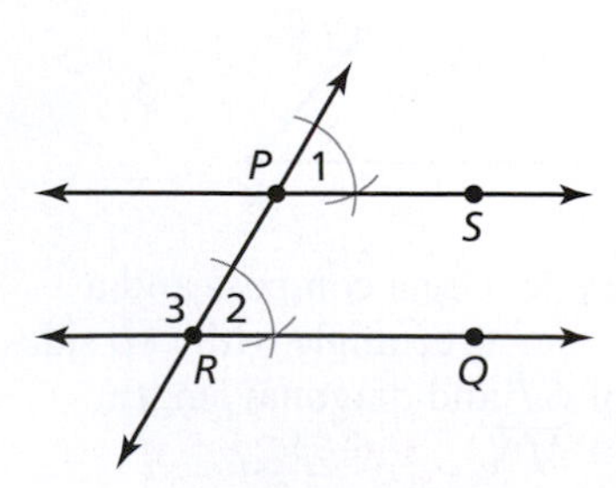

7. The diagram at the left illustrates the construction of $\overline{PS}$ parallel to $\overline{RQ}$ through point P. Which statement justifies this construction?

A. $m\angle 1 = m\angle 2$ **B.** $m\angle 1 = m\angle 3$ **C.** $\overline{PR} \cong \overline{RQ}$

D. $\overline{PS} \cong \overline{RQ}$ **E.** None of the above

14 Transformations

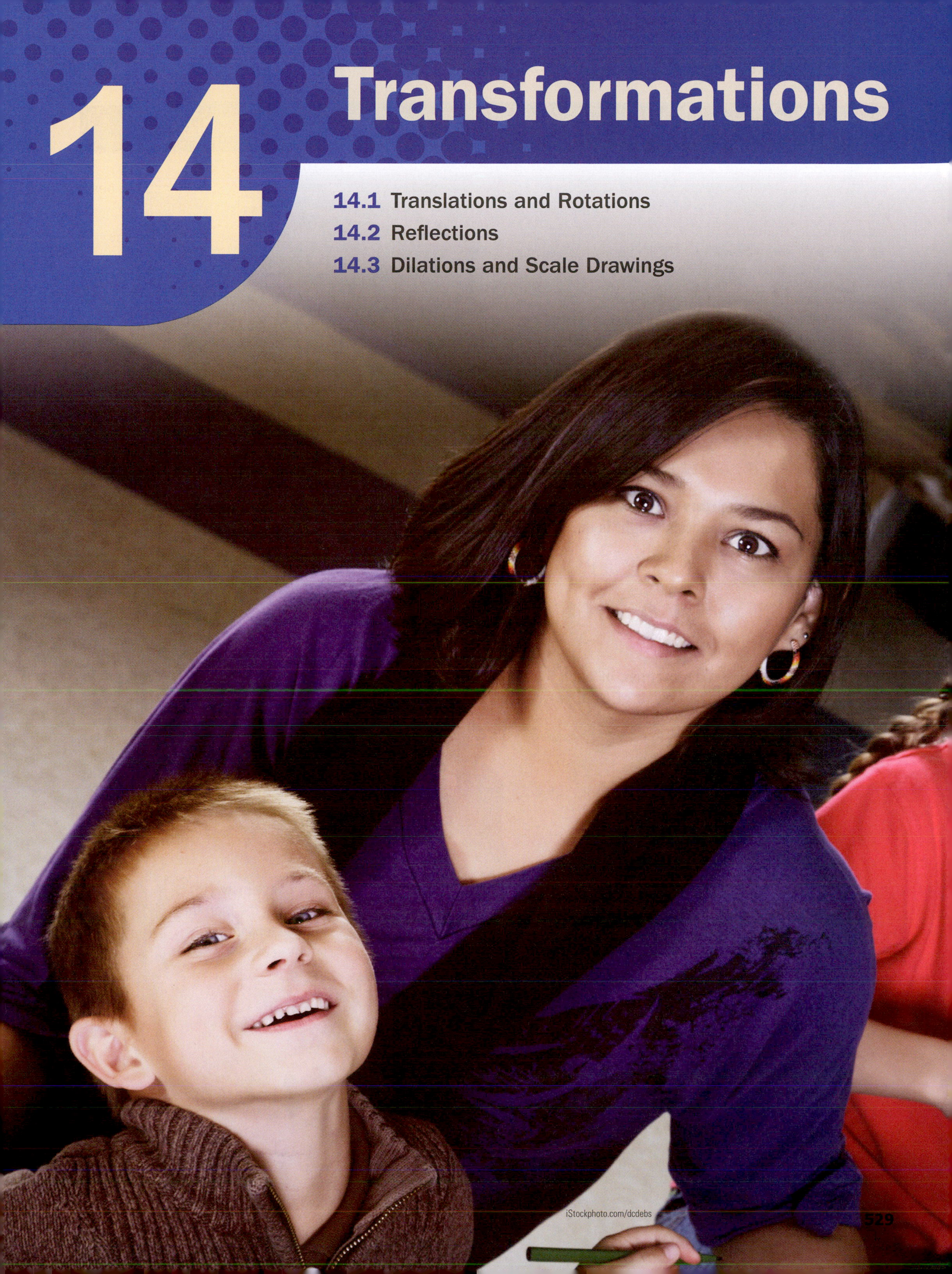

iStockphoto.com/dcdebs

Activity: Describing Translations and Rotations as Dance Steps

Materials:

- Pencil

Learning Objective:

Understand and describe the concept of translations and rotations in the plane. Use translations and rotations to describe dance steps.

Name ______________________________

The footprints show the dance steps for the salsa. The yellow glow shows the foot that the dancer's weight is on.

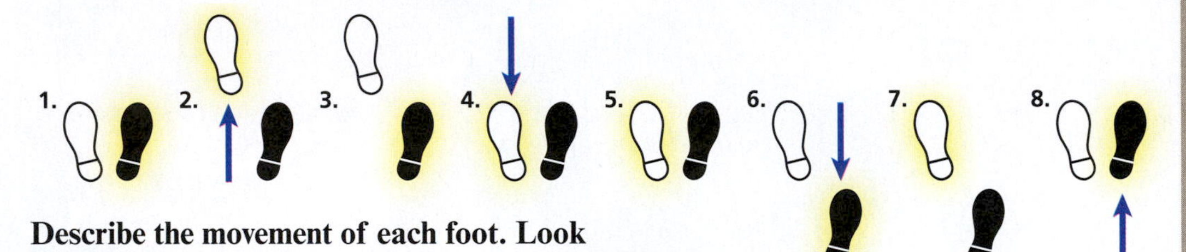

Describe the movement of each foot. Look only at the end result of each movement.

1. Hold the beat.
2.
3. Rock back onto right foot.
4.
5. Hold the beat.
6.
7. Rock forward onto left foot.
8.

The dance instructor is teaching the man the basic steps in the dance called the *salsa*.

A *translation* of a figure in the plane is a transformation in which the figure *slides*, but does not turn. A *rotation* is a transformation in which the figure *turns*.

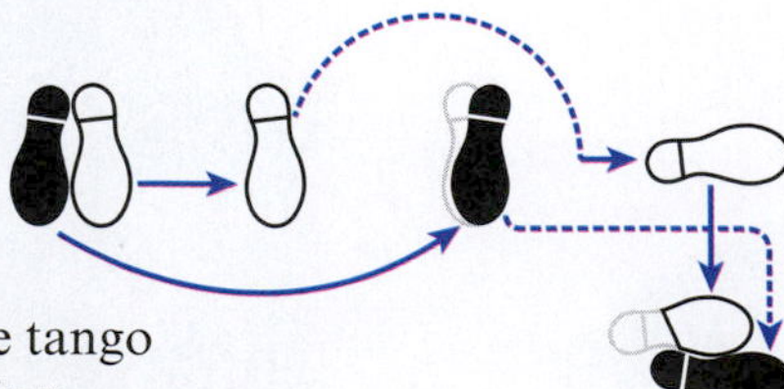

9. Are the steps in the salsa translations or rotations? Explain your reasoning.
10. As shown at the right, are the basic steps in the tango translations or rotations? Explain your reasoning.

14.1 Translations and Rotations

Standards

Grades 6–8
Geometry
Students should understand congruence and similarity using physical models, transparencies, or geometry software.

- Understand and identify the four basic types of transformations.
- Understand and use the concept of a translation.
- Understand and use the concept of a rotation.

Basic Types of Transformations

A **transformation** changes a figure into another figure. The new figure is called the **image**. There are four basic types of transformations.

1. Translation (Section 14.1)

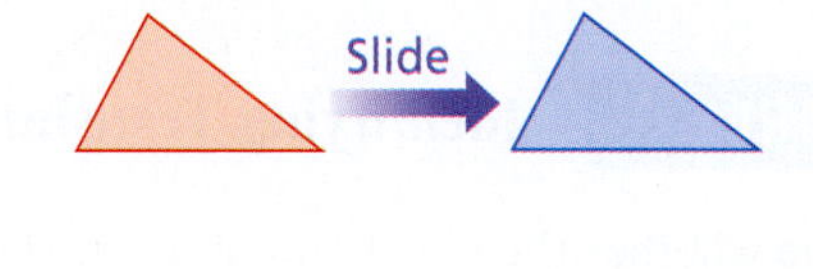

2. Rotation (Section 14.1)

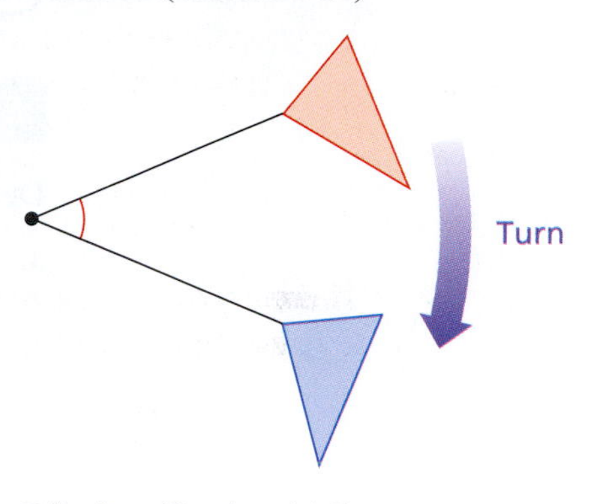

3. Reflection (Section 14.2)

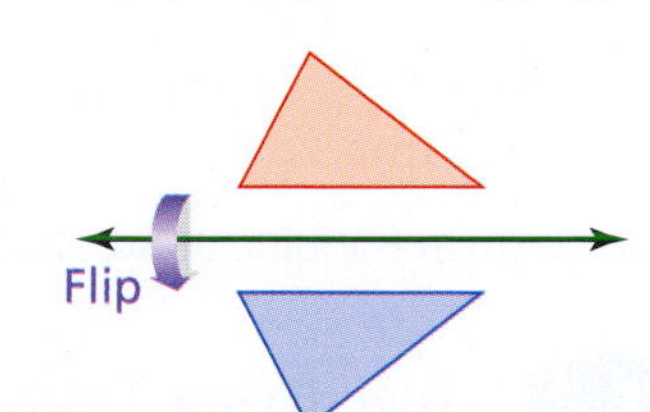

4. Dilation (Section 14.3)

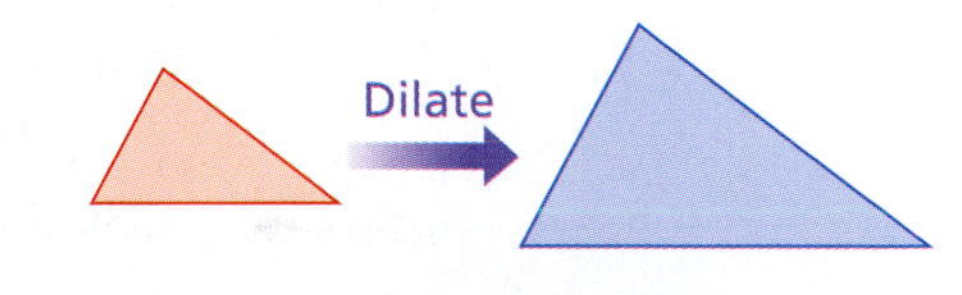

The first three types of transformations above (translations, rotations, reflections) are **rigid** transformations because they preserve the size and shape of the original figure. In other words, the image is *congruent* to the original figure.

Classroom Tip

For the tessellation shown in Example 1, you can convince your students that there are no dilations because all of the lizards are congruent. To convince your students that there are no reflections, make a copy of the print, cut out one of the lizards, and reflect it. The reflected image does not occur in the print.

Identifying Transformations in Art

What types of transformations are shown in the tessellation? Explain your reasoning.

There are no reflections or dilations in this tessellation.

SOLUTION

There are translations and rotations.

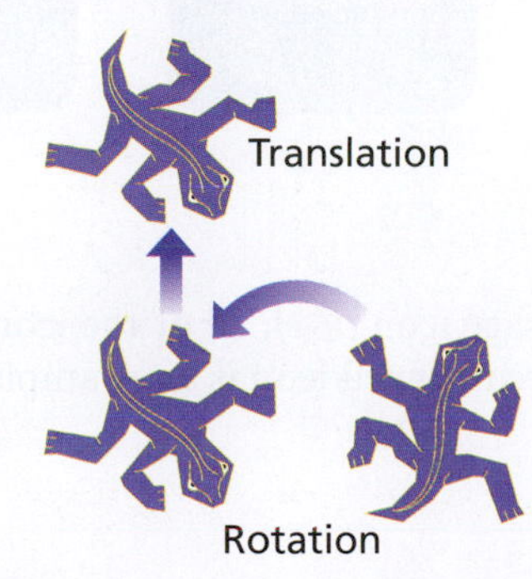

Translations

Definition of Translation

A **translation** is a transformation in which a figure *slides* but does not turn. Every point of the figure moves the same distance and in the same direction.

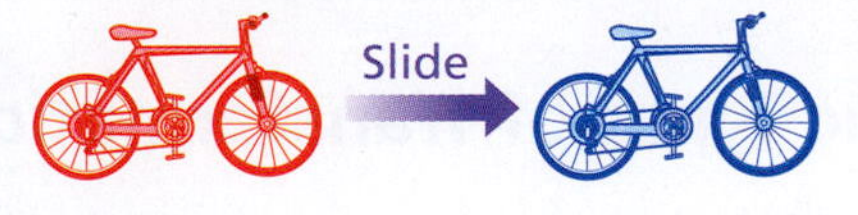

The original figure and its image are congruent.

EXAMPLE 2 Identifying Translations

Determine whether the blue figure is a translation of the red figure.

a.

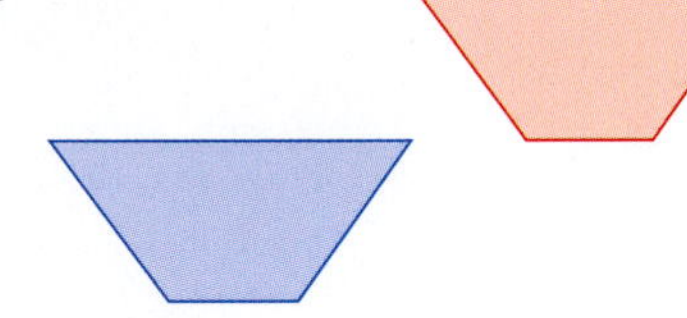

b.

SOLUTION

a. The red trapezoid *slides* to form the blue trapezoid. So, the blue trapezoid *is* a translation of the red trapezoid.

b. The red 3 *turns* to form the blue 3. So, the blue 3 is *not* a translation of the red 3.

EXAMPLE 3 Identifying Translations in Real Life

You rearrange the icons on your smartphone. Do the icons represent examples of translations? Explain your reasoning.

Original

Rearranged

Classroom Tip

Ask your students to make a list of real-life examples of translations that they have seen during the past week. You can mention a few examples, such as a magnet sliding down a refrigerator or a car backing out of a garage.

SOLUTION.

Except for the message icon ( ), all of the icons slide without changing their size or shape. So, each rearranged icon is an example of a translation.

Oleksiy Mark/Shutterstock.com

EXAMPLE 4 Counting Possible Translations

Classroom Tip

A translation cannot be flipped, enlarged, or rotated. Notice that the triangle below is not a translation of the original triangle in Example 4 because it is not a slide of the original triangle.

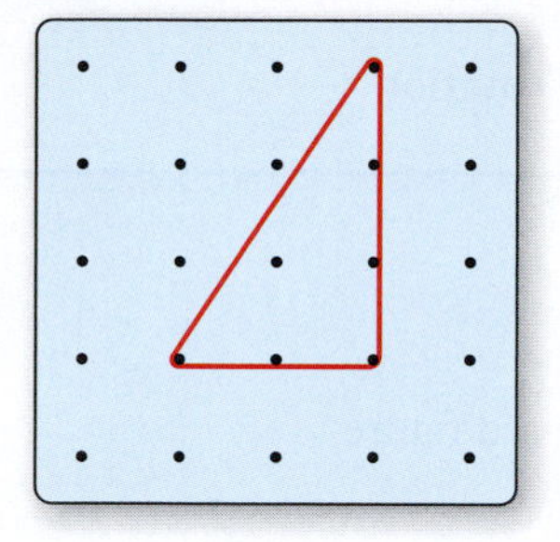

A right triangle is formed on the geoboard with a rubber band. How many different translations of the triangle are possible? The entire triangle must remain on the geoboard.

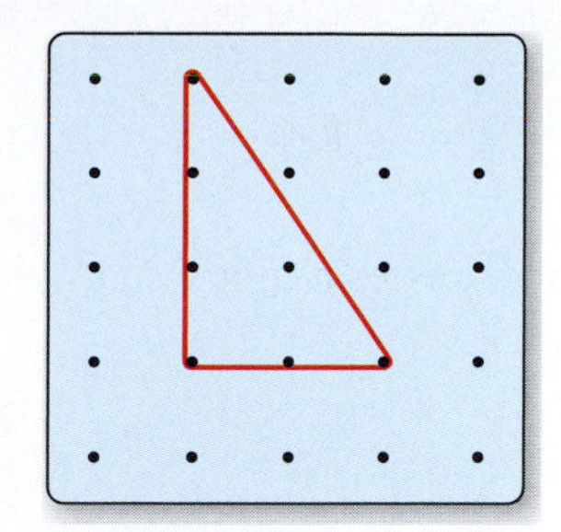

SOLUTION

There are 5 possible translations as shown below.

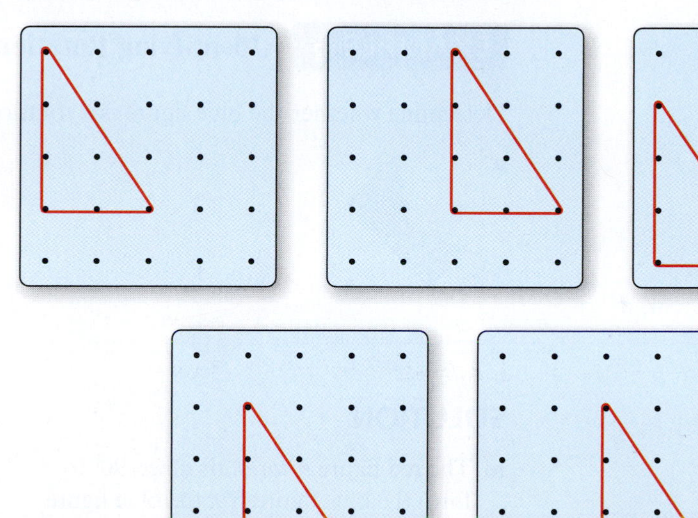

EXAMPLE 5 Identifying Translations in Real Life

The game board at the right shows the locations of 7 ships. Are any of the ships translations of each other?

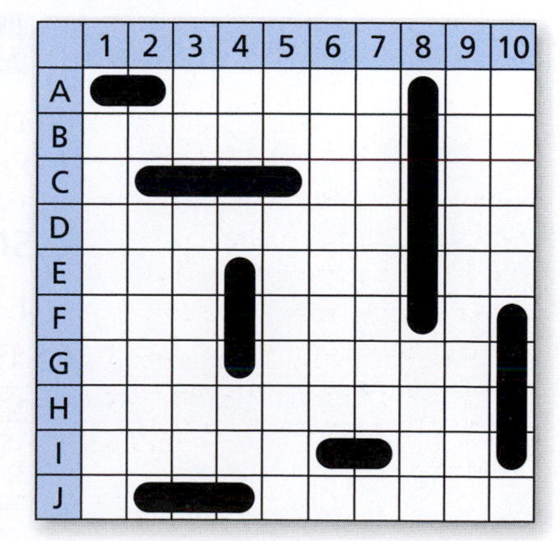

SOLUTION

There are two ships that are 2 units long, two that are 3 units long, two that are 4 units long, and one that is 6 units long. The only ships that you can slide, without turning, to coincide with another ship are the 2-unit ships.

So, the 2-unit ships are translations of each other.

Rotations

Definition of Rotation

A **rotation** is a transformation in which a figure *turns* about a point called the **center of rotation**. The number of degrees a figure rotates is the **angle of rotation**. The original figure and its image are congruent.

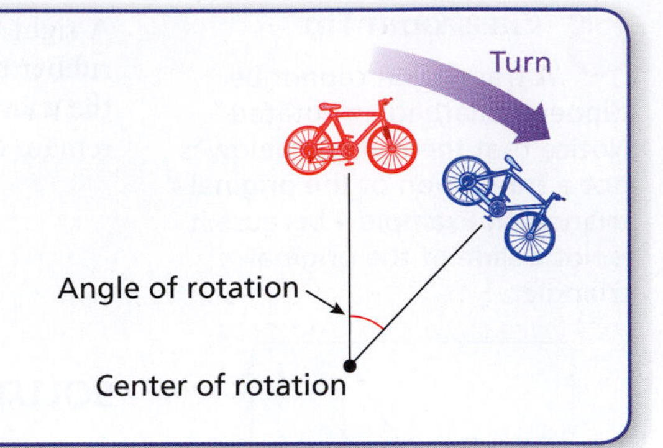

EXAMPLE 6 Identifying Rotations

Determine whether the blue figure is a rotation of the red figure.

a.

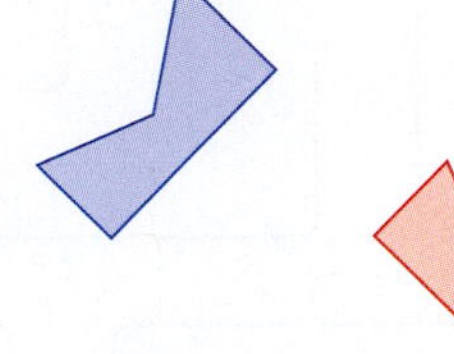

b.

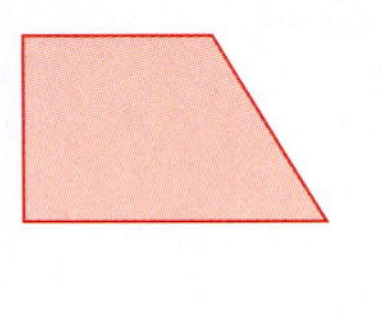

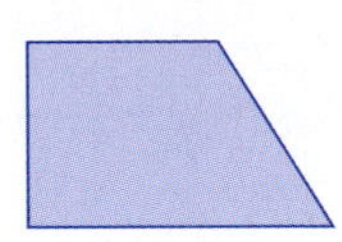

SOLUTION

a. The red figure *rotates*, or *turns*, 90° to form the blue figure. So, the blue figure *is* a rotation of the red figure.

b. The red trapezoid *slides* to form the blue trapezoid. So, the blue trapezoid *is not* a translation of the red trapezoid.

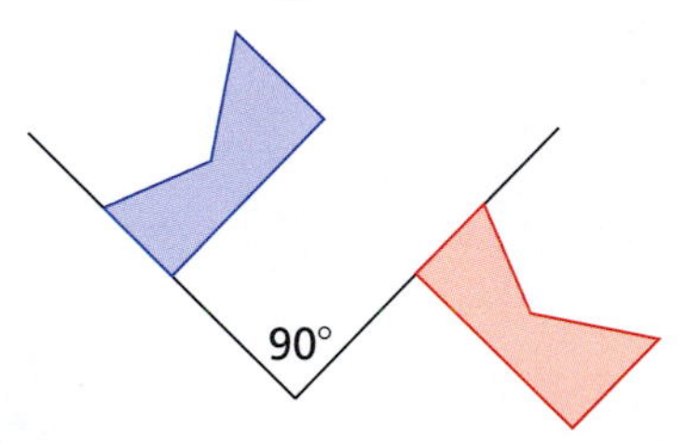

EXAMPLE 7 Finding the Center of Rotation

The red triangle rotates to form the blue triangle. Find the center of rotation.

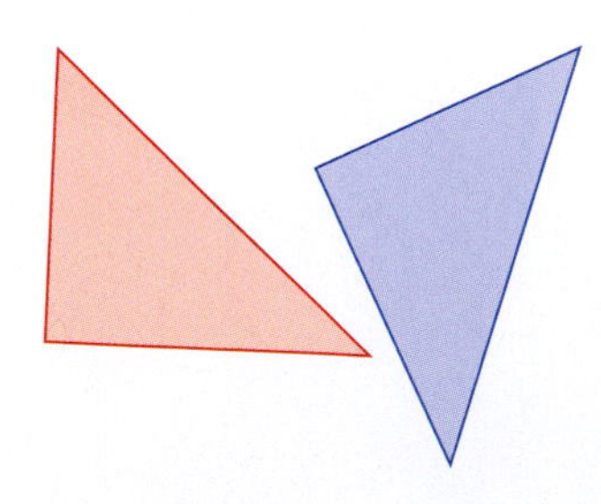

SOLUTION

1. Identify a pair of corresponding points (one point on each triangle).
2. Connect the points with a line segment.
3. Draw the perpendicular bisector of the line segment.
4. Repeat steps 1 through 3 with another set of corresponding points.
5. The point of intersection of these bisectors is the center of rotation.

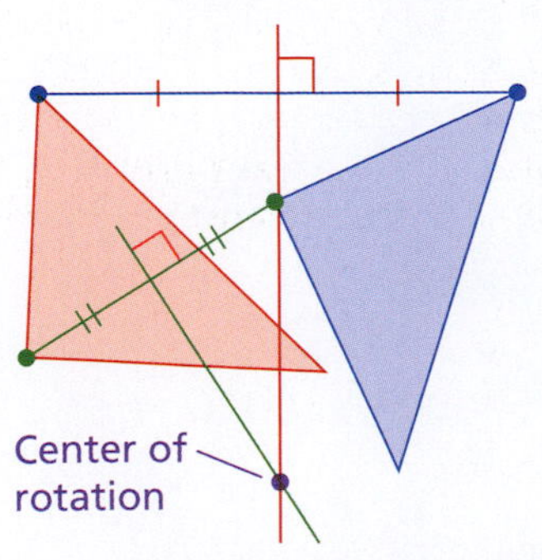

Classroom Tip

Be sure your students understand that a figure has rotational symmetry when a rotation of 180° or less produces an identical image. A figure that coincides with itself only after a 360° rotation (a *full turn*) does *not* have rotational symmetry.

Definition of Rotational Symmetry

A figure has **rotational symmetry** when a rotation of 180° (a *half-turn*) or less around a point in the figure produces an image that fits exactly on the original figure.

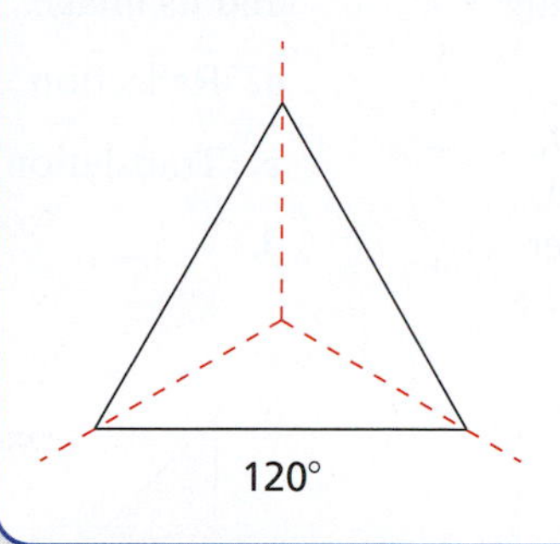
120°

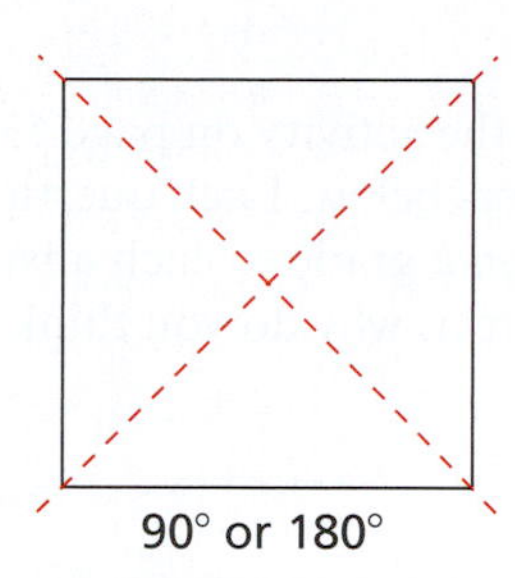
90° or 180°

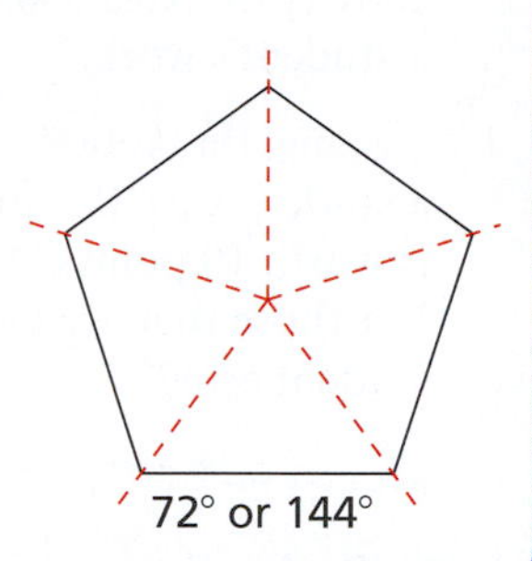
72° or 144°

Mathematical Practices

Model with Mathematics

MP4 Mathematically proficient students apply what they know and are comfortable making assumptions to simplify a complicated situation.

EXAMPLE 8 Identifying Rotational Symmetry in Real Life

Which of the stained glass window designs has rotational symmetry? Explain your reasoning.

a.

b.

c.

SOLUTION

a. The design will coincide with itself when it rotates 90° and 180°. So, it has rotational symmetry.

b. The design will coincide with itself when it rotates 60°, 120°, and 180°. So, it has rotational symmetry.

c. The only way the figure will coincide with itself is when it rotates 360°. So, it does *not* have rotational symmetry.

EXAMPLE 9 Identifying Rotational Symmetry in Real Life

Classroom Tip

Every figure that rotates 360° using any point as the center of rotation moves onto itself. Such a transformation is an *identity transformation.*

A rotational *ambigram* is a word that has a rotational symmetry of 180°. Identify the names. Do they have rotational symmetry?

SOLUTION

The names are "katie," "michelle," and "matthew." Rotate the book and read the words upside down. You can see that each has a rotational symmetry of 180°.

14.1 Exercises

Free step-by-step solutions to odd-numbered exercises at *MathematicalPractices.com.*

1. **Writing a Solution Key** Write a solution key for the activity on page 530. Describe a rubric for grading a student's work.

2. **Grading the Activity** In the activity on page 530, a student gave the answers below. Each question is worth 10 points. Assign a grade to each answer. For those that are incorrect, why do you think the student erred?

Sample Student Work

> **9.** Are the steps in the salsa translations or rotations? Explain your reasoning.
>
> *Rotations because the dancer is rotating which foot he/she uses to step.*
>
> **10.** As shown at the right, are the basic steps in the tangotranslations or rotations? Explain your reasoning.
>
> *There is a combination of rotations and translations. For instance, the white foot moves over which is a translation and then turns to point right which is a rotation.*

3. **Matching** Match each transformation to the figure and its image.

   a. Reflection

   b. Dilation

   c. Translation

   d. Rotation

   i.

   ii.

   iii.

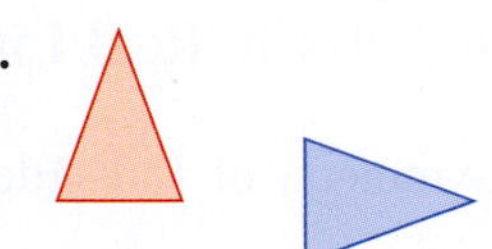

   iv.

4. **Matching** Match each transformation to the figure and its image.

   a. Reflection **b.** Dilation

   c. Translation **d.** Rotation

   i.

   ii.

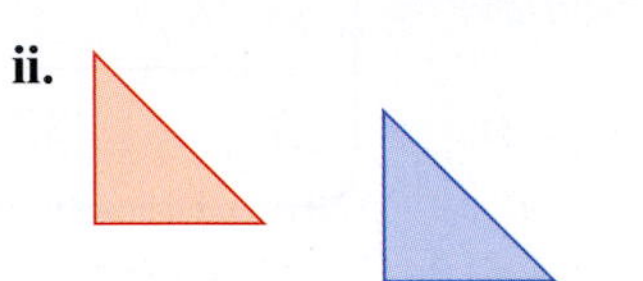

   iii.

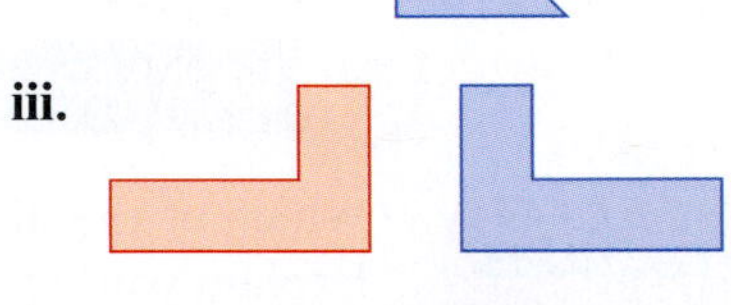

   iv.

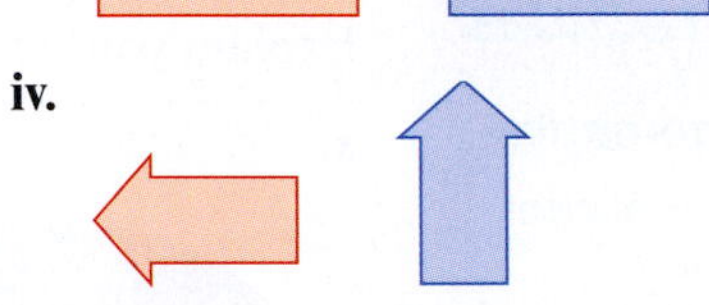

5. **Transformations in Real Life** Name the type of transformation modeled by the action.

   a. Riding down an escalator

   b. Blowing on a pinwheel

   c. Looking in a mirror

6. **Transformations in Real Life** Name the type of transformation modeled by the action.

   a. Making a handprint in clay

   b. Going up in an elevator

   c. Riding a carousel

7. **Identifying Translations** Determine whether the blue figure is a translation of the red figure. Explain your reasoning.

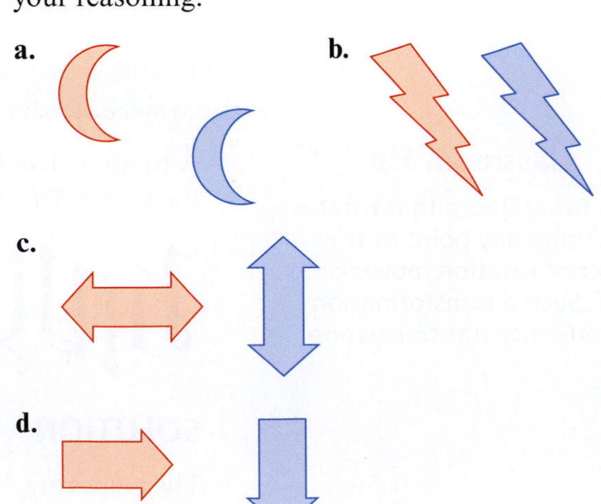

8. **Identifying Translations** Determine whether the blue figure is a translation of the red figure. Explain your reasoning.

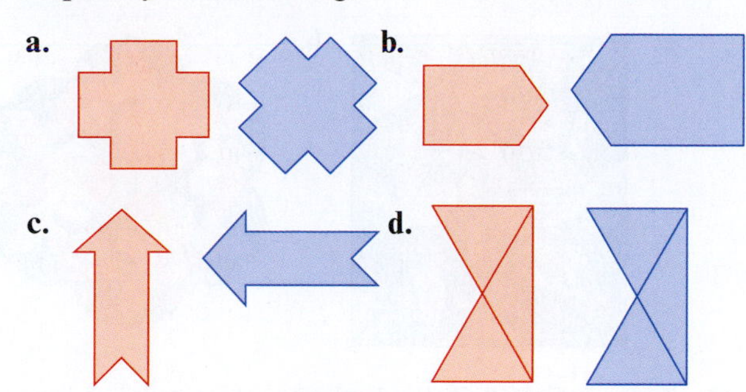

9. **Counting Possible Translations** A triangle is formed on the geoboard with a rubber band. How many different translations of the triangle are possible? The entire triangle must remain on the geoboard.

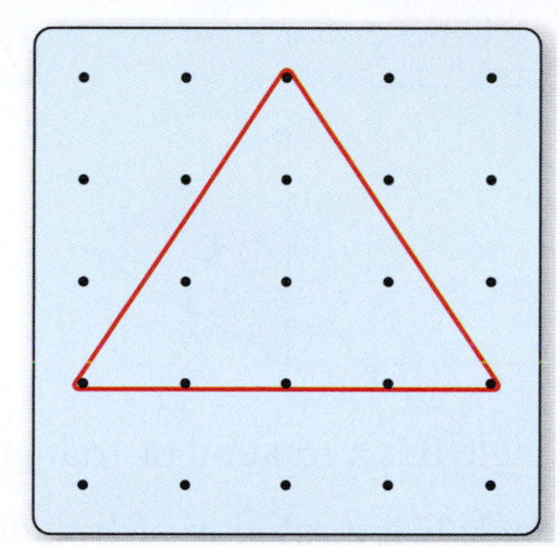

10. **Counting Possible Translations** A square is formed on the geoboard with a rubber band. How many different translations of the square are possible? The entire square must remain on the geoboard.

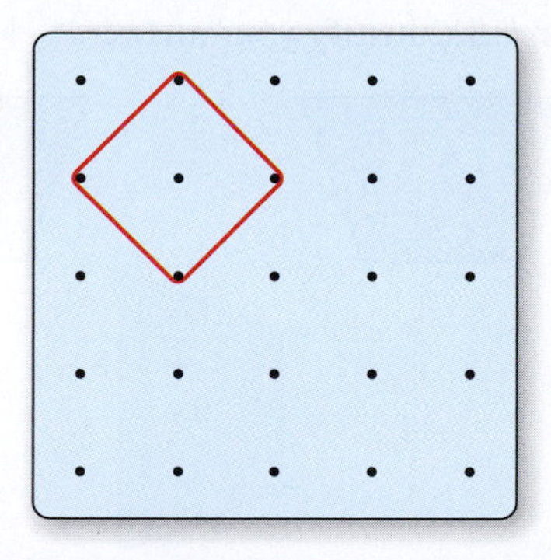

11. **Identifying Rotations** Determine whether the blue figure is a rotation of the red figure. Explain your reasoning.

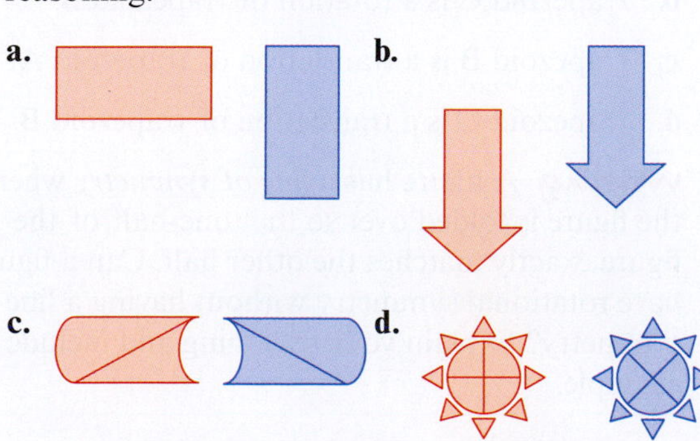

12. **Identifying Rotations** Determine whether the blue figure is a rotation of the red figure. Explain your reasoning.

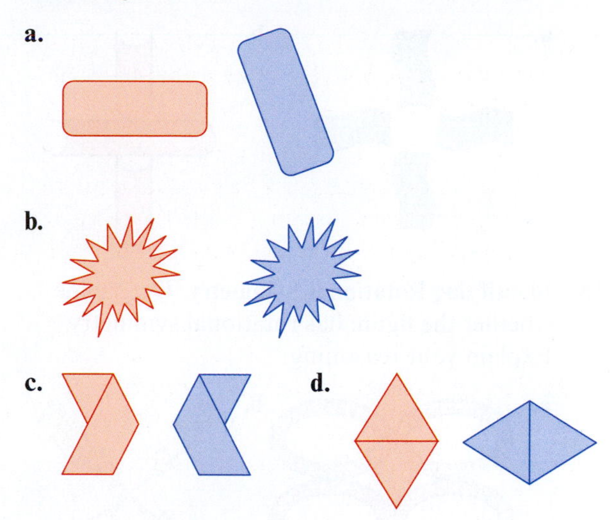

13. **Finding the Angle of Rotation** Find the angle of rotation of the minute hand of an analog clock over 20 minutes.

14. **Finding the Angle of Rotation** Find the angle of rotation of the minute hand of an analog clock over 15 minutes.

15. **Finding Centers of Rotation** Find the center of rotation of the figure and its image.

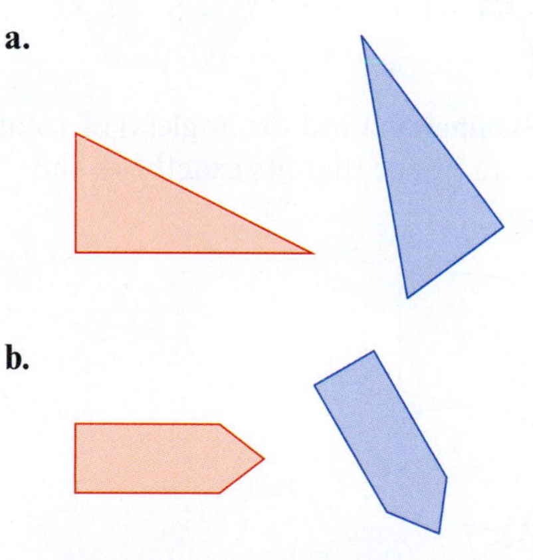

16. **Finding Centers of Rotation** Find the center of rotation of the figure and its image.

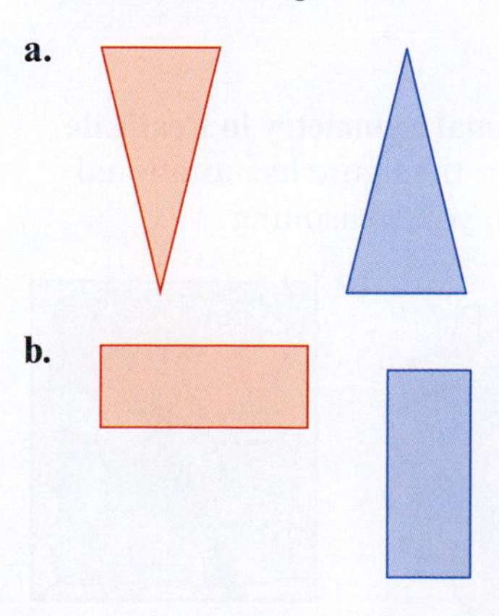

17. **Identifying Rotational Symmetry** Determine whether the figure has rotational symmetry. Explain your reasoning.

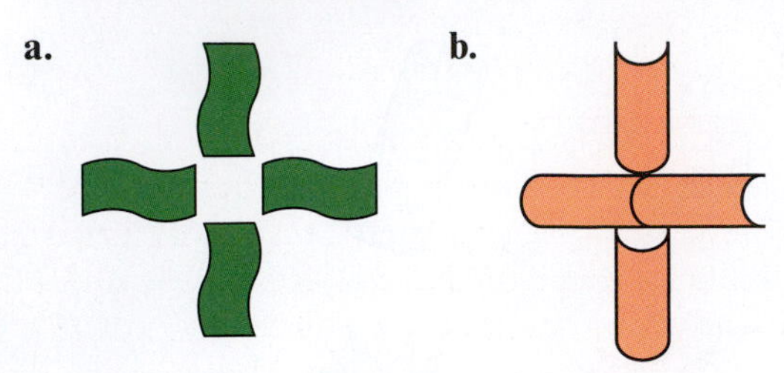

18. **Identifying Rotational Symmetry** Determine whether the figure has rotational symmetry. Explain your reasoning.

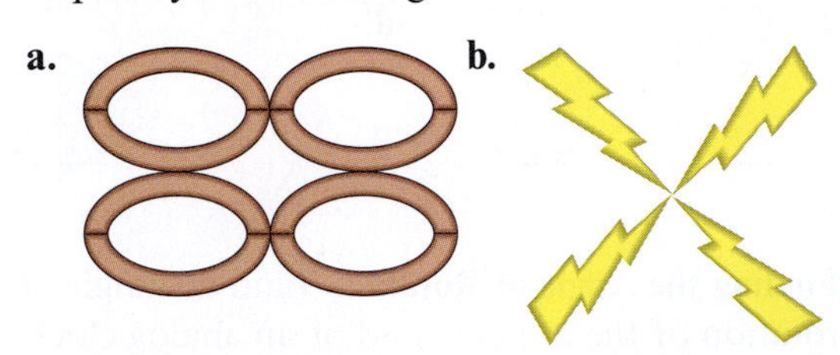

19. **Rotational Symmetry** Find the angle(s) of rotation that produce an image that fits exactly on the original figure.

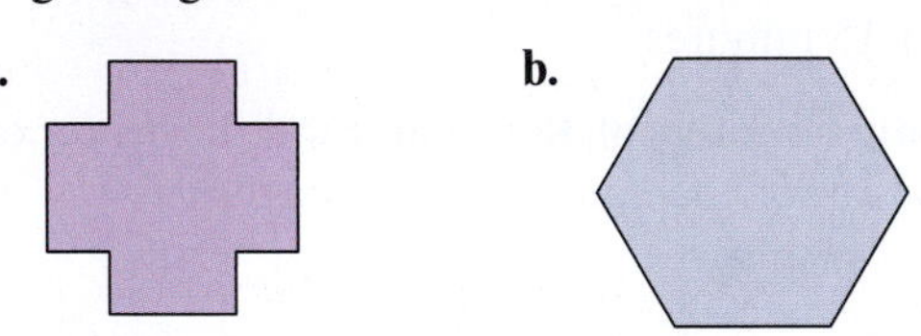

20. **Rotational Symmetry** Find the angle(s) of rotation that produce an image that fits exactly on the original figure.

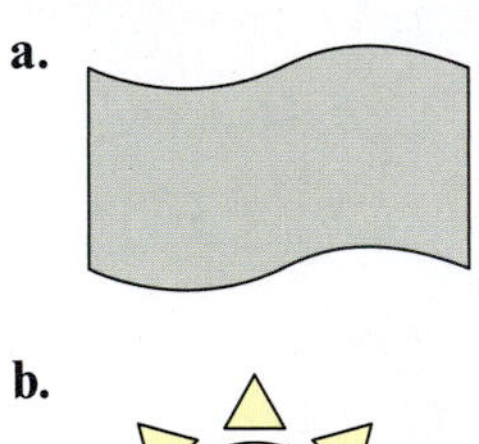

21. **Identifying Rotational Symmetry in Real Life** Determine whether the figure has rotational symmetry. Explain your reasoning.

22. **Identifying Rotational Symmetry in Real Life** Determine whether the figure has rotational symmetry. Explain your reasoning.

23. **True or False?** Tell whether each statement is *true* or *false*. Justify your answer.

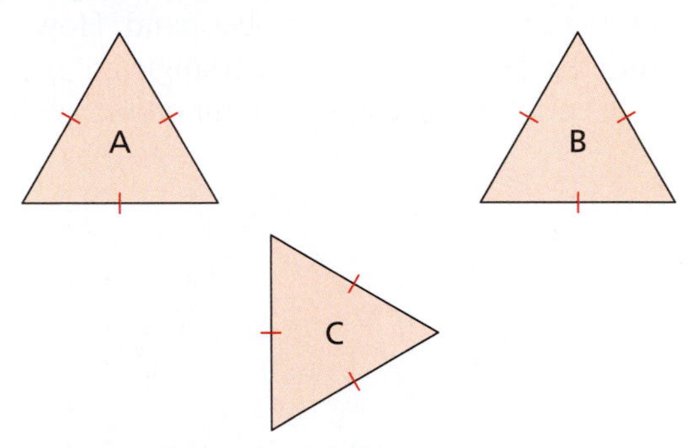

a. Triangle B is a rotation of triangle A.

b. Triangle C is a rotation of triangle B.

c. Triangle B is a translation of triangle A.

d. Triangle C is a translation of triangle B.

24. **True or False?** Tell whether the statement below is *true* or *false*. Justify your answer.

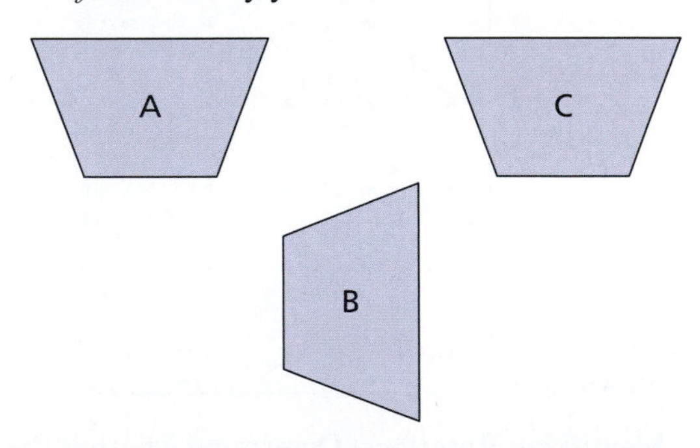

a. Trapezoid B is a rotation of trapezoid A.

b. Trapezoid C is a rotation of trapezoid B.

c. Trapezoid B is a translation of trapezoid A.

d. Trapezoid C is a translation of trapezoid B.

25. *Writing* A figure has a *line of symmetry* when the figure is folded over so that one-half of the figure exactly matches the other half. Can a figure have rotational symmetry without having a line of symmetry? Explain your reasoning and include an example.

26. *Writing* A student in your class does not understand the difference between a translation and a rotation. Describe a real-life example that you could use to explain the difference.

27. **Identifying the Center of Rotation** Rectangle $ABCD$ can be rotated onto rectangle $EFGD$. Which point is the center of rotation?

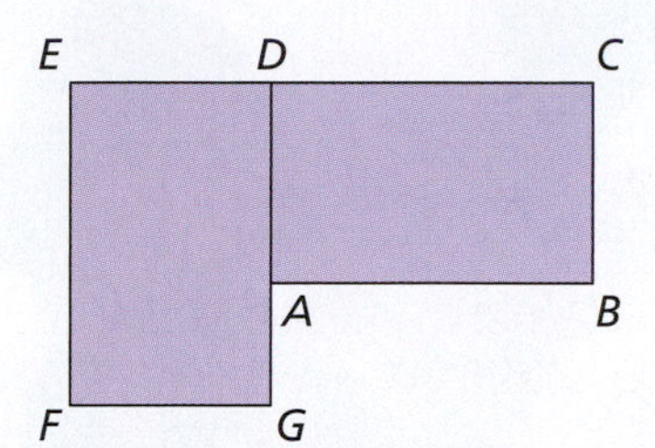

28. **Identifying the Center of Rotation** Triangle TPR can be rotated onto triangle TRS. Which point could be the center of rotation?

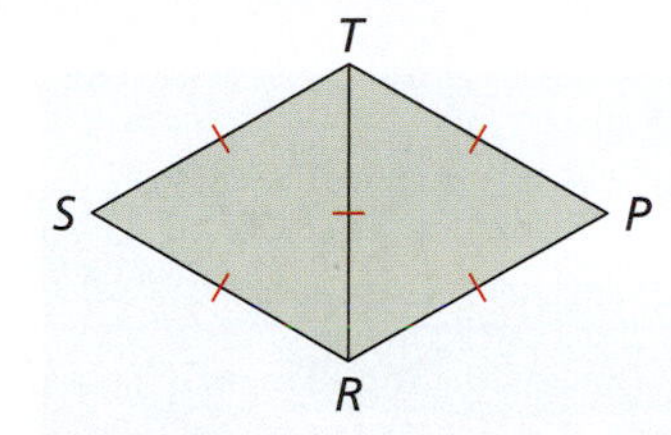

29. **Grading Student Work** On a diagnostic test, one of your students does the following work. Explain what the student did wrong. Which topics would you encourage the student to review?

True or False?

A. **B.** **C.**

1. Figure B is a rotation of figure A. True
2. Figure C is a rotation of figure B. False
3. Figure B is a translation of figure A. True
4. Figure C is a translation of figure B. True

30. **Grading Student Work** On a diagnostic test, one of your students does the following work. Explain what the student did wrong. Which topics would you encourage the student to review?

Draw an example of a translation.

31. **Activity: In Your Classroom** Design an activity you could use to show your students that direction (clockwise or counterclockwise) does not matter when rotating 180°.

32. **Creating a Visual: In Your Classroom** In Exercises 27 and 28, what do you observe about these rotations? Describe how you could make a visual representation to assist your class in identifying the center of rotation.

33. **Tessellation** In Section 10.3, translating a cutout from one side of a square to the opposite side produced an irregular shape that tessellated a plane. Create an irregular shape by rotating a cutout from one side of a square to an adjacent side. Use this shape to tessellate a plane. How does this tessellation differ from the tessellation in Section 10.3?

34. **Quilting** Describe the transformations that can be seen in the quilt block.

35. **Identifying Rotational Symmetry** Name three regular polygons that have 180° rotational symmetry. Name three regular polygons that do *not* have 180° rotational symmetry.

36. **Chess** In chess, the knight can only move in an L-shaped pattern.

- Move two vertical squares then one horizontal square
- Move two horizontal squares then one vertical square
- Move one vertical square then two horizontal squares
- Move one horizontal square then two vertical squares

Write a series of translations to move the knight from g8 to g5.

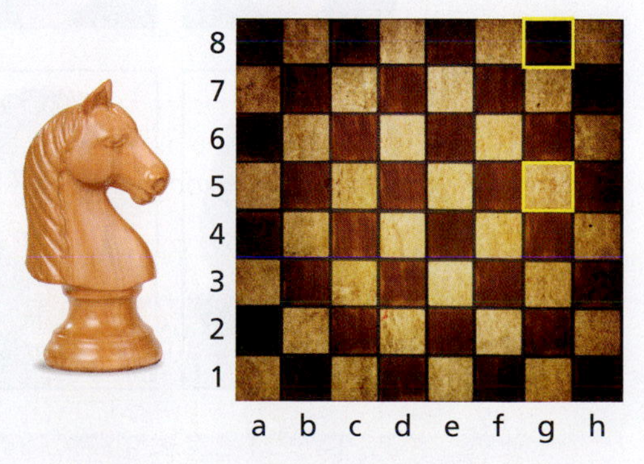

Maryna Gviazdovska/Shutterstock.com; Triff/Shutterstock.com

Activity: Recognizing Reflections and Symmetry

Materials:

- Pencil

Learning Objective:

Understand the concept of reflections. Understand the definition of symmetrical. Recognize when a photograph is symmetrical.

Name ______________________________

When you are looking at the photo that shows a mountain lake, you can see the *reflection*, or mirror image, of the mountain in the lake. When you fold the photo on the line shown, the mountain and its reflection will coincide.

Fold line

A figure is *symmetrical* when it is a reflection of itself in a line.

Are people's faces symmetrical? Which of the following photographs is the real person and which is a symmetrical creation?

1.

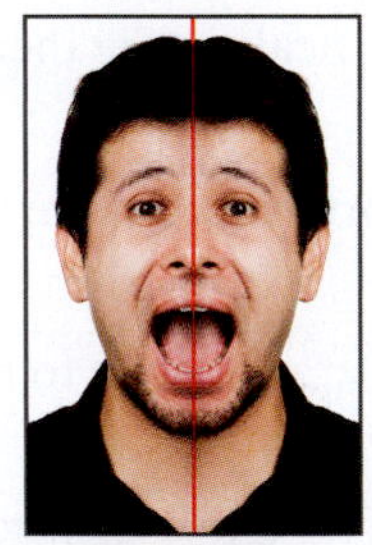

2.

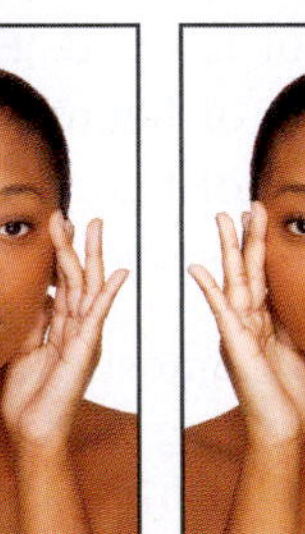

3.

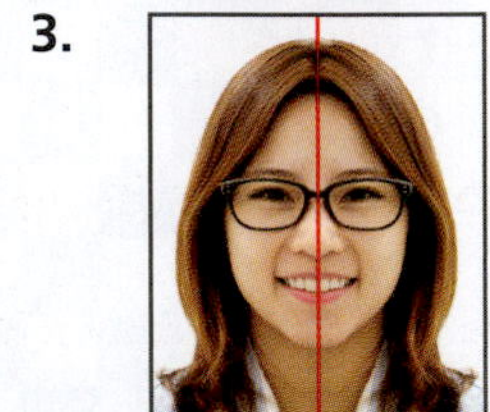

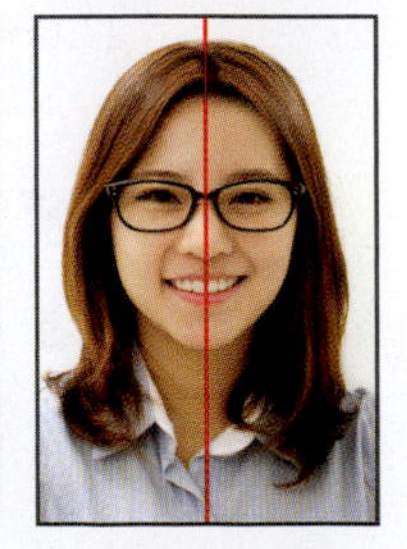

4.

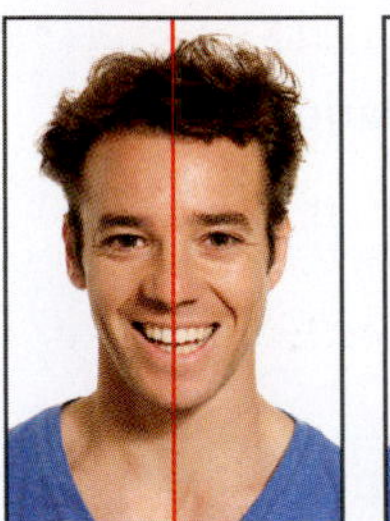

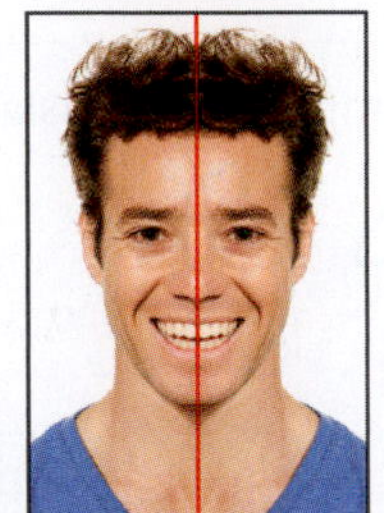

14.2 Reflections

Standards

Grades 3–5
Geometry
Students should draw and identify lines and angles, and classify shapes by properties of their lines and angles.

Grades 6–8
Geometry
Students should understand congruence and similarity using physical models, transparencies, or geometry software.

- Understand and use the concept of a reflection in a line.
- Understand and use the definition of a glide reflection.
- Understand and use the definition of a frieze pattern.

Reflections

Definition of Reflection

A **reflection** is a transformation in which a figure *flips* in a line called the **line of reflection**. A reflection creates a mirror image of the original figure.

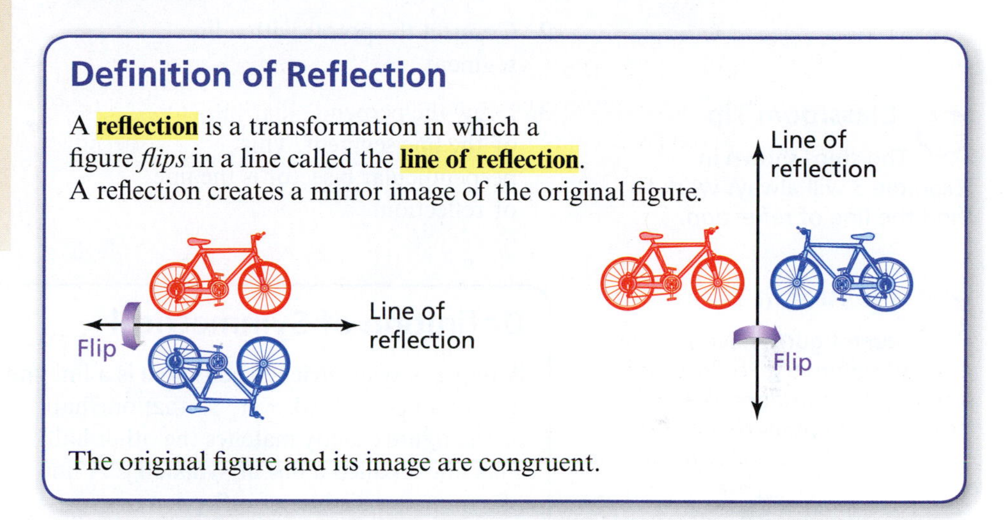

The original figure and its image are congruent.

EXAMPLE 1 Identifying Reflections

Determine whether the blue figure is a reflection of the red figure.

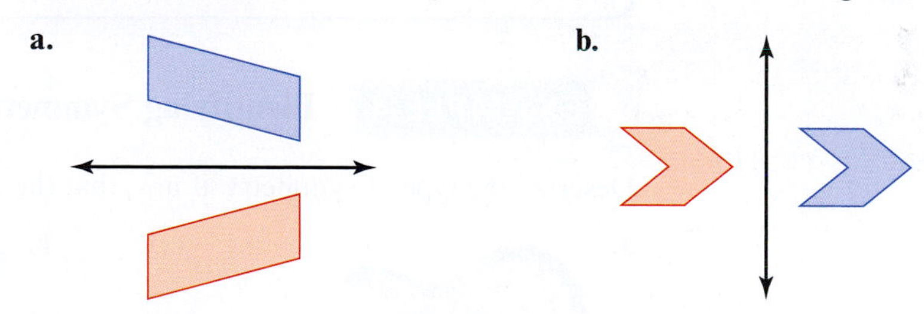

SOLUTION

a. The red parallelogram *flips* to form the blue parallelogram. So, the blue parallelogram *is* a reflection of the red parallelogram.

b. The red arrow cannot be flipped in the line to form the blue arrow. When you flip the red arrow in the line, it will point to the left. So, the blue arrow is *not* a reflection of the red arrow. It is a translation.

EXAMPLE 2 Identifying Reflections in Typesetting

Which of the letters of the alphabet are the same when reflected in a horizontal line?

SOLUTION

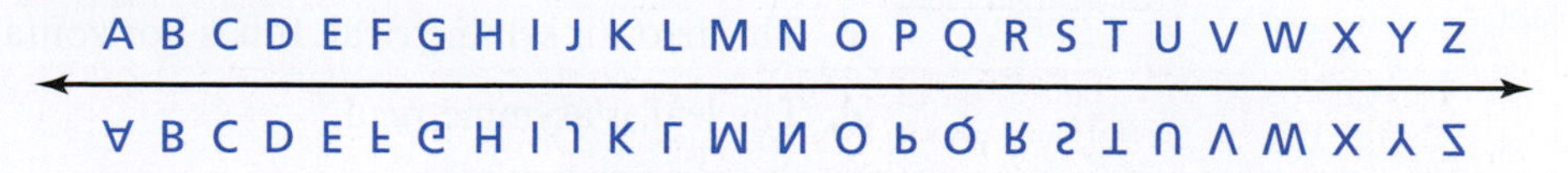

For the font shown above, the letters B, C, D, E, H, I, K, O, and X are the same.

EXAMPLE 3 Finding a Line of Reflection

The red triangle reflects to form the blue triangle. Find the line of reflection.

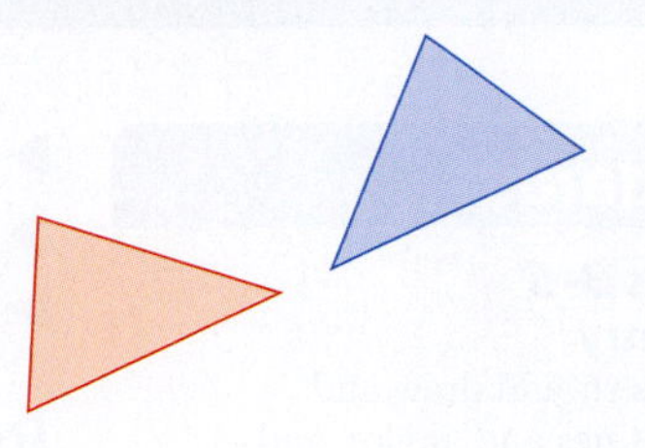

SOLUTION

1. Identify a pair of corresponding points (one point on each triangle).
2. Connect the points with a line segment.
3. Draw the perpendicular bisector of the line segment. This perpendicular bisector is the line of reflection.

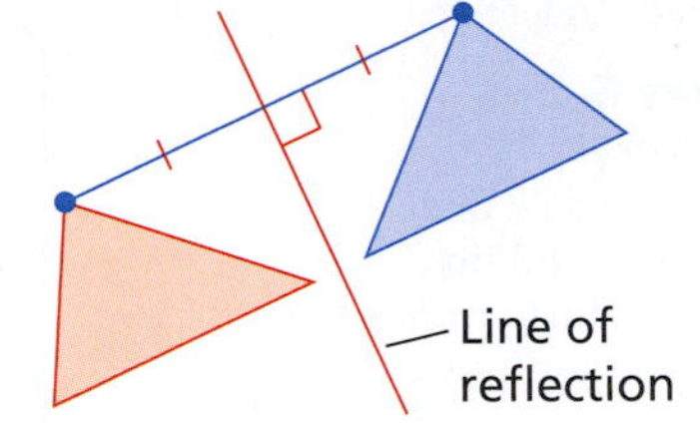

Classroom Tip

The steps shown in Example 3 will always work to find the line of reflection.

Classroom Tip

Some figures have several lines of symmetry. For instance, the regular pentagon below has five lines of symmetry.

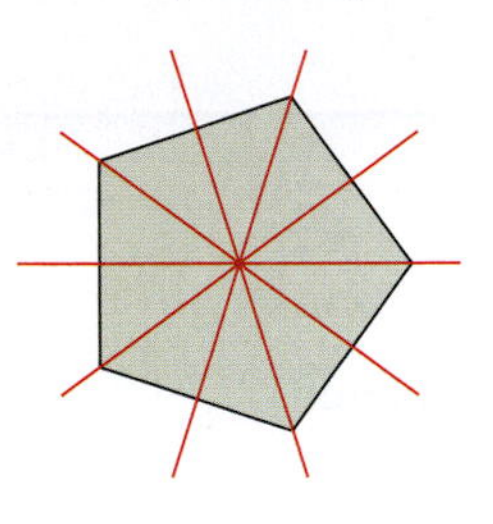

Definition of Symmetrical

A figure is **symmetrical** when there is a line the figure can be "folded over" so that one half of the figure exactly matches the other half. This line is called a **line of symmetry**. A line of symmetry is a line of reflection.

When the line of symmetry is vertical, the figure has **vertical symmetry**. When the line of symmetry is horizontal, the figure has **horizontal symmetry**. A figure that has no line of symmetry is called **asymmetrical**.

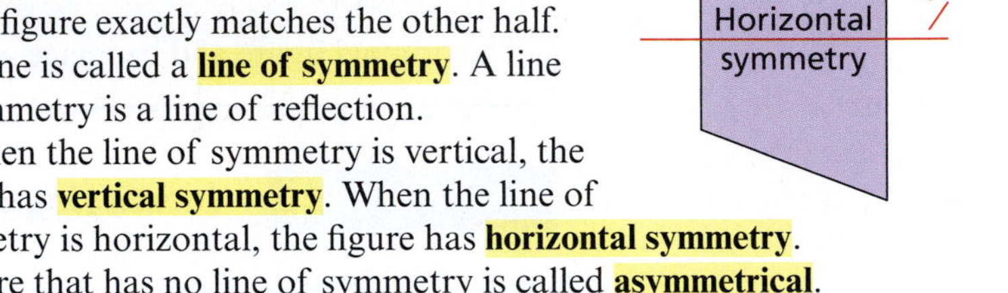

EXAMPLE 4 Identifying Symmetry in Real Life

Describe the type of symmetry, if any, that the object has.

a.

b.

c.

d.

Mathematical Practices

Attend to Precision

MP6 Mathematically proficient students use clear definitions in discussions with others and in their own reasoning.

SOLUTION

a. The electric guitar is asymmetrical.

b. The butterfly is symmetrical. It has vertical symmetry.

c. The arrow is symmetrical. It has horizontal symmetry.

d. The leaf is asymmetrical.

Glide Reflections

Definition of Glide Reflection

A **glide reflection** is a transformation that is composed of a translation (the glide) followed by a reflection in a line that is parallel to the direction of the translation.

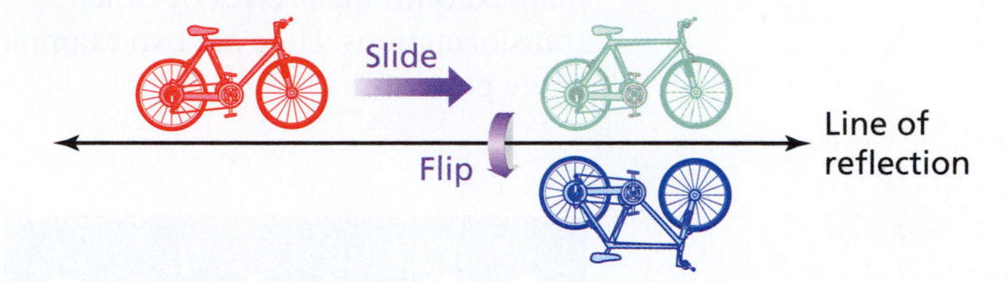

The order of the two transformations is not important. You obtain the same result regardless of which transformation you perform first.

EXAMPLE 5 Identifying Glide Reflections in Real Life

Tell whether the objects represent glide reflections. If they do, then show that the results are the same whether the reflection is performed first or second.

a.

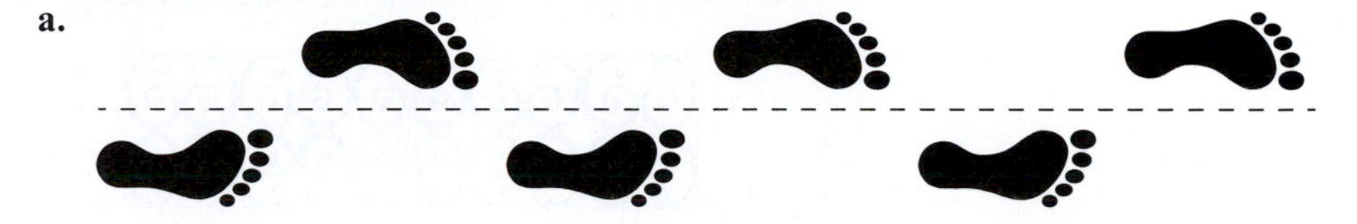

b.

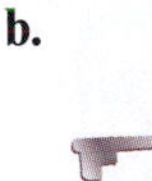

SOLUTION

a. Yes, the footprints do represent a glide reflection. Each footprint can glide forward, then reflect about a horizontal line to create the next footprint, or each footprint can reflect about a horizontal line, then glide forward.

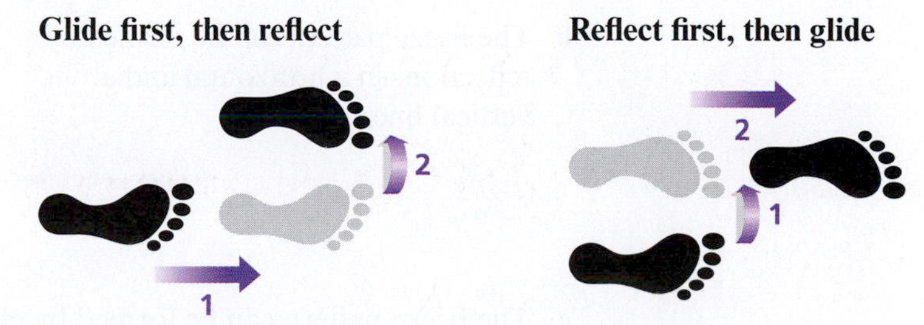

b. No, the pendants of the necklace do *not* represent a glide reflection. You cannot translate and reflect the pendant to coincide with the other pendant. They must rotate to coincide with one another.

Frieze Patterns

A **frieze** is a horizontal band that runs at the top of a building. A frieze is often decorated with a design that repeats. A **frieze pattern** extends to the left and right so that the pattern can be mapped onto itself. Besides horizontal translations, some frieze patterns can be mapped onto themselves by other transformations. Here are two examples of frieze patterns.

EXAMPLE 6 **Identifying Transformations in Frieze Patterns**

Describe the transformation(s) that form each frieze pattern.

a.

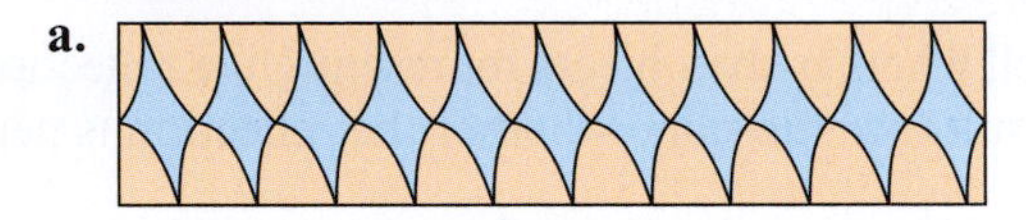

b.

c.

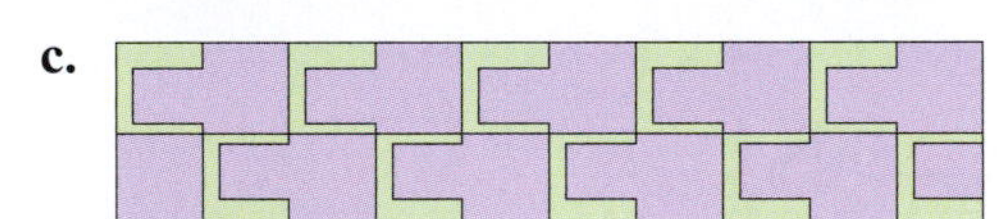

Classroom Tip

A nice project for students is to have them design their own frieze patterns. Then ask each student whether their pattern has horizontal or vertical symmetry, or both.

SOLUTION

a. The frieze pattern can be formed by translations.

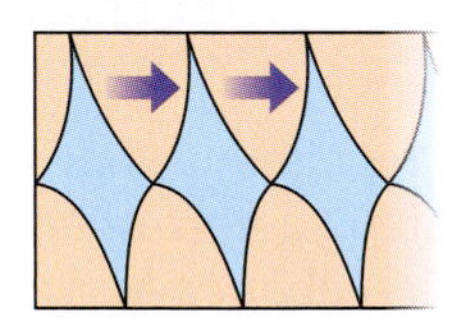

b. The frieze pattern can be formed by reflections in a horizontal and a vertical line.

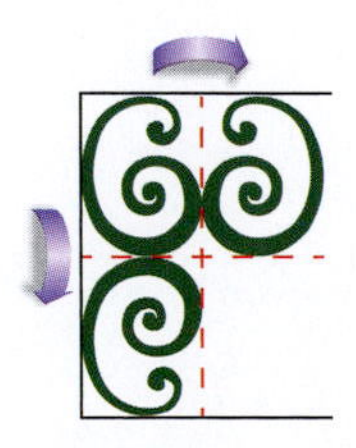

c. The frieze pattern can be formed by glide reflections.

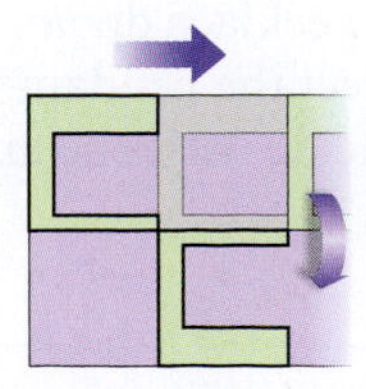

EXAMPLE 7 Identifying Symmetry in a Frieze Pattern

a. Does the frieze pattern have horizontal symmetry? Explain.

b. Does the frieze pattern have vertical symmetry? Explain.

SOLUTION

a. Fold the pattern horizontally. The top and bottom halves of the pattern coincide.

So, the frieze pattern has horizontal symmetry.

b. Fold the pattern vertically. The left and right halves of the pattern do *not* coincide. So, the frieze pattern does not have vertical symmetry.

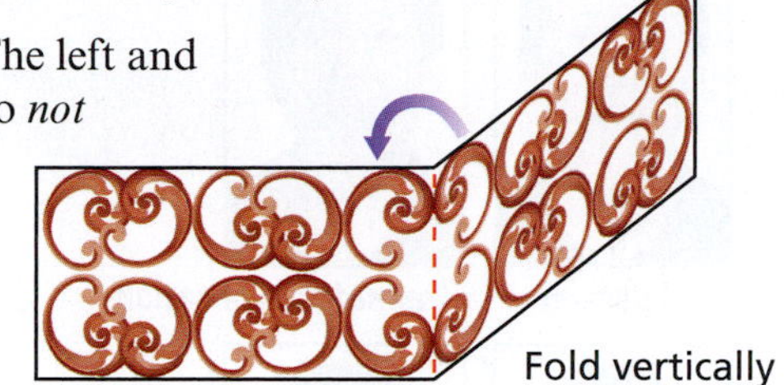

EXAMPLE 8 Analyzing a Frieze Pattern in Real Life

Wallpaper is usually hung vertically using vertical lines as guides. What characteristic of the wallpaper design shown at the left makes it possible to be used as wallpaper?

SOLUTION

The wallpaper design represents a frieze pattern. The pattern along the edge repeats the pattern down the middle of the roll. When hung side-by-side, the edges align to repeat the pattern that is shown down the middle of the roll.

14.2 Exercises

Free step-by-step solutions to odd-numbered exercises at *MathematicalPractices.com.*

1. **Writing a Solution Key** Write a solution key for the activity on page 540. Describe a rubric for grading a student's work.

2. **Grading the Activity** In the activity on page 540, a student gave the answers below. Each question is worth 10 points. Assign a grade to each answer. For those that are incorrect, why do you think the student erred?

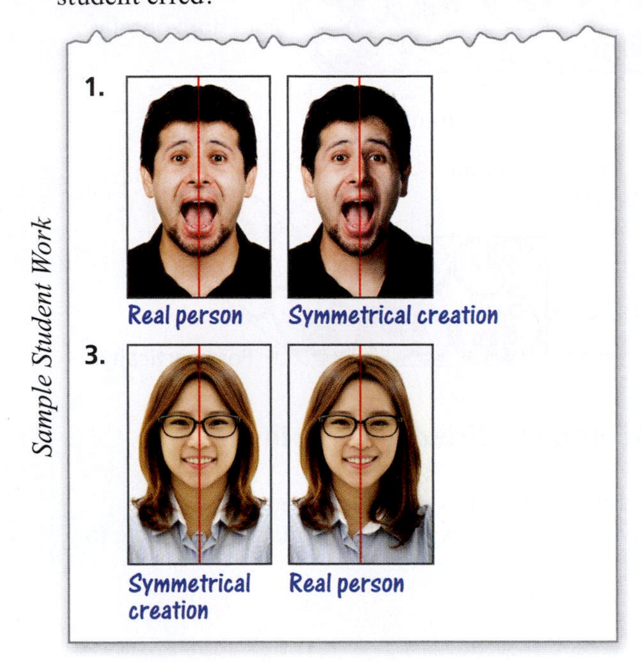

3. **Identifying Reflections** Determine whether each blue figure is a reflection of the red figure. Explain your reasoning.

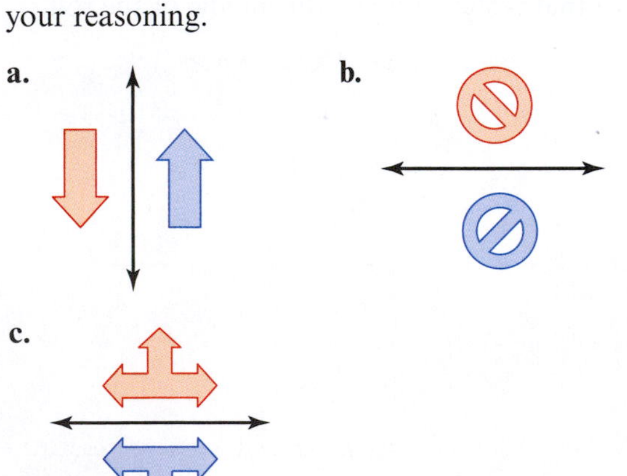

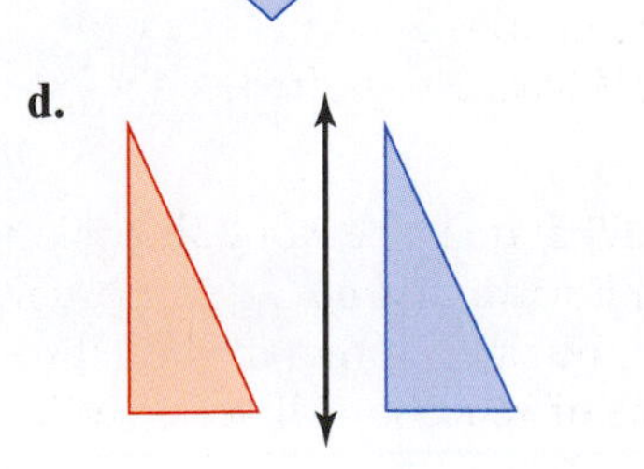

4. **Identifying Reflections** Determine whether each blue figure is a reflection of the red figure. Explain your reasoning.

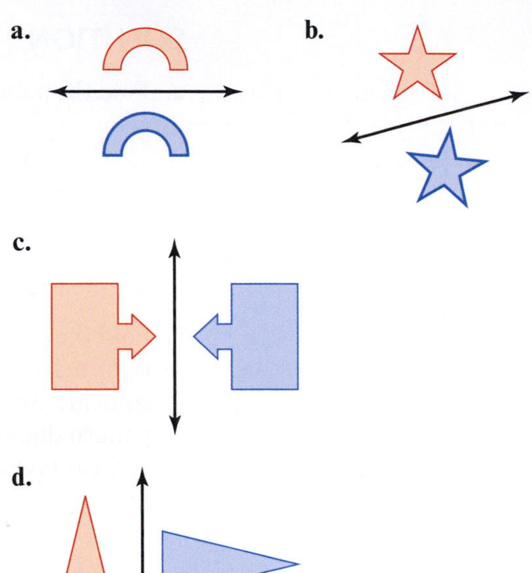

5. **Finding Lines of Reflection** Each blue figure is a reflection of the red figure. Draw the line of reflection.

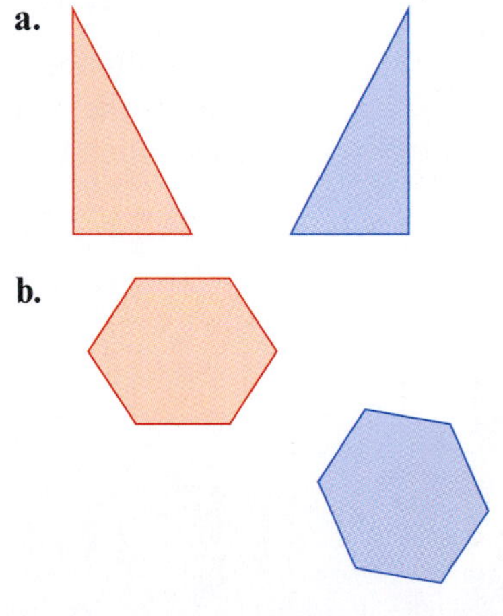

6. **Finding Lines of Reflection** Each blue figure is a reflection of the red figure. Draw the line of reflection.

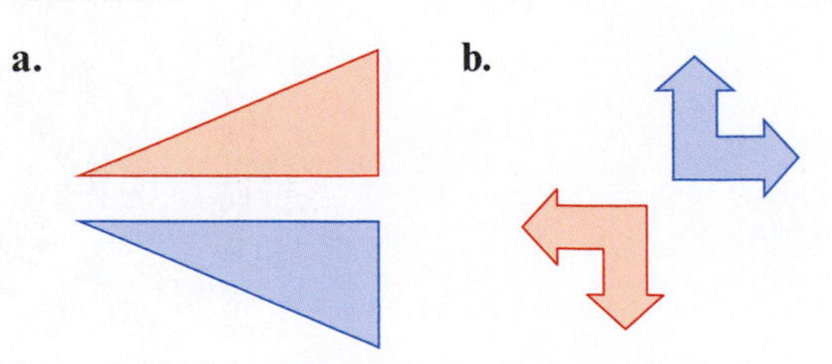

7. Identifying Symmetry Describe the type of symmetry, if any, that each figure has.

a.

b.
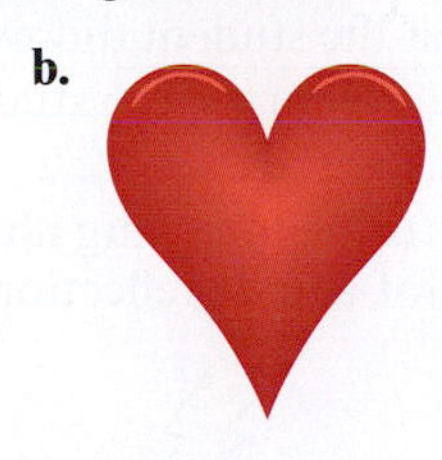

8. Identifying Symmetry Describe the type of symmetry, if any, that each figure has.

a.
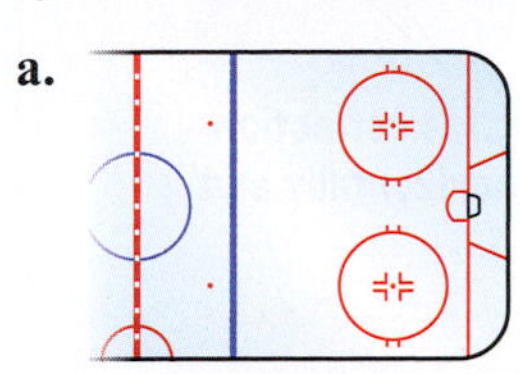

b.

9. Identifying Lines of Symmetry Determine the number of lines of symmetry in each figure.

a.
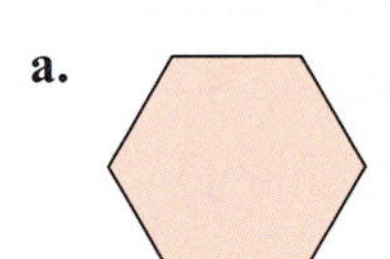

b.
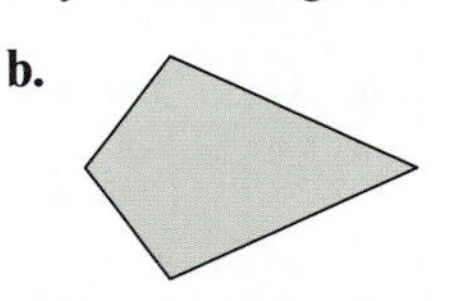

10. Identifying Lines of Symmetry Determine the number of lines of symmetry in each figure.

a.
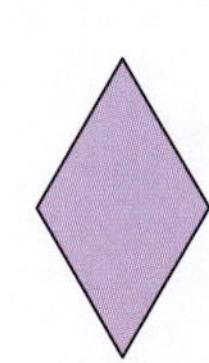

b.
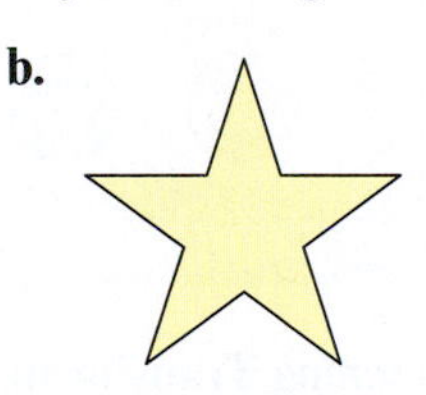

11. Drawing Reflections Complete each image by drawing the reflection in the given line of symmetry.

a.

b.

12. Drawing Reflections Complete each image by drawing the reflection in the given line of symmetry.

a.
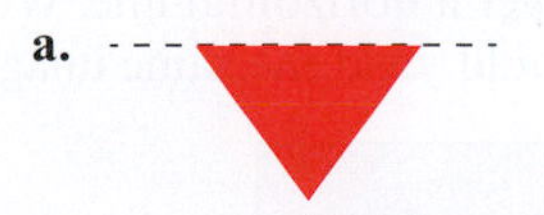

b.

13. Identifying Glide Reflections Determine whether each pair of figures is a glide reflection. Explain your reasoning.

a.

b.

14. Identifying Glide Reflections Determine whether each pair of figures is a glide reflection. Explain your reasoning.

a.

b.
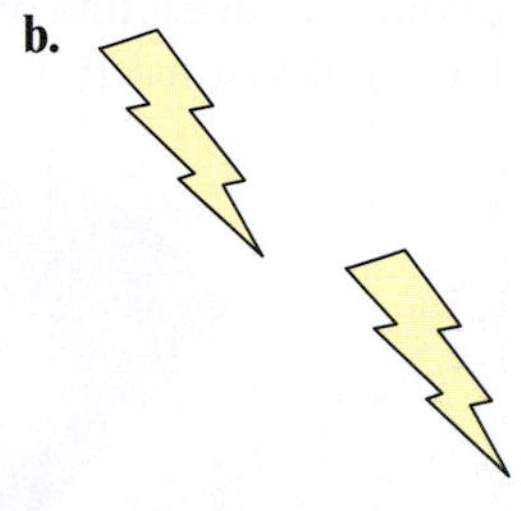

15. Identifying Transformations in Frieze Patterns Describe the transformation(s) that form each frieze pattern.

a.

b.

16. Identifying Symmetry in Frieze Patterns Describe any symmetry in each frieze pattern.

a.

b.

17. Word Reflections What does the word MOM look like when reflected in a horizontal line under the word? in a vertical line to the right of the word? Does the word MOM have a line of reflection? If so, describe the line.

18. Word Reflections What does the word BIB look like when reflected in a horizontal line under the word? in a vertical line to the right of the word? Does the word BIB have a line of reflection? If so, describe the line.

19. **Transformations in Real Life** Does the soccer field have a line of symmetry? If so, draw the line(s) of symmetry.

20. **Transformations in Real Life** Does the baseball diamond have a line of symmetry? If so, draw the line(s) of symmetry.

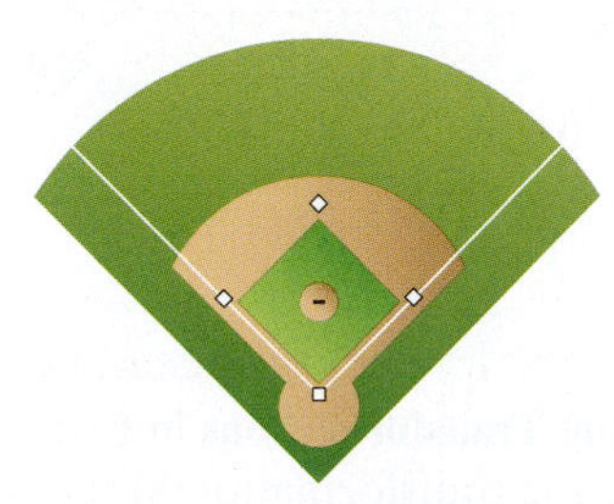

21. **Activity: In Your Classroom** Design an activity that students can complete with partners to determine when a frieze pattern has horizontal symmetry, vertical symmetry, or neither. Then have students create their own frieze pattern to share.

22. **Artwork: In Your Classroom** To review transformations, have students create their own artwork by tracing their handprints and arranging them on a poster to demonstrate each transformation. Create an example to demonstrate the activity to your students.

23. **Review Activity: In Your Classroom** To review transformations, have students create a study tool, such as a graphic organizer that reviews the definition of each transformation and includes examples. Create a sample of this activity.

24. **Reflection Art: In Your Classroom** Fold a piece of paper in half. Unfold the paper and sign your name on one half of the paper with chalk or paint. Refold the paper and rub hard to create the reflection of your name on the other half of the paper. What mathematical term could be assigned to the fold line in the paper? Repeat the activity creating a design that also has rotational symmetry.

Dave King/Dorling Kindersley/Getty Images

25. **Grading Student Work** On a diagnostic test, one of your students does the following work. Explain what the student did wrong. Which topics would you encourage the student to review?

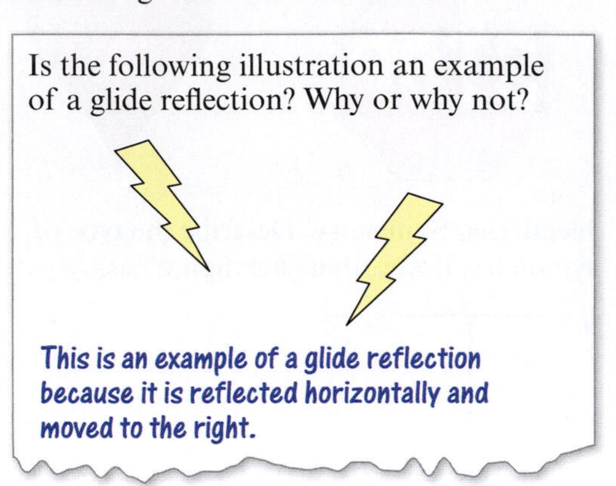

Is the following illustration an example of a glide reflection? Why or why not?

This is an example of a glide reflection because it is reflected horizontally and moved to the right.

26. **Grading Student Work** On a diagnostic test, one of your students does the following work. Explain what the student did wrong. Which topics would you encourage the student to review?

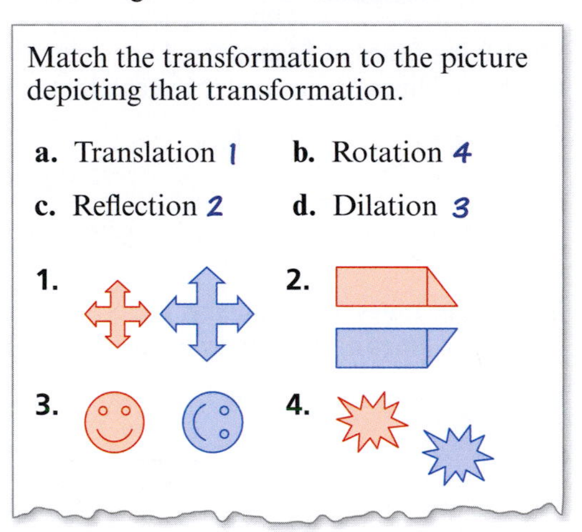

Match the transformation to the picture depicting that transformation.

a. Translation *1* **b.** Rotation *4*

c. Reflection *2* **d.** Dilation *3*

1. **2.**

3. **4.**

27. **Sketching Transformations** Sketch a reflection of the figure over a vertical line. Then sketch a reflection of the image over a different vertical line. What single transformation would yield the second image?

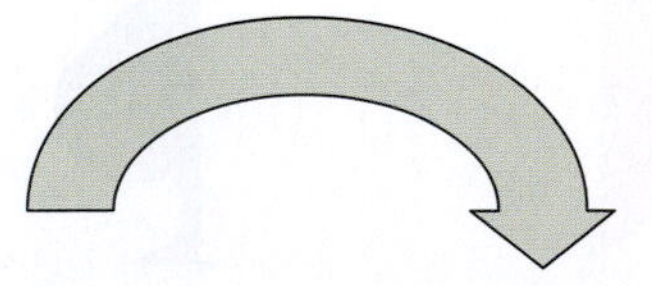

28. **Sketching Transformations** Sketch a reflection of the figure over a vertical line. Then sketch a reflection of the image over a horizontal line. What single transformation would yield the same image?

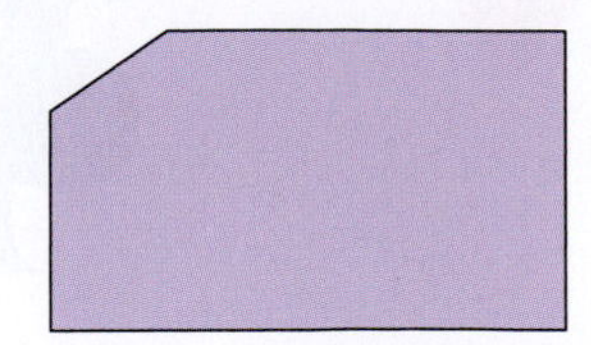

29. *Writing* Determine how many lines of symmetry a square has. How many different ways can you fold a square so that the edges of the figure match up perfectly? Try other regular polygons. What do you observe about the number of sides and the number of lines of symmetry? Predict what would happen when doing this activity with a circle.

30. *Writing* Look up the definitions of translation, rotation, and reflection in a dictionary. How do those definitions relate to the definitions for translation, rotation, and reflection studied in this chapter?

31. Combinations of Transformations The blue figure is a glide reflection of the red figure.

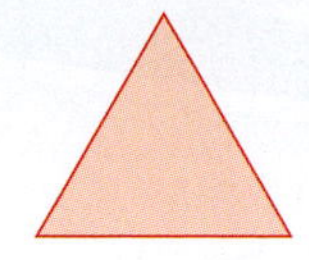

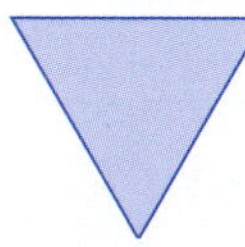

a. What other single transformation would produce the same image?

b. Give an example of a pair of transformations that would produce the same image.

32. Combinations of Transformations The blue figure is a translation of the red figure.

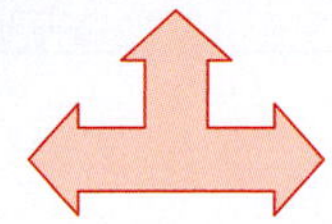

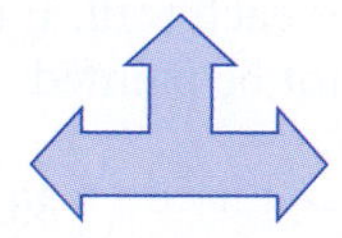

a. What other single transformation would produce the same image?

b. Give an example of a pair of transformations that would produce the same image.

33. Determining Lines of Symmetry Determine the lines of symmetry, if any, in the symbol for recycling.

34. Determining Lines of Symmetry Draw the lines of symmetry for the helm of a ship.

35. Symmetry Triangle ABD is a reflection of triangle CBD. The length of $\overline{AB}$ is 10 inches and the length of $\overline{AD}$ is 8 inches.

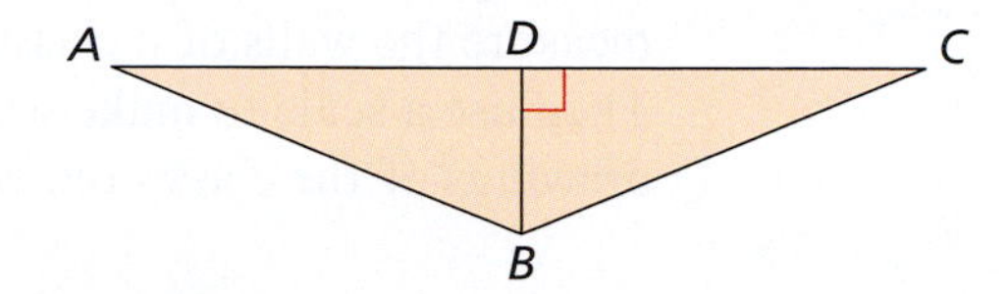

a. Find the perimeter of triangle ABC. Explain how you determined the measurements needed to find the perimeter.

b. Find the area of triangle ABC. Explain how you determined the measurements needed to find the area.

36. Symmetry Trapezoid $ABCD$ is a reflection of trapezoid $AFED$ and $\overline{AB} \cong \overline{CD}$. The length of $\overline{AB}$ is 5 meters and the length of $\overline{BC}$ is 8 meters.

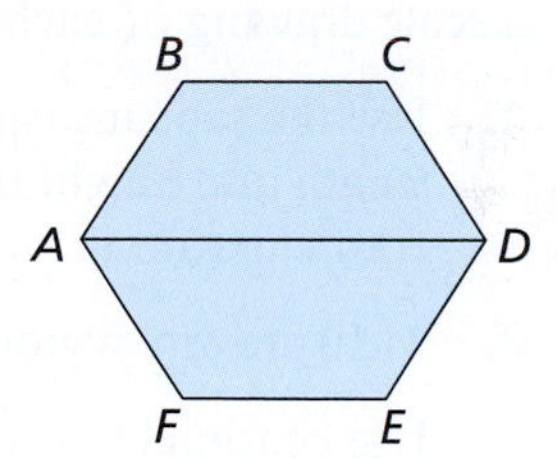

a. Find the perimeter of hexagon $ABCDEF$. Explain how you determined the measurements needed to find the perimeter.

b. Line segment $\overline{BF}$ is drawn and has a length of 8 meters. $\overline{BF}$ is perpendicular to $\overline{AD}$. Find the area of hexagon $ABCDEF$. Explain how you determined the measurements needed to find the area.

37. Wallpaper Border You are hanging wallpaper border at the top of your living room walls. What transformation is taking place from wall to wall when the corner of the wall is at line A? line B? Explain your reasoning.

Activity: Creating a Scale Drawing

Materials:

- Pencil
- Tape measure
- Centimeter grid paper

Learning Objective:

Understand the concept of a scale drawing. Use a tape measure to measure the walls of a classroom. Then use a scale to make scale drawings of the classroom walls.

Name ______________________

A *scale drawing* is a representation of a real-life object that is not the same size as the actual object. The ratio of the dimensions of the scale drawing to the dimensions of the actual object is called the *scale*.

You want to paint the four walls of your classroom. To begin, you make a scale drawing of each wall.

1. Use the tape measure to find the length and height of each wall of your classroom.
2. Measure any regions that will *not* be painted.
3. Use centimeter grid paper to make a scale drawing of each wall. Each centimeter corresponds to 1 foot. Include the regions that will not be painted. A sample is shown.

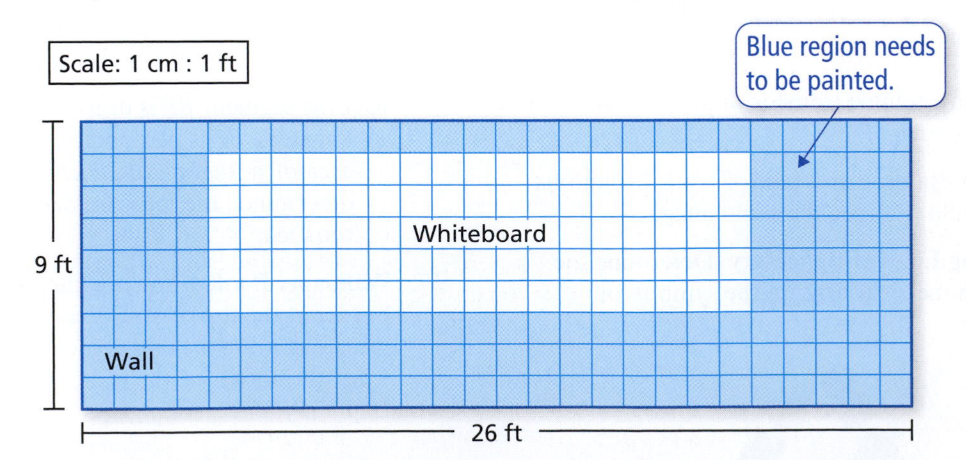

4. One gallon of paint covers about 350 square feet. How many gallons of paint do you need to cover the walls with one coat of paint?

14.3 Dilations and Scale Drawings

Standards

Grades 6–8
Geometry
Students should draw, construct, and describe geometrical figures and describe the relationships between them.

- Understand and use the concept of dilation.
- Find the scale factor of a dilation. Use a scale factor to draw a dilation.
- Use the concept of a dilation to read and create scale drawings.

Dilations

Definition of Dilation

A **dilation** is a transformation in which a figure is made larger or smaller with respect to a fixed point called the **center of dilation**.

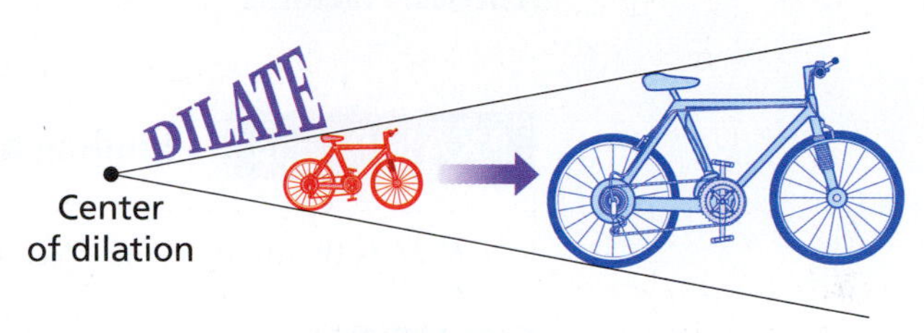

The original figure and its image are similar. The ratio of the dimensions of the image to the dimensions of the original figure is the **scale factor** of the dilation. When the scale factor is greater than 1, the dilation is an **enlargement**. When the scale factor is between 0 and 1, the dilation is a **reduction**.

EXAMPLE 1 **Identifying Dilations**

Determine whether the blue figure is a dilation of the red figure. If it is, tell whether it is a reduction or an enlargement. Then estimate the scale factor.

a.

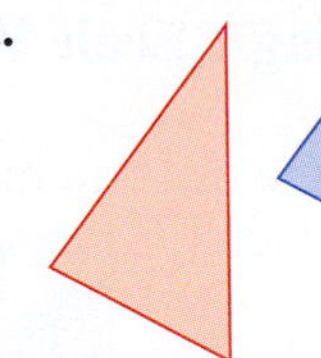

b.

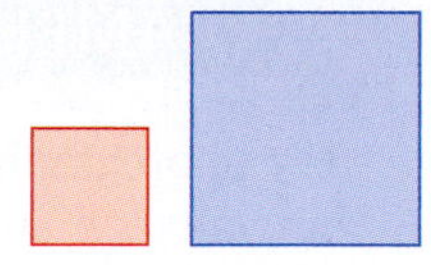

c.

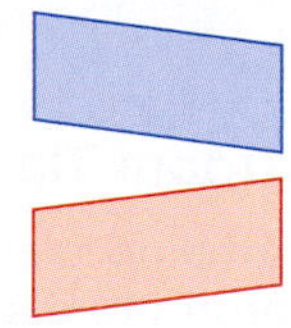

Classroom Tip

When an ophthalmologist examines your eyes, you may be given eye drops that cause your pupils to dilate (get larger). This allows the ophthalmologist to look way into the back of your eyes.

SOLUTION

a. Lines connecting corresponding vertices meet at a point. So, the blue figure is a dilation.

Because the blue triangle is smaller, the dilation is a *reduction*. The dimensions of the blue triangle are about one-half the dimensions of the red triangle. So, the scale factor is about $\frac{1}{2}$.

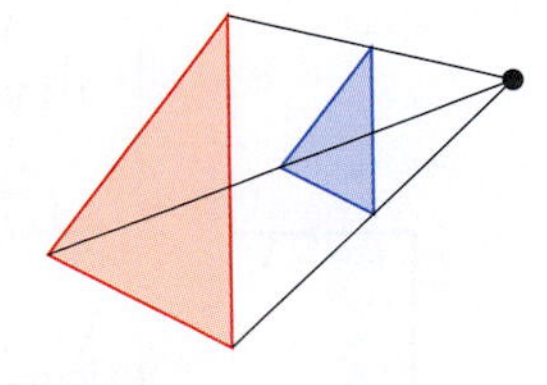

b. Lines connecting corresponding vertices meet at a point. So, the blue figure is a dilation.

Because the blue square is larger, the dilation is an *enlargement*. The dimensions of the blue square are about twice the dimensions of the red square. So, the scale factor is about 2.

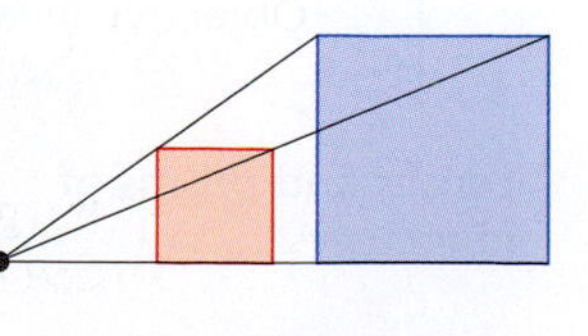

c. The red figure *reflects* to form the blue figure. So, the blue figure is not a dilation.

Finding and Using Scale Factors

The distances from the center of dilation to the corresponding points of the original figure and its image can also be used to find the scale factor.

EXAMPLE 2 Finding a Scale Factor

Find the scale factor of the dilation.

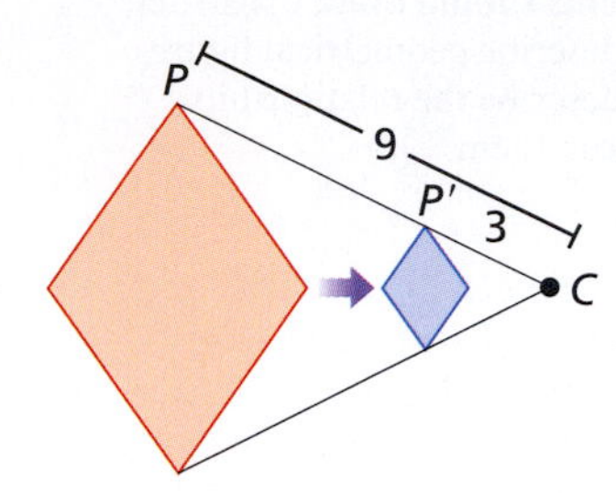

SOLUTION

Find the ratio of CP' to CP.

$$\frac{CP'}{CP} = \frac{3}{9} = \frac{1}{3}$$

The scale factor is $\frac{1}{3}$.

EXAMPLE 3 Finding a Side Length

$\triangle J'K'L'$ is the image of $\triangle JKL$. Find the value of x.

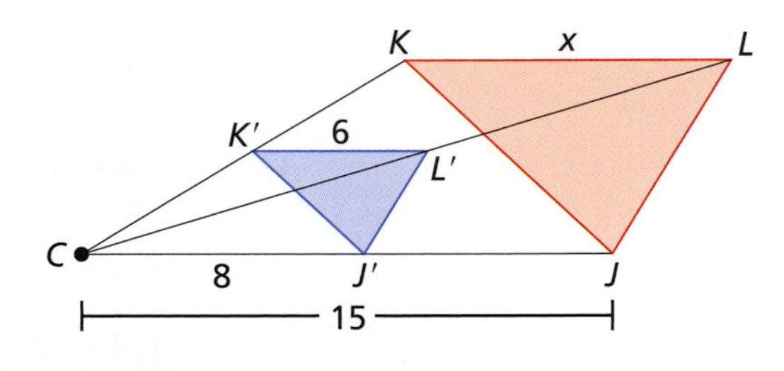

SOLUTION

Write and solve a proportion.

$$\frac{CJ'}{CJ} = \frac{K'L'}{KL}$$

$$\frac{8}{15} = \frac{6}{x}$$

$$8x = 90$$

$$x = 11.25$$

EXAMPLE 4 Finding a Scale Factor in Real Life

The photographs below show the size of a bird with and without the use of binoculars. Estimate the scale factor.

Without magnification

With magnification

1 2 3 4 5 6 7

The magnification of the binoculars is 7X.

SOLUTION

Compare the dimensions of the two photographs. You can see that with magnification, the image is about 7 times greater. So, the scale factor is

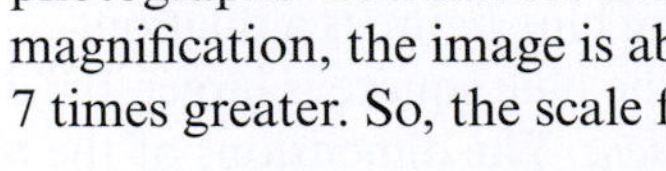

$$\frac{\text{Size with magnification}}{\text{Size without magnification}} = 7.$$

Kletr/Shutterstock.com

Classroom Tip

A pair of binoculars uses lenses and prisms to magnify an object. The magnification of a pair of binoculars is specified as kX, where k is the scale factor.

Eyepiece

Prisms

Objective lens

Lenses and prisms of binoculars

EXAMPLE 5 Drawing a Dilation

For the red parallelogram shown, draw an enlargement using a scale factor of 3.

SOLUTION

Begin by locating a center for the dilation.

Center
of dilation

Next, draw a line segment from the center of dilation to each of the four vertices of the parallelogram.

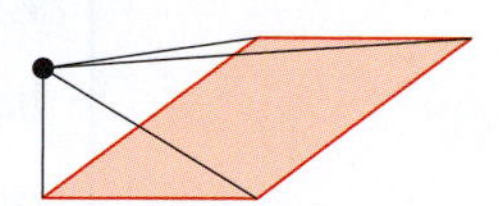

Because the scale factor is 3, triple the length of each line segment. Connect the ends of the longer line segments to form the image.

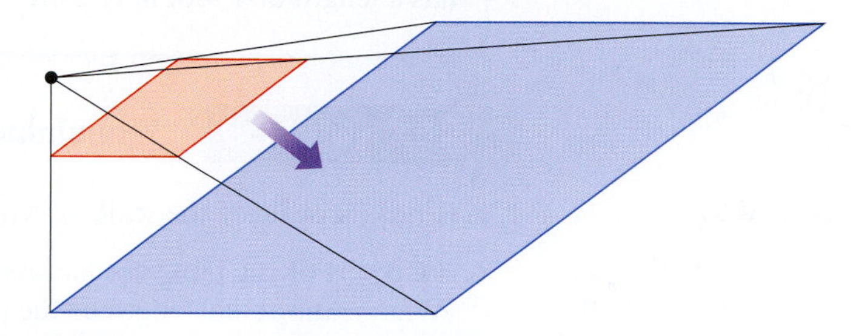

Classroom Tip

The location of the center of dilation determines the location of the image. It does not, however, determine the size of the image. To better understand this, trace the red parallelogram in Example 5. Then use a different center of dilation. You will discover that the blue image is the same size but has a different location.

EXAMPLE 6 Comparing Perimeter and Area

The dimensions of the red parallelogram in Example 5 are shown at the right. Compare (a) the perimeters and (b) the areas of the original figure and its image.

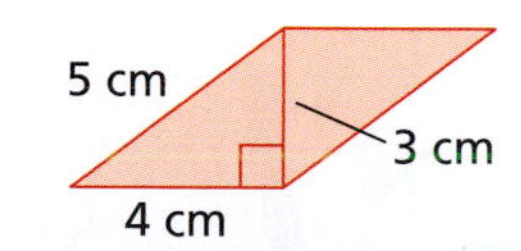

SOLUTION

a. The perimeter of the original (red) figure is

$5 + 4 + 5 + 4 = 18$ centimeters. Perimeter of original figure

The length of each line segment is tripled in Example 5, so the dimensions are tripled. The perimeter of the image (blue figure) is

$15 + 12 + 15 + 12 = 54$ centimeters. Perimeter of image

Because $\frac{54}{18} = 3$, you know that the perimeter of the blue parallelogram is three times the perimeter of the red parallelogram.

b. The area of the original (red) figure is

$4 \cdot 3 = 12$ square centimeters. Area of original figure

The area of the image (blue figure) is

$12 \cdot 9 = 108$ square centimeters. Perimeter of image

Because $\frac{108}{12} = 9$, you know that the area of the blue parallelogram is nine times the area of the red parallelogram.

Mathematical Practices

Make Sense of Problems and Persevere in Solving Them

MP1 Mathematically proficient students consider analogous problems.

Classroom Tip

When a figure is dilated by a scale factor of k, the perimeter is multiplied by a factor of k, and the area is multiplied by a factor of k^2. In Example 6, the ratio of the perimeters is equal to the scale factor of 3. The ratio of the areas is equal to $3^2 = 9$.

Scale Drawings

Definition of a Scale Drawing

A **scale drawing** is a proportional, two-dimensional drawing of an object. The **scale** gives the ratio that compares the measurements of the drawing with the actual measurements of the object. This ratio is usually written as drawing length : actual length.

Example The scale drawing has a scale of $\frac{1}{8}$ inch to 1 foot, or $\frac{1}{8}$ in. : 1 ft. So, anything with a length of $\frac{1}{8}$ inch in the drawing has a length of 1 foot in real life.

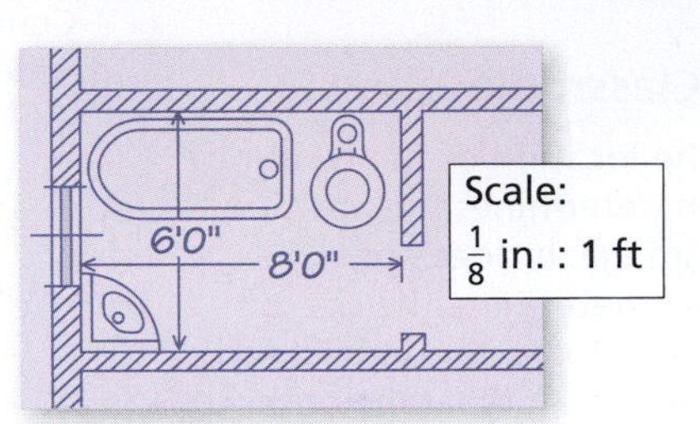

EXAMPLE 7 Examining a Scale Drawing

a. Find the scale of the scale drawing of a junkyard hammer.

b. One part of the junkyard hammer measures 0.5 inch long in the scale drawing. What is the actual length of the part?

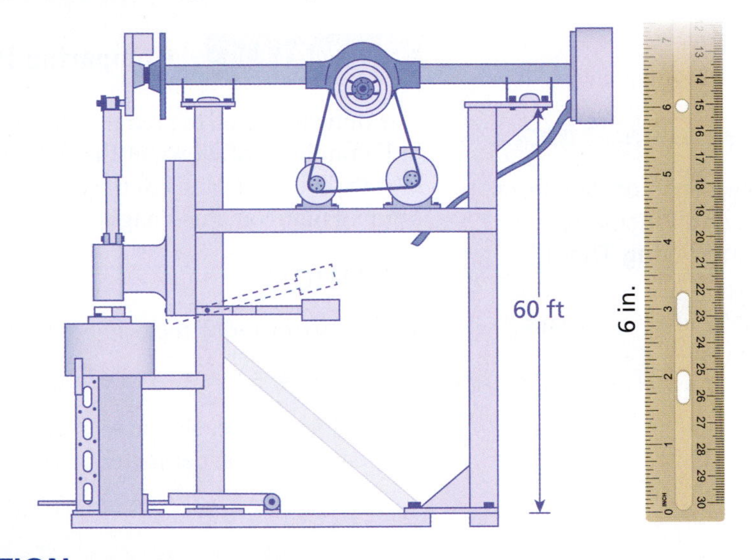

Classroom Tip

Scale drawings used to be drawn by hand in engineering and architecture. Now, however, the process is done using a Computer Aided Design (CAD) system.

SOLUTION

a. Using the ruler, the given length of 60 feet in the drawing is 6 inches long. Write the ratio of these two lengths.

$$\frac{\text{Length in drawing}}{\text{Actual length}} = \frac{6 \text{ in.}}{60 \text{ ft}} = \frac{1 \text{ in.}}{10 \text{ ft}}$$

So, the scale is 1 in. : 10 ft.

b. Use the scale from part (a) to write and solve a proportion.

$$\frac{1 \text{ in.}}{10 \text{ ft}} = \frac{0.5 \text{ in.}}{x \text{ ft}}$$

$$x = 10 \cdot 0.5$$

$$x = 5$$

The actual length of the part is 5 feet.

Classroom Tip

A scale can be written without units when the units are the same. So, the scale of a scale drawing can be written as a scale factor. In Example 7, you can write the following.

$$\frac{6 \text{ in.}}{60 \text{ ft}} = \frac{6 \text{ in.}}{720 \text{ in.}} = \frac{1 \text{ in.}}{120 \text{ in.}}$$

The scale factor is 1 : 120.

EXAMPLE 8 Using a Grid to Create a Scale Drawing

Use a grid to create a scale drawing of the horse. Use a scale of 2 : 1.

SOLUTION

Begin by placing a grid over the drawing of the horse.

Next, draw a grid whose dimensions are doubled. Finally, use the grid to copy each square section of the drawing.

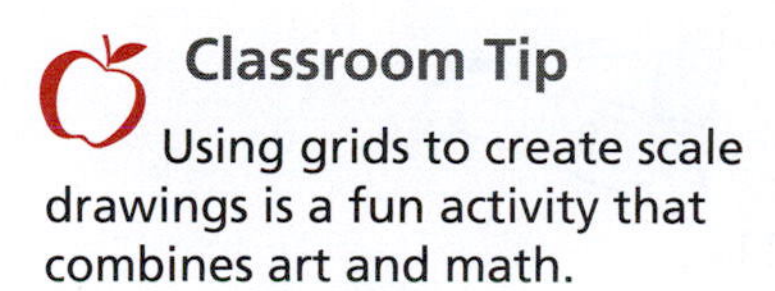

Classroom Tip

Using grids to create scale drawings is a fun activity that combines art and math.

Canicula/Shutterstock.com

14.3 Exercises

Free step-by-step solutions to odd-numbered exercises at *MathematicalPractices.com.*

1. Writing a Solution Key Write a solution key for the activity on page 550. Describe a rubric for grading a student's work.

2. Grading the Activity In the activity on page 550, a student gave the answers below. Each question is worth 10 points. Assign a grade to each answer. For those that are incorrect, why do you think the student erred?

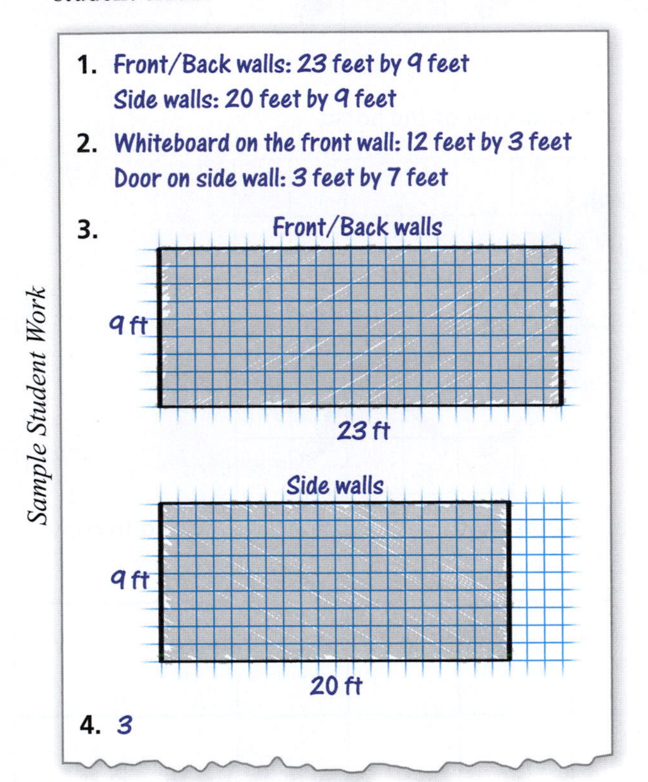
Sample Student Work

1. Front/Back walls: 23 feet by 9 feet
   Side walls: 20 feet by 9 feet
2. Whiteboard on the front wall: 12 feet by 3 feet
   Door on side wall: 3 feet by 7 feet
3. Front/Back walls
   9 ft
   23 ft
   Side walls
   9 ft
   20 ft
4. 3

3. Identifying Dilations Determine whether the blue figure is an enlargement or a reduction of the red figure. Explain your reasoning.

a.

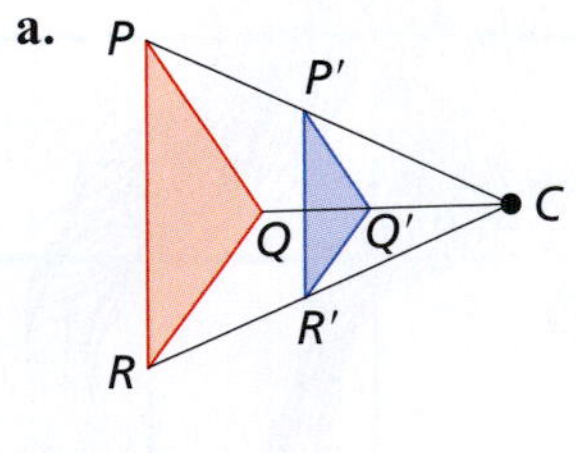

b.

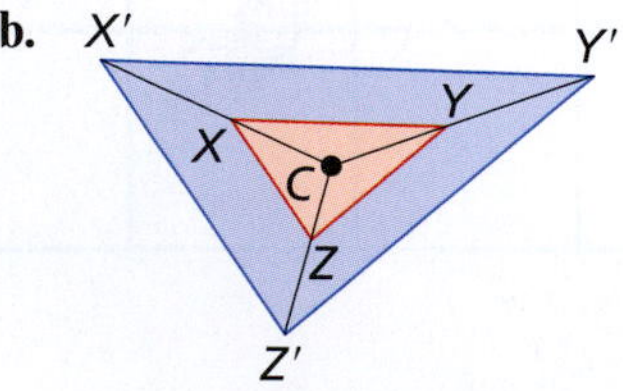

4. Identifying Dilations Determine whether the blue figure is an enlargement or a reduction of the red figure. Explain your reasoning.

a.

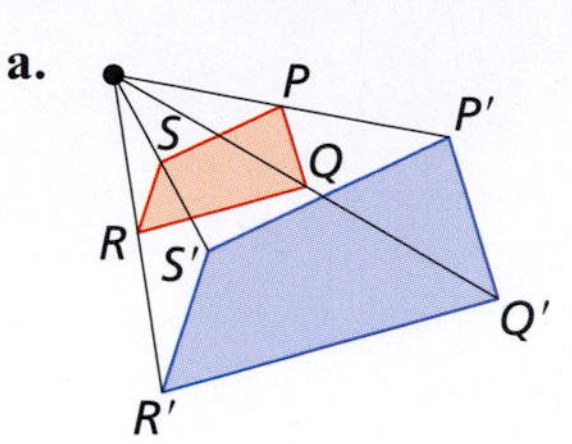

b.

5. Finding Scale Factors Determine whether the blue figure is an enlargement or a reduction of the red figure. Find the scale factor of the dilation.

a.

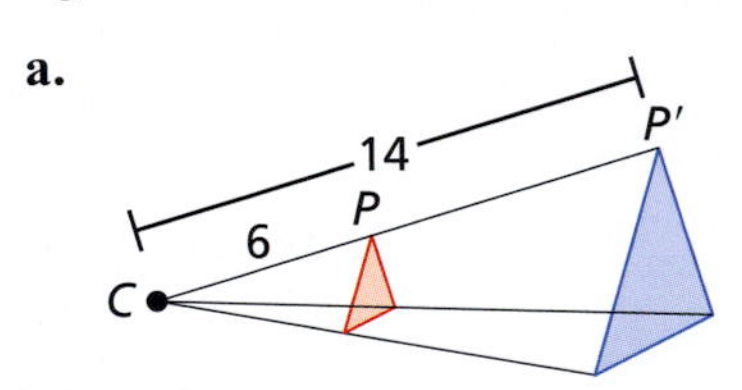

b.

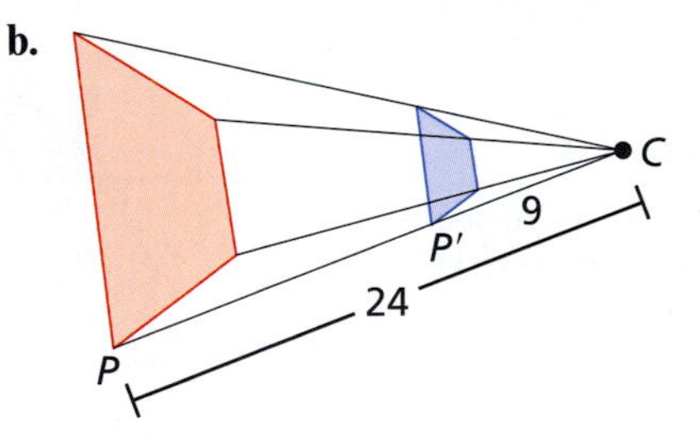

c.

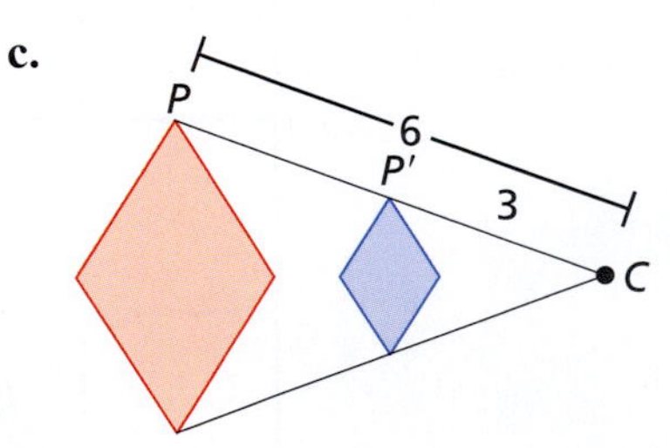

d.

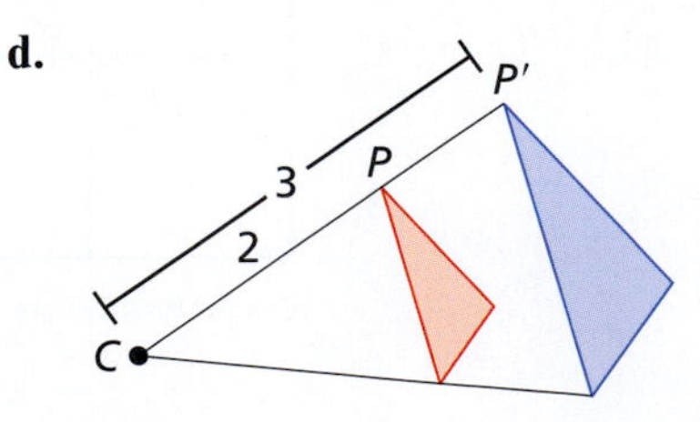

6. Finding Scale Factors Determine whether the blue figure is an enlargement or a reduction of the red figure. Find the scale factor of the dilation.

a.

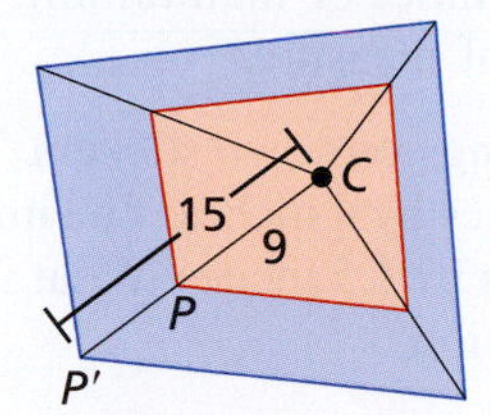

b.

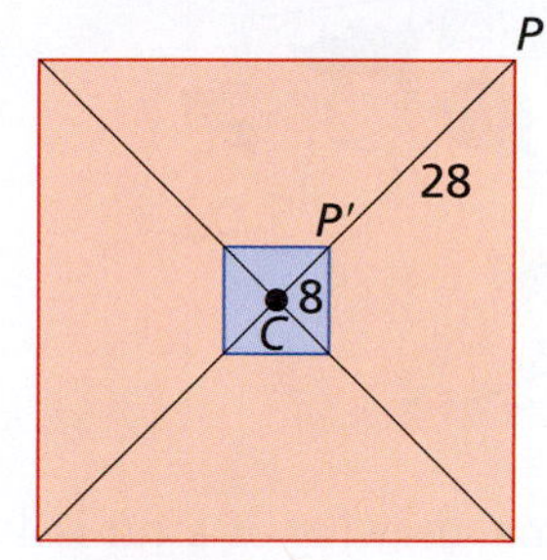

c.

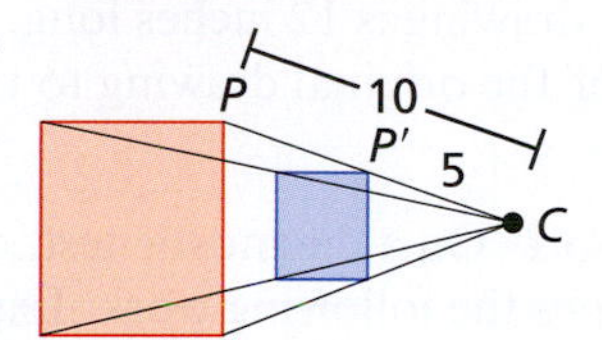

d.

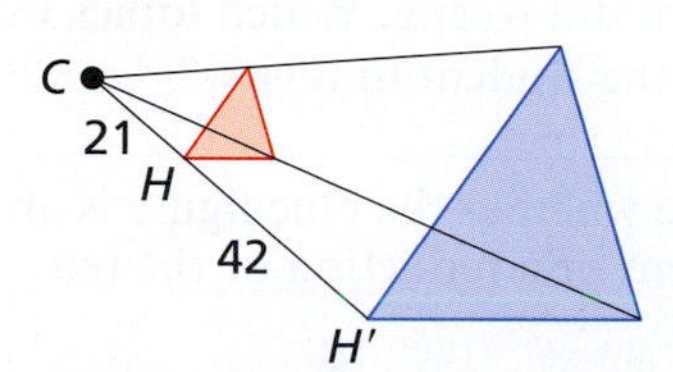

7. Solving Proportions Solve each proportion.

a. $\frac{5}{9} = \frac{x}{45}$ **b.** $\frac{6}{7} = \frac{48}{x}$

8. Solving Proportions Solve each proportion.

a. $\frac{x}{25} = \frac{8}{100}$ **b.** $\frac{12}{x} = \frac{2}{7}$

9. Finding Side Lengths Write and solve a proportion to find each missing length. Tell whether the dilation is a *reduction* or an *enlargement*.

a.

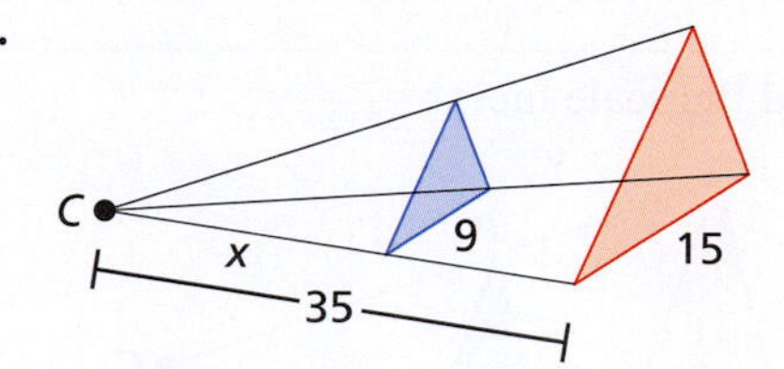

b.

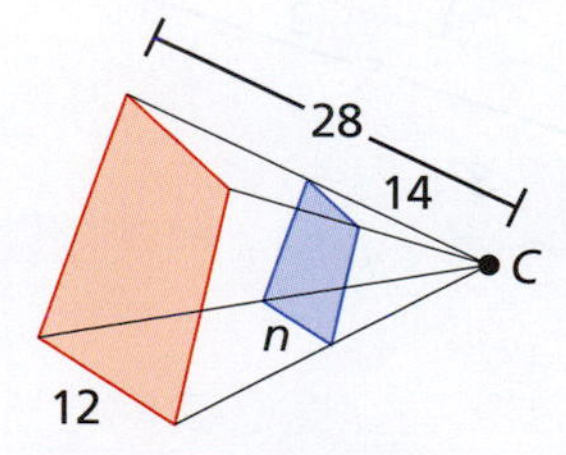

10. Finding Side Lengths Write and solve a proportion to find each missing length. Tell whether the dilation is a *reduction* or an *enlargement*.

a.

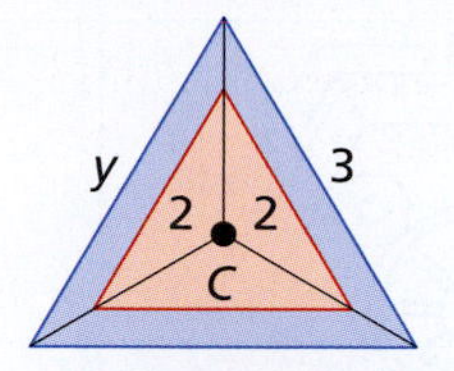

b.

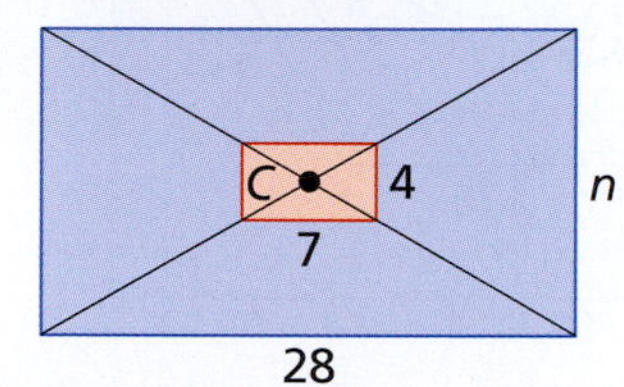

11. Drawing Dilations Draw the quadrilateral on a separate piece of paper. Then draw the specified dilation.

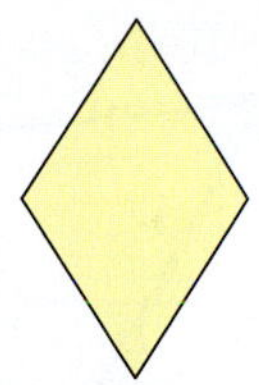

a. Enlargement using a scale factor of 4

b. Reduction using a scale factor of $\frac{1}{2}$

12. Drawing Dilations Draw the pentagon on a separate piece of paper. Then draw the specified dilation.

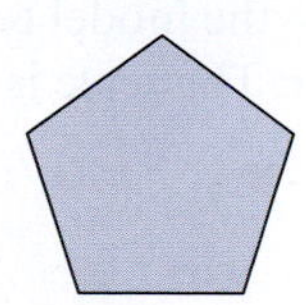

a. Enlargement using a scale factor of 2

b. Reduction using a scale factor of $\frac{1}{4}$

13. Blueprint The blueprint has a scale of 1 in. : 8 ft. Find the width of the actual bedroom.

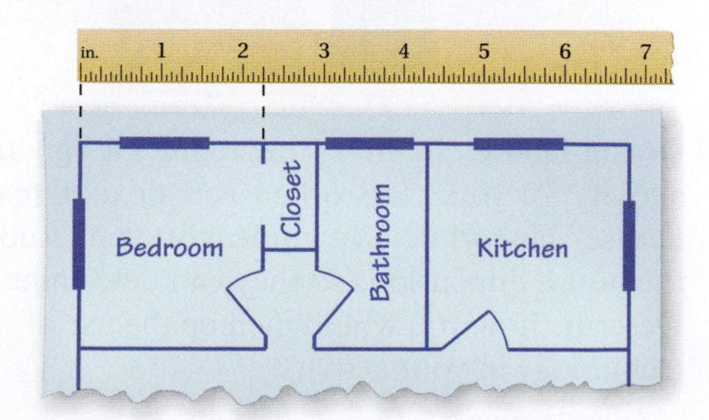

14. **Blueprint** The blueprint has a scale of 1 in. : 3 ft. Find the perimeter of the actual object.

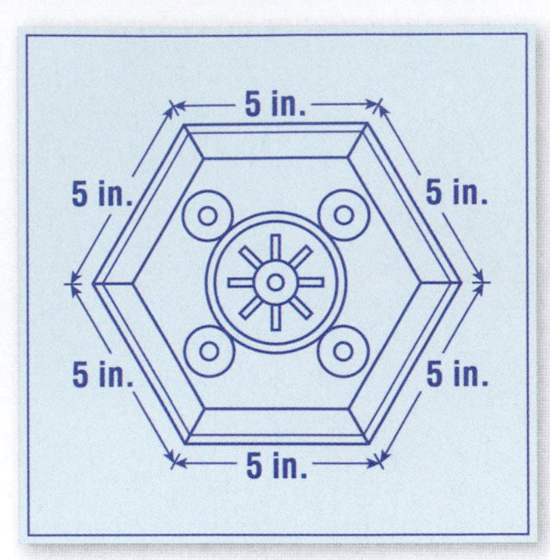

15. **Doll House** An original drawing of a doll house is shown below. The enlargement of the drawing is 36 inches high. What is the scale factor of original drawing to the enlargement?

16. **Architecture** Use the model to find the actual height of the skyscraper. The scale is 1 in. : 100 ft.

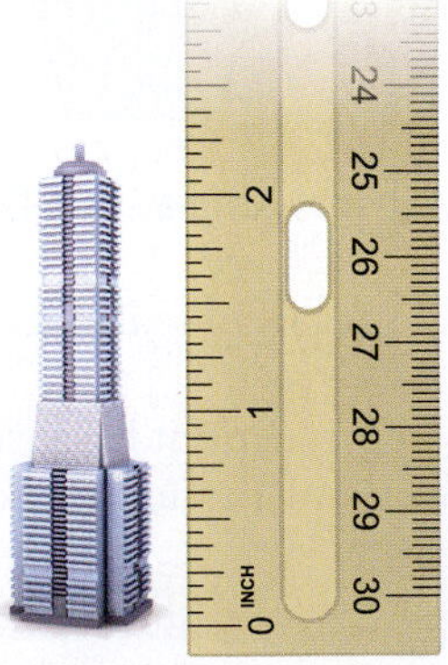

17. **Dream House: In Your Classroom** Design an activity for your classroom to create a "Dream House" blueprint. As a guide, give your students the house dimensions so they can determine a scale. Describe how you would include the use of technology in your activity.

18. **Activity: In Your Classroom** Design an activity for your students to explore why a scale factor of $a : b$ will not yield the same results as a scale factor of $b : a$. Include the importance of maintaining consistency with units of measure.

19. **Stickers** You are creating your own stickers. The original design is 4 inches by 4 inches. The image on the stickers is 1.5 inches by 1.5 inches. What is the scale factor of this dilation?

20. **Model Car** An original drawing of a car is 4 inches long. The enlarged drawing is 12 inches long. What is the scale factor of the original drawing to the enlargement?

21. **Grading Student Work** On a diagnostic test, one of your students does the following work. Explain what the student did wrong. Which topics would you encourage the student to review?

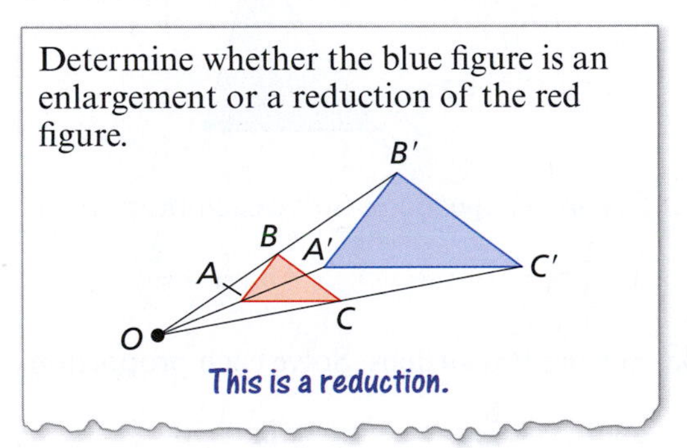

22. **Grading Student Work** On a diagnostic test, one of your students does the following work. Explain what the student did wrong. Which topics would you encourage the student to review?

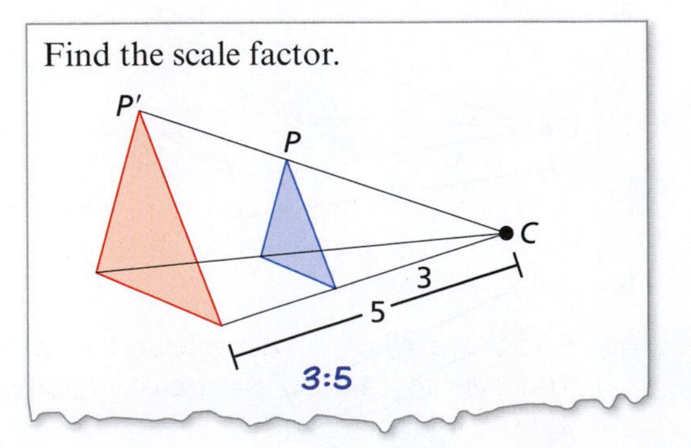

23. *Writing* Describe the difference between a scale factor of 1 : 10 and a scale of 1 in. : 10 ft. Explain when you would use a scale factor and when you would use a scale.

24. *Writing* Give an example of how an architect or an engineer would use dilation. Why is correctly calculating the scale factor imperative?

25. **Drawing a Dilation** Sketch a quadrilateral $ABCD$.

a. Draw point E inside $ABCD$. Then sketch a dilation of $ABCD$ with center E and a scale factor of 3.

b. Draw point F outside $ABCD$. Then sketch a dilation of $ABCD$ with center F and a scale factor of 3.

c. Sketch a dilation of $ABCD$ with center A and a scale factor of 3.

d. Describe the similarities and differences in the dilations in parts (a) – (c).

26. **Drawing a Dilation** Sketch a triangle XYZ.

a. Draw point W inside XYZ. Then sketch a dilation of XYZ with center W and a scale factor of $\frac{1}{2}$.

b. Draw point V outside XYZ. Then sketch a dilation of XYZ with center V and a scale factor of $\frac{1}{2}$.

c. Sketch a dilation of XYZ with center X and a scale factor of $\frac{1}{2}$.

d. Describe the similarities and differences in the dilations in parts (a) – (c).

27. **Comparing Perimeters and Areas** Use the triangle below.

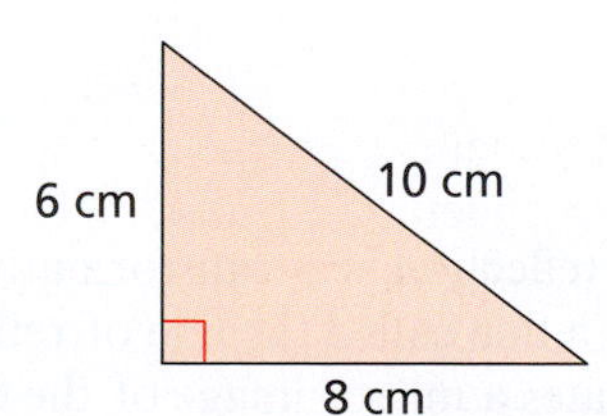

a. Draw a reduction using a scale factor of $\frac{1}{2}$.

b. Find the perimeters of the triangle and its image. How does the ratio of the perimeters compare to the scale factor?

c. Find the areas of the triangle and its image. How does the ratio of the areas compare to the scale factor?

28. **Comparing Perimeters and Areas** Use the parallelogram below.

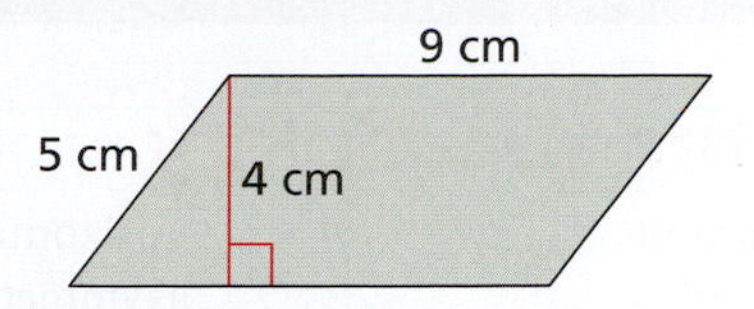

a. Draw an enlargement using a scale factor of 3.

b. Find the perimeters of the parallelogram and its image. How does the ratio of the perimeters compare to the scale factor?

c. Find the areas of the parallelogram and its image. How does the ratio of the areas compare to the scale factor?

29. **Planting a Garden** There is 50 feet of fencing around your rectangular vegetable garden. The width of the garden is 11 feet. You want to enlarge your garden by a scale factor of 2.

a. Draw a sketch of your current garden. Choose one corner of the garden as the center of dilation and sketch the new garden.

b. How much more fencing do you need to enclose the new garden?

c. Last year, you planted the garden in 90 minutes. How many hours will it take to plant the garden this year?

30. **Neighborhood Mural** You are designing a mural for the outside wall of the neighborhood art house. You do a rough sketch of your design on an 8.5-inch by 11-inch piece of paper. The scale of your sketch to the mural is 1 in. : 2 ft. The outside wall of the neighborhood art house is 25 feet by 40 feet. What is the area of the wall that is not covered by the mural?

31. **Dilating Areas** A rectangle with an area of 8 square inches is dilated by a scale factor of two. What is the new area?

32. **Wall Hanging** A 4-inch by 6-inch postcard is enlarged to a wall hanging. The scale factor is 5 : 1. What is the area of the wall hanging?

Clarsen55 | Dreamstime.com

Chapter Summary

Chapter Vocabulary

transformation *(p. 531)*
image *(p. 531)*
rigid *(p. 531)*
translation *(p. 532)*
rotation *(p. 534)*
center of rotation *(p. 534)*
angle of rotation *(p. 534)*
rotational symmetry *(p. 535)*
reflection *(p. 541)*
line of reflection *(p. 541)*
symmetrical *(p. 542)*
line of symmetry *(p. 542)*
vertical symmetry *(p. 542)*
horizontal symmetry *(p. 542)*
asymmetrical *(p. 542)*
glide reflection *(p. 543)*
frieze *(p. 544)*
frieze pattern *(p. 544)*
dilation *(p. 551)*
center of dilation *(p. 551)*
scale factor *(p. 551)*
enlargement *(p. 551)*
reduction *(p. 551)*
scale drawing *(p. 554)*
scale *(p. 554)*

Chapter Learning Objectives — Review Exercises

14.1 Translations and Rotations *(page 531)* — 1–20

- Understand and identify the four basic types of transformations.
- Understand and use the concept of a translation.
- Understand and use the concept of a rotation.

14.2 Reflections *(page 541)* — 21–38

- Understand and use the concept of a reflection in a line.
- Understand and use the definition of a glide reflection.
- Understand and use the definition of a frieze pattern.

14.3 Dilations and Scale Drawings *(page 551)* — 39–49

- Understand and use the concept of dilation.
- Find the scale factor of a dilation. Use a scale factor to draw a dilation.
- Use the concept of a dilation to read and create scale drawings.

Important Concepts and Formulas

Translation A translation is a transformation in which a figure slides but does not turn. Every point of the figure moves the same distance and in the same direction.

Rotation A rotation is a transformation in which a figure turns about a point called the center of rotation. The number of degrees a figure rotates is the angle of rotation.

Rotational Symmetry A figure has rotational symmetry when a rotation of 180° or less produces an image that fits exactly on the original figure.

Reflection A reflection is a transformation in which a figure flips in a line called the line of reflection. A reflection creates a mirror image of the original figure.

Glide Reflection A glide reflection is a transformation that is composed of a translation followed by a reflection in a line that is parallel to the direction of the translation.

Dilation A dilation is a transformation in which a figure is made larger or smaller with respect to a fixed point called the center of dilation.

Review Exercises

Free step-by-step solutions to odd-numbered exercises at MathematicalPractices.com.

14.1 Translations and Rotations

Identifying a Translation Determine whether the blue figure is a translation of the red figure. Explain your reasoning.

1.

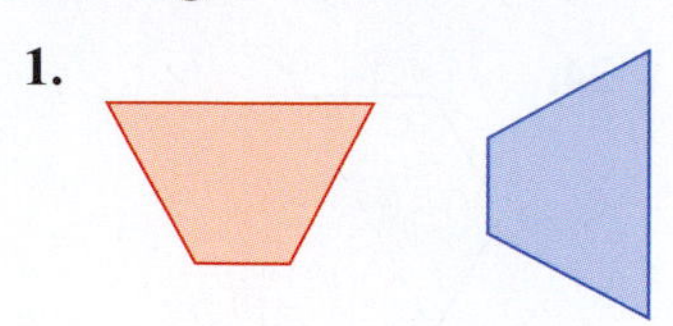

2.

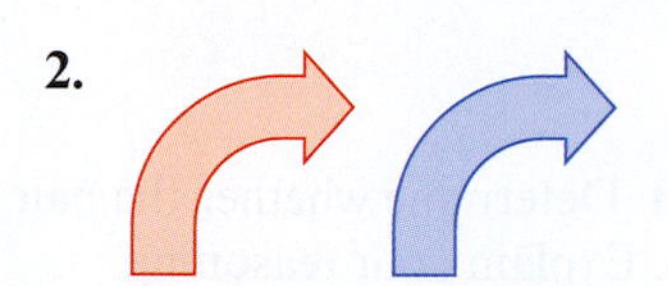

3.

4.

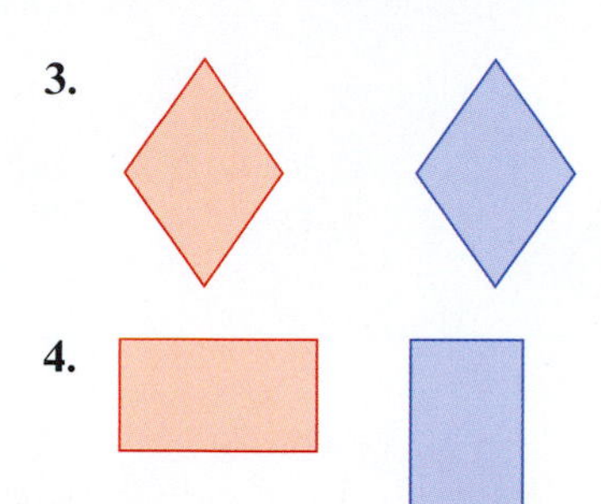

Counting Possible Translations A figure is formed on the geoboard with a rubber band. How many different translations of the figure are possible? The entire figure must remain on the geoboard.

5.

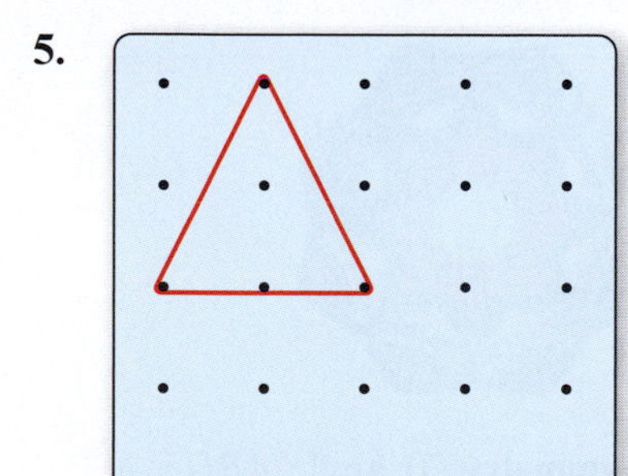

6.

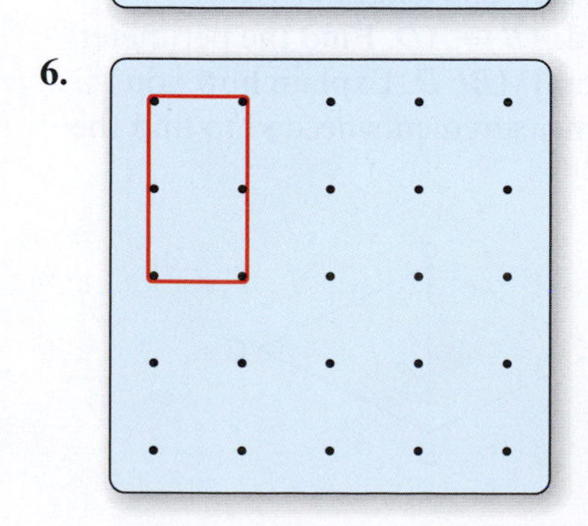

Identifying a Rotation Determine whether the blue figure is a rotation of the red figure. Explain your reasoning.

7.

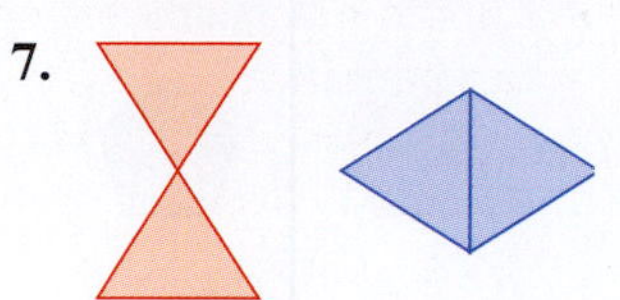

8.

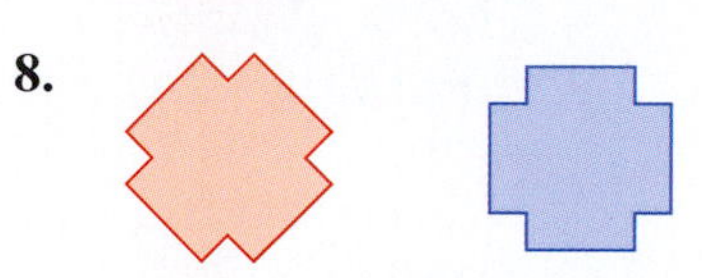

9.

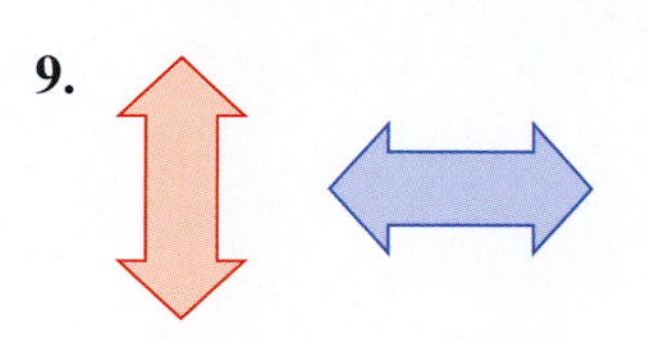

10.

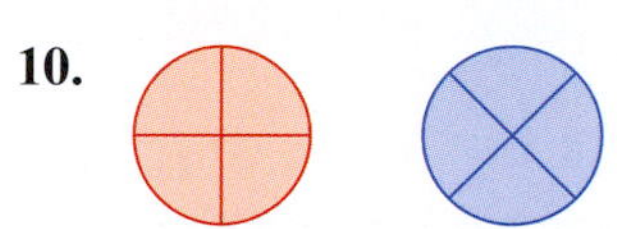

Finding the Center of Rotation Find the center of rotation of the figure and its image.

11. 12.

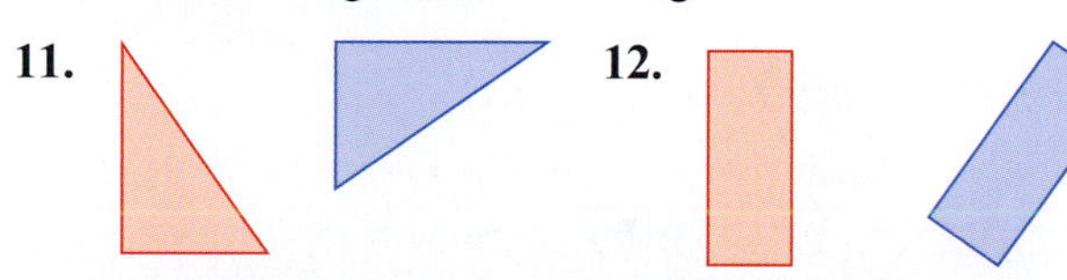

Identifying Rotational Symmetry Determine whether the figure has rotational symmetry. Explain your reasoning.

13. 14.

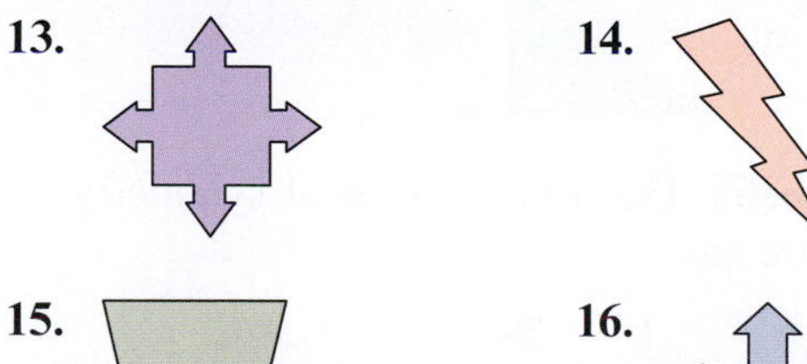

15. 16.

True or False? Tell whether the statement below is *true* or *false*. Justify your answer.

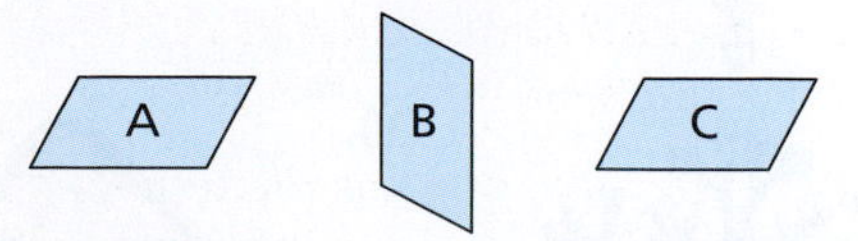

17. Parallelogram C is a rotation of parallelogram B.

18. Parallelogram B is a translation of parallelogram A.

19. Parallelogram C is a translation of parallelogram B.

20. Parallelogram B is a rotation of parallelogram A.

14.2 Reflections

Identifying a Reflection Determine whether the blue figure is a reflection of the red figure. Explain your reasoning.

21.

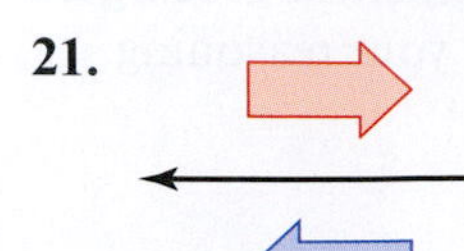

22.

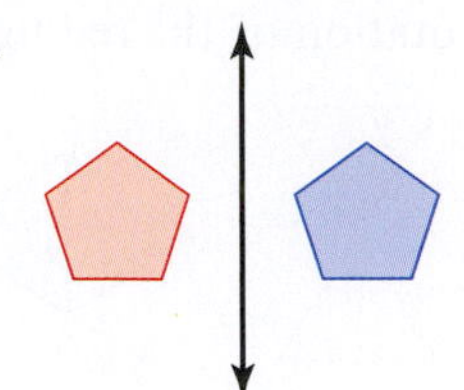

23.

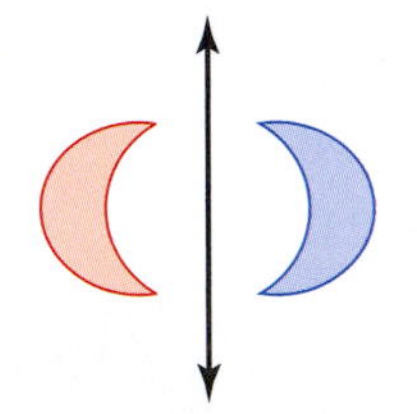

24.

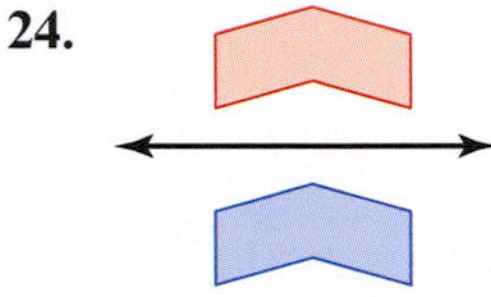

Finding a Line of Reflection The blue figure is a reflection of the red figure. Draw the line of reflection.

25.

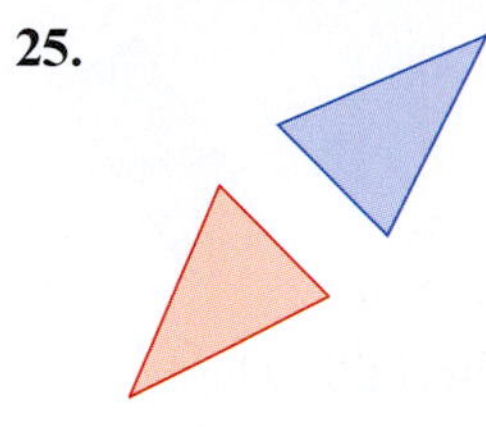

26.

Identifying Symmetry Describe the type of symmetry, if any, that the figure has.

27.

28.

29.

30.

Kichigin/Shutterstock.com; Nella/Shutterstock.com
Loskutnikov/Shutterstock.com

Identifying Lines of Symmetry Determine the number of lines of symmetry in the figure.

31.

32.

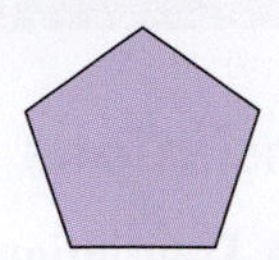

Drawing a Reflection Complete the picture by drawing the reflection in the given line of symmetry.

33.

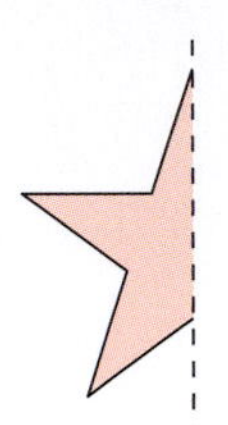

34.

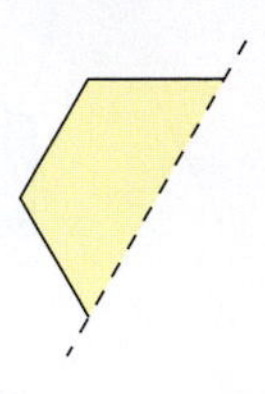

Identifying a Glide Reflection Determine whether the pair of figures is a glide reflection. Explain your reasoning.

35.

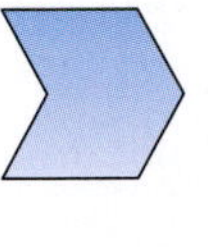
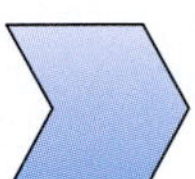

36.

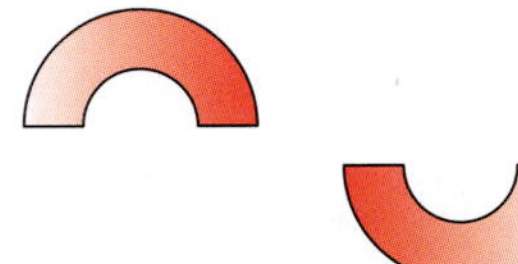

37. **Determining Lines of Symmetry** Find the number of lines of symmetry in the image from a kaleidoscope.

38. **Calculating with Symmetry** Triangle ABD is a reflection of $\triangle CBD$. The length of $\overline{AB}$ is 5 centimeters and $\overline{AB} \cong \overline{AD}$. Find the perimeter of the quadrilateral $ABCD$. Explain how you determined the measurements needed to find the perimeter.

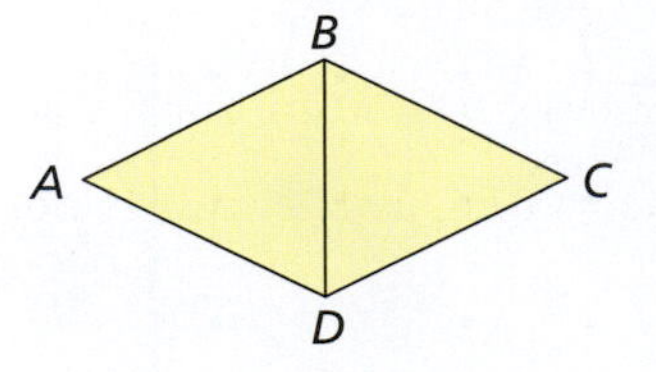

Mastering_Microstock/Shutterstock.com

14.3 Dilations and Scale Drawings

Identifying a Dilation Determine whether the blue figure is an enlargement or a reduction of the red figure. Explain your reasoning.

39.

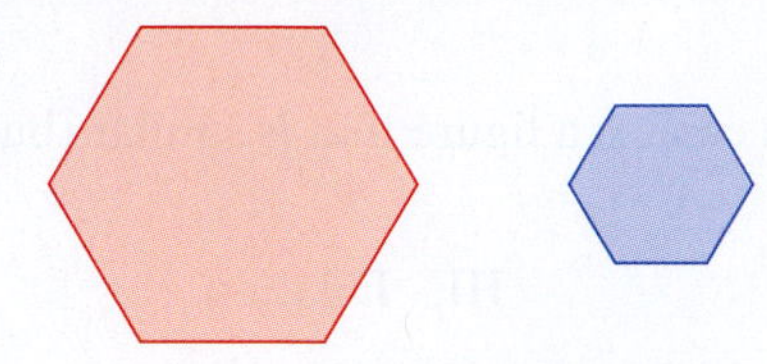

40.

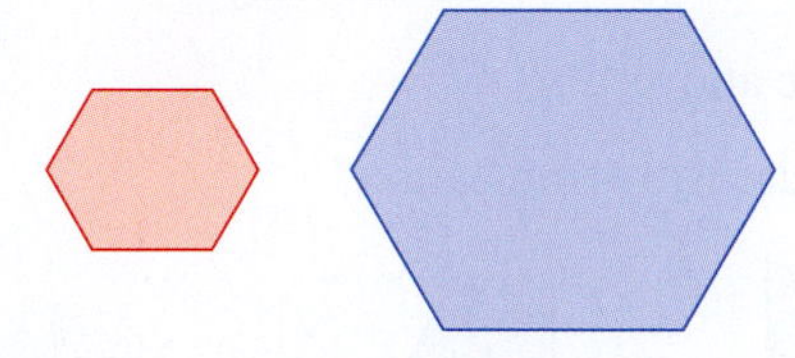

Finding a Scale Factor Determine whether the blue figure is an enlargement or a reduction of the red figure. Find the scale factor of the dilation.

41.

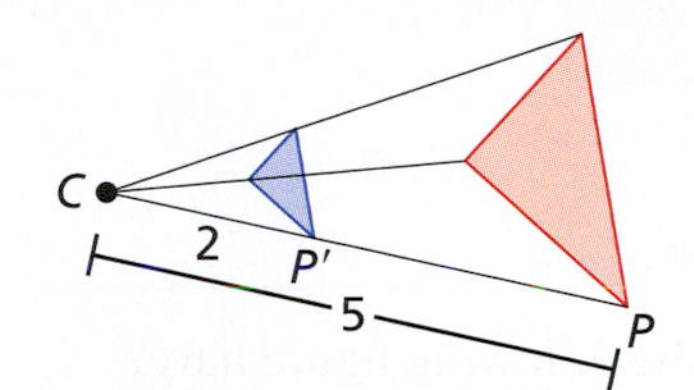

42.

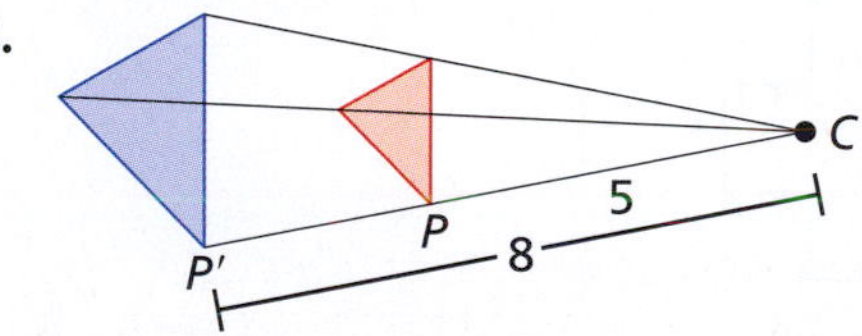

43. **Finding a Side Length** Write and solve a proportion to find the missing length. Tell whether the dilation is a *reduction* or an *enlargement*.

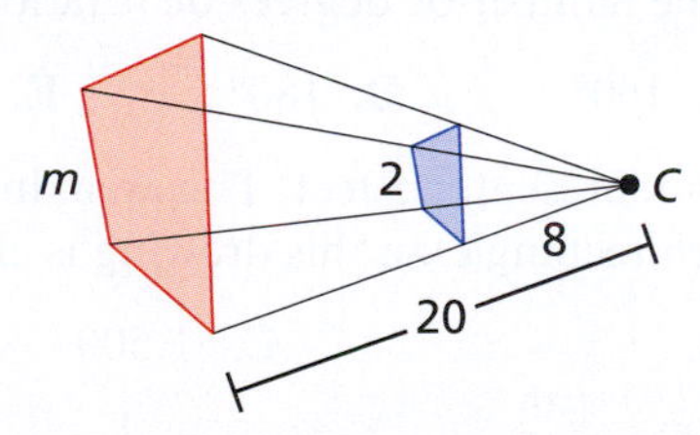

44. **Drawing Dilations** Draw the figure on separate paper. Then draw the specified dilation.

a. Enlargement using a scale factor of 3

b. Reduction using a scale factor of $\frac{1}{2}$

Using a Scale Drawing Use the scale drawing to find the actual length of the object given the scale.

45. 1 cm : 2 ft

46. 1 cm : 1 ft

47. **Replica Prints** A gift shop sells replica prints of a painting that is 30 inches by 21 inches. The ratio of the dimensions of the painting to the dimensions of a replica is 1 : 3. What are the dimensions of a replica print?

48. **Architecture** The height of a wall in a blueprint is 5 inches. The actual wall is 100 inches high. Find the scale of the blueprint.

49. **Comparing Perimeters and Areas** Use the parallelogram below.

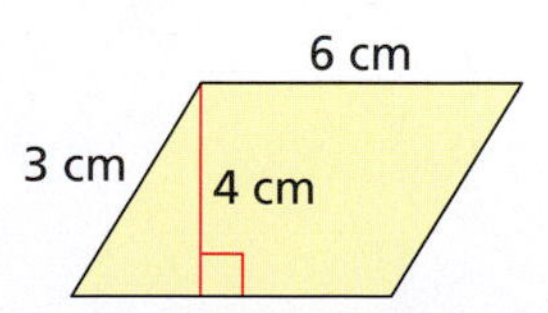

a. Draw an enlargement using a scale factor of 4.

b. Find the perimeters of the parallelogram and its image. How does the ratio of the perimeters compare to the scale factor?

c. Find the areas of the parallelogram and its image. How does the ratio of the areas compare to the scale factor?

Chapter Test

The questions on this practice test are modeled after questions on state certification exams for K–8 teaching.

Test Directions: Each of the questions is followed by five choices. Choose the *best* response to each question.

1. Which of the following transformations creates a figure that is similar (but not congruent) to the original figure?

I. Translation II. Rotation III. Dilation

A. I only **B.** II only **C.** III only
D. II and III **E.** All of the above

2. Which figure can be obtained from figure X by translation?

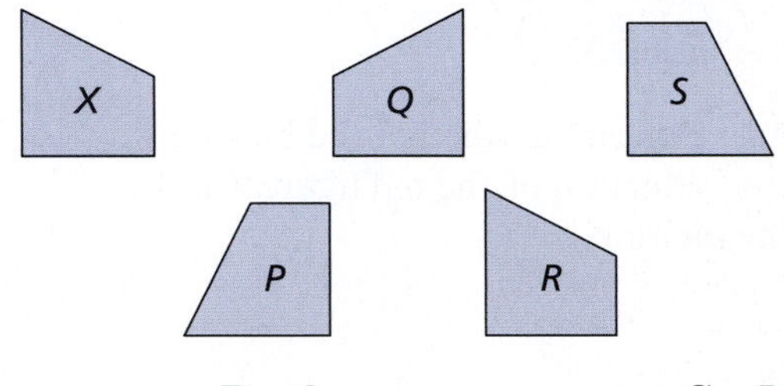

A. P **B.** Q **C.** R
D. S **E.** None of the above

3. How many lines of symmetry, if any, does the following figure have?

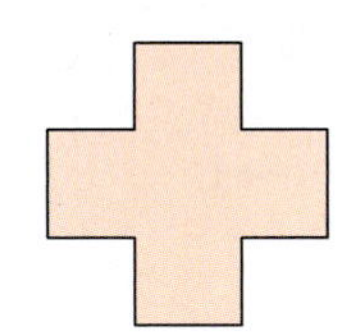

A. 0 **B.** 1 **C.** 2 **D.** 4 **E.** 8

A B C

4. If the equilateral triangle at the left is rotated clockwise about its center so that point B in the new image corresponds to point C in the original image, which of the following could be the number of degrees of rotation?

A. 110° **B.** 120° **C.** 160° **D.** 180° **E.** 240°

5. On a certain scale drawing, 1 inch represents 15 feet. The area, in square feet, represented by a 2-inch by 8.5-inch rectangle on this drawing is closest to

A. 200 **B.** 300 **C.** 1,500
D. 2,600 **E.** 3,800

6. The Statue of Liberty casts a shadow 37 meters long at the same time that a nearby vertical 5-meter pole casts a shadow that is 2 meters long. The height, in meters, of the Statue of Liberty is within which of the following ranges?

A. 10 m to 15 m **B.** 60 m to 65 m **C.** 80 m to 85 m
D. 90 m to 95 m **E.** 105 m to 110 m

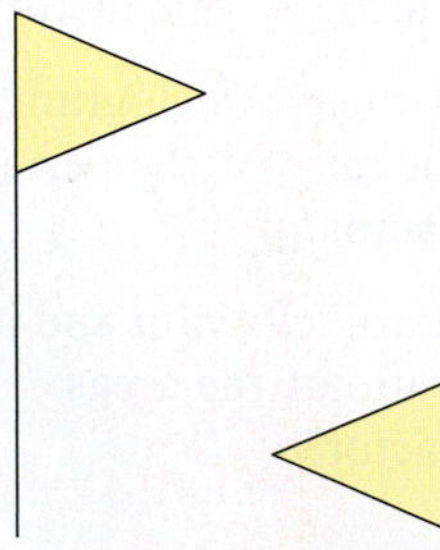

Figure I Figure II

7. Study the figures at the left. Which transformation, if any, of Figure I is shown in Figure II?

A. No transformation **B.** Reflection **C.** Rotation
D. Translation **E.** Dilation

15 Coordinate Geometry

Activity: Plotting Points in the Coordinate Plane

Materials:

- Pencil

Learning Objective:

Understand how to plot points in the coordinate plane, including points in all four quadrants.

Name ______________________

Plot the ordered pairs in the coordinate plane. Connect the points in numerical order.

1(1, 11)	2(−4, 9)	3(−5, 7)	4(−6, 6)	5(−6, 5)	6(−3, 2)
7(−4, 1)	8(−8, 3)	9(−9, 2)	10(−9, 1)	11(−5, −1)	12(−4, −4)
13(−5, −6)	14(−7 −5)	15(−8, −6)	16(−8, −7)	17(−4, −10)	18(−2, −6)
19(−1, −6)	20(0, −11)	21(5, −9)	22(5, −8)	23(4, −7)	24(2, −8)
25(2, −5)	26(4, −2)	27(9, −1)	28(9, 0)	29(8, 1)	30(4, 0)
31(3, 1)	32(7, 4)	33(7, 5)	34(6, 6)	35(5, 8)	

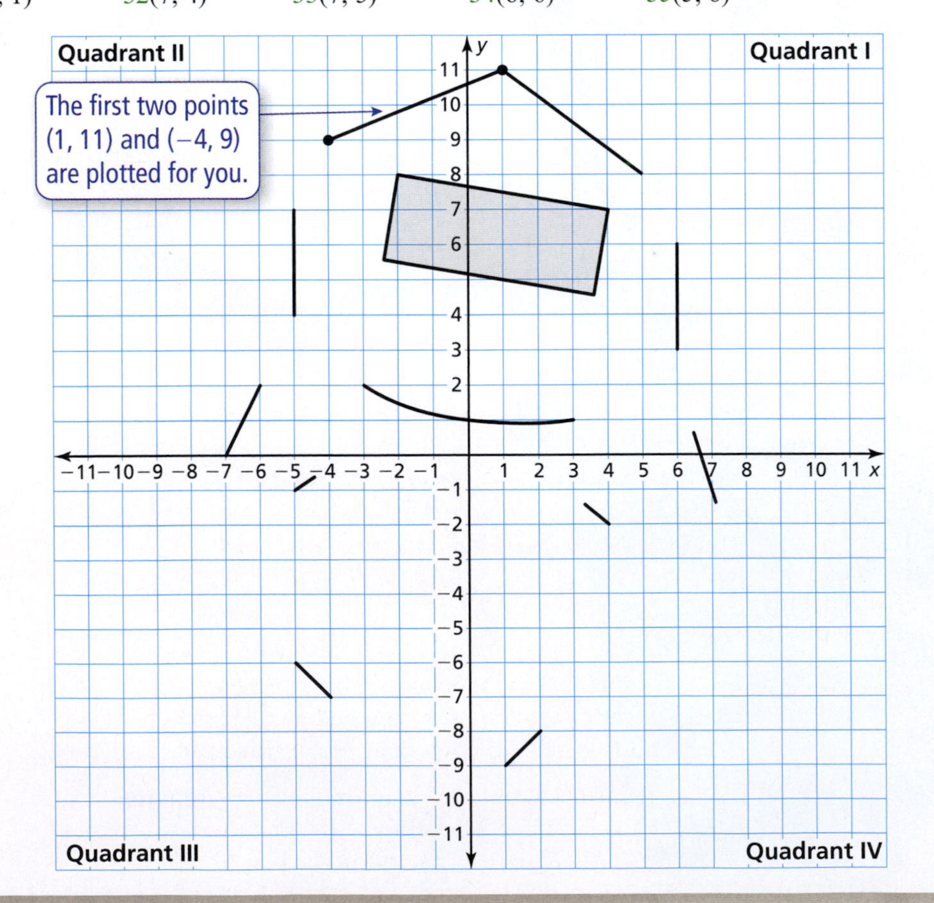

Hurst Photo/Shutterstock.com

15.1 The Coordinate Plane and Distance

Standards

Grades 3–5
Geometry
Students should graph points on the coordinate plane to solve real-world and mathematical problems.

Grades 6–8
Geometry
Students should understand and apply the Pythagorean Theorem.

- Identify points in a coordinate plane using ordered pairs.
- Represent transformations in a coordinate plane.
- Use the Pythagorean Theorem to find the distance between two points.
- Find the midpoint of a line segment.

The Coordinate Plane

A **coordinate plane**, or **Cartesian plane** (named after the French mathematician René Descartes), is formed by the intersection of a horizontal number line and a vertical number line. The number lines intersect at the **origin** and separate the coordinate plane into four regions called **quadrants**.

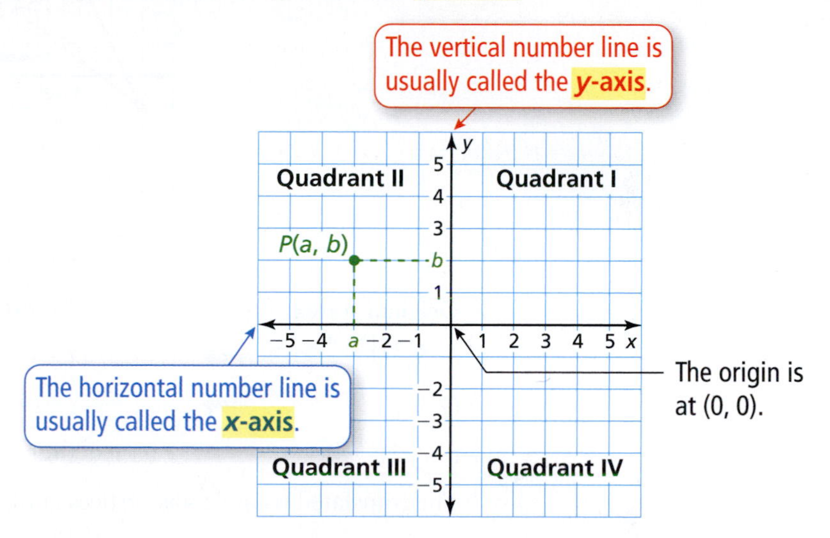

An **ordered pair** (a, b) locates a point P in a coordinate plane. A line from P perpendicular to the x-axis intersects at a point with coordinate a and a line from P perpendicular to the y-axis intersects at a point with coordinate b.

The ***x*-coordinate** corresponds to a number on the x-axis. → **Ordered pair** $(-3, 2)$ ← The ***y*-coordinate** corresponds to a number on the y-axis.

EXAMPLE 1 Identifying Ordered Pairs

Write the ordered pair that corresponds to (a) point B and (b) point D. If possible, identify the quadrant where each point lies.

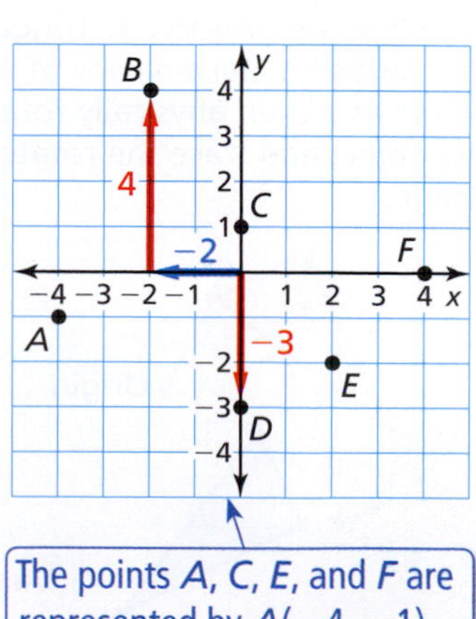

SOLUTION

a. Point B is 2 units to the left of the origin and 4 units up. So, the x-coordinate is -2 and the y-coordinate is 4. The ordered pair $(-2, 4)$ corresponds to point B. It lies in Quadrant II.

b. Point D lies on the y-axis 3 units down from the origin. So, the x-coordinate is 0 and the y-coordinate is -3. The ordered pair $(0, -3)$ corresponds to point D. Because point D lies on an axis, it does not lie in a quadrant.

Classroom Tip
There is a one-to-one correspondence between all the points in the coordinate plane and all the ordered pairs of real numbers.

Transformations in a Coordinate Plane

EXAMPLE 2 Translating a Triangle

Translate the triangle whose vertices are $(-5, -1)$, $(-1, 2)$, and $(-3, 3)$ five units right.

SOLUTION

Plot and label the vertices. Then connect the points to form the triangle. To translate the triangle five units right, add 5 to each of the x-coordinates of the vertices.

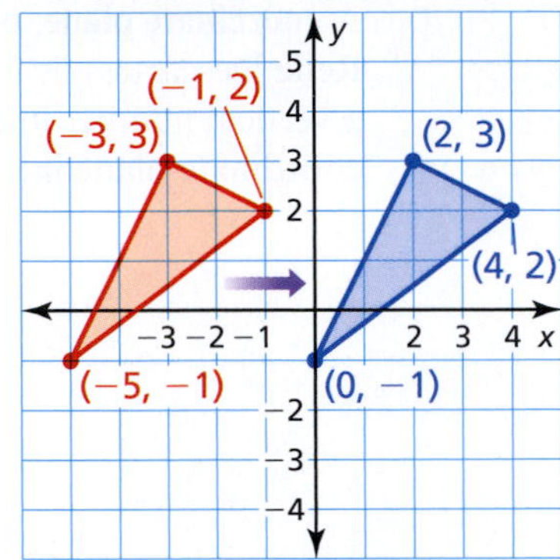

Translation (5 units right)

Original vertex		Translated vertex
$(-5, -1)$	→	$(-5 + 5, -1) = (0, -1)$
$(-1, 2)$	→	$(-1 + 5, 2) = (4, 2)$
$(-3, 3)$	→	$(-3 + 5, 3) = (2, 3)$

The translated triangle has vertices at $(0, -1)$, $(4, 2)$ and $(2, 3)$.

EXAMPLE 3 Rotating a Triangle

Rotate the triangle whose vertices are $(-5, -1)$, $(-1, 2)$, and $(-3, 3)$ 90° counterclockwise about the origin.

SOLUTION

Plot and label the vertices. Then connect the points to form the triangle. It can be shown that to rotate a figure in a coordinate plane 90° counterclockwise about the origin, each point (x, y) transforms to the point $(-y, x)$.

Classroom Tip

Rotations are typically difficult for students to visualize, and rules vary depending on the degree of rotation. So, rather than presenting rules for how each point (x, y) transforms, it is usually more effective to suggest that students make a copy of the object and then physically rotate the object and trace the rotated image.

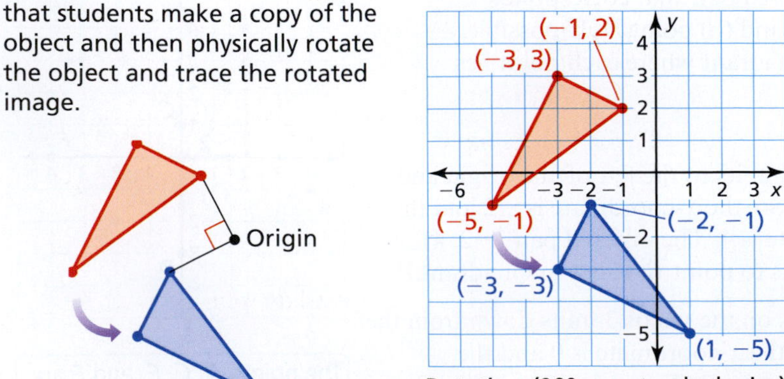

Rotation (90° counterclockwise)

Original vertex		Rotated vertex
$(-5, -1)$	→	$(1, -5)$
$(-1, 2)$	→	$(-2, -1)$
$(-3, 3)$	→	$(-3, -3)$

The rotated triangle has vertices at $(1, -5)$, $(-2, -1)$, and $(-3, -3)$.

EXAMPLE 4 Reflecting a Triangle

Reflect the triangle whose vertices are $(-5, -1)$, $(-1, 2)$, and $(-3, 3)$ in the y-axis.

SOLUTION

Plot and label the vertices. Then connect the points to form the triangle. Imagine reflecting each vertex of the triangle in the y-axis so that you create its mirror image. The result is that each vertex of the image has the same y-coordinate as its corresponding vertex. The x-coordinate of each vertex of the image is the opposite of the x-coordinate of its corresponding vertex.

> **Classroom Tip**
> The four transformations shown in Examples 2-5 were discussed in Chapter 14. To review the transformations, refer to the following sections.
>
> 14.1 Translation
>
> 14.1 Rotation
>
> 14.2 Reflection
>
> 14.2 Glide Reflection

> **Classroom Tip**
> Recall from Section 8.3 that to find the opposite of a number, you can multiply by -1.

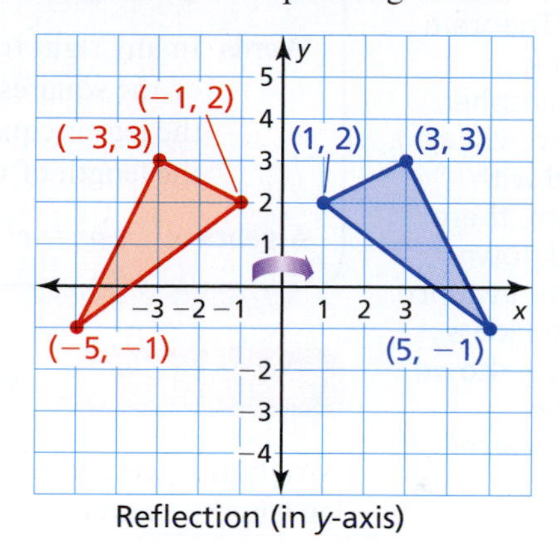

Reflection (in y-axis)

Original vertex		Translated vertex
$(-5, -1)$		$(-5 \cdot (-1), -1) = (5, -1)$
$(-1, 2)$		$(-1 \cdot (-1), 2) = (1, 2)$
$(-3, 3)$		$(-3 \cdot (-1), 3) = (3, 3)$

The reflected triangle has vertices at $(5, -1)$, $(1, 2)$ and $(3, 3)$.

EXAMPLE 5 Glide Reflecting a Triangle

Glide reflect the triangle whose vertices are $(-5, -1)$, $(-1, 2)$, and $(-3, 3)$ five units right, and then reflect it in the x-axis.

SOLUTION

Plot and label the vertices. Then connect the points to form the triangle. This transformation requires two steps. The first step is to slide the triangle 5 units right. The second step is to reflect the triangle in the x-axis.

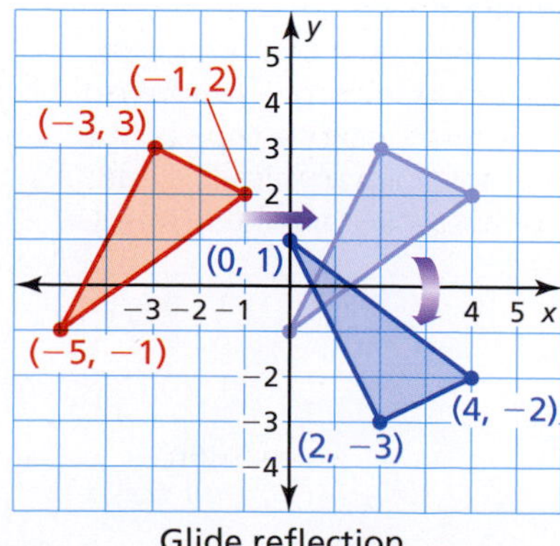

Glide reflection

Original vertex		Translated vertex
$(-5, -1)$		$(0, 1)$
$(-1, 2)$		$(4, -2)$
$(-3, 3)$		$(2, -3)$

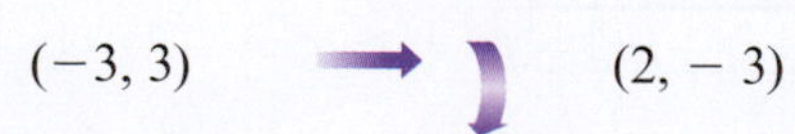

The glide reflected triangle has vertices at $(0, 1)$, $(4, -2)$, and $(2, -3)$. Note that it does not matter which step you perform first. You obtain the same result either way.

The Pythagorean Theorem and Distance

The sides of a right triangle have special names. The **legs** are the two sides that form the right angle. The **hypotenuse** is the side opposite the right angle.

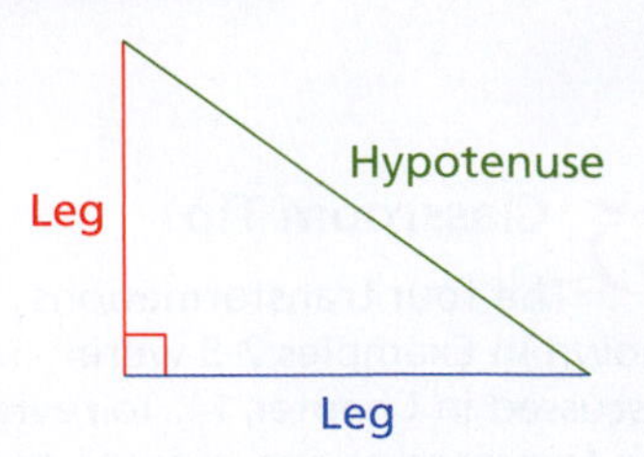

The Pythagorean Theorem

Words In any right triangle, the sum of the squares of the lengths of the legs is equal to the square of the length of the hypotenuse.

Algebra $a^2 + b^2 = c^2$

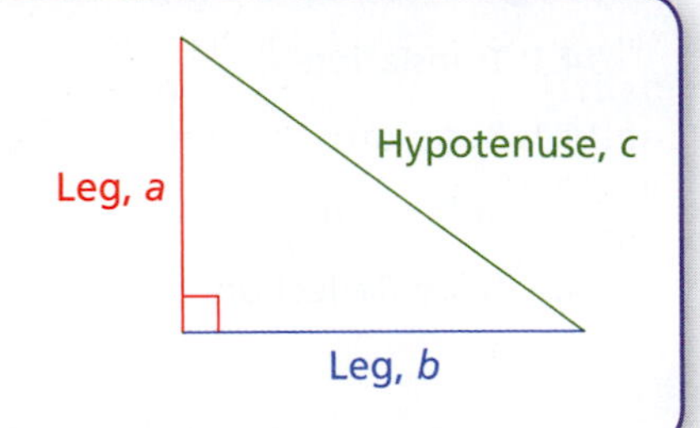

Classroom Tip

The Pythagorean Theorem is named after the Greek mathematician and philospher, Pythagoras (c. 570 B.C.–c. 490 B.C.). Although he is credited with discovering this theorem, there is evidence that it was known long before he lived. For instance, the Egyptian pyramid builders used ropes with knots (as shown below) to verify that the corners of the pyramids were right angles.

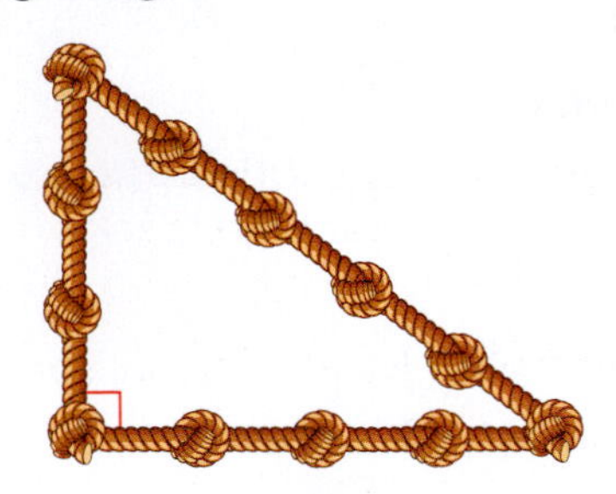

$3^2 + 4^2 = 5^2$

This particular application is actually the "Converse of the Pythagorean Theorem," which states that if a triangle has side lengths such that $a^2 + b^2 = c^2$, then the triangle must be a right triangle.

EXAMPLE 6 **Verifying the Pythagorean Theorem**

A right triangle is drawn with leg lengths a and b, and a hypotenuse of length c. Squares are drawn along the triangle's three sides as shown. The areas of the squares are labeled a^2, b^2, and c^2. Explain how this figure demonstrates that the Pythagorean Theorem is valid.

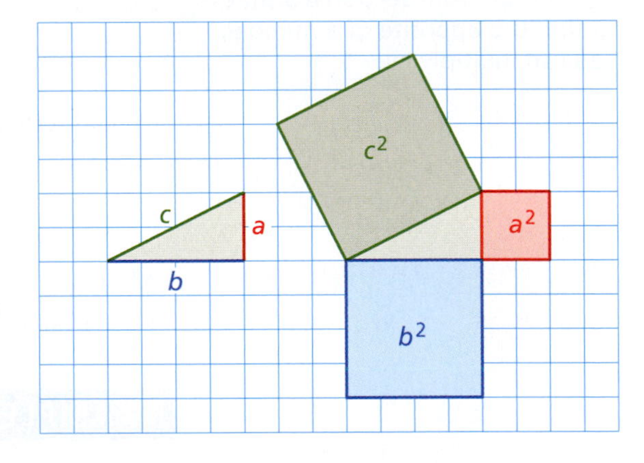

SOLUTION

Here is one way to use the figure to verify the Pythagorean Theorem.

1. Cut out the 3 squares whose areas are labeled a^2, b^2, and c^2.
2. Make eight copies of the right triangle and cut them out.
3. Arrange 4 copies of the right triangle with c^2 to form a large square as shown.
4. Arrange the other 4 copies of the right triangle with a^2 and b^2 to form another large square as shown.

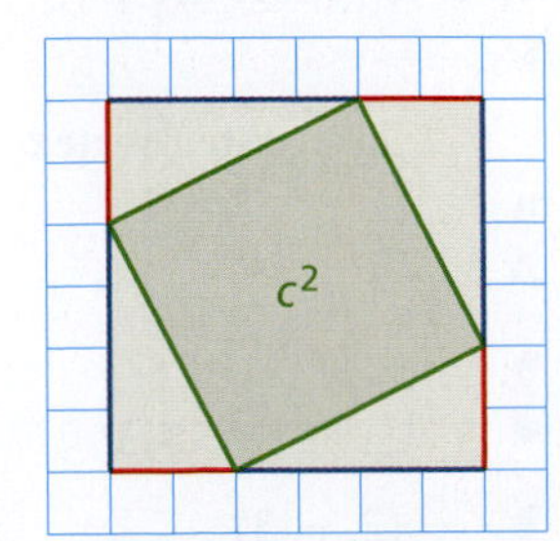

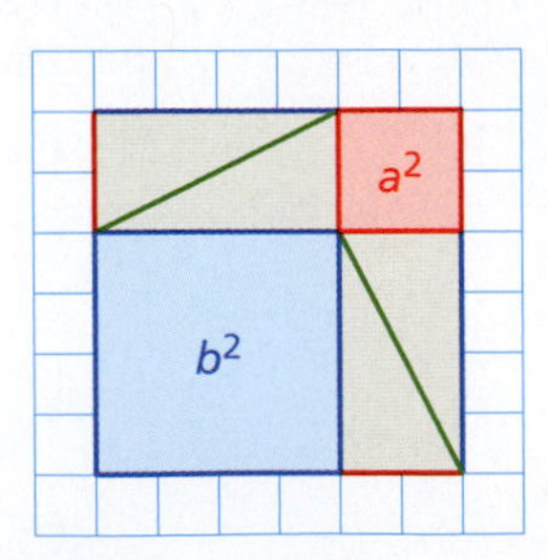

Each large square has a side length of $a + b$. So, their areas are equal. If you remove the four congruent triangles from each large square, then the remaining areas are c^2 and $a^2 + b^2$. So, it follows that $a^2 + b^2 = c^2$.

EXAMPLE 7 Finding the Length of a Hypotenuse

Find the length of the hypotenuse of the triangle.

SOLUTION

Use the Pythagorean Theorem.

$a^2 + b^2 = c^2$	Write the Pythagorean Theorem.
$5^2 + 12^2 = c^2$	Substitute 5 for a and 12 for b.
$25 + 144 = c^2$	Evaluate powers.
$169 = c^2$	Add.
$\sqrt{169} = \sqrt{c^2}$	Take positive square root of each side.
$13 = c$	Simplify.

The length of the hypotenuse is 13 meters.

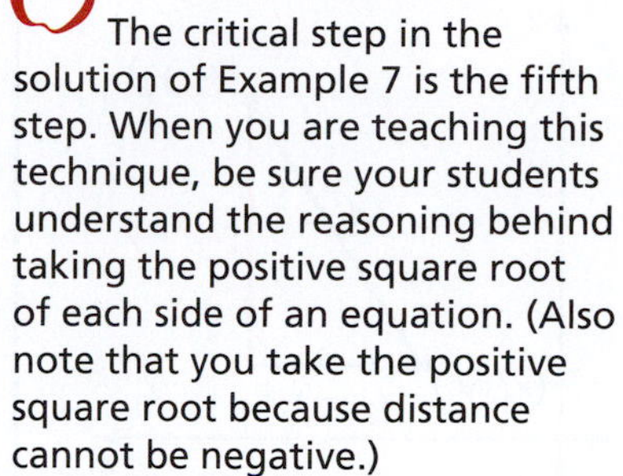

Classroom Tip

The critical step in the solution of Example 7 is the fifth step. When you are teaching this technique, be sure your students understand the reasoning behind taking the positive square root of each side of an equation. (Also note that you take the positive square root because distance cannot be negative.)

EXAMPLE 8 Using the Pythagorean Theorem in Real Life

Televisions are advertised by the lengths of the diagonals of their screens. A store has a sale on televisions 40 inches and larger. Is the television shown at the left on sale? Explain.

SOLUTION

1. **Understand the Problem** You are given the length and width of the television screen. You need to determine whether the screen has a diagonal that is 40 inches long or longer.
2. **Make a Plan** Use the Pythagorean Theorem to find the length of the diagonal.
3. **Solve the Problem**

$a^2 + b^2 = c^2$	Write the Pythagorean Theorem.
$24^2 + 32^2 = c^2$	Substitute 24 for a and 32 for b.
$576^2 + 1024^2 = c^2$	Evaluate powers.
$1600 = c^2$	Add.
$\sqrt{1600} = \sqrt{c^2}$	Take positive square root of each side.
$40 = c$	Simplify.

The length of the diagonal is 40 inches. So, the television is on sale.

4. **Look Back** One way to check that your answer is reasonable is to make a scale drawing of the television screen. Using centimeter grid paper and a scale of 1 cm : 8 in., you can make the scale drawing shown. By the Pythagorean Theorem, you know that the length of the diagonal in your drawing is 5 centimeters. You can then write and solve the proportion

$$\frac{1 \text{ cm}}{8 \text{ in.}} = \frac{5 \text{ cm}}{x \text{ in.}}$$

to find that $x = 40$.

Mathematical Practices

Construct Viable Arguments and Critique the Reasoning of Others

MP3 Mathematically proficient students make conjectures and build a logical progression of statements to explore the truth of their conjectures.

iStockphoto.com/JanPietruszka; iStockphoto.com/alexsl

The Pythagorean Theorem can be used to derive a formula for the distance between any two points in a coordinate plane.

Classroom Tip

In the Distance Formula, x_1 is read as "x sub one," and y_2 is read as "y sub two." The numbers 1 and 2 in x_1 and y_2 are called *subscripts*. Unlike exponents, subscripts are simply a convention for specifying a value.

The Distance Formula

The distance d between any two points (x_1, y_1) and (x_2, y_2) is given by the formula

$$d = \sqrt{(x_2 - x_1)^2 + (y_2 - y_1)^2}.$$

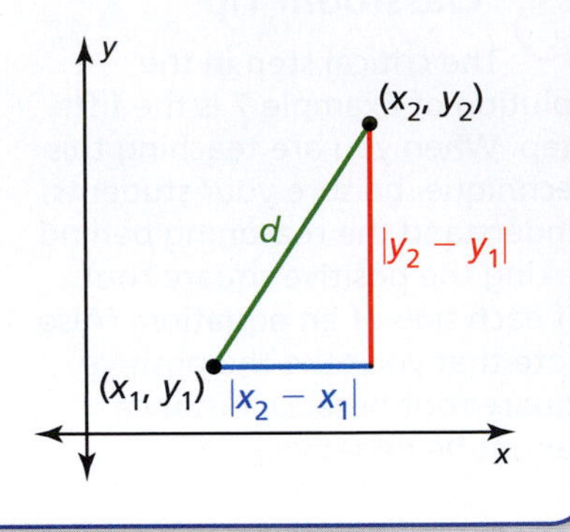

EXAMPLE 9 Finding the Distance Between Two Points

Find the distance d between the two points shown.

SOLUTION

The two plotted points are $(-1, 2)$ and $(2, 4)$.
So, let $(x_1, y_1) = (-1, 2)$ and $(x_2, y_2) = (2, 4)$.
Then apply the Distance Formula.

$d = \sqrt{(x_2 - x_1)^2 + (y_2 - y_1)^2}$ Distance Formula

$= \sqrt{[2 - (-1)]^2 + (4 - 2)^2}$ Substitute.

$= \sqrt{3^2 + 2^2}$ Simplify.

$= \sqrt{9 + 4}$ Evaluate powers.

$= \sqrt{13}$ Add.

≈ 3.61 Use a calculator.

The distance is about 3.61 units.

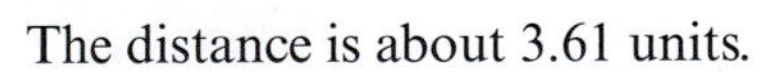

EXAMPLE 10 Using the Distance Formula in Real Life

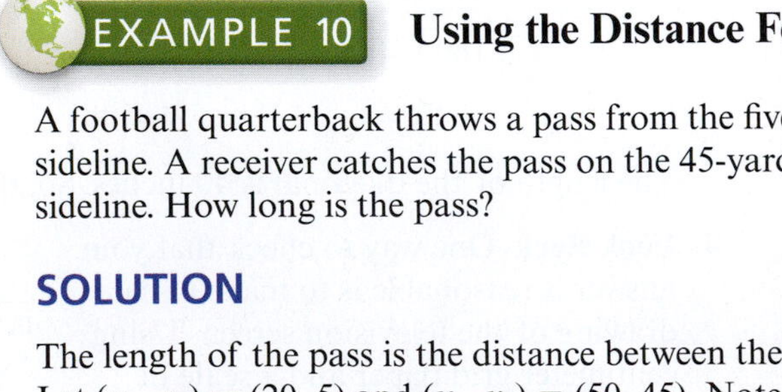

A football quarterback throws a pass from the five-yard line, 20 yards from the sideline. A receiver catches the pass on the 45-yard line, 50 yards from the same sideline. How long is the pass?

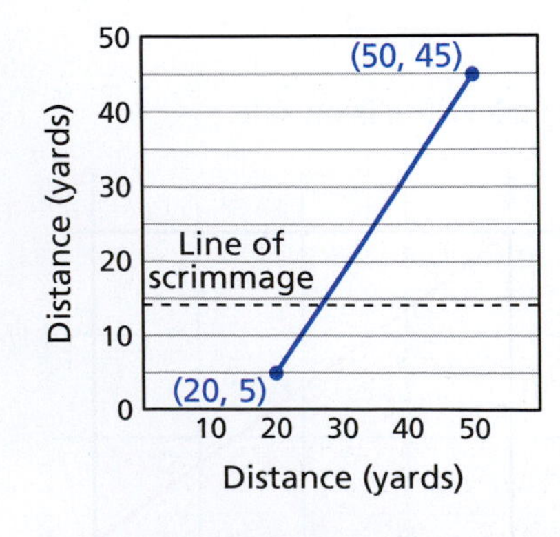

SOLUTION

The length of the pass is the distance between the points (20, 5) and (50, 45). Let $(x_1, y_1) = (20, 5)$ and $(x_2, y_2) = (50, 45)$. Note that you can label either point as (x_1, y_1) and the other (x_2, y_2).

$d = \sqrt{(x_2 - x_1)^2 + (y_2 - y_1)^2}$ Distance Formula

$= \sqrt{(50 - 20)^2 + (45 - 5)^2}$ Substitute.

$= \sqrt{2500}$ Simplify.

$= 50$ Evaluate the square root.

The pass is 50 yards long.

The Midpoint Formula

The **midpoint** of a line segment that joins two points is the point that divides the segment into two congruent segments. To find the coordinates of the midpoint of a line segment, find the averages of the x-coordinates and of the y-coordinates of the endpoints.

The Midpoint Formula

The midpoint of the line segment joining the two points (x_1, y_1) and (x_2, y_2) is

$$\text{Midpoint} = \left(\frac{x_1 + x_2}{2}, \frac{y_1 + y_2}{2}\right)$$

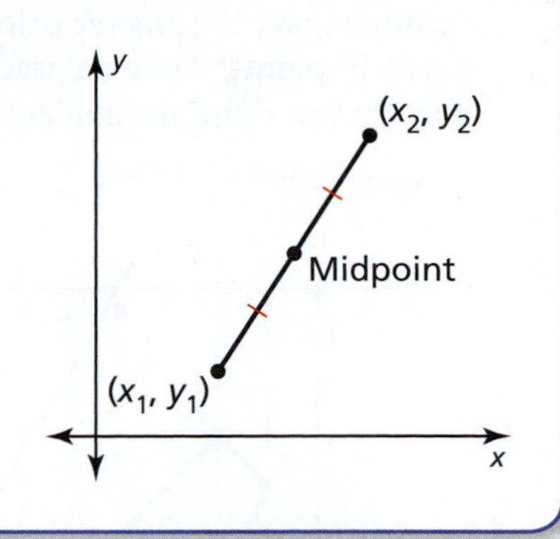

EXAMPLE 11 **Finding the Midpoint of a Line Segment**

a. Find the midpoint of the line segment joining the two points shown.

b. Verify that the midpoint divides the line segment into two congruent segments.

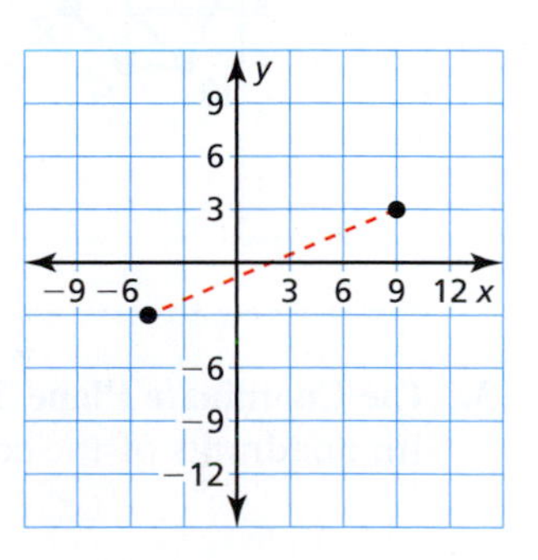

SOLUTION

a. The two plotted points are $(-5, -3)$ and $(9, 3)$. So, let $(x_1, y_1) = (-5, -3)$ and $(x_2, y_2) = (9, 3)$. Then apply the Midpoint Formula.

$$\text{Midpoint} = \left(\frac{x_1 + x_2}{2}, \frac{y_1 + y_2}{2}\right) \quad \text{Midpoint Formula}$$

$$= \left(\frac{-5 + 9}{2}, \frac{-3 + 3}{2}\right) \quad \text{Substitute.}$$

$$= \left(\frac{4}{2}, \frac{0}{2}\right) \quad \text{Add.}$$

$$= (2, 0) \quad \text{Divide.}$$

The midpoint is $(2, 0)$.

b. To verify that the point $(2, 0)$ is the midpoint, apply the Distance Formula for each set of points.

Distance between $(-5, -3)$ and $(2, 0)$

$$d = \sqrt{(x_2 - x_1)^2 + (y_2 - y_1)^2}$$
$$= \sqrt{(-5 - 2)^2 + (-3 - 0)^2}$$
$$= \sqrt{49 + 9}$$
$$= \sqrt{58}$$

Distance between $(9, 3)$ and $(2, 0)$

$$d = \sqrt{(x_2 - x_1)^2 + (y_2 - y_1)^2}$$
$$= \sqrt{(9 - 2)^2 + (3 - 0)^2}$$
$$= \sqrt{49 + 9}$$
$$= \sqrt{58}$$

The two line segments are congruent.

15.1 Exercises

Free step-by-step solutions to odd-numbered exercises at *MathematicalPractices.com*.

1. **Writing a Solution Key** Write a solution key for the activity on page 566. Describe a rubric for grading a student's work.

2. **Grading the Activity** In the activity on page 566, a student gave the answer below. The question is worth 10 points. Assign a grade to the answer. Why do you think the student erred?

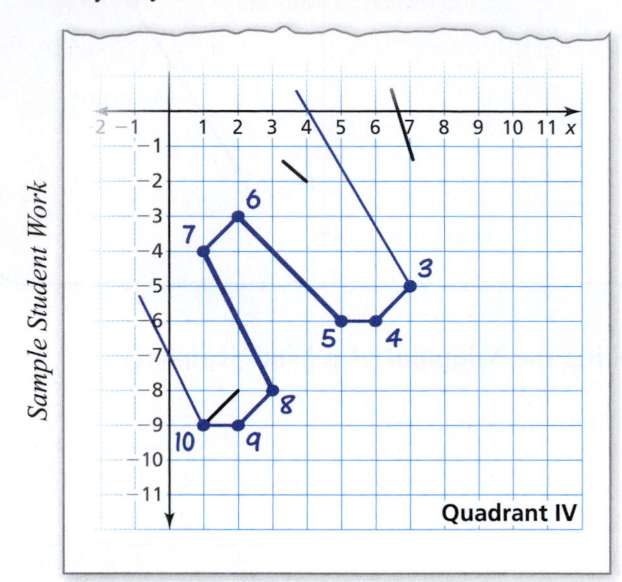

3. **The Coordinate Plane** Label the 1st, 2nd, 3rd, and 4th quadrants of the coordinate plane.

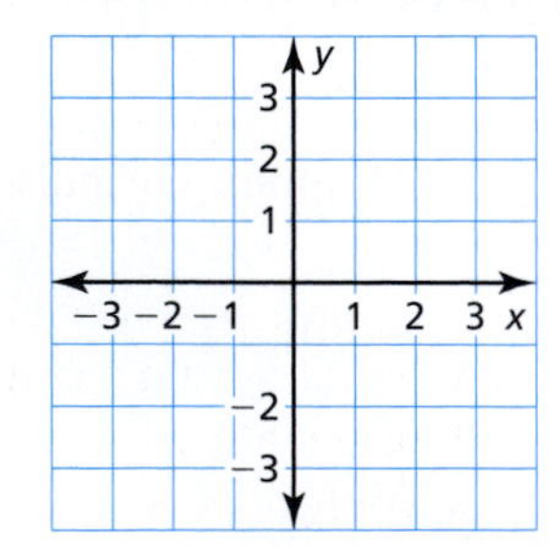

4. **The Coordinate Plane** Label the x-axis and y-axis of the coordinate plane. Label the origin as an ordered pair.

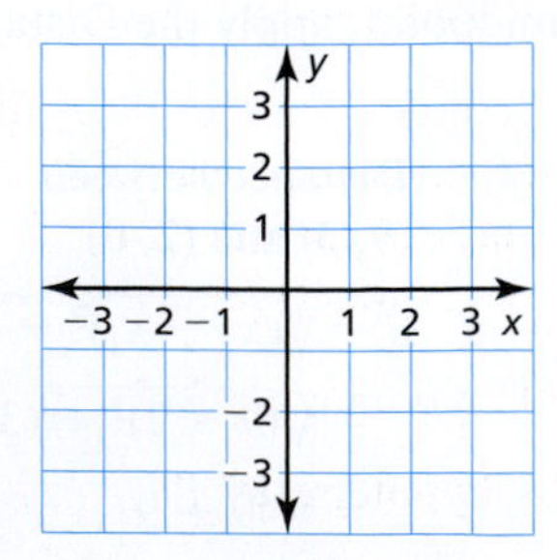

5. **Plotting Points** Plot each point in a coordinate plane.

a. $A(3, 5)$ **b.** $B(-2, -4)$
c. $C(5, -1)$ **d.** $D(0, 7)$
e. $E\left(7.2, \frac{3}{2}\right)$ **f.** $F\left(-\frac{7}{4}, \sqrt{7}\right)$
g. $G(\pi, 6.8)$ **h.** $H(\sqrt{3}, 2.2)$

6. **Plotting Points** Plot each point in a coordinate plane.

a. $A(-2, 5)$ **b.** $B(4, 7)$
c. $C(6, 0)$ **d.** $D(-3, -5)$
e. $E\left(\sqrt{101}, \frac{6}{12}\right)$ **f.** $F\left(4.4, \frac{7}{3}\right)$
g. $G\left(-5.75, -\frac{19}{5}\right)$ **h.** $H(5, \sqrt{85})$

7. **Identifying Ordered Pairs** Write the ordered pair that corresponds to each point.

a. Point A **b.** Point B
c. Point C **d.** Point D

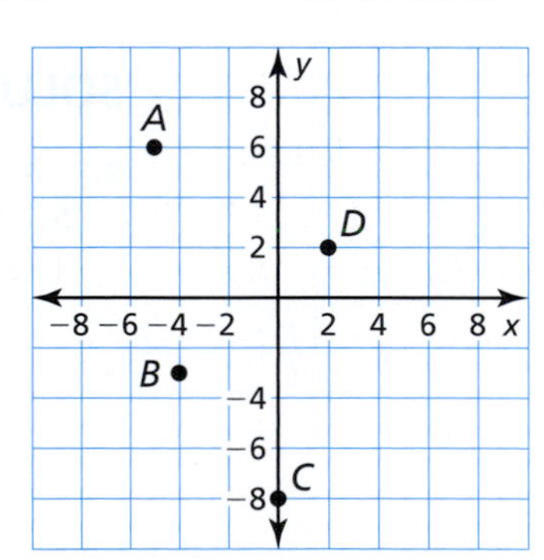

8. **Identifying Ordered Pairs** Write the ordered pair that corresponds to each point.

a. Point E **b.** Point F
c. Point G **d.** Point H

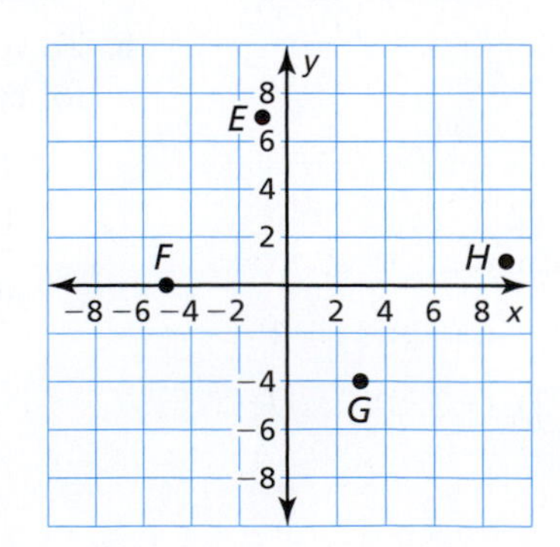

9. Translating Triangles Translate each triangle.

a. Translate the triangle whose vertices are $(1, 6)$, $(4, 7)$, and $(7, 1)$ four units left.

b. Translate the triangle whose vertices are $(-7, 7)$, $(-5, 3)$, and $(-2, 7)$ five units down.

c. Translate the triangle whose vertices are $(-4, -2)$, $(-4, -6)$, and $(-1, -6)$ eight units right.

d. Translate the triangle whose vertices are $(1, -7)$, $(6, -10)$, and $(11, -3)$ six units up.

10. Translating Triangles Translate each triangle.

a. Translate the triangle whose vertices are $(-3, 1)$, $(1, 3)$, and $(2, -2)$ two units left.

b. Translate the triangle whose vertices are $(-4, -2)$, $(0, 0)$, and $(-1, -4)$ four units down.

c. Translate the triangle whose vertices are $(-7, 3)$, $(-6, 8)$, and $(-5, 6)$ three units right.

d. Translate the triangle whose vertices are $(4, -8)$, $(6, -7)$, and $(7, -8)$ eight units up.

11. Rotating Triangles Rotate each triangle with the given vertices 90° counterclockwise about the origin.

a. $(-8, -2)$, $(-3, -5)$, and $(2, 2)$

b. $(2, -3)$, $(5, -3)$, and $(4, -7)$

c. $(2, 2)$, $(7, 4)$, and $(3, 7)$

d. $(-3, 1)$, $(-1, 1)$, and $(-1, 3)$

12. Rotating Triangles Rotate each triangle with the given vertices 90° clockwise about the origin.

a. $(1, 5)$, $(4, 7)$, and $(7, 0)$

b. $(-4, 2)$, $(-3, 8)$, and $(-1, 3)$

c. $(-4, -4)$, $(-7, -5)$, and $(-5, -7)$

d. $(-1, 1)$, $(1, -3)$, and $(5, -2)$

13. Reflecting Triangles Reflect each triangle with the given vertices in the y-axis.

a. $(-6, -5)$, $(-3, -1)$, and $(-3, -6)$

b. $(1, -1)$, $(1, -4)$, and $(3, -1)$

c. $(3, 4)$, $(7, 2)$, and $(4, 7)$

d. $(-4, 2)$, $(-4, 5)$, and $(0, 5)$

14. Reflecting Triangles Reflect each triangle with the given vertices in the x-axis.

a. $(3, -2)$, $(5, -3)$, and $(6, 2)$

b. $(5, 2)$, $(4, 5)$, and $(6, 4)$

c. $(-4, 2)$, $(-3, 0)$, and $(-1, 4)$

d. $(-3, -1)$, $(-1, 2)$, and $(2, -2)$

15. Glide Reflecting Triangles Translate each triangle and then reflect it in the y-axis.

a. Translate the triangle whose vertices are $(1, 1)$, $(5, -3)$, and $(9, -2)$ four units up.

b. Translate the triangle whose vertices are $(-7, 4)$, $(-3, 3)$, and $(-3, 5)$ three units down.

c. Translate the triangle whose vertices are $(-1, -1)$, $(0, -4)$, and $(2, -3)$ five units down.

d. Translate the triangle whose vertices are $(2, 8)$, $(3, 5)$, and $(4, 8)$ one unit up.

16. Glide Reflecting Triangles Translate each triangle and then reflect it in the x-axis.

a. Translate the triangle whose vertices are $(-7, 3)$, $(-3, 4)$, and $(-1, 2)$ five units right.

b. Translate the triangle whose vertices are $(3, -2)$, $(2, -4)$, and $(4, -5)$ six units left.

c. Translate the triangle whose vertices are $(-6, -1)$, $(-4, -1)$, and $(-2, -3)$ three units left.

d. Translate the triangle whose vertices are $(0, 4)$, $(1, 7)$, and $(2, 3)$ four units right.

17. The Pythagorean Theorem The lengths of the legs of a right triangle are given. Find the length of the hypotenuse.

a. 6 and 8 **b.** 9 and 3

c. 8 and 15 **d.** 2 and $2\sqrt{3}$

18. The Pythagorean Theorem The lengths of the legs of a right triangle are given. Find the length of the hypotenuse.

a. 2 and 2 **b.** 5 and 10

c. 12 and 9 **d.** 7 and 24

19. The Pythagorean Theorem The lengths of one leg and the hypotenuse of a right triangle are given. Find the length of the other leg.

a. Leg: 5, hypotenuse: 13

b. Leg: 7, hypotenuse: 25

c. Leg: 8, hypotenuse: 17

d. Leg: 15, hypotenuse: 25

20. The Pythagorean Theorem The lengths of one leg and the hypotenuse of a right triangle are given. Find the length of the other leg.

a. Leg: 21, hypotenuse: 29

b. Leg: 30, hypotenuse: 34

c. Leg: 24, hypotenuse: 30

d. Leg: 10, hypotenuse: 26

21. The Distance Formula Find the distance between the two points.

a. (6, 1), (4, 4) **b.** (−4, −1), (9, −3)

c. (8, 7), (10, 2) **d.** (10, −4), (7, −8)

22. The Distance Formula Find the distance between the two points.

a. (−3, 11), (5, 5) **b.** (15, −7), (5, −3)

c. (21, 43), (32, 23) **d.** (−22, −10), (−32, −10)

23. The Midpoint Formula Find the midpoint of the line segment joining the two points.

a. (2, 5), (4, 3) **b.** (−1, 3), (4, 8)

c. (0, −2), (4, 0) **d.** (−1, 2), (−4, 1)

24. The Midpoint Formula Find the midpoint of the line segment joining the two points.

a. (0, 0), (4, 6) **b.** (3, 8), (7, 6)

c. (−5, 6), (9, 7) **d.** (−12, 0), (6, 1)

25. Determining Right Triangles The coordinates of the vertices of a triangle are given. Determine whether the triangle is a right triangle. Explain your reasoning. (*Hint:* The hypotenuse of a right triangle is always the longest side.)

a. (12, 3), (0, −1), (−2, 5)

b. (2, 3), (−6, 1), (−2, −3)

26. Determining Right Triangles The coordinates of the vertices of a triangle are given. Determine whether the triangle is a right triangle. Explain your reasoning. (*Hint:* The hypotenuse of a right triangle is always the longest side.)

a. (1, 2), (5, −2), (−3, −2)

b. (4, 0), (5, 2), (1, 4)

27. Writing James Garfield, the 20th President of the United States, developed a proof of the Pythagorean Theorem using the area of a right trapezoid. Use the figure and write a paragraph showing $a^2 + b^2 = c^2$. (*Hint:* Write an expression for the area of the trapezoid and set this equal to the sum of the areas of the three triangles.)

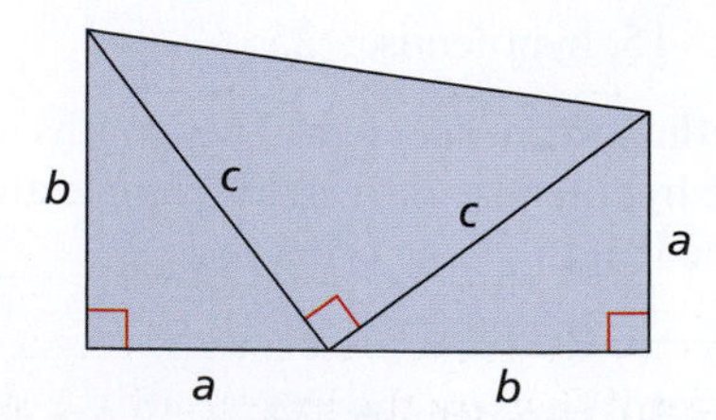

28. Writing A glide reflection is a combination of a translation and a reflection where the order of the transformations does not matter. Explain why order matters when you translate and rotate a figure about the origin in a coordinate plane.

29. Activity: In Your Classroom Fold a piece of paper in half twice so that when it is unfolded it represents a coordinate plane. Label the x- and y-axes and the four quadrants. Fold the paper on the axes and cut a scalene triangle out of the middle of the four layers. Unfold the paper again.

a. Which transformation of the figure in Quadrant I has the figure in Quadrant II as its image?

b. Which transformation of the figure in Quadrant II has the figure in Quadrant IV as its image?

c. Describe the concepts your students will practice and discover by doing this activity.

30. Activity: In Your Classroom The vertices of a triangle are (−3, 5), (4, −2), and (−7, −8).

a. Find the lengths of the sides of the triangle.

b. Find the midpoint of each side. What are the side lengths of the triangle that has the midpoints as its vertices?

c. Are the two triangles similar? Explain.

d. Describe the concepts your students will practice and discover by doing this activity.

31. Grading Student Work On a diagnostic test, one of your students does the following work. Explain what the student did wrong. Which topics would you encourage the student to review?

Find the length of the hypotenuse of the triangle. Show all of your work.

3 c 4

$a^2 + b^2 = c^2$
$3^2 + 4^2 = c^2$
$9 + 16 = c^2$
$25 = c^2$
$\pm 5 = c$

32. Grading Student Work On a diagnostic test, one of your students does the following work. Explain what the student did wrong. Which topics would you encourage the student to review?

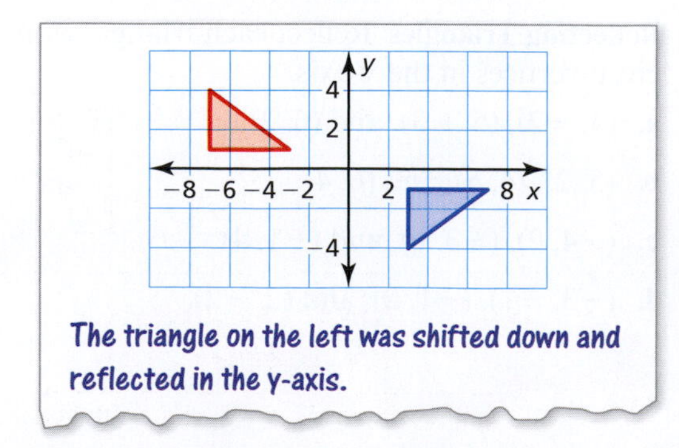

33. Baseball The bases on a baseball field are 90 feet apart. The pitcher's mound is between second base and home plate, 60.5 feet from home plate. What is the distance from the pitcher's mound to second base?

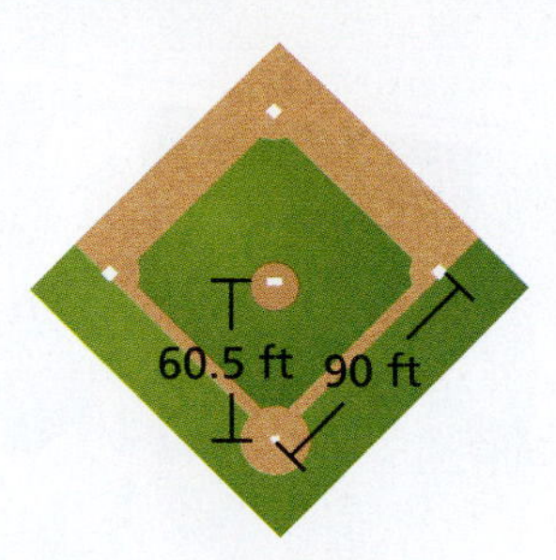

34. Baseball The bases on a baseball field are 60 feet apart. The pitcher's mound is between second base and home plate, 46 feet from home plate. What is the distance from the pitcher's mound to second base?

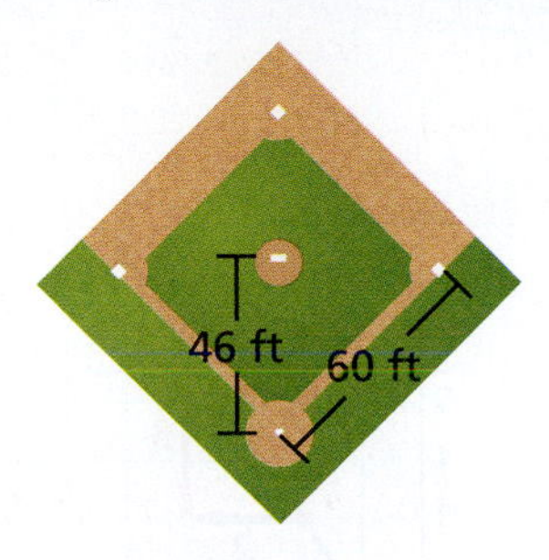

35. Describe the Transformation Describe the transformations that will move the five smaller figures to form the rectangle.

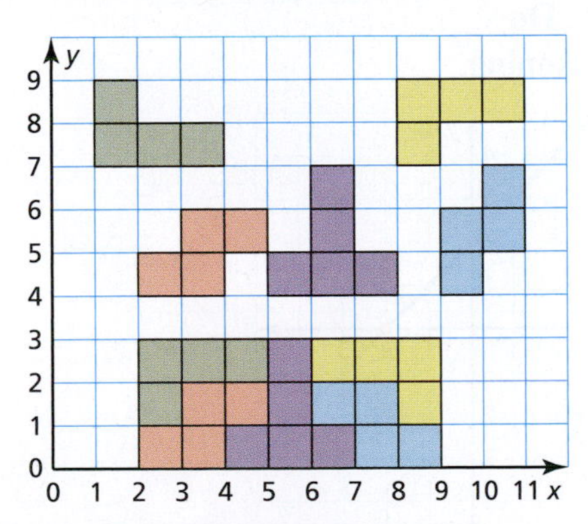

36. Describe the Transformation Describe the transformations that will move the five smaller figures to form the rectangle.

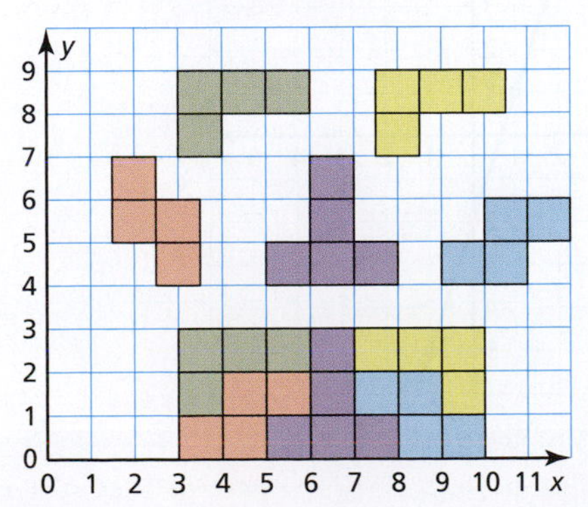

37. Diagonals in a Prism Find the length of the longest line segment in the rectangular prism.

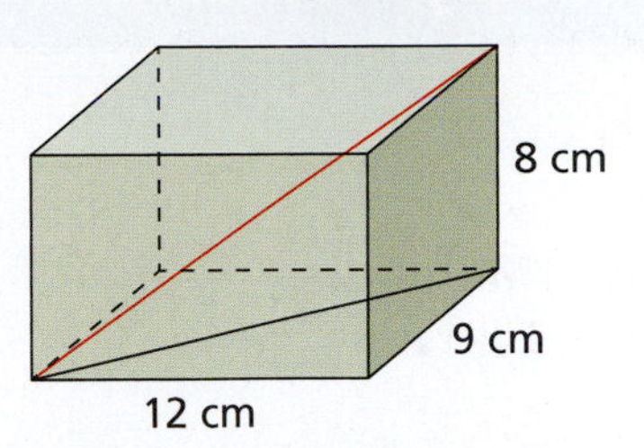

38. Diagonals in a Prism Find the length of the longest line segment in the rectangular prism.

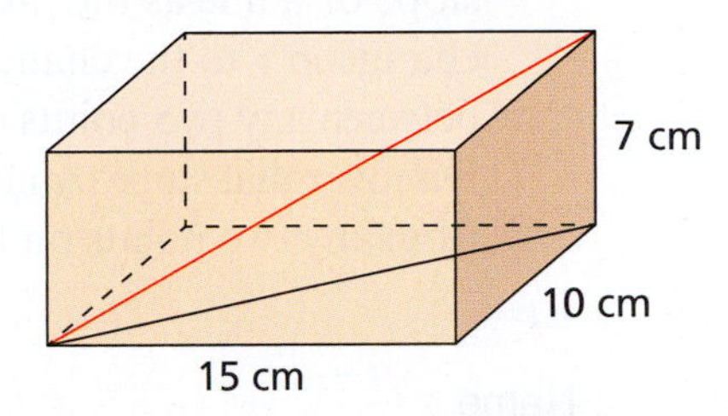

39. Television Set You buy a 48-inch-wide by 30-inch-high television set. The television will be hung on a wall so that the bottom is 6 feet from the floor and the left side is 5 feet from the corner of the wall.

a. What is the length of the diagonal of the television?

b. Draw a coordinate plane to represent the wall using the origin as the bottom left corner. Identify the coordinates of the vertices of the television when it is placed at the bottom left corner. Then, draw the translation of the television on the wall and identify the coordinates of the vertices.

40. School Flag You are designing a rectangular school flag that is 4 feet long and 2 feet wide as shown in the figure.

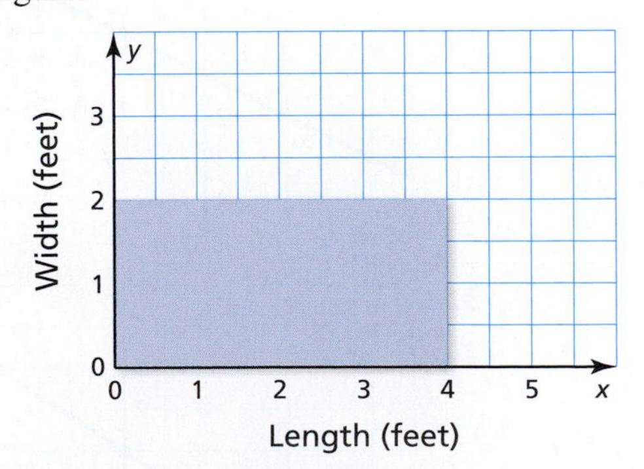

a. Draw an isosceles triangle on the flag using the 4-foot side as the base and having a height of 2 feet. What are the lengths of the sides of the triangle?

b. Draw another triangle by connecting the midpoints of the 3 sides of the first triangle. What are the coordinates of the vertices of this triangle?

Activity: Finding the Slope of a Line

Materials:

- Pencil

Learning Objective:

Understand the definition of the slope of a line as the ratio of the change in y to the change in x between any two points on the line. Discover that slope can be determined from any two points on the line.

Name ______________________________

Slope is the rate of change between any two points on a line. It is the measure of the steepness of the line. To find the slope of a line, find the ratio of the directed change in y (vertical change) to the directed change in x (horizontal change) from one point to another.

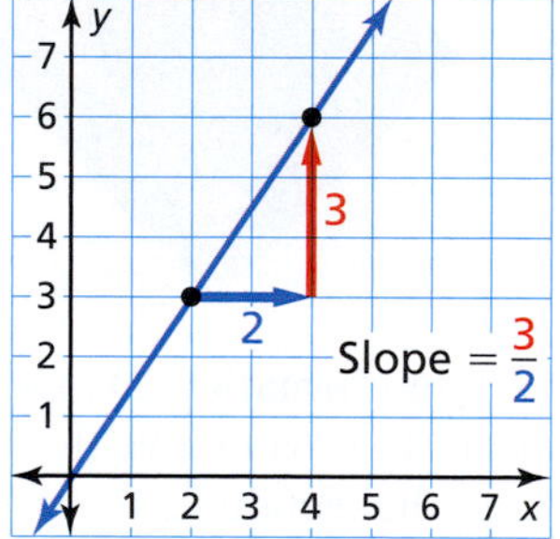

$$\text{Slope} = \frac{\text{change in } y}{\text{change in } x}$$

Find the slope of the line using the two red points. Then find the slope of the line using the two black points. Do you get the same slope each time? Explain your reasoning.

1.

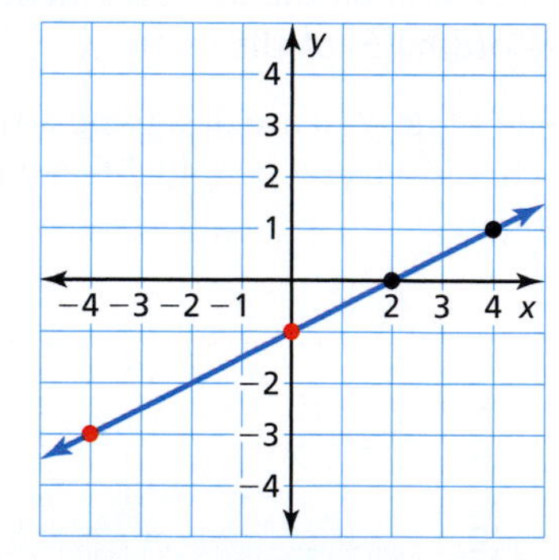

2.

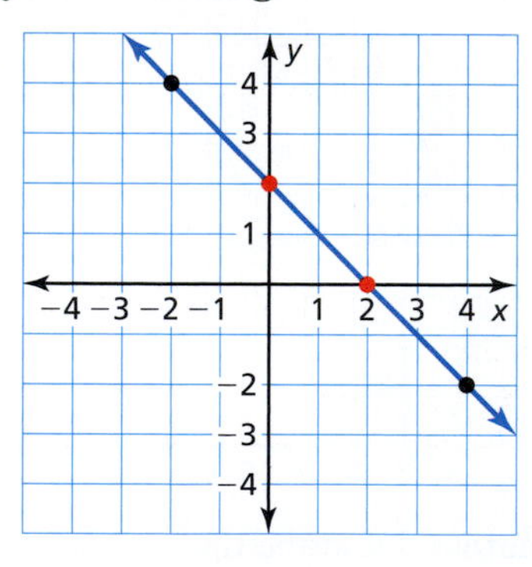

3.

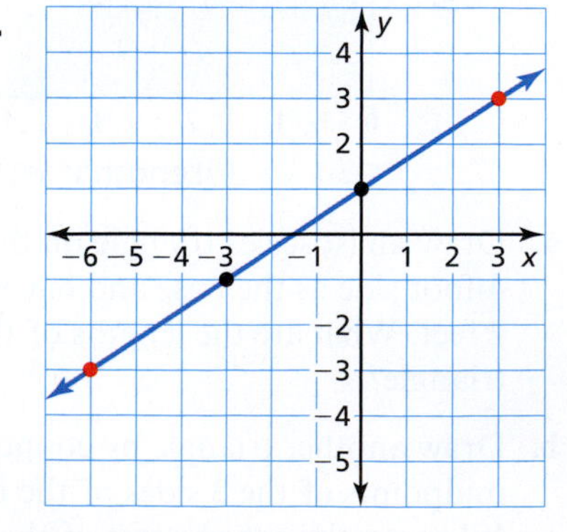

4.

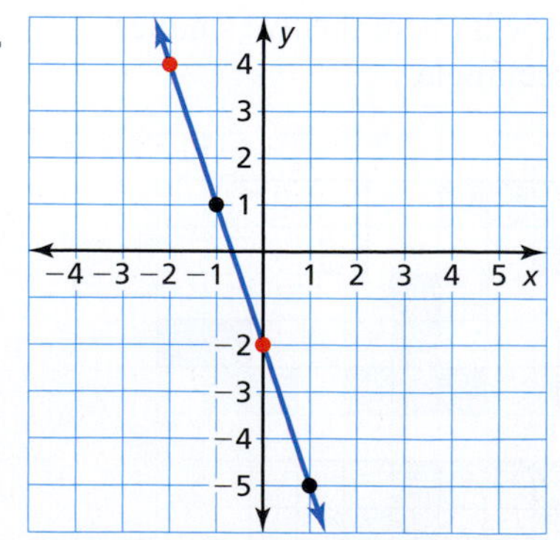

15.2 The Slope of a Line

Standards

Grades 3–5
Operations and Algebraic Thinking
Students should analyze patterns and relationships.

Grades 6–8
Expressions and Equations
Students should understand the connections between proportional relationships, lines, and linear equations.

- Find the slope of a line.
- Interpret the slope of a line as a rate of change.
- Find the slopes of parallel and perpendicular lines.

The Slope of a Line

Definition of the Slope of a Line

The **slope** of a line is the ratio of the change in y (the **rise**) to the change in x (the **run**) between any two points (x_1, y_1) and (x_2, y_2), on the line.

$$\text{Slope} = \frac{\text{rise}}{\text{run}} = \frac{\text{change in } y}{\text{change in } x} = \frac{y_2 - y_1}{x_2 - x_1}$$

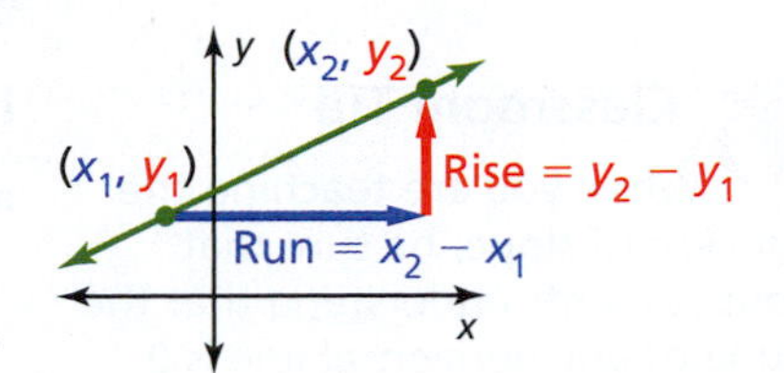

Positive Slope

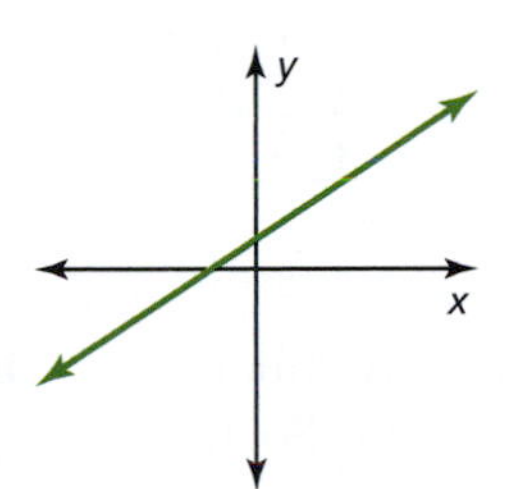

The line rises from left to right.

Negative Slope

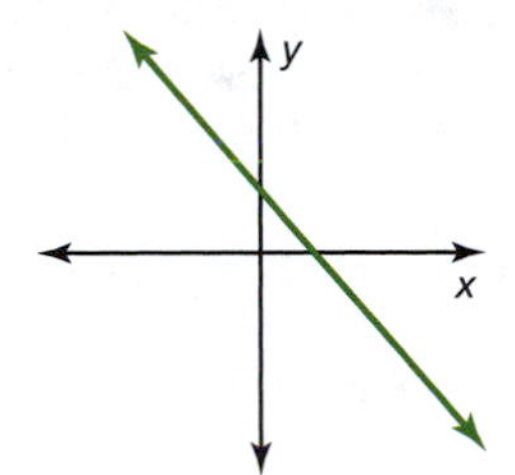

The line falls from left to right.

EXAMPLE 1 **Finding the Slopes of Lines**

a.

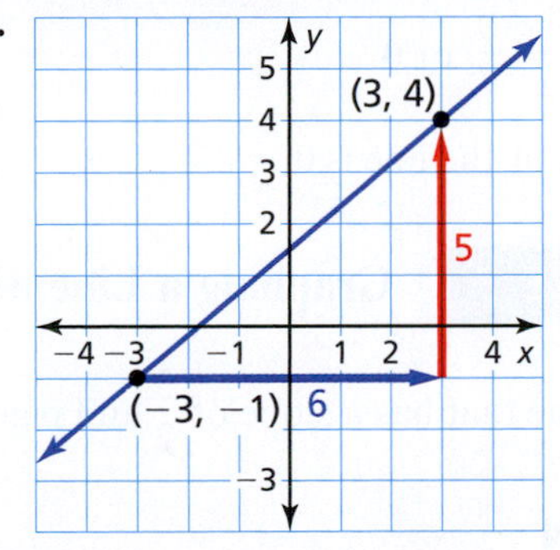

The line rises from left to right. So, the slope is positive. Let $(x_1, y_1) = (-3, -1)$ and $(x_2, y_2) = (3, 4)$.

$$\text{Slope} = \frac{y_2 - y_1}{x_2 - x_1} = \frac{4 - (-1)}{3 - (-3)} = \frac{5}{6}$$

b.

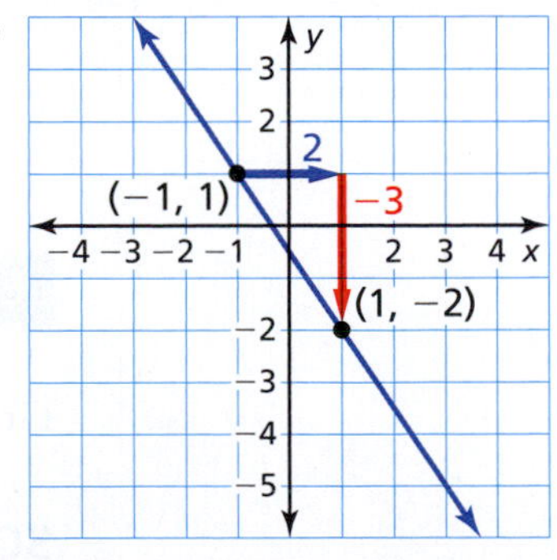

The line falls from left to right. So, the slope is negative. Let $(x_1, y_1) = (-1, 1)$ and $(x_2, y_2) = (1, -2)$.

$$\text{Slope} = \frac{y_2 - y_1}{x_2 - x_1} = \frac{-2 - 1}{1 - (-1)} = \frac{-3}{2}, \text{ or } -\frac{3}{2}$$

Classroom Tip

To help your students gain an understanding of rise and run, try having them measure some real-life slopes. For instance, you can use a stack of books and two rulers. Use one ruler to create the line. Then use the other ruler to measure the rise (the height of the books) and the run (the distance from the books to the end of the ruler).

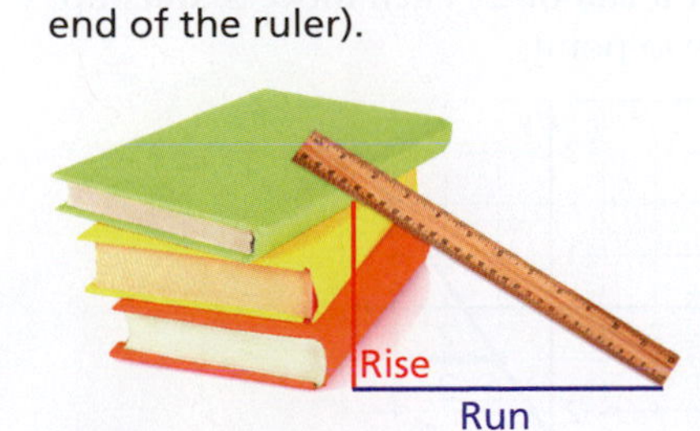

Africa Studio/Shutterstock.com

Notice that slope allows you to compare the steepness of lines. For instance, of the three lines shown at the right, Line 3 has the greatest slope and it is the steepest. Similarly, Line 1 has the least slope and it is less steep than the other two lines.

Be careful when comparing the steepness of lines with negative slopes. A line with slope -5 is steeper than the lines shown, even though its slope is negative.

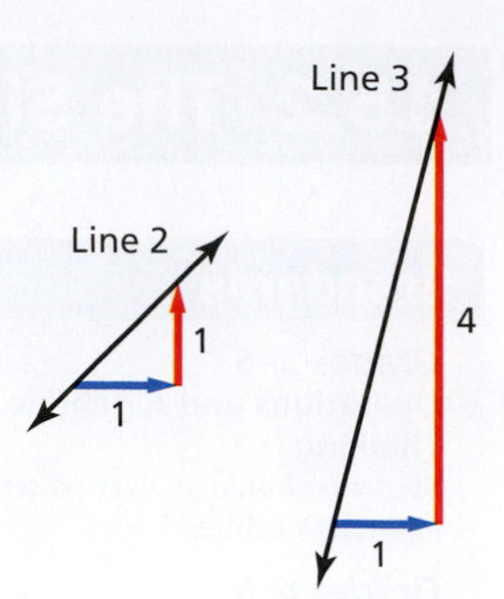

EXAMPLE 2 Horizontal and Vertical Lines

Classroom Tip

When you are teaching the concept of slope, be sure that your students understand that the slope of *any* horizontal line is 0 and the slope of *any* vertical line is undefined.

Find the slope of each line.

a.

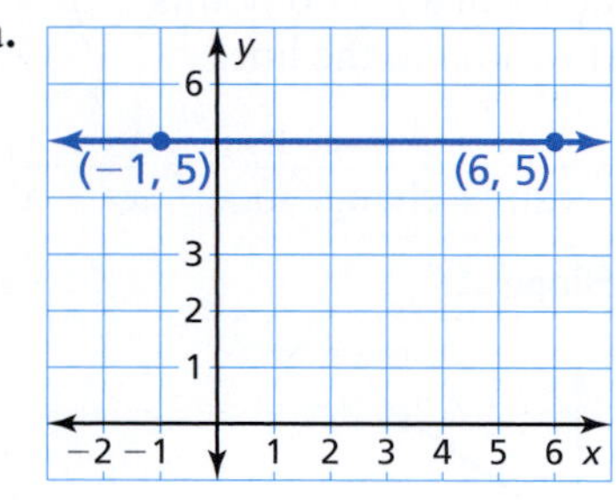

b.

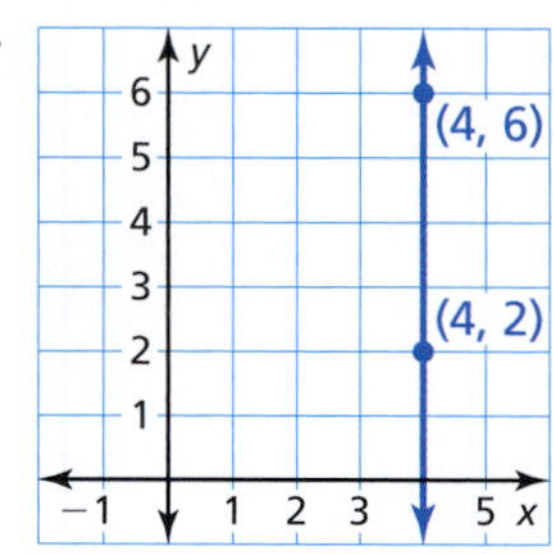

SOLUTION

a. The line is not rising or falling. There is no change in y. So, the change in y is zero.

$$\text{Slope} = \frac{y_2 - y_1}{x_2 - x_1} = \frac{5 - 5}{6 - (-1)} = \frac{0}{7}, \text{ or } 0$$

The slope of the line is 0.

b. The line is vertical. There is no change in x. So, the change in x is zero.

$$\text{Slope} = \frac{y_2 - y_1}{x_2 - x_1} = \frac{6 - 2}{4 - 4} = \frac{4}{0} \quad \text{Undefined}$$

Because division by zero is undefined, the slope of the line is undefined.

EXAMPLE 3 Graphing a Line with a Given Slope

Graph the line that has a slope of $\frac{3}{2}$ and passes through the point $(1, -1)$.

SOLUTION

Start at the point $(1, -1)$. Move 2 units right for a run of 2. Then move 3 units up for a rise of 3. Finally, draw a line through the two points.

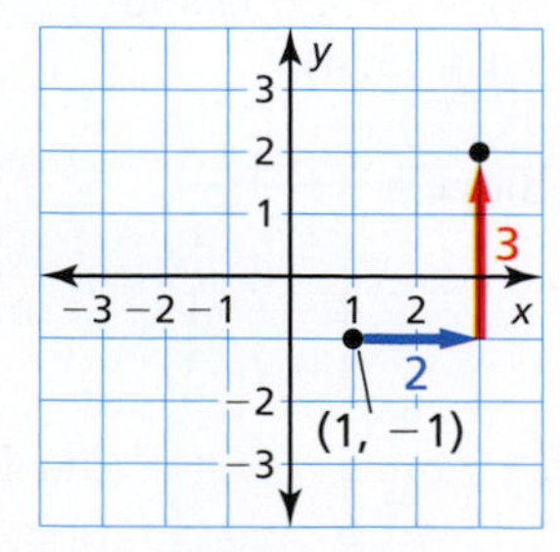

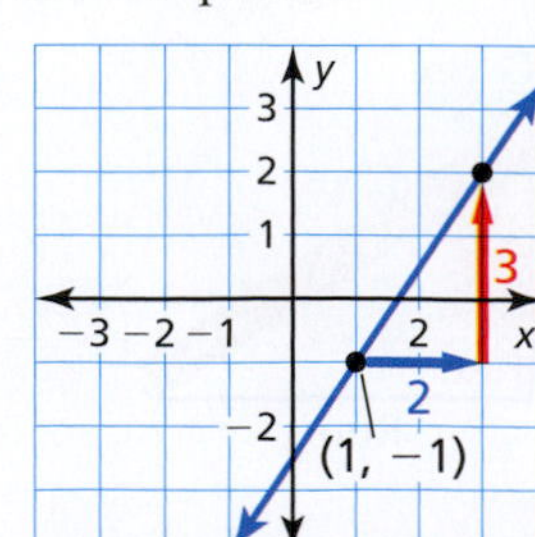

Slope as a Rate of Change

In Section 6.4, you learned that a rate is a ratio in which the units for the numerator and denominator are *different*. Because the slope of a line is a ratio, many real-life applications of slope involve rates. Below are two examples.

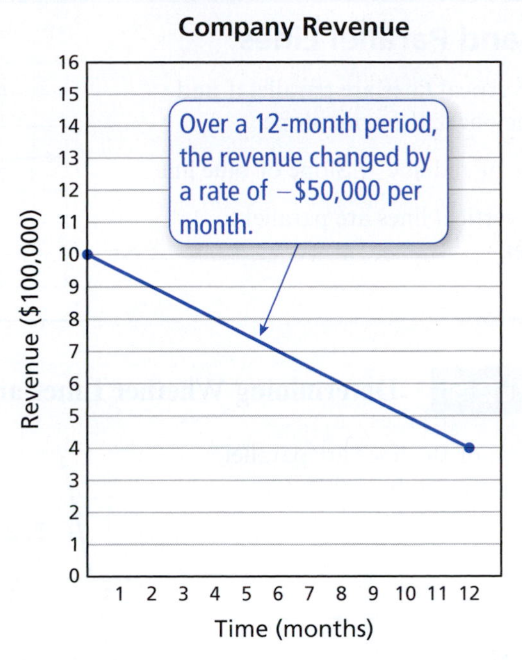

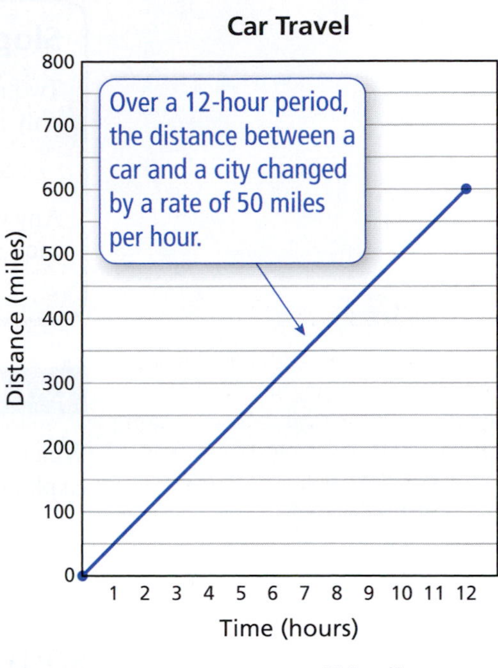

$$\text{Rate of change} = \frac{-\$600{,}000}{12 \text{ months}} \qquad \text{Rate of change} = \frac{600 \text{ miles}}{12 \text{ hours}}$$

Classroom Tip

When you are interpreting slope as a rate of change, look at the units on the vertical and horizontal axes. The rate of change is "the vertical units per the horizontal units."

EXAMPLE 4 Using Slope in Real Life

The graph shows the cost of traveling by car on a turnpike.

a. Find the slope of the line.

b. Explain the meaning of the slope as a rate of change.

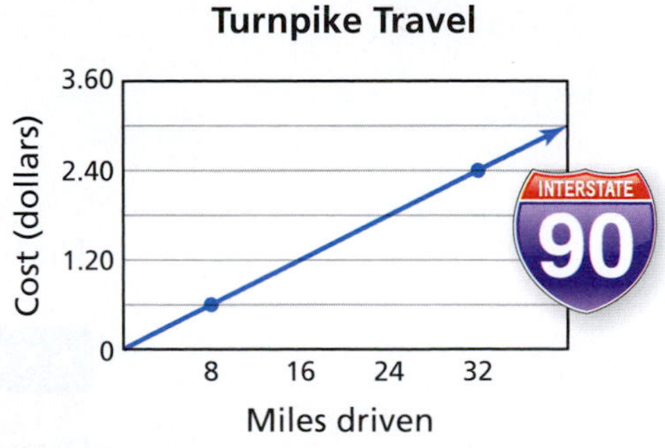

SOLUTION

a. The two points shown on the graph are (8, 0.60) and (32, 2.40). Use these two points to find the slope of the line.

$$\text{Slope} = \frac{2.40 - 0.60}{32 - 8}$$

$$= \frac{1.8}{24}$$

$$= 0.075$$

b. Because the vertical axis is measured in dollars and the horizontal axis is measured in miles, the rate of change is

$$\text{Rate of change} = \frac{\$0.075}{1 \text{ mi}}.$$

This means that the cost of driving on the turnpike is $0.075 per mile. In other words, it costs 7 and 1/2 cents for every mile driven on the turnpike.

Mathematical Practices

Attend to Precision

MP6 Mathematically proficient students use clear definitions in discussion with others and in their own reasoning.

Slopes of Parallel and Perpendicular Lines

Lines in the same plane that do not intersect are *parallel* lines. Slope can be used to determine whether two lines are parallel.

Slope and Parallel Lines

Two nonvertical lines are parallel if and only if they have the same slope.

Slope of red line = Slope of blue line

Any two vertical lines are parallel to each other.

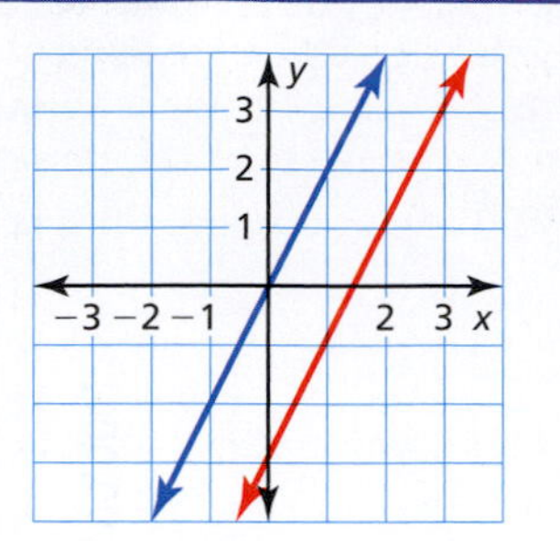

EXAMPLE 5 **Determining Whether Lines are Parallel**

Which, if any, of the lines are parallel? Explain your reasoning.

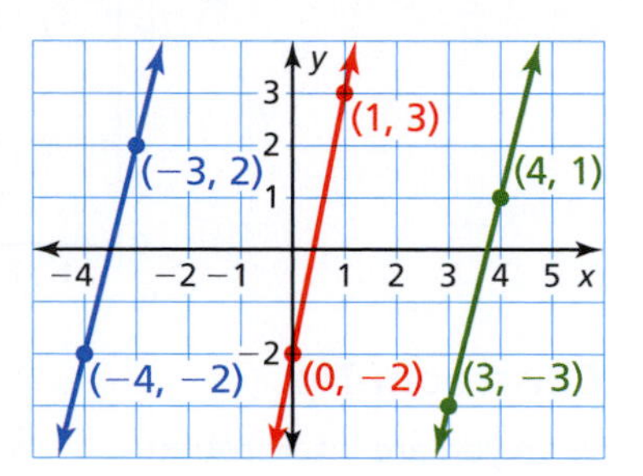

SOLUTION

Find the slope of each line.

Blue line

$$\text{Slope} = \frac{y_2 - y_1}{x_2 - x_1} = \frac{-2 - 2}{-4 - (-3)} = \frac{-4}{-1}, \text{ or } 4$$

Red line

$$\text{Slope} = \frac{y_2 - y_1}{x_2 - x_1} = \frac{-2 - 3}{0 - 1} = \frac{-5}{-1}, \text{ or } 5$$

Green line

$$\text{Slope} = \frac{y_2 - y_1}{x_2 - x_1} = \frac{-3 - 1}{3 - 4} = \frac{-4}{-1}, \text{ or } 4$$

The blue line and the green line each have a slope of 4. So, they are parallel.

EXAMPLE 6 **Classifying a Quadrilateral**

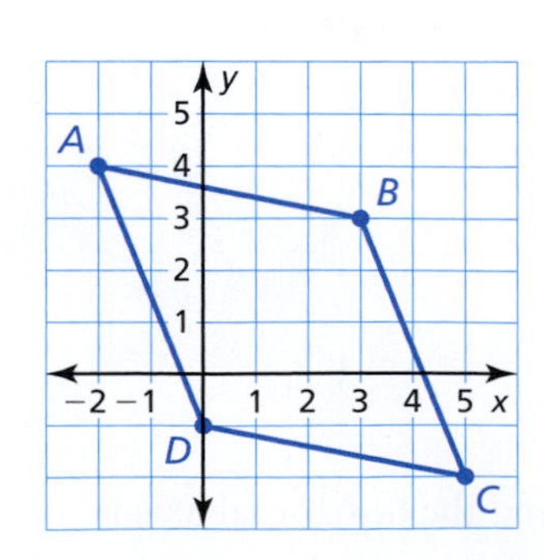

Determine whether quadrilateral $ABCD$ is a parallelogram. Explain your reasoning.

SOLUTION

In Section 10.2, you learned that opposite sides of a parallelogram are parallel. So, find the slopes of lines that include the four sides of $ABCD$.

$\overleftrightarrow{AB}$ $\dfrac{3 - 4}{3 - (-2)} = -\dfrac{1}{5}$ $\qquad$ $\overleftrightarrow{BC}$ $\dfrac{-2 - 3}{5 - 3} = -\dfrac{5}{2}$

$\overleftrightarrow{CD}$ $\dfrac{-1 - (-2)}{0 - 5} = -\dfrac{1}{5}$ $\qquad$ $\overleftrightarrow{DA}$ $\dfrac{4 - (-1)}{-2 - 0} = -\dfrac{5}{2}$

$\overleftrightarrow{AB}$ and $\overleftrightarrow{CD}$ are parallel. $\overleftrightarrow{BC}$ and $\overleftrightarrow{DA}$ are parallel. Because the quadrilateral has two pairs of opposite sides that are parallel, it is a parallelogram.

Lines in the same plane that intersect at right angles are *perpendicular* lines. Slope can be used to determine whether two lines are perpendicular.

Classroom Tip

In algebra, it is common to identify the slope of a line using the letter m. Using this notation, you can write the result shown at the right as

$$m_1 = \frac{-1}{m_2}$$

where m_1 and m_2 are the slopes of the two perpendicular lines.

Slope and Perpendicular Lines

Two nonvertical lines are perpendicular to each other if and only if their slopes are negative reciprocals.

$$\text{Slope of red line} = \frac{-1}{\text{Slope of blue line}}$$

Vertical lines are perpendicular to horizontal lines.

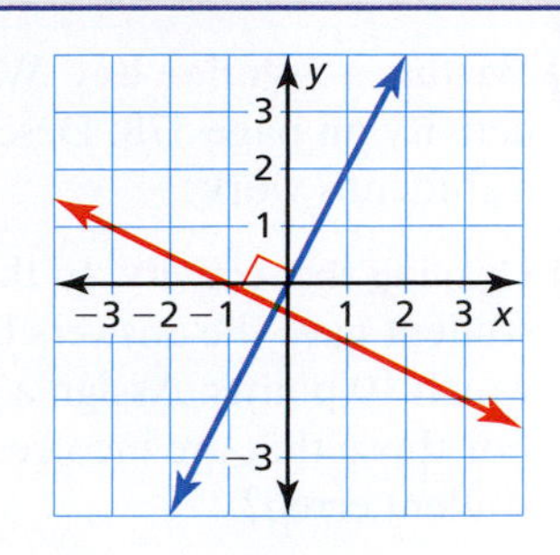

EXAMPLE 7 Drawing Parallel and Perpendicular Lines

a. Draw a line that is parallel to the line shown and passes through the point (−2, 1).

b. Draw a line that is perpendicular to the line shown and passes through the point (−2, 1).

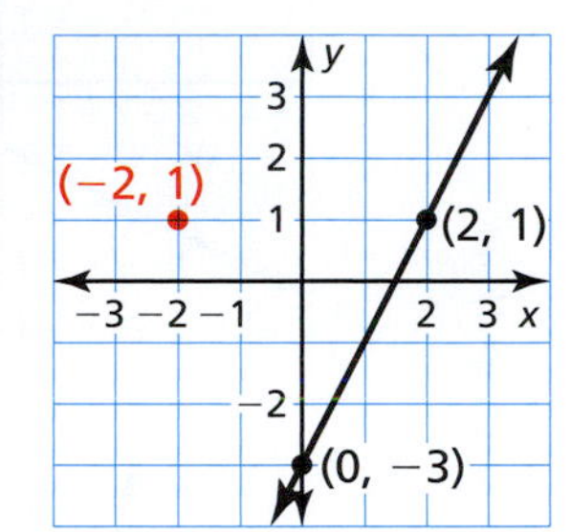

SOLUTION

Begin by finding the slope of the given line.

$$\text{Slope} = \frac{1-(-3)}{2-0}$$

$$= \frac{4}{2}$$

$$= 2 \qquad \text{Slope of line}$$

a. To draw a line that is parallel to the given line, you need to draw a line that has the same slope, 2.

Start at the point (−2, 1). Move 1 unit right and 2 units up. Plot a point at this location. Draw a line through the two points. The line has a slope of 2, so it is parallel to the given line.

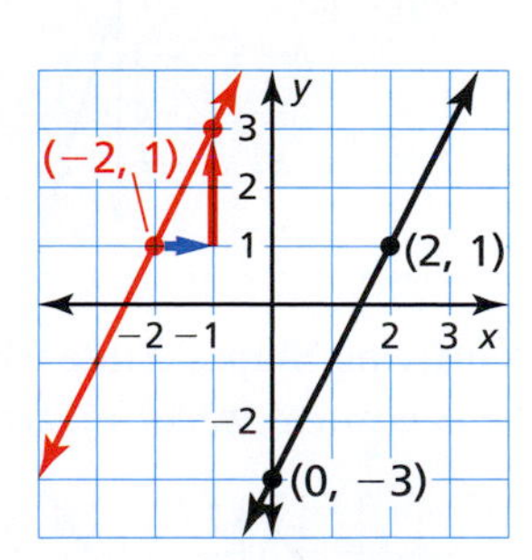

b. To draw a line that is perpendicular to the given line, you need to draw a line that has a slope that is the negative reciprocal of 2.

The reciprocal of 2 is $\frac{1}{2}$. So, the slope of the perpendicular line should be $-\frac{1}{2}$.

Start at the point (−2, 1). Move 2 units right and 1 unit down. Plot a point at this location. Draw a line through the two points. The line has a slope of $-\frac{1}{2}$, so it is perpendicular to the given line.

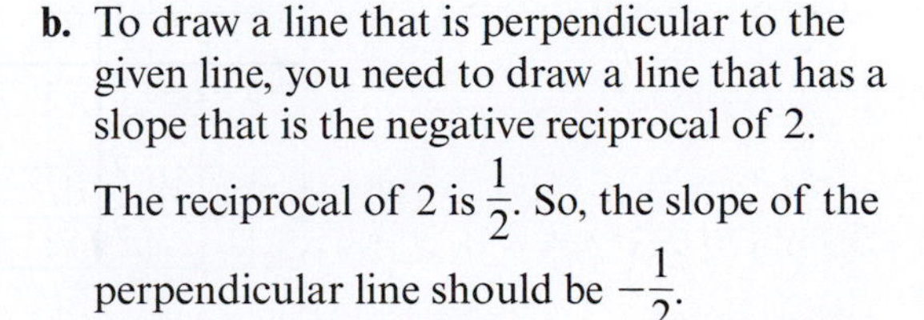

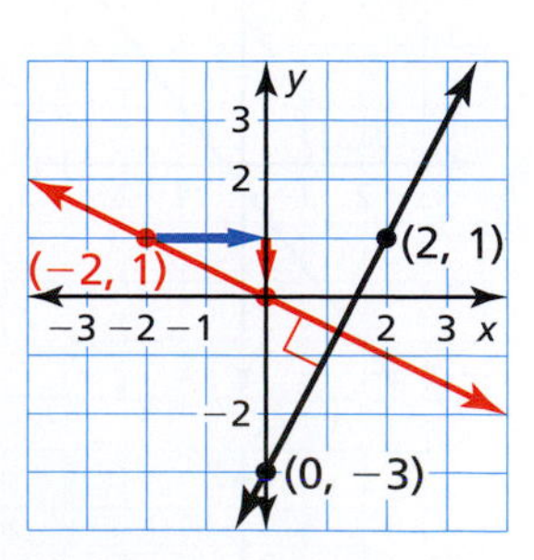

15.2 Exercises

Free step-by-step solutions to odd-numbered exercises at *MathematicalPractices.com.*

1. Writing a Solution Key Write a solution key for the activity on page 578. Describe a rubric for grading a student's work.

2. Grading the Activity In the activity on page 578, a student gave the answers below. Each question is worth 10 points. Assign a grade to each answer. For those that are incorrect, why do you think the student erred?

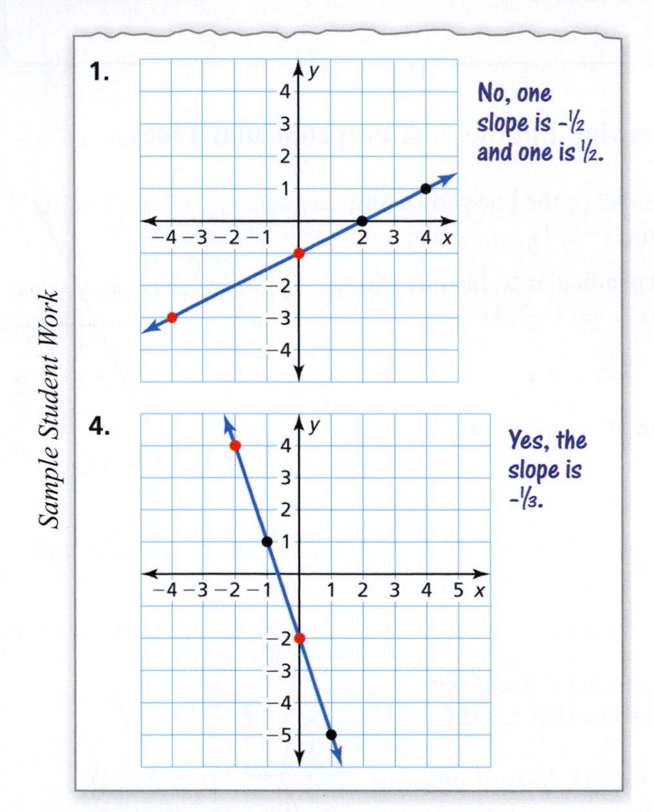

3. Identifying Slopes State whether the slope of each line is *positive*, *negative*, *zero*, or *undefined*.

a.

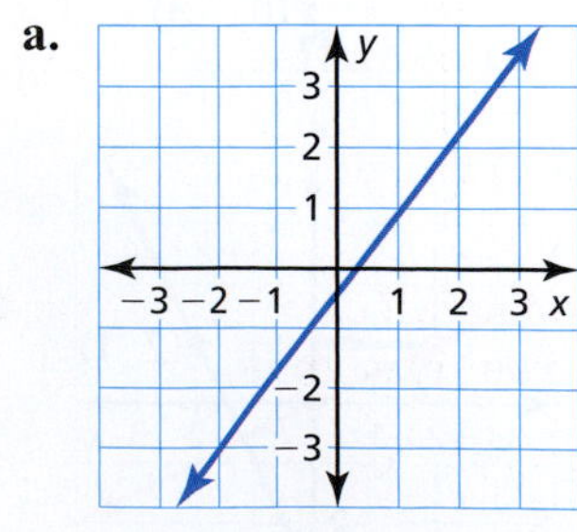

b.

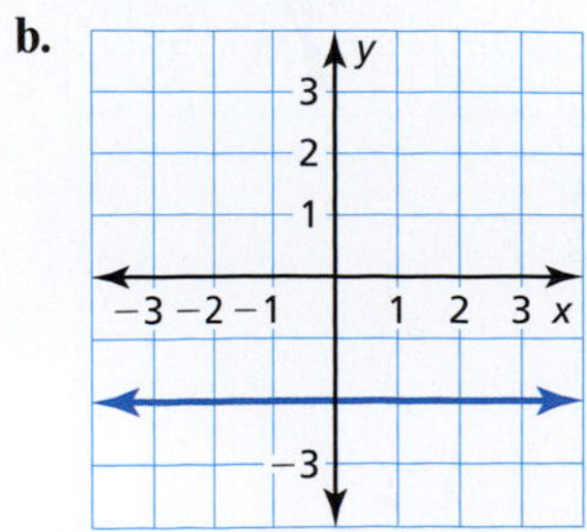

4. Identifying Slopes State whether the slope of each line is *positive*, *negative*, *zero*, or *undefined*.

a.

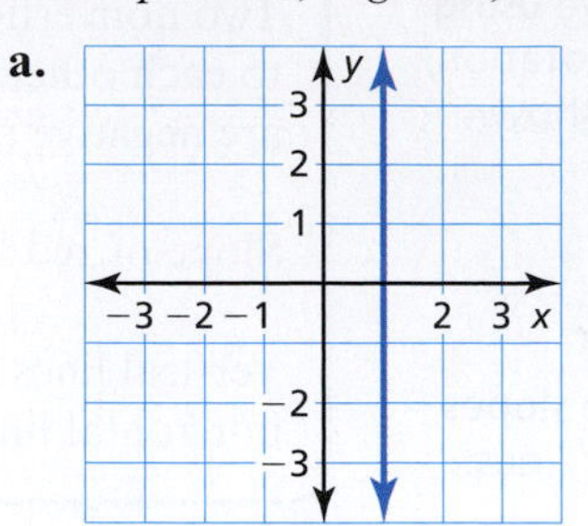

b.

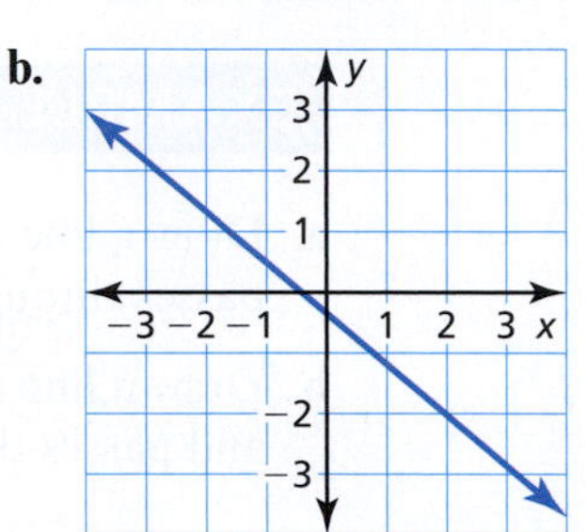

5. Finding Slopes Find the slope of each line.

a. **b.**

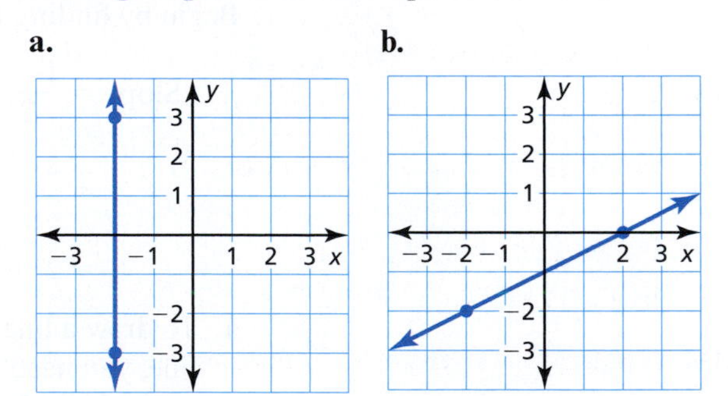

6. Finding Slopes Find the slope of each line.

a. **b.**

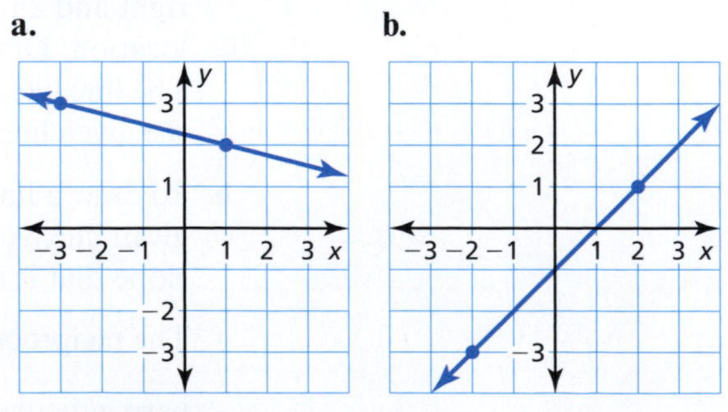

7. Finding Slopes The two points lie on a line. Find the slope of each line.

a. $(3, 5), (4, -7)$

b. $(6, -3), (6, 2)$

8. Finding Slopes The two points lie on a line. Find the slope of each line.

a. $(-1, 4), (5, 4)$

b. $(-5, 2), (-1, 8)$

9. Finding Slope The table shows the points that lie on a line. Find the slope of the line.

x	6	6	6	6
y	−3	0	3	6

10. Finding Slope The table shows the points that lie on a line. Find the slope of the line.

x	−4	−2	0	2
y	7	5	3	1

11. Graphing Lines Graph the line with the given slope that passes through the given point.

a. Slope: -2, point: $(0, 3)$

b. Slope: $\frac{5}{4}$, point: $(-2, -3)$

12. Graphing Lines Graph the line with the given slope that passes through the given point.

a. Slope: $\frac{3}{4}$, point: $(1, 3)$

b. Slope: 0, point: $(0, -3)$

13. Finding Points on Lines Determine whether the two points lie on a line with a slope of $\frac{1}{2}$.

a. $(1, 2)$, $(4, 1)$

b. $(3, 1)$, $(5, 2)$

14. Finding Points on Lines Determine whether the two points lie on a line with a slope of $-\frac{4}{3}$.

a. $(-5, 5)$, $(-2, 1)$

b. $(0, 2)$, $(6, -4)$

15. Rate of Change The graph shows the distance a cruise ship travels in relation to time.

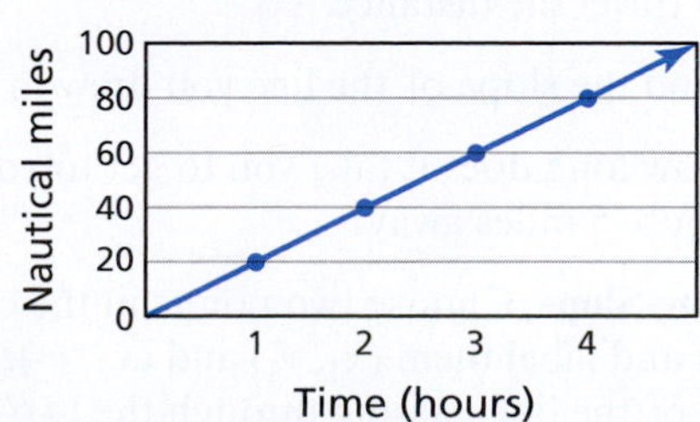

a. Find the slope of the line.

b. Explain the meaning of the slope as a rate of change.

16. Rate of Change The graph shows the distance a train travels in relation to time.

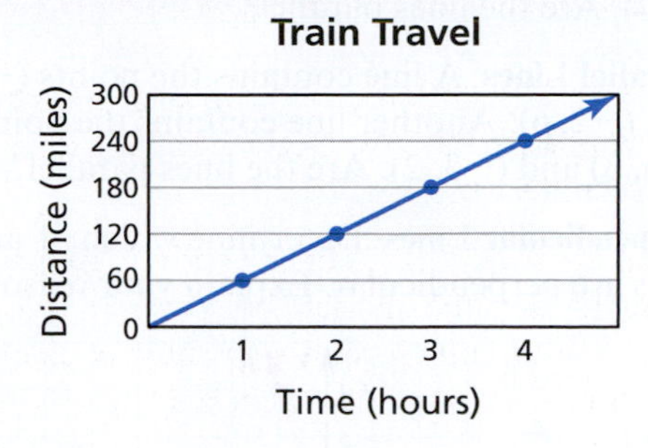

a. Find the slope of the line.

b. Explain the meaning of the slope as a rate of change.

17. Parallel and Perpendicular Lines A line has a slope of 1.

a. What is the slope of a line that is parallel to the given line?

b. What is the slope of a line that is perpendicular to the given line?

18. Parallel and Perpendicular Lines A line has a slope of $-\frac{2}{5}$.

a. What is the slope of a line that is parallel to the given line?

b. What is the slope of a line that is perpendicular to the given line?

19. Parallel Lines Determine which, if any, of the lines are parallel. Explain your reasoning.

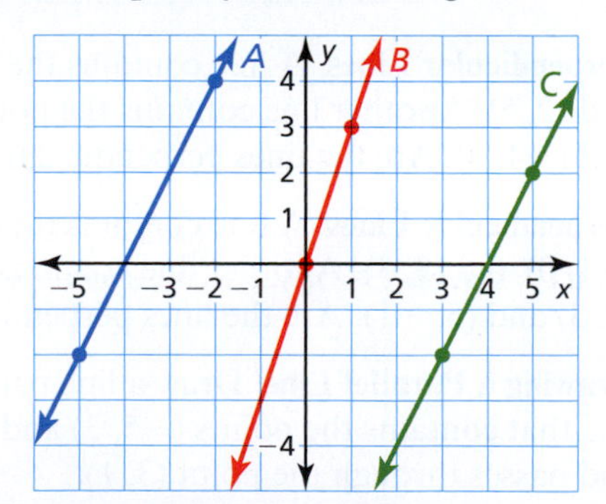

20. Parallel Lines Determine which, if any, of the lines are parallel. Explain your reasoning.

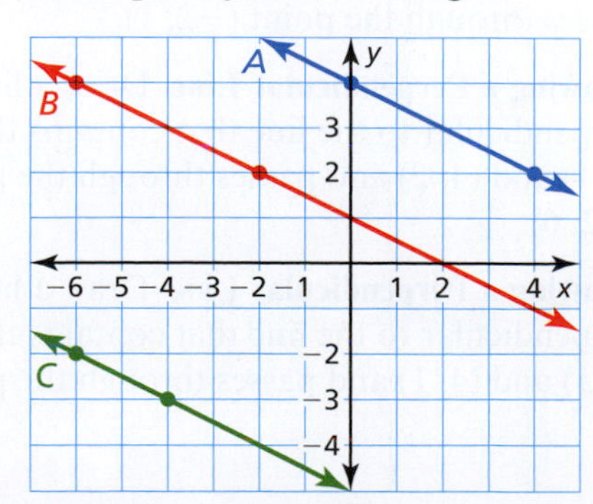

21. Parallel Lines A line contains the points (2, 1) and (6, 5). Another line contains the points (−3, 2) and (2, 7). Are the lines parallel?

22. Parallel Lines A line contains the points (−4, 5) and (−2, 6). Another line contains the points (−6, 3) and (−3, 2). Are the lines parallel?

23. Perpendicular Lines Determine which, if any, of the lines are perpendicular. Explain your reasoning.

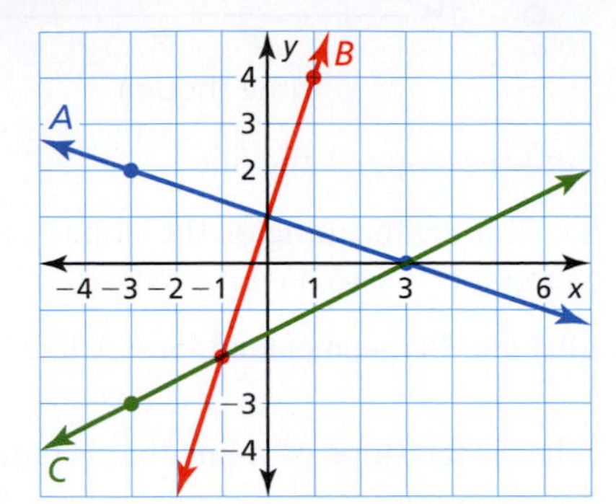

24. Perpendicular Lines Determine which, if any, of the lines are perpendicular. Explain your reasoning.

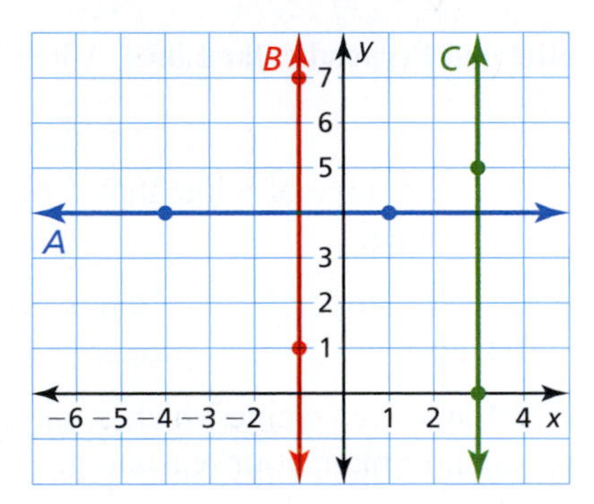

25. Perpendicular Lines A line contains the points (1, 2) and (2, 5). Another line contains the points (− 5, 3) and (−4, 0). Are the lines perpendicular?

26. Perpendicular Lines A line contains the points (−1, 0) and (3, 3). Another line contains the points (2, 3) and (5, −1). Are the lines perpendicular?

27. Drawing a Parallel Line Draw a line parallel to the line that contains the points (−5, 3) and (−5, −2) and passes through the point (3, 1).

28. Drawing a Parallel Line Draw a line parallel to the line that contains the points (−1, −3) and (2, 0) and passes through the point (−2, 1).

29. Drawing a Perpendicular Line Draw a line perpendicular to the line that contains the points (1, 4) and (4, 2) and passes through the point (−2, 6).

30. Drawing a Perpendicular Line Draw a line perpendicular to the line that contains the points (0, 1) and (4, 1) and passes through the point (2, 1).

31. Geometry Determine whether the quadrilateral is a *parallelogram*, *rectangle*, or *neither*. Explain your reasoning.

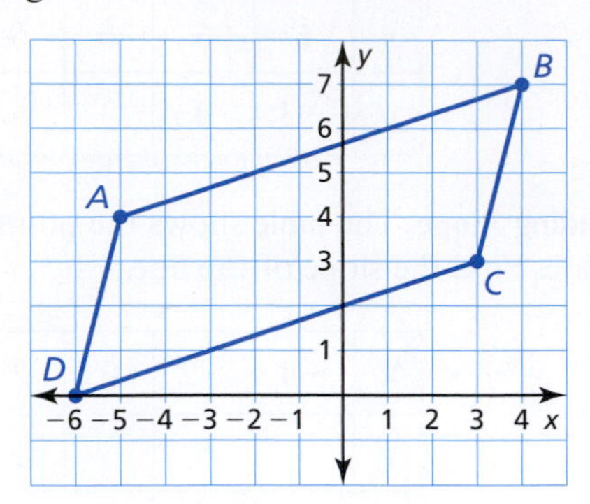

32. Geometry Determine whether the quadrilateral is a *parallelogram*, *rectangle*, or *neither*. Explain your reasoning.

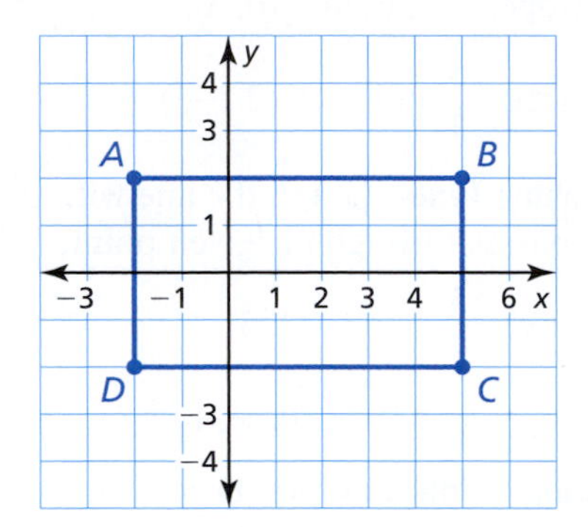

33. Rate of Change You are skateboarding at a constant rate of 5 miles per hour.

a. Draw a graph showing the relation between the distance traveled and the amount of time it takes to travel the distance.

b. Find the slope of the line you drew in part (a).

c. How long does it take you to get to your friend's house 2 miles away?

34. Rate of Change You are biking at a constant rate of 10 miles per hour.

a. Draw a graph showing the relation between the distance traveled and the amount of time it takes to travel the distance.

b. Find the slope of the line you drew in part (a).

c. How long does it take you to get to your friend's house 5 miles away?

35. Finding Slope Choose two points in the coordinate plane and label them (x_1, y_1) and (x_2, y_2). Find the slope of the line passing through the two points using the slope formula. Then evaluate the expression $\frac{y_1 - y_2}{x_1 - x_2}$. Repeat this process for several pairs of points. Are your results the same for each pair? Explain.

36. Finding Slope Choose two points in the coordinate plane and label them (x_1, y_1) and (x_2, y_2). Find the slope of the line passing through the two points using the slope formula. Then evaluate the expression $\frac{x_2 - x_1}{y_2 - y_1}$. Repeat this process for several pairs of points. Are your results the same for each pair? Explain.

37. Determining Slope Without graphing, determine whether the points $(-3, -3)$, $(-2, -1)$, and $(0, 3)$ lie on the same line. Explain your reasoning.

38. Determining Slope Without graphing, determine whether the points $(-4, 6)$, $(0, 4)$, and $(2, 2)$ lie on the same line. Explain your reasoning.

39. Finding Points on Lines Find the coordinates of two other points collinear (on the same line) with each of the following given pairs of points.

a. $(-1, -3)$, $(1, 1)$

b. $(1, 6)$, $(1, -5)$

40. Finding Points on Lines Find the coordinates of two other points collinear (on the same line) with each of the following given pairs of points.

a. $(2, 4)$, $(5, 3)$

b. $(-6, -3)$, $(4, -3)$

41. Discussion: In Your Classroom Lines A, B, and C are graphed in a coordinate plane. Line A is parallel to line B and line B is perpendicular to line C. Discuss whether line A is parallel or perpendicular to line C.

42. Discussion: In Your Classroom Point (x, y) is on a line with a slope of $\frac{5}{3}$. Discuss whether each statement will determine another point on the line. Discuss how to improve the statement.

a. Move up 5, then move over 3

b. Move up 5, then move right 3

43. Problem Solving Two of your students are sitting on a seesaw. Use the picture below to calculate the slope of the seesaw.

Sergey Lavrentev/Shutterstock.com

44. Problem Solving You build a tower of cards and measure the length of the base of the cards on the floor to be 12 inches. The height of the tower of cards is 15 inches. What is the slope of the line of cards on the left side of the tower? What is the slope of the line of cards on the right side of the tower? Explain your reasoning.

45. Grading Student Work On a diagnostic test, one of your students does the following work. Explain what the student did wrong. Which topics would you encourage the student to review?

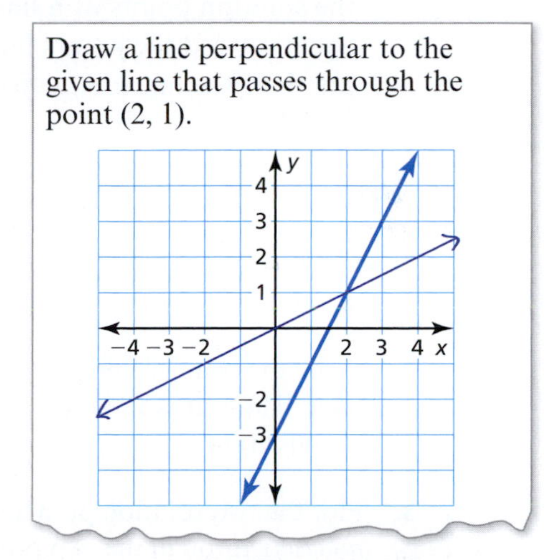

46. Grading Student Work On a diagnostic test, one of your students does the following work. Explain what the student did wrong. Which topics would you encourage the student to review?

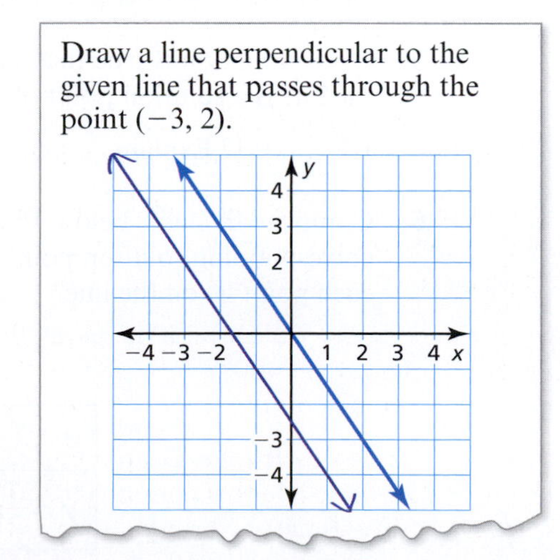

47. Writing Describe the difference between a line that has a slope of zero and a line that has an undefined slope.

48. Writing Can a line have two different slopes? Explain.

Zerbor/Shutterstock.com

Activity: Discovering the Graph of a Linear Equation

Materials:

- Pencil

Learning Objective:

Understand the concept of solution points of a linear equation in two variables. Discover that there is a one-to-one correspondence between the solution points of a linear equation and the points on the graph of the linear equation.

Name ______________________________

1. Use the equation $y = \frac{1}{2}x + 1$ to complete the table.

	Solution points	
x	4	−4
$y = \frac{1}{2}x + 1$		

2. Write the two ordered pairs given by the table. These are called *solution points* of the equation.

3. Plot the two solution points. Draw a line *precisely* through the two points.

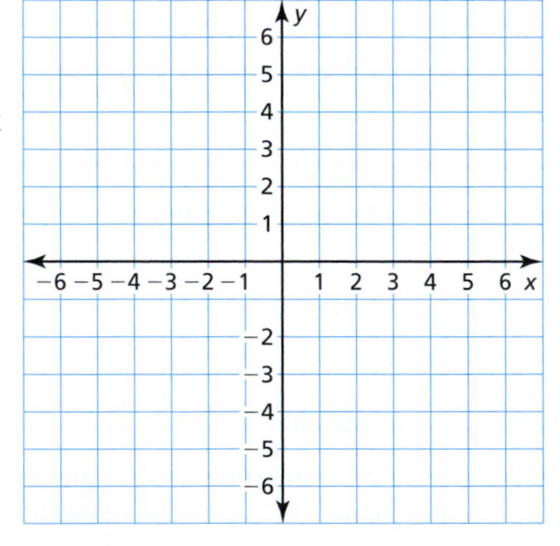

4. Choose a different point on the line. Check that this point is a solution point of the equation $y = \frac{1}{2}x + 1$.

5. Do you think it is true that *any* point on the line is a solution point of the equation $y = \frac{1}{2}x + 1$? Explain.

6. Complete the table below. Plot the five corresponding solution points. Does each point lie on the line?

	Solution points				
x	−6	−2	0	1.5	5
$y = \frac{1}{2}x + 1$					

7. Do you think it is true that *any* solution point of the equation $y = \frac{1}{2}x + 1$ is a point on the line? Explain.

iStockphoto.com/kali9

15.3 Linear Equations in Two Variables

Standards

Grades 3–5
Operations and Algebraic Thinking
Students should analyze patterns and relationships.

Grades 6–8
Expressions and Equations
Students should represent and analyze quantitative relationships between dependent and independent variables.

- Sketch the graphs of linear equations.
- Recognize the equations of horizontal and vertical lines.
- Find and interpret the slopes and *y*-intercepts of the graphs of linear equations.

The Graph of a Linear Equation

Graph of a Linear Equation

A **linear equation in two variables** is an equation whose graph is a line. The points on the line are **solutions** of the equation. You can use a graph to show the solutions of a linear equation.

Example The graph below represents the linear equation $y = x + 1$.

x	*y*	(*x*, *y*)
−1	0	(−1, 0)
0	1	(0, 1)
2	3	(2, 3)

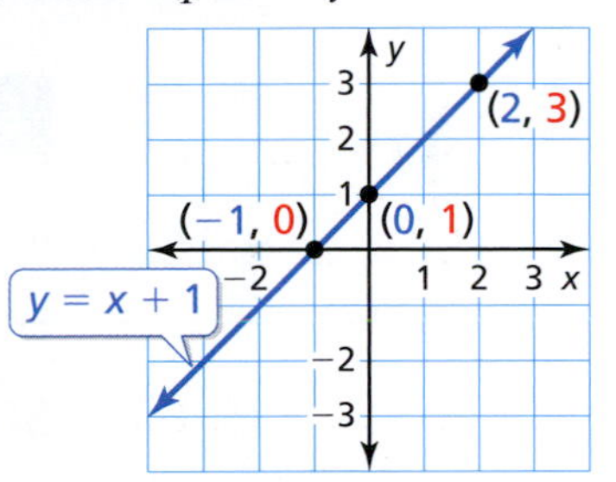

EXAMPLE 1 **Sketching the Graph of a Linear Equation**

Sketch the graph of $y = -2x + 1$.

SOLUTION

Begin by making a table of values. Choose *x*-values −1, 0, and 2.

x	$y = -2x + 1$	*y*	(*x*, *y*)
−1	$y = -2(-1) + 1$	3	(−1, 3)
0	$y = -2(0) + 1$	1	(0, 1)
2	$y = -2(2) + 1$	−3	(2, −3)

Next, plot the ordered pairs. Finally, draw a line through the points.

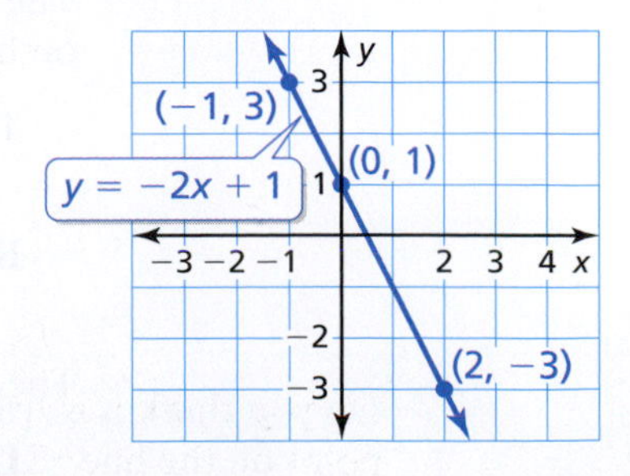

Although you only need two points to determine a line, plot at least three points. This will serve as a check that all of the points lie on the line.

Classroom Tip

Up until the time of René Descartes (1596–1650), algebra and geometry were separate fields of mathematics. Descartes's invention of the coordinate plane was of great importance to mathematics. For the first time, people could "see" solutions of equations. No longer did people have to work with algebra from a purely symbolic point of view. Descartes's combination of geometry and algebra is called analytic (or coordinate) geometry. One of the main discoveries in analytic geometry is that all of the important types of graphs (lines, parabolas, circles, ellipses, and so on) can be represented by straightforward algebraic equations.

Horizontal and Vertical Lines

Classroom Tip

In the equations shown at the right, notice that x and y are variables, while a and b are constants.

Equations of Horizontal and Vertical Lines

The equation of any **horizontal line** is of the form $y = b$.

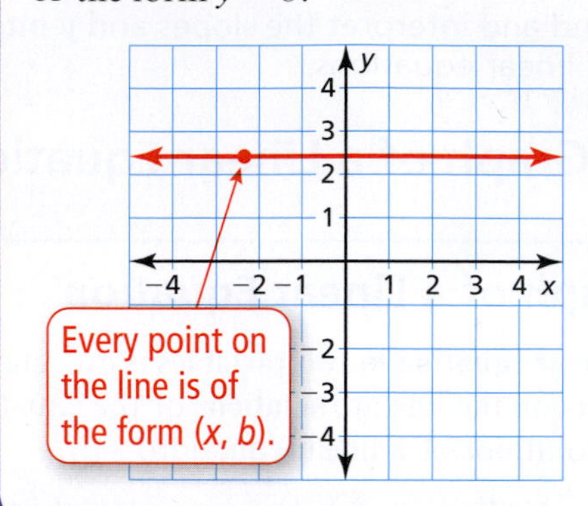

The equation of any **vertical line** is of the form $x = a$.

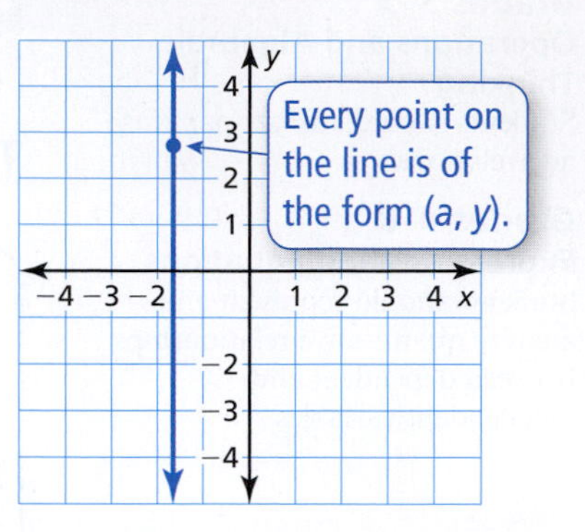

EXAMPLE 2 **Writing Equations of Horizontal and Vertical Lines**

a. Write the equation of the horizontal line.

b. Write the equation of the vertical line.

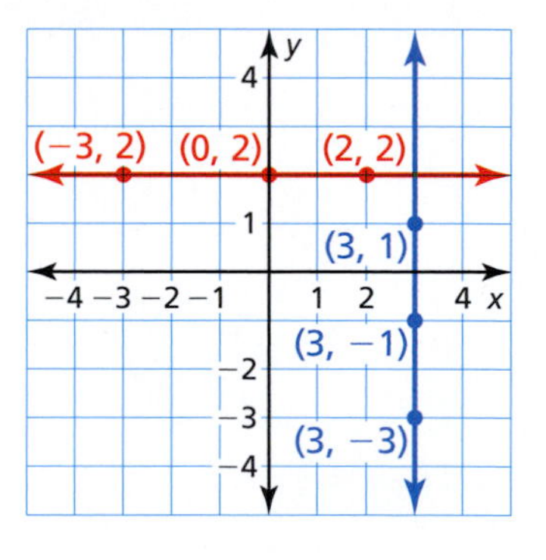

SOLUTION

a. Every point on the horizontal line is of the form $(x, 2)$. So, the equation of the horizontal line is

$y = 2$. Horizontal line

b. Every point on the vertical line is of the form $(3, y)$. So, the equation of the vertical line is

$x = 3$. Vertical line

EXAMPLE 3 **Writing Equations of Lines that Form a Rectangle**

Write the equations of the lines that form the four sides of the rectangle.

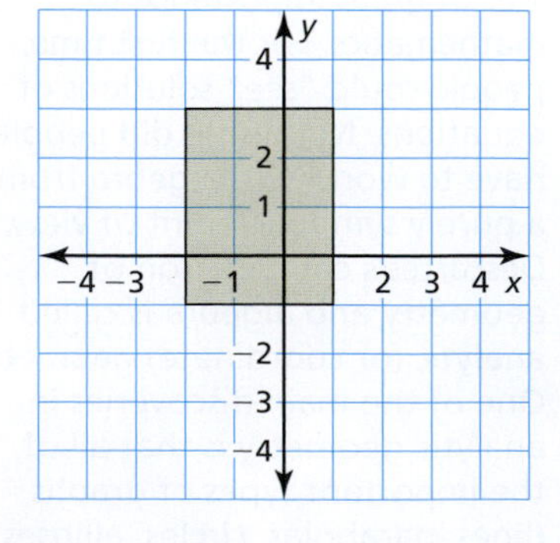

SOLUTION

The top and bottom sides of the rectangle lie on horizontal lines.

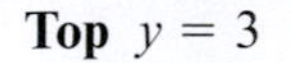

Top $y = 3$ Each point on the top side has the form $(x, 3)$.

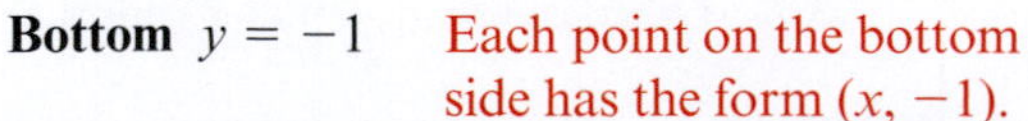

Bottom $y = -1$ Each point on the bottom side has the form $(x, -1)$.

The left and right sides of the rectangle lie on vertical lines.

Left $x = -2$ Each point on the left side has the form $(-2, y)$.

Right $x = 1$ Each point on the right side has the form $(1, y)$.

Equations of Lines and Slope

EXAMPLE 4 **Finding the Slope of a Line**

Mathematical Practices

Construct Viable Arguments and Critique the Reasoning of Others

MP1 Mathematically proficient students look for entry points to a solution.

Sketch the graph of $y = 3x + 1$. Determine the slope of the line.

SOLUTION

Begin by making a table of values and finding two solution points.

x	$y = 3x + 1$	y	(x, y)
0	$y = 3(0) + 1$	1	(0, 1)
2	$y = 3(2) + 1$	7	(2, 7)

Plot the points (0, 1) and (2, 7). Then draw a line through them. Find the slope.

$$\text{Slope} = \frac{y_2 - y_1}{x_2 - x_1} = \frac{7 - 1}{2 - 0} = \frac{6}{2} = 3$$

The slope of the line is 3.

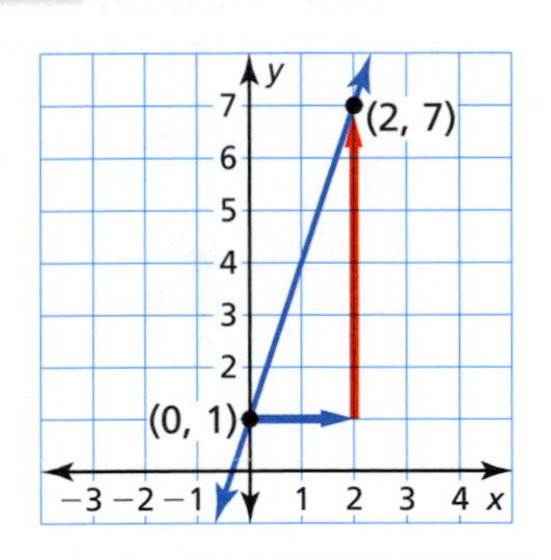

In Example 4, notice that the slope of the line is the coefficient of x and the constant term represents the y-coordinate of the point where the line crosses the y-axis.

$y = 3x + 1$

Slope = 3 y-intercept = 1

You can generalize these observations to describe the equation of any nonvertical line.

Intercepts of a Line

The **x-intercept** of a line is the x-coordinate of the point where the line crosses the x-axis. It occurs when $y = 0$. The **y-intercept** of a line is the y-coordinate of the point where the line crosses the y-axis. It occurs when $x = 0$.

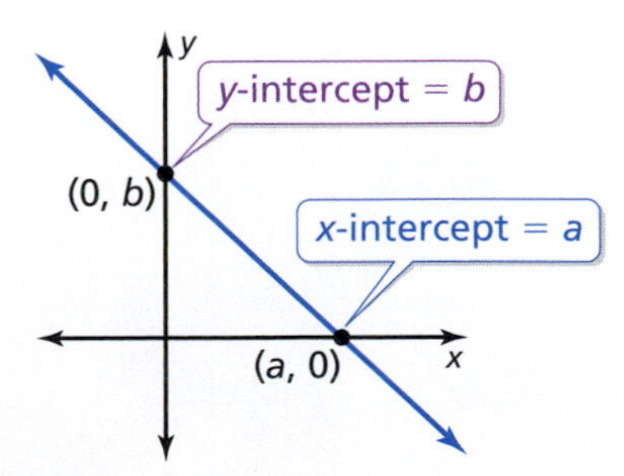

Slope-Intercept Form of a Linear Equation

Words An equation written in the form $y = mx + b$ is in **slope-intercept form**. The slope of the line is m and the y-intercept of the line is b.

Algebra $y = mx + b$

Slope y-intercept

Classroom Tip

Math historians have tried to determine why the letter m is used for slope. No one has been able to come up with a definitive answer. So, when you are teaching the slope-intercept form of a linear equation, suggest that m and b represent

b = begin

and

m = move.

You can use the b-value to plot the "beginning" point $(0, b)$. Then use the m-value to "move" from point $(0, b)$ to plot the next point.

EXAMPLE 5 Using m and b to Sketch a Graph

Sketch the graph of $y = -3x + 3$. Identify the x-intercept.

SOLUTION

Identify the slope and the y-intercept from the equation.

$$y = -3x + 3$$

Slope: -3; y-intercept: 3

Begin The y-intercept is 3. So, *begin* by plotting the point (0, 3).

Move The slope is -3. So,

$$\text{Slope} = \frac{\text{rise}}{\text{run}} = \frac{-3}{1}.$$

Move 1 unit right for a run of 1.
Then *move* 3 units down for a rise of -3.

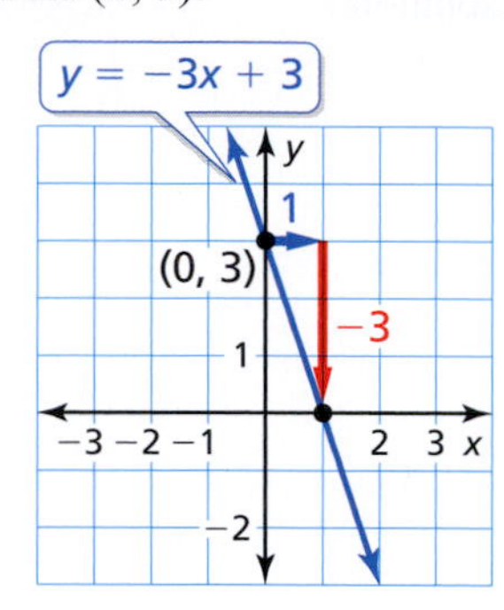

Draw a line through the points. It crosses the x-axis at (1, 0). So, the x-intercept is 1.

Check $y = -3x + 3$

$0 = -3x + 3$

$1 = x$ ✓

EXAMPLE 6 Interpreting Slope and y-Intercept in Real Life

The cost y (in dollars) of taking a taxi x miles is $y = 2.5x + 2$. (a) Interpret the slope and the y-intercept. (b) Graph the equation. (c) What is the cost of a 3-mile taxi ride?

SOLUTION

a. The slope is 2.5. So, the rate is \$2.50 per mile. The y-intercept is 2. So, there is an initial fee of \$2 to take the taxi.

b. The slope of the line is 2.5, or $\frac{5}{2}$. The y-intercept is 2.
Use the slope and y-intercept to sketch the graph of the equation.

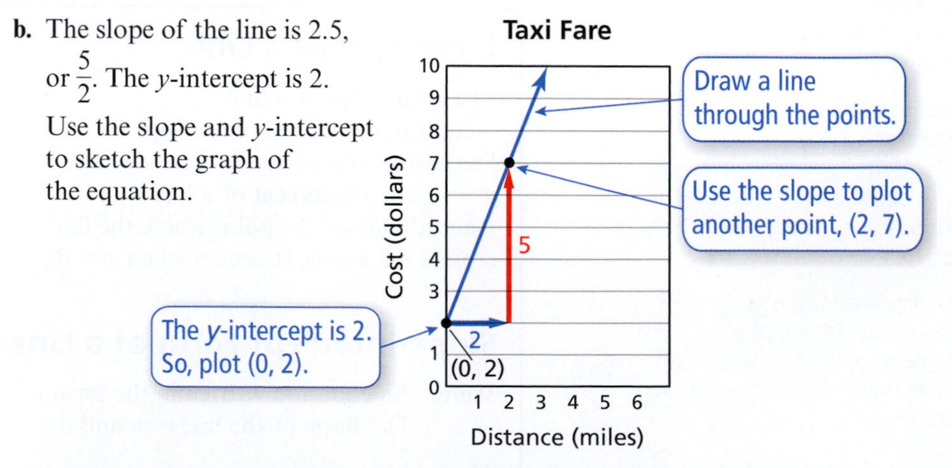

c. For a 3-mile taxi ride, the value of x is 3. Substitute 3 for x and evaluate.

$$y = 2.5(3) + 2 = 9.5$$

So, the cost is \$9.50.

When you are given the equation of a line in a form other than $y = mx + b$, you can use the properties of equality to rewrite the equation in slope-intercept form.

EXAMPLE 7 Rewriting an Equation in Slope-Intercept Form

Sketch the graph of $-2x + 3y = -6$.

SOLUTION

Begin by rewriting the equation in slope-intercept form.

$-2x + 3y = -6$ Write the equation.

$3y = 2x - 6$ Add $2x$ to each side.

$y = \frac{2}{3}x - 2$ Divide each side by 3.

Next, use the slope, $\frac{2}{3}$, and y-intercept, -2, to sketch the graph of the equation.

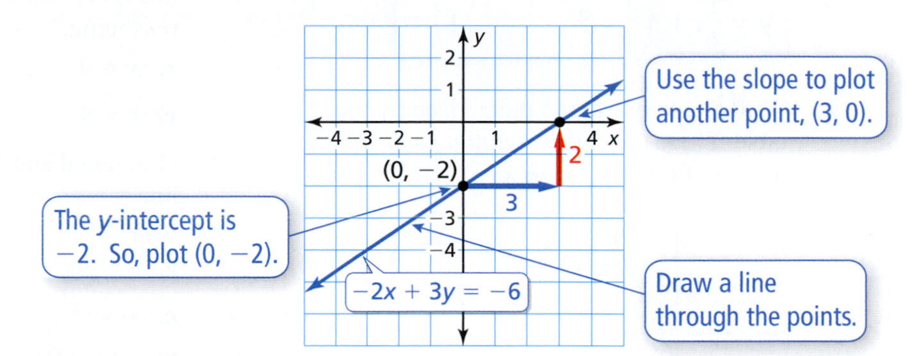

Classroom Tip There are 3 basic ways to sketch the graph of a linear equation.

Plotting points (Example 1) Find two or more solution points of the equation. Plot the points, and draw the line that passes through them.

Using Slope-Intercept Form (Examples 5 and 7) Identify the slope and y-intercept. Then use the slope and y-intercept to plot two points and draw the line that passes through them.

Using Intercepts (Example 8) Find and plot the x-intercept and the y-intercept. Then draw the line that passes through the two intercepts.

EXAMPLE 8 Using Intercepts to Graph an Equation

Sketch the graph of $x + 3y = -3$ using the x- and y-intercepts.

SOLUTION

x-intercept To find the x-intercept, substitute 0 for y.

$x + 3y = -3$ Write the equation.

$x + 3(0) = -3$ Substitute 0 for y.

$x = -3$ Solve for x.

y-intercept To find the y-intercept, substitute 0 for x.

$x + 3y = -3$ Write the equation.

$0 + 3y = -3$ Substitute 0 for x.

$y = -1$ Solve for y.

Plot the intercepts and draw the line that passes through them.

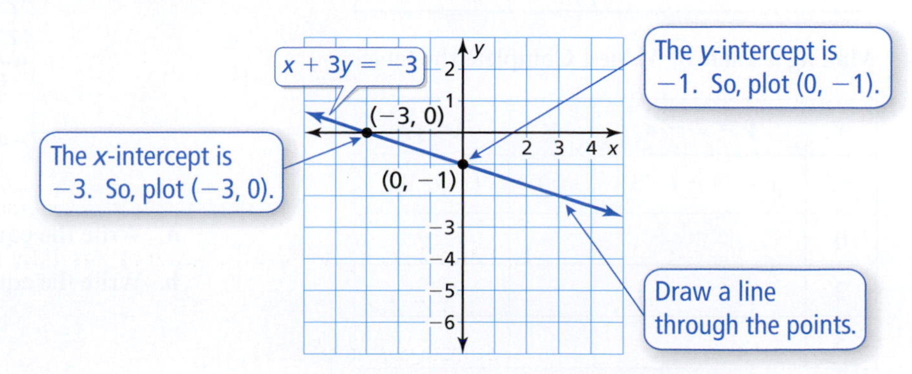

15.3 Exercises

Free step-by-step solutions to odd-numbered exercises at *MathematicalPractices.com.*

1. **Writing a Solution Key** Write a solution key for the activity on page 588. Describe a rubric for grading a student's work.

2. **Grading the Activity** In the activity on page 588, a student gave the answers below. Each question is worth 10 points. Assign a grade to each answer. For those that are incorrect, why do you think the student erred?

Sample Student Work

1.

	Solution points	
x	4	-4
$y = \frac{1}{2}x + 1$	3	-1

2. Write the two ordered pairs given by the table. These are called *solution points* of the equation. (-1, -4) (3,4)

3. (3,4) (-1, -4)

3. **Making a Table of Values** Complete the table.

x	$y = 4x - 5$	y	(x, y)
-5	$y = 4(-5) - 5$	-25	$(-5, -25)$
-3			
1			
4			

4. **Making a Table of Values** Complete the table.

x	$y = 7 - x$	y	(x, y)
-3	$y = 7 - (-3)$	10	$(-3, 10)$
0			
2			
5			

5. **Sketching Graphs of Linear Equations** Sketch the graph of each equation.

a. $y = 5x + 1$ **b.** $y = 6 - 3x$

c. $y = \frac{1}{3}x + 2$ **d.** $y = -2(x - 4)$

6. **Sketching Graphs of Linear Equations** Sketch the graph of each equation.

a. $y = -x + 8$ **b.** $y = 9 - 2x$

c. $y = -\frac{3}{4}x + 1$ **d.** $y = 5\left(x + \frac{1}{5}\right)$

7. **Horizontal and Vertical Lines** Decide whether each line is *horizontal*, *vertical*, or *neither*. Explain your reasoning.

a. $x = 7$ **b.** $y = 3x$

c. $y = 9$ **d.** $x = -2$

8. **Horizontal and Vertical Lines** Decide whether each line is *horizontal*, *vertical*, or *neither*. Explain your reasoning.

a. $x = 6y$ **b.** $y = -5$

c. $x = 0$ **d.** $y = 4$

9. **Sketching Horizontal and Vertical Lines** Sketch the graph of each equation.

a. $x = 8$ **b.** $y = 0$

c. $x = -10$ **d.** $y = 7$

10. **Sketching Horizontal and Vertical Lines** Sketch the graph of each equation.

a. $x = -3$ **b.** $y = 12$

c. $x = 1$ **d.** $y = -9$

11. **Writing Equations of Horizontal and Vertical Lines** Refer to the graph below.

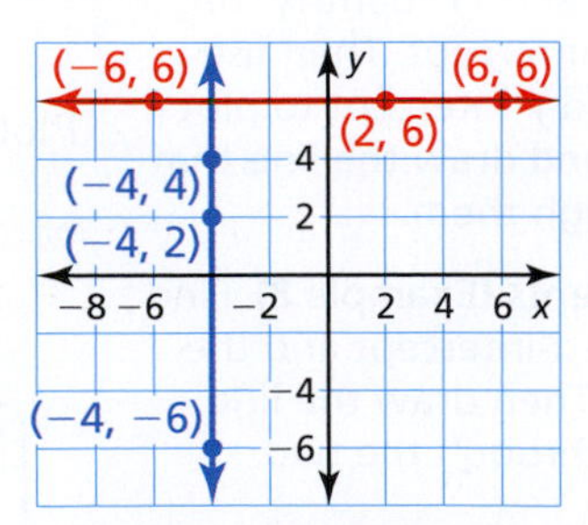

a. Write the equation of the horizontal line.

b. Write the equation of the vertical line.

12. Writing Equations of Horizontal and Vertical Lines Refer to the graph below.

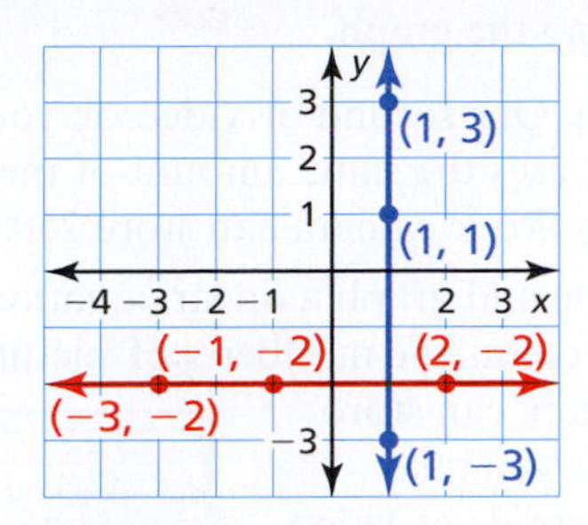

a. Write the equation of the horizontal line.

b. Write the equation of the vertical line.

13. Finding Slopes of Lines Find the slope of the graph of each equation, if possible.

a. $y = 7x + 2$　　b. $y = -13x$

c. $y = 3\left(x - \frac{1}{2}\right)$　　d. $y = 8$

14. Finding Slopes of Lines Find the slope of the graph of each equation, if possible.

a. $y = x$　　b. $y = -\frac{2}{5}x + 9$

c. $y = -(x + 4)$　　d. $x = 10$

15. Finding the *x*- and *y*-Intercepts of Lines Find the intercept(s) of the graph of each equation.

a. $y = 8x + 4$　　b. $y = -3x$

c. $2y - x = 6$　　d. $x = 7$

16. Finding the *x*- and *y*-Intercepts of Lines Find the intercept(s) of the graph of each equation.

a. $y = x - 9$　　b. $y = -2(x - 3)$

c. $-y + 6x = 4$　　d. $y = 1$

17. Using Intercepts to Sketch Graphs Graph each equation using the *x*- and *y*-intercepts.

a. $y = 5x - 10$　　b. $y = -4\left(2x + \frac{1}{2}\right)$

c. $y = -3x - 9$　　d. $4y + 6x = 24$

18. Using Intercepts to Sketch Graphs Graph each equation using the *x*- and *y*-intercepts.

a. $y = -x + 6$　　b. $y = -(x - 7)$

c. $y = \frac{1}{3}x + 1$　　d. $4y - 3x = 12$

19. Using *m* and *b* to Sketch Graphs of Lines Use the slope and *y*-intercept to sketch the graph of each equation. Identify the *x*-intercept.

a. $y = 3x - 5$　　b. $y = -x$

c. $y = \frac{5}{2}x + 1$　　d. $y = -4x + 7$

20. Using *m* and *b* to Sketch Graphs of Lines Use the slope and *y*-intercept to sketch the graph of each equation. Identify the *x*-intercept.

a. $y = 2x + 1$　　b. $y = -\frac{1}{4}x$

c. $y = \frac{4}{3}x - 6$　　d. $y = -5x + 8$

21. Rewriting Equations in Slope-Intercept Form Write each equation in slope-intercept form. Then sketch the graph of the equation.

a. $y = -2(3x + 4)$　　b. $5y - x = 10$

c. $y = -0.6(5x + 15)$　　d. $16x + 4y = -28$

22. Rewriting Equations in Slope-Intercept Form Write each equation in slope-intercept form. Then sketch the graph of the equation.

a. $y = -(x - 1)$　　b. $y + 2x = 6$

c. $-1.5(8y - 2x) = 0$　　d. $-3y + 9x = 6$

23. Writing Equations of Lines Write the equation of each line in slope-intercept form.

a.

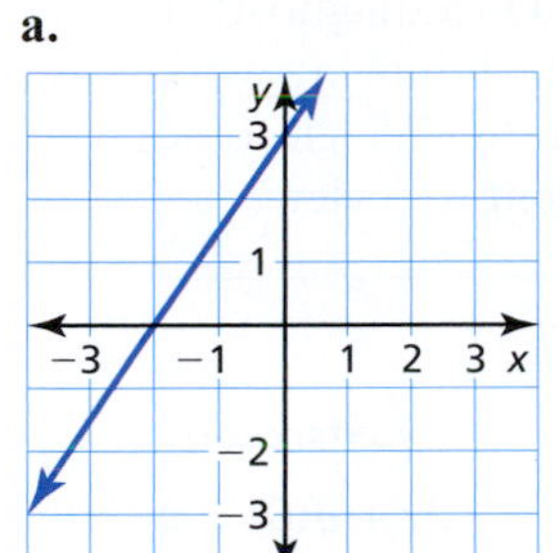

b.

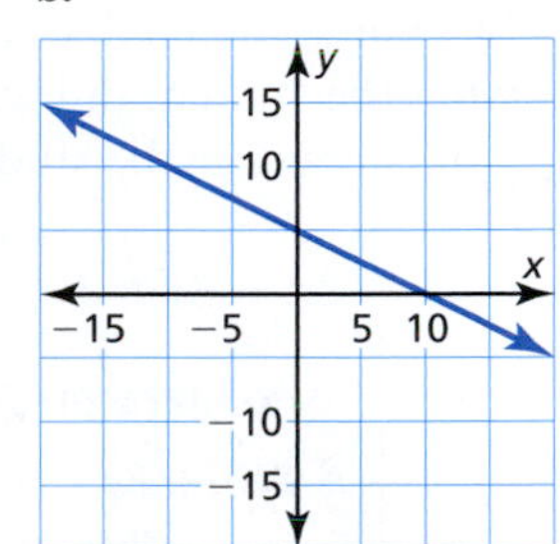

24. Writing Equations of Lines Write the equation of each line in slope-intercept form.

a.

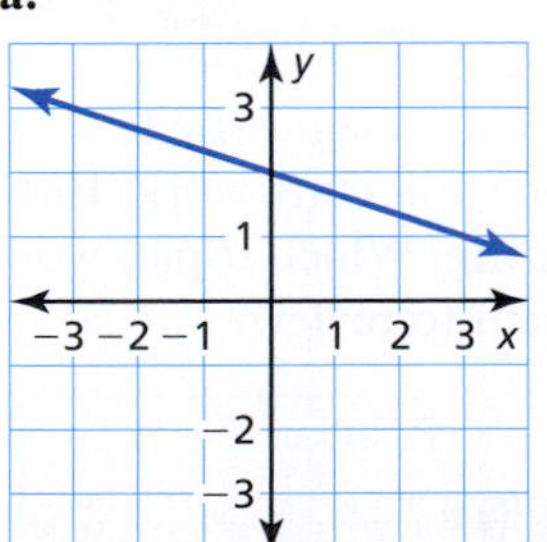

b.

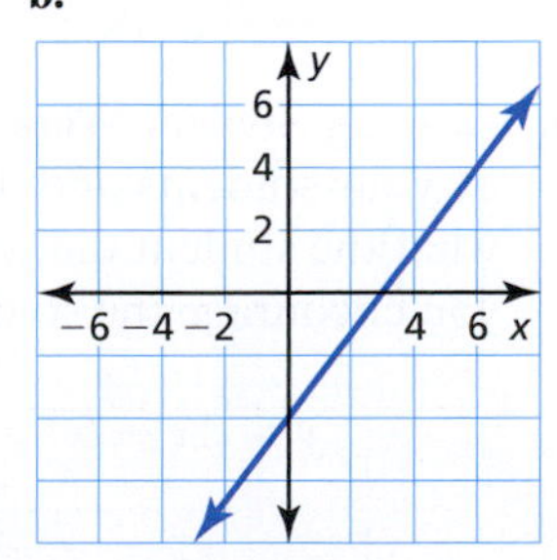

25. Writing Equations of Lines that Form a Polygon Write the equations of the lines that form the three sides of the triangle.

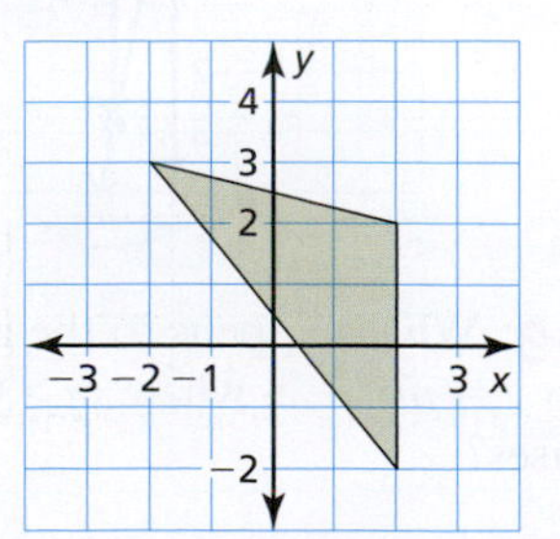

26. Writing Equations of Lines that Form a Polygon Write the equations of the lines that form the four sides of the right trapezoid with vertices at $(-12, 6)$, $(-12, -4)$, $(2, -4)$, and $(12, 6)$.

27. Interpreting Slope and y-Intercept in Real Life The speed y (in feet per second) of a car x seconds after the driver begins to apply the brakes can be represented by the equation $y = 90 - 15x$.

a. Identify and interpret the slope and y-intercept.

b. Graph the equation.

c. After how many seconds does the car come to a stop?

28. Interpreting Slope and y-Intercept in Real Life The value y (in dollars) of your computer after x years can be approximated using the equation $y = 1300 - 250x$.

a. Identify and interpret the slope and y-intercept.

b. Graph the equation.

c. What is the value of your computer after 3 years?

29. Grading Student Work On a diagnostic test, one of your students does the following work. Explain what the student did wrong. Which topics would you encourage the student to review?

$2x + 3y = 6$

x-intercept:	y-intercept:
$2(0) + 3y = 6$	$2x + 3(0) = 6$
$3y = 6$	$2x = 6$
$y = 2$	$x = 3$

The x-intercept is (0, 2) and the y-intercept is (3, 0).

30. Grading Student Work On a diagnostic test, one of your students does the following work. Explain what the student did wrong. Which topics would you encourage the student to review?

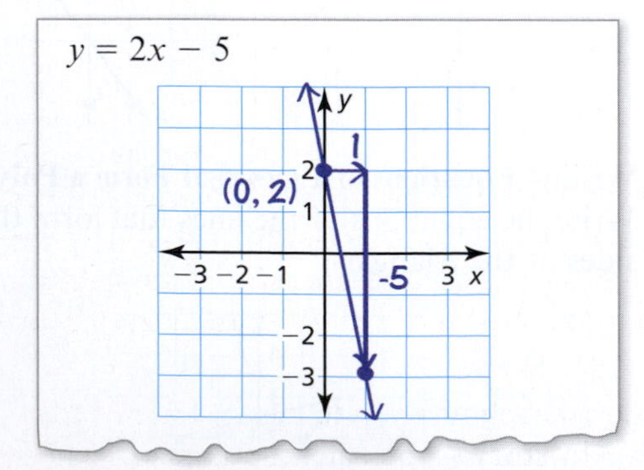

31. Writing What happens to the graph of the equation $y = mx + 3$, where $m > 0$ as the value of m increases?

32. Writing Explain how to show that the point $(3, 5)$ lies on the graph of $y = -2x + 11$ without sketching the graph.

33. Camera One second of video on your digital camera uses the same amount of memory as two pictures. Your camera can store 250 pictures.

a. Write and graph a linear equation that represents the number y of pictures your camera can store after you take x seconds of video.

b. How many pictures can your camera store after you take the video shown?

34. Walkathon One of your friends gives you \$15 for a charity walkathon. Another friend gives you a set amount for each mile you walk. After 4 miles, you have raised a total of \$25.

a. Write and graph a linear equation that represents the amount of money y you have raised after x miles.

b. Use the graph to estimate how much money you raise after 6 miles.

c. Use the equation to find exactly how much you raise after 6 miles.

35. Break-Even Point A manufacturer produces items at a cost of \$0.50 each after an initial setup cost of \$550. The manufacturer sells the items for \$1 each.

a. Write an equation that represents the total cost y_1 (in dollars) of manufacturing x items.

b. Write an equation that represents the total revenue y_2 (in dollars) earned for selling x items.

c. Sketch the graphs of the equations in parts (a) and (b) on the same coordinate plane.

d. How many items must the company sell to break even?

36. Motor Boats A motor boat is traveling at a speed of 30 miles per hour. A second motor boat is 1 mile in front of the first motor boat and is traveling in the same direction at a speed of 20 miles per hour.

a. Write an equation that represents the position s_1 (in miles) of the first boat after t minutes.

b. Write an equation that represents the position s_2 (in miles) of the second boat after t minutes.

c. Sketch the graphs of the equations in parts (a) and (b) on the same coordinate plane.

d. After how many minutes will the first boat catch up to the second boat?

37. Writing an Equation Using a Translation Write the equation of a line that is a translation 3 units to the left of the graph of $y = x + 11$.

38. Writing an Equation Using a Translation Write the equation of the line that is a translation 7 units to the right of the graph of $y = 4x + 9$.

39. Sketching the Graphs of Linear Equations The graph of $y = mx$ is shown below. Sketch the graph of each linear equation.

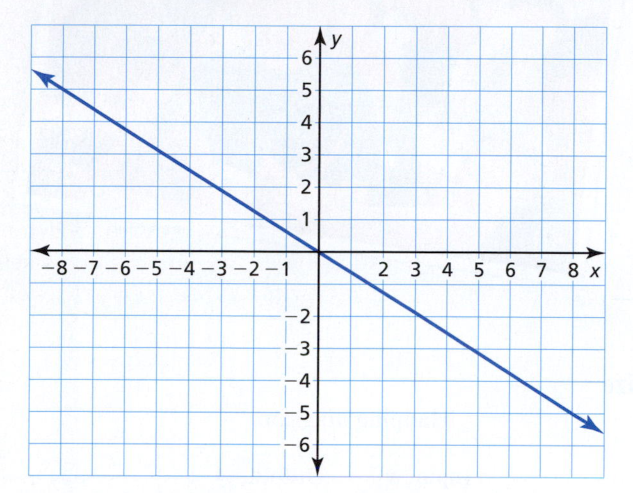

a. $y = mx + 4$

b. $y = mx - 4$

40. Sketching the Graphs of Linear Equations The graph of $y = mx - 2$ is shown below. Sketch the graph of each linear equation.

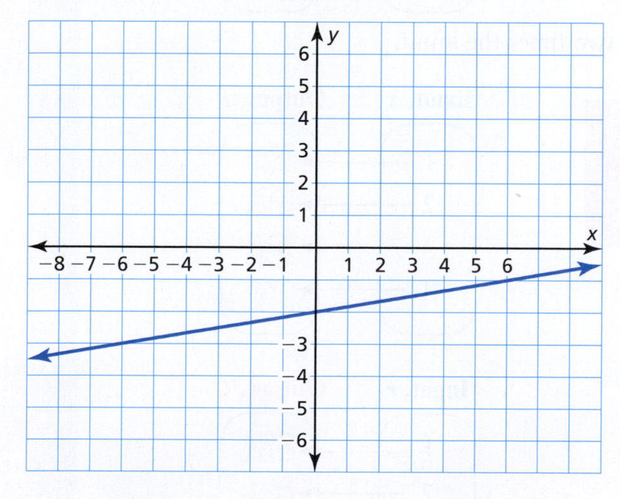

a. $y = mx$

b. $y = mx + 2$

41. Writing an Equation of a Line The *point-slope form* of the equation of a line is $y - y_1 = m(x - x_1)$, where m is the slope and (x_1, y_1) is a point on the line. Write the equation of a line in point-slope form parallel to the graph of $y = \frac{3}{4}x + 1$ passing through the point (12, 7).

42. Writing an Equation of a Line The *point-slope form* of the equation of a line is $y - y_1 = m(x - x_1)$, where m is the slope and (x_1, y_1) is a point on the line. Write the equation of the line in point-slope form perpendicular to the graph of $y = \frac{1}{2}x - 9$ passing through the point (8, 3).

43. Writing an Equation of a Perpendicular Bisector Write an equation of the perpendicular bisector of $\overline{EF}$, where $E = (3, 2)$ and $F = (3, 6)$.

44. Writing an Equation of a Perpendicular Bisector Write an equation of the perpendicular bisector of $\overline{JK}$, where $J = (-4, 2)$ and $K = (2, 2)$.

45. Activity: In Your Classroom The graph below shows three lines.

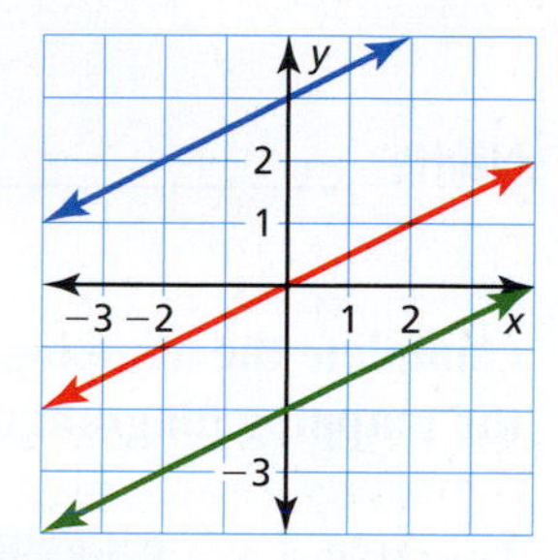

a. Find the slope of each line.

b. Find the y-intercept of each line.

c. Write an equation of each line.

d. Determine what all three lines have in common.

e. Repeat parts (a)–(d) for the graph below.

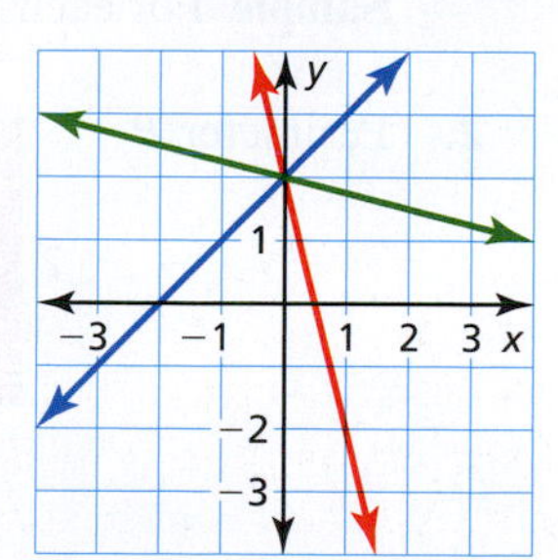

f. Describe the concepts your students will practice and discover by doing this kind of activity.

46. Activity: In Your Classroom You are teaching your students how to use a linear equation in two variables to model and solve a real-life problem. You have your students use the graph of a linear equation to write a story about a train approaching a train station.

a. What should you tell your students to include in the story to interpret the slope of the line? the y-intercept? the x-intercept?

b. Why is it helpful to have your students label the axes of the graph that represents the story?

Activity: Understanding Mapping Diagrams

Materials:

- Pencil

Learning Objective:

Understand the concept of a mapping diagram. Be able to use formulas for perimeter, area, and circumference to complete mapping diagrams.

Name ______________________________

Complete the mapping diagram. Summarize the mapping diagram in words.

Mapping diagram

1. Area A

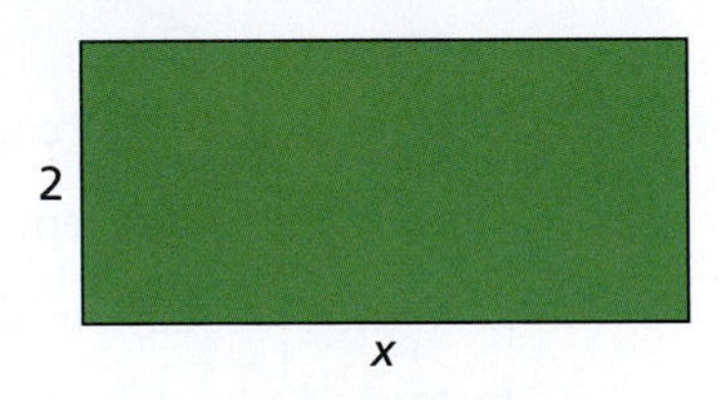

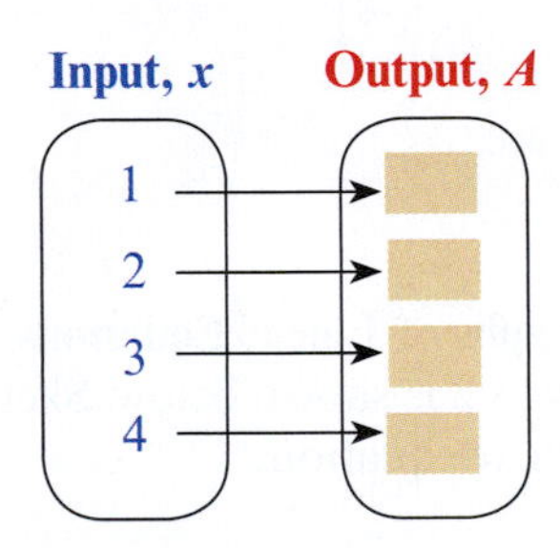

Sample For each input x, the output A is two times the input.

2. Perimeter P

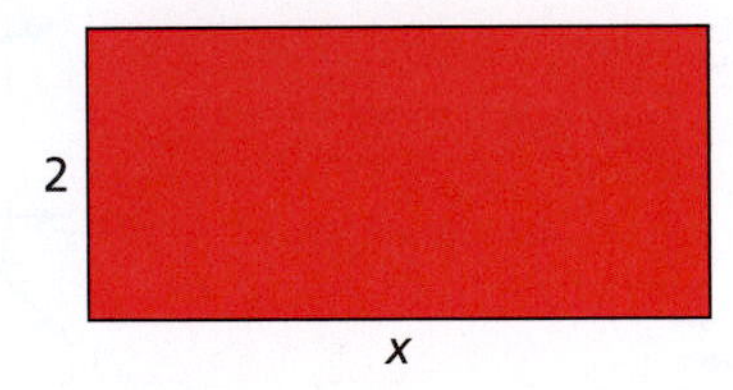

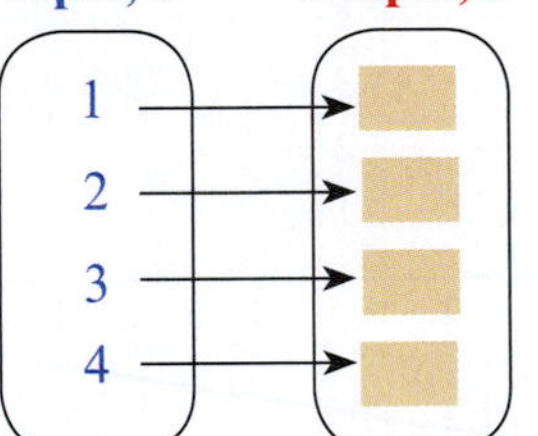

3. Circumference C

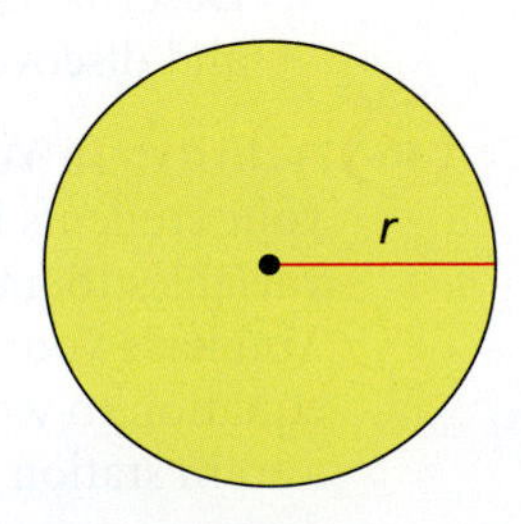

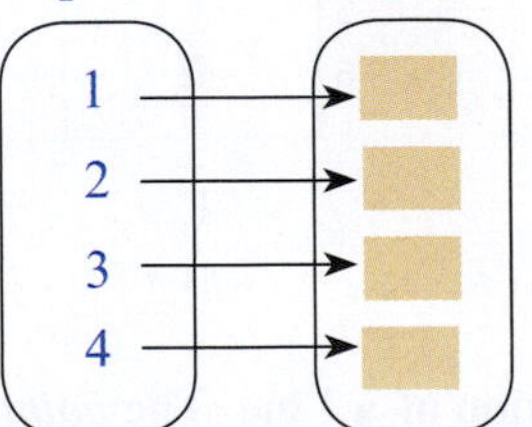

15.4 Functions

Standards

Grades 3–5
Operations and Algebraic Thinking
Students should analyze patterns and relationships.

Grades 6–8
Functions
Students should define, evaluate, and compare functions.

- Represent functions using ordered pairs or mapping diagrams.
- Represent functions using words or equations.
- Represent functions using tables or graphs.

Functions as Ordered Pairs or Mapping Diagrams

Definition of a Function

A **relation** pairs **inputs** with **outputs**. When a relation pairs each input with *exactly one* output, it is a **function**.

Two ways that a function can be represented are by ordered pairs and by a **mapping diagram**. The x-values are the inputs and the y-values are the outputs.

Ordered pairs

Inputs → (0, 1), (1, 2), (2, 4) ← Outputs

Mapping diagram

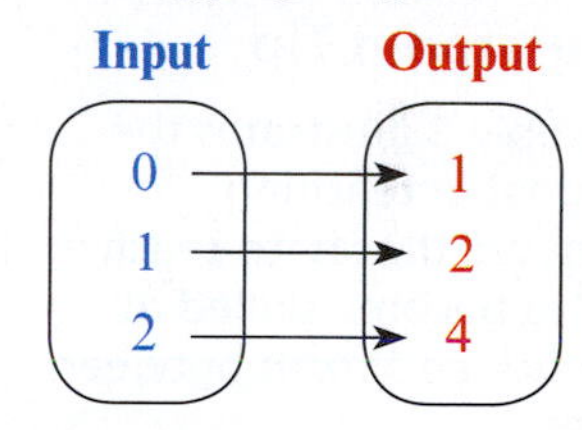

EXAMPLE 1 **Representing Functions as Sets of Ordered Pairs**

a.

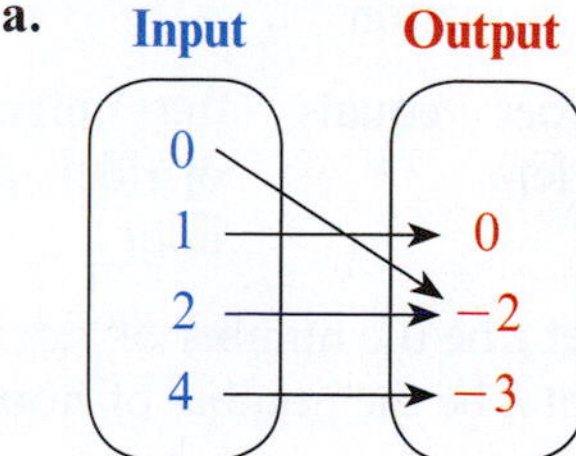

The ordered pairs are (0, −2), (1, 0), (2, −2), and (4, −3). Each input has exactly one output. So, the relation is a function.

b.

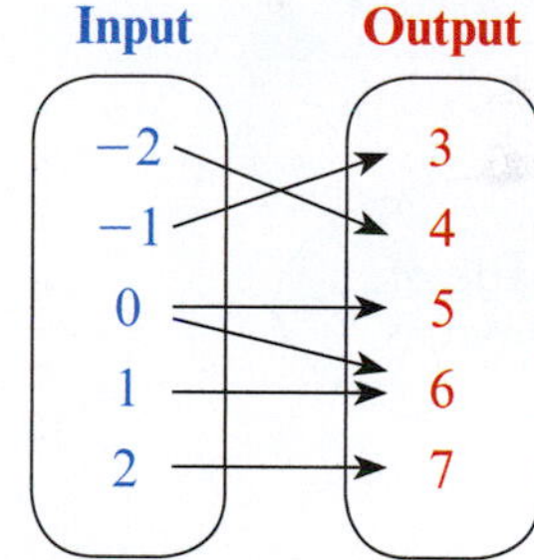

The ordered pairs are (−2, 4), (−1, 3), (0, 5), (0, 6), (1, 6) and (2, 7). The input 0 has two outputs, 5 and 6. So, the relation is *not* a function.

Classroom Tip

The concept of a function is considered one of the most important in mathematics. The study of functions forms the core of both precalculus and calculus. In these courses, functions are classified into five types.

1. Polynomial functions
2. Rational functions
3. Radical functions
4. Exponential and logarithmic functions
5. Trigonometric functions

EXAMPLE 2 **Representing a Function as a Mapping Diagram**

Draw a mapping diagram of (2, 6), (0, 4), (5, 9), and (3, 5). Determine whether the relation is a function.

SOLUTION

Order the inputs and the outputs from least to greatest.

Inputs 0, 2, 3, 5 **Outputs** 4, 5, 6, 9

Create a mapping diagram by drawing arrows from the inputs to their outputs. Each input has exactly one output. So, the relation is a function.

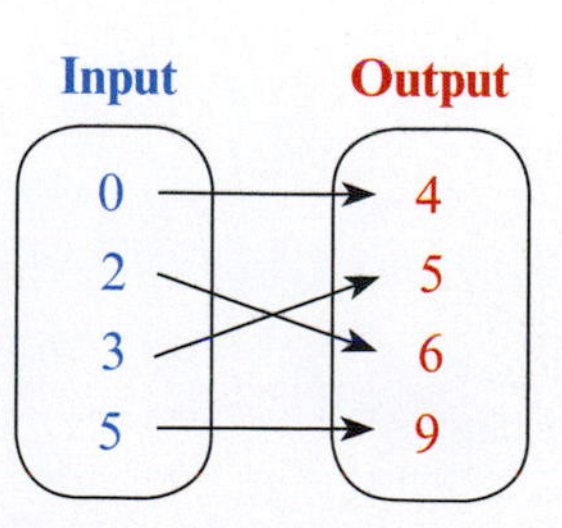

Functions as Words or Equations

Representing a Function as Words or as an Equation

A **function rule** is an equation that describes the relationship between inputs and outputs. The variable that represents input values is the **independent variable**. The variable that represents output values is the **dependent variable**. A dependent variable *depends* on the value of the independent variable.

Example **Words** The output is five less than the input.

Equation $y = x - 5$

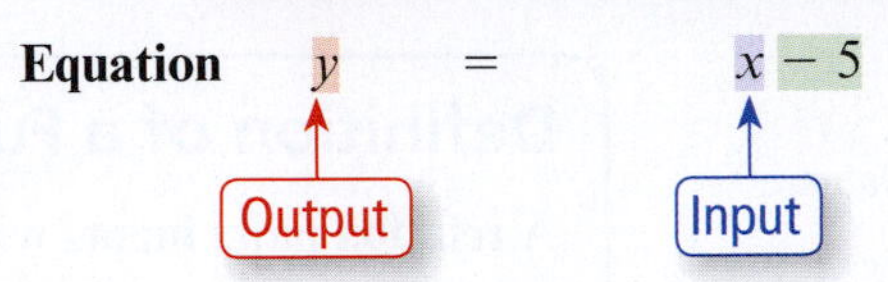

Writing a Function Rule

Classroom Tip

Example 3 illustrates the primary goal in teaching mathematics: that is, to teach students to become skilled at moving back and forth between languages.

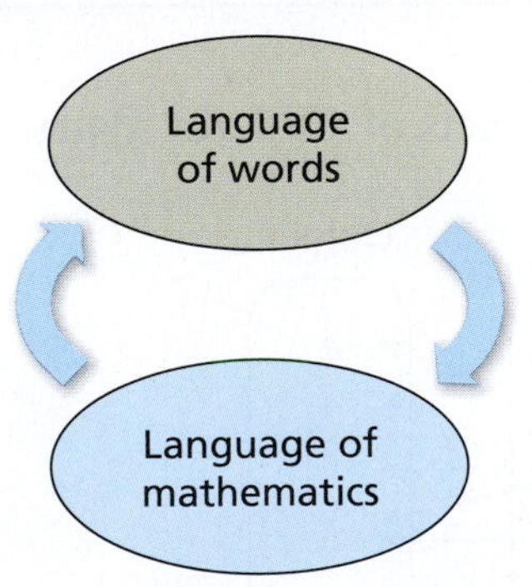

A ride at an amusement park can accommodate 1000 riders in an hour.

a. Write an equation that relates riders and time.

b. Identify the independent variable and the dependent variable in the equation from part (a).

c. The park is open 12 hours each day. How many people can go on the ride each day?

SOLUTION

a. Begin by writing the function in words. Then assign variables. Finally, translate the words into an equation.

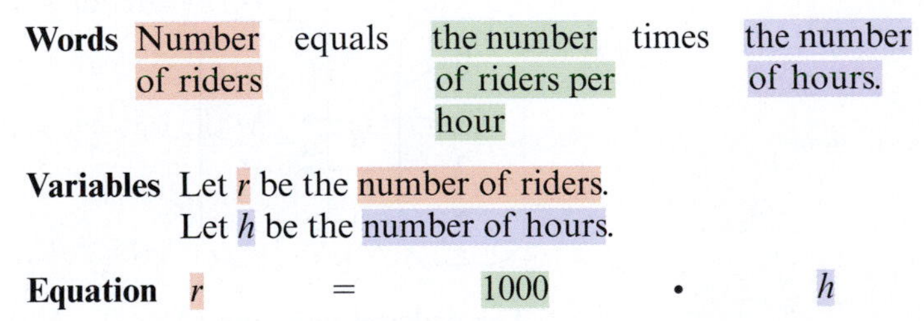

Words Number of riders equals the number of riders per hour times the number of hours.

Variables Let r be the number of riders.
Let h be the number of hours.

Equation $r = 1000 \cdot h$

An equation that represents this function is $r = 1000h$.

b. The number of hours is a quantity that can change freely. The number of riders *depends* on the number of hours. So, h is the independent variable and r is the dependent variable.

$r = 1000h$

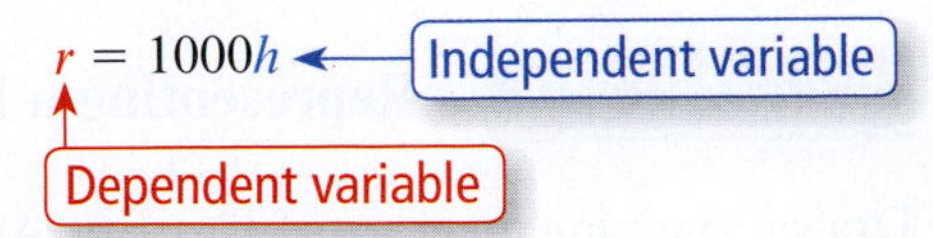

c. Substitute 12 for h in the equation from part (a).

$r = 1000h$ Write the equation.

$= 1000(12)$ Substitute 12 for h.

$= 12{,}000$ Multiply.

Each day, 12,000 people can go on the ride.

Functions as Tables or Graphs

Representing a Function as an Input-Output Table

A function can be represented by an **input-output table**.

Example The table below represents the function $y = x + 2$.

Input, x	Output, y	
		← $y = x + 2$
1	3	← $3 = 1 + 2$
2	4	← $4 = 2 + 2$
3	5	← $5 = 3 + 2$
4	6	← $6 = 4 + 2$

EXAMPLE 4 **Finding a Missing Input**

Each output in the table is 5 more than twice the input. Find the missing input.

Input, x	1	5	10	20	?
Output, y	7	15	25	45	53

Mathematical Practices

Reason Abstractly and Quantitatively

MP2 Mathematically proficient students bring two complementary abilities to bear on problems involving quantitative relationships: the ability to *decontextualize* and the ability to *contextualize*.

SOLUTION

Step 1

Write an equation for the function shown by the table.

Words Output is five more than twice the input.

Variables Let y be the output value and x be the input value.

Equation $y \quad = \quad 5 + \quad 2 \cdot \quad x$

An equation that represents this function is $y = 5 + 2x$.

Step 2

Substitute 53 for y. Then solve for x.

$y = 5 + 2x$ Write the equation.

$53 = 5 + 2x$ Substitute 53 for y.

$48 = 2x$ Subtraction Property of Equality

$24 = x$ Division Property of Equality

The missing input is $x = 24$.

Step 3

Check your solution.

$y = 5 + 2x$ Write the equation.

$53 \stackrel{?}{=} 5 + 2(24)$ Substitute 53 for y and 24 for x.

$53 \stackrel{?}{=} 5 + 48$ Multiply.

$53 = 53$ ✓ Add.

Representing a Function as a Graph

A function can be represented by a **graph**.

Example The graph below represents the function $y = x + 2$.

Input, x	Output, y	Ordered pair (x, y)
1	3	(1, 3)
2	4	(2, 4)
3	5	(3, 5)

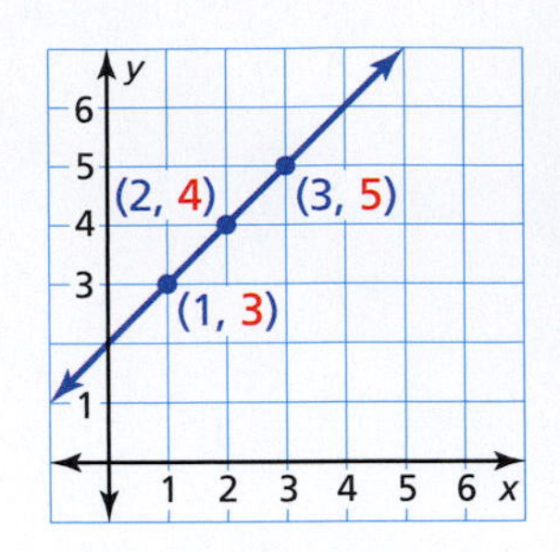

By drawing a line through the points, you graph *all* of the solutions of $y = x + 2$.

EXAMPLE 5 Sketching the Graph of a Function

The function $p = 20g$ gives the number of pounds p of carbon dioxide produced when burning g gallons of gasoline that does not contain ethanol. Graph this function.

SOLUTION

Step 1 Begin by making an input-output table.

Input, g	$p = 20g$	Output, p	Ordered pair, (g, p)
0	$p = 20(0)$	0	(0, 0)
1	$p = 20(1)$	20	(1, 20)
2	$p = 20(2)$	40	(2, 40)
3	$p = 20(3)$	60	(3, 60)

Step 2 Next, plot the ordered pairs from the table.

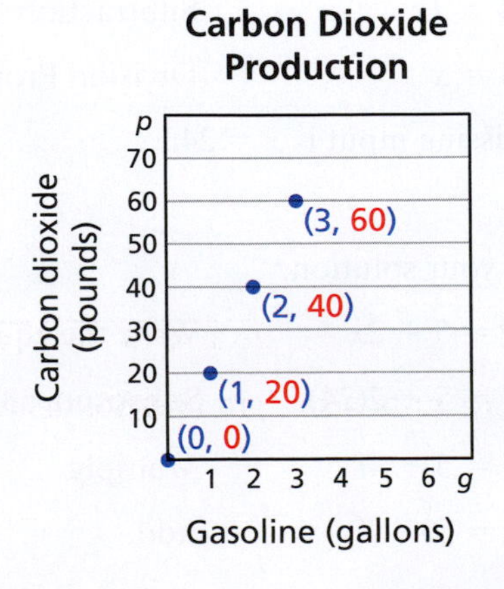

Step 3 The equation $p = 20g$ is of the form $y = mx + b$. So, it is linear. Sketch its graph by drawing a line through the points.

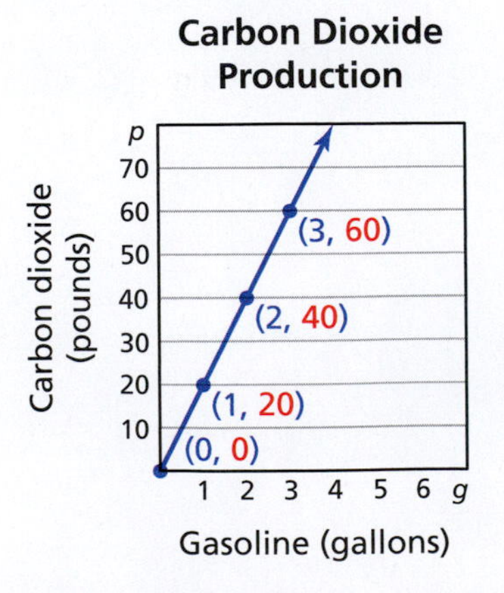

In this section, you have seen that there are several ways to represent a function.

Words Each output is 2 more than the input.

Equation $y = x + 2$

Input-output table

Input, x	Output, y
1	3
2	4
3	5
4	6

Mapping diagram

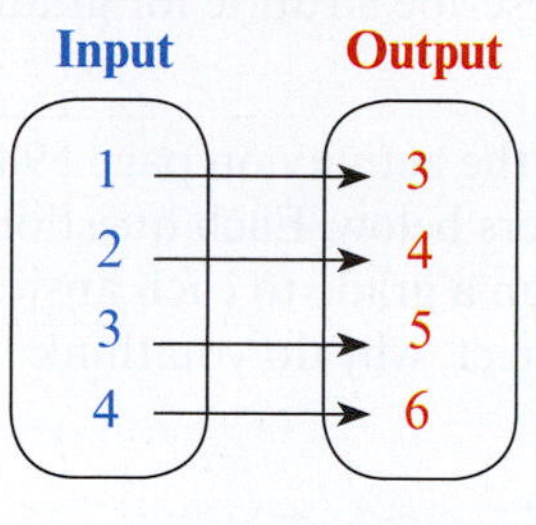

Graph

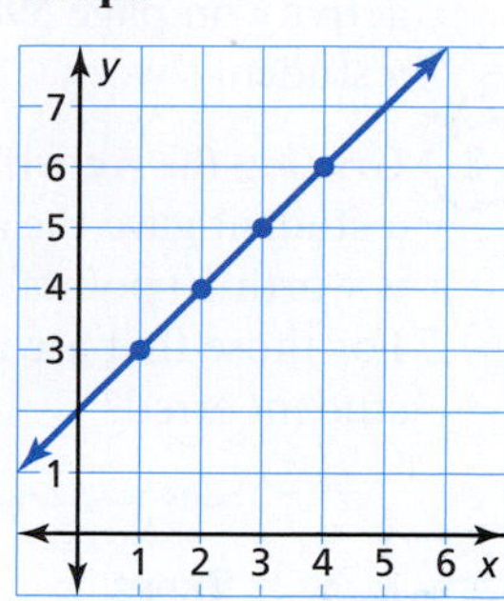

Classroom Tip

In Section 15.3, you learned that a linear equation in x and y can be written in the form $y = mx + b$. In the context of functions, this type of equation defines y as a linear function of x. The graph of a linear function is a straight line.

EXAMPLE 6 Identifying a Linear Function

Does the graph represent a linear function? Explain your reasoning.

a.

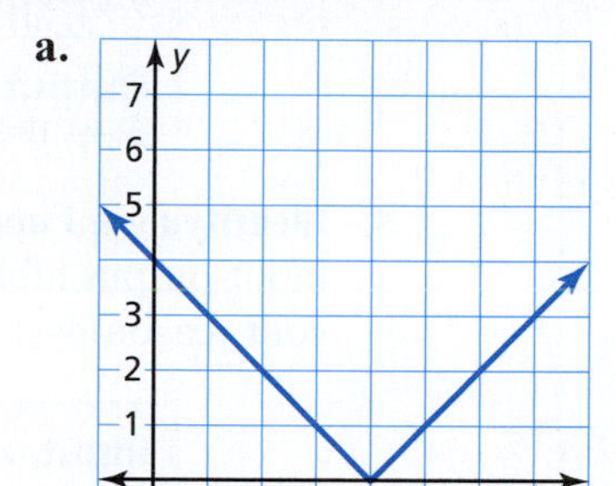

b.

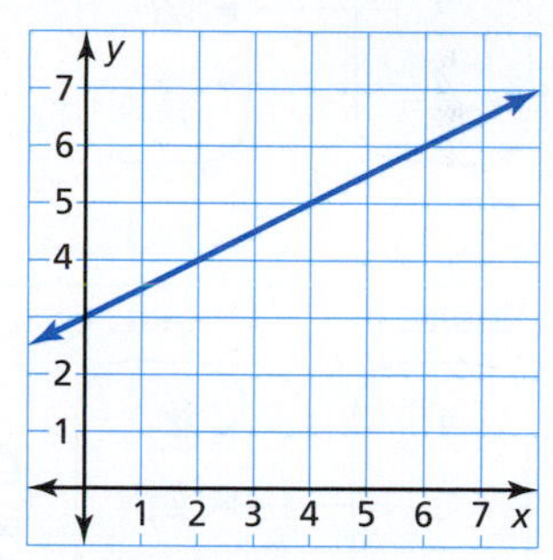

SOLUTION

a. The graph is not a straight line. So, the graph does not represent a linear function.

b. The graph is a straight line. So, the graph does represent a linear function.

EXAMPLE 7 Identifying a Linear Function

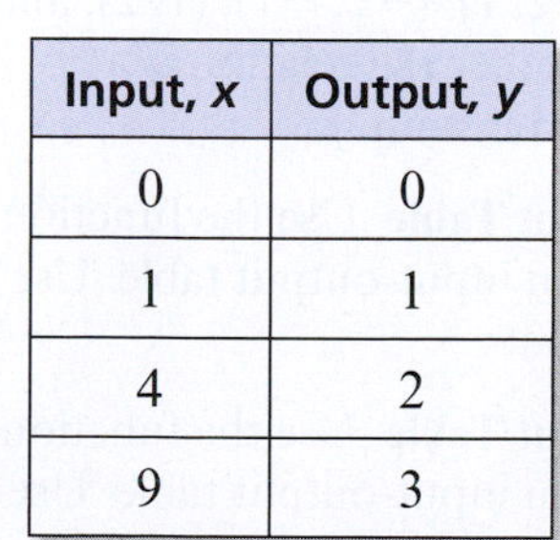

Input, x	Output, y
0	0
1	1
4	2
9	3

Does the input-output table at the left represent a linear function? Explain your reasoning.

SOLUTION

The ordered pairs in the table are (0, 0), (1, 1), (4, 2), and (9, 3). Plot these ordered pairs and draw a graph through the points.

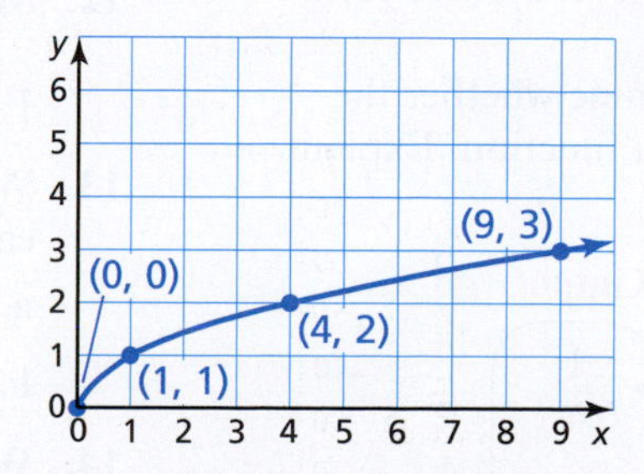

The graph is not a straight line. So, the function is not linear.

15.4 Exercises

Free step-by-step solutions to odd-numbered exercises at *MathematicalPractices.com.*

1. **Writing a Solution Key** Write a solution key for the activity on page 598. Describe a rubric for grading a student's work.

2. **Grading the Activity** In the activity on page 598, a student gave the answers below. Each question is worth 10 points. Assign a grade to each answer. For those that are incorrect, why do you think the student erred?

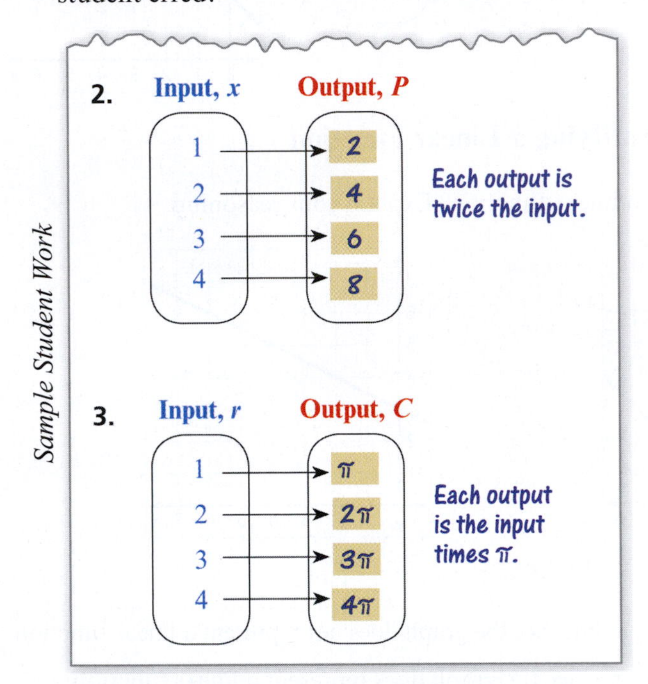

3. **Identifying Functions** Determine whether each relation is a function. Explain your reasoning.
   a. (1, 2), (2, 3), (3, 4), (4, 5), (5, 6), (6, 7)
   b. (−2, 3), (0, 7), (−2, 4), (1, 6), (3, 8), (8, 9)

4. **Identifying Functions** Determine whether each relation is a function. Explain your reasoning.
   a. (−5, 0), (0, 0), (5, 0), (5, 10), (10, 10), (−1, 0)
   b. $(\sqrt{8}, 6)$, $\left(\frac{1}{2}, 2\right)$, (8, 9), (5, 6), (1, 2), (11, 11)

5. **Identifying a Function** Determine whether the mapping diagram represents a function. Explain your reasoning.

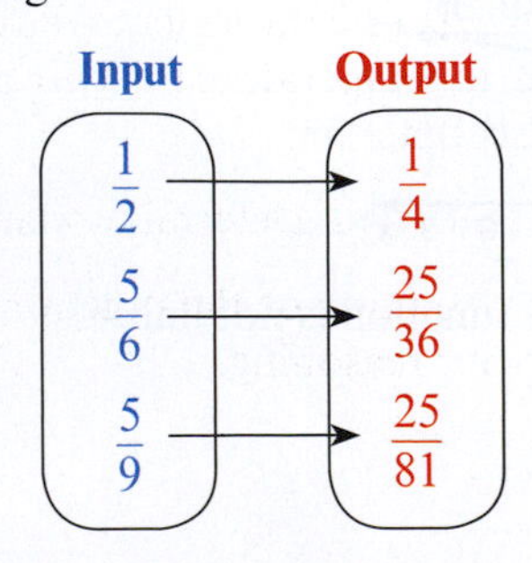

6. **Identifying a Function** Determine whether the mapping diagram represents a function. Explain your reasoning.

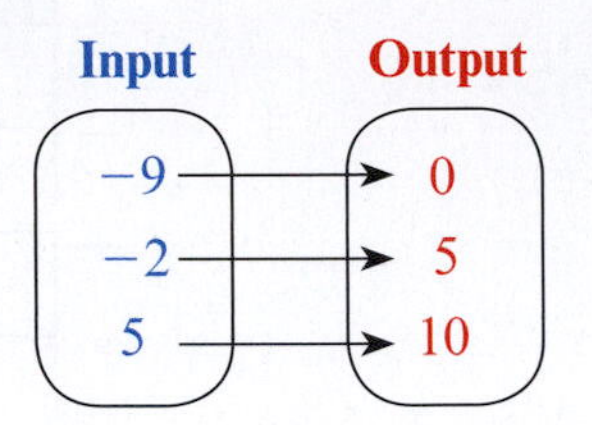

7. **Identifying a Function** Determine whether the input-output table represents a function. Explain your reasoning.

Input, *x*	0	1	2	3
Output, *y*	2	4	6	8

8. **Identifying a Function** Determine whether the input-output table represents a function. Explain your reasoning.

Input, *x*	1	3	3	7
Output, *y*	1	2	3	1

9. **Drawing a Mapping Diagram** Draw a mapping diagram of (4, 12), (0, 8), (10, 18), and (6, 10). Determine whether the relation is a function. Explain your reasoning.

10. **Drawing a Mapping Diagram** Draw a mapping diagram of (−3, 0), (−2, 1), (−2, −1), (1, 2), and (1, −2). Determine whether the relation is a function. Explain your reasoning.

11. **Making an Input-Output Table** Use the function $y = 10x + 2$ to make an input-output table. Use 0, 1, 2, 3, and 4 as the inputs.

12. **Making an Input-Output Table** Use the function $y = 4x^2 - 3$ to make an input-output table. Use 0, 1, 2, 3, and 4 as the inputs.

13. **Writing Function Rules** Write a function rule for each statement.
    a. The output is four more than the input.
    b. The output is three more than twice the input.

14. **Writing Function Rules** Write a function rule for each statement.
    a. The output is three times the input.
    b. The output is seven less than one-half the input.

15. **Using Tables to Write Function Rules** Write a function rule for each input-output table.

a.

Input, x	1	2	3	4
Output, y	2	4	6	8

b.

Input, x	1	2	3	4
Output, y	1	4	9	16

16. **Using Tables to Write Function Rules** Write a function rule for each input-output table.

a.

Input, x	1	2	3	4
Output, y	4	7	10	13

b.

Input, x	1	2	3	4
Output, y	−0.5	0	0.5	1

17. **Writing a Function Rule** A restaurant serves 50 customers per hour.

a. Write an equation that relates the number of customers C and time t.

b. Identify the independent and dependent variables in the equation in part (a).

c. The restaurant is open 15 hours each day. How many people does the restaurant serve each day?

18. **Writing a Function Rule** A campground charges \$15 per car plus \$2 for each person inside the car to camp overnight.

a. Write an equation that relates the cost of camping for one night C and the number of people in a car p.

b. Identify the independent and dependent variables in the equation in part (a).

c. Five people are in a car. What is the total cost to camp for the night?

19. **Writing a Function Rule** A cell phone company charges a \$25 monthly service charge and \$0.05 a minute for every minute used.

a. Write an equation that relates the cost of the monthly phone bill C and the number of minutes used m.

b. Identify the independent and dependent variables in the equation in part (a).

c. You use 300 minutes in a month. How much is your cell phone bill for the month?

20. **Writing a Function Rule** You decide to start a dog-sitting business. You charge a \$25 fee plus \$2 per hour.

a. Write an equation that relates the cost for watching a dog C and the number of hours watched h.

b. Identify the independent and dependent variables in the equation in part (a).

c. You watch a dog for six and a half hours. How much do you charge the owner?

21. **Finding a Missing Input** Each output in the table is one more than twice the input. Find the missing input.

Input, x	1	5	15	?	25
Output, y	3	11	31	41	51

22. **Finding a Missing Input** Each output in the table is two less than five times the input. Find the missing input.

Input, x	1	2	4	?	8
Output, y	3	8	18	28	38

23. **Graphing a Function** Graph the function using the input-output table.

Input, x	1	2	3	4
Output, y	3	4	5	6

24. **Graphing a Function** Graph the function using the input-output table.

Input, x	1	2	3	4
Output, y	6	4	2	0

25. **Graphing a Function** The function $d = 65t$ gives the distance d in miles a car travels in t hours. Graph this function.

26. **Graphing a Function** The function $F = \frac{9}{5}C + 32$ is the formula to convert temperature in degrees Celsius C to temperature in degrees Fahrenheit F. Graph this function.

27. **Finding a Solution** Determine whether the point (6, 10) is a solution of the function $y = \frac{2}{3}x + 6$. Explain your reasoning.

28. Finding a Solution Determine whether the point (20, 67) is a solution of the function $y = \frac{7}{4}x + 33$. Explain your reasoning.

29. Finding a Solution Is the point (6, 5) a solution of the function represented by the graph? Explain your reasoning.

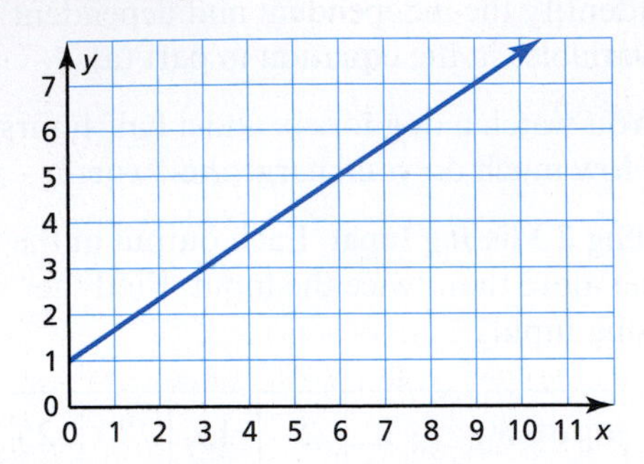

30. Finding a Solution Is the point (2, 3) a solution of the function represented by the graph? Explain your reasoning.

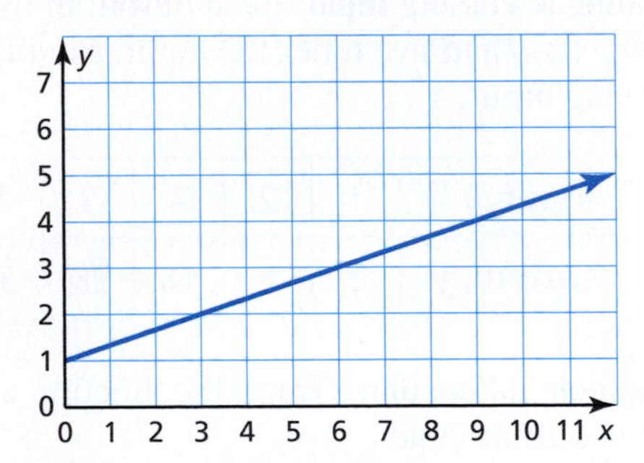

31. Identifying Linear Functions Determine whether each graph represents a linear function. Explain your reasoning.

a.

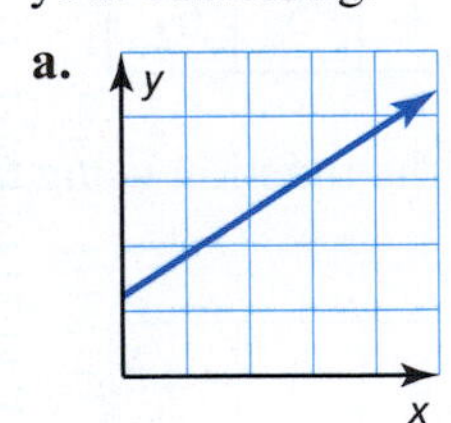

b.

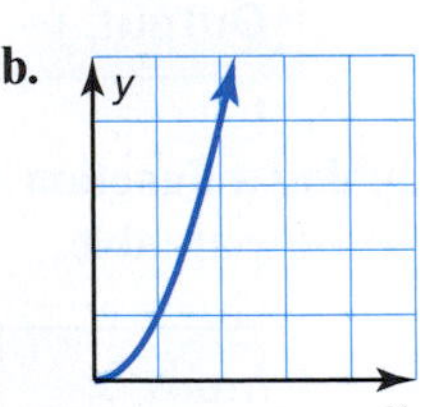

32. Identifying Linear Functions Determine whether each graph represents a linear function. Explain your reasoning.

a.

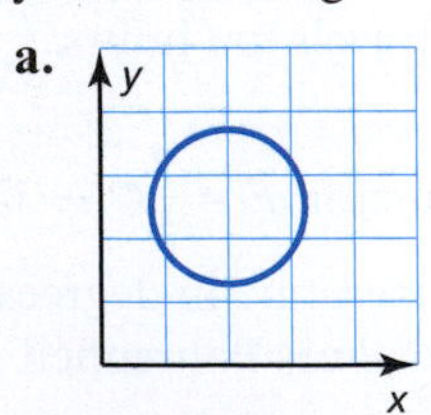

b.

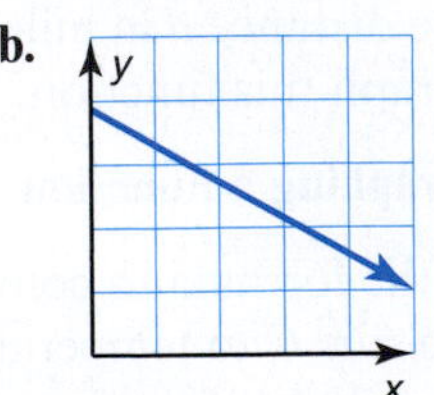

33. Identifying Linear Functions Determine whether each equation represents a linear function. Explain your reasoning.

a. $y = -x + 2$

b. $y = x^2 + 2$

34. Identifying Linear Functions Determine whether each equation represents a linear function. Explain your reasoning.

a. $x - y = 6$ **b.** $3 = x^3 + y$

35. Identifying Linear Functions Determine whether the input-output table represents a linear function. Explain your reasoning.

Input, *x*	1	2	3	4
Output, *y*	2	5	10	17

36. Identifying Linear Functions Determine whether the input-output table represents a linear function. Explain your reasoning.

Input, *x*	1	2	3	4
Output, *y*	8	6	4	2

37. Activity: In Your Classroom You create a life-size coordinate plane on your classroom floor with masking tape.

a. Four students representing a linear equation stand in the coordinate plane. What information can be determined by the line formed by the students?

b. Give students an equation in slope-intercept form. Describe how three students can use the slope-intercept form to "graph" the equation in the coordinate plane.

38. Activity: In Your Classroom Set your mathematics book on the floor and measure the distance from the floor to the top of the book. Place another copy of the book on top and measure the distance from the floor to the top of the books. Repeat several times.

a. Write an equation that relates the distance d from the floor to the top of your books and the number of books b.

b. Set the books on top of a table. Write a new equation that represents the distance d from the floor to the top of the books and the number of books b.

c. Describe the concepts your students will practice and discover by doing this activity.

39. Writing Describe an example of a real-life situation that represents a function.

40. Writing Use a newspaper, magazine, or some other type of print to search for an example of a function. Explain why your example represents a function and graph the function in a coordinate plane.

41. **Grading Student Work** On a diagnostic test, one of your students does the following work. Explain what the student did wrong. Which topics would you encourage the student to review?

Input, x	1	2	3	4	5
Output, y	2	2	2	2	2

The table does not represent a function because all of the x's have the same y value.

42. **Grading Student Work** On a diagnostic test, one of your students does the following work. Explain what the student did wrong. Which topics would you encourage the student to review?

Does the graph represent a linear function? Explain your reasoning.

The graph represents a linear function. The lines are straight.

43. **Business** A calculator costs \$25.00 to manufacture and sells for \$110.00.

a. Write a function that calculates the revenue R from selling n calculators.

b. The cost to set up the production line is \$750.00. Write a function for the cost C of setting up the production line and manufacturing the calculators.

Nikola Kikovic/Shutterstock.com

c. Write a function for the profit P. Then find the profit for selling 50 calculators. (*Hint:* Profit = Revenue − Cost)

d. The cost of the material to make calculator screens increases by \$4.00. Write a new cost function.

e. Using the new cost function, how many calculators must be sold for the business to break even? (*Hint:* Revenue = Cost)

44. **Business** An electronics company makes a computer tablet that sells for \$499.00. The cost, including parts and labor, to make one tablet is \$259.60.

a. Write a function that calculates the revenue R the company makes from selling t tablets.

b. The company spends \$500 to store the unsold tablets. Write a function for the cost C of parts, labor, and storage.

c. Write a function for the profit P. Then find the profit for selling 1000 tablets. (*Hint:* Profit = Revenue − Cost)

d. The cost of material to make the tablet increases by \$40 per tablet. Write a new cost function.

e. Using the new cost function, how many tablets will the company have to sell in order to make a \$1,000,000 profit?

45. **Analyzing Graphs** A math tutoring company offers two fee schedules. Schedule 1 charges a one-time service fee of \$50 and a monthly fee of \$10 for tutoring services. Schedule 2 charges a \$100 one-time service fee and a monthly fee of \$5.

a. Write a function for each schedule that relates the cost C for tutoring services for m months.

b. Graph each function in part (a) in the same coordinate plane.

c. When would a person choose schedule 1? When would a person choose schedule 2? Explain your reasoning.

Monkey Business Images/Shutterstock.com

46. **Analyzing Graphs** A cell phone company offers two plans. Plan A charges a monthly fee of \$15 and \$0.05 a minute. Plan B charges no monthly fee and \$0.15 a minute.

a. Write a function for each plan that relates the monthly cost C for m minutes of use.

b. Graph each function in part (a) in the same coordinate plane.

c. When would a customer choose plan A? When would a customer choose plan B? Explain your reasoning.

Chapter Summary

Chapter Vocabulary

coordinate plane *(p. 567)*
Cartesian plane *(p. 567)*
origin *(p. 567)*
quadrants *(p. 567)*
x-axis *(p. 567)*
y-axis *(p. 567)*
ordered pair *(p. 567)*
x-coordinate *(p. 567)*
y-coordinate *(p. 567)*
legs *(p. 570)*
hypotenuse *(p. 570)*
midpoint *(p. 573)*
slope *(p. 579)*
rise *(p. 579)*
run *(p. 579)*
linear equation in two variables *(p. 589)*
solutions *(p. 589)*
horizontal line *(p. 590)*
vertical line *(p. 590)*
x-intercept *(p. 591)*)
y-intercept *(p. 591)*
slope-intercept form *(p. 591)*
relation *(p. 599)*
inputs *(p. 599)*
outputs *(p. 599)*
function *(p. 599)*
mapping diagram *(p. 599)*
function rule *(p. 600)*
independent variable *(p. 600)*
dependent variable *(p. 600)*
input-output table *(p. 601)*
graph *(p. 602)*

Chapter Learning Objectives — Review Exercises

15.1 The Coordinate Plane and Distance *(page 567)* — 1–26

- Identify points in a coordinate plane using ordered pairs.
- Represent transformations in a coordinate plane.
- Use the Pythagorean Theorem to find the distance between two points.
- Find the midpoint of a line segment.

15.2 The Slope of a Line *(page 579)* — 27–48

- Find the slope of the line.
- Interpret the slope of a line as a rate of change.
- Find the slopes of parallel and perpendicular lines.

15.3 Linear Equations in Two Variables *(page 589)* — 49–88

- Sketch the graphs of linear equations.
- Recognize the equations of horizontal and vertical lines.
- Find and interpret the slopes and y-intercepts of the graphs of linear equations.

15.4 Functions *(page 599)* — 89–104

- Represent functions using ordered pairs or mapping diagrams.
- Represent functions using words or equations.
- Represent functions using tables or graphs.

Important Concepts and Formulas

Pythagorean Theorem $a^2 + b^2 = c^2$

Distance Formula $d = \sqrt{(x_2 - x_1) + (y_2 - y_1)}$

Midpoint Formula $\left(\dfrac{x_1 + x_2}{2}, \dfrac{y_1 + y_2}{2}\right)$

Slope $\dfrac{\text{rise}}{\text{run}} = \dfrac{y_2 - y_1}{x_2 - x_1}$

Slope-Intercept Form $y = mx + b$

Definition of a Function When a relation pairs each input with exactly one output, the relation is a function.

Review Exercises

Free step-by-step solutions to odd-numbered exercises at MathematicalPractices.com.

15.1 The Coordinate Plane and Distance

Plotting Points Plot the point in a coordinate plane.

1. $A(3, 4)$
2. $B(-6, 5)$
3. $C(8, -9)$
4. $D(-5, -3)$
5. $E(\sqrt{5}, 8)$
6. $F(-7.4, 3.6)$

Identifying Ordered Pairs Write the ordered pair that corresponds to the point.

7. Point A
8. Point B
9. Point C
10. Point D
11. Point E
12. Point F

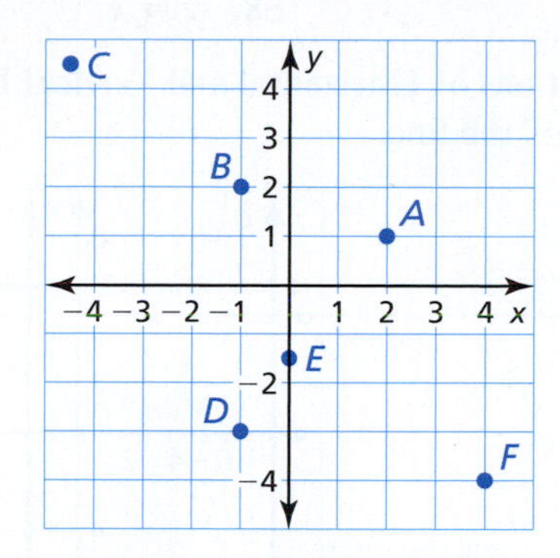

Translating Triangles Translate the triangle.

13. Translate the triangle whose vertices are $(-1, 2)$, $(-2, 5)$, and $(-5, 1)$ seven units right.

14. Translate the triangle whose vertices are $(1, 2)$, $(3, 1)$, and $(4, 6)$ three units down.

Rotating Triangles Rotate the triangle with the given vertices 90° counterclockwise about the origin.

15. $(-1, -5)$, $(-2, -1)$, and $(-7, -5)$

16. $(1, -3)$, $(3, -1)$, and $(3, -5)$

Glide Reflecting Triangles Translate the triangle and then reflect it in the given line.

17. Translate the triangle whose vertices are $(1, 2)$, $(5, 5)$, and $(7, 1)$ three units down, then reflect it in the y-axis.

18. Translate the triangle whose vertices are $(1, 1)$, $(2, 5)$, and $(5, 1)$ four units left, then reflect it in the x-axis.

The Pythagorean Theorem The lengths of the legs of a right triangle are given. Find the length of the hypotenuse.

19. 4 and 7
20. $\sqrt{3}$ and $\sqrt{3}$

The Pythagorean Theorem The lengths of one leg and the hypotenuse of a right triangle are given. Find the length of the other leg.

21. Leg: 21, hypotenuse: 29

22. Leg: 36, hypotenuse: 60

The Distance and Midpoint Formulas Find the distance between the two points. Then find the coordinates of the midpoint of the line segment joining the two points.

23. $(-2, 3)$, $(3, 1)$

24. $(5, 8)$, $(-1, -4)$

25. $(0, 4)$, $(3, 0)$

26. $(-5, -2)$, $(-1, -6)$

15.2 The Slope of a Line

Identifying Slope State whether the slope of the line is *positive*, *negative*, *zero*, or *undefined*.

27. Line a
28. Line b
29. Line c
30. Line d

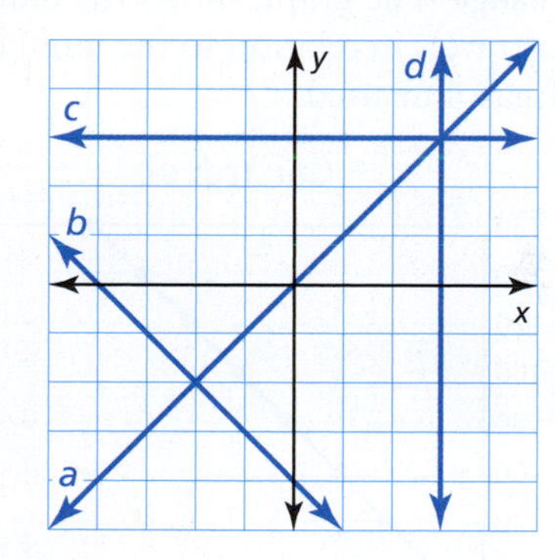

Finding Slope Find the slope of each line.

31. **a.** $\overleftrightarrow{AB}$
b. $\overleftrightarrow{CD}$

32. **a.** $\overleftrightarrow{EF}$
b. $\overleftrightarrow{GH}$

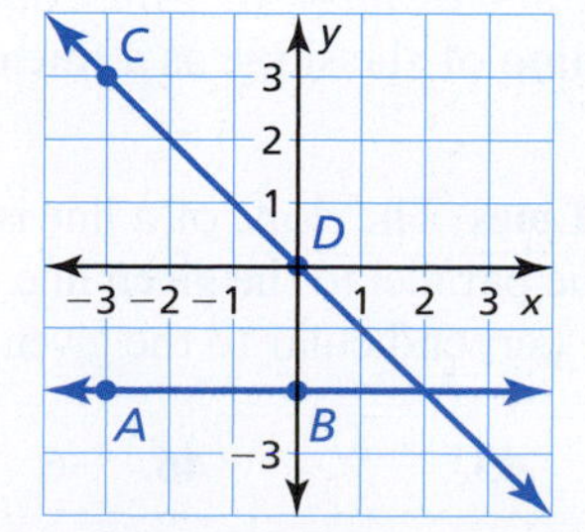

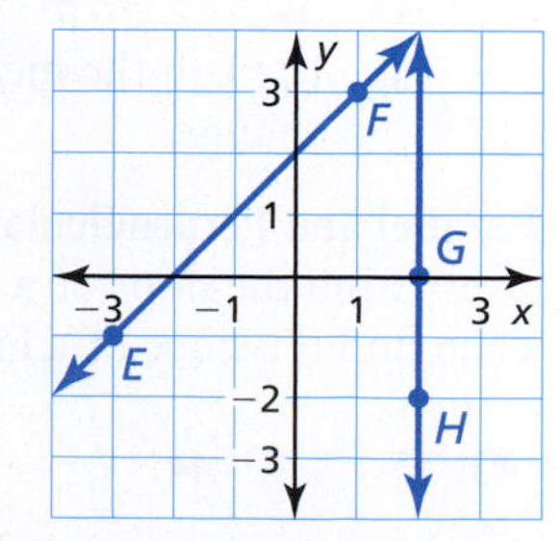

Finding Slope The two points lie on a line. Find the slope of the line.

33. $(-2, -1)$, $(2, 2)$
34. $(1, 5)$, $(4, 5)$
35. $(-4, 6)$, $(-4, 1)$
36. $(-6, -1)$, $(-5, -5)$

Graphing a Line Graph the line with the given slope that passes through the given point.

37. Slope: 3, point: (4, 1)

38. Slope: $-\frac{4}{3}$, point: (−3, 2)

39. Slope: undefined, point: (−1, 7)

40. Slope: −1, point: (0, 4)

41. Rate of Change The graph shows the distance you traveled in relation to time on a business trip.

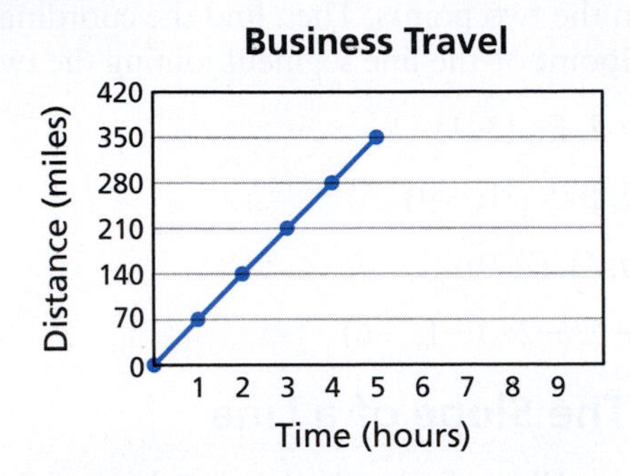

a. Find the slope of the line.

b. Explain the meaning of the slope as a rate of change.

42. Rate of Change The graph shows the number of miles a car travels in relation to the number of gallons of gasoline used.

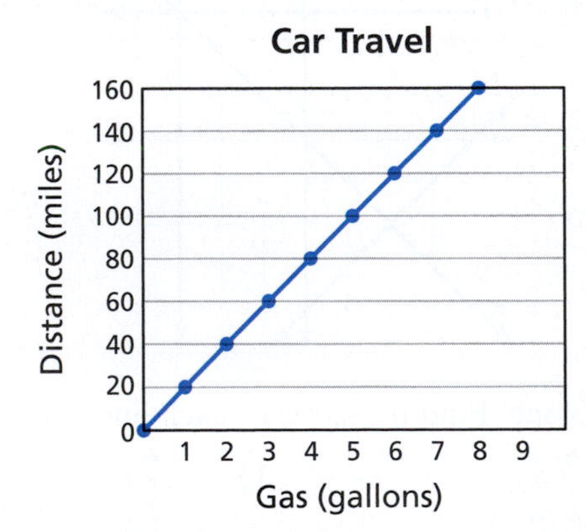

a. Find the slope of the line.

b. Explain the meaning of the slope as a rate of change.

Parallel and Perpendicular Lines The slope of a line is given. Find the slope of a line parallel to the given line. Then find the slope of a line perpendicular to the given line.

43. 7 **44.** $\frac{5}{2}$ **45.** $-\frac{2}{3}$ **46.** −6

Parallel and Perpendicular Lines A line contains the points (3, 2) and (4, 5).

47. Draw a line parallel to the given line that passes through the point (0, 5).

48. Draw a line perpendicular to the given line that passes through the point (−4, −2).

15.3 Linear Equations in Two Variables

Sketching the Graph of a Linear Equation Sketch the graph of the equation.

49. $y = 3x + 4$ **50.** $y = \frac{2}{3}x - 1$

51. $y = -3\left(x - \frac{4}{3}\right)$ **52.** $y = -\frac{2}{5}(x - 10)$

Horizontal and Vertical Lines Decide whether the line is *horizontal*, *vertical*, or *neither*. Explain your reasoning.

53. $x = 3$ **54.** $y = 2$

55. $y = -5$ **56.** $x = \frac{2}{3}$

57. $y = 4x$ **58.** $y = x$

Writing Equations of Horizontal and Vertical Lines Write the equation of the line.

59. Line *a*

60. Line *b*

61. Line *c*

62. Line *d*

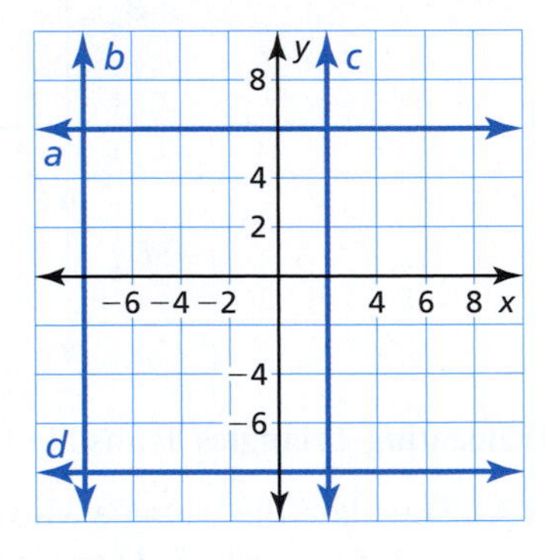

Finding the Slope of a Line Find the slope of the graph of the equation if possible.

63. $y = 3x + 2$ **64.** $y = -x$

65. $4y + 2x = 1$ **66.** $y = 6$

67. $y = -\frac{4}{3}x + 2$ **68.** $x = 1$

Using Intercepts to Sketch a Graph Graph the equation using the *x*- and *y*-intercepts.

69. $y = x + 3$ **70.** $y = x - 4$

71. $-y - 3x = 2$ **72.** $y + x = -\frac{1}{2}$

73. $y = 5(2x - 1)$ **74.** $3y + 2x = 6$

Using *m* and *b* to Sketch a Graph Use the slope and *y*-intercept to sketch the graph of the equation. Identify the *x*-intercept.

75. $-3y + 4x = 9$ **76.** $y = -4$

77. $8y + 2x = 16$ **78.** $-y = x$

79. $y = \frac{5}{4}x - 5$ **80.** $y = \frac{1}{3}(9x + 15)$

Rewriting an Equation in Slope-Intercept Form Write the equation in slope-intercept form. Then sketch the graph of the equation.

81. $3y + 2x = 3$

82. $-4y = x + 48$

83. $-2x = 15 - 3y$

84. $-x = \frac{1}{2}(4 - 2y)$

Writing the Equation of a Line Write the equation of the line in slope-intercept form.

85. Line *a*

86. Line *b*

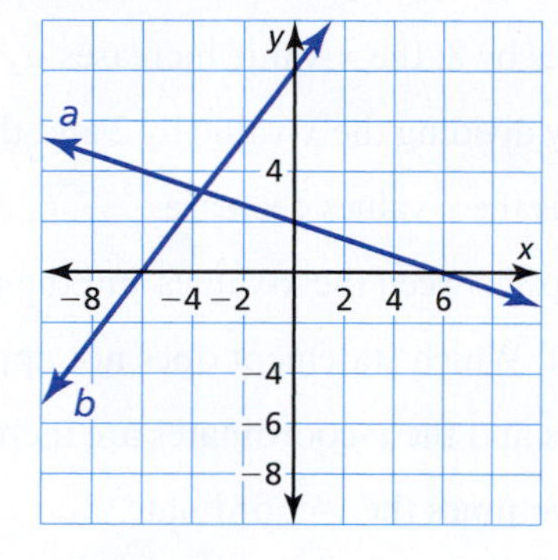

87. Car Rental You can determine the cost *y* (in dollars) of renting a car after driving *x* miles with the equation $y = 0.3x + 105$.

a. Graph the equation. Identify and interpret the slope and *y*-intercept.

b. You drive a distance of 130 miles. How much do you pay for the rental car?

88. Phone Bill You can determine the cost *y* (in dollars) of a prepaid phone when purchasing *x* minutes with the equation $y = 0.15x + 25$.

a. Graph the equation. Identify and interpret the slope and *y*-intercept.

b. You purchase 200 minutes. How much is your phone bill?

15.4 Functions

Identifying Functions Determine whether the relation is a function. Explain your reasoning.

89. (5, 3), (6, 4), (7, 5), (8, 6)

90. (9, 7), (8, 5), (7, 3), (9, 1)

Identifying Functions Determine whether the mapping diagram represents a function. Explain your reasoning.

91.

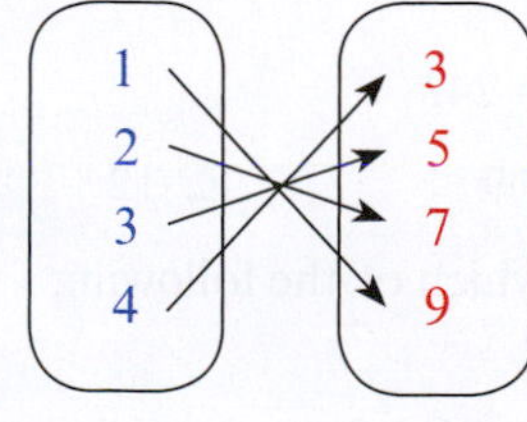

92.

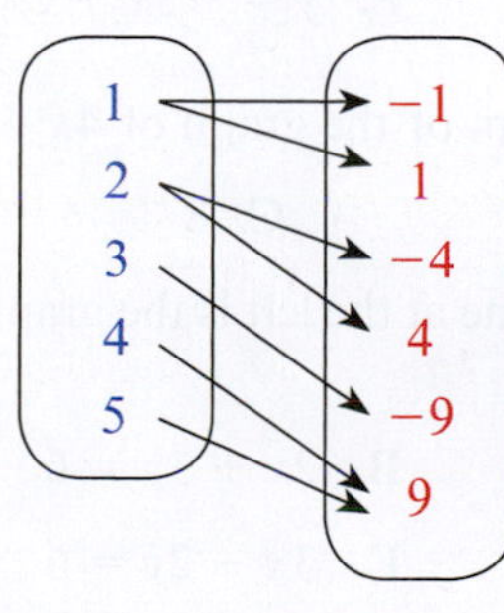

Identifying Functions Determine whether the table represents a function. Explain your reasoning.

93.

Input, *x*	1	2	3	4	5
Output, *y*	2	4	6	8	10

94.

Input, *x*	1	2	3	4	5
Output, *y*	2	2	2	2	2

Drawing Mapping Diagrams Draw a mapping diagram of the relation. Determine whether the relation is a function. Explain your reasoning.

95. (1, 2), (3, 4), (5, 6), (7, 8)

96. (1, 1), (2, 4), (3, 9), (4, 16)

Writing Function Rules Write a function rule for the statement.

97. The output is three more than half the input.

98. The output is four less than three times the input.

Using Tables to Write Function Rules Write a function rule for the input-output table.

99.

Input, *x*	1	2	3	4	5
Output, *y*	8	6	4	2	0

100.

Input, *x*	1	2	3	4	5
Output, *y*	5	8	11	14	17

Finding a Solution Determine whether the point (2, 4) is a solution of the function represented by the graph. Explain your reasoning.

101.

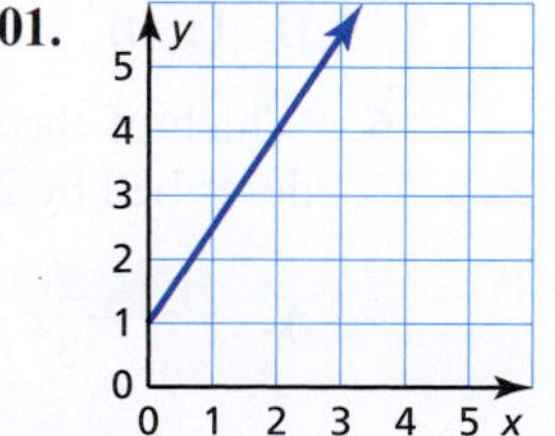

102.

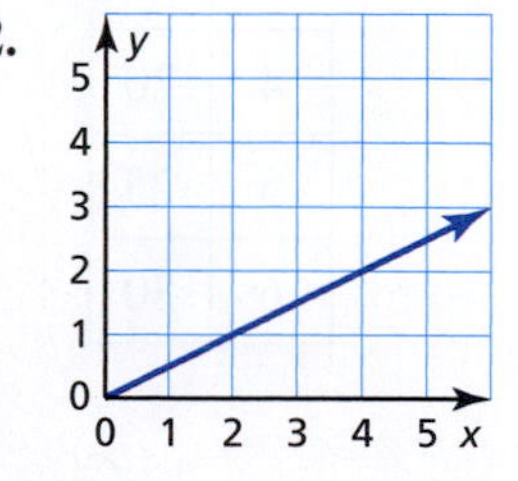

Identifying Linear Functions Determine whether the graph represents a linear function. Explain your reasoning.

103.

104.

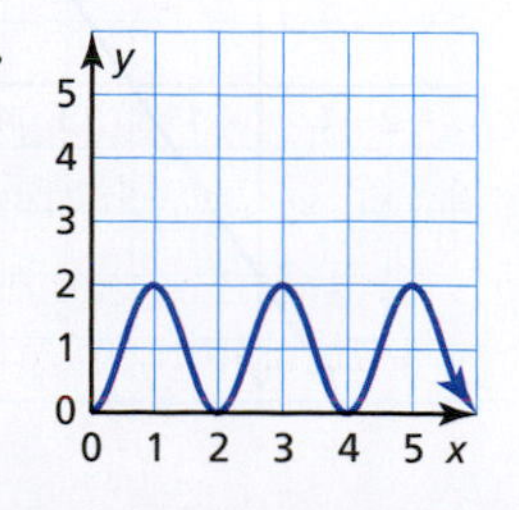

Chapter Test

The questions on this practice test are modeled after questions on state certification exams for K–8 teaching.

Test Directions: Each of the questions is followed by five choices. Choose the *best* response to each question.

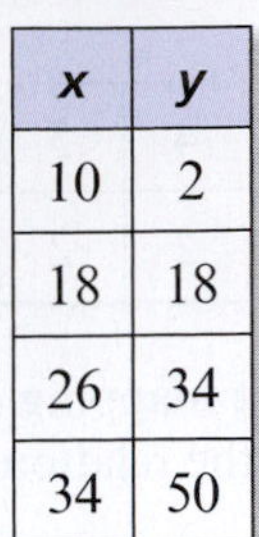

x	y
10	2
18	18
26	34
34	50

1. Which statement correctly describes the relationship shown in the table at the left?

A. As the x-value increases by 8, the y-value increases by 16.

B. As the y-value increases by 8, the x-value increases by 16.

C. The y-value is found by dividing the x-value by 2 and then subtracting 3.

D. The x-values increase as the y-values decrease.

E. There is no relationships between the x-values and the y-values.

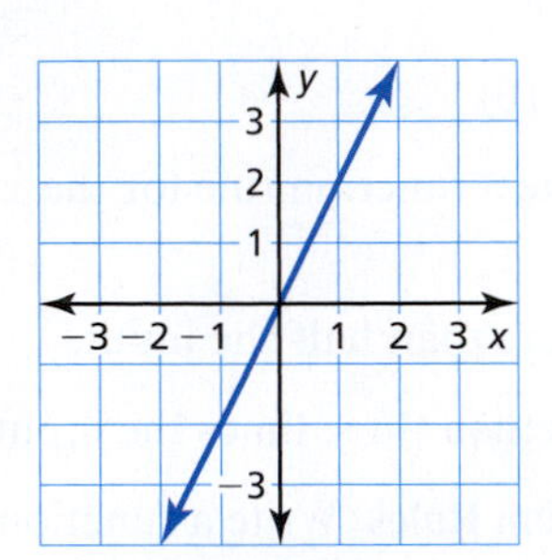

2. Examine the line at the left. Which statement does not apply?

A. Both the x-coordinates and the y-coordinates are increasing.

B. The x-coordinate is two times the y-coordinate.

C. The y-coordinate is two times the x-coordinate.

D. The point (9, 18) will lie on the graph.

E. The x- and y-intercepts have the same value.

x	y
0	5
2	13
6	29
7	33

3. Which of the following equations expresses the relationship between x and y in the table at the left?

A. $y = x + 5$ **B.** $y = x + 6$ **C.** $y = 4x + 5$

D. $y = 4x - 1$ **E.** $y = 4x - 5$

4. Find the coordinates (x, y) of the midpoint of the line segment on a graph that connects the points $(-5, 3)$ and $(3, -5)$.

A. $(-1, -1)$ **B.** $(1, -1)$ **C.** $(-1, 1)$

D. $(1, 1)$ **E.** $(4, 4)$

x	y
3	15
4	20
5	25
6	30

5. Which ordered pair could also be included in the table at the left?

A. (7, 42) **B.** (8, 50) **C.** (2, 10)

D. (1, 0) **E.** (0, 5)

6. Which of the following equations describes a line parallel to the line described by $2y + 3x = 8$?

A. $y = -\frac{2}{3}x - 2$ **B.** $y = \frac{2}{3}x + 3$ **C.** $y = -\frac{3}{2}x + 1$

D. $y = \frac{3}{2}x - 1$ **E.** $y = -3x + 2$

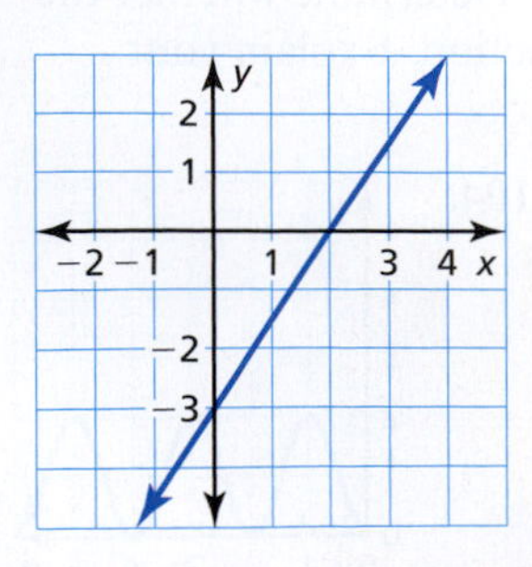

7. What is the x-intercept of the graph of $4x + 2y = 24$?

A. 2 **B.** 6 **C.** 8 **D.** 9 **E.** 12

8. The line in the xy-plane at the left is the graph of which of the following equations?

A. $2x + 3y = 4$ **B.** $2x + 3y = 6$ **C.** $3x + 2y = 4$

D. $3x + 2y = 6$ **E.** $3x - 2y = 6$

16 Probability

Activity: Comparing the Likelihood of Two Events

Materials:

- Pencil
- Marbles
- Bag

Learning Objective:

Perform experiments in probability, and record and graphically display the results of the experiments. Become familiar with the concept of likelihood, based on experimental probability.

Name ______________________________

Put 2 green marbles and 1 blue marble in a bag. Perform each experiment 36 times. Record each result as GG (green, green), GB (green, blue), or BB (blue, blue) using tally marks. A result of GB or BG represents the same result.

1. Randomly draw two marbles from the bag.

GG	
GB	

2. Randomly draw one marble from the bag. Then put the marble back in the bag and draw a second marble.

GG	
GB	
BB	

3. Make a bar graph of your results.

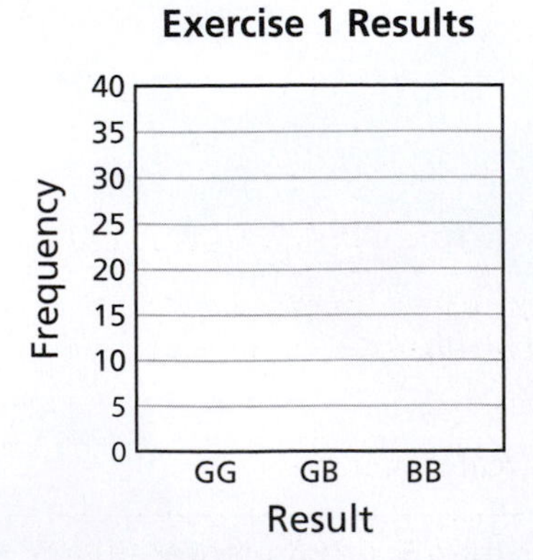

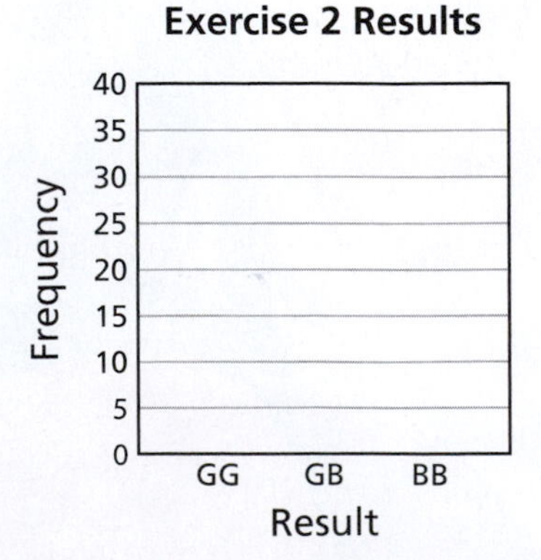

4. Do you think you are more likely to draw 2 green marbles in the first or second exercise? Explain.

16.1 Introduction to Probability

Standards

Grades 6–8

Statistics and Probability
Students should investigate chance processes and develop, use, and evaluate probability models.

- Identify events and outcomes in an experiment.
- Understand that probability is a measure of the likelihood of an event.
- Compare theoretical, experimental, and subjective probabilities.

Outcomes and Events

Definition of Outcomes and Events

An **experiment** is an investigation or procedure that has varying results. The possible results of an experiment are called **outcomes**. A collection of one or more outcomes is an **event**. The outcomes of a specific event are called **favorable outcomes**.

Example Randomly selecting a marble from a group of marbles is an experiment. Each marble in the group is an outcome. Selecting a green marble from the group is an event. When you perform an experiment *at random* or *randomly*, all of the possible outcomes are equally likely.

Possible outcomes

Event Selecting a green marble

Number of favorable outcomes 2

EXAMPLE 1 **Identifying Outcomes and Events**

You roll the number cube shown.

a. What are the possible outcomes?

b. What are the favorable outcomes of rolling an even number?

c. What are the favorable outcomes of rolling a number greater than 5?

Classroom Tip
When performing an experiment, assume that the faces of an n-sided polyhedron represent the numbers 1 through n. In Example 1, the faces of the number cube are numbered 1 through 6.

SOLUTION

a. The six possible outcomes are rolling a 1, 2, 3, 4, 5, and 6.

b. There are 3 even numbers and 3 odd numbers.
So, the favorable outcomes of the event "rolling an even number" are rolling a 2, 4, or 6.

Even	*Not* even
2, 4, 6	1, 3, 5

c. There is only one number greater than 5.

Greater than 5	*Not* greater than 5
6	1, 2, 3, 4, 5

So, the only favorable outcome of the event "rolling a number greater than 5" is rolling a 6.

iStockphoto.com/sweetym

Probability and Likelihood

Classroom Tip

In the definition at the right, notice that the more likely an event is to occur, the greater its probability. For instance, if two events have probabilities of $\frac{1}{3}$ and $\frac{1}{2}$, the event that has a probability of $\frac{1}{2}$ is more likely to occur because $\frac{1}{2}$ is greater than $\frac{1}{3}$.

Definition of Probability as a Measure of Likelihood

The **probability** of an event is a number that measures the likelihood that the event will occur. Probabilities are between 0 and 1, including 0 and 1. The diagram relates likelihoods (above the diagram) and probabilities (below the diagram).

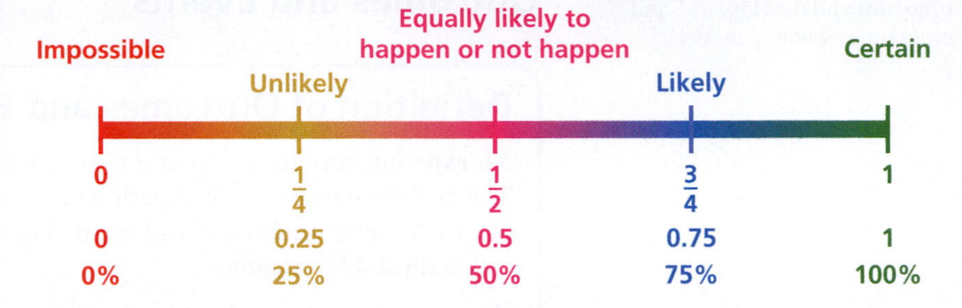

Probabilities can be written as fractions, decimals, or percents.

Describing the Likelihood of an Event

The table below gives the probabilities of three types of space objects impacting Earth or entering Earth's atmosphere. Describe the likelihood of each event.

NASA says there is no chance of the 885-foot asteroid, Apophis, smashing into Earth in its 2029 flyby, and less than 1 in 1,000,000 chance of a collision in 2036.

	Probability of a space object impacting Earth			
	Space object	**Diameter**	**Probability of impact**	**Date(s) of impact**
a.	2000 SG344	121 ft	$\frac{1}{1000}$	2071
b.	Meteor	Varying sizes	1	Every day
c.	Apophis	885 ft	0	2029

SOLUTION

a. With a probability of $\frac{1}{1000}$, or 0.1%, this event is *very unlikely*. Of 1000 identical space objects, you would expect only one of them to hit Earth.

b. With a probability of 1, this event is *certain*. Meteors enter Earth's atmosphere every day.

c. With a probability of 0, this event is impossible.

Mopic/Shutterstock.com

Finding Probabilities

There are three basic ways to estimate (or find exactly) the probability that an event will occur.

1. **Theoretical probability** Used in cases where all possible outcomes of an experiment are known and can be counted. Gambling is a common example.
2. **Experimental probability** Used in cases where a representative sample can be taken and counted. Quality control is a common example.
3. **Subjective probability** Used in cases where personal judgement or intuition determines the likelihood of an event to occur. Recovering from a surgery is a common example.

Definition of Theoretical Probability

Suppose an event can occur in n ways out of a total possible number T of equally likely outcomes. The **theoretical probability** that the event will occur is the ratio

$$\text{Theoretical probability} = \frac{\text{Number of favorable outcomes}}{\text{Number of possible outcomes}} = \frac{n}{T}.$$

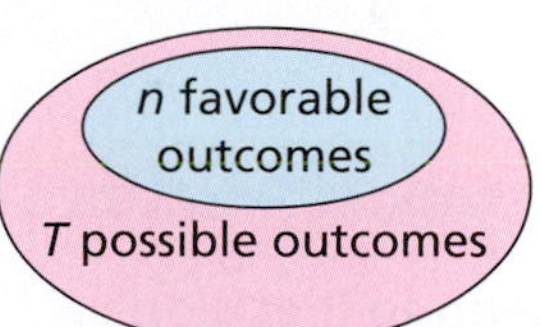

EXAMPLE 3 **Finding Theoretical Probabilities**

You roll the number cube shown.

a. What is the theoretical probability of rolling an even number?

b. What is the theoretical probability of rolling a number greater than 5?

Classroom Tip When you are teaching probability, be sure your students understand the difference between what is *possible* and what is *probable*. For instance, in the example at the right, if you roll a number cube many times, it is *possible* that you will never roll an even number. Although this is possible, it is unlikely. The likelihood is that you will roll an even number about half of the time.

SOLUTION

The number of possible outcomes $\{1, 2, 3, 4, 5, 6\}$ and the numbers of favorable outcomes were counted in Example 1.

a. There are 3 even numbers and 3 odd numbers.

$$\text{Theoretical probability} = \frac{\text{Number of favorable outcomes}}{\text{Number of possible outcomes}} = \frac{3}{6} = \frac{1}{2}$$

The theoretical probability of rolling an even number is $\frac{1}{2}$, or 50%.

b. There is only one number greater than 5.

$$\text{Theoretical probability} = \frac{\text{Number of favorable outcomes}}{\text{Number of possible outcomes}} = \frac{1}{6}$$

The theoretical probability of rolling a number greater than 5 is $\frac{1}{6}$, or about 17%.

EXAMPLE 4 Comparing Theoretical and Experimental Probabilities

You spin the spinner 100 times and obtain the following results.

Yellow	Red	Blue	White
38	31	19	12

a. Use the results to find the *experimental* probability of landing on each color.

b. Find the *theoretical* probability of landing on each color.

c. Compare the experimental and theoretical probabilities.

d. Conduct a simulation for this experiment 10,000 times using the spinner at *MathematicalPractices.com*. Then compare the experimental and theoretical probabilities.

SOLUTION

a. To find each *experimental* probability, divide the number of times each color is spun by the total number of spins, 100.

Yellow $\frac{38}{100} = 38\%$ Blue $\frac{19}{100} = 19\%$

Red $\frac{31}{100} = 31\%$ White $\frac{12}{100} = 12\%$

Mathematical Practices

Attend to Precision

MP6 Mathematically proficient students try to communicate precisely to others.

b. To find each *theoretical* probability, divide the number of favorable outcomes on the spinner by the total number of possible outcomes on the spinner, 10.

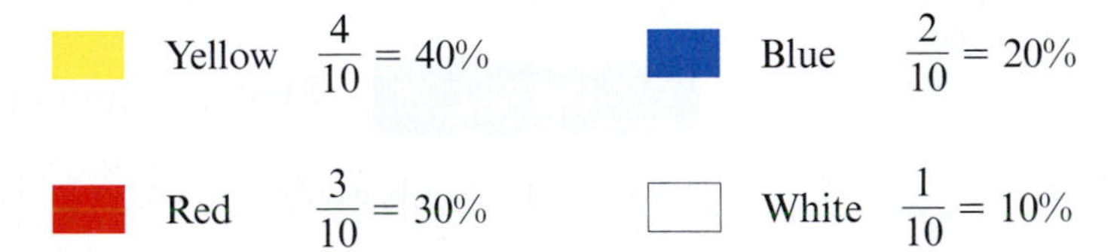

Yellow $\frac{4}{10} = 40\%$ Blue $\frac{2}{10} = 20\%$

Red $\frac{3}{10} = 30\%$ White $\frac{1}{10} = 10\%$

c. The experimental probabilities are about the same as the theoretical probabilities.

d. Using the simulator at *MathematicalPractices.com*, set the number of trials to 10,000. The results of one such simulation are shown below.

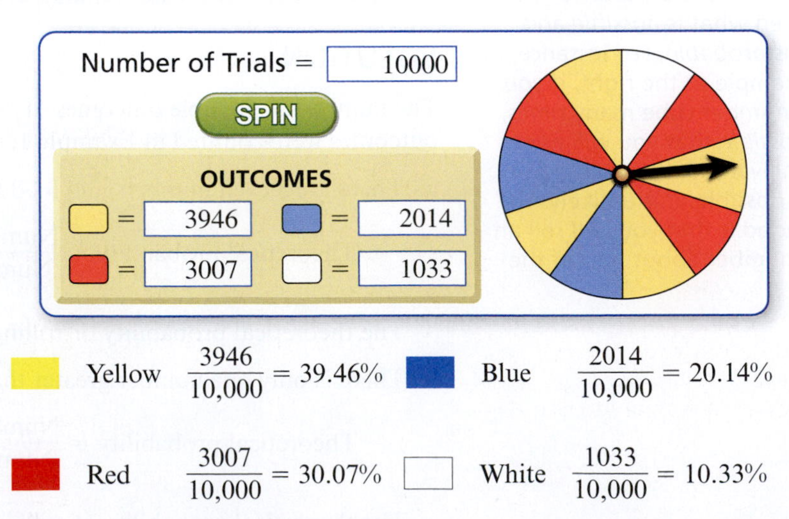

Yellow $\frac{3946}{10{,}000} = 39.46\%$ Blue $\frac{2014}{10{,}000} = 20.14\%$

Red $\frac{3007}{10{,}000} = 30.07\%$ White $\frac{1033}{10{,}000} = 10.33\%$

Notice that these experimental probabilities are very close to the theoretical probabilities from part (b). As the total number of spins increases, the closer the experimental probabilities will be to the theoretical probabilities.

Classroom Tip

In Example 4, notice that the sum of all probabilities equals 1.

Finding a Theoretical Probability

The NBA Draft Lottery is an annual event held by the National Basketball Association. The 14 teams that missed the playoffs in the previous season participate in a lottery to determine the order for the first 3 picks of the draft.

The NBA Draft Lottery

1. 250 combinations
2. 199 combinations
3. 156 combinations
4. 119 combinations
5. 88 combinations
6. 63 combinations
7. 43 combinations
8. 28 combinations
9. 17 combinations
10. 11 combinations
11. 8 combinations
12. 7 combinations
13. 6 combinations
14. 5 combinations

To conduct the lottery, each of the numbers from 1 through 14 are put into a lottery machine. Four of the numbers are drawn. The team that has that combination of numbers (order is not counted) wins the first pick. There are 1001 possible combinations of 4 numbers (from 1 through 14). However, one combination is not used, leaving only 1000 valid combinations.

The 14 teams are ordered (from worst record to best record) and are randomly given 4-number combinations, as shown at the left. The lottery is weighted so that the team with the worst record has the best chance to obtain one of the first three picks.

a. What is the probability that the team with the worst record wins the first pick of the draft?

b. What is the probability that the team in the 14th spot wins the first pick of the draft?

SOLUTION

a. There are 250 combinations given to the team with the worst record. So, the theoretical probability that one of its combinations is drawn is

$$\text{Theoretical probability} = \frac{\text{Number of favorable outcomes}}{\text{Number of possible outcomes}}$$

$$= \frac{250}{1000}$$

$$= \frac{1}{4}.$$

So, the theoretical probability of the team with the worst record winning the first pick is $\frac{1}{4}$, or 25%.

Classroom Tip

The occurrence of unlikely events is both interesting and amazing. Ask your students to find examples of unlikely events that actually happened in history.

b. There are only 5 combinations given to the team in the 14th spot. So, the theoretical probability that one of its combinations is drawn is

$$\text{Theoretical probability} = \frac{\text{Number of favorable outcomes}}{\text{Number of possible outcomes}}$$

$$= \frac{5}{1000}$$

$$= \frac{1}{200}.$$

So, the theoretical probability of the team in the 14th spot winning the first pick is $\frac{1}{200}$, or 0.5%.

16.1 Exercises

Free step-by-step solutions to odd-numbered exercises at *MathematicalPractices.com.*

1. Writing a Solution Key Write a solution key for the activity on page 614. Describe a rubric for grading a student's work.

2. Grading the Activity In the activity on page 614, a student gave the answers below. Each question is worth 10 points. Assign a grade to each answer. For those that are incorrect, why do you think the student erred?

Sample Student Work

1.

GG	𝍸 𝍸 ||||
GB	𝍸 𝍸 𝍸 𝍸 ||

2.

GG	𝍸 𝍸 𝍸 ||
GB	𝍸 𝍸 𝍸
BB	||||

4. First, because there are only two possible outcomes.

3. Identifying Possible Outcomes Determine the number of possible outcomes for each experiment.

a. Flipping a coin

b. Rolling a number cube

c. Choosing a letter from the alphabet

4. Identifying Possible Outcomes Determine the number of possible outcomes for each experiment.

a. Choosing a letter from the word PROBABILITY

b. Rolling an octahedron

c. Choosing a multiple choice answer *a–e*

5. Identifying Outcomes You are given a bag with 6 green marbles, 10 blue marbles, 4 yellow marbles, and 8 red marbles. You randomly draw one marble.

a. What are the possible outcomes?

b. How many favorable outcomes are there for the event of drawing a blue marble?

c. How many favorable outcomes are there for the event of drawing a yellow marble?

6. Identifying Outcomes You roll a number cube.

a. What are the possible outcomes?

b. What are the favorable outcomes for the event of rolling an odd number?

c. What are the favorable outcomes for the event of rolling a number less than 3?

7. Identifying Outcomes and Events You randomly draw one card from a standard deck of 52 playing cards.

a. Determine the number of possible outcomes.

b. How many favorable outcomes are there for the event of drawing a red card?

c. How many favorable outcomes are there for the event of drawing a spade?

d. How many favorable outcomes are there for the event of drawing a black jack?

8. Identifying Outcomes and Events You roll a dodecahedron with the numbers 1 through 12.

a. Determine the number of possible outcomes.

b. What are the favorable outcomes for the event of rolling an even number?

c. What are the favorable outcomes for the event of rolling a number less than 5?

d. What are the favorable outcomes for the event of rolling a number with 2 digits?

9. Describing Likelihood Describe the likelihood of each event as *impossible*, *unlikely*, *equally likely*, *likely*, or *certain*.

	Event	Probability
a.	Having a male child	0.5
b.	Drawing the ace of spades	$\frac{1}{52}$
c.	Drawing a card that is not a 2	$\frac{12}{13}$

10. Describing Likelihood Describe the likelihood of each event as *impossible*, *unlikely*, *equally likely*, *likely*, or *certain*.

	Event	Probability
a.	Rolling a number greater than 0	1
b.	Rolling a 2.4	0
c.	Correctly guessing the answer on a true/false question	0.5

11. Finding Theoretical Probabilities You are given a bag with 8 green marbles, 4 blue marbles, 12 yellow marbles, and 10 red marbles. Find the theoretical probability of each random event.

a. Drawing a green marble

b. Drawing a red marble

c. Drawing a marble that is not yellow

12. Finding Theoretical Probabilities You are given a standard deck of 52 playing cards. Find the theoretical probability of each random event.

a. Drawing a black card

b. Drawing a face card (jack, queen, or king)

c. Drawing a diamond

13. Finding Theoretical Probabilities A roulette wheel has 38 slots, 36 of which are numbered 1–36. Half of these 36 slots are red and half are black. The remaining two slots are green and are numbered 0 and 00. A small white ball is tossed into the spinning wheel and has an equally likely chance of landing in each slot. Find the theoretical probability of each event.

a. Landing in a red slot

b. Landing in a green slot

c. Landing in an odd-numbered slot

14. Finding Theoretical Probabilities You roll an icosahedron that contains each letter of the alphabet *except Q, U, V, X, Y,* and *Z*. Find the theoretical probability of each event.

a. Rolling a vowel

b. Rolling a consonant

c. Rolling a letter in the word MATH

15. Theoretical Probability Four bouncy-ball machines each contain 100 bouncyballs. The theoretical probability of randomly dispensing an orange ball is given in simplified form. Find the number of orange balls in each bouncy-ball machine.

a. $\frac{1}{4}$ **b.** $\frac{1}{20}$

c. $\frac{2}{5}$ **d.** $\frac{3}{4}$

16. Theoretical Probability Four jars of coins each contain 240 coins. The theoretical probability of randomly drawing a quarter is given in simplified form. Find the number of quarters in each jar.

a. $\frac{1}{3}$ **b.** $\frac{1}{6}$

c. $\frac{3}{8}$ **d.** $\frac{2}{3}$

17. Finding Experimental Probabilities The bar graph shows the results of rolling a number cube 120 times. Find the experimental probability of each event.

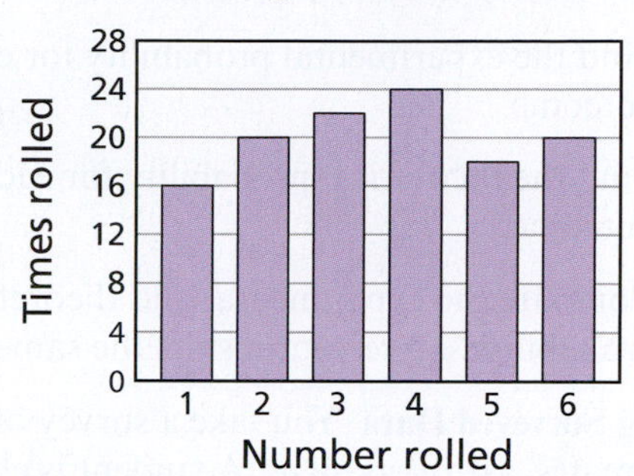

a. Rolling a 1

b. Rolling a 5

c. Rolling an even number

18. Finding Experimental Probabilities The bar graph shows the results of spinning a spinner 60 times. Find the experimental probability of each event.

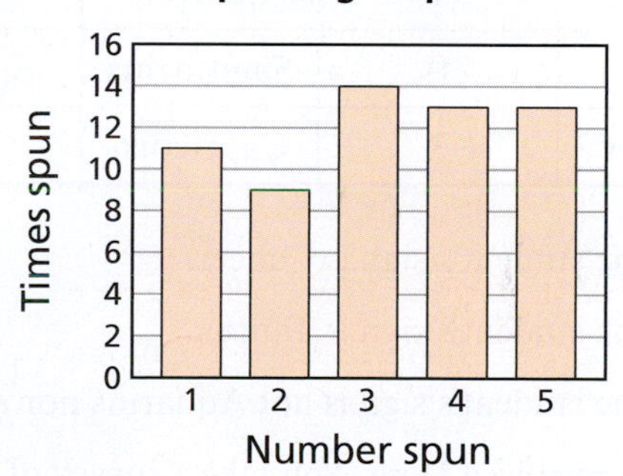

a. Spinning a 2

b. Spinning an odd number

c. Spinning a number greater than 3

19. Comparing Probabilities Conduct your own experiment by flipping a coin.

a. List the possible outcomes.

b. Perform the experiment 30 times. Make a table to record your results.

c. Find the experimental probability for each outcome.

d. Find the theoretical probability for each outcome.

e. Compare the experimental and theoretical probabilities. Are your results the same? Explain.

20. Comparing Probabilities Conduct your own experiment by rolling a number cube.

a. List the possible outcomes.

b. Perform the experiment 36 times. Make a table to record your results.

c. Find the experimental probability for each outcome.

d. Find the theoretical probability for each outcome.

e. Compare the experimental and theoretical probabilities. Are your results the same? Explain.

21. Using Surveyed Data You take a survey of your classmates' astrology signs. A student is chosen at random. Find the probability of each event.

Sign	Number of students	Sign	Number of students
Aquarius	3	Leo	1
Pisces	1	Virgo	3
Aries	5	Libra	4
Taurus	1	Scorpio	3
Gemini	0	Sagittarius	2
Cancer	2	Capricorn	5

a. The student's sign is Cancer.

b. The student's sign is Taurus.

c. The student's sign is not Aquarius nor Aries.

22. Using Surveyed Data You take a survey of your classmates' birthday months. A student is chosen at random. Find the probability of each event.

Month	Number of students	Month	Number of students
January	0	July	4
February	3	August	2
March	5	September	0
April	2	October	3
May	5	November	4
June	1	December	6

a. The student was born in August.

b. The student was born in May.

c. The student was born in one of the last 4 months of the year.

23. Digital Numbers Digital clocks, calculators, scoreboards, and many other devices use a 7-segment display like the one shown below. The device lights up different segments (labeled *a*–*g*) to display a number 0–9. A number is chosen at random. Find the probability of each event.

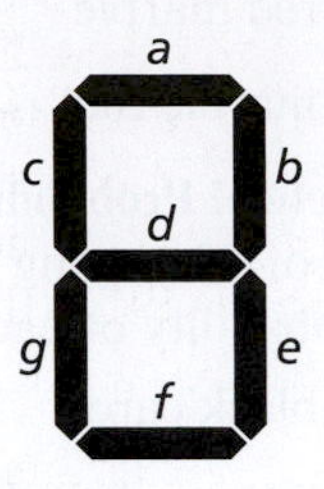

a. Segment *c* will be lit.

b. Segment *d* will be lit.

c. Segments *d* and *e* will be lit.

24. Board Games The board game Scrabble® contains 100 tiles, 98 of which are labeled with a letter and a point value, and 2 blank tiles. The distribution of tiles and point values is shown in the table. A tile is drawn at random. Find the probability of each event.

Point value	Number of tiles
0 points	2
1 point	68
2 points	7
3 points	8
4 points	10
5 points	1
8 points	2
10 points	2

a. Drawing a tile that is worth 1 point

b. Drawing a tile that is worth 10 points

c. Drawing a tile that is worth 0 points

d. Drawing a tile that is worth at least 5 points

25. Making a Prediction You have 500 songs on your MP3 player. Seven of the last 10 songs played were rock songs. How many of the 500 songs do you expect to be rock songs?

26. Making a Prediction Jim has the best batting average on the school baseball team. In his last 25 at bats, he had 14 hits. How many hits do you expect him to get in his next 150 at bats?

27. **Problem Solving** A high school consists of 195 freshman, 224 sophomores, 179 juniors, and 202 seniors. A student is chosen at random to win $100. What is the probability that the student is a senior?

28. **Problem Solving** A college calculus class consists of 26 females and 14 males. There are 8 biology majors, 18 mathematics majors, and 14 computer science majors. If a student is chosen at random, what is the probability that the student is a female? What is the probability that the student is a mathematics major?

29. **Produce** The local farmer's market sells baskets of different types of fruit. There are 2 baskets of apples, 4 baskets of strawberries, 2 baskets of raspberries, 1 basket of blackberries, 1 basket of cherries, 3 baskets of blueberries, and 2 baskets of mixed berries. You choose a basket at random. What is the probability that you choose a basket of blueberries?

30. **Crayons** Each student in your class has a box of 16 crayons. Your students like to trade crayons while coloring. One student ends up with 2 blue, 3 green, 1 black, 1 brown, 2 red, 1 yellow, 3 orange, 1 yellow-orange, and 2 blue-green crayons. If the student randomly chooses one crayon to use, what is the probability that the student will choose a red crayon?

31. **Grading Student Work** On a diagnostic test, one of your students does the following work. Explain what the student did wrong. Which topics would you encourage the student to review?

> Calculate the probability of rolling a number greater than 4 on a standard number cube.
>
> $$\frac{3}{6} = \frac{1}{2}$$
>
> The probability of rolling a number greater than 4 is 1/2.

32. **Grading Student Work** On a diagnostic test, one of your students does the following work. Explain what the student did wrong. Which topics would you encourage the student to review?

> You randomly draw one card form a standard deck of 52 playing cards. Calculate the probability of drawing a red 10.
>
> $$\frac{4}{52} = \frac{1}{13}$$
>
> The probability of drawing a red 10 is 1/13.

33. *Writing* Describe the difference between experimental probability and theoretical probability.

34. *Writing* Describe why the sum of all probabilities for the outcomes of one experiment must equal 1.

35. **Activity: In Your Classroom** Have your students conduct an in-class experiment.

    a. Have each student write his or her full name on a piece of paper and place it in a hat. Does each student have the same probability of his or her name being drawn? Calculate the probability of drawing one particular student's name.

    b. Now have each student write his or her full name on 3 pieces of paper and place them in a hat. Does each student have the same probability of his or her name being drawn? Calculate the probability of drawing one particular student's name.

    c. Are the probabilities calculated in part (a) and part (b) the same? Explain your reasoning.

36. **Activity: In Your Classroom** Have your students conduct an in-class experiment.

    a. Give each student a coin to flip. What is the probability of landing on heads? Does each student have the same probability of his or her coin landing on heads?

    b. Now, give each student 2 coins to flip. What is the probability of both coins landing on heads? Does each student have the same probability of the coins both landing on heads?

    c. Are the probabilities calculated in part (a) and part (b) the same? Explain your reasoning.

37. **Find the Missing Value** An experiment has 5 different possible outcomes. The probabilities of the first 4 outcomes are $\frac{1}{5}, \frac{1}{4}, \frac{1}{20}$, and $\frac{2}{5}$. What is the probability of the last outcome?

38. **Find the Missing Value** An experiment has 6 different possible outcomes. The probabilities of the first 5 outcomes are $\frac{3}{20}, \frac{7}{20}, \frac{1}{10}, \frac{1}{25}$, and $\frac{4}{25}$. What is the probability of the last outcome?

39. **Critical Thinking** In general, the probability of an event is $\frac{1}{n}$, where n is the number of outcomes. Using this, create a formula to calculate the probability that the event does *not* occur. Explain.

40. **Critical Thinking** If the probability of an event is 0, what is the probability that the event will *not* occur? Explain.

Activity: Using a Tree Diagram to List Outcomes

Materials:

- Pencil
- Coins

Learning Objective:

Understand how a tree diagram can be used to list the possible outcomes of an experiment. Be able to use the results to find the probabilities of various events.

Name ______________________

The tree diagram shows the possible outcomes when two coins are flipped.

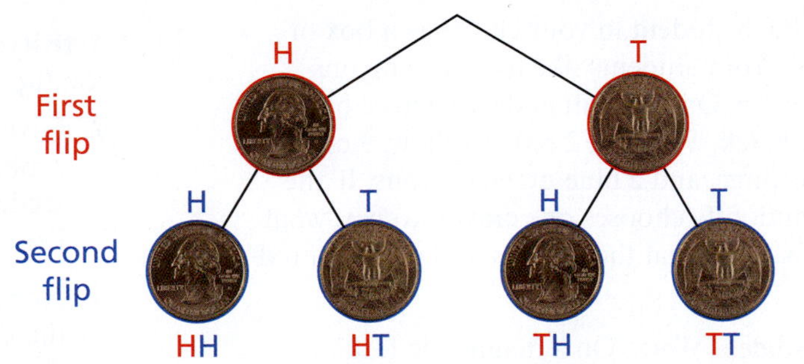

1. How many outcomes are possible? ▭
   List them. ______

2. Find the following probabilities when 2 coins are flipped.

   Probability of 2 heads = ▭

   Probability of 1 head and 1 tail = ▭

   Probability of 2 tails = ▭

3. Complete the tree diagram at the right for flipping 3 coins.

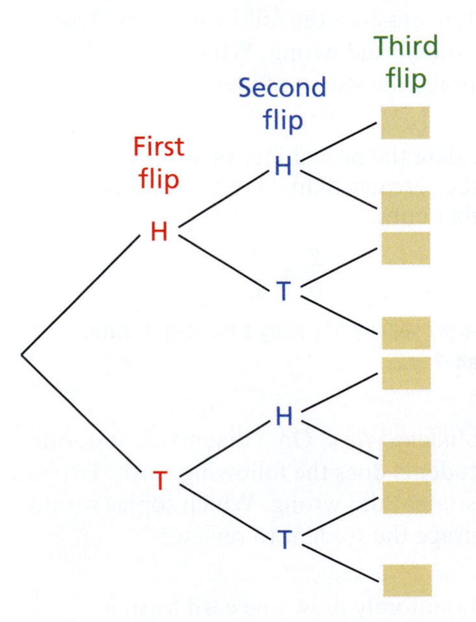

4. How many outcomes are possible? ▭
   List them. ______________________

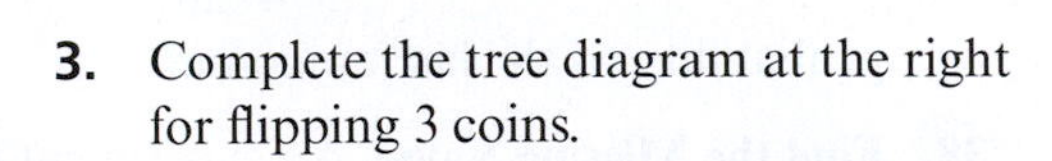

5. Find the following probabilities when 3 coins are flipped.

   Probability of 3 heads = ▭

   Probability of 2 heads and 1 tail = ▭

   Probability of 1 head and 2 tails = ▭

   Probability of 3 tails = ▭

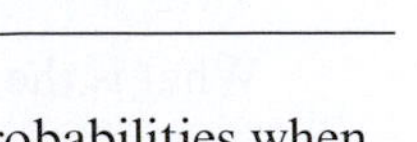

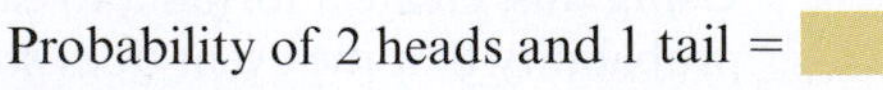

16.2 Counting Techniques

Standards

Grades 6–8

Statistics and Probability

Students should investigate chance processes and develop, use, and evaluate probability models.

- Use tree diagrams, tables, or the Fundamental Counting Principle to find the possible outcomes of experiments.
- Use permutations to find probabilities.
- Use combinations to find probabilities.

Tree Diagrams and Tables

Definition of a Tree Diagram

A **tree diagram** is a diagram that displays all possible outcomes of a probability experiment by using branches that originate from a starting point.

Example A bag contains 3 marbles: 1 red, 1 blue, and 1 green. One marble is randomly drawn from the bag. Then, without replacement, another marble is randomly drawn. The tree diagram shows the six possible outcomes.

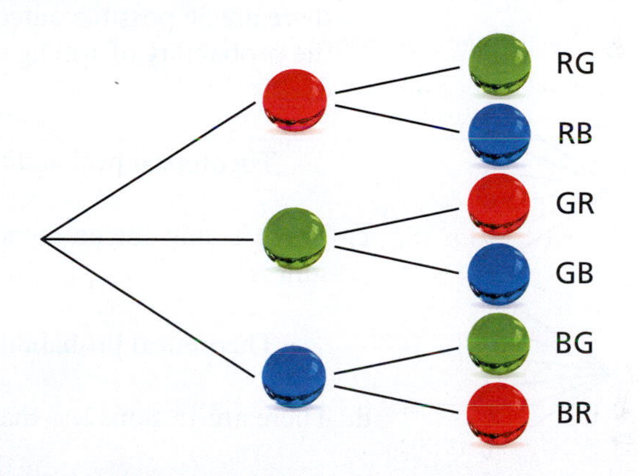

Classroom Tip

Tree diagrams can be used to find probabilities. For instance, you can use the tree diagram at the right to find the probability of randomly drawing a red and a green marble. This outcome occurs twice. There are 6 possible outcomes. So, the probability is $\frac{2}{6} = \frac{1}{3}$.

EXAMPLE 1 **Finding Probability Using a Tree Diagram**

You flip the coin and roll the number cube shown. What is the probability that you flip a head and roll a 3?

SOLUTION

Begin by drawing a tree diagram.

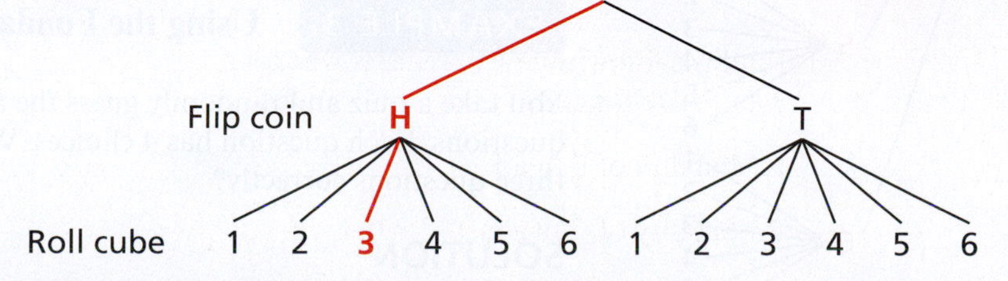

There are 12 possible outcomes. Of these, only one is favorable. So, the probability of flipping heads and rolling a 3 is

$$\text{Theoretical probability} = \frac{\text{Number of favorable outcomes}}{\text{Number of possible outcomes}} = \frac{1}{12}.$$

iStockphoto.com/sweetym; Robert Pernell/Shutterstock.com

EXAMPLE 2 Using a Table to Count Outcomes

You roll the two dice shown.

a. What are the possible sums of the two dice?

b. Which sum occurs most often? What is the probability of rolling this sum?

c. What is the probability of rolling a sum of 2 (or "snake eyes"), as shown above?

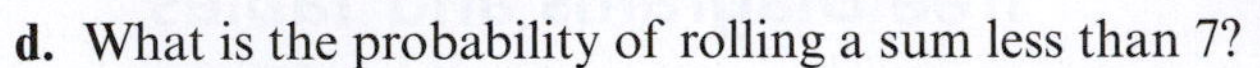

d. What is the probability of rolling a sum less than 7?

Classroom Tip

Tree diagrams and tables serve essentially the same purpose. They are convenient ways to list the possible outcomes of an experiment. The table shows the sums of two dice. Below is a tree diagram that shows the possible outcomes when rolling two dice.

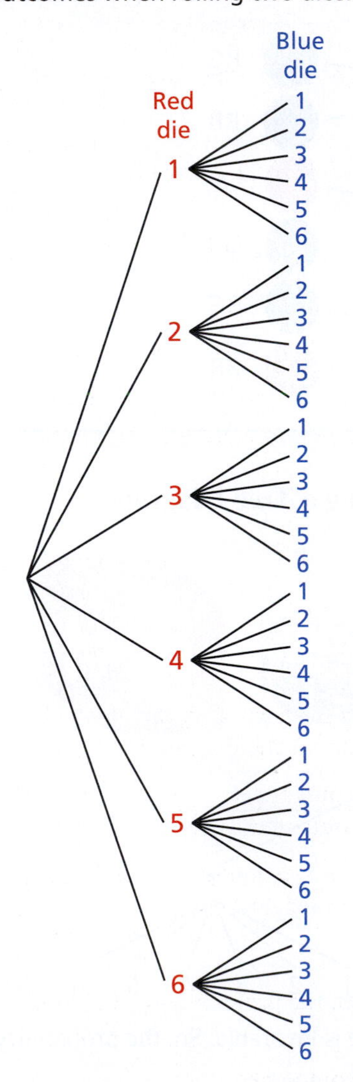

SOLUTION

Begin by making a table that shows the possible sums.

		Red die					
		1	2	3	4	5	6
Blue die	1	2	3	4	5	6	7
	2	3	4	5	6	7	8
	3	4	5	6	7	8	9
	4	5	6	7	8	9	10
	5	6	7	8	9	10	11
	6	7	8	9	10	11	12

a. The possible sums are 2, 3, 4, 5, 6, 7, 8, 9, 10, 11, and 12.

b. The sum that occurs most often is 7. It occurs 6 times. Because there are 36 possible outcomes, the probability of rolling a sum of 7 is

$$\text{Theoretical probability} = \frac{\text{Number of favorable outcomes}}{\text{Number of possible outcomes}} = \frac{6}{36} = \frac{1}{6}.$$

c. There is only one outcome that has a sum of 2. The probability of rolling this sum is

$$\text{Theoretical probability} = \frac{\text{Number of favorable outcomes}}{\text{Number of possible outcomes}} = \frac{1}{36}.$$

d. There are 15 sums less than 7. The probability of rolling a sum less than 7 is

$$\text{Theoretical probability} = \frac{\text{Number of favorable outcomes}}{\text{Number of possible outcomes}} = \frac{15}{36} = \frac{5}{12}.$$

Fundamental Counting Principle

An event M has m possible outcomes. An event N has n possible outcomes. The total number of outcomes of event M followed by event N is $m \cdot n$. The Fundamental Counting Principle can be extended to more than two events.

EXAMPLE 3 Using the Fundamental Counting Principle

You take a quiz and randomly guess the answer for three multiple-choice questions. Each question has 4 choices. What is the probability you answer all three questions correctly?

SOLUTION

There are 4 choices for each question. So, there are $4 \cdot 4 \cdot 4 = 64$ possible outcomes. There is only 1 way to answer all three questions correctly.

$$\text{Theoretical probability} = \frac{\text{Number of favorable outcomes}}{\text{Number of possible outcomes}} = \frac{1}{64}$$

Chris Sargent/Shutterstock.com

Permutations

Definition of Permutation

A **permutation** is an ordered arrangement of all or part of a set of objects. A **permutation of n objects taken all at a time** is an ordered arrangement of all the members of a set. If n is the number of members of a set, then the total possible number of permutations is **n factorial**, written as $n!$.

$n! = n \cdot (n-1) \cdot (n-2) \cdot \ldots \cdot 3 \cdot 2 \cdot 1$ n factorial

Example There are six permutations of 3 different colored marbles.

$3! = 3 \cdot 2 \cdot 1$

$= 6$ 6 permutations of 3 marbles

Classroom Tip

Zero factorial is defined to be 1. So, $0! = 1$. To further understand why this is true, first look at a set that has 2 elements, A and B. This set has two permutations: AB and BA.

Next, look at a set that has only 1 element, such as A. This set has 1 single permutation: A.

You have already learned that a set with zero elements is called the *empty set*. So, how many ways can you order a set that has no elements? Even though there are no elements, an empty set can only be ordered one way. So, $0! = 1$.

EXAMPLE 4 Finding a Probability Using a Permutation

You randomly draw the four alphabet blocks shown out of a bag one at a time. What is the probability that you draw the blocks in alphabetical order?

SOLUTION

Method 1 One way to find the probability is to first list the possible orders in which you can draw the four blocks.

ABCD	ABDC	ACBD	ACDB	ADBC	ADCB
BACD	BADC	BCAD	BCDA	BDAC	BDCA
CABD	CADB	CBAD	CBDA	CDAB	CDBA
DABC	DACB	DBAC	DBCA	DCAB	DCBA

There are 24 possible orders. Of these, only the first one listed is in alphabetical order. So, the probability of drawing the blocks in alphabetical order is

$$\text{Theoretical probability} = \frac{\text{Number of favorable outcomes}}{\text{Number of possible outcomes}} = \frac{1}{24}.$$

Method 2 Another way to find the probability is to use permutations. Because there are four blocks, there are 4! ways that you can order them.

$4! = 4 \cdot 3 \cdot 2 \cdot 1 = 24$ 4 factorial

Because there is only one that is in alphabetical order, the probability of drawing the blocks in alphabetical order is

$$\text{Theoretical probability} = \frac{\text{Number of favorable outcomes}}{\text{Number of possible outcomes}} = \frac{1}{24}.$$

Classroom Tip

When you are introducing students to permutations and factorials, ask them to write out the possible orders, as shown in Method 1 of Example 4. They can then use a factorial as a check.

Permutations of n Objects Taken r at a Time

A **permutation of n objects taken r at a time** is a permutation containing only r members of a set. The number of permutations of n objects taken r at a time is

$$P(n, r) = \frac{n!}{(n - r)!}.$$

Example There are six permutations of 3 different colored marbles taken 2 at a time.

$$P(3, 2) = \frac{3!}{(3 - 2)!} = \frac{3!}{1!} = \frac{6}{1} = 6$$

Count by using formula

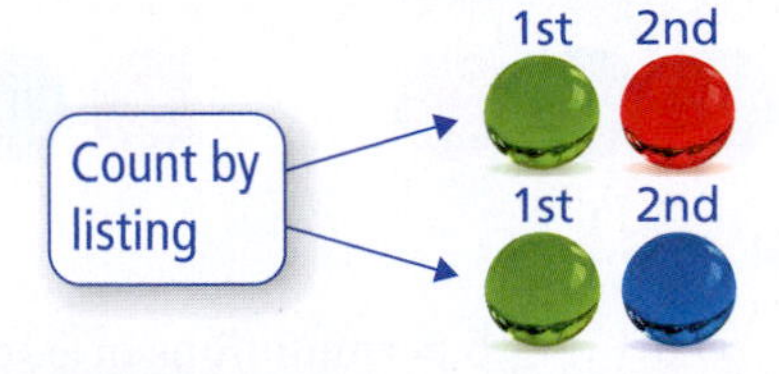

Classroom Tip

Be sure you see that order is important when counting permutations. For instance, the following two lists each have 1 green marble and 1 red marble, but they are considered different.

When counting "combinations" (as seen on page 629), order is not important.

EXAMPLE 5 Finding a Probability Using a Permutation

You randomly draw two of the four alphabet blocks shown out of a bag one at a time. What is the probability that you draw two blocks in alphabetical order?

SOLUTION

Method 1 One way to find the probability is to first list the possible orders in which you can draw two blocks.

AB AC AD BA BC BD

CA CB CD DA DB DC

There are 12 possible orders. Of these, six are in alphabetical order. So, the probability of drawing two blocks in alphabetical order is

$$\text{Theoretical probability} = \frac{\text{Number of favorable outcomes}}{\text{Number of possible outcomes}} = \frac{6}{12} = \frac{1}{2}.$$

Method 2 Another way to find the probability is to use the formula for the number of permutations of 4 objects taken 2 at a time.

$$P(4, 2) = \frac{4!}{(4 - 2)!} = \frac{4!}{2!} = \frac{24}{2} = 12$$

So, the number of possible outcomes is 12. To find the number of favorable outcomes, list the outcomes that are in alphabetical order: AB, AC, AD, BC, BD, and CD. There are 6 favorable outcomes. So, the probability of drawing two blocks in alphabetical order is

$$\text{Theoretical probability} = \frac{\text{Number of favorable outcomes}}{\text{Number of possible outcomes}} = \frac{6}{12} = \frac{1}{2}.$$

Mathematical Practices

Make Sense of Problems and Persevere in Solving Them

MP1 Mathematically proficient students plan a solution pathway rather than simply jumping into a solution attempt.

Combinations

Combinations of n Objects Taken r at a Time

A **combination** is a selection of several objects from a larger group of objects, where (unlike permutations) the order of selecting the objects does not matter. A **combination of n objects taken r at a time** is a combination containing only r members of a set. The number of combinations of n objects taken r at a time is

$$C(n, r) = \frac{n!}{r!(n-r)!}.$$

Example There are three combinations of 3 different colored marbles taken 2 at a time.

$$C(3, 2) = \frac{3!}{2!(3-2)!} = \frac{3!}{2! \cdot 1!} = \frac{6}{2} = 3$$

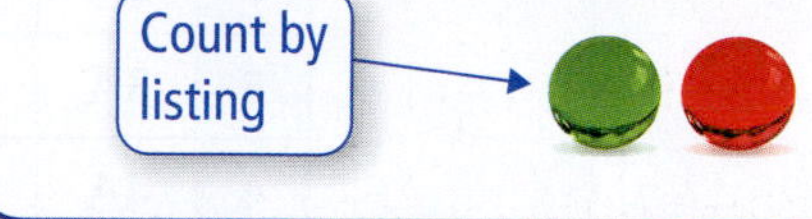

EXAMPLE 6 Using Combinations in Real Life

Your instructor randomly chooses four students from your class of 20 students to work on a project. Your three best friends are in the class.

a. What is the probability that you and your three friends are chosen for the project?

b. How likely is it that you and your three friends are chosen for the project?

Classroom Tip

Notice in Example 6 that as n gets larger, $n!$ gets extremely large. For instance,

$20! = 2,432,902,008,176,640,000.$

For this reason, it is essential that you use the cancellation technique shown when working with factorials of large numbers.

SOLUTION

a. Begin by finding the number of combinations of 20 students taken 4 at a time.

$$C(20, 4) = \frac{20!}{4!(20-4)!}$$

$$= \frac{20!}{4! \cdot 16!}$$

$$= \frac{20 \cdot 19 \cdot 18 \cdot 17 \cdot \cancel{16!}^{1}}{(4 \cdot 3 \cdot 2 \cdot 1)(\cancel{16!})_{1}}$$

Note that 20! can be written as 20 • 19 • 18 • 17 • 16!.

$$= \frac{\overset{5}{\cancel{20}} \cdot 19 \cdot \overset{3}{\cancel{18}} \cdot 17}{\underset{1}{\cancel{4}} \cdot \underset{1}{\cancel{3}} \cdot \underset{1}{\cancel{2}} \cdot 1}$$

$$= 4845$$

Because only one of the groups of 4 students consists of you and your three friends, the probability is

$$\text{Theoretical probability} = \frac{\text{Number of favorable outcomes}}{\text{Number of possible outcomes}} = \frac{1}{4845}.$$

b. A probability of 1 out of 4845 is very unlikely.

16.2 Exercises

Free step-by-step solutions to odd-numbered exercises at *MathematicalPractices.com*.

1. **Writing a Solution Key** Write a solution key for the activity on page 624. Describe a rubric for grading a student's work.

2. **Grading the Activity** In the activity on page 624, a student gave the answers below. Each question is worth 10 points. Assign a grade to each answer. For those that are incorrect, why do you think the student erred?

Sample Student Work

> **1.** How many outcomes are possible? 6
>
> List them. H, T, H, T, H, T
>
> **2.** Probability of 2 heads = $2/6 = 1/3$
>
> Probability of 1 head and 1 tail =
>
> $1/6 + 1/6 = 1/3$
>
> Probability of 2 tails = $2/6 = 1/3$

3. **Finding Outcomes Using a Tree Diagram** Complete the tree diagram for rolling a number cube and flipping a coin. How many outcomes are there?

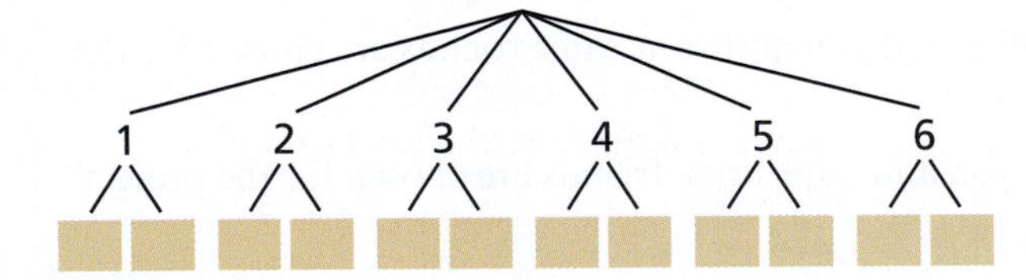

4. **Finding Outcomes Using a Tree Diagram** Complete the tree diagram for rolling a number cube and choosing a red, green, or blue marble. How many outcomes are there?

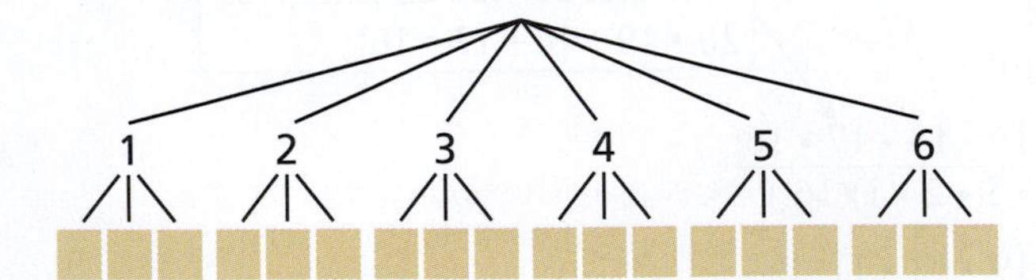

5. **Finding Probability Using a Tree Diagram** You flip a coin and roll a number cube. What is the probability that you flip a tail and roll a four?

6. **Finding Probability Using a Tree Diagram** You flip a coin and roll a number cube. What is the probability that you flip a tail and roll an even number?

7. **Finding Probability Using a Tree Diagram** You flip two coins and roll a number cube. What is the probability that you flip two heads and roll a 6?

8. **Finding Probability Using a Tree Diagram** A bag contains 1 yellow, 1 green, 1 red, and 1 blue marble. You randomly draw one marble. Without replacing the marble, you draw another marble, and then another. What is the probability that the marbles drawn (in any order) are the red, blue, and yellow marbles?

9. **Finding Probability Using a Table** The table shows the outcomes of rolling a number cube and spinning a spinner with three sections: A, B, and C.

	A	B	C
1	1A	1B	1C
2	2A	2B	2C
3	3A	3B	3C
4	4A	4B	4C
5	5A	5B	5C
6	6A	6B	6C

a. What is the total number of outcomes?

b. What is the probability of rolling a 6 and spinning A?

c. Draw a tree diagram to show the possible outcomes.

10. **Finding Probability Using a Table** A bag contains 1 red, 1 blue, 1 green, 1 orange, and 1 purple marble. The table shows the outcomes of drawing a marble from the bag and flipping a coin.

	Heads	Tails
Red	RH	RT
Blue	BH	BT
Green	GH	GT
Orange	OH	OT
Purple	PH	PT

a. What is the total number of outcomes?

b. What is the probability of drawing a red marble and flipping a head?

c. Draw a tree diagram to show the possible outcomes.

11. Finding Probability Using a Table A bag contains 1 red marble, 1 green marble, and 1 blue marble.

a. Create a table to show the outcomes of drawing a marble from the bag and spinning the spinner shown.

b. How many outcomes are there?

c. What is the probability of randomly drawing the blue marble and spinning red?

12. Finding Probability Using a Table

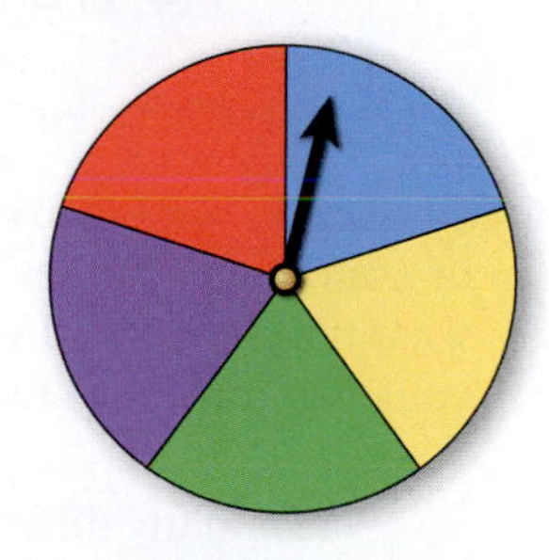

a. Create a table to show the outcomes of rolling the number cube and spinning the spinner.

b. How many outcomes are there?

c. What is the probability of rolling a 5 and spinning green?

13. Fundamental Counting Principle How many different two-digit numbers can be formed when each digit is one of the numbers 2, 4, 6, or 8?

14. Fundamental Counting Principle How many different pairs of letters of the alphabet can be formed?

15. Fundamental Counting Principle How many three-digit numbers can be formed under each condition?

a. The last digit cannot be 0 or 1.

b. The leading digit cannot be an odd number.

16. Fundamental Counting Principle How many three-digit numbers can be formed under each condition?

a. None of the digits can be 0.

b. The leading digit is one of the digits 0 through 5.

HomeStudio/Shutterstock.com

17. Finding a Probability You flip a coin four times. What is the probability of flipping heads all 4 times?

18. Finding a Probability You roll a pair of six-sided number cubes, numbered 1 through 6. What is the probability of rolling two even numbers?

19. Permutation or Combination? Determine whether each situation is a *permutation* or *combination*. Explain your reasoning.

a. Choosing four out of nine books for leisure reading

b. Displaying seven mannequins in a store window

20. Permutation or Combination? Determine whether each situation is a *permutation* or *combination*. Explain your reasoning.

a. 7 people passing through a turnstile

b. Drawing 5 cards from a standard 52-card deck

21. Finding Numbers of Permutations Find the number of permutations of each group of letters.

a. S, N, A, P

b. H, Y, P, E, R, B, O, L, A

22. Finding Numbers of Permutations Find the number of permutations of each group of letters.

a. G, E, C, K, O

b. P, Y, T, H, A, G, O, R, U, S

23. Permutations and Combinations Use the appropriate formula to calculate each permutation or combination.

a. $P(6, 3)$ b. $C(5, 3)$

c. $P(6, 6)$ d. $C(9, 2)$

24. Permutations and Combinations Use the appropriate formula to calculate each permutation or combination.

a. $P(15, 4)$ b. $C(9, 4)$

c. $P(100, 2)$ d. $C(7, 4)$

25. Finding Numbers of Combinations Write and evaluate an expression of the form $C(n, r)$ that represents each number of combinations of r marbles that can be taken from the six marbles shown.

a. $r = 1$ b. $r = 2$

c. $r = 3$ d. $r = 4$

e. $r = 5$ f. $r = 6$

26. Finding Numbers of Combinations Write and evaluate an expression of the form $C(n, r)$ that represents each number of combinations of r cubes that can be taken from the eight cubes shown.

a. $r = 1$ b. $r = 2$
c. $r = 3$ d. $r = 4$
e. $r = 5$ f. $r = 6$
g. $r = 7$ h. $r = 8$

27. Fundraising A fundraising committee selects four of its twelve members to serve as ushers at a fundraising event.

a. Is the possible number of outcomes found using permutations or combinations? Explain your reasoning.

b. How many different groups of ushers are possible?

28. Batting Order The manager of a baseball team must determine the batting order of the nine starting players.

a. Is the possible number of outcomes found using permutations or combinations? Explain your reasoning.

b. How many different batting orders are possible?

29. Sudoku Puzzle A 9 x 9 Sudoku puzzle must contain the digits 1 through 9 in each row, column, and 3 x 3 grid. How many different ways can one row be filled with these digits?

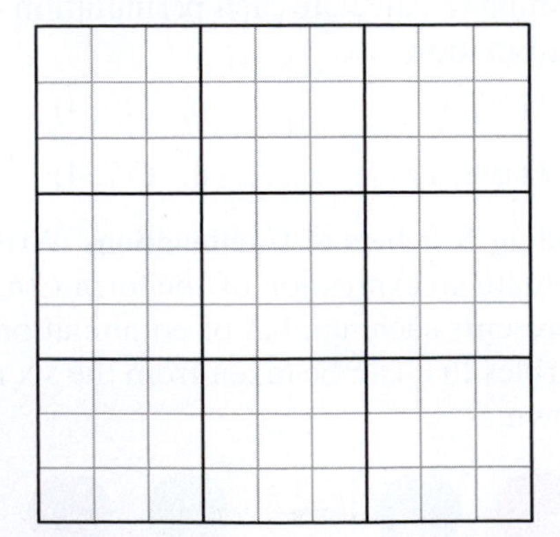

30. Transportation Your field experience instructor wants to line up 8 cars for transportation to a grade school. Between you and your classmates, there are 25 cars available. How many different groups of 8 cars are possible?

31. Teams How many different five-player starting teams for a basketball game can be formed from a team of fourteen players?

32. Combination Lock The dial of a combination lock has the numbers 0 through 39. The combination to open the lock is actually a permutation of 3 numbers. How many combinations are possible for this lock?

33. Quiz In how many ways can a 10-question true-false quiz be answered?

34. Quiz A multiple choice quiz has five questions and four answers per question. In how many ways can the quiz be answered?

35. Grading Student Work On a diagnostic test, one of your students does the following work. Explain what the student did wrong. Which topics would you encourage the student to review?

$$P(4, 2) = \frac{4!}{2!(4-2)!} = 6$$

36. Grading Student Work On a diagnostic test, one of your students does the following work. Explain what the student did wrong. Which topics would you encourage the student to review?

How many different 6-player starting squads can be formed from a volleyball team of 11 players?

$$C(11, 6) = \frac{11!}{6(11-6)!} = 55{,}440$$

37. Writing Explain how the Fundamental Counting Principle can be used to find the number of permutations of n objects.

38. Writing Is the number of permutations of m letters that can be formed from the 26 letters of the alphabet greater than the number of combinations of m letters that can be formed from the 26 letters of the alphabet? Explain.

39. License Plate Numbers All standard car license plate numbers in a state consist of three letters followed by a four-digit number. How many different standard license plate numbers can be assigned in this state?

40. License Plate Numbers All standard car license plate numbers in a state consist of two letters followed by a four-digit number. The letters I and O are not used. How many different standard license plate numbers can be assigned in this state?

41. *Writing* How is $C(n, r)$ related to $C(n, n - r)$? Explain.

42. *Writing* Explain how to use the numbers of possible outcomes of events A, B, and C to find the possible number of outcomes of event A followed by event B followed by event C. What counting technique does this involve?

43. **Counting Techniques: In Your Classroom** A diner advertises that it offers more than 10,000 different dinners. Each dinner includes choices of 8 drinks, 12 entrees, 10 sides, and a varying number of desserts.

a. Explain how to explore with your students the least number of desserts the diner must have for the advertisement to be true. Then find the number.

b. The diner begins to advertise that they offer more than 50,000 different dinners after adding a choice of dinner breads to each dinner. How many dinner bread choices do they offer? Explain.

44. **Probability: In Your Classroom** The prize winners for a lottery are determined by drawing 5 balls from a drum of 53 white numbered balls and 1 ball from a drum of 42 black numbered balls. A lottery ticket that matches the numbers of all 6 balls wins the jackpot. A ticket that matches only the number of the black ball wins \$3.

a. Explain how to show your students the use of counting techniques to find the probability of winning the jackpot. Then find the probability.

b. The lottery organizers state that the approximate probability of matching only the black ball is $\frac{1}{70}$. One of your students says it should be $\frac{1}{42}$. Explain what your student is missing. (Keep in mind what has to happen to match *only* the black ball.) Then show how to use the Fundamental Counting Principle to find the probability.

45. **Grab Bags** A teacher has 10 grab bags. Seven of the grab bags contain toys. The other three grab bags contain snacks. The teacher invites three students to each randomly select a grab bag as a reward for good behavior.

a. What is the probability that all three grab bags contain snacks?

b. What is the probability that all three grab bags contain toys?

c. What is the probability that at least one of the grab bags contains a toy?

46. **Defective Units** A batch of 16 stereos coming off the production line contains 4 stereos with defects that cannot be detected visually. An inspector randomly checks 4 stereos from the entire batch.

a. What is the probability that all four stereos have defects?

b. What is the probability that none of the four stereos has a defect?

c. What is the probability that at least one of the stereos is without defects?

47. **Probability** You roll the two number cubes shown. Use the table in Example 2 to find each answer.

a. What is the probability that the result of rolling the two number cubes is a sum of either 7 or 10?

b. Describe three different events for rolling the two number cubes that have probabilities of $\frac{1}{2}$.

48. **Telephone Numbers** Telephone numbers within the same area code in North America consist of seven-digit numbers of the form XXX-XXXX, subject to the following restrictions.

- The first digit may only begin with 2–9.
- The second and third digits cannot both be 1.
- 555-0100 through 555-0199 cannot be used because they are reserved for fictional numbers.

Considering these restrictions, how many telephone numbers are possible within the same area code?

49. **Roads** The triangles represent the relative locations of seven towns.

a. How can you use combinations to find the number of roads needed to connect each town directly to every other town? Explain.

b. How many roads are needed?

c. The roads are built in random order. What is the probability that the roads connecting the towns shown by the red triangles are the first roads built?

Activity: Comparing Independent and Dependent Events

Materials:

- Pencil
- Bag
- Marbles

Learning Objective:

Understand how an experiment where you draw marbles from a bag can yield different results when the marbles are *replaced* or *not replaced* in the bag.

Name ______________________________

Put two green marbles and one purple marble in a bag. Draw one marble. Then replace the marble and draw a second marble.

1. Complete the tree diagram. Let G = green and P = purple. What is the probability that both marbles are green? ____

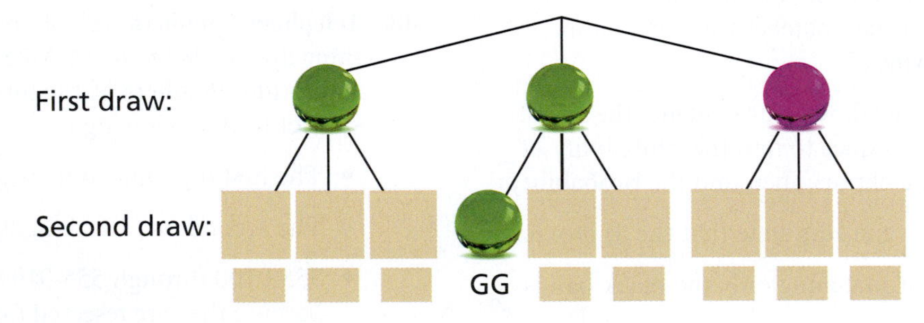

2. Use the tree diagram to determine whether the probability of drawing a green marble on the second draw *depends* on the color of the first marble drawn. Explain.

Use the same bag with two green marbles and one purple marble. Draw one marble. Do not replace the marble. Then draw a second marble.

3. Complete the tree diagram. What is the probability that both marbles are green? ____

4. Use the tree diagram to determine whether the probability of drawing a green marble on the second draw *depends* on the color of the first marble drawn. Explain.

16.3 Independent and Dependent Events

Standards

Grades 6–8

Statistics and Probability

Students should investigate chance processes and develop, use, and evaluate probability models.

- Determine whether two events are independent or dependent.
- Find the probability of two independent events.
- Find the probability of two dependent events.

Independent and Dependent Events

A **compound event** consists of two or more events. Compound events may be *independent events* or *dependent events.*

Definition of Independent and Dependent Events

Events are **independent events** when the occurrence of one event *does not* affect the likelihood that the other event(s) will occur. Events are **dependent events** when the occurrence of one event *does* affect the likelihood that the other event(s) will occur.

EXAMPLE 1 **Identifying Independent and Dependent Events**

a. You flip heads on a penny. You then flip tails on a quarter. Are these events independent or dependent? What is the probability of this event?

b. You randomly choose one of 10 students to lead a group. Then you randomly choose another student to lead another group. Are these events independent or dependent? What is the probability of randomly choosing a girl to lead the first group?

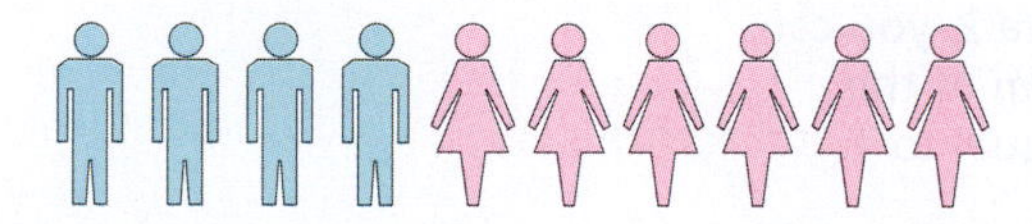

SOLUTION

a. The outcome of flipping the penny does not affect the outcome of flipping the quarter. These events are independent. Use a tree diagram to find all possible outcomes. The probability of a compound event is the ratio of the number of favorable outcomes to the number of possible outcomes.

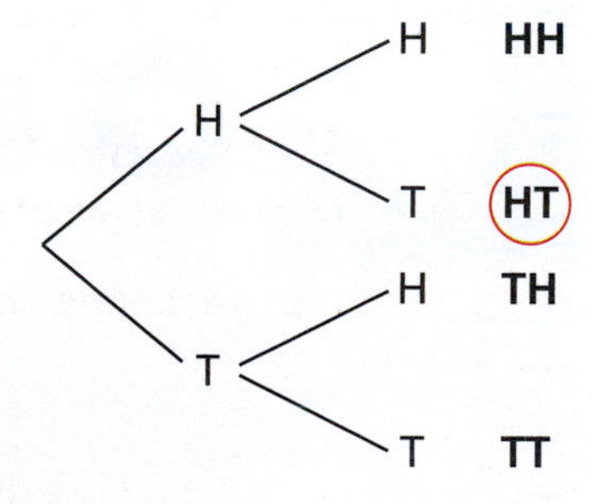

Of the four possible outcomes, one is favorable. So, the probability of flipping heads on a penny and tails on a quarter is

$$\text{Theoretical probability} = \frac{\text{Number of favorable outcomes}}{\text{Number of possible outcomes}} = \frac{1}{4}.$$

b. You cannot choose the same student to lead both groups. There are fewer students to choose from to lead the second group. These events are dependent.

Of the ten possible outcomes, six are favorable. So, the probability of randomly choosing a girl to lead the first group is

$$\text{Theoretical probability} = \frac{\text{Number of favorable outcomes}}{\text{Number of possible outcomes}} = \frac{6}{10} = \frac{3}{5}.$$

Classroom Tip

In Example 1(b), notice that you can determine the probability that the first choice is a girl. However, without knowing what the first choice is, you cannot determine the probability of choosing another girl or a boy to lead the second group. In other words, the probability of the second outcome depends on the first outcome. You will look at this example further on page 639.

Finding the Probability of Independent Events

Classroom Tip

The probability of an event can be written as

P(event).

The Probability of Independent Events

Words The probability of two or more independent events is the product of the probabilities of the events.

Algebra $P(A \text{ and } B) = P(A) \cdot P(B)$

Probability of both events — $P(A \text{ and } B)$; Probability of first event — $P(A)$; Probability of second event — $P(B)$

EXAMPLE 2 Finding the Probability of Independent Events

You are throwing 3 darts at the dartboard shown. The darts are equally likely to hit any point on the dartboard. What is the probability that your first dart hits the blue region, your second dart hits the green region, and your third dart hits the red region?

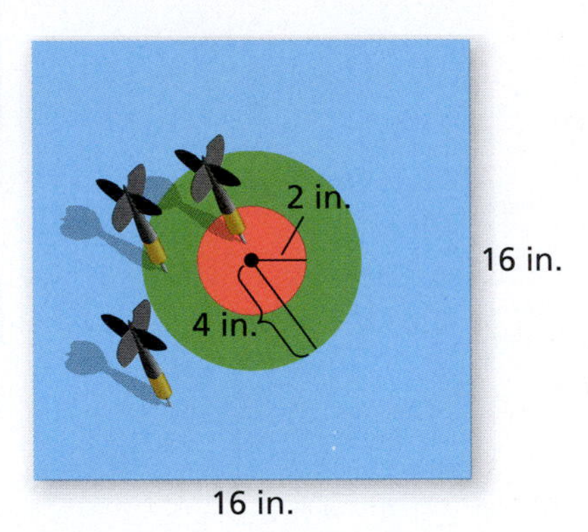

Classroom Tip

The dartboard problem at the right is an example of geometric probability. In geometric probability, probabilities are found using the ratios of favorable outcome regions to the area of the entire region. In Example 2, you can check that the sum of the probabilities is equal to 1.

$0.80 + 0.15 + 0.05 = 1.00$ ✓

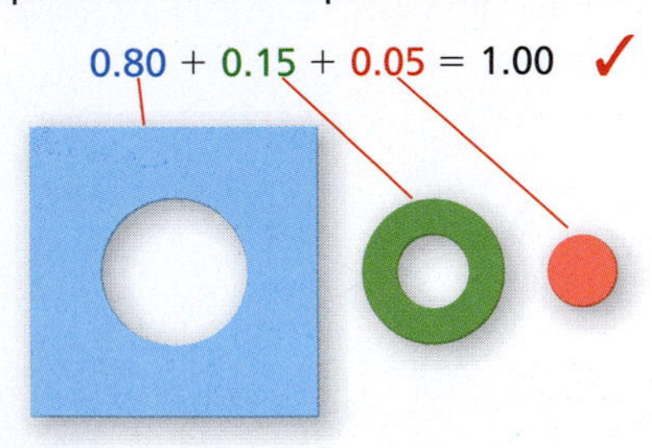

SOLUTION

The probability of hitting a particular region is given by the ratio of its area to the area of the entire dartboard.

$$P(\text{hitting blue region}) = \frac{\text{Area of blue region}}{\text{Area of dartboard}} = \frac{16^2 - \pi(4)^2}{16^2}$$

$$\approx \frac{256 - (3.14)(16)}{256}$$

$$\approx 0.80$$

$$P(\text{hitting green region}) = \frac{\text{Area of green region}}{\text{Area of dartboard}} = \frac{\pi(4)^2 - \pi(2)^2}{16^2}$$

$$= \frac{16\pi - 4\pi}{16^2}$$

$$\approx \frac{12(3.14)}{256}$$

$$\approx 0.15$$

$$P(\text{hitting red region}) = \frac{\text{Area of red region}}{\text{Area of dartboard}} = \frac{\pi(2)^2}{16^2}$$

$$\approx \frac{3.14(4)}{256}$$

$$\approx 0.05$$

Because these events are independent, the probability is given by the product of the three probabilities.

$$P(\text{blue, green, red}) \approx 0.80 \cdot 0.15 \cdot 0.05 = 0.006$$

So, the probability is about 0.006, or 0.6%. In other words, you can expect this compound event to occur about 6 times out of every 1000 throws (of the 3 darts).

In Example 4 on page 618, you used a *simulation*. Simulations can be programmed to run on a computer. The advantage of this is that you can perform experiments thousands, or even millions, of times to find the experimental probabilities of events.

EXAMPLE 3 Using a Simulation

You run a simulator for the dartboard problem described in Example 2. (See *MathematicalPractices.com* to run the simulator.) When you set the number of throws to 10,000, you obtain the results shown below.

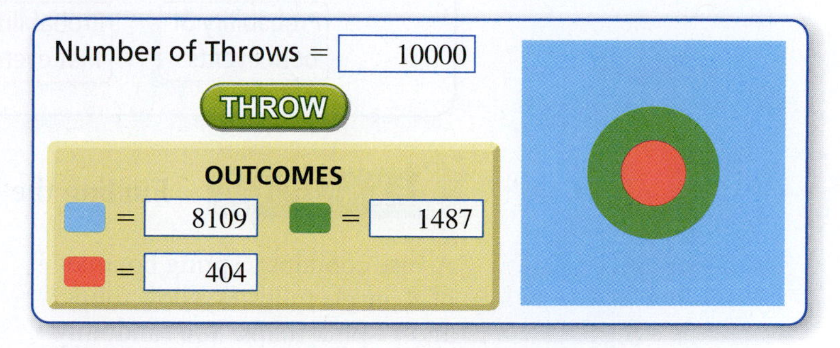

a. Find the experimental probability of a dart hitting each region.

b. Find the experimental probability that your first dart hits the blue region, your second dart hits the green region, and your third dart hits the red region.

Mathematical Practices

Use Appropriate Tools Strategically

MP5 Mathematically proficient students use technological tools to explore and deepen their understanding for concepts.

Classroom Tip

Remember that when comparing experimental and theoretical probabilities, do not expect to obtain results that are exactly the same. You should, however, obtain results that are similar, as shown in Examples 2 and 3.

SOLUTION

a. $$P(\text{hitting blue region}) = \frac{\text{Number of hits in blue region}}{\text{Total number of throws}} = \frac{8109}{10{,}000} \approx 0.81$$

$$P(\text{hitting green region}) = \frac{\text{Number of hits in green region}}{\text{Total number of throws}} = \frac{1487}{10{,}000} \approx 0.15$$

$$P(\text{hitting red region}) = \frac{\text{Number of hits in red region}}{\text{Total number of throws}} = \frac{404}{10{,}000} \approx 0.04$$

b. Using the results above, the experimental probability is given by the product of the three probabilities.

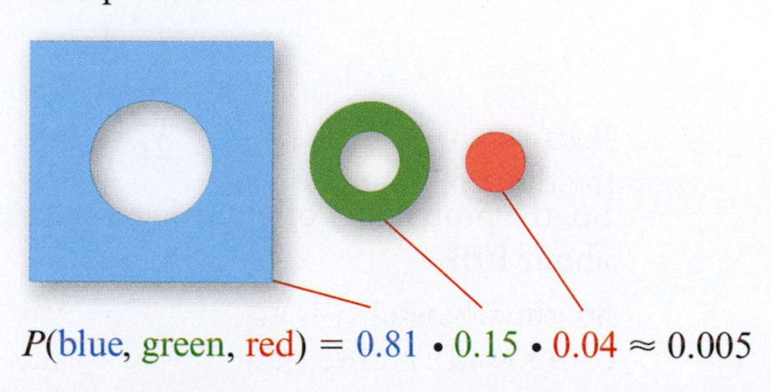

$$P(\text{blue, green, red}) = 0.81 \cdot 0.15 \cdot 0.04 \approx 0.005$$

Finding the Probability of Dependent Events

The Probability of Dependent Events

Words The probability of two dependent events A and B is the probability of A times the probability of B after A occurs.

Algebra $P(A \text{ and } B) = P(A) \cdot P(B \text{ after } A)$

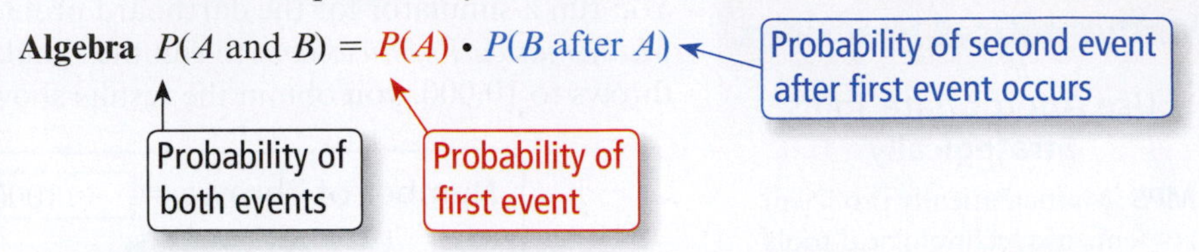

EXAMPLE 4 Finding the Probability of Dependent Events

A vase contains a spring bouquet of 8 purple tulips, 9 yellow tulips, and 11 pink tulips. You randomly choose a tulip from the vase. Then your friend randomly chooses another tulip from the vase. What is the probability that you choose a purple tulip and your friend chooses a yellow tulip?

SOLUTION

Choosing the first tulip changes the number of tulips left in the vase. So, the two events are *dependent*.

$$P(\text{purple}) = \frac{8}{28} = \frac{2}{7}$$

There are 8 purple tulips. There is a total of 28 tulips.

$$P(\text{yellow after purple}) = \frac{9}{27} = \frac{1}{3}$$

There are 9 yellow tulips. There are 27 tulips left.

Use the formula to find the probability.

$$P(A \text{ and } B) = P(A) \cdot P(B \text{ after } A)$$

$$P(\text{purple and yellow}) = P(\text{purple}) \cdot P(\text{yellow after purple})$$

$$= \frac{2}{7} \cdot \frac{1}{3}$$

$$= \frac{2}{21}$$

So, the probability of choosing a purple tulip and then a yellow tulip is $\frac{2}{21}$, or about 10%.

EXAMPLE 5 Finding the Probability of Dependent Events

In Example 1 on page 635, you looked at the following problem.

> You randomly choose one of 10 students to lead a group. Then you randomly choose another student to lead another group.

What is the probability that you choose each of the following?

a. 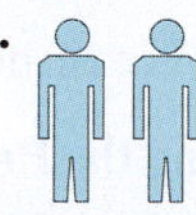

b. 

c. 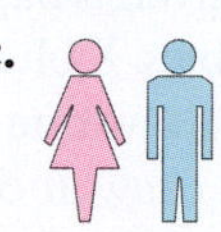

d.

SOLUTION

a. $P(\text{first is boy}) = \dfrac{4}{10} = \dfrac{2}{5}$ $\quad P(\text{second is boy}) = \dfrac{3}{9} = \dfrac{1}{3}$

The probability of choosing a boy to lead each group is

$$P(\text{boy and boy}) = \frac{2}{5} \cdot \frac{1}{3} = \frac{2}{15}.$$

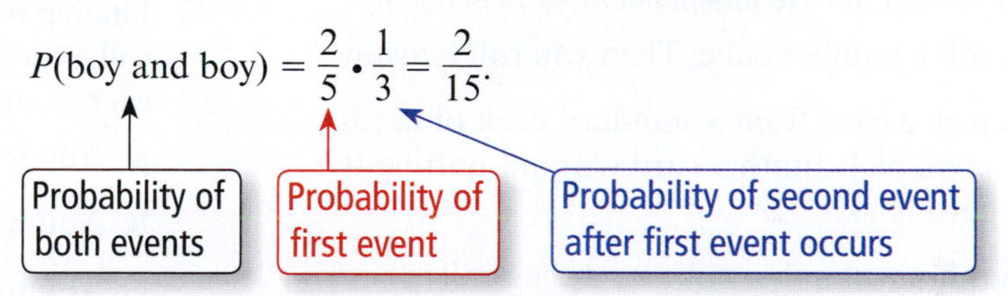

b. $P(\text{first is boy}) = \dfrac{4}{10} = \dfrac{2}{5}$ $\quad P(\text{second is girl}) = \dfrac{6}{9} = \dfrac{2}{3}$

The probability of choosing a boy first and a girl second is

$$P(\text{boy and girl}) = \frac{2}{5} \cdot \frac{2}{3} = \frac{4}{15}.$$

c. $P(\text{first is girl}) = \dfrac{6}{10} = \dfrac{3}{5}$ $\quad P(\text{second is boy}) = \dfrac{4}{9}$

The probability of choosing a girl first and a boy second is

$$P(\text{girl and boy}) = \frac{\overset{1}{\cancel{3}}}{5} \cdot \frac{4}{\underset{3}{\cancel{9}}} = \frac{4}{15}.$$

d. $P(\text{first is girl}) = \dfrac{6}{10} = \dfrac{3}{5}$ $\quad P(\text{second is girl}) = \dfrac{5}{9}$

The probability of choosing a girl first and a girl second is

$$P(\text{girl and girl}) = \frac{\overset{1}{\cancel{3}}}{\underset{1}{\cancel{5}}} \cdot \frac{\overset{1}{\cancel{5}}}{\underset{3}{\cancel{9}}} = \frac{1}{3}.$$

Notice that parts (a)–(d) represent all of the possible outcomes. In cases like this, you can check your results by checking that the four probabilities add up to 1.

$$\frac{2}{15} + \frac{4}{15} + \frac{4}{15} + \frac{1}{3} = \frac{2}{15} + \frac{4}{15} + \frac{4}{15} + \frac{5}{15} = \frac{15}{15} = 1 \checkmark$$

16.3 Exercises

Free step-by-step solutions to odd-numbered exercises at MathematicalPractices.com.

1. **Writing a Solution Key** Write a solution key for the activity on page 634. Describe a rubric for grading a student's work.

2. **Grading the Activity** In the activity on page 634, a student gave the answers below. Each question is worth 10 points. Assign a grade to each answer. For those that are incorrect, why do you think the student erred?

Sample Student Work

3\.

G	P	G		G	P
GG	GP	GG	GP	GP	PP

The probability that both marbles are green is $2/6 = 1/3$.

4\. It doesn't depend on the first draw because you're more likely to draw a green marble no matter what because there are 2 green marbles and only 1 purple.

3. **Identifying Independent and Dependent Events** State whether the events are *independent* or *dependent*.
   - **a.** You roll a number cube. Then you roll it again.
   - **b.** You pick a card from a standard deck of cards. Then you pick another card without putting the first card back.
   - **c.** You roll one number cube, and then roll a second number cube.
   - **d.** You choose one crayon from a box of 8 crayons. Without putting the first crayon back, you choose another crayon.

4. **Identifying Independent and Dependent Events** State whether the events are *independent* or *dependent*.
   - **a.** You roll an icosahedron. Then you roll it again.
   - **b.** You pick a card from a standard deck of cards. Then you put that card back in the deck and pick another card.
   - **c.** You pick a marble from a bag of 4 green and 6 red marbles. Without putting the first marble back, you pick a second marble.
   - **d.** You and two friends flip 3 coins at the same time.

5. **Finding the Probability of Independent Events** You roll a number cube. Then you roll it again. Find each probability.
   - **a.** P(even and then odd)
   - **b.** P(one and then two)
   - **c.** P(six and then even)
   - **d.** P(four and then prime)

iStockphoto.com/sweetym

6. **Finding the Probability of Independent Events** You draw a card from a standard deck of 52 cards, replace it, and draw another card. Find each probability.
   - **a.** P(jack and then black)
   - **b.** P(red and then diamond)
   - **c.** P(four and then face card)
   - **d.** P(ace of hearts and then two)

7. **Finding the Probability of Independent Events** You are given a bag with 6 red marbles, 12 green marbles, 8 blue marbles, and 10 yellow marbles. You draw one marble, put it back in the bag, and then draw another marble. Find the probability of each event. Describe the likelihood of each event.
   - **a.** You draw 2 green marbles.
   - **b.** You draw a red marble and then a blue marble.
   - **c.** You draw 2 marbles that are not red.

8. **Finding the Probability of Independent Events** You roll a number cube 3 times. Find the probability of each event. Describe the likelihood of each event.
   - **a.** You roll a 1, then a 2, and then a 3.
   - **b.** You roll three odd numbers.
   - **c.** You roll three numbers greater than 1.

9. **Finding the Probability of Independent Events** You flip 2 coins. What is the probability that you flip tails on both coins?

10. **Finding the Probability of Independent Events** You cannot remember your 3-digit bike-lock combination. What is the probability that you guess the combination correctly?

11. **Finding the Probability of Independent Events** When a baby is born, the probability of having a girl is one-half and the probability of having a boy is one-half. What is the probability that a family will have 2 boys and then 2 girls?

12. **Finding the Probability of Independent Events** You are taking a multiple-choice quiz in which each question has the choice of *a*, *b*, *c*, or *d* as the answer.
    - **a.** The quiz has one question. What is the probability that you will guess the correct answer?
    - **b.** The quiz has five questions. What is the probability that you will guess the correct answers to all five questions?

Tatiana Popova/Shutterstock.com

13. **Geometric Probability** You throw three pennies into the pool shown. The pennies are equally likely to land in any section of the pool. What is the probability that the first penny lands in the deep section, the second penny lands in the medium section, and the third penny lands in the shallow section?

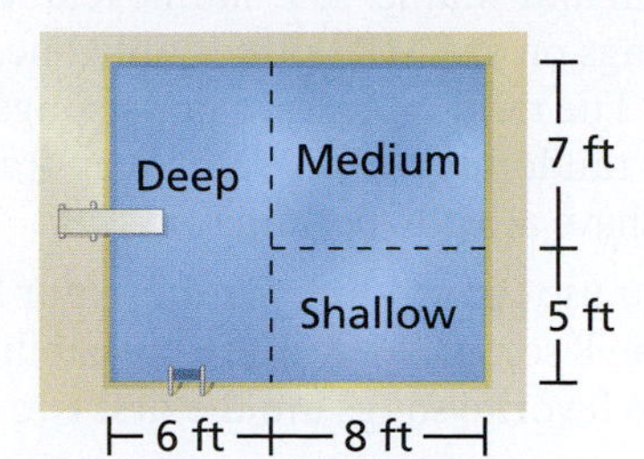

14. **Geometric Probability** You throw four softballs at the strike-zone target shown. The softballs are equally likely to hit any point of the strike-zone target. What is the probability that the first softball hits zone 1, the second softball hits zone 2, the third softball hits zone 3, and the fourth softball hits zone 4?

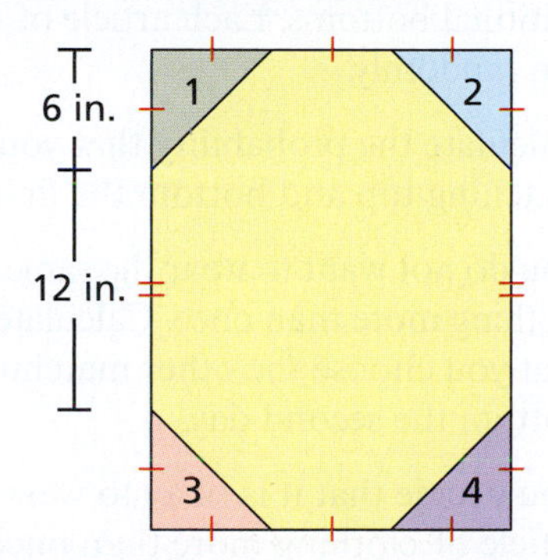

15. **Experimental Probability** Flip 3 coins a total of 48 times and record the results. Find the experimental probability for each event.
    - **a.** Flipping 2 heads and 1 tail
    - **b.** Flipping 2 tails and 1 head
    - **c.** Flipping 3 heads

16. **Experimental Probability** Roll two number cubes a total of 36 times and record the results in a table. Find the experimental probability for each event.
    - **a.** Rolling a 2 and a 5
    - **b.** Rolling two 4s
    - **c.** Rolling two odd numbers

17. **Finding the Probability of Dependent Events** You randomly choose a crayon from the box, keep it, and then randomly choose another crayon. Find each probability.
    - **a.** *P*(yellow and then orange)
    - **b.** *P*(green and then purple)
    - **c.** *P*(blue and then green)
    - **d.** *P*(two blue)

18. **Finding the Probability of Dependent Events** Each letter in the word INDEPENDENT is written on a separate piece of paper and placed in a hat. You randomly draw one piece of paper, keep it, and then randomly draw another piece of paper. Find each probability.
    - **a.** *P*(I and then D)
    - **b.** *P*(E and then E)
    - **c.** *P*(D and then E)
    - **d.** *P*(N and then T)

19. **Finding the Probability of Dependent Events** The table shows the socks that are in your sock drawer. You randomly choose one sock, put it on your left foot, then choose another. Find the probability of each compound event. Then, describe the likelihood.

Socks in a drawer	
White	
Black	
Green	
Blue	
Yellow	

   - **a.** Choosing a black sock and then a white sock
   - **b.** Choosing a blue sock and then a yellow sock
   - **c.** Choosing a pair of green socks

20. **Finding the Probability of Dependent Events** The table shows the types of children's books on a shelf. The room is dark. You randomly choose three books to read, one at a time, without replacement. Find the probability of each compound event. Then, describe the likelihood.

Topic	Quantity	Topic	Quantity
Manners	2	Insects	3
Sharing	4	Birds	3
Farm animals	6	Rainbows	2

   - **a.** Choosing a book about manners and then two books about sharing
   - **b.** Choosing a book about insects, then a book about birds, and then a book about rainbows
   - **c.** Choosing three books about farm animals

21. Finding the Probability of Dependent Events There are 2 five-dollar bills and 6 one-dollar bills in your wallet. You randomly select one bill, and then select another from your wallet. What is the probability that you select a five-dollar bill and then a one-dollar bill?

22. Finding the Probability of Dependent Events There are 10 girls and 9 boys in your class. You randomly assign each student a seat in a row. What is the probability that you assign two boys to the first two seats in the row?

23. Finding the Probability of Dependent Events The table shows tickets in an envelope to different concerts. You randomly select two tickets, without replacement. What is the probability that you select two rock concert tickets?

Rock	Pop	Country	Hip-hop
6	8	6	4

24. Finding the Probability of Dependent Events A restaurant is handing out coupons to customers at random. Coupons include 10% off entire bill, 25% off entire bill, and buy one entree, get one free. There are 50 of each type of coupon. You and your friend are the first two customers to receive coupons. What is the probability that you get a 10% off coupon and then your friend gets a 25% off coupon?

25. Charity Raffle The poster shows the information for a local charity raffle. A husband and wife decide to each enter the raffle. The total number of entries is 500. What is the probability that the husband and wife win both the grand prize and the runner-up prize?

26. Batteries A company has determined that 1% of the batteries it sells are defective. You recently purchased 2 batteries from the company. What is the probability that both batteries are defective?

27. Activity: In Your Classroom You randomly arrange your students into a straight line. Which concepts could you teach your students with this activity? Explain.

28. Activity: In Your Classroom You have students write their names on pieces of paper and place them in a bag for a prize drawing. Which concepts could you teach your students with this activity? Explain.

29. Music An MP3 player has two similar features, random and shuffle. The shuffle feature plays all the songs on the MP3 player only once in random order. The random feature plays songs on the MP3 player randomly but does not keep track of which songs have already been played.

a. You use the random feature. Your MP3 player has 98 songs. What is the probability that your two favorite songs are the first two songs played?

b. You use the shuffle feature. Your MP3 player has 350 songs. What is the probability that the newest song and the oldest song are the first two songs played?

30. Outfits Your wardrobe consists of 2 pairs of matching tops and bottoms, 7 additional tops, and 4 additional bottoms. Each article of clothing is chosen randomly.

a. Calculate the probability that you choose a matching top and bottom the first day.

b. You do not want to wear the same article of clothing more than once. Calculate the probability that you choose the other matching top and bottom the second day.

c. You decide that it is okay to wear the same article of clothing more than once. Calculate the probability that you choose the other matching top and bottom the second day.

31. Sports Your friend is the starting first baseman for the school baseball team. He has a batting average of .350. (In other words, he gets a hit 35% of the time he is up to bat.)

a. Your friend usually gets up to bat 3 times each game. Is each at bat independent of the previous at bat?

b. Calculate the probability that your friend will get a hit in his next 2 at bats.

c. Calculate the probability that your friend will get a hit in his next 3 at bats.

Karkas/Shutterstock.com

32. **Pin Number** Your bank assigns a 4-digit PIN to your debit card. You wonder if a random number generator application on your smartphone can guess your PIN.

 a. What is the probability that the application generates the first digit correctly?

 b. What is the probability that the application generates the first two digits correctly?

 c. What is the probability that the application generates the correct PIN?

33. **Grading Student Work** On a diagnostic test, one of your students does the following work. Explain what the student did wrong. Which topics would you encourage the student to review?

> A bag contains 5 red marbles, 6 blue marbles, 7 green marbles, and 8 orange marbles. You choose three marbles, one at a time, and replace each marble after it is drawn. Find the probability that you choose 1 orange, then 1 green, and then 1 blue marble.
>
> $$P(\text{orange, green, blue}) = \frac{8}{26} \cdot \frac{7}{25} \cdot \frac{6}{24}$$
> $$= \frac{4}{13} \cdot \frac{7}{25} \cdot \frac{1}{4}$$
> $$= \frac{7}{13 \cdot 25}$$
> $$= \frac{7}{325}$$

34. **Grading Student Work** On a diagnostic test, one of your students does the following work. Explain what the student did wrong. Which topics would you encourage the student to review?

> You choose two cards from a standard deck of 52 cards. Find the probability of drawing a king and an ace.
>
> $$P = (\text{king and ace}) = \frac{4}{52} \cdot \frac{4}{52} = \frac{1}{13} \cdot \frac{1}{13} = \frac{1}{169}$$

35. **Writing** Describe a real-life example of 3 independent events. Explain why the events are independent.

ALMAGAMI/Shutterstock.com

36. **Writing** Describe a real-life example of 2 dependent events. Explain why the events are dependent.

37. **Reasoning** A bag contains 9 marbles. Some are red, some are blue, and some are yellow. The probability of selecting a red marble, a blue marble, and then a yellow marble without replacement is $\frac{1}{21}$. How many of each color marble are in the bag? Is there more than one correct answer? Explain your reasoning.

38. **Reasoning** A bag contains 6 sports balls. Some are basketballs, some are volleyballs, and some are soccer balls. There is a large hole in the bag. The probability of a basketball falling out of the hole, then a volleyball falling out of the hole, and then a soccer ball falling out of the hole is $\frac{1}{20}$. How many of each type of ball are in the bag? Is there more than one correct answer? Explain your reasoning.

39. **Reasoning** A bag contains a total of n marbles, k of which are blue. You draw three marbles, with replacement. Write an expression for the probability that you draw three blue marbles.

40. **Reasoning** A jar contains a total of n bouncy balls, k of which are tie-dye. You draw four bouncy balls, without replacement. Write an expression for the probability that you draw four tie-dye bouncy balls.

41. **Writing** Describe the difference between finding the probability of an independent event and a dependent event.

42. **Writing** Explain how compound events are related to simple events. (A simple event is an event in which only one experiment takes place.)

43. **Reasoning** Write an expression for the probability that a coin shows heads each time when it is flipped n times in a row.

44. **Reasoning** Write an expression for the probability that a number cube shows 4 each time when it is rolled n times in a row.

45. **Geography** You randomly order the tiles below. What is the probability that the tiles spell the name of a windy city in Illinois?

Activity: Discovering an Expected Value

Materials:

- Pencil
- Number cube

Learning Objective:

Determine whether a game is fair, over time, based on the sum of the payoffs.

Name 

You are playing a game in which you roll the number cube shown one time. The game has the following rules.

- **When you roll a 6, you win five points.**
- **When you roll a 1, 2, 3, 4, or 5, you lose one point.**

1. Play the game 30 times. Keep track of the number you roll and the points won or lost.

Game	Number	Payoff
1		
2		
3		
4		
5		
6		
7		
8		
9		
10		

Game	Number	Payoff
11		
12		
13		
14		
15		
16		
17		
18		
19		
20		

Game	Number	Payoff
21		
22		
23		
24		
25		
26		
27		
28		
29		
30		

Find the sum of your 30 payoffs.

Sum =

2. Is this a "good" game? In other words, if given the chance to play this game, would you play it? Explain your reasoning.

3. If you play this game 300 times, what would you expect the sum of your payoffs to be? Explain your reasoning.

Sum =

16.4 Expected Value

Standards

Grades 6–8
Statistics and Probability
Students should investigate chance processes and develop, use, and evaluate probability models.

- Find the expected value of an experiment.
- Use expected value to make decisions.
- Use expected value to decide whether a game is fair.

Expected Value

Definition of Expected Value

Consider an experiment that has two possible outcomes. Each outcome has a payoff (which can be positive, negative, or zero). The **expected value** of the experiment is as follows.

$$\text{Expected value} = \begin{pmatrix}\text{Probability}\\ \text{outcome 1}\end{pmatrix} \cdot \begin{pmatrix}\text{Payoff for}\\ \text{outcome 1}\end{pmatrix} + \begin{pmatrix}\text{Probability}\\ \text{outcome 2}\end{pmatrix} \cdot \begin{pmatrix}\text{Payoff for}\\ \text{outcome 2}\end{pmatrix}$$

This definition can be extended to experiments with 3 or more outcomes.

EXAMPLE 1 **Comparing Expected Values**

A child asks his parents for money. Whose offer has the greater expected value?

Father The child flips a coin. If the child flips heads, then the father will give the child \$20. If the child flips tails, then the father gives the child nothing.

Mother The child rolls a die. The mother will give the child \$3 for each dot that appears on the side that lands up.

SOLUTION

Father Each outcome has a probability of one-half.

$$\text{Expected value} = \left(\frac{1}{2}\right)(\$20) + \left(\frac{1}{2}\right)(\$0) = \$10$$

The father's offer has an expected value of \$10.

Mother There are six possible outcomes. With this many outcomes, a spreadsheet is convenient.

	A	B	C	D
1	Number	Payoff	Probability	Expected Value
2	1	\$3.00	1/6	\$0.50
3	2	\$6.00	1/6	\$1.00
4	3	\$9.00	1/6	\$1.50
5	4	\$12.00	1/6	\$2.00
6	5	\$15.00	1/6	\$2.50
7	6	\$18.00	1/6	\$3.00
8				**\$10.50**

The mother's offer has an expected value of \$10.50. Because \$10.50 > \$10.00, the mother's offer has the greater expected value.

Classroom Tip

The expected value of an experiment represents the "average payoff" when the experiment is conducted many times. For instance, if you buy a state lottery ticket for \$1 and the expected value is −\$0.50, then you should expect an average loss of −\$0.50 for each ticket you purchase. As a general rule, all commercial forms of gambling have negative expected values. Casinos and state lotteries make a profit because this is true.

Expected Value and Decision Making

The most common use of expected value is to make decisions between two or more courses of action. For instance, the child from Example 1 can use expected value to compare the offers from his father and mother, and determine that his mother's offer is slightly better.

EXAMPLE 2 Comparing Two Expected Values

Daniel Kahneman, a professor at Princeton University, became the first psychologist to win the Nobel Prize in Economic Sciences (2002). He and a colleague developed a theory, called *Prospect Theory*, to help explain people's attitudes toward risks involving gains and losses.

Kahneman and his colleague found that most people incorrectly estimate probabilities in predictable ways. As part of their work, they used the two examples below. Compare the expected values of the two options in each experiment.

Mathematical Practices

Construct Viable Arguments and Critique the Reasoning of Others

MP3 Mathematically proficient students listen to or read the arguments of others, decide whether they make sense, and ask useful questions to clarify or improve the arguments.

a. In one experiment, subjects were given two options.

(1) An 80% probability of winning $4000 and a 20% probability of not winning anything

(2) A 100% probability of winning $3000

b. In another experiment, subjects were given two options.

(1) An 80% probability of losing $4000 and a 20% probability of not losing anything

(2) A 100% probability of losing $3000

SOLUTION

a. Here are the first two options the subjects were given.

Greater expected value

	Expected value
(1) 80% chance win $4000 20% chance win $0	$(0.8)(4000) + (0.2)(0) = \$3200$
(2) 100% chance win $3000	$(1.0)(3000) = \$3000$

Preferred by 80% of subjects

b. Here are the second two options the subjects were given.

Preferred by 92% of subjects

	Expected value
(1) 80% chance lose $4000 20% chance lose $0	$(0.8)(-4000) + (0.2)(0) = -\$3200$
(2) 100% chance lose $3000	$(1.0)(-3000) = -\$3000$

Greater expected value

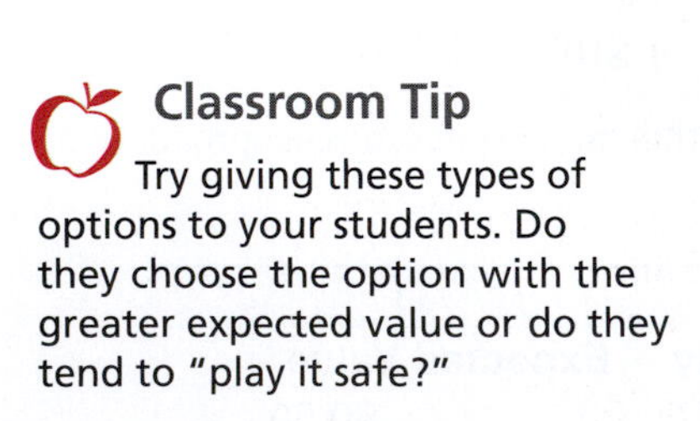

Classroom Tip

Try giving these types of options to your students. Do they choose the option with the greater expected value or do they tend to "play it safe?"

What Kahneman found surprising is that in neither case did the majority of the subjects intuitively choose the option with the greater expected value. As described in *Prospect Theory*, this occurs because individuals overweigh losses when they are described as definitive, as opposed to situations where they are described as possible. In other words, people tend to fear losses more than they value gains. A $1 loss is more painful than the pleasure of a $1 gain.

Comparing Two Expected Values

Your company is developing a smartphone. Based on reports from your development and market research departments, you have the following projections.

Smartphone Alpha

Cost of development: $2,500,000
Projected sales: 50% chance of net sales of $5,000,000
30% chance of net sales of $3,000,000
20% chance of net sales of $1,500,000

Smartphone Beta

Cost of development: $1,500,000
Projected sales: 30% chance of net sales of $4,000,000
60% chance of net sales of $2,000,000
10% chance of net sales of $500,000

Which smartphone should your company develop? Explain your reasoning.

SOLUTION

A *decision tree* can help organize your thinking. A **decision tree** is a tree diagram that includes expected values.

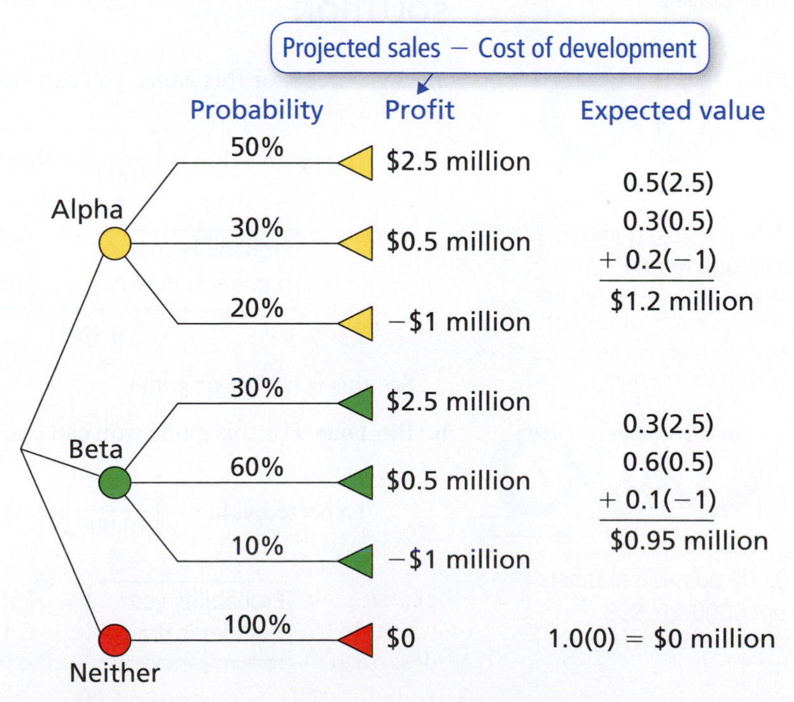

From the decision tree, you can see that the Alpha model has twice the probability of losing $1 million, but it has the greater expected value. So, using expected value, your company should develop the Alpha model smartphone.

Fair Games

Definition of Fair Game

A game is called a **fair game** when its expected value is 0.

Example The game described in the activity on page 644 where you are rolling a number cube is fair because its expected value is 0.

- When you roll a 6, you win five points.
- When you roll a 1, 2, 3, 4, or 5, you lose one point.

$$\text{Expected value} = \left(\frac{1}{6}\right)(5) + \left(\frac{5}{6}\right)(-1) = \frac{5}{6} + \left(-\frac{5}{6}\right) = 0$$

Probability of rolling a 6 — Probability of rolling a 1, 2, 3, 4, or 5

EXAMPLE 4 Identifying Fair Games

Consider the following state lottery games. Are they fair games? Explain your reasoning.

a. Big Three You choose a 3-digit number. The cost of a lottery ticket is \$1. If your number is chosen, then you win \$500. If your number is not chosen, then you lose the cost of your ticket.

b. Big Four You choose a 4-digit number. The cost of a lottery ticket is \$2. If your number is chosen, then you win \$10,000. If your number is not chosen, then you lose the cost of your ticket.

Classroom Tip

In Example 4, notice that neither of the state lottery games are fair. In fact, both are shockingly unfair. It is typical of state lottery systems for a player to lose about \$0.50 of each dollar played. This is a contrast to casinos, in which a player typically loses between \$0.10 and \$0.02 of each dollar played.

1000 possible numbers from 000 to 999.

SOLUTION

a. Big Three For this game, you can choose from 1000 possible numbers.

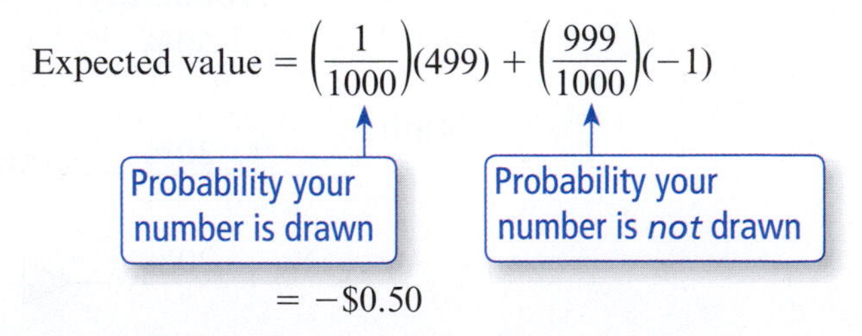

$$\text{Expected value} = \left(\frac{1}{1000}\right)(499) + \left(\frac{999}{1000}\right)(-1)$$

$$= -\$0.50$$

So, this is not a fair game.

10,000 possible numbers from 0000 to 9999.

b. Big Four For this game, you can choose from 10,000 possible numbers.

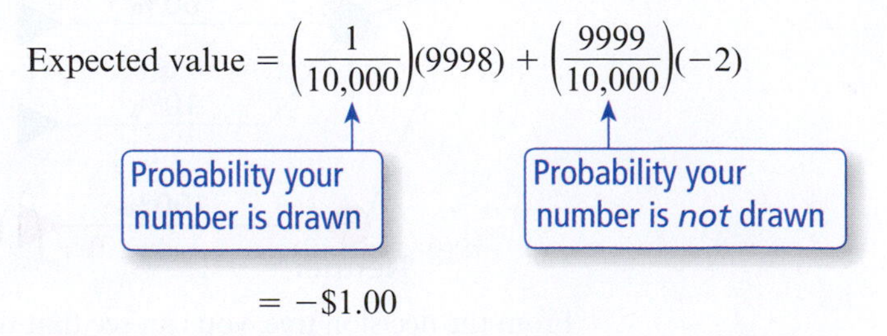

$$\text{Expected value} = \left(\frac{1}{10{,}000}\right)(9998) + \left(\frac{9999}{10{,}000}\right)(-2)$$

$$= -\$1.00$$

So, this is not a fair game.

EXAMPLE 5 Creating a Fair Game and an Unfair Game

Use the spinner at the right to create (a) a fair game and (b) an unfair game.

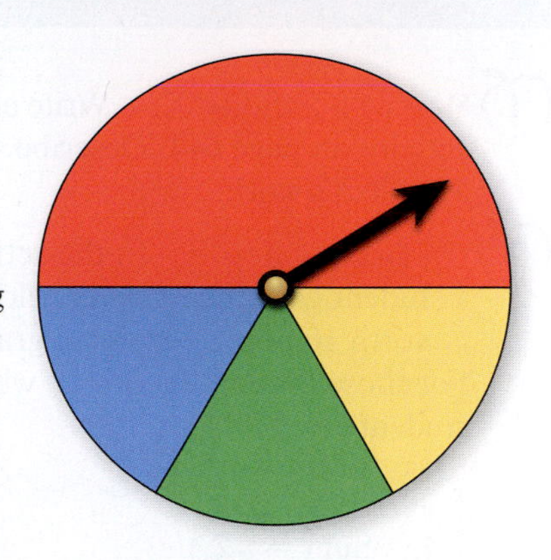

SOLUTION

In both cases, there are many possible correct answers. First, find the probability of spinning each color.

$$P(\text{spinning red}) = \frac{1}{2}$$

$$P(\text{spinning blue}) = \frac{1}{6}$$

$$P(\text{spinning green}) = \frac{1}{6}$$

$$P(\text{spinning yellow}) = \frac{1}{6}$$

a. Fair game Your goal is to create payoffs for which the expected value is 0. Here is an example.

Lose 1 point | Win 1 point | Win 1 point | Win 1 point

$$\text{Expected value} = \left(\frac{1}{2}\right)(-1) + \left(\frac{1}{6}\right)(1) + \left(\frac{1}{6}\right)(1) + \left(\frac{1}{6}\right)(1)$$

$$= 0$$

Here is a more interesting example.

Lose 5 points | Win 2 points | Win 3 points | Win 10 points

$$\text{Expected value} = \left(\frac{1}{2}\right)(-5) + \left(\frac{1}{6}\right)(2) + \left(\frac{1}{6}\right)(3) + \left(\frac{1}{6}\right)(10)$$

$$= 0$$

b. Unfair game Your goal is to create payoffs for which the expected value is not 0. Typically, unfair games have negative expected value, which means that they favor the "house." However, it is possible that an unfair game can have positive expected values, which means that they favor the player. Here is an unfair game that favors the "house."

Lose 2 points | Win 1 point | Win 2 points | Win 2 points

$$\text{Expected value} = \left(\frac{1}{2}\right)(-2) + \left(\frac{1}{6}\right)(1) + \left(\frac{1}{6}\right)(2) + \left(\frac{1}{6}\right)(2)$$

$$\approx -0.17$$

Here is an unfair game that favors the player.

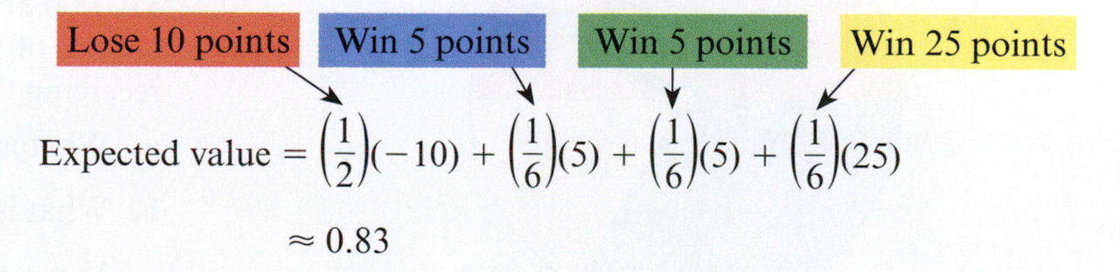

$$\text{Expected value} = \left(\frac{1}{2}\right)(-10) + \left(\frac{1}{6}\right)(5) + \left(\frac{1}{6}\right)(5) + \left(\frac{1}{6}\right)(25)$$

$$\approx 0.83$$

16.4 Exercises

Free step-by-step solutions to odd-numbered exercises at MathematicalPractices.com.

1. **Writing a Solution Key** Write a solution key for the activity on page 644. Describe a rubric for grading a student's work.

2. **Grading the Activity** In the activity on page 644, a student gave the answers below. Each question is worth 10 points. Assign a grade to each answer. For those that are incorrect, why do you think the student erred?

Sample Student Work

1. Sum = 2
2. Is this a "good" game? In other words, if given the chance to play this game, would you play it? Explain your reasoning.
   No this isn't a good game because the payoff is not high enough.
3. If you play this game 300 times, what would you expect the sum of your payoffs to be? Explain your reasoning.
   Sum = 250
   In 300 rolls, about 50 would be 6.

3. **Identifying Probability and Payoff** Identify the probability and payoff for each outcome in the experiment.
   a. You flip one coin. You earn $5 if you flip heads and lose $5 if you flip tails.
   b. You roll a number cube. You earn $3 if you roll an even number and lose $1 if you roll an odd number.

4. **Identifying Probability and Payoff** Identify the probability and payoff for each outcome in the experiment.
   a. You draw a card from a standard deck of 52 cards. You earn 26 points if it is a face card and lose 13 points if it is not a face card.
   b. You spin the spinner shown. You lose 6 points if the arrow lands on blue, green or yellow and earn 20 points if the arrow lands on red.

5. **Finding Expected Value** You draw a card from a standard deck of 52 cards. Find the expected value for each experiment.
   a. You earn 16 points if it is a diamond and lose 4 points if it is not a diamond.
   b. You earn 52 points if it is an ace and lose 13 points if it is not an ace.

6. **Finding Expected Value** You roll a standard 6-sided die. Find the expected value for each experiment.
   a. You earn 6 points for each dot on the side that lands up.
   b. You earn 6 points if it lands on a prime number and lose 6 points if it does not land on a prime number.

7. **Comparing Expected Values** Option 1: You have an 80% chance of winning $2000 and a 20% chance of winning nothing. Option 2: You have a 100% chance of winning $1000.
   a. Predict which option has a greater expected value.
   b. Calculate the expected value for each option. Was your prediction correct?

8. **Comparing Expected Values** Option 1: You have a 60% chance of losing $5000 and a 40% chance of losing nothing. Option 2: You have a 100% chance of losing $2000.
   a. Predict which option has a greater expected value.
   b. Calculate the expected value for each option. Was your prediction correct?

9. **Comparing Expected Values** Scholarship A is worth $1000 and costs $10 to apply. Scholarship B is worth $5000 and costs $25 to apply. You have a 5% chance of receiving Scholarship A and a 1% chance of receiving the Scholarship B.
   a. What is your expected value for Scholarship A?
   b. What is your expected value for Scholarship B?
   c. You can only apply for one scholarship. Which one should you choose?

10. **Comparing Expected Values** Grant A is worth $4000 and costs $50 to apply. Grant B is worth $10,000 and costs $100 to apply. You have a 15% chance of receiving Grant A and a 4% chance of receiving Grant B.
    a. What is the expected value for Grant A?
    b. What is your expected value for Grant B?
    c. You only have $125 to apply for a grant. Which grant should you choose?

11. **Using a Decision Tree** An electronics company is developing a new digital camera. Projections for two possible models are shown in the table. Use a decision tree to decide which camera they should produce.

Snapshot camera: cost of development: $3 million	
Probability	**Net sales**
40%	$6 million
40%	$5 million
20%	$1 million

Zoomax camera: cost of development: $5 million	
Probability	**Net sales**
20%	$10 million
60%	$8 million
20%	$3 million

12. **Using a Decision Tree** An electronics company is developing a new MP3 player. Projections for two possible models are shown in the table. Use a decision tree to decide which MP3 player they should produce.

Zuke MP3 player: cost of development: $3 million	
Probability	**Net sales**
10%	$9 million
55%	$5 million
20%	$4 million
15%	$1 million

Juke MP3 player: cost of development: $5 million	
Probability	**Net sales**
20%	$10 million
35%	$7 million
25%	$5 million
20%	$2 million

13. **Using a Decision Tree** An electronics company is developing a new laptop computer. Projections for two possible models are shown in the table. Use a decision tree to decide which laptop they should produce.

Stylist laptop: cost of development: $8 million	
Probability	**Net sales**
10%	$16 million
70%	$12 million
20%	$6 million

Practix laptop: cost of development: $10 million	
Probability	**Net sales**
30%	$18 million
50%	$14 million
20%	$8 million

14. **Using a Decision Tree** An electronics company is developing a new tablet. Projections for two possible models are shown in the table. Use a decision tree to decide which tablet they should produce.

Tabatha tablet: cost of development: $3 million	
Probability	**Net sales**
20%	$10 million
45%	$6 million
25%	$5 million
10%	$0.5 million

Tobias tablet: cost of development: $4 million	
Probability	**Net sales**
10%	$12 million
40%	$10 million
30%	$4 million
20%	$1 million

15. Calculating Expected Value There are 10 quarters, 10 dimes, 10 nickels, and 20 pennies in a jar. One coin is randomly dumped out of the jar. What is the expected value for this experiment?

16. Calculating Expected Value There are four $20 bills, six $10 bills, eight $5 bills, and two $1 bills in your wallet. You select one bill at random. What is the expected value for this experiment?

17. Calculating Expected Value A student on the newspaper committee recorded attendance and weather conditions at the high school football games for one season. The table shows the data collected. What is the expected attendance for a game next season?

Weather	Probability	Attendance
Cold	0.38	300
Mild	0.45	800
Warm	0.12	1250
Hot	0.05	900

18. Calculating Expected Value A music artist has released numerous CDs over the past ten years. An assistant has made a table to show the projected profits of the next CD based on previous releases. Use the table to calculate the expected income of the next CD.

Probability	Profit
10%	−$10,000
15%	−$2000
10%	0
35%	$15,000
25%	$20,000
5%	$35,000

19. Identifying Fair Games Your friend asks if you would like to play a game in which you roll a number cube. If the number cube lands on an even number, then you win 1 point. If the number cube lands on an odd number, then you lose 1 point. Is this a fair game? Explain your reasoning.

20. Identifying Fair Games Your friend asks if you would like to play a game in which you toss two coins. If the coins land on the same side, then you win 2 points. If the coins land on different sides, then you lose 1 point. Is this game fair? Explain your reasoning.

21. *Writing* An experiment has 4 possible outcomes. A student states that the expected value of the experiment is

$$\text{Expected value} = \left(\frac{1}{4}\right)(2) + \left(\frac{1}{4}\right)(-1) + \left(\frac{1}{4}\right)(1) = \frac{1}{2}.$$

Explain what error the student made and what you would tell the student to check when setting up expected value equations.

22. *Writing* An experiment has 3 possible outcomes. A student states that the expected value of the experiment is

$$\text{Expected value} = \left(\frac{1}{2}\right)(3) + \left(\frac{1}{2}\right)(-2) + \left(\frac{1}{2}\right)(1) = 1.$$

Explain what error the student made and what you would tell the student to check when setting up expected value equations.

23. Movie Production A production company is developing a new movie. The cost of development for the movie will be $2 million. The projected sales are shown in the table. Should the company produce the movie? Explain your reasoning.

Probability	Net sales
35%	$0.5 million
30%	$2 million
20%	$3 million
15%	$4 million

24. Sports Car A car designer has ideas for a new sports car. The cost of development is $45 million. The table below shows the projected sales. Should the designer develop the car? Explain your reasoning.

Probability	Net sales
15%	$30 million
20%	$40 million
30%	$50 million
20%	$55 million
15%	$70 million

25. Identifying Fair Games The state lottery offers a game called the Easy Three. You choose a 3-digit number. The cost of a ticket is $3. If your number is chosen, then you win $1000. If your number is not chosen, then you lose the cost of your ticket. What is the expected value of the game? Is this a fair game?

26. Identifying Fair Games The state lottery offers a game called Lucky Sevens. The cost of a ticket is \$10. It includes a 5-digit scratch off number. If your number contains three 7s, then you win \$1000. If your number contains four 7s, then you win \$2500. If your number contains five 7s, then you win \$10,000. If your number contains zero to two sevens, you lose the cost of your ticket. What is the expected value of the game? Is this a fair game?

27. Investment Comparison You want to invest \$1000 and have two investment options.

a. The speculative investment has a 40% chance of complete loss, 25% chance of no loss or gain, 15% chance of 100% gain, and 20% chance of 250% gain. Find the expected value.

b. The conservative investment has a 10% chance of complete loss, 35% chance of no loss or gain, 30% chance of 50% gain, and 25% chance of 100% gain. Find the expected value.

c. Do you think most investors would choose the option with the greater expected value? Explain your reasoning.

28. Investment Comparison You want to invest \$5000 and have two investment options.

a. The speculative investment has a 30% chance of complete loss, 25% chance of no loss or gain, 35% chance of 250% gain, and 10% chance of 750% gain. Find the expected value.

b. The conservative investment has a 5% chance of complete loss, 15% chance of no loss or gain, 60% chance of 30% gain, and 20% chance of 60% gain. Find the expected value.

c. Do you think most investors would choose the option with the greater expected value? Explain your reasoning.

29. Grading Student Work On a diagnostic test, one of your students does the following work. Explain what the student did wrong. Which topics would you encourage the student to review?

> The lottery game Sure Four is played by purchasing a \$2 ticket and choosing a 4-digit number. If your number is chosen, you win \$5000. If not, you lose the price of the ticket. What is the expected value for this game?
>
> $\left(\frac{1}{10{,}000}\right)(4998) \approx 0.5$
>
> The expected value is about \$0.50.

30. Grading Student Work On a diagnostic test, one of your students does the following work. Explain what the student did wrong. Which topics would you encourage the student to review?

> You play a game in which you roll two number cubes. The game has the following rules.
> - If the sum of the numbers is less than 7, then you win 1 point.
> - If the sum of the numbers is 7 or greater, then you lose 1 point.
> - Is this game fair?
>
> Yes, it is a fair game because the sum of the payoffs is 0.

31. Activity: In Your Classroom Design an unfair game that involves drawing marbles from a bag. Explain how you would demonstrate to your students that the game is unfair.

32. Discussion: In Your Classroom One of your students does not understand why the expected value of an experiment is not one of the outcomes. How could you explain this to your student?

33. Championship Playoffs At the end of the season, 4 volleyball teams are tied for first place with the same record. The athletic director designs this playoff structure to determine the league champions. Each team has the same probability for winning an individual game.

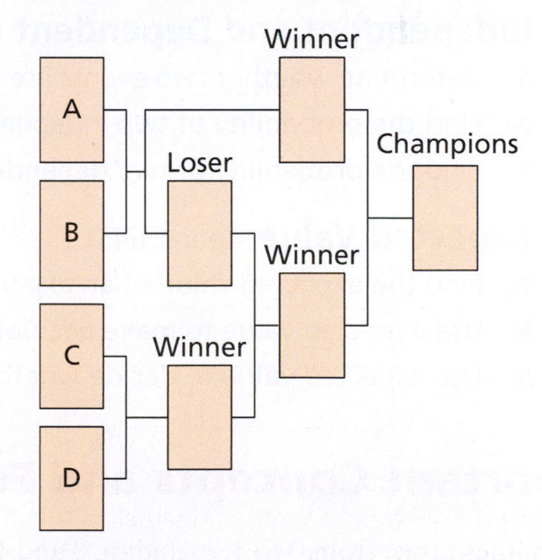

a. Would this playoff structure be considered fair? Explain your reasoning.

b. Why do you think anyone would use this structure?

c. Create an alternate playoff structure that would be considered fair.

Chapter Summary

Chapter Vocabulary

experiment (*p. 615*)
outcomes (*p. 615*)
event (*p. 615*)
favorable outcomes (*p. 615*)
probability (*p. 616*)
theoretical probability (*p. 617*)
tree diagram (*p. 625*)
permutation (*p. 627*)
permutation of n objects taken all at a time (*p. 627*)
n factorial (*p. 627*)
permutation of n objects taken r at a time (*p. 628*)
combination (*p. 629*)
combination of n objects taken r at a time (*p. 629*)
compound event (*p. 635*)
independent events (*p. 635*)
dependent events (*p. 635*)
expected value (*p. 645*)
decision tree (*p. 647*)
fair game (*p. 648*)

Chapter Learning Objectives — Review Exercises

16.1 Introduction to Probability (*page 615*) — 1–18

- Identify events and outcomes in an experiment.
- Understand that probability is a measure of the likelihood of an event.
- Compare theoretical, experimental, and subjective probabilities.

16.2 Counting Techniques (*page 625*) — 19–50

- Use tree diagrams, tables, or the Fundamental Counting Principle to find the possible outcomes of experiments.
- Use permutations to find probabilities.
- Use combinations to find probabilities.

16.3 Independent and Dependent Events (*page 635*) — 51–66

- Determine whether two events are independent or dependent.
- Find the probability of two independent events.
- Find the probability of two dependent events.

16.4 Expected Value (*page 645*) — 67–75

- Find the expected value of an experiment.
- Use expected value to make decisions.
- Use expected value to decide whether a game is fair.

Important Concepts and Formulas

Probabilities range from 0 to 1, including 0 and 1.

$$\text{Theoretical probability} = \frac{\text{Number of favorable outcomes}}{\text{Number of possible outcomes}}$$

Permutation $P(n, r) = \dfrac{n!}{(n-r)!}$

Combination $C(n, r) = \dfrac{n!}{r!(n-r)!}$

Probability of independent events

$P(A \text{ and } B) = P(A) \cdot P(B)$

Probability of dependent events

$P(A \text{ and } B) = P(A) \cdot P(B \text{ after } A)$

$$\text{Expected value} = \left(\begin{matrix}\text{Probability}\\ \text{outcome 1}\end{matrix}\right) \cdot \left(\begin{matrix}\text{Payoff for}\\ \text{outcome 1}\end{matrix}\right) + \left(\begin{matrix}\text{Probability}\\ \text{outcome 2}\end{matrix}\right) \cdot \left(\begin{matrix}\text{Payoff for}\\ \text{outcome 2}\end{matrix}\right)$$

Monkey Business Images/Shutterstock.com

Review Exercises

Free step-by-step solutions to odd-numbered exercises at MathematicalPractices.com.

16.1 Introduction to Probability

Identifying Possible Outcomes Determine the number of possible outcomes for the experiment.

1. Rolling a dodecahedron
2. Choosing a letter from the word EXPERIMENT

Identifying Outcomes and Events List the favorable outcomes for the experiment.

3. Rolling an even number on a number cube
4. Drawing a 10 from a standard deck of 52 cards
5. Drawing a heart from a standard deck of 52 cards
6. Choosing a letter from the word MATHEMATICS

Describing Likelihood Describe the likelihood of the event as *impossible*, *unlikely*, *equally likely*, *likely*, or *certain*.

7. Rolling a number greater than 1 on a number cube
8. Choosing the letter P from the word ALGEBRA

Finding Theoretical Probability Find the theoretical probability of the event.

9. Drawing a card that is *not* a face card from a standard deck of 52 cards
10. Rolling an odd number on an octahedron
11. Drawing a red face card from a standard deck of cards
12. Choosing the letter E from the word THEORETICAL

Finding Experimental Probability The bar graph shows the results of spinning a spinner 100 times. Find the experimental probability of the event.

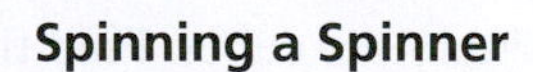

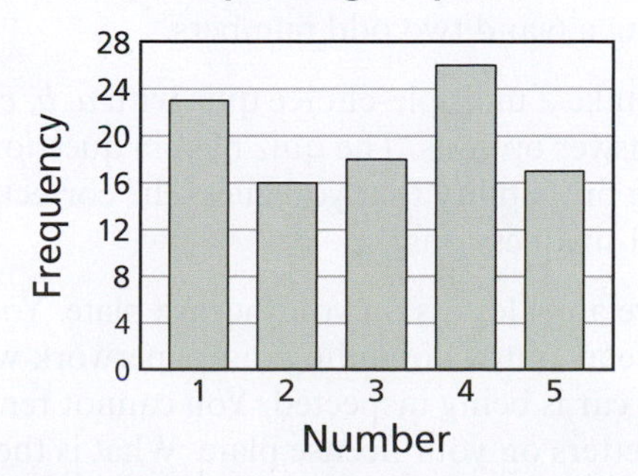

13. Spinning a 1
14. Spinning an even number
15. Spinning a number less than 3
16. Spinning a prime number
17. **Race for a Cure** There are 237 females and 342 males signed up for a 5k charity run. Each runner is given a 3-digit number ranging from 101 through 679. A runner is chosen at random to win a race-day hooded sweatshirt. What is the probability that the runner is female? What is the probability that the runner is wearing a number that starts with a 3?
18. **Chinese Auction** A Chinese auction is held to raise money for a little league association. You put 15 tickets in a container to win a basket filled with children's books and movies. The total number of tickets for the basket is 210. What is the probability that you will win the basket?

16.2 Counting Techniques

Finding Outcomes Using a Tree Diagram Draw a tree diagram to show the possible outcomes.

19. Flipping 2 coins and rolling a number cube
20. Choosing 2 marbles (without replacement) from a bag with 1 green, 1 blue, 1 yellow, 1 red, and 1 orange marble

Finding Probability Using a Tree Diagram The tree diagram shows the possible outcomes for flipping a coin and spinning a spinner with 5 equal sections. Find the probability of the event.

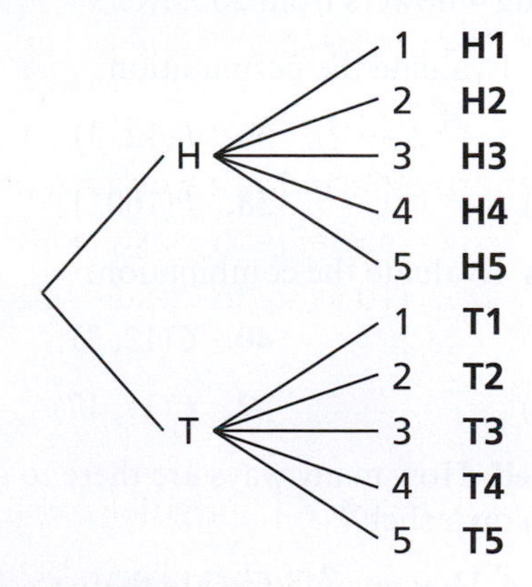

21. Flipping a head
22. Spinning a 3
23. Flipping a head and spinning an even number
24. Flipping a tail and spinning a 4

Finding Probability Using a Table The table shows the possible outcomes for rolling 2 number cubes. Find the probability of the event.

	1	2	3	4	5	6
1	1–1	1–2	1–3	1–4	1–5	1–6
2	2–1	2–2	2–3	2–4	2–5	2–6
3	3–1	3–2	3–3	3–4	3–5	3–6
4	4–1	4–2	4–3	4–4	4–5	4–6
5	5–1	5–2	5–3	5–4	5–5	5–6
6	6–1	6–2	6–3	6–4	6–5	6–6

25. Rolling one 3

26. Rolling two 6s

27. Rolling a 1 and a 2

28. Rolling two even numbers

Permuation or Combination? Determine whether the situation is a *permutation* or a *combination*. Explain your reasoning.

29. Choosing 3 crayons from a box of 16 crayons

30. Selecting the batting order for a 10-player softball team

31. Placing 3 books out of 12 in a book bag

32. Drawing 5 cards from a standard deck of 52 cards

33. Choosing an 8-character password

34. Choosing 4 flowers from 20 flowers

Permutations Evaluate the permutation.

35. $P(5, 4)$

36. $P(12, 3)$

37. $P(24, 4)$

38. $P(100, 1)$

Combinations Evaluate the combination.

39. $C(8, 1)$

40. $C(12, 3)$

41. $C(48, 6)$

42. $C(17, 17)$

43. Bookshelf How many ways are there to arrange 6 books on a shelf?

44. Password How many ways are there to choose a 5-digit personal identification number?

45. Student Assembly How many ways are there to choose 4 students to represent a class of 24 students at an assembly?

46. Burger Toppings How many ways are there to choose 5 burger toppings from 15 toppings?

47. Race Day In a 5k race, there are a total of 579 runners. How many different possibilities are there for first, second, and third place?

48. Lottery A lottery game chooses a 6-digit number using the digits 0 through 9. How many possible 6-digit numbers are there? What is the probability that the 6-digit number you pick will be the winning number?

49. Try Outs Twenty-three girls try out for a 14-player volleyball team. How many ways are there to choose 14 girls?

50. Essay Questions You are given a list of 5 prompts to study, 2 of which will be on the English exam. How many ways are there to choose 2 of these prompts? If you only study 2 of the prompts, what is the probability that you study the same 2 that appear on the exam?

16.3 Independent and Dependent Events

Identifying Independent and Dependent Events State whether the events are *independent* or *dependent*.

51. You roll two number cubes.

52. The gender of each child in a 3-child family

53. You choose a card from a standard deck of 52 cards without replacement and then choose another card.

54. You choose 2 socks from a drawer, in the dark, without replacement.

55. You and three friends flip 4 coins.

56. A teacher chooses 2 student representatives, one at a time, from a class of 24 students.

Finding the Probability of Independent Events Find the probability of the event.

57. You roll 2 number cubes. Find the probability of rolling a two and an even number.

58. You roll 3 number cubes. Find the probability of rolling a 6 and two odd numbers.

59. You take a multiple-choice quiz with a, b, c, and d as answer options. The quiz has six questions. What is the probability that you guess the correct answer to all six questions?

60. There are 4 letters on your license plate. You are inside an auto shop filling out paperwork while your car is being inspected. You cannot remember the letters on your license plate. What is the probability that you guess all 4 letters correctly?

Finding the Probability of Dependent Events Find the probability of the event.

61. You randomly choose 2 socks from a drawer of 20 socks in the dark. There are 6 white, 8 black, and 6 tie-dye. Find the probability of choosing a black sock and then a tie-dye sock.

62. You randomly draw 3 straws without replacement from a jar of 20 pink, 20 orange, 30 green, and 20 yellow straws. Find the probability of drawing 1 pink, then 1 orange, and then 1 yellow straw.

63. Each letter in the word TEACHER is written on a piece of paper and placed in a hat. Two pieces are randomly drawn without replacement. What is the probability of drawing both Es?

64. You randomly draw 4 cards without replacement from a standard deck of 52 cards. Find the probability of drawing 4 kings.

65. **Problem Solving** You have 12 shirts and 7 pairs of pants. You randomly choose one of each to wear each day of the week, Monday through Friday. There is one matching top and bottom.

a. You do not want to wear any article of clothing more than once. What is the probability that you choose the matching top and bottom on the first day? What is the probability that you choose a top and bottom that do not match on the first day?

b. You are okay with wearing any article of clothing more than once. What is the probability that you choose the matching top the first three days? What is the probability that you choose the matching bottom the first three days?

66. **Problem Solving** A beverage company offers a promotional code under the cap of each bottle. The code is to be typed into the company's website for a chance to win prizes. The advertisement claims that 1 in 12 wins. You buy 24 bottles of the beverage.

a. How many bottles do you expect to have a winning code?

b. None of your bottles have a winning code. Is the company's advertisement false? Explain your reasoning.

16.4 Expected Value

Calculating Expected Value for an Event Calculate the expected value for the experiment.

67. You flip two coins. You earn $16 if you flip two heads and lose $4 if you do not flip two heads.

68. You roll a number cube. You earn $12 if it lands on a number greater than 2 and lose $3 if it lands on 2 or less.

Comparing Expected Values Predict which option has a greater expected value. Calculate the expected value for each option. Then decide which option has a greater expected value.

69. Option 1: You flip a coin. You earn $4 if you flip tails and lose $2 if you flip heads. Option 2: You draw a card from a standard deck of 52 cards. You earn $8 if it is a club and lose $2 if it is not a club.

70. You roll two number cubes. Option 1: You earn $9 if only one cube lands on 1 and lose $3 if neither or both cubes land on 1. Option 2: You earn $12 if both cubes land on the same number and lose $2 if both cubes do not land on the same number.

71. Option 1: You have a 100% chance of winning $500. Option 2: You have a 70% chance of winning $1500 and a 30% chance of losing $200.

72. Option 1: You have an 80% chance of losing $500 and a 20% chance of losing nothing. Option 2: You have a 100% chance of losing $250.

Identifying Fair Games Calculate the expected value for the game. Then, state whether the game is fair or not. Explain your reasoning.

73. You roll a number cube 20 times. If it lands on an even number, you earn 2 points. If it lands on an odd number, you lose 2 points.

74. You choose a 3-digit number at random. You pay $1 to enter your number into the lottery. If your number is chosen, then you win $1000. If not, then you lose the cost of your ticket.

75. **Using a Decision Tree** The projections for two possible electronic models are shown in the tables. Decide which model should be produced.

Cali calculator: cost of development: $4 million	
Probability	**Net sales**
50%	$6 million
30%	$5 million
20%	$1 million

Florida calculator: cost of development: $2 million	
Probability	**Net sales**
20%	$4 million
60%	$3 million
20%	$1 million

Chapter Test

The questions on this practice test are modeled after questions on state certification exams for K–8 teaching.

Test Directions: Each of the questions is followed by five choices. Choose the *best* response to each question.

1. A jar contains 20 balls. These balls are numbered 1 through 20. What is the probability that a ball chosen randomly from the jar has a number on it which is divisible by 5?

A. $\frac{1}{20}$ **B.** $\frac{1}{5}$ **C.** $\frac{1}{4}$ **D.** 4 **E.** 5

2. Sal packed two dozen sandwiches. There were 6 ham, 6 turkey, and 12 salami sandwiches. If Sal picks a sandwich at random, what is the probability of selecting a ham sandwich?

A. $\frac{6}{12}$ **B.** $\frac{12}{18}$ **C.** $\frac{6}{24}$ **D.** $\frac{1}{24}$ **E.** $\frac{6}{18}$

3. Randi tossed a six-sided number cube 25 times and got the results shown in the table below. What percent of the time did Randi get a 4?

Number on cube	1	2	3	4	5	6
Frequency	6	3	6	1	2	7

A. 4% **B.** 7% **C.** 21% **D.** 28% **E.** 33%

4. A bag contains four coins, the state quarters for NY, SC, PA, and AZ. Two quarters are drawn from the bag simultaneously. What is the probability that the two quarters drawn from the bag will be the NY and PA state quarters?

A. $\frac{1}{12}$ **B.** $\frac{1}{8}$ **C.** $\frac{1}{6}$ **D.** $\frac{1}{4}$ **E.** $\frac{1}{2}$

5. A bag contains 6 red marbles and 4 blue marbles. A person reaches into the bag twice and removes a marble each time without returning it to the bag. What is the probability that the first marble will be red and the second marble will be blue?

A. $\frac{1}{6} \times \frac{1}{4}$ **B.** $\frac{5}{10} \times \frac{3}{9}$ **C.** $\frac{6}{10} \times \frac{4}{9}$

D. $\frac{6}{10} \times \frac{4}{8}$ **E.** $\frac{6}{10} \times \frac{4}{10}$

6. A lunch special includes a sandwich, soup, snack, and drink. There are 4 sandwiches, 4 soups, 2 snacks, and 3 drinks to choose from. How many different lunch specials are possible?

A. $4 \times 4 \times 2 \times 3$ **B.** $4 \times 3 \times 2 \times 1$ **C.** $(4 + 4 + 2 + 3) \div 2$

D. $(4 + 4 + 2 + 3) \times 2$ **E.** $(4 + 4 + 2 + 3) \div 4$

7. Brandon has eight new shirts and four new pairs of pants. How many possible new outfit combinations of a shirt and pants does this give him?

A. 8 **B.** 16 **C.** 18 **D.** 24 **E.** 32

8. A company makes electronic gadgets. One out of every 50 gadgets is faulty, but the company doesn't know which ones are faulty until a buyer complains. Suppose the company makes a $3 profit on the sale of any working gadget, but it suffers a loss of $80 for every faulty gadget because they have to repair the unit. What is the expected value?

A. −$7.78 **B.** −$2.68 **C.** −$1.34 **D.** $1.34 **E.** $2.68

17 Statistics

Activity: Reading Data from Two Types of Plots

Materials:

- Pencil

Learning Objective:

Determine frequencies from a dot plot and from a stem-and-leaf plot. Read data from a stem-and-leaf plot.

Name ________________________

1. You intercept a secret message. Use the secret code shown at the right to decode the message below.

32 44 15   2 32 15 22 2   2 44 9 12

32 44 15   32 15 3 2,   32 44 15

26 15 29 35 15 2   2 44 9 12

32 44 15   9 3 15 2.

Secret Code

A = 29	J = 11	S = 2
B = 33	K = 18	T = 32
C = 7	L = 26	U = 19
D = 20	M = 22	V = 35
E = 15	N = 3	W = 12
F = 31	O = 9	X = 25
G = 8	P = 4	Y = 13
H = 44	Q = 10	Z = 1
I = 5	R = 21	

2. Two ways to organize the numbers in the coded message are shown. How many numbers are there in all? How does each display show this?

Dot plot

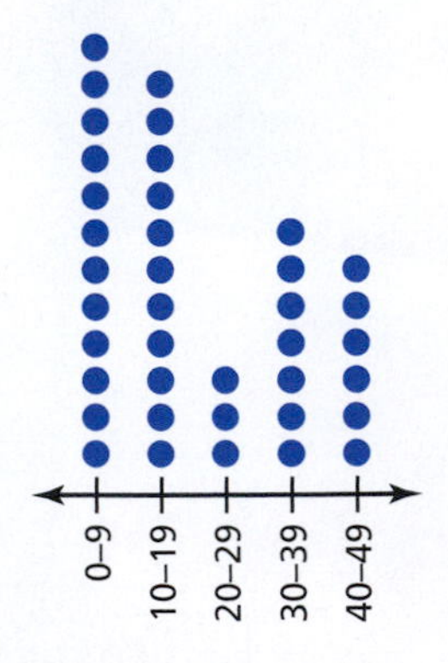

Stem-and-leaf plot

Stem	Leaf
0	2 2 2 2 2 2 2 3 3 9 9 9
1	2 2 5 5 5 5 5 5 5 5 5
2	2 6 9
3	2 2 2 2 2 2 5
4	4 4 4 4 4 4

Key: 2 | 6 = 26

3. How many numbers are greater than or equal to 40? How does each display show this?

4. Which display is the secret message about? Explain.

5. Which display shows that 29 is one of the numbers? How can you tell?

6. Use the stem-and-leaf plot to make an ordered list of all the numbers in the data set.

Pressmaster/Shutterstock.com

17.1 Graphs of Data

Standards

Grades K–2

Measurement and Data
Students should represent and interpret data.

Grades 3–5

Measurement and Data
Students should represent and interpret data.

Grades 6–8

Statistics and Probability
Students should summarize and describe distributions and investigate patterns of association in bivariate data.

- Organize data using stem-and-leaf plots.
- Organize data using bar graphs, frequency tables, and histograms.
- Organize data using pictographs and circle graphs.
- Read and use scatter plots.

Stem-and-Leaf Plots

Stem-and-Leaf Plots

A **stem-and-leaf plot** uses the digits of data values to organize a data set. Each data value is broken into a **stem** (digit or digits on the left) and a **leaf** (digit or digits on the right). A stem-and-leaf plot shows how data are distributed.

Stem	Leaf
2	0 0 1 2 5 7
3	1 4 8
4	2
5	8 9

Key: 2 | 0 = 20

The *key* explains what the stems and leaves represent.

Making a Stem-and-Leaf Plot

Make a stem-and-leaf plot of the lengths of 12 phone calls.

DATE	MINUTES
JULY 9	55
JULY 9	3
JULY 9	6
JULY 10	14
JULY 10	18
JULY 10	5
JULY 10	23
JULY 11	30
JULY 11	23
JULY 11	10
JULY 11	2
JULY 11	36

SOLUTION

Step 1 Order the data.

2, 3, 5, 6, 10, 14, 18, 23, 23, 30, 36, 55

Step 2 Choose the stems and leaves. Because the data values range from 2 to 55, use the *tens* digits for the stems and the *ones* digits for the leaves.

Step 3 Write the stems to the left of the vertical line.

Step 4 Write the leaves for each stem to the right of the vertical line.

Phone Call Lengths

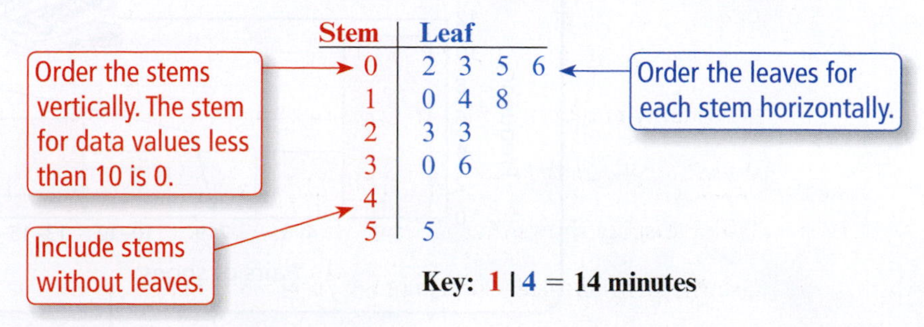

Classroom Tip

Example 1 shows that data can be organized in many ways. Choose a way to organize a data set based on the information you want to show. Here are 3 ways the data in Example 1 are organized.

1. **Table** The table shows the date and the number of minutes of each call.
2. **Ordered list** The list shows the lengths of the calls from the shortest time to the longest time.
3. **Stem-and-leaf plot** The stem-and-leaf plot shows how the data are distributed. You can see that seven calls were less than 20 minutes long.

Bar Graphs, Frequency Tables, and Histograms

A **bar graph** is used to compare and display data grouped in categories. Bar graphs have been used throughout this text. For instance, look back at some of the exercises on pages 8–11. There are many types of bar graphs. A few examples are shown below.

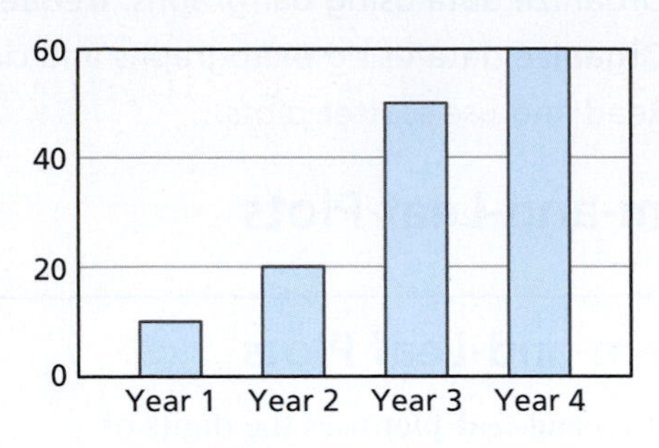

Simple bar graph

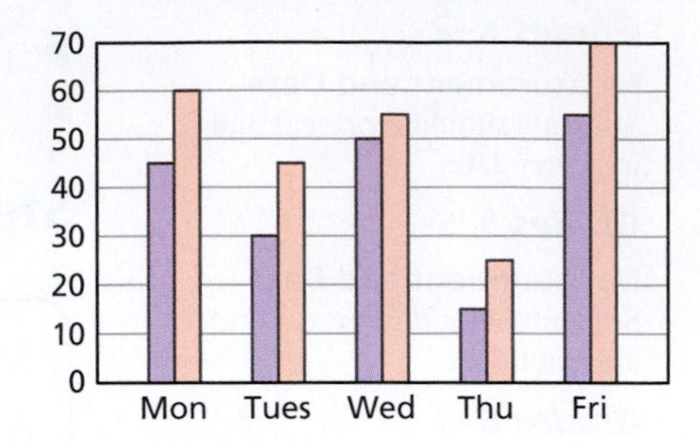

Double bar graph

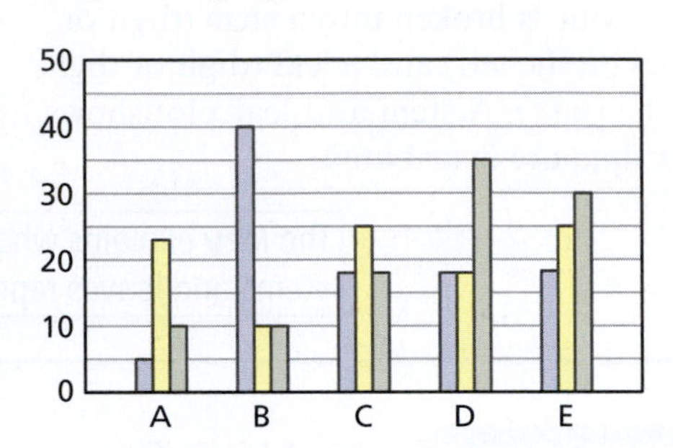

Triple bar graph

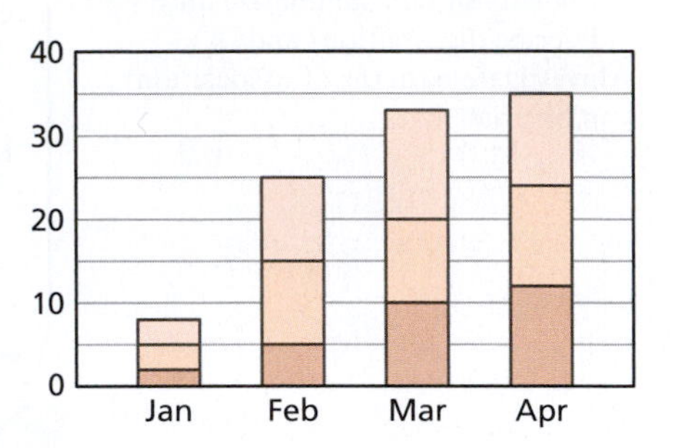

Stacked bar graph

Classroom Tip

Students learn to make and read bar graphs in early grades. Here is an example of a bar graph made by a student in Grade 4.

Frequency Tables and Histograms

A **frequency table** groups data values into intervals of the same size. The **frequency** is the number of data values in an interval. The frequency table at the right shows the pairs of shoes owned by students in a class.

A **histogram** is a bar graph that shows the frequency of data values in intervals of the same size. The height of a bar represents the frequency of the data values in the interval. The histogram below represents the data in the frequency table.

Pairs of shoes	Frequency
1–3	11
4–6	4
7–9	0
10–12	3
13–15	6

Shoes Owned

Frequency: 0, 2, 4, 6, 8, 10, 12

Pairs of shoes: 1–3, 4–6, 7–9, 10–12, 13–15

Include any interval with a frequency of 0. The bar height is 0.

There is no space between the bars of a histogram.

EXAMPLE 2 Making and Using a Histogram

The average speeds (in miles per hour) of the winners of the Daytona 500 from 1959 through 2013 are listed below.

136, 125, 150, 153, 152, 154, 142, 161, 147, 143, 158, 150, 144, 162, 157, 141, 154, 152, 153, 160, 144, 178, 170, 154, 156, 151, 172, 148, 176, 138, 148, 166, 148, 160, 155, 157, 142, 154, 148, 173, 162, 156, 162, 143, 134, 156, 135, 143, 149, 153, 133, 137, 130, 140, 159

a. Make a frequency table for the data starting with the interval 120–129 miles per hour.

b. Use the frequency table to make a histogram for the data.

c. How many of the winning average speeds are less than 140 miles per hour?

d. How many of the winning average speeds are at least 160 miles per hour?

Mathematical Practices

Reason Abstractly and Quantitatively

MP2 Mathematically proficient students create a coherent representation of the problem at hand, attending to the meaning of quantities, not just how to compute them, and knowing and flexibly using different properties of operations and objects.

SOLUTION

a. Form intervals of the same size that contain all the data starting with the interval 120–129. Tally the data to find the frequency for each interval.

Average speed (miles per hour)	Tally	Frequency
120–129	I	1
130–139	𝍸 II	7
140–149	𝍸 𝍸 𝍸	15
150–159	𝍸 𝍸 𝍸 𝍸	20
160–169	𝍸 II	7
170–179	𝍸	5

b. Make a histogram with intervals of average speeds along the horizontal axis and frequencies along the vertical axis.

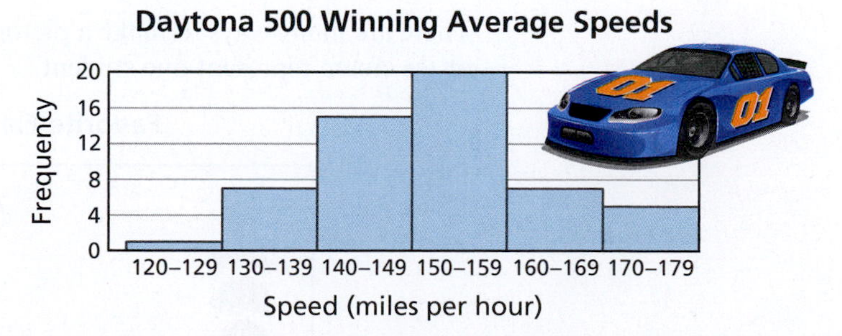

c. One winning average speed is in the 120–129 miles per hour interval and seven winning average speeds are in the 130–139 miles per hour interval. So, $1 + 7 = 8$ winning average speeds are less than 140 miles per hour.

d. Seven winning average speeds are in the 160–169 miles per hour interval and five winning average speeds are in the 170–179 miles per hour interval. So, $7 + 5 = 12$ winning average speeds are at least 160 miles per hour.

Pictographs and Circle Graphs

Pictographs are used by the media to create visual interest in a data display. The form of a pictograph may be like that of a bar graph or some other type of graph.

Pictographs

A **pictograph** uses images or symbols to display data. A pictograph includes a key that explains what each image or symbol represents.

Example The pictograph at the right shows the approximate numbers of apples available at small food store.

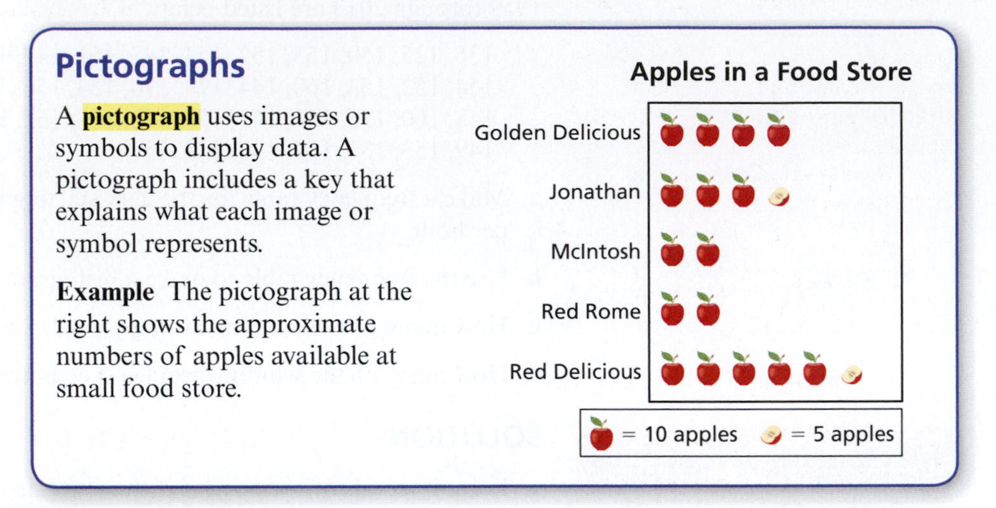

EXAMPLE 3 **Making a Pictograph**

You conduct a survey asking the students in your class to name their favorite flavors of ice cream. Make a pictograph of the results shown in the frequency table.

Sum	Tally	Frequency
Chocolate	𝍸 I	6
Vanilla	III	3
Strawberry	I	1
Chocolate chip	II	2
Other	𝍸 I	6

SOLUTION

There are many ways to make a pictograph. One way is to let a picture of a scoop of ice cream represent one student.

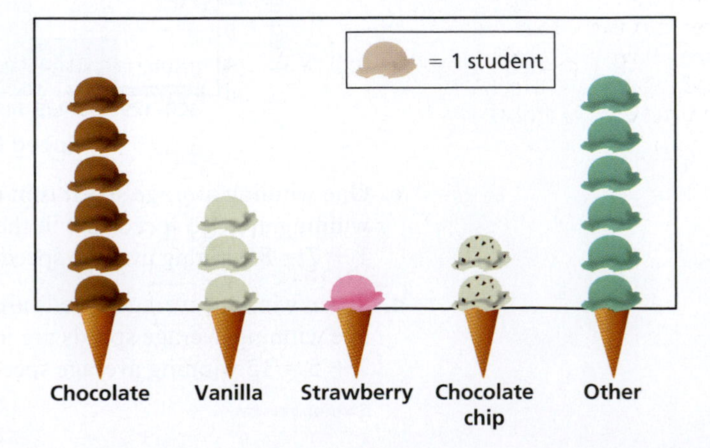

Classroom Tip

Circle graphs have been used throughout this text. Accurate circle graphs are difficult to make by hand. It is still a good idea, however, to ask students to sketch rough circle graphs. It gives them practice with estimating percents and degrees.

Here is an example of a circle graph made by a student.

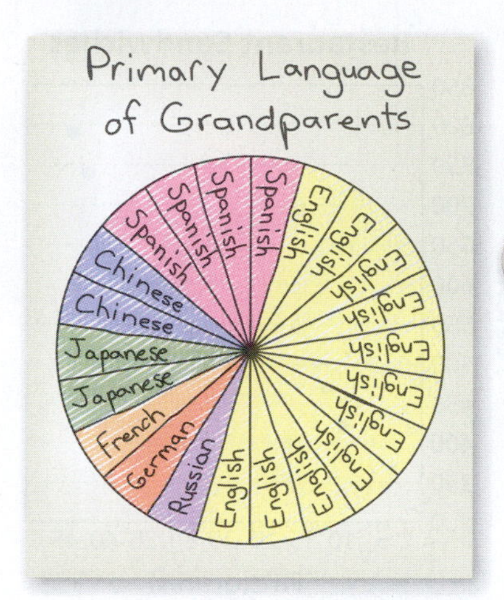

Circle Graphs

A **circle graph** (or *pie chart*) displays data as parts of a whole using sectors of a circle. The sum of the central angle measures in a circle graph is 360°.

To find the central angle measure for each sector, multiply the fraction of the whole (or the decimal form of the percent) that the sector represents by 360°.

EXAMPLE 4 Making a Circle Graph

The table shows the results of a survey. Display the data in a circle graph.

Favorite amusement park	People
Disney World	25
Busch Gardens	15
Universal Studios	12
Marineland	8

SOLUTION

Step 1 Find the total number of people surveyed.

$$25 + 15 + 12 + 8 = 60$$

Step 2 To find the central angle measure for each sector of the circle graph, multiply the fraction of people who chose each park by 360°.

Disney World

$$\frac{25}{60} \cdot 360^\circ = 150^\circ$$

Busch Gardens

$$\frac{15}{60} \cdot 360^\circ = 90^\circ$$

Universal Studios

$$\frac{12}{60} \cdot 360^\circ = 72^\circ$$

Marineland

$$\frac{8}{60} \cdot 360^\circ = 48^\circ$$

Step 3 Use a protractor to draw the angle measures found in Step 2 as central angles of a circle. Then label the sectors.

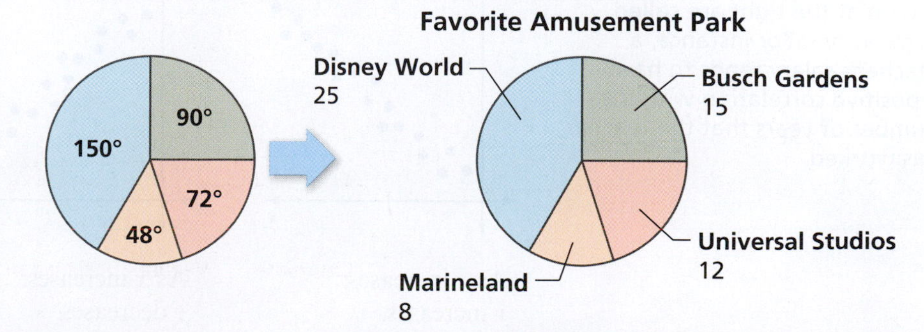

Classroom Tip

When you teach your students to check that the sum of the percents in a circle graph is 100%, remind them that the percents may have been rounded. In such cases, *round-off error* can produce a total that is slightly different from 100%.

Scatter Plots

Scatter Plots

A **scatter plot** is a graph that shows the relationship between two data sets. The two data sets are graphed as ordered pairs in a coordinate plane.

EXAMPLE 5 **Reading a Scatter Plot**

The scatter plot at the right shows the fat x (in grams) and the number y of calories in 12 restaurant sandwiches.

a. How many calories are in the sandwich that contains 17 grams of fat?

b. How many grams of fat are in the sandwich that contains 600 calories?

c. What tends to happen to the number of calories as the number of grams of fat increases?

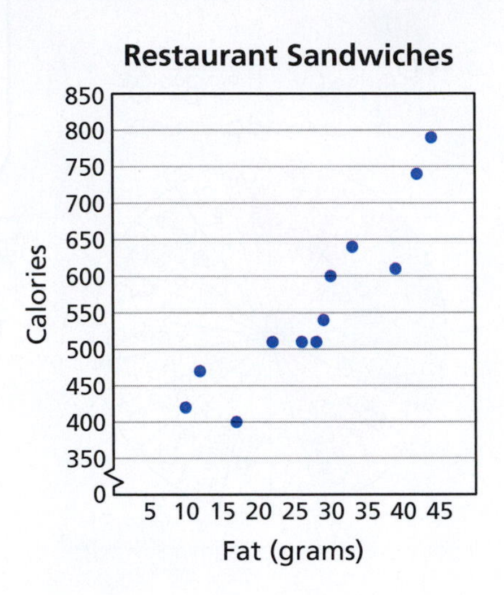

SOLUTION

a. Draw a horizontal line from the point that has an x-value of 17. It crosses the y-axis at 400. So, the sandwich has 400 calories.

b. Draw a vertical line from the point that has a y-value of 600. It crosses the x-axis at 30. So, the sandwich has 30 grams of fat.

c. The plotted points tend to go up from left to right. So, as the number of grams of fat increases, the number of calories increases.

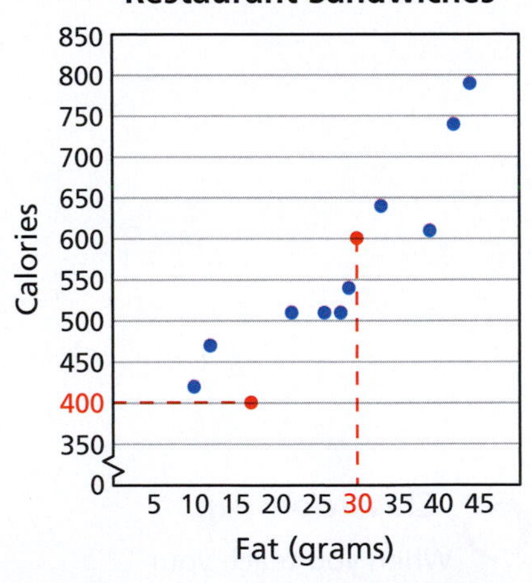

You can see from Example 5 that a scatter plot can show trends in the data.

Classroom Tip
In statistics, the relationships shown at the right are called *correlations*. For instance, a teacher's salary tends to have a positive correlation with the number of years that the teacher has worked.

Positive relationship

As x increases, y increases.

Negative relationship

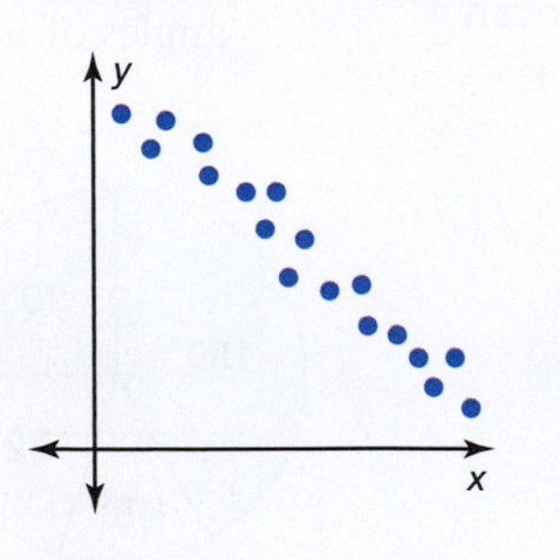

As x increases, y decreases.

No relationship

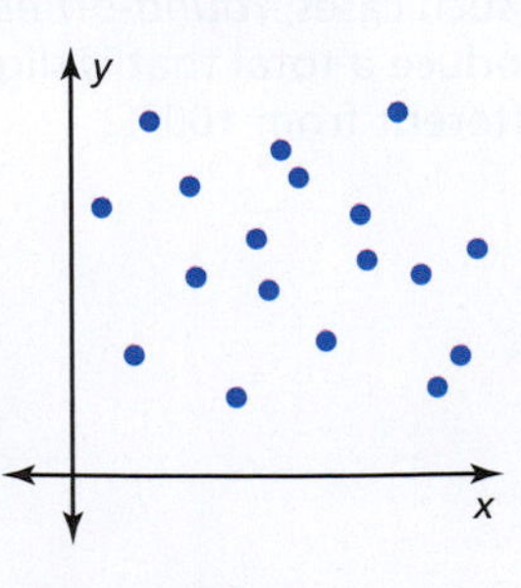

The points show no pattern.

EXAMPLE 6 Using a Scatter Plot

Decide whether the data points in each scatter plot show a *positive*, a *negative*, or *no* relationship.

a. **Television Size and Price**

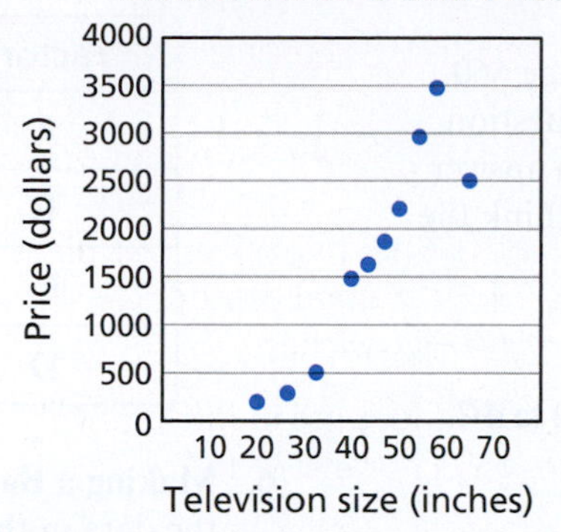

b. **Age and Pets Owned**

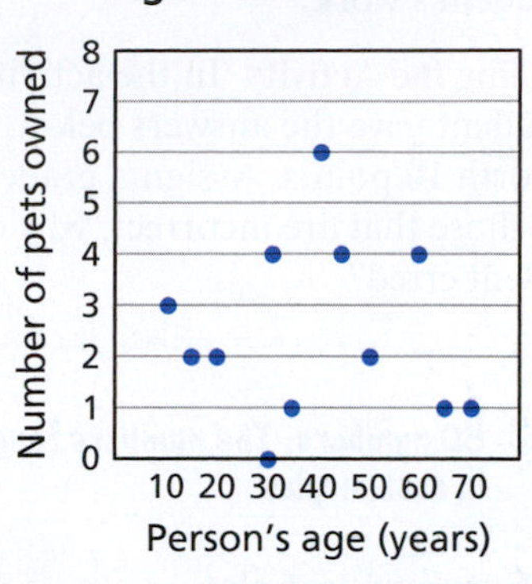

SOLUTION

a. As the sizes x of the televisions increase, the prices y increase. So, the scatter plot shows a positive relationship.

b. The number y of pets owned does not depend on the age x of the pet owner. So, the scatter plot shows no relationship.

A **line of fit** is a line drawn close to most of the data points on a scatter plot. A line of fit can be used to estimate data on a graph. There should be approximately as many points above the line as below it.

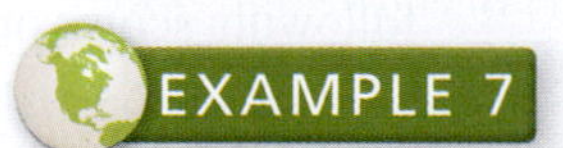

Using a Scatter Plot

The scatter plot at the right shows the first 8 weeks of DVD sales for a movie. A line of fit has been drawn close to the data points.

a. What is the slope of the line of fit?

b. Interpret the slope of the line of fit in the context of the data.

SOLUTION

a. The line passes through (5, 10) and (6, 8).

$$\text{Slope} = \frac{\text{rise}}{\text{run}} = \frac{-2}{1} = -2$$

So, the line has a slope of -2.

b. In the context of the data, a slope of -2 implies that DVD sales are decreasing about $2 million per week.

17.1 Exercises

Free step-by-step solutions to odd-numbered exercises at MathematicalPractices.com.

1. **Writing a Solution Key** Write a solution key for the activity on page 660. Describe a rubric for grading a student's work.

2. **Grading the Activity** In the activity on page 660, a student gave the answers below. Each question is worth 10 points. Assign a grade to each answer. For those that are incorrect, why do you think the student erred?

Sample Student Work

2. 50 numbers; The numbers range from 0 to 49 in each display.

Stem-and-leaf plot

Stem	Leaf
0	2 2 2 2 2 2 2 2 3 3 9 9 9
1	2 2 5 5 5 5 5 5 5 5 5 5
2	2 6 9
3	2 2 2 2 2 2 5
4	4 4 4 4 4 4

Key: 2 | 6 = 26

3. 10 numbers; Each display includes 40-49.

4. The stem-and-leaf plot; The message talks about the stems and leaves.

6. 0, 1, 2, 2, 2, 2, 2, 2, 2, 2, 2, 2, 2, 2, 2, 2, 2, 2, 2, 3, 3, 3, 4, 4, 4, 4, 4, 4, 4, 5, 5, 5, 5, 5, 5, 5, 5, 5, 5, 6, 9, 9, 9, 9

3. **Making a Stem-and-Leaf Plot** The different prices (in cents) of postage stamps from 1974 through 2013 are given below.

   10, 13, 15, 18, 20, 22, 25, 29, 32, 33, 34, 37, 39, 41, 42, 44, 45, 46

   a. Make a stem-and-leaf plot of the data.

   b. How many prices are less than \$0.30?

   c. What percent of the prices are greater than \$0.32?

4. **Making a Stem-and-Leaf Plot** The ages (in years) at inauguration of the first 44 United States presidents are given below.

   57, 61, 57, 57, 58, 57, 61, 54, 68, 51, 49, 64, 50, 48, 65, 52, 56, 46, 54, 49, 51, 47, 55, 55, 54, 42, 51, 56, 55, 51, 54, 51, 60, 62, 43, 55, 56, 61, 52, 69, 64, 46, 54, 47

   a. Make a stem-and-leaf plot of the data.

   b. How many presidents were in their 60s when they were inaugurated?

   c. What was the age of the youngest president?

5. **Making a Bar Graph** Make a bar graph to display the data in the table.

Factory	Profit (millions of dollars)
A	12.5
B	11.3
C	21.6
D	17.4

6. **Making a Bar Graph** Make a bar graph to display the data in the table.

Day	Study hours
Monday	3
Tuesday	1
Wednesday	4
Thursday	2.5

7. **Making a Histogram** A class of 30 students has the following scores on a 100-point exam.

   100, 68, 93, 82, 71, 77, 70, 79, 69, 64, 67, 79, 84, 91, 100, 75, 91, 70, 55, 85, 68, 91, 64, 64, 61, 89, 96, 67, 81, 68

   a. Make a frequency table for the data starting with the interval 51–60.

   b. Use the frequency table to make a histogram for the data.

   c. How many scores are greater than 80?

   d. How many scores are 60 or less?

8. **Making a Histogram** The heights (in inches) of the 32 members of a track team are shown.

   64, 68, 69, 72, 71, 67, 70, 74, 69, 75, 67, 63, 74, 72, 72, 73, 76, 71, 70, 68, 72, 72, 71, 67, 64, 71, 70, 68, 69, 71, 70, 73

   a. Make a frequency table for the data starting with the interval 63–64.

   b. Use the frequency table to make a histogram for the data.

   c. How many team members are 6 feet tall or less?

   d. How many team members are taller than a height of 6 feet, 2 inches?

9. **Making a Pictograph** The table shows the numbers of three types of trees planted by a landscaper in one day. Make a pictograph to show the data. Include a key.

Tree type	Tally	Frequency
Maple	卌 卌 II	12
Dogwood	卌 III	8
Willow	卌 I	6

10. **Making a Pictograph** You ask your students what drink they consume most often. Make a pictograph of the results shown in the table. Include a key.

Drink	Tally	Frequency
Water	卌 IIII	9
Milk	卌 I	6
Fruit juice	卌 卌 I	11
Soda	卌 II	7

11. **Making a Circle Graph** The table shows the numbers of students with perfect attendance in each grade. Display the data in a circle graph.

Grade	6	7	8
Number	10	9	17

12. **Making a Circle Graph** A school booster club consists of 4 school administrators, 11 teachers, 25 students, and 32 parents. Display this information in a circle graph.

13. **Reading a Scatter Plot** The scatter plot compares the body weights and shoe sizes of 12 soccer players.

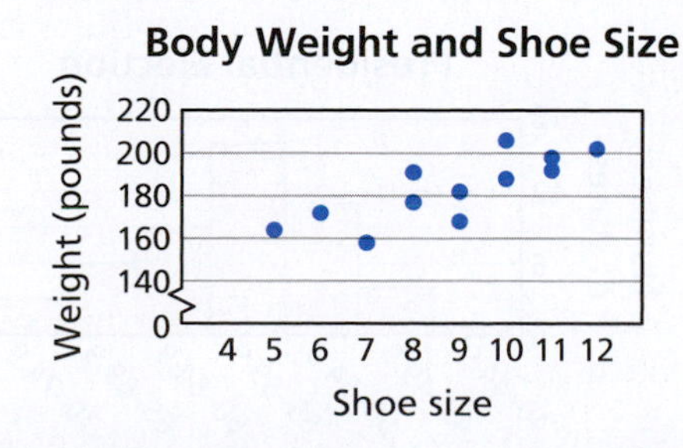

a. What is the weight of the player with size 12 shoes?

b. What is the shoe size of the 158-pound player?

c. Do the body weights and shoe sizes show a *positive*, a *negative*, or *no* relationship? Explain.

14. **Reading a Scatter Plot** The scatter plot shows how long it took different numbers of workers to do the same job.

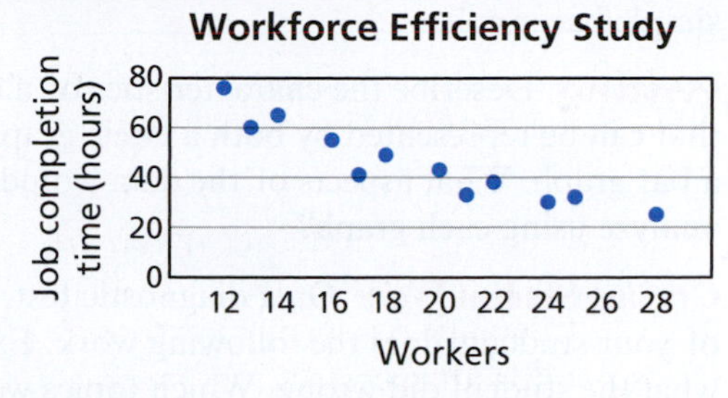

a. How many hours did it take the crew of 13 workers to complete the job?

b. How many workers were in the crew that completed the job in 25 hours?

c. Do the numbers of workers and job completion times show a *positive*, a *negative*, or *no* relationship? Explain.

15. **Using a Scatter Plot** The scatter plot shows the study time and exam scores for several students for the same exam. A line of fit is drawn close to the data points.

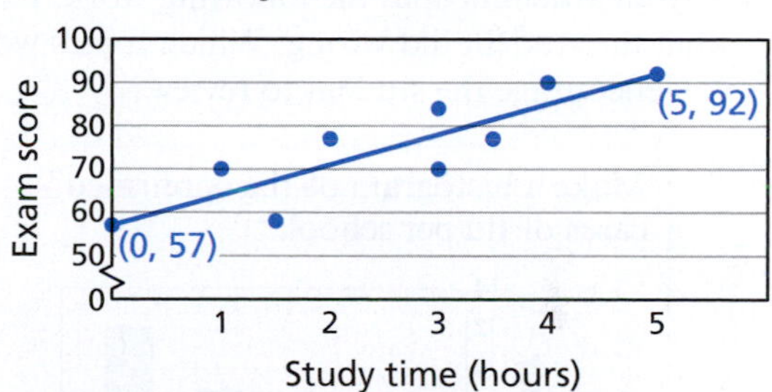

a. What is the slope of the line of fit?

b. Interpret the slope of the line of fit in the context of the data.

16. **Using a Scatter Plot** The scatter plot shows the yearly profits of a company during its first nine years. A line of fit is drawn close to the data points.

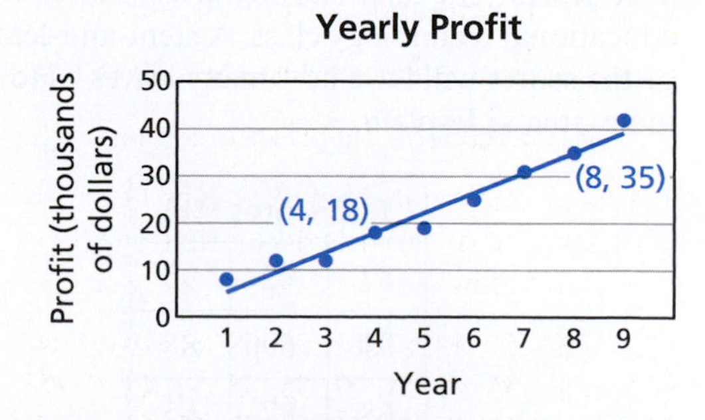

a. What is the slope of the line of fit?

b. Interpret the slope of the line of fit in the context of the data.

17. *Writing* Describe the types of data that you can represent with a bar graph. Then explain when a histogram is a better choice of data display than a simple bar graph.

18. *Writing* Describe the characteristics of a data set that can be represented by both a circle graph and a bar graph. What aspects of the data would you analyze using each graph?

19. Grading Student Work On a diagnostic test, one of your students does the following work. Explain what the student did wrong. Which topics would you encourage the student to review?

Make a stem-and-leaf plot for the data.
51, 25, 47, 42, 55, 26, 50, 44, 55

Stem	Leaf
2	5 6
4	2 4 7
5	0 1 5 5

Key: 4 | 2 = 42

20. Grading Student Work On a diagnostic test, one of your students does the following work. Explain what the student did wrong. Which topics would you encourage the student to review?

Make a histogram of the confirmed cases of flu per school.

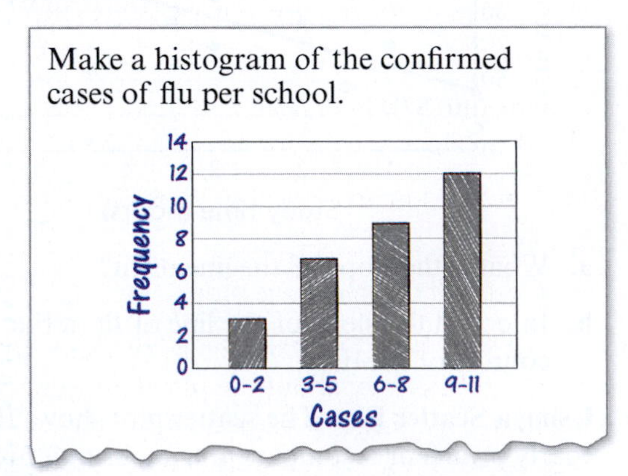

21. Test Scores The table shows the test scores of an educational technology class. A stem-and-leaf plot of the scores will have how many leaves? How many stems? Explain.

Test scores (%)		
87	82	95
91	69	88
68	87	65
81	97	85
80	90	62

22. Stem-and-Leaf Plot The table shows the numbers of bicycles sold at a bike shop each month for one year. Make a stem-and-leaf plot of the data. Explain your choice of stems and leaves.

Bikes sold			
78	112	105	99
86	96	115	100
79	81	99	108

23. Circle Graph The table shows the seasonal amounts of rainfall a city had in a recent year. Display the data in a circle graph.

Season	Rainfall (inches)
Spring	9
Summer	18
Fall	6
Winter	3

24. Retail The circle graph shows how a store collected $7200 in sales.

Department Store Sales

80°
140°
Other 75°

a. Find the sales for each category shown in the circle graph.

b. One-fourth of shirt sales were long sleeve shirts. Find the central angle measure of the section that would represent long sleeve shirts on the circle graph.

25. Voting The histogram shows the percent of the voting-age population that voted in a recent presidential election. Explain whether each statement is supported by the graph.

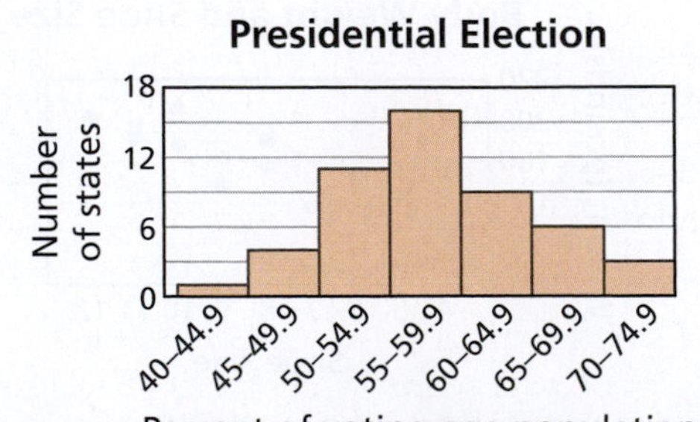

a. Only 40% of one state voted.

b. Well over half of the states had between 49.9% and 65% that voted.

26. Histograms The table shows the weights of guide dogs enrolled in a training program.

Weight (pounds)					
81	88	57	82	70	85
71	51	82	77	79	77
83	80	54	80	81	73
59	84	75	76	68	78
83	78	55	67	85	79

a. Make a histogram of the data starting with the interval 51–55.

b. Make another histogram of the data using a different-sized interval.

c. Compare and contrast the two histograms.

27. Drawing a Line of Fit: In Your Classroom The graph from Example 7 is shown.

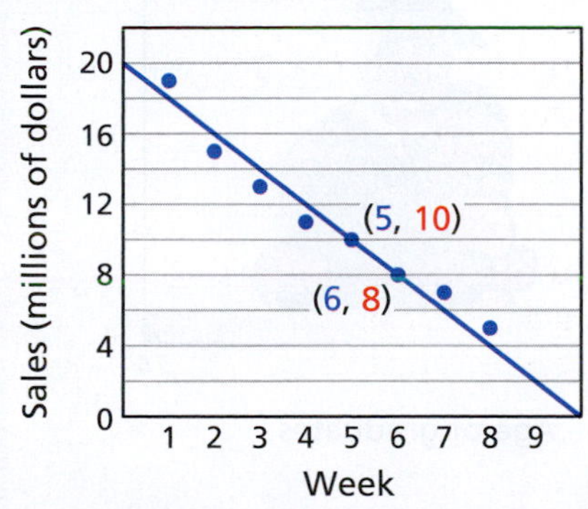

a. A common way to draw a line of fit on a scatter plot is to draw a line through two of the points. Explain why the points (5, 10) and (6, 8) were used to draw the line of fit.

b. Explain how you would demonstrate the process of using the line of fit to predict the DVD sales in week 9 for your students.

28. Analyzing Lines of Fit: In Your Classroom Consider lines of fit for various scatter plots.

a. Describe the slope of a line of fit for a scatter plot of data points with a negative relationship. How can you demonstrate this for your students?

b. Describe the slope of a line of fit for a scatter plot of data points with a positive relationship. How can you demonstrate this for your students?

c. Does it make sense to draw a line of fit for a scatter plot of data points with no relationship? How can you demonstrate this for your students?

29. Using a Line of Fit The table shows the daily high temperature x (in degrees Fahrenheit) and the number y of hot chocolates sold at a cafe on eight random days.

Temperature, x	30	36	44	51	60	68
Number sold, y	45	43	36	35	30	27

a. Make a scatter plot of the data.

b. What type of relationship does the scatter plot show?

c. Use two points to draw a line of fit for the scatter plot.

d. Find and interpret the slope of the line of fit.

e. Predict the number of hot chocolates sold when the high temperature is 40°F.

30. Using a Line of Fit The table shows the weights y (in pounds) of x pints of blueberries.

Pints, x	0	1	2	3	4	5
Weight, y	0	0.8	1.5	2.2	3.0	3.75

a. Make a scatter plot of the data.

b. What type of relationship does the scatter plot show?

c. Use two points to draw a line of fit for the scatter plot.

d. Find and interpret the slope of the line of fit.

e. Blueberries cost $2.75 per pound. How much do 1.5 pints of blueberries cost?

31. Choosing a Data Display The table shows the results of a survey about the plans of graduating seniors at a college. What is an appropriate data display for these results? Explain your reasoning.

Plans after graduation	Percent
Graduate school	18
Job in field of major	62
Job outside field of major	14
Travel	24

32. Choosing a Data Display What is another type of data display that can be used to show the information in the dot plot in the activity on page 660? Draw the data display and explain your reasoning.

Activity: Understanding the Effect of an Outlier on the Mean

Materials:
- Pencil

Learning Objective:

Understand that the mean of a set of numbers can be a misleading average when there is an outlier. Think of other ways to represent an "average" of a data set that has an outlier.

Name ____________________

Read each paragraph. Think of a better way to represent the "average" so that the statement is not so misleading.

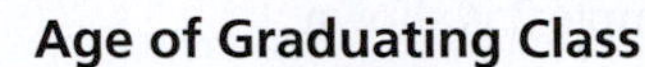

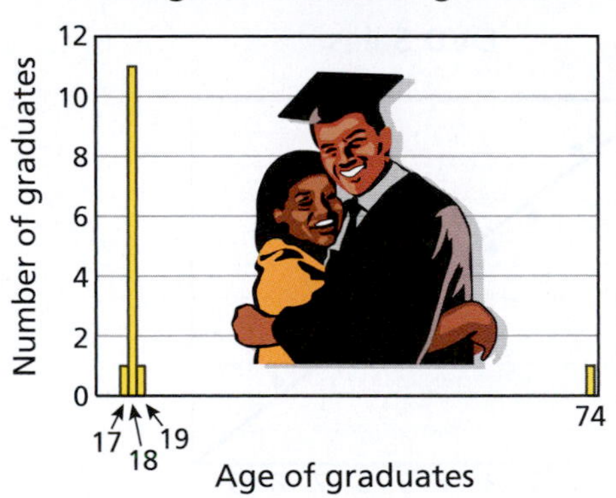

1. Someone criticizes a small high school by saying "Last year, the average age of the graduating class was 22 years old."

You later find out that one member of the class was a senior citizen who went back to school to earn a diploma. The ages of the class are shown in the bar graph.

a. What percent of the ages are below this average?

b. What number is a better "average"?

c. How did you find this average?

Families' Yearly Income

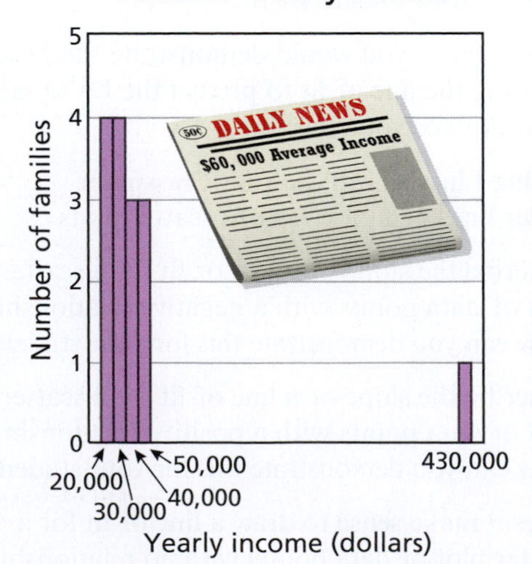

2. Most of the people in a small town struggle with the cost of living. A nearby newspaper columnist makes the comment "That town is not so bad off. The average annual income for a family is $60,000."

The incomes are shown in the bar graph.

a. What percent of the incomes are below this average?

b. What number is a better "average"?

c. How did you find this average?

17.2 Measures of Central Tendency

Standards

Grades 6–8

Statistics and Probability

Students should develop understanding of statistical variability and investigate patterns of association in bivariate data.

- Find and use the mean of a data set.
- Find and use the median and mode of a data set.
- Understand the effects of outliers on measures of central tendency.

Mean

A **measure of central tendency** is a number that describes the "center" or "middle" of a data set when the numbers are written in order. The most common measures of central tendency are mean, median, and mode.

Definition of Mean

Words The **mean** of a data set is the sum of the data divided by the number of data values.

Example Data: 8, 5, 6, 9 Mean: $\dfrac{8 + 5 + 6 + 9}{4} = \dfrac{28}{4} = 7$

4 data values

Balancing point model You can think of the mean of a data set as the point on a scale where the data balance.

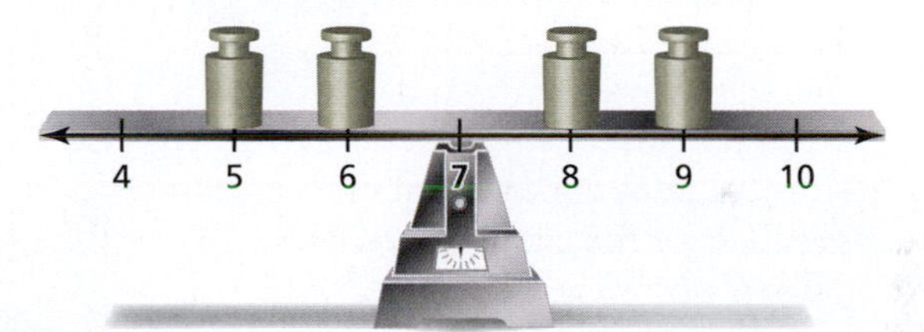

EXAMPLE 1 **Comparing Two Means**

The double bar graph shows the monthly rainfall amounts (in inches) for two cities over a six-month period. Compare the mean monthly rainfall amounts of the two cities for these months.

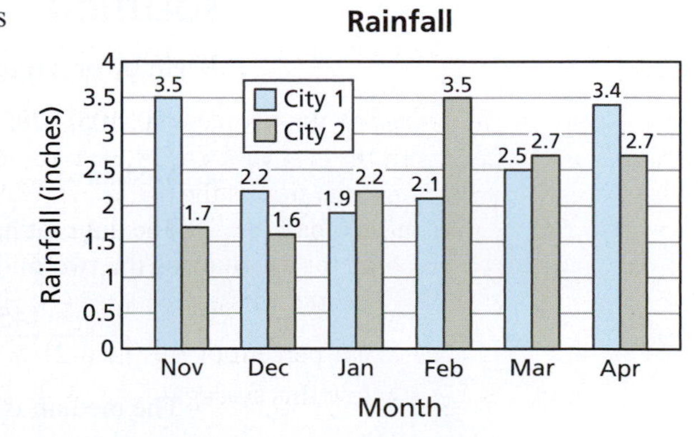

SOLUTION

Find the mean monthly rainfall in each city for the six-month period.

Mean of City 1 $\dfrac{3.5 + 2.2 + 1.9 + 2.1 + 2.5 + 3.4}{6} = \dfrac{15.6}{6} = 2.6$

Mean of City 2 $\dfrac{1.7 + 1.6 + 2.2 + 3.5 + 2.7 + 2.7}{6} = \dfrac{14.4}{6} = 2.4$

City 1 had a greater mean monthly rainfall amount than City 2.

Median and Mode

Definition of Median

Words Order the data. For a data set with an odd number of values, the **median** is the middle value. For a data set with an even number of values, the **median** is the mean of the two middle values.

Examples Data: 5, 8, 9, 12, 14 The median is 9.

Data: 2, 3, 5, 7, 10, 11 The median is $\frac{5+7}{2}$, or 6.

Definition of Mode

Words The **mode** of a data set is the value or values that occur most often. Data can have one mode, more than one mode, or no mode. When all values occur only once, there is no mode.

Example Data: 11, 13, 15, 15, 18, 21, 24, 24 The modes are 15 and 24.

To find either the median or the mode, it is helpful to begin by ordering the data.

Finding the Median and Mode

Find the (a) median and (b) mode of the bowling scores.

Bowling scores	
120	135
160	125
90	205
160	175
105	145

SOLUTION

Begin by ordering the data.

90, 105, 120, 125, 135, 145, 160, 160, 175, 205

a. Median

The data set has an even number of values. So, the median is the mean of the two middle values.

$$\frac{135+145}{2} = \frac{280}{2}, \text{ or } 140$$

The median is 140.

b. Mode

The value 160 occurs twice. Each of the other scores occurs only once.

90, 105, 120, 125, 135, 145, 160, 160, 175, 205

The value 160 occurs most often.

So, the mode is 160.

EXAMPLE 3 Finding a Mode

Each student in a class voted for his or her favorite type of movie. The list shows the results. Organize the data in a frequency table. Then find the mode.

Favorite Types of Movies		
Comedy	Drama	Horror
Horror	Drama	Horror
Comedy	Comedy	Action
Action	Comedy	Action
Horror	Drama	Comedy
Comedy	Comedy	Horror
Horror	Comedy	Action
Horror	Action	Drama

Classroom Tip

Notice in Example 3 that the definition of mode can be applied to a data set that consists of data other than numbers.

SOLUTION

Type	Tally	Frequency
Action	𝍸	5
Comedy	𝍸 III	8
Drama	IIII	4
Horror	𝍸 II	7

Make a tally for each vote.

Comedy received the most votes. So, the mode is comedy.

EXAMPLE 4 Changing the Values of a Data Set

Students' hourly wages	
\$3.87	\$7.25
\$8.25	\$8.45
\$8.25	\$8.04
\$6.99	

The table shows the hourly wages of seven students working at an amusement park for the summer. At midseason each hourly wage increases \$0.40. How does this increase affect the mean, median, and mode of the hourly wages?

SOLUTION

Begin by finding the mean, median, and mode of the data.

Mean $\dfrac{3.87 + 7.25 + 8.25 + 8.45 + 8.25 + 8.04 + 6.99}{7} = \dfrac{51.10}{7} = 7.30$

Median 3.87, 6.99, 7.25, 8.04, 8.25, 8.25, 8.45 Order the data.

Because the data set has an odd number of values, the median is the middle value, 8.04.

Mode 3.87, 6.99, 7.25, 8.04, 8.25, 8.25, 8.45

The mode is the data value that occurs most often, 8.25.

Make a new table by adding \$0.40 to each hourly wage.

Students' hourly wages	
\$4.27	\$7.65
\$8.65	\$8.85
\$8.65	\$8.44
\$7.39	

Mean $\dfrac{4.27 + 7.65 + 8.65 + 8.85 + 8.65 + 8.44 + 7.39}{7} = \dfrac{53.90}{7} = 7.70$

Median 4.27, 7.29, 7.65, 8.44, 8.65, 8.65, 8.85 Order the data.

The median is 8.44.

Mode 4.27, 7.29, 7.65, 8.44, 8.65, 8.65, 8.85

The mode is 8.65.

By increasing each hourly wage \$0.40, the mean, median, and mode all increase \$0.40.

Effects of Outliers

An **outlier** in a data set is a number that is much greater or much less than the other numbers in the set. A data set can have more than one outlier, as shown below.

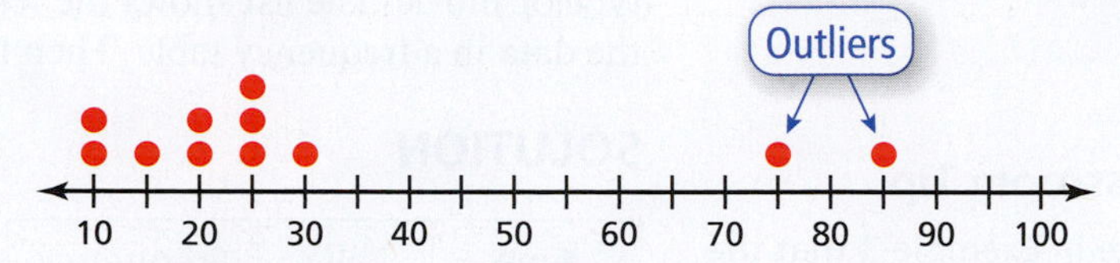

The mean is the most commonly used average, but it can be greatly affected by outliers in the data set.

EXAMPLE 5 Finding the Best "Average"

Find the mean, median, and mode of the shoe prices. Which measure best represents the data?

SOLUTION

Mean $\frac{20 + 31 + 37 + 20 + 122 + 48 + 45 + 65}{8} = \frac{388}{8}$, or 48.5

The mean is 48.5.

Median 20, 20, 31, 37, 45, 48, 65, 122 — Order the data.

$\frac{37 + 45}{2} = \frac{82}{2}$, or 41 — Mean of the middle two values

The median is 41.

Mode 20, 20, 31, 37, 45, 48, 65, 122 — The value 20 occurs most often.

The mode is 20.

To compare these three measures of central tendency, plot them with the data along a number line.

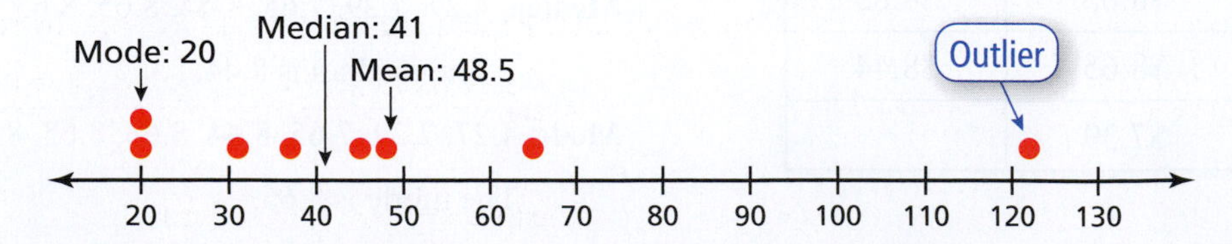

You can see that the median is central to and close to most of the data, whereas the mode is the least data value, and the mean is greater than most of the data. Of these three "averages," the median best represents the data.

Classroom Tip

In choosing the measure of central tendency that best describes a set of data, remember that the goal is to communicate a representative "average." When you teach students about measures of central tendency, make sure they understand that the mean is not always a representative average.

EXAMPLE 6 Removing an Outlier

Remove the outlier from the data in Example 5 and find the mean, median, and mode of the resulting data set. Which measure of central tendency is most affected by the outlier?

SOLUTION

The price \$122 is much greater than the other prices, so it is the outlier.

	Mean	Median	Mode
With outlier (Example 5)	48.5	41	20
Without outlier	38	37	20

Notice that the mode is not affected by the outlier. The median is somewhat affected. The measure of central tendency that is the most affected by the outlier, however, is the mean.

EXAMPLE 7 Analyzing the "Average" for Team GPA

The data show the GPAs of the students on a college baseball team.

1.35, 1.65, 2.08, 2.12, 2.12, 2.40, 2.44, 2.56, 2.68, 2.73, 2.79, 2.80, 2.81, 2.95, 3.00, 3.04, 3.16, 3.18, 3.23, 3.26, 3.37, 3.54, 3.54, 3.55, 3.68, 3.76, 3.92, 3.92, 4.04, 4.08

a. Explain why the mode is not useful for describing the team GPA.

b. Which average represents a more impressive team GPA, the *mean* or the *median*? Explain.

c. The players with the two lowest GPAs are released from the team for having GPAs below 2.0. Compare how the mean and the median GPAs change for the team.

Mathematical Practices

Construct Viable Arguments and Critique the Reasoning of Others

MP3 Mathematically proficient students construct arguments using concrete referents such as objects, drawings, diagrams, and actions.

SOLUTION

a. There are three modes, 2.12, 3.54, and 3.92, and they do not describe the typical GPA for the team well.

b. **Mean** $\dfrac{1.35 + 1.65 + \ldots + 4.04 + 4.08}{30} = \dfrac{89.75}{30} \approx 2.99$

Median $\dfrac{3.00 + 3.04}{2} = 3.02$

The median represents a more impressive team GPA, both because it is greater than the mean and because it is greater than 3.0.

c. Find the mean and median GPAs for the remaining students.

Mean $\dfrac{2.08 + 2.12 + \ldots + 4.04 + 4.08}{30} = \dfrac{86.75}{28} \approx 3.10$

Median $\dfrac{3.04 + 3.16}{2} = 3.10$

Releasing the players increases the mean GPA by $3.10 - 2.99 = 0.11$ and the median GPA by $3.10 - 3.02 = 0.08$. The mean increased slightly more than the median. The mean and median are now both 3.10.

17.2 Exercises

Free step-by-step solutions to odd-numbered exercises at MathematicalPractices.com.

1. **Writing a Solution Key** Write a solution key for the activity on page 672. Describe a rubric for grading a student's work.

2. **Grading the Activity** In the activity on page 672, a student gave the answers below. Each question is worth 10 points. Assign a grade to each answer. For those that are incorrect, why do you think the student erred?

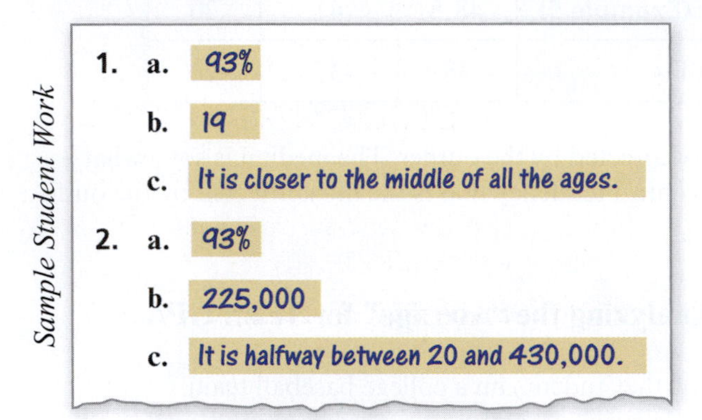

3. **Comparing Two Means** The double bar graph shows the monthly snowfall amounts (in inches) for two cities over a six-month period. Compare the mean monthly snowfall amounts of the two cities for these months.

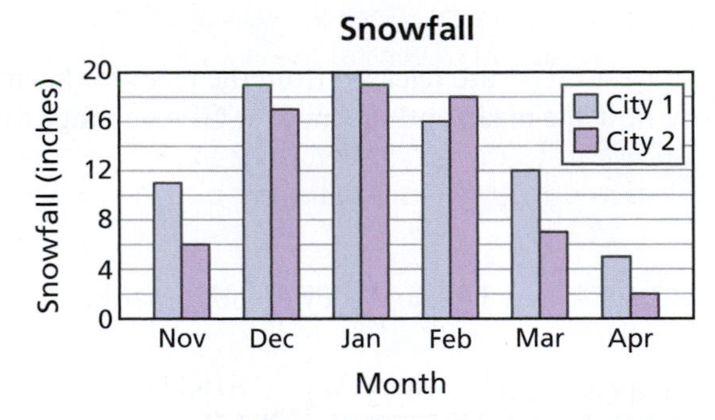

4. **Comparing Two Means** The double bar graph shows the heights (in inches) of both the women's and men's basketball teams at a university. Compare the mean height of these two teams.

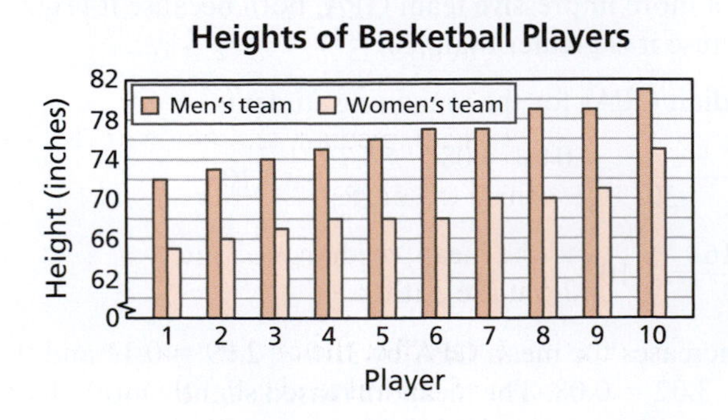

5. **Finding the Median and Mode** Find the median and mode for each set of numbers.
   - **a.** 8, 12, 9, 13, 17, 16, 8, 10, 11, 7, 6, 13, 12, 12
   - **b.** 79, 88, 67, 92, 94, 86, 75, 98, 68, 79, 83, 69

6. **Finding the Median and Mode** Find the median and mode for each set of numbers.
   - **a.** 41, 52, 34, 56, 61, 44, 49, 58, 56, 43, 37, 40, 51
   - **b.** 108, 102, 112, 126, 119, 111, 123, 106, 113

7. **Finding the Mean, Median, and Mode** Find the mean, median, and mode for each set of numbers.
   - **a.** 10, 3, 10, 5, 2, 1, 6, 5, 2, 4, 8, 10, 4
   - **b.** 22, 27, 28, 23, 20, 27, 26, 28, 23, 29, 23

8. **Finding the Mean, Median, and Mode** Find the mean, median, and mode for each set of numbers.
   - **a.** 69, 68, 55, 69, 54, 51, 64, 57, 69
   - **b.** 67, 41, 48, 27, 41, 25, 68, 62, 40

9. **Finding the Mean, Median, and Mode** A class's pre-chapter and post-chapter test scores are given. Find the class mean, median, and mode for each set of data.

   Pre-chapter: 72, 69, 78, 76, 62, 83, 73, 62, 82, 79

   Post-chapter: 67, 82, 97, 89, 75, 81, 92, 94, 86, 89

10. **Finding the Mean, Median, and Mode** A baseball team's pre-season and post-season batting averages were recorded. Find the team mean, median, and mode for each set of data.

    Pre-season: .154, .305, .356, .498, .365, .276, .295, .421, .372

    Post-season: .375, .458, .245, .304, .420, .513, .176, .297, .420

11. **Finding the Mean, Median, and Mode** An 8th grade class was surveyed on their shoe sizes. The results are shown in the bar graph below. Use the data to find the mean, median, and mode.

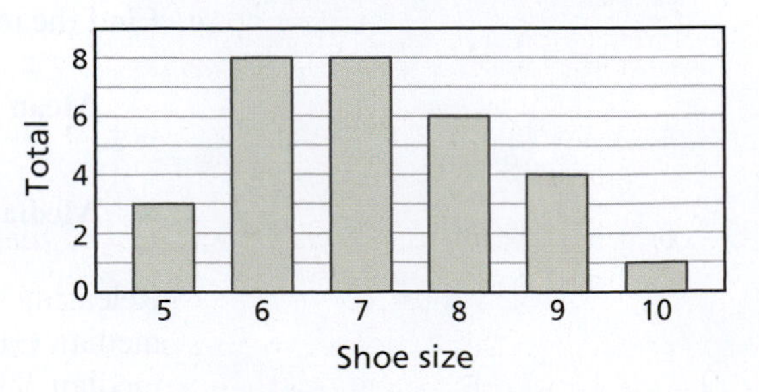

12. **Finding the Mean, Median, and Mode** The circle graph shows the results of a survey of a college class. Find the mean, median, and mode of the data.

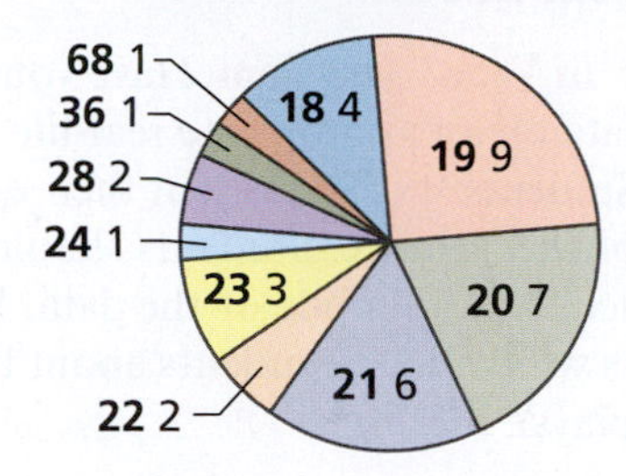

13. **Changing the Values of a Data Set** The table below shows the yearly salaries for the employees of a company.

Department salaries	
$45,000	$42,000
$54,000	$39,000
$39,000	$49,000
$47,000	$41,000
$51,000	$50,000

a. Find the mean, median, and mode of the data.

b. After one year, each employee's salary increases by $1,000. Find the mean, median, and mode of the data after the increase.

c. How are the mean, median, and mode affected by the increase?

14. **Changing the Values of a Data Set** The time spent traveling to a meeting for store managers of a company is shown in the table below. The company decides to change the meeting location. The new travel times are shown in a separate column.

Manager	Travel time	New travel time
Manager 1	20 min	7 min
Manager 2	11 min	15 min
Manager 3	16 min	12 min
Manager 4	23 min	9 min

a. Find the mean, median, and mode of the travel times.

b. Find the mean, median, and mode of the new travel times.

c. Do you think the managers agree with the new location? Explain your reasoning.

15. **Finding the Best Representation** The IQ scores of a high school academic decathlon team are shown below. Find the mean, median, and mode for the data. Which measure of central tendency do you think best represents the team's IQ scores? Explain.

IQ: 108, 99, 92, 123, 104, 100, 99, 116, 132, 118

16. **Finding the Best Representation** The judge's scores for an Olympic gymnast's routine are shown below. Find the mean, median, and mode for the data. Which measure of central tendency do you think best represents the gymnast's score? Explain.

Judge's scores: 10.0, 8.0, 8.5, 5.0, 5.0, 9.0, 7.5, 9.5

17. **Examining Outliers** Use the data below.

19, 16, 23, 35, 28, 20, 16, 36, 98, 13, 26, 29, 31

a. Find the mean, median, and mode of the data.

b. Find the mean, median, and mode of the data without the outlier 98.

c. Describe the effect of the outlier on the measures of central tendency.

18. **Examining Outliers** Use the data below.

58, 67, 94, 85, 78, 76, 6, 99, 100, 88, 76, 2, 82, 91, 98

a. Find the mean, median, and mode of the data.

b. Find the mean, median, and mode of the data without the outliers 2 and 6.

c. Describe the effect of the outlier on the measures of central tendency.

19. **Examining Outliers** You want to know what the "average" cost of tuition is for schools in your area. You do some research and record the results in the table below.

School	Yearly tuition
School A	$16,000
School B	$23,000
School C	$11,000
School D	$39,000
School E	$22,000

a. Find the mean, median, and mode of the data.

b. Determine the outlier.

c. Find the mean, median, and mode of the data without the outlier.

d. Describe the effect the outlier has on the measures of central tendency for tuition costs in your area.

20. Examining Outliers You want to know what the "average" price of gas is in your area. You do some research and record the results in the table below.

Gas station	Gas price/gallon
Sunset	$3.49
Circular O	$3.69
City Square	$4.29
Quick Stop	$3.35
Chase	$3.58

a. Find the mean, median, and mode of the data.

b. Determine the outlier.

c. Find the mean, median, and mode of the data without the outlier.

d. Describe the effect the outlier has on the measures of central tendency for gas prices in your area.

21. Describing Data Sets Which data set has a mean of 14.2, a median of 17.5, and a mode of 19?

Set A 20, 17, 19, 4, 17, 18, 19, 6, 3, 19

Set B 19, 6, 12, 5, 21, 16, 19, 14, 20, 2

Set C 18, 9, 16, 17, 22, 20, 4, 17, 18, 10

22. Describing Data Sets Which data set has a mean of 83, a median of 86, and a mode of 94?

Set A 94, 94, 68, 71, 88, 89, 96, 59, 66, 77, 93

Set B 92, 64, 75, 95, 99, 86, 63, 67, 91, 80, 64

Set C 86, 95, 85, 59, 70, 94, 99, 69, 90, 94, 72

23. Writing Describe a real-life situation where the mean is not a good indicator of central tendency.

24. Writing Write a real-life data set where the median is the best measure of central tendency. Explain your reasoning.

25. Problem Solving The mean of the data set is 82. Find the missing value.

Data set: 66, 80, 64, 100, ___, 99

26. Problem Solving The median of the data set is 53. Find the missing value.

Data set: 43, 48, 62, 64, ___, 45, 58

27. Problem Solving The mode of the data set is 27. Find the missing value.

Data set: 27, 20, 13, 29, 14, ___, 23

28. Problem Solving The mean of the data set is 34. Find the missing value

Data set: 40, ___, 27, 40, 41, 37, 27

29. Questions: In Your Classroom You assign your students an exercise to find the mean number of students per classroom. One student comes to you and asks, "How can the mean be 17.5?" What would you tell your student?

30. Activity: In Your Classroom Have your students collect data on an appropriate real-life topic of their choice. Students should present their data in a data display of their choice. Students should find the mean, median, and mode of the data. What questions will you ask students about their data and data displays?

31. Finding the Best Representation A stadium vendor decides to sell only one size hat this year. Last year's sales are recorded according to size. Which measure of central tendency would be useful? Explain your reasoning.

32. Finding the Best Representation A popcorn vendor decides to sell only one size of popcorn this year. Which measure of central tendency best represents the data? Explain your reasoning.

33. Temperatures The bar graph shows the high temperature in a city each day for six days. Find the mean, median, and mode of the data. Which measure of central tendency best represents the data? Explain.

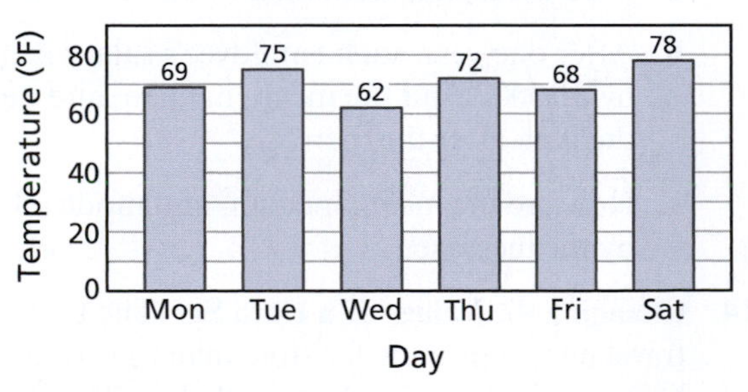

34. Gas Mileage The results of a survey of drivers on what type of gas mileage their vehicles get when traveling on the highway are shown in the bar graph. Find the mean, median, and mode of the data. Which measure of central tendency best represents the data? Explain.

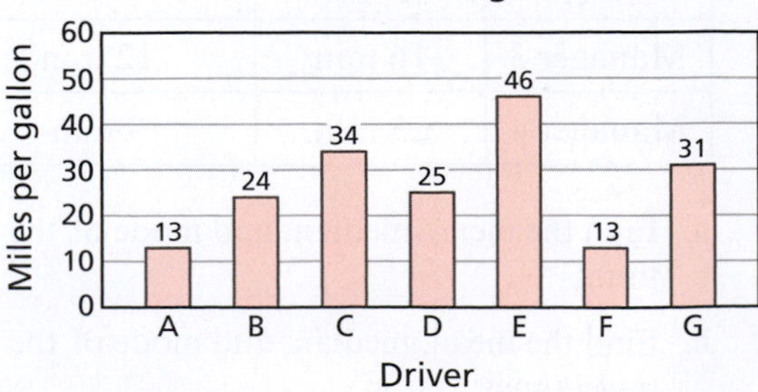

35. Reasoning The mean of five numbers is 6. When one of the numbers is removed, the mean is still 6. What is the value of the number that was removed?

36. Reasoning The mean of six numbers is 11. When one of the numbers is removed, the mean becomes 10. What is the value of the number that was removed?

37. Texting The results of a survey of a college class on how often they send text messages each day are shown in the frequency table. Which measure of central tendency can you find from the table? Do you think this best represents the whole population of cell phone owners? Explain your reasoning.

Texts sent per day	Frequency
Never	3
Rarely	7
Sometimes	16
Hourly	11
More often	2

38. Internet Connection Students were surveyed about the type of technology they use most to access the internet. The results are shown in the pictograph. Which measure(s) of central tendency can be determined from the pictograph? Find the measure(s).

Type of Technology Used Most

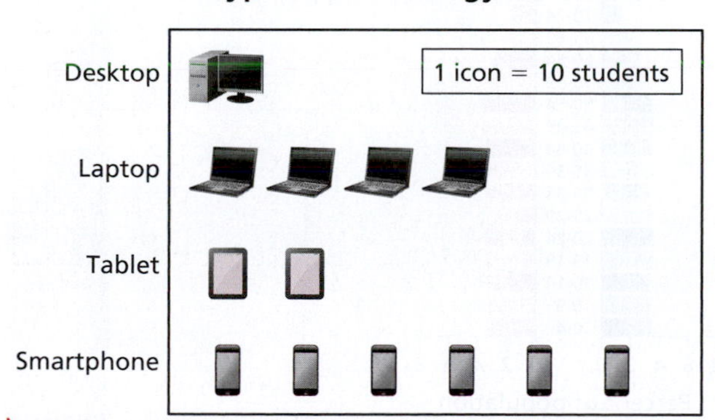

39. Grading Student Work On a diagnostic test, one of your students does the following work. Explain what the student did wrong. Which topics would you encourage the student to review?

> Find the mean, median, and mode of the set of data.
> Movie ratings (stars): 3.5, 4.5, 5.0, 4.5, 4.0, 3.0, 4.5, 3.0
>
> Mean
> $\frac{3.5 + 4.5 + 5.0 + 4.5 + 4.0 + 3.0 + 4.5 + 3.0}{8} = 4.0$
>
> Median 3.0, 3.0, 3.5, 4.0, 4.5, 4.5, 4.5, 5.0
> The median is 4.5.
>
> Mode $\frac{4.0 + 4.5}{2} = 4.25$

40. Grading Student Work On a diagnostic test, one of your students does the following work. Explain what the student did wrong. Which topics would you encourage the student to review?

> Find the mean, median, and mode of the set of data.
> Golf Scores:
> 56, 48, 61, 42, 65, 47, 52, 54, 60
>
> Mean
> $\frac{56 + 48 + 61 + 42 + 65 + 47 + 52 + 54 + 60}{9} \approx 53.9$
>
> Median 42, 47, 48, 52, 54, 56, 60, 61, 65
> The median is 54.
>
> Mode 54

41. Corrective Lenses A group of 100 adults was asked what type of corrective lenses they wear daily. The results are shown in the circle graph. Which measure(s) of central tendency can be determined from the circle graph? Find the measure(s).

Corrective Lenses

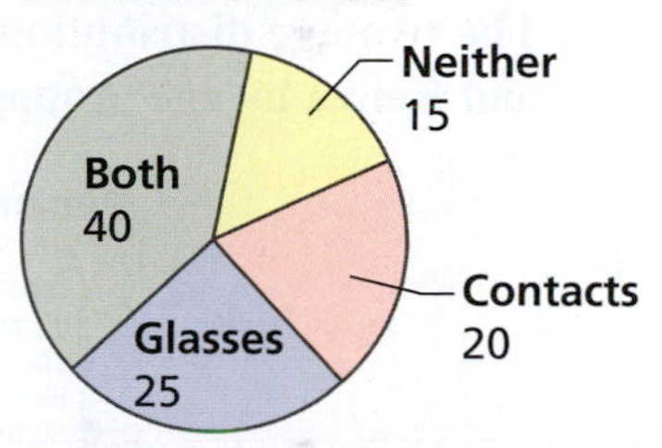

42. Cause of Death The table shows the top 5 causes of death in 2011. Which measure(s) of central tendency can be determined from the table? Find the measure(s). What information can you gather from these measures?

Cause	Percentage
Heart disease	12.9%
Stroke	11.4%
Lower respiratory infections	5.9%
Chronic obstructive pulmonary disease	5.4%
Diarrhoeal diseases	3.5%

43. Problem Solving In your class, eight students do not receive a weekly allowance, five students receive \$3, six students receive \$5, four students receive \$6, and two students receive \$8. Is the mean weekly allowance \$3.40 or \$5.00? Explain.

44. Problem Solving A set of 8 backpacks has a mean weight of 14 pounds. A different set of 12 backpacks has a mean weight of 9 pounds. What is the mean weight of all 20 backpacks? Explain your reasoning.

Activity: Analyzing the Age Distributions of Two Populations

Materials:

- Pencil

Learning Objective:

Read and analyze an age distribution pyramid for a population. Use an age distribution pyramid to estimate the median age of a population.

Name ______________________________

An *age distribution pyramid* is a graph that is used to analyze the ages of males and females in a population.

The two age distribution pyramids below show the populations of the United States and Kenya by age groups.

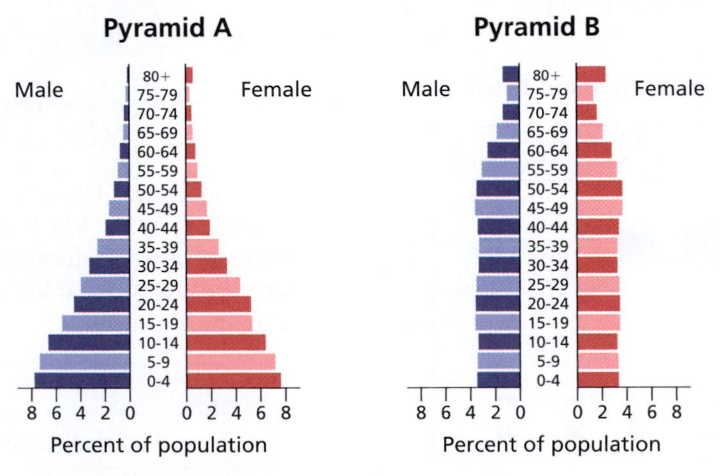

1. Which age distribution pyramid represents Kenya? Explain your reasoning.

____

2. Estimate the median age of the population of Kenya. Explain your reasoning.

____

3. Estimate the median age of the population of the United States. Explain your reasoning.

____

4. The birth rate in a country is a ratio of the number of births to the total population. Which country has the greater birth rate? Explain your reasoning.

____

Zou Zou/Shutterstock.com

17.3 Variation

Standards

Grades 6–8

Statistics and Probability

Students should develop an understanding of statistical variability and summarize and describe distributions.

- Find the range of a data set.
- Use standard deviation to describe a distribution.
- Describe the variation in a distribution, including a normal distribution.
- Compare different types of distributions.

Range

Range

The **range** of a data set is the difference of the greatest value and the least value. The range describes how spread out the data are.

Example Ordered data: 3, 4, 6, 8, 10

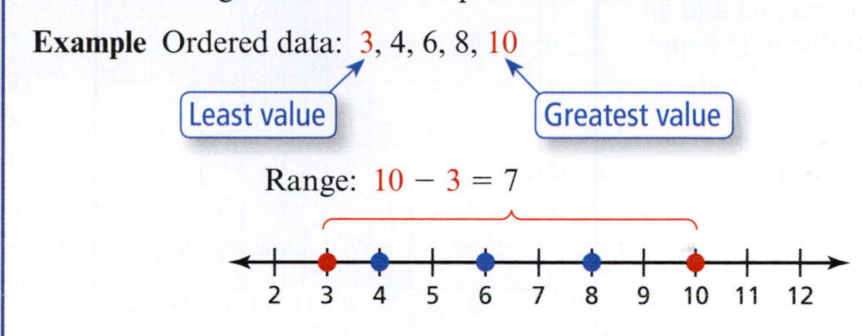

EXAMPLE 1 Finding the Range of a Data Set

The table shows the record lengths (in inches) of six venomous snakes. What is the range of the lengths?

Snake	Record length (inches)
Copperhead	53
Cottonmouth	74.5
Diamondback rattlesnake	96
Timber rattlesnake	74.5
Pygmy rattlesnake	31
Coral snake	47.5

SOLUTION

Begin by ordering the data from least to greatest.

31, 47.5, 53, 74.5, 74.5, 96

The least value is 31. The greatest value is 96. The range is the difference of the greatest value and the least value.

$$96 - 31 = 65$$

The range of the lengths is 65 inches.

Standard Deviation and Variation

Standard Deviation and Variation

Standard deviation is a measurement that shows how much **variation** or *dispersion* there is from the mean. A small standard deviation indicates the data are clustered tightly around the mean. A large standard deviation indicates the data are spread out over a large range of values.

Example In each histogram below, the mean is 11. The data set on the left has a smaller standard deviation because its data values are clustered more tightly around the mean. The data set on the right has a greater standard deviation because its data values are more spread out.

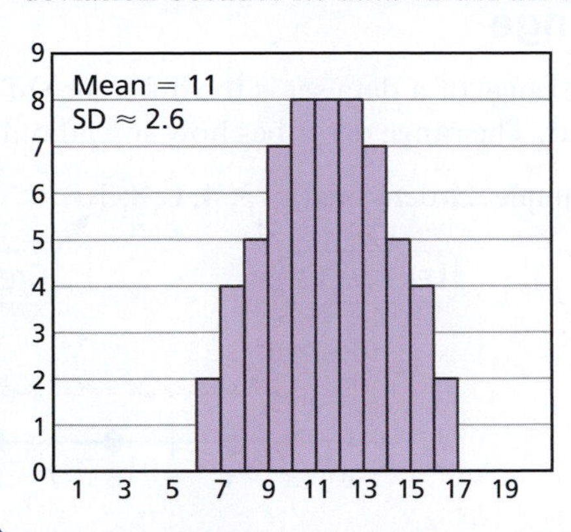

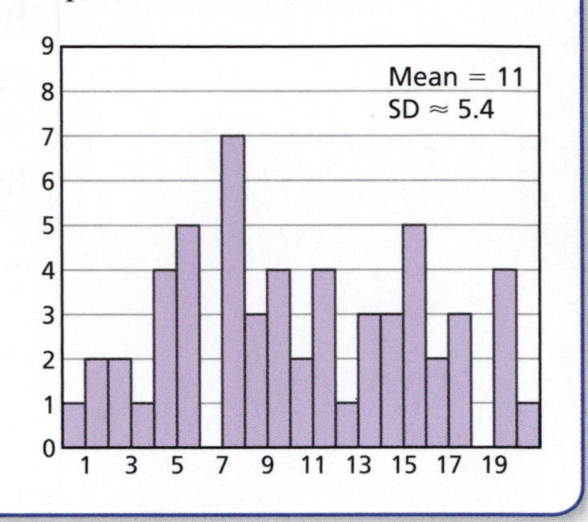

Classroom Tip

The standard deviation of a data set can be thought of as an "average deviation from the mean." To calculate the standard deviation, (a) find the difference of each value and the mean, (b) square each difference, (c) add all of the squares, (d) divide the sum by the number of data values, and (e) take the square root of the result.

EXAMPLE 2 Comparing Standard Deviations

The histograms below show the distributions of samples of heights of adult American men and women. There were 250 people in each sample. What do the standard deviations tell you about the heights in the samples?

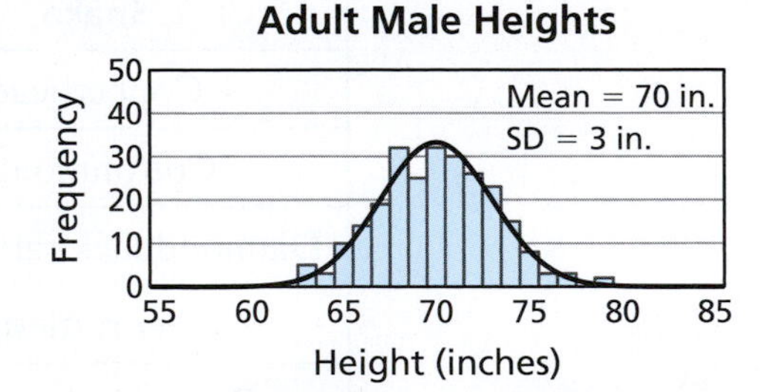

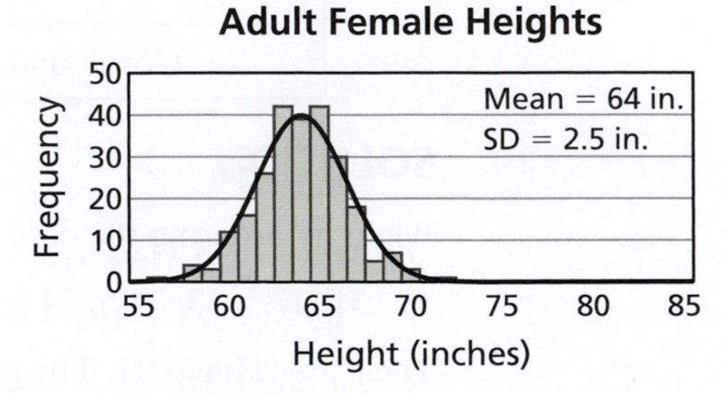

Mathematical Practices

Model with Mathematics

MP4 Mathematically proficient students routinely interpret their mathematical results in the context of the situation and the results make sense.

SOLUTION

Standard deviation is a measure of variation. The greater the standard deviation, the more the data are spread out. So, there is greater variation among the male heights than among the female heights.

Normal Distribution

Normal Distribution

A **normal distribution** is a distribution whose histogram is bell-shaped. It is symmetrical about its mean. In a normal distribution, approximately 68.2% of the numbers are within 1 standard deviation of the mean, 95.4% of the numbers are within 2 standard deviations of the mean, and 99.8% of the numbers are within 3 standard deviations of the mean.

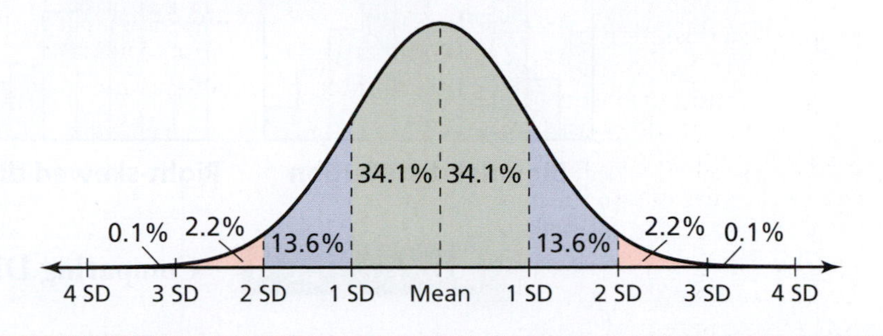

EXAMPLE 3 A Famous Normal Distribution

A famous data set was collected in Scotland in the mid-1800s. It contains the chest sizes (in inches) of 5738 men in the Scottish Militia. What percent of the chest sizes lie within 1 standard deviation of the mean?

Chest size	Number of men
33	3
34	18
35	81
36	185
37	420
38	749
39	1073
40	1079
41	934
42	658
43	370
44	92
45	50
46	21
47	4
48	1

Scottish Militiamen

Mean ≈ 40 in.
SD ≈ 2 in.

Frequency: 0, 200, 400, 600, 800, 1000, 1200

Chest size (inches): 33, 35, 37, 39, 41, 43, 45, 47

SOLUTION

The number of chest sizes within 1 standard deviation of the mean is

$$749 + 1073 + 1079 + 934 + 658 = 4493.$$

This is about 78.3% of the total number of chest sizes, which is somewhat greater than the percent predicted by a normal distribution.

The data set shown in Example 3 became famous during the 19th century because it helped mathematicians realize that many data sets involving quantities in nature have bell-shaped distributions.

Types of Distributions

Although many naturally occurring data sets have normal (bell-shaped) distributions, there are other types of distributions. Here are three of them.

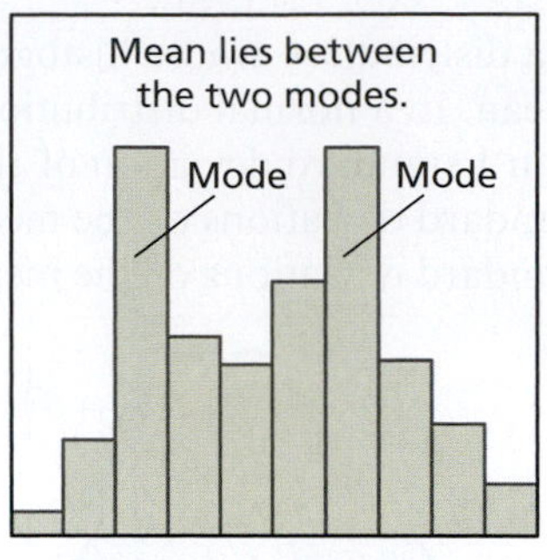

Bimodal distribution

Mean lies to the right of the median.

Right-skewed distribution

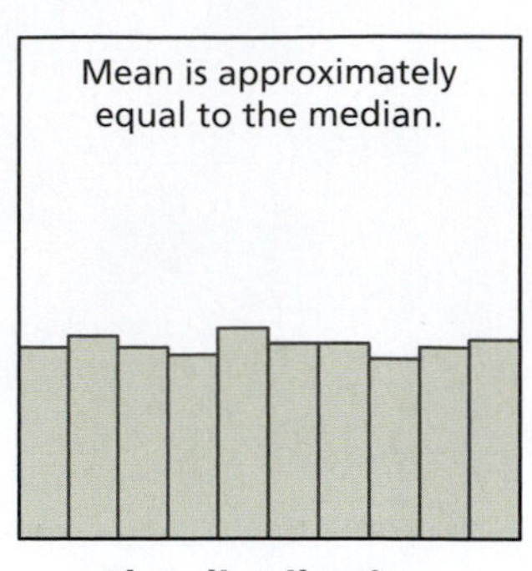

Flat distribution

Comparing Distributions

The histograms below show the distributions of game lengths (in numbers of turns) when there are two players.

a. Describe the differences in the distributions.

b. *Cootie* is recommended for players 3–6 years old. *Chutes and Ladders* is recommended for players 3–7 years old. *Candyland* is recommended for players 3–6 years old. Do these recommendations agree with the data shown in the histograms?

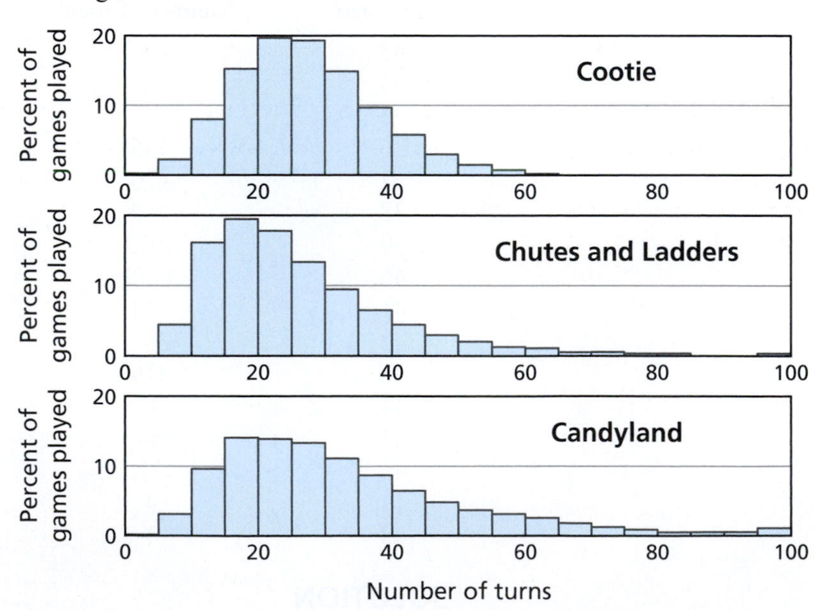

SOLUTION

a. The distribution for *Cootie* is nearly bell-shaped like a normal distribution, although it is slightly right-skewed. The distributions for *Chutes and Ladders* and *Candyland* are more strongly right-skewed.

b. The recommendations are not what you might expect. The histograms show that *Candyland* tends to take more turns than *Chutes and Ladders*. So, you would think that 7-year-olds would be more likely to be included in the recommendations for *Candyland* than for *Chutes and Ladders*.

EXAMPLE 5 Comparing Flat Distributions

The number cube shown is rolled 60 times, 600 times, and 6000 times. The bar graphs show the results in mixed order. Which bar graph shows the results for each number of rolls? Explain.

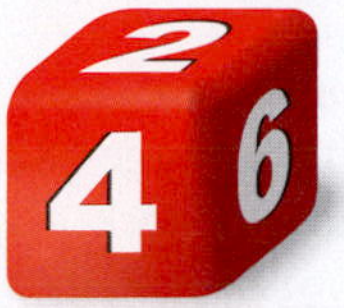

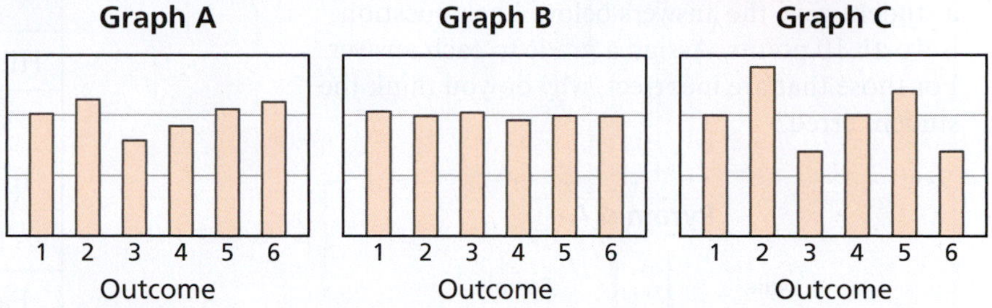

SOLUTION

When a number cube is rolled, each outcome (1, 2, 3, 4, 5, 6) is likely to occur about $\frac{1}{6}$ of the time. In fact, the greater the number of rolls, the more likely it is that each outcome will occur about $\frac{1}{6}$ of the time. So, a greater number of rolls tends to result in a flatter distribution. Based on the flatness of the distributions, you can assume that the results of 60 rolls are shown by Graph C, the results of 600 rolls are shown by Graph A, and the results of 6000 rolls are shown by Graph B.

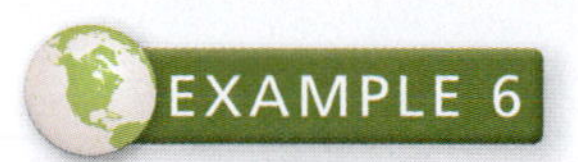

EXAMPLE 6 Describing a Distribution

Classroom Tip

Examples 4 and 6 both show how distributions can be used by researchers in business and manufacturing. By drawing a histogram for a data set, you are often able to see patterns that would be difficult to see by simply looking at the set of numbers.

The graph shows the full-time starting salaries of 18,630 law school graduates from the class of 2011. What type of distribution is shown by the graph? What does the graph show about the salaries of these graduates?

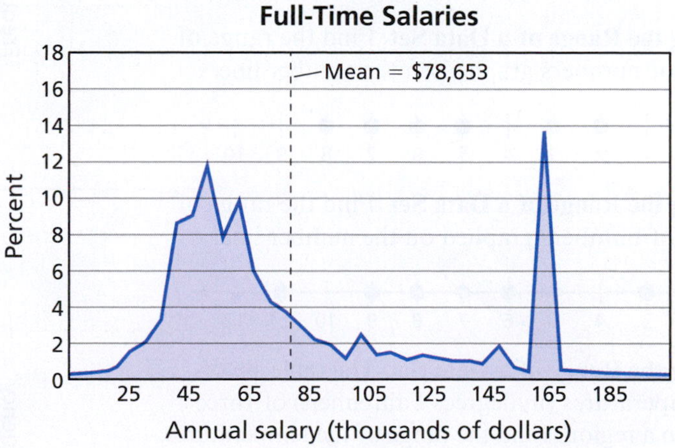

SOLUTION

The graph has two major peaks, so the distribution is approximately bimodal. Most of the salaries were close to the modes, shown by the peaks. The peak at the right shows that 14% of the salaries were \$160,000. The peak at the left involves a wider interval and shows that many (52%) of the salaries were between \$40,000 and \$65,000. Notice that the mean, \$78,653, is between the two peaks, and few of the salaries were close to the mean.

17.3 Exercises

Free step-by-step solutions to odd-numbered exercises at MathematicalPractices.com.

1. **Writing a Solution Key** Write a solution key for the activity on page 682. Describe a rubric for grading a student's work.

2. **Grading the Activity** In the activity on page 682, a student gave the answers below. Each question is worth 10 points. Assign a grade to each answer. For those that are incorrect, why do you think the student erred?

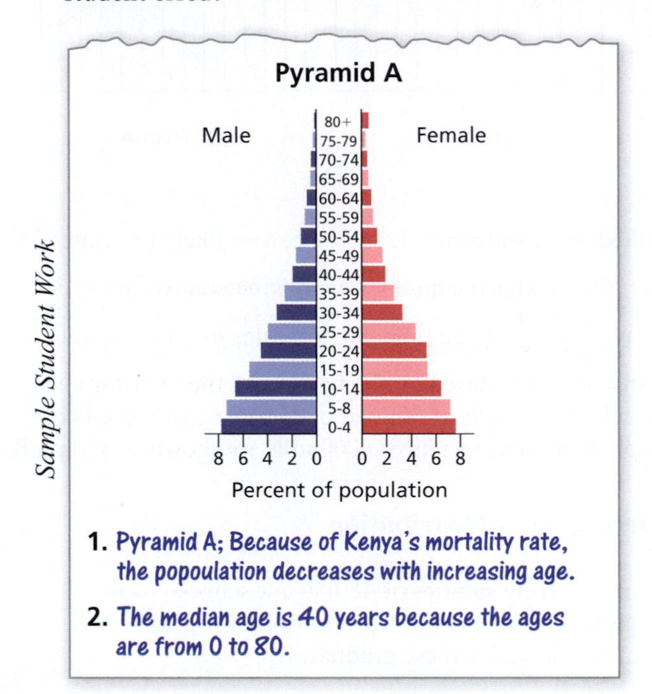

3. **Finding the Range of a Data Set** Find the range of the set of numbers graphed on the number line.

4. **Finding the Range of a Data Set** Find the range of the set of numbers graphed on the number line.

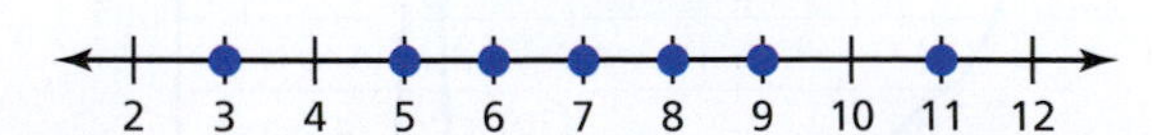

5. **Finding the Range of a Data Set** The table shows the temperatures (in degrees Fahrenheit) of three towns in a region. Find the range of the temperatures.

Town	Temperature
Mount Hope	72°F
Grove Valley	83°F
Daintree	76°F

6. **Finding the Range of a Data Set** The table shows the prices of a gallon of skim milk at seven stores. Find the range of the prices.

Store	Price
Hillside Grocery	$3.49
Value Foods	$3.37
In & Out Shop	$4.11
Dairy Stop	$3.52
Farmer's Mart	$3.26
Lee's Market	$3.42
One Stop Shop	$3.66

7. **Comparing Standard Deviations** The histograms below represent data sets with the same mean (15) but different standard deviations (1.58 and 2.41). For which histogram is the standard deviation 1.58? Explain.

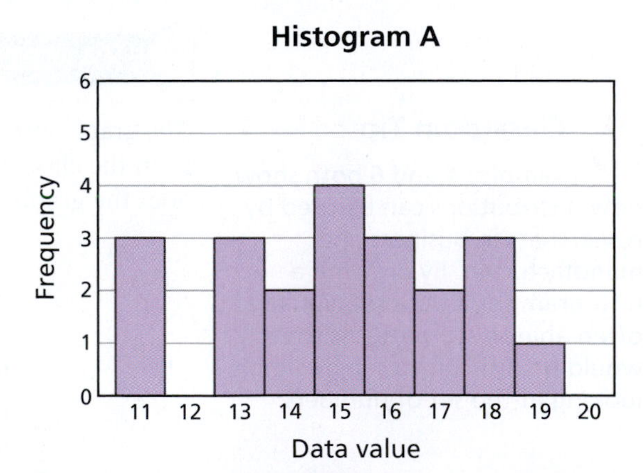

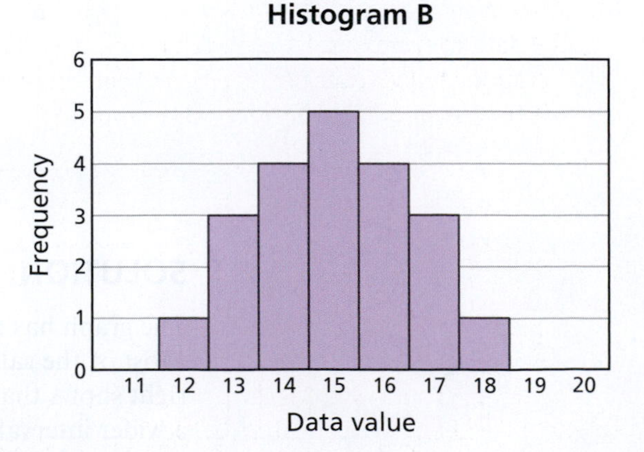

8. Comparing Standard Deviations The histograms below represent data sets with the same mean (9) but different standard deviations (3.06 and 4.24). For which histogram is the standard deviation 3.06? Explain.

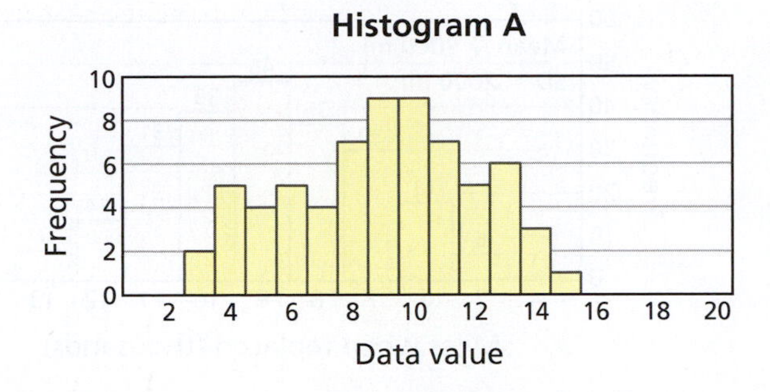

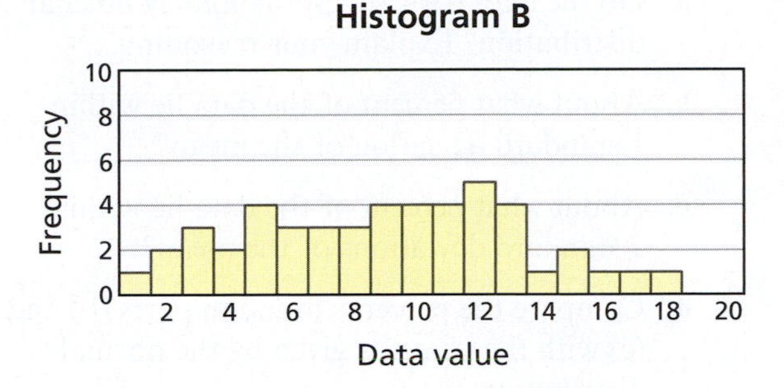

9. Analyzing Standard Deviation What percent of the data represented by the bar graph lie within 1 standard deviation of the mean? What percent lie within 2 standard deviations of the mean?

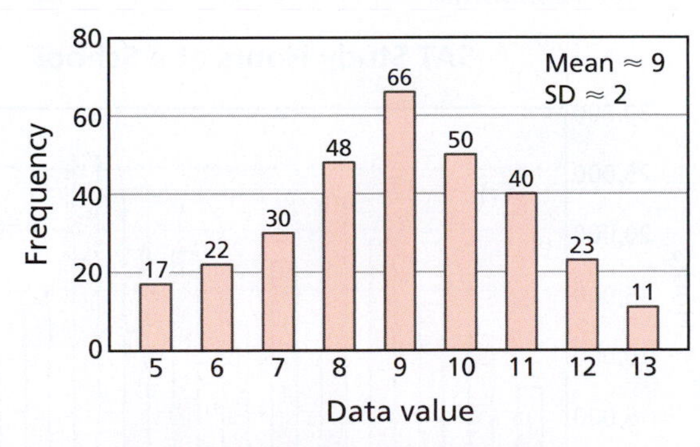

10. Analyzing Standard Deviation What percent of the data represented by the bar graph lie within 1 standard deviation of the mean? What percent lie within 2 standard deviations of the mean?

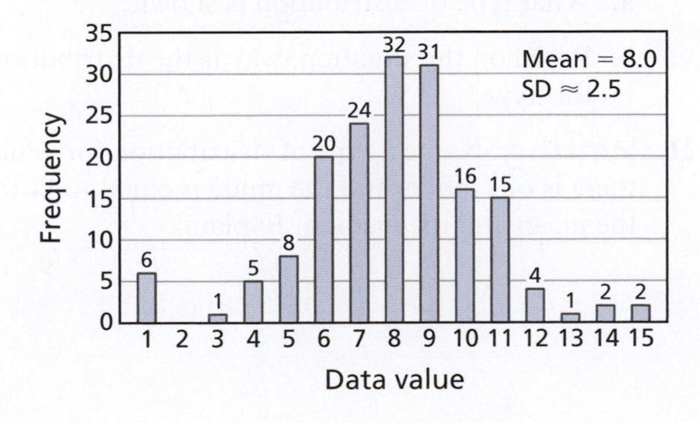

11. Analyzing a Normal Distribution The data in a data set are normally distributed with a mean of 34 and a standard deviation of 11. Estimate the percent of the data that are between 12 and 56.

12. Analyzing a Normal Distribution The data in a data set are normally distributed with a mean of 150 and a standard deviation of 30. Estimate the percent of the data that are less than 120 or greater than 180.

13. Describing a Distribution Is the data set shown best described as a *normal*, *bimodal*, *right-skewed*, *left-skewed*, or *flat* distribution? How do the mean and median of the data set compare?

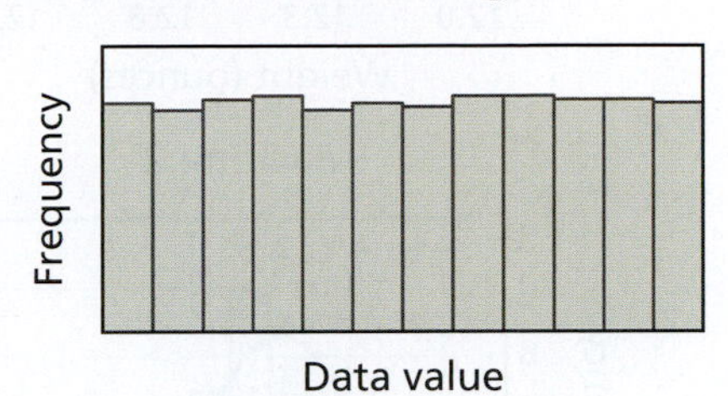

14. Describing a Distribution Is the data set shown best described as a *normal*, *bimodal*, *right-skewed*, *left-skewed*, or *flat* distribution? How do the mean, median, and mode(s) of the data set compare?

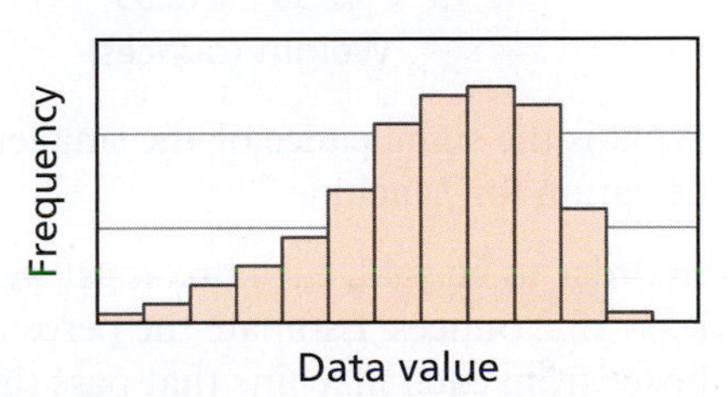

15. Describing a Distribution Is the data set shown best described as a *normal*, *bimodal*, *right-skewed*, *left-skewed*, or *flat* distribution? How do the mean and mode(s) of the data set compare?

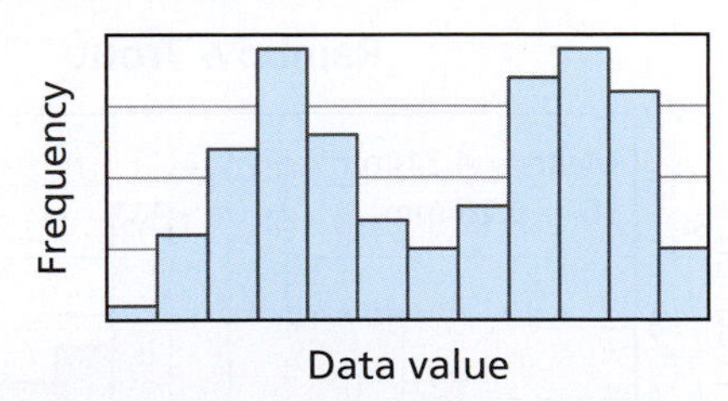

16. Describing a Distribution Is the data set shown best described as a *normal*, *bimodal*, *right-skewed*, *left-skewed*, or *flat* distribution? How do the mean, median, and mode(s) of the data set compare?

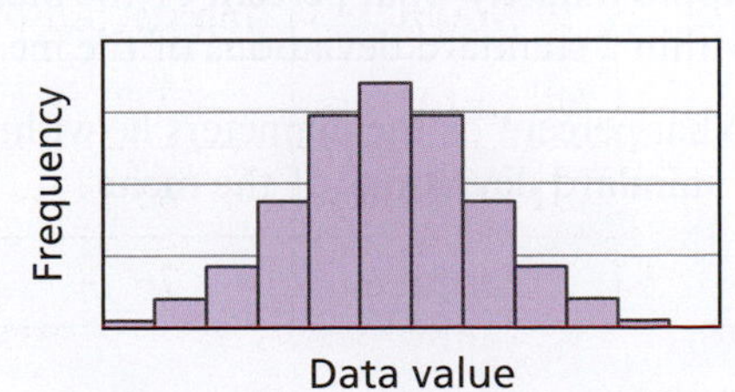

17. Comparing Standard Deviations The graphs show the distributions of weights of boxes of cereal filled by two machines.

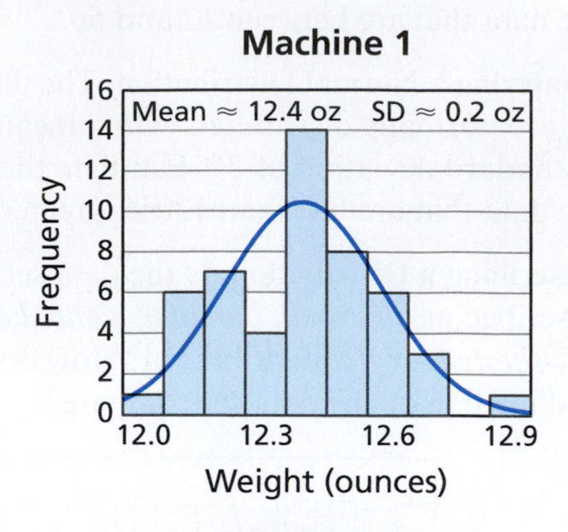

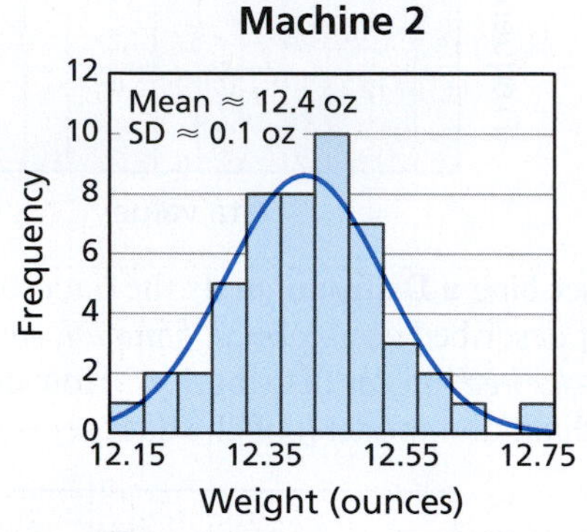

a. What is the significance of the smaller standard deviation for Machine 2?

b. In order to be sold, each box must weigh at least 12.2 ounces. Estimate the percent of boxes from each machine that pass this requirement.

18. Using Standard Deviations The histogram shows the distribution of the diameters of 100 rainbow trout eggs.

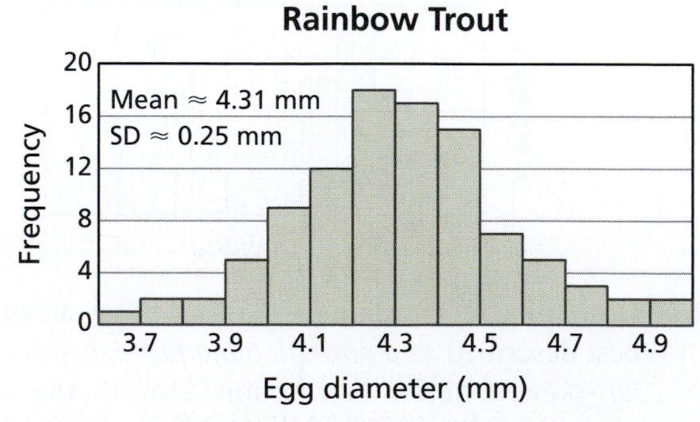

a. Approximately what percent of the diameters lie within 2 standard deviations of the mean?

b. What percent of the diameters lie within 3 standard deviations of the mean?

19. Using Standard Deviations The histogram shows the distribution of the numbers of miles of wear on 230 motorcycle tires of the same model when they were replaced.

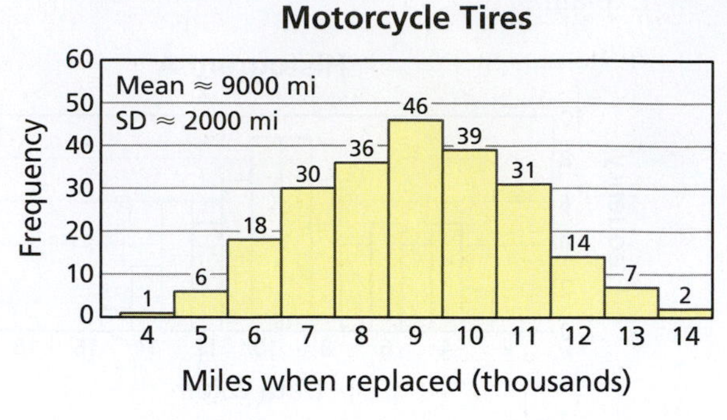

a. Do the data have an approximately normal distribution? Explain your reasoning.

b. About what percent of the data lie within 1 standard deviation of the mean?

c. About what percent of the data lie within 2 standard deviations of the mean?

d. Compare the percents found in parts (b) and (c) with the percents given by the normal distribution.

20. Describing a Distribution The bar graph shows the distribution of the total numbers of hours students at a school spent studying for the SAT over 12 months.

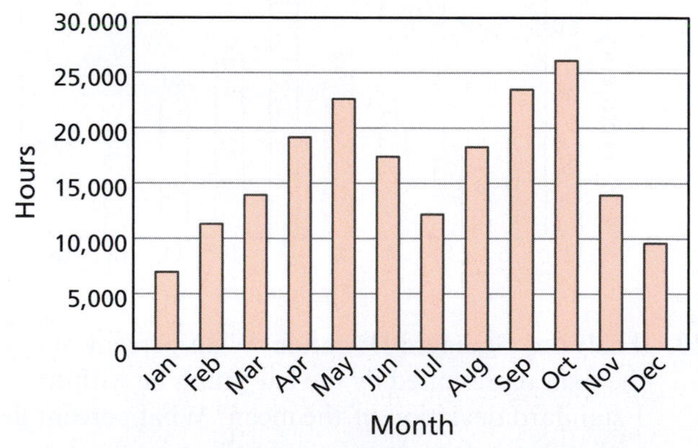

a. What type of distribution is shown?

b. Based on the situation, why is the distribution of this type?

21. *Writing* Name a type of distribution for which there is one mode, and the mode is equal to both the mean and the median. Explain.

22. *Writing* In a right-skewed distribution, are more of the data likely to be *greater* than the mode or *less* than the mode? Explain.

23. **Grading Student Work** On a diagnostic test, one of your students does the following work. Explain what the student did wrong. Which topics would you encourage the student to review?

Species	Adult length
Blue whale	29 m
Fin whale	26.8 m
Gray whale	13.8 m
Killer whale	9.6 m
Sperm whale	16.5 m

Range = 29 − 16.5 = 12.5

24. **Grading Student Work** On a diagnostic test, one of your students does the following work. Explain what the student did wrong. Which topics would you encourage the student to review?

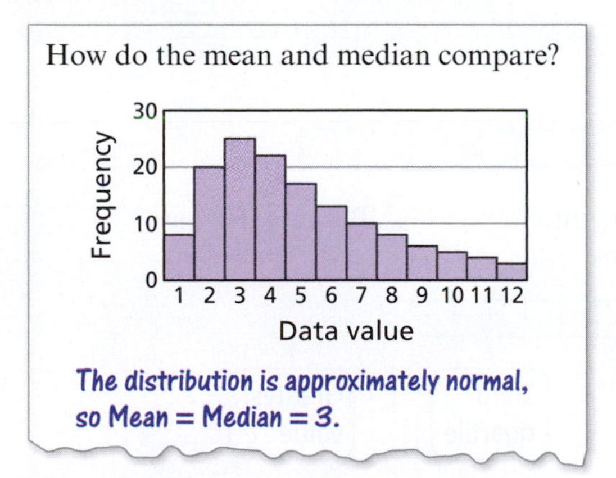

25. **Activity: In Your Classroom** Design an activity that you can use with your students to analyze a real-life distribution by finding the totals of two number cubes rolled together 100 (or more) times.

a. Specify how you will tell your students to display their results to determine what type of distribution the results show.

b. What type of distribution do you expect the results to show?

c. Explain how you might use the table in Example 2 on page 626 to discuss with students how their distributions compare with the expected results.

26. **Activity: In Your Classroom** Roll a number cube 4 times and record the numbers rolled. Find the mean and standard deviation of the numbers. (A method for finding the standard deviation is given in the classroom tip on page 684.) Repeat as necessary to answer the following questions.

a. What is the smallest possible standard deviation of the four numbers resulting from 4 rolls of a number cube? Describe a set of four numbers that has the smallest possible standard deviation.

b. Rolling what four numbers produces the greatest possible standard deviation?

c. What concepts might this activity reinforce for your students?

27. **Comparing Distributions** The histograms show the distributions of male mathematics and writing SAT scores in a recent year.

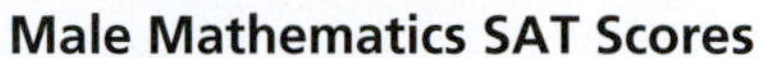

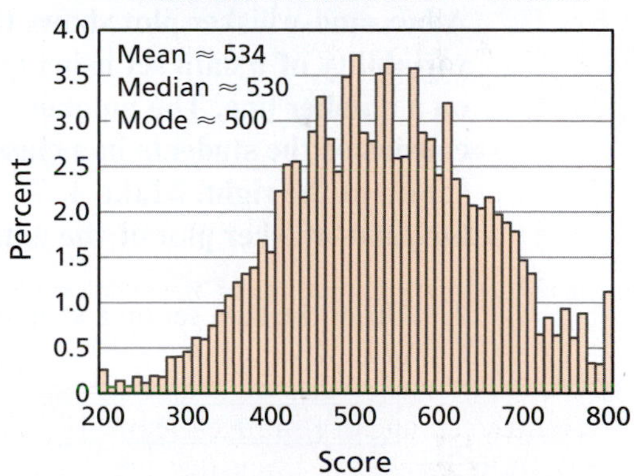

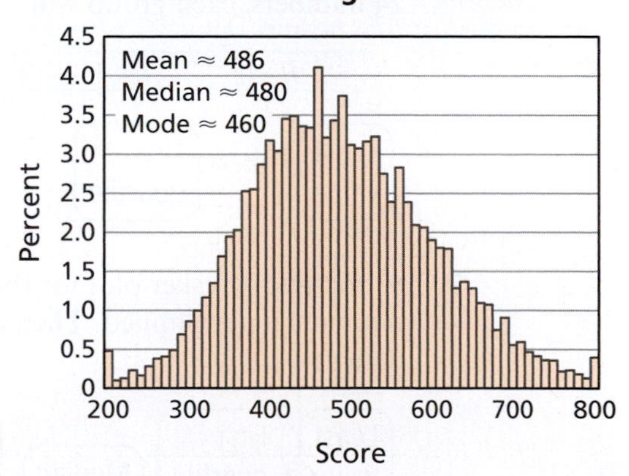

a. Describe the differences in the distributions of the male mathematics and writing SAT scores.

b. Does the distribution of male writing SAT scores appear to be skewed? If so, is this supported by the values of the mean and the median? Explain.

Activity: Drawing and Reading a Box-and-Whisker Plot

Materials:

- Pencil
- Grid paper

Learning Objective:

Discover how to find the first and third quartiles for a data set. Discover how a box-and-whisker plot represents a data set.

Name ______________________________

A box-and-whisker plot shows the variability of a data set using quartiles on a number line. The numbers of first cousins of the students in a class are shown at the right. Make a box-and-whisker plot of the data.

Numbers of First Cousins

3	10	18	8
9	3	0	32
23	19	13	8
6	3	3	10
12	45	1	5
13	24	16	14

1. Order the data set on a strip of grid paper.

2. Fold the paper in half twice to divide the data into four groups. Because there are 24 numbers, each group will have six numbers.

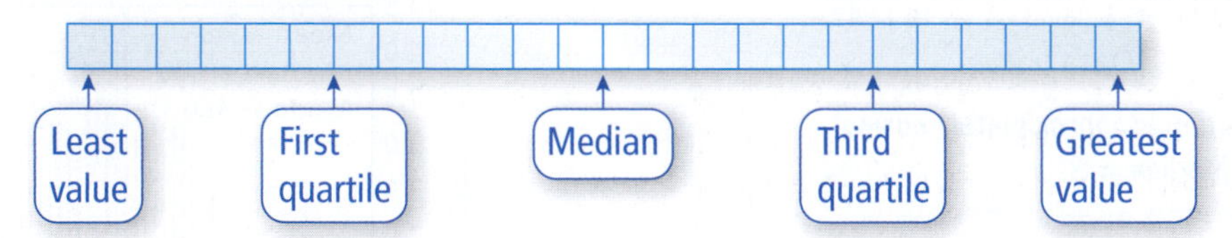

3. A box-and-whisker plot for the data is shown below. Explain how each of the five numbers is determined. Then explain how the box-and-whisker plot represents the data set.

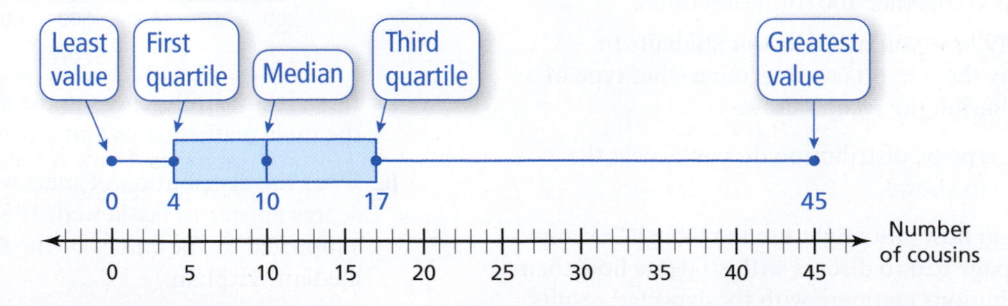

17.4 Other Types of Graphs

Standards

Grades K–2
Measurement and Data
Students should represent and interpret data.

Grades 3–5
Measurement and Data
Students should represent and interpret data.

Grades 6–8
Statistics and Probability
Students should develop an understanding of statistical variability and summarize and describe distributions.

- Read and make box-and-whisker plots and line plots.
- Choose an appropriate type of graph for a data set.
- Recognize when a graph is misleading.

Box-and-Whisker Plots and Line Plots

Box-and-Whisker Plot

A **box-and-whisker plot** displays a data set along a number line using medians. **Quartiles** divide the data set into four equal parts. The median (second quartile) divides the data set into two halves. The median of the lower half is the first quartile. The median of the upper half is the third quartile.

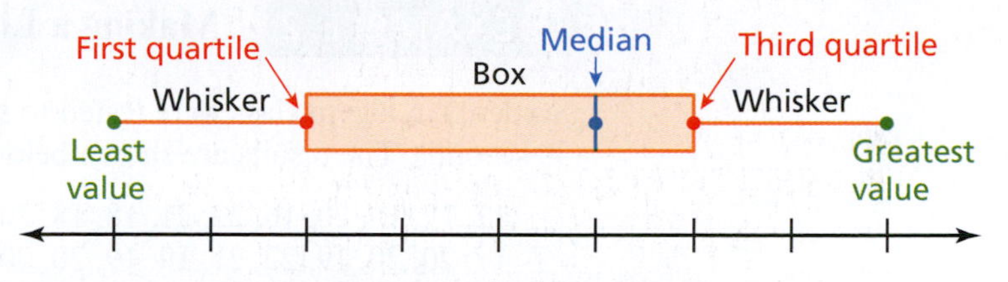

EXAMPLE 1 Making a Box-and-Whisker Plot

Make a box-and-whisker plot for the ages of the players on a basketball team.

24, 30, 30, 22, 25, 22, 18, 25, 28, 30, 25, 27

SOLUTION

Step 1 Order the data. Find the median and the quartiles.

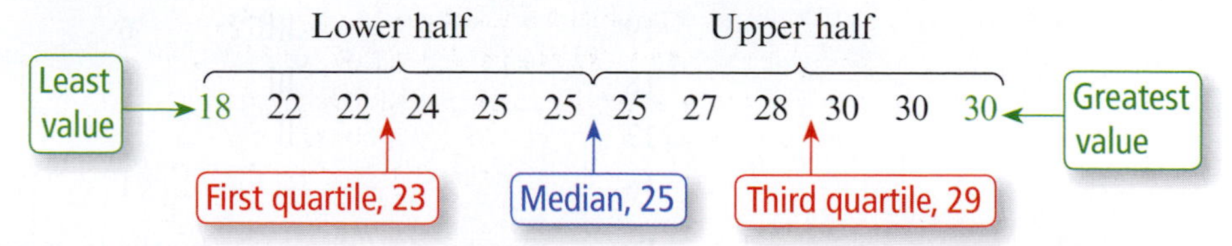

Step 2 Draw a number line that includes the least and greatest values. Graph points above the number line for the least value, greatest value, median, first quartile, and third quartile.

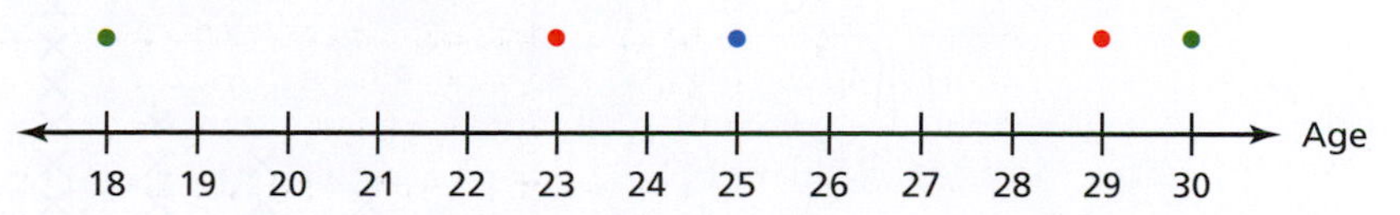

Step 3 Draw a box using the first and third quartiles. Draw a vertical line through the median. Draw whiskers from the box to the least and greatest values. Add a title to the graph.

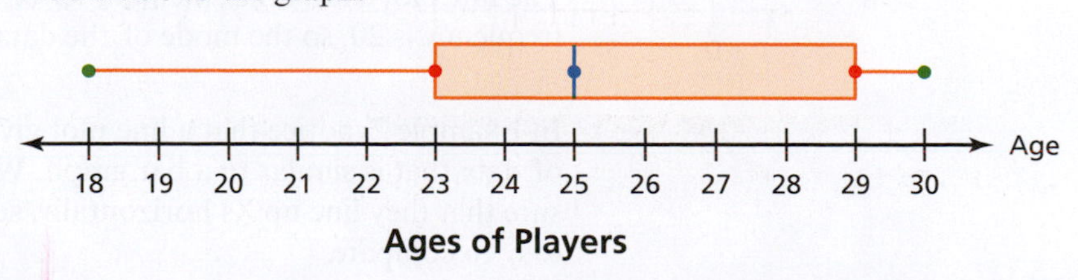

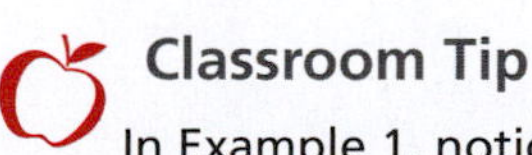

Classroom Tip

In Example 1, notice that the box-and-whisker plot gives a quick overview of how the data are distributed. By looking at the box-and-whisker plot, you can quickly observe the following.

1. The ages are from 18 through 30.
2. Half of the ages are from 23 through 29.
3. The median age is 25.

Line Plot

A **line plot** shows data on a number line using Xs (or other marks) to show frequencies.

Example The number of coins in each stack is represented in the line plot by an X above the number line.

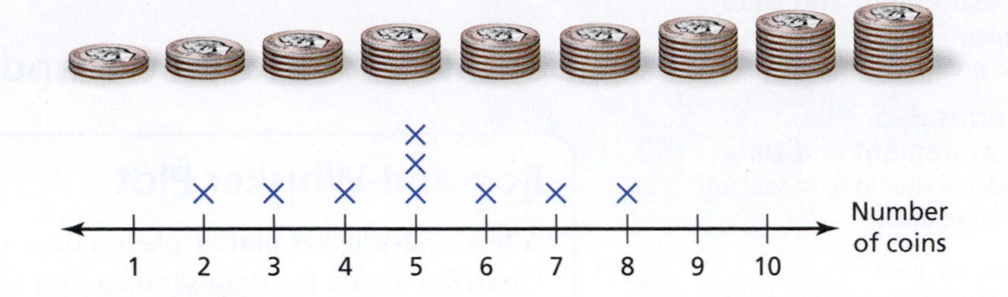

EXAMPLE 2 Making a Line Plot

Students in a gym class were tested to see how many sit-ups they could do in 30 seconds. The results are shown below.

17, 19, 19, 20, 16, 22, 21, 18, 18, 20, 18, 21,
17, 20, 20, 19, 22, 21, 19, 19, 20, 20, 19, 20

Make a line plot for the data. What is the mode of the data?

SOLUTION

Begin by making a frequency table for the data.

Number of sit-ups	Tally	Frequency
22	II	2
21	III	3
20	~~IIII~~ II	7
19	~~IIII~~ I	6
18	III	3
17	II	2
16	I	1

Draw a number line that includes the least and greatest values. Draw Xs above the number line to represent the frequencies.

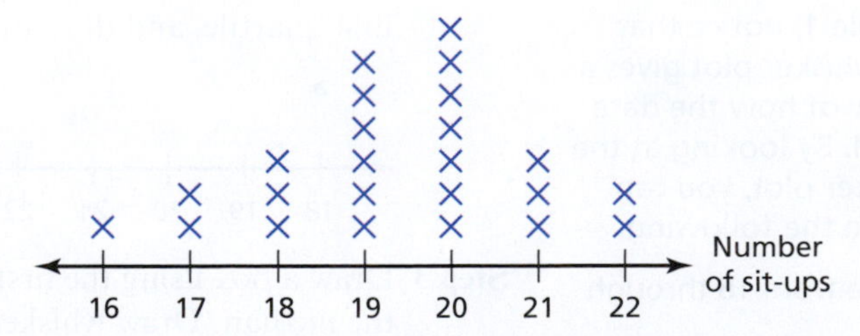

The line plot shows that the number of sit-ups that occurred with the greatest frequency is 20, so the mode of the data is 20 sit-ups.

In Example 2, notice that a line plot gives a visual overview of the distribution of data that is similar to a bar graph. When your students make line plots, be sure that they line up Xs horizontally, so that the frequencies in the columns are easy to compare.

Summary of Types of Statistical Graphs

Here is a summary of several of the more common types of data displays.

Pictograph — Shows data using pictures

Bar graph — Shows data in specific categories

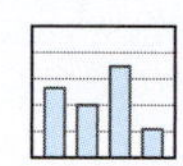

Circle graph — Shows data as parts of a whole

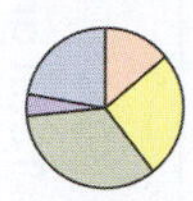

Line graph — Shows how data change over time

Histogram — Shows frequencies of data values in intervals of the same size

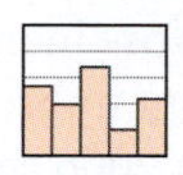

Stem-and-leaf plot — Orders numerical data and shows how they are distributed

1	0 2 3 6
2	1 1 5
3	9
4	0 6

Box-and-whisker plot — Shows the variability of a data set using quartiles

Line plot — Shows the number of times each value occurs in a data set

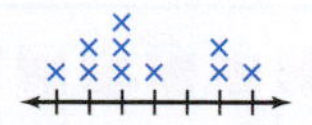

Scatter plot — Shows the relationship between two data sets using ordered pairs in a coordinate plane

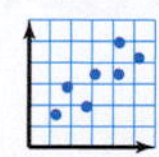

Choosing an Appropriate Data Display

Choose an appropriate data display for the situation. Explain your reasoning.

a. The number of students in a marching band each year

b. A comparison of the shoe sizes and heights of several people

SOLUTION

a. You want to show change over time. A line graph is an appropriate data display.

b. You want to compare two data sets. A scatter plot is an appropriate data display.

Misleading Graphs

Although graphs of data sets can help you see the big picture and recognize trends, graphs can also be misleading. Sadly, sometimes this is intentional. Because of this, when you use a graph to form or support a conclusion, be sure that the graph represents the data well.

A Misleading Line Graph

Which line graph is misleading? Why?

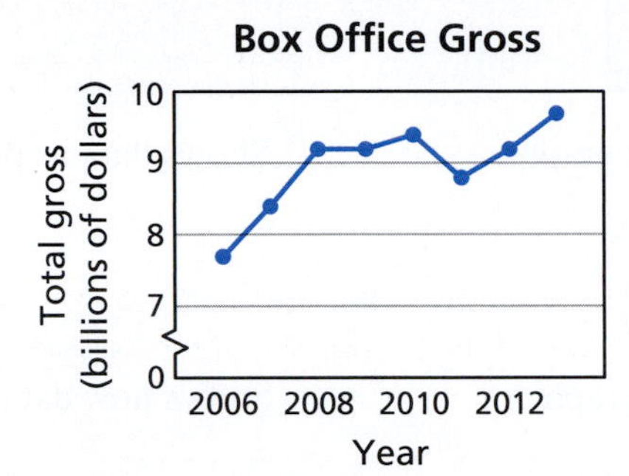

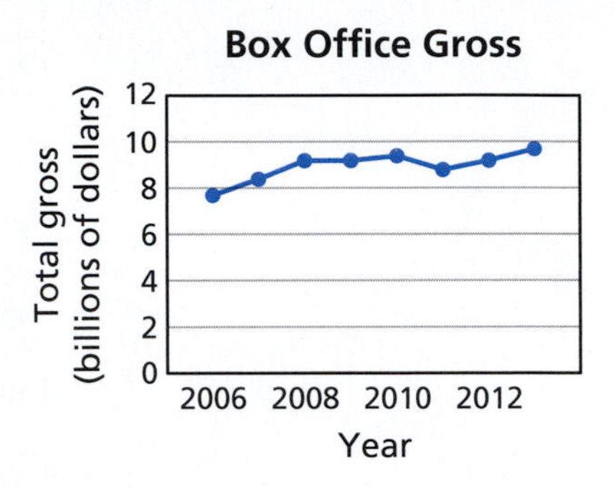

Classroom Tip

Using a broken vertical axis is the most common way that the media uses to create misleading graphs. A break in the vertical or horizontal axis, however, can sometimes help to make a graph more readable. Ask your students to find examples of graphs with broken vertical or horizontal axes in magazines, newspapers, or on the Internet. Have them analyze whether the breaks in the axes make the graphs misleading or more readable.

SOLUTION

The vertical axis of the line graph on the left has a break and begins at 7. This graph makes it appear that the total gross increased rapidly from 2006 through 2008. The line graph on the right has an unbroken vertical axis. This graph shows the relatively slow rate of growth from 2006 through 2008 more accurately. So, the graph on the left is misleading.

A Misleading Pictograph

From the pictograph, a person concludes that the numbers of cans and boxes of food donated are about the same. Is this conclusion accurate? Explain your reasoning.

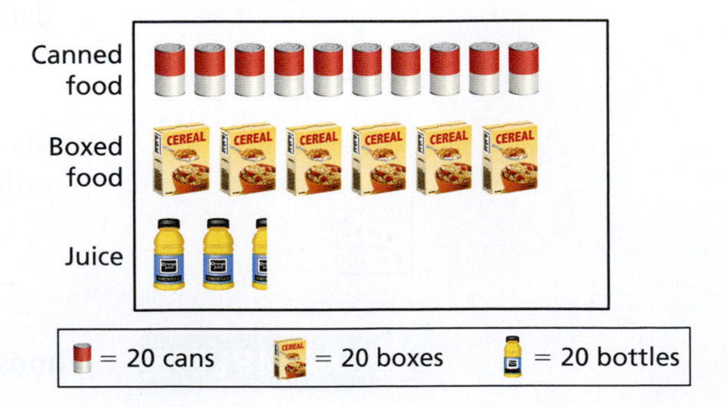

SOLUTION

The person assumed that the numbers of cans and boxes of food donated were about the same because the lengths of their rows in the pictograph are about the same. The key of the pictograph, however, shows that each symbol represents 20 items. By counting the symbols, you can see that about 200 cans and 120 boxes of food were donated. So, the conclusion is not accurate.

EXAMPLE 6 A Misleading Bar Graph

Why is the bar graph misleading?

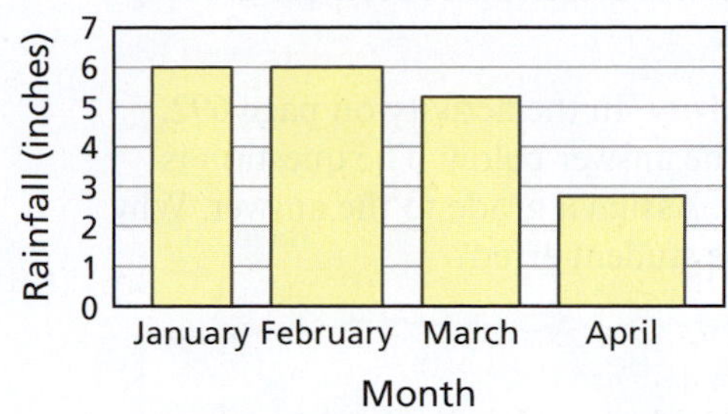

SOLUTION

The bars have different widths. This may give the perception that the wider bars represent greater rainfall amounts. In a bar graph, each bar should have the same width.

EXAMPLE 7 A Misleading Histogram

Mathematical Practices

Construct Viable Arguments and Critique the Reasoning of Others

MP3 Mathematically proficient students read the arguments of others, decide whether they make sense, and ask useful questions to clarify or improve the arguments.

The following data represent the number of miles each student in a gym class walked during a week.

1, 1, 2, 2, 2, 2, 2, 3, 3, 3, 3, 3, 3,

4, 4, 4, 4, 4, 4, 4, 5, 5, 5, 5, 5, 5, 5, 5,

6, 6, 6, 6, 6, 7, 7, 7, 8, 8, 8, 9, 9, 9, 9, 10, 11, 12

One of the students made the following histogram of the data. Is the histogram misleading? Explain your reasoning.

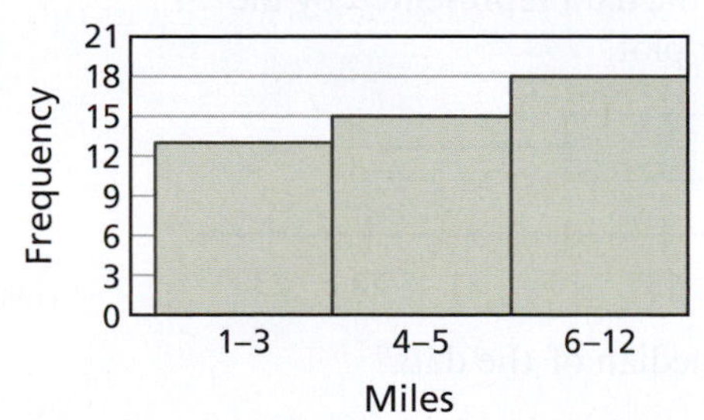

SOLUTION

The histogram is misleading. The intervals 1–3, 4–5, and 6–12 all have different sizes. In a histogram, the intervals on the horizontal axis must be the same size. Here is an accurate histogram of the data.

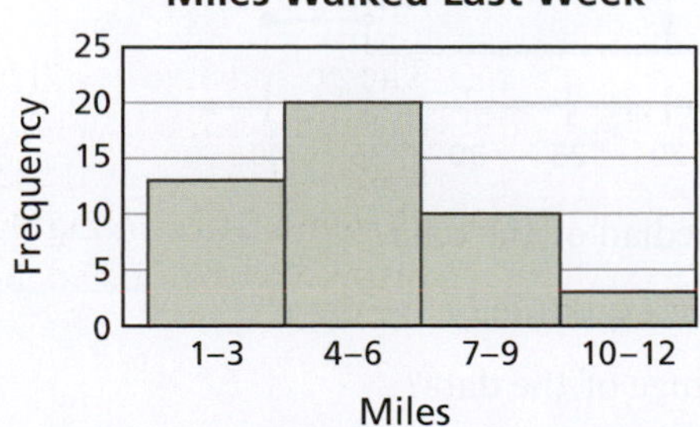

17.4 Exercises

Free step-by-step solutions to odd-numbered exercises at MathematicalPractices.com.

1. **Writing a Solution Key** Write a solution key for the activity on page 692. Describe a rubric for grading a student's work.

2. **Grading the Activity** In the activity on page 692, a student gave the answer below. The question is worth 10 points. Assign a grade to the answer. Why do you think the student erred?

Sample Student Work

> 3. A box-and-whisker plot for the data is shown below. Explain how each of the five numbers is determined. Then explain how the box-and-whisker plot represents the data set.
>
> The least of the data values is 0.
>
> The first quartile, 4, is the mean of the lower half of the data.
>
> The median 10 is the value closest to the middle.
>
> The third quartile, 17, is the mean of the upper half of the data.
>
> The greatest of the data values is 45.
>
> The box-and-whisker plot shows the number of data values in each quarter of the data.

3. **Interpreting a Box-and-Whisker Plot** Answer the questions about the data represented by the box-and-whisker plot.

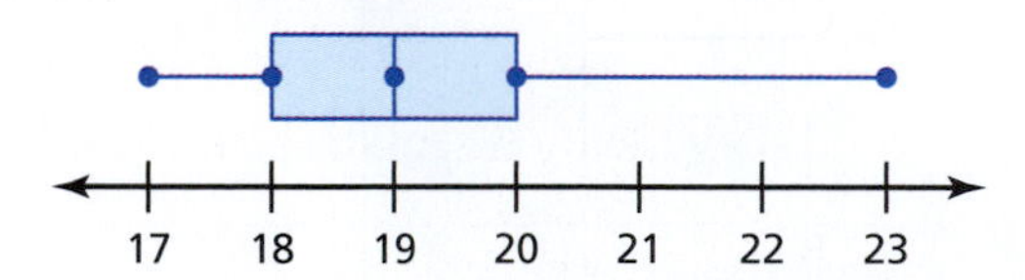

a. What is the median of the data?

b. What is the first quartile of the data?

c. What is the range of the data?

4. **Interpreting a Box-and-Whisker Plot** Answer the questions about the data represented by the box-and-whisker plot.

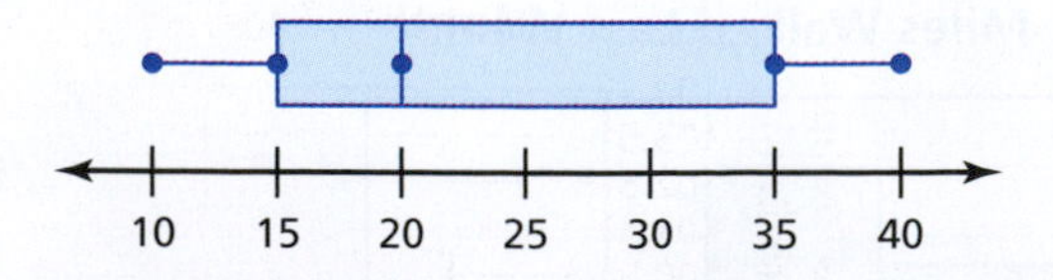

a. What is the median of the data?

b. What is the third quartile of the data?

c. What is the range of the data?

5. **Making a Box-and-Whisker Plot** Make a box-and-whisker plot for the data.

a. Hours studying: 2, 1, 3, 1, 4, 0, 5, 1, 1, 2, 3, 2

b. Donations (dollars): 10, 30, 5, 15, 50, 25, 5, 20, 15, 35, 10, 30, 20, 10, 20, 5

6. **Making a Box-and-Whisker Plot** Make a box-and-whisker plot for the data.

a. Quiz scores: 8, 12, 9, 10, 12, 8, 5, 9, 10, 8, 9, 11

b. Ski lengths (centimeters): 180, 175, 205, 165, 160, 210, 175, 180, 190, 205, 190, 160, 165, 195, 190, 200

7. **Making a Line Plot** Make a line plot for the data.

a. Credit hour load: 15, 14, 17, 12, 16, 17, 15, 14, 18, 17, 14, 15, 16, 17, 14, 12, 13, 17, 15

b. Checkout time (minutes): 3, 2, 3, 4, 5, 5, 2, 1, 3, 6, 4, 5, 2, 1, 3, 2, 1, 1, 1, 2, 3, 2, 4

8. **Making a Line Plot** Make a line plot for the data.

a. Team errors by game: 4, 3, 4, 2, 5, 1, 7, 3, 4, 5, 1, 0, 1, 2, 2, 1, 3, 2

b. Clearance shoe sizes: 11, 8.5, 11.5, 11, 7, 8, 7.5, 8.5, 8.5, 11.5, 11.5, 11.5, 11, 7, 7, 11.5, 8.5, 8.5, 11.5, 11, 11.5, 11

9. **Choosing a Data Display** Display the data in a way that best describes the data. Explain your choice of display.

Notebooks sold in one week		
192 red	170 green	203 black
183 pink	448 blue	165 yellow
210 purple	250 orange	179 white

10. **Choosing a Data Display** Display the data in a way that best shows the numerical order of the data and the distribution of the data by decade. Explain your choice of display.

Players' birth years					
1994	1995	1979	1992	1989	1974
1995	1984	1983	1976	1984	1993
1976	1995	1989	1982	1991	1980

11. **Choosing a Data Display** Choose an appropriate data display for the situation. Explain your reasoning.
    a. The variability of the test scores of a class
    b. The amount in a savings fund from month to month
    c. The outcomes of rolling a number cube 60 times
    d. The fraction of the student body in each grade

12. **Choosing a Data Display** Choose an appropriate data display for the situation. Explain your reasoning.
    a. The heights of classmates by intervals
    b. The number of birds observed by species using pictures
    c. The finish times of marathoners by quartiles
    d. The relationship between wage and years of experience for several employees

13. **A Misleading Line Graph** Explain why the line graph is misleading.

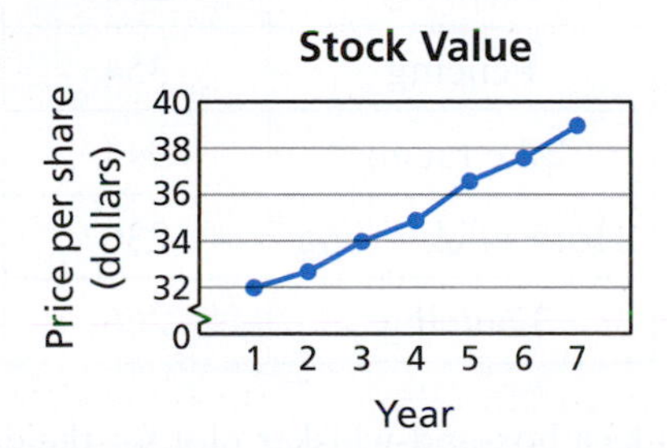

14. **A Misleading Line Graph** Explain why the line graph is misleading.

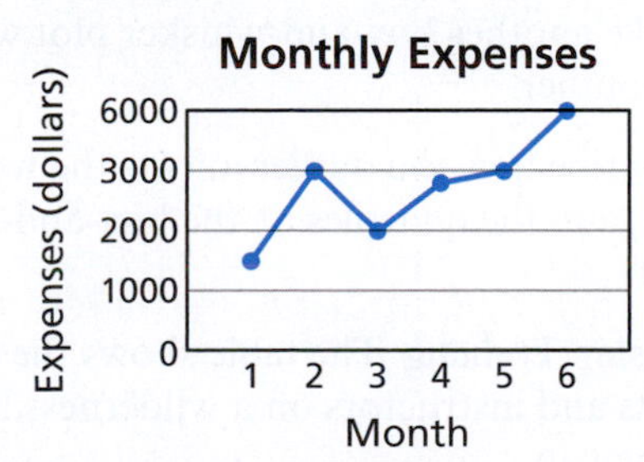

15. **A Misleading Pictograph** Explain why the pictograph is misleading.

16. **A Misleading Pictograph** Explain why the pictograph is misleading.

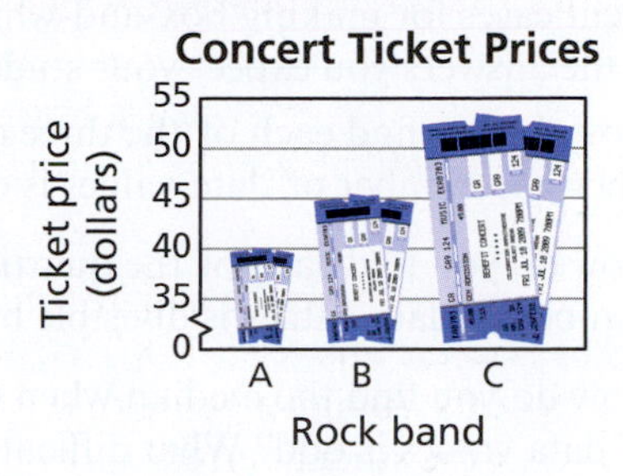

17. **A Misleading Bar Graph** Explain why the bar graph is misleading.

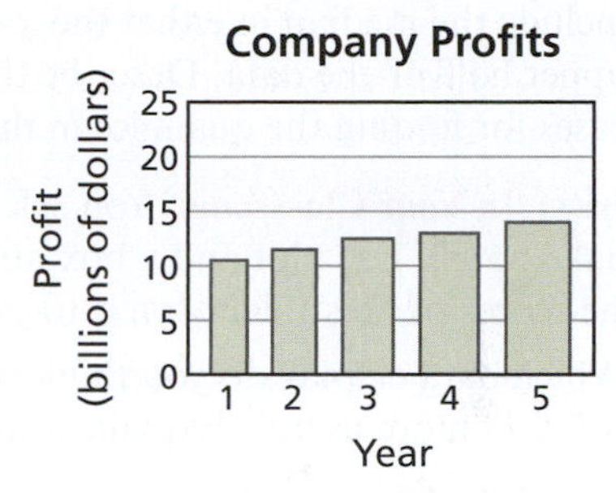

18. **A Misleading Bar Graph** Explain why the bar graph is misleading.

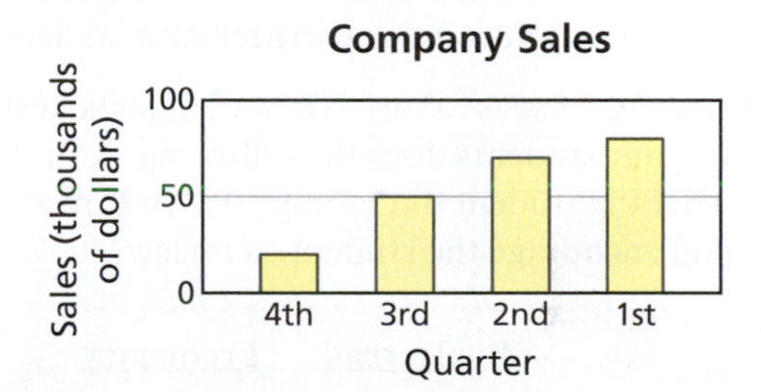

19. **A Misleading Histogram** Explain why the histogram is misleading.

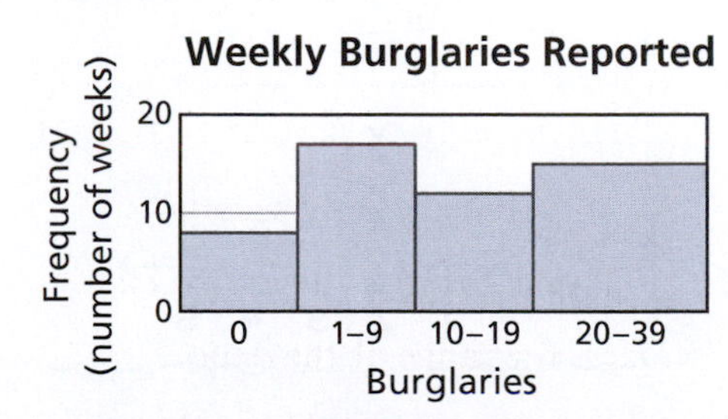

20. **A Misleading Histogram** Explain why the histogram is misleading.

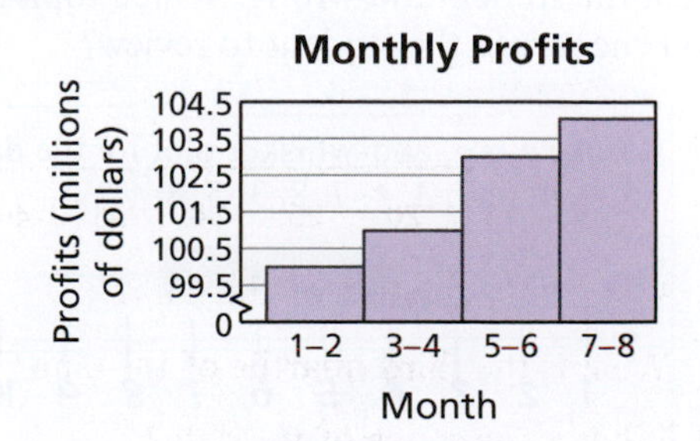

21. **Activity: In Your Classroom** You ask your students the following questions to help them discover the different cases for making box-and-whisker plots. State the answers you expect your students to give.

a. How do you find each of the three quartiles when the number of data values is divisible by 4?

b. How do you find each of the quartiles when the number of data values is divisible by 2 but not 4?

c. How do you find the median when the number of data values is odd? What difficulty do you have in finding the first and third quartiles?

d. When the number of data values is odd, do not include the median in either the lower half or the upper half of the data. Describe the two possible cases for finding the quartiles in this situation.

22. **Activity: In Your Classroom** You ask your students to make both a line plot and a box-and-whisker plot of the scores of 10 students on a 10-point quiz.

a. Which data display do you think your students will find more useful? Explain your reasoning.

b. As a follow-up question, you ask your students which of the two types of data displays is more useful for other types of scores. Describe the types of answers you expect your students to give.

23. **Grading Student Work** On a diagnostic test, one of your students does the following work. Explain what the student did wrong. Which topics would you encourage the student to review?

Books read	Frequency
1	2
2	1
3	2
4	3
5	2

X X X X X X
1 2 3 4 5

24. **Grading Student Work** On a diagnostic test, one of your students does the following work. Explain what the student did wrong. Which topics would you encourage the student to review?

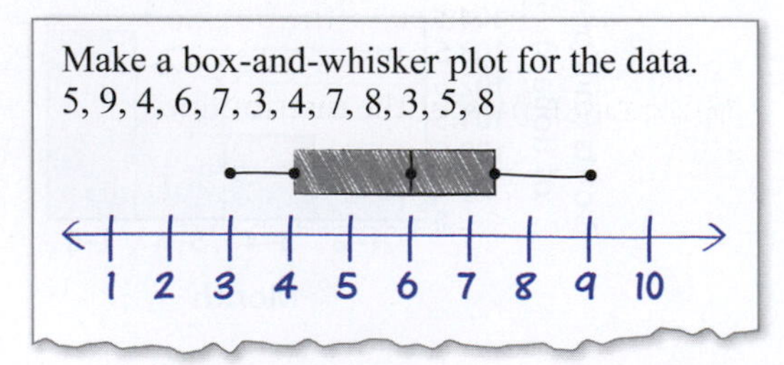
Make a box-and-whisker plot for the data.
5, 9, 4, 6, 7, 3, 4, 7, 8, 3, 5, 8

25. **Comparing Box-and-Whisker Plots** The box-and-whisker plots show the monthly car sales for two sales representatives. Compare and contrast the sales for the two representatives.

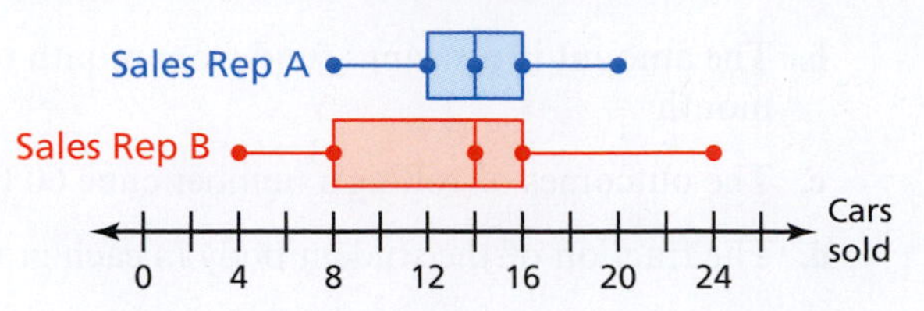

26. **Using Box-and-Whisker Plots** The table shows the numbers of calories burned per hour for nine activities.

Calories burned per hour	
Fishing	207
Mowing the lawn	325
Canoeing	236
Bowling	177
Hunting	295
Fencing	354
Bike racing	944
Horseback riding	236
Dancing	266

a. Make a box-and-whisker plot for the data.

b. Identify the outlier.

c. Make another box-and-whisker plot without the outlier.

d. Describe how the outlier affects the whiskers, the box, and the quartiles of the box-and-whisker plot.

27. **Leadership Training** The table shows the ages of the students and instructors on a wilderness leadership camping trip.

Ages				
14	18	13	15	15
24	16	17	18	15
15	19	16	24	14
15	14	16	17	16

a. Display the data in a line plot.

b. Describe the distribution of the data.

28. Comparing Data Displays The stem-and-leaf plot shows the numbers of students located in classrooms of an elementary school.

Students by Classroom

Stem	Leaf
1	8 8 8 9 9 9
2	1 1 2 3 3 4 5 5 5 8
3	1 1 2 2 2

Key: 1 | 8 = 18 students

a. Make a line plot for the data.

b. Which is the better data display for showing how many times each number occurs in the data set?

c. Which is the better data display for showing how the data are distributed by groups of 10?

29. Comparing Data Displays Mathematicians have computed and analyzed billions of digits of the decimal representation of the irrational number π. One thing that they analyze about the digits is the frequency of each of the numbers 0 through 9. The table shows the frequency of each number in the first 100,000 digits of π.

Number	Frequency
0	9999
1	10,137
2	9908
3	10,025
4	9971
5	10,026
6	10,029
7	10,025
8	9978
9	9902

a. Display the data in a bar graph.

b. Display the data in a circle graph.

c. Which data display is more appropriate? Explain.

d. Describe the distribution.

30. Choosing a Data Display A nutritionist wants to use a data display to show the favorite vegetables of the students at a school. Choose an appropriate data display for the situation. Explain your reasoning.

31. *Writing* When will a box-and-whisker plot *not* have one or both whiskers?

32. *Writing* Look at the line plot in Example 2. Without doing any calculations, will the box for a box-and-whisker plot for the data be longer or shorter than the combined length of the whiskers? Explain your reasoning.

33. Sports A survey asked 100 students what sports they play. The results are shown in the circle graph.

Sports Played

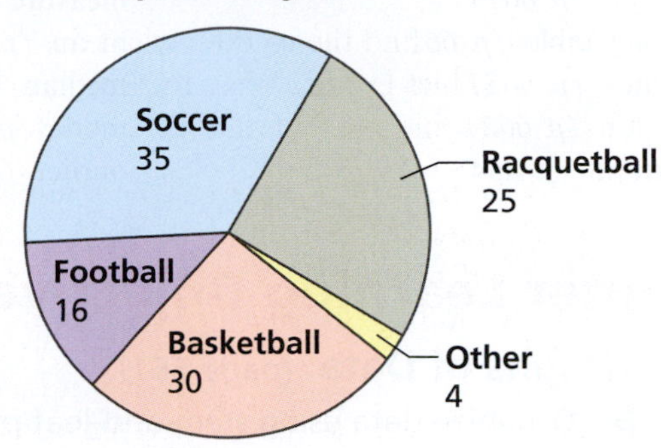

a. Explain why the graph is misleading.

b. What type of data display would be more appropriate for the data? Explain.

34. Chemicals A scientist gathers data about a decaying chemical compound. The results are shown in the line graph. Is the data display misleading? Explain.

Decaying Chemical Compound

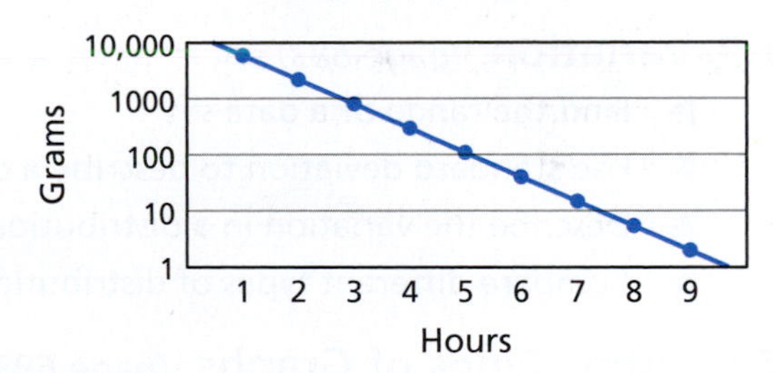

35. Choosing a Data Display A market research company wants to make a data display to summarize the variability of the SAT scores of graduating seniors in the United States. Is a stem-and-leaf plot, a histogram, or a box-and-whisker plot the most appropriate data display to use? Explain.

36. Reasoning You made a histogram of the ages of all your relatives at a family reunion one year ago. The histogram still represents the correct information today. How can this be true?

37. Reasoning The shape of the box-and-whisker plot for a distribution is shown below. Draw a histogram that could represent the distribution.

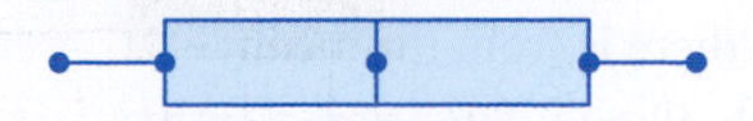

Chapter Summary

Chapter Vocabulary

stem-and-leaf plot *(p. 661)*
stem *(p. 661)*
leaf *(p. 661)*
bar graph *(p. 662)*
frequency table *(p. 662)*
frequency *(p. 662)*
histogram *(p. 662)*
pictograph *(p. 664)*
circle graph *(p. 665)*
scatter plot *(p. 666)*
line of fit *(p. 667)*
measure of central tendency *(p. 673)*
mean *(p. 673)*
median *(p. 674)*
mode *(p. 674)*
outlier *(p. 676)*
range *(p. 683)*
standard deviation *(p. 684)*
variation *(p. 684)*
normal distribution *(p. 685)*
box-and-whisker plot *(p. 693)*
quartiles *(p. 693)*
line plot *(p. 694)*

Chapter Learning Objectives — Review Exercises

17.1 Graphs of Data *(page 661)* — 1–10

- Organize data using stem-and-leaf plots.
- Organize data using bar graphs, frequency tables, and histograms.
- Organize data using pictographs and circle graphs.
- Read and use scatter plots.

17.2 Measures of Central Tendency *(page 673)* — 11–22

- Find and use the mean of a data set.
- Find and use the median and mode of a data set.
- Understand the effects of outliers on measures of central tendency.

17.3 Variation *(page 683)* — 23–30

- Find the range of a data set.
- Use standard deviation to describe a distribution.
- Describe the variation in a distribution, including a normal distribution.
- Compare different types of distributions.

17.4 Other Types of Graphs *(page 693)* — 31–44

- Read and make box-and-whisker plots and line plots.
- Choose an appropriate type of graph for a data set.
- Recognize when a graph is misleading.

Important Concepts and Formulas

Mean The sum of the data divided by the number of data values

Median The middle data value or the mean of the middle two data values of an ordered data set

Mode The value or values that occur most often

Range The difference of the greatest and least data values

Standard deviation Measurement that shows how much deviation there is from the mean

Pictograph Shows data using pictures

Bar graph Shows data in specific categories

Circle graph Shows data as parts of a whole

Line graph Shows how data change over time

Histogram Shows frequencies of data values in intervals of the same size

Stem-and-leaf plot Orders numerical data and shows how they are distributed

Box-and-whisker plot Shows variability using quartiles

Line plot Shows the number of times each data value occurs

Scatter plot Shows the relationship between two data sets using ordered pairs in a coordinate plane

Review Exercises

Free step-by-step solutions to odd-numbered exercises at MathematicalPractices.com.

17.1 Graphs of Data

Making a Stem-and-Leaf Plot Make a stem-and-leaf plot of the data.

1. Speeds of cars (miles per hour): 54, 55, 55, 58, 58, 58, 60, 60, 63, 64, 66, 66, 66, 68, 70, 72
2. Words per paragraph: 32, 55, 46, 65, 34, 48, 35, 51, 56, 37, 51, 64, 42, 56, 31, 44, 46, 51, 18

Making a Bar Graph Make a bar graph to display the data in the table.

3.

Week	Knee range of motion (degrees)
1	60
2	85
3	95
4	100

4.

Part-time credit load	Cost per credit (dollars)
1	556
3	407
6	369
9	357

5. **Making a Histogram** The numbers of goals scored by a soccer team in 10 games are shown.

   1, 0, 1, 2, 6, 3, 0, 2, 1, 3

   a. Make a frequency table for the data starting with the interval 0–1.

   b. Use the frequency table to make a histogram for the data.

6. **Making a Histogram** The numbers of minutes you spent on homework are shown for 20 school days.

   75, 40, 55, 26, 65, 35, 35, 14, 5, 50, 45, 8, 35, 0, 42, 5, 70, 60, 12, 52

   a. Make a frequency table for the data starting with the interval 0–19.

   b. Use the frequency table to make a histogram for the data.

   c. How many times did you spend at least 40 minutes on homework?

Making Data Displays Use the data in the table.

Favorite book genre	Students
Fantasy	20
Historical fiction	10
Mystery	15
Nonfiction	5

7. Make a pictograph to show the data. Include a key.
8. Display the data in a circle graph.
9. **Reading a Scatter Plot** The scatter plot shows the amount of money donated to a charity.

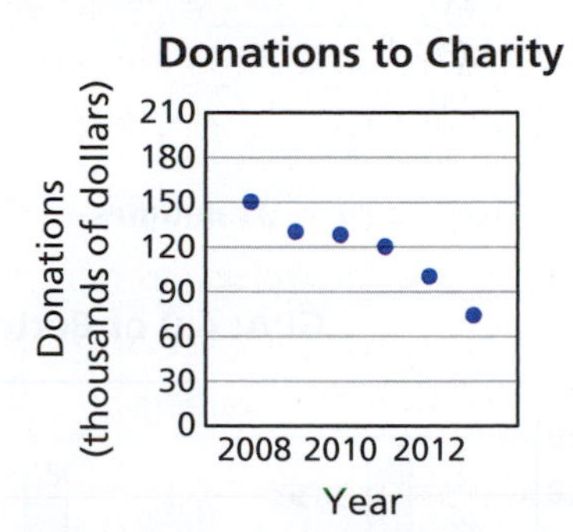

   a. How much money was donated in 2008?

   b. In what year did the charity receive $120,000?

   c. Do the donations and years show a *positive*, a *negative*, or *no* relationship? Explain.

10. **Using a Scatter Plot** The scatter plot shows the number of geese that migrated to a park each season. A line of fit is drawn close to the data points.

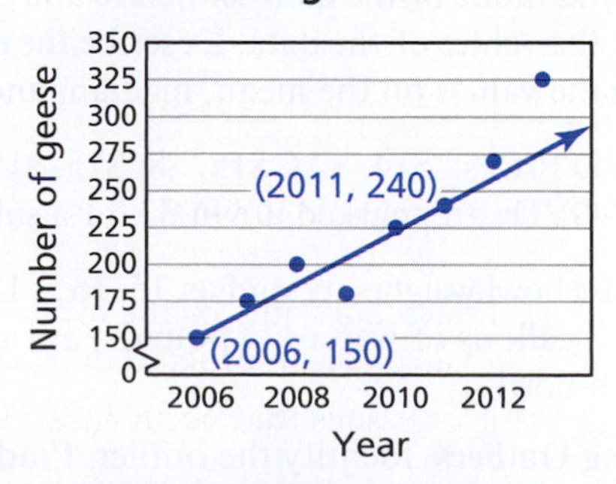

   a. What is the slope of the line of fit?

   b. Interpret the slope of the line of fit in the context of the data.

17.2 Measures of Central Tendency

Finding the Mean, Median, and Mode Find the mean, median, and mode of the data.

11. Pages in a book: 156, 212, 538, 77, 388, 419, 212

12. Track team mile times (minutes): 7.0, 6.5, 5.4, 6.2, 7.8, 6.8

Comparing the Means Compare the means of the two sets of data.

13. **Women's Swim Race Times**

Stem	Leaf
2	8 9
3	0 1 1 4 5 8
4	1 2

Key: 2 | 8 = 28 minutes

Men's Swim Race Times

Stem	Leaf
2	7 8 8
3	0 0 1 2 3 7
4	1

Key: 2 | 7 = 27 minutes

14.

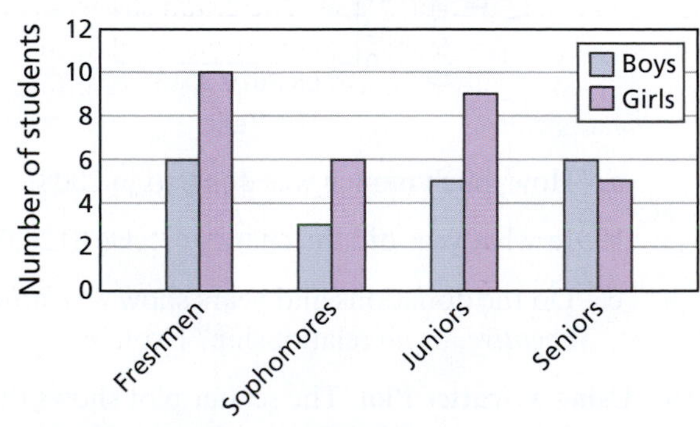

Changing the Values of a Data Set Find the mean, median, and mode of the data set before and after changing the values of the data. Describe the effects of changing the values on the mean, median, and mode.

15. DVD Prices: \$14, \$23, \$15, \$8, \$25, \$18, \$12, \$10
All DVDs are marked down \$3 for a sale.

16. Glass bowl weights (pounds): 15, 16.5, 12, 13, 11.5, 13; 3 gallons of water (25 pounds) are added to each bowl.

Examining Outliers Identify the outlier. Find the mean, median, and mode of the data with and without the outlier. Describe the effect of the outlier on the measures of central tendency.

17. 47, 56, 57, 63, 89, 44, 56, 49, 52, 50, 59, 46

18. 27, 36, 25, 26, 22, 28, 24, 2, 22, 37, 25, 34, 30

Finding the Best Representation Find the mean, median, and mode of the data. Decide which measure of central tendency best represents the data. Explain your reasoning.

19. Distances of kicked footballs (yards): 35, 52, 43, 65, 47, 39, 24, 27, 43, 29, 51

20. Sizes of shirts sold: 10, 14, 18, 14, 16, 12, 14, 10, 14

21. **Team Roster** The circle graph shows the number of players on each team in a basketball league. Find the mean, median, and mode of the data and decide which measure of central tendency best represents the data.

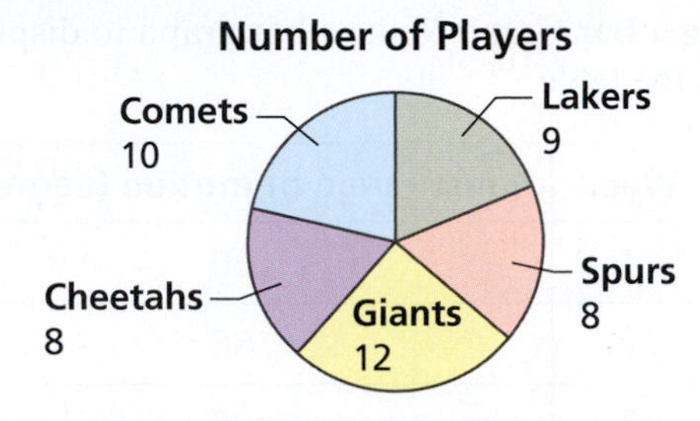

22. **Pleasure Reading** A class of students was asked how many books they read for pleasure in the past month. The results are shown in the bar graph. Find the mean, median, and mode of the data and decide which measure of central tendency best represents the data.

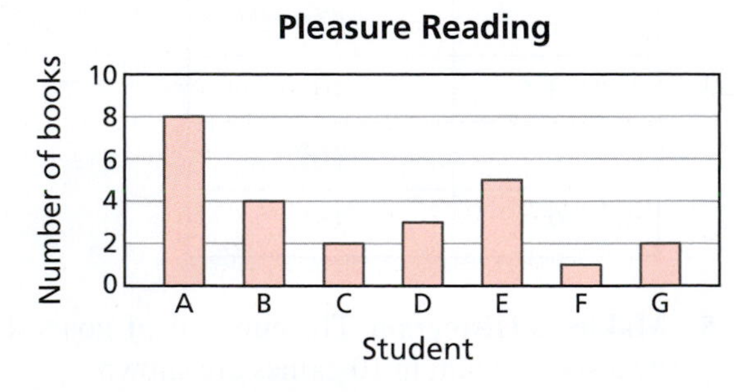

17.3 Variation

Finding the Range Find the range of the data.

23. Speech length (minutes): 22, 18, 19, 23, 17, 24, 26, 19, 22, 19, 18, 21

24. Prices: \$1.37, \$1.42, \$1.29, \$1.72, \$1.18, \$1.33, \$1.42

25. **Analyzing Standard Deviation** What percent of the data represented by the bar graph lie within 1 standard deviation of the mean? What percent lie within 2 standard deviations of the mean?

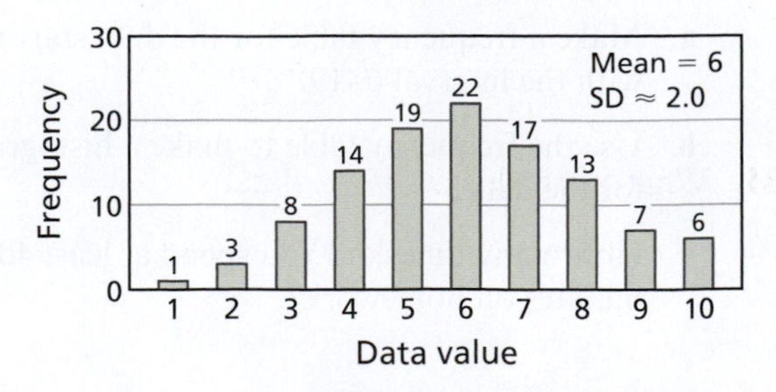

26. **Comparing Standard Deviations** Two data sets have the same mean (20) but different standard deviations (2.5 and 3.8). The histogram for one data set is more bell-shaped than the histogram for the other data set, which is flatter. Which is the standard deviation of the bell-shaped distribution? Explain.

27. **Analyzing a Normal Distribution** The data in a data set are normally distributed with a mean of 50 and a standard deviation of 15. Estimate the percent of the data that are between 35 and 65.

28. **Analyzing a Normal Distribution** The data in a data set are normally distributed with a mean of 220 and a standard deviation of 45. Estimate the percent of the data that are less than 130 or greater than 310.

29. **Describing a Distribution** Is the data set shown best described as a *normal*, *bimodal*, *right-skewed*, *left-skewed*, or *flat* distribution? How do the mean, median, and mode(s) of the data set compare?

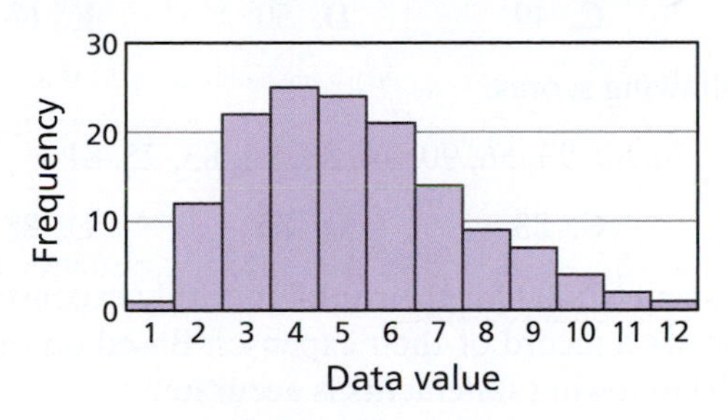

30. **Describing a Distribution** Is the data set shown best described as a *normal*, *bimodal*, *right-skewed*, *left-skewed*, or *flat* distribution? How do the mean and mode(s) of the data set compare?

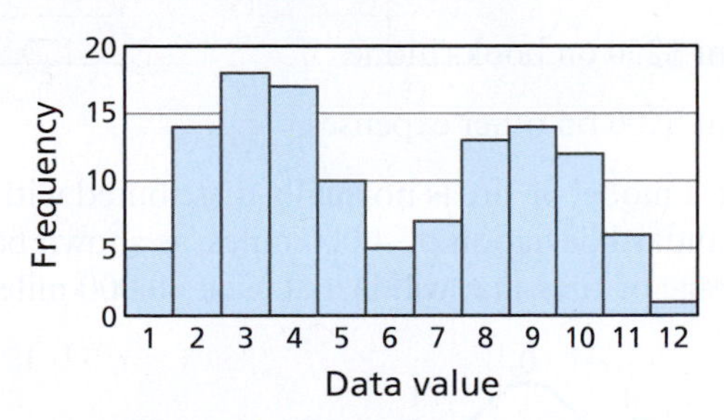

17.4 Other Types of Graphs

Interpreting a Box-and-Whisker Plot Answer the questions about the data represented by the box-and-whisker plot.

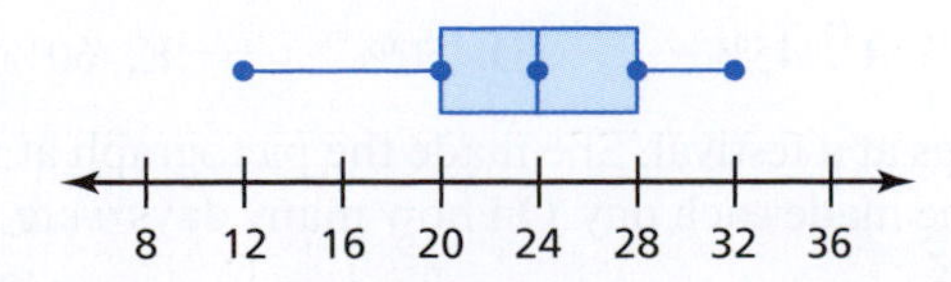

31. What is the median of the data?

32. What is the first quartile of the data?

33. What is the third quartile of the data?

34. What is the range of the data?

Making a Box-and-Whisker Plot Make a box-and-whisker plot for the data.

35. Team arm spans (inches): 70, 68, 74, 71, 73, 71, 76, 67, 75, 72, 73, 72

36. Experience (years): 1, 8, 5, 15, 11, 4, 5, 21, 12, 29, 10, 16, 3, 7, 11, 5, 32, 16, 3, 0

Making a Line Plot Make a line plot for the data.

37. Litter sizes (number of puppies): 6, 8, 5, 7, 9, 3, 11, 5, 6, 8, 4, 2, 5, 7, 6, 7, 8

38. Lengths of newborns (inches): 19.5, 21, 20.5, 18.5, 20, 20.5, 21, 19, 19.5, 21, 20, 19.5

Choosing a Data Display Choose an appropriate data display for the situation. Explain your reasoning.

39. A person's height from year to year

40. The profits of six companies for the past year

41. Commuting distances by intervals of 90 employees

42. The fraction of total income from each of four sources

43. **Analyzing a Line Graph** The graph shows the amount of money raised by students each month to pay for a class trip. Is the data display misleading? Explain.

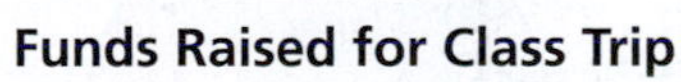

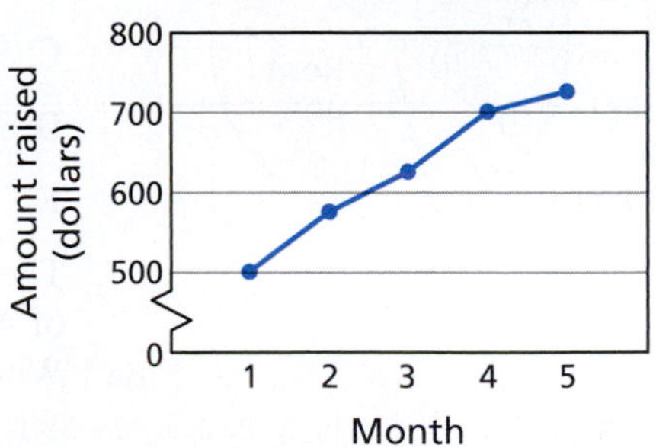

44. **A Misleading Histogram** Explain why the histogram is misleading.

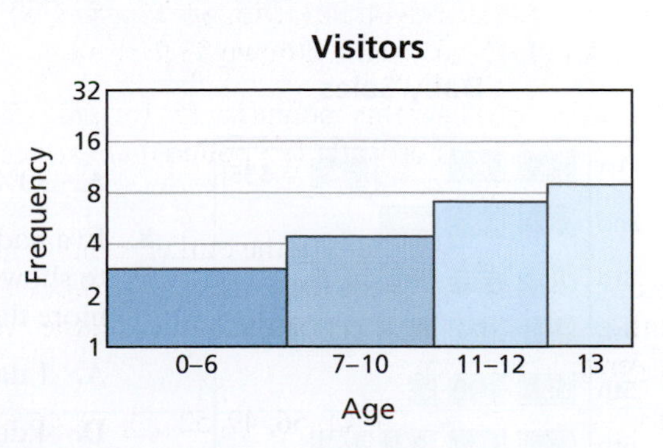

Chapter Test

The questions on this practice test are modeled after questions on state certification exams for K–8 teaching.

Test Directions: Each of the questions is followed by five choices. Choose the *best* response to each question.

Acme Calculators

Sales (millions of dollars); Scientific; Graphing; Year: 2010, 2011, 2012, 2013, 2014

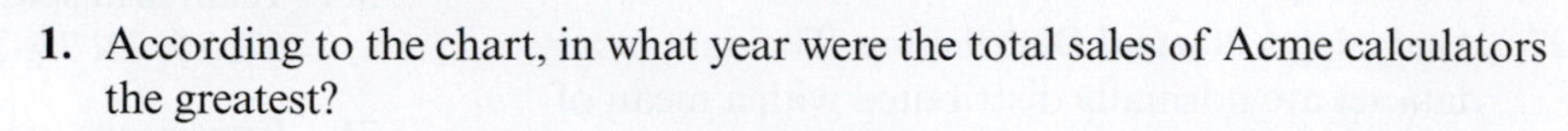

1. According to the chart, in what year were the total sales of Acme calculators the greatest?

   A. 2010 **B.** 2011 **C.** 2012 **D.** 2013 **E.** 2014

2. Which set of data below does not have a mean of 50?

   A. 40, 50, 50, 60 **B.** 10, 20, 20 **C.** 30, 40, 45, 55, 60, 70
   D. 25, 75 **E.** 35, 35, 35, 70, 75

3. Neil had scores of 86, 77, 98, and 91 on four history tests. What is the lowest grade that he can get on the next test to guarantee an average of at least 86?

   A. 78 **B.** 80 **C.** 83 **D.** 86 **E.** 89

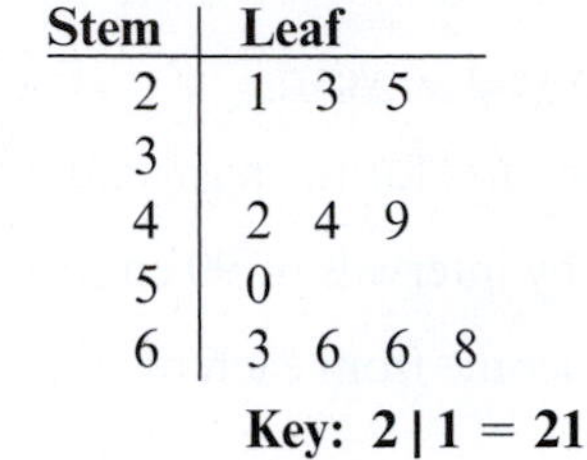

Stem	Leaf
2	1 3 5
3	
4	2 4 9
5	0
6	3 6 6 8

Key: 2 | 1 = 21

4. Examine the stem-and-leaf plot at the left. What is the median of the data?

   A. 44 **B.** 47 **C.** 49 **D.** 50 **E.** 66

5. Find the mode of the following scores:

   92, 83, 66, 83, 52, 70, 80, 93, 56, 90, 44, 88, 96, 83, 75, 81

   A. 77 **B.** 82 **C.** 83 **D.** 85 **E.** 88

Monthly Expenses

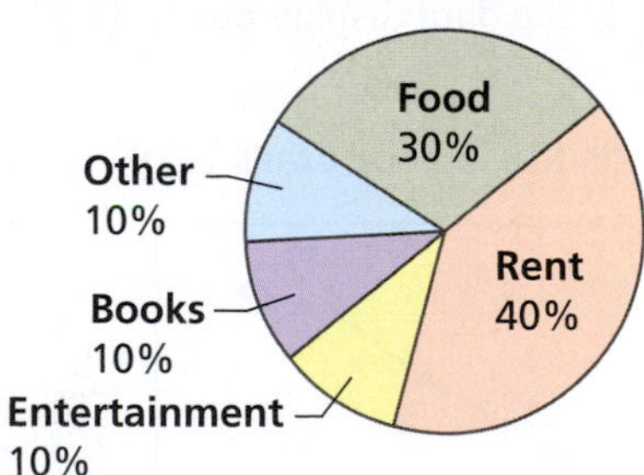

6. Two college roommates spent $2,000 for their total monthly expenses. The pie graph at the left indicates a record of their expenses. Based on the given information, which of the following statements is accurate?

   A. The roommates spent $700 on food alone.
   B. The roommates spent $600 on rent alone.
   C. The roommates spent $100 on entertainment alone.
   D. The roommates spent $250 on books alone.
   E. The roommates spent $200 on other expenses.

7. The wear-out mileage of a model of tire is normally distributed with a mean of 40,000 miles and a standard deviation of 4,000 miles, as shown below. What will be the percentage of tires that will last at least 40,000 miles?

   −1 0 1; 36,000 40,000 44,000

   A. 30% **B.** 40% **C.** 45% **D.** 50% **E.** 60%

Daily Sales

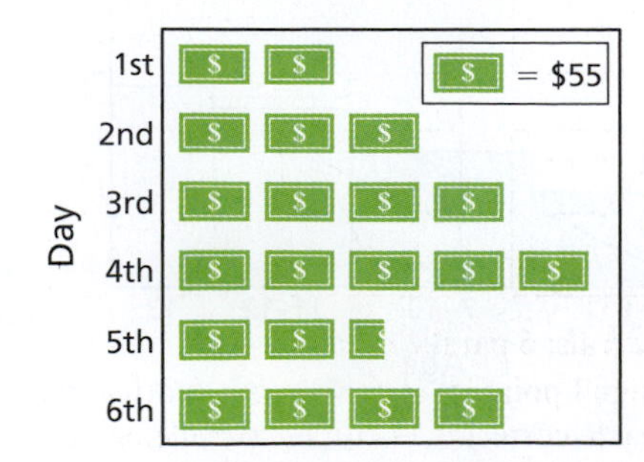

8. Eva had been selling paintings at a festival. She made the pictograph at the left to show how much money she made each day. On how many days were sales more than $150?

   A. 1 day **B.** 2 days **C.** 3 days
   D. 4 days **E.** 5 days

Answers to Odd-Numbered Exercises and Tests

CHAPTER 1

Section 1.1 (page 8)

1. Solution key:

1. Odd integers; 9, 11, 13
2. Increase by consecutive integers; 17, 23, 30
3. Perfect squares; 25, 36, 49 4. Decrease by 3; 11, 8, 5
5. Upper and lower case of every other letter; E, e, G
6. Every other upper case letter in reverse order; R, P, N
7. Alternate between square and circle, and change color from square to circle; green, blue, blue
8. Rotate 90° clockwise; green blue, green blue, blue green
9. Two possible answers: large circle followed by two small circles, alternate colors; green, red, green; or two large circles followed by two small circles, alternate colors; green, red, green
10. Multiply by 3, alternate colors; 81 (blue), 243 (red), 729 (blue)

Sample rubric: Assign 1 point to each correct pattern description, and 1 point to each correct number, letter, or figure based on the pattern. There is a total of 40 points.

3. a. Decrease by 8; 66, 58, 50
b. Increase by consecutive integers; 22, 29, 37
c. Multiply by 2; 80, 160, 320
d. Perfect squares in decreasing order; 36, 25, 16

5. a. Halve the object;
b. Move raised square one space to the right;

7. The number is the sum of the two numbers above it. Row 9: 1, 8, 28, 56, 70, 56, 28, 8, 1

9. a. *Sample answer:* Sales increased and then decreased.
b. *Sample answer:* The temperature was fairly constant, increased, and then decreased.
c. *Sample answer:* Younger age groups are larger.

11. *Sample answer:* Most fish are caught early in the day.

13.

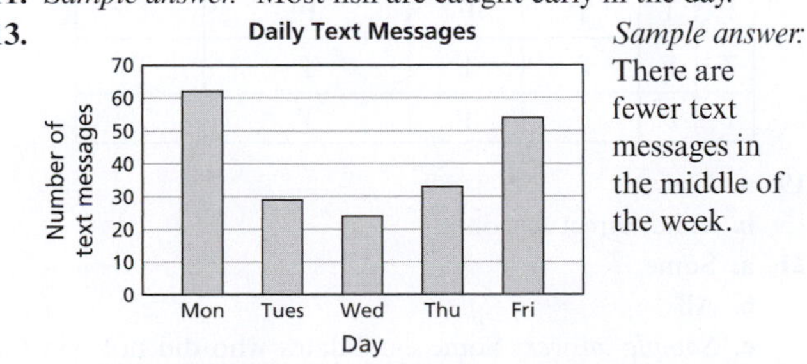

Sample answer: There are fewer text messages in the middle of the week.

15.

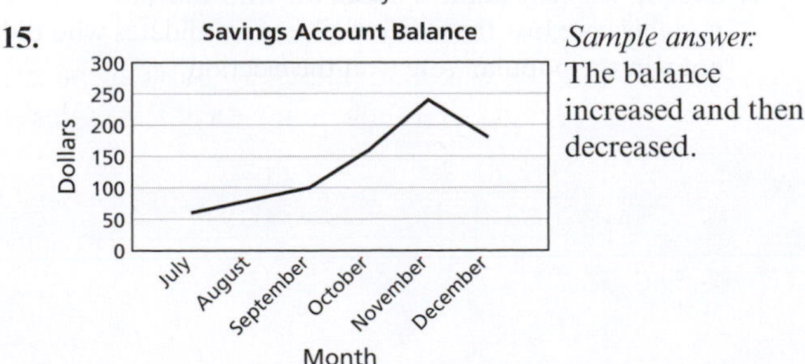

Sample answer: The balance increased and then decreased.

17.

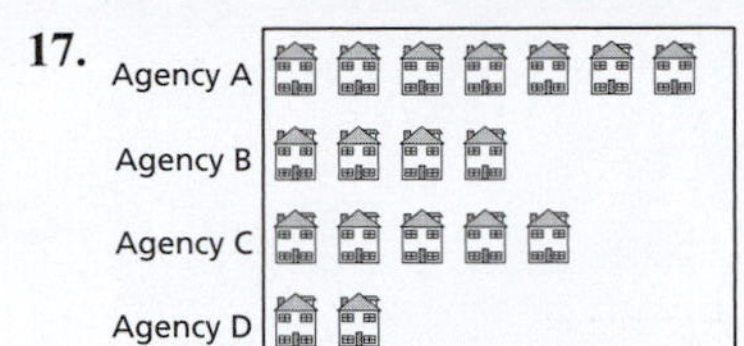

Sample answer: The data do not show a particular pattern.

19. F, F, F, F, C; *Sample answer:* The number of fastballs increases by 1 but the number of curve balls is constant.

21. *Sample answer:* You have a part-time job that pays $8 per hour. What is the total amount you make after each hour in a work day?

23. *Sample answer:* You tell a joke to three friends. Each of your friends tells the joke to three different friends, who then tell the joke to three more friends. How many people heard the joke without telling another person?

25. a. Each number is the sum of the previous two numbers.
b. 144, 233, 377 **c.** Answers will vary.

27. *Sample answer:* The student only looked at 20 and 40, and should review looking for a pattern.

29. a. The number of parts needed to earn $140 for the day
b. and c.

Parts produced	0	1	2	3	4	5
Earnings	$80	$92	$104	$116	$128	$140

31. *Sample answer:* The profits increased, decreased, and then increased.

33. a; *Sample answer:* The bars increase from left to right.

35. *Sample answer:* The forwards and centers tend to be taller than the guards.

Section 1.2 (page 18)

1. Solution key:

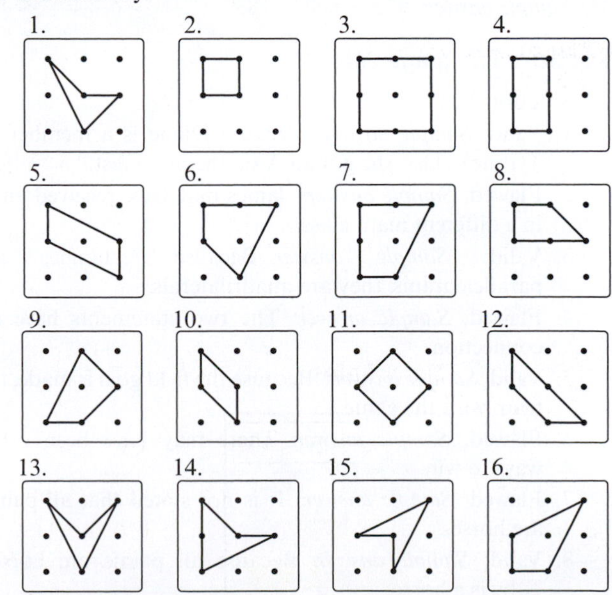

16 distinct quadrilaterals; 6 parallelograms

Sample rubric: Assign 1 point for each distinct quadrilateral and 1 point for each correctly identified parallelogram. There is a total of 22 points.

3. a. $4 + 3 - 2 = 5$ **b.** $5 + 4 + 3 = 12$
c. $2 \times 5 - 4 = 6$

5. a. Yes **b.** Yes **7. a.** 11 **b.** 21

9. *Sample answer:*

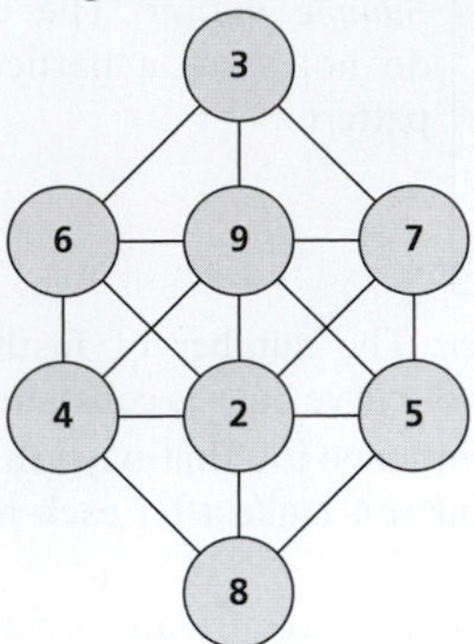

11. a. *Sample answer:* 75 cm × 40 cm
b. *Sample answer:* 60 cm × 100 cm

13. a. 14 **b.** 84

15. Answers will vary.

17. $29

19. a. 304 pages **b.** 190 pages

21. a. $28.06 **b.** $28.11

23. *Sample answer:* In one week, you spend $63 on groceries, $27 on movie tickets, and $48 on clothes. You have $80 left at the end of the week. How much did you have at the beginning of the week?

25. 40; *Sample answer:* Add diameters one at a time and count the sections.

27. 22; *Sample answer:* Each of the two end tables can seat 3 people and the eight tables in between can each seat 2 people.

29. *Sample answer:* The student miscalculated counting to the seventh entry, and should review the strategy of solving a simpler problem.

31. 26.0 ft **33.** 12 weeks

35. *Sample answer:* $975 + 864 = 1839$

Section 1.3 *(page 28)*

1. Solution key:

1. Valid; *Sample answer:* Because Diane is a member of Tyrone's class, she got an A on the math test.
2. Flawed; *Sample answer:* James may have received an A in a different math class.
3. Valid; *Sample answer:* Because rectangles are parallelograms, they are quadrilaterals.
4. Flawed; *Sample answer:* The two statements have no connection.
5. Valid; *Sample answer:* Because the field goal is made, the team wins the game.
6. Flawed; *Sample answer:* There may have been other ways to win.
7. Flawed; *Sample answer:* It is not stated that all paints are horses.
8. Valid; *Sample answer:* Because all paints are horses, Toby is a horse.

Sample rubric: Assign 1 point for each correct response and 2 points for each correct reasoning. There is a total of 24 points.

3. a. Non-statement; A phrase is not a statement.
b. Statement; George Washington was not a United States president.
c. Statement; $3 + 5 \neq 8$
d. Non-statement; The value of x is unknown.

5. a. True **b.** False **c.** False **d.** True

7. a. Patience is not important in teaching.
b. The gold medalist did not have her fastest time ever.
c. The shoes are comfortable.
d. $4 + 1 \neq 5$

9. a. All piano players are musicians.
b. Some trumpet players are piano players.
c. Some musicians are trumpet players.
d. No trumpet players are not musicians.

11.

p	Not(*p*)
T	F
F	T

Not(*p*): "The Reds did not win."

13.

p	*q*	*p* and *q*
T	T	T
F	T	F
T	F	F
F	F	F

15.

p	*q*	*p* and *q*	Not(*p* and *q*)	Not(*p*)	Not(*q*)	Not(*p*) or Not(*q*)
T	T	T	F	F	F	F
F	T	F	T	T	F	T
T	F	F	T	F	T	T
F	F	F	T	T	T	T

17.

p	*q*	Not(*p*)	Not(*q*)	If Not(*p*), then Not(*q*).	If *q*, then *p*.
T	T	F	F	T	T
F	T	T	F	F	F
T	F	F	T	T	T
F	F	T	T	T	T

19. a. Flawed
b. Valid; direct reasoning

21. a. Some
b. All
c. *Sample answer:* Some candidates who did not win the popular vote lost the election. Some candidates who did not win the popular vote won the election.

23. *Sample answer:* Next semester you will take a foreign language class and you will take a math class.

25. *Sample answer:* If you join a study group, then you will focus better.

27. a. *Sample answer:* The student completed the table for (p and q), and should review conditional statements.

b. *Sample answer:* The student completed the table for [q and Not(p)], and should review compound statements.

29. a. *Sample answer:* All politicians are honest people. Some honest people are politicians.

b. *Sample answer:* No politicians are honest people. No honest people are politicians.

31. a. The amphibian is not a frog or the amphibian is not green.

b. The amphibian is not a frog and the amphibian is not green.

33. a. The policy is considered to be unconstitutional.

b. Direct **c.** Answers will vary.

35. a. No; The person has not been a permanent resident for at least 3 years.

b. No; The conditions require that the immigrant be married.

Chapter 1 Review *(page 33)*

1. Increase by 10; 52, 62, 72 **3.** Decrease by 7; 48, 41, 34

5. Multiply by 3; 243, 729, 2187

7. The number is the sum of the two numbers above it. Row 6: 2, 10, 20, 20, 10, 2

9. *Sample answer:* Production increased and then decreased.

11. a. and b.

Hours	40	41	42	43	44	45	46
Earnings	\$480	\$498	\$516	\$534	\$552	\$570	\$588

13. *Sample answer:* More 9th graders had lower grades.

15. a. 7 **b.** 5 **17.** \$298 **19.** 2:45 P.M.

21. 4 burgers, 3 salads

23. Non-statement; A question is not a statement.

25. Non-statement; The value of x is unknown.

27. I am not home. **29.** I do not enjoy science class.

31. All basketball players are students.

33. All panda bears share fur colors with polar bears and black bears.

35. No polar bears share fur color with brown bears.

37.

p	Not(p)
T	F
F	T

Not(p): "Dubstep is not a genre of music."

39. a. Valid; direct **b.** Flawed **c.** Valid; indirect

Chapter 1 Test *(page 36)*

1. D **2.** C **3.** C **4.** D

5. A **6.** D **7.** A **8.** C

CHAPTER 2

Section 2.1 *(page 44)*

1. Solution key:

1. 1 2. 3 3. 4 4. 13 5. 8 6. 5
7. 7 8. 6 9. 0 10. 9 11. 12 12. 10
13. 18 14. 2 15. 14 16. 16
17. 11, 15, 17, 19, 20

Sample rubric: Assign 1 point to each correct response. There is a total of 17 points.

3. a. 1, 2, 3, 4, 5 **b.** 0

5. a. $\{1, 2, 3, 4, 5, 6, 7, 8, 9\}$ **b.** $\{s, e, t\}$

7. a. *Sample answer:* $\{x \mid x$ is an odd integer from 1 through 7.$\}$

b. *Sample answer*: $\{x \mid x$ is a letter from the word *math*.$\}$

9. $A = B$; The elements are the same.

11. a. $B = \{2, 4, 6, 8, 10\}$; $C = \{2, 4, 6, 8, 10\}$; $B = C$; The elements are the same.

b. $A = \{10, 12, 14, 15, 16, 18, 20\}$
$B = \{10, 12, 14, 15, 16, 18, 20\}$
$C = \{10, 12, 14, 16, 18, 20\}$;
$A = B$; The elements are the same.

13. a. 6 **b.** 5

15. a. Finite; 3 **b.** Finite; 5 **c.** Infinite **d.** Finite; 1

17. a. Yes; The sets have a cardinality of 3.

b. No; The cardinalities are different.

19. a. $C = \{2, 3, 5, 7\} = B$; $B \sim C$ because $B = C$. $A \sim D$ because the sets have a cardinality of 5.

b. $A = \{\ \} = D$; $A \sim D$ because $A = D$.

21. 2

23. a. 1; { }; no proper subsets **b.** 2; { }, {vase}; { }

c. 16; { }, {north}, {south}, {east}, {west}, {north, south}, {north, east}, {north, west}, {south, east}, {south, west}, {east, west}, {north, south, east}, {north, south, west}, {north, east, west}, {south, east, west}, {north, south, east, west};
{ }, {north}, {south}, {east}, {west}, {north, south}, {north, east}, {north, west}, {south, east}, {south, west}, {east, west}, {north, south, east}, {north, south, west}, {north, east, west}, {south, east, west}

25. a. False; $\{\ldots -3, -2, -1, 0, 1, 2, 3, \ldots\} \not\subseteq \{0, 1, 2, 3 \ldots\}$

b. True; $\{1, 2, 3, \ldots\} \subseteq \{0, 1, 2, 3 \ldots\}$

c. False; $\{1, 2, 3, \ldots\} \not\subset \{1, 2, 3 \ldots\}$

d. True; $\{1, 2, 3, \ldots\} \subset \{\ldots -3, -2, -1, 0, 1, 2, 3 \ldots\}$

27. $4 > 3$

29. a. *Sample diagram:*

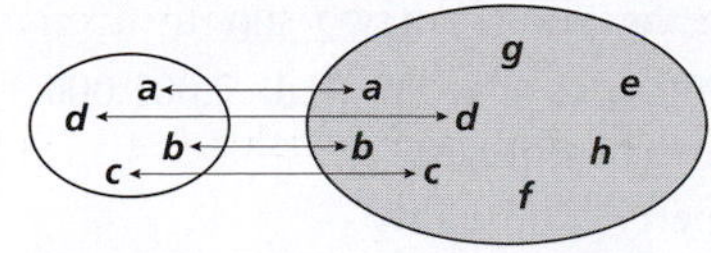

$4 < 8$

b. *Sample diagram:*

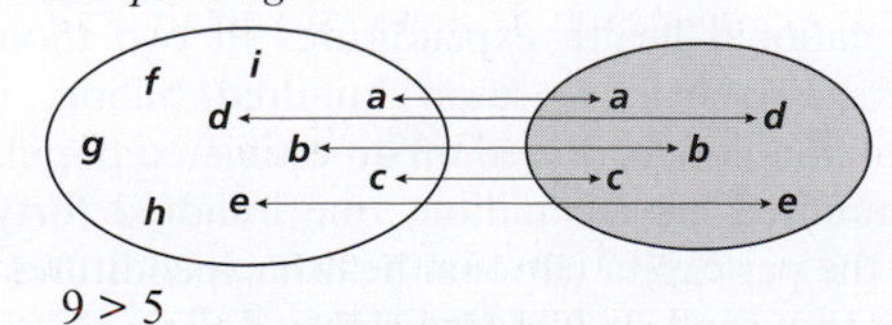

$9 > 5$

31. a. $A = \{1, 2, 3, 4, 5, 6, 7, 8, 9, 10, 11\}$
b. $B = \{1, 2, 3, 4, 5, 6, 7, 8, 9\}$
c. $C = \{-4, -3, -2, -1\}$ **d.** $D = \{8, 9, 10, \ldots\}$

33. a. *Sample answer:* $A = \{x \mid x$ is a whole number and $0 < x.\}$
b. *Sample answer:* $B = \{y \mid y$ is an even integer, and $1 < y$ and $y < 21.\}$
c. *Sample answer:* $C = \{z \mid z$ is a negative integer, and $-5 \leq z$ and $z \leq -1.\}$

35. *Sample answer:* The student mixed up sets A and B, and should review subsets.

37. a. 16 **b.** 15 **c.** 4 **d.** Answers will vary.

39. a. False; G might be a proper subset of H.
b. True; A set is a subset of itself.

41. *Sample answer:* Because all of the elements of set E are within set F, and all the elements of set F are within set G, it follows that all the elements in set E are in set G, so $E \subseteq G$.

43. 6

45. Yes; *Sample answer:* The null set cannot equal a non-empty set, so it is a proper subset of every non-empty set.

47. 20

Section 2.2 *(page 54)*

1. Solution key:
1. Twenty-three thousand, five hundred
2. Six hundred fifty thousand
3. Five hundred forty 4. Three hundred eighty-one
5. Ten thousand, nine hundred forty-seven
6. 922,004 7. 537 8. 10,400 9. 306,017
10. 419,059 11. 613,042 12. 584,270

Sample rubric: Assign 3 points to each correct word name in problems 1 through 5. Assign 1 point to each correct digit in problems 6 through 12. There is a total of 53 points.

3. a. 16 **b.** $x = 23$ **c.** $a = 7$

5. a. Ones **b.** Tens **c.** Thousands
d. Hundred thousands **e.** Ten millions **f.** Hundreds

7. a. Four hundred **b.** Thirty
c. Twenty thousand **d.** Seven hundred thousand
e. Sixty thousand **f.** Eight hundred million

9. a. 248 **b.** 2024 **c.** 3457

11. a. Six hundred ninety-one
b. Ninety-eight thousand, four hundred thirty-five
c. Six hundred six thousand, one
d. Two million, nine hundred thousand, four hundred fifty

13. a. 723 **b.** 8604 **c.** 200,076 **d.** 16,824,786

15. a. 80 **b.** 410 **c.** 1060 **d.** 12,670

17. a. 700 **b.** 2100 **c.** 4400 **d.** 22,500

19. a. 9000 **b.** 79,000 **c.** 113,000 **d.** 2,001,000

21. a. Reflexive Property of Equality
b. Symmetric Property of Equality
c. *Sample answer:* Use the Transitive Property of Equality to write $n = 100$ and $100 = r$, so $n = r$.

23. The total national health expenditures in two thousand eleven were two trillion, seven hundred billion, seven hundred million dollars. Based on an estimated population of three hundred eleven million, one hundred forty-one thousand, the per capita national health expenditures were about eight thousand six hundred eighty dollars.

25. The International Space Station weighs 419,600 kilograms and contains 935 cubic meters of habitable space. It orbits at a distance of 386 kilometers above the Earth, moving at a speed of 32,410 kilometers per hour.

27. *Sample answer:* The student did not recognize that the leading digit is in the millions place, and should review the base-ten system for whole numbers.

29. *Sample answer:* An employee earns $28,831.57 per year. To the nearest thousand, what is the employee's annual salary?

31. a. 8 small cubes **b.** 1 long
c. *Sample answer:* Count the number of small cubes. If there are less than five, then remove them. If there are five or more, remove them and add one long.

33. a. 801,500 through 802,499
b. 1,836,500 through 1,837,499
c. 249,500 through 250,499
d. 1,252,500 through 1,253,499

35. a. 390
b. *Sample answer:* Yes; the estimates are not close.

Section 2.3 *(page 64)*

1. Solution key:
1. 4 2. 39 3. 15 4. 50 5. 108
6. 10; 20; 40; 50; 60 7. 5; 10; 20; 25; 30
8. 4; 6; 8; 12; 14 9. 3; 6; 9; 15; 18
10. 25; 50; 75; 125; 150

Sample rubric: Assign 1 point to each correct number. There is a total of 30 points.

3. a. $1 < 4$ **b.** $15 < 20$ **c.** $28 < 44$

5. a. 0 1 2 3 4 5
$4 > 2$
b. 3 4 5 6 7 8
$3 < 8$
c. 10 12 14 16 18 20
$20 > 10$
d. 21 22 23 24 25 26
$22 < 26$

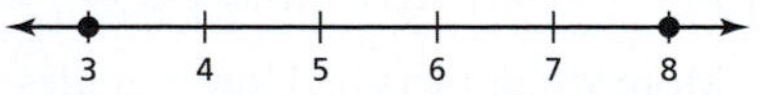

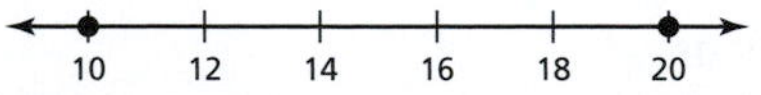

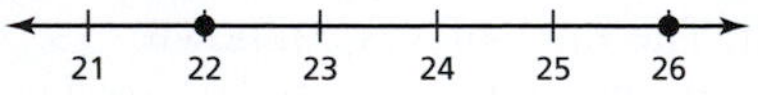

7. a. 12 **b.** 59 **c.** 443 **d.** 8005

9. a. 5 **b.** 13 **c.** 17 **d.** 63

11. a. 11 **b.** 28 **c.** 4150 **d.** 8623

13. a. 3435 **b.** 125,326 **c.** 1,362,000 **d.** 800,539

15. a. The numerals for 10 and 8 are multiplied.
b. The numerals for 30 and 3 are added.
c. The numerals for 200, 20, and 1 are added.

17. a.  **b.** 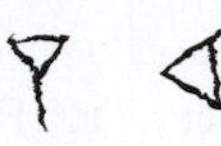
c.  **d.**

19. a. 17 **b.** 576 **c.** 99 **d.** 1174

21. a. XIX **b** XLIII **c.** CCCXXXIV **d.** MDLXVII

23. a. 9 **b.** 77 **25. a.** 23 **b.** 106

27. a. *Sample answer:* A graph shows that a is to the left of c, so $a < c$.
b. *Sample answer:* No, $a \nless a$.

29. **a.** Ten
b. Yes; The value of a digit depends on the position or place it occupies in the numeral.
c. A blank space **d.** 36,994

31. **a.** 1101 **b.** 13 **c.** 1111; 15
d. *Sample answer:* The set of whole numbers from 1 through 15.

33. No; *Sample answer:* The value of a digit does not depend on the position or place it occupies.

35. **a.** *Sample answer:* The student confused place value and base, and should review bases for numeral systems.
b. *Sample answer:* The student added 10 instead of subtracting 10, and should review the Roman numeral system.

37. **a.** 26BF **b.** 9919

Chapter 2 Review *(page 69)*

1. {21, 23, 25, 27, 29}

3. *Sample answer:* $\{x \mid x$ is an odd integer, and $16 < x$ and $x < 22.\}$

5. $A = C$; The elements are the same.

7. Finite; 4 **9.** Finite; 3

11. Yes; the sets have a cardinality of 3.

13. $C = E$ because the elements are the same. $A \sim D$ because the sets have a cardinality of 3. $B \sim C \sim E$ because the sets have a cardinality of 4.

15. True; An integer is a real number, and the sets are not equal.

17. False; 0 is not a positive integer.

19. *Sample diagram:*

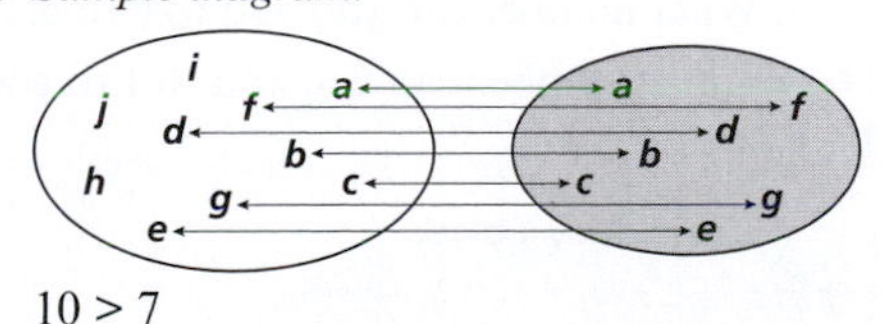

$10 > 7$

21. $A = \{5, 6, 7, 8, 9, 10\}$ **23.** $C = \{1, 2, 3\}$

25. 15 **27.** 21 **29.** $a = 5$ **31.** $x = 9$ **33.** Ones

35. Hundreds **37.** Tens **39.** Four thousand

41. Ten **43.** 1124 **45.** Four hundred thirty-one

47. One hundred twenty-three thousand, four hundred sixty two

49. 327 **51.** 18,214 **53.** 2,551,165 **55.** 90

57. 2510 **59.** 500 **61.** 78,300 **63.** 1000

65. 56,000

67. A school's French club would like to help raise one thousand, five hundred sixty-five dollars to help pay for a class trip overseas. They plan to sell magazine subscriptions for twenty-two dollars, coupon booklets for fifteen dollars, and raise at least three hundred fifty dollars at a car wash.

69. $3 < 5$ **71.** $19 < 39$ **73.** 17 **75.** 820

77. 6 **79.** 21 **81.** 15 **83.** 325 **85.** 1213

87. 1,310,116

89. The numerals for 10 and 6 are multiplied.

91. The numerals for 9 and 10 are added.

93. **95.**

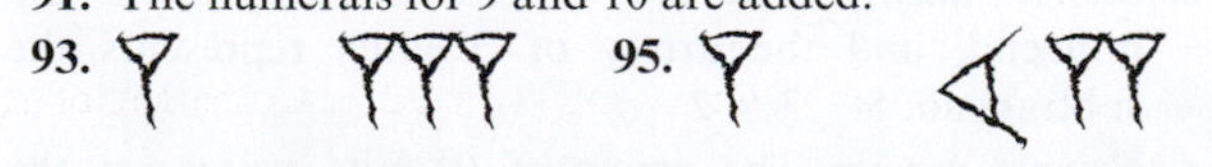

97. 16 **99.** 1560 **101.** XI **103.** CML **105.** 10

107. 30 **109.** 24 **111.** 120

Chapter 2 Test *(page 72)*

1. D **2.** C **3.** D **4.** A
5. D **6.** E **7.** E **8.** A

CHAPTER 3

Section 3.1 *(page 82)*

1. Solution key:
1. $336 + 224 = 560$ 2. $57 + 55 = 112$
3. $134 + 38 = 172$ 4. $68 + 47 = 115$
5. $47 + 126 = 173$

Sample rubric: Assign 1 point to each correct digit in the addition problem. There is a total of 39 points.

3. **a.** 6, 5; 11 **b.** 7, 9; 16

5. $4 + 3 = 7$; *Sample answer:* The cardinalities of A and B are the addends. The cardinality of $A \cup B$ is the sum.

7. **a.** Commutative **b.** Additive Identity **c.** Associative

9. **a.** $8 + 7$ **b.** $(4 + 6) + 7$ **c.** 4

11. **a.** *Sample answer:* Use the Additive Identity Property.
b. *Sample answer:* Count on by 1.
c. *Sample answer:* Put a 1 in front of a.

13. **a.** *Sample answer:* The arrows of 4 units and 3 units represent the addends 4 and 3; $4 + 3 = 7$
b. *Sample answer:* The arrows of 2 units and 6 units represent the addends 2 and 6; $2 + 6 = 8$

15. **a.**

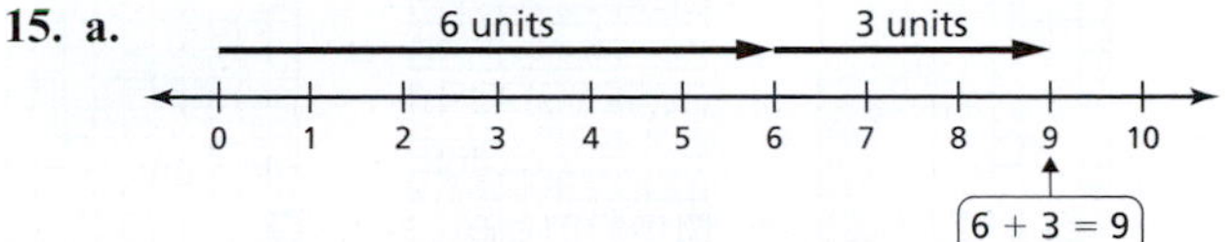

b.

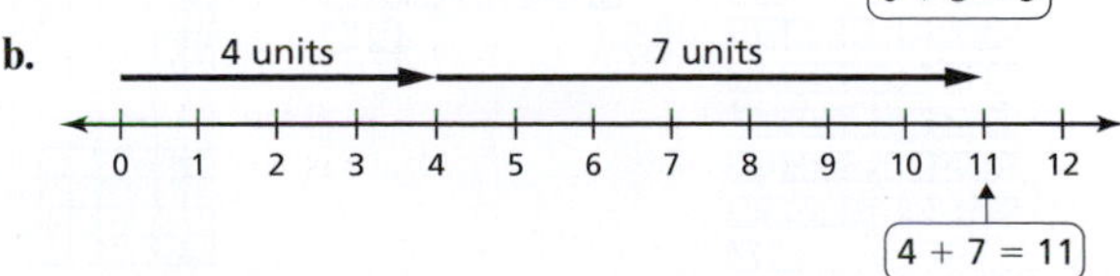

c.

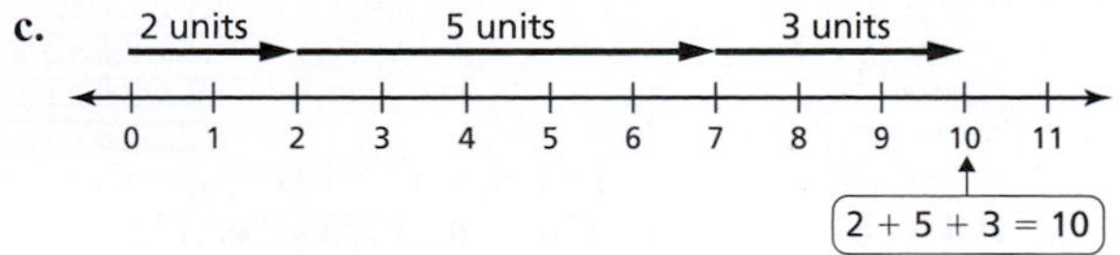

d.

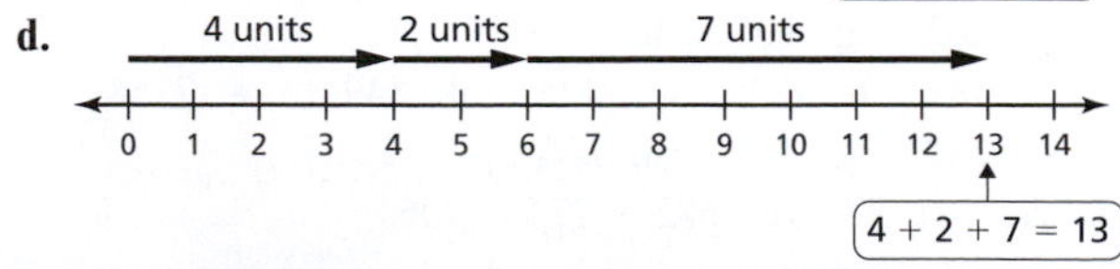

17. *Sample answer:* Join all of the blocks and regroup as needed.

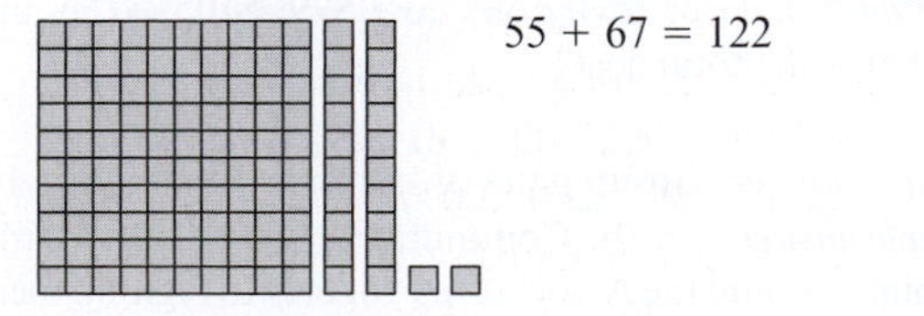

19. a.

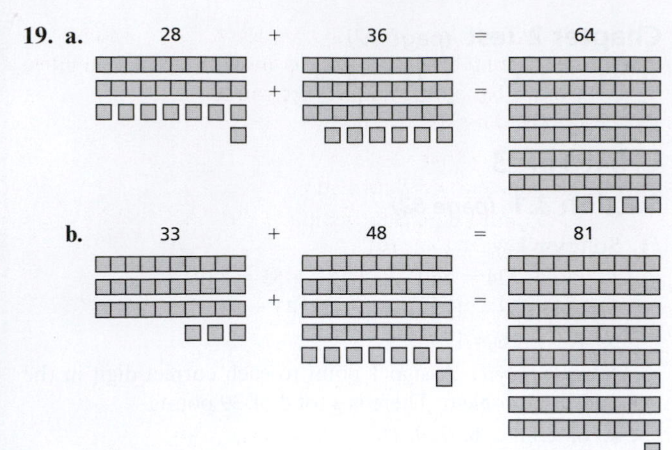

c.

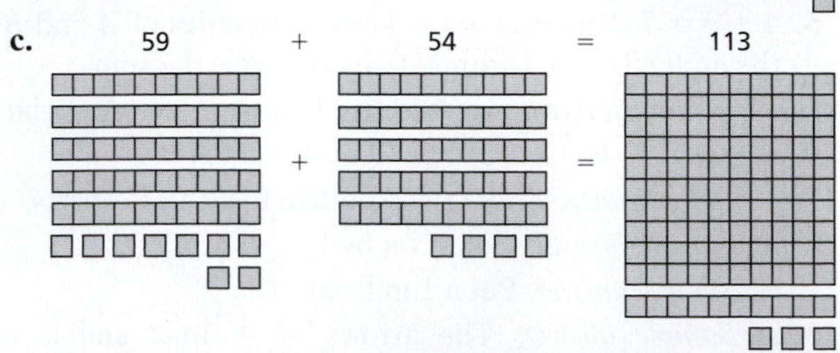

d.

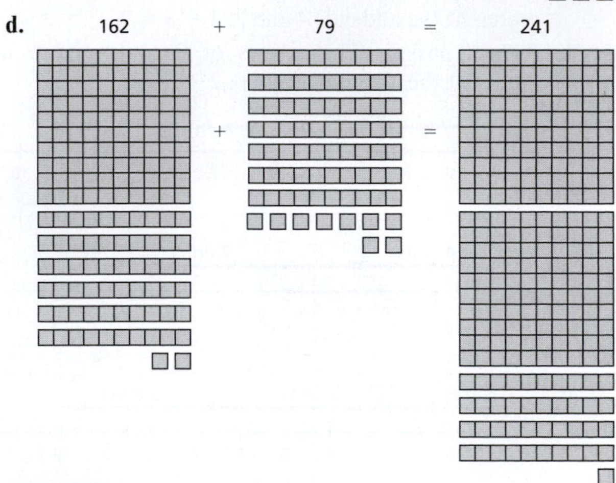

21. a. 104 **b.** 1131 **c.** 570 **d.** 1353 **e.** 132
f. 1262 **g.** 105 **h.** 417

23. a. 106 **b.** 922 **c.** 5425 **d.** 5494 **e.** 7154
f. 7232 **g.** 168 **h.** 744

25. \$93 **27.** Yes; 682 + 224 = 906

29. *Sample answer:* 1 + 6 = 3 + 4

31. *Sample answer:* Four textbooks cost \$79, \$105, \$156, and \$88. What is the total cost?

33. a. 30

b. *Sample answer:* Group pairs of numbers that sum to 10.

c. *Sample answer:* Use the Commutative Property to reorder the numbers and the Associative Property to regroup them.

35.

49 + 45 = 49 + (1 + 44)	Think of 45 as 1 + 44.
= (49 + 1) + 44	Regroup using the Associative Property.
= 50 + 44	Add 49 and 1.
= 94	Add 50 and 44.

37. a. *Sample answer:* The student changed 24 to 1 + 25 instead of 1 + 23, and should review the properties of addition for whole numbers.

b. *Sample answer:* The student did not regroup the sum of the ones as 14 = 1(10) + 4, and should review using algorithms to add whole numbers.

39.

8	1	6
3	5	7
4	9	2

Strategies will vary.

41. a. 161

b. *Sample answer:* Subtract 43 from 161, or add the years without 43.

Section 3.2 *(page 92)*

1. Solution key:

1. 336 − 224 = 112 2. 57 − 55 = 2
3. 134 − 38 = 96 4. 68 − 47 = 21
5. 126 − 47 = 79

Sample rubric: Assign 1 point to each correct digit in the subtraction problem. There is a total of 34 points.

3. a. Minuend: 10; subtrahend: 7; difference: 3

b. Minuend: 9; subtrahend: 2; difference: 7

5. a. What number can you add to 9 to get 11? 11 − 9

b. What number can you add to 4 to get 7? 7 − 4

c. What number can you add to 2 to get 8? 8 − 2

7. a. 7 + ☐ = 9; What number can you add to 7 to get 9?

b. 1 + ☐ = 11; What number can you add to 1 to get 11?

c. 6 + ☐ = 8; What number can you add to 6 to get 8?

d. 1 + ☐ = 6; What number can you add to 1 to get 6?

9. a. 9 − 2 **b.** 11 − 5

11. a.

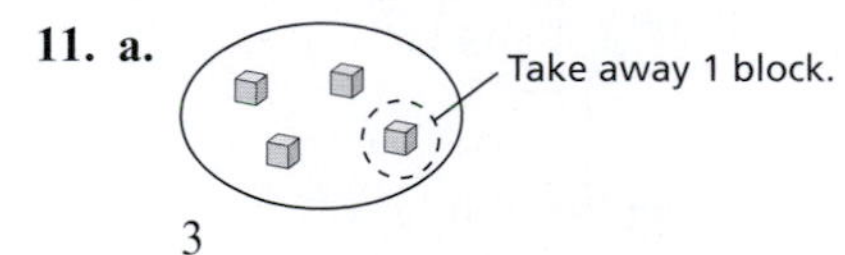

3

b.

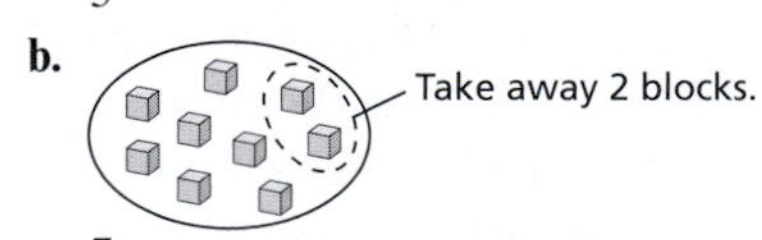

7

c.

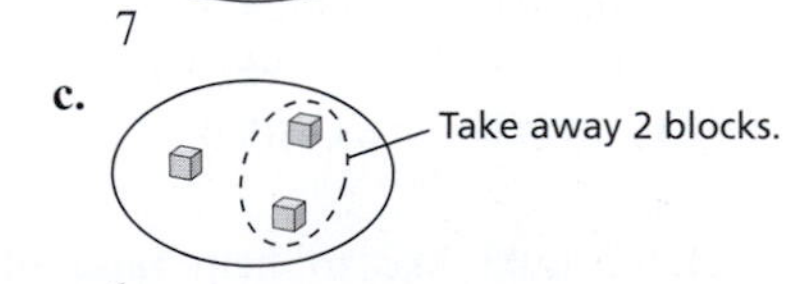

1

d.

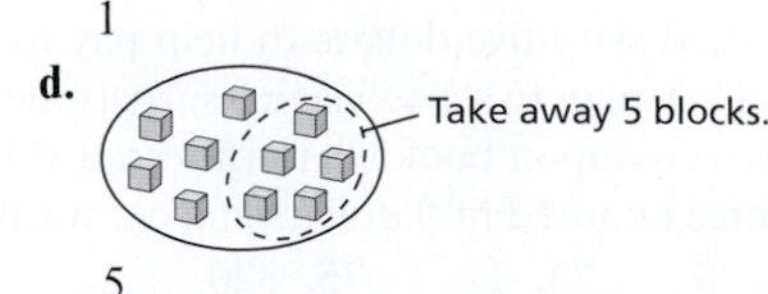

5

13. a. *Sample answer:* The arrow of 7 units represents the minuend, and the arrow of 4 units represents the subtrahend; 7 − 4 = 3

b. *Sample answer:* The arrow of 5 units represents the minuend, and the arrow of 3 units represents the subtrahend; 5 − 3 = 2

c. *Sample answer:* The arrow of 9 units represents the minuend, and the arrow of 1 unit represents the subtrahend; 9 − 1 = 8

15. a.

12 units
2 units
0 1 2 3 4 5 6 7 8 9 10 11 12
12 − 2 = 10

b.

7 units
1 unit
0 1 2 3 4 5 6 7
7 − 1 = 6

c.

2 units
2 units
0 1 2 3
2 − 2 = 0

d.

9 units
4 units
0 1 2 3 4 5 6 7 8 9
9 − 4 = 5

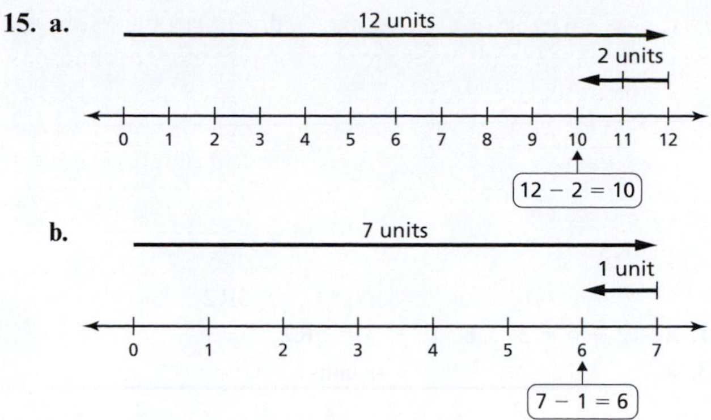

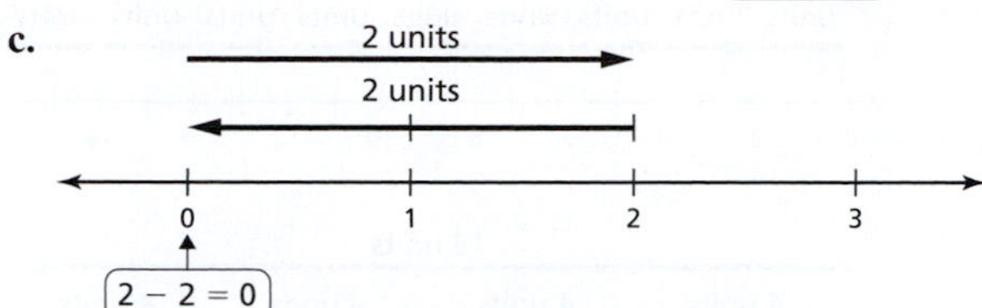

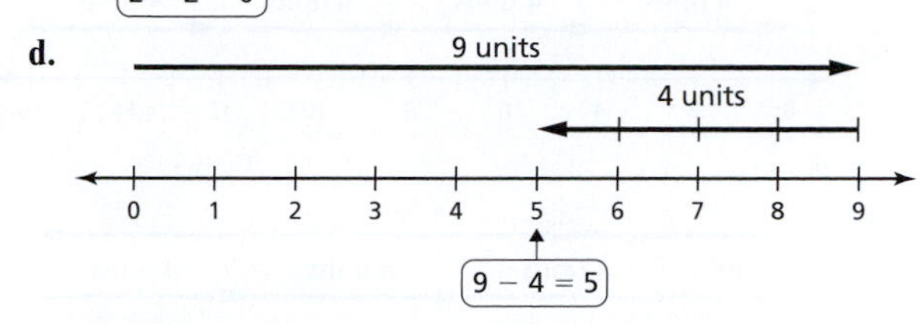

17. Start with 7 tens and 7 ones, then take away 3 tens and 6 ones; 77 − 36 = 41

19. a.

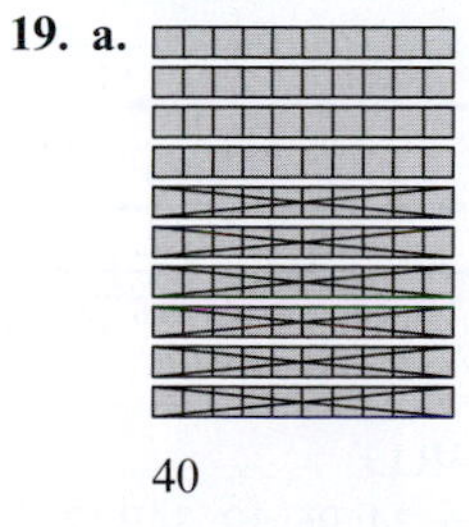

40

b.

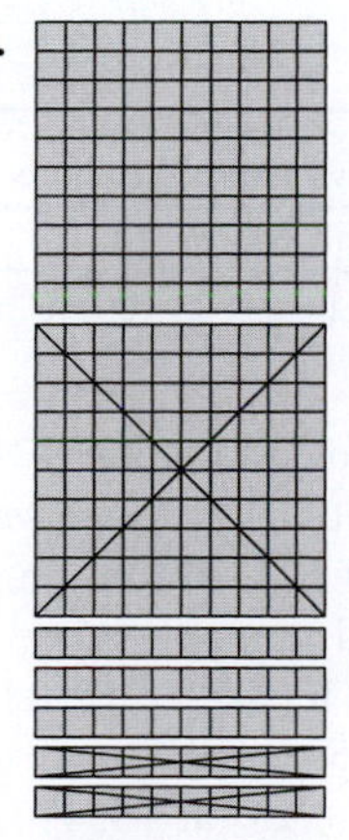

130

c.

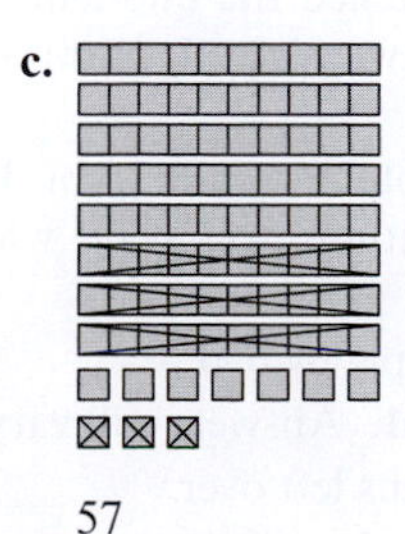

57

d.

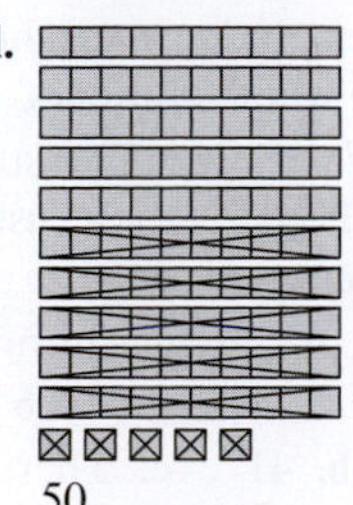

50

21. a. 31 **b.** 44 **c.** 4909 **d.** 6906

23. a. 2220 **b.** 1133 **c.** 855 **d.** 1686

25. *Sample answer:* 6 − 2 = 9 − 5 **27.** 60 mm **29.** 132

31. *Sample answer:* The student did not change the tens place, and should review using algorithms to subtract whole numbers.

33. Answers will vary.

35. a. 1150

b. *Sample answer:* Consume only 400 calories at breakfast.

37. *Sample answer:* You are in an exercise program where you jog for 20 minutes each day. You just finished 12 minutes. How many remaining minutes are you to jog?

39. a. 9; *Sample answer:* Addition

b. 9; *Sample answer:* Addition

c. 8; *Sample answer:* Subtraction

d. 4; *Sample answer:* Subtraction

41. a. \$155 **b.** You; \$1191

43. a-d. Answers will vary.

Section 3.3 *(page 104)*

1. Solution key:

1. 6 × 14 = 84 2. 5 × 16 = 80
3. 14 × 14 = 196 4. 18 × 16 = 288
5.

13 units
7 units

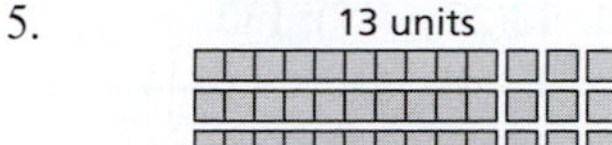

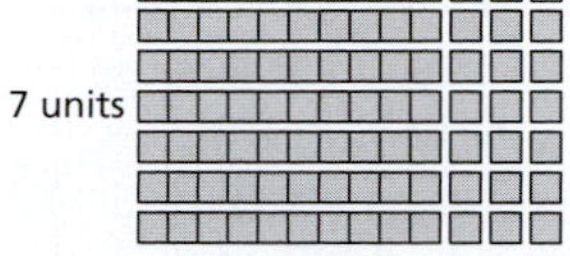

91; *Sample answer:* 70 + 21 = 91

6.

17 units
9 units

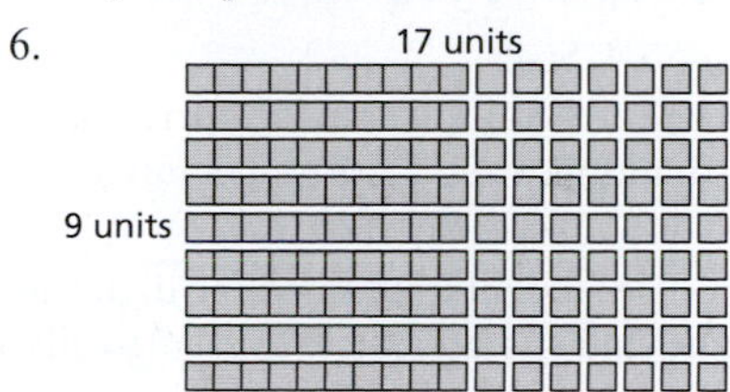

153; *Sample answer:* 90 + 63 = 153

7.

15 units
12 units

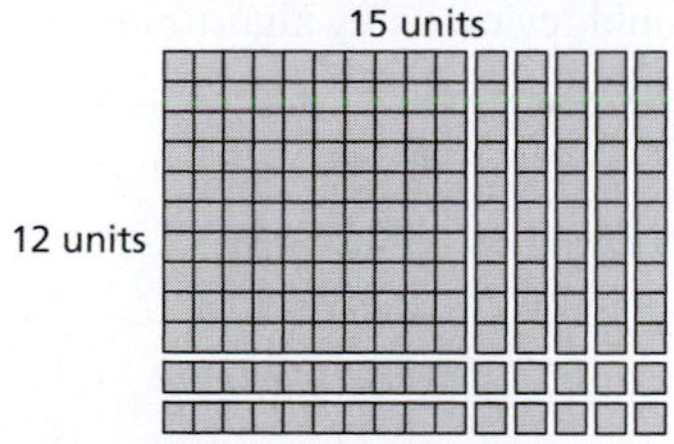

180; *Sample answer:* 100 + 70 + 10 = 180

8.

13 units
18 units

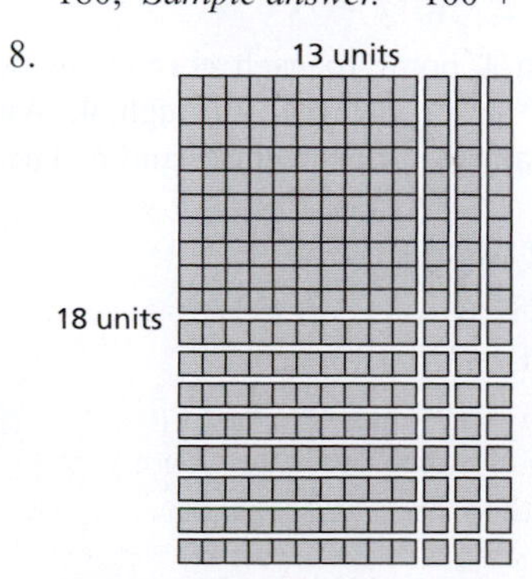

234; *Sample answer:* 100 + 110 + 24 = 234

Sample rubric: Assign 1 point to each correct factor and product in problems 1 through 4. Assign 2 points for a correct model and 2 points for a correct explanation in problems 5 through 8. There is a total of 28 points.

3. a. 4 × 3 **b.** 7 × 5 **5. a.** 4, 5; 20 **b.** 7, 2; 14

7. a. 30 **b.** 32 **c.** 35 **d.** 27

9. a. 9 • 7 **b.** (4 • 5) • 7 **c.** 9 **d.** 0

11. a. *Sample answer:* Use the Multiplicative Identity Property.
b. *Sample answer:* Find $a + a$.
c. *Sample answer:* Put a 0 after a.

13. a. *Sample answer:* The rectangle is 3 units by 8 units; $3 \times 8 = 24$
b. *Sample answer:* The rectangle is 12 units by 13 units; $12 \times 13 = 156$

15. a. 6×40; 10×7; 10×40
b. 2×6; 2×20; 2×300; 40×6; 40×20; 40×300

17. a. 3120 **b.** 2925 **c.** 2016 **d.** 29,260 **e.** 8712 **f.** 68,100

19. a. 1302 **b.** 5544 **c.** 5960 **d.** 13,912 **e.** 95,200 **f.** 42,460 **g.** 74,538 **h.** 259,501

21. a. $6 \cdot 9 + 6 \cdot 5$; 84 **b.** $7 \cdot (13 + 7)$; 140
c. $9 \cdot 20 - 9 \cdot 3$; 153 **d.** $13(27 - 17)$; 130
e. $4(9) + 4(7) + 4(6)$; 88 **f.** $8(6 + 3 + 9)$; 144

23. a. 752 **b.** 984 **25.** 511

27. a. Associative Property **b.** $(2 \cdot 3) \cdot 4$
c. Answers will vary.

29. 2280 **31.** *Sample answer:* $1 \cdot 8 = 2 \cdot 4$

33. *Sample answer:* $1 \cdot 3 \cdot 8 = 2 \cdot 3 \cdot 4$

35. a. \$128 **b.** $32 \cdot 5 \cdot 4$; \$640

37. *Sample answer:* The Zero Multiplication Property cuts down on the number of multiplication facts to memorize.

39. a. 2592 in.^{2} **b.** $36 \cdot (72 - 18)$; 1944 in.^{2}

41. a. *Sample answer:* The student multiplied the digits instead of by place value, and should review using algorithms to multiply whole numbers.
b. *Sample answer:* The student did not align the digits properly, and should review using algorithms to multiply whole numbers.

43. \$36,000 to \$276,000

Section 3.4 *(page 114)*

1. Solution key:

1. 5; $20 \div 4 = 5$
2. 4; $20 \div 5 = 4$
3. 3; $18 \div 6 = 3$
4. 6; $18 \div 3 = 6$
5. 8
6. 3

Sample rubric: Assign 1 point to each correct dividend, divisor, and quotient in problems 1 through 4. Assign 2 points for a correct answer in problems 5 and 6. There is a total of 16 points.

3. Divisor: 2; dividend: 8; quotient: 4

5. a. 6; $30 \div 5$ **b.** 7; $28 \div 4$

7. a. $25 \div 5 = 5$ **b.** $31 \div 7 = 4R3$

9. a.

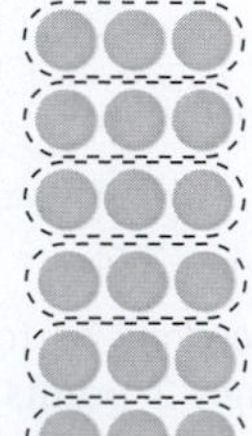

6

b.

5

c.

4R3

d.

5R2

11. a. $12 \div 4 = 3$ **b.** $17 \div 3 = 5R2$

13. a.

18 units
2 units 2 units 2 units 2 units 2 units 2 units 2 units 2 units 2 units
0 2 4 6 8 10 12 14 16 18

9

b.

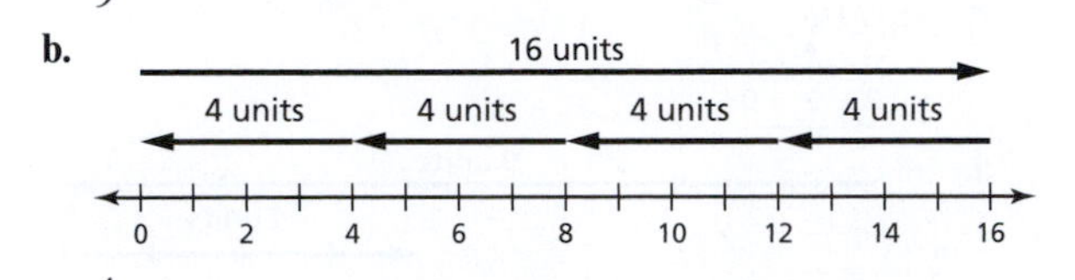

4

c.

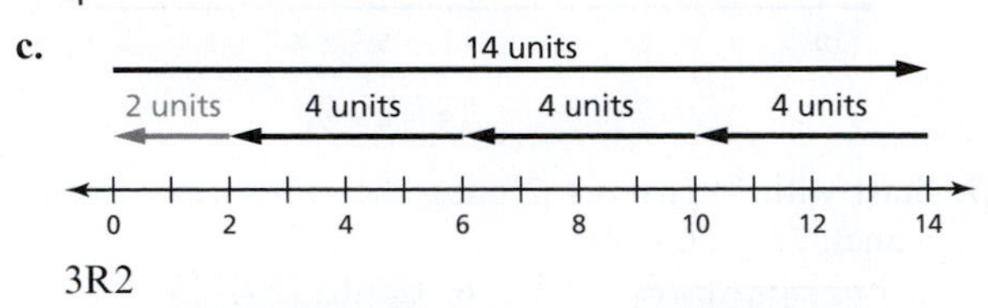

3R2

d.

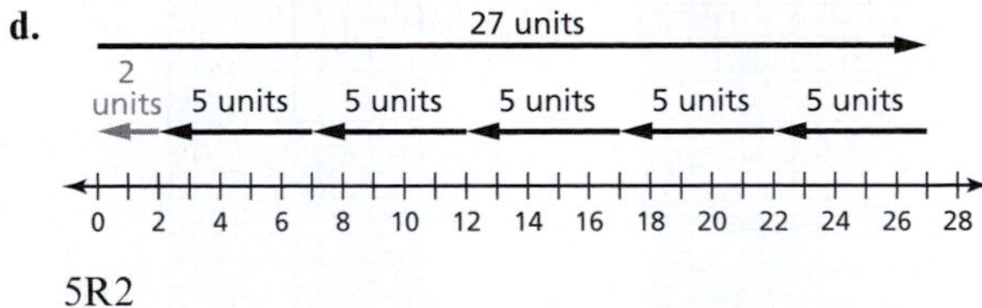

5R2

15. a. Scaffold algorithm; 124; 13; 23; 23R13
b. Traditional, compressed algorithm; 34; 96; 12; 34R12

17. a. 47R4 **b.** 32R9 **c.** 57 **d.** 23 **e.** 56R8 **f.** 89R21

19. a. 47 **b.** 34 **c.** 29R17 **d.** 62 **e.** 52R22 **f.** 19R15

21. 145 lb **23. a.** 3254 **b.** 2993

25. a. *Sample answer:* The student switched the quotient and the remainder, and should review using algorithms to divide whole numbers.
b. *Sample answer:* The student wrote 150 instead of 157, and should review using algorithms to divide whole numbers.

27. a. \$14 **b.** *Sample answer:* Multiply by 650.

29. *Sample answer:* $2 \div 1 = 6 \div 3$ **31.** Answers will vary.

33. a. 45 **b.** 41 **c.** There are 16 cans left over.

35. 66 ft^3

37. a. 25 calories
b. *Sample answer:* $500 \div (60 \div 3)$
c. Answers will vary.

Chapter 3 Review *(page 119)*

1. $9 + 6$ **3.** $67 + (9 + 1)$ **5.** 12

7. *Sample answer:* The arrows of 2 units and 5 units represent the addends 2 and 5; $2 + 5 = 7$

9.

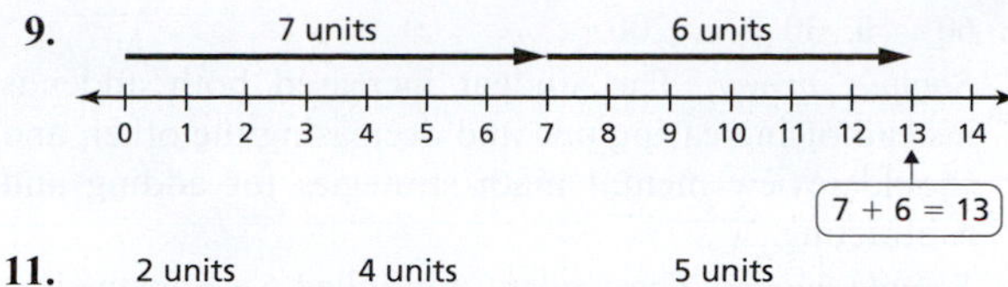

11.

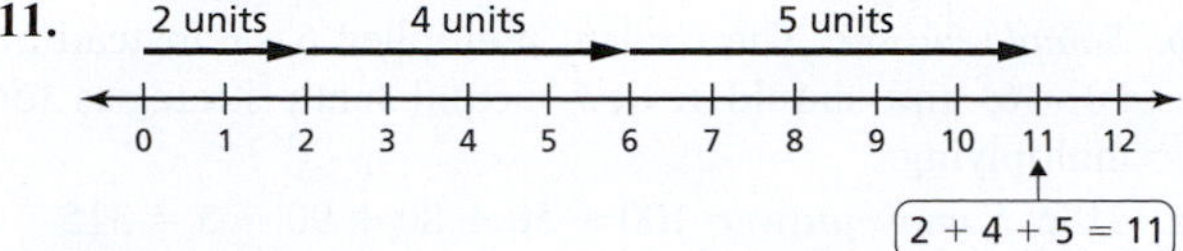

13. *Sample answer:* Join all of the blocks and regroup as needed.

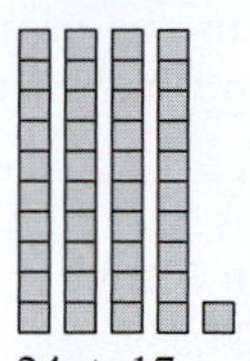

24 + 17 = 41

15.

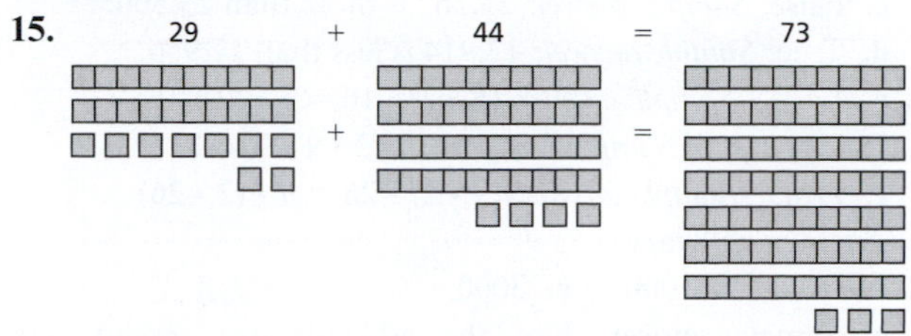

17.

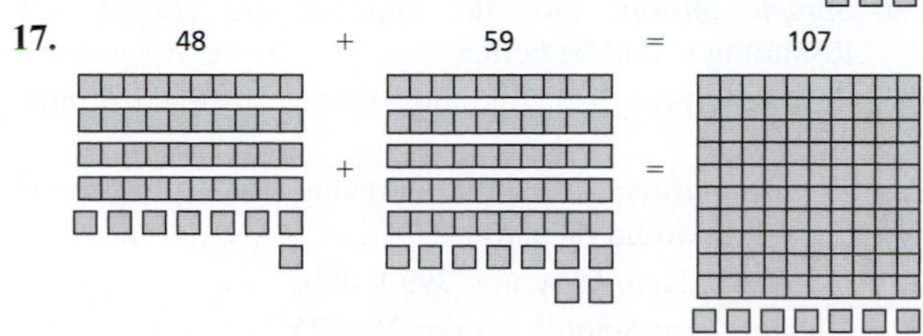

19. 161 **21.** 948 **23.** 96 **25.** 6604 **27.** 2550

29. 667

31. \$62,834; *Sample answer:* The standard algorithm because there are fewer steps.

33. What number can you add to 8 to get 19? 19 − 8

35. What number can you add to 6 to get 23? 23 − 6

37. 22 + ☐ = 24; What number can you add to 22 to get 24?

39. 6 + ☐ = 18; What number can you add to 6 to get 18?

41. 12 − 5

43.

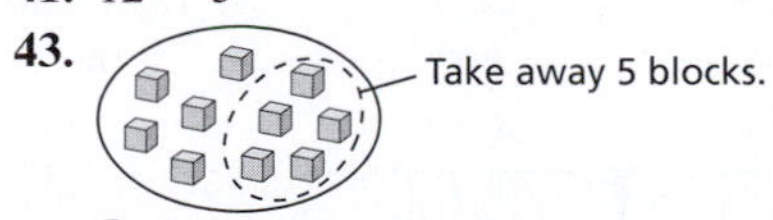

5

45.

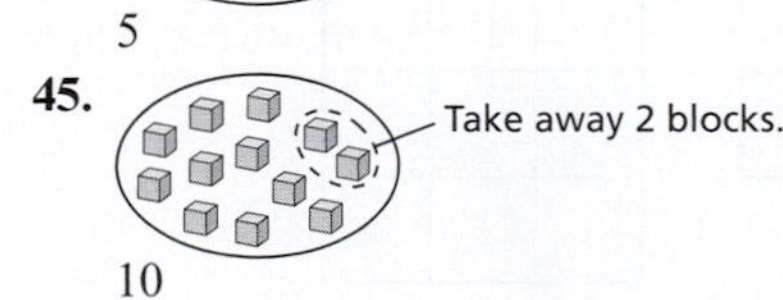

10

47.

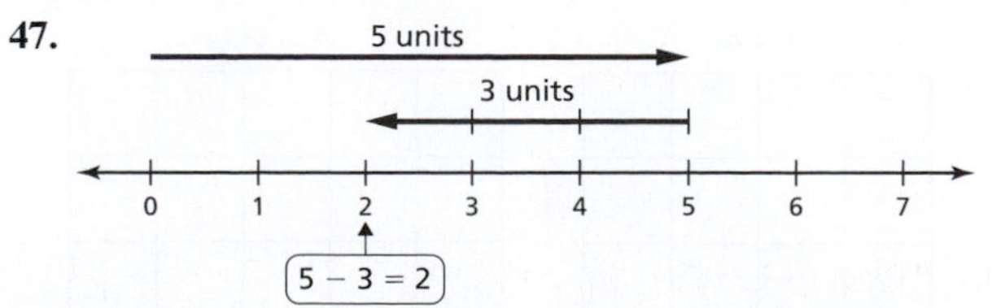

49.

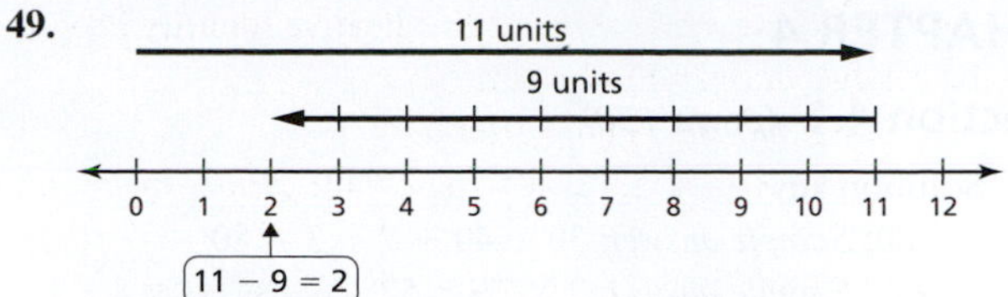

51.

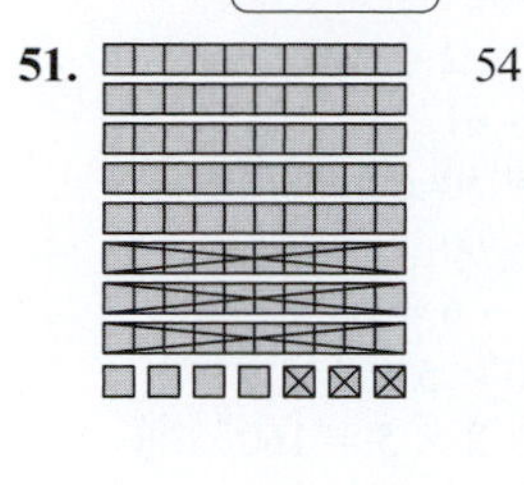

54

53.

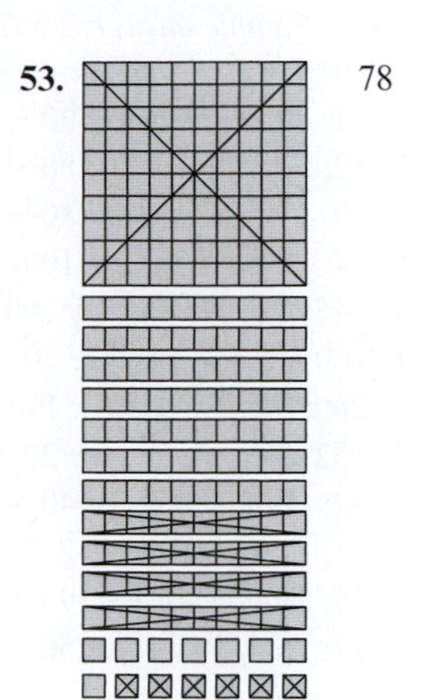

78

55. 207 **57.** 758 **59.** 81 **61.** 981 **63.** 19

65. 63 **67.** 52 **69.** 9 • 6 **71.** 7 • (16 • 4) **73.** 22

75. *Sample answer:* The rectangle is 5 units by 9 units; 5 × 9 = 45

77. 2655 **79.** 3496 **81.** 7720 **83.** 26,961

85. 6 • 10 + 6 • 8; 108 **87.** 9 • 40 − 9 • 7; 297

89. 4(7) + 4(8) + 4(9); 96 **91.** 696 **93.** 2079

95. 448 in. **97. a.** \$168 **b.** 6 • 4 • 28; \$672

99. 7; 42 ÷ 6 **101.** 4; 32 ÷ 8 **103.** 18 ÷ 6 = 3

105.

4

107.

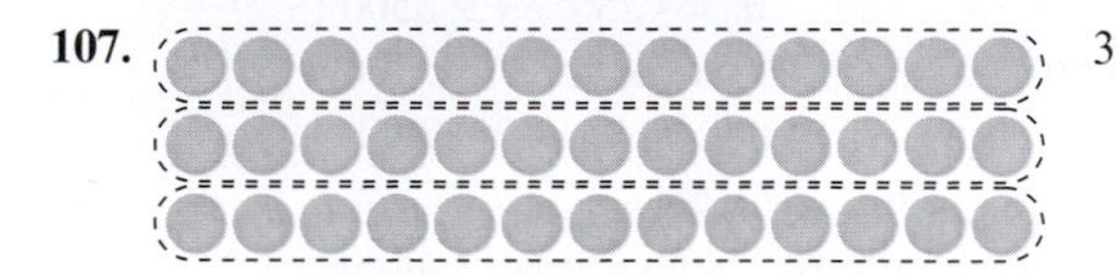

3

109.

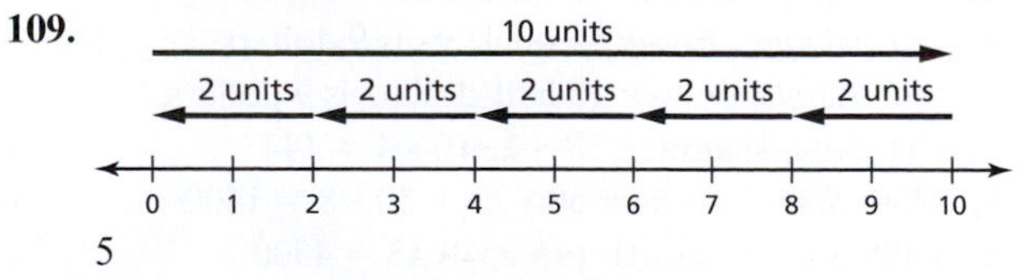

5

111.

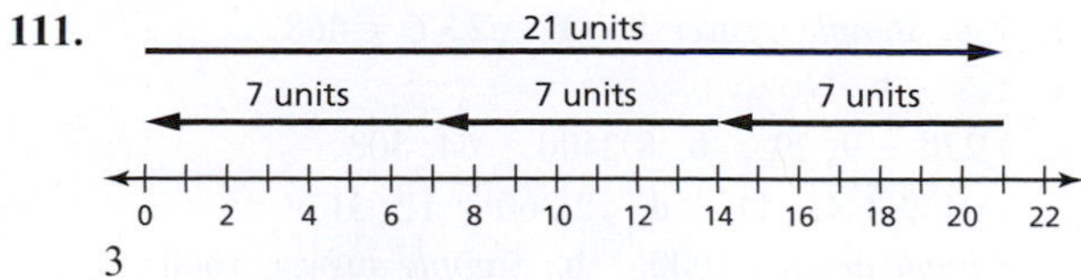

3

113. 42 **115.** 37R3 **117.** 66R15 **119.** 44

121. 59R9 **123.** 65R5 **125.** About 37 ft^3

Chapter 3 Test *(page 122)*

1. B **2.** E **3.** C **4.** D

5. E **6.** B **7.** C **8.** D

CHAPTER 4

Section 4.1 *(page 130)*

1. Solution key:

1. 80; *Sample answer:* $30 + 40 + 7 + 3 = 80$
2. 52; *Sample answer:* $2 \times 26 = 52$
3. 102; *Sample answer:* $75 + 25 + 2 = 102$
4. 61; *Sample answer:* $30 + 31 = 61$
5. 52; *Sample answer:* $80 - 30 = 50$ and $7 - 5 = 2$
6. 50; *Sample answer:* $80 - 30 = 50$
7. 64; *Sample answer:* $100 - 30 - 6 = 64$
8. 145; *Sample answer:* $200 - 50 - 5 = 145$
9. 160; *Sample answer:* $30 \times 5 + 2 \times 5 = 160$
10. 260; *Sample answer:* Put a zero at the end of 26.
11. 152; *Sample answer:* $20 \times 8 - 1 \times 8 = 152$
12. 92; *Sample answer:* $40 \times 2 + 6 \times 2 = 92$
13. 21; *Sample answer:* $210 = 21 \times 10$
14. 35; *Sample answer:* $90 \div 3 + 15 \div 3 = 35$
15. 17; *Sample answer:* $50 \div 5 + 35 \div 5 = 17$
16. 219; *Sample answer:* $440 \div 2 - 2 \div 2 = 219$

Sample rubric: Assign 1 point to each correct sum, difference, product, and quotient. Assign 2 points to each correct reasoning. There is a total of 48 points.

3. a. No **b.** Yes **c.** Yes **d.** Yes **e.** Yes **f.** No

5. a. 81; *Sample answer:* $50 + 31 = 81$
b. 90; *Sample answer:* $60 + 20 + 7 + 3 = 90$
c. 360; *Sample answer:* $200 + 100 + 20 + 40 = 360$
d. 43; *Sample answer:* $73 - 30 = 43$
e. 25; *Sample answer:* $100 - 75 = 25$
f. 140; *Sample answer:* $440 - 300 = 140$

7. a. 87¢; *Sample answer:* $27 + 60 = 87$
b. \$133; *Sample answer:* $60 + 73 = 133$
c. 94¢; *Sample answer:* $20 + 30 + 50 - 6 = 94$
d. \$125; *Sample answer:* $50 + 40 + 30 + 5 = 125$

9. a. $2 \cdot 5 \cdot 6 \cdot 3$; 180 **b.** $4 \cdot 25 \cdot 7 \cdot 5$; 3500
c. $2 \cdot 50 \cdot 13 \cdot 3$; 3900 **d.** $250 \cdot 4 \cdot 7$; 7000
e. $5 \cdot 20 \cdot 17$; 1700 **f.** $200 \cdot 5 \cdot 23$; 23,000

11. a. 265 **b.** 276 **c.** 4368

13. a. 170; *Sample answer:* 34 nickels is 17 dimes.
b. 900; *Sample answer:* 36 quarters is 9 dollars.
c. 900; *Sample answer:* 18 half dollars is 9 dollars.

15. a. 144; *Sample answer:* $30 \cdot 4 + 6 \cdot 4 = 144$
b. 1400; *Sample answer:* $50 \cdot 20 + 50 \cdot 8 = 1400$
c. 4300; *Sample answer:* $(4 \cdot 25) \cdot 43 = 4300$
d. 468; *Sample answer:* $80 \cdot 6 - 2 \cdot 6 = 468$

17. a. Yes **b.** No

19. a. $9\overline{)270 - 9}$; 29 **b.** $8\overline{)2400 + 64}$; 308
c. $45\overline{)450 + 45}$; 11 **d.** $12\overline{)360 + 12}$; 31

21. a. *Sample answer:* 1000 **b.** *Sample answer:* 1600

23. a. *Sample answer:* 300 **b.** *Sample answer:* 100
c. *Sample answer:* 400

25. a. *Sample answer:* 8000 **b.** *Sample answer:* 8000
c. *Sample answer:* 33,000

27. a. *Sample answer:* 6 **b.** *Sample answer:* 3
c. *Sample answer:* 31 **d.** *Sample answer:* 225

29. a. *Sample answer:* \$1800 **b.** \$1600
c. \$2000 **d.** \$1780

31. a. 60 **b.** 30 **c.** 200

33. a. *Sample answer:* The student increased both addends instead of increasing one and decreasing the other, and should review mental math strategies for adding and subtracting.
b. *Sample answer:* The student multiplied $6 \cdot 6$ instead of $60 \cdot 6$, and should review mental math strategies for multiplying.

35. a. 315¢; *Sample answer:* $100 + 50 + 80 + 90 - 5 = 315$
b. 801¢; *Sample answer:* $90 \times 9 - 1 \times 9 = 801$
c. \$240; *Sample answer:* $(5 \cdot 2) \cdot 24 = 240$

37. *Sample answer:* To add two numbers, add to one number and subtract from the other. To subtract two numbers, add or subtract to both numbers.

39. 22; $352 = 320 + 32$

41. a. False; *Sample answer:* 24,828 is more than 24,535.
b. True; *Sample answer:* 24,500 is less than 24,535.
c. False; *Sample answer:* 23,562 is more than 23,500.
d. True; *Sample answer:* 17,914 is less than 17,950.

43. a. Equal; *Sample answer:* $(6 \cdot 2) \cdot 16 = 6 \cdot (2 \cdot 16)$
b. Not equal; *Sample answer:* $12 \cdot 2 \cdot 9 \cdot 2 \neq 12 \cdot 9$
c. Equal; *Sample answer:* $(9 \cdot 2) \cdot 26 = 9 \cdot (2 \cdot 26)$
Answers will vary.

45. a. 400 **b.** 1000 **c.** 3000

47. a. *Sample answer:* No; the addends are spread out. Rounding would be better.
b. *Sample answer:* Yes; the addends are clustered around 2500.
c. *Sample answer:* No; the addends are spread out. Rounding would be better.

49. a. Less than; *Sample answer:* $399 < 400$
b. Greater than; *Sample answer:* $26 > 25$
c. Less than; *Sample answer:* $30 \cdot 29 < 899$

51. *Sample answer:* When all of the addends round down; $1003 + 2007 + 310 + 904 \approx 4200$

53. Yes; $10 \cdot 40 \cdot 40 = 16{,}000$

55. Yes; *Sample answer:* Rounding down produces $1400 + 400 + 1200 = 3000$, which is a low estimate.

Section 4.2 *(page 140)*

1. Solution key:

1. 10
2. $A = 1^2$ $A = 2^2$ $A = 3^2$

$A = 4^2$

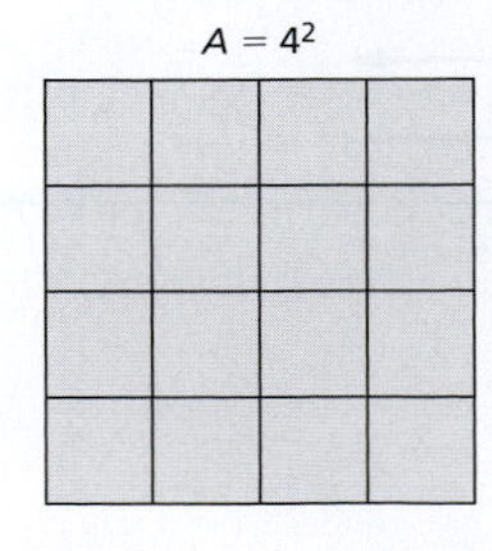

$A = 5^2$

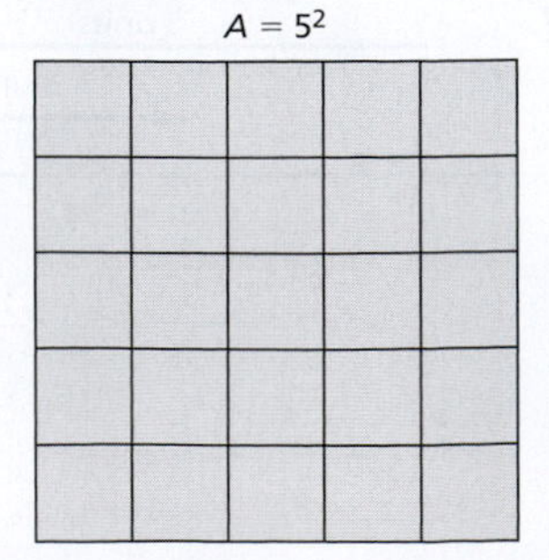

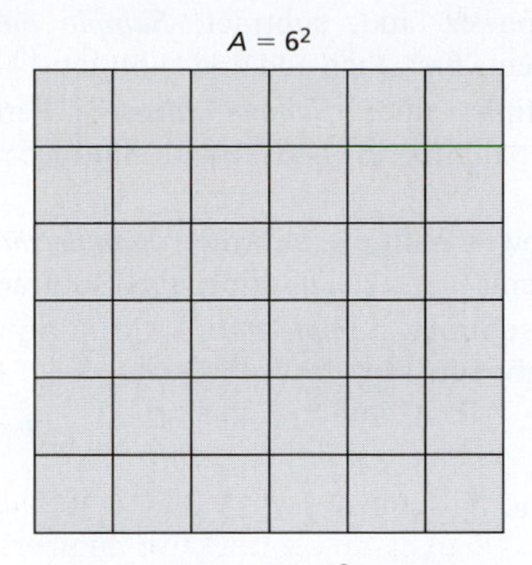

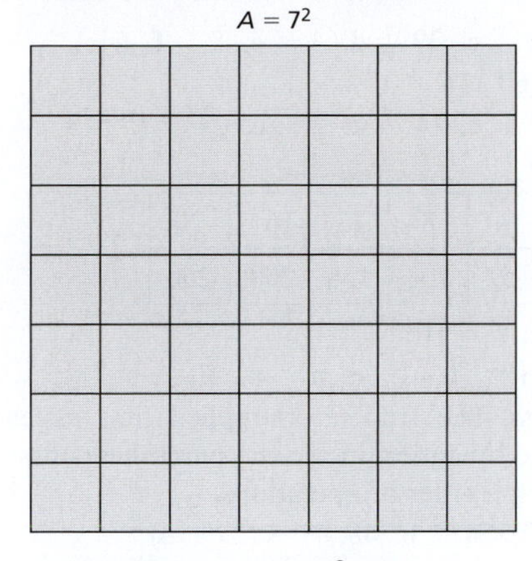

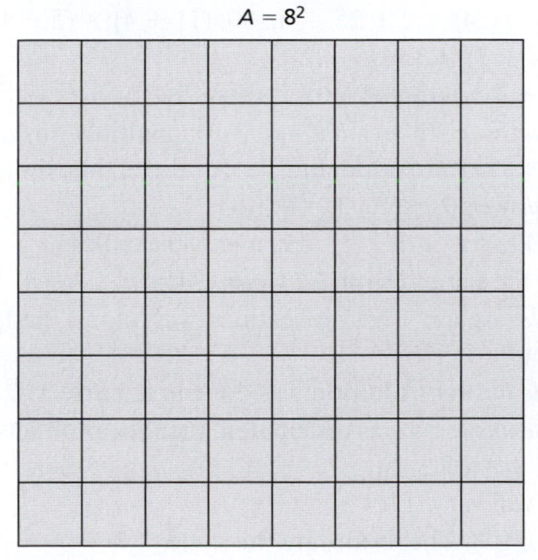

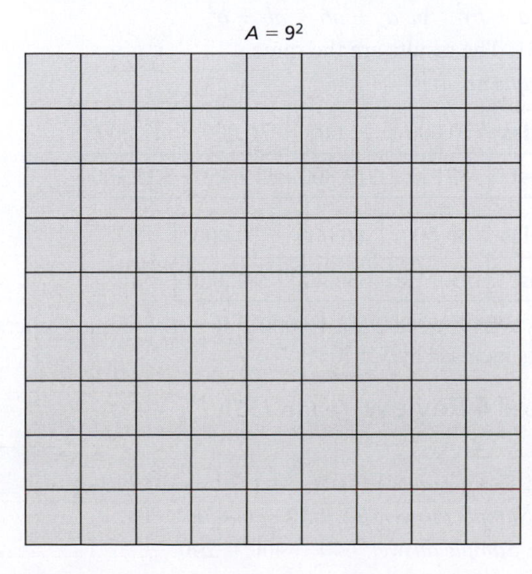

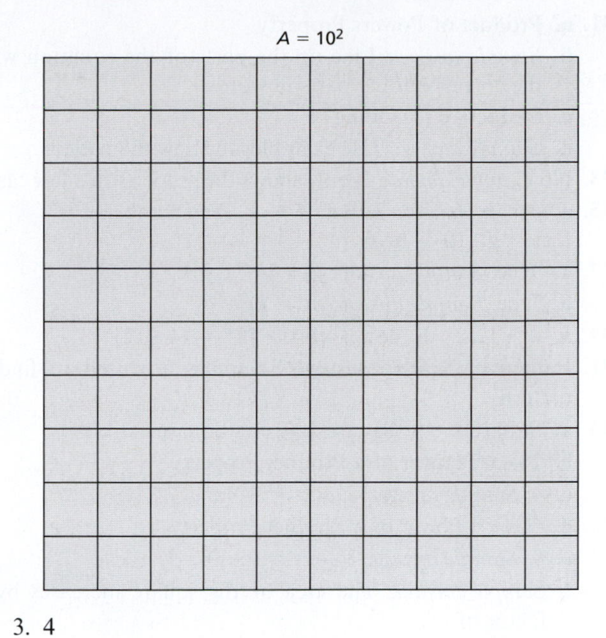

3. 4
4.

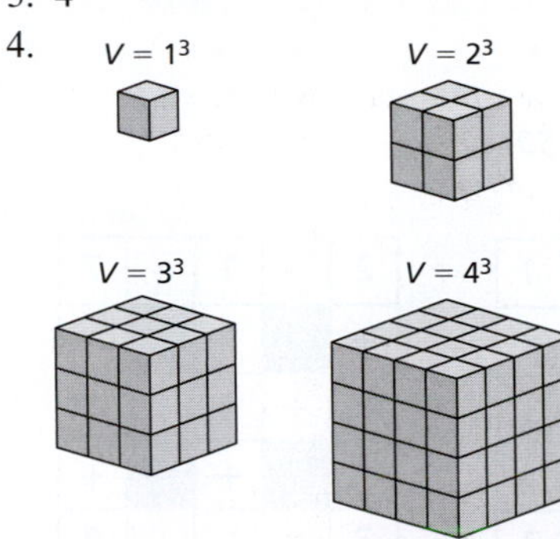

Sample rubric: Assign 1 point to each correct response in problems 1 and 3. Assign 1 point to each correct sketch and area or volume in exponential form in problems 2 and 4. There is a total of 30 points.

3. a. 17^2 **b.** 31^3 **c.** 4^5 **d.** x^6

5. a. $6^5 \cdot x^2$ **b.** $15^3 \cdot 7^2 \cdot y^3$

7. a. $13 \cdot 13 \cdot 13 \cdot 13$ **b.** $9 \cdot 9 \cdot 9 \cdot 9 \cdot 9 \cdot 9$

c. $y \cdot y \cdot y \cdot y \cdot y \cdot y \cdot y$ **d.** $b \cdot b \cdot b \cdot b \cdot b \cdot b \cdot b \cdot b$

9. 4^3; 64 units^3 **11.** 20^2; 400 in.^{2}

13. a. 32 **b.** 81 **c.** 1 **d.** 1

15. a. $3^0, 3^1, 3^2, 3^3, 3^4, 3^5, \ldots$ **b.** $10^0, 10^1, 10^2, 10^3, 10^4, 10^5, \ldots$

17. a. 2^7; $2^3 \cdot 2^4 = 2^{3+4}$ **b.** 6^5; $\frac{6^7}{6^2} = 6^{7-2}$

19. a. 11^9; Product of Powers Property

b. 23^3; Quotient of Powers Property

c. 19^4; Quotient of Powers Property

d. x^{15}; Product of Powers Property

21. a. 12^5 **b.** 30^3 **c.** 35^4 **d.** 42^9 **e.** 120^7 **f.** 108^3

23. a. $3^6 \cdot x^6$ **b.** $5^8 \cdot y^8$ **c.** $4^3 \cdot x^3 \cdot y^3$ **d.** $6^2 \cdot x^2 \cdot y^2$

25. a. 2^6; 64 **b.** 5^6; 15,625 **c.** 11^0; 1

d. 10^9; 1,000,000,000

27. a. $2^6 \cdot a^2$ **b.** $5^4 \cdot c^{12}$

29. a. *Sample answer:* The student multiplied the exponents instead of adding them, and should review the properties of exponents.

b. *Sample answer:* The student added the exponents instead of using 3, and should review the properties of exponents.

31. a. Product of Powers Property
b. *Sample answer:* Line up the parts of the equation with the corresponding parts in words.
c. Deductive reasoning
d. *Sample answer:* Use examples to show the pattern.
33. No; *Sample answer:* It only shows the property in a few cases.
35. a. 6 **b.** 6 **c.** 3 **d.** 4 **e.** Any number x
f. 5 **g.** 10 **h.** 6
37. a. True. *Sample answer:* $k + 1 - 1 = k$
b. True. *Sample answer:* $nk = kn$
39. a. $4^2 \cdot 12^2$ **b.** 48^2 **c.** $4^2 \cdot 12^2 = (4 \cdot 12)^2$
41. Inductive; *Sample answer:* Examples are used to find a pattern.
43. a. Square 1: x^2; Square 2: $(2x)^2$; Square 3: $(3x)^2$
b. $2^2 \cdot x^2$; Power of a Product Property
c. 4; *Sample answer:* $2^2 \cdot x^2 = 4 \cdot x^2$
d. $3^2 \cdot x^2$; Power of a Product Property
e. 9; *Sample answer:* $3^2 \cdot x^2 = 9 \cdot x^2$
f. *Sample answer:* The area of the square increases by a factor of n^2.
45. 24 ft **47.** 2^8; 256

Section 4.3 (page 150)

1. Solution key:

1.

9	−	8	+	1
+		×		×
4	÷	2	+	3
−		−		+
6	×	5	−	7

2.

2	−	1	+	7
+		×		×
8	÷	4	+	6
−		+		+
3	×	5	−	9

3.

2	×	4	+	5
×		×		×
9	−	6	+	1
−		÷		+
8	+	3	−	7

4.

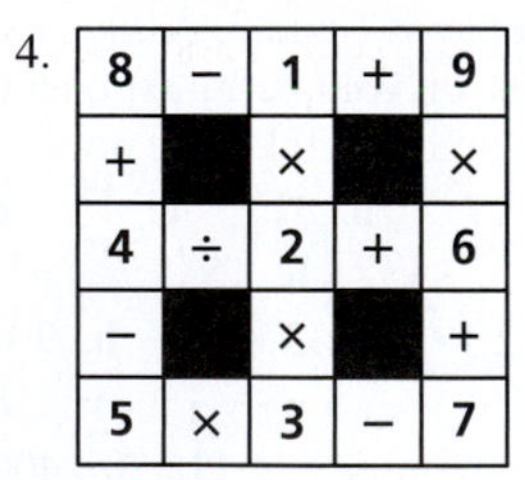

8	−	1	+	9
+		×		×
4	÷	2	+	6
−		×		+
5	×	3	−	7

5. a. 60 b. 0
(b); *Sample answer:* Multiply first.
6. a. 9 b. 21
(b); *Sample answer:* Divide first.
7. a. 63 b. 39
(b); *Sample answer:* Multiply first.
8. a. 6 b. 2
(b); *Sample answer:* Divide first.
Sample rubric: Assign 1 point for each correct entry in problems 1 through 4. Assign 1 point for each correct calculation and 2 points for each correct reasoning in problems 5 through 8. There is a total of 36 points.

3. a. ×; *Sample answer:* Multiply first; 2
b. ×; *Sample answer:* Multiply first; 19
c. ÷; *Sample answer:* Divide first; 13
d. ÷; *Sample answer:* Divide first; 2
5. a. Evaluate the power, add, subtract; *Sample answer:* Evaluate exponents first, then add and subtract; 11
b. Subtract, multiply, add; *Sample answer:* Perform operations in parentheses first, then multiply, then add; 10
c. Evaluate the power, multiply, subtract; *Sample answer:* Evaluate exponents first, then multiply, then subtract; 65
d. Multiply, add, subtract; *Sample answer:* Multiply first, then add and subtract; 14
7. a. 12 **b.** 43 **c.** 2 **d.** 79 **e.** 40 **f.** 24 **g.** 2
9. a. $(3 + 4) \cdot 5$ **b.** $23 - 3 \times 6$ **c.** $12 - (6 - 4)$
d. $14 \div 7 + 10$ **e.** $7^2 + 4 + 6$ **f.** $33 \div (8 + 3)$
11. $18 - 9$
13. a. 36 **b.** 32 **c.** 39 **d.** 4 **e.** 8 **f.** 64
15. a. $22h$ **b.** $160 - w$ **c.** $n \div 8$ **d.** s^2
17. a. 27 **b.** 42 **c.** 14 **d.** 2 **e.** 24 **f.** 144
g. 13 **h.** 13
19. a.

n	0	4	8	12	16	20
$13n$	0	52	104	156	208	260

b. $n = 16$ **c.** $n = 16$
21. 17 **23. a.** iv **b.** i **c.** ii **d.** iii
25. *Sample answer:* The student multiplied first instead of performing the operation inside the parentheses first, and should review the order of operations.
27. a. 30; $(10 + 5) \times 4$ **b.** 48; $83 - (29 - 6)$
c. 83; $(46 + 24) \div 2 + 25$ **d.** 30; $(11 + 4) \times (5 - 1)$
e. 28; $(14 + 7) \times 3 - (5 - 2)$
f. 30; $(2 + 4)^2 + (6 + 2) \times 3$
29. *Sample answer:* In evaluating $(3x)^2$, multiply inside the parentheses first. In evaluating $3x^2$, evaluate the power first.
31. *Sample answer:* $9 - 5 - 3 = 8 - 6 - 1$
33. $6 \div 2 + 4 = 5 \times 1 + 2$ **35. a–d.** 80 candles
37. They are the same. $(6 \cdot n + 6) \div 6 - 1 = n$
39. a. *Sample answer:* Multiplication was done first; not correct
b. *Sample answer:* Addition was done first; correct
c. *Sample answer:* 4 was distributed, then the products were added; correct
Answers will vary.
41. $12(x + 5)$; 108; The results are the same.
43. a. $(a + b)^2$ **b.** $a^2 + ab + ab + b^2$
c. 81; The results are the same.
45. *Sample answer:*

Miles	10,000	20,000	30,000	40,000
Cost	\$8150	\$13,800	\$19,450	\$25,100

Miles	50,000	60,000	70,000
Cost	\$30,750	\$36,400	\$42,050

For every increase of 10,000 miles, the annual operating costs increase by \$5650.

Chapter 4 Review (page 155)

1. No **3.** No
5. 113; *Sample answer:* $10 + 30 + 70 + 3 = 113$
7. 80; *Sample answer:* $60 + 10 + 6 + 4 = 80$
9. 240; *Sample answer:* $640 - 400 = 240$

11. $2 \cdot 5 \cdot 3 \cdot 7 = 210$ **13.** $4 \cdot 250 \cdot 13 = 13{,}000$
15. $20 \cdot 5 \cdot 7 \cdot 8 = 5600$ **17.** 282 **19.** 2580
21. 190; *Sample answer:* 38 nickels is 19 dimes.
23. 3800; *Sample answer:* 76 half dollars is 38 dollars.
25. 329; *Sample answer:* $40 \cdot 7 + 7 \cdot 7 = 329$
27. 2700; *Sample answer:* $50 \cdot 50 + 4 \cdot 50 = 2700$
29. $9\overline{)360 + 9}$; 41 **31.** $35\overline{)3500 + 70}$; 102
33. *Sample answer:* 900 **35.** *Sample answer:* 1500
37. *Sample answer:* 300 **39.** *Sample answer:* 100
41. *Sample answer:* 4000 **43.** *Sample answer:* 24,000
45. *Sample answer:* 4 **47.** *Sample answer:* 70
49. a. *Sample answer:* \$1600 **b.** \$1400
c. \$1800 **d.** \$1630
51. a. *Sample answer:* \$1900 **b.** \$1800
c. \$2200 **d.** \$2000
53. 4000
55. Yes; *Sample answer:* Rounding down produces $300 + 1200 + 400 = 1900$, which is a low estimate.
57. 12^3 **59.** 6^5 **61.** $2^3 \cdot x^2$ **63.** $14 \cdot 14 \cdot 14$
65. $x \cdot x \cdot x \cdot x \cdot x$ **67.** 11^2; 121 in.^{2} **69.** 243
71. 1 **73.** $7^0, 7^1, 7^2, 7^3, 7^4, \ldots$
75. 11^9; Product of Powers Property
77. 5^{12}; Product of Powers Property
79. 60^3; Power of a Product Property
81. 5^3; Quotient of Powers Property
83. 3^9; Quotient of Powers Property
85. $10^6 = 1{,}000{,}000$ **87.** $2^8 = 256$ **89.** $3^2 \cdot a^2$
91. $7^3 \cdot a^6 \cdot b^3$ **93.** 4 **95.** 2 **97.** 7
99. 25; *Sample answer:* $(5x)^2 = 25x^2$
101. Evaluate the power, subtract, add; *Sample answer:* Evaluate exponents first, then subtract and add; 19
103. Evaluate the power, multiply, add; *Sample answer:* Evaluate exponents first, then multiply, then add; 40
105. 13 **107.** 30 **109.** 28 **111.** $(6 - 2)3$
113. $11 + (18 - 14)$ **115.** $18 + 7 + 5^2$ **117.** 42
119. 30 **121.** 8 **123.** $10{,}000h$ **125.** $x \div 11$
127. 37 **129.** 11 **131.** 100 **133.** 19
135. 60; $4 \times (16 - 4)$ **137.** 83; $72 \div (8 + 28) + 46$
139. 90; $70 - (4 + 44) \div 2 + 2$
141. $9(8 + x)$; 117; The results are the same.
143. a–d. 33 gal

Chapter 4 Test *(page 158)*

1. D **2.** C **3.** B **4.** B
5. C **6.** D **7.** B **8.** E

CHAPTER 5

Section 5.1 *(page 166)*

1. Solution key:
1. 6, 14, 18, 32, 50; *Sample answer:* A whole number is divisible by 2 when its ones digit is 0, 2, 4, 6, or 8.
2. 15, 30, 45, 75, 90; *Sample answer:* A whole number is divisible by 5 when its ones digit is 0 or 5.
3. 30, 90; *Sample answer:* A whole number is divisible by 10 when its ones digit is 0.
4. 24, 76, 148, 220, 504; *Sample answer:* A whole number is divisible by 4 when the number represented by its last two digits is divisible by 4.
5. 6, 18, 27, 30, 72; *Sample answer:* A whole number is divisible by 3 when the sum of its digits is divisible by 3.

Sample rubric: Assign 1 point for each correct number and 2 points for each correct rule. There is a total of 32 points.

3. a. 1, 2, 3, 4, 5, 6, 10, 12, 15, 20, 30, 60
b. 1, 2, 4, 5, 10, 20, 25, 50, 100
5. a. *Sample answer:* 6, 9 **b.** *Sample answer:* 10, 15
c. *Sample answer:* 12, 18 **d.** *Sample answer:* 20, 30
7. a. True. 20 is divisible by 4.
b. False. 4 is not divisible by 20.
c. True. 20 is divisible by 4.
d. False. 20 is divisible by 4.
e. True. 20 is divisible by 4.
9. a. Yes; 6 is an even number.
b. Yes; 10 is an even number.
c. No; 61 is not an even number.
d. No; 97 is not an even number.
e. Yes; 254 is an even number.
f. Yes; 782 is an even number.
11. a. Divisible by 5 and 10; the ones digit is 0.
b. Divisible by 5; the ones digit is 5.
c. Not divisible by 5 or 10; the ones digit is not 0 or 5.
d. Not divisible by 5 or 10; the ones digit is not 0 or 5.
e. Divisible by 5 and 10; the ones digit is 0.
f. Divisible by 5; the ones digit is 5.
13. a. 18 **b.** 40 **c.** 32
15. a. Divisible by 4 and 8; $4 \mid 56$ and $8 \mid 56$
b. Divisible by 4; $4 \mid 60$ and $8 \nmid 460$
c. Not divisible by 4 or 8; $4 \nmid 19$ and $8 \nmid 519$
d. Divisible by 4 and 8; $4 \mid 36$ and $8 \mid 536$
e. Not divisible by 4 or 8; $4 \nmid 30$ and $8 \nmid 530$
f. Divisible by 4; $4 \mid 76$ and $8 \nmid 676$
17. a. Divisible by 3 and 9; $3 \mid (2 + 4 + 3)$ and $9 \mid (2 + 4 + 3)$
b. Divisible by 3 and 9; $3 \mid (7 + 8 + 3)$ and $9 \mid (7 + 8 + 3)$
c. Divisible by 3; $3 \mid (1 + 8 + 5 + 7)$ and $9 \nmid (1 + 8 + 5 + 7)$
d. Divisible by 3; $3 \mid (2 + 0 + 5 + 8)$ and $9 \nmid (2 + 0 + 5 + 8)$
e. Not divisible by 3 or 9; $3 \nmid (3 + 2 + 5 + 3)$ and $9 \nmid (3 + 2 + 5 + 3)$
f. Not divisible by 3 or 9; $3 \nmid (4 + 2 + 5 + 9)$ and $9 \nmid (4 + 2 + 5 + 9)$
19. a. Yes; 474 is even and $3 \mid (4 + 7 + 4)$.
b. No; 547 is not even.
c. Yes; 864 is even and $3 \mid (8 + 6 + 4)$.
d. No; 913 is not even.
e. Yes; 2460 is even and $3 \mid (2 + 4 + 6 + 0)$.
f. No; 3289 is not even.
21. a. 8 **b.** 8 **c.** 9 **d.** 4
23. a. True; *Sample answer:* $8n = 4(2n)$
b. False; *Sample answer:* $3 \mid 12$ but $9 \nmid 12$.
25. 1, 2, 4, 8, 16
27. a. Yes **b.** No **c.** Yes **d.** No **29.** Yes
31. *Sample answer:* The student only used the last digit to test for divisibility by 4, and should review the divisibility test for 4.
33. No

35. No; *Sample answer:* $3 \mid (5 + 7)$ but $3 \nmid 5$ and $3 \nmid 7$

37. No

39. a. No

b. Yes; *Sample answer:* 371 yards is 1113 feet, and 1113 is divisible by 1, 3, 7, 21, 53, 159, 371, and 1113.

41. 9

43. Answers will vary.

45. *Sample answer:* The cost of the limes is \$0.73 and $3 \nmid 73$.

Section 5.2 *(page 176)*

1. Solution key:

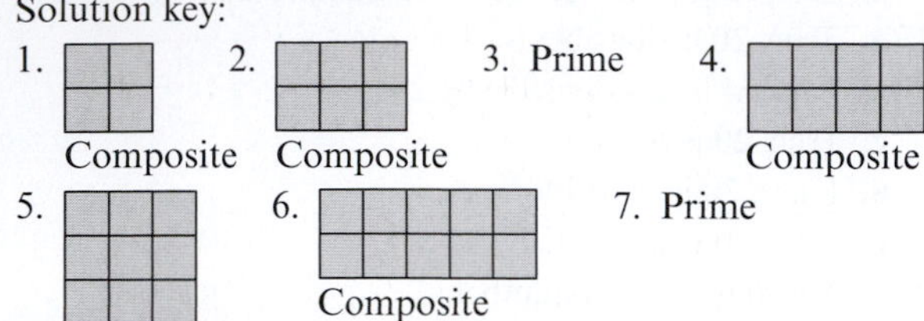

1\. Composite 2. Composite 3. Prime 4. Composite

5\. Composite 6. Composite 7. Prime

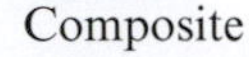

8\.

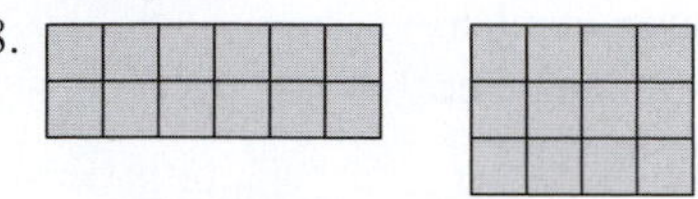

9\. Prime

10\.

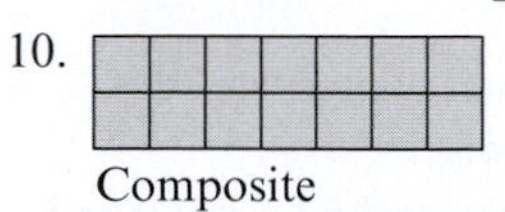

Composite

Sample rubric: Assign 1 point for each correct rectangle and 1 point for each correct label of prime or composite. There is a total of 18 points.

3. a. 1, 3, 9 **b.** 1, 2, 4, 8, 16 **c.** 1, 3, 17, 51
d. 1, 3, 7, 9, 21, 63 **e.** 1, 5, 25, 125
f. 1, 2, 3, 4, 6, 8, 9, 12, 16, 18, 24, 36, 48, 72, 144

5. a. Prime **b.** Prime **c.** Composite
d. Composite **e.** Composite **f.** Prime

7. 30: 1, 2, 3, 5, 6, 10, 15, 30; 31: 1, 31; 32: 1, 2, 4, 8, 16, 32; 33: 1, 3, 11, 33; 34: 1, 2, 17, 34; 35: 1, 5, 7, 35; 36: 1, 2, 3, 4, 6, 9, 12, 18, 36; 37: 1, 37; 38: 1, 2, 19, 38; 39: 1, 3, 13, 39; primes: 31, 37

9.

101	~~102~~	103	~~104~~	~~105~~	~~106~~	107	~~108~~	109	~~110~~
~~111~~	~~112~~	113	~~114~~	~~115~~	~~116~~	~~117~~	~~118~~	~~119~~	~~120~~

101, 103, 107, 109, 113

11. a. $2 \cdot 2 \cdot 3$ or $2^2 \cdot 3$ **b.** $5 \cdot 5$ or 5^2
c. $2 \cdot 3 \cdot 11$ **d.** $2 \cdot 3 \cdot 3$ or $2 \cdot 3^2$
e. $5 \cdot 29$ **f.** $2 \cdot 2 \cdot 5 \cdot 5$ or $2^2 \cdot 5^2$

13. 210, 105, 35; yes; *Sample answer:* Multiply the factors.

15. a. 2 **b.** 5 **c.** 11 **d.** 13

17. a. Composite **b.** Prime **c.** Prime
d. Prime **e.** Composite **f.** Composite

19. a. Prime **b.** Composite; $3 \cdot 11 \cdot 11$ or $3 \cdot 11^2$
c. Composite; $37 \cdot 37$ or 37^2 **d.** Prime
e. Composite; $7 \cdot 11 \cdot 13 \cdot 19$
f. Composite; $7 \cdot 7 \cdot 17 \cdot 19$ or $7^2 \cdot 17 \cdot 19$

21. a. $2^5 \cdot 11$ **b.** $2^4 \cdot 3^3 \cdot 5$ **c.** $7^3 \cdot 11$
d. $7^2 \cdot 11^2$ **e.** $2^2 \cdot 5^3 \cdot 7$ **f.** $3^4 \cdot 5^2$

23. a. False; 2 is an even prime.
b. False; 5 is a multiple of 5, but not composite.
c. True; $30 = 2 \cdot 3 \cdot 5$
d. False; you need to test for prime factors less than or equal to 11.

25. 2, 4, 7, 14 **27.** Yes; 133 is composite.

29. a. 5
b. 1 unit: 30; 2 units: 15; 3 units: 10; 5 units: 6; 6 units: 5; 10 units: 3
c. Answers will vary.

31. a. *Sample answer:*

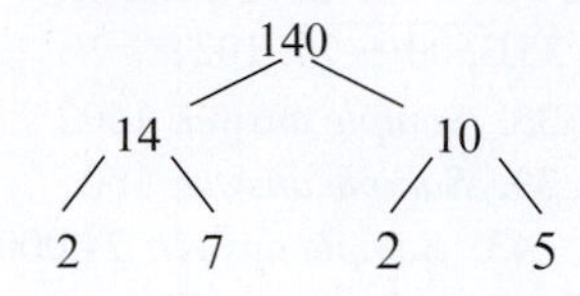

b. No; *Sample answer:* 140 can be factored as $2 \cdot 70$ first.
c. Answers will vary.

33. 31; *Sample answer:* $1367 < 37^2$

35. No; 3; *Sample answer:* $3 \mid (1 + 7 + 4 + 9)$, so $3 \mid 1749$.

37. 8 is not prime; $2^{49} \cdot 3 \cdot 5^{12}$

39. a. $2^{66} \cdot 11^8$ **b.** $2^{12} \cdot 3^{30} \cdot 7^{16}$

41. *Sample answer:* The student thought 1 was a prime, and should review the definitions of prime and composite numbers.

43. $48 = 2^4 \cdot 3$; $4 = 2^2$, $6 = 2 \cdot 3$, $8 = 2^3$, $12 = 2^2 \cdot 3$, $16 = 2^4$, $24 = 2^3 \cdot 3$, $48 = 2^4 \cdot 3$; *Sample answer:* The sets of prime factors for each composite factor of 48 are subsets of the prime factors of 48.

45. a. Perfect **b.** Not perfect **c.** Perfect

47. 6, 9, 10, 14, 15, 21, 22, 25, 26

49. *Sample answer:* The next consecutive prime is added to the next consecutive even number; $37 + 12 = 49$

51. 7 is prime; $9 = 3 + 3 + 3$; 11 is prime; 13 is prime; $15 = 5 + 5 + 5$; 17 is prime, 19 is prime; $21 = 2 + 2 + 17$; 23 is prime, $25 = 3 + 3 + 19$; $27 = 2 + 2 + 23$; 29 is prime; 31 is prime.

Section 5.3 *(page 186)*

1. Solution key:
1\. $a = 12$, $b = 30$, GCF = 6, LCM = 60
2\. $a = 60$, $b = 18$, GCF = 6, LCM = 180
3\. $a = 6$, $b = 30$, GCF = 6, LCM = 30
4\. $a = 70$, $b = 252$, GCF = 14, LCM = 1260
Sample rubric: Assign 1 point for each correct number. There is a total of 16 points.

3. a. 2 **b.** 1 **c.** 6 **d.** 3 **e.** 15 **f.** 50

5. a. 14 **b.** 12 **c.** 10 **d.** 12 **e.** 2 **f.** 1

7. a. 30 **b.** 18 **c.** 13 **d.** 18 **e.** 8 **f.** 120
g. 15 **h.** 34

9. a. 60 **b.** 12 **c.** 21 **d.** 56 **e.** 160 **f.** 140

11. a. 4 **b.** 30 **c.** 21

13. a.

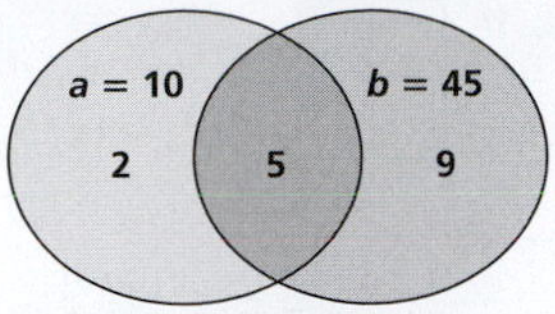

GCF(10, 45) = 5;
LCM(10, 45) = 90

b.

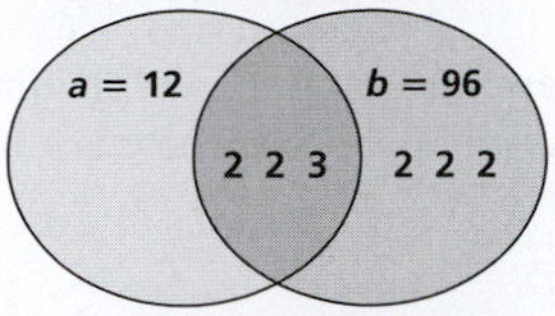

GCF(12, 96) = 12;
LCM(12, 96) = 96

c.

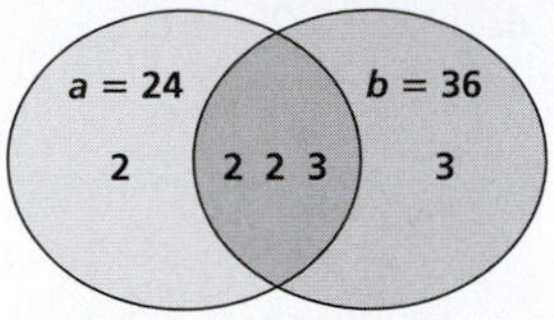

GCF(24, 36) = 12;
LCM(24, 36) = 72

15. LCM(48, 72); *Sample answer:* GCF(48, 72) = 24 and LCM(48, 72) = 144

17. a. $ab = 75$, GCF(5, 15) • LCM(5, 15) = 5 • 15 = 75
b. $ab = 147$, GCF(7, 21) • LCM(7, 21) = 7 • 21 = 147
c. $ab = 160$, GCF(8, 20) • LCM(8, 20) = 4 • 40 = 160

19. 4 **21. a.** 2 **b.** 15 **c.** 6

23. a. 6 **b.** 30 **c.** 120

25. a. 390 mi **b.** 10 gal, 13 gal

27. *Sample answer:* The student used 36 as the GCF instead of 18, and should review the subtraction algorithm for finding the GCF.

29. Answers will vary. **31.** 42 min **33.** 9 in.

35. Gear A: 2 revolutions; Gear B: 5 revolutions

37. Incorrect; *Sample answer:* With 5 school days in one week, the next math class on a Friday will be in 3 weeks.

39. 3 packs of napkins, 6 packs of paper plates, and 25 packs of plastic cups

41. False; *Sample answer:* GCF(3, 10) = 1

43. Not true; *Sample answer:* 2 • 3 • 12 = 72 but GCF(2, 3, 12) • LCM(2, 3, 12) = 1 • 12 = 12.

45. 61

Chapter 5 Review *(page 191)*

1. Yes; 328 is even. **3.** No; 5793 is not even.
5. Yes; 9412 is even. **7.** Divisible by 5; the ones digit is 5.
9. Divisible by 5; the ones digit is 5.
11. Divisible by 5 and 10; the ones digit is 0.
13. Divisible by 4; $4 \mid 72$ and $8 \nmid 972$
15. Not divisible by 4 or 8; $4 \nmid 23$ and $8 \nmid 223$
17. Divisible by 4 and 8; $4 \mid 84$ and $8 \mid 384$
19. Not divisible by 3 or 9; $3 \nmid (1 + 7 + 3)$ and $9 \nmid (1 + 7 + 3)$
21. Divisible by 3 and 9; $3 \mid (7 + 2 + 5 + 4)$ and $9 \mid (7 + 2 + 5 + 4)$
23. Divisible by 3 and 9; $3 \mid (9 + 0 + 7 + 2)$ and $9 \mid (9 + 0 + 7 + 2)$
25. Yes; 24 is even and $3 \mid (2 + 4)$. **27.** No; 1839 is not even.
29. Yes; 3012 is even and $3 \mid (3 + 0 + 1 + 2)$.
31. Difference Property; 96 − 12 = 84
33. Product Property; 336 is a multiple of 48.
35. Difference Property; 572 − 112 = 460
37. 1, 2, 13 **39.** Yes

41. a. Yes
b. Yes; *Sample answer:* 621 is divisible by 1, 3, 9, 23, 27, 69, 207, and 621.

43. 4 **45.** Composite **47.** Prime **49.** Composite
51. 4, 3, 2, 2, 2, 2 **53.** 47; *Sample answer:* $2371 < 49^2$
55. Composite; $2^3 \cdot 7$ **57.** Composite; $5 \cdot 19$
59. Composite; $2^2 \cdot 7^2$ **61.** Composite; $3 \cdot 419$
63. Composite; $19 \cdot 89$ **65.** $3^{13} \cdot 5^{11} \cdot 11^2$
67. 2, 3, 5, 10, 15 **69.** No; *Sample answer:* 139 is prime.
71. 2, 3, 5, 7, 11, 13 **73.** 9 **75.** 4 **77.** 24 **79.** 6
81. 2 **83.** 19 **85.** 45 **87.** 48

89.

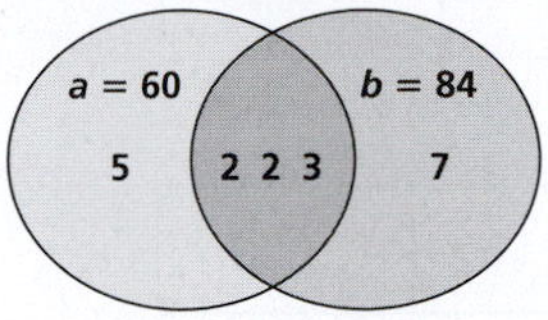

GCF(60, 84) = 12;
LCM(60, 84) = 420

91. 4 **93.** 9 **95.** 17 **97.** 741 **99.** 162
101. 840
103. $ab = 243$, GCF(9, 27) • LCM(9, 27) = 9 • 27 = 243
105. $ab = 80$, GCF(8, 10) • LCM(8, 10) = 2 • 40 = 80
107. 72 min **109.** 26 **111.** 80

Chapter Test *(page 194)*

1. D **2.** C **3.** E **4.** A
5. D **6.** B **7.** C **8.** B

CHAPTER 6

Section 6.1 *(page 204)*

1. Solution key:
1. 2; $\frac{1}{2}$ 2. 3; $\frac{1}{3}$ 3. 2; $\frac{1}{2}$ 4. 6; $\frac{1}{6}$ 5. 3; $\frac{1}{3}$
6. 1; $\frac{1}{1}$ 7. 6; $\frac{1}{6}$ 8. 2; $\frac{1}{2}$ 9. 4; $\frac{1}{4}$ 10. 5; $\frac{1}{5}$

Sample rubric: Assign 1 point to each correct number and fraction. There is a total of 20 points.

3. a. $\frac{3}{7}$; numerator: 3; denominator: 7
b. $\frac{5}{9}$; numerator: 5; denominator: 9
c. $\frac{2}{5}$; numerator: 2; denominator: 5
d. $\frac{7}{10}$; numerator: 7; denominator: 10

5. a. $\frac{2}{12}$ **b.** $\frac{6}{12}$ **c.** $\frac{6}{12}$
7. a. $\frac{20}{50}$ **b.** $\frac{10}{50}$ **c.** $\frac{6}{50}$

9. a.

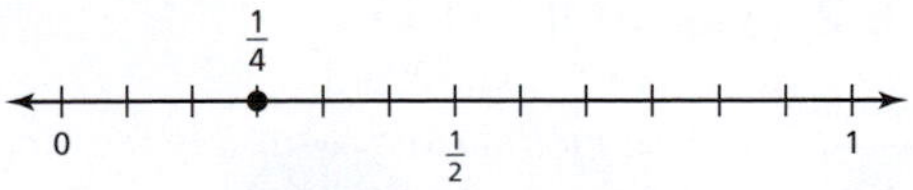

b.

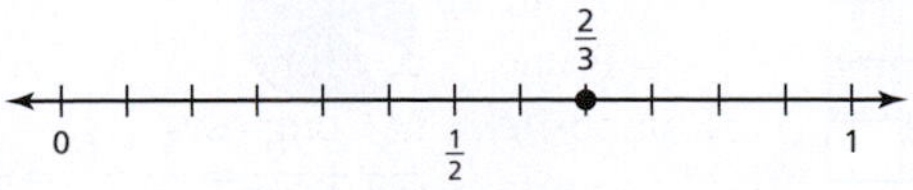

c.

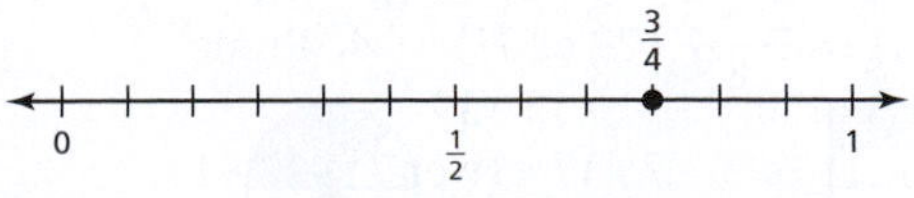

d.

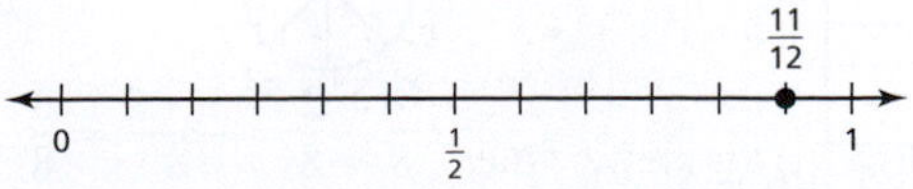

11. a. $\frac{5}{4}$
b. $\frac{5}{2}$
c. $3\frac{1}{8}$
d. $2\frac{3}{4}$

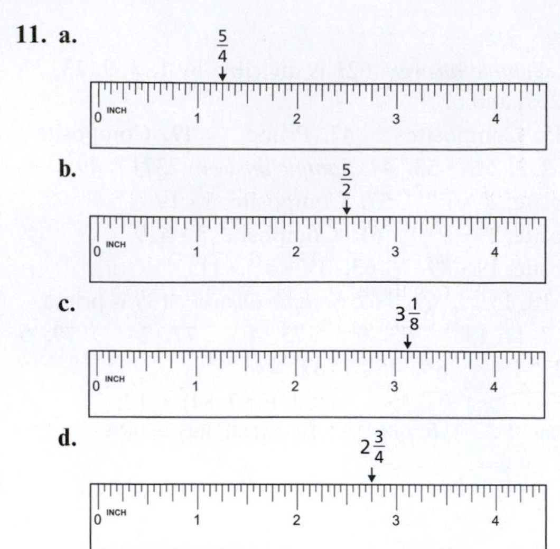

13. *Sample answer:* $\frac{35}{60} = \frac{7}{12}$

15. a. $\frac{1}{3}$ **b.** $\frac{1}{4}$ **c.** $\frac{1}{2}$ **d.** $\frac{2}{3}$

17. a. Equivalent **b.** Not equivalent **c.** Not equivalent
d. Equivalent

19. a. Equivalent **b.** Not equivalent **c.** Not equivalent
d. Equivalent

21. a. $>$ **b.** $<$ **c.** $=$ **d.** $>$

23. $\frac{1}{16}, \frac{8}{17}, \frac{5}{9}, \frac{11}{12}$; *Sample answer:* $\frac{1}{16}$ is close to 0, $\frac{8}{17} < \frac{1}{2}, \frac{5}{9} > \frac{1}{2}$, and $\frac{11}{12}$ is close to 1.

25. a. $\frac{72}{118}; \frac{28}{118}; \frac{13}{118}; \frac{5}{118}$ **b.** Answers will vary.

27. *Sample answer:* Divide out common factors.

29. a. *Sample answer:* $\frac{25}{100}$ **b.** *Sample answer:* $\frac{21}{100}$

31. *Sample answer:* The student factored 118 incorrectly, and should review finding the prime factorization of a number.

33. $\frac{7}{20}$; *Sample answer:* The other sections are $\frac{13}{20}$ of the graph.

35. Answers will vary. **37.** *Sample answer:* $\frac{1}{101}$

39. a. Yes; *Sample answer:* Shade 12 parts.
b. Yes; *Sample answer:* Shade 6 parts.
c. Yes; *Sample answer:* Shade 4 parts.
d. Yes; *Sample answer:* Shade 3 parts.
e. No; *Sample answer:* Not possible.
f. Yes; *Sample answer:* Shade 2 parts.

41. Shilling; *Sample answer:* It is easiest to read.

Section 6.2 *(page 214)*

1. Solution key:

1.

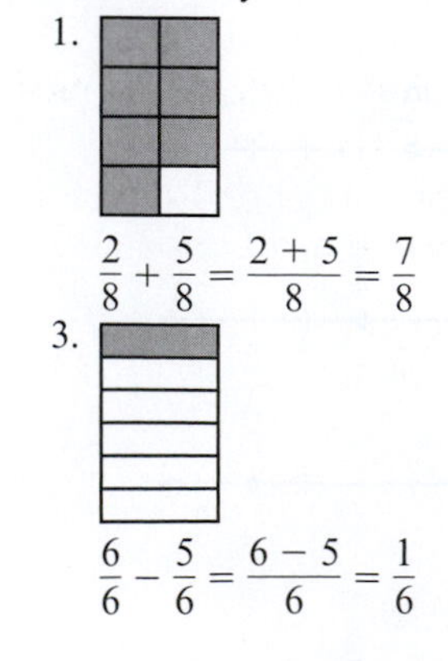

$\frac{2}{8} + \frac{5}{8} = \frac{2+5}{8} = \frac{7}{8}$

2.

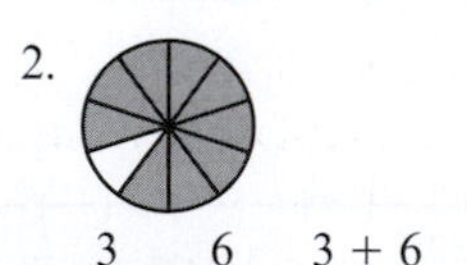

$\frac{3}{10} + \frac{6}{10} = \frac{3+6}{10} = \frac{9}{10}$

3.

$\frac{6}{6} - \frac{5}{6} = \frac{6-5}{6} = \frac{1}{6}$

4.

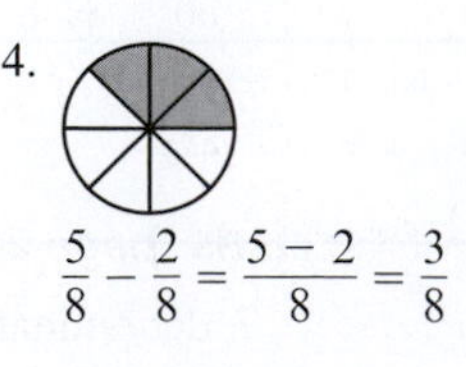

$\frac{5}{8} - \frac{2}{8} = \frac{5-2}{8} = \frac{3}{8}$

5.

$\frac{1}{2} + \frac{1}{8} = \frac{4}{8} + \frac{1}{8} = \frac{5}{8}$

6.

$\frac{1}{3} - \frac{1}{6} = \frac{2}{6} - \frac{1}{6} = \frac{1}{6}$

Sample rubric: Assign 1 point to each correct shading and each correct number in the work shown. There is a total of 62 points.

3. a. $\frac{1}{4} + \frac{3}{4}; 1$ **b.** $\frac{5}{12} + \frac{8}{12}; \frac{13}{12}$

5. a. $1\frac{1}{3}$ **b.** $1\frac{4}{5}$ **c.** $1\frac{4}{9}$ **d.** $1\frac{1}{5}$ **e.** $2\frac{1}{4}$ **f.** $2\frac{3}{7}$

7. a. 6 **b.** $2\frac{1}{4}$ **c.** $6\frac{3}{5}$

9. a. $\frac{1}{5} + \frac{2}{5}; \frac{3}{5}$ **b.** $\frac{2}{4} + \frac{1}{4}; \frac{3}{4}$ **c.** $\frac{2}{6} + \frac{3}{6}; \frac{5}{6}$

11. a. $\frac{3}{4} - \frac{2}{4}; \frac{1}{4}$ **b.** $\frac{2}{3} - \frac{1}{3}; \frac{1}{3}$

13. a. $\frac{1}{7}$ **b.** $\frac{1}{3}$ **c.** $\frac{1}{2}$ **d.** 0

15. a. $1\frac{1}{2}$ **b.** $1\frac{2}{5}$ **c.** $1\frac{3}{5}$

17. a. $\frac{3}{4}$ **b.** $\frac{2}{3}$
c. $\frac{3}{4}$ **d.** $\frac{5}{7}$

19. *Sample answer:* Square; divides into 9 equal sections easily.

21. a. $\frac{6}{12} - \frac{1}{4}; \frac{1}{4}$ **b.** $\frac{2}{4} + \frac{2}{3}; 1\frac{1}{6}$

23. a. $\frac{1}{3}$ **b.** $1\frac{1}{3}$ **c.** $1\frac{1}{14}$ **d.** $\frac{1}{5}$

25. a. $\frac{1}{6}$ **b.** $1\frac{5}{8}$ **c.** $1\frac{3}{10}$

27. a. $\frac{1}{16}$ **b.** $\frac{4}{15}$ **c.** $\frac{1}{9}$ **d.** $\frac{1}{2}$ **29. a.** $\frac{1}{6}$ **b.** $1\frac{3}{20}$ **c.** $\frac{3}{26}$

31. *Sample answer:* $\frac{1}{4} + \frac{1}{2} = \frac{3}{8} + \frac{3}{8}$

33. a. 4 km
b. *Sample answer:* Group together days that add up to whole numbers.
c. *Sample answer:* Use the Commutative Property to reorder the numbers. Use the Associative Property to group the numbers.

35. *Sample answer:* Group together pairs that add up to whole numbers. 6

37. a. Commutative Property **b.** Associative Property
c. Additive Identity Property

39. a. *Sample answer:* $\left(\frac{3}{8} + \frac{5}{8}\right) + \left(\frac{4}{3} + \frac{2}{3}\right) + \left(\frac{4}{7} + \frac{10}{7}\right)$ **b.** 5

41. *Sample answer:* The student added the numerators and the denominators, and should review adding fractions with unlike denominators.

43. *Sample answer:* The denominators may be different.

45. Answers will vary. **47. a.** $\frac{5}{18}$ **b.** Answers will vary.

49. a. With preparation: $4\frac{19}{20}$; Without preparation: 4
b. *Sample answer:* Round $4\frac{19}{20}$ up to 5, and find $5 \div 6$ and $4 \div 6$.

Section 6.3 *(page 226)*

1. Solution key:

1. 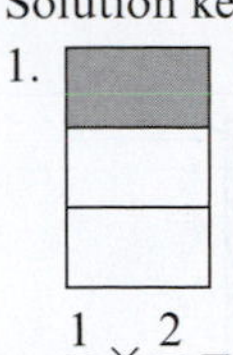
$\frac{1}{2} \times \frac{2}{3} = \frac{1}{3}$

2. $\frac{2}{3} \times \frac{6}{10} = \frac{4}{10}$

3. 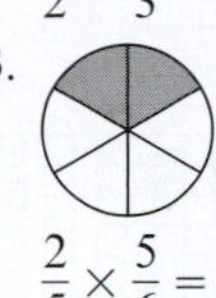
$\frac{2}{5} \times \frac{5}{6} = \frac{2}{6}$

4. 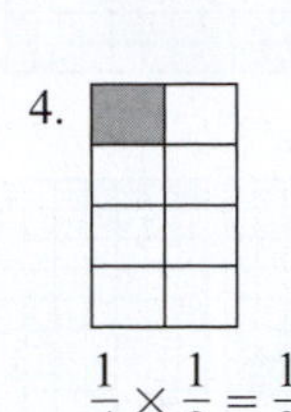
$\frac{1}{4} \times \frac{1}{2} = \frac{1}{8}$

5. $\frac{1}{4}$

Sample rubric: Assign 1 point to each correct shading and each correct number in the work shown. There is a total of 30 points.

3. a. $6 \cdot \frac{2}{5}; \frac{12}{5}$ **b.** $3 \cdot \frac{1}{8}; \frac{3}{8}$ **c.** $7 \cdot \frac{3}{4}; \frac{21}{4}$

5. a.–c.

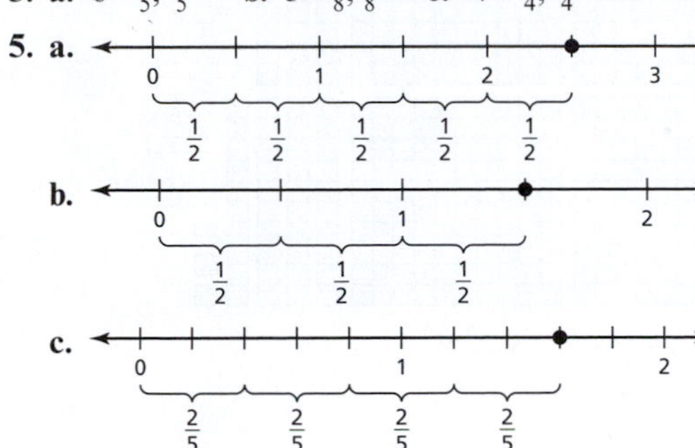

7. a. $\frac{1}{2} \cdot \frac{4}{5}$ **b.** $\frac{2}{3} \cdot \frac{3}{4}$ **c.** $\frac{2}{3} \cdot \frac{1}{4}$

9. a. $\frac{4}{5}$ **b.** $\frac{5}{2}$ **c.** $\frac{2}{7}$ **d.** $\frac{1}{6}$ **e.** $\frac{3}{2}$ **f.** $\frac{5}{3}$ **g.** $\frac{9}{4}$ **h.** 0

11. a. $\frac{3}{2}$; 1 **b.** $\frac{4}{5}$; 1 **c.** $\frac{2}{7}$; 1 **13. a.** $6\frac{1}{2}$ **b.** 1 **c.** $2\frac{1}{4}$

15.

$\frac{3}{2}\left(\frac{2}{3}x\right) = \frac{3}{2} \cdot \frac{4}{3}$	Multiply by the reciprocal of $\frac{2}{3}$.
$\left(\frac{3}{2} \cdot \frac{2}{3}\right)x = \frac{3}{2} \cdot \frac{4}{3}$	Associative Property of Multiplication
$(1)x = \frac{3}{2} \cdot \frac{4}{3}$	Use the Multiplicative Inverse Property.
$x = 2$	Use the Multiplicative Identity Property. Multiply $\frac{3}{2} \cdot \frac{4}{3}$.

17. a. $\frac{12}{15} = \frac{12}{15}$ **b.** $\frac{4}{15} = \frac{4}{15}$ **19. a.** $\frac{9}{2}$ **b.** $\frac{2}{9}$ **c.** $\frac{3}{2}$

21. $4\frac{1}{2} \div \frac{1}{2}$; 9 **23. a.** $\frac{4}{5}$ **b.** 2 **c.** $\frac{7}{4}$

25. $\frac{2}{5} \cdot \frac{1}{4}$; *Sample answer:* $\frac{2}{5} \div 4, \frac{1}{4} \div \frac{5}{2}$

27. a. $\frac{9}{5}$ **b.** $\frac{1}{3}$ **c.** $\frac{3}{40}$ **d.** 5

29. a. 60 **b.** 2 **c.** 4

31. a. 14

b.

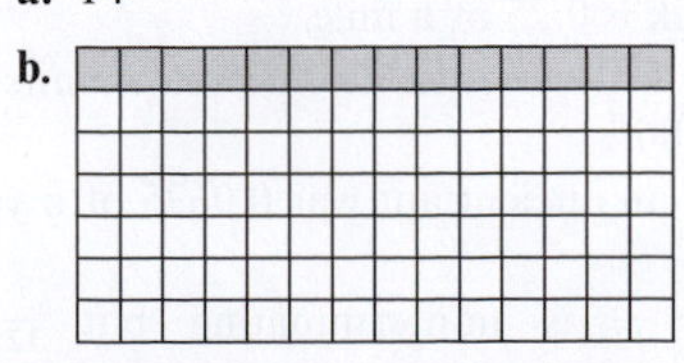

33. a. 6 cups

b.

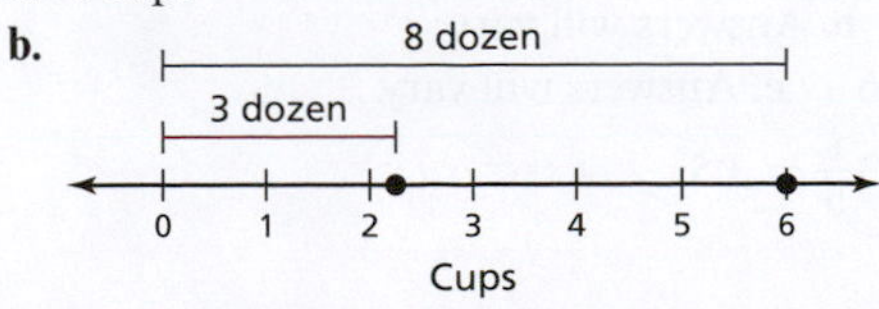

35. *Sample answer:* When the fraction is improper.

37. a. $\frac{13}{100}$ **b.** 52 **c.** 234 **d.** Answers will vary.

39. *Sample answer:* $\frac{2}{9} \times \frac{3}{7} = \frac{1}{3} \times \frac{2}{7}$

41. a. *Sample answer:* The student did not multiply the denominators, and should review multiplying fractions.

b. *Sample answer:* The student did not use the reciprocal of the divisor, and should review dividing fractions.

43. a. 14 **b.** Answers will vary. **c.** Answers will vary.

45. *Sample answer:* Multiply by the inverse. **47.** 24

49. a. About \$54 **b.** *Sample answer:* Find the exact answer.

51. a. About 20 mpg

b. $18\frac{24}{25}$ mpg; *Sample answer:* The estimate is close to the actual mileage.

Section 6.4 *(page 236)*

1. Solution key:

1. $\frac{3}{6}; \frac{1}{2}$ 2. $\frac{6}{10}; \frac{3}{5}$

3. $\neq$; Not a proportion 4. $=$; Proportion

Sample rubric: Assign 1 point to each correct numerator and denominator in problems 1 and 2. Assign 1 point to each correct symbol and answer in problems 3 and 4. There is a total of 12 points.

3. a. $\frac{5}{2}$ **b.** $\frac{2}{5}$ **c.** $\frac{5}{7}$ **d.** $\frac{2}{7}$ **5. a.** $\frac{1}{1}$ **b.** $\frac{7}{10}$

7. $\frac{377}{120}; \frac{754}{240}$; *Sample answer:* The ratios are proportional.

9. a. $\frac{55\text{ mi}}{1\text{ h}}$ **b.** $\frac{90\text{ mi}}{1\text{ h}}$ **c.** $\frac{500\text{ mi}}{1\text{ h}}$

11. $\frac{1320\text{ ft}}{1\text{ min}}$; *Sample answer:* The speed of a riding mower is 15 miles per hour. What is the speed in feet per minute?

13. $\frac{15\text{ mi}}{1\text{ h}}$

15. a. Proportional **b.** Not proportional **c.** Proportional

17. $\frac{3}{4} = \frac{18}{24}$ **19.** No; *Sample answer:* $\frac{14}{16} \neq \frac{13}{15}$

21. \$728,000 per year

23. a. *Sample answer:* The student miscalculated the product, and should review multiplying fractions.

b. *Sample answer:* The student ignored the $\frac{1}{2}$ in the denominator, and should review how to simplify complex fractions.

25. a. 6 : 4

b. *Sample answer:* i; It describes the ratio when simplified to 3 : 2.

c. Answers will vary.

27. None are proportional; *Sample answer:* None of the ratios are equivalent.

29. *Sample answer:* Write a ratio that compares how much you have in savings to how much you have in debt.

31. 1 : 48 **33. a.** 60 **b.** 8

35. a. *Sample answer:* 6 ft **b.** $\frac{\text{Giant height}}{6\text{ ft}} = \frac{20}{9}$; $13\frac{1}{3}$ ft

c. Answers will vary.

Review Exercises *(page 241)*

1. $\frac{7}{10}$; numerator: 7; denominator: 10

3. $\frac{11}{20}$; numerator: 11; denominator: 20

5. $\frac{3}{5}$; numerator: 3; denominator: 5

7. $\frac{1}{6}$

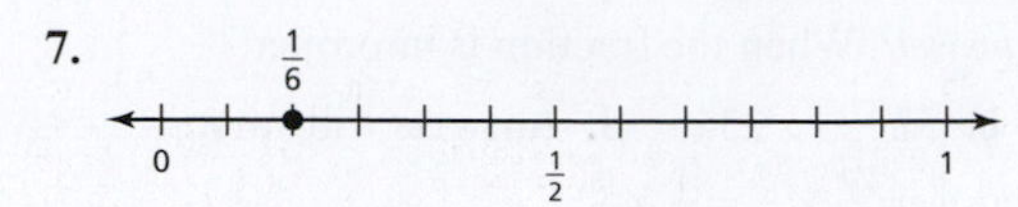

9. $\frac{5}{6}$

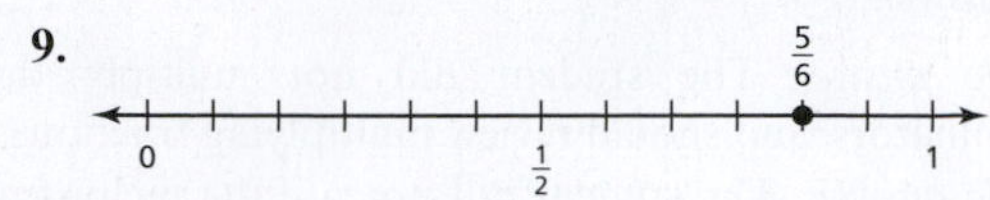

11. $\frac{7}{2}$

13. $2\frac{3}{8}$

15. Equivalent **17.** Equivalent **19.** Equivalent
21. Equivalent **23.** Equivalent **25.** Equivalent
27. > **29.** > **31.** =

33. $\frac{1}{17}, \frac{7}{15}, \frac{5}{8}, \frac{9}{10}$, *Sample answer:* $\frac{1}{17}$ is close to 0, $\frac{7}{15} < \frac{1}{2}, \frac{5}{8} > \frac{1}{2}$, and $\frac{9}{10}$ is close to 1.

35. $\frac{13}{51}; \frac{12}{51}$ **37.** $\frac{25}{51}$ **39.** $\frac{3}{5}$ **41.** 1 **43.** $3\frac{1}{2}$

45. $1\frac{1}{5}$ **47.** 6 **49.** $6\frac{1}{10}$ **51.** $5\frac{1}{2}$ **53.** $\frac{1}{4}$ **55.** 0

57. 2 **59.** 2 **61.** $2\frac{3}{4}$ **63.** 2 **65.** $2\frac{4}{7}$ **67.** $\frac{5}{12}$

69. $\frac{17}{10}$ **71.** $\frac{1}{6}$ **73.** $\frac{17}{14}$ **75.** 1 **77.** $3\frac{17}{24}$ **79.** $2\frac{5}{8}$

81. $\frac{13}{20}$

83. *Sample answer:* Reorder and group together pairs that add up to whole numbers. 5

85. $\frac{3}{4} \cdot \frac{1}{3}$ **87.** $\frac{1}{4} \cdot \frac{1}{3}$ **89.** $\frac{2}{3} \cdot \frac{1}{3}$ **91.** $\frac{10}{3}$ **93.** $\frac{5}{6}$

95. 0 **97.** $\frac{3}{4}$ **99.** $\frac{1}{2}$ **101.** $\frac{1}{12}$ **103.** $\frac{5}{36}$ **105.** $\frac{3}{14}$

107. $\frac{3}{16}$ **109.** 18 **111.** $\frac{5}{12}$ **113.** $\frac{3}{2}$ **115.** $\frac{3}{10}$

117. $\frac{3}{14}$ **119.** $\frac{1}{20}$ **121.** 8 **123.** 12 **125.** 40

127. 7 **129.** 18

131. a. About 130 ft^2

b. $130\frac{27}{32}$; *Sample answer:* The estimate is close to the actual area.

133. $\frac{1}{1}$ **135.** $\frac{5}{16}$ **137.** $\frac{45 \text{ mi}}{1 \text{ h}}$ **139.** $\frac{700 \text{ mi}}{1 \text{ h}}$

141. \$936,000 **143.** $\frac{3}{4} = \frac{18}{24}$

145. a. 6 : 3 **b.** 20 **c.** 50; *Sample answer:* $\frac{5}{20} = \frac{50}{200}$

Chapter Test *(page 244)*

1. A **2.** B **3.** E **4.** B
5. E **6.** A **7.** D **8.** D

CHAPTER 7

Section 7.1 *(page 252)*

1. Solution key:

1. \$3.32; $\$3\frac{32}{100}$
2. \$2.13; $\$2\frac{13}{100}$
3. \$4.60; $\$4\frac{60}{100}$
4. \$6.81; $\$6\frac{81}{100}$

Sample rubric: Assign 1 point to each correct number of dollars and number of cents. There is a total of 8 points.

3. a. Tenths **b.** Hundreds **c.** Hundredths

5. a. 0.3 **b.** 8.29 **c.** 13.09 **d.** 0.741

7. a. $9\frac{4}{10}$ **b.** $3\frac{612}{1000}$ **c.** $2\frac{1}{100}$ **9.** 2.15

11. a. $1 + \frac{8}{10}$

b. $3 + \frac{1}{10} + \frac{9}{100}$

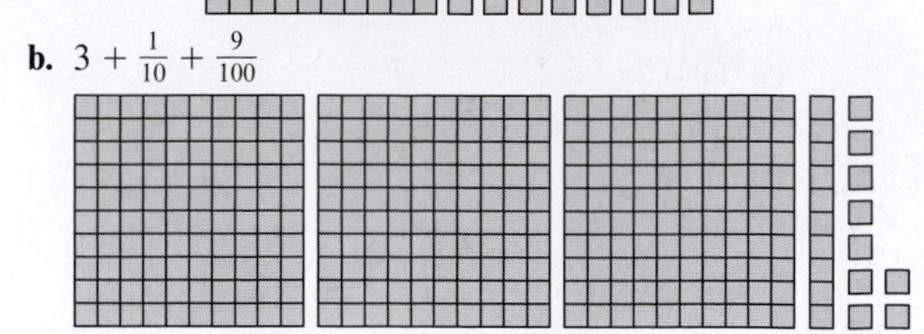

c. $4 + \frac{7}{10} + \frac{3}{100}$

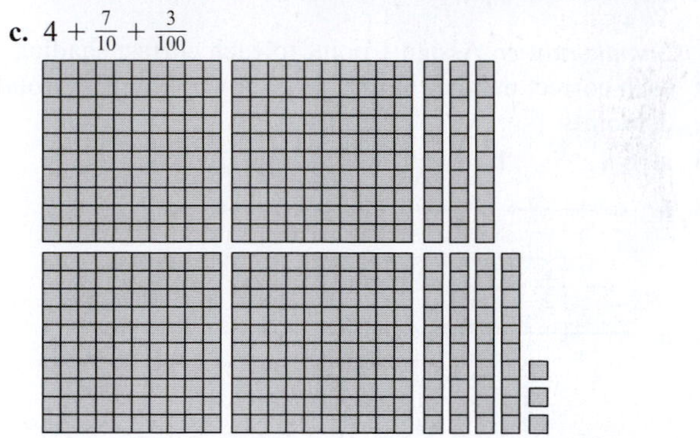

13. a. 45.8 **b.** 9.27 **c.** 3.07

15. a. $3 + \frac{3}{10} + \frac{8}{100}$

b.

c. Three and thirty-eight hundredths

17. a. Yes; 0.04 **b.** No; *Sample answer:* $9 = 3 \cdot 3$

c. No; *Sample answer:* $14 = 2 \cdot 7$ **d.** Yes; 0.7

19.

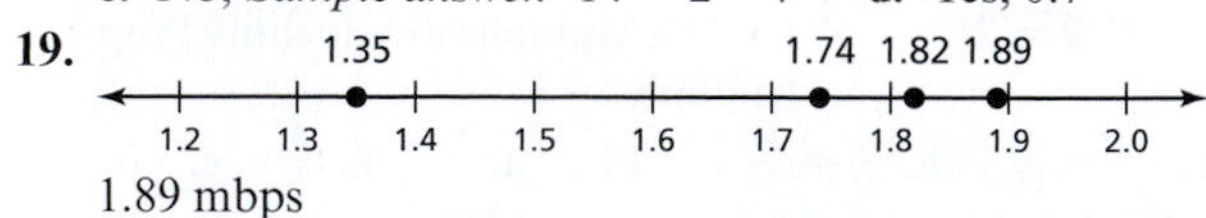

1.89 mbps

21. 0.498, 0.508, 0.55, 0.6 **23.** $\frac{13}{16}$, 0.835, $\frac{17}{20}$, 0.9

25. Valedictorian: Alyssa; salutatorian: Oscar

27. *Sample answer:* The student thought the second place to the right was the tenths place, and should review writing decimal notation.

29. a. A lap around a track is 0.25 of a mile.

b. The atomic mass of Silicon (in atomic mass units) is approximately 28.086.

c. A fossil shows that an ancient ant was 0.0625 of a yard long.

31. Yes; *Sample answer:* $\frac{1}{12}$ is nonterminating but $\frac{3}{12}$ is terminating.

33. a. Class 2 **b.** Answers will vary.

35. a and b. 0.16 **c.** Answers will vary.

37. $5\frac{9}{12} = 5.75$; $3\frac{6}{12} = 3.5$

39. a. *Sample answer:* 1.85 **b.** *Sample answer:* 1.88
c. *Sample answer:* 1.865
41. Answers will vary. **43.** 0.4321; 0.1234
45. a. 1, 2, 4, 5, 7, or 8 **b.** 6

Section 7.2 *(page 262)*

1. Solution key:
1. \$2.31 + \$1.64 = \$3.95; *Sample answer:* Put all the money in one group.
2. \$1.13 + \$3.35 = \$4.48; *Sample answer:* Put all the money in one group.
3. \$3.77 − \$2.27 = \$1.50; *Sample answer:* Remove 2 dollars, 1 quarter, and 2 pennies from the first group.
4. \$5.28 − \$1.25 = \$4.03; *Sample answer:* Change the five to 5 ones, then remove 1 dollar and 1 quarter.

Sample rubric: Assign 1 point to each correct quantity and 2 points to a correct reasoning. There is a total of 20 points.

3. a. 1.16 + 0.43; 1.59 **b.** 1.68 + 1.34; 3.02
5. a. 3.56 − 1.32; 2.24 **b.** 0.79 − 0.27; 0.52
7. a. 2 + 1.58 = 3.58

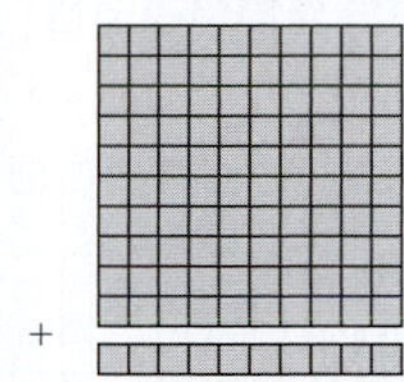
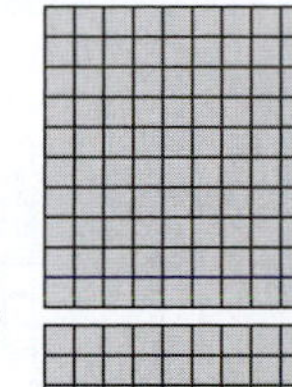
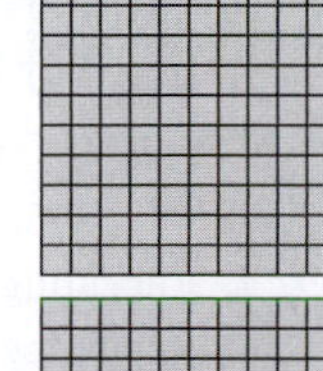
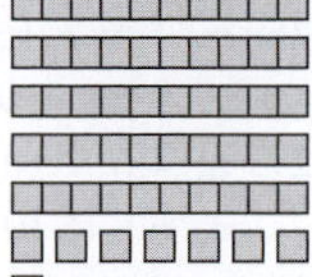

b. 1.82 + 1.4 = 3.22

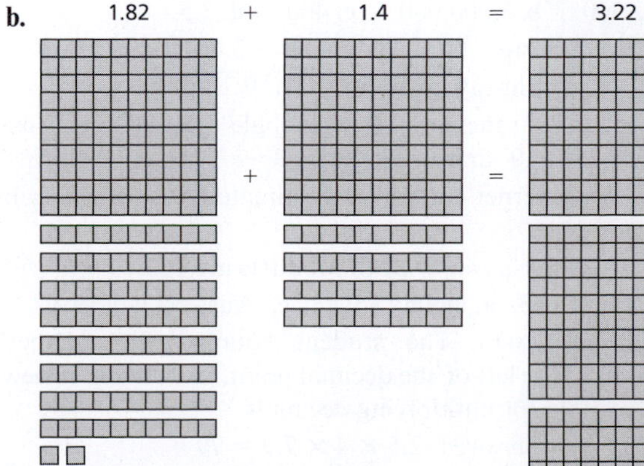
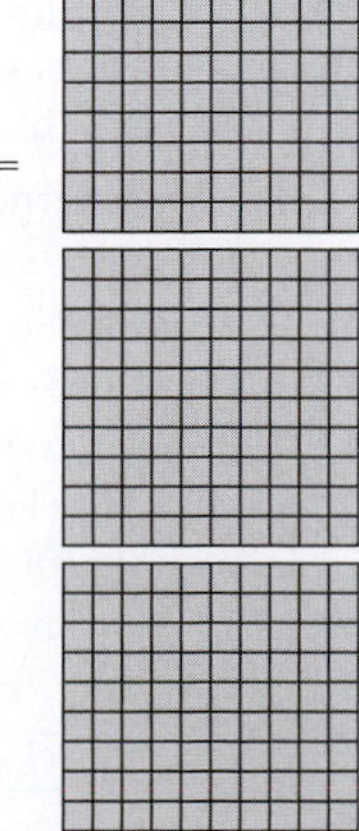

c. 1.45 + 2.08 = 3.53

d.

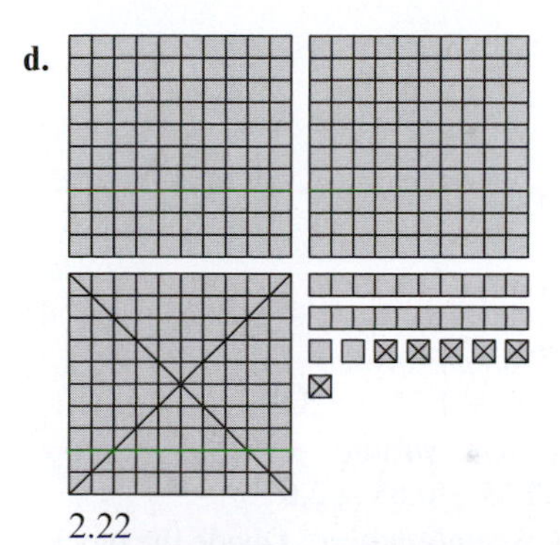

2.22

e.

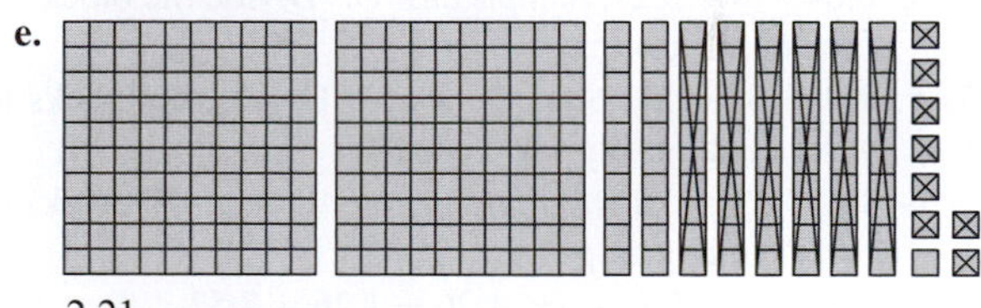

2.21

f.

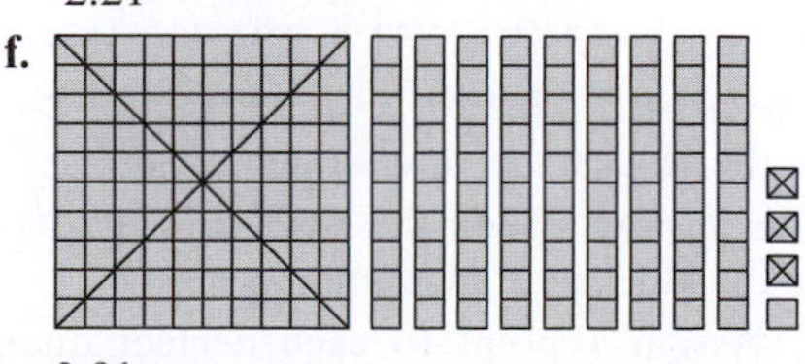

0.91

9. a. 6.3; 6.25; 6.254 **b.** 0.1; 0.09; 0.086
c. 44.7; 44.70; 44.698 **d.** 5.0; 5.00; 5.001
11. a. 9.2 **b.** 6.53 **c.** 5.293 **d.** 15.991
e. 28.042 **f.** 6.897
13. a. 8.11 **b.** 2.76 **c.** 7.39 **d.** 9.914
e. 6.632 **f.** 35.92
15. a. 4.25 **b.** 3.355 **c.** 8.006 **d.** 15.112
17. a. 13.1; *Sample answer:* 5.1 + 8.0 = 13.1
b. 2.8; *Sample answer:* 6.8 − 4.0 = 2.8
c. 5.7; *Sample answer:* 10.7 − 5.0 = 5.7
d. 8.87; *Sample answer:* 2.00 + 4.00 + 3.00 − 0.13 = 8.87

19. *Sample answer:* It helps keep place-value positions lined up.

21. *Sample answer:* $0.2 + 0.8 = 0.4 + 0.6$

23. 1.14 m **25. a.** 213.2 mi **b.** 66.8 mi

27. The typical towing capacity for this truck brand is about 7400 pounds. *Sample answer:* Towing capacities are usually rounded to the nearest hundred.

29. a. 1.4 h **b.** 3.6 h

31. *Sample answer:* $0.1 + 0.3 - 0.2 = 0.5 + 0.6 - 0.9$

33. *Sample answer:* The subtraction is the same, the difference is the placement of the decimal point.

35. *Sample answer:* You run 1.7 miles one day and 2.3 miles the next. How far did you run both days?

37. *Sample answer:* The student did not line up the decimal points, and should review the algorithm for adding decimals.

39. a. Location 1: 54.84 m; Location 2: 53.09 m
b. Location 2; 1.75 m
c. Answers will vary.

41.

1.8	8.1	3.6
6.3	4.5	2.7
5.4	0.9	7.2

Section 7.3 *(page 272)*

1. Solution key:
1. $3 \cdot 1.4 = 4.2$; *Sample answer:* $1.4 + 1.4 + 1.4 = 4.2$
2. $4 \cdot 1.33 = 5.32$; *Sample answer:* $1.33 + 1.33 + 1.33 + 1.33 = 5.32$
3. $4 \cdot 0.65 = 2.6$; *Sample answer:* $0.65 + 0.65 + 0.65 + 0.65 = 2.6$
4. $0.84 \div 4 = 0.21$; *Sample answer:* Divide the blocks into four equal groups.
5. $1.2 \div 3 = 0.4$; *Sample answer:* Divide the blocks into three equal groups.
6. $1.32 \div 3 = 0.44$; *Sample answer:* Divide the blocks into three equal groups.
7. 2.52; *Sample answer:* $1.26 + 1.26 = 2.52$
8. 9.24; *Sample answer:* $2.31 + 2.31 + 2.31 + 2.31 = 9.24$
9. 2.13; *Sample answer:* Model 6.39 with base-ten blocks, then divide the blocks into three equal groups.
10. 1.22; *Sample answer:* Model 6.1 with base-ten blocks, then divide the blocks into five equal groups.

Sample rubric: Assign 1 point to each correct quantity and 2 points to each correct reasoning. There is a total of 36 points.

3. a. 20.7 **b.** 2.38 **c.** 19.5 **d.** 16.2064

5. $3 \cdot 0.43 = 1.29$

7. a.

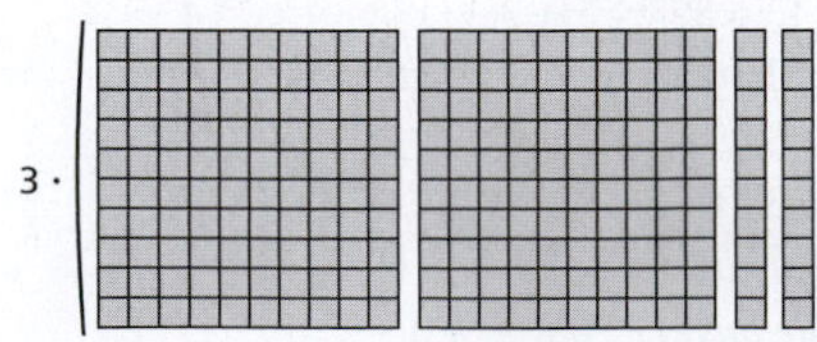

6.6

b.

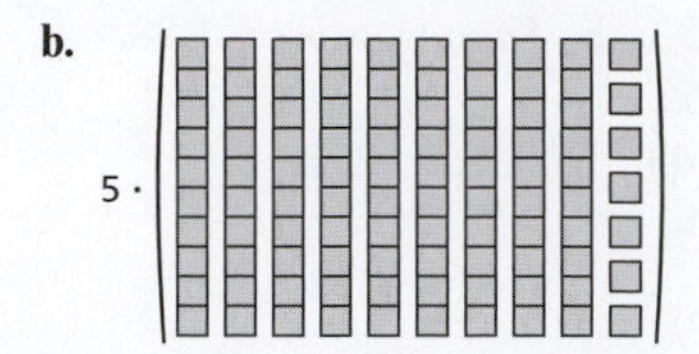

4.85

9. a. 12.6 **b.** 25 **c.** 15.604 **d.** 8.1238

11. a. 6.25 **b.** 1.48 **c.** 3.147

13. $0.72 \div 6 = 0.12$

15. a.

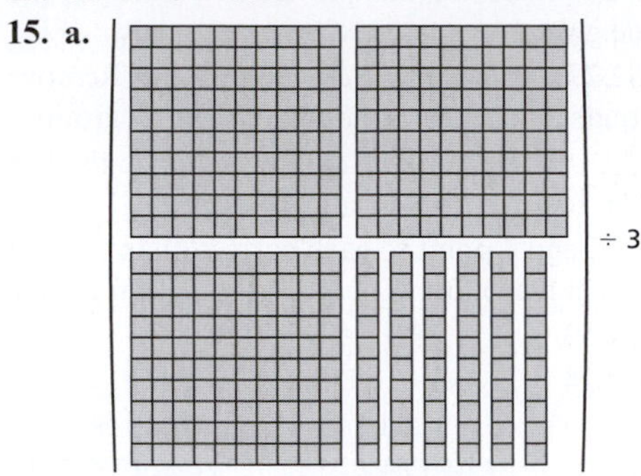

1.2

b.

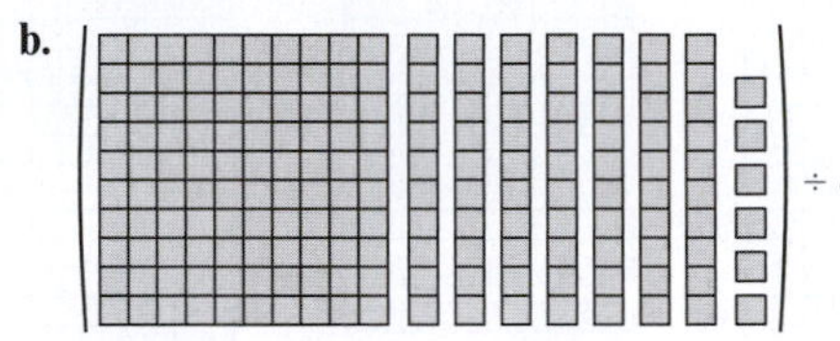

0.44

17. a. 9 **b.** 6.7 **c.** 4.8 **d.** 2.75

19. a. Repeating **b.** Terminating **c.** Neither
d. Repeating

21. a. 0.875 **b.** $0.\overline{5}$ **c.** 0.04 **d.** $0.2\overline{6}$

23. a. Yes **b.** No; $26.5 > 10$; 2.65×10^9
c. No; The base of the power is not 10; 7.92×10^2
d. Yes

25. a. 1.5; 7; 15,000,000 **b.** 3.28; 2; 328
c. 9.96; 6; 9,960,000 **d.** 8.875; 10; 88,750,000,000

27. a. 4.6×10^1 **b.** 9.5×10^3 **c.** 2.0425×10^5
d. 6.8×10^{10}

29. a. 92,000,000 **b.** 405,000,000,000

31. a. 300 **b.** 6,000,000 **c.** 400 **d.** 3.5

33. *Sample answer:* $0.6 \div 0.4 = 0.9 \div 0.6$

35. $9.98 per Blu-ray Disc™ **37.** $0.\overline{714285}$

39. a. During one month, Google had approximately 1.75×10^8 unique visitors.
b. The Internet contains an estimated 5×10^{12} megabytes of data.

41. Yes; *Sample answer:* Both products are 162.

43. 6.7 h **45. a. and b.** 7.02 **c.** Answers will vary.

47. *Sample answer:* The student counted the number of digits to the left of the decimal point, and should review the algorithm for multiplying decimals.

49. 79; *Sample answer:* $2.5 \times 4 \times 7.9 = 79$

51. a. $42.60 **b.** $42.60 **c.** Answers will vary.

53. 439.425 in.^{2}
55. *Sample answer:* $0.7 \times 2.5 \div 0.5 = 1.5 \div 0.6 \times 1.4$
57. 0.21×0.3; 0.23×0.1
59. *Sample answer:* Because $1 \le a < 10$ and $1 \le b < 10$, $1 \le ab < 100$.

Section 7.4 *(page 282)*

1. Solution key:
1. 30; 30 2. 35; 35 3. 60; 60 4. 80; 80
5. *Sample answer:* 25; 15; 10; 50
Sample rubric: Assign 1 point to each correct quantity. There is a total of 12 points.

3. a. 0.4 **b.** 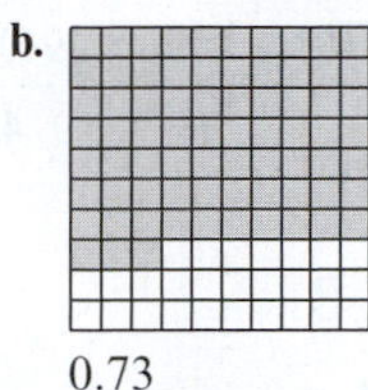
0.73

5. a. 14% **b.** 92% **c.** 130% **d.** 0.004%
7. a. 0.30; $\frac{3}{10}$ **b.** 0.06; $\frac{3}{50}$ **c.** 2.24; $\frac{56}{25}$ **d.** 0.045; $\frac{9}{200}$
9.

Fraction	Decimal	Percent
4/5	0.8	80%
1/4	0.25	25%
19/50	0.38	38%
1/200	0.005	0.5%
1/100	0.01	1%

11. a. $\frac{21}{50}$, 43.6%, $\frac{7}{16}$, 0.45 **b.** 1.225, 1.23, $\frac{31}{25}$, 125%
13. a. 64%; 21,760 **b.** 50%; 1300
15. a. 7 **b.** 714 **c.** 40,950 **d.** 8592
17. a. iii **b.** i **c.** ii **19. a.** 108 **b.** 55.8
21. a. 80% **b.** 37.5% **23. a.** 200 **b.** 350
25. a. 12.5% **b.** 212.5 **c.** 1435 **d.** 145
27. a. \$1468.80 **b.** \$350
29. Greater than; *Sample answer:* 45% < 100%
31. Yes; *Sample answer:* $0.25 \cdot 800 = 200$ and $0.80 \cdot 50 = 40$.
33. \$56,000
35. *Sample answer:* The student divided 180 by 9 and wrote 20 as 20%, and should review solving percent problems.
37. a. 40% **b.** 11.5%
39. Yes; yes; *Sample answer:* An amount could triple, or decrease to a negative number.
41. a. \$80 **b.** \$3480
43. a. $0.30 \cdot 25 = 7.5$; \$7.50 **b.** \$32.50
c. *Sample answer:* The selling price is $a = (1 + p)b$.
45. a. $x = 3$ **b.** $x = 6$

Review Exercises *(page 287)*

1. Hundredths **3.** Tenths **5.** Hundreds **7.** 4.3
9. 82.09 **11.** 19.007 **13.** $15\frac{8}{10}$ **15.** $2\frac{73}{100}$
17. $7\frac{32}{1000}$
19. $4 + \frac{8}{10} + \frac{9}{100}$; four and eighty-nine hundredths
21. $13 + \frac{7}{10} + \frac{6}{100}$; thirteen and seventy-six hundredths
23. $2 + \frac{0}{10} + \frac{2}{100} + \frac{9}{1000}$; two and twenty-nine thousandths
25. Yes; 0.3 **27.** No; *Sample answer:* $70 = 2 \cdot 5 \cdot 7$
29. Yes; 42.8 **31.** 0.1, 0.139, 0.1392, 0.14, 0.145
33. A fluid ounce is equal to 0.125 of a cup.
35. Earth's distance from the Sun is about 0.192 of Jupiter's distance from the Sun.
37.

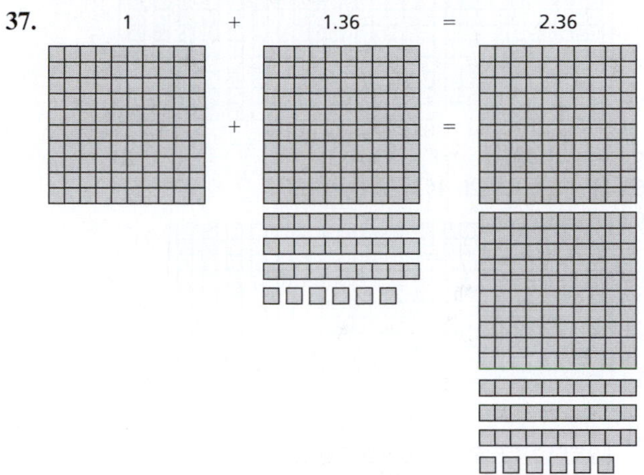

39.

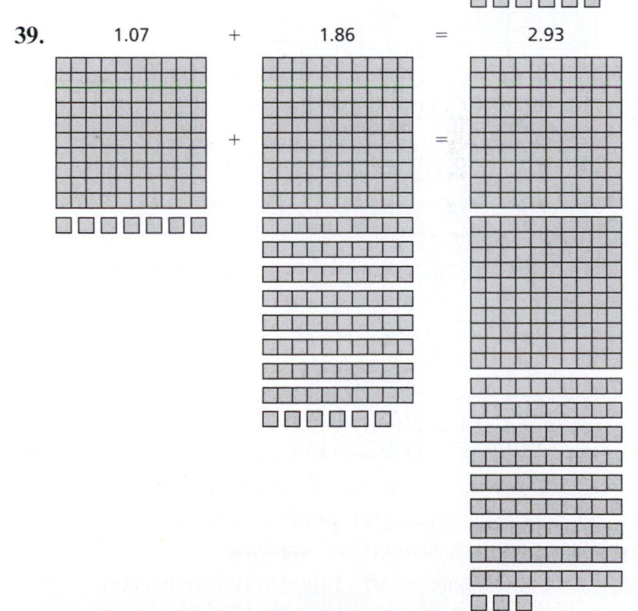

41.

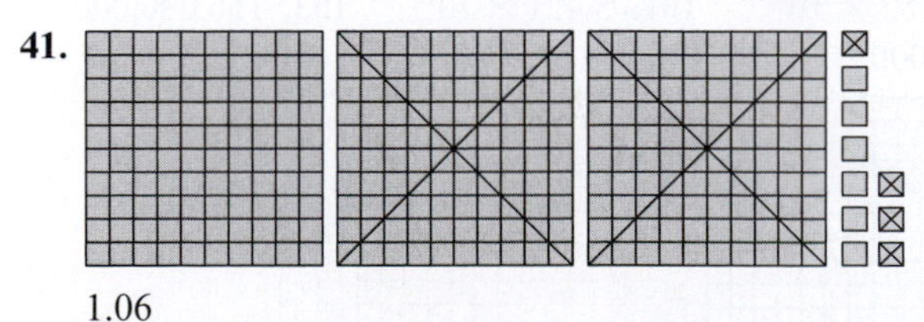
1.06

43. 2.2; 2.25; 2.248 **45.** 0.4; 0.40; 0.397 **47.** 7.15
49. 19.498 **51.** 1.81 **53.** 14.535 **55.** 2.43
57. 35.459 **59.** 13.2; *Sample answer:* $8.2 + 5.0 = 13.2$
61. 2.75; *Sample answer:* $5.75 - 3.0 = 2.75$
63. 21.38 lb **65.** \$385.76 **67.** 9.18 **69.** 6.316
71.

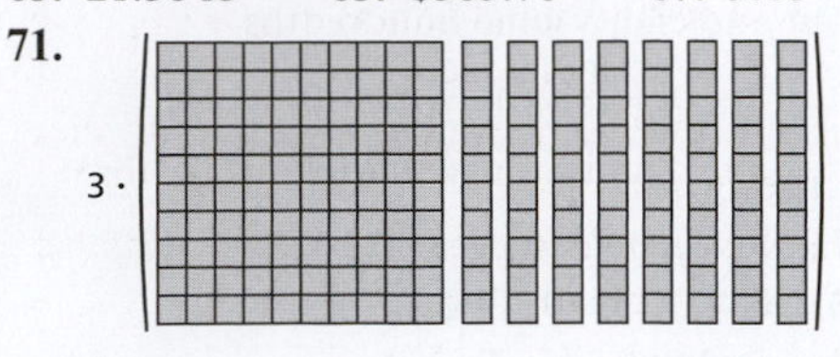

5.4

73.

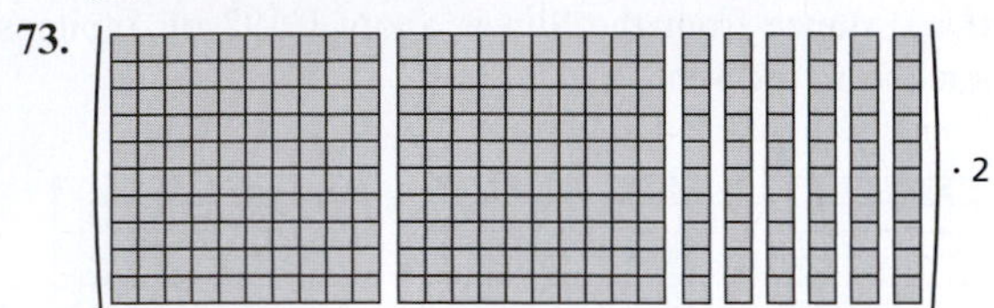

5.2

75. 119 **77.** 20.4516
79.

÷ 6

0.4

81.

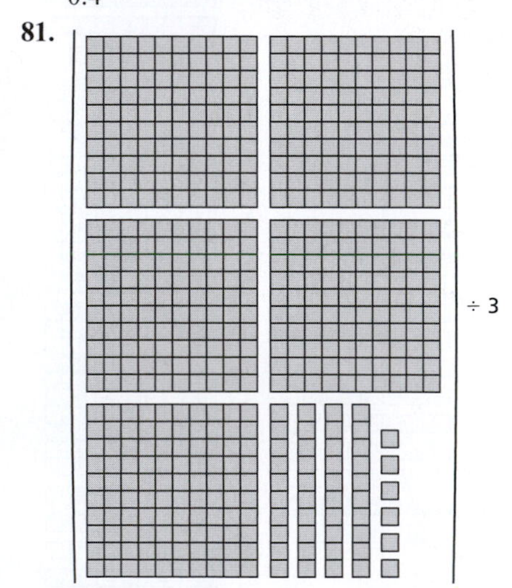

1.82

83. 22.4 **85.** 2.56 **87.** $0.\overline{72}$ **89.** 0.125
91. No; $0.45 < 1$; 4.5×10^{13} **93.** Yes
95. 2.9; 6; 2,900,000 **97.** 1.08; 10; 10,800,000,000
99. 7.82×10^3 **101.** 3.421×10^9 **103.** 156,000,000
105. 3000 **107.** 14.85
109.

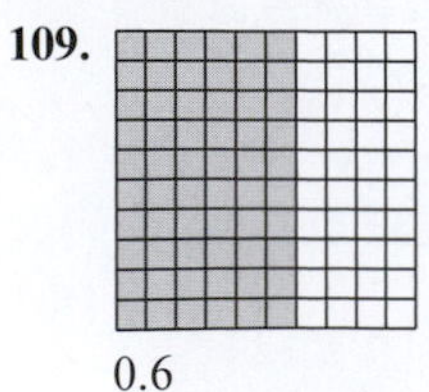

0.6

111.

Fraction	Decimal	Percent
$\frac{21}{50}$	0.42	42%
$\frac{5}{8}$	0.625	62.5%
$\frac{9}{100}$	0.09	9%
$\frac{1}{1250}$	0.0008	0.08%

113. $\frac{17}{20}$, 87.7%, $\frac{22}{25}$, 0.884 **115.** 26%; 20,644 **117.** 75%
119. 45% **121.** 161 **123.** 60 **125.** 30,000
127. 39.2%

Chapter 7 Test *(page 290)*

1. C **2.** C **3.** A **4.** A **5.** C **6.** C
7. D **8.** B

CHAPTER 8

Section 8.1 *(page 298)*

1. Solution key:
1. $-4 < 3$ 2. $-2 > -5$ 3. $2 > 0$
4. $-6 < 0$ 5. 0 6. -3
7. Anchorage; *Sample answer:* -3 is to the left of 0 on a number line.

Sample rubric: Assign 1 point to each correct integer and city. Assign 2 points to a correct reasoning. There is a total of 13 points.

3. a. 5 6 7 8 9 10
b. −18 −16 −14 −12 −10 −8
c. −15 −14 −13 −12 −11 −10
d. 16 18 20 22 24 26

5. a. 64 **b.** 18 **c.** -25 **d.** 29

7. a. −3 −2 −1 0 1 2 3
b. −12 12; −15 −10 −5 0 5 10 15
c. −42 42; −50 −40 −30 −20 −10 0 10 20 30 40 50
d. −9 9; −10 −8 −6 −4 −2 0 2 4 6 8 10
e. −16 16; −20 −15 −10 −5 0 5 10 15 20
f. −20 −15 −10 −5 0 5 10 15 20

9. a. < **b.** > **c.** < **d.** < **e.** > **f.** >
11. a. $-11, -7, 4, 15, 22$ **b.** $-15, -3, 0, 5, 23$
c. $-12, -10, -9, 2, 18$ **d.** $-30, -21, -1, 16, 32$
13. a. 5 **b.** 0 **c.** -9 **d.** -25 **e.** 17 **f.** -21
15. a. < **b.** < **c.** > **d.** = **e.** = **f.** >
17. a. $-|-13|, -11, -9, 3, |-17|$ **b.** $-7, -|5|, 0, 4, |-8|$
c. $-22, -|-2|, |1|, |-10|, 14$
d. $-24, 0, |-6|, 13, |29|$ **e.** $-16, -|5|, 0, |-11|, 12$

19. **a.** True; *Sample answer:* The opposite of a positive integer is a negative integer.
b. False; *Sample answer:* The absolute value of a positive integer is itself.
c. False; *Sample answer:* Zero is neither positive nor negative.

21. **a.** -3 **b.** $-5, 3$

23. **a.** 750 ft/sec **b.** 672 ft/sec
c. 543 ft/sec **d.** 495 ft/sec

25. -50 430
0 100 200 300 400

27. 29; -7; Juneau

29. **a.** Launch control system enabled; launch verification; main engines start; boosters ignite; rocket clears launch pad tower
b. Rocket clears launch pad tower

31. -19; -23; The diver at -23 m

33. Answers will vary.

35. **a.** Methanol, ethanol, propanol, water
b. Water; *Sample answer:* Water has the highest boiling point.

37. *Sample answer:* $-6, -4, 0, 2, 8, 10$; $-10, -8, -2, 0, 4, 6$; The order reverses.

39. *d, c, b, a, e; b, a, e; d, c*

41. *Sample answer:* The student listed the negative integers by absolute value, and should review comparing and ordering integers.

43. -19; -14

Section 8.2 *(page 308)*

1. Solution key:
1. ⊖⊖⊖⊖⊖⊖⊖ -7
2. 2; ⊖ -1
3. 2 4. -1

Sample rubric: Assign 1 point to each correct model and integer. There is a total of 9 points.

3. **a.** Positive; *Sample answer:* $|3| > |-1|$
b. Negative; *Sample answer:* Both numbers are negative.
c. Negative; *Sample answer:* $|-9| > |6|$
d. Negative; *Sample answer:* $|-7| > |4|$

5. **a.** -16 **b.** 8 **c.** -7 **d.** 12

7. **a.** *Sample answer:* The negative counters represent -3 and the positive counters represent 2. $-3 + 2$; -1
b. *Sample answer:* The positive counters represent 2 and the negative counter represents -1. $2 + (-1)$; 1

9. **a.** 2 **b.** 6 **c.** -5 **d.** 7 **e.** -12 **f.** -11

11. *Sample answer:* The blue arrow represents 3 and the red arrow represents -7. $3 + (-7)$; -4

13. **a.** 3 **b.** -3 **c.** -6 **d.** 0

15. **a.** $-2 + 3$ **b.** $-3 + (2 + 1)$ **c.** -5

17. **a.** -11 **b.** -15 **c.** 5 **d.** 4 **e.** 1 **f.** -6

19. ii

21. **a.** -7 **b.** -5 **c.** -9 **d.** -4 **e.** 7 **f.** -3

23. *Sample answer:* The blue arrow represents 2 and the red arrow represents 5. $2 - 5$; -3

25. **a.** -6 **b.** 3 **c.** -28 **d.** -11

27. **a.** 1 **b.** 5 **c.** -9 **d.** -6 **e.** 4 **f.** -7

29. **a.** -3 **b.** 18 **c.** -18 **d.** 2 **e.** -14 **f.** -1

31. **a.** *Sample answer:* $-7 + 1 = -9 + 3$
b. *Sample answer:* $-2 - (-5) = -4 - (-7)$

33. **a.** $-10 + 5$; -5; $-5°C$
b. $-3 + 8 + 0 + (-7)$; -2; loss of 2 yd

35. 3 points **37.** 20 ft below sea level

39. **a.** *Sample answer:* The student added instead of subtracted, and should review adding integers.
b. *Sample answer:* The student evaluated $3 - 6$, and should review subtracting integers.

41. **a.** True; *Sample answer:* Subtracting a negative integer is the same as adding a positive integer.
b. False; *Sample answer:* $-3 - (-5) = 2$
c. False; *Sample answer:* Subtract the absolute values.

43. 5 over; *Sample answer:*
$-1 + 0 + 0 + 0 + 1 + 1 + 1 + 1 + 2 = 5$

45. 5 **47.** Answers will vary. **49.** **a.** 11 **b.** 22

51.

-5	25	-17
-11	1	13
19	-23	7

Section 8.3 *(page 318)*

1. Solution key:
1. -3; ⊖⊖⊖⊖⊖⊖⊖⊖⊖ -9
2. 3; ⊕⊕⊕⊕⊕⊕ ⊕⊕⊕⊕⊕⊕ 12
3. -6; ⊖⊖⊖ -3
4. 12; ⊕⊕⊕⊕⊕⊕ 6
5. 4; -8 6. 5; -15

Sample rubric: Assign 1 point to each correct model and integer. There is a total of 16 points.

3. *Sample answer:* Form groups of the same size to model repeated addition.
a. -10 **b.** -42 **c.** -9 **d.** -12

5. **a.** 60 **b.** 9 **c.** -22 **d.** -160

7. **a.** $8 \cdot (-2)$ **b.** $[4 \cdot (-5)] \cdot 7$ **c.** -9 **d.** 0

9. **a.** -90 **b.** -504

11. **a.** -21; *Sample answer:* $-1 \cdot a = -a$
b. -15; *Sample answer:* $-a \cdot b = -(ab)$

13. **a.** 0 **b.** -90 **c.** 42 **d.** -20

15. **a.** -2 **b.** -4 **c.** -4 **d.** -10

17. **a.** 18 **b.** 9 **c.** 18 **d.** -3 **e.** 11 **f.** -3

19. **a.** -8 **b.** -7 **c.** 6

21. **a.** $\frac{1}{4}$ **b.** $\frac{1}{9}$ **c.** $\frac{1}{216}$ **d.** -1

23. **a.** $\frac{1}{9}$ **b.** 1 **c.** $\frac{1}{4096}$ **d.** $\frac{1}{16}$ **e.** $\frac{1}{216}$ **f.** 4

25. **a.** 0.0235 **b.** 0.000111
c. 0.000000751 **d.** 0.00000000057

27. **a.** 2.64×10^{-2} **b.** 5.87×10^{-4}
c. 1.2×10^{-6} **d.** 8.95×10^{-8}

29. **a.** 3 **b.** -16 **c.** -11 **d.** -64 **e.** -4

31. $-2 \cdot 30$; -60 **33.** $-3°F$ **35.** 2.66×10^{-23} g

37. 9.9×10^{-5} m

39. a. *Sample answer:* The student used the common sign, and should review multiplying integers.

b. *Sample answer:* The student evaluated $(-3)^3$, and should review negative integer exponents.

41. *Sample answer:* $-2 \times 9 \div (-3) = 6 \times 2 \div (2)$

43. a. $x < 0$; $x > 0$; *Sample answer:* The negative changes the sign of x.

b. $x \neq 0$; cannot be negative; *Sample answer:* The square of a nonzero number is positive.

c. $x > 0$; $x < 0$; *Sample answer:* x^3 has the same sign as x.

45. a. $35 - 3(10) - 3(5)$; *Sample answer:* You have three face cards and three other cards.

b. -10 points

47. a. No

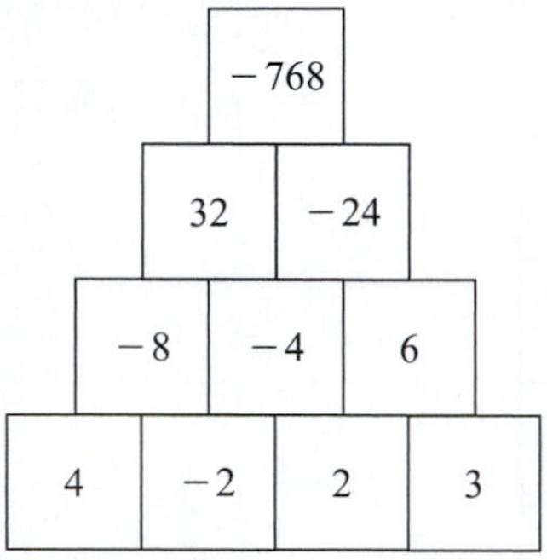

b. Answers will vary.

49. -2 in. **51. a.** 243 **b.** 9

53. a. $37 - 7t$ **b.** $-23 + 5t$

c.

Minute	First solution	Second solution
1	30	−18
2	23	−13
3	16	−8
4	9	−3
5	2	2
6	−5	7
7	−12	12
8	−19	17
9	−26	22
10	−33	27

d. 5

55. *Sample answer:* $-7 + (-7) + (-7) + (-7) + (-7)$

57. No; *Sample answer:* $-3 < 2$ but $(-3)^2 > 2^2$

Chapter 8 Review *(page 323)*

1.
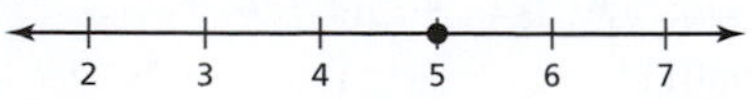

3.
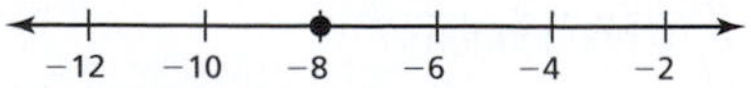

5. 19 **7.** -83

9.
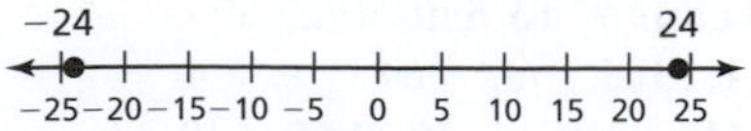

11.
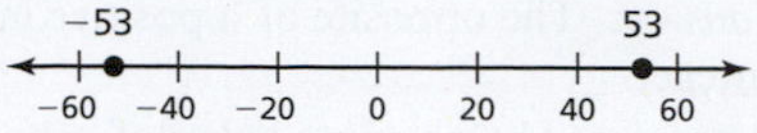

13. $<$ **15.** $>$ **17.** $-13, -7, 0, 5, 19$

19. $-30, -12, -8, 0, 15$ **21.** 51 **23.** -16 **25.** $>$

27. $=$ **29.** $-|9|, -|-7|, 0, 13, |-28|$

31. $-22, -8, -|5|, |-19|, |34|$ **33.** -72; -80; January

35. 76 under **37.** -15 **39.** 2 **41.** -4

43. 2 **45.** 0 **47.** -22

49. Commutative Property of Addition; *Sample answer:* The addends were switched.

51. Associative Property of Addition; *Sample answer:* The addends were regrouped.

53. 9 **55.** 4 **57.** -2 **59.** -7 **61.** -16

63. 11 **65.** -19 **67.** 0 **69.** 10 **71.** 4 **73.** 1

75. -12 **77.** 18 **79.** -22

81. a. $5 - 2 + 4 - 3 + 1$ **b.** 5 cards **83.** -7°C

85. $-\$12$; *Sample answer:* The teacher spent \$12 out of pocket.

87. -8 **89.** -13 **91.** -210 **93.** 84 **95.** 96

97. -54

99. Multiplicative Identity Property; *Sample answer:* 1 was multiplied.

101. Associative Property of Multiplication; *Sample answer:* The factors were regrouped.

103. 330 **105.** -1680 **107.** -5 **109.** -1

111. 9 **113.** -8 **115.** -3 **117.** -9 **119.** $\frac{1}{81}$

121. $\frac{1}{625}$ **123.** 81 **125.** $\frac{1}{64}$ **127.** -125 **129.** 64

131. 0.00000157 **133.** 0.064 **135.** 2.69×10^{-3}

137. 7.3×10^{-8} **139.** -16 **141.** 11 **143.** 5°F

145. -4

Chapter 8 Test *(page 326)*

1. C **2.** C **3.** E **4.** A **5.** D **6.** D
7. D **8.** D

CHAPTER 9

Section 9.1 *(page 334)*

1. Solution key:

1. $-\frac{3}{6}$; *Sample answer:* -1 is six tick marks from 0 and a is three tick marks from 0.
2. $-\frac{3}{8}$; *Sample answer:* -1 is eight tick marks from 0 and b is three tick marks from 0.
3. $\frac{5}{12}$; *Sample answer:* 1 is twelve tick marks from 0 and c is five tick marks from 0.
4. $-\frac{4}{9}$; *Sample answer:* -1 is nine tick marks from 0 and d is four tick marks from 0.
5. $-\frac{4}{6} < \frac{1}{6}$ 6. $-\frac{1}{6} > -\frac{5}{6}$
7. $-\frac{2}{6}, -\frac{2}{6}, \frac{2}{6}$ 8. $-\frac{3}{6}, -\frac{3}{6}, \frac{3}{6}$

Sample rubric: Assign 1 point to each correct rational number and 2 points to each correct reasoning. There is a total of 22 points.

3. a. $-\frac{5}{3}$ **b.** $-\frac{1}{3}$ **c.** 1 **d.** $\frac{4}{3}$

5. a.

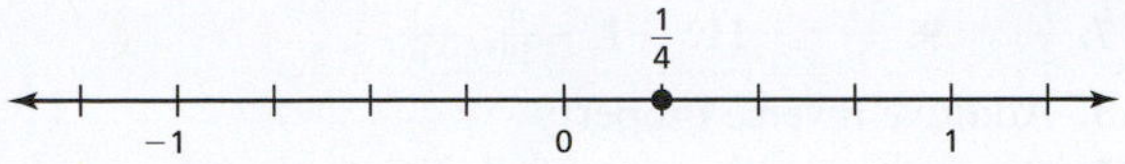

b.

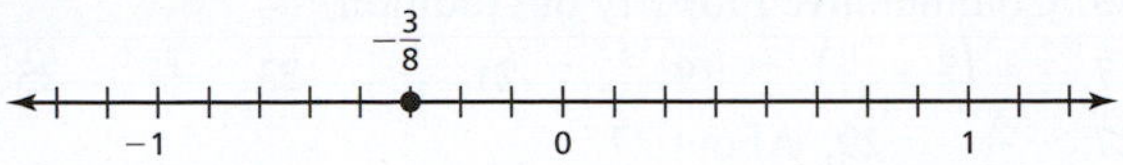

c.

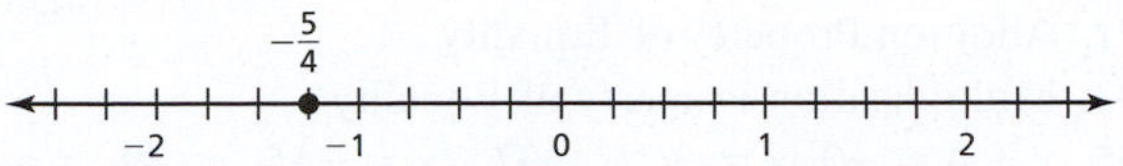

d.

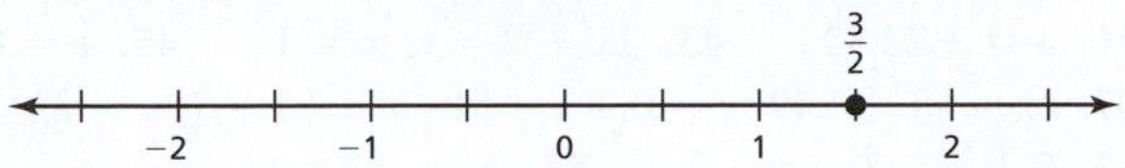

7. a. $\frac{3}{2}$ **b.** $\frac{-4}{1}$ **c.** $\frac{7}{4}$ **d.** $\frac{-1}{3}$ **9.** $\frac{-5}{1}, \frac{5}{-1}, -\frac{5}{1}, -\frac{-5}{-1}$

11. a. Not equal **b.** Equal **c.** Equal **d.** Not equal

13. a. $<$ **b.** $<$ **c.** $=$ **d.** $>$ **15.** $-\frac{5}{2}, -2, 0, \frac{5}{4}, \frac{3}{2}$

17. a. $-\frac{1}{5}$; 5 **b.** $\frac{7}{3}$; $-\frac{3}{7}$ **c.** -6; $\frac{1}{6}$ **d.** $\frac{13}{16}$; $-\frac{16}{13}$

19. a. Additive Identity Property
b. Multiplicative Inverse Property
c. Associative Property of Multiplication
d. Distributive Property of Multiplication Over Addition

21. a. $\frac{1}{3} + \left(\frac{1}{5} + \frac{3}{13}\right)$ **b.** 0 **c.** $\frac{2}{5} + \left(-\frac{4}{6}\right)$
d. $-\frac{10}{11}$ **e.** $\frac{4}{3} \cdot \left(-\frac{6}{12}\right)$

23. a. 2 **b.** $-\frac{7}{5}$ **c.** $-\frac{5}{9}$ **d.** $\frac{7}{4}$
e. $\frac{5}{12}$ **f.** $\frac{2}{3}$ **g.** $-\frac{10}{7}$ **h.** $-\frac{1}{10}$

25. a. True; *Sample answer:* Every natural number is an integer and every integer is a rational number.
b. False; *Sample answer:* Repeating decimals are rational.
c. True; *Sample answer:* True by definition.

27. $\frac{5}{6} \cdot \left(\frac{6}{5}x\right) = \frac{5}{6} \cdot (-3)$
$\left(\frac{5}{6} \cdot \frac{6}{5}\right)x = \frac{5}{6} \cdot (-3)$ Associative Property of Multiplication
$(1)x = \frac{5}{6} \cdot (-3)$ Multiplicative Inverse Property
$x = -\frac{5}{2}$ Multiplicative Identity Property and multiply $\frac{5}{6}$ and -3.

29. *Sample answer:* $6 - 10 = -4$; integers, rational numbers

31. *Sample answer:* $2 \div 10 = 0.2$; rational numbers

33. a. Week 1 **b.** Week 2 **c.** Answers will vary.

35. $-5\frac{3}{10}$ m

37. a. $\frac{3}{4}$; *Sample answer:* Commutative Property of Addition, Associative Property of Addition
b. $-\frac{9}{8}$; *Sample answer:* Commutative Property of Multiplication, Associative Property of Multiplication, Multiplicative Inverse Property, Multiplicative Identity Property
c. $\frac{1}{6}$; *Sample answer:* Distributive Property of Multiplication Over Addition, Multiplicative Identity Property
d. 1; *Sample answer:* Associative Property of Addition

39. $\frac{ad + bc}{bd}$; *Sample answer:* bd and $(ad + bc)$ are integers, so $\frac{ad + bc}{bd}$ is rational.

41. *Sample answer:* The student multiplied across, and should review equality of rational numbers.

43. 0; *Sample answer:* $0 + 0 = 0$

45. a. $\$1\frac{1}{2}$ hundred thousand **b.** $22\frac{2}{3}$ times greater
c. $-\$1\frac{9}{10}$ hundred thousand

47. 72 years old

Section 9.2 *(page 344)*

1. Solution key:
1. $b - 9 = 2$; 11 2. $5n = 30$; 6
3. $x + 10 = 24$; 14 4. $\frac{y}{4} = 12$; 48
5. *Sample answer:* m decreased by $\frac{1}{2}$ is 0; $\frac{1}{2}$
6. *Sample answer:* The product of 2 and a number x is -1; $-\frac{1}{2}$
7. *Sample answer:* The quotient of a number n and 4 is $\frac{3}{4}$; 3
8. *Sample answer:* The sum of a number d and $\frac{1}{4}$ is $\frac{5}{4}$; 1
9. $x - 2 = 3$; 5 10. $-2 = 2x$; -1

Sample rubric: Assign 1 point to each correct equation or sentence and solution. There is a total of 20 points.

3. a. Expression **b.** Equation
c. Equation **d.** Expression

5. a. -1 **b.** 13 **c.** -7 **d.** 21 **e.** 3 **f.** 18

7. a. $x - 4 = -4$; $x = 0$ **b.** $x - 3 = 2$; $x = 5$

9. a. $x = 7$ **b.** $x = 3$ **c.** $y = 15$
d. $y = 17$ **e.** $z = 4$ **f.** $z = 7$

11. a. Subtraction **b.** Addition
c. Division **d.** Multiplication

13. a. Solution **b.** Not a solution **c.** Solution
d. Not a solution **e.** Solution **f.** Not a solution

15. a. $x = 17$ **b.** $a = \frac{7}{8}$ **c.** $y = 108$ **d.** $b = 7$
e. $y = 3.3$ **f.** $u = 13.5$

17. $-13 = 3x - 1$; $x = -4$

19. a. $x = -2$ **b.** $x = 2$ **c.** $y = 10$ **d.** $y = -1$

21. a. $x = 2$ **b.** $d = 2$ **c.** $m = -3$ **d.** $a = 6$
e. $d = 10$ **f.** $c = 15.9$ **g.** $g = \frac{7}{20}$ **h.** $c = 18$

23. a. $n + 14 = 24$; $n = 10$ **b.** $n - 3 = -15$; $n = -12$
c. $\frac{n}{4} = -2$; $n = -8$ **d.** $12n = 36$; $n = 3$
e. $3n - 9 = 15$; $n = 8$

25. $675 + h = 1500$; 825 ft **27.** $y + 14 = 105$; 91 pins

29. $5.5r = 45.65$; \$8.30 **31.** $15 + 11c = 92$; 7 classes

33. Answers will vary.

35. a. *Sample answer:* The student added -1.5 instead of 1.5, and should review solving one-step equations.
b. *Sample answer:* The student subtracted 3 from 7, and should review solving two-step equations.

37. *Sample answer:* You spend \$6 on supplies to paint mugs. The mugs sell for \$5. How many mugs do you need to sell to make a profit of \$9?

39. a. $3x - 8 = 10$; $x = 6$; $2y - 15 = 5$; $y = 10$
b. $\frac{2x}{5} = 4$; $x = 10$; $7 - 5y = 8$; $y = -\frac{1}{5}$

41. The amount of increase of running time

43. *Sample answer:* It costs \$3 to rent a video and \$4 to buy snacks; $C = 3x + 4$; \$31

45. 120

47. a. $2(24.95) + 3(12.99) + 3x = 132.07$; $x = 14.4$
b. $0.80x = 14.40$; $x = 18$

49. 40; *Sample answer:* There is a total of 240 tokens, so you had 160 and your colleague had 80.

Section 9.3 *(page 354)*

1. Solution key:
1. =; *Sample answer:* $1^2 = 1$
2. ≈; *Sample answer:* $1.41421^2 \neq 2$
3. =; *Sample answer:* $2^2 = 4$
4. =; *Sample answer:* $1.5^2 = 2.25$
5. ≈; *Sample answer:* $1.73205^2 \neq 3$
6. =; *Sample answer:* $10^2 = 100$
7. ≈; *Sample answer:* $3.16227^2 \neq 10$
8. =; *Sample answer:* $0.6^2 = 0.36$
9. No

Sample rubric: Assign 1 point to each correct symbol and 2 points to each correct reasoning in problems 1 through 8. Assign 1 point to a correct response in problem 9. There is a total of 25 points.

3. a. Rational, real **b.** Integer, rational, real **c.** Irrational, real **d.** Rational, real **e.** Natural, integer, rational, real

5. a. True; *Sample answer:* Terminating decimals are rational numbers.
b. False; *Sample answer:* No rational numbers are irrational numbers.

7. a. Rational **b.** Rational **c.** Irrational **d.** 8; rational **e.** Irrational **f.** 7; rational

9. a. $0.\overline{454}, \frac{1}{2}, 0.5\overline{4}, 0.\overline{54}, 0.\overline{554}, 0.\overline{5}, \sqrt{0.54}$
b. $0.7, \frac{3}{4}, 0.7\overline{5}, 0.\overline{75}, 0.\overline{755}, \sqrt{0.7}, \sqrt{0.75}$

11. a. *Sample answer:* $\sqrt{5}, \sqrt{6}$
b. *Sample answer:* $\sqrt{\frac{53}{100}}, \sqrt{\frac{67}{100}}$
c. *Sample answer:* $\sqrt{0.02}, \sqrt{0.03}$
d. *Sample answer:* $\sqrt{3}, \sqrt{5}$

13. a. 5 m **b.** 8.6 cm **15. a.** 110 m **b.** 286 ft

17. a. Yes **b.** No **19. a.** Yes **b.** No

21. a. 9 **b.** 4 **c.** 256 **d.** $\frac{1}{16}$

23. a. About 2996.07 **b.** About 0.135

25. About $59,187.20

27. *Sample answer:* The student used the wrong formula, and should review the definition of π.

29. 9.42 in.

31. No; *Sample answer:* $\sqrt{9 + 16} = 5$ but $\sqrt{9} + \sqrt{16} = 7$

33. *Sample answer:* Any irrational number and its opposite sum to 0.

35. a. $\sqrt{6}$ **b.** $\sqrt{8}$

37. Answers will vary. **39.** About 13.9 in.

41. a. $\sqrt{98} \approx 9.9$ ft **b.** $\sqrt{147} \approx 12.1$ ft

Chapter 9 Review *(page 359)*

1. a. $\frac{4}{3}$ **b.** $-\frac{2}{3}$ **c.** -1 **d.** $\frac{5}{3}$

3. $\frac{1}{2}$

−2 −1 0 1 2

5. $\frac{2}{7}$

−1 0 1

7. $\frac{1}{2}$ **9.** $\frac{13}{6}$ **11.** $-1, -\frac{6}{10}, -\frac{1}{10}, \frac{1}{3}, \frac{3}{5}$

13. Additive Inverse Property

15. Commutative Property of Addition

17. $\frac{2}{5} + \left(\frac{4}{7} + \frac{1}{15}\right)$ **19.** $\frac{4}{5}$ **21.** $\frac{1}{24}$ **23.** $-\frac{1}{8}$ **25.** $\frac{3}{16}$

27. $-\frac{5}{12}$ **29.** About 27

31. Addition Property of Equality

33. Multiplication Property of Equality

35. $x + 4 = -3; x = -7$ **37.** $x = -55$ **39.** $z = -17$

41. $u = -22.32$ **43.** $2x - 2 = 6; x = 4$ **45.** $y = 0$

47. $m = 18$ **49.** $j = 0.9$ **51.** $n - 15 = 7; n = 22$

53. $5 + n = 3; n = -2$ **55.** $5n - 12 = 68; n = 16$

57. $3x - 5 = 4; x = 3; 2y - 4 = 10; y = 7$

59. $27.5 + h = 40; h = 12.5$ **61.** $17.5w = 196; w = 11.2$

63. Irrational, real **65.** Integer, rational, real

67. Rational **69.** $\frac{1}{5}$; rational **71.** Irrational

73. $0.3, \frac{3}{8}, 0.38, 0.3\overline{8}, 0.\overline{38}, \sqrt{0.3}, \sqrt{0.38}$

75. 9.4 cm **77.** 220 yd **79.** Yes **81.** Yes **83.** 12

85. 27 **87.** About 0.099 **89.** About $20,400

Chapter 9 Test *(page 362)*

1. C **2.** C **3.** B **4.** C **5.** B
6. C **7.** D **8.** E

CHAPTER 10

Section 10.1 *(page 370)*

1. Solution key:
Steps 1–3: Check students' work.
Step 4: The angle measures sum to 180°.
Step 5: Yes; *Sample answer:* The angle measures sum to 180° each time.
Step 6: *Sample answer:* The sum of the angle measures of a triangle is 180°.
Sample rubric: Assign 1 point to correct work in Steps 1 through 3. Assign 1 point to each correct response in Steps 4 through 6. There is a total of 4 points.

3. a. Obtuse; *Sample answer:* The measure is greater than 90° and less than 180°.
b. Acute; *Sample answer:* The measure is less than 90°.
c. Straight; *Sample answer:* The measure is 180°.
d. Right; *Sample answer:* The measure is 90°.

5. 90°; right

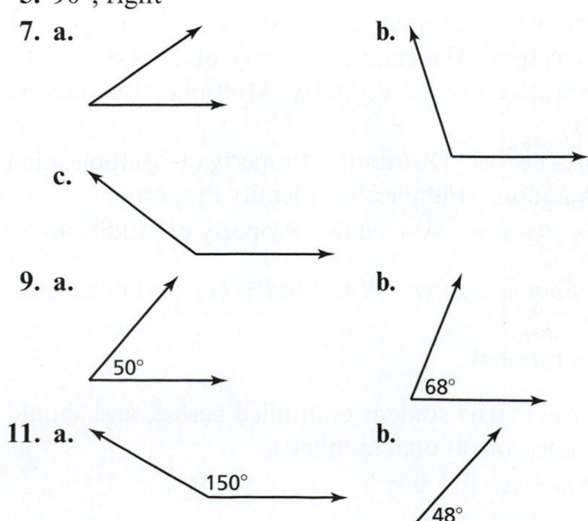

13. **a.** No; *Sample answer:* The figure is not closed.
b. No; *Sample answer:* The figure has 4 sides.
c. Yes; *Sample answer:* The figure is closed and has 3 straight sides.

15. **a.** *Sample answer:* **b.** *Sample answer:*

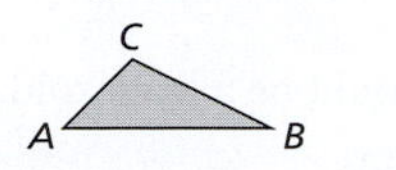

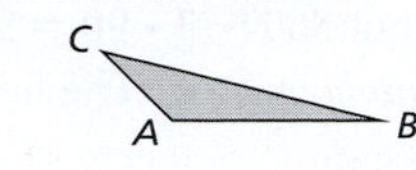

c. *Sample answer:*

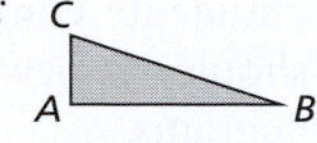

17. **a.** 80; acute **b.** 90; right
c. 94; obtuse **d.** 76.3; acute

19. **a.** Yes **b.** No; 63°

21. *Sample answer:*

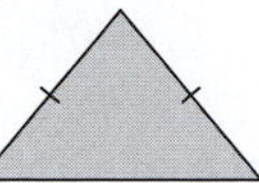

23. **a.** 72; acute, isosceles **b.** 32; obtuse, isosceles
c. 21; obtuse, scalene
d. 60; acute, equiangular, isosceles, equilateral

25. **a.** Complementary **b.** Neither **c.** Supplementary

27. **a.** *Sample answer:* Obtuse, acute, and straight
b. *Sample answer:* The obtuse and acute angles are supplementary.

29. *Sample answer:* The student subtracted from 180° instead of 90°, and should review classifying angles.

31. **a.** Always; *Sample answer:* $90 + 90 = 180$
b. Sometimes; *Sample answer:* $30 + 60 = 90$ but $50 + 45 \neq 90$
c. Sometimes; *Sample answer:* The triangle can be isosceles or scalene..
d. Never; *Sample answer:* The sum of the angles would be greater than 180.

33. Isosceles, scalene; *Sample answer:* The three sides cannot be equal.

35. Right isosceles, right scalene, obtuse scalene; *Sample answer:*

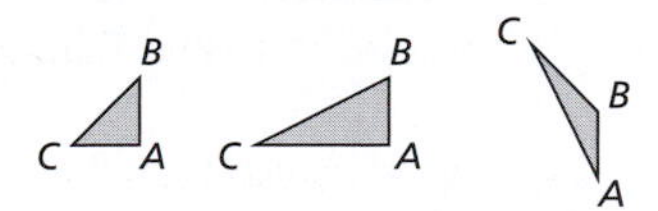

37. **a.** *Sample answer:* Draw the graph and measure the angle.
b. For the tax **c.** Against the tax
d. Don't care, Undecided
e.

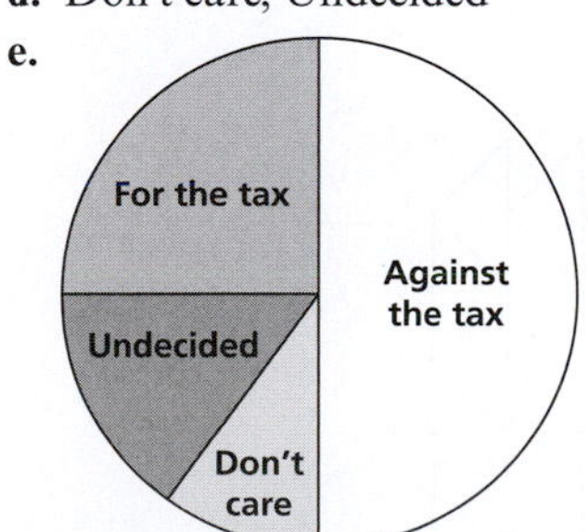

f. Right

39. **a.** E **b.** Median; D is the midpoint.
c. and d.

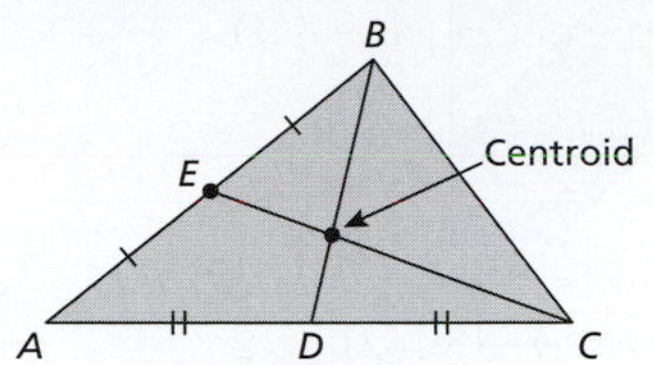

Sample answer: The centroid is where the medians intersect.

e. *Sample answer:* Draw a segment through A and the centroid.

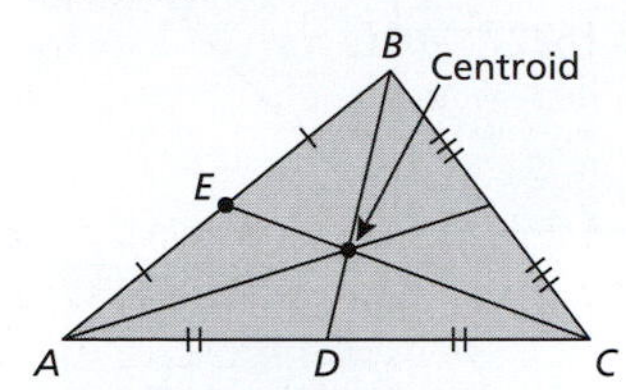

f. Answers will vary.

Section 10.2 *(page 382)*

1. Solution key:

1.

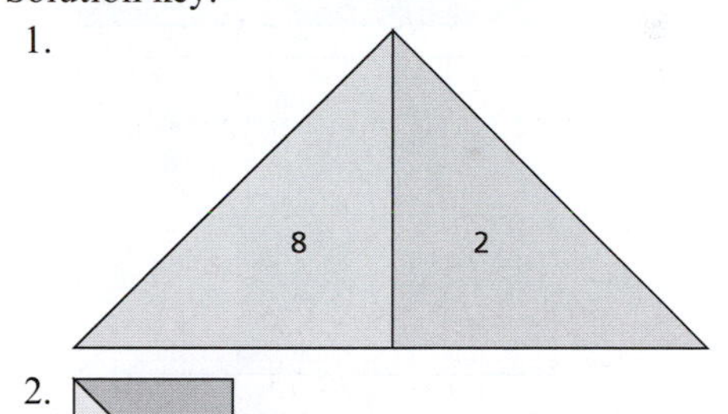

2.

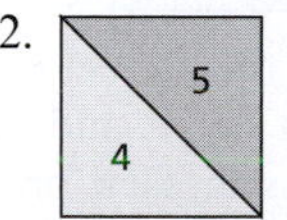

3.

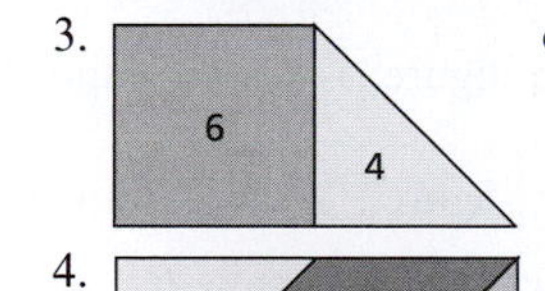

or

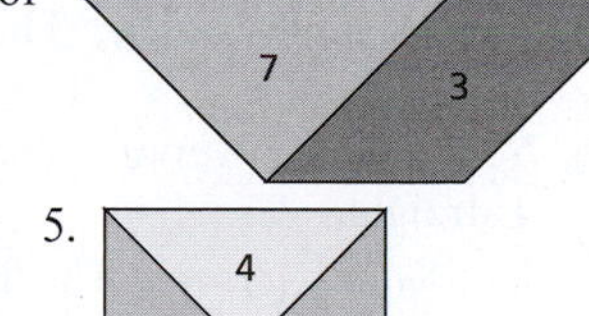

4. 5.

6.

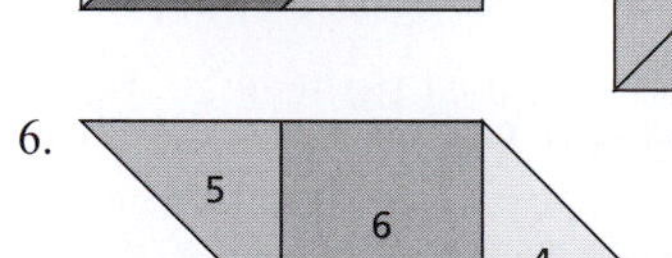

or

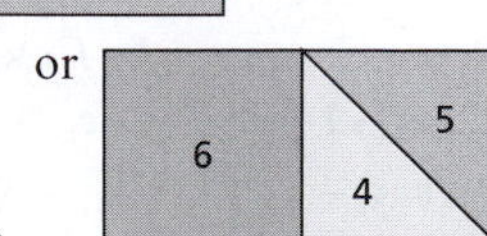

7.

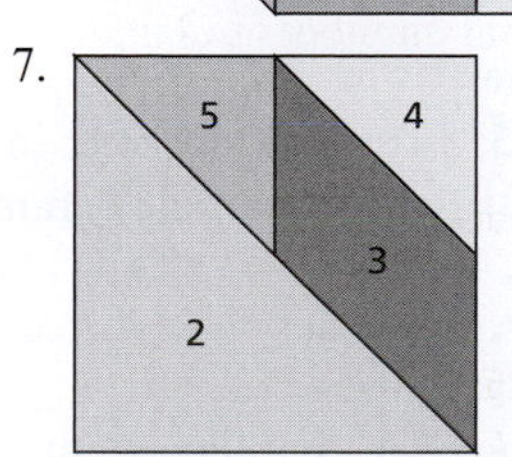

8. *Sample answer:*

9. 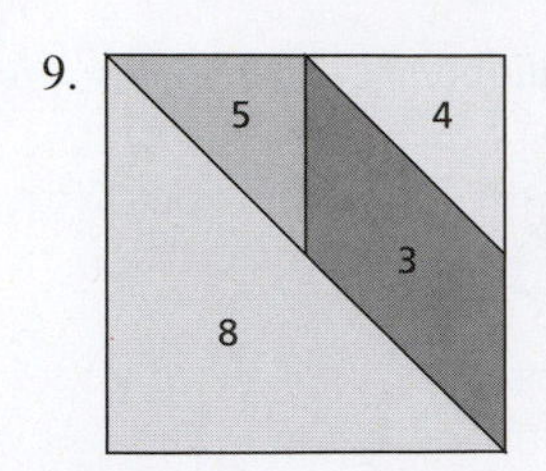

10. *Sample answer:*

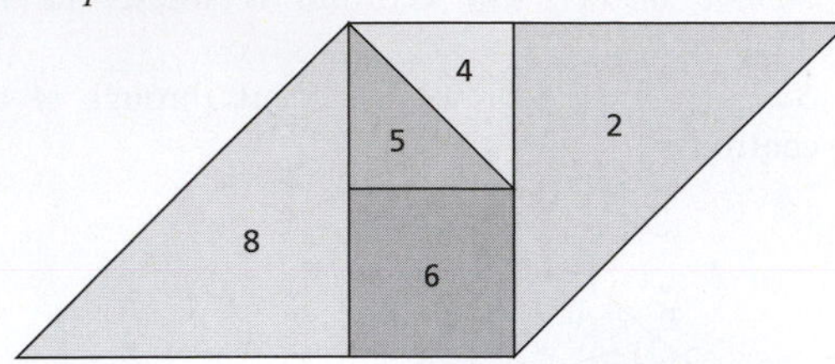

11. *Sample answer:*

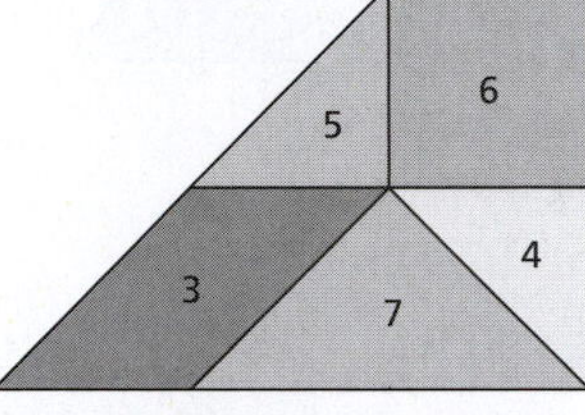

12. *Sample answer:*

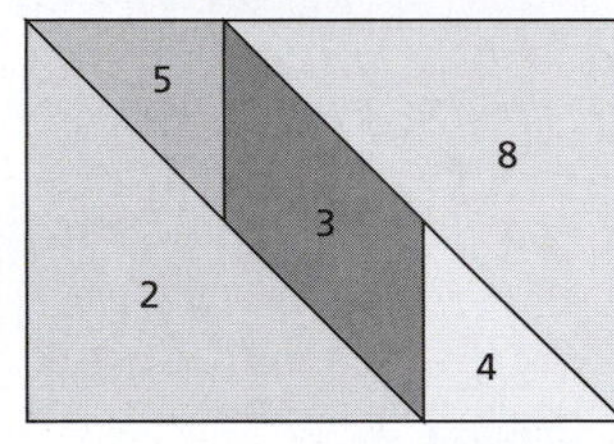

Sample rubric: Assign 1 point to using the correct number of pieces, 1 point to forming the correct shape, and 1 point to having the correct number of points in each problem. There is a total of 36 points.

3. a. Yes; *Sample answer:* The figure is closed and has 4 straight sides.
b. Yes; *Sample answer:* The figure is closed and has 4 straight sides.
c. No; *Sample answer:* The figure has 5 sides.
d. No; *Sample answer:* One side is curved.

5. a. Parallelogram **b.** Isosceles trapezoid, trapezoid
c. Kite **d.** Trapezoid

7. a. Convex **b.** Concave **c.** Concave **d.** Convex

9. a. 12 **b.** 3 **c.** 5

11. a. 80° **b.** 120° **c.** 113° **d.** 82°

13. a. 47 **b.** 70 **c.** 95

15. a. Parallelogram **b.** Square

17. 1: parallelogram; 2: square; 3: isosceles trapezoid; square

19. a. Always; *Sample answer:* A square is a parallelogram and has 4 congruent sides.
b. Never; *Sample answer:* A rhombus has 4 pairs of adjacent sides that are congruent.
c. Sometimes; *Sample answer:* A parallelogram is a rectangle when the angle measures are 90°.
d. Never; *Sample answer:* Parallel lines do not intersect.

21. a. Yes; *Sample answer:* The two parts line up.
b. No **c.** Answers will vary.

23. a. 120° **b.** Isosceles

25. a. *Sample answer:* 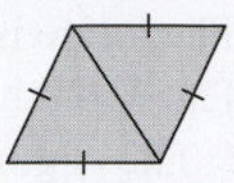 **b.** Isosceles

27. No; *Sample answer:* The fourth angle would be 180°.

29. Yes; no; *Sample answer:* Right trapezoids have two right angles, but $360 - 3 \cdot 90 = 90$.

31. No; *Sample answer:* The figure could be a trapezoid.

33. 6

35. *Sample answer:* The student confused rhombus and parallelogram, and should review the definitions of rhombuses and parallelograms.

Section 10.3 *(page 392)*

1. Solution key:
1. $180° + 180° = 360°$ 2. 540° 3. 720°
4. 900° 5. 1080°
6. *Sample answer:* The sum increases by 180 each time.

Sample rubric: Assign 1 point to a correct sum and 2 points to a correct pattern. There is a total of 7 points.

3. a. Yes; *Sample answer:* The figure is closed and has straight sides.
b. No; *Sample answer:* The figure is not closed.
c. No; *Sample answer:* One side is curved.

5. a. 720° **b.** 540° **7. a.** 540° **b.** 900°

9. a. Yes; *Sample answer:* The sides and angles are congruent.
b. No; *Sample answer:* The angles are not congruent.
c. No; *Sample answer:* The sides are not congruent.
d. No; *Sample answer:* The angles are not congruent.

11. a. Interior: 60°; central: 120°
b. Interior: 135°; central: 45°

13. a. 118 **b.** 80

15. a. 60° 60° 60°
b. 120° 120° 120° 120° 120° 120°
c. 144° 144° 144° 144° 144° 144° 144° 144° 144° 144°

17. *Sample answer:*

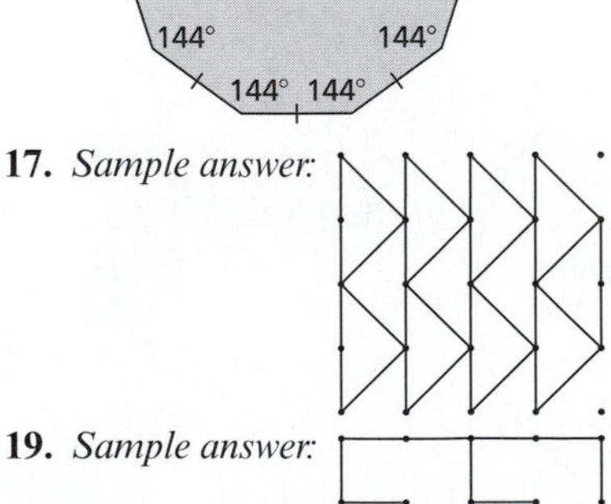

19. *Sample answer:*

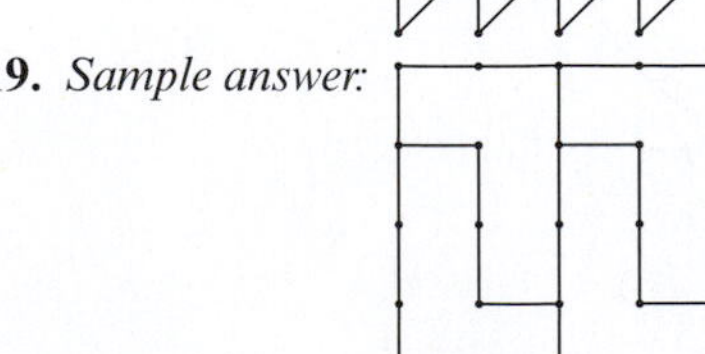

21. *Sample answer:* The student switched n and $n - 2$, and should review regular polygons.

23. a. Obtuse, straight **b.** 110°; 290°
c. *Sample answer:* The sum is 360°.
d. Answers will vary. *Sample answer:* Subtract from 360°.

25. *Sample answer:*

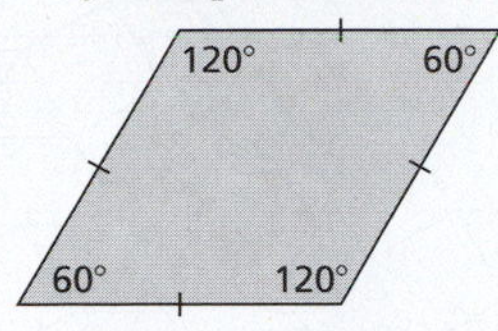

27. *Sample answer:* Evaluate $180° \times 6 - 360°$; $180° \cdot n - 360°$

29. 240°

31. Rectangle; *Sample answer:* To make the wall more stable.

33. No; *Sample answer:* The heptagons overlap.

35. a. *Sample answer:*

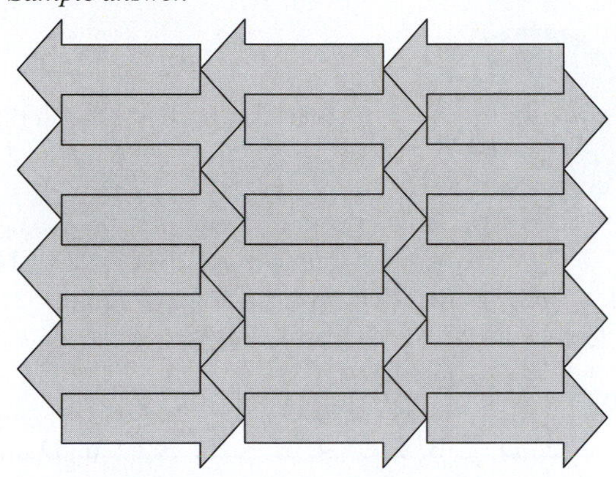

b. No; *Sample answer:* There will be gaps or overlaps.
c. No; *Sample answer:* There will be gaps or overlaps.
d.

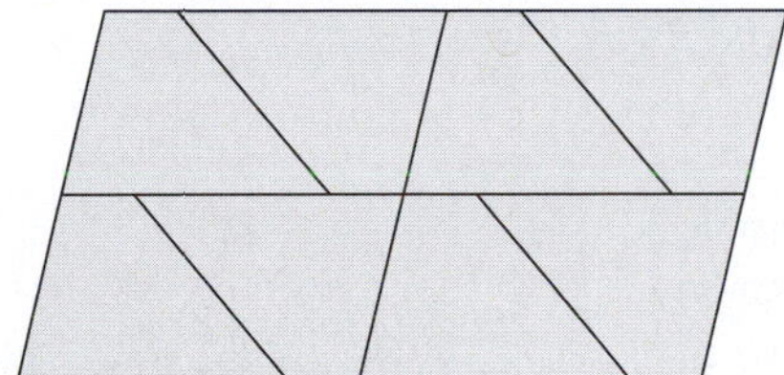

37. Tessellations will vary. *Sample answer:* Cut a shape out of one congruent side, then flip the shape vertically and tape to the other congruent side.

Section 10.4 *(page 402)*

1. Solution key:

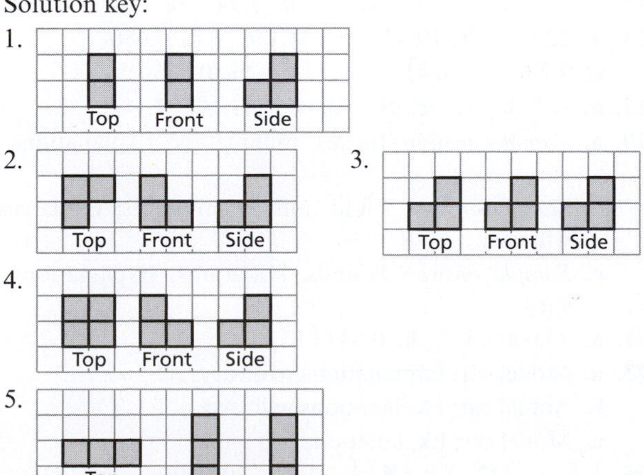

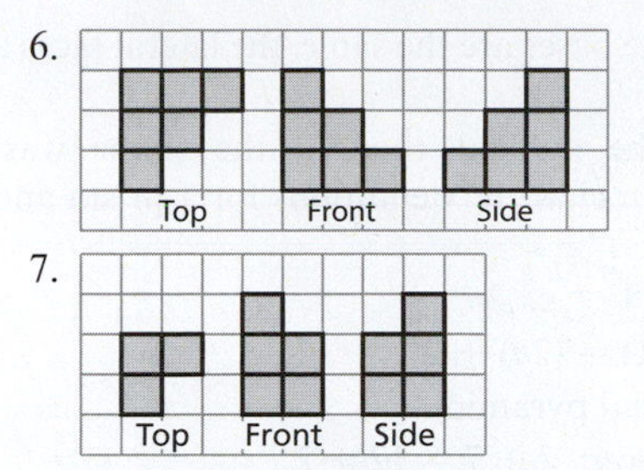

8.

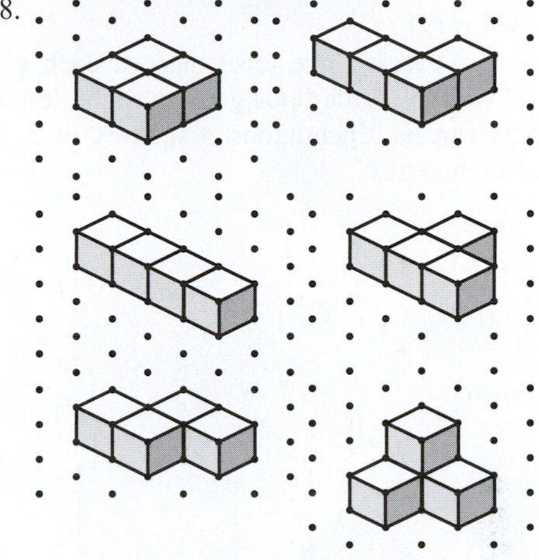

Sample rubric: Assign 1 point to each correct view in problems 1 through 7. Assign 1 point to each correct solid in problem 8. There is a total of 27 points.

3. a. Yes; *Sample answer:* The figure is a solid and the faces are polygons.
b. No; *Sample answer:* The faces are not polygons.
c. No; *Sample answer:* Two faces are a semicircle.
d. Yes; *Sample answer:* The figure is a solid and the faces are polygons.

5. a. Neither **b.** Pyramid **c.** Prism **d.** Neither

7. a. Hexagonal pyramid **b.** Pentagonal prism

9. *Sample answer:*

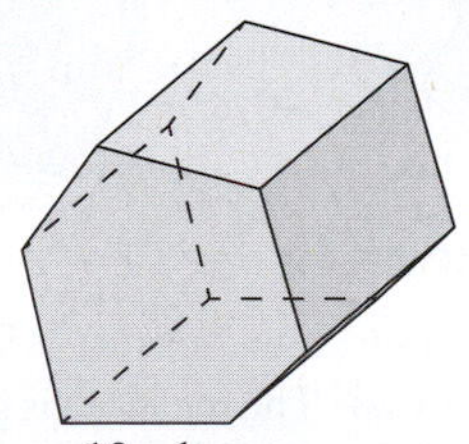

12 vertices, 8 faces, 18 edges

11. a. 10 **b.** 16 **c.** 12

13. a. 9
b. Truncated tetrahedron, cuboctahedron, icosidodecahedron, snub dodecahedron, small rhombicosidodecahedron, small rhombicuboctahedron, snub cube, truncated cube, truncated dodecahedron

15. a. Regular; tetrahedron
b. Semiregular; truncated tetrahedron

17. a. False; *Sample answer:* A cube has 6 congruent faces.
b. True; *Sample answer:* There are 5 faces and 6 vertices.
c. True; *Sample answer:* The pattern in Example 6(b) can be tessellated.
d. False; *Sample answer:* The lateral faces may not be parallel.

19. Neither; *Sample answer:* There are not two congruent, parallel bases, and the lateral faces do not meet at a point.

21. a. Snub dodecahedron **b.** 92

23. *Sample answer:* The bases are the same, the lateral faces are different.

25. *Sample answer:* The student thought the shape was a prism, and should review the definitions for a prism and a pyramid.

27. a. $n + 1$ **b.** $n + 1$ **c.** $2n$
d. $(n + 1) + (n + 1) = (2n) + 2$

29. a. 7 **b.** Hexagonal pyramid
c. Yes; *Sample answer:* $7 + 7 = 12 + 2$
d. Answers will vary.

31. *Sample answer:* At least three faces meet at each vertex, so the angles of the regular polygons must be less than 120°. So, there can be 3 pentagons, 3 squares, or 3, 4, or 5 triangles at each vertex.

33. iii

35.

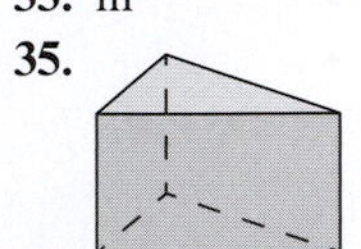

Triangular prism

37. a.

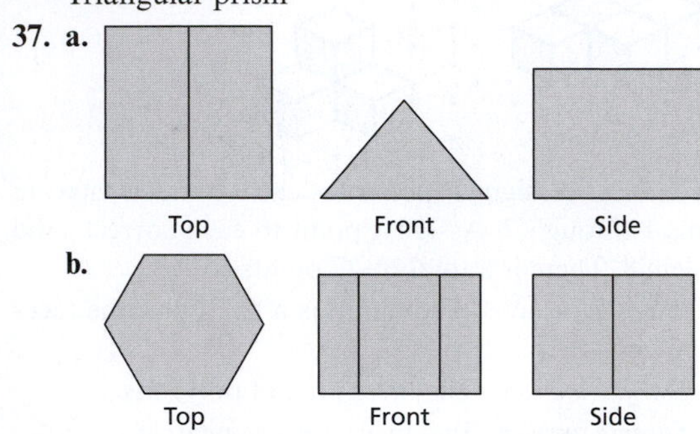

39. a. Rectangle **b.** Triangle

41. a.

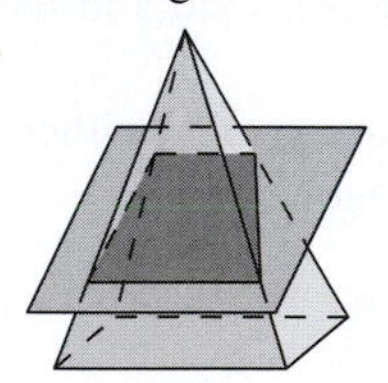

b.

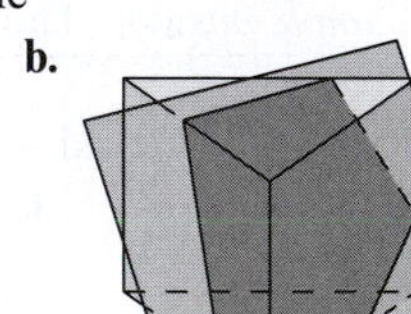

Chapter 10 Review *(page 407)*

1. Right; *Sample answer:* The measure is 90°.

3. Obtuse; *Sample answer:* The measure is greater than 90° and less than 180°.

5. **7.**

9. 85°; 5° **11.** 90; right, scalene

13. 60; acute, equiangular, isosceles, equilateral

15. a. Obtuse **b.** Acute **c.** They are supplementary.

17. Square, rhombus, rectangle, parallelogram

19. Rhombus, parallelogram **21.** Concave

23. 6 **25.** 69 **27.** 116 **29.** 63

31. a. Kite **b.** Square **33.** 720°

35. No; *Sample answer:* The sides are not congruent.

37. Yes; *Sample answer:* The sides and angles are congruent.

39. Interior: 108°; central: 72°

41. *Sample answer:*

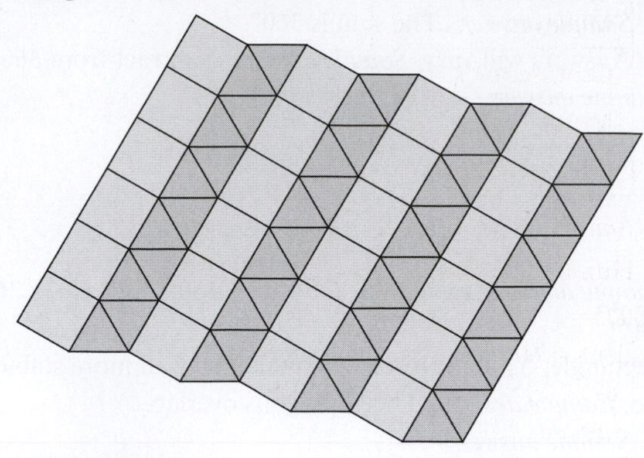

43.

45. iv **47.** Decagonal pyramid **49.** 18 **51.** 14

53. 11 **55.** Neither; regular hexagonal pyramid

57. a. Octahedron **b.** 8 faces, 12 edges, 6 vertices

Chapter 10 Test *(page 410)*

1. C **2.** D **3.** E **4.** E **5.** A **6.** D **7.** B

CHAPTER 11

Section 11.1 *(page 418)*

1. Solution key:
1. 3 2. 2.5 3. 2; 1 4. 4; 1.5
5. *Sample answer:* Pen
6. *Sample answer:* Pencil box or crayon box
Sample rubric: Assign 1 point to each correct measurement or object. There is a total of 8 points.

3. a. Length **b.** Weight **c.** Weight

5. 5.5 in. **7.** 53 lb **9.** 28 fl oz

11. a. 36 **b.** 44,352 **c.** 1.375
d. 6400 **e.** 2.375 **f.** 9.6

13. a. 1.2 **b.** 9250 **c.** 3 **d.** 7.24 **e.** 7200 **f.** 0.573

15. a. 22.86 **b.** 19.32 **c.** 30.176 **d.** 0.686
e. 9.281 **f.** 0.418 **g.** 7.8 **h.** 0.666

17. a. $<$ **b.** $>$ **c.** $>$ **d.** $<$ **e.** $>$ **f.** $>$

19. a. *Sample answer:* Inches, millimeters; Explanations will vary.
b. *Sample answer:* Fluid ounces, milliliters; Explanations will vary.
c. *Sample answer:* Pounds, kilograms; Explanations will vary.

21. a. 33,000 mL **b.** 0.033 kL

23. a. Model car; Explanations will vary.
b. Actual car; Explanations will vary.
c. Model car; Explanations will vary.

25. 1.5 c **27.** 7 ft 1.5 in. **29.** 20 times

31. About 500 mph **33.** About 0.426 L

35. *Sample answer:* The student chose a U.S. customary unit and should review the metric system of measure.

37. a.

Millimeter	1000
Centimeter	100
Meter	1
Kilometer	0.001

Sample answer: Each value is a power of 10.

b. Hundred or hundredth **c.** $\frac{1}{1000}$ sec

39. 1.6975

41. 1 talent = 60 mina

43. a. 5.87×10^{12} mi **b.** 2.48×10^{13} mi

45. a. Answers will vary.

b. *Sample answer:* The measurements may be different.

c. Answers will vary.

47. Answers will vary.

Section 11.2 *(page 428)*

1. Solution key:

1. 6 units 2. 10 units 3. 14 units

4. *Sample answer:* 5. *Sample answer:*

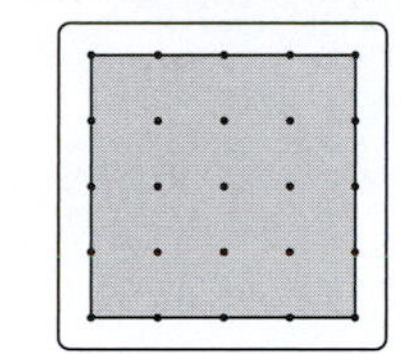

6. *Sample answer:*

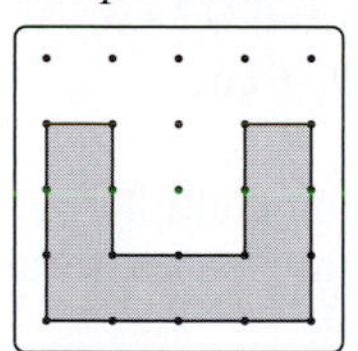

7. 14 units 8. 12 units 9. 14 units

Sample rubric: Assign 1 point to each correct perimeter and 2 points to each correct polygon. There is a total of 12 points.

3. a. Inches **b.** Square miles **c.** Square feet **d.** Miles

5. a. 66 cm **b.** 73 ft **7.** 12 units; 8 units^2

9. 210 ft; 2000 ft^2 **11.** 10 units^2 **13.** 80 cm^2

15. 59.5 m^2 **17.** 24 ft^2 **19.** 70 in.^2 **21.** 28 in.^2

23. 192 in.; 2160 in.^2 **25.** 20 units

27. 6 units^2; *Sample answer:* Make the width as small as possible.

29. 5 in. **31.** 248.5 ft^2

33. *Sample answer:* The student did not multiply by 2 and should review finding the perimeter of a polygon.

35. *Sample answer:* Move triangles 1 and 2 to create a rectangle that is b units by $\frac{a}{2}$ units.

37. a.

Length, ℓ	9	8	7	6	5
Width, w	1	2	3	4	5
Area, A	9	16	21	24	25

b. *Sample answer:* Make a square. **c.** Answers will vary.

39. 64 cm^2 **41. a.** 75.55 in.^2 **b.** 84.125 in.^2

Section 11.3 *(page 438)*

1. Solution key:

1. 166 units^2

2.

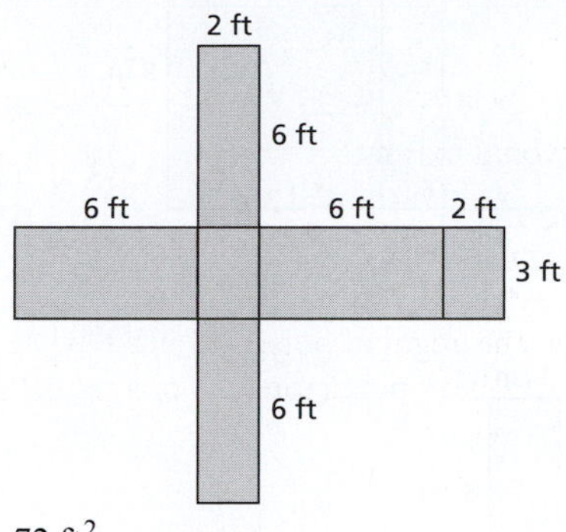

72 ft^2

3.

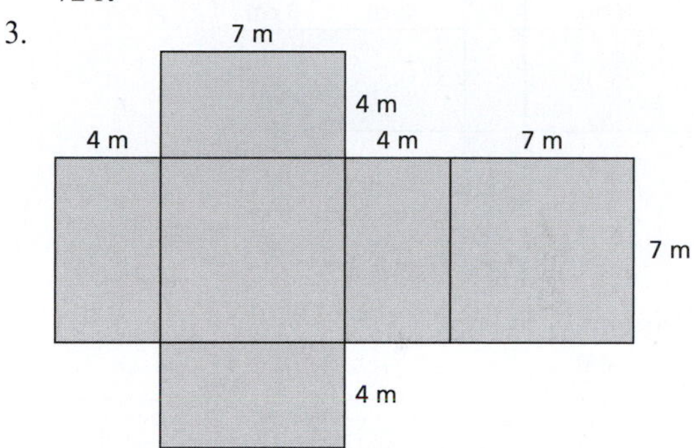

210 m^2

4.

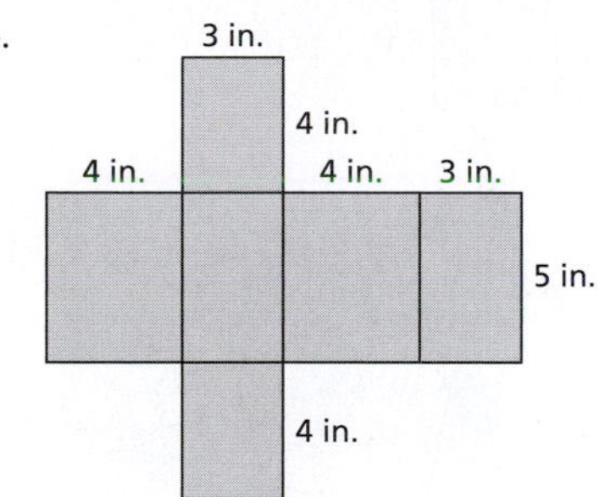

94 in.^2

5.

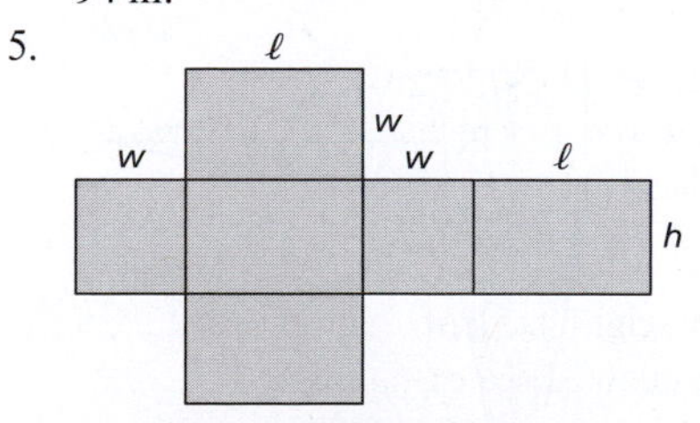

$S = 2\ell w + 2\ell h + 2wh$

Sample rubric: Assign 1 point to each correct net and 1 point to each correct surface area or formula. There is a total of 9 points.

3.

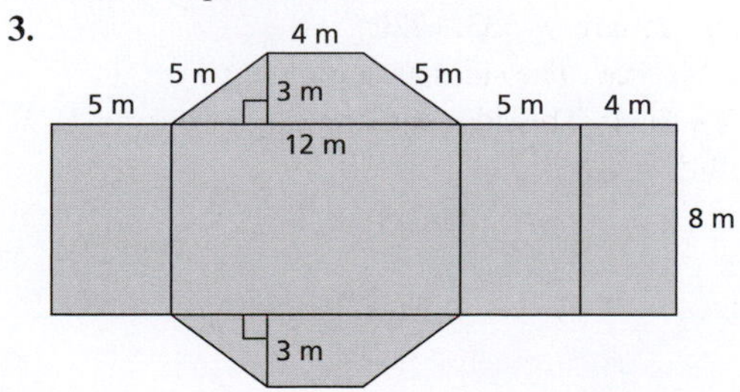

5. a.

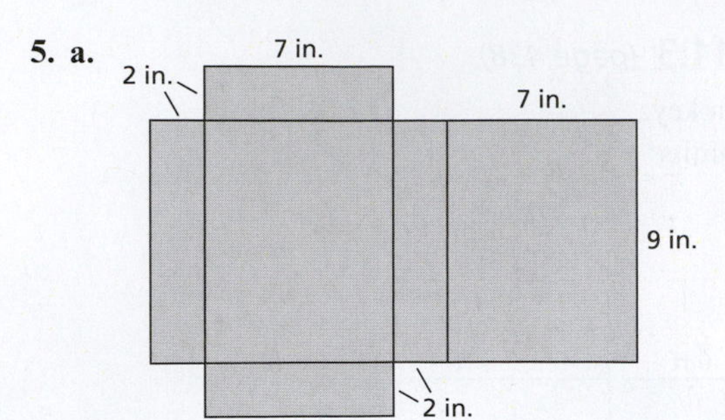

190 in.^{2}

b.

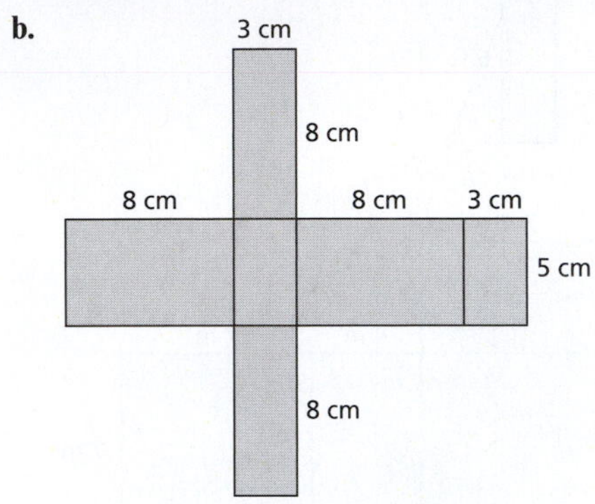

158 cm^2

7.

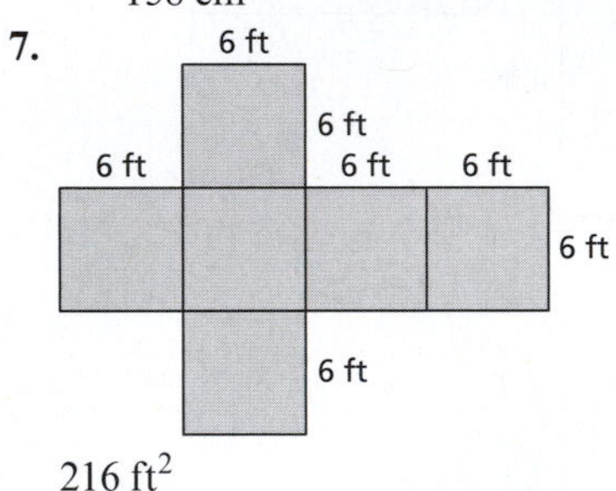

216 ft^2

9. a.

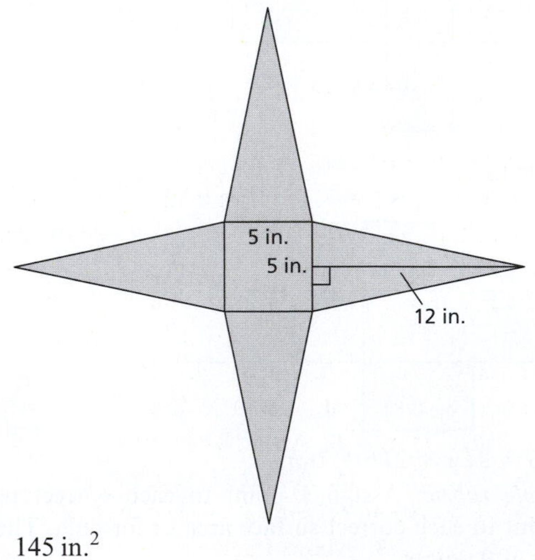

145 in.^{2}

b.

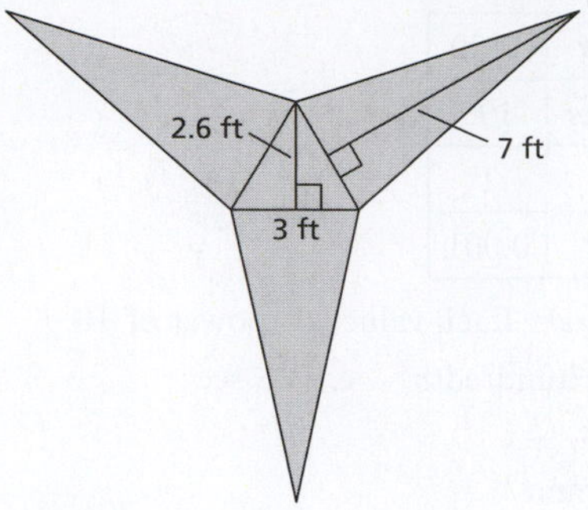

35.4 ft^2

11.

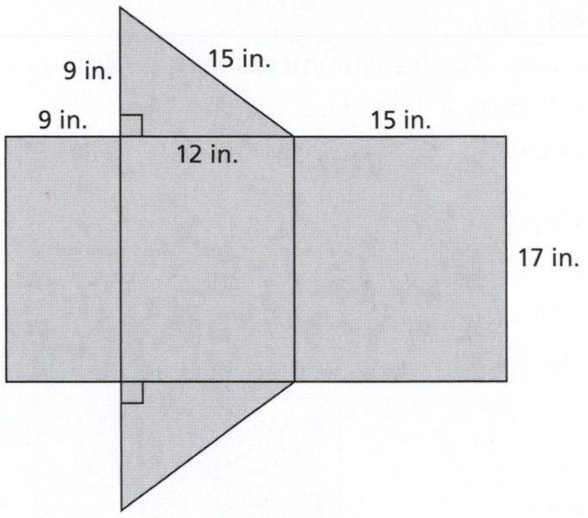

720 in.^{2}

13. a. 96 m^2 **b.** 26 in.^{2}

15. a. True; *Sample answer:* The net contains all the faces.

b. False; *Sample answer:* $6(2)^3 = 48$ and $6(2 \cdot 3)^3 = 1296$

17. a. 166 units^2 **b.** 116 units^2 **19.** 6 cm

21. 13.5 ft^2 **23.** 1 **25.** 105 in.^{2}

27. *Sample answer:* The student did not multiply by 2, and should review finding the surface area of a rectangular prism.

29. 36 : 64 **31.** 540 in.^{2}

33. a. None; *Sample answer:* The area removed equals the area added.

b. 96 in.^{2}

c. Yes; *Sample answer:* The surface areas are equal.

35. 129.5 ft^2

37. *Sample answer:* Multiply the height by the perimeter.

39. a. 98 units^2

b. *Sample answer:* Make a 1 by 3 by 8 prism.

c. 2 by 3 by 4; 52 units^2 **d.** Answers will vary.

Section 11.4 *(page 448)*

1. Solution key:

1. 12 units^3
12 units^3
+ 12 units^3
36 units^3

2. 16 units^3
16 units^3
16 units^3
16 units^3
+ 16 units^3
80 units^3

3.

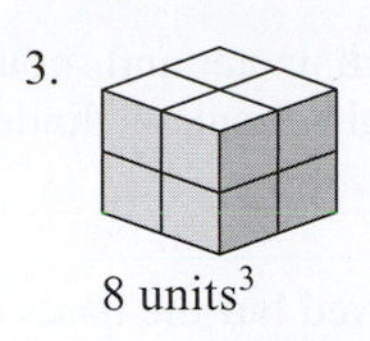

8 units^3

4.

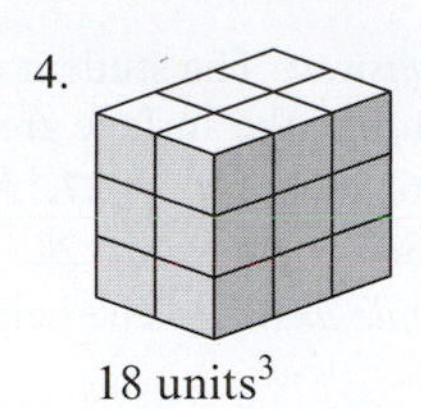

18 units^3

5.

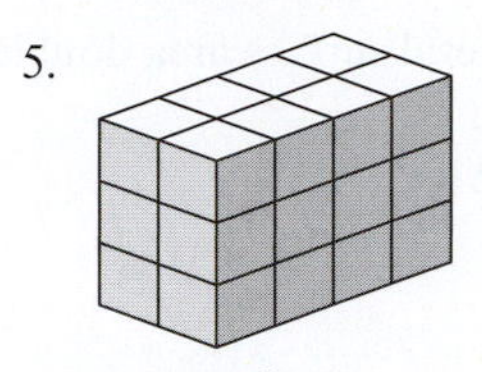

24 units^3

6. $V = \ell wh$

Sample rubric: Assign 1 point to each correct volume or formula and 1 point to each correct prism. There is a total of 9 points.

3. 1 by 1 by 8, 1 by 2 by 4, 2 by 2 by 2; 8 units^3

5. 24 units^3 **7. a.** 280 in.^{3} **b.** 108 ft^3 **9.** 5 in.

11. 918 cm^3 **13. a.** 128 ft^3 **b.** 105 cm^3 **c.** 33 mm^3

15. a. 714 ft^3 **b.** 4576 ft^3 **c.** 2950 mm^3 **17.** 60 ft^3

19.

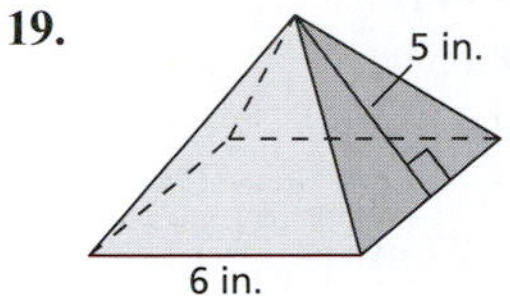

48 in.^{3}

21. Answers will vary.

23. *Sample answer:* The student substituted the wrong values into the formula, and should review finding the volume of a prism.

25. *Sample answer:* Double the length.

27. 360 in.^{3} **29.** Box 3; Box 1

31. 9324.028 cm^3 **33.** 900 ft^2

Chapter 11 Review *(page 453)*

1. $2\frac{2}{3}$ **3.** 4800 **5.** 14 **7.** 150,000 **9.** 1.8

11. 7.2 **13.** 1.8 **15.** 0.731 **17.** 17.529 **19.** <

21. > **23.** >

25. *Sample answer:* Ounces, grams; Explanations will vary.

27. *Sample answer:* Gallons, liters; Explanations will vary.

29. *Sample answer:* Pounds, kilograms; Explanations will vary.

31. About 422 g **33.** No; *Sample answer:* 2 L ≈ 2.08 qt

35. Mix B **37.** 70 ft **39.** 21.5 cm

41. 15 ft; 13.5 ft^2 **43.** 36 in.^{2} **45.** 209 cm^2

47. 29 in.^{2} **49.** 131.2 ft

51.

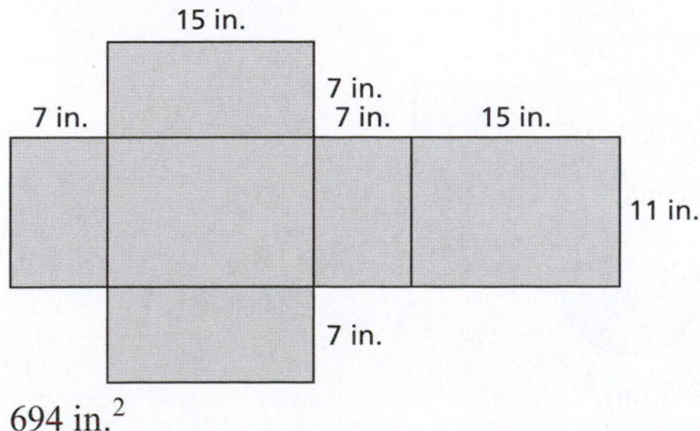

694 in.^{2}

53.

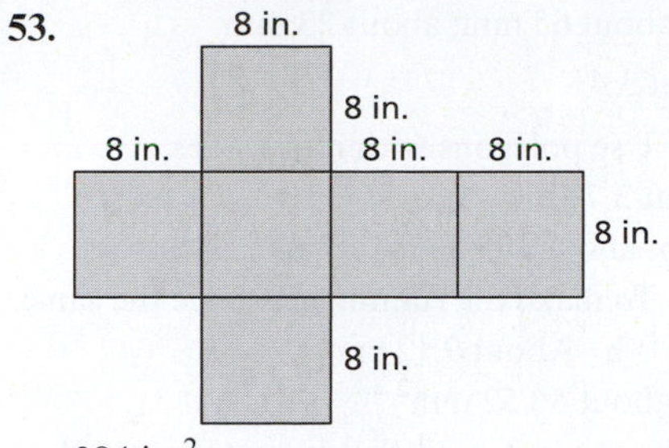

384 in.^{2}

55.

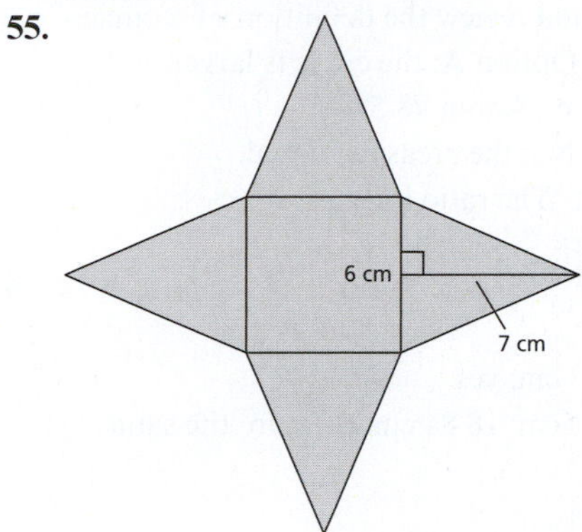

120 cm^2

57. 486 in.^{2} **59.** 1400 mm^2 **61.** 5 ft **63.** 16 : 25

65. 832 in.^{2} **67.** 63 in.^{3} **69.** 7.5 ft^3 **71.** 72 in.^{3}

73. 1074 in.^{3} **75.** 1671 cm^3 **77.** 260 ft^3

Chapter 11 Test *(page 456)*

1. C **2.** D **3.** E **4.** A **5.** C

6. D **7.** D **8.** D

CHAPTER 12

Section 12.1 *(page 464)*

1. Solution key:

1. Check students' work 2. $2r$ 3. $2\pi r$

4. Check students' work.

5. Check students' work. 6. r; πr

7. πr^2; *Sample answer:* The area of the circle is πr^2.

Sample rubric: Assign 1 point to correct work in problems 1, 4, and 5. Assign 1 point to each correct expression in problems 2, 3, 6, and 7. There is a total of 7 points.

3. a. 6 m **b.** 7 in.

5. a. About 47.1 mm **b.** About 34.54 ft **c.** About 28.26 in. **d.** About 25.12 m

7. a. About 12.85 m **b.** About 84.81 mm **c.** About 10.28 ft **d.** About 51.4 cm

9. a. 4 in. **b.** About 6.28 m **c.** 16.625 mm **d.** About 12.56 cm

11. a. About 452.16 ft^2 **b.** About 530.66 in.^{2} **c.** About 154 m^2 **d.** About 803.84 cm^2

13. a. About 57.75 cm^2 **b.** About 79.13 in.^{2} **c.** About 17.27 in.^{2} **d.** About 8.79 m^2

15. a. About 8.39 m^2 **b.** About 34.70 in.^{2}

17. a. About 38.56 mm **b.** About 6.28 m

19. 38 in. **21.** 50 yd^2

23. **a.** About 92 mm, about 65 mm; about 23 mm
b. $\frac{4}{1}; \frac{65}{23}$ **c.** About 3.41
d. *Sample answer:* Use polygons with more sides.
25. 5040 **27.** About 3.77 in.
29. **a.** About 398.02 m, about 405.56 m; about 7.54 m
b. *Sample answer:* To make the running distance the same.
31. **a.** About 308 in.^{2} **b.** About 9.12 cm^2
33. About 18.84 mm; about 56.52 mm^2
35. *Sample answer:* The student used the diameter instead of the radius, and should review the definition of a circle.
37. **a.** *Sample answer:* Option A; the circle is larger.
b. About 78.5% **c.** About 78.5%
d. *Sample answer:* No; the areas are equal.
39. Yes; *Sample answer:* The ratio is 2π.

Section 12.2 *(page 474)*

1. Solution key:
1. *Sample answer:* 3 cm; yes
2. *Sample answer:* 5cm; 18.84 cm; they are the same.
3.

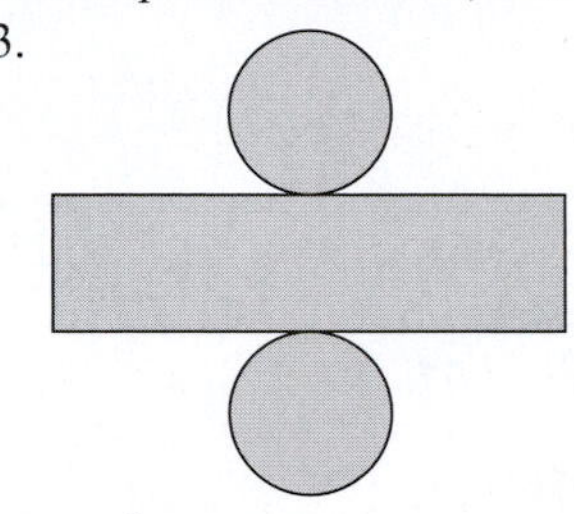

4. $2\pi r^2 + 2\pi rh$
5. *Sample answer:* About 150.72 cm^2

Sample rubric: Assign 1 point to each correct quantity, relationship, formula, or net. There is a total of 8 points.

3. About 75.36 in.^{2}
5. **a.**

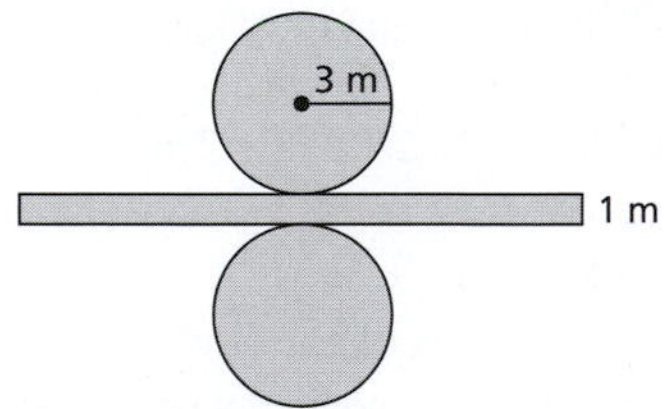

About 75.36 m^2

b.

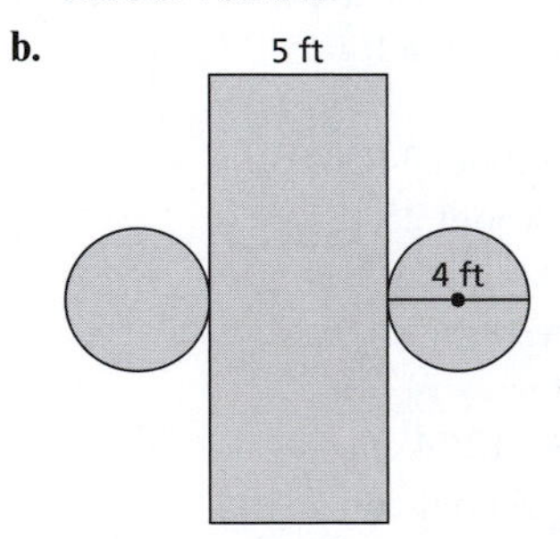

About 87.92 ft^2

7. 25 ft
9. **a.** About 43.96 in.^{2} **b.** About 452.16 cm^2
c. About 418 ft^2 **d.** About 898.04 m^2
11. 20 in. **13.** $r = 4$ m; $\ell = 6$ m **15.** 80 cm^2
17. **a.** About 113.04 ft^2 **b.** About 452.16 cm^2
19. About 78.5 in.^{2} **21.** About 19,899.75 in.^{2}
23. *Sample answer:* The student used the diameter and should review finding the surface area of a right circular cylinder.
25. About 263.76 mm^2 **27.** Answers will vary.
29. About 1848 cm^2 **31.** 22
33. No; *Sample answer:* The height is halved but the bases are the same.
35. No; *Sample answer:* Only the lateral surface area doubles.
37. **a.** $2r; r$ **b.** $4\pi r^2; 6\pi r^2$ **c.** 2 : 3
39. **a.** About 81,000,000 mi^2 **b.** About 41%

Section 12.3 *(page 484)*

1. Solution key:
1. 1000 2. 1000 3. 50.24 cm^2
4. Yes; *Sample answer:* $20 \cdot 50.24 \approx 1000$

Sample rubric: Assign 1 point to a correct quantity and 2 points to a correct reasoning. There is a total of 5 points.

3. **a.** About 3.14 cm^2 **b.** About 3.14 cm^3
c. About 9.42 cm^3
5. **a.** About 301.44 ft^3 **b.** About 314 in.^{3}
c. About 84.78 mm^3 **d.** About 57.88 cm^3
7. **a.** About 183.33 in.^{3} **b.** About 314 ft^3
c. About 34.65 yd^3 **d.** About 18.65 yd^3
9. **a.** About 4186.67 cm^3 **b.** About 179.67 mm^3
11. **a.** About 8 ft **b.** About 11 cm
13. About 1232 in.^{3} **15.** About 482.27 m^3
17. About 0.2826 m^3
19. The volume of the cone is $\frac{2}{3}$ the volume of the cylinder.
21. **a.** *Sample answer:* The student multiplied by 2, and should review finding the volume of a cylinder.
b. *Sample answer:* The student used the diameter, and should review finding the volume of a cone.
23. About 3.59 m^3
25. Cylinder; *Sample answer:* Three cones cost more than one cylinder.
27. About 3477 km **29.** Answers will vary.
31. *Sample answer:* Prisms and cylinders; they are congruent.
33. About 186 lb

Chapter 12 Review *(page 489)*

1. 12 mm **3.** 18 in. **5.** About 50.24 yd
7. About 40.82 mm **9.** About 36 m **11.** 20 mm
13. About 32.97 in. **15.** About 706.5 cm^2
17. About 616 mm^2 **19.** About 314 m^2
21. About 82.13 in.^{2} **23.** About 8.71 ft^2
25. 12 in.; *Sample answer:* 14 in.^{2} > 13 in.^{2}
27.

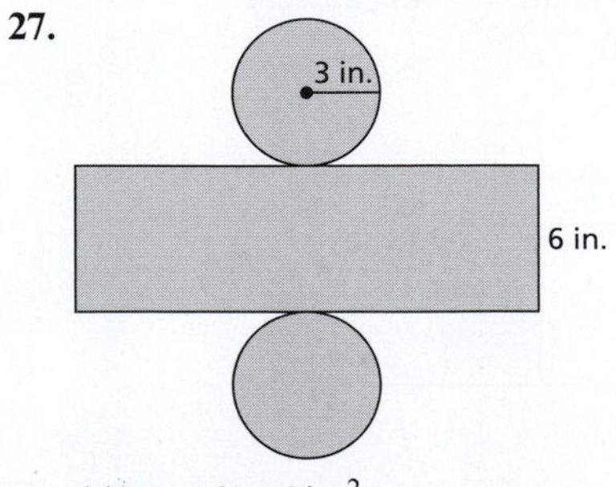

About 169.56 in.^{2}

29. 38 m **31.** About 84.78 mm^2 **33.** 8 ft **35.** 56 ft^2
37. About 31,400 mm^2 **39.** About 314 cm^2
41. About 2816 m^2

43. *Sample answer:* The area is half. **45.** About 198 cm^3
47. About 452.16 ft^3 **49.** About 267.95 yd^3
51. About 4.19 km^3 **53.** About 20 in.
55. About 469.33 ft^3
57. a. About 7597.33 ft^3 **b.** About 27%

Chapter 12 Test *(page 492)*

1. A **2.** B **3.** C **4.** D **5.** C **6.** A **7.** C

CHAPTER 13

Section 13.1 *(page 500)*

1. Solution key:
1. Check students' work.
2. Students' measurements will vary. $A \approx 24.6°$, $B \approx 41.8°$, $C \approx 113.6°$
3. No; *Sample answer:* The angles are always the same.
4. Yes; *Sample answer:* When the side lengths are the same, the angles are the same.

Sample rubric: Assign 1 point to correct work in problem 1. Assign 1 point to each correct angle measure in problem 2. Assign 1 point to each correct answer and 2 points to each correct reasoning in problems 3 and 4. There is a total of 10 points.

3. a. $\triangle EFD$ **b.** $\triangle FDE$ **c.** $\triangle CBA$
5. a. $\overline{DE}$ **b.** $\overline{AC}$ **c.** $\overline{FD}$ **d.** $\angle A$ **e.** $\angle B$ **f.** $\angle F$
7. a. 22.1 cm **b.** 14.1 cm **c.** 26.1 cm
d. 89° **e.** 33° **f.** 58°
9. a. $\angle B$ **b.** $\angle F$ **c.** $\angle A$
11. a. Congruent; *Sample answer:* SSS Congruence Postulate
b. Congruent; *Sample answer:* SAS Congruence Postulate
c. Not congruent; *Sample answer:* Corresponding angles are not congruent.
13. a. *Sample answer:* $\overline{AB} \cong \overline{DE}$, $\overline{BC} \cong \overline{EF}$, $\angle B \cong \angle E$
b. $\overline{DF} \cong \overline{AC}$, $\overline{DE} \cong \overline{AB}$, $\overline{EF} \cong \overline{BC}$
15. a. Yes; *Sample answer:* Use the SAS Congruence Postulate.
b. Yes; *Sample answer:* Use the ASA Congruence Postulate.
17. a. Yes; *Sample answer:* ASA Congruence Postulate
b. Yes; *Sample answer:* Measure the distance from the tent to the tree.
19. a. *Sample answer:* $\overline{AC} \cong \overline{XZ}$; SSS Congruence Postulate
b. *Sample answer:* $\angle BAC \cong \angle DAC$; ASA Congruence Postulate
21. *Sample answer:*

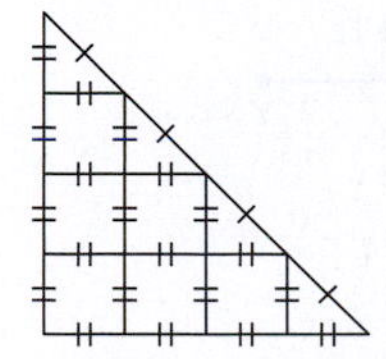

SSS Congruence Postulate

23. *Sample answer:* The student did not realize the angle must be between the two sides and should review the SAS Congruence Postulate.
25. Answers will vary.
27. Yes; *Sample answer:*
$\angle E \cong \angle G$ and $m\angle HFG + m\angle EFH = m\angle FHE + m\angle GHF$
29. Yes; *Sample answer:* ASA Congruence Postulate
31. They are equal; *Sample answer:* SAS Congruence Postulate

Section 13.2 *(page 510)*

1. Solution key:
1. a. Yes; *Sample answer:* $\frac{5}{5} = \frac{6}{6}$
   b. No; *Sample answer:* $\frac{5}{6} \neq \frac{6}{6}$
   c. Yes; *Sample answer:* $\frac{8}{8} = \frac{6}{6}$
2. a. No; *Sample answer:* $\frac{5}{3} \neq \frac{6}{4}$
   b. Yes; *Sample answer:* $\frac{3}{2} = \frac{6}{4}$
   c. Yes; *Sample answer:* $\frac{9}{6} = \frac{6}{4}$
3. a. Yes; *Sample answer:* $\frac{3}{4} = \frac{6}{8}$
   b. No; *Sample answer:* $\frac{4}{6} \neq \frac{6}{8}$
   c. No; *Sample answer:* $\frac{8}{10} \neq \frac{6}{8}$

Sample rubric: Assign 1 point to each correct answer and 2 points to each correct reasoning. There is a total of 27 points.

3. a. 70° **b.** 20° **c.** 70° **5. a.** 12 **b.** 1
7. $x = 6$; $y = 8$
9. a. Yes; *Sample answer:* AA Similarity Postulate
b. No; *Sample answer:* Corresponding angles are not congruent.
c. Yes; *Sample answer:* SSS Similarity Theorem
11. Yes; *Sample answer:* Corresponding sides are proportional and corresponding angles are congruent.
13. a. 22 **b.** 33 **c.** 39 **15.** 12
17. Similar; *Sample answer:* AA Similarity Postulate; $\triangle ABE \sim \triangle ACD$
19. Not similar; *Sample answer:* Corresponding sides are not proportional.
21. *Sample answer:* In real life, similar can refer to measurements that are approximately equal, such as a pen and a pencil having a similar size.
23. *Sample answer:* The student only compared one pair of angles and should review the definition of similar triangles.
25. a. Answers will vary. **b.** They are equal.
c. Answers will vary.
27. a. Yes; *Sample answer:* AA Similarity Postulate
b. 28.5 ft
29. a. Yes; *Sample answer:* AA Similarity Postulate
b. 50 ft
31. a.

b. *Sample answer:* The two roads are parallel, so the angles are congruent.
c. 1 mi; 2 mi **d.** About 2.3 mi

Section 13.3 *(page 520)*

1. Solution key: 1–8. Check students' work.
Sample rubric: Assign 1 point to each correct construction in problems 1 through 5 and 2 points to each correct construction in problems 6 through 8. There is a total of 11 points.

3. a. A B **b.** C D

5. 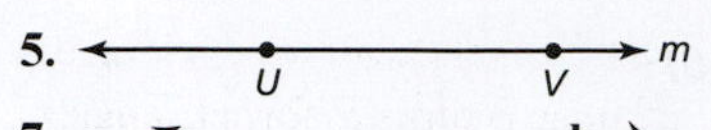

7. a. **b.**

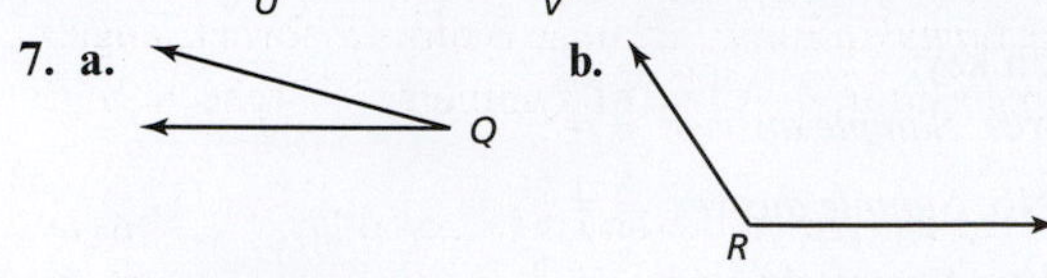

9.

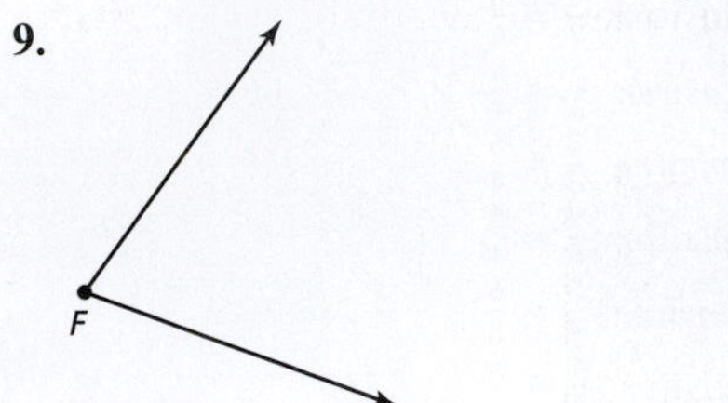

11. a. P J K **b.**

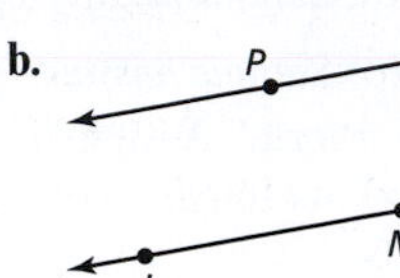

13.

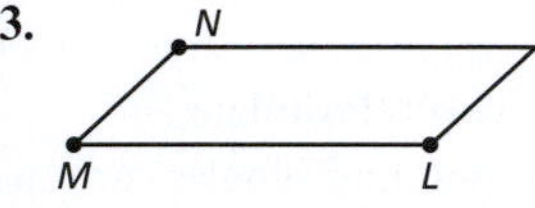

15. a.

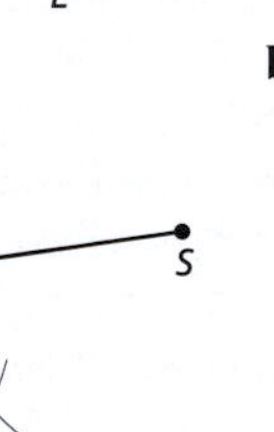

b.

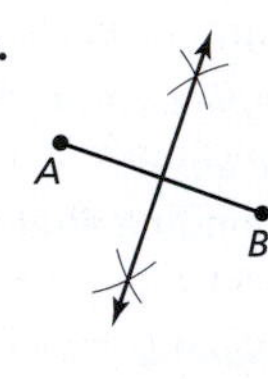

17. a. **b.**

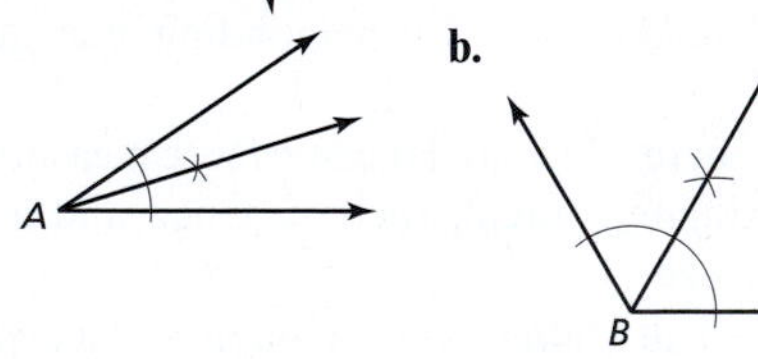

19. L N

21.

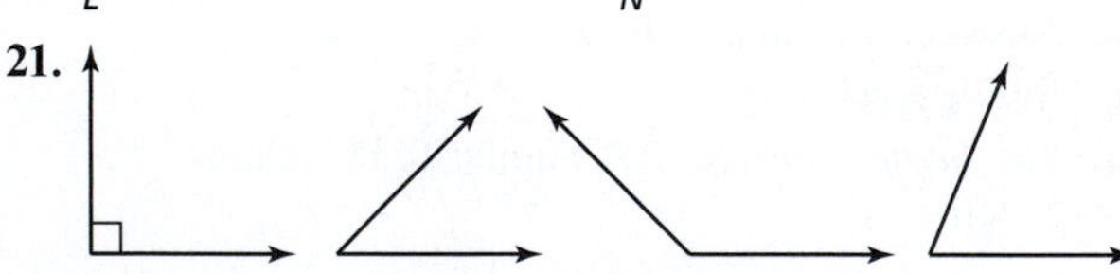

Sample answer: Draw two perpendicular lines; bisect the 90° angle; construct a 45° angle adjacent to a 90° angle; bisect a 135° angle.

23. *Sample answer:* The student constructed the sides from the wrong vertices and should review constructing a parallelogram.

25. a. **b.**

27.

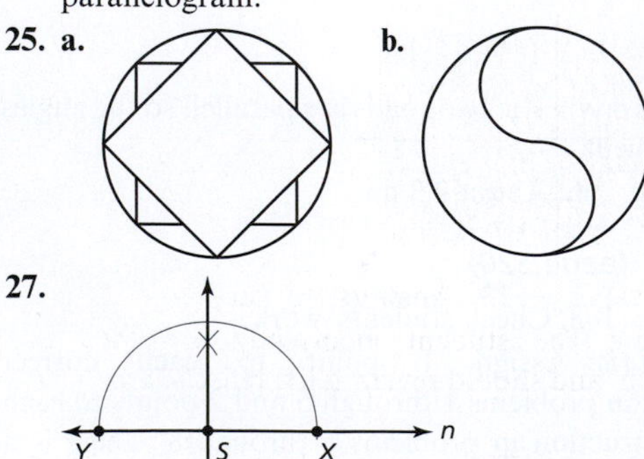

29. *Sample answer:* Let A be the point of intersection of the arcs. By the SSS Congruence Postulate, $\triangle XSA \cong \triangle YSA$, so $m\angle XSA = m\angle YSA$. $m\angle XSY = 180°$, so each is 90°.

31. a. and b. **c.** They are perpendicular.

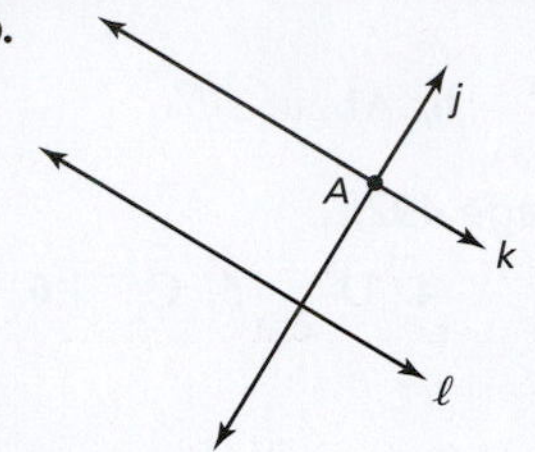

33.

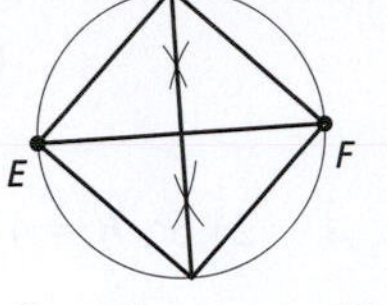

Sample answer: Construct a perpendicular bisector of $\overline{EF}$, then draw a circle with center at the intersection and radius half the length of $\overline{EF}$. Connect the points on the circle that intersect the lines.

35. *Sample answer:* Construct two sides of an equilateral triangle, as in Exercise 34.

37. a.

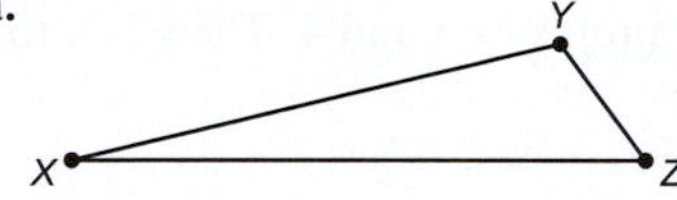

b. SSS Congruence Postulate **c.** Answers will vary.

Chapter 13 Review *(page 525)*

1. $\overline{KL}$ **3.** $\overline{JL}$ **5.** $\angle J$ **7.** 1.6 in. **9.** 3 in.

11. 119°

13. Congruent; *Sample answer:* SAS Congruence Postulate

15. Not congruent; *Sample answer:* Corresponding sides are not congruent.

17. Congruent; *Sample answer:* SSS Congruence Postulate

19. No; *Sample answer:* $\overline{QT}$ and $\overline{ST}$ may not be congruent.

21. Yes; *Sample answer:* Use the SSS Congruence Postulate.

23. *Sample answer:* Two pairs of corresponding sides are $\overline{VX}$, $\overline{ZX}$ and $\overline{WX}$, $\overline{YX}$.

25. 63° **27.** $a = 4$; $b = 3.5$

29. Not similar; *Sample answer:* Corresponding side lengths are not proportional.

31. Similar; *Sample answer:* AA Similarity Postulate

33. Not similar; *Sample answer:* Corresponding angles are not congruent.

35. 7 **37.** 4.2 **39.** 74° **41.** 554 ft

43. X Y

45.

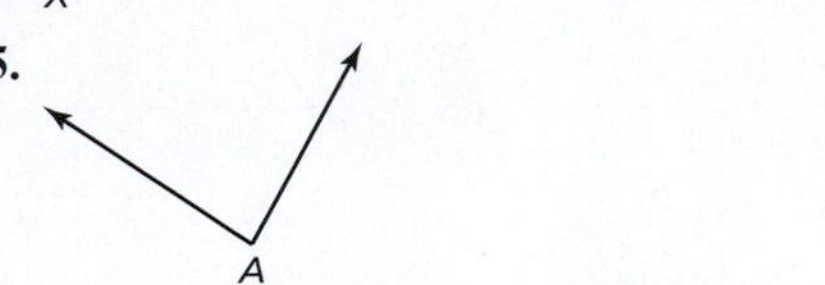

47. **49.**

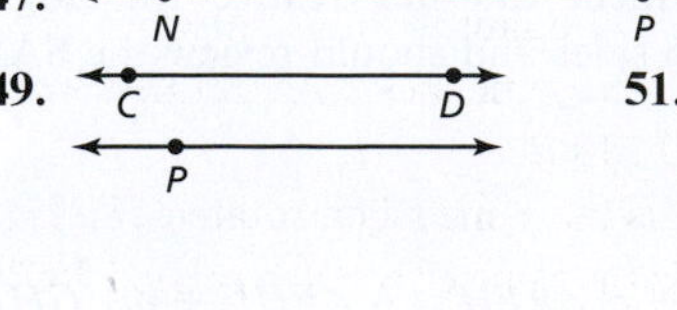

51.

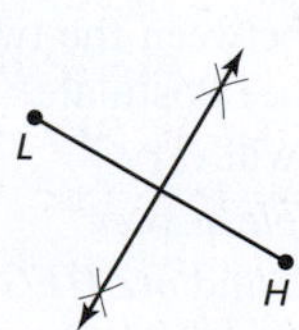

53.

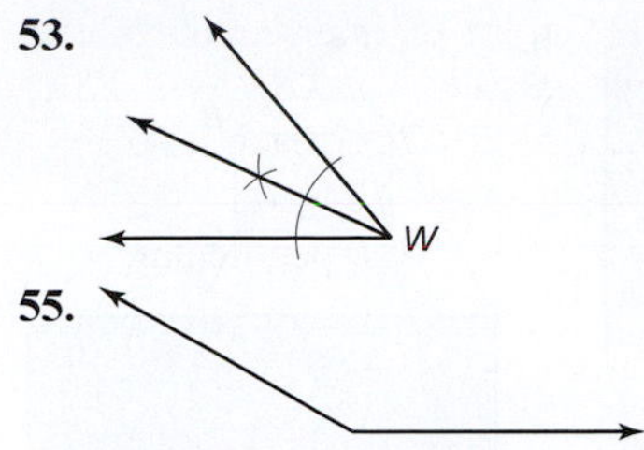

55.

Sample answer: Construct adjacent copies of the angle.

57.

Sample answer: Construct adjacent copies of the angles.

59.

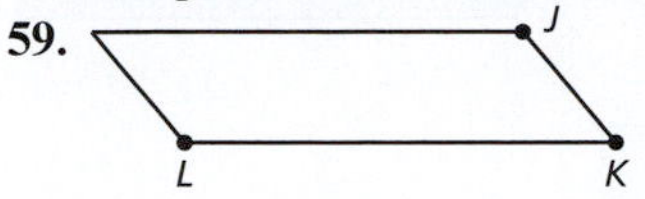

Chapter 13 Test *(page 528)*

1. B **2.** E **3.** C **4.** A **5.** D **6.** E **7.** A

CHAPTER 14

Section 14.1 *(page 536)*

1. Solution key:
  2. Step forward with left foot.
  4. Step back with left foot.
  6. Step back with right foot.
  8. Step forward with right foot.
  9. Translations; *Sample answer:* The steps are slides.
  10. Both; *Sample answer:* Some steps are slides and some are turns.

  Sample rubric: Assign 1 point to each correct response in problems 2, 4, 6, and 8. Assign 1 point to each correct answer and 2 points to each correct reasoning in problems 9 and 10. There is a total of 10 points.

3. a. ii **b.** iv **c.** i **d.** iii

5. a. Translation **b.** Rotation **c.** Reflection

7. a. Yes; *Sample answer:* The figure slides.
b. Yes; *Sample answer:* The figure slides.
c. No; *Sample answer:* The figure turns.
d. No; *Sample answer:* The figure turns.

9. 1

11. a. Yes; *Sample answer:* The figure turns.
b. No; *Sample answer:* The figure slides.
c. No; *Sample answer:* The figure flips.
d. Yes; *Sample answer:* The figure turns.

13. 120°

15. a. **b.**

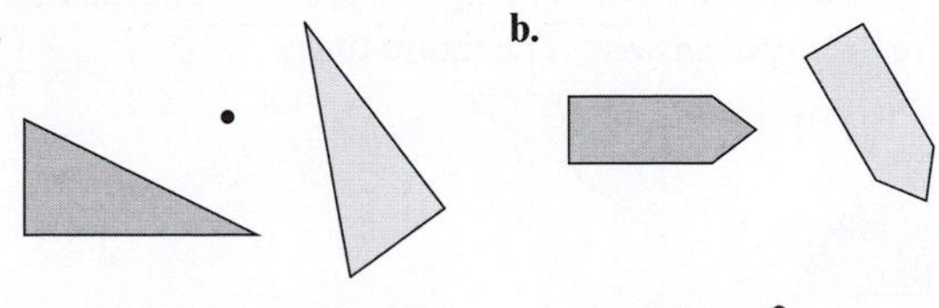

17. a. Yes; *Sample answer:* It is the same when rotated 90°.
b. No; *Sample answer:* It is the same only when rotated 360°.

19. a. 90°, 180° **b.** 60°, 120°, 180°

21. a. Yes; *Sample answer:* It is the same when rotated 180°.
b. No; *Sample answer:* It is the same only when rotated 360°.

23. a. True; *Sample answer:* Triangle A turns to form triangle B.
b. True; *Sample answer:* Triangle B turns to form triangle C.
c. True; *Sample answer:* Triangle A slides to form triangle B.
d. False; *Sample answer:* Triangle B turns to form triangle C.

25. Yes; *Sample answer:* A line of symmetry is a reflection; Z

27. *D*

29. *Sample answer:* The student misunderstood translations and rotations, and should review the definitions of translation and rotation.

31. Answers will vary

33. *Sample answer:*

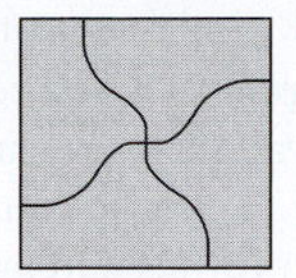

The tessellation involves rotations.

35. *Sample answer:* Square, hexagon, octagon; triangle, pentagon, heptagon

Section 14.2 *(page 546)*

1. Solution key:
  1. Symmetrical; real 2. Real; symmetrical
  3. Symmetrical; real 4. Real; symmetrical

  Assign 1 point to each correct label. There is a total of 8 points.

3. a. No; *Sample answer:* The figure turns.
b. Yes; *Sample answer:* The figure flips.
c. Yes; *Sample answer:* The figure flips.
d. No; *Sample answer:* The figure slides.

5. a.

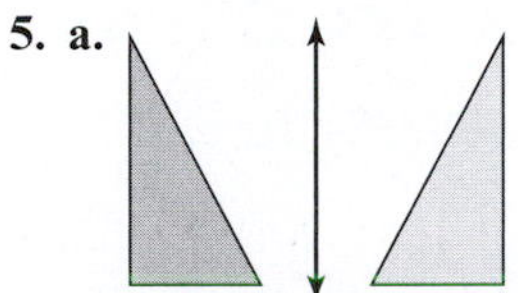

b.

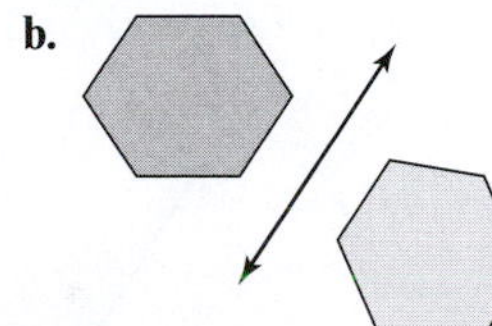

7. a. Asymmetrical **b.** Vertical

9. a. 6 **b.** 1

11. a.

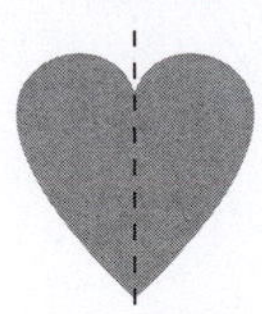

b.

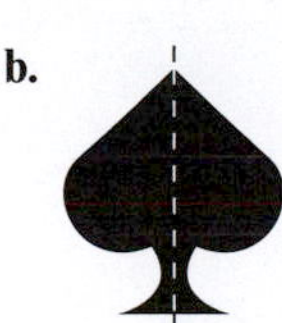

13. a. Yes; *Sample answer:* Slide down, reflect across
b. Yes; *Sample answer:* Slide right, reflect down

15. a. *Sample answer:* Reflections
b. *Sample answer:* Translations

17. WOW; MOM; yes; vertical line through O

19. Yes

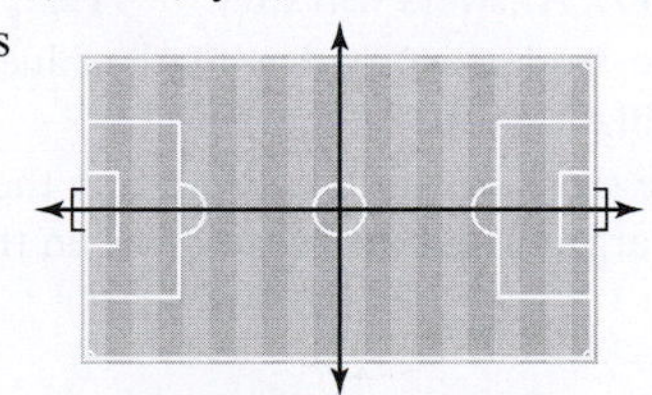

21. Answers will vary. **23.** Answers will vary.

25. *Sample answer:* The student incorrectly described the transformation, and should review glide reflections.

27. Translation

29. 4; 4; *Sample answer:* They are equal; there are infinitely many lines.

31. a. *Sample answer:* 120° rotation
b. *Sample answer:* Translate right, rotate 180°

33. Asymmetrical

35. a. 36 in.; *Sample answer:* $AB = BC$ and $AD = DC$.

b. 48 in.^{2}; *Sample answer:* $AB = BC$, and use the Pythagorean Theorem to find DB.

37. Translations; reflections; *Sample answer:* B is a line of reflection.

Section 14.3 *(page 556)*

1. Solution key: 1–4. Check students' work.
Sample rubric: Assign 1 point to each correct measurement, 2 points to each correct wall scale drawing, and 2 points to a correct amount of paint. The total number of points will vary based on the number of measurements.

3. a. Reduction; *Sample answer:* It is smaller.

b. Enlargement; *Sample answer:* It is larger.

5. a. Enlargement; $\frac{7}{3}$ **b.** Reduction; $\frac{3}{8}$

c. Reduction; $\frac{1}{2}$ **d.** Enlargement; $\frac{3}{2}$

7. a. 25 **b.** 56

9. a. $\frac{x}{9} = \frac{35}{15}$; 21; reduction **b.** $\frac{n}{12} = \frac{14}{28}$; 6; reduction

11. a.

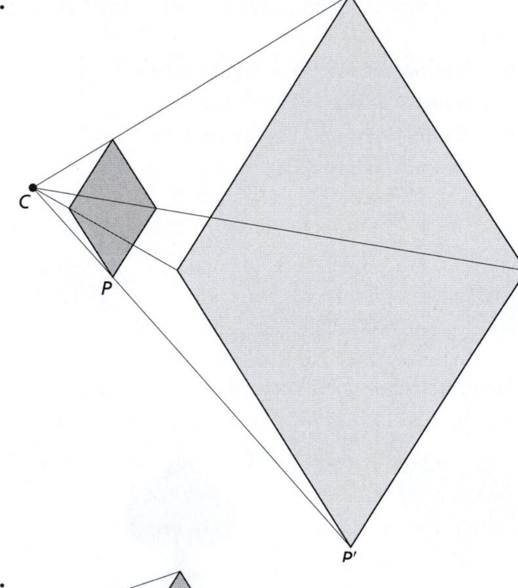

b.

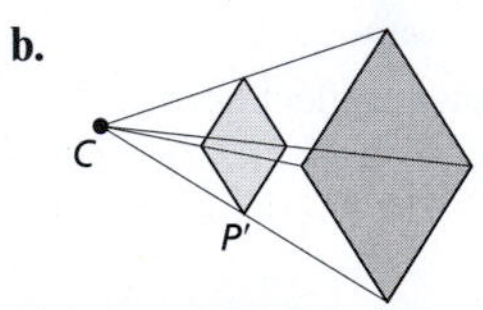

13. 18 ft **15.** 3 **17.** Answers will vary. **19.** $\frac{3}{8}$

21. *Sample answer:* The student started with the blue figure, and should review dilations.

23. *Sample answer:* The scale has different units, so the actual object is 120 times larger; Use a scale factor when the units are the same.

25. a.

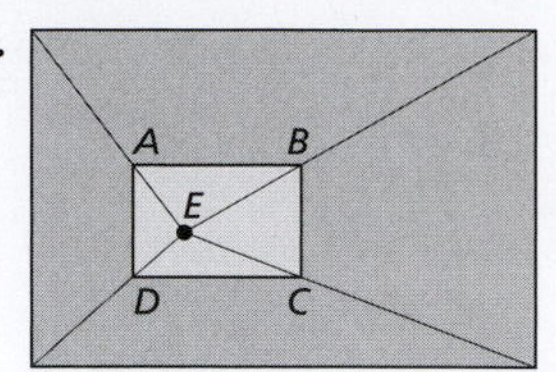

b.

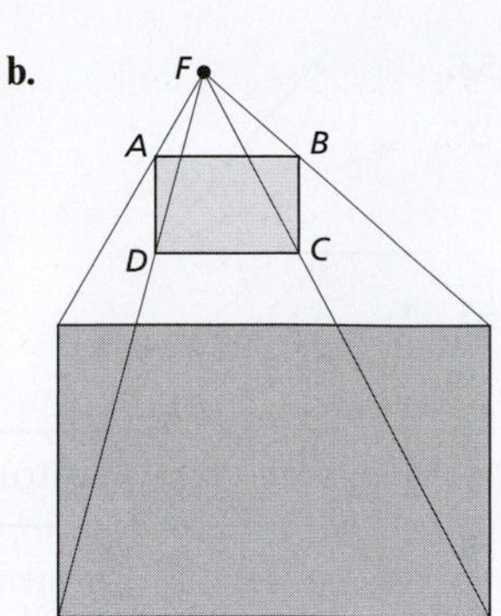

c.

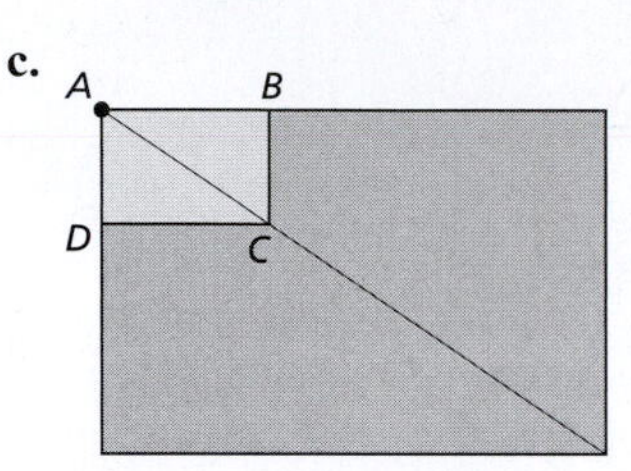

d. *Sample answer:* The images are translations of each other.

27. a.

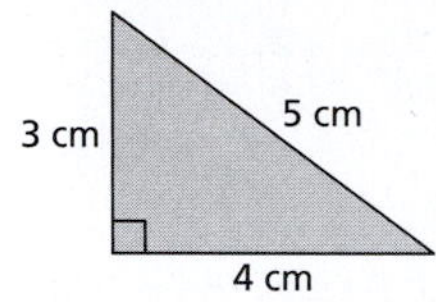

b. 24 cm; 12 cm; They are equal.

c. 24 cm^2; 6 cm^2; The ratio is the square of the scale factor.

29. a.

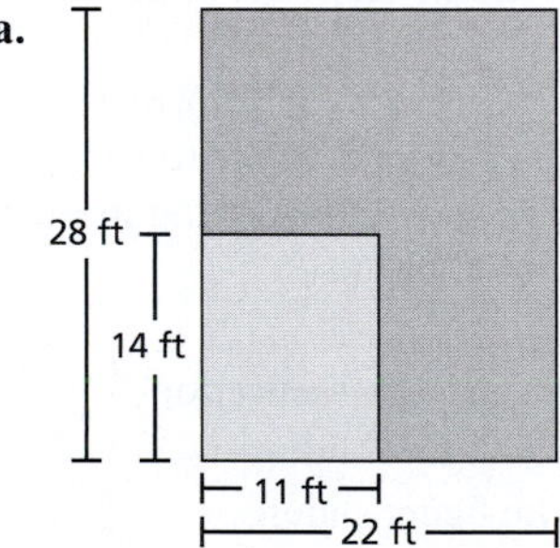

b. 50 ft **c.** 3 h

31. 32 in.^{2}

Chapter 14 Review *(page 561)*

1. No; *Sample answer:* The figure turns.

3. Yes; *Sample answer:* The figure slides.

5. 8

7. No; *Sample answer:* The figures are not congruent.

9. Yes; *Sample answer:* The figure turns.

11.

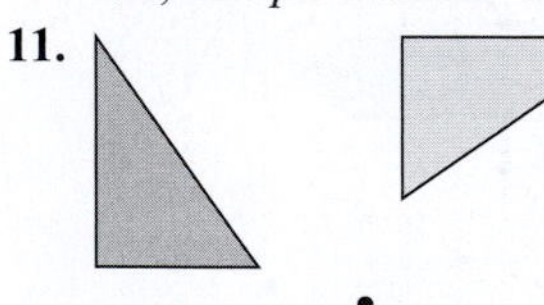

13. Yes; *Sample answer:* It is the same when rotated 90°.

15. No; *Sample answer:* It is the same only when rotated 360°.

17. True; *Sample answer:* Parallelogram B turns to form parallelogram C.

19. False; *Sample answer:* Parallelogram B turns to form parallelogram C.

21. No; *Sample answer:* The figure turns.

23. Yes; *Sample answer:* The figure flips.

25.

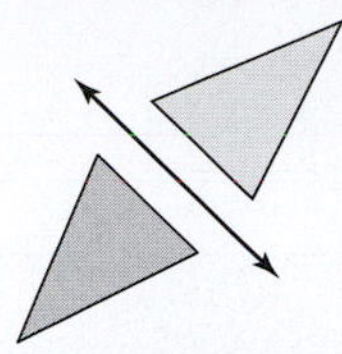

27. Asymmetrical **29.** Horizontal, vertical **31.** 2

33.

35. Yes; *Sample answer:* Slide right, reflect down **37.** 8

39. Reduction; *Sample answer:* It is smaller.

41. Reduction; $\frac{2}{5}$ **43.** $\frac{m}{2} = \frac{20}{8}$; 5; reduction

45. 8 ft **47.** 10 in. by 7 in.

49. a.

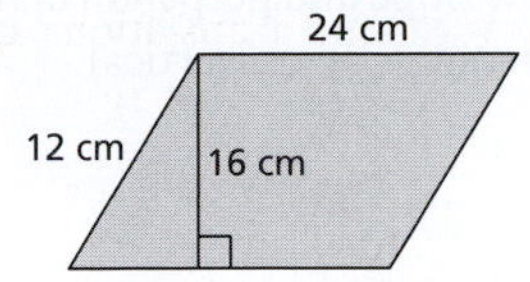

b. 18 cm; 72 cm; They are the same.

c. 24 cm^2; 384 cm^2; The ratio is the square of the scale factor.

Chapter 14 Test *(page 564)*

1. C **2.** C **3.** D **4.** B **5.** E **6.** D **7.** C

CHAPTER 15

Section 15.1 *(page 574)*

1. Solution key:

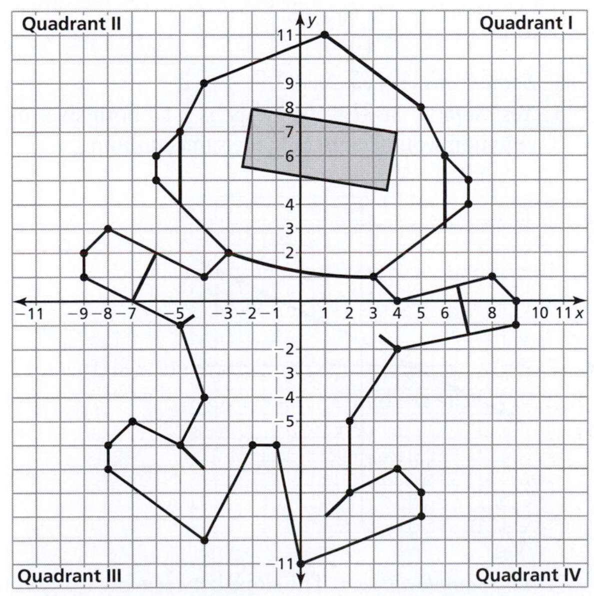

Sample rubric: Assign 1 point to each correctly plotted point and 5 points to correctly connecting the points. There is a total of 40 points.

3.

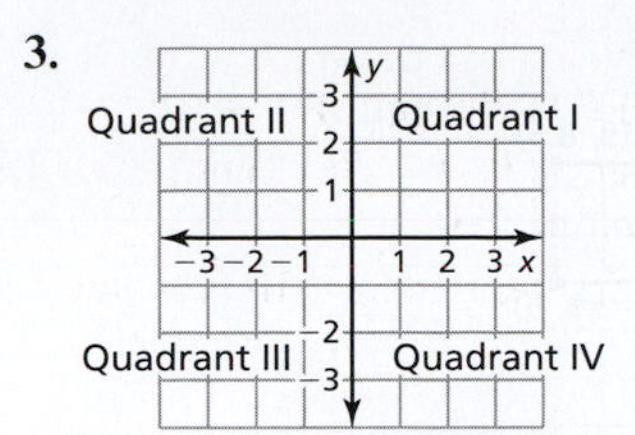

5.

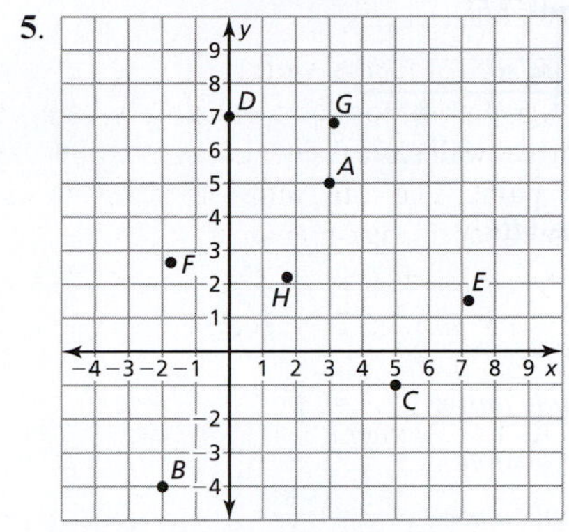

7. a. $(-5, 6)$ **b.** $(-4, -3)$ **c.** $(0, -8)$ **d.** $(2, 2)$

9. a. $(-3, 6), (0, 7), (3, 1)$ **b.** $(-7, 2), (-5, -2), (-2, 2)$
c. $(4, -2), (4, -6), (7, -6)$ **d.** $(1, -1), (6, -4), (11, 3)$

11. a. $(2, -8), (5, -3), (-2, 2)$ **b.** $(3, 2), (3, 5), (7, 4)$
c. $(-2, 2), (-4, 7), (-7, 3)$
d. $(-1, -3), (-1, -1), (-3, -1)$

13. a $(6, -5), (3, -1), (3, -6)$
b. $(-1, -1), (-1, -4), (-3, -1)$
c. $(-3, 4), (-7, 2), (-4, 7)$ **d.** $(4, 2), (4, 5), (0, 5)$

15. a. $(-1, 5), (-5, 1), (-9, 2)$ **b.** $(7, 1), (3, 0), (3, 2)$
c. $(1, -6), (0, -9), (-2, -8)$ **d.** $(-2, 9), (-3, 6), (-4, 9)$

17. a. 10 **b.** About 9.49 **c.** 17 **d.** 4

19. a. 12 **b.** 24 **c.** 15 **d.** 20

21. a. About 3.61 **b.** About 13.15 **c.** About 5.39 **d.** 5

23. a. $(3, 4)$ **b.** $(1.5, 5.5)$ **c.** $(2, -1)$ **d.** $(-2.5, 1.5)$

25. a. Yes; *Sample answer:* $(\sqrt{160})^2 + (\sqrt{40})^2 = (\sqrt{200})^2$
b. No; *Sample answer:* $(\sqrt{32})^2 + (\sqrt{52})^2 \neq (\sqrt{68})^2$

27. $\frac{1}{2}(a + b)(a + b) = \frac{1}{2}ab + \frac{1}{2}ab + \frac{1}{2}c^2$ implies $a^2 + b^2 = c^2$.

29. a. Reflection in the y-axis
b. 180° rotation about the origin **c.** Answers will vary.

31. *Sample answer:* The student found both square roots of 25, and should review using the Pythagorean Theorem.

33. About 66.78 ft

35. *Sample answer:* Red: translate 4 units down; purple: translate 4 units down and 1 unit left; blue: rotate 90° clockwise about the origin, then translate 11 units up and 2 units right; green: reflect in x-axis, then translate 10 units up and 1 unit right; yellow: reflect in y-axis, then translate 6 units down and 17 units right

37. 17 cm

39. a. About 56.6 in.

b.

(5, 8.5) (9, 8.5) (5, 6) (9, 6) (0, 2.5) (4, 2.5) (0, 0) (4, 0)

Section 15.2 *(page 584)*

1. Solution key:

1. $\frac{2}{4}; \frac{1}{2}$; Yes; *Sample answer:* $\frac{2}{4} = \frac{1}{2}$
2. $\frac{-2}{2}; \frac{-6}{6}$; Yes; *Sample answer:* $\frac{-2}{2} = \frac{-6}{6}$
3. $\frac{6}{9}; \frac{2}{3}$; Yes; *Sample answer:* $\frac{6}{9} = \frac{2}{3}$
4. $\frac{-6}{2}; \frac{-6}{2}$; Yes; *Sample answer:* $\frac{-6}{2} = \frac{-6}{2}$

Sample rubric: Assign 1 point to each correct slope and 2 points to each correct reasoning. There is a total of 16 points.

3. a. Positive **b.** Zero **5. a.** Undefined **b.** $\frac{1}{2}$

7. a. -12 **b.** Undefined **9.** Undefined

11. a.

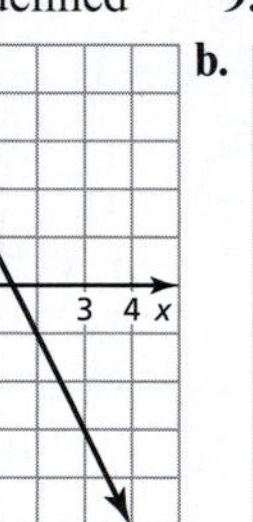

b.

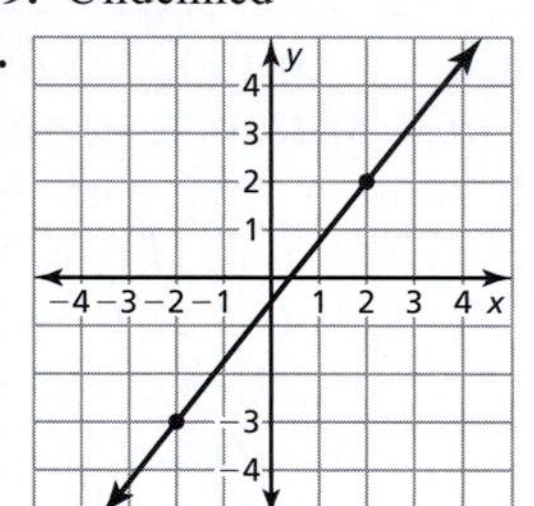

13. a. No **b.** Yes

15. a. 20

b. *Sample answer:* The ship travels 20 nautical miles each hour.

17. a. 1 **b.** -1

19. *A* and *C*; *Sample answer:* The slopes are equal.

21. Yes

23. *A* and *B*; *Sample answer:* The slopes are negative reciprocals of each other.

25. No

27.

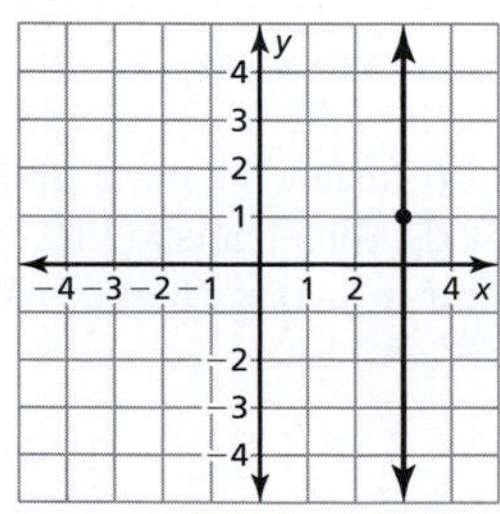

29.

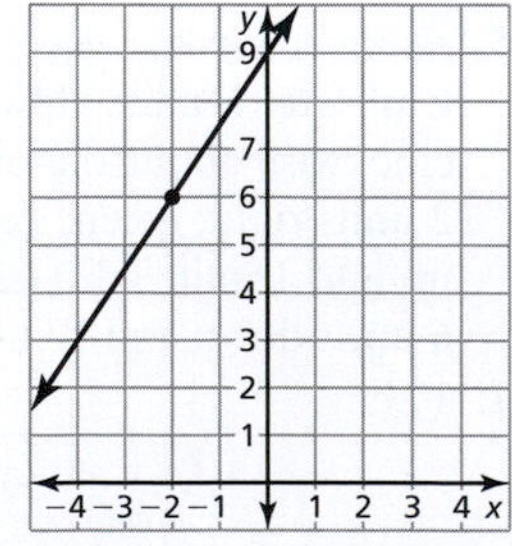

31. Parallelogram; *Sample answer:* Opposite sides are parallel.

33. a.

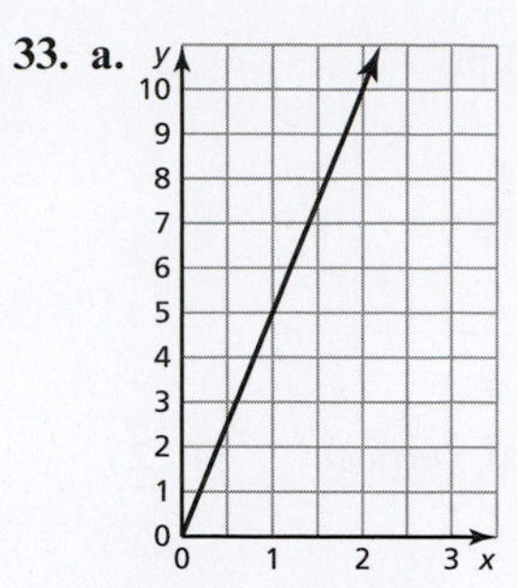

b. 5 **c.** 24 min

35. Points will vary. Yes; *Sample answer:* The sign changed in both differences.

37. Yes; *Sample answer:* The slopes between the points are equal.

39. a. *Sample answer:* (2, 3), $(-2, -5)$

b. *Sample answer:* (1, 2), (1, 0)

41. Perpendicular **43.** $\frac{1}{2}$

45. *Sample answer:* The student did not use the negative reciprocal, and should review slope and perpendicular lines.

47. *Sample answer:* One is horizontal, one is vertical.

Section 15.3 *(page 594)*

1. Solution key:

1. 3; -1 2. (4, 3), $(-4, -1)$

3.

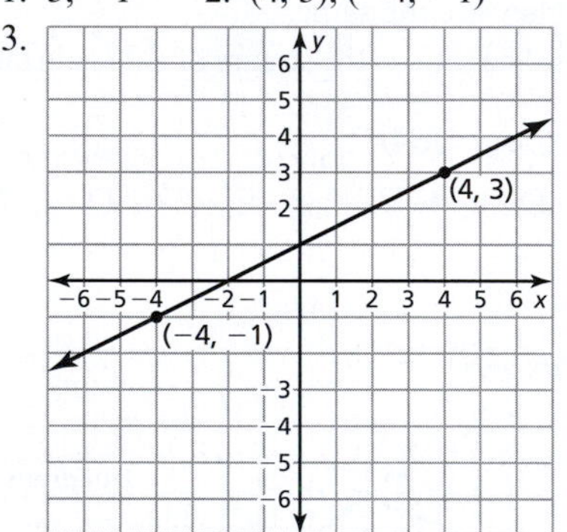

4. *Sample answer:* (2, 2); $\frac{1}{2}(2) + 1 = 2$
5. Yes; *Sample answer:* The slope is $\frac{1}{2}$, so both sides of the equation would change by the same amount.
6. -2; 0; 1; 1.75; 3.5

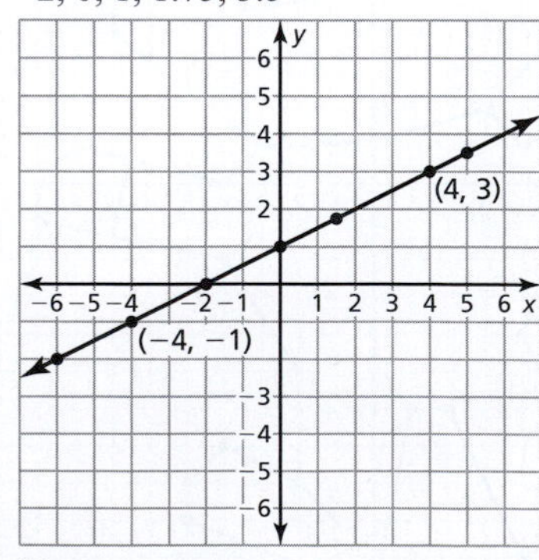

Yes

7. Yes; *Sample answer:* The slope between any two solution points is $\frac{1}{2}$.

Sample rubric: Assign 1 point to each correct *y*-value, ordered pair, plotted point, line, or check and 2 points to each correct reasoning. There is a total of 22 points.

3.

x	$y = 4x - 5$	y	(x, y)
-5	$y = 4(-5) - 5$	-25	$(-5, -25)$
-3	$y = 4(-3) - 5$	-17	$(-3, -17)$
1	$y = 4(1) - 5$	-1	$(1, -1)$
4	$y = 4(4) - 5$	11	$(4, 11)$

5. a. 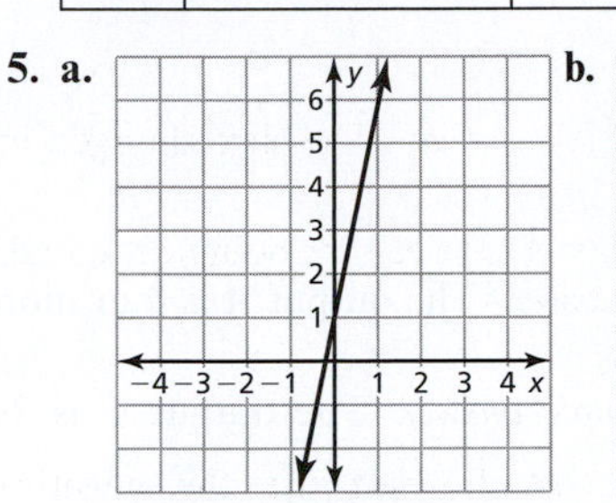 **b.** 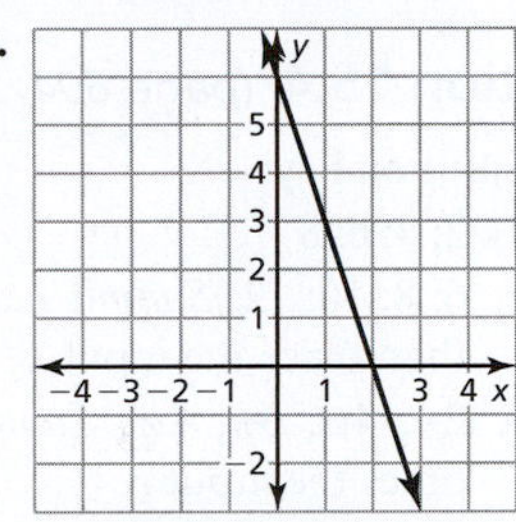

c. 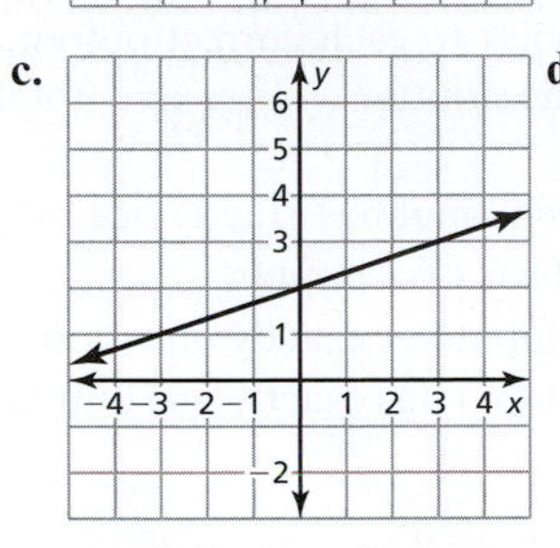 **d.** 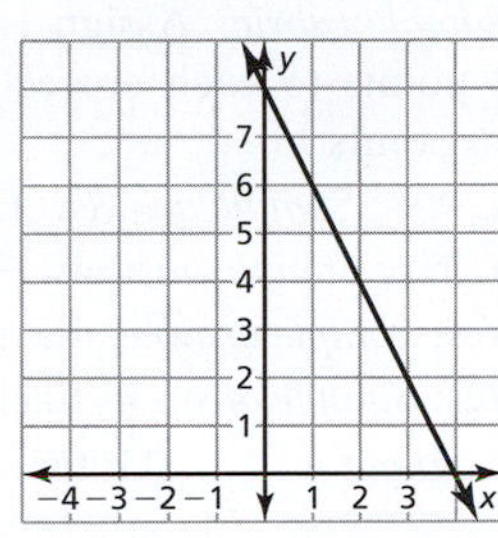

7. a. Vertical; *Sample answer:* x is constant.
b. Neither; *Sample answer:* Neither x nor y is constant.
c. Horizontal; *Sample answer:* y is constant.
d. Vertical; *Sample answer:* x is constant.

9. a. 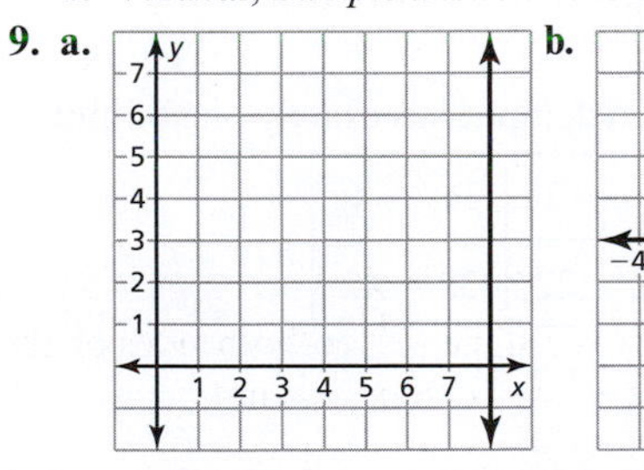 **b.** 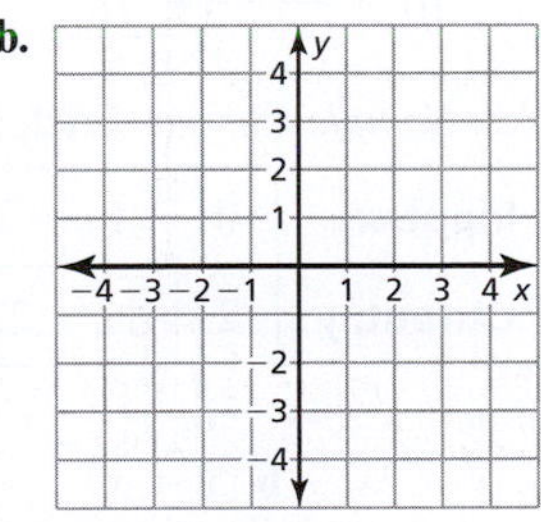

c. 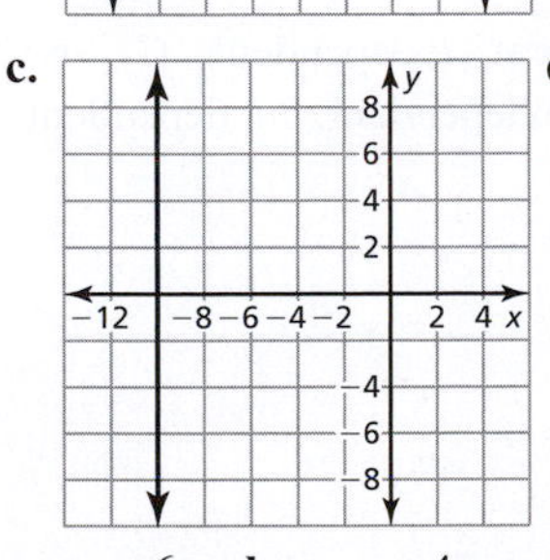 **d.** 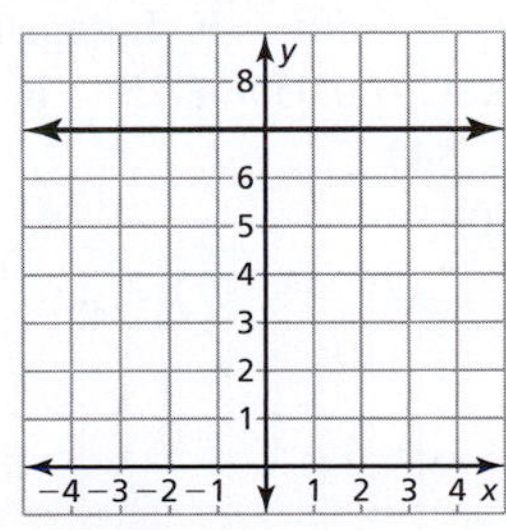

11. a. $y = 6$ **b.** $x = -4$

13. a. 7 **b.** -13 **c.** 3 **d.** 0

15. a. x-intercept: $-\frac{1}{2}$; y-intercept: 4
b. x-intercept: 0; y-intercept: 0
c. x-intercept: -6; y-intercept: 3
d. x-intercept: 7; y-intercept: none

17. a. 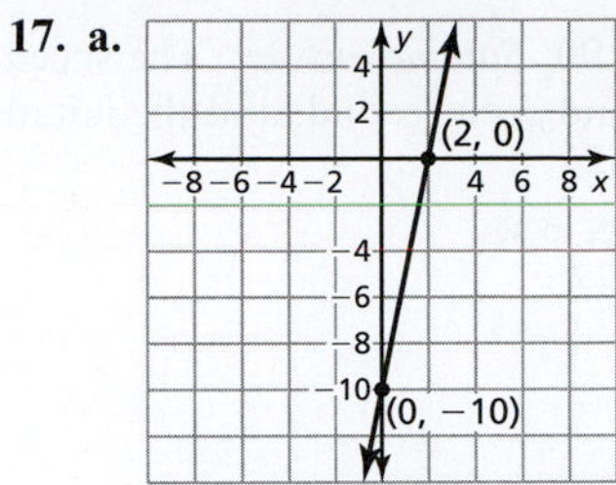

b. 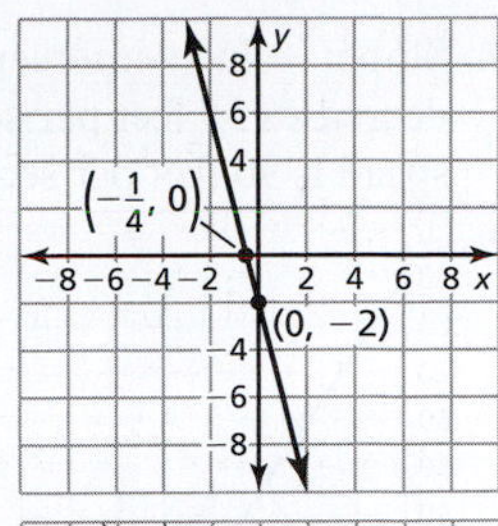

c. 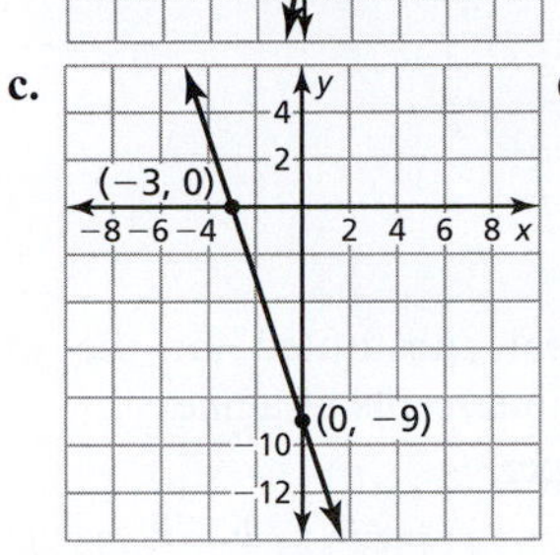

d. 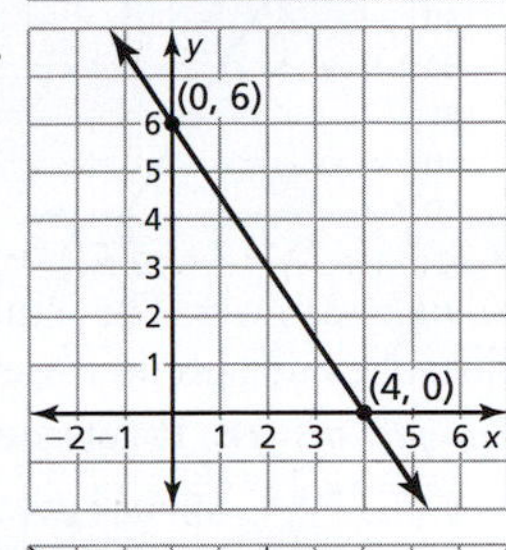

19. a. 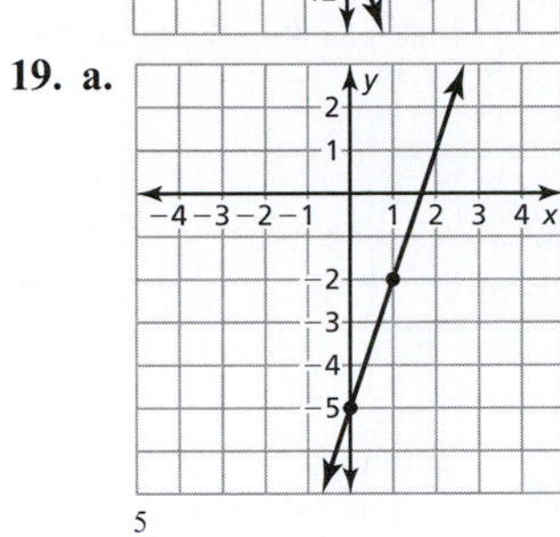
$\frac{5}{3}$

b. 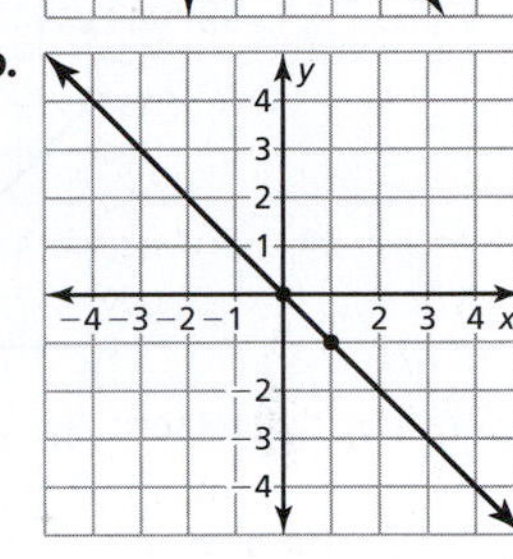
0

c. 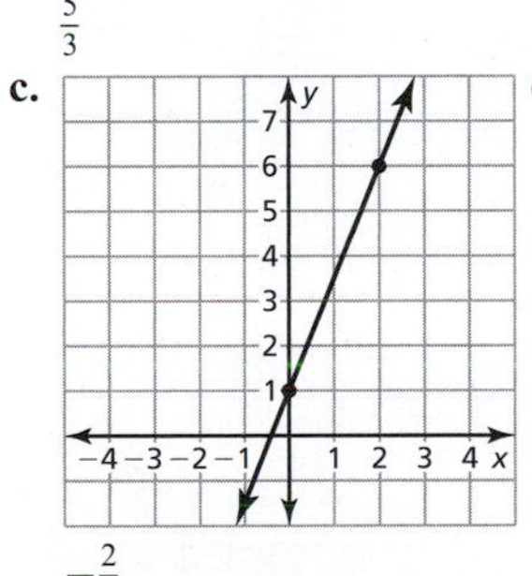
$-\frac{2}{5}$

d. 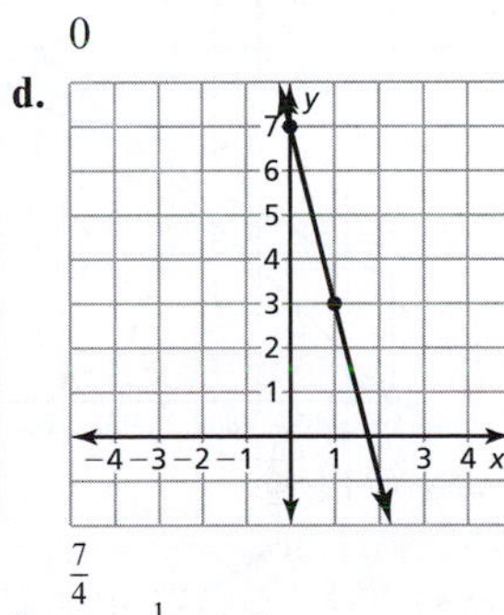
$\frac{7}{4}$

21. a. $y = -6x - 8$
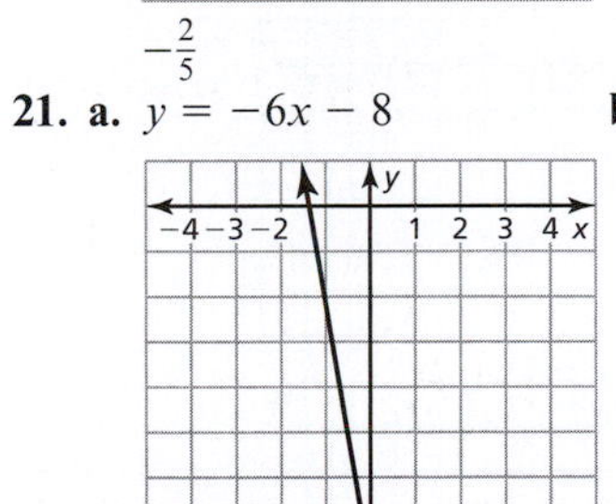

b. $y = \frac{1}{5}x + 2$
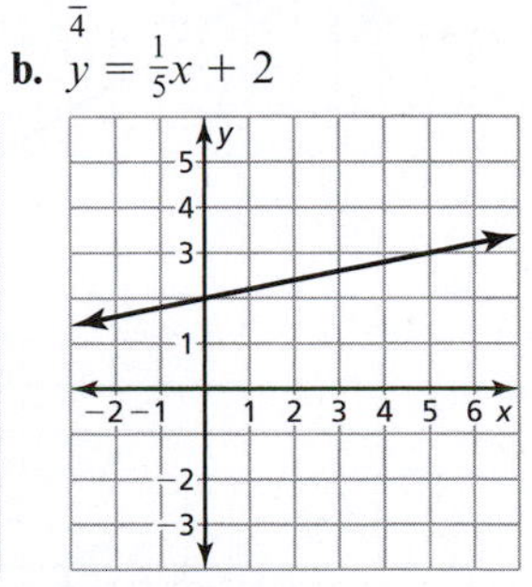

c. $y = -3x - 9$
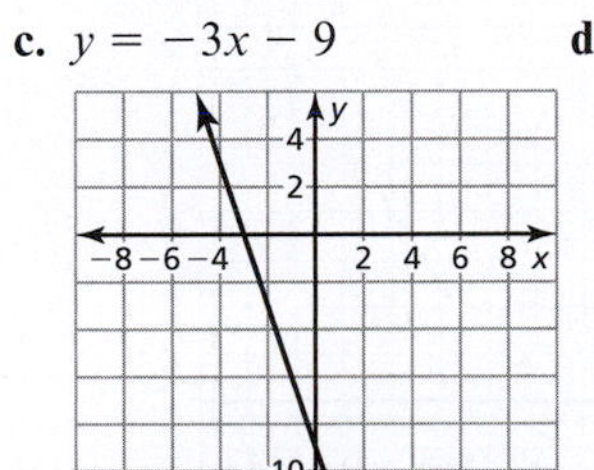

d. $y = -4x - 7$
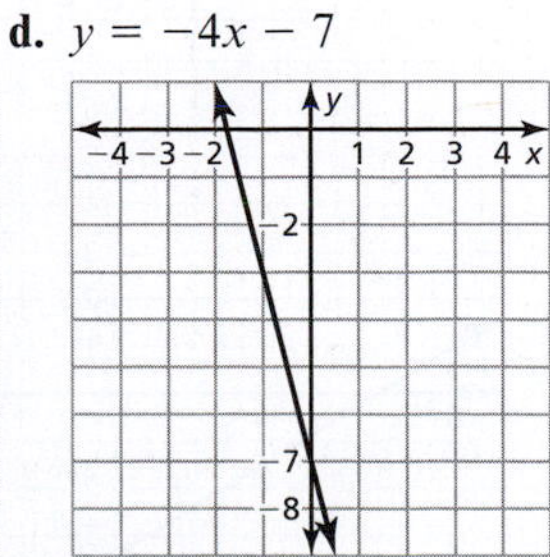

23. a. $y = \frac{3}{2}x + 3$ **b.** $y = -\frac{1}{2}x + 5$

25. $x = 2$; $y = -\frac{1}{4}x + \frac{5}{2}$; $y = -\frac{5}{4}x + \frac{1}{2}$

27. a. Slope: -15; y-intercept: 90; *Sample answer:* The speed decreases 15 feet per second each second, and the initial speed is 90 feet per second.

b.

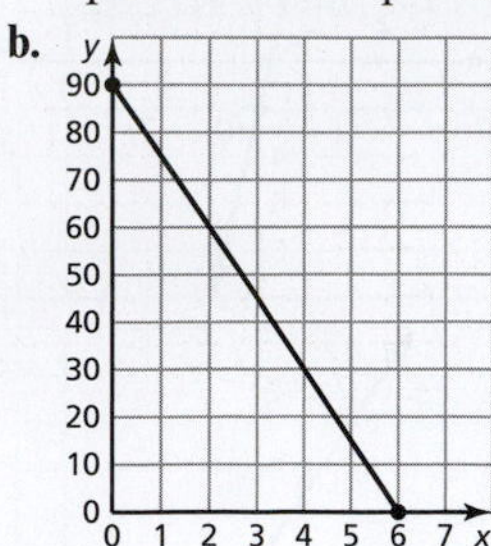

c. 6 sec

29. *Sample answer:* The student substituted for the wrong variable, and should review intercepts of a line.

31. *Sample answer:* It gets steeper.

33. a. $y = 250 - 2x$

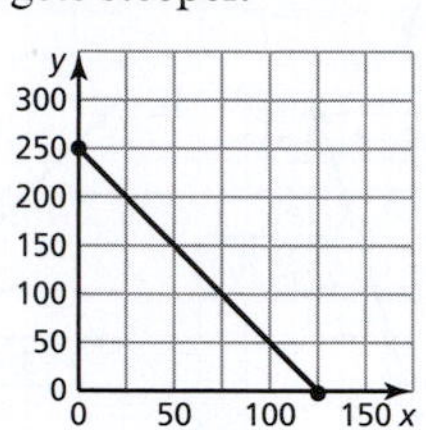

b. 70

35. a. $y_1 = 0.5x + 550$ **b.** $y_2 = x$

c.

d. 1100

37. $y = x + 14$

39. a.

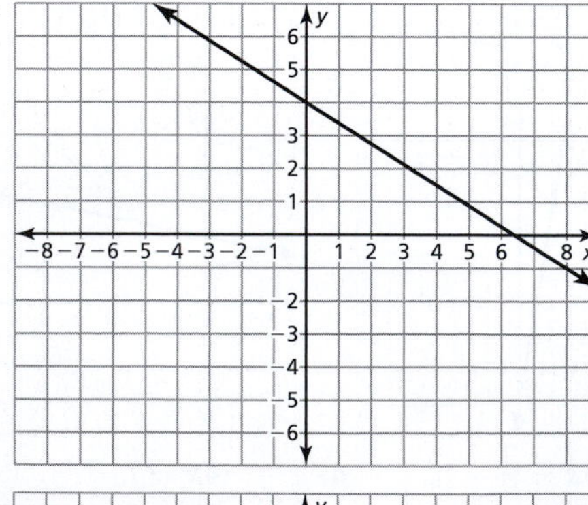

b.

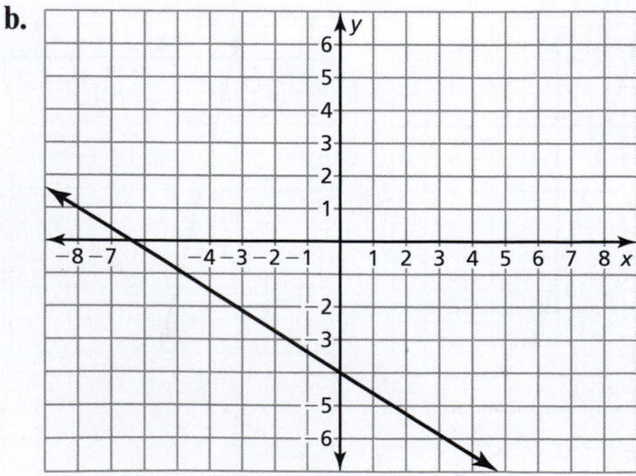

41. $y - 7 = \frac{3}{4}(x - 12)$ **43.** $y = 4$

45. a. Blue: $\frac{1}{2}$; red: $\frac{1}{2}$; green: $\frac{1}{2}$

b. Blue: 3; red: 0; green: -2

c. Blue: $y = \frac{1}{2}x + 3$; red: $y = \frac{1}{2}x$; green: $y = \frac{1}{2}x - 2$

d. Same slope

e. Blue: slope is 1, y-intercept is 2, equation is $y = x + 2$; red: slope is -4, y-intercept is 2, equation is $y = -4x + 2$; green: slope is $-\frac{1}{4}$, y-intercept is 2, equation is $y = -\frac{1}{4}x + 2$; same y-intercept

f. Answers will vary.

Section 15.4 *(page 604)*

1. Solution key:

1. 2; 4; 6; 8
2. 6; 8; 10; 12; *Sample answer:* The output A is four more than twice the input x.
3. 2π; 4π; 6π; 8π; *Sample answer:* The output C is 2π times the input r.

Sample rubric: Assign 1 point to each correct output and 2 points to each correct description. There is a total of 16 points.

3. a. Yes; *Sample answer:* Each input has exactly one output.

b. No; *Sample answer:* -2 has two outputs.

5. Yes; *Sample answer:* Each input has exactly one output.

7. Yes; *Sample answer:* Each input has exactly one output.

9.

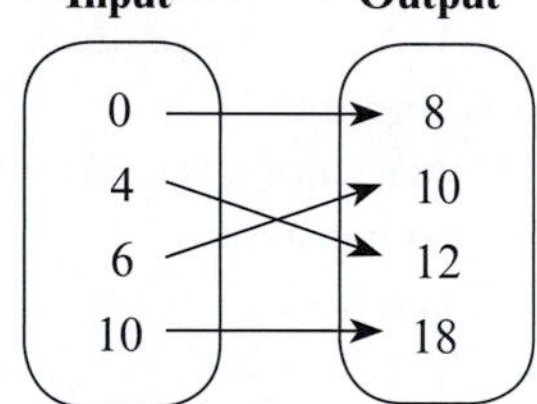

Yes; *Sample answer:* Each input has exactly one output.

11.

Input, x	0	1	2	3	4
Output, y	2	12	22	32	42

13. a. $y = x + 4$ **b.** $y = 2x + 3$

15. a. $y = 2x$ **b.** $y = x^2$

17. a. $C = 50t$ **b.** Independent: t; dependent: C **c.** 750

19. a. $C = 0.05m + 25$ **b.** Independent: m; dependent: C

c. \$40

21. 20

23.

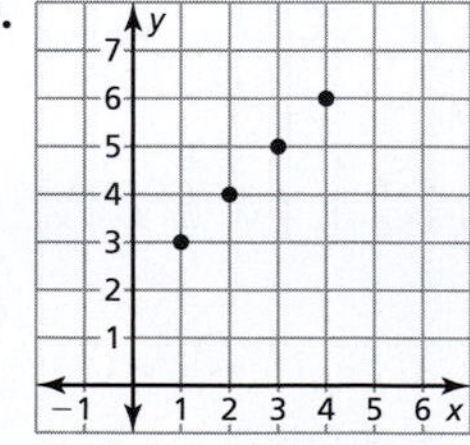

25.

27. Yes; *Sample answer:* $\frac{2}{3}(6) + 6 = 10$

29. Yes; *Sample answer:* (6, 5) is on the line.

31. a. Yes; *Sample answer:* The graph is a straight line.
b. No; *Sample answer:* The graph is not a straight line.

33. a. Yes; *Sample answer:* The graph is a straight line.
b. No; *Sample answer:* The graph is not a straight line.

35. No; *Sample answer:* The points do not lie on a straight line.

37. a. Answers will vary. **b.** Answers will vary.

39. *Sample answer:* Height of a plant over time

41. *Sample answer:* The student thought outputs needed to be unique, and should review the definition of a function.

43. a. $R = 110n$ **b.** $C = 25n + 750$
c. $P = 85n - 750$; \$3500 **d.** $C = 29n + 750$ **e.** 10

45. a. Schedule 1: $C = 10m + 50$; schedule 2: $C = 5m + 100$
b.

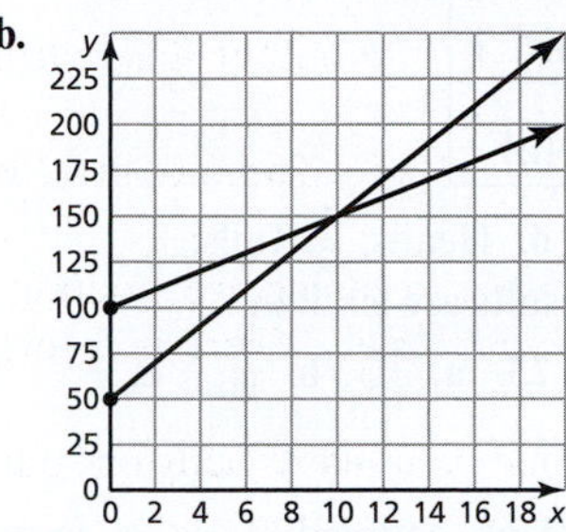

c. When $m < 10$; when $m > 10$; *Sample answer:* At $m = 10$, the schedules cost the same.

Chapter 15 Review *(page 609)*

1–5.

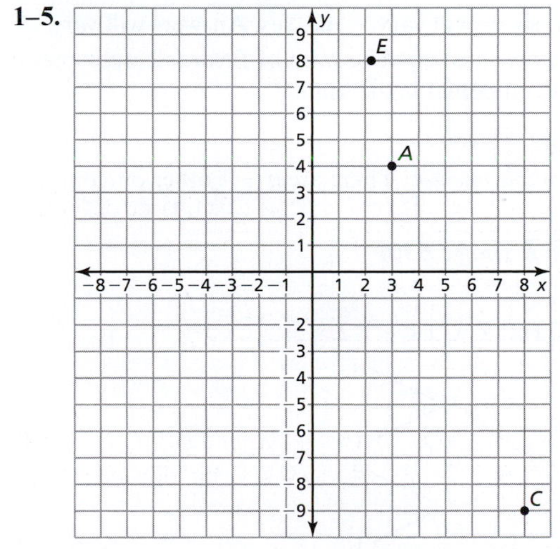

7. (2, 1) **9.** (−4.5, 4.5) **11.** (0, −1.5)

13. (6, 2), (5, 5), (2, 1) **15.** (5, −1), (1, −2), (5, −7)

17. (−5, 2), (−7, −2), (−1, −1) **19.** About 8.06

21. 20 **23.** About 5.39; (0.5, 2) **25.** 5; (1.5, 2)

27. Positive **29.** Zero **31. a.** 0 **b.** −1

33. $\frac{3}{4}$ **35.** Undefined

37.

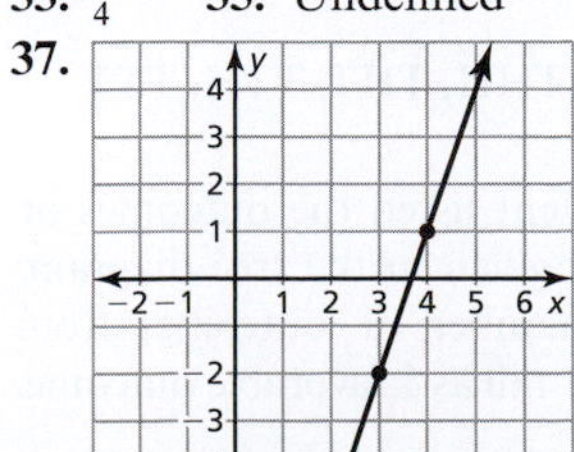

39.

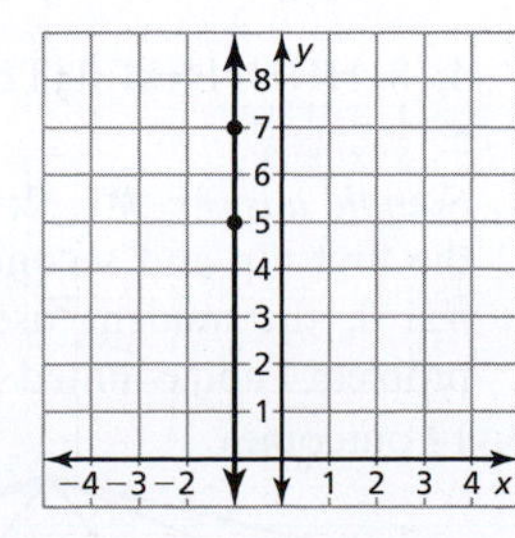

41. a. 70 **b.** *Sample answer:* You travel 70 miles each hour.

43. $7; -\frac{1}{7}$ **45.** $-\frac{2}{3}; \frac{3}{2}$

47.

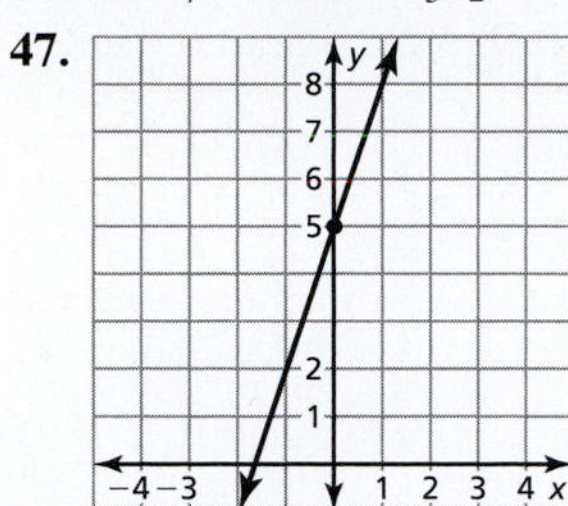

49.

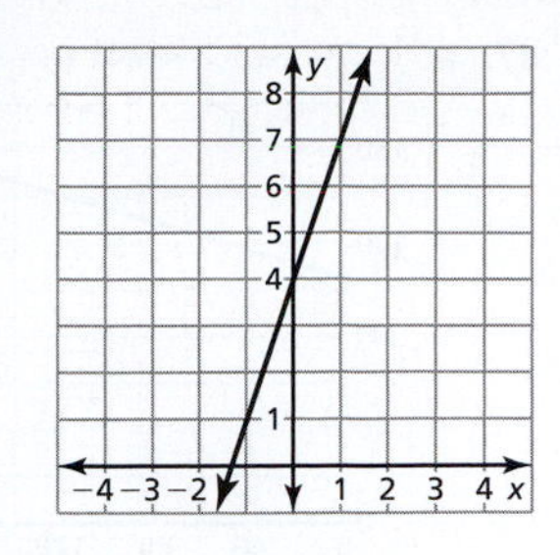

51.

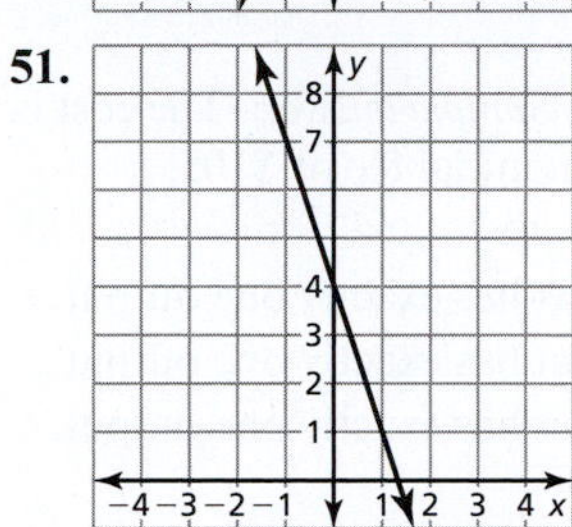

53. Vertical; *Sample answer:* x is constant.

55. Horizontal; *Sample answer:* y is constant.

57. Neither; *Sample answer:* Neither x nor y is constant.

59. $y = 6$ **61.** $x = 2$ **63.** 3 **65.** $-\frac{1}{2}$ **67.** $-\frac{4}{3}$

69.

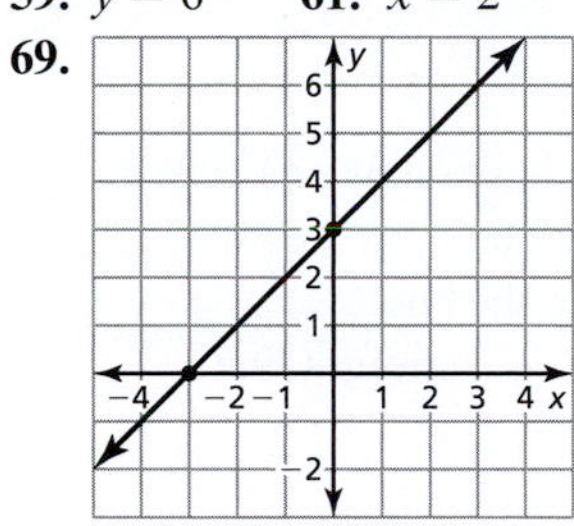

71.

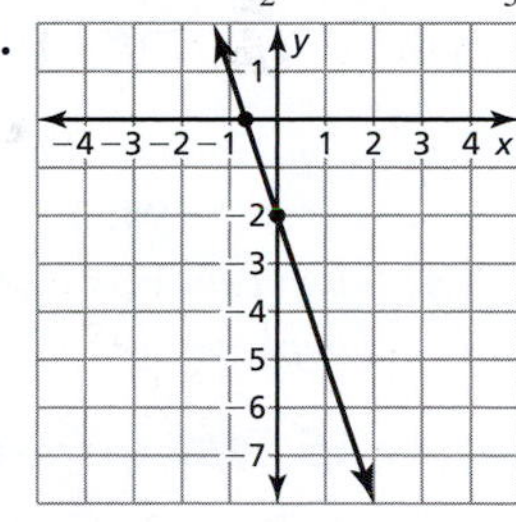

73.

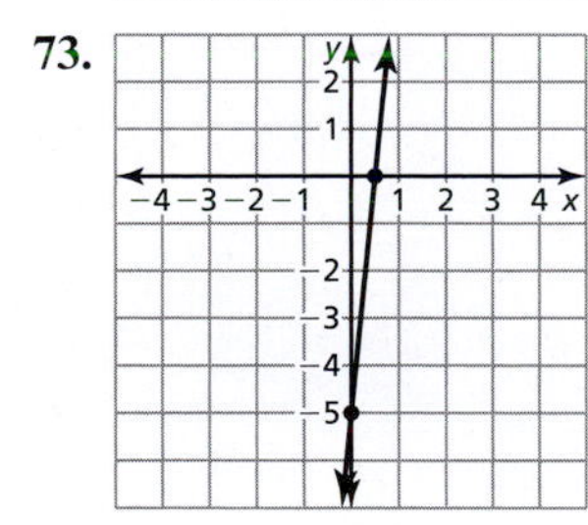

75.

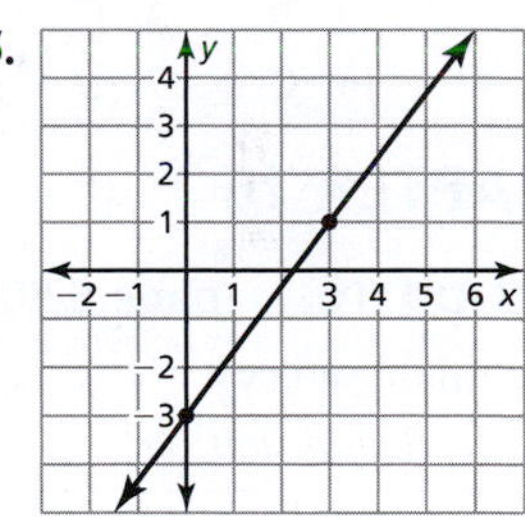

$\frac{9}{4}$

77.

79.

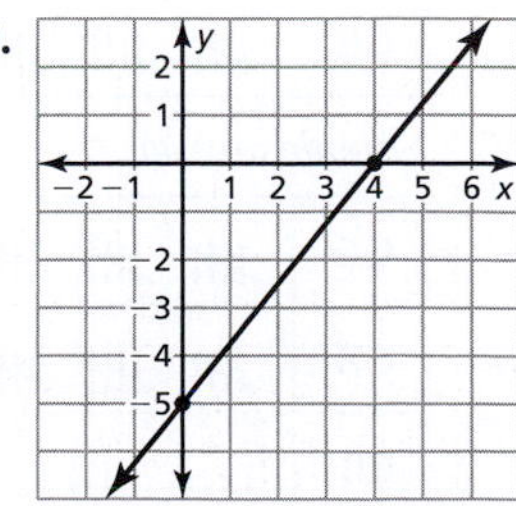

8

4

81. $y = -\frac{2}{3}x + 1$

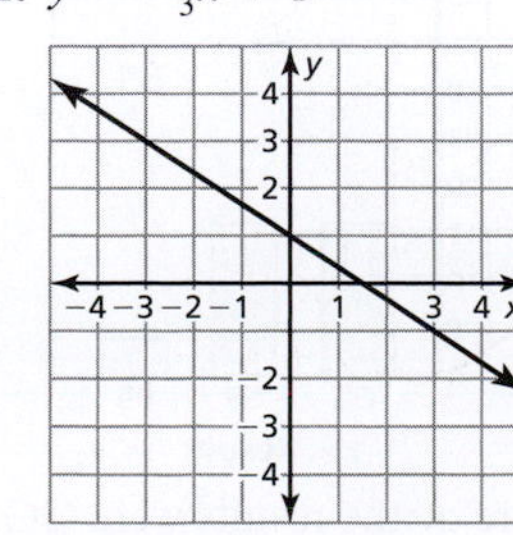

83. $y = \frac{2}{3}x + 5$

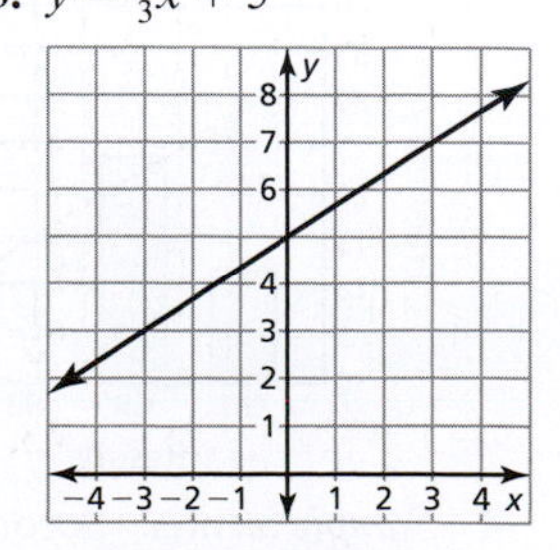

85. $y = -\frac{1}{3}x + 2$

87. a.

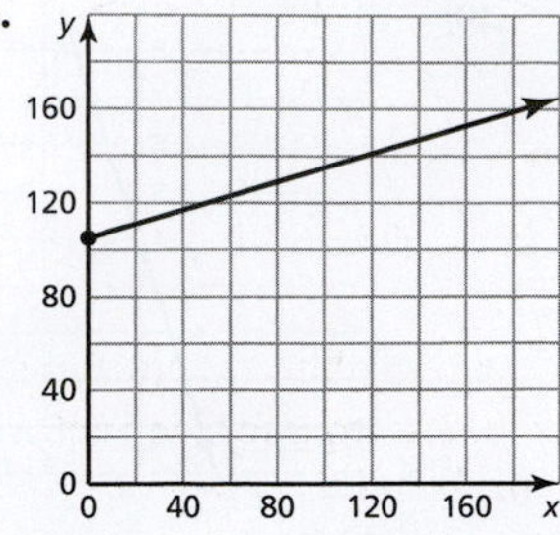

Slope: 0.3; y-intercept: 105; *Sample answer:* The cost is \$0.30 per mile and there is an initial fee of \$105.

b. \$144

89. Yes; *Sample answer:* Each input has exactly one output.

91. Yes; *Sample answer:* Each input has exactly one output.

93. Yes; *Sample answer:* Each input has exactly one output.

95.

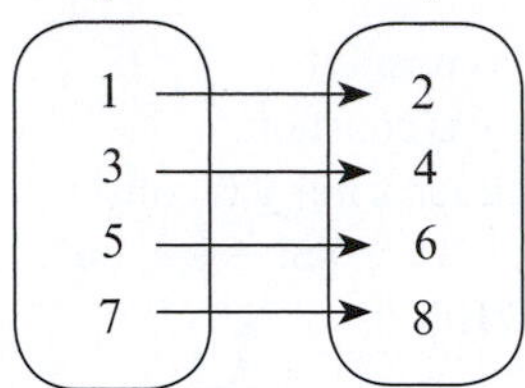

Yes; *Sample answer:* Each input has exactly one output.

97. $y = \frac{1}{2}x + 3$ **99.** $y = -2x + 10$

101. Yes; *Sample answer:* (2, 4) is on the line.

103. Yes; *Sample answer:* The graph is a straight line.

Chapter 15 Test *(page 612)*

1. A **2.** B **3.** C **4.** A

5. C **6.** C **7.** B **8.** E

CHAPTER 16

Section 16.1 *(page 620)*

1. Solution key:

1. *Sample answer:*

GG	𝍸 𝍸 II
GB	𝍸 𝍸 𝍸 𝍸 IIII

2. *Sample answer:*

GG	𝍸 𝍸 𝍸 I
GB	𝍸 𝍸 𝍸 I
BB	IIII

3.

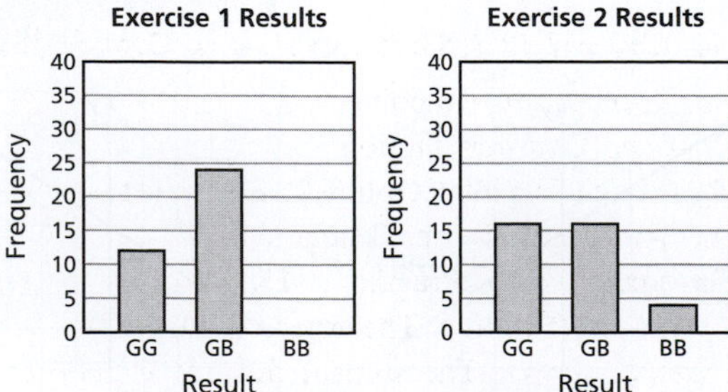

4. *Sample answer:* Second; more of the results were GG.

Sample rubric: Assign 1 point to each completed table, 1 point to each correct graph, and 2 points to a correct reasoning. There is a total of 6 points.

3. a. 2 **b.** 6 **c.** 26

5. a. Green, blue, yellow, red **b.** 10 **c.** 4

7. a. 52 **b.** 26 **c.** 13 **d.** 2

9. a. Equally likely **b.** Unlikely **c.** Likely

11. a. $\frac{4}{17}$ **b.** $\frac{5}{17}$ **c.** $\frac{11}{17}$ **13. a.** $\frac{9}{19}$ **b.** $\frac{1}{19}$ **c.** $\frac{9}{19}$

15. a. 25 **b.** 5 **c.** 40 **d.** 75

17. a. $\frac{2}{15}$ **b.** $\frac{3}{20}$ **c.** $\frac{8}{15}$

19. a. Heads, tails

b. *Sample answer:*

Heads	𝍸 𝍸 𝍸 I
Tails	𝍸 𝍸 IIII

c. Heads: $\frac{8}{15}$; Tails: $\frac{7}{15}$ **d.** Heads: $\frac{1}{2}$; Tails: $\frac{1}{2}$

e. *Sample answer:* The results are similar.

21. a. $\frac{1}{15}$ **b.** $\frac{1}{30}$ **c.** $\frac{11}{15}$ **23. a.** $\frac{3}{5}$ **b.** $\frac{7}{10}$ **c.** $\frac{3}{5}$

25. 350 **27.** $\frac{101}{400}$ **29.** $\frac{1}{5}$

31. *Sample answer:* The student counted 4 as a favorable outcome, and should review identifying outcomes and events.

33. *Sample answer:* Experimental probability is based on results and theoretical probability is based on the possible outcomes.

35. a. Yes; Answers will vary. **b.** Yes; Answers will vary.

c. Yes; *Sample answer:* The ratio of favorable outcomes to possible outcomes is the same.

37. $\frac{1}{10}$

39. $\frac{n-1}{n}$; *Sample answer:* There are $n - 1$ other outcomes.

Section 16.2 *(page 630)*

1. Solution key:

1. 4; HH, HT, TH, TT 2. $\frac{1}{4}$; $\frac{1}{2}$; $\frac{1}{4}$

3.

H — H — H, T; H — T — H, T; T — H — H, T; T — T — H, T

4. 8; HHH, HHT, HTH, HTT, THH, THT, TTH, TTT

5. $\frac{1}{8}$; $\frac{3}{8}$; $\frac{3}{8}$; $\frac{1}{8}$

Sample answer: #1: 2; the student listed the outcomes of the first flip and second flip as given in the tree diagram; #2: 0; the student used the number of outcomes from problem 1 and counted 1 head or tail as 1 favorable outcome.

3. 12 outcomes

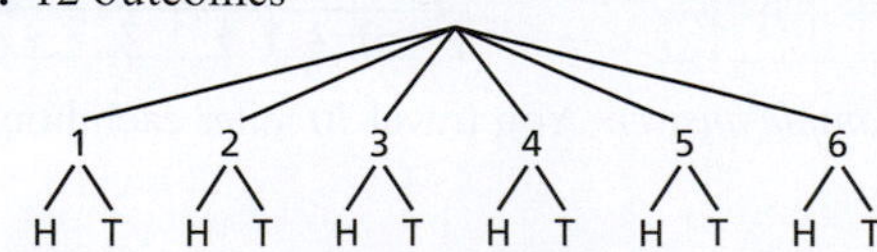

5. $\frac{1}{12}$ **7.** $\frac{1}{24}$

9. a. 18 **b.** $\frac{1}{18}$

c.

1 2 3 4 5 6

A B C A B C A B C A B C A B C A B C

11. a.

	Red	Blue	Green
Red	RR	RB	RG
Yellow	YR	YB	YG
Blue	BR	BB	BG
Green	GR	GB	GG

b. 12 **c.** $\frac{1}{12}$

13. 16 **15. a.** 800 **b.** 500 **17.** $\frac{1}{16}$

19. a. Combination; *Sample answer:* Order does not matter.

b. Permutation; *Sample answer:* Order matters.

21. a. 24 **b.** 362,880

23. a. 120 **b.** 10 **c.** 720 **d.** 36

25. a. $C(6, 1) = 6$ **b.** $C(6, 2) = 15$ **c.** $C(6, 3) = 20$

d. $C(6, 4) = 15$ **e.** $C(6, 5) = 6$ **f.** $C(6, 6) = 1$

27. a. Combinations; *Sample answer:* Order does not matter.

b. 495

29. 362,880 **31.** 2002 **33.** 1024

35. *Sample answer:* The student used the formula for a combination, and should review permutations.

37. *Sample answer:* The events are selecting the first object, second object, and so on, so the total number of outcomes is $n \times (n-1) \times (n-2) \times \ldots \times 3 \times 2 \times 1 = n!$

39. 175,760,000

41. They are equal. *Sample answer:* Both expressions are equal to $\frac{n!}{(n-r)!r!}$.

43. a. Answers will vary. 11

b. 5; *Sample answer:* There are 5 times as many combinations.

45. a. $\frac{1}{120}$ **b.** $\frac{7}{24}$ **c.** $\frac{119}{120}$

47. a. $\frac{1}{4}$

b. *Sample answer:* Rolling an odd sum; rolling a sum of 2, 6, 7, 8, or 12; rolling a sum of 4, 6, 9, 10, 11, or 12

49. a. $C(7, 2)$; *Sample answer:* Each road connects 2 towns.

b. 21 **c.** $\frac{1}{1330}$

Section 16.3 *(page 640)*

1. Solution key:

1.

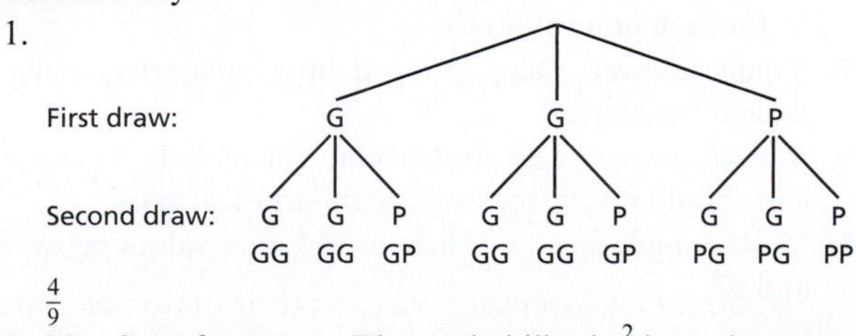

$\frac{4}{9}$

2. No; *Sample answer:* The probability is $\frac{2}{3}$ in each case.

3.

First draw: G G P

Second draw: G P G P G G

GG GP GG GP PG PG

$\frac{1}{3}$

4. Yes; *Sample answer:* The probability is either $\frac{1}{2}$ or 1.

Sample rubric: Assign 2 points to each correct diagram, 1 point to each correct probability, and 2 points to each correct reasoning. There is a total of 10 points.

3. a. Independent **b.** Dependent

c. Independent **d.** Dependent

5. a. $\frac{1}{4}$ **b.** $\frac{1}{36}$ **c.** $\frac{1}{12}$ **d.** $\frac{1}{12}$

7. a. $\frac{1}{9}$; unlikely **b.** $\frac{1}{27}$; unlikely **c.** $\frac{25}{36}$; likely

9. $\frac{1}{4}$ **11.** $\frac{1}{16}$ **13.** $\frac{5}{147}$ **15.** Answers will vary.

17. a. $\frac{1}{132}$ **b.** $\frac{1}{66}$ **c.** $\frac{1}{33}$ **d.** $\frac{1}{66}$

19. a. $\frac{2}{31}$; unlikely **b.** $\frac{9}{248}$; unlikely **c.** $\frac{3}{248}$; unlikely

21. $\frac{3}{14}$ **23.** $\frac{5}{92}$ **25.** $\frac{1}{124{,}750}$ **27.** Answers will vary.

29. a. $\frac{1}{4802}$ **b.** $\frac{1}{61{,}075}$

31. a. Yes **b.** About 0.123 **c.** About 0.043

33. *Sample answer:* The student thought the events were dependent, and should review independent and dependent events.

35. *Sample answer:* Three teams winning three separate quarterfinal matches in a tournament. The outcome of one match does not affect the outcome of another.

37. 2, 3, 4, in some order; *Sample answer:* The product is 24 and the sum is 9.

39. $\left(\frac{k}{n}\right)^3$

41. *Sample answer:* The number of outcomes changes for dependent events.

43. $\left(\frac{1}{2}\right)^n$ **45.** $\frac{1}{2520}$

Section 16.4 *(page 650)*

1. Solution key:

1. Check students' work.
2. *Sample answer:* No; there are more ways to lose than win.
3. 0; *Sample answer:* The sum of the payoffs is 0, and they have the same probabilities.

Sample rubric: Assign 1 point to each correct sum and 2 points to each correct reasoning. There is a total of 6 points.

3. a. Heads: $\frac{1}{2}$, \$5; tails: $\frac{1}{2}$, −\$5

b. 1: $\frac{1}{6}$, −\$1; 2: $\frac{1}{6}$, \$3; 3: $\frac{1}{6}$, −\$1; 4: $\frac{1}{6}$, \$3; 5: $\frac{1}{6}$, −\$1; 6: $\frac{1}{6}$, \$3

5. a. 1 point **b.** −8 points

7. a. *Sample answer:* Option 2

b. Option 1: \$1600; Option 2: \$1000; yes

9. a. \$40 **b.** \$25 **c.** Scholarship A

11. Zoomax **13.** Practix **15.** 8.4 cents **17.** 669

19. Yes; *Sample answer:* The expected value is 0.

21. *Sample answer:* The student did not include one of the outcomes; check that the probabilities sum to 1.

23. No; *Sample answer:* The expected value is –\$25,000.
25. −\$2; no
27. a. \$250 **b.** \$300 **c.** Answers will vary.
29. *Sample answer:* The student only considered one outcome, and should review the definition of expected value.
31. Answers will vary.
33. a. No; *Sample answer:* There's a $\frac{3}{4}$ chance of A or B winning, and a $\frac{1}{4}$ chance of C or D winning.
b. Answers will vary. **c.** Answers will vary.

Chapter 16 Review *(page 655)*

1. 12 **3.** 2, 4, 6
5. 2, 3, 4, 5, 6, 7, 8, 9, 10, jack, queen, king, and ace of hearts
7. Likely **9.** $\frac{10}{13}$ **11.** $\frac{3}{26}$ **13.** $\frac{23}{100}$
15. $\frac{39}{100}$ **17.** $\frac{79}{193}$; $\frac{100}{579}$
19.

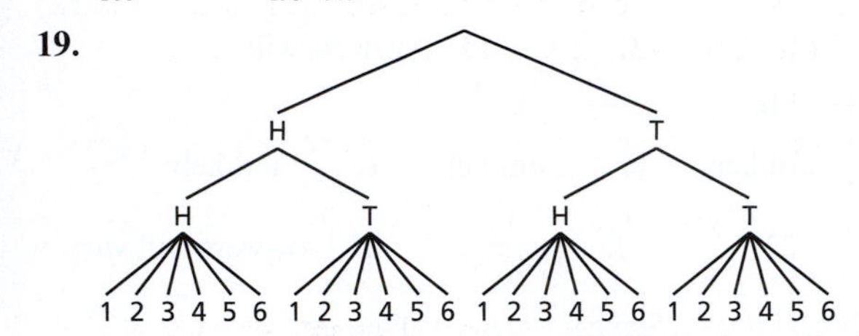

21. $\frac{1}{2}$ **23.** $\frac{1}{5}$ **25.** $\frac{5}{18}$ **27.** $\frac{1}{18}$
29. Combination; *Sample answer:* Order does not matter.
31. Combination; *Sample answer:* Order does not matter.
33. Permutation; *Sample answer:* Order matters.
35. 120 **37.** 255,024 **39.** 8 **41.** 12,271,512
43. 720 **45.** 10,626 **47.** 193,099,974 **49.** 817,190
51. Independent **53.** Dependent **55.** Independent
57. $\frac{1}{12}$ **59.** $\frac{1}{4096}$ **61.** $\frac{12}{95}$ **63.** $\frac{1}{21}$
65. a. $\frac{1}{84}$; $\frac{83}{84}$ **b.** $\frac{1}{1728}$; $\frac{1}{343}$ **67.** \$1
69. Option 1: \$1; Option 2: \$0.50; Option 1
71. Option 1: \$500; Option 2: \$990; Option 2
73. \$0; yes; *Sample answer:* The expected value is 0.
75. Florida

Chapter 16 Test *(page 658)*

1. B **2.** C **3.** A **4.** C **5.** C
6. A **7.** E **8.** D

CHAPTER 17

Section 17.1 *(page 668)*

1. Solution key:
1. The stems show the tens. The leaves show the ones.
2. 39; *Sample answer:* 39 dots; 39 leaves
3. 6; *Sample answer:* 6 dots in 40–49, 6 leaves after 4
4. Stem-and-leaf plot; *Sample answer:* It mentions stems and leaves.
5. Stem-and-leaf plot; *Sample answer:* The stem of 2 has a leaf of 9.
6. 2, 2, 2, 2, 2, 2, 2, 3, 3, 9, 9, 9, 12, 12, 15, 15, 15, 15, 15, 15, 15, 15, 15, 22, 26, 29, 32, 32, 32, 32, 32, 32, 35, 44, 44, 44, 44, 44, 44

Sample rubric: Assign 10 points to a correct message and 3 points to each correct response in problems 2 through 5. There is a total of 25 points.

3. a. **Postage Stamp Prices**

Stem	Leaf
1	0 3 5 8
2	0 2 5 9
3	2 3 4 7 9
4	1 2 4 5 6

Key: 1 | 0 = 10 cents

b. 8 **c.** 50%
5.

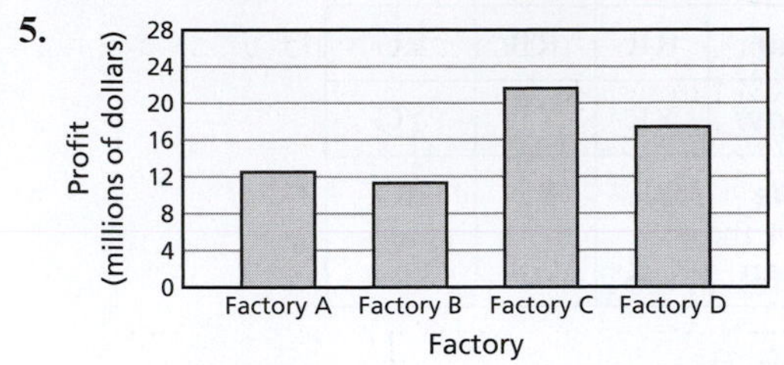

7. a.

Exam scores	51–60	61–70	71–80	81–90	91–100
Frequency	1	12	5	5	7

b.

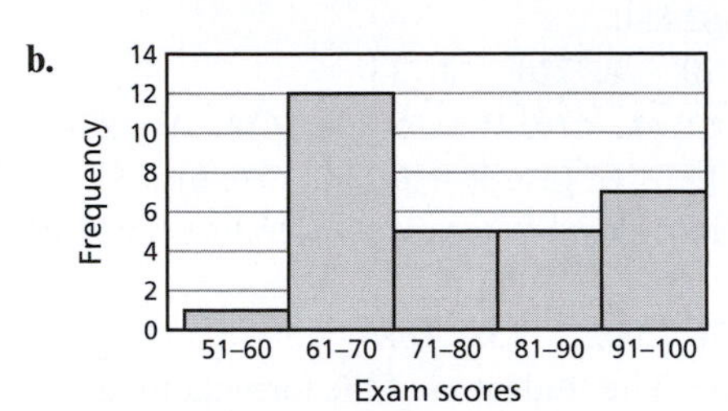

c. 12 **d.** 1
9.

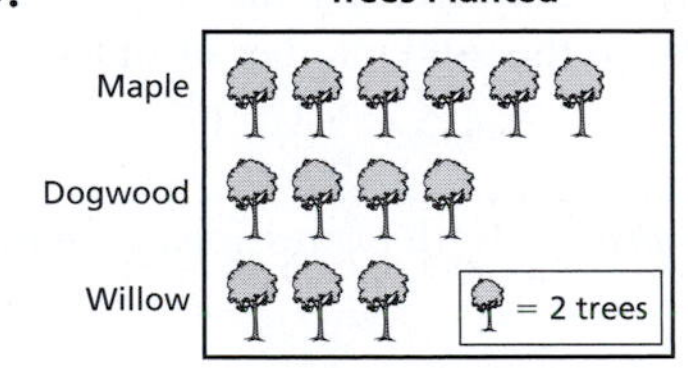

11.

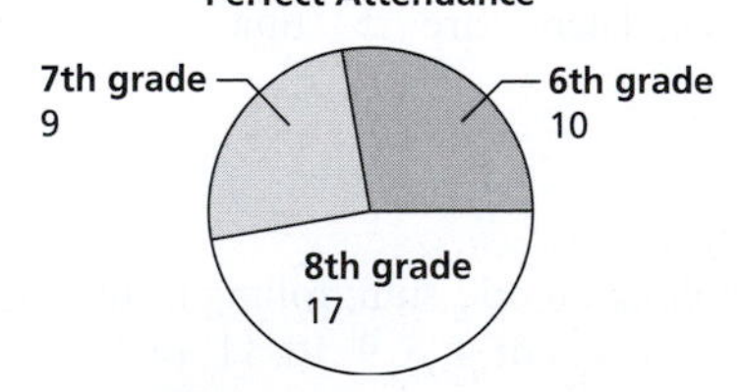

13. a. 200 lb **b.** 7
c. Positive; *Sample answer:* As shoe size increases, the weight increases.
15. a. 7
b. *Sample answer:* The exam score increased about 7 points for each hour of study.
17. *Sample answer:* Data grouped into categories; when the data are numerical.
19. *Sample answer:* The student did not include a stem of 3, and should review making a stem-and-leaf plot.
21. 15; 4; *Sample answer:* There are 15 data values between 62 and 97.

23. **Seasonal Rainfall Amounts**

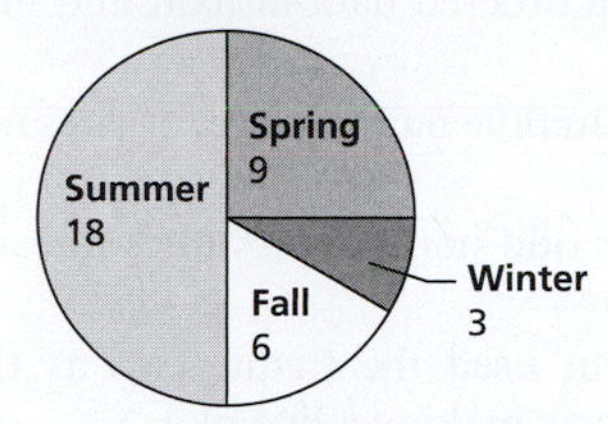

25. a. No; *Sample answer:* The value could be any number from 40% through 44.9%.

b. Yes; *Sample answer:* 36 states are in that range.

27. a. *Sample answer:* The line through the points is close to most of the data values.

b. Answers will vary.

29. a.

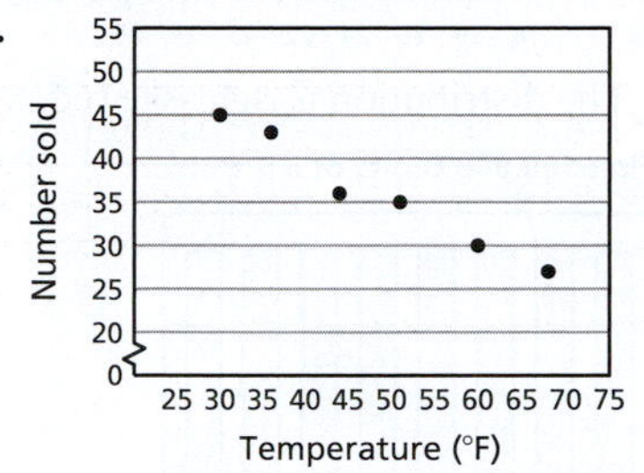

b. Negative

c. *Sample answer:*

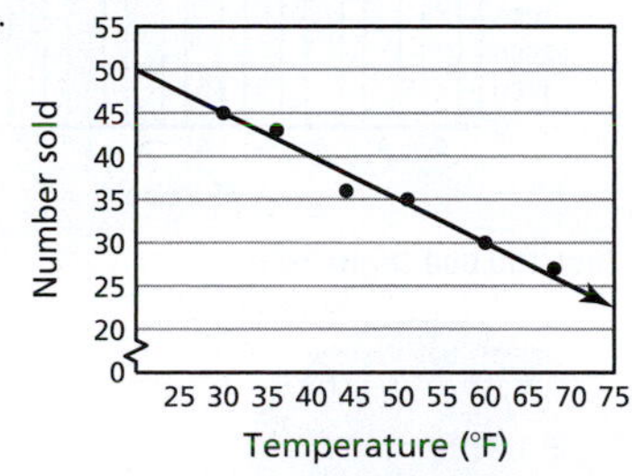

d. $-\frac{1}{2}$; *Sample answer:* The number of hot chocolates sold decreases by 1 for each 2° increase in temperature.

e. 40

31. *Sample answer:* Circle graph; the data are percents in categories.

Section 17.2 *(page 678)*

1. Solution key:

1. a. About 93% b. *Sample answer:* 18
   c. *Sample answer:* Most graduates were 18.
2. a. About 93% b. *Sample answer:* $35,000
   c. *Sample answer:* Most incomes are between $20,000 and $50,000.

Sample rubric: Assign 1 point to each correct response. There is a total of 6 points.

3. City 1: about 13.8 in.; City 2: 11.5 in.; City 1 had a greater mean.

5. a. 11.5; 12 **b.** 81; 79

7. a. About 5.4; 5; 10 **b.** About 25.1; 26; 23

9. 73.6, 74.5, 62; 85.2, 87.5, 89 **11.** 7.1; 7; 6 and 7

13. a. $45,700, $46,000, $39,000 **b.** $46,700, $47,000, $40,000

c. They increase by $1000.

15. 109.1; 106; 99; *Sample answer:* Median; it is close to most of the data.

17. a. 30, 26, 16 **b.** 24.3, 24.5, 16

c. *Sample answer:* The outlier increased the mean and median.

19. a. $22,200, $22,000, no mode **b.** School D

c. $18,000, $19,000, no mode

d. *Sample answer:* The outlier increased the mean and median.

21. A

23. *Sample answer:* Average age of workers and residents in a nursing home

25. 83 **27.** 27 **29.** Answers will vary.

31. Mode; *Sample answer:* The mode would be the most popular size.

33. About 70.7, 70.5, no mode; mean or median; *Sample answer:* They are close to most of the data.

35. 6

37. Mode; *Sample answer:* No; the sample is very small compared to the whole population.

39. *Sample answer:* The student confused median and mode, and should review the definitions.

41. Mode; both

43. $3.40; *Sample answer:* The 25 students receive $85 in a week.

Section 17.3 *(page 688)*

1. Solution key:

1. Pyramid A; *Sample answer:* The age distribution in Pyramid B is approximately uniform, so it represents the United States.
2. *Sample answer:* 17; The middle of the data is in the 15–19 interval.
3. *Sample answer:* 37; The middle of the data is in the 35–39 interval.
4. Kenya; *Sample answer:* More of the population is in the 0–4 interval.

Sample rubric: Assign 1 point to each correct response and 2 points to each correct reasoning. There is a total of 12 points.

3. 6 **5.** 11°F

7. B; *Sample answer:* The data are less spread out.

9. About 76.2%; 100% **11.** About 95.4%

13. Flat; about equal

15. Bimodal; The mean is between the two modes.

17. a. *Sample answer:* There is less variation in the weights.

b. Machine 1: 84.1%; Machine 2: 97.7%

19. a. Yes; *Sample answer:* The histogram is bell-shaped.

b. About 79.1% **c.** About 98.7%

d. *Sample answer:* The percents in parts (b) and (c) are greater.

21. *Sample answer:* Normal; the measures of central tendency are approximately equal.

23. *Sample answer:* The student used the first and last numbers instead of least and greatest, and should review finding the range of a data set.

25. a. Answers will vary. **b.** Normal

c. Answers will vary.

27. a. *Sample answer:* The mean, median, and mode are greater for the mathematics scores.

b. Slightly; yes; *Sample answer:* The mean is greater than the median.

Section 17.4 *(page 698)*

1. Solution key:

1. 0, 1, 3, 3, 3, 3, 5, 6, 8, 8, 9, 10, 10, 12, 13, 13, 14, 16, 18, 19, 23, 24, 32, 45
2. 0, 1, 3, 3, 3, 3; 5, 6, 8, 8, 9, 10; 10, 12, 13, 13, 14, 16; 18, 19, 23, 24, 32, 45
3. *Sample answer:* The least and greatest values are the first and last in the ordered list. The others are the averages of the two numbers they are between in the ordered list. The box-and-whisker plot shows the data by quarters.

Sample rubric: Assign 6 points to correctly ordering the data set and dividing it into four groups, 1 point to each correct explanation of the numbers, and 2 points to a correct explanation of the box-and-whisker plot. There is a total of 13 points.

3. a. 19 **b.** 18 **c.** 6

5. a.

0 1 2 3 5

Hours studying (0 1 2 3 4 5 6)

b.

5 10 17.5 27.5 50

Donations (dollars) (0 10 20 30 40 50 60)

7. a.

Credit hour load (11 12 13 14 15 16 17 18 19)

b.

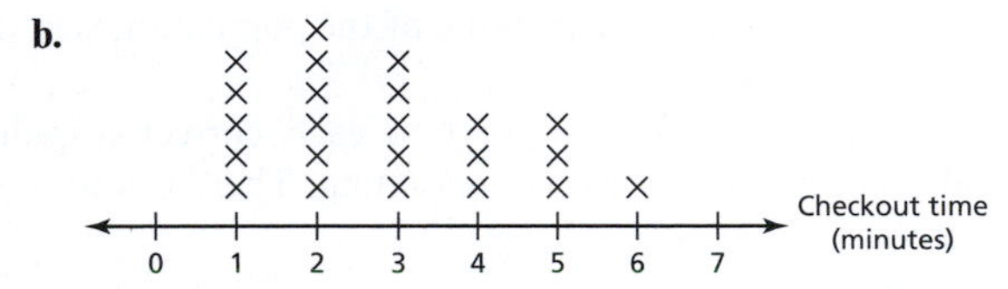

9. *Sample answer:*

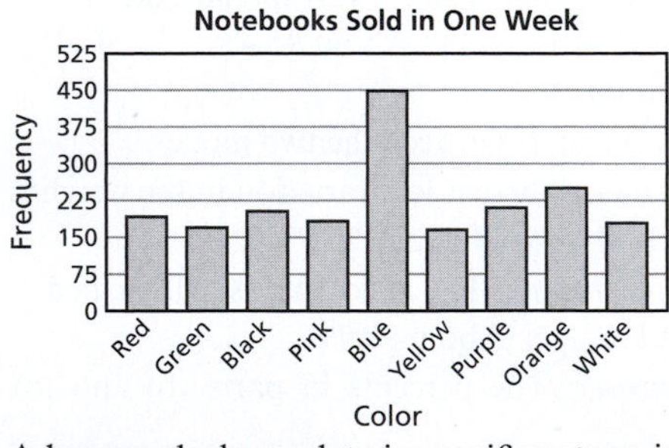

A bar graph shows data in specific categories.

11. a. *Sample answer:* Box-and-whisker plot; shows the variability of data.

b. *Sample answer:* Line graph; shows change over time.

c. *Sample answer:* Line plot; shows the number of times each value occurs.

d. *Sample answer:* Circle graph; shows data as parts of a whole.

13. *Sample answer:* There is a break in the vertical axis.

15. *Sample answer:* The icons are different sizes.

17. *Sample answer:* The width changes.

19. *Sample answer:* The intervals are different sizes.

21. a. *Sample answer:* Split the ordered data into equal groups.

b. *Sample answer:* Split the ordered data in half, and find the median of each half.

c. *Sample answer:* Use the middle number; the set does not split evenly.

d. *Sample answer:* Even or odd number of values in each half

23. *Sample answer:* The student used the frequencies as the data values, and should review making a line plot.

25. *Sample answer:* Both have the same median, Sales Rep B has more variability.

27. a.

Ages (12 13 14 15 16 17 18 19 20 21 22 23 24 25)

b. *Sample answer:* The distribution is right-skewed.

29. a. First 100,000 Digits of π

Frequency (0 1000 2000 3000 4000 5000 6000 7000 8000 9000 10,000 11,000); Number (0 1 2 3 4 5 6 7 8 9)

b. First 100,000 Digits of π

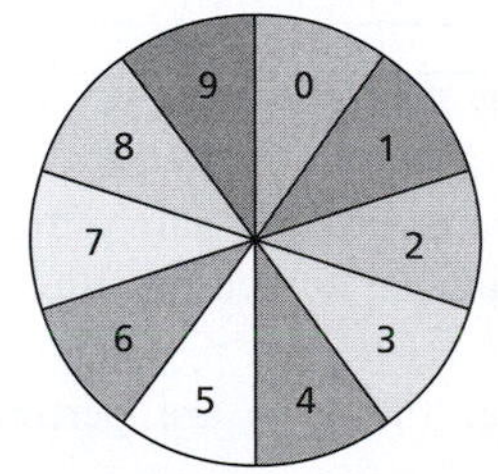

c. *Sample answer:* Bar graph; it is easier to see the differences in frequencies.

d. Flat

31. *Sample answer:* When the first or last quarter of the data values are all the same value.

33. a. *Sample answer:* The numbers add to 110, so the categories overlap.

b. *Sample answer:* Bar graph; shows the data in specific categories.

35. Box-and-whisker plot; *Sample answer:* Shows the variability of the data

37. Answers will vary.

Chapter 17 Review *(page 703)*

1. Speeds of Cars

Stem	Leaf
5	4 5 5 8 8 8
6	0 0 3 4 6 6 6 8
7	0 2

Key: 5 | 4 = 54 mph

3.

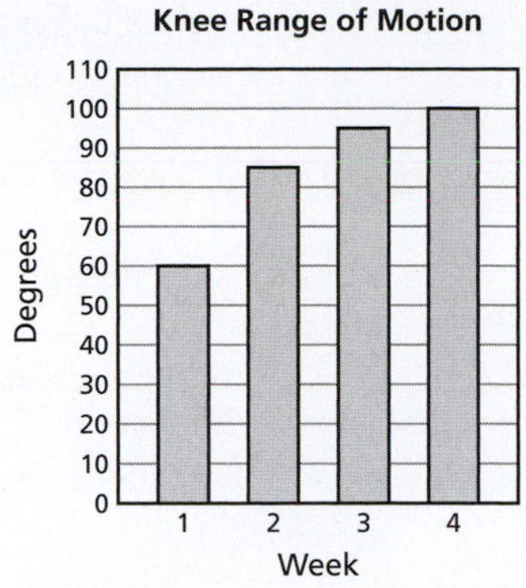
Knee Range of Motion
Degrees
110
100
90
80
70
60
50
40
30
20
10
0
1
2
3
4
Week

5. a.

Goals scored	Frequency
0–1	5
2–3	4
4–5	0
6–7	1

b.

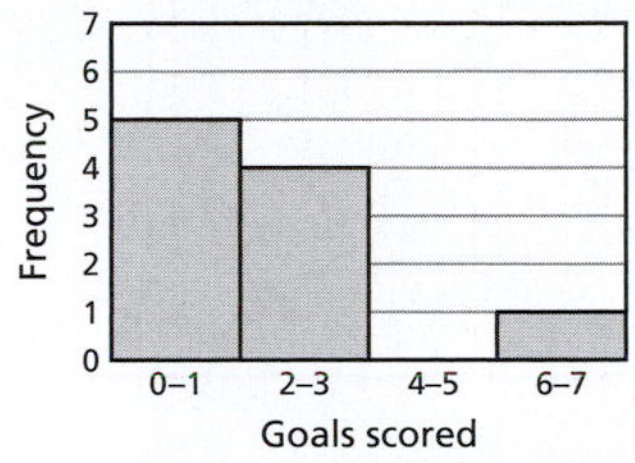
Frequency
7
6
5
4
3
2
1
0
0–1
2–3
4–5
6–7
Goals scored

7.

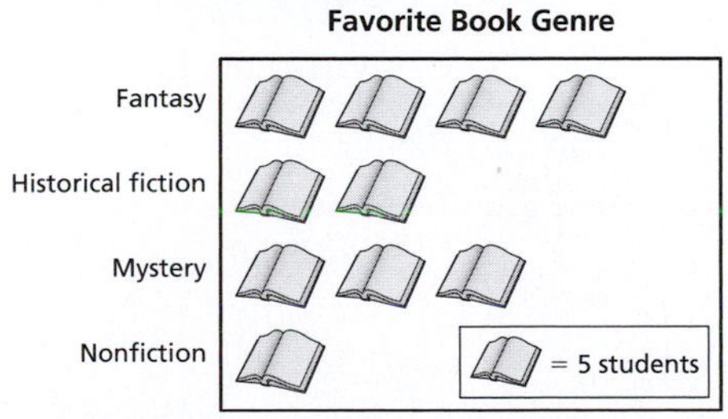
Favorite Book Genre
Fantasy
Historical fiction
Mystery
Nonfiction
= 5 students

9. a. \$150,000 **b.** 2011
c. Negative; *Sample answer:* As the years increase, the donations decrease.

11. 286; 212; 212

13. Women's: 33.9; men's: 31.7; The women's mean is greater.

15. \$15.63, \$14.50, no mode; \$12.63, \$11.50, no mode; The mean and median decrease.

17. About 55.7, 54, 56; about 52.6, 52, 56; The outlier increased mean and the median.

19. About 41.4, 43, 43; *Sample answer:* Median or mode; Both are 43.

21. About 9.4, 9, 8; *Sample answer:* Median

23. 9 **25.** About 77.3%; about 99.1%

27. 68.2%

29. Right-skewed; The mean lies to the right of the median and mode.

31. 24 **33.** 28

35.

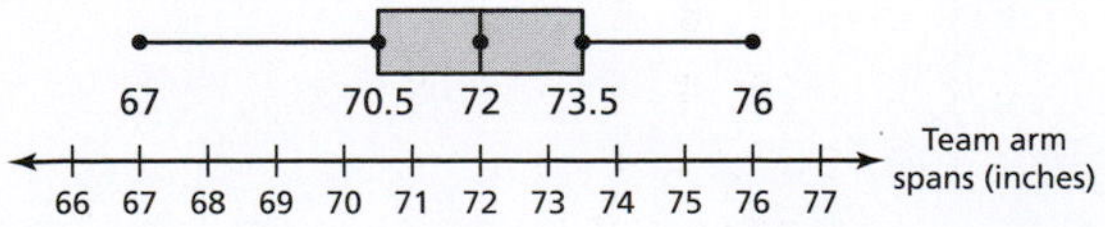
67
70.5
72
73.5
76
66 67 68 69 70 71 72 73 74 75 76 77
Team arm spans (inches)

37.

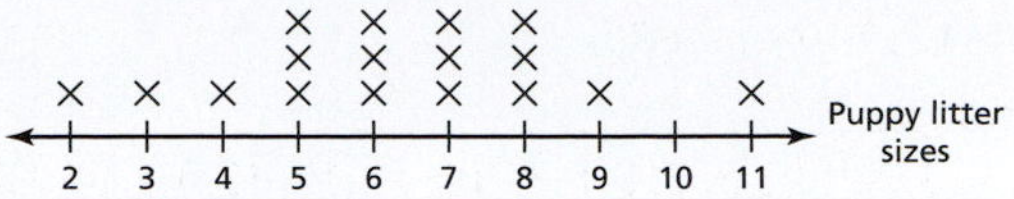
2 3 4 5 6 7 8 9 10 11
Puppy litter sizes

39. *Sample answer:* Line graph; shows how data change over time.

41. *Sample answer:* Histogram; shows data in intervals.

43. *Sample answer:* There is a break in the vertical axis.

Chapter 17 Test *(page 706)*

1. D **2.** B **3.** A **4.** C **5.** C
6. E **7.** D **8.** D

CHAPTER 17

Index

E

H

I

K

L

Q

R

S

Coordinate Plane

Midpoint Formula

Midpoint of line segment joining (x_1, y_1) and (x_2, y_2) $\left(\frac{x_1 + x_2}{2}, \frac{y_1 + y_2}{2}\right)$

Distance Formula

d = distance between points (x_1, y_1) and (x_2, y_2)

$$d = \sqrt{(x_2 - x_1)^2 + (y_2 - y_1)^2}$$

Rules of Exponents

Let m and n be integers and assume $a \neq 0$ and $b \neq 0$.

$a^0 = 1$

$a^m \cdot a^n = a^{m+n}$

$\frac{a^m}{a^n} = a^{m-n}$

$(a^m)^n = a^{mn}$

$(ab)^m = a^m \cdot b^m$

$a^{-n} = \frac{1}{a^n}$

Properties of Equality

Addition Property of Equality

If $a = b$, then $a + c = b + c$.

Subtraction Property of Equality

If $a = b$, then $a - c = b - c$.

Multiplication Property of Equality

If $a = b$, then $ac = bc$.

Division Property of Equality

If $a = b$, and $c \neq 0$, then $\frac{a}{c} = \frac{b}{c}$.

Cancellation Property of Multiplication

If $ac = bc$, and $c \neq 0$, then $a = b$.

Properties of Rational Numbers

Let $\frac{a}{b}$, $\frac{c}{d}$, and $\frac{e}{f}$ be any rational numbers.

Commutative Property of Addition

$$\frac{a}{b} + \frac{c}{d} = \frac{c}{d} + \frac{a}{b}$$

Commutative Property of Multiplication

$$\frac{a}{b} \cdot \frac{c}{d} = \frac{c}{d} \cdot \frac{a}{b}$$

Associative Property of Addition

$$\left(\frac{a}{b} + \frac{c}{d}\right) + \frac{e}{f} = \frac{a}{b} + \left(\frac{c}{d} + \frac{e}{f}\right)$$

Associative Property of Multiplication

$$\left(\frac{a}{b} \cdot \frac{c}{d}\right) \cdot \frac{e}{f} = \frac{a}{b} \cdot \left(\frac{c}{d} \cdot \frac{e}{f}\right)$$

Distributive Property of Multiplication Over Addition and Subtraction

$$\frac{a}{b}\left(\frac{c}{d} + \frac{e}{f}\right) = \frac{a}{b} \cdot \frac{c}{d} + \frac{a}{b} \cdot \frac{e}{f}$$

$$\frac{a}{b}\left(\frac{c}{d} - \frac{e}{f}\right) = \frac{a}{b} \cdot \frac{c}{d} - \frac{a}{b} \cdot \frac{e}{f}$$

Additive Identity Property

$$\frac{a}{b} + 0 = \frac{a}{b}, 0 + \frac{a}{b} = \frac{a}{b}$$

Additive Inverse Property

$$\frac{a}{b} + \left(-\frac{a}{b}\right) = 0$$

Multiplicative Identity Property

$$\frac{a}{b} \cdot 1 = \frac{a}{b}, 1 \cdot \frac{a}{b} = \frac{a}{b}$$

Multiplicative Inverse Property

$$\frac{a}{b} \cdot \frac{b}{a} = 1, a \neq 0, b \neq 0$$

Zero Multiplication Property

$$\frac{a}{b} \cdot 0 = 0, 0 \cdot \frac{a}{b} = 0$$